Concrete & Masonry Costs with RSMeans data

Scott Keller, Senior Editor

2021
39th annual edition

Vice President, Data
Tim Duggan

Vice President, Product
Ted Kail

Product Manager
Jason Jordan

Principal Engineer
Bob Mewis

Engineers: Architectural Divisions
1, 3, 4, 5, 6, 7, 8, 9, 10, 11, 12, 13, 41
Matthew Doheny, Manager
Sam Babbitt
Sergeleen Edouard
Richard Goldin
Scott Keller
Thomas Lane

Engineers: Civil Divisions and Wages
2, 31, 32, 33, 34, 35, 44, 46
Derrick Hale, PE, Manager
Christopher Babbitt
Stephen Bell
Michael Lynch
Elisa Mello
David Yazbek

Engineers: Mechanical, Electrical, Plumbing & Conveying Divisions
14, 21, 22, 23, 25, 26, 27, 28, 48
Joseph Kelble, Manager
Brian Adams
Michelle Curran
Antonio D'Aulerio
Thomas Lyons
Jake MacDonald

Contributing Engineers
John Melin, PE, Manager
Paul Cowan
Barry Hutchinson
Gerard Lafond, PE
Matthew Sorrentino

Technical Project Lead
Kevin Souza

Production
Debra Panarelli, Manager
Jonathan Forgit
Sharon Larsen
Sheryl Rose
Janice Thalin

Data Quality
Joseph Ingargiola, Manager
David Byars
Audrey Considine
Ellen D'Amico

Cover Design
Blaire Collins

Innovation
Ray Diwakar, Vice President
Kedar Gaikwad
Todd Glowac
Srini Narla
Joseph Woughter

RSMeans data from Gordian
Construction Publishers & Consultants
1099 Hingham Street, Suite 201
Rockland, MA 02370
United States of America
1.800.448.8182
RSMeans.com

Printed in the United States of America
ISSN 1075-0274
ISBN 978-1-950656-53-0

0111 $319.00 per copy (in United States)
Price is subject to change without prior notice.

Related Data and Services

Our engineers recommend the following products and services to complement *Concrete & Masonry Costs with RSMeans data:*

Annual Cost Data Books
2021 Building Construction Costs with RSMeans data
2021 Heavy Construction Costs with RSMeans data

Reference Books
Concrete Repair & Maintenance Illustrated
Unit Price Estimating Methods
Building Security: Strategies & Costs
Designing & Building with the IBC
Estimating Building Costs
RSMeans Estimating Handbook
Green Building: Project Planning & Estimating
How to Estimate with RSMeans data
Plan Reading & Material Takeoff
Project Scheduling & Management for Construction

Virtual, Instructor-led & On-site Training Offerings
Site Work & Heavy Construction Estimating
Unit Price Estimating
Training for our estimating solution on CD
Plan Reading & Material Takeoff

RSMeans data
For access to the latest cost data, an intuitive search, and an easy-to-use estimate builder, take advantage of the time savings available from our online application.

To learn more visit: **RSMeans.com/online**

Enterprise Solutions
Building owners, facility managers, building product manufacturers, and attorneys across the public and private sectors engage with RSMeans data Enterprise to solve unique challenges where trusted construction cost data is critical.

To learn more visit: **RSMeans.com/Enterprise**

Custom Built Data Sets
Building and Space Models: Quickly plan construction costs across multiple locations based on geography, project size, building system component, product options, and other variables for precise budgeting and cost control.

Predictive Analytics: Accurately plan future builds with custom graphical interactive dashboards, negotiate future costs of tenant build-outs, and identify and compare national account pricing.

Consulting
Building Product Manufacturing Analytics: Validate your claims and assist with new product launches.

Third-Party Legal Resources: Used in cases of construction cost or estimate disputes, construction product failure vs. installation failure, eminent domain, class action construction product liability, and more.

API
For resellers or internal application integration, RSMeans data is offered via API. Deliver Unit, Assembly, and Square Foot Model data within your interface. To learn more about how you can provide your customers with the latest in localized construction cost data visit:
RSMeans.com/API

Table of Contents

Foreword

The Value of RSMeans data from Gordian

Since 1942, RSMeans data has been the industry-standard materials, labor, and equipment cost information database for contractors, facility owners and managers, architects, engineers, and anyone else that requires the latest localized construction cost information. More than 75 years later, the objective remains the same: to provide facility and construction professionals with the most current and comprehensive construction cost database possible.

With the constant influx of new construction methods and materials, in addition to ever-changing labor and material costs, last year's cost data is not reliable for today's designs, estimates, or budgets. Gordian's cost engineers apply real-world construction experience to identify and quantify new building products and methodologies, adjust productivity rates, and adjust costs to local market conditions across the nation. This adds up to more than 22,000 hours in cost research annually. This unparalleled construction cost expertise is why so many facility and construction professionals rely on RSMeans data year over year.

About Gordian

Gordian originated in the spirit of innovation and a strong commitment to helping clients reach and exceed their construction goals. In 1982, Gordian's chairman and founder, Harry H. Mellon, created Job Order Contracting while serving as chief engineer at the Supreme Headquarters Allied Powers Europe. Job Order Contracting is a unique indefinite delivery/indefinite quantity (IDIQ) process, which enables facility owners to complete a substantial number of repair, maintenance, and construction projects with a single, competitively awarded contract. Realizing facility and infrastructure owners across various industries could greatly benefit from the time and cost saving advantages of this innovative construction procurement solution, he established Gordian in 1990.

Continuing the commitment to provide the most relevant and accurate facility and construction data, software, and expertise in the industry, Gordian enhanced the fortitude of its data with the acquisition of RSMeans in 2014. And in an effort to expand its facility management capabilities, Gordian acquired Sightlines, the leading provider of facilities benchmarking data and analysis, in 2015.

Our Offerings

Gordian is the leader in facility and construction cost data, software, and expertise for all phases of the building life cycle. From planning to design, procurement, construction, and operations, Gordian's solutions help clients maximize efficiency, optimize cost savings, and increase building quality with its highly specialized data engineers, software, and unique proprietary data sets.

Our Commitment

At Gordian, we do more than talk about the quality of our data and the usefulness of its application. We stand behind all of our RSMeans data—from historical cost indexes to construction materials and techniques—to craft current costs and predict future trends. If you have any questions about our products or services, please call us toll-free at 800.448.8182 or visit our website at gordian.com.

How the Cost Data Is Built: An Overview

Unit Prices*
All cost data have been divided into 50 divisions according to the MasterFormat® system of classification and numbering.

Assemblies*
The cost data in this section have been organized in an "Assemblies" format. These assemblies are the functional elements of a building and are arranged according to the 7 elements of the UNIFORMAT II classification system. For a complete explanation of a typical "Assembly", see "RSMeans data: Assemblies—How They Work."

Residential Models*
Model buildings for four classes of construction—economy, average, custom, and luxury—are developed and shown with complete costs per square foot.

Commercial/Industrial/ Institutional Models*
This section contains complete costs for 77 typical model buildings expressed as costs per square foot.

Green Commercial/Industrial/ Institutional Models*
This section contains complete costs for 25 green model buildings expressed as costs per square foot.

References*
This section includes information on Equipment Rental Costs, Crew Listings, Historical Cost Indexes, City Cost Indexes, Location Factors, Reference Tables, and Change Orders, as well as a listing of abbreviations.

- **Equipment Rental Costs:** Included are the average costs to rent and operate hundreds of pieces of construction equipment.
- **Crew Listings:** This section lists all the crews referenced in the cost data. A crew is composed of more than one trade classification and/or the addition of power equipment to any trade classification. Power equipment is included in the cost of the crew. Costs are shown both with bare labor rates and with the installing contractor's overhead and profit added. For each, the total crew cost per eight-hour day and the composite cost per labor-hour are listed.

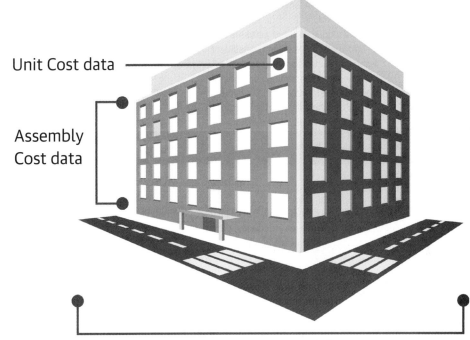

Unit Cost data

Assembly Cost data

Square Foot Models

- **Historical Cost Indexes:** These indexes provide you with data to adjust construction costs over time.
- **City Cost Indexes:** All costs in this data set are U.S. national averages. Costs vary by region. You can adjust for this by CSI Division to over 730 cities in 900+ 3-digit zip codes throughout the U.S. and Canada by using this data.
- **Location Factors:** You can adjust total project costs to over 730 cities in 900+ 3-digit zip codes throughout the U.S. and Canada by using the weighted number, which applies across all divisions.
- **Reference Tables:** At the beginning of selected major classifications in the Unit Prices are reference numbers indicators. These numbers refer you to related information in the Reference Section. In this section, you'll find reference tables, explanations, and estimating information that support how we develop the unit price data, technical data, and estimating procedures.
- **Change Orders:** This section includes information on the factors that influence the pricing of change orders.

- **Abbreviations:** A listing of abbreviations used throughout this information, along with the terms they represent, is included.

Index (printed versions only)
A comprehensive listing of all terms and subjects will help you quickly find what you need when you are not sure where it occurs in MasterFormat®.

Conclusion
This information is designed to be as comprehensive and easy to use as possible.

The Construction Specifications Institute (CSI) and Construction Specifications Canada (CSC) have produced the 2018 edition of MasterFormat®, a system of titles and numbers used extensively to organize construction information.

All unit prices in the RSMeans cost data are now arranged in the 50-division MasterFormat® 2018 system.

---

*   Not all information is available in all data sets

Note: The material prices in RSMeans cost data are "contractor's prices." They are the prices that contractors can expect to pay at the lumberyards, suppliers'/distributors' warehouses, etc. Small orders of specialty items would be higher than the costs shown, while very large orders, such as truckload lots, would be less. The variation would depend on the size, timing, and negotiating power of the contractor. The labor costs are primarily for new construction or major renovation rather than repairs or minor alterations. With reasonable exercise of judgment, the figures can be used for any building work.

Estimating with RSMeans data: Unit Prices

Following these steps will allow you to complete an accurate estimate using RSMeans data: Unit Prices.

1. Scope Out the Project

- Think through the project and identify the CSI divisions needed in your estimate.
- Identify the individual work tasks that will need to be covered in your estimate.
- The Unit Price data have been divided into 50 divisions according to CSI MasterFormat® 2018.
- In printed versions, the Unit Price Section Table of Contents on page 1 may also be helpful when scoping out your project.
- Experienced estimators find it helpful to begin with Division 2 and continue through completion. Division 1 can be estimated after the full project scope is known.

2. Quantify

- Determine the number of units required for each work task that you identified.
- Experienced estimators include an allowance for waste in their quantities. (Waste is not included in our Unit Price line items unless otherwise stated.)

3. Price the Quantities

- Use the search tools available to locate individual Unit Price line items for your estimate.
- Reference Numbers indicated within a Unit Price section refer to additional information that you may find useful.
- The crew indicates who is performing the work for that task. Crew codes are expanded in the Crew Listings in the Reference Section to include all trades and equipment that comprise the crew.
- The Daily Output is the amount of work the crew is expected to complete in one day.
- The Labor-Hours value is the amount of time it will take for the crew to install one unit of work.
- The abbreviated Unit designation indicates the unit of measure upon which the crew, productivity, and prices are based.
- Bare Costs are shown for materials, labor, and equipment needed to complete the Unit Price line item. Bare costs do not include waste, project overhead, payroll insurance, payroll taxes, main office overhead, or profit.
- The Total Incl O&P cost is the billing rate or invoice amount of the installing contractor or subcontractor who performs the work for the Unit Price line item.

4. Multiply

- Multiply the total number of units needed for your project by the Total Incl O&P cost for each Unit Price line item.
- Be careful that your take off unit of measure matches the unit of measure in the Unit column.
- The price you calculate is an estimate for a completed item of work.
- Keep scoping individual tasks, determining the number of units required for those tasks, matching each task with individual Unit Price line items, and multiplying quantities by Total Incl O&P costs.
- An estimate completed in this manner is priced as if a subcontractor, or set of subcontractors, is performing the work. The estimate does not yet include Project Overhead or Estimate Summary components such as general contractor markups on subcontracted work, general contractor office overhead and profit, contingency, and location factors.

5. Project Overhead

- Include project overhead items from Division 1–General Requirements.
- These items are needed to make the job run. They are typically, but not always, provided by the general contractor. Items include, but are not limited to, field personnel, insurance, performance bond, permits, testing, temporary utilities, field office and storage facilities, temporary scaffolding and platforms, equipment mobilization and demobilization, temporary roads and sidewalks, winter protection, temporary barricades and fencing, temporary security, temporary signs, field engineering and layout, final cleaning, and commissioning.
- Each item should be quantified and matched to individual Unit Price line items in Division 1, then priced and added to your estimate.
- An alternate method of estimating project overhead costs is to apply a percentage of the total project cost—usually 5% to 15% with an average of 10% (see General Conditions).
- Include other project related expenses in your estimate such as:
  - Rented equipment not itemized in the Crew Listings
  - Rubbish handling throughout the project (see section 02 41 19.19)

6. Estimate Summary

- Include sales tax as required by laws of your state or county.
- Include the general contractor's markup on self-performed work, usually 5% to 15% with an average of 10%.
- Include the general contractor's markup on subcontracted work, usually 5% to 15% with an average of 10%.
- Include the general contractor's main office overhead and profit:
  - RSMeans data provides general guidelines on the general contractor's main office overhead (see section 01 31 13.60 and Reference Number R013113-50).
  - Markups will depend on the size of the general contractor's operations, projected annual revenue, the level of risk, and the level of competition in the local area and for this project in particular.
- Include a contingency, usually 3% to 5%, if appropriate.
- Adjust your estimate to the project's location by using the City Cost Indexes or the Location Factors in the Reference Section:
  - Look at the rules in "How to Use the City Cost Indexes" to see how to apply the Indexes for your location.
  - When the proper Index or Factor has been identified for the project's location, convert it to a multiplier by dividing it by 100, then multiply that multiplier by your estimated total cost. The original estimated total cost will now be adjusted up or down from the national average to a total that is appropriate for your location.

Editors' Note:

We urge you to spend time reading and understanding the supporting material. An accurate estimate requires experience, knowledge, and careful calculation. The more you know about how we at RSMeans developed the data, the more accurate your estimate will be. In addition, it is important to take into consideration the reference material such as Equipment Listings, Crew Listings, City Cost Indexes, Location Factors, and Reference Tables.

How to Use the Cost Data: The Details

What's Behind the Numbers? The Development of Cost Data

RSMeans data engineers continually monitor developments in the construction industry in order to ensure reliable, thorough, and up-to-date cost information. While overall construction costs may vary relative to general economic conditions, price fluctuations within the industry are dependent upon many factors. Individual price variations may, in fact, be opposite to overall economic trends. Therefore, costs are constantly tracked and complete updates are performed yearly. Also, new items are frequently added in response to changes in materials and methods.

Costs in U.S. Dollars

All costs represent U.S. national averages and are given in U.S. dollars. The City Cost Index (CCI) with RSMeans data can be used to adjust costs to a particular location. The CCI for Canada can be used to adjust U.S. national averages to local costs in Canadian dollars. No exchange rate conversion is necessary because it has already been factored in.

G The processes or products identified by the green symbol in our publications have been determined to be environmentally responsible and/or resource-efficient solely by RSMeans data engineering staff. The inclusion of the green symbol does not represent compliance with any specific industry association or standard.

Material Costs

RSMeans data engineers contact manufacturers, dealers, distributors, and contractors all across the U.S. and Canada to determine national average material costs. If you have access to current material costs for your specific location, you may wish to make adjustments to reflect differences from the national average. Included within material costs are fasteners for a normal installation. RSMeans data engineers use manufacturers' recommendations, written specifications, and/or standard construction practices for the sizing and spacing of fasteners. Adjustments to material costs may be required for your specific application or location. The manufacturer's warranty is assumed. Extended warranties are not included in the material costs. **Material costs do not include sales tax.**

Labor Costs

Labor costs are based upon a mathematical average of trade-specific wages in 30 major U.S. cities. The type of wage (union, open shop, or residential) is identified on the inside back cover of printed publications or selected by the estimator when using the electronic products. Markups for the wages can also be found on the inside back cover of printed publications and/or under the labor references found in the electronic products.

- If wage rates in your area vary from those used, or if rate increases are expected within a given year, labor costs should be adjusted accordingly.

Labor costs reflect productivity based on actual working conditions. In addition to actual installation, these figures include time spent during a normal weekday on tasks, such as material receiving and handling, mobilization at the site, site movement, breaks, and cleanup.

Productivity data is developed over an extended period so as not to be influenced by abnormal variations and reflects a typical average.

Equipment Costs

Equipment costs include not only rental but also operating costs for equipment under normal use. The operating costs include parts and labor for routine servicing, such as the repair and replacement of pumps, filters, and worn lines. Normal operating expendables, such as fuel, lubricants, tires, and electricity (where applicable), are also included. Extraordinary operating expendables with highly variable wear patterns, such as diamond bits and blades, are excluded. These costs are included under materials. Equipment rental rates are obtained from industry sources throughout North America—contractors, suppliers, dealers, manufacturers, and distributors.

Rental rates can also be treated as reimbursement costs for contractor-owned equipment. Owned equipment costs include depreciation, loan payments, interest, taxes, insurance, storage, and major repairs.

Equipment costs do not include operators' wages.

Equipment Cost/Day—The cost of equipment required for each crew is included in the Crew Listings in the Reference Section (small tools that are considered essential everyday tools are not listed out separately). The Crew Listings itemize specialized tools and heavy equipment along with labor trades. The daily cost of itemized equipment included in a crew is based on dividing the weekly bare rental rate by 5 (number of working days per week), then adding the hourly operating cost times 8 (the number of hours per day). This Equipment Cost/Day is shown in the last column of the Equipment Rental Costs in the Reference Section.

Mobilization, Demobilization—The cost to move construction equipment from an equipment yard or rental company to the job site and back again is not included in equipment costs. Mobilization (to the site) and demobilization (from the site) costs can be found in the Unit Price Section. If a piece of equipment is already at the job site, it is not appropriate to utilize mobilization or demobilization costs again in an estimate.

Overhead and Profit

Total Cost including O&P for the installing contractor is shown in the last column of the Unit Price and/or Assemblies. This figure is the sum of the bare material cost plus 10% for profit, the bare labor cost plus total overhead and profit, and the bare equipment cost plus 10% for profit. Details for the calculation of overhead and profit on labor are shown on the inside back cover of the printed product and in the Reference Section of the electronic product.

General Conditions

Cost data in this data set are presented in two ways: Bare Costs and Total Cost including O&P (Overhead and Profit). General Conditions, or General Requirements, of the contract should also be added to the Total Cost including O&P when applicable. Costs for General Conditions are listed in Division 1 of the Unit Price Section and in the Reference Section.

General Conditions for the installing contractor may range from 0% to 10% of the Total Cost including O&P. For the general or prime contractor, costs for General Conditions may range from 5% to 15% of the Total Cost including O&P, with a figure of 10% as the most typical allowance. If applicable, the Assemblies and Models sections use costs that include the installing contractor's overhead and profit (O&P).

Factors Affecting Costs

Costs can vary depending upon a number of variables. Here's a listing of some factors that affect costs and points to consider.

Quality—The prices for materials and the workmanship upon which productivity is based represent sound construction work. They are also in line with industry standard and manufacturer specifications and are frequently used by federal, state, and local governments.

Overtime—We have made no allowance for overtime. If you anticipate premium time or work beyond normal working hours, be sure to make an appropriate adjustment to your labor costs.

Productivity—The productivity, daily output, and labor-hour figures for each line item are based on an eight-hour work day in daylight hours in moderate temperatures and up to a 14' working height unless otherwise indicated. For work that extends beyond normal work hours or is performed under adverse conditions, productivity may decrease.

Size of Project—The size, scope of work, and type of construction project will have a significant impact on cost. Economies of scale can reduce costs for large projects. Unit costs can often run higher for small projects.

Location—Material prices are for metropolitan areas. However, in dense urban areas, traffic and site storage limitations may increase costs. Beyond a 20-mile radius of metropolitan areas, extra trucking or transportation charges may also increase the material costs slightly. On the other hand, lower wage rates may be in effect. Be sure to consider both of these factors when preparing an estimate, particularly if the job site is located in a central city or remote rural location. In addition, highly specialized subcontract items may require travel and per-diem expenses for mechanics.

Other Factors—

- season of year
- contractor management
- weather conditions
- local union restrictions
- building code requirements
- availability of:
  - adequate energy
  - skilled labor
  - building materials
- owner's special requirements/restrictions
- safety requirements
- environmental considerations
- access

Unpredictable Factors—General business conditions influence "in-place" costs of all items. Substitute materials and construction methods may have to be employed. These may affect the installed cost and/or life cycle costs. Such factors may be difficult to evaluate and cannot necessarily be predicted on the basis of the job's location in a particular section of the country. Thus, where these factors apply, you may find significant but unavoidable cost variations for which you will have to apply a measure of judgment to your estimate.

Rounding of Costs

In printed publications only, all unit prices in excess of $5.00 have been rounded to make them easier to use and still maintain adequate precision of the results.

How Subcontracted Items Affect Costs

A considerable portion of all large construction jobs is usually subcontracted. In fact, the percentage done by subcontractors is constantly increasing and may run over 90%. Since the workers employed by these companies do nothing else but install their particular products, they soon become experts in that line. As a result, installation by these firms is accomplished so efficiently that the total in-place cost, even with the general contractor's overhead and profit, is no more, and often less, than if the principal contractor had handled the installation. Companies that deal with construction specialties are anxious to have their products perform well and, consequently, the installation will be the best possible.

Contingencies

The allowance for contingencies generally provides for unforeseen construction difficulties. On alterations or repair jobs, 20% is not too much. If drawings are final and only field contingencies are being considered, 2% or 3% is probably sufficient and often nothing needs to be added. Contractually, changes in plans will be covered by extras. The contractor should consider inflationary price trends and possible material shortages during the course of the job. These escalation factors are dependent upon both economic conditions and the anticipated time between the estimate and actual construction. If drawings are not complete or approved, or a budget cost is wanted, it is wise to add 5% to 10%. Contingencies, then, are a matter of judgment.

Important Estimating Considerations

The productivity, or daily output, of each craftsman or crew assumes a well-managed job where tradesmen with the proper tools and equipment, along with the appropriate construction materials, are present. Included are daily set-up and cleanup time, break time, and plan layout time. Unless otherwise indicated, time for material movement on site (for items

that can be transported by hand) of up to 200' into the building and to the first or second floor is also included. If material has to be transported by other means, over greater distances, or to higher floors, an additional allowance should be considered by the estimator.

While horizontal movement is typically a sole function of distances, vertical transport introduces other variables that can significantly impact productivity. In an occupied building, the use of elevators (assuming access, size, and required protective measures are acceptable) must be understood at the time of the estimate. For new construction, hoist wait and cycle times can easily be 15 minutes and may result in scheduled access extending beyond the normal work day. Finally, all vertical transport will impose strict weight limits likely to preclude the use of any motorized material handling.

The productivity, or daily output, also assumes installation that meets manufacturer/designer/ standard specifications. A time allowance for quality control checks, minor adjustments, and any task required to ensure proper function or operation is also included. For items that require connections to services, time is included for positioning, leveling, securing the unit, and making all the necessary connections (and start up where applicable) to ensure a complete installation. Estimating of the services themselves (electrical, plumbing, water, steam, hydraulics, dust collection, etc.) is separate.

In some cases, the estimator must consider the use of a crane and an appropriate crew for the installation of large or heavy items. For those situations where a crane is not included in the assigned crew and as part of the line item cost, then equipment rental costs, mobilization and demobilization costs, and operator and support personnel costs must be considered.

Labor-Hours

The labor-hours expressed in this publication are derived by dividing the total daily labor-hours for the crew by the daily output. Based on average installation time and the assumptions listed above, the labor-hours include: direct labor, indirect labor, and nonproductive time. A typical day for a craftsman might include but is not limited to:

- Direct Work
  - ☐ Measuring and layout
  - ☐ Preparing materials
  - ☐ Actual installation
  - ☐ Quality assurance/quality control
- Indirect Work
  - ☐ Reading plans or specifications
  - ☐ Preparing space
  - ☐ Receiving materials
  - ☐ Material movement
  - ☐ Giving or receiving instruction
  - ☐ Miscellaneous
- Non-Work
  - ☐ Chatting
  - ☐ Personal issues
  - ☐ Breaks
  - ☐ Interruptions (i.e., sickness, weather, material or equipment shortages, etc.)

If any of the items for a typical day do not apply to the particular work or project situation, the estimator should make any necessary adjustments.

Final Checklist

Estimating can be a straightforward process provided you remember the basics. Here's a checklist of some of the steps you should remember to complete before finalizing your estimate.

Did you remember to:

- factor in the City Cost Index for your locale?
- take into consideration which items have been marked up and by how much?
- mark up the entire estimate sufficiently for your purposes?
- read the background information on techniques and technical matters that could impact your project time span and cost?
- include all components of your project in the final estimate?
- double check your figures for accuracy?
- call RSMeans data engineers if you have any questions about your estimate or the data you've used? Remember, Gordian stands behind all of our products, including our extensive RSMeans data solutions. If you have any questions about your estimate, about the costs you've used from our data, or even about the technical aspects of the job that may affect your estimate, feel free to call the Gordian RSMeans editors at 1.800.448.8182.

Unit Price Section

Table of Contents

Table of Contents (cont.)

RSMeans data: Unit Prices— How They Work

All RSMeans data: Unit Prices are organized in the same way.

03 30 Cast-In-Place Concrete

03 30 53 – Miscellaneous Cast-In-Place Concrete

03 30 53.40 Concrete In Place		Crew	Daily Output	Labor-Hours	Unit	Material	2021 Bare Costs Labor	Equipment	Total	Total Incl O&P
0010	**CONCRETE IN PLACE**	R033053-10								
0020	Including forms (4 uses), Grade 60 rebar, concrete (Portland cement	R033053-60								
0050	Type I), placement and finishing unless otherwise indicated	R033105-10								
0300	Beams (3500 psi), 5 kip/L.F., 10' span	R033105-20	C-14A	15.62	12.804	C.Y.	420	700	29	1,149	1,550
0350	25' span	R033105-50	"	18.55	10.782		440	590	24.50	1,054.50	1,375
0500	Chimney foundations (5000 psi), over 5 C.Y.	R033105-65	C-14C	32.22	3.476		187	182	.83	369.83	480
0510	(3500 psi), under 5 C.Y.	R033105-70	"	23.71	4.724		223	247	1.13	471.13	615
0700	Columns, square (4000 psi), 12" x 12", up to 1% reinforcing by area	R033105-80	C-14A	11.96	16.722		475	915	37.50	1,427.50	1,950
3540	Equipment pad (3000 psi), 3' x 3' x 6" thick		C-14H	45	1.067	Ea.	55.50	57	.61	113.11	147
3550	4' x 4' x 6" thick			30	1.600		84	85.50	.91	170.41	220
3560	5' x 5' x 8" thick			18	2.667		150	143	1.52	294.52	380
3570	6' x 6' x 8" thick			14	3.429		204	184	1.95	389.95	500
3580	8' x 8' x 10" thick			8	6		415	320	3.42	738.42	945
3590	10' x 10' x 12" thick			5	9.600		720	515	5.45	1,240.45	1,550

It is important to understand the structure of RSMeans data: Unit Prices so that you can find information easily and use it correctly.

---

1 Line Numbers

Line Numbers consist of 12 characters, which identify a unique location in the database for each task. The first 6 or 8 digits conform to the Construction Specifications Institute MasterFormat® 2018. The remainder of the digits are a further breakdown in order to arrange items in understandable groups of similar tasks. Line numbers are consistent across all of our publications, so a line number in any of our products will always refer to the same item of work.

2 Descriptions

Descriptions are shown in a hierarchical structure to make them readable. In order to read a complete description, read up through the indents to the top of the section. Include everything that is above and to the left that is not contradicted by information below. For instance, the complete description for line 03 30 53.40 3550 is "Concrete in place, including forms (4 uses), Grade 60 rebar, concrete (Portland cement Type 1), placement and finishing unless otherwise indicated; Equipment pad (3,000 psi), 4' × 4' × 6" thick."

3 RSMeans data

When using **RSMeans data**, it is important to read through an entire section to ensure that you use the data that most closely matches your work. Note that sometimes there is additional information shown in the section that may improve your price. There are frequently lines that further describe, add to, or adjust data for specific situations.

4 Reference Information

Gordian's RSMeans engineers have created **reference** information to assist you in your estimate. **If** there is information that applies to a section, it will be indicated at the start of the section. The Reference Section is located in the back of the data set.

5 Crews

Crews include labor and/or equipment necessary to accomplish each task. In this case, Crew C-14H is used. Gordian's RSMeans staff selects a crew to represent the workers and equipment that are

typically used for that task. In this case, Crew C-14H consists of one carpenter foreman (outside), two carpenters, one rodman, one laborer, one cement finisher, and one gas engine vibrator. Details of all crews can be found in the Reference Section.

Crews - Standard

Crew C-14H	Bare Costs		Incl. Subs O & P		Cost Per Labor-Hour	
	Hr.	Daily	Hr.	Daily	Bare Costs	Incl. O&P
1 Carpenter Foreman (outside)	$56.70	$453.60	$84.60	$676.80	$53.53	$79.68
2 Carpenters	54.70	875.20	81.65	1306.40		
1 Rodman (reinf.)	58.90	471.20	88.05	704.40		
1 Laborer	44.40	355.20	66.25	530.00		
1 Cement Finisher	51.80	414.40	75.90	607.20		
1 Gas Engine Vibrator		27.15		29.86	.57	.62
48 L.H., Daily Totals		$2596.75		$3854.67	$54.10	$80.31

6 Daily Output

The **Daily Output** is the amount of work that the crew can do in a normal 8-hour workday, including mobilization, layout, movement of materials, and cleanup. In this case, crew C-14H can install thirty 4' × 4' × 6" thick concrete pads in a day. Daily output is variable and based on many factors, including the size of the job, location, and environmental conditions. RSMeans data represents work done in daylight (or adequate lighting) and temperate conditions.

7 Labor-Hours

The figure in the **Labor-Hours** column is the amount of labor required to perform one unit of work–in this case the amount of labor required to construct one 4' × 4' equipment pad. This figure is calculated by dividing the number of hours of labor in the crew by the daily output (48 labor-hours divided by 30 pads = 1.6 hours of labor per pad). Multiply 1.6 times 60 to see the value in minutes: 60 × 1.6 = 96

minutes. Note: the labor-hour figure is not dependent on the crew size. A change in crew size will result in a corresponding change in daily output, but the labor-hours per unit of work will not change.

8 Unit of Measure

All RSMeans data: Unit Prices include the typical **Unit of Measure** used for estimating that item. For concrete-in-place the typical unit is cubic yards (C.Y.) or each (Ea.). For installing broadloom carpet it is square yard and for gypsum board it is square foot. The estimator needs to take special care that the unit in the data matches the unit in the take-off. Unit conversions may be found in the Reference Section.

9 Bare Costs

Bare Costs are the costs of materials, labor, and equipment that the installing contractor pays. They represent the cost, in U.S. dollars, for one unit of work. They do not include any markups for profit or labor burden.

10 Bare Total

The **Total column** represents the total bare cost for the installing contractor in U.S. dollars. In this case, the sum of $84 for material + $85.50 for labor + $.91 for equipment is $170.41.

11 Total Incl O&P

The **Total Incl O&P column** is the total cost, including overhead and profit, that the installing contractor will charge the customer. This represents the cost of materials plus 10% profit, the cost of labor plus labor burden and 10% profit, and the cost of equipment plus 10% profit. It does not include the general contractor's overhead and profit. Note: See the inside back cover of the printed product or the Reference Section of the electronic product for details on how the labor burden is calculated.

National Average

*The RSMeans data in our print publications represent a "national average" cost. This data should be modified to the project location using the **City Cost Indexes** or **Location Factors** tables found in the Reference Section. Use the Location Factors to adjust estimate totals if the project covers multiple trades. Use the City Cost Indexes (CCI) for single trade*

projects or projects where a more detailed analysis is required. All figures in the two tables are derived from the same research. The last row of data in the CCI—the weighted average—is the same as the numbers reported for each location in the location factor table.

Project Name: Pre-Engineered Steel Building		Architect: As Shown						
Location:	**Anywhere, USA**						**01/01/21**	**STD**
Line Number	**Description**	**Qty**	**Unit**	**Material**	**Labor**	**Equipment**	**SubContract**	**Estimate Total**
03 30 53.40 3940	Strip footing, 12" x 24", reinforced	15	C.Y.	$2,625.00	$1,830.00	$8.40	$0.00	
03 30 53.40 3950	Strip footing, 12" x 36", reinforced	34	C.Y.	$5,644.00	$3,315.00	$15.30	$0.00	
03 11 13.65 3000	Concrete slab edge forms	500	L.F.	$225.00	$1,390.00	$0.00	$0.00	
03 22 11.10 0200	Welded wire fabric reinforcing	150	C.S.F.	$3,150.00	$4,575.00	$0.00	$0.00	
03 31 13.35 0300	Ready mix concrete, 4000 psi for slab on grade	278	C.Y.	$35,862.00	$0.00	$0.00	$0.00	
03 31 13.70 4300	Place, strike off & consolidate concrete slab	278	C.Y.	$0.00	$5,560.00	$136.22	$0.00	
03 35 13.30 0250	Machine float & trowel concrete slab	15,000	S.F.	$0.00	$10,350.00	$750.00	$0.00	
03 15 16.20 0140	Cut control joints in concrete slab	950	L.F.	$38.00	$437.00	$57.00	$0.00	
03 39 23.13 0300	Sprayed concrete curing membrane	150	C.S.F.	$1,830.00	$1,125.00	$0.00	$0.00	
Division 03	**Subtotal**			**$49,374.00**	**$28,582.00**	**$966.92**	**$0.00**	**$78,922.92**
08 36 13.10 2650	Manual 10' x 10' steel sectional overhead door	8	Ea.	$9,800.00	$3,880.00	$0.00	$0.00	
08 36 13.10 2860	Insulation and steel back panel for OH door	800	S.F.	$4,760.00	$0.00	$0.00	$0.00	
Division 08	**Subtotal**			**$14,560.00**	**$3,880.00**	**$0.00**	**$0.00**	**$18,440.00**
13 34 19.50 1100	Pre-Engineered Steel Building, 100' x 150' x 24'	15,000	SF Flr.	$0.00	$0.00	$0.00	$367,500.00	
13 34 19.50 6050	Framing for PESB door opening, 3' x 7'	4	Opng.	$0.00	$0.00	$0.00	$2,300.00	
13 34 19.50 6100	Framing for PESB door opening, 10' x 10'	8	Opng.	$0.00	$0.00	$0.00	$9,400.00	
13 34 19.50 6200	Framing for PESB window opening, 4' x 3'	6	Opng.	$0.00	$0.00	$0.00	$3,360.00	
13 34 19.50 5750	PESB door, 3' x 7', single leaf	4	Opng.	$2,740.00	$772.00	$0.00	$0.00	
13 34 19.50 7750	PESB sliding window, 4' x 3' with screen	6	Opng.	$2,940.00	$660.00	$68.70	$0.00	
13 34 19.50 6550	PESB gutter, eave type, 26 ga., painted	300	L.F.	$2,595.00	$906.00	$0.00	$0.00	
13 34 19.50 8650	PESB roof vent, 12" wide x 10' long	15	Ea.	$705.00	$3,615.00	$0.00	$0.00	
13 34 19.50 6900	PESB insulation, vinyl faced, 4" thick	27,400	S.F.	$14,522.00	$10,412.00	$0.00	$0.00	
Division 13	**Subtotal**			**$23,502.00**	**$16,365.00**	**$68.70**	**$382,560.00**	**$422,495.70**
	Subtotal			**$87,436.00**	**$48,827.00**	**$1,035.62**	**$382,560.00**	**$519,858.62**
Division 01	**General Requirements @ 7%**			6,120.52	3,417.89	72.49	26,779.20	
	Estimate Subtotal			$93,556.52	$52,244.89	$1,108.11	$409,339.20	$519,858.62
	Sales Tax @ 5%			4,677.83		55.41	10,233.48	
	Subtotal A			98,234.35	52,244.89	1,163.52	419,572.68	
	GC O & P			9,823.43	26,331.42	116.35	41,957.27	
	Subtotal B			108,057.78	78,576.31	1,279.87	461,529.95	$649,443.91
	Contingency @ 5%							32,472.20
	Subtotal C							$681,916.11
	Bond @ $12/1000 +10% O&P							9,001.29
	Subtotal D							$690,917.40
	Location Adjustment Factor				114.20			98,110.27
	Grand Total							**$789,027.67**

This estimate is based on an interactive spreadsheet. You are free to download it and adjust it to your methodology.
A copy of this spreadsheet is available at **RSMeans.com/2021books.**

Sample Estimate

This sample demonstrates the elements of an estimate, including a tally of the RSMeans data lines and a summary of the markups on a contractor's work to arrive at a total cost to the owner. The Location Factor with RSMeans data is added at the bottom of the estimate to adjust the cost of the work to a specific location.

---

1 Work Performed

The body of the estimate shows the RSMeans data selected, including the line number, a brief description of each item, its take-off unit and quantity, and the bare costs of materials, labor, and equipment. This estimate also includes a column titled "SubContract." This data is taken from the column "Total Incl O&P" and represents the total that a subcontractor would charge a general contractor for the work, including the sub's markup for overhead and profit.

2 Division 1, General Requirements

This is the first division numerically but the last division estimated. Division 1 includes project-wide needs provided by the general contractor. These requirements vary by project but may include temporary facilities and utilities, security, testing, project cleanup, etc. For small projects a percentage can be used—typically between 5% and 15% of project cost. For large projects the costs may be itemized and priced individually.

3 Sales Tax

If the work is subject to state or local sales taxes, the amount must be added to the estimate. Sales tax may be added to material costs, equipment costs, and subcontracted work. In this case, sales tax was added in all three categories. It was assumed that approximately half the subcontracted work would be material cost, so the tax was applied to 50% of the subcontract total.

4 GC O&P

This entry represents the general contractor's markup on material, labor, equipment, and subcontractor costs. Our standard markup on materials, equipment, and subcontracted work is 10%. In this estimate, the markup on the labor performed by the GC's workers uses "Skilled Workers Average" shown in Column F on the table "Installing Contractor's Overhead & Profit," which can be found on the inside back cover of the printed product or in the Reference Section of the electronic product.

5 Contingency

A factor for contingency may be added to any estimate to represent the cost of unknowns that may occur between the time that the estimate is performed and the time the project is constructed. The amount of the allowance will depend on the stage of design at which the estimate is done and the contractor's assessment of the risk involved. Refer to section 01 21 16.50 for contingency allowances.

6 Bonds

Bond costs should be added to the estimate. The figures here represent a typical performance bond, ensuring the owner that if the general contractor does not complete the obligations in the construction contract the bonding company will pay the cost for completion of the work.

7 Location Adjustment

Published prices are based on national average costs. If necessary, adjust the total cost of the project using a location factor from the "Location Factor" table or the "City Cost Index" table. Use location factors if the work is general, covering multiple trades. If the work is by a single trade (e.g., masonry) use the more specific data found in the "City Cost Indexes."

Estimating Tips
01 20 00 Price and Payment Procedures

- Allowances that should be added to estimates to cover contingencies and job conditions that are not included in the national average material and labor costs are shown in Section 01 21.

- When estimating historic preservation projects (depending on the condition of the existing structure and the owner's requirements), a 15–20% contingency or allowance is recommended, regardless of the stage of the drawings.

01 30 00 Administrative Requirements

- Before determining a final cost estimate, it is good practice to review all the items listed in Subdivisions 01 31 and 01 32 to make final adjustments for items that may need customizing to specific job conditions.

- Requirements for initial and periodic submittals can represent a significant cost to the General Requirements of a job. Thoroughly check the submittal specifications when estimating a project to determine any costs that should be included.

01 40 00 Quality Requirements

- All projects will require some degree of quality control. This cost is not included in the unit cost of construction listed in each division. Depending upon the terms of the contract, the various costs of inspection and testing can be the responsibility of either the owner or the contractor. Be sure to include the required costs in your estimate.

01 50 00 Temporary Facilities and Controls

- Barricades, access roads, safety nets, scaffolding, security, and many more requirements for the execution of a safe project are elements of direct cost. These costs can easily be overlooked when preparing an estimate. When looking through the major classifications of this subdivision, determine which items apply to each division in your estimate.

- Construction equipment rental costs can be found in the Reference Section in Section 01 54 33. Operators' wages are not included in equipment rental costs.

- Equipment mobilization and demobilization costs are not included in equipment rental costs and must be considered separately.

- The cost of small tools provided by the installing contractor for his workers is covered in the "Overhead" column on the "Installing Contractor's Overhead and Profit" table that lists labor trades, base rates, and markups. Therefore, it is included in the "Total Incl. O&P" cost of any unit price line item.

01 70 00 Execution and Closeout Requirements

- When preparing an estimate, thoroughly read the specifications to determine the requirements for Contract Closeout. Final cleaning, record documentation, operation and maintenance data, warranties and bonds, and spare parts and maintenance materials can all be elements of cost for the completion of a contract. Do not overlook these in your estimate.

Reference Numbers

Reference numbers are shown at the beginning of some major classifications. These numbers refer to related items in the Reference Section. The reference information may be an estimating procedure, an alternate pricing method, or technical information.

Note: Not all subdivisions listed here necessarily appear. ■

Same Data. Simplified.

Enjoy the convenience and efficiency of accessing your costs anywhere:

- **Skip the multiplier** by setting your location
- **Quickly search,** edit, favorite and share costs
- **Stay on top of price changes** with automatic updates

Discover more at rsmeans.com/online

01 11 Summary of Work

01 11 31 – Professional Consultants

01 11 31.10 Architectural Fees		Crew	Daily Output	Labor-Hours	Unit	Material	2021 Bare Costs Labor	Equipment	Total	Total Incl O&P
0010	**ARCHITECTURAL FEES** R011110-10									
0020	For new construction									
0060	Minimum				Project				4.90%	4.90%
0090	Maximum				"				16%	16%

01 11 31.20 Construction Management Fees

0010	**CONSTRUCTION MANAGEMENT FEES**									
0020	$1,000,000 job, minimum				Project				4.50%	4.50%
0050	Maximum								7.50%	7.50%
0300	$50,000,000 job, minimum								2.50%	2.50%
0350	Maximum				▼				4%	4%

01 11 31.30 Engineering Fees

0010	**ENGINEERING FEES** R011110-30									
1200	Structural, minimum				Project				1%	1%
1300	Maximum				"				2.50%	2.50%

01 11 31.50 Models

0010	**MODELS**									
0500	2 story building, scaled 100' x 200', simple materials and details				Ea.	4,775			4,775	5,250
0510	Elaborate materials and details				"	31,400			31,400	34,500

01 11 31.75 Renderings

0010	**RENDERINGS** Color, matted, 20" x 30", eye level,									
0020	1 building, minimum				Ea.	2,375			2,375	2,600
0050	Average					3,425			3,425	3,775
0100	Maximum					5,600			5,600	6,150
1000	5 buildings, minimum					4,125			4,125	4,550
1100	Maximum					8,250			8,250	9,075
2000	Aerial perspective, color, 1 building, minimum					3,100			3,100	3,425
2100	Maximum					8,550			8,550	9,400
3000	5 buildings, minimum					6,650			6,650	7,300
3100	Maximum				▼	12,400			12,400	13,600

01 21 Allowances

01 21 16 – Contingency Allowances

01 21 16.50 Contingencies

0010	**CONTINGENCIES**, Add to estimate									
0020	Conceptual stage				Project				20%	20%
0050	Schematic stage								15%	15%
0100	Preliminary working drawing stage (Design Dev.)								10%	10%
0150	Final working drawing stage				▼				3%	3%

01 21 53 – Factors Allowance

01 21 53.60 Security Factors

0010	**SECURITY FACTORS** R012153-60									
0100	Additional costs due to security requirements									
0110	Daily search of personnel, supplies, equipment and vehicles									
0120	Physical search, inventory and doc of assets, at entry				Costs	30%				
0130	At entry and exit					50%				
0140	Physical search, at entry					6.25%				
0150	At entry and exit					12.50%				
0160	Electronic scan search, at entry					2%				
0170	At entry and exit				▼	4%				

01 21 Allowances

01 21 53 – Factors Allowance

01 21 53.60 Security Factors

		Crew	Daily Output	Labor-Hours	Unit	Material	2021 Bare Costs Labor	Equipment	Total	Total Incl O&P
0180	Visual inspection only, at entry				Costs		.25%			
0190	At entry and exit						.50%			
0200	ID card or display sticker only, at entry						.12%			
0210	At entry and exit				▼		.25%			
0220	Day 1 as described below, then visual only for up to 5 day job duration									
0230	Physical search, inventory and doc of assets, at entry				Costs		5%			
0240	At entry and exit						10%			
0250	Physical search, at entry						1.25%			
0260	At entry and exit						2.50%			
0270	Electronic scan search, at entry						.42%			
0280	At entry and exit				▼		.83%			
0290	Day 1 as described below, then visual only for 6-10 day job duration									
0300	Physical search, inventory and doc of assets, at entry				Costs		2.50%			
0310	At entry and exit						5%			
0320	Physical search, at entry						.63%			
0330	At entry and exit						1.25%			
0340	Electronic scan search, at entry						.21%			
0350	At entry and exit				▼		.42%			
0360	Day 1 as described below, then visual only for 11-20 day job duration									
0370	Physical search, inventory and doc of assets, at entry				Costs		1.25%			
0380	At entry and exit						2.50%			
0390	Physical search, at entry						.31%			
0400	At entry and exit						.63%			
0410	Electronic scan search, at entry						.10%			
0420	At entry and exit				▼		.21%			
0430	Beyond 20 days, costs are negligible									
0440	Escort required to be with tradesperson during work effort				Costs		6.25%			

01 21 53.65 Infectious Disease Precautions

		Crew	Daily Output	Labor-Hours	Unit	Material	2021 Bare Costs Labor	Equipment	Total	Total Incl O&P
0010	**INFECTIOUS DISEASE PRECAUTIONS** R012153-65									
0100	Additional costs due to infectious disease precautions									
0120	Daily temperature checks				Costs		1%			
0130	Donning & doffing masks and gloves						1%			
0140	Washing hands						1%			
0150	Informational meetings						2%			
0160	Maintaining social distance						1%			
0170	Disinfecting tools or equipment				▼		3%			
0200	N95 rated masks				Ea.	1.40			1.40	1.54
0210	Surgical masks				"	.16			.16	.18
0220	Black nitrile disposable gloves				Pair	.18			.18	.20
0250	Hand washing station & 2 x weekly service				Week				158	174

01 21 55 – Job Conditions Allowance

01 21 55.50 Job Conditions

		Crew	Daily Output	Labor-Hours	Unit	Material	2021 Bare Costs Labor	Equipment	Total	Total Incl O&P
0010	**JOB CONDITIONS** Modifications to applicable									
0020	cost summaries									
0100	Economic conditions, favorable, deduct				Project				2%	2%
0200	Unfavorable, add								5%	5%
0300	Hoisting conditions, favorable, deduct								2%	2%
0400	Unfavorable, add								5%	5%
0700	Labor availability, surplus, deduct								1%	1%
0800	Shortage, add								10%	10%
0900	Material storage area, available, deduct								1%	1%
1000	Not available, add				▼				2%	2%

For customer support on your Concrete & Masonry Costs with RSMeans Data, call 800.448.8182.

11

01 21 Allowances

01 21 55 – Job Conditions Allowance

01 21 55.50 Job Conditions	Crew	Daily Output	Labor-Hours	Unit	Material	2021 Bare Costs Labor	Equipment	Total	Total Incl O&P
1100	Subcontractor availability, surplus, deduct				Project				5%	5%
1200	Shortage, add								12%	12%
1300	Work space, available, deduct								2%	2%
1400	Not available, add								5%	5%

01 21 57 – Overtime Allowance

01 21 57.50 Overtime

| 0010 | **OVERTIME** for early completion of projects or where | R012909-90 | | | | | | | |
| 0020 | labor shortages exist, add to usual labor, up to | | | | Costs | | 100% | | | |

01 21 63 – Taxes

01 21 63.10 Taxes

0010	**TAXES**	R012909-80								
0020	Sales tax, State, average				%	5.08%				
0050	Maximum	R012909-85				7.50%				
0200	Social Security, on first $118,500 of wages						7.65%			
0300	Unemployment, combined Federal and State, minimum	R012909-86					.60%			
0350	Average						9.60%			
0400	Maximum						12%			

01 31 Project Management and Coordination

01 31 13 – Project Coordination

01 31 13.20 Field Personnel

		Crew	Daily Output	Labor-Hours	Unit	Material	Labor	Equipment	Total	Total Incl O&P
0010	**FIELD PERSONNEL**									
0020	Clerk, average				Week		495		495	750
0100	Field engineer, junior engineer						1,241		1,241	1,877
0120	Engineer						1,825		1,825	2,775
0140	Senior engineer						2,400		2,400	3,625
0160	General purpose laborer, average	1 Clab	.20	40			1,775		1,775	2,650
0180	Project manager, minimum						2,175		2,175	3,300
0200	Average						2,500		2,500	3,800
0220	Maximum						2,850		2,850	4,325
0240	Superintendent, minimum						2,125		2,125	3,225
0260	Average						2,325		2,325	3,525
0280	Maximum						2,650		2,650	4,025
0290	Timekeeper, average						1,350		1,350	2,050

01 31 13.30 Insurance

0010	**INSURANCE**	R013113-40								
0020	Builders risk, standard, minimum				Job				.24%	.24%
0050	Maximum	R013113-50							.80%	.80%
0200	All-risk type, minimum								.25%	.25%
0250	Maximum	R013113-60							.62%	.62%
0400	Contractor's equipment floater, minimum				Value				.50%	.50%
0450	Maximum				"				1.50%	1.50%
0600	Public liability, average				Job				2.02%	2.02%
0800	Workers' compensation & employer's liability, average									
0850	by trade, carpentry, general				Payroll		11.97%			
0900	Clerical						.38%			
0950	Concrete						10.84%			
1000	Electrical						4.91%			
1050	Excavation						7.81%			
1100	Glazing						11.29%			

12

01 31 13.30 Insurance

		Crew	Daily Output	Labor-Hours	Unit	Material	2021 Bare Costs Labor	2021 Bare Costs Equipment	Total	Total Incl O&P
1150	Insulation				Payroll		10.07%			
1200	Lathing						7.58%			
1250	Masonry						13.40%			
1300	Painting & decorating						10.44%			
1350	Pile driving						12.06%			
1400	Plastering						10.24%			
1450	Plumbing						5.77%			
1500	Roofing						27.34%			
1550	Sheet metal work (HVAC)						7.56%			
1600	Steel erection, structural						17.21%			
1650	Tile work, interior ceramic						8.10%			
1700	Waterproofing, brush or hand caulking						6.05%			
1800	Wrecking						15.48%			
2000	Range of 35 trades in 50 states, excl. wrecking & clerical, min.						1.37%			
2100	Average						10.60%			
2200	Maximum						120.29%			

01 31 13.40 Main Office Expense

		Crew	Daily Output	Labor-Hours	Unit	Material	2021 Bare Costs Labor	2021 Bare Costs Equipment	Total	Total Incl O&P
0010	**MAIN OFFICE EXPENSE** Average for General Contractors R013113-50									
0020	As a percentage of their annual volume									
0125	Annual volume under $1,000,000				% Vol.				17.50%	
0145	Up to $2,500,000								8%	
0150	Up to $4,000,000								6.80%	
0200	Up to $7,000,000								5.60%	
0250	Up to $10,000,000								5.10%	
0300	Over $10,000,000								3.90%	

01 31 13.50 General Contractor's Mark-Up

		Crew	Daily Output	Labor-Hours	Unit	Material	2021 Bare Costs Labor	2021 Bare Costs Equipment	Total	Total Incl O&P
0010	**GENERAL CONTRACTOR'S MARK-UP** on Change Orders									
0200	Extra work, by subcontractors, add				%				10%	10%
0250	By General Contractor, add								15%	15%
0400	Omitted work, by subcontractors, deduct all but								5%	5%
0450	By General Contractor, deduct all but								7.50%	7.50%
0600	Overtime work, by subcontractors, add								15%	15%
0650	By General Contractor, add								10%	10%

01 31 13.80 Overhead and Profit

		Crew	Daily Output	Labor-Hours	Unit	Material	2021 Bare Costs Labor	2021 Bare Costs Equipment	Total	Total Incl O&P
0010	**OVERHEAD & PROFIT** Allowance to add to items in this R013113-50									
0020	book that do not include Subs O&P, average				%				25%	
0100	Allowance to add to items in this book that R013113-55									
0110	do include Subs O&P, minimum				%				5%	5%
0150	Average								10%	10%
0200	Maximum								15%	15%
0300	Typical, by size of project, under $100,000								30%	
0350	$500,000 project								25%	
0400	$2,000,000 project								20%	
0450	Over $10,000,000 project								15%	

01 31 13.90 Performance Bond

		Crew	Daily Output	Labor-Hours	Unit	Material	2021 Bare Costs Labor	2021 Bare Costs Equipment	Total	Total Incl O&P
0010	**PERFORMANCE BOND** R013113-80									
0020	For buildings, minimum				Job				.60%	.60%
0100	Maximum				"				2.50%	2.50%

For customer support on your Concrete & Masonry Costs with RSMeans Data, call 800.448.8182.

13

01 31 Project Management and Coordination

01 31 14 – Facilities Services Coordination

01 31 14.20 Lock Out/Tag Out		Crew	Daily Output	Labor-Hours	Unit	Material	2021 Bare Costs Labor	Equipment	Total	Total Incl O&P
0010	**LOCK OUT/TAG OUT**									
0020	Miniature circuit breaker lock out device	1 Elec	220	.036	Ea.	24	2.32		26.32	30
0030	Miniature pin circuit breaker lock out device		220	.036		19.90	2.32		22.22	25.50
0040	Single circuit breaker lock out device		220	.036		21	2.32		23.32	26.50
0050	Multi-pole circuit breaker lock out device (15 to 225 Amp)		210	.038		19.90	2.43		22.33	25.50
0060	Large 3 pole circuit breaker lock out device (over 225 Amp)		210	.038		21	2.43		23.43	27
0080	Square D I-Line circuit breaker lock out device		210	.038		33	2.43		35.43	39.50
0090	Lock out disconnect switch, 30 to 100 Amp		330	.024		11	1.54		12.54	14.40
0100	100 to 400 Amp		330	.024		11	1.54		12.54	14.40
0110	Over 400 Amp		330	.024		11	1.54		12.54	14.40
0120	Lock out hasp for multiple lockout tags		200	.040		5.60	2.55		8.15	9.95
0130	Electrical cord plug lock out device		220	.036		10.50	2.32		12.82	15
0140	Electrical plug prong lock out device (3-wire grounding plug)		220	.036		6.95	2.32		9.27	11.10
0150	Wall switch lock out		200	.040		25	2.55		27.55	32
0160	Fire alarm pull station lock out	1 Stpi	200	.040		18.75	2.73		21.48	24.50
0170	Sprinkler valve tamper and flow switch lock out device	1 Skwk	220	.036		16.60	2.08		18.68	21.50
0180	Lock out sign		330	.024		18.90	1.38		20.28	23
0190	Lock out tag		440	.018		5	1.04		6.04	7.05

01 32 Construction Progress Documentation

01 32 13 – Scheduling of Work

01 32 13.50 Scheduling of Work

0010	**SCHEDULING**									
0025	Scheduling, critical path, $50 million project, initial schedule				Ea.				25,750	28,325
0030	Monthly updates				"				3,348	3,682

01 32 33 – Photographic Documentation

01 32 33.50 Photographs

0010	**PHOTOGRAPHS**									
0020	8" x 10", 4 shots, 2 prints ea., std. mounting				Set	545			545	600
0100	Hinged linen mounts					550			550	605
0200	8" x 10", 4 shots, 2 prints each, in color					575			575	630
0300	For I.D. slugs, add to all above					5.10			5.10	5.60
0500	Aerial photos, initial fly-over, 5 shots, digital images					350			350	385
0550	10 shots, digital images, 1 print					595			595	655
0600	For each additional print from fly-over					310			310	340
0700	For full color prints, add					40%				
0750	Add for traffic control area				Ea.	360			360	395
0900	For over 30 miles from airport, add per				Mile	7.20			7.20	7.90
1500	Time lapse equipment, camera and projector, buy				Ea.	2,600			2,600	2,850
1550	Rent per month				"	1,325			1,325	1,450
1700	Cameraman and processing, black & white				Day	1,300			1,300	1,425
1720	Color				"	1,525			1,525	1,675

01 41 Regulatory Requirements

01 41 26 – Permit Requirements

01 41 26.50 Permits		Crew	Daily Output	Labor-Hours	Unit	Material	2021 Bare Costs Labor	Equipment	Total	Total Incl O&P
0010	**PERMITS**									
0020	Rule of thumb, most cities, minimum				Job				.50%	.50%
0100	Maximum				"				2%	2%

01 45 Quality Control

01 45 23 – Testing and Inspecting Services

01 45 23.50 Testing		Crew	Daily Output	Labor-Hours	Unit	Material	2021 Bare Costs Labor	Equipment	Total	Total Incl O&P
0010	**TESTING** and Inspecting Services									
0012	Testing, for concrete or steel building, minimum				Week				4,585	5,045
0025	Maximum				"				6,005	6,605
0600	Concrete testing, aggregates, abrasion, ASTM C 131				Ea.				205	225
0650	Absorption, ASTM C 127								77	85
0800	Petrographic analysis, ASTM C 295								910	1,000
0900	Specific gravity, ASTM C 127								77	85
1000	Sieve analysis, washed, ASTM C 136								130	140
1050	Unwashed								130	140
1200	Sulfate soundness								182	200
1300	Weight per cubic foot								80	88
1500	Cement, physical tests, ASTM C 150								320	350
1600	Chemical tests, ASTM C 150								245	270
1800	Compressive test, cylinder, delivered to lab, ASTM C 39								36	40
1900	Picked up by lab, 30 minute round trip, add to above	1 Skwk	16	.500			28.50		28.50	43
1950	1 hour round trip, add to above		8	1			57		57	86
2000	2 hour round trip, add to above		4	2			114		114	172
2200	Compressive strength, cores (not incl. drilling), ASTM C 42								95	105
2300	Patching core holes, to 12" diameter	1 Cefi	22	.364		35	18.85		53.85	66
2400	Drying shrinkage at 28 days								236	260
2500	Flexural test beams, ASTM C 78								136	150
2600	Mix design, one batch mix								259	285
2650	Added trial batches								120	132
2800	Modulus of elasticity, ASTM C 469								195	215
2900	Tensile test, cylinders, ASTM C 496								52	58
3000	Water-Cement ratio curve, 3 batches								141	155
3100	4 batches								186	205
3300	Masonry testing, absorption, per 5 brick, ASTM C 67								72	80
3350	Chemical resistance, per 2 brick								50	55
3400	Compressive strength, per 5 brick, ASTM C 67								95	105
3420	Efflorescence, per 5 brick, ASTM C 67								101	112
3440	Imperviousness, per 5 brick								87	96
3470	Modulus of rupture, per 5 brick								86	95
3500	Moisture, block only								68	75
3550	Mortar, compressive strength, set of 3								32	35
4100	Reinforcing steel, bend test								68	75
4200	Tensile test, up to #8 bar								68	75
4220	#9 to #11 bar								114	125
4240	#14 bar and larger								182	200
4400	Soil testing, Atterberg limits, liquid and plastic limits								91	100
4510	Hydrometer analysis								155	170
4600	Sieve analysis, washed, ASTM D 422								136	150
4710	Consolidation test (ASTM D 2435), minimum								455	500
4730	Soil testing									

For customer support on your Concrete & Masonry Costs with RSMeans Data, call 800.448.8182.

15

01 45 Quality Control

01 45 23 – Testing and Inspecting Services

01 45 23.50 Testing	Crew	Daily Output	Labor-Hours	Unit	Material	2021 Bare Costs Labor	Equipment	Total	Total Incl O&P
4750	Moisture content, ASTM D 2216				Ea.				18	20
4780	Permeability test, double ring infiltrometer								500	550
4800	Permeability, var. or constant head, undist., ASTM D 2434								264	290
4850	Recompacted								250	275
4900	Proctor compaction, 4" standard mold, ASTM D 698								230	253
4950	6" modified mold								110	120
5100	Shear tests, triaxial, minimum								410	450
5150	Maximum								545	600
5300	Direct shear, minimum, ASTM D 3080								320	350
5350	Maximum								410	450
5550	Technician for inspection, per day, earthwork								556	612
5570	Concrete								556	612
5790	Welding								556	612
6000	Welding certification, minimum								91	100
6100	Maximum								364	400
7000	Underground storage tank									
7520	Volumetric tightness, <= 12,000 gal.				Ea.	470			470	515
7530	12,000 – 29,999 gal.					585			585	645
7540	>= 30,000 gal.					665			665	730
7600	Vadose zone (soil gas) sampling, 10-40 samples, min.				Day				1,375	1,500
7610	Maximum				"				2,275	2,500
7700	Ground water monitoring incl. drilling 3 wells, min.				Total				4,550	5,000
7710	Maximum				"				6,375	7,000
8000	X-ray concrete slabs				Ea.				182	200

01 51 Temporary Utilities

01 51 13 – Temporary Electricity

01 51 13.80 Temporary Utilities

		Crew	Daily Output	Labor-Hours	Unit	Material	Labor	Equipment	Total	Total Incl O&P
0010	**TEMPORARY UTILITIES**									
0350	Lighting, lamps, wiring, outlets, 40,000 S.F. building, 8 strings	1 Elec	34	.235	CSF Flr	3.55	15		18.55	26.50
0360	16 strings	"	17	.471		7.10	30		37.10	52.50
0400	Power for temp lighting only, 6.6 KWH, per month								.92	1.01
0430	11.8 KWH, per month								1.65	1.82
0450	23.6 KWH, per month								3.30	3.63
0600	Power for job duration incl. elevator, etc., minimum								53	58
0650	Maximum								110	121
0675	Temporary cooling				Ea.	1,050			1,050	1,150

01 52 Construction Facilities

01 52 13 – Field Offices and Sheds

01 52 13.20 Office and Storage Space

		Crew	Daily Output	Labor-Hours	Unit	Material	Labor	Equipment	Total	Total Incl O&P
0010	**OFFICE AND STORAGE SPACE**									
0020	Office trailer, furnished, no hookups, 20' x 8', buy	2 Skwk	1	16	Ea.	10,000	915		10,915	12,400
0250	Rent per month					192			192	211
0300	32' x 8', buy	2 Skwk	.70	22.857		15,000	1,300		16,300	18,500
0350	Rent per month					243			243	267
0400	50' x 10', buy	2 Skwk	.60	26.667		30,500	1,525		32,025	35,800
0450	Rent per month					355			355	390
0500	50' x 12', buy	2 Skwk	.50	32		25,800	1,825		27,625	31,200

01 52 Construction Facilities

01 52 13 – Field Offices and Sheds

01 52 13.20 Office and Storage Space	Crew	Daily Output	Labor-Hours	Unit	Material	2021 Bare Costs Labor	Equipment	Total	Total Incl O&P
0550	Rent per month				Ea.	450			450	500
0700	For air conditioning, rent per month, add					53			53	58.50
0800	For delivery, add per mile				Mile	11.95			11.95	13.15
0900	Bunk house trailer, 8' x 40' duplex dorm with kitchen, no hookups, buy	2 Carp	1	16	Ea.	89,000	875		89,875	99,500
0910	9 man with kitchen and bath, no hookups, buy		1	16		91,500	875		92,375	102,000
0920	18 man sleeper with bath, no hookups, buy		1	16		98,500	875		99,375	110,000
1000	Portable buildings, prefab, on skids, economy, 8' x 8'		265	.060	S.F.	29.50	3.30		32.80	37.50
1100	Deluxe, 8' x 12'		150	.107	"	33.50	5.85		39.35	45.50
1200	Storage boxes, 20' x 8', buy	2 Skwk	1.80	8.889	Ea.	3,100	510		3,610	4,200
1250	Rent per month					88			88	96.50
1300	40' x 8', buy	2 Skwk	1.40	11.429		4,600	655		5,255	6,050
1350	Rent per month					135			135	148

01 52 13.40 Field Office Expense

		Crew	Daily Output	Labor-Hours	Unit	Material	2021 Bare Costs Labor	Equipment	Total	Total Incl O&P
0010	**FIELD OFFICE EXPENSE**									
0100	Office equipment rental average				Month	228			228	251
0120	Office supplies, average				"	90.50			90.50	99.50
0125	Office trailer rental, see Section 01 52 13.20									
0140	Telephone bill; avg. bill/month incl. long dist.				Month	87			87	96
0160	Lights & HVAC				"	163			163	179

01 54 Construction Aids

01 54 09 – Protection Equipment

01 54 09.60 Safety Nets

		Crew	Daily Output	Labor-Hours	Unit	Material	2021 Bare Costs Labor	Equipment	Total	Total Incl O&P
0010	**SAFETY NETS**									
0020	No supports, stock sizes, nylon, 3-1/2" mesh				S.F.	3.16			3.16	3.48
0100	Polypropylene, 6" mesh					1.63			1.63	1.79
0200	Small mesh debris nets, 1/4" mesh, stock sizes					.57			.57	.63
0220	Combined 3-1/2" mesh and 1/4" mesh, stock sizes					4.92			4.92	5.40
0300	Rental, 4" mesh, stock sizes, 3 months					.88			.88	.97
0320	6 month rental					1.03			1.03	1.13
0340	12 months					1.42			1.42	1.56

01 54 16 – Temporary Hoists

01 54 16.50 Weekly Forklift Crew

		Crew	Daily Output	Labor-Hours	Unit	Material	2021 Bare Costs Labor	Equipment	Total	Total Incl O&P
0010	**WEEKLY FORKLIFT CREW**									
0100	All-terrain forklift, 45' lift, 35' reach, 9000 lb. capacity	A-3P	.20	40	Week		2,225	1,725	3,950	5,200

01 54 19 – Temporary Cranes

01 54 19.50 Daily Crane Crews

		Crew	Daily Output	Labor-Hours	Unit	Material	2021 Bare Costs Labor	Equipment	Total	Total Incl O&P
0010	**DAILY CRANE CREWS** for small jobs, portal to portal	R015433-15								
0100	12-ton truck-mounted hydraulic crane	A-3H	1	8	Day		490	735	1,225	1,525
0200	25-ton	A-3I	1	8			490	810	1,300	1,625
0300	40-ton	A-3J	1	8			490	1,275	1,765	2,150
0400	55-ton	A-3K	1	16			910	1,525	2,435	3,025
0500	80-ton	A-3L	1	16			910	2,200	3,110	3,775
0900	If crane is needed on a Saturday, Sunday or Holiday									
0910	At time-and-a-half, add				Day		50%			
0920	At double time, add				"		100%			

For customer support on your Concrete & Masonry Costs with RSMeans Data, call 800.448.8182.

17

01 54 Construction Aids

01 54 19 – Temporary Cranes

01 54 19.60 Monthly Tower Crane Crew	Crew	Daily Output	Labor-Hours	Unit	Material	2021 Bare Costs Labor	Equipment	Total	Total Incl O&P
0010	**MONTHLY TOWER CRANE CREW**, excludes concrete footing									
0100	Static tower crane, 130' high, 106' jib, 6200 lb. capacity	A-3N	.05	176	Month		10,800	38,200	49,000	58,000

01 54 23 – Temporary Scaffolding and Platforms

01 54 23.70 Scaffolding

		Crew	Daily Output	Labor-Hours	Unit	Material	Labor	Equipment	Total	Total Incl O&P
0010	**SCAFFOLDING** R015423-10									
0906	Complete system for face of walls, no plank, material only rent/mo				C.S.F.	81.50			81.50	90
0910	Steel tubular, heavy duty shoring, buy									
0920	Frames 5' high 2' wide				Ea.	92.50			92.50	102
0925	5' high 4' wide					118			118	130
0930	6' high 2' wide					120			120	132
0935	6' high 4' wide					119			119	131
0940	Accessories									
0945	Cross braces				Ea.	19.20			19.20	21
0950	U-head, 8" x 8"					23			23	25
0955	J-head, 4" x 8"					16.70			16.70	18.35
0960	Base plate, 8" x 8"					18.05			18.05	19.90
0965	Leveling jack					36			36	39.50
1000	Steel tubular, regular, buy									
1100	Frames 3' high 5' wide				Ea.	54			54	59.50
1150	5' high 5' wide					71			71	78
1200	6'-4" high 5' wide					83.50			83.50	92
1350	7'-6" high 6' wide					156			156	171
1500	Accessories, cross braces					19.20			19.20	21
1550	Guardrail post					24			24	26.50
1600	Guardrail 7' section					8.15			8.15	9
1650	Screw jacks & plates					26			26	28.50
1700	Sidearm brackets					26			26	29
1750	8" casters					37.50			37.50	41
1800	Plank 2" x 10" x 16'-0"					66.50			66.50	73.50
1900	Stairway section					298			298	330
1910	Stairway starter bar					32.50			32.50	35.50
1920	Stairway inside handrail					57.50			57.50	63.50
1930	Stairway outside handrail					91.50			91.50	101
1940	Walk-thru frame guardrail					42			42	46.50
2000	Steel tubular, regular, rent/mo.									
2100	Frames 3' high 5' wide				Ea.	9.90			9.90	10.90
2150	5' high 5' wide					9.90			9.90	10.90
2200	6'-4" high 5' wide					5.75			5.75	6.35
2250	7'-6" high 6' wide					9.90			9.90	10.90
2500	Accessories, cross braces					3.35			3.35	3.69
2550	Guardrail post					3.25			3.25	3.58
2600	Guardrail 7' section					11.10			11.10	12.20
2650	Screw jacks & plates					4.35			4.35	4.79
2700	Sidearm brackets					5.20			5.20	5.70
2750	8" casters					7.85			7.85	8.65
2800	Outrigger for rolling tower					9.10			9.10	10
2850	Plank 2" x 10" x 16'-0"					9.90			9.90	10.90
2900	Stairway section					32			32	35.50
2940	Walk-thru frame guardrail					8.25			8.25	9.10
3000	Steel tubular, heavy duty shoring, rent/mo.									
3250	5' high 2' & 4' wide				Ea.	8.45			8.45	9.25
3300	6' high 2' & 4' wide					8.45			8.45	9.25

01 54 Construction Aids

01 54 23 – Temporary Scaffolding and Platforms

01 54 23.70 Scaffolding

		Crew	Daily Output	Labor-Hours	Unit	Material	2021 Bare Costs Labor	2021 Bare Costs Equipment	Total	Total Incl O&P
3500	Accessories, cross braces				Ea.	.91			.91	1
3600	U-head, 8" x 8"					2.49			2.49	2.74
3650	J-head, 4" x 8"					2.49			2.49	2.74
3700	Base plate, 8" x 8"					.91			.91	1
3750	Leveling jack					2.47			2.47	2.72
4000	Scaffolding, stl. tubular, reg., no plank, labor only to erect & dismantle									
4100	Building exterior 2 stories	3 Carp	8	3	C.S.F.		164		164	245
4150	4 stories	"	8	3			164		164	245
4200	6 stories	4 Carp	8	4			219		219	325
4250	8 stories		8	4			219		219	325
4300	10 stories		7.50	4.267			233		233	350
4350	12 stories		7.50	4.267			233		233	350
5700	Planks, 2" x 10" x 16'-0", labor only to erect & remove to 50' H	3 Carp	72	.333	Ea.		18.25		18.25	27
5800	Over 50' high	4 Carp	80	.400	"		22		22	32.50
6000	Heavy duty shoring for elevated slab forms to 8'-2" high, floor area									
6100	Labor only to erect & dismantle	4 Carp	16	2	C.S.F.		109		109	163
6110	Materials only, rent/mo.				"	43.50			43.50	48
6500	To 14'-8" high									
6600	Labor only to erect & dismantle	4 Carp	10	3.200	C.S.F.		175		175	261
6610	Materials only, rent/mo				"	64			64	70

01 54 23.75 Scaffolding Specialties

		Crew	Daily Output	Labor-Hours	Unit	Material	2021 Bare Costs Labor	2021 Bare Costs Equipment	Total	Total Incl O&P
0010	**SCAFFOLDING SPECIALTIES**									
1200	Sidewalk bridge, heavy duty steel posts & beams, including									
1210	parapet protection & waterproofing (material cost is rent/month)									
1220	8' to 10' wide, 2 posts	3 Carp	15	1.600	L.F.	65.50	87.50		153	203
1230	3 posts		10	2.400		96.50	131		227.50	300
1510	planking (material cost is rent/month)		45	.533		9.80	29		38.80	54.50
1600	For 2 uses per month, deduct from all above					50%				
1700	For 1 use every 2 months, add to all above					100%				
1900	Catwalks, 20" wide, no guardrails, 7' span, buy				Ea.	148			148	163
2000	10' span, buy					207			207	228
3720	Putlog, standard, 8' span, with hangers, buy					78.50			78.50	86
3730	Rent per month					16.30			16.30	17.90
3750	12' span, buy					100			100	110
3755	Rent per month					20.50			20.50	22.50
3760	Trussed type, 16' span, buy					254			254	280
3770	Rent per month					24.50			24.50	27
3790	22' span, buy					287			287	315
3795	Rent per month					32.50			32.50	35.50
3800	Rolling ladders with handrails, 30" wide, buy, 2 step					289			289	320
4000	7 step					1,125			1,125	1,225
4050	10 step					1,375			1,375	1,500
4100	Rolling towers, buy, 5' wide, 7' long, 10' high					1,250			1,250	1,375
4200	For additional 5' high sections, to buy					180			180	198
4300	Complete incl. wheels, railings, outriggers,									
4350	21' high, to buy				Ea.	2,050			2,050	2,250
4400	Rent/month = 5% of purchase cost				"	103			103	113

01 54 26 – Temporary Swing Staging

01 54 26.50 Swing Staging

		Crew	Daily Output	Labor-Hours	Unit	Material	2021 Bare Costs Labor	2021 Bare Costs Equipment	Total	Total Incl O&P
0010	**SWING STAGING**, 500 lb. cap., 2' wide to 24' long, hand operated									
0020	Steel cable type, with 60' cables, buy				Ea.	6,425			6,425	7,075
0030	Rent per month				"	645			645	705

01 54 Construction Aids

01 54 26 – Temporary Swing Staging

01 54 26.50 Swing Staging	Crew	Daily Output	Labor-Hours	Unit	Material	2021 Bare Costs Labor	2021 Bare Costs Equipment	Total	Total Incl O&P
0600	Lightweight (not for masons) 24' long for 150' height,									
0610	manual type, buy				Ea.	13,200			13,200	14,500
0620	Rent per month					1,325			1,325	1,450
0700	Powered, electric or air, to 150' high, buy					34,200			34,200	37,600
0710	Rent per month					2,400			2,400	2,625
0780	To 300' high, buy					34,700			34,700	38,200
0800	Rent per month					2,425			2,425	2,675
1000	Bosun's chair or work basket 3' x 3'6", to 300' high, electric, buy					13,600			13,600	15,000
1010	Rent per month					950			950	1,050
2200	Move swing staging (setup and remove)	E-4	2	16	Move		975	74.50	1,049.50	1,575

01 54 36 – Equipment Mobilization

01 54 36.50 Mobilization

		Crew	Daily Output	Labor-Hours	Unit	Material	2021 Bare Costs Labor	2021 Bare Costs Equipment	Total	Total Incl O&P
0010	**MOBILIZATION** (Use line item again for demobilization)	R015436-50								
0015	Up to 25 mi. haul dist. (50 mi. RT for mob/demob crew)									
1200	Small equipment, placed in rear of, or towed by pickup truck	A-3A	4	2	Ea.		111	44	155	214
1300	Equipment hauled on 3-ton capacity towed trailer	A-3Q	2.67	3			167	93	260	350
1400	20-ton capacity	B-34U	2	8			425	224	649	885
1500	40-ton capacity	B-34N	2	8			440	345	785	1,050
1600	50-ton capacity	B-34V	1	24			1,350	995	2,345	3,100
1700	Crane, truck-mounted, up to 75 ton (driver only)	1 Eqhv	4	2			123		123	183
1800	Over 75 ton (with chase vehicle)	A-3E	2.50	6.400			360	70.50	430.50	620
2400	Crane, large lattice boom, requiring assembly	B-34W	.50	144			7,775	7,075	14,850	19,400
2500	For each additional 5 miles haul distance, add						10%	10%		
3000	For large pieces of equipment, allow for assembly/knockdown									
3001	For mob/demob of vibroflotation equip, see Section 31 45 13.10									
3100	For mob/demob of micro-tunneling equip, see Section 33 05 07.36									
3200	For mob/demob of pile driving equip, see Section 31 62 19.10									
3300	For mob/demob of caisson drilling equip, see Section 31 63 26.13									

01 55 Vehicular Access and Parking

01 55 23 – Temporary Roads

01 55 23.50 Roads and Sidewalks

		Crew	Daily Output	Labor-Hours	Unit	Material	2021 Bare Costs Labor	2021 Bare Costs Equipment	Total	Total Incl O&P
0010	**ROADS AND SIDEWALKS** Temporary									
0050	Roads, gravel fill, no surfacing, 4" gravel depth	B-14	715	.067	S.Y.	5.35	3.13	.30	8.78	10.85
0100	8" gravel depth	"	615	.078	"	10.65	3.64	.35	14.64	17.55
1000	Ramp, 3/4" plywood on 2" x 6" joists, 16" OC	2 Carp	300	.053	S.F.	2.44	2.92		5.36	7.05
1100	On 2" x 10" joists, 16" OC	"	275	.058	"	3.39	3.18		6.57	8.50

01 56 Temporary Barriers and Enclosures

01 56 13 – Temporary Air Barriers

01 56 13.60 Tarpaulins

		Crew	Daily Output	Labor-Hours	Unit	Material	2021 Bare Costs Labor	2021 Bare Costs Equipment	Total	Total Incl O&P
0010	**TARPAULINS**									
0020	Cotton duck, 10-13.13 oz./S.Y., 6' x 8'				S.F.	.74			.74	.81
0050	30' x 30'					.62			.62	.68
0100	Polyvinyl coated nylon, 14-18 oz., minimum					1.46			1.46	1.61
0150	Maximum					1.46			1.46	1.61
0200	Reinforced polyethylene 3 mils thick, white					.06			.06	.07
0300	4 mils thick, white, clear or black					.17			.17	.19
0400	5.5 mils thick, clear					.16			.16	.18

01 56 Temporary Barriers and Enclosures

01 56 13 – Temporary Air Barriers

01 56 13.60 Tarpaulins

		Crew	Daily Output	Labor-Hours	Unit	Material	2021 Bare Costs Labor	Equipment	Total	Total Incl O&P
0500	White, fire retardant				S.F.	.56			.56	.62
0600	12 mils, oil resistant, fire retardant					.51			.51	.56
0700	8.5 mils, black					.31			.31	.34
0710	Woven polyethylene, 6 mils thick					.06			.06	.07
0730	Polyester reinforced w/integral fastening system, 11 mils thick					.23			.23	.25
0740	Polyethylene, reflective, 23 mils thick					1.34			1.34	1.47

01 56 13.90 Winter Protection

		Crew	Daily Output	Labor-Hours	Unit	Material	2021 Bare Costs Labor	Equipment	Total	Total Incl O&P
0010	**WINTER PROTECTION**									
0100	Framing to close openings	2 Clab	500	.032	S.F.	.77	1.42		2.19	2.97
0200	Tarpaulins hung over scaffolding, 8 uses, not incl. scaffolding		1500	.011		.25	.47		.72	.99
0250	Tarpaulin polyester reinf. w/integral fastening system, 11 mils thick		1600	.010		.25	.44		.69	.94
0300	Prefab fiberglass panels, steel frame, 8 uses		1200	.013		3.06	.59		3.65	4.25

01 56 16 – Temporary Dust Barriers

01 56 16.10 Dust Barriers, Temporary

		Crew	Daily Output	Labor-Hours	Unit	Material	2021 Bare Costs Labor	Equipment	Total	Total Incl O&P
0010	**DUST BARRIERS, TEMPORARY**, erect and dismantle									
0020	Spring loaded telescoping pole & head, to 12', erect and dismantle	1 Clab	240	.033	Ea.		1.48		1.48	2.21
0025	Cost per day (based upon 250 days)				Day	.22			.22	.25
0030	To 21', erect and dismantle	1 Clab	240	.033	Ea.		1.48		1.48	2.21
0035	Cost per day (based upon 250 days)				Day	.73			.73	.81
0040	Accessories, caution tape reel, erect and dismantle	1 Clab	480	.017	Ea.		.74		.74	1.10
0045	Cost per day (based upon 250 days)				Day	.34			.34	.38
0060	Foam rail and connector, erect and dismantle	1 Clab	240	.033	Ea.		1.48		1.48	2.21
0065	Cost per day (based upon 250 days)				Day	.10			.10	.11
0070	Caution tape	1 Clab	384	.021	C.L.F.	2.55	.93		3.48	4.18
0080	Zipper, standard duty		60	.133	Ea.	10.75	5.90		16.65	20.50
0090	Heavy duty		48	.167	"	10.85	7.40		18.25	23
0100	Polyethylene sheet, 4 mil		37	.216	Sq.	2.70	9.60		12.30	17.25
0110	6 mil		37	.216	"	7.65	9.60		17.25	23
1000	Dust partition, 6 mil polyethylene, 1" x 3" frame	2 Carp	2000	.008	S.F.	.49	.44		.93	1.19
1080	2" x 4" frame		2000	.008	"	.60	.44		1.04	1.31
4000	Dust & infectious control partition, adj. to 10' high, obscured, 4' panel		90	.178	Ea.	580	9.70		589.70	655
4010	3' panel		90	.178		550	9.70		559.70	620
4020	2' panel		90	.178		430	9.70		439.70	490
4030	1' panel	1 Carp	90	.089		285	4.86		289.86	320
4040	6" panel	"	90	.089		265	4.86		269.86	299
4050	2' panel with HEPA filtered discharge port	2 Carp	90	.178		500	9.70		509.70	565
4060	3' panel with 32" door		90	.178		895	9.70		904.70	1,000
4070	4' panel with 36" door		90	.178		995	9.70		1,004.70	1,125
4080	4'-6" panel with 44" door		90	.178		1,200	9.70		1,209.70	1,350
4090	Hinged corner		80	.200		185	10.95		195.95	220
4100	Outside corner		80	.200		150	10.95		160.95	181
4110	T post		80	.200		150	10.95		160.95	181
4120	Accessories, ceiling grid clip		360	.044		7.45	2.43		9.88	11.85
4130	Panel locking clip		360	.044		5.15	2.43		7.58	9.30
4140	Panel joint closure strip		360	.044		8	2.43		10.43	12.45
4150	Screw jack		360	.044		6.65	2.43		9.08	10.95
4160	Digital pressure difference gauge					275			275	305
4180	Combination lockset	1 Carp	13	.615		200	33.50		233.50	271
4185	Sealant tape, 2" wide	1 Clab	192	.042	C.L.F.	10.65	1.85		12.50	14.50
4190	System in place, including door and accessories									
4200	Based upon 25 uses	2 Carp	51	.314	L.F.	9.85	17.15		27	36.50
4210	Based upon 50 uses		51	.314		4.94	17.15		22.09	31

01 56 Temporary Barriers and Enclosures

01 56 16 – Temporary Dust Barriers

01 56 16.10 Dust Barriers, Temporary	Crew	Daily Output	Labor-Hours	Unit	Material	2021 Bare Costs Labor	2021 Bare Costs Equipment	Total	Total Incl O&P
4230 Based upon 100 uses	2 Carp	51	.314	L.F.	2.47	17.15		19.62	28

01 56 23 – Temporary Barricades

01 56 23.10 Barricades

		Crew	Daily Output	Labor-Hours	Unit	Material	Labor	Equipment	Total	Total Incl O&P
0010	**BARRICADES**									
0020	5' high, 3 rail @ 2" x 8", fixed	2 Carp	20	.800	L.F.	9.45	44		53.45	76
0150	Movable	"	30	.533	"	8.15	29		37.15	52.50
0300	Stock units, 58' high, 8' wide, reflective, buy				Ea.	211			211	232
0350	With reflective tape, buy				"	410			410	455
0400	Break-a-way 3" PVC pipe barricade									
0410	with 3 ea. 1' x 4' reflectorized panels, buy				Ea.	146			146	161
0500	Barricades, plastic, 8" x 24" wide, foldable					57.50			57.50	63.50
0800	Traffic cones, PVC, 18" high					11.60			11.60	12.75
0850	28" high				▼	26			26	28.50
1000	Guardrail, wooden, 3' high, 1" x 6" on 2" x 4" posts	2 Carp	200	.080	L.F.	2.05	4.38		6.43	8.80
1100	2" x 6" on 4" x 4" posts	"	165	.097		3.46	5.30		8.76	11.70
1200	Portable metal with base pads, buy				▼	14.35			14.35	15.80
1250	Typical installation, assume 10 reuses	2 Carp	600	.027		2.19	1.46		3.65	4.59
1300	Barricade tape, polyethylene, 7 mil, 3" wide x 300' long roll	1 Clab	128	.063	Ea.	7.65	2.78		10.43	12.55
3000	Detour signs, set up and remove									
3010	Reflective aluminum, MUTCD, 24" x 24", post mounted	1 Clab	20	.400	Ea.	3.46	17.75		21.21	30.50
4000	Roof edge portable barrier stands and warning flags, 50 uses	1 Rohe	9100	.001	L.F.	.08	.03		.11	.14
4010	100 uses	"	9100	.001	"	.04	.03		.07	.09

01 56 26 – Temporary Fencing

01 56 26.50 Temporary Fencing

		Crew	Daily Output	Labor-Hours	Unit	Material	Labor	Equipment	Total	Total Incl O&P
0010	**TEMPORARY FENCING**									
0020	Chain link, 11 ga., 4' high	2 Clab	400	.040	L.F.	3.05	1.78		4.83	6
0100	6' high		300	.053		5.80	2.37		8.17	9.90
0200	Rented chain link, 6' high, to 1000' (up to 12 mo.)		400	.040		3.44	1.78		5.22	6.45
0250	Over 1000' (up to 12 mo.)	▼	300	.053		4.07	2.37		6.44	8
0350	Plywood, painted, 2" x 4" frame, 4' high	A-4	135	.178		9.45	9.25		18.70	24
0400	4" x 4" frame, 8' high	"	110	.218		17.05	11.35		28.40	35.50
0500	Wire mesh on 4" x 4" posts, 4' high	2 Carp	100	.160		12.25	8.75		21	26.50
0550	8' high	"	80	.200	▼	18	10.95		28.95	36

01 56 29 – Temporary Protective Walkways

01 56 29.50 Protection

		Crew	Daily Output	Labor-Hours	Unit	Material	Labor	Equipment	Total	Total Incl O&P
0010	**PROTECTION**									
0020	Stair tread, 2" x 12" planks, 1 use	1 Carp	75	.107	Tread	7.50	5.85		13.35	16.95
0100	Exterior plywood, 1/2" thick, 1 use		65	.123		3.01	6.75		9.76	13.35
0200	3/4" thick, 1 use		60	.133	▼	4.15	7.30		11.45	15.45
2200	Sidewalks, 2" x 12" planks, 2 uses		350	.023	S.F.	1.25	1.25		2.50	3.24
2300	Exterior plywood, 2 uses, 1/2" thick		750	.011		.50	.58		1.08	1.42
2400	5/8" thick		650	.012		.58	.67		1.25	1.64
2500	3/4" thick	▼	600	.013	▼	.69	.73		1.42	1.85

01 57 Temporary Controls

01 57 33 – Temporary Security

01 57 33.50 Watchman		Crew	Daily Output	Labor-Hours	Unit	Material	2021 Bare Costs Labor	Equipment	Total	Total Incl O&P
0010	**WATCHMAN**									
0020	Service, monthly basis, uniformed person, minimum				Hr.				27.65	30.40
0100	Maximum								56	61.62
0200	Person and command dog, minimum								31	34
0300	Maximum								60	65

01 58 Project Identification

01 58 13 – Temporary Project Signage

01 58 13.50 Signs		Crew	Daily Output	Labor-Hours	Unit	Material	2021 Bare Costs Labor	Equipment	Total	Total Incl O&P
0010	**SIGNS**									
0020	High intensity reflectorized, no posts, buy				Ea.	22			22	24

01 66 Product Storage and Handling Requirements

01 66 19 – Material Handling

01 66 19.10 Material Handling		Crew	Daily Output	Labor-Hours	Unit	Material	2021 Bare Costs Labor	Equipment	Total	Total Incl O&P
0010	**MATERIAL HANDLING**									
0020	Above 2nd story, via stairs, per C.Y. of material per floor	2 Clab	145	.110	C.Y.		4.90		4.90	7.30
0030	Via elevator, per C.Y. of material		240	.067			2.96		2.96	4.42
0050	Distances greater than 200', per C.Y. of material per each addl 200'		300	.053			2.37		2.37	3.53

01 71 Examination and Preparation

01 71 23 – Field Engineering

01 71 23.13 Construction Layout		Crew	Daily Output	Labor-Hours	Unit	Material	2021 Bare Costs Labor	Equipment	Total	Total Incl O&P
0010	**CONSTRUCTION LAYOUT**									
1100	Crew for layout of building, trenching or pipe laying, 2 person crew	A-6	1	16	Day		880	34.50	914.50	1,375
1200	3 person crew	A-7	1	24			1,425	34.50	1,459.50	2,175
1400	Crew for roadway layout, 4 person crew	A-8	1	32			1,850	34.50	1,884.50	2,800

01 71 23.19 Surveyor Stakes		Crew	Daily Output	Labor-Hours	Unit	Material	2021 Bare Costs Labor	Equipment	Total	Total Incl O&P
0010	**SURVEYOR STAKES**									
0020	Hardwood, 1" x 1" x 48" long				C	70			70	77
0100	2" x 2" x 18" long					78			78	86
0150	2" x 2" x 24" long					150			150	165

01 74 Cleaning and Waste Management

01 74 13 – Progress Cleaning

01 74 13.20 Cleaning Up		Crew	Daily Output	Labor-Hours	Unit	Material	2021 Bare Costs Labor	Equipment	Total	Total Incl O&P
0010	**CLEANING UP**									
0020	After job completion, allow, minimum				Job				.30%	.30%
0040	Maximum				"				1%	1%
0050	Cleanup of floor area, continuous, per day, during const.	A-5	24	.750	M.S.F.	2.77	33.50	2.07	38.34	56
0100	Final by GC at end of job	"	11.50	1.565	"	2.93	70.50	4.32	77.75	113

For customer support on your Concrete & Masonry Costs with RSMeans Data, call 800.448.8182.

23

01 76 Protecting Installed Construction

01 76 13 – Temporary Protection of Installed Construction

01 76 13.20 Temporary Protection	Crew	Daily Output	Labor-Hours	Unit	Material	2021 Bare Costs Labor	Equipment	Total	Total Incl O&P
0010 **TEMPORARY PROTECTION**									
0020 Flooring, 1/8" tempered hardboard, taped seams	2 Carp	1500	.011	S.F.	.51	.58		1.09	1.43
0030 Peel away carpet protection	1 Clab	3200	.003	"	.11	.11		.22	.29

01 91 Commissioning

01 91 13 – General Commissioning Requirements

01 91 13.50 Building Commissioning

	Crew	Daily Output	Labor-Hours	Unit	Material	2021 Bare Costs Labor	Equipment	Total	Total Incl O&P
0010 **BUILDING COMMISSIONING**									
0100 Systems operation and verification during turnover				%				.25%	.25%
0150 Including all systems subcontractors								.50%	.50%
0200 Systems design assistance, operation, verification and training								.50%	.50%
0250 Including all systems subcontractors								1%	1%

Estimating Tips
02 30 00 Subsurface Investigation
In preparing estimates on structures involving earthwork or foundations, all information concerning soil characteristics should be obtained. Look particularly for hazardous waste, evidence of prior dumping of debris, and previous stream beds.

02 40 00 Demolition and Structure Moving
The costs shown for selective demolition do not include rubbish handling or disposal. These items should be estimated separately using RSMeans data or other sources.
- Historic preservation often requires that the contractor remove materials from the existing structure, rehab them, and replace them. The estimator must be aware of any related measures and precautions that must be taken when doing selective demolition and cutting and patching. Requirements may include special handling and storage, as well as security.

- In addition to Subdivision 02 41 00, you can find selective demolition items in each division. Example: Roofing demolition is in Division 7.
- Absent of any other specific reference, an approximate demolish-in-place cost can be obtained by halving the new-install labor cost. To remove for reuse, allow the entire new-install labor figure.

02 40 00 Building Deconstruction
This section provides costs for the careful dismantling and recycling of most low-rise building materials.

02 50 00 Containment of Hazardous Waste
This section addresses on-site hazardous waste disposal costs.

02 80 00 Hazardous Material Disposal/Remediation
This subdivision includes information on hazardous waste handling, asbestos remediation, lead remediation, and mold remediation. See reference numbers

RO28213-20 and RO28319-60 for further guidance in using these unit price lines.

02 90 00 Monitoring Chemical Sampling, Testing Analysis
This section provides costs for on-site sampling and testing hazardous waste.

Reference Numbers
Reference numbers are shown at the beginning of some major classifications. These numbers refer to related items in the Reference Section. The reference information may be an estimating procedure, an alternate pricing method, or technical information.

Note: Not all subdivisions listed here necessarily appear. ■

Same Data. Simplified.
Enjoy the convenience and efficiency of accessing your costs anywhere:
- **Skip the multiplier** by setting your location
- **Quickly search,** edit, favorite and share costs
- **Stay on top of price changes** with automatic updates

Discover more at rsmeans.com/online

02 21 Surveys

02 21 13 – Site Surveys

02 21 13.09 Topographical Surveys	Crew	Daily Output	Labor-Hours	Unit	Material	2021 Bare Costs Labor	Equipment	Total	Total Incl O&P
0010 **TOPOGRAPHICAL SURVEYS**									
0020 Topographical surveying, conventional, minimum	A-7	3.30	7.273	Acre	35	430	10.40	475.40	695
0050 Average	"	1.95	12.308		52.50	730	17.60	800.10	1,175
0100 Maximum	A-8	.60	53.333		70	3,075	57.50	3,202.50	4,750

02 21 13.13 Boundary and Survey Markers	Crew	Daily Output	Labor-Hours	Unit	Material	2021 Bare Costs Labor	Equipment	Total	Total Incl O&P
0010 **BOUNDARY AND SURVEY MARKERS**									
0300 Lot location and lines, large quantities, minimum	A-7	2	12	Acre	34.50	715	17.15	766.65	1,125
0320 Average	"	1.25	19.200		61	1,150	27.50	1,238.50	1,800
0400 Small quantities, maximum	A-8	1	32		77	1,850	34.50	1,961.50	2,875
0600 Monuments, 3' long	A-7	10	2.400	Ea.	37.50	143	3.43	183.93	258
0800 Property lines, perimeter, cleared land	"	1000	.024	L.F.	.08	1.43	.03	1.54	2.26
0900 Wooded land	A-8	875	.037	"	.10	2.11	.04	2.25	3.30

02 21 13.16 Aerial Surveys	Crew	Daily Output	Labor-Hours	Unit	Material	2021 Bare Costs Labor	Equipment	Total	Total Incl O&P
0010 **AERIAL SURVEYS**									
1500 Aerial surveying, including ground control, minimum fee, 10 acres				Total				4,700	4,700
1510 100 acres								9,400	9,400
1550 From existing photography, deduct								1,625	1,625
1600 2' contours, 10 acres				Acre				470	470
1850 100 acres								94	94
2000 1000 acres								90	90
2050 10,000 acres								85	85

02 31 Geophysical Investigations

02 31 23 – Electromagnetic Investigations

02 31 23.10 Ground Penetrating Radar

	Crew	Daily Output	Labor-Hours	Unit	Material	2021 Bare Costs Labor	Equipment	Total	Total Incl O&P
0010 **GROUND PENETRATING RADAR**									
0100 Ground Penetrating Radar				Hr.		210		210	235

02 32 Geotechnical Investigations

02 32 13 – Subsurface Drilling and Sampling

02 32 13.10 Boring and Exploratory Drilling

	Crew	Daily Output	Labor-Hours	Unit	Material	2021 Bare Costs Labor	Equipment	Total	Total Incl O&P
0010 **BORING AND EXPLORATORY DRILLING**									
0020 Borings, initial field stake out & determination of elevations	A-6	1	16	Day		880	34.50	914.50	1,375
0100 Drawings showing boring details				Total		335		335	425
0200 Report and recommendations from P.E.						775		775	970
0300 Mobilization and demobilization	B-55	4	6			275	310	585	750
0350 For over 100 miles, per added mile		450	.053	Mile		2.45	2.76	5.21	6.70
0600 Auger holes in earth, no samples, 2-1/2" diameter		78.60	.305	L.F.		14	15.85	29.85	38.50
0650 4" diameter		67.50	.356			16.30	18.45	34.75	45
0800 Cased borings in earth, with samples, 2-1/2" diameter		55.50	.432		25	19.85	22.50	67.35	81.50
0850 4" diameter		32.60	.736		28.50	34	38	100.50	124
1000 Drilling in rock, "BX" core, no sampling	B-56	34.90	.458			23	45.50	68.50	84
1050 With casing & sampling		31.70	.505		25	25	50	100	120
1200 "NX" core, no sampling		25.92	.617			31	61.50	92.50	114
1250 With casing and sampling		25	.640		26	32	63.50	121.50	146
1400 Borings, earth, drill rig and crew with truck mounted auger	B-55	1	24	Day		1,100	1,250	2,350	3,025
1450 Rock using crawler type drill	B-56	1	16	"		800	1,600	2,400	2,950
1500 For inner city borings, add, minimum								10%	10%
1510 Maximum								20%	20%

26

02 32 Geotechnical Investigations

02 32 19 – Exploratory Excavations

02 32 19.10 Test Pits	Crew	Daily Output	Labor-Hours	Unit	Material	2021 Bare Costs Labor	Equipment	Total	Total Incl O&P
0010 **TEST PITS**									
0020 Hand digging, light soil	1 Clab	4.50	1.778	C.Y.		79		79	118
0100 Heavy soil	"	2.50	3.200			142		142	212
0120 Loader-backhoe, light soil	B-11M	28	.571			29.50	8.40	37.90	53.50
0130 Heavy soil	"	20	.800	↓		41.50	11.75	53.25	74.50
1000 Subsurface exploration, mobilization				Mile				6.75	8.40
1010 Difficult access for rig, add				Hr.				260	320
1020 Auger borings, drill rig, incl. samples				L.F.				26.50	33
1030 Hand auger								31.50	40
1050 Drill and sample every 5', split spoon				↓				31.50	40
1060 Extra samples				Ea.				36	45.50

02 41 Demolition

02 41 13 – Selective Site Demolition

02 41 13.15 Hydrodemolition

	Crew	Daily Output	Labor-Hours	Unit	Material	Labor	Equipment	Total	Total Incl O&P
0010 **HYDRODEMOLITION**									
0015 Hydrodemolition, concrete pavement									
0120 Includes removal but excludes disposal of concrete									
0130 2" depth	B-5E	1000	.064	S.F.		3.15	2.75	5.90	7.70
0410 4" depth		800	.080			3.93	3.43	7.36	9.65
0420 6" depth		600	.107			5.25	4.58	9.83	12.85
0520 8" depth	↓	300	.213	↓		10.50	9.15	19.65	25.50

02 41 13.16 Hydroexcavation

	Crew	Daily Output	Labor-Hours	Unit	Material	Labor	Equipment	Total	Total Incl O&P
0010 **HYDROEXCAVATION**									
0015 Hydroexcavation									
0105 Assumes onsite disposal									
0110 Mobilization or Demobilization	B-6D	8	2.500	Ea.		127	160	287	365
0130 Normal Conditions		48	.417	B.C.Y.		21	26.50	47.50	61
0140 Adverse Conditions		30	.667	"		34	42.50	76.50	97.50
0160 Minimum labor/equipment charge	↓	4	5	Ea.		253	320	573	730

02 41 13.17 Demolish, Remove Pavement and Curb

	Crew	Daily Output	Labor-Hours	Unit	Material	Labor	Equipment	Total	Total Incl O&P
0010 **DEMOLISH, REMOVE PAVEMENT AND CURB** R024119-10									
5010 Pavement removal, bituminous roads, up to 3" thick	B-38	690	.058	S.Y.		2.90	1.76	4.66	6.25
5050 4"-6" thick		420	.095			4.76	2.90	7.66	10.30
5200 Concrete to 6" thick, hydraulic hammer, mesh reinforced		255	.157			7.85	4.77	12.62	16.95
5300 Rod reinforced		200	.200	↓		10	6.10	16.10	21.50
5400 Concrete, 7"-24" thick, plain		33	1.212	C.Y.		60.50	37	97.50	131
5500 Reinforced	↓	24	1.667	"		83	50.50	133.50	180
5600 With hand held air equipment, bituminous, to 6" thick	B-39	1900	.025	S.F.		1.18	.19	1.37	1.97
5700 Concrete to 6" thick, no reinforcing		1600	.030			1.40	.22	1.62	2.32
5800 Mesh reinforced		1400	.034			1.60	.25	1.85	2.66
5900 Rod reinforced	↓	765	.063	↓		2.92	.46	3.38	4.87
6000 Curbs, concrete, plain	B-6	360	.067	L.F.		3.21	.60	3.81	5.45
6100 Reinforced	"	275	.087	"		4.20	.79	4.99	7.10

02 41 13.23 Utility Line Removal

	Crew	Daily Output	Labor-Hours	Unit	Material	Labor	Equipment	Total	Total Incl O&P
0010 **UTILITY LINE REMOVAL**									
0015 No hauling, abandon catch basin or manhole	B-6	7	3.429	Ea.		165	31	196	280
0020 Remove existing catch basin or manhole, masonry		4	6			289	54	343	490
0030 Catch basin or manhole frames and covers, stored		13	1.846			89	16.65	105.65	150
0040 Remove and reset	↓	7	3.429	↓		165	31	196	280

02 41 Demolition

02 41 13 – Selective Site Demolition

02 41 13.30 Minor Site Demolition

02 41 13.30 Minor Site Demolition	Crew	Daily Output	Labor-Hours	Unit	Material	2021 Bare Costs Labor	Equipment	Total	Total Incl O&P
0010 **MINOR SITE DEMOLITION** R024119-10									
1000 Masonry walls, block, solid	B-5	1800	.031	C.F.		1.52	.83	2.35	3.19
1200 Brick, solid		900	.062			3.04	1.67	4.71	6.35
1400 Stone, with mortar		900	.062			3.04	1.67	4.71	6.35
1500 Dry set	↓	1500	.037	↓		1.82	1	2.82	3.82
4050 Brick, set in mortar	B-6	185	.130	S.Y.		6.25	1.17	7.42	10.60
4100 Concrete, plain, 4"		160	.150			7.20	1.35	8.55	12.25
4110 Plain, 5"		140	.171			8.25	1.54	9.79	14
4120 Plain, 6"		120	.200			9.60	1.80	11.40	16.35
4200 Mesh reinforced, concrete, 4"		150	.160			7.70	1.44	9.14	13.10
4210 5" thick		131	.183			8.80	1.65	10.45	14.95
4220 6" thick	↓	112	.214	↓		10.30	1.93	12.23	17.45
4300 Slab on grade removal, plain	B-5	45	1.244	C.Y.		61	33.50	94.50	127
4310 Mesh reinforced		33	1.697			83	45.50	128.50	174
4320 Rod reinforced	↓	25	2.240			109	60	169	229
4400 For congested sites or small quantities, add up to								200%	200%
4450 For disposal on site, add	B-11A	232	.069			3.57	6.55	10.12	12.50
4500 To 5 miles, add	B-34D	76	.105	↓		5.40	8.60	14	17.50

02 41 16 – Structure Demolition

02 41 16.13 Building Demolition

	Crew	Daily Output	Labor-Hours	Unit	Material	2021 Bare Costs Labor	Equipment	Total	Total Incl O&P
0010 **BUILDING DEMOLITION** Large urban projects, incl. 20 mi. haul R024119-10									
0011 No foundation or dump fees, C.F. is vol. of building standing									
0020 Steel	B-8	21500	.003	C.F.		.15	.13	.28	.38
0050 Concrete		15300	.004			.21	.19	.40	.53
0080 Masonry		20100	.003			.16	.14	.30	.40
0100 Mixture of types	↓	20100	.003			.16	.14	.30	.40
0500 Small bldgs, or single bldgs, no salvage included, steel	B-3	14800	.003			.16	.16	.32	.41
0600 Concrete		11300	.004			.21	.20	.41	.53
0650 Masonry	↓	14800	.003	↓		.16	.16	.32	.41
0750 For buildings with no interior walls, deduct								30%	30%
1000 Demolition single family house, one story, wood 1600 S.F.	B-3	1	48	Ea.		2,375	2,300	4,675	6,075
1020 3200 S.F.		.50	96			4,750	4,600	9,350	12,200
1200 Demolition two family house, two story, wood 2400 S.F.		.67	71.964			3,550	3,450	7,000	9,125
1220 4200 S.F.		.38	128			6,325	6,150	12,475	16,200
1300 Demolition three family house, three story, wood 3200 S.F.		.50	96			4,750	4,600	9,350	12,200
1320 5400 S.F.	↓	.30	160			7,925	7,675	15,600	20,300
5000 For buildings with no interior walls, deduct								30%	30%

02 41 16.15 Explosive/Implosive Demolition

	Crew	Daily Output	Labor-Hours	Unit	Material	2021 Bare Costs Labor	Equipment	Total	Total Incl O&P
0010 **EXPLOSIVE/IMPLOSIVE DEMOLITION** R024119-10									
0011 Large projects,									
0020 No disposal fee based on building volume, steel building	B-5B	16900	.003	C.F.		.16	.16	.32	.41
0100 Concrete building		16900	.003			.16	.16	.32	.41
0200 Masonry building	↓	16900	.003	↓		.16	.16	.32	.41
0400 Disposal of material, minimum	B-3	445	.108	C.Y.		5.35	5.20	10.55	13.65
0500 Maximum	"	365	.132	"		6.50	6.30	12.80	16.65

02 41 16.17 Building Demolition Footings and Foundations

	Crew	Daily Output	Labor-Hours	Unit	Material	2021 Bare Costs Labor	Equipment	Total	Total Incl O&P
0010 **BUILDING DEMOLITION FOOTINGS AND FOUNDATIONS** R024119-10									
0200 Floors, concrete slab on grade,									
0240 4" thick, plain concrete	B-13L	5000	.003	S.F.		.20	.42	.62	.75
0280 Reinforced, wire mesh		4000	.004			.25	.53	.78	.95
0300 Rods	↓	4500	.004	↓		.22	.47	.69	.84

02 41 Demolition

02 41 16 – Structure Demolition

02 41 16.17 Building Demolition Footings and Foundations

		Crew	Daily Output	Labor-Hours	Unit	Material	2021 Bare Costs Labor	2021 Bare Costs Equipment	Total	Total Incl O&P
0400	6" thick, plain concrete	B-13L	4000	.004	S.F.		.25	.53	.78	.95
0420	Reinforced, wire mesh		3200	.005			.31	.66	.97	1.18
0440	Rods		3600	.004	↓		.27	.58	.85	1.05
1000	Footings, concrete, 1' thick, 2' wide		300	.053	L.F.		3.28	7	10.28	12.60
1080	1'-6" thick, 2' wide		250	.064			3.93	8.40	12.33	15.10
1120	3' wide		200	.080			4.92	10.50	15.42	18.85
1140	2' thick, 3' wide	↓	175	.091			5.60	12	17.60	21.50
1200	Average reinforcing, add								10%	10%
1220	Heavy reinforcing, add				↓				20%	20%
2000	Walls, block, 4" thick	B-13L	8000	.002	S.F.		.12	.26	.38	.47
2040	6" thick		6000	.003			.16	.35	.51	.63
2080	8" thick		4000	.004			.25	.53	.78	.95
2100	12" thick	↓	3000	.005			.33	.70	1.03	1.26
2200	For horizontal reinforcing, add								10%	10%
2220	For vertical reinforcing, add								20%	20%
2400	Concrete, plain concrete, 6" thick	B-13L	4000	.004			.25	.53	.78	.95
2420	8" thick		3500	.005			.28	.60	.88	1.08
2440	10" thick		3000	.005			.33	.70	1.03	1.26
2500	12" thick	↓	2500	.006			.39	.84	1.23	1.52
2600	For average reinforcing, add								10%	10%
2620	For heavy reinforcing, add								20%	20%
4000	For congested sites or small quantities, add up to				↓				200%	200%
4200	Add for disposal, on site	B-11A	232	.069	C.Y.		3.57	6.55	10.12	12.50
4250	To five miles	B-30	220	.109	"		5.90	8.45	14.35	18

02 41 19 – Selective Demolition

02 41 19.13 Selective Building Demolition

		Crew	Daily Output	Labor-Hours	Unit	Material	2021 Bare Costs Labor	2021 Bare Costs Equipment	Total	Total Incl O&P
0010	**SELECTIVE BUILDING DEMOLITION**									
0020	Costs related to selective demolition of specific building components									
0025	are included under Common Work Results (XX 05)									
0030	in the component's appropriate division.									

02 41 19.16 Selective Demolition, Cutout

		Crew	Daily Output	Labor-Hours	Unit	Material	2021 Bare Costs Labor	2021 Bare Costs Equipment	Total	Total Incl O&P
0010	**SELECTIVE DEMOLITION, CUTOUT** R024119-10									
0020	Concrete, elev. slab, light reinforcement, under 6 C.F.	B-9	65	.615	C.F.		27.50	5.45	32.95	47
0050	Light reinforcing, over 6 C.F.		75	.533	"		24	4.74	28.74	40.50
0200	Slab on grade to 6" thick, not reinforced, under 8 S.F.		85	.471	S.F.		21	4.18	25.18	36
0250	8-16 S.F.	↓	175	.229	"		10.25	2.03	12.28	17.55
0255	For over 16 S.F. see Line 02 41 16.17 0400									
0600	Walls, not reinforced, under 6 C.F.	B-9	60	.667	C.F.		30	5.95	35.95	51
0650	6-12 C.F.	"	80	.500	"		22.50	4.45	26.95	38.50
0655	For over 12 C.F. see Line 02 41 16.17 2500									
1000	Concrete, elevated slab, bar reinforced, under 6 C.F.	B-9	45	.889	C.F.		40	7.90	47.90	68
1050	Bar reinforced, over 6 C.F.		50	.800	"		36	7.10	43.10	61.50
1200	Slab on grade to 6" thick, bar reinforced, under 8 S.F.		75	.533	S.F.		24	4.74	28.74	40.50
1250	8-16 S.F.	↓	150	.267	"		11.95	2.37	14.32	20.50
1255	For over 16 S.F. see Line 02 41 16.17 0440									
1400	Walls, bar reinforced, under 6 C.F.	B-9	50	.800	C.F.		36	7.10	43.10	61.50
1450	6-12 C.F.	"	70	.571	"		25.50	5.10	30.60	43.50
1455	For over 12 C.F. see Lines 02 41 16.17 2500 and 2600									
2000	Brick, to 4 S.F. opening, not including toothing									
2040	4" thick	B-9	30	1.333	Ea.		59.50	11.85	71.35	102
2060	8" thick		18	2.222			99.50	19.75	119.25	171
2080	12" thick	↓	10	4	↓		179	35.50	214.50	305

For customer support on your Concrete & Masonry Costs with RSMeans Data, call 800.448.8182.

02 41 Demolition

02 41 19 – Selective Demolition

02 41 19.16 Selective Demolition, Cutout

		Crew	Daily Output	Labor-Hours	Unit	Material	2021 Bare Costs Labor	Equipment	Total	Total Incl O&P
2400	Concrete block, to 4 S.F. opening, 2" thick	B-9	35	1.143	Ea.		51	10.15	61.15	87.50
2420	4" thick		30	1.333			59.50	11.85	71.35	102
2440	8" thick		27	1.481			66.50	13.15	79.65	114
2460	12" thick		24	1.667			74.50	14.80	89.30	127
2600	Gypsum block, to 4 S.F. opening, 2" thick		80	.500			22.50	4.45	26.95	38.50
2620	4" thick		70	.571			25.50	5.10	30.60	43.50
2640	8" thick		55	.727			32.50	6.45	38.95	55.50
2800	Terra cotta, to 4 S.F. opening, 4" thick		70	.571			25.50	5.10	30.60	43.50
2840	8" thick		65	.615			27.50	5.45	32.95	47
2880	12" thick		50	.800			36	7.10	43.10	61.50
3000	Toothing masonry cutouts, brick, soft old mortar	1 Brhe	40	.200	V.L.F.		8.75		8.75	13.15
3100	Hard mortar		30	.267			11.65		11.65	17.55
3200	Block, soft old mortar		70	.114			4.99		4.99	7.55
3400	Hard mortar		50	.160			7		7	10.55

02 41 19.18 Selective Demolition, Disposal Only

		Crew	Daily Output	Labor-Hours	Unit	Material	2021 Bare Costs Labor	Equipment	Total	Total Incl O&P
0010	**SELECTIVE DEMOLITION, DISPOSAL ONLY** R024119-10									
0015	Urban bldg w/salvage value allowed									
0020	Including loading and 5 mile haul to dump									
0200	Steel frame	B-3	430	.112	C.Y.		5.50	5.35	10.85	14.15
0300	Concrete frame		365	.132			6.50	6.30	12.80	16.65
0400	Masonry construction		445	.108			5.35	5.20	10.55	13.65
0500	Wood frame		247	.194			9.60	9.35	18.95	24.50

02 41 19.19 Selective Demolition

		Crew	Daily Output	Labor-Hours	Unit	Material	2021 Bare Costs Labor	Equipment	Total	Total Incl O&P
0010	**SELECTIVE DEMOLITION,** Rubbish Handling R024119-10									
0020	The following are to be added to the demolition prices									
0600	Dumpster, weekly rental, 1 dump/week, 6 C.Y. capacity (2 tons)				Week	415			415	455
0700	10 C.Y. capacity (3 tons)					480			480	530
0725	20 C.Y. capacity (5 tons) R024119-20					565			565	625
0800	30 C.Y. capacity (7 tons)					730			730	800
0840	40 C.Y. capacity (10 tons)					775			775	850
2000	Load, haul, dump and return, 0'-50' haul, hand carried	2 Clab	24	.667	C.Y.		29.50		29.50	44
2005	Wheeled		37	.432			19.20		19.20	28.50
2040	0'-100' haul, hand carried		16.50	.970			43		43	64
2045	Wheeled		25	.640			28.50		28.50	42.50
2050	Forklift	A-3R	25	.320			17.75	11.35	29.10	39
2080	Haul and return, add per each extra 100' haul, hand carried	2 Clab	35.50	.451			20		20	30
2085	Wheeled		54	.296			13.15		13.15	19.65
2120	For travel in elevators, up to 10 floors, add		140	.114			5.05		5.05	7.55
2130	0'-50' haul, incl. up to 5 riser stairs, hand carried		23	.696			31		31	46
2135	Wheeled		35	.457			20.50		20.50	30.50
2140	6-10 riser stairs, hand carried		22	.727			32.50		32.50	48
2145	Wheeled		34	.471			21		21	31
2150	11-20 riser stairs, hand carried		20	.800			35.50		35.50	53
2155	Wheeled		31	.516			23		23	34
2160	21-40 riser stairs, hand carried		16	1			44.50		44.50	66.50
2165	Wheeled		24	.667			29.50		29.50	44
2170	0'-100' haul, incl. 5 riser stairs, hand carried		15	1.067			47.50		47.50	70.50
2175	Wheeled		23	.696			31		31	46
2180	6-10 riser stairs, hand carried		14	1.143			50.50		50.50	75.50
2185	Wheeled		21	.762			34		34	50.50
2190	11-20 riser stairs, hand carried		12	1.333			59		59	88.50
2195	Wheeled		18	.889			39.50		39.50	59

02 41 Demolition

02 41 19 – Selective Demolition

02 41 19.19 Selective Demolition

		Crew	Daily Output	Labor-Hours	Unit	Material	2021 Bare Costs Labor	Equipment	Total	Total Incl O&P
2200	21-40 riser stairs, hand carried	2 Clab	8	2	C.Y.		89		89	133
2205	Wheeled		12	1.333			59		59	88.50
2210	Haul and return, add per each extra 100' haul, hand carried		35.50	.451			20		20	30
2215	Wheeled		54	.296	▼		13.15		13.15	19.65
2220	For each additional flight of stairs, up to 5 risers, add		550	.029	Flight		1.29		1.29	1.93
2225	6-10 risers, add		275	.058			2.58		2.58	3.85
2230	11-20 risers, add		138	.116			5.15		5.15	7.70
2235	21-40 risers, add	▼	69	.232	▼		10.30		10.30	15.35
3000	Loading & trucking, including 2 mile haul, chute loaded	B-16	45	.711	C.Y.		33	12.85	45.85	63.50
3040	Hand loading truck, 50' haul	"	48	.667			31	12.05	43.05	60
3080	Machine loading truck	B-17	120	.267			13.05	5.20	18.25	25
5000	Haul, per mile, up to 8 C.Y. truck	B-34B	1165	.007			.35	.50	.85	1.08
5100	Over 8 C.Y. truck	"	1550	.005	▼		.26	.37	.63	.81

02 41 19.20 Selective Demolition, Dump Charges

		Crew	Daily Output	Labor-Hours	Unit	Material	2021 Bare Costs Labor	Equipment	Total	Total Incl O&P
0010	**SELECTIVE DEMOLITION, DUMP CHARGES** R024119-10									
0020	Dump charges, typical urban city, tipping fees only									
0100	Building construction materials				Ton	74			74	81
0200	Trees, brush, lumber					63			63	69.50
0300	Rubbish only					63			63	69.50
0500	Reclamation station, usual charge				▼	74			74	81

02 41 19.25 Selective Demolition, Saw Cutting

		Crew	Daily Output	Labor-Hours	Unit	Material	2021 Bare Costs Labor	Equipment	Total	Total Incl O&P
0010	**SELECTIVE DEMOLITION, SAW CUTTING** R024119-10									
0015	Asphalt, up to 3" deep	B-89	1050	.015	L.F.	.11	.79	1.02	1.92	2.43
0020	Each additional inch of depth	"	1800	.009		.04	.46	.59	1.09	1.38
1200	Masonry walls, hydraulic saw, brick, per inch of depth	B-89B	300	.053		.04	2.78	5.20	8.02	9.90
1220	Block walls, solid, per inch of depth	"	250	.064		.04	3.34	6.25	9.63	11.85
2000	Brick or masonry w/hand held saw, per inch of depth	A-1	125	.064		.05	2.84	.91	3.80	5.30
5000	Wood sheathing to 1" thick, on walls	1 Carp	200	.040			2.19		2.19	3.27
5020	On roof	"	250	.032	▼		1.75		1.75	2.61

02 41 19.27 Selective Demolition, Torch Cutting

		Crew	Daily Output	Labor-Hours	Unit	Material	2021 Bare Costs Labor	Equipment	Total	Total Incl O&P
0010	**SELECTIVE DEMOLITION, TORCH CUTTING** R024119-10									
0020	Steel, 1" thick plate	E-25	333	.024	L.F.	2.23	1.50	.04	3.77	4.81
0040	1" diameter bar	"	600	.013	Ea.	.37	.83	.02	1.22	1.72
1000	Oxygen lance cutting, reinforced concrete walls									
1040	12"-16" thick walls	1 Clab	10	.800	L.F.		35.50		35.50	53
1080	24" thick walls	"	6	1.333	"		59		59	88.50

02 58 Snow Control

02 58 13 – Snow Fencing

02 58 13.10 Snow Fencing System

		Crew	Daily Output	Labor-Hours	Unit	Material	2021 Bare Costs Labor	Equipment	Total	Total Incl O&P
0010	**SNOW FENCING SYSTEM**									
7001	Snow fence on steel posts 10' OC, 4' high	B-1	500	.048	L.F.	.39	2.16		2.55	3.66

Division Notes

	CREW	DAILY OUTPUT	LABOR-HOURS	UNIT	BARE COSTS				TOTAL INCL O&P
					MAT.	LABOR	EQUIP.	TOTAL	

Estimating Tips
General
- Carefully check all the plans and specifications. Concrete often appears on drawings other than structural drawings, including mechanical and electrical drawings for equipment pads. The cost of cutting and patching is often difficult to estimate. See Subdivision 03 81 for Concrete Cutting, Subdivision 02 41 19.16 for Cutout Demolition, Subdivision 03 05 05.10 for Concrete Demolition, and Subdivision 02 41 19.19 for Rubbish Handling (handling, loading, and hauling of debris).
- Always obtain concrete prices from suppliers near the job site. A volume discount can often be negotiated, depending upon competition in the area. Remember to add for waste, particularly for slabs and footings on grade.

03 10 00 Concrete Forming and Accessories
- A primary cost for concrete construction is forming. Most jobs today are constructed with prefabricated forms. The selection of the forms best suited for the job and the total square feet of forms required for efficient concrete forming and placing are key elements in estimating concrete construction. Enough forms must be available for erection to make efficient use of the concrete placing equipment and crew.
- Concrete accessories for forming and placing depend upon the systems used. Study the plans and specifications to ensure that all special accessory requirements have been included in the cost estimate, such as anchor bolts, inserts, and hangers.
- Included within costs for forms-in-place are all necessary bracing and shoring.

03 20 00 Concrete Reinforcing
- Ascertain that the reinforcing steel supplier has included all accessories, cutting, bending, and an allowance for lapping, splicing, and waste. A good rule of thumb is 10% for lapping, splicing, and waste. Also, 10% waste should be allowed for welded wire fabric.
- The unit price items in the subdivisions for Reinforcing In Place, Glass Fiber Reinforcing, and Welded Wire Fabric include the labor to install accessories such as beam and slab bolsters, high chairs, and bar ties and tie wire. The material cost for these accessories is not included; they may be obtained from the Accessories Subdivisions.

03 30 00 Cast-In-Place Concrete
- When estimating structural concrete, pay particular attention to requirements for concrete additives, curing methods, and surface treatments. Special consideration for climate, hot or cold, must be included in your estimate. Be sure to include requirements for concrete placing equipment and concrete finishing.
- For accurate concrete estimating, the estimator must consider each of the following major components individually: forms, reinforcing steel, ready-mix concrete, placement of the concrete, and finishing of the top surface. For faster estimating, Subdivision 03 30 53.40 for Concrete-In-Place can be used; here, various items of concrete work are presented that include the costs of all five major components (unless specifically stated otherwise).

03 40 00 Precast Concrete
03 50 00 Cast Decks and Underlayment
- The cost of hauling precast concrete structural members is often an important factor. For this reason, it is important to get a quote from the nearest supplier. It may become economically feasible to set up precasting beds on the site if the hauling costs are prohibitive.

Reference Numbers
Reference numbers are shown at the beginning of some major classifications. These numbers refer to related items in the Reference Section. The reference information may be an estimating procedure, an alternate pricing method, or technical information.

Note: Not all subdivisions listed here necessarily appear. ∎

Same Data. Simplified.

Enjoy the convenience and efficiency of accessing your costs anywhere:

- **Skip the multiplier** by setting your location
- **Quickly search,** edit, favorite and share costs
- **Stay on top of price changes** with automatic updates

Discover more at rsmeans.com/online

03 01 30.62 Concrete Patching

		Crew	Daily Output	Labor-Hours	Unit	Material	2021 Bare Costs Labor	Equipment	Total	Total Incl O&P
0010	**CONCRETE PATCHING**									
0100	Floors, 1/4" thick, small areas, regular grout	1 Cefi	170	.047	S.F.	1.53	2.44		3.97	5.25
0150	Epoxy grout		100	.080		11.15	4.14		15.29	18.30
0160	Two part polymer grout mix		150	.053		1.74	2.76		4.50	5.95
0200	Overhead, including chipping or sand blasting,									
0210	Priming, and two part polymer mix, 1/4" deep	1 Cefi	45	.178	S.F.	1.62	9.20		10.82	15.30
0220	1/2" deep		35	.229		3.24	11.85		15.09	21
0230	3/4" deep		25	.320		4.87	16.60		21.47	30
2000	Walls, including chipping, cleaning and epoxy grout									
2100	1/4" deep	1 Cefi	65	.123	S.F.	6.55	6.40		12.95	16.55
2150	1/2" deep		50	.160		13.10	8.30		21.40	26.50
2200	3/4" deep		40	.200		19.65	10.35		30	36.50
2400	Walls, including chipping or sand blasting,									
2410	Priming, and two part polymer mix, 1/4" deep	1 Cefi	80	.100	S.F.	1.62	5.20		6.82	9.40
2420	1/2" deep		60	.133		3.24	6.90		10.14	13.65
2430	3/4" deep		40	.200		4.87	10.35		15.22	20.50

03 01 30.71 Concrete Crack Repair

		Crew	Daily Output	Labor-Hours	Unit	Material	2021 Bare Costs Labor	Equipment	Total	Total Incl O&P
0010	**CONCRETE CRACK REPAIR**									
1000	Structural repair of concrete cracks by epoxy injection (ACI RAP-1)									
1001	suitable for horizontal, vertical and overhead repairs									
1010	Clean/grind concrete surface(s) free of contaminants	1 Cefi	400	.020	L.F.		1.04		1.04	1.52
1015	Rout crack with v-notch crack chaser, if needed	C-32	600	.027		.08	1.28	.25	1.61	2.26
1020	Blow out crack with oil-free dry compressed air (1 pass)	C-28	3000	.003			.14	.01	.15	.21
1030	Install surface-mounted entry ports (spacing = concrete depth)	1 Cefi	400	.020	Ea.	.62	1.04		1.66	2.20
1040	Cap crack at surface with epoxy gel (per side/face)		400	.020	L.F.	.69	1.04		1.73	2.28
1050	Snap off ports, grind off epoxy cap residue after injection		200	.040	"		2.07		2.07	3.04
1100	Manual injection with 2-part epoxy cartridge, excludes prep									
1110	Up to 1/32" (0.03125") wide x 4" deep	1 Cefi	160	.050	L.F.	.35	2.59		2.94	4.18
1120	6" deep		120	.067		.52	3.45		3.97	5.60
1130	8" deep		107	.075		.69	3.87		4.56	6.40
1140	10" deep		100	.080		.87	4.14		5.01	7
1150	12" deep		96	.083		1.04	4.32		5.36	7.50
1210	Up to 1/16" (0.0625") wide x 4" deep		160	.050		.69	2.59		3.28	4.56
1220	6" deep		120	.067		1.04	3.45		4.49	6.20
1230	8" deep		107	.075		1.39	3.87		5.26	7.20
1240	10" deep		100	.080		1.74	4.14		5.88	7.95
1250	12" deep		96	.083		2.08	4.32		6.40	8.65
1310	Up to 3/32" (0.09375") wide x 4" deep		160	.050		1.04	2.59		3.63	4.95
1320	6" deep		120	.067		1.56	3.45		5.01	6.75
1330	8" deep		107	.075		2.08	3.87		5.95	7.95
1340	10" deep		100	.080		2.60	4.14		6.74	8.90
1350	12" deep		96	.083		3.13	4.32		7.45	9.80
1410	Up to 1/8" (0.125") wide x 4" deep		160	.050		1.39	2.59		3.98	5.35
1420	6" deep		120	.067		2.08	3.45		5.53	7.35
1430	8" deep		107	.075		2.78	3.87		6.65	8.70
1440	10" deep		100	.080		3.47	4.14		7.61	9.85
1450	12" deep		96	.083		4.17	4.32		8.49	10.95
1500	Pneumatic injection with 2-part bulk epoxy, excludes prep									
1510	Up to 5/32" (0.15625") wide x 4" deep	C-31	240	.033	L.F.	.32	1.73	1.34	3.39	4.35
1520	6" deep		180	.044		.48	2.30	1.79	4.57	5.85
1530	8" deep		160	.050		.63	2.59	2.01	5.23	6.70
1540	10" deep		150	.053		.79	2.76	2.15	5.70	7.30

34

For customer support on your Concrete & Masonry Costs with RSMeans Data, call 800.448.8182.

03 01 Maintenance of Concrete

03 01 30 – Maintenance of Cast-In-Place Concrete

03 01 30.71 Concrete Crack Repair	Crew	Daily Output	Labor-Hours	Unit	Material	2021 Bare Costs Labor	2021 Bare Costs Equipment	Total	Total Incl O&P
1550	12" deep	C-31	144	.056	L.F.	.95	2.88	2.23	6.06	7.75
1610	Up to 3/16" (0.1875") wide x 4" deep		240	.033		.38	1.73	1.34	3.45	4.42
1620	6" deep		180	.044		.57	2.30	1.79	4.66	5.95
1630	8" deep		160	.050		.76	2.59	2.01	5.36	6.85
1640	10" deep		150	.053		.95	2.76	2.15	5.86	7.45
1650	12" deep		144	.056		1.14	2.88	2.23	6.25	7.95
1710	Up to 1/4" (0.25") wide x 4" deep		240	.033		.51	1.73	1.34	3.58	4.56
1720	6" deep		180	.044		.76	2.30	1.79	4.85	6.20
1730	8" deep		160	.050		1.02	2.59	2.01	5.62	7.15
1740	10" deep		150	.053		1.27	2.76	2.15	6.18	7.80
1750	12" deep	▼	144	.056	▼	1.52	2.88	2.23	6.63	8.35
2000	Non-structural filling of concrete cracks by gravity-fed resin (ACI RAP-2)									
2001	suitable for individual cracks in stable horizontal surfaces only									
2010	Clean/grind concrete surface(s) free of contaminants	1 Cefi	400	.020	L.F.		1.04		1.04	1.52
2020	Rout crack with v-notch crack chaser, if needed	C-32	600	.027		.08	1.28	.25	1.61	2.26
2030	Blow out crack with oil-free dry compressed air (1 pass)	C-28	3000	.003			.14	.01	.15	.21
2040	Cap crack with epoxy gel at underside of elevated slabs, if needed	1 Cefi	400	.020		.69	1.04		1.73	2.28
2050	Insert backer rod into crack, if needed		400	.020		.03	1.04		1.07	1.55
2060	Partially fill crack with fine dry sand, if needed		800	.010		.02	.52		.54	.78
2070	Apply two beads of sealant alongside crack to form reservoir, if needed	▼	400	.020	▼	.27	1.04		1.31	1.81
2100	Manual filling with squeeze bottle of 2-part epoxy resin									
2110	Full depth crack up to 1/16" (0.0625") wide x 4" deep	1 Cefi	300	.027	L.F.	.07	1.38		1.45	2.10
2120	6" deep		240	.033		.11	1.73		1.84	2.65
2130	8" deep		200	.040		.14	2.07		2.21	3.19
2140	10" deep		185	.043		.18	2.24		2.42	3.47
2150	12" deep		175	.046		.21	2.37		2.58	3.70
2210	Full depth crack up to 1/8" (0.125") wide x 4" deep		270	.030		.14	1.53		1.67	2.40
2220	6" deep		215	.037		.21	1.93		2.14	3.05
2230	8" deep		180	.044		.28	2.30		2.58	3.68
2240	10" deep		165	.048		.35	2.51		2.86	4.07
2250	12" deep		155	.052		.42	2.67		3.09	4.38
2310	Partial depth crack up to 3/16" (0.1875") wide x 1" deep		300	.027		.05	1.38		1.43	2.08
2410	Up to 1/4" (0.250") wide x 1" deep		270	.030		.07	1.53		1.60	2.33
2510	Up to 5/16" (0.3125") wide x 1" deep		250	.032		.09	1.66		1.75	2.53
2610	Up to 3/8" (0.375") wide x 1" deep	▼	240	.033	▼	.11	1.73		1.84	2.65

03 01 30.72 Concrete Surface Repairs

		Crew	Daily Output	Labor-Hours	Unit	Material	2021 Bare Costs Labor	2021 Bare Costs Equipment	Total	Total Incl O&P
0010	**CONCRETE SURFACE REPAIRS**									
1000	Spall repairs using low-pressure spraying (ACI RAP-3)									
1001	Suitable for vertical and overhead repairs up to 3" thick									
1010	Sound the concrete surface to locate delaminated areas	1 Cefi	2000	.004	S.F.		.21		.21	.30
1020	Remove concrete in repair areas to fully expose reinforcing bars	B-9E	160	.100			4.81	.21	5.02	7.35
1030	Mark the perimeter of each repair area	1 Cefi	2000	.004	▼		.21		.21	.30
1040	Saw cut the perimeter of each repair area down to reinforcing bars	B-89C	160	.050	L.F.	.03	2.59	.36	2.98	4.24
1050	If corroded rebar is exposed, remove concrete to 3/4" below corroded rebars									
1051	Single layer of #4 rebar	B-9E	64	.250	S.F.		12.05	.51	12.56	18.30
1052	#5 rebar		53.30	.300			14.45	.62	15.07	22
1053	#6 rebar		45.70	.350			16.85	.72	17.57	26
1054	#7 rebar		40	.400			19.25	.82	20.07	29.50
1055	#8 rebar		35.60	.449			21.50	.92	22.42	33
1056	Double layer of #4 rebar		35.60	.449			21.50	.92	22.42	33
1057	#5 rebar		32	.500			24	1.03	25.03	36.50
1058	#6 rebar		29.10	.550	▼		26.50	1.13	27.63	40

03 01 30.72 Concrete Surface Repairs		Crew	Daily Output	Labor-Hours	Unit	Material	2021 Bare Costs Labor	Equipment	Total	Total Incl O&P
1059	#7 rebar	B-9E	26.70	.599	S.F.		29	1.23	30.23	44
1060	#8 rebar	↓	24.60	.650	↓		31.50	1.33	32.83	47.50
1100	See section 03 21 11.60 for placement of supplemental reinforcement bars									
1104	See section 02 41 19.19 for handling & removal of debris									
1106	See section 02 41 19.20 for disposal of debris									
1110	Final cleaning by sandblasting	E-11	1100	.029	S.F.	.52	1.41	.27	2.20	3.06
1120	By high pressure water	C-29	500	.016	↓		.71	.19	.90	1.27
1130	Blow off dust and debris with oil-free dry compressed air	C-28	16000	.001	↓		.03		.03	.04
1140	Mix up repair material with concrete mixer	C-30	200	.040	C.F.	12.50	1.78	.74	15.02	17.20
1150	Place repair material by low-pressure spray in lifts up to 3" thick	C-9	200	.160	"		7.85	2.38	10.23	14.25
1160	Screed and initial float of repair material by hand	C-10	2000	.012	S.F.		.59		.59	.87
1170	Float and steel trowel finish repair material by hand	"	1265	.019			.94		.94	1.38
1180	Cure with sprayed membrane curing compound	2 Cefi	6200	.003	↓	.12	.13		.25	.33
2000	Surface repairs using form-and-pour techniques (ACI RAP-4)									
2001	Suitable for walls, columns, beam sides/bottoms, and slab soffits									
2010	Sound the concrete surface to locate delaminated areas	1 Cefi	2000	.004	S.F.		.21		.21	.30
2020	Remove concrete in repair areas to fully expose reinforcing bars	B-9E	160	.100			4.81	.21	5.02	7.35
2030	Mark the perimeter of each repair area	1 Cefi	2000	.004	↓		.21		.21	.30
2040	Saw cut the perimeter of each repair area down to reinforcing bars	B-89C	160	.050	L.F.	.03	2.59	.36	2.98	4.24
2050	If corroded rebar is exposed, remove concrete to 3/4" below corroded rebars									
2051	Single layer of #4 rebar	B-9E	64	.250	S.F.		12.05	.51	12.56	18.30
2052	#5 rebar		53.30	.300			14.45	.62	15.07	22
2053	#6 rebar		45.70	.350			16.85	.72	17.57	26
2054	#7 rebar		40	.400			19.25	.82	20.07	29.50
2055	#8 rebar		35.60	.449			21.50	.92	22.42	33
2056	Double layer of #4 rebar		35.60	.449			21.50	.92	22.42	33
2057	#5 rebar		32	.500			24	1.03	25.03	36.50
2058	#6 rebar		29.10	.550			26.50	1.13	27.63	40
2059	#7 rebar		26.70	.599			29	1.23	30.23	44
2060	#8 rebar	↓	24.60	.650	↓		31.50	1.33	32.83	47.50
2100	See section 03 21 11.60 for placement of supplemental reinforcement bars									
2105	See section 02 41 19.19 for handling & removal of debris									
2110	See section 02 41 19.20 for disposal of debris									
2120	For slab soffit repairs, drill 2" diam. holes to 6" deep for conc placement	B-89A	16.50	.970	Ea.	.27	49	7.35	56.62	82.50
2130	For each additional inch of slab thickness in same hole, add	"	1080	.015		.04	.75	.11	.90	1.30
2140	Drill holes for expansion shields, 3/4" diameter x 4" deep	1 Carp	45	.178	↓		9.70		9.70	14.50
2150	Final cleaning by sandblasting	E-11	1100	.029	S.F.	.52	1.41	.27	2.20	3.06
2160	By high pressure water	C-29	500	.016	↓		.71	.19	.90	1.27
2170	Blow off dust and debris with oil-free dry compressed air	C-28	16000	.001	↓		.03		.03	.04
2180	Build wood forms including chutes where needed for concrete placement	C-2	960	.050	SFCA	4.63	2.67		7.30	9.10
2190	Insert expansion shield, coil rod, coil tie and plastic cone in each hole	1 Carp	32	.250	Ea.	5.50	13.70		19.20	26.50
2200	Install wood forms and fasten with coil rod and coil nut	C-2	960	.050	SFCA	.58	2.67		3.25	4.62
2210	Mix up bagged repair material by hand with a mixing paddle	1 Cefi	30	.267	C.F.	12.50	13.80		26.30	34
2220	By concrete mixer	C-30	200	.040		12.50	1.78	.74	15.02	17.20
2230	Place repair material by hand and consolidate	C-6A	200	.080	↓		4.14	.23	4.37	6.30
2240	Remove wood forms	C-2	1600	.030	SFCA		1.60		1.60	2.39
2250	Chip/grind off extra concrete at fill chutes	B-9E	16	1	Ea.	.12	48	2.05	50.17	73.50
2260	Break ties and patch voids	1 Cefi	270	.030	"	.08	1.53		1.61	2.34
2270	Cure with sprayed membrane curing compound	2 Cefi	6200	.003	S.F.	.12	.13		.25	.33
3000	Surface repairs using form-and-pump techniques (ACI RAP-5)									
3001	Suitable for walls, columns, beam sides/bottoms, and slab soffits									
3010	Sound the concrete surface to locate delaminated areas	1 Cefi	2000	.004	S.F.		.21		.21	.30
3020	Remove concrete in repair areas to fully expose reinforcing bars	B-9E	160	.100	↓		4.81	.21	5.02	7.35

36

For customer support on your Concrete & Masonry Costs with RSMeans Data, call 800.448.8182.

03 01 30 – Maintenance of Cast-In-Place Concrete

03 01 30.72 Concrete Surface Repairs	Crew	Daily Output	Labor-Hours	Unit	Material	2021 Bare Costs Labor	Equipment	Total	Total Incl O&P
3030	Mark the perimeter of each repair area	1 Cefi	2000	.004	S.F.		.21		.21	.30
3040	Saw cut the perimeter of each repair area down to reinforcing bars	B-89C	160	.050	L.F.	.03	2.59	.36	2.98	4.24
3050	If corroded rebar is exposed, remove concrete to 3/4" below corroded rebars									
3051	Single layer of #4 rebar	B-9E	64	.250	S.F.		12.05	.51	12.56	18.30
3052	#5 rebar		53.30	.300			14.45	.62	15.07	22
3053	#6 rebar		45.70	.350			16.85	.72	17.57	26
3054	#7 rebar		40	.400			19.25	.82	20.07	29.50
3055	#8 rebar		35.60	.449			21.50	.92	22.42	33
3056	Double layer of #4 rebar		35.60	.449			21.50	.92	22.42	33
3057	#5 rebar		32	.500			24	1.03	25.03	36.50
3058	#6 rebar		29.10	.550			26.50	1.13	27.63	40
3059	#7 rebar		26.70	.599			29	1.23	30.23	44
3060	#8 rebar		24.60	.650			31.50	1.33	32.83	47.50
3100	See section 03 21 11.60 for placement of supplemental reinforcement bars									
3104	See section 02 41 19.19 for handling & removal of debris									
3106	See section 02 41 19.20 for disposal of debris									
3110	Drill holes for expansion shields, 3/4" diameter x 4" deep	1 Carp	45	.178	Ea.		9.70		9.70	14.50
3120	Final cleaning by sandblasting	E-11	1100	.029	S.F.	.52	1.41	.27	2.20	3.06
3130	By high pressure water	C-29	500	.016			.71	.19	.90	1.27
3140	Blow off dust and debris with oil-free dry compressed air	C-28	16000	.001			.03		.03	.04
3150	Build wood forms and drill holes for injection ports	C-2	960	.050	SFCA	4.63	2.67		7.30	9.10
3160	Assemble PVC injection ports with ball valves and install in forms	1 Plum	8	1	Ea.	90.50	67.50		158	201
3170	Insert expansion shield, coil rod, coil tie and plastic cone in each hole	1 Carp	32	.250	"	5.50	13.70		19.20	26.50
3180	Install wood forms and fasten with coil rod and coil nut	C-2	960	.050	SFCA	.58	2.67		3.25	4.62
3190	Mix up self-consolidating repair material with concrete mixer	C-30	200	.040	C.F.	12.75	1.78	.74	15.27	17.45
3200	Place repair material by pump and pressurize	C-9	200	.160	"		7.85	2.38	10.23	14.25
3210	Remove wood forms	C-2	1600	.030	SFCA		1.60		1.60	2.39
3220	Chip/grind off extra concrete at injection ports	B-9E	64	.250	Ea.	.12	12.05	.51	12.68	18.45
3230	Break ties and patch voids	1 Cefi	270	.030	"	.08	1.53		1.61	2.34
3240	Cure with sprayed membrane curing compound	2 Cefi	6200	.003	S.F.	.12	.13		.25	.33
4000	Vertical & overhead spall repair by hand application (ACI RAP-6)									
4001	Generally recommended for thin repairs that are cosmetic in nature									
4010	Remove concrete in repair areas to fully expose reinforcing bars	B-9E	160	.100	S.F.		4.81	.21	5.02	7.35
4020	Saw cut the perimeter of each repair area to 1/2" depth	B-89C	160	.050	L.F.	.02	2.59	.36	2.97	4.22
4030	If corroded rebar is exposed, remove concrete to 3/4" below corroded rebars									
4031	Single layer of #4 rebar	B-9E	64	.250	S.F.		12.05	.51	12.56	18.30
4032	#5 rebar		53.30	.300			14.45	.62	15.07	22
4033	#6 rebar		45.70	.350			16.85	.72	17.57	26
4034	#7 rebar		40	.400			19.25	.82	20.07	29.50
4035	#8 rebar		35.60	.449			21.50	.92	22.42	33
4036	Double layer of #4 rebar		35.60	.449			21.50	.92	22.42	33
4037	#5 rebar		32	.500			24	1.03	25.03	36.50
4038	#6 rebar		29.10	.550			26.50	1.13	27.63	40
4039	#7 rebar		26.70	.599			29	1.23	30.23	44
4040	#8 rebar		24.60	.650			31.50	1.33	32.83	47.50
4080	See section 03 21 11.60 for placement of supplemental reinforcement bars									
4090	See section 02 41 19.19 for handling & removal of debris									
4095	See section 02 41 19.20 for disposal of debris									
4100	Final cleaning by sandblasting	E-11	1100	.029	S.F.	.52	1.41	.27	2.20	3.06
4110	By high pressure water	C-29	500	.016			.71	.19	.90	1.27
4120	Blow off dust and debris with oil-free dry compressed air	C-28	16000	.001			.03		.03	.04
4130	Brush on bonding material	1 Cefi	320	.025		.11	1.30		1.41	2.02
4140	Mix up bagged repair material by hand with a mixing paddle	"	30	.267	C.F.	73.50	13.80		87.30	101

03 01 30.72 Concrete Surface Repairs		Crew	Daily Output	Labor-Hours	Unit	Material	2021 Bare Costs Labor	2021 Bare Costs Equipment	Total	Total Incl O&P
4150	By concrete mixer	C-30	200	.040	C.F.	73.50	1.78	.74	76.02	84
4160	Place repair material by hand in lifts up to 2" deep for vertical repairs	C-10	48	.500			24.50		24.50	36.50
4170	Up to 1" deep for overhead repairs		36	.667			33		33	48.50
4180	Screed and initial float of repair material by hand		2000	.012	S.F.		.59		.59	.87
4190	Float and steel trowel finish repair material by hand		1265	.019			.94		.94	1.38
4200	Cure with sprayed membrane curing compound	2 Cefi	6200	.003		.12	.13		.25	.33
5000	Spall repair of horizontal concrete surfaces (ACI RAP-7)									
5001	Suitable for structural slabs, slabs-on-grade, balconies, interior floors									
5010	Sound the concrete surface to locate delaminated areas	1 Cefi	2000	.004	SF Flr.		.21		.21	.30
5020	Mark the perimeter of each repair area	"	2000	.004	"		.21		.21	.30
5030	Saw cut the perimeter of each repair area down to reinforcing bars	B-89	1060	.015	L.F.	.03	.79	1.01	1.83	2.33
5040	Remove concrete in repair areas to fully expose reinforcing bars	B-9E	160	.100	S.F.		4.81	.21	5.02	7.35
5050	If corroded rebar is exposed, remove concrete to 3/4" below corroded rebars									
5051	Single layer of #4 rebar	B-9E	64	.250	S.F.		12.05	.51	12.56	18.30
5052	#5 rebar		53.30	.300			14.45	.62	15.07	22
5053	#6 rebar		45.70	.350			16.85	.72	17.57	26
5054	#7 rebar		40	.400			19.25	.82	20.07	29.50
5055	#8 rebar		35.60	.449			21.50	.92	22.42	33
5056	Double layer of #4 rebar		35.60	.449			21.50	.92	22.42	33
5057	#5 rebar		32	.500			24	1.03	25.03	36.50
5058	#6 rebar		29.10	.550			26.50	1.13	27.63	40
5059	#7 rebar		26.70	.599			29	1.23	30.23	44
5060	#8 rebar		24.60	.650			31.50	1.33	32.83	47.50
5100	See section 03 21 11.60 for placement of supplemental reinforcement bars									
5110	See section 02 41 19.19 for handling & removal of debris									
5115	See section 02 41 19.20 for disposal of debris									
5120	Final cleaning by sandblasting	E-11	1100	.029	S.F.	.52	1.41	.27	2.20	3.06
5130	By high pressure water	C-29	500	.016			.71	.19	.90	1.27
5140	Blow off dust and debris with oil-free dry compressed air	C-28	16000	.001			.03		.03	.04
5160	Brush on bonding material	1 Cefi	320	.025		.11	1.30		1.41	2.02
5170	Mix up bagged repair material by hand with a mixing paddle	"	30	.267	C.F.	73.50	13.80		87.30	101
5180	By concrete mixer	C-30	200	.040		73.50	1.78	.74	76.02	84
5190	Place repair material by hand, consolidate and screed	C-6A	200	.080			4.14	.23	4.37	6.30
5200	Bull float, manual float & broom finish	C-10	1850	.013	S.F.		.64		.64	.94
5210	Bull float, manual float & trowel finish	"	1265	.019			.94		.94	1.38
5220	Cure with sprayed membrane curing compound	2 Cefi	6200	.003		.12	.13		.25	.33
6000	Spall repairs by the Preplaced Aggregate method (ACI RAP-9)									
6001	Suitable for vertical, overhead, and underwater repairs									
6010	Sound the concrete surface to locate delaminated areas	1 Cefi	2000	.004	S.F.		.21		.21	.30
6020	Remove concrete in repair areas to fully expose reinforcing bars	B-9E	160	.100			4.81	.21	5.02	7.35
6030	Mark the perimeter of each repair area	1 Cefi	2000	.004			.21		.21	.30
6040	Saw cut the perimeter of each repair area down to reinforcing bars	B-89C	160	.050	L.F.	.03	2.59	.36	2.98	4.24
6050	If corroded rebar is exposed, remove concrete to 3/4" below corroded rebars									
6051	Single layer of #4 rebar	B-9E	64	.250	S.F.		12.05	.51	12.56	18.30
6052	#5 rebar		53.30	.300			14.45	.62	15.07	22
6053	#6 rebar		45.70	.350			16.85	.72	17.57	26
6054	#7 rebar		40	.400			19.25	.82	20.07	29.50
6055	#8 rebar		35.60	.449			21.50	.92	22.42	33
6056	Double layer of #4 rebar		35.60	.449			21.50	.92	22.42	33
6057	#5 rebar		32	.500			24	1.03	25.03	36.50
6058	#6 rebar		29.10	.550			26.50	1.13	27.63	40
6059	#7 rebar		26.70	.599			29	1.23	30.23	44
6060	#8 rebar		24.60	.650			31.50	1.33	32.83	47.50

03 01 30 – Maintenance of Cast-In-Place Concrete

03 01 30.72 Concrete Surface Repairs	Crew	Daily Output	Labor-Hours	Unit	Material	2021 Bare Costs Labor	2021 Bare Costs Equipment	Total	Total Incl O&P
6100	See section 03 21 11.60 for placement of supplemental reinforcement bars								
6104	See section 02 41 19.19 for handling & removal of debris								
6106	See section 02 41 19.20 for disposal of debris								
6110	Drill holes for expansion shields, 3/4" diameter x 4" deep	1 Carp	45	.178	Ea.		9.70		9.70	14.50
6120	Final cleaning by sandblasting	E-11	1100	.029	S.F.	.52	1.41	.27	2.20	3.06
6130	By high pressure water	C-29	500	.016			.71	.19	.90	1.27
6140	Blow off dust and debris with oil-free dry compressed air	C-28	16000	.001			.03		.03	.04
6150	Build wood forms and drill holes for injection ports	C-2	960	.050	SFCA	4.63	2.67		7.30	9.10
6160	Assemble PVC injection ports with ball valves and install in forms	1 Plum	8	1	Ea.	90.50	67.50		158	201
6170	Insert expansion shield, coil rod, coil tie and plastic cone in each hole	1 Carp	32	.250	"	5.50	13.70		19.20	26.50
6180	Install wood forms and fasten with coil rod and coil nut	C-2	960	.050	SFCA	.58	2.67		3.25	4.62
6185	Install pre-washed gap-graded aggregate as forms are installed	B-24	160	.150	C.F.	1.11	7.55		8.66	12.40
6190	Mix up pre-bagged grout material with concrete mixer	C-30	200	.040		15.40	1.78	.74	17.92	20.50
6200	Place grout by pump from bottom up	C-9	200	.160			7.85	2.38	10.23	14.25
6210	Remove wood forms	C-2	1600	.030	SFCA		1.60		1.60	2.39
6220	Chip/grind off extra concrete at injection ports	B-9E	64	.250	Ea.	.12	12.05	.51	12.68	18.45
6230	Break ties and patch voids	1 Cefi	270	.030	"	.08	1.53		1.61	2.34
6240	Cure with sprayed membrane curing compound	2 Cefi	6200	.003	S.F.	.12	.13		.25	.33
7000	Leveling & reprofiling of vertical & overhead surfaces (ACI RAP-10)									
7001	Used to provide an acceptable surface for aesthetic or protective coatings									
7010	Clean the surface by sandblasting, leaving an open profile	E-11	1100	.029	S.F.	.52	1.41	.27	2.20	3.06
7020	By high pressure water	C-29	500	.016			.71	.19	.90	1.27
7030	Blow off dust and debris with oil-free dry compressed air	C-28	16000	.001			.03		.03	.04
7040	Mix up bagged repair material by hand with a mixing paddle	1 Cefi	30	.267	C.F.	73.50	13.80		87.30	101
7050	Place repair material by hand with steel trowel to fill pores & honeycombs	C-10	1265	.019	S.F.		.94		.94	1.38
7060	Cure with sprayed membrane curing compound	2 Cefi	6200	.003	"	.12	.13		.25	.33
8000	Concrete repair by shotcrete application (ACI RAP-12)									
8001	Suitable for vertical and overhead surfaces									
8010	Sound the concrete surface to locate delaminated areas	1 Cefi	2000	.004	S.F.		.21		.21	.30
8020	Remove concrete in repair areas to fully expose reinforcing bars	B-9E	160	.100			4.81	.21	5.02	7.35
8030	Mark the perimeter of each repair area	1 Cefi	2000	.004			.21		.21	.30
8040	Saw cut the perimeter of each repair area down to reinforcing bars	B-89C	160	.050	L.F.	.03	2.59	.36	2.98	4.24
8050	If corroded rebar is exposed, remove concrete to 3/4" below corroded rebars									
8051	Single layer of #4 rebar	B-9E	64	.250	S.F.		12.05	.51	12.56	18.30
8052	#5 rebar		53.30	.300			14.45	.62	15.07	22
8053	#6 rebar		45.70	.350			16.85	.72	17.57	26
8054	#7 rebar		40	.400			19.25	.82	20.07	29.50
8055	#8 rebar		35.60	.449			21.50	.92	22.42	33
8056	Double layer of #4 rebar		35.60	.449			21.50	.92	22.42	33
8057	#5 rebar		32	.500			24	1.03	25.03	36.50
8058	#6 rebar		29.10	.550			26.50	1.13	27.63	40
8059	#7 rebar		26.70	.599			29	1.23	30.23	44
8060	#8 rebar		24.60	.650			31.50	1.33	32.83	47.50
8100	See section 03 21 11.60 for placement of supplemental reinforcement bars									
8110	Final cleaning by sandblasting	E-11	1100	.029	S.F.	.52	1.41	.27	2.20	3.06
8120	By high pressure water	C-29	500	.016			.71	.19	.90	1.27
8130	Blow off dust and debris with oil-free dry compressed air	C-28	16000	.001			.03		.03	.04
8140	Mix up repair material with concrete mixer	C-30	200	.040	C.F.	12.50	1.78	.74	15.02	17.20
8150	Place repair material by wet shotcrete process to full depth of repair	C-8C	200	.240	"		11.60	3.15	14.75	20.50
8160	Carefully trim excess material and steel trowel finish without overworking	C-10	1265	.019	S.F.		.94		.94	1.38
8170	Cure with sprayed membrane curing compound	2 Cefi	6200	.003	"	.12	.13		.25	.33
9000	Surface repair by Methacrylate flood coat (ACI RAP-13)									
9010	Suitable for healing and sealing horizontal surfaces only									

03 01 Maintenance of Concrete

03 01 30 – Maintenance of Cast-In-Place Concrete

03 01 30.72 Concrete Surface Repairs	Crew	Daily Output	Labor-Hours	Unit	Material	2021 Bare Costs Labor	Equipment	Total	Total Incl O&P
9020	Large cracks must previously have been repaired or filled									
9100	Shotblast entire surface to remove contaminants	A-1A	4000	.002	S.F.		.11	.05	.16	.23
9200	Blow off dust and debris with oil-free dry compressed air	C-28	16000	.001			.03		.03	.04
9300	Flood coat surface w/Methacrylate, distribute w/broom/squeegee, no prep.	3 Cefi	8000	.003		.97	.16		1.13	1.30
9400	Lightly broadcast even coat of dry silica sand while sealer coat is tacky	1 Cefi	8000	.001	▼	.01	.05		.06	.10

03 05 Common Work Results for Concrete

03 05 05 – Selective Demolition for Concrete

03 05 05.10 Selective Demolition, Concrete

		Crew	Daily Output	Labor-Hours	Unit	Material	2021 Bare Costs Labor	Equipment	Total	Total Incl O&P
0010	**SELECTIVE DEMOLITION, CONCRETE** R024119-10									
0012	Excludes saw cutting, torch cutting, loading or hauling									
0050	Break into small pieces, reinf. less than 1% of cross-sectional area	B-9	24	1.667	C.Y.		74.50	14.80	89.30	127
0060	Reinforcing 1% to 2% of cross-sectional area		16	2.500			112	22	134	192
0070	Reinforcing more than 2% of cross-sectional area	▼	8	5	▼		224	44.50	268.50	385
0150	Remove whole pieces, up to 2 tons per piece	E-18	36	1.111	Ea.		67	42.50	109.50	150
0160	2-5 tons per piece		30	1.333			80.50	51	131.50	180
0170	5-10 tons per piece		24	1.667			101	63.50	164.50	225
0180	10-15 tons per piece	▼	18	2.222			134	85	219	300
0250	Precast unit embedded in masonry, up to 1 C.F.	D-1	16	1			48.50		48.50	73.50
0260	1-2 C.F.		12	1.333			65		65	98
0270	2-5 C.F.		10	1.600			78		78	117
0280	5-10 C.F.	▼	8	2	▼		97.50		97.50	147
0990	For hydrodemolition see Section 02 41 13.15									

03 05 13 – Basic Concrete Materials

03 05 13.20 Concrete Admixtures and Surface Treatments

		Crew	Daily Output	Labor-Hours	Unit	Material	2021 Bare Costs Labor	Equipment	Total	Total Incl O&P
0010	**CONCRETE ADMIXTURES AND SURFACE TREATMENTS**									
0040	Abrasives, aluminum oxide, over 20 tons				Lb.	1.90			1.90	2.09
0050	1 to 20 tons					1.98			1.98	2.17
0070	Under 1 ton					2.05			2.05	2.26
0100	Silicon carbide, black, over 20 tons					2.97			2.97	3.27
0110	1 to 20 tons					3.09			3.09	3.40
0120	Under 1 ton				▼	3.21			3.21	3.53
0200	Air entraining agent, .7 to 1.5 oz. per bag, 55 gallon drum				Gal.	22			22	24.50
0220	5 gallon pail					34.50			34.50	38
0300	Bonding agent, acrylic latex, 250 S.F./gallon, 5 gallon pail					27			27	29.50
0320	Epoxy resin, 80 S.F./gallon, 4 gallon case				▼	62.50			62.50	69
0400	Calcium chloride, 50 lb. bags, T.L. lots				Ton	1,450			1,450	1,575
0420	Less than truckload lots				Bag	24			24	26.50
0500	Carbon black, liquid, 2 to 8 lb. per bag of cement				Lb.	16.05			16.05	17.65
0600	Concrete admixture, integral colors, dry pigment, 5 lb. bag				Ea.	31			31	34
0610	10 lb. bag					47			47	51.50
0620	25 lb. bag				▼	82			82	90.50
0920	Dustproofing compound, 250 S.F./gal., 5 gallon pail				Gal.	8			8	8.80
1010	Epoxy based, 125 S.F./gal., 5 gallon pail				"	65.50			65.50	72
1100	Hardeners, metallic, 55 lb. bags, natural (grey)				Lb.	1.04			1.04	1.15
1200	Colors					1.53			1.53	1.68
1300	Non-metallic, 55 lb. bags, natural grey					.60			.60	.66
1320	Colors				▼	.85			.85	.93
1550	Release agent, for tilt slabs, 5 gallon pail				Gal.	18.55			18.55	20.50
1570	For forms, 5 gallon pail				▼	13.90			13.90	15.30

03 05 13 – Basic Concrete Materials

	03 05 13.20 Concrete Admixtures and Surface Treatments	Crew	Daily Output	Labor-Hours	Unit	Material	2021 Bare Costs Labor	Equipment	Total	Total Incl O&P
1590	Concrete release agent for forms, 100% biodegradable, zero VOC, 5 gal. pail **G**				Gal.	21.50			21.50	24
1595	55 gallon drum **G**					19.90			19.90	22
1600	Sealer, hardener and dustproofer, epoxy-based, 125 S.F./gal., 5 gallon unit					65.50			65.50	72
1620	3 gallon unit					72			72	79
1630	Sealer, solvent-based, 250 S.F./gal., 55 gallon drum					21			21	23
1640	5 gallon pail					41			41	45
1650	Sealer, water based, 350 S.F., 55 gallon drum					26.50			26.50	29
1660	5 gallon pail					36			36	39.50
1900	Set retarder, 100 S.F./gal., 1 gallon pail				▼	6.35			6.35	6.95
2000	Waterproofing, integral 1 lb. per bag of cement				Lb.	1.04			1.04	1.14
2100	Powdered metallic, 40 lbs. per 100 S.F., standard colors					4.50			4.50	4.95
2120	Premium colors				▼	6.30			6.30	6.95
3000	For colored ready-mix concrete, add to prices in section 03 31 13.35									
3100	Subtle shades, 5 lb. dry pigment per C.Y., add				C.Y.	31			31	34
3400	Medium shades, 10 lb. dry pigment per C.Y., add					47			47	51.50
3700	Deep shades, 25 lb. dry pigment per C.Y., add				▼	82			82	90.50
6000	Concrete ready mix additives, recycled coal fly ash, mixed at plant **G**				Ton	58.50			58.50	64.50
6010	Recycled blast furnace slag, mixed at plant **G**				"	88			88	96.50

03 05 13.25 Aggregate

0010	**AGGREGATE** R033105-20									
0100	Lightweight vermiculite or perlite, 4 C.F. bag, C.L. lots **G**				Bag	28			28	30.50
0150	L.C.L. lots **G**				"	31			31	34
0250	Sand & stone, loaded at pit, crushed bank gravel				Ton	21			21	23
0350	Sand, washed, for concrete R033105-50					21.50			21.50	23.50
0400	For plaster or brick					21.50			21.50	23.50
0450	Stone, 3/4" to 1-1/2"					20			20	22
0470	Round, river stone					43.50			43.50	48
0500	3/8" roofing stone & 1/2" pea stone					32			32	35
0550	For trucking 10-mile round trip, add to the above	B-34B	117	.068			3.51	4.95	8.46	10.70
0600	For trucking 30-mile round trip, add to the above	"	72	.111	▼		5.70	8.05	13.75	17.35
0850	Sand & stone, loaded at pit, crushed bank gravel				C.Y.	29.50			29.50	32.50
0950	Sand, washed, for concrete					30			30	33
1000	For plaster or brick					30			30	33
1050	Stone, 3/4" to 1-1/2"					38.50			38.50	42.50
1055	Round, river stone					45			45	49.50
1100	3/8" roofing stone & 1/2" pea stone					31			31	34
1150	For trucking 10-mile round trip, add to the above	B-34B	78	.103			5.25	7.45	12.70	16
1200	For trucking 30-mile round trip, add to the above	"	48	.167	▼		8.55	12.05	20.60	26
1310	Onyx chips, 50 lb. bags				Cwt.	80			80	88
1330	Quartz chips, 50 lb. bags					34.50			34.50	37.50
1410	White marble, 3/8" to 1/2", 50 lb. bags				▼	10.25			10.25	11.30
1430	3/4", bulk				Ton	179			179	197

03 05 13.30 Cement

0010	**CEMENT** R033105-20									
0240	Portland, Type I/II, T.L. lots, 94 lb. bags				Bag	14.70			14.70	16.15
0250	L.T.L./L.C.L. lots R033105-30				"	19.95			19.95	22
0300	Trucked in bulk, per cwt.				Cwt.	9.10			9.10	10
0400	Type III, high early strength, T.L. lots, 94 lb. bags R033105-40				Bag	11.80			11.80	13
0420	L.T.L. or L.C.L. lots					21.50			21.50	23.50
0500	White, type III, high early strength, T.L. or C.L. lots, bags R033105-50					23.50			23.50	26
0520	L.T.L. or L.C.L. lots					43			43	47.50
0600	White, type I, T.L. or C.L. lots, bags				▼	29.50			29.50	32.50

03 05 Common Work Results for Concrete

03 05 13 – Basic Concrete Materials

03 05 13.30 Cement

		Crew	Daily Output	Labor-Hours	Unit	Material	2021 Bare Costs Labor	Equipment	Total	Total Incl O&P
0620	L.T.L. or L.C.L. lots				Bag	31			31	34

03 05 13.80 Waterproofing and Dampproofing

		Crew	Daily Output	Labor-Hours	Unit	Material	2021 Bare Costs Labor	Equipment	Total	Total Incl O&P
0010	**WATERPROOFING AND DAMPPROOFING**									
0050	Integral waterproofing, add to cost of regular concrete				C.Y.	6.25			6.25	6.85

03 05 13.85 Winter Protection

		Crew	Daily Output	Labor-Hours	Unit	Material	2021 Bare Costs Labor	Equipment	Total	Total Incl O&P
0010	**WINTER PROTECTION**									
0012	For heated ready mix, add				C.Y.	5.85			5.85	6.40
0100	Temporary heat to protect concrete, 24 hours	2 Clab	50	.320	M.S.F.	181	14.20		195.20	220
0200	Temporary shelter for slab on grade, wood frame/polyethylene sheeting									
0201	Build or remove, light framing for short spans	2 Carp	10	1.600	M.S.F.	530	87.50		617.50	710
0210	Large framing for long spans	"	3	5.333	"	620	292		912	1,125
0500	Electrically heated pads, 110 volts, 15 watts/S.F., buy				S.F.	14.50			14.50	15.95
0600	20 watts/S.F., buy					19.25			19.25	21
0710	Electrically heated pads, 15 watts/S.F., 20 uses				↓	.72			.72	.80

03 11 Concrete Forming

03 11 13 – Structural Cast-In-Place Concrete Forming

03 11 13.05 Forms, Buy or Rent

			Crew	Daily Output	Labor-Hours	Unit	Material	2021 Bare Costs Labor	Equipment	Total	Total Incl O&P
0010	**FORMS, BUY OR RENT**										
0015	Aluminum, smooth face, 3' x 8', buy	G				SFCA	13.25			13.25	14.55
0020	2' x 8'	G					17.90			17.90	19.70
0050	12" x 8'	G					22.50			22.50	25
0100	6" x 8'	G					29.50			29.50	32.50
0150	3' x 4'	G					16.30			16.30	17.95
0200	2' x 4'	G					20.50			20.50	22.50
0250	12" x 4'	G					26.50			26.50	29
0300	6" x 4'	G					41			41	45
0500	Textured brick face, 3' x 8', buy	G					17.65			17.65	19.45
0550	2' x 8'	G					25.50			25.50	28
0600	12" x 8'	G					28.50			28.50	31.50
0650	6" x 8'	G					43			43	47.50
0700	3' x 4'	G					21.50			21.50	23.50
0750	2' x 4'	G					25			25	27.50
0800	12" x 4'	G					36.50			36.50	40
0850	6" x 4'	G					67.50			67.50	74.50
1000	Average cost incl. accessories but not incl. ties, buy	G					24			24	26.50
1100	Rent per month	G					1.21			1.21	1.33
2000	Metal framed plywood 2' x 8', buy						8			8	8.80
2050	2' x 3'						8.80			8.80	9.65
2100	1' x 3'						12			12	13.20
2200	Average cost incl. accessories but not incl. ties, buy						10.80			10.80	11.85
2300	Rent per month						.54			.54	.59
3000	Plywood modular prefabricated 2' x 8', buy						7.45			7.45	8.20
3050	Average cost incl. accessories but not incl. ties, buy						10.10			10.10	11.10
3200	Rent per month					↓	1.01			1.01	1.11
3900	Wood, 5/8" exterior plyform, buy					S.F.	2.32			2.32	2.55
3950	3/4" exterior plyform					"	2.12			2.12	2.33
4100	Lumber, framing, average					M.B.F.	1,200			1,200	1,325

03 11 Concrete Forming

03 11 13 – Structural Cast-In-Place Concrete Forming

03 11 13.20 Forms In Place, Beams and Girders

		Crew	Daily Output	Labor-Hours	Unit	Material	2021 Bare Costs Labor	Equipment	Total	Total Incl O&P
0010	**FORMS IN PLACE, BEAMS AND GIRDERS** R031113-40									
0500	Exterior spandrel, job-built plywood, 12" wide, 1 use R031113-60	C-2	225	.213	SFCA	4.66	11.35		16.01	22
0550	2 use		275	.175		2.43	9.30		11.73	16.60
0600	3 use		295	.163		1.86	8.70		10.56	15
0650	4 use		310	.155		1.51	8.25		9.76	13.95
1000	18" wide, 1 use		250	.192		4.22	10.25		14.47	19.95
1050	2 use		275	.175		2.32	9.30		11.62	16.45
1100	3 use		305	.157		1.69	8.40		10.09	14.35
1150	4 use		315	.152		1.39	8.10		9.49	13.70
1500	24" wide, 1 use		265	.181		3.86	9.65		13.51	18.65
1550	2 use		290	.166		2.18	8.85		11.03	15.55
1600	3 use		315	.152		1.54	8.10		9.64	13.85
1650	4 use		325	.148		1.25	7.85		9.10	13.15
2000	Interior beam, job-built plywood, 12" wide, 1 use		300	.160		5.45	8.55		14	18.75
2050	2 use		340	.141		2.62	7.55		10.17	14.15
2100	3 use		364	.132		2.18	7.05		9.23	12.90
2150	4 use		377	.127		1.78	6.80		8.58	12.10
2500	24" wide, 1 use		320	.150		3.96	8		11.96	16.30
2550	2 use		365	.132		2.23	7		9.23	12.90
2600	3 use		385	.125		1.57	6.65		8.22	11.65
2650	4 use		395	.122		1.28	6.50		7.78	11.05
3000	Encasing steel beam, hung, job-built plywood, 1 use		325	.148		4.48	7.85		12.33	16.70
3050	2 use		390	.123		2.46	6.55		9.01	12.50
3100	3 use		415	.116		1.79	6.15		7.94	11.15
3150	4 use		430	.112		1.46	5.95		7.41	10.50
3500	Bottoms only, to 30" wide, job-built plywood, 1 use		230	.209		6.60	11.15		17.75	24
3550	2 use		265	.181		3.70	9.65		13.35	18.45
3600	3 use		280	.171		2.64	9.15		11.79	16.55
3650	4 use		290	.166		2.15	8.85		11	15.50
4000	Sides only, vertical, 36" high, job-built plywood, 1 use		335	.143		6.60	7.65		14.25	18.70
4050	2 use		405	.119		3.64	6.30		9.94	13.45
4100	3 use		430	.112		2.65	5.95		8.60	11.80
4150	4 use		445	.108		2.15	5.75		7.90	10.95
4500	Sloped sides, 36" high, 1 use		305	.157		6.60	8.40		15	19.75
4550	2 use		370	.130		3.69	6.90		10.59	14.35
4600	3 use		405	.119		2.64	6.30		8.94	12.35
4650	4 use		425	.113		2.14	6		8.14	11.35
5000	Upstanding beams, 36" high, 1 use		225	.213		7.65	11.35		19	25.50
5050	2 use		255	.188		4.28	10.05		14.33	19.70
5100	3 use		275	.175		3.09	9.30		12.39	17.30
5150	4 use		280	.171		2.51	9.15		11.66	16.40

03 11 13.25 Forms In Place, Columns

		Crew	Daily Output	Labor-Hours	Unit	Material	2021 Bare Costs Labor	Equipment	Total	Total Incl O&P
0010	**FORMS IN PLACE, COLUMNS** R031113-40									
0500	Round fiberglass, 4 use per mo., rent, 12" diameter	C-1	160	.200	L.F.	22.50	10.45		32.95	40.50
0550	16" diameter R031113-60		150	.213		26	11.10		37.10	45
0600	18" diameter		140	.229		28.50	11.90		40.40	49.50
0650	24" diameter		135	.237		35.50	12.35		47.85	57.50
0700	28" diameter		130	.246		40	12.85		52.85	63
0800	30" diameter		125	.256		42	13.35		55.35	66
0850	36" diameter		120	.267		55	13.90		68.90	81.50
1500	Round fiber tube, recycled paper, 1 use, 8" diameter [G]		155	.206		2.94	10.75		13.69	19.30
1550	10" diameter [G]		155	.206		3.51	10.75		14.26	19.90

03 11 13 – Structural Cast-In-Place Concrete Forming

03 11 13.25 Forms In Place, Columns		Crew	Daily Output	Labor-Hours	Unit	Material	2021 Bare Costs Labor	Equipment	Total	Total Incl O&P
1600	12" diameter	G	C-1	150	.213	L.F.	4.14	11.10		15.24	21
1650	14" diameter	G		145	.221		5.85	11.50		17.35	23.50
1700	16" diameter	G		140	.229		6.30	11.90		18.20	25
1720	18" diameter	G		140	.229		7.40	11.90		19.30	26
1750	20" diameter	G		135	.237		7.65	12.35		20	27
1800	24" diameter	G		130	.246		12.70	12.85		25.55	33
1850	30" diameter	G		125	.256		17.60	13.35		30.95	39.50
1900	36" diameter	G		115	.278		21.50	14.50		36	45
1950	42" diameter	G		100	.320		43	16.70		59.70	72.50
2000	48" diameter	G	▼	85	.376	▼	89	19.65		108.65	128
2200	For seamless type, add						15%				
3000	Round, steel, 4 use per mo., rent, regular duty, 14" diameter	G	C-1	145	.221	L.F.	18.50	11.50		30	37.50
3050	16" diameter	G		125	.256		18.85	13.35		32.20	41
3100	Heavy duty, 20" diameter	G		105	.305		20.50	15.90		36.40	46
3150	24" diameter	G		85	.376		22.50	19.65		42.15	54
3200	30" diameter	G		70	.457		25.50	24		49.50	63.50
3250	36" diameter	G		60	.533		27.50	28		55.50	71.50
3300	48" diameter	G		50	.640		41.50	33.50		75	96
3350	60" diameter	G	▼	45	.711	▼	50.50	37		87.50	111
4500	For second and succeeding months, deduct						50%				
5000	Job-built plywood, 8" x 8" columns, 1 use		C-1	165	.194	SFCA	3.73	10.10		13.83	19.20
5050	2 use			195	.164		2.14	8.55		10.69	15.10
5100	3 use			210	.152		1.49	7.95		9.44	13.50
5150	4 use			215	.149		1.22	7.75		8.97	12.95
5500	12" x 12" columns, 1 use			180	.178		3.76	9.25		13.01	18
5550	2 use			210	.152		2.07	7.95		10.02	14.15
5600	3 use			220	.145		1.50	7.60		9.10	12.95
5650	4 use			225	.142		1.22	7.40		8.62	12.40
6000	16" x 16" columns, 1 use			185	.173		3.85	9		12.85	17.70
6050	2 use			215	.149		2.04	7.75		9.79	13.85
6100	3 use			230	.139		1.55	7.25		8.80	12.50
6150	4 use			235	.136		1.26	7.10		8.36	12
6500	24" x 24" columns, 1 use			190	.168		4.43	8.80		13.23	17.95
6550	2 use			216	.148		2.44	7.70		10.14	14.25
6600	3 use			230	.139		1.77	7.25		9.02	12.75
6650	4 use			238	.134		1.44	7		8.44	12.05
7000	36" x 36" columns, 1 use			200	.160		3.55	8.35		11.90	16.35
7050	2 use			230	.139		1.99	7.25		9.24	13
7100	3 use			245	.131		1.42	6.80		8.22	11.70
7150	4 use		▼	250	.128	▼	1.15	6.65		7.80	11.20
7400	Steel framed plywood, based on 50 uses of purchased										
7420	forms, and 4 uses of bracing lumber										
7500	8" x 8" column		C-1	340	.094	SFCA	2.29	4.91		7.20	9.80
7550	10" x 10"			350	.091		2.03	4.77		6.80	9.35
7600	12" x 12"			370	.086		1.72	4.51		6.23	8.65
7650	16" x 16"			400	.080		1.33	4.17		5.50	7.65
7700	20" x 20"			420	.076		1.18	3.97		5.15	7.25
7750	24" x 24"			440	.073		.84	3.79		4.63	6.55
7755	30" x 30"			440	.073		1.04	3.79		4.83	6.80
7760	36" x 36"		▼	460	.070		.92	3.63		4.55	6.40

03 11 Concrete Forming

03 11 13 – Structural Cast-In-Place Concrete Forming

03 11 13.30 Forms In Place, Culvert

		Crew	Daily Output	Labor-Hours	Unit	Material	2021 Bare Costs Labor	Equipment	Total	Total Incl O&P
0010	**FORMS IN PLACE, CULVERT** R031113-40									
0015	5' to 8' square or rectangular, 1 use	C-1	170	.188	SFCA	5.65	9.80		15.45	21
0050	2 use R031113-60		180	.178		3.33	9.25		12.58	17.50
0100	3 use		190	.168		2.55	8.80		11.35	15.90
0150	4 use		200	.160		2.16	8.35		10.51	14.85

03 11 13.35 Forms In Place, Elevated Slabs

		Crew	Daily Output	Labor-Hours	Unit	Material	2021 Bare Costs Labor	Equipment	Total	Total Incl O&P
0010	**FORMS IN PLACE, ELEVATED SLABS** R031113-40									
1000	Flat plate, job-built plywood, to 15' high, 1 use R031113-60	C-2	470	.102	S.F.	5.75	5.45		11.20	14.50
1050	2 use		520	.092		3.16	4.92		8.08	10.85
1100	3 use		545	.088		2.30	4.70		7	9.55
1150	4 use		560	.086		1.87	4.57		6.44	8.85
1500	15' to 20' high ceilings, 4 use		495	.097		1.90	5.15		7.05	9.80
1600	21' to 35' high ceilings, 4 use		450	.107		2.23	5.70		7.93	10.95
2000	Flat slab, drop panels, job-built plywood, to 15' high, 1 use		449	.107		6.75	5.70		12.45	15.95
2050	2 use		509	.094		3.72	5.05		8.77	11.60
2100	3 use		532	.090		2.70	4.81		7.51	10.15
2150	4 use		544	.088		2.20	4.70		6.90	9.40
2250	15' to 20' high ceilings, 4 use		480	.100		3.15	5.35		8.50	11.40
2350	20' to 35' high ceilings, 4 use		435	.110		3.47	5.90		9.37	12.60
3000	Floor slab hung from steel beams, 1 use		485	.099		3.61	5.30		8.91	11.85
3050	2 use		535	.090		2.66	4.78		7.44	10.05
3100	3 use		550	.087		2.34	4.65		6.99	9.50
3150	4 use		565	.085		2.18	4.53		6.71	9.15
3500	Floor slab, with 1-way joist pans, 1 use		415	.116		10.20	6.15		16.35	20.50
3550	2 use		445	.108		7.15	5.75		12.90	16.50
3600	3 use		475	.101		6.15	5.40		11.55	14.80
3650	4 use		500	.096		5.65	5.10		10.75	13.85
4500	With 2-way waffle domes, 1 use		405	.119		10.75	6.30		17.05	21.50
4520	2 use		450	.107		7.70	5.70		13.40	16.95
4530	3 use		460	.104		6.65	5.55		12.20	15.65
4550	4 use		470	.102		6.15	5.45		11.60	14.95
5000	Box out for slab openings, over 16" deep, 1 use		190	.253	SFCA	4.83	13.45		18.28	25.50
5050	2 use		240	.200	"	2.66	10.65		13.31	18.80
5500	Shallow slab box outs, to 10 S.F.		42	1.143	Ea.	19.20	61		80.20	112
5550	Over 10 S.F. (use perimeter)		600	.080	L.F.	2.56	4.27		6.83	9.15
6000	Bulkhead forms for slab, with keyway, 1 use, 2 piece		500	.096		2.35	5.10		7.45	10.25
6100	3 piece (see also edge forms)		460	.104		2.68	5.55		8.23	11.25
6200	Slab bulkhead form, 4-1/2" high, exp metal, w/keyway & stakes G	C-1	1200	.027		.94	1.39		2.33	3.10
6210	5-1/2" high G		1100	.029		1.19	1.52		2.71	3.57
6215	7-1/2" high G		960	.033		1.40	1.74		3.14	4.13
6220	9-1/2" high G		840	.038		1.56	1.99		3.55	4.68
6500	Curb forms, wood, 6" to 12" high, on elevated slabs, 1 use		180	.178	SFCA	2.63	9.25		11.88	16.75
6550	2 use		205	.156		1.45	8.15		9.60	13.75
6600	3 use		220	.145		1.05	7.60		8.65	12.45
6650	4 use		225	.142		.86	7.40		8.26	12
7000	Edge forms to 6" high, on elevated slab, 4 use		500	.064	L.F.	.34	3.34		3.68	5.35
7070	7" to 12" high, 1 use		162	.198	SFCA	1.94	10.30		12.24	17.50
7080	2 use		198	.162		1.07	8.45		9.52	13.70
7090	3 use		222	.144		.78	7.50		8.28	12.05
7101	4 use		350	.091		.63	4.77		5.40	7.80
7500	Depressed area forms to 12" high, 4 use		300	.107	L.F.	1.07	5.55		6.62	9.50
7550	12" to 24" high, 4 use		175	.183		1.46	9.55		11.01	15.85

For customer support on your Concrete & Masonry Costs with RSMeans Data, call 800.448.8182.

45

03 11 13 – Structural Cast-In-Place Concrete Forming

03 11 13.35 Forms In Place, Elevated Slabs

		Crew	Daily Output	Labor-Hours	Unit	Material	2021 Bare Costs Labor	Equipment	Total	Total Incl O&P
8000	Perimeter deck and rail for elevated slabs, straight	C-1	90	.356	L.F.	18.85	18.55		37.40	48
8050	Curved		65	.492	↓	26	25.50		51.50	67
8500	Void forms, round plastic, 8" high x 3" diameter Ⓖ		450	.071	Ea.	1.62	3.71		5.33	7.35
8550	4" diameter Ⓖ		425	.075		2.41	3.93		6.34	8.50
8600	6" diameter Ⓖ		400	.080		4	4.17		8.17	10.60
8650	8" diameter Ⓖ	↓	375	.085	↓	7.10	4.45		11.55	14.45

03 11 13.40 Forms In Place, Equipment Foundations

		Crew	Daily Output	Labor-Hours	Unit	Material	2021 Bare Costs Labor	Equipment	Total	Total Incl O&P
0010	**FORMS IN PLACE, EQUIPMENT FOUNDATIONS** R031113-40									
0020	1 use	C-2	160	.300	SFCA	3.96	16		19.96	28.50
0050	2 use R031113-60		190	.253		2.18	13.45		15.63	22.50
0100	3 use		200	.240		1.59	12.80		14.39	21
0150	4 use	↓	205	.234	↓	1.30	12.50		13.80	20

03 11 13.45 Forms In Place, Footings

		Crew	Daily Output	Labor-Hours	Unit	Material	2021 Bare Costs Labor	Equipment	Total	Total Incl O&P
0010	**FORMS IN PLACE, FOOTINGS** R031113-40									
0020	Continuous wall, plywood, 1 use	C-1	375	.085	SFCA	7.85	4.45		12.30	15.30
0050	2 use R031113-60		440	.073		4.32	3.79		8.11	10.40
0100	3 use		470	.068		3.14	3.55		6.69	8.75
0150	4 use		485	.066	↓	2.56	3.44		6	7.95
0500	Dowel supports for footings or beams, 1 use		500	.064	L.F.	1.56	3.34		4.90	6.70
1000	Integral starter wall, to 4" high, 1 use	↓	400	.080		1.58	4.17		5.75	7.95
1500	Keyway, 4 use, tapered wood, 2" x 4"	1 Carp	530	.015		.37	.83		1.20	1.64
1550	2" x 6"		500	.016		.53	.88		1.41	1.89
2000	Tapered plastic		530	.015		1.58	.83		2.41	2.97
2250	For keyway hung from supports, add	↓	150	.053	↓	1.56	2.92		4.48	6.05
3000	Pile cap, square or rectangular, job-built plywood, 1 use	C-1	290	.110	SFCA	4.48	5.75		10.23	13.55
3050	2 use		346	.092		2.46	4.82		7.28	9.90
3100	3 use		371	.086		1.79	4.50		6.29	8.65
3150	4 use		383	.084		1.46	4.36		5.82	8.10
4000	Triangular or hexagonal, 1 use		225	.142		5.25	7.40		12.65	16.85
4050	2 use		280	.114		2.89	5.95		8.84	12.10
4100	3 use		305	.105		2.10	5.45		7.55	10.45
4150	4 use		315	.102		1.71	5.30		7.01	9.80
5000	Spread footings, job-built lumber, 1 use		305	.105		3.28	5.45		8.73	11.75
5050	2 use		371	.086		1.82	4.50		6.32	8.70
5100	3 use		401	.080		1.31	4.16		5.47	7.65
5150	4 use		414	.077	↓	1.06	4.03		5.09	7.15
6000	Supports for dowels, plinths or templates, 2' x 2' footing		25	1.280	Ea.	9.30	66.50		75.80	110
6050	4' x 4' footing		22	1.455		18.55	76		94.55	134
6100	8' x 8' footing		20	1.600		37	83.50		120.50	165
6150	12' x 12' footing		17	1.882	↓	51	98		149	202
7000	Plinths, job-built plywood, 1 use		250	.128	SFCA	5.30	6.65		11.95	15.75
7100	4 use	↓	270	.119	"	1.73	6.20		7.93	11.10

03 11 13.47 Forms In Place, Gas Station Forms

		Crew	Daily Output	Labor-Hours	Unit	Material	2021 Bare Costs Labor	Equipment	Total	Total Incl O&P
0010	**FORMS IN PLACE, GAS STATION FORMS**									
0050	Curb fascia, with template, 12 ga. steel, left in place, 9" high Ⓖ	1 Carp	50	.160	L.F.	13.80	8.75		22.55	28.50
1000	Sign or light bases, 18" diameter, 9" high Ⓖ		9	.889	Ea.	87	48.50		135.50	169
1050	30" diameter, 13" high Ⓖ	↓	8	1		138	54.50		192.50	234
2000	Island forms, 10' long, 9" high, 3'-6" wide Ⓖ	C-1	10	3.200		385	167		552	675
2050	4' wide Ⓖ		9	3.556		400	185		585	715
2500	20' long, 9" high, 4' wide Ⓖ		6	5.333		640	278		918	1,125
2550	5' wide Ⓖ	↓	5	6.400	↓	670	335		1,005	1,225

03 11 Concrete Forming

03 11 13 – Structural Cast-In-Place Concrete Forming

03 11 13.50 Forms In Place, Grade Beam

		Crew	Daily Output	Labor-Hours	Unit	Material	2021 Bare Costs Labor	Equipment	Total	Total Incl O&P
0010	**FORMS IN PLACE, GRADE BEAM** R031113-40									
0020	Job-built plywood, 1 use	C-2	530	.091	SFCA	3.97	4.83		8.80	11.55
0050	2 use R031113-60		580	.083		2.19	4.41		6.60	9
0100	3 use		600	.080		1.59	4.27		5.86	8.10
0150	4 use	↓	605	.079	↓	1.29	4.23		5.52	7.70

03 11 13.55 Forms In Place, Mat Foundation

		Crew	Daily Output	Labor-Hours	Unit	Material	2021 Bare Costs Labor	Equipment	Total	Total Incl O&P
0010	**FORMS IN PLACE, MAT FOUNDATION** R031113-40									
0020	Job-built plywood, 1 use	C-2	290	.166	SFCA	4.43	8.85		13.28	18.05
0050	2 use R031113-60		310	.155		1.80	8.25		10.05	14.30
0100	3 use		330	.145		1.20	7.75		8.95	12.90
0120	4 use	↓	350	.137	↓	1.06	7.30		8.36	12.05

03 11 13.65 Forms In Place, Slab On Grade

		Crew	Daily Output	Labor-Hours	Unit	Material	2021 Bare Costs Labor	Equipment	Total	Total Incl O&P
0010	**FORMS IN PLACE, SLAB ON GRADE** R031113-40									
1000	Bulkhead forms w/keyway, wood, 6" high, 1 use	C-1	510	.063	L.F.	1.65	3.27		4.92	6.70
1050	2 uses R031113-60		400	.080		.91	4.17		5.08	7.20
1100	4 uses		350	.091		.54	4.77		5.31	7.70
1400	Bulkhead form for slab, 4-1/2" high, exp metal, incl. keyway & stakes [G]		1200	.027		.94	1.39		2.33	3.10
1410	5-1/2" high [G]		1100	.029		1.19	1.52		2.71	3.57
1420	7-1/2" high [G]		960	.033		1.40	1.74		3.14	4.13
1430	9-1/2" high [G]		840	.038	↓	1.56	1.99		3.55	4.68
2000	Curb forms, wood, 6" to 12" high, on grade, 1 use		215	.149	SFCA	2.99	7.75		10.74	14.90
2050	2 use		250	.128		1.65	6.65		8.30	11.75
2100	3 use		265	.121		1.20	6.30		7.50	10.70
2150	4 use		275	.116	↓	.97	6.05		7.02	10.10
3000	Edge forms, wood, 4 use, on grade, to 6" high		600	.053	L.F.	.45	2.78		3.23	4.65
3050	7" to 12" high		435	.074	SFCA	1.14	3.83		4.97	6.95
3060	Over 12"		350	.091	"	1.43	4.77		6.20	8.65
3500	For depressed slabs, 4 use, to 12" high		300	.107	L.F.	1.18	5.55		6.73	9.60
3550	To 24" high		175	.183		1.53	9.55		11.08	15.95
4000	For slab blockouts, to 12" high, 1 use		200	.160		1.20	8.35		9.55	13.75
4050	To 24" high, 1 use		120	.267		1.51	13.90		15.41	22.50
4100	Plastic (extruded), to 6" high, multiple use, on grade	↓	800	.040	↓	6.50	2.09		8.59	10.25
5000	Screed, 24 ga. metal key joint, see Section 03 15 16.30									
5020	Wood, incl. wood stakes, 1" x 3"	C-1	900	.036	L.F.	1.23	1.85		3.08	4.13
5050	2" x 4"		900	.036	"	1.31	1.85		3.16	4.21
6000	Trench forms in floor, wood, 1 use		160	.200	SFCA	3.15	10.45		13.60	19
6050	2 use		175	.183		1.73	9.55		11.28	16.15
6100	3 use		180	.178		1.26	9.25		10.51	15.25
6150	4 use		185	.173	↓	1.02	9		10.02	14.60
8760	Void form, corrugated fiberboard, 4" x 12", 4' long [G]		3000	.011	S.F.	3.85	.56		4.41	5.05
8770	6" x 12", 4' long	↓	3000	.011		4.64	.56		5.20	5.95
8780	1/4" thick hardboard protective cover for void form	2 Carp	1500	.011	↓	.94	.58		1.52	1.90

03 11 13.85 Forms In Place, Walls

		Crew	Daily Output	Labor-Hours	Unit	Material	2021 Bare Costs Labor	Equipment	Total	Total Incl O&P
0010	**FORMS IN PLACE, WALLS** R031113-10									
0100	Box out for wall openings, to 16" thick, to 10 S.F.	C-2	24	2	Ea.	42	107		149	205
0150	Over 10 S.F. (use perimeter) R031113-40	"	280	.171	L.F.	3.66	9.15		12.81	17.65
0250	Brick shelf, 4" W, add to wall forms, use wall area above shelf									
0260	1 use R031113-60	C-2	240	.200	SFCA	4.02	10.65		14.67	20.50
0300	2 use		275	.175		2.21	9.30		11.51	16.35
0350	4 use		300	.160	↓	1.61	8.55		10.16	14.50
0500	Bulkhead, wood with keyway, 1 use, 2 piece	↓	265	.181	L.F.	3.19	9.65		12.84	17.90

For customer support on your Concrete & Masonry Costs with RSMeans Data, call 800.448.8182.

47

03 11 13 – Structural Cast-In-Place Concrete Forming

03 11 13.85 Forms In Place, Walls		Crew	Daily Output	Labor-Hours	Unit	Material	2021 Bare Costs Labor	Equipment	Total	Total Incl O&P
0600	Bulkhead forms with keyway, 1 piece expanded metal, 8" wall [G]	C-1	1000	.032	L.F.	1.40	1.67		3.07	4.03
0610	10" wall [G]		800	.040		1.56	2.09		3.65	4.83
0620	12" wall [G]		525	.061		1.87	3.18		5.05	6.80
0700	Buttress, to 8' high, 1 use	C-2	350	.137	SFCA	5.60	7.30		12.90	17.05
0750	2 use		430	.112		3.07	5.95		9.02	12.30
0800	3 use		460	.104		2.24	5.55		7.79	10.75
0850	4 use		480	.100		1.84	5.35		7.19	10
1000	Corbel or haunch, to 12" wide, add to wall forms, 1 use		150	.320	L.F.	3.75	17.05		20.80	29.50
1050	2 use		170	.282		2.06	15.05		17.11	25
1100	3 use		175	.274		1.50	14.60		16.10	23.50
1150	4 use		180	.267		1.22	14.20		15.42	22.50
2000	Wall, job-built plywood, to 8' high, 1 use		370	.130	SFCA	4.22	6.90		11.12	14.95
2050	2 use		435	.110		2.72	5.90		8.62	11.80
2100	3 use		495	.097		1.98	5.15		7.13	9.90
2150	4 use		505	.095		1.61	5.05		6.66	9.30
2400	Over 8' to 16' high, 1 use		280	.171		4.66	9.15		13.81	18.75
2450	2 use		345	.139		2.03	7.40		9.43	13.30
2500	3 use		375	.128		1.45	6.80		8.25	11.80
2550	4 use		395	.122		1.19	6.50		7.69	10.95
2700	Over 16' high, 1 use		235	.204		4.14	10.90		15.04	21
2750	2 use		290	.166		2.28	8.85		11.13	15.65
2800	3 use		315	.152		1.66	8.10		9.76	13.95
2850	4 use		330	.145		1.35	7.75		9.10	13.10
4000	Radial, smooth curved, job-built plywood, 1 use		245	.196		3.77	10.45		14.22	19.75
4050	2 use		300	.160		2.07	8.55		10.62	15.05
4100	3 use		325	.148		1.51	7.85		9.36	13.40
4150	4 use		335	.143		1.23	7.65		8.88	12.75
4200	Below grade, job-built plywood, 1 use		225	.213		4.23	11.35		15.58	21.50
4210	2 use		225	.213		2.33	11.35		13.68	19.55
4220	3 use		225	.213		1.96	11.35		13.31	19.15
4230	4 use		225	.213		1.38	11.35		12.73	18.50
4300	Curved, 2' chords, job-built plywood, to 8' high, 1 use		290	.166		3.29	8.85		12.14	16.75
4350	2 use		355	.135		1.81	7.20		9.01	12.75
4400	3 use		385	.125		1.32	6.65		7.97	11.35
4450	4 use		400	.120		1.07	6.40		7.47	10.75
4500	Over 8' to 16' high, 1 use		290	.166		1.38	8.85		10.23	14.65
4525	2 use		355	.135		.76	7.20		7.96	11.60
4550	3 use		385	.125		.55	6.65		7.20	10.50
4575	4 use		400	.120		.46	6.40		6.86	10.05
4600	Retaining wall, battered, job-built plyw'd, to 8' high, 1 use		300	.160		3.08	8.55		11.63	16.15
4650	2 use		355	.135		1.69	7.20		8.89	12.60
4700	3 use		375	.128		1.23	6.80		8.03	11.55
4750	4 use		390	.123		1	6.55		7.55	10.90
4900	Over 8' to 16' high, 1 use		240	.200		3.39	10.65		14.04	19.65
4950	2 use		295	.163		1.87	8.70		10.57	15
5000	3 use		305	.157		1.36	8.40		9.76	14
5050	4 use		320	.150		1.10	8		9.10	13.15
5100	Retaining wall form, plywood, smooth curve, 1 use		200	.240		4.94	12.80		17.74	24.50
5120	2 use		235	.204		2.72	10.90		13.62	19.25
5130	3 use		250	.192		1.98	10.25		12.23	17.45
5140	4 use		260	.185		1.62	9.85		11.47	16.50
5500	For gang wall forming, 192 S.F. sections, deduct					10%	10%			
5550	384 S.F. sections, deduct					20%	20%			

03 11 Concrete Forming

03 11 13 – Structural Cast-In-Place Concrete Forming

03 11 13.85 Forms In Place, Walls

		Crew	Daily Output	Labor-Hours	Unit	Material	2021 Bare Costs Labor	Equipment	Total	Total Incl O&P
7500	Lintel or sill forms, 1 use	1 Carp	30	.267	SFCA	5.15	14.60		19.75	27.50
7520	2 use		34	.235		2.84	12.85		15.69	22.50
7540	3 use		36	.222		2.07	12.15		14.22	20.50
7560	4 use		37	.216		1.68	11.85		13.53	19.50
7800	Modular prefabricated plywood, based on 20 uses of purchased									
7820	forms, and 4 uses of bracing lumber									
7860	To 8' high	C-2	800	.060	SFCA	1.31	3.20		4.51	6.20
8060	Over 8' to 16' high		600	.080		1.40	4.27		5.67	7.90
8600	Pilasters, 1 use		270	.178		4.92	9.50		14.42	19.55
8620	2 use		330	.145		2.70	7.75		10.45	14.55
8640	3 use		370	.130		1.97	6.90		8.87	12.45
8660	4 use		385	.125		1.60	6.65		8.25	11.65
9010	Steel framed plywood, based on 50 uses of purchased									
9020	forms, and 4 uses of bracing lumber									
9060	To 8' high	C-2	600	.080	SFCA	.74	4.27		5.01	7.15
9260	Over 8' to 16' high		450	.107		.74	5.70		6.44	9.30
9460	Over 16' to 20' high		400	.120		.74	6.40		7.14	10.35
9475	For elevated walls, add						10%			
9480	For battered walls, 1 side battered, add					10%	10%			
9485	For battered walls, 2 sides battered, add					15%	15%			

03 11 16 – Architectural Cast-in-Place Concrete Forming

03 11 16.13 Concrete Form Liners

		Crew	Daily Output	Labor-Hours	Unit	Material	2021 Bare Costs Labor	Equipment	Total	Total Incl O&P
0010	**CONCRETE FORM LINERS**									
5750	Liners for forms (add to wall forms), ABS plastic									
5800	Aged wood, 4" wide, 1 use	1 Carp	256	.031	SFCA	3.11	1.71		4.82	5.95
5820	2 use		256	.031		1.71	1.71		3.42	4.43
5830	3 use		256	.031		1.24	1.71		2.95	3.92
5840	4 use		256	.031		1.01	1.71		2.72	3.66
5900	Fractured rope rib, 1 use		192	.042		4.57	2.28		6.85	8.45
5925	2 use		192	.042		2.51	2.28		4.79	6.15
5950	3 use		192	.042		1.83	2.28		4.11	5.40
6000	4 use		192	.042		1.49	2.28		3.77	5.05
6100	Ribbed, 3/4" deep x 1-1/2" OC, 1 use		224	.036		5.05	1.95		7	8.45
6125	2 use		224	.036		2.77	1.95		4.72	5.95
6150	3 use		224	.036		2.01	1.95		3.96	5.15
6200	4 use		224	.036		1.63	1.95		3.58	4.72
6300	Rustic brick pattern, 1 use		224	.036		3.23	1.95		5.18	6.45
6325	2 use		224	.036		1.78	1.95		3.73	4.87
6350	3 use		224	.036		1.29	1.95		3.24	4.34
6400	4 use		224	.036		1.05	1.95		3	4.07
6500	3/8" striated, random, 1 use		224	.036		3.52	1.95		5.47	6.80
6525	2 use		224	.036		1.94	1.95		3.89	5.05
6550	3 use		224	.036		1.41	1.95		3.36	4.47
6600	4 use		224	.036		1.14	1.95		3.09	4.18
6850	Random vertical rustication, 1 use		384	.021		6.10	1.14		7.24	8.45
6900	2 use		384	.021		3.37	1.14		4.51	5.40
6925	3 use		384	.021		2.45	1.14		3.59	4.39
6950	4 use		384	.021		1.99	1.14		3.13	3.89
7050	Wood, beveled edge, 3/4" deep, 1 use		384	.021	L.F.	.14	1.14		1.28	1.85
7100	1" deep, 1 use		384	.021	"	.24	1.14		1.38	1.96
7200	4" wide aged cedar, 1 use		256	.031	SFCA	3.23	1.71		4.94	6.10
7300	4" variable depth rough cedar		224	.036	"	4.52	1.95		6.47	7.90

For customer support on your Concrete & Masonry Costs with RSMeans Data, call 800.448.8182.

49

03 11 19 – Insulating Concrete Forming

03 11 19.10 Insulating Forms, Left In Place	Crew	Daily Output	Labor-Hours	Unit	Material	2021 Bare Costs Labor	2021 Bare Costs Equipment	Total	Total Incl O&P
0010 INSULATING FORMS, LEFT IN PLACE									
0020 Forms include layout, exclude rebar, embedments, bucks for openings,									
0030 scaffolding, wall bracing, concrete, and concrete placing.									
0040 S.F. is for exterior face but includes forms for both faces of wall									
0100 Straight blocks or panels, molded, walls up to 4' high									
0110 4" core wall	4 Carp	1984	.016	S.F.	3.64	.88		4.52	5.30
0120 6" core wall		1808	.018		3.66	.97		4.63	5.50
0130 8" core wall		1536	.021		3.79	1.14		4.93	5.85
0140 10" core wall		1152	.028		4.32	1.52		5.84	7
0150 12" core wall		992	.032		4.91	1.76		6.67	8.05
0200 90 degree corner blocks or panels, molded, walls up to 4' high									
0210 4" core wall	4 Carp	1880	.017	S.F.	3.75	.93		4.68	5.50
0220 6" core wall		1708	.019		3.77	1.02		4.79	5.70
0230 8" core wall		1324	.024		3.90	1.32		5.22	6.25
0240 10" core wall		987	.032		4.51	1.77		6.28	7.60
0250 12" core wall		884	.036		4.83	1.98		6.81	8.25
0300 45 degree corner blocks or panels, molded, walls up to 4' high									
0310 4" core wall	4 Carp	1880	.017	S.F.	3.95	.93		4.88	5.75
0320 6" core wall		1712	.019		4.08	1.02		5.10	6
0330 8" core wall		1324	.024		4.19	1.32		5.51	6.60
0400 T blocks or panels, molded, walls up to 4' high									
0420 6" core wall	4 Carp	1540	.021	S.F.	4.81	1.14		5.95	7
0430 8" core wall		1325	.024		4.86	1.32		6.18	7.30
0440 Non-standard corners or Ts requiring trimming & strapping		192	.167		4.91	9.10		14.01	19
0500 Radius blocks or panels, molded, walls up to 4' high, 6" core wall									
0520 5' to 10' diameter, molded blocks or panels	4 Carp	2400	.013	S.F.	5.95	.73		6.68	7.60
0530 10' to 15' diameter, requiring trimming and strapping, add		500	.064		4.47	3.50		7.97	10.15
0540 15'-1" to 30' diameter, requiring trimming and strapping, add		1200	.027		4.47	1.46		5.93	7.10
0550 30'-1" to 60' diameter, requiring trimming and strapping, add		1600	.020		4.47	1.09		5.56	6.55
0560 60'-1" to 100' diameter, requiring trimming and strapping, add		2800	.011		4.47	.63		5.10	5.85
0600 Additional labor for blocks/panels in higher walls (excludes scaffolding)									
0610 4'-1" to 9'-4" high, add						10%			
0620 9'-5" to 12'-0" high, add						20%			
0630 12'-1" to 20'-0" high, add						35%			
0640 Over 20'-0" high, add						55%			
0700 Taper block or panels, molded, single course									
0720 6" core wall	4 Carp	1600	.020	S.F.	3.93	1.09		5.02	5.95
0730 8" core wall	"	1392	.023	"	3.98	1.26		5.24	6.25
0800 ICF brick ledge (corbel) block or panels, molded, single course									
0820 6" core wall	4 Carp	1200	.027	S.F.	4.33	1.46		5.79	6.95
0830 8" core wall	"	1152	.028	"	4.43	1.52		5.95	7.15
0900 ICF curb (shelf) block or panels, molded, single course									
0930 8" core wall	4 Carp	688	.047	S.F.	3.74	2.54		6.28	7.90
0940 10" core wall	"	544	.059	"	4.27	3.22		7.49	9.50
0950 Wood form to hold back concrete to form shelf, 8" high	2 Carp	400	.040	L.F.	1.41	2.19		3.60	4.82
1000 ICF half height block or panels, molded, single course									
1010 4" core wall	4 Carp	1248	.026	S.F.	4.74	1.40		6.14	7.30
1020 6" core wall		1152	.028		4.76	1.52		6.28	7.50
1030 8" core wall		942	.034		4.81	1.86		6.67	8.05
1040 10" core wall		752	.043		5.30	2.33		7.63	9.30
1050 12" core wall		648	.049		5.25	2.70		7.95	9.80
1100 ICF half height block/panels, made by field sawing full height block/panels									

03 11 Concrete Forming

03 11 19 – Insulating Concrete Forming

03 11 19.10 Insulating Forms, Left In Place	Crew	Daily Output	Labor-Hours	Unit	Material	2021 Bare Costs Labor	Equipment	Total	Total Incl O&P
1110	4" core wall	4 Carp	800	.040	S.F.	1.82	2.19		4.01	5.25
1120	6" core wall		752	.043		1.83	2.33		4.16	5.50
1130	8" core wall		600	.053		1.90	2.92		4.82	6.45
1140	10" core wall		496	.065		2.16	3.53		5.69	7.65
1150	12" core wall	▼	400	.080	▼	2.46	4.38		6.84	9.25
1200	Additional insulation inserted into forms between ties									
1210	1 layer (2" thick)	4 Carp	14000	.002	S.F.	1.04	.13		1.17	1.33
1220	2 layers (4" thick)		7000	.005		2.08	.25		2.33	2.66
1230	3 layers (6" thick)	▼	4622	.007	▼	3.12	.38		3.50	4
1300	EPS window/door bucks, molded, permanent									
1310	4" core wall (9" wide)	2 Carp	200	.080	L.F.	3.49	4.38		7.87	10.40
1320	6" core wall (11" wide)		200	.080		3.58	4.38		7.96	10.50
1330	8" core wall (13" wide)		176	.091		3.75	4.97		8.72	11.55
1340	10" core wall (15" wide)		152	.105		4.88	5.75		10.63	13.95
1350	12" core wall (17" wide)		152	.105		5.45	5.75		11.20	14.60
1360	2" x 6" temporary buck bracing (includes installing and removing)	▼	400	.040	▼	1.06	2.19		3.25	4.43
1400	Wood window/door bucks (instead of EPS bucks), permanent									
1410	4" core wall (9" wide)	2 Carp	400	.040	L.F.	2.02	2.19		4.21	5.50
1420	6" core wall (11" wide)		400	.040		2.50	2.19		4.69	6
1430	8" core wall (13" wide)		350	.046		8.15	2.50		10.65	12.75
1440	10" core wall (15" wide)		300	.053		11.30	2.92		14.22	16.75
1450	12" core wall (17" wide)		300	.053		11.30	2.92		14.22	16.75
1460	2" x 6" temporary buck bracing (includes installing and removing)	▼	800	.020	▼	1.06	1.09		2.15	2.79
1500	ICF alignment brace (incl. stiff-back, diagonal kick-back, work platform									
1501	bracket & guard rail post), fastened to one face of wall forms @ 6' O.C.									
1510	1st tier up to 10' tall									
1520	Rental of ICF alignment brace set, per set				Week	10.10			10.10	11.10
1530	Labor (includes installing & removing)	2 Carp	30	.533	Ea.		29		29	43.50
1560	2nd tier from 10' to 20' tall (excludes mason's scaffolding up to 10' high)									
1570	Rental of ICF alignment brace set, per set				Week	10.10			10.10	11.10
1580	Labor (includes installing & removing)	4 Carp	30	1.067	Ea.		58.50		58.50	87
1600	2" x 10" wood plank for work platform, 16' long									
1610	Plank material cost pro-rated over 20 uses				Ea.	1.61			1.61	1.78
1620	Labor (includes installing & removing)	2 Carp	48	.333	"		18.25		18.25	27
1700	2" x 4" lumber for top & middle rails for work platform									
1710	Railing material cost pro-rated over 20 uses				Ea.	.03			.03	.04
1720	Labor (includes installing & removing)	2 Carp	2400	.007	L.F.		.36		.36	.54
1800	ICF accessories									
1810	Wire clip to secure forms in place	2 Carp	2100	.008	Ea.	.39	.42		.81	1.05
1820	Masonry anchor embedment (excludes ties by mason)		1600	.010		4.26	.55		4.81	5.50
1830	Ledger anchor embedment (excludes timber hanger & screws)	▼	128	.125	▼	7.65	6.85		14.50	18.60
1900	See section 01 54 23.70 for mason's scaffolding components									
1910	See section 03 15 19.05 for anchor bolt sleeves									
1920	See section 03 15 19.10 for anchor bolts									
1930	See section 03 15 19.20 for dovetail anchor components									
1940	See section 03 15 19.30 for embedded inserts									
1950	See section 03 21 05.10 for rebar accessories									
1960	See section 03 21 11.60 for reinforcing bars in place									
1970	See section 03 31 13.35 for ready-mix concrete material									
1980	See section 03 31 13.70 for placement and consolidation of concrete									
1990	See section 06 05 23.60 for timber connectors									

For customer support on your Concrete & Masonry Costs with RSMeans Data, call 800.448.8182.

51

03 11 Concrete Forming

03 11 19 – Insulating Concrete Forming

03 11 19.60 Roof Deck Form Boards

		Crew	Daily Output	Labor-Hours	Unit	Material	2021 Bare Costs Labor	2021 Bare Costs Equipment	Total	Total Incl O&P
0010	**ROOF DECK FORM BOARDS** R051223-50									
0050	Includes bulb tee sub-purlins @ 32-5/8" OC									
0070	Non-asbestos fiber cement, 5/16" thick	C-13	2950	.008	S.F.	3.38	.48	.05	3.91	4.51
0100	Fiberglass, 1" thick		2700	.009		3.94	.52	.06	4.52	5.20
0500	Wood fiber, 1" thick G		2700	.009		2.42	.52	.06	3	3.51

03 11 23 – Permanent Stair Forming

03 11 23.75 Forms In Place, Stairs

		Crew	Daily Output	Labor-Hours	Unit	Material	2021 Bare Costs Labor	2021 Bare Costs Equipment	Total	Total Incl O&P
0010	**FORMS IN PLACE, STAIRS** R031113-40									
0015	(Slant length x width), 1 use	C-2	165	.291	S.F.	9.45	15.50		24.95	33.50
0050	2 use R031113-60		170	.282		5.30	15.05		20.35	28.50
0100	3 use		180	.267		3.92	14.20		18.12	25.50
0150	4 use		190	.253		3.24	13.45		16.69	23.50
1000	Alternate pricing method (1.0 L.F./S.F.), 1 use		100	.480	LF Rsr	9.45	25.50		34.95	48.50
1050	2 use		105	.457		5.30	24.50		29.80	42.50
1100	3 use		110	.436		3.92	23.50		27.42	39
1150	4 use		115	.417		3.24	22.50		25.74	36.50
2000	Stairs, cast on sloping ground (length x width), 1 use		220	.218	S.F.	3.91	11.65		15.56	21.50
2025	2 use		232	.207		2.15	11.05		13.20	18.80
2050	3 use		244	.197		1.56	10.50		12.06	17.35
2100	4 use		256	.188		1.27	10		11.27	16.30

03 15 Concrete Accessories

03 15 05 – Concrete Forming Accessories

03 15 05.12 Chamfer Strips

		Crew	Daily Output	Labor-Hours	Unit	Material	2021 Bare Costs Labor	2021 Bare Costs Equipment	Total	Total Incl O&P
0010	**CHAMFER STRIPS**									
2000	Polyvinyl chloride, 1/2" wide with leg	1 Carp	535	.015	L.F.	.76	.82		1.58	2.06
2200	3/4" wide with leg		525	.015		.84	.83		1.67	2.16
2400	1" radius with leg		515	.016		1.04	.85		1.89	2.41
2800	2" radius with leg		500	.016		1.78	.88		2.66	3.27
5000	Wood, 1/2" wide		535	.015		.12	.82		.94	1.35
5200	3/4" wide		525	.015		.14	.83		.97	1.39
5400	1" wide		515	.016		.24	.85		1.09	1.53

03 15 05.15 Column Form Accessories

		Crew	Daily Output	Labor-Hours	Unit	Material	2021 Bare Costs Labor	2021 Bare Costs Equipment	Total	Total Incl O&P
0010	**COLUMN FORM ACCESSORIES**									
1000	Column clamps, adjustable to 24" x 24", buy G				Set	182			182	200
1100	Rent per month G					13.70			13.70	15.10
1300	For sizes to 30" x 30", buy G					244			244	268
1400	Rent per month G					15.80			15.80	17.35
1600	For sizes to 36" x 36", buy G					275			275	305
1700	Rent per month G					17.90			17.90	19.70
2000	Bar type with wedges, 36" x 36", buy G					193			193	212
2100	Rent per month G					14.75			14.75	16.25
2300	48" x 48", buy G					256			256	282
2400	Rent per month G					19.30			19.30	21
3000	Scissor type with wedges, 36" x 36", buy G					172			172	189
3100	Rent per month G					15.55			15.55	17.10
3300	60" x 60", buy G					260			260	286
3400	Rent per month G					21.50			21.50	24
4000	Friction collars 2'-6" diam., buy G					2,075			2,075	2,275
4100	Rent per month G					174			174	191

03 15 Concrete Accessories

03 15 05 – Concrete Forming Accessories

03 15 05.15 Column Form Accessories		Crew	Daily Output	Labor-Hours	Unit	Material	2021 Bare Costs Labor	2021 Bare Costs Equipment	Total	Total Incl O&P
4300	4'-0" diam., buy	G			Set	2,450			2,450	2,700
4400	Rent per month	G			▼	214			214	236

03 15 05.30 Hangers

		Crew	Daily Output	Labor-Hours	Unit	Material	Labor	Equipment	Total	Total Incl O&P
0010	**HANGERS**									
0020	Slab and beam form									
0500	Banding iron									
0550	3/4" x 22 ga., 14 L.F./lb. or 1/2" x 14 ga., 7 L.F./lb.	G			Lb.	1.46			1.46	1.61
1000	Fascia ties, coil type, to 24" long	G			C	490			490	540
1500	Frame ties to 8-1/8"	G				570			570	625
1550	8-1/8" to 10-1/8"	G				600			600	655
1600	10-1/8" to 12-1/8"	G				610			610	675
1650	12-1/8" to 14-1/8"	G				635			635	700
1700	14-1/8" to 16-1/8"	G				655			655	725
2000	Half hanger	G			▼	830			830	915
2500	Haunch hanger, for 1" haunch									
2600	Flange to 8-1/8"	G			C	615			615	680
2650	8-1/8" to 10-1/8"	G				655			655	725
2700	10-1/8" to 12-1/8"	G				690			690	755
2750	12-1/8" to 14-1/8"	G				710			710	780
2800	14-1/8" to 16-1/8"	G				750			750	825
3000	Haunch half hanger, for 1" haunch	G				410			410	450
5000	Snap tie hanger, to 30" overall length, 3000#	G				490			490	540
5050	To 36" overall length	G				545			545	600
5100	To 48" overall length	G			▼	700			700	775
5500	Steel beam hanger									
5600	Flange to 8-1/8"	G			C	570			570	625
5650	8-1/8" to 10-1/8"	G				600			600	655
5700	10-1/8" to 12-1/8"	G				610			610	675
5750	12-1/8" to 14-1/8"	G				635			635	700
5800	14-1/8" to 16-1/8"	G			▼	655			655	725
5900	Coil threaded rods, continuous, 1/2" diameter	G			L.F.	1.54			1.54	1.69
6000	Tie hangers to 30" overall length, 4000#	G			C	555			555	610
6100	To 36" overall length	G				610			610	670
6150	To 48" overall length	G				770			770	845
6500	Tie back hanger, up to 12-1/8" flange	G			▼	1,525			1,525	1,675
8000	Wire beam saddles, to 18" overall length									
8100	1-1/2" joist, 7 gauge	G			C	630			630	690
8200	4 gauge	G			"	660			660	725
8500	Wire, black annealed, 15 gauge	G			Cwt.	150			150	165
8600	16 gauge	G			"	153			153	168

03 15 05.70 Shores

		Crew	Daily Output	Labor-Hours	Unit	Material	Labor	Equipment	Total	Total Incl O&P
0010	**SHORES**									
0020	Erect and strip, by hand, horizontal members									
0500	Aluminum joists and stringers	G	2 Carp	60	.267	Ea.		14.60		14.60	22
0600	Steel, adjustable beams	G		45	.356			19.45		19.45	29
0700	Wood joists			50	.320			17.50		17.50	26
0800	Wood stringers			30	.533			29		29	43.50
1000	Vertical members to 10' high	G		55	.291			15.90		15.90	24
1050	To 13' high	G		50	.320			17.50		17.50	26
1100	To 16' high	G		45	.356	▼		19.45		19.45	29
1500	Reshoring	G	▼	1400	.011	S.F.	.62	.63		1.25	1.61
1600	Flying truss system	G	C-17D	9600	.009	SFCA		.50	.08	.58	.84

03 15 05 – Concrete Forming Accessories

03 15 05.70 Shores

			Crew	Daily Output	Labor-Hours	Unit	Material	2021 Bare Costs Labor	Equipment	Total	Total Incl O&P
1760	Horizontal, aluminum joists, 6-1/4" high x 5' to 21' span, buy	G				L.F.	16.05			16.05	17.65
1770	Beams, 7-1/4" high x 4' to 30' span	G				"	19.25			19.25	21
1810	Horizontal, steel beam, W8x10, 7' span, buy	G				Ea.	63			63	69
1830	10' span	G					73			73	80
1920	15' span	G					125			125	138
1940	20' span	G				↓	176			176	194
1970	Steel stringer, W8x10, 4' to 16' span, buy	G				L.F.	7.30			7.30	8
3000	Rent for job duration, aluminum joist @ 2' OC, per mo.	G				SF Flr.	.40			.40	.44
3050	Steel W8x10	G					.18			.18	.20
3060	Steel adjustable	G				↓	.18			.18	.20
3500	#1 post shore, steel, 5'-7" to 9'-6" high, 10,000# cap., buy	G				Ea.	158			158	174
3550	#2 post shore, 7'-3" to 12'-10" high, 7800# capacity	G					182			182	200
3600	#3 post shore, 8'-10" to 16'-1" high, 3800# capacity	G				↓	199			199	219
5010	Frame shoring systems, steel, 12,000#/leg, buy										
5040	Frame, 2' wide x 6' high	G				Ea.	120			120	132
5250	X-brace	G					19.20			19.20	21
5550	Base plate	G					18.05			18.05	19.90
5600	Screw jack	G					36			36	39.50
5650	U-head, 8" x 8"	G				↓	23			23	25

03 15 05.75 Sleeves and Chases

			Crew	Daily Output	Labor-Hours	Unit	Material	2021 Bare Costs Labor	Equipment	Total	Total Incl O&P
0010	**SLEEVES AND CHASES**										
0100	Plastic, 1 use, 12" long, 2" diameter		1 Carp	100	.080	Ea.	1.27	4.38		5.65	7.95
0150	4" diameter			90	.089		2.99	4.86		7.85	10.55
0200	6" diameter			75	.107		11.30	5.85		17.15	21
0250	12" diameter			60	.133		18.05	7.30		25.35	31
5000	Sheet metal, 2" diameter	G		100	.080		1.61	4.38		5.99	8.30
5100	4" diameter	G		90	.089		2.01	4.86		6.87	9.45
5150	6" diameter	G		75	.107		1.74	5.85		7.59	10.60
5200	12" diameter	G		60	.133		3.67	7.30		10.97	14.95
6000	Steel pipe, 2" diameter	G		100	.080		4.35	4.38		8.73	11.35
6100	4" diameter	G		90	.089		17.20	4.86		22.06	26
6150	6" diameter	G		75	.107		19.45	5.85		25.30	30
6200	12" diameter	G	↓	60	.133	↓	87	7.30		94.30	107

03 15 05.80 Snap Ties

			Crew	Daily Output	Labor-Hours	Unit	Material	2021 Bare Costs Labor	Equipment	Total	Total Incl O&P
0010	**SNAP TIES**, 8-1/4" L&W (Lumber and wedge)										
0100	2250 lb., w/flat washer, 8" wall	G				C	75			75	82.50
0150	10" wall	G					132			132	145
0200	12" wall	G					189			189	208
0250	16" wall	G					167			167	184
0300	18" wall	G					174			174	191
0500	With plastic cone, 8" wall	G					64.50			64.50	71
0550	10" wall	G					67			67	73.50
0600	12" wall	G					72.50			72.50	79.50
0650	16" wall	G					79.50			79.50	87.50
0700	18" wall	G					81.50			81.50	90
1000	3350 lb., w/flat washer, 8" wall	G					163			163	179
1100	10" wall	G					178			178	196
1150	12" wall	G					189			189	208
1200	16" wall	G					209			209	230
1250	18" wall	G					218			218	240
1500	With plastic cone, 8" wall	G					129			129	142
1550	10" wall	G					141			141	155

03 15 Concrete Accessories

03 15 05 – Concrete Forming Accessories

	03 15 05.80 Snap Ties		Crew	Daily Output	Labor-Hours	Unit	Material	2021 Bare Costs Labor	Equipment	Total	Total Incl O&P
1600	12" wall	G				C	152			152	167
1650	16" wall	G					165			165	182
1700	18" wall	G				▼	167			167	184

03 15 05.85 Stair Tread Inserts

		Crew	Daily Output	Labor-Hours	Unit	Material	Labor	Equipment	Total	Total Incl O&P
0010	**STAIR TREAD INSERTS**									
0105	Cast nosing insert, abrasive surface, pre-drilled, includes screws									
0110	Aluminum, 3" wide x 3' long	1 Cefi	32	.250	Ea.	57.50	12.95		70.45	82.50
0120	4' long		31	.258		74.50	13.35		87.85	102
0130	5' long	▼	30	.267	▼	89.50	13.80		103.30	119
0135	Extruded nosing insert, black abrasive strips, continuous anchor									
0140	Aluminum, 3" wide x 3' long	1 Cefi	64	.125	Ea.	35.50	6.50		42	48.50
0150	4' long		60	.133		48	6.90		54.90	63
0160	5' long	▼	56	.143	▼	62.50	7.40		69.90	79.50
0165	Extruded nosing insert, black abrasive strips, pre-drilled, incl. screws									
0170	Aluminum, 3" wide x 3' long	1 Cefi	32	.250	Ea.	45	12.95		57.95	68.50
0180	4' long		31	.258		59	13.35		72.35	84.50
0190	5' long	▼	30	.267	▼	83.50	13.80		97.30	112

03 15 05.95 Wall and Foundation Form Accessories

			Crew	Daily Output	Labor-Hours	Unit	Material	Labor	Equipment	Total	Total Incl O&P
0010	**WALL AND FOUNDATION FORM ACCESSORIES**										
0020	Coil tie system										
0050	Coil ties 1/2", 6000 lb., to 8"	G				C	281			281	310
0100	10" to 12"	G					340			340	370
0120	18"	G					400			400	440
0150	24"	G					475			475	520
0200	36"	G					610			610	670
0220	48"	G					760			760	835
0300	3/4", 12,000 lb., to 8"	G					545			545	600
0320	10" to 12"	G					605			605	665
0350	18"	G					725			725	795
0400	24"	G					805			805	885
0420	36"	G					1,000			1,000	1,100
0450	48"	G					1,525			1,525	1,675
0500	1", 24,000 lb., to 8"	G					955			955	1,050
0520	10" to 12"	G					1,150			1,150	1,250
0550	18"	G					1,300			1,300	1,425
0600	24"	G					1,550			1,550	1,725
0620	36"	G					2,025			2,025	2,225
0650	48"	G					2,500			2,500	2,750
0700	1-1/4", 36,000 lb., to 8"	G					1,400			1,400	1,550
0720	10" to 12"	G					1,700			1,700	1,875
0750	18"	G					2,225			2,225	2,425
0800	24"	G					2,550			2,550	2,800
0820	36"	G					3,475			3,475	3,825
0850	48"	G					4,400			4,400	4,850
0900	Coil bolts, 1/2" diameter x 3" long	G					144			144	158
0920	6" long	G					234			234	257
0940	12" long	G					410			410	455
0960	18" long	G					540			540	595
1000	3/4" diameter x 3" long	G					315			315	345
1020	6" long	G					580			580	640
1040	12" long	G					865			865	950
1060	18" long	G					1,125			1,125	1,250

03 15 05 – Concrete Forming Accessories

03 15 05.95 Wall and Foundation Form Accessories		Crew	Daily Output	Labor-Hours	Unit	Material	2021 Bare Costs Labor	Equipment	Total	Total Incl O&P
1100	1" diameter x 3" long	G				C	1,525			1,525	1,675
1120	6" long	G					1,550			1,550	1,725
1140	12" long	G					2,375			2,375	2,625
1160	18" long	G					3,175			3,175	3,475
1200	1-1/4" diameter x 3" long	G					2,400			2,400	2,650
1220	6" long	G					2,450			2,450	2,700
1240	12" long	G					3,775			3,775	4,150
1260	18" long	G					4,850			4,850	5,325
1300	Adjustable coil tie, 3/4" diameter, 20" long	G					1,200			1,200	1,325
1350	3/4" diameter, 36" long	G					1,425			1,425	1,550
1400	Tie cones, plastic, 1" setback length, 1/2" bolt diameter						87			87	95.50
1420	3/4" bolt diameter						171			171	188
1440	2" setback length, 1" bolt diameter						465			465	515
1460	1-1/4" bolt diameter						560			560	620
1500	Welding coil tie, 1/2" diameter x 4" long	G					270			270	297
1550	3/4" diameter x 6" long	G					350			350	385
1600	1" diameter x 8" long	G					560			560	620
1700	Waler holders, 1/2" diameter	G					266			266	292
1750	3/4" diameter	G					296			296	325
1900	Flat washers, 4" x 5" x 1/4" for 3/4" diameter	G					1,275			1,275	1,400
1950	5" x 5" x 7/16" for 1" diameter	G				▼	1,600			1,600	1,750
2000	Footings, turnbuckle form aligner	G				Ea.	13.50			13.50	14.85
2050	Spreaders for footer, adjustable	G				"	26.50			26.50	29.50
2100	Lagstud, threaded, 1/2" diameter	G				C.L.F.	149			149	164
2150	3/4" diameter	G					275			275	305
2200	1" diameter	G					545			545	600
2250	1-1/4" diameter	G				▼	1,450			1,450	1,575
2300	Lagnuts, 1/2" diameter	G				C	39			39	43
2350	3/4" diameter	G					94			94	103
2400	1" diameter	G					270			270	297
2450	1-1/4" diameter	G					375			375	415
2600	Lagnuts with handle, 1/2" diameter	G					365			365	400
2650	3/4" diameter	G					490			490	540
2700	1" diameter	G					1,025			1,025	1,125
2750	Plastic set back plugs for 3/4" diameter						99.50			99.50	110
2800	Rock anchors, 1/2" diameter	G					1,325			1,325	1,475
2850	3/4" diameter	G					1,775			1,775	1,950
2900	1" diameter	G					1,925			1,925	2,100
2950	Batter washer, 1/2" diameter	G				▼	665			665	730
3000	Form oil, up to 1200 S.F./gallon coverage					Gal.	17.15			17.15	18.90
3050	Up to 800 S.F./gallon					"	21.50			21.50	23.50
3500	Form patches, 1-3/4" diameter					C	23.50			23.50	26
3550	2-3/4" diameter					"	40			40	44
4000	Nail stakes, 3/4" diameter, 18" long	G				Ea.	2.66			2.66	2.93
4050	24" long	G					2.64			2.64	2.90
4200	30" long	G					3.33			3.33	3.66
4250	36" long	G				▼	4.20			4.20	4.62
5000	Pencil rods, 1/4" diameter	G				C.L.F.	65.50			65.50	72
5200	Clamps for 1/4" pencil rods	G				Ea.	6.25			6.25	6.85
5300	Clamping jacks for 1/4" pencil rod clamp	G				"	101			101	111
6000	She-bolts, 7/8" x 20"	G				C	6,100			6,100	6,700
6150	1-1/4" x 20"	G					6,950			6,950	7,650
6200	1-1/2" x 20"	G				▼	9,550			9,550	10,500

03 15 Concrete Accessories

03 15 05 – Concrete Forming Accessories

03 15 05.95 Wall and Foundation Form Accessories		Crew	Daily Output	Labor-Hours	Unit	Material	2021 Bare Costs Labor	Equipment	Total	Total Incl O&P
6300	Inside rods, threaded, 1/2" diameter x 6" G				C	175			175	193
6350	1/2" diameter x 12" G					310			310	340
6400	5/8" diameter x 6" G					415			415	455
6450	5/8" diameter x 12" G					605			605	665
6500	3/4" diameter x 6" G					460			460	505
6550	3/4" diameter x 12" G				↓	655			655	720
6700	For wing nuts, see taper ties									
7000	Taper tie system									
7100	Taper ties, 3/4" to 1/2" diameter x 30" G				Ea.	59.50			59.50	65
7150	1" to 3/4" diameter x 30" G					82.50			82.50	91
7200	1-1/4" to 1" diameter x 30" G				↓	96			96	106
7300	Wing nuts, 1/2" diameter G				C	695			695	765
7350	3/4" diameter G					740			740	815
7400	7/8" diameter G					815			815	900
7450	1" diameter G					815			815	900
7500	1-1/8" diameter G					760			760	835
7550	1-1/4" diameter G					1,050			1,050	1,150
7600	1-1/2" diameter G					1,225			1,225	1,350
7700	Flat washers, 1/4" x 3" x 4", for 1/2" diam. bolt G					310			310	340
7750	1/4" x 4" x 5", for 3/4" diam. bolt G					1,275			1,275	1,400
7800	7/16" x 5" x 5", for 1" diam. bolt G					1,600			1,600	1,750
7850	7/16" x 5" x 5", for 1-1/4" diam. bolt G				↓	1,450			1,450	1,600
9000	Wood form accessories									
9100	Tie plate G				C	640			640	700
9200	Corner washer G					1,400			1,400	1,550
9300	Panel bolt G					965			965	1,050
9400	Panel wedge G					161			161	177
9500	Stud clamp G				↓	690			690	760

03 15 13 – Waterstops

03 15 13.50 Waterstops		Crew	Daily Output	Labor-Hours	Unit	Material	2021 Bare Costs Labor	Equipment	Total	Total Incl O&P
0010	**WATERSTOPS**, PVC and Rubber									
0020	PVC, ribbed 3/16" thick, 4" wide	1 Carp	155	.052	L.F.	1.66	2.82		4.48	6.05
0050	6" wide		145	.055		2.62	3.02		5.64	7.40
0500	With center bulb, 6" wide, 3/16" thick		135	.059		2.73	3.24		5.97	7.85
0550	3/8" thick		130	.062		3.16	3.37		6.53	8.50
0600	9" wide x 3/8" thick		125	.064		4.17	3.50		7.67	9.85
0800	Dumbbell type, 6" wide, 3/16" thick		150	.053		4.87	2.92		7.79	9.70
0850	3/8" thick		145	.055		4.30	3.02		7.32	9.25
1000	9" wide, 3/8" thick, plain		130	.062		6.60	3.37		9.97	12.25
1050	Center bulb		130	.062		9.95	3.37		13.32	15.95
1250	Ribbed type, split, 3/16" thick, 6" wide		145	.055		2.73	3.02		5.75	7.50
1300	3/8" thick		130	.062		5.35	3.37		8.72	10.90
2000	Rubber, flat dumbbell, 3/8" thick, 6" wide		145	.055		5.60	3.02		8.62	10.65
2050	9" wide		135	.059		11.35	3.24		14.59	17.35
2500	Flat dumbbell split, 3/8" thick, 6" wide		145	.055		2.73	3.02		5.75	7.50
2550	9" wide		135	.059		5.35	3.24		8.59	10.75
3000	Center bulb, 1/4" thick, 6" wide		145	.055		6.95	3.02		9.97	12.15
3050	9" wide		135	.059		14.50	3.24		17.74	21
3500	Center bulb split, 3/8" thick, 6" wide		145	.055		5.40	3.02		8.42	10.45
3550	9" wide	↓	135	.059	↓	9	3.24		12.24	14.75
5000	Waterstop fittings, rubber, flat									
5010	Dumbbell or center bulb, 3/8" thick,									

03 15 Concrete Accessories

03 15 13 – Waterstops

03 15 13.50 Waterstops

		Crew	Daily Output	Labor-Hours	Unit	Material	2021 Bare Costs Labor	Equipment	Total	Total Incl O&P
5200	Field union, 6" wide	1 Carp	50	.160	Ea.	29	8.75		37.75	44.50
5250	9" wide		50	.160		32.50	8.75		41.25	49
5500	Flat cross, 6" wide		30	.267		43.50	14.60		58.10	69.50
5550	9" wide		30	.267		64.50	14.60		79.10	93
6000	Flat tee, 6" wide		30	.267		47.50	14.60		62.10	74
6050	9" wide		30	.267		62	14.60		76.60	90.50
6500	Flat ell, 6" wide		40	.200		46.50	10.95		57.45	68
6550	9" wide		40	.200		57.50	10.95		68.45	79.50
7000	Vertical tee, 6" wide		25	.320		18.25	17.50		35.75	46
7050	9" wide		25	.320		30.50	17.50		48	60
7500	Vertical ell, 6" wide		35	.229		22	12.50		34.50	43
7550	9" wide		35	.229		42.50	12.50		55	65

03 15 16 – Concrete Construction Joints

03 15 16.20 Control Joints, Saw Cut

		Crew	Daily Output	Labor-Hours	Unit	Material	2021 Bare Costs Labor	Equipment	Total	Total Incl O&P
0010	**CONTROL JOINTS, SAW CUT**									
0100	Sawcut control joints in green concrete									
0120	1" depth	C-27	2000	.008	L.F.	.03	.41	.06	.50	.70
0140	1-1/2" depth		1800	.009		.04	.46	.06	.56	.79
0160	2" depth		1600	.010		.06	.52	.07	.65	.90
0180	Sawcut joint reservoir in cured concrete									
0182	3/8" wide x 3/4" deep, with single saw blade	C-27	1000	.016	L.F.	.04	.83	.11	.98	1.38
0184	1/2" wide x 1" deep, with double saw blades		900	.018		.08	.92	.13	1.13	1.58
0186	3/4" wide x 1-1/2" deep, with double saw blades		800	.020		.17	1.04	.14	1.35	1.87
0190	Water blast joint to wash away laitance, 2 passes	C-29	2500	.003			.14	.04	.18	.25
0200	Air blast joint to blow out debris and air dry, 2 passes	C-28	2000	.004			.21	.02	.23	.32
0300	For backer rod, see Section 07 91 23.10									
0340	For joint sealant, see Section 03 15 16.30 or 07 92 13.20									
0900	For replacement of joint sealant, see Section 07 01 90.81									

03 15 16.30 Expansion Joints

		Crew	Daily Output	Labor-Hours	Unit	Material	2021 Bare Costs Labor	Equipment	Total	Total Incl O&P
0010	**EXPANSION JOINTS**									
0020	Keyed, cold, 24 ga., incl. stakes, 3-1/2" high [G]	1 Carp	200	.040	L.F.	.89	2.19		3.08	4.25
0050	4-1/2" high [G]		200	.040		.94	2.19		3.13	4.30
0100	5-1/2" high [G]		195	.041		1.19	2.24		3.43	4.66
0150	7-1/2" high [G]		190	.042		1.40	2.30		3.70	4.98
0160	9-1/2" high [G]		185	.043		1.56	2.37		3.93	5.25
0300	Poured asphalt, plain, 1/2" x 1"	1 Clab	450	.018		.94	.79		1.73	2.21
0350	1" x 2"		400	.020		3.77	.89		4.66	5.50
0500	Neoprene, liquid, cold applied, 1/2" x 1"		450	.018		2.72	.79		3.51	4.17
0550	1" x 2"		400	.020		10.75	.89		11.64	13.15
0700	Polyurethane, poured, 2 part, 1/2" x 1"		400	.020		1.42	.89		2.31	2.89
0750	1" x 2"		350	.023		5.65	1.01		6.66	7.75
0900	Rubberized asphalt, hot or cold applied, 1/2" x 1"		450	.018		.34	.79		1.13	1.55
0950	1" x 2"		400	.020		1.29	.89		2.18	2.75
1100	Hot applied, fuel resistant, 1/2" x 1"		450	.018		.51	.79		1.30	1.74
1150	1" x 2"		400	.020		1.94	.89		2.83	3.46
2000	Premolded, bituminous fiber, 1/2" x 6"	1 Carp	375	.021		.41	1.17		1.58	2.19
2050	1" x 12"		300	.027		1.63	1.46		3.09	3.97
2140	Concrete expansion joint, recycled paper and fiber, 1/2" x 6" [G]		390	.021		.43	1.12		1.55	2.14
2150	1/2" x 12" [G]		360	.022		.86	1.22		2.08	2.76
2250	Cork with resin binder, 1/2" x 6"		375	.021		1.49	1.17		2.66	3.38
2300	1" x 12"		300	.027		3.65	1.46		5.11	6.20
2500	Neoprene sponge, closed cell, 1/2" x 6"		375	.021		2.41	1.17		3.58	4.39

03 15 16 – Concrete Construction Joints

03 15 16.30 Expansion Joints

		Crew	Daily Output	Labor-Hours	Unit	Material	2021 Bare Costs Labor	2021 Bare Costs Equipment	Total	Total Incl O&P
2550	1" x 12"	1 Carp	300	.027	L.F.	7.85	1.46		9.31	10.80
2750	Polyethylene foam, 1/2" x 6"		375	.021		.65	1.17		1.82	2.46
2800	1" x 12"		300	.027		3.44	1.46		4.90	5.95
3000	Polyethylene backer rod, 3/8" diameter		460	.017		.16	.95		1.11	1.60
3050	3/4" diameter		460	.017		.08	.95		1.03	1.50
3100	1" diameter		460	.017		.15	.95		1.10	1.59
3500	Polyurethane foam, with polybutylene, 1/2" x 1/2"		475	.017		1.25	.92		2.17	2.76
3550	1" x 1"		450	.018		3.18	.97		4.15	4.95
3750	Polyurethane foam, regular, closed cell, 1/2" x 6"		375	.021		.93	1.17		2.10	2.76
3800	1" x 12"		300	.027		3.31	1.46		4.77	5.80
4000	Polyvinyl chloride foam, closed cell, 1/2" x 6"		375	.021		2.52	1.17		3.69	4.51
4050	1" x 12"		300	.027		8.70	1.46		10.16	11.75
4250	Rubber, gray sponge, 1/2" x 6"		375	.021		2.12	1.17		3.29	4.07
4300	1" x 12"		300	.027		7.60	1.46		9.06	10.60
4400	Redwood heartwood, 1" x 4"		400	.020		2.97	1.09		4.06	4.90
4450	1" x 6"	↓	375	.021	↓	2.70	1.17		3.87	4.71
5000	For installation in walls, add						75%			
5250	For installation in boxouts, add						25%			

03 15 19 – Cast-In Concrete Anchors

03 15 19.05 Anchor Bolt Accessories

			Crew	Daily Output	Labor-Hours	Unit	Material	2021 Bare Costs Labor	2021 Bare Costs Equipment	Total	Total Incl O&P
0010	**ANCHOR BOLT ACCESSORIES**										
0015	For anchor bolts set in fresh concrete, see Section 03 15 19.10										
8150	Anchor bolt sleeve, plastic, 1" diameter bolts		1 Carp	60	.133	Ea.	10.45	7.30		17.75	22.50
8500	1-1/2" diameter			28	.286		13.10	15.65		28.75	38
8600	2" diameter			24	.333		15.05	18.25		33.30	43.50
8650	3" diameter		↓	20	.400		27.50	22		49.50	63
8800	Templates, steel, 8" bolt spacing	G	2 Carp	16	1		10.75	54.50		65.25	93.50
8850	12" bolt spacing	G		15	1.067		11.20	58.50		69.70	99.50
8900	16" bolt spacing	G		14	1.143		13.45	62.50		75.95	108
8950	24" bolt spacing	G		12	1.333		17.90	73		90.90	129
9100	Wood, 8" bolt spacing			16	1		1.40	54.50		55.90	83
9150	12" bolt spacing			15	1.067		1.85	58.50		60.35	89
9200	16" bolt spacing			14	1.143		2.32	62.50		64.82	96
9250	24" bolt spacing		↓	16	1	↓	3.24	54.50		57.74	85

03 15 19.10 Anchor Bolts

			Crew	Daily Output	Labor-Hours	Unit	Material	2021 Bare Costs Labor	2021 Bare Costs Equipment	Total	Total Incl O&P
0010	**ANCHOR BOLTS**										
0015	Made from recycled materials										
0025	Single bolts installed in fresh concrete, no templates										
0030	Hooked w/nut and washer, 1/2" diameter, 8" long	G	1 Carp	132	.061	Ea.	1.55	3.32		4.87	6.65
0040	12" long	G		131	.061		1.72	3.34		5.06	6.90
0050	5/8" diameter, 8" long	G		129	.062		3.40	3.39		6.79	8.80
0060	12" long	G		127	.063		4.18	3.45		7.63	9.75
0070	3/4" diameter, 8" long	G		127	.063		4.18	3.45		7.63	9.75
0080	12" long	G	↓	125	.064	↓	5.25	3.50		8.75	11
0090	2-bolt pattern, including job-built 2-hole template, per set										
0100	J-type, incl. hex nut & washer, 1/2" diameter x 6" long	G	1 Carp	21	.381	Set	7.20	21		28.20	39
0110	12" long	G		21	.381		7.90	21		28.90	39.50
0120	18" long	G		21	.381		8.90	21		29.90	41
0130	3/4" diameter x 8" long	G		20	.400		12.80	22		34.80	46.50
0140	12" long	G		20	.400		14.90	22		36.90	49
0150	18" long	G		20	.400		18.05	22		40.05	52.50
0160	1" diameter x 12" long	G		19	.421		28	23		51	65.50

03 15 19.10 Anchor Bolts		Crew	Daily Output	Labor-Hours	Unit	Material	2021 Bare Costs Labor	Equipment	Total	Total Incl O&P
0170	18" long	G 1 Carp	19	.421	Set	33	23		56	70.50
0180	24" long	G	19	.421		39	23		62	77.50
0190	36" long	G	18	.444		52	24.50		76.50	93.50
0200	1-1/2" diameter x 18" long	G	17	.471		49	25.50		74.50	92.50
0210	24" long	G	16	.500		57.50	27.50		85	105
0300	L-type, incl. hex nut & washer, 3/4" diameter x 12" long	G	20	.400		21.50	22		43.50	56
0310	18" long	G	20	.400		26.50	22		48.50	61.50
0320	24" long	G	20	.400		31	22		53	67
0330	30" long	G	20	.400		38.50	22		60.50	74.50
0340	36" long	G	20	.400		43	22		65	80
0350	1" diameter x 12" long	G	19	.421		34	23		57	72
0360	18" long	G	19	.421		41.50	23		64.50	80
0370	24" long	G	19	.421		50	23		73	89.50
0380	30" long	G	19	.421		58.50	23		81.50	98.50
0390	36" long	G	18	.444		66	24.50		90.50	109
0400	42" long	G	18	.444		79	24.50		103.50	124
0410	48" long	G	18	.444		88.50	24.50		113	134
0420	1-1/4" diameter x 18" long	G	18	.444		43	24.50		67.50	84
0430	24" long	G	18	.444		50	24.50		74.50	92
0440	30" long	G	17	.471		57.50	25.50		83	102
0450	36" long	G	17	.471		64.50	25.50		90	110
0460	42" long	G 2 Carp	32	.500		72.50	27.50		100	121
0470	48" long	G	32	.500		82	27.50		109.50	132
0480	54" long	G	31	.516		96	28		124	148
0490	60" long	G	31	.516		105	28		133	158
0500	1-1/2" diameter x 18" long	G	33	.485		48	26.50		74.50	92.50
0510	24" long	G	32	.500		55.50	27.50		83	102
0520	30" long	G	31	.516		62	28		90	110
0530	36" long	G	30	.533		70.50	29		99.50	121
0540	42" long	G	30	.533		79.50	29		108.50	131
0550	48" long	G	29	.552		89	30		119	143
0560	54" long	G	28	.571		107	31.50		138.50	165
0570	60" long	G	28	.571		117	31.50		148.50	176
0580	1-3/4" diameter x 18" long	G	31	.516		80.50	28		108.50	131
0590	24" long	G	30	.533		93.50	29		122.50	147
0600	30" long	G	29	.552		108	30		138	164
0610	36" long	G	28	.571		123	31.50		154.50	182
0620	42" long	G	27	.593		137	32.50		169.50	200
0630	48" long	G	26	.615		150	33.50		183.50	217
0640	54" long	G	26	.615		186	33.50		219.50	255
0650	60" long	G	25	.640		200	35		235	273
0660	2" diameter x 24" long	G	27	.593		147	32.50		179.50	211
0670	30" long	G	27	.593		166	32.50		198.50	231
0680	36" long	G	26	.615		181	33.50		214.50	251
0690	42" long	G	25	.640		202	35		237	275
0700	48" long	G	24	.667		231	36.50		267.50	310
0710	54" long	G	23	.696		274	38		312	355
0720	60" long	G	23	.696		295	38		333	380
0730	66" long	G	22	.727		315	40		355	405
0740	72" long	G	21	.762		345	41.50		386.50	440
1000	4-bolt pattern, including job-built 4-hole template, per set									
1100	J-type, incl. hex nut & washer, 1/2" diameter x 6" long	G 1 Carp	19	.421	Set	9.95	23		32.95	45.50
1110	12" long	G	19	.421		11.30	23		34.30	47

03 15 19.10 Anchor Bolts		Crew	Daily Output	Labor-Hours	Unit	Material	2021 Bare Costs Labor	Equipment	Total	Total Incl O&P
1120	18" long	G	1 Carp	18	.444	Set	13.40	24.50		37.90	51
1130	3/4" diameter x 8" long	G		17	.471		21	25.50		46.50	62
1140	12" long	G		17	.471		25.50	25.50		51	66.50
1150	18" long	G		17	.471		31.50	25.50		57	73.50
1160	1" diameter x 12" long	G		16	.500		51.50	27.50		79	97.50
1170	18" long	G		15	.533		61	29		90	111
1180	24" long	G		15	.533		73.50	29		102.50	125
1190	36" long	G		15	.533		99	29		128	153
1200	1-1/2" diameter x 18" long	G		13	.615		94	33.50		127.50	154
1210	24" long	G		12	.667		111	36.50		147.50	177
1300	L-type, incl. hex nut & washer, 3/4" diameter x 12" long	G		17	.471		38.50	25.50		64	81
1310	18" long	G		17	.471		48.50	25.50		74	91.50
1320	24" long	G		17	.471		58	25.50		83.50	102
1330	30" long	G		16	.500		72.50	27.50		100	121
1340	36" long	G		16	.500		82	27.50		109.50	131
1350	1" diameter x 12" long	G		16	.500		63.50	27.50		91	111
1360	18" long	G		15	.533		78.50	29		107.50	130
1370	24" long	G		15	.533		95.50	29		124.50	149
1380	30" long	G		15	.533		112	29		141	167
1390	36" long	G		15	.533		128	29		157	184
1400	42" long	G		14	.571		154	31.50		185.50	216
1410	48" long	G		14	.571		172	31.50		203.50	236
1420	1-1/4" diameter x 18" long	G		14	.571		81.50	31.50		113	136
1430	24" long	G		14	.571		96	31.50		127.50	153
1440	30" long	G		13	.615		110	33.50		143.50	173
1450	36" long	G		13	.615		125	33.50		158.50	188
1460	42" long	G	2 Carp	25	.640		141	35		176	208
1470	48" long	G		24	.667		160	36.50		196.50	231
1480	54" long	G		23	.696		188	38		226	263
1490	60" long	G		23	.696		206	38		244	283
1500	1-1/2" diameter x 18" long	G		25	.640		91.50	35		126.50	154
1510	24" long	G		24	.667		106	36.50		142.50	172
1520	30" long	G		23	.696		119	38		157	188
1530	36" long	G		22	.727		136	40		176	210
1540	42" long	G		22	.727		155	40		195	230
1550	48" long	G		21	.762		173	41.50		214.50	253
1560	54" long	G		20	.800		210	44		254	297
1570	60" long	G		20	.800		230	44		274	320
1580	1-3/4" diameter x 18" long	G		22	.727		156	40		196	232
1590	24" long	G		21	.762		183	41.50		224.50	263
1600	30" long	G		21	.762		212	41.50		253.50	295
1610	36" long	G		20	.800		241	44		285	330
1620	42" long	G		19	.842		270	46		316	365
1630	48" long	G		18	.889		297	48.50		345.50	400
1640	54" long	G		18	.889		365	48.50		413.50	480
1650	60" long	G		17	.941		395	51.50		446.50	510
1660	2" diameter x 24" long	G		19	.842		290	46		336	390
1670	30" long	G		18	.889		325	48.50		373.50	435
1680	36" long	G		18	.889		360	48.50		408.50	470
1690	42" long	G		17	.941		400	51.50		451.50	515
1700	48" long	G		16	1		460	54.50		514.50	585
1710	54" long	G		15	1.067		545	58.50		603.50	685
1720	60" long	G		15	1.067		585	58.50		643.50	730

03 15 19.10 Anchor Bolts

		Crew	Daily Output	Labor-Hours	Unit	Material	2021 Bare Costs Labor	Equipment	Total	Total Incl O&P
1730	66" long	G 2 Carp	14	1.143	Set	625	62.50		687.50	785
1740	72" long	G ↓	14	1.143	↓	685	62.50		747.50	850
1990	For galvanized, add				Ea.	75%				

03 15 19.20 Dovetail Anchor System

		Crew	Daily Output	Labor-Hours	Unit	Material	2021 Bare Costs Labor	Equipment	Total	Total Incl O&P
0010	**DOVETAIL ANCHOR SYSTEM**									
0500	Dovetail anchor slot, galvanized, foam-filled, 26 ga.	G 1 Carp	425	.019	L.F.	1.38	1.03		2.41	3.06
0600	24 ga.	G	400	.020		.30	1.09		1.39	1.96
0625	22 ga.	G	400	.020		.45	1.09		1.54	2.13
0900	Stainless steel, foam-filled, 26 ga.	G ↓	375	.021	↓	2.21	1.17		3.38	4.17
1200	Dovetail brick anchor, corrugated, galvanized, 3-1/2" long, 16 ga.	G 1 Bric	10.50	.762	C	31.50	41		72.50	96
1300	12 ga.	G	10.50	.762		40	41		81	106
1500	Seismic, galvanized, 3-1/2" long, 16 ga.	G	10.50	.762		74.50	41		115.50	144
1600	12 ga.	G	10.50	.762		101	41		142	173
2000	Dovetail cavity wall, corrugated, galvanized, 5-1/2" long, 16 ga.	G	10.50	.762		49.50	41		90.50	116
2100	12 ga.	G	10.50	.762		61.50	41		102.50	129
3000	Dovetail furring anchors, corrugated, galvanized, 1-1/2" long, 16 ga.	G	10.50	.762		28	41		69	92
3100	12 ga.	G	10.50	.762		32	41		73	97
6000	Dovetail stone panel anchors, galvanized, 1/8" x 1" wide, 3-1/2" long	G	10.50	.762		90.50	41		131.50	162
6100	1/4" x 1" wide	G ↓	10.50	.762	↓	177	41		218	257

03 15 19.30 Inserts

		Crew	Daily Output	Labor-Hours	Unit	Material	2021 Bare Costs Labor	Equipment	Total	Total Incl O&P
0010	**INSERTS**									
1000	Inserts, slotted nut type for 3/4" bolts, 4" long	G 1 Carp	84	.095	Ea.	22.50	5.20		27.70	32.50
2100	6" long	G	84	.095		26.50	5.20		31.70	37.50
2150	8" long	G	84	.095		36.50	5.20		41.70	48.50
2200	Slotted, strap type, 4" long	G	84	.095		22.50	5.20		27.70	32.50
2250	6" long	G	84	.095		27	5.20		32.20	37.50
2300	8" long	G	84	.095		35.50	5.20		40.70	47.50
2350	Strap for slotted insert, 4" long	G	84	.095		13.65	5.20		18.85	23
4100	6" long	G	84	.095		14.80	5.20		20	24
4150	8" long	G	84	.095		18.30	5.20		23.50	28
4200	10" long	G ↓	84	.095	↓	22.50	5.20		27.70	32.50
6000	Thin slab, ferrule type									
6100	1/2" diameter bolt	G 1 Carp	60	.133	Ea.	3.83	7.30		11.13	15.10
6150	5/8" diameter bolt	G	60	.133		4.73	7.30		12.03	16.10
6200	3/4" diameter bolt	G	60	.133		5.50	7.30		12.80	16.95
6250	1" diameter bolt	G ↓	60	.133	↓	8.85	7.30		16.15	20.50
7000	Loop ferrule type									
7100	1/4" diameter bolt	G 1 Carp	84	.095	Ea.	2.11	5.20		7.31	10.10
7150	3/8" diameter bolt	G	84	.095		2.11	5.20		7.31	10.10
7200	1/2" diameter bolt	G	84	.095		2.54	5.20		7.74	10.60
7250	5/8" diameter bolt	G	84	.095		3.39	5.20		8.59	11.55
7300	3/4" diameter bolt	G	84	.095		3.80	5.20		9	12
7350	7/8" diameter bolt	G	84	.095		9.85	5.20		15.05	18.60
7400	1" diameter bolt	G ↓	84	.095	↓	13.70	5.20		18.90	23
7500	Threaded coil type with filler plug									
7600	3/4" diameter bolt	G 1 Carp	60	.133	Ea.	5.70	7.30		13	17.20
7650	1" diameter bolt	G	60	.133		10.75	7.30		18.05	22.50
7700	1-1/4" diameter bolt	G	60	.133		20.50	7.30		27.80	33.50
7750	1-1/2" diameter bolt	G ↓	60	.133		32.50	7.30		39.80	47
9000	Wedge type									
9100	For 3/4" diameter bolt	G 1 Carp	60	.133	Ea.	10.10	7.30		17.40	22
9200	Askew head bolts, black									

03 15 Concrete Accessories

03 15 19 – Cast-In Concrete Anchors

03 15 19.30 Inserts

			Crew	Daily Output	Labor-Hours	Unit	Material	2021 Bare Costs Labor	Equipment	Total	Total Incl O&P
9400	3/4" diameter, 1-1/2" long	G				Ea.	4.76			4.76	5.25
9450	2" long	G					3.79			3.79	4.17
9500	3" long	G					9.45			9.45	10.40
9600	Nuts, black										
9700	3/4" bolt	G				Ea.	2.09			2.09	2.30
9800	Cut washers, black										
9900	3/4" bolt	G				Ea.	1.36			1.36	1.50
9950	For galvanized inserts, add						30%				

03 15 19.45 Machinery Anchors

			Crew	Daily Output	Labor-Hours	Unit	Material	2021 Bare Costs Labor	Equipment	Total	Total Incl O&P
0010	**MACHINERY ANCHORS**, heavy duty, incl. sleeve, floating base nut,										
0020	lower stud & coupling nut, fiber plug, connecting stud, washer & nut.										
0030	For flush mounted embedment in poured concrete heavy equip. pads.										
0200	Stud & bolt, 1/2" diameter	G	E-16	40	.400	Ea.	60.50	24.50	3.72	88.72	109
0300	5/8" diameter	G		35	.457		67	28	4.25	99.25	122
0500	3/4" diameter	G		30	.533		80.50	32.50	4.96	117.96	145
0600	7/8" diameter	G		25	.640		95	39	5.95	139.95	171
0800	1" diameter	G		20	.800		105	49	7.45	161.45	199
0900	1-1/4" diameter	G		15	1.067		135	65.50	9.90	210.40	260

03 21 Reinforcement Bars

03 21 05 – Reinforcing Steel Accessories

03 21 05.10 Rebar Accessories

			Crew	Daily Output	Labor-Hours	Unit	Material	2021 Bare Costs Labor	Equipment	Total	Total Incl O&P
0010	**REBAR ACCESSORIES**										
0030	Steel & plastic made from recycled materials										
0100	Beam bolsters (BB), lower, 1-1/2" high, plain steel	G				C.L.F.	22			22	24
0102	Galvanized	G					50.50			50.50	55.50
0104	Stainless tipped legs	G					420			420	460
0106	Plastic tipped legs	G					38			38	42
0108	Epoxy dipped	G					56			56	61.50
0110	2" high, plain	G					27			27	29.50
0120	Galvanized	G					64.50			64.50	71
0140	Stainless tipped legs	G					510			510	565
0160	Plastic tipped legs	G					43			43	47.50
0162	Epoxy dipped	G					64			64	70.50
0200	Upper (BBU), 1-1/2" high, plain steel	G					98			98	108
0210	3" high	G					110			110	121
0500	Slab bolsters, continuous (SB), 1" high, plain steel	G					19			19	21
0502	Galvanized	G					40			40	44
0504	Stainless tipped legs	G					410			410	455
0506	Plastic tipped legs	G					34			34	37.50
0510	2" high, plain steel	G					24			24	26.50
0515	Galvanized	G					50.50			50.50	55.50
0520	Stainless tipped legs	G					490			490	535
0525	Plastic tipped legs	G					41			41	45
0530	For bolsters with wire runners (SBR), add	G					39.50			39.50	43.50
0540	For bolsters with plates (SBP), add	G					109			109	119
0700	Bag ties, 16 ga., plain, 4" long	G				C	.71			.71	.78
0710	5" long	G					.82			.82	.90
0720	6" long	G					.93			.93	1.02
0730	7" long	G					1.02			1.02	1.12
1200	High chairs, individual (HC), 3" high, plain steel	G					50			50	55

For customer support on your Concrete & Masonry Costs with RSMeans Data, call 800.448.8182.

63

03 21 05 – Reinforcing Steel Accessories

03 21 05.10 Rebar Accessories		Crew	Daily Output	Labor-Hours	Unit	Material	2021 Bare Costs Labor	Equipment	Total	Total Incl O&P
1202	Galvanized	G				C	114			114	126
1204	Stainless tipped legs	G					470			470	515
1206	Plastic tipped legs	G					66			66	72.50
1210	5" high, plain	G					60			60	66
1212	Galvanized	G					147			147	162
1214	Stainless tipped legs	G					590			590	650
1216	Plastic tipped legs	G					76			76	83.50
1220	8" high, plain	G					83			83	91.50
1222	Galvanized	G					180			180	198
1224	Stainless tipped legs	G					635			635	695
1226	Plastic tipped legs	G					99			99	109
1230	12" high, plain	G					139			139	153
1232	Galvanized	G					360			360	395
1234	Stainless tipped legs	G					690			690	760
1236	Plastic tipped legs	G					154			154	169
1400	Individual high chairs, with plate (HCP), 5" high	G					139			139	153
1410	8" high	G					162			162	178
1500	Bar chair (BC), 1-1/2" high, plain steel	G					46			46	50.50
1520	Galvanized	G					52			52	57
1530	Stainless tipped legs	G					420			420	465
1540	Plastic tipped legs	G					49			49	54
1700	Continuous high chairs (CHC), legs 8" OC, 4" high, plain steel	G				C.L.F.	62			62	68
1705	Galvanized	G					62.50			62.50	69
1710	Stainless tipped legs	G					560			560	615
1715	Plastic tipped legs	G					78			78	86
1718	Epoxy dipped	G					93			93	102
1720	6" high, plain	G					85			85	93.50
1725	Galvanized	G					100			100	110
1730	Stainless tipped legs	G					580			580	640
1735	Plastic tipped legs	G					101			101	111
1738	Epoxy dipped	G					117			117	129
1740	8" high, plain	G					122			122	134
1745	Galvanized	G					154			154	169
1750	Stainless tipped legs	G					620			620	685
1755	Plastic tipped legs	G					138			138	152
1758	Epoxy dipped	G					153			153	168
1900	For continuous bottom wire runners, add	G					37			37	40.50
1940	For continuous bottom plate, add	G					216			216	238
2200	Screed chair base, 1/2" coil thread diam., 2-1/2" high, plain steel	G				C	385			385	425
2210	Galvanized	G					430			430	475
2220	5-1/2" high, plain	G					465			465	510
2250	Galvanized	G					510			510	560
2300	3/4" coil thread diam., 2-1/2" high, plain steel	G					465			465	515
2310	Galvanized	G					535			535	590
2320	5-1/2" high, plain steel	G					585			585	645
2350	Galvanized	G					725			725	800
2400	Screed holder, 1/2" coil thread diam. for pipe screed, plain steel, 6" long	G					465			465	510
2420	12" long	G					710			710	785
2500	3/4" coil thread diam. for pipe screed, plain steel, 6" long	G					580			580	640
2520	12" long	G					925			925	1,025
2700	Screw anchor for bolts, plain steel, 3/4" diameter x 4" long	G					530			530	585
2720	1" diameter x 6" long	G					1,000			1,000	1,100
2740	1-1/2" diameter x 8" long	G					1,350			1,350	1,500

03 21 05 – Reinforcing Steel Accessories

03 21 05.10 Rebar Accessories

		Crew	Daily Output	Labor-Hours	Unit	Material	2021 Bare Costs Labor	Equipment	Total	Total Incl O&P
2800	Screw anchor eye bolts, 3/4" x 3" long	G				C	3,125			3,125	3,450
2820	1" x 3-1/2" long	G					4,050			4,050	4,450
2840	1-1/2" x 6" long	G					12,400			12,400	13,600
2900	Screw anchor bolts, 3/4" x 9" long	G					1,975			1,975	2,175
2920	1" x 12" long	G					3,825			3,825	4,200
3001	Slab lifting inserts, single pickup, galv, 3/4" diam., 5" high	G					2,100			2,100	2,300
3010	6" high	G					2,075			2,075	2,275
3030	7" high	G					2,100			2,100	2,325
3100	1" diameter, 5-1/2" high	G					2,150			2,150	2,350
3120	7" high	G					2,225			2,225	2,450
3200	Double pickup lifting inserts, 1" diameter, 5-1/2" high	G					4,150			4,150	4,550
3220	7" high	G					4,600			4,600	5,050
3330	1-1/2" diameter, 8" high	G					5,700			5,700	6,275
3800	Subgrade chairs, #4 bar head, 3-1/2" high	G					48			48	53
3850	12" high	G					54.50			54.50	59.50
3900	#6 bar head, 3-1/2" high	G					48			48	53
3950	12" high	G					54.50			54.50	59.50
4200	Subgrade stakes, no nail holes, 3/4" diameter, 12" long	G					370			370	410
4250	24" long	G					490			490	540
4300	7/8" diameter, 12" long	G					415			415	455
4350	24" long	G					710			710	780
4500	Tie wire, 16 ga. annealed steel	G				Cwt.	153			153	168

03 21 05.75 Splicing Reinforcing Bars

			Crew	Daily Output	Labor-Hours	Unit	Material	2021 Bare Costs Labor	Equipment	Total	Total Incl O&P
0010	**SPLICING REINFORCING BARS**	R032110-70									
0020	Including holding bars in place while splicing										
0100	Standard, self-aligning type, taper threaded, #4 bars	G	C-25	190	.168	Ea.	8.30	8		16.30	21.50
0105	#5 bars	G		170	.188		10.35	8.95		19.30	25
0110	#6 bars	G		150	.213		11.65	10.10		21.75	28.50
0120	#7 bars	G		130	.246		13.45	11.70		25.15	33
0300	#8 bars	G		115	.278		22.50	13.20		35.70	45
0305	#9 bars	G	C-5	105	.533		26	31.50	5.60	63.10	81
0310	#10 bars	G		95	.589		29	34.50	6.20	69.70	90.50
0320	#11 bars	G		85	.659		31	38.50	6.90	76.40	99
0330	#14 bars	G		65	.862		49	50.50	9.05	108.55	139
0340	#18 bars	G		45	1.244		74.50	73	13.05	160.55	205
0500	Transition self-aligning, taper threaded, #18-14	G		45	1.244		82.50	73	13.05	168.55	214
0510	#18-11	G		45	1.244		84	73	13.05	170.05	216
0520	#14-11	G		65	.862		55.50	50.50	9.05	115.05	146
0540	#11-10	G		85	.659		38.50	38.50	6.90	83.90	108
0550	#10-9	G		95	.589		36.50	34.50	6.20	77.20	98.50
0560	#9-8	G	C-25	105	.305		33	14.45		47.45	59
0580	#8-7	G		115	.278		30.50	13.20		43.70	54
0590	#7-6	G		130	.246		23	11.70		34.70	43.50
0600	Position coupler for curved bars, taper threaded, #4 bars	G		160	.200		39.50	9.50		49	58
0610	#5 bars	G		145	.221		41.50	10.45		51.95	61.50
0620	#6 bars	G		130	.246		58	11.70		69.70	82
0630	#7 bars	G		110	.291		62	13.80		75.80	89.50
0640	#8 bars	G		100	.320		64	15.20		79.20	94
0650	#9 bars	G	C-5	90	.622		69.50	36.50	6.50	112.50	138
0660	#10 bars	G		80	.700		75.50	41	7.35	123.85	153
0670	#11 bars	G		70	.800		78.50	47	8.40	133.90	166
0680	#14 bars	G		55	1.018		106	59.50	10.65	176.15	217

03 21 Reinforcement Bars

03 21 05 – Reinforcing Steel Accessories

03 21 05.75 Splicing Reinforcing Bars

			Crew	Daily Output	Labor-Hours	Unit	Material	2021 Bare Costs Labor	Equipment	Total	Total Incl O&P
0690	#18 bars	G	C-5	40	1.400	Ea.	163	82	14.65	259.65	320
0700	Transition position coupler for curved bars, taper threaded, #18-14	G		40	1.400		161	82	14.65	257.65	315
0710	#18-11	G		40	1.400		151	82	14.65	247.65	305
0720	#14-11	G		55	1.018		105	59.50	10.65	175.15	216
0730	#11-10	G		70	.800		84.50	47	8.40	139.90	172
0740	#10-9	G	↓	80	.700		81.50	41	7.35	129.85	159
0750	#9-8	G	C-25	90	.356		76	16.85		92.85	110
0760	#8-7	G		100	.320		69.50	15.20		84.70	100
0770	#7-6	G		110	.291		67.50	13.80		81.30	96
0800	Sleeve type w/grout filler, for precast concrete, #6 bars	G		72	.444		33	21		54	68.50
0802	#7 bars	G		64	.500		39	23.50		62.50	79.50
0805	#8 bars	G	↓	56	.571		46.50	27		73.50	93.50
0807	#9 bars	G		48	.667		54	31.50		85.50	109
0810	#10 bars	G	C-5	40	1.400		65.50	82	14.65	162.15	212
0900	#11 bars	G		32	1.750		72.50	103	18.35	193.85	253
0920	#14 bars	G	↓	24	2.333		113	137	24.50	274.50	355
1000	Sleeve type w/ferrous filler, for critical structures, #6 bars	G	C-25	72	.444		108	21		129	151
1210	#7 bars	G		64	.500		109	23.50		132.50	157
1220	#8 bars	G	↓	56	.571		115	27		142	168
1230	#9 bars	G	C-5	48	1.167		118	68.50	12.25	198.75	245
1240	#10 bars	G		40	1.400		126	82	14.65	222.65	277
1250	#11 bars	G		32	1.750		152	103	18.35	273.35	340
1260	#14 bars	G		24	2.333		190	137	24.50	351.50	440
1270	#18 bars	G	↓	16	3.500		276	205	36.50	517.50	650
2000	Weldable half coupler, taper threaded, #4 bars	G	E-16	120	.133		12.95	8.15	1.24	22.34	28.50
2100	#5 bars	G		112	.143		15.25	8.75	1.33	25.33	32
2200	#6 bars	G		104	.154		24	9.45	1.43	34.88	42.50
2300	#7 bars	G		96	.167		28	10.20	1.55	39.75	48.50
2400	#8 bars	G		88	.182		29.50	11.15	1.69	42.34	51.50
2500	#9 bars	G		80	.200		32.50	12.25	1.86	46.61	56.50
2600	#10 bars	G		72	.222		33	13.60	2.07	48.67	60
2700	#11 bars	G		64	.250		35.50	15.35	2.33	53.18	65
2800	#14 bars	G		56	.286		41	17.50	2.66	61.16	75
2900	#18 bars	G	↓	48	.333	↓	66.50	20.50	3.10	90.10	108

03 21 11 – Plain Steel Reinforcement Bars

03 21 11.50 Reinforcing Steel, Mill Base Plus Extras

						Unit	Material	Labor	Equipment	Total	Total Incl O&P
0010	**REINFORCING STEEL, MILL BASE PLUS EXTRAS** R032110-10										
0150	Reinforcing, A615 grade 40, mill base	G				Ton	635			635	700
0200	Detailed, cut, bent, and delivered R032110-20	G					905			905	995
0650	Reinforcing steel, A615 grade 60, mill base	G					645			645	705
0700	Detailed, cut, bent, and delivered	G				↓	910			910	1,000
1000	Reinforcing steel, extras, included in delivered price										
1005	Mill extra, added for delivery to shop R032110-40					Ton	42.50			42.50	46.50
1010	Shop extra, added for handling & storage						48.50			48.50	53.50
1020	Shop extra, added for bending, limited percent of bars R032110-50						39.50			39.50	43
1030	Average percent of bars						78.50			78.50	86.50
1050	Large percent of bars R032110-70						157			157	173
1200	Shop extra, added for detailing, under 50 tons						58.50			58.50	64.50
1250	50 to 150 tons R032110-80						44			44	48.50
1300	150 to 500 tons						42			42	46
1350	Over 500 tons						39.50			39.50	43.50
1700	Shop extra, added for listing						6.10			6.10	6.70

03 21 Reinforcement Bars

03 21 11 – Plain Steel Reinforcement Bars

03 21 11.50 Reinforcing Steel, Mill Base Plus Extras

		Crew	Daily Output	Labor-Hours	Unit	Material	2021 Bare Costs Labor	Equipment	Total	Total Incl O&P
2000	Mill extra, added for quantity, under 20 tons				Ton	26			26	28.50
2100	Shop extra, added for quantity, under 20 tons					39			39	43
2200	20 to 49 tons					29.50			29.50	32
2250	50 to 99 tons					19.55			19.55	21.50
2300	100 to 300 tons					11.70			11.70	12.90
2500	Shop extra, added for size, #3					163			163	180
2550	#4					81.50			81.50	90
2600	#5					41			41	45
2650	#6					36.50			36.50	40.50
2700	#7 to #11					49			49	54
2750	#14					61			61	67.50
2800	#18					69.50			69.50	76.50
2900	Shop extra, added for delivery to job					16.15			16.15	17.75

03 21 11.60 Reinforcing In Place

			Crew	Daily Output	Labor-Hours	Unit	Material	2021 Bare Costs Labor	Equipment	Total	Total Incl O&P
0010	**REINFORCING IN PLACE**, 50-60 ton lots, A615 Grade 60	R032110-10									
0020	Includes labor, but not material cost, to install accessories										
0030	Made from recycled materials										
0100	Beams & Girders, #3 to #7	G	4 Rodm	1.60	20	Ton	1,250	1,175		2,425	3,125
0150	#8 to #18	R032110-20 G		2.70	11.852		1,250	700		1,950	2,425
0200	Columns, #3 to #7	G		1.50	21.333		1,250	1,250		2,500	3,250
0210	#3 to #7, alternate method	G		3000	.011	Lb.	.65	.63		1.28	1.65
0250	#8 to #18	G		2.30	13.913	Ton	1,250	820		2,070	2,600
0260	#8 to #18, alternate method	R032110-40 G		4600	.007	Lb.	.65	.41		1.06	1.32
0300	Spirals, hot rolled, 8" to 15" diameter	G		2.20	14.545	Ton	1,575	855		2,430	3,025
0320	15" to 24" diameter	R032110-50 G		2.20	14.545		1,525	855		2,380	2,950
0330	24" to 36" diameter	G		2.30	13.913		1,450	820		2,270	2,800
0340	36" to 48" diameter	R032110-70 G		2.40	13.333		1,375	785		2,160	2,675
0360	48" to 64" diameter	G		2.50	12.800		1,525	755		2,280	2,800
0380	64" to 84" diameter	R032110-80 G		2.60	12.308		1,575	725		2,300	2,825
0390	84" to 96" diameter	G		2.70	11.852		1,650	700		2,350	2,875
0400	Elevated slabs, #4 to #7	G		2.90	11.034		1,250	650		1,900	2,350
0500	Footings, #4 to #7	G		2.10	15.238		1,250	900		2,150	2,725
0550	#8 to #18	G		3.60	8.889		1,250	525		1,775	2,150
0600	Slab on grade, #3 to #7	G		2.30	13.913		1,250	820		2,070	2,600
0700	Walls, #3 to #7	G		3	10.667		1,250	630		1,880	2,325
0750	#8 to #18	G		4	8		1,250	470		1,720	2,075
0900	For other than 50-60 ton lots										
1000	Under 10 ton job, #3 to #7, add						25%	10%			
1010	#8 to #18, add						20%	10%			
1050	10-50 ton job, #3 to #7, add						10%				
1060	#8 to #18, add						5%				
1100	60-100 ton job, #3 to #7, deduct						5%				
1110	#8 to #18, deduct						10%				
1150	Over 100 ton job, #3 to #7, deduct						10%				
1160	#8 to #18, deduct						15%				
1200	Reinforcing in place, A615 Grade 75, add	G				Ton	129			129	142
1220	Grade 90, add						145			145	160
2000	Unloading & sorting, add to above		C-5	100	.560			33	5.85	38.85	55.50
2200	Crane cost for handling, 90 picks/day, up to 1.5 tons/bundle, add to above			135	.415			24.50	4.35	28.85	41.50
2210	1.0 ton/bundle			92	.609			35.50	6.40	41.90	60.50
2220	0.5 ton/bundle			35	1.600			94	16.75	110.75	158
2400	Dowels, 2 feet long, deformed, #3	G	2 Rodm	520	.031	Ea.	.51	1.81		2.32	3.27

For customer support on your Concrete & Masonry Costs with RSMeans Data, call 800.448.8182.

67

03 21 Reinforcement Bars

03 21 11 – Plain Steel Reinforcement Bars

03 21 11.60 Reinforcing In Place

			Crew	Daily Output	Labor-Hours	Unit	Material	2021 Bare Costs Labor	Equipment	Total	Total Incl O&P
2410	#4	G	2 Rodm	480	.033	Ea.	.91	1.96		2.87	3.94
2420	#5	G		435	.037		1.42	2.17		3.59	4.81
2430	#6	G		360	.044	▼	2.05	2.62		4.67	6.15
2450	Longer and heavier dowels, add	G		725	.022	Lb.	.68	1.30		1.98	2.69
2500	Smooth dowels, 12" long, 1/4" or 3/8" diameter	G		140	.114	Ea.	.87	6.75		7.62	11
2520	5/8" diameter	G		125	.128		1.53	7.55		9.08	12.95
2530	3/4" diameter	G	▼	110	.145	▼	1.89	8.55		10.44	14.90
2600	Dowel sleeves for CIP concrete, 2-part system										
2610	Sleeve base, plastic, for 5/8" smooth dowel sleeve, fasten to edge form		1 Rodm	200	.040	Ea.	.53	2.36		2.89	4.10
2615	Sleeve, plastic, 12" long, for 5/8" smooth dowel, snap onto base			400	.020		1.54	1.18		2.72	3.45
2620	Sleeve base, for 3/4" smooth dowel sleeve			175	.046		.64	2.69		3.33	4.73
2625	Sleeve, 12" long, for 3/4" smooth dowel			350	.023		1.42	1.35		2.77	3.57
2630	Sleeve base, for 1" smooth dowel sleeve			150	.053		.68	3.14		3.82	5.45
2635	Sleeve, 12" long, for 1" smooth dowel		▼	300	.027		1.37	1.57		2.94	3.86
2700	Dowel caps, visual warning only, plastic, #3 to #8		2 Rodm	800	.020		.39	1.18		1.57	2.19
2720	#8 to #18			750	.021		.89	1.26		2.15	2.85
2750	Impalement protective, plastic, #4 to #9		▼	800	.020	▼	1.22	1.18		2.40	3.10

03 21 13 – Galvanized Reinforcement Steel Bars

03 21 13.10 Galvanized Reinforcing

			Crew	Daily Output	Labor-Hours	Unit	Material	Labor	Equipment	Total	Total Incl O&P
0010	**GALVANIZED REINFORCING**										
0150	Add to plain steel rebar pricing for galvanized rebar					Ton	485			485	535

03 21 16 – Epoxy-Coated Reinforcement Steel Bars

03 21 16.10 Epoxy-Coated Reinforcing

			Crew	Daily Output	Labor-Hours	Unit	Material	Labor	Equipment	Total	Total Incl O&P
0010	**EPOXY-COATED REINFORCING**										
0100	Add to plain steel rebar pricing for epoxy-coated rebar					Ton	970			970	1,075

03 21 19 – Stainless Steel Reinforcement Bars

03 21 19.10 Stainless Steel Reinforcing

			Crew	Daily Output	Labor-Hours	Unit	Material	Labor	Equipment	Total	Total Incl O&P
0010	**STAINLESS STEEL REINFORCING**										
0100	Add to plain steel rebar pricing for stainless steel rebar						300%				

03 21 21 – Composite Reinforcement Bars

03 21 21.11 Glass Fiber-Reinforced Polymer Reinf. Bars

			Crew	Daily Output	Labor-Hours	Unit	Material	Labor	Equipment	Total	Total Incl O&P
0010	**GLASS FIBER-REINFORCED POLYMER REINFORCEMENT BARS**										
0020	Includes labor, but not material cost, to install accessories										
0050	#2 bar, .043 lb./L.F.		4 Rodm	9500	.003	L.F.	.41	.20		.61	.75
0100	#3 bar, .092 lb./L.F.			9300	.003		.66	.20		.86	1.03
0150	#4 bar, .160 lb./L.F.			9100	.004		.95	.21		1.16	1.36
0200	#5 bar, .258 lb./L.F.			8700	.004		1.17	.22		1.39	1.61
0250	#6 bar, .372 lb./L.F.			8300	.004		2.04	.23		2.27	2.58
0300	#7 bar, .497 lb./L.F.			7900	.004		2.52	.24		2.76	3.13
0350	#8 bar, .620 lb./L.F.			7400	.004		2.29	.25		2.54	2.90
0400	#9 bar, .800 lb./L.F.			6800	.005		4.15	.28		4.43	4.98
0450	#10 bar, 1.08 lb./L.F.		▼	5800	.006	▼	3.77	.32		4.09	4.64
0500	For bends, add per bend					Ea.	1.59			1.59	1.75

For customer support on your Concrete & Masonry Costs with RSMeans Data, call 800.448.8182.

03 22 Fabric and Grid Reinforcing

03 22 11 – Plain Welded Wire Fabric Reinforcing

03 22 11.10 Plain Welded Wire Fabric		Crew	Daily Output	Labor-Hours	Unit	Material	2021 Bare Costs Labor	Equipment	Total	Total Incl O&P
0010	**PLAIN WELDED WIRE FABRIC** ASTM A185 R032205-30									
0020	Includes labor, but not material cost, to install accessories									
0030	Made from recycled materials									
0050	Sheets									
0100	6 x 6 - W1.4 x W1.4 (10 x 10) 21 lb./C.S.F.	G	2 Rodm	35	.457	C.S.F.	14.95	27		41.95	57
0200	6 x 6 - W2.1 x W2.1 (8 x 8) 30 lb./C.S.F.	G		31	.516		21	30.50		51.50	69
0300	6 x 6 - W2.9 x W2.9 (6 x 6) 42 lb./C.S.F.	G		29	.552		33.50	32.50		66	85.50
0400	6 x 6 - W4 x W4 (4 x 4) 58 lb./C.S.F.	G		27	.593		39	35		74	95
0500	4 x 4 - W1.4 x W1.4 (10 x 10) 31 lb./C.S.F.	G		31	.516		24.50	30.50		55	72.50
0600	4 x 4 - W2.1 x W2.1 (8 x 8) 44 lb./C.S.F.	G		29	.552		30.50	32.50		63	82
0650	4 x 4 - W2.9 x W2.9 (6 x 6) 61 lb./C.S.F.	G		27	.593		49	35		84	106
0700	4 x 4 - W4 x W4 (4 x 4) 85 lb./C.S.F.	G		25	.640		61	37.50		98.50	124
0750	Rolls									
0800	2 x 2 - #14 galv., 21 lb./C.S.F., beam & column wrap	G	2 Rodm	6.50	2.462	C.S.F.	45.50	145		190.50	267
0900	2 x 2 - #12 galv. for gunite reinforcing	G	"	6.50	2.462	"	63	145		208	287

03 22 13 – Galvanized Welded Wire Fabric Reinforcing

03 22 13.10 Galvanized Welded Wire Fabric

0010	**GALVANIZED WELDED WIRE FABRIC**									
0100	Add to plain welded wire pricing for galvanized welded wire					Lb.	.24			.24	.27

03 22 16 – Epoxy-Coated Welded Wire Fabric Reinforcing

03 22 16.10 Epoxy-Coated Welded Wire Fabric

0010	**EPOXY-COATED WELDED WIRE FABRIC**									
0100	Add to plain welded wire pricing for epoxy-coated welded wire					Lb.	.49			.49	.53

03 23 Stressed Tendon Reinforcing

03 23 05 – Prestressing Tendons

03 23 05.50 Prestressing Steel

03 23 05.50 Prestressing Steel		Crew	Daily Output	Labor-Hours	Unit	Material	2021 Bare Costs Labor	Equipment	Total	Total Incl O&P
0010	**PRESTRESSING STEEL** R034136-90									
0100	Grouted strand, in beams, post-tensioned in field, 50' span, 100 kip	G	C-3	1200	.053	Lb.	2.89	2.94	.15	5.98	7.75
0150	300 kip	G		2700	.024		1.22	1.31	.07	2.60	3.36
0300	100' span, 100 kip	G		1700	.038		2.89	2.07	.11	5.07	6.40
0350	300 kip	G		3200	.020		2.48	1.10	.06	3.64	4.44
0500	200' span, 100 kip	G		2700	.024		2.89	1.31	.07	4.27	5.20
0550	300 kip	G		3500	.018		2.48	1.01	.05	3.54	4.29
0800	Grouted bars, in beams, 50' span, 42 kip	G		2600	.025		1.25	1.36	.07	2.68	3.48
0850	143 kip	G		3200	.020		1.20	1.10	.06	2.36	3.02
1000	75' span, 42 kip	G		3200	.020		1.26	1.10	.06	2.42	3.10
1050	143 kip	G		4200	.015		1.07	.84	.04	1.95	2.48
1200	Ungrouted strand, in beams, 50' span, 100 kip	G	C-4	1275	.025		.65	1.49	.04	2.18	3
1250	300 kip	G		1475	.022		.65	1.29	.04	1.98	2.69
1400	100' span, 100 kip	G		1500	.021		.65	1.27	.04	1.96	2.65
1450	300 kip	G		1650	.019		.65	1.15	.03	1.83	2.48
1600	200' span, 100 kip	G		1500	.021		.65	1.27	.04	1.96	2.65
1650	300 kip	G		1700	.019		.65	1.12	.03	1.80	2.43
1800	Ungrouted bars, in beams, 50' span, 42 kip	G		1400	.023		.62	1.36	.04	2.02	2.75
1850	143 kip	G		1700	.019		.62	1.12	.03	1.77	2.39
2000	75' span, 42 kip	G		1800	.018		.62	1.06	.03	1.71	2.29
2050	143 kip	G		2200	.015		.62	.86	.03	1.51	2
2220	Ungrouted single strand, 100' elevated slab, 25 kip	G		1200	.027		.65	1.58	.05	2.28	3.14
2250	35 kip	G		1475	.022		.65	1.29	.04	1.98	2.69

For customer support on your Concrete & Masonry Costs with RSMeans Data, call 800.448.8182.

69

03 23 05 – Prestressing Tendons

03 23 05.50 Prestressing Steel	Crew	Daily Output	Labor-Hours	Unit	Material	2021 Bare Costs Labor	Equipment	Total	Total Incl O&P
3000	Slabs on grade, 0.5-inch diam. non-bonded strands, HDPE sheathed,								
3050	attached dead-end anchors, loose stressing-end anchors								
3100	25' x 30' slab, strands @ 36" OC, placing	2 Rodm	2940	.005	S.F.	.69	.32		1.01	1.23
3105	Stressing	C-4A	3750	.004			.25	.02	.27	.40
3110	42" OC, placing	2 Rodm	3200	.005		.61	.29		.90	1.11
3115	Stressing	C-4A	4040	.004			.23	.02	.25	.37
3120	48" OC, placing	2 Rodm	3510	.005		.53	.27		.80	.99
3125	Stressing	C-4A	4390	.004			.21	.02	.23	.34
3150	25' x 40' slab, strands @ 36" OC, placing	2 Rodm	3370	.005		.66	.28		.94	1.15
3155	Stressing	C-4A	4360	.004			.22	.02	.24	.34
3160	42" OC, placing	2 Rodm	3760	.004		.57	.25		.82	1
3165	Stressing	C-4A	4820	.003			.20	.02	.22	.31
3170	48" OC, placing	2 Rodm	4090	.004		.51	.23		.74	.90
3175	Stressing	C-4A	5190	.003			.18	.01	.19	.29
3200	30' x 30' slab, strands @ 36" OC, placing	2 Rodm	3260	.005		.66	.29		.95	1.16
3205	Stressing	C-4A	4190	.004			.22	.02	.24	.36
3210	42" OC, placing	2 Rodm	3530	.005		.60	.27		.87	1.05
3215	Stressing	C-4A	4500	.004			.21	.02	.23	.33
3220	48" OC, placing	2 Rodm	3840	.004		.53	.25		.78	.95
3225	Stressing	C-4A	4850	.003			.19	.02	.21	.31
3230	30' x 40' slab, strands @ 36" OC, placing	2 Rodm	3780	.004		.64	.25		.89	1.07
3235	Stressing	C-4A	4920	.003			.19	.02	.21	.31
3240	42" OC, placing	2 Rodm	4190	.004		.56	.22		.78	.95
3245	Stressing	C-4A	5410	.003			.17	.01	.18	.28
3250	48" OC, placing	2 Rodm	4520	.004		.50	.21		.71	.86
3255	Stressing	C-4A	5790	.003			.16	.01	.17	.25
3260	30' x 50' slab, strands @ 36" OC, placing	2 Rodm	4300	.004		.60	.22		.82	.99
3265	Stressing	C-4A	5650	.003			.17	.01	.18	.26
3270	42" OC, placing	2 Rodm	4720	.003		.54	.20		.74	.89
3275	Stressing	C-4A	6150	.003			.15	.01	.16	.24
3280	48" OC, placing	2 Rodm	5240	.003		.47	.18		.65	.79
3285	Stressing	C-4A	6760	.002			.14	.01	.15	.22

03 24 05 – Reinforcing Fibers

03 24 05.30 Synthetic Fibers

		Crew	Daily Output	Labor-Hours	Unit	Material	2021 Bare Costs Labor	Equipment	Total	Total Incl O&P
0010	**SYNTHETIC FIBERS**									
0100	Synthetic fibers, add to concrete					Lb.	6.15			6.15	6.75
0110	1-1/2 lb./C.Y.					C.Y.	9.50			9.50	10.45

03 24 05.70 Steel Fibers

			Crew	Daily Output	Labor-Hours	Unit	Material	2021 Bare Costs Labor	Equipment	Total	Total Incl O&P
0010	**STEEL FIBERS**										
0140	ASTM A850, Type V, continuously deformed, 1-1/2" long x 0.045" diam.										
0150	Add to price of ready mix concrete	G				Lb.	1.20			1.20	1.32
0205	Alternate pricing, dosing at 5 lb./C.Y., add to price of RMC	G				C.Y.	6			6	6.60
0210	10 lb./C.Y.	G					12			12	13.20
0215	15 lb./C.Y.	G					18			18	19.80
0220	20 lb./C.Y.	G					24			24	26.50
0225	25 lb./C.Y.	G					30			30	33
0230	30 lb./C.Y.	G					36			36	39.50
0235	35 lb./C.Y.	G					42			42	46
0240	40 lb./C.Y.	G					48			48	53

70

For customer support on your Concrete & Masonry Costs with RSMeans Data, call 800.448.8182.

03 24 Fibrous Reinforcing

03 24 05 – Reinforcing Fibers

03 24 05.70 Steel Fibers		Crew	Daily Output	Labor-Hours	Unit	Material	2021 Bare Costs Labor	Equipment	Total	Total Incl O&P
0250	50 lb./C.Y.	G			C.Y.	60			60	66
0275	75 lb./C.Y.	G				90			90	99
0300	100 lb./C.Y.	G				120			120	132

03 30 Cast-In-Place Concrete

03 30 53 – Miscellaneous Cast-In-Place Concrete

03 30 53.40 Concrete In Place

			Crew	Daily Output	Labor-Hours	Unit	Material	2021 Bare Costs Labor	Equipment	Total	Total Incl O&P
0010	**CONCRETE IN PLACE**	R033053-10									
0020	Including forms (4 uses), Grade 60 rebar, concrete (Portland cement	R033053-60									
0050	Type I), placement and finishing unless otherwise indicated	R033105-10									
0300	Beams (3500 psi), 5 kip/L.F., 10' span	R033105-20	C-14A	15.62	12.804	C.Y.	420	700	29	1,149	1,550
0350	25' span	R033105-50	"	18.55	10.782		440	590	24.50	1,054.50	1,375
0500	Chimney foundations (5000 psi), over 5 C.Y.	R033105-65	C-14C	32.22	3.476		187	182	.83	369.83	480
0510	(3500 psi), under 5 C.Y.	R033105-70	"	23.71	4.724		223	247	1.13	471.13	615
0700	Columns, square (4000 psi), 12" x 12", up to 1% reinforcing by area	R033105-80	C-14A	11.96	16.722		475	915	37.50	1,427.50	1,950
0720	Up to 2% reinforcing by area	R033105-85		10.13	19.743		735	1,075	44.50	1,854.50	2,450
0740	Up to 3% reinforcing by area	R033053-50		9.03	22.148		1,075	1,200	50	2,325	3,050
0800	16" x 16", up to 1% reinforcing by area			16.22	12.330		375	675	27.50	1,077.50	1,450
0820	Up to 2% reinforcing by area			12.57	15.911		620	870	36	1,526	2,025
0840	Up to 3% reinforcing by area			10.25	19.512		945	1,075	44	2,064	2,700
0900	24" x 24", up to 1% reinforcing by area			23.66	8.453		310	460	19	789	1,050
0920	Up to 2% reinforcing by area			17.71	11.293		545	620	25.50	1,190.50	1,550
0940	Up to 3% reinforcing by area			14.15	14.134		860	775	32	1,667	2,125
1000	36" x 36", up to 1% reinforcing by area			33.69	5.936		268	325	13.35	606.35	795
1020	Up to 2% reinforcing by area			23.32	8.576		475	470	19.30	964.30	1,250
1040	Up to 3% reinforcing by area			17.82	11.223		795	615	25.50	1,435.50	1,825
1100	Columns, round (4000 psi), tied, 12" diameter, up to 1% reinforcing by area			20.97	9.537		420	520	21.50	961.50	1,275
1120	Up to 2% reinforcing by area			15.27	13.098		675	715	29.50	1,419.50	1,850
1140	Up to 3% reinforcing by area			12.11	16.515		1,025	905	37	1,967	2,525
1200	16" diameter, up to 1% reinforcing by area			31.49	6.351		365	345	14.30	724.30	935
1220	Up to 2% reinforcing by area			19.12	10.460		625	570	23.50	1,218.50	1,575
1240	Up to 3% reinforcing by area			13.77	14.524		935	795	32.50	1,762.50	2,225
1300	20" diameter, up to 1% reinforcing by area			41.04	4.873		335	266	10.95	611.95	780
1320	Up to 2% reinforcing by area			24.05	8.316		575	455	18.70	1,048.70	1,325
1340	Up to 3% reinforcing by area			17.01	11.758		905	645	26.50	1,576.50	2,000
1400	24" diameter, up to 1% reinforcing by area			51.85	3.857		340	211	8.70	559.70	700
1420	Up to 2% reinforcing by area			27.06	7.391		600	405	16.65	1,021.65	1,275
1440	Up to 3% reinforcing by area			18.29	10.935		915	600	24.50	1,539.50	1,925
1500	36" diameter, up to 1% reinforcing by area			75.04	2.665		330	146	6	482	590
1520	Up to 2% reinforcing by area			37.49	5.335		560	292	12	864	1,075
1540	Up to 3% reinforcing by area			22.84	8.757		880	480	19.70	1,379.70	1,700
1900	Elevated slab (4000 psi), flat slab with drops, 125 psf Sup. Load, 20' span		C-14B	38.45	5.410		345	295	11.75	651.75	835
1950	30' span			50.99	4.079		360	223	8.85	591.85	735
2100	Flat plate, 125 psf Sup. Load, 15' span			30.24	6.878		310	375	14.95	699.95	920
2150	25' span			49.60	4.194		320	229	9.10	558.10	700
2300	Waffle const., 30" domes, 125 psf Sup. Load, 20' span			37.07	5.611		340	305	12.20	657.20	845
2350	30' span			44.07	4.720		315	258	10.25	583.25	740
2500	One way joists, 30" pans, 125 psf Sup. Load, 15' span			27.38	7.597		440	415	16.50	871.50	1,125
2550	25' span			31.15	6.677		405	365	14.50	784.50	1,000
2700	One way beam & slab, 125 psf Sup. Load, 15' span			20.59	10.102		340	550	22	912	1,225
2750	25' span			28.36	7.334		315	400	15.90	730.90	960

71

03 30 53.40 Concrete In Place	Crew	Daily Output	Labor-Hours	Unit	Material	2021 Bare Costs Labor	Equipment	Total	Total Incl O&P
2900 Two way beam & slab, 125 psf Sup. Load, 15' span	C-14B	24.04	8.652	C.Y.	325	470	18.80	813.80	1,075
2950 25' span	↓	35.87	5.799	↓	276	315	12.60	603.60	790
3100 Elevated slabs, flat plate, including finish, not									
3110 including forms or reinforcing									
3150 Regular concrete (4000 psi), 4" slab	C-8	2613	.021	S.F.	1.72	1.05	.16	2.93	3.62
3200 6" slab		2585	.022		2.52	1.06	.16	3.74	4.52
3250 2-1/2" thick floor fill		2685	.021		1.12	1.02	.16	2.30	2.91
3300 Lightweight, 110 #/C.F., 2-1/2" thick floor fill		2585	.022		1.31	1.06	.16	2.53	3.19
3400 Cellular concrete, 1-5/8" fill, under 5000 S.F.		2000	.028		.89	1.37	.21	2.47	3.24
3450 Over 10,000 S.F.		2200	.025		.85	1.24	.19	2.28	3
3500 Add per floor for 3 to 6 stories high		31800	.002			.09	.01	.10	.14
3520 For 7 to 20 stories high	↓	21200	.003	↓		.13	.02	.15	.21
3540 Equipment pad (3000 psi), 3' x 3' x 6" thick	C-14H	45	1.067	Ea.	55.50	57	.61	113.11	147
3550 4' x 4' x 6" thick		30	1.600		84	85.50	.91	170.41	220
3560 5' x 5' x 8" thick		18	2.667		150	143	1.52	294.52	380
3570 6' x 6' x 8" thick		14	3.429		204	184	1.95	389.95	500
3580 8' x 8' x 10" thick		8	6		415	320	3.42	738.42	945
3590 10' x 10' x 12" thick	↓	5	9.600	↓	720	515	5.45	1,240.45	1,550
3600 Flexural concrete on grade, direct chute, 500 psi, no forms, reinf, finish	C-8A	150	.320	C.Y.	124	15.10		139.10	160
3610 650 psi		150	.320		137	15.10		152.10	174
3620 750 psi		150	.320		206	15.10		221.10	250
3650 Pumped, 500 psi	C-8	70	.800		124	39	6.05	169.05	202
3660 650 psi		70	.800		137	39	6.05	182.05	216
3670 750 psi	↓	70	.800		206	39	6.05	251.05	292
3800 Footings (3000 psi), spread under 1 C.Y.	C-14C	28	4		208	209	.96	417.96	540
3813 Install new concrete (3000 psi) light pole base, 24" diam. x 8'	C-1	2.66	12.030		345	625		970	1,325
3825 1 C.Y. to 5 C.Y.	C-14C	43	2.605		266	136	.63	402.63	495
3850 Over 5 C.Y.	"	75	1.493		245	78	.36	323.36	385
3900 Footings, strip (3000 psi), 18" x 9", unreinforced	C-14L	40	2.400		159	123	.67	282.67	360
3920 18" x 9", reinforced	C-14C	35	3.200		189	167	.77	356.77	460
3925 20" x 10", unreinforced	C-14L	45	2.133		153	109	.60	262.60	335
3930 20" x 10", reinforced	C-14C	40	2.800		178	146	.67	324.67	415
3935 24" x 12", unreinforced	C-14L	55	1.745		150	89.50	.49	239.99	299
3940 24" x 12", reinforced	C-14C	48	2.333		175	122	.56	297.56	375
3945 36" x 12", unreinforced	C-14L	70	1.371		144	70	.38	214.38	263
3950 36" x 12", reinforced	C-14C	60	1.867		166	97.50	.45	263.95	330
4000 Foundation mat (3000 psi), under 10 C.Y.		38.67	2.896		259	151	.70	410.70	510
4050 Over 20 C.Y.	↓	56.40	1.986		225	104	.48	329.48	405
4200 Wall, free-standing (3000 psi), 8" thick, 8' high	C-14D	45.83	4.364		199	237	9.80	445.80	585
4250 14' high		27.26	7.337		237	400	16.50	653.50	875
4260 12" thick, 8' high		64.32	3.109		179	169	7	355	455
4270 14' high		40.01	4.999		191	272	11.25	474.25	630
4300 15" thick, 8' high		80.02	2.499		172	136	5.60	313.60	400
4350 12' high		51.26	3.902		172	212	8.80	392.80	515
4500 18' high	↓	48.85	4.094	↓	195	223	9.20	427.20	555
4520 Handicap access ramp (4000 psi), railing both sides, 3' wide	C-14H	14.58	3.292	L.F.	405	176	1.88	582.88	710
4525 5' wide		12.22	3.928		420	210	2.24	632.24	775
4530 With 6" curb and rails both sides, 3' wide		8.55	5.614		415	300	3.20	718.20	910
4535 5' wide	↓	7.31	6.566	↓	425	350	3.74	778.74	995
4650 Slab on grade (3500 psi), not including finish, 4" thick	C-14E	60.75	1.449	C.Y.	149	77.50	.45	226.95	279
4700 6" thick	"	92	.957	"	144	51	.30	195.30	234
4701 Thickened slab edge (3500 psi), for slab on grade poured									
4702 monolithically with slab; depth is in addition to slab thickness;									

03 30 53 – Miscellaneous Cast-In-Place Concrete

03 30 53.40 Concrete In Place	Crew	Daily Output	Labor-Hours	Unit	Material	2021 Bare Costs Labor	Equipment	Total	Total Incl O&P
4703	formed vertical outside edge, earthen bottom and inside slope									
4705	8" deep x 8" wide bottom, unreinforced	C-14L	2190	.044	L.F.	4.58	2.24	.01	6.83	8.40
4710	8" x 8", reinforced	C-14C	1670	.067		7.50	3.51	.02	11.03	13.50
4715	12" deep x 12" wide bottom, unreinforced	C-14L	1800	.053		9.05	2.73	.01	11.79	14.05
4720	12" x 12", reinforced	C-14C	1310	.086		14.40	4.47	.02	18.89	22.50
4725	16" deep x 16" wide bottom, unreinforced	C-14L	1440	.067		15	3.41	.02	18.43	21.50
4730	16" x 16", reinforced	C-14C	1120	.100		21.50	5.25	.02	26.77	31.50
4735	20" deep x 20" wide bottom, unreinforced	C-14L	1150	.083		22.50	4.27	.02	26.79	31
4740	20" x 20", reinforced	C-14C	920	.122		30.50	6.35	.03	36.88	43
4745	24" deep x 24" wide bottom, unreinforced	C-14L	930	.103		31.50	5.30	.03	36.83	42.50
4750	24" x 24", reinforced	C-14C	740	.151	▼	42	7.90	.04	49.94	58
4751	Slab on grade (3500 psi), incl. troweled finish, not incl. forms									
4760	or reinforcing, over 10,000 S.F., 4" thick	C-14F	3425	.021	S.F.	1.65	1.04	.01	2.70	3.36
4820	6" thick		3350	.021		2.41	1.07	.01	3.49	4.23
4840	8" thick		3184	.023		3.29	1.12	.01	4.42	5.30
4900	12" thick		2734	.026		4.94	1.31	.01	6.26	7.40
4950	15" thick	▼	2505	.029	▼	6.20	1.42	.01	7.63	8.95
5000	Slab on grade (3000 psi), incl. broom finish, not incl. forms									
5001	or reinforcing, 4" thick	C-14G	2873	.019	S.F.	1.62	.95	.01	2.58	3.20
5010	6" thick		2590	.022		2.53	1.06	.01	3.60	4.35
5020	8" thick	▼	2320	.024	▼	3.30	1.18	.01	4.49	5.40
5200	Lift slab in place above the foundation, incl. forms, reinforcing,									
5210	concrete (4000 psi) and columns, over 20,000 S.F./floor	C-14B	2113	.098	S.F.	7.50	5.35	.21	13.06	16.50
5250	10,000 S.F. to 20,000 S.F./floor		1650	.126		8.20	6.90	.27	15.37	19.55
5300	Under 10,000 S.F./floor	▼	1500	.139	▼	8.85	7.55	.30	16.70	21.50
5500	Lightweight, ready mix, including screed finish only,									
5510	not including forms or reinforcing									
5550	1:4 (2500 psi) for structural roof decks	C-14B	260	.800	C.Y.	126	43.50	1.74	171.24	205
5600	1:6 (3000 psi) for ground slab with radiant heat	C-14F	92	.783		118	39	.30	157.30	187
5650	1:3:2 (2000 psi) with sand aggregate, roof deck	C-14B	260	.800		112	43.50	1.74	157.24	190
5700	Ground slab (2000 psi)	C-14F	107	.673		112	33.50	.26	145.76	172
5900	Pile caps (3000 psi), incl. forms and reinf., sq. or rect., under 10 C.Y.	C-14C	54.14	2.069		221	108	.50	329.50	405
5950	Over 10 C.Y.		75	1.493		202	78	.36	280.36	340
6000	Triangular or hexagonal, under 10 C.Y.		53	2.113		153	111	.51	264.51	335
6050	Over 10 C.Y.	▼	85	1.318		173	69	.32	242.32	293
6200	Retaining walls (3000 psi), gravity, 4' high see Section 32 32	C-14D	66.20	3.021		176	164	6.80	346.80	445
6250	10' high		125	1.600		166	87	3.60	256.60	315
6300	Cantilever, level backfill loading, 8' high		70	2.857		185	155	6.45	346.45	440
6350	16' high	▼	91	2.198	▼	179	119	4.95	302.95	380
6800	Stairs (3500 psi), not including safety treads, free standing, 3'-6" wide	C-14H	83	.578	LF Nose	7.60	31	.33	38.93	54.50
6850	Cast on ground		125	.384	"	6.05	20.50	.22	26.77	37.50
7000	Stair landings, free standing		200	.240	S.F.	6.10	12.85	.14	19.09	26
7050	Cast on ground	▼	475	.101	"	4.50	5.40	.06	9.96	13.05

03 31 Structural Concrete

03 31 13 – Heavyweight Structural Concrete

03 31 13.25 Concrete, Hand Mix

		Crew	Daily Output	Labor-Hours	Unit	Material	2021 Bare Costs Labor	Equipment	Total	Total Incl O&P
0010	**CONCRETE, HAND MIX** for small quantities or remote areas									
0050	Includes bulk local aggregate, bulk sand, bagged Portland									
0060	cement (Type I) and water, using gas powered cement mixer									
0125	2500 psi	C-30	135	.059	C.F.	4.35	2.63	1.09	8.07	9.90
0130	3000 psi		135	.059		4.71	2.63	1.09	8.43	10.35
0135	3500 psi		135	.059		4.92	2.63	1.09	8.64	10.55
0140	4000 psi		135	.059		5.15	2.63	1.09	8.87	10.85
0145	4500 psi		135	.059		5.45	2.63	1.09	9.17	11.15
0150	5000 psi		135	.059		5.85	2.63	1.09	9.57	11.55
0300	Using pre-bagged dry mix and wheelbarrow (80-lb. bag = 0.6 C.F.)									
0340	4000 psi	1 Clab	48	.167	C.F.	13.90	7.40		21.30	26.50

03 31 13.30 Concrete, Volumetric Site-Mixed

		Crew	Daily Output	Labor-Hours	Unit	Material	2021 Bare Costs Labor	Equipment	Total	Total Incl O&P
0010	**CONCRETE, VOLUMETRIC SITE-MIXED**									
0015	Mixed on-site in volumetric truck									
0020	Includes local aggregate, sand, Portland cement (Type I) and water									
0025	Excludes all additives and treatments									
0100	3000 psi, 1 C.Y. mixed and discharged				C.Y.	205			205	226
0110	2 C.Y.					160			160	176
0120	3 C.Y.					145			145	160
0130	4 C.Y.					128			128	140
0140	5 C.Y.					125			125	138
0200	For truck holding/waiting time past first 2 on-site hours, add				Hr.	93			93	102
0210	For trip charge beyond first 20 miles, each way, add				Mile	3.68			3.68	4.05
0220	For each additional increase of 500 psi, add				Ea.	4.72			4.72	5.20

03 31 13.35 Heavyweight Concrete, Ready Mix

			Crew	Daily Output	Labor-Hours	Unit	Material	2021 Bare Costs Labor	Equipment	Total	Total Incl O&P
0010	**HEAVYWEIGHT CONCRETE, READY MIX**, delivered	R033105-10									
0012	Includes local aggregate, sand, Portland cement (Type I) and water										
0015	Excludes all additives and treatments	R033105-20									
0020	2000 psi					C.Y.	109			109	120
0100	2500 psi	R033105-30					112			112	124
0150	3000 psi						124			124	137
0200	3500 psi	R033105-40					127			127	139
0300	4000 psi						129			129	142
0350	4500 psi	R033105-50					133			133	146
0400	5000 psi						137			137	151
0411	6000 psi						141			141	155
0412	8000 psi						149			149	163
0413	10,000 psi						156			156	172
0414	12,000 psi						164			164	180
0416	1:4 topping						112			112	123
0417	1:3 topping						115			115	126
0418	1:2 topping						135			135	148
0420	High early strength, 2000 psi						115			115	127
0440	2500 psi						119			119	131
0460	3000 psi						134			134	148
0480	3500 psi						137			137	151
0520	4000 psi						140			140	154
0540	4500 psi						144			144	158
0560	5000 psi						148			148	163
0561	6000 psi						154			154	169
0562	8000 psi						165			165	181
0563	10,000 psi						176			176	193

03 31 Structural Concrete

03 31 13 – Heavyweight Structural Concrete

03 31 13.35 Heavyweight Concrete, Ready Mix

		Crew	Daily Output	Labor-Hours	Unit	Material	2021 Bare Costs Labor	Equipment	Total	Total Incl O&P
0564	12,000 psi				C.Y.	187			187	205
1300	For winter concrete (hot water), add					5.85			5.85	6.40
1410	For mid-range water reducer, add					4.22			4.22	4.64
1420	For high-range water reducer/superplasticizer, add					6.65			6.65	7.30
1430	For retarder, add					4.22			4.22	4.64
1440	For non-Chloride accelerator, add					8.70			8.70	9.60
1450	For Chloride accelerator, per 1%, add					3.69			3.69	4.06
1460	For fiber reinforcing, synthetic (1 lb./C.Y.), add					8.30			8.30	9.15
1500	For Saturday delivery, add					7.70			7.70	8.45
1510	For truck holding/waiting time past 1st hour per load, add				Hr.	114			114	125
1520	For short load (less than 4 C.Y.), add per load				Ea.	97			97	107
4000	Flowable fill: ash, cement, aggregate, water									
4100	40-80 psi				C.Y.	77			77	84.50
4150	Structural: ash, cement, aggregate, water & sand									
4200	50 psi				C.Y.	77			77	85
4250	140 psi					78			78	85.50
4300	500 psi					80			80	88
4350	1000 psi					83.50			83.50	91.50

03 31 13.70 Placing Concrete

		Crew	Daily Output	Labor-Hours	Unit	Material	2021 Bare Costs Labor	Equipment	Total	Total Incl O&P
0010	**PLACING CONCRETE** R033105-70									
0020	Includes labor and equipment to place, level (strike off) and consolidate									
0050	Beams, elevated, small beams, pumped	C-20	60	1.067	C.Y.		50.50	7.95	58.45	84
0100	With crane and bucket	C-7	45	1.600			77	24	101	141
0200	Large beams, pumped	C-20	90	.711			33.50	5.30	38.80	56
0250	With crane and bucket	C-7	65	1.108			53	16.75	69.75	97.50
0400	Columns, square or round, 12" thick, pumped	C-20	60	1.067			50.50	7.95	58.45	84
0450	With crane and bucket	C-7	40	1.800			86.50	27.50	114	159
0600	18" thick, pumped	C-20	90	.711			33.50	5.30	38.80	56
0650	With crane and bucket	C-7	55	1.309			63	19.80	82.80	116
0800	24" thick, pumped	C-20	92	.696			33	5.20	38.20	54.50
0850	With crane and bucket	C-7	70	1.029			49.50	15.55	65.05	90.50
1000	36" thick, pumped	C-20	140	.457			21.50	3.41	24.91	36.50
1050	With crane and bucket	C-7	100	.720			34.50	10.90	45.40	63.50
1200	Duct bank, direct chute	C-6	155	.310			14.25	.35	14.60	21.50
1400	Elevated slabs, less than 6" thick, pumped	C-20	140	.457			21.50	3.41	24.91	36.50
1450	With crane and bucket	C-7	95	.758			36.50	11.45	47.95	66.50
1500	6" to 10" thick, pumped	C-20	160	.400			18.95	2.99	21.94	31.50
1550	With crane and bucket	C-7	110	.655			31.50	9.90	41.40	57.50
1600	Slabs over 10" thick, pumped	C-20	180	.356			16.85	2.66	19.51	28
1650	With crane and bucket	C-7	130	.554			26.50	8.40	34.90	49
1900	Footings, continuous, shallow, direct chute	C-6	120	.400			18.40	.45	18.85	28
1950	Pumped	C-20	150	.427			20	3.19	23.19	33.50
2000	With crane and bucket	C-7	90	.800			38.50	12.10	50.60	70.50
2100	Footings, continuous, deep, direct chute	C-6	140	.343			15.75	.39	16.14	24
2150	Pumped	C-20	160	.400			18.95	2.99	21.94	31.50
2200	With crane and bucket	C-7	110	.655			31.50	9.90	41.40	57.50
2400	Footings, spread, under 1 C.Y., direct chute	C-6	55	.873			40	.99	40.99	60.50
2450	Pumped	C-20	65	.985			46.50	7.35	53.85	77.50
2500	With crane and bucket	C-7	45	1.600			77	24	101	141
2600	Over 5 C.Y., direct chute	C-6	120	.400			18.40	.45	18.85	28
2650	Pumped	C-20	150	.427			20	3.19	23.19	33.50
2700	With crane and bucket	C-7	100	.720			34.50	10.90	45.40	63.50

For customer support on your Concrete & Masonry Costs with RSMeans Data, call 800.448.8182.

75

03 31 Structural Concrete

03 31 13 – Heavyweight Structural Concrete

03 31 13.70 Placing Concrete	Crew	Daily Output	Labor-Hours	Unit	Material	2021 Bare Costs Labor	2021 Bare Costs Equipment	Total	Total Incl O&P
2900	Foundation mats, over 20 C.Y., direct chute	C-6	350	.137	C.Y.		6.30	.15	6.45	9.55
2950	Pumped	C-20	400	.160			7.60	1.20	8.80	12.60
3000	With crane and bucket	C-7	300	.240			11.50	3.63	15.13	21
3200	Grade beams, direct chute	C-6	150	.320			14.70	.36	15.06	22.50
3250	Pumped	C-20	180	.356			16.85	2.66	19.51	28
3300	With crane and bucket	C-7	120	.600			29	9.10	38.10	53
3500	High rise, for more than 5 stories, pumped, add per story	C-20	2100	.030			1.44	.23	1.67	2.40
3510	With crane and bucket, add per story	C-7	2100	.034			1.64	.52	2.16	3.02
3700	Pile caps, under 5 C.Y., direct chute	C-6	90	.533			24.50	.60	25.10	37
3750	Pumped	C-20	110	.582			27.50	4.35	31.85	46
3800	With crane and bucket	C-7	80	.900			43	13.65	56.65	79.50
3850	Pile cap, 5 C.Y. to 10 C.Y., direct chute	C-6	175	.274			12.60	.31	12.91	19.10
3900	Pumped	C-20	200	.320			15.15	2.39	17.54	25
3950	With crane and bucket	C-7	150	.480			23	7.25	30.25	42.50
4000	Over 10 C.Y., direct chute	C-6	215	.223			10.25	.25	10.50	15.55
4050	Pumped	C-20	240	.267			12.65	1.99	14.64	21
4100	With crane and bucket	C-7	185	.389			18.65	5.90	24.55	34.50
4300	Slab on grade, up to 6" thick, direct chute	C-6	110	.436			20	.49	20.49	30.50
4350	Pumped	C-20	130	.492			23.50	3.68	27.18	38.50
4400	With crane and bucket	C-7	110	.655			31.50	9.90	41.40	57.50
4600	Over 6" thick, direct chute	C-6	165	.291			13.35	.33	13.68	20.50
4650	Pumped	C-20	185	.346			16.40	2.58	18.98	27.50
4700	With crane and bucket	C-7	145	.497			24	7.50	31.50	44
4900	Walls, 8" thick, direct chute	C-6	90	.533			24.50	.60	25.10	37
4950	Pumped	C-20	100	.640			30.50	4.78	35.28	50.50
5000	With crane and bucket	C-7	80	.900			43	13.65	56.65	79.50
5050	12" thick, direct chute	C-6	100	.480			22	.54	22.54	33.50
5100	Pumped	C-20	110	.582			27.50	4.35	31.85	46
5200	With crane and bucket	C-7	90	.800			38.50	12.10	50.60	70.50
5300	15" thick, direct chute	C-6	105	.457			21	.52	21.52	32
5350	Pumped	C-20	120	.533			25.50	3.98	29.48	42
5400	With crane and bucket	C-7	95	.758			36.50	11.45	47.95	66.50
5600	Wheeled concrete dumping, add to placing costs above									
5610	Walking cart, 50' haul, add	C-18	32	.281	C.Y.		12.55	3.65	16.20	23
5620	150' haul, add		24	.375			16.75	4.87	21.62	30.50
5700	250' haul, add		18	.500			22.50	6.50	29	40.50
5800	Riding cart, 50' haul, add	C-19	80	.113			5	1.74	6.74	9.40
5810	150' haul, add		60	.150			6.70	2.32	9.02	12.55
5900	250' haul, add		45	.200			8.90	3.09	11.99	16.70
6000	Concrete in-fill for pan-type metal stairs and landings. Manual placement									
6010	includes up to 50' horizontal haul from point of concrete discharge.									
6100	Stair pan treads, 2" deep									
6110	Flights in 1st floor level up/down from discharge point	C-8A	3200	.015	S.F.		.71		.71	1.05
6120	2nd floor level		2500	.019			.91		.91	1.34
6130	3rd floor level		2000	.024			1.13		1.13	1.68
6140	4th floor level		1800	.027			1.26		1.26	1.87
6200	Intermediate stair landings, pan-type 4" deep									
6210	Flights in 1st floor level up/down from discharge point	C-8A	2000	.024	S.F.		1.13		1.13	1.68
6220	2nd floor level		1500	.032			1.51		1.51	2.24
6230	3rd floor level		1200	.040			1.89		1.89	2.80
6240	4th floor level		1000	.048			2.27		2.27	3.36

03 31 Structural Concrete

03 31 16 – Lightweight Structural Concrete

03 31 16.10 Lightweight Concrete, Ready Mix	Crew	Daily Output	Labor-Hours	Unit	Material	2021 Bare Costs Labor	Equipment	Total	Total Incl O&P
0010 **LIGHTWEIGHT CONCRETE, READY MIX**									
0700 Lightweight, 110#/C.F.									
0740 2500 psi				C.Y.	126			126	138
0760 3000 psi					118			118	130
0780 3500 psi					135			135	148
0820 4000 psi					154			154	169
0840 4500 psi					141			141	156
0860 5000 psi					144			144	158

03 35 Concrete Finishing

03 35 13 – High-Tolerance Concrete Floor Finishing

03 35 13.30 Finishing Floors, High Tolerance

	Crew	Daily Output	Labor-Hours	Unit	Material	Labor	Equipment	Total	Total Incl O&P
0010 **FINISHING FLOORS, HIGH TOLERANCE**									
0012 Finishing of fresh concrete flatwork requires that concrete									
0013 first be placed, struck off & consolidated									
0015 Basic finishing for various unspecified flatwork									
0100 Bull float only	C-10	4000	.006	S.F.		.30		.30	.44
0125 Bull float & manual float		2000	.012			.59		.59	.87
0150 Bull float, manual float & broom finish, w/edging & joints		1850	.013			.64		.64	.94
0200 Bull float, manual float & manual steel trowel		1265	.019			.94		.94	1.38
0210 For specified Random Access Floors in ACI Classes 1, 2, 3 and 4 to achieve									
0215 Composite Overall Floor Flatness and Levelness values up to FF35/FL25									
0250 Bull float, machine float & machine trowel (walk-behind)	C-10C	1715	.014	S.F.		.69	.05	.74	1.08
0300 Power screed, bull float, machine float & trowel (walk-behind)	C-10D	2400	.010			.49	.08	.57	.81
0350 Power screed, bull float, machine float & trowel (ride-on)	C-10E	4000	.006			.30	.06	.36	.51
0352 For specified Random Access Floors in ACI Classes 5, 6, 7 and 8 to achieve									
0354 Composite Overall Floor Flatness and Levelness values up to FF50/FL50									
0356 Add for two-dimensional restraightening after power float	C-10	6000	.004	S.F.		.20		.20	.29
0358 For specified Random or Defined Access Floors in ACI Class 9 to achieve									
0360 Composite Overall Floor Flatness and Levelness values up to FF100/FL100									
0362 Add for two-dimensional restraightening after bull float & power float	C-10	3000	.008	S.F.		.39		.39	.58
0364 For specified Superflat Defined Access Floors in ACI Class 9 to achieve									
0366 Minimum Floor Flatness and Levelness values of FF100/FL100									
0368 Add for 2-dim'l restraightening after bull float, power float, power trowel	C-10	2000	.012	S.F.		.59		.59	.87

03 35 16 – Heavy-Duty Concrete Floor Finishing

03 35 16.30 Finishing Floors, Heavy-Duty

	Crew	Daily Output	Labor-Hours	Unit	Material	Labor	Equipment	Total	Total Incl O&P
0010 **FINISHING FLOORS, HEAVY-DUTY**									
1800 Floor abrasives, dry shake on fresh concrete, .25 psf, aluminum oxide	1 Cefi	850	.009	S.F.	.48	.49		.97	1.23
1850 Silicon carbide		850	.009		.74	.49		1.23	1.53
2000 Floor hardeners, dry shake, metallic, light service, .50 psf		850	.009		.56	.49		1.05	1.32
2050 Medium service, .75 psf		750	.011		.84	.55		1.39	1.73
2100 Heavy service, 1.0 psf		650	.012		1.12	.64		1.76	2.16
2150 Extra heavy, 1.5 psf		575	.014		1.68	.72		2.40	2.90
2300 Non-metallic, light service, .50 psf		850	.009		.14	.49		.63	.87
2350 Medium service, .75 psf		750	.011		.21	.55		.76	1.04
2400 Heavy service, 1.0 psf		650	.012		.28	.64		.92	1.24
2450 Extra heavy, 1.5 psf		575	.014		.43	.72		1.15	1.53
2800 Trap rock wearing surface, dry shake, for monolithic floors									
2810 2.0 psf	C-10B	1250	.032	S.F.	.02	1.52	.27	1.81	2.57
3800 Dustproofing, liquid, for cured concrete, solvent-based, 1 coat	1 Cefi	1900	.004		.19	.22		.41	.53

03 35 Concrete Finishing

03 35 16 – Heavy-Duty Concrete Floor Finishing

03 35 16.30 Finishing Floors, Heavy-Duty		Crew	Daily Output	Labor-Hours	Unit	Material	2021 Bare Costs Labor	Equipment	Total	Total Incl O&P
3850	2 coats	1 Cefi	1300	.006	S.F.	.69	.32		1.01	1.23
4000	Epoxy-based, 1 coat		1500	.005		.17	.28		.45	.58
4050	2 coats		1500	.005		.34	.28		.62	.77

03 35 19 – Colored Concrete Finishing

03 35 19.30 Finishing Floors, Colored

		Crew	Daily Output	Labor-Hours	Unit	Material	Labor	Equipment	Total	Total Incl O&P
0010	**FINISHING FLOORS, COLORED**									
3000	Floor coloring, dry shake on fresh concrete (0.6 psf)	1 Cefi	1300	.006	S.F.	.48	.32		.80	1
3050	(1.0 psf)	"	625	.013	"	.80	.66		1.46	1.85
3100	Colored dry shake powder only				Lb.	.80			.80	.88
3600	1/2" topping using 0.6 psf dry shake powdered color	C-10B	590	.068	S.F.	7.30	3.21	.57	11.08	13.40
3650	1.0 psf dry shake powdered color	"	590	.068	"	7.60	3.21	.57	11.38	13.75

03 35 23 – Exposed Aggregate Concrete Finishing

03 35 23.30 Finishing Floors, Exposed Aggregate

		Crew	Daily Output	Labor-Hours	Unit	Material	Labor	Equipment	Total	Total Incl O&P
0010	**FINISHING FLOORS, EXPOSED AGGREGATE**									
1600	Exposed local aggregate finish, seeded on fresh concrete, 3 lb./S.F.	1 Cefi	625	.013	S.F.	.20	.66		.86	1.19
1650	4 lb./S.F.	"	465	.017	"	.34	.89		1.23	1.68

03 35 29 – Tooled Concrete Finishing

03 35 29.30 Finishing Floors, Tooled

		Crew	Daily Output	Labor-Hours	Unit	Material	Labor	Equipment	Total	Total Incl O&P
0010	**FINISHING FLOORS, TOOLED**									
4400	Stair finish, fresh concrete, float finish	1 Cefi	275	.029	S.F.		1.51		1.51	2.21
4500	Steel trowel finish		200	.040			2.07		2.07	3.04
4600	Silicon carbide finish, dry shake on fresh concrete, .25 psf		150	.053		.48	2.76		3.24	4.57

03 35 29.60 Finishing Walls

		Crew	Daily Output	Labor-Hours	Unit	Material	Labor	Equipment	Total	Total Incl O&P
0010	**FINISHING WALLS**									
0020	Break ties and patch voids	1 Cefi	540	.015	S.F.	.04	.77		.81	1.17
0050	Burlap rub with grout		450	.018		.04	.92		.96	1.40
0100	Carborundum rub, dry		270	.030			1.53		1.53	2.25
0150	Wet rub		175	.046			2.37		2.37	3.47
0300	Bush hammer, green concrete	B-39	1000	.048			2.24	.36	2.60	3.73
0350	Cured concrete	"	650	.074			3.44	.55	3.99	5.75
0500	Acid etch	1 Cefi	575	.014		.13	.72		.85	1.21
0600	Float finish, 1/16" thick	"	300	.027		.38	1.38		1.76	2.44
0700	Sandblast, light penetration	E-11	1100	.029		.52	1.41	.27	2.20	3.06
0750	Heavy penetration	"	375	.085		1.04	4.15	.81	6	8.45
0850	Grind form fins flush	1 Clab	700	.011	L.F.		.51		.51	.76

03 35 33 – Stamped Concrete Finishing

03 35 33.50 Slab Texture Stamping

		Crew	Daily Output	Labor-Hours	Unit	Material	Labor	Equipment	Total	Total Incl O&P
0010	**SLAB TEXTURE STAMPING**									
0050	Stamping requires that concrete first be placed, struck off, consolidated,									
0060	bull floated and free of bleed water. Decorative stamping tasks include:									
0100	Step 1 - first application of dry shake colored hardener	1 Cefi	6400	.001	S.F.	.43	.06		.49	.56
0110	Step 2 - bull float		6400	.001			.06		.06	.09
0130	Step 3 - second application of dry shake colored hardener		6400	.001		.21	.06		.27	.32
0140	Step 4 - bull float, manual float & steel trowel	3 Cefi	1280	.019			.97		.97	1.42
0150	Step 5 - application of dry shake colored release agent	1 Cefi	6400	.001		.10	.06		.16	.20
0160	Step 6 - place, tamp & remove mats	3 Cefi	2400	.010		.85	.52		1.37	1.69
0170	Step 7 - touch up edges, mat joints & simulated grout lines	1 Cefi	1280	.006			.32		.32	.47
0300	Alternate stamping estimating method includes all tasks above	4 Cefi	800	.040		1.58	2.07		3.65	4.78
0400	Step 8 - pressure wash @ 3000 psi after 24 hours	1 Cefi	1600	.005			.26		.26	.38
0500	Step 9 - roll 2 coats cure/seal compound when dry	"	800	.010		.81	.52		1.33	1.65

03 35 Concrete Finishing

03 35 43 – Polished Concrete Finishing

03 35 43.10 Polished Concrete Floors	Crew	Daily Output	Labor-Hours	Unit	Material	2021 Bare Costs Labor	Equipment	Total	Total Incl O&P
0010 **POLISHED CONCRETE FLOORS** R033543-10									
0015 Processing of cured concrete to include grinding, honing,									
0020 and polishing of interior floors with 22" segmented diamond									
0025 planetary floor grinder (2 passes in different directions per grit)									
0100 Removal of pre-existing coatings, dry, with carbide discs using									
0105 dry vacuum pick-up system, final hand sweeping									
0110 Glue, adhesive or tar	J-4	1.60	15	M.S.F.	23	740	134	897	1,275
0120 Paint, epoxy, 1 coat		3.60	6.667		23	330	59.50	412.50	575
0130 2 coats		1.80	13.333		23	660	119	802	1,125
0200 Grinding and edging, wet, including wet vac pick-up and auto									
0205 scrubbing between grit changes									
0210 40-grit diamond/metal matrix	J-4A	1.60	20	M.S.F.	42.50	960	247	1,249.50	1,750
0220 80-grit diamond/metal matrix		2	16		42.50	770	197	1,009.50	1,400
0230 120-grit diamond/metal matrix		2.40	13.333		42.50	640	164	846.50	1,175
0240 200-grit diamond/metal matrix		2.80	11.429		42.50	550	141	733.50	1,000
0300 Spray on dye or stain (1 coat)	1 Cefi	16	.500		215	26		241	275
0400 Spray on densifier/hardener (2 coats)	"	8	1		305	52		357	410
0410 Auto scrubbing after 2nd coat, when dry	J-4B	16	.500			22	11.20	33.20	45.50
0500 Honing and edging, wet, including wet vac pick-up and auto									
0505 scrubbing between grit changes									
0510 100-grit diamond/resin matrix	J-4A	2.80	11.429	M.S.F.	42.50	550	141	733.50	1,000
0520 200-grit diamond/resin matrix	"	2.80	11.429	"	42.50	550	141	733.50	1,000
0530 Dry, including dry vacuum pick-up system, final hand sweeping									
0540 400-grit diamond/resin matrix	J-4A	2.80	11.429	M.S.F.	42.50	550	141	733.50	1,000
0600 Polishing and edging, dry, including dry vac pick-up and hand									
0605 sweeping between grit changes									
0610 800-grit diamond/resin matrix	J-4A	2.80	11.429	M.S.F.	42.50	550	141	733.50	1,000
0620 1500-grit diamond/resin matrix		2.80	11.429		42.50	550	141	733.50	1,000
0630 3000-grit diamond/resin matrix		2.80	11.429		42.50	550	141	733.50	1,000
0700 Auto scrubbing after final polishing step	J-4B	16	.500			22	11.20	33.20	45.50

03 37 Specialty Placed Concrete

03 37 13 – Shotcrete

03 37 13.30 Gunite (Dry-Mix)

	Crew	Daily Output	Labor-Hours	Unit	Material	Labor	Equipment	Total	Total Incl O&P
0010 **GUNITE (DRY-MIX)**									
0020 Typical in place, 1" layers, no mesh included	C-16	2000	.028	S.F.	.52	1.37	.20	2.09	2.82
0100 Mesh for gunite 2 x 2, #12	2 Rodm	800	.020		.63	1.18		1.81	2.45
0150 #4 reinforcing bars @ 6" each way	"	500	.032		2.07	1.88		3.95	5.10
0300 Typical in place, including mesh, 2" thick, flat surfaces	C-16	1000	.056		1.66	2.74	.39	4.79	6.30
0350 Curved surfaces		500	.112		1.66	5.50	.79	7.95	10.80
0500 4" thick, flat surfaces		750	.075		2.70	3.65	.52	6.87	8.95
0550 Curved surfaces		350	.160		2.70	7.80	1.12	11.62	15.80
0900 Prepare old walls, no scaffolding, good condition	C-10	1000	.024			1.18		1.18	1.74
0950 Poor condition	"	275	.087			4.31		4.31	6.35
1100 For high finish requirement or close tolerance, add						50%			
1150 Very high						110%			

03 37 13.60 Shotcrete (Wet-Mix)

	Crew	Daily Output	Labor-Hours	Unit	Material	Labor	Equipment	Total	Total Incl O&P
0010 **SHOTCRETE (WET-MIX)**									
0020 Wet mix, placed @ up to 12 C.Y./hour, 3000 psi	C-8C	80	.600	C.Y.	167	29	7.85	203.85	236
0100 Up to 35 C.Y./hour	C-8E	240	.200	"	151	9.55	2.72	163.27	183

For customer support on your Concrete & Masonry Costs with RSMeans Data, call 800.448.8182.

03 37 Specialty Placed Concrete

03 37 13 – Shotcrete

03 37 13.60 Shotcrete (Wet-Mix)

		Crew	Daily Output	Labor-Hours	Unit	Material	2021 Bare Costs Labor	Equipment	Total	Total Incl O&P
1010	Fiber reinforced, 1" thick	C-8C	1740	.028	S.F.	1.23	1.34	.36	2.93	3.74
1020	2" thick		900	.053		2.45	2.58	.70	5.73	7.30
1030	3" thick		825	.058		3.68	2.82	.76	7.26	9.10
1040	4" thick	↓	750	.064	↓	4.91	3.10	.84	8.85	10.95

03 37 23 – Roller-Compacted Concrete

03 37 23.50 Concrete, Roller-Compacted

		Crew	Daily Output	Labor-Hours	Unit	Material	2021 Bare Costs Labor	Equipment	Total	Total Incl O&P
0010	**CONCRETE, ROLLER-COMPACTED**									
0100	Mass placement, 1' lift, 1' layer	B-10C	1280	.009	C.Y.		.51	1.59	2.10	2.51
0200	2' lift, 6" layer	"	1600	.008			.41	1.28	1.69	2.01
0210	Vertical face, formed, 1' lift	B-11V	400	.060			2.66	.42	3.08	4.44
0220	6" lift	"	200	.120			5.35	.83	6.18	8.85
0300	Sloped face, nonformed, 1' lift	B-11L	384	.042			2.15	2.79	4.94	6.30
0360	6" lift	"	192	.083	↓		4.31	5.60	9.91	12.55
0400	Surface preparation, vacuum truck	B-6A	3280	.006	S.Y.		.31	.11	.42	.58
0450	Water clean	B-9A	3000	.008			.37	.16	.53	.73
0460	Water blast	B-9B	800	.030			1.40	.71	2.11	2.88
0500	Joint bedding placement, 1" thick	B-11C	975	.016	↓		.85	.22	1.07	1.50
0510	Conveyance of materials, 18 C.Y. truck, 5 min. cycle	B-34F	2048	.004	C.Y.		.20	.46	.66	.81
0520	10 min. cycle		1024	.008			.40	.92	1.32	1.62
0540	15 min. cycle	↓	680	.012			.60	1.39	1.99	2.43
0550	With crane and bucket	C-23A	1600	.025			1.25	1.66	2.91	3.68
0560	With 4 C.Y. loader, 4 min. cycle	B-10U	480	.025			1.35	2.02	3.37	4.24
0570	8 min. cycle		240	.050			2.71	4.03	6.74	8.45
0580	12 min. cycle	↓	160	.075			4.06	6.05	10.11	12.70
0590	With belt conveyor	C-7D	600	.093			4.37	.34	4.71	6.90
0600	With 17 C.Y. scraper, 5 min. cycle	B-33J	1440	.006			.33	1.68	2.01	2.34
0610	10 min. cycle		720	.011			.66	3.37	4.03	4.68
0620	15 min. cycle		480	.017			.98	5.05	6.03	7
0630	20 min. cycle	↓	360	.022	↓		1.31	6.75	8.06	9.35
0640	Water cure, small job, < 500 C.Y.	B-94C	8	1	Hr.		44.50	11.50	56	79
0650	Large job, over 500 C.Y.	B-59A	8	3	"		140	58	198	273
0660	RCC paving, with asphalt paver including material	B-25C	1000	.048	C.Y.	77	2.38	2.39	81.77	90.50
0670	8" thick layers		4200	.011	S.Y.	20	.57	.57	21.14	23.50
0680	12" thick layers	↓	2800	.017	"	30	.85	.85	31.70	35

03 39 Concrete Curing

03 39 13 – Water Concrete Curing

03 39 13.50 Water Curing

		Crew	Daily Output	Labor-Hours	Unit	Material	2021 Bare Costs Labor	Equipment	Total	Total Incl O&P
0010	**WATER CURING**									
0015	With burlap, 4 uses assumed, 7.5 oz.	2 Clab	55	.291	C.S.F.	13.75	12.90		26.65	34.50
0100	10 oz.	"	55	.291	"	25	12.90		37.90	46.50
0400	Curing blankets, 1" to 2" thick, buy				S.F.	.64			.64	.70

03 39 23 – Membrane Concrete Curing

03 39 23.13 Chemical Compound Membrane Concrete Curing

		Crew	Daily Output	Labor-Hours	Unit	Material	2021 Bare Costs Labor	Equipment	Total	Total Incl O&P
0010	**CHEMICAL COMPOUND MEMBRANE CONCRETE CURING**									
0300	Sprayed membrane curing compound	2 Clab	95	.168	C.S.F.	12.20	7.50		19.70	24.50
0700	Curing compound, solvent based, 400 S.F./gal., 55 gallon lots				Gal.	21			21	23
0720	5 gallon lots					41			41	45
0800	Curing compound, water based, 250 S.F./gal., 55 gallon lots					26.50			26.50	29
0820	5 gallon lots				↓	36			36	39.50

For customer support on your Concrete & Masonry Costs with RSMeans Data, call 800.448.8182.

03 39 Concrete Curing

03 39 23 – Membrane Concrete Curing

03 39 23.23 Sheet Membrane Concrete Curing	Crew	Daily Output	Labor- Hours	Unit	Material	2021 Bare Costs Labor	Equipment	Total	Total Incl O&P
0010 **SHEET MEMBRANE CONCRETE CURING**									
0200 Curing blanket, burlap/poly, 2-ply	2 Clab	70	.229	C.S.F.	20.50	10.15		30.65	37.50

03 41 Precast Structural Concrete

03 41 13 – Precast Concrete Hollow Core Planks

03 41 13.50 Precast Slab Planks

		Crew	Daily Output	Labor- Hours	Unit	Material	2021 Bare Costs Labor	Equipment	Total	Total Incl O&P
0010	**PRECAST SLAB PLANKS** R034105-30									
0020	Prestressed roof/floor members, grouted, solid, 4" thick	C-11	2400	.030	S.F.	8.40	1.79	.97	11.16	13.05
0050	6" thick		2800	.026		8.55	1.54	.83	10.92	12.65
0100	Hollow, 8" thick		3200	.023		12.20	1.35	.73	14.28	16.30
0150	10" thick		3600	.020		9.90	1.20	.65	11.75	13.45
0200	12" thick		4000	.018		11.70	1.08	.58	13.36	15.15

03 41 16 – Precast Concrete Slabs

03 41 16.20 Precast Concrete Channel Slabs

		Crew	Daily Output	Labor- Hours	Unit	Material	2021 Bare Costs Labor	Equipment	Total	Total Incl O&P
0010	**PRECAST CONCRETE CHANNEL SLABS**									
0335	Lightweight concrete channel slab, long runs, 2-3/4" thick	C-12	1575	.030	S.F.	12.75	1.66	.30	14.71	16.85
0375	3-3/4" thick		1550	.031		14	1.69	.31	16	18.25
0475	4-3/4" thick		1525	.031		12.80	1.71	.31	14.82	17
1275	Short pieces, 2-3/4" thick		785	.061		19.15	3.33	.61	23.09	26.50
1375	3-3/4" thick		770	.062		21	3.39	.62	25.01	28.50
1475	4-3/4" thick		762	.063		19.20	3.43	.62	23.25	27

03 41 16.50 Precast Lightweight Concrete Plank

		Crew	Daily Output	Labor- Hours	Unit	Material	2021 Bare Costs Labor	Equipment	Total	Total Incl O&P
0010	**PRECAST LIGHTWEIGHT CONCRETE PLANK**									
0015	Lightweight plank, nailable, T&G, 2" thick	C-12	1800	.027	S.F.	9.95	1.45	.26	11.66	13.40
0150	For premium ceiling finish, add				"	10%				
0200	For sloping roofs, slope over 4 in 12, add						25%			
0250	Slope over 6 in 12, add						150%			

03 41 23 – Precast Concrete Stairs

03 41 23.50 Precast Stairs

		Crew	Daily Output	Labor- Hours	Unit	Material	2021 Bare Costs Labor	Equipment	Total	Total Incl O&P
0010	**PRECAST STAIRS**									
0020	Precast concrete treads on steel stringers, 3' wide	C-12	75	.640	Riser	135	35	6.35	176.35	207
0300	Front entrance, 5' wide with 48" platform, 2 risers		16	3	Flight	690	163	29.50	882.50	1,025
0350	5 risers		12	4		1,275	218	39.50	1,532.50	1,775
0500	6' wide, 2 risers		15	3.200		735	174	31.50	940.50	1,100
0550	5 risers		11	4.364		1,400	238	43	1,681	1,950
0700	7' wide, 2 risers		14	3.429		1,075	187	34	1,296	1,500
0750	5 risers		10	4.800		1,700	261	47.50	2,008.50	2,325
1200	Basement entrance stairwell, 6 steps, incl. steel bulkhead door	B-51	22	2.182		1,900	99	9.05	2,008.05	2,250
1250	14 steps	"	11	4.364		3,425	198	18.05	3,641.05	4,075

03 41 33 – Precast Structural Pretensioned Concrete

03 41 33.10 Precast Beams

		Crew	Daily Output	Labor- Hours	Unit	Material	2021 Bare Costs Labor	Equipment	Total	Total Incl O&P
0010	**PRECAST BEAMS** R034105-30									
0011	L-shaped, 20' span, 12" x 20"	C-11	32	2.250	Ea.	4,800	135	72.50	5,007.50	5,550
0060	18" x 36"		24	3		6,625	179	97	6,901	7,650
0100	24" x 44"		22	3.273		7,925	196	106	8,227	9,150
0150	30' span, 12" x 36"		24	3		9,350	179	97	9,626	10,700
0200	18" x 44"		20	3.600		11,100	215	116	11,431	12,800
0250	24" x 52"		16	4.500		13,400	269	145	13,814	15,400

For customer support on your Concrete & Masonry Costs with RSMeans Data, call 800.448.8182.

81

03 41 Precast Structural Concrete

03 41 33 – Precast Structural Pretensioned Concrete

03 41 33.10 Precast Beams

		Crew	Daily Output	Labor-Hours	Unit	Material	2021 Bare Costs Labor	2021 Bare Costs Equipment	Total	Total Incl O&P
0400	40' span, 12" x 52"	C-11	20	3.600	Ea.	14,100	215	116	14,431	16,000
0450	18" x 52"		16	4.500		15,800	269	145	16,214	18,000
0500	24" x 52"		12	6		17,900	360	194	18,454	20,500
1200	Rectangular, 20' span, 12" x 20"		32	2.250		4,150	135	72.50	4,357.50	4,850
1250	18" x 36"		24	3		6,050	179	97	6,326	7,025
1300	24" x 44"		22	3.273		7,050	196	106	7,352	8,175
1400	30' span, 12" x 36"		24	3		6,875	179	97	7,151	7,950
1450	18" x 44"		20	3.600		9,550	215	116	9,881	11,000
1500	24" x 52"		16	4.500		11,600	269	145	12,014	13,400
1600	40' span, 12" x 52"		20	3.600		11,400	215	116	11,731	13,100
1650	18" x 52"		16	4.500		13,500	269	145	13,914	15,500
1700	24" x 52"		12	6		15,500	360	194	16,054	17,900
2000	"T" shaped, 20' span, 12" x 20"		32	2.250		5,775	135	72.50	5,982.50	6,625
2050	18" x 36"		24	3		7,600	179	97	7,876	8,750
2100	24" x 44"		22	3.273		7,825	196	106	8,127	9,025
2200	30' span, 12" x 36"		24	3		10,400	179	97	10,676	11,800
2250	18" x 44"		20	3.600		13,100	215	116	13,431	14,900
2300	24" x 52"		16	4.500		12,800	269	145	13,214	14,600
2500	40' span, 12" x 52"		20	3.600		16,800	215	116	17,131	19,000
2550	18" x 52"		16	4.500		15,900	269	145	16,314	18,100
2600	24" x 52"		12	6		17,000	360	194	17,554	19,500

03 41 33.15 Precast Columns

		Crew	Daily Output	Labor-Hours	Unit	Material	2021 Bare Costs Labor	2021 Bare Costs Equipment	Total	Total Incl O&P
0010	**PRECAST COLUMNS** R034105-30									
0020	Rectangular to 12' high, 16" x 16"	C-11	120	.600	L.F.	272	36	19.35	327.35	375
0050	24" x 24"		96	.750		370	45	24	439	505
0300	24' high, 28" x 28"		192	.375		420	22.50	12.10	454.60	515
0350	36" x 36"		144	.500		565	30	16.15	611.15	685
0700	24' high, 1 haunch, 12" x 12"		32	2.250	Ea.	5,775	135	72.50	5,982.50	6,625
0800	20" x 20"		28	2.571	"	7,750	154	83	7,987	8,850

03 41 33.25 Precast Joists

		Crew	Daily Output	Labor-Hours	Unit	Material	2021 Bare Costs Labor	2021 Bare Costs Equipment	Total	Total Incl O&P
0010	**PRECAST JOISTS** R034105-30									
0015	40 psf L.L., 6" deep for 12' spans	C-12	600	.080	L.F.	38.50	4.36	.79	43.65	49.50
0050	8" deep for 16' spans		575	.083		63.50	4.54	.83	68.87	77.50
0100	10" deep for 20' spans		550	.087		111	4.75	.86	116.61	131
0150	12" deep for 24' spans		525	.091		153	4.98	.91	158.89	176

03 41 33.60 Precast Tees

		Crew	Daily Output	Labor-Hours	Unit	Material	2021 Bare Costs Labor	2021 Bare Costs Equipment	Total	Total Incl O&P
0010	**PRECAST TEES** R034105-30									
0020	Quad tee, short spans, roof	C-11	7200	.010	S.F.	10.65	.60	.32	11.57	13.05
0050	Floor		7200	.010		10.65	.60	.32	11.57	13.05
0200	Double tee, floor members, 60' span		8400	.009		11.90	.51	.28	12.69	14.20
0250	80' span		8000	.009		16.95	.54	.29	17.78	19.80
0300	Roof members, 30' span		4800	.015		14.55	.90	.48	15.93	17.95
0350	50' span		6400	.011		12.25	.67	.36	13.28	14.90
0400	Wall members, up to 55' high		3600	.020		16.65	1.20	.65	18.50	21
0500	Single tee roof members, 40' span		3200	.023		17.55	1.35	.73	19.63	22
0550	80' span		5120	.014		19.80	.84	.45	21.09	24
0600	100' span		6000	.012		30.50	.72	.39	31.61	35
0650	120' span		6000	.012		31.50	.72	.39	32.61	36
1000	Double tees, floor members									
1100	Lightweight, 20" x 8' wide, 45' span	C-11	20	3.600	Ea.	4,225	215	116	4,556	5,100
1150	24" x 8' wide, 50' span		18	4		4,700	239	129	5,068	5,650
1200	32" x 10' wide, 60' span		16	4.500		7,025	269	145	7,439	8,325

03 41 Precast Structural Concrete

03 41 33 – Precast Structural Pretensioned Concrete

03 41 33.60 Precast Tees

		Crew	Daily Output	Labor-Hours	Unit	Material	2021 Bare Costs Labor	2021 Bare Costs Equipment	Total	Total Incl O&P
1250	Standard weight, 12" x 8' wide, 20' span	C-11	22	3.273	Ea.	1,700	196	106	2,002	2,300
1300	16" x 8' wide, 25' span		20	3.600		2,125	215	116	2,456	2,800
1350	18" x 8' wide, 30' span		20	3.600		2,550	215	116	2,881	3,275
1400	20" x 8' wide, 45' span		18	4		3,850	239	129	4,218	4,725
1450	24" x 8' wide, 50' span		16	4.500		4,275	269	145	4,689	5,275
1500	32" x 10' wide, 60' span	▼	14	5.143	▼	6,400	305	166	6,871	7,675
2000	Roof members									
2050	Lightweight, 20" x 8' wide, 40' span	C-11	20	3.600	Ea.	3,750	215	116	4,081	4,575
2100	24" x 8' wide, 50' span		18	4		4,700	239	129	5,068	5,650
2150	32" x 10' wide, 60' span		16	4.500		7,025	269	145	7,439	8,325
2200	Standard weight, 12" x 8' wide, 30' span		22	3.273		2,550	196	106	2,852	3,250
2250	16" x 8' wide, 30' span		20	3.600		2,675	215	116	3,006	3,400
2300	18" x 8' wide, 30' span		20	3.600		2,825	215	116	3,156	3,550
2350	20" x 8' wide, 40' span		18	4		3,400	239	129	3,768	4,250
2400	24" x 8' wide, 50' span		16	4.500		4,275	269	145	4,689	5,275
2450	32" x 10' wide, 60' span	▼	14	5.143	▼	6,400	305	166	6,871	7,675

03 45 Precast Architectural Concrete

03 45 13 – Faced Architectural Precast Concrete

03 45 13.50 Precast Wall Panels

		Crew	Daily Output	Labor-Hours	Unit	Material	2021 Bare Costs Labor	2021 Bare Costs Equipment	Total	Total Incl O&P
0010	**PRECAST WALL PANELS** R034513-10									
0050	Uninsulated, smooth gray									
0150	Low rise, 4' x 8' x 4" thick	C-11	320	.225	S.F.	33	13.45	7.25	53.70	64.50
0210	8' x 8', 4" thick		576	.125		32.50	7.45	4.04	43.99	52
0250	8' x 16' x 4" thick		1024	.070		32	4.20	2.27	38.47	44.50
0600	High rise, 4' x 8' x 4" thick		288	.250		33	14.95	8.05	56	68
0650	8' x 8' x 4" thick		512	.141		32.50	8.40	4.54	45.44	54
0700	8' x 16' x 4" thick		768	.094		32	5.60	3.03	40.63	47.50
0750	10' x 20', 6" thick	▼	1400	.051		55	3.07	1.66	59.73	67
0800	Insulated panel, 2" polystyrene, add					1.23			1.23	1.35
0850	2" urethane, add					.78			.78	.86
1200	Finishes, white, add					3.43			3.43	3.77
1250	Exposed aggregate, add					.71			.71	.78
1300	Granite faced, domestic, add					28.50			28.50	31.50
1350	Brick faced, modular, red, add				▼	8.95			8.95	9.85
2200	Fiberglass reinforced cement with urethane core									
2210	R20, 8' x 8', 5" plain finish	E-2	750	.075	S.F.	30	4.45	2.28	36.73	42.50
2220	Exposed aggregate or brick finish	"	600	.093	"	49.50	5.55	2.86	57.91	66

03 47 Site-Cast Concrete

03 47 13 – Tilt-Up Concrete

03 47 13.50 Tilt-Up Wall Panels

		Crew	Daily Output	Labor-Hours	Unit	Material	2021 Bare Costs Labor	2021 Bare Costs Equipment	Total	Total Incl O&P
0010	**TILT-UP WALL PANELS** R034713-20									
0015	Wall panel construction, walls only, 5-1/2" thick	C-14	1600	.090	S.F.	6.15	4.81	.91	11.87	14.95
0100	7-1/2" thick		1550	.093		7.85	4.96	.94	13.75	17.10
0500	Walls and columns, 5-1/2" thick walls, 12" x 12" columns		1565	.092		9.25	4.91	.93	15.09	18.55
0550	7-1/2" thick wall, 12" x 12" columns		1370	.105	▼	11.60	5.60	1.06	18.26	22.50
0800	Columns only, site precast, 12" x 12"		200	.720	L.F.	24.50	38.50	7.30	70.30	92.50
0850	16" x 16"	▼	105	1.371	"	36	73	13.90	122.90	164

For customer support on your Concrete & Masonry Costs with RSMeans Data, call 800.448.8182.

83

03 48 Precast Concrete Specialties

03 48 43 – Precast Concrete Trim

03 48 43.40 Precast Lintels

		Crew	Daily Output	Labor-Hours	Unit	Material	2021 Bare Costs Labor	2021 Bare Costs Equipment	Total	Total Incl O&P
0010	**PRECAST LINTELS**, smooth gray, prestressed, stock units only									
0800	4" wide x 8" high x 4' long	D-10	28	1.143	Ea.	35.50	61.50	15.45	112.45	148
0850	8' long		24	1.333		81	71.50	18.05	170.55	216
1000	6" wide x 8" high x 4' long		26	1.231		60	66	16.65	142.65	183
1050	10' long		22	1.455		156	78	19.65	253.65	310
1200	8" wide x 8" high x 4' long		24	1.333		58	71.50	18.05	147.55	191
1250	12' long		20	1.600		201	86	21.50	308.50	375
1275	For custom sizes, types, colors, or finishes of precast lintels, add					150%				

03 48 43.90 Precast Window Sills

		Crew	Daily Output	Labor-Hours	Unit	Material	2021 Bare Costs Labor	2021 Bare Costs Equipment	Total	Total Incl O&P
0010	**PRECAST WINDOW SILLS**									
0600	Precast concrete, 4" tapers to 3", 9" wide	D-1	70	.229	L.F.	22.50	11.15		33.65	41.50
0650	11" wide	"	60	.267	"	40.50	13		53.50	64

03 51 Cast Roof Decks

03 51 13 – Cementitious Wood Fiber Decks

03 51 13.50 Cementitious/Wood Fiber Planks

		Crew	Daily Output	Labor-Hours	Unit	Material	2021 Bare Costs Labor	2021 Bare Costs Equipment	Total	Total Incl O&P
0010	**CEMENTITIOUS/WOOD FIBER PLANKS** R051223-50									
0050	Plank, beveled edge, 1" thick	2 Carp	1000	.016	S.F.	3.66	.88		4.54	5.35
0100	1-1/2" thick		975	.016		5.55	.90		6.45	7.45
0150	T&G, 2" thick		950	.017		3.83	.92		4.75	5.60
0200	2-1/2" thick		925	.017		4.34	.95		5.29	6.20
0250	3" thick		900	.018		4.86	.97		5.83	6.80
1000	Bulb tee, sub-purlin and grout, 6' span, add	E-1	5000	.005		2.20	.28	.03	2.51	2.89
1100	8' span	"	4200	.006		2.20	.34	.04	2.58	2.98

03 51 16 – Gypsum Concrete Roof Decks

03 51 16.50 Gypsum Roof Deck

		Crew	Daily Output	Labor-Hours	Unit	Material	2021 Bare Costs Labor	2021 Bare Costs Equipment	Total	Total Incl O&P
0010	**GYPSUM ROOF DECK**									
1000	Poured gypsum, 2" thick	C-8	6000	.009	S.F.	.80	.46	.07	1.33	1.64
1100	3" thick	"	4800	.012	"	1.20	.57	.09	1.86	2.27

03 52 Lightweight Concrete Roof Insulation

03 52 16 – Lightweight Insulating Concrete

03 52 16.13 Lightweight Cellular Insulating Concrete

		Crew	Daily Output	Labor-Hours	Unit	Material	2021 Bare Costs Labor	2021 Bare Costs Equipment	Total	Total Incl O&P
0010	**LIGHTWEIGHT CELLULAR INSULATING CONCRETE** R035216-10									
0020	Portland cement and foaming agent [G]	C-8	50	1.120	C.Y.	134	55	8.50	197.50	238

03 52 16.16 Lightweight Aggregate Insulating Concrete

		Crew	Daily Output	Labor-Hours	Unit	Material	2021 Bare Costs Labor	2021 Bare Costs Equipment	Total	Total Incl O&P
0010	**LIGHTWEIGHT AGGREGATE INSULATING CONCRETE** R035216-10									
0100	Poured vermiculite or perlite, field mix,									
0110	1:6 field mix [G]	C-8	50	1.120	C.Y.	330	55	8.50	393.50	455
0200	Ready mix, 1:6 mix, roof fill, 2" thick [G]		10000	.006	S.F.	1.85	.27	.04	2.16	2.49
0250	3" thick [G]		7700	.007		2.77	.36	.06	3.19	3.64
0400	Expanded volcanic glass rock, 1" thick [G]	2 Carp	1500	.011		.61	.58		1.19	1.54
0450	3" thick [G]	"	1200	.013		1.83	.73		2.56	3.10
1020	Lightweight insulating fill									
1040	1000 psi [G]				C.Y.	126			126	139
1200	1:6 mix perlite				"	110			110	121

03 53 Concrete Topping

03 53 16 – Iron-Aggregate Concrete Topping

03 53 16.50 Floor Topping		Crew	Daily Output	Labor-Hours	Unit	Material	2021 Bare Costs Labor	Equipment	Total	Total Incl O&P
0010	**FLOOR TOPPING**									
0400	Integral topping/finish, on fresh concrete, using 1:1:2 mix, 3/16" thick	C-10B	1000	.040	S.F.	.17	1.89	.34	2.40	3.35
0450	1/2" thick		950	.042		.44	1.99	.35	2.78	3.82
0500	3/4" thick		850	.047		.66	2.23	.39	3.28	4.46
0600	1" thick		750	.053		.88	2.53	.45	3.86	5.20
0800	Granolithic topping, on fresh or cured concrete, 1:1:1-1/2 mix, 1/2" thick		590	.068		.49	3.21	.57	4.27	5.90
0820	3/4" thick		580	.069		.74	3.27	.58	4.59	6.30
0850	1" thick		575	.070		.99	3.29	.58	4.86	6.60
0950	2" thick		500	.080		1.97	3.79	.67	6.43	8.50
1200	Heavy duty, 1:1:2, 3/4" thick, preshrunk, gray, 20 M.S.F.		320	.125		.93	5.90	1.05	7.88	10.95
1300	100 M.S.F.		380	.105		.66	4.99	.88	6.53	9.10

03 54 Cast Underlayment

03 54 13 – Gypsum Cement Underlayment

03 54 13.50 Poured Gypsum Underlayment		Crew	Daily Output	Labor-Hours	Unit	Material	2021 Bare Costs Labor	Equipment	Total	Total Incl O&P
0010	**POURED GYPSUM UNDERLAYMENT**									
0400	Underlayment, gypsum based, self-leveling 2500 psi, pumped, 1/2" thick	C-8	24000	.002	S.F.	.20	.11	.02	.33	.41
0500	3/4" thick		20000	.003		.30	.14	.02	.46	.55
0600	1" thick		16000	.004		.40	.17	.03	.60	.72
1400	Hand placed, 1/2" thick	C-18	450	.020		.20	.89	.26	1.35	1.84
1500	3/4" thick	"	300	.030		.30	1.34	.39	2.03	2.76

03 54 16 – Hydraulic Cement Underlayment

03 54 16.50 Cement Underlayment		Crew	Daily Output	Labor-Hours	Unit	Material	2021 Bare Costs Labor	Equipment	Total	Total Incl O&P
0010	**CEMENT UNDERLAYMENT**									
2510	Underlayment, P.C. based, self-leveling, 4100 psi, pumped, 1/4" thick	C-8	20000	.003	S.F.	2.09	.14	.02	2.25	2.52
2520	1/2" thick		19000	.003		4.19	.14	.02	4.35	4.84
2530	3/4" thick		18000	.003		6.30	.15	.02	6.47	7.15
2540	1" thick		17000	.003		8.35	.16	.02	8.53	9.45
2550	1-1/2" thick		15000	.004		12.55	.18	.03	12.76	14.10
2560	Hand placed, 1/2" thick	C-18	450	.020		4.19	.89	.26	5.34	6.25
2610	Topping, P.C. based, self-leveling, 6100 psi, pumped, 1/4" thick	C-8	20000	.003		3.40	.14	.02	3.56	3.96
2620	1/2" thick		19000	.003		6.80	.14	.02	6.96	7.75
2630	3/4" thick		18000	.003		10.20	.15	.02	10.37	11.50
2660	1" thick		17000	.003		13.60	.16	.02	13.78	15.20
2670	1-1/2" thick		15000	.004		20.50	.18	.03	20.71	23
2680	Hand placed, 1/2" thick	C-18	450	.020		6.80	.89	.26	7.95	9.10

03 61 Cementitious Grouting

03 61 13 – Dry-Pack Grouting

03 61 13.50 Grout, Dry-Pack		Crew	Daily Output	Labor-Hours	Unit	Material	2021 Bare Costs Labor	Equipment	Total	Total Incl O&P
0010	**GROUT, DRY-PACK**									
4000	Dry-pack, under beams or walls, cement & sand, 1" thick	D-3	170	.247	S.F.	1.02	12.35		13.37	19.70
4020	2" thick	"	150	.280	"	2.03	14		16.03	23

For customer support on your Concrete & Masonry Costs with RSMeans Data, call 800.448.8182.

85

03 62 Non-Shrink Grouting

03 62 13 – Non-Metallic Non-Shrink Grouting

03 62 13.50 Grout, Non-Metallic Non-Shrink	Crew	Daily Output	Labor-Hours	Unit	Material	2021 Bare Costs Labor	Equipment	Total	Total Incl O&P
0010 **GROUT, NON-METALLIC NON-SHRINK**									
0300 Non-shrink, non-metallic, 1" deep	1 Cefi	35	.229	S.F.	6.80	11.85		18.65	25
0350 2" deep	"	25	.320	"	13.60	16.60		30.20	39.50

03 62 16 – Metallic Non-Shrink Grouting

03 62 16.50 Grout, Metallic Non-Shrink

	Crew	Daily Output	Labor-Hours	Unit	Material	Labor	Equipment	Total	Total Incl O&P
0010 **GROUT, METALLIC NON-SHRINK**									
0020 Column & machine bases, non-shrink, metallic, 1" deep	1 Cefi	35	.229	S.F.	9.10	11.85		20.95	27.50
0050 2" deep	"	25	.320	"	18.25	16.60		34.85	44.50

03 63 Epoxy Grouting

03 63 05 – Grouting of Dowels and Fasteners

03 63 05.10 Epoxy Only

	Crew	Daily Output	Labor-Hours	Unit	Material	Labor	Equipment	Total	Total Incl O&P
0010 **EPOXY ONLY**									
1500 Chemical anchoring, epoxy cartridge, excludes layout, drilling, fastener									
1530 For fastener 3/4" diam. x 6" embedment	2 Skwk	72	.222	Ea.	4.89	12.70		17.59	24.50
1535 1" diam. x 8" embedment		66	.242		7.35	13.85		21.20	29
1540 1-1/4" diam. x 10" embedment		60	.267		14.70	15.25		29.95	39
1545 1-3/4" diam. x 12" embedment		54	.296		24.50	16.90		41.40	52.50
1550 14" embedment		48	.333		29.50	19.05		48.55	61
1555 2" diam. x 12" embedment		42	.381		39	22		61	75.50
1560 18" embedment		32	.500		49	28.50		77.50	97

03 81 Concrete Cutting

03 81 13 – Flat Concrete Sawing

03 81 13.50 Concrete Floor/Slab Cutting

	Crew	Daily Output	Labor-Hours	Unit	Material	Labor	Equipment	Total	Total Incl O&P
0010 **CONCRETE FLOOR/SLAB CUTTING**									
0050 Includes blade cost, layout and set-up time									
0300 Saw cut concrete slabs, plain, up to 3" deep	B-89	1060	.015	L.F.	.10	.79	1.01	1.90	2.40
0320 Each additional inch of depth		3180	.005		.03	.26	.34	.63	.80
0400 Mesh reinforced, up to 3" deep		980	.016		.12	.85	1.09	2.06	2.60
0420 Each additional inch of depth		2940	.005		.04	.28	.36	.68	.86
0500 Rod reinforced, up to 3" deep		800	.020		.14	1.04	1.33	2.51	3.19
0520 Each additional inch of depth		2400	.007		.05	.35	.44	.84	1.06

03 81 13.75 Concrete Saw Blades

	Crew	Daily Output	Labor-Hours	Unit	Material	Labor	Equipment	Total	Total Incl O&P
0010 **CONCRETE SAW BLADES**									
3000 Blades for saw cutting, included in cutting line items									
3020 Diamond, 12" diameter				Ea.	219			219	241
3040 18" diameter					345			345	380
3080 24" diameter					490			490	535
3120 30" diameter					865			865	950
3160 36" diameter					1,225			1,225	1,350
3200 42" diameter					1,850			1,850	2,050

03 81 16 – Track Mounted Concrete Wall Sawing

03 81 16.50 Concrete Wall Cutting

	Crew	Daily Output	Labor-Hours	Unit	Material	Labor	Equipment	Total	Total Incl O&P
0010 **CONCRETE WALL CUTTING**									
0750 Includes blade cost, layout and set-up time									
0800 Concrete walls, hydraulic saw, plain, per inch of depth	B-89B	250	.064	L.F.	.03	3.34	6.25	9.62	11.85
0820 Rod reinforcing, per inch of depth	"	150	.107	"	.05	5.55	10.40	16	19.80

03 82 Concrete Boring

03 82 13 – Concrete Core Drilling

03 82 13.10 Core Drilling	Crew	Daily Output	Labor-Hours	Unit	Material	2021 Bare Costs Labor	Equipment	Total	Total Incl O&P
0010	**CORE DRILLING**								
0015	Includes bit cost, layout and set-up time								
0020	Reinforced concrete slab, up to 6" thick								
0100	1" diameter core	B-89A	17	.941	Ea.	.20	48	7.15	55.35	79.50
0150	For each additional inch of slab thickness in same hole, add		1440	.011		.03	.56	.08	.67	.98
0200	2" diameter core		16.50	.970		.27	49	7.35	56.62	82.50
0250	For each additional inch of slab thickness in same hole, add		1080	.015		.04	.75	.11	.90	1.30
0300	3" diameter core		16	1		.41	51	7.60	59.01	85
0350	For each additional inch of slab thickness in same hole, add		720	.022		.07	1.13	.17	1.37	1.96
0500	4" diameter core		15	1.067		.50	54	8.10	62.60	90.50
0550	For each additional inch of slab thickness in same hole, add		480	.033		.08	1.69	.25	2.02	2.91
0700	6" diameter core		14	1.143		.80	58	8.70	67.50	97.50
0750	For each additional inch of slab thickness in same hole, add		360	.044		.13	2.26	.34	2.73	3.90
0900	8" diameter core		13	1.231		1.16	62.50	9.35	73.01	105
0950	For each additional inch of slab thickness in same hole, add		288	.056		.19	2.82	.42	3.43	4.90
1100	10" diameter core		12	1.333		1.43	67.50	10.15	79.08	114
1150	For each additional inch of slab thickness in same hole, add		240	.067		.24	3.38	.51	4.13	5.85
1300	12" diameter core		11	1.455		1.63	74	11.05	86.68	125
1350	For each additional inch of slab thickness in same hole, add		206	.078		.27	3.94	.59	4.80	6.85
1500	14" diameter core		10	1.600		2.02	81	12.15	95.17	138
1550	For each additional inch of slab thickness in same hole, add		180	.089		.34	4.51	.68	5.53	7.85
1700	18" diameter core		9	1.778		2.80	90	13.50	106.30	153
1750	For each additional inch of slab thickness in same hole, add		144	.111		.47	5.65	.84	6.96	9.90
1754	24" diameter core		8	2		4.07	102	15.20	121.27	173
1756	For each additional inch of slab thickness in same hole, add		120	.133		.68	6.75	1.01	8.44	12
1760	For horizontal holes, add to above						20%	20%		
1770	Prestressed hollow core plank, 8" thick									
1780	1" diameter core	B-89A	17.50	.914	Ea.	.26	46.50	6.95	53.71	77.50
1790	For each additional inch of plank thickness in same hole, add		3840	.004		.03	.21	.03	.27	.39
1794	2" diameter core		17.25	.928		.36	47	7.05	54.41	78.50
1796	For each additional inch of plank thickness in same hole, add		2880	.006		.04	.28	.04	.36	.52
1800	3" diameter core		17	.941		.55	48	7.15	55.70	80
1810	For each additional inch of plank thickness in same hole, add		1920	.008		.07	.42	.06	.55	.78
1820	4" diameter core		16.50	.970		.67	49	7.35	57.02	83
1830	For each additional inch of plank thickness in same hole, add		1280	.013		.08	.63	.10	.81	1.14
1840	6" diameter core		15.50	1.032		1.06	52.50	7.85	61.41	88.50
1850	For each additional inch of plank thickness in same hole, add		960	.017		.13	.85	.13	1.11	1.56
1860	8" diameter core		15	1.067		1.55	54	8.10	63.65	91.50
1870	For each additional inch of plank thickness in same hole, add		768	.021		.19	1.06	.16	1.41	1.97
1880	10" diameter core		14	1.143		1.90	58	8.70	68.60	98.50
1890	For each additional inch of plank thickness in same hole, add		640	.025		.24	1.27	.19	1.70	2.37
1900	12" diameter core		13.50	1.185		2.18	60	9	71.18	102
1910	For each additional inch of plank thickness in same hole, add		548	.029		.27	1.48	.22	1.97	2.76
3000	Bits for core drilling, included in drilling line items									
3010	Diamond, premium, 1" diameter				Ea.	78			78	86
3020	2" diameter					107			107	118
3030	3" diameter					165			165	182
3040	4" diameter					202			202	222
3060	6" diameter					320			320	350
3080	8" diameter					465			465	510
3110	10" diameter					570			570	630
3120	12" diameter					655			655	720
3140	14" diameter					805			805	890

03 82 Concrete Boring

03 82 13 – Concrete Core Drilling

03 82 13.10 Core Drilling	Crew	Daily Output	Labor-Hours	Unit	Material	2021 Bare Costs Labor	Equipment	Total	Total Incl O&P
3180	18" diameter				Ea.	1,125			1,125	1,225
3240	24" diameter				↓	1,625			1,625	1,800

03 82 16 – Concrete Drilling

03 82 16.10 Concrete Impact Drilling

		Crew	Daily Output	Labor-Hours	Unit	Material	Labor	Equipment	Total	Total Incl O&P
0010	**CONCRETE IMPACT DRILLING**									
0020	Includes bit cost, layout and set-up time, no anchors									
0050	Up to 4" deep in concrete/brick floors/walls									
0100	Holes, 1/4" diameter	1 Carp	75	.107	Ea.	.05	5.85		5.90	8.75
0150	For each additional inch of depth in same hole, add		430	.019		.01	1.02		1.03	1.53
0200	3/8" diameter		63	.127		.05	6.95		7	10.40
0250	For each additional inch of depth in same hole, add		340	.024		.01	1.29		1.30	1.93
0300	1/2" diameter		50	.160		.05	8.75		8.80	13.10
0350	For each additional inch of depth in same hole, add		250	.032		.01	1.75		1.76	2.62
0400	5/8" diameter		48	.167		.08	9.10		9.18	13.70
0450	For each additional inch of depth in same hole, add		240	.033		.02	1.82		1.84	2.74
0500	3/4" diameter		45	.178		.11	9.70		9.81	14.60
0550	For each additional inch of depth in same hole, add		220	.036		.03	1.99		2.02	3
0600	7/8" diameter		43	.186		.13	10.20		10.33	15.35
0650	For each additional inch of depth in same hole, add		210	.038		.03	2.08		2.11	3.15
0700	1" diameter		40	.200		.21	10.95		11.16	16.60
0750	For each additional inch of depth in same hole, add		190	.042		.05	2.30		2.35	3.50
0800	1-1/4" diameter		38	.211		.35	11.50		11.85	17.60
0850	For each additional inch of depth in same hole, add		180	.044		.09	2.43		2.52	3.73
0900	1-1/2" diameter		35	.229		.37	12.50		12.87	19.05
0950	For each additional inch of depth in same hole, add	↓	165	.048	↓	.09	2.65		2.74	4.06
1000	For ceiling installations, add						40%			

Estimating Tips
04 05 00 Common Work Results for Masonry

- The terms mortar and grout are often used interchangeably—and incorrectly. Mortar is used to bed masonry units, seal the entry of air and moisture, provide architectural appearance, and allow for size variations in the units. Grout is used primarily in reinforced masonry construction and to bond the masonry to the reinforcing steel. Common mortar types are M (2500 psi), S (1800 psi), N (750 psi), and O (350 psi), and they conform to ASTM C270. Grout is either fine or coarse and conforms to ASTM C476, and in-place strengths generally exceed 2500 psi. Mortar and grout are different components of masonry construction and are placed by entirely different methods. An estimator should be aware of their unique uses and costs.

- Mortar is included in all assembled masonry line items. The mortar cost, part of the assembled masonry material cost, includes all ingredients, all labor, and all equipment required. Please see reference number R040513-10.

- Waste, specifically the loss/droppings of mortar and the breakage of brick and block, is included in all unit cost lines that include mortar and masonry units in this division. A factor of 25% is added for mortar and 3% for brick and concrete masonry units.

- Scaffolding or staging is not included in any of the Division 4 costs. Refer to Subdivision 01 54 23 for scaffolding and staging costs.

04 20 00 Unit Masonry

- The most common types of unit masonry are brick and concrete masonry. The major classifications of brick are building brick (ASTM C62), facing brick (ASTM C216), glazed brick, fire brick, and pavers. Many varieties of texture and appearance can exist within these classifications, and the estimator would be wise to check local custom and availability within the project area. For repair and remodeling jobs, matching the existing brick may be the most important criteria.

- Brick and concrete block are priced by the piece and then converted into a price per square foot of wall. Openings less than two square feet are generally ignored by the estimator because any savings in units used are offset by the cutting and trimming required.

- It is often difficult and expensive to find and purchase small lots of historic brick. Costs can vary widely. Many design issues affect costs, selection of mortar mix, and repairs or replacement of masonry materials. Cleaning techniques must be reflected in the estimate.

- All masonry walls, whether interior or exterior, require bracing. The cost of bracing walls during construction should be included by the estimator, and this bracing must remain in place until permanent bracing is complete. Permanent bracing of masonry walls is accomplished by masonry itself, in the form of pilasters or abutting wall corners, or by anchoring the walls to the structural frame. Accessories in the form of anchors, anchor slots, and ties are used, but their supply and installation can be by different trades. For instance, anchor slots on spandrel beams and columns are supplied and welded in place by the steel fabricator, but the ties from the slots into the masonry are installed by the bricklayer. Regardless of the installation method, the estimator must be certain that these accessories are accounted for in pricing.

Reference Numbers

Reference numbers are shown at the beginning of some major classifications. These numbers refer to related items in the Reference Section. The reference information may be an estimating procedure, an alternate pricing method, or technical information.

Note: Not all subdivisions listed here necessarily appear. ■

Same Data. Simplified.

Enjoy the convenience and efficiency of accessing your costs anywhere:

- **Skip the multiplier** by setting your location
- **Quickly search,** edit, favorite and share costs
- **Stay on top of price changes** with automatic updates

Discover more at rsmeans.com/online

04 01 Maintenance of Masonry

04 01 20 – Maintenance of Unit Masonry

04 01 20.10 Patching Masonry	Crew	Daily Output	Labor-Hours	Unit	Material	2021 Bare Costs Labor	Equipment	Total	Total Incl O&P
0010 PATCHING MASONRY									
0500 CMU patching, includes chipping, cleaning and epoxy									
0520 1/4" deep	1 Cefi	65	.123	S.F.	4.23	6.40		10.63	14
0540 3/8" deep		50	.160		6.35	8.30		14.65	19.15
0580 1/2" deep	↓	40	.200	↓	8.45	10.35		18.80	24.50

04 01 20.20 Pointing Masonry	Crew	Daily Output	Labor-Hours	Unit	Material	2021 Bare Costs Labor	Equipment	Total	Total Incl O&P
0010 POINTING MASONRY									
0300 Cut and repoint brick, hard mortar, running bond	1 Bric	80	.100	S.F.	.57	5.35		5.92	8.75
0320 Common bond		77	.104		.57	5.60		6.17	9.05
0360 Flemish bond		70	.114		.60	6.15		6.75	9.90
0400 English bond		65	.123		.60	6.60		7.20	10.60
0600 Soft old mortar, running bond		100	.080		.57	4.30		4.87	7.10
0620 Common bond		96	.083		.57	4.48		5.05	7.40
0640 Flemish bond		90	.089		.60	4.77		5.37	7.85
0680 English bond		82	.098		.60	5.25		5.85	8.55
0700 Stonework, hard mortar		140	.057	L.F.	.76	3.07		3.83	5.45
0720 Soft old mortar		160	.050	"	.76	2.69		3.45	4.89
1000 Repoint, mask and grout method, running bond		95	.084	S.F.	.76	4.52		5.28	7.65
1020 Common bond		90	.089		.76	4.77		5.53	8.05
1040 Flemish bond		86	.093		.80	5		5.80	8.45
1060 English bond		77	.104		.80	5.60		6.40	9.30
2000 Scrub coat, sand grout on walls, thin mix, brushed		120	.067		3.89	3.58		7.47	9.65
2020 Troweled	↓	98	.082	↓	5.40	4.38		9.78	12.50

04 01 20.30 Pointing CMU	Crew	Daily Output	Labor-Hours	Unit	Material	2021 Bare Costs Labor	Equipment	Total	Total Incl O&P
0010 POINTING CMU									
0300 Cut and repoint block, hard mortar, running bond	1 Bric	190	.042	S.F.	.23	2.26		2.49	3.67
0310 Stacked bond		200	.040		.23	2.15		2.38	3.50
0600 Soft old mortar, running bond		230	.035		.23	1.87		2.10	3.07
0610 Stacked bond	↓	245	.033	↓	.23	1.75		1.98	2.90

04 01 20.40 Sawing Masonry	Crew	Daily Output	Labor-Hours	Unit	Material	2021 Bare Costs Labor	Equipment	Total	Total Incl O&P
0010 SAWING MASONRY									
0050 Brick or block by hand, per inch depth	A-1	125	.064	L.F.	.05	2.84	.91	3.80	5.30

04 01 20.41 Unit Masonry Stabilization	Crew	Daily Output	Labor-Hours	Unit	Material	2021 Bare Costs Labor	Equipment	Total	Total Incl O&P
0010 UNIT MASONRY STABILIZATION									
0100 Structural repointing method									
0110 Cut/grind mortar joint	1 Bric	240	.033	L.F.		1.79		1.79	2.70
0120 Clean and mask joint		2500	.003		.13	.17		.30	.40
0130 Epoxy paste and 1/4" FRP rod		240	.033		2.06	1.79		3.85	4.97
0132 3/8" FRP rod		160	.050		2.92	2.69		5.61	7.25
0140 Remove masking	↓	14400	.001	↓		.03		.03	.04
0300 Structural fabric method									
0310 Primer	1 Bric	600	.013	S.F.	1.01	.72		1.73	2.19
0320 Apply filling/leveling paste		720	.011		.88	.60		1.48	1.86
0330 Epoxy, glass fiber fabric		720	.011		9.25	.60		9.85	11.10
0340 Carbon fiber fabric	↓	720	.011	↓	19.70	.60		20.30	22.50

04 01 20.50 Toothing Masonry	Crew	Daily Output	Labor-Hours	Unit	Material	2021 Bare Costs Labor	Equipment	Total	Total Incl O&P
0010 TOOTHING MASONRY									
0500 Brickwork, soft old mortar	1 Clab	40	.200	V.L.F.		8.90		8.90	13.25
0520 Hard mortar		30	.267			11.85		11.85	17.65
0700 Blockwork, soft old mortar		70	.114			5.05		5.05	7.55
0720 Hard mortar	↓	50	.160	↓		7.10		7.10	10.60

04 01 Maintenance of Masonry

04 01 20 – Maintenance of Unit Masonry

04 01 20.52 Cleaning Masonry

		Crew	Daily Output	Labor-Hours	Unit	Material	2021 Bare Costs Labor	2021 Bare Costs Equipment	Total	Total Incl O&P
0010	**CLEANING MASONRY**									
0200	By chemical, brush and rinse, new work, light construction dust	D-1	1000	.016	S.F.	.07	.78		.85	1.25
0220	Medium construction dust		800	.020		.10	.97		1.07	1.58
0240	Heavy construction dust, drips or stains		600	.027		.14	1.30		1.44	2.11
0260	Low pressure wash and rinse, light restoration, light soil		800	.020		.16	.97		1.13	1.65
0270	Average soil, biological staining		400	.040		.24	1.95		2.19	3.21
0280	Heavy soil, biological and mineral staining, paint		330	.048		.32	2.36		2.68	3.92
0300	High pressure wash and rinse, heavy restoration, light soil		600	.027		.09	1.30		1.39	2.06
0310	Average soil, biological staining		400	.040		.14	1.95		2.09	3.09
0320	Heavy soil, biological and mineral staining, paint	▼	250	.064		.19	3.12		3.31	4.91
0400	High pressure wash, water only, light soil	C-29	500	.016			.71	.19	.90	1.27
0420	Average soil, biological staining		375	.021			.95	.26	1.21	1.70
0440	Heavy soil, biological and mineral staining, paint		250	.032			1.42	.39	1.81	2.55
0800	High pressure water and chemical, light soil		450	.018		.19	.79	.22	1.20	1.63
0820	Average soil, biological staining		300	.027		.29	1.18	.32	1.79	2.45
0840	Heavy soil, biological and mineral staining, paint	▼	200	.040		.39	1.78	.49	2.66	3.62
1200	Sandblast, wet system, light soil	J-6	1750	.018		.35	.88	.17	1.40	1.88
1220	Average soil, biological staining		1100	.029		.52	1.40	.27	2.19	2.96
1240	Heavy soil, biological and mineral staining, paint		700	.046		.69	2.20	.43	3.32	4.51
1400	Dry system, light soil		2500	.013		.35	.62	.12	1.09	1.43
1420	Average soil, biological staining		1750	.018		.52	.88	.17	1.57	2.07
1440	Heavy soil, biological and mineral staining, paint	▼	1000	.032		.69	1.54	.30	2.53	3.38
1800	For walnut shells, add					.95			.95	1.05
1820	For corn chips, add					.99			.99	1.09
2000	Steam cleaning, light soil	A-1H	750	.011			.47	.10	.57	.82
2020	Average soil, biological staining		625	.013			.57	.12	.69	.99
2040	Heavy soil, biological and mineral staining	▼	375	.021			.95	.21	1.16	1.64
4000	Add for masking doors and windows	1 Clab	800	.010	▼	.19	.44		.63	.87
4200	Add for pedestrian protection				Job				10%	10%

04 01 20.70 Brick Washing

		Crew	Daily Output	Labor-Hours	Unit	Material	2021 Bare Costs Labor	2021 Bare Costs Equipment	Total	Total Incl O&P
0010	**BRICK WASHING** R040130-10									
0012	Acid cleanser, smooth brick surface	1 Bric	560	.014	S.F.	.06	.77		.83	1.22
0050	Rough brick		400	.020		.07	1.07		1.14	1.70
0060	Stone, acid wash	▼	600	.013	▼	.09	.72		.81	1.18
1000	Muriatic acid, price per gallon in 5 gallon lots				Gal.	11.15			11.15	12.30

04 05 Common Work Results for Masonry

04 05 05 – Selective Demolition for Masonry

04 05 05.10 Selective Demolition

		Crew	Daily Output	Labor-Hours	Unit	Material	2021 Bare Costs Labor	2021 Bare Costs Equipment	Total	Total Incl O&P
0010	**SELECTIVE DEMOLITION** R024119-10									
0200	Bond beams, 8" block with #4 bar	2 Clab	32	.500	L.F.		22		22	33
0300	Concrete block walls, unreinforced, 2" thick		1200	.013	S.F.		.59		.59	.88
0310	4" thick		1150	.014			.62		.62	.92
0320	6" thick		1100	.015			.65		.65	.96
0330	8" thick		1050	.015			.68		.68	1.01
0340	10" thick		1000	.016			.71		.71	1.06
0360	12" thick		950	.017			.75		.75	1.12
0380	Reinforced alternate courses, 2" thick		1130	.014			.63		.63	.94
0390	4" thick		1080	.015			.66		.66	.98
0400	6" thick	▼	1035	.015			.69		.69	1.02

04 05 05.10 Selective Demolition	Crew	Daily Output	Labor-Hours	Unit	Material	2021 Bare Costs Labor	2021 Bare Costs Equipment	Total	Total Incl O&P
0410	8" thick	2 Clab	990	.016	S.F.		.72		.72	1.07
0420	10" thick		940	.017			.76		.76	1.13
0430	12" thick		890	.018			.80		.80	1.19
0440	Reinforced alternate courses & vertically 48" OC, 4" thick		900	.018			.79		.79	1.18
0450	6" thick		850	.019			.84		.84	1.25
0460	8" thick		800	.020			.89		.89	1.33
0480	10" thick		750	.021			.95		.95	1.41
0490	12" thick		700	.023			1.01		1.01	1.51
1000	Chimney, 16" x 16", soft old mortar	1 Clab	55	.145	C.F.		6.45		6.45	9.65
1020	Hard mortar		40	.200			8.90		8.90	13.25
1030	16" x 20", soft old mortar		55	.145			6.45		6.45	9.65
1040	Hard mortar		40	.200			8.90		8.90	13.25
1050	16" x 24", soft old mortar		55	.145			6.45		6.45	9.65
1060	Hard mortar		40	.200			8.90		8.90	13.25
1080	20" x 20", soft old mortar		55	.145			6.45		6.45	9.65
1100	Hard mortar		40	.200			8.90		8.90	13.25
1110	20" x 24", soft old mortar		55	.145			6.45		6.45	9.65
1120	Hard mortar		40	.200			8.90		8.90	13.25
1140	20" x 32", soft old mortar		55	.145			6.45		6.45	9.65
1160	Hard mortar		40	.200			8.90		8.90	13.25
1200	48" x 48", soft old mortar		55	.145			6.45		6.45	9.65
1220	Hard mortar		40	.200			8.90		8.90	13.25
1250	Metal, high temp steel jacket, 24" diameter	E-2	130	.431	V.L.F.		25.50	13.20	38.70	54
1260	60" diameter	"	60	.933			55.50	28.50	84	117
1280	Flue lining, up to 12" x 12"	1 Clab	200	.040			1.78		1.78	2.65
1282	Up to 24" x 24"		150	.053			2.37		2.37	3.53
2000	Columns, 8" x 8", soft old mortar		48	.167			7.40		7.40	11.05
2020	Hard mortar		40	.200			8.90		8.90	13.25
2060	16" x 16", soft old mortar		16	.500			22		22	33
2100	Hard mortar		14	.571			25.50		25.50	38
2140	24" x 24", soft old mortar		8	1			44.50		44.50	66.50
2160	Hard mortar		6	1.333			59		59	88.50
2200	36" x 36", soft old mortar		4	2			89		89	133
2220	Hard mortar		3	2.667			118		118	177
2230	Alternate pricing method, soft old mortar		30	.267	C.F.		11.85		11.85	17.65
2240	Hard mortar		23	.348	"		15.45		15.45	23
3000	Copings, precast or masonry, to 8" wide									
3020	Soft old mortar	1 Clab	180	.044	L.F.		1.97		1.97	2.94
3040	Hard mortar	"	160	.050	"		2.22		2.22	3.31
3100	To 12" wide									
3120	Soft old mortar	1 Clab	160	.050	L.F.		2.22		2.22	3.31
3140	Hard mortar	"	140	.057	"		2.54		2.54	3.79
4000	Fireplace, brick, 30" x 24" opening									
4020	Soft old mortar	1 Clab	2	4	Ea.		178		178	265
4040	Hard mortar		1.25	6.400			284		284	425
4100	Stone, soft old mortar		1.50	5.333			237		237	355
4120	Hard mortar		1	8			355		355	530
4150	Up to 48" fireplace, 15' chimney and foundation		.28	28.571			1,275		1,275	1,900
4400	Premanufactured, up to 48"	2 Clab	14	1.143			50.50		50.50	75.50
5000	Veneers, brick, soft old mortar	1 Clab	140	.057	S.F.		2.54		2.54	3.79
5020	Hard mortar		125	.064			2.84		2.84	4.24
5050	Glass block, up to 4" thick		500	.016			.71		.71	1.06
5100	Granite and marble, 2" thick		180	.044			1.97		1.97	2.94

04 05 Common Work Results for Masonry

04 05 05 – Selective Demolition for Masonry

04 05 05.10 Selective Demolition

04 05 05.10 Selective Demolition	Crew	Daily Output	Labor-Hours	Unit	Material	2021 Bare Costs Labor	Equipment	Total	Total Incl O&P
5120	4" thick	1 Clab	170	.047	S.F.		2.09		2.09	3.12
5140	Stone, 4" thick		180	.044			1.97		1.97	2.94
5160	8" thick	↓	175	.046	↓		2.03		2.03	3.03
5400	Alternate pricing method, stone, 4" thick		60	.133	C.F.		5.90		5.90	8.85
5420	8" thick		85	.094			4.18		4.18	6.25
5450	Solid masonry		130	.062			2.73		2.73	4.08
5460	Stone or precast sills, treads, copings		130	.062			2.73		2.73	4.08
5470	Solid stone or precast	↓	110	.073	↓		3.23		3.23	4.82
5500	Remove and reset steel lintel	1 Bric	40	.200	L.F.		10.75		10.75	16.20
5600	Vent box removal	1 Clab	50	.160	S.F.		7.10		7.10	10.60
5700	Remove block pilaster for fence, 6' high		2.33	3.433	Ea.		152		152	227
5800	Remove 12" x 12" step flashing from mortar joints	↓	240	.033	C.F.		1.48		1.48	2.21

04 05 13 – Masonry Mortaring

04 05 13.10 Cement

04 05 13.10 Cement	Crew	Daily Output	Labor-Hours	Unit	Material	2021 Bare Costs Labor	Equipment	Total	Total Incl O&P
0010	**CEMENT**									
0100	Masonry, 70 lb. bag, T.L. lots				Bag	11.75			11.75	12.90
0150	L.T.L. lots					12.45			12.45	13.70
0200	White, 70 lb. bag, T.L. lots					20.50			20.50	22.50
0250	L.T.L. lots				↓	17.10			17.10	18.80

04 05 13.20 Lime

04 05 13.20 Lime	Crew	Daily Output	Labor-Hours	Unit	Material	2021 Bare Costs Labor	Equipment	Total	Total Incl O&P
0010	**LIME**									
0020	Masons, hydrated, 50 lb. bag, T.L. lots				Bag	11.05			11.05	12.15
0050	L.T.L. lots					12.15			12.15	13.40
0200	Finish, double hydrated, 50 lb. bag, T.L. lots					12.05			12.05	13.25
0250	L.T.L. lots				↓	13.25			13.25	14.55

04 05 13.23 Surface Bonding Masonry Mortaring

04 05 13.23 Surface Bonding Masonry Mortaring	Crew	Daily Output	Labor-Hours	Unit	Material	2021 Bare Costs Labor	Equipment	Total	Total Incl O&P
0010	**SURFACE BONDING MASONRY MORTARING**									
0020	Gray or white colors, not incl. block work	1 Bric	540	.015	S.F.	.20	.80		1	1.42

04 05 13.30 Mortar

04 05 13.30 Mortar	Crew	Daily Output	Labor-Hours	Unit	Material	2021 Bare Costs Labor	Equipment	Total	Total Incl O&P
0010	**MORTAR** R040513-10									
0020	With masonry cement									
0100	Type M, 1:1:6 mix	1 Brhe	143	.056	C.F.	6.85	2.44		9.29	11.25
0200	Type N, 1:3 mix		143	.056		5.45	2.44		7.89	9.70
0300	Type O, 1:3 mix		143	.056		4.82	2.44		7.26	9
0400	Type PM, 1:1:6 mix, 2500 psi		143	.056		6.80	2.44		9.24	11.15
0500	Type S, 1/2:1:4 mix	↓	143	.056	↓	6.60	2.44		9.04	10.95
2000	With Portland cement and lime									
2100	Type M, 1:1/4:3 mix	1 Brhe	143	.056	C.F.	9.45	2.44		11.89	14.10
2200	Type N, 1:1:6 mix, 750 psi		143	.056		7.30	2.44		9.74	11.75
2300	Type O, 1:2:9 mix (Pointing Mortar)		143	.056		8.85	2.44		11.29	13.40
2400	Type PL, 1:1/2:4 mix, 2500 psi		143	.056		6.10	2.44		8.54	10.40
2600	Type S, 1:1/2:4 mix, 1800 psi	↓	143	.056		9.50	2.44		11.94	14.15
2650	Pre-mixed, type S or N					6.40			6.40	7.05
2700	Mortar for glass block	1 Brhe	143	.056	↓	13	2.44		15.44	18
2900	Mortar for fire brick, dry mix, 10 lb. pail				Ea.	25			25	27.50

04 05 13.91 Masonry Restoration Mortaring

04 05 13.91 Masonry Restoration Mortaring	Crew	Daily Output	Labor-Hours	Unit	Material	2021 Bare Costs Labor	Equipment	Total	Total Incl O&P
0010	**MASONRY RESTORATION MORTARING**									
0020	Masonry restoration mix				Lb.	.73			.73	.80
0050	White				"	1.20			1.20	1.32

For customer support on your Concrete & Masonry Costs with RSMeans Data, call 800.448.8182.

93

04 05 13 – Masonry Mortaring

04 05 13.93 Mortar Pigments

		Crew	Daily Output	Labor-Hours	Unit	Material	2021 Bare Costs Labor	Equipment	Total	Total Incl O&P
0010	**MORTAR PIGMENTS**, 50 lb. bags (2 bags per M bricks) R040513-10									
0020	Color admixture, range 2 to 10 lb. per bag of cement, light colors				Lb.	6.65			6.65	7.30
0050	Medium colors					7.65			7.65	8.40
0100	Dark colors					15.85			15.85	17.40

04 05 13.95 Sand

		Crew	Daily Output	Labor-Hours	Unit	Material	2021 Bare Costs Labor	Equipment	Total	Total Incl O&P
0010	**SAND**, screened and washed at pit									
0020	For mortar, per ton				Ton	21.50			21.50	23.50
0050	With 10 mile haul					43			43	47
0100	With 30 mile haul					71.50			71.50	78.50
0200	Screened and washed, at the pit				C.Y.	30			30	33
0250	With 10 mile haul					59.50			59.50	65.50
0300	With 30 mile haul					99			99	109

04 05 13.98 Mortar Admixtures

		Crew	Daily Output	Labor-Hours	Unit	Material	2021 Bare Costs Labor	Equipment	Total	Total Incl O&P
0010	**MORTAR ADMIXTURES**									
0020	Waterproofing admixture, per quart (1 qt. to 2 bags of masonry cement)				Qt.	3.66			3.66	4.03

04 05 16 – Masonry Grouting

04 05 16.30 Grouting

		Crew	Daily Output	Labor-Hours	Unit	Material	2021 Bare Costs Labor	Equipment	Total	Total Incl O&P
0010	**GROUTING**									
0011	Bond beams & lintels, 8" deep, 6" thick, 0.15 C.F./L.F.	D-4	1480	.022	L.F.	.83	1.06	.13	2.02	2.65
0020	8" thick, 0.2 C.F./L.F.		1400	.023		1.33	1.12	.14	2.59	3.30
0050	10" thick, 0.25 C.F./L.F.		1200	.027		1.39	1.31	.16	2.86	3.66
0060	12" thick, 0.3 C.F./L.F.		1040	.031		1.66	1.51	.18	3.35	4.30
0200	Concrete block cores, solid, 4" thk., by hand, 0.067 C.F./S.F. of wall	D-8	1100	.036	S.F.	.37	1.81		2.18	3.13
0210	6" thick, pumped, 0.175 C.F./S.F.	D-4	720	.044		.97	2.18	.26	3.41	4.64
0250	8" thick, pumped, 0.258 C.F./S.F.		680	.047		1.43	2.31	.28	4.02	5.35
0300	10" thick, pumped, 0.340 C.F./S.F.		660	.048		1.88	2.38	.29	4.55	5.95
0350	12" thick, pumped, 0.422 C.F./S.F.		640	.050		2.34	2.46	.30	5.10	6.60
0500	Cavity walls, 2" space, pumped, 0.167 C.F./S.F. of wall		1700	.019		.93	.93	.11	1.97	2.53
0550	3" space, 0.250 C.F./S.F.		1200	.027		1.39	1.31	.16	2.86	3.66
0600	4" space, 0.333 C.F./S.F.		1150	.028		1.85	1.37	.17	3.39	4.26
0700	6" space, 0.500 C.F./S.F.		800	.040		2.77	1.97	.24	4.98	6.25
0800	Door frames, 3' x 7' opening, 2.5 C.F. per opening		60	.533	Opng.	13.85	26	3.17	43.02	58
0850	6' x 7' opening, 3.5 C.F. per opening		45	.711	"	19.40	35	4.23	58.63	78.50
2000	Grout, C476, for bond beams, lintels and CMU cores		350	.091	C.F.	5.55	4.49	.54	10.58	13.45

04 05 19 – Masonry Anchorage and Reinforcing

04 05 19.05 Anchor Bolts

		Crew	Daily Output	Labor-Hours	Unit	Material	2021 Bare Costs Labor	Equipment	Total	Total Incl O&P
0010	**ANCHOR BOLTS**									
0015	Installed in fresh grout in CMU bond beams or filled cores, no templates									
0020	Hooked, with nut and washer, 1/2" diameter, 8" long	1 Bric	132	.061	Ea.	1.55	3.25		4.80	6.60
0030	12" long		131	.061		1.72	3.28		5	6.85
0040	5/8" diameter, 8" long		129	.062		3.40	3.33		6.73	8.75
0050	12" long		127	.063		4.18	3.38		7.56	9.70
0060	3/4" diameter, 8" long		127	.063		4.18	3.38		7.56	9.70
0070	12" long		125	.064		5.25	3.44		8.69	10.95

04 05 19.16 Masonry Anchors

		Crew	Daily Output	Labor-Hours	Unit	Material	2021 Bare Costs Labor	Equipment	Total	Total Incl O&P
0010	**MASONRY ANCHORS**									
0020	For brick veneer, galv., corrugated, 7/8" x 7", 22 ga.	1 Bric	10.50	.762	C	15.95	41		56.95	79
0100	24 ga.		10.50	.762		10.40	41		51.40	73
0150	16 ga.		10.50	.762		33	41		74	97.50
0200	Buck anchors, galv., corrugated, 16 ga., 2" bend, 8" x 2"		10.50	.762		58.50	41		99.50	126

04 05 19.16 Masonry Anchors		Crew	Daily Output	Labor-Hours	Unit	Material	2021 Bare Costs Labor	Equipment	Total	Total Incl O&P
0250	8" x 3"	1 Bric	10.50	.762	C	63.50	41		104.50	131
0300	Adjustable, rectangular, 4-1/8" wide									
0350	Anchor and tie, 3/16" wire, mill galv.									
0400	2-3/4" eye, 3-1/4" tie	1 Bric	1.05	7.619	M	480	410		890	1,150
0500	4-3/4" tie		1.05	7.619		520	410		930	1,200
0520	5-1/2" tie		1.05	7.619		565	410		975	1,225
0550	4-3/4" eye, 3-1/4" tie		1.05	7.619		525	410		935	1,200
0570	4-3/4" tie		1.05	7.619		580	410		990	1,250
0580	5-1/2" tie		1.05	7.619		680	410		1,090	1,375
0660	Cavity wall, Z-type, galvanized, 6" long, 1/8" diam.		10.50	.762	C	24.50	41		65.50	88.50
0670	3/16" diameter		10.50	.762		30.50	41		71.50	95
0680	1/4" diameter		10.50	.762		40	41		81	106
0850	8" long, 3/16" diameter		10.50	.762		28	41		69	92
0855	1/4" diameter		10.50	.762		50.50	41		91.50	117
1000	Rectangular type, galvanized, 1/4" diameter, 2" x 6"		10.50	.762		80	41		121	150
1050	4" x 6"		10.50	.762		91	41		132	162
1100	3/16" diameter, 2" x 6"		10.50	.762		54	41		95	121
1150	4" x 6"		10.50	.762		55.50	41		96.50	123
1200	Mesh wall tie, 1/2" mesh, hot dip galvanized									
1400	16 ga., 12" long, 3" wide	1 Bric	9	.889	C	91.50	47.50		139	173
1420	6" wide		9	.889		134	47.50		181.50	220
1440	12" wide		8.50	.941		212	50.50		262.50	310
1500	Rigid partition anchors, plain, 8" long, 1" x 1/8"		10.50	.762		191	41		232	273
1550	1" x 1/4"		10.50	.762		325	41		366	420
1580	1-1/2" x 1/8"		10.50	.762		227	41		268	310
1600	1-1/2" x 1/4"		10.50	.762		360	41		401	455
1650	2" x 1/8"		10.50	.762		335	41		376	425
1700	2" x 1/4"		10.50	.762		395	41		436	495
2000	Column flange ties, wire, galvanized									
2300	3/16" diameter, up to 3" wide	1 Bric	10.50	.762	C	80.50	41		121.50	150
2350	To 5" wide		10.50	.762		87.50	41		128.50	158
2400	To 7" wide		10.50	.762		94.50	41		135.50	166
2600	To 9" wide		10.50	.762		100	41		141	172
2650	1/4" diameter, up to 3" wide		10.50	.762		112	41		153	185
2700	To 5" wide		10.50	.762		133	41		174	208
2800	To 7" wide		10.50	.762		149	41		190	226
2850	To 9" wide		10.50	.762		162	41		203	240
2900	For hot dip galvanized, add					35%				
4000	Channel slots, 1-3/8" x 1/2" x 8"									
4100	12 ga., plain	1 Bric	10.50	.762	C	254	41		295	340
4150	16 ga., galvanized	"	10.50	.762	"	143	41		184	219
4200	Channel slot anchors									
4300	16 ga., galvanized, 1-1/4" x 3-1/2"				C	64			64	70.50
4350	1-1/4" x 5-1/2"					74.50			74.50	82
4400	1-1/4" x 7-1/2"					68.50			68.50	75.50
4500	1/8" plain, 1-1/4" x 3-1/2"					143			143	158
4550	1-1/4" x 5-1/2"					153			153	168
4600	1-1/4" x 7-1/2"					167			167	184
4700	For corrugation, add					77			77	85
4750	For hot dip galvanized, add					35%				
5000	Dowels									
5100	Plain, 1/4" diameter, 3" long				C	49			49	54
5150	4" long					54.50			54.50	59.50

For customer support on your Concrete & Masonry Costs with RSMeans Data, call 800.448.8182.

95

04 05 19.16 Masonry Anchors

		Crew	Daily Output	Labor-Hours	Unit	Material	2021 Bare Costs Labor	Equipment	Total	Total Incl O&P
5200	6" long				C	66.50			66.50	73
5300	3/8" diameter, 3" long					65			65	71.50
5350	4" long					80.50			80.50	88.50
5400	6" long					93			93	102
5500	1/2" diameter, 3" long					94.50			94.50	104
5550	4" long					112			112	123
5600	6" long					146			146	161
5700	5/8" diameter, 3" long					130			130	143
5750	4" long					159			159	175
5800	6" long					219			219	241
6000	3/4" diameter, 3" long					161			161	177
6100	4" long					205			205	225
6150	6" long					293			293	320
6300	For hot dip galvanized, add					35%				

04 05 19.26 Masonry Reinforcing Bars

		Crew	Daily Output	Labor-Hours	Unit	Material	2021 Bare Costs Labor	Equipment	Total	Total Incl O&P
0010	**MASONRY REINFORCING BARS** R040519-50									
0015	Steel bars A615, placed horiz., #3 & #4 bars	1 Bric	450	.018	Lb.	.62	.95		1.57	2.12
0020	#5 & #6 bars		800	.010		.62	.54		1.16	1.49
0050	Placed vertical, #3 & #4 bars		350	.023		.62	1.23		1.85	2.53
0060	#5 & #6 bars		650	.012		.62	.66		1.28	1.68
0200	Joint reinforcing, regular truss, to 6" wide, mill std galvanized		30	.267	C.L.F.	23	14.30		37.30	47
0250	12" wide		20	.400		29.50	21.50		51	65
0400	Cavity truss with drip section, to 6" wide		30	.267		23.50	14.30		37.80	47
0450	12" wide		20	.400		26.50	21.50		48	62
0500	Joint reinforcing, ladder type, mill std galvanized									
0600	9 ga. sides, 9 ga. ties, 4" wall	1 Bric	30	.267	C.L.F.	22	14.30		36.30	45.50
0650	6" wall		30	.267		22	14.30		36.30	45.50
0700	8" wall		25	.320		26.50	17.20		43.70	55
0750	10" wall		20	.400		26	21.50		47.50	61
0800	12" wall		20	.400		27.50	21.50		49	62.50
1000	Truss type									
1100	9 ga. sides, 9 ga. ties, 4" wall	1 Bric	30	.267	C.L.F.	24.50	14.30		38.80	48.50
1150	6" wall		30	.267		26	14.30		40.30	50
1200	8" wall		25	.320		27.50	17.20		44.70	56.50
1250	10" wall		20	.400		23.50	21.50		45	58.50
1300	12" wall		20	.400		25	21.50		46.50	60
1500	3/16" sides, 9 ga. ties, 4" wall		30	.267		28.50	14.30		42.80	53
1550	6" wall		30	.267		36	14.30		50.30	61.50
1600	8" wall		25	.320		32.50	17.20		49.70	62
1650	10" wall		20	.400		39	21.50		60.50	75.50
1700	12" wall		20	.400		38	21.50		59.50	74.50
2000	3/16" sides, 3/16" ties, 4" wall		30	.267		40.50	14.30		54.80	66
2050	6" wall		30	.267		41.50	14.30		55.80	67.50
2100	8" wall		25	.320		40.50	17.20		57.70	70.50
2150	10" wall		20	.400		45	21.50		66.50	82
2200	12" wall		20	.400		46.50	21.50		68	84
2500	Cavity truss type, galvanized									
2600	9 ga. sides, 9 ga. ties, 4" wall	1 Bric	25	.320	C.L.F.	35.50	17.20		52.70	65
2650	6" wall		25	.320		40.50	17.20		57.70	70.50
2700	8" wall		20	.400		48	21.50		69.50	85
2750	10" wall		15	.533		43	28.50		71.50	90.50
2800	12" wall		15	.533		52	28.50		80.50	100

04 05 Common Work Results for Masonry

04 05 19 – Masonry Anchorage and Reinforcing

04 05 19.26 Masonry Reinforcing Bars	Crew	Daily Output	Labor-Hours	Unit	Material	2021 Bare Costs Labor	Equipment	Total	Total Incl O&P
3000	3/16" sides, 9 ga. ties, 4" wall	1 Bric	25	.320	C.L.F.	45	17.20		62.20	75.50
3050	6" wall		25	.320		33	17.20		50.20	62
3100	8" wall		20	.400		60.50	21.50		82	99
3150	10" wall		15	.533		43	28.50		71.50	90
3200	12" wall	↓	15	.533	↓	43	28.50		71.50	90
3500	For hot dip galvanizing, add				Ton	485			485	535

04 05 23 – Masonry Accessories

04 05 23.13 Masonry Control and Expansion Joints

		Crew	Daily Output	Labor-Hours	Unit	Material	Labor	Equipment	Total	Total Incl O&P
0010	**MASONRY CONTROL AND EXPANSION JOINTS**									
0020	Rubber, for double wythe 8" minimum wall (Brick/CMU)	1 Bric	400	.020	L.F.	2.55	1.07		3.62	4.43
0025	"T" shaped		320	.025		1.37	1.34		2.71	3.53
0030	Cross-shaped for CMU units		280	.029		1.68	1.53		3.21	4.16
0050	PVC, for double wythe 8" minimum wall (Brick/CMU)		400	.020		1.64	1.07		2.71	3.42
0120	"T" shaped		320	.025		1.08	1.34		2.42	3.21
0160	Cross-shaped for CMU units	↓	280	.029	↓	1.28	1.53		2.81	3.72

04 05 23.19 Masonry Cavity Drainage, Weepholes, and Vents

		Crew	Daily Output	Labor-Hours	Unit	Material	Labor	Equipment	Total	Total Incl O&P
0010	**MASONRY CAVITY DRAINAGE, WEEPHOLES, AND VENTS**									
0020	Extruded aluminum, 4" deep, 2-3/8" x 8-1/8"	1 Bric	30	.267	Ea.	41	14.30		55.30	67
0050	5" x 8-1/8"		25	.320		50	17.20		67.20	81
0100	2-1/4" x 25"		25	.320		84.50	17.20		101.70	119
0150	5" x 16-1/2"		22	.364		62	19.55		81.55	98
0175	5" x 24"		22	.364		94	19.55		113.55	133
0200	6" x 16-1/2"		22	.364		101	19.55		120.55	141
0250	7-3/4" x 16-1/2"	↓	20	.400		93.50	21.50		115	136
0400	For baked enamel finish, add					35%				
0500	For cast aluminum, painted, add					60%				
1000	Stainless steel ventilators, 6" x 6"	1 Bric	25	.320		227	17.20		244.20	276
1050	8" x 8"		24	.333		254	17.90		271.90	305
1100	12" x 12"		23	.348		274	18.70		292.70	330
1150	12" x 6"		24	.333		275	17.90		292.90	325
1200	Foundation block vent, galv., 1-1/4" thk, 8" high, 16" long, no damper	↓	30	.267		18.55	14.30		32.85	42
1250	For damper, add				↓	3.74			3.74	4.11
1450	Drainage and ventilation fabric	2 Bric	1450	.011	S.F.	.82	.59		1.41	1.79

04 05 23.95 Wall Plugs

		Crew	Daily Output	Labor-Hours	Unit	Material	Labor	Equipment	Total	Total Incl O&P
0010	**WALL PLUGS** (for nailing to brickwork)									
0020	25 ga., galvanized, plain	1 Bric	10.50	.762	C	52	41		93	119
0050	Wood filled	"	10.50	.762	"	121	41		162	196

04 21 Clay Unit Masonry

04 21 13 – Brick Masonry

04 21 13.13 Brick Veneer Masonry

		Crew	Daily Output	Labor-Hours	Unit	Material	Labor	Equipment	Total	Total Incl O&P
0010	**BRICK VENEER MASONRY**, T.L. lots, excl. scaff., grout & reinforcing R042110-10									
0015	Material costs incl. 3% brick and 25% mortar waste									
0020	Standard, select common, 4" x 2-2/3" x 8" (6.75/S.F.) R042110-20	D-8	1.50	26.667	M	705	1,325		2,030	2,775
0050	Red, 4" x 2-2/3" x 8", running bond		1.50	26.667		750	1,325		2,075	2,825
0100	Full header every 6th course (7.88/S.F.) R042110-50		1.45	27.586		750	1,375		2,125	2,900
0150	English, full header every 2nd course (10.13/S.F.)		1.40	28.571		750	1,425		2,175	2,975
0200	Flemish, alternate header every course (9.00/S.F.)		1.40	28.571		750	1,425		2,175	2,975
0250	Flemish, alt. header every 6th course (7.13/S.F.)		1.45	27.586		750	1,375		2,125	2,900
0300	Full headers throughout (13.50/S.F.)	↓	1.40	28.571	↓	745	1,425		2,170	2,975

For customer support on your Concrete & Masonry Costs with RSMeans Data, call 800.448.8182.

97

04 21 13.13 Brick Veneer Masonry	Crew	Daily Output	Labor-Hours	Unit	Material	2021 Bare Costs Labor	2021 Bare Costs Equipment	Total	Total Incl O&P
0350	Rowlock course (13.50/S.F.)	D-8	1.35	29.630	M	745	1,475		2,220	3,050
0400	Rowlock stretcher (4.50/S.F.)		1.40	28.571		760	1,425		2,185	2,975
0450	Soldier course (6.75/S.F.)		1.40	28.571		750	1,425		2,175	2,975
0500	Sailor course (4.50/S.F.)		1.30	30.769		760	1,525		2,285	3,125
0601	Buff or gray face, running bond (6.75/S.F.)		1.50	26.667		750	1,325		2,075	2,825
0700	Glazed face, 4" x 2-2/3" x 8", running bond		1.40	28.571		2,475	1,425		3,900	4,875
0750	Full header every 6th course (7.88/S.F.)		1.35	29.630		2,375	1,475		3,850	4,825
1000	Jumbo, 6" x 4" x 12" (3.00/S.F.)		1.30	30.769		1,975	1,525		3,500	4,475
1051	Norman, 4" x 2-2/3" x 12" (4.50/S.F.)		1.45	27.586		1,225	1,375		2,600	3,425
1100	Norwegian, 4" x 3-1/5" x 12" (3.75/S.F.)		1.40	28.571		2,075	1,425		3,500	4,450
1150	Economy, 4" x 4" x 8" (4.50/S.F.)		1.40	28.571		1,050	1,425		2,475	3,325
1201	Engineer, 4" x 3-1/5" x 8" (5.63/S.F.)		1.45	27.586		710	1,375		2,085	2,850
1251	Roman, 4" x 2" x 12" (6.00/S.F.)		1.50	26.667		1,375	1,325		2,700	3,525
1300	S.C.R., 6" x 2-2/3" x 12" (4.50/S.F.)		1.40	28.571		1,375	1,425		2,800	3,675
1350	Utility, 4" x 4" x 12" (3.00/S.F.)		1.08	37.037		1,900	1,850		3,750	4,850
1360	For less than truck load lots, add					15%				
1400	For battered walls, add						30%			
1450	For corbels, add						75%			
1500	For curved walls, add						30%			
1550	For pits and trenches, deduct						20%			
1999	Alternate method of figuring by square foot									
2000	Standard, sel. common, 4" x 2-2/3" x 8" (6.75/S.F.)	D-8	230	.174	S.F.	4.75	8.65		13.40	18.20
2020	Red, 4" x 2-2/3" x 8", running bond		220	.182		5.05	9.05		14.10	19.20
2050	Full header every 6th course (7.88/S.F.)		185	.216		5.90	10.75		16.65	22.50
2100	English, full header every 2nd course (10.13/S.F.)		140	.286		7.55	14.20		21.75	30
2150	Flemish, alternate header every course (9.00/S.F.)		150	.267		6.75	13.25		20	27.50
2200	Flemish, alt. header every 6th course (7.13/S.F.)		205	.195		5.35	9.70		15.05	20.50
2250	Full headers throughout (13.50/S.F.)		105	.381		10.05	18.95		29	39.50
2300	Rowlock course (13.50/S.F.)		100	.400		10.05	19.90		29.95	41
2350	Rowlock stretcher (4.50/S.F.)		310	.129		3.41	6.40		9.81	13.40
2400	Soldier course (6.75/S.F.)		200	.200		5.05	9.95		15	20.50
2450	Sailor course (4.50/S.F.)		290	.138		3.41	6.85		10.26	14.10
2600	Buff or gray face, running bond (6.75/S.F.)		220	.182		5.35	9.05		14.40	19.50
2700	Glazed face brick, running bond		210	.190		16	9.45		25.45	32
2750	Full header every 6th course (7.88/S.F.)		170	.235		18.70	11.70		30.40	38
3000	Jumbo, 6" x 4" x 12" running bond (3.00/S.F.)		435	.092		5.40	4.57		9.97	12.85
3050	Norman, 4" x 2-2/3" x 12" running bond (4.5/S.F.)		320	.125		6.70	6.20		12.90	16.75
3100	Norwegian, 4" x 3-1/5" x 12" (3.75/S.F.)		375	.107		7.70	5.30		13	16.45
3150	Economy, 4" x 4" x 8" (4.50/S.F.)		310	.129		4.72	6.40		11.12	14.85
3200	Engineer, 4" x 3-1/5" x 8" (5.63/S.F.)		260	.154		3.99	7.65		11.64	15.90
3250	Roman, 4" x 2" x 12" (6.00/S.F.)		250	.160		8.10	7.95		16.05	21
3300	S.C.R., 6" x 2-2/3" x 12" (4.50/S.F.)		310	.129		6.15	6.40		12.55	16.40
3350	Utility, 4" x 4" x 12" (3.00/S.F.)		360	.111		5.55	5.50		11.05	14.40
3360	For less than truck load lots, add					.10%				
3370	For battered walls, add						30%			
3380	For corbels, add						75%			
3400	For cavity wall construction, add						15%			
3450	For stacked bond, add						10%			
3500	For interior veneer construction, add						15%			
3510	For pits and trenches, deduct						20%			
3550	For curved walls, add						30%			

04 21 Clay Unit Masonry

04 21 13 – Brick Masonry

04 21 13.14 Thin Brick Veneer

		Crew	Daily Output	Labor-Hours	Unit	Material	2021 Bare Costs Labor	2021 Bare Costs Equipment	Total	Total Incl O&P
0010	**THIN BRICK VENEER**									
0015	Material costs incl. 3% brick and 25% mortar waste									
0020	On & incl. metal panel support sys, modular, 2-2/3" x 5/8" x 8", red	D-7	92	.174	S.F.	13.85	8.10		21.95	27
0100	Closure, 4" x 5/8" x 8"		110	.145		9.45	6.80		16.25	20.50
0110	Norman, 2-2/3" x 5/8" x 12"		110	.145		9.85	6.80		16.65	20.50
0120	Utility, 4" x 5/8" x 12"		125	.128		9.20	5.95		15.15	18.85
0130	Emperor, 4" x 3/4" x 16"		175	.091		10.40	4.27		14.67	17.70
0140	Super emperor, 8" x 3/4" x 16"		195	.082		11.40	3.83		15.23	18.15
0150	For L shaped corners with 4" return, add				L.F.	9.25			9.25	10.20
0200	On masonry/plaster back-up, modular, 2-2/3" x 5/8" x 8", red	D-7	137	.117	S.F.	8.80	5.45		14.25	17.65
0210	Closure, 4" x 5/8" x 8"		165	.097		4.40	4.52		8.92	11.45
0220	Norman, 2-2/3" x 5/8" x 12"		165	.097		4.77	4.52		9.29	11.85
0230	Utility, 4" x 5/8" x 12"		185	.086		4.12	4.03		8.15	10.45
0240	Emperor, 4" x 3/4" x 16"		260	.062		5.30	2.87		8.17	10.05
0250	Super emperor, 8" x 3/4" x 16"		285	.056		6.35	2.62		8.97	10.80
0260	For L shaped corners with 4" return, add				L.F.	10			10	11
0270	For embedment into pre-cast concrete panels, add				S.F.	14.40			14.40	15.85

04 21 13.15 Chimney

		Crew	Daily Output	Labor-Hours	Unit	Material	2021 Bare Costs Labor	2021 Bare Costs Equipment	Total	Total Incl O&P
0010	**CHIMNEY**, excludes foundation, scaffolding, grout and reinforcing									
0100	Brick, 16" x 16", 8" flue	D-1	18.20	.879	V.L.F.	26	43		69	93.50
0150	16" x 20" with one 8" x 12" flue		16	1		44	48.50		92.50	122
0200	16" x 24" with two 8" x 8" flues		14	1.143		65	55.50		120.50	156
0250	20" x 20" with one 12" x 12" flue		13.70	1.168		51.50	57		108.50	143
0300	20" x 24" with two 8" x 12" flues		12	1.333		73	65		138	178
0350	20" x 32" with two 12" x 12" flues		10	1.600		91	78		169	217

04 21 13.18 Columns

		Crew	Daily Output	Labor-Hours	Unit	Material	2021 Bare Costs Labor	2021 Bare Costs Equipment	Total	Total Incl O&P
0010	**COLUMNS**, solid, excludes scaffolding, grout and reinforcing									
0050	Brick, 8" x 8", 9 brick/V.L.F.	D-1	56	.286	V.L.F.	6.50	13.90		20.40	28
0100	12" x 8", 13.5 brick/V.L.F.		37	.432		9.75	21		30.75	42.50
0200	12" x 12", 20 brick/V.L.F.		25	.640		14.50	31		45.50	63
0300	16" x 12", 27 brick/V.L.F.		19	.842		19.55	41		60.55	83.50
0400	16" x 16", 36 brick/V.L.F.		14	1.143		26	55.50		81.50	113
0500	20" x 16", 45 brick/V.L.F.		11	1.455		32.50	71		103.50	143
0600	20" x 20", 56 brick/V.L.F.		9	1.778		40.50	86.50		127	175
0700	24" x 20", 68 brick/V.L.F.		7	2.286		49	111		160	222
0800	24" x 24", 81 brick/V.L.F.		6	2.667		58.50	130		188.50	261
1000	36" x 36", 182 brick/V.L.F.		3	5.333		132	260		392	535

04 21 13.30 Oversized Brick

		Crew	Daily Output	Labor-Hours	Unit	Material	2021 Bare Costs Labor	2021 Bare Costs Equipment	Total	Total Incl O&P
0010	**OVERSIZED BRICK**, excludes scaffolding, grout and reinforcing									
0100	Veneer, 4" x 2.25" x 16"	D-8	387	.103	S.F.	4.90	5.15		10.05	13.15
0102	8" x 2.25" x 16", multicell		265	.151		17.75	7.50		25.25	31
0105	4" x 2.75" x 16"		412	.097		5.85	4.83		10.68	13.70
0107	8" x 2.75" x 16", multicell		295	.136		17.75	6.75		24.50	29.50
0110	4" x 4" x 16"		460	.087		3.90	4.32		8.22	10.80
0120	4" x 8" x 16"		533	.075		5.60	3.73		9.33	11.80
0122	4" x 8" x 16" multicell		327	.122		15.90	6.10		22	26.50
0125	Loadbearing, 6" x 4" x 16", grouted and reinforced		387	.103		12.05	5.15		17.20	21
0130	8" x 4" x 16", grouted and reinforced		327	.122		12.20	6.10		18.30	22.50
0132	10" x 4" x 16", grouted and reinforced		327	.122		28.50	6.10		34.60	40.50
0135	6" x 8" x 16", grouted and reinforced		440	.091		15.60	4.52		20.12	24
0140	8" x 8" x 16", grouted and reinforced		400	.100		16.45	4.97		21.42	25.50

04 21 13 – Brick Masonry

04 21 13.30 Oversized Brick

		Crew	Daily Output	Labor-Hours	Unit	Material	2021 Bare Costs Labor	Equipment	Total	Total Incl O&P
0145	Curtainwall/reinforced veneer, 6" x 4" x 16"	D-8	387	.103	S.F.	17.45	5.15		22.60	27
0150	8" x 4" x 16"		327	.122		20.50	6.10		26.60	31.50
0152	10" x 4" x 16"		327	.122		31.50	6.10		37.60	44
0155	6" x 8" x 16"		440	.091		21.50	4.52		26.02	31
0160	8" x 8" x 16"		400	.100		29	4.97		33.97	39.50
0200	For 1 to 3 slots in face, add					15%				
0210	For 4 to 7 slots in face, add					25%				
0220	For bond beams, add					20%				
0230	For bullnose shapes, add					20%				
0240	For open end knockout, add					10%				
0250	For white or gray color group, add					10%				
0260	For 135 degree corner, add					250%				

04 21 13.35 Common Building Brick

		Crew	Daily Output	Labor-Hours	Unit	Material	2021 Bare Costs Labor	Equipment	Total	Total Incl O&P
0010	**COMMON BUILDING BRICK**, C62, T.L. lots, material only R042110-20									
0020	Standard				M	610			610	670
0050	Select				"	580			580	640

04 21 13.40 Structural Brick

		Crew	Daily Output	Labor-Hours	Unit	Material	2021 Bare Costs Labor	Equipment	Total	Total Incl O&P
0010	**STRUCTURAL BRICK** C652, Grade SW, incl. mortar, scaffolding not incl.									
0100	Standard unit, 4-5/8" x 2-3/4" x 9-5/8"	D-8	245	.163	S.F.	4.33	8.10		12.43	17
0120	Bond beam		225	.178		4.45	8.85		13.30	18.20
0140	V cut bond beam		225	.178		4.34	8.85		13.19	18.10
0160	Stretcher quoin, 5-5/8" x 2-3/4" x 9-5/8"		245	.163		7.35	8.10		15.45	20.50
0180	Corner quoin		245	.163		8	8.10		16.10	21
0200	Corner, 45 degree, 4-5/8" x 2-3/4" x 10-7/16"		235	.170		9.05	8.45		17.50	22.50

04 21 13.45 Face Brick

		Crew	Daily Output	Labor-Hours	Unit	Material	2021 Bare Costs Labor	Equipment	Total	Total Incl O&P
0010	**FACE BRICK** Material Only, C216, T.L. lots R042110-20									
0300	Standard modular, 4" x 2-2/3" x 8"				M	625			625	690
0450	Economy, 4" x 4" x 8"					910			910	1,000
0510	Economy, 4" x 4" x 12"					1,450			1,450	1,575
0550	Jumbo, 6" x 4" x 12"					1,625			1,625	1,800
0610	Jumbo, 8" x 4" x 12"					1,625			1,625	1,800
0650	Norwegian, 4" x 3-1/5" x 12"					1,875			1,875	2,075
0710	Norwegian, 6" x 3-1/5" x 12"					1,575			1,575	1,750
0850	Standard glazed, plain colors, 4" x 2-2/3" x 8"					2,200			2,200	2,425
1000	Deep trim shades, 4" x 2-2/3" x 8"					2,375			2,375	2,625
1080	Jumbo utility, 4" x 4" x 12"					1,675			1,675	1,850
1120	4" x 8" x 8"					1,975			1,975	2,150
1140	4" x 8" x 16"					6,875			6,875	7,550
1260	Engineer, 4" x 3-1/5" x 8"					580			580	640
1350	King, 4" x 2-3/4" x 10"					575			575	635
1770	Standard modular, double glazed, 4" x 2-2/3" x 8"					3,000			3,000	3,300
1850	Jumbo, colored glazed ceramic, 6" x 4" x 12"					2,800			2,800	3,075
2050	Jumbo utility, glazed, 4" x 4" x 12"					5,750			5,750	6,325
2100	4" x 8" x 8"					6,775			6,775	7,450
2150	4" x 16" x 8"					7,225			7,225	7,950

04 21 26 – Glazed Structural Clay Tile Masonry

04 21 26.10 Structural Facing Tile

		Crew	Daily Output	Labor-Hours	Unit	Material	2021 Bare Costs Labor	Equipment	Total	Total Incl O&P
0010	**STRUCTURAL FACING TILE**, std. colors, excl. scaffolding, grout, reinforcing									
0020	6T series, 5-1/3" x 12", 2.3 pieces per S.F., glazed 1 side, 2" thick	D-8	225	.178	S.F.	9.90	8.85		18.75	24
0100	4" thick		220	.182		13.10	9.05		22.15	28
0150	Glazed 2 sides		195	.205		17.45	10.20		27.65	34.50

04 21 Clay Unit Masonry

04 21 26 – Glazed Structural Clay Tile Masonry

04 21 26.10 Structural Facing Tile		Crew	Daily Output	Labor-Hours	Unit	Material	2021 Bare Costs Labor	Equipment	Total	Total Incl O&P
0250	6" thick	D-8	210	.190	S.F.	18.70	9.45		28.15	35
0300	Glazed 2 sides		185	.216		22.50	10.75		33.25	40.50
0400	8" thick		180	.222	↓	23	11.05		34.05	41.50
0500	Special shapes, group 1		400	.100	Ea.	7.55	4.97		12.52	15.80
0550	Group 2		375	.107		13.55	5.30		18.85	23
0600	Group 3		350	.114		17.05	5.70		22.75	27.50
0650	Group 4		325	.123		35.50	6.10		41.60	48
0700	Group 5		300	.133		42	6.65		48.65	56.50
0750	Group 6		275	.145	↓	62.50	7.25		69.75	80
1000	Fire rated, 4" thick, 1 hr. rating		210	.190	S.F.	18.45	9.45		27.90	35
1300	Acoustic, 4" thick	↓	210	.190	"	40	9.45		49.45	58
2000	8W series, 8" x 16", 1.125 pieces per S.F.									
2050	2" thick, glazed 1 side	D-8	360	.111	S.F.	9.55	5.50		15.05	18.80
2100	4" thick, glazed 1 side		345	.116		13.50	5.75		19.25	23.50
2150	Glazed 2 sides		325	.123		14.40	6.10		20.50	25
2200	6" thick, glazed 1 side		330	.121		31	6		37	43.50
2250	8" thick, glazed 1 side		310	.129	↓	29.50	6.40		35.90	42
2500	Special shapes, group 1		300	.133	Ea.	15.90	6.65		22.55	27.50
2550	Group 2		280	.143		21	7.10		28.10	33.50
2600	Group 3		260	.154		22.50	7.65		30.15	36.50
2650	Group 4		250	.160		54	7.95		61.95	71
2700	Group 5		240	.167		47	8.30		55.30	64
2750	Group 6		230	.174	↓	103	8.65		111.65	126
3000	4" thick, glazed 1 side		345	.116	S.F.	15.80	5.75		21.55	26
3100	Acoustic, 4" thick	↓	345	.116	"	22.50	5.75		28.25	33
3120	4W series, 8" x 8", 2.25 pieces per S.F.									
3125	2" thick, glazed 1 side	D-8	360	.111	S.F.	9.20	5.50		14.70	18.40
3130	4" thick, glazed 1 side		345	.116		12.65	5.75		18.40	22.50
3135	Glazed 2 sides		325	.123		15.75	6.10		21.85	26.50
3140	6" thick, glazed 1 side		330	.121		16.15	6		22.15	27
3150	8" thick, glazed 1 side		310	.129	↓	24.50	6.40		30.90	36.50
3155	Special shapes, group 1		300	.133	Ea.	7.60	6.65		14.25	18.40
3160	Group 2	↓	280	.143	"	8.85	7.10		15.95	20.50
3200	For designer colors, add					25%				
3300	For epoxy mortar joints, add				S.F.	1.88			1.88	2.07

04 21 29 – Terra Cotta Masonry

04 21 29.10 Terra Cotta Masonry Components

		Crew	Daily Output	Labor-Hours	Unit	Material	2021 Bare Costs Labor	Equipment	Total	Total Incl O&P
0010	**TERRA COTTA MASONRY COMPONENTS**									
0020	Coping, split type, not glazed, 9" wide	D-1	90	.178	L.F.	11.60	8.65		20.25	26
0100	13" wide		80	.200		16.70	9.75		26.45	33
0200	Coping, split type, glazed, 9" wide		90	.178		20.50	8.65		29.15	35.50
0250	13" wide	↓	80	.200	↓	24.50	9.75		34.25	41.50
0500	Partition or back-up blocks, scored, in C.L. lots									
0700	Non-load bearing 12" x 12", 3" thick, special order	D-8	550	.073	S.F.	20.50	3.61		24.11	28
0750	4" thick, standard		500	.080		6.95	3.98		10.93	13.65
0800	6" thick		450	.089		8.40	4.42		12.82	15.90
0850	8" thick		400	.100		10.35	4.97		15.32	18.85
1000	Load bearing, 12" x 12", 4" thick, in walls		500	.080		6.45	3.98		10.43	13.10
1050	In floors		750	.053		6.45	2.65		9.10	11.10
1200	6" thick, in walls		450	.089		9.80	4.42		14.22	17.40
1250	In floors		675	.059		9.80	2.95		12.75	15.20
1400	8" thick, in walls	↓	400	.100	↓	10.85	4.97		15.82	19.45

04 21 Clay Unit Masonry

04 21 29 – Terra Cotta Masonry

04 21 29.10 Terra Cotta Masonry Components	Crew	Daily Output	Labor-Hours	Unit	Material	2021 Bare Costs Labor	Equipment	Total	Total Incl O&P
1450	In floors	D-8	575	.070	S.F.	10.85	3.46		14.31	17.15
1600	10" thick, in walls, special order		350	.114		28.50	5.70		34.20	39.50
1650	In floors, special order		500	.080		28.50	3.98		32.48	37
1800	12" thick, in walls, special order		300	.133		26.50	6.65		33.15	39
1850	In floors, special order		450	.089		26.50	4.42		30.92	35.50
2000	For reinforcing with steel rods, add to above					15%	5%			
2100	For smooth tile instead of scored, add					2.94			2.94	3.23
2200	For L.C.L. quantities, add					10%	10%			

04 21 29.20 Terra Cotta Tile

		Crew	Daily Output	Labor-Hours	Unit	Material	2021 Bare Costs Labor	Equipment	Total	Total Incl O&P
0010	**TERRA COTTA TILE**, on walls, dry set, 1/2" thick									
0100	Square, hexagonal or lattice shapes, unglazed	1 Tilf	135	.059	S.F.	5.20	3.06		8.26	10.20
0300	Glazed, plain colors		130	.062		8.10	3.18		11.28	13.55
0400	Intense colors		125	.064		9.45	3.31		12.76	15.25

04 22 Concrete Unit Masonry

04 22 10 – Concrete Masonry Units

04 22 10.10 Concrete Block

		Crew	Daily Output	Labor-Hours	Unit	Material	2021 Bare Costs Labor	Equipment	Total	Total Incl O&P
0010	**CONCRETE BLOCK** Material Only R042210-20									
0020	2" x 8" x 16" solid, normal-weight, 2,000 psi				Ea.	1.21			1.21	1.33
0050	3,500 psi					1.40			1.40	1.54
0100	5,000 psi					1.50			1.50	1.65
0150	Lightweight, std.					1.49			1.49	1.64
0300	3" x 8" x 16" solid, normal-weight, 2,000 psi					.96			.96	1.06
0350	3,500 psi					1.40			1.40	1.54
0400	5,000 psi					1.64			1.64	1.80
0450	Lightweight, std.					1.47			1.47	1.62
0600	4" x 8" x 16" hollow, normal-weight, 2,000 psi					1.51			1.51	1.66
0650	3,500 psi					1.43			1.43	1.57
0700	5,000 psi					1.82			1.82	2
0750	Lightweight, std.					1.43			1.43	1.57
1300	Solid, normal-weight, 2,000 psi					1.89			1.89	2.08
1350	3,500 psi					1.67			1.67	1.84
1400	5,000 psi					1.65			1.65	1.82
1450	Lightweight, std.					1.54			1.54	1.69
1600	6" x 8" x 16" hollow, normal-weight, 2,000 psi					1.93			1.93	2.12
1650	3,500 psi					2.08			2.08	2.29
1700	5,000 psi					2.24			2.24	2.46
1750	Lightweight, std.					2.27			2.27	2.50
2300	Solid, normal-weight, 2,000 psi					2.67			2.67	2.94
2350	3,500 psi					1.65			1.65	1.82
2400	5,000 psi					2.25			2.25	2.48
2450	Lightweight, std.					2.60			2.60	2.86
2600	8" x 8" x 16" hollow, normal-weight, 2,000 psi					1.73			1.73	1.90
2650	3,500 psi					2.49			2.49	2.74
2700	5,000 psi					2.64			2.64	2.90
2750	Lightweight, std.					2.91			2.91	3.20
3200	Solid, normal-weight, 2,000 psi					2.77			2.77	3.05
3250	3,500 psi					2.98			2.98	3.28
3300	5,000 psi					3.25			3.25	3.58
3350	Lightweight, std.					2.36			2.36	2.60
3400	10" x 8" x 16" hollow, normal-weight, 2,000 psi					1.73			1.73	1.90

04 22 Concrete Unit Masonry

04 22 10 – Concrete Masonry Units

04 22 10.10 Concrete Block

		Crew	Daily Output	Labor-Hours	Unit	Material	2021 Bare Costs Labor	Equipment	Total	Total Incl O&P
3410	3,500 psi				Ea.	2.49			2.49	2.74
3420	5,000 psi					2.64			2.64	2.90
3430	Lightweight, std.					2.91			2.91	3.20
3480	Solid, normal-weight, 2,000 psi					2.77			2.77	3.05
3490	3,500 psi					2.98			2.98	3.28
3500	5,000 psi					3.25			3.25	3.58
3510	Lightweight, std.					2.36			2.36	2.60
3600	12" x 8" x 16" hollow, normal-weight, 2,000 psi					3.25			3.25	3.58
3650	3,500 psi					3.06			3.06	3.37
3700	5,000 psi					3.66			3.66	4.03
3750	Lightweight, std.					2.70			2.70	2.97
4300	Solid, normal-weight, 2,000 psi					4.13			4.13	4.54
4350	3,500 psi					3.71			3.71	4.08
4400	5,000 psi					3.57			3.57	3.93
4500	Lightweight, std.					3.44			3.44	3.78

04 22 10.11 Autoclave Aerated Concrete Block

			Crew	Daily Output	Labor-Hours	Unit	Material	2021 Bare Costs Labor	Equipment	Total	Total Incl O&P
0010	**AUTOCLAVE AERATED CONCRETE BLOCK**, excl. scaffolding, grout & reinforcing										
0050	Solid, 4" x 8" x 24", incl. mortar	G	D-8	600	.067	S.F.	1.59	3.31		4.90	6.75
0060	6" x 8" x 24"	G		600	.067		2.56	3.31		5.87	7.80
0070	8" x 8" x 24"	G		575	.070		3.40	3.46		6.86	8.95
0080	10" x 8" x 24"	G		575	.070		4.08	3.46		7.54	9.70
0090	12" x 8" x 24"	G		550	.073		4.89	3.61		8.50	10.85

04 22 10.12 Chimney Block

		Crew	Daily Output	Labor-Hours	Unit	Material	2021 Bare Costs Labor	Equipment	Total	Total Incl O&P
0010	**CHIMNEY BLOCK**, excludes scaffolding, grout and reinforcing									
0220	1 piece, with 8" x 8" flue, 16" x 16"	D-1	28	.571	V.L.F.	19.55	28		47.55	63.50
0230	2 piece, 16" x 16"		26	.615		20	30		50	67
0240	2 piece, with 8" x 12" flue, 16" x 20"		24	.667		36	32.50		68.50	88.50

04 22 10.14 Concrete Block, Back-Up

		Crew	Daily Output	Labor-Hours	Unit	Material	2021 Bare Costs Labor	Equipment	Total	Total Incl O&P
0010	**CONCRETE BLOCK, BACK-UP**, C90, 2000 psi R042210-20									
0020	Normal weight, 8" x 16" units, tooled joint 1 side									
0050	Not-reinforced, 2000 psi, 2" thick	D-8	475	.084	S.F.	1.72	4.19		5.91	8.20
0200	4" thick		460	.087		2.20	4.32		6.52	8.90
0300	6" thick		440	.091		2.81	4.52		7.33	9.90
0350	8" thick		400	.100		2.74	4.97		7.71	10.50
0400	10" thick		330	.121		3.29	6		9.29	12.70
0450	12" thick	D-9	310	.155		4.74	7.55		12.29	16.55
1000	Reinforced, alternate courses, 4" thick	D-8	450	.089		2.38	4.42		6.80	9.25
1100	6" thick		430	.093		3.01	4.62		7.63	10.25
1150	8" thick		395	.101		2.95	5.05		8	10.85
1200	10" thick		320	.125		3.47	6.20		9.67	13.15
1250	12" thick	D-9	300	.160		4.93	7.80		12.73	17.20
2000	Lightweight, not reinforced, 4" thick	D-8	460	.087		2.11	4.32		6.43	8.80
2100	6" thick		445	.090		3.20	4.47		7.67	10.25
2150	8" thick		435	.092		4.07	4.57		8.64	11.40
2200	10" thick		410	.098		4.32	4.85		9.17	12.05
2250	12" thick	D-9	390	.123		4.12	6		10.12	13.60
3000	Reinforced, alternate courses, 4" thick	D-8	450	.089		2.29	4.42		6.71	9.15
3100	6" thick		430	.093		3.39	4.62		8.01	10.70
3150	8" thick		420	.095		4.28	4.73		9.01	11.85
3200	10" thick		400	.100		4.50	4.97		9.47	12.45
3250	12" thick	D-9	380	.126		4.31	6.15		10.46	14

For customer support on your Concrete & Masonry Costs with RSMeans Data, call 800.448.8182.

103

04 22 10.16 Concrete Block, Bond Beam	Crew	Daily Output	Labor-Hours	Unit	Material	2021 Bare Costs Labor	Equipment	Total	Total Incl O&P
0010 **CONCRETE BLOCK, BOND BEAM**, C90, 2000 psi									
0020 Not including grout or reinforcing									
0125 Regular block, 6" thick	D-8	584	.068	L.F.	2.76	3.40		6.16	8.20
0130 8" high, 8" thick	"	565	.071		2.93	3.52		6.45	8.55
0150 12" thick	D-9	510	.094		4.35	4.58		8.93	11.70
0525 Lightweight, 6" thick	D-8	592	.068		3.11	3.36		6.47	8.45
0530 8" high, 8" thick	"	575	.070		3.66	3.46		7.12	9.20
0550 12" thick	D-9	520	.092		4.91	4.50		9.41	12.15
2000 Including grout and 2 #5 bars									
2100 Regular block, 8" high, 8" thick	D-8	300	.133	L.F.	5.55	6.65		12.20	16.10
2150 12" thick	D-9	250	.192		7.70	9.35		17.05	22.50
2500 Lightweight, 8" high, 8" thick	D-8	305	.131		6.30	6.50		12.80	16.70
2550 12" thick	D-9	255	.188		8.25	9.15		17.40	23

04 22 10.18 Concrete Block, Column	Crew	Daily Output	Labor-Hours	Unit	Material	2021 Bare Costs Labor	Equipment	Total	Total Incl O&P
0010 **CONCRETE BLOCK, COLUMN** or pilaster									
0050 Including vertical reinforcing (4-#4 bars) and grout									
0160 1 piece unit, 16" x 16"	D-1	26	.615	V.L.F.	16.60	30		46.60	63.50
0170 2 piece units, 16" x 20"		24	.667		21.50	32.50		54	72.50
0180 20" x 20"		22	.727		33	35.50		68.50	89.50
0190 22" x 24"		18	.889		48.50	43.50		92	119
0200 20" x 32"		14	1.143		57.50	55.50		113	147

04 22 10.19 Concrete Block, Insulation Inserts	Crew	Daily Output	Labor-Hours	Unit	Material	2021 Bare Costs Labor	Equipment	Total	Total Incl O&P
0010 **CONCRETE BLOCK, INSULATION INSERTS**									
0100 Styrofoam, plant installed, add to block prices									
0200 8" x 16" units, 6" thick				S.F.	1.28			1.28	1.41
0250 8" thick					1.65			1.65	1.82
0300 10" thick					1.34			1.34	1.47
0350 12" thick					1.71			1.71	1.88
0500 8" x 8" units, 8" thick					1.63			1.63	1.79
0550 12" thick					1.62			1.62	1.78

04 22 10.23 Concrete Block, Decorative	Crew	Daily Output	Labor-Hours	Unit	Material	2021 Bare Costs Labor	Equipment	Total	Total Incl O&P
0010 **CONCRETE BLOCK, DECORATIVE**, C90, 2000 psi									
0020 Embossed, simulated brick face									
0100 8" x 16" units, 4" thick	D-8	400	.100	S.F.	3.17	4.97		8.14	11
0200 8" thick		340	.118		3.29	5.85		9.14	12.40
0250 12" thick		300	.133		4.82	6.65		11.47	15.30
0400 Embossed both sides									
0500 8" thick	D-8	300	.133	S.F.	5.55	6.65		12.20	16.10
0550 12" thick	"	275	.145	"	6.75	7.25		14	18.30
1000 Fluted high strength									
1100 8" x 16" x 4" thick, flutes 1 side,	D-8	345	.116	S.F.	4.17	5.75		9.92	13.30
1150 Flutes 2 sides		335	.119		4.72	5.95		10.67	14.15
1200 8" thick		300	.133		5.30	6.65		11.95	15.80
1250 For special colors, add					.66			.66	.72
1400 Deep grooved, smooth face									
1450 8" x 16" x 4" thick	D-8	345	.116	S.F.	2.64	5.75		8.39	11.60
1500 8" thick	"	300	.133	"	4.54	6.65		11.19	15
2000 Formblock, incl. inserts & reinforcing									
2100 8" x 16" x 8" thick	D-8	345	.116	S.F.	4.49	5.75		10.24	13.65
2150 12" thick	"	310	.129	"	5.85	6.40		12.25	16.05
2500 Ground face									

04 22 Concrete Unit Masonry

04 22 10 – Concrete Masonry Units

04 22 10.23 Concrete Block, Decorative	Crew	Daily Output	Labor-Hours	Unit	Material	2021 Bare Costs Labor	2021 Bare Costs Equipment	Total	Total Incl O&P
2600 8" x 16" x 4" thick	D-8	345	.116	S.F.	4.30	5.75		10.05	13.45
2650 6" thick		325	.123		4.92	6.10		11.02	14.60
2700 8" thick		300	.133		5.70	6.65		12.35	16.25
2750 12" thick	D-9	265	.181		7.30	8.80		16.10	21.50
2900 For special colors, add, minimum					15%				
2950 For special colors, add, maximum					45%				
4000 Slump block									
4100 4" face height x 16" x 4" thick	D-1	165	.097	S.F.	5.40	4.72		10.12	13
4150 6" thick		160	.100		6.65	4.87		11.52	14.65
4200 8" thick		155	.103		9.50	5.05		14.55	18
4250 10" thick		140	.114		12.15	5.55		17.70	22
4300 12" thick		130	.123		13.45	6		19.45	24
4400 6" face height x 16" x 6" thick		155	.103		6.20	5.05		11.25	14.35
4450 8" thick		150	.107		7.65	5.20		12.85	16.25
4500 10" thick		130	.123		14.55	6		20.55	25
4550 12" thick		120	.133		14.55	6.50		21.05	26
5000 Split rib profile units, 1" deep ribs, 8 ribs									
5100 8" x 16" x 4" thick	D-8	345	.116	S.F.	4.90	5.75		10.65	14.10
5150 6" thick		325	.123		5.65	6.10		11.75	15.40
5200 8" thick		300	.133		6.85	6.65		13.50	17.55
5250 12" thick	D-9	275	.175		7.85	8.50		16.35	21.50
5400 For special deeper colors, 4" thick, add					1.42			1.42	1.57
5450 12" thick, add					1.38			1.38	1.52
5600 For white, 4" thick, add					1.42			1.42	1.57
5650 6" thick, add					1.38			1.38	1.52
5700 8" thick, add					1.48			1.48	1.63
5750 12" thick, add					1.42			1.42	1.57
6000 Split face									
6100 8" x 16" x 4" thick	D-8	350	.114	S.F.	3.86	5.70		9.56	12.80
6150 6" thick		325	.123		4.38	6.10		10.48	14
6200 8" thick		300	.133		5.60	6.65		12.25	16.15
6250 12" thick	D-9	270	.178		6.25	8.65		14.90	19.95
6300 For scored, add					.39			.39	.43
6400 For special deeper colors, 4" thick, add					.66			.66	.72
6450 6" thick, add					.84			.84	.92
6500 8" thick, add					.71			.71	.78
6550 12" thick, add					.84			.84	.92
6650 For white, 4" thick, add					1.42			1.42	1.57
6700 6" thick, add					1.29			1.29	1.42
6750 8" thick, add					1.41			1.41	1.55
6800 12" thick, add					1.38			1.38	1.52
7000 Scored ground face, 2 to 5 scores									
7100 8" x 16" x 4" thick	D-8	340	.118	S.F.	8.85	5.85		14.70	18.55
7150 6" thick		310	.129		10.15	6.40		16.55	21
7200 8" thick		290	.138		11.50	6.85		18.35	23
7250 12" thick	D-9	265	.181		17.10	8.80		25.90	32
8000 Hexagonal face profile units, 8" x 16" units									
8100 4" thick, hollow	D-8	340	.118	S.F.	4.18	5.85		10.03	13.40
8200 Solid		340	.118		5.30	5.85		11.15	14.60
8300 6" thick, hollow		310	.129		4.19	6.40		10.59	14.25
8350 8" thick, hollow		290	.138		4.17	6.85		11.02	14.95
8500 For stacked bond, add					26%				
8550 For high rise construction, add per story	D-8	67.80	.590	M.S.F.		29.50		29.50	44

105

04 22 Concrete Unit Masonry

04 22 10 – Concrete Masonry Units

04 22 10.23 Concrete Block, Decorative	Crew	Daily Output	Labor-Hours	Unit	Material	2021 Bare Costs Labor	Equipment	Total	Total Incl O&P
8600	For scored block, add					10%				
8650	For honed or ground face, per face, add				Ea.	1.19			1.19	1.31
8700	For honed or ground end, per end, add				"	1.54			1.54	1.69
8750	For bullnose block, add					10%				
8800	For special color, add					13%				

04 22 10.24 Concrete Block, Exterior

		Crew	Daily Output	Labor-Hours	Unit	Material	Labor	Equipment	Total	Total Incl O&P
0010	**CONCRETE BLOCK, EXTERIOR**, C90, 2000 psi R042210-20									
0020	Reinforced alt courses, tooled joints 2 sides									
0100	Normal weight, 8" x 16" x 6" thick	D-8	395	.101	S.F.	2.56	5.05		7.61	10.40
0200	8" thick		360	.111		4.25	5.50		9.75	12.95
0250	10" thick	↓	290	.138		4.58	6.85		11.43	15.40
0300	12" thick	D-9	250	.192		5.35	9.35		14.70	20
0500	Lightweight, 8" x 16" x 6" thick	D-8	450	.089		3.64	4.42		8.06	10.65
0600	8" thick		430	.093		3.55	4.62		8.17	10.85
0650	10" thick	↓	395	.101		4.60	5.05		9.65	12.65
0700	12" thick	D-9	350	.137	↓	5.05	6.70		11.75	15.60

04 22 10.26 Concrete Block Foundation Wall

		Crew	Daily Output	Labor-Hours	Unit	Material	Labor	Equipment	Total	Total Incl O&P
0010	**CONCRETE BLOCK FOUNDATION WALL**, C90/C145									
0050	Normal-weight, cut joints, horiz joint reinf, no vert reinf.									
0200	Hollow, 8" x 16" x 6" thick	D-8	455	.088	S.F.	3.38	4.37		7.75	10.30
0250	8" thick		425	.094		3.33	4.68		8.01	10.70
0300	10" thick	↓	350	.114		3.87	5.70		9.57	12.80
0350	12" thick	D-9	300	.160		5.35	7.80		13.15	17.65
0500	Solid, 8" x 16" block, 6" thick	D-8	440	.091		4.22	4.52		8.74	11.45
0550	8" thick	"	415	.096		4.50	4.79		9.29	12.15
0600	12" thick	D-9	350	.137	↓	6.35	6.70		13.05	17
1000	Reinforced, #4 vert @ 48"									
1100	Hollow, 8" x 16" block, 4" thick	D-8	455	.088	S.F.	3.42	4.37		7.79	10.35
1125	6" thick		445	.090		4.47	4.47		8.94	11.65
1150	8" thick		415	.096		4.88	4.79		9.67	12.55
1200	10" thick	↓	340	.118		5.90	5.85		11.75	15.25
1250	12" thick	D-9	290	.166		7.80	8.05		15.85	21
1500	Solid, 8" x 16" block, 6" thick	D-8	430	.093		4.22	4.62		8.84	11.60
1600	8" thick	"	405	.099		4.50	4.91		9.41	12.35
1650	12" thick	D-9	340	.141	↓	6.35	6.90		13.25	17.30

04 22 10.28 Concrete Block, High Strength

		Crew	Daily Output	Labor-Hours	Unit	Material	Labor	Equipment	Total	Total Incl O&P
0010	**CONCRETE BLOCK, HIGH STRENGTH**									
0050	Hollow, reinforced alternate courses, 8" x 16" units									
0200	3500 psi, 4" thick	D-8	440	.091	S.F.	2.39	4.52		6.91	9.45
0250	6" thick		395	.101		2.47	5.05		7.52	10.30
0300	8" thick	↓	360	.111		4.16	5.50		9.66	12.85
0350	12" thick	D-9	250	.192		5.25	9.35		14.60	19.85
0500	5000 psi, 4" thick	D-8	440	.091		2.42	4.52		6.94	9.45
0550	6" thick		395	.101		3.24	5.05		8.29	11.15
0600	8" thick	↓	360	.111		4.55	5.50		10.05	13.30
0650	12" thick	D-9	300	.160	↓	5.20	7.80		13	17.50
1000	For 75% solid block, add					30%				
1050	For 100% solid block, add					50%				

For customer support on your Concrete & Masonry Costs with RSMeans Data, call 800.448.8182.

04 22 10.30 Concrete Block, Interlocking

		Crew	Daily Output	Labor-Hours	Unit	Material	2021 Bare Costs Labor	2021 Bare Costs Equipment	Total	Total Incl O&P
0010	**CONCRETE BLOCK, INTERLOCKING**									
0100	Not including grout or reinforcing									
0200	8" x 16" units, 2,000 psi, 8" thick	D-1	245	.065	S.F.	3.31	3.18		6.49	8.45
0300	12" thick		220	.073		4.90	3.54		8.44	10.75
0350	16" thick	↓	185	.086		7.15	4.21		11.36	14.25
0400	Including grout & reinforcing, 8" thick	D-4	245	.131		9.35	6.40	.78	16.53	21
0450	12" thick		220	.145		11.25	7.15	.87	19.27	24
0500	16" thick	↓	185	.173	↓	13.75	8.50	1.03	23.28	29

04 22 10.32 Concrete Block, Lintels

		Crew	Daily Output	Labor-Hours	Unit	Material	2021 Bare Costs Labor	2021 Bare Costs Equipment	Total	Total Incl O&P
0010	**CONCRETE BLOCK, LINTELS**, C90, normal weight									
0100	Including grout and horizontal reinforcing									
0200	8" x 8" x 8", 1 #4 bar	D-4	300	.107	L.F.	4.90	5.25	.63	10.78	14
0250	2 #4 bars		295	.108		5.20	5.35	.65	11.20	14.40
0400	8" x 16" x 8", 1 #4 bar		275	.116		4.31	5.70	.69	10.70	14.10
0450	2 #4 bars		270	.119		4.59	5.85	.71	11.15	14.60
1000	12" x 8" x 8", 1 #4 bar		275	.116		6.45	5.70	.69	12.84	16.45
1100	2 #4 bars		270	.119		6.75	5.85	.71	13.31	16.95
1150	2 #5 bars		270	.119		7.05	5.85	.71	13.61	17.30
1200	2 #6 bars		265	.121		7.40	5.95	.72	14.07	17.85
1500	12" x 16" x 8", 1 #4 bar		250	.128		7.60	6.30	.76	14.66	18.65
1600	2 #3 bars		245	.131		7.60	6.40	.78	14.78	18.90
1650	2 #4 bars		245	.131		7.85	6.40	.78	15.03	19.15
1700	2 #5 bars	↓	240	.133	↓	8.20	6.55	.79	15.54	19.70

04 22 10.33 Lintel Block

		Crew	Daily Output	Labor-Hours	Unit	Material	2021 Bare Costs Labor	2021 Bare Costs Equipment	Total	Total Incl O&P
0010	**LINTEL BLOCK**									
3481	Lintel block 6" x 8" x 8"	D-1	300	.053	Ea.	1.45	2.60		4.05	5.50
3501	6" x 16" x 8"		275	.058		2.02	2.83		4.85	6.50
3521	8" x 8" x 8"		275	.058		1.40	2.83		4.23	5.80
3561	8" x 16" x 8"	↓	250	.064	↓	1.55	3.12		4.67	6.40

04 22 10.34 Concrete Block, Partitions

		Crew	Daily Output	Labor-Hours	Unit	Material	2021 Bare Costs Labor	2021 Bare Costs Equipment	Total	Total Incl O&P
0010	**CONCRETE BLOCK, PARTITIONS**, excludes scaffolding R042210-20									
0100	Acoustical slotted block									
0200	4" thick, type A-1	D-8	315	.127	S.F.	6.65	6.30		12.95	16.80
0210	8" thick		275	.145		7.55	7.25		14.80	19.20
0250	8" thick, type Q		275	.145		14.70	7.25		21.95	27
0260	4" thick, type RSC		315	.127		11.20	6.30		17.50	22
0270	6" thick		295	.136		11.70	6.75		18.45	23
0280	8" thick		275	.145		12.25	7.25		19.50	24.50
0290	12" thick		250	.160		12.75	7.95		20.70	26
0300	8" thick, type RSR		275	.145		12.25	7.25		19.50	24.50
0400	8" thick, type RSC/RF		275	.145		9.75	7.25		17	21.50
0410	10" thick		260	.154		12	7.65		19.65	24.50
0420	12" thick		250	.160		11.25	7.95		19.20	24.50
0430	12" thick, type RSC/RF-4		250	.160		16.30	7.95		24.25	30
0500	NRC .60 type R, 8" thick		265	.151		13.05	7.50		20.55	25.50
0600	NRC .65 type RR, 8" thick		265	.151		8.75	7.50		16.25	21
0700	NRC .65 type 4R-RF, 8" thick		265	.151		13.50	7.50		21	26
0710	NRC .70 type R, 12" thick	↓	245	.163	↓	14	8.10		22.10	27.50
1000	Lightweight block, tooled joints, 2 sides, hollow									
1100	Not reinforced, 8" x 16" x 4" thick	D-8	440	.091	S.F.	2	4.52		6.52	9
1150	6" thick	↓	410	.098	↓	3.10	4.85		7.95	10.70

04 22 10.34 Concrete Block, Partitions	Crew	Daily Output	Labor-Hours	Unit	Material	2021 Bare Costs Labor	Equipment	Total	Total Incl O&P
1200	8" thick	D-8	385	.104	S.F.	3.97	5.15		9.12	12.15
1250	10" thick	↓	370	.108		4.22	5.35		9.57	12.75
1300	12" thick	D-9	350	.137		4.02	6.70		10.72	14.45
1500	Reinforced alternate courses, 4" thick	D-8	435	.092		2.17	4.57		6.74	9.30
1600	6" thick		405	.099		3.26	4.91		8.17	11
1650	8" thick		380	.105		4.17	5.25		9.42	12.50
1700	10" thick	↓	365	.110		4.41	5.45		9.86	13.05
1750	12" thick	D-9	345	.139		4.23	6.80		11.03	14.85
2000	Not reinforced, 8" x 24" x 4" thick, hollow		460	.104		1.46	5.10		6.56	9.25
2100	6" thick		440	.109		2.21	5.30		7.51	10.45
2150	8" thick		415	.116		2.87	5.65		8.52	11.65
2200	10" thick		385	.125		3.08	6.05		9.13	12.55
2250	12" thick		365	.132		3	6.40		9.40	12.95
2400	Reinforced alternate courses, 4" thick		455	.105		1.62	5.15		6.77	9.55
2500	6" thick		435	.110		2.40	5.35		7.75	10.75
2550	8" thick		410	.117		3.06	5.70		8.76	11.95
2600	10" thick		380	.126		3.27	6.15		9.42	12.85
2650	12" thick	↓	360	.133		3.20	6.50		9.70	13.30
2800	Solid, not reinforced, 8" x 16" x 2" thick	D-8	440	.091		1.94	4.52		6.46	8.95
2900	4" thick		420	.095		2.13	4.73		6.86	9.50
2950	6" thick		390	.103		3.47	5.10		8.57	11.50
3000	8" thick		365	.110		3.35	5.45		8.80	11.90
3050	10" thick	↓	350	.114		4.41	5.70		10.11	13.40
3100	12" thick	D-9	330	.145		4.86	7.10		11.96	16
3300	Solid, reinforced alternate courses, 4" thick	D-8	415	.096		2.29	4.79		7.08	9.70
3400	6" thick		385	.104		3.85	5.15		9	12.05
3450	8" thick		360	.111		3.55	5.50		9.05	12.20
3500	10" thick	↓	345	.116		4.60	5.75		10.35	13.75
3550	12" thick	D-9	325	.148	↓	5.05	7.20		12.25	16.40
4000	Regular block, tooled joints, 2 sides, hollow									
4100	Not reinforced, 8" x 16" x 4" thick	D-8	430	.093	S.F.	2.09	4.62		6.71	9.25
4150	6" thick		400	.100		2.71	4.97		7.68	10.50
4200	8" thick		375	.107		2.64	5.30		7.94	10.90
4250	10" thick	↓	360	.111		3.19	5.50		8.69	11.80
4300	12" thick	D-9	340	.141		4.64	6.90		11.54	15.45
4500	Reinforced alternate courses, 8" x 16" x 4" thick	D-8	425	.094		2.28	4.68		6.96	9.55
4550	6" thick		395	.101		2.91	5.05		7.96	10.80
4600	8" thick		370	.108		2.85	5.35		8.20	11.25
4650	10" thick	↓	355	.113		4.68	5.60		10.28	13.60
4700	12" thick	D-9	335	.143		4.83	7		11.83	15.80
4900	Solid, not reinforced, 2" thick	D-8	435	.092		1.66	4.57		6.23	8.75
5000	3" thick		430	.093		1.42	4.62		6.04	8.50
5050	4" thick		415	.096		2.52	4.79		7.31	10
5100	6" thick		385	.104		3.55	5.15		8.70	11.70
5150	8" thick	↓	360	.111		3.81	5.50		9.31	12.50
5200	12" thick	D-9	325	.148		5.65	7.20		12.85	17.05
5500	Solid, reinforced alternate courses, 4" thick	D-8	420	.095		2.69	4.73		7.42	10.10
5550	6" thick		380	.105		3.71	5.25		8.96	12
5600	8" thick	↓	355	.113		4.01	5.60		9.61	12.85
5650	12" thick	D-9	320	.150	↓	4.74	7.30		12.04	16.20

04 22 Concrete Unit Masonry

04 22 10 – Concrete Masonry Units

04 22 10.38 Concrete Brick	Crew	Daily Output	Labor-Hours	Unit	Material	2021 Bare Costs Labor	2021 Bare Costs Equipment	Total	Total Incl O&P
0010 **CONCRETE BRICK**, C55, grade N, type 1									
0100 Regular, 4" x 2-1/4" x 8"	D-8	660	.061	Ea.	.66	3.01		3.67	5.25
0125 Rusticated, 4" x 2-1/4" x 8"		660	.061		.74	3.01		3.75	5.35
0150 Frog, 4" x 2-1/4" x 8"		660	.061		.71	3.01		3.72	5.30
0200 Double, 4" x 4-7/8" x 8"	↓	535	.075	↓	1.16	3.72		4.88	6.85

04 22 10.42 Concrete Block, Screen Block

	Crew	Daily Output	Labor-Hours	Unit	Material	Labor	Equipment	Total	Total Incl O&P
0010 **CONCRETE BLOCK, SCREEN BLOCK**									
0200 8" x 16", 4" thick	D-8	330	.121	S.F.	11.55	6		17.55	22
0300 8" thick		270	.148		13.35	7.35		20.70	26
0350 12" x 12", 4" thick		290	.138		7.50	6.85		14.35	18.60
0500 8" thick	↓	250	.160	↓	7.30	7.95		15.25	20

04 22 10.44 Glazed Concrete Block

	Crew	Daily Output	Labor-Hours	Unit	Material	Labor	Equipment	Total	Total Incl O&P
0010 **GLAZED CONCRETE BLOCK** C744									
0100 Single face, 8" x 16" units, 2" thick	D-8	360	.111	S.F.	12.25	5.50		17.75	22
0200 4" thick		345	.116		12.60	5.75		18.35	22.50
0250 6" thick		330	.121		14.85	6		20.85	25.50
0300 8" thick		310	.129		16.50	6.40		22.90	28
0350 10" thick	↓	295	.136		16.95	6.75		23.70	29
0400 12" thick	D-9	280	.171		18.20	8.35		26.55	32.50
0700 Double face, 8" x 16" units, 4" thick	D-8	340	.118		17.10	5.85		22.95	27.50
0750 6" thick		320	.125		21.50	6.20		27.70	33.50
0800 8" thick		300	.133	↓	23	6.65		29.65	35.50
1000 Jambs, bullnose or square, single face, 8" x 16", 2" thick		315	.127	Ea.	20.50	6.30		26.80	32
1050 4" thick		285	.140	"	21.50	7		28.50	34
1200 Caps, bullnose or square, 8" x 16", 2" thick		420	.095	L.F.	20.50	4.73		25.23	29.50
1250 4" thick		380	.105	"	23.50	5.25		28.75	33.50
1256 Corner, bullnose or square, 2" thick		280	.143	Ea.	23.50	7.10		30.60	36.50
1258 4" thick		270	.148		27	7.35		34.35	41
1260 6" thick		260	.154		30	7.65		37.65	44.50
1270 8" thick		250	.160		39.50	7.95		47.45	55.50
1280 10" thick		240	.167		32	8.30		40.30	48
1290 12" thick		230	.174	↓	31.50	8.65		40.15	47.50
1500 Cove base, 8" x 16", 2" thick		315	.127	L.F.	10.80	6.30		17.10	21.50
1550 4" thick		285	.140		10.90	7		17.90	22.50
1600 6" thick		265	.151		11.60	7.50		19.10	24
1650 8" thick	↓	245	.163	↓	12.15	8.10		20.25	25.50

04 23 Glass Unit Masonry

04 23 13 – Vertical Glass Unit Masonry

04 23 13.10 Glass Block

	Crew	Daily Output	Labor-Hours	Unit	Material	Labor	Equipment	Total	Total Incl O&P
0010 **GLASS BLOCK**									
0100 Plain, 4" thick, under 1,000 S.F., 6" x 6"	D-8	115	.348	S.F.	26.50	17.30		43.80	55
0150 8" x 8"		160	.250		16.15	12.45		28.60	36.50
0160 end block		160	.250		54	12.45		66.45	78
0170 90 degree corner		160	.250		68	12.45		80.45	93.50
0180 45 degree corner		160	.250		56	12.45		68.45	80.50
0200 12" x 12"		175	.229		27.50	11.35		38.85	47
0210 4" x 8"		160	.250		35.50	12.45		47.95	57.50
0220 6" x 8"		160	.250		21	12.45		33.45	41.50
0300 1,000 to 5,000 S.F., 6" x 6"	↓	135	.296		26	14.75		40.75	50.50

For customer support on your Concrete & Masonry Costs with RSMeans Data, call 800.448.8182.

109

04 23 13 – Vertical Glass Unit Masonry

04 23 13.10 Glass Block		Crew	Daily Output	Labor-Hours	Unit	Material	2021 Bare Costs Labor	Equipment	Total	Total Incl O&P
0350	8" x 8"	D-8	190	.211	S.F.	15.85	10.45		26.30	33
0400	12" x 12"		215	.186		27	9.25		36.25	43.50
0410	4" x 8"		215	.186		34.50	9.25		43.75	52
0420	6" x 8"		215	.186		20.50	9.25		29.75	36.50
0500	Over 5,000 S.F., 6" x 6"		145	.276		25	13.70		38.70	48
0550	8" x 8"		215	.186		15.35	9.25		24.60	31
0600	12" x 12"		240	.167		26	8.30		34.30	41
0610	4" x 8"		240	.167		33.50	8.30		41.80	49.50
0620	6" x 8"	▼	240	.167	▼	19.90	8.30		28.20	34.50
0700	For solar reflective blocks, add					100%				
1000	Thinline, plain, 3-1/8" thick, under 1,000 S.F., 6" x 6"	D-8	115	.348	S.F.	24.50	17.30		41.80	53
1050	8" x 8"		160	.250		15.75	12.45		28.20	36
1200	Over 5,000 S.F., 6" x 6"		145	.276		28.50	13.70		42.20	51.50
1250	8" x 8"		215	.186		15.80	9.25		25.05	31.50
1400	For cleaning block after installation (both sides), add	▼	1000	.040	▼	.16	1.99		2.15	3.18
4000	Accessories									
4100	Anchors, 20 ga. galv., 1-3/4" wide x 24" long				Ea.	6.25			6.25	6.90
4200	Emulsion asphalt				Gal.	11.30			11.30	12.45
4300	Expansion joint, fiberglass				L.F.	.76			.76	.84
4400	Steel mesh, double galvanized				"	1.06			1.06	1.17

04 24 16 – Manufactured Adobe Unit Masonry

04 24 16.06 Adobe Brick

	04 24 16.06 Adobe Brick		Crew	Daily Output	Labor-Hours	Unit	Material	2021 Bare Costs Labor	Equipment	Total	Total Incl O&P
0010	**ADOBE BRICK**, Semi-stabilized, with cement mortar										
0060	Brick, 10" x 4" x 14", 2.6/S.F.	G	D-8	560	.071	S.F.	6.85	3.55		10.40	12.90
0080	12" x 4" x 16", 2.3/S.F.	G		580	.069		7.40	3.43		10.83	13.30
0100	10" x 4" x 16", 2.3/S.F.	G		590	.068		7.60	3.37		10.97	13.45
0120	8" x 4" x 16", 2.3/S.F.	G		560	.071		5.30	3.55		8.85	11.20
0140	4" x 4" x 16", 2.3/S.F.	G		540	.074		4.31	3.68		7.99	10.30
0160	6" x 4" x 16", 2.3/S.F.	G		540	.074		3.97	3.68		7.65	9.90
0180	4" x 4" x 12", 3.0/S.F.	G		520	.077		5.50	3.82		9.32	11.80
0200	8" x 4" x 12", 3.0/S.F.	G	▼	520	.077	▼	4.45	3.82		8.27	10.65

04 25 20 – Pre-Fabricated Masonry Panels

04 25 20.10 Brick and Epoxy Mortar Panels

	04 25 20.10 Brick and Epoxy Mortar Panels	Crew	Daily Output	Labor-Hours	Unit	Material	2021 Bare Costs Labor	Equipment	Total	Total Incl O&P
0010	**BRICK AND EPOXY MORTAR PANELS**									
0020	Prefabricated brick & epoxy mortar, 4" thick, minimum	C-11	775	.093	S.F.	8.40	5.55	3	16.95	21
0100	Maximum	"	500	.144		9.85	8.60	4.65	23.10	29
0200	For 2" concrete back-up, add					50%				
0300	For 1" urethane & 3" concrete back-up, add				▼	70%				

04 27 Multiple-Wythe Unit Masonry

04 27 10 – Multiple-Wythe Masonry

04 27 10.20 Cavity Walls

	Crew	Daily Output	Labor-Hours	Unit	Material	2021 Bare Costs Labor	Equipment	Total	Total Incl O&P
0010 **CAVITY WALLS**, brick and CMU, includes joint reinforcing and ties									
0200 4" face brick, 4" block	D-8	165	.242	S.F.	7.35	12.05		19.40	26.50
0400 6" block		145	.276		7.80	13.70		21.50	29
0600 8" block	↓	125	.320	↓	7.60	15.90		23.50	32.50

04 27 10.30 Brick Walls

	Crew	Daily Output	Labor-Hours	Unit	Material	2021 Bare Costs Labor	Equipment	Total	Total Incl O&P
0010 **BRICK WALLS**, including mortar, excludes scaffolding R042110-20									
0020 Estimating by number of brick									
0140 Face brick, 4" thick wall, 6.75 brick/S.F.	D-8	1.45	27.586	M	740	1,375		2,115	2,900
0150 Common brick, 4" thick wall, 6.75 brick/S.F. R042110-50		1.60	25		720	1,250		1,970	2,675
0204 8" thick, 13.50 brick/S.F.		1.80	22.222		740	1,100		1,840	2,500
0250 12" thick, 20.25 brick/S.F.		1.90	21.053		745	1,050		1,795	2,400
0304 16" thick, 27.00 brick/S.F.		2	20		755	995		1,750	2,325
0500 Reinforced, face brick, 4" thick wall, 6.75 brick/S.F.		1.40	28.571		760	1,425		2,185	3,000
0520 Common brick, 4" thick wall, 6.75 brick/S.F.		1.55	25.806		745	1,275		2,020	2,750
0550 8" thick, 13.50 brick/S.F.		1.75	22.857		765	1,125		1,890	2,550
0600 12" thick, 20.25 brick/S.F.		1.85	21.622		770	1,075		1,845	2,475
0650 16" thick, 27.00 brick/S.F.	↓	1.95	20.513	↓	775	1,025		1,800	2,375
0790 Alternate method of figuring by square foot									
0800 Face brick, 4" thick wall, 6.75 brick/S.F.	D-8	215	.186	S.F.	4.99	9.25		14.24	19.45
0850 Common brick, 4" thick wall, 6.75 brick/S.F.		240	.167		4.87	8.30		13.17	17.85
0900 8" thick, 13.50 brick/S.F.		135	.296		10	14.75		24.75	33
1000 12" thick, 20.25 brick/S.F.		95	.421		15.05	21		36.05	48
1050 16" thick, 27.00 brick/S.F.		75	.533		20.50	26.50		47	62.50
1200 Reinforced, face brick, 4" thick wall, 6.75 brick/S.F.		210	.190		5.15	9.45		14.60	19.90
1220 Common brick, 4" thick wall, 6.75 brick/S.F.		235	.170		5	8.45		13.45	18.25
1250 8" thick, 13.50 brick/S.F.		130	.308		10.30	15.30		25.60	34.50
1260 8" thick, 2.25 brick/S.F.		130	.308		1.89	15.30		17.19	25
1300 12" thick, 20.25 brick/S.F.		90	.444		15.55	22		37.55	50.50
1350 16" thick, 27.00 brick/S.F.	↓	70	.571	↓	21	28.50		49.50	66

04 27 10.40 Steps

	Crew	Daily Output	Labor-Hours	Unit	Material	2021 Bare Costs Labor	Equipment	Total	Total Incl O&P
0010 **STEPS**									
0012 Entry steps, select common brick	D-1	.30	53.333	M	580	2,600		3,180	4,575

04 41 Dry-Placed Stone

04 41 10 – Dry Placed Stone

04 41 10.10 Rough Stone Wall

		Crew	Daily Output	Labor-Hours	Unit	Material	2021 Bare Costs Labor	Equipment	Total	Total Incl O&P
0011 **ROUGH STONE WALL**, Dry										
0012 Dry laid (no mortar), under 18" thick	G	D-1	60	.267	C.F.	14.75	13		27.75	36
0100 Random fieldstone, under 18" thick	G	D-12	60	.533		14.75	26		40.75	56
0150 Over 18" thick	G	"	63	.508	↓	17.70	25		42.70	57
0500 Field stone veneer	G	D-8	120	.333	S.F.	14	16.55		30.55	40.50
0510 Valley stone veneer	G		120	.333		14	16.55		30.55	40.50
0520 River stone veneer	G	↓	120	.333	↓	14	16.55		30.55	40.50
0600 Rubble stone walls, in mortar bed, up to 18" thick	G	D-11	75	.320	C.F.	17.80	16.35		34.15	44

04 43 10 – Masonry with Natural and Processed Stone

04 43 10.05 Ashlar Veneer	Crew	Daily Output	Labor-Hours	Unit	Material	2021 Bare Costs Labor	Equipment	Total	Total Incl O&P
0011	**ASHLAR VENEER** +/- 4" thk, random or random rectangular								
0150	Sawn face, split joints, low priced stone	D-8	140	.286	S.F.	13	14.20		27.20	36
0200	Medium priced stone		130	.308		15.55	15.30		30.85	40
0300	High priced stone		120	.333		18.75	16.55		35.30	45.50
0600	Seam face, split joints, medium price stone		125	.320		18.45	15.90		34.35	44.50
0700	High price stone		120	.333		17.95	16.55		34.50	45
1000	Split or rock face, split joints, medium price stone		125	.320		10.45	15.90		26.35	35.50
1100	High price stone		120	.333		18.30	16.55		34.85	45

04 43 10.10 Bluestone

		Crew	Daily Output	Labor-Hours	Unit	Material	Labor	Equipment	Total	Total Incl O&P
0010	**BLUESTONE**, cut to size									
0100	Paving, natural cleft, to 4', 1" thick	D-8	150	.267	S.F.	8.15	13.25		21.40	29
0150	1-1/2" thick		145	.276		8.10	13.70		21.80	29.50
0200	Smooth finish, 1" thick		150	.267		7.85	13.25		21.10	28.50
0250	1-1/2" thick		145	.276		9.15	13.70		22.85	30.50
0300	Thermal finish, 1" thick		150	.267		7.60	13.25		20.85	28.50
0350	1-1/2" thick		145	.276		8.10	13.70		21.80	29.50
0500	Sills, natural cleft, 10" wide to 6' long, 1-1/2" thick	D-11	70	.343	L.F.	13.50	17.50		31	41.50
0550	2" thick		63	.381		14.75	19.45		34.20	46
0600	Smooth finish, 1-1/2" thick		70	.343		14.45	17.50		31.95	42.50
0650	2" thick		63	.381		17.10	19.45		36.55	48.50
0800	Thermal finish, 1-1/2" thick		70	.343		12.75	17.50		30.25	40.50
0850	2" thick		63	.381		10.95	19.45		30.40	41.50
1000	Stair treads, natural cleft, 12" wide, 6' long, 1-1/2" thick	D-10	115	.278		15.60	14.95	3.76	34.31	44
1050	2" thick		105	.305		17	16.35	4.12	37.47	47.50
1100	Smooth finish, 1-1/2" thick		115	.278		14.60	14.95	3.76	33.31	42.50
1150	2" thick		105	.305		15.45	16.35	4.12	35.92	46
1300	Thermal finish, 1-1/2" thick		115	.278		15	14.95	3.76	33.71	43
1350	2" thick		105	.305		14.30	16.35	4.12	34.77	45
2000	Coping, finished top & 2 sides, 12" to 6'									
2100	Natural cleft, 1-1/2" thick	D-10	115	.278	L.F.	12.95	14.95	3.76	31.66	41
2150	2" thick		105	.305		15.05	16.35	4.12	35.52	45.50
2200	Smooth finish, 1-1/2" thick		115	.278		12.90	14.95	3.76	31.61	41
2250	2" thick		105	.305		15.05	16.35	4.12	35.52	45.50
2300	Thermal finish, 1-1/2" thick		115	.278		12.95	14.95	3.76	31.66	41
2350	2" thick		105	.305		15.05	16.35	4.12	35.52	45.50

04 43 10.45 Granite

		Crew	Daily Output	Labor-Hours	Unit	Material	Labor	Equipment	Total	Total Incl O&P
0010	**GRANITE**, cut to size									
0050	Veneer, polished face, 3/4" to 1-1/2" thick									
0150	Low price, gray, light gray, etc.	D-10	130	.246	S.F.	25	13.20	3.33	41.53	51
0180	Medium price, pink, brown, etc.		130	.246		28.50	13.20	3.33	45.03	55
0220	High price, red, black, etc.		130	.246		43	13.20	3.33	59.53	70.50
0300	1-1/2" to 2-1/2" thick, veneer									
0350	Low price, gray, light gray, etc.	D-10	130	.246	S.F.	29	13.20	3.33	45.53	55.50
0500	Medium price, pink, brown, etc.		130	.246		34	13.20	3.33	50.53	61
0550	High price, red, black, etc.		130	.246		53	13.20	3.33	69.53	81.50
0700	2-1/2" to 4" thick, veneer									
0750	Low price, gray, light gray, etc.	D-10	110	.291	S.F.	39	15.60	3.93	58.53	71
0850	Medium price, pink, brown, etc.		110	.291		44.50	15.60	3.93	64.03	77
0950	High price, red, black, etc.		110	.291		58.50	15.60	3.93	78.03	92
1000	For bush hammered finish, deduct					5%				
1050	Coarse rubbed finish, deduct					10%				
1100	Honed finish, deduct					5%				

04 43 10.45 Granite

		Crew	Daily Output	Labor-Hours	Unit	Material	2021 Bare Costs Labor	Equipment	Total	Total Incl O&P
1150	Thermal finish, deduct				S.F.	18%				
1800	Carving or bas-relief, from templates or plaster molds									
1850	Low price, gray, light gray, etc.	D-10	80	.400	C.F.	164	21.50	5.40	190.90	219
1875	Medium price, pink, brown, etc.		80	.400		345	21.50	5.40	371.90	420
1900	High price, red, black, etc.		80	.400		530	21.50	5.40	556.90	620
2000	Intricate or hand finished pieces									
2010	Mouldings, radius cuts, bullnose edges, etc.									
2050	Add for low price gray, light gray, etc.					30%				
2075	Add for medium price, pink, brown, etc.					165%				
2100	Add for high price red, black, etc.					300%				
2450	For radius under 5', add				L.F.	100%				
2500	Steps, copings, etc., finished on more than one surface									
2550	Low price, gray, light gray, etc.	D-10	50	.640	C.F.	92	34.50	8.65	135.15	162
2575	Medium price, pink, brown, etc.		50	.640		120	34.50	8.65	163.15	193
2600	High price, red, black, etc.		50	.640		147	34.50	8.65	190.15	223
2700	Pavers, Belgian block, 8"-13" long, 4"-6" wide, 4"-6" deep	D-11	120	.200	S.F.	25	10.20		35.20	42.50
2800	Pavers, 4" x 4" x 4" blocks, split face and joints									
2850	Low price, gray, light gray, etc.	D-11	80	.300	S.F.	12.10	15.30		27.40	36.50
2875	Medium price, pink, brown, etc.		80	.300		19.45	15.30		34.75	44.50
2900	High price, red, black, etc.		80	.300		27	15.30		42.30	52.50
3000	Pavers, 4" x 4" x 4", thermal face, sawn joints									
3050	Low price, gray, light gray, etc.	D-11	65	.369	S.F.	24.50	18.85		43.35	55.50
3075	Medium price, pink, brown, etc.		65	.369		28.50	18.85		47.35	60
3100	High price, red, black, etc.		65	.369		32.50	18.85		51.35	64
4000	Soffits, 2" thick, low price, gray, light gray	D-13	35	1.371		38	71.50	12.35	121.85	163
4050	Medium price, pink, brown, etc.		35	1.371		61	71.50	12.35	144.85	188
4100	High price, red, black, etc.		35	1.371		84	71.50	12.35	167.85	213
4200	Low price, gray, light gray, etc.		35	1.371		63.50	71.50	12.35	147.35	191
4250	Medium price, pink, brown, etc.		35	1.371		86.50	71.50	12.35	170.35	216
4300	High price, red, black, etc.		35	1.371		109	71.50	12.35	192.85	241
5000	Reclaimed or antique									
5010	Treads, up to 12" wide	D-10	100	.320	L.F.	24	17.15	4.33	45.48	57.50
5020	Up to 18" wide		100	.320		45.50	17.15	4.33	66.98	81
5030	Capstone, size varies		50	.640		26.50	34.50	8.65	69.65	90
5040	Posts		30	1.067	V.L.F.	31.50	57	14.40	102.90	136

04 43 10.50 Lightweight Natural Stone

		Crew	Daily Output	Labor-Hours	Unit	Material	2021 Bare Costs Labor	Equipment	Total	Total Incl O&P
0011	**LIGHTWEIGHT NATURAL STONE** Lava type									
0100	Veneer, rubble face, sawed back, irregular shapes [G]	D-10	130	.246	S.F.	9.15	13.20	3.33	25.68	33.50
0200	Sawed face and back, irregular shapes [G]		130	.246	"	9.15	13.20	3.33	25.68	33.50
1000	Reclaimed or antique, barn or foundation stone		1	32	Ton	180	1,725	435	2,340	3,250

04 43 10.55 Limestone

		Crew	Daily Output	Labor-Hours	Unit	Material	2021 Bare Costs Labor	Equipment	Total	Total Incl O&P
0010	**LIMESTONE**, cut to size									
0020	Veneer facing panels									
0500	Texture finish, light stick, 4-1/2" thick, 5' x 12'	D-4	300	.107	S.F.	26	5.25	.63	31.88	37.50
0750	5" thick, 5' x 14' panels	D-10	275	.116		28	6.25	1.57	35.82	41.50
1000	Sugarcube finish, 2" thick, 3' x 5' panels		275	.116		31	6.25	1.57	38.82	45
1050	3" thick, 4' x 9' panels		275	.116		32	6.25	1.57	39.82	46
1200	4" thick, 5' x 11' panels		275	.116		35	6.25	1.57	42.82	49.50
1400	Sugarcube, textured finish, 4-1/2" thick, 5' x 12'		275	.116		38.50	6.25	1.57	46.32	53
1450	5" thick, 5' x 14' panels		275	.116		39	6.25	1.57	46.82	54
2000	Coping, sugarcube finish, top & 2 sides		30	1.067	C.F.	66.50	57	14.40	137.90	175
2100	Sills, lintels, jambs, trim, stops, sugarcube finish, simple		20	1.600		66.50	86	21.50	174	226

04 43 10.55 Limestone

		Crew	Daily Output	Labor-Hours	Unit	Material	2021 Bare Costs Labor	Equipment	Total	Total Incl O&P
2150	Detailed	D-10	20	1.600	C.F.	66.50	86	21.50	174	226
2300	Steps, extra hard, 14" wide, 6" rise	↓	50	.640	L.F.	29	34.50	8.65	72.15	92.50
3000	Quoins, plain finish, 6" x 12" x 12"	D-12	25	1.280	Ea.	54.50	63		117.50	155
3050	6" x 16" x 24"	"	25	1.280	"	73	63		136	175

04 43 10.60 Marble

		Crew	Daily Output	Labor-Hours	Unit	Material	2021 Bare Costs Labor	Equipment	Total	Total Incl O&P
0011	**MARBLE**, ashlar, split face, +/- 4" thick, random									
0040	Lengths 1' to 4' & heights 2" to 7-1/2", average	D-8	175	.229	S.F.	16.85	11.35		28.20	35.50
0100	Base, polished, 3/4" or 7/8" thick, polished, 6" high	D-10	65	.492	L.F.	11.25	26.50	6.65	44.40	59
0300	Carvings or bas-relief, from templates, simple design		80	.400	S.F.	157	21.50	5.40	183.90	210
0350	Intricate design	↓	80	.400	"	345	21.50	5.40	371.90	420
0600	Columns, cornices, mouldings, etc.									
0650	Hand or special machine cut, simple design	D-10	35	.914	C.F.	71	49	12.35	132.35	165
0700	Intricate design	"	35	.914	"	330	49	12.35	391.35	445
1000	Facing, polished finish, cut to size, 3/4" to 7/8" thick									
1050	Carrara or equal	D-10	130	.246	S.F.	35	13.20	3.33	51.53	62
1100	Arabescato or equal		130	.246		37	13.20	3.33	53.53	64
1300	1-1/4" thick, Botticino Classico or equal		125	.256		22.50	13.75	3.46	39.71	49.50
1350	Statuarietto or equal		125	.256		37.50	13.75	3.46	54.71	66
1500	2" thick, Crema Marfil or equal		120	.267		48	14.30	3.61	65.91	78.50
1550	Cafe Pinta or equal	↓	120	.267	↓	66.50	14.30	3.61	84.41	98.50
1700	Rubbed finish, cut to size, 4" thick									
1740	Average	D-10	100	.320	S.F.	41.50	17.15	4.33	62.98	76.50
1780	Maximum	"	100	.320	"	77.50	17.15	4.33	98.98	116
2200	Window sills, 6" x 3/4" thick	D-1	85	.188	L.F.	13.10	9.15		22.25	28
2500	Flooring, polished tiles, 12" x 12" x 3/8" thick									
2510	Thin set, Giallo Solare or equal	D-11	90	.267	S.F.	14.75	13.60		28.35	36.50
2600	Sky Blue or equal		90	.267		11.35	13.60		24.95	33
2700	Mortar bed, Giallo Solare or equal		65	.369		14.85	18.85		33.70	45
2740	Sky Blue or equal	↓	65	.369		11.35	18.85		30.20	41
2780	Travertine, 3/8" thick, Sierra or equal	D-10	130	.246		11.50	13.20	3.33	28.03	36
2790	Silver or equal	"	130	.246		30.50	13.20	3.33	47.03	57
2800	Patio tile, non-slip, 1/2" thick, flame finish	D-11	75	.320	↓	13.45	16.35		29.80	39.50
2900	Shower or toilet partitions, 7/8" thick partitions									
3050	3/4" or 1-1/4" thick stiles, polished 2 sides, average	D-11	75	.320	S.F.	48.50	16.35		64.85	77.50
3201	Soffits, add to above prices				"	20%	100%			
3210	Stairs, risers, 7/8" thick x 6" high	D-10	115	.278	L.F.	14.95	14.95	3.76	33.66	43
3360	Treads, 12" wide x 1-1/4" thick	"	115	.278	"	46	14.95	3.76	64.71	77
3500	Thresholds, 3' long, 7/8" thick, 4" to 5" wide, plain	D-12	24	1.333	Ea.	34.50	65.50		100	137
3550	Beveled		24	1.333	"	76	65.50		141.50	183
3700	Window stools, polished, 7/8" thick, 5" wide	↓	85	.376	L.F.	19.65	18.50		38.15	49.50

04 43 10.75 Sandstone or Brownstone

		Crew	Daily Output	Labor-Hours	Unit	Material	2021 Bare Costs Labor	Equipment	Total	Total Incl O&P
0011	**SANDSTONE OR BROWNSTONE**									
0100	Sawed face veneer, 2-1/2" thick, to 2' x 4' panels	D-10	130	.246	S.F.	21.50	13.20	3.33	38.03	47.50
0150	4" thick, to 3'-6" x 8' panels		100	.320		21.50	17.15	4.33	42.98	55
0300	Split face, random sizes	↓	100	.320	↓	15	17.15	4.33	36.48	47.50
0350	Cut stone trim (limestone)									
0360	Ribbon stone, 4" thick, 5' pieces	D-8	120	.333	Ea.	159	16.55		175.55	200
0370	Cove stone, 4" thick, 5' pieces		105	.381		160	18.95		178.95	205
0380	Cornice stone, 10" to 12" wide		90	.444		198	22		220	251
0390	Band stone, 4" thick, 5' pieces		145	.276		107	13.70		120.70	139
0410	Window and door trim, 3" to 4" wide		160	.250		91	12.45		103.45	119
0420	Key stone, 18" long	↓	60	.667	↓	91.50	33		124.50	150

114

For customer support on your Concrete & Masonry Costs with RSMeans Data, call 800.448.8182.

04 43 10.80 Slate	Crew	Daily Output	Labor-Hours	Unit	Material	2021 Bare Costs Labor	Equipment	Total	Total Incl O&P
0010 **SLATE**									
0040 Pennsylvania - blue gray to black									
0050 Vermont - unfading green, mottled green & purple, gray & purple									
0100 Virginia - blue black									
0200 Exterior paving, natural cleft, 1" thick									
0250 6" x 6" Pennsylvania	D-12	100	.320	S.F.	7.05	15.75		22.80	31.50
0300 Vermont		100	.320		11.40	15.75		27.15	36
0350 Virginia		100	.320		14.70	15.75		30.45	39.50
0500 24" x 24", Pennsylvania		120	.267		13.55	13.10		26.65	34.50
0550 Vermont		120	.267		28.50	13.10		41.60	51
0600 Virginia		120	.267		21.50	13.10		34.60	43.50
0700 18" x 30" Pennsylvania		120	.267		15.35	13.10		28.45	36.50
0750 Vermont		120	.267		28.50	13.10		41.60	51
0800 Virginia		120	.267		19	13.10		32.10	41
1000 Interior flooring, natural cleft, 1/2" thick									
1100 6" x 6" Pennsylvania	D-12	100	.320	S.F.	4.16	15.75		19.91	28
1150 Vermont		100	.320		10.05	15.75		25.80	34.50
1200 Virginia		100	.320		11.60	15.75		27.35	36.50
1300 24" x 24" Pennsylvania		120	.267		8.10	13.10		21.20	28.50
1350 Vermont		120	.267		23	13.10		36.10	45
1400 Virginia		120	.267		15.35	13.10		28.45	36.50
1500 18" x 24" Pennsylvania		120	.267		8.10	13.10		21.20	28.50
1550 Vermont		120	.267		16.85	13.10		29.95	38.50
1600 Virginia		120	.267		15.55	13.10		28.65	37
2000 Facing panels, 1-1/4" thick, to 4' x 4' panels									
2100 Natural cleft finish, Pennsylvania	D-10	180	.178	S.F.	35	9.55	2.40	46.95	55.50
2110 Vermont		180	.178		28.50	9.55	2.40	40.45	48
2120 Virginia		180	.178		34.50	9.55	2.40	46.45	55
2150 Sand rubbed finish, surface, add					10.60			10.60	11.65
2200 Honed finish, add					7.65			7.65	8.45
2500 Ribbon, natural cleft finish, 1" thick, to 9 S.F.	D-10	80	.400		13.65	21.50	5.40	40.55	53
2550 Sand rubbed finish		80	.400		18.50	21.50	5.40	45.40	58.50
2600 Honed finish		80	.400		17.20	21.50	5.40	44.10	57
2700 1-1/2" thick		78	.410		17.75	22	5.55	45.30	58.50
2750 Sand rubbed finish		78	.410		23.50	22	5.55	51.05	65
2800 Honed finish		78	.410		22	22	5.55	49.55	63.50
2850 2" thick		76	.421		21.50	22.50	5.70	49.70	64
2900 Sand rubbed finish		76	.421		29.50	22.50	5.70	57.70	73
2950 Honed finish		76	.421		27	22.50	5.70	55.20	70.50
3100 Stair landings, 1" thick, black, clear	D-1	65	.246		21	12		33	41
3200 Ribbon	"	65	.246		23	12		35	43
3500 Stair treads, sand finish, 1" thick x 12" wide									
3550 Under 3 L.F.	D-10	85	.376	L.F.	23	20	5.10	48.10	61
3600 3 L.F. to 6 L.F.	"	120	.267	"	25	14.30	3.61	42.91	52.50
3700 Ribbon, sand finish, 1" thick x 12" wide									
3750 To 6 L.F.	D-10	120	.267	L.F.	21	14.30	3.61	38.91	48.50
4000 Stools or sills, sand finish, 1" thick, 6" wide	D-12	160	.200		12	9.85		21.85	28
4100 Honed finish		160	.200		11.45	9.85		21.30	27.50
4200 10" wide		90	.356		18.50	17.50		36	47
4250 Honed finish		90	.356		17.20	17.50		34.70	45.50
4400 2" thick, 6" wide		140	.229		19.30	11.25		30.55	38
4450 Honed finish		140	.229		18.40	11.25		29.65	37.50

04 43 Stone Masonry

04 43 10 – Masonry with Natural and Processed Stone

04 43 10.80 Slate

		Crew	Daily Output	Labor-Hours	Unit	Material	2021 Bare Costs Labor	Equipment	Total	Total Incl O&P
4600	10" wide	D-12	90	.356	L.F.	30	17.50		47.50	59.50
4650	Honed finish	↓	90	.356	↓	28.50	17.50		46	58
4800	For lengths over 3', add					25%				

04 43 10.85 Window Sill

		Crew	Daily Output	Labor-Hours	Unit	Material	2021 Bare Costs Labor	Equipment	Total	Total Incl O&P
0010	**WINDOW SILL**									
0020	Bluestone, thermal top, 10" wide, 1-1/2" thick	D-1	85	.188	S.F.	8.40	9.15		17.55	23
0050	2" thick		75	.213	"	12.95	10.40		23.35	30
0100	Cut stone, 5" x 8" plain		48	.333	L.F.	12.80	16.25		29.05	38.50
0200	Face brick on edge, brick, 8" wide		80	.200		3.53	9.75		13.28	18.60
0400	Marble, 9" wide, 1" thick		85	.188		9.10	9.15		18.25	24
0900	Slate, colored, unfading, honed, 12" wide, 1" thick		85	.188		9.30	9.15		18.45	24
0950	2" thick	↓	70	.229	↓	11.30	11.15		22.45	29

04 51 Flue Liner Masonry

04 51 10 – Clay Flue Lining

04 51 10.10 Flue Lining

		Crew	Daily Output	Labor-Hours	Unit	Material	2021 Bare Costs Labor	Equipment	Total	Total Incl O&P
0010	**FLUE LINING**, including mortar									
0020	Clay, 8" x 8"	D-1	125	.128	V.L.F.	5.25	6.25		11.50	15.20
0100	8" x 12"		103	.155		9.80	7.55		17.35	22
0200	12" x 12"		93	.172		12.30	8.40		20.70	26
0300	12" x 18"		84	.190		26	9.30		35.30	42.50
0400	18" x 18"		75	.213		29.50	10.40		39.90	48
0500	20" x 20"		66	.242		46	11.80		57.80	69
0600	24" x 24"		56	.286		68	13.90		81.90	96
1000	Round, 18" diameter		66	.242		43	11.80		54.80	65.50
1100	24" diameter	↓	47	.340	↓	89	16.60		105.60	123

04 54 Refractory Brick Masonry

04 54 10 – Refractory Brick Work

04 54 10.10 Fire Brick

		Crew	Daily Output	Labor-Hours	Unit	Material	2021 Bare Costs Labor	Equipment	Total	Total Incl O&P
0010	**FIRE BRICK**									
0012	Low duty, 2000°F, 9" x 2-1/2" x 4-1/2"	D-1	.60	26.667	M	1,750	1,300		3,050	3,875
0050	High duty, 3000°F	"	.60	26.667	"	3,150	1,300		4,450	5,400

04 54 10.20 Fire Clay

		Crew	Daily Output	Labor-Hours	Unit	Material	2021 Bare Costs Labor	Equipment	Total	Total Incl O&P
0010	**FIRE CLAY**									
0020	Gray, high duty, 100 lb. bag				Bag	28.50			28.50	31
0050	100 lb. drum, premixed (400 brick per drum)				Drum	44			44	48.50

04 57 Masonry Fireplaces

04 57 10 – Brick or Stone Fireplaces

04 57 10.10 Fireplace

		Crew	Daily Output	Labor-Hours	Unit	Material	2021 Bare Costs Labor	Equipment	Total	Total Incl O&P
0010	**FIREPLACE**									
0100	Brick fireplace, not incl. foundations or chimneys									
0110	30" x 29" opening, incl. chamber, plain brickwork	D-1	.40	40	Ea.	630	1,950		2,580	3,625
0200	Fireplace box only (110 brick)	"	2	8	"	162	390		552	765
0300	For elaborate brickwork and details, add					35%	35%			
0400	For hearth, brick & stone, add	D-1	2	8	Ea.	214	390		604	820
0410	For steel, damper, cleanouts, add		4	4		18.10	195		213.10	315
0600	Plain brickwork, incl. metal circulator		.50	32	↓	1,075	1,550		2,625	3,525
0800	Face brick only, standard size, 8" x 2-2/3" x 4"	↓	.30	53.333	M	615	2,600		3,215	4,600

04 71 Manufactured Brick Masonry

04 71 10 – Simulated or Manufactured Brick

04 71 10.10 Simulated Brick

		Crew	Daily Output	Labor-Hours	Unit	Material	2021 Bare Costs Labor	Equipment	Total	Total Incl O&P
0010	**SIMULATED BRICK**									
0020	Aluminum, baked on colors	1 Carp	200	.040	S.F.	4.53	2.19		6.72	8.25
0050	Fiberglass panels		200	.040		9.90	2.19		12.09	14.15
0100	Urethane pieces cemented in mastic		150	.053		8.30	2.92		11.22	13.45
0150	Vinyl siding panels	↓	200	.040		11.10	2.19		13.29	15.45
0160	Cement base, brick, incl. mastic	D-1	100	.160	↓	9	7.80		16.80	21.50
0170	Corner		50	.320	V.L.F.	22	15.60		37.60	48
0180	Stone face, incl. mastic		100	.160	S.F.	10.05	7.80		17.85	23
0190	Corner	↓	50	.320	V.L.F.	3.50	15.60		19.10	27.50

04 72 Cast Stone Masonry

04 72 10 – Cast Stone Masonry Features

04 72 10.10 Coping

		Crew	Daily Output	Labor-Hours	Unit	Material	2021 Bare Costs Labor	Equipment	Total	Total Incl O&P
0010	**COPING**, stock units									
0050	Precast concrete, 10" wide, 4" tapers to 3-1/2", 8" wall	D-1	75	.213	L.F.	25	10.40		35.40	43
0100	12" wide, 3-1/2" tapers to 3", 10" wall		70	.229		27	11.15		38.15	46.50
0110	14" wide, 4" tapers to 3-1/2", 12" wall		65	.246		36.50	12		48.50	58
0150	16" wide, 4" tapers to 3-1/2", 14" wall		60	.267	↓	39	13		52	62.50
0250	Precast concrete corners		40	.400	Ea.	51	19.50		70.50	85.50
0300	Limestone for 12" wall, 4" thick		90	.178	L.F.	16.90	8.65		25.55	31.50
0350	6" thick		80	.200		23.50	9.75		33.25	40.50
0500	Marble, to 4" thick, no wash, 9" wide		90	.178		10.40	8.65		19.05	24.50
0550	12" wide		80	.200		18.50	9.75		28.25	35
0700	Terra cotta, 9" wide		90	.178		8.10	8.65		16.75	22
0750	12" wide		80	.200		8.20	9.75		17.95	23.50
0800	Aluminum, for 12" wall	↓	80	.200	↓	9.45	9.75		19.20	25

04 72 20 – Cultured Stone Veneer

04 72 20.10 Cultured Stone Veneer Components

		Crew	Daily Output	Labor-Hours	Unit	Material	2021 Bare Costs Labor	Equipment	Total	Total Incl O&P
0010	**CULTURED STONE VENEER COMPONENTS**									
0110	On wood frame and sheathing substrate, random sized cobbles, corner stones	D-8	70	.571	V.L.F.	8.45	28.50		36.95	52.50
0120	Field stones		140	.286	S.F.	5.90	14.20		20.10	28
0130	Random sized flats, corner stones		70	.571	V.L.F.	8.65	28.50		37.15	52.50
0140	Field stones		140	.286	S.F.	9.20	14.20		23.40	31.50
0150	Horizontal lined ledgestones, corner stones		75	.533	V.L.F.	8.35	26.50		34.85	49
0160	Field stones		150	.267	S.F.	6.30	13.25		19.55	27
0170	Random shaped flats, corner stones		65	.615	V.L.F.	8.40	30.50		38.90	55

04 72 Cast Stone Masonry

04 72 20 – Cultured Stone Veneer

04 72 20.10 Cultured Stone Veneer Components	Crew	Daily Output	Labor-Hours	Unit	Material	2021 Bare Costs Labor	Equipment	Total	Total Incl O&P
0180	Field stones	D-8	150	.267	S.F.	6.15	13.25		19.40	27
0190	Random shaped/textured face, corner stones		65	.615	V.L.F.	8.45	30.50		38.95	55.50
0200	Field stones		130	.308	S.F.	6.10	15.30		21.40	30
0210	Random shaped river rock, corner stones		65	.615	V.L.F.	8.45	30.50		38.95	55.50
0220	Field stones		130	.308	S.F.	6.10	15.30		21.40	30
0240	On concrete or CMU substrate, random sized cobbles, corner stones		70	.571	V.L.F.	7.60	28.50		36.10	51.50
0250	Field stones		140	.286	S.F.	5.45	14.20		19.65	27.50
0260	Random sized flats, corner stones		70	.571	V.L.F.	7.80	28.50		36.30	51.50
0270	Field stones		140	.286	S.F.	8.75	14.20		22.95	31
0280	Horizontal lined ledgestones, corner stones		75	.533	V.L.F.	7.50	26.50		34	48.50
0290	Field stones		150	.267	S.F.	5.85	13.25		19.10	26.50
0300	Random shaped flats, corner stones		70	.571	V.L.F.	7.50	28.50		36	51.50
0310	Field stones		140	.286	S.F.	5.75	14.20		19.95	28
0320	Random shaped/textured face, corner stones		65	.615	V.L.F.	7.60	30.50		38.10	54.50
0330	Field stones		130	.308	S.F.	5.70	15.30		21	29.50
0340	Random shaped river rock, corner stones		65	.615	V.L.F.	7.60	30.50		38.10	54.50
0350	Field stones		130	.308	S.F.	5.70	15.30		21	29.50
0360	Cultured stone veneer, #15 felt weather resistant barrier	1 Clab	3700	.002	Sq.	5.70	.10		5.80	6.40
0370	Expanded metal lath, diamond, 2.5 lb./S.Y., galvanized	1 Lath	85	.094	S.Y.	3.87	5.10		8.97	11.70
0390	Water table or window sill, 18" long	1 Bric	80	.100	Ea.	9.25	5.35		14.60	18.25

04 73 Manufactured Stone Masonry

04 73 20 – Simulated or Manufactured Stone

04 73 20.10 Simulated Stone

		Crew	Daily Output	Labor-Hours	Unit	Material	2021 Bare Costs Labor	Equipment	Total	Total Incl O&P
0010	**SIMULATED STONE**									
0100	Insulated fiberglass panels, 5/8" ply backer	L-4	200	.120	S.F.	11.60	6.25		17.85	22

Estimating Tips

05 05 00 Common Work Results for Metals

- Nuts, bolts, washers, connection angles, and plates can add a significant amount to both the tonnage of a structural steel job and the estimated cost. As a rule of thumb, add 10% to the total weight to account for these accessories.

- Type 2 steel construction, commonly referred to as "simple construction," consists generally of field-bolted connections with lateral bracing supplied by other elements of the building, such as masonry walls or x-bracing. The estimator should be aware, however, that shop connections may be accomplished by welding or bolting. The method may be particular to the fabrication shop and may have an impact on the estimated cost.

05 10 00 Structural Steel

- Steel items can be obtained from two sources: a fabrication shop or a metals service center. Fabrication shops can fabricate items under more controlled conditions than crews in the field can. They are also more efficient and can produce items more economically. Metal service centers serve as a source of long mill shapes to both fabrication shops and contractors.

- Most line items in this structural steel subdivision, and most items in 05 50 00 Metal Fabrications, are indicated as being shop fabricated. The bare material cost for these shop fabricated items is the "Invoice Cost" from the shop and includes the mill base price of steel plus mill extras, transportation to the shop, shop drawings and detailing where warranted, shop fabrication and handling, sandblasting and a shop coat of primer paint, all necessary structural bolts, and delivery to the job site. The bare labor cost and bare equipment cost for these shop fabricated items are for field installation or erection.

- Line items in Subdivision 05 12 23.40 Lightweight Framing, and other items scattered in Division 5, are indicated as being field fabricated. The bare material cost for these field fabricated items is the "Invoice Cost" from the metals service center and includes the mill base price of steel plus mill extras, transportation to the metals service center, material handling, and delivery of long lengths of mill shapes to the job site. Material costs for structural bolts and welding rods should be added to the estimate. The bare labor cost and bare equipment cost for these items are for both field fabrication and field installation or erection, and include time for cutting, welding, and drilling in the fabricated metal items. Drilling into concrete and fasteners to fasten field fabricated items to other work is not included and should be added to the estimate.

05 20 00 Steel Joist Framing

- In any given project the total weight of open web steel joists is determined by the loads to be supported and the design. However, economies can be realized in minimizing the amount of labor used to place the joists. This is done by maximizing the joist spacing and therefore minimizing the number of joists required to be installed on the job. Certain spacings and locations may be required by the design, but in other cases maximizing the spacing and keeping it as uniform as possible will keep the costs down.

05 30 00 Steel Decking

- The takeoff and estimating of a metal deck involve more than the area of the floor or roof and the type of deck specified or shown on the drawings. Many different sizes and types of openings may exist. Small openings

for individual pipes or conduits may be drilled after the floor/roof is installed, but larger openings may require special deck lengths as well as reinforcing or structural support. The estimator should determine who will be supplying this reinforcing. Additionally, some deck terminations are part of the deck package, such as screed angles and pour stops, and others will be part of the steel contract, such as angles attached to structural members and cast-in-place angles and plates. The estimator must ensure that all pieces are accounted for in the complete estimate.

05 50 00 Metal Fabrications

- The most economical steel stairs are those that use common materials, standard details, and most importantly, a uniform and relatively simple method of field assembly. Commonly available A36/A992 channels and plates are very good choices for the main stringers of the stairs, as are angles and tees for the carrier members. Risers and treads are usually made by specialty shops, and it is most economical to use a typical detail in as many places as possible. The stairs should be pre-assembled and shipped directly to the site. The field connections should be simple and straightforward enough to be accomplished efficiently, and with minimum equipment and labor.

Reference Numbers

Reference numbers are shown at the beginning of some major classifications. These numbers refer to related items in the Reference Section. The reference information may be an estimating procedure, an alternate pricing method, or technical information.

Note: Not all subdivisions listed here necessarily appear. ■

Same Data. Simplified.

Enjoy the convenience and efficiency of accessing your costs anywhere:

- **Skip the multiplier** by setting your location
- **Quickly search,** edit, favorite and share costs
- **Stay on top of price changes** with automatic updates

Discover more at rsmeans.com/online

05 05 05 – Selective Demolition for Metals

05 05 05.10 Selective Demolition, Metals

		Crew	Daily Output	Labor-Hours	Unit	Material	2021 Bare Costs Labor	2021 Bare Costs Equipment	Total	Total Incl O&P
0010	**SELECTIVE DEMOLITION, METALS** R024119-10									
0015	Excludes shores, bracing, cutting, loading, hauling, dumping									
0020	Remove nuts only up to 3/4" diameter	1 Sswk	480	.017	Ea.		1.01		1.01	1.56
0030	7/8" to 1-1/4" diameter		240	.033			2.01		2.01	3.11
0040	1-3/8" to 2" diameter		160	.050			3.02		3.02	4.67
0060	Unbolt and remove structural bolts up to 3/4" diameter		240	.033			2.01		2.01	3.11
0070	7/8" to 2" diameter		160	.050			3.02		3.02	4.67
0140	Light weight framing members, remove whole or cut up, up to 20 lb.		240	.033			2.01		2.01	3.11
0150	21-40 lb.	2 Sswk	210	.076			4.59		4.59	7.10
0160	41-80 lb.	3 Sswk	180	.133			8.05		8.05	12.45
0170	81-120 lb.	4 Sswk	150	.213			12.85		12.85	19.90
0230	Structural members, remove whole or cut up, up to 500 lb.	E-19	48	.500			29.50	32	61.50	80.50
0240	1/4-2 tons	E-18	36	1.111			67	42.50	109.50	150
0250	2-5 tons	E-24	30	1.067			64	19.55	83.55	120
0260	5-10 tons	E-20	24	2.667			159	88.50	247.50	340
0270	10-15 tons	E-2	18	3.111			186	95	281	390
0340	Fabricated item, remove whole or cut up, up to 20 lb.	1 Sswk	96	.083			5.05		5.05	7.80
0350	21-40 lb.	2 Sswk	84	.190			11.50		11.50	17.75
0360	41-80 lb.	3 Sswk	72	.333			20		20	31
0370	81-120 lb.	4 Sswk	60	.533			32		32	50
0380	121-500 lb.	E-19	48	.500			29.50	32	61.50	80.50
0390	501-1000 lb.	"	36	.667			39.50	42.50	82	107
0500	Steel roof decking, uncovered, bare	B-2	5000	.008	S.F.		.36		.36	.53

05 05 19 – Post-Installed Concrete Anchors

05 05 19.10 Chemical Anchors

		Crew	Daily Output	Labor-Hours	Unit	Material	2021 Bare Costs Labor	2021 Bare Costs Equipment	Total	Total Incl O&P
0010	**CHEMICAL ANCHORS**									
0020	Includes layout & drilling									
1430	Chemical anchor, w/rod & epoxy cartridge, 3/4" diameter x 9-1/2" long	B-89A	27	.593	Ea.	11.90	30	4.50	46.40	63
1435	1" diameter x 11-3/4" long		24	.667		24	34	5.05	63.05	82.50
1440	1-1/4" diameter x 14" long		21	.762		38.50	38.50	5.80	82.80	106
1445	1-3/4" diameter x 15" long		20	.800		70.50	40.50	6.10	117.10	145
1450	18" long		17	.941		84.50	48	7.15	139.65	172
1455	2" diameter x 18" long		16	1		148	51	7.60	206.60	247
1460	24" long		15	1.067		194	54	8.10	256.10	305

05 05 19.20 Expansion Anchors

		Crew	Daily Output	Labor-Hours	Unit	Material	2021 Bare Costs Labor	2021 Bare Costs Equipment	Total	Total Incl O&P
0010	**EXPANSION ANCHORS**									
0100	Anchors for concrete, brick or stone, no layout and drilling									
0200	Expansion shields, zinc, 1/4" diameter, 1-5/16" long, single G	1 Carp	90	.089	Ea.	.31	4.86		5.17	7.60
0300	1-3/8" long, double G		85	.094		.42	5.15		5.57	8.15
0400	3/8" diameter, 1-1/2" long, single G		85	.094		.51	5.15		5.66	8.25
0500	2" long, double G		80	.100		1.22	5.45		6.67	9.50
0600	1/2" diameter, 2-1/16" long, single G		80	.100		1.23	5.45		6.68	9.50
0700	2-1/2" long, double G		75	.107		2.25	5.85		8.10	11.20
0800	5/8" diameter, 2-5/8" long, single G		75	.107		1.97	5.85		7.82	10.85
0900	2-3/4" long, double G		70	.114		3.80	6.25		10.05	13.55
1000	3/4" diameter, 2-3/4" long, single G		70	.114		3.75	6.25		10	13.50
1100	3-15/16" long, double G		65	.123		5.55	6.75		12.30	16.15
2100	Hollow wall anchors for gypsum wall board, plaster or tile									
2300	1/8" diameter, short G	1 Carp	160	.050	Ea.	.31	2.74		3.05	4.42
2400	Long G		150	.053		.30	2.92		3.22	4.68
2500	3/16" diameter, short G		150	.053		.61	2.92		3.53	5
2600	Long G		140	.057		.71	3.13		3.84	5.45

05 05 19 – Post-Installed Concrete Anchors

05 05 19.20 Expansion Anchors

		Crew	Daily Output	Labor-Hours	Unit	Material	2021 Bare Costs Labor	2021 Bare Costs Equipment	Total	Total Incl O&P
2700	1/4" diameter, short	G 1 Carp	140	.057	Ea.	.58	3.13		3.71	5.30
2800	Long	G	130	.062		1.04	3.37		4.41	6.15
3000	Toggle bolts, bright steel, 1/8" diameter, 2" long	G	85	.094		.33	5.15		5.48	8.05
3100	4" long	G	80	.100		.38	5.45		5.83	8.55
3400	1/4" diameter, 3" long	G	75	.107		.44	5.85		6.29	9.20
3500	6" long	G	70	.114		.59	6.25		6.84	10
3600	3/8" diameter, 3" long	G	70	.114		.73	6.25		6.98	10.15
3700	6" long	G	60	.133		1.62	7.30		8.92	12.70
5000	Screw anchors for concrete, masonry,									
5100	stone & tile, no layout or drilling included									
5700	Lag screw shields, 1/4" diameter, short	G 1 Carp	90	.089	Ea.	.28	4.86		5.14	7.55
5900	3/8" diameter, short	G	85	.094		.55	5.15		5.70	8.30
6100	1/2" diameter, short	G	80	.100		1.02	5.45		6.47	9.25
6300	5/8" diameter, short	G	70	.114		1.60	6.25		7.85	11.10
6600	Lead, #6 & #8, 3/4" long	G	260	.031		.22	1.68		1.90	2.75
6700	#10 - #14, 1-1/2" long	G	200	.040		.52	2.19		2.71	3.84
6800	#16 & #18, 1-1/2" long	G	160	.050		.70	2.74		3.44	4.85
6900	Plastic, #6 & #8, 3/4" long		260	.031		.05	1.68		1.73	2.57
7100	#10 & #12, 1" long		220	.036		.07	1.99		2.06	3.05
8000	Wedge anchors, not including layout or drilling									
8050	Carbon steel, 1/4" diameter, 1-3/4" long	G 1 Carp	150	.053	Ea.	.76	2.92		3.68	5.20
8100	3-1/4" long	G	140	.057		1	3.13		4.13	5.75
8150	3/8" diameter, 2-1/4" long	G	145	.055		.43	3.02		3.45	4.98
8200	5" long	G	140	.057		.76	3.13		3.89	5.50
8250	1/2" diameter, 2-3/4" long	G	140	.057		1.14	3.13		4.27	5.95
8300	7" long	G	125	.064		1.96	3.50		5.46	7.40
8350	5/8" diameter, 3-1/2" long	G	130	.062		2.13	3.37		5.50	7.35
8400	8-1/2" long	G	115	.070		4.54	3.81		8.35	10.70
8450	3/4" diameter, 4-1/4" long	G	115	.070		3.45	3.81		7.26	9.50
8500	10" long	G	95	.084		7.85	4.61		12.46	15.55
8550	1" diameter, 6" long	G	100	.080		5.65	4.38		10.03	12.75
8575	9" long	G	85	.094		7.35	5.15		12.50	15.75
8600	12" long	G	75	.107		7.95	5.85		13.80	17.40
8650	1-1/4" diameter, 9" long	G	70	.114		33.50	6.25		39.75	46.50
8700	12" long	G	60	.133		43	7.30		50.30	58.50
8750	For type 303 stainless steel, add					350%				
8800	For type 316 stainless steel, add					450%				
8950	Self-drilling concrete screw, hex washer head, 3/16" diam. x 1-3/4" long	G 1 Carp	300	.027	Ea.	.21	1.46		1.67	2.41
8960	2-1/4" long	G	250	.032		.29	1.75		2.04	2.93
8970	Phillips flat head, 3/16" diam. x 1-3/4" long	G	300	.027		.20	1.46		1.66	2.40
8980	2-1/4" long	G	250	.032		.31	1.75		2.06	2.95

05 05 21 – Fastening Methods for Metal

05 05 21.10 Cutting Steel

		Crew	Daily Output	Labor-Hours	Unit	Material	2021 Bare Costs Labor	2021 Bare Costs Equipment	Total	Total Incl O&P
0010	**CUTTING STEEL**									
0020	Hand burning, incl. preparation, torch cutting & grinding, no staging									
0050	Steel to 1/4" thick	E-25	400	.020	L.F.	.64	1.25	.03	1.92	2.68
0100	1/2" thick		320	.025		1.04	1.56	.04	2.64	3.60
0150	3/4" thick		260	.031		1.61	1.92	.05	3.58	4.80
0200	1" thick		200	.040		2.23	2.49	.06	4.78	6.40

05 05 21 – Fastening Methods for Metal

05 05 21.15 Drilling Steel

		Crew	Daily Output	Labor-Hours	Unit	Material	2021 Bare Costs Labor	2021 Bare Costs Equipment	Total	Total Incl O&P
0010	**DRILLING STEEL**									
1910	Drilling & layout for steel, up to 1/4" deep, no anchor									
1920	Holes, 1/4" diameter	1 Sswk	112	.071	Ea.	.06	4.31		4.37	6.70
1925	For each additional 1/4" depth, add		336	.024		.06	1.44		1.50	2.29
1930	3/8" diameter		104	.077		.08	4.64		4.72	7.30
1935	For each additional 1/4" depth, add		312	.026		.08	1.55		1.63	2.48
1940	1/2" diameter		96	.083		.08	5.05		5.13	7.90
1945	For each additional 1/4" depth, add		288	.028		.08	1.68		1.76	2.68
1950	5/8" diameter		88	.091		.12	5.50		5.62	8.65
1955	For each additional 1/4" depth, add		264	.030		.12	1.83		1.95	2.96
1960	3/4" diameter		80	.100		.16	6.05		6.21	9.55
1965	For each additional 1/4" depth, add		240	.033		.16	2.01		2.17	3.29
1970	7/8" diameter		72	.111		.19	6.70		6.89	10.55
1975	For each additional 1/4" depth, add		216	.037		.19	2.23		2.42	3.67
1980	1" diameter		64	.125		.30	7.55		7.85	12
1985	For each additional 1/4" depth, add		192	.042		.30	2.51		2.81	4.22
1990	For drilling up, add						40%			

05 05 21.90 Welding Steel

		Crew	Daily Output	Labor-Hours	Unit	Material	2021 Bare Costs Labor	2021 Bare Costs Equipment	Total	Total Incl O&P
0010	**WELDING STEEL**, Structural R050521-20									
0020	Field welding, 1/8" E6011, cost per welder, no operating engineer	E-14	8	1	Hr.	5.45	62.50	18.60	86.55	123
0200	With 1/2 operating engineer	E-13	8	1.500		5.45	90	18.60	114.05	165
0300	With 1 operating engineer	E-12	8	2		5.45	118	18.60	142.05	206
0500	With no operating engineer, 2# weld rod per ton	E-14	8	1	Ton	5.45	62.50	18.60	86.55	123
0600	8# E6011 per ton	"	2	4		22	249	74.50	345.50	490
0800	With one operating engineer per welder, 2# E6011 per ton	E-12	8	2		5.45	118	18.60	142.05	206
0900	8# E6011 per ton	"	2	8		22	470	74.50	566.50	820
1200	Continuous fillet, down welding									
1300	Single pass, 1/8" thick, 0.1#/L.F.	E-14	150	.053	L.F.	.27	3.32	.99	4.58	6.55
1400	3/16" thick, 0.2#/L.F.		75	.107		.54	6.65	1.98	9.17	13.10
1500	1/4" thick, 0.3#/L.F.		50	.160		.82	9.95	2.97	13.74	19.55
1610	5/16" thick, 0.4#/L.F.		38	.211		1.09	13.10	3.91	18.10	26
1800	3 passes, 3/8" thick, 0.5#/L.F.		30	.267		1.36	16.60	4.96	22.92	32.50
2010	4 passes, 1/2" thick, 0.7#/L.F.		22	.364		1.90	22.50	6.75	31.15	44.50
2200	5 to 6 passes, 3/4" thick, 1.3#/L.F.		12	.667		3.54	41.50	12.40	57.44	82
2400	8 to 11 passes, 1" thick, 2.4#/L.F.		6	1.333		6.55	83	25	114.55	164
2600	For vertical joint welding, add						20%			
2700	Overhead joint welding, add						300%			
2900	For semi-automatic welding, obstructed joints, deduct						5%			
3000	Exposed joints, deduct						15%			
4000	Cleaning and welding plates, bars, or rods									
4010	to existing beams, columns, or trusses	E-14	12	.667	L.F.	1.36	41.50	12.40	55.26	79.50

05 05 23 – Metal Fastenings

05 05 23.30 Lag Screws

			Crew	Daily Output	Labor-Hours	Unit	Material	2021 Bare Costs Labor	2021 Bare Costs Equipment	Total	Total Incl O&P
0010	**LAG SCREWS**										
0020	Steel, 1/4" diameter, 2" long	G	1 Carp	200	.040	Ea.	.13	2.19		2.32	3.41
0100	3/8" diameter, 3" long	G		150	.053		.44	2.92		3.36	4.83
0200	1/2" diameter, 3" long	G		130	.062		1.08	3.37		4.45	6.20
0300	5/8" diameter, 3" long	G		120	.067		1.72	3.65		5.37	7.35

05 05 23 – Metal Fastenings

05 05 23.35 Machine Screws

		Crew	Daily Output	Labor-Hours	Unit	Material	2021 Bare Costs Labor	Equipment	Total	Total Incl O&P
0010	**MACHINE SCREWS**									
0020	Steel, round head, #8 x 1" long	G 1 Carp	4.80	1.667	C	3.98	91		94.98	140
0110	#8 x 2" long	G	2.40	3.333		14.70	182		196.70	288
0200	#10 x 1" long	G	4	2		6.25	109		115.25	170
0300	#10 x 2" long	G	2	4		10.90	219		229.90	335

05 05 23.50 Powder Actuated Tools and Fasteners

		Crew	Daily Output	Labor-Hours	Unit	Material	2021 Bare Costs Labor	Equipment	Total	Total Incl O&P
0010	**POWDER ACTUATED TOOLS & FASTENERS**									
0020	Stud driver, .22 caliber, single shot				Ea.	146			146	160
0100	.27 caliber, semi automatic, strip				"	625			625	685
0300	Powder load, single shot, .22 cal, power level 2, brown				C	7.85			7.85	8.60
0400	Strip, .27 cal, power level 4, red					12			12	13.20
0600	Drive pin, .300 x 3/4" long	G 1 Carp	4.80	1.667		16.95	91		107.95	155
0700	.300 x 3" long with washer	G "	4	2		22	109		131	187

05 05 23.70 Structural Blind Bolts

		Crew	Daily Output	Labor-Hours	Unit	Material	2021 Bare Costs Labor	Equipment	Total	Total Incl O&P
0010	**STRUCTURAL BLIND BOLTS**									
0100	1/4" diameter x 1/4" grip	G 1 Sswk	240	.033	Ea.	1.27	2.01		3.28	4.51
0150	1/2" grip	G	216	.037		1.01	2.23		3.24	4.57
0200	3/8" diameter x 1/2" grip	G	232	.034		1.69	2.08		3.77	5.10
0250	3/4" grip	G	208	.038		1.96	2.32		4.28	5.75
0300	1/2" diameter x 1/2" grip	G	224	.036		3.82	2.15		5.97	7.55
0350	3/4" grip	G	200	.040		3.82	2.41		6.23	7.95
0400	5/8" diameter x 3/4" grip	G	216	.037		6	2.23		8.23	10.05
0450	1" grip	G	192	.042		6	2.51		8.51	10.50

05 05 23.85 Weld Shear Connectors

		Crew	Daily Output	Labor-Hours	Unit	Material	2021 Bare Costs Labor	Equipment	Total	Total Incl O&P
0010	**WELD SHEAR CONNECTORS**									
0020	3/4" diameter, 3-3/16" long	G E-10	960	.017	Ea.	.58	1.02	1.04	2.64	3.36
0030	3-3/8" long	G	950	.017		.60	1.03	1.05	2.68	3.42
0200	3-7/8" long	G	945	.017		.65	1.04	1.06	2.75	3.48
0300	4-3/16" long	G	935	.017		.68	1.05	1.07	2.80	3.55
0500	4-7/8" long	G	930	.017		.76	1.05	1.07	2.88	3.65
0600	5-3/16" long	G	920	.017		.79	1.07	1.09	2.95	3.71
0800	5-3/8" long	G	910	.018		.80	1.08	1.10	2.98	3.76
0900	6-3/16" long	G	905	.018		.87	1.08	1.10	3.05	3.85
1000	7-3/16" long	G	895	.018		1.09	1.10	1.12	3.31	4.13
1100	8-3/16" long	G	890	.018		1.19	1.10	1.12	3.41	4.25
1500	7/8" diameter, 3-11/16" long	G	920	.017		.95	1.07	1.09	3.11	3.88
1600	4-3/16" long	G	910	.018		1.01	1.08	1.10	3.19	4
1700	5-3/16" long	G	905	.018		1.15	1.08	1.10	3.33	4.16
1800	6-3/16" long	G	895	.018		1.29	1.10	1.12	3.51	4.35
1900	7-3/16" long	G	890	.018		1.43	1.10	1.12	3.65	4.51
2000	8-3/16" long	G	880	.018		1.56	1.11	1.14	3.81	4.69

05 05 23.87 Weld Studs

		Crew	Daily Output	Labor-Hours	Unit	Material	2021 Bare Costs Labor	Equipment	Total	Total Incl O&P
0010	**WELD STUDS**									
0020	1/4" diameter, 2-11/16" long	G E-10	1120	.014	Ea.	.39	.88	.89	2.16	2.77
0100	4-1/8" long	G	1080	.015		.37	.91	.92	2.20	2.84
0200	3/8" diameter, 4-1/8" long	G	1080	.015		.42	.91	.92	2.25	2.89
0300	6-1/8" long	G	1040	.015		.55	.94	.96	2.45	3.12
0400	1/2" diameter, 2-1/8" long	G	1040	.015		.39	.94	.96	2.29	2.95
0500	3-1/8" long	G	1025	.016		.47	.96	.97	2.40	3.07
0600	4-1/8" long	G	1010	.016		.55	.97	.99	2.51	3.20
0700	5-5/16" long	G	990	.016		.68	.99	1.01	2.68	3.39

123

For customer support on your Concrete & Masonry Costs with RSMeans Data, call 800.448.8182.

05 05 Common Work Results for Metals

05 05 23 – Metal Fastenings

05 05 23.87 Weld Studs		Crew	Daily Output	Labor-Hours	Unit	Material	2021 Bare Costs Labor	Equipment	Total	Total Incl O&P
0800	6-1/8" long	G	E-10	975	.016	Ea.	.74	1.01	1.02	2.77	3.50
0900	8-1/8" long	G		960	.017		1.04	1.02	1.04	3.10	3.86
1000	5/8" diameter, 2-11/16" long	G		1000	.016		.68	.98	1	2.66	3.36
1010	4-3/16" long	G		990	.016		.84	.99	1.01	2.84	3.56
1100	6-9/16" long	G		975	.016		1.09	1.01	1.02	3.12	3.89
1200	8-3/16" long	G		960	.017		1.46	1.02	1.04	3.52	4.33

05 05 23.90 Welding Rod		Crew	Daily Output	Labor-Hours	Unit	Material	2021 Bare Costs Labor	Equipment	Total	Total Incl O&P
0010	**WELDING ROD**									
0020	Steel, type 6011, 1/8" diam., less than 500#				Lb.	2.72			2.72	2.99
0100	500# to 2,000#					2.45			2.45	2.70
0200	2,000# to 5,000#					2.30			2.30	2.53
0300	5/32" diam., less than 500#					3.94			3.94	4.33
0310	500# to 2,000#					3.55			3.55	3.91
0320	2,000# to 5,000#					3.34			3.34	3.67
0400	3/16" diam., less than 500#					2.87			2.87	3.16
0500	500# to 2,000#					2.59			2.59	2.85
0600	2,000# to 5,000#					2.43			2.43	2.68
0620	Steel, type 6010, 1/8" diam., less than 500#					5.70			5.70	6.30
0630	500# to 2,000#					5.15			5.15	5.65
0640	2,000# to 5,000#					4.83			4.83	5.30
0650	Steel, type 7018 Low Hydrogen, 1/8" diam., less than 500#					2.99			2.99	3.28
0660	500# to 2,000#					2.69			2.69	2.96
0670	2,000# to 5,000#					2.53			2.53	2.78
0700	Steel, type 7024 Jet Weld, 1/8" diam., less than 500#					3.72			3.72	4.09
0710	500# to 2,000#					3.35			3.35	3.69
0720	2,000# to 5,000#					3.15			3.15	3.46
1550	Aluminum, type 4043 TIG, 1/8" diam., less than 10#					7			7	7.70
1560	10# to 60#					6.30			6.30	6.90
1570	Over 60#					5.90			5.90	6.50
1600	Aluminum, type 5356 TIG, 1/8" diam., less than 10#					6.55			6.55	7.20
1610	10# to 60#					5.90			5.90	6.50
1620	Over 60#					5.55			5.55	6.10
1900	Cast iron, type 8 Nickel, 1/8" diam., less than 500#					46.50			46.50	51.50
1910	500# to 1,000#					42			42	46
1920	Over 1,000#					39.50			39.50	43.50
2000	Stainless steel, type 316/316L, 1/8" diam., less than 500#					7.20			7.20	7.90
2100	500# to 1,000#					6.50			6.50	7.15
2220	Over 1,000#					6.10			6.10	6.70

05 12 Structural Steel Framing

05 12 23 – Structural Steel for Buildings

05 12 23.10 Ceiling Supports		Crew	Daily Output	Labor-Hours	Unit	Material	2021 Bare Costs Labor	Equipment	Total	Total Incl O&P
0010	**CEILING SUPPORTS**									
1000	Entrance door/folding partition supports, shop fabricated	G	E-4	60	.533	L.F.	26.50	32.50	2.48	61.48	81.50
1100	Linear accelerator door supports	G		14	2.286		120	139	10.65	269.65	360
1200	Lintels or shelf angles, hung, exterior hot dipped galv.	G		267	.120		18	7.30	.56	25.86	31.50
1250	Two coats primer paint instead of galv.	G		267	.120		15.60	7.30	.56	23.46	29
1400	Monitor support, ceiling hung, expansion bolted	G		4	8	Ea.	420	485	37	942	1,250
1450	Hung from pre-set inserts	G		6	5.333		450	325	25	800	1,025
1600	Motor supports for overhead doors	G		4	8		213	485	37	735	1,025
1700	Partition support for heavy folding partitions, without pocket	G		24	1.333	L.F.	60	81	6.20	147.20	198

05 12 23.10 Ceiling Supports

		Crew	Daily Output	Labor-Hours	Unit	Material	2021 Bare Costs Labor	Equipment	Total	Total Incl O&P
1750	Supports at pocket only	G E-4	12	2.667	L.F.	120	162	12.40	294.40	395
2000	Rolling grilles & fire door supports	G	34	.941		51.50	57	4.38	112.88	150
2100	Spider-leg light supports, expansion bolted to ceiling slab	G	8	4	Ea.	171	243	18.60	432.60	585
2150	Hung from pre-set inserts	G	12	2.667	"	185	162	12.40	359.40	470
2400	Toilet partition support	G	36	.889	L.F.	60	54	4.13	118.13	154
2500	X-ray travel gantry support	G	12	2.667	"	206	162	12.40	380.40	490

05 12 23.15 Columns, Lightweight

		Crew	Daily Output	Labor-Hours	Unit	Material	2021 Bare Costs Labor	Equipment	Total	Total Incl O&P
0010	**COLUMNS, LIGHTWEIGHT**									
1000	Lightweight units (lally), 3-1/2" diameter	E-2	780	.072	L.F.	5.40	4.28	2.20	11.88	14.90
1050	4" diameter	"	900	.062	"	9.25	3.71	1.90	14.86	18

05 12 23.17 Columns, Structural

		Crew	Daily Output	Labor-Hours	Unit	Material	2021 Bare Costs Labor	Equipment	Total	Total Incl O&P
0010	**COLUMNS, STRUCTURAL** R051223-10									
0015	Made from recycled materials									
0020	Shop fab'd for 100-ton, 1-2 story project, bolted connections									
0800	Steel, concrete filled, extra strong pipe, 3-1/2" diameter	E-2	660	.085	L.F.	43.50	5.05	2.60	51.15	58.50
0830	4" diameter		780	.072		48.50	4.28	2.20	54.98	62.50
0890	5" diameter		1020	.055		58	3.27	1.68	62.95	70.50
0930	6" diameter		1200	.047		76.50	2.78	1.43	80.71	90
0940	8" diameter		1100	.051		76.50	3.04	1.56	81.10	90.50
1100	For galvanizing, add				Lb.	.33			.33	.37
8090	For projects 75 to 99 tons, add				%	10%				
8092	50 to 74 tons, add					20%				
8094	25 to 49 tons, add					30%	10%			
8096	10 to 24 tons, add					50%	25%			
8098	2 to 9 tons, add					75%	50%			
8099	Less than 2 tons, add					100%	100%			

05 12 23.18 Corner Guards

		Crew	Daily Output	Labor-Hours	Unit	Material	2021 Bare Costs Labor	Equipment	Total	Total Incl O&P
0010	**CORNER GUARDS**									
0020	Steel angle w/anchors, 1" x 1" x 1/4", 1.5#/L.F.	2 Carp	160	.100	L.F.	7.45	5.45		12.90	16.35
0100	2" x 2" x 1/4" angles, 3.2#/L.F.		150	.107		11.50	5.85		17.35	21.50
0200	3" x 3" x 5/16" angles, 6.1#/L.F.		140	.114		19	6.25		25.25	30.50
0300	4" x 4" x 5/16" angles, 8.2#/L.F.		120	.133		15.75	7.30		23.05	28
0350	For angles drilled and anchored to masonry, add					15%	120%			
0370	Drilled and anchored to concrete, add					20%	170%			
0400	For galvanized angles, add					35%				
0450	For stainless steel angles, add					100%				

05 12 23.20 Curb Edging

		Crew	Daily Output	Labor-Hours	Unit	Material	2021 Bare Costs Labor	Equipment	Total	Total Incl O&P
0010	**CURB EDGING**									
0020	Steel angle w/anchors, shop fabricated, on forms, 1" x 1", 0.8#/L.F.	G E-4	350	.091	L.F.	1.69	5.55	.43	7.67	10.95
0100	2" x 2" angles, 3.92#/L.F.	G	330	.097		6.65	5.90	.45	13	16.90
0200	3" x 3" angles, 6.1#/L.F.	G	300	.107		10.90	6.50	.50	17.90	22.50
0300	4" x 4" angles, 8.2#/L.F.	G	275	.116		14.20	7.05	.54	21.79	27
1000	6" x 4" angles, 12.3#/L.F.	G	250	.128		20.50	7.80	.60	28.90	35.50
1050	Steel channels with anchors, on forms, 3" channel, 5#/L.F.	G	290	.110		8.35	6.70	.51	15.56	20
1100	4" channel, 5.4#/L.F.	G	270	.119		8.95	7.20	.55	16.70	21.50
1200	6" channel, 8.2#/L.F.	G	255	.125		14.20	7.65	.58	22.43	28
1300	8" channel, 11.5#/L.F.	G	225	.142		19.40	8.65	.66	28.71	35.50
1400	10" channel, 15.3#/L.F.	G	180	.178		25.50	10.80	.83	37.13	45.50
1500	12" channel, 20.7#/L.F.	G	140	.229		34	13.90	1.06	48.96	60
2000	For curved edging, add					35%	10%			

05 12 23.40 Lightweight Framing

		Crew	Daily Output	Labor-Hours	Unit	Material	2021 Bare Costs Labor	Equipment	Total	Total Incl O&P
0010	**LIGHTWEIGHT FRAMING** R051223-35									
0015	Made from recycled materials									
0400	Angle framing, field fabricated, 4" and larger G	E-3	440	.055	Lb.	.77	3.33	.34	4.44	6.35
0450	Less than 4" angles R051223-45 G		265	.091		.79	5.50	.56	6.85	10.05
0600	Channel framing, field fabricated, 8" and larger G		500	.048		.79	2.93	.30	4.02	5.75
0650	Less than 8" channels G		335	.072		.79	4.37	.44	5.60	8.10
1000	Continuous slotted channel framing system, shop fab, simple framing G	2 Sswk	2400	.007		4.09	.40		4.49	5.10
1200	Complex framing G	"	1600	.010		4.62	.60		5.22	6.05
1300	Cross bracing, rods, shop fabricated, 3/4" diameter G	E-3	700	.034		1.58	2.09	.21	3.88	5.20
1310	7/8" diameter G		850	.028		1.58	1.72	.18	3.48	4.59
1320	1" diameter G		1000	.024		1.58	1.46	.15	3.19	4.16
1330	Angle, 5" x 5" x 3/8" G		2800	.009		1.58	.52	.05	2.15	2.61
1350	Hanging lintels, shop fabricated G		850	.028		1.58	1.72	.18	3.48	4.59

05 12 23.45 Lintels

		Crew	Daily Output	Labor-Hours	Unit	Material	2021 Bare Costs Labor	Equipment	Total	Total Incl O&P
0010	**LINTELS**									
0015	Made from recycled materials									
0020	Plain steel angles, shop fabricated, under 500 lb. G	1 Bric	550	.015	Lb.	1.02	.78		1.80	2.30
0100	500 to 1,000 lb. G		640	.013		.99	.67		1.66	2.10
0200	1,000 to 2,000 lb. G		640	.013		.96	.67		1.63	2.07
0300	2,000 to 4,000 lb. G		640	.013		.94	.67		1.61	2.04
0500	For built-up angles and plates, add to above G					1.32			1.32	1.45
0700	For engineering, add to above					.13			.13	.15
0900	For galvanizing, add to above, under 500 lb.					.39			.39	.42
0950	500 to 2,000 lb.					.35			.35	.39
1000	Over 2,000 lb.					.33			.33	.37
2000	Steel angles, 3-1/2" x 3", 1/4" thick, 2'-6" long G	1 Bric	47	.170	Ea.	14.25	9.15		23.40	29.50
2100	4'-6" long G		26	.308		25.50	16.50		42	53
2500	3-1/2" x 3-1/2" x 5/16", 5'-0" long G		18	.444		38	24		62	78
2600	4" x 3-1/2", 1/4" thick, 5'-0" long G		21	.381		32.50	20.50		53	67
2700	9'-0" long G		12	.667		59	36		95	119
2800	4" x 3-1/2" x 5/16", 7'-0" long G		12	.667		57	36		93	117
2900	5" x 3-1/2" x 5/16", 10'-0" long G		8	1		92	53.50		145.50	182

05 12 23.65 Plates

		Crew	Daily Output	Labor-Hours	Unit	Material	2021 Bare Costs Labor	Equipment	Total	Total Incl O&P
0010	**PLATES** R051223-80									
0015	Made from recycled materials									
0020	For connections & stiffener plates, shop fabricated									
0050	1/8" thick (5.1 lb./S.F.) G				S.F.	6.75			6.75	7.40
0100	1/4" thick (10.2 lb./S.F.) G					13.45			13.45	14.80
0300	3/8" thick (15.3 lb./S.F.) G					20			20	22
0400	1/2" thick (20.4 lb./S.F.) G					27			27	29.50
0450	3/4" thick (30.6 lb./S.F.) G					40.50			40.50	44.50
0500	1" thick (40.8 lb./S.F.) G					54			54	59
2000	Steel plate, warehouse prices, no shop fabrication									
2100	1/4" thick (10.2 lb./S.F.) G				S.F.	8.10			8.10	8.90
2210	1/4" steel plate, welded in place	E-18	528	.076		8.10	4.58	2.89	15.57	19.15
2220	1/2" steel plate, welded in place		480	.083		16.15	5.05	3.18	24.38	29
2230	3/4" steel plate, welded in place		384	.104		24	6.30	3.97	34.27	40.50
2240	1" steel plate, welded in place		320	.125		32.50	7.55	4.77	44.82	52.50
2250	1-1/2" steel plate, welded in place		256	.156		48.50	9.45	5.95	63.90	74.50
2260	2" steel plate, welded in place		192	.208		64.50	12.60	7.95	85.05	99

05 12 23 – Structural Steel for Buildings

05 12 23.75 Structural Steel Members		Crew	Daily Output	Labor-Hours	Unit	Material	2021 Bare Costs Labor	2021 Bare Costs Equipment	Total	Total Incl O&P
0010	**STRUCTURAL STEEL MEMBERS** R051223-10									
0015	Made from recycled materials									
0020	Shop fab'd for 100-ton, 1-2 story project, bolted connections									
0100	Beam or girder, W 6 x 9 G	E-2	600	.093	L.F.	13.05	5.55	2.86	21.46	26
0120	x 15 G		600	.093		22	5.55	2.86	30.41	35.50
0140	x 20 G		600	.093		29	5.55	2.86	37.41	43.50
0300	W 8 x 10 G		600	.093		14.50	5.55	2.86	22.91	27.50
0320	x 15 G		600	.093		22	5.55	2.86	30.41	35.50
0350	x 21 G		600	.093		30.50	5.55	2.86	38.91	45
0360	x 24 G		550	.102		35	6.05	3.11	44.16	51
0370	x 28 G		550	.102		40.50	6.05	3.11	49.66	57
0500	x 31 G		550	.102		45	6.05	3.11	54.16	62
0520	x 35 G		550	.102		51	6.05	3.11	60.16	68.50
0540	x 48 G		550	.102		69.50	6.05	3.11	78.66	89
0600	W 10 x 12 G		600	.093		17.40	5.55	2.86	25.81	31
0620	x 15 G		600	.093		22	5.55	2.86	30.41	35.50
0700	x 22 G		600	.093		32	5.55	2.86	40.41	46.50
0720	x 26 G		600	.093		37.50	5.55	2.86	45.91	53
0740	x 33 G		550	.102		48	6.05	3.11	57.16	65
0900	x 49 G		550	.102		71	6.05	3.11	80.16	90.50
1100	W 12 x 16 G		880	.064		23	3.80	1.95	28.75	33.50
1300	x 22 G		880	.064		32	3.80	1.95	37.75	43
1500	x 26 G		880	.064		37.50	3.80	1.95	43.25	49.50
1520	x 35 G		810	.069		51	4.12	2.11	57.23	64.50
1560	x 50 G		750	.075		72.50	4.45	2.28	79.23	89.50
1580	x 58 G		750	.075		84	4.45	2.28	90.73	102
1700	x 72 G		640	.088		104	5.20	2.68	111.88	126
1740	x 87 G		640	.088		126	5.20	2.68	133.88	150
1900	W 14 x 26 G		990	.057		37.50	3.37	1.73	42.60	48.50
2100	x 30 G		900	.062		43.50	3.71	1.90	49.11	56
2300	x 34 G		810	.069		49.50	4.12	2.11	55.73	63
2320	x 43 G		810	.069		62.50	4.12	2.11	68.73	77
2340	x 53 G		800	.070		77	4.17	2.14	83.31	93.50
2360	x 74 G		760	.074		107	4.39	2.25	113.64	127
2380	x 90 G		740	.076		131	4.51	2.31	137.82	153
2500	x 120 G		720	.078		174	4.64	2.38	181.02	202
2700	W 16 x 26 G		1000	.056		37.50	3.34	1.71	42.55	48.50
2900	x 31 G		900	.062		45	3.71	1.90	50.61	57.50
3100	x 40 G		800	.070		58	4.17	2.14	64.31	73
3120	x 50 G		800	.070		72.50	4.17	2.14	78.81	89
3140	x 67 G		760	.074		97	4.39	2.25	103.64	116
3300	W 18 x 35 G	E-5	960	.083		51	5	1.94	57.94	66
3920	x 65 G		900	.089		94.50	5.35	2.07	101.92	114
3940	x 76 G		900	.089		110	5.35	2.07	117.42	131
3960	x 86 G		900	.089		125	5.35	2.07	132.42	147
3980	x 106 G		900	.089		154	5.35	2.07	161.42	179
8490	For projects 75 to 99 tons, add					10%				
8492	50 to 74 tons, add					20%				
8494	25 to 49 tons, add					30%	10%			
8496	10 to 24 tons, add					50%	25%			
8498	2 to 9 tons, add					75%	50%			
8499	Less than 2 tons, add					100%	100%			

05 12 Structural Steel Framing

05 12 23 – Structural Steel for Buildings

05 12 23.80 Subpurlins		Crew	Daily Output	Labor-Hours	Unit	Material	2021 Bare Costs Labor	Equipment	Total	Total Incl O&P
0010	**SUBPURLINS** R051223-50									
0015	Made from recycled materials									
0020	Bulb tees, shop fabricated, painted, 32-5/8" OC, 40 psf L.L.									
0200	Type 218, max 10'-2" span, 3.19 plf, 2-1/8" high x 2-1/8" wide [G]	E-1	3100	.008	S.F.	1.77	.46	.05	2.28	2.70
1420	For 24-5/8" spacing, add					33%	33%			
1430	For 48-5/8" spacing, deduct					33%	33%			

05 14 Structural Aluminum Framing

05 14 23 – Non-Exposed Structural Aluminum Framing

05 14 23.05 Aluminum Shapes

		Crew	Daily Output	Labor-Hours	Unit	Material	2021 Bare Costs Labor	Equipment	Total	Total Incl O&P
0010	**ALUMINUM SHAPES**									
0015	Made from recycled materials									
0020	Structural shapes, 1" to 10" members, under 1 ton [G]	E-2	4000	.014	Lb.	5.20	.83	.43	6.46	7.45
0050	1 to 5 tons [G]		4300	.013		4.38	.78	.40	5.56	6.45
0100	Over 5 tons [G]		4600	.012		4.10	.73	.37	5.20	6.05
0300	Extrusions, over 5 tons, stock shapes [G]		1330	.042		3.48	2.51	1.29	7.28	9.10
0400	Custom shapes [G]		1330	.042		4.38	2.51	1.29	8.18	10.10

05 15 Wire Rope Assemblies

05 15 16 – Steel Wire Rope Assemblies

05 15 16.05 Accessories for Steel Wire Rope

		Crew	Daily Output	Labor-Hours	Unit	Material	2021 Bare Costs Labor	Equipment	Total	Total Incl O&P
0010	**ACCESSORIES FOR STEEL WIRE ROPE**									
0015	Made from recycled materials									
1500	Thimbles, heavy duty, 1/4" [G]	E-17	160	.100	Ea.	.45	6.15		6.60	10
1510	1/2" [G]		160	.100		1.98	6.15		8.13	11.70
1520	3/4" [G]		105	.152		4.50	9.35		13.85	19.40
1530	1" [G]		52	.308		9	18.85		27.85	39
1540	1-1/4" [G]		38	.421		13.85	26		39.85	55.50
1550	1-1/2" [G]		13	1.231		39	75.50		114.50	160
1560	1-3/4" [G]		8	2		80.50	123		203.50	279
1570	2" [G]		6	2.667		117	163		280	380
1580	2-1/4" [G]		4	4		158	245		403	555
1600	Clips, 1/4" diameter [G]		160	.100		1.97	6.15		8.12	11.65
1610	3/8" diameter [G]		160	.100		2.16	6.15		8.31	11.90
1620	1/2" diameter [G]		160	.100		3.47	6.15		9.62	13.30
1630	3/4" diameter [G]		102	.157		5.65	9.60		15.25	21
1640	1" diameter [G]		64	.250		9.40	15.35		24.75	34
1650	1-1/4" diameter [G]		35	.457		15.40	28		43.40	60.50
1670	1-1/2" diameter [G]		26	.615		21	37.50		58.50	81.50
1680	1-3/4" diameter [G]		16	1		48.50	61.50		110	148
1690	2" diameter [G]		12	1.333		54	81.50		135.50	185
1700	2-1/4" diameter [G]		10	1.600		79	98		177	239
1800	Sockets, open swage, 1/4" diameter [G]		160	.100		32	6.15		38.15	44.50
1810	1/2" diameter [G]		77	.208		46	12.75		58.75	70
1820	3/4" diameter [G]		19	.842		71	51.50		122.50	159
1830	1" diameter [G]		9	1.778		127	109		236	310
1840	1-1/4" diameter [G]		5	3.200		177	196		373	500
1850	1-1/2" diameter [G]		3	5.333		390	325		715	930
1860	1-3/4" diameter [G]		3	5.333		690	325		1,015	1,250

05 15 Wire Rope Assemblies

05 15 16 – Steel Wire Rope Assemblies

05 15 16.05 Accessories for Steel Wire Rope

		Crew	Daily Output	Labor-Hours	Unit	Material	2021 Bare Costs Labor	Equipment	Total	Total Incl O&P
1870	2" diameter	G E-17	1.50	10.667	Ea.	1,050	655		1,705	2,150
1900	Closed swage, 1/4" diameter	G	160	.100		19.10	6.15		25.25	30.50
1910	1/2" diameter	G	104	.154		33	9.45		42.45	51
1920	3/4" diameter	G	32	.500		49.50	30.50		80	102
1930	1" diameter	G	15	1.067		86.50	65.50		152	197
1940	1-1/4" diameter	G	7	2.286		130	140		270	360
1950	1-1/2" diameter	G	4	4		236	245		481	640
1960	1-3/4" diameter	G	3	5.333		350	325		675	890
1970	2" diameter	G	2	8		675	490		1,165	1,500
2000	Open spelter, galv., 1/4" diameter	G	160	.100		75	6.15		81.15	91.50
2010	1/2" diameter	G	70	.229		78	14		92	108
2020	3/4" diameter	G	26	.615		117	37.50		154.50	188
2030	1" diameter	G	10	1.600		325	98		423	510
2040	1-1/4" diameter	G	5	3.200		465	196		661	815
2050	1-1/2" diameter	G	4	4		985	245		1,230	1,450
2060	1-3/4" diameter	G	2	8		1,725	490		2,215	2,650
2070	2" diameter	G	1.20	13.333		1,975	815		2,790	3,450
2080	2-1/2" diameter	G	1	16		3,650	980		4,630	5,525
2100	Closed spelter, galv., 1/4" diameter	G	160	.100		32.50	6.15		38.65	45.50
2110	1/2" diameter	G	88	.182		35	11.15		46.15	55.50
2120	3/4" diameter	G	30	.533		52.50	32.50		85	109
2130	1" diameter	G	13	1.231		112	75.50		187.50	240
2140	1-1/4" diameter	G	7	2.286		179	140		319	415
2150	1-1/2" diameter	G	6	2.667		385	163		548	680
2160	1-3/4" diameter	G	2.80	5.714		515	350		865	1,100
2170	2" diameter	G	2	8		635	490		1,125	1,450
2200	Jaw & jaw turnbuckles, 1/4" x 4"	G	160	.100		8.40	6.15		14.55	18.75
2250	1/2" x 6"	G	96	.167		10.65	10.20		20.85	27.50
2260	1/2" x 9"	G	77	.208		14.20	12.75		26.95	35.50
2270	1/2" x 12"	G	66	.242		15.95	14.85		30.80	40.50
2300	3/4" x 6"	G	38	.421		21	26		47	63
2310	3/4" x 9"	G	30	.533		23	32.50		55.50	76
2320	3/4" x 12"	G	28	.571		29.50	35		64.50	86.50
2330	3/4" x 18"	G	23	.696		35.50	42.50		78	105
2350	1" x 6"	G	17	.941		40.50	57.50		98	134
2360	1" x 12"	G	13	1.231		44.50	75.50		120	166
2370	1" x 18"	G	10	1.600		66.50	98		164.50	225
2380	1" x 24"	G	9	1.778		73	109		182	250
2400	1-1/4" x 12"	G	7	2.286		74.50	140		214.50	299
2410	1-1/4" x 18"	G	6.50	2.462		92	151		243	335
2420	1-1/4" x 24"	G	5.60	2.857		124	175		299	410
2450	1-1/2" x 12"	G	5.20	3.077		305	189		494	630
2460	1-1/2" x 18"	G	4	4		325	245		570	740
2470	1-1/2" x 24"	G	3.20	5		440	305		745	960
2500	1-3/4" x 18"	G	3.20	5		665	305		970	1,200
2510	1-3/4" x 24"	G	2.80	5.714		755	350		1,105	1,375
2550	2" x 24"	G	1.60	10		1,025	615		1,640	2,075

05 15 16.50 Steel Wire Rope

0010	**STEEL WIRE ROPE**									
0015	Made from recycled materials									
0020	6 x 19, bright, fiber core, 5000' rolls, 1/2" diameter	G			L.F.	.77			.77	.85
0050	Steel core	G				1.01			1.01	1.11

For customer support on your Concrete & Masonry Costs with RSMeans Data, call 800.448.8182.

129

05 15 16 – Steel Wire Rope Assemblies

05 15 16.50 Steel Wire Rope		Crew	Daily Output	Labor-Hours	Unit	Material	2021 Bare Costs Labor	Equipment	Total	Total Incl O&P
0100	Fiber core, 1" diameter	G			L.F.	2.59			2.59	2.85
0150	Steel core	G				2.96			2.96	3.25
0300	6 x 19, galvanized, fiber core, 1/2" diameter	G				1.13			1.13	1.25
0350	Steel core	G				1.30			1.30	1.43
0400	Fiber core, 1" diameter	G				3.32			3.32	3.65
0450	Steel core	G				3.48			3.48	3.83
0500	6 x 7, bright, IPS, fiber core, <500 L.F. w/acc., 1/4" diameter	G	E-17	6400	.003		.46	.15		.61	.75
0510	1/2" diameter	G		2100	.008		1.12	.47		1.59	1.95
0520	3/4" diameter	G		960	.017		2.03	1.02		3.05	3.81
0550	6 x 19, bright, IPS, IWRC, <500 L.F. w/acc., 1/4" diameter	G		5760	.003		.85	.17		1.02	1.20
0560	1/2" diameter	G		1730	.009		1.38	.57		1.95	2.39
0570	3/4" diameter	G		770	.021		2.39	1.27		3.66	4.60
0580	1" diameter	G		420	.038		4.05	2.34		6.39	8.05
0590	1-1/4" diameter	G		290	.055		6.70	3.38		10.08	12.65
0600	1-1/2" diameter	G		192	.083		8.25	5.10		13.35	17
0610	1-3/4" diameter	G	E-18	240	.167		13.15	10.05	6.35	29.55	37
0620	2" diameter	G		160	.250		16.90	15.10	9.55	41.55	52
0630	2-1/4" diameter	G		160	.250		22.50	15.10	9.55	47.15	58.50
0650	6 x 37, bright, IPS, IWRC, <500 L.F. w/acc., 1/4" diameter	G	E-17	6400	.003		1.03	.15		1.18	1.38
0660	1/2" diameter	G		1730	.009		1.75	.57		2.32	2.81
0670	3/4" diameter	G		770	.021		2.83	1.27		4.10	5.10
0680	1" diameter	G		430	.037		4.49	2.28		6.77	8.45
0690	1-1/4" diameter	G		290	.055		6.80	3.38		10.18	12.70
0700	1-1/2" diameter	G		190	.084		9.70	5.15		14.85	18.65
0710	1-3/4" diameter	G	E-18	260	.154		15.40	9.30	5.85	30.55	37.50
0720	2" diameter	G		200	.200		20	12.10	7.65	39.75	49
0730	2-1/4" diameter	G		160	.250		26.50	15.10	9.55	51.15	62.50
0800	6 x 19 & 6 x 37, swaged, 1/2" diameter	G	E-17	1220	.013		4.34	.80		5.14	6
0810	9/16" diameter	G		1120	.014		5.05	.88		5.93	6.90
0820	5/8" diameter	G		930	.017		6	1.05		7.05	8.20
0830	3/4" diameter	G		640	.025		7.60	1.53		9.13	10.75
0840	7/8" diameter	G		480	.033		9.60	2.04		11.64	13.70
0850	1" diameter	G		350	.046		11.70	2.80		14.50	17.25
0860	1-1/8" diameter	G		288	.056		14.40	3.41		17.81	21
0870	1-1/4" diameter	G		230	.070		17.45	4.26		21.71	26
0880	1-3/8" diameter	G		192	.083		20	5.10		25.10	30
0890	1-1/2" diameter	G	E-18	300	.133		24.50	8.05	5.10	37.65	45

05 15 16.60 Galvanized Steel Wire Rope and Accessories

05 15 16.60 Galvanized Steel Wire Rope and Accessories		Crew	Daily Output	Labor-Hours	Unit	Material	Labor	Equipment	Total	Total Incl O&P
0010	**GALVANIZED STEEL WIRE ROPE & ACCESSORIES**									
0015	Made from recycled materials									
3000	Aircraft cable, galvanized, 7 x 7 x 1/8"	G	E-17	5000	.003	L.F.	.16	.20		.36	.48
3100	Clamps, 1/8"	G	"	125	.128	Ea.	1.41	7.85		9.26	13.70

05 15 16.70 Temporary Cable Safety Railing

05 15 16.70 Temporary Cable Safety Railing		Crew	Daily Output	Labor-Hours	Unit	Material	Labor	Equipment	Total	Total Incl O&P
0010	**TEMPORARY CABLE SAFETY RAILING**, Each 100' strand incl.									
0020	2 eyebolts, 1 turnbuckle, 100' cable, 2 thimbles, 6 clips									
0025	Made from recycled materials									
0100	One strand using 1/4" cable & accessories	G	2 Sswk	4	4	C.L.F.	79.50	241		320.50	460
0200	1/2" cable & accessories	G	"	2	8	"	157	480		637	920

05 21 Steel Joist Framing

05 21 13 – Deep Longspan Steel Joist Framing

05 21 13.50 Deep Longspan Joists

		Crew	Daily Output	Labor-Hours	Unit	Material	2021 Bare Costs Labor	Equipment	Total	Total Incl O&P
0010	**DEEP LONGSPAN JOISTS**									
3010	DLH series, 40-ton job lots, bolted cross bridging, shop primer									
3015	Made from recycled materials									
3040	Spans to 144' (shipped in 2 pieces) [G]	E-7	13	6.154	Ton	2,475	370	155	3,000	3,475
3500	For less than 40-ton job lots									
3502	For 30 to 39 tons, add				%	10%				
3504	20 to 29 tons, add					20%				
3506	10 to 19 tons, add					30%				
3507	5 to 9 tons, add					50%	25%			
3508	1 to 4 tons, add					75%	50%			
3509	Less than 1 ton, add					100%	100%			
4010	SLH series, 40-ton job lots, bolted cross bridging, shop primer									
4040	Spans to 200' (shipped in 3 pieces) [G]	E-7	13	6.154	Ton	2,550	370	155	3,075	3,550
6100	For less than 40-ton job lots									
6102	For 30 to 39 tons, add				%	10%				
6104	20 to 29 tons, add					20%				
6106	10 to 19 tons, add					30%				
6107	5 to 9 tons, add					50%	25%			
6108	1 to 4 tons, add					75%	50%			
6109	Less than 1 ton, add					100%	100%			

05 21 16 – Longspan Steel Joist Framing

05 21 16.50 Longspan Joists

		Crew	Daily Output	Labor-Hours	Unit	Material	2021 Bare Costs Labor	Equipment	Total	Total Incl O&P
0010	**LONGSPAN JOISTS**									
2000	LH series, 40-ton job lots, bolted cross bridging, shop primer									
2015	Made from recycled materials									
2040	Longspan joists, LH series, up to 96' [G]	E-7	13	6.154	Ton	2,150	370	155	2,675	3,125
2600	For less than 40-ton job lots									
2602	For 30 to 39 tons, add				%	10%				
2604	20 to 29 tons, add					20%				
2606	10 to 19 tons, add					30%				
2607	5 to 9 tons, add					50%	25%			
2608	1 to 4 tons, add					75%	50%			
2609	Less than 1 ton, add					100%	100%			
6000	For welded cross bridging, add						30%			

05 21 19 – Open Web Steel Joist Framing

05 21 19.10 Open Web Joists

		Crew	Daily Output	Labor-Hours	Unit	Material	2021 Bare Costs Labor	Equipment	Total	Total Incl O&P
0010	**OPEN WEB JOISTS**									
0015	Made from recycled materials									
0050	K series, 40-ton lots, horiz. bridging, spans to 30', shop primer [G]	E-7	12	6.667	Ton	1,900	400	168	2,468	2,900
0440	K series, 30' to 50' spans [G]	"	17	4.706	"	1,950	283	118	2,351	2,725
0800	For less than 40-ton job lots									
0802	For 30 to 39 tons, add				%	10%				
0804	20 to 29 tons, add					20%				
0806	10 to 19 tons, add					30%				
0807	5 to 9 tons, add					50%	25%			
0808	1 to 4 tons, add					75%	50%			
0809	Less than 1 ton, add					100%	100%			
1010	CS series, 40-ton job lots, horizontal bridging, shop primer									
1040	Spans to 30' [G]	E-7	12	6.667	Ton	1,975	400	168	2,543	2,975
1500	For less than 40-ton job lots									
1502	For 30 to 39 tons, add				%	10%				
1504	20 to 29 tons, add					20%				

05 21 Steel Joist Framing

05 21 19 - Open Web Steel Joist Framing

05 21 19.10 Open Web Joists

		Crew	Daily Output	Labor-Hours	Unit	Material	2021 Bare Costs Labor	Equipment	Total	Total Incl O&P
1506	10 to 19 tons, add				%	30%				
1507	5 to 9 tons, add					50%	25%			
1508	1 to 4 tons, add					75%	50%			
1509	Less than 1 ton, add				↓	100%	100%			
6200	For shop prime paint other than mfrs. standard, add					20%				
6300	For bottom chord extensions, add per chord	G			Ea.	42.50			42.50	46.50
6400	Individual steel bearing plate, 6" x 6" x 1/4" with J-hook	G	1 Bric	160	.050	"	7.90	2.69		10.59	12.75

05 21 23 - Steel Joist Girder Framing

05 21 23.50 Joist Girders

		Crew	Daily Output	Labor-Hours	Unit	Material	2021 Bare Costs Labor	Equipment	Total	Total Incl O&P
0010	**JOIST GIRDERS**									
0015	Made from recycled materials									
7020	Joist girders, 40-ton job lots, shop primer	G	E-5	13	6.154	Ton	1,925	370	143	2,438	2,825
7100	For less than 40-ton job lots									
7102	For 30 to 39 tons, add					Ton	10%				
7104	20 to 29 tons, add						20%				
7106	10 to 19 tons, add						30%				
7107	5 to 9 tons, add						50%	25%			
7108	1 to 4 tons, add						75%	50%			
7109	Less than 1 ton, add						100%	100%			
8000	Trusses, 40-ton job lots, shop fabricated WT chords, shop primer	G	E-5	11	7.273	↓	6,300	435	169	6,904	7,775
8100	For less than 40-ton job lots									
8102	For 30 to 39 tons, add					Ton	10%				
8104	20 to 29 tons, add						20%				
8106	10 to 19 tons, add						30%				
8107	5 to 9 tons, add						50%	25%			
8108	1 to 4 tons, add						75%	50%			
8109	Less than 1 ton, add					↓	100%	100%			

05 31 Steel Decking

05 31 13 - Steel Floor Decking

05 31 13.50 Floor Decking

			Crew	Daily Output	Labor-Hours	Unit	Material	2021 Bare Costs Labor	Equipment	Total	Total Incl O&P
0010	**FLOOR DECKING**	R053100-10									
0015	Made from recycled materials										
5100	Non-cellular composite decking, galvanized, 1-1/2" deep, 16 ga.	G	E-4	3500	.009	S.F.	4.90	.56	.04	5.50	6.30
5120	18 ga.	G		3650	.009		3.02	.53	.04	3.59	4.19
5140	20 ga.	G		3800	.008		3.57	.51	.04	4.12	4.75
5200	2" deep, 22 ga.	G		3860	.008		2.41	.50	.04	2.95	3.47
5300	20 ga.	G		3600	.009		3.43	.54	.04	4.01	4.66
5400	18 ga.	G		3380	.009		3.15	.58	.04	3.77	4.41
5500	16 ga.	G		3200	.010		4.84	.61	.05	5.50	6.30
5700	3" deep, 22 ga.	G		3200	.010		2.63	.61	.05	3.29	3.88
5800	20 ga.	G		3000	.011		3.77	.65	.05	4.47	5.20
5900	18 ga.	G		2850	.011		3.15	.68	.05	3.88	4.59
6000	16 ga.	G	↓	2700	.012	↓	5.50	.72	.06	6.28	7.25

05 31 23 - Steel Roof Decking

05 31 23.50 Roof Decking

			Crew	Daily Output	Labor-Hours	Unit	Material	2021 Bare Costs Labor	Equipment	Total	Total Incl O&P
0010	**ROOF DECKING**										
0015	Made from recycled materials										
2100	Open type, 1-1/2" deep, Type B, wide rib, galv., 22 ga., under 50 sq.	G	E-4	4500	.007	S.F.	2.72	.43	.03	3.18	3.70
2400	Over 500 squares	G	↓	5100	.006	↓	1.96	.38	.03	2.37	2.77

132

05 31 Steel Decking

05 31 23 – Steel Roof Decking

05 31 23.50 Roof Decking

		Crew	Daily Output	Labor-Hours	Unit	Material	2021 Bare Costs Labor	Equipment	Total	Total Incl O&P
2600	20 ga., under 50 squares	G E-4	3865	.008	S.F.	3.03	.50	.04	3.57	4.15
2700	Over 500 squares	G	4300	.007		2.18	.45	.03	2.66	3.14
2900	18 ga., under 50 squares	G	3800	.008		3.89	.51	.04	4.44	5.10
3000	Over 500 squares	G	4300	.007		2.80	.45	.03	3.28	3.82
3050	16 ga., under 50 squares	G	3700	.009		5.25	.53	.04	5.82	6.65
3100	Over 500 squares	G	4200	.008		3.79	.46	.04	4.29	4.93
3200	3" deep, Type N, 22 ga., under 50 squares	G	3600	.009		3.75	.54	.04	4.33	5
3300	20 ga., under 50 squares	G	3400	.009		4.09	.57	.04	4.70	5.45
3400	18 ga., under 50 squares	G	3200	.010		5.30	.61	.05	5.96	6.85
3500	16 ga., under 50 squares	G	3000	.011		7.05	.65	.05	7.75	8.80
3700	4-1/2" deep, Type J, 20 ga., over 50 squares	G	2700	.012		3.99	.72	.06	4.77	5.55
3800	18 ga.	G	2460	.013		5.25	.79	.06	6.10	7.10
3900	16 ga.	G	2350	.014		6.95	.83	.06	7.84	9
4100	6" deep, Type H, 18 ga., over 50 squares	G	2000	.016		6.40	.97	.07	7.44	8.65
4200	16 ga.	G	1930	.017		9.10	1.01	.08	10.19	11.65
4300	14 ga.	G	1860	.017		11.65	1.05	.08	12.78	14.55
4500	7-1/2" deep, Type H, 18 ga., over 50 squares	G	1690	.019		7.60	1.15	.09	8.84	10.25
4600	16 ga.	G	1590	.020		9.45	1.22	.09	10.76	12.40
4700	14 ga.	G	1490	.021		11.80	1.31	.10	13.21	15.10
4800	For painted instead of galvanized, deduct					5%				
5000	For acoustical perforated with fiberglass insulation, add				S.F.	25%				
5100	For type F intermediate rib instead of type B wide rib, add	G				25%				
5150	For type A narrow rib instead of type B wide rib, add	G				25%				

05 31 33 – Steel Form Decking

05 31 33.50 Form Decking

		Crew	Daily Output	Labor-Hours	Unit	Material	2021 Bare Costs Labor	Equipment	Total	Total Incl O&P
0010	**FORM DECKING**									
0015	Made from recycled materials									
6100	Slab form, steel, 28 ga., 9/16" deep, Type UFS, uncoated	G E-4	4000	.008	S.F.	1.91	.49	.04	2.44	2.89
6200	Galvanized	G	4000	.008		1.69	.49	.04	2.22	2.65
6220	24 ga., 1" deep, Type UF1X, uncoated	G	3900	.008		1.87	.50	.04	2.41	2.87
6240	Galvanized	G	3900	.008		2.20	.50	.04	2.74	3.23
6300	24 ga., 1-5/16" deep, Type UFX, uncoated	G	3800	.008		1.99	.51	.04	2.54	3.02
6400	Galvanized	G	3800	.008		2.34	.51	.04	2.89	3.40
6500	22 ga., 1-5/16" deep, uncoated	G	3700	.009		2.52	.53	.04	3.09	3.62
6600	Galvanized	G	3700	.009		2.57	.53	.04	3.14	3.68
6700	22 ga., 2" deep, uncoated	G	3600	.009		3.27	.54	.04	3.85	4.49
6800	Galvanized	G	3600	.009		3.21	.54	.04	3.79	4.42
7000	Sheet metal edge closure form, 12" wide with 2 bends, galvanized									
7100	18 ga.	G E-14	360	.022	L.F.	5.30	1.38	.41	7.09	8.45
7200	16 ga.	G "	360	.022	"	7.20	1.38	.41	8.99	10.50

For customer support on your Concrete & Masonry Costs with RSMeans Data, call 800.448.8182.

133

05 35 Raceway Decking Assemblies

05 35 13 – Steel Cellular Decking

05 35 13.50 Cellular Decking

		Crew	Daily Output	Labor-Hours	Unit	Material	2021 Bare Costs Labor	Equipment	Total	Total Incl O&P
0010	**CELLULAR DECKING**									
0015	Made from recycled materials									
0200	Cellular units, galv, 1-1/2" deep, Type BC, 20-20 ga., over 15 squares [G]	E-4	1460	.022	S.F.	11.70	1.33	.10	13.13	15.05
0250	18-20 ga. [G]		1420	.023		9.20	1.37	.10	10.67	12.35
0300	18-18 ga. [G]		1390	.023		9.40	1.40	.11	10.91	12.65
0320	16-18 ga. [G]		1360	.024		16.25	1.43	.11	17.79	20
0340	16-16 ga. [G]		1330	.024		18.10	1.46	.11	19.67	22.50
0400	3" deep, Type NC, galvanized, 20-20 ga. [G]		1375	.023		12.90	1.41	.11	14.42	16.45
0500	18-20 ga. [G]		1350	.024		10.75	1.44	.11	12.30	14.15
0600	18-18 ga. [G]		1290	.025		10.70	1.51	.12	12.33	14.20
0700	16-18 ga. [G]		1230	.026		17.50	1.58	.12	19.20	22
0800	16-16 ga. [G]		1150	.028		19.05	1.69	.13	20.87	24
1000	4-1/2" deep, Type JC, galvanized, 18-20 ga. [G]		1100	.029		12.40	1.77	.14	14.31	16.55
1100	18-18 ga. [G]		1040	.031		12.30	1.87	.14	14.31	16.60
1200	16-18 ga. [G]		980	.033		20	1.99	.15	22.14	25
1300	16-16 ga. [G]		935	.034		22	2.08	.16	24.24	27.50
1500	For acoustical deck, add					15%				
1700	For cells used for ventilation, add					15%				
1900	For multi-story or congested site, add						50%			
8000	Metal deck and trench, 2" thick, 20 ga., combination									
8010	60% cellular, 40% non-cellular, inserts and trench [G]	R-4	1100	.036	S.F.	23.50	2.23	.14	25.87	29.50

05 53 Metal Gratings

05 53 13 – Bar Gratings

05 53 13.10 Floor Grating, Aluminum

		Crew	Daily Output	Labor-Hours	Unit	Material	2021 Bare Costs Labor	Equipment	Total	Total Incl O&P
0010	**FLOOR GRATING, ALUMINUM**, field fabricated from panels									
0015	Made from recycled materials									
0110	Bearing bars @ 1-3/16" OC, cross bars @ 4" OC,									
0111	Up to 300 S.F., 1" x 1/8" bar [G]	E-4	900	.036	S.F.	24.50	2.16	.17	26.83	30.50
0112	Over 300 S.F. [G]		850	.038		22.50	2.29	.18	24.97	28
0113	1-1/4" x 1/8" bar, up to 300 S.F. [G]		800	.040		21.50	2.43	.19	24.12	27.50
0114	Over 300 S.F. [G]		1000	.032		19.45	1.95	.15	21.55	24.50
0122	1-1/4" x 3/16" bar, up to 300 S.F. [G]		750	.043		30.50	2.59	.20	33.29	37.50
0124	Over 300 S.F. [G]		1000	.032		28	1.95	.15	30.10	33.50
0132	1-1/2" x 3/16" bar, up to 300 S.F. [G]		700	.046		44.50	2.78	.21	47.49	53.50
0134	Over 300 S.F. [G]		1000	.032		40.50	1.95	.15	42.60	47.50
0136	1-3/4" x 3/16" bar, up to 300 S.F. [G]		500	.064		48	3.89	.30	52.19	59.50
0138	Over 300 S.F. [G]		1000	.032		43.50	1.95	.15	45.60	51
0146	2-1/4" x 3/16" bar, up to 300 S.F. [G]		600	.053		66	3.24	.25	69.49	78
0148	Over 300 S.F. [G]		1000	.032		60	1.95	.15	62.10	69
0162	Cross bars @ 2" OC, 1" x 1/8", up to 300 S.F. [G]		600	.053		29	3.24	.25	32.49	37.50
0164	Over 300 S.F. [G]		1000	.032		26.50	1.95	.15	28.60	32
0172	1-1/4" x 3/16" bar, up to 300 S.F. [G]		600	.053		46.50	3.24	.25	49.99	56.50
0174	Over 300 S.F. [G]		1000	.032		42.50	1.95	.15	44.60	49.50
0182	1-1/2" x 3/16" bar, up to 300 S.F. [G]		600	.053		56	3.24	.25	59.49	67
0184	Over 300 S.F. [G]		1000	.032		50.50	1.95	.15	52.60	59
0186	1-3/4" x 3/16" bar, up to 300 S.F. [G]		600	.053		72	3.24	.25	75.49	84.50
0188	Over 300 S.F. [G]		1000	.032		65	1.95	.15	67.10	75
0200	For straight cuts, add				L.F.	4.56			4.56	5
0300	For curved cuts, add					5.60			5.60	6.15
0400	For straight banding, add [G]					5.80			5.80	6.40

05 53 13 – Bar Gratings

05 53 13.10 Floor Grating, Aluminum		Crew	Daily Output	Labor-Hours	Unit	Material	2021 Bare Costs Labor	Equipment	Total	Total Incl O&P
0500	For curved banding, add	G			L.F.	7.05			7.05	7.75
0600	For aluminum checkered plate nosings, add	G				7.55			7.55	8.35
0700	For straight toe plate, add	G				11.45			11.45	12.55
0800	For curved toe plate, add	G				13.35			13.35	14.65
1000	For cast aluminum abrasive nosings, add	G				11.10			11.10	12.20
1400	Extruded I bars are 10% less than 3/16" bars									
1600	Heavy duty, all extruded plank, 3/4" deep, 1.8 #/S.F.	G	E-4	1100	.029	S.F.	28.50	1.77	.14	30.41	34.50
1700	1-1/4" deep, 2.9 #/S.F.	G		1000	.032		37.50	1.95	.15	39.60	44
1800	1-3/4" deep, 4.2 #/S.F.	G		925	.035		43.50	2.10	.16	45.76	51.50
1900	2-1/4" deep, 5.0 #/S.F.	G		875	.037		70	2.22	.17	72.39	80.50
2100	For safety serrated surface, add						15%				

05 53 13.70 Floor Grating, Steel		Crew	Daily Output	Labor-Hours	Unit	Material	2021 Bare Costs Labor	Equipment	Total	Total Incl O&P
0010	**FLOOR GRATING, STEEL**, field fabricated from panels										
0015	Made from recycled materials										
0300	Platforms, to 12' high, rectangular	G	E-4	3150	.010	Lb.	3.29	.62	.05	3.96	4.63
0400	Circular	G	"	2300	.014	"	4.11	.85	.06	5.02	5.90
0410	Painted bearing bars @ 1-3/16"										
0412	Cross bars @ 4" OC, 3/4" x 1/8" bar, up to 300 S.F.	G	E-2	500	.112	S.F.	8.65	6.70	3.43	18.78	23.50
0414	Over 300 S.F.	G		750	.075		7.85	4.45	2.28	14.58	17.95
0422	1-1/4" x 3/16", up to 300 S.F.	G		400	.140		13.85	8.35	4.28	26.48	33
0424	Over 300 S.F.	G		600	.093		12.60	5.55	2.86	21.01	25.50
0432	1-1/2" x 3/16", up to 300 S.F.	G		400	.140		16	8.35	4.28	28.63	35
0434	Over 300 S.F.	G		600	.093		14.55	5.55	2.86	22.96	27.50
0436	1-3/4" x 3/16", up to 300 S.F.	G		400	.140		22	8.35	4.28	34.63	42
0438	Over 300 S.F.	G		600	.093		20	5.55	2.86	28.41	33.50
0452	2-1/4" x 3/16", up to 300 S.F.	G		300	.187		26.50	11.15	5.70	43.35	52.50
0454	Over 300 S.F.	G		450	.124		24	7.40	3.81	35.21	42
0462	Cross bars @ 2" OC, 3/4" x 1/8", up to 300 S.F.	G		500	.112		16.30	6.70	3.43	26.43	32
0464	Over 300 S.F.	G		750	.075		13.60	4.45	2.28	20.33	24.50
0472	1-1/4" x 3/16", up to 300 S.F.	G		400	.140		19.80	8.35	4.28	32.43	39.50
0474	Over 300 S.F.	G		600	.093		16.50	5.55	2.86	24.91	30
0482	1-1/2" x 3/16", up to 300 S.F.	G		400	.140		22.50	8.35	4.28	35.13	42
0484	Over 300 S.F.	G		600	.093		18.55	5.55	2.86	26.96	32
0486	1-3/4" x 3/16", up to 300 S.F.	G		400	.140		33.50	8.35	4.28	46.13	54.50
0488	Over 300 S.F.	G		600	.093		28	5.55	2.86	36.41	42.50
0502	2-1/4" x 3/16", up to 300 S.F.	G		300	.187		33	11.15	5.70	49.85	59.50
0504	Over 300 S.F.	G		450	.124		27.50	7.40	3.81	38.71	45.50
0690	For galvanized grating, add						25%				
0800	For straight cuts, add					L.F.	6.50			6.50	7.15
0900	For curved cuts, add						8.25			8.25	9.10
1000	For straight banding, add	G					7.05			7.05	7.75
1100	For curved banding, add	G					9.20			9.20	10.15
1200	For checkered plate nosings, add	G					8.10			8.10	8.90
1300	For straight toe or kick plate, add	G					14.50			14.50	15.95
1400	For curved toe or kick plate, add	G					16.45			16.45	18.05
1500	For abrasive nosings, add	G					10.75			10.75	11.85
1600	For safety serrated surface, bearing bars @ 1-3/16" OC, add						15%				
1700	Bearing bars @ 15/16" OC, add						25%				
2000	Stainless steel gratings, close spaced, 1" x 1/8" bars, up to 300 S.F.	G	E-4	450	.071	S.F.	53.50	4.32	.33	58.15	66
2100	Standard spacing, 3/4" x 1/8" bars	G		500	.064		104	3.89	.30	108.19	120
2200	1-1/4" x 3/16" bars	G		400	.080		130	4.86	.37	135.23	151

For customer support on your Concrete & Masonry Costs with RSMeans Data, call 800.448.8182.

135

05 53 16 – Plank Gratings

05 53 16.50 Grating Planks

		Crew	Daily Output	Labor-Hours	Unit	Material	2021 Bare Costs Labor	Equipment	Total	Total Incl O&P
0010	**GRATING PLANKS**, field fabricated from planks									
0020	Aluminum, 9-1/2" wide, 14 ga., 2" rib	[G] E-4	950	.034	L.F.	36	2.05	.16	38.21	43
0200	Galvanized steel, 9-1/2" wide, 14 ga., 2-1/2" rib	[G]	950	.034		19.55	2.05	.16	21.76	25
0300	4" rib	[G]	950	.034		21	2.05	.16	23.21	26.50
0500	12 ga., 2-1/2" rib	[G]	950	.034		22	2.05	.16	24.21	27.50
0600	3" rib	[G]	950	.034		26.50	2.05	.16	28.71	32.50
0800	Stainless steel, type 304, 16 ga., 2" rib	[G]	950	.034		40.50	2.05	.16	42.71	48
0900	Type 316	[G]	950	.034		56	2.05	.16	58.21	65.50

05 53 19 – Expanded Metal Gratings

05 53 19.10 Expanded Grating, Aluminum

		Crew	Daily Output	Labor-Hours	Unit	Material	2021 Bare Costs Labor	Equipment	Total	Total Incl O&P
0010	**EXPANDED GRATING, ALUMINUM**									
1200	Expanded aluminum, .65 #/S.F.	[G] E-4	1050	.030	S.F.	22	1.85	.14	23.99	27

05 53 19.20 Expanded Grating, Steel

		Crew	Daily Output	Labor-Hours	Unit	Material	2021 Bare Costs Labor	Equipment	Total	Total Incl O&P
0010	**EXPANDED GRATING, STEEL**									
2400	Expanded steel grating, at ground, 3.0 #/S.F.	[G] E-4	900	.036	S.F.	8.70	2.16	.17	11.03	13.10
2500	3.14 #/S.F.	[G]	900	.036		6.50	2.16	.17	8.83	10.70
2600	4.0 #/S.F.	[G]	850	.038		7.95	2.29	.18	10.42	12.50
2650	4.27 #/S.F.	[G]	850	.038		8.70	2.29	.18	11.17	13.30
2700	5.0 #/S.F.	[G]	800	.040		14.95	2.43	.19	17.57	20.50
2800	6.25 #/S.F.	[G]	750	.043		19.20	2.59	.20	21.99	25
2900	7.0 #/S.F.	[G]	700	.046		21	2.78	.21	23.99	28
3100	For flattened expanded steel grating, add					8%				
3300	For elevated installation above 15', add						15%			

05 53 19.30 Grating Frame

		Crew	Daily Output	Labor-Hours	Unit	Material	2021 Bare Costs Labor	Equipment	Total	Total Incl O&P
0010	**GRATING FRAME**, field fabricated									
0020	Aluminum, for gratings 1" to 1-1/2" deep	[G] 1 Sswk	70	.114	L.F.	3.28	6.90		10.18	14.25
0100	For each corner, add	[G]			Ea.	4.89			4.89	5.40

05 54 13 – Floor Plates

05 54 13.20 Checkered Plates

		Crew	Daily Output	Labor-Hours	Unit	Material	2021 Bare Costs Labor	Equipment	Total	Total Incl O&P
0010	**CHECKERED PLATES**, steel, field fabricated									
0015	Made from recycled materials									
0020	1/4" & 3/8", 2000 to 5000 S.F., bolted	[G] E-4	2900	.011	Lb.	1.05	.67	.05	1.77	2.26
0100	Welded	[G]	4400	.007	"	1.02	.44	.03	1.49	1.84
0300	Pit or trench cover and frame, 1/4" plate, 2' to 3' wide	[G]	100	.320	S.F.	12.65	19.45	1.49	33.59	45.50
0400	For galvanizing, add	[G]			Lb.	.30			.30	.33
0500	Platforms, 1/4" plate, no handrails included, rectangular	[G] E-4	4200	.008		3.49	.46	.04	3.99	4.60
0600	Circular	"	2500	.013		4.37	.78	.06	5.21	6.10

05 54 13.70 Trench Covers

		Crew	Daily Output	Labor-Hours	Unit	Material	2021 Bare Costs Labor	Equipment	Total	Total Incl O&P
0010	**TRENCH COVERS**, field fabricated									
0020	Cast iron grating with bar stops and angle frame, to 18" wide	[G] 1 Sswk	20	.400	L.F.	198	24		222	256
0100	Frame only (both sides of trench), 1" grating	[G]	45	.178		1.22	10.70		11.92	17.95
0150	2" grating	[G]	35	.229		2.58	13.80		16.38	24.50
0200	Aluminum, stock units, including frames and									
0210	3/8" plain cover plate, 4" opening	[G] E-4	205	.156	L.F.	15.15	9.50	.73	25.38	32
0300	6" opening	[G]	185	.173		19.05	10.50	.80	30.35	38
0400	10" opening	[G]	170	.188		33.50	11.45	.88	45.83	55.50
0500	16" opening	[G]	155	.206		38.50	12.55	.96	52.01	62.50

136

For customer support on your Concrete & Masonry Costs with RSMeans Data, call 800.448.8182.

05 54 Metal Floor Plates

05 54 13 – Floor Plates

05 54 13.70 Trench Covers		Crew	Daily Output	Labor-Hours	Unit	Material	2021 Bare Costs Labor	Equipment	Total	Total Incl O&P
0700	Add per inch for additional widths to 24"	G				L.F.	1.68			1.68	1.85
0900	For custom fabrication, add						50%				
1100	For 1/4" plain cover plate, deduct						12%				
1500	For cover recessed for tile, 1/4" thick, deduct						12%				
1600	3/8" thick, add						5%				
1800	For checkered plate cover, 1/4" thick, deduct						12%				
1900	3/8" thick, add						2%				
2100	For slotted or round holes in cover, 1/4" thick, add						3%				
2200	3/8" thick, add						4%				
2300	For abrasive cover, add						12%				

05 55 Metal Stair Treads and Nosings

05 55 19 – Metal Stair Tread Covers

05 55 19.50 Stair Tread Covers for Renovation

		Crew	Daily Output	Labor-Hours	Unit	Material	2021 Bare Costs Labor	Equipment	Total	Total Incl O&P
0010	**STAIR TREAD COVERS FOR RENOVATION**									
0205	Extruded tread cover with nosing, pre-drilled, includes screws									
0210	Aluminum with black abrasive strips, 9" wide x 3' long	1 Carp	24	.333	Ea.	105	18.25		123.25	142
0220	4' long		22	.364		137	19.90		156.90	181
0230	5' long		20	.400		186	22		208	238
0240	11" wide x 3' long		24	.333		147	18.25		165.25	189
0250	4' long		22	.364		201	19.90		220.90	251
0260	5' long		20	.400		240	22		262	297
0305	Black abrasive strips with yellow front strips									
0310	Aluminum, 9" wide x 3' long	1 Carp	24	.333	Ea.	127	18.25		145.25	167
0320	4' long		22	.364		170	19.90		189.90	217
0330	5' long		20	.400		218	22		240	272
0340	11" wide x 3' long		24	.333		162	18.25		180.25	205
0350	4' long		22	.364		206	19.90		225.90	257
0360	5' long		20	.400		269	22		291	330
0405	Black abrasive strips with photoluminescent front strips									
0410	Aluminum, 9" wide x 3' long	1 Carp	24	.333	Ea.	161	18.25		179.25	204
0420	4' long		22	.364		182	19.90		201.90	231
0430	5' long		20	.400		256	22		278	315
0440	11" wide x 3' long		24	.333		182	18.25		200.25	228
0450	4' long		22	.364		243	19.90		262.90	297
0460	5' long		20	.400		305	22		327	370

05 56 Metal Castings

05 56 13 – Metal Construction Castings

05 56 13.50 Construction Castings

			Crew	Daily Output	Labor-Hours	Unit	Material	2021 Bare Costs Labor	Equipment	Total	Total Incl O&P
0010	**CONSTRUCTION CASTINGS**										
0020	Manhole covers and frames, see Section 33 44 13.13										
0100	Column bases, cast iron, 16" x 16", approx. 65 lb.	G	E-4	46	.696	Ea.	139	42.50	3.23	184.73	222
0200	32" x 32", approx. 256 lb.	G	"	23	1.391		505	84.50	6.45	595.95	695
0400	Cast aluminum for wood columns, 8" x 8"	G	1 Carp	32	.250		34	13.70		47.70	58
0500	12" x 12"	G	"	32	.250		60.50	13.70		74.20	87
0600	Miscellaneous C.I. castings, light sections, less than 150 lb.	G	E-4	3200	.010	Lb.	9.05	.61	.05	9.71	11
1100	Heavy sections, more than 150 lb.	G		4200	.008		5.25	.46	.04	5.75	6.55
1300	Special low volume items	G		3200	.010		11.25	.61	.05	11.91	13.40

137

05 56 Metal Castings

05 56 13 – Metal Construction Castings

05 56 13.50 Construction Castings		Crew	Daily Output	Labor-Hours	Unit	Material	2021 Bare Costs Labor	Equipment	Total	Total Incl O&P
1500	For ductile iron, add				Lb.	100%				

05 58 Formed Metal Fabrications

05 58 21 – Formed Chain

05 58 21.05 Alloy Steel Chain

	05 58 21.05 Alloy Steel Chain		Crew	Daily Output	Labor-Hours	Unit	Material	2021 Bare Costs Labor	Equipment	Total	Total Incl O&P
0010	**ALLOY STEEL CHAIN**, Grade 80, for lifting										
0015	Self-colored, cut lengths, 1/4"	G	E-17	4	4	C.L.F.	294	245		539	705
0020	3/8"	G		2	8		1,350	490		1,840	2,225
0030	1/2"	G		1.20	13.333		2,425	815		3,240	3,925
0040	5/8"	G		.72	22.222		3,175	1,350		4,525	5,575
0050	3/4"	G	E-18	.48	83.333		1,750	5,025	3,175	9,950	13,200
0060	7/8"	G		.40	100		2,950	6,050	3,825	12,825	16,700
0070	1"	G		.35	114		5,450	6,900	4,350	16,700	21,400
0080	1-1/4"	G		.24	167		13,500	10,100	6,350	29,950	37,300
0110	Hook, Grade 80, Clevis slip, 1/4"	G				Ea.	30			30	33
0120	3/8"	G					48			48	53
0130	1/2"	G					74.50			74.50	82
0140	5/8"	G					101			101	112
0150	3/4"	G					129			129	142
0160	Hook, Grade 80, eye/sling w/hammerlock coupling, 15 ton	G					430			430	475
0170	22 ton	G					1,975			1,975	2,175
0180	37 ton	G					3,575			3,575	3,925

Estimating Tips
06 05 00 Common Work Results for Wood, Plastics, and Composites

- Common to any wood-framed structure are the accessory connector items such as screws, nails, adhesives, hangers, connector plates, straps, angles, and hold-downs. For typical wood-framed buildings, such as residential projects, the aggregate total for these items can be significant, especially in areas where seismic loading is a concern. For floor and wall framing, the material cost is based on 10 to 25 lbs. of accessory connectors per MBF. Hold-downs, hangers, and other connectors should be taken off by the piece.

  Included with material costs are fasteners for a normal installation. Gordian's RSMeans engineers use manufacturers' recommendations, written specifications, and/or standard construction practice for the sizing and spacing of fasteners. Prices for various fasteners are shown for informational purposes only. Adjustments should be made if unusual fastening conditions exist.

06 10 00 Carpentry

- Lumber is a traded commodity and therefore sensitive to supply and demand in the marketplace. Even with "budgetary" estimating of wood-framed projects, it is advisable to call local suppliers for the latest market pricing.

- The common quantity unit for wood-framed projects is "thousand board feet" (MBF). A board foot is a volume of wood—1" x 1' x 1' or 144 cubic inches. Board-foot quantities are generally calculated using nominal material dimensions—dressed sizes are ignored. Board foot per lineal foot of any stick of lumber can be calculated by dividing the nominal cross-sectional area by 12. As an example, 2,000 lineal feet of 2 x 12 equates to 4 MBF by dividing the nominal area, 2 x 12, by 12, which equals 2, and multiplying that by 2,000 to give 4,000 board feet. This simple rule applies to all nominal dimensioned lumber.

- Waste is an issue of concern at the quantity takeoff for any area of construction. Framing lumber is sold in even foot lengths, i.e., 8', 10', 12', 14', 16', and depending on spans, wall heights, and the grade of lumber, waste is inevitable. A rule of thumb for lumber waste is 5–10% depending on material quality and the complexity of the framing.

- Wood in various forms and shapes is used in many projects, even where the main structural framing is steel, concrete, or masonry. Plywood as a back-up partition material and 2x boards used as blocking and cant strips around roof edges are two common examples. The estimator should ensure that the costs of all wood materials are included in the final estimate.

06 20 00 Finish Carpentry

- It is necessary to consider the grade of workmanship when estimating labor costs for erecting millwork and an interior finish. In practice, there are three grades: premium, custom, and economy. The RSMeans daily output for base and case moldings is in the range of 200 to 250 L.F. per carpenter per day. This is appropriate for most average custom-grade projects. For premium projects, an adjustment to productivity of 25–50% should be made, depending on the complexity of the job.

Reference Numbers

Reference numbers are shown at the beginning of some major classifications. These numbers refer to related items in the Reference Section. The reference information may be an estimating procedure, an alternate pricing method, or technical information.

Note: Not all subdivisions listed here necessarily appear. ∎

Same Data. Simplified.

Enjoy the convenience and efficiency of accessing your costs anywhere:

- **Skip the multiplier** by setting your location
- **Quickly search,** edit, favorite and share costs
- **Stay on top of price changes** with automatic updates

Discover more at rsmeans.com/online

06 05 05 – Selective Demolition for Wood, Plastics, and Composites

06 05 05.10 Selective Demolition Wood Framing	Crew	Daily Output	Labor-Hours	Unit	Material	2021 Bare Costs Labor	Equipment	Total	Total Incl O&P
0010 **SELECTIVE DEMOLITION WOOD FRAMING** R024119-10									
0100 Timber connector, nailed, small	1 Clab	96	.083	Ea.		3.70		3.70	5.50
0110 Medium		60	.133			5.90		5.90	8.85
0120 Large		48	.167			7.40		7.40	11.05
0130 Bolted, small		48	.167			7.40		7.40	11.05
0140 Medium		32	.250			11.10		11.10	16.55
0150 Large		24	.333			14.80		14.80	22
3162 Alternate pricing method	B-1	1.10	21.818	M.B.F.		985		985	1,475

06 11 Wood Framing

06 11 10 – Framing with Dimensional, Engineered or Composite Lumber

06 11 10.02 Blocking

06 11 10.02 Blocking	Crew	Daily Output	Labor-Hours	Unit	Material	2021 Bare Costs Labor	Equipment	Total	Total Incl O&P
0010 **BLOCKING**									
2600 Miscellaneous, to wood construction									
2620 2" x 4"	1 Carp	.17	47.059	M.B.F.	1,050	2,575		3,625	5,000
2625 Pneumatic nailed		.21	38.095		1,050	2,075		3,125	4,275
2660 2" x 8"		.27	29.630		1,200	1,625		2,825	3,750
2665 Pneumatic nailed		.33	24.242		1,225	1,325		2,550	3,325
2720 To steel construction									
2740 2" x 4"	1 Carp	.14	57.143	M.B.F.	1,050	3,125		4,175	5,825
2780 2" x 8"	"	.21	38.095	"	1,200	2,075		3,275	4,425

06 11 10.04 Wood Bracing

06 11 10.04 Wood Bracing	Crew	Daily Output	Labor-Hours	Unit	Material	2021 Bare Costs Labor	Equipment	Total	Total Incl O&P
0010 **WOOD BRACING**									
0012 Let-in, with 1" x 6" boards, studs @ 16" OC	1 Carp	150	.053	L.F.	1.15	2.92		4.07	5.60
0202 Studs @ 24" OC	"	230	.035	"	1.15	1.90		3.05	4.10

06 11 10.06 Bridging

06 11 10.06 Bridging	Crew	Daily Output	Labor-Hours	Unit	Material	2021 Bare Costs Labor	Equipment	Total	Total Incl O&P
0010 **BRIDGING**									
0012 Wood, for joists 16" OC, 1" x 3"	1 Carp	130	.062	Pr.	.98	3.37		4.35	6.05
0017 Pneumatic nailed		170	.047		1.06	2.57		3.63	5
0102 2" x 3" bridging		130	.062		.98	3.37		4.35	6.10
0107 Pneumatic nailed		170	.047		1.02	2.57		3.59	4.97
0302 Steel, galvanized, 18 ga., for 2" x 10" joists at 12" OC		130	.062		1.77	3.37		5.14	6.95
0352 16" OC		135	.059		.03	3.24		3.27	4.87
0402 24" OC		140	.057		2.62	3.13		5.75	7.55
0602 For 2" x 14" joists at 16" OC		130	.062		2.26	3.37		5.63	7.50
0902 Compression type, 16" OC, 2" x 8" joists		200	.040		1.34	2.19		3.53	4.75
1002 2" x 12" joists		200	.040		1.34	2.19		3.53	4.74

06 11 10.10 Beam and Girder Framing

06 11 10.10 Beam and Girder Framing	Crew	Daily Output	Labor-Hours	Unit	Material	2021 Bare Costs Labor	Equipment	Total	Total Incl O&P
0010 **BEAM AND GIRDER FRAMING** R061110-30									
3500 Single, 2" x 6"	2 Carp	.70	22.857	M.B.F.	1,075	1,250		2,325	3,050
3505 Pneumatic nailed		.81	19.704		1,075	1,075		2,150	2,775
3520 2" x 8"		.86	18.605		1,200	1,025		2,225	2,850
3525 Pneumatic nailed		1	16.048		1,225	880		2,105	2,650
3540 2" x 10"		1	16		1,225	875		2,100	2,650
3545 Pneumatic nailed		1.16	13.793		1,225	755		1,980	2,475
3560 2" x 12"		1.10	14.545		1,250	795		2,045	2,575
3565 Pneumatic nailed		1.28	12.539		1,275	685		1,960	2,425
3580 2" x 14"		1.17	13.675		1,525	750		2,275	2,800
3585 Pneumatic nailed		1.36	11.791		1,525	645		2,170	2,675
3600 3" x 8"		1.10	14.545		2,050	795		2,845	3,450

06 11 Wood Framing

06 11 10 – Framing with Dimensional, Engineered or Composite Lumber

06 11 10.10 Beam and Girder Framing

		Crew	Daily Output	Labor-Hours	Unit	Material	2021 Bare Costs Labor	Equipment	Total	Total Incl O&P
3620	3" x 10"	2 Carp	1.25	12.800	M.B.F.	1,975	700		2,675	3,225
3640	3" x 12"		1.35	11.852		2,350	650		3,000	3,550
3660	3" x 14"	↓	1.40	11.429		1,850	625		2,475	2,950
3680	4" x 8"	F-3	2.66	15.038		2,425	845	179	3,449	4,125
3700	4" x 10"		3.16	12.658		2,400	710	151	3,261	3,875
3720	4" x 12"		3.60	11.111		1,825	625	132	2,582	3,075
3740	4" x 14"	↓	3.96	10.101		2,000	565	120	2,685	3,175
4000	Double, 2" x 6"	2 Carp	1.25	12.800		1,075	700		1,775	2,225
4005	Pneumatic nailed		1.45	11.034		1,075	605		1,680	2,075
4020	2" x 8"		1.60	10		1,200	545		1,745	2,150
4025	Pneumatic nailed		1.86	8.621		1,225	470		1,695	2,050
4040	2" x 10"		1.92	8.333		1,225	455		1,680	2,025
4045	Pneumatic nailed		2.23	7.185		1,225	395		1,620	1,925
4060	2" x 12"		2.20	7.273		1,250	400		1,650	1,975
4065	Pneumatic nailed		2.55	6.275		1,275	345		1,620	1,900
4080	2" x 14"		2.45	6.531		1,525	355		1,880	2,200
4085	Pneumatic nailed		2.84	5.634		1,525	310		1,835	2,150
5000	Triple, 2" x 6"		1.65	9.697		1,075	530		1,605	1,975
5005	Pneumatic nailed		1.91	8.377		1,075	460		1,535	1,850
5020	2" x 8"		2.10	7.619		1,200	415		1,615	1,950
5025	Pneumatic nailed		2.44	6.568		1,225	360		1,585	1,875
5040	2" x 10"		2.50	6.400		1,225	350		1,575	1,875
5045	Pneumatic nailed		2.90	5.517		1,225	300		1,525	1,800
5060	2" x 12"		2.85	5.614		1,250	305		1,555	1,825
5065	Pneumatic nailed		3.31	4.840		1,275	265		1,540	1,800
5080	2" x 14"		3.15	5.079		1,525	278		1,803	2,100
5085	Pneumatic nailed	↓	3.35	4.770	↓	1,525	261		1,786	2,100

06 11 10.12 Ceiling Framing

		Crew	Daily Output	Labor-Hours	Unit	Material	2021 Bare Costs Labor	Equipment	Total	Total Incl O&P
0010	**CEILING FRAMING**									
6400	Suspended, 2" x 3"	2 Carp	.50	32	M.B.F.	1,200	1,750		2,950	3,950
6450	2" x 4"		.59	27.119		1,050	1,475		2,525	3,375
6500	2" x 6"		.80	20		1,075	1,100		2,175	2,800
6550	2" x 8"	↓	.86	18.605	↓	1,200	1,025		2,225	2,850

06 11 10.14 Posts and Columns

		Crew	Daily Output	Labor-Hours	Unit	Material	2021 Bare Costs Labor	Equipment	Total	Total Incl O&P
0010	**POSTS AND COLUMNS**									
0400	4" x 4"	2 Carp	.52	30.769	M.B.F.	1,725	1,675		3,400	4,400
0420	4" x 6"		.55	29.091		1,450	1,600		3,050	3,975
0440	4" x 8"		.59	27.119		2,450	1,475		3,925	4,900
0460	6" x 6"		.65	24.615		2,300	1,350		3,650	4,525
0480	6" x 8"		.70	22.857		2,425	1,250		3,675	4,525
0500	6" x 10"	↓	.75	21.333	↓	1,975	1,175		3,150	3,925

06 11 10.18 Joist Framing

		Crew	Daily Output	Labor-Hours	Unit	Material	2021 Bare Costs Labor	Equipment	Total	Total Incl O&P
0010	**JOIST FRAMING** R061110-30									
2650	Joists, 2" x 4"	2 Carp	.83	19.277	M.B.F.	1,050	1,050		2,100	2,725
2655	Pneumatic nailed		.96	16.667		1,050	910		1,960	2,525
2680	2" x 6"		1.25	12.800		1,075	700		1,775	2,225
2685	Pneumatic nailed		1.44	11.111		1,075	610		1,685	2,075
2700	2" x 8"		1.46	10.959		1,200	600		1,800	2,225
2705	Pneumatic nailed		1.68	9.524		1,225	520		1,745	2,125
2720	2" x 10"		1.49	10.738		1,225	585		1,810	2,225
2725	Pneumatic nailed		1.71	9.357		1,225	510		1,735	2,125
2740	2" x 12"		1.75	9.143		1,250	500		1,750	2,125

06 11 Wood Framing

06 11 10 – Framing with Dimensional, Engineered or Composite Lumber

06 11 10.18 Joist Framing		Crew	Daily Output	Labor-Hours	Unit	Material	2021 Bare Costs Labor	Equipment	Total	Total Incl O&P
2745	Pneumatic nailed	2 Carp	2.01	7.960	M.B.F.	1,275	435		1,710	2,050
2760	2" x 14"		1.79	8.939		1,525	490		2,015	2,400
2765	Pneumatic nailed		2.06	7.767		1,525	425		1,950	2,325
2780	3" x 6"		1.39	11.511		2,025	630		2,655	3,175
2790	3" x 8"		1.90	8.421		2,050	460		2,510	2,950
2800	3" x 10"		1.95	8.205		1,975	450		2,425	2,850
2820	3" x 12"		1.80	8.889		2,350	485		2,835	3,300
2840	4" x 6"		1.60	10		1,450	545		1,995	2,425
2860	4" x 10"		2	8		2,400	440		2,840	3,300
2880	4" x 12"		1.80	8.889		1,825	485		2,310	2,725

06 11 10.24 Miscellaneous Framing

		Crew	Daily Output	Labor-Hours	Unit	Material	2021 Bare Costs Labor	Equipment	Total	Total Incl O&P
0010	**MISCELLANEOUS FRAMING**									
8500	Firestops, 2" x 4"	2 Carp	.51	31.373	M.B.F.	1,050	1,725		2,775	3,700
8505	Pneumatic nailed		.62	25.806		1,050	1,400		2,450	3,275
8520	2" x 6"		.60	26.667		1,075	1,450		2,525	3,350
8525	Pneumatic nailed		.73	21.858		1,075	1,200		2,275	2,950
8540	2" x 8"		.60	26.667		1,200	1,450		2,650	3,500
8560	2" x 12"		.70	22.857		1,250	1,250		2,500	3,250
8600	Nailers, treated, wood construction, 2" x 4"		.53	30.189		1,350	1,650		3,000	3,950
8605	Pneumatic nailed		.64	25.157		1,350	1,375		2,725	3,550
8620	2" x 6"		.75	21.333		1,250	1,175		2,425	3,150
8625	Pneumatic nailed		.90	17.778		1,275	970		2,245	2,850
8640	2" x 8"		.93	17.204		1,425	940		2,365	2,975
8645	Pneumatic nailed		1.12	14.337		1,450	785		2,235	2,750
8660	Steel construction, 2" x 4"		.50	32		1,350	1,750		3,100	4,100
8680	2" x 6"		.70	22.857		1,250	1,250		2,500	3,250
8700	2" x 8"		.87	18.391		1,425	1,000		2,425	3,075
8760	Rough bucks, treated, for doors or windows, 2" x 6"		.40	40		1,250	2,200		3,450	4,675
8765	Pneumatic nailed		.48	33.333		1,275	1,825		3,100	4,125
8780	2" x 8"		.51	31.373		1,425	1,725		3,150	4,125
8785	Pneumatic nailed		.61	26.144		1,450	1,425		2,875	3,700
8800	Stair stringers, 2" x 10"		.22	72.727		1,225	3,975		5,200	7,300
8820	2" x 12"		.26	61.538		1,250	3,375		4,625	6,400
8840	3" x 10"		.31	51.613		1,975	2,825		4,800	6,400
8860	3" x 12"		.38	42.105		2,350	2,300		4,650	6,025

06 11 10.30 Roof Framing

		Crew	Daily Output	Labor-Hours	Unit	Material	2021 Bare Costs Labor	Equipment	Total	Total Incl O&P
0010	**ROOF FRAMING** R061110-30									
6070	Fascia boards, 2" x 8"	2 Carp	.30	53.333	M.B.F.	1,200	2,925		4,125	5,675
6080	2" x 10"		.30	53.333		1,225	2,925		4,150	5,700
7000	Rafters, to 4 in 12 pitch, 2" x 6"		1	16		1,075	875		1,950	2,475
7060	2" x 8"		1.26	12.698		1,200	695		1,895	2,350
7300	Hip and valley rafters, 2" x 6"		.76	21.053		1,075	1,150		2,225	2,900
7360	2" x 8"		.96	16.667		1,200	910		2,110	2,675
7540	Hip and valley jacks, 2" x 6"		.60	26.667		1,075	1,450		2,525	3,350
7600	2" x 8"		.65	24.615		1,200	1,350		2,550	3,325
7780	For slopes steeper than 4 in 12, add						30%			
7790	For dormers or complex roofs, add						50%			
7800	Rafter tie, 1" x 4", #3	2 Carp	.27	59.259	M.B.F.	2,475	3,250		5,725	7,575
7820	Ridge board, #2 or better, 1" x 6"		.30	53.333		2,300	2,925		5,225	6,875
7840	1" x 8"		.37	43.243		2,525	2,375		4,900	6,300
7860	1" x 10"		.42	38.095		3,425	2,075		5,500	6,850
7880	2" x 6"		.50	32		1,075	1,750		2,825	3,800

06 11 Wood Framing

06 11 10 – Framing with Dimensional, Engineered or Composite Lumber

06 11 10.30 Roof Framing		Crew	Daily Output	Labor-Hours	Unit	Material	2021 Bare Costs Labor	Equipment	Total	Total Incl O&P
7900	2" x 8"	2 Carp	.60	26.667	M.B.F.	1,200	1,450		2,650	3,500
7920	2" x 10"		.66	24.242		1,225	1,325		2,550	3,325
7940	Roof cants, split, 4" x 4"		.86	18.605		1,725	1,025		2,750	3,425
7960	6" x 6"		1.80	8.889		2,275	485		2,760	3,250
7980	Roof curbs, untreated, 2" x 6"		.52	30.769		1,075	1,675		2,750	3,675
8000	2" x 12"		.80	20		1,250	1,100		2,350	3,000

06 11 10.32 Sill and Ledger Framing										
0010	**SILL AND LEDGER FRAMING**									
4482	Ledgers, nailed, 2" x 4"	2 Carp	.50	32	M.B.F.	1,050	1,750		2,800	3,775
4484	2" x 6"		.60	26.667		1,075	1,450		2,525	3,350
4486	Bolted, not including bolts, 3" x 8"		.65	24.615		2,050	1,350		3,400	4,250
4488	3" x 12"		.70	22.857		2,325	1,250		3,575	4,450
4490	Mud sills, redwood, construction grade, 2" x 4"		.59	27.119		8,200	1,475		9,675	11,300
4492	2" x 6"		.78	20.513		9,350	1,125		10,475	12,000
4500	Sills, 2" x 4"		.40	40		1,050	2,200		3,250	4,425
4520	2" x 6"		.55	29.091		1,050	1,600		2,650	3,525
4540	2" x 8"		.67	23.881		1,200	1,300		2,500	3,275
4600	Treated, 2" x 4"		.36	44.444		1,350	2,425		3,775	5,100
4620	2" x 6"		.50	32		1,250	1,750		3,000	4,000
4640	2" x 8"		.60	26.667		1,425	1,450		2,875	3,750
4700	4" x 4"		.60	26.667		1,675	1,450		3,125	4,025
4720	4" x 6"		.70	22.857		1,450	1,250		2,700	3,475
4740	4" x 8"		.80	20		2,300	1,100		3,400	4,175
4760	4" x 10"		.87	18.391		2,150	1,000		3,150	3,850

06 11 10.34 Sleepers										
0010	**SLEEPERS**									
0300	On concrete, treated, 1" x 2"	2 Carp	.39	41.026	M.B.F.	2,700	2,250		4,950	6,300
0320	1" x 3"		.50	32		2,925	1,750		4,675	5,850
0340	2" x 4"		.99	16.162		1,575	885		2,460	3,050
0360	2" x 6"		1.30	12.308		1,475	675		2,150	2,625

06 11 10.38 Treated Lumber Framing Material										
0010	**TREATED LUMBER FRAMING MATERIAL**									
0100	2" x 4"				M.B.F.	1,350			1,350	1,475
0110	2" x 6"					1,250			1,250	1,375
0120	2" x 8"					1,425			1,425	1,575
0130	2" x 10"					1,200			1,200	1,325
0140	2" x 12"					1,825			1,825	2,025
0200	4" x 4"					1,675			1,675	1,850
0210	4" x 6"					1,450			1,450	1,600
0220	4" x 8"					2,300			2,300	2,550

06 11 10.40 Wall Framing										
0010	**WALL FRAMING** R061110-30									
5860	Headers over openings, 2" x 6"	2 Carp	.36	44.444	M.B.F.	1,075	2,425		3,500	4,800
5865	2" x 6", pneumatic nailed		.43	37.209		1,075	2,025		3,100	4,225
5880	2" x 8"		.45	35.556		1,200	1,950		3,150	4,225
5885	2" x 8", pneumatic nailed		.54	29.630		1,225	1,625		2,850	3,775
5900	2" x 10"		.53	30.189		1,225	1,650		2,875	3,825
5905	2" x 10", pneumatic nailed		.67	23.881		1,225	1,300		2,525	3,300

For customer support on your Concrete & Masonry Costs with RSMeans Data, call 800.448.8182.

143

06 11 Wood Framing

06 11 10 – Framing with Dimensional, Engineered or Composite Lumber

06 11 10.42 Furring

		Crew	Daily Output	Labor-Hours	Unit	Material	2021 Bare Costs Labor	2021 Bare Costs Equipment	Total	Total Incl O&P
0010	**FURRING**									
0012	Wood strips, 1" x 2", on walls, on wood	1 Carp	550	.015	L.F.	.42	.80		1.22	1.65
0015	On wood, pneumatic nailed		710	.011		.42	.62		1.04	1.38
0300	On masonry		495	.016		.45	.88		1.33	1.82
0400	On concrete		260	.031		.45	1.68		2.13	3.01
0600	1" x 3", on walls, on wood		550	.015		.62	.80		1.42	1.87
0605	On wood, pneumatic nailed		710	.011		.62	.62		1.24	1.60
0700	On masonry		495	.016		.67	.88		1.55	2.06
0800	On concrete		260	.031		.67	1.68		2.35	3.25
0850	On ceilings, on wood		350	.023		.62	1.25		1.87	2.55
0855	On wood, pneumatic nailed		450	.018		.62	.97		1.59	2.13
0900	On masonry		320	.025		.67	1.37		2.04	2.78
0950	On concrete	↓	210	.038	↓	.67	2.08		2.75	3.85

06 13 Heavy Timber Construction

06 13 23 – Heavy Timber Framing

06 13 23.10 Heavy Framing

		Crew	Daily Output	Labor-Hours	Unit	Material	2021 Bare Costs Labor	2021 Bare Costs Equipment	Total	Total Incl O&P
0010	**HEAVY FRAMING**									
0020	Beams, single 6" x 10"	2 Carp	1.10	14.545	M.B.F.	2,150	795		2,945	3,575
0100	Single 8" x 16"		1.20	13.333		2,675	730		3,405	4,025
0200	Built from 2" lumber, multiple 2" x 14"		.90	17.778		1,525	970		2,495	3,125
0210	Built from 3" lumber, multiple 3" x 6"		.70	22.857		2,025	1,250		3,275	4,100
0220	Multiple 3" x 8"		.80	20		2,050	1,100		3,150	3,875
0230	Multiple 3" x 10"		.90	17.778		1,975	970		2,945	3,625
0240	Multiple 3" x 12"		1	16		2,325	875		3,200	3,875
0250	Built from 4" lumber, multiple 4" x 6"		.80	20		1,425	1,100		2,525	3,200
0260	Multiple 4" x 8"		.90	17.778		2,425	970		3,395	4,125
0270	Multiple 4" x 10"		1	16		2,400	875		3,275	3,925
0280	Multiple 4" x 12"		1.10	14.545		1,800	795		2,595	3,200
0290	Columns, structural grade, 1500 fb, 4" x 4"		.60	26.667		2,425	1,450		3,875	4,850
0300	6" x 6"		.65	24.615		2,325	1,350		3,675	4,550
0400	8" x 8"		.70	22.857		2,050	1,250		3,300	4,125
0500	10" x 10"		.75	21.333		2,475	1,175		3,650	4,475
0600	12" x 12"		.80	20		2,150	1,100		3,250	4,000
0800	Floor planks, 2" thick, T&G, 2" x 6"		1.05	15.238		1,900	835		2,735	3,350
0900	2" x 10"		1.10	14.545		1,925	795		2,720	3,300
1100	3" thick, 3" x 6"		1.05	15.238		2,275	835		3,110	3,750
1200	3" x 10"		1.10	14.545		2,300	795		3,095	3,750
1400	Girders, structural grade, 12" x 12"		.80	20		2,150	1,100		3,250	4,000
1500	10" x 16"	↓	1	16	↓	3,600	875		4,475	5,250

06 16 Sheathing

06 16 13 – Insulating Sheathing

06 16 13.10 Insulating Sheathing

		Crew	Daily Output	Labor-Hours	Unit	Material	2021 Bare Costs Labor	Equipment	Total	Total Incl O&P
0010	**INSULATING SHEATHING**									
0020	Expanded polystyrene, 1#/C.F. density, 3/4" thick, R2.89	G 2 Carp	1400	.011	S.F.	.40	.63		1.03	1.37
0030	1" thick, R3.85	G	1300	.012		.48	.67		1.15	1.53
0040	2" thick, R7.69	G	1200	.013		.80	.73		1.53	1.97
0050	Extruded polystyrene, 15 psi compressive strength, 1" thick, R5	G	1300	.012		.68	.67		1.35	1.75
0060	2" thick, R10	G	1200	.013		.83	.73		1.56	2.01
0070	Polyisocyanurate, 2#/C.F. density, 3/4" thick	G	1400	.011		.65	.63		1.28	1.64
0080	1" thick	G	1300	.012		.62	.67		1.29	1.68
0090	1-1/2" thick	G	1250	.013		1.06	.70		1.76	2.21
0100	2" thick	G	1200	.013		1.11	.73		1.84	2.31

06 16 23 – Subflooring

06 16 23.10 Subfloor

		Crew	Daily Output	Labor-Hours	Unit	Material	2021 Bare Costs Labor	Equipment	Total	Total Incl O&P
0010	**SUBFLOOR** R061636-20									
0011	Plywood, CDX, 1/2" thick	2 Carp	1500	.011	SF Flr.	1	.58		1.58	1.97
0015	Pneumatic nailed		1860	.009		1	.47		1.47	1.80
0100	5/8" thick		1350	.012		1.16	.65		1.81	2.24
0105	Pneumatic nailed		1674	.010		1.16	.52		1.68	2.05
0200	3/4" thick		1250	.013		1.38	.70		2.08	2.57
0205	Pneumatic nailed		1550	.010		1.38	.56		1.94	2.36
0300	1-1/8" thick, 2-4-1 including underlayment		1050	.015		2.88	.83		3.71	4.41
0450	1" x 8", laid regular		1000	.016		2.69	.88		3.57	4.27
0460	Laid diagonal		850	.019		2.69	1.03		3.72	4.50
0500	1" x 10", laid regular		1100	.015		3.58	.80		4.38	5.10
0600	Laid diagonal		900	.018		3.58	.97		4.55	5.40
8990	Subfloor adhesive, 3/8" bead	1 Carp	2300	.003	L.F.	.10	.19		.29	.39

06 16 26 – Underlayment

06 16 26.10 Wood Product Underlayment

		Crew	Daily Output	Labor-Hours	Unit	Material	2021 Bare Costs Labor	Equipment	Total	Total Incl O&P
0010	**WOOD PRODUCT UNDERLAYMENT**									
0015	Plywood, underlayment grade, 1/4" thick	2 Carp	1500	.011	S.F.	1.34	.58		1.92	2.34
0018	Pneumatic nailed		1860	.009		1.34	.47		1.81	2.17
0030	3/8" thick R061636-20		1500	.011		1.48	.58		2.06	2.50
0070	Pneumatic nailed		1860	.009		1.48	.47		1.95	2.33
0100	1/2" thick		1450	.011		1.83	.60		2.43	2.92
0105	Pneumatic nailed		1798	.009		1.83	.49		2.32	2.75
0200	5/8" thick		1400	.011		2.16	.63		2.79	3.30
0205	Pneumatic nailed		1736	.009		2.16	.50		2.66	3.12
0300	3/4" thick		1300	.012		2.21	.67		2.88	3.43
0305	Pneumatic nailed		1612	.010		2.21	.54		2.75	3.24
0500	Particle board, 3/8" thick	G	1500	.011		.45	.58		1.03	1.36
0505	Pneumatic nailed	G	1860	.009		.45	.47		.92	1.19
0600	1/2" thick	G	1450	.011		.71	.60		1.31	1.69
0605	Pneumatic nailed	G	1798	.009		.71	.49		1.20	1.52
0800	5/8" thick	G	1400	.011		.98	.63		1.61	2.01
0805	Pneumatic nailed	G	1736	.009		.98	.50		1.48	1.83
0900	3/4" thick	G	1300	.012		1.06	.67		1.73	2.17
0905	Pneumatic nailed	G	1612	.010		1.06	.54		1.60	1.98
1100	Hardboard, underlayment grade, 4' x 4', .215" thick	G	1500	.011		.94	.58		1.52	1.90

For customer support on your Concrete & Masonry Costs with RSMeans Data, call 800.448.8182.

145

06 16 Sheathing

06 16 33 – Wood Board Sheathing

06 16 33.10 Board Sheathing	Crew	Daily Output	Labor-Hours	Unit	Material	2021 Bare Costs Labor	Equipment	Total	Total Incl O&P
0009	**BOARD SHEATHING**									
0010	Roof, 1" x 6" boards, laid horizontal	2 Carp	725	.022	S.F.	2.49	1.21		3.70	4.54
0020	On steep roof		520	.031		2.49	1.68		4.17	5.25
0040	On dormers, hips, & valleys		480	.033		2.49	1.82		4.31	5.45
0050	Laid diagonal		650	.025		2.49	1.35		3.84	4.75
0070	1" x 8" boards, laid horizontal		875	.018		2.69	1		3.69	4.45
0080	On steep roof		635	.025		2.69	1.38		4.07	5
0090	On dormers, hips, & valleys		580	.028		2.75	1.51		4.26	5.30
0100	Laid diagonal		725	.022		2.69	1.21		3.90	4.76
0110	Skip sheathing, 1" x 4", 7" OC	1 Carp	1200	.007		1.03	.36		1.39	1.68
0120	1" x 6", 9" OC		1450	.006		1.14	.30		1.44	1.71
0180	T&G sheathing/decking, 1" x 6"		1000	.008		2.44	.44		2.88	3.33
0190	2" x 6"		1000	.008		5.60	.44		6.04	6.85
0200	Walls, 1" x 6" boards, laid regular	2 Carp	650	.025		2.49	1.35		3.84	4.75
0210	Laid diagonal		585	.027		2.49	1.50		3.99	4.97
0220	1" x 8" boards, laid regular		765	.021		2.69	1.14		3.83	4.67
0230	Laid diagonal		650	.025		2.69	1.35		4.04	4.97

06 16 36 – Wood Panel Product Sheathing

06 16 36.10 Sheathing

		Crew	Daily Output	Labor-Hours	Unit	Material	2021 Bare Costs Labor	Equipment	Total	Total Incl O&P
0010	**SHEATHING** R061110-30									
0012	Plywood on roofs, CDX									
0030	5/16" thick	2 Carp	1600	.010	S.F.	.90	.55		1.45	1.81
0035	Pneumatic nailed R061636-20		1952	.008		.90	.45		1.35	1.66
0050	3/8" thick		1525	.010		.78	.57		1.35	1.72
0055	Pneumatic nailed		1860	.009		.78	.47		1.25	1.56
0100	1/2" thick		1400	.011		1	.63		1.63	2.03
0105	Pneumatic nailed		1708	.009		1	.51		1.51	1.86
0200	5/8" thick		1300	.012		1.16	.67		1.83	2.27
0205	Pneumatic nailed		1586	.010		1.16	.55		1.71	2.09
0300	3/4" thick		1200	.013		1.38	.73		2.11	2.61
0305	Pneumatic nailed		1464	.011		1.38	.60		1.98	2.41
0500	Plywood on walls, with exterior CDX, 3/8" thick		1200	.013		.78	.73		1.51	1.95
0505	Pneumatic nailed		1488	.011		.78	.59		1.37	1.74
0600	1/2" thick		1125	.014		1	.78		1.78	2.26
0605	Pneumatic nailed		1395	.011		1	.63		1.63	2.04
0700	5/8" thick		1050	.015		1.16	.83		1.99	2.51
0705	Pneumatic nailed		1302	.012		1.16	.67		1.83	2.27
0800	3/4" thick		975	.016		1.38	.90		2.28	2.86
0805	Pneumatic nailed		1209	.013		1.38	.72		2.10	2.60

06 52 Plastic Structural Assemblies

06 52 10 – Fiberglass Structural Assemblies

06 52 10.40 Fiberglass Floor Grating

		Crew	Daily Output	Labor-Hours	Unit	Material	2021 Bare Costs Labor	Equipment	Total	Total Incl O&P
0010	**FIBERGLASS FLOOR GRATING**									
0100	Reinforced polyester, fire retardant, 1" x 4" grid, 1" thick	E-4	510	.063	S.F.	13.10	3.81	.29	17.20	20.50
0200	1-1/2" x 6" mesh, 1-1/2" thick		500	.064		15.15	3.89	.30	19.34	23
0300	With grit surface, 1-1/2" x 6" grid, 1-1/2" thick		500	.064		15.45	3.89	.30	19.64	23.50

Estimating Tips

07 10 00 Dampproofing and Waterproofing

- Be sure of the job specifications before pricing this subdivision. The difference in cost between waterproofing and dampproofing can be great. Waterproofing will hold back standing water. Dampproofing prevents the transmission of water vapor. Also included in this section are vapor retarding membranes.

07 20 00 Thermal Protection

- Insulation and fireproofing products are measured by area, thickness, volume, or R-value. Specifications may give only what the specific R-value should be in a certain situation. The estimator may need to choose the type of insulation to meet that R-value.

07 30 00 Steep Slope Roofing
07 40 00 Roofing and Siding Panels

- Many roofing and siding products are bought and sold by the square. One square is equal to an area that measures 100 square feet.

  This simple change in unit of measure could create a large error if the estimator is not observant. Accessories necessary for a complete installation must be figured into any calculations for both material and labor.

07 50 00 Membrane Roofing
07 60 00 Flashing and Sheet Metal
07 70 00 Roofing and Wall Specialties and Accessories

- The items in these subdivisions compose a roofing system. No one component completes the installation, and all must be estimated. Built-up or single-ply membrane roofing systems are made up of many products and installation trades. Wood blocking at roof perimeters or penetrations, parapet coverings, reglets, roof drains, gutters, downspouts, sheet metal flashing, skylights, smoke vents, and roof hatches all need to be considered along with the roofing material. Several different installation trades will need to work together on the roofing system. Inherent difficulties in the scheduling and coordination of various trades must be accounted for when estimating labor costs.

07 90 00 Joint Protection

- To complete the weather-tight shell, the sealants and caulkings must be estimated. Where different materials meet—at expansion joints, at flashing penetrations, and at hundreds of other locations throughout a construction project—caulking and sealants provide another line of defense against water penetration. Often, an entire system is based on the proper location and placement of caulking or sealants. The detailed drawings that are included as part of a set of architectural plans show typical locations for these materials. When caulking or sealants are shown at typical locations, this means the estimator must include them for all the locations where this detail is applicable. Be careful to keep different types of sealants separate, and remember to consider backer rods and primers if necessary.

Reference Numbers

Reference numbers are shown at the beginning of some major classifications. These numbers refer to related items in the Reference Section. The reference information may be an estimating procedure, an alternate pricing method, or technical information.

Note: Not all subdivisions listed here necessarily appear. ■

---

07 01 50 – Maintenance of Membrane Roofing

07 01 50.10 Roof Coatings		Crew	Daily Output	Labor-Hours	Unit	Material	2021 Bare Costs Labor	Equipment	Total	Total Incl O&P
0010	**ROOF COATINGS**									
0012	Asphalt, brush grade, material only				Gal.	8.80			8.80	9.65
0200	Asphalt base, fibered aluminum coating [G]					9.55			9.55	10.50
0300	Asphalt primer, 5 gal.				↓	5.50			5.50	6.05
0600	Coal tar pitch, 200 lb. barrels				Ton	1,200			1,200	1,325
0700	Tar roof cement, 5 gal. lots				Gal.	14.35			14.35	15.80
0800	Glass fibered roof & patching cement, 5 gal.				"	9.55			9.55	10.50
0900	Reinforcing glass membrane, 450 S.F./roll				Ea.	59.50			59.50	65.50
1000	Neoprene roof coating, 5 gal., 2 gal./sq.				Gal.	30.50			30.50	33.50
1100	Roof patch & flashing cement, 5 gal.					9.75			9.75	10.70
1200	Roof resurtant, glass fibered, 3 gal./sq.				↓	8.85			8.85	9.75

07 01 90 – Maintenance of Joint Protection

07 01 90.81 Joint Sealant Replacement

		Crew	Daily Output	Labor-Hours	Unit	Material	2021 Bare Costs Labor	Equipment	Total	Total Incl O&P
0010	**JOINT SEALANT REPLACEMENT**									
0050	Control joints in concrete floors/slabs									
0100	Option 1 for joints with hard dry sealant									
0110	Step 1: Sawcut to remove 95% of old sealant									
0112	1/4" wide x 1/2" deep, with single saw blade	C-27	4800	.003	L.F.	.01	.17	.02	.20	.29
0114	3/8" wide x 3/4" deep, with single saw blade		4000	.004		.02	.21	.03	.26	.36
0116	1/2" wide x 1" deep, with double saw blades		3600	.004		.05	.23	.03	.31	.42
0118	3/4" wide x 1-1/2" deep, with double saw blades	↓	3200	.005		.10	.26	.04	.40	.53
0120	Step 2: Water blast joint faces and edges	C-29	2500	.003			.14	.04	.18	.25
0130	Step 3: Air blast joint faces and edges	C-28	2000	.004			.21	.02	.23	.32
0140	Step 4: Sand blast joint faces and edges	E-11	2000	.016			.78	.15	.93	1.37
0150	Step 5: Air blast joint faces and edges	C-28	2000	.004	↓		.21	.02	.23	.32
0200	Option 2 for joints with soft pliable sealant									
0210	Step 1: Plow joint with rectangular blade	B-62	2600	.009	L.F.		.44	.07	.51	.74
0220	Step 2: Sawcut to re-face joint faces									
0222	1/4" wide x 1/2" deep, with single saw blade	C-27	2400	.007	L.F.	.02	.35	.05	.42	.58
0224	3/8" wide x 3/4" deep, with single saw blade		2000	.008		.03	.41	.06	.50	.71
0226	1/2" wide x 1" deep, with double saw blades		1800	.009		.06	.46	.06	.58	.81
0228	3/4" wide x 1-1/2" deep, with double saw blades	↓	1600	.010		.13	.52	.07	.72	.98
0230	Step 3: Water blast joint faces and edges	C-29	2500	.003			.14	.04	.18	.25
0240	Step 4: Air blast joint faces and edges	C-28	2000	.004			.21	.02	.23	.32
0250	Step 5: Sand blast joint faces and edges	E-11	2000	.016			.78	.15	.93	1.37
0260	Step 6: Air blast joint faces and edges	C-28	2000	.004	↓		.21	.02	.23	.32
0290	For saw cutting new control joints, see Section 03 15 16.20									
8910	For backer rod, see Section 07 91 23.10									
8920	For joint sealant, see Section 03 15 16.30 or 07 92 13.20									

07 11 Dampproofing

07 11 13 – Bituminous Dampproofing

07 11 13.10 Bituminous Asphalt Coating

		Crew	Daily Output	Labor-Hours	Unit	Material	2021 Bare Costs Labor	Equipment	Total	Total Incl O&P
0010	**BITUMINOUS ASPHALT COATING**									
0030	Brushed on, below grade, 1 coat	1 Rofc	665	.012	S.F.	.22	.58		.80	1.18
0100	2 coat		500	.016		.44	.77		1.21	1.73
0300	Sprayed on, below grade, 1 coat		830	.010		.22	.46		.68	.99
0400	2 coat	↓	500	.016	↓	.43	.77		1.20	1.72
0500	Asphalt coating, with fibers				Gal.	9.55			9.55	10.50
0600	Troweled on, asphalt with fibers, 1/16" thick	1 Rofc	500	.016	S.F.	.42	.77		1.19	1.71

07 11 Dampproofing

07 11 13 – Bituminous Dampproofing

07 11 13.10 Bituminous Asphalt Coating	Crew	Daily Output	Labor-Hours	Unit	Material	2021 Bare Costs Labor	Equipment	Total	Total Incl O&P
0700	1/8" thick	1 Rofc	400	.020	S.F.	.73	.96		1.69	2.37
1000	1/2" thick	▼	350	.023	▼	2.39	1.10		3.49	4.42

07 11 16 – Cementitious Dampproofing

07 11 16.20 Cementitious Parging

		Crew	Daily Output	Labor-Hours	Unit	Material	2021 Bare Costs Labor	Equipment	Total	Total Incl O&P
0010	**CEMENTITIOUS PARGING**									
0020	Portland cement, 2 coats, 1/2" thick	D-1	250	.064	S.F.	.49	3.12		3.61	5.25
0100	Waterproofed Portland cement, 1/2" thick, 2 coats	"	250	.064	"	2.26	3.12		5.38	7.20

07 12 Built-up Bituminous Waterproofing

07 12 13 – Built-Up Asphalt Waterproofing

07 12 13.20 Membrane Waterproofing

		Crew	Daily Output	Labor-Hours	Unit	Material	2021 Bare Costs Labor	Equipment	Total	Total Incl O&P
0010	**MEMBRANE WATERPROOFING**									
0012	On slabs, 1 ply, felt, mopped	G-1	3000	.019	S.F.	.48	.84	.19	1.51	2.10
0015	On walls, 1 ply, felt, mopped		3000	.019		.48	.84	.19	1.51	2.10
0100	On slabs, 1 ply, glass fiber fabric, mopped		2100	.027		.53	1.20	.27	2	2.82
0105	On walls, 1 ply, glass fiber fabric, mopped		2100	.027		.53	1.20	.27	2	2.82
0300	On slabs, 2 ply, felt, mopped		2500	.022		.96	1.01	.23	2.20	2.93
0305	On walls, 2 ply, felt, mopped		2500	.022		.96	1.01	.23	2.20	2.93
0400	On slabs, 2 ply, glass fiber fabric, mopped		1650	.034		1.13	1.53	.35	3.01	4.09
0405	On walls, 2 ply, glass fiber fabric, mopped		1650	.034		1.13	1.53	.35	3.01	4.09
0600	On slabs, 3 ply, felt, mopped		2100	.027		1.44	1.20	.27	2.91	3.82
0605	On walls, 3 ply, felt, mopped		2100	.027		1.44	1.20	.27	2.91	3.82
0700	On slabs, 3 ply, glass fiber fabric, mopped		1550	.036		1.59	1.63	.37	3.59	4.79
0705	On walls, 3 ply, glass fiber fabric, mopped	▼	1550	.036		1.59	1.63	.37	3.59	4.79
0710	Asphaltic hardboard protection board, 1/8" thick	2 Rofc	500	.032		.77	1.54		2.31	3.35
1000	EPS membrane protection board, 1/4" thick		3500	.005		.34	.22		.56	.73
1050	3/8" thick		3500	.005		.38	.22		.60	.77
1060	1/2" thick		3500	.005	▼	.42	.22		.64	.82
1070	Fiberglass fabric, black, 20/10 mesh	▼	116	.138	Sq.	15.45	6.65		22.10	28

07 13 Sheet Waterproofing

07 13 53 – Elastomeric Sheet Waterproofing

07 13 53.10 Elastomeric Sheet Waterproofing and Access.

		Crew	Daily Output	Labor-Hours	Unit	Material	2021 Bare Costs Labor	Equipment	Total	Total Incl O&P
0010	**ELASTOMERIC SHEET WATERPROOFING AND ACCESS.**									
0090	EPDM, plain, 45 mils thick	2 Rofc	580	.028	S.F.	.74	1.33		2.07	2.98
0100	60 mils thick		570	.028		.95	1.35		2.30	3.23
0300	Nylon reinforced sheets, 45 mils thick		580	.028		.87	1.33		2.20	3.11
0400	60 mils thick	▼	570	.028	▼	1.01	1.35		2.36	3.30
0600	Vulcanizing splicing tape for above, 2" wide				C.L.F.	61.50			61.50	67.50
0700	4" wide				"	320			320	350
0900	Adhesive, bonding, 60 S.F./gal.				Gal.	30.50			30.50	33.50
1000	Splicing, 75 S.F./gal.				"	37.50			37.50	41.50
1200	Neoprene sheets, plain, 45 mils thick	2 Rofc	580	.028	S.F.	2.25	1.33		3.58	4.64
1300	60 mils thick		570	.028		1.59	1.35		2.94	3.94
1500	Nylon reinforced, 45 mils thick		580	.028		2.36	1.33		3.69	4.76
1600	60 mils thick		570	.028		2.64	1.35		3.99	5.10
1800	120 mils thick	▼	500	.032	▼	5.70	1.54		7.24	8.80
1900	Adhesive, splicing, 150 S.F./gal. per coat				Gal.	37.50			37.50	41.50
2100	Fiberglass reinforced, fluid applied, 1/8" thick	2 Rofc	500	.032	S.F.	1.63	1.54		3.17	4.29

149

07 13 Sheet Waterproofing

07 13 53 – Elastomeric Sheet Waterproofing

07 13 53.10 Elastomeric Sheet Waterproofing and Access.

	07 13 53.10 Elastomeric Sheet Waterproofing and Access.	Crew	Daily Output	Labor-Hours	Unit	Material	2021 Bare Costs Labor	Equipment	Total	Total Incl O&P
2200	Polyethylene and rubberized asphalt sheets, 60 mils thick	2 Rofc	550	.029	S.F.	1.26	1.40		2.66	3.66
2400	Polyvinyl chloride sheets, plain, 10 mils thick		580	.028		.15	1.33		1.48	2.33
2500	20 mils thick		570	.028		.26	1.35		1.61	2.48
2700	30 mils thick		560	.029		.21	1.38		1.59	2.46
3000	Adhesives, trowel grade, 40-100 S.F./gal.				Gal.	25.50			25.50	28
3100	Brush grade, 100-250 S.F./gal.				"	23			23	25.50
3300	Bitumen modified polyurethane, fluid applied, 55 mils thick	2 Rofc	665	.024	S.F.	1.18	1.16		2.34	3.18

07 16 Cementitious and Reactive Waterproofing

07 16 16 – Crystalline Waterproofing

07 16 16.20 Cementitious Waterproofing

	07 16 16.20 Cementitious Waterproofing	Crew	Daily Output	Labor-Hours	Unit	Material	2021 Bare Costs Labor	Equipment	Total	Total Incl O&P
0010	**CEMENTITIOUS WATERPROOFING**									
0020	1/8" application, sprayed on	G-2A	1000	.024	S.F.	.58	1.03	.64	2.25	2.97
0050	4 coat cementitious metallic slurry	1 Cefi	1.20	6.667	C.S.F.	72	345		417	585

07 17 Bentonite Waterproofing

07 17 13 – Bentonite Panel Waterproofing

07 17 13.10 Bentonite

	07 17 13.10 Bentonite	Crew	Daily Output	Labor-Hours	Unit	Material	2021 Bare Costs Labor	Equipment	Total	Total Incl O&P
0010	**BENTONITE**									
0020	Panels, 4' x 4', 3/16" thick	1 Rofc	625	.013	S.F.	2.22	.62		2.84	3.44
0100	Rolls, 3/8" thick, with geotextile fabric both sides	"	550	.015	"	1.51	.70		2.21	2.80
0300	Granular bentonite, 50 lb. bags (.625 C.F.)				Bag	16.60			16.60	18.25
0400	3/8" thick, troweled on	1 Rofc	475	.017	S.F.	.83	.81		1.64	2.23
0500	Drain board, expanded polystyrene, 1-1/2" thick	1 Rohe	1600	.005		.48	.18		.66	.82
0510	2" thick		1600	.005		.64	.18		.82	.99
0520	3" thick		1600	.005		.96	.18		1.14	1.35
0530	4" thick		1600	.005		1.28	.18		1.46	1.70
0600	With filter fabric, 1-1/2" thick		1600	.005		.55	.18		.73	.89
0625	2" thick		1600	.005		.71	.18		.89	1.07
0650	3" thick		1600	.005		1.03	.18		1.21	1.42
0675	4" thick		1600	.005		1.35	.18		1.53	1.77

07 19 Water Repellents

07 19 19 – Silicone Water Repellents

07 19 19.10 Silicone Based Water Repellents

	07 19 19.10 Silicone Based Water Repellents	Crew	Daily Output	Labor-Hours	Unit	Material	2021 Bare Costs Labor	Equipment	Total	Total Incl O&P
0010	**SILICONE BASED WATER REPELLENTS**									
0020	Water base liquid, roller applied	2 Rofc	7000	.002	S.F.	.43	.11		.54	.65
0200	Silicone or stearate, sprayed on CMU, 1 coat	1 Rofc	4000	.002		.40	.10		.50	.61
0300	2 coats	"	3000	.003		.81	.13		.94	1.10

07 21 Thermal Insulation

07 21 13 – Board Insulation

07 21 13.10 Rigid Insulation

	Crew	Daily Output	Labor-Hours	Unit	Material	2021 Bare Costs Labor	Equipment	Total	Total Incl O&P
0010 **RIGID INSULATION**, for walls									
0040 Fiberglass, 1.5#/C.F., unfaced, 1" thick, R4.1	G 1 Carp	1000	.008	S.F.	.36	.44		.80	1.05
0060 1-1/2" thick, R6.2	G	1000	.008		.44	.44		.88	1.13
0080 2" thick, R8.3	G	1000	.008		.49	.44		.93	1.19
0120 3" thick, R12.4	G	800	.010		.59	.55		1.14	1.47
0370 3#/C.F., unfaced, 1" thick, R4.3	G	1000	.008		.57	.44		1.01	1.28
0390 1-1/2" thick, R6.5	G	1000	.008		.73	.44		1.17	1.45
0400 2" thick, R8.7	G	890	.009		1.25	.49		1.74	2.11
0420 2-1/2" thick, R10.9	G	800	.010		1.04	.55		1.59	1.96
0440 3" thick, R13	G	800	.010		1.63	.55		2.18	2.61
0520 Foil faced, 1" thick, R4.3	G	1000	.008		.80	.44		1.24	1.53
0540 1-1/2" thick, R6.5	G	1000	.008		1.18	.44		1.62	1.95
0560 2" thick, R8.7	G	890	.009		1.49	.49		1.98	2.37
0580 2-1/2" thick, R10.9	G	800	.010		1.75	.55		2.30	2.75
0600 3" thick, R13	G	800	.010		1.85	.55		2.40	2.86
1600 Isocyanurate, 4' x 8' sheet, foil faced, both sides									
1610 1/2" thick	G 1 Carp	800	.010	S.F.	.29	.55		.84	1.14
1620 5/8" thick	G	800	.010		.62	.55		1.17	1.50
1630 3/4" thick	G	800	.010		.44	.55		.99	1.30
1640 1" thick	G	800	.010		.62	.55		1.17	1.50
1650 1-1/2" thick	G	730	.011		.76	.60		1.36	1.73
1660 2" thick	G	730	.011		.78	.60		1.38	1.75
1670 3" thick	G	730	.011		3.13	.60		3.73	4.33
1680 4" thick	G	730	.011		2.60	.60		3.20	3.75
1700 Perlite, 1" thick, R2.77	G	800	.010		.46	.55		1.01	1.33
1750 2" thick, R5.55	G	730	.011		.83	.60		1.43	1.80
1900 Extruded polystyrene, 25 psi compressive strength, 1" thick, R5	G	800	.010		.61	.55		1.16	1.49
1940 2" thick, R10	G	730	.011		1.23	.60		1.83	2.24
1960 3" thick, R15	G	730	.011		1.75	.60		2.35	2.82
2100 Expanded polystyrene, 1" thick, R3.85	G	800	.010		.32	.55		.87	1.17
2120 2" thick, R7.69	G	730	.011		.64	.60		1.24	1.59
2140 3" thick, R11.49	G	730	.011		.96	.60		1.56	1.95

07 21 13.13 Foam Board Insulation

	Crew	Daily Output	Labor-Hours	Unit	Material	2021 Bare Costs Labor	Equipment	Total	Total Incl O&P
0010 **FOAM BOARD INSULATION**									
0600 Polystyrene, expanded, 1" thick, R4	G 1 Carp	680	.012	S.F.	.32	.64		.96	1.31
0700 2" thick, R8	G "	675	.012	"	.64	.65		1.29	1.67

07 21 16 – Blanket Insulation

07 21 16.20 Blanket Insulation for Walls

	Crew	Daily Output	Labor-Hours	Unit	Material	2021 Bare Costs Labor	Equipment	Total	Total Incl O&P
0010 **BLANKET INSULATION FOR WALLS**									
0410 Foil faced fiberglass, 3-1/2" thick, R13, 11" wide	G 1 Carp	1150	.007	S.F.	.55	.38		.93	1.18
0420 15" wide	G	1350	.006		.55	.32		.87	1.09
0440 23" wide	G	1600	.005		.55	.27		.82	1.02
0442 R15, 11" wide	G	1150	.007		.47	.38		.85	1.09
0444 15" wide	G	1350	.006		.47	.32		.79	1
0446 23" wide	G	1600	.005		.47	.27		.74	.93
0448 6" thick, R19, 11" wide	G	1150	.007		.76	.38		1.14	1.41
0460 15" wide	G	1350	.006		.76	.32		1.08	1.32
0480 23" wide	G	1600	.005		.76	.27		1.03	1.25
0482 R21, 11" wide	G	1150	.007		.66	.38		1.04	1.30
0484 15" wide	G	1350	.006		.66	.32		.98	1.21
0486 23" wide	G	1600	.005		.66	.27		.93	1.14
0488 9" thick, R30, 11" wide	G	985	.008		1.19	.44		1.63	1.97

07 21 16 – Blanket Insulation

07 21 16.20 Blanket Insulation for Walls		Crew	Daily Output	Labor-Hours	Unit	Material	2021 Bare Costs Labor	Equipment	Total	Total Incl O&P
0500	15" wide	G 1 Carp	1150	.007	S.F.	1.19	.38		1.57	1.88
0550	23" wide	G	1350	.006		1.19	.32		1.51	1.79
0560	12" thick, R38, 11" wide	G	985	.008		1.13	.44		1.57	1.90
0570	15" wide	G	1150	.007		1.13	.38		1.51	1.81
0580	23" wide	G	1350	.006		1.13	.32		1.45	1.72
0620	Unfaced fiberglass, 3-1/2" thick, R13, 11" wide	G	1150	.007		.33	.38		.71	.93
0832	R15, 11" wide	G	1150	.007		.41	.38		.79	1.02

07 21 19 – Foamed In Place Insulation

07 21 19.10 Masonry Foamed In Place Insulation

		Crew	Daily Output	Labor-Hours	Unit	Material	2021 Bare Costs Labor	Equipment	Total	Total Incl O&P
0010	**MASONRY FOAMED IN PLACE INSULATION**									
0100	Amino-plast foam, injected into block core, 6" block	G G-2A	6000	.004	Ea.	.17	.17	.11	.45	.58
0110	8" block	G	5000	.005		.21	.21	.13	.55	.69
0120	10" block	G	4000	.006		.26	.26	.16	.68	.87
0130	12" block	G	3000	.008		.35	.34	.21	.90	1.16
0140	Injected into cavity wall	G	13000	.002	B.F.	.06	.08	.05	.19	.24
0150	Preparation, drill holes into mortar joint every 4 V.L.F., 5/8" diameter	1 Clab	960	.008	Ea.		.37		.37	.55
0160	7/8" diameter		680	.012			.52		.52	.78
0170	Patch drilled holes, 5/8" diameter		1800	.004		.04	.20		.24	.33
0180	7/8" diameter		1200	.007		.05	.30		.35	.50

07 21 23 – Loose-Fill Insulation

07 21 23.10 Poured Loose-Fill Insulation

		Crew	Daily Output	Labor-Hours	Unit	Material	2021 Bare Costs Labor	Equipment	Total	Total Incl O&P
0010	**POURED LOOSE-FILL INSULATION**									
0020	Cellulose fiber, R3.8 per inch	G 1 Carp	200	.040	C.F.	.70	2.19		2.89	4.04
0021	4" thick	G	1000	.008	S.F.	.17	.44		.61	.83
0022	6" thick	G	800	.010	"	.28	.55		.83	1.13
0080	Fiberglass wool, R4 per inch	G	200	.040	C.F.	.55	2.19		2.74	3.88
0081	4" thick	G	600	.013	S.F.	.19	.73		.92	1.30
0082	6" thick	G	400	.020	"	.27	1.09		1.36	1.92
0100	Mineral wool, R3 per inch	G	200	.040	C.F.	.56	2.19		2.75	3.89
0101	4" thick	G	600	.013	S.F.	.18	.73		.91	1.29
0102	6" thick	G	400	.020	"	.28	1.09		1.37	1.94
0300	Polystyrene, R4 per inch	G	200	.040	C.F.	1.56	2.19		3.75	4.99
0301	4" thick	G	600	.013	S.F.	.51	.73		1.24	1.66
0302	6" thick	G	400	.020	"	.78	1.09		1.87	2.49
0400	Perlite, R2.78 per inch	G	200	.040	C.F.	6.55	2.19		8.74	10.50
0401	4" thick	G	1000	.008	S.F.	2.19	.44		2.63	3.06
0402	6" thick	G	800	.010	"	3.29	.55		3.84	4.43

07 21 23.20 Masonry Loose-Fill Insulation

		Crew	Daily Output	Labor-Hours	Unit	Material	2021 Bare Costs Labor	Equipment	Total	Total Incl O&P
0010	**MASONRY LOOSE-FILL INSULATION**, vermiculite or perlite									
0100	In cores of concrete block, 4" thick wall, .115 C.F./S.F.	G D-1	4800	.003	S.F.	.76	.16		.92	1.07
0200	6" thick wall, .175 C.F./S.F.	G	3000	.005		1.15	.26		1.41	1.65
0300	8" thick wall, .258 C.F./S.F.	G	2400	.007		1.70	.32		2.02	2.35
0400	10" thick wall, .340 C.F./S.F.	G	1850	.009		2.23	.42		2.65	3.09
0500	12" thick wall, .422 C.F./S.F.	G	1200	.013		2.77	.65		3.42	4.03
0600	Poured cavity wall, vermiculite or perlite, water repellent	G	250	.064	C.F.	6.55	3.12		9.67	11.95
0700	Foamed in place, urethane in 2-5/8" cavity	G G-2A	1035	.023	S.F.	1.45	.99	.62	3.06	3.84
0800	For each 1" added thickness, add	G	"	2372	.010	"	.55	.43	.27	1.25	1.59

07 21 Thermal Insulation

07 21 29 – Sprayed Insulation

07 21 29.10 Sprayed-On Insulation		Crew	Daily Output	Labor- Hours	Unit	Material	2021 Bare Costs Labor	Equipment	Total	Total Incl O&P
0010	**SPRAYED-ON INSULATION**									
0020	Fibrous/cementitious, finished wall, 1" thick, R3.7	G	G-2	2050	.012	S.F.	.36	.54	.09	.99	1.31
0100	Attic, 5.2" thick, R19	G		1550	.015	"	.42	.72	.12	1.26	1.67
0200	Fiberglass, R4 per inch, vertical	G		1600	.015	B.F.	.20	.69	.12	1.01	1.38
0210	Horizontal	G		1200	.020	"	.20	.93	.16	1.29	1.77

07 25 Weather Barriers

07 25 10 – Weather Barriers or Wraps

07 25 10.10 Weather Barriers

		Crew	Daily Output	Labor- Hours	Unit	Material	2021 Bare Costs Labor	Equipment	Total	Total Incl O&P
0010	**WEATHER BARRIERS**									
0400	Asphalt felt paper, #15	1 Carp	37	.216	Sq.	5.70	11.85		17.55	24
0401	Per square foot	"	3700	.002	S.F.	.06	.12		.18	.24
0450	Housewrap, exterior, spun bonded polypropylene									
0470	Small roll	1 Carp	3800	.002	S.F.	.15	.12		.27	.33
0480	Large roll	"	4000	.002	"	.18	.11		.29	.35
2100	Asphalt felt roof deck vapor barrier, class 1 metal decks	1 Rofc	37	.216	Sq.	24.50	10.40		34.90	44
2200	For all other decks	"	37	.216		18.65	10.40		29.05	37.50
2800	Asphalt felt, 50% recycled content, 15 lb., 4 sq./roll	1 Carp	36	.222		5.95	12.15		18.10	24.50
2810	30 lb., 2 sq./roll	"	36	.222		10.20	12.15		22.35	29.50
3000	Building wrap, spun bonded polyethylene	2 Carp	8000	.002	S.F.	.14	.11		.25	.31

07 26 Vapor Retarders

07 26 13 – Above-Grade Vapor Retarders

07 26 13.10 Vapor Retarders

			Crew	Daily Output	Labor- Hours	Unit	Material	2021 Bare Costs Labor	Equipment	Total	Total Incl O&P
0010	**VAPOR RETARDERS**										
0020	Aluminum and kraft laminated, foil 1 side	G	1 Carp	37	.216	Sq.	17.65	11.85		29.50	37
0100	Foil 2 sides	G		37	.216		18.40	11.85		30.25	37.50
0600	Polyethylene vapor barrier, standard, 2 mil	G		37	.216		3.03	11.85		14.88	21
0700	4 mil	G		37	.216		2.70	11.85		14.55	20.50
0900	6 mil	G		37	.216		7.65	11.85		19.50	26
1200	10 mil	G		37	.216		8.15	11.85		20	26.50
1300	Clear reinforced, fire retardant, 8 mil	G		37	.216		26.50	11.85		38.35	46.50
1350	Cross laminated type, 3 mil	G		37	.216		12.85	11.85		24.70	32
1400	4 mil	G		37	.216		14.25	11.85		26.10	33.50
1800	Reinf. waterproof, 2 mil polyethylene backing, 1 side			37	.216		12	11.85		23.85	31
1900	2 sides			37	.216		15.15	11.85		27	34.50

For customer support on your Concrete & Masonry Costs with RSMeans Data, call 800.448.8182.

153

07 31 Shingles and Shakes

07 31 26 – Slate Shingles

07 31 26.10 Slate Roof Shingles		Crew	Daily Output	Labor-Hours	Unit	Material	2021 Bare Costs Labor	Equipment	Total	Total Incl O&P
0010	**SLATE ROOF SHINGLES**									
0100	Buckingham Virginia black, 3/16" - 1/4" thick [G]	1 Rots	1.75	4.571	Sq.	555	220		775	965
0200	1/4" thick [G]		1.75	4.571		555	220		775	965
0900	Pennsylvania black, Bangor, #1 clear [G]		1.75	4.571		500	220		720	905
1200	Vermont, unfading, green, mottled green [G]		1.75	4.571		505	220		725	910
1300	Semi-weathering green & gray [G]		1.75	4.571		405	220		625	800
1400	Purple [G]		1.75	4.571		450	220		670	845
1500	Black or gray [G]		1.75	4.571		485	220		705	890
1600	Red [G]		1.75	4.571		1,175	220		1,395	1,650
1700	Variegated purple		1.75	4.571		460	220		680	860
2700	Ridge shingles, slate		200	.040	L.F.	10.20	1.93		12.13	14.35

07 32 Roof Tiles

07 32 13 – Clay Roof Tiles

07 32 13.10 Clay Tiles		Crew	Daily Output	Labor-Hours	Unit	Material	2021 Bare Costs Labor	Equipment	Total	Total Incl O&P
0010	**CLAY TILES**, including accessories									
0300	Flat shingle, interlocking, 15", 166 pcs./sq., fireflashed blend	3 Rots	6	4	Sq.	555	193		748	925
0500	Terra cotta red		6	4		655	193		848	1,025
0600	Roman pan and top, 18", 102 pcs./sq., fireflashed blend		5.50	4.364		515	210		725	905
1100	Barrel mission tile, 18", 166 pcs./sq., fireflashed blend		5.50	4.364		430	210		640	810
1140	Terra cotta red		5.50	4.364		435	210		645	820
1700	Scalloped edge flat shingle, 14", 145 pcs./sq., fireflashed blend		6	4		1,125	193		1,318	1,550
1800	Terra cotta red		6	4		1,100	193		1,293	1,550
3010	#15 felt underlayment	1 Rofc	64	.125		5.70	6.05		11.75	16
3020	#30 felt underlayment		58	.138		9.60	6.65		16.25	21.50
3040	Polyethylene and rubberized asph. underlayment		22	.364		68.50	17.55		86.05	104

07 32 16 – Concrete Roof Tiles

07 32 16.10 Concrete Tiles		Crew	Daily Output	Labor-Hours	Unit	Material	2021 Bare Costs Labor	Equipment	Total	Total Incl O&P
0010	**CONCRETE TILES**									
0020	Corrugated, 13" x 16-1/2", 90 per sq., 950 lb./sq.									
0050	Earthtone colors, nailed to wood deck	1 Rots	1.35	5.926	Sq.	118	286		404	595
0150	Blues		1.35	5.926		106	286		392	580
0200	Greens		1.35	5.926		117	286		403	595
0250	Premium colors		1.35	5.926		117	286		403	595
0500	Shakes, 13" x 16-1/2", 90 per sq., 950 lb./sq.									
0600	All colors, nailed to wood deck	1 Rots	1.50	5.333	Sq.	172	257		429	605
1500	Accessory pieces, ridge & hip, 10" x 16-1/2", 8 lb. each	"	120	.067	Ea.	3.74	3.21		6.95	9.30
1700	Rake, 6-1/2" x 16-3/4", 9 lb. each					3.74			3.74	4.11
1800	Mansard hip, 10" x 16-1/2", 9.2 lb. each					3.74			3.74	4.11
1900	Hip starter, 10" x 16-1/2", 10.5 lb. each					10.35			10.35	11.40
2000	3 or 4 way apex, 10" each side, 11.5 lb. each					11.85			11.85	13.05

07 46 Siding

07 46 23 – Wood Siding

07 46 23.10 Wood Board Siding

		Crew	Daily Output	Labor-Hours	Unit	Material	2021 Bare Costs Labor	2021 Bare Costs Equipment	Total	Total Incl O&P
0010	**WOOD BOARD SIDING**									
3200	Wood, cedar bevel, A grade, 1/2" x 6"	1 Carp	295	.027	S.F.	4.42	1.48		5.90	7.05
3300	1/2" x 8"		330	.024		7.50	1.33		8.83	10.25
3500	3/4" x 10", clear grade		375	.021		7.35	1.17		8.52	9.80
3600	"B" grade		375	.021		4.38	1.17		5.55	6.55
3800	Cedar, rough sawn, 1" x 4", A grade, natural		220	.036		8.80	1.99		10.79	12.60
3900	Stained		220	.036		8.75	1.99		10.74	12.55
4100	1" x 12", board & batten, #3 & Btr., natural		420	.019		4.83	1.04		5.87	6.85
4200	Stained		420	.019		5.25	1.04		6.29	7.35
4400	1" x 8" channel siding, #3 & Btr., natural		330	.024		5.25	1.33		6.58	7.80
4500	Stained		330	.024		6	1.33		7.33	8.60
4700	Redwood, clear, beveled, vertical grain, 1/2" x 4"		220	.036		4.99	1.99		6.98	8.45
4750	1/2" x 6"		295	.027		5.40	1.48		6.88	8.15
4800	1/2" x 8"		330	.024		5.55	1.33		6.88	8.10
5000	3/4" x 10"		375	.021		5.35	1.17		6.52	7.65
5200	Channel siding, 1" x 10", B grade		375	.021		4.90	1.17		6.07	7.15
5250	Redwood, T&G boards, B grade, 1" x 4"		220	.036		7	1.99		8.99	10.65
5270	1" x 8"		330	.024		9	1.33		10.33	11.90
5400	White pine, rough sawn, 1" x 8", natural		330	.024		2.41	1.33		3.74	4.63
5500	Stained		330	.024		2.41	1.33		3.74	4.63

07 46 29 – Plywood Siding

07 46 29.10 Plywood Siding Options

		Crew	Daily Output	Labor-Hours	Unit	Material	2021 Bare Costs Labor	2021 Bare Costs Equipment	Total	Total Incl O&P
0010	**PLYWOOD SIDING OPTIONS**									
0900	Plywood, medium density overlaid, 3/8" thick	2 Carp	750	.021	S.F.	1.20	1.17		2.37	3.06
1000	1/2" thick		700	.023		1.59	1.25		2.84	3.62
1100	3/4" thick		650	.025		2.62	1.35		3.97	4.89
1600	Texture 1-11, cedar, 5/8" thick, natural		675	.024		2.58	1.30		3.88	4.78
1700	Factory stained		675	.024		2.86	1.30		4.16	5.10
1900	Texture 1-11, fir, 5/8" thick, natural		675	.024		1.38	1.30		2.68	3.46
2000	Factory stained		675	.024		2.02	1.30		3.32	4.16
2050	Texture 1-11, S.Y.P., 5/8" thick, natural		675	.024		1.44	1.30		2.74	3.52
2100	Factory stained		675	.024		1.54	1.30		2.84	3.63
2200	Rough sawn cedar, 3/8" thick, natural		675	.024		1.28	1.30		2.58	3.35
2300	Factory stained		675	.024		1.54	1.30		2.84	3.63
2500	Rough sawn fir, 3/8" thick, natural		675	.024		1.02	1.30		2.32	3.06
2600	Factory stained		675	.024		1.13	1.30		2.43	3.18
2800	Redwood, textured siding, 5/8" thick		675	.024		2.05	1.30		3.35	4.20
3000	Polyvinyl chloride coated, 3/8" thick		750	.021		1.18	1.17		2.35	3.04

07 51 Built-Up Bituminous Roofing

07 51 13 – Built-Up Asphalt Roofing

07 51 13.50 Walkways for Built-Up Roofs

		Crew	Daily Output	Labor-Hours	Unit	Material	2021 Bare Costs Labor	2021 Bare Costs Equipment	Total	Total Incl O&P
0010	**WALKWAYS FOR BUILT-UP ROOFS**									
0020	Asphalt impregnated, 3' x 6' x 1/2" thick	1 Rofc	400	.020	S.F.	2.11	.96		3.07	3.88
0100	3' x 3' x 3/4" thick	"	400	.020		6.10	.96		7.06	8.25
0300	Concrete patio blocks, 2" thick, natural	1 Clab	115	.070		3.57	3.09		6.66	8.55
0400	Colors	"	115	.070		3.56	3.09		6.65	8.55

For customer support on your Concrete & Masonry Costs with RSMeans Data, call 800.448.8182.

155

07 65 Flexible Flashing

07 65 10 – Sheet Metal Flashing

07 65 10.10 Sheet Metal Flashing and Counter Flashing

	07 65 10.10 Sheet Metal Flashing and Counter Flashing	Crew	Daily Output	Labor-Hours	Unit	Material	2021 Bare Costs Labor	Equipment	Total	Total Incl O&P
0010	**SHEET METAL FLASHING AND COUNTER FLASHING**									
0011	Including up to 4 bends									
0020	Aluminum, mill finish, .013" thick	1 Rofc	145	.055	S.F.	1.30	2.66		3.96	5.75
0030	.016" thick		145	.055		1.16	2.66		3.82	5.60
0060	.019" thick		145	.055		1.56	2.66		4.22	6.05
0100	.032" thick		145	.055		1.53	2.66		4.19	6
0200	.040" thick		145	.055		2.38	2.66		5.04	6.95
0300	.050" thick		145	.055		3.25	2.66		5.91	7.90
0325	Mill finish 5" x 7" step flashing, .016" thick		1920	.004	Ea.	.17	.20		.37	.52
0350	Mill finish 12" x 12" step flashing, .016" thick		1600	.005	"	.61	.24		.85	1.06
0400	Painted finish, add				S.F.	.31			.31	.34
1000	Mastic-coated 2 sides, .005" thick	1 Rofc	330	.024		1.87	1.17		3.04	3.95
1100	.016" thick		330	.024		2.06	1.17		3.23	4.16
1600	Copper, 16 oz. sheets, under 1000 lb.		115	.070		7.90	3.35		11.25	14.15
1700	Over 4000 lb.		155	.052		7.90	2.49		10.39	12.75
1900	20 oz. sheets, under 1000 lb.		110	.073		11.20	3.51		14.71	18.05
2000	Over 4000 lb.		145	.055		10.65	2.66		13.31	16
2200	24 oz. sheets, under 1000 lb.		105	.076		15.60	3.67		19.27	23
2300	Over 4000 lb.		135	.059		14.80	2.86		17.66	21
2500	32 oz. sheets, under 1000 lb.		100	.080		22.50	3.86		26.36	31.50
2600	Over 4000 lb.		130	.062		21.50	2.97		24.47	29
5800	Lead, 2.5 lb./S.F., up to 12" wide		135	.059		6.30	2.86		9.16	11.60
5900	Over 12" wide		135	.059		4.34	2.86		7.20	9.40
8900	Stainless steel sheets, 32 ga.		155	.052		3.55	2.49		6.04	7.95
9000	28 ga.		155	.052		5.10	2.49		7.59	9.65
9100	26 ga.		155	.052		5.20	2.49		7.69	9.80
9200	24 ga.		155	.052		5.95	2.49		8.44	10.60
9290	For mechanically keyed flashing, add					40%				
9400	Terne coated stainless steel, .015" thick, 28 ga.	1 Rofc	155	.052	S.F.	8.15	2.49		10.64	13
9500	.018" thick, 26 ga.		155	.052		9.10	2.49		11.59	14.05
9600	Zinc and copper alloy (brass), .020" thick		155	.052		9.05	2.49		11.54	14
9700	.027" thick		155	.052		10.65	2.49		13.14	15.75
9800	.032" thick		155	.052		13.30	2.49		15.79	18.70
9900	.040" thick		155	.052		19.45	2.49		21.94	25.50

07 65 12 – Fabric and Mastic Flashings

07 65 12.10 Fabric and Mastic Flashing and Counter Flashing

		Crew	Daily Output	Labor-Hours	Unit	Material	2021 Bare Costs Labor	Equipment	Total	Total Incl O&P
0010	**FABRIC AND MASTIC FLASHING AND COUNTER FLASHING**									
1300	Asphalt flashing cement, 5 gallon				Gal.	9.60			9.60	10.55
4900	Fabric, asphalt-saturated cotton, specification grade	1 Rofc	35	.229	S.Y.	3.32	11		14.32	21.50
5000	Utility grade		35	.229		1.45	11		12.45	19.45
5300	Close-mesh fabric, saturated, 17 oz./S.Y.		35	.229		2.21	11		13.21	20.50
5500	Fiberglass, resin-coated		35	.229		1.14	11		12.14	19.10

07 65 13 – Laminated Sheet Flashing

07 65 13.10 Laminated Sheet Flashing

		Crew	Daily Output	Labor-Hours	Unit	Material	2021 Bare Costs Labor	Equipment	Total	Total Incl O&P
0010	**LAMINATED SHEET FLASHING**, Including up to 4 bends									
0500	Aluminum, fabric-backed 2 sides, mill finish, .004" thick	1 Rofc	330	.024	S.F.	1.58	1.17		2.75	3.63
0700	.005" thick		330	.024		1.86	1.17		3.03	3.94
0750	Mastic-backed, self adhesive		460	.017		3.41	.84		4.25	5.10
0800	Mastic-coated 2 sides, .004" thick		330	.024		1.58	1.17		2.75	3.63
2800	Copper, paperbacked 1 side, 2 oz.		330	.024		2.15	1.17		3.32	4.26
2900	3 oz.		330	.024		3.14	1.17		4.31	5.35
3100	Paperbacked 2 sides, 2 oz.		330	.024		2.38	1.17		3.55	4.51

07 65 Flexible Flashing

07 65 13 – Laminated Sheet Flashing

07 65 13.10 Laminated Sheet Flashing

		Crew	Daily Output	Labor-Hours	Unit	Material	2021 Bare Costs Labor	Equipment	Total	Total Incl O&P
3150	3 oz.	1 Rofc	330	.024	S.F.	2.23	1.17		3.40	4.34
3200	5 oz.		330	.024		3.88	1.17		5.05	6.15
3250	7 oz.		330	.024		6.10	1.17		7.27	8.65
3400	Mastic-backed 2 sides, copper, 2 oz.		330	.024		2.27	1.17		3.44	4.39
3500	3 oz.		330	.024		2.57	1.17		3.74	4.72
3700	5 oz.		330	.024		3.91	1.17		5.08	6.20
3800	Fabric-backed 2 sides, copper, 2 oz.		330	.024		2.04	1.17		3.21	4.13
4000	3 oz.		330	.024		2.88	1.17		4.05	5.05
4100	5 oz.		330	.024		4.04	1.17		5.21	6.35
4300	Copper-clad stainless steel, .015" thick, under 500 lb.		115	.070		6.90	3.35		10.25	13.05
4400	Over 2000 lb.		155	.052		6.80	2.49		9.29	11.50
4600	.018" thick, under 500 lb.		100	.080		7.90	3.86		11.76	14.95
4700	Over 2000 lb.		145	.055		8.15	2.66		10.81	13.25
8550	Shower pan, 3 ply copper and fabric, 3 oz.		155	.052		4.12	2.49		6.61	8.55
8600	7 oz.		155	.052		4.98	2.49		7.47	9.55
9300	Stainless steel, paperbacked 2 sides, .005" thick	▼	330	.024	▼	4.05	1.17		5.22	6.35

07 65 19 – Plastic Sheet Flashing

07 65 19.10 Plastic Sheet Flashing and Counter Flashing

		Crew	Daily Output	Labor-Hours	Unit	Material	2021 Bare Costs Labor	Equipment	Total	Total Incl O&P
0010	**PLASTIC SHEET FLASHING AND COUNTER FLASHING**									
7300	Polyvinyl chloride, black, 10 mil	1 Rofc	285	.028	S.F.	.27	1.35		1.62	2.49
7400	20 mil		285	.028		.26	1.35		1.61	2.48
7600	30 mil		285	.028		.33	1.35		1.68	2.55
7700	60 mil		285	.028		.88	1.35		2.23	3.16
7900	Black or white for exposed roofs, 60 mil	▼	285	.028	▼	4.78	1.35		6.13	7.45
8060	PVC tape, 5" x 45 mils, for joint covers, 100 L.F./roll				Ea.	185			185	203
8850	Polyvinyl chloride, 30 mil	1 Rofc	160	.050	S.F.	1.90	2.41		4.31	6

07 65 23 – Rubber Sheet Flashing

07 65 23.10 Rubber Sheet Flashing and Counter Flashing

		Crew	Daily Output	Labor-Hours	Unit	Material	2021 Bare Costs Labor	Equipment	Total	Total Incl O&P
0010	**RUBBER SHEET FLASHING AND COUNTER FLASHING**									
8100	Rubber, butyl, 1/32" thick	1 Rofc	285	.028	S.F.	2.30	1.35		3.65	4.72
8200	1/16" thick		285	.028		4.80	1.35		6.15	7.50
8300	Neoprene, cured, 1/16" thick		285	.028		2.30	1.35		3.65	4.72
8400	1/8" thick	▼	285	.028	▼	6.05	1.35		7.40	8.85

07 71 Roof Specialties

07 71 26 – Reglets

07 71 26.10 Reglets and Accessories

		Crew	Daily Output	Labor-Hours	Unit	Material	2021 Bare Costs Labor	Equipment	Total	Total Incl O&P
0010	**REGLETS AND ACCESSORIES**									
0020	Reglet, aluminum, .025" thick, in parapet	1 Carp	225	.036	L.F.	1.90	1.94		3.84	4.99
0300	16 oz. copper		225	.036		7.30	1.94		9.24	10.95
0400	Galvanized steel, 24 ga.		225	.036		1.51	1.94		3.45	4.56
0600	Stainless steel, .020" thick	▼	225	.036		4.12	1.94		6.06	7.45
0900	Counter flashing for above, 12" wide, .032" aluminum	1 Shee	150	.053		2.01	3.49		5.50	7.45
1200	16 oz. copper		150	.053		7.30	3.49		10.79	13.30
1300	Galvanized steel, 26 ga.		150	.053		1.40	3.49		4.89	6.80
1500	Stainless steel, .020" thick	▼	150	.053	▼	6.85	3.49		10.34	12.80

07 71 Roof Specialties

07 71 29 – Manufactured Roof Expansion Joints

07 71 29.10 Expansion Joints	Crew	Daily Output	Labor-Hours	Unit	Material	2021 Bare Costs Labor	Equipment	Total	Total Incl O&P
0010 **EXPANSION JOINTS**									
0300 Butyl or neoprene center with foam insulation, metal flanges									
0400 Aluminum, .032" thick for openings to 2-1/2"	1 Rofc	165	.048	L.F.	14.90	2.34		17.24	20
0600 For joint openings to 3-1/2"		165	.048		12.25	2.34		14.59	17.30
0610 For joint openings to 5"		165	.048		14.60	2.34		16.94	19.85
0620 For joint openings to 8"		165	.048		17.85	2.34		20.19	23.50
0700 Copper, 16 oz. for openings to 2-1/2"		165	.048		25	2.34		27.34	31.50
0900 For joint openings to 3-1/2"		165	.048		19.45	2.34		21.79	25.50
0910 For joint openings to 5"		165	.048		24.50	2.34		26.84	31
0920 For joint openings to 8"		165	.048		26.50	2.34		28.84	33
1000 Galvanized steel, 26 ga. for openings to 2-1/2"		165	.048		12.35	2.34		14.69	17.40
1200 For joint openings to 3-1/2"		165	.048		10.40	2.34		12.74	15.25
1210 For joint openings to 5"		165	.048		15.05	2.34		17.39	20.50
1220 For joint openings to 8"		165	.048		16.20	2.34		18.54	21.50
1300 Lead-coated copper, 16 oz. for openings to 2-1/2"		165	.048		36	2.34		38.34	43.50
1500 For joint openings to 3-1/2"		165	.048		39.50	2.34		41.84	47
1600 Stainless steel, .018", for openings to 2-1/2"		165	.048		12.30	2.34		14.64	17.30
1800 For joint openings to 3-1/2"		165	.048		14.15	2.34		16.49	19.35
1810 For joint openings to 5"		165	.048		13.85	2.34		16.19	19
1820 For joint openings to 8"		165	.048		22	2.34		24.34	28.50
1900 Neoprene, double-seal type with thick center, 4-1/2" wide		125	.064		14.45	3.08		17.53	21
1950 Polyethylene bellows, with galv steel flat flanges		100	.080		7.75	3.86		11.61	14.75
1960 With galvanized angle flanges		100	.080		7.80	3.86		11.66	14.85
2000 Roof joint with extruded aluminum cover, 2"	1 Shee	115	.070		33.50	4.55		38.05	44
2100 Expansion, stainless flange, foam center, standard	1 Rofc	100	.080		13.75	3.86		17.61	21.50
2200 Large	"	100	.080		18.50	3.86		22.36	27
2500 Roof to wall joint with extruded aluminum cover	1 Shee	115	.070		30.50	4.55		35.05	40.50
2700 Wall joint, closed cell foam on PVC cover, 9" wide	1 Rofc	125	.064		6.25	3.08		9.33	11.85
2800 12" wide	"	115	.070		7.05	3.35		10.40	13.20

07 72 Roof Accessories

07 72 26 – Ridge Vents

07 72 26.10 Ridge Vents and Accessories

	Crew	Daily Output	Labor-Hours	Unit	Material	Labor	Equipment	Total	Total Incl O&P
0010 **RIDGE VENTS AND ACCESSORIES**									
2300 Ridge vent strip, mill finish	1 Shee	155	.052	L.F.	4.17	3.38		7.55	9.70

07 72 53 – Snow Guards

07 72 53.10 Snow Guard Options

	Crew	Daily Output	Labor-Hours	Unit	Material	Labor	Equipment	Total	Total Incl O&P
0010 **SNOW GUARD OPTIONS**									
0100 Slate & asphalt shingle roofs, fastened with nails	1 Rofc	160	.050	Ea.	12.45	2.41		14.86	17.60
0200 Standing seam metal roofs, fastened with set screws		48	.167		17.80	8.05		25.85	32.50
0300 Surface mount for metal roofs, fastened with solder		48	.167		7.75	8.05		15.80	21.50
0400 Double rail pipe type, including pipe		130	.062	L.F.	36	2.97		38.97	44.50

07 72 80 – Vent Options

07 72 80.30 Vent Options

	Crew	Daily Output	Labor-Hours	Unit	Material	Labor	Equipment	Total	Total Incl O&P
0010 **VENT OPTIONS**									
0020 Plastic, for insulated decks, 1 per M.S.F.	1 Rofc	40	.200	Ea.	25	9.65		34.65	43
0100 Heavy duty		20	.400		61.50	19.30		80.80	99.50
0300 Aluminum		30	.267		23.50	12.85		36.35	47
0800 Polystyrene baffles, 12" wide for 16" OC rafter spacing	1 Carp	90	.089		.52	4.86		5.38	7.80

07 72 Roof Accessories

07 72 80 – Vent Options

07 72 80.30 Vent Options	Crew	Daily Output	Labor-Hours	Unit	Material	2021 Bare Costs Labor	Equipment	Total	Total Incl O&P
0900 For 24" OC rafter spacing	1 Carp	110	.073	Ea.	.90	3.98		4.88	6.95

07 76 Roof Pavers

07 76 16 – Roof Decking Pavers

07 76 16.10 Roof Pavers and Supports

		Crew	Daily Output	Labor-Hours	Unit	Material	2021 Bare Costs Labor	Equipment	Total	Total Incl O&P
0010	**ROOF PAVERS AND SUPPORTS**									
1000	Roof decking pavers, concrete blocks, 2" thick, natural	1 Clab	115	.070	S.F.	3.57	3.09		6.66	8.55
1100	Colors		115	.070	"	3.56	3.09		6.65	8.55
1200	Support pedestal, bottom cap		960	.008	Ea.	2.65	.37		3.02	3.47
1300	Top cap		960	.008		4.86	.37		5.23	5.90
1400	Leveling shims, 1/16"		1920	.004		1.21	.19		1.40	1.61
1500	1/8"		1920	.004		1.99	.19		2.18	2.47
1600	Buffer pad		960	.008		4.50	.37		4.87	5.50
1700	PVC legs (4" SDR 35)		2880	.003	Inch	.18	.12		.30	.38
2000	Alternate pricing method, system in place		101	.079	S.F.	8	3.52		11.52	14

07 81 Applied Fireproofing

07 81 16 – Cementitious Fireproofing

07 81 16.10 Sprayed Cementitious Fireproofing

		Crew	Daily Output	Labor-Hours	Unit	Material	2021 Bare Costs Labor	Equipment	Total	Total Incl O&P
0010	**SPRAYED CEMENTITIOUS FIREPROOFING**									
0050	Not including canvas protection, normal density									
0100	Per 1" thick, on flat plate steel	G-2	3000	.008	S.F.	.59	.37	.06	1.02	1.27
0200	Flat decking		2400	.010		.59	.46	.08	1.13	1.43
0400	Beams		1500	.016		.59	.74	.13	1.46	1.89
0500	Corrugated or fluted decks		1250	.019		.88	.89	.15	1.92	2.46
0700	Columns, 1-1/8" thick		1100	.022		.66	1.01	.17	1.84	2.42
0800	2-3/16" thick		700	.034		1.63	1.59	.27	3.49	4.45
0900	For canvas protection, add		5000	.005		.11	.22	.04	.37	.49
1000	Not including canvas protection, high density									
1100	Per 1" thick, on flat plate steel	G-2	3000	.008	S.F.	2.30	.37	.06	2.73	3.15
1110	On flat decking		2400	.010		2.30	.46	.08	2.84	3.31
1120	On beams		1500	.016		2.30	.74	.13	3.17	3.77
1130	Corrugated or fluted decks		1250	.019		2.30	.89	.15	3.34	4.02
1140	Columns, 1-1/8" thick		1100	.022		2.59	1.01	.17	3.77	4.54
1150	2-3/16" thick		1100	.022		5.20	1.01	.17	6.38	7.40
1170	For canvas protection, add		5000	.005		.11	.22	.04	.37	.49
1200	Not including canvas protection, retrofitting									
1210	Per 1" thick, on flat plate steel	G-2	1500	.016	S.F.	.52	.74	.13	1.39	1.81
1220	On flat decking		1200	.020		.52	.93	.16	1.61	2.12
1230	On beams		750	.032		.52	1.48	.25	2.25	3.06
1240	Corrugated or fluted decks		625	.038		.78	1.78	.30	2.86	3.84
1250	Columns, 1-1/8" thick		550	.044		.59	2.02	.35	2.96	4.04
1260	2-3/16" thick		500	.048		1.17	2.22	.38	3.77	5
1400	Accessories, preliminary spattered texture coat		4500	.005		.05	.25	.04	.34	.47
1410	Bonding agent	1 Plas	1000	.008		.11	.40		.51	.71

07 91 Preformed Joint Seals

07 91 13 – Compression Seals

07 91 13.10 Compression Seals

		Crew	Daily Output	Labor-Hours	Unit	Material	2021 Bare Costs Labor	Equipment	Total	Total Incl O&P
0010	**COMPRESSION SEALS**									
4900	O-ring type cord, 1/4"	1 Bric	472	.017	L.F.	.42	.91		1.33	1.83
4910	1/2"		440	.018		1.11	.98		2.09	2.69
4920	3/4"		424	.019		2.23	1.01		3.24	3.98
4930	1"		408	.020		4.48	1.05		5.53	6.50
4940	1-1/4"		384	.021		10.80	1.12		11.92	13.55
4950	1-1/2"		368	.022		13.25	1.17		14.42	16.30
4960	1-3/4"		352	.023		15.45	1.22		16.67	18.85
4970	2"		344	.023		27	1.25		28.25	31.50

07 91 16 – Joint Gaskets

07 91 16.10 Joint Gaskets

		Crew	Daily Output	Labor-Hours	Unit	Material	2021 Bare Costs Labor	Equipment	Total	Total Incl O&P
0010	**JOINT GASKETS**									
4400	Joint gaskets, neoprene, closed cell w/adh, 1/8" x 3/8"	1 Bric	240	.033	L.F.	.35	1.79		2.14	3.09
4500	1/4" x 3/4"		215	.037		.74	2		2.74	3.82
4700	1/2" x 1"		200	.040		1.79	2.15		3.94	5.20
4800	3/4" x 1-1/2"		165	.048		1.88	2.60		4.48	6

07 91 23 – Backer Rods

07 91 23.10 Backer Rods

		Crew	Daily Output	Labor-Hours	Unit	Material	2021 Bare Costs Labor	Equipment	Total	Total Incl O&P
0010	**BACKER RODS**									
0030	Backer rod, polyethylene, 1/4" diameter	1 Bric	4.60	1.739	C.L.F.	2.77	93.50		96.27	144
0050	1/2" diameter		4.60	1.739		19.90	93.50		113.40	163
0070	3/4" diameter		4.60	1.739		7.70	93.50		101.20	149
0090	1" diameter		4.60	1.739		15.45	93.50		108.95	158

07 91 26 – Joint Fillers

07 91 26.10 Joint Fillers

		Crew	Daily Output	Labor-Hours	Unit	Material	2021 Bare Costs Labor	Equipment	Total	Total Incl O&P
0010	**JOINT FILLERS**									
4360	Butyl rubber filler, 1/4" x 1/4"	1 Bric	290	.028	L.F.	.23	1.48		1.71	2.48
4365	1/2" x 1/2"		250	.032		.92	1.72		2.64	3.60
4370	1/2" x 3/4"		210	.038		1.38	2.05		3.43	4.60
4375	3/4" x 3/4"		230	.035		2.07	1.87		3.94	5.10
4380	1" x 1"		180	.044		2.76	2.39		5.15	6.65
4390	For coloring, add					12%				
4980	Polyethylene joint backing, 1/4" x 2"	1 Bric	2.08	3.846	C.L.F.	14.85	207		221.85	325
4990	1/4" x 6"		1.28	6.250	"	33.50	335		368.50	540
5600	Silicone, room temp vulcanizing foam seal, 1/4" x 1/2"		1312	.006	L.F.	.45	.33		.78	.99
5610	1/2" x 1/2"		656	.012		.91	.65		1.56	1.99
5620	1/2" x 3/4"		442	.018		1.36	.97		2.33	2.95
5630	3/4" x 3/4"		328	.024		2.04	1.31		3.35	4.21
5640	1/8" x 1"		1312	.006		.45	.33		.78	.99
5650	1/8" x 3"		442	.018		1.36	.97		2.33	2.95
5670	1/4" x 3"		295	.027		2.72	1.46		4.18	5.20
5680	1/4" x 6"		148	.054		5.45	2.90		8.35	10.35
5690	1/2" x 6"		82	.098		10.85	5.25		16.10	19.85
5700	1/2" x 9"		52.50	.152		16.30	8.20		24.50	30.50
5710	1/2" x 12"		33	.242		21.50	13		34.50	43.50

07 92 Joint Sealants

07 92 13 – Elastomeric Joint Sealants

07 92 13.10 Masonry Joint Sealants

		Crew	Daily Output	Labor-Hours	Unit	Material	2021 Bare Costs Labor	Equipment	Total	Total Incl O&P
0010	**MASONRY JOINT SEALANTS**, 1/2" x 1/2" joint									
0050	Re-caulk only, oil base	1 Bric	225	.036	L.F.	.53	1.91		2.44	3.46
0100	Acrylic latex		205	.039		.30	2.10		2.40	3.49
0200	Polyurethane		200	.040		.73	2.15		2.88	4.04
0300	Silicone		195	.041		.59	2.20		2.79	3.97
1000	Cut out and re-caulk, oil base		145	.055		.53	2.96		3.49	5.05
1050	Acrylic latex		130	.062		.30	3.30		3.60	5.30
1100	Polyurethane		125	.064		.73	3.44		4.17	6
1150	Silicone	▼	120	.067	▼	.59	3.58		4.17	6.05

07 92 13.20 Caulking and Sealant Options

		Crew	Daily Output	Labor-Hours	Unit	Material	2021 Bare Costs Labor	Equipment	Total	Total Incl O&P
0010	**CAULKING AND SEALANT OPTIONS**									
0050	Latex acrylic based, bulk				Gal.	30.50			30.50	33.50
0055	Bulk in place 1/4" x 1/4" bead	1 Bric	300	.027	L.F.	.10	1.43		1.53	2.27
0060	1/4" x 3/8"		294	.027		.16	1.46		1.62	2.38
0065	1/4" x 1/2"		288	.028		.21	1.49		1.70	2.48
0075	3/8" x 3/8"		284	.028		.24	1.51		1.75	2.54
0080	3/8" x 1/2"		280	.029		.32	1.53		1.85	2.66
0085	3/8" x 5/8"		276	.029		.40	1.56		1.96	2.78
0095	3/8" x 3/4"		272	.029		.48	1.58		2.06	2.90
0100	1/2" x 1/2"		275	.029		.42	1.56		1.98	2.82
0105	1/2" x 5/8"		269	.030		.53	1.60		2.13	2.99
0110	1/2" x 3/4"		263	.030		.64	1.63		2.27	3.16
0115	1/2" x 7/8"		256	.031		.74	1.68		2.42	3.35
0120	1/2" x 1"		250	.032		.85	1.72		2.57	3.52
0125	3/4" x 3/4"		244	.033		.96	1.76		2.72	3.70
0130	3/4" x 1"		225	.036		1.28	1.91		3.19	4.28
0135	1" x 1"	▼	200	.040	▼	1.70	2.15		3.85	5.10
0190	Cartridges				Gal.	41.50			41.50	45.50
0200	11 fl. oz. cartridge				Ea.	3.57			3.57	3.93
0500	1/4" x 1/2"	1 Bric	288	.028	L.F.	.29	1.49		1.78	2.57
0600	1/2" x 1/2"		275	.029		.58	1.56		2.14	2.99
0800	3/4" x 3/4"		244	.033		1.31	1.76		3.07	4.09
0900	3/4" x 1"		225	.036		1.75	1.91		3.66	4.81
1000	1" x 1"	▼	200	.040	▼	2.19	2.15		4.34	5.65
1400	Butyl based, bulk				Gal.	41			41	45.50
1500	Cartridges				"	41			41	45.50
1700	1/4" x 1/2", 154 L.F./gal.	1 Bric	288	.028	L.F.	.27	1.49		1.76	2.54
1800	1/2" x 1/2", 77 L.F./gal.	"	275	.029	"	.53	1.56		2.09	2.94
2300	Polysulfide compounds, 1 component, bulk				Gal.	84			84	92
2600	1 or 2 component, in place, 1/4" x 1/4", 308 L.F./gal.	1 Bric	300	.027	L.F.	.27	1.43		1.70	2.46
2700	1/2" x 1/4", 154 L.F./gal.		288	.028		.54	1.49		2.03	2.85
2900	3/4" x 3/8", 68 L.F./gal.		272	.029		1.23	1.58		2.81	3.74
3000	1" x 1/2", 38 L.F./gal.	▼	250	.032	▼	2.21	1.72		3.93	5
3200	Polyurethane, 1 or 2 component				Gal.	54.50			54.50	60
3500	Bulk, in place, 1/4" x 1/4"	1 Bric	300	.027	L.F.	.18	1.43		1.61	2.36
3655	1/2" x 1/4"		288	.028		.35	1.49		1.84	2.64
3800	3/4" x 3/8"		272	.029		.80	1.58		2.38	3.26
3900	1" x 1/2"	▼	250	.032	▼	1.42	1.72		3.14	4.15
4100	Silicone rubber, bulk				Gal.	65.50			65.50	72.50
4200	Cartridges				"	44.50			44.50	49

For customer support on your Concrete & Masonry Costs with RSMeans Data, call 800.448.8182.

161

07 92 Joint Sealants

07 92 16 – Rigid Joint Sealants

07 92 16.10 Rigid Joint Sealants	Crew	Daily Output	Labor-Hours	Unit	Material	2021 Bare Costs Labor	Equipment	Total	Total Incl O&P
0010 **RIGID JOINT SEALANTS**									
5800 Tapes, sealant, PVC foam adhesive, 1/16" x 1/4"				C.L.F.	8.95			8.95	9.85
5900 1/16" x 1/2"					9.05			9.05	9.95
5950 1/16" x 1"					15.05			15.05	16.55
6000 1/8" x 1/2"					9.20			9.20	10.10

07 92 19 – Acoustical Joint Sealants

07 92 19.10 Acoustical Sealant	Crew	Daily Output	Labor-Hours	Unit	Material	2021 Bare Costs Labor	Equipment	Total	Total Incl O&P
0010 **ACOUSTICAL SEALANT**									
0020 Acoustical sealant, elastomeric, cartridges				Ea.	8.05			8.05	8.85
0025 In place, 1/4" x 1/4"	1 Bric	300	.027	L.F.	.33	1.43		1.76	2.52
0030 1/4" x 1/2"		288	.028		.66	1.49		2.15	2.97
0035 1/2" x 1/2"		275	.029		1.31	1.56		2.87	3.80
0040 1/2" x 3/4"		263	.030		1.97	1.63		3.60	4.63
0045 3/4" x 3/4"		244	.033		2.96	1.76		4.72	5.90
0050 1" x 1"		200	.040		5.25	2.15		7.40	9.05

07 95 Expansion Control

07 95 13 – Expansion Joint Cover Assemblies

07 95 13.50 Expansion Joint Assemblies	Crew	Daily Output	Labor-Hours	Unit	Material	2021 Bare Costs Labor	Equipment	Total	Total Incl O&P
0010 **EXPANSION JOINT ASSEMBLIES**									
0200 Floor cover assemblies, 1" space, aluminum	1 Sswk	38	.211	L.F.	22.50	12.70		35.20	44.50
0300 Bronze		38	.211		53	12.70		65.70	78
0500 2" space, aluminum		38	.211		14.90	12.70		27.60	36
0600 Bronze		38	.211		61.50	12.70		74.20	87.50
0800 Wall and ceiling assemblies, 1" space, aluminum		38	.211		17.20	12.70		29.90	38.50
0900 Bronze		38	.211		54.50	12.70		67.20	79
1100 2" space, aluminum		38	.211		18.05	12.70		30.75	39.50
1200 Bronze		38	.211		52	12.70		64.70	77
1400 Floor to wall assemblies, 1" space, aluminum		38	.211		21	12.70		33.70	42.50
1500 Bronze or stainless		38	.211		63.50	12.70		76.20	89.50
1700 Gym floor angle covers, aluminum, 3" x 3" angle		46	.174		21	10.50		31.50	39.50
1800 3" x 4" angle		46	.174		25	10.50		35.50	44
2300 Roof to wall, low profile, 1" space		57	.140		25	8.45		33.45	40
2400 High profile		57	.140		32.50	8.45		40.95	48.50

Estimating Tips

08 10 00 Doors and Frames

All exterior doors should be addressed for their energy conservation (insulation and seals).

- Most metal doors and frames look alike, but there may be significant differences among them. When estimating these items, be sure to choose the line item that most closely compares to the specification or door schedule requirements regarding:
  - □ type of metal
  - □ metal gauge
  - □ door core material
  - □ fire rating
  - □ finish

- Wood and plastic doors vary considerably in price. The primary determinant is the veneer material. Lauan, birch, and oak are the most common veneers. Other variables include the following:
  - □ hollow or solid core
  - □ fire rating
  - □ flush or raised panel
  - □ finish

- Door pricing includes bore for cylindrical locksets and mortise for hinges.

08 30 00 Specialty Doors and Frames

- There are many varieties of special doors, and they are usually priced per each. Add frames, hardware, or operators required for a complete installation.

08 40 00 Entrances, Storefronts, and Curtain Walls

- Glazed curtain walls consist of the metal tube framing and the glazing material. The cost data in this subdivision is presented for the metal tube framing alone or the composite wall. If your estimate requires a detailed takeoff of the framing, be sure to add the glazing cost and any tints.

08 50 00 Windows

- Steel windows are unglazed and aluminum can be glazed or unglazed. Some metal windows are priced without glass. Refer to 08 80 00 Glazing for glass pricing. The grade C indicates commercial grade windows, usually ASTM C-35.

- All wood windows and vinyl are priced preglazed. The glazing is insulating glass. Add the cost of screens and grills if required and not already included.

08 70 00 Hardware

- Hardware costs add considerably to the cost of a door. The most efficient method to determine the hardware requirements for a project is to review the door and hardware schedule together. One type of door may have different hardware, depending on the door usage.

- Door hinges are priced by the pair, with most doors requiring 1-1/2 pairs per door. The hinge prices do not include installation labor because it is included in door installation.

Hinges are classified according to the frequency of use, base material, and finish.

08 80 00 Glazing

- Different openings require different types of glass. The most common types are:
  - □ float
  - □ tempered
  - □ insulating
  - □ impact-resistant
  - □ ballistic-resistant

- Most exterior windows are glazed with insulating glass. Entrance doors and window walls, where the glass is less than 18" from the floor, are generally glazed with tempered glass. Interior windows and some residential windows are glazed with float glass.

- Coastal communities require the use of impact-resistant glass, dependent on wind speed.

- The insulation or 'u' value is a strong consideration, along with solar heat gain, to determine total energy efficiency.

Reference Numbers

Reference numbers are shown at the beginning of some major classifications. These numbers refer to related items in the Reference Section. The reference information may be an estimating procedure, an alternate pricing method, or technical information.

Note: Not all subdivisions listed here necessarily appear. ■

Same Data. Simplified.

Enjoy the convenience and efficiency of accessing your costs anywhere:

- **Skip the multiplier** by setting your location
- **Quickly search,** edit, favorite and share costs
- **Stay on top of price changes** with automatic updates

Discover more at rsmeans.com/online

08 12 Metal Frames

08 12 13 – Hollow Metal Frames

08 12 13.13 Standard Hollow Metal Frames		Crew	Daily Output	Labor-Hours	Unit	Material	2021 Bare Costs Labor	Equipment	Total	Total Incl O&P
0010	**STANDARD HOLLOW METAL FRAMES**									
0020	16 ga., up to 5-3/4" jamb depth									
0025	3'-0" x 6'-8" single	G	2 Carp	16	1	Ea.	228	54.50		282.50	335
0028	3'-6" wide, single	G		16	1		269	54.50		323.50	380
0030	4'-0" wide, single	G		16	1		305	54.50		359.50	415
0040	6'-0" wide, double	G		14	1.143		299	62.50		361.50	425
0045	8'-0" wide, double	G		14	1.143		245	62.50		307.50	365
0100	3'-0" x 7'-0" single	G		16	1		217	54.50		271.50	320
0110	3'-6" wide, single	G		16	1		228	54.50		282.50	335
0112	4'-0" wide, single	G		16	1		253	54.50		307.50	360
0140	6'-0" wide, double	G		14	1.143		296	62.50		358.50	420
0145	8'-0" wide, double	G		14	1.143		231	62.50		293.50	350
1000	16 ga., up to 4-7/8" deep, 3'-0" x 7'-0" single	G		16	1		174	54.50		228.50	273
1140	6'-0" wide, double	G		14	1.143		212	62.50		274.50	330
1200	16 ga., 8-3/4" deep, 3'-0" x 7'-0" single	G		16	1		231	54.50		285.50	335
1240	6'-0" wide, double	G		14	1.143		275	62.50		337.50	395
2800	14 ga., up to 3-7/8" deep, 3'-0" x 7'-0" single	G		16	1		230	54.50		284.50	335
2840	6'-0" wide, double	G		14	1.143		350	62.50		412.50	480
3000	14 ga., up to 5-3/4" deep, 3'-0" x 6'-8" single	G		16	1		156	54.50		210.50	253
3002	3'-6" wide, single	G		16	1		246	54.50		300.50	355
3005	4'-0" wide, single	G		16	1		133	54.50		187.50	228
3600	up to 5-3/4" jamb depth, 4'-0" x 7'-0" single	G		15	1.067		216	58.50		274.50	325
3620	6'-0" wide, double	G		12	1.333		209	73		282	340
3640	8'-0" wide, double	G		12	1.333		315	73		388	460
3700	8'-0" high, 4'-0" wide, single	G		15	1.067		298	58.50		356.50	415
3740	8'-0" wide, double	G		12	1.333		345	73		418	490
4000	6-3/4" deep, 4'-0" x 7'-0" single	G		15	1.067		288	58.50		346.50	400
4020	6'-0" wide, double	G		12	1.333		325	73		398	470
4040	8'-0" wide, double	G		12	1.333		243	73		316	375
4100	8'-0" high, 4'-0" wide, single	G		15	1.067		160	58.50		218.50	263
4140	8'-0" wide, double	G		12	1.333		450	73		523	605
4400	8-3/4" deep, 4'-0" x 7'-0", single	G		15	1.067		460	58.50		518.50	590
4440	8'-0" wide, double	G		12	1.333		440	73		513	590
4500	4'-0" x 8'-0", single	G		15	1.067		480	58.50		538.50	615
4540	8'-0" wide, double	G		12	1.333		475	73		548	630
4900	For welded frames, add						69.50			69.50	76.50
5400	14 ga., "B" label, up to 5-3/4" deep, 4'-0" x 7'-0" single	G	2 Carp	15	1.067		183	58.50		241.50	288
5440	8'-0" wide, double	G		12	1.333		230	73		303	360
5800	6-3/4" deep, 7'-0" high, 4'-0" wide, single	G		15	1.067		166	58.50		224.50	269
5840	8'-0" wide, double	G		12	1.333		345	73		418	490
6200	8-3/4" deep, 4'-0" x 7'-0" single	G		15	1.067		271	58.50		329.50	385
6240	8'-0" wide, double	G		12	1.333		395	73		468	545
6300	For "A" label use same price as "B" label										
6400	For baked enamel finish, add						30%	15%			
6500	For galvanizing, add						20%				
7900	Transom lite frames, fixed, add		2 Carp	155	.103	S.F.	53	5.65		58.65	67
8000	Movable, add		"	130	.123	"	71.50	6.75		78.25	88.50

08 12 13.25 Channel Metal Frames

		Crew	Daily Output	Labor-Hours	Unit	Material	2021 Bare Costs Labor	Equipment	Total	Total Incl O&P
0010	**CHANNEL METAL FRAMES**									
0020	Steel channels with anchors and bar stops									
0100	6" channel @ 8.2#/L.F., 3' x 7' door, weighs 150#	G	E-4	13	2.462	Ea.	237	150	11.45	398.45	505
0200	8" channel @ 11.5#/L.F., 6' x 8' door, weighs 275#	G	"	9	3.556		435	216	16.55	667.55	835

08 12 Metal Frames

08 12 13 – Hollow Metal Frames

	08 12 13.25 Channel Metal Frames		Crew	Daily Output	Labor-Hours	Unit	Material	2021 Bare Costs Labor	Equipment	Total	Total Incl O&P
0300	8' x 12' door, weighs 400#	G	E-4	6.50	4.923	Ea.	635	299	23	957	1,175
0400	10" channel @ 15.3#/L.F., 10' x 10' door, weighs 500#	G		6	5.333		790	325	25	1,140	1,400
0500	12' x 12' door, weighs 600#	G		5.50	5.818		950	355	27	1,332	1,625
0600	12" channel @ 20.7#/L.F., 12' x 12' door, weighs 825#	G		4.50	7.111		1,300	430	33	1,763	2,125
0700	12' x 16' door, weighs 1000#	G		4	8		1,575	485	37	2,097	2,550
0800	For frames without bar stops, light sections, deduct						15%				
0900	Heavy sections, deduct						10%				

08 31 Access Doors and Panels

08 31 13 – Access Doors and Frames

08 31 13.20 Bulkhead/Cellar Doors

		Crew	Daily Output	Labor-Hours	Unit	Material	Labor	Equipment	Total	Total Incl O&P
0010	**BULKHEAD/CELLAR DOORS**									
0020	Steel, not incl. sides, 44" x 62"	1 Carp	5.50	1.455	Ea.	720	79.50		799.50	910
0100	52" x 73"		5.10	1.569		900	86		986	1,125
0500	With sides and foundation plates, 57" x 45" x 24"		4.70	1.702		895	93		988	1,125
0600	42" x 49" x 51"		4.30	1.860		610	102		712	820

08 31 13.30 Commercial Floor Doors

		Crew	Daily Output	Labor-Hours	Unit	Material	Labor	Equipment	Total	Total Incl O&P
0010	**COMMERCIAL FLOOR DOORS**									
0020	Aluminum tile, steel frame, one leaf, 2' x 2' opng.	2 Sswk	3.50	4.571	Opng.	740	276		1,016	1,225
0050	3'-6" x 3'-6" opening		3.50	4.571		1,575	276		1,851	2,150
0500	Double leaf, 4' x 4' opening		3	5.333		1,600	320		1,920	2,275
0550	5' x 5' opening		3	5.333		2,675	320		2,995	3,425

08 31 13.35 Industrial Floor Doors

		Crew	Daily Output	Labor-Hours	Unit	Material	Labor	Equipment	Total	Total Incl O&P
0010	**INDUSTRIAL FLOOR DOORS**									
0020	Steel 300 psf L.L., single leaf, 2' x 2', 175#	2 Sswk	6	2.667	Opng.	835	161		996	1,175
0050	3' x 3' opening, 300#		5.50	2.909		1,400	175		1,575	1,825
0300	Double leaf, 4' x 4' opening, 455#		5	3.200		2,675	193		2,868	3,250
0350	5' x 5' opening, 645#		4.50	3.556		3,175	214		3,389	3,800
1000	Aluminum, 300 psf L.L., single leaf, 2' x 2', 60#		6	2.667		735	161		896	1,050
1050	3' x 3' opening, 100#		5.50	2.909		890	175		1,065	1,250
1500	Double leaf, 4' x 4' opening, 160#		5	3.200		3,200	193		3,393	3,825
1550	5' x 5' opening, 235#		4.50	3.556		4,125	214		4,339	4,875
2000	Aluminum, 150 psf L.L., single leaf, 2' x 2', 60#		6	2.667		765	161		926	1,100
2050	3' x 3' opening, 95#		5.50	2.909		1,150	175		1,325	1,525
2500	Double leaf, 4' x 4' opening, 150#		5	3.200		1,475	193		1,668	1,925
2550	5' x 5' opening, 230#		4.50	3.556		1,975	214		2,189	2,475

08 34 Special Function Doors

08 34 59 – Vault Doors and Day Gates

08 34 59.10 Secure Storage Doors

		Crew	Daily Output	Labor-Hours	Unit	Material	Labor	Equipment	Total	Total Incl O&P
0010	**SECURE STORAGE DOORS**									
0020	Door and frame, 32" x 78", clear opening									
0100	1 hour test, 32" door, weighs 750 lb.	2 Sswk	1.50	10.667	Opng.	8,325	645		8,970	10,200
0200	2 hour test, 32" door, weighs 950 lb.		1.30	12.308		7,825	740		8,565	9,775
0250	40" door, weighs 1130 lb.		1	16		8,525	965		9,490	10,900
0300	4 hour test, 32" door, weighs 1025 lb.		1.20	13.333		10,300	805		11,105	12,700
0350	40" door, weighs 1140 lb.		.90	17.778		10,800	1,075		11,875	13,600
0600	For time lock, two movement, add	1 Elec	2	4	Ea.	1,575	255		1,830	2,100
0800	Day gate, painted, steel, 32" wide	2 Sswk	1.50	10.667		2,025	645		2,670	3,250

08 34 Special Function Doors

08 34 59 – Vault Doors and Day Gates

08 34 59.10 Secure Storage Doors	Crew	Daily Output	Labor-Hours	Unit	Material	2021 Bare Costs Labor	Equipment	Total	Total Incl O&P
0850	40" wide	2 Sswk	1.40	11.429	Ea.	2,575	690		3,265	3,900
0900	Aluminum, 32" wide		1.50	10.667		2,900	645		3,545	4,175
0950	40" wide		1.40	11.429		3,150	690		3,840	4,550
2050	Security vault door, class I, 3' wide, 3-1/2" thick	E-24	.19	167	Opng.	14,600	10,000	3,050	27,650	34,800
2100	Class II, 3' wide, 7" thick		.19	167		17,700	10,000	3,050	30,750	38,200
2150	Class III, 9R, 3' wide, 10" thick		.13	250		21,900	15,000	4,575	41,475	52,000

08 51 Metal Windows

08 51 23 – Steel Windows

08 51 23.10 Steel Sash

08 51 23.10 Steel Sash	Crew	Daily Output	Labor-Hours	Unit	Material	2021 Bare Costs Labor	Equipment	Total	Total Incl O&P
0010	**STEEL SASH** Custom units, glazing and trim not included									
0100	Casement, 100% vented	2 Sswk	200	.080	S.F.	70	4.82		74.82	84.50
0200	50% vented		200	.080		57	4.82		61.82	70.50
0300	Fixed		200	.080		29.50	4.82		34.32	40
1000	Projected, commercial, 40% vented		200	.080		54	4.82		58.82	67
1100	Intermediate, 50% vented		200	.080		64.50	4.82		69.32	78.50
1500	Industrial, horizontally pivoted		200	.080		56.50	4.82		61.32	69.50
1600	Fixed		200	.080		32.50	4.82		37.32	43.50
2000	Industrial security sash, 50% vented		200	.080		62	4.82		66.82	75.50
2100	Fixed		200	.080		49.50	4.82		54.32	62
2500	Picture window		200	.080		33	4.82		37.82	43.50
3000	Double-hung		200	.080		62.50	4.82		67.32	76.50
5000	Mullions for above, open interior face		240	.067	L.F.	11	4.02		15.02	18.30
5100	With interior cover		240	.067	"	18.25	4.02		22.27	26

08 51 23.20 Steel Windows

08 51 23.20 Steel Windows	Crew	Daily Output	Labor-Hours	Unit	Material	2021 Bare Costs Labor	Equipment	Total	Total Incl O&P
0010	**STEEL WINDOWS** Stock, including frame, trim and insul. glass									
1000	Custom units, double-hung, 2'-8" x 4'-6" opening	2 Sswk	12	1.333	Ea.	725	80.50		805.50	920
1100	2'-4" x 3'-9" opening		12	1.333		595	80.50		675.50	780
1500	Commercial projected, 3'-9" x 5'-5" opening		10	1.600		1,250	96.50		1,346.50	1,550
1600	6'-9" x 4'-1" opening		7	2.286		1,675	138		1,813	2,050
2000	Intermediate projected, 2'-9" x 4'-1" opening		12	1.333		710	80.50		790.50	905
2100	4'-1" x 5'-5" opening		10	1.600		1,450	96.50		1,546.50	1,725

Estimating Tips
General

- Room Finish Schedule: A complete set of plans should contain a room finish schedule. If one is not available, it would be well worth the time and effort to obtain one.

09 20 00 Plaster and Gypsum Board

- Lath is estimated by the square yard plus a 5% allowance for waste. Furring, channels, and accessories are measured by the linear foot. An extra foot should be allowed for each accessory miter or stop.

- Plaster is also estimated by the square yard. Deductions for openings vary by preference, from zero deduction to 50% of all openings over 2 feet in width. The estimator should allow one extra square foot for each linear foot of horizontal interior or exterior angle located below the ceiling level. Also, double the areas of small radius work.

- Drywall accessories, studs, track, and acoustical caulking are all measured by the linear foot. Drywall taping is figured by the square foot. Gypsum wallboard is estimated by the square foot. No material deductions should be made for door or window openings under 32 S.F.

09 60 00 Flooring

- Tile and terrazzo areas are taken off on a square foot basis. Trim and base materials are measured by the linear foot. Accent tiles are listed per each. Two basic methods of installation are used. Mud set is approximately 30% more expensive than thin set.

The cost of grout is included with tile unit price lines unless otherwise noted. In terrazzo work, be sure to include the linear footage of embedded decorative strips, grounds, machine rubbing, and power cleanup.

- Wood flooring is available in strip, parquet, or block configuration. The latter two types are set in adhesives with quantities estimated by the square foot. The laying pattern will influence labor costs and material waste. In addition to the material and labor for laying wood floors, the estimator must make allowances for sanding and finishing these areas, unless the flooring is prefinished.

- Sheet flooring is measured by the square yard. Roll widths vary, so consideration should be given to use the most economical width, as waste must be figured into the total quantity. Consider also the installation methods available—direct glue down or stretched. Direct glue-down installation is assumed with sheet carpet unit price lines unless otherwise noted.

09 70 00 Wall Finishes

- Wall coverings are estimated by the square foot. The area to be covered is measured—length by height of the wall above the baseboards—to calculate the square footage of each wall. This figure is divided by the number of square feet in the single roll which is being used. Deduct, in full, the areas of openings such as doors and windows. Where a pattern match is required allow 25–30% waste.

09 80 00 Acoustic Treatment

- Acoustical systems fall into several categories. The takeoff of these materials should be by the square foot of area with a 5% allowance for waste. Do not forget about scaffolding, if applicable, when estimating these systems.

09 90 00 Painting and Coating

- New line items created for cut-ins with reference diagram.

- A major portion of the work in painting involves surface preparation. Be sure to include cleaning, sanding, filling, and masking costs in the estimate.

- Protection of adjacent surfaces is not included in painting costs. When considering the method of paint application, an important factor is the amount of protection and masking required. These must be estimated separately and may be the determining factor in choosing the method of application.

Reference Numbers

Reference numbers are shown at the beginning of some major classifications. These numbers refer to related items in the Reference Section. The reference information may be an estimating procedure, an alternate pricing method, or technical information.

Note: Not all subdivisions listed here necessarily appear. ■

---

09 05 05 – Selective Demolition for Finishes

09 05 05.20 Selective Demolition, Flooring	Crew	Daily Output	Labor-Hours	Unit	Material	2021 Bare Costs Labor	Equipment	Total	Total Incl O&P
0010 **SELECTIVE DEMOLITION, FLOORING** R024119-10									
0200 Brick with mortar	2 Clab	475	.034	S.F.		1.50		1.50	2.23
0400 Carpet, bonded, including surface scraping		2000	.008			.36		.36	.53
0440 Scrim applied		8000	.002			.09		.09	.13
0480 Tackless		9000	.002			.08		.08	.12
0550 Carpet tile, releasable adhesive		5000	.003			.14		.14	.21
0560 Permanent adhesive		1850	.009			.38		.38	.57
0600 Composition, acrylic or epoxy		400	.040			1.78		1.78	2.65
0700 Concrete, scarify skin	A-1A	225	.036			2.03	.93	2.96	4.07
0800 Resilient, sheet goods	2 Clab	1400	.011			.51		.51	.76
0820 For gym floors	"	900	.018			.79		.79	1.18
0850 Vinyl or rubber cove base	1 Clab	1000	.008	L.F.		.36		.36	.53
0860 Vinyl or rubber cove base, molded corner	"	1000	.008	Ea.		.36		.36	.53
0870 For glued and caulked installation, add to labor						50%			
0900 Vinyl composition tile, 12" x 12"	2 Clab	1000	.016	S.F.		.71		.71	1.06
2000 Tile, ceramic, thin set		675	.024			1.05		1.05	1.57
2020 Mud set		625	.026			1.14		1.14	1.70
2200 Marble, slate, thin set		675	.024			1.05		1.05	1.57
2220 Mud set		625	.026			1.14		1.14	1.70
2600 Terrazzo, thin set		450	.036			1.58		1.58	2.36
2620 Mud set		425	.038			1.67		1.67	2.49
2640 Terrazzo, cast in place		300	.053			2.37		2.37	3.53
3000 Wood, block, on end	1 Carp	400	.020			1.09		1.09	1.63
3200 Parquet		450	.018			.97		.97	1.45
3400 Strip flooring, interior, 2-1/4" x 25/32" thick		325	.025			1.35		1.35	2.01
3500 Exterior, porch flooring, 1" x 4"		220	.036			1.99		1.99	2.97
3800 Subfloor, tongue and groove, 1" x 6"		325	.025			1.35		1.35	2.01
3820 1" x 8"		430	.019			1.02		1.02	1.52
3840 1" x 10"		520	.015			.84		.84	1.26
4000 Plywood, nailed		600	.013			.73		.73	1.09
4100 Glued and nailed		400	.020			1.09		1.09	1.63
4200 Hardboard, 1/4" thick		760	.011			.58		.58	.86

09 05 05.30 Selective Demolition, Walls and Partitions

	Crew	Daily Output	Labor-Hours	Unit	Material	2021 Bare Costs Labor	Equipment	Total	Total Incl O&P
0010 **SELECTIVE DEMOLITION, WALLS AND PARTITIONS** R024119-10									
0100 Brick, 4" to 12" thick	B-9	220	.182	C.F.		8.15	1.62	9.77	13.95
0200 Concrete block, 4" thick		1150	.035	S.F.		1.56	.31	1.87	2.67
0280 8" thick		1050	.038			1.71	.34	2.05	2.92
0300 Exterior stucco 1" thick over mesh		3200	.013			.56	.11	.67	.96
3750 Terra cotta block and plaster, to 6" thick	B-1	175	.137			6.20		6.20	9.20
3800 Toilet partitions, slate or marble	1 Clab	5	1.600	Ea.		71		71	106

168

For customer support on your Concrete & Masonry Costs with RSMeans Data, call 800.448.8182.

09 22 Supports for Plaster and Gypsum Board

09 22 03 – Fastening Methods for Finishes

09 22 03.20 Drilling Plaster/Drywall

		Crew	Daily Output	Labor-Hours	Unit	Material	2021 Bare Costs Labor	2021 Bare Costs Equipment	Total	Total Incl O&P
0010	**DRILLING PLASTER/DRYWALL**									
1100	Drilling & layout for drywall/plaster walls, up to 1" deep, no anchor									
1200	Holes, 1/4" diameter	1 Carp	150	.053	Ea.	.01	2.92		2.93	4.36
1300	3/8" diameter		140	.057		.01	3.13		3.14	4.68
1400	1/2" diameter		130	.062		.01	3.37		3.38	5
1500	3/4" diameter		120	.067		.01	3.65		3.66	5.45
1600	1" diameter		110	.073		.03	3.98		4.01	6
1700	1-1/4" diameter		100	.080		.04	4.38		4.42	6.60
1800	1-1/2" diameter	▼	90	.089	▼	.05	4.86		4.91	7.30
1900	For ceiling installations, add						40%			

09 22 36 – Lath

09 22 36.23 Metal Lath

		Crew	Daily Output	Labor-Hours	Unit	Material	2021 Bare Costs Labor	2021 Bare Costs Equipment	Total	Total Incl O&P
0010	**METAL LATH** R092000-50									
0020	Diamond, expanded, 2.5 lb./S.Y., painted				S.Y.	5.45			5.45	6
0100	Galvanized					3.87			3.87	4.26
0300	3.4 lb./S.Y., painted					4.72			4.72	5.20
0400	Galvanized					4.99			4.99	5.50
0600	For #15 asphalt sheathing paper, add					.51			.51	.56
0900	Flat rib, 1/8" high, 2.75 lb., painted					3.87			3.87	4.26
1000	Foil backed					3.63			3.63	3.99
1200	3.4 lb./S.Y., painted					4.07			4.07	4.48
1300	Galvanized					4.24			4.24	4.66
1500	For #15 asphalt sheathing paper, add					.51			.51	.56
1800	High rib, 3/8" high, 3.4 lb./S.Y., painted					4.82			4.82	5.30
1900	Galvanized				▼	4			4	4.40
2400	3/4" high, painted, .60 lb./S.F.				S.F.	.82			.82	.90
2500	.75 lb./S.F.				"	1.72			1.72	1.89
2800	Stucco mesh, painted, 3.6 lb.				S.Y.	4.09			4.09	4.50
3000	K-lath, perforated, absorbent paper, regular					4.39			4.39	4.83
3100	Heavy duty					5.20			5.20	5.70
3300	Waterproof, heavy duty, grade B backing					5.05			5.05	5.55
3400	Fire resistant backing					5.60			5.60	6.15
3600	2.5 lb. diamond painted, on wood framing, on walls	1 Lath	85	.094		5.45	5.10		10.55	13.45
3700	On ceilings		75	.107		5.45	5.75		11.20	14.45
3900	3.4 lb. diamond painted, on wood framing, on walls		80	.100		4.07	5.40		9.47	12.40
4000	On ceilings		70	.114		4.07	6.20		10.27	13.55
4200	3.4 lb. diamond painted, wired to steel framing		75	.107		4.07	5.75		9.82	12.95
4300	On ceilings		60	.133		4.07	7.20		11.27	15.05
4500	Columns and beams, wired to steel		40	.200		4.07	10.80		14.87	20.50
4600	Cornices, wired to steel		35	.229		4.07	12.35		16.42	22.50
4800	Screwed to steel studs, 2.5 lb.		80	.100		5.45	5.40		10.85	13.90
4900	3.4 lb.		75	.107		4.72	5.75		10.47	13.65
5100	Rib lath, painted, wired to steel, on walls, 2.5 lb.		75	.107		3.87	5.75		9.62	12.70
5200	3.4 lb.		70	.114		4.82	6.20		11.02	14.35
5400	4.0 lb.	▼	65	.123		5.75	6.65		12.40	16
5500	For self-furring lath, add					.12			.12	.13
5700	Suspended ceiling system, incl. 3.4 lb. diamond lath, painted	1 Lath	15	.533		4.20	29		33.20	46.50
5800	Galvanized	"	15	.533	▼	3.84	29		32.84	46

For customer support on your Concrete & Masonry Costs with RSMeans Data, call 800.448.8182.

169

09 24 Cement Plastering

09 24 23 – Cement Stucco

09 24 23.40 Stucco

		Crew	Daily Output	Labor-Hours	Unit	Material	2021 Bare Costs Labor	Equipment	Total	Total Incl O&P
0010	**STUCCO** R092000-50									
0015	3 coats 7/8" thick, float finish, with mesh, on wood frame	J-2	63	.762	S.Y.	7.70	37	1.80	46.50	65.50
0100	On masonry construction, no mesh incl.	J-1	67	.597		3.10	28.50	1.69	33.29	48
0300	For trowel finish, add	1 Plas	170	.047			2.35		2.35	3.49
0600	For coloring, add	J-1	685	.058		.43	2.79	.17	3.39	4.80
0700	For special texture, add	"	200	.200		1.50	9.55	.57	11.62	16.45
0900	For soffits, add	J-2	155	.310		2.29	15.10	.73	18.12	26
1000	Stucco, with bonding agent, 3 coats, on walls, no mesh incl.	J-1	200	.200		4.31	9.55	.57	14.43	19.55
1200	Ceilings		180	.222		3.83	10.60	.63	15.06	20.50
1300	Beams		80	.500		3.83	24	1.42	29.25	41.50
1500	Columns		100	.400		3.83	19.10	1.13	24.06	34
1600	Mesh, galvanized, nailed to wood, 1.8 lb.	1 Lath	60	.133		8	7.20		15.20	19.35
1800	3.6 lb.		55	.145		4.09	7.85		11.94	16
1900	Wired to steel, galvanized, 1.8 lb.		53	.151		8	8.15		16.15	20.50
2100	3.6 lb.		50	.160		4.09	8.65		12.74	17.15

09 28 Backing Boards and Underlayments

09 28 13 – Cementitious Backing Boards

09 28 13.10 Cementitious Backerboard

		Crew	Daily Output	Labor-Hours	Unit	Material	2021 Bare Costs Labor	Equipment	Total	Total Incl O&P
0010	**CEMENTITIOUS BACKERBOARD**									
0070	Cementitious backerboard, on floor, 3' x 4' x 1/2" sheets	2 Carp	525	.030	S.F.	.99	1.67		2.66	3.58
0080	3' x 5' x 1/2" sheets		525	.030		.81	1.67		2.48	3.38
0090	3' x 6' x 1/2" sheets		525	.030		.72	1.67		2.39	3.28
0100	3' x 4' x 5/8" sheets		525	.030		.92	1.67		2.59	3.50
0110	3' x 5' x 5/8" sheets		525	.030		.99	1.67		2.66	3.58
0120	3' x 6' x 5/8" sheets		525	.030		.98	1.67		2.65	3.56
0150	On wall, 3' x 4' x 1/2" sheets		350	.046		.99	2.50		3.49	4.82
0160	3' x 5' x 1/2" sheets		350	.046		.81	2.50		3.31	4.62
0170	3' x 6' x 1/2" sheets		350	.046		.72	2.50		3.22	4.52
0180	3' x 4' x 5/8" sheets		350	.046		.92	2.50		3.42	4.74
0190	3' x 5' x 5/8" sheets		350	.046		.99	2.50		3.49	4.82
0200	3' x 6' x 5/8" sheets		350	.046		.98	2.50		3.48	4.80
0250	On counter, 3' x 4' x 1/2" sheets		180	.089		.99	4.86		5.85	8.35
0260	3' x 5' x 1/2" sheets		180	.089		.81	4.86		5.67	8.15
0270	3' x 6' x 1/2" sheets		180	.089		.72	4.86		5.58	8.05
0300	3' x 4' x 5/8" sheets		180	.089		.92	4.86		5.78	8.25
0310	3' x 5' x 5/8" sheets		180	.089		.99	4.86		5.85	8.35
0320	3' x 6' x 5/8" sheets		180	.089		.98	4.86		5.84	8.30

09 30 Tiling

09 30 29 – Metal Tiling

09 30 29.10 Metal Tile

		Crew	Daily Output	Labor-Hours	Unit	Material	2021 Bare Costs Labor	Equipment	Total	Total Incl O&P
0010	**METAL TILE** 4' x 4' sheet, 24 ga., tile pattern, nailed									
0200	Stainless steel	2 Carp	512	.031	S.F.	28	1.71		29.71	33.50
0400	Aluminized steel	"	512	.031	"	20.50	1.71		22.21	25

09 30 Tiling

09 30 95 - Tile & Stone Setting Materials and Specialties

	09 30 95.10 Moisture Resistant, Anti-Fracture Membrane	Crew	Daily Output	Labor-Hours	Unit	Material	2021 Bare Costs Labor	Equipment	Total	Total Incl O&P
0010	**MOISTURE RESISTANT, ANTI-FRACTURE MEMBRANE**									
0200	Elastomeric membrane, 1/16" thick	D-7	275	.058	S.F.	1.53	2.71		4.24	5.65

09 31 Thin-Set Tiling

09 31 13 - Thin-Set Ceramic Tiling

09 31 13.10 Thin-Set Ceramic Tile

		Crew	Daily Output	Labor-Hours	Unit	Material	2021 Bare Costs Labor	Equipment	Total	Total Incl O&P
0010	**THIN-SET CERAMIC TILE**									
0020	Backsplash, average grade tiles	1 Tilf	50	.160	S.F.	3.20	8.25		11.45	15.60
0022	Custom grade tiles		50	.160		6.40	8.25		14.65	19.15
0024	Luxury grade tiles		50	.160		12.80	8.25		21.05	26
0026	Economy grade tiles	↓	50	.160	↓	2.92	8.25		11.17	15.30
0100	Base, using 1' x 4" high piece with 1" x 1" tiles	D-7	128	.125	L.F.	4.90	5.85		10.75	13.90
0300	For 6" high base, 1" x 1" tile face, add					1.26			1.26	1.38
0400	For 2" x 2" tile face, add to above					.76			.76	.84
0700	Cove base, 4-1/4" x 4-1/4"	D-7	128	.125		4.09	5.85		9.94	13
1000	6" x 4-1/4" high		137	.117		4.24	5.45		9.69	12.60
1300	Sanitary cove base, 6" x 4-1/4" high		124	.129		4.86	6		10.86	14.15
1600	6" x 6" high		117	.137	↓	4.92	6.40		11.32	14.70
1800	Bathroom accessories, average (soap dish, toothbrush holder)		82	.195	Ea.	10.30	9.10		19.40	24.50
1900	Bathtub, 5', rec. 4-1/4" x 4-1/4" tile wainscot, adhesive set 6' high		2.90	5.517		203	257		460	600
2100	7' high wainscot		2.50	6.400		237	299		536	695
2200	8' high wainscot		2.20	7.273	↓	271	340		611	795
2500	Bullnose trim, 4-1/4" x 4-1/4"		128	.125	L.F.	4.18	5.85		10.03	13.10
2800	2" x 6"		124	.129	"	4	6		10	13.20
3300	Ceramic tile, porcelain type, 1 color, color group 2, 1" x 1"		183	.087	S.F.	5.75	4.08		9.83	12.30
3310	2" x 2" or 2" x 1"	↓	190	.084		4.04	3.93		7.97	10.20
3350	For random blend, 2 colors, add					1			1	1.10
3360	4 colors, add					1.43			1.43	1.57
3370	For color group 3, add					.65			.65	.72
3380	For abrasive non-slip tile, add					.41			.41	.45
4300	Specialty tile, 4-1/4" x 4-1/4" x 1/2", decorator finish	D-7	183	.087		13.75	4.08		17.83	21
4500	Add for epoxy grout, 1/16" joint, 1" x 1" tile		800	.020		.66	.93		1.59	2.09
4600	2" x 2" tile		820	.020		.62	.91		1.53	2.01
4610	Add for epoxy grout, 1/8" joint, 8" x 8" x 3/8" tile, add	↓	900	.018	↓	1.56	.83		2.39	2.93
4800	Pregrouted sheets, walls, 4-1/4" x 4-1/4", 6" x 4-1/4"									
4810	and 8-1/2" x 4-1/4", 4 S.F. sheets, silicone grout	D-7	240	.067	S.F.	5.80	3.11		8.91	10.95
5100	Floors, unglazed, 2 S.F. sheets,									
5110	urethane adhesive	D-7	180	.089	S.F.	2.07	4.15		6.22	8.35
5400	Walls, interior, 4-1/4" x 4-1/4" tile		190	.084		3.02	3.93		6.95	9.05
5500	6" x 4-1/4" tile		190	.084		3.17	3.93		7.10	9.25
5700	8-1/2" x 4-1/4" tile		190	.084		6.40	3.93		10.33	12.75
5800	6" x 6" tile		175	.091		4.22	4.27		8.49	10.90
5810	8" x 8" tile		170	.094		4.86	4.39		9.25	11.75
5820	12" x 12" tile		160	.100		4.82	4.67		9.49	12.10
5830	16" x 16" tile		150	.107		5.50	4.98		10.48	13.30
6000	Decorated wall tile, 4-1/4" x 4-1/4", color group 1		270	.059		3.73	2.76		6.49	8.15
6100	Color group 4	↓	180	.089	↓	69	4.15		73.15	81.50

For customer support on your Concrete & Masonry Costs with RSMeans Data, call 800.448.8182.

171

09 31 Thin-Set Tiling

09 31 33 – Thin-Set Stone Tiling

09 31 33.10 Tiling, Thin-Set Stone	Crew	Daily Output	Labor-Hours	Unit	Material	2021 Bare Costs Labor	2021 Bare Costs Equipment	Total	Total Incl O&P
0010 **TILING, THIN-SET STONE**									
3000 Floors, natural clay, random or uniform, color group 1	D-7	183	.087	S.F.	4.98	4.08		9.06	11.40
3100 Color group 2	"	183	.087	"	5.70	4.08		9.78	12.20

09 32 Mortar-Bed Tiling

09 32 13 – Mortar-Bed Ceramic Tiling

09 32 13.10 Ceramic Tile

	Crew	Daily Output	Labor-Hours	Unit	Material	2021 Bare Costs Labor	2021 Bare Costs Equipment	Total	Total Incl O&P
0010 **CERAMIC TILE**									
0050 Base, using 1' x 4" high pc. with 1" x 1" tiles	D-7	82	.195	L.F.	5.45	9.10		14.55	19.30
0600 Cove base, 4-1/4" x 4-1/4" high		91	.176		4.28	8.20		12.48	16.70
0900 6" x 4-1/4" high		100	.160		4.43	7.45		11.88	15.75
1200 Sanitary cove base, 6" x 4-1/4" high		93	.172		5.05	8.05		13.10	17.30
1500 6" x 6" high		84	.190		5.10	8.90		14	18.60
2400 Bullnose trim, 4-1/4" x 4-1/4"		82	.195		4.31	9.10		13.41	18.05
2700 2" x 6" bullnose trim		84	.190		4.09	8.90		12.99	17.50
6210 Wall tile, 4-1/4" x 4-1/4", better grade	1 Tilf	50	.160	S.F.	9.60	8.25		17.85	22.50
6240 2" x 2"		50	.160		7.85	8.25		16.10	21
6250 6" x 6"		55	.145		9.95	7.50		17.45	22
6260 8" x 8"		60	.133		10.25	6.90		17.15	21.50
6300 Exterior walls, frostproof, 4-1/4" x 4-1/4"	D-7	102	.157		7.45	7.30		14.75	18.90
6400 1-3/8" x 1-3/8"		93	.172		6.95	8.05		15	19.40
6600 Crystalline glazed, 4-1/4" x 4-1/4", plain		100	.160		4.86	7.45		12.31	16.25
6700 4-1/4" x 4-1/4", scored tile		100	.160		6.70	7.45		14.15	18.25
6900 6" x 6" plain		93	.172		5.25	8.05		13.30	17.55
7000 For epoxy grout, 1/16" joints, 4-1/4" tile, add		800	.020		.43	.93		1.36	1.83
7200 For tile set in dry mortar, add		1735	.009			.43		.43	.63
7300 For tile set in Portland cement mortar, add		290	.055		.22	2.57		2.79	4

09 32 16 – Mortar-Bed Quarry Tiling

09 32 16.10 Quarry Tile

	Crew	Daily Output	Labor-Hours	Unit	Material	2021 Bare Costs Labor	2021 Bare Costs Equipment	Total	Total Incl O&P
0010 **QUARRY TILE**									
0100 Base, cove or sanitary, to 5" high, 1/2" thick	D-7	110	.145	L.F.	6.65	6.80		13.45	17.20
0300 Bullnose trim, red, 6" x 6" x 1/2" thick		120	.133		5.20	6.20		11.40	14.85
0400 4" x 4" x 1/2" thick		110	.145		4.23	6.80		11.03	14.55
0600 4" x 8" x 1/2" thick, using 8" as edge		130	.123		5.45	5.75		11.20	14.40
0700 Floors, 1,000 S.F. lots, red, 4" x 4" x 1/2" thick		120	.133	S.F.	9.40	6.20		15.60	19.40
0900 6" x 6" x 1/2" thick		140	.114		10.15	5.35		15.50	18.95
1000 4" x 8" x 1/2" thick		130	.123		6.65	5.75		12.40	15.70
1300 For waxed coating, add					.76			.76	.84
1500 For non-standard colors, add					.50			.50	.55
1600 For abrasive surface, add					.54			.54	.59
1800 Brown tile, imported, 6" x 6" x 3/4"	D-7	120	.133		7.80	6.20		14	17.65
1900 8" x 8" x 1"		110	.145		10.25	6.80		17.05	21
2100 For thin set mortar application, deduct		700	.023			1.07		1.07	1.56
2200 For epoxy grout & mortar, 6" x 6" x 1/2", add		350	.046		2.43	2.13		4.56	5.80
2700 Stair tread, 6" x 6" x 3/4", plain		50	.320		8.55	14.95		23.50	31.50
2800 Abrasive		47	.340		8.75	15.90		24.65	32.50
3000 Wainscot, 6" x 6" x 1/2", thin set, red		105	.152		3.31	7.10		10.41	14.05
3100 Non-standard colors		105	.152		3.79	7.10		10.89	14.55
3300 Window sill, 6" wide, 3/4" thick		90	.178	L.F.	8.50	8.30		16.80	21.50
3400 Corners		80	.200	Ea.	6.20	9.35		15.55	20.50

09 32 Mortar-Bed Tiling

09 32 23 – Mortar-Bed Glass Mosaic Tiling

09 32 23.10 Glass Mosaics		Crew	Daily Output	Labor-Hours	Unit	Material	2021 Bare Costs Labor	Equipment	Total	Total Incl O&P
0010	**GLASS MOSAICS** 3/4" tile on 12" sheets, standard grout									
0300	Color group 1 & 2	D-7	73	.219	S.F.	19.50	10.20		29.70	36.50
0350	Color group 3		73	.219		5.55	10.20		15.75	21
0400	Color group 4		73	.219		29.50	10.20		39.70	47.50
0450	Color group 5		73	.219		30	10.20		40.20	48
0500	Color group 6		73	.219		39.50	10.20		49.70	58
0600	Color group 7		73	.219		41.50	10.20		51.70	60.50
0700	Color group 8, golds, silvers & specialties		64	.250		6.60	11.65		18.25	24.50
1720	For glass mosaic tiles set in Portland cement mortar, add	▼	290	.055	▼	.01	2.57		2.58	3.77

09 34 Waterproofing-Membrane Tiling

09 34 13 – Waterproofing-Membrane Ceramic Tiling

09 34 13.10 Ceramic Tile Waterproofing Membrane

		Crew	Daily Output	Labor-Hours	Unit	Material	2021 Bare Costs Labor	Equipment	Total	Total Incl O&P
0010	**CERAMIC TILE WATERPROOFING MEMBRANE**									
0020	On floors, including thinset									
0030	Fleece laminated polyethylene grid, 1/8" thick	D-7	250	.064	S.F.	2.13	2.99		5.12	6.70
0040	5/16" thick	"	250	.064	"	2.63	2.99		5.62	7.25
0050	On walls, including thinset									
0060	Fleece laminated polyethylene sheet, 8 mil thick	D-7	480	.033	S.F.	2.43	1.56		3.99	4.94
0070	Accessories, including thinset									
0080	Joint and corner sheet, 4 mils thick, 5" wide	1 Tilf	240	.033	L.F.	1.34	1.72		3.06	4
0090	7-1/4" wide		180	.044		1.72	2.30		4.02	5.25
0100	10" wide		120	.067	▼	2.09	3.45		5.54	7.35
0110	Pre-formed corners, inside		32	.250	Ea.	9.15	12.95		22.10	29
0120	Outside		32	.250		7.65	12.95		20.60	27.50
0130	2" flanged floor drain with 6" stainless steel grate		16	.500	▼	390	26		416	470
0140	EPS, sloped shower floor		480	.017	S.F.	11.60	.86		12.46	14
0150	Curb	▼	32	.250	L.F.	17.55	12.95		30.50	38

09 35 Chemical-Resistant Tiling

09 35 13 – Chemical-Resistant Ceramic Tiling

09 35 13.10 Chemical-Resistant Ceramic Tiling

		Crew	Daily Output	Labor-Hours	Unit	Material	2021 Bare Costs Labor	Equipment	Total	Total Incl O&P
0010	**CHEMICAL-RESISTANT CERAMIC TILING**									
0100	4-1/4" x 4-1/4" x 1/4", 1/8" joint	D-7	130	.123	S.F.	10.80	5.75		16.55	20.50
0200	6" x 6" x 1/2" thick		120	.133		8.40	6.20		14.60	18.35
0300	8" x 8" x 1/2" thick		110	.145		9.95	6.80		16.75	21
0400	4-1/4" x 4-1/4" x 1/4", 1/4" joint		130	.123		11.60	5.75		17.35	21
0500	6" x 6" x 1/2" thick		120	.133		9.60	6.20		15.80	19.65
0600	8" x 8" x 1/2" thick		110	.145		10.65	6.80		17.45	21.50
0700	4-1/4" x 4-1/4" x 1/4", 3/8" joint		130	.123		12.35	5.75		18.10	22
0800	6" x 6" x 1/2" thick		120	.133		10.60	6.20		16.80	21
0900	8" x 8" x 1/2" thick		110	.145	▼	11.90	6.80		18.70	23

09 35 16 – Chemical-Resistant Quarry Tiling

09 35 16.10 Chemical-Resistant Quarry Tiling

		Crew	Daily Output	Labor-Hours	Unit	Material	2021 Bare Costs Labor	Equipment	Total	Total Incl O&P
0010	**CHEMICAL-RESISTANT QUARRY TILING**									
0100	4" x 8" x 1/2" thick, 1/8" joint	D-7	130	.123	S.F.	10.10	5.75		15.85	19.50
0200	6" x 6" x 1/2" thick		120	.133		10.20	6.20		16.40	20.50
0300	8" x 8" x 1/2" thick	▼	110	.145	▼	7.60	6.80		14.40	18.30

173

For customer support on your Concrete & Masonry Costs with RSMeans Data, call 800.448.8182.

09 35 Chemical-Resistant Tiling

09 35 16 – Chemical-Resistant Quarry Tiling

09 35 16.10 Chemical-Resistant Quarry Tiling	Crew	Daily Output	Labor-Hours	Unit	Material	2021 Bare Costs Labor	2021 Bare Costs Equipment	Total	Total Incl O&P
0400	4" x 8" x 1/2" thick, 1/4" joint	D-7	130	.123	S.F.	11.40	5.75		17.15	21
0500	6" x 6" x 1/2" thick		120	.133		11.35	6.20		17.55	21.50
0600	8" x 8" x 1/2" thick		110	.145		8.30	6.80		15.10	19.05
0700	4" x 8" x 1/2" thick, 3/8" joint		130	.123		12.55	5.75		18.30	22
0800	6" x 6" x 1/2" thick		120	.133		12.40	6.20		18.60	23
0900	8" x 8" x 1/2" thick		110	.145		9.55	6.80		16.35	20.50

09 63 Masonry Flooring

09 63 13 – Brick Flooring

09 63 13.10 Miscellaneous Brick Flooring

		Crew	Daily Output	Labor-Hours	Unit	Material	2021 Bare Costs Labor	2021 Bare Costs Equipment	Total	Total Incl O&P
0010	**MISCELLANEOUS BRICK FLOORING**									
0020	Acid-proof shales, red, 8" x 3-3/4" x 1-1/4" thick	D-7	.43	37.209	M	615	1,725		2,340	3,200
0050	2-1/4" thick	D-1	.40	40		1,000	1,950		2,950	4,025
0200	Acid-proof clay brick, 8" x 3-3/4" x 2-1/4" thick [G]	"	.40	40		1,025	1,950		2,975	4,050
0260	Cast ceramic, pressed, 4" x 8" x 1/2", unglazed	D-7	100	.160	S.F.	7.40	7.45		14.85	19.05
0270	Glazed		100	.160		9.85	7.45		17.30	22
0280	Hand molded flooring, 4" x 8" x 3/4", unglazed		95	.168		9.75	7.85		17.60	22.50
0290	Glazed		95	.168		10.75	7.85		18.60	23.50
0300	8" hexagonal, 3/4" thick, unglazed		85	.188		9.35	8.80		18.15	23
0310	Glazed		85	.188		16.90	8.80		25.70	31.50
0400	Heavy duty industrial, cement mortar bed, 2" thick, not incl. brick	D-1	80	.200		.90	9.75		10.65	15.70
0450	Acid-proof joints, 1/4" wide	"	65	.246		1.70	12		13.70	19.90
0500	Pavers, 8" x 4", 1" to 1-1/4" thick, red	D-7	95	.168		4.30	7.85		12.15	16.25
0510	Ironspot	"	95	.168		5.30	7.85		13.15	17.35
0540	1-3/8" to 1-3/4" thick, red	D-1	95	.168		4.15	8.20		12.35	16.90
0560	Ironspot		95	.168		5.25	8.20		13.45	18.15
0580	2-1/4" thick, red		90	.178		4.23	8.65		12.88	17.70
0590	Ironspot		90	.178		6.55	8.65		15.20	20.50
0800	For sidewalks and patios with pavers, see Section 32 14 16.10									
0870	For epoxy joints, add	D-1	600	.027	S.F.	3.22	1.30		4.52	5.50
0880	For Furan underlayment, add	"	600	.027		2.66	1.30		3.96	4.89
0890	For waxed surface, steam cleaned, add	A-1H	1000	.008		.21	.36	.08	.65	.84

09 63 40 – Stone Flooring

09 63 40.10 Marble

		Crew	Daily Output	Labor-Hours	Unit	Material	2021 Bare Costs Labor	2021 Bare Costs Equipment	Total	Total Incl O&P
0010	**MARBLE**									
0020	Thin gauge tile, 12" x 6", 3/8", white Carara	D-7	60	.267	S.F.	16.90	12.45		29.35	37
0100	Travertine		60	.267		9.55	12.45		22	28.50
0200	12" x 12" x 3/8", thin set, floors		60	.267		12.40	12.45		24.85	32
0300	On walls		52	.308		9.95	14.35		24.30	32
1000	Marble threshold, 4" wide x 36" long x 5/8" thick, white		60	.267	Ea.	12.40	12.45		24.85	32

09 63 40.20 Slate Tile

		Crew	Daily Output	Labor-Hours	Unit	Material	2021 Bare Costs Labor	2021 Bare Costs Equipment	Total	Total Incl O&P
0010	**SLATE TILE**									
0020	Vermont, 6" x 6" x 1/4" thick, thin set	D-7	180	.089	S.F.	7.65	4.15		11.80	14.50
0200	See also Section 32 14 40.10									

09 66 Terrazzo Flooring

09 66 13 – Portland Cement Terrazzo Flooring

09 66 13.10 Portland Cement Terrazzo	Crew	Daily Output	Labor-Hours	Unit	Material	2021 Bare Costs Labor	Equipment	Total	Total Incl O&P
0010	**PORTLAND CEMENT TERRAZZO**, cast-in-place									
0020	Cove base, 6" high, 16 ga. zinc	1 Mstz	20	.400	L.F.	3.85	20.50		24.35	34
0100	Curb, 6" high and 6" wide		6	1.333		6.35	69		75.35	108
0300	Divider strip for floors, 14 ga., 1-1/4" deep, zinc		375	.021		1.40	1.10		2.50	3.15
0400	Brass		375	.021		2.82	1.10		3.92	4.71
0600	Heavy top strip 1/4" thick, 1-1/4" deep, zinc		300	.027		2.14	1.38		3.52	4.37
0900	Galv. bottoms, brass		300	.027		2.72	1.38		4.10	5
1200	For thin set floors, 16 ga., 1/2" x 1/2", zinc		350	.023		1.46	1.18		2.64	3.34
1300	Brass	▼	350	.023	▼	2.92	1.18		4.10	4.94
1500	Floor, bonded to concrete, 1-3/4" thick, gray cement	J-3	75	.213	S.F.	3.86	10.15	3.45	17.46	23
1600	White cement, mud set		75	.213		4.55	10.15	3.45	18.15	23.50
1800	Not bonded, 3" total thickness, gray cement		70	.229		4.75	10.90	3.70	19.35	25
1900	White cement, mud set	▼	70	.229	▼	5.50	10.90	3.70	20.10	26
2100	For Venetian terrazzo, 1" topping, add					50%	50%			
2200	For heavy duty abrasive terrazzo, add					50%	50%			
2700	Monolithic terrazzo, 1/2" thick									
2710	10' panels	J-3	125	.128	S.F.	3.42	6.10	2.07	11.59	14.95
3000	Stairs, cast-in-place, pan filled treads		30	.533	L.F.	4.40	25.50	8.65	38.55	51.50
3100	Treads and risers	▼	14	1.143	"	6.50	54.50	18.50	79.50	107
3300	For stair landings, add to floor prices						50%			
3400	Stair stringers and fascia	J-3	30	.533	S.F.	5.25	25.50	8.65	39.40	52.50
3600	For abrasive metal nosings on stairs, add		150	.107	L.F.	9.95	5.10	1.73	16.78	20.50
3700	For abrasive surface finish, add		600	.027	S.F.	1.84	1.27	.43	3.54	4.35
3900	For raised abrasive strips, add		150	.107	L.F.	1.36	5.10	1.73	8.19	10.85
4000	Wainscot, bonded, 1-1/2" thick		30	.533	S.F.	4.03	25.50	8.65	38.18	51
4200	1/4" thick	▼	40	.400	"	6.55	19.05	6.45	32.05	42.50

09 66 16 – Terrazzo Floor Tile

09 66 16.10 Tile or Terrazzo Base

		Crew	Daily Output	Labor-Hours	Unit	Material	2021 Bare Costs Labor	Equipment	Total	Total Incl O&P
0010	**TILE OR TERRAZZO BASE**	R096613-10								
0020	Scratch coat only	1 Mstz	150	.053	S.F.	.48	2.76		3.24	4.56
0500	Scratch and brown coat only	"	75	.107	"	.96	5.50		6.46	9.10

09 66 16.13 Portland Cement Terrazzo Floor Tile

		Crew	Daily Output	Labor-Hours	Unit	Material	2021 Bare Costs Labor	Equipment	Total	Total Incl O&P
0010	**PORTLAND CEMENT TERRAZZO FLOOR TILE**									
1200	Floor tiles, non-slip, 1" thick, 12" x 12"	D-1	60	.267	S.F.	27.50	13		40.50	49.50
1300	1-1/4" thick, 12" x 12"		60	.267		25.50	13		38.50	47.50
1500	16" x 16"		50	.320		28	15.60		43.60	54.50
1600	1-1/2" thick, 16" x 16"	▼	45	.356		26	17.30		43.30	54.50
1800	For Venetian terrazzo, add					7.75			7.75	8.55
1900	For white cement, add	▼			▼	.71			.71	.78

09 66 16.30 Terrazzo, Precast

		Crew	Daily Output	Labor-Hours	Unit	Material	2021 Bare Costs Labor	Equipment	Total	Total Incl O&P
0010	**TERRAZZO, PRECAST**									
0020	Base, 6" high, straight	1 Mstz	70	.114	L.F.	12.50	5.90		18.40	22.50
0100	Cove		60	.133		16.95	6.90		23.85	29
0300	8" high, straight		60	.133		15.95	6.90		22.85	27.50
0400	Cove	▼	50	.160		24.50	8.30		32.80	39
0600	For white cement, add					.60			.60	.66
0700	For 16 ga. zinc toe strip, add					2.35			2.35	2.59
0900	Curbs, 4" x 4" high	1 Mstz	40	.200		43.50	10.35		53.85	62.50
1000	8" x 8" high	"	30	.267		47.50	13.80		61.30	72.50
2400	Stair treads, 1-1/2" thick, non-slip, three line pattern	2 Mstz	70	.229		54	11.85		65.85	77
2500	Nosing and two lines		70	.229		54	11.85		65.85	77
2700	2" thick treads, straight	▼	60	.267	▼	62.50	13.80		76.30	89

09 66 Terrazzo Flooring

09 66 16 – Terrazzo Floor Tile

09 66 16.30 Terrazzo, Precast		Crew	Daily Output	Labor-Hours	Unit	Material	2021 Bare Costs Labor	Equipment	Total	Total Incl O&P
2800	Curved	2 Mstz	50	.320	L.F.	84.50	16.55		101.05	117
3000	Stair risers, 1" thick, to 6" high, straight sections		60	.267		16.25	13.80		30.05	38
3100	Cove		50	.320		20.50	16.55		37.05	46.50
3300	Curved, 1" thick, to 6" high, vertical		48	.333		28.50	17.25		45.75	56
3400	Cove		38	.421		42	22		64	78
3600	Stair tread and riser, single piece, straight, smooth surface		60	.267		66.50	13.80		80.30	93.50
3700	Non skid surface		40	.400		86	20.50		106.50	125
3900	Curved tread and riser, smooth surface		40	.400		95.50	20.50		116	135
4000	Non skid surface		32	.500		117	26		143	167
4200	Stair stringers, notched, 1" thick		25	.640		38	33		71	90.50
4300	2" thick		22	.727	▼	45	37.50		82.50	105
4500	Stair landings, structural, non-slip, 1-1/2" thick		85	.188	S.F.	42.50	9.75		52.25	61.50
4600	3" thick	▼	75	.213		59	11.05		70.05	81
4800	Wainscot, 12" x 12" x 1" tiles	1 Mstz	12	.667		9.50	34.50		44	61
4900	16" x 16" x 1-1/2" tiles	"	8	1	▼	18.75	52		70.75	96

09 66 23 – Resinous Matrix Terrazzo Flooring

09 66 23.16 Epoxy-Resin Terrazzo Flooring

0010	**EPOXY-RESIN TERRAZZO FLOORING**									
2500	Epoxy terrazzo, 1/4" thick, granite chips	J-3	200	.080	S.F.	5.95	3.81	1.29	11.05	13.50
2550	Average		175	.091		5.75	4.36	1.48	11.59	14.30
2600	Recycled aggregate		150	.107		5.85	5.10	1.73	12.68	15.75
2650	Epoxy terrazzo, 3/8" thick, marble chips		200	.080		5.85	3.81	1.29	10.95	13.35
2675	Glass or mother of pearl	▼	200	.080	▼	7.30	3.81	1.29	12.40	14.95

09 66 33 – Conductive Terrazzo Flooring

09 66 33.10 Conductive Terrazzo

0010	**CONDUCTIVE TERRAZZO**									
2400	Bonded conductive floor for hospitals	J-3	90	.178	S.F.	5.55	8.45	2.88	16.88	21.50

09 91 Painting

09 91 23 – Interior Painting

09 91 23.72 Walls and Ceilings, Interior

0010	**WALLS AND CEILINGS, INTERIOR**									
0100	Concrete, drywall or plaster, latex, primer or sealer coat									
0150	Smooth finish, cut-in by brush	1 Pord	1150	.007	L.F.	.02	.32		.34	.50
0200	Brushwork		1150	.007	S.F.	.07	.32		.39	.55
0240	Roller		1350	.006		.07	.28		.35	.48
0280	Spray		2750	.003	▼	.06	.14		.20	.26
0290	Sand finish, cut-in by brush		975	.008	L.F.	.02	.38		.40	.59
0300	Brushwork		975	.008	S.F.	.07	.38		.45	.64
0340	Roller		1150	.007		.07	.32		.39	.55
0380	Spray		2275	.004	▼	.06	.16		.22	.30
0590	Paint 2 coats, smooth finish, cut-in by brush		680	.012	L.F.	.04	.55		.59	.85
0800	Brushwork		680	.012	S.F.	.16	.55		.71	.99
0840	Roller		800	.010		.16	.46		.62	.87
0880	Spray		1625	.005	▼	.15	.23		.38	.51
0890	Sand finish, cut-in by brush		605	.013	L.F.	.04	.61		.65	.95
0900	Brushwork		605	.013	S.F.	.16	.61		.77	1.09
0940	Roller		1020	.008		.16	.36		.52	.72
0980	Spray		1700	.005		.15	.22		.37	.49
1600	Glaze coating, 2 coats, spray, clear		1200	.007	▼	.56	.31		.87	1.08

09 91 Painting

09 91 23.72 Walls and Ceilings, Interior

		Crew	Daily Output	Labor-Hours	Unit	Material	2021 Bare Costs Labor	Equipment	Total	Total Incl O&P
1640	Multicolor	1 Pord	1200	.007	S.F.	.92	.31		1.23	1.47
1660	Painting walls, complete, including surface prep, primer &									
1670	2 coats finish, on drywall or plaster, with roller	1 Pord	325	.025	S.F.	.24	1.14		1.38	1.96
1700	For oil base paint, add					10%				
1800	For ceiling installations, add						25%			
2000	Masonry or concrete block, primer/sealer, latex paint									
2090	Primer, smooth finish, cut-in by brush	1 Pord	1000	.008	L.F.	.04	.37		.41	.59
2100	Brushwork		1000	.008	S.F.	.15	.37		.52	.72
2110	Roller		1150	.007		.12	.32		.44	.61
2180	Spray		2400	.003		.10	.15		.25	.34
2190	Sand finish, cut-in by brush		850	.009	L.F.	.03	.44		.47	.68
2200	Brushwork		850	.009	S.F.	.12	.44		.56	.78
2210	Roller		975	.008		.12	.38		.50	.70
2280	Spray		2050	.004		.10	.18		.28	.38
2590	Primer plus one finish coat, smooth cut-in by brush		525	.015	L.F.	.08	.71		.79	1.14
2800	Brushwork		525	.015	S.F.	.32	.71		1.03	1.40
2810	Roller		615	.013		.20	.60		.80	1.12
2880	Spray		1200	.007		.18	.31		.49	.65
2890	Sand finish, cut-in by brush		450	.018	L.F.	.05	.83		.88	1.27
2900	Brushwork		450	.018	S.F.	.20	.83		1.03	1.44
2910	Roller		515	.016		.20	.72		.92	1.29
2980	Spray		1025	.008		.18	.36		.54	.73
3600	Glaze coating, 3 coats, spray, clear		900	.009		.80	.41		1.21	1.49
3620	Multicolor		900	.009		1.12	.41		1.53	1.84
4000	Block filler, 1 coat, brushwork		425	.019		.14	.87		1.01	1.45
4100	Silicone, water repellent, 2 coats, spray		2000	.004		.47	.19		.66	.79
4120	For oil base paint, add					10%				
8200	For work 8'-15' H, add						10%			
8300	For work over 15' H, add						20%			
8400	For light textured surfaces, add						10%			
8410	Heavy textured, add						25%			

09 91 23.74 Walls and Ceilings, Interior, Zero VOC Latex

		Crew	Daily Output	Labor-Hours	Unit	Material	2021 Bare Costs Labor	Equipment	Total	Total Incl O&P
0010	**WALLS AND CEILINGS, INTERIOR, ZERO VOC LATEX**									
0100	Concrete, dry wall or plaster, latex, primer or sealer coat									
0190	Smooth finish, cut-in by brush	1 Pord	1150	.007	L.F.	.02	.32		.34	.50
0200	Brushwork	G	1150	.007	S.F.	.08	.32		.40	.56
0240	Roller	G	1350	.006		.07	.28		.35	.49
0280	Spray	G	2750	.003		.06	.14		.20	.26
0290	Sand finish, cut-in by brush		975	.008	L.F.	.02	.38		.40	.59
0300	Brushwork	G	975	.008	S.F.	.07	.38		.45	.65
0340	Roller	G	1150	.007		.08	.32		.40	.57
0380	Spray	G	2275	.004		.06	.16		.22	.31
0790	Paint 2 coats, smooth finish, cut-in by brush		680	.012	L.F.	.04	.55		.59	.85
0800	Brushwork	G	680	.012	S.F.	.04	.55		.59	.85
0840	Roller	G	800	.010		.16	.46		.62	.86
0880	Spray	G	1625	.005		.14	.23		.37	.49
0890	Sand finish, cut-in by brush		605	.013	L.F.	.04	.61		.65	.95
0900	Brushwork	G	605	.013	S.F.	.15	.61		.76	1.07
0940	Roller	G	1020	.008		.16	.36		.52	.71
0980	Spray	G	1700	.005		.14	.22		.36	.47
1800	For ceiling installations, add	G					25%			
8200	For work 8' - 15' H, add						10%			

For customer support on your Concrete & Masonry Costs with RSMeans Data, call 800.448.8182.

177

09 91 Painting

09 91 23 – Interior Painting

09 91 23.74 Walls and Ceilings, Interior, Zero VOC Latex	Crew	Daily Output	Labor-Hours	Unit	Material	2021 Bare Costs Labor	2021 Bare Costs Equipment	Total	Total Incl O&P
8300	For work over 15' H, add				S.F.		20%			

09 91 23.75 Dry Fall Painting

		Crew	Daily Output	Labor-Hours	Unit	Material	Labor	Equipment	Total	Total Incl O&P
0010	**DRY FALL PAINTING**									
0100	Sprayed on walls, gypsum board or plaster									
0220	One coat	1 Pord	2600	.003	S.F.	.09	.14		.23	.30
0250	Two coats		1560	.005		.17	.24		.41	.54
0280	Concrete or textured plaster, one coat		1560	.005		.09	.24		.33	.44
0310	Two coats		1300	.006		.17	.29		.46	.61
0340	Concrete block, one coat		1560	.005		.09	.24		.33	.44
0370	Two coats		1300	.006		.17	.29		.46	.61
0400	Wood, one coat		877	.009		.09	.42		.51	.72
0430	Two coats		650	.012		.17	.57		.74	1.04
0440	On ceilings, gypsum board or plaster									
0470	One coat	1 Pord	1560	.005	S.F.	.09	.24		.33	.44
0500	Two coats		1300	.006		.17	.29		.46	.61
0530	Concrete or textured plaster, one coat		1560	.005		.09	.24		.33	.44
0560	Two coats		1300	.006		.17	.29		.46	.61
0570	Structural steel, bar joists or metal deck, one coat		1560	.005		.09	.24		.33	.44
0580	Two coats		1040	.008		.17	.36		.53	.72

09 96 High-Performance Coatings

09 96 56 – Epoxy Coatings

09 96 56.20 Wall Coatings

		Crew	Daily Output	Labor-Hours	Unit	Material	Labor	Equipment	Total	Total Incl O&P
0010	**WALL COATINGS**									
0100	Acrylic glazed coatings, matte	1 Pord	525	.015	S.F.	.38	.71		1.09	1.47
0200	Gloss		305	.026		.77	1.22		1.99	2.66
0300	Epoxy coatings, solvent based		525	.015		.48	.71		1.19	1.58
0400	Water based		170	.047		.38	2.19		2.57	3.66
0600	Exposed aggregate, troweled on, 1/16" to 1/4", solvent based		235	.034		.78	1.58		2.36	3.21
0700	Water based (epoxy or polyacrylate)		130	.062		1.68	2.86		4.54	6.10
0900	1/2" to 5/8" aggregate, solvent based		130	.062		1.46	2.86		4.32	5.85
1000	Water based		80	.100		2.54	4.65		7.19	9.70
1500	Exposed aggregate, sprayed on, 1/8" aggregate, solvent based		295	.027		.59	1.26		1.85	2.52
1600	Water based		145	.055		1.27	2.56		3.83	5.20
1800	High build epoxy, 50 mil, solvent based		390	.021		.78	.95		1.73	2.27
1900	Water based		95	.084		1.40	3.91		5.31	7.35
2100	Laminated epoxy with fiberglass, solvent based		295	.027		.87	1.26		2.13	2.83
2200	Water based		145	.055		1.66	2.56		4.22	5.65
2400	Sprayed perlite or vermiculite, 1/16" thick, solvent based		2935	.003		.28	.13		.41	.50
2500	Water based		640	.013		.88	.58		1.46	1.83
2700	Vinyl plastic wall coating, solvent based		735	.011		.41	.51		.92	1.20
2800	Water based		240	.033		1.03	1.55		2.58	3.43
3000	Urethane on smooth surface, 2 coats, solvent based		1135	.007		.35	.33		.68	.88
3100	Water based		665	.012		.63	.56		1.19	1.52
3600	Ceramic-like glazed coating, cementitious, solvent based		440	.018		.50	.84		1.34	1.80
3700	Water based		345	.023		1.02	1.08		2.10	2.72
3900	Resin base, solvent based		640	.013		.35	.58		.93	1.25
4000	Water based		330	.024		.68	1.13		1.81	2.42

178

For customer support on your Concrete & Masonry Costs with RSMeans Data, call 800.448.8182.

Estimating Tips

General

- The items in this division are usually priced per square foot or each.

- Many items in Division 10 require some type of support system or special anchors that are not usually furnished with the item. The required anchors must be added to the estimate in the appropriate division.

- Some items in Division 10, such as lockers, may require assembly before installation. Verify the amount of assembly required. Assembly can often exceed installation time.

10 20 00 Interior Specialties

- Support angles and blocking are not included in the installation of toilet compartments, shower/ dressing compartments, or cubicles. Appropriate line items from Division 5 or 6 may need to be added to support the installations.

- Toilet partitions are priced by the stall. A stall consists of a side wall, pilaster, and door with hardware. Toilet tissue holders and grab bars are extra.

- The required acoustical rating of a folding partition can have a significant impact on costs. Verify the sound transmission coefficient rating of the panel priced against the specification requirements.

- Grab bar installation does not include supplemental blocking or backing to support the required load. When grab bars are installed at an existing facility, provisions must be made to attach the grab bars to a solid structure.

Reference Numbers

Reference numbers are shown at the beginning of some major classifications. These numbers refer to related items in the Reference Section. The reference information may be an estimating procedure, an alternate pricing method, or technical information.

Note: Not all subdivisions listed here necessarily appear. ■

Same Data. Simplified.

Enjoy the convenience and efficiency of accessing your costs anywhere:

- **Skip the multiplier** by setting your location
- **Quickly search,** edit, favorite and share costs
- **Stay on top of price changes** with automatic updates

Discover more at rsmeans.com/online

10 21 Compartments and Cubicles

10 21 13 – Toilet Compartments

10 21 13.40 Stone Toilet Compartments

		Crew	Daily Output	Labor-Hours	Unit	Material	2021 Bare Costs Labor	Equipment	Total	Total Incl O&P
0010	**STONE TOILET COMPARTMENTS**									
0100	Cubicles, ceiling hung, marble	2 Marb	2	8	Ea.	1,650	425		2,075	2,475
0600	For handicap units, add					480			480	530
0800	Floor & ceiling anchored, marble	2 Marb	2.50	6.400		1,675	340		2,015	2,375
1400	For handicap units, add					380			380	415
1600	Floor mounted, marble	2 Marb	3	5.333		1,050	284		1,334	1,575
2400	Floor mounted, headrail braced, marble	"	3	5.333		1,200	284		1,484	1,750
2900	For handicap units, add					330			330	360
4100	Entrance screen, floor mounted marble, 58" high, 48" wide	2 Marb	9	1.778		725	94.50		819.50	935
4600	Urinal screen, 18" wide, ceiling braced, marble	D-1	6	2.667	↓	760	130		890	1,025
5100	Floor mounted, headrail braced									
5200	Marble	D-1	6	2.667	Ea.	650	130		780	910
5700	Pilaster, flush, marble		9	1.778		850	86.50		936.50	1,075
6200	Post braced, marble	↓	9	1.778	↓	830	86.50		916.50	1,050

10 28 Toilet, Bath, and Laundry Accessories

10 28 13 – Toilet Accessories

10 28 13.13 Commercial Toilet Accessories

		Crew	Daily Output	Labor-Hours	Unit	Material	2021 Bare Costs Labor	Equipment	Total	Total Incl O&P
0010	**COMMERCIAL TOILET ACCESSORIES**									
0200	Curtain rod, stainless steel, 5' long, 1" diameter	1 Carp	13	.615	Ea.	26.50	33.50		60	79.50
0300	1-1/4" diameter		13	.615		18.30	33.50		51.80	70.50
0400	Diaper changing station, horizontal, wall mounted, plastic	↓	10	.800	↓	315	44		359	410
0500	Dispenser units, combined soap & towel dispensers,									
0510	Mirror and shelf, flush mounted	1 Carp	10	.800	Ea.	355	44		399	455
0600	Towel dispenser and waste receptacle,									
0610	18 gallon capacity	1 Carp	10	.800	Ea.	385	44		429	490
0800	Grab bar, straight, 1-1/4" diameter, stainless steel, 18" long		24	.333		10.90	18.25		29.15	39
0900	24" long		23	.348		13.30	19.05		32.35	43
1000	30" long		22	.364		27.50	19.90		47.40	59.50
1100	36" long		20	.400		13.30	22		35.30	47
1105	42" long		20	.400		41.50	22		63.50	78
1120	Corner, 36" long		20	.400		95	22		117	137
1200	1-1/2" diameter, 24" long		23	.348		37.50	19.05		56.55	70
1300	36" long		20	.400		43	22		65	79.50
1310	42" long		18	.444		34	24.50		58.50	74
1500	Tub bar, 1-1/4" diameter, 24" x 36"		14	.571		110	31.50		141.50	168
1600	Plus vertical arm		12	.667		108	36.50		144.50	174
1900	End tub bar, 1" diameter, 90° angle, 16" x 32"		12	.667		102	36.50		138.50	167
2300	Hand dryer, surface mounted, electric, 115 volt, 20 amp		4	2		425	109		534	630
2400	230 volt, 10 amp		4	2		585	109		694	810
2450	Hand dryer, touch free, 1400 watt, 81,000 rpm		4	2		1,350	109		1,459	1,675
2600	Hat and coat strip, stainless steel, 4 hook, 36" long		24	.333		36.50	18.25		54.75	67
2700	6 hook, 60" long		20	.400		106	22		128	149
3000	Mirror, with stainless steel 3/4" square frame, 18" x 24"		20	.400		50	22		72	87.50
3100	36" x 24"		15	.533		63.50	29		92.50	114
3200	48" x 24"		10	.800		168	44		212	251
3300	72" x 24"		6	1.333		385	73		458	535
3500	With 5" stainless steel shelf, 18" x 24"		20	.400		83	22		105	124
3600	36" x 24"		15	.533		290	29		319	365
3700	48" x 24"		10	.800		285	44		329	380
3800	72" x 24"	↓	6	1.333	↓	241	73		314	375

10 28 Toilet, Bath, and Laundry Accessories

10 28 13 – Toilet Accessories

10 28 13.13 Commercial Toilet Accessories

		Crew	Daily Output	Labor-Hours	Unit	Material	2021 Bare Costs Labor	Equipment	Total	Total Incl O&P
4100	Mop holder strip, stainless steel, 5 holders, 48" long	1 Carp	20	.400	Ea.	72	22		94	112
4200	Napkin/tampon dispenser, recessed		15	.533		635	29		664	745
4220	Semi-recessed		6.50	1.231		345	67.50		412.50	480
4250	Napkin receptacle, recessed		6.50	1.231		170	67.50		237.50	287
4300	Robe hook, single, regular		96	.083		15.30	4.56		19.86	23.50
4400	Heavy duty, concealed mounting		56	.143		25.50	7.80		33.30	39.50
4600	Soap dispenser, chrome, surface mounted, liquid		20	.400		67.50	22		89.50	107
5000	Recessed stainless steel, liquid		10	.800		172	44		216	255
5600	Shelf, stainless steel, 5" wide, 18 ga., 24" long		24	.333		66.50	18.25		84.75	100
5700	48" long		16	.500		212	27.50		239.50	274
5800	8" wide shelf, 18 ga., 24" long		22	.364		85	19.90		104.90	123
5900	48" long		14	.571		146	31.50		177.50	207
6000	Toilet seat cover dispenser, stainless steel, recessed		20	.400		181	22		203	232
6050	Surface mounted		15	.533		34	29		63	81
6100	Toilet tissue dispenser, surface mounted, SS, single roll		30	.267		16	14.60		30.60	39.50
6200	Double roll		24	.333		23.50	18.25		41.75	53
6240	Plastic, twin/jumbo dbl. roll		24	.333		30.50	18.25		48.75	60.50
6400	Towel bar, stainless steel, 18" long		23	.348		38.50	19.05		57.55	70.50
6500	30" long		21	.381		46	21		67	82
6700	Towel dispenser, stainless steel, surface mounted		16	.500		38.50	27.50		66	83.50
6800	Flush mounted, recessed		10	.800		67.50	44		111.50	140
6900	Plastic, touchless, battery operated		16	.500		79	27.50		106.50	128
7000	Towel holder, hotel type, 2 guest size		20	.400		54	22		76	92
7200	Towel shelf, stainless steel, 24" long, 8" wide		20	.400		85	22		107	126
7400	Tumbler holder, for tumbler only		30	.267		37.50	14.60		52.10	63
7410	Tumbler holder, recessed		20	.400		8.25	22		30.25	41.50
7500	Soap, tumbler & toothbrush		30	.267		17.40	14.60		32	41
7510	Tumbler & toothbrush holder		20	.400		11.15	22		33.15	45
8000	Waste receptacles, stainless steel, with top, 13 gallon		10	.800		315	44		359	415
8100	36 gallon		8	1		535	54.50		589.50	670
9996	Bathroom access., grab bar, straight, 1-1/2" diam., SS, 42" L install only	↓	18	.444	↓		24.50		24.50	36.50

10 28 19 – Tub and Shower Enclosures

10 28 19.10 Partitions, Shower

		Crew	Daily Output	Labor-Hours	Unit	Material	2021 Bare Costs Labor	Equipment	Total	Total Incl O&P
0010	**PARTITIONS, SHOWER** floor mounted, no plumbing									
2900	Marble shower stall, stock design, with shower door	2 Marb	1.20	13.333	Ea.	2,275	710		2,985	3,600
3000	With curtain		1.30	12.308		2,025	655		2,680	3,200
3200	Receptors, precast terrazzo, 32" x 32"		14	1.143		288	61		349	405
3300	48" x 34"		9.50	1.684		410	89.50		499.50	585
3500	Plastic, simulated terrazzo receptor, 32" x 32"		14	1.143		171	61		232	280
3600	32" x 48"		12	1.333		305	71		376	440
3800	Precast concrete, colors, 32" x 32"		14	1.143		235	61		296	350
3900	48" x 48"	↓	8	2	↓	325	106		431	515

For customer support on your Concrete & Masonry Costs with RSMeans Data, call 800.448.8182.

181

10 31 Manufactured Fireplaces

10 31 13 – Manufactured Fireplace Chimneys

10 31 13.10 Fireplace Chimneys

		Crew	Daily Output	Labor-Hours	Unit	Material	2021 Bare Costs Labor	Equipment	Total	Total Incl O&P
0010	**FIREPLACE CHIMNEYS**									
0500	Chimney dbl. wall, all stainless, over 8'-6", 7" diam., add to fireplace	1 Carp	33	.242	V.L.F.	97	13.25		110.25	127
0600	10" diameter, add to fireplace		32	.250		102	13.70		115.70	134
0700	12" diameter, add to fireplace		31	.258		204	14.10		218.10	245
0800	14" diameter, add to fireplace		30	.267	↓	212	14.60		226.60	255
1000	Simulated brick chimney top, 4' high, 16" x 16"		10	.800	Ea.	485	44		529	600
1100	24" x 24"	↓	7	1.143	"	575	62.50		637.50	730

10 31 13.20 Chimney Accessories

		Crew	Daily Output	Labor-Hours	Unit	Material	2021 Bare Costs Labor	Equipment	Total	Total Incl O&P
0010	**CHIMNEY ACCESSORIES**									
0020	Chimney screens, galv., 13" x 13" flue	1 Bric	8	1	Ea.	58	53.50		111.50	145
0050	24" x 24" flue		5	1.600		124	86		210	266
0200	Stainless steel, 13" x 13" flue		8	1		106	53.50		159.50	197
0250	20" x 20" flue		5	1.600		148	86		234	292
2400	Squirrel and bird screens, galvanized, 8" x 8" flue		16	.500		57	27		84	103
2450	13" x 13" flue	↓	12	.667	↓	148	36		184	217

10 31 16 – Manufactured Fireplace Forms

10 31 16.10 Fireplace Forms

		Crew	Daily Output	Labor-Hours	Unit	Material	2021 Bare Costs Labor	Equipment	Total	Total Incl O&P
0010	**FIREPLACE FORMS**									
1800	Fireplace forms, no accessories, 32" opening	1 Bric	3	2.667	Ea.	700	143		843	980
1900	36" opening		2.50	3.200		890	172		1,062	1,225
2000	40" opening		2	4		1,175	215		1,390	1,625
2100	78" opening	↓	1.50	5.333	↓	1,725	286		2,011	2,325

10 31 23 – Prefabricated Fireplaces

10 31 23.10 Fireplace, Prefabricated

		Crew	Daily Output	Labor-Hours	Unit	Material	2021 Bare Costs Labor	Equipment	Total	Total Incl O&P
0010	**FIREPLACE, PREFABRICATED**, free standing or wall hung									
0100	With hood & screen, painted	1 Carp	1.30	6.154	Ea.	1,650	335		1,985	2,300
0150	Average		1	8		1,525	440		1,965	2,325
0200	Stainless steel		.90	8.889	↓	2,800	485		3,285	3,800
1500	Simulated logs, gas fired, 40,000 BTU, 2' long, manual safety pilot		7	1.143	Set	575	62.50		637.50	725
1600	Adjustable flame remote pilot		6	1.333		2,425	73		2,498	2,750
1700	Electric, 1,500 BTU, 1'-6" long, incandescent flame		7	1.143		284	62.50		346.50	405
1800	1,500 BTU, LED flame	↓	6	1.333	↓	345	73		418	485

10 32 Fireplace Specialties

10 32 13 – Fireplace Dampers

10 32 13.10 Dampers

		Crew	Daily Output	Labor-Hours	Unit	Material	2021 Bare Costs Labor	Equipment	Total	Total Incl O&P
0010	**DAMPERS**									
0800	Damper, rotary control, steel, 30" opening	1 Bric	6	1.333	Ea.	125	71.50		196.50	246
0850	Cast iron, 30" opening		6	1.333		124	71.50		195.50	245
0880	36" opening		6	1.333		128	71.50		199.50	248
0900	48" opening		6	1.333		150	71.50		221.50	273
0920	60" opening		6	1.333		360	71.50		431.50	505
0950	72" opening		5	1.600		430	86		516	605
1000	84" opening, special order		5	1.600		925	86		1,011	1,150
1050	96" opening, special order		4	2		940	107		1,047	1,175
1200	Steel plate, poker control, 60" opening		8	1		330	53.50		383.50	440
1250	84" opening, special order		5	1.600		600	86		686	790
1400	"Universal" type, chain operated, 32" x 20" opening		8	1		255	53.50		308.50	360
1450	48" x 24" opening	↓	5	1.600	↓	380	86		466	550

10 32 Fireplace Specialties

10 32 23 – Fireplace Doors

10 32 23.10 Doors		Crew	Daily Output	Labor-Hours	Unit	Material	2021 Bare Costs Labor	Equipment	Total	Total Incl O&P
0010	**DOORS**									
0400	Cleanout doors and frames, cast iron, 8" x 8"	1 Bric	12	.667	Ea.	53	36		89	113
0450	12" x 12"		10	.800		103	43		146	178
0500	18" x 24"		8	1		153	53.50		206.50	249
0550	Cast iron frame, steel door, 24" x 30"		5	1.600		380	86		466	550
1600	Dutch oven door and frame, cast iron, 12" x 15" opening		13	.615		134	33		167	197
1650	Copper plated, 12" x 15" opening	↓	13	.615	↓	262	33		295	340

10 35 Stoves

10 35 13 – Heating Stoves

10 35 13.10 Wood Burning Stoves		Crew	Daily Output	Labor-Hours	Unit	Material	2021 Bare Costs Labor	Equipment	Total	Total Incl O&P
0010	**WOOD BURNING STOVES**									
0015	Cast iron, less than 1,500 S.F.	2 Carp	1.30	12.308	Ea.	1,450	675		2,125	2,600
0020	1,500 to 2,000 S.F.		1	16		2,550	875		3,425	4,100
0030	greater than 2,000 S.F.	↓	.80	20		3,200	1,100		4,300	5,125
0050	For gas log lighter, add				↓	53.50			53.50	59

10 74 Manufactured Exterior Specialties

10 74 46 – Window Wells

10 74 46.10 Area Window Wells		Crew	Daily Output	Labor-Hours	Unit	Material	2021 Bare Costs Labor	Equipment	Total	Total Incl O&P
0010	**AREA WINDOW WELLS**, Galvanized steel									
0020	20 ga., 3'-2" wide, 1' deep	1 Sswk	29	.276	Ea.	17.95	16.65		34.60	45
0100	2' deep		23	.348		29.50	21		50.50	65
0300	16 ga., 3'-2" wide, 1' deep		29	.276		27.50	16.65		44.15	55.50
0400	3' deep		23	.348		48	21		69	85.50
0600	Welded grating for above, 15 lb., painted		45	.178		103	10.70		113.70	130
0700	Galvanized		45	.178		126	10.70		136.70	155
0900	Translucent plastic cap for above	↓	60	.133	↓	20.50	8.05		28.55	35

10 75 Flagpoles

10 75 16 – Ground-Set Flagpoles

10 75 16.10 Flagpoles		Crew	Daily Output	Labor-Hours	Unit	Material	2021 Bare Costs Labor	Equipment	Total	Total Incl O&P
0010	**FLAGPOLES**, ground set									
7300	Foundations for flagpoles, including									
7400	excavation and concrete, to 35' high poles	C-1	10	3.200	Ea.	725	167		892	1,050
7600	40' to 50' high		3.50	9.143		1,350	475		1,825	2,175
7700	Over 60' high	↓	2	16	↓	1,675	835		2,510	3,100

For customer support on your Concrete & Masonry Costs with RSMeans Data, call 800.448.8182.

183

Division Notes

		CREW	DAILY OUTPUT	LABOR-HOURS	UNIT	BARE COSTS				TOTAL INCL O&P
						MAT.	LABOR	EQUIP.	TOTAL	

Estimating Tips
General

- The items in this division are usually priced per square foot or each. Many of these items are purchased by the owner for installation by the contractor. Check the specifications for responsibilities and include time for receiving, storage, installation, and mechanical and electrical hookups in the appropriate divisions.

- Many items in Division 11 require some type of support system that is not usually furnished with the item. Examples of these systems include blocking for the attachment of casework and support angles for ceiling-hung projection screens. The required blocking or supports must be added to the estimate in the appropriate division.

- Some items in Division 11 may require assembly or electrical hookups. Verify the amount of assembly required or the need for a hard electrical connection and add the appropriate costs.

Reference Numbers

Reference numbers are shown at the beginning of some major classifications. These numbers refer to related items in the Reference Section. The reference information may be an estimating procedure, an alternate pricing method, or technical information.

Same Data. Simplified.

Enjoy the convenience and efficiency of accessing your costs anywhere:

- **Skip the multiplier** by setting your location
- **Quickly search,** edit, favorite and share costs
- **Stay on top of price changes** with automatic updates

Discover more at rsmeans.com/online

11 13 Loading Dock Equipment

11 13 13 – Loading Dock Bumpers

11 13 13.10 Dock Bumpers

		Crew	Daily Output	Labor-Hours	Unit	Material	2021 Bare Costs Labor	Equipment	Total	Total Incl O&P
0010	**DOCK BUMPERS** Bolts not included									
0020	2" x 6" to 4" x 8", average	1 Carp	.30	26.667	M.B.F.	1,425	1,450		2,875	3,750
0050	Bumpers, lam. rubber blocks 4-1/2" thick, 10" high, 14" long		26	.308	Ea.	70	16.85		86.85	103
0200	24" long		22	.364		103	19.90		122.90	143
0300	36" long		17	.471		365	25.50		390.50	440
0500	12" high, 14" long		25	.320		69.50	17.50		87	103
0550	24" long		20	.400		85.50	22		107.50	127
0600	36" long		15	.533		111	29		140	166
0800	Laminated rubber blocks 6" thick, 10" high, 14" long		22	.364		79.50	19.90		99.40	117
0850	24" long		18	.444		142	24.50		166.50	193
0900	36" long		13	.615		208	33.50		241.50	280
0910	20" high, 11" long		13	.615		139	33.50		172.50	204
0920	Extruded rubber bumpers, T section, 22" x 22" x 3" thick		41	.195		75.50	10.65		86.15	99
0940	Molded rubber bumpers, 24" x 12" x 3" thick	↓	20	.400		71.50	22		93.50	111
1000	Welded installation of above bumpers	E-14	8	1		3.99	62.50	18.60	85.09	121
1100	For drilled anchors, add per anchor	1 Carp	36	.222	↓	23.50	12.15		35.65	44
1301	Steel bumpers, see Section 10 26									

11 13 16 – Loading Dock Seals and Shelters

11 13 16.10 Dock Seals and Shelters

		Crew	Daily Output	Labor-Hours	Unit	Material	2021 Bare Costs Labor	Equipment	Total	Total Incl O&P
0010	**DOCK SEALS AND SHELTERS**									
3600	Door seal for door perimeter, 12" x 12", vinyl covered	1 Carp	26	.308	L.F.	54	16.85		70.85	84.50
3700	Loading dock, seal for perimeter, 9' x 8', with 12" vinyl	2 Carp	6	2.667	Ea.	1,500	146		1,646	1,875

11 13 19 – Stationary Loading Dock Equipment

11 13 19.10 Dock Equipment

		Crew	Daily Output	Labor-Hours	Unit	Material	2021 Bare Costs Labor	Equipment	Total	Total Incl O&P
0010	**DOCK EQUIPMENT**									
2200	Dock boards, heavy duty, 60" x 60", aluminum, 5,000 lb. capacity				Ea.	1,050			1,050	1,150
2700	9,000 lb. capacity					1,325			1,325	1,450
3200	15,000 lb. capacity					1,550			1,550	1,700
4200	Platform lifter, 6' x 6', portable, 3,000 lb. capacity					9,775			9,775	10,700
4250	4,000 lb. capacity					12,000			12,000	13,200
4400	Fixed, 6' x 8', 5,000 lb. capacity	E-16	.70	22.857		9,450	1,400	213	11,063	12,800
4500	Levelers, hinged for trucks, 10 ton capacity, 6' x 8'		1.08	14.815		4,675	910	138	5,723	6,675
4650	7' x 8'		1.08	14.815		6,600	910	138	7,648	8,800
4670	Air bag power operated, 10 ton cap., 6' x 8'		1.08	14.815		5,175	910	138	6,223	7,250
4680	7' x 8'		1.08	14.815		5,600	910	138	6,648	7,700
4700	Hydraulic, 10 ton capacity, 6' x 8'		1.08	14.815		5,025	910	138	6,073	7,075
4800	7' x 8'	↓	1.08	14.815	↓	5,325	910	138	6,373	7,400

Estimating Tips
General

- The items in this division are usually priced per square foot or each. Most of these items are purchased by the owner and installed by the contractor. Do not assume the items in Division 12 will be purchased and installed by the contractor. Check the specifications for responsibilities and include receiving, storage, installation, and mechanical and electrical hookups in the appropriate divisions.

- Some items in this division require some type of support system that is not usually furnished with the item. Examples of these systems include blocking for the attachment of casework and heavy drapery rods. The required blocking must be added to the estimate in the appropriate division.

Reference Numbers

Reference numbers are shown at the beginning of some major classifications. These numbers refer to related items in the Reference Section. The reference information may be an estimating procedure, an alternate pricing method, or technical information.

Same Data. Simplified.

Enjoy the convenience and efficiency of accessing your costs anywhere:

- **Skip the multiplier** by setting your location
- **Quickly search,** edit, favorite and share costs
- **Stay on top of price changes** with automatic updates

Discover more at rsmeans.com/online

---

12 48 Rugs and Mats

12 48 13 – Entrance Floor Mats and Frames

12 48 13.13 Entrance Floor Mats	Crew	Daily Output	Labor-Hours	Unit	Material	2021 Bare Costs Labor	Equipment	Total	Total Incl O&P
0010 **ENTRANCE FLOOR MATS**									
0020 Recessed, black rubber, 3/8" thick, solid	1 Clab	155	.052	S.F.	2.42	2.29		4.71	6.10
0050 Perforated		155	.052		24	2.29		26.29	30
0100 1/2" thick, solid		155	.052		29	2.29		31.29	35.50
0150 Perforated		155	.052		9.80	2.29		12.09	14.20
0200 In colors, 3/8" thick, solid		155	.052		27.50	2.29		29.79	34
0250 Perforated		155	.052		25.50	2.29		27.79	32
0300 1/2" thick, solid		155	.052		40.50	2.29		42.79	48
0350 Perforated	↓	155	.052	↓	5.35	2.29		7.64	9.25
1225 Recessed, alum. rail, hinged mat, 7/16" thick									
1250 Carpet insert	1 Clab	360	.022	S.F.	71	.99		71.99	80
1275 Vinyl insert		360	.022		71	.99		71.99	80
1300 Abrasive insert	↓	360	.022	↓	71	.99		71.99	80
1325 Recessed, vinyl rail, hinged mat, 7/16" thick									
1350 Carpet insert	1 Clab	360	.022	S.F.	79	.99		79.99	88
1375 Vinyl insert		360	.022		79	.99		79.99	88
1400 Abrasive insert	↓	360	.022	↓	79	.99		79.99	88

Estimating Tips
General

- The items and systems in this division are usually estimated, purchased, supplied, and installed as a unit by one or more subcontractors. The estimator must ensure that all parties are operating from the same set of specifications and assumptions, and that all necessary items are estimated and will be provided. Many times the complex items and systems are covered, but the more common ones, such as excavation or a crane, are overlooked for the very reason that everyone assumes nobody could miss them. The estimator should be the central focus and be able to ensure that all systems are complete.

- It is important to consider factors such as site conditions, weather, shape and size of building, as well as labor availability as they may impact the overall cost of erecting special structures and systems included in this division.

- Another area where problems can develop in this division is at the interface between systems.

The estimator must ensure, for instance, that anchor bolts, nuts, and washers are estimated and included for the air-supported structures and pre-engineered buildings to be bolted to their foundations. Utility supply is a common area where essential items or pieces of equipment can be missed or overlooked because each subcontractor may feel it is another's responsibility. The estimator should also be aware of certain items which may be supplied as part of a package but installed by others, and ensure that the installing contractor's estimate includes the cost of installation. Conversely, the estimator must also ensure that items are not costed by two different subcontractors, resulting in an inflated overall estimate.

13 30 00 Special Structures

- The foundations and floor slab, as well as rough mechanical and electrical, should be estimated, as this work is required for the assembly and erection of the structure. Generally, as noted in the data set, the pre-engineered building comes

as a shell. Pricing is based on the size and structural design parameters stated in the reference section. Additional features, such as windows and doors with their related structural framing, must also be included by the estimator. Here again, the estimator must have a clear understanding of the scope of each portion of the work and all the necessary interfaces.

Reference Numbers

Reference numbers are shown at the beginning of some major classifications. These numbers refer to related items in the Reference Section. The reference information may be an estimating procedure, an alternate pricing method, or technical information.

Note: Not all subdivisions listed here necessarily appear. ■

Same Data. Simplified.

Enjoy the convenience and efficiency of accessing your costs anywhere:

- **Skip the multiplier** by setting your location
- **Quickly search,** edit, favorite and share costs
- **Stay on top of price changes** with automatic updates

Discover more at rsmeans.com/online

13 11 Swimming Pools

13 11 13 – Below-Grade Swimming Pools

13 11 13.50 Swimming Pools

		Crew	Daily Output	Labor-Hours	Unit	Material	2021 Bare Costs Labor	Equipment	Total	Total Incl O&P
0010	**SWIMMING POOLS** Residential in-ground, vinyl lined									
0020	Concrete sides, w/equip, sand bottom	B-52	300	.187	SF Surf	29.50	9.50	1.91	40.91	48.50
0100	Metal or polystyrene sides	R131113-20 B-14	410	.117		24.50	5.45	.53	30.48	35.50
0200	Add for vermiculite bottom					1.87			1.87	2.06
0500	Gunite bottom and sides, white plaster finish									
0600	12' x 30' pool	B-52	145	.386	SF Surf	54.50	19.60	3.95	78.05	93.50
0720	16' x 32' pool		155	.361		49	18.35	3.70	71.05	85.50
0750	20' x 40' pool		250	.224		44	11.35	2.29	57.64	68
0810	Concrete bottom and sides, tile finish									
0820	12' x 30' pool	B-52	80	.700	SF Surf	55	35.50	7.15	97.65	121
0830	16' x 32' pool		95	.589		45.50	30	6.05	81.55	101
0840	20' x 40' pool		130	.431		36.50	22	4.41	62.91	77.50
1100	Motel, gunite with plaster finish, incl. medium									
1150	capacity filtration & chlorination	B-52	115	.487	SF Surf	67.50	24.50	4.98	96.98	117
1200	Municipal, gunite with plaster finish, incl. high									
1250	capacity filtration & chlorination	B-52	100	.560	SF Surf	87	28.50	5.75	121.25	144
1350	Add for formed gutters				L.F.	128			128	141
1360	Add for stainless steel gutters				"	380			380	415
1700	Filtration and deck equipment only, as % of total				Total				20%	20%
1800	Automatic vacuum, hand tools, etc., 20' x 40' pool				SF Pool				.56	.62
1900	5,000 S.F. pool				"				.11	.12
3000	Painting pools, preparation + 3 coats, 20' x 40' pool, epoxy	2 Pord	.33	48.485	Total	1,300	2,250		3,550	4,775
3100	Rubber base paint, 18 gallons	"	.33	48.485		1,250	2,250		3,500	4,725
3500	42' x 82' pool, 75 gallons, epoxy paint	3 Pord	.14	171		5,500	7,975		13,475	17,900
3600	Rubber base paint	"	.14	171		5,150	7,975		13,125	17,500

13 12 Fountains

13 12 13 – Exterior Fountains

13 12 13.10 Outdoor Fountains

		Crew	Daily Output	Labor-Hours	Unit	Material	2021 Bare Costs Labor	Equipment	Total	Total Incl O&P
0010	**OUTDOOR FOUNTAINS**									
0100	Outdoor fountain, 48" high with bowl and figures	2 Clab	2	8	Ea.	385	355		740	955
0200	Commercial, concrete or cast stone, 40-60" H, simple		2	8		975	355		1,330	1,600
0220	Average		2	8		1,900	355		2,255	2,600
0240	Ornate		2	8		4,300	355		4,655	5,275
0260	Metal, 72" high		2	8		1,525	355		1,880	2,200
0280	90" high		2	8		2,200	355		2,555	2,950
0300	120" high		2	8		5,275	355		5,630	6,325
0320	Resin or fiberglass, 40-60" H, wall type		2	8		690	355		1,045	1,300
0340	Waterfall type		2	8		1,175	355		1,530	1,800

13 12 23 – Interior Fountains

13 12 23.10 Indoor Fountains

		Crew	Daily Output	Labor-Hours	Unit	Material	2021 Bare Costs Labor	Equipment	Total	Total Incl O&P
0010	**INDOOR FOUNTAINS**									
0100	Commercial, floor type, resin or fiberglass, lighted, cascade type	2 Clab	2	8	Ea.	435	355		790	1,000
0120	Tiered type		2	8		680	355		1,035	1,275
0140	Waterfall type		2	8		485	355		840	1,050

13 49 Radiation Protection

13 49 13 – Integrated X-Ray Shielding Assemblies

13 49 13.50 Lead Sheets

		Crew	Daily Output	Labor-Hours	Unit	Material	2021 Bare Costs Labor	2021 Bare Costs Equipment	Total	Total Incl O&P
0010	**LEAD SHEETS**									
0300	Lead sheets, 1/16" thick	2 Lath	135	.119	S.F.	11.10	6.40		17.50	21.50
0400	1/8" thick		120	.133		25	7.20		32.20	38
0500	Lead shielding, 1/4" thick		135	.119		47	6.40		53.40	61
0550	1/2" thick	↓	120	.133	↓	87.50	7.20		94.70	107
0950	Lead headed nails (average 1 lb. per sheet)				Lb.	8.85			8.85	9.70
1000	Butt joints in 1/8" lead or thicker, 2" batten strip x 7' long	2 Lath	240	.067	Ea.	34.50	3.61		38.11	43.50
1200	X-ray protection, average radiography or fluoroscopy									
1210	room, up to 300 S.F. floor, 1/16" lead, economy	2 Lath	.25	64	Total	13,300	3,450		16,750	19,700
1500	7'-0" walls, deluxe	"	.15	107	"	15,900	5,775		21,675	25,900
1600	Deep therapy X-ray room, 250 kV capacity,									
1800	up to 300 S.F. floor, 1/4" lead, economy	2 Lath	.08	200	Total	37,000	10,800		47,800	56,500
1900	7'-0" walls, deluxe	"	.06	267	"	45,700	14,400		60,100	71,000

13 49 19 – Lead-Lined Materials

13 49 19.50 Shielding Lead

		Crew	Daily Output	Labor-Hours	Unit	Material	2021 Bare Costs Labor	2021 Bare Costs Equipment	Total	Total Incl O&P
0010	**SHIELDING LEAD**									
0100	Laminated lead in wood doors, 1/16" thick, no hardware				S.F.	52.50			52.50	57.50
0200	Lead lined door frame, not incl. hardware,									
0210	1/16" thick lead, butt prepared for hardware	1 Lath	2.40	3.333	Ea.	905	180		1,085	1,250
0850	Window frame with 1/16" lead and voice passage, 36" x 60"	2 Glaz	2	8		4,300	420		4,720	5,350
0870	24" x 36" frame		4	4	↓	2,325	211		2,536	2,875
0900	Lead gypsum board, 5/8" thick with 1/16" lead		160	.100	S.F.	16	5.25		21.25	25.50
0910	1/8" lead	↓	140	.114		27	6		33	39
0930	1/32" lead	2 Lath	200	.080	↓	12.05	4.33		16.38	19.60

13 49 21 – Lead Glazing

13 49 21.50 Lead Glazing

		Crew	Daily Output	Labor-Hours	Unit	Material	2021 Bare Costs Labor	2021 Bare Costs Equipment	Total	Total Incl O&P
0010	**LEAD GLAZING**									
0600	Lead glass, 1/4" thick, 2.0 mm LE, 12" x 16"	2 Glaz	13	1.231	Ea.	495	65		560	635
0700	24" x 36"		8	2		1,675	105		1,780	1,975
0800	36" x 60"	↓	2	8	↓	4,025	420		4,445	5,050
2000	X-ray viewing panels, clear lead plastic									
2010	7 mm thick, 0.3 mm LE, 2.3 lb./S.F.	H-3	139	.115	S.F.	256	5.45		261.45	290
2020	12 mm thick, 0.5 mm LE, 3.9 lb./S.F.		82	.195		345	9.25		354.25	395
2030	18 mm thick, 0.8 mm LE, 5.9 lb./S.F.		54	.296		420	14.05		434.05	480
2040	22 mm thick, 1.0 mm LE, 7.2 lb./S.F.		44	.364		690	17.20		707.20	785
2050	35 mm thick, 1.5 mm LE, 11.5 lb./S.F.		28	.571		915	27		942	1,050
2060	46 mm thick, 2.0 mm LE, 15.0 lb./S.F.	↓	21	.762	↓	1,200	36		1,236	1,350

13 49 23 – Integrated RFI/EMI Shielding Assemblies

13 49 23.50 Modular Shielding Partitions

		Crew	Daily Output	Labor-Hours	Unit	Material	2021 Bare Costs Labor	2021 Bare Costs Equipment	Total	Total Incl O&P
0010	**MODULAR SHIELDING PARTITIONS**									
4000	X-ray barriers, modular, panels mounted within framework for									
4002	attaching to floor, wall or ceiling, upper portion is clear lead									
4005	plastic window panels 48"H, lower portion is opaque leaded									
4008	steel panels 36"H, structural supports not incl.									
4010	1-section barrier, 36"W x 84"H overall									
4020	0.5 mm LE panels	H-3	6.40	2.500	Ea.	9,250	118		9,368	10,400
4030	0.8 mm LE panels		6.40	2.500		9,975	118		10,093	11,200
4040	1.0 mm LE panels		5.33	3.002		13,800	142		13,942	15,400
4050	1.5 mm LE panels	↓	5.33	3.002	↓	16,300	142		16,442	18,100
4060	2-section barrier, 72"W x 84"H overall									
4070	0.5 mm LE panels	H-3	4	4	Ea.	14,000	189		14,189	15,700

13 49 Radiation Protection

13 49 23 – Integrated RFI/EMI Shielding Assemblies

13 49 23.50 Modular Shielding Partitions	Crew	Daily Output	Labor-Hours	Unit	Material	2021 Bare Costs Labor	Equipment	Total	Total Incl O&P
4080	0.8 mm LE panels	H-3	4	4	Ea.	14,400	189		14,589	16,200
4090	1.0 mm LE panels		3.56	4.494		15,700	213		15,913	17,600
5000	1.5 mm LE panels		3.20	5		28,100	237		28,337	31,400
5010	3-section barrier, 108"W x 84"H overall									
5020	0.5 mm LE panels	H-3	3.20	5	Ea.	21,000	237		21,237	23,500
5030	0.8 mm LE panels		3.20	5		21,700	237		21,937	24,200
5040	1.0 mm LE panels		2.67	5.993		23,600	284		23,884	26,300
5050	1.5 mm LE panels		2.46	6.504		42,700	310		43,010	47,400
7000	X-ray barriers, mobile, mounted within framework w/casters on									
7005	bottom, clear lead plastic window panels on upper portion,									
7010	opaque on lower, 30"W x 75"H overall, incl. framework									
7020	24"H upper w/0.5 mm LE, 48"H lower w/0.8 mm LE	1 Carp	16	.500	Ea.	4,000	27.50		4,027.50	4,475
7030	48"W x 75"H overall, incl. framework									
7040	36"H upper w/0.5 mm LE, 36"H lower w/0.8 mm LE	1 Carp	16	.500	Ea.	6,325	27.50		6,352.50	7,000
7050	36"H upper w/1.0 mm LE, 36"H lower w/1.5 mm LE	"	16	.500	"	7,500	27.50		7,527.50	8,275
7060	72"W x 75"H overall, incl. framework									
7070	36"H upper w/0.5 mm LE, 36"H lower w/0.8 mm LE	1 Carp	16	.500	Ea.	7,500	27.50		7,527.50	8,275
7080	36"H upper w/1.0 mm LE, 36"H lower w/1.5 mm LE	"	16	.500	"	9,375	27.50		9,402.50	10,300

13 53 Meteorological Instrumentation

13 53 09 – Weather Instrumentation

13 53 09.50 Weather Station

		Crew	Daily Output	Labor-Hours	Unit	Material	2021 Bare Costs Labor	Equipment	Total	Total Incl O&P
0010	**WEATHER STATION**									
0020	Remote recording, solar powered, with rain gauge & display, 400' range				Ea.	890			890	980
0100	1 mile range				"	1,700			1,700	1,875

Estimating Tips

22 10 00 Plumbing Piping and Pumps

This subdivision is primarily basic pipe and related materials. The pipe may be used by any of the mechanical disciplines, i.e., plumbing, fire protection, heating, and air conditioning.

Note: CPVC plastic piping approved for fire protection is located in 21 11 13.

- The labor adjustment factors listed in Subdivision 22 01 02.20 apply throughout Divisions 21, 22, and 23. CAUTION: the correct percentage may vary for the same items. For example, the percentage add for the basic pipe installation should be based on the maximum height that the installer must install for that particular section. If the pipe is to be located 14' above the floor but it is suspended on threaded rod from beams, the bottom flange of which is 18' high (4' rods), then the height is actually 18' and the add is 20%. The pipe cover, however, does not have to go above the 14' and so the add should be 10%.

- Most pipe is priced first as straight pipe with a joint (coupling, weld, etc.) every 10' and a hanger usually every 10'. There are exceptions with hanger spacing such as for cast iron pipe (5')

and plastic pipe (3 per 10'). Following each type of pipe there are several lines listing sizes and the amount to be subtracted to delete couplings and hangers. This is for pipe that is to be buried or supported together on trapeze hangers. The reason that the couplings are deleted is that these runs are usually long, and frequently longer lengths of pipe are used. By deleting the couplings, the estimator is expected to look up and add back the correct reduced number of couplings.

- When preparing an estimate, it may be necessary to approximate the fittings. Fittings usually run between 25% and 50% of the cost of the pipe. The lower percentage is for simpler runs, and the higher number is for complex areas, such as mechanical rooms.

- For historic restoration projects, the systems must be as invisible as possible, and pathways must be sought for pipes, conduit, and ductwork. While installations in accessible spaces (such as basements and attics) are relatively straightforward to estimate, labor costs may be more difficult to determine when delivery systems must be concealed.

22 40 00 Plumbing Fixtures

- Plumbing fixture costs usually require two lines: the fixture itself and its "rough-in, supply, and waste."

- In the Assemblies Section (Plumbing D2010) for the desired fixture, the System Components Group at the center of the page shows the fixture on the first line. The rest of the list (fittings, pipe, tubing, etc.) will total up to what we refer to in the Unit Price section as "Rough-in, supply, waste, and vent." Note that for most fixtures we allow a nominal 5' of tubing to reach from the fixture to a main or riser.

- Remember that gas- and oil-fired units need venting.

Reference Numbers

Reference numbers are shown at the beginning of some major classifications. These numbers refer to related items in the Reference Section. The reference information may be an estimating procedure, an alternate pricing method, or technical information.

Note: Not all subdivisions listed here necessarily appear. ∎

Same Data. Simplified.

Enjoy the convenience and efficiency of accessing your costs anywhere:

- **Skip the multiplier** by setting your location
- **Quickly search,** edit, favorite and share costs
- **Stay on top of price changes** with automatic updates

Discover more at rsmeans.com/online

Note: "Powered in part by CINX™, based on licensed proprietary information of Harrison Publishing House, Inc."

Note: Trade Service, in part, has been used as a reference source for some of the material prices used in Division 22.

22 05 Common Work Results for Plumbing

22 05 76 – Facility Drainage Piping Cleanouts

22 05 76.10 Cleanouts	Crew	Daily Output	Labor-Hours	Unit	Material	2021 Bare Costs Labor	Equipment	Total	Total Incl O&P
0010 **CLEANOUTS**									
0060 Floor type									
0080 Round or square, scoriated nickel bronze top									
0100 2" pipe size	1 Plum	10	.800	Ea.	440	54		494	560
0120 3" pipe size		8	1		460	67.50		527.50	605
0140 4" pipe size	↓	6	1.333	↓	615	90.50		705.50	815
0980 Round top, recessed for terrazzo									
1000 2" pipe size	1 Plum	9	.889	Ea.	630	60		690	785
1080 3" pipe size		6	1.333		675	90.50		765.50	875
1100 4" pipe size	↓	4	2		805	135		940	1,075
1120 5" pipe size	Q-1	6	2.667	↓	1,275	162		1,437	1,650

22 13 Facility Sanitary Sewerage

22 13 19 – Sanitary Waste Piping Specialties

22 13 19.13 Sanitary Drains

	Crew	Daily Output	Labor-Hours	Unit	Material	2021 Bare Costs Labor	Equipment	Total	Total Incl O&P
0010 **SANITARY DRAINS**									
0400 Deck, auto park, CI, 13" top									
0440 3", 4", 5", and 6" pipe size	Q-1	8	2	Ea.	2,100	122		2,222	2,500
0480 For galvanized body, add				"	1,300			1,300	1,425
2000 Floor, medium duty, CI, deep flange, 7" diam. top									
2040 2" and 3" pipe size	Q-1	12	1.333	Ea.	355	81		436	510
2080 For galvanized body, add					162			162	178
2120 With polished bronze top				↓	440			440	485
2400 Heavy duty, with sediment bucket, CI, 12" diam. loose grate									
2420 2", 3", 4", 5", and 6" pipe size	Q-1	9	1.778	Ea.	1,050	108		1,158	1,300
2460 With polished bronze top				"	1,575			1,575	1,725
2500 Heavy duty, cleanout & trap w/bucket, CI, 15" top									
2540 2", 3", and 4" pipe size	Q-1	6	2.667	Ea.	6,950	162		7,112	7,875
2560 For galvanized body, add					2,800			2,800	3,100
2580 With polished bronze top				↓	11,200			11,200	12,300

22 14 Facility Storm Drainage

22 14 26 – Facility Storm Drains

22 14 26.13 Roof Drains

	Crew	Daily Output	Labor-Hours	Unit	Material	2021 Bare Costs Labor	Equipment	Total	Total Incl O&P
0010 **ROOF DRAINS**									
0140 Cornice, CI, 45° or 90° outlet									
0200 3" and 4" pipe size	Q-1	12	1.333	Ea.	400	81		481	555
0260 For galvanized body, add					128			128	141
0280 For polished bronze dome, add				↓	95.50			95.50	105

Estimating Tips

The labor adjustment factors listed in Subdivision 22 01 02.20 also apply to Division 23.

23 10 00 Facility Fuel Systems

- The prices in this subdivision for above- and below-ground storage tanks do not include foundations or hold-down slabs, unless noted. The estimator should refer to Divisions 3 and 31 for foundation system pricing. In addition to the foundations, required tank accessories, such as tank gauges, leak detection devices, and additional manholes and piping, must be added to the tank prices.

23 50 00 Central Heating Equipment

- When estimating the cost of an HVAC system, check to see who is responsible for providing and installing the temperature control system. It is possible to overlook controls, assuming that they would be included in the electrical estimate.

- When looking up a boiler, be careful on specified capacity. Some manufacturers rate their products on output while others use input.

- Include HVAC insulation for pipe, boiler, and duct (wrap and liner).

- Be careful when looking up mechanical items to get the correct pressure rating and connection type (thread, weld, flange).

23 70 00 Central HVAC Equipment

- Combination heating and cooling units are sized by the air conditioning requirements. (See Reference No. R236000-20 for the preliminary sizing guide.)

- A ton of air conditioning is nominally 400 CFM.

- Rectangular duct is taken off by the linear foot for each size, but its cost is usually estimated by the pound. Remember that SMACNA standards now base duct on internal pressure.

- Prefabricated duct is estimated and purchased like pipe: straight sections and fittings.

- Note that cranes or other lifting equipment are not included on any lines in Division 23. For example, if a crane is required to lift a heavy piece of pipe into place high above a gym floor, or to put a rooftop unit on the roof of a four-story building, etc., it must be added. Due to the potential for extreme variation—from nothing additional required to a major crane or helicopter—we feel that including a nominal amount for "lifting contingency" would be useless and detract from the accuracy of the estimate. When using equipment rental cost data from RSMeans, do not forget to include the cost of the operator(s).

Reference Numbers

Reference numbers are shown at the beginning of some major classifications. These numbers refer to related items in the Reference Section. The reference information may be an estimating procedure, an alternate pricing method, or technical information.

Note: Not all subdivisions listed here necessarily appear. ■

Same Data. Simplified.

Enjoy the convenience and efficiency of accessing your costs anywhere:

- **Skip the multiplier** by setting your location
- **Quickly search,** edit, favorite and share costs
- **Stay on top of price changes** with automatic updates

Discover more at rsmeans.com/online

23 37 Air Outlets and Inlets

23 37 13 – Diffusers, Registers, and Grilles

23 37 13.30 Grilles

		Crew	Daily Output	Labor-Hours	Unit	Material	2021 Bare Costs Labor	Equipment	Total	Total Incl O&P
0010	**GRILLES**									
0020	Aluminum, unless noted otherwise									
1000	Air return, steel, 6" x 6"	1 Shee	26	.308	Ea.	24	20		44	57
1020	10" x 6"		24	.333		24	22		46	59.50
1080	16" x 8"		22	.364		35	24		59	74.50
1100	12" x 12"		22	.364		35	24		59	74.50
1120	24" x 12"		18	.444		45.50	29		74.50	94
1220	24" x 18"		16	.500		56.50	32.50		89	112
1280	36" x 24"		14	.571		97.50	37.50		135	164
3000	Filter grille with filter, 12" x 12"		24	.333		68.50	22		90.50	109
3020	18" x 12"		20	.400		85.50	26		111.50	134
3040	24" x 18"		18	.444		110	29		139	165
3060	24" x 24"		16	.500		133	32.50		165.50	196
6000	For steel grilles instead of aluminum in above, deduct					10%				

23 37 15 – Air Outlets and Inlets, HVAC Louvers

23 37 15.40 HVAC Louvers

		Crew	Daily Output	Labor-Hours	Unit	Material	2021 Bare Costs Labor	Equipment	Total	Total Incl O&P
0010	**HVAC LOUVERS**									
0100	Aluminum, extruded, with screen, mill finish									
1002	Brick vent, see also Section 04 05 23.19									
1100	Standard, 4" deep, 8" wide, 5" high	1 Shee	24	.333	Ea.	38	22		60	74.50
1200	Modular, 4" deep, 7-3/4" wide, 5" high		24	.333		39	22		61	76
1300	Speed brick, 4" deep, 11-5/8" wide, 3-7/8" high		24	.333		39	22		61	76
1400	Fuel oil brick, 4" deep, 8" wide, 5" high		24	.333		67.50	22		89.50	108
2000	Cooling tower and mechanical equip., screens, light weight		40	.200	S.F.	16.85	13.10		29.95	38.50
2020	Standard weight		35	.229		45	14.95		59.95	72
2500	Dual combination, automatic, intake or exhaust		20	.400		61.50	26		87.50	107
2520	Manual operation		20	.400		45.50	26		71.50	89.50
2540	Electric or pneumatic operation		20	.400		45.50	26		71.50	89.50
2560	Motor, for electric or pneumatic		14	.571	Ea.	525	37.50		562.50	635
3000	Fixed blade, continuous line									
3100	Mullion type, stormproof	1 Shee	28	.286	S.F.	45.50	18.70		64.20	78
3200	Stormproof		28	.286		45.50	18.70		64.20	78
3300	Vertical line		28	.286		54	18.70		72.70	87.50
3500	For damper to use with above, add					50%	30%			
3520	Motor, for damper, electric or pneumatic	1 Shee	14	.571	Ea.	525	37.50		562.50	635
4000	Operating, 45°, manual, electric or pneumatic		24	.333	S.F.	61	22		83	100
4100	Motor, for electric or pneumatic		14	.571	Ea.	525	37.50		562.50	635
4200	Penthouse, roof		56	.143	S.F.	27	9.35		36.35	43.50
4300	Walls		40	.200		63.50	13.10		76.60	90
5000	Thinline, under 4" thick, fixed blade		40	.200		26.50	13.10		39.60	49
5010	Finishes, applied by mfr. at additional cost, available in colors									
5020	Prime coat only, add				S.F.	3.62			3.62	3.98
5040	Baked enamel finish coating, add					6.65			6.65	7.35
5060	Anodized finish, add					7.20			7.20	7.95
5080	Duranodic finish, add					13.15			13.15	14.45
5100	Fluoropolymer finish coating, add					20.50			20.50	23
9980	For small orders (under 10 pieces), add					25%				

Estimating Tips
26 05 00 Common Work Results for Electrical

- Conduit should be taken off in three main categories—power distribution, branch power, and branch lighting—so the estimator can concentrate on systems and components, therefore making it easier to ensure all items have been accounted for.

- For cost modifications for elevated conduit installation, add the percentages to labor according to the height of installation and only to the quantities exceeding the different height levels, not to the total conduit quantities. Refer to subdivision 26 01 02.20 for labor adjustment factors.

- Remember that aluminum wiring of equal ampacity is larger in diameter than copper and may require larger conduit.

- If more than three wires at a time are being pulled, deduct percentages from the labor hours of that grouping of wires.

- When taking off grounding systems, identify separately the type and size of wire, and list each unique type of ground connection.

- The estimator should take the weights of materials into consideration when completing a takeoff. Topics to consider include: How will the materials be supported? What methods of support are available? How high will the support structure have to reach? Will the final support structure be able to withstand the total burden? Is the support material included or separate from the fixture, equipment, and material specified?

- Do not overlook the costs for equipment used in the installation. If scaffolding or highlifts are available in the field, contractors may use them in lieu of the proposed ladders and rolling staging.

26 20 00 Low-Voltage Electrical Transmission

- Supports and concrete pads may be shown on drawings for the larger equipment, or the support system may be only a piece of plywood for the back of a panelboard. In either case, they must be included in the costs.

26 40 00 Electrical and Cathodic Protection

- When taking off cathodic protection systems, identify the type and size of cable, and list each unique type of anode connection.

26 50 00 Lighting

- Fixtures should be taken off room by room using the fixture schedule, specifications, and the ceiling plan. For large concentrations of lighting fixtures in the same area, deduct the percentages from labor hours.

Reference Numbers

Reference numbers are shown at the beginning of some major classifications. These numbers refer to related items in the Reference Section. The reference information may be an estimating procedure, an alternate pricing method, or technical information.

Note: Not all subdivisions listed here necessarily appear. ∎

Same Data. Simplified.

Enjoy the convenience and efficiency of accessing your costs anywhere:

- **Skip the multiplier** by setting your location
- **Quickly search,** edit, favorite and share costs
- **Stay on top of price changes** with automatic updates

Discover more at rsmeans.com/online

Note: "Powered in part by CINX™, based on licensed proprietary information of Harrison Publishing House, Inc."

Note: The following companies, in part, have been used as a reference source for some of the material prices used in Division 26:

Electriflex

Trade Service

26 05 39.30 Conduit In Concrete Slab	Crew	Daily Output	Labor-Hours	Unit	Material	2021 Bare Costs Labor	Equipment	Total	Total Incl O&P
0010 **CONDUIT IN CONCRETE SLAB** Including terminations,									
0020 fittings and supports									
3230 PVC, schedule 40, 1/2" diameter	1 Elec	270	.030	L.F.	.36	1.89		2.25	3.19
3250 3/4" diameter		230	.035		.37	2.22		2.59	3.70
3270 1" diameter		200	.040		.42	2.55		2.97	4.25
3300 1-1/4" diameter		170	.047		.73	3		3.73	5.25
3330 1-1/2" diameter		140	.057		.76	3.64		4.40	6.25
3350 2" diameter		120	.067		.90	4.25		5.15	7.30
4350 Rigid galvanized steel, 1/2" diameter		200	.040		2.32	2.55		4.87	6.35
4400 3/4" diameter		170	.047		2.48	3		5.48	7.15
4450 1" diameter		130	.062		3.73	3.92		7.65	9.90
4500 1-1/4" diameter		110	.073		5.25	4.63		9.88	12.70
4600 1-1/2" diameter		100	.080		5.90	5.10		11	14.05
4800 2" diameter		90	.089		7.20	5.65		12.85	16.35

26 05 43 – Underground Ducts and Raceways for Electrical Systems

26 05 43.10 Trench Duct	Crew	Daily Output	Labor-Hours	Unit	Material	2021 Bare Costs Labor	Equipment	Total	Total Incl O&P
0010 **TRENCH DUCT** Steel with cover									
0020 Standard adjustable, depths to 4"									
0100 Straight, single compartment, 9" wide	2 Elec	40	.400	L.F.	400	25.50		425.50	480
0200 12" wide		32	.500		530	32		562	630
0400 18" wide		26	.615		500	39		539	610
0600 24" wide		22	.727		595	46.50		641.50	725
0800 30" wide		20	.800		231	51		282	330
1000 36" wide		16	1		256	63.50		319.50	375
1200 Horizontal elbow, 9" wide		5.40	2.963	Ea.	415	189		604	740
1400 12" wide		4.60	3.478		440	222		662	810
1600 18" wide		4	4		590	255		845	1,025
1800 24" wide		3.20	5		835	320		1,155	1,400
2000 30" wide		2.60	6.154		1,225	390		1,615	1,925
2200 36" wide		2.40	6.667		1,450	425		1,875	2,225
2400 Vertical elbow, 9" wide		5.40	2.963		186	189		375	485
2600 12" wide		4.60	3.478		158	222		380	505
2800 18" wide		4	4		181	255		436	580
3000 24" wide		3.20	5		225	320		545	725
3200 30" wide		2.60	6.154		249	390		639	855
3400 36" wide		2.40	6.667		274	425		699	930
3600 Cross, 9" wide		4	4		685	255		940	1,125
3800 12" wide		3.20	5		725	320		1,045	1,275
4000 18" wide		2.60	6.154		865	390		1,255	1,525
4200 24" wide		2.20	7.273		1,075	465		1,540	1,875
4400 30" wide		2	8		1,500	510		2,010	2,400
4600 36" wide		1.80	8.889		1,875	565		2,440	2,925
4800 End closure, 9" wide		14.40	1.111		45.50	71		116.50	155
5000 12" wide		12	1.333		52	85		137	183
5200 18" wide		10	1.600		79.50	102		181.50	239
5400 24" wide		8	2		105	127		232	305
5600 30" wide		6.60	2.424		162	154		316	410
5800 36" wide		5.80	2.759		156	176		332	430
6000 Tees, 9" wide		4	4		440	255		695	865
6200 12" wide		3.60	4.444		510	283		793	980
6400 18" wide		3.20	5		655	320		975	1,200
6600 24" wide		3	5.333		940	340		1,280	1,525

26 05 43 – Underground Ducts and Raceways for Electrical Systems

26 05 43.10 Trench Duct	Crew	Daily Output	Labor-Hours	Unit	Material	2021 Bare Costs Labor	Equipment	Total	Total Incl O&P
6800	30" wide	2 Elec	2.60	6.154	Ea.	1,225	390		1,615	1,925
7000	36" wide		2	8		1,600	510		2,110	2,525
7200	Riser, and cabinet connector, 9" wide		5.40	2.963		192	189		381	490
7400	12" wide		4.60	3.478		224	222		446	575
7600	18" wide		4	4		238	255		493	640
7800	24" wide		3.20	5		335	320		655	840
8000	30" wide		2.60	6.154		385	390		775	1,000
8200	36" wide		2	8		445	510		955	1,250
8400	Insert assembly, cell to conduit adapter, 1-1/4"	1 Elec	16	.500		69.50	32		101.50	124

For customer support on your Concrete & Masonry Costs with RSMeans Data, call 800.448.8182.

199

Division Notes

		CREW	DAILY OUTPUT	LABOR-HOURS	UNIT	BARE COSTS				TOTAL INCL O&P
						MAT.	LABOR	EQUIP.	TOTAL	

Estimating Tips

31 05 00 Common Work Results for Earthwork

- Estimating the actual cost of performing earthwork requires careful consideration of the variables involved. This includes items such as type of soil, whether water will be encountered, dewatering, whether banks need bracing, disposal of excavated earth, and length of haul to fill or spoil sites, etc. If the project has large quantities of cut or fill, consider raising or lowering the site to reduce costs, while paying close attention to the effect on site drainage and utilities.

- If the project has large quantities of fill, creating a borrow pit on the site can significantly lower the costs.

- It is very important to consider what time of year the project is scheduled for completion. Bad weather can create large cost overruns from dewatering, site repair, and lost productivity from cold weather.

Reference Numbers

Reference numbers are shown at the beginning of some major classifications. These numbers refer to related items in the Reference Section. The reference information may be an estimating procedure, an alternate pricing method, or technical information.

Note: Not all subdivisions listed here necessarily appear. ■

Same Data. Simplified.

Enjoy the convenience and efficiency of accessing your costs anywhere:

- **Skip the multiplier** by setting your location
- **Quickly search,** edit, favorite and share costs
- **Stay on top of price changes** with automatic updates

Discover more at rsmeans.com/online

31 05 Common Work Results for Earthwork

31 05 13 – Soils for Earthwork

31 05 13.10 Borrow		Crew	Daily Output	Labor-Hours	Unit	Material	2021 Bare Costs Labor	Equipment	Total	Total Incl O&P
0010	**BORROW**									
0020	Spread, 200 HP dozer, no compaction, 2 mile RT haul									
0200	Common borrow	B-15	600	.047	C.Y.	13.50	2.45	4.46	20.41	23.50
0700	Screened loam		600	.047		29.50	2.45	4.46	36.41	41
0800	Topsoil, weed free	↓	600	.047		27	2.45	4.46	33.91	38.50
0900	For 5 mile haul, add	B-34B	200	.040	↓		2.05	2.90	4.95	6.25

31 05 16 – Aggregates for Earthwork

31 05 16.10 Borrow		Crew	Daily Output	Labor-Hours	Unit	Material	2021 Bare Costs Labor	Equipment	Total	Total Incl O&P
0010	**BORROW**									
0020	Spread, with 200 HP dozer, no compaction, 2 mile RT haul									
0100	Bank run gravel	B-15	600	.047	L.C.Y.	29.50	2.45	4.46	36.41	41
0300	Crushed stone (1.40 tons per C.Y.), 1-1/2"		600	.047		25.50	2.45	4.46	32.41	37
0320	3/4"		600	.047		25.50	2.45	4.46	32.41	37
0340	1/2"		600	.047		19.60	2.45	4.46	26.51	30
0360	3/8"		600	.047		33	2.45	4.46	39.91	45
0400	Sand, washed, concrete		600	.047		33	2.45	4.46	39.91	44.50
0500	Dead or bank sand		600	.047		18.75	2.45	4.46	25.66	29
0600	Select structural fill	↓	600	.047		30	2.45	4.46	36.91	41.50
0900	For 5 mile haul, add	B-34B	200	.040	↓		2.05	2.90	4.95	6.25
1000	For flowable fill, see Section 03 31 13.35									

31 06 Schedules for Earthwork

31 06 60 – Schedules for Special Foundations and Load Bearing Elements

31 06 60.14 Piling Special Costs		Crew	Daily Output	Labor-Hours	Unit	Material	2021 Bare Costs Labor	Equipment	Total	Total Incl O&P
0010	**PILING SPECIAL COSTS**									
0011	Piling special costs, pile caps, see Section 03 30 53.40									
0500	Cutoffs, concrete piles, plain	1 Pile	5.50	1.455	Ea.		81.50		81.50	126
0600	With steel thin shell, add		38	.211			11.75		11.75	18.20
0700	Steel pile or "H" piles		19	.421			23.50		23.50	36.50
0800	Wood piles	↓	38	.211	↓		11.75		11.75	18.20
0900	Pre-augering up to 30' deep, average soil, 24" diameter	B-43	180	.267	L.F.		13.05	4.27	17.32	24
0920	36" diameter		115	.417			20.50	6.70	27.20	38
0960	48" diameter		70	.686			33.50	11	44.50	62
0980	60" diameter	↓	50	.960	↓		47	15.35	62.35	87
1000	Testing, any type piles, test load is twice the design load									
1050	50 ton design load, 100 ton test				Ea.				14,000	15,500
1100	100 ton design load, 200 ton test								20,000	22,000
1150	150 ton design load, 300 ton test								26,000	28,500
1200	200 ton design load, 400 ton test								28,000	31,000
1250	400 ton design load, 800 ton test				↓				32,000	35,000
1500	Wet conditions, soft damp ground									
1600	Requiring mats for crane, add								40%	40%
1700	Barge mounted driving rig, add								30%	30%

31 06 60.15 Mobilization		Crew	Daily Output	Labor-Hours	Unit	Material	2021 Bare Costs Labor	Equipment	Total	Total Incl O&P
0010	**MOBILIZATION**									
0020	Set up & remove, air compressor, 600 CFM	A-5	3.30	5.455	Ea.		245	15.05	260.05	380
0100	1,200 CFM	"	2.20	8.182			365	22.50	387.50	575
0200	Crane, with pile leads and pile hammer, 75 ton	B-19	.60	107			6,100	3,400	9,500	13,000
0300	150 ton	"	.36	178			10,200	5,675	15,875	21,800
0500	Drill rig, for caissons, to 36", minimum	B-43	2	24	↓		1,175	385	1,560	2,175

31 06 Schedules for Earthwork

31 06 60 – Schedules for Special Foundations and Load Bearing Elements

31 06 60.15 Mobilization	Crew	Daily Output	Labor-Hours	Unit	Material	2021 Bare Costs Labor	Equipment	Total	Total Incl O&P
0520	Maximum	B-43	.50	96	Ea.		4,700	1,525	6,225	8,700
0600	Up to 84"	↓	1	48			2,350	770	3,120	4,350
0800	Auxiliary boiler, for steam small	A-5	1.66	10.843			485	30	515	760
0900	Large	"	.83	21.687			975	60	1,035	1,525
1100	Rule of thumb: complete pile driving set up, small	B-19	.45	142			8,125	4,550	12,675	17,400
1200	Large	"	.27	237	↓		13,500	7,575	21,075	28,900
1500	Mobilization, barge, by tug boat	B-83	25	.640	Mile		33	29	62	81.50

31 22 Grading

31 22 16 – Fine Grading

31 22 16.10 Finish Grading

		Crew	Daily Output	Labor-Hours	Unit	Material	2021 Bare Costs Labor	Equipment	Total	Total Incl O&P
0010	**FINISH GRADING**									
0012	Finish grading area to be paved with grader, small area	B-11L	400	.040	S.Y.		2.07	2.68	4.75	6.05
0100	Large area		2000	.008	"		.41	.54	.95	1.21
3500	Finish grading lagoon bottoms	↓	4	4	M.S.F.		207	268	475	605

31 23 Excavation and Fill

31 23 16 – Excavation

31 23 16.13 Excavating, Trench

		Crew	Daily Output	Labor-Hours	Unit	Material	2021 Bare Costs Labor	Equipment	Total	Total Incl O&P
0010	**EXCAVATING, TRENCH**									
0011	Or continuous footing									
0020	Common earth with no sheeting or dewatering included									
0050	1' to 4' deep, 3/8 C.Y. excavator	B-11C	150	.107	B.C.Y.		5.50	1.44	6.94	9.80
0060	1/2 C.Y. excavator	B-11M	200	.080			4.14	1.18	5.32	7.45
0090	4' to 6' deep, 1/2 C.Y. excavator	"	200	.080			4.14	1.18	5.32	7.45
0100	5/8 C.Y. excavator	B-12Q	250	.064			3.39	2.42	5.81	7.70
0300	1/2 C.Y. excavator, truck mounted	B-12J	200	.080			4.23	4.25	8.48	11
0500	6' to 10' deep, 3/4 C.Y. excavator	B-12F	225	.071			3.76	3.12	6.88	9.05
0600	1 C.Y. excavator, truck mounted	B-12K	400	.040			2.12	2.46	4.58	5.85
0900	10' to 14' deep, 3/4 C.Y. excavator	B-12F	200	.080			4.23	3.51	7.74	10.15
1000	1-1/2 C.Y. excavator	B-12B	540	.030			1.57	1.29	2.86	3.76
1300	14' to 20' deep, 1 C.Y. excavator	B-12A	320	.050			2.65	2.60	5.25	6.80
1340	20' to 24' deep, 1 C.Y. excavator	"	288	.056			2.94	2.89	5.83	7.55
1352	4' to 6' deep, 1/2 C.Y. excavator w/trench box	B-13H	188	.085			4.50	5.15	9.65	12.35
1354	5/8 C.Y. excavator	"	235	.068			3.60	4.13	7.73	9.90
1362	6' to 10' deep, 3/4 C.Y. excavator w/trench box	B-13G	212	.075			3.99	3.87	7.86	10.20
1374	10' to 14' deep, 3/4 C.Y. excavator w/trench box	"	188	.085			4.50	4.37	8.87	11.50
1376	1-1/2 C.Y. excavator	B-13E	508	.032			1.67	1.60	3.27	4.25
1381	14' to 20' deep, 1 C.Y. excavator w/trench box	B-13D	301	.053			2.81	3.16	5.97	7.65
1386	20' to 24' deep, 1 C.Y. excavator w/trench box	"	271	.059			3.13	3.51	6.64	8.50
1400	By hand with pick and shovel 2' to 6' deep, light soil	1 Clab	8	1			44.50		44.50	66.50
1500	Heavy soil	"	4	2			89		89	133
1700	For tamping backfilled trenches, air tamp, add	A-1G	100	.080	E.C.Y.		3.55	.55	4.10	5.90
1900	Vibrating plate, add	B-18	180	.133	"		6	.92	6.92	9.95
2100	Trim sides and bottom for concrete pours, common earth		1500	.016	S.F.		.72	.11	.83	1.20
2300	Hardpan	↓	600	.040	"		1.80	.28	2.08	2.99
2400	Pier and spread footing excavation, add to above				B.C.Y.				30%	30%
5020	Loam & sandy clay with no sheeting or dewatering included									
5050	1' to 4' deep, 3/8 C.Y. tractor loader/backhoe	B-11C	162	.099	B.C.Y.		5.10	1.33	6.43	9.05

31 23 16.13 Excavating, Trench

		Crew	Daily Output	Labor-Hours	Unit	Material	2021 Bare Costs Labor	2021 Bare Costs Equipment	Total	Total Incl O&P
5060	1/2 C.Y. excavator	B-11M	216	.074	B.C.Y.		3.83	1.09	4.92	6.90
5080	4' to 6' deep, 1/2 C.Y. excavator	"	216	.074			3.83	1.09	4.92	6.90
5090	5/8 C.Y. excavator	B-12Q	276	.058			3.07	2.19	5.26	7
5130	1/2 C.Y. excavator, truck mounted	B-12J	216	.074			3.92	3.94	7.86	10.20
5140	6' to 10' deep, 3/4 C.Y. excavator	B-12F	243	.066			3.49	2.89	6.38	8.40
5160	1 C.Y. excavator, truck mounted	B-12K	432	.037			1.96	2.28	4.24	5.45
5190	10' to 14' deep, 3/4 C.Y. excavator	B-12F	216	.074			3.92	3.25	7.17	9.40
5210	1-1/2 C.Y. excavator	B-12B	583	.027			1.45	1.19	2.64	3.48
5250	14' to 20' deep, 1 C.Y. excavator	B-12A	346	.046			2.45	2.41	4.86	6.30
5300	20' to 24' deep, 1 C.Y. excavator	"	311	.051			2.72	2.68	5.40	7
5352	4' to 6' deep, 1/2 C.Y. excavator w/trench box	B-13H	205	.078			4.13	4.73	8.86	11.35
5354	5/8 C.Y. excavator	"	257	.062			3.30	3.77	7.07	9.05
5362	6' to 10' deep, 3/4 C.Y. excavator w/trench box	B-13G	231	.069			3.67	3.55	7.22	9.35
5370	10' to 14' deep, 3/4 C.Y. excavator w/trench box	"	205	.078			4.13	4	8.13	10.55
5374	1-1/2 C.Y. excavator	B-13E	554	.029			1.53	1.47	3	3.90
5382	14' to 20' deep, 1 C.Y. excavator w/trench box	B-13D	329	.049			2.57	2.89	5.46	7
5392	20' to 24' deep, 1 C.Y. excavator w/trench box	"	295	.054	▼		2.87	3.23	6.10	7.85
6020	Sand & gravel with no sheeting or dewatering included									
6050	1' to 4' deep, 3/8 C.Y. excavator	B-11C	165	.097	B.C.Y.		5	1.31	6.31	8.90
6060	1/2 C.Y. excavator	B-11M	220	.073			3.76	1.07	4.83	6.80
6080	4' to 6' deep, 1/2 C.Y. excavator	"	220	.073			3.76	1.07	4.83	6.80
6090	5/8 C.Y. excavator	B-12Q	275	.058			3.08	2.20	5.28	7
6130	1/2 C.Y. excavator, truck mounted	B-12J	220	.073			3.85	3.87	7.72	10
6140	6' to 10' deep, 3/4 C.Y. excavator	B-12F	248	.065			3.41	2.83	6.24	8.20
6160	1 C.Y. excavator, truck mounted	B-12K	440	.036			1.92	2.24	4.16	5.35
6190	10' to 14' deep, 3/4 C.Y. excavator	B-12F	220	.073			3.85	3.19	7.04	9.25
6210	1-1/2 C.Y. excavator	B-12B	594	.027			1.43	1.17	2.60	3.42
6250	14' to 20' deep, 1 C.Y. excavator	B-12A	352	.045			2.41	2.37	4.78	6.20
6300	20' to 24' deep, 1 C.Y. excavator	"	317	.050			2.67	2.63	5.30	6.85
6352	4' to 6' deep, 1/2 C.Y. excavator w/trench box	B-13H	209	.077			4.05	4.64	8.69	11.15
6354	5/8 C.Y. excavator	"	261	.061			3.24	3.72	6.96	8.95
6362	6' to 10' deep, 3/4 C.Y. excavator w/trench box	B-13G	236	.068			3.59	3.48	7.07	9.20
6370	10' to 14' deep, 3/4 C.Y. excavator w/trench box	"	209	.077			4.05	3.93	7.98	10.35
6374	1-1/2 C.Y. excavator	B-13E	564	.028			1.50	1.44	2.94	3.83
6382	14' to 20' deep, 1 C.Y. excavator w/trench box	B-13D	334	.048			2.54	2.85	5.39	6.90
6392	20' to 24' deep, 1 C.Y. excavator w/trench box	"	301	.053	▼		2.81	3.16	5.97	7.65
7020	Dense hard clay with no sheeting or dewatering included									
7050	1' to 4' deep, 3/8 C.Y. excavator	B-11C	132	.121	B.C.Y.		6.25	1.64	7.89	11.15
7060	1/2 C.Y. excavator	B-11M	176	.091			4.70	1.34	6.04	8.45
7080	4' to 6' deep, 1/2 C.Y. excavator	"	176	.091			4.70	1.34	6.04	8.45
7090	5/8 C.Y. excavator	B-12Q	220	.073			3.85	2.75	6.60	8.75
7130	1/2 C.Y. excavator, truck mounted	B-12J	176	.091			4.81	4.83	9.64	12.45
7140	6' to 10' deep, 3/4 C.Y. excavator	B-12F	198	.081			4.28	3.54	7.82	10.30
7160	1 C.Y. excavator, truck mounted	B-12K	352	.045			2.41	2.80	5.21	6.65
7190	10' to 14' deep, 3/4 C.Y. excavator	B-12F	176	.091			4.81	3.99	8.80	11.55
7210	1-1/2 C.Y. excavator	B-12B	475	.034			1.78	1.46	3.24	4.27
7250	14' to 20' deep, 1 C.Y. excavator	B-12A	282	.057			3	2.95	5.95	7.75
7300	20' to 24' deep, 1 C.Y. excavator	"	254	.063	▼		3.33	3.28	6.61	8.60

31 23 Excavation and Fill

31 23 16 – Excavation

31 23 16.14 Excavating, Utility Trench	Crew	Daily Output	Labor- Hours	Unit	Material	2021 Bare Costs Labor	2021 Bare Costs Equipment	Total	Total Incl O&P
0010 EXCAVATING, UTILITY TRENCH									
0011 Common earth									
0050 Trenching with chain trencher, 12 HP, operator walking									
0100 4" wide trench, 12" deep	B-53	800	.010	L.F.		.56	.20	.76	1.05
0150 18" deep		750	.011			.59	.21	.80	1.11
0200 24" deep		700	.011			.63	.23	.86	1.20
0300 6" wide trench, 12" deep		650	.012			.68	.24	.92	1.29
0350 18" deep		600	.013			.74	.26	1	1.39
0400 24" deep		550	.015			.81	.29	1.10	1.52
0450 36" deep		450	.018			.99	.35	1.34	1.86
0600 8" wide trench, 12" deep		475	.017			.93	.33	1.26	1.76
0650 18" deep		400	.020			1.11	.40	1.51	2.09
0700 24" deep		350	.023			1.27	.45	1.72	2.39
0750 36" deep	▼	300	.027	▼		1.48	.53	2.01	2.79
1000 Backfill by hand including compaction, add									
1050 4" wide trench, 12" deep	A-1G	800	.010	L.F.		.44	.07	.51	.74
1100 18" deep		530	.015			.67	.10	.77	1.11
1150 24" deep		400	.020			.89	.14	1.03	1.48
1300 6" wide trench, 12" deep		540	.015			.66	.10	.76	1.09
1350 18" deep		405	.020			.88	.13	1.01	1.46
1400 24" deep		270	.030			1.32	.20	1.52	2.18
1450 36" deep		180	.044			1.97	.30	2.27	3.27
1600 8" wide trench, 12" deep		400	.020			.89	.14	1.03	1.48
1650 18" deep		265	.030			1.34	.21	1.55	2.23
1700 24" deep		200	.040			1.78	.27	2.05	2.95
1750 36" deep	▼	135	.059	▼		2.63	.40	3.03	4.38
2000 Chain trencher, 40 HP operator riding									
2050 6" wide trench and backfill, 12" deep	B-54	1200	.007	L.F.		.37	.38	.75	.96
2100 18" deep		1000	.008			.44	.45	.89	1.16
2150 24" deep		975	.008			.46	.46	.92	1.19
2200 36" deep		900	.009			.49	.50	.99	1.29
2250 48" deep		750	.011			.59	.60	1.19	1.54
2300 60" deep		650	.012			.68	.69	1.37	1.78
2400 8" wide trench and backfill, 12" deep		1000	.008			.44	.45	.89	1.16
2450 18" deep		950	.008			.47	.47	.94	1.22
2500 24" deep		900	.009			.49	.50	.99	1.29
2550 36" deep		800	.010			.56	.56	1.12	1.45
2600 48" deep		650	.012			.68	.69	1.37	1.78
2700 12" wide trench and backfill, 12" deep		975	.008			.46	.46	.92	1.19
2750 18" deep		860	.009			.52	.52	1.04	1.35
2800 24" deep		800	.010			.56	.56	1.12	1.45
2850 36" deep		725	.011			.61	.62	1.23	1.59
3000 16" wide trench and backfill, 12" deep		835	.010			.53	.54	1.07	1.38
3050 18" deep		750	.011			.59	.60	1.19	1.54
3100 24" deep	▼	700	.011	▼		.63	.64	1.27	1.66
3200 Compaction with vibratory plate, add								35%	35%
5100 Hand excavate and trim for pipe bells after trench excavation									
5200 8" pipe	1 Clab	155	.052	L.F.		2.29		2.29	3.42
5300 18" pipe	"	130	.062	"		2.73		2.73	4.08

31 23 16.16 Structural Excavation for Minor Structures

		Crew	Daily Output	Labor-Hours	Unit	Material	2021 Bare Costs Labor	2021 Bare Costs Equipment	Total	Total Incl O&P
0010	**STRUCTURAL EXCAVATION FOR MINOR STRUCTURES**									
0015	Hand, pits to 6' deep, sandy soil	1 Clab	8	1	B.C.Y.		44.50		44.50	66.50
0100	Heavy soil or clay		4	2			89		89	133
0300	Pits 6' to 12' deep, sandy soil		5	1.600			71		71	106
0500	Heavy soil or clay		3	2.667			118		118	177
0700	Pits 12' to 18' deep, sandy soil		4	2			89		89	133
0900	Heavy soil or clay		2	4			178		178	265
1100	Hand loading trucks from stock pile, sandy soil		12	.667			29.50		29.50	44
1300	Heavy soil or clay	▼	8	1	▼		44.50		44.50	66.50
1500	For wet or muck hand excavation, add to above								50%	50%
6000	Machine excavation, for spread and mat footings, elevator pits,									
6001	and small building foundations									
6030	Common earth, hydraulic backhoe, 1/2 C.Y. bucket	B-12E	55	.291	B.C.Y.		15.40	8.30	23.70	32
6035	3/4 C.Y. bucket	B-12F	90	.178			9.40	7.80	17.20	22.50
6040	1 C.Y. bucket	B-12A	108	.148			7.85	7.70	15.55	20
6050	1-1/2 C.Y. bucket	B-12B	144	.111			5.90	4.83	10.73	14.05
6060	2 C.Y. bucket	B-12C	200	.080			4.23	4.71	8.94	11.50
6070	Sand and gravel, 3/4 C.Y. bucket	B-12F	100	.160			8.45	7	15.45	20.50
6080	1 C.Y. bucket	B-12A	120	.133			7.05	6.95	14	18.15
6090	1-1/2 C.Y. bucket	B-12B	160	.100			5.30	4.35	9.65	12.70
6100	2 C.Y. bucket	B-12C	220	.073			3.85	4.29	8.14	10.45
6110	Clay, till, or blasted rock, 3/4 C.Y. bucket	B-12F	80	.200			10.60	8.75	19.35	25.50
6120	1 C.Y. bucket	B-12A	95	.168			8.90	8.75	17.65	23
6130	1-1/2 C.Y. bucket	B-12B	130	.123			6.50	5.35	11.85	15.60
6140	2 C.Y. bucket	B-12C	175	.091			4.84	5.40	10.24	13.15
6230	Sandy clay & loam, hydraulic backhoe, 1/2 C.Y. bucket	B-12E	60	.267			14.10	7.60	21.70	29.50
6235	3/4 C.Y. bucket	B-12F	98	.163			8.65	7.15	15.80	21
6240	1 C.Y. bucket	B-12A	116	.138			7.30	7.20	14.50	18.80
6250	1-1/2 C.Y. bucket	B-12B	156	.103	▼		5.45	4.46	9.91	13
9010	For mobilization or demobilization, see Section 01 54 36.50									
9020	For dewatering, see Section 31 23 19.20									
9022	For larger structures, see Bulk Excavation, Section 31 23 16.42									
9024	For loading onto trucks, add								15%	15%
9026	For hauling, see Section 31 23 23.20									
9030	For sheeting or soldier bms/lagging, see Section 31 52 16.10									

31 23 16.26 Rock Removal

		Crew	Daily Output	Labor-Hours	Unit	Material	2021 Bare Costs Labor	2021 Bare Costs Equipment	Total	Total Incl O&P
0010	**ROCK REMOVAL**									
0015	Drilling only rock, 2" hole for rock bolts	B-47	316	.076	L.F.		3.70	5.15	8.85	11.15
0800	2-1/2" hole for pre-splitting		250	.096			4.68	6.50	11.18	14.15
4600	Quarry operations, 2-1/2" to 3-1/2" diameter	▼	240	.100	▼		4.88	6.80	11.68	14.70

31 23 16.30 Drilling and Blasting Rock

		Crew	Daily Output	Labor-Hours	Unit	Material	2021 Bare Costs Labor	2021 Bare Costs Equipment	Total	Total Incl O&P
0010	**DRILLING AND BLASTING ROCK**									
0020	Rock, open face, under 1,500 C.Y.	B-47	225	.107	B.C.Y.	4.69	5.20	7.25	17.14	21
0100	Over 1,500 C.Y.		300	.080		4.69	3.90	5.45	14.04	16.95
0200	Areas where blasting mats are required, under 1,500 C.Y.		175	.137		4.69	6.70	9.30	20.69	25.50
0250	Over 1,500 C.Y.	▼	250	.096		4.69	4.68	6.50	15.87	19.30
0300	Bulk drilling and blasting, can vary greatly, average								9.65	12.20
0500	Pits, average								25.50	31.50
1300	Deep hole method, up to 1,500 C.Y.	B-47	50	.480		4.69	23.50	32.50	60.69	76
1400	Over 1,500 C.Y.		66	.364		4.69	17.75	24.50	46.94	58.50
1900	Restricted areas, up to 1,500 C.Y.		13	1.846		4.69	90	125	219.69	277
2000	Over 1,500 C.Y.	▼	20	1.200	▼	4.69	58.50	81.50	144.69	182

31 23 Excavation and Fill

31 23 16 – Excavation

31 23 16.30 Drilling and Blasting Rock

		Crew	Daily Output	Labor-Hours	Unit	Material	2021 Bare Costs Labor	Equipment	Total	Total Incl O&P
2200	Trenches, up to 1,500 C.Y.	B-47	22	1.091	B.C.Y.	13.60	53	74	140.60	176
2300	Over 1,500 C.Y.		26	.923		13.60	45	62.50	121.10	151
2500	Pier holes, up to 1,500 C.Y.		22	1.091		4.69	53	74	131.69	166
2600	Over 1,500 C.Y.	↓	31	.774		4.69	38	52.50	95.19	120
2800	Boulders under 1/2 C.Y., loaded on truck, no hauling	B-100	80	.150			8.10	11.55	19.65	25
2900	Boulders, drilled, blasted	B-47	100	.240	▽	4.69	11.70	16.30	32.69	40.50
3100	Jackhammer operators with foreman compressor, air tools	B-9	1	40	Day		1,800	355	2,155	3,075
3300	Track drill, compressor, operator and foreman	B-47	1	24	"		1,175	1,625	2,800	3,550
3500	Blasting caps				Ea.	6.70			6.70	7.35
3700	Explosives					.55			.55	.61
3800	Blasting mats, for purchase, no mobilization, 10' x 15' x 12"					1,275			1,275	1,400
3900	Blasting mats, rent, for first day					205			205	225
4000	Per added day					68.50			68.50	75.50
4200	Preblast survey for 6 room house, individual lot, minimum	A-6	2.40	6.667			365	14.35	379.35	565
4300	Maximum	"	1.35	11.852	▽		650	25.50	675.50	1,000
4500	City block within zone of influence, minimum	A-8	25200	.001	S.F.		.07		.07	.11
4600	Maximum	"	15100	.002	"		.12		.12	.18

31 23 16.42 Excavating, Bulk Bank Measure

		Crew	Daily Output	Labor-Hours	Unit	Material	2021 Bare Costs Labor	Equipment	Total	Total Incl O&P
0010	**EXCAVATING, BULK BANK MEASURE** R312316-40									
0011	Common earth piled									
0020	For loading onto trucks, add								15%	15%
0100	For hauling, see Section 31 23 23.20 R312316-45									
0200	Excavator, hydraulic, crawler mtd., 1 C.Y. cap. = 100 C.Y./hr.	B-12A	800	.020	B.C.Y.		1.06	1.04	2.10	2.72
0250	1-1/2 C.Y. cap. = 125 C.Y./hr.	B-12B	1000	.016			.85	.70	1.55	2.03
0260	2 C.Y. cap. = 165 C.Y./hr.	B-12C	1320	.012			.64	.71	1.35	1.75
0300	3 C.Y. cap. = 260 C.Y./hr.	B-12D	2080	.008			.41	1.05	1.46	1.77
0305	3-1/2 C.Y. cap. = 300 C.Y./hr.	"	2400	.007			.35	.91	1.26	1.53
0310	Wheel mounted, 1/2 C.Y. cap. = 40 C.Y./hr.	B-12E	320	.050			2.65	1.43	4.08	5.50
0360	3/4 C.Y. cap. = 60 C.Y./hr.	B-12F	480	.033			1.76	1.46	3.22	4.24
0500	Clamshell 1/2 C.Y. cap. = 20 C.Y./hr.	B-12G	160	.100			5.30	5.50	10.80	13.95
0550	1 C.Y. cap. = 35 C.Y./hr.	B-12H	280	.057			3.02	4.36	7.38	9.30
0950	Dragline, 1/2 C.Y. cap. = 30 C.Y./hr.	B-12I	240	.067			3.53	4.48	8.01	10.20
1000	3/4 C.Y. cap. = 35 C.Y./hr.	"	280	.057			3.02	3.84	6.86	8.75
1050	1-1/2 C.Y. cap. = 65 C.Y./hr.	B-12P	520	.031			1.63	2.49	4.12	5.15
1200	Front end loader, track mtd., 1-1/2 C.Y. cap. = 70 C.Y./hr.	B-10N	560	.021			1.16	1.02	2.18	2.85
1250	2-1/2 C.Y. cap. = 95 C.Y./hr.	B-100	760	.016			.85	1.22	2.07	2.61
1300	3 C.Y. cap. = 130 C.Y./hr.	B-10P	1040	.012			.62	1.10	1.72	2.14
1350	5 C.Y. cap. = 160 C.Y./hr.	B-10Q	1280	.009			.51	1.14	1.65	2.01
1500	Wheel mounted, 3/4 C.Y. cap. = 45 C.Y./hr.	B-10R	360	.033			1.80	.85	2.65	3.62
1550	1-1/2 C.Y. cap. = 80 C.Y./hr.	B-10S	640	.019			1.01	.69	1.70	2.27
1600	2-1/4 C.Y. cap. = 100 C.Y./hr.	B-10T	800	.015			.81	.80	1.61	2.09
1650	5 C.Y. cap. = 185 C.Y./hr.	B-10U	1480	.008			.44	.65	1.09	1.37
1800	Hydraulic excavator, truck mtd. 1/2 C.Y. = 30 C.Y./hr.	B-12J	240	.067			3.53	3.54	7.07	9.15
1850	48" bucket, 1 C.Y. = 45 C.Y./hr.	B-12K	360	.044			2.35	2.73	5.08	6.50
3700	Shovel, 1/2 C.Y. cap. = 55 C.Y./hr.	B-12L	440	.036			1.92	1.99	3.91	5.05
3750	3/4 C.Y. cap. = 85 C.Y./hr.	B-12M	680	.024			1.25	1.59	2.84	3.61
3800	1 C.Y. cap. = 120 C.Y./hr.	B-12N	960	.017			.88	1.28	2.16	2.73
3850	1-1/2 C.Y. cap. = 160 C.Y./hr.	B-120	1280	.013			.66	1.03	1.69	2.12
4000	For soft soil or sand, deduct								15%	15%
4100	For heavy soil or stiff clay, add								60%	60%
4200	For wet excavation with clamshell or dragline, add								100%	100%
4250	All other equipment, add								50%	50%

For customer support on your Concrete & Masonry Costs with RSMeans Data, call 800.448.8182.

207

31 23 16.42 Excavating, Bulk Bank Measure

		Crew	Daily Output	Labor-Hours	Unit	Material	2021 Bare Costs Labor	2021 Bare Costs Equipment	Total	Total Incl O&P
4400	Clamshell in sheeting or cofferdam, minimum	B-12H	160	.100	B.C.Y.		5.30	7.65	12.95	16.30
4450	Maximum	"	60	.267	▼		14.10	20.50	34.60	43.50
5000	Excavating, bulk bank measure, sandy clay & loam piled									
5020	For loading onto trucks, add								15%	15%
5100	Excavator, hydraulic, crawler mtd., 1 C.Y. cap. = 120 C.Y./hr.	B-12A	960	.017	B.C.Y.		.88	.87	1.75	2.27
5150	1-1/2 C.Y. cap. = 150 C.Y./hr.	B-12B	1200	.013			.71	.58	1.29	1.69
5300	2 C.Y. cap. = 195 C.Y./hr.	B-12C	1560	.010			.54	.60	1.14	1.47
5400	3 C.Y. cap. = 300 C.Y./hr.	B-12D	2400	.007			.35	.91	1.26	1.53
5500	3-1/2 C.Y. cap. = 350 C.Y./hr.	"	2800	.006			.30	.78	1.08	1.31
5610	Wheel mounted, 1/2 C.Y. cap. = 44 C.Y./hr.	B-12E	352	.045			2.41	1.30	3.71	5
5660	3/4 C.Y. cap. = 66 C.Y./hr.	B-12F	528	.030	▼		1.60	1.33	2.93	3.85
8000	For hauling excavated material, see Section 31 23 23.20									

31 23 16.46 Excavating, Bulk, Dozer

		Crew	Daily Output	Labor-Hours	Unit	Material	2021 Bare Costs Labor	2021 Bare Costs Equipment	Total	Total Incl O&P
0010	**EXCAVATING, BULK, DOZER**									
0011	Open site									
2000	80 HP, 50' haul, sand & gravel	B-10L	460	.026	B.C.Y.		1.41	.88	2.29	3.07
2200	150' haul, sand & gravel		230	.052			2.82	1.76	4.58	6.15
2400	300' haul, sand & gravel	▼	120	.100			5.40	3.38	8.78	11.75
3000	105 HP, 50' haul, sand & gravel	B-10W	700	.017			.93	.92	1.85	2.39
3200	150' haul, sand & gravel		310	.039			2.10	2.07	4.17	5.40
3300	300' haul, sand & gravel	▼	140	.086			4.64	4.58	9.22	11.95
4000	200 HP, 50' haul, sand & gravel	B-10B	1400	.009			.46	1.09	1.55	1.88
4200	150' haul, sand & gravel		595	.020			1.09	2.55	3.64	4.44
4400	300' haul, sand & gravel	▼	310	.039	▼		2.10	4.90	7	8.50

31 23 16.50 Excavation, Bulk, Scrapers

		Crew	Daily Output	Labor-Hours	Unit	Material	2021 Bare Costs Labor	2021 Bare Costs Equipment	Total	Total Incl O&P
0010	**EXCAVATION, BULK, SCRAPERS**									
0100	Elev. scraper 11 C.Y., sand & gravel 1,500' haul, 1/4 dozer	B-33F	690	.020	B.C.Y.		1.11	2.18	3.29	4.06
0150	3,000' haul		610	.023			1.26	2.47	3.73	4.59
0200	5,000' haul		505	.028			1.52	2.98	4.50	5.55
0300	Common earth, 1,500' haul		600	.023			1.28	2.51	3.79	4.67
0350	3,000' haul		530	.026			1.45	2.84	4.29	5.30
0400	5,000' haul		440	.032			1.74	3.42	5.16	6.35
0410	Sandy clay & loam, 1,500' haul		648	.022			1.18	2.32	3.50	4.33
0420	3,000' haul		572	.024			1.34	2.63	3.97	4.89
0430	5,000' haul		475	.029			1.62	3.17	4.79	5.90
0500	Clay, 1,500' haul		375	.037			2.05	4.01	6.06	7.45
0550	3,000' haul		330	.042			2.33	4.56	6.89	8.45
0600	5,000' haul	▼	275	.051	▼		2.79	5.45	8.24	10.15
1000	Self propelled scraper, 14 C.Y., 1/4 push dozer									
1050	Sand and gravel, 1,500' haul	B-33D	920	.015	B.C.Y.		.83	3.12	3.95	4.67
1100	3,000' haul		805	.017			.95	3.57	4.52	5.35
1200	5,000' haul		645	.022			1.19	4.45	5.64	6.65
1300	Common earth, 1,500' haul		800	.018			.96	3.59	4.55	5.40
1350	3,000' haul		700	.020			1.10	4.10	5.20	6.15
1400	5,000' haul		560	.025			1.37	5.15	6.52	7.70
1420	Sandy clay & loam, 1,500' haul		864	.016			.89	3.32	4.21	4.97
1430	3,000' haul		786	.018			.98	3.65	4.63	5.50
1440	5,000' haul		605	.023			1.27	4.74	6.01	7.10
1500	Clay, 1,500' haul		500	.028			1.54	5.75	7.29	8.60
1550	3,000' haul		440	.032			1.74	6.50	8.24	9.80
1600	5,000' haul	▼	350	.040			2.19	8.20	10.39	12.25
2000	21 C.Y., 1/4 push dozer, sand & gravel, 1,500' haul	B-33E	1180	.012	▼		.65	2.63	3.28	3.86

31 23 16 – Excavation

31 23 16.50 Excavation, Bulk, Scrapers	Crew	Daily Output	Labor-Hours	Unit	Material	2021 Bare Costs Labor	2021 Bare Costs Equipment	Total	Total Incl O&P
2100	3,000' haul	B-33E	910	.015	B.C.Y.		.84	3.41	4.25	5
2200	5,000' haul		750	.019			1.02	4.14	5.16	6.10
2300	Common earth, 1,500' haul		1030	.014			.75	3.01	3.76	4.42
2350	3,000' haul		790	.018			.97	3.93	4.90	5.75
2400	5,000' haul		650	.022			1.18	4.77	5.95	7
2420	Sandy clay & loam, 1,500' haul		1112	.013			.69	2.79	3.48	4.10
2430	3,000' haul		854	.016			.90	3.63	4.53	5.35
2440	5,000' haul		702	.020			1.09	4.42	5.51	6.50
2500	Clay, 1,500' haul		645	.022			1.19	4.81	6	7.05
2550	3,000' haul		495	.028			1.55	6.25	7.80	9.20
2600	5,000' haul		405	.035			1.90	7.65	9.55	11.25
2700	Towed, 10 C.Y., 1/4 push dozer, sand & gravel, 1,500' haul	B-33B	560	.025			1.37	4.27	5.64	6.75
2720	3,000' haul		450	.031			1.71	5.30	7.01	8.40
2730	5,000' haul		365	.038			2.10	6.55	8.65	10.35
2750	Common earth, 1,500' haul		420	.033			1.83	5.70	7.53	8.95
2770	3,000' haul		400	.035			1.92	6	7.92	9.45
2780	5,000' haul		310	.045			2.48	7.70	10.18	12.20
2785	Sandy clay & loam, 1,500' haul		454	.031			1.69	5.25	6.94	8.30
2790	3,000' haul		432	.032			1.78	5.55	7.33	8.75
2795	5,000' haul		340	.041			2.26	7.05	9.31	11.10
2800	Clay, 1,500' haul		315	.044			2.44	7.60	10.04	12
2820	3,000' haul		300	.047			2.56	8	10.56	12.55
2840	5,000' haul		225	.062			3.41	10.65	14.06	16.80
2900	15 C.Y., 1/4 push dozer, sand & gravel, 1,500' haul	B-33C	800	.018			.96	3.01	3.97	4.74
2920	3,000' haul		640	.022			1.20	3.77	4.97	5.95
2940	5,000' haul		520	.027			1.48	4.63	6.11	7.30
2960	Common earth, 1,500' haul		600	.023			1.28	4.02	5.30	6.35
2980	3,000' haul		560	.025			1.37	4.30	5.67	6.75
3000	5,000' haul		440	.032			1.74	5.50	7.24	8.65
3005	Sandy clay & loam, 1,500' haul		648	.022			1.18	3.72	4.90	5.85
3010	3,000' haul		605	.023			1.27	3.98	5.25	6.25
3015	5,000' haul		475	.029			1.62	5.05	6.67	8
3020	Clay, 1,500' haul		450	.031			1.71	5.35	7.06	8.45
3040	3,000' haul		420	.033			1.83	5.75	7.58	9
3060	5,000' haul		320	.044			2.40	7.55	9.95	11.85

31 23 19 – Dewatering

31 23 19.20 Dewatering Systems	Crew	Daily Output	Labor-Hours	Unit	Material	2021 Bare Costs Labor	2021 Bare Costs Equipment	Total	Total Incl O&P
0010	**DEWATERING SYSTEMS**									
0020	Excavate drainage trench, 2' wide, 2' deep	B-11C	90	.178	C.Y.		9.20	2.40	11.60	16.35
0100	2' wide, 3' deep, with backhoe loader	"	135	.119			6.15	1.60	7.75	10.90
0200	Excavate sump pits by hand, light soil	1 Clab	7.10	1.127			50		50	74.50
0300	Heavy soil	"	3.50	2.286			101		101	151
0500	Pumping 8 hrs., attended 2 hrs./day, incl. 20 L.F.									
0550	of suction hose & 100 L.F. discharge hose									
0600	2" diaphragm pump used for 8 hrs.	B-10H	4	3	Day		162	25	187	270
0650	4" diaphragm pump used for 8 hrs.	B-10I	4	3			162	37.50	199.50	283
0800	8 hrs. attended, 2" diaphragm pump	B-10H	1	12			650	99	749	1,075
0900	3" centrifugal pump	B-10J	1	12			650	92	742	1,075
1000	4" diaphragm pump	B-10I	1	12			650	149	799	1,125
1100	6" centrifugal pump	B-10K	1	12			650	297	947	1,300
1300	CMP, incl. excavation 3' deep, 12" diameter	B-6	115	.209	L.F.	12	10.05	1.88	23.93	30
1400	18" diameter		100	.240	"	18.75	11.55	2.16	32.46	40

For customer support on your Concrete & Masonry Costs with RSMeans Data, call 800.448.8182.

209

31 23 Excavation and Fill

31 23 19 – Dewatering

31 23 19.20 Dewatering Systems

		Crew	Daily Output	Labor-Hours	Unit	Material	2021 Bare Costs Labor	Equipment	Total	Total Incl O&P
1600	Sump hole construction, incl. excavation and gravel, pit	B-6	1250	.019	C.F.	1.12	.92	.17	2.21	2.81
1700	With 12" gravel collar, 12" pipe, corrugated, 16 ga.		70	.343	L.F.	23.50	16.50	3.09	43.09	54
1800	15" pipe, corrugated, 16 ga.		55	.436		31	21	3.93	55.93	70
1900	18" pipe, corrugated, 16 ga.		50	.480		35.50	23	4.32	62.82	79
2000	24" pipe, corrugated, 14 ga.		40	.600		43	29	5.40	77.40	96
2200	Wood lining, up to 4' x 4', add		300	.080	SFCA	15.95	3.85	.72	20.52	24

31 23 19.40 Wellpoints

		Crew	Daily Output	Labor-Hours	Unit	Material	2021 Bare Costs Labor	Equipment	Total	Total Incl O&P
0010	**WELLPOINTS**									
0011	For equipment rental, see 01 54 33 in Reference Section									
0100	Installation and removal of single stage system									
0110	Labor only, 0.75 labor-hours per L.F.	1 Clab	10.70	.748	LF Hdr		33		33	49.50
0200	2.0 labor-hours per L.F.	"	4	2	"		89		89	133
0400	Pump operation, 4 @ 6 hr. shifts									
0410	Per 24 hr. day	4 Eqlt	1.27	25.197	Day		1,400		1,400	2,075
0500	Per 168 hr. week, 160 hr. straight, 8 hr. double time		.18	178	Week		9,875		9,875	14,700
0550	Per 4.3 week month		.04	800	Month		44,400		44,400	66,000
0600	Complete installation, operation, equipment rental, fuel &									
0610	removal of system with 2" wellpoints 5' OC									
0700	100' long header, 6" diameter, first month	4 Eqlt	3.23	9.907	LF Hdr	160	550		710	995
0800	Thereafter, per month		4.13	7.748		128	430		558	780
1000	200' long header, 8" diameter, first month		6	5.333		153	296		449	610
1100	Thereafter, per month		8.39	3.814		72	212		284	395
1300	500' long header, 8" diameter, first month		10.63	3.010		56	167		223	310
1400	Thereafter, per month		20.91	1.530		40	85		125	171
1600	1,000' long header, 10" diameter, first month		11.62	2.754		48	153		201	281
1700	Thereafter, per month		41.81	.765		24	42.50		66.50	90
1900	Note: above figures include pumping 168 hrs. per week,									
1910	the pump operator, and one stand-by pump.									

31 23 23 – Fill

31 23 23.13 Backfill

		Crew	Daily Output	Labor-Hours	Unit	Material	2021 Bare Costs Labor	Equipment	Total	Total Incl O&P
0010	**BACKFILL** R312323-30									
0015	By hand, no compaction, light soil	1 Clab	14	.571	L.C.Y.		25.50		25.50	38
0100	Heavy soil		11	.727	"		32.50		32.50	48
0300	Compaction in 6" layers, hand tamp, add to above		20.60	.388	E.C.Y.		17.25		17.25	25.50
0400	Roller compaction operator walking, add	B-10A	100	.120			6.50	1.67	8.17	11.55
0500	Air tamp, add	B-9D	190	.211			9.45	1.73	11.18	15.95
0600	Vibrating plate, add	A-1D	60	.133			5.90	.53	6.43	9.45
0800	Compaction in 12" layers, hand tamp, add to above	1 Clab	34	.235			10.45		10.45	15.60
0900	Roller compaction operator walking, add	B-10A	150	.080			4.33	1.11	5.44	7.65
1000	Air tamp, add	B-9	285	.140			6.30	1.25	7.55	10.75
1100	Vibrating plate, add	A-1E	90	.089			3.95	1.84	5.79	7.90
1300	Dozer backfilling, bulk, up to 300' haul, no compaction	B-10B	1200	.010	L.C.Y.		.54	1.27	1.81	2.20
1400	Air tamped, add	B-11B	80	.200	E.C.Y.		10	5.35	15.35	21
1600	Compacting backfill, 6" to 12" lifts, vibrating roller	B-10C	800	.015			.81	2.55	3.36	4.02
1700	Sheepsfoot roller	B-10D	750	.016			.87	2.60	3.47	4.15
1900	Dozer backfilling, trench, up to 300' haul, no compaction	B-10B	900	.013	L.C.Y.		.72	1.69	2.41	2.94
2000	Air tamped, add	B-11B	80	.200	E.C.Y.		10	5.35	15.35	21
2200	Compacting backfill, 6" to 12" lifts, vibrating roller	B-10C	700	.017			.93	2.91	3.84	4.59
2300	Sheepsfoot roller	B-10D	650	.018			1	3	4	4.78
3000	For flowable fill, see Section 03 31 13.35									

31 23 23 – Fill

31 23 23.16 Fill By Borrow and Utility Bedding	Crew	Daily Output	Labor-Hours	Unit	Material	2021 Bare Costs Labor	2021 Bare Costs Equipment	Total	Total Incl O&P
0010 **FILL BY BORROW AND UTILITY BEDDING**									
0015 Fill by borrow, load, 1 mile haul, spread with dozer									
0020 for embankments	B-15	1200	.023	L.C.Y.	13.50	1.23	2.23	16.96	19.15
0035 Select fill for shoulders & embankments	"	1200	.023	"	30	1.23	2.23	33.46	37.50
0040 Fill,for hauling over 1 mile,add to above per C.Y., see Section 31 23 23.20				Mile				1.41	1.73

31 23 23.17 General Fill

	Crew	Daily Output	Labor-Hours	Unit	Material	Labor	Equipment	Total	Total Incl O&P
0010 **GENERAL FILL**									
0011 Spread dumped material, no compaction									
0020 By dozer	B-10B	1000	.012	L.C.Y.		.65	1.52	2.17	2.64
0100 By hand	1 Clab	12	.667	"		29.50		29.50	44
0500 Gravel fill, compacted, under floor slabs, 4" deep	B-37	10000	.005	S.F.	.60	.22	.03	.85	1.01
0600 6" deep		8600	.006		.89	.26	.03	1.18	1.40
0700 9" deep		7200	.007		1.49	.31	.04	1.84	2.14
0800 12" deep		6000	.008		2.08	.37	.04	2.49	2.90
1000 Alternate pricing method, 4" deep		120	.400	E.C.Y.	44.50	18.65	2.16	65.31	79.50
1100 6" deep		160	.300		44.50	13.95	1.62	60.07	72
1200 9" deep		200	.240		44.50	11.20	1.29	56.99	67
1300 12" deep		220	.218		44.50	10.15	1.18	55.83	65.50
1500 For fill under exterior paving, see Section 32 11 23.23									
1600 For flowable fill, see Section 03 31 13.35									

31 23 23.20 Hauling

	Crew	Daily Output	Labor-Hours	Unit	Material	Labor	Equipment	Total	Total Incl O&P
0010 **HAULING**									
0011 Excavated or borrow, loose cubic yards									
0012 no loading equipment, including hauling, waiting, loading/dumping									
0013 time per cycle (wait, load, travel, unload or dump & return)									
0014 8 C.Y. truck, 15 MPH avg., cycle 0.5 miles, 10 min. wait/ld./uld.	B-34A	320	.025	L.C.Y.		1.28	1.27	2.55	3.32
0016 cycle 1 mile		272	.029			1.51	1.50	3.01	3.91
0018 cycle 2 miles		208	.038			1.97	1.96	3.93	5.10
0020 cycle 4 miles		144	.056			2.85	2.83	5.68	7.35
0022 cycle 6 miles		112	.071			3.66	3.64	7.30	9.50
0024 cycle 8 miles		88	.091			4.66	4.63	9.29	12.05
0026 20 MPH avg., cycle 0.5 mile		336	.024			1.22	1.21	2.43	3.16
0028 cycle 1 mile		296	.027			1.39	1.38	2.77	3.58
0030 cycle 2 miles		240	.033			1.71	1.70	3.41	4.43
0032 cycle 4 miles		176	.045			2.33	2.32	4.65	6.05
0034 cycle 6 miles		136	.059			3.02	3	6.02	7.80
0036 cycle 8 miles		112	.071			3.66	3.64	7.30	9.50
0044 25 MPH avg., cycle 4 miles		192	.042			2.14	2.12	4.26	5.55
0046 cycle 6 miles		160	.050			2.57	2.55	5.12	6.65
0048 cycle 8 miles		128	.063			3.21	3.18	6.39	8.30
0050 30 MPH avg., cycle 4 miles		216	.037			1.90	1.89	3.79	4.92
0052 cycle 6 miles		176	.045			2.33	2.32	4.65	6.05
0054 cycle 8 miles		144	.056			2.85	2.83	5.68	7.35
0114 15 MPH avg., cycle 0.5 mile, 15 min. wait/ld./uld.		224	.036			1.83	1.82	3.65	4.74
0116 cycle 1 mile		200	.040			2.05	2.04	4.09	5.30
0118 cycle 2 miles		168	.048			2.44	2.43	4.87	6.30
0120 cycle 4 miles		120	.067			3.42	3.40	6.82	8.85
0122 cycle 6 miles		96	.083			4.28	4.25	8.53	11.05
0124 cycle 8 miles		80	.100			5.15	5.10	10.25	13.25
0126 20 MPH avg., cycle 0.5 mile		232	.034			1.77	1.76	3.53	4.57
0128 cycle 1 mile		208	.038			1.97	1.96	3.93	5.10
0130 cycle 2 miles		184	.043			2.23	2.22	4.45	5.75

For customer support on your Concrete & Masonry Costs with RSMeans Data, call 800.448.8182.

211

31 23 23.20 Hauling		Crew	Daily Output	Labor-Hours	Unit	Material	2021 Bare Costs		Total	Total Incl O&P
							Labor	Equipment		
0132	cycle 4 miles	B-34A	144	.056	L.C.Y.		2.85	2.83	5.68	7.35
0134	cycle 6 miles		112	.071			3.66	3.64	7.30	9.50
0136	cycle 8 miles		96	.083			4.28	4.25	8.53	11.05
0144	25 MPH avg., cycle 4 miles		152	.053			2.70	2.68	5.38	7
0146	cycle 6 miles		128	.063			3.21	3.18	6.39	8.30
0148	cycle 8 miles		112	.071			3.66	3.64	7.30	9.50
0150	30 MPH avg., cycle 4 miles		168	.048			2.44	2.43	4.87	6.30
0152	cycle 6 miles		144	.056			2.85	2.83	5.68	7.35
0154	cycle 8 miles		120	.067			3.42	3.40	6.82	8.85
0214	15 MPH avg., cycle 0.5 mile, 20 min. wait/ld./uld.		176	.045			2.33	2.32	4.65	6.05
0216	cycle 1 mile		160	.050			2.57	2.55	5.12	6.65
0218	cycle 2 miles		136	.059			3.02	3	6.02	7.80
0220	cycle 4 miles		104	.077			3.95	3.92	7.87	10.20
0222	cycle 6 miles		88	.091			4.66	4.63	9.29	12.05
0224	cycle 8 miles		72	.111			5.70	5.65	11.35	14.75
0226	20 MPH avg., cycle 0.5 mile		176	.045			2.33	2.32	4.65	6.05
0228	cycle 1 mile		168	.048			2.44	2.43	4.87	6.30
0230	cycle 2 miles		144	.056			2.85	2.83	5.68	7.35
0232	cycle 4 miles		120	.067			3.42	3.40	6.82	8.85
0234	cycle 6 miles		96	.083			4.28	4.25	8.53	11.05
0236	cycle 8 miles		88	.091			4.66	4.63	9.29	12.05
0244	25 MPH avg., cycle 4 miles		128	.063			3.21	3.18	6.39	8.30
0246	cycle 6 miles		112	.071			3.66	3.64	7.30	9.50
0248	cycle 8 miles		96	.083			4.28	4.25	8.53	11.05
0250	30 MPH avg., cycle 4 miles		136	.059			3.02	3	6.02	7.80
0252	cycle 6 miles		120	.067			3.42	3.40	6.82	8.85
0254	cycle 8 miles		104	.077			3.95	3.92	7.87	10.20
0314	15 MPH avg., cycle 0.5 mile, 25 min. wait/ld./uld.		144	.056			2.85	2.83	5.68	7.35
0316	cycle 1 mile		128	.063			3.21	3.18	6.39	8.30
0318	cycle 2 miles		112	.071			3.66	3.64	7.30	9.50
0320	cycle 4 miles		96	.083			4.28	4.25	8.53	11.05
0322	cycle 6 miles		80	.100			5.15	5.10	10.25	13.25
0324	cycle 8 miles		64	.125			6.40	6.35	12.75	16.60
0326	20 MPH avg., cycle 0.5 mile		144	.056			2.85	2.83	5.68	7.35
0328	cycle 1 mile		136	.059			3.02	3	6.02	7.80
0330	cycle 2 miles		120	.067			3.42	3.40	6.82	8.85
0332	cycle 4 miles		104	.077			3.95	3.92	7.87	10.20
0334	cycle 6 miles		88	.091			4.66	4.63	9.29	12.05
0336	cycle 8 miles		80	.100			5.15	5.10	10.25	13.25
0344	25 MPH avg., cycle 4 miles		112	.071			3.66	3.64	7.30	9.50
0346	cycle 6 miles		96	.083			4.28	4.25	8.53	11.05
0348	cycle 8 miles		88	.091			4.66	4.63	9.29	12.05
0350	30 MPH avg., cycle 4 miles		112	.071			3.66	3.64	7.30	9.50
0352	cycle 6 miles		104	.077			3.95	3.92	7.87	10.20
0354	cycle 8 miles		96	.083			4.28	4.25	8.53	11.05
0414	15 MPH avg., cycle 0.5 mile, 30 min. wait/ld./uld.		120	.067			3.42	3.40	6.82	8.85
0416	cycle 1 mile		112	.071			3.66	3.64	7.30	9.50
0418	cycle 2 miles		96	.083			4.28	4.25	8.53	11.05
0420	cycle 4 miles		80	.100			5.15	5.10	10.25	13.25
0422	cycle 6 miles		72	.111			5.70	5.65	11.35	14.75
0424	cycle 8 miles		64	.125			6.40	6.35	12.75	16.60
0426	20 MPH avg., cycle 0.5 mile		120	.067			3.42	3.40	6.82	8.85
0428	cycle 1 mile		112	.071			3.66	3.64	7.30	9.50

31 23 Excavation and Fill

31 23 23 – Fill

31 23 23.20 Hauling		Crew	Daily Output	Labor-Hours	Unit	Material	2021 Bare Costs Labor	2021 Bare Costs Equipment	Total	Total Incl O&P
0430	cycle 2 miles	B-34A	104	.077	L.C.Y.		3.95	3.92	7.87	10.20
0432	cycle 4 miles		88	.091			4.66	4.63	9.29	12.05
0434	cycle 6 miles		80	.100			5.15	5.10	10.25	13.25
0436	cycle 8 miles		72	.111			5.70	5.65	11.35	14.75
0444	25 MPH avg., cycle 4 miles		96	.083			4.28	4.25	8.53	11.05
0446	cycle 6 miles		88	.091			4.66	4.63	9.29	12.05
0448	cycle 8 miles		80	.100			5.15	5.10	10.25	13.25
0450	30 MPH avg., cycle 4 miles		96	.083			4.28	4.25	8.53	11.05
0452	cycle 6 miles		88	.091			4.66	4.63	9.29	12.05
0454	cycle 8 miles		80	.100			5.15	5.10	10.25	13.25
0514	15 MPH avg., cycle 0.5 mile, 35 min. wait/ld./uld.		104	.077			3.95	3.92	7.87	10.20
0516	cycle 1 mile		96	.083			4.28	4.25	8.53	11.05
0518	cycle 2 miles		88	.091			4.66	4.63	9.29	12.05
0520	cycle 4 miles		72	.111			5.70	5.65	11.35	14.75
0522	cycle 6 miles		64	.125			6.40	6.35	12.75	16.60
0524	cycle 8 miles		56	.143			7.35	7.30	14.65	18.95
0526	20 MPH avg., cycle 0.5 mile		104	.077			3.95	3.92	7.87	10.20
0528	cycle 1 mile		96	.083			4.28	4.25	8.53	11.05
0530	cycle 2 miles		96	.083			4.28	4.25	8.53	11.05
0532	cycle 4 miles		80	.100			5.15	5.10	10.25	13.25
0534	cycle 6 miles		72	.111			5.70	5.65	11.35	14.75
0536	cycle 8 miles		64	.125			6.40	6.35	12.75	16.60
0544	25 MPH avg., cycle 4 miles		88	.091			4.66	4.63	9.29	12.05
0546	cycle 6 miles		80	.100			5.15	5.10	10.25	13.25
0548	cycle 8 miles		72	.111			5.70	5.65	11.35	14.75
0550	30 MPH avg., cycle 4 miles		88	.091			4.66	4.63	9.29	12.05
0552	cycle 6 miles		80	.100			5.15	5.10	10.25	13.25
0554	cycle 8 miles		72	.111			5.70	5.65	11.35	14.75
1014	12 C.Y. truck, cycle 0.5 mile, 15 MPH avg., 15 min. wait/ld./uld.	B-34B	336	.024			1.22	1.72	2.94	3.73
1016	cycle 1 mile		300	.027			1.37	1.93	3.30	4.17
1018	cycle 2 miles		252	.032			1.63	2.30	3.93	4.96
1020	cycle 4 miles		180	.044			2.28	3.22	5.50	6.95
1022	cycle 6 miles		144	.056			2.85	4.02	6.87	8.70
1024	cycle 8 miles		120	.067			3.42	4.83	8.25	10.40
1025	cycle 10 miles		96	.083			4.28	6.05	10.33	13.05
1026	20 MPH avg., cycle 0.5 mile		348	.023			1.18	1.66	2.84	3.59
1028	cycle 1 mile		312	.026			1.32	1.86	3.18	4.01
1030	cycle 2 miles		276	.029			1.49	2.10	3.59	4.53
1032	cycle 4 miles		216	.037			1.90	2.68	4.58	5.80
1034	cycle 6 miles		168	.048			2.44	3.45	5.89	7.45
1036	cycle 8 miles		144	.056			2.85	4.02	6.87	8.70
1038	cycle 10 miles		120	.067			3.42	4.83	8.25	10.40
1040	25 MPH avg., cycle 4 miles		228	.035			1.80	2.54	4.34	5.50
1042	cycle 6 miles		192	.042			2.14	3.02	5.16	6.50
1044	cycle 8 miles		168	.048			2.44	3.45	5.89	7.45
1046	cycle 10 miles		144	.056			2.85	4.02	6.87	8.70
1050	30 MPH avg., cycle 4 miles		252	.032			1.63	2.30	3.93	4.96
1052	cycle 6 miles		216	.037			1.90	2.68	4.58	5.80
1054	cycle 8 miles		180	.044			2.28	3.22	5.50	6.95
1056	cycle 10 miles		156	.051			2.63	3.71	6.34	8
1060	35 MPH avg., cycle 4 miles		264	.030			1.55	2.19	3.74	4.73
1062	cycle 6 miles		228	.035			1.80	2.54	4.34	5.50
1064	cycle 8 miles		204	.039			2.01	2.84	4.85	6.15

31 23 23.20 Hauling		Crew	Daily Output	Labor-Hours	Unit	Material	2021 Bare Costs Labor	Equipment	Total	Total Incl O&P
1066	cycle 10 miles	B-34B	180	.044	L.C.Y.		2.28	3.22	5.50	6.95
1068	cycle 20 miles		120	.067			3.42	4.83	8.25	10.40
1069	cycle 30 miles		84	.095			4.89	6.90	11.79	14.90
1070	cycle 40 miles		72	.111			5.70	8.05	13.75	17.35
1072	40 MPH avg., cycle 6 miles		240	.033			1.71	2.41	4.12	5.20
1074	cycle 8 miles		216	.037			1.90	2.68	4.58	5.80
1076	cycle 10 miles		192	.042			2.14	3.02	5.16	6.50
1078	cycle 20 miles		120	.067			3.42	4.83	8.25	10.40
1080	cycle 30 miles		96	.083			4.28	6.05	10.33	13.05
1082	cycle 40 miles		72	.111			5.70	8.05	13.75	17.35
1084	cycle 50 miles		60	.133			6.85	9.65	16.50	21
1094	45 MPH avg., cycle 8 miles		216	.037			1.90	2.68	4.58	5.80
1096	cycle 10 miles		204	.039			2.01	2.84	4.85	6.15
1098	cycle 20 miles		132	.061			3.11	4.39	7.50	9.50
1100	cycle 30 miles		108	.074			3.80	5.35	9.15	11.60
1102	cycle 40 miles		84	.095			4.89	6.90	11.79	14.90
1104	cycle 50 miles		72	.111			5.70	8.05	13.75	17.35
1106	50 MPH avg., cycle 10 miles		216	.037			1.90	2.68	4.58	5.80
1108	cycle 20 miles		144	.056			2.85	4.02	6.87	8.70
1110	cycle 30 miles		108	.074			3.80	5.35	9.15	11.60
1112	cycle 40 miles		84	.095			4.89	6.90	11.79	14.90
1114	cycle 50 miles		72	.111			5.70	8.05	13.75	17.35
1214	15 MPH avg., cycle 0.5 mile, 20 min. wait/ld./uld.		264	.030			1.55	2.19	3.74	4.73
1216	cycle 1 mile		240	.033			1.71	2.41	4.12	5.20
1218	cycle 2 miles		204	.039			2.01	2.84	4.85	6.15
1220	cycle 4 miles		156	.051			2.63	3.71	6.34	8
1222	cycle 6 miles		132	.061			3.11	4.39	7.50	9.50
1224	cycle 8 miles		108	.074			3.80	5.35	9.15	11.60
1225	cycle 10 miles		96	.083			4.28	6.05	10.33	13.05
1226	20 MPH avg., cycle 0.5 mile		264	.030			1.55	2.19	3.74	4.73
1228	cycle 1 mile		252	.032			1.63	2.30	3.93	4.96
1230	cycle 2 miles		216	.037			1.90	2.68	4.58	5.80
1232	cycle 4 miles		180	.044			2.28	3.22	5.50	6.95
1234	cycle 6 miles		144	.056			2.85	4.02	6.87	8.70
1236	cycle 8 miles		132	.061			3.11	4.39	7.50	9.50
1238	cycle 10 miles		108	.074			3.80	5.35	9.15	11.60
1240	25 MPH avg., cycle 4 miles		192	.042			2.14	3.02	5.16	6.50
1242	cycle 6 miles		168	.048			2.44	3.45	5.89	7.45
1244	cycle 8 miles		144	.056			2.85	4.02	6.87	8.70
1246	cycle 10 miles		132	.061			3.11	4.39	7.50	9.50
1250	30 MPH avg., cycle 4 miles		204	.039			2.01	2.84	4.85	6.15
1252	cycle 6 miles		180	.044			2.28	3.22	5.50	6.95
1254	cycle 8 miles		156	.051			2.63	3.71	6.34	8
1256	cycle 10 miles		144	.056			2.85	4.02	6.87	8.70
1260	35 MPH avg., cycle 4 miles		216	.037			1.90	2.68	4.58	5.80
1262	cycle 6 miles		192	.042			2.14	3.02	5.16	6.50
1264	cycle 8 miles		168	.048			2.44	3.45	5.89	7.45
1266	cycle 10 miles		156	.051			2.63	3.71	6.34	8
1268	cycle 20 miles		108	.074			3.80	5.35	9.15	11.60
1269	cycle 30 miles		72	.111			5.70	8.05	13.75	17.35
1270	cycle 40 miles		60	.133			6.85	9.65	16.50	21
1272	40 MPH avg., cycle 6 miles		192	.042			2.14	3.02	5.16	6.50
1274	cycle 8 miles		180	.044			2.28	3.22	5.50	6.95

31 23 23 – Fill

31 23 23.20 Hauling		Crew	Daily Output	Labor-Hours	Unit	Material	2021 Bare Costs Labor	Equipment	Total	Total Incl O&P
1276	cycle 10 miles	B-34B	156	.051	L.C.Y.		2.63	3.71	6.34	8
1278	cycle 20 miles		108	.074			3.80	5.35	9.15	11.60
1280	cycle 30 miles		84	.095			4.89	6.90	11.79	14.90
1282	cycle 40 miles		72	.111			5.70	8.05	13.75	17.35
1284	cycle 50 miles		60	.133			6.85	9.65	16.50	21
1294	45 MPH avg., cycle 8 miles		180	.044			2.28	3.22	5.50	6.95
1296	cycle 10 miles		168	.048			2.44	3.45	5.89	7.45
1298	cycle 20 miles		120	.067			3.42	4.83	8.25	10.40
1300	cycle 30 miles		96	.083			4.28	6.05	10.33	13.05
1302	cycle 40 miles		72	.111			5.70	8.05	13.75	17.35
1304	cycle 50 miles		60	.133			6.85	9.65	16.50	21
1306	50 MPH avg., cycle 10 miles		180	.044			2.28	3.22	5.50	6.95
1308	cycle 20 miles		132	.061			3.11	4.39	7.50	9.50
1310	cycle 30 miles		96	.083			4.28	6.05	10.33	13.05
1312	cycle 40 miles		84	.095			4.89	6.90	11.79	14.90
1314	cycle 50 miles		72	.111			5.70	8.05	13.75	17.35
1414	15 MPH avg., cycle 0.5 mile, 25 min. wait/ld./uld.		204	.039			2.01	2.84	4.85	6.15
1416	cycle 1 mile		192	.042			2.14	3.02	5.16	6.50
1418	cycle 2 miles		168	.048			2.44	3.45	5.89	7.45
1420	cycle 4 miles		132	.061			3.11	4.39	7.50	9.50
1422	cycle 6 miles		120	.067			3.42	4.83	8.25	10.40
1424	cycle 8 miles		96	.083			4.28	6.05	10.33	13.05
1425	cycle 10 miles		84	.095			4.89	6.90	11.79	14.90
1426	20 MPH avg., cycle 0.5 mile		216	.037			1.90	2.68	4.58	5.80
1428	cycle 1 mile		204	.039			2.01	2.84	4.85	6.15
1430	cycle 2 miles		180	.044			2.28	3.22	5.50	6.95
1432	cycle 4 miles		156	.051			2.63	3.71	6.34	8
1434	cycle 6 miles		132	.061			3.11	4.39	7.50	9.50
1436	cycle 8 miles		120	.067			3.42	4.83	8.25	10.40
1438	cycle 10 miles		96	.083			4.28	6.05	10.33	13.05
1440	25 MPH avg., cycle 4 miles		168	.048			2.44	3.45	5.89	7.45
1442	cycle 6 miles		144	.056			2.85	4.02	6.87	8.70
1444	cycle 8 miles		132	.061			3.11	4.39	7.50	9.50
1446	cycle 10 miles		108	.074			3.80	5.35	9.15	11.60
1450	30 MPH avg., cycle 4 miles		168	.048			2.44	3.45	5.89	7.45
1452	cycle 6 miles		156	.051			2.63	3.71	6.34	8
1454	cycle 8 miles		132	.061			3.11	4.39	7.50	9.50
1456	cycle 10 miles		120	.067			3.42	4.83	8.25	10.40
1460	35 MPH avg., cycle 4 miles		180	.044			2.28	3.22	5.50	6.95
1462	cycle 6 miles		156	.051			2.63	3.71	6.34	8
1464	cycle 8 miles		144	.056			2.85	4.02	6.87	8.70
1466	cycle 10 miles		132	.061			3.11	4.39	7.50	9.50
1468	cycle 20 miles		96	.083			4.28	6.05	10.33	13.05
1469	cycle 30 miles		72	.111			5.70	8.05	13.75	17.35
1470	cycle 40 miles		60	.133			6.85	9.65	16.50	21
1472	40 MPH avg., cycle 6 miles		168	.048			2.44	3.45	5.89	7.45
1474	cycle 8 miles		156	.051			2.63	3.71	6.34	8
1476	cycle 10 miles		144	.056			2.85	4.02	6.87	8.70
1478	cycle 20 miles		96	.083			4.28	6.05	10.33	13.05
1480	cycle 30 miles		84	.095			4.89	6.90	11.79	14.90
1482	cycle 40 miles		60	.133			6.85	9.65	16.50	21
1484	cycle 50 miles		60	.133			6.85	9.65	16.50	21
1494	45 MPH avg., cycle 8 miles		156	.051			2.63	3.71	6.34	8

31 23 23.20 Hauling		Crew	Daily Output	Labor-Hours	Unit	Material	2021 Bare Costs Labor	2021 Bare Costs Equipment	Total	Total Incl O&P
1496	cycle 10 miles	B-34B	144	.056	L.C.Y.		2.85	4.02	6.87	8.70
1498	cycle 20 miles		108	.074			3.80	5.35	9.15	11.60
1500	cycle 30 miles		84	.095			4.89	6.90	11.79	14.90
1502	cycle 40 miles		72	.111			5.70	8.05	13.75	17.35
1504	cycle 50 miles		60	.133			6.85	9.65	16.50	21
1506	50 MPH avg., cycle 10 miles		156	.051			2.63	3.71	6.34	8
1508	cycle 20 miles		120	.067			3.42	4.83	8.25	10.40
1510	cycle 30 miles		96	.083			4.28	6.05	10.33	13.05
1512	cycle 40 miles		72	.111			5.70	8.05	13.75	17.35
1514	cycle 50 miles		60	.133			6.85	9.65	16.50	21
1614	15 MPH avg., cycle 0.5 mile, 30 min. wait/ld./uld.		180	.044			2.28	3.22	5.50	6.95
1616	cycle 1 mile		168	.048			2.44	3.45	5.89	7.45
1618	cycle 2 miles		144	.056			2.85	4.02	6.87	8.70
1620	cycle 4 miles		120	.067			3.42	4.83	8.25	10.40
1622	cycle 6 miles		108	.074			3.80	5.35	9.15	11.60
1624	cycle 8 miles		84	.095			4.89	6.90	11.79	14.90
1625	cycle 10 miles		84	.095			4.89	6.90	11.79	14.90
1626	20 MPH avg., cycle 0.5 mile		180	.044			2.28	3.22	5.50	6.95
1628	cycle 1 mile		168	.048			2.44	3.45	5.89	7.45
1630	cycle 2 miles		156	.051			2.63	3.71	6.34	8
1632	cycle 4 miles		132	.061			3.11	4.39	7.50	9.50
1634	cycle 6 miles		120	.067			3.42	4.83	8.25	10.40
1636	cycle 8 miles		108	.074			3.80	5.35	9.15	11.60
1638	cycle 10 miles		96	.083			4.28	6.05	10.33	13.05
1640	25 MPH avg., cycle 4 miles		144	.056			2.85	4.02	6.87	8.70
1642	cycle 6 miles		132	.061			3.11	4.39	7.50	9.50
1644	cycle 8 miles		108	.074			3.80	5.35	9.15	11.60
1646	cycle 10 miles		108	.074			3.80	5.35	9.15	11.60
1650	30 MPH avg., cycle 4 miles		144	.056			2.85	4.02	6.87	8.70
1652	cycle 6 miles		132	.061			3.11	4.39	7.50	9.50
1654	cycle 8 miles		120	.067			3.42	4.83	8.25	10.40
1656	cycle 10 miles		108	.074			3.80	5.35	9.15	11.60
1660	35 MPH avg., cycle 4 miles		156	.051			2.63	3.71	6.34	8
1662	cycle 6 miles		144	.056			2.85	4.02	6.87	8.70
1664	cycle 8 miles		132	.061			3.11	4.39	7.50	9.50
1666	cycle 10 miles		120	.067			3.42	4.83	8.25	10.40
1668	cycle 20 miles		84	.095			4.89	6.90	11.79	14.90
1669	cycle 30 miles		72	.111			5.70	8.05	13.75	17.35
1670	cycle 40 miles		60	.133			6.85	9.65	16.50	21
1672	40 MPH avg., cycle 6 miles		144	.056			2.85	4.02	6.87	8.70
1674	cycle 8 miles		132	.061			3.11	4.39	7.50	9.50
1676	cycle 10 miles		120	.067			3.42	4.83	8.25	10.40
1678	cycle 20 miles		96	.083			4.28	6.05	10.33	13.05
1680	cycle 30 miles		72	.111			5.70	8.05	13.75	17.35
1682	cycle 40 miles		60	.133			6.85	9.65	16.50	21
1684	cycle 50 miles		48	.167			8.55	12.05	20.60	26
1694	45 MPH avg., cycle 8 miles		144	.056			2.85	4.02	6.87	8.70
1696	cycle 10 miles		132	.061			3.11	4.39	7.50	9.50
1698	cycle 20 miles		96	.083			4.28	6.05	10.33	13.05
1700	cycle 30 miles		84	.095			4.89	6.90	11.79	14.90
1702	cycle 40 miles		60	.133			6.85	9.65	16.50	21
1704	cycle 50 miles		60	.133			6.85	9.65	16.50	21
1706	50 MPH avg., cycle 10 miles		132	.061			3.11	4.39	7.50	9.50

31 23 23.20 Hauling		Crew	Daily Output	Labor-Hours	Unit	Material	2021 Bare Costs Labor	Equipment	Total	Total Incl O&P
1708	cycle 20 miles	B-34B	108	.074	L.C.Y.		3.80	5.35	9.15	11.60
1710	cycle 30 miles		84	.095			4.89	6.90	11.79	14.90
1712	cycle 40 miles		72	.111			5.70	8.05	13.75	17.35
1714	cycle 50 miles		60	.133			6.85	9.65	16.50	21
2000	Hauling, 8 C.Y. truck, small project cost per hour	B-34A	8	1	Hr.		51.50	51	102.50	133
2100	12 C.Y. truck	B-34B	8	1			51.50	72.50	124	156
2150	16.5 C.Y. truck	B-34C	8	1			51.50	79.50	131	164
2175	18 C.Y. 8 wheel truck	B-34I	8	1			51.50	94	145.50	181
2200	20 C.Y. truck	B-34D	8	1			51.50	81.50	133	167
2300	Grading at dump, or embankment if required, by dozer	B-10B	1000	.012	L.C.Y.		.65	1.52	2.17	2.64
2310	Spotter at fill or cut, if required	1 Clab	8	1	Hr.		44.50		44.50	66.50
9014	18 C.Y. truck, 8 wheels,15 min. wait/ld./uld.,15 MPH, cycle 0.5 mi.	B-34I	504	.016	L.C.Y.		.81	1.50	2.31	2.86
9016	cycle 1 mile		450	.018			.91	1.67	2.58	3.20
9018	cycle 2 miles		378	.021			1.09	1.99	3.08	3.81
9020	cycle 4 miles		270	.030			1.52	2.79	4.31	5.35
9022	cycle 6 miles		216	.037			1.90	3.49	5.39	6.70
9024	cycle 8 miles		180	.044			2.28	4.19	6.47	8
9025	cycle 10 miles		144	.056			2.85	5.25	8.10	10
9026	20 MPH avg., cycle 0.5 mile		522	.015			.79	1.44	2.23	2.77
9028	cycle 1 mile		468	.017			.88	1.61	2.49	3.08
9030	cycle 2 miles		414	.019			.99	1.82	2.81	3.48
9032	cycle 4 miles		324	.025			1.27	2.33	3.60	4.45
9034	cycle 6 miles		252	.032			1.63	2.99	4.62	5.70
9036	cycle 8 miles		216	.037			1.90	3.49	5.39	6.70
9038	cycle 10 miles		180	.044			2.28	4.19	6.47	8
9040	25 MPH avg., cycle 4 miles		342	.023			1.20	2.20	3.40	4.21
9042	cycle 6 miles		288	.028			1.43	2.62	4.05	5
9044	cycle 8 miles		252	.032			1.63	2.99	4.62	5.70
9046	cycle 10 miles		216	.037			1.90	3.49	5.39	6.70
9050	30 MPH avg., cycle 4 miles		378	.021			1.09	1.99	3.08	3.81
9052	cycle 6 miles		324	.025			1.27	2.33	3.60	4.45
9054	cycle 8 miles		270	.030			1.52	2.79	4.31	5.35
9056	cycle 10 miles		234	.034			1.75	3.22	4.97	6.15
9060	35 MPH avg., cycle 4 miles		396	.020			1.04	1.90	2.94	3.64
9062	cycle 6 miles		342	.023			1.20	2.20	3.40	4.21
9064	cycle 8 miles		288	.028			1.43	2.62	4.05	5
9066	cycle 10 miles		270	.030			1.52	2.79	4.31	5.35
9068	cycle 20 miles		162	.049			2.53	4.65	7.18	8.90
9070	cycle 30 miles		126	.063			3.26	6	9.26	11.45
9072	cycle 40 miles		90	.089			4.56	8.35	12.91	16
9074	40 MPH avg., cycle 6 miles		360	.022			1.14	2.09	3.23	4
9076	cycle 8 miles		324	.025			1.27	2.33	3.60	4.45
9078	cycle 10 miles		288	.028			1.43	2.62	4.05	5
9080	cycle 20 miles		180	.044			2.28	4.19	6.47	8
9082	cycle 30 miles		144	.056			2.85	5.25	8.10	10
9084	cycle 40 miles		108	.074			3.80	7	10.80	13.40
9086	cycle 50 miles		90	.089			4.56	8.35	12.91	16
9094	45 MPH avg., cycle 8 miles		324	.025			1.27	2.33	3.60	4.45
9096	cycle 10 miles		306	.026			1.34	2.46	3.80	4.72
9098	cycle 20 miles		198	.040			2.07	3.81	5.88	7.30
9100	cycle 30 miles		144	.056			2.85	5.25	8.10	10
9102	cycle 40 miles		126	.063			3.26	6	9.26	11.45
9104	cycle 50 miles		108	.074			3.80	7	10.80	13.40

217

31 23 23 – Fill

31 23 23.20 Hauling		Crew	Daily Output	Labor-Hours	Unit	Material	2021 Bare Costs Labor	Equipment	Total	Total Incl O&P
9106	50 MPH avg., cycle 10 miles	B-34I	324	.025	L.C.Y.		1.27	2.33	3.60	4.45
9108	cycle 20 miles		216	.037			1.90	3.49	5.39	6.70
9110	cycle 30 miles		162	.049			2.53	4.65	7.18	8.90
9112	cycle 40 miles		126	.063			3.26	6	9.26	11.45
9114	cycle 50 miles		108	.074			3.80	7	10.80	13.40
9214	20 min. wait/ld./uld.,15 MPH, cycle 0.5 mi.		396	.020			1.04	1.90	2.94	3.64
9216	cycle 1 mile		360	.022			1.14	2.09	3.23	4
9218	cycle 2 miles		306	.026			1.34	2.46	3.80	4.72
9220	cycle 4 miles		234	.034			1.75	3.22	4.97	6.15
9222	cycle 6 miles		198	.040			2.07	3.81	5.88	7.30
9224	cycle 8 miles		162	.049			2.53	4.65	7.18	8.90
9225	cycle 10 miles		144	.056			2.85	5.25	8.10	10
9226	20 MPH avg., cycle 0.5 mile		396	.020			1.04	1.90	2.94	3.64
9228	cycle 1 mile		378	.021			1.09	1.99	3.08	3.81
9230	cycle 2 miles		324	.025			1.27	2.33	3.60	4.45
9232	cycle 4 miles		270	.030			1.52	2.79	4.31	5.35
9234	cycle 6 miles		216	.037			1.90	3.49	5.39	6.70
9236	cycle 8 miles		198	.040			2.07	3.81	5.88	7.30
9238	cycle 10 miles		162	.049			2.53	4.65	7.18	8.90
9240	25 MPH avg., cycle 4 miles		288	.028			1.43	2.62	4.05	5
9242	cycle 6 miles		252	.032			1.63	2.99	4.62	5.70
9244	cycle 8 miles		216	.037			1.90	3.49	5.39	6.70
9246	cycle 10 miles		198	.040			2.07	3.81	5.88	7.30
9250	30 MPH avg., cycle 4 miles		306	.026			1.34	2.46	3.80	4.72
9252	cycle 6 miles		270	.030			1.52	2.79	4.31	5.35
9254	cycle 8 miles		234	.034			1.75	3.22	4.97	6.15
9256	cycle 10 miles		216	.037			1.90	3.49	5.39	6.70
9260	35 MPH avg., cycle 4 miles		324	.025			1.27	2.33	3.60	4.45
9262	cycle 6 miles		288	.028			1.43	2.62	4.05	5
9264	cycle 8 miles		252	.032			1.63	2.99	4.62	5.70
9266	cycle 10 miles		234	.034			1.75	3.22	4.97	6.15
9268	cycle 20 miles		162	.049			2.53	4.65	7.18	8.90
9270	cycle 30 miles		108	.074			3.80	7	10.80	13.40
9272	cycle 40 miles		90	.089			4.56	8.35	12.91	16
9274	40 MPH avg., cycle 6 miles		288	.028			1.43	2.62	4.05	5
9276	cycle 8 miles		270	.030			1.52	2.79	4.31	5.35
9278	cycle 10 miles		234	.034			1.75	3.22	4.97	6.15
9280	cycle 20 miles		162	.049			2.53	4.65	7.18	8.90
9282	cycle 30 miles		126	.063			3.26	6	9.26	11.45
9284	cycle 40 miles		108	.074			3.80	7	10.80	13.40
9286	cycle 50 miles		90	.089			4.56	8.35	12.91	16
9294	45 MPH avg., cycle 8 miles		270	.030			1.52	2.79	4.31	5.35
9296	cycle 10 miles		252	.032			1.63	2.99	4.62	5.70
9298	cycle 20 miles		180	.044			2.28	4.19	6.47	8
9300	cycle 30 miles		144	.056			2.85	5.25	8.10	10
9302	cycle 40 miles		108	.074			3.80	7	10.80	13.40
9304	cycle 50 miles		90	.089			4.56	8.35	12.91	16
9306	50 MPH avg., cycle 10 miles		270	.030			1.52	2.79	4.31	5.35
9308	cycle 20 miles		198	.040			2.07	3.81	5.88	7.30
9310	cycle 30 miles		144	.056			2.85	5.25	8.10	10
9312	cycle 40 miles		126	.063			3.26	6	9.26	11.45
9314	cycle 50 miles		108	.074			3.80	7	10.80	13.40
9414	25 min. wait/ld./uld.,15 MPH, cycle 0.5 mi.		306	.026			1.34	2.46	3.80	4.72

31 23 23 – Fill

31 23 23.20 Hauling		Crew	Daily Output	Labor-Hours	Unit	Material	2021 Bare Costs Labor	2021 Bare Costs Equipment	Total	Total Incl O&P
9416	cycle 1 mile	B-34I	288	.028	L.C.Y.		1.43	2.62	4.05	5
9418	cycle 2 miles		252	.032			1.63	2.99	4.62	5.70
9420	cycle 4 miles		198	.040			2.07	3.81	5.88	7.30
9422	cycle 6 miles		180	.044			2.28	4.19	6.47	8
9424	cycle 8 miles		144	.056			2.85	5.25	8.10	10
9425	cycle 10 miles		126	.063			3.26	6	9.26	11.45
9426	20 MPH avg., cycle 0.5 mile		324	.025			1.27	2.33	3.60	4.45
9428	cycle 1 mile		306	.026			1.34	2.46	3.80	4.72
9430	cycle 2 miles		270	.030			1.52	2.79	4.31	5.35
9432	cycle 4 miles		234	.034			1.75	3.22	4.97	6.15
9434	cycle 6 miles		198	.040			2.07	3.81	5.88	7.30
9436	cycle 8 miles		180	.044			2.28	4.19	6.47	8
9438	cycle 10 miles		144	.056			2.85	5.25	8.10	10
9440	25 MPH avg., cycle 4 miles		252	.032			1.63	2.99	4.62	5.70
9442	cycle 6 miles		216	.037			1.90	3.49	5.39	6.70
9444	cycle 8 miles		198	.040			2.07	3.81	5.88	7.30
9446	cycle 10 miles		180	.044			2.28	4.19	6.47	8
9450	30 MPH avg., cycle 4 miles		252	.032			1.63	2.99	4.62	5.70
9452	cycle 6 miles		234	.034			1.75	3.22	4.97	6.15
9454	cycle 8 miles		198	.040			2.07	3.81	5.88	7.30
9456	cycle 10 miles		180	.044			2.28	4.19	6.47	8
9460	35 MPH avg., cycle 4 miles		270	.030			1.52	2.79	4.31	5.35
9462	cycle 6 miles		234	.034			1.75	3.22	4.97	6.15
9464	cycle 8 miles		216	.037			1.90	3.49	5.39	6.70
9466	cycle 10 miles		198	.040			2.07	3.81	5.88	7.30
9468	cycle 20 miles		144	.056			2.85	5.25	8.10	10
9470	cycle 30 miles		108	.074			3.80	7	10.80	13.40
9472	cycle 40 miles		90	.089			4.56	8.35	12.91	16
9474	40 MPH avg., cycle 6 miles		252	.032			1.63	2.99	4.62	5.70
9476	cycle 8 miles		234	.034			1.75	3.22	4.97	6.15
9478	cycle 10 miles		216	.037			1.90	3.49	5.39	6.70
9480	cycle 20 miles		144	.056			2.85	5.25	8.10	10
9482	cycle 30 miles		126	.063			3.26	6	9.26	11.45
9484	cycle 40 miles		90	.089			4.56	8.35	12.91	16
9486	cycle 50 miles		90	.089			4.56	8.35	12.91	16
9494	45 MPH avg., cycle 8 miles		234	.034			1.75	3.22	4.97	6.15
9496	cycle 10 miles		216	.037			1.90	3.49	5.39	6.70
9498	cycle 20 miles		162	.049			2.53	4.65	7.18	8.90
9500	cycle 30 miles		126	.063			3.26	6	9.26	11.45
9502	cycle 40 miles		108	.074			3.80	7	10.80	13.40
9504	cycle 50 miles		90	.089			4.56	8.35	12.91	16
9506	50 MPH avg., cycle 10 miles		234	.034			1.75	3.22	4.97	6.15
9508	cycle 20 miles		180	.044			2.28	4.19	6.47	8
9510	cycle 30 miles		144	.056			2.85	5.25	8.10	10
9512	cycle 40 miles		108	.074			3.80	7	10.80	13.40
9514	cycle 50 miles		90	.089			4.56	8.35	12.91	16
9614	30 min. wait/ld./uld.,15 MPH, cycle 0.5 mi.		270	.030			1.52	2.79	4.31	5.35
9616	cycle 1 mile		252	.032			1.63	2.99	4.62	5.70
9618	cycle 2 miles		216	.037			1.90	3.49	5.39	6.70
9620	cycle 4 miles		180	.044			2.28	4.19	6.47	8
9622	cycle 6 miles		162	.049			2.53	4.65	7.18	8.90
9624	cycle 8 miles		126	.063			3.26	6	9.26	11.45
9625	cycle 10 miles		126	.063			3.26	6	9.26	11.45

31 23 23.20 Hauling

	Crew	Daily Output	Labor-Hours	Unit	Material	2021 Bare Costs Labor	Equipment	Total	Total Incl O&P
9626	20 MPH avg., cycle 0.5 mile	B-34I	270	.030	L.C.Y.		1.52	2.79	4.31	5.35
9628	cycle 1 mile		252	.032			1.63	2.99	4.62	5.70
9630	cycle 2 miles		234	.034			1.75	3.22	4.97	6.15
9632	cycle 4 miles		198	.040			2.07	3.81	5.88	7.30
9634	cycle 6 miles		180	.044			2.28	4.19	6.47	8
9636	cycle 8 miles		162	.049			2.53	4.65	7.18	8.90
9638	cycle 10 miles		144	.056			2.85	5.25	8.10	10
9640	25 MPH avg., cycle 4 miles		216	.037			1.90	3.49	5.39	6.70
9642	cycle 6 miles		198	.040			2.07	3.81	5.88	7.30
9644	cycle 8 miles		180	.044			2.28	4.19	6.47	8
9646	cycle 10 miles		162	.049			2.53	4.65	7.18	8.90
9650	30 MPH avg., cycle 4 miles		216	.037			1.90	3.49	5.39	6.70
9652	cycle 6 miles		198	.040			2.07	3.81	5.88	7.30
9654	cycle 8 miles		180	.044			2.28	4.19	6.47	8
9656	cycle 10 miles		162	.049			2.53	4.65	7.18	8.90
9660	35 MPH avg., cycle 4 miles		234	.034			1.75	3.22	4.97	6.15
9662	cycle 6 miles		216	.037			1.90	3.49	5.39	6.70
9664	cycle 8 miles		198	.040			2.07	3.81	5.88	7.30
9666	cycle 10 miles		180	.044			2.28	4.19	6.47	8
9668	cycle 20 miles		126	.063			3.26	6	9.26	11.45
9670	cycle 30 miles		108	.074			3.80	7	10.80	13.40
9672	cycle 40 miles		90	.089			4.56	8.35	12.91	16
9674	40 MPH avg., cycle 6 miles		216	.037			1.90	3.49	5.39	6.70
9676	cycle 8 miles		198	.040			2.07	3.81	5.88	7.30
9678	cycle 10 miles		180	.044			2.28	4.19	6.47	8
9680	cycle 20 miles		144	.056			2.85	5.25	8.10	10
9682	cycle 30 miles		108	.074			3.80	7	10.80	13.40
9684	cycle 40 miles		90	.089			4.56	8.35	12.91	16
9686	cycle 50 miles		72	.111			5.70	10.45	16.15	20
9694	45 MPH avg., cycle 8 miles		216	.037			1.90	3.49	5.39	6.70
9696	cycle 10 miles		198	.040			2.07	3.81	5.88	7.30
9698	cycle 20 miles		144	.056			2.85	5.25	8.10	10
9700	cycle 30 miles		126	.063			3.26	6	9.26	11.45
9702	cycle 40 miles		108	.074			3.80	7	10.80	13.40
9704	cycle 50 miles		90	.089			4.56	8.35	12.91	16
9706	50 MPH avg., cycle 10 miles		198	.040			2.07	3.81	5.88	7.30
9708	cycle 20 miles		162	.049			2.53	4.65	7.18	8.90
9710	cycle 30 miles		126	.063			3.26	6	9.26	11.45
9712	cycle 40 miles		108	.074			3.80	7	10.80	13.40
9714	cycle 50 miles		90	.089			4.56	8.35	12.91	16

31 23 23.24 Compaction, Structural

		Crew	Daily Output	Labor-Hours	Unit	Material	2021 Bare Costs Labor	Equipment	Total	Total Incl O&P
0010	**COMPACTION, STRUCTURAL** R312323-30									
0020	Steel wheel tandem roller, 5 tons	B-10E	8	1.500	Hr.		81	32.50	113.50	157
0100	10 tons	B-10F	8	1.500	"		81	31	112	155
0300	Sheepsfoot or wobbly wheel roller, 8" lifts, common fill	B-10G	1300	.009	E.C.Y.		.50	1.05	1.55	1.89
0400	Select fill	"	1500	.008			.43	.91	1.34	1.65
0600	Vibratory plate, 8" lifts, common fill	A-1D	200	.040			1.78	.16	1.94	2.83
0700	Select fill	"	216	.037			1.64	.15	1.79	2.61

220

For customer support on your Concrete & Masonry Costs with RSMeans Data, call 800.448.8182.

31 25 Erosion and Sedimentation Controls

31 25 14 – Stabilization Measures for Erosion and Sedimentation Control

31 25 14.16 Rolled Erosion Control Mats and Blankets

		Crew	Daily Output	Labor-Hours	Unit	Material	2021 Bare Costs Labor	Equipment	Total	Total Incl O&P
0010	**ROLLED EROSION CONTROL MATS AND BLANKETS**									
0020	Jute mesh, 100 S.Y. per roll, 4' wide, stapled [G]	B-80A	2400	.010	S.Y.	1	.44	.35	1.79	2.15
0060	Polyethylene 3 dimensional geomatrix, 50 mil thick [G]		700	.034		3.47	1.52	1.21	6.20	7.45
0062	120 mil thick [G]		515	.047		7.45	2.07	1.65	11.17	13.10
0070	Paper biodegradable mesh [G]	B-1	2500	.010		.14	.43		.57	.80
0080	Paper mulch [G]	B-64	20000	.001		.14	.04	.02	.20	.23
0100	Plastic netting, stapled, 2" x 1" mesh, 20 mil [G]	B-1	2500	.010		.31	.43		.74	.99
0120	Revegetation mat, webbed [G]	2 Clab	1000	.016		2.72	.71		3.43	4.05
0200	Polypropylene mesh, stapled, 6.5 oz./S.Y. [G]	B-1	2500	.010		1.83	.43		2.26	2.66
0300	Tobacco netting, or jute mesh #2, stapled [G]	"	2500	.010		.26	.43		.69	.94
0400	Soil sealant, liquid sprayed from truck [G]	B-81	5000	.005		.35	.25	.11	.71	.88
0600	Straw in polymeric netting, biodegradable log	A-2	1000	.024	L.F.	2.25	1.10	.20	3.55	4.34
0705	Sediment Log, Filter Sock, 9"		1000	.024		2.25	1.10	.20	3.55	4.34
0710	Sediment Log, Filter Sock, 12"		1000	.024		3.50	1.10	.20	4.80	5.70
1000	Silt fence, install and remove [G]	B-62	650	.037		.44	1.78	.28	2.50	3.43

31 32 Soil Stabilization

31 32 13 – Soil Mixing Stabilization

31 32 13.30 Calcium Chloride

		Crew	Daily Output	Labor-Hours	Unit	Material	2021 Bare Costs Labor	Equipment	Total	Total Incl O&P
0010	**CALCIUM CHLORIDE**									
0020	Calcium chloride, delivered, 100 lb. bags, truckload lots				Ton	705			705	775
0030	Solution, 4 lb. flake per gallon, tank truck delivery				Gal.	1.69			1.69	1.86

31 36 Gabions

31 36 13 – Gabion Boxes

31 36 13.10 Gabion Box Systems

		Crew	Daily Output	Labor-Hours	Unit	Material	2021 Bare Costs Labor	Equipment	Total	Total Incl O&P
0010	**GABION BOX SYSTEMS**									
0400	Gabions, galvanized steel mesh mats or boxes, stone filled, 6" deep	B-13	200	.280	S.Y.	17.45	13.50	2.93	33.88	42.50
0500	9" deep		163	.344		26	16.60	3.60	46.20	57
0600	12" deep		153	.366		35.50	17.65	3.84	56.99	69.50
0700	18" deep		102	.549		45	26.50	5.75	77.25	95.50
0800	36" deep		60	.933		73	45	9.80	127.80	158

31 37 Riprap

31 37 13 – Machined Riprap

31 37 13.10 Riprap and Rock Lining

		Crew	Daily Output	Labor-Hours	Unit	Material	2021 Bare Costs Labor	Equipment	Total	Total Incl O&P
0010	**RIPRAP AND ROCK LINING**									
0011	Random, broken stone									
0100	Machine placed for slope protection	B-12G	62	.258	L.C.Y.	32	13.65	14.15	59.80	71
0110	3/8 to 1/4 C.Y. pieces, grouted	B-13	80	.700	S.Y.	67.50	34	7.35	108.85	133
0200	18" minimum thickness, not grouted	"	53	1.057	"	19.85	51	11.05	81.90	110
0300	Dumped, 50 lb. average	B-11A	800	.020	Ton	28.50	1.03	1.90	31.43	35
0350	100 lb. average		700	.023		28.50	1.18	2.17	31.85	35.50
0370	300 lb. average		600	.027		28.50	1.38	2.53	32.41	36.50

For customer support on your Concrete & Masonry Costs with RSMeans Data, call 800.448.8182.

221

31 41 Shoring

31 41 13 – Timber Shoring

31 41 13.10 Building Shoring

		Crew	Daily Output	Labor-Hours	Unit	Material	2021 Bare Costs Labor	Equipment	Total	Total Incl O&P
0010	**BUILDING SHORING**									
0020	Shoring, existing building, with timber, no salvage allowance	B-51	2.20	21.818	M.B.F.	1,250	990	90.50	2,330.50	2,950
1000	On cribbing with 35 ton screw jacks, per box and jack	"	3.60	13.333	Jack	65	605	55	725	1,025

31 41 16 – Sheet Piling

31 41 16.10 Sheet Piling Systems

		Crew	Daily Output	Labor-Hours	Unit	Material	2021 Bare Costs Labor	Equipment	Total	Total Incl O&P
0010	**SHEET PILING SYSTEMS**									
0020	Sheet piling, 50,000 psi steel, not incl. wales, 22 psf, left in place	B-40	10.81	5.920	Ton	1,900	340	325	2,565	2,950
0100	Drive, extract & salvage		6	10.667		530	610	590	1,730	2,150
0300	20' deep excavation, 27 psf, left in place		12.95	4.942		1,900	282	273	2,455	2,800
0400	Drive, extract & salvage		6.55	9.771		530	560	540	1,630	2,025
0600	25' deep excavation, 38 psf, left in place		19	3.368		1,900	192	186	2,278	2,575
0700	Drive, extract & salvage		10.50	6.095		530	350	335	1,215	1,475
0900	40' deep excavation, 38 psf, left in place		21.20	3.019		1,900	172	166	2,238	2,525
1000	Drive, extract & salvage		12.25	5.224		530	298	288	1,116	1,350
1200	15' deep excavation, 22 psf, left in place		983	.065	S.F.	22	3.72	3.59	29.31	33.50
1300	Drive, extract & salvage		545	.117		5.95	6.70	6.50	19.15	24
1500	20' deep excavation, 27 psf, left in place		960	.067		27.50	3.81	3.68	34.99	40.50
1600	Drive, extract & salvage		485	.132		7.70	7.55	7.30	22.55	28
1800	25' deep excavation, 38 psf, left in place		1000	.064		41	3.66	3.53	48.19	54.50
1900	Drive, extract & salvage		553	.116		10.60	6.60	6.40	23.60	29
2100	Rent steel sheet piling and wales, first month				Ton	330			330	365
2200	Per added month					33			33	36.50
2300	Rental piling left in place, add to rental					1,225			1,225	1,350
2500	Wales, connections & struts, 2/3 salvage					530			530	585
2700	High strength piling, 60,000 psi, add					190			190	208
2800	65,000 psi, add					284			284	315
3000	Tie rod, not upset, 1-1/2" to 4" diameter with turnbuckle					2,375			2,375	2,600
3100	No turnbuckle					1,675			1,675	1,850
3300	Upset, 1-3/4" to 4" diameter with turnbuckle					2,500			2,500	2,750
3400	No turnbuckle					2,125			2,125	2,325
3600	Lightweight, 18" to 28" wide, 7 ga., 9.22 psf, and									
3610	9 ga., 8.6 psf, minimum				Lb.	.72			.72	.79
3700	Average					.92			.92	1.01
3750	Maximum					1.01			1.01	1.11
3900	Wood, solid sheeting, incl. wales, braces and spacers, *R314116-40*									
3910	drive, extract & salvage, 8' deep excavation	B-31	330	.121	S.F.	3.34	5.70	.78	9.82	13.05
4000	10' deep, 50 S.F./hr. in & 150 S.F./hr. out		300	.133		3.44	6.25	.85	10.54	14
4100	12' deep, 45 S.F./hr. in & 135 S.F./hr. out		270	.148		3.54	6.95	.95	11.44	15.30
4200	14' deep, 42 S.F./hr. in & 126 S.F./hr. out		250	.160		3.64	7.50	1.02	12.16	16.35
4300	16' deep, 40 S.F./hr. in & 120 S.F./hr. out		240	.167		3.76	7.80	1.07	12.63	16.95
4400	18' deep, 38 S.F./hr. in & 114 S.F./hr. out		230	.174		3.88	8.15	1.11	13.14	17.65
4500	20' deep, 35 S.F./hr. in & 105 S.F./hr. out		210	.190		4.01	8.95	1.22	14.18	19.05
4520	Left in place, 8' deep, 55 S.F./hr.		440	.091		6	4.26	.58	10.84	13.60
4540	10' deep, 50 S.F./hr.		400	.100		6.35	4.69	.64	11.68	14.65
4560	12' deep, 45 S.F./hr.		360	.111		6.70	5.20	.71	12.61	15.90
4565	14' deep, 42 S.F./hr.		335	.119		7.10	5.60	.76	13.46	17
4570	16' deep, 40 S.F./hr.		320	.125		7.50	5.85	.80	14.15	17.90
4580	18' deep, 38 S.F./hr.		305	.131		8	6.15	.84	14.99	18.85
4590	20' deep, 35 S.F./hr.		280	.143		8.60	6.70	.91	16.21	20.50
4700	Alternate pricing, left in place, 8' deep		1.76	22.727	M.B.F.	1,350	1,075	145	2,570	3,225
4800	Drive, extract and salvage, 8' deep		1.32	30.303	"	1,200	1,425	194	2,819	3,675
5000	For treated lumber add cost of treatment to lumber									

31 43 Concrete Raising

31 43 13 – Pressure Grouting

31 43 13.13 Concrete Pressure Grouting

		Crew	Daily Output	Labor-Hours	Unit	Material	2021 Bare Costs Labor	Equipment	Total	Total Incl O&P
0010	**CONCRETE PRESSURE GROUTING**									
0020	Grouting, pressure, cement & sand, 1:1 mix, minimum	B-61	124	.323	Bag	21.50	15.15	2.62	39.27	49
0100	Maximum		51	.784	"	21.50	37	6.35	64.85	85.50
0200	Cement and sand, 1:1 mix, minimum		250	.160	C.F.	43	7.50	1.30	51.80	60
0300	Maximum		100	.400		64.50	18.80	3.25	86.55	103
0400	Epoxy cement grout, minimum		137	.292		805	13.75	2.37	821.12	910
0500	Maximum	↓	57	.702	↓	805	33	5.70	843.70	940
0700	Alternate pricing method: (Add for materials)									
0710	5 person crew and equipment	B-61	1	40	Day		1,875	325	2,200	3,150

31 43 19 – Mechanical Jacking

31 43 19.10 Slabjacking

		Crew	Daily Output	Labor-Hours	Unit	Material	2021 Bare Costs Labor	Equipment	Total	Total Incl O&P
0010	**SLABJACKING**									
0100	4" thick slab	D-4	1500	.021	S.F.	.39	1.05	.13	1.57	2.15
0150	6" thick slab		1200	.027		.55	1.31	.16	2.02	2.74
0200	8" thick slab		1000	.032		.65	1.57	.19	2.41	3.29
0250	10" thick slab		900	.036		.70	1.75	.21	2.66	3.63
0300	12" thick slab	↓	850	.038	↓	.76	1.85	.22	2.83	3.87

31 45 Vibroflotation and Densification

31 45 13 – Vibroflotation

31 45 13.10 Vibroflotation Densification

		Crew	Daily Output	Labor-Hours	Unit	Material	2021 Bare Costs Labor	Equipment	Total	Total Incl O&P
0010	**VIBROFLOTATION DENSIFICATION** R314513-90									
0900	Vibroflotation compacted sand cylinder, minimum	B-60	750	.075	V.L.F.		3.84	3.04	6.88	9.10
0950	Maximum		325	.172			8.85	7.05	15.90	21
1100	Vibro replacement compacted stone cylinder, minimum		500	.112			5.75	4.57	10.32	13.60
1150	Maximum		250	.224	↓		11.50	9.15	20.65	27.50
1300	Mobilization and demobilization, minimum		.47	119	Total		6,125	4,850	10,975	14,500
1400	Maximum	↓	.14	400	"		20,600	16,300	36,900	48,600

31 46 Needle Beams

31 46 13 – Cantilever Needle Beams

31 46 13.10 Needle Beams

		Crew	Daily Output	Labor-Hours	Unit	Material	2021 Bare Costs Labor	Equipment	Total	Total Incl O&P
0010	**NEEDLE BEAMS**									
0011	Incl. wood shoring 10' x 10' opening									
0400	Block, concrete, 8" thick	B-9	7.10	5.634	Ea.	67.50	252	50	369.50	505
0420	12" thick		6.70	5.970		73	267	53	393	540
0800	Brick, 4" thick with 8" backup block		5.70	7.018		73	315	62.50	450.50	620
1000	Brick, solid, 8" thick		6.20	6.452		67.50	289	57.50	414	570
1040	12" thick		4.90	8.163		73	365	72.50	510.50	705
1080	16" thick	↓	4.50	8.889		84	400	79	563	775
2000	Add for additional floors of shoring	B-1	6	4	↓	67.50	180		247.50	345

223

For customer support on your Concrete & Masonry Costs with RSMeans Data, call 800.448.8182.

31 48 Underpinning

31 48 13 – Underpinning Piers

31 48 13.10 Underpinning Foundations	Crew	Daily Output	Labor-Hours	Unit	Material	2021 Bare Costs Labor	Equipment	Total	Total Incl O&P
0010 **UNDERPINNING FOUNDATIONS**									
0011 Including excavation,									
0020 forming, reinforcing, concrete and equipment									
0100 5' to 16' below grade, 100 to 500 C.Y.	B-52	2.30	24.348	C.Y.	345	1,225	249	1,819	2,500
0200 Over 500 C.Y.		2.50	22.400		310	1,125	229	1,664	2,300
0400 16' to 25' below grade, 100 to 500 C.Y.		2	28		380	1,425	286	2,091	2,850
0500 Over 500 C.Y.		2.10	26.667		355	1,350	273	1,978	2,725
0700 26' to 40' below grade, 100 to 500 C.Y.		1.60	35		410	1,775	360	2,545	3,500
0800 Over 500 C.Y.	▼	1.80	31.111	▼	380	1,575	320	2,275	3,125
0900 For under 50 C.Y., add					10%	40%			
1000 For 50 C.Y. to 100 C.Y., add				▼	5%	20%			

31 52 Cofferdams

31 52 16 – Timber Cofferdams

31 52 16.10 Cofferdams

	Crew	Daily Output	Labor-Hours	Unit	Material	2021 Bare Costs Labor	Equipment	Total	Total Incl O&P
0010 **COFFERDAMS**									
0011 Incl. mobilization and temporary sheeting									
0080 Soldier beams & lagging H-piles with 3" wood sheeting									
0090 horizontal between piles, including removal of wales & braces									
0100 No hydrostatic head, 15' deep, 1 line of braces, minimum	B-50	545	.206	S.F.	12.50	11.15	4.79	28.44	36
0200 Maximum		495	.226		13.85	12.30	5.30	31.45	40
0400 15' to 22' deep with 2 lines of braces, 10" H, minimum		360	.311		14.70	16.90	7.25	38.85	49.50
0500 Maximum		330	.339		16.65	18.40	7.90	42.95	55
0700 23' to 35' deep with 3 lines of braces, 12" H, minimum		325	.345		19.20	18.70	8.05	45.95	58.50
0800 Maximum		295	.380		21	20.50	8.85	50.35	64.50
1000 36' to 45' deep with 4 lines of braces, 14" H, minimum		290	.386		21.50	21	9	51.50	65.50
1100 Maximum		265	.423		22.50	23	9.85	55.35	71
1300 No hydrostatic head, left in place, 15' deep, 1 line of braces, min.		635	.176		16.65	9.55	4.11	30.31	37.50
1400 Maximum		575	.195		17.85	10.55	4.54	32.94	40.50
1600 15' to 22' deep with 2 lines of braces, minimum		455	.246		25	13.35	5.75	44.10	54.50
1700 Maximum		415	.270		27.50	14.65	6.30	48.45	60
1900 23' to 35' deep with 3 lines of braces, minimum		420	.267		29.50	14.45	6.20	50.15	61.50
2000 Maximum		380	.295		33	16	6.85	55.85	68
2200 36' to 45' deep with 4 lines of braces, minimum		385	.291		35.50	15.80	6.80	58.10	70.50
2300 Maximum	▼	350	.320		41.50	17.35	7.45	66.30	80.50
2350 Lagging only, 3" thick wood between piles 8' OC, minimum	B-46	400	.120		2.77	6.05	.10	8.92	12.35
2370 Maximum		250	.192		4.16	9.70	.17	14.03	19.50
2400 Open sheeting no bracing, for trenches to 10' deep, min.		1736	.028		1.25	1.40	.02	2.67	3.52
2450 Maximum	▼	1510	.032		1.39	1.60	.03	3.02	4
2500 Tie-back method, add to open sheeting, add, minimum								20%	20%
2550 Maximum				▼				60%	60%
2700 Tie-backs only, based on tie-backs total length, minimum	B-46	86.80	.553	L.F.	19.55	28	.48	48.03	64.50
2750 Maximum		38.50	1.247	"	34.50	63	1.08	98.58	135
3500 Tie-backs only, typical average, 25' long		2	24	Ea.	860	1,200	21	2,081	2,825
3600 35' long	▼	1.58	30.380	"	1,150	1,525	26.50	2,701.50	3,600
6000 See also Section 31 41 16.10									

224

For customer support on your Concrete & Masonry Costs with RSMeans Data, call 800.448.8182.

31 56 Slurry Walls

31 56 23 – Lean Concrete Slurry Walls

31 56 23.20 Slurry Trench

		Crew	Daily Output	Labor-Hours	Unit	Material	2021 Bare Costs Labor	2021 Bare Costs Equipment	Total	Total Incl O&P
0010	**SLURRY TRENCH**									
0011	Excavated slurry trench in wet soils									
0020	backfilled with 3,000 psi concrete, no reinforcing steel									
0050	Minimum	C-7	333	.216	C.F.	9.20	10.35	3.27	22.82	29
0100	Maximum		200	.360	"	15.45	17.25	5.45	38.15	48.50
0200	Alternate pricing method, minimum		150	.480	S.F.	18.45	23	7.25	48.70	63
0300	Maximum		120	.600		27.50	29	9.10	65.60	83.50
0500	Reinforced slurry trench, minimum	B-48	177	.316		13.85	15.75	5.90	35.50	45
0600	Maximum	"	69	.812		46	40.50	15.20	101.70	128
0800	Haul for disposal, 2 mile haul, excavated material, add	B-34B	99	.081	C.Y.		4.15	5.85	10	12.65
0900	Haul bentonite castings for disposal, add	"	40	.200	"		10.25	14.50	24.75	31.50

31 62 Driven Piles

31 62 13 – Concrete Piles

31 62 13.23 Prestressed Concrete Piles

		Crew	Daily Output	Labor-Hours	Unit	Material	2021 Bare Costs Labor	2021 Bare Costs Equipment	Total	Total Incl O&P
0010	**PRESTRESSED CONCRETE PILES**, 200 piles									
0020	Unless specified otherwise, not incl. pile caps or mobilization									
2200	Precast, prestressed, 50' long, cylinder, 12" diam., 2-3/8" wall	B-19	720	.089	V.L.F.	36.50	5.10	2.84	44.44	51
2300	14" diameter, 2-1/2" wall		680	.094		38	5.40	3.01	46.41	53.50
2500	16" diameter, 3" wall		640	.100		51.50	5.70	3.20	60.40	68.50
2600	18" diameter, 3-1/2" wall	B-19A	600	.107		65	6.10	4.87	75.97	86
2800	20" diameter, 4" wall		560	.114		56.50	6.55	5.20	68.25	77.50
2900	24" diameter, 5" wall		520	.123		80.50	7.05	5.60	93.15	105
3100	Precast, prestressed, 40' long, 10" thick, square	B-19	700	.091		14.50	5.20	2.92	22.62	27
3200	12" thick, square		680	.094		26	5.40	3.01	34.41	40
3400	14" thick, square		600	.107		29.50	6.10	3.41	39.01	45.50
3500	Octagonal		640	.100		28.50	5.70	3.20	37.40	43.50
3700	16" thick, square		560	.114		31	6.55	3.65	41.20	48
3800	Octagonal		600	.107		34.50	6.10	3.41	44.01	51
4000	18" thick, square	B-19A	520	.123		53	7.05	5.60	65.65	75.50
4100	Octagonal	B-19	560	.114		46	6.55	3.65	56.20	64.50
4300	20" thick, square	B-19A	480	.133		49	7.60	6.10	62.70	72.50
4400	Octagonal	B-19	520	.123		51.50	7.05	3.93	62.48	71.50
4600	24" thick, square	B-19A	440	.145		70	8.30	6.65	84.95	97
4700	Octagonal	B-19	480	.133		60.50	7.60	4.26	72.36	83.50
4730	Precast, prestressed, 60' long, 10" thick, square		700	.091		15.25	5.20	2.92	23.37	28
4740	12" thick, square (60' long)		680	.094		27	5.40	3.01	35.41	41
4750	Mobilization for 10,000 L.F. pile job, add		3300	.019			1.11	.62	1.73	2.37
4800	25,000 L.F. pile job, add		8500	.008			.43	.24	.67	.92

31 62 16 – Steel Piles

31 62 16.13 Steel Piles

		Crew	Daily Output	Labor-Hours	Unit	Material	2021 Bare Costs Labor	2021 Bare Costs Equipment	Total	Total Incl O&P
0010	**STEEL PILES**									
0100	Step tapered, round, concrete filled									
0110	8" tip, 12" butt, 60 ton capacity, 30' depth	B-19	760	.084	V.L.F.	17.75	4.81	2.69	25.25	30
0120	60' depth with extension		740	.086		37.50	4.94	2.76	45.20	52
0130	80' depth with extensions		700	.091		58.50	5.20	2.92	66.62	75.50
0250	"H" Sections, 50' long, HP8 x 36		640	.100		17.75	5.70	3.20	26.65	31.50
0400	HP10 x 42		610	.105		21	6	3.35	30.35	36
0500	HP10 x 57		610	.105		27	6	3.35	36.35	43
0700	HP12 x 53		590	.108		28	6.20	3.47	37.67	44.50

For customer support on your Concrete & Masonry Costs with RSMeans Data, call 800.448.8182.

225

31 62 Driven Piles

31 62 16 – Steel Piles

31 62 16.13 Steel Piles

31 62 16.13 Steel Piles		Crew	Daily Output	Labor-Hours	Unit	Material	2021 Bare Costs Labor	Equipment	Total	Total Incl O&P
0800	HP12 x 74	B-19A	590	.108	V.L.F.	34.50	6.20	4.95	45.65	53
1000	HP14 x 73		540	.119		36.50	6.75	5.40	48.65	56.50
1100	HP14 x 89		540	.119		46.50	6.75	5.40	58.65	67.50
1300	HP14 x 102		510	.125		53	7.15	5.75	65.90	76
1400	HP14 x 117		510	.125		61.50	7.15	5.75	74.40	85
1600	Splice on standard points, not in leads, 8" or 10"	1 Sswl	5	1.600	Ea.	117	96.50		213.50	278
1700	12" or 14"		4	2		165	121		286	370
1900	Heavy duty points, not in leads, 10" wide		4	2		187	121		308	395
2100	14" wide		3.50	2.286		231	138		369	465

31 62 19 – Timber Piles

31 62 19.10 Wood Piles

31 62 19.10 Wood Piles		Crew	Daily Output	Labor-Hours	Unit	Material	2021 Bare Costs Labor	Equipment	Total	Total Incl O&P
0010	**WOOD PILES**									
0011	Friction or end bearing, not including									
0050	mobilization or demobilization									
0100	ACZA treated piles, 1.0 lb/C.F., up to 30' long, 12" butts, 8" points	B-19	625	.102	V.L.F.	15.05	5.85	3.27	24.17	29
0200	30' to 39' long, 12" butts, 8" points		700	.091		15.90	5.20	2.92	24.02	28.50
0300	40' to 49' long, 12" butts, 7" points		720	.089		16.75	5.10	2.84	24.69	29.50
0400	50' to 59' long, 13" butts, 7" points		800	.080		23.50	4.57	2.56	30.63	36
0500	60' to 69' long, 13" butts, 7" points		840	.076		27	4.35	2.43	33.78	39
0600	70' to 80' long, 13" butts, 6" points		840	.076		27	4.35	2.43	33.78	39.50
0800	ACZA Treated piles, 1.5 lb./C.F.									
0810	friction or end bearing, ASTM class B									
1000	Up to 30' long, 12" butts, 8" points	B-19	625	.102	V.L.F.	17.85	5.85	3.27	26.97	32
1100	30' to 39' long, 12" butts, 8" points		700	.091		18.50	5.20	2.92	26.62	31.50
1200	40' to 49' long, 12" butts, 7" points		720	.089		25.50	5.10	2.84	33.44	39.50
1300	50' to 59' long, 13" butts, 7" points		800	.080		27.50	4.57	2.56	34.63	40
1400	60' to 69' long, 13" butts, 6" points	B-19A	840	.076		25	4.35	3.48	32.83	38
1500	70' to 80' long, 13" butts, 6" points	"	840	.076		26	4.35	3.48	33.83	39
1600	ACZA treated piles, 2.5 lb./C.F.									
1610	8" butts, 10' long	B-19	400	.160	V.L.F.	8.05	9.15	5.10	22.30	28.50
1620	11' to 16' long		500	.128		8.05	7.30	4.09	19.44	24.50
1630	17' to 20' long		575	.111		8.05	6.35	3.56	17.96	22.50
1640	10" butts, 10' to 16' long		500	.128		23.50	7.30	4.09	34.89	41.50
1650	17' to 20' long		575	.111		23.50	6.35	3.56	33.41	39.50
1660	21' to 40' long		700	.091		23.50	5.20	2.92	31.62	37
1670	12" butts, 10' to 20' long		575	.111		29	6.35	3.56	38.91	45.50
1680	21' to 35' long		650	.098		29.50	5.60	3.15	38.25	44
1690	36' to 40' long		700	.091		30	5.20	2.92	38.12	44
1695	14" butts, to 40' long		700	.091		16.70	5.20	2.92	24.82	29.50
1700	Boot for pile tip, minimum	1 Pile	27	.296	Ea.	47	16.55		63.55	77.50
1800	Maximum		21	.381		142	21.50		163.50	189
2000	Point for pile tip, minimum		20	.400		47	22.50		69.50	86.50
2100	Maximum		15	.533		170	30		200	233
2300	Splice for piles over 50' long, minimum	B-46	35	1.371		59	69	1.19	129.19	171
2400	Maximum		20	2.400		77.50	121	2.09	200.59	272
2600	Concrete encasement with wire mesh and tube		331	.145	V.L.F.	77.50	7.30	.13	84.93	97
2700	Mobilization for 10,000 L.F. pile job, add	B-19	3300	.019			1.11	.62	1.73	2.37
2800	25,000 L.F. pile job, add	"	8500	.008			.43	.24	.67	.92

31 62 Driven Piles

31 62 23 – Composite Piles

31 62 23.13 Concrete-Filled Steel Piles

		Crew	Daily Output	Labor-Hours	Unit	Material	2021 Bare Costs Labor	Equipment	Total	Total Incl O&P
0010	**CONCRETE-FILLED STEEL PILES** no mobilization or demobilization									
2600	Pipe piles, 50' L, 8" diam., 29 lb./L.F., no concrete	B-19	500	.128	V.L.F.	21	7.30	4.09	32.39	38.50
2700	Concrete filled		460	.139		26	7.95	4.45	38.40	45.50
2900	10" diameter, 34 lb./L.F., no concrete		500	.128		25.50	7.30	4.09	36.89	43.50
3000	Concrete filled		450	.142		31	8.10	4.54	43.64	51.50
3200	12" diameter, 44 lb./L.F., no concrete		475	.135		33	7.70	4.30	45	53
3300	Concrete filled		415	.154		37.50	8.80	4.93	51.23	60
3500	14" diameter, 46 lb./L.F., no concrete		430	.149		35.50	8.50	4.76	48.76	57
3600	Concrete filled		355	.180		40.50	10.30	5.75	56.55	66.50
3800	16" diameter, 52 lb./L.F., no concrete		385	.166		43	9.50	5.30	57.80	68
3900	Concrete filled		335	.191		55	10.90	6.10	72	84
4100	18" diameter, 59 lb./L.F., no concrete		355	.180		49.50	10.30	5.75	65.55	76.50
4200	Concrete filled	↓	310	.206	↓	63	11.80	6.60	81.40	94
4400	Splices for pipe piles, stl., not in leads, 8" diameter	1 Sswl	5	1.600	Ea.	90	96.50		186.50	248
4410	10" diameter		4.75	1.684		88.50	102		190.50	255
4430	12" diameter		4.50	1.778		111	107		218	288
4500	14" diameter		4.25	1.882		157	114		271	350
4600	16" diameter		4	2		177	121		298	380
4650	18" diameter		3.75	2.133		293	129		422	520
4710	Steel pipe pile backing rings, w/spacer, 8" diameter		12	.667		9.30	40		49.30	72.50
4720	10" diameter		12	.667		12.25	40		52.25	75.50
4730	12" diameter		10	.800		14.75	48		62.75	90.50
4740	14" diameter		9	.889		18.85	53.50		72.35	104
4750	16" diameter		8	1		22.50	60.50		83	118
4760	18" diameter		6	1.333		24.50	80.50		105	151
4800	Points, standard, 8" diameter		4.61	1.735		102	105		207	275
4840	10" diameter		4.45	1.798		134	108		242	315
4880	12" diameter		4.25	1.882		167	114		281	360
4900	14" diameter		4.05	1.975		224	119		343	430
5000	16" diameter		3.37	2.374		315	143		458	565
5050	18" diameter		3.50	2.286		400	138		538	655
5200	Points, heavy duty, 10" diameter		2.90	2.759		256	166		422	540
5240	12" diameter		2.95	2.712		315	164		479	600
5260	14" diameter		2.95	2.712		310	164		474	600
5280	16" diameter		2.95	2.712		375	164		539	665
5290	18" diameter		2.80	2.857	↓	445	172		617	755
5500	For reinforcing steel, add	↓	1150	.007	Lb.	.80	.42		1.22	1.53
5700	For thick wall sections, add				"	.70			.70	.77

31 63 Bored Piles

31 63 26 – Drilled Caissons

31 63 26.13 Fixed End Caisson Piles

		Crew	Daily Output	Labor-Hours	Unit	Material	2021 Bare Costs Labor	Equipment	Total	Total Incl O&P
0010	**FIXED END CAISSON PILES** R316326-60									
0015	Including excavation, concrete, 50 lb. reinforcing									
0020	per C.Y., not incl. mobilization, boulder removal, disposal									
0100	Open style, machine drilled, to 50' deep, in stable ground, no									
0110	casings or ground water, 18" diam., 0.065 C.Y./L.F.	B-43	200	.240	V.L.F.	10.10	11.75	3.84	25.69	33
0200	24" diameter, 0.116 C.Y./L.F.		190	.253		18.05	12.35	4.04	34.44	43
0300	30" diameter, 0.182 C.Y./L.F.		150	.320		28	15.65	5.10	48.75	60
0400	36" diameter, 0.262 C.Y./L.F.	↓	125	.384	↓	40.50	18.80	6.15	65.45	80

31 63 26 – Drilled Caissons

31 63 26.13 Fixed End Caisson Piles		Crew	Daily Output	Labor-Hours	Unit	Material	2021 Bare Costs Labor	Equipment	Total	Total Incl O&P
0500	48" diameter, 0.465 C.Y./L.F.	B-43	100	.480	V.L.F.	72.50	23.50	7.70	103.70	123
0600	60" diameter, 0.727 C.Y./L.F.		90	.533		113	26	8.55	147.55	172
0700	72" diameter, 1.05 C.Y./L.F.		80	.600		163	29.50	9.60	202.10	235
0800	84" diameter, 1.43 C.Y./L.F.		75	.640		222	31.50	10.25	263.75	305
1000	For bell excavation and concrete, add									
1020	4' bell diameter, 24" shaft, 0.444 C.Y.	B-43	20	2.400	Ea.	55.50	117	38.50	211	279
1040	6' bell diameter, 30" shaft, 1.57 C.Y.		5.70	8.421		195	410	135	740	980
1060	8' bell diameter, 36" shaft, 3.72 C.Y.		2.40	20		465	980	320	1,765	2,300
1080	9' bell diameter, 48" shaft, 4.48 C.Y.		2	24		560	1,175	385	2,120	2,800
1100	10' bell diameter, 60" shaft, 5.24 C.Y.		1.70	28.235		650	1,375	450	2,475	3,250
1120	12' bell diameter, 72" shaft, 8.74 C.Y.		1	48		1,100	2,350	770	4,220	5,550
1140	14' bell diameter, 84" shaft, 13.6 C.Y.		.70	68.571		1,700	3,350	1,100	6,150	8,050
1200	Open style, machine drilled, to 50' deep, in wet ground, pulled									
1300	casing and pumping, 18" diameter, 0.065 C.Y./L.F.	B-48	160	.350	V.L.F.	10.10	17.45	6.55	34.10	44.50
1400	24" diameter, 0.116 C.Y./L.F.		125	.448		18.05	22.50	8.40	48.95	62.50
1500	30" diameter, 0.182 C.Y./L.F.		85	.659		28	33	12.30	73.30	93.50
1600	36" diameter, 0.262 C.Y./L.F.		60	.933		40.50	46.50	17.45	104.45	134
1700	48" diameter, 0.465 C.Y./L.F.	B-49	55	1.600		72.50	83.50	29.50	185.50	238
1800	60" diameter, 0.727 C.Y./L.F.		35	2.514		113	131	46.50	290.50	375
1900	72" diameter, 1.05 C.Y./L.F.		30	2.933		163	153	54.50	370.50	470
2000	84" diameter, 1.43 C.Y./L.F.		25	3.520		222	184	65.50	471.50	595
2100	For bell excavation and concrete, add									
2120	4' bell diameter, 24" shaft, 0.444 C.Y.	B-48	19.80	2.828	Ea.	55.50	141	53	249.50	330
2140	6' bell diameter, 30" shaft, 1.57 C.Y.		5.70	9.825		195	490	184	869	1,150
2160	8' bell diameter, 36" shaft, 3.72 C.Y.		2.40	23.333		465	1,175	435	2,075	2,725
2180	9' bell diameter, 48" shaft, 4.48 C.Y.	B-49	3.30	26.667		560	1,400	495	2,455	3,250
2200	10' bell diameter, 60" shaft, 5.24 C.Y.		2.80	31.429		650	1,650	585	2,885	3,825
2220	12' bell diameter, 72" shaft, 8.74 C.Y.		1.60	55		1,100	2,875	1,025	5,000	6,650
2240	14' bell diameter, 84" shaft, 13.6 C.Y.		1	88		1,700	4,600	1,625	7,925	10,600
2300	Open style, machine drilled, to 50' deep, in soft rocks and									
2400	medium hard shales, 18" diameter, 0.065 C.Y./L.F.	B-49	50	1.760	V.L.F.	10.10	92	32.50	134.60	185
2500	24" diameter, 0.116 C.Y./L.F.		30	2.933		18.05	153	54.50	225.55	310
2600	30" diameter, 0.182 C.Y./L.F.		20	4.400		28	230	81.50	339.50	465
2700	36" diameter, 0.262 C.Y./L.F.		15	5.867		40.50	305	109	454.50	625
2800	48" diameter, 0.465 C.Y./L.F.		10	8.800		72.50	460	163	695.50	950
2900	60" diameter, 0.727 C.Y./L.F.		7	12.571		113	655	233	1,001	1,375
3000	72" diameter, 1.05 C.Y./L.F.		6	14.667		163	765	272	1,200	1,625
3100	84" diameter, 1.43 C.Y./L.F.		5	17.600		222	920	325	1,467	1,975
3200	For bell excavation and concrete, add									
3220	4' bell diameter, 24" shaft, 0.444 C.Y.	B-49	10.90	8.073	Ea.	55.50	420	150	625.50	860
3240	6' bell diameter, 30" shaft, 1.57 C.Y.		3.10	28.387		195	1,475	525	2,195	3,025
3260	8' bell diameter, 36" shaft, 3.72 C.Y.		1.30	67.692		465	3,525	1,250	5,240	7,175
3280	9' bell diameter, 48" shaft, 4.48 C.Y.		1.10	80		560	4,175	1,475	6,210	8,525
3300	10' bell diameter, 60" shaft, 5.24 C.Y.		.90	97.778		650	5,100	1,825	7,575	10,400
3320	12' bell diameter, 72" shaft, 8.74 C.Y.		.60	147		1,100	7,675	2,725	11,500	15,700
3340	14' bell diameter, 84" shaft, 13.6 C.Y.		.40	220		1,700	11,500	4,075	17,275	23,700
3600	For rock excavation, sockets, add, minimum		120	.733	C.F.		38.50	13.60	52.10	72.50
3650	Average		95	.926			48.50	17.20	65.70	91.50
3700	Maximum		48	1.833			96	34	130	182
3900	For 50' to 100' deep, add				V.L.F.				7%	7%
4000	For 100' to 150' deep, add								25%	25%
4100	For 150' to 200' deep, add								30%	30%
4200	For casings left in place, add				Lb.	1.39			1.39	1.53

31 63 26 – Drilled Caissons

31 63 26.13 Fixed End Caisson Piles

		Crew	Daily Output	Labor-Hours	Unit	Material	2021 Bare Costs Labor	Equipment	Total	Total Incl O&P
4300	For other than 50 lb. reinf. per C.Y., add or deduct				Lb.	1.31			1.31	1.44
4400	For steel I-beam cores, add	B-49	8.30	10.602	Ton	2,100	555	197	2,852	3,375
4500	Load and haul excess excavation, 2 miles	B-34B	178	.045	L.C.Y.		2.31	3.25	5.56	7.05
5000	Bottom inspection	1 Skwk	1.20	6.667	Ea.		380		380	575

31 63 26.16 Concrete Caissons for Marine Construction

		Crew	Daily Output	Labor-Hours	Unit	Material	2021 Bare Costs Labor	Equipment	Total	Total Incl O&P
0010	**CONCRETE CAISSONS FOR MARINE CONSTRUCTION**									
0100	Caissons, incl. mobilization and demobilization, up to 50 miles									
0200	Uncased shafts, 30 to 80 tons cap., 17" diam., 10' depth	B-44	88	.727	V.L.F.	25	41	23.50	89.50	116
0300	25' depth		165	.388		17.80	22	12.55	52.35	66.50
0400	80 to 150 ton capacity, 22" diameter, 10' depth		80	.800		31	45	26	102	131
0500	20' depth		130	.492		25	27.50	15.90	68.40	87
0700	Cased shafts, 10 to 30 ton capacity, 10-5/8" diam., 20' depth		175	.366		17.80	20.50	11.80	50.10	64
0800	30' depth		240	.267		16.60	14.95	8.60	40.15	50.50
0850	30 to 60 ton capacity, 12" diameter, 20' depth		160	.400		25	22.50	12.90	60.40	75.50
0900	40' depth		230	.278		19.15	15.60	9	43.75	55
1000	80 to 100 ton capacity, 16" diameter, 20' depth		160	.400		35.50	22.50	12.90	70.90	87
1100	40' depth		230	.278		33	15.60	9	57.60	70.50
1200	110 to 140 ton capacity, 17-5/8" diameter, 20' depth		160	.400		38.50	22.50	12.90	73.90	90
1300	40' depth		230	.278		35.50	15.60	9	60.10	73
1400	140 to 175 ton capacity, 19" diameter, 20' depth		130	.492		41.50	27.50	15.90	84.90	105
1500	40' depth		210	.305		38.50	17.10	9.85	65.45	79

31 63 29 – Drilled Concrete Piers and Shafts

31 63 29.13 Uncased Drilled Concrete Piers

		Crew	Daily Output	Labor-Hours	Unit	Material	2021 Bare Costs Labor	Equipment	Total	Total Incl O&P
0010	**UNCASED DRILLED CONCRETE PIERS**									
0020	Unless specified otherwise, not incl. pile caps or mobilization									
0100	Cast in place, thin wall shell pile, straight sided,									
0110	not incl. reinforcing, 8" diam., 16 ga., 5.8 lb./L.F.	B-19	700	.091	V.L.F.	9.60	5.20	2.92	17.72	21.50
0200	10" diameter, 16 ga. corrugated, 7.3 lb./L.F.		650	.098		12.60	5.60	3.15	21.35	26
0300	12" diameter, 16 ga. corrugated, 8.7 lb./L.F.		600	.107		16.30	6.10	3.41	25.81	31
0400	14" diameter, 16 ga. corrugated, 10.0 lb./L.F.		550	.116		19.20	6.65	3.72	29.57	35
0500	16" diameter, 16 ga. corrugated, 11.6 lb./L.F.		500	.128		23.50	7.30	4.09	34.89	41.50
0800	Cast in place friction pile, 50' long, fluted,									
0810	tapered steel, 4,000 psi concrete, no reinforcing									
0900	12" diameter, 7 ga.	B-19	600	.107	V.L.F.	29.50	6.10	3.41	39.01	45.50
1000	14" diameter, 7 ga.		560	.114		32	6.55	3.65	42.20	49.50
1100	16" diameter, 7 ga.		520	.123		38	7.05	3.93	48.98	56.50
1200	18" diameter, 7 ga.		480	.133		44	7.60	4.26	55.86	65
1300	End bearing, fluted, constant diameter,									
1320	4,000 psi concrete, no reinforcing									
1340	12" diameter, 7 ga.	B-19	600	.107	V.L.F.	31	6.10	3.41	40.51	47
1360	14" diameter, 7 ga.		560	.114		39	6.55	3.65	49.20	56.50
1380	16" diameter, 7 ga.		520	.123		45	7.05	3.93	55.98	64.50
1400	18" diameter, 7 ga.		480	.133		49.50	7.60	4.26	61.36	71

31 63 29.20 Cast In Place Piles, Adds

		Crew	Daily Output	Labor-Hours	Unit	Material	2021 Bare Costs Labor	Equipment	Total	Total Incl O&P
0010	**CAST IN PLACE PILES, ADDS**									
1500	For reinforcing steel, add				Lb.	1.24			1.24	1.37
1700	For ball or pedestal end, add	B-19	11	5.818	C.Y.	151	330	186	667	875
1900	For lengths above 60', concrete, add	"	11	5.818	"	158	330	186	674	885
2000	For steel thin shell, pipe only				Lb.	1.54			1.54	1.69

For customer support on your Concrete & Masonry Costs with RSMeans Data, call 800.448.8182.

229

Division Notes

	CREW	DAILY OUTPUT	LABOR-HOURS	UNIT	BARE COSTS				TOTAL INCL O&P
					MAT.	LABOR	EQUIP.	TOTAL	

Estimating Tips

32 01 00 Operations and Maintenance of Exterior Improvements

- Recycling of asphalt pavement is becoming very popular and is an alternative to removal and replacement. It can be a good value engineering proposal if removed pavement can be recycled, either at the project site or at another site that is reasonably close to the project site. Sections on repair of flexible and rigid pavement are included.

32 10 00 Bases, Ballasts, and Paving

- When estimating paving, keep in mind the project schedule. Also note that prices for asphalt and concrete are generally higher in the cold seasons. Lines for pavement markings, including tactile warning systems and fence lines, are included.

32 90 00 Planting

- The timing of planting and guarantee specifications often dictate the costs for establishing tree and shrub growth and a stand of grass or ground cover. Establish the work performance schedule to coincide with the local planting season. Maintenance and growth guarantees can add 20–100% to the total landscaping cost and can be contractually cumbersome. The cost to replace trees and shrubs can be as high as 5% of the total cost, depending on the planting zone, soil conditions, and time of year.

Reference Numbers

Reference numbers are shown at the beginning of some major classifications. These numbers refer to related items in the Reference Section. The reference information may be an estimating procedure, an alternate pricing method, or technical information.

Note: Not all subdivisions listed here necessarily appear. ∎

Same Data. Simplified.

Enjoy the convenience and efficiency of accessing your costs anywhere:

- **Skip the multiplier** by setting your location
- **Quickly search,** edit, favorite and share costs
- **Stay on top of price changes** with automatic updates

Discover more at rsmeans.com/online

---

32 01 Operation and Maintenance of Exterior Improvements

32 01 13 – Flexible Paving Surface Treatment

32 01 13.61 Slurry Seal (Latex Modified)

		Crew	Daily Output	Labor-Hours	Unit	Material	2021 Bare Costs Labor	Equipment	Total	Total Incl O&P
0010	**SLURRY SEAL (LATEX MODIFIED)**									
0011	Chip seal, slurry seal, and microsurfacing, see section 32 12 36									
3780	Rubberized asphalt (latex) seal	B-45	5000	.003	S.Y.	1.44	.18	.17	1.79	2.02
5400	Thermoplastic coal-tar, Type I, small or irregular area	B-90	2400	.027		4.23	1.31	.93	6.47	7.65
5450	Roadway or large area		8000	.008		4.23	.39	.28	4.90	5.55
5500	Type II, small or irregular area		2400	.027		5.40	1.31	.93	7.64	8.95
6000	Gravel surfacing on asphalt, screened and rolled	B-11L	160	.100	C.Y.	23.50	5.15	6.70	35.35	41
7000	For subbase treatment, see Section 31 32 13									

32 01 19 – Rigid Paving Surface Treatment

32 01 19.61 Sealing of Joints In Rigid Paving

		Crew	Daily Output	Labor-Hours	Unit	Material	2021 Bare Costs Labor	Equipment	Total	Total Incl O&P
0010	**SEALING OF JOINTS IN RIGID PAVING**									
1000	Sealing, concrete pavement, preformed elastomeric	B-77	500	.080	L.F.	3.58	3.65	2.15	9.38	11.75
2000	Waterproofing, membrane, tar and fabric, small area	B-63	233	.172	S.Y.	15.10	8	.77	23.87	29.50
2500	Large area		1435	.028		13.30	1.30	.13	14.73	16.75
3000	Preformed rubberized asphalt, small area		100	.400		19.50	18.65	1.80	39.95	51.50
3500	Large area		367	.109		17.75	5.10	.49	23.34	27.50

32 06 Schedules for Exterior Improvements

32 06 10 – Schedules for Bases, Ballasts, and Paving

32 06 10.10 Sidewalks, Driveways and Patios

		Crew	Daily Output	Labor-Hours	Unit	Material	2021 Bare Costs Labor	Equipment	Total	Total Incl O&P
0010	**SIDEWALKS, DRIVEWAYS AND PATIOS** No base									
0020	Asphaltic concrete, 2" thick	B-37	720	.067	S.Y.	7.25	3.11	.36	10.72	13
0100	2-1/2" thick	"	660	.073	"	9.15	3.39	.39	12.93	15.60
0300	Concrete, 3,000 psi, CIP, 6 x 6 - W1.4 x W1.4 mesh,									
0310	broomed finish, no base, 4" thick	B-24	600	.040	S.F.	2.22	2.01		4.23	5.45
0350	5" thick		545	.044		2.77	2.22		4.99	6.35
0400	6" thick		510	.047		3.23	2.37		5.60	7.05
0450	For bank run gravel base, 4" thick, add	B-18	2500	.010		.68	.43	.07	1.18	1.47
0520	8" thick, add	"	1600	.015		1.38	.68	.10	2.16	2.64
0550	Exposed aggregate finish, add to above, minimum	B-24	1875	.013		.13	.64		.77	1.09
0600	Maximum	"	455	.053		.44	2.65		3.09	4.41
1000	Crushed stone, 1" thick, white marble	2 Clab	1700	.009		.50	.42		.92	1.17
1050	Bluestone	"	1700	.009		.19	.42		.61	.83
1700	Redwood, prefabricated, 4' x 4' sections	2 Carp	316	.051		7.35	2.77		10.12	12.25
1750	Redwood planks, 1" thick, on sleepers	"	240	.067		7.35	3.65		11	13.55
2250	Stone dust, 4" thick	B-62	900	.027	S.Y.	5.95	1.28	.20	7.43	8.70

32 06 10.20 Steps

		Crew	Daily Output	Labor-Hours	Unit	Material	2021 Bare Costs Labor	Equipment	Total	Total Incl O&P
0010	**STEPS**									
0011	Incl. excav., borrow & concrete base as required									
0100	Brick steps	B-24	35	.686	LF Riser	18.55	34.50		53.05	71.50
0200	Railroad ties	2 Clab	25	.640		3.86	28.50		32.36	47
0300	Bluestone treads, 12" x 2" or 12" x 1-1/2"	B-24	30	.800		45	40		85	109
0600	Precast concrete, see Section 03 41 23.50									

32 11 Base Courses

32 11 23 – Aggregate Base Courses

32 11 23.23 Base Course Drainage Layers

		Crew	Daily Output	Labor-Hours	Unit	Material	2021 Bare Costs Labor	Equipment	Total	Total Incl O&P
0010	**BASE COURSE DRAINAGE LAYERS**									
0011	For Soil Stabilization, see Section 31 32									
0012	For roadways and large areas									
0050	Crushed 3/4" stone base, compacted, 3" deep	B-36C	5200	.008	S.Y.	2.50	.42	.83	3.75	4.29
0100	6" deep		5000	.008		5	.44	.86	6.30	7.10
0200	9" deep		4600	.009		7.50	.48	.93	8.91	10
0300	12" deep	▼	4200	.010		10	.52	1.02	11.54	12.95
0301	Crushed 1-1/2" stone base, compacted to 4" deep	B-36B	6000	.011		5.90	.56	.83	7.29	8.25
0302	6" deep		5400	.012		8.85	.63	.92	10.40	11.70
0303	8" deep		4500	.014		11.85	.75	1.10	13.70	15.35
0304	12" deep	▼	3800	.017	▼	17.75	.89	1.31	19.95	22.50
0350	Bank run gravel, spread and compacted									
0370	6" deep	B-32	6000	.005	S.Y.	5.70	.30	.47	6.47	7.25
0390	9" deep		4900	.007		8.55	.36	.58	9.49	10.65
0400	12" deep	▼	4200	.008	▼	11.45	.42	.68	12.55	13.95
6900	For small and irregular areas, add						50%	50%		
7000	Prepare and roll sub-base, small areas to 2,500 S.Y.	B-32A	1500	.016	S.Y.		.87	1.16	2.03	2.57
8000	Large areas over 2,500 S.Y.	"	3500	.007			.37	.50	.87	1.10
8050	For roadways	B-32	4000	.008	▼		.44	.71	1.15	1.44

32 11 26 – Asphaltic Base Courses

32 11 26.19 Bituminous-Stabilized Base Courses

		Crew	Daily Output	Labor-Hours	Unit	Material	2021 Bare Costs Labor	Equipment	Total	Total Incl O&P
0010	**BITUMINOUS-STABILIZED BASE COURSES**									
0020	For roadways and large paved areas									
0700	Liquid application to gravel base, asphalt emulsion	B-45	6000	.003	Gal.	5.10	.15	.14	5.39	5.95
0800	Prime and seal, cut back asphalt		6000	.003	"	6.05	.15	.14	6.34	7
1000	Macadam penetration crushed stone, 2 gal./S.Y., 4" thick		6000	.003	S.Y.	10.20	.15	.14	10.49	11.60
1100	6" thick, 3 gal./S.Y.		4000	.004		15.35	.22	.21	15.78	17.40
1200	8" thick, 4 gal./S.Y.	▼	3000	.005	▼	20.50	.29	.28	21.07	23.50
8900	For small and irregular areas, add						50%	50%		

32 12 36 – Seal Coats

32 12 36.13 Chip Seal

		Crew	Daily Output	Labor-Hours	Unit	Material	2021 Bare Costs Labor	Equipment	Total	Total Incl O&P
0010	**CHIP SEAL**									
0011	Excludes crack repair and flush coat									
1000	Fine - PMCRS-2h (20lbs/sy, 1/4" (No.10), .30gal/sy app. rate)									
1010	Small, irregular areas	B-91	5000	.013	S.Y.	1.10	.68	.46	2.24	2.72
1020	Parking Lot	B-91D	15000	.007		1.10	.36	.23	1.69	1.99
1030	Roadway	"	30000	.003		1.10	.18	.11	1.39	1.59
1090	For Each .5% Latex Additive, Add				▼	.23			.23	.25
1100	Medium Fine - PMCRS-2h (25lbs/sy, 5/16" (No.8), .35gal/sy app. rate)									
1110	Small, irregular areas	B-91	4000	.016	S.Y.	1.29	.84	.57	2.70	3.30
1120	Parking Lot	B-91D	12000	.009		1.29	.44	.28	2.01	2.38
1130	Roadway	"	24000	.004		1.29	.22	.14	1.65	1.90
1190	For Each .5% Latex Additive, Add				▼	.28			.28	.31
1200	Medium - PMCRS-2h (30lbs/sy, 3/8" (No.6), .40gal/sy app. rate)									
1210	Small, irregular areas	B-91	3330	.019	S.Y.	1.41	1.01	.69	3.11	3.82
1220	Parking Lot	B-91D	10000	.010		1.41	.53	.34	2.28	2.71
1230	Roadway	"	20000	.005		1.41	.27	.17	1.85	2.14
1290	For Each .5% Latex Additive, Add				▼	.34			.34	.37
1300	Course - PMCRS-2h (30lbs/sy, 1/2" (No.4), .40gal/sy app. rate)									
1310	Small, irregular areas	B-91	2500	.026	S.Y.	1.39	1.35	.92	3.66	4.55
1320	Parking Lot	B-91D	7500	.014		1.39	.71	.45	2.55	3.08
1330	Roadway	"	15000	.007	▼	1.39	.36	.23	1.98	2.30

233

32 12 36.13 Chip Seal

		Crew	Daily Output	Labor-Hours	Unit	Material	2021 Bare Costs Labor	Equipment	Total	Total Incl O&P
1390	For Each .5% Latex Additive, Add				S.Y.	.34			.34	.37
1400	Double - PMCRS-2h (Course Base with Fine Top)									
1410	Small, irregular areas	B-91	2000	.032	S.Y.	2.36	1.69	1.14	5.19	6.40
1420	Parking Lot	B-91D	6000	.017		2.36	.89	.57	3.82	4.54
1430	Roadway	"	12000	.009		2.36	.44	.28	3.08	3.57
1490	For Each .5% Latex Additive, Add					.56			.56	.62

32 12 36.14 Flush Coat

		Crew	Daily Output	Labor-Hours	Unit	Material	2021 Bare Costs Labor	Equipment	Total	Total Incl O&P
0010	**FLUSH COAT**									
0011	Fog Seal with Sand Cover (18gal/sy, 6lbs/sy)									
1010	Small, irregular areas	B-91	2000	.032	S.Y.	.52	1.69	1.14	3.35	4.36
1020	Parking lot		6000	.011		.52	.56	.38	1.46	1.84
1030	Roadway		12000	.005		.52	.28	.19	.99	1.21

32 12 36.33 Slurry Seal

		Crew	Daily Output	Labor-Hours	Unit	Material	2021 Bare Costs Labor	Equipment	Total	Total Incl O&P
0010	**SLURRY SEAL**									
0011	Includes sweeping and cleaning of area									
1000	Type I-PMCQS-1h-EAS (12lbs/sy, 1/8", 20% asphalt emulsion)									
1010	Small, irregular areas	B-90	8000	.008	S.Y.	.98	.39	.28	1.65	1.97
1020	Parking Lot		25000	.003		.98	.13	.09	1.20	1.36
1030	Roadway		50000	.001		.98	.06	.04	1.08	1.21
1090	For Each .5% Latex Additive, Add					.15			.15	.16
1100	Type II-PMCQS-1h-EAS (15lbs/sy, 1/4", 18% asphalt emulsion)									
1110	Small, irregular areas	B-90	6000	.011	S.Y.	1.10	.52	.37	1.99	2.40
1120	Parking lot		20000	.003		1.10	.16	.11	1.37	1.56
1130	Roadway		40000	.002		1.10	.08	.06	1.24	1.39
1190	For Each .5% Latex Additive, Add					.18			.18	.20
1200	Type III-PMCQS-1h-EAS (25lbs/sy, 3/8", 15% asphalt emulsion)									
1210	Small, irregular areas	B-90	4000	.016	S.Y.	1.51	.79	.56	2.86	3.44
1220	Parking lot		12000	.005		1.51	.26	.19	1.96	2.25
1230	Roadway		24000	.003		1.51	.13	.09	1.73	1.96
1290	For Each .5% Latex Additive, Add					.29			.29	.32

32 12 36.36 Microsurfacing

		Crew	Daily Output	Labor-Hours	Unit	Material	2021 Bare Costs Labor	Equipment	Total	Total Incl O&P
0010	**MICROSURFACING**									
1100	Type II-MSE (20lbs/sy, 1/4", 18% microsurfacing emulsion)									
1110	Small, irregular areas	B-90	5000	.013	S.Y.	1.65	.63	.44	2.72	3.24
1120	Parking lot		15000	.004		1.65	.21	.15	2.01	2.28
1130	Roadway		30000	.002		1.65	.10	.07	1.82	2.05
1200	Type IIIa-MSE (32lbs/sy, 3/8", 15% microsurfacing emulsion)									
1210	Small, irregular areas	B-90	3000	.021	S.Y.	2.29	1.05	.74	4.08	4.89
1220	Parking lot		9000	.007		2.29	.35	.25	2.89	3.31
1230	Roadway		18000	.004		2.29	.17	.12	2.58	2.92

234

For customer support on your Concrete & Masonry Costs with RSMeans Data, call 800.448.8182.

32 13 Rigid Paving

32 13 13 – Concrete Paving

32 13 13.25 Concrete Pavement, Highways

		Crew	Daily Output	Labor-Hours	Unit	Material	2021 Bare Costs Labor	Equipment	Total	Total Incl O&P
0010	**CONCRETE PAVEMENT, HIGHWAYS**									
0015	Including joints, finishing and curing									
0020	Fixed form, 12' pass, unreinforced, 6" thick	B-26	3000	.029	S.Y.	26.50	1.44	1.19	29.13	33
0100	8" thick		2750	.032		36.50	1.58	1.30	39.38	44
0110	8" thick, small area		1375	.064		36.50	3.15	2.60	42.25	47.50
0200	9" thick		2500	.035		41.50	1.73	1.43	44.66	49.50
0300	10" thick		2100	.042		45.50	2.06	1.70	49.26	55
0310	10" thick, small area		1050	.084		45.50	4.13	3.41	53.04	60
0400	12" thick		1800	.049		52	2.41	1.99	56.40	63.50
0410	Conc. pavement, w/jt.& curing, fix form, 24' pass, unreinforced, 6"T		6000	.015		25	.72	.60	26.32	29
0430	8" thick		5500	.016		34.50	.79	.65	35.94	40
0440	9" thick		5000	.018		39.50	.87	.72	41.09	45
0450	10" thick		4200	.021		43.50	1.03	.85	45.38	50
0460	12" thick		3600	.024		50	1.20	.99	52.19	58
0470	15" thick		3000	.029		66.50	1.44	1.19	69.13	76.50
0500	Fixed form 12' pass, 15" thick		1500	.059		66.50	2.89	2.38	71.77	80
0510	For small irregular areas, add				%	10%	100%	100%		
0520	Welded wire fabric, sheets for rigid paving 2.33 lb./S.Y.	2 Rodm	389	.041	S.Y.	1.34	2.42		3.76	5.10
0530	Reinforcing steel for rigid paving 12 lb./S.Y.		666	.024		7.85	1.42		9.27	10.70
0540	Reinforcing steel for rigid paving 18 lb./S.Y.		444	.036		11.75	2.12		13.87	16.05
0620	Slip form, 12' pass, unreinforced, 6" thick	B-26A	5600	.016		25.50	.77	.66	26.93	30
0624	8" thick		5300	.017		35	.82	.70	36.52	40.50
0626	9" thick		4820	.018		40	.90	.77	41.67	46
0628	10" thick		4050	.022		44	1.07	.91	45.98	50.50
0630	12" thick		3470	.025		50.50	1.25	1.06	52.81	58.50
0640	Slip form, 24' pass, unreinforced, 6" thick		11200	.008		25	.39	.33	25.72	28.50
0644	8" thick		10600	.008		33.50	.41	.35	34.26	38
0646	9" thick		9640	.009		38.50	.45	.38	39.33	43.50
0648	10" thick		8100	.011		42.50	.53	.46	43.49	48
0650	12" thick		6940	.013		49	.62	.53	50.15	55.50
0652	15" thick		5780	.015		61.50	.75	.64	62.89	70
0700	Finishing, broom finish small areas	2 Cefi	120	.133			6.90		6.90	10.10
1000	Curing, with sprayed membrane by hand	2 Clab	1500	.011		1.10	.47		1.57	1.92
1650	For integral coloring, see Section 03 05 13.20									

32 14 Unit Paving

32 14 13 – Precast Concrete Unit Paving

32 14 13.13 Interlocking Precast Concrete Unit Paving

		Crew	Daily Output	Labor-Hours	Unit	Material	2021 Bare Costs Labor	Equipment	Total	Total Incl O&P
0010	**INTERLOCKING PRECAST CONCRETE UNIT PAVING**									
0020	"V" blocks for retaining soil	D-1	205	.078	S.F.	11.20	3.80		15	18.10

32 14 13.16 Precast Concrete Unit Paving Slabs

		Crew	Daily Output	Labor-Hours	Unit	Material	2021 Bare Costs Labor	Equipment	Total	Total Incl O&P
0010	**PRECAST CONCRETE UNIT PAVING SLABS**									
0750	Exposed local aggregate, natural	2 Bric	250	.064	S.F.	8.80	3.44		12.24	14.90
0800	Colors		250	.064		9.90	3.44		13.34	16.10
0850	Exposed granite or limestone aggregate		250	.064		9.10	3.44		12.54	15.20
0900	Exposed white tumblestone aggregate		250	.064		11.15	3.44		14.59	17.45

32 14 Unit Paving

32 14 16 – Brick Unit Paving

32 14 16.10 Brick Paving

		Crew	Daily Output	Labor-Hours	Unit	Material	2021 Bare Costs Labor	Equipment	Total	Total Incl O&P
0010	**BRICK PAVING**									
0012	4" x 8" x 1-1/2", without joints (4.5 bricks/S.F.)	D-1	110	.145	S.F.	2.74	7.10		9.84	13.65
0100	Grouted, 3/8" joint (3.9 bricks/S.F.)		90	.178		2.26	8.65		10.91	15.55
0200	4" x 8" x 2-1/4", without joints (4.5 bricks/S.F.)		110	.145		2.61	7.10		9.71	13.50
0300	Grouted, 3/8" joint (3.9 bricks/S.F.)		90	.178		2.26	8.65		10.91	15.55
0455	Pervious brick paving, 4" x 8" x 3-1/4", without joints (4.5 bricks/S.F.)		110	.145		2.83	7.10		9.93	13.75
0500	Bedding, asphalt, 3/4" thick	B-25	5130	.017		.68	.83	.53	2.04	2.58
0540	Course washed sand bed, 1" thick	B-18	5000	.005		.37	.22	.03	.62	.77
0580	Mortar, 1" thick	D-1	300	.053		.76	2.60		3.36	4.75
0620	2" thick		200	.080		1.52	3.90		5.42	7.55
1500	Brick on 1" thick sand bed laid flat, 4.5/S.F.		100	.160		3.08	7.80		10.88	15.15
2000	Brick pavers, laid on edge, 7.2/S.F.		70	.229		4.74	11.15		15.89	22
2500	For 4" thick concrete bed and joints, add		595	.027		1.34	1.31		2.65	3.44
2800	For steam cleaning, add	A-1H	950	.008		.10	.37	.08	.55	.76

32 14 23 – Asphalt Unit Paving

32 14 23.10 Asphalt Blocks

		Crew	Daily Output	Labor-Hours	Unit	Material	2021 Bare Costs Labor	Equipment	Total	Total Incl O&P
0010	**ASPHALT BLOCKS**									
0020	Rectangular, 6" x 12" x 1-1/4", w/bed & neopr. adhesive	D-1	135	.119	S.F.	10.10	5.75		15.85	19.80
0100	3" thick		130	.123		14.15	6		20.15	24.50
0300	Hexagonal tile, 8" wide, 1-1/4" thick		135	.119		10.10	5.75		15.85	19.80
0400	2" thick		130	.123		14.15	6		20.15	24.50
0500	Square, 8" x 8", 1-1/4" thick		135	.119		10.10	5.75		15.85	19.80
0600	2" thick		130	.123		14.15	6		20.15	24.50
0900	For exposed aggregate (ground finish), add					.61			.61	.67
0910	For colors, add					.61			.61	.67

32 14 40 – Stone Paving

32 14 40.10 Stone Pavers

		Crew	Daily Output	Labor-Hours	Unit	Material	2021 Bare Costs Labor	Equipment	Total	Total Incl O&P
0010	**STONE PAVERS**									
1100	Flagging, bluestone, irregular, 1" thick,	D-1	81	.198	S.F.	9.65	9.60		19.25	25
1150	Snapped random rectangular, 1" thick		92	.174		14.65	8.45		23.10	29
1200	1-1/2" thick		85	.188		17.60	9.15		26.75	33
1250	2" thick		83	.193		20.50	9.40		29.90	36.50
1300	Slate, natural cleft, irregular, 3/4" thick		92	.174		9.70	8.45		18.15	23.50
1350	Random rectangular, gauged, 1/2" thick		105	.152		21	7.40		28.40	34
1400	Random rectangular, butt joint, gauged, 1/4" thick		150	.107		22.50	5.20		27.70	33
1450	For sand rubbed finish, add					10			10	11
1500	For interior setting, add					25%			25%	25%
1550	Granite blocks, 3-1/2" x 3-1/2" x 3-1/2"	D-1	92	.174	S.F.	21.50	8.45		29.95	37
1600	4" to 12" long, 3" to 5" wide, 3" to 5" thick		98	.163		18.05	7.95		26	32
1650	6" to 15" long, 3" to 6" wide, 3" to 5" thick		105	.152		9.65	7.40		17.05	22

32 16 13 – Curbs and Gutters

32 16 13.13 Cast-in-Place Concrete Curbs and Gutters

		Crew	Daily Output	Labor-Hours	Unit	Material	2021 Bare Costs Labor	2021 Bare Costs Equipment	Total	Total Incl O&P
0010	**CAST-IN-PLACE CONCRETE CURBS AND GUTTERS**									
0290	Forms only, no concrete									
0300	Concrete, wood forms, 6" x 18", straight	C-2	500	.096	L.F.	4	5.10		9.10	12.05
0400	6" x 18", radius	"	200	.240	"	4.21	12.80		17.01	23.50
0402	Forms and concrete complete									
0404	Concrete, wood forms, 6" x 18", straight & concrete	C-2A	500	.096	L.F.	7.50	5.05		12.55	15.80
0406	6" x 18", radius		200	.240		7.70	12.70		20.40	27.50
0410	Steel forms, 6" x 18", straight		700	.069		7.30	3.62		10.92	13.40
0411	6" x 18", radius	↓	400	.120	↓	7.15	6.35		13.50	17.30
0415	Machine formed, 6" x 18", straight	B-69A	2000	.024		5.15	1.16	.62	6.93	8.05
0416	6" x 18", radius	"	900	.053		5.20	2.58	1.37	9.15	11.10
0421	Curb and gutter, straight									
0422	with 6" high curb and 6" thick gutter, wood forms									
0430	24" wide, 0.055 C.Y./L.F.	C-2A	375	.128	L.F.	21	6.75		27.75	33
0435	30" wide, 0.066 C.Y./L.F.		340	.141		23	7.45		30.45	36
0440	Steel forms, 24" wide, straight		700	.069		9.70	3.62		13.32	16.05
0441	Radius		500	.096		8.55	5.05		13.60	17
0442	30" wide, straight		700	.069		11.35	3.62		14.97	17.90
0443	Radius	↓	500	.096		9.70	5.05		14.75	18.20
0445	Machine formed, 24" wide, straight	B-69A	2000	.024		6.95	1.16	.62	8.73	10.05
0446	Radius		900	.053		6.95	2.58	1.37	10.90	13
0447	30" wide, straight		2000	.024		8.10	1.16	.62	9.88	11.30
0448	Radius		900	.053		8.10	2.58	1.37	12.05	14.25
0451	Median mall, machine formed, 2' x 9" high, straight	↓	2200	.022	↓	6.95	1.06	.56	8.57	9.85
0452	Radius	B-69B	900	.053		6.95	2.58	.89	10.42	12.45
0453	4' x 9" high, straight		2000	.024		13.95	1.16	.40	15.51	17.50
0454	Radius	↓	800	.060	↓	13.95	2.90	1	17.85	21

32 16 13.23 Precast Concrete Curbs and Gutters

		Crew	Daily Output	Labor-Hours	Unit	Material	2021 Bare Costs Labor	2021 Bare Costs Equipment	Total	Total Incl O&P
0010	**PRECAST CONCRETE CURBS AND GUTTERS**									
0550	Precast, 6" x 18", straight	B-29	700	.080	L.F.	9.05	3.86	1.22	14.13	17.05
0600	6" x 18", radius	"	325	.172	"	9.65	8.30	2.62	20.57	26

32 16 13.33 Asphalt Curbs

		Crew	Daily Output	Labor-Hours	Unit	Material	2021 Bare Costs Labor	2021 Bare Costs Equipment	Total	Total Incl O&P
0010	**ASPHALT CURBS**									
0012	Curbs, asphaltic, machine formed, 8" wide, 6" high, 40 L.F./ton	B-27	1000	.032	L.F.	1.84	1.44	.25	3.53	4.45
0100	8" wide, 8" high, 30 L.F./ton		900	.036		2.46	1.60	.28	4.34	5.40
0150	Asphaltic berm, 12" W, 3" to 6" H, 35 L.F./ton, before pavement	↓	700	.046		.04	2.05	.36	2.45	3.51
0200	12" W, 1-1/2" to 4" H, 60 L.F./ton, laid with pavement	B-2	1050	.038	↓	.02	1.71		1.73	2.58

32 16 13.43 Stone Curbs

		Crew	Daily Output	Labor-Hours	Unit	Material	2021 Bare Costs Labor	2021 Bare Costs Equipment	Total	Total Incl O&P
0010	**STONE CURBS**									
1000	Granite, split face, straight, 5" x 16"	D-13	275	.175	L.F.	16.45	9.10	1.57	27.12	33.50
1100	6" x 18"	"	250	.192		21.50	10	1.73	33.23	41
1300	Radius curbing, 6" x 18", over 10' radius	B-29	260	.215	↓	26.50	10.40	3.27	40.17	48
1400	Corners, 2' radius	"	80	.700	Ea.	89	34	10.65	133.65	160
1600	Edging, 4-1/2" x 12", straight	D-13	300	.160	L.F.	8.25	8.35	1.44	18.04	23
1800	Curb inlets (guttermouth) straight	B-29	41	1.366	Ea.	198	66	21	285	340
2000	Indian granite (Belgian block)									
2100	Jumbo, 10-1/2" x 7-1/2" x 4", grey	D-1	150	.107	L.F.	8.30	5.20		13.50	16.95
2150	Pink		150	.107		9.80	5.20		15	18.60
2200	Regular, 9" x 4-1/2" x 4-1/2", grey		160	.100		4.77	4.87		9.64	12.60
2250	Pink		160	.100		6.50	4.87		11.37	14.50
2300	Cubes, 4" x 4" x 4", grey		175	.091		2.62	4.45		7.07	9.60
2350	Pink	↓	175	.091	↓	4.04	4.45		8.49	11.15

237

For customer support on your Concrete & Masonry Costs with RSMeans Data, call 800.448.8182.

32 16 Curbs, Gutters, Sidewalks, and Driveways

32 16 13 – Curbs and Gutters

32 16 13.43 Stone Curbs		Crew	Daily Output	Labor-Hours	Unit	Material	2021 Bare Costs Labor	Equipment	Total	Total Incl O&P
2400	6" x 6" x 6", pink	D-1	155	.103	L.F.	13.50	5.05		18.55	22.50
2500	Alternate pricing method for Indian granite									
2550	Jumbo, 10-1/2" x 7-1/2" x 4" (30 lb.), grey				Ton	470			470	520
2600	Pink					570			570	625
2650	Regular, 9" x 4-1/2" x 4-1/2" (20 lb.), grey					335			335	370
2700	Pink					455			455	500
2750	Cubes, 4" x 4" x 4" (5 lb.), grey					320			320	350
2800	Pink					520			520	570
2850	6" x 6" x 6" (25 lb.), pink					525			525	580
2900	For pallets, add					22			22	24

32 17 Paving Specialties

32 17 13 – Parking Bumpers

32 17 13.13 Metal Parking Bumpers

	32 17 13.13 Metal Parking Bumpers	Crew	Daily Output	Labor-Hours	Unit	Material	Labor	Equipment	Total	Total Incl O&P
0010	**METAL PARKING BUMPERS**									
0015	Bumper rails for garages, 12 ga. rail, 6" wide, with steel									
0020	posts 12'-6" OC, minimum	E-4	190	.168	L.F.	19.75	10.25	.78	30.78	38
0030	Average		165	.194		24.50	11.80	.90	37.20	46
0100	Maximum		140	.229		29.50	13.90	1.06	44.46	55
0300	12" channel rail, minimum		160	.200		24.50	12.15	.93	37.58	47
0400	Maximum		120	.267		37	16.20	1.24	54.44	67
1300	Pipe bollards, conc. filled/paint, 8' L x 4' D hole, 6" diam.	B-6	20	1.200	Ea.	680	57.50	10.80	748.30	845
1400	8" diam.		15	1.600		745	77	14.40	836.40	950
1500	12" diam.		12	2		1,025	96	18	1,139	1,300
2030	Folding with individual padlocks	B-2	50	.800		182	36		218	254

32 17 13.16 Plastic Parking Bumpers

	32 17 13.16 Plastic Parking Bumpers	Crew	Daily Output	Labor-Hours	Unit	Material	Labor	Equipment	Total	Total Incl O&P
0010	**PLASTIC PARKING BUMPERS**									
1200	Thermoplastic, 6" x 10" x 6'-0"	B-2	120	.333	Ea.	46	14.95		60.95	73

32 17 13.19 Precast Concrete Parking Bumpers

	32 17 13.19 Precast Concrete Parking Bumpers	Crew	Daily Output	Labor-Hours	Unit	Material	Labor	Equipment	Total	Total Incl O&P
0010	**PRECAST CONCRETE PARKING BUMPERS**									
1000	Wheel stops, precast concrete incl. dowels, 6" x 10" x 6'-0"	B-2	120	.333	Ea.	58	14.95		72.95	86
1100	8" x 13" x 6'-0"	"	120	.333	"	86.50	14.95		101.45	118

32 17 13.26 Wood Parking Bumpers

	32 17 13.26 Wood Parking Bumpers	Crew	Daily Output	Labor-Hours	Unit	Material	Labor	Equipment	Total	Total Incl O&P
0010	**WOOD PARKING BUMPERS**									
0020	Parking barriers, timber w/saddles, treated type									
0100	4" x 4" for cars	B-2	520	.077	L.F.	3.91	3.45		7.36	9.45
0200	6" x 6" for trucks		520	.077	"	8.20	3.45		11.65	14.15
0600	Flexible fixed stanchion, 2' high, 3" diameter		100	.400	Ea.	44.50	17.90		62.40	75.50

32 31 Fences and Gates

32 31 13 – Chain Link Fences and Gates

32 31 13.20 Fence, Chain Link Industrial

		Crew	Daily Output	Labor-Hours	Unit	Material	2021 Bare Costs Labor	2021 Bare Costs Equipment	Total	Total Incl O&P
0010	**FENCE, CHAIN LINK INDUSTRIAL**									
0011	Schedule 40, including concrete									
0020	3 strands barb wire, 2" post @ 10' OC, set in concrete, 6' H									
0200	9 ga. wire, galv. steel, in concrete	B-80C	240	.100	L.F.	19.40	4.59	1.05	25.04	29.50
0300	Aluminized steel		240	.100		26.50	4.59	1.05	32.14	37
0301	Fence, wrought iron		240	.100		43	4.59	1.05	48.64	55
0303	Fence, commercial 4' high		240	.100		32.50	4.59	1.05	38.14	43.50
0304	Fence, commercial 6' high		240	.100		58	4.59	1.05	63.64	72
0500	6 ga. wire, galv. steel		240	.100		25	4.59	1.05	30.64	35.50
0600	Aluminized steel		240	.100		34	4.59	1.05	39.64	45.50
0800	6 ga. wire, 6' high but omit barbed wire, galv. steel		250	.096		22.50	4.40	1.01	27.91	32
0900	Aluminized steel, in concrete		250	.096		30.50	4.40	1.01	35.91	41.50
0920	8' H, 6 ga. wire, 2-1/2" line post, galv. steel, in concrete		180	.133		35.50	6.10	1.41	43.01	49.50
0940	Aluminized steel, in concrete		180	.133	↓	43.50	6.10	1.41	51.01	58.50
1100	Add for corner posts, 3" diam., galv. steel, in concrete		40	.600	Ea.	89	27.50	6.30	122.80	146
1200	Aluminized steel, in concrete		40	.600		99.50	27.50	6.30	133.30	157
1300	Add for braces, galv. steel		80	.300		49	13.75	3.16	65.91	78
1350	Aluminized steel		80	.300		52	13.75	3.16	68.91	81.50
1400	Gate for 6' high fence, 1-5/8" frame, 3' wide, galv. steel		10	2.400		208	110	25.50	343.50	420
1500	Aluminized steel, in concrete	↓	10	2.400	↓	224	110	25.50	359.50	440
2000	5'-0" high fence, 9 ga., no barbed wire, 2" line post, in concrete									
2010	10' OC, 1-5/8" top rail, in concrete									
2100	Galvanized steel, in concrete	B-80C	300	.080	L.F.	20.50	3.67	.84	25.01	29
2200	Aluminized steel, in concrete		300	.080	"	20	3.67	.84	24.51	28.50
2400	Gate, 4' wide, 5' high, 2" frame, galv. steel, in concrete		10	2.400	Ea.	229	110	25.50	364.50	445
2500	Aluminized steel, in concrete		10	2.400	"	183	110	25.50	318.50	395
3100	Overhead slide gate, chain link, 6' high, to 18' wide, in concrete	↓	38	.632	L.F.	97	29	6.65	132.65	158
3110	Cantilever type, in concrete	B-80	48	.667		189	32.50	22	243.50	281
3120	8' high, in concrete		24	1.333		182	65	44	291	345
3130	10' high, in concrete	↓	18	1.778	↓	211	86.50	58.50	356	425

32 31 13.80 Residential Chain Link Gate

		Crew	Daily Output	Labor-Hours	Unit	Material	2021 Bare Costs Labor	2021 Bare Costs Equipment	Total	Total Incl O&P
0010	**RESIDENTIAL CHAIN LINK GATE**									
0110	Residential 4' gate, single incl. hardware and concrete	B-80C	10	2.400	Ea.	191	110	25.50	326.50	400
0120	5'		10	2.400		201	110	25.50	336.50	415
0130	6'		10	2.400		213	110	25.50	348.50	425
0510	Residential 4' gate, double incl. hardware and concrete		10	2.400		290	110	25.50	425.50	510
0520	5'		10	2.400		305	110	25.50	440.50	525
0530	6'	↓	10	2.400	↓	350	110	25.50	485.50	575

32 31 13.82 Internal Chain Link Gate

		Crew	Daily Output	Labor-Hours	Unit	Material	2021 Bare Costs Labor	2021 Bare Costs Equipment	Total	Total Incl O&P
0010	**INTERNAL CHAIN LINK GATE**									
0110	Internal 6' gate, single incl. post flange, hardware and concrete	B-80C	10	2.400	Ea.	340	110	25.50	475.50	565
0120	8'		10	2.400		380	110	25.50	515.50	605
0130	10'		10	2.400		505	110	25.50	640.50	745
0510	Internal 6' gate, double incl. post flange, hardware and concrete		10	2.400		575	110	25.50	710.50	820
0520	8'		10	2.400		645	110	25.50	780.50	900
0530	10'	↓	10	2.400	↓	805	110	25.50	940.50	1,075

32 31 13.84 Industrial Chain Link Gate

		Crew	Daily Output	Labor-Hours	Unit	Material	2021 Bare Costs Labor	2021 Bare Costs Equipment	Total	Total Incl O&P
0010	**INDUSTRIAL CHAIN LINK GATE**									
0110	Industrial 8' gate, single incl. hardware and concrete	B-80C	10	2.400	Ea.	525	110	25.50	660.50	765
0120	10'		10	2.400		590	110	25.50	725.50	840
0510	Industrial 8' gate, double incl. hardware and concrete		10	2.400		790	110	25.50	925.50	1,050
0520	10'	↓	10	2.400		895	110	25.50	1,030.50	1,175

239

For customer support on your Concrete & Masonry Costs with RSMeans Data, call 800.448.8182.

32 31 Fences and Gates

32 31 13 – Chain Link Fences and Gates

32 31 13.88 Chain Link Transom	Crew	Daily Output	Labor-Hours	Unit	Material	2021 Bare Costs Labor	Equipment	Total	Total Incl O&P
0010	**CHAIN LINK TRANSOM**									
0110	Add for, single transom, 3' wide, incl. components & hardware	B-80C	10	2.400	Ea.	119	110	25.50	254.50	325
0120	Add for, double transom, 6' wide, incl. components & hardware	"	10	2.400	"	127	110	25.50	262.50	330

32 31 19 – Decorative Metal Fences and Gates

32 31 19.10 Decorative Fence

		Crew	Daily Output	Labor-Hours	Unit	Material	Labor	Equipment	Total	Total Incl O&P
0010	**DECORATIVE FENCE**									
5300	Tubular picket, steel, 6' sections, 1-9/16" posts, 4' high	B-80C	300	.080	L.F.	38.50	3.67	.84	43.01	49
5400	2" posts, 5' high		240	.100		44	4.59	1.05	49.64	56
5600	2" posts, 6' high		200	.120		53	5.50	1.26	59.76	68
5700	Staggered picket, 1-9/16" posts, 4' high		300	.080		45	3.67	.84	49.51	56
5800	2" posts, 5' high		240	.100		45	4.59	1.05	50.64	57.50
5900	2" posts, 6' high		200	.120		59	5.50	1.26	65.76	74.50
6200	Gates, 4' high, 3' wide	B-1	10	2.400	Ea.	335	108		443	530
6300	5' high, 3' wide		10	2.400		246	108		354	430
6400	6' high, 3' wide		10	2.400		370	108		478	565
6500	4' wide		10	2.400		330	108		438	520

32 31 26 – Wire Fences and Gates

32 31 26.10 Fences, Misc. Metal

		Crew	Daily Output	Labor-Hours	Unit	Material	Labor	Equipment	Total	Total Incl O&P
0010	**FENCES, MISC. METAL**									
0012	Chicken wire, posts @ 4', 1" mesh, 4' high	B-80C	410	.059	L.F.	4.32	2.69	.62	7.63	9.45
0100	2" mesh, 6' high		350	.069		3.75	3.15	.72	7.62	9.60
0200	Galv. steel, 12 ga., 2" x 4" mesh, posts 5' OC, 3' high		300	.080		2.79	3.67	.84	7.30	9.50
0300	5' high		300	.080		3.30	3.67	.84	7.81	10.05
0400	14 ga., 1" x 2" mesh, 3' high		300	.080		3.10	3.67	.84	7.61	9.85
0500	5' high		300	.080		4.34	3.67	.84	8.85	11.20
1000	Kennel fencing, 1-1/2" mesh, 6' long, 3'-6" wide, 6'-2" high	2 Clab	4	4	Ea.	505	178		683	820
1050	12' long		4	4		715	178		893	1,050
1200	Top covers, 1-1/2" mesh, 6' long		15	1.067		146	47.50		193.50	232
1250	12' long		12	1.333		219	59		278	330
4492	Security fence, prison grade, barbed wire, set in concrete, 10' high	B-80	22	1.455	L.F.	64.50	71	48	183.50	230
4494	Security fence, prison grade, razor wire, set in concrete, 10' high		18	1.778		65.50	86.50	58.50	210.50	266
4500	Security fence, prison grade, set in concrete, 12' high		25	1.280		56	62.50	42	160.50	201
4600	16' high		20	1.600		87	78	52.50	217.50	270
4990	Security fence, prison grade, set in concrete, 10' high		25	1.280		56	62.50	42	160.50	201

32 32 Retaining Walls

32 32 13 – Cast-in-Place Concrete Retaining Walls

32 32 13.10 Retaining Walls, Cast Concrete

		Crew	Daily Output	Labor-Hours	Unit	Material	Labor	Equipment	Total	Total Incl O&P
0010	**RETAINING WALLS, CAST CONCRETE**									
1800	Concrete gravity wall with vertical face including excavation & backfill									
1850	No reinforcing									
1900	6' high, level embankment	C-17C	36	2.306	L.F.	84	133	15.20	232.20	310
2000	33° slope embankment		32	2.594		106	150	17.10	273.10	360
2200	8' high, no surcharge		27	3.074		113	177	20.50	310.50	415
2300	33° slope embankment		24	3.458		137	199	23	359	475
2500	10' high, level embankment		19	4.368		162	252	29	443	590
2600	33° slope embankment		18	4.611		224	266	30.50	520.50	680
2800	Reinforced concrete cantilever, incl. excavation, backfill & reinf.									
2900	6' high, 33° slope embankment	C-17C	35	2.371	L.F.	83	137	15.65	235.65	315

32 32 Retaining Walls

32 32 13 – Cast-in-Place Concrete Retaining Walls

32 32 13.10 Retaining Walls, Cast Concrete

		Crew	Daily Output	Labor-Hours	Unit	Material	2021 Bare Costs Labor	2021 Bare Costs Equipment	Total	Total Incl O&P
3000	8' high, 33° slope embankment	C-17C	29	2.862	L.F.	96	165	18.85	279.85	375
3100	10' high, 33° slope embankment		20	4.150		124	239	27.50	390.50	525
3200	20' high, 500 lb./L.F. surcharge	↓	7.50	11.067	↓	375	640	73	1,088	1,450
3500	Concrete cribbing, incl. excavation and backfill									
3700	12' high, open face	B-13	210	.267	S.F.	41.50	12.85	2.79	57.14	68
3900	Closed face	"	210	.267	"	39	12.85	2.79	54.64	65.50
4100	Concrete filled slurry trench, see Section 31 56 23.20									

32 32 23 – Segmental Retaining Walls

32 32 23.13 Segmental Conc. Unit Masonry Retaining Walls

		Crew	Daily Output	Labor-Hours	Unit	Material	2021 Bare Costs Labor	2021 Bare Costs Equipment	Total	Total Incl O&P
0010	**SEGMENTAL CONC. UNIT MASONRY RETAINING WALLS**									
7100	Segmental retaining wall system, incl. pins and void fill									
7120	base and backfill not included									
7140	Large unit, 8" high x 18" wide x 20" deep, 3 plane split	B-62	300	.080	S.F.	15.15	3.85	.60	19.60	23
7150	Straight split		300	.080		15.10	3.85	.60	19.55	23
7160	Medium, lt. wt., 8" high x 18" wide x 12" deep, 3 plane split		400	.060		7.80	2.89	.45	11.14	13.40
7170	Straight split		400	.060		10.85	2.89	.45	14.19	16.70
7180	Small unit, 4" x 18" x 10" deep, 3 plane split		400	.060		16.90	2.89	.45	20.24	23.50
7190	Straight split		400	.060		13.05	2.89	.45	16.39	19.15
7200	Cap unit, 3 plane split		300	.080		15.30	3.85	.60	19.75	23.50
7210	Cap unit, straight split	↓	300	.080		15.30	3.85	.60	19.75	23.50
7250	Geo-grid soil reinforcement 4' x 50'	2 Clab	22500	.001		.80	.03		.83	.93
7255	Geo-grid soil reinforcement 6' x 150'	"	22500	.001	↓	.62	.03		.65	.73

32 32 26 – Metal Crib Retaining Walls

32 32 26.10 Metal Bin Retaining Walls

		Crew	Daily Output	Labor-Hours	Unit	Material	2021 Bare Costs Labor	2021 Bare Costs Equipment	Total	Total Incl O&P
0010	**METAL BIN RETAINING WALLS**									
0011	Aluminized steel bin, excavation									
0020	and backfill not included, 10' wide									
0100	4' high, 5.5' deep	B-13	650	.086	S.F.	28.50	4.16	.90	33.56	38.50
0200	8' high, 5.5' deep		615	.091		33	4.40	.95	38.35	43.50
0300	10' high, 7.7' deep		580	.097		36.50	4.66	1.01	42.17	48.50
0400	12' high, 7.7' deep		530	.106		39.50	5.10	1.11	45.71	52.50
0500	16' high, 7.7' deep		515	.109		42	5.25	1.14	48.39	55
0600	16' high, 9.9' deep		500	.112		46.50	5.40	1.17	53.07	61
0700	20' high, 9.9' deep		470	.119		52.50	5.75	1.25	59.50	67.50
0800	20' high, 12.1' deep		460	.122		45.50	5.90	1.28	52.68	60
0900	24' high, 12.1' deep		455	.123		48.50	5.95	1.29	55.74	64
1000	24' high, 14.3' deep		450	.124		68.50	6	1.30	75.80	86
1100	28' high, 14.3' deep	↓	440	.127	↓	71.50	6.15	1.33	78.98	89
1300	For plain galvanized bin type walls, deduct					10%				

32 32 29 – Timber Retaining Walls

32 32 29.10 Landscape Timber Retaining Walls

		Crew	Daily Output	Labor-Hours	Unit	Material	2021 Bare Costs Labor	2021 Bare Costs Equipment	Total	Total Incl O&P
0010	**LANDSCAPE TIMBER RETAINING WALLS**									
0100	Treated timbers, 6" x 6"	1 Clab	265	.030	L.F.	2.84	1.34		4.18	5.10
0110	6" x 8"	"	200	.040	"	7.40	1.78		9.18	10.80
0120	Drilling holes in timbers for fastening, 1/2"	1 Carp	450	.018	Inch		.97		.97	1.45
0130	5/8"	"	450	.018	"		.97		.97	1.45
0140	Reinforcing rods for fastening, 1/2"	1 Clab	312	.026	L.F.	.46	1.14		1.60	2.20
0150	5/8"	"	312	.026	"	.71	1.14		1.85	2.48
0160	Reinforcing fabric	2 Clab	2500	.006	S.Y.	2.25	.28		2.53	2.90
0170	Gravel backfill		28	.571	C.Y.	26.50	25.50		52	67.50
0180	Perforated pipe, 4" diameter with silt sock	↓	1200	.013	L.F.	1.26	.59		1.85	2.27

32 32 Retaining Walls

32 32 29 – Timber Retaining Walls

32 32 29.10 Landscape Timber Retaining Walls	Crew	Daily Output	Labor-Hours	Unit	Material	2021 Bare Costs Labor	2021 Bare Costs Equipment	Total	Total Incl O&P
0190	Galvanized 60d common nails	1 Clab	625	.013	Ea.	.15	.57		.72	1.01
0200	20d common nails	"	3800	.002	"	.03	.09		.12	.18

32 32 36 – Gabion Retaining Walls

32 32 36.10 Stone Gabion Retaining Walls

		Crew	Daily Output	Labor-Hours	Unit	Material	Labor	Equipment	Total	Total Incl O&P
0010	**STONE GABION RETAINING WALLS**									
4300	Stone filled gabions, not incl. excavation,									
4350	Galvanized, 6' high, 33° slope embankment	B-13	49	1.143	L.F.	56	55	12	123	157
4500	Highway surcharge		27	2.074		96.50	100	21.50	218	279
4600	9' high, up to 33° slope embankment		24	2.333		126	113	24.50	263.50	335
4700	Highway surcharge		16	3.500		169	169	36.50	374.50	480
4900	12' high, up to 33° slope embankment		14	4		196	193	42	431	550
5000	Highway surcharge		11	5.091		241	246	53.50	540.50	690
5950	For PVC coating, add					12%				

32 32 53 – Stone Retaining Walls

32 32 53.10 Retaining Walls, Stone

		Crew	Daily Output	Labor-Hours	Unit	Material	Labor	Equipment	Total	Total Incl O&P
0010	**RETAINING WALLS, STONE**									
0015	Including excavation, concrete footing and									
0020	stone 3' below grade. Price is exposed face area.									
0200	Decorative random stone, to 6' high, 1'-6" thick, dry set	D-1	35	.457	S.F.	71	22.50		93.50	112
0300	Mortar set		40	.400		73	19.50		92.50	110
0500	Cut stone, to 6' high, 1'-6" thick, dry set		35	.457		73.50	22.50		96	115
0600	Mortar set		40	.400		74	19.50		93.50	111
0800	Random stone, 6' to 10' high, 2' thick, dry set		45	.356		75.50	17.30		92.80	109
0900	Mortar set		50	.320		90	15.60		105.60	122
1100	Cut stone, 6' to 10' high, 2' thick, dry set		45	.356		75.50	17.30		92.80	109
1200	Mortar set		50	.320		90	15.60		105.60	123

32 33 Site Furnishings

32 33 33 – Site Manufactured Planters

32 33 33.10 Planters

		Crew	Daily Output	Labor-Hours	Unit	Material	Labor	Equipment	Total	Total Incl O&P
0010	**PLANTERS**									
0012	Concrete, sandblasted, precast, 48" diameter, 24" high	2 Clab	15	1.067	Ea.	665	47.50		712.50	800
0100	Fluted, precast, 7' diameter, 36" high	"	10	1.600	"	1,700	71		1,771	1,950

32 33 43 – Site Seating and Tables

32 33 43.13 Site Seating

		Crew	Daily Output	Labor-Hours	Unit	Material	Labor	Equipment	Total	Total Incl O&P
0010	**SITE SEATING**									
0012	Seating, benches, park, precast conc., w/backs, wood rails, 4' long	2 Clab	5	3.200	Ea.	670	142		812	950
0100	8' long	"	4	4	"	1,175	178		1,353	1,550

32 34 Fabricated Bridges

32 34 13 – Fabricated Pedestrian Bridges

32 34 13.10 Bridges, Pedestrian

		Crew	Daily Output	Labor-Hours	Unit	Material	2021 Bare Costs Labor	Equipment	Total	Total Incl O&P
0010	**BRIDGES, PEDESTRIAN**									
0011	Spans over streams, roadways, etc.									
0020	including erection, not including foundations									
0050	Precast concrete, complete in place, 8' wide, 60' span	E-2	215	.260	S.F.	146	15.55	7.95	169.50	193
0100	100' span		185	.303		159	18.05	9.25	186.30	213
0150	120' span		160	.350		173	21	10.70	204.70	235
0200	150' span		145	.386		180	23	11.80	214.80	247

32 35 Screening Devices

32 35 16 – Sound Barriers

32 35 16.10 Traffic Barriers, Highway Sound Barriers

		Crew	Daily Output	Labor-Hours	Unit	Material	2021 Bare Costs Labor	Equipment	Total	Total Incl O&P
0010	**TRAFFIC BARRIERS, HIGHWAY SOUND BARRIERS**									
0020	Highway sound barriers, not including footing									
0100	Precast concrete, concrete columns @ 30' OC, 8" T, 8' H	C-12	400	.120	L.F.	175	6.55	1.19	182.74	204
0110	12' H		265	.181		263	9.85	1.80	274.65	305
0120	16' H		200	.240		350	13.05	2.38	365.43	405
0130	20' H		160	.300		440	16.35	2.97	459.32	510
0400	Lt. wt. composite panel, cementitious face, st. posts @ 12' OC, 8' H	B-80B	190	.168		182	7.95	1.26	191.21	213
0410	12' H		125	.256		272	12.10	1.91	286.01	320
0420	16' H		95	.337		365	15.90	2.51	383.41	425
0430	20' H		75	.427		455	20	3.18	478.18	535

Division Notes

		CREW	DAILY OUTPUT	LABOR-HOURS	UNIT	BARE COSTS				TOTAL INCL O&P
						MAT.	LABOR	EQUIP.	TOTAL	

Estimating Tips
33 10 00 Water Utilities
33 30 00 Sanitary Sewerage Utilities
33 40 00 Storm Drainage Utilities

- Never assume that the water, sewer, and drainage lines will go in at the early stages of the project. Consider the site access needs before dividing the site in half with open trenches, loose pipe, and machinery obstructions. Always inspect the site to establish that the site drawings are complete. Check off all existing utilities on your drawings as you locate them. Be especially careful with underground utilities because appurtenances are sometimes buried during regrading or repaving operations. If you find any discrepancies, mark up the site plan for further research. Differing site conditions can be very costly if discovered later in the project.

- See also Section 33 01 00 for restoration of pipe where removal/replacement may be undesirable. Use of new types of piping materials can reduce the overall project cost. Owners/design engineers should consider the installing contractor as a valuable source of current information on utility products and local conditions that could lead to significant cost savings.

Reference Numbers
Reference numbers are shown at the beginning of some major classifications. These numbers refer to related items in the Reference Section. The reference information may be an estimating procedure, an alternate pricing method, or technical information.

Note: Not all subdivisions listed here necessarily appear. ■

Same Data. Simplified.

Enjoy the convenience and efficiency of accessing your costs anywhere:

- **Skip the multiplier** by setting your location
- **Quickly search,** edit, favorite and share costs
- **Stay on top of price changes** with automatic updates

Discover more at rsmeans.com/online

33 01 Operation and Maintenance of Utilities

33 01 10 – Operation and Maintenance of Water Utilities

33 01 10.10 Corrosion Resistance	Crew	Daily Output	Labor-Hours	Unit	Material	2021 Bare Costs Labor	2021 Bare Costs Equipment	Total	Total Incl O&P
0010 **CORROSION RESISTANCE**									
0012 Wrap & coat, add to pipe, 4" diameter				L.F.	2.50			2.50	2.75
0040 6" diameter					3.71			3.71	4.08
0060 8" diameter					5.05			5.05	5.55
0100 12" diameter					6.65			6.65	7.30
0200 24" diameter					14.85			14.85	16.30
0500 Coating, bituminous, per diameter inch, 1 coat, add					.19			.19	.21
0540 3 coat					.62			.62	.68
0560 Coal tar epoxy, per diameter inch, 1 coat, add					.25			.25	.28
0600 3 coat					.69			.69	.76

33 05 Common Work Results for Utilities

33 05 07 – Trenchless Installation of Utility Piping

33 05 07.36 Microtunneling

		Crew	Daily Output	Labor-Hours	Unit	Material	2021 Bare Costs Labor	2021 Bare Costs Equipment	Total	Total Incl O&P
0010	**MICROTUNNELING**									
0011	Not including excavation, backfill, shoring,									
0020	or dewatering, average 50'/day, slurry method									
0100	24" to 48" outside diameter, minimum				L.F.				965	965
0110	Adverse conditions, add				"				500	500
1000	Rent microtunneling machine, average monthly lease				Month				97,500	107,000
1010	Operating technician				Day				630	705
1100	Mobilization and demobilization, minimum				Job				41,200	45,900
1110	Maximum				"				445,500	490,500

33 05 61 – Concrete Manholes

33 05 61.10 Storm Drainage Manholes, Frames and Covers

		Crew	Daily Output	Labor-Hours	Unit	Material	2021 Bare Costs Labor	2021 Bare Costs Equipment	Total	Total Incl O&P
0010	**STORM DRAINAGE MANHOLES, FRAMES & COVERS**									
0020	Excludes footing, excavation, backfill (See line items for frame & cover)									
0050	Brick, 4' inside diameter, 4' deep	D-1	1	16	Ea.	700	780		1,480	1,950
0100	6' deep		.70	22.857		1,000	1,125		2,125	2,775
0150	8' deep		.50	32		1,300	1,550		2,850	3,775
0200	Add for 1' depth increase		4	4	V.L.F.	176	195		371	485
0400	Concrete blocks (radial), 4' ID, 4' deep		1.50	10.667	Ea.	430	520		950	1,250
0500	6' deep		1	16		570	780		1,350	1,800
0600	8' deep		.70	22.857		710	1,125		1,835	2,450
0700	For depths over 8', add		5.50	2.909	V.L.F.	73	142		215	293
0800	Concrete, cast in place, 4' x 4', 8" thick, 4' deep	C-14H	2	24	Ea.	680	1,275	13.70	1,968.70	2,650
0900	6' deep		1.50	32		985	1,725	18.25	2,728.25	3,650
1000	8' deep		1	48		1,400	2,575	27.50	4,002.50	5,400
1100	For depths over 8', add		8	6	V.L.F.	166	320	3.42	489.42	665
1110	Precast, 4' ID, 4' deep	B-22	4.10	7.317	Ea.	905	380	70	1,355	1,625
1120	6' deep		3	10		1,075	515	95.50	1,685.50	2,050
1130	8' deep		2	15		1,900	775	143	2,818	3,400

33 05 Common Work Results for Utilities

33 05 61 – Concrete Manholes

33 05 61.10 Storm Drainage Manholes, Frames and Covers

		Crew	Daily Output	Labor-Hours	Unit	Material	2021 Bare Costs Labor	Equipment	Total	Total Incl O&P
1140	For depths over 8', add	B-22	16	1.875	V.L.F.	107	97	17.90	221.90	283
1150	5' ID, 4' deep	B-6	3	8	Ea.	1,875	385	72	2,332	2,725
1160	6' deep		2	12		3,250	575	108	3,933	4,550
1170	8' deep		1.50	16	▼	2,750	770	144	3,664	4,325
1180	For depths over 8', add		12	2	V.L.F.	470	96	18	584	680
1190	6' ID, 4' deep		2	12	Ea.	2,100	575	108	2,783	3,275
1200	6' deep		1.50	16		4,600	770	144	5,514	6,350
1210	8' deep		1	24	▼	5,375	1,150	216	6,741	7,900
1220	For depths over 8', add	▼	8	3	V.L.F.	425	144	27	596	715
1250	Slab tops, precast, 8" thick									
1300	4' diameter manhole	B-6	8	3	Ea.	288	144	27	459	560
1400	5' diameter manhole		7.50	3.200		685	154	29	868	1,000
1500	6' diameter manhole	▼	7	3.429		715	165	31	911	1,075
3800	Steps, heavyweight cast iron, 7" x 9"	1 Bric	40	.200		17.25	10.75		28	35
3900	8" x 9"		40	.200		20.50	10.75		31.25	39
3928	12" x 10-1/2"		40	.200		31	10.75		41.75	50
4000	Standard sizes, galvanized steel		40	.200		24	10.75		34.75	42.50
4100	Aluminum		40	.200		31	10.75		41.75	50
4150	Polyethylene	▼	40	.200	▼	32.50	10.75		43.25	52

33 05 63 – Concrete Vaults and Chambers

33 05 63.13 Precast Concrete Utility Structures

		Crew	Daily Output	Labor-Hours	Unit	Material	2021 Bare Costs Labor	Equipment	Total	Total Incl O&P
0010	**PRECAST CONCRETE UTILITY STRUCTURES**, 6" thick									
0050	5' x 10' x 6' high, ID	B-13	2	28	Ea.	1,825	1,350	293	3,468	4,350
0100	6' x 10' x 6' high, ID		2	28		1,900	1,350	293	3,543	4,450
0150	5' x 12' x 6' high, ID		2	28		2,000	1,350	293	3,643	4,550
0200	6' x 12' x 6' high, ID		1.80	31.111		2,250	1,500	325	4,075	5,075
0250	6' x 13' x 6' high, ID		1.50	37.333		2,950	1,800	390	5,140	6,375
0300	8' x 14' x 7' high, ID	▼	1	56	▼	3,200	2,700	585	6,485	8,175
0350	Hand hole, precast concrete, 1-1/2" thick									
0400	1'-0" x 2'-0" x 1'-9", ID, light duty	B-1	4	6	Ea.	525	270		795	985
0450	4'-6" x 3'-2" x 2'-0", OD, heavy duty	B-6	3	8	"	1,600	385	72	2,057	2,400

33 05 97 – Identification and Signage for Utilities

33 05 97.05 Utility Connection

		Crew	Daily Output	Labor-Hours	Unit	Material	2021 Bare Costs Labor	Equipment	Total	Total Incl O&P
0010	**UTILITY CONNECTION**									
0020	Water, sanitary, stormwater, gas, single connection	B-14	1	48	Ea.	4,925	2,225	216	7,366	9,000
0030	Telecommunication	"	3	16	"	420	745	72	1,237	1,650

33 05 97.10 Utility Accessories

		Crew	Daily Output	Labor-Hours	Unit	Material	2021 Bare Costs Labor	Equipment	Total	Total Incl O&P
0010	**UTILITY ACCESSORIES**									
0400	Underground tape, detectable, reinforced, alum. foil core, 2"	1 Clab	150	.053	C.L.F.	3	2.37		5.37	6.85

33 16 Water Utility Storage Tanks

33 16 36 – Ground-Level Reinforced Concrete Water Storage Tanks

33 16 36.16 Prestressed Conc. Water Storage Tanks	Crew	Daily Output	Labor-Hours	Unit	Material	2021 Bare Costs Labor	Equipment	Total	Total Incl O&P
0010 **PRESTRESSED CONC. WATER STORAGE TANKS**									
0020 Not including fdn., pipe or pumps, 250,000 gallons				Ea.				299,000	329,500
0100 500,000 gallons								487,000	536,000
0300 1,000,000 gallons								707,000	807,500
0400 2,000,000 gallons								1,072,000	1,179,000
0600 4,000,000 gallons								1,706,000	1,877,000
0700 6,000,000 gallons								2,266,000	2,493,000
0750 8,000,000 gallons								2,924,000	3,216,000
0800 10,000,000 gallons								3,533,000	3,886,000

33 31 Sanitary Sewerage Piping

33 31 11 – Public Sanitary Sewerage Gravity Piping

		Crew	Daily Output	Labor-Hours	Unit	Material	2021 Bare Costs Labor	Equipment	Total	Total Incl O&P
3416	16" diameter	B-21	80	.350	L.F.	218	17.85	2.39	238.24	269
3418	18" diameter	B-13B	80	.700		380	34	12.40	426.40	485
3420	20" diameter		60	.933		415	45	16.50	476.50	540
3424	24" diameter		60	.933		565	45	16.50	626.50	705
3436	36" diameter		40	1.400		875	67.50	25	967.50	1,100

33 31 11.25 Sewage Collection, Polyvinyl Chloride Pipe

		Crew	Daily Output	Labor-Hours	Unit	Material	2021 Bare Costs Labor	Equipment	Total	Total Incl O&P
0010	**SEWAGE COLLECTION, POLYVINYL CHLORIDE PIPE**									
0020	Not including excavation or backfill									
2000	20' lengths, SDR 35, B&S, 4" diameter	B-20	375	.064	L.F.	2.21	3.16		5.37	7.15
2040	6" diameter		350	.069		4.26	3.38		7.64	9.75
2080	13' lengths, SDR 35, B&S, 8" diameter		335	.072		8.80	3.53		12.33	15
2120	10" diameter	B-21	330	.085		12.05	4.33	.58	16.96	20.50
2160	12" diameter		320	.088		17.05	4.47	.60	22.12	26
2170	14" diameter		280	.100		19.45	5.10	.68	25.23	30
2200	15" diameter		240	.117		22	5.95	.80	28.75	34
2250	16" diameter		220	.127		29	6.50	.87	36.37	42
4000	Piping, DWV PVC, no exc./bkfill., 10' L, Sch 40, 4" diameter	B-20	375	.064		4.78	3.16		7.94	9.95
4010	6" diameter		350	.069		11.20	3.38		14.58	17.35
4020	8" diameter		335	.072		17.85	3.53		21.38	25

33 31 11.37 Centrif. Cst. Fbgs-Reinf. Polymer Mort. Util. Pipe

		Crew	Daily Output	Labor-Hours	Unit	Material	2021 Bare Costs Labor	Equipment	Total	Total Incl O&P
0010	**CENTRIF. CST. FBGS-REINF. POLYMER MORT. UTIL. PIPE**									
0020	Not including excavation or backfill									
3348	48" diameter	B-13B	40	1.400	L.F.	181	67.50	25	273.50	325
3354	54" diameter		40	1.400		295	67.50	25	387.50	455
3363	63" diameter		20	2.800		365	135	49.50	549.50	655
3418	18" diameter		60	.933		128	45	16.50	189.50	226
3424	24" diameter		60	.933		141	45	16.50	202.50	240
3436	36" diameter		40	1.400		159	67.50	25	251.50	305
3444	44" diameter		40	1.400		175	67.50	25	267.50	320
3448	48" diameter		40	1.400		195	67.50	25	287.50	345

33 34 Onsite Wastewater Disposal

33 34 13 – Septic Tanks

33 34 13.13 Concrete Septic Tanks

	33 34 13.13 Concrete Septic Tanks	Crew	Daily Output	Labor-Hours	Unit	Material	2021 Bare Costs Labor	Equipment	Total	Total Incl O&P
0010	**CONCRETE SEPTIC TANKS**									
0011	Not including excavation or piping									
0015	Septic tanks, precast, 1,000 gallon	B-21	8	3.500	Ea.	1,000	179	24	1,203	1,400
0100	2,000 gallon	"	5	5.600		2,550	286	38	2,874	3,300
0200	5,000 gallon	B-13	3.50	16		7,350	770	168	8,288	9,425
0300	15,000 gallon, 4 piece	B-13B	1.70	32.941		25,700	1,600	580	27,880	31,300
0400	25,000 gallon, 4 piece		1.10	50.909		48,400	2,450	900	51,750	57,500
0500	40,000 gallon, 4 piece	↓	.80	70		53,500	3,375	1,250	58,125	65,500
0520	50,000 gallon, 5 piece	B-13C	.60	93.333		61,500	4,500	3,850	69,850	79,000
0640	75,000 gallon, cast in place	C-14C	.25	448		75,000	23,400	108	98,508	117,500
0660	100,000 gallon	"	.15	747		93,000	39,000	179	132,179	160,000
1150	Leaching field chambers, 13' x 3'-7" x 1'-4", standard	B-13	16	3.500		475	169	36.50	680.50	820
1200	Heavy duty, 8' x 4' x 1'-6"		14	4		325	193	42	560	690
1300	13' x 3'-9" x 1'-6"		12	4.667		1,400	225	49	1,674	1,925
1350	20' x 4' x 1'-6"	↓	5	11.200		1,250	540	117	1,907	2,300
1400	Leaching pit, precast concrete, 3' diameter, 3' deep	B-21	8	3.500		710	179	24	913	1,075
1500	6' diameter, 3' section		4.70	5.957		1,100	305	40.50	1,445.50	1,725
2000	Velocity reducing pit, precast conc., 6' diameter, 3' deep	↓	4.70	5.957	↓	1,875	305	40.50	2,220.50	2,550

33 34 13.33 Polyethylene Septic Tanks

	33 34 13.33 Polyethylene Septic Tanks	Crew	Daily Output	Labor-Hours	Unit	Material	2021 Bare Costs Labor	Equipment	Total	Total Incl O&P
0010	**POLYETHYLENE SEPTIC TANKS**									
0015	High density polyethylene, 1,000 gallon	B-21	8	3.500	Ea.	1,600	179	24	1,803	2,050
0020	1,250 gallon		8	3.500		1,325	179	24	1,528	1,750
0025	1,500 gallon	↓	7	4	↓	1,400	204	27.50	1,631.50	1,850

33 34 51 – Drainage Field Systems

33 34 51.10 Drainage Field Excavation and Fill

	33 34 51.10 Drainage Field Excavation and Fill	Crew	Daily Output	Labor-Hours	Unit	Material	2021 Bare Costs Labor	Equipment	Total	Total Incl O&P
0010	**DRAINAGE FIELD EXCAVATION AND FILL**									
2200	Septic tank & drainage field excavation with 3/4 C.Y. backhoe	B-12F	145	.110	C.Y.		5.85	4.84	10.69	14
2400	4' trench for disposal field, 3/4 C.Y. backhoe	"	335	.048	L.F.		2.53	2.09	4.62	6.05
2600	Gravel fill, run of bank	B-6	150	.160	C.Y.	26.50	7.70	1.44	35.64	42.50
2800	Crushed stone, 3/4"	"	150	.160	"	41	7.70	1.44	50.14	58

33 34 51.13 Utility Septic Tank Tile Drainage Field

	33 34 51.13 Utility Septic Tank Tile Drainage Field	Crew	Daily Output	Labor-Hours	Unit	Material	2021 Bare Costs Labor	Equipment	Total	Total Incl O&P
0010	**UTILITY SEPTIC TANK TILE DRAINAGE FIELD**									
0015	Distribution box, concrete, 5 outlets	2 Clab	20	.800	Ea.	75	35.50		110.50	136
0020	7 outlets		16	1		93.50	44.50		138	170
0025	9 outlets		8	2		530	89		619	720
0115	Distribution boxes, HDPE, 5 outlets		20	.800		75	35.50		110.50	136
0117	6 outlets		15	1.067		75	47.50		122.50	153
0118	7 outlets		15	1.067		74.50	47.50		122	153
0120	8 outlets	↓	10	1.600		79	71		150	193
0240	Distribution boxes, outlet flow leveler	1 Clab	50	.160	↓	2.99	7.10		10.09	13.90
0365	chamber 16" H x 34" W	2 Clab	300	.053	L.F.	19.40	2.37		21.77	25

33 42 11.40 Piping, Storm Drainage, Corrugated Metal

	33 42 11.40 Piping, Storm Drainage, Corrugated Metal	Crew	Daily Output	Labor-Hours	Unit	Material	2021 Bare Costs Labor	Equipment	Total	Total Incl O&P
0010	**PIPING, STORM DRAINAGE, CORRUGATED METAL**									
0020	Not including excavation or backfill									
2000	Corrugated metal pipe, galvanized									
2020	Bituminous coated with paved invert, 20' lengths									
2040	8" diameter, 16 ga.	B-14	330	.145	L.F.	7.25	6.80	.65	14.70	18.80
2060	10" diameter, 16 ga.		260	.185		7.55	8.60	.83	16.98	22
2080	12" diameter, 16 ga.		210	.229		12	10.65	1.03	23.68	30
2100	15" diameter, 16 ga.		200	.240		12.70	11.20	1.08	24.98	32
2120	18" diameter, 16 ga.		190	.253		18.75	11.75	1.14	31.64	39.50
2140	24" diameter, 14 ga.		160	.300		20.50	13.95	1.35	35.80	45.50
2160	30" diameter, 14 ga.	B-13	120	.467		26.50	22.50	4.89	53.89	68
2180	36" diameter, 12 ga.		120	.467		29.50	22.50	4.89	56.89	71.50
2200	48" diameter, 12 ga.		100	.560		40	27	5.85	72.85	91
2220	60" diameter, 10 ga.	B-13B	75	.747		70	36	13.20	119.20	146
2240	72" diameter, 8 ga.	"	45	1.244		73	60	22	155	194
2500	Galvanized, uncoated, 20' lengths									
2520	8" diameter, 16 ga.	B-14	355	.135	L.F.	8.20	6.30	.61	15.11	19.05
2540	10" diameter, 16 ga.		280	.171		8.15	8	.77	16.92	22
2560	12" diameter, 16 ga.		220	.218		8.90	10.15	.98	20.03	26
2580	15" diameter, 16 ga.		220	.218		11.60	10.15	.98	22.73	29
2600	18" diameter, 16 ga.		205	.234		12.90	10.90	1.05	24.85	31.50
2620	24" diameter, 14 ga.		175	.274		24.50	12.80	1.23	38.53	47
2640	30" diameter, 14 ga.	B-13	130	.431		28	21	4.51	53.51	66.50
2660	36" diameter, 12 ga.		130	.431		32	21	4.51	57.51	71
2680	48" diameter, 12 ga.		110	.509		47.50	24.50	5.35	77.35	94.50
2690	60" diameter, 10 ga.	B-13B	78	.718		78	34.50	12.70	125.20	151
2780	End sections, 8" diameter	B-14	35	1.371	Ea.	102	64	6.15	172.15	214
2785	10" diameter		35	1.371		80.50	64	6.15	150.65	191
2790	12" diameter		35	1.371		96	64	6.15	166.15	207
2800	18" diameter		30	1.600		112	74.50	7.20	193.70	242
2810	24" diameter	B-13	25	2.240		189	108	23.50	320.50	395
2820	30" diameter		25	2.240		340	108	23.50	471.50	560
2825	36" diameter		20	2.800		455	135	29.50	619.50	735
2830	48" diameter		10	5.600		990	270	58.50	1,318.50	1,575
2835	60" diameter	B-13B	5	11.200		1,950	540	198	2,688	3,150
2840	72" diameter	"	4	14		2,250	675	248	3,173	3,750

33 42 11.60 Sewage/Drainage Collection, Concrete Pipe

	33 42 11.60 Sewage/Drainage Collection, Concrete Pipe	Crew	Daily Output	Labor-Hours	Unit	Material	2021 Bare Costs Labor	Equipment	Total	Total Incl O&P
0010	**SEWAGE/DRAINAGE COLLECTION, CONCRETE PIPE**									
1000	Non-reinforced pipe, extra strength, B&S or T&G joints									
1010	6" diameter	B-14	265.04	.181	L.F.	8.05	8.45	.81	17.31	22.50
1020	8" diameter		224	.214		8.90	10	.96	19.86	25.50
1030	10" diameter		216	.222		9.85	10.35	1	21.20	27.50
1040	12" diameter		200	.240		10.25	11.20	1.08	22.53	29
1050	15" diameter		180	.267		14.70	12.40	1.20	28.30	36
1060	18" diameter		144	.333		18.45	15.55	1.50	35.50	45
1070	21" diameter		112	.429		19.30	19.95	1.93	41.18	53.50
1080	24" diameter		100	.480		22.50	22.50	2.16	47.16	60.50
2000	Reinforced culvert, class 3, no gaskets									
2010	12" diameter	B-14	150	.320	L.F.	16.50	14.90	1.44	32.84	41.50
2020	15" diameter		150	.320		20.50	14.90	1.44	36.84	46
2030	18" diameter		132	.364		25	16.95	1.64	43.59	55.50
2035	21" diameter		120	.400		28.50	18.65	1.80	48.95	61

33 42 11 – Stormwater Gravity Piping

33 42 11.60 Sewage/Drainage Collection, Concrete Pipe		Crew	Daily Output	Labor-Hours	Unit	Material	2021 Bare Costs Labor	Equipment	Total	Total Incl O&P
2040	24" diameter	B-14	100	.480	L.F.	27.50	22.50	2.16	52.16	66
2045	27" diameter	B-13	92	.609		46.50	29.50	6.40	82.40	103
2050	30" diameter		88	.636		44	30.50	6.65	81.15	102
2060	36" diameter	▼	72	.778		80	37.50	8.15	125.65	153
2070	42" diameter	B-13B	68	.824		104	40	14.55	158.55	190
2080	48" diameter		64	.875		108	42.50	15.45	165.95	199
2085	54" diameter		56	1		148	48.50	17.70	214.20	254
2090	60" diameter		48	1.167		229	56.50	20.50	306	360
2100	72" diameter		40	1.400		254	67.50	25	346.50	410
2120	84" diameter	▼	32	1.750		410	84.50	31	525.50	615
2140	96" diameter		24	2.333		495	113	41.50	649.50	760
2200	With gaskets, class 3, 12" diameter	B-21	168	.167		18.15	8.50	1.14	27.79	34
2220	15" diameter		160	.175		22.50	8.95	1.19	32.64	39.50
2230	18" diameter		152	.184		28	9.40	1.26	38.66	46
2235	21" diameter		152	.184		31	9.40	1.26	41.66	50
2240	24" diameter	▼	136	.206		34	10.50	1.40	45.90	54.50
2260	30" diameter	B-13	88	.636		52.50	30.50	6.65	89.65	111
2270	36" diameter	"	72	.778		90	37.50	8.15	135.65	163
2290	48" diameter	B-13B	64	.875		121	42.50	15.45	178.95	214
2310	72" diameter	"	40	1.400	▼	275	67.50	25	367.50	430
2330	Flared ends, 12" diameter	B-21	31	.903	Ea.	340	46	6.15	392.15	450
2340	15" diameter		25	1.120		395	57	7.65	459.65	530
2400	18" diameter		20	1.400		445	71.50	9.55	526.05	610
2420	24" diameter	▼	14	2		515	102	13.65	630.65	740
2440	36" diameter	B-13	10	5.600	▼	1,125	270	58.50	1,453.50	1,725
3080	Radius pipe, add to pipe prices, 12" to 60" diameter				L.F.	50%				
3090	Over 60" diameter, add				"	20%				
3500	Reinforced elliptical, 8' lengths, C507 class 3									
3520	14" x 23" inside, round equivalent 18" diameter	B-21	82	.341	L.F.	41	17.45	2.33	60.78	73.50
3530	24" x 38" inside, round equivalent 30" diameter	B-13	58	.966		70	46.50	10.10	126.60	158
3540	29" x 45" inside, round equivalent 36" diameter		52	1.077		110	52	11.30	173.30	211
3550	38" x 60" inside, round equivalent 48" diameter		38	1.474		178	71	15.45	264.45	320
3560	48" x 76" inside, round equivalent 60" diameter		26	2.154		203	104	22.50	329.50	405
3570	58" x 91" inside, round equivalent 72" diameter	▼	22	2.545	▼	345	123	26.50	494.50	590

33 42 13 – Stormwater Culverts

33 42 13.13 Public Pipe Culverts

		Crew	Daily Output	Labor-Hours	Unit	Material	Labor	Equipment	Total	Total Incl O&P
0010	**PUBLIC PIPE CULVERTS**									
0020	Headwall, concrete									
0100	C.I.P., 30 degree skewed wingwall, 12" diameter pipe	C-14H	3.20	15	Ea.	231	805	8.55	1,044.55	1,475
0110	18" diameter pipe		2.29	20.957		325	1,125	11.95	1,461.95	2,050
0120	24" diameter pipe		1.60	30		530	1,600	17.10	2,147.10	3,000
0130	30" diameter pipe		1.20	40		690	2,150	23	2,863	3,950
0140	36" diameter pipe		.91	52.724		820	2,825	30	3,675	5,125
0150	48" diameter pipe		.56	85.714		1,350	4,600	49	5,999	8,350
0160	60" diameter pipe	▼	.38	126		2,975	6,750	72	9,797	13,500
0520	Precast, 12" diameter pipe	B-6	12	2		2,350	96	18	2,464	2,750
0530	18" diameter pipe		10	2.400		2,350	115	21.50	2,486.50	2,775
0540	24" diameter pipe		10	2.400		3,175	115	21.50	3,311.50	3,700
0550	30" diameter pipe		8	3		2,475	144	27	2,646	2,950
0560	36" diameter pipe	▼	8	3		3,275	144	27	3,446	3,850
0570	48" diameter pipe	B-69	7	6.857		6,925	335	208	7,468	8,350
0580	60" diameter pipe	"	6	8	▼	11,500	390	243	12,133	13,600

For customer support on your Concrete & Masonry Costs with RSMeans Data, call 800.448.8182.

251

33 42 13 – Stormwater Culverts

33 42 13.15 Oval Arch Culverts

		Crew	Daily Output	Labor-Hours	Unit	Material	2021 Bare Costs Labor	Equipment	Total	Total Incl O&P
0010	**OVAL ARCH CULVERTS**									
3000	Corrugated galvanized or aluminum, coated & paved									
3020	17" x 13", 16 ga., 15" equivalent	B-14	200	.240	L.F.	16.90	11.20	1.08	29.18	36.50
3040	21" x 15", 16 ga., 18" equivalent		150	.320		17.30	14.90	1.44	33.64	42.50
3060	28" x 20", 14 ga., 24" equivalent		125	.384		26.50	17.90	1.73	46.13	57.50
3080	35" x 24", 14 ga., 30" equivalent		100	.480		32.50	22.50	2.16	57.16	72
3100	42" x 29", 12 ga., 36" equivalent	B-13	100	.560		30	27	5.85	62.85	80
3120	49" x 33", 12 ga., 42" equivalent		90	.622		39	30	6.50	75.50	94.50
3140	57" x 38", 12 ga., 48" equivalent		75	.747		51.50	36	7.85	95.35	119
3160	Steel, plain oval arch culverts, plain									
3180	17" x 13", 16 ga., 15" equivalent	B-14	225	.213	L.F.	11.05	9.95	.96	21.96	28
3200	21" x 15", 16 ga., 18" equivalent		175	.274		14.40	12.80	1.23	28.43	36.50
3220	28" x 20", 14 ga., 24" equivalent		150	.320		20	14.90	1.44	36.34	45.50
3240	35" x 24", 14 ga., 30" equivalent	B-13	108	.519		26	25	5.45	56.45	72.50
3260	42" x 29", 12 ga., 36" equivalent		108	.519		46.50	25	5.45	76.95	94.50
3280	49" x 33", 12 ga., 42" equivalent		92	.609		56	29.50	6.40	91.90	113
3300	57" x 38", 12 ga., 48" equivalent		75	.747		56.50	36	7.85	100.35	125
3320	End sections, 17" x 13"		22	2.545	Ea.	132	123	26.50	281.50	360
3340	42" x 29"		17	3.294	"	545	159	34.50	738.50	875
3360	Multi-plate arch, steel	B-20	1690	.014	Lb.	1.29	.70		1.99	2.47

33 42 33 – Stormwater Curbside Drains and Inlets

33 42 33.13 Catch Basins

		Crew	Daily Output	Labor-Hours	Unit	Material	2021 Bare Costs Labor	Equipment	Total	Total Incl O&P
0010	**CATCH BASINS**									
0011	Not including footing & excavation									
1600	Frames & grates, C.I., 24" square, 500 lb.	B-6	7.80	3.077	Ea.	395	148	27.50	570.50	685
1700	26" D shape, 600 lb.		7	3.429		735	165	31	931	1,075
1800	Light traffic, 18" diameter, 100 lb.		10	2.400		216	115	21.50	352.50	435
1900	24" diameter, 300 lb.		8.70	2.759		209	133	25	367	455
2000	36" diameter, 900 lb.		5.80	4.138		720	199	37.50	956.50	1,125
2100	Heavy traffic, 24" diameter, 400 lb.		7.80	3.077		291	148	27.50	466.50	570
2200	36" diameter, 1,150 lb.		3	8		955	385	72	1,412	1,700
2300	Mass. State standard, 26" diameter, 475 lb.		7	3.429		695	165	31	891	1,050
2400	30" diameter, 620 lb.		7	3.429		380	165	31	576	695
2500	Watertight, 24" diameter, 350 lb.		7.80	3.077		430	148	27.50	605.50	725
2600	26" diameter, 500 lb.		7	3.429		445	165	31	641	770
2700	32" diameter, 575 lb.		6	4		915	192	36	1,143	1,325
2800	3 piece cover & frame, 10" deep,									
2900	1,200 lb., for heavy equipment	B-6	3	8	Ea.	1,075	385	72	1,532	1,850
3000	Raised for paving 1-1/4" to 2" high									
3100	4 piece expansion ring									
3200	20" to 26" diameter	1 Clab	3	2.667	Ea.	223	118		341	425
3300	30" to 36" diameter	"	3	2.667	"	310	118		428	515
3320	Frames and covers, existing, raised for paving, 2", including									
3340	row of brick, concrete collar, up to 12" wide frame	B-6	18	1.333	Ea.	52.50	64	12	128.50	166
3360	20" to 26" wide frame		11	2.182		79	105	19.65	203.65	266
3380	30" to 36" wide frame		9	2.667		98	128	24	250	325
3400	Inverts, single channel brick	D-1	3	5.333		113	260		373	515
3500	Concrete		5	3.200		137	156		293	385
3600	Triple channel, brick		2	8		186	390		576	790
3700	Concrete		3	5.333		156	260		416	560

33 42 Stormwater Conveyance

33 42 33 – Stormwater Curbside Drains and Inlets

33 42 33.50 Stormwater Management	Crew	Daily Output	Labor-Hours	Unit	Material	2021 Bare Costs Labor	Equipment	Total	Total Incl O&P
0010 **STORMWATER MANAGEMENT**									
0030 Add per S.F. of impervious surface	B-37	6000	.008	S.F.	3.19	.37	.04	3.60	4.12

For customer support on your Concrete & Masonry Costs with RSMeans Data, call 800.448.8182.

253

Division Notes

		CREW	DAILY OUTPUT	LABOR-HOURS	UNIT	BARE COSTS				TOTAL INCL O&P
						MAT.	LABOR	EQUIP.	TOTAL	

Estimating Tips
34 11 00 Rail Tracks
This subdivision includes items that may involve either repair of existing or construction of new railroad tracks. Additional preparation work, such as the roadbed earthwork, would be found in Division 31. Additional new construction siding and turnouts are found in Subdivision 34 72. Maintenance of railroads is found under 34 01 23 Operation and Maintenance of Railways.

34 40 00 Traffic Signals
This subdivision includes traffic signal systems. Other traffic control devices such as traffic signs are found in Subdivision 10 14 53 Traffic Signage.

34 70 00 Vehicle Barriers
This subdivision includes security vehicle barriers, guide and guard rails, crash barriers, and delineators. The actual maintenance and construction of concrete and asphalt pavement are found in Division 32.

Reference Numbers
Reference numbers are shown at the beginning of some major classifications. These numbers refer to related items in the Reference Section. The reference information may be an estimating procedure, an alternate pricing method, or technical information.

Note: Not all subdivisions listed here necessarily appear. ∎

Same Data. Simplified.
Enjoy the convenience and efficiency of accessing your costs anywhere:
- **Skip the multiplier** by setting your location
- **Quickly search,** edit, favorite and share costs
- **Stay on top of price changes** with automatic updates

Discover more at rsmeans.com/online

34 11 Rail Tracks

34 11 33 – Track Cross Ties

34 11 33.13 Concrete Track Cross Ties	Crew	Daily Output	Labor-Hours	Unit	Material	2021 Bare Costs Labor	2021 Bare Costs Equipment	Total	Total Incl O&P
0010 **CONCRETE TRACK CROSS TIES**									
1400 Ties, concrete, 8'-6" long, 30" OC	B-14	80	.600	Ea.	132	28	2.70	162.70	189

34 11 93 – Track Appurtenances and Accessories

34 11 93.50 Track Accessories

	Crew	Daily Output	Labor-Hours	Unit	Material	Labor	Equipment	Total	Total Incl O&P
0010 **TRACK ACCESSORIES**									
0020 Car bumpers, test	B-14	2	24	Ea.	4,150	1,125	108	5,383	6,350
0100 Heavy duty	"	2	24	"	7,850	1,125	108	9,083	10,400

34 71 Roadway Construction

34 71 13 – Vehicle Barriers

34 71 13.26 Vehicle Guide Rails

	Crew	Daily Output	Labor-Hours	Unit	Material	Labor	Equipment	Total	Total Incl O&P
0010 **VEHICLE GUIDE RAILS**									
0012 Corrugated stl., galv. stl. posts, 6'-3" OC - "W-Beam Guiderail"	B-80	850	.038	L.F.	12.10	1.84	1.24	15.18	17.40
0100 Double face		570	.056	"	36.50	2.74	1.85	41.09	46
0200 End sections, galvanized, flared		50	.640	Ea.	98.50	31	21	150.50	178
0300 Wrap around end		50	.640		141	31	21	193	225
0325 W-Beam, stl., galv. Terminal Connector, Concrete Mounted		8	4		249	195	132	576	710
0350 Anchorage units		15	2.133		1,200	104	70	1,374	1,550
0410 Thrie Beam, stl., galv. stl. posts, 6'-3" OC		850	.038	L.F.	23.50	1.84	1.24	26.58	29.50
0440 Thrie Beam, stl., galv, Wrap Around End Section		50	.640	Ea.	90	31	21	142	168
0450 Thrie Beam, stl., galv. Terminal Connector, Concrete Mounted		8	4	"	274	195	132	601	735
0510 Rub Rail, stl, galv, steel posts (not included), 6'-3" OC		850	.038	L.F.	1.70	1.84	1.24	4.78	5.95
0600 Cable guide rail, 3 at 3/4" cables, steel posts, single face		900	.036		12.60	1.73	1.17	15.50	17.75
0650 Double face		635	.050		23.50	2.46	1.66	27.62	31.50
0700 Wood posts		950	.034		13.30	1.64	1.11	16.05	18.25
0750 Double face		650	.049		23.50	2.40	1.62	27.52	31.50
0800 Anchorage units, breakaway		15	2.133	Ea.	1,550	104	70	1,724	1,925
0900 Guide rail, steel box beam, 6" x 6"		120	.267	L.F.	38.50	13	8.75	60.25	71
0950 End assembly	B-80A	48	.500	Ea.	1,550	22	17.70	1,589.70	1,750
1100 Median barrier, steel box beam, 6" x 8"	B-80	215	.149	L.F.	52.50	7.25	4.90	64.65	74
1120 Shop curved	B-80A	92	.261	"	71	11.60	9.25	91.85	105
1140 End assembly		48	.500	Ea.	2,575	22	17.70	2,614.70	2,875
1150 Corrugated beam		400	.060	L.F.	59	2.66	2.13	63.79	71.50
1400 Resilient guide fence and light shield, 6' high	B-2	130	.308	"	25	13.80		38.80	48
1500 Concrete posts, individual, 6'-5", triangular	B-80	110	.291	Ea.	68	14.20	9.55	91.75	107
1550 Square		110	.291		74.50	14.20	9.55	98.25	114
1600 Wood guide posts		150	.213		45	10.40	7	62.40	73
1650 Timber guide rail, 4" x 8" with 6" x 8" wood posts, treated		960	.033	L.F.	13.90	1.63	1.10	16.63	18.95
2000 Median, precast concrete, 3'-6" high, 2' wide, single face	B-29	380	.147		60.50	7.10	2.24	69.84	79.50
2200 Double face	"	340	.165		69.50	7.95	2.50	79.95	91
2300 Cast in place, steel forms	C-2	170	.282		49	15.05		64.05	76.50
2320 Slipformed	C-7	352	.205		43	9.80	3.10	55.90	65.50
2400 Speed bumps, thermoplastic, 10-1/2" x 2-1/4" x 48" long	B-2	120	.333	Ea.	144	14.95		158.95	182
3030 Impact barrier, MUTCD, barrel type	B-16	30	1.067	"	395	49.50	19.30	463.80	530

34 72 Railway Construction

34 72 16 – Railway Siding

34 72 16.50 Railroad Sidings		Crew	Daily Output	Labor-Hours	Unit	Material	2021 Bare Costs Labor	Equipment	Total	Total Incl O&P
0010	**RAILROAD SIDINGS**	R347216-10									
1020	100 lb. new rail	R347216-20	B-14	22	2.182	L.F.	194	102	9.80	305.80	375

For customer support on your Concrete & Masonry Costs with RSMeans Data, call 800.448.8182.

257

Assemblies Section

Table of Contents

Table of Contents

RSMeans data: Assemblies— How They Work

Assemblies estimating provides a fast and reasonably accurate way to develop construction costs. An assembly is the grouping of individual work items—with appropriate quantities— to provide a cost for a major construction component in a convenient unit of measure.

An assemblies estimate is often used during early stages of design development to compare the cost impact of various design alternatives on total building cost.

Assemblies estimates are also used as an efficient tool to verify construction estimates.

Assemblies estimates do not require a completed design or detailed drawings. Instead, they are based on the general size of the structure and other known parameters of the project. The degree of accuracy of an assemblies estimate is generally within +/- 15%.

Most assemblies consist of three major elements: a graphic, the system components, and the cost data itself. The **Graphic** is a visual representation showing the typical appearance of the assembly

① Unique 12-character Identifier

Our assemblies are identified by a **unique 12-character identifier**. The assemblies are numbered using UNIFORMAT II, ASTM Standard E1557. The first 5 characters represent this system to Level 3. The last 7 characters represent further breakdown in order to arrange items in understandable groups of similar tasks. Line numbers are consistent across all of our publications, so a line number in any assemblies data set will always refer to the same item.

② Narrative Descriptions

Our assemblies descriptions appear in two formats: narrative and table. **Narrative descriptions** are shown in a hierarchical structure to make them readable. In order to read a complete description, read up through the indents to the top of the section. Include everything that is above and to the left that is not contradicted by information below.

Narrative Format

G10 Site Preparation

G1030 Site Earthwork

Trenching Systems are shown on a cost per linear foot basis. The systems include: excavation; backfill and removal of spoil; and compaction for various depths and trench bottom widths. The backfill has been reduced to accommodate a pipe of suitable diameter and bedding.

The slope for trench sides varies from none to 1:1.

The Expanded System Listing shows Trenching Systems that range from 2' to 12' in width. Depths range from 2' to 25'.

System Components		QUANTITY	UNIT	COST PER L.F.		
				EQUIP.	LABOR	TOTAL
① SYSTEM G1030 805 1310 TRENCHING, COMMON EARTH, NO SLOPE, 2' WIDE, 2' DP, 3/8 C.Y. BUCKET						
	Excavation, trench, hyd. backhoe, track mtd., 3/8 C.Y. bucket	.148	B.C.Y.	.24	1.21	1.45
	Backfill and load spoil, from stockpile	.153	L.C.Y.	.13	.37	.50
	Compaction by vibrating plate, 6" lifts, 4 passes	.118	E.C.Y.	.03	.45	.48
	Remove excess spoil, 8 C.Y. dump truck, 2 mile roundtrip	.040	L.C.Y.	.15	.20	.35
	TOTAL			.55	2.23	2.78

G1030 805	Trenching Common Earth	COST PER L.F.		
		EQUIP.	LABOR	TOTAL
1310	Trenching, common earth, no slope, 2' wide, 2' deep, 3/8 C.Y. bucket	.54	2.24	2.78
1330	4' deep, 3/8 C.Y. bucket	.98	4.48	5.46
1400	② 4' wide, 2' deep, 3/8 C.Y. bucket	1.28	4.45	5.73
1420	4' deep, 1/2 C.Y. bucket	1.94	7.45	9.39
1800	1/2 to 1 slope, 2' wide, 2' deep, 3/8 C.Y. bucket	.76	3.36	4.12
1820	4' deep, 3/8 C.Y. bucket	1.85	8.95	10.80
1860	8' deep, 1/2 C.Y. bucket	4.78	23	27.78
2300	4' wide, 2' deep, 3/8 C.Y. bucket	1.49	5.55	7.04
2320	4' deep, 1/2 C.Y. bucket	2.72	11.35	14.07
2360	8' deep, 1/2 C.Y. bucket	13.10	31.50	44.60
2400	12' deep, 1 C.Y. bucket	24	59.50	83.50
2840	6' wide, 6' deep, 5/8 C.Y. bucket w/trench box	13.15	26.50	39.65
2900	12' deep, 1-1/2 C.Y. bucket	19.75	53	72.75
3020	24' deep, 3-1/2 C.Y. bucket	73	143	216
3100	8' wide, 12' deep, 1-1/2 C.Y. bucket w/trench box	24.50	60.50	85
3500	1 to 1 slope, 2' wide, 2' deep, 3/8 C.Y. bucket	.98	4.48	5.46
3540	4' deep, 3/8 C.Y. bucket	2.72	13.45	16.17
3580	8' deep, 1/2 C.Y. bucket	5.95	29	34.95
3800	4' wide, 2' deep, 3/8 C.Y. bucket	1.71	6.70	8.41
3840	4' deep, 1/2 C.Y. bucket	3.50	15.20	18.70
3880	8' deep, 1/2 C.Y. bucket	19.50	47	66.50
3920	12' deep, 1 C.Y. bucket	41.50	99.50	141
4030	6' wide, 6' deep, 5/8 C.Y. bucket w/trench box	17.30	35.50	52.80
4060	12' deep, 1-1/2 C.Y. bucket	30	81	111
4090	24' deep, 3-1/2 C.Y. bucket	121	240	361
4500	8' wide, 12' deep, 1-1/2 C.Y. bucket w/trench box	34.50	88	122.50
4950	12' wide, 20' deep, 3-1/2 C.Y. bucket w/trench box	109	207	316

For supplemental customizable square foot estimating forms, visit: **RSMeans.com/2021books**

in question. It is frequently accompanied by additional explanatory technical information describing the class of items. The **System Components** is a listing of the individual tasks that make up the assembly, including the quantity and unit of measure for each item, along with the cost of material and installation. The **Assemblies**

data below lists prices for other similar systems with dimensional and/or size variations.

All of our assemblies costs represent the cost for the installing contractor. An allowance for profit has been added to all material, labor, and equipment rental costs. A markup for labor burdens, including workers' compensation, fixed

overhead, and business overhead, is included with installation costs.

The information in RSMeans cost data represents a "national average" cost. This data should be modified to the project location using the **City Cost Indexes** or **Location Factors** tables found in the Reference Section.

Table Format

A10 Foundations

A1010 Standard Foundations

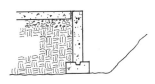

The Foundation Bearing Wall System includes: forms up to 6' high (four uses); 3,000 p.s.i. concrete placed and vibrated; and form removal with breaking form ties and patching walls. The wall systems list walls from 6" to 16" thick and are designed with minimum reinforcement.

Excavation and backfill are not included.

Please see the reference section for further design and cost information.

System Components **4**			COST PER L.F.		
	QUANTITY	UNIT	MAT.	INST.	TOTAL
SYSTEM A1010 105 1500					
FOUNDATION WALL, CAST IN PLACE, DIRECT CHUTE, 4' HIGH, 6" THICK					
Formwork	8.000	SFCA	6.48	50.80	57.28
Reinforcing	3.300	Lb.	2.27	1.55	3.82
Unloading & sorting reinforcing	3.300	Lb.		.09	.09
Concrete, 3,000 psi	.074	C.Y.	10.14		10.14
Place concrete, direct chute	.074	C.Y.		2.75	2.75
Finish walls, break ties and patch voids, one side	4.000	S.F.	.20	4.48	4.68
TOTAL			19.09	59.67	78.76

A1010 105			**Wall Foundations**					
	WALL HEIGHT (FT.)	PLACING METHOD	CONCRETE (C.Y. per L.F.)	REINFORCING (LBS. per L.F.)	WALL THICKNESS (IN.)	COST PER L.F.		
						MAT.	INST.	TOTAL
1500	4'	direct chute	.074	3.3	6	19.10	59.50	78.60
1520			.099	4.8	8	23.50	61.50	85
1540			.123	6.0	10	27.50	62.50	90
1560			.148	7.2	12	32	64	96
1580			.173	8.1	14	36	65	101
1600			.197	9.44	16	40	66.50	106.50
1700	4'	pumped	.074	3.3	6	19.10	60.50	79.60
1720			.099	4.8	8	23.50	62.50	86
1740			.123	6.0	10	27.50	64	91.50
1760			.148	7.2	12	32	65.50	97.50
1780			.173	8.1	14	36	66.50	102.50
1800			.197	9.44	16	40	68	108
3000	6'	direct chute	.111	4.95	6	28.50	89.50	118
3020			.149	7.20	8	35.50	92	127.50
3040			.184	9.00	10	41.50	93.50	135
3060			.222	10.8	12	48	96	144
3080			.260	12.15	14	54	97.50	151.50
3100			.300	14.39	16	61	99.50	160.50
3200	6'	pumped	.111	4.95	6	28.50	91	119.50
3220			.149	7.20	8	35.50	94	129.50
3240			.184	9.00	10	41.50	96	137.50
3260			.222	10.8	12	48	98.50	146.50
3280			.260	12.15	14	54	100	154
3300			.300	14.39	16	61	103	164

3 Unit of Measure

All RSMeans data: Assemblies include a typical **Unit of Measure** used for estimating that item. For instance, for continuous footings or foundation walls the unit is linear feet (L.F.). For spread footings the unit is each (Ea.). The estimator needs to take special care that the unit in the data matches the unit in the takeoff. Abbreviations and unit conversions can be found in the Reference Section.

4 System Components

System components are listed separately to detail what is included in the development of the total system price.

5 Table Descriptions

Table descriptions work similar to Narrative Descriptions, except that if there is a blank in the column at a particular line number, read up to the description above in the same column.

Sample Estimate

This sample demonstrates the elements of an estimate, including a tally of the RSMeans data lines. Published assemblies costs include all markups for labor burden and profit for the installing contractor. This estimate adds a summary of the markups applied by a general contractor on the installing contractor's work. These figures represent the total cost to the owner. The location factor with RSMeans data is applied at the bottom of the estimate to adjust the cost of the work to a specific location.

Project Name:	Interior Fit-out, ABC Office			
Location:	**Anywhere, USA**		Date: 1/1/2021	STD
Assembly Number	**Description**	**Qty.**	**Unit**	**Subtotal**
C1010 124 1200	Wood partition, 2 x 4 @ 16" OC w/5/8" FR gypsum board	560	S.F.	$3,169.60
C1020 114 1800	Metal door & frame, flush hollow core, 3'-0" x 7'-0"	2	Ea.	$2,580.00
C3010 230 0080	Painting, brushwork, primer & 2 coats	1,120	S.F.	$1,512.00
C3020 410 0140	Carpet, tufted, nylon, roll goods, 12' wide, 26 oz	240	S.F.	$684.00
C3030 210 6000	Acoustic ceilings, 24" x 48" tile, tee grid suspension	200	S.F.	$1,812.00
D5020 125 0560	Receptacles incl plate, box, conduit, wire, 20 A duplex	8	Ea.	$2,448.00
D5020 125 0720	Light switch incl plate, box, conduit, wire, 20 A single pole	2	Ea.	$598.00
D5020 210 0560	Fluorescent fixtures, recess mounted, 20 per 1000 SF	200	S.F.	$2,414.00
	Assembly Subtotal			**$15,217.60**
	Sales Tax @ ②		5 %	$ 380.44
	General Requirements @ ③		7 %	$ 1,065.23
	Subtotal A			**$16,663.27**
	GC Overhead @ ④		5 %	$ 833.16
	Subtotal B			**$17,496.44**
	GC Profit @ ⑤		5 %	$ 874.82
	Subtotal C			**$18,371.26**
	Adjusted by Location Factor ⑥	114.2		$ 20,979.98
	Architects Fee @ ⑦		8 %	$ 1,678.40
	Contingency @ ⑧		15 %	$ 3,147.00
	Project Total Cost			**$ 25,805.38**

This estimate is based on an interactive spreadsheet. You are free to download it and adjust it to your methodology. A copy of this spreadsheet is available at **RSMeans.com/2021books.**

① Work Performed

The body of the estimate shows the RSMeans data selected, including line numbers, a brief description of each item, its takeoff quantity and unit, and the total installed cost, including the installing contractor's overhead and profit.

② Sales Tax

If the work is subject to state or local sales taxes, the amount must be added to the estimate. In a conceptual estimate it can be assumed that one half of the total represents material costs. Therefore, apply the sales tax rate to 50% of the assembly subtotal.

③ General Requirements

This item covers project-wide needs provided by the general contractor. These items vary by project but may include temporary facilities and utilities, security, testing, project cleanup, etc. In assemblies estimates a percentage is used—typically between 5% and 15% of project cost.

④ General Contractor Overhead

This entry represents the general contractor's markup on all work to cover project administration costs.

⑤ General Contractor Profit

This entry represents the GC's profit on all work performed. The value included here can vary widely by project and is influenced by the GC's perception of the project's financial risk and market conditions.

⑥ Location Factor

RSMeans published data are based on national average costs. If necessary, adjust the total cost of the project using a location factor from the "Location Factor" table or the "City Cost Indexes" table found in the Reference Section. Use location factors if the work is general, covering the work of multiple trades. If the work is by a single trade (e.g., masonry) use the more specific data found in the City Cost Indexes.

To adjust costs by location factors, multiply the base cost by the factor and divide by 100.

⑦ Architect's Fee

If appropriate, add the design cost to the project estimate. These fees vary based on project complexity and size. Typical design and engineering fees can be found in the Reference Section.

⑧ Contingency

A factor for contingency may be added to any estimate to represent the cost of unknowns that may occur between the time that the estimate is performed and the time the project is constructed. The amount of the allowance will depend on the stage of design at which the estimate is done, as well as the contractor's assessment of the risk involved.

A1010 Standard Foundations

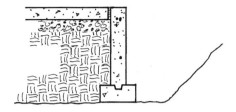

The Foundation Bearing Wall System includes: forms up to 6' high (four uses); 3,000 p.s.i. concrete placed and vibrated; and form removal with breaking form ties and patching walls. The wall systems list walls from 6" to 16" thick and are designed with minimum reinforcement.

Excavation and backfill are not included.

Please see the reference section for further design and cost information.

System Components				COST PER L.F.		
	QUANTITY	UNIT		MAT.	INST.	TOTAL
SYSTEM A1010 105 1500						
FOUNDATION WALL, CAST IN PLACE, DIRECT CHUTE, 4' HIGH, 6" THICK						
Formwork	8.000	SFCA		6.48	50.80	57.28
Reinforcing	3.300	Lb.		2.27	1.55	3.82
Unloading & sorting reinforcing	3.300	Lb.			.09	.09
Concrete, 3,000 psi	.074	C.Y.		10.14		10.14
Place concrete, direct chute	.074	C.Y.			2.75	2.75
Finish walls, break ties and patch voids, one side	4.000	S.F.		.20	4.48	4.68
			TOTAL	19.09	59.67	78.76

A1010 105			Wall Foundations						
	WALL HEIGHT (FT.)	PLACING METHOD	CONCRETE (C.Y. per L.F.)	REINFORCING (LBS. per L.F.)	WALL THICKNESS (IN.)	COST PER L.F.			
						MAT.	INST.	TOTAL	
1500	4'	direct chute	.074	3.3	6	19.10	59.50	78.60	
1520			.099	4.8	8	23.50	61.50	85	
1540			.123	6.0	10	27.50	62.50	90	
1560			.148	7.2	12	32	64	96	
1580			.173	8.1	14	36	65	101	
1600			.197	9.44	16	40	66.50	106.50	
1700	4'	pumped	.074	3.3	6	19.10	60.50	79.60	
1720			.099	4.8	8	23.50	62.50	86	
1740			.123	6.0	10	27.50	64	91.50	
1760			.148	7.2	12	32	65.50	97.50	
1780			.173	8.1	14	36	66.50	102.50	
1800			.197	9.44	16	40	68	108	
3000	6'	direct chute	.111	4.95	6	28.50	89.50	118	
3020			.149	7.20	8	35.50	92	127.50	
3040			.184	9.00	10	41.50	93.50	135	
3060			.222	10.8	12	48	96	144	
3080			.260	12.15	14	54	97.50	151.50	
3100			.300	14.39	16	61	99.50	160.50	
3200	6'	pumped	.111	4.95	6	28.50	91	119.50	
3220			.149	7.20	8	35.50	94	129.50	
3240			.184	9.00	10	41.50	96	137.50	
3260			.222	10.8	12	48	98.50	146.50	
3280			.260	12.15	14	54	100	154	
3300			.300	14.39	16	61	103	164	

A10 Foundations

A1010 Standard Foundations

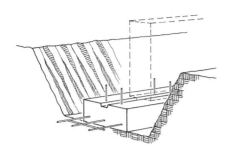

The Strip Footing System includes: excavation; hand trim; all forms needed for footing placement; forms for 2" x 6" keyway (four uses); dowels; and 3,000 p.s.i. concrete.

The footing size required varies for different soils. Soil bearing capacities are listed for 3 KSF and 6 KSF. Depths of the system range from 8" and deeper. Widths range from 16" and wider. Smaller strip footings may not require reinforcement.

Please see the reference section for further design and cost information.

System Components	QUANTITY	UNIT	COST PER L.F.		
			MAT.	INST.	TOTAL
SYSTEM A1010 110 2500					
STRIP FOOTING, LOAD 5.1 KLF, SOIL CAP. 3 KSF, 24" WIDE X 12" DEEP, REINF.					
Trench excavation	.148	C.Y.		1.45	1.45
Hand trim	2.000	S.F.		2.40	2.40
Compacted backfill	.074	C.Y.		.34	.34
Formwork, 4 uses	2.000	S.F.	5.64	10.30	15.94
Keyway form, 4 uses	1.000	L.F.	.58	1.31	1.89
Reinforcing, fy = 60000 psi	3.000	Lb.	2.13	2.01	4.14
Dowels	2.000	Ea.	2	5.88	7.88
Concrete, f'c = 3000 psi	.074	C.Y.	10.14		10.14
Place concrete, direct chute	.074	C.Y.		2.08	2.08
Screed finish	2.000	S.F.		.88	.88
TOTAL			20.49	26.65	47.14

A1010 110	Strip Footings	COST PER L.F.		
		MAT.	INST.	TOTAL
2100	Strip footing, load 2.6 KLF, soil capacity 3 KSF, 16" wide x 8" deep, plain	8.90	12.45	21.35
2300	Load 3.9 KLF, soil capacity 3 KSF, 24" wide x 8" deep, plain	11.20	14.10	25.30
2500	Load 5.1 KLF, soil capacity 3 KSF, 24" wide x 12" deep, reinf.	20.50	26.50	47
2700	Load 11.1 KLF, soil capacity 6 KSF, 24" wide x 12" deep, reinf.	20.50	26.50	47
2900	Load 6.8 KLF, soil capacity 3 KSF, 32" wide x 12" deep, reinf.	24.50	29	53.50
3100	Load 14.8 KLF, soil capacity 6 KSF, 32" wide x 12" deep, reinf.	24.50	29	53.50
3300	Load 9.3 KLF, soil capacity 3 KSF, 40" wide x 12" deep, reinf.	29	31.50	60.50
3500	Load 18.4 KLF, soil capacity 6 KSF, 40" wide x 12" deep, reinf.	29	31.50	60.50
3700	Load 10.1 KLF, soil capacity 3 KSF, 48" wide x 12" deep, reinf.	32	34	66
3900	Load 22.1 KLF, soil capacity 6 KSF, 48" wide x 12" deep, reinf.	34	36	70
4100	Load 11.8 KLF, soil capacity 3 KSF, 56" wide x 12" deep, reinf.	37.50	37.50	75
4300	Load 25.8 KLF, soil capacity 6 KSF, 56" wide x 12" deep, reinf.	40.50	40.50	81
4500	Load 10 KLF, soil capacity 3 KSF, 48" wide x 16" deep, reinf.	40.50	39.50	80
4700	Load 22 KLF, soil capacity 6 KSF, 48" wide, 16" deep, reinf.	41.50	40.50	82
4900	Load 11.6 KLF, soil capacity 3 KSF, 56" wide x 16" deep, reinf.	46	56.50	102.50
5100	Load 25.6 KLF, soil capacity 6 KSF, 56" wide x 16" deep, reinf.	48.50	58.50	107
5300	Load 13.3 KLF, soil capacity 3 KSF, 64" wide x 16" deep, reinf.	52.50	47	99.50
5500	Load 29.3 KLF, soil capacity 6 KSF, 64" wide x 16" deep, reinf.	56	50.50	106.50
5700	Load 15 KLF, soil capacity 3 KSF, 72" wide x 20" deep, reinf.	69.50	56.50	126
5900	Load 33 KLF, soil capacity 6 KSF, 72" wide x 20" deep, reinf.	73.50	60.50	134
6100	Load 18.3 KLF, soil capacity 3 KSF, 88" wide x 24" deep, reinf.	98	71.50	169.50
6300	Load 40.3 KLF, soil capacity 6 KSF, 88" wide x 24" deep, reinf.	106	79.50	185.50
6500	Load 20 KLF, soil capacity 3 KSF, 96" wide x 24" deep, reinf.	106	76	182
6700	Load 44 KLF, soil capacity 6 KSF, 96" wide x 24" deep, reinf.	113	82	195

A1010 Standard Foundations

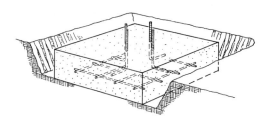

The Spread Footing System includes: excavation; backfill; forms (four uses); all reinforcement; 3,000 p.s.i. concrete (chute placed); and float finish.

Footing systems are priced per individual unit. The Expanded System Listing at the bottom shows various footing sizes. It is assumed that excavation is done by a truck mounted hydraulic excavator with an operator and oiler.

Backfill is with a dozer, and compaction by air tamp. The excavation and backfill equipment is assumed to operate at 30 C.Y. per hour.

Please see the reference section for further design and cost information.

System Components	QUANTITY	UNIT	COST EACH MAT.	COST EACH INST.	COST EACH TOTAL
SYSTEM A1010 210 7100					
SPREAD FOOTINGS, LOAD 25K, SOIL CAPACITY 3 KSF, 3' SQ X 12" DEEP					
Bulk excavation	.590	C.Y.		5.40	5.40
Hand trim	9.000	S.F.		10.80	10.80
Compacted backfill	.260	C.Y.		1.19	1.19
Formwork, 4 uses	12.000	S.F.	14.04	72	86.04
Reinforcing, fy = 60,000 psi	.006	Ton	8.25	8.10	16.35
Dowel or anchor bolt templates	6.000	L.F.	10.32	29.88	40.20
Concrete, f'c = 3,000 psi	.330	C.Y.	45.21		45.21
Place concrete, direct chute	.330	C.Y.		9.25	9.25
Float finish	9.000	S.F.		3.96	3.96
TOTAL			77.82	140.58	218.40

A1010 210	Spread Footings	COST EACH MAT.	COST EACH INST.	COST EACH TOTAL
7090	Spread footings, 3000 psi concrete, chute delivered			
7100	Load 25K, soil capacity 3 KSF, 3'-0" sq. x 12" deep	78	141	219
7150	Load 50K, soil capacity 3 KSF, 4'-6" sq. x 12" deep	163	242	405
7200	Load 50K, soil capacity 6 KSF, 3'-0" sq. x 12" deep	78	141	219
7250	Load 75K, soil capacity 3 KSF, 5'-6" sq. x 13" deep	256	340	596
7300	Load 75K, soil capacity 6 KSF, 4'-0" sq. x 12" deep	133	208	341
7350	Load 100K, soil capacity 3 KSF, 6'-0" sq. x 14" deep	320	410	730
7410	Load 100K, soil capacity 6 KSF, 4'-6" sq. x 15" deep	199	284	483
7450	Load 125K, soil capacity 3 KSF, 7'-0" sq. x 17" deep	510	585	1,095
7500	Load 125K, soil capacity 6 KSF, 5'-0" sq. x 16" deep	255	340	595
7550	Load 150K, soil capacity 3 KSF 7'-6" sq. x 18" deep	610	680	1,290
7610	Load 150K, soil capacity 6 KSF, 5'-6" sq. x 18" deep	340	430	770
7650	Load 200K, soil capacity 3 KSF, 8'-6" sq. x 20" deep	870	900	1,770
7700	Load 200K, soil capacity 6 KSF, 6'-0" sq. x 20" deep	440	530	970
7750	Load 300K, soil capacity 3 KSF, 10'-6" sq. x 25" deep	1,600	1,475	3,075
7810	Load 300K, soil capacity 6 KSF, 7'-6" sq. x 25" deep	835	885	1,720
7850	Load 400K, soil capacity 3 KSF, 12'-6" sq. x 28" deep	2,525	2,175	4,700
7900	Load 400K, soil capacity 6 KSF, 8'-6" sq. x 27" deep	1,150	1,150	2,300
7950	Load 500K, soil capacity 3 KSF, 14'-0" sq. x 31" deep	3,475	2,825	6,300
8010	Load 500K, soil capacity 6 KSF, 9'-6" sq. x 30" deep	1,575	1,500	3,075
8050	Load 600K, soil capacity 3 KSF, 16'-0" sq. x 35" deep	5,075	3,900	8,975
8100	Load 600K, soil capacity 6 KSF, 10'-6" sq. x 33" deep	2,125	1,925	4,050

A1010 Standard Foundations

A1010 210	Spread Footings	COST EACH		
		MAT.	INST.	TOTAL
8150	Load 700K, soil capacity 3 KSF, 17'-0" sq. x 37" deep	5,975	4,450	10,425
8200	Load 700K, soil capacity 6 KSF, 11'-6" sq. x 36" deep	2,725	2,350	5,075
8250	Load 800K, soil capacity 3 KSF, 18'-0" sq. x 39" deep	7,075	5,150	12,225
8300	Load 800K, soil capacity 6 KSF, 12'-0" sq. x 37" deep	3,050	2,575	5,625
8350	Load 900K, soil capacity 3 KSF, 19'-0" sq. x 40" deep	8,225	5,900	14,125
8400	Load 900K, soil capacity 6 KSF, 13'-0" sq. x 39" deep	3,775	3,050	6,825
8450	Load 1000K, soil capacity 3 KSF, 20'-0" sq. x 42" deep	9,475	6,625	16,100
8500	Load 1000K, soil capacity 6 KSF, 13'-6" sq. x 41" deep	4,275	3,400	7,675
8550	Load 1200K, soil capacity 6 KSF, 15'-0" sq. x 48" deep	5,675	4,325	10,000
8600	Load 1400K, soil capacity 6 KSF, 16'-0" sq. x 47" deep	6,925	5,125	12,050
8650	Load 1600K, soil capacity 6 KSF, 18'-0" sq. x 52" deep	9,600	6,800	16,400

A1010 Standard Foundations

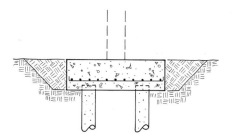

These pile cap systems include excavation with a truck mounted hydraulic excavator, hand trimming, compacted backfill, forms for concrete, templates for dowels or anchor bolts, reinforcing steel and concrete placed and floated.

Pile embedment is assumed as 6". Design is consistent with the Concrete Reinforcing Steel Institute Handbook f'c = 3000 psi, fy = 60,000.

Please see the reference section for further design and cost information.

System Components	QUANTITY	UNIT	COST EACH		
			MAT.	INST.	TOTAL
SYSTEM A1010 250 5100					
CAP FOR 2 PILES, 6'-6"X3'-6"X20", 15 TON PILE, 8" MIN. COL., 45K COL. LOAD					
Excavation, bulk, hyd excavator, truck mtd. 30" bucket 1/2 CY	2.890	C.Y.		26.44	26.44
Trim sides and bottom of trench, regular soil	23.000	S.F.		27.60	27.60
Dozer backfill & roller compaction	1.500	C.Y.		6.89	6.89
Forms in place pile cap, square or rectangular, 4 uses	33.000	SFCA	52.80	214.50	267.30
Templates for dowels or anchor bolts	8.000	Ea.	13.76	39.84	53.60
Reinforcing in place footings, #8 to #14	.025	Ton	34.38	19.63	54.01
Concrete ready mix, regular weight, 3000 psi	1.400	C.Y.	191.80		191.80
Place and vibrate concrete for pile caps, under 5 CY, direct chute	1.400	C.Y.		52.02	52.02
Float finish	23.000	S.F.		10.12	10.12
TOTAL			292.74	397.04	689.78

A1010 250		Pile Caps						
	NO. PILES	SIZE FT-IN X FT-IN X IN	PILE CAPACITY (TON)	COLUMN SIZE (IN)	COLUMN LOAD (K)	COST EACH		
						MAT.	INST.	TOTAL
5100	2	6-6x3-6x20	15	8	45	293	395	688
5150		26	40	8	155	360	485	845
5200		34	80	11	314	485	625	1,110
5250		37	120	14	473	520	665	1,185
5300	3	5-6x5-1x23	15	8	75	350	455	805
5350		28	40	10	232	390	515	905
5400		32	80	14	471	450	585	1,035
5450		38	120	17	709	520	670	1,190
5500	4	5-6x5-6x18	15	10	103	395	455	850
5550		30	40	11	308	570	655	1,225
5600		36	80	16	626	670	765	1,435
5650		38	120	19	945	705	800	1,505
5700	6	8-6x5-6x18	15	12	156	655	660	1,315
5750		37	40	14	458	1,050	975	2,025
5800		40	80	19	936	1,200	1,075	2,275
5850		45	120	24	1413	1,350	1,200	2,550
5900	8	8-6x7-9x19	15	12	205	985	890	1,875
5950		36	40	16	610	1,400	1,175	2,575
6000		44	80	22	1243	1,725	1,425	3,150
6050		47	120	27	1881	1,875	1,525	3,400

A1010 Standard Foundations

A1010 250	Pile Caps

	NO. PILES	SIZE FT-IN X FT-IN X IN	PILE CAPACITY (TON)	COLUMN SIZE (IN)	COLUMN LOAD (K)	COST EACH		
						MAT.	INST.	TOTAL
6100	10	11-6x7-9x21	15	14	250	1,375	1,100	2,475
6150		39	40	17	756	2,100	1,575	3,675
6200		47	80	25	1547	2,550	1,875	4,425
6250		49	120	31	2345	2,700	1,975	4,675
6300	12	11-6x8-6x22	15	15	316	1,750	1,325	3,075
6350		49	40	19	900	2,750	1,975	4,725
6400		52	80	27	1856	3,050	2,175	5,225
6450		55	120	34	2812	3,300	2,325	5,625
6500	14	11-6x10-9x24	15	16	345	2,275	1,625	3,900
6550		41	40	21	1056	2,950	2,025	4,975
6600		55	80	29	2155	3,850	2,575	6,425
6700	16	11-6x11-6x26	15	18	400	2,750	1,875	4,625
6750		48	40	22	1200	3,625	2,400	6,025
6800		60	80	31	2460	4,500	2,925	7,425
6900	18	13-0x11-6x28	15	20	450	3,125	2,050	5,175
6950		49	40	23	1349	4,225	2,675	6,900
7000		56	80	33	2776	4,975	3,150	8,125
7100	20	14-6x11-6x30	15	20	510	3,825	2,450	6,275
7150		52	40	24	1491	5,075	3,150	8,225

A1010 Standard Foundations

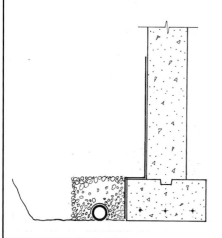

General: Footing drains can be placed either inside or outside of foundation walls depending upon the source of water to be intercepted. If the source of subsurface water is principally from grade or a subsurface stream above the bottom of the footing, outside drains should be used. For high water tables, use inside drains or both inside and outside.

The effectiveness of underdrains depends on good waterproofing. This must be carefully installed and protected during construction.

Costs below include the labor and materials for the pipe and 6″ of crushed stone around pipe. Excavation and backfill are not included.

System Components			QUANTITY	UNIT	COST PER L.F.		
					MAT.	INST.	TOTAL
SYSTEM A1010 310 1000							
FOUNDATION UNDERDRAIN, OUTSIDE ONLY, PVC, 4″ DIAM.							
PVC pipe 4″ diam. S.D.R. 35			1.000	L.F.	2.43	4.72	7.15
Crushed stone 3/4″ to 1/2″			.070	C.Y.	2	.92	2.92
		TOTAL			4.43	5.64	10.07

A1010 310	Foundation Underdrain	COST PER L.F.		
		MAT.	INST.	TOTAL
1000	Foundation underdrain, outside only, PVC, 4″ diameter	4.43	5.65	10.08
1100	6″ diameter	7.25	6.25	13.50
1400	Perforated HDPE, 6″ diameter	4.57	2.36	6.93
1450	8″ diameter	8.95	2.95	11.90
1500	12″ diameter	11.30	7.30	18.60
1600	Corrugated metal, 16 ga. asphalt coated, 6″ diameter	9.55	5.85	15.40
1650	8″ diameter	14.65	6.25	20.90
1700	10″ diameter	18.05	8.25	26.30
3000	Outside and inside, PVC, 4″ diameter	8.85	11.25	20.10
3100	6″ diameter	14.50	12.45	26.95
3400	Perforated HDPE, 6″ diameter	9.15	4.72	13.87
3450	8″ diameter	17.85	5.90	23.75
3500	12″ diameter	22.50	14.60	37.10
3600	Corrugated metal, 16 ga., asphalt coated, 6″ diameter	19.15	11.70	30.85
3650	8″ diameter	29.50	12.45	41.95
3700	10″ diameter	36	16.45	52.45

A1010 Standard Foundations

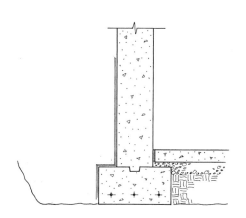

General: Apply foundation wall dampproofing over clean concrete giving particular attention to the joint between the wall and the footing. Use care in backfilling to prevent damage to the dampproofing.

Costs for four types of dampproofing are listed below.

System Components	QUANTITY	UNIT	COST PER L.F. MAT.	COST PER L.F. INST.	COST PER L.F. TOTAL
SYSTEM A1010 320 1000					
FOUNDATION DAMPPROOFING, BITUMINOUS, 1 COAT, 4' HIGH					
Bituminous asphalt dampproofing brushed on below grade, 1 coat	4.000	S.F.	.96	3.76	4.72
Labor for protection of dampproofing during backfilling	4.000	S.F.		1.59	1.59
TOTAL			.96	5.35	6.31

A1010 320	Foundation Dampproofing	COST PER L.F. MAT.	COST PER L.F. INST.	COST PER L.F. TOTAL
1000	Foundation dampproofing, bituminous, 1 coat, 4' high	.96	5.35	6.31
1400	8' high	1.92	10.70	12.62
1800	12' high	2.88	16.60	19.48
2000	2 coats, 4' high	1.92	6.60	8.52
2400	8' high	3.84	13.20	17.04
2800	12' high	5.75	20.50	26.25
3000	Asphalt with fibers, 1/16" thick, 4' high	1.84	6.60	8.44
3400	8' high	3.68	13.20	16.88
3800	12' high	5.50	20.50	26
4000	1/8" thick, 4' high	3.24	7.85	11.09
4400	8' high	6.50	15.65	22.15
4800	12' high	9.70	24	33.70
5000	Asphalt coated board and mastic, 1/4" thick, 4' high	5.70	7.15	12.85
5400	8' high	11.35	14.30	25.65
5800	12' high	17.05	22	39.05
6000	1/2" thick, 4' high	8.50	9.95	18.45
6400	8' high	17.05	19.85	36.90
6800	12' high	25.50	30.50	56
7000	Cementitious coating, on walls, 1/8" thick coating, 4' high	2.56	9.30	11.86
7400	8' high	5.10	18.65	23.75
7800	12' high	7.70	28	35.70
8000	Cementitious/metallic slurry, 4 coat, 1/2"thick, 2' high	1.59	10.10	11.69
8400	4' high	3.18	20	23.18
8800	6' high	4.77	30.50	35.27

A1020 Special Foundations

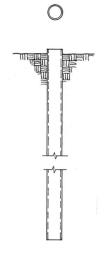

The Cast-in-Place Concrete Pile System includes: a defined number of 4,000 p.s.i. concrete piles with thin-wall, straight-sided, steel shells that have a standard steel plate driving point. An allowance for cutoffs is included.

The Expanded System Listing shows costs per cluster of piles. Clusters range from one pile to twenty piles. Loads vary from 50 Kips to 1,600 Kips. Both end-bearing and friction-type piles are shown.

Please see the reference section for cost of mobilization of the pile driving equipment and other design and cost information.

System Components	QUANTITY	UNIT	COST EACH		
			MAT.	INST.	TOTAL
SYSTEM A1020 110 2220					
CIP SHELL CONCRETE PILE, 25' LONG, 50K LOAD, END BEARING, 1 PILE					
7 Ga. shell, 12" diam.	27.000	V.L.F.	918	352.35	1,270.35
Steel pipe pile standard point, 12" or 14" diameter pile	1.000	Ea.	246	184	430
Pile cutoff, conc. pile with thin steel shell	1.000	Ea.		18.20	18.20
TOTAL			1,164	554.55	1,718.55

A1020 110	C.I.P. Concrete Piles	COST EACH		
		MAT.	INST.	TOTAL
2220	CIP shell concrete pile, 25' long, 50K load, end bearing, 1 pile	1,175	555	1,730
2240	100K load, end bearing, 2 pile cluster	2,075	925	3,000
2260	200K load, end bearing, 4 pile cluster	4,175	1,850	6,025
2280	400K load, end bearing, 7 pile cluster	7,275	3,250	10,525
2300	10 pile cluster	10,400	4,625	15,025
2320	800K load, end bearing, 13 pile cluster	21,600	7,600	29,200
2340	17 pile cluster	28,200	9,925	38,125
2360	1200K load, end bearing, 14 pile cluster	23,200	8,175	31,375
2380	19 pile cluster	31,500	11,100	42,600
2400	1600K load, end bearing, 19 pile cluster	31,500	11,100	42,600
2420	50' long, 50K load, end bearing, 1 pile	1,925	800	2,725
2440	Friction type, 2 pile cluster	3,700	1,600	5,300
2460	3 pile cluster	5,525	2,400	7,925
2480	100K load, end bearing, 2 pile cluster	3,850	1,600	5,450
2500	Friction type, 4 pile cluster	7,375	3,200	10,575
2520	6 pile cluster	11,100	4,800	15,900
2540	200K load, end bearing, 4 pile cluster	7,700	3,200	10,900
2560	Friction type, 8 pile cluster	14,800	6,425	21,225
2580	10 pile cluster	18,500	8,025	26,525
2600	400K load, end bearing, 7 pile cluster	13,500	5,625	19,125
2620	Friction type, 16 pile cluster	29,500	12,800	42,300
2640	19 pile cluster	35,100	15,200	50,300
2660	800K load, end bearing, 14 pile cluster	43,100	14,100	57,200
2680	20 pile cluster	61,500	20,200	81,700
2700	1200K load, end bearing, 15 pile cluster	46,100	15,100	61,200
2720	1600K load, end bearing, 20 pile cluster	61,500	20,200	81,700

A1020 Special Foundations

A1020 110	C.I.P. Concrete Piles	COST EACH		
		MAT.	INST.	TOTAL
3740	75' long, 50K load, end bearing, 1 pile	2,975	1,325	4,300
3760	Friction type, 2 pile cluster	5,725	2,625	8,350
3780	3 pile cluster	8,600	3,950	12,550
3800	100K load, end bearing, 2 pile cluster	5,950	2,625	8,575
3820	Friction type, 3 pile cluster	8,600	3,950	12,550
3840	5 pile cluster	14,300	6,575	20,875
3860	200K load, end bearing, 4 pile cluster	11,900	5,275	17,175
3880	6 pile cluster	17,900	7,900	25,800
3900	Friction type, 6 pile cluster	17,200	7,900	25,100
3910	7 pile cluster	20,000	9,225	29,225
3920	400K load, end bearing, 7 pile cluster	20,900	9,225	30,125
3930	11 pile cluster	32,800	14,500	47,300
3940	Friction type, 12 pile cluster	34,400	15,800	50,200
3950	14 pile cluster	40,100	18,400	58,500
3960	800K load, end bearing, 15 pile cluster	70,500	24,300	94,800
3970	20 pile cluster	93,500	32,400	125,900
3980	1200K load, end bearing, 17 pile cluster	79,500	27,500	107,000

The Precast Concrete Pile System includes: pre-stressed concrete piles; standard steel driving point; and an allowance for cutoffs.

The Expanded System Listing shows costs per cluster of piles. Clusters range from one pile to twenty piles. Loads vary from 50 Kips to 1,600 Kips. Both end-bearing and friction type piles are listed.

Please see the reference section for cost of mobilization of the pile driving equipment and other design and cost information.

System Components	QUANTITY	UNIT	COST EACH		
			MAT.	INST.	TOTAL
SYSTEM A1020 120 2220					
PRECAST CONCRETE PILE, 50' LONG, 50K LOAD, END BEARING, 1 PILE					
Precast, prestressed conc. piles, 10" square, no mobil.	53.000	V.L.F.	845.35	591.48	1,436.83
Steel pipe pile standard point, 8" to 10" diameter	1.000	Ea.	56.50	81	137.50
Piling special costs cutoffs concrete piles plain	1.000	Ea.		126	126
TOTAL			901.85	798.48	1,700.33

A1020 120	Precast Concrete Piles	COST EACH		
		MAT.	INST.	TOTAL
2220	Precast conc pile, 50' long, 50K load, end bearing, 1 pile	900	800	1,700
2240	Friction type, 2 pile cluster	3,275	1,650	4,925
2260	4 pile cluster	6,525	3,300	9,825
2280	100K load, end bearing, 2 pile cluster	1,800	1,600	3,400
2300	Friction type, 2 pile cluster	3,275	1,650	4,925
2320	4 pile cluster	6,525	3,300	9,825
2340	7 pile cluster	11,400	5,800	17,200
2360	200K load, end bearing, 3 pile cluster	2,700	2,400	5,100
2380	4 pile cluster	3,600	3,200	6,800
2400	Friction type, 8 pile cluster	13,100	6,625	19,725
2420	9 pile cluster	14,700	7,450	22,150
2440	14 pile cluster	22,900	11,600	34,500
2460	400K load, end bearing, 6 pile cluster	5,400	4,800	10,200
2480	8 pile cluster	7,225	6,400	13,625
2500	Friction type, 14 pile cluster	22,900	11,600	34,500
2520	16 pile cluster	26,100	13,200	39,300
2540	18 pile cluster	29,400	14,900	44,300
2560	800K load, end bearing, 12 pile cluster	11,700	10,400	22,100
2580	16 pile cluster	14,400	12,800	27,200
2600	1200K load, end bearing, 19 pile cluster	54,000	18,600	72,600
2620	20 pile cluster	57,000	19,500	76,500
2640	1600K load, end bearing, 19 pile cluster	54,000	18,600	72,600
4660	100' long, 50K load, end bearing, 1 pile	1,725	1,375	3,100
4680	Friction type, 1 pile	3,125	1,425	4,550

A1020 Special Foundations

A1020 120	Precast Concrete Piles	COST EACH		
		MAT.	INST.	TOTAL
4700	2 pile cluster	6,225	2,850	9,075
4720	100K load, end bearing, 2 pile cluster	3,475	2,750	6,225
4740	Friction type, 2 pile cluster	6,225	2,850	9,075
4760	3 pile cluster	9,350	4,275	13,625
4780	4 pile cluster	12,500	5,700	18,200
4800	200K load, end bearing, 3 pile cluster	5,200	4,125	9,325
4820	4 pile cluster	6,925	5,525	12,450
4840	Friction type, 3 pile cluster	9,350	4,275	13,625
4860	5 pile cluster	15,600	7,125	22,725
4880	400K load, end bearing, 6 pile cluster	10,400	8,275	18,675
4900	8 pile cluster	13,900	11,000	24,900
4910	Friction type, 8 pile cluster	24,900	11,400	36,300
4920	10 pile cluster	31,200	14,300	45,500
4930	800K load, end bearing, 13 pile cluster	22,500	17,900	40,400
4940	16 pile cluster	27,700	22,100	49,800
4950	1200K load, end bearing, 19 pile cluster	104,000	32,400	136,400
4960	20 pile cluster	109,500	34,100	143,600
4970	1600K load, end bearing, 19 pile cluster	104,000	32,400	136,400

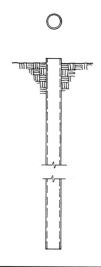

The Steel Pipe Pile System includes: steel pipe sections filled with 4,000 p.s.i. concrete; a standard steel driving point; splices when required and an allowance for cutoffs.

The Expanded System Listing shows costs per cluster of piles. Clusters range from one pile to twenty piles. Loads vary from 50 Kips to 1,600 Kips. Both end-bearing and friction-type piles are shown.

Please see the reference section for cost of mobilization of the pile driving equipment and other design and cost information.

System Components	QUANTITY	UNIT	COST EACH		
			MAT.	INST.	TOTAL
SYSTEM A1020 130 2220					
CONC. FILL STEEL PIPE PILE, 50' LONG, 50K LOAD, END BEARING, 1 PILE					
Piles, steel, pipe, conc. filled, 12" diameter	53.000	V.L.F.	2,173	999.05	3,172.05
Steel pipe pile, standard point, for 12" or 14" diameter pipe	1.000	Ea.	246	184	430
Pile cut off, concrete pile, thin steel shell	1.000	Ea.		18.20	18.20
TOTAL			2,419	1,201.25	3,620.25

A1020 130	Steel Pipe Piles	COST EACH		
		MAT.	INST.	TOTAL
2220	Conc. fill steel pipe pile, 50' long, 50K load, end bearing, 1 pile	2,425	1,200	3,625
2240	Friction type, 2 pile cluster	4,850	2,400	7,250
2250	100K load, end bearing, 2 pile cluster	4,850	2,400	7,250
2260	3 pile cluster	7,250	3,600	10,850
2300	Friction type, 4 pile cluster	9,675	4,800	14,475
2320	5 pile cluster	12,100	6,000	18,100
2340	10 pile cluster	24,200	12,000	36,200
2360	200K load, end bearing, 3 pile cluster	7,250	3,600	10,850
2380	4 pile cluster	9,675	4,800	14,475
2400	Friction type, 4 pile cluster	9,675	4,800	14,475
2420	8 pile cluster	19,400	9,600	29,000
2440	9 pile cluster	21,800	10,800	32,600
2460	400K load, end bearing, 6 pile cluster	14,500	7,200	21,700
2480	7 pile cluster	16,900	8,400	25,300
2500	Friction type, 9 pile cluster	21,800	10,800	32,600
2520	16 pile cluster	38,700	19,200	57,900
2540	19 pile cluster	46,000	22,800	68,800
2560	800K load, end bearing, 11 pile cluster	26,600	13,200	39,800
2580	14 pile cluster	33,900	16,800	50,700
2600	15 pile cluster	36,300	18,000	54,300
2620	Friction type, 17 pile cluster	41,100	20,400	61,500
2640	1200K load, end bearing, 16 pile cluster	38,700	19,200	57,900
2660	20 pile cluster	48,400	24,000	72,400
2680	1600K load, end bearing, 17 pile cluster	41,100	20,400	61,500
3700	100' long, 50K load, end bearing, 1 pile	4,725	2,350	7,075
3720	Friction type, 1 pile	4,725	2,350	7,075

For customer support on your Concrete & Masonry Costs with RSMeans Data, call 800.448.8182.

A10 Foundations

A1020 Special Foundations

A1020 130	Steel Pipe Piles	COST EACH		
		MAT.	INST.	TOTAL
3740	2 pile cluster	9,450	4,725	14,175
3760	100K load, end bearing, 2 pile cluster	9,450	4,725	14,175
3780	Friction type, 2 pile cluster	9,450	4,725	14,175
3800	3 pile cluster	14,200	7,075	21,275
3820	200K load, end bearing, 3 pile cluster	14,200	7,075	21,275
3840	4 pile cluster	18,900	9,425	28,325
3860	Friction type, 3 pile cluster	14,200	7,075	21,275
3880	4 pile cluster	18,900	9,425	28,325
3900	400K load, end bearing, 6 pile cluster	28,300	14,100	42,400
3910	7 pile cluster	33,100	16,500	49,600
3920	Friction type, 5 pile cluster	23,600	11,800	35,400
3930	8 pile cluster	37,800	18,900	56,700
3940	800K load, end bearing, 11 pile cluster	52,000	25,900	77,900
3950	14 pile cluster	66,000	33,000	99,000
3960	15 pile cluster	71,000	35,400	106,400
3970	1200K load, end bearing, 16 pile cluster	75,500	37,700	113,200
3980	20 pile cluster	94,500	47,100	141,600
3990	1600K load, end bearing, 17 pile cluster	80,500	40,100	120,600

A1020 Special Foundations

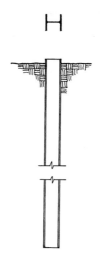

A Steel "H" Pile System includes: steel H sections; heavy duty driving point; splices where applicable and allowance for cutoffs.

The Expanded System Listing shows costs per cluster of piles. Clusters range from one pile to eighteen piles. Loads vary from 50 Kips to 2,000 Kips. All loads for Steel H Pile systems are given in terms of end bearing capacity.

Steel sections range from 10" x 10" to 14" x 14" in the Expanded System Listing. The 14" x 14" steel section is used for all H piles used in applications requiring a working load over 800 Kips.

Please see the reference section for cost of mobilization of the pile driving equipment and other design and cost information.

System Components	QUANTITY	UNIT	COST EACH		
			MAT.	INST.	TOTAL
SYSTEM A1020 140 2220					
STEEL H PILES, 50' LONG, 100K LOAD, END BEARING, 1 PILE					
Steel H piles 10" x 10", 42 #/L.F.	53.000	V.L.F.	1,219	680.52	1,899.52
Heavy duty point, 10"	1.000	Ea.	206	187	393
Pile cut off, steel pipe or H piles	1.000	Ea.		36.50	36.50
TOTAL			1,425	904.02	2,329.02

A1020 140	Steel H Piles	COST EACH		
		MAT.	INST.	TOTAL
2220	Steel H piles, 50' long, 100K load, end bearing, 1 pile	1,425	905	2,330
2260	2 pile cluster	2,850	1,800	4,650
2280	200K load, end bearing, 2 pile cluster	2,850	1,800	4,650
2300	3 pile cluster	4,275	2,700	6,975
2320	400K load, end bearing, 3 pile cluster	4,275	2,700	6,975
2340	4 pile cluster	5,700	3,625	9,325
2360	6 pile cluster	8,550	5,425	13,975
2380	800K load, end bearing, 5 pile cluster	7,125	4,525	11,650
2400	7 pile cluster	9,975	6,325	16,300
2420	12 pile cluster	17,100	10,800	27,900
2440	1200K load, end bearing, 8 pile cluster	11,400	7,225	18,625
2460	11 pile cluster	15,700	9,950	25,650
2480	17 pile cluster	24,200	15,400	39,600
2500	1600K load, end bearing, 10 pile cluster	18,700	9,450	28,150
2520	14 pile cluster	26,200	13,200	39,400
2540	2000K load, end bearing, 12 pile cluster	22,400	11,300	33,700
2560	18 pile cluster	33,700	17,000	50,700
3580	100' long, 50K load, end bearing, 1 pile	4,625	2,150	6,775
3600	100K load, end bearing, 1 pile	4,625	2,150	6,775
3620	2 pile cluster	9,275	4,275	13,550
3640	200K load, end bearing, 2 pile cluster	9,275	4,275	13,550
3660	3 pile cluster	13,900	6,425	20,325
3680	400K load, end bearing, 3 pile cluster	13,900	6,425	20,325
3700	4 pile cluster	18,500	8,575	27,075
3720	6 pile cluster	27,800	12,900	40,700
3740	800K load, end bearing, 5 pile cluster	23,200	10,700	33,900

A10 Foundations

A1020 Special Foundations

A1020 140	Steel H Piles	COST EACH		
		MAT.	INST.	TOTAL
3760	7 pile cluster	32,500	15,000	47,500
3780	12 pile cluster	55,500	25,700	81,200
3800	1200K load, end bearing, 8 pile cluster	37,100	17,100	54,200
3820	11 pile cluster	51,000	23,600	74,600
3840	17 pile cluster	79,000	36,400	115,400
3860	1600K load, end bearing, 10 pile cluster	46,400	21,400	67,800
3880	14 pile cluster	65,000	30,000	95,000
3900	2000K load, end bearing, 12 pile cluster	55,500	25,700	81,200
3920	18 pile cluster	83,500	38,600	122,100

For customer support on your Concrete & Masonry Costs with RSMeans Data, call 800.448.8182.

A10 Foundations

A1020 Special Foundations

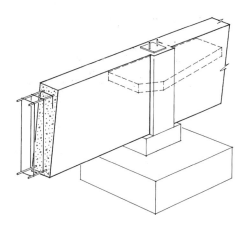

The Grade Beam System includes: excavation with a truck mounted backhoe; hand trim; backfill; forms (four uses); reinforcing steel; and 3,000 p.s.i. concrete placed from chute.

Superimposed loads vary in the listing from 1 Kip per linear foot (KLF) and above. In the Expanded System Listing, the span of the beams varies from 15' to 40'. Depth varies from 28" to 52". Width varies from 12" and wider.

Please see the reference section for further design and cost information.

System Components	QUANTITY	UNIT	COST PER L.F.		
			MAT.	INST.	TOTAL
SYSTEM A1020 210 2220					
GRADE BEAM, 15' SPAN, 28" DEEP, 12" WIDE, 8 KLF LOAD					
Excavation, trench, hydraulic backhoe, 3/8 CY bucket	.260	C.Y.		2.54	2.54
Trim sides and bottom of trench, regular soil	2.000	S.F.		2.40	2.40
Backfill, by hand, compaction in 6" layers, using vibrating plate	.170	C.Y.		1.60	1.60
Forms in place, grade beam, 4 uses	4.700	SFCA	6.67	29.61	36.28
Reinforcing in place, beams & girders, #8 to #14	.019	Ton	26.13	19.95	46.08
Concrete ready mix, regular weight, 3000 psi	.090	C.Y.	12.33		12.33
Place and vibrate conc. for grade beam, direct chute	.090	C.Y.		2.02	2.02
TOTAL			45.13	58.12	103.25

A1020 210	Grade Beams	COST PER L.F.		
		MAT.	INST.	TOTAL
2220	Grade beam, 15' span, 28" deep, 12" wide, 8 KLF load	45	58	103
2240	14" wide, 12 KLF load	46.50	58.50	105
2260	40" deep, 12" wide, 16 KLF load	46.50	72.50	119
2280	20 KLF load	53.50	77.50	131
2300	52" deep, 12" wide, 30 KLF load	62	98	160
2320	40 KLF load	75.50	108	183.50
2340	50 KLF load	89.50	118	207.50
3360	20' span, 28" deep, 12" wide, 2 KLF load	28.50	45.50	74
3380	16" wide, 4 KLF load	37.50	50.50	88
3400	40" deep, 12" wide, 8 KLF load	46.50	72.50	119
3420	12 KLF load	59	80.50	139.50
3440	14" wide, 16 KLF load	71.50	90	161.50
3460	52" deep, 12" wide, 20 KLF load	77	109	186
3480	14" wide, 30 KLF load	106	129	235
3500	20" wide, 40 KLF load	123	136	259
3520	24" wide, 50 KLF load	152	155	307
4540	30' span, 28" deep, 12" wide, 1 KLF load	30	46.50	76.50
4560	14" wide, 2 KLF load	47.50	63.50	111
4580	40" deep, 12" wide, 4 KLF load	56	76.50	132.50
4600	18" wide, 8 KLF load	79.50	92.50	172
4620	52" deep, 14" wide, 12 KLF load	102	126	228
4640	20" wide, 16 KLF load	124	137	261
4660	24" wide, 20 KLF load	152	156	308
4680	36" wide, 30 KLF load	218	194	412
4700	48" wide, 40 KLF load	287	235	522
5720	40' span, 40" deep, 12" wide, 1 KLF load	41	68.50	109.50

A1020 Special Foundations

A1020 210	Grade Beams	COST PER L.F.		
		MAT.	INST.	TOTAL
5740	2 KLF load	50.50	75.50	126
5760	52" deep, 12" wide, 4 KLF load	75.50	108	183.50
5780	20" wide, 8 KLF load	119	134	253
5800	28" wide, 12 KLF load	163	161	324
5820	38" wide, 16 KLF load	224	196	420
5840	46" wide, 20 KLF load	293	241	534

A1020 Special Foundations

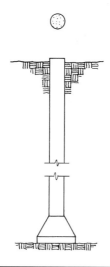

Caisson Systems are listed for three applications: stable ground, wet ground and soft rock. Concrete used is 3,000 p.s.i. placed from chute. Included are a bell at the bottom of the caisson shaft (if applicable) along with required excavation and disposal of excess excavated material up to two miles from job site.

The Expanded System lists cost per caisson. End-bearing loads vary from 200 Kips to 3,200 Kips. The dimensions of the caissons range from 2' x 50' to 7' x 200'.

Please see the reference section for further design and cost information.

System Components	QUANTITY	UNIT	COST EACH		
			MAT.	INST.	TOTAL
SYSTEM A1020 310 2200					
CAISSON, STABLE GROUND, 3000 PSI CONC., 10 KSF BRNG, 200K LOAD, 2'X50'					
Caissons, drilled, to 50', 24" shaft diameter, .116 C.Y./L.F.	50.000	V.L.F.	992.50	1,145	2,137.50
Reinforcing in place, columns, #3 to #7	.060	Ton	82.50	112.50	195
4' bell diameter, 24" shaft, 0.444 C.Y.	1.000	Ea.	61	217.50	278.50
Load & haul excess excavation, 2 miles	6.240	C.Y.		43.87	43.87
TOTAL			1,136	1,518.87	2,654.87

A1020 310	Caissons	COST EACH		
		MAT.	INST.	TOTAL
2200	Caisson, stable ground, 3000 PSI conc, 10 KSF brng, 200K load, 2'-0"x50'-0	1,125	1,525	2,650
2400	400K load, 2'-6"x50'-0"	1,850	2,425	4,275
2600	800K load, 3'-0"x100'-0"	5,150	5,700	10,850
2800	1200K load, 4'-0"x100'-0"	8,850	7,275	16,125
3000	1600K load, 5'-0"x150'-0"	19,900	11,400	31,300
3200	2400K load, 6'-0"x150'-0"	29,200	15,000	44,200
3400	3200K load, 7'-0"x200'-0"	52,500	21,900	74,400
5000	Wet ground, 3000 PSI conc., 10 KSF brng, 200K load, 2'-0"x50'-0"	985	2,250	3,235
5200	400K load, 2'-6"x50'-0"	1,625	3,800	5,425
5400	800K load, 3'-0"x100'-0"	4,475	10,200	14,675
5600	1200K load, 4'-0"x100'-0"	7,675	16,900	24,575
5800	1600K load, 5'-0"x150'-0"	17,100	36,400	53,500
6000	2400K load, 6'-0"x150'-0"	25,100	44,900	70,000
6200	3200K load, 7'-0"x200'-0"	45,000	72,000	117,000
7800	Soft rock, 3000 PSI conc., 10 KSF brng, 200K load, 2'-0"x50'-0"	985	18,900	19,885
8000	400K load, 2'-6"x50'-0"	1,625	21,500	23,125
8200	800K load, 3'-0"x100'-0"	4,475	56,500	60,975
8400	1200K load, 4'-0"x100'-0"	7,675	82,500	90,175
8600	1600K load, 5'-0"x150'-0"	17,100	169,500	186,600
8800	2400K load, 6'-0"x150'-0"	25,100	201,500	226,600
9000	3200K load, 7'-0"x200'-0"	45,000	321,000	366,000

A1020 Special Foundations

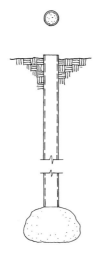

Pressure Injected Piles are usually uncased up to 25' and cased over 25' depending on soil conditions.

These costs include excavation and hauling of excess materials; steel casing over 25'; reinforcement; 3,000 p.s.i. concrete; plus mobilization and demobilization of equipment for a distance of up to fifty miles to and from the job site.

The Expanded System lists cost per cluster of piles. Clusters range from one pile to eight piles. End-bearing loads range from 50 Kips to 1,600 Kips.

Please see the reference section for further design and cost information.

System Components	QUANTITY	UNIT	COST EACH MAT.	COST EACH INST.	COST EACH TOTAL
SYSTEM A1020 710 4200					
PRESSURE INJECTED FOOTING, END BEARING, 50' LONG, 50K LOAD, 1 PILE					
Pressure injected footings, cased, 30-60 ton cap., 12" diameter	50.000	V.L.F.	1,050	1,695	2,745
Pile cutoff, concrete pile with thin steel shell	1.000	Ea.		18.20	18.20
TOTAL			1,050	1,713.20	2,763.20

A1020 710	Pressure Injected Footings	COST EACH MAT.	COST EACH INST.	COST EACH TOTAL
2200	Pressure injected footing, end bearing, 25' long, 50K load, 1 pile	490	1,200	1,690
2400	100K load, 1 pile	490	1,200	1,690
2600	2 pile cluster	980	2,375	3,355
2800	200K load, 2 pile cluster	980	2,375	3,355
3200	400K load, 4 pile cluster	1,950	4,750	6,700
3400	7 pile cluster	3,425	8,325	11,750
3800	1200K load, 6 pile cluster	4,125	9,025	13,150
4000	1600K load, 7 pile cluster	4,825	10,500	15,325
4200	50' long, 50K load, 1 pile	1,050	1,725	2,775
4400	100K load, 1 pile	1,825	1,725	3,550
4600	2 pile cluster	3,650	3,425	7,075
4800	200K load, 2 pile cluster	3,650	3,425	7,075
5000	4 pile cluster	7,300	6,850	14,150
5200	400K load, 4 pile cluster	7,300	6,850	14,150
5400	8 pile cluster	14,600	13,700	28,300
5600	800K load, 7 pile cluster	12,800	12,000	24,800
5800	1200K load, 6 pile cluster	11,700	10,300	22,000
6000	1600K load, 7 pile cluster	13,700	12,000	25,700

A1030 Slab on Grade

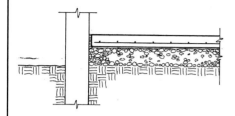

There are four types of Slab on Grade Systems listed: Non-industrial, Light industrial, Industrial and Heavy industrial. Each type is listed two ways: reinforced and non-reinforced. A Slab on Grade system includes three passes with a grader; 6" of compacted gravel fill; polyethylene vapor barrier; 3500 p.s.i. concrete placed by chute; bituminous fibre expansion joint; all necessary edge forms (4 uses); steel trowel finish; and sprayed-on membrane curing compound.

The Expanded System Listing shows costs on a per square foot basis. Thicknesses of the slabs range from 4" and above. Non-industrial applications are for foot traffic only with negligible abrasion. Light industrial applications are for pneumatic wheels and light abrasion. Industrial applications are for solid rubber wheels and moderate abrasion. Heavy industrial applications are for steel wheels and severe abrasion. All slabs are either shown unreinforced or reinforced with welded wire fabric.

System Components	QUANTITY	UNIT	COST PER S.F.		
			MAT.	INST.	TOTAL
SYSTEM A1030 120 2220					
SLAB ON GRADE, 4" THICK, NON INDUSTRIAL, NON REINFORCED					
Fine grade, 3 passes with grader and roller	.110	S.Y.		.66	.66
Gravel under floor slab, 4" deep, compacted	1.000	S.F.	.65	.36	1.01
Polyethylene vapor barrier, standard, 6 mil	1.000	S.F.	.08	.18	.26
Concrete ready mix, regular weight, 3500 psi	.012	C.Y.	1.67		1.67
Place and vibrate concrete for slab on grade, 4" thick, direct chute	.012	C.Y.		.37	.37
Expansion joint, premolded bituminous fiber, 1/2" x 6"	.220	L.F.	.10	.38	.48
Edge forms in place for slab on grade to 6" high, 4 uses	.030	L.F.	.02	.12	.14
Cure with sprayed membrane curing compound	1.000	S.F.	.13	.11	.24
Finishing floor, monolithic steel trowel	1.000	S.F.		1.08	1.08
TOTAL			2.65	3.26	5.91

A1030 120	Plain & Reinforced	COST PER S.F.		
		MAT.	INST.	TOTAL
2220	Slab on grade, 4" thick, non industrial, non reinforced	2.65	3.27	5.92
2240	Reinforced	2.75	3.56	6.31
2260	Light industrial, non reinforced	3.30	3.98	7.28
2280	Reinforced	3.46	4.38	7.84
2300	Industrial, non reinforced	4.31	8.20	12.51
2320	Reinforced	4.48	8.60	13.08
3340	5" thick, non industrial, non reinforced	3.07	3.36	6.43
3360	Reinforced	3.23	3.76	6.99
3380	Light industrial, non reinforced	3.72	4.07	7.79
3400	Reinforced	3.89	4.47	8.36
3420	Heavy industrial, non reinforced	5.50	9.80	15.30
3440	Reinforced	5.60	10.25	15.85
4460	6" thick, non industrial, non reinforced	3.62	3.28	6.90
4480	Reinforced	3.91	3.83	7.74
4500	Light industrial, non reinforced	4.29	3.99	8.28
4520	Reinforced	4.66	4.42	9.08
4540	Heavy industrial, non reinforced	6.05	9.95	16
4560	Reinforced	6.35	10.45	16.80
5580	7" thick, non industrial, non reinforced	4.06	3.39	7.45
5600	Reinforced	4.51	3.97	8.48
5620	Light industrial, non reinforced	4.74	4.10	8.84
5640	Reinforced	5.20	4.68	9.88
5660	Heavy industrial, non reinforced	6.50	9.80	16.30
5680	Reinforced	6.80	10.35	17.15

A10 Foundations

A1030 Slab on Grade

A1030 120	Plain & Reinforced	COST PER S.F.		
		MAT.	INST.	TOTAL
6700	8″ thick, non industrial, non reinforced	4.48	3.45	7.93
6740	Light industrial, non reinforced	5.15	4.16	9.31
6760	Reinforced	5.55	4.65	10.20
6780	Heavy industrial, non reinforced	6.95	9.90	16.85
6800	Reinforced	7.40	10.40	17.80

For customer support on your Concrete & Masonry Costs with RSMeans Data, call 800.448.8182.

A2010 Basement Excavation

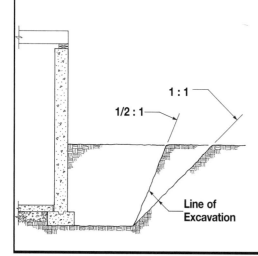

1 : 1

1/2 : 1

Line of Excavation

Pricing Assumptions: Two-thirds of excavation is by 2-1/2 C.Y. wheel mounted front end loader and one-third by 1-1/2 C.Y. hydraulic excavator.

Two-mile round trip haul by 12 C.Y. tandem trucks is included for excavation wasted and storage of suitable fill from excavated soil. For excavation in clay, all is wasted and the cost of suitable backfill with two-mile haul is included.

Sand and gravel assumes 15% swell and compaction; common earth assumes 25% swell and 15% compaction; clay assumes 40% swell and 15% compaction (non-clay).

In general, the following items are accounted for in the costs in the table below.

1. Excavation for building or other structure to depth and extent indicated.
2. Backfill compacted in place.
3. Haul of excavated waste.
4. Replacement of unsuitable material with bank run gravel.

Note: Additional excavation and fill beyond this line of general excavation for the building (as required for isolated spread footings, strip footings, etc.) are included in the cost of the appropriate component systems.

System Components	QUANTITY	UNIT	COST PER S.F.		
			MAT.	INST.	TOTAL
SYSTEM A2010 110 2280					
EXCAVATE & FILL, 1000 S.F., 8′ DEEP, SAND, ON SITE STORAGE					
Excavating bulk shovel, 1.5 C.Y. bucket, 150 cy/hr	.262	C.Y.		.56	.56
Excavation, front end loader, 2-1/2 C.Y.	.523	C.Y.		1.36	1.36
Haul earth, 12 C.Y. dump truck	.341	L.C.Y.		2.37	2.37
Backfill, dozer bulk push 300′, including compaction	.562	C.Y.		2.58	2.58
TOTAL				6.87	6.87

A2010 110	Building Excavation & Backfill	COST PER S.F.		
		MAT.	INST.	TOTAL
2220	Excav & fill, 1000 S.F., 4′ sand, gravel, or common earth, on site storage		1.17	1.17
2240	Off site storage		1.72	1.72
2260	Clay excavation, bank run gravel borrow for backfill	3.25	2.59	5.84
2280	8′ deep, sand, gravel, or common earth, on site storage		6.85	6.85
2300	Off site storage		14.65	14.65
2320	Clay excavation, bank run gravel borrow for backfill	13.20	12.10	25.30
2340	16′ deep, sand, gravel, or common earth, on site storage		18.70	18.70
2350	Off site storage		36.50	36.50
2360	Clay excavation, bank run gravel borrow for backfill	36.50	31	67.50
3380	4000 S.F., 4′ deep, sand, gravel, or common earth, on site storage		.63	.63
3400	Off site storage		1.18	1.18
3420	Clay excavation, bank run gravel borrow for backfill	1.66	1.32	2.98
3440	8′ deep, sand, gravel, or common earth, on site storage		4.76	4.76
3460	Off site storage		8.20	8.20
3480	Clay excavation, bank run gravel borrow for backfill	6	7.20	13.20
3500	16′ deep, sand, gravel, or common earth, on site storage		11.40	11.40
3520	Off site storage		22	22
3540	Clay, excavation, bank run gravel borrow for backfill	16.10	17.20	33.30
4560	10,000 S.F., 4′ deep, sand, gravel, or common earth, on site storage		.36	.36
4580	Off site storage		.70	.70
4600	Clay excavation, bank run gravel borrow for backfill	1.01	.81	1.82
4620	8′ deep, sand, gravel, or common earth, on site storage		4.09	4.09
4640	Off site storage		6.15	6.15
4660	Clay excavation, bank run gravel borrow for backfill	3.64	5.60	9.24
4680	16′ deep, sand, gravel, or common earth, on site storage		9.20	9.20
4700	Off site storage		15.50	15.50
4720	Clay excavation, bank run gravel borrow for backfill	9.65	12.80	22.45

A2010 Basement Excavation

A2010 110	Building Excavation & Backfill	COST PER S.F.		
		MAT.	INST.	TOTAL
5740	30,000 S.F., 4' deep, sand, gravel, or common earth, on site storage		.20	.20
5760	Off site storage		.41	.41
5780	Clay excavation, bank run gravel borrow for backfill	.59	.47	1.06
5800	8' deep, sand, gravel, or common earth, on site storage		3.65	3.65
5820	Off site storage		4.79	4.79
5840	Clay excavation, bank run gravel borrow for backfill	2.05	4.50	6.55
5860	16' deep, sand, gravel, or common earth, on site storage		7.80	7.80
5880	Off site storage		11.25	11.25
5900	Clay excavation, bank run gravel borrow for backfill	5.35	9.85	15.20
6910	100,000 S.F., 4' deep, sand, gravel, or common earth, on site storage		.11	.11
6920	Off site storage		.23	.23
6930	Clay excavation, bank run gravel borrow for backfill	.29	.25	.54
6940	8' deep, sand, gravel, or common earth, on site storage		3.39	3.39
6950	Off site storage		4	4
6960	Clay excavation, bank run gravel borrow for backfill	1.11	3.85	4.96
6970	16' deep, sand, gravel, or common earth, on site storage		7.05	7.05
6980	Off site storage		8.85	8.85
6990	Clay excavation, bank run gravel borrow for backfill	2.89	8.15	11.04

A2020 Basement Walls

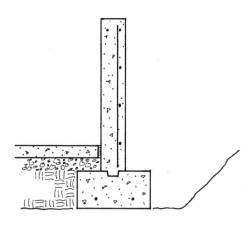

The Foundation Bearing Wall System includes: forms up to 16' high (four uses); 3,000 p.s.i. concrete placed and vibrated; and form removal with breaking form ties and patching walls. The wall systems list walls from 6" to 16" thick and are designed with minimum reinforcement.

Excavation and backfill are not included.

Please see the reference section for further design and cost information.

A2020 110		Walls, Cast in Place						
	WALL HEIGHT (FT.)	PLACING METHOD	CONCRETE (C.Y. per L.F.)	REINFORCING (LBS. per L.F.)	WALL THICKNESS (IN.)	COST PER L.F.		
						MAT.	INST.	TOTAL
5000	8'	direct chute	.148	6.6	6	38	119	157
5020			.199	9.6	8	47	125	172
5040			.250	12	10	56	125	181
5060			.296	14.39	12	64	128	192
5080			.347	16.19	14	66.50	128	194.50
5100			.394	19.19	16	80.50	133	213.50
5200	8'	pumped	.148	6.6	6	38	121	159
5220			.199	9.6	8	47	125	172
5240			.250	12	10	56	128	184
5260			.296	14.39	12	64	131	195
5280			.347	16.19	14	66.50	131	197.50
5300			.394	19.19	16	80.50	137	217.50
6020	10'	direct chute	.248	12	8	59	153	212
6040			.307	14.99	10	69	156	225
6060			.370	17.99	12	80	160	240
6080			.433	20.24	14	90	162	252
6100			.493	23.99	16	101	166	267
6220	10'	pumped	.248	12	8	59	157	216
6240			.307	14.99	10	69	160	229
6260			.370	17.99	12	80	164	244
6280			.433	20.24	14	90	166	256
6300			.493	23.99	16	101	171	272
7220	12'	pumped	.298	14.39	8	71	188	259
7240			.369	17.99	10	83	192	275
7260			.444	21.59	12	95.50	197	292.50
7280			.52	24.29	14	108	200	308
7300			.591	28.79	16	121	205	326
7420	12'	crane & bucket	.298	14.39	8	71	197	268
7440			.369	17.99	10	83	201	284
7460			.444	21.59	12	95.50	208	303.50
7480			.52	24.29	14	108	213	321
7500			.591	28.79	16	121	222	343
8220	14'	pumped	.347	16.79	8	82.50	219	301.50
8240			.43	20.99	10	96.50	224	320.50
8260			.519	25.19	12	112	230	342
8280			.607	28.33	14	126	233	359
8300			.69	33.59	16	141	239	380

A20 Basement Construction

A2020 Basement Walls

A2020 110	Walls, Cast in Place						

	WALL HEIGHT (FT.)	PLACING METHOD	CONCRETE (C.Y. per L.F.)	REINFORCING (LBS. per L.F.)	WALL THICKNESS (IN.)	COST PER L.F.		
						MAT.	INST.	TOTAL
8420	14'	crane & bucket	.347	16.79	8	82.50	229	311.50
8440			.43	20.99	10	96.50	234	330.50
8460			.519	25.19	12	112	243	355
8480			.607	28.33	14	126	248	374
8500			.69	33.59	16	141	256	397
9220	16'	pumped	.397	19.19	8	94.50	251	345.50
9240			.492	23.99	10	111	256	367
9260			.593	28.79	12	128	263	391
9280			.693	32.39	14	144	266	410
9300			.788	38.38	16	161	273	434
9420	16'	crane & bucket	.397	19.19	8	94.50	262	356.50
9440			.492	23.99	10	111	268	379
9460			.593	28.79	12	128	277	405
9480			.693	32.39	14	144	283	427
9500			.788	38.38	16	161	293	454

For customer support on your Concrete & Masonry Costs with RSMeans Data, call 800.448.8182.

A2020 Basement Walls

A2020 220	Subdrainage Piping	COST PER L.F.		
		MAT.	INST.	TOTAL
2000	Piping, excavation & backfill excluded, PVC, perforated			
2110	3" diameter	2.43	4.72	7.15
2130	4" diameter	2.43	4.72	7.15
2140	5" diameter	4.69	5.05	9.74
2150	6" diameter	4.69	5.05	9.74
3000	Metal alum. or steel, perforated asphalt coated			
3150	6" diameter	7	4.66	11.66
3160	8" diameter	11.50	4.79	16.29
3170	10" diameter	14.35	6.55	20.90
3180	12" diameter	16.10	8.25	24.35
3220	18" diameter	24	11.45	35.45
4000	Corrugated HDPE tubing			
4130	4" diameter	.89	.88	1.77
4150	6" diameter	2	1.18	3.18
4160	8" diameter	5.80	1.51	7.31
4180	12" diameter	6.75	5.20	11.95
4200	15" diameter	10.75	5.90	16.65
4220	18" diameter	14.15	8.50	22.65

Same Data. Simplified.

Enjoy the convenience and efficiency of accessing your costs anywhere:

- **Skip the multiplier** by setting your location
- **Quickly search,** edit, favorite and share costs
- **Stay on top of price changes** with automatic updates

Discover more at rsmeans.com/online

CONCRETE COLUMNS

General: It is desirable for purposes of consistency and simplicity to maintain constant column sizes throughout the building height. To do this, concrete strength may be varied (higher strength concrete at lower stories and lower strength concrete at upper stories), as well as varying the amount of reinforcing.

The first portion of the table provides probable minimum column sizes with related costs and weights per lineal foot of story height for bottom level columns.

The second portion of the table provides costs by column size for top level columns with minimum code reinforcement. Probable maximum loads for these columns are also given.

How to Use Table:

1. Enter the second portion (minimum reinforcing) of the table with the minimum allowable column size from the selected cast in place floor system.

   If the total load on the column does not exceed the allowable working load shown, use the cost per L.F. multiplied by the length of columns required to obtain the column cost.

2. If the total load on the column exceeds the allowable working load shown in the second portion of the table, enter the first portion of the

table with the total load on the column and the minimum allowable column size from the selected cast in place floor system.

Select a cost per L.F. for bottom level columns by total load or minimum allowable column size.

Select a cost per L.F. for top level columns using the column size required for bottom level columns from the second portion of the table.

$$\frac{\text{Btm. + Top Col. Costs/L.F.}}{2} = \text{Avg. Col. Cost/L.F.}$$

Column Cost = Average Col. Cost/L.F. x Length of Cols. Required.

See reference section to determine total loads.

Design and Pricing Assumptions:
Normal wt. concrete, f'c = 4 or 6 KSI, placed by pump.
Steel, fy = 60 KSI, spliced every other level.
Minimum design eccentricity of 0.1t.
Assumed load level depth is 8″ (weights prorated to full story basis).
Gravity loads only (no frame or lateral loads included).

Please see the reference section for further design and cost information.

System Components			COST PER V.L.F.		
	QUANTITY	UNIT	MAT.	INST.	TOTAL
SYSTEM B1010 201 1050					
ROUND TIED COLUMNS, 4 KSI CONCRETE, 100K MAX. LOAD, 10′ STORY, 12″ SIZE					
Forms in place, columns, round fiber tube, 12″ diam., 1 use	1.000	L.F.	4.55	16.60	21.15
Reinforcing in place, column ties	1.393	Lb.	5.46	7.23	12.69
Concrete ready mix, regular weight, 4000 psi	.029	C.Y.	4.12		4.12
Placing concrete, incl. vibrating, 12″ sq./round columns, pumped	.029	C.Y.		2.43	2.43
Finish, burlap rub w/grout	3.140	S.F.	.16	4.24	4.40
TOTAL			14.29	30.50	44.79

B1010 201		C.I.P. Column - Round Tied						
	LOAD (KIPS)	STORY HEIGHT (FT.)	COLUMN SIZE (IN.)	COLUMN WEIGHT (P.L.F.)	CONCRETE STRENGTH (PSI)	COST PER V.L.F.		
						MAT.	INST.	TOTAL
1050	100	10	12	110	4000	14.30	30.50	44.80
1060		12	12	111	4000	14.55	31	45.55
1070		14	12	112	4000	14.85	31	45.85
1080	150	10	12	110	4000	15.65	32.50	48.15
1090		12	12	111	4000	15.95	32.50	48.45
1100		14	14	153	4000	19.75	35.50	55.25
1120	200	10	14	150	4000	20	36	56
1140		12	14	152	4000	20.50	36.50	57
1160		14	14	153	4000	21	37	58

B1010 Floor Construction

B1010 201	C.I.P. Column - Round Tied

	LOAD (KIPS)	STORY HEIGHT (FT.)	COLUMN SIZE (IN.)	COLUMN WEIGHT (P.L.F.)	CONCRETE STRENGTH (PSI)	COST PER V.L.F.		
						MAT.	INST.	TOTAL
1180	300	10	16	194	4000	24	39	63
1190		12	18	250	4000	30	44.50	74.50
1200		14	18	252	4000	31	45.50	76.50
1220	400	10	20	306	4000	35.50	50	85.50
1230		12	20	310	4000	36	51	87
1260		14	20	313	4000	37	52	89
1280	500	10	22	368	4000	46.50	56.50	103
1300		12	22	372	4000	47.50	57.50	105
1325		14	22	375	4000	48.50	59	107.50
1350	600	10	24	439	4000	52.50	63	115.50
1375		12	24	445	4000	54	64.50	118.50
1400		14	24	448	4000	55	66	121
1420	700	10	26	517	4000	65	70.50	135.50
1430		12	26	524	4000	66	72	138
1450		14	26	528	4000	67.50	74	141.50
1460	800	10	28	596	4000	72	78	150
1480		12	28	604	4000	73.50	80	153.50
1490		14	28	609	4000	75	82	157
1500	900	10	28	596	4000	76.50	83.50	160
1510		12	28	604	4000	78	85.50	163.50
1520		14	28	609	4000	79.50	87.50	167
1530	1000	10	30	687	4000	83	87	170
1540		12	30	695	4000	84.50	89	173.50
1620		14	30	701	4000	86.50	91.50	178
1640	100	10	12	110	6000	14.65	30.50	45.15
1660		12	12	111	6000	14.95	31	45.95
1680		14	12	112	6000	15.20	31	46.20
1700	150	10	12	110	6000	15.75	32	47.75
1710		12	12	111	6000	16.05	32.50	48.55
1720		14	12	112	6000	16.30	32.50	48.80
1730	200	10	12	110	6000	16.55	33	49.55
1740		12	12	111	6000	16.85	33.50	50.35
1760		14	12	112	6000	17.10	34	51.10
1780	300	10	14	150	6000	21	36.50	57.50
1790		12	14	152	6000	21.50	37	58.50
1800		14	16	153	6000	24.50	38.50	63
1810	400	10	16	194	6000	25.50	40.50	66
1820		12	16	196	6000	26.50	41	67.50
1830		14	18	252	6000	31	44.50	75.50
1850	500	10	18	247	6000	32.50	46.50	79
1870		12	18	250	6000	33	47	80
1880		14	20	252	6000	36.50	50	86.50
1890	600	10	20	306	6000	38	52	90
1900		12	20	310	6000	38.50	53	91.50
1905		14	20	313	6000	39.50	54	93.50
1910	700	10	22	368	6000	47.50	56.50	104
1915		12	22	372	6000	48.50	57.50	106
1920		14	22	375	6000	49.50	59	108.50
1925	800	10	24	439	6000	54	63	117
1930		12	24	445	6000	55.50	64.50	120
1935		14	24	448	6000	56.50	66	122.50
1940	900	10	24	439	6000	57.50	67.50	125
1945		12	24	445	6000	58.50	68.50	127

297

B1010 Floor Construction

B1010 201		C.I.P. Column - Round Tied						
	LOAD (KIPS)	STORY HEIGHT (FT.)	COLUMN SIZE (IN.)	COLUMN WEIGHT (P.L.F.)	CONCRETE STRENGTH (PSI)	COST PER V.L.F.		
						MAT.	INST.	TOTAL
1970	1000	10	26	517	6000	68	72	140
1980		12	26	524	6000	69	74	143
1995		14	26	528	6000	70.50	75.50	146

B1010 202		C.I.P. Columns, Round Tied - Minimum Reinforcing						
	LOAD (KIPS)	STORY HEIGHT (FT.)	COLUMN SIZE (IN.)	COLUMN WEIGHT (P.L.F.)	CONCRETE STRENGTH (PSI)	COST PER V.L.F.		
						MAT.	INST.	TOTAL
2500	100	10-14	12	107	4000	12.90	28.50	41.40
2510	200	10-14	16	190	4000	21.50	36	57.50
2520	400	10-14	20	295	4000	31.50	45	76.50
2530	600	10-14	24	425	4000	47.50	55.50	103
2540	800	10-14	28	580	4000	64.50	68	132.50
2550	1100	10-14	32	755	4000	82.50	79	161.50
2560	1400	10-14	36	960	4000	98	92.50	190.50
2570								

CONCRETE COLUMNS

General: It is desirable for purposes of consistency and simplicity to maintain constant column sizes throughout the building height. To do this, concrete strength may be varied (higher strength concrete at lower stories and lower strength concrete at upper stories), as well as varying the amount of reinforcing.

The first portion of the table provides probable minimum column sizes with related costs and weights per lineal foot of story height for bottom level columns.

The second portion of the table provides costs by column size for top level columns with minimum code reinforcement. Probable maximum loads for these columns are also given.

How to Use Table:

1. Enter the second portion (minimum reinforcing) of the table with the minimum allowable column size from the selected cast in place floor system.

   If the total load on the column does not exceed the allowable working load shown, use the cost per L.F. multiplied by the length of columns required to obtain the column cost.

2. If the total load on the column exceeds the allowable working load shown in the second portion of the table, enter the first portion of the table with the total load on the column and the minimum allowable column size from the selected cast in place floor system.

Select a cost per L.F. for bottom level columns by total load or minimum allowable column size.

Select a cost per L.F. for top level columns using the column size required for bottom level columns from the second portion of the table.

$$\frac{\text{Btm.} + \text{Top Col. Costs/L.F.}}{2} = \text{Avg. Col. Cost/L.F.}$$

Column Cost = Average Col. Cost/L.F. x Length of Cols. Required.

See reference section in back of book to determine total loads.

Design and Pricing Assumptions:
Normal wt. concrete, f'c = 4 or 6 KSI, placed by pump.
Steel, fy = 60 KSI, spliced every other level.
Minimum design eccentricity of 0.1t.
Assumed load level depth is 8″ (weights prorated to full story basis).
Gravity loads only (no frame or lateral loads included).

Please see the reference section for further design and cost information.

System Components	QUANTITY	UNIT	COST PER V.L.F.		
			MAT.	INST.	TOTAL
SYSTEM B1010 203 0640					
SQUARE COLUMNS, 100K LOAD, 10' STORY, 10″ SQUARE					
Forms in place, columns, plywood, 10″ x 10″, 4 uses	3.323	SFCA	4.47	36.83	41.30
Chamfer strip, wood, 3/4″ wide	4.000	L.F.	.60	4.96	5.56
Reinforcing in place, column ties	1.405	Lb.	4.83	6.39	11.22
Concrete ready mix, regular weight, 4000 psi	.026	C.Y.	3.69		3.69
Placing concrete, incl. vibrating, 12″ sq./round columns, pumped	.026	C.Y.		2.18	2.18
Finish, break ties, patch voids, burlap rub w/grout	3.333	S.F.	.17	4.50	4.67
TOTAL			13.76	54.86	68.62

B1010 203		C.I.P. Column, Square Tied						
	LOAD (KIPS)	STORY HEIGHT (FT.)	COLUMN SIZE (IN.)	COLUMN WEIGHT (P.L.F.)	CONCRETE STRENGTH (PSI)	COST PER V.L.F.		
						MAT.	INST.	TOTAL
0640	100	10	10	96	4000	13.75	55	68.75
0680		12	10	97	4000	14	55	69
0700		14	12	142	4000	18.35	67	85.35
0710								

299

B1010 203		C.I.P. Column, Square Tied					

	LOAD (KIPS)	STORY HEIGHT (FT.)	COLUMN SIZE (IN.)	COLUMN WEIGHT (P.L.F.)	CONCRETE STRENGTH (PSI)	COST PER V.L.F.		
						MAT.	INST.	TOTAL
0740	150	10	10	96	4000	16.15	58	74.15
0780		12	12	142	4000	19.05	68	87.05
0800		14	12	143	4000	19.40	68	87.40
0840	200	10	12	140	4000	20	69	89
0860		12	12	142	4000	20.50	69.50	90
0900		14	14	196	4000	24	77.50	101.50
0920	300	10	14	192	4000	25.50	79.50	105
0960		12	14	194	4000	26	80	106
0980		14	16	253	4000	30	88.50	118.50
1020	400	10	16	248	4000	32	91	123
1060		12	16	251	4000	32.50	92	124.50
1080		14	16	253	4000	33	92.50	125.50
1200	500	10	18	315	4000	38	101	139
1250		12	20	394	4000	45.50	115	160.50
1300		14	20	397	4000	46.50	116	162.50
1350	600	10	20	388	4000	47.50	118	165.50
1400		12	20	394	4000	48.50	119	167.50
1600		14	20	397	4000	49.50	120	169.50
1900	700	10	20	388	4000	54.50	130	184.50
2100		12	22	474	4000	54.50	131	185.50
2300		14	22	478	4000	56	132	188
2600	800	10	22	388	4000	57	134	191
2900		12	22	474	4000	58	135	193
3200		14	22	478	4000	59.50	137	196.50
3400	900	10	24	560	4000	64	146	210
3800		12	24	567	4000	65	148	213
4000		14	24	571	4000	66.50	150	216.50
4250	1000	10	24	560	4000	69.50	153	222.50
4500		12	26	667	4000	72.50	160	232.50
4750		14	26	673	4000	74.50	162	236.50
5600	100	10	10	96	6000	14.05	54.50	68.55
5800		12	10	97	6000	14.30	55	69.30
6000		14	12	142	6000	18.85	67	85.85
6200	150	10	10	96	6000	16.50	58	74.50
6400		12	12	98	6000	19.55	68	87.55
6600		14	12	143	6000	19.90	68	87.90
6800	200	10	12	140	6000	20.50	69	89.50
7000		12	12	142	6000	21	69.50	90.50
7100		14	14	196	6000	24.50	77.50	102
7300	300	10	14	192	6000	25.50	79	104.50
7500		12	14	194	6000	26	79.50	105.50
7600		14	14	196	6000	26.50	80	106.50
7700	400	10	14	192	6000	27.50	81.50	109
7800		12	14	194	6000	28	82	110
7900		14	16	253	6000	31.50	88.50	120
8000	500	10	16	248	6000	32.50	91	123.50
8050		12	16	251	6000	33.50	92	125.50
8100		14	16	253	6000	34	92.50	126.50
8200	600	10	18	315	6000	37.50	102	139.50
8300		12	18	319	6000	38.50	103	141.50
8400		14	18	321	6000	39.50	104	143.50

B1010 Floor Construction

B1010 203		C.I.P. Column, Square Tied						
	LOAD (KIPS)	STORY HEIGHT (FT.)	COLUMN SIZE (IN.)	COLUMN WEIGHT (P.L.F.)	CONCRETE STRENGTH (PSI)	COST PER V.L.F.		
						MAT.	INST.	TOTAL
8500	700	10	18	315	6000	40	105	145
8600		12	18	319	6000	41	106	147
8700		14	18	321	6000	41.50	107	148.50
8800	800	10	20	388	6000	47	115	162
8900		12	20	394	6000	47.50	116	163.50
9000		14	20	397	6000	48.50	118	166.50
9100	900	10	20	388	6000	50.50	120	170.50
9300		12	20	394	6000	51.50	122	173.50
9600		14	20	397	6000	52.50	123	175.50
9800	1000	10	22	469	6000	57.50	132	189.50
9840		12	22	474	6000	58.50	134	192.50
9900		14	22	478	6000	60	135	195

B1010 204		C.I.P. Column, Square Tied-Minimum Reinforcing						
	LOAD (KIPS)	STORY HEIGHT (FT.)	COLUMN SIZE (IN.)	COLUMN WEIGHT (P.L.F.)	CONCRETE STRENGTH (PSI)	COST PER V.L.F.		
						MAT.	INST.	TOTAL
9913	150	10-14	12	135	4000	16.45	59	75.45
9918	300	10-14	16	240	4000	26.50	77.50	104
9924	500	10-14	20	375	4000	40.50	109	149.50
9930	700	10-14	24	540	4000	55.50	135	190.50
9936	1000	10-14	28	740	4000	73	164	237
9942	1400	10-14	32	965	4000	89	184	273
9948	1800	10-14	36	1220	4000	111	215	326
9954	2300	10-14	40	1505	4000	135	247	382

B1010 Floor Construction

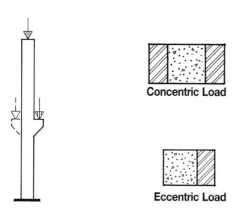

Concentric Load

Eccentric Load

General: Data presented here is for plant produced members transported 50 miles to 100 miles to the site and erected.

Design and pricing assumptions:
Normal wt. concrete, f'c = 5 KSI

Main reinforcement, fy = 60 KSI
Ties, fy = 40 KSI

Minimum design eccentricity, 0.1t.

Concrete encased structural steel haunches are assumed where practical; otherwise galvanized rebar haunches are assumed.

Base plates are integral with columns.

Foundation anchor bolts, nuts and washers are included in price.

System Components	QUANTITY	UNIT	COST PER V.L.F.		
			MAT.	INST.	TOTAL
SYSTEM B1010 206 0560					
PRECAST TIED COLUMN, CONCENTRIC LOADING, 100 K MAX.					
TWO STORY - 10′ PER STORY, 5 KSI CONCRETE, 12″X12″					
Precast column, two story-10′/story, 5 KSI conc., 12″x12″	1.000	Ea.	252		252
Anchor bolts	.050	Set	3.18	1.93	5.11
Steel bearing plates; top, bottom, haunches	3.250	Lb.	5.20		5.20
Erection crew	.012	Hr.		13.74	13.74
TOTAL			260.38	15.67	276.05

B1010 206		Tied, Concentric Loaded Precast Concrete Columns					
	LOAD (KIPS)	STORY HEIGHT (FT.)	COLUMN SIZE (IN.)	COLUMN WEIGHT (P.L.F.)	LOAD LEVELS	COST PER V.L.F.		
						MAT.	INST.	TOTAL
0560	100	10	12x12	164	2	260	15.65	275.65
0570		12	12x12	162	2	255	13.05	268.05
0580		14	12x12	161	2	254	12.85	266.85
0590	150	10	12x12	166	3	255	12.70	267.70
0600		12	12x12	169	3	254	11.40	265.40
0610		14	12x12	162	3	254	11.25	265.25
0620	200	10	12x12	168	4	254	13.55	267.55
0630		12	12x12	170	4	253	12.25	265.25
0640		14	12x12	220	4	253	12.15	265.15
0680	300	10	14x14	225	3	345	12.70	357.70
0690		12	14x14	225	3	345	11.40	356.40
0700		14	14x14	250	3	345	11.25	356.25
0710	400	10	16x16	255	4	300	13.55	313.55
0720		12	16x16	295	4	300	12.25	312.25
0750		14	16x16	305	4	242	12.15	254.15
0790	450	10	16x16	320	3	300	12.70	312.70
0800		12	16x16	315	3	300	11.40	311.40
0810		14	16x16	330	3	300	11.25	311.25
0820	600	10	18x18	405	4	380	13.55	393.55
0830		12	18x18	395	4	380	12.25	392.25
0840		14	18x18	410	4	380	12.15	392.15
0910	800	10	20x20	495	4	340	13.55	353.55
0920		12	20x20	505	4	340	12.25	352.25
0930		14	20x20	510	4	340	12.15	352.15

B1010 Floor Construction

B1010 206	Tied, Concentric Loaded Precast Concrete Columns						
	LOAD (KIPS)	STORY HEIGHT (FT.)	COLUMN SIZE (IN.)	COLUMN WEIGHT (P.L.F.)	LOAD LEVELS	COST PER V.L.F.	
						MAT.	INST.	TOTAL

	LOAD (KIPS)	STORY HEIGHT (FT.)	COLUMN SIZE (IN.)	COLUMN WEIGHT (P.L.F.)	LOAD LEVELS	MAT.	INST.	TOTAL
0970	900	10	22x22	625	3	410	12.70	422.70
0980		12	22x22	610	3	410	11.40	421.40
0990		14	22x22	605	3	410	11.25	421.25

B1010 207	Tied, Eccentric Loaded Precast Concrete Columns

	LOAD (KIPS)	STORY HEIGHT (FT.)	COLUMN SIZE (IN.)	COLUMN WEIGHT (P.L.F.)	LOAD LEVELS	MAT.	INST.	TOTAL
1130	100	10	12x12	161	2	251	15.65	266.65
1140		12	12x12	159	2	250	13.05	263.05
1150		14	12x12	159	2	250	12.85	262.85
1160	150	10	12x12	161	3	246	12.70	258.70
1170		12	12x12	160	3	246	11.40	257.40
1180		14	12x12	159	3	247	11.25	258.25
1190	200	10	12x12	161	4	243	13.55	256.55
1200		12	12x12	160	4	243	12.25	255.25
1210		14	12x12	177	4	244	12.15	256.15
1250	300	10	12x12	185	3	335	12.70	347.70
1260		12	14x14	215	3	335	11.40	346.40
1270		14	14x14	215	3	335	11.25	346.25
1280	400	10	14x14	235	4	330	13.55	343.55
1290		12	16x16	285	4	330	12.25	342.25
1300		14	16x16	295	4	330	12.15	342.15
1360	450	10	14x14	245	3	292	12.70	304.70
1370		12	16x16	285	3	293	11.40	304.40
1380		14	16x16	290	3	292	11.25	303.25
1390	600	10	18x18	385	4	365	13.55	378.55
1400		12	18x18	380	4	365	12.25	377.25
1410		14	18x18	375	4	365	12.15	377.15
1480	800	10	20x20	490	4	325	13.55	338.55
1490		12	20x20	480	4	325	12.25	337.25
1500		14	20x20	475	4	325	12.15	337.15

B1010   Floor Construction

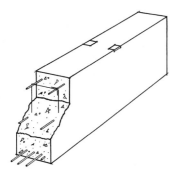

General: Beams priced in the following table are plant produced prestressed members transported to the site and erected.

Pricing Assumptions: Prices are based upon 10,000 S.F. to 20,000 S.F. projects and 50 mile to 100 mile transport.

Normal steel for connections is included in price. Deduct 20% from prices for Southern states. Add 10% to prices for Western states.

Design Assumptions: Normal weight concrete to 150 lbs./C.F. f'c = 5 KSI. Prestressing Steel: 250 KSI straight strand. Non-Prestressing Steel: fy = 60 KSI.

B1010 213		**Rectangular Precast Beams**						
	SPAN (FT.)	**SUPERIMPOSED LOAD (K.L.F.)**	**SIZE W X D (IN.)**	**BEAM WEIGHT (P.L.F.)**	**TOTAL LOAD (K.L.F.)**	**COST PER L.F.**		
						MAT.	**INST.**	**TOTAL**
2200	15	2.32	12x16	200	2.52	211	22	233
2250		3.80	12x20	250	4.05	229	22	251
2300		5.60	12x24	300	5.90	236	22.50	258.50
2350		7.65	12x28	350	8.00	258	22.50	280.50
2400		5.85	18x20	375	6.73	282	22.50	304.50
2450		8.26	18x24	450	8.71	287	24.50	311.50
2500		11.39	18x28	525	11.91	300	24.50	324.50
2550		18.90	18x36	675	19.58	335	24.50	359.50
2600		10.78	24x24	600	11.38	310	24.50	334.50
2700		15.12	24x28	700	15.82	335	24.50	359.50
2750		25.23	24x36	900	26.13	375	25.50	400.50
2800	20	1.22	12x16	200	1.44	211	16.35	227.35
2850		2.03	12x20	250	2.28	229	17	246
2900		3.02	12x24	300	3.32	236	17	253
2950		4.15	12x28	350	4.50	258	17	275
3000		6.85	12x36	450	7.30	252	18.30	270.30
3050		3.13	18x20	375	3.50	282	17	299
3100		4.45	18x24	450	4.90	287	18.30	305.30
3150		6.18	18x28	525	6.70	300	18.30	318.30
3200		10.33	18x36	675	11.00	335	19.10	354.10
3400		15.48	18x44	825	16.30	350	21	371
3450		21.63	18x52	975	22.60	370	21	391
3500		5.80	24x24	600	6.40	310	18.30	328.30
3600		8.20	24x28	700	8.90	335	19.10	354.10
3850		29.20	24x52	1300	30.50	425	28.50	453.50
3900	25	1.82	12x24	300	2.12	236	13.60	249.60
4000		2.53	12x28	350	2.88	258	14.65	272.65
4050		5.18	12x36	450	5.63	252	14.65	266.65
4100		6.55	12x44	550	7.10	288	14.65	302.65
4150		9.08	12x52	650	9.73	315	15.25	330.25
4200		1.86	18x20	375	2.24	282	14.65	296.65
4250		2.69	18x24	450	3.14	287	14.65	301.65
4300		3.76	18x28	525	4.29	300	14.65	314.65
4350		6.37	18x36	675	7.05	335	16.65	351.65
4400		9.60	18x44	872	10.43	350	18.30	368.30
4450		13.49	18x52	975	14.47	370	18.30	388.30

For customer support on your Concrete & Masonry Costs with RSMeans Data, call 800.448.8182.

B1010 Floor Construction

B1010 213		Rectangular Precast Beams						

	SPAN (FT.)	SUPERIMPOSED LOAD (K.L.F.)	SIZE W X D (IN.)	BEAM WEIGHT (P.L.F.)	TOTAL LOAD (K.L.F.)	COST PER L.F.		
						MAT.	INST.	TOTAL
4500	25	18.40	18x60	1125	19.53	395	20.50	415.50
4600		3.50	24x24	600	4.10	310	16.65	326.65
4700		5.00	24x28	700	5.70	335	16.65	351.65
4800		8.50	24x36	900	9.40	375	18.30	393.30
4900		12.85	24x44	1100	13.95	385	20.50	405.50
5000		18.22	24x52	1300	19.52	425	23	448
5050		24.48	24x60	1500	26.00	455	26	481
5100	30	1.65	12x28	350	2.00	258	12.75	270.75
5150		2.79	12x36	450	3.24	252	12.75	264.75
5200		4.38	12x44	550	4.93	288	13.85	301.85
5250		6.10	12x52	650	6.75	315	13.85	328.85
5300		8.36	12x60	750	9.11	340	15.25	355.25
5400		1.72	18x24	450	2.17	287	12.75	299.75
5450		2.45	18x28	525	2.98	300	13.85	313.85
5750		4.21	18x36	675	4.89	335	13.85	348.85
6000		6.42	18x44	825	7.25	350	15.25	365.25
6250		9.07	18x52	975	10.05	370	17	387
6500		12.43	18x60	1125	13.56	395	19.10	414.10
6750		2.24	24x24	600	2.84	310	13.85	323.85
7000		3.26	24x28	700	3.96	335	15.25	350.25
7100		5.63	24x36	900	6.53	375	17	392
7200	35	8.59	24x44	1100	9.69	385	18.70	403.70
7300		12.25	24x52	1300	13.55	425	18.70	443.70
7400		16.54	24x60	1500	18.04	455	18.70	473.70
7500	40	2.23	12x44	550	2.78	288	11.45	299.45
7600		3.15	12x52	650	3.80	315	11.45	326.45
7700		4.38	12x60	750	5.13	340	12.75	352.75
7800		2.08	18x36	675	2.76	335	12.75	347.75
7900		3.25	18x44	825	4.08	350	12.75	362.75
8000		4.68	18x52	975	5.66	370	14.35	384.35
8100		6.50	18x60	1175	7.63	395	14.35	409.35
8200		2.78	24x36	900	3.60	375	17.60	392.60
8300		4.35	24x44	1100	5.45	385	17.60	402.60
8400		6.35	24x52	1300	7.65	425	19.10	444.10
8500		8.65	24x60	1500	10.15	455	19.10	474.10
8600	45	2.35	12x52	650	3.00	315	11.30	326.30
8800		3.29	12x60	750	4.04	340	11.30	351.30
9000		2.39	18x44	825	3.22	350	14.55	364.55
9020		3.49	18x52	975	4.47	370	14.55	384.55
9200		4.90	18x60	1125	6.03	395	17	412
9250		3.21	24x44	1100	4.31	385	18.50	403.50
9300		4.72	24x52	1300	6.02	425	18.50	443.50
9500		6.52	24x60	1500	8.02	455	18.50	473.50
9600	50	2.64	18x52	975	3.62	370	15.25	385.25
9700		3.75	18x60	1125	4.88	395	18.30	413.30
9750		3.58	24x52	1300	4.88	425	18.30	443.30
9800		5.00	24x60	1500	6.50	455	20.50	475.50
9950	55	3.86	24x60	1500	5.36	455	18.50	473.50

305

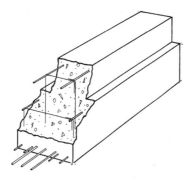

General: Beams priced in the following table are plant produced prestressed members transported to the site and erected.

Pricing Assumptions: Prices are based upon 10,000 S.F. to 20,000 S.F. projects and 50 mile to 100 mile transport.

Normal steel for connections is included in price. Deduct 20% from prices for Southern states. Add 10% to prices for Western states.

Design Assumptions: Normal weight concrete to 150 lbs./C.F. f'c = 5 KSI. Prestressing Steel: 250 KSI straight strand. Non-Prestressing Steel: fy = 60 KSI.

B1010 214		"T" Shaped Precast Beams						
	SPAN (FT.)	SUPERIMPOSED LOAD (K.L.F.)	SIZE W X D (IN.)	BEAM WEIGHT (P.L.F.)	TOTAL LOAD (K.L.F.)	COST PER L.F.		
						MAT.	INST.	TOTAL
2300	15	2.8	12x16	260	3.06	285	22.50	307.50
2350		4.33	12x20	355	4.69	320	22.50	342.50
2400		6.17	12x24	445	6.62	340	24.50	364.50
2500		8.37	12x28	515	8.89	355	24.50	379.50
2550		13.54	12x36	680	14.22	380	24.50	404.50
2600		8.83	18x24	595	9.43	345	24.50	369.50
2650		12.11	18x28	690	12.8	375	24.50	399.50
2700		19.72	18x36	905	24.63	420	25.50	445.50
2800		11.52	24x24	745	12.27	370	24.50	394.50
2900		15.85	24x28	865	16.72	390	25.50	415.50
3000		26.07	24x36	1130	27.20	465	28	493
3100	20	1.46	12x16	260	1.72	285	17	302
3200		2.28	12x20	355	2.64	320	17	337
3300		3.28	12x24	445	3.73	340	18.30	358.30
3400		4.49	12x28	515	5.00	355	18.30	373.30
3500		7.32	12x36	680	8.00	380	19.10	399.10
3600		11.26	12x44	840	12.10	390	21	411
3700		4.70	18x24	595	5.30	345	18.30	363.30
3800		6.51	18x28	690	7.20	375	19.10	394.10
3900		10.7	18x36	905	11.61	420	21	441
4300		16.19	18x44	1115	17.31	480	23	503
4400		22.77	18x52	1330	24.10	440	23	463
4500		6.15	24x24	745	6.90	370	19.10	389.10
4600		8.54	24x28	865	9.41	390	19.10	409.10
4700		14.17	24x36	1130	15.30	465	23	488
4800		21.41	24x44	1390	22.80	430	28.50	458.50
4900		30.25	24x52	1655	31.91	470	28.50	498.50

B1010 Floor Construction

B1010 214	"T" Shaped Precast Beams

	SPAN (FT.)	SUPERIMPOSED LOAD (K.L.F.)	SIZE W X D (IN.)	BEAM WEIGHT (P.L.F.)	TOTAL LOAD (K.L.F.)	COST PER L.F.		
						MAT.	INST.	TOTAL
5000	25	2.68	12x28	515	3.2	355	14.65	369.65
5050		4.44	12x36	680	5.12	380	16.65	396.65
5100		6.90	12x44	840	7.74	390	16.65	406.65
5200		9.75	12x52	1005	10.76	465	17.45	482.45
5300		13.43	12x60	1165	14.60	550	18.30	568.30
5350		3.92	18x28	690	4.61	375	16.65	391.65
5400		6.52	18x36	905	7.43	420	17.45	437.45
5500		9.96	18x44	1115	11.08	480	19.30	499.30
5600		14.09	18x52	1330	15.42	440	20.50	460.50
5650		19.39	18x60	1540	20.93	460	23	483
5700		3.67	24x24	745	4.42	370	16.65	386.65
5750		5.15	24x28	865	6.02	390	18.30	408.30
5800		8.66	24x36	1130	9.79	465	20.50	485.50
5850		13.20	24x44	1390	14.59	430	23	453
5900		18.76	24x52	1655	20.42	470	24.50	494.50
5950		25.35	24x60	1916	27.27	510	26	536
6000	30	2.88	12x36	680	3.56	380	13.85	393.85
6100		4.54	12x44	840	5.38	390	15.25	405.25
6200		6.46	12x52	1005	7.47	465	15.25	480.25
6250		8.97	12x60	1165	10.14	550	17	567
6300		4.25	18x36	905	5.16	420	15.25	435.25
6350		6.57	18x44	1115	7.69	480	17	497
6400		9.38	18x52	1330	10.71	440	19.10	459.10
6500		13.00	18x60	1540	14.54	460	22	482
6700		3.31	24x28	865	4.18	390	17	407
6750		5.67	24x36	1130	6.80	465	19.10	484.10
6800		8.74	24x44	1390	10.13	430	20.50	450.50
6850		12.52	24x52	1655	14.18	470	22	492
6900		17.00	24x60	1215	18.92	510	23.50	533.50
7000	35	3.11	12x44	840	3.95	390	13.10	403.10
7100		4.48	12x52	1005	5.49	465	13.80	478.80
7200		6.28	12x60	1165	7.45	550	15.40	565.40
7300		4.53	18x44	1115	5.65	480	15.40	495.40
7500		6.54	18x52	1330	7.87	440	17.45	457.45
7600		9.14	18x60	1540	10.68	460	18.70	478.70
7700		3.87	24x36	1130	5.00	465	18.70	483.70
7800		6.05	24x44	1390	7.44	430	20	450
7900		8.76	24x52	1655	10.42	470	20	490
8000		12.00	24x60	1915	13.92	510	22	532
8100	40	3.19	12x52	1005	4.2	465	12.75	477.75
8200		4.53	12x60	1165	5.7	550	14.35	564.35
8300		4.70	18x52	1330	6.03	440	16.35	456.35
8400		6.64	18x60	1540	8.18	460	16.35	476.35
8600		4.31	24x44	1390	5.7	430	19.10	449.10
8700		6.32	24x52	1655	7.98	470	21	491
8800		8.74	24x60	1915	10.66	510	21	531
8900	45	3.34	12x60	1165	4.51	550	14.55	564.55
9000		4.92	18x60	1540	6.46	460	17	477
9250		4.64	24x52	1655	6.30	470	20.50	490.50
9900		6.5	24x60	1915	8.42	510	20.50	530.50
9950	50	4.9	24x60	1915	6.82	510	23	533

B1010 Floor Construction

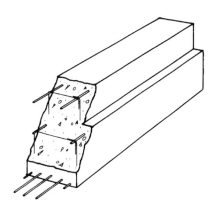

General: Beams priced in the following table are plant produced prestressed members transported to the site and erected.

Pricing Assumptions: Prices are based upon 10,000 S.F. to 20,000 S.F. projects and 50 mile to 100 mile transport.

Normal steel for connections is included in price. Deduct 20% from prices for Southern states. Add 10% to prices for Western states.

Design Assumptions: Normal weight concrete to 150 lbs./C.F. f'c = 5 KSI. Prestressing Steel: 250 KSI straight strand. Non-Prestressing Steel: fy = 60 KSI.

B1010 215		"L" Shaped Precast Beams						
	SPAN (FT.)	SUPERIMPOSED LOAD (K.L.F.)	SIZE W X D (IN.)	BEAM WEIGHT (P.L.F.)	TOTAL LOAD (K.L.F.)	COST PER L.F.		
						MAT.	INST.	TOTAL
2250	15	2.58	12x16	230	2.81	269	22.50	291.50
2300		4.10	12x20	300	4.40	263	22.50	285.50
2400		5.92	12x24	370	6.29	283	22.50	305.50
2450		8.09	12x28	435	8.53	297	22.50	319.50
2500		12.95	12x36	565	13.52	345	24.50	369.50
2600		8.55	18x24	520	9.07	294	24.50	318.50
2650		11.83	18x28	610	12.44	345	24.50	369.50
2700		19.30	18x36	790	20.09	365	24.50	389.50
2750		11.24	24x24	670	11.91	350	24.50	374.50
2800		15.40	24x28	780	16.18	355	25.50	380.50
2850		25.65	24x36	1015	26.67	410	26.50	436.50
2900	20	2.18	12x20	300	2.48	263	17	280
2950		3.17	12x24	370	3.54	283	17.60	300.60
3000		4.37	12x28	435	4.81	297	17.60	314.60
3100		7.04	12x36	565	7.60	345	19.10	364.10
3150		10.80	12x44	695	11.50	355	19.90	374.90
3200		15.08	12x52	825	15.91	390	21	411
3250		4.58	18x24	520	5.10	294	17.60	311.60
3300		6.39	18x28	610	7.00	345	18.30	363.30
3350		10.51	18x36	790	11.30	365	19.10	384.10
3400		15.73	18x44	970	16.70	410	21	431
3550		22.15	18x52	1150	23.30	435	22	457
3600		6.03	24x24	370	6.40	350	18.30	368.30
3650		8.32	24x28	780	9.10	355	21	376
3700		13.98	24x36	1015	15.00	410	22	432
3800		21.05	24x44	1245	22.30	435	25.50	460.50
3900		29.73	24x52	1475	31.21	495	28.50	523.50

B1010 Floor Construction

B1010 215				"L" Shaped Precast Beams			

	SPAN (FT.)	SUPERIMPOSED LOAD (K.L.F.)	SIZE W X D (IN.)	BEAM WEIGHT (P.L.F.)	TOTAL LOAD (K.L.F.)	COST PER L.F.		
						MAT.	INST.	TOTAL
4000	25	2.64	12x28	435	3.08	297	14.65	311.65
4100		4.30	12x36	565	4.87	345	15.25	360.25
4200		6.65	12x44	695	7.35	355	15.95	370.95
4250		9.35	12x52	875	10.18	390	16.65	406.65
4300		12.81	12x60	950	13.76	435	17.45	452.45
4350		2.74	18x24	570	3.76	294	15.25	309.25
4400		3.67	18x28	610	4.28	345	15.95	360.95
4450		6.44	18x36	790	7.23	365	17.45	382.45
4500		9.72	18x44	970	10.69	410	19.30	429.30
4600		13.76	18x52	1150	14.91	435	19.30	454.30
4700		18.33	18x60	1330	20.16	470	21.50	491.50
4800		3.62	24x24	370	3.99	350	16.65	366.65
4900		5.04	24x28	780	5.82	355	17.45	372.45
5000		8.58	24x36	1015	9.60	410	18.30	428.30
5100		13.00	24x44	1245	14.25	435	21.50	456.50
5200		18.50	24x52	1475	19.98	495	24.50	519.50
5300	30	2.80	12x36	565	3.37	345	13.30	358.30
5400		4.42	12x44	695	5.12	355	14.55	369.55
5500		6.24	12x52	875	7.07	390	14.55	404.55
5600		8.60	12x60	950	9.55	435	16.10	451.10
5700		2.50	18x28	610	3.11	345	14.55	359.55
5800		4.23	18x36	790	5.02	365	14.55	379.55
5900		6.45	18x44	970	7.42	410	16.10	426.10
6000		9.20	18x52	1150	10.35	435	17.95	452.95
6100		12.67	18x60	1330	14.00	470	20.50	490.50
6200		3.26	24x28	780	4.04	355	16.10	371.10
6300		5.65	24x36	1015	6.67	410	17.95	427.95
6400		8.66	24x44	1245	9.90	435	19.10	454.10
6500	35	3.00	12x44	695	3.70	355	13.10	368.10
6600		4.27	12x52	825	5.20	390	13.10	403.10
6700		6.00	12x60	950	6.95	435	14.55	449.55
6800		2.90	18x36	790	3.69	365	13.10	378.10
6900		4.48	18x44	970	5.45	410	14.55	424.55
7000		6.45	18x52	1150	7.60	435	15.40	450.40
7100		8.95	18x60	1330	10.28	470	16.35	486.35
7200		3.88	24x36	1015	4.90	410	18.70	428.70
7300		6.03	24x44	1245	7.28	435	20	455
7400		8.71	24x52	1475	10.19	495	20	515
7500	40	3.15	12x52	825	3.98	390	12.05	402.05
7600		4.43	12x60	950	5.38	435	13.45	448.45
7700		3.20	18x44	970	4.17	410	14.35	424.35
7800		4.68	18x52	1150	5.83	435	15.25	450.25
7900		6.55	18x60	1330	7.88	470	15.25	485.25
8000		4.43	24x44	1245	5.68	435	21	456
8100		6.33	24x52	1475	7.81	495	21	516
8200	45	3.30	12x60	950	4.25	435	12.75	447.75
8300		3.45	18x52	1150	4.60	435	15.65	450.65
8500		4.44	18x60	1330	5.77	470	18.50	488.50
8750		3.16	24x44	1245	4.41	435	20.50	455.50
9000		4.68	24x52	1475	6.16	495	20.50	515.50
9900	50	3.71	18x60	1330	5.04	470	18.30	488.30
9950		3.82	24x52	1475	5.30	495	20.50	515.50

For customer support on your Concrete & Masonry Costs with RSMeans Data, call 800.448.8182.

B1010 Floor Construction

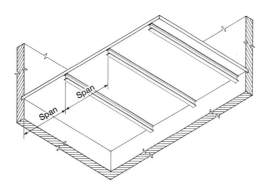

General: Solid concrete slabs of uniform depth reinforced for flexure in one direction and for temperature and shrinkage in the other direction.

Design and Pricing Assumptions:
Concrete f'c = 3 KSI normal weight placed by pump.
Reinforcement fy = 60 KSI.
Deflection ≤ span/360.
Forms, four uses hung from steel beams plus edge forms.
Steel trowel finish for finish floor and cure.

System Components	QUANTITY	UNIT	COST PER S.F. MAT.	COST PER S.F. INST.	COST PER S.F. TOTAL
SYSTEM B1010 217 2000					
6'-0" SINGLE SPAN, 4" SLAB DEPTH, 40 PSF SUPERIMPOSED LOAD					
Forms in place, floor slab forms hung from steel beams, 4 uses	1.000	S.F.	2.40	6.75	9.15
Edge forms to 6" high on elevated slab, 4 uses	.333	L.F.	.13	1.66	1.79
Reinforcing in place, elevated slabs #4 to #7	1.030	Lb.	.73	.50	1.23
Concrete, ready mix, regular weight, 3000 psi	.333	C.F.	1.69		1.69
Place and vibrate concrete, elevated slab less than 6", pumped	.333	C.F.		.57	.57
Finishing floor, monolithic steel trowel finish for finish floor	1.000	S.F.		1.08	1.08
Curing with sprayed membrane curing compound	1.000	S.F.	.13	.11	.24
TOTAL			5.08	10.67	15.75

B1010 217		Cast in Place Slabs, One Way						
	SLAB DESIGN & SPAN (FT.)	SUPERIMPOSED LOAD (P.S.F.)	THICKNESS (IN.)	TOTAL LOAD (P.S.F.)		COST PER S.F. MAT.	COST PER S.F. INST.	COST PER S.F. TOTAL
2000	Single 6	40	4	90		5.10	10.65	15.75
2100		75	4	125		5.10	10.65	15.75
2200		125	4	175		5.10	10.65	15.75
2300		200	4	250		5.35	10.85	16.20
2500	Single 8	40	4	90		4.99	9.40	14.39
2600		75	4	125		5.05	10.25	15.30
2700		125	4-1/2	181		5.25	10.35	15.60
2800		200	5	262		5.85	10.65	16.50
3000	Single 10	40	4	90		5.25	10.15	15.40
3100		75	4	125		5.25	10.15	15.40
3200		125	5	188		5.80	10.40	16.20
3300		200	7-1/2	293		7.75	11.65	19.40
3500	Single 15	40	5-1/2	90		6.25	10.30	16.55
3600		75	6-1/2	156		7	10.85	17.85
3700		125	7-1/2	219		7.60	11.05	18.65
3800		200	8-1/2	306		8.20	11.35	19.55
4000	Single 20	40	7-1/2	115		7.55	10.80	18.35
4100		75	9	200		8.20	10.80	19
4200		125	10	250		9.05	11.20	20.25
4300		200	10	324		9.70	11.65	21.35

B1010 Floor Construction

B1010 217	Cast in Place Slabs, One Way

	SLAB DESIGN & SPAN (FT.)	SUPERIMPOSED LOAD (P.S.F.)	THICKNESS (IN.)	TOTAL LOAD (P.S.F.)		COST PER S.F.		
						MAT.	INST.	TOTAL
4500	Multi 6	40	4	90		5.15	10.10	15.25
4600		75	4	125		5.15	10.10	15.25
4700		125	4	175		5.15	10.10	15.25
4800		200	4	250		5.15	10.10	15.25
5000	Multi 8	40	4	90		5.15	9.85	15
5100		75	4	125		5.15	9.85	15
5200		125	4	175		5.15	9.85	15
5300		200	4	250		5.15	9.85	15
5500	Multi 10	40	4	90		5.15	9.70	14.85
5600		75	4	125		5.15	9.75	14.90
5700		125	4	175		5.35	9.90	15.25
5800		200	5	263		5.95	10.30	16.25
6600	Multi 15	40	5	103		5.90	10.05	15.95
6800		75	5	138		6.20	10.25	16.45
7000		125	6-1/2	206		6.90	10.50	17.40
7100		200	6-1/2	281		7.10	10.65	17.75
7500	Multi 20	40	5-1/2	109		6.80	10.40	17.20
7600		75	6-1/2	156		7.35	10.70	18.05
7800		125	9	238		9.25	11.60	20.85
8000		200	9	313		9.25	11.60	20.85

B1010 Floor Construction

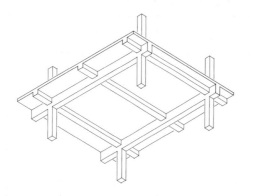

General: Solid concrete one-way slab cast monolithically with reinforced concrete support beams and girders.

Design and Pricing Assumptions:
Concrete f'c = 3 KSI, normal weight, placed by concrete pump.
Reinforcement, fy = 60 KSI.
Forms, four use.
Finish, steel trowel.
Curing, spray on membrane.
Based on 4 bay x 4 bay structure.

System Components	QUANTITY	UNIT	MAT.	INST.	TOTAL
SYSTEM B1010 219 3000					
BM. & SLAB ONE WAY 15′ X 15′ BAY, 40 PSF S.LOAD, 12″ MIN. COL.					
Forms in place, flat plate to 15′ high, 4 uses	.858	S.F.	1.77	5.83	7.60
Forms in place, exterior spandrel, 12″ wide, 4 uses	.142	SFCA	.24	1.75	1.99
Forms in place, interior beam. 12″ wide, 4 uses	.306	SFCA	.60	3.11	3.71
Reinforcing in place, elevated slabs #4 to #7	1.600	Lb.	1.14	.78	1.92
Concrete, ready mix, regular weight, 3000 psi	.410	C.F.	2.08		2.08
Place and vibrate concrete, elevated slab less than 6″, pump	.410	C.F.		.69	.69
Finish floor, monolithic steel trowel finish for finish floor	1.000	S.F.		1.08	1.08
Cure with sprayed membrane curing compound	.010	C.S.F.	.13	.11	.24
TOTAL			5.96	13.35	19.31

B1010 219	Cast in Place Beam & Slab, One Way							
	BAY SIZE (FT.)	SUPERIMPOSED LOAD (P.S.F.)	MINIMUM COL. SIZE (IN.)	SLAB THICKNESS (IN.)	TOTAL LOAD (P.S.F.)	MAT.	INST.	TOTAL
3000	15x15	40	12	4	120	5.95	13.35	19.30
3100		75	12	4	138	6.05	13.45	19.50
3200		125	12	4	188	6.20	13.55	19.75
3300		200	14	4	266	6.60	14.05	20.65
3600	15x20	40	12	4	102	6.10	13.25	19.35
3700		75	12	4	140	6.35	13.65	20
3800		125	14	4	192	6.70	14.10	20.80
3900		200	16	4	272	7.40	15	22.40
4200	20x20	40	12	5	115	6.60	12.90	19.50
4300		75	14	5	154	7.15	13.90	21.05
4400		125	16	5	206	7.45	14.60	22.05
4500		200	18	5	287	8.35	15.65	24
5000	20x25	40	12	5-1/2	121	6.85	12.90	19.75
5100		75	14	5-1/2	160	7.50	14	21.50
5200		125	16	5-1/2	215	8.05	14.80	22.85
5300		200	18	5-1/2	294	8.75	15.85	24.60
5500	25x25	40	12	6	129	7.20	12.65	19.85
5600		75	16	6	171	7.85	13.60	21.45
5700		125	18	6	227	9.15	15.55	24.70
5800		200	2	6	300	10.10	16.70	26.80
6500	25x30	40	14	6-1/2	132	7.35	12.90	20.25
6600		75	16	6-1/2	172	7.95	13.70	21.65
6700		125	18	6-1/2	231	9.35	15.50	24.85
6800		200	20	6-1/2	312	10.15	16.85	27

B1010 Floor Construction

| B1010 219 | Cast in Place Beam & Slab, One Way |

	BAY SIZE (FT.)	SUPERIMPOSED LOAD (P.S.F.)	MINIMUM COL. SIZE (IN.)	SLAB THICKNESS (IN.)	TOTAL LOAD (P.S.F.)	COST PER S.F.		
						MAT.	INST.	TOTAL
7000	30x30	40	14	7-1/2	150	8.45	13.90	22.35
7100		75	18	7-1/2	191	9.35	14.65	24
7300		125	20	7-1/2	245	9.95	15.50	25.45
7400		200	24	7-1/2	328	11.05	17.15	28.20
7500	30x35	40	16	8	158	8.90	14.30	23.20
7600		75	18	8	196	9.50	14.75	24.25
7700		125	22	8	254	10.65	16.40	27.05
7800		200	26	8	332	11.55	16.95	28.50
8000	35x35	40	16	9	169	9.95	14.75	24.70
8200		75	20	9	213	10.75	16.05	26.80
8400		125	24	9	272	11.80	16.65	28.45
8600		200	26	9	355	12.95	17.75	30.70
9000	35x40	40	18	9	174	10.20	15	25.20
9300		75	22	9	214	11	16.20	27.20
9400		125	26	9	273	11.95	16.75	28.70
9600		200	30	9	355	13.10	17.85	30.95

B1010 Floor Construction

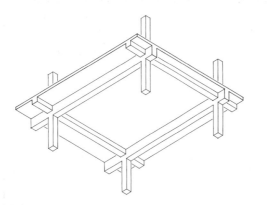

General: Solid concrete two-way slab cast monolithically with reinforced concrete support beams and girders.

Design and Pricing Assumptions:
Concrete f'c = 3 KSI, normal weight, placed by concrete pump.
Reinforcement, fy = 60 KSI.
Forms, four use.
Finish, steel trowel.
Curing, spray on membrane.
Based on 4 bay x 4 bay structure.

System Components				COST PER S.F.		
	QUANTITY	UNIT		MAT.	INST.	TOTAL
SYSTEM B1010 220 2000						
15' X 15' BAY, 40 PSF S. LOAD, 12" MIN. COL						
Forms in place, flat plate to 15' high, 4 uses	.894	S.F.		1.84	6.08	7.92
Forms in place, exterior spandrel, 12" wide, 4 uses	.142	SFCA		.24	1.75	1.99
Forms in place, interior beam. 12" wide, 4 uses	.178	SFCA		.35	1.81	2.16
Reinforcing in place, elevated slabs #4 to #7	1.598	Lb.		1.13	.78	1.91
Concrete, ready mix, regular weight, 3000 psi	.437	C.F.		2.22		2.22
Place and vibrate concrete, elevated slab less than 6", pump	.437	C.F.			.74	.74
Finish floor, monolithic steel trowel finish for finish floor	1.000	S.F.			1.08	1.08
Cure with sprayed membrane curing compound	.010	C.S.F.		.13	.11	.24
TOTAL				**5.91**	**12.35**	**18.26**

B1010 220		Cast in Place Beam & Slab, Two Way						
	BAY SIZE (FT.)	SUPERIMPOSED LOAD (P.S.F.)	MINIMUM COL. SIZE (IN.)	SLAB THICKNESS (IN.)	TOTAL LOAD (P.S.F.)	COST PER S.F.		
						MAT.	INST.	TOTAL
2000	15x15	40	12	4-1/2	106	5.90	12.35	18.25
2200		75	12	4-1/2	143	6.10	12.50	18.60
2250		125	12	4-1/2	195	6.50	12.85	19.35
2300		200	14	4-1/2	274	7.05	13.60	20.65
2400	15 x 20	40	12	5-1/2	120	6.65	12.25	18.90
2600		75	12	5-1/2	159	7.25	13.10	20.35
2800		125	14	5-1/2	213	7.80	14.20	22
2900		200	16	5-1/2	294	8.55	15.10	23.65
3100	20 x 20	40	12	6	132	7.35	12.45	19.80
3300		75	14	6	167	7.80	13.60	21.40
3400		125	16	6	220	8.10	14.10	22.20
3600		200	18	6	301	8.95	15.05	24
4000	20 x 25	40	12	7	141	7.85	13	20.85
4300		75	14	7	181	9	14.20	23.20
4500		125	16	7	236	9.15	14.60	23.75
4700		200	18	7	317	10.10	15.70	25.80
5100	25 x 25	40	12	7-1/2	149	8.20	12.10	20.30
5200		75	16	7-1/2	185	8.95	14.10	23.05
5300		125	18	7-1/2	250	9.70	15.30	25
5400		200	20	7-1/2	332	11.50	16.50	28
6000	25 x 30	40	14	8-1/2	165	9.05	13.15	22.20
6200		75	16	8-1/2	201	9.70	14.30	24
6400		125	18	8-1/2	259	10.60	15.40	26
6600		200	20	8-1/2	341	11.95	16.35	28.30

B10 Superstructure

B1010 Floor Construction

B1010 220	Cast in Place Beam & Slab, Two Way

	BAY SIZE (FT.)	SUPERIMPOSED LOAD (P.S.F.)	MINIMUM COL. SIZE (IN.)	SLAB THICKNESS (IN.)	TOTAL LOAD (P.S.F.)	COST PER S.F.		
						MAT.	INST.	TOTAL
7000	30 x 30	40	14	9	170	9.85	14.05	23.90
7100		75	18	9	212	10.80	15.10	25.90
7300		125	20	9	267	11.55	15.65	27.20
7400		200	24	9	351	12.60	16.50	29.10
7600	30 x 35	40	16	10	188	10.80	14.85	25.65
7700		75	18	10	225	11.45	15.40	26.85
8000		125	22	10	282	12.70	16.55	29.25
8200		200	26	10	371	13.70	17.05	30.75
8500	35 x 35	40	16	10-1/2	193	11.55	15.10	26.65
8600		75	20	10-1/2	233	12.10	15.70	27.80
9000		125	24	10-1/2	287	13.50	16.85	30.35
9100		200	26	10-1/2	370	14	17.90	31.90
9300	35 x 40	40	18	11-1/2	208	12.45	15.65	28.10
9500		75	22	11-1/2	247	13.05	16.05	29.10
9800		125	24	11-1/2	302	13.90	16.60	30.50
9900		200	26	11-1/2	383	14.95	17.65	32.60

B1010 Floor Construction

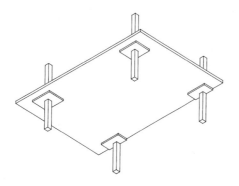

General: Flat Slab: Solid uniform depth concrete two-way slabs with drop panels at columns and no column capitals.

Design and Pricing Assumptions:
Concrete f'c = 3 KSI, placed by concrete pump.
Reinforcement, fy = 60 KSI.
Forms, four use.
Finish, steel trowel.
Curing, spray on membrane.
Based on 4 bay x 4 bay structure.

System Components	QUANTITY	UNIT	COST PER S.F.		
			MAT.	INST.	TOTAL
SYSTEM B1010 222 1700					
15' X 15' BAY, 40 PSF S. LOAD, 12" MIN. COL., 6" SLAB, 1-1/2" DROP, 117 PSF					
Forms in place, flat slab with drop panels, to 15' high, 4 uses	.993	S.F.	2.40	6.95	9.35
Forms in place, exterior spandrel, 12" wide, 4 uses	.034	SFCA	.06	.42	.48
Reinforcing in place, elevated slabs #4 to #7	1.588	Lb.	1.13	.78	1.91
Concrete, ready mix, regular weight, 3000 psi	.513	C.F.	2.60		2.60
Place and vibrate concrete, elevated slab, 6" to 10" pump	.513	C.F.		.74	.74
Finish floor, monolithic steel trowel finish for finish floor	1.000	S.F.		1.08	1.08
Cure with sprayed membrane curing compound	.010	C.S.F.	.13	.11	.24
TOTAL			6.32	10.08	16.40

B1010 222			Cast in Place Flat Slab with Drop Panels					
	BAY SIZE (FT.)	SUPERIMPOSED LOAD (P.S.F.)	MINIMUM COL. SIZE (IN.)	SLAB & DROP (IN.)	TOTAL LOAD (P.S.F.)	COST PER S.F.		
						MAT.	INST.	TOTAL
1700	15 x 15	40	12	6 - 1-1/2	117	6.30	10.10	16.40
1720		75	12	6 - 2-1/2	153	6.45	10.15	16.60
1760		125	14	6 - 3-1/2	205	6.75	10.35	17.10
1780		200	16	6 - 4-1/2	281	7.10	10.60	17.70
1840	15 x 20	40	12	6-1/2 - 2	124	6.75	10.20	16.95
1860		75	14	6-1/2 - 4	162	7.05	10.40	17.45
1880		125	16	6-1/2 - 5	213	7.45	10.65	18.10
1900		200	18	6-1/2 - 6	293	7.65	10.85	18.50
1960	20 x 20	40	12	7 - 3	132	7.05	10.35	17.40
1980		75	16	7 - 4	168	7.50	10.60	18.10
2000		125	18	7 - 6	221	8.30	11	19.30
2100		200	20	8 - 6-1/2	309	8.45	11.15	19.60
2300	20 x 25	40	12	8 - 5	147	7.90	10.75	18.65
2400		75	18	8 - 6-1/2	184	8.50	11.15	19.65
2600		125	20	8 - 8	236	9.20	11.55	20.75
2800		200	22	8-1/2 - 8-1/2	323	9.60	11.80	21.40
3200	25 x 25	40	12	8-1/2 - 5-1/2	154	8.25	10.90	19.15
3400		75	18	8-1/2 - 7	191	8.70	11.15	19.85
4000		125	20	8-1/2 - 8-1/2	243	9.35	11.60	20.95
4400		200	24	9 - 8-1/2	329	9.80	11.85	21.65
5000	25 x 30	40	14	9-1/2 - 7	168	8.95	11.25	20.20
5200		75	18	9-1/2 - 7	203	9.55	11.65	21.20
5600		125	22	9-1/2 - 8	256	10	11.90	21.90
5800		200	24	10 - 10	342	10.65	12.30	22.95

B10 Superstructure

B1010 Floor Construction

B1010 222				Cast in Place Flat Slab with Drop Panels				
	BAY SIZE (FT.)	SUPERIMPOSED LOAD (P.S.F.)	MINIMUM COL. SIZE (IN.)	SLAB & DROP (IN.)	TOTAL LOAD (P.S.F.)	COST PER S.F.		
						MAT.	INST.	TOTAL
6400	30 x 30	40	14	10-1/2 - 7-1/2	182	9.65	11.55	21.20
6600		75	18	10-1/2 - 7-1/2	217	10.30	11.95	22.25
6800		125	22	10-1/2 - 9	269	10.75	12.25	23
7000		200	26	11 - 11	359	11.55	12.70	24.25
7400	30 x 35	40	16	11-1/2 - 9	196	10.50	11.95	22.45
7900		75	20	11-1/2 - 9	231	11.20	12.40	23.60
8000		125	24	11-1/2 - 11	284	11.70	12.70	24.40
9000	35 x 35	40	16	12 - 9	202	10.80	12.05	22.85
9400		75	20	12 - 11	240	11.60	12.60	24.20
9600		125	24	12 - 11	290	12	12.80	24.80

B1010   Floor Construction

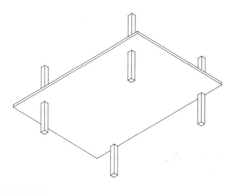

General: Flat Plates: Solid uniform depth concrete two-way slab without drops or interior beams. Primary design limit is shear at columns.

Design and Pricing Assumptions:
Concrete f'c to 4 KSI, placed by concrete pump.
Reinforcement, fy = 60 KSI.
Forms, four use.
Finish, steel trowel.
Curing, spray on membrane.
Based on 4 bay x 4 bay structure.

System Components	QUANTITY	UNIT	COST PER S.F.		
			MAT.	INST.	TOTAL
SYSTEM B1010 223 2000					
15'X15' BAY, 40 PSF S. LOAD, 12" MIN. COL.					
Forms in place, flat plate to 15' high, 4 uses	.992	S.F.	2.04	6.75	8.79
Edge forms to 6" high on elevated slab, 4 uses	.065	L.F.	.02	.32	.34
Reinforcing in place, elevated slabs #4 to #7	1.706	Lb.	1.21	.84	2.05
Concrete, ready mix, regular weight, 3000 psi	.459	C.F.	2.33		2.33
Place and vibrate concrete, elevated slab less than 6", pump	.459	C.F.		.78	.78
Finish floor, monolithic steel trowel finish for finish floor	1.000	S.F.		1.08	1.08
Cure with sprayed membrane curing compound	.010	C.S.F.	.13	.11	.24
TOTAL			5.73	9.88	15.61

B1010 223		Cast in Place Flat Plate						
	BAY SIZE (FT.)	SUPERIMPOSED LOAD (P.S.F.)	MINIMUM COL. SIZE (IN.)	SLAB THICKNESS (IN.)	TOTAL LOAD (P.S.F.)	COST PER S.F.		
						MAT.	INST.	TOTAL
2000	15 x 15	40	12	5-1/2	109	5.75	9.90	15.65
2200		75	14	5-1/2	144	5.80	9.85	15.65
2400		125	20	5-1/2	194	6	9.95	15.95
2600		175	22	5-1/2	244	6.15	10	16.15
3000	15 x 20	40	14	7	127	6.60	10	16.60
3400		75	16	7-1/2	169	7	10.15	17.15
3600		125	22	8-1/2	231	7.70	10.45	18.15
3800		175	24	8-1/2	281	7.75	10.45	18.20
4200	20 x 20	40	16	7	127	6.60	9.95	16.55
4400		75	20	7-1/2	175	7.05	10.20	17.25
4600		125	24	8-1/2	231	7.70	10.40	18.10
5000		175	24	8-1/2	281	7.75	10.50	18.25
5600	20 x 25	40	18	8-1/2	146	7.65	10.45	18.10
6000		75	20	9	188	7.90	10.55	18.45
6400		125	26	9-1/2	244	8.55	10.85	19.40
6600		175	30	10	300	8.85	10.95	19.80
7000	25 x 25	40	20	9	152	7.90	10.50	18.40
7400		75	24	9-1/2	194	8.35	10.75	19.10
7600		125	30	10	250	8.90	11	19.90
8000								

B10 Superstructure

B1010 Floor Construction

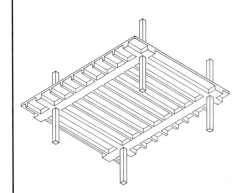

General: Combination of thin concrete slab and monolithic ribs at uniform spacing to reduce dead weight and increase rigidity. The ribs (or joists) are arranged parallel in one direction between supports.

Square end joists simplify forming. Tapered ends can increase span or provide for heavy load.

Costs for multiple span joists are provided in this section. Single span joist costs are not provided here.

Design and Pricing Assumptions:
Concrete f'c = 4 KSI, normal weight placed by concrete pump.
Reinforcement, fy = 60 KSI.
Forms, four use.
  4-1/2" slab.
  30" pans, sq. ends (except for shear req.).
  6" rib thickness.
  Distribution ribs as required.
Finish, steel trowel.
Curing, spray on membrane.
Based on 4 bay x 4 bay structure.

System Components	QUANTITY	UNIT	COST PER S.F. MAT.	COST PER S.F. INST.	COST PER S.F. TOTAL
SYSTEM B1010 226 2000					
15' X 15' BAY, 40 PSF S. LOAD, 12" MIN. COLUMN					
Forms in place, floor slab, with 1-way joist pans, 4 use	.905	S.F.	5.61	6.92	12.53
Forms in place, exterior spandrel, 12" wide, 4 uses	.170	SFCA	.28	2.09	2.37
Forms in place, interior beam. 12" wide, 4 uses	.095	SFCA	.19	.96	1.15
Edge forms, 7"-12" high on elevated slab, 4 uses	.010	L.F.	.01	.08	.09
Reinforcing in place, elevated slabs #4 to #7	.628	Lb.	.45	.31	.76
Concrete ready mix, regular weight, 4000 psi	.555	C.F.	2.92		2.92
Place and vibrate concrete, elevated slab, 6" to 10" pump	.555	C.F.		.80	.80
Finish floor, monolithic steel trowel finish for finish floor	1.000	S.F.		1.08	1.08
Cure with sprayed membrane curing compound	.010	S.F.	.13	.11	.24
TOTAL			9.59	12.35	21.94

B1010 226		Cast in Place Multispan Joist Slab						
	BAY SIZE (FT.)	SUPERIMPOSED LOAD (P.S.F.)	MINIMUM COL. SIZE (IN.)	RIB DEPTH (IN.)	TOTAL LOAD (P.S.F.)	COST PER S.F. MAT.	COST PER S.F. INST.	COST PER S.F. TOTAL
2000	15 x 15	40	12	8	115	9.60	12.35	21.95
2100		75	12	8	150	9.65	12.40	22.05
2200		125	12	8	200	9.80	12.50	22.30
2300		200	14	8	275	10.05	13	23.05
2600	15 x 20	40	12	8	115	9.75	12.35	22.10
2800		75	12	8	150	9.95	13	22.95
3000		125	14	8	200	10.20	13.25	23.45
3300		200	16	8	275	10.65	13.45	24.10
3600	20 x 20	40	12	10	120	9.95	12.20	22.15
3900		75	14	10	155	10.30	12.90	23.20
4000		125	16	10	205	10.35	13.10	23.45
4100		200	18	10	280	10.75	13.65	24.40
4300	20 x 25	40	12	10	120	9.85	12.35	22.20
4400		75	14	10	155	10.25	12.95	23.20
4500		125	16	10	205	10.80	13.65	24.45
4600		200	18	12	280	11.20	14.30	25.50
4700	25 x 25	40	12	12	125	10.05	12.05	22.10
4800		75	16	12	160	10.50	12.70	23.20
4900		125	18	12	210	11.45	13.95	25.40
5000		200	20	14	291	11.95	14.25	26.20

319

B1010 Floor Construction

B1010 226	Cast in Place Multispan Joist Slab

	BAY SIZE (FT.)	SUPERIMPOSED LOAD (P.S.F.)	MINIMUM COL. SIZE (IN.)	RIB DEPTH (IN.)	TOTAL LOAD (P.S.F.)	COST PER S.F.		
						MAT.	INST.	TOTAL
5400	25 x 30	40	14	12	125	10.40	12.60	23
5600		75	16	12	160	10.70	13	23.70
5800		125	18	12	210	11.35	13.90	25.25
6000		200	20	14	291	12.05	14.50	26.55
6200	30 x 30	40	14	14	131	10.85	12.75	23.60
6400		75	18	14	166	11.10	13.15	24.25
6600		125	20	14	216	11.75	13.80	25.55
6700		200	24	16	297	12.40	14.30	26.70
6900	30 x 35	40	16	14	131	11.10	13.30	24.40
7000		75	18	14	166	11.30	13.35	24.65
7100		125	22	14	216	11.40	14.05	25.45
7200		200	26	16	297	12.45	14.80	27.25
7400	35 x 35	40	16	16	137	11.35	13.15	24.50
7500		75	20	16	172	11.85	13.60	25.45
7600		125	24	16	222	11.90	13.65	25.55
7700		200	26	20	309	12.85	14.55	27.40
8000	35 x 40	40	18	16	137	11.70	13.60	25.30
8100		75	22	16	172	12.25	14.15	26.40
8300		125	26	16	222	12.15	14	26.15
8400		200	30	20	309	13.15	14.65	27.80
8750	40 x 40	40	18	20	149	12.35	13.60	25.95
8800		75	24	20	184	12.65	13.85	26.50
8900		125	26	20	234	13.15	14.45	27.60
9100	40 x 45	40	20	20	149	12.80	14.05	26.85
9500		75	24	20	184	12.90	14.15	27.05
9800		125	28	20	234	13.25	14.65	27.90

B1010   Floor Construction

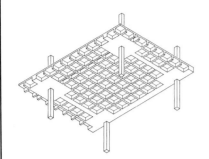

General: Waffle slabs are basically flat slabs with hollowed out domes on bottom side to reduce weight. Solid concrete heads at columns function as drops without increasing depth. The concrete ribs function as two-way right angle joist.

Joists are formed with standard sized domes. Thin slabs cover domes and are usually reinforced with welded wire fabric. Ribs have bottom steel and may have stirrups for shear.

Design and Pricing Assumptions:
Concrete f'c = 4 KSI, normal weight
   placed by concrete pump.
Reinforcement, fy = 60 KSI.
Forms, four use.
   4-1/2″ slab.
   30″ x 30″ voids.
   6″ wide ribs.
   (ribs @ 36″ O.C.).
   Rib depth filler beams as required.
   Solid concrete heads at columns.
Finish, steel trowel.
Curing, spray on membrane.
Based on 4 bay x 4 bay structure.

System Components	QUANTITY	UNIT	COST PER S.F. MAT.	COST PER S.F. INST.	COST PER S.F. TOTAL
SYSTEM B1010 227 3900					
20′ X 20′ BAY, 40 PSF S. LOAD, 12″ MIN. COLUMN					
Forms in place, floor slab with 2-way waffle domes, 4 use	1.000	S.F.	6.80	8.15	14.95
Edge forms, 7″-12″ high on elevated slab, 4 uses	.052	SFCA	.06	.41	.47
Forms in place, bulkhead for slab with keyway, 1 use, 3 piece	.010	L.F.	.03	.08	.11
Reinforcing in place, elevated slabs #4 to #7	1.580	Lb.	1.12	.77	1.89
Welded wire fabric rolls, 6 x 6 - W4 x W4 (4 x 4) 58 lb./c.s.f	1.000	S.F.	.43	.52	.95
Concrete ready mix, regular weight, 4000 psi	.690	C.F.	3.63		3.63
Place and vibrate concrete, elevated slab, over 10″, pump	.690	C.F.		1.09	1.09
Finish floor, monolithic steel trowel finish for finish floor	1.000	S.F.		1.08	1.08
Cure with sprayed membrane curing compound	.010	C.S.F.	.13	.11	.24
TOTAL			12.20	12.21	24.41

B1010 227			Cast in Place Waffle Slab					
	BAY SIZE (FT.)	SUPERIMPOSED LOAD (P.S.F.)	MINIMUM COL. SIZE (IN.)	RIB DEPTH (IN.)	TOTAL LOAD (P.S.F.)	COST PER S.F. MAT.	COST PER S.F. INST.	COST PER S.F. TOTAL
3900	20 x 20	40	12	8	144	12.20	12.20	24.40
4000		75	12	8	179	12.40	12.35	24.75
4100		125	16	8	229	12.60	12.50	25.10
4200		200	18	8	304	13.20	12.90	26.10
4400	20 x 25	40	12	8	146	12.40	12.30	24.70
4500		75	14	8	181	12.65	12.45	25.10
4600		125	16	8	231	12.90	12.60	25.50
4700		200	18	8	306	13.40	12.95	26.35
4900	25 x 25	40	12	10	150	12.70	12.40	25.10
5000		75	16	10	185	13.05	12.65	25.70
5300		125	18	10	235	13.35	12.85	26.20
5500		200	20	10	310	13.70	13.10	26.80
5700	25 x 30	40	14	10	154	12.90	12.50	25.40
5800		75	16	10	189	13.20	12.70	25.90
5900		125	18	10	239	13.55	12.90	26.45
6000		200	20	12	329	14.65	13.50	28.15
6400	30 x 30	40	14	12	169	13.60	12.75	26.35
6500		75	18	12	204	13.90	12.95	26.85
6600		125	20	12	254	14.10	13.10	27.20
6700		200	24	12	329	15.20	13.80	29

B1010 Floor Construction

| B1010 227 | Cast in Place Waffle Slab |

	BAY SIZE (FT.)	SUPERIMPOSED LOAD (P.S.F.)	MINIMUM COL. SIZE (IN.)	RIB DEPTH (IN.)	TOTAL LOAD (P.S.F.)	COST PER S.F.		
						MAT.	INST.	TOTAL
6900	30 x 35	40	16	12	169	13.85	12.90	26.75
7000		75	18	12	204	13.85	12.90	26.75
7100		125	22	12	254	14.45	13.30	27.75
7200		200	26	14	334	15.85	14.10	29.95
7400	35 x 35	40	16	14	174	14.55	13.20	27.75
7500		75	20	14	209	14.85	13.40	28.25
7600		125	24	14	259	15.10	13.60	28.70
7700		200	26	16	346	16.10	14.30	30.40
8000	35 x 40	40	18	14	176	14.85	13.45	28.30
8300		75	22	14	211	15.30	13.75	29.05
8500		125	26	16	271	15.95	14	29.95
8750		200	30	20	372	17.30	14.80	32.10
9200	40 x 40	40	18	14	176	15.30	13.75	29.05
9400		75	24	14	211	15.85	14.10	29.95
9500		125	26	16	271	16.20	14.15	30.35
9700	40 x 45	40	20	16	186	15.80	13.90	29.70
9800		75	24	16	221	16.35	14.25	30.60
9900		125	28	16	271	16.60	14.45	31.05

B10 Superstructure

B1010 Floor Construction

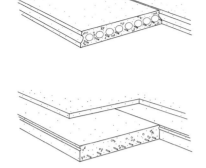

General: Units priced here are for plant produced prestressed members, transported to site and erected.

Normal weight concrete is most frequently used. Lightweight concrete may be used to reduce dead weight.

Structural topping is sometimes used on floors: insulating concrete or rigid insulation on roofs.

Camber and deflection may limit use by depth considerations.

Prices are based upon 10,000 S.F. to 20,000 S.F. projects, and 50 mile to 100 mile transport.

Concrete is f'c = 5 KSI and Steel is fy = 250 or 300 KSI

Note: Deduct from prices 20% for Southern states. Add to prices 10% for Western states.

Description of Table: Enter table at span and load. Most economical sections will generally consist of normal weight concrete without topping. If acceptable, note this price, depth and weight. For topping and/or lightweight concrete, note appropriate data.

Generally used on masonry and concrete bearing or reinforced concrete and steel framed structures.

The solid 4″ slabs are used for light loads and short spans. The 6″ to 12″ thick hollow core units are used for longer spans and heavier loads. Cores may carry utilities.

Topping is used structurally for loads or rigidity and architecturally to level or slope surface.

Camber and deflection and change in direction of spans must be considered (door openings, etc.), especially untopped.

System Components	QUANTITY	UNIT	COST PER S.F.		
			MAT.	INST.	TOTAL
SYSTEM B1010 230 2000					
10′ SPAN, 40 LBS S.F. WORKING LOAD, 2″ TOPPING					
Precast prestressed concrete roof/floor slabs 4″ thick, grouted	1.000	S.F.	9.25	3.82	13.07
Edge forms to 6″ high on elevated slab, 4 uses	.100	L.F.	.04	.50	.54
Welded wire fabric 6 x 6 - W1.4 x W1.4 (10 x 10), 21 lb/csf, 10% lap	.010	C.S.F.	.16	.41	.57
Concrete, ready mix, regular weight, 3000 psi	.170	C.F.	.86		.86
Place and vibrate concrete, elevated slab less than 6″, pumped	.170	C.F.		.29	.29
Finishing floor, monolithic steel trowel finish for resilient tile	1.000	S.F.		1.38	1.38
Curing with sprayed membrane curing compound	.010	C.S.F.	.13	.11	.24
TOTAL			10.44	6.51	16.95

B1010 229		Precast Plank with No Topping						
	SPAN (FT.)	SUPERIMPOSED LOAD (P.S.F.)	TOTAL DEPTH (IN.)	DEAD LOAD (P.S.F.)	TOTAL LOAD (P.S.F.)	COST PER S.F.		
						MAT.	INST.	TOTAL
0720	10	40	4	50	90	9.25	3.82	13.07
0750		75	6	50	125	9.40	3.27	12.67
0770		100	6	50	150	9.40	3.27	12.67
0800	15	40	6	50	90	9.40	3.27	12.67
0820		75	6	50	125	9.40	3.27	12.67
0850		100	6	50	150	9.40	3.27	12.67
0875	20	40	6	50	90	9.40	3.27	12.67
0900		75	6	50	125	9.40	3.27	12.67
0920		100	6	50	150	9.40	3.27	12.67
0950	25	40	6	50	90	9.40	3.27	12.67
0970		75	8	55	130	13.45	2.87	16.32
1000		100	8	55	155	13.45	2.87	16.32
1200	30	40	8	55	95	13.45	2.87	16.32
1300		75	8	55	130	13.45	2.87	16.32
1400		100	10	70	170	10.90	2.55	13.45
1500	40	40	10	70	110	10.90	2.55	13.45
1600		75	12	70	145	12.85	2.29	15.14

323

B1010 Floor Construction

B1010 229		Precast Plank with No Topping						
	SPAN (FT.)	SUPERIMPOSED LOAD (P.S.F.)	TOTAL DEPTH (IN.)	DEAD LOAD (P.S.F.)	TOTAL LOAD (P.S.F.)	COST PER S.F.		
						MAT.	INST.	TOTAL
1700	45	40	12	70	110	12.85	2.29	15.14

B1010 230		Precast Plank with 2″ Concrete Topping						
	SPAN (FT.)	SUPERIMPOSED LOAD (P.S.F.)	TOTAL DEPTH (IN.)	DEAD LOAD (P.S.F.)	TOTAL LOAD (P.S.F.)	COST PER S.F.		
						MAT.	INST.	TOTAL
2000	10	40	6	75	115	10.45	6.50	16.95
2100		75	8	75	150	10.60	5.95	16.55
2200		100	8	75	175	10.60	5.95	16.55
2500	15	40	8	75	115	10.60	5.95	16.55
2600		75	8	75	150	10.60	5.95	16.55
2700		100	8	75	175	10.60	5.95	16.55
2800	20	40	8	75	115	10.60	5.95	16.55
2900		75	8	75	150	10.60	5.95	16.55
3000		100	8	75	175	10.60	5.95	16.55
3100	25	40	8	75	115	10.60	5.95	16.55
3200		75	8	75	150	10.60	5.95	16.55
3300		100	10	80	180	14.65	5.55	20.20
3400	30	40	10	80	120	14.65	5.55	20.20
3500		75	10	80	155	14.65	5.55	20.20
3600		100	10	80	180	14.65	5.55	20.20
3700	35	40	12	95	135	12.10	5.25	17.35
3800		75	12	95	170	12.10	5.25	17.35
3900		100	14	95	195	14.05	4.97	19.02
4000	40	40	12	95	135	12.10	5.25	17.35
4500		75	14	95	170	14.05	4.97	19.02
5000	45	40	14	95	135	14.05	4.97	19.02

B1010 Floor Construction

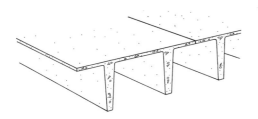

Most widely used for moderate span floors and roofs. At shorter spans, they tend to be competitive with hollow core slabs. They are also used as wall panels.

System Components	QUANTITY	UNIT	COST PER S.F.		
			MAT.	INST.	TOTAL
SYSTEM B1010 235 6700					
PRECAST, DOUBLE "T", 2" TOPPING, 30' SPAN, 30 PSF SUP. LOAD, 18" X 8'					
Double "T" beams, reg. wt, 18" x 8' w, 30' span	1.000	S.F.	12.92	1.91	14.83
Edge forms to 6" high on elevated slab, 4 uses	.050	L.F.	.02	.25	.27
Concrete, ready mix, regular weight, 3000 psi	.250	C.F.	1.27		1.27
Place and vibrate concrete, elevated slab less than 6", pumped	.250	C.F.		.43	.43
Finishing floor, monolithic steel trowel finish for finish floor	1.000	S.F.		1.08	1.08
Curing with sprayed membrane curing compound	.010	C.S.F.	.13	.11	.24
TOTAL			14.34	3.78	18.12

B1010 234		Precast Double "T" Beams with No Topping					
	SPAN (FT.)	SUPERIMPOSED LOAD (P.S.F.)	DBL. "T" SIZE D (IN.) W (FT.)	CONCRETE "T" TYPE	TOTAL LOAD (P.S.F.)	COST PER S.F.		
						MAT.	INST.	TOTAL
1500	30	30	18x8	Reg. Wt.	92	12.90	1.91	14.81
1600		40	18x8	Reg. Wt.	102	13.10	2.46	15.56
1700		50	18x8	Reg. Wt	112	13.10	2.46	15.56
1800		75	18x8	Reg. Wt.	137	13.10	2.55	15.65
1900		100	18x8	Reg. Wt.	162	13.10	2.55	15.65
2000	40	30	20x8	Reg. Wt.	87	11.70	1.58	13.28
2100		40	20x8	Reg. Wt.	97	11.80	1.90	13.70
2200		50	20x8	Reg. Wt.	107	11.80	1.90	13.70
2300		75	20x8	Reg. Wt.	132	11.85	2.04	13.89
2400		100	20x8	Reg. Wt.	157	12	2.50	14.50
2500	50	30	24x8	Reg. Wt.	103	11.75	1.44	13.19
2600		40	24x8	Reg. Wt.	113	11.85	1.75	13.60
2700		50	24x8	Reg. Wt.	123	11.90	1.86	13.76
2800		75	24x8	Reg. Wt.	148	11.90	1.89	13.79
2900		100	24x8	Reg. Wt.	173	12.05	2.34	14.39
3000	60	30	24x8	Reg. Wt.	82	11.90	1.89	13.79
3100		40	32x10	Reg. Wt.	104	11.85	1.54	13.39
3150		50	32x10	Reg. Wt.	114	11.80	1.34	13.14
3200		75	32x10	Reg. Wt.	139	11.80	1.45	13.25
3250		100	32x10	Reg. Wt.	164	11.95	1.81	13.76
3300	70	30	32x10	Reg. Wt.	94	11.80	1.43	13.23
3350		40	32x10	Reg. Wt.	104	11.80	1.45	13.25
3400		50	32x10	Reg. Wt.	114	11.95	1.80	13.75
3450		75	32x10	Reg. Wt.	139	12.05	2.16	14.21
3500		100	32x10	Reg. Wt.	164	12.25	2.88	15.13
3550	80	30	32x10	Reg. Wt.	94	11.95	1.80	13.75
3600		40	32x10	Reg. Wt.	104	12.15	2.52	14.67
3900		50	32x10	Reg. Wt.	114	12.25	2.87	15.12

B1010 Floor Construction

| B1010 234 | | | Precast Double "T" Beams with No Topping | | | | | |

	SPAN (FT.)	SUPERIMPOSED LOAD (P.S.F.)	DBL. "T" SIZE D (IN.) W (FT.)	CONCRETE "T" TYPE	TOTAL LOAD (P.S.F.)	COST PER S.F.		
						MAT.	INST.	TOTAL
4300	50	30	20x8	Lt. Wt.	66	13	1.75	14.75
4400		40	20x8	Lt. Wt.	76	12.95	1.58	14.53
4500		50	20x8	Lt. Wt.	86	13.05	1.90	14.95
4600		75	20x8	Lt. Wt.	111	13.15	2.17	15.32
4700		100	20x8	Lt. Wt.	136	13.30	2.62	15.92
4800	60	30	24x8	Lt. Wt.	70	13	1.72	14.72
4900		40	32x10	Lt. Wt.	88	13	1.21	14.21
5000		50	32x10	Lt. Wt.	98	13.05	1.46	14.51
5200		75	32x10	Lt. Wt.	123	13.15	1.68	14.83
5400		100	32x10	Lt. Wt.	148	13.25	2.04	15.29
5600	70	30	32x10	Lt. Wt.	78	13	1.21	14.21
5750		40	32x10	Lt. Wt.	88	13.05	1.46	14.51
5900		50	32x10	Lt. Wt.	98	13.15	1.67	14.82
6000		75	32x10	Lt. Wt.	123	13.25	2.03	15.28
6100		100	32x10	Lt. Wt.	148	13.50	2.75	16.25
6200	80	30	32x10	Lt. Wt.	78	13.15	1.67	14.82
6300		40	32x10	Lt. Wt.	88	13.25	2.03	15.28
6400		50	32x10	Lt. Wt.	98	13.35	2.39	15.74

| B1010 235 | | | Precast Double "T" Beams With 2″ Topping | | | | | |

	SPAN (FT.)	SUPERIMPOSED LOAD (P.S.F.)	DBL. "T" SIZE D (IN.) W (FT.)	CONCRETE "T" TYPE	TOTAL LOAD (P.S.F.)	COST PER S.F.		
						MAT.	INST.	TOTAL
6700	30	30	18x8	Reg. Wt.	117	14.35	3.77	18.12
6750		40	18x8	Reg. Wt.	127	14.35	3.77	18.12
6800		50	18x8	Reg. Wt.	137	14.45	4.10	18.55
6900		75	18x8	Reg. Wt.	162	14.45	4.10	18.55
7000		100	18x8	Reg. Wt.	187	14.50	4.24	18.74
7100	40	30	18x8	Reg. Wt.	120	11.20	3.61	14.81
7200		40	20x8	Reg. Wt.	130	13.15	3.45	16.60
7300		50	20x8	Reg. Wt.	140	13.25	3.77	17.02
7400		75	20x8	Reg. Wt.	165	13.30	3.91	17.21
7500		100	20x8	Reg. Wt.	190	13.45	4.36	17.81
7550	50	30	24x8	Reg. Wt.	120	13.25	3.62	16.87
7600		40	24x8	Reg. Wt.	130	13.30	3.73	17.03
7750		50	24x8	Reg. Wt.	140	13.30	3.75	17.05
7800		75	24x8	Reg. Wt.	165	13.45	4.21	17.66
7900		100	32x10	Reg. Wt.	189	13.30	3.46	16.76
8000	60	30	32x10	Reg. Wt.	118	13.15	2.95	16.10
8100		40	32x10	Reg. Wt.	129	13.20	3.20	16.40
8200		50	32x10	Reg. Wt.	139	13.30	3.45	16.75
8300		75	32x10	Reg. Wt.	164	13.35	3.67	17.02
8350		100	32x10	Reg. Wt.	189	13.45	4.03	17.48
8400	70	30	32x10	Reg. Wt.	119	13.25	3.31	16.56
8450		40	32x10	Reg. Wt.	129	13.35	3.67	17.02
8500		50	32x10	Reg. Wt.	139	13.45	4.03	17.48
8550		75	32x10	Reg. Wt.	164	13.70	4.74	18.44
8600	80	30	32x10	Reg. Wt.	119	13.70	4.74	18.44
8800	50	30	20x8	Lt. Wt.	105	14.45	3.75	18.20
8850		40	24x8	Lt. Wt.	121	14.50	3.92	18.42
8900		50	24x8	Lt. Wt.	131	14.60	4.20	18.80
8950		75	24x8	Lt. Wt.	156	14.60	4.20	18.80
9000		100	24x8	Lt. Wt.	181	14.75	4.65	19.40

B1010 Floor Construction

B1010 235	Precast Double "T" Beams With 2" Topping

	SPAN (FT.)	SUPERIMPOSED LOAD (P.S.F.)	DBL. "T" SIZE D (IN.) W (FT.)	CONCRETE "T" TYPE	TOTAL LOAD (P.S.F.)	COST PER S.F.		
						MAT.	INST.	TOTAL
9200	60	30	32x10	Lt. Wt.	103	14.40	3.07	17.47
9300		40	32x10	Lt. Wt.	113	14.50	3.32	17.82
9350		50	32x10	Lt. Wt.	123	14.50	3.32	17.82
9400		75	32x10	Lt. Wt.	148	14.55	3.54	18.09
9450		100	32x10	Lt. Wt.	173	14.70	3.90	18.60
9500	70	30	32x10	Lt. Wt.	103	14.55	3.54	18.09
9550		40	32x10	Lt. Wt.	113	14.55	3.54	18.09
9600		50	32x10	Lt. Wt.	123	14.70	3.90	18.60
9650		75	32x10	Lt. Wt.	148	14.80	4.26	19.06
9700	80	30	32x10	Lt. Wt.	103	14.70	3.89	18.59
9800		40	32x10	Lt. Wt.	113	14.80	4.25	19.05
9900		50	32x10	Lt. Wt.	123	14.90	4.61	19.51

For customer support on your Concrete & Masonry Costs with RSMeans Data, call 800.448.8182.

B1010 Floor Construction

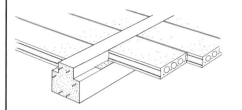

General: Units priced here are for plant produced prestressed members transported to the site and erected.

System has precast prestressed concrete beams and precast, prestressed hollow core slabs spanning the longer direction when applicable.

Camber and deflection must be considered when using untopped hollow core slabs.

Design and Pricing Assumptions:
Prices are based on 10,000 S.F. to 20,000 S.F. projects and 50 mile to 100 mile transport.

Concrete is f'c 5 KSI and prestressing steel is fy = 250 or 300 KSI.

System Components				COST PER S.F.		
	QUANTITY	UNIT		MAT.	INST.	TOTAL
SYSTEM B1010 236 5200						
20' X 20' BAY, 6" DEPTH, 40 PSF S. LOAD, 110 PSF TOTAL LOAD						
Precast concrete beam, T-shaped, 20' span, 12" x 20"	.038	L.F.		15.36	.82	16.18
Precast concrete beam, L-shaped, 20' span, 12" x 20"	.025	L.F.		8.42	.54	8.96
Precast prestressed concrete roof/floor slabs 6" deep, grouted	1.000	S.F.		9.40	3.27	12.67
TOTAL				33.18	4.63	37.81

B1010 236		Precast Beam & Plank with No Topping						
	BAY SIZE (FT.)	SUPERIMPOSED LOAD (P.S.F.)	PLANK THICKNESS (IN.)	TOTAL DEPTH (IN.)	TOTAL LOAD (P.S.F.)	COST PER S.F.		
						MAT.	INST.	TOTAL
5200	20x20	40	6	20	110	33	4.63	37.63
5400		75	6	24	148	25.50	4.15	29.65
5600		100	6	24	173	34	4.69	38.69
5800	20x25	40	8	24	113	33.50	4.03	37.53
6000		75	8	24	149	33.50	4.03	37.53
6100		100	8	36	183	33.50	3.97	37.47
6200	25x25	40	8	28	118	35	3.82	38.82
6400		75	8	36	158	34.50	3.82	38.32
6500		100	8	36	183	34.50	3.82	38.32
7000	25x30	40	8	36	110	31.50	3.69	35.19
7200		75	10	36	159	30	3.39	33.39
7400		100	12	36	188	31	3.08	34.08
7600	30x30	40	8	36	121	32.50	3.58	36.08
8000		75	10	44	140	32	3.59	35.59
8250		100	12	52	206	33.50	3.02	36.52
8500	30x35	40	12	44	135	29	2.93	31.93
8750		75	12	52	176	30.50	2.92	33.42
9000	35x35	40	12	52	141	30.50	2.86	33.36
9250		75	12	60	181	31.50	2.85	34.35
9500	35x40	40	12	52	137	30.50	2.86	33.36
9750	40x40	40	12	60	141	31.50	2.83	34.33

B1010 Floor Construction

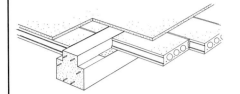

General: Beams and hollow core slabs priced here are for plant produced prestressed members transported to the site and erected.

The 2″ structural topping is applied after the beams and hollow core slabs are in place and is reinforced with W.W.F.

Design and Pricing Assumptions:

Prices are based on 10,000 S.F. to 20,000 S.F. projects and 50 mile to 100 mile transport.

Concrete for prestressed members is f'c 5 KSI.

Concrete for topping is f'c 3000 PSI and placed by pump.

Prestressing steel is fy = 250 or 300 KSI.

W.W.F. is 6 x 6 – W1.4 x W1.4 (10 x 10).

System Components	QUANTITY	UNIT	COST PER S.F. MAT.	COST PER S.F. INST.	COST PER S.F. TOTAL
SYSTEM B1010 238 4300					
20′ X 20′ BAY, 6″ PLANK, 40 PSF S. LOAD, 135 PSF TOTAL LOAD					
Precast concrete beam, T-shaped, 20′ span, 12″ x 20″	.038	L.F.	15.04	.80	15.84
Precast concrete beam, L-shaped, 20′ span, 12″ x 20″	.025	L.F.	8.42	.54	8.96
Precast prestressed concrete roof/floor slabs 6″ deep, grouted	1.000	S.F.	9.40	3.27	12.67
Edge forms to 6″ high on elevated slab, 4 uses	.050	L.F.	.02	.25	.27
Forms in place, bulkhead for slab with keyway, 1 use, 2 piece	.013	L.F.	.03	.10	.13
Welded wire fabric rolls, 6 x 6 - W1.4 x W1.4 (10 x 10), 21 lb/csf	.010	C.S.F.	.16	.41	.57
Concrete, ready mix, regular weight, 3000 psi	.170	C.F.	.86		.86
Place and vibrate concrete, elevated slab less than 6″, pump	.170	C.F.		.29	.29
Finish floor, monolithic steel trowel finish for finish floor	1.000	S.F.		1.08	1.08
Cure with sprayed membrane curing compound	.010	C.S.F.	.13	.11	.24
TOTAL			34.06	6.85	40.91

B1010 238		Precast Beam & Plank with 2″ Topping						
	BAY SIZE (FT.)	SUPERIMPOSED LOAD (P.S.F.)	PLANK THICKNESS (IN.)	TOTAL DEPTH (IN.)	TOTAL LOAD (P.S.F.)	COST PER S.F. MAT.	COST PER S.F. INST.	COST PER S.F. TOTAL
4300	20x20	40	6	22	135	34	6.85	40.85
4400		75	6	24	173	35.50	6.95	42.45
4500		100	6	28	200	35.50	6.90	42.40
4600	20x25	40	6	26	134	32	6.70	38.70
5000		75	8	30	177	35	6.20	41.20
5200		100	8	30	202	35	6.20	41.20
5400	25x25	40	6	38	143	32	6.40	38.40
5600		75	8	38	183	25	6.05	31.05
6000		100	8	46	216	28	5.60	33.60
6200	25x30	40	8	38	144	32.50	5.80	38.30
6400		75	10	46	200	30.50	5.45	35.95
6600		100	10	46	225	30.50	5.50	36
7000	30x30	40	8	46	150	33.50	5.70	39.20
7200		75	10	54	181	35	5.70	40.70
7600		100	10	54	231	35	5.70	40.70
7800	30x35	40	10	54	166	30	5.25	35.25
8000		75	12	54	200	31.50	4.99	36.49
8200	35x35	40	10	62	170	31	5.20	36.20
9300		75	12	62	206	34.50	4.98	39.48
9500	35x40	40	12	62	167	33	4.92	37.92
9600	40x40	40	12	62	173	30.50	4.90	35.40

B1010 Floor Construction

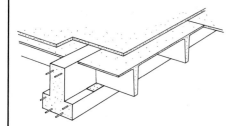

General: Beams and double tees priced here are for plant produced prestressed members transported to the site and erected.

The 2″ structural topping is applied after the beams and double tees are in place and is reinforced with W.W.F.

Design and Pricing Assumptions: Prices are based on 10,000 S.F. to 20,000 S.F. projects and 50 mile to 100 mile transport.

Concrete for prestressed members is f'c 5 KSI.

Concrete for topping is f'c 3000 PSI and placed by pump.

Prestressing steel is fy = 250 or 300 KSI.

W.W.F. is 6 x 6 – W1.4 x W1.4 (10x10).

System Components	QUANTITY	UNIT	COST PER S.F.		
			MAT.	INST.	TOTAL
SYSTEM B1010 239 3000					
25′ X 30′ BAY, 38″ DEPTH, 130 PSF T.L., 2″ TOPPING					
Precast concrete beam, T-shaped, 25′ span, 12″ x 36″	.025	L.F.	11.78	.51	12.29
Precast concrete beam, L-shaped, 25′ span, 12″ x 28″	.017	L.F.	6.24	.31	6.55
Double T, standard weight, 16″ x 8′ w, 25′ span	.989	S.F.	12.16	1.89	14.05
Edge forms to 6″ high on elevated slab, 4 uses	.037	L.F.	.01	.18	.19
Forms in place, bulkhead for slab with keyway, 1 use, 2 piece	.010	L.F.	.03	.08	.11
Welded wire fabric rolls, 6 x 6 - W1.4 x W1.4 (10 x 10), 21 lb/csf	1.000	S.F.	.16	.41	.57
Concrete, ready mix, regular weight, 3000 psi	.170	C.F.	.86		.86
Place and vibrate concrete, elevated slab less than 6″, pumped	.170	C.F.		.29	.29
Finishing floor, monolithic steel trowel finish for finish floor	1.000	S.F.		1.08	1.08
Curing with sprayed membrane curing compound	.010	S.F.	.13	.11	.24
TOTAL			31.37	4.86	36.23

B1010 239	Precast Double "T" & 2" Topping on Precast Beams						
	BAY SIZE (FT.)	SUPERIMPOSED LOAD (P.S.F.)	DEPTH (IN.)		TOTAL LOAD (P.S.F.)	COST PER S.F.		

	BAY SIZE (FT.)	SUPERIMPOSED LOAD (P.S.F.)	DEPTH (IN.)		TOTAL LOAD (P.S.F.)	MAT.	INST.	TOTAL
3000	25x30	40	38		130	31.50	4.86	36.36
3100		75	38		168	31.50	4.86	36.36
3300		100	46		196	31	4.83	35.83
3600	30x30	40	46		150	32	4.71	36.71
3750		75	46		174	32	4.71	36.71
4000		100	54		203	33.50	4.70	38.20
4100	30x40	40	46		136	28	4.26	32.26
4300		75	54		173	29	4.23	33.23
4400		100	62		204	30	4.24	34.24
4600	30x50	40	54		138	26	3.95	29.95
4800		75	54		181	24.50	3.99	28.49
5000		100	54		219	25	3.64	28.64
5200	30x60	40	62		151	25	3.57	28.57
5400		75	62		192	23.50	3.63	27.13
5600		100	62		215	23.50	3.63	27.13
5800	35x40	40	54		139	29	4.17	33.17
6000		75	62		179	30	4.01	34.01
6250		100	62		212	30.50	4.16	34.66
6500	35x50	40	62		142	27	3.91	30.91
6750		75	62		186	27.50	4.04	31.54
7300		100	62		231	26	3.69	29.69

For customer support on your Concrete & Masonry Costs with RSMeans Data, call 800.448.8182.

B10 Superstructure

B1010 Floor Construction

B1010 239		Precast Double "T" & 2" Topping on Precast Beams						

	BAY SIZE (FT.)	SUPERIMPOSED LOAD (P.S.F.)	DEPTH (IN.)		TOTAL LOAD (P.S.F.)	COST PER S.F.		
						MAT.	INST.	TOTAL
7600	35x60	40	54		154	25	3.42	28.42
7750		75	54		179	25	3.44	28.44
8000		100	62		224	25.50	3.49	28.99
8250	40x40	40	62		145	32.50	4.19	36.69
8400		75	62		187	29.50	4.20	33.70
8750		100	62		223	30	4.29	34.29
9000	40x50	40	62		151	27	3.94	30.94
9300		75	62		193	26	2.87	28.87
9800	40x60	40	62		164	30	2.75	32.75

For customer support on your Concrete & Masonry Costs with RSMeans Data, call 800.448.8182.

B1010 Floor Construction

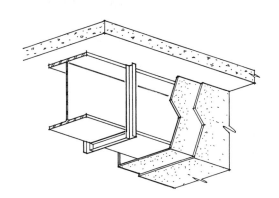

The table below lists fireproofing costs for steel beams by type, beam size, thickness and fire rating. Weights listed are for the fireproofing material only.

System Components	QUANTITY	UNIT	COST PER L.F.		
			MAT.	INST.	TOTAL
SYSTEM B1010 710 0400					
FIREPROOFING, 3000 P.S.I. CONC., 12" X 4" BEAM, 1 HR. RATING					
Formwork, wood stud and plywood, 4 uses	2.670	SFCA	4.27	23.76	28.03
Welded wire fabric, 2 x 2 #14 galv. 21 lb/C.S.F. beam wrap	2.700	S.F.	1.35	5.86	7.21
Concrete ready mix, 3000 p.s.i.	.513	C.F.	2.60		2.60
Place concrete, elevated beams, pump	.513	C.F.		1.60	1.60
TOTAL			8.22	31.22	39.44

B1010 710 — Steel Beam Fireproofing

	ENCASEMENT SYSTEM	BEAM SIZE (IN.)	THICKNESS (IN.)	FIRE RATING (HRS.)	WEIGHT (P.L.F.)	COST PER L.F.		
						MAT.	INST.	TOTAL
0400	Concrete	12x4	1	1	77	8.25	31	39.25
0450	3000 PSI		1-1/2	2	93	9.50	34	43.50
0500			2	3	121	10.55	37	47.55
0550		14x5	1	1	100	10.65	38.50	49.15
0600			1-1/2	2	122	12.15	42.50	54.65
0650			2	3	142	13.65	47	60.65
0700		16x7	1	1	147	12.45	41	53.45
0750			1-1/2	2	169	13.40	42.50	55.90
0800			2	3	195	15	46.50	61.50
0850		18x7-1/2	1	1	172	14.35	47.50	61.85
0900			1-1/2	2	196	16.05	52	68.05
0950			2	3	225	17.80	56.50	74.30
1000		24x9	1	1	264	19.35	57.50	76.85
1050			1-1/2	2	295	21	60.50	81.50
1100			2	3	328	22.50	64	86.50
1150		30x10-1/2	1	1	366	25	72.50	97.50
1200			1-1/2	2	404	27	77	104
1250			2	3	449	28.50	81	109.50

B1010 Floor Construction

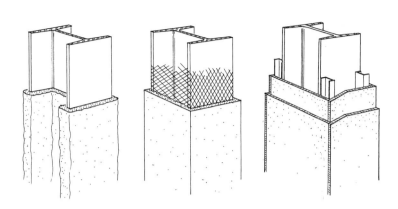

Listed below are costs per V.L.F. for fireproofing by material, column size, thickness and fire rating. Weights listed are for the fireproofing material only.

System Components	QUANTITY	UNIT	COST PER V.L.F.		
			MAT.	INST.	TOTAL
SYSTEM B1010 720 3000					
CONCRETE FIREPROOFING, 8″ STEEL COLUMN, 1″ THICK, 1 HR. FIRE RATING					
Forms in place, columns, plywood, 4 uses	3.330	SFCA	4.23	33.13	37.36
Welded wire fabric, 2 x 2 #14 galv. 21 lb./C.S.F., column wrap	2.700	S.F.	1.35	5.86	7.21
Concrete ready mix, regular weight, 3000 psi	.621	C.F.	3.15		3.15
Place and vibrate concrete, 12″ sq./round columns, pumped	.621	C.F.		1.93	1.93
TOTAL			8.73	40.92	49.65

B1010 720			Steel Column Fireproofing					
	ENCASEMENT SYSTEM	COLUMN SIZE (IN.)	THICKNESS (IN.)	FIRE RATING (HRS.)	WEIGHT (P.L.F.)	COST PER V.L.F.		
						MAT.	INST.	TOTAL
3000	Concrete	8	1	1	110	8.75	41	49.75
3050			1-1/2	2	133	10	45.50	55.50
3100			2	3	145	11.25	49	60.25
3150		10	1	1	145	11.30	50.50	61.80
3200			1-1/2	2	168	12.55	54.50	67.05
3250			2	3	196	13.95	58.50	72.45
3300		14	1	1	258	14.25	59.50	73.75
3350			1-1/2	2	294	15.50	63.50	79
3400			2	3	325	17	67.50	84.50

B2010 Exterior Walls

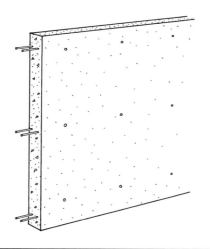

The table below describes a concrete wall system for exterior closure. There are several types of wall finishes priced from plain finish to a finish with 3/4″ rustication strip.

Design Assumptions:
Conc. f'c = 3000 to 5000 psi
Reinf. fy = 60,000 psi

System Components	QUANTITY	UNIT	COST PER S.F.		
			MAT.	INST.	TOTAL
SYSTEM B2010 101 2100					
CONC. WALL, REINFORCED, 8′ HIGH, 6″ THICK, PLAIN FINISH, 3,000 PSI					
Forms in place, wall, job built plyform to 8′ high, 4 uses	2.000	SFCA	3.54	15.10	18.64
Reinforcing in place, walls, #3 to #7	.752	Lb.	.53	.35	.88
Concrete ready mix, regular weight, 3000 psi	.018	C.Y.	2.47		2.47
Place and vibrate concrete, walls 6″ thick, pump	.018	C.Y.		.90	.90
Finish wall, break ties, patch voids	2.000	S.F.	.10	2.24	2.34
TOTAL			6.64	18.59	25.23

B2010 101	Cast In Place Concrete	COST PER S.F.		
		MAT.	INST.	TOTAL
2100	Conc wall reinforced, 8′ high, 6″ thick, plain finish, 3000 PSI	6.65	18.60	25.25
2200	4000 PSI	6.75	18.60	25.35
2300	5000 PSI	6.90	18.60	25.50
2400	Rub concrete 1 side, 3000 PSI	6.65	22	28.65
2500	4000 PSI	6.75	22	28.75
2600	5000 PSI	6.90	22	28.90
2700	Aged wood liner, 3000 PSI	7.75	21	28.75
2800	4000 PSI	7.85	21	28.85
2900	5000 PSI	8	21	29
3000	Sand blast light 1 side, 3000 PSI	7.20	21	28.20
3100	4000 PSI	7.30	21	28.30
3300	5000 PSI	7.45	21	28.45
3400	Sand blast heavy 1 side, 3000 PSI	7.80	26	33.80
3500	4000 PSI	7.85	26	33.85
3600	5000 PSI	8.05	26	34.05
3700	3/4″ bevel rustication strip, 3000 PSI	6.75	19.70	26.45
3800	4000 PSI	6.85	19.70	26.55
3900	5000 PSI	7	19.70	26.70
4000	8″ thick, plain finish, 3000 PSI	7.70	19.05	26.75
4100	4000 PSI	7.80	19.05	26.85
4200	5000 PSI	8.05	19.05	27.10
4300	Rub concrete 1 side, 3000 PSI	7.70	22.50	30.20
4400	4000 PSI	7.80	22.50	30.30
4500	5000 PSI	8.05	22.50	30.55
4550	8″ thick, aged wood liner, 3000 PSI	8.80	21.50	30.30
4600	4000 PSI	8.90	21.50	30.40

B2010 Exterior Walls

B2010 101	Cast In Place Concrete	COST PER S.F.		
		MAT.	INST.	TOTAL
4700	5000 PSI	9.15	21.50	30.65
4750	Sand blast light 1 side, 3000 PSI	8.25	21.50	29.75
4800	4000 PSI	8.40	21.50	29.90
4900	5000 PSI	8.60	21.50	30.10
5000	Sand blast heavy 1 side, 3000 PSI	8.85	26.50	35.35
5100	4000 PSI	8.95	26.50	35.45
5200	5000 PSI	9.15	26.50	35.65
5300	3/4" bevel rustication strip, 3000 PSI	7.80	20	27.80
5400	4000 PSI	7.90	20	27.90
5500	5000 PSI	8.15	20	28.15
5600	10" thick, plain finish, 3000 PSI	8.70	19.50	28.20
5700	4000 PSI	8.85	19.50	28.35
5800	5000 PSI	9.10	19.50	28.60
5900	Rub concrete 1 side, 3000 PSI	8.70	23	31.70
6000	4000 PSI	8.85	23	31.85
6100	5000 PSI	9.10	23	32.10
6200	Aged wood liner, 3000 PSI	9.80	22	31.80
6300	4000 PSI	9.95	22	31.95
6400	5000 PSI	10.25	22	32.25
6500	Sand blast light 1 side, 3000 PSI	9.25	22	31.25
6600	4000 PSI	9.40	22	31.40
6700	5000 PSI	9.70	22	31.70
6800	Sand blast heavy 1 side, 3000 PSI	9.85	27	36.85
6900	4000 PSI	10	27	37
7000	5000 PSI	10.25	27	37.25
7100	3/4" bevel rustication strip, 3000 PSI	8.80	20.50	29.30
7200	4000 PSI	8.95	20.50	29.45
7300	5000 PSI	9.20	20.50	29.70
7400	12" thick, plain finish, 3000 PSI	9.90	20	29.90
7500	4000 PSI	10.05	20	30.05
7600	5000 PSI	10.40	20	30.40
7700	Rub concrete 1 side, 3000 PSI	9.90	23.50	33.40
7800	4000 PSI	10.05	23.50	33.55
7900	5000 PSI	10.40	23.50	33.90
8000	Aged wood liner, 3000 PSI	11	22.50	33.50
8100	4000 PSI	11.20	22.50	33.70
8200	5000 PSI	11.50	22.50	34
8300	Sand blast light 1 side, 3000 PSI	10.45	22.50	32.95
8400	4000 PSI	10.65	22.50	33.15
8500	5000 PSI	11	22.50	33.50
8600	Sand blast heavy 1 side, 3000 PSI	11.05	27.50	38.55
8700	4000 PSI	11.20	27.50	38.70
8800	5000 PSI	11.55	27.50	39.05
8900	3/4" bevel rustication strip, 3000 PSI	10	21	31
9000	4000 PSI	10.15	21	31.15
9500	5000 PSI	10.50	21	31.50

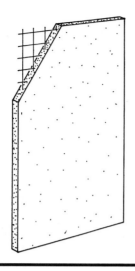

Precast concrete wall panels are either solid or insulated with plain, colored or textured finishes. Transportation is an important cost factor. Prices below are based on delivery within fifty miles of a plant. Engineering data is available from fabricators to assist with construction details. Usual minimum job size for economical use of panels is about 5000 S.F. Small jobs can double the prices below. For large, highly repetitive jobs, deduct up to 15% from the prices below.

B2010 102			Flat Precast Concrete					
	THICKNESS (IN.)	PANEL SIZE (FT.)	FINISHES	RIGID INSULATION (IN)	TYPE	COST PER S.F.		
						MAT.	INST.	TOTAL
3000	4	5x18	smooth gray	none	low rise	19.80	8.75	28.55
3050		6x18				16.55	7.30	23.85
3100		8x20				32.50	3.54	36.04
3150		12x20				31	3.34	34.34
3200	6	5x18	smooth gray	2	low rise	21	9.40	30.40
3250		6x18				17.75	7.95	25.70
3300		8x20				34.50	4.30	38.80
3350		12x20				31	3.94	34.94
3400	8	5x18	smooth gray	2	low rise	44.50	5.40	49.90
3450		6x18				42.50	5.15	47.65
3500		8x20				38.50	4.75	43.25
3550		12x20				35	4.37	39.37
3600	4	4x8	white face	none	low rise	74	5.05	79.05
3650		8x8				56	5.55	61.55
3700		10x10				49	3.33	52.33
3750		20x10				44	3.02	47.02
3800	5	4x8	white face	none	low rise	76	5.15	81.15
3850		8x8				57	3.91	60.91
3900		10x10				50.50	3.46	53.96
3950		20x20				46.50	3.17	49.67
4000	6	4x8	white face	none	low rise	78.50	5.35	83.85
4050		8x8				59.50	4.07	63.57
4100		10x10				52.50	3.59	56.09
4150		20x10				48.50	3.30	51.80
4200	6	4x8	white face	2	low rise	80	6.10	86.10
4250		8x8				61	4.80	65.80
4300		10x10				54	4.32	58.32
4350		20x10				48.50	3.30	51.80
4400	7	4x8	white face	none	low rise	80.50	5.50	86
4450		8x8				62	4.22	66.22
4500		10x10				55.50	3.78	59.28
4550		20x10				50.50	3.46	53.96
4600	7	4x8	white face	2	low rise	82	6.20	88.20
4650		8x8				63	4.95	67.95
4700		10x10				56.50	4.51	61.01

B2010 Exterior Walls

B2010 102					Flat Precast Concrete			
	THICKNESS (IN.)	PANEL SIZE (FT.)	FINISHES	RIGID INSULATION (IN)	TYPE	COST PER S.F.		
						MAT.	INST.	TOTAL
4750	7	20x10				52	4.19	56.19
4800	8	4x8	white face	none	low rise	82	5.60	87.60
4850		8x8				63.50	4.33	67.83
4900		10x10				57	3.89	60.89
4950		20x10				52.50	3.58	56.08
5000	8	4x8	white face	2	low rise	83.50	6.35	89.85
5100		10x10				65.50	5.35	70.85
5150		20x10				65.50	5.35	70.85

B2010 103					Fluted Window or Mullion Precast Concrete			
	THICKNESS (IN.)	PANEL SIZE (FT.)	FINISHES	RIGID INSULATION (IN)	TYPE	COST PER S.F.		
						MAT.	INST.	TOTAL
5200	4	4x8	smooth gray	none	high rise	45.50	20	65.50
5250		8x8				32.50	14.30	46.80
5300		10x10				62.50	6.75	69.25
5350		20x10				54.50	5.90	60.40
5400	5	4x8	smooth gray	none	high rise	46	20.50	66.50
5450		8x8				33.50	14.80	48.30
5500		10x10				65	7.05	72.05
5550		20x10				57.50	6.25	63.75
5600	6	4x8	smooth gray	none	high rise	47.50	21	68.50
5650		8x8				34.50	15.30	49.80
5700		10x10				67	7.25	74.25
5750		20x10				59.50	6.45	65.95
5800	6	4x8	smooth gray	2	high rise	48.50	21.50	70
5850		8x8				36	16.05	52.05
5900		10x10				68.50	7.95	76.45
5950		20x10				61	7.15	68.15
6000	7	4x8	smooth gray	none	high rise	48.50	21.50	70
6050		8x8				35.50	15.75	51.25
6100		10x10				70	7.55	77.55
6150		20x10				61.50	6.65	68.15
6200	7	4x8	smooth gray	2	high rise	49.50	22	71.50
6250		8x8				37	16.45	53.45
6300		10x10				71	8.30	79.30
6350		20x10				62.50	7.35	69.85
6400	8	4x8	smooth gray	none	high rise	49	21.50	70.50
6450		8x8				36.50	16.15	52.65
6500		10x10				72	7.80	79.80
6550		20x10				64.50	6.95	71.45
6600	8	4x8	smooth gray	2	high rise	50.50	22.50	73
6650		8x8				38	16.85	54.85
6700		10x10				73.50	8.50	82
6750		20x10				65.50	7.70	73.20

B2010 104					Ribbed Precast Concrete			
	THICKNESS (IN.)	PANEL SIZE (FT.)	FINISHES	RIGID INSULATION(IN.)	TYPE	COST PER S.F.		
						MAT.	INST.	TOTAL
6800	4	4x8	aggregate	none	high rise	63.50	20	83.50
6850		8x8				46.50	14.60	61.10
6900		10x10				78.50	5.30	83.80
6950		20x10				69.50	4.68	74.18

B2010 Exterior Walls

| B2010 104 | | | | | Ribbed Precast Concrete | | | |

	THICKNESS (IN.)	PANEL SIZE (FT.)	FINISHES	RIGID INSULATION(IN.)	TYPE	COST PER S.F.		
						MAT.	INST.	TOTAL
7000	5	4x8	aggregate	none	high rise	64.50	20.50	85
7050		8x8				48	15.05	63.05
7100		10x10				81.50	5.50	87
7150		20x10				73	4.91	77.91
7200	6	4x8	aggregate	none	high rise	66	21	87
7250		8x8				49	15.45	64.45
7300		10x10				84	5.70	89.70
7350		20x10				75.50	5.10	80.60
7400	6	4x8	aggregate	2	high rise	67.50	21.50	89
7450		8x8				50.50	16.20	66.70
7500		10x10				85.50	6.40	91.90
7550		20x10				77	5.80	82.80
7600	7	4x8	aggregate	none	high rise	67.50	21	88.50
7650		8x8				50.50	15.80	66.30
7700		10x10				87	5.85	92.85
7750		20x10				77.50	5.25	82.75
7800	7	4x8	aggregate	2	high rise	69	22	91
7850		8x8				51.50	16.55	68.05
7900		10x10				88	6.60	94.60
7950		20x10				79	5.95	84.95
8000	8	4x8	aggregate	none	high rise	68.50	21.50	90
8050		8x8				52	16.30	68.30
8100		10x10				90	6.05	96.05
8150		20x10				81	5.45	86.45
8200	8	4x8	aggregate	2	high rise	70	22.50	92.50
8250		8x8				53	17	70
8300		10x10				91.50	8.55	100.05
8350		20x10				82.50	6.20	88.70

| B2010 105 | | | Precast Concrete Specialties | | | | | |

	TYPE	SIZE				COST PER L.F.		
						MAT.	INST.	TOTAL
8400	Coping, precast	6" wide				27.50	15.65	43.15
8450	Stock units	10" wide				29.50	16.75	46.25
8460		12" wide				40	18.05	58.05
8480		14" wide				43	19.55	62.55
8500	Window sills	6" wide				24.50	16.75	41.25
8550	Precast	10" wide				44.50	19.55	64.05
8600		14" wide				40.50	23.50	64
8610								

B2010 Exterior Walls

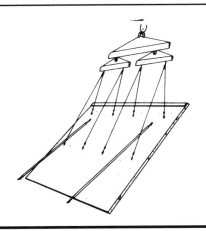

The advantage of tilt up construction is in the low cost of forms and placing of concrete and reinforcing. Tilt up has been used for several types of buildings, including warehouses, stores, offices, and schools. The panels are cast in forms on the ground, or floor slab. Most jobs use 5-1/2" thick solid reinforced concrete panels.

Design Assumptions:
Conc. f'c = 3000 psi
Reinf. fy = 60,000

System Components	QUANTITY	UNIT	COST PER S.F. MAT.	COST PER S.F. INST.	COST PER S.F. TOTAL
SYSTEM B2010 106 3200					
TILT-UP PANELS, 20' X 25', BROOM FINISH, 5-1/2" THICK, 3000 PSI					
Apply liquid bond release agent	500.000	S.F.	.02	.16	.18
Edge forms in place for slab on grade	120.000	L.F.	.12	1	1.12
Reinforcing in place	.350	Ton	.96	.86	1.82
Footings, form braces, steel	1.000	Set	.83		.83
Slab lifting inserts	1.000	Set	.19		.19
Framing, less than 4" angles	1.000	Set	.17	1.74	1.91
Concrete, ready mix, regular weight, 3000 psi	8.550	C.Y.	2.34		2.34
Place and vibrate concrete for slab on grade, 4" thick, direct chute	8.550	C.Y.		.52	.52
Finish floor, monolithic broom finish	500.000	S.F.		.94	.94
Cure with curing compound, sprayed	500.000	S.F.	.13	.11	.24
Erection crew	.058	Day		1.52	1.52
TOTAL			4.76	6.85	11.61

B2010 106	Tilt-Up Concrete Panel	COST PER S.F. MAT.	COST PER S.F. INST.	COST PER S.F. TOTAL
3200	Tilt-up conc panels, broom finish, 5-1/2" thick, 3000 PSI	4.77	6.85	11.62
3250	5000 PSI	4.87	6.75	11.62
3300	6" thick, 3000 PSI	5.25	7	12.25
3350	5000 PSI	5.35	6.90	12.25
3400	7-1/2" thick, 3000 PSI	6.60	7.30	13.90
3450	5000 PSI	6.80	7.15	13.95
3500	8" thick, 3000 PSI	7.10	7.45	14.55
3550	5000 PSI	7.30	7.35	14.65
3700	Steel trowel finish, 5-1/2" thick, 3000 PSI	4.77	7	11.77
3750	5000 PSI	4.87	6.85	11.72
3800	6" thick, 3000 PSI	5.25	7.15	12.40
3850	5000 PSI	5.35	7.05	12.40
3900	7-1/2" thick, 3000 PSI	6.60	7.40	14
3950	5000 PSI	6.80	7.30	14.10
4000	8" thick, 3000 PSI	7.10	7.60	14.70
4050	5000 PSI	7.30	7.45	14.75
4200	Exp. aggregate finish, 5-1/2" thick, 3000 PSI	5.10	7	12.10
4250	5000 PSI	5.25	6.90	12.15
4300	6" thick, 3000 PSI	5.60	7.15	12.75
4350	5000 PSI	5.70	7.05	12.75
4400	7-1/2" thick, 3000 PSI	6.95	7.45	14.40
4450	5000 PSI	7.15	7.30	14.45

B2010 Exterior Walls

B2010 106	Tilt-Up Concrete Panel	COST PER S.F.		
		MAT.	INST.	TOTAL
4500	8" thick, 3000 PSI	7.45	7.60	15.05
4550	5000 PSI	7.65	7.50	15.15
4600	Exposed aggregate & vert. rustication 5-1/2" thick, 3000 PSI	7.30	8.70	16
4650	5000 PSI	7.40	8.60	16
4700	6" thick, 3000 PSI	7.80	8.85	16.65
4750	5000 PSI	7.90	8.75	16.65
4800	7-1/2" thick, 3000 PSI	9.15	9.15	18.30
4850	5000 PSI	9.35	9	18.35
4900	8" thick, 3000 PSI	9.60	9.30	18.90
4950	5000 PSI	9.85	9.20	19.05
5000	Vertical rib & light sandblast, 5-1/2" thick, 3000 PSI	7.25	11.45	18.70
5050	5000 PSI	7.40	11.30	18.70
5100	6" thick, 3000 PSI	7.75	11.60	19.35
5150	5000 PSI	7.85	11.50	19.35
5200	7-1/2" thick, 3000 PSI	9.10	11.85	20.95
5250	5000 PSI	9.30	11.75	21.05
5300	8" thick, 3000 PSI	9.60	12.05	21.65
5350	5000 PSI	9.80	11.90	21.70
6000	Broom finish w/2" polystyrene insulation, 6" thick, 3000 PSI	4.58	8.50	13.08
6050	5000 PSI	4.76	8.50	13.26
6100	Broom finish 2" fiberplank insulation, 6" thick, 3000 PSI	5.30	8.40	13.70
6150	5000 PSI	5.45	8.40	13.85
6200	Exposed aggregate w/2" polystyrene insulation, 6" thick, 3000 PSI	4.80	8.50	13.30
6250	5000 PSI	4.98	8.50	13.48
6300	Exposed aggregate 2" fiberplank insulation, 6" thick, 3000 PSI	5.50	8.40	13.90
6350	5000 PSI	5.70	8.40	14.10

B2010 Exterior Walls

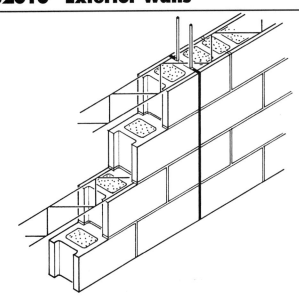

Exterior concrete block walls are defined in the following terms; structural reinforcement, weight, percent solid, size, strength and insulation. Within each of these categories, two to four variations are shown. No costs are included for brick shelf or relieving angles.

System Components	QUANTITY	UNIT	COST PER S.F. MAT.	COST PER S.F. INST.	COST PER S.F. TOTAL
SYSTEM B2010 109 1400					
UNREINFORCED CONCRETE BLOCK WALL, 8″ X 8″ X 16″, PERLITE CORE FILL					
Concrete block wall, 8″ thick	1.000	S.F.	2.90	8	10.90
Perlite insulation	1.000	S.F.	1.86	.49	2.35
Horizontal joint reinforcing, alternate courses	.800	S.F.	.23	.21	.44
Control joint	.050	L.F.	.09	.08	.17
TOTAL			5.08	8.78	13.86

B2010 109			Concrete Block Wall - Regular Weight			COST PER S.F.		
	TYPE	SIZE (IN.)	STRENGTH (P.S.I.)	CORE FILL		MAT.	INST.	TOTAL
1200	Hollow	4x8x16	2,000	none		2.58	7.20	9.78
1250			4,500	none		2.96	7.20	10.16
1300		6x8x16	2,000	perlite		4.52	8.15	12.67
1310				styrofoam		4.67	7.75	12.42
1340				none		3.26	7.75	11.01
1350			4,500	perlite		4.90	8.15	13.05
1360				styrofoam		5.05	7.75	12.80
1390				none		3.64	7.75	11.39
1400		8x8x16	2,000	perlite		5.10	8.80	13.90
1410				styrofoam		5.05	8.30	13.35
1440				none		3.22	8.30	11.52
1450			4,500	perlite		6.20	8.80	15
1460				styrofoam		6.15	8.30	14.45
1490				none		4.34	8.30	12.64
1500		12x8x16	2,000	perlite		8.50	11.65	20.15
1510				styrofoam		7.30	10.70	18
1540				none		5.45	10.70	16.15
1550			4,500	perlite		9	11.65	20.65
1560				styrofoam		7.80	10.70	18.50
1590				none		5.95	10.70	16.65
2000	75% solid	4x8x16	2,000	none		3.14	7.30	10.44
2050			4,500	none		3.63	7.30	10.93

341

B2010 Exterior Walls

| B2010 109 | | | Concrete Block Wall - Regular Weight | | | | |

	TYPE	SIZE (IN.)	STRENGTH (P.S.I.)	CORE FILL		COST PER S.F.		
						MAT.	INST.	TOTAL
2100	75% solid	6x8x16	2,000	perlite		4.60	8.05	12.65
2140				none		3.97	7.85	11.82
2150			4,500	perlite		5.10	8.05	13.15
2190				none		4.47	7.85	12.32
2200		8x8x16	2,000	perlite		4.79	8.65	13.44
2240				none		3.86	8.40	12.26
2250			4,500	perlite		6.25	8.65	14.90
2290				none		5.30	8.40	13.70
2300		12x8x16	2,000	perlite		8.15	11.35	19.50
2340				none		6.65	10.85	17.50
2350			4,500	perlite		8.80	11.35	20.15
2390				none		7.30	10.85	18.15
2500	Solid	4x8x16	2,000	none		3.06	7.45	10.51
2550			4,500	none		4.09	7.40	11.49
2600		6x8x16	2,000	none		4.18	8.05	12.23
2650			4,500	none		5	7.95	12.95
2700		8x8x16	2,000	none		4.52	8.60	13.12
2750			4,500	none		5.95	8.50	14.45
2800		12x8x16	2,000	none		6.55	11.20	17.75
2850			4,500	none		8.20	11	19.20

| B2010 110 | | | Concrete Block Wall - Lightweight | | | | |

	TYPE	SIZE (IN.)	WEIGHT (P.C.F.)	CORE FILL		COST PER S.F.		
						MAT.	INST.	TOTAL
3100	Hollow	8x4x16	105	perlite		5	8.30	13.30
3110				styrofoam		4.98	7.80	12.78
3140				none		3.16	7.80	10.96
3150			85	perlite		7.25	8.10	15.35
3160				styrofoam		7.20	7.60	14.80
3190				none		5.35	7.60	12.95
3200		4x8x16	105	none		2.48	7.05	9.53
3250			85	none		3.51	6.90	10.41
3300		6x8x16	105	perlite		4.95	7.95	12.90
3310				styrofoam		5.10	7.55	12.65
3340				none		3.69	7.55	11.24
3350			85	perlite		5.70	7.80	13.50
3360				styrofoam		5.85	7.40	13.25
3390				none		4.45	7.40	11.85
3400		8x8x16	105	perlite		6.55	8.60	15.15
3410				styrofoam		6.50	8.10	14.60
3440				none		4.69	8.10	12.79
3450			85	perlite		6.15	8.40	14.55
3460				styrofoam		6.10	7.90	14
3490				none		4.28	7.90	12.18
3500		12x8x16	105	perlite		7.80	11.35	19.15
3510				styrofoam		6.65	10.40	17.05
3540				none		4.75	10.40	15.15
3550			85	perlite		10.50	11.10	21.60
3560				styrofoam		9.30	10.15	19.45
3590				none		7.45	10.15	17.60
3600		4x8x24	105	none		1.89	7.90	9.79
3650			85	none		4.38	6.60	10.98
3690	For stacked bond add						.40	.40

B2010 Exterior Walls

B2010 110	Concrete Block Wall - Lightweight

	TYPE	SIZE (IN.)	WEIGHT (P.C.F.)	CORE FILL		COST PER S.F. MAT.	COST PER S.F. INST.	COST PER S.F. TOTAL
3700	Hollow	6x8x24	105	perlite		3.97	8.65	12.62
3710				styrofoam		4.12	8.25	12.37
3740				none		2.71	8.25	10.96
3750			85	perlite		7.90	7.30	15.20
3760				styrofoam		8.05	6.90	14.95
3790				none		6.65	6.90	13.55
3800		8x8x24	105	perlite		5.35	9.30	14.65
3810				styrofoam		5.30	8.80	14.10
3840				none		3.47	8.80	12.27
3850			85	perlite		10.35	7.85	18.20
3860				styrofoam		10.30	7.35	17.65
3890				none		8.45	7.35	15.80
3900		12x8x24	105	perlite		6.70	10.95	17.65
3910				styrofoam		5.50	10	15.50
3940				none		3.63	10	13.63
3950			85	perlite		11.55	10.70	22.25
3960				styrofoam		10.35	9.75	20.10
3990				none		8.50	9.75	18.25
4000	75% solid	4x8x16	105	none		3.58	7.15	10.73
4050			85	none		4.35	7	11.35
4100		6x8x16	105	perlite		6.05	7.85	13.90
4140				none		5.45	7.65	13.10
4150			85	perlite		6.15	7.65	13.80
4190				none		5.55	7.45	13
4200		8x8x16	105	perlite		7.85	8.45	16.30
4240				none		6.90	8.20	15.10
4250			85	perlite		6.20	8.25	14.45
4290				none		5.25	8	13.25
4300		12x8x16	105	perlite		8.35	11.05	19.40
4340				none		6.85	10.55	17.40
4350			85	perlite		10.75	10.75	21.50
4390				none		9.25	10.25	19.50
4500	Solid	4x8x16	105	none		2.62	7.40	10.02
4550			85	none		6	7.05	13.05
4600		6x8x16	105	none		4.10	7.95	12.05
4650			85	none		8.30	7.55	15.85
4700		8x8x16	105	none		4.01	8.50	12.51
4750			85	none		8.75	8.10	16.85
4800		12x8x16	105	none		5.70	11	16.70
4850			85	none		12.90	10.40	23.30
4900	For stacked bond, add						.40	.40

343

B2010 Exterior Walls

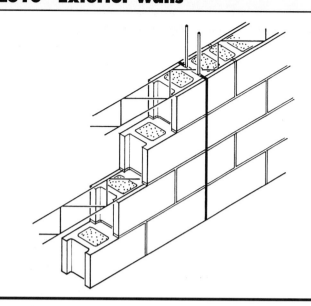

Exterior concrete block walls are defined in the following terms; structural reinforcement, weight, percent solid, size, strength and insulation. Within each of these categories, two to four variations are shown. No costs are included for brick shelf or relieving angles.

System Components	QUANTITY	UNIT	COST PER S.F.		
			MAT.	INST.	TOTAL
SYSTEM B2010 111 5200					
REINFORCED CONCRETE BLOCK WALL, 4″X8″X16″, #4 VERT. REINF. AT 48″ O.C.					
Concrete block wall	1.000	S.F.	2.30	6.95	9.25
Horizontal joint reinforcing, 9 ga. ladder type	.800	L.F.	.19	.17	.36
Vertical joint reinforcing, #4 steel rods	.180	Lb.	.12	.33	.45
Grout solid, 4″ cores	.167	S.F.	.07	.45	.52
Control joint, 4″ P.V.C.	.050	L.F.	.09	.08	.17
TOTAL			2.77	7.98	10.75

B2010 111		Reinforced Concrete Block Wall - Regular Weight				COST PER S.F.		
	TYPE	SIZE (IN.)	STRENGTH (P.S.I.)	VERT. REINF & GROUT SPACING		MAT.	INST.	TOTAL
5200	Hollow	4x8x16	2,000	#4 @ 48″		2.77	8	10.77
5250			4,500	#4 @ 48″		3.15	8	11.15
5300		6x8x16	2,000	#4 @ 48″		3.58	8.55	12.13
5330				#5 @ 32″		3.85	8.90	12.75
5340				#5 @ 16″		4.43	10	14.43
5350			4,500	#4 @ 28″		3.96	8.55	12.51
5380				#5 @ 32″		4.23	8.90	13.13
5390				#5 @ 16″		4.81	10	14.81
5400		8x8x16	2,000	#4 @ 48″		3.72	9.15	12.87
5430				#5 @ 32″		3.91	9.65	13.56
5440				#5 @ 16″		4.59	11.05	15.64
5450		8x8x16	4,500	#4 @ 48″		4.73	9.25	13.98
5480				#5 @ 32″		5.05	9.65	14.70
5490				#5 @ 16″		5.70	11.05	16.75
5500		12x8x16	2,000	#4 @ 48″		6	11.70	17.70
5530				#5 @ 32″		6.35	12.15	18.50
5540				#5 @ 16″		7.30	13.55	20.85
5550			4,500	#4 @ 48″		6.50	11.70	18.20
5580				#5 @ 32″		6.85	12.15	19
5590				#5 @ 16″		7.80	13.55	21.35

B2010 Exterior Walls

B2010 111	Reinforced Concrete Block Wall - Regular Weight

	TYPE	SIZE (IN.)	STRENGTH (P.S.I.)	VERT. REINF & GROUT SPACING		COST PER S.F. MAT.	INST.	TOTAL
6100	75% solid	6x8x16	2,000	#4 @ 48"		4.15	8.55	12.70
6130				#5 @ 32"		4.34	8.80	13.14
6140				#5 @ 16"		4.70	9.70	14.40
6150			4,500	#4 @ 48"		4.65	8.55	13.20
6180				#5 @ 32"		4.84	8.80	13.64
6190				#5 @ 16"		5.20	9.70	14.90
6200		8x8x16	2,000	#4 @ 48"		4.05	9.20	13.25
6230				#5 @ 32"		4.26	9.50	13.76
6240				#5 @ 16"		4.65	10.60	15.25
6250			4,500	#4 @ 48"		5.50	9.20	14.70
6280				#5 @ 32"		5.70	9.50	15.20
6290				#5 @ 16"		6.10	10.60	16.70
6300		12x8x16	2,000	#4 @ 48"		6.95	11.65	18.60
6330				#5 @ 32"		7.20	12	19.20
6340				#5 @ 16"		7.80	13.10	20.90
6350			4,500	#4 @ 48"		7.60	11.65	19.25
6380				#4 @ 32"		7.85	12	19.85
6390				#5 @ 16"		8.45	13.10	21.55
6500	Solid-double Wythe	2-4x8x16	2,000	#4 @ 48" E.W.		7.15	16.85	24
6530				#5 @ 16" E.W.		8.05	17.90	25.95
6550			4,500	#4 @ 48" E.W.		9.20	16.75	25.95
6580				#5 @ 16" E.W.		10.15	17.80	27.95
6600		2-6x8x16	2,000	#4 @ 48" E.W.		9.35	18.10	27.45
6630				#5 @ 16" E.W.		10.30	19.15	29.45
6650			4,000	#4 @ 48" E.W.		11.05	17.90	28.95
6680				#5 @ 16" E.W.		11.95	18.95	30.90

B2010 112	Reinforced Concrete Block Wall - Lightweight

	TYPE	SIZE (IN.)	WEIGHT (P.C.F.)	VERT REINF. & GROUT SPACING		COST PER S.F. MAT.	INST.	TOTAL
7100	Hollow	8x4x16	105	#4 @ 48"		3.55	8.75	12.30
7130				#5 @ 32"		3.85	9.15	13
7140				#5 @ 16"		4.53	10.55	15.08
7150			85	#4 @ 48"		5.75	8.55	14.30
7180				#5 @ 32"		6.05	8.95	15
7190				#5 @ 16"		6.75	10.35	17.10
7200		4x8x16	105	#4 @ 48"		2.67	7.85	10.52
7250			85	#4 @ 48"		3.70	7.70	11.40
7300		6x8x16	105	#4 @ 48"		4.01	8.35	12.36
7330				#5 @ 32"		4.28	8.70	12.98
7340				#5 @ 16"		4.86	9.80	14.66
7350			85	#4 @ 48"		4.77	8.20	12.97
7380				#5 @ 32"		5.05	8.55	13.60
7390				#5 @ 16"		5.60	9.65	15.25
7400		8x8x16	105	#4 @ 48"		5.10	9.05	14.15
7430				#5 @ 32"		5.40	9.45	14.85
7440				#5 @ 16"		6.05	10.85	16.90
7450		8x8x16	85	#4 @ 48"		4.67	8.85	13.52
7480				#5 @ 32"		4.97	9.25	14.22
7490				#5 @ 16"		5.65	10.65	16.30

For customer support on your Concrete & Masonry Costs with RSMeans Data, call 800.448.8182.

B2010 Exterior Walls

B2010 112	Reinforced Concrete Block Wall - Lightweight

	TYPE	SIZE (IN.)	WEIGHT (P.C.F.)	VERT REINF. & GROUT SPACING		COST PER S.F. MAT.	COST PER S.F. INST.	COST PER S.F. TOTAL
7510	Hollow	12x8x16		#4 @ 48"		5.30	11.40	16.70
7530				#5 @ 32"		5.70	11.85	17.55
7540				#5 @ 16"		6.60	13.25	19.85
7550			85	#4 @ 48"		8	11.15	19.15
7580				#5 @ 32"		8.35	11.60	19.95
7590				#5 @ 16"		9.30	13	22.30
7600		4x8x24	105	#4 @ 48"		2.08	8.70	10.78
7650			85	#4 @ 48"		4.57	7.40	11.97
7700		6x8x24	105	#4 @ 48"		3.03	9.05	12.08
7730				#5 @ 32"		3.30	9.40	12.70
7740				#5 @ 16"		3.88	10.50	14.38
7750			85	#4 @ 48"		6.95	7.70	14.65
7780				#5 @ 32"		7.20	8.05	15.25
7790				#5 @ 16"		7.80	9.15	16.95
7800		8x8x24	105	#4 @ 48"		3.86	9.75	13.61
7840				#5 @ 16"		4.84	11.55	16.39
7850			85	#4 @ 48"		8.85	8.30	17.15
7880				#5 @ 32"		9.15	8.70	17.85
7890				#5 @ 16"		9.85	10.10	19.95
7900		12x8x24	105	#4 @ 48"		4.18	11	15.18
7930				#5 @ 32"		4.56	11.45	16.01
7940				#5 @ 16"		5.50	12.85	18.35
7950			85	#4 @ 48"		9.05	10.75	19.80
7980				#5 @ 32"		9.40	11.20	20.60
7990				#5 @ 16"		10.35	12.60	22.95
8100	75% solid	6x8x16	105	#4 @ 48"		5.60	8.35	13.95
8130				#5 @ 32"		5.80	8.60	14.40
8140				#5 @ 16"		6.15	9.50	15.65
8150			85	#4 @ 48"		5.70	8.15	13.85
8180				#5 @ 32"		5.90	8.40	14.30
8190				#5 @ 16"		6.25	9.30	15.55
8200		8x8x16	105	#4 @ 48"		7.10	9	16.10
8230				#5 @ 32"		7.30	9.30	16.60
8240				#5 @ 16"		7.70	10.40	18.10
8250			85	#4 @ 48"		5.45	8.80	14.25
8280				#5 @ 32"		5.65	9.10	14.75
8290				#5 @ 16"		6.05	10.20	16.25
8300		12x8x16	105	#4 @ 48"		7.15	11.35	18.50
8330				#5 @ 32"		7.40	11.70	19.10
8340				#5 @ 16"		8	12.80	20.80
8350			85	#4 @ 48"		9.55	11.05	20.60
8380				#5 @ 32"		9.80	11.40	21.20
8390				#5 @ 16"		10.40	12.50	22.90
8500	Solid-double	2-4x8x16	105	#4 @ 48"		6.30	16.75	23.05
8530	Wythe		105	#5 @ 16"		7.20	17.80	25
8550			85	#4 @ 48"		13	16.05	29.05
8580				#5 @ 16"		13.90	17.10	31
8600		2-6x8x16	105	#4 @ 48"		9.20	17.90	27.10
8630				#5 @ 16"		10.15	18.95	29.10
8650			85	#4 @ 48"		17.55	17.10	34.65
8680				#5 @ 16"		18.50	18.15	36.65
8900	For stacked bond add						.40	.40

B2010 Exterior Walls

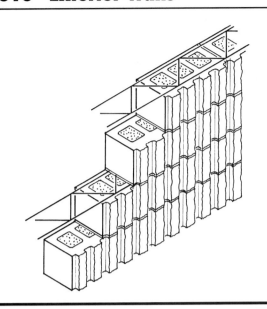

Exterior split ribbed block walls are defined in the following terms; structural reinforcement, weight, percent solid, size, number of ribs and insulation. Within each of these categories two to four variations are shown. No costs are included for brick shelf or relieving angles. Costs include control joints every 20' and horizontal reinforcing.

System Components	QUANTITY	UNIT	COST PER S.F. MAT.	COST PER S.F. INST.	COST PER S.F. TOTAL
SYSTEM B2010 113 1430					
UNREINFORCED SPLIT RIB BLOCK WALL, 8"X8"X16", 8 RIBS(HEX), PERLITE FILL					
Split ribbed block wall, 8" thick	1.000	S.F.	7.90	9.80	17.70
Perlite insulation	1.000	S.F.	1.86	.49	2.35
Horizontal joint reinforcing, alternate courses	.800	L.F.	.23	.21	.44
Control joint	.050	L.F.	.09	.08	.17
TOTAL			10.08	10.58	20.66

B2010 113			Split Ribbed Block Wall - Regular Weight			COST PER S.F.		
	TYPE	SIZE (IN.)	RIBS	CORE FILL		MAT.	INST.	TOTAL
1220	Hollow	4x8x16	4	none		5.45	8.95	14.40
1250			8	none		5.95	8.95	14.90
1280			16	none		6.40	9.05	15.45
1430		8x8x16	8	perlite		10.10	10.60	20.70
1440				styrofoam		9.65	10.10	19.75
1450				none		8.20	10.10	18.30
1530		12x8x16	8	perlite		12.50	14.10	26.60
1550				none		9.45	13.15	22.60
2520	Solid	4x8x16	4	none		7.80	9.20	17
2580			16	none		9.25	9.35	18.60
2680		8x8x16	8	none		12.75	10.45	23.20
2750		12x8x16	8	none		14.55	13.65	28.20

B2010 115			Reinforced Split Ribbed Block Wall - Regular Weight			COST PER S.F.		
	TYPE	SIZE (IN.)	RIBS	VERT. REINF. & GROUT SPACING		MAT.	INST.	TOTAL
5200	Hollow	4x8x16	4	#4 @ 48"		5.60	9.75	15.35
5260			16	#4 @ 48"		6.55	9.85	16.40
5430		8x8x16	8	#4 @ 48"		8.60	11.05	19.65
5450				#5 @ 16"		9.60	12.85	22.45
5530		12x8x16	8	#4 @ 48"		10	14.15	24.15
5550				#5 @ 16"		11.30	16	27.30

347

B2010 Exterior Walls

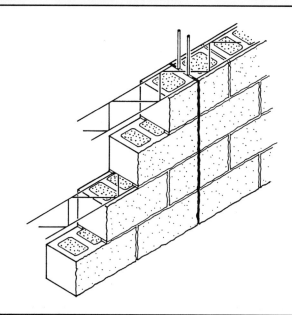

Exterior split face block walls are defined in the following terms; structural reinforcement, weight, percent solid, size, scores and insulation. Within each of these categories two to four variations are shown. No costs are included for brick shelf or relieving angles. Costs include control joints every 20′ and horizontal reinforcing.

System Components	QUANTITY	UNIT	COST PER S.F. MAT.	COST PER S.F. INST.	COST PER S.F. TOTAL
SYSTEM B2010 117 1600					
UNREINFORCED SPLIT FACE BLOCK WALL, 8″X8″X16″, 0 SCORES, PERLITE FILL					
Split face block wall, 8″ thick	1.000	S.F.	5.90	10.15	16.05
Perlite insulation	1.000	S.F.	1.86	.49	2.35
Horizontal joint reinforcing, alternate course	.800	L.F.	.23	.21	.44
Control joint	.050	L.F.	.09	.08	.17
TOTAL			8.08	10.93	19.01

B2010 117 — Split Face Block Wall - Regular Weight

	TYPE	SIZE (IN.)	SCORES	CORE FILL		MAT.	INST.	TOTAL
1200	Hollow	8x4x16	0	perlite		8.85	11.70	20.55
1400		4x8x16	0	none		4.33	8.80	13.13
1450			1	none		4.70	8.95	13.65
1600		8x8x16	0	perlite		8.10	10.95	19.05
1610				styrofoam		8.05	10.45	18.50
1640				none		6.20	10.45	16.65
1700		12x8x16	0	perlite		10.05	14.35	24.40
1740				none		7	13.40	20.40
2400	Solid	8x4x16	0	none		9.55	11.60	21.15
2600		4x8x16	0	none		6.15	9.05	15.20
2800		8x8x16	0	none		8.75	10.80	19.55
2900		12x8x16	0	none		9.75	13.90	23.65

B2010 119 — Reinforced Split Face Block Wall - Regular Weight

	TYPE	SIZE (IN.)	SCORES	VERT. REINF. & GROUT SPACING		MAT.	INST.	TOTAL
5200	Hollow	8x4x16	0	#4 @ 48″		7.35	12.15	19.50
5400		4x8x16	0	#4 @ 48″		4.52	9.60	14.12
5600		8x8x16	0	#4 @ 48″		6.60	11.40	18
5640				#5 @ 16″		7.15	12.50	19.65
5700		12x8x16	0	#4 @ 48″		7.55	14.40	21.95
5740				#5 @ 16″		8.85	16.25	25.10

B2010 Exterior Walls

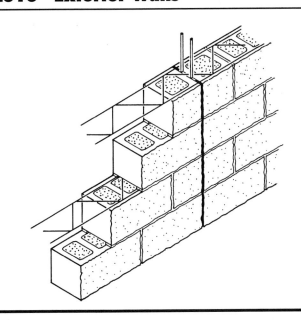

Exterior ground face block walls are defined in the following terms; structural reinforcement, weight, percent solid, size, scores and insulation. Within each of these categories two to four variations are shown. No costs are included for brick shelf or relieving angles. Costs include control joints every 20' and horizontal reinforcing.

System Components	QUANTITY	UNIT	COST PER S.F.		
			MAT.	INST.	TOTAL
SYSTEM B2010 121 1600					
UNREINF. GROUND FACE BLOCK WALL, 8"X8"X16", 0 SCORES, PERLITE FILL					
Ground face block wall, 8" thick	1.000	S.F.	12.35	10.35	22.70
Perlite insulation	1.000	S.F.	1.86	.49	2.35
Horizontal joint reinforcing, alternate course	.800	L.F.	.23	.21	.44
Control joint	.050	L.F.	.09	.08	.17
TOTAL			14.53	11.13	25.66

B2010 121 — Ground Face Block Wall

	TYPE	SIZE (IN.)	SCORES	CORE FILL		COST PER S.F.		
						MAT.	INST.	TOTAL
1200	Hollow	4x8x16	0	none		9.75	8.95	18.70
1250			1	none		10.70	9.05	19.75
1300			2 to 5	none		11.60	9.20	20.80
1600		8x8x16	0	perlite		14.55	11.15	25.70
1610				styrofoam		14.50	10.65	25.15
1640				none		12.65	10.65	23.30
1800		12x8x16	0	perlite		22	14.60	36.60
1840				none		18.85	13.65	32.50
3200	Solid	4x8x16	0	none		14.30	9.20	23.50
3300			2 to 5	none		17	9.45	26.45
3600		8x8x16	0	none		18.45	11	29.45
3800		12x8x16	0	none		27.50	14.15	41.65

B2010 122 — Reinforced Ground Face Block Wall

	TYPE	SIZE (IN.)	SCORES	VERT. REINF. & GROUT SPACING		COST PER S.F.		
						MAT.	INST.	TOTAL
5200	Hollow	4x8x16	0	#4 @ 48"		9.90	9.75	19.65
5300			2 to 5	#4 @ 48"		11.75	10	21.75
5600		8x8x16	0	#4 @ 48"		13.05	11.60	24.65
5630				#5 @ 16"		14.05	13.40	27.45
5800		12x8x16	0	#4 @ 48"		19.40	14.65	34.05
5830				#5 @ 16"		20.50	16.50	37

For customer support on your Concrete & Masonry Costs with RSMeans Data, call 800.448.8182.

B2010 Exterior Walls

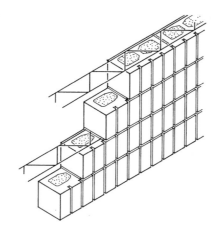

Exterior miscellaneous block walls are defined in the following terms: structural reinforcement, finish texture, percent solid, size, weight and insulation. Within each of these categories two to four variations are shown. No costs are included for brick shelf or relieving angles. Costs include control joints every 20′ and horizontal reinforcing.

System Components	QUANTITY	UNIT	COST PER S.F.		
			MAT.	INST.	TOTAL
SYSTEM B2010 123 1300					
UNREINFORCED DEEP GROOVE BLOCK WALL, 8″ X 8″ X 16″, PERLITE FILL					
Deep groove block partition, 8″ thick	1.000	S.F.	4.33	10	14.33
Perlite insulation	1.000	S.F.	1.86	.49	2.35
Horizontal joint reforcing, alternate courses	.800	S.F.	.23	.21	.44
Control joint	.050	L.F.	.09	.08	.17
TOTAL			6.51	10.78	17.29

B2010 123		Miscellaneous Block Wall						
	TYPE	SIZE (IN.)	WEIGHT (P.C.F.)	CORE FILL		COST PER S.F.		
						MAT.	INST.	TOTAL
1100	Deep groove-hollow	4x8x16	125	none		2.89	8.95	11.84
1200		6x8x16	125	perlite		5.40	9.85	15.25
1300		8x8x16	125	perlite		6.50	10.80	17.30
2100	Fluted	4x8x16	125	none		4.54	8.95	13.49
2200		6x8x16	125	perlite		7.90	9.85	17.75
2240				none		6.65	9.45	16.10
2500	Hex-hollow	4x8x16	125	none		4.68	8.95	13.63
2600		6x8x16	125	perlite		5.95	10.30	16.25
2640				none		4.69	9.90	14.59
3100	Slump block	4x4x16	125	none		6.20	7.35	13.55
3200		6x4x16	125	perlite		8.85	8	16.85
3240				none		7.60	7.60	15.20

B2010 124		Reinforced Misc. Block Walls						
	TYPE	SIZE (IN.)	WEIGHT (P.C.F.)	VERT. REINF. & GROUT SPACING		COST PER S.F.		
						MAT.	INST.	TOTAL
5100	Deep groove-hollow	4x8x16	125	#4 @ 48″		3.08	9.75	12.83
5200		6x8x16	125	#4 @ 48″		4.45	10.25	14.70
6100	Fluted	4x8x16	125	#4 @ 48″		4.73	9.75	14.48
6200		6x8x16	125	#4 @ 48″		6.95	10.25	17.20
6500	Hex-hollow	4x8x16	125	#4 @ 48″		4.87	9.75	14.62
6650		6x8x16	105	#4 @ 48″		6.15	10.40	16.55
7100	Slump block	4x4x16	125	#4 @ 48″		6.35	8.15	14.50
7200		6x4x16	125	#4 @ 48″		7.90	8.40	16.30

For customer support on your Concrete & Masonry Costs with RSMeans Data, call 800.448.8182.

B2010 Exterior Walls

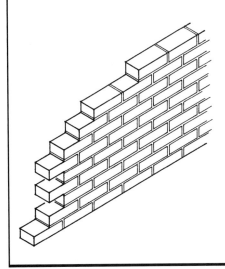

Exterior solid brick walls are defined in the following terms: structural reinforcement, type of brick, thickness, and bond. Six different types of face brick are presented, with single wythes shown in four different bonds. Shelf angles are included at every floor with an average 12' floor to floor height and control joints are included every 20'.

System Components	QUANTITY	UNIT	COST PER S.F.		
			MAT.	INST.	TOTAL
SYSTEM B2010 125 1000					
BRICK WALL, COMMON, SINGLE WYTHE, 4″ THICK, RUNNING BOND					
Common brick wall, running bond	1.000	S.F.	5.20	13	18.20
Wash brick	1.000	S.F.	.06	1.16	1.22
Control joint, backer rod	.100	L.F.		.14	.14
Control joint, sealant	.125	L.F.	.04	.28	.32
Shelf angle	1.000	Lb.	1.12	1.18	2.30
Flashing	.100	S.F.	.17	.43	.60
TOTAL			6.59	16.19	22.78

B2010 125			Solid Brick Walls - Single Wythe			COST PER S.F.		
	TYPE	THICKNESS (IN.)	BOND			MAT.	INST.	TOTAL
1000	Common	4	running			6.60	16.20	22.80
1010			common			8	18.55	26.55
1050			Flemish			8.85	22.50	31.35
1100			English			9.95	23.50	33.45
1150	Standard	4	running			6.90	16.80	23.70
1160			common			8.15	19.40	27.55
1200			Flemish			9.05	23	32.05
1250			English			10.15	24.50	34.65
1450	Engineer	4	running			5.70	14.70	20.40
1460			common			6.70	16.80	23.50
1500			Flemish			7.40	20.50	27.90
1550			English			8.30	21.50	29.80
2950	SCR	6	running			8.20	12.85	21.05
2960			common			9.75	14.70	24.45
3000			Flemish			10.85	17.45	28.30
3050			English			12.30	18.55	30.85
3100	Norwegian	6	running			7	11.40	18.40
3110			common			8.25	13	21.25
3150			Flemish			9.15	15.45	24.60
3200			English			10.25	16.20	26.45

B20 Exterior Enclosure

B2010 Exterior Walls

B2010 125			Solid Brick Walls - Single Wythe					
	TYPE	THICKNESS (IN.)	BOND			COST PER S.F.		
						MAT.	INST.	TOTAL
3250	Jumbo	6	running			6.40	10.10	16.50
3260			common			7.55	11.40	18.95
3300			Flemish			8.35	13.55	21.90
3350			English			9.45	14.30	23.75

For customer support on your Concrete & Masonry Costs with RSMeans Data, call 800.448.8182.

B2010 Exterior Walls

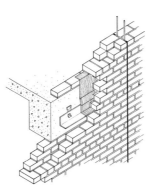

Exterior solid brick walls are defined in the following terms; structural reinforcement, type of brick, thickness, and bond. Sixteen different types of face bricks are presented, with single wythes shown in four different bonds. Six types of reinforced single wythe walls are also given, and twice that many for reinforced double wythe. Shelf angles are included in system components. These walls do not include ties and as such are not tied to a backup wall.

System Components	QUANTITY	UNIT	COST PER S.F. MAT.	COST PER S.F. INST.	COST PER S.F. TOTAL
SYSTEM B2010 126 4100					
DOUBLE WYTHE, COMMON BRICK, 8″ THICK					
Common brick, double wythe	1.000	S.F.	10.70	23	33.70
Wash smooth brick	1.000	S.F.	.06	1.16	1.22
Control joint, 4″ P.V.C.	.050	L.F.	.09	.08	.17
Steel lintels	1.000	Lb.	1.12	1.18	2.30
Caulking backer rod, polyethylene, 1/2″ diameter	.100	L.F.		.14	.14
Caulking, butyl base, 1/4″x1/2″	.125	L.F.	.04	.28	.32
Aluminum flashing, mill finish, .019″	.100	S.F.	.17	.43	.60
Grout cavity wall	.500	S.F.	.51	.76	1.27
Horizontal joint reinforcing, 9 ga. truss type	.800	L.F.	.24	.21	.45
TOTAL			12.93	27.24	40.17

B2010 126		Solid Brick Walls - Double Wythe			COST PER S.F.		
	TYPE	THICKNESS (IN.)	COLLAR JOINT THICKNESS (IN.)		MAT.	INST.	TOTAL
4100	Common	8	3/4		12.95	27	39.95
4150	Standard	8	1/2		13.25	28	41.25
4200	Glazed	8	1/2		49	29	78
4250	Engineer	8	1/2		10.90	24.50	35.40
4300	Economy	8	3/4		17.40	21.50	38.90
4350	Double	8	3/4		12.20	16.50	28.70
4400	Fire	8	3/4		27	24.50	51.50
4450	King	7	3/4		9.50	22	31.50
4500	Roman	8	1		19.25	25.50	44.75
4550	Norman	8	3/4		13.50	21	34.50
4600	Norwegian	8	3/4		11.45	18.50	29.95
4650	Utility	8	3/4		13.20	16	29.20
4700	Triple	8	3/4		9.40	14.40	23.80
4750	SCR	12	3/4		15.75	21.50	37.25
4800	Norwegian	12	3/4		13.45	18.90	32.35

B2010 Exterior Walls

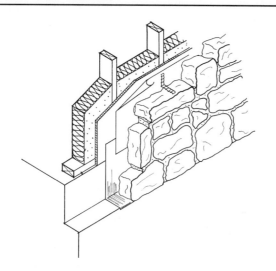

The table below lists costs per S.F. for stone veneer walls on various backup using different stone. Typical components for a system are shown in the component block below.

System Components	QUANTITY	UNIT	COST PER S.F.		
			MAT.	INST.	TOTAL
SYSTEM B2010 128 2000					
ASHLAR STONE VENEER, 4″, 2″ X 4″ STUD 16″ O.C. BACK UP, 8′ HIGH					
Sawn face, split joints, low priced stone	1.000	S.F.	14.30	21.50	35.80
Framing, 2″ x 4″ studs 8′ high 16″ O.C.	.083	B.F.	.86	1.63	2.49
Wall ties for stone veneer, galv. corrg 7/8″ x 7″, 22 gauge	.700	Ea.	.12	.43	.55
Asphalt felt sheathing paper, 15 lb.	1.000	S.F.	.06	.18	.24
Sheathing plywood on wall CDX 1/2″	1.000	S.F.	1.10	1.16	2.26
Fiberglass insulation, 3-1/2″, cr-11	1.000	S.F.	.61	.48	1.09
Flashing, copper, paperback 1 side, 3 oz	.125	S.F.	.43	.24	.67
TOTAL			17.48	25.62	43.10

B2010 128	Stone Veneer	COST PER S.F.		
		MAT.	INST.	TOTAL
2000	Ashlar veneer, 4″, 2″ x 4″ stud backup, 16″ O.C., 8′ high, low priced stone	17.50	25.50	43
2050	2″ x 6″ stud backup, 16″ O.C.	20	30	50
2100	Metal stud backup, 8′ high, 16″ O.C.	17.95	26	43.95
2150	24″ O.C.	17.60	25.50	43.10
2200	Conc. block backup, 4″ thick	19.05	30.50	49.55
2300	6″ thick	19.75	31	50.75
2350	8″ thick	19.65	31.50	51.15
2400	10″ thick	20.50	33	53.50
2500	12″ thick	22	35.50	57.50
3100	High priced stone, wood stud backup, 10′ high, 16″ O.C.	24	29	53
3200	Metal stud backup, 10′ high, 16″ O.C.	24	29.50	53.50
3250	24″ O.C.	24	29	53
3300	Conc. block backup, 10′ high, 4″ thick	25.50	34	59.50
3350	6″ thick	26	34	60
3400	8″ thick	26.50	36.50	63
3450	10″ thick	26.50	36.50	63
3500	12″ thick	28	39	67
4000	Indiana limestone 2″ thk., sawn finish, wood stud backup, 10′ high, 16″ O.C.	37	15.20	52.20
4100	Metal stud backup, 10′ high, 16″ O.C.	39.50	19.25	58.75
4150	24″ O.C.	39.50	19.25	58.75
4200	Conc. block backup, 4″ thick	39	20	59
4250	6″ thick	39.50	20.50	60
4300	8″ thick	39.50	21	60.50

B2010 Exterior Walls

B2010 128	Stone Veneer	COST PER S.F.		
		MAT.	INST.	TOTAL
4350	10" thick	40	22.50	62.50
4400	12" thick	41.50	25	66.50
4450	2" thick, smooth finish, wood stud backup, 8' high, 16" O.C.	37	15.20	52.20
4550	Metal stud backup, 8' high, 16" O.C.	37.50	15.30	52.80
4600	24" O.C.	37.50	14.75	52.25
4650	Conc. block backup, 4" thick	39	19.85	58.85
4700	6" thick	39.50	20	59.50
4750	8" thick	39.50	21	60.50
4800	10" thick	40	22.50	62.50
4850	12" thick	41.50	25	66.50
5350	4" thick, smooth finish, wood stud backup, 8' high, 16" O.C.	41.50	15.20	56.70
5450	Metal stud backup, 8' high, 16" O.C.	42	15.55	57.55
5500	24" O.C.	42	15	57
5550	Conc. block backup, 4" thick	43.50	20	63.50
5600	6" thick	44	20.50	64.50
5650	8" thick	48	26	74
5700	10" thick	44.50	22.50	67
5750	12" thick	46	25	71
6000	Granite, gray or pink, 2" thick, wood stud backup, 8' high, 16" O.C.	35	27.50	62.50
6100	Metal studs, 8' high, 16" O.C.	35.50	28	63.50
6150	24" O.C.	35.50	27.50	63
6200	Conc. block backup, 4" thick	37	32.50	69.50
6250	6" thick	37.50	33	70.50
6300	8" thick	37.50	33.50	71
6350	10" thick	38	35	73
6400	12" thick	39.50	37.50	77
6900	4" thick, wood stud backup, 8' high, 16" O.C.	46	32	78
7000	Metal studs, 8' high, 16" O.C.	46.50	32.50	79
7050	24" O.C.	46.50	31.50	78
7100	Conc. block backup, 4" thick	48	37	85
7150	6" thick	48.50	37	85.50
7200	8" thick	48.50	38	86.50
7250	10" thick	49	39.50	88.50
7300	12" thick	50.50	41.50	92

For customer support on your Concrete & Masonry Costs with RSMeans Data, call 800.448.8182.

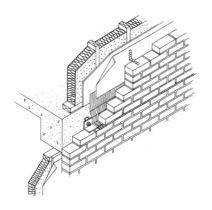

Exterior brick veneer/stud backup walls are defined in the following terms: type of brick and studs, stud spacing and bond. All systems include a back-up wall, a control joint every 20', a brick shelf every 12' of height, ties to the backup and the necessary dampproofing, flashing and insulation.

System Components	QUANTITY	UNIT	COST PER S.F.		
			MAT.	INST.	TOTAL
SYSTEM B2010 129 1100					
STANDARD BRICK VENEER, 2″ X 4″ STUD BACKUP @ 16″ O.C., RUNNING BOND					
Standard brick wall, 4″ thick, running bond	1.000	S.F.	5.50	13.60	19.10
Wash smooth brick	1.000	S.F.	.06	1.16	1.22
Joint backer rod	.100	L.F.		.14	.14
Sealant	.125	L.F.	.04	.28	.32
Wall ties, corrugated, 7/8″ x 7″, 22 gauge	.003	Ea.	.05	.18	.23
Shelf angle	1.000	Lb.	1.12	1.18	2.30
Wood stud partition, backup, 2″ x 4″ @ 16″ O.C.	1.000	S.F.	.81	1.31	2.12
Sheathing, plywood, CDX, 1/2″	1.000	S.F.	1.10	.93	2.03
Building paper, asphalt felt, 15 lb.	1.000	S.F.	.06	.18	.24
Fiberglass insulation, batts, 3-1/2″ thick paper backing	1.000	S.F.	.61	.48	1.09
Flashing, copper, paperbacked	.100	S.F.	.17	.43	.60
TOTAL			9.52	19.87	29.39

B2010 129		Brick Veneer/Wood Stud Backup						
	FACE BRICK	STUD BACKUP	STUD SPACING (IN.)	BOND	FACE	COST PER S.F.		
						MAT.	INST.	TOTAL
1100	Standard	2x4-wood	16	running		9.55	19.85	29.40
1120				common		10.80	22.50	33.30
1140				Flemish		11.70	26	37.70
1160				English		12.80	28	40.80
1400		2x6-wood	16	running		10.20	19.80	30
1420				common		11.20	22.50	33.70
1440				Flemish		12.10	26.50	38.60
1460				English		13.20	28	41.20
1500			24	running		9.65	19.70	29.35
1520				common		10.90	22.50	33.40
1540				Flemish		11.80	26	37.80
1560				English		12.90	27.50	40.40
1700	Glazed	2x4-wood	16	running		27.50	20.50	48
1720				common		32.50	23.50	56
1740				Flemish		35.50	28	63.50
1760				English		40	29.50	69.50

B2010 Exterior Walls

B2010 129			Brick Veneer/Wood Stud Backup				

	FACE BRICK	STUD BACKUP	STUD SPACING (IN.)	BOND	FACE	COST PER S.F.		
						MAT.	INST.	TOTAL
2000	Glazed	2x6-wood	16	running		28	20.50	48.50
2020				common		33	23.50	56.50
2040				Flemish		36	28	64
2060				English		40.50	29.50	70
2100			24	running		27.50	20.50	48
2120				common		32.50	23	55.50
2140				Flemish		35.50	27.50	63
2160				English		40	29	69
2300	Engineer	2x4-wood	16	running		8.35	17.75	26.10
2320				common		9.35	19.85	29.20
2340				Flemish		10.05	23.50	33.55
2360				English		10.95	24.50	35.45
2600		2x6-wood	16	running		8.75	17.90	26.65
2620				common		9.75	20	29.75
2640				Flemish		10.45	23.50	33.95
2660				English		11.35	24.50	35.85
2700			24	running		8.45	17.60	26.05
2720				common		9.45	19.70	29.15
2740				Flemish		10.15	23	33.15
2760				English		11.05	24.50	35.55
2900	Roman	2x4-wood	16	running		12.55	18.25	30.80
2920				common		14.45	20.50	34.95
2940				Flemish		15.70	24	39.70
2960				English		17.40	25.50	42.90
3200		2x6-wood	16	running		12.95	18.40	31.35
3220				common		14.85	20.50	35.35
3240				Flemish		16.10	24	40.10
3260				English		17.80	25.50	43.30
3300			24	running		12.65	18.10	30.75
3320				common		14.55	20.50	35.05
3340				Flemish		15.80	23.50	39.30
3360				English		17.50	25.50	43
3500	Norman	2x4-wood	16	running		9.65	15.60	25.25
3520				common		10.95	17.35	28.30
3540				Flemish		22	20	42
3560				English		12.95	21.50	34.45
3800		2x6-wood	16	running		10.05	15.75	25.80
3820				common		11.35	17.50	28.85
3840				Flemish		21.50	19.55	41.05
3860				English		13.35	21.50	34.85
3900			24	running		9.75	15.45	25.20
3920				common		11.05	17.20	28.25
3940				Flemish		22	20	42
3960				English		13.05	21	34.05
4100	Norwegian	2x4-wood	16	running		8.65	14.25	22.90
4120				common		9.70	15.75	25.45
4140				Flemish		10.45	18.25	28.70
4160				English		11.35	19	30.35
4400		2x6-wood	16	running		9.05	14.40	23.45
4420				common		10.10	15.90	26
4440				Flemish		10.85	18.40	29.25
4460				English		11.75	19.15	30.90

For customer support on your Concrete & Masonry Costs with RSMeans Data, call 800.448.8182.

B2010 Exterior Walls

B2010 129	Brick Veneer/Wood Stud Backup

	FACE BRICK	STUD BACKUP	STUD SPACING (IN.)	BOND	FACE	COST PER S.F.		
						MAT.	INST.	TOTAL
4500	Norwegian	2x6-wood	24	running		8.75	14.10	22.85
4520				common		9.80	15.60	25.40
4540				Flemish		10.55	18.10	28.65
4560				English		11.45	18.85	30.30
4600	Oversized, 4" x 2-1/4" x 16"	2x4-wood	16	running	plain face	9.45	14	23.45
4610					1 to 3 slot face	9.85	14.60	24.45
4620					4 to 7 slot face	10.15	15.10	25.25
4630	Oversized, 4" x 2-3/4" x 16"	2x4-wood	16	running	plain face	10.50	13.50	24
4640					1 to 3 slot face	10.95	14.05	25
4650					4 to 7 slot face	11.35	14.50	25.85
4660	Oversized, 4" x 4" x 16"	2x4-wood	16	running	plain face	8.30	12.75	21.05
4670					1 to 3 slot face	8.65	13.25	21.90
4680					4 to 7 slot face	8.90	13.65	22.55
4690	Oversized, 4" x 8" x 16"	2x4-wood	16	running	plain face	10.25	11.85	22.10
4700					1 to 3 slot face	10.70	12.30	23
4710					4 to 7 slot face	11.05	12.65	23.70
4720	Oversized, 4" x 2-1/4" x 16"	2x6-wood	16	running	plain face	9.85	14.15	24
4730					1 to 3 slot face	10.25	14.75	25
4740					4 to 7 slot face	10.60	15.20	25.80
4750	Oversized, 4" x 2-3/4" x 16"	2x6-wood	16	running	plain face	10.90	13.65	24.55
4760					1 to 3 slot face	11.40	14.20	25.60
4770					4 to 7 slot face	11.80	14.65	26.45
4780	Oversized, 4" x 4" x 16"	2x6-wood	16	running	plain face	8.75	12.90	21.65
4790					1 to 3 slot face	9.05	13.40	22.45
4800					4 to 7 slot face	9.30	13.80	23.10
4810	Oversized, 4" x 8" x 16"	2x6-wood	16	running	plain face	10.65	12	22.65
4820					1 to 3 slot face	11.10	12.45	23.55
4830					4 to 7 slot face	11.50	12.80	24.30

B2010 130	Brick Veneer/Metal Stud Backup

	FACE BRICK	STUD BACKUP	STUD SPACING (IN.)	BOND	FACE	COST PER S.F.		
						MAT.	INST.	TOTAL
5050	Standard	16 ga x 6"LB	16	running		10.85	21	31.85
5100		25ga.x6"NLB	24	running		8.50	19.45	27.95
5120				common		9.75	22	31.75
5140				Flemish		10.25	25	35.25
5160				English		11.75	27.50	39.25
5200		20ga.x3-5/8"NLB	16	running		8.75	20.50	29.25
5220				common		10	23	33
5240				Flemish		10.90	26.50	37.40
5260				English		12	28	40
5300			24	running		8.55	19.65	28.20
5320				common		9.80	22.50	32.30
5340				Flemish		10.70	26	36.70
5360				English		11.80	27.50	39.30
5400		16ga.x3-5/8"LB	16	running		9.35	20.50	29.85
5420				common		10.60	23	33.60
5440				Flemish		11.50	27	38.50
5460				English		12.60	28.50	41.10
5500			24	running		9	19.95	28.95
5520				common		10.25	22.50	32.75
5540				Flemish		11.15	26.50	37.65
5560				English		12.25	28	40.25

B2010 Exterior Walls

B2010 130	Brick Veneer/Metal Stud Backup

	FACE BRICK	STUD BACKUP	STUD SPACING (IN.)	BOND	FACE	COST PER S.F.		
						MAT.	INST.	TOTAL
5700	Glazed	25ga.x6"NLB	24	running		26.50	20	46.50
5720				common		31.50	23	54.50
5740				Flemish		34.50	27.50	62
5760				English		39	29	68
5800		20ga.x3-5/8"NLB	24	running		26.50	20.50	47
5820				common		31.50	23	54.50
5840				Flemish		34.50	27.50	62
5860				English		39	29	68
6000		16ga.x3-5/8"LB	16	running		27.50	21	48.50
6020				common		32.50	24	56.50
6040				Flemish		35.50	28.50	64
6060				English		40	30	70
6100			24	running		27	20.50	47.50
6120				common		32	23.50	55.50
6140				Flemish		35	28	63
6160				English		39.50	29.50	69
6300	Engineer	25ga.x6"NLB	24	running		7.30	17.35	24.65
6320				common		8.30	19.45	27.75
6340				Flemish		9	23	32
6360				English		9.90	24	33.90
6400		20ga.x3-5/8"NLB	16	running		7.55	18.15	25.70
6420				common		8.55	20.50	29.05
6440				Flemish		9.25	24	33.25
6460				English		10.15	25	35.15
6500	Engineer	20ga.x3-5/8"NLB	24	running		6.90	16.55	23.45
6520				common		8.35	19.65	28
6540				Flemish		9.05	23	32.05
6560				English		9.95	24	33.95
6600		16ga.x3-5/8"LB	16	running		8.20	18.45	26.65
6620				common		9.15	20.50	29.65
6640				Flemish		9.85	24	33.85
6660				English		10.75	25	35.75
6700			24	running		7.80	17.85	25.65
6720				common		8.80	19.95	28.75
6740				Flemish		9.50	23.50	33
6760				English		10.40	24.50	34.90
6900	Roman	25ga.x6"NLB	24	running		11.50	17.85	29.35
6920				common		13.40	20	33.40
6940				Flemish		14.65	23.50	38.15
6960				English		16.35	25	41.35
7000		20ga.x3-5/8"NLB	16	running		11.75	18.65	30.40
7020				common		13.65	21	34.65
7040				Flemish		14.90	24.50	39.40
7060				English		16.60	26	42.60
7100			24	running		11.55	18.05	29.60
7120				common		13.45	20.50	33.95
7140				Flemish		14.70	23.50	38.20
7160				English		16.40	25.50	41.90
7200		16ga.x3-5/8"LB	16	running		12.35	18.95	31.30
7220				common		14.25	21	35.25
7240				Flemish		15.50	24.50	40
7260				English		17.20	26	43.20

B2010 130			Brick Veneer/Metal Stud Backup				

	FACE BRICK	STUD BACKUP	STUD SPACING (IN.)	BOND	FACE	COST PER S.F.		
						MAT.	INST.	TOTAL
7300	Roman		24	running		12	18.35	30.35
7320				common		13.90	20.50	34.40
7340				Flemish		15.15	24	39.15
7360				English		16.85	25.50	42.35
7500	Norman	25ga.x6"NLB	24	running		8.60	15.20	23.80
7520				common		9.90	16.95	26.85
7540				Flemish		21	19.80	40.80
7560				English		11.90	21	32.90
7600		20ga.x3-5/8"NLB	24	running		8.65	15.40	24.05
7620				common		9.95	17.15	27.10
7640				Flemish		21	20	41
7660				English		11.95	21	32.95
7800		16ga.x3-5/8"LB	16	running		9.45	16.30	25.75
7820				common		10.75	18.05	28.80
7840				Flemish		22	21	43
7860				English		12.75	22	34.75
7900			24	running		9.10	15.70	24.80
7920				common		10.40	17.45	27.85
7940				Flemish		21.50	20.50	42
7960				English		12.40	21.50	33.90
8100	Norwegian	25ga.x6"NLB	24	running		7.60	13.85	21.45
8120				common		8.65	15.35	24
8140				Flemish		9.40	17.85	27.25
8160				English		10.30	18.60	28.90
8200		20ga.x3-5/8"NLB	16	running		7.85	14.65	22.50
8220				common		8.90	16.15	25.05
8240				Flemish		9.65	18.65	28.30
8260				English		10.55	19.40	29.95
8300			24	running		7.65	14.05	21.70
8320				common		8.70	15.55	24.25
8340				Flemish		9.45	18.05	27.50
8360				English		10.35	18.80	29.15
8400		16ga.x3-5/8"LB	16	running		8.45	14.95	23.40
8420				common		9.50	16.45	25.95
8440				Flemish		10.25	18.95	29.20
8460				English		11.15	19.70	30.85
8500			24	running		8.10	14.35	22.45
8520				common		9.15	15.85	25
8540				Flemish		9.90	18.35	28.25
8560				English		10.80	19.10	29.90
8600	Oversized, 4" x 2-1/4" x 16"	25 ga. x 6" NLB	24	running	plain face	9.10	13.60	22.70
8610					1 to 3 slot face	9.50	14.20	23.70
8620					4 to 7 slot face	9.80	14.65	24.45
8630	Oversized, 4" x 2-3/4" x 16"	25 ga. x 6" NLB	24	running	plain face	10.15	13.10	23.25
8640					1 to 3 slot face	10.60	13.65	24.25
8650					4 to 7 slot face	11	14.10	25.10
8660	Oversized, 4" x 4" x 16"	25 ga. x 6" NLB	24	running	plain face	7.95	12.35	20.30
8670					1 to 3 slot face	8.30	12.85	21.15
8680					4 to 7 slot face	8.55	13.25	21.80
8690	Oversized, 4" x 8" x 16"	25 ga x 6" NLB	24	running	plain face	9.90	11.45	21.35
8700					1 to 3 slot face	10.35	11.90	22.25
8710					4 to 7 slot face	10.70	12.25	22.95

B2010 Exterior Walls

B2010 130	Brick Veneer/Metal Stud Backup

	FACE BRICK	STUD BACKUP	STUD SPACING (IN.)	BOND	FACE	COST PER S.F.		
						MAT.	INST.	TOTAL
8720	Oversized, 4" x 2-1/4" x 16"	20 ga. x 3-5/8" NLB	24	running	plain face	9.15	13.80	22.95
8730					1 to 3 slot face	9.55	14.40	23.95
8740					4 to 7 slot face	9.90	14.90	24.80
8750	Oversized, 4" x 2-3/4" x 16"	20 ga. x 3-5/8" NLB	24	running	plain face	10.20	13.30	23.50
8760					1 to 3 slot face	10.70	13.85	24.55
8770					4 to 7 slot face	11.10	14.30	25.40
8780	Oversized, 4" x 4" x 16"	20 ga. x 3-5/8" NLB	24	running	plain face	8.05	12.55	20.60
8790					1 to 3 slot face	8.35	13.05	21.40
8800					4 to 7 slot face	8.65	13.45	22.10
8810	Oversized, 4" x 8" x 16"	20 ga x 3-5/8" NLB	24	running	plain face	9.95	11.65	21.60
8820					1 to 3 slot face	10.40	12.10	22.50
8830					4 to 7 slot face	10.80	12.45	23.25

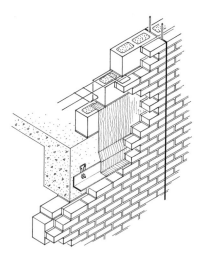

Exterior brick face composite walls are defined in the following terms: type of face brick and backup masonry, thickness of backup masonry and insulation. A special section is included on triple wythe construction at the back. Seven types of face brick are shown with various thicknesses of seven types of backup. All systems include a brick shelf, ties to the backup and necessary dampproofing, flashing, and control joints every 20'.

System Components	QUANTITY	UNIT	COST PER S.F.		
			MAT.	INST.	TOTAL
SYSTEM B2010 132 1120					
COMPOSITE WALL, STANDARD BRICK FACE, 6" C.M.U. BACKUP, PERLITE FILL					
Face brick veneer, standard, running bond	1.000	S.F.	5.50	13.60	19.10
Wash brick	1.000	S.F.	.06	1.16	1.22
Concrete block backup, 6" thick	1.000	S.F.	2.98	7.50	10.48
Wall ties	.300	Ea.	.11	.18	.29
Perlite insulation, poured	1.000	S.F.	1.26	.39	1.65
Flashing, aluminum	.100	S.F.	.17	.43	.60
Shelf angle	1.000	Lb.	1.12	1.18	2.30
Control joint	.050	L.F.	.09	.08	.17
Backer rod	.100	L.F.		.14	.14
Sealant	.125	L.F.	.04	.28	.32
Collar joint	1.000	S.F.	.51	.76	1.27
TOTAL			11.84	25.70	37.54

B2010 132		**Brick Face Composite Wall - Double Wythe**				COST PER S.F.		
	FACE BRICK	BACKUP MASONRY	BACKUP THICKNESS (IN.)	BACKUP CORE FILL		MAT.	INST.	TOTAL
1000	Standard	common brick	4	none		12.80	31	43.80
1040		SCR brick	6	none		14.40	27.50	41.90
1080		conc. block	4	none		9.90	25	34.90
1120			6	perlite		11.85	25.50	37.35
1160				styrofoam		12	25.50	37.50
1200			8	perlite		12.35	26.50	38.85
1240				styrofoam		12.30	26	38.30
1280		L.W. block	4	none		9.80	24.50	34.30
1320			6	perlite		12.25	25.50	37.75
1360				styrofoam		12.40	25	37.40
1400			8	perlite		13.85	26	39.85
1440				styrofoam		13.80	25.50	39.30
1520		glazed block	4	none		21.50	26.50	48
1560			6	perlite		25	27.50	52.50
1600				styrofoam		25.50	27	52.50
1640			8	perlite		27.50	28	55.50
1680				styrofoam		27.50	27.50	55

B2010 Exterior Walls

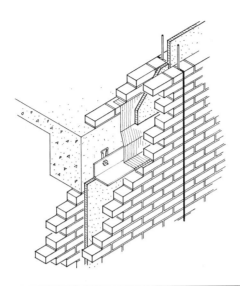

Exterior brick face cavity walls are defined in the following terms: cavity treatment, type of face brick, backup masonry, total thickness and insulation. Seven types of face brick are shown with various types of backup. All systems include a brick shelf, ties to the backups and necessary dampproofing, flashing, and control joints every 20'.

System Components	QUANTITY	UNIT	COST PER S.F.		
			MAT.	INST.	TOTAL
SYSTEM B2010 134 1000					
CAVITY WALL, STANDARD BRICK FACE, COMMON BRICK BACKUP, POLYSTYRENE					
Face brick veneer, standard, running bond	1.000	S.F.	5.50	13.60	19.10
Wash brick	1.000	S.F.	.06	1.16	1.22
Common brick wall backup, 4" thick	1.000	S.F.	5.20	13	18.20
Wall ties	.300	L.F.	.13	.18	.31
Polystyrene insulation board, 1" thick	1.000	S.F.	.35	.82	1.17
Flashing, aluminum	.100	S.F.	.17	.43	.60
Shelf angle	1.000	Lb.	1.12	1.18	2.30
Control joint	.050	L.F.	.09	.08	.17
Backer rod	.100	L.F.		.14	.14
Sealant	.125	L.F.	.04	.28	.32
TOTAL			12.66	30.87	43.53

B2010 134		Brick Face Cavity Wall							
	FACE BRICK	BACKUP MASONRY	TOTAL THICKNESS (IN.)	CAVITY INSULATION		COST PER S.F.			
						MAT.	INST.	TOTAL	
1000	Standard	4" common brick	10	polystyrene		12.65	31	43.65	
1020				none		12.30	30	42.30	
1040		6" SCR brick	12	polystyrene		14.30	27.50	41.80	
1060				none		13.95	26.50	40.45	
1080		4" conc. block	10	polystyrene		9.75	25	34.75	
1100				none		9.40	24	33.40	
1120		6" conc. block	12	polystyrene		10.50	25.50	36	
1140				none		10.15	24.50	34.65	
1160		4" L.W. block	10	polystyrene		9.65	24.50	34.15	
1180				none		9.30	24	33.30	
1200		6" L.W. block	12	polystyrene		10.90	25	35.90	
1220				none		10.55	24.50	35.05	
1240		4" glazed block	10	polystyrene		21.50	26.50	48	
1260				none		21	26	47	
1280		6" glazed block	12	polystyrene		21.50	26.50	48	
1300				none		21	26	47	

B2010 Exterior Walls

B2010 134			Brick Face Cavity Wall				

	FACE BRICK	BACKUP MASONRY	TOTAL THICKNESS (IN.)	CAVITY INSULATION		COST PER S.F.		
						MAT.	INST.	TOTAL
1320	Standard	4" clay tile	10	polystyrene		14.55	24	38.55
1340				none		14.20	23	37.20
1500	Glazed	4" common brick	10	polystyrene		30.50	31.50	62
1580		4" conc. block	10	polystyrene		28	25.50	53.50
1620		6" conc. block	12	polystyrene		28.50	26	54.50
1660		4" L.W. block	10	polystyrene		27.50	25.50	53
1700		6" L.W. block	12	polystyrene		29	26	55
2000	Engineer	4" common brick	10	polystyrene		11.50	29	40.50
2080		4" conc. block	10	polystyrene		8.60	22.50	31.10
2120		6" conc. block	12	polystyrene		9.30	23.50	32.80
2160		4" L.W. block	10	polystyrene		8.50	22.50	31
2200		6" L.W. block	12	polystyrene		9.75	23	32.75
2500	Roman	4" common brick	10	polystyrene		15.65	29.50	45.15
2580		4" conc. block	10	polystyrene		12.75	23	35.75
2620		6" conc. block	12	polystyrene		13.50	24	37.50
2660		4" L.W. block	10	polystyrene		12.65	23	35.65
2700		6" L.W. block	12	polystyrene		13.90	23.50	37.40
3000	Norman	4" common brick	10	polystyrene		12.75	26.50	39.25
3080		4" conc. block	10	polystyrene		9.85	20.50	30.35
3120		6" conc. block	12	polystyrene		10.60	21	31.60
3160		4" L.W. block	10	polystyrene		9.75	20.50	30.25
3200		6" L.W. block	12	polystyrene		11	21	32
3500	Norwegian	4" common brick	10	polystyrene		11.75	25.50	37.25
3580		4" conc. block	10	polystyrene		8.85	19.25	28.10
3620		6" conc. block	12	polystyrene		9.60	19.80	29.40
3660		4" L.W. block	10	polystyrene		8.75	19.10	27.85
3700		6" L.W. block	12	polystyrene		10	19.60	29.60
4000	Utility	4" common brick	10	polystyrene		12.60	24	36.60
4080		4" conc. block	10	polystyrene		9.70	17.90	27.60
4120		6" conc. block	12	polystyrene		10.45	18.45	28.90
4160		4" L.W. block	10	polystyrene		9.60	17.75	27.35
4200		6" L.W. block	12	polystyrene		10.85	18.25	29.10

B2010 Exterior Walls

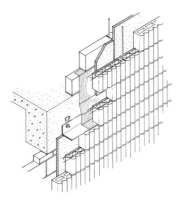

Exterior block face cavity walls are defined in the following terms: cavity treatment, type of face block, backup masonry, total thickness and insulation. Multiple types of face block are shown with various types of backup. All systems include a brick shelf and necessary dampproofing, flashing, and control joints every 20'.

System Components	QUANTITY	UNIT	COST PER S.F.		
			MAT.	INST.	TOTAL
SYSTEM B2010 137 1000					
CAVITY WALL, DEEP GROOVE BLOCK FACE, 4″ BLOCK BACKUP, RIGID INSUL.					
Deep groove block veneer, 4″ thick	1.000	S.F.	3.69	8.95	12.64
Concrete block wall backup, 4″ thick	1.000	S.F.	2.30	6.95	9.25
Horizontal joint reinforcing	.800	L.F.	.23	.26	.49
Polystyrene insulation board, 1″ thick	1.000	S.F.	.35	.82	1.17
Flashing, aluminum	.100	S.F.	.17	.43	.60
Shelf angle	1.000	Lb.	1.12	1.18	2.30
Control joint	.050	L.F.	.09	.08	.17
Backer rod	.100	L.F.		.14	.14
Sealant	.125	L.F.	.04	.28	.32
TOTAL			7.99	19.09	27.08

B2010 137		Block Face Cavity Wall						
	FACE BLOCK	BACKUP MASONRY	TOTAL THICKNESS (IN.)	CAVITY INSULATION		COST PER S.F.		
						MAT.	INST.	TOTAL
1000	Deep groove	4″ conc. block	10	polystyrene		8	19.10	27.10
1060		6″ conc. block	12	polystyrene		8.70	19.65	28.35
1600	Fluted	4″ conc. block	10	polystyrene		8	19.10	27.10
2200	Ground face	4″ conc. block	10	polystyrene		19.65	19.25	38.90
2250	1 Score			none		19.30	18.40	37.70
2500	Hexagonal	4″ conc. block	10	polystyrene		8.70	18.85	27.55
2560		6″ conc. block	12	polystyrene		9.40	19.40	28.80
3100	Slump block	4″ conc. block	10	polystyrene		10.20	17.25	27.45
3150	4x16			none		9.85	16.40	26.25
4000	Split rib	4″ conc. block	10	polystyrene		13.25	19.10	32.35

B20 Exterior Enclosure

B2010 Exterior Walls

B2010 138				Block Face Cavity Wall - Insulated Backup				

	FACE BLOCK	BACKUP MASONRY	TOTAL THICKNESS (IN.)	BACKUP CORE INSULATION		COST PER S.F.		
						MAT.	INST.	TOTAL
5010	Deep groove	6" conc. block	10	perlite		9.60	19.20	28.80
5060		8" conc. block	12	perlite		10.10	19.80	29.90
5600	Fluted	6" conc. block	10	perlite		10.15	18.95	29.10
6200	Ground face	6" conc. block	10	perlite		21	19.35	40.35
6250	1 Score			styrofoam		21.50	18.95	40.45
6500	Hexagonal	6" conc. block	10	perlite		10.30	18.95	29.25
6560		8" conc. block	12	perlite		10.80	19.55	30.35
7100	Slump block	6" conc. block	10	perlite		11.80	17.35	29.15
7150	4x16			styrofoam		11.95	16.95	28.90
7700	Split rib	6" conc. block	10	perlite		14.85	19.20	34.05

B2010 Exterior Walls

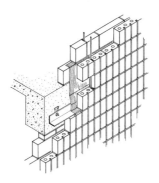

Exterior block face composite walls are defined in the following terms: type of face block and backup masonry, total thickness and insulation. All systems include shelf angles and necessary dampproofing, flashing, and control joints every 20'.

System Components	QUANTITY	UNIT	COST PER S.F.		
			MAT.	INST.	TOTAL
SYSTEM B2010 139 1000					
COMPOSITE WALL, GROOVED BLOCK FACE, 4″ C.M.U. BACKUP, PERLITE FILL					
Deep groove block veneer, 4″ thick	1.000	S.F.	3.69	8.95	12.64
Concrete block wall, backup, 4″ thick	1.000	S.F.	2.30	6.95	9.25
Horizontal joint reinforcing	.800	L.F.	.24	.21	.45
Perlite insulation, poured	1.000	S.F.	.83	.24	1.07
Flashing, aluminum	.100	S.F.	.17	.43	.60
Shelf angle	1.000	Lb.	1.12	1.18	2.30
Control joint	.050	L.F.	.09	.08	.17
Backer rod	.100	L.F.		.14	.14
Sealant	.125	L.F.	.04	.28	.32
Collar joint	1.000	S.F.	.51	.76	1.27
TOTAL			8.99	19.22	28.21

B2010 139		Block Face Composite Wall						
	FACE BLOCK	BACKUP MASONRY	TOTAL THICKNESS (IN.)	BACKUP CORE INSULATION		COST PER S.F.		
						MAT.	INST.	TOTAL
1000	Deep groove	4″ conc. block	8	perlite		9	19.20	28.20
1050	Reg. wt.			none		8.15	19	27.15
1120		8″ conc. block	12	perlite		10.60	20.50	31.10
1170				styrofoam		10.55	20	30.55
3400	Ground face	4″ conc. block	8	perlite		20.50	19.35	39.85
3450	1 Score			none		19.85	19.15	39
3520		8″ conc. block	12	perlite		22.50	21	43.50
3570				styrofoam		22	20	42
4000	Hexagonal	4″ conc. block	8	perlite		9.70	18.95	28.65
4050	Reg. wt.			none		8.90	18.75	27.65
4120		8″ conc. block	12	perlite		11.40	20.50	31.90
4170				styrofoam		11.25	19.85	31.10
5200	Slump block	4″ conc. block	8	perlite		11.20	17.35	28.55
5250	4x16			none		10.40	17.15	27.55
5320		8″ conc. block	12	perlite		12.80	18.70	31.50
5370				styrofoam		12.75	18.25	31
5800	Split face	4″ conc. block	8	perlite		11.70	19.20	30.90
5850	1 Score			none		10.90	19	29.90
5920		8″ conc. block	12	perlite		13.30	20.50	33.80
5970				styrofoam		13.25	20	33.25

B2010 Exterior Walls

B2010 139			Block Face Composite Wall					
	FACE BLOCK	BACKUP MASONRY	TOTAL THICKNESS (IN.)	BACKUP CORE INSULATION		COST PER S.F.		
						MAT.	INST.	TOTAL
7000	Split rib	4″ conc. block	8	perlite		14.25	19.20	33.45
7050	8 Rib			none		13.45	19	32.45
7120		8″ conc. block	12	perlite		15.85	20.50	36.35
7170				styrofoam		15.80	20	35.80

B2010 Exterior Walls

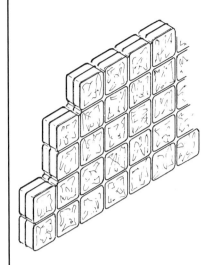

The table below lists costs per S.F. for glass block walls. Included in the costs are the following special accessories required for glass block walls.

Glass block accessories required for proper installation.

Wall ties: Galvanized double steel mesh full length of joint.

Fiberglass expansion joint at sides and top.

Silicone caulking: One gallon does 95 L.F.

Oakum: One lb. does 30 L.F.

Asphalt emulsion: One gallon does 600 L.F.

If block are not set in wall chase, use 2'-0" long wall anchors at 2'-0" O.C.

System Components	QUANTITY	UNIT	COST PER S.F. MAT.	COST PER S.F. INST.	COST PER S.F. TOTAL
SYSTEM B2010 140 2300					
GLASS BLOCK, 4" THICK, 6" X 6" PLAIN, UNDER 1,000 S.F.					
Glass block, 4" thick, 6" x 6" plain, under 1000 S.F.	4.100	Ea.	29	26	55
Glass block, cleaning blocks after installation, both sides add	2.000	S.F.	.18	3	3.18
TOTAL			29.18	29	58.18

B2010 140	Glass Block	COST PER S.F. MAT.	COST PER S.F. INST.	COST PER S.F. TOTAL
2300	Glass block 4" thick, 6"x6" plain, under 1,000 S.F.	29	29	58
2400	1,000 to 5,000 S.F.	28.50	25	53.50
2500	Over 5,000 S.F.	27.50	23.50	51
2600	Solar reflective, under 1,000 S.F.	41	39.50	80.50
2700	1,000 to 5,000 S.F.	40	34	74
2800	Over 5,000 S.F.	38.50	31.50	70
3500	8"x8" plain, under 1,000 S.F.	17.95	21.50	39.45
3600	1,000 to 5,000 S.F.	17.60	18.75	36.35
3700	Over 5,000 S.F.	17.10	16.95	34.05
3800	Solar reflective, under 1,000 S.F.	25	29	54
3900	1,000 to 5,000 S.F.	24.50	25	49.50
4000	Over 5,000 S.F.	24	22.50	46.50
5000	12"x12" plain, under 1,000 S.F.	30	20	50
5100	1,000 to 5,000 S.F.	29.50	16.95	46.45
5200	Over 5,000 S.F.	28.50	15.50	44
5300	Solar reflective, under 1,000 S.F.	42	27	69
5400	1,000 to 5,000 S.F.	41.50	22.50	64
5600	Over 5,000 S.F.	40	20.50	60.50
5800	3" thinline, 6"x6" plain, under 1,000 S.F.	27	29	56
5900	Over 5,000 S.F.	31	23.50	54.50
6000	Solar reflective, under 1,000 S.F.	38	39.50	77.50
6100	Over 5,000 S.F.	43.50	31.50	75
6200	8"x8" plain, under 1,000 S.F.	17.55	21.50	39.05
6300	Over 5,000 S.F.	17.55	16.95	34.50
6400	Solar reflective, under 1,000 S.F.	24.50	29	53.50
6500	Over 5,000 S.F.	24.50	22.50	47

B2010 Exterior Walls

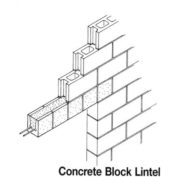

Concrete Block Lintel

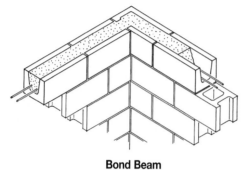

Bond Beam

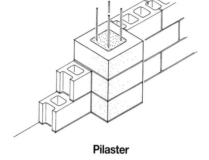

Pilaster

Concrete block specialties are divided into lintels, pilasters, and bond beams. Lintels are defined by span, thickness, height and wall loading. Span refers to the clear opening but the cost includes 8″ of bearing at both ends.

Bond beams and pilasters are defined by height thickness and weight of the masonry unit itself. Components for bond beams also include grout and reinforcing. For pilasters, components include four #5 reinforcing bars and type N mortar.

System Components	QUANTITY	UNIT	COST PER LINTEL		
			MAT.	INST.	TOTAL
SYSTEM B2010 144 3100					
CONCRETE BLOCK LINTEL, 6″ X 8″, LOAD 300 LB/L.F. 3′-4″ SPAN					
Lintel blocks, 6″ x 8″ x 8″	7.000	Ea.	11.20	27.37	38.57
Joint reinforcing, #3 & #4 steel bars, horizontal	3.120	Lb.	2.12	4.49	6.61
Grouting, 8″ deep, 6″ thick, .15 C.F./L.F.	.700	C.F.	5.11	6.45	11.56
Temporary shoring, lintel forms	1.000	Set	1.85	17.65	19.50
Temporary shoring, wood joists	1.000	Set		26	26
Temporary shoring, vertical members	1.000	Set		48	48
Mortar, masonry cement, 1:3 mix, type N	.270	C.F.	1.62	.99	2.61
TOTAL			21.90	130.95	152.85

B2010 144		Concrete Block Lintel						
	SPAN	THICKNESS (IN.)	HEIGHT (IN.)	WALL LOADING P.L.F.		COST PER LINTEL		
						MAT.	INST.	TOTAL
3100	3′-4″	6	8	300		22	131	153
3150		6	16	1,000		33	141	174
3200		8	8	300		23.50	138	161.50
3300		8	8	1,000		26	142	168
3400		8	16	1,000		33.50	151	184.50
3450	4′-0″	6	8	300		25.50	142	167.50
3500		6	16	1,000		44	161	205
3600		8	8	300		27.50	149	176.50
3700		8	16	1,000		38.50	163	201.50

B2010 Exterior Walls

B2010 144 — Concrete Block Lintel

	SPAN	THICKNESS (IN.)	HEIGHT (IN.)	WALL LOADING P.L.F.		COST PER LINTEL		
						MAT.	INST.	TOTAL
3800	4'-8"	6	8	300		28	144	172
3900		6	16	1,000		43	159	202
4000		8	8	300		31.50	159	190.50
4100		8	16	1,000		43.50	175	218.50
4200	5'-4"	6	8	300		34.50	160	194.50
4300		6	16	1,000		49	167	216
4400		8	8	300		37.50	175	212.50
4500		8	16	1,000		50	191	241
4600	6'-0"	6	8	300		41.50	194	235.50
4700		6	16	300		52.50	204	256.50
4800		6	16	1,000		55.50	211	266.50
4900		8	8	300		41.50	212	253.50
5000		8	16	1,000		56.50	232	288.50
5100	6'-8"	6	16	300		50	205	255
5200		6	16	1,000		53.50	213	266.50
5300		8	8	300		49	221	270
5400		8	16	1,000		62	245	307
5500	7'-4"	6	16	300		55	213	268
5600		6	16	1,000		71	230	301
5700		8	8	300		61	240	301
5800		8	16	300		75	258	333
5900		8	16	1,000		75	258	333
6000	8'-0"	6	16	300		59.50	223	282.50
6100		6	16	1,000		77	241	318
6200		8	16	300		84	272	356
6500		8	16	1,000		84	272	356

B2010 145 — Concrete Block Specialties

	TYPE	HEIGHT	THICKNESS	WEIGHT (P.C.F.)		COST PER L.F.		
						MAT.	INST.	TOTAL
7100	Bond beam	8	8	125		6.10	8.85	14.95
7200				105		6.90	8.75	15.65
7300			12	125		8.05	11.05	19.10
7400				105		8.65	10.90	19.55
7500	Pilaster	8	16	125		14.80	29	43.80
7600			20	125		17.80	35	52.80

B2010 Exterior Walls

B2010 153	Oversized Brick Curtain Wall	COST PER S.F.		
		MAT.	INST.	TOTAL
1000	Oversized brk curtain wall, 6"x4"x16", incl rebar, insul, bond beam, plain	25	11.80	36.80
1010	1 to 3 slots in face	31	14.35	45.35
1020	4 to 7 slots in face	35.50	16.05	51.55
1030	8" x 4" x 16", incl rebar, insul & bond beam, plain face	29	13.50	42.50
1040	1 to 3 slots in face	36.50	16.50	53
1050	4 to 7 slots in face	41	18.50	59.50
1060	6" x 8" x 16", incl rebar, insul & bond beam, plain face	30.50	10.65	41.15
1070	1 to 3 slots in face	38.50	12.90	51.40
1080	4 to 7 slots in face	44	14.40	58.40
1090	8" x 8" x 16", incl rebar, insul & bond beam, plain face	40.50	11.50	52
1100	1 to 3 slots in face	51	13.95	64.95
1110	4 to 7 slots in face	58	15.60	73.60

Same Data. Simplified.

Enjoy the convenience and efficiency of accessing your costs anywhere:

- **Skip the multiplier** by setting your location
- **Quickly search,** edit, favorite and share costs
- **Stay on top of price changes** with automatic updates

Discover more at rsmeans.com/online

C1010   Partitions

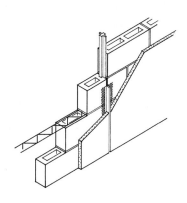

The Concrete Block Partition Systems are defined by weight and type of block, thickness, type of finish and number of sides finished. System components include joint reinforcing on alternate courses and vertical control joints.

System Components	QUANTITY	UNIT	COST PER S.F.		
			MAT.	INST.	TOTAL
SYSTEM C1010 102 1020					
CONC. BLOCK PARTITION, 8″ X 16″, 4″ TK., 2 CT. GYP. PLASTER 2 SIDES					
Conc. block partition, 4″ thick	1.000	S.F.	2.30	6.95	9.25
Control joint	.050	L.F.	.19	.17	.36
Horizontal joint reinforcing	.800	L.F.	.09	.08	.17
Gypsum plaster, 2 coat, on masonry	2.000	S.F.	1.48	6.61	8.09
TOTAL			4.06	13.81	17.87

C1010 102		Concrete Block Partitions - Regular Weight				COST PER S.F.		
	TYPE	THICKNESS (IN.)	TYPE FINISH	SIDES FINISHED		MAT.	INST.	TOTAL
1000	Hollow	4	none	0		2.58	7.20	9.78
1010			gyp. plaster 2 coat	1		3.32	10.50	13.82
1020				2		4.06	13.80	17.86
1100			lime plaster - 2 coat	1		3.04	10.50	13.54
1150			lime portland - 2 coat	1		3.06	10.50	13.56
1200			portland - 3 coat	1		3.10	10.95	14.05
1400			5/8″ drywall	1		3.28	9.65	12.93
1410				2		3.98	12.10	16.08
1500		6	none	0		3.26	7.75	11.01
1510			gyp. plaster 2 coat	1		4	11.05	15.05
1520				2		4.74	14.35	19.09
1600			lime plaster - 2 coat	1		3.72	11.05	14.77
1650			lime portland - 2 coat	1		3.74	11.05	14.79
1700			portland - 3 coat	1		3.78	11.50	15.28
1900			5/8″ drywall	1		3.96	10.20	14.16
1910				2		4.66	12.65	17.31
2000		8	none	0		3.22	8.30	11.52
2010			gyp. plaster 2 coat	1		3.96	11.60	15.56
2020			gyp. plaster 2 coat	2		4.70	14.90	19.60
2100			lime plaster - 2 coat	1		3.68	11.60	15.28
2150			lime portland - 2 coat	1		3.70	11.60	15.30
2200			portland - 3 coat	1		3.74	12.05	15.79
2400			5/8″ drywall	1		3.92	10.75	14.67
2410				2		4.62	13.20	17.82

C10 Interior Construction

C1010 Partitions

C1010 102	Concrete Block Partitions - Regular Weight							

	TYPE	THICKNESS (IN.)	TYPE FINISH	SIDES FINISHED		COST PER S.F.		
						MAT.	INST.	TOTAL
2500	Hollow	10	none	0		3.83	8.65	12.48
2510			gyp. plaster 2 coat	1		4.57	11.95	16.52
2520				2		5.30	15.25	20.55
2600			lime plaster - 2 coat	1		4.28	11.95	16.23
2650			lime portland - 2 coat	1		4.31	11.95	16.26
2700			portland - 3 coat	1		4.35	12.40	16.75
2900			5/8" drywall	1		4.53	11.10	15.63
2910				2		5.25	13.55	18.80
3000	Solid	2	none	0		2.11	7.15	9.26
3010			gyp. plaster	1		2.85	10.45	13.30
3020				2		3.59	13.75	17.34
3100			lime plaster - 2 coat	1		2.57	10.45	13.02
3150			lime portland - 2 coat	1		2.59	10.45	13.04
3200			portland - 3 coat	1		2.63	10.90	13.53
3400			5/8" drywall	1		2.81	9.60	12.41
3410				2		3.51	12.05	15.56
3500		4	none	0		3.06	7.45	10.51
3510			gyp. plaster	1		3.90	10.80	14.70
3520				2		4.54	14.05	18.59
3600			lime plaster - 2 coat	1		3.52	10.75	14.27
3650			lime portland - 2 coat	1		3.54	10.75	14.29
3700			portland - 3 coat	1		3.58	11.20	14.78
3900			5/8" drywall	1		3.76	9.90	13.66
3910				2		4.46	12.35	16.81
4000		6	none	0		4.18	8.05	12.23
4010			gyp. plaster	1		4.92	11.35	16.27
4020				2		5.65	14.65	20.30
4100			lime plaster - 2 coat	1		4.64	11.35	15.99
4150			lime portland - 2 coat	1		4.66	11.35	16.01
4200			portland - 3 coat	1		4.70	11.80	16.50
4400			5/8" drywall	1		4.88	10.50	15.38
4410				2		5.60	12.95	18.55

C1010 104	Concrete Block Partitions - Lightweight							

	TYPE	THICKNESS (IN.)	TYPE FINISH	SIDES FINISHED		COST PER S.F.		
						MAT.	INST.	TOTAL
5000	Hollow	4	none	0		2.48	7.05	9.53
5010			gyp. plaster	1		3.22	10.35	13.57
5020				2		3.96	13.65	17.61
5100			lime plaster - 2 coat	1		2.94	10.35	13.29
5150			lime portland - 2 coat	1		2.96	10.35	13.31
5200			portland - 3 coat	1		3	10.80	13.80
5400			5/8" drywall	1		3.18	9.50	12.68
5410				2		3.88	11.95	15.83
5500		6	none	0		3.69	7.55	11.24
5510			gyp. plaster	1		4.43	10.85	15.28
5520			gyp. plaster	2		5.15	14.15	19.30
5600			lime plaster - 2 coat	1		4.15	10.85	15
5650			lime portland - 2 coat	1		4.17	10.85	15.02
5700			portland - 3 coat	1		4.21	11.30	15.51
5900			5/8" drywall	1		4.39	10	14.39
5910				2		5.10	12.45	17.55

For customer support on your Concrete & Masonry Costs with RSMeans Data, call 800.448.8182.

C1010 Partitions

C1010 104	Concrete Block Partitions - Lightweight

	TYPE	THICKNESS (IN.)	TYPE FINISH	SIDES FINISHED		COST PER S.F.		
						MAT.	INST.	TOTAL
6000	Hollow	8	none	0		4.69	8.10	12.79
6010			gyp. plaster	1		5.45	11.40	16.85
6020				2		6.15	14.70	20.85
6100			lime plaster - 2 coat	1		5.15	11.40	16.55
6150			lime portland - 2 coat	1		5.15	11.40	16.55
6200			portland - 3 coat	1		5.20	11.85	17.05
6400			5/8" drywall	1		5.40	10.55	15.95
6410				2		6.10	13	19.10
6500		10	none	0		4.96	8.45	13.41
6510			gyp. plaster	1		5.70	11.75	17.45
6520				2		6.45	15.05	21.50
6600			lime plaster - 2 coat	1		5.40	11.75	17.15
6650			lime portland - 2 coat	1		5.45	11.75	17.20
6700			portland - 3 coat	1		5.50	12.20	17.70
6900			5/8" drywall	1		5.65	10.90	16.55
6910				2		6.35	13.35	19.70
7000	Solid	4	none	0		2.62	7.40	10.02
7010			gyp. plaster	1		3.36	10.70	14.06
7020				2		4.10	14	18.10
7100			lime plaster - 2 coat	1		3.08	10.70	13.78
7150			lime portland - 2 coat	1		3.10	10.70	13.80
7200			portland - 3 coat	1		3.14	11.15	14.29
7400			5/8" drywall	1		3.32	9.85	13.17
7410				2		4.02	12.30	16.32
7500		6	none	0		4.10	7.95	12.05
7510			gyp. plaster	1		4.99	11.35	16.34
7520				2		5.60	14.55	20.15
7600			lime plaster - 2 coat	1		4.56	11.25	15.81
7650			lime portland - 2 coat	1		4.58	11.25	15.83
7700			portland - 3 coat	1		4.62	11.70	16.32
7900			5/8" drywall	1		4.80	10.40	15.20
7910				2		5.50	12.85	18.35
8000		8	none	0		4.01	8.50	12.51
8010			gyp. plaster	1		4.75	11.80	16.55
8020				2		5.50	15.10	20.60
8100			lime plaster - 2 coat	1		4.47	11.80	16.27
8150			lime portland - 2 coat	1		4.49	11.80	16.29
8200			portland - 3 coat	1		4.53	12.25	16.78
8400			5/8" drywall	1		4.71	10.95	15.66
8410				2		5.40	13.40	18.80

C1010 Partitions

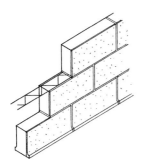

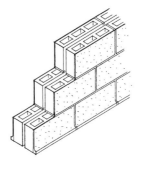

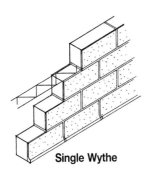

Single Wythe

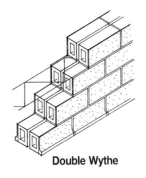

Double Wythe

C1010 120	Tile Partitions	COST PER S.F.		
		MAT.	INST.	TOTAL
1000	8W series 8"x16", 4" thick wall, reinf every 2 courses, glazed 1 side	14.85	8.70	23.55
1100	Glazed 2 sides	15.85	9.20	25.05
1200	Glazed 2 sides, using 2 wythes of 2" thick tile	21	16.60	37.60
1300	6" thick wall, horizontal reinf every 2 courses, glazed 1 side	34.50	9.10	43.60
1400	Glazed 2 sides, each face different color, 2" and 4" tile	25.50	17	42.50
1500	8" thick wall, glazed 2 sides using 2 wythes of 4" thick tile	29.50	17.40	46.90
1600	10" thick wall, glazed 2 sides using 1 wythe of 4" & 1 wythe of 6" tile	49.50	17.80	67.30
1700	Glazed 2 sides cavity wall, using 2 wythes of 4" thick tile	29.50	17.40	46.90
1800	12" thick wall, glazed 2 sides using 2 wythes of 6" thick tile	69	18.20	87.20
1900	Glazed 2 sides cavity wall, using 2 wythes of 4" thick tile	29.50	17.40	46.90
2100	6T series 5-1/3"x12" tile, 4" thick, non load bearing glazed one side	14.40	13.60	28
2200	Glazed two sides	19.20	15.35	34.55
2300	Glazed two sides, using two wythes of 2" thick tile	22	26.50	48.50
2400	6" thick, glazed one side	20.50	14.25	34.75
2500	Glazed two sides	24.50	16.20	40.70
2600	Glazed two sides using 2" thick tile and 4" thick tile	25.50	27	52.50
2700	8" thick, glazed one side	25	16.65	41.65
2800	Glazed two sides using two wythes of 4" thick tile	29	27	56
2900	Glazed two sides using 6" thick tile and 2" thick tile	31.50	27.50	59
3000	10" thick cavity wall, glazed two sides using two wythes of 4" tile	29	27	56
3100	12" thick, glazed two sides using 4" thick tile and 8" thick tile	39.50	30.50	70
3200	2" thick facing tile, glazed one side, on 6" concrete block	13.50	15.80	29.30
3300	On 8" concrete block	13.40	16.30	29.70
3400	On 10" concrete block	14	16.60	30.60

377

C2010 Stair Construction

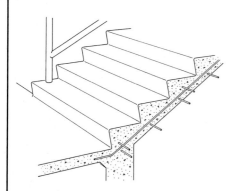

General Design: See reference section for code requirements. Maximum height between landings is 12'; usual stair angle is 20° to 50° with 30° to 35° best. Usual relation of riser to treads is:

Riser + tread = 17.5.
2x (Riser) + tread = 25.
Riser x tread = 70 or 75.

Maximum riser height is 7" for commercial, 8-1/4" for residential.
Usual riser height is 6-1/2" to 7-1/4".

Minimum tread width is 11" for commercial and 9" for residential.

For additional information please see reference section.

Cost Per Flight: Table below lists the cost per flight for 4'-0" wide stairs.
Side walls are not included.
Railings are included.

System Components	QUANTITY	UNIT	COST PER FLIGHT MAT.	COST PER FLIGHT INST.	COST PER FLIGHT TOTAL
SYSTEM C2010 110 0560					
STAIRS, C.I.P. CONCRETE WITH LANDING, 12 RISERS					
Concrete in place, free standing stairs not incl. safety treads	48.000	L.F.	400.80	2,225.28	2,626.08
Concrete in place, free standing stair landing	32.000	S.F.	214.40	616	830.40
Cast alum nosing insert, abr surface, pre-drilled, 3" wide x 4' long	12.000	Ea.	984	235.20	1,219.20
Industrial railing, welded, 2 rail 3'-6" high 1-1/2" pipe	18.000	L.F.	801	223.92	1,024.92
Wall railing with returns, steel pipe	17.000	L.F.	328.95	211.48	540.43
TOTAL			2,729.15	3,511.88	6,241.03

C2010 110	Stairs	COST PER FLIGHT MAT.	COST PER FLIGHT INST.	COST PER FLIGHT TOTAL
0470	Stairs, C.I.P. concrete, w/o landing, 12 risers, w/o nosing	1,525	2,650	4,175
0480	With nosing	2,525	2,900	5,425
0550	W/landing, 12 risers, w/o nosing	1,750	3,275	5,025
0560	With nosing	2,725	3,500	6,225
0570	16 risers, w/o nosing	2,125	4,125	6,250
0580	With nosing	3,450	4,425	7,875
0590	20 risers, w/o nosing	2,525	4,950	7,475
0600	With nosing	4,175	5,350	9,525
0610	24 risers, w/o nosing	2,900	5,800	8,700
0620	With nosing	4,875	6,275	11,150

G1030 Site Earthwork

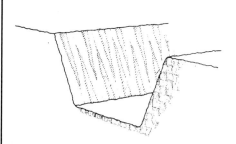

Trenching Systems are shown on a cost per linear foot basis. The systems include: excavation; backfill and removal of spoil; and compaction for various depths and trench bottom widths. The backfill has been reduced to accommodate a pipe of suitable diameter and bedding.

The slope for trench sides varies from none to 1:1.

The Expanded System Listing shows Trenching Systems that range from 2' to 12' in width. Depths range from 2' to 25'.

System Components			COST PER L.F.		
	QUANTITY	UNIT	EQUIP.	LABOR	TOTAL
SYSTEM G1030 805 1310					
TRENCHING, COMMON EARTH, NO SLOPE, 2' WIDE, 2' DP, 3/8 C.Y. BUCKET					
Excavation, trench, hyd. backhoe, track mtd., 3/8 C.Y. bucket	.148	B.C.Y.	.24	1.21	1.45
Backfill and load spoil, from stockpile	.153	L.C.Y.	.13	.37	.50
Compaction by vibrating plate, 6" lifts, 4 passes	.118	E.C.Y.	.03	.45	.48
Remove excess spoil, 8 C.Y. dump truck, 2 mile roundtrip	.040	L.C.Y.	.15	.20	.35
TOTAL			.55	2.23	2.78

G1030 805	Trenching Common Earth	COST PER L.F.		
		EQUIP.	LABOR	TOTAL
1310	Trenching, common earth, no slope, 2' wide, 2' deep, 3/8 C.Y. bucket	.54	2.24	2.78
1330	4' deep, 3/8 C.Y. bucket	.98	4.48	5.46
1400	4' wide, 2' deep, 3/8 C.Y. bucket	1.28	4.45	5.73
1420	4' deep, 1/2 C.Y. bucket	1.94	7.45	9.39
1800	1/2 to 1 slope, 2' wide, 2' deep, 3/8 C.Y. bucket	.76	3.36	4.12
1820	4' deep, 3/8 C.Y. bucket	1.85	8.95	10.80
1860	8' deep, 1/2 C.Y. bucket	4.78	23	27.78
2300	4' wide, 2' deep, 3/8 C.Y. bucket	1.49	5.55	7.04
2320	4' deep, 1/2 C.Y. bucket	2.72	11.35	14.07
2360	8' deep, 1/2 C.Y. bucket	13.10	31.50	44.60
2400	12' deep, 1 C.Y. bucket	24	59.50	83.50
2840	6' wide, 6' deep, 5/8 C.Y. bucket w/trench box	13.15	26.50	39.65
2900	12' deep, 1-1/2 C.Y. bucket	19.75	53	72.75
3020	24' deep, 3-1/2 C.Y. bucket	73	143	216
3100	8' wide, 12' deep, 1-1/2 C.Y. bucket w/trench box	24.50	60.50	85
3500	1 to 1 slope, 2' wide, 2' deep, 3/8 C.Y. bucket	.98	4.48	5.46
3540	4' deep, 3/8 C.Y. bucket	2.72	13.45	16.17
3580	8' deep, 1/2 C.Y. bucket	5.95	29	34.95
3800	4' wide, 2' deep, 3/8 C.Y. bucket	1.71	6.70	8.41
3840	4' deep, 1/2 C.Y. bucket	3.50	15.20	18.70
3880	8' deep, 1/2 C.Y. bucket	19.50	47	66.50
3920	12' deep, 1 C.Y. bucket	41.50	99.50	141
4030	6' wide, 6' deep, 5/8 C.Y. bucket w/trench box	17.30	35.50	52.80
4060	12' deep, 1-1/2 C.Y. bucket	30	81	111
4090	24' deep, 3-1/2 C.Y. bucket	121	240	361
4500	8' wide, 12' deep, 1-1/2 C.Y. bucket w/trench box	34.50	88	122.50
4950	12' wide, 20' deep, 3-1/2 C.Y. bucket w/trench box	109	207	316

G10 Site Preparation

G1030 Site Earthwork

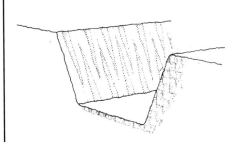

Trenching Systems are shown on a cost per linear foot basis. The systems include: excavation; backfill and removal of spoil; and compaction for various depths and trench bottom widths. The backfill has been reduced to accommodate a pipe of suitable diameter and bedding.

The slope for trench sides varies from none to 1:1.

The Expanded System Listing shows Trenching Systems that range from 2' to 12' in width. Depths range from 2' to 25'.

System Components	QUANTITY	UNIT	COST PER L.F.		
			EQUIP.	LABOR	TOTAL
SYSTEM G1030 807 1310					
TRENCHING, SAND & GRAVEL, NO SLOPE, 2' WIDE, 2' DEEP, 3/8 C.Y. BUCKET					
Excavation, trench, hyd. backhoe, track mtd., 3/8 C.Y. bucket	.148	B.C.Y.	.21	1.10	1.31
Backfill and load spoil, from stockpile	.140	L.C.Y.	.12	.34	.46
Compaction by vibrating plate 18" wide, 6" lifts, 4 passes	.118	E.C.Y.	.03	.45	.48
Remove excess spoil, 8 C.Y. dump truck, 2 mile roundtrip	.035	L.C.Y.	.13	.18	.31
TOTAL			.49	2.07	2.56

G1030 807	Trenching Sand & Gravel	COST PER L.F.		
		EQUIP.	LABOR	TOTAL
1310	Trenching, sand & gravel, no slope, 2' wide, 2' deep, 3/8 C.Y. bucket	.49	2.07	2.56
1320	3' deep, 3/8 C.Y. bucket	.75	3.45	4.20
1330	4' deep, 3/8 C.Y. bucket	.89	4.15	5.05
1340	6' deep, 3/8 C.Y. bucket	1.72	4.97	6.70
1350	8' deep, 1/2 C.Y. bucket	2.26	6.60	8.85
1360	10' deep, 1 C.Y. bucket	2.52	6.95	9.45
1400	4' wide, 2' deep, 3/8 C.Y. bucket	1.16	4.10	5.25
1410	3' deep, 3/8 C.Y. bucket	1.56	6.20	7.75
1420	4' deep, 1/2 C.Y. bucket	1.78	6.95	8.75
1430	6' deep, 1/2 C.Y. bucket	4.07	10.30	14.40
1440	8' deep, 1/2 C.Y. bucket	6.20	14.55	20.50
1450	10' deep, 1 C.Y. bucket	6	14.75	21
1460	12' deep, 1 C.Y. bucket	7.60	18.45	26
1470	15' deep, 1-1/2 C.Y. bucket	7.50	21.50	29
1480	18' deep, 2-1/2 C.Y. bucket	10.55	23.50	34
1520	6' wide, 6' deep, 5/8 C.Y. bucket w/trench box	8.25	16.65	25
1530	8' deep, 3/4 C.Y. bucket	10.75	22	32.50
1540	10' deep, 1 C.Y. bucket	10.05	22	32
1550	12' deep, 1-1/2 C.Y. bucket	9.60	24.50	34.50
1560	16' deep, 2 C.Y. bucket	15.50	31	46.50
1570	20' deep, 3-1/2 C.Y. bucket	17.90	36.50	54.50
1580	24' deep, 3-1/2 C.Y. bucket	22.50	44.50	67
1640	8' wide, 12' deep, 1-1/2 C.Y. bucket w/trench box	13.45	31.50	45
1650	15' deep, 1-1/2 C.Y. bucket	16.40	40	56
1660	18' deep, 2-1/2 C.Y. bucket	23	44.50	68
1680	24' deep, 3-1/2 C.Y. bucket	30.50	57.50	88
1730	10' wide, 20' deep, 3-1/2 C.Y. bucket w/trench box	30.50	58	88
1740	24' deep, 3-1/2 C.Y. bucket	38	71.50	109
1780	12' wide, 20' deep, 3-1/2 C.Y. bucket w/trench box	36.50	68.50	105
1790	25' deep, 3-1/2 C.Y. bucket	47.50	88.50	136
1800	1/2:1 slope, 2' wide, 2' deep, 3/8 C.Y. bucket	.69	3.11	3.80
1810	3' deep, 3/8 C.Y. bucket	1.14	5.45	6.60
1820	4' deep, 3/8 C.Y. bucket	1.68	8.35	10
1840	6' deep, 3/8 C.Y. bucket	4.16	12.45	16.60

G1030 Site Earthwork

G1030 807	Trenching Sand & Gravel	COST PER L.F.		
		EQUIP.	LABOR	TOTAL
1860	8' deep, 1/2 C.Y. bucket	6.60	19.90	26.50
1880	10' deep, 1 C.Y. bucket	8.65	24.50	33
2300	4' wide, 2' deep, 3/8 C.Y. bucket	1.36	5.15	6.50
2310	3' deep, 3/8 C.Y. bucket	2.01	8.55	10.55
2320	4' deep, 1/2 C.Y. bucket	2.49	10.60	13.05
2340	6' deep, 1/2 C.Y. bucket	6.95	18.35	25.50
2360	8' deep, 1/2 C.Y. bucket	12.05	29.50	41.50
2380	10' deep, 1 C.Y. bucket	13.30	34	47
2400	12' deep, 1 C.Y. bucket	22.50	56.50	78.50
2430	15' deep, 1-1/2 C.Y. bucket	21.50	62.50	84
2460	18' deep, 2-1/2 C.Y. bucket	42	95	137
2840	6' wide, 6' deep, 5/8 C.Y. bucket w/trench box	12.05	25.50	37.50
2860	8' deep, 3/4 C.Y. bucket	17.45	37	54.50
2880	10' deep, 1 C.Y. bucket	18.15	41.50	59.50
2900	12' deep, 1-1/2 C.Y. bucket	18.80	50.50	69.50
2940	16' deep, 2 C.Y. bucket	35.50	74	110
2980	20' deep, 3-1/2 C.Y. bucket	47.50	99.50	147
3020	24' deep, 3-1/2 C.Y. bucket	67	137	204
3100	8' wide, 12' deep, 1-1/4 C.Y. bucket w/trench box	22.50	57.50	80
3120	15' deep, 1-1/2 C.Y. bucket	31	80.50	111
3140	18' deep, 2-1/2 C.Y. bucket	48.50	99	148
3180	24' deep, 3-1/2 C.Y. bucket	75	150	224
3270	10' wide, 20' deep, 3-1/2 C.Y. bucket w/trench box	60	121	181
3280	24' deep, 3-1/2 C.Y. bucket	82.50	163	246
3370	12' wide, 20' deep, 3-1/2 C.Y. bucket w/trench box	67	134	201
3380	25' deep, 3-1/2 C.Y. bucket	96	188	284
3500	1:1 slope, 2' wide, 2' deep, 3/8 C.Y. bucket	1.76	5.25	7
3520	3' deep, 3/8 C.Y. bucket	1.58	7.80	9.40
3540	4' deep, 3/8 C.Y. bucket	2.48	12.50	15
3560	6' deep, 3/8 C.Y. bucket	4.16	12.45	16.65
3580	8' deep, 1/2 C.Y. bucket	10.95	33	44
3600	10' deep, 1 C.Y. bucket	14.80	42	57
3800	4' wide, 2' deep, 3/8 C.Y. bucket	1.56	6.20	7.75
3820	3' deep, 3/8 C.Y. bucket	2.45	10.90	13.35
3840	4' deep, 1/2 C.Y. bucket	3.21	14.20	17.40
3860	6' deep, 1/2 C.Y. bucket	9.80	26.50	36
3880	8' deep, 1/2 C.Y. bucket	17.95	44.50	62
3900	10' deep, 1 C.Y. bucket	20.50	53	73.50
3920	12' deep, 1 C.Y. bucket	30	75	105
3940	15' deep, 1-1/2 C.Y. bucket	35	104	139
3960	18' deep, 2-1/2 C.Y. bucket	57.50	131	188
4030	6' wide, 6' deep, 5/8 C.Y. bucket w/trench box	15.85	34	49.50
4040	8' deep, 3/4 C.Y. bucket	24	52.50	76.50
4050	10' deep, 1 C.Y. bucket	26	61	87
4060	12' deep, 1-1/2 C.Y. bucket	28	76.50	104
4070	16' deep, 2 C.Y. bucket	55.50	117	173
4080	20' deep, 3-1/2 C.Y. bucket	76.50	163	239
4090	24' deep, 3-1/2 C.Y. bucket	111	228	340
4500	8' wide, 12' deep, 1-1/2 C.Y. bucket w/trench box	32	83.50	115
4550	15' deep, 1-1/2 C.Y. bucket	45	121	166
4600	18' deep, 2-1/2 C.Y. bucket	73.50	153	227
4650	24' deep, 3-1/2 C.Y. bucket	119	241	360
4800	10' wide, 20' deep, 3-1/2 C.Y. bucket w/trench box	89.50	184	273
4850	24' deep, 3-1/2 C.Y. bucket	127	255	380
4950	12' wide, 20' deep, 3-1/2 C.Y. bucket w/trench box	95.50	195	290
4980	25' deep, 3-1/2 C.Y. bucket	144	288	430

G1030 Site Earthwork

The Pipe Bedding System is shown for various pipe diameters. Compacted bank sand is used for pipe bedding and to fill 12″ over the pipe. No backfill is included. Various side slopes are shown to accommodate different soil conditions. Pipe sizes vary from 6″ to 84″ diameter.

System Components	QUANTITY	UNIT	COST PER L.F.		
			MAT.	INST.	TOTAL
SYSTEM G1030 815 1440					
PIPE BEDDING, SIDE SLOPE 0 TO 1, 1′ WIDE, PIPE SIZE 6″ DIAMETER					
Borrow, bank sand, 2 mile haul, machine spread	.086	C.Y.	1.75	.73	2.48
Compaction, vibrating plate	.086	C.Y.		.25	.25
TOTAL			1.75	.98	2.73

G1030 815	Pipe Bedding	COST PER L.F.		
		MAT.	INST.	TOTAL
1440	Pipe bedding, side slope 0 to 1, 1′ wide, pipe size 6″ diameter	1.75	.97	2.72
1460	2′ wide, pipe size 8″ diameter	3.79	2.11	5.90
1480	Pipe size 10″ diameter	3.87	2.15	6.02
1500	Pipe size 12″ diameter	3.96	2.20	6.16
1520	3′ wide, pipe size 14″ diameter	6.40	3.57	9.97
1540	Pipe size 15″ diameter	6.50	3.60	10.10
1560	Pipe size 16″ diameter	6.55	3.64	10.19
1580	Pipe size 18″ diameter	6.65	3.71	10.36
1600	4′ wide, pipe size 20″ diameter	9.45	5.25	14.70
1620	Pipe size 21″ diameter	9.55	5.30	14.85
1640	Pipe size 24″ diameter	9.75	5.45	15.20
1660	Pipe size 30″ diameter	9.95	5.55	15.50
1680	6′ wide, pipe size 32″ diameter	17	9.45	26.45
1700	Pipe size 36″ diameter	17.40	9.70	27.10
1720	7′ wide, pipe size 48″ diameter	27.50	15.40	42.90
1740	8′ wide, pipe size 60″ diameter	33.50	18.75	52.25
1760	10′ wide, pipe size 72″ diameter	47	26	73
1780	12′ wide, pipe size 84″ diameter	62	34.50	96.50
2140	Side slope 1/2 to 1, 1′ wide, pipe size 6″ diameter	3.27	1.82	5.09
2160	2′ wide, pipe size 8″ diameter	5.55	3.09	8.64
2180	Pipe size 10″ diameter	5.95	3.32	9.27
2200	Pipe size 12″ diameter	6.30	3.51	9.81
2220	3′ wide, pipe size 14″ diameter	9.10	5.05	14.15
2240	Pipe size 15″ diameter	9.35	5.20	14.55
2260	Pipe size 16″ diameter	9.55	5.30	14.85
2280	Pipe size 18″ diameter	10.05	5.60	15.65
2300	4′ wide, pipe size 20″ diameter	13.30	7.40	20.70
2320	Pipe size 21″ diameter	13.55	7.55	21.10
2340	Pipe size 24″ diameter	14.45	8.05	22.50
2360	Pipe size 30″ diameter	16	8.90	24.90
2380	6′ wide, pipe size 32″ diameter	23.50	13.15	36.65
2400	Pipe size 36″ diameter	25	13.95	38.95
2420	7′ wide, pipe size 48″ diameter	39	22	61
2440	8′ wide, pipe size 60″ diameter	49.50	27.50	77
2460	10′ wide, pipe size 72″ diameter	68	38	106
2480	12′ wide, pipe size 84″ diameter	89.50	49.50	139
2620	Side slope 1 to 1, 1′ wide, pipe size 6″ diameter	4.79	2.66	7.45
2640	2′ wide, pipe size 8″ diameter	7.40	4.10	11.50

383

G1030 Site Earthwork

G1030 815	Pipe Bedding	COST PER L.F.		
		MAT.	INST.	TOTAL
2660	Pipe size 10" diameter	8	4.46	12.46
2680	Pipe size 12" diameter	8.75	4.86	13.61
2700	3' wide, pipe size 14" diameter	11.85	6.60	18.45
2720	Pipe size 15" diameter	12.20	6.80	19
2740	Pipe size 16" diameter	12.60	7	19.60
2760	Pipe size 18" diameter	13.45	7.50	20.95
2780	4' wide, pipe size 20" diameter	17.10	9.50	26.60
2800	Pipe size 21" diameter	17.55	9.75	27.30
2820	Pipe size 24" diameter	19.05	10.60	29.65
2840	Pipe size 30" diameter	22	12.25	34.25
2860	6' wide, pipe size 32" diameter	30	16.80	46.80
2880	Pipe size 36" diameter	33	18.25	51.25
2900	7' wide, pipe size 48" diameter	50.50	28	78.50
2920	8' wide, pipe size 60" diameter	66	36.50	102.50
2940	10' wide, pipe size 72" diameter	89.50	50	139.50
2960	12' wide, pipe size 84" diameter	117	65	182

G2030 Pedestrian Paving

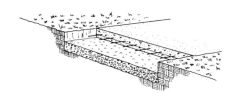

The Concrete Sidewalk System includes: excavation; compacted gravel base (hand graded); forms; welded wire fabric; and 3,000 p.s.i. air-entrained concrete (broom finish).

The Expanded System Listing shows Concrete Sidewalk systems with wearing course depths ranging from 4″ to 6″. The gravel base ranges from 4″ to 8″. Sidewalk widths are shown ranging from 3′ to 5′. Costs are on a linear foot basis.

System Components	QUANTITY	UNIT	COST PER L.F.		
			MAT.	INST.	TOTAL
SYSTEM G2030 120 1580					
CONCRETE SIDEWALK 4″ THICK, 4″ GRAVEL BASE, 3′ WIDE					
Excavation, box out with dozer	.100	B.C.Y.		.24	.24
Gravel base, haul 2 miles, spread with dozer	.037	L.C.Y.	1.20	.32	1.52
Compaction with vibrating plate	.037	E.C.Y.		.11	.11
Fine grade by hand	.333	S.Y.		4.23	4.23
Concrete in place including forms and reinforcing	.037	C.Y.	7.35	8.94	16.29
Backfill edges by hand	.010	L.C.Y.		.38	.38
TOTAL			8.55	14.22	22.77

G2030 120	Concrete Sidewalks	COST PER L.F.		
		MAT.	INST.	TOTAL
1580	Concrete sidewalk, 4″ thick, 4″ gravel base, 3′ wide	8.55	14.20	22.75
1600	4′ wide	11.40	17.40	28.80
1620	5′ wide	14.25	20.50	34.75
1640	6″ gravel base, 3′ wide	9.15	14.45	23.60
1660	4′ wide	12.20	17.70	29.90
1680	5′ wide	15.25	21	36.25
1700	8″ gravel base, 3′ wide	9.75	14.70	24.45
1720	4′ wide	13	18.05	31.05
1740	5′ wide	16.25	21.50	37.75
1800	5″ thick concrete, 4″ gravel base, 3′ wide	10.30	15.35	25.65
1820	4′ wide	13.75	18.85	32.60
1840	5′ wide	17.20	22.50	39.70
1860	6″ gravel base, 3′ wide	10.95	15.60	26.55
1880	4′ wide	14.55	19.20	33.75
1900	5′ wide	18.20	22	40.20
1920	8″ gravel base, 3′ wide	11.55	15.85	27.40
1940	4′ wide	15.40	20.50	35.90
1960	5′ wide	19.20	22.50	41.70
2120	6″ thick concrete, 4″ gravel base, 3′ wide	11.85	16.15	28
2140	4′ wide	15.80	19.90	35.70
2160	5′ wide	19.75	23.50	43.25
2180	6″ gravel base, 3′ wide	12.45	16.45	28.90
2200	4′ wide	16.60	20	36.60
2220	5′ wide	21	24	45
2240	8″ gravel base, 3′ wide	13.05	16.70	29.75
2260	4′ wide	17.40	20.50	37.90
2280	5′ wide	22	24.50	46.50

G2030 Pedestrian Paving

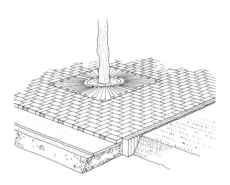

The Plaza Systems listed include several brick and tile paving surfaces on three different bases: gravel, slab on grade and suspended slab. The system cost includes this base cost with the exception of the suspended slab. The type of bedding for the pavers depends on the base being used, and alternate bedding may be desirable. Also included in the paving costs are edging and precast grating costs and where concrete bases are involved, expansion joints.

System Components	QUANTITY	UNIT	COST PER S.F. MAT.	COST PER S.F. INST.	COST PER S.F. TOTAL
SYSTEM G2030 150 2050					
PLAZA, BRICK PAVERS, 4" X 8" X 1-3/4", GRAVEL BASE, STONE DUST BED					
Compact subgrade, static roller, 4 passes	.111	S.Y.		.05	.05
Bank gravel, 2 mi haul, dozer spread	.012	L.C.Y.	.40	.11	.51
Compact gravel bedding or base, vibrating plate	.012	E.C.Y.		.04	.04
Grading fine grade, 3 passes with grader	.111	S.Y.		.67	.67
Coarse washed sand bed, 1" thick	.003	C.Y.	.06	.08	.14
Brick paver, 4" x 8" x 1-3/4"	4.150	Ea.	4.59	4.89	9.48
Brick edging, stood on end, 3 per L.F.	.060	L.F.	.37	.73	1.10
Concrete tree grate, 5' square	.004	Ea.	1.83	.31	2.14
TOTAL			7.25	6.88	14.13

G2030 150	Brick & Tile Plazas	COST PER S.F. MAT.	COST PER S.F. INST.	COST PER S.F. TOTAL
1050	Plaza, asphalt pavers, 6" x 12" x 1-1/4", gravel base, asphalt bedding	13.80	12.60	26.40
1100	Slab on grade, asphalt bedding	16.25	13.65	29.90
1150	Suspended slab, insulated & mastic bedding	18.45	16.75	35.20
1300	6" x 12" x 3", gravel base, asphalt, bedding	19	13.05	32.05
1350	Slab on grade, asphalt bedding	20.50	14	34.50
1400	Suspended slab, insulated & mastic bedding	23	17.10	40.10
2050	Brick pavers, 4" x 8" x 1-3/4", gravel base, stone dust bedding	7.25	6.85	14.10
2100	Slab on grade, asphalt bedding	9.80	9.85	19.65
2150	Suspended slab, insulated & no bedding	11.95	12.95	24.90
2300	4" x 8" x 2-1/4", gravel base, stone dust bedding	6	6.85	12.85
2350	Slab on grade, asphalt bedding	8.50	9.85	18.35
2400	Suspended slab, insulated & no bedding	10.70	12.95	23.65
2550	Shale pavers, 4" x 8" x 2-1/4", gravel base, stone dust bedding	6.80	7.80	14.60
2600	Slab on grade, asphalt bedding	9.30	10.80	20.10
2650	Suspended slab, insulated & no bedding	6.25	9.05	15.30
3050	Thin set tile, 4" x 4" x 3/8", slab on grade	9.60	10.95	20.55
3300	4" x 4" x 3/4", slab on grade	11.05	7.30	18.35
3550	Concrete paving stone, 4" x 8" x 2-1/2", gravel base, sand bedding	5.30	4.85	10.15
3600	Slab on grade, asphalt bedding	7.25	7.20	14.45
3650	Suspended slab, insulated & no bedding	4.19	5.45	9.64
3800	4" x 8" x 3-1/4", gravel base, sand bedding	5.30	4.85	10.15
3850	Slab on grade, asphalt bedding	6.55	5.60	12.15
3900	Suspended slab, insulated & no bedding	4.19	5.45	9.64
4050	Concrete patio blocks, 8" x 16" x 2", gravel base, sand bedding	15.10	6.30	21.40
4100	Slab on grade, asphalt bedding	17.35	9.10	26.45
4150	Suspended slab, insulated & no bedding	14	6.90	20.90

G2030 Pedestrian Paving

G2030 150	Brick & Tile Plazas	COST PER S.F.		
		MAT.	INST.	TOTAL
4300	16" x 16" x 2", gravel base, sand bedding	20	5.05	25.05
4350	Slab on grade, asphalt bedding	22	6.30	28.30
4400	Suspended slab, insulated & no bedding	19.40	6.15	25.55
4550	24" x 24" x 2", gravel base, sand bedding	24.50	3.87	28.37
4600	Slab on grade, asphalt bedding	27	6.70	33.70
4650	Suspended slab, insulated & no bedding	24	4.95	28.95
5050	Bluestone flagging, 3/4" thick irregular, gravel base, sand bedding	15.40	14.90	30.30
5100	Slab on grade, mastic bedding	22	20.50	42.50
5300	1" thick, irregular, gravel base, sand bedding	13.50	16.35	29.85
5350	Slab on grade, mastic bedding	19.95	21.50	41.45
5550	Flagstone, 3/4" thick irregular, gravel base, sand bedding	25.50	14.20	39.70
5600	Slab on grade, mastic bedding	32	19.60	51.60
5800	1-1/2" thick, random rectangular, gravel base, sand bedding	27	16.35	43.35
5850	Slab on grade, mastic bedding	33.50	21.50	55
6050	Granite pavers, 3-1/2" x 3-1/2" x 3-1/2", gravel base, sand bedding	27	15	42
6100	Slab on grade, mortar bedding	30.50	24	54.50
6300	4" x 4" x 4", gravel base, sand bedding	28	14.60	42.60
6350	Slab on grade, mortar bedding	30.50	19.50	50
6550	4" x 12" x 4", gravel base, sand bedding	23	14.25	37.25
6600	Slab on grade, mortar bedding	25.50	19.15	44.65
6800	6" x 15" x 4", gravel base, sand bedding	13.75	13.45	27.20
6850	Slab on grade, mortar bedding	16.05	18.35	34.40
7050	Limestone, 3" thick, gravel base, sand bedding	14.05	18.55	32.60
7100	Slab on grade, mortar bedding	16.35	23.50	39.85
7300	4" thick, gravel base, sand bedding	17.60	19	36.60
7350	Slab on grade, mortar bedding	19.90	24	43.90
7550	5" thick, gravel base, sand bedding	21	19.50	40.50
7600	Slab on grade, mortar bedding	23.50	24.50	48
8050	Slate flagging, 3/4" thick, gravel base, sand bedding	13.85	15	28.85
8100	Slab on grade, mastic bedding	29	21.50	50.50
8300	1" thick, gravel base, sand bedding	15.60	16.05	31.65
8350	Slab on grade, mastic bedding	22.50	22	44.50

G2030 Pedestrian Paving

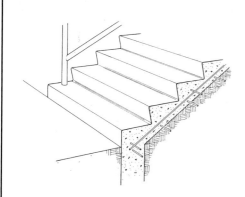

The Step System has three basic types: railroad tie, cast-in-place concrete or brick with a concrete base. System elements include: gravel base compaction; and backfill.

Wood Step Systems use 6″ x 6″ railroad ties that produce 3′ to 6′ wide steps that range from 2-riser to 5-riser configurations. Cast in Place Concrete Step Systems are either monolithic or aggregate finish. They range from 3′ to 6′ in width. Concrete Steps Systems are either in a 2-riser or 5-riser configuration. Precast Concrete Step Systems are listed for 4′ to 7′ widths with both 2-riser and 5-riser models. Brick Step Systems are placed on a 12″ concrete base. The size range is the same as cost in place concrete. Costs are on a per unit basis. All systems are assumed to include a full landing at the top 4′ long.

System Components	QUANTITY	UNIT	COST PER EACH		
			MAT.	INST.	TOTAL
SYSTEM G2030 310 2520					
STAIRS; CONCRETE, CAST IN PLACE, 3′ WIDE, 2 RISERS					
Excavate footing trench by hand	1.278	C.Y.		84.99	84.99
Bank run gravel, material only	.426	C.Y.	12.57		12.57
Haul gravel from pit	.426	C.Y.		12.67	12.67
Place gravel with hydraulic backhoe	.426	C.Y.		3.90	3.90
Backfill compaction, 6″ layers, hand tamp	.426	C.Y.		10.86	10.86
Forms in place	18.000	SFCA	25.20	268.20	293.40
Reinforcing in place	52.000	Lb.	35.75	35.10	70.85
Cast in place concrete, materials and installation	1.104	C.Y.	174.43	84.26	258.69
Finish concrete, broom finish	15.600	S.F.		14.66	14.66
TOTAL			247.95	514.64	762.59

G2030 310	Stairs; Concrete, Cast in Place	COST PER EACH		
		MAT.	INST.	TOTAL
2520	Concrete, cast in place, 3′ wide, 2 risers	248	515	763
2540	5 risers	330	730	1,060
2560	4′ wide, 2 risers	305	675	980
2580	5 risers	405	940	1,345
2600	5′ wide, 2 risers	365	840	1,205
2620	5 risers	470	1,125	1,595
2640	6′ wide, 2 risers	420	980	1,400
2660	5 risers	545	1,300	1,845
2800	Exposed aggregate finish, 3′ wide, 2 risers	251	530	781
2820	5 risers	335	730	1,065
2840	4′ wide, 2 risers	310	675	985
2860	5 risers	410	940	1,350
2880	5′ wide, 2 risers	375	840	1,215
2900	5 risers	475	1,125	1,600
2920	6′ wide, 2 risers	425	980	1,405
2940	5 risers	555	1,300	1,855
3200	Precast, 4′ wide, 2 risers	690	375	1,065
3220	5 risers	1,100	555	1,655
3240	5′ wide, 2 risers	780	450	1,230
3260	5 risers	1,425	600	2,025
3280	6′ wide, 2 risers	835	475	1,310
3300	5 risers	1,575	645	2,220
3320	7′ wide, 2 risers	1,200	505	1,705
3340	5 risers	1,900	690	2,590

G20 Site Improvements

G2030 Pedestrian Paving

G2030 310	Stairs; Concrete, Cast in Place	COST PER EACH		
		MAT.	INST.	TOTAL
4100	Brick, incl 12" conc base, 3' wide, 2 risers	415	1,325	1,740
4120	5 risers	595	1,975	2,570
4140	4' wide, 2 risers	520	1,625	2,145
4160	5 risers	740	2,425	3,165
4180	5' wide, 2 risers	625	1,925	2,550
4200	5 risers	875	2,850	3,725
4220	6' wide, 2 risers	720	2,225	2,945
4240	5 risers	1,025	3,275	4,300

For customer support on your Concrete & Masonry Costs with RSMeans Data, call 800.448.8182.

G2040 Site Development

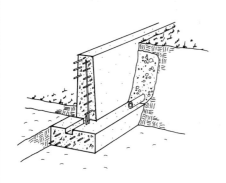

There are four basic types of Concrete Retaining Wall Systems: reinforced concrete with level backfill; reinforced concrete with sloped backfill or surcharge; unreinforced with level backfill; and unreinforced with sloped backfill or surcharge. System elements include: all necessary forms (4 uses); 3,000 p.s.i. concrete with an 8" chute; all necessary reinforcing steel; and underdrain. Exposed concrete is patched and rubbed.

The Expanded System Listing shows walls that range in thickness from 10" to 18" for reinforced concrete walls with level backfill and 12" to 24" for reinforced walls with sloped backfill. Walls range from a height of 4' to 20'. Unreinforced level and sloped backfill walls range from a height of 3' to 10'.

System Components	QUANTITY	UNIT	COST PER L.F.		
			MAT.	INST.	TOTAL
SYSTEM G2040 210 1000					
CONC. RETAIN. WALL REINFORCED, LEVEL BACKFILL, 4' HIGH					
Forms in place, cont. wall footing & keyway, 4 uses	2.000	S.F.	5.64	10.30	15.94
Forms in place, retaining wall forms, battered to 8' high, 4 uses	8.000	SFCA	8.80	78.40	87.20
Reinforcing in place, walls, #3 to #7	.004	Ton	5.50	3.76	9.26
Concrete ready mix, regular weight, 3000 psi	.204	C.Y.	27.95		27.95
Placing concrete and vibrating footing con., shallow direct chute	.074	C.Y.		2.08	2.08
Placing concrete and vibrating walls, 8" thick, direct chute	.130	C.Y.		4.84	4.84
Pipe bedding, crushed or screened bank run gravel	1.000	L.F.	4.02	1.44	5.46
Pipe, subdrainage, corrugated plastic, 4" diameter	1.000	L.F.	.89	.88	1.77
Finish walls and break ties, patch walls	4.000	S.F.	.20	4.48	4.68
TOTAL			53	106.18	159.18

G2040 210	Concrete Retaining Walls	COST PER L.F.		
		MAT.	INST.	TOTAL
1000	Conc. retain. wall, reinforced, level backfill, 4' high x 2'-2" base, 10" th	53	106	159
1200	6' high x 3'-3" base, 10" thick	76	154	230
1400	8' high x 4'-3" base, 10" thick	99	201	300
1600	10' high x 5'-4" base, 13" thick	129	295	424
2200	16' high x 8'-6" base, 16" thick	241	480	721
2600	20' high x 10'-5" base, 18" thick	350	620	970
3000	Sloped backfill, 4' high x 3'-2" base, 12" thick	64	110	174
3200	6' high x 4'-6" base, 12" thick	90	158	248
3400	8' high x 5'-11" base, 12" thick	119	208	327
3600	10' high x 7'-5" base, 16" thick	170	310	480
3800	12' high x 8'-10" base, 18" thick	221	375	596
4200	16' high x 11'-10" base, 21" thick	370	530	900
4600	20' high x 15'-0" base, 24" thick	570	715	1,285
5000	Unreinforced, level backfill, 3'-0" high x 1'-6" base	28	72	100
5200	4'-0" high x 2'-0" base	43	95	138
5400	6'-0" high x 3'-0" base	80	147	227
5600	8'-0" high x 4'-0" base	126	196	322
5800	10'-0" high x 5'-0" base	186	305	491
7000	Sloped backfill, 3'-0" high x 2'-0" base	33.50	74.50	108
7200	4'-0" high x 3'-0" base	55	99	154
7400	6'-0" high x 5'-0" base	114	156	270
7600	8'-0" high x 7'-0" base	191	213	404
7800	10'-0" high x 9'-0" base	292	330	622

For customer support on your Concrete & Masonry Costs with RSMeans Data, call 800.448.8182.

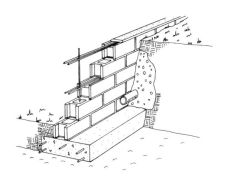

The Masonry Retaining Wall System includes: a reinforced concrete foundation footing for the wall with all necessary forms and 3,000 p.s.i. concrete; sand aggregate concrete blocks with steel reinforcing; solid grouting and underdrain.

The Expanded System Listing shows walls that range in thickness from 8″ to 12″ and in height from 3′-4″ to 8′.

System Components	QUANTITY	UNIT	COST PER L.F.		
			MAT.	INST.	TOTAL
SYSTEM G2040 220 1000					
REINF. CMU WALL, LEVEL FILL, 8″ THICK, 3′-4″ HIGH, 2′-4″ CONC. FTG.					
Forms in place, continuous wall footings 4 uses	1.670	S.F.	4.71	8.60	13.31
Joint reinforcing, #5 & #6 steel bars, vertical	5.590	Lb.	3.80	5.59	9.39
Concrete block, found. wall, cut joints, 8″ x 16″ x 8″ block	3.330	S.F.	12.19	23.48	35.67
Grout concrete block cores solid, 8″ thick, .258 CF/SF, pumped	3.330	S.F.	5.23	12.59	17.82
Concrete ready mix, regular wt., 3000 psi	.070	C.Y.	8.45		8.45
Pipe bedding, crushed or screened bank run gravel	1.000	L.F.	4.02	1.44	5.46
Pipe, subdrainage, corrugated plastic, 4″ diameter	1.000	L.F.	.89	.88	1.77
TOTAL			39.29	52.58	91.87

G2040 220	Masonry Retaining Walls	COST PER L.F.		
		MAT.	INST.	TOTAL
1000	Wall, reinforced CMU, level fill, 8″ thick, 3′-4″ high, 2′-4″ base	39.50	52.50	92
1200	4′-0″ high x 2′-9″ base	46	63.50	109.50
1400	4′-8″ high x 3′-3″ base	55	75	130
1600	5′-4″ high x 3′-8″ base	63.50	85.50	149
1800	6′-0″ high x 4′-2″ base	75	97.50	172.50
1840	10″ thick, 4′-0″ high x 2′-10″ base	47	64.50	111.50
1860	4′-8″ high x 3′-4″ base	56	76	132
1880	5′-4″ high x 3′-10″ base	65	86.50	151.50
1900	6′-0″ high x 4′-4″ base	76.50	99	175.50
1901	6′-0″ high x 4′-4″ base	76.50	99	175.50
1920	6′-8″ high x 4′-10″ base	84	109	193
1940	7′-4″ high x 5′-4″ base	100	123	223
2000	12″ thick, 5′-4″ high x 4′-0″ base	77	111	188
2200	6′-0″ high x 4′-6″ base	91	127	218
2400	6′-8″ high x 5′-0″ base	100	140	240
2600	7′-4″ high x 5′-6″ base	117	158	275
2800	8′-0″ high x 5′-11″ base	126	171	297

G2040 Site Development

The Stone Retaining Wall System is constructed of one of four types of stone. Each of the four types is listed in terms of cost per ton. Construction is either dry set or mortar set. System elements include excavation; concrete base; crushed stone; underdrain; and backfill.

The Expanded System Listing shows five heights above grade for each type, ranging from 3' above grade to 12' above grade.

System Components	QUANTITY	UNIT	COST PER L.F.		
			MAT.	INST.	TOTAL
SYSTEM G2040 260 2400					
STONE RETAINING WALL, DRY SET, STONE AT $250/TON, 3' ABOVE GRADE					
Excavation, trench, hyd backhoe	.880	C.Y.		6.55	6.55
Strip footing, 36" x 12", reinforced	.111	C.Y.	20.31	16.27	36.58
Stone, wall material, type 1	6.550	C.F.	59.28		59.28
Setting stone wall, dry	6.550	C.F.		76.96	76.96
Stone borrow, delivered, 3/8", machine spread	.320	C.Y.	11.68	2.74	14.42
Piping, subdrainage, perforated PVC, 4" diameter	1.000	L.F.	2.43	4.72	7.15
Backfill with dozer, trench, up to 300' haul, no compaction	1.019	C.Y.		3	3
TOTAL			93.70	110.24	203.94

G2040 260	Stone Retaining Walls	COST PER L.F.		
		MAT.	INST.	TOTAL
2400	Stone retaining wall, dry set, stone at $250/ton, height above grade 3'	93.50	110	203.50
2420	Height above grade 4'	112	131	243
2440	Height above grade 6'	146	176	322
2460	Height above grade 8'	205	276	481
2480	Height above grade 10'	263	360	623
2500	Height above grade 12'	325	450	775
2600	$350/ton stone, height above grade 3'	118	110	228
2620	Height above grade 4'	143	131	274
2640	Height above grade 6'	191	176	367
2660	Height above grade 8'	272	276	548
2680	Height above grade 10'	355	360	715
2700	Height above grade 12'	440	450	890
2800	$450/ton stone, height above grade 3'	141	110	251
2820	Height above grade 4'	174	131	305
2840	Height above grade 6'	235	176	411
2860	Height above grade 8'	340	276	616
2880	Height above grade 10'	445	360	805
2900	Height above grade 12'	555	450	1,005
3000	$650/ton stone, height above grade 3'	188	110	298
3020	Height above grade 4'	235	131	366
3040	Height above grade 6'	325	176	501
3060	Height above grade 8'	470	276	746
3080	Height above grade 10'	620	360	980
3100	Height above grade 12'	780	450	1,230
5020	Mortar set, stone at $250/ton, height above grade 3'	93.50	97.50	191
5040	Height above grade 4'	112	115	227

G2040 Site Development

G2040 260	Stone Retaining Walls	COST PER L.F.		
		MAT.	INST.	TOTAL
5060	Height above grade 6'	146	153	299
5080	Height above grade 8'	205	237	442
5100	Height above grade 10'	263	310	573
5120	Height above grade 12'	325	385	710
5300	$350/ton stone, height above grade 3'	118	97.50	215.50
5320	Height above grade 4'	143	115	258
5340	Height above grade 6'	191	153	344
5360	Height above grade 8'	272	237	509
5380	Height above grade 10'	355	310	665
5400	Height above grade 12'	440	385	825
5500	$450/ton stone, height above grade 3'	141	97.50	238.50
5520	Height above grade 4'	174	115	289
5540	Height above grade 6'	235	153	388
5560	Height above grade 8'	340	237	577
5580	Height above grade 10'	445	310	755
5600	Height above grade 12'	555	385	940
5700	$650/ton stone, height above grade 3'	188	97.50	285.50
5720	Height above grade 4'	235	115	350
5740	Height above grade 6'	325	153	478
5760	Height above grade 8'	470	237	707
5780	Height above grade 10'	620	310	930
5800	Height above grade 12'	780	385	1,165

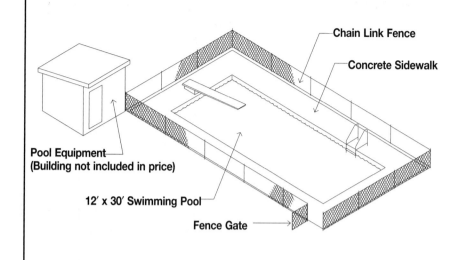

Chain Link Fence

Concrete Sidewalk

Pool Equipment
(Building not included in price)

12' x 30' Swimming Pool

Fence Gate

The Swimming Pool System is a complete package. Everything from excavation to deck hardware is included in system costs. Below are three basic types of pool systems: residential, motel, and municipal. Systems elements include: excavation, pool materials, installation, deck hardware, pumps and filters, sidewalk, and fencing.

The Expanded System Listing shows three basic types of pools with a variety of finishes and basic materials. These systems are either vinyl lined with metal sides; gunite shell with a cement plaster finish or tile finish; or concrete sided with vinyl lining. Pool sizes listed here vary from 12' x 30' to 60' x 82.5'. All costs are on a per unit basis.

System Components			COST EACH		
	QUANTITY	UNIT	MAT.	INST.	TOTAL
SYSTEM G2040 920 1000					
SWIMMING POOL, RESIDENTIAL, CONC. SIDES, VINYL LINED, 12' X 30'					
Swimming pool, residential, in-ground including equipment	360.000	S.F.	11,700	5,832	17,532
4" thick reinforced concrete sidewalk, broom finish, no base	400.000	S.F.	980	1,192	2,172
Chain link fence, residential, 4' high	124.000	L.F.	762.60	596.44	1,359.04
Fence gate, chain link, 4' high	1.000	Ea.	104	192	296
TOTAL			13,546.60	7,812.44	21,359.04

G2040 920	Swimming Pools	COST EACH		
		MAT.	INST.	TOTAL
1000	Swimming pool, residential class, concrete sides, vinyl lined, 12' x 30'	13,500	7,800	21,300
1100	16' x 32'	15,400	8,825	24,225
1200	20' x 40'	19,800	11,300	31,100
1250				
1500	Tile finish, 12' x 30'	29,600	33,100	62,700
1600	16' x 32'	30,100	34,800	64,900
1700	20' x 40'	45,100	48,900	94,000
3000	Gunite shell, cement plaster finish, 12' x 30'	23,400	14,000	37,400
3100	16' x 32'	29,700	18,300	48,000
3200	20' x 40'	41,200	18,100	59,300
3210				
3750				
4000	Motel class, concrete sides, vinyl lined, 20' x 40'	28,400	15,500	43,900
4100	28' x 60'	39,900	21,700	61,600
4500	Tile finish, 20' x 40'	53,000	56,500	109,500
4600	28' x 60'	84,500	91,500	176,000
6000	Gunite shell, cement plaster finish, 20' x 40'	76,500	45,100	121,600
6100	28' x 60'	114,500	67,500	182,000
7000	Municipal class, gunite shell, cement plaster finish, 42' x 75'	305,000	158,000	463,000
7100	60' x 82.5'	392,000	203,000	595,000
7500	Concrete walls, tile finish, 42' x 75'	341,500	214,500	556,000
7600	60' x 82.5'	445,000	284,500	729,500
7700	Tile finish and concrete gutter, 42' x 75'	375,000	214,500	589,500
7800	60' x 82.5'	485,000	284,500	769,500
7900	Tile finish and stainless gutter, 42' x 75'	439,000	214,500	653,500
8000	60' x 82.5'	649,000	328,500	977,500

G3020 Sanitary Sewer

G3020 110	Drainage & Sewage Piping	COST PER L.F.		
		MAT.	INST.	TOTAL
2000	Piping, excavation & backfill excluded, PVC, plain			
2130	4" diameter	2.43	4.72	7.15
2150	6" diameter	4.69	5.05	9.74
2160	8" diameter	9.70	5.30	15
2900	Box culvert, precast, 8' long			
3000	6' x 3'	239	36.50	275.50
3020	6' x 7'	325	41	366
3040	8' x 3'	360	45	405
3060	8' x 8'	420	51	471
3080	10' x 3'	410	46.50	456.50
3100	10' x 8'	575	64	639
3120	12' x 3'	990	51	1,041
3140	12' x 8'	985	76.50	1,061.50
4000	Concrete, nonreinforced			
4150	6" diameter	8.90	13.50	22.40
4160	8" diameter	9.75	15.95	25.70
4170	10" diameter	10.85	16.55	27.40
4180	12" diameter	11.30	17.90	29.20
4200	15" diameter	16.20	19.85	36.05
4220	18" diameter	20.50	24.50	45
4250	24" diameter	24.50	36	60.50
4400	Reinforced, no gasket			
4580	12" diameter	18.15	23.50	41.65
4600	15" diameter	22.50	23.50	46
4620	18" diameter	28	27.50	55.50
4650	24" diameter	30	36	66
4670	30" diameter	48.50	53.50	102
4680	36" diameter	88	65	153
4690	42" diameter	114	75.50	189.50
4700	48" diameter	119	80	199
4720	60" diameter	251	107	358
4730	72" diameter	280	128	408
4740	84" diameter	455	160	615
4800	With gasket			
4980	12" diameter	19.95	13.95	33.90
5000	15" diameter	25	14.65	39.65
5020	18" diameter	30.50	15.45	45.95
5050	24" diameter	37.50	17.25	54.75
5070	30" diameter	57.50	53.50	111
5080	36" diameter	98.50	65	163.50
5090	42" diameter	107	64	171
5100	48" diameter	134	80	214
5120	60" diameter	224	102	326
5130	72" diameter	300	128	428
5140	84" diameter	545	214	759
5700	Corrugated metal, alum. or galv. bit. coated			
5760	8" diameter	8	10.80	18.80
5770	10" diameter	8.30	13.75	22.05
5780	12" diameter	13.20	17.05	30.25
5800	15" diameter	14	17.90	31.90
5820	18" diameter	20.50	18.80	39.30
5850	24" diameter	23	22.50	45.50
5870	30" diameter	29	39	68
5880	36" diameter	32.50	39	71.50
5900	48" diameter	44	47	91
5920	60" diameter	77	68.50	145.50
5930	72" diameter	80	114	194
6000	Plain			

G3020 Sanitary Sewer

G3020 110	Drainage & Sewage Piping	COST PER L.F.		
		MAT.	INST.	TOTAL
6060	8" diameter	9	10.05	19.05
6070	10" diameter	9	12.75	21.75
6080	12" diameter	9.80	16.25	26.05
6100	15" diameter	12.75	16.25	29
6120	18" diameter	14.20	17.40	31.60
6140	24" diameter	26.50	20.50	47
6170	30" diameter	30.50	36	66.50
6180	36" diameter	35	36	71
6200	48" diameter	52	42.50	94.50
6220	60" diameter	85.50	65.50	151
6230	72" diameter	172	166	338
6300	Steel or alum. oval arch, coated & paved invert			
6400	15" equivalent diameter	18.60	17.90	36.50
6420	18" equivalent diameter	19.05	23.50	42.55
6450	24" equivalent diameter	29	28.50	57.50
6470	30" equivalent diameter	36	36	72
6480	36" equivalent diameter	33	47	80
6490	42" equivalent diameter	42.50	52	94.50
6500	48" equivalent diameter	56.50	62.50	119
6600	Plain			
6700	15" equivalent diameter	12.15	15.85	28
6720	18" equivalent diameter	15.85	20.50	36.35
6750	24" equivalent diameter	22	23.50	45.50
6770	30" equivalent diameter	29	43.50	72.50
6780	36" equivalent diameter	51	43.50	94.50
6790	42" equivalent diameter	62	51	113
6800	48" equivalent diameter	62	62.50	124.50
8000	Polyvinyl chloride SDR 35			
8130	4" diameter	2.43	4.72	7.15
8150	6" diameter	4.69	5.05	9.74
8160	8" diameter	9.70	5.30	15
8170	10" diameter	13.25	7.15	20.40
8180	12" diameter	18.75	7.35	26.10
8200	15" diameter	24	9.80	33.80

G3030 Storm Sewer

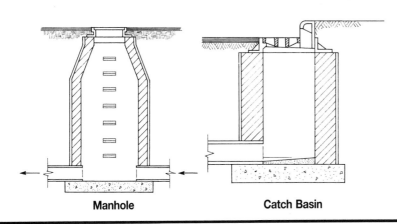

Manhole

Catch Basin

The Manhole and Catch Basin System includes: excavation with a backhoe; a formed concrete footing; frame and cover; cast iron steps and compacted backfill.

The Expanded System Listing shows manholes that have a 4', 5' and 6' inside diameter riser. Depths range from 4' to 14'. Construction material shown is either concrete, concrete block, precast concrete, or brick.

System Components	QUANTITY	UNIT	COST PER EACH		
			MAT.	INST.	TOTAL
SYSTEM G3030 210 1920					
MANHOLE/CATCH BASIN, BRICK, 4' I.D. RISER, 4' DEEP					
Excavation, hydraulic backhoe, 3/8 C.Y. bucket	14.815	B.C.Y.		133.78	133.78
Trim sides and bottom of excavation	64.000	S.F.		76.80	76.80
Forms in place, manhole base, 4 uses	20.000	SFCA	23.40	120	143.40
Reinforcing in place footings, #4 to #7	.019	Ton	26.13	25.65	51.78
Concrete, 3000 psi	.925	C.Y.	126.73		126.73
Place and vibrate concrete, footing, direct chute	.925	C.Y.		56.04	56.04
Catch basin or MH, brick, 4' ID, 4' deep	1.000	Ea.	770	1,175	1,945
Catch basin or MH steps; heavy galvanized cast iron	1.000	Ea.	19	16.20	35.20
Catch basin or MH frame and cover	1.000	Ea.	320	251.50	571.50
Fill, granular	12.954	L.C.Y.	427.48		427.48
Backfill, spread with wheeled front end loader	12.954	L.C.Y.		38.08	38.08
Backfill compaction, 12" lifts, air tamp	12.954	E.C.Y.		139.52	139.52
TOTAL			1,712.74	2,032.57	3,745.31

G3030 210	Manholes & Catch Basins	COST PER EACH		
		MAT.	INST.	TOTAL
1920	Manhole/catch basin, brick, 4' I.D. riser, 4' deep	1,725	2,025	3,750
1940	6' deep	2,450	2,825	5,275
1960	8' deep	3,300	3,875	7,175
1980	10' deep	4,025	4,800	8,825
3000	12' deep	5,150	5,250	10,400
3020	14' deep	6,475	7,325	13,800
3200	Block, 4' I.D. riser, 4' deep	1,425	1,650	3,075
3220	6' deep	1,975	2,325	4,300
3240	8' deep	2,650	3,200	5,850
3260	10' deep	3,150	3,975	7,125
3280	12' deep	4,075	5,050	9,125
3300	14' deep	5,150	6,175	11,325
4620	Concrete, cast-in-place, 4' I.D. riser, 4' deep	1,700	2,775	4,475
4640	6' deep	2,425	3,725	6,150
4660	8' deep	3,425	5,375	8,800
4680	10' deep	4,125	6,700	10,825
4700	12' deep	5,250	8,300	13,550
4720	14' deep	6,525	9,975	16,500
5820	Concrete, precast, 4' I.D. riser, 4' deep	1,950	1,500	3,450
5840	6' deep	2,525	2,025	4,550
5860	8' deep	3,975	2,850	6,825
5880	10' deep	4,550	3,500	8,050

G3030 Storm Sewer

G3030 210	Manholes & Catch Basins	COST PER EACH		
		MAT.	INST.	TOTAL
5900	12' deep	5,525	4,250	9,775
5920	14' deep	6,700	5,500	12,200
6000	5' I.D. riser, 4' deep	3,100	1,625	4,725
6020	6' deep	5,025	2,275	7,300
6040	8' deep	5,050	3,025	8,075
6060	10' deep	6,775	3,850	10,625
6080	12' deep	8,675	4,925	13,600
6100	14' deep	10,700	6,025	16,725
6200	6' I.D. riser, 4' deep	3,525	2,125	5,650
6220	6' deep	6,775	2,825	9,600
6240	8' deep	8,300	3,950	12,250
6260	10' deep	10,000	5,000	15,000
6280	12' deep	12,000	6,325	18,325
6300	14' deep	14,100	7,675	21,775

G3030 Storm Sewer

The Headwall Systems are listed in concrete and different stone wall materials for two different backfill slope conditions. The backfill slope directly affects the length of the wing walls. Walls are listed for different culvert sizes starting at 30″ diameter. Excavation and backfill are included in the system components, and are figured from an elevation 2′ below the bottom of the pipe.

System Components	QUANTITY	UNIT	COST PER EACH		
			MAT.	INST.	TOTAL
SYSTEM G3030 310 2000					
HEADWALL, C.I.P. CONCRETE FOR 30″ PIPE, 3′ LONG WING WALLS					
Excavation, hydraulic backhoe, 3/8 C.Y. bucket	28.000	B.C.Y.		252.84	252.84
Formwork, 2 uses	157.000	SFCA	573.71	2,170.99	2,744.70
Reinforcing in place, A615 Gr 60, longer and heavier dowels, add	45.000	Lb.	33.75	87.30	121.05
Concrete, 3000 psi	2.600	C.Y.	356.20		356.20
Place concrete, spread footings, direct chute	2.600	C.Y.		157.51	157.51
Backfill, dozer	28.000	L.C.Y.		61.60	61.60
TOTAL			963.66	2,730.24	3,693.90

G3030 310	Headwalls	COST PER EACH		
		MAT.	INST.	TOTAL
2000	Headwall, 1-1/2 to 1 slope soil, C.I.P. conc, 30″ pipe, 3′ long wing walls	965	2,725	3,690
2060	Pipe size 48″, 4′-6″ long wing walls	1,750	4,475	6,225
2100	Pipe size 60″, 5′-6″ long wing walls	2,400	5,925	8,325
2150				
2500	$250/ton stone, pipe size 30″, 3′ long wing walls	595	1,000	1,595
2560	Pipe size 48″, 4′-6″ long wing walls	1,175	1,700	2,875
2600	Pipe size 60″, 5′-6″ long wing walls	1,700	2,275	3,975
2650				
3000	$350/ton stone, pipe size 30″, 3′ long wing walls	830	1,000	1,830
3060	Pipe size 48″, 4′-6″ long wing walls	1,650	1,700	3,350
3100	Pipe size 60″, 5′-6″ long wing walls	2,375	2,275	4,650
3150				
4500	2 to 1 slope soil, C.I.P. concrete, pipe size 30″, 4′-3″ long wing walls	1,150	3,175	4,325
4560	Pipe size 48″, 6′-6″ long wing walls	2,150	5,375	7,525
4600	Pipe size 60″, 8′-0″ long wing walls	2,975	7,175	10,150
4650				
5000	$250/ton stone, pipe size 30″, 4′-3″ long wing walls	720	1,150	1,870
5060	Pipe size 48″, 6′-6″ long wing walls	1,475	2,025	3,500
5100	Pipe size 60″, 8′-0″ long wing walls	2,125	2,800	4,925
5150				
5500	$350/ton stone, pipe size 30″, 4′-3″ long wing walls	1,000	1,150	2,150
5560	Pipe size 48″, 6′-6″ long wing walls	2,075	2,025	4,100
5600	Pipe size 60″, 8′-0″ long wing walls	3,000	2,800	5,800
5650				

Reference Section

All the reference information is in one section, making it easy to find what you need to know ... and easy to use the data set on a daily basis. This section is visually identified by a vertical black bar on the page edges.

In this Reference Section, we've included Equipment Rental Costs, a listing of rental and operating costs; Crew Listings, a full listing of all crews and equipment, and their costs; Historical Cost Indexes for cost comparisons over time; City Cost Indexes and Location Factors for adjusting costs to the region you are in; Reference Tables, where you will find explanations, estimating information and procedures, or technical data; Change Orders, information on pricing changes to contract documents; and an explanation of all the Abbreviations in the data set.

Table of Contents

Estimating Tips

- This section contains the average costs to rent and operate hundreds of pieces of construction equipment. This is useful information when one is estimating the time and material requirements of any particular operation in order to establish a unit or total cost. Bare equipment costs shown on a unit cost line include, not only rental, but also operating costs for equipment under normal use.

Rental Costs

- Equipment rental rates are obtained from the following industry sources throughout North America: contractors, suppliers, dealers, manufacturers, and distributors.

- Rental rates vary throughout the country, with larger cities generally having lower rates. Lease plans for new equipment are available for periods in excess of six months, with a percentage of payments applying toward purchase.

- Monthly rental rates vary from 2% to 5% of the purchase price of the equipment depending on the anticipated life of the equipment and its wearing parts.

- Weekly rental rates are about 1/3 of the monthly rates, and daily rental rates are about 1/3 of the weekly rate.

- Rental rates can also be treated as reimbursement costs for contractor-owned equipment. Owned equipment costs include depreciation, loan payments, interest, taxes, insurance, storage, and major repairs.

Operating Costs

- The operating costs include parts and labor for routine servicing, such as the repair and replacement of pumps, filters, and worn lines. Normal operating expendables, such as fuel, lubricants, tires, and electricity (where applicable), are also included.

- Extraordinary operating expendables with highly variable wear patterns, such as diamond bits and blades, are excluded. These costs can be found as material costs in the Unit Price section.

- The hourly operating costs listed do not include the operator's wages.

Equipment Cost/Day

- Any power equipment required by a crew is shown in the Crew Listings with a daily cost.

- This daily cost of equipment needed by a crew includes both the rental cost and the operating cost and is based on dividing the weekly rental rate by 5 (the number of working days in the week), then adding the hourly operating cost multiplied by 8 (the number of hours in a day). This "Equipment Cost/Day" is shown in the far right column of the Equipment Rental section.

- If equipment is needed for only one or two days, it is best to develop your own cost by including components for daily rent and hourly operating costs. This is important when the listed Crew for a task does not contain the equipment needed, such as a crane for lifting mechanical heating/cooling equipment up onto a roof.

- If the quantity of work is less than the crew's Daily Output shown for a Unit Price line item that includes a bare unit equipment cost, the recommendation is to estimate one day's rental cost and operating cost for equipment shown in the Crew Listing for that line item.

- Please note, in some cases the equipment description in the crew is followed by a time period in parenthesis. For example: (daily) or (monthly). In these cases the equipment cost/day is calculated by adding the rental cost per time period to the hourly operating cost multiplied by 8.

Mobilization, Demobilization Costs

- The cost to move construction equipment from an equipment yard or rental company to the job site and back again is not included in equipment rental costs listed in the Reference Section. It is also not included in the bare equipment cost of any Unit Price line item or in any equipment costs shown in the Crew Listings.

- Mobilization (to the site) and demobilization (from the site) costs can be found in the Unit Price section.

- If a piece of equipment is already at the job site, it is not appropriate to utilize mobilization or demobilization costs again in an estimate. ■

---

01 54 33 | Equipment Rental

		UNIT	HOURLY OPER. COST	RENT PER DAY	RENT PER WEEK	RENT PER MONTH	EQUIPMENT COST/DAY	
10	0010	**CONCRETE EQUIPMENT RENTAL** without operators R015433 -10						**10**
	0200	Bucket, concrete lightweight, 1/2 C.Y.	Ea.	.76	38.50	116.16	350	29.30	
	0300	1 C.Y.		.99	63.50	190	570	45.90	
	0400	1-1/2 C.Y.		1.29	60.50	181.82	545	46.70	
	0500	2 C.Y.		1.40	74	222.23	665	55.65	
	0580	8 C.Y.		7.01	94.50	282.83	850	112.65	
	0600	Cart, concrete, self-propelled, operator walking, 10 C.F.		2.88	157	469.50	1,400	116.95	
	0700	Operator riding, 18 C.F.		4.85	167	500.50	1,500	138.95	
	0800	Conveyer for concrete, portable, gas, 16" wide, 26' long		10.71	162	484.86	1,450	182.70	
	0900	46' long		11.09	177	530.31	1,600	194.80	
	1000	56' long		11.26	194	580.82	1,750	206.25	
	1100	Core drill, electric, 2-1/2 H.P., 1" to 8" bit diameter		1.58	83.50	250	750	62.65	
	1150	11 H.P., 8" to 18" cores		5.45	130	390	1,175	121.60	
	1200	Finisher, concrete floor, gas, riding trowel, 96" wide		9.75	155	465.11	1,400	171.05	
	1300	Gas, walk-behind, 3 blade, 36" trowel		2.06	101	302.50	910	76.95	
	1400	4 blade, 48" trowel		3.10	116	347.50	1,050	94.30	
	1500	Float, hand-operated (Bull float), 48" wide		.08	13	39	117	8.45	
	1570	Curb builder, 14 H.P., gas, single screw		14.14	256	767.69	2,300	266.65	
	1590	Double screw		12.50	256	767.69	2,300	253.55	
	1600	Floor grinder, concrete and terrazzo, electric, 22" path		3.07	119	357.50	1,075	96.05	
	1700	Edger, concrete, electric, 7" path		1.19	57.50	172.50	520	44.05	
	1750	Vacuum pick-up system for floor grinders, wet/dry		1.63	103	309.38	930	74.95	
	1800	Mixer, powered, mortar and concrete, gas, 6 C.F., 18 H.P.		7.48	89	267.50	805	113.35	
	1900	10 C.F., 25 H.P.		9.08	124	372.50	1,125	147.15	
	2000	16 C.F.		9.44	146	436.88	1,300	162.90	
	2100	Concrete, stationary, tilt drum, 2 C.Y.		7.98	81	242.43	725	112.35	
	2120	Pump, concrete, truck mounted, 4" line, 80' boom		31.24	290	868.70	2,600	423.65	
	2140	5" line, 110' boom		39.99	290	868.70	2,600	493.65	
	2160	Mud jack, 50 C.F. per hr.		6.50	231	691.93	2,075	190.35	
	2180	225 C.F. per hr.		8.61	296	888.91	2,675	246.65	
	2190	Shotcrete pump rig, 12 C.Y./hr.		16.73	226	676.78	2,025	269.20	
	2200	35 C.Y./hr.		15.91	290	868.70	2,600	301.05	
	2600	Saw, concrete, manual, gas, 18 H.P.		5.59	115	345	1,025	113.70	
	2650	Self-propelled, gas, 30 H.P.		7.97	82	245.62	735	112.85	
	2675	V-groove crack chaser, manual, gas, 6 H.P.		1.65	100	300	900	73.25	
	2700	Vibrators, concrete, electric, 60 cycle, 2 H.P.		.47	70	210	630	45.80	
	2800	3 H.P.		.57	81	242.50	730	53.05	
	2900	Gas engine, 5 H.P.		1.56	17.05	51.21	154	22.70	
	3000	8 H.P.		2.10	17.25	51.74	155	27.15	
	3050	Vibrating screed, gas engine, 8 H.P.		2.83	108	325	975	87.65	
	3120	Concrete transit mixer, 6 x 4, 250 H.P., 8 C.Y., rear discharge		51.08	70.50	212.13	635	451.10	
	3200	Front discharge		59.30	136	409.10	1,225	556.20	
	3300	6 x 6, 285 H.P., 12 C.Y., rear discharge		58.56	167	500	1,500	568.50	
	3400	Front discharge		61.02	167	500	1,500	588.15	
20	0010	**EARTHWORK EQUIPMENT RENTAL** without operators R015433 -10							**20**
	0040	Aggregate spreader, push type, 8' to 12' wide	Ea.	2.62	65	195	585	59.95	
	0045	Tailgate type, 8' wide		2.56	64	191.92	575	58.90	
	0055	Earth auger, truck mounted, for fence & sign posts, utility poles		13.95	152	454.55	1,375	202.55	
	0060	For borings and monitoring wells		42.95	84	252.53	760	394.15	
	0070	Portable, trailer mounted		2.32	92.50	277.50	835	74.05	
	0075	Truck mounted, for caissons, water wells		86.00	134	402	1,200	768.40	
	0080	Horizontal boring machine, 12" to 36" diameter, 45 H.P.		22.93	110	330	990	249.45	
	0090	12" to 48" diameter, 65 H.P.		31.47	130	390	1,175	329.75	
	0095	Auger, for fence posts, gas engine, hand held		.45	84.50	254.05	760	54.40	
	0100	Excavator, diesel hydraulic, crawler mounted, 1/2 C.Y. cap.		21.92	470	1,408.39	4,225	457	
	0120	5/8 C.Y. capacity		29.30	615	1,851.77	5,550	604.75	
	0140	3/4 C.Y. capacity		32.95	730	2,190.83	6,575	701.80	
	0150	1 C.Y. capacity		41.58	835	2,500	7,500	832.65	

01 54 33 | Equipment Rental

		UNIT	HOURLY OPER. COST	RENT PER DAY	RENT PER WEEK	RENT PER MONTH	EQUIPMENT COST/DAY	
0200	1-1/2 C.Y. capacity	Ea.	49.02	505	1,518	4,550	695.80	20
0300	2 C.Y. capacity		57.09	810	2,430	7,300	942.70	
0320	2-1/2 C.Y. capacity		83.38	1,500	4,500	13,500	1,567	
0325	3-1/2 C.Y. capacity		121.20	2,025	6,072	18,200	2,184	
0330	4-1/2 C.Y. capacity		152.99	3,700	11,132	33,400	3,450	
0335	6 C.Y. capacity		194.11	3,250	9,765.80	29,300	3,506	
0340	7 C.Y. capacity		176.77	3,425	10,303.22	30,900	3,475	
0342	Excavator attachments, bucket thumbs		3.43	261	783.90	2,350	184.20	
0345	Grapples		3.17	225	674.15	2,025	160.20	
0346	Hydraulic hammer for boom mounting, 4000 ft lb.		13.60	900	2,702.04	8,100	649.20	
0347	5000 ft lb.		16.10	960	2,884.20	8,650	705.60	
0348	8000 ft lb.		23.75	1,225	3,643.20	10,900	918.65	
0349	12,000 ft lb.		25.95	1,125	3,373	10,100	882.20	
0350	Gradall type, truck mounted, 3 ton @ 15' radius, 5/8 C.Y.		43.83	835	2,500	7,500	850.65	
0370	1 C.Y. capacity		59.93	840	2,525.30	7,575	984.55	
0400	Backhoe-loader, 40 to 45 H.P., 5/8 C.Y. capacity		12.01	286	857.50	2,575	267.55	
0450	45 H.P. to 60 H.P., 3/4 C.Y. capacity		18.19	118	353.54	1,050	216.20	
0460	80 H.P., 1-1/4 C.Y. capacity		20.54	118	353.54	1,050	235.05	
0470	112 H.P., 1-1/2 C.Y. capacity		33.28	620	1,855.22	5,575	637.30	
0482	Backhoe-loader attachment, compactor, 20,000 lb.		6.50	157	470.34	1,400	146.05	
0485	Hydraulic hammer, 750 ft lb.		3.71	108	324.01	970	94.50	
0486	Hydraulic hammer, 1200 ft lb.		6.61	207	621.89	1,875	177.25	
0500	Brush chipper, gas engine, 6" cutter head, 35 H.P.		9.25	212	635	1,900	201	
0550	Diesel engine, 12" cutter head, 130 H.P.		23.88	292	875	2,625	366.05	
0600	15" cutter head, 165 H.P.		26.82	515	1,545	4,625	523.60	
0750	Bucket, clamshell, general purpose, 3/8 C.Y.		1.41	92.50	277.78	835	66.85	
0800	1/2 C.Y.		1.53	92.50	277.78	835	67.80	
0850	3/4 C.Y.		1.65	92.50	277.78	835	68.80	
0900	1 C.Y.		1.71	92.50	277.78	835	69.25	
0950	1-1/2 C.Y.		2.81	92.50	277.78	835	78.05	
1000	2 C.Y.		2.94	95	285	855	80.55	
1010	Bucket, dragline, medium duty, 1/2 C.Y.		.83	92.50	277.78	835	62.15	
1020	3/4 C.Y.		.79	92.50	277.78	835	61.85	
1030	1 C.Y.		.80	92.50	277.78	835	62	
1040	1-1/2 C.Y.		1.27	92.50	277.78	835	65.70	
1050	2 C.Y.		1.30	92.50	277.78	835	65.95	
1070	3 C.Y.		2.10	92.50	277.78	835	72.30	
1200	Compactor, manually guided 2-drum vibratory smooth roller, 7.5 H.P.		7.28	181	542.50	1,625	166.75	
1250	Rammer/tamper, gas, 8"		2.23	48.50	146.06	440	47	
1260	15"		2.65	55.50	167.23	500	54.65	
1300	Vibratory plate, gas, 18" plate, 3000 lb. blow		2.15	24.50	73.55	221	31.90	
1350	21" plate, 5000 lb. blow		2.64	241	722.50	2,175	165.60	
1370	Curb builder/extruder, 14 H.P., gas, single screw		12.50	256	767.69	2,300	253.55	
1390	Double screw		12.50	256	767.69	2,300	253.55	
1500	Disc harrow attachment, for tractor		.48	83.50	249.80	750	53.80	
1810	Feller buncher, shearing & accumulating trees, 100 H.P.		43.75	465	1,393.97	4,175	628.80	
1860	Grader, self-propelled, 25,000 lb.		33.65	1,125	3,366.73	10,100	942.50	
1910	30,000 lb.		33.15	1,350	4,040.48	12,100	1,073	
1920	40,000 lb.		52.35	1,325	4,000	12,000	1,219	
1930	55,000 lb.		67.53	1,675	5,000	15,000	1,540	
1950	Hammer, pavement breaker, self-propelled, diesel, 1000 to 1250 lb.		28.65	600	1,800	5,400	589.25	
2000	1300 to 1500 lb.		43.19	1,025	3,051.51	9,150	955.80	
2050	Pile driving hammer, steam or air, 4150 ft lb. @ 225 bpm		12.26	505	1,518	4,550	401.70	
2100	8750 ft lb. @ 145 bpm		14.47	710	2,125.20	6,375	540.85	
2150	15,000 ft lb. @ 60 bpm		14.81	845	2,530	7,600	624.45	
2200	24,450 ft lb. @ 111 bpm		15.83	980	2,934.80	8,800	713.60	
2250	Leads, 60' high for pile driving hammers up to 20,000 ft lb.		3.70	305	910.80	2,725	211.80	
2300	90' high for hammers over 20,000 ft lb.		5.50	545	1,639.44	4,925	371.85	

01 54 33 | Equipment Rental

		UNIT	HOURLY OPER. COST	RENT PER DAY	RENT PER WEEK	RENT PER MONTH	EQUIPMENT COST/DAY	
20	2350	Diesel type hammer, 22,400 ft lb.	Ea.	17.98	495	1,489.40	4,475	441.70	20
	2400	41,300 ft lb.		25.91	625	1,881.35	5,650	583.55	
	2450	141,000 ft lb.		41.70	995	2,978.80	8,925	929.35	
	2500	Vib. elec. hammer/extractor, 200 kW diesel generator, 34 H.P.		41.74	725	2,168.78	6,500	767.70	
	2550	80 H.P.		73.68	1,050	3,135.58	9,400	1,217	
	2600	150 H.P.		136.35	2,000	6,035.99	18,100	2,298	
	2700	Hydro excavator w/ext boom 12 C.Y., 1200 gallons		38.15	1,625	4,857.60	14,600	1,277	
	2800	Log chipper, up to 22" diameter, 600 H.P.		45.00	325	975	2,925	555	
	2850	Logger, for skidding & stacking logs, 150 H.P.		43.84	940	2,813.18	8,450	913.35	
	2860	Mulcher, diesel powered, trailer mounted		20.06	310	924.26	2,775	345.35	
	2900	Rake, spring tooth, with tractor		14.85	375	1,123.58	3,375	343.50	
	3000	Roller, vibratory, tandem, smooth drum, 20 H.P.		7.88	325	978.60	2,925	258.75	
	3050	35 H.P.		10.22	275	825	2,475	246.80	
	3100	Towed type vibratory compactor, smooth drum, 50 H.P.		25.50	525	1,581.85	4,750	520.35	
	3150	Sheepsfoot, 50 H.P.		25.87	365	1,100	3,300	426.95	
	3170	Landfill compactor, 220 H.P.		70.51	1,675	5,035.45	15,100	1,571	
	3200	Pneumatic tire roller, 80 H.P.		13.04	410	1,228.10	3,675	349.90	
	3250	120 H.P.		19.56	565	1,700	5,100	496.50	
	3300	Sheepsfoot vibratory roller, 240 H.P.		62.77	1,425	4,303.41	12,900	1,363	
	3320	340 H.P.		84.58	2,200	6,565.78	19,700	1,990	
	3350	Smooth drum vibratory roller, 75 H.P.		23.55	660	1,982.18	5,950	584.80	
	3400	125 H.P.		27.86	750	2,247.17	6,750	672.35	
	3410	Rotary mower, brush, 60", with tractor		18.96	365	1,097.45	3,300	371.15	
	3420	Rototiller, walk-behind, gas, 5 H.P.		2.15	60	180	540	53.25	
	3422	8 H.P.		2.84	116	347.50	1,050	92.20	
	3440	Scrapers, towed type, 7 C.Y. capacity		6.50	129	386.72	1,150	129.30	
	3450	10 C.Y. capacity		7.27	172	517.37	1,550	161.65	
	3500	15 C.Y. capacity		7.46	199	595.76	1,775	178.85	
	3525	Self-propelled, single engine, 14 C.Y. capacity		134.48	2,250	6,739.92	20,200	2,424	
	3550	Dual engine, 21 C.Y. capacity		142.64	2,525	7,575.90	22,700	2,656	
	3600	31 C.Y. capacity		189.53	3,650	10,954.15	32,900	3,707	
	3640	44 C.Y. capacity		234.76	4,700	14,083.90	42,300	4,695	
	3650	Elevating type, single engine, 11 C.Y. capacity		62.42	935	2,800	8,400	1,059	
	3700	22 C.Y. capacity		115.65	1,625	4,850	14,600	1,895	
	3710	Screening plant, 110 H.P. w/5' x 10' screen		21.32	650	1,952.56	5,850	561.10	
	3720	5' x 16' screen		26.92	1,225	3,666.67	11,000	948.65	
	3850	Shovel, crawler-mounted, front-loading, 7 C.Y. capacity		220.61	3,975	11,915.21	35,700	4,148	
	3855	12 C.Y. capacity		339.92	5,500	16,514.06	49,500	6,022	
	3860	Shovel/backhoe bucket, 1/2 C.Y.		2.72	74	221.58	665	66.05	
	3870	3/4 C.Y.		2.69	83	248.76	745	71.25	
	3880	1 C.Y.		2.78	92	275.42	825	77.35	
	3890	1-1/2 C.Y.		2.98	108	324.01	970	88.65	
	3910	3 C.Y.		3.47	146	438.98	1,325	115.55	
	3950	Stump chipper, 18" deep, 30 H.P.		6.95	223	668	2,000	189.20	
	4110	Dozer, crawler, torque converter, diesel 80 H.P.		25.48	335	1,010.12	3,025	405.85	
	4150	105 H.P.		34.64	605	1,818.22	5,450	640.80	
	4200	140 H.P.		41.64	660	1,980	5,950	729.15	
	4260	200 H.P.		63.72	1,675	5,050.60	15,200	1,520	
	4310	300 H.P.		81.45	1,900	5,667.20	17,000	1,785	
	4360	410 H.P.		107.70	3,250	9,727.46	29,200	2,807	
	4370	500 H.P.		134.58	3,925	11,788.10	35,400	3,434	
	4380	700 H.P.		232.23	5,525	16,587.71	49,800	5,175	
	4400	Loader, crawler, torque conv., diesel, 1-1/2 C.Y., 80 H.P.		29.80	555	1,667.71	5,000	572	
	4450	1-1/2 to 1-3/4 C.Y., 95 H.P.		30.54	705	2,120	6,350	668.35	
	4510	1-3/4 to 2-1/4 C.Y., 130 H.P.		48.19	900	2,700	8,100	925.50	
	4530	2-1/2 to 3-1/4 C.Y., 190 H.P.		58.30	1,125	3,400	10,200	1,146	
	4560	3-1/2 to 5 C.Y., 275 H.P.		72.05	1,475	4,400	13,200	1,456	
	4610	Front end loader, 4WD, articulated frame, diesel, 1 to 1-1/4 C.Y., 70 H.P.		16.77	286	857.06	2,575	305.60	

01 54 33 | Equipment Rental

		UNIT	HOURLY OPER. COST	RENT PER DAY	RENT PER WEEK	RENT PER MONTH	EQUIPMENT COST/DAY	
20	4620	1-1/2 to 1-3/4 C.Y., 95 H.P.	Ea.	20.17	465	1,400	4,200	441.40	**20**
	4650	1-3/4 to 2 C.Y., 130 H.P.		21.26	380	1,140	3,425	398.05	
	4710	2-1/2 to 3-1/2 C.Y., 145 H.P.		29.79	665	2,000	6,000	638.30	
	4730	3 to 4-1/2 C.Y., 185 H.P.		32.38	835	2,500	7,500	759	
	4760	5-1/4 to 5-3/4 C.Y., 270 H.P.		53.67	900	2,692.98	8,075	967.95	
	4810	7 to 9 C.Y., 475 H.P.		91.99	2,575	7,744.59	23,200	2,285	
	4870	9 to 11 C.Y., 620 H.P.		133.10	2,725	8,204.77	24,600	2,706	
	4880	Skid-steer loader, wheeled, 10 C.F., 30 H.P. gas		9.66	170	511.19	1,525	179.50	
	4890	1 C.Y., 78 H.P., diesel		18.60	495	1,487.50	4,475	446.30	
	4892	Skid-steer attachment, auger		.75	149	447.50	1,350	95.50	
	4893	Backhoe		.75	123	370.35	1,100	80.05	
	4894	Broom		.71	136	407.50	1,225	87.20	
	4895	Forks		.16	32	96	288	20.45	
	4896	Grapple		.73	89.50	267.93	805	59.45	
	4897	Concrete hammer		1.06	183	550	1,650	118.50	
	4898	Tree spade		.61	105	313.56	940	67.55	
	4899	Trencher		.66	102	305	915	66.25	
	4900	Trencher, chain, boom type, gas, operator walking, 12 H.P.		4.21	208	624.51	1,875	158.60	
	4910	Operator riding, 40 H.P.		16.83	525	1,580	4,750	450.70	
	5000	Wheel type, diesel, 4' deep, 12" wide		69.33	975	2,926.54	8,775	1,140	
	5100	6' deep, 20" wide		88.37	885	2,655	7,975	1,238	
	5150	Chain type, diesel, 5' deep, 8" wide		16.45	365	1,097.45	3,300	351.05	
	5200	Diesel, 8' deep, 16" wide		90.47	1,950	5,853.08	17,600	1,894	
	5202	Rock trencher, wheel type, 6" wide x 18" deep		47.46	91	272.73	820	434.20	
	5206	Chain type, 18" wide x 7' deep		105.44	286	858.60	2,575	1,015	
	5210	Tree spade, self-propelled		13.79	232	696.98	2,100	249.70	
	5250	Truck, dump, 2-axle, 12 ton, 8 C.Y. payload, 220 H.P.		24.16	355	1,071.50	3,225	407.60	
	5300	Three axle dump, 16 ton, 12 C.Y. payload, 400 H.P.		45.03	365	1,095.41	3,275	579.35	
	5310	Four axle dump, 25 ton, 18 C.Y. payload, 450 H.P.		50.45	585	1,750	5,250	753.60	
	5350	Dump trailer only, rear dump, 16-1/2 C.Y.		5.80	153	459.89	1,375	138.35	
	5400	20 C.Y.		6.26	172	517.37	1,550	153.55	
	5450	Flatbed, single axle, 1-1/2 ton rating		19.22	74.50	223.67	670	198.50	
	5500	3 ton rating		25.95	1,075	3,212.18	9,625	850.05	
	5550	Off highway rear dump, 25 ton capacity		63.42	1,525	4,600	13,800	1,427	
	5600	35 ton capacity		67.70	675	2,020.24	6,050	945.65	
	5610	50 ton capacity		84.88	2,175	6,500	19,500	1,979	
	5620	65 ton capacity		90.64	2,025	6,062.12	18,200	1,938	
	5630	100 ton capacity		122.70	2,975	8,936.41	26,800	2,769	
	6000	Vibratory plow, 25 H.P., walking	▼	6.84	305	909.11	2,725	236.55	
40	0010	**GENERAL EQUIPMENT RENTAL** without operators R015433 -10							**40**
	0020	Aerial lift, scissor type, to 20' high, 1200 lb. capacity, electric	Ea.	3.52	154	462.50	1,400	120.70	
	0030	To 30' high, 1200 lb. capacity		3.82	235	705	2,125	171.55	
	0040	Over 30' high, 1500 lb. capacity		5.20	288	865	2,600	214.55	
	0070	Articulating boom, to 45' high, 500 lb. capacity, diesel R015433 -15		10.04	250	750	2,250	230.30	
	0075	To 60' high, 500 lb. capacity		13.82	305	909.11	2,725	292.40	
	0080	To 80' high, 500 lb. capacity		16.24	995	2,987.50	8,975	727.45	
	0085	To 125' high, 500 lb. capacity		18.56	1,675	4,997.50	15,000	1,148	
	0100	Telescoping boom to 40' high, 500 lb. capacity, diesel		11.37	320	954.56	2,875	281.90	
	0105	To 45' high, 500 lb. capacity		12.66	325	974.77	2,925	296.20	
	0110	To 60' high, 500 lb. capacity		16.56	267	800	2,400	292.45	
	0115	To 80' high, 500 lb. capacity		21.52	360	1,077.80	3,225	387.75	
	0120	To 100' high, 500 lb. capacity		29.06	985	2,950	8,850	822.45	
	0125	To 120' high, 500 lb. capacity		29.51	1,525	4,545	13,600	1,145	
	0195	Air compressor, portable, 6.5 CFM, electric		.91	53.50	160	480	39.30	
	0196	Gasoline		.66	56	167.50	505	38.80	
	0200	Towed type, gas engine, 60 CFM		9.55	129	387.50	1,175	153.85	
	0300	160 CFM	▼	10.60	213	637.50	1,925	212.30	

For customer support on your Concrete & Masonry Costs with RSMeans Data, call 800.448.8182.

407

		UNIT	HOURLY OPER. COST	RENT PER DAY	RENT PER WEEK	RENT PER MONTH	EQUIPMENT COST/DAY
01 54 33	**Equipment Rental**						
0400	Diesel engine, rotary screw, 250 CFM	Ea.	12.23	175	525	1,575	202.85
0500	365 CFM		16.19	355	1,070	3,200	343.55
0550	450 CFM		20.19	300	900	2,700	341.50
0600	600 CFM		34.51	251	752.54	2,250	426.55
0700	750 CFM		35.04	525	1,580	4,750	596.30
0930	Air tools, breaker, pavement, 60 lb.		.57	81.50	245	735	53.60
0940	80 lb.		.57	81	242.50	730	53.05
0950	Drills, hand (jackhammer), 65 lb.		.68	64	192.50	580	43.90
0960	Track or wagon, swing boom, 4" drifter		62.50	1,050	3,135.41	9,400	1,127
0970	5" drifter		62.50	1,050	3,135.41	9,400	1,127
0975	Track mounted quarry drill, 6" diameter drill		102.94	1,900	5,722.33	17,200	1,968
0980	Dust control per drill		1.05	25.50	76.26	229	23.65
0990	Hammer, chipping, 12 lb.		.61	46.50	140	420	32.85
1000	Hose, air with couplings, 50' long, 3/4" diameter		.07	11	33	99	7.15
1100	1" diameter		.08	12.35	37	111	8.05
1200	1-1/2" diameter		.22	35	105	315	22.80
1300	2" diameter		.24	45	135	405	28.95
1400	2-1/2" diameter		.36	57.50	172.50	520	37.40
1410	3" diameter		.42	58.50	175	525	38.35
1450	Drill, steel, 7/8" x 2'		.09	13.05	39.12	117	8.50
1460	7/8" x 6'		.12	19.85	59.58	179	12.85
1520	Moil points		.03	7	21	63	4.40
1525	Pneumatic nailer w/accessories		.48	39.50	119.19	360	27.70
1530	Sheeting driver for 60 lb. breaker		.04	7.85	23.52	70.50	5.05
1540	For 90 lb. breaker		.13	10.65	31.88	95.50	7.45
1550	Spade, 25 lb.		.51	7.50	22.47	67.50	8.55
1560	Tamper, single, 35 lb.		.60	58.50	175	525	39.75
1570	Triple, 140 lb.		.90	62.50	187.09	560	44.60
1580	Wrenches, impact, air powered, up to 3/4" bolt		.43	50	149.75	450	33.40
1590	Up to 1-1/4" bolt		.58	67.50	202.50	610	45.15
1600	Barricades, barrels, reflectorized, 1 to 99 barrels		.03	4	12	36	2.65
1610	100 to 200 barrels		.03	4.46	13.38	40	2.90
1620	Barrels with flashers, 1 to 99 barrels		.04	6.45	19.39	58	4.15
1630	100 to 200 barrels		.03	5.15	15.52	46.50	3.35
1640	Barrels with steady burn type C lights		.05	8.55	25.61	77	5.50
1650	Illuminated board, trailer mounted, with generator		3.32	141	423.30	1,275	111.20
1670	Portable barricade, stock, with flashers, 1 to 6 units		.04	6.45	19.34	58	4.15
1680	25 to 50 units		.03	6	18.03	54	3.85
1685	Butt fusion machine, wheeled, 1.5 HP electric, 2" - 8" diameter pipe		2.66	190	568.50	1,700	134.95
1690	Tracked, 20 HP diesel, 4" - 12" diameter pipe		10.04	565	1,702.05	5,100	420.75
1695	83 HP diesel, 8" - 24" diameter pipe		51.84	1,125	3,358.65	10,100	1,086
1700	Carts, brick, gas engine, 1000 lb. capacity		2.98	65.50	196.97	590	63.20
1800	1500 lb., 7-1/2' lift		2.95	70	210.50	630	65.70
1822	Dehumidifier, medium, 6 lb./hr., 150 CFM		1.20	77.50	232.03	695	56.05
1824	Large, 18 lb./hr., 600 CFM		2.22	585	1,750	5,250	367.75
1830	Distributor, asphalt, trailer mounted, 2000 gal., 38 H.P. diesel		11.13	355	1,071.32	3,225	303.25
1840	3000 gal., 38 H.P. diesel		13.02	385	1,149.71	3,450	334.10
1850	Drill, rotary hammer, electric		1.13	72	216.17	650	52.25
1860	Carbide bit, 1-1/2" diameter, add to electric rotary hammer		.03	41.50	125	375	25.25
1865	Rotary, crawler, 250 H.P.		137.40	2,450	7,375	22,100	2,574
1870	Emulsion sprayer, 65 gal., 5 H.P. gas engine		2.80	108	324.01	970	87.20
1880	200 gal., 5 H.P. engine		7.31	181	543.50	1,625	167.20
1900	Floor auto-scrubbing machine, walk-behind, 28" path		5.69	223	670	2,000	179.55
1930	Floodlight, mercury vapor, or quartz, on tripod, 1000 watt		.46	36.50	110	330	25.70
1940	2000 watt		.60	28.50	85.55	257	21.90
1950	Floodlights, trailer mounted with generator, 1 - 300 watt light		3.59	79.50	238.30	715	76.35
1960	2 - 1000 watt lights		4.54	88.50	265.48	795	89.40
2000	4 - 300 watt lights		4.29	101	303.11	910	95

01 54 33 | Equipment Rental

		UNIT	HOURLY OPER. COST	RENT PER DAY	RENT PER WEEK	RENT PER MONTH	EQUIPMENT COST/DAY	
40	2005	Foam spray rig, incl. box trailer, compressor, generator, proportioner	Ea.	25.76	540	1,620.05	4,850	530.15	40
	2015	Forklift, pneumatic tire, rough terr, straight mast, 5000 lb, 12' lift, gas		18.82	221	663.70	2,000	283.30	
	2025	8000 lb, 12' lift		22.95	365	1,097.45	3,300	403.10	
	2030	5000 lb, 12' lift, diesel		15.59	247	740.71	2,225	272.85	
	2035	8000 lb, 12' lift, diesel		16.90	280	839.82	2,525	303.15	
	2045	All terrain, telescoping boom, diesel, 5000 lb, 10' reach, 19' lift		17.40	200	600	1,800	259.25	
	2055	6600 lb, 29' reach, 42' lift		21.29	236	707.08	2,125	311.75	
	2065	10,000 lb, 31' reach, 45' lift		23.30	267	800	2,400	346.45	
	2070	Cushion tire, smooth floor, gas, 5000 lb capacity		8.32	262	785	2,350	223.60	
	2075	8000 lb capacity		11.47	320	957.50	2,875	283.25	
	2085	Diesel, 5000 lb capacity		7.82	257	770	2,300	216.55	
	2090	12,000 lb capacity		12.16	400	1,206.59	3,625	338.60	
	2095	20,000 lb capacity		17.41	800	2,405	7,225	620.30	
	2100	Generator, electric, gas engine, 1.5 kW to 3 kW		2.60	46.50	140	420	48.80	
	2200	5 kW		3.25	91	272.50	820	80.45	
	2300	10 kW		5.98	108	322.50	970	112.35	
	2400	25 kW		7.47	405	1,222.25	3,675	304.25	
	2500	Diesel engine, 20 kW		9.29	222	665	2,000	207.35	
	2600	50 kW		16.09	360	1,085	3,250	345.70	
	2700	100 kW		28.85	485	1,455	4,375	521.85	
	2800	250 kW		54.84	825	2,472.50	7,425	933.25	
	2850	Hammer, hydraulic, for mounting on boom, to 500 ft lb.		2.93	94.50	283.25	850	80.05	
	2860	1000 ft lb.		4.65	141	423.30	1,275	121.85	
	2900	Heaters, space, oil or electric, 50 MBH		1.48	47	141.42	425	40.10	
	3000	100 MBH		2.74	47	141.42	425	50.25	
	3100	300 MBH		8.00	136	409.10	1,225	145.80	
	3150	500 MBH		13.28	208	625	1,875	231.20	
	3200	Hose, water, suction with coupling, 20' long, 2" diameter		.02	5.65	17	51	3.55	
	3210	3" diameter		.03	14.15	42.50	128	8.75	
	3220	4" diameter		.03	28.50	85.01	255	17.25	
	3230	6" diameter		.11	41	122.96	370	25.50	
	3240	8" diameter		.27	54	161.92	485	34.55	
	3250	Discharge hose with coupling, 50' long, 2" diameter		.01	6.50	19.50	58.50	4	
	3260	3" diameter		.01	7.35	22	66	4.50	
	3270	4" diameter		.02	21	63.25	190	12.80	
	3280	6" diameter		.06	29.50	88.04	264	18.10	
	3290	8" diameter		.24	38	113.85	340	24.70	
	3295	Insulation blower		.84	118	353.54	1,050	77.40	
	3300	Ladders, extension type, 16' to 36' long		.18	41.50	125	375	26.45	
	3400	40' to 60' long		.64	123	368	1,100	78.75	
	3405	Lance for cutting concrete		2.22	66.50	200	600	57.80	
	3407	Lawn mower, rotary, 22", 5 H.P.		1.06	38.50	115	345	31.50	
	3408	48" self-propelled		2.92	160	481	1,450	119.60	
	3410	Level, electronic, automatic, with tripod and leveling rod		1.06	43	129.50	390	34.40	
	3430	Laser type, for pipe and sewer line and grade		2.20	117	350	1,050	87.55	
	3440	Rotating beam for interior control		.91	64	192.50	580	45.80	
	3460	Builder's optical transit, with tripod and rod		.10	43	129.50	390	26.70	
	3500	Light towers, towable, with diesel generator, 2000 watt		4.31	102	307.29	920	95.90	
	3600	4000 watt		4.55	142	425	1,275	121.40	
	3700	Mixer, powered, plaster and mortar, 6 C.F., 7 H.P.		2.08	86.50	260	780	68.60	
	3800	10 C.F., 9 H.P.		2.26	124	372.50	1,125	92.60	
	3850	Nailer, pneumatic		.48	30.50	91	273	22.10	
	3900	Paint sprayers complete, 8 CFM		.86	62.50	186.74	560	44.20	
	4000	17 CFM		1.61	111	333.84	1,000	79.70	
	4020	Pavers, bituminous, rubber tires, 8' wide, 50 H.P., diesel		32.31	575	1,724.57	5,175	603.40	
	4030	10' wide, 150 H.P.		96.77	1,975	5,905.34	17,700	1,955	
	4050	Crawler, 8' wide, 100 H.P., diesel		88.64	2,075	6,245.03	18,700	1,958	
	4060	10' wide, 150 H.P.		105.21	2,375	7,133.45	21,400	2,268	

01 54 33 | Equipment Rental

		UNIT	HOURLY OPER. COST	RENT PER DAY	RENT PER WEEK	RENT PER MONTH	EQUIPMENT COST/DAY	
40								**40**
4070	Concrete paver, 12' to 24' wide, 250 H.P.	Ea.	88.67	1,700	5,121.45	15,400	1,734	
4080	Placer-spreader-trimmer, 24' wide, 300 H.P.		118.92	2,575	7,760.56	23,300	2,503	
4100	Pump, centrifugal gas pump, 1-1/2" diam., 65 GPM		3.97	54.50	164.10	490	64.60	
4200	2" diameter, 130 GPM		5.04	44	132.50	400	66.80	
4300	3" diameter, 250 GPM		5.18	55	165	495	74.40	
4400	6" diameter, 1500 GPM		22.46	92.50	277.78	835	235.25	
4500	Submersible electric pump, 1-1/4" diameter, 55 GPM		.40	35	105	315	24.25	
4600	1-1/2" diameter, 83 GPM		.45	43.50	130	390	29.60	
4700	2" diameter, 120 GPM		1.66	55	165	495	46.30	
4800	3" diameter, 300 GPM		3.07	109	327.50	985	90.05	
4900	4" diameter, 560 GPM		14.90	76	227.50	685	164.70	
5000	6" diameter, 1590 GPM		22.30	65.50	196.97	590	217.80	
5100	Diaphragm pump, gas, single, 1-1/2" diameter		1.14	38.50	116.16	350	32.35	
5200	2" diameter		4.02	92.50	277.78	835	87.70	
5300	3" diameter		4.09	86	258	775	84.35	
5400	Double, 4" diameter		6.10	96	287.88	865	106.35	
5450	Pressure washer 5 GPM, 3000 psi		3.92	110	330	990	97.35	
5460	7 GPM, 3000 psi		5.00	90	270	810	93.95	
5470	High pressure water jet 10 KSI		40.02	730	2,185.92	6,550	757.40	
5480	40 KSI		28.21	990	2,975.28	8,925	820.75	
5500	Trash pump, self-priming, gas, 2" diameter		3.85	109	328.29	985	96.50	
5600	Diesel, 4" diameter		6.74	163	489.91	1,475	151.95	
5650	Diesel, 6" diameter		17.02	163	489.91	1,475	234.15	
5655	Grout pump		18.92	284	851.83	2,550	321.75	
5700	Salamanders, L.P. gas fired, 100,000 Btu		2.92	57.50	172.50	520	57.85	
5705	50,000 Btu		1.68	23	69	207	27.25	
5720	Sandblaster, portable, open top, 3 C.F. capacity		.61	132	395	1,175	83.85	
5730	6 C.F. capacity		1.02	133	399	1,200	87.90	
5740	Accessories for above		.14	24	72.12	216	15.55	
5750	Sander, floor		.78	73.50	220	660	50.20	
5760	Edger		.52	35	105	315	25.20	
5800	Saw, chain, gas engine, 18" long		1.77	63.50	190	570	52.20	
5900	Hydraulic powered, 36" long		.79	59	176.77	530	41.65	
5950	60" long		.79	65.50	196.97	590	45.70	
6000	Masonry, table mounted, 14" diameter, 5 H.P.		1.34	76.50	230	690	56.70	
6050	Portable cut-off, 8 H.P.		1.83	72.50	217.50	655	58.15	
6100	Circular, hand held, electric, 7-1/4" diameter		.23	13.85	41.50	125	10.15	
6200	12" diameter		.24	41	122.50	370	26.40	
6250	Wall saw, w/hydraulic power, 10 H.P.		3.32	99.50	299	895	86.40	
6275	Shot blaster, walk-behind, 20" wide		4.79	284	851.83	2,550	208.70	
6280	Sidewalk broom, walk-behind		2.27	83.50	250.85	755	68.30	
6300	Steam cleaner, 100 gallons per hour		3.38	83.50	250.85	755	77.20	
6310	200 gallons per hour		4.38	101	303.11	910	95.70	
6340	Tar kettle/pot, 400 gallons		16.65	128	383.85	1,150	209.95	
6350	Torch, cutting, acetylene-oxygen, 150' hose, excludes gases		.46	15.50	46.51	140	12.95	
6360	Hourly operating cost includes tips and gas		21.17	7.10	21.23	63.50	173.60	
6410	Toilet, portable chemical		.13	23.50	70.03	210	15.05	
6420	Recycle flush type		.16	29	86.75	260	18.65	
6430	Toilet, fresh water flush, garden hose,		.20	34.50	103.47	310	22.25	
6440	Hoisted, non-flush, for high rise		.16	28	84.66	254	18.20	
6465	Tractor, farm with attachment		17.57	395	1,184.37	3,550	377.45	
6480	Trailers, platform, flush deck, 2 axle, 3 ton capacity		1.71	96	287.41	860	71.15	
6500	25 ton capacity		6.31	145	433.76	1,300	137.20	
6600	40 ton capacity		8.14	206	616.66	1,850	188.45	
6700	3 axle, 50 ton capacity		8.82	228	683.33	2,050	207.25	
6800	75 ton capacity		11.21	305	907.63	2,725	271.20	
6810	Trailer mounted cable reel for high voltage line work		5.94	28.50	85.86	258	64.75	
6820	Trailer mounted cable tensioning rig		11.79	28.50	85.86	258	111.50	

01 54 33 | Equipment Rental

		UNIT	HOURLY OPER. COST	RENT PER DAY	RENT PER WEEK	RENT PER MONTH	EQUIPMENT COST/DAY	
40	6830	Cable pulling rig	Ea.	74.51	28.50	85.86	258	613.30	**40**
	6850	Portable cable/wire puller, 8000 lb max pulling capacity		3.74	121	363.64	1,100	102.65	
	6900	Water tank trailer, engine driven discharge, 5000 gallons		7.24	160	480.79	1,450	154.10	
	6925	10,000 gallons		9.87	218	653.25	1,950	209.60	
	6950	Water truck, off highway, 6000 gallons		72.61	845	2,529.89	7,600	1,087	
	7010	Tram car for high voltage line work, powered, 2 conductor		6.95	29.50	88	264	73.20	
	7020	Transit (builder's level) with tripod		.10	17.75	53.30	160	11.45	
	7030	Trench box, 3000 lb., 6' x 8'		.57	97.50	293.15	880	63.15	
	7040	7200 lb., 6' x 20'		.73	189	566.72	1,700	119.15	
	7050	8000 lb., 8' x 16'		1.09	205	615	1,850	131.70	
	7060	9500 lb., 8' x 20'		1.22	235	705.51	2,125	150.85	
	7065	11,000 lb., 8' x 24'		1.28	221	663.70	2,000	142.95	
	7070	12,000 lb., 10' x 20'		1.51	267	799.57	2,400	172	
	7100	Truck, pickup, 3/4 ton, 2 wheel drive		9.35	62.50	187.09	560	112.20	
	7200	4 wheel drive		9.60	167	500	1,500	176.75	
	7250	Crew carrier, 9 passenger		12.81	109	328.29	985	168.15	
	7290	Flat bed truck, 20,000 lb. GVW		15.44	134	402.40	1,200	204.05	
	7300	Tractor, 4 x 2, 220 H.P.		22.52	218	653.25	1,950	310.80	
	7410	330 H.P.		32.72	298	893.64	2,675	440.45	
	7500	6 x 4, 380 H.P.		36.53	345	1,034.74	3,100	499.15	
	7600	450 H.P.		44.76	420	1,254.23	3,775	608.95	
	7610	Tractor, with A frame, boom and winch, 225 H.P.		25.04	296	886.76	2,650	377.65	
	7620	Vacuum truck, hazardous material, 2500 gallons		12.94	315	940.67	2,825	291.70	
	7625	5,000 gallons		13.18	445	1,332.62	4,000	371.95	
	7650	Vacuum, HEPA, 16 gallon, wet/dry		.86	122	365	1,100	79.90	
	7655	55 gallon, wet/dry		.79	25.50	76.50	230	21.60	
	7660	Water tank, portable		.74	162	485.11	1,450	102.90	
	7690	Sewer/catch basin vacuum, 14 C.Y., 1500 gallons		17.52	670	2,012	6,025	542.60	
	7700	Welder, electric, 200 amp		3.86	33.50	101.01	305	51.10	
	7800	300 amp		5.61	104	313.14	940	107.55	
	7900	Gas engine, 200 amp		9.05	59	176.77	530	107.80	
	8000	300 amp		10.25	111	333.96	1,000	148.75	
	8100	Wheelbarrow, any size		.07	11.15	33.50	101	7.20	
	8200	Wrecking ball, 4000 lb.		2.53	62	186	560	57.45	
50	0010	**HIGHWAY EQUIPMENT RENTAL** without operators [R015433-10]							**50**
	0050	Asphalt batch plant, portable drum mixer, 100 ton/hr.	Ea.	89.47	1,550	4,677.24	14,000	1,651	
	0060	200 ton/hr.		103.21	1,675	4,990.80	15,000	1,824	
	0070	300 ton/hr.		121.29	1,950	5,853.08	17,600	2,141	
	0100	Backhoe attachment, long stick, up to 185 H.P., 10-1/2' long		.37	26	77.34	232	18.45	
	0140	Up to 250 H.P., 12' long		.42	29	86.75	260	20.70	
	0180	Over 250 H.P., 15' long		.57	39.50	118.11	355	28.20	
	0200	Special dipper arm, up to 100 H.P., 32' long		1.17	80.50	241.44	725	57.65	
	0240	Over 100 H.P., 33' long		1.46	101	303.11	910	72.30	
	0280	Catch basin/sewer cleaning truck, 3 ton, 9 C.Y., 1000 gal.		35.81	425	1,280.36	3,850	542.60	
	0300	Concrete batch plant, portable, electric, 200 C.Y./hr.		24.47	565	1,698.44	5,100	535.50	
	0520	Grader/dozer attachment, ripper/scarifier, rear mounted, up to 135 H.P.		3.19	64	192.32	575	63.95	
	0540	Up to 180 H.P.		4.18	97	290.56	870	91.60	
	0580	Up to 250 H.P.		5.92	155	465.11	1,400	140.40	
	0700	Pvmt. removal bucket, for hyd. excavator, up to 90 H.P.		2.18	59	176.64	530	52.80	
	0740	Up to 200 H.P.		2.33	75.50	225.76	675	63.80	
	0780	Over 200 H.P.		2.55	92.50	276.98	830	75.80	
	0900	Aggregate spreader, self-propelled, 187 H.P.		51.21	750	2,247.17	6,750	859.10	
	1000	Chemical spreader, 3 C.Y.		3.20	99.50	299	895	85.40	
	1900	Hammermill, traveling, 250 H.P.		68.03	520	1,565.69	4,700	857.40	
	2000	Horizontal borer, 3" diameter, 13 H.P. gas driven		5.47	234	702.03	2,100	184.15	
	2150	Horizontal directional drill, 20,000 lb. thrust, 78 H.P. diesel		27.86	535	1,606.09	4,825	544.10	
	2160	30,000 lb. thrust, 115 H.P.		34.24	625	1,868.72	5,600	647.65	
	2170	50,000 lb. thrust, 170 H.P.		49.09	720	2,156.61	6,475	824.05	

01 54 33 | Equipment Rental

		UNIT	HOURLY OPER. COST	RENT PER DAY	RENT PER WEEK	RENT PER MONTH	EQUIPMENT COST/DAY	
50	2190	Mud trailer for HDD, 1500 gallons, 175 H.P., gas	Ea.	25.76	177	530.31	1,600	312.15	**50**
	2200	Hydromulcher, diesel, 3000 gallon, for truck mounting		17.61	193	580	1,750	256.85	
	2300	Gas, 600 gallon		7.57	96	287.88	865	118.15	
	2400	Joint & crack cleaner, walk behind, 25 H.P.		3.19	46	137.38	410	53	
	2500	Filler, trailer mounted, 400 gallons, 20 H.P.		8.43	173	517.50	1,550	170.95	
	3000	Paint striper, self-propelled, 40 gallon, 22 H.P.		6.83	123	368.69	1,100	128.35	
	3100	120 gallon, 120 H.P.		17.83	385	1,151.54	3,450	372.90	
	3200	Post drivers, 6" I-Beam frame, for truck mounting		12.54	325	969.72	2,900	294.25	
	3400	Road sweeper, self-propelled, 8' wide, 90 H.P.		36.34	720	2,164.75	6,500	723.65	
	3450	Road sweeper, vacuum assisted, 4 C.Y., 220 gallons		58.98	680	2,038.13	6,125	879.45	
	4000	Road mixer, self-propelled, 130 H.P.		46.79	835	2,508.46	7,525	876	
	4100	310 H.P.		75.91	2,175	6,558.59	19,700	1,919	
	4220	Cold mix paver, incl. pug mill and bitumen tank, 165 H.P.		96.11	2,350	7,015.87	21,000	2,172	
	4240	Pavement brush, towed		3.47	101	303.11	910	88.35	
	4250	Paver, asphalt, wheel or crawler, 130 H.P., diesel		95.37	2,300	6,898.28	20,700	2,143	
	4300	Paver, road widener, gas, 1' to 6', 67 H.P.		47.22	985	2,952.67	8,850	968.30	
	4400	Diesel, 2' to 14', 88 H.P.		57.06	1,175	3,501.40	10,500	1,157	
	4600	Slipform pavers, curb and gutter, 2 track, 75 H.P.		58.53	1,275	3,814.96	11,400	1,231	
	4700	4 track, 165 H.P.		36.11	855	2,560.72	7,675	801.05	
	4800	Median barrier, 215 H.P.		59.13	1,350	4,076.26	12,200	1,288	
	4901	Trailer, low bed, 75 ton capacity		10.84	286	857.06	2,575	258.10	
	5000	Road planer, walk behind, 10" cutting width, 10 H.P.		2.48	258	773	2,325	174.45	
	5100	Self-propelled, 12" cutting width, 64 H.P.		8.34	192	575.77	1,725	181.85	
	5120	Traffic line remover, metal ball blaster, truck mounted, 115 H.P.		47.03	1,025	3,100	9,300	996.25	
	5140	Grinder, truck mounted, 115 H.P.		51.41	1,025	3,100	9,300	1,031	
	5160	Walk-behind, 11 H.P.		3.59	143	430	1,300	114.75	
	5200	Pavement profiler, 4' to 6' wide, 450 H.P.		218.77	1,275	3,838.46	11,500	2,518	
	5300	8' to 10' wide, 750 H.P.		334.94	1,350	4,015.23	12,000	3,483	
	5400	Roadway plate, steel, 1" x 8' x 20'		.09	61.50	184.69	555	37.65	
	5600	Stabilizer, self-propelled, 150 H.P.		41.55	1,050	3,131.37	9,400	958.70	
	5700	310 H.P.		76.95	1,325	3,939.47	11,800	1,404	
	5800	Striper, truck mounted, 120 gallon paint, 460 H.P.		49.24	350	1,046	3,150	603.10	
	5900	Thermal paint heating kettle, 115 gallons		7.78	75	225	675	107.25	
	6000	Tar kettle, 330 gallon, trailer mounted		12.40	96	287.50	865	156.70	
	7000	Tunnel locomotive, diesel, 8 to 12 ton		30.11	625	1,881.35	5,650	617.20	
	7005	Electric, 10 ton		29.60	715	2,142.65	6,425	665.30	
	7010	Muck cars, 1/2 C.Y. capacity		2.33	27	81	243	34.80	
	7020	1 C.Y. capacity		2.54	35	105.56	315	41.45	
	7030	2 C.Y. capacity		2.69	39.50	118.11	355	45.10	
	7040	Side dump, 2 C.Y. capacity		2.90	49	146.33	440	52.50	
	7050	3 C.Y. capacity		3.90	53.50	160.96	485	63.40	
	7060	5 C.Y. capacity		5.69	69.50	207.99	625	87.10	
	7100	Ventilating blower for tunnel, 7-1/2 H.P.		2.16	53.50	159.91	480	49.30	
	7110	10 H.P.		2.45	55.50	167.23	500	53.05	
	7120	20 H.P.		3.58	72.50	217.40	650	72.15	
	7140	40 H.P.		6.21	96	287.43	860	107.20	
	7160	60 H.P.		8.79	103	308.33	925	132	
	7175	75 H.P.		10.49	160	480.79	1,450	180.10	
	7180	200 H.P.		21.03	315	945.90	2,850	357.45	
	7800	Windrow loader, elevating		54.49	1,700	5,125	15,400	1,461	
60	0010	**LIFTING AND HOISTING EQUIPMENT RENTAL** without operators							**60**
	0150	Crane, flatbed mounted, 3 ton capacity	Ea.	14.58	203	610.30	1,825	238.75	
	0200	Crane, climbing, 106' jib, 6000 lb. capacity, 410 fpm		40.13	2,675	8,034	24,100	1,928	
	0300	101' jib, 10,250 lb. capacity, 270 fpm		46.90	2,300	6,868.82	20,600	1,749	
	0500	Tower, static, 130' high, 106' jib, 6200 lb. capacity at 400 fpm		45.61	2,300	6,916	20,700	1,748	
	0520	Mini crawler spider crane, up to 24" wide, 1990 lb. lifting capacity		12.65	555	1,672.31	5,025	435.70	
	0525	Up to 30" wide, 6450 lb. lifting capacity		14.70	660	1,985.87	5,950	514.75	
	0530	Up to 52" wide, 6680 lb. lifting capacity		23.38	810	2,430.08	7,300	673.05	

Reference boxes in row 0010: R015433 -10

Reference box in row 0200: R312316 -45

01 54 33 | Equipment Rental

		UNIT	HOURLY OPER. COST	RENT PER DAY	RENT PER WEEK	RENT PER MONTH	EQUIPMENT COST/DAY
0535	Up to 55" wide, 8920 lb. lifting capacity	Ea.	26.10	895	2,691.37	8,075	747.10
0540	Up to 66" wide, 13,350 lb. lifting capacity		35.34	1,400	4,180.77	12,500	1,119
0600	Crawler mounted, lattice boom, 1/2 C.Y., 15 tons at 12' radius		37.34	855	2,558	7,675	810.30
0700	3/4 C.Y., 20 tons at 12' radius		54.72	960	2,874	8,625	1,013
0800	1 C.Y., 25 tons at 12' radius		68.10	1,025	3,038	9,125	1,152
0900	1-1/2 C.Y., 40 tons at 12' radius		66.98	1,150	3,476	10,400	1,231
1000	2 C.Y., 50 tons at 12' radius		89.66	1,375	4,120	12,400	1,541
1100	3 C.Y., 75 tons at 12' radius		65.65	2,400	7,210	21,600	1,967
1200	100 ton capacity, 60' boom		86.78	2,700	8,080.96	24,200	2,310
1300	165 ton capacity, 60' boom		107.18	3,025	9,091.08	27,300	2,676
1400	200 ton capacity, 70' boom		139.61	3,875	11,616.38	34,800	3,440
1500	350 ton capacity, 80' boom		184.04	4,200	12,626.50	37,900	3,998
1600	Truck mounted, lattice boom, 6 x 4, 20 tons at 10' radius		40.16	2,000	6,026	18,100	1,526
1700	25 tons at 10' radius		43.16	2,400	7,210	21,600	1,787
1800	8 x 4, 30 tons at 10' radius		54.71	2,575	7,725	23,200	1,983
1900	40 tons at 12' radius		54.71	2,825	8,446	25,300	2,127
2000	60 tons at 15' radius		54.23	1,675	5,000.09	15,000	1,434
2050	82 tons at 15' radius		60.03	1,800	5,404.14	16,200	1,561
2100	90 tons at 15' radius		67.06	1,950	5,883.95	17,700	1,713
2200	115 tons at 15' radius		75.66	2,200	6,591.03	19,800	1,923
2300	150 tons at 18' radius		81.91	2,775	8,343	25,000	2,324
2350	165 tons at 18' radius		87.91	2,825	8,498	25,500	2,403
2400	Truck mounted, hydraulic, 12 ton capacity		29.80	395	1,186.89	3,550	475.80
2500	25 ton capacity		36.72	490	1,464.67	4,400	586.70
2550	33 ton capacity		54.71	910	2,727.32	8,175	983.15
2560	40 ton capacity		54.71	910	2,727.32	8,175	983.15
2600	55 ton capacity		54.32	925	2,777.83	8,325	990.15
2700	80 ton capacity		71.13	1,475	4,444.53	13,300	1,458
2720	100 ton capacity		75.72	1,575	4,722.31	14,200	1,550
2740	120 ton capacity		106.34	1,850	5,555.66	16,700	1,962
2760	150 ton capacity		113.80	2,050	6,186.99	18,600	2,148
2800	Self-propelled, 4 x 4, with telescoping boom, 5 ton		15.29	435	1,298	3,900	381.95
2900	12-1/2 ton capacity		21.63	435	1,298	3,900	432.65
3000	15 ton capacity		34.77	455	1,363.66	4,100	550.85
3050	20 ton capacity		22.55	655	1,969.73	5,900	574.35
3100	25 ton capacity		37.06	1,425	4,293.01	12,900	1,155
3150	40 ton capacity		45.35	665	1,994.99	5,975	761.80
3200	Derricks, guy, 20 ton capacity, 60' boom, 75' mast		22.97	1,450	4,378	13,100	1,059
3300	100' boom, 115' mast		36.41	2,050	6,180	18,500	1,527
3400	Stiffleg, 20 ton capacity, 70' boom, 37' mast		25.66	635	1,906	5,725	586.50
3500	100' boom, 47' mast		39.71	685	2,060	6,175	729.70
3550	Helicopter, small, lift to 1250 lb. maximum, w/pilot		100.14	2,175	6,500.12	19,500	2,101
3600	Hoists, chain type, overhead, manual, 3/4 ton		.15	10.25	30.70	92	7.30
3900	10 ton		.80	6.25	18.78	56.50	10.10
4000	Hoist and tower, 5000 lb. cap., portable electric, 40' high		4.82	146	438	1,325	126.15
4100	For each added 10' section, add		.12	32.50	98	294	20.55
4200	Hoist and single tubular tower, 5000 lb. electric, 100' high		7.03	108	325	975	121.20
4300	For each added 6'-6" section, add		.21	39.50	118	355	25.25
4400	Hoist and double tubular tower, 5000 lb., 100' high		7.65	108	325	975	126.20
4500	For each added 6'-6" section, add		.23	42.50	128	385	27.45
4550	Hoist and tower, mast type, 6000 lb., 100' high		9.32	95.50	286.87	860	131.95
4570	For each added 10' section, add		.13	32.50	98	294	20.65
4600	Hoist and tower, personnel, electric, 2000 lb., 100' @ 125 fpm		17.67	25.50	75.76	227	156.55
4700	3000 lb., 100' @ 200 fpm		20.22	25.50	75.76	227	176.95
4800	3000 lb., 150' @ 300 fpm		22.44	25.50	75.76	227	194.70
4900	4000 lb., 100' @ 300 fpm		23.22	25.50	75.76	227	200.90
5000	6000 lb., 100' @ 275 fpm		24.95	25.50	75.76	227	214.75
5100	For added heights up to 500', add	L.F.	.01	3.33	10	30	2.10

60

01 54 33 | Equipment Rental

			UNIT	HOURLY OPER. COST	RENT PER DAY	RENT PER WEEK	RENT PER MONTH	EQUIPMENT COST/DAY	
60	5200	Jacks, hydraulic, 20 ton	Ea.	.05	19.85	59.60	179	12.30	60
	5500	100 ton		.40	26	78.50	236	18.95	
	6100	Jacks, hydraulic, climbing w/50' jackrods, control console, 30 ton cap.		2.19	32	96	288	36.70	
	6150	For each added 10' jackrod section, add		.05	5	15	45	3.40	
	6300	50 ton capacity		3.52	34.50	103	310	48.75	
	6350	For each added 10' jackrod section, add		.06	5	15	45	3.50	
	6500	125 ton capacity		9.20	53.50	160	480	105.55	
	6550	For each added 10' jackrod section, add		.62	5	15	45	8	
	6600	Cable jack, 10 ton capacity with 200' cable		1.60	36.50	110	330	34.80	
	6650	For each added 50' of cable, add		.22	15.35	46	138	11	
70	0010	**WELLPOINT EQUIPMENT RENTAL** without operators [R015433 -10]							70
	0020	Based on 2 months rental							
	0100	Combination jetting & wellpoint pump, 60 H.P. diesel	Ea.	15.83	305	922	2,775	311.05	
	0200	High pressure gas jet pump, 200 H.P., 300 psi	"	34.17	278	833.35	2,500	440.05	
	0300	Discharge pipe, 8" diameter	L.F.	.01	1.41	4.24	12.75	.90	
	0350	12" diameter		.01	2.09	6.26	18.80	1.35	
	0400	Header pipe, flows up to 150 GPM, 4" diameter		.01	.74	2.22	6.65	.50	
	0500	400 GPM, 6" diameter		.01	1.08	3.23	9.70	.70	
	0600	800 GPM, 8" diameter		.01	1.41	4.24	12.75	.95	
	0700	1500 GPM, 10" diameter		.01	1.75	5.25	15.75	1.15	
	0800	2500 GPM, 12" diameter		.03	2.09	6.26	18.80	1.45	
	0900	4500 GPM, 16" diameter		.03	2.42	7.27	22	1.70	
	0950	For quick coupling aluminum and plastic pipe, add		.03	9.60	28.84	86.50	6.05	
	1100	Wellpoint, 25' long, with fittings & riser pipe, 1-1/2" or 2" diameter	Ea.	.07	133	399	1,200	80.35	
	1200	Wellpoint pump, diesel powered, 4" suction, 20 H.P.		7.07	152	454.55	1,375	147.50	
	1300	6" suction, 30 H.P.		9.49	140	420	1,250	159.90	
	1400	8" suction, 40 H.P.		12.85	253	757.59	2,275	254.35	
	1500	10" suction, 75 H.P.		18.75	268	803.05	2,400	310.60	
	1600	12" suction, 100 H.P.		27.51	300	904.06	2,700	400.90	
	1700	12" suction, 175 H.P.		39.37	320	959.61	2,875	506.90	
80	0010	**MARINE EQUIPMENT RENTAL** without operators [R015433 -10]	Ea.						80
	0200	Barge, 400 Ton, 30' wide x 90' long		17.84	1,200	3,632.05	10,900	869.15	
	0240	800 Ton, 45' wide x 90' long		22.41	1,500	4,468.20	13,400	1,073	
	2000	Tugboat, diesel, 100 H.P.		29.92	240	721.18	2,175	383.65	
	2040	250 H.P.		58.10	435	1,306.49	3,925	726.10	
	2080	380 H.P.		126.49	1,300	3,919.48	11,800	1,796	
	3000	Small work boat, gas, 16-foot, 50 H.P.		11.48	48.50	145.01	435	120.85	
	4000	Large, diesel, 48-foot, 200 H.P.		75.57	1,375	4,154.64	12,500	1,436	

Crews - Standard

Crew No.	Bare Costs Hr.	Daily	Incl. Subs O&P Hr.	Daily	Cost Per Labor-Hour Bare Costs	Incl. O&P
Crew A-1	Hr.	Daily	Hr.	Daily	Bare Costs	Incl. O&P
1 Building Laborer	$44.40	$355.20	$66.25	$530.00	$44.40	$66.25
1 Concrete Saw, Gas Manual		113.70		125.07	14.21	15.63
8 L.H., Daily Totals		$468.90		$655.07	$58.61	$81.88
Crew A-1A	Hr.	Daily	Hr.	Daily	Bare Costs	Incl. O&P
1 Skilled Worker	$57.10	$456.80	$85.90	$687.20	$57.10	$85.90
1 Shot Blaster, 20"		208.70		229.57	26.09	28.70
8 L.H., Daily Totals		$665.50		$916.77	$83.19	$114.60
Crew A-1B	Hr.	Daily	Hr.	Daily	Bare Costs	Incl. O&P
1 Building Laborer	$44.40	$355.20	$66.25	$530.00	$44.40	$66.25
1 Concrete Saw		112.85		124.14	14.11	15.52
8 L.H., Daily Totals		$468.05		$654.13	$58.51	$81.77
Crew A-1C	Hr.	Daily	Hr.	Daily	Bare Costs	Incl. O&P
1 Building Laborer	$44.40	$355.20	$66.25	$530.00	$44.40	$66.25
1 Chain Saw, Gas, 18"		52.20		57.42	6.53	7.18
8 L.H., Daily Totals		$407.40		$587.42	$50.92	$73.43
Crew A-1D	Hr.	Daily	Hr.	Daily	Bare Costs	Incl. O&P
1 Building Laborer	$44.40	$355.20	$66.25	$530.00	$44.40	$66.25
1 Vibrating Plate, Gas, 18"		31.90		35.09	3.99	4.39
8 L.H., Daily Totals		$387.10		$565.09	$48.39	$70.64
Crew A-1E	Hr.	Daily	Hr.	Daily	Bare Costs	Incl. O&P
1 Building Laborer	$44.40	$355.20	$66.25	$530.00	$44.40	$66.25
1 Vibrating Plate, Gas, 21"		165.60		182.16	20.70	22.77
8 L.H., Daily Totals		$520.80		$712.16	$65.10	$89.02
Crew A-1F	Hr.	Daily	Hr.	Daily	Bare Costs	Incl. O&P
1 Building Laborer	$44.40	$355.20	$66.25	$530.00	$44.40	$66.25
1 Rammer/Tamper, Gas, 8"		47.00		51.70	5.88	6.46
8 L.H., Daily Totals		$402.20		$581.70	$50.27	$72.71
Crew A-1G	Hr.	Daily	Hr.	Daily	Bare Costs	Incl. O&P
1 Building Laborer	$44.40	$355.20	$66.25	$530.00	$44.40	$66.25
1 Rammer/Tamper, Gas, 15"		54.65		60.12	6.83	7.51
8 L.H., Daily Totals		$409.85		$590.12	$51.23	$73.76
Crew A-1H	Hr.	Daily	Hr.	Daily	Bare Costs	Incl. O&P
1 Building Laborer	$44.40	$355.20	$66.25	$530.00	$44.40	$66.25
1 Exterior Steam Cleaner		77.20		84.92	9.65	10.62
8 L.H., Daily Totals		$432.40		$614.92	$54.05	$76.86
Crew A-1J	Hr.	Daily	Hr.	Daily	Bare Costs	Incl. O&P
1 Building Laborer	$44.40	$355.20	$66.25	$530.00	$44.40	$66.25
1 Cultivator, Walk-Behind, 5 H.P.		53.25		58.58	6.66	7.32
8 L.H., Daily Totals		$408.45		$588.58	$51.06	$73.57
Crew A-1K	Hr.	Daily	Hr.	Daily	Bare Costs	Incl. O&P
1 Building Laborer	$44.40	$355.20	$66.25	$530.00	$44.40	$66.25
1 Cultivator, Walk-Behind, 8 H.P.		92.20		101.42	11.53	12.68
8 L.H., Daily Totals		$447.40		$631.42	$55.92	$78.93
Crew A-1M	Hr.	Daily	Hr.	Daily	Bare Costs	Incl. O&P
1 Building Laborer	$44.40	$355.20	$66.25	$530.00	$44.40	$66.25
1 Snow Blower, Walk-Behind		68.30		75.13	8.54	9.39
8 L.H., Daily Totals		$423.50		$605.13	$52.94	$75.64

Crew No.	Bare Costs Hr.	Daily	Incl. Subs O&P Hr.	Daily	Cost Per Labor-Hour Bare Costs	Incl. O&P
Crew A-2	Hr.	Daily	Hr.	Daily	Bare Costs	Incl. O&P
2 Laborers	$44.40	$710.40	$66.25	$1060.00	$45.87	$68.50
1 Truck Driver (light)	48.80	390.40	73.00	584.00		
1 Flatbed Truck, Gas, 1.5 Ton		198.50		218.35	8.27	9.10
24 L.H., Daily Totals		$1299.30		$1862.35	$54.14	$77.60
Crew A-2A	Hr.	Daily	Hr.	Daily	Bare Costs	Incl. O&P
2 Laborers	$44.40	$710.40	$66.25	$1060.00	$45.87	$68.50
1 Truck Driver (light)	48.80	390.40	73.00	584.00		
1 Flatbed Truck, Gas, 1.5 Ton		198.50		218.35		
1 Concrete Saw		112.85		124.14	12.97	14.27
24 L.H., Daily Totals		$1412.15		$1986.48	$58.84	$82.77
Crew A-2B	Hr.	Daily	Hr.	Daily	Bare Costs	Incl. O&P
1 Truck Driver (light)	$48.80	$390.40	$73.00	$584.00	$48.80	$73.00
1 Flatbed Truck, Gas, 1.5 Ton		198.50		218.35	24.81	27.29
8 L.H., Daily Totals		$588.90		$802.35	$73.61	$100.29
Crew A-3A	Hr.	Daily	Hr.	Daily	Bare Costs	Incl. O&P
1 Equip. Oper. (light)	$55.50	$444.00	$82.70	$661.60	$55.50	$82.70
1 Pickup Truck, 4x4, 3/4 Ton		176.75		194.43	22.09	24.30
8 L.H., Daily Totals		$620.75		$856.02	$77.59	$107.00
Crew A-3B	Hr.	Daily	Hr.	Daily	Bare Costs	Incl. O&P
1 Equip. Oper. (medium)	$59.00	$472.00	$87.90	$703.20	$55.15	$82.30
1 Truck Driver (heavy)	51.30	410.40	76.70	613.60		
1 Dump Truck, 12 C.Y., 400 H.P.		579.35		637.28		
1 F.E. Loader, W.M., 2.5 C.Y.		638.30		702.13	76.10	83.71
16 L.H., Daily Totals		$2100.05		$2656.22	$131.25	$166.01
Crew A-3C	Hr.	Daily	Hr.	Daily	Bare Costs	Incl. O&P
1 Equip. Oper. (light)	$55.50	$444.00	$82.70	$661.60	$55.50	$82.70
1 Loader, Skid Steer, 78 H.P.		446.30		490.93	55.79	61.37
8 L.H., Daily Totals		$890.30		$1152.53	$111.29	$144.07
Crew A-3D	Hr.	Daily	Hr.	Daily	Bare Costs	Incl. O&P
1 Truck Driver (light)	$48.80	$390.40	$73.00	$584.00	$48.80	$73.00
1 Pickup Truck, 4x4, 3/4 Ton		176.75		194.43		
1 Flatbed Trailer, 25 Ton		137.20		150.92	39.24	43.17
8 L.H., Daily Totals		$704.35		$929.35	$88.04	$116.17
Crew A-3E	Hr.	Daily	Hr.	Daily	Bare Costs	Incl. O&P
1 Equip. Oper. (crane)	$61.45	$491.60	$91.55	$732.40	$56.38	$84.13
1 Truck Driver (heavy)	51.30	410.40	76.70	613.60		
1 Pickup Truck, 4x4, 3/4 Ton		176.75		194.43	11.05	12.15
16 L.H., Daily Totals		$1078.75		$1540.43	$67.42	$96.28
Crew A-3F	Hr.	Daily	Hr.	Daily	Bare Costs	Incl. O&P
1 Equip. Oper. (crane)	$61.45	$491.60	$91.55	$732.40	$56.38	$84.13
1 Truck Driver (heavy)	51.30	410.40	76.70	613.60		
1 Pickup Truck, 4x4, 3/4 Ton		176.75		194.43		
1 Truck Tractor, 6x4, 380 H.P.		499.15		549.07		
1 Lowbed Trailer, 75 Ton		258.10		283.91	58.38	64.21
16 L.H., Daily Totals		$1836.00		$2373.40	$114.75	$148.34

415

Crew No.	Bare Costs		Incl. Subs O&P		Cost Per Labor-Hour	
	Hr.	Daily	Hr.	Daily	Bare Costs	Incl. O&P
Crew A-3G	Hr.	Daily	Hr.	Daily	Bare Costs	Incl. O&P
1 Equip. Oper. (crane)	$61.45	$491.60	$91.55	$732.40	$56.38	$84.13
1 Truck Driver (heavy)	51.30	410.40	76.70	613.60		
1 Pickup Truck, 4x4, 3/4 Ton		176.75		194.43		
1 Truck Tractor, 6x4, 450 H.P.		608.95		669.85		
1 Lowbed Trailer, 75 Ton		258.10		283.91	65.24	71.76
16 L.H., Daily Totals		$1945.80		$2494.18	$121.61	$155.89
Crew A-3H	Hr.	Daily	Hr.	Daily	Bare Costs	Incl. O&P
1 Equip. Oper. (crane)	$61.45	$491.60	$91.55	$732.40	$61.45	$91.55
1 Hyd. Crane, 12 Ton (Daily)		733.15		806.47	91.64	100.81
8 L.H., Daily Totals		$1224.75		$1538.87	$153.09	$192.36
Crew A-3I	Hr.	Daily	Hr.	Daily	Bare Costs	Incl. O&P
1 Equip. Oper. (crane)	$61.45	$491.60	$91.55	$732.40	$61.45	$91.55
1 Hyd. Crane, 25 Ton (Daily)		810.50		891.55	101.31	111.44
8 L.H., Daily Totals		$1302.10		$1623.95	$162.76	$202.99
Crew A-3J	Hr.	Daily	Hr.	Daily	Bare Costs	Incl. O&P
1 Equip. Oper. (crane)	$61.45	$491.60	$91.55	$732.40	$61.45	$91.55
1 Hyd. Crane, 40 Ton (Daily)		1287.00		1415.70	160.88	176.96
8 L.H., Daily Totals		$1778.60		$2148.10	$222.32	$268.51
Crew A-3K	Hr.	Daily	Hr.	Daily	Bare Costs	Incl. O&P
1 Equip. Oper. (crane)	$61.45	$491.60	$91.55	$732.40	$56.98	$84.90
1 Equip. Oper. (oiler)	52.50	420.00	78.25	626.00		
1 Hyd. Crane, 55 Ton (Daily)		1377.00		1514.70		
1 P/U Truck, 3/4 Ton (Daily)		143.85		158.24	95.05	104.56
16 L.H., Daily Totals		$2432.45		$3031.34	$152.03	$189.46
Crew A-3L	Hr.	Daily	Hr.	Daily	Bare Costs	Incl. O&P
1 Equip. Oper. (crane)	$61.45	$491.60	$91.55	$732.40	$56.98	$84.90
1 Equip. Oper. (oiler)	52.50	420.00	78.25	626.00		
1 Hyd. Crane, 80 Ton (Daily)		2058.00		2263.80		
1 P/U Truck, 3/4 Ton (Daily)		143.85		158.24	137.62	151.38
16 L.H., Daily Totals		$3113.45		$3780.43	$194.59	$236.28
Crew A-3M	Hr.	Daily	Hr.	Daily	Bare Costs	Incl. O&P
1 Equip. Oper. (crane)	$61.45	$491.60	$91.55	$732.40	$56.98	$84.90
1 Equip. Oper. (oiler)	52.50	420.00	78.25	626.00		
1 Hyd. Crane, 100 Ton (Daily)		2253.00		2478.30		
1 P/U Truck, 3/4 Ton (Daily)		143.85		158.24	149.80	164.78
16 L.H., Daily Totals		$3308.45		$3994.93	$206.78	$249.68
Crew A-3N	Hr.	Daily	Hr.	Daily	Bare Costs	Incl. O&P
1 Equip. Oper. (crane)	$61.45	$491.60	$91.55	$732.40	$61.45	$91.55
1 Tower Crane (monthly)		1737.00		1910.70	217.13	238.84
8 L.H., Daily Totals		$2228.60		$2643.10	$278.57	$330.39
Crew A-3P	Hr.	Daily	Hr.	Daily	Bare Costs	Incl. O&P
1 Equip. Oper. (light)	$55.50	$444.00	$82.70	$661.60	$55.50	$82.70
1 A.T. Forklift, 31' reach, 45' lift		346.45		381.10	43.31	47.64
8 L.H., Daily Totals		$790.45		$1042.69	$98.81	$130.34
Crew A-3Q	Hr.	Daily	Hr.	Daily	Bare Costs	Incl. O&P
1 Equip. Oper. (light)	$55.50	$444.00	$82.70	$661.60	$55.50	$82.70
1 Pickup Truck, 4x4, 3/4 Ton		176.75		194.43		
1 Flatbed Trailer, 3 Ton		71.15		78.27	30.99	34.09
8 L.H., Daily Totals		$691.90		$934.29	$86.49	$116.79

Crew No.	Bare Costs		Incl. Subs O&P		Cost Per Labor-Hour	
Crew A-3R	Hr.	Daily	Hr.	Daily	Bare Costs	Incl. O&P
1 Equip. Oper. (light)	$55.50	$444.00	$82.70	$661.60	$55.50	$82.70
1 Forklift, Smooth Floor, 8,000 Lb.		283.25		311.57	35.41	38.95
8 L.H., Daily Totals		$727.25		$973.17	$90.91	$121.65
Crew A-4	Hr.	Daily	Hr.	Daily	Bare Costs	Incl. O&P
2 Carpenters	$54.70	$875.20	$81.65	$1306.40	$51.95	$77.40
1 Painter, Ordinary	46.45	371.60	68.90	551.20		
24 L.H., Daily Totals		$1246.80		$1857.60	$51.95	$77.40
Crew A-5	Hr.	Daily	Hr.	Daily	Bare Costs	Incl. O&P
2 Laborers	$44.40	$710.40	$66.25	$1060.00	$44.89	$67.00
.25 Truck Driver (light)	48.80	97.60	73.00	146.00		
.25 Flatbed Truck, Gas, 1.5 Ton		49.63		54.59	2.76	3.03
18 L.H., Daily Totals		$857.63		$1260.59	$47.65	$70.03
Crew A-6	Hr.	Daily	Hr.	Daily	Bare Costs	Incl. O&P
1 Instrument Man	$57.10	$456.80	$85.90	$687.20	$54.88	$82.15
1 Rodman/Chainman	52.65	421.20	78.40	627.20		
1 Level, Electronic		34.40		37.84	2.15	2.37
16 L.H., Daily Totals		$912.40		$1352.24	$57.02	$84.52
Crew A-7	Hr.	Daily	Hr.	Daily	Bare Costs	Incl. O&P
1 Chief of Party	$68.50	$548.00	$102.35	$818.80	$59.42	$88.88
1 Instrument Man	57.10	456.80	85.90	687.20		
1 Rodman/Chainman	52.65	421.20	78.40	627.20		
1 Level, Electronic		34.40		37.84	1.43	1.58
24 L.H., Daily Totals		$1460.40		$2171.04	$60.85	$90.46
Crew A-8	Hr.	Daily	Hr.	Daily	Bare Costs	Incl. O&P
1 Chief of Party	$68.50	$548.00	$102.35	$818.80	$57.73	$86.26
1 Instrument Man	57.10	456.80	85.90	687.20		
2 Rodmen/Chainmen	52.65	842.40	78.40	1254.40		
1 Level, Electronic		34.40		37.84	1.08	1.18
32 L.H., Daily Totals		$1881.60		$2798.24	$58.80	$87.44
Crew A-9	Hr.	Daily	Hr.	Daily	Bare Costs	Incl. O&P
1 Asbestos Foreman	$61.45	$491.60	$94.00	$752.00	$61.01	$93.34
7 Asbestos Workers	60.95	3413.20	93.25	5222.00		
64 L.H., Daily Totals		$3904.80		$5974.00	$61.01	$93.34
Crew A-10A	Hr.	Daily	Hr.	Daily	Bare Costs	Incl. O&P
1 Asbestos Foreman	$61.45	$491.60	$94.00	$752.00	$61.12	$93.50
2 Asbestos Workers	60.95	975.20	93.25	1492.00		
24 L.H., Daily Totals		$1466.80		$2244.00	$61.12	$93.50
Crew A-10B	Hr.	Daily	Hr.	Daily	Bare Costs	Incl. O&P
1 Asbestos Foreman	$61.45	$491.60	$94.00	$752.00	$61.08	$93.44
3 Asbestos Workers	60.95	1462.80	93.25	2238.00		
32 L.H., Daily Totals		$1954.40		$2990.00	$61.08	$93.44
Crew A-10C	Hr.	Daily	Hr.	Daily	Bare Costs	Incl. O&P
3 Asbestos Workers	$60.95	$1462.80	$93.25	$2238.00	$60.95	$93.25
1 Flatbed Truck, Gas, 1.5 Ton		198.50		218.35	8.27	9.10
24 L.H., Daily Totals		$1661.30		$2456.35	$69.22	$102.35

Crew No.	Bare Costs		Incl. Subs O&P		Cost Per Labor-Hour	
Crew A-10D	Hr.	Daily	Hr.	Daily	Bare Costs	Incl. O&P
2 Asbestos Workers	$60.95	$975.20	$93.25	$1492.00	$58.96	$89.08
1 Equip. Oper. (crane)	61.45	491.60	91.55	732.40		
1 Equip. Oper. (oiler)	52.50	420.00	78.25	626.00		
1 Hydraulic Crane, 33 Ton		983.15		1081.46	30.72	33.80
32 L.H., Daily Totals		$2869.95		$3931.86	$89.69	$122.87

Crew A-11	Hr.	Daily	Hr.	Daily	Bare Costs	Incl. O&P
1 Asbestos Foreman	$61.45	$491.60	$94.00	$752.00	$61.01	$93.34
7 Asbestos Workers	60.95	3413.20	93.25	5222.00		
2 Chip. Hammers, 12 Lb., Elec.		65.70		72.27	1.03	1.13
64 L.H., Daily Totals		$3970.50		$6046.27	$62.04	$94.47

Crew A-12	Hr.	Daily	Hr.	Daily	Bare Costs	Incl. O&P
1 Asbestos Foreman	$61.45	$491.60	$94.00	$752.00	$61.01	$93.34
7 Asbestos Workers	60.95	3413.20	93.25	5222.00		
1 Trk-Mtd Vac, 14 CY, 1500 Gal.		542.60		596.86		
1 Flatbed Truck, 20,000 GVW		204.05		224.46	11.67	12.83
64 L.H., Daily Totals		$4651.45		$6795.31	$72.68	$106.18

Crew A-13	Hr.	Daily	Hr.	Daily	Bare Costs	Incl. O&P
1 Equip. Oper. (light)	$55.50	$444.00	$82.70	$661.60	$55.50	$82.70
1 Trk-Mtd Vac, 14 CY, 1500 Gal.		542.60		596.86		
1 Flatbed Truck, 20,000 GVW		204.05		224.46	93.33	102.66
8 L.H., Daily Totals		$1190.65		$1482.92	$148.83	$185.36

Crew B-1	Hr.	Daily	Hr.	Daily	Bare Costs	Incl. O&P
1 Labor Foreman (outside)	$46.40	$371.20	$69.25	$554.00	$45.07	$67.25
2 Laborers	44.40	710.40	66.25	1060.00		
24 L.H., Daily Totals		$1081.60		$1614.00	$45.07	$67.25

Crew B-1A	Hr.	Daily	Hr.	Daily	Bare Costs	Incl. O&P
1 Labor Foreman (outside)	$46.40	$371.20	$69.25	$554.00	$45.07	$67.25
2 Laborers	44.40	710.40	66.25	1060.00		
2 Cutting Torches		25.90		28.49		
2 Sets of Gases		347.20		381.92	15.55	17.10
24 L.H., Daily Totals		$1454.70		$2024.41	$60.61	$84.35

Crew B-1B	Hr.	Daily	Hr.	Daily	Bare Costs	Incl. O&P
1 Labor Foreman (outside)	$46.40	$371.20	$69.25	$554.00	$49.16	$73.33
2 Laborers	44.40	710.40	66.25	1060.00		
1 Equip. Oper. (crane)	61.45	491.60	91.55	732.40		
2 Cutting Torches		25.90		28.49		
2 Sets of Gases		347.20		381.92		
1 Hyd. Crane, 12 Ton		475.80		523.38	26.53	29.18
32 L.H., Daily Totals		$2422.10		$3280.19	$75.69	$102.51

Crew B-1C	Hr.	Daily	Hr.	Daily	Bare Costs	Incl. O&P
1 Labor Foreman (outside)	$46.40	$371.20	$69.25	$554.00	$45.07	$67.25
2 Laborers	44.40	710.40	66.25	1060.00		
1 Telescoping Boom Lift, to 60'		292.45		321.69	12.19	13.40
24 L.H., Daily Totals		$1374.05		$1935.69	$57.25	$80.65

Crew B-1D	Hr.	Daily	Hr.	Daily	Bare Costs	Incl. O&P
2 Laborers	$44.40	$710.40	$66.25	$1060.00	$44.40	$66.25
1 Small Work Boat, Gas, 50 H.P.		120.85		132.94		
1 Pressure Washer, 7 GPM		93.95		103.35	13.43	14.77
16 L.H., Daily Totals		$925.20		$1296.28	$57.83	$81.02

Crew No.	Bare Costs		Incl. Subs O&P		Cost Per Labor-Hour	
Crew B-1E	Hr.	Daily	Hr.	Daily	Bare Costs	Incl. O&P
1 Labor Foreman (outside)	$46.40	$371.20	$69.25	$554.00	$44.90	$67.00
3 Laborers	44.40	1065.60	66.25	1590.00		
1 Work Boat, Diesel, 200 H.P.		1436.00		1579.60		
2 Pressure Washers, 7 GPM		187.90		206.69	50.75	55.82
32 L.H., Daily Totals		$3060.70		$3930.29	$95.65	$122.82

Crew B-1F	Hr.	Daily	Hr.	Daily	Bare Costs	Incl. O&P
2 Skilled Workers	$57.10	$913.60	$85.90	$1374.40	$52.87	$79.35
1 Laborer	44.40	355.20	66.25	530.00		
1 Small Work Boat, Gas, 50 H.P.		120.85		132.94		
1 Pressure Washer, 7 GPM		93.95		103.35	8.95	9.85
24 L.H., Daily Totals		$1483.60		$2140.68	$61.82	$89.19

Crew B-1G	Hr.	Daily	Hr.	Daily	Bare Costs	Incl. O&P
2 Laborers	$44.40	$710.40	$66.25	$1060.00	$44.40	$66.25
1 Small Work Boat, Gas, 50 H.P.		120.85		132.94	7.55	8.31
16 L.H., Daily Totals		$831.25		$1192.93	$51.95	$74.56

Crew B-1H	Hr.	Daily	Hr.	Daily	Bare Costs	Incl. O&P
2 Skilled Workers	$57.10	$913.60	$85.90	$1374.40	$52.87	$79.35
1 Laborer	44.40	355.20	66.25	530.00		
1 Small Work Boat, Gas, 50 H.P.		120.85		132.94	5.04	5.54
24 L.H., Daily Totals		$1389.65		$2037.34	$57.90	$84.89

Crew B-1J	Hr.	Daily	Hr.	Daily	Bare Costs	Incl. O&P
1 Labor Foreman (inside)	$44.90	$359.20	$67.00	$536.00	$44.65	$66.63
1 Laborer	44.40	355.20	66.25	530.00		
16 L.H., Daily Totals		$714.40		$1066.00	$44.65	$66.63

Crew B-1K	Hr.	Daily	Hr.	Daily	Bare Costs	Incl. O&P
1 Carpenter Foreman (inside)	$55.20	$441.60	$82.40	$659.20	$54.95	$82.03
1 Carpenter	54.70	437.60	81.65	653.20		
16 L.H., Daily Totals		$879.20		$1312.40	$54.95	$82.03

Crew B-2	Hr.	Daily	Hr.	Daily	Bare Costs	Incl. O&P
1 Labor Foreman (outside)	$46.40	$371.20	$69.25	$554.00	$44.80	$66.85
4 Laborers	44.40	1420.80	66.25	2120.00		
40 L.H., Daily Totals		$1792.00		$2674.00	$44.80	$66.85

Crew B-2A	Hr.	Daily	Hr.	Daily	Bare Costs	Incl. O&P
1 Labor Foreman (outside)	$46.40	$371.20	$69.25	$554.00	$45.07	$67.25
2 Laborers	44.40	710.40	66.25	1060.00		
1 Telescoping Boom Lift, to 60'		292.45		321.69	12.19	13.40
24 L.H., Daily Totals		$1374.05		$1935.69	$57.25	$80.65

Crew B-3	Hr.	Daily	Hr.	Daily	Bare Costs	Incl. O&P
1 Labor Foreman (outside)	$46.40	$371.20	$69.25	$554.00	$49.47	$73.84
2 Laborers	44.40	710.40	66.25	1060.00		
1 Equip. Oper. (medium)	59.00	472.00	87.90	703.20		
2 Truck Drivers (heavy)	51.30	820.80	76.70	1227.20		
1 Crawler Loader, 3 C.Y.		1146.00		1260.60		
2 Dump Trucks, 12 C.Y., 400 H.P.		1158.70		1274.57	48.01	52.82
48 L.H., Daily Totals		$4679.10		$6079.57	$97.48	$126.66

Crew B-3A	Hr.	Daily	Hr.	Daily	Bare Costs	Incl. O&P
4 Laborers	$44.40	$1420.80	$66.25	$2120.00	$47.32	$70.58
1 Equip. Oper. (medium)	59.00	472.00	87.90	703.20		
1 Hyd. Excavator, 1.5 C.Y.		695.80		765.38	17.40	19.13
40 L.H., Daily Totals		$2588.60		$3588.58	$64.72	$89.71

Crew No.	Bare Costs Hr.	Daily	Incl. Subs O&P Hr.	Daily	Cost Per Labor-Hour Bare Costs	Incl. O&P
Crew B-3B						
2 Laborers	$44.40	$710.40	$66.25	$1060.00	$49.77	$74.28
1 Equip. Oper. (medium)	59.00	472.00	87.90	703.20		
1 Truck Driver (heavy)	51.30	410.40	76.70	613.60		
1 Backhoe Loader, 80 H.P.		235.05		258.56		
1 Dump Truck, 12 C.Y., 400 H.P.		579.35		637.28	25.45	28.00
32 L.H., Daily Totals		$2407.20		$3272.64	$75.22	$102.27
Crew B-3C						
3 Laborers	$44.40	$1065.60	$66.25	$1590.00	$48.05	$71.66
1 Equip. Oper. (medium)	59.00	472.00	87.90	703.20		
1 Crawler Loader, 4 C.Y.		1456.00		1601.60	45.50	50.05
32 L.H., Daily Totals		$2993.60		$3894.80	$93.55	$121.71
Crew B-4						
1 Labor Foreman (outside)	$46.40	$371.20	$69.25	$554.00	$45.88	$68.49
4 Laborers	44.40	1420.80	66.25	2120.00		
1 Truck Driver (heavy)	51.30	410.40	76.70	613.60		
1 Truck Tractor, 220 H.P.		310.80		341.88		
1 Flatbed Trailer, 40 Ton		188.45		207.29	10.40	11.44
48 L.H., Daily Totals		$2701.65		$3836.78	$56.28	$79.93
Crew B-5						
1 Labor Foreman (outside)	$46.40	$371.20	$69.25	$554.00	$48.86	$72.86
4 Laborers	44.40	1420.80	66.25	2120.00		
2 Equip. Oper. (medium)	59.00	944.00	87.90	1406.40		
1 Air Compressor, 250 cfm		202.85		223.13		
2 Breakers, Pavement, 60 lb.		107.20		117.92		
2 -50' Air Hoses, 1.5"		45.60		50.16		
1 Crawler Loader, 3 C.Y.		1146.00		1260.60	26.82	29.50
56 L.H., Daily Totals		$4237.65		$5732.22	$75.67	$102.36
Crew B-5A						
1 Labor Foreman (outside)	$46.40	$371.20	$69.25	$554.00	$49.08	$73.22
6 Laborers	44.40	2131.20	66.25	3180.00		
2 Equip. Oper. (medium)	59.00	944.00	87.90	1406.40		
1 Equip. Oper. (light)	55.50	444.00	82.70	661.60		
2 Truck Drivers (heavy)	51.30	820.80	76.70	1227.20		
1 Air Compressor, 365 cfm		343.55		377.90		
2 Breakers, Pavement, 60 lb.		107.20		117.92		
8 -50' Air Hoses, 1"		64.40		70.84		
2 Dump Trucks, 8 C.Y., 220 H.P.		815.20		896.72	13.86	15.24
96 L.H., Daily Totals		$6041.55		$8492.58	$62.93	$88.46
Crew B-5B						
1 Powderman	$57.10	$456.80	$85.90	$687.20	$54.83	$81.97
2 Equip. Oper. (medium)	59.00	944.00	87.90	1406.40		
3 Truck Drivers (heavy)	51.30	1231.20	76.70	1840.80		
1 F.E. Loader, W.M., 2.5 C.Y.		638.30		702.13		
3 Dump Trucks, 12 C.Y., 400 H.P.		1738.05		1911.86		
1 Air Compressor, 365 cfm		343.55		377.90	56.66	62.33
48 L.H., Daily Totals		$5351.90		$6926.29	$111.50	$144.30

Crew No.	Bare Costs Hr.	Daily	Incl. Subs O&P Hr.	Daily	Cost Per Labor-Hour Bare Costs	Incl. O&P
Crew B-5C						
3 Laborers	$44.40	$1065.60	$66.25	$1590.00	$51.09	$76.23
1 Equip. Oper. (medium)	59.00	472.00	87.90	703.20		
2 Truck Drivers (heavy)	51.30	820.80	76.70	1227.20		
1 Equip. Oper. (crane)	61.45	491.60	91.55	732.40		
1 Equip. Oper. (oiler)	52.50	420.00	78.25	626.00		
2 Dump Trucks, 12 C.Y., 400 H.P.		1158.70		1274.57		
1 Crawler Loader, 4 C.Y.		1456.00		1601.60		
1 S.P. Crane, 4x4, 25 Ton		1155.00		1270.50	58.90	64.79
64 L.H., Daily Totals		$7039.70		$9025.47	$110.00	$141.02
Crew B-5D						
1 Labor Foreman (outside)	$46.40	$371.20	$69.25	$554.00	$49.16	$73.34
4 Laborers	44.40	1420.80	66.25	2120.00		
2 Equip. Oper. (medium)	59.00	944.00	87.90	1406.40		
1 Truck Driver (heavy)	51.30	410.40	76.70	613.60		
1 Air Compressor, 250 cfm		202.85		223.13		
2 Breakers, Pavement, 60 lb.		107.20		117.92		
2 -50' Air Hoses, 1.5"		45.60		50.16		
1 Crawler Loader, 3 C.Y.		1146.00		1260.60		
1 Dump Truck, 12 C.Y., 400 H.P.		579.35		637.28	32.52	35.77
64 L.H., Daily Totals		$5227.40		$6983.10	$81.68	$109.11
Crew B-5E						
1 Labor Foreman (outside)	$46.40	$371.20	$69.25	$554.00	$49.16	$73.34
4 Laborers	44.40	1420.80	66.25	2120.00		
2 Equip. Oper. (medium)	59.00	944.00	87.90	1406.40		
1 Truck Driver (heavy)	51.30	410.40	76.70	613.60		
1 Water Tank Trailer, 5000 Gal.		154.10		169.51		
1 High Pressure Water Jet 40 KSI		820.75		902.83		
2 -50' Air Hoses, 1.5"		45.60		50.16		
1 Crawler Loader, 3 C.Y.		1146.00		1260.60		
1 Dump Truck, 12 C.Y., 400 H.P.		579.35		637.28	42.90	47.19
64 L.H., Daily Totals		$5892.20		$7714.38	$92.07	$120.54
Crew B-6						
2 Laborers	$44.40	$710.40	$66.25	$1060.00	$48.10	$71.73
1 Equip. Oper. (light)	55.50	444.00	82.70	661.60		
1 Backhoe Loader, 48 H.P.		216.20		237.82	9.01	9.91
24 L.H., Daily Totals		$1370.60		$1959.42	$57.11	$81.64
Crew B-6A						
.5 Labor Foreman (outside)	$46.40	$185.60	$69.25	$277.00	$50.64	$75.51
1 Laborer	44.40	355.20	66.25	530.00		
1 Equip. Oper. (medium)	59.00	472.00	87.90	703.20		
1 Vacuum Truck, 5000 Gal.		371.95		409.14	18.60	20.46
20 L.H., Daily Totals		$1384.75		$1919.35	$69.24	$95.97
Crew B-6B						
2 Labor Foremen (outside)	$46.40	$742.40	$69.25	$1108.00	$45.07	$67.25
4 Laborers	44.40	1420.80	66.25	2120.00		
1 S.P. Crane, 4x4, 5 Ton		381.95		420.14		
1 Flatbed Truck, Gas, 1.5 Ton		198.50		218.35		
1 Butt Fusion Mach., 4"-12" diam.		420.75		462.82	20.86	22.94
48 L.H., Daily Totals		$3164.40		$4329.32	$65.92	$90.19

Crew No.	Bare Costs		Incl. Subs O&P		Cost Per Labor-Hour	
Crew B-6C	Hr.	Daily	Hr.	Daily	Bare Costs	Incl. O&P
2 Labor Foremen (outside)	$46.40	$742.40	$69.25	$1108.00	$45.07	$67.25
4 Laborers	44.40	1420.80	66.25	2120.00		
1 S.P. Crane, 4x4, 12 Ton		432.65		475.92		
1 Flatbed Truck, Gas, 3 Ton		850.05		935.05		
1 Butt Fusion Mach., 8"-24" diam.		1086.00		1194.60	49.35	54.28
48 L.H., Daily Totals		$4531.90		$5833.57	$94.41	$121.53

Crew B-6D	Hr.	Daily	Hr.	Daily	Bare Costs	Incl. O&P
.5 Labor Foreman (outside)	$46.40	$185.60	$69.25	$277.00	$50.64	$75.51
1 Laborer	44.40	355.20	66.25	530.00		
1 Equip. Oper. (medium)	59.00	472.00	87.90	703.20		
1 Hydro Excavator, 12 C.Y.		1277.00		1404.70	63.85	70.23
20 L.H., Daily Totals		$2289.80		$2914.90	$114.49	$145.75

Crew B-7	Hr.	Daily	Hr.	Daily	Bare Costs	Incl. O&P
1 Labor Foreman (outside)	$46.40	$371.20	$69.25	$554.00	$47.17	$70.36
4 Laborers	44.40	1420.80	66.25	2120.00		
1 Equip. Oper. (medium)	59.00	472.00	87.90	703.20		
1 Brush Chipper, 12", 130 H.P.		366.05		402.65		
1 Crawler Loader, 3 C.Y.		1146.00		1260.60		
2 Chain Saws, Gas, 36" Long		83.30		91.63	33.24	36.56
48 L.H., Daily Totals		$3859.35		$5132.09	$80.40	$106.92

Crew B-7A	Hr.	Daily	Hr.	Daily	Bare Costs	Incl. O&P
2 Laborers	$44.40	$710.40	$66.25	$1060.00	$48.10	$71.73
1 Equip. Oper. (light)	55.50	444.00	82.70	661.60		
1 Rake w/Tractor		343.50		377.85		
2 Chain Saws, Gas, 18"		104.40		114.84	18.66	20.53
24 L.H., Daily Totals		$1602.30		$2214.29	$66.76	$92.26

Crew B-7B	Hr.	Daily	Hr.	Daily	Bare Costs	Incl. O&P
1 Labor Foreman (outside)	$46.40	$371.20	$69.25	$554.00	$47.76	$71.26
4 Laborers	44.40	1420.80	66.25	2120.00		
1 Equip. Oper. (medium)	59.00	472.00	87.90	703.20		
1 Truck Driver (heavy)	51.30	410.40	76.70	613.60		
1 Brush Chipper, 12", 130 H.P.		366.05		402.65		
1 Crawler Loader, 3 C.Y.		1146.00		1260.60		
2 Chain Saws, Gas, 36" Long		83.30		91.63		
1 Dump Truck, 8 C.Y., 220 H.P.		407.60		448.36	35.77	39.34
56 L.H., Daily Totals		$4677.35		$6194.05	$83.52	$110.61

Crew B-7C	Hr.	Daily	Hr.	Daily	Bare Costs	Incl. O&P
1 Labor Foreman (outside)	$46.40	$371.20	$69.25	$554.00	$47.76	$71.26
4 Laborers	44.40	1420.80	66.25	2120.00		
1 Equip. Oper. (medium)	59.00	472.00	87.90	703.20		
1 Truck Driver (heavy)	51.30	410.40	76.70	613.60		
1 Brush Chipper, 12", 130 H.P.		366.05		402.65		
1 Crawler Loader, 3 C.Y.		1146.00		1260.60		
2 Chain Saws, Gas, 36" Long		83.30		91.63		
1 Dump Truck, 12 C.Y., 400 H.P.		579.35		637.28	38.83	42.72
56 L.H., Daily Totals		$4849.10		$6382.97	$86.59	$113.98

Crew No.	Bare Costs		Incl. Subs O&P		Cost Per Labor-Hour	
Crew B-8	Hr.	Daily	Hr.	Daily	Bare Costs	Incl. O&P
1 Labor Foreman (outside)	$46.40	$371.20	$69.25	$554.00	$51.04	$76.15
2 Laborers	44.40	710.40	66.25	1060.00		
2 Equip. Oper. (medium)	59.00	944.00	87.90	1406.40		
1 Equip. Oper. (oiler)	52.50	420.00	78.25	626.00		
2 Truck Drivers (heavy)	51.30	820.80	76.70	1227.20		
1 Hyd. Crane, 25 Ton		586.70		645.37		
1 Crawler Loader, 3 C.Y.		1146.00		1260.60		
2 Dump Trucks, 12 C.Y., 400 H.P.		1158.70		1274.57	45.18	49.70
64 L.H., Daily Totals		$6157.80		$8054.14	$96.22	$125.85

Crew B-9	Hr.	Daily	Hr.	Daily	Bare Costs	Incl. O&P
1 Labor Foreman (outside)	$46.40	$371.20	$69.25	$554.00	$44.80	$66.85
4 Laborers	44.40	1420.80	66.25	2120.00		
1 Air Compressor, 250 cfm		202.85		223.13		
2 Breakers, Pavement, 60 lb.		107.20		117.92		
2 -50' Air Hoses, 1.5"		45.60		50.16	8.89	9.78
40 L.H., Daily Totals		$2147.65		$3065.22	$53.69	$76.63

Crew B-9A	Hr.	Daily	Hr.	Daily	Bare Costs	Incl. O&P
2 Laborers	$44.40	$710.40	$66.25	$1060.00	$46.70	$69.73
1 Truck Driver (heavy)	51.30	410.40	76.70	613.60		
1 Water Tank Trailer, 5000 Gal.		154.10		169.51		
1 Truck Tractor, 220 H.P.		310.80		341.88		
2 -50' Discharge Hoses, 3"		9.00		9.90	19.75	21.72
24 L.H., Daily Totals		$1594.70		$2194.89	$66.45	$91.45

Crew B-9B	Hr.	Daily	Hr.	Daily	Bare Costs	Incl. O&P
2 Laborers	$44.40	$710.40	$66.25	$1060.00	$46.70	$69.73
1 Truck Driver (heavy)	51.30	410.40	76.70	613.60		
2 -50' Discharge Hoses, 3"		9.00		9.90		
1 Water Tank Trailer, 5000 Gal.		154.10		169.51		
1 Truck Tractor, 220 H.P.		310.80		341.88		
1 Pressure Washer		97.35		107.08	23.80	26.18
24 L.H., Daily Totals		$1692.05		$2301.97	$70.50	$95.92

Crew B-9D	Hr.	Daily	Hr.	Daily	Bare Costs	Incl. O&P
1 Labor Foreman (outside)	$46.40	$371.20	$69.25	$554.00	$44.80	$66.85
4 Common Laborers	44.40	1420.80	66.25	2120.00		
1 Air Compressor, 250 cfm		202.85		223.13		
2 -50' Air Hoses, 1.5"		45.60		50.16		
2 Air Powered Tampers		79.50		87.45	8.20	9.02
40 L.H., Daily Totals		$2119.95		$3034.74	$53.00	$75.87

Crew B-9E	Hr.	Daily	Hr.	Daily	Bare Costs	Incl. O&P
1 Cement Finisher	$51.80	$414.40	$75.90	$607.20	$48.10	$71.08
1 Laborer	44.40	355.20	66.25	530.00		
1 Chip. Hammers, 12 Lb., Elec.		32.85		36.13	2.05	2.26
16 L.H., Daily Totals		$802.45		$1173.34	$50.15	$73.33

Crew B-10	Hr.	Daily	Hr.	Daily	Bare Costs	Incl. O&P
1 Equip. Oper. (medium)	$59.00	$472.00	$87.90	$703.20	$54.13	$80.68
.5 Laborer	44.40	177.60	66.25	265.00		
12 L.H., Daily Totals		$649.60		$968.20	$54.13	$80.68

Crew B-10A	Hr.	Daily	Hr.	Daily	Bare Costs	Incl. O&P
1 Equip. Oper. (medium)	$59.00	$472.00	$87.90	$703.20	$54.13	$80.68
.5 Laborer	44.40	177.60	66.25	265.00		
1 Roller, 2-Drum, W.B., 7.5 H.P.		166.75		183.43	13.90	15.29
12 L.H., Daily Totals		$816.35		$1151.63	$68.03	$95.97

Crew B-10B

Crew No.	Bare Costs Hr.	Daily	Incl. Subs O&P Hr.	Daily	Cost Per Labor-Hour Bare Costs	Incl. O&P
1 Equip. Oper. (medium)	$59.00	$472.00	$87.90	$703.20	$54.13	$80.68
.5 Laborer	44.40	177.60	66.25	265.00		
1 Dozer, 200 H.P.		1520.00		1672.00	126.67	139.33
12 L.H., Daily Totals		$2169.60		$2640.20	$180.80	$220.02

Crew B-10C

Crew No.	Bare Costs Hr.	Daily	Incl. Subs O&P Hr.	Daily	Cost Per Labor-Hour Bare Costs	Incl. O&P
1 Equip. Oper. (medium)	$59.00	$472.00	$87.90	$703.20	$54.13	$80.68
.5 Laborer	44.40	177.60	66.25	265.00		
1 Dozer, 200 H.P.		1520.00		1672.00		
1 Vibratory Roller, Towed, 23 Ton		520.35		572.38	170.03	187.03
12 L.H., Daily Totals		$2689.95		$3212.59	$224.16	$267.72

Crew B-10D

Crew No.	Bare Costs Hr.	Daily	Incl. Subs O&P Hr.	Daily	Cost Per Labor-Hour Bare Costs	Incl. O&P
1 Equip. Oper. (medium)	$59.00	$472.00	$87.90	$703.20	$54.13	$80.68
.5 Laborer	44.40	177.60	66.25	265.00		
1 Dozer, 200 H.P.		1520.00		1672.00		
1 Sheepsft. Roller, Towed		426.95		469.64	162.25	178.47
12 L.H., Daily Totals		$2596.55		$3109.84	$216.38	$259.15

Crew B-10E

Crew No.	Bare Costs Hr.	Daily	Incl. Subs O&P Hr.	Daily	Cost Per Labor-Hour Bare Costs	Incl. O&P
1 Equip. Oper. (medium)	$59.00	$472.00	$87.90	$703.20	$54.13	$80.68
.5 Laborer	44.40	177.60	66.25	265.00		
1 Tandem Roller, 5 Ton		258.75		284.63	21.56	23.72
12 L.H., Daily Totals		$908.35		$1252.83	$75.70	$104.40

Crew B-10F

Crew No.	Bare Costs Hr.	Daily	Incl. Subs O&P Hr.	Daily	Cost Per Labor-Hour Bare Costs	Incl. O&P
1 Equip. Oper. (medium)	$59.00	$472.00	$87.90	$703.20	$54.13	$80.68
.5 Laborer	44.40	177.60	66.25	265.00		
1 Tandem Roller, 10 Ton		246.80		271.48	20.57	22.62
12 L.H., Daily Totals		$896.40		$1239.68	$74.70	$103.31

Crew B-10G

Crew No.	Bare Costs Hr.	Daily	Incl. Subs O&P Hr.	Daily	Cost Per Labor-Hour Bare Costs	Incl. O&P
1 Equip. Oper. (medium)	$59.00	$472.00	$87.90	$703.20	$54.13	$80.68
.5 Laborer	44.40	177.60	66.25	265.00		
1 Sheepsfoot Roller, 240 H.P.		1363.00		1499.30	113.58	124.94
12 L.H., Daily Totals		$2012.60		$2467.50	$167.72	$205.63

Crew B-10H

Crew No.	Bare Costs Hr.	Daily	Incl. Subs O&P Hr.	Daily	Cost Per Labor-Hour Bare Costs	Incl. O&P
1 Equip. Oper. (medium)	$59.00	$472.00	$87.90	$703.20	$54.13	$80.68
.5 Laborer	44.40	177.60	66.25	265.00		
1 Diaphragm Water Pump, 2"		87.70		96.47		
1 -20' Suction Hose, 2"		3.55		3.90		
2 -50' Discharge Hoses, 2"		8.00		8.80	8.27	9.10
12 L.H., Daily Totals		$748.85		$1077.38	$62.40	$89.78

Crew B-10I

Crew No.	Bare Costs Hr.	Daily	Incl. Subs O&P Hr.	Daily	Cost Per Labor-Hour Bare Costs	Incl. O&P
1 Equip. Oper. (medium)	$59.00	$472.00	$87.90	$703.20	$54.13	$80.68
.5 Laborer	44.40	177.60	66.25	265.00		
1 Diaphragm Water Pump, 4"		106.35		116.99		
1 -20' Suction Hose, 4"		17.25		18.98		
2 -50' Discharge Hoses, 4"		25.60		28.16	12.43	13.68
12 L.H., Daily Totals		$798.80		$1132.32	$66.57	$94.36

Crew B-10J

Crew No.	Bare Costs Hr.	Daily	Incl. Subs O&P Hr.	Daily	Cost Per Labor-Hour Bare Costs	Incl. O&P
1 Equip. Oper. (medium)	$59.00	$472.00	$87.90	$703.20	$54.13	$80.68
.5 Laborer	44.40	177.60	66.25	265.00		
1 Centrifugal Water Pump, 3"		74.40		81.84		
1 -20' Suction Hose, 3"		8.75		9.63		
2 -50' Discharge Hoses, 3"		9.00		9.90	7.68	8.45
12 L.H., Daily Totals		$741.75		$1069.57	$61.81	$89.13

Crew B-10K

Crew No.	Bare Costs Hr.	Daily	Incl. Subs O&P Hr.	Daily	Cost Per Labor-Hour Bare Costs	Incl. O&P
1 Equip. Oper. (medium)	$59.00	$472.00	$87.90	$703.20	$54.13	$80.68
.5 Laborer	44.40	177.60	66.25	265.00		
1 Centr. Water Pump, 6"		235.25		258.77		
1 -20' Suction Hose, 6"		25.50		28.05		
2 -50' Discharge Hoses, 6"		36.20		39.82	24.75	27.22
12 L.H., Daily Totals		$946.55		$1294.85	$78.88	$107.90

Crew B-10L

Crew No.	Bare Costs Hr.	Daily	Incl. Subs O&P Hr.	Daily	Cost Per Labor-Hour Bare Costs	Incl. O&P
1 Equip. Oper. (medium)	$59.00	$472.00	$87.90	$703.20	$54.13	$80.68
.5 Laborer	44.40	177.60	66.25	265.00		
1 Dozer, 80 H.P.		405.85		446.44	33.82	37.20
12 L.H., Daily Totals		$1055.45		$1414.64	$87.95	$117.89

Crew B-10M

Crew No.	Bare Costs Hr.	Daily	Incl. Subs O&P Hr.	Daily	Cost Per Labor-Hour Bare Costs	Incl. O&P
1 Equip. Oper. (medium)	$59.00	$472.00	$87.90	$703.20	$54.13	$80.68
.5 Laborer	44.40	177.60	66.25	265.00		
1 Dozer, 300 H.P.		1785.00		1963.50	148.75	163.63
12 L.H., Daily Totals		$2434.60		$2931.70	$202.88	$244.31

Crew B-10N

Crew No.	Bare Costs Hr.	Daily	Incl. Subs O&P Hr.	Daily	Cost Per Labor-Hour Bare Costs	Incl. O&P
1 Equip. Oper. (medium)	$59.00	$472.00	$87.90	$703.20	$54.13	$80.68
.5 Laborer	44.40	177.60	66.25	265.00		
1 F.E. Loader, T.M., 1.5 C.Y.		572.00		629.20	47.67	52.43
12 L.H., Daily Totals		$1221.60		$1597.40	$101.80	$133.12

Crew B-10O

Crew No.	Bare Costs Hr.	Daily	Incl. Subs O&P Hr.	Daily	Cost Per Labor-Hour Bare Costs	Incl. O&P
1 Equip. Oper. (medium)	$59.00	$472.00	$87.90	$703.20	$54.13	$80.68
.5 Laborer	44.40	177.60	66.25	265.00		
1 F.E. Loader, T.M., 2.25 C.Y.		925.50		1018.05	77.13	84.84
12 L.H., Daily Totals		$1575.10		$1986.25	$131.26	$165.52

Crew B-10P

Crew No.	Bare Costs Hr.	Daily	Incl. Subs O&P Hr.	Daily	Cost Per Labor-Hour Bare Costs	Incl. O&P
1 Equip. Oper. (medium)	$59.00	$472.00	$87.90	$703.20	$54.13	$80.68
.5 Laborer	44.40	177.60	66.25	265.00		
1 Crawler Loader, 3 C.Y.		1146.00		1260.60	95.50	105.05
12 L.H., Daily Totals		$1795.60		$2228.80	$149.63	$185.73

Crew B-10Q

Crew No.	Bare Costs Hr.	Daily	Incl. Subs O&P Hr.	Daily	Cost Per Labor-Hour Bare Costs	Incl. O&P
1 Equip. Oper. (medium)	$59.00	$472.00	$87.90	$703.20	$54.13	$80.68
.5 Laborer	44.40	177.60	66.25	265.00		
1 Crawler Loader, 4 C.Y.		1456.00		1601.60	121.33	133.47
12 L.H., Daily Totals		$2105.60		$2569.80	$175.47	$214.15

Crew B-10R

Crew No.	Bare Costs Hr.	Daily	Incl. Subs O&P Hr.	Daily	Cost Per Labor-Hour Bare Costs	Incl. O&P
1 Equip. Oper. (medium)	$59.00	$472.00	$87.90	$703.20	$54.13	$80.68
.5 Laborer	44.40	177.60	66.25	265.00		
1 F.E. Loader, W.M., 1 C.Y.		305.60		336.16	25.47	28.01
12 L.H., Daily Totals		$955.20		$1304.36	$79.60	$108.70

Crew B-10S

Crew No.	Bare Costs Hr.	Daily	Incl. Subs O&P Hr.	Daily	Cost Per Labor-Hour Bare Costs	Incl. O&P
1 Equip. Oper. (medium)	$59.00	$472.00	$87.90	$703.20	$54.13	$80.68
.5 Laborer	44.40	177.60	66.25	265.00		
1 F.E. Loader, W.M., 1.5 C.Y.		441.40		485.54	36.78	40.46
12 L.H., Daily Totals		$1091.00		$1453.74	$90.92	$121.15

Crew B-10T

Crew No.	Bare Costs Hr.	Daily	Incl. Subs O&P Hr.	Daily	Cost Per Labor-Hour Bare Costs	Incl. O&P
1 Equip. Oper. (medium)	$59.00	$472.00	$87.90	$703.20	$54.13	$80.68
.5 Laborer	44.40	177.60	66.25	265.00		
1 F.E. Loader, W.M., 2.5 C.Y.		638.30		702.13	53.19	58.51
12 L.H., Daily Totals		$1287.90		$1670.33	$107.33	$139.19

Crew B-10U	Hr.	Daily	Hr.	Daily	Bare Costs	Incl. O&P
1 Equip. Oper. (medium)	$59.00	$472.00	$87.90	$703.20	$54.13	$80.68
.5 Laborer	44.40	177.60	66.25	265.00		
1 F.E. Loader, W.M., 5.5 C.Y.		967.95		1064.74	80.66	88.73
12 L.H., Daily Totals		$1617.55		$2032.94	$134.80	$169.41

Crew B-10V	Hr.	Daily	Hr.	Daily	Bare Costs	Incl. O&P
1 Equip. Oper. (medium)	$59.00	$472.00	$87.90	$703.20	$54.13	$80.68
.5 Laborer	44.40	177.60	66.25	265.00		
1 Dozer, 700 H.P.		5175.00		5692.50	431.25	474.38
12 L.H., Daily Totals		$5824.60		$6660.70	$485.38	$555.06

Crew B-10W	Hr.	Daily	Hr.	Daily	Bare Costs	Incl. O&P
1 Equip. Oper. (medium)	$59.00	$472.00	$87.90	$703.20	$54.13	$80.68
.5 Laborer	44.40	177.60	66.25	265.00		
1 Dozer, 105 H.P.		640.80		704.88	53.40	58.74
12 L.H., Daily Totals		$1290.40		$1673.08	$107.53	$139.42

Crew B-10X	Hr.	Daily	Hr.	Daily	Bare Costs	Incl. O&P
1 Equip. Oper. (medium)	$59.00	$472.00	$87.90	$703.20	$54.13	$80.68
.5 Laborer	44.40	177.60	66.25	265.00		
1 Dozer, 410 H.P.		2807.00		3087.70	233.92	257.31
12 L.H., Daily Totals		$3456.60		$4055.90	$288.05	$337.99

Crew B-10Y	Hr.	Daily	Hr.	Daily	Bare Costs	Incl. O&P
1 Equip. Oper. (medium)	$59.00	$472.00	$87.90	$703.20	$54.13	$80.68
.5 Laborer	44.40	177.60	66.25	265.00		
1 Vibr. Roller, Towed, 12 Ton		584.80		643.28	48.73	53.61
12 L.H., Daily Totals		$1234.40		$1611.48	$102.87	$134.29

Crew B-11A	Hr.	Daily	Hr.	Daily	Bare Costs	Incl. O&P
1 Equipment Oper. (med.)	$59.00	$472.00	$87.90	$703.20	$51.70	$77.08
1 Laborer	44.40	355.20	66.25	530.00		
1 Dozer, 200 H.P.		1520.00		1672.00	95.00	104.50
16 L.H., Daily Totals		$2347.20		$2905.20	$146.70	$181.57

Crew B-11B	Hr.	Daily	Hr.	Daily	Bare Costs	Incl. O&P
1 Equipment Oper. (light)	$55.50	$444.00	$82.70	$661.60	$49.95	$74.47
1 Laborer	44.40	355.20	66.25	530.00		
1 Air Powered Tamper		39.75		43.73		
1 Air Compressor, 365 cfm		343.55		377.90		
2 -50' Air Hoses, 1.5"		45.60		50.16	26.81	29.49
16 L.H., Daily Totals		$1228.10		$1663.39	$76.76	$103.96

Crew B-11C	Hr.	Daily	Hr.	Daily	Bare Costs	Incl. O&P
1 Equipment Oper. (med.)	$59.00	$472.00	$87.90	$703.20	$51.70	$77.08
1 Laborer	44.40	355.20	66.25	530.00		
1 Backhoe Loader, 48 H.P.		216.20		237.82	13.51	14.86
16 L.H., Daily Totals		$1043.40		$1471.02	$65.21	$91.94

Crew B-11J	Hr.	Daily	Hr.	Daily	Bare Costs	Incl. O&P
1 Equipment Oper. (med.)	$59.00	$472.00	$87.90	$703.20	$51.70	$77.08
1 Laborer	44.40	355.20	66.25	530.00		
1 Grader, 30,000 Lbs.		1073.00		1180.30		
1 Ripper, Beam & 1 Shank		91.60		100.76	72.79	80.07
16 L.H., Daily Totals		$1991.80		$2514.26	$124.49	$157.14

Crew B-11K	Hr.	Daily	Hr.	Daily	Bare Costs	Incl. O&P
1 Equipment Oper. (med.)	$59.00	$472.00	$87.90	$703.20	$51.70	$77.08
1 Laborer	44.40	355.20	66.25	530.00		
1 Trencher, Chain Type, 8' D		1894.00		2083.40	118.38	130.21
16 L.H., Daily Totals		$2721.20		$3316.60	$170.07	$207.29

Crew B-11L	Hr.	Daily	Hr.	Daily	Bare Costs	Incl. O&P
1 Equipment Oper. (med.)	$59.00	$472.00	$87.90	$703.20	$51.70	$77.08
1 Laborer	44.40	355.20	66.25	530.00		
1 Grader, 30,000 Lbs.		1073.00		1180.30	67.06	73.77
16 L.H., Daily Totals		$1900.20		$2413.50	$118.76	$150.84

Crew B-11M	Hr.	Daily	Hr.	Daily	Bare Costs	Incl. O&P
1 Equipment Oper. (med.)	$59.00	$472.00	$87.90	$703.20	$51.70	$77.08
1 Laborer	44.40	355.20	66.25	530.00		
1 Backhoe Loader, 80 H.P.		235.05		258.56	14.69	16.16
16 L.H., Daily Totals		$1062.25		$1491.76	$66.39	$93.23

Crew B-11N	Hr.	Daily	Hr.	Daily	Bare Costs	Incl. O&P
1 Labor Foreman (outside)	$46.40	$371.20	$69.25	$554.00	$52.47	$78.36
2 Equipment Operators (med.)	59.00	944.00	87.90	1406.40		
6 Truck Drivers (heavy)	51.30	2462.40	76.70	3681.60		
1 F.E. Loader, W.M., 5.5 C.Y.		967.95		1064.74		
1 Dozer, 410 H.P.		2807.00		3087.70		
6 Dump Trucks, Off Hwy., 50 Ton		11874.00		13061.40	217.35	239.08
72 L.H., Daily Totals		$19426.55		$22855.85	$269.81	$317.44

Crew B-11Q	Hr.	Daily	Hr.	Daily	Bare Costs	Incl. O&P
1 Equipment Operator (med.)	$59.00	$472.00	$87.90	$703.20	$54.13	$80.68
.5 Laborer	44.40	177.60	66.25	265.00		
1 Dozer, 140 H.P.		729.15		802.07	60.76	66.84
12 L.H., Daily Totals		$1378.75		$1770.27	$114.90	$147.52

Crew B-11R	Hr.	Daily	Hr.	Daily	Bare Costs	Incl. O&P
1 Equipment Operator (med.)	$59.00	$472.00	$87.90	$703.20	$54.13	$80.68
.5 Laborer	44.40	177.60	66.25	265.00		
1 Dozer, 200 H.P.		1520.00		1672.00	126.67	139.33
12 L.H., Daily Totals		$2169.60		$2640.20	$180.80	$220.02

Crew B-11S	Hr.	Daily	Hr.	Daily	Bare Costs	Incl. O&P
1 Equipment Operator (med.)	$59.00	$472.00	$87.90	$703.20	$54.13	$80.68
.5 Laborer	44.40	177.60	66.25	265.00		
1 Dozer, 300 H.P.		1785.00		1963.50		
1 Ripper, Beam & 1 Shank		91.60		100.76	156.38	172.02
12 L.H., Daily Totals		$2526.20		$3032.46	$210.52	$252.71

Crew B-11T	Hr.	Daily	Hr.	Daily	Bare Costs	Incl. O&P
1 Equipment Operator (med.)	$59.00	$472.00	$87.90	$703.20	$54.13	$80.68
.5 Laborer	44.40	177.60	66.25	265.00		
1 Dozer, 410 H.P.		2807.00		3087.70		
1 Ripper, Beam & 2 Shanks		140.40		154.44	245.62	270.18
12 L.H., Daily Totals		$3597.00		$4210.34	$299.75	$350.86

Crew B-11U	Hr.	Daily	Hr.	Daily	Bare Costs	Incl. O&P
1 Equipment Operator (med.)	$59.00	$472.00	$87.90	$703.20	$54.13	$80.68
.5 Laborer	44.40	177.60	66.25	265.00		
1 Dozer, 520 H.P.		3434.00		3777.40	286.17	314.78
12 L.H., Daily Totals		$4083.60		$4745.60	$340.30	$395.47

For customer support on your Concrete & Masonry Costs with RSMeans Data, call 800.448.8182.

421

Crew No.	Bare Costs		Incl. Subs O&P		Cost Per Labor-Hour	

Crew B-11V

	Hr.	Daily	Hr.	Daily	Bare Costs	Incl. O&P
3 Laborers	$44.40	$1065.60	$66.25	$1590.00	$44.40	$66.25
1 Roller, 2-Drum, W.B., 7.5 H.P.		166.75		183.43	6.95	7.64
24 L.H., Daily Totals		$1232.35		$1773.43	$51.35	$73.89

Crew B-11W

	Hr.	Daily	Hr.	Daily	Bare Costs	Incl. O&P
1 Equipment Operator (med.)	$59.00	$472.00	$87.90	$703.20	$51.37	$76.76
1 Common Laborer	44.40	355.20	66.25	530.00		
10 Truck Drivers (heavy)	51.30	4104.00	76.70	6136.00		
1 Dozer, 200 H.P.		1520.00		1672.00		
1 Vibratory Roller, Towed, 23 Ton		520.35		572.38		
10 Dump Trucks, 8 C.Y., 220 H.P.		4076.00		4483.60	63.71	70.08
96 L.H., Daily Totals		$11047.55		$14097.18	$115.08	$146.85

Crew B-11Y

	Hr.	Daily	Hr.	Daily	Bare Costs	Incl. O&P
1 Labor Foreman (outside)	$46.40	$371.20	$69.25	$554.00	$49.49	$73.80
5 Common Laborers	44.40	1776.00	66.25	2650.00		
3 Equipment Operators (med.)	59.00	1416.00	87.90	2109.60		
1 Dozer, 80 H.P.		405.85		446.44		
2 Rollers, 2-Drum, W.B., 7.5 H.P.		333.50		366.85		
4 Vibrating Plates, Gas, 21"		662.40		728.64	19.47	21.42
72 L.H., Daily Totals		$4964.95		$6855.52	$68.96	$95.22

Crew B-12A

	Hr.	Daily	Hr.	Daily	Bare Costs	Incl. O&P
1 Equip. Oper. (crane)	$61.45	$491.60	$91.55	$732.40	$52.92	$78.90
1 Laborer	44.40	355.20	66.25	530.00		
1 Hyd. Excavator, 1 C.Y.		832.65		915.91	52.04	57.24
16 L.H., Daily Totals		$1679.45		$2178.32	$104.97	$136.14

Crew B-12B

	Hr.	Daily	Hr.	Daily	Bare Costs	Incl. O&P
1 Equip. Oper. (crane)	$61.45	$491.60	$91.55	$732.40	$52.92	$78.90
1 Laborer	44.40	355.20	66.25	530.00		
1 Hyd. Excavator, 1.5 C.Y.		695.80		765.38	43.49	47.84
16 L.H., Daily Totals		$1542.60		$2027.78	$96.41	$126.74

Crew B-12C

	Hr.	Daily	Hr.	Daily	Bare Costs	Incl. O&P
1 Equip. Oper. (crane)	$61.45	$491.60	$91.55	$732.40	$52.92	$78.90
1 Laborer	44.40	355.20	66.25	530.00		
1 Hyd. Excavator, 2 C.Y.		942.70		1036.97	58.92	64.81
16 L.H., Daily Totals		$1789.50		$2299.37	$111.84	$143.71

Crew B-12D

	Hr.	Daily	Hr.	Daily	Bare Costs	Incl. O&P
1 Equip. Oper. (crane)	$61.45	$491.60	$91.55	$732.40	$52.92	$78.90
1 Laborer	44.40	355.20	66.25	530.00		
1 Hyd. Excavator, 3.5 C.Y.		2184.00		2402.40	136.50	150.15
16 L.H., Daily Totals		$3030.80		$3664.80	$189.43	$229.05

Crew B-12E

	Hr.	Daily	Hr.	Daily	Bare Costs	Incl. O&P
1 Equip. Oper. (crane)	$61.45	$491.60	$91.55	$732.40	$52.92	$78.90
1 Laborer	44.40	355.20	66.25	530.00		
1 Hyd. Excavator, .5 C.Y.		457.00		502.70	28.56	31.42
16 L.H., Daily Totals		$1303.80		$1765.10	$81.49	$110.32

Crew B-12F

	Hr.	Daily	Hr.	Daily	Bare Costs	Incl. O&P
1 Equip. Oper. (crane)	$61.45	$491.60	$91.55	$732.40	$52.92	$78.90
1 Laborer	44.40	355.20	66.25	530.00		
1 Hyd. Excavator, .75 C.Y.		701.80		771.98	43.86	48.25
16 L.H., Daily Totals		$1548.60		$2034.38	$96.79	$127.15

Crew B-12G

	Hr.	Daily	Hr.	Daily	Bare Costs	Incl. O&P
1 Equip. Oper. (crane)	$61.45	$491.60	$91.55	$732.40	$52.92	$78.90
1 Laborer	44.40	355.20	66.25	530.00		
1 Crawler Crane, 15 Ton		810.30		891.33		
1 Clamshell Bucket, .5 C.Y.		67.80		74.58	54.88	60.37
16 L.H., Daily Totals		$1724.90		$2228.31	$107.81	$139.27

Crew B-12H

	Hr.	Daily	Hr.	Daily	Bare Costs	Incl. O&P
1 Equip. Oper. (crane)	$61.45	$491.60	$91.55	$732.40	$52.92	$78.90
1 Laborer	44.40	355.20	66.25	530.00		
1 Crawler Crane, 25 Ton		1152.00		1267.20		
1 Clamshell Bucket, 1 C.Y.		69.25		76.17	76.33	83.96
16 L.H., Daily Totals		$2068.05		$2605.78	$129.25	$162.86

Crew B-12I

	Hr.	Daily	Hr.	Daily	Bare Costs	Incl. O&P
1 Equip. Oper. (crane)	$61.45	$491.60	$91.55	$732.40	$52.92	$78.90
1 Laborer	44.40	355.20	66.25	530.00		
1 Crawler Crane, 20 Ton		1013.00		1114.30		
1 Dragline Bucket, .75 C.Y.		61.85		68.03	67.18	73.90
16 L.H., Daily Totals		$1921.65		$2444.74	$120.10	$152.80

Crew B-12J

	Hr.	Daily	Hr.	Daily	Bare Costs	Incl. O&P
1 Equip. Oper. (crane)	$61.45	$491.60	$91.55	$732.40	$52.92	$78.90
1 Laborer	44.40	355.20	66.25	530.00		
1 Gradall, 5/8 C.Y.		850.65		935.72	53.17	58.48
16 L.H., Daily Totals		$1697.45		$2198.11	$106.09	$137.38

Crew B-12K

	Hr.	Daily	Hr.	Daily	Bare Costs	Incl. O&P
1 Equip. Oper. (crane)	$61.45	$491.60	$91.55	$732.40	$52.92	$78.90
1 Laborer	44.40	355.20	66.25	530.00		
1 Gradall, 3 Ton, 1 C.Y.		984.55		1083.01	61.53	67.69
16 L.H., Daily Totals		$1831.35		$2345.41	$114.46	$146.59

Crew B-12L

	Hr.	Daily	Hr.	Daily	Bare Costs	Incl. O&P
1 Equip. Oper. (crane)	$61.45	$491.60	$91.55	$732.40	$52.92	$78.90
1 Laborer	44.40	355.20	66.25	530.00		
1 Crawler Crane, 15 Ton		810.30		891.33		
1 F.E. Attachment, .5 C.Y.		66.05		72.66	54.77	60.25
16 L.H., Daily Totals		$1723.15		$2226.39	$107.70	$139.16

Crew B-12M

	Hr.	Daily	Hr.	Daily	Bare Costs	Incl. O&P
1 Equip. Oper. (crane)	$61.45	$491.60	$91.55	$732.40	$52.92	$78.90
1 Laborer	44.40	355.20	66.25	530.00		
1 Crawler Crane, 20 Ton		1013.00		1114.30		
1 F.E. Attachment, .75 C.Y.		71.25		78.38	67.77	74.54
16 L.H., Daily Totals		$1931.05		$2455.07	$120.69	$153.44

Crew B-12N

	Hr.	Daily	Hr.	Daily	Bare Costs	Incl. O&P
1 Equip. Oper. (crane)	$61.45	$491.60	$91.55	$732.40	$52.92	$78.90
1 Laborer	44.40	355.20	66.25	530.00		
1 Crawler Crane, 25 Ton		1152.00		1267.20		
1 F.E. Attachment, 1 C.Y.		77.35		85.08	76.83	84.52
16 L.H., Daily Totals		$2076.15		$2614.68	$129.76	$163.42

Crew B-12O

	Hr.	Daily	Hr.	Daily	Bare Costs	Incl. O&P
1 Equip. Oper. (crane)	$61.45	$491.60	$91.55	$732.40	$52.92	$78.90
1 Laborer	44.40	355.20	66.25	530.00		
1 Crawler Crane, 40 Ton		1231.00		1354.10		
1 F.E. Attachment, 1.5 C.Y.		88.65		97.52	82.48	90.73
16 L.H., Daily Totals		$2166.45		$2714.01	$135.40	$169.63

For customer support on your Concrete & Masonry Costs with RSMeans Data, call 800.448.8182.

Crew No.	Bare Costs		Incl. Subs O&P		Cost Per Labor-Hour	
Crew B-12P	Hr.	Daily	Hr.	Daily	Bare Costs	Incl. O&P
1 Equip. Oper. (crane)	$61.45	$491.60	$91.55	$732.40	$52.92	$78.90
1 Laborer	44.40	355.20	66.25	530.00		
1 Crawler Crane, 40 Ton		1231.00		1354.10		
1 Dragline Bucket, 1.5 C.Y.		65.70		72.27	81.04	89.15
16 L.H., Daily Totals		$2143.50		$2688.77	$133.97	$168.05
Crew B-12Q	Hr.	Daily	Hr.	Daily	Bare Costs	Incl. O&P
1 Equip. Oper. (crane)	$61.45	$491.60	$91.55	$732.40	$52.92	$78.90
1 Laborer	44.40	355.20	66.25	530.00		
1 Hyd. Excavator, 5/8 C.Y.		604.75		665.23	37.80	41.58
16 L.H., Daily Totals		$1451.55		$1927.63	$90.72	$120.48
Crew B-12S	Hr.	Daily	Hr.	Daily	Bare Costs	Incl. O&P
1 Equip. Oper. (crane)	$61.45	$491.60	$91.55	$732.40	$52.92	$78.90
1 Laborer	44.40	355.20	66.25	530.00		
1 Hyd. Excavator, 2.5 C.Y.		1567.00		1723.70	97.94	107.73
16 L.H., Daily Totals		$2413.80		$2986.10	$150.86	$186.63
Crew B-12T	Hr.	Daily	Hr.	Daily	Bare Costs	Incl. O&P
1 Equip. Oper. (crane)	$61.45	$491.60	$91.55	$732.40	$52.92	$78.90
1 Laborer	44.40	355.20	66.25	530.00		
1 Crawler Crane, 75 Ton		1967.00		2163.70		
1 F.E. Attachment, 3 C.Y.		115.55		127.11	130.16	143.18
16 L.H., Daily Totals		$2929.35		$3553.20	$183.08	$222.08
Crew B-12V	Hr.	Daily	Hr.	Daily	Bare Costs	Incl. O&P
1 Equip. Oper. (crane)	$61.45	$491.60	$91.55	$732.40	$52.92	$78.90
1 Laborer	44.40	355.20	66.25	530.00		
1 Crawler Crane, 75 Ton		1967.00		2163.70		
1 Dragline Bucket, 3 C.Y.		72.30		79.53	127.46	140.20
16 L.H., Daily Totals		$2886.10		$3505.63	$180.38	$219.10
Crew B-12Y	Hr.	Daily	Hr.	Daily	Bare Costs	Incl. O&P
1 Equip. Oper. (crane)	$61.45	$491.60	$91.55	$732.40	$50.08	$74.68
2 Laborers	44.40	710.40	66.25	1060.00		
1 Hyd. Excavator, 3.5 C.Y.		2184.00		2402.40	91.00	100.10
24 L.H., Daily Totals		$3386.00		$4194.80	$141.08	$174.78
Crew B-12Z	Hr.	Daily	Hr.	Daily	Bare Costs	Incl. O&P
1 Equip. Oper. (crane)	$61.45	$491.60	$91.55	$732.40	$50.08	$74.68
2 Laborers	44.40	710.40	66.25	1060.00		
1 Hyd. Excavator, 2.5 C.Y.		1567.00		1723.70	65.29	71.82
24 L.H., Daily Totals		$2769.00		$3516.10	$115.38	$146.50
Crew B-13	Hr.	Daily	Hr.	Daily	Bare Costs	Incl. O&P
1 Labor Foreman (outside)	$46.40	$371.20	$69.25	$554.00	$48.28	$72.01
4 Laborers	44.40	1420.80	66.25	2120.00		
1 Equip. Oper. (crane)	61.45	491.60	91.55	732.40		
1 Equip. Oper. (oiler)	52.50	420.00	78.25	626.00		
1 Hyd. Crane, 25 Ton		586.70		645.37	10.48	11.52
56 L.H., Daily Totals		$3290.30		$4677.77	$58.76	$83.53

Crew No.	Bare Costs		Incl. Subs O&P		Cost Per Labor-Hour	
Crew B-13A	Hr.	Daily	Hr.	Daily	Bare Costs	Incl. O&P
1 Labor Foreman (outside)	$46.40	$371.20	$69.25	$554.00	$50.83	$75.85
2 Laborers	44.40	710.40	66.25	1060.00		
2 Equipment Operators (med.)	59.00	944.00	87.90	1406.40		
2 Truck Drivers (heavy)	51.30	820.80	76.70	1227.20		
1 Crawler Crane, 75 Ton		1967.00		2163.70		
1 Crawler Loader, 4 C.Y.		1456.00		1601.60		
2 Dump Trucks, 8 C.Y., 220 H.P.		815.20		896.72	75.68	83.25
56 L.H., Daily Totals		$7084.60		$8909.62	$126.51	$159.10
Crew B-13B	Hr.	Daily	Hr.	Daily	Bare Costs	Incl. O&P
1 Labor Foreman (outside)	$46.40	$371.20	$69.25	$554.00	$48.28	$72.01
4 Laborers	44.40	1420.80	66.25	2120.00		
1 Equip. Oper. (crane)	61.45	491.60	91.55	732.40		
1 Equip. Oper. (oiler)	52.50	420.00	78.25	626.00		
1 Hyd. Crane, 55 Ton		990.15		1089.17	17.68	19.45
56 L.H., Daily Totals		$3693.75		$5121.56	$65.96	$91.46
Crew B-13C	Hr.	Daily	Hr.	Daily	Bare Costs	Incl. O&P
1 Labor Foreman (outside)	$46.40	$371.20	$69.25	$554.00	$48.28	$72.01
4 Laborers	44.40	1420.80	66.25	2120.00		
1 Equip. Oper. (crane)	61.45	491.60	91.55	732.40		
1 Equip. Oper. (oiler)	52.50	420.00	78.25	626.00		
1 Crawler Crane, 100 Ton		2310.00		2541.00	41.25	45.38
56 L.H., Daily Totals		$5013.60		$6573.40	$89.53	$117.38
Crew B-13D	Hr.	Daily	Hr.	Daily	Bare Costs	Incl. O&P
1 Laborer	$44.40	$355.20	$66.25	$530.00	$52.92	$78.90
1 Equip. Oper. (crane)	61.45	491.60	91.55	732.40		
1 Hyd. Excavator, 1 C.Y.		832.65		915.91		
1 Trench Box		119.15		131.07	59.49	65.44
16 L.H., Daily Totals		$1798.60		$2309.38	$112.41	$144.34
Crew B-13E	Hr.	Daily	Hr.	Daily	Bare Costs	Incl. O&P
1 Laborer	$44.40	$355.20	$66.25	$530.00	$52.92	$78.90
1 Equip. Oper. (crane)	61.45	491.60	91.55	732.40		
1 Hyd. Excavator, 1.5 C.Y.		695.80		765.38		
1 Trench Box		119.15		131.07	50.93	56.03
16 L.H., Daily Totals		$1661.75		$2158.84	$103.86	$134.93
Crew B-13F	Hr.	Daily	Hr.	Daily	Bare Costs	Incl. O&P
1 Laborer	$44.40	$355.20	$66.25	$530.00	$52.92	$78.90
1 Equip. Oper. (crane)	61.45	491.60	91.55	732.40		
1 Hyd. Excavator, 3.5 C.Y.		2184.00		2402.40		
1 Trench Box		119.15		131.07	143.95	158.34
16 L.H., Daily Totals		$3149.95		$3795.86	$196.87	$237.24
Crew B-13G	Hr.	Daily	Hr.	Daily	Bare Costs	Incl. O&P
1 Laborer	$44.40	$355.20	$66.25	$530.00	$52.92	$78.90
1 Equip. Oper. (crane)	61.45	491.60	91.55	732.40		
1 Hyd. Excavator, .75 C.Y.		701.80		771.98		
1 Trench Box		119.15		131.07	51.31	56.44
16 L.H., Daily Totals		$1667.75		$2165.45	$104.23	$135.34
Crew B-13H	Hr.	Daily	Hr.	Daily	Bare Costs	Incl. O&P
1 Laborer	$44.40	$355.20	$66.25	$530.00	$52.92	$78.90
1 Equip. Oper. (crane)	61.45	491.60	91.55	732.40		
1 Gradall, 5/8 C.Y.		850.65		935.72		
1 Trench Box		119.15		131.07	60.61	66.67
16 L.H., Daily Totals		$1816.60		$2329.18	$113.54	$145.57

Crew B-13I

	Bare Costs		Incl. Subs O&P		Cost Per Labor-Hour	
Crew B-13I	Hr.	Daily	Hr.	Daily	Bare Costs	Incl. O&P
1 Laborer	$44.40	$355.20	$66.25	$530.00	$52.92	$78.90
1 Equip. Oper. (crane)	61.45	491.60	91.55	732.40		
1 Gradall, 3 Ton, 1 C.Y.		984.55		1083.01		
1 Trench Box		119.15		131.07	68.98	75.88
16 L.H., Daily Totals		$1950.50		$2476.47	$121.91	$154.78

Crew B-13J

Crew B-13J	Hr.	Daily	Hr.	Daily	Bare Costs	Incl. O&P
1 Laborer	$44.40	$355.20	$66.25	$530.00	$52.92	$78.90
1 Equip. Oper. (crane)	61.45	491.60	91.55	732.40		
1 Hyd. Excavator, 2.5 C.Y.		1567.00		1723.70		
1 Trench Box		119.15		131.07	105.38	115.92
16 L.H., Daily Totals		$2532.95		$3117.17	$158.31	$194.82

Crew B-13K

Crew B-13K	Hr.	Daily	Hr.	Daily	Bare Costs	Incl. O&P
2 Equip. Opers. (crane)	$61.45	$983.20	$91.55	$1464.80	$61.45	$91.55
1 Hyd. Excavator, .75 C.Y.		701.80		771.98		
1 Hyd. Hammer, 4000 ft-lb		649.20		714.12		
1 Hyd. Excavator, .75 C.Y.		701.80		771.98	128.30	141.13
16 L.H., Daily Totals		$3036.00		$3722.88	$189.75	$232.68

Crew B-13L

Crew B-13L	Hr.	Daily	Hr.	Daily	Bare Costs	Incl. O&P
2 Equip. Opers. (crane)	$61.45	$983.20	$91.55	$1464.80	$61.45	$91.55
1 Hyd. Excavator, 1.5 C.Y.		695.80		765.38		
1 Hyd. Hammer, 5000 ft-lb		705.60		776.16		
1 Hyd. Excavator, .75 C.Y.		701.80		771.98	131.45	144.60
16 L.H., Daily Totals		$3086.40		$3778.32	$192.90	$236.15

Crew B-13M

Crew B-13M	Hr.	Daily	Hr.	Daily	Bare Costs	Incl. O&P
2 Equip. Opers. (crane)	$61.45	$983.20	$91.55	$1464.80	$61.45	$91.55
1 Hyd. Excavator, 2.5 C.Y.		1567.00		1723.70		
1 Hyd. Hammer, 8000 ft-lb		918.65		1010.52		
1 Hyd. Excavator, 1.5 C.Y.		695.80		765.38	198.84	218.72
16 L.H., Daily Totals		$4164.65		$4964.40	$260.29	$310.27

Crew B-13N

Crew B-13N	Hr.	Daily	Hr.	Daily	Bare Costs	Incl. O&P
2 Equip. Opers. (crane)	$61.45	$983.20	$91.55	$1464.80	$61.45	$91.55
1 Hyd. Excavator, 3.5 C.Y.		2184.00		2402.40		
1 Hyd. Hammer, 12,000 ft-lb		882.20		970.42		
1 Hyd. Excavator, 1.5 C.Y.		695.80		765.38	235.13	258.64
16 L.H., Daily Totals		$4745.20		$5603.00	$296.57	$350.19

Crew B-14

Crew B-14	Hr.	Daily	Hr.	Daily	Bare Costs	Incl. O&P
1 Labor Foreman (outside)	$46.40	$371.20	$69.25	$554.00	$46.58	$69.49
4 Laborers	44.40	1420.80	66.25	2120.00		
1 Equip. Oper. (light)	55.50	444.00	82.70	661.60		
1 Backhoe Loader, 48 H.P.		216.20		237.82	4.50	4.95
48 L.H., Daily Totals		$2452.20		$3573.42	$51.09	$74.45

Crew B-14A

Crew B-14A	Hr.	Daily	Hr.	Daily	Bare Costs	Incl. O&P
1 Equip. Oper. (crane)	$61.45	$491.60	$91.55	$732.40	$55.77	$83.12
.5 Laborer	44.40	177.60	66.25	265.00		
1 Hyd. Excavator, 4.5 C.Y.		3450.00		3795.00	287.50	316.25
12 L.H., Daily Totals		$4119.20		$4792.40	$343.27	$399.37

Crew B-14B

Crew B-14B	Hr.	Daily	Hr.	Daily	Bare Costs	Incl. O&P
1 Equip. Oper. (crane)	$61.45	$491.60	$91.55	$732.40	$55.77	$83.12
.5 Laborer	44.40	177.60	66.25	265.00		
1 Hyd. Excavator, 6 C.Y.		3506.00		3856.60	292.17	321.38
12 L.H., Daily Totals		$4175.20		$4854.00	$347.93	$404.50

Crew B-14C

Crew B-14C	Hr.	Daily	Hr.	Daily	Bare Costs	Incl. O&P
1 Equip. Oper. (crane)	$61.45	$491.60	$91.55	$732.40	$55.77	$83.12
.5 Laborer	44.40	177.60	66.25	265.00		
1 Hyd. Excavator, 7 C.Y.		3475.00		3822.50	289.58	318.54
12 L.H., Daily Totals		$4144.20		$4819.90	$345.35	$401.66

Crew B-14F

Crew B-14F	Hr.	Daily	Hr.	Daily	Bare Costs	Incl. O&P
1 Equip. Oper. (crane)	$61.45	$491.60	$91.55	$732.40	$55.77	$83.12
.5 Laborer	44.40	177.60	66.25	265.00		
1 Hyd. Shovel, 7 C.Y.		4148.00		4562.80	345.67	380.23
12 L.H., Daily Totals		$4817.20		$5560.20	$401.43	$463.35

Crew B-14G

Crew B-14G	Hr.	Daily	Hr.	Daily	Bare Costs	Incl. O&P
1 Equip. Oper. (crane)	$61.45	$491.60	$91.55	$732.40	$55.77	$83.12
.5 Laborer	44.40	177.60	66.25	265.00		
1 Hyd. Shovel, 12 C.Y.		6022.00		6624.20	501.83	552.02
12 L.H., Daily Totals		$6691.20		$7621.60	$557.60	$635.13

Crew B-14J

Crew B-14J	Hr.	Daily	Hr.	Daily	Bare Costs	Incl. O&P
1 Equip. Oper. (medium)	$59.00	$472.00	$87.90	$703.20	$54.13	$80.68
.5 Laborer	44.40	177.60	66.25	265.00		
1 F.E. Loader, 8 C.Y.		2285.00		2513.50	190.42	209.46
12 L.H., Daily Totals		$2934.60		$3481.70	$244.55	$290.14

Crew B-14K

Crew B-14K	Hr.	Daily	Hr.	Daily	Bare Costs	Incl. O&P
1 Equip. Oper. (medium)	$59.00	$472.00	$87.90	$703.20	$54.13	$80.68
.5 Laborer	44.40	177.60	66.25	265.00		
1 F.E. Loader, 10 C.Y.		2706.00		2976.60	225.50	248.05
12 L.H., Daily Totals		$3355.60		$3944.80	$279.63	$328.73

Crew B-15

Crew B-15	Hr.	Daily	Hr.	Daily	Bare Costs	Incl. O&P
1 Equipment Oper. (med.)	$59.00	$472.00	$87.90	$703.20	$52.51	$78.41
.5 Laborer	44.40	177.60	66.25	265.00		
2 Truck Drivers (heavy)	51.30	820.80	76.70	1227.20		
2 Dump Trucks, 12 C.Y., 400 H.P.		1158.70		1274.57		
1 Dozer, 200 H.P.		1520.00		1672.00	95.67	105.23
28 L.H., Daily Totals		$4149.10		$5141.97	$148.18	$183.64

Crew B-16

Crew B-16	Hr.	Daily	Hr.	Daily	Bare Costs	Incl. O&P
1 Labor Foreman (outside)	$46.40	$371.20	$69.25	$554.00	$46.63	$69.61
2 Laborers	44.40	710.40	66.25	1060.00		
1 Truck Driver (heavy)	51.30	410.40	76.70	613.60		
1 Dump Truck, 12 C.Y., 400 H.P.		579.35		637.28	18.10	19.92
32 L.H., Daily Totals		$2071.35		$2864.89	$64.73	$89.53

Crew B-17

Crew B-17	Hr.	Daily	Hr.	Daily	Bare Costs	Incl. O&P
2 Laborers	$44.40	$710.40	$66.25	$1060.00	$48.90	$72.97
1 Equip. Oper. (light)	55.50	444.00	82.70	661.60		
1 Truck Driver (heavy)	51.30	410.40	76.70	613.60		
1 Backhoe Loader, 48 H.P.		216.20		237.82		
1 Dump Truck, 8 C.Y., 220 H.P.		407.60		448.36	19.49	21.44
32 L.H., Daily Totals		$2188.60		$3021.38	$68.39	$94.42

Crew B-17A

Crew B-17A	Hr.	Daily	Hr.	Daily	Bare Costs	Incl. O&P
2 Labor Foremen (outside)	$46.40	$742.40	$69.25	$1108.00	$47.54	$71.08
6 Laborers	44.40	2131.20	66.25	3180.00		
1 Skilled Worker Foreman (out)	59.10	472.80	88.90	711.20		
1 Skilled Worker	57.10	456.80	85.90	687.20		
80 L.H., Daily Totals		$3803.20		$5686.40	$47.54	$71.08

424

For customer support on your Concrete & Masonry Costs with RSMeans Data, call 800.448.8182.

Crew No.	Bare Costs		Incl. Subs O&P		Cost Per Labor-Hour	

Crew B-17B	Hr.	Daily	Hr.	Daily	Bare Costs	Incl. O&P
2 Laborers	$44.40	$710.40	$66.25	$1060.00	$48.90	$72.97
1 Equip. Oper. (light)	55.50	444.00	82.70	661.60		
1 Truck Driver (heavy)	51.30	410.40	76.70	613.60		
1 Backhoe Loader, 48 H.P.		216.20		237.82		
1 Dump Truck, 12 C.Y., 400 H.P.		579.35		637.28	24.86	27.35
32 L.H., Daily Totals		$2360.35		$3210.30	$73.76	$100.32

Crew B-18	Hr.	Daily	Hr.	Daily	Bare Costs	Incl. O&P
1 Labor Foreman (outside)	$46.40	$371.20	$69.25	$554.00	$45.07	$67.25
2 Laborers	44.40	710.40	66.25	1060.00		
1 Vibrating Plate, Gas, 21"		165.60		182.16	6.90	7.59
24 L.H., Daily Totals		$1247.20		$1796.16	$51.97	$74.84

Crew B-19	Hr.	Daily	Hr.	Daily	Bare Costs	Incl. O&P
1 Pile Driver Foreman (outside)	$57.90	$463.20	$89.50	$716.00	$57.11	$87.06
4 Pile Drivers	55.90	1788.80	86.40	2764.80		
2 Equip. Oper. (crane)	61.45	983.20	91.55	1464.80		
1 Equip. Oper. (oiler)	52.50	420.00	78.25	626.00		
1 Crawler Crane, 40 Ton		1231.00		1354.10		
1 Lead, 90' High		371.85		409.04		
1 Hammer, Diesel, 22k ft-lb		441.70		485.87	31.95	35.14
64 L.H., Daily Totals		$5699.75		$7820.60	$89.06	$122.20

Crew B-19A	Hr.	Daily	Hr.	Daily	Bare Costs	Incl. O&P
1 Pile Driver Foreman (outside)	$57.90	$463.20	$89.50	$716.00	$57.11	$87.06
4 Pile Drivers	55.90	1788.80	86.40	2764.80		
2 Equip. Oper. (crane)	61.45	983.20	91.55	1464.80		
1 Equip. Oper. (oiler)	52.50	420.00	78.25	626.00		
1 Crawler Crane, 75 Ton		1967.00		2163.70		
1 Lead, 90' High		371.85		409.04		
1 Hammer, Diesel, 41k ft-lb		583.55		641.90	45.66	50.23
64 L.H., Daily Totals		$6577.60		$8786.24	$102.78	$137.29

Crew B-19B	Hr.	Daily	Hr.	Daily	Bare Costs	Incl. O&P
1 Pile Driver Foreman (outside)	$57.90	$463.20	$89.50	$716.00	$57.11	$87.06
4 Pile Drivers	55.90	1788.80	86.40	2764.80		
2 Equip. Oper. (crane)	61.45	983.20	91.55	1464.80		
1 Equip. Oper. (oiler)	52.50	420.00	78.25	626.00		
1 Crawler Crane, 40 Ton		1231.00		1354.10		
1 Lead, 90' High		371.85		409.04		
1 Hammer, Diesel, 22k ft-lb		441.70		485.87		
1 Barge, 400 Ton		869.15		956.07	45.53	50.08
64 L.H., Daily Totals		$6568.90		$8776.67	$102.64	$137.14

Crew B-19C	Hr.	Daily	Hr.	Daily	Bare Costs	Incl. O&P
1 Pile Driver Foreman (outside)	$57.90	$463.20	$89.50	$716.00	$57.11	$87.06
4 Pile Drivers	55.90	1788.80	86.40	2764.80		
2 Equip. Oper. (crane)	61.45	983.20	91.55	1464.80		
1 Equip. Oper. (oiler)	52.50	420.00	78.25	626.00		
1 Crawler Crane, 75 Ton		1967.00		2163.70		
1 Lead, 90' High		371.85		409.04		
1 Hammer, Diesel, 41k ft-lb		583.55		641.90		
1 Barge, 400 Ton		869.15		956.07	59.24	65.17
64 L.H., Daily Totals		$7446.75		$9742.31	$116.36	$152.22

Crew B-20	Hr.	Daily	Hr.	Daily	Bare Costs	Incl. O&P
1 Labor Foreman (outside)	$46.40	$371.20	$69.25	$554.00	$49.30	$73.80
1 Skilled Worker	57.10	456.80	85.90	687.20		
1 Laborer	44.40	355.20	66.25	530.00		
24 L.H., Daily Totals		$1183.20		$1771.20	$49.30	$73.80

Crew B-20A	Hr.	Daily	Hr.	Daily	Bare Costs	Incl. O&P
1 Labor Foreman (outside)	$46.40	$371.20	$69.25	$554.00	$53.16	$79.34
1 Laborer	44.40	355.20	66.25	530.00		
1 Plumber	67.70	541.60	101.05	808.40		
1 Plumber Apprentice	54.15	433.20	80.80	646.40		
32 L.H., Daily Totals		$1701.20		$2538.80	$53.16	$79.34

Crew B-21	Hr.	Daily	Hr.	Daily	Bare Costs	Incl. O&P
1 Labor Foreman (outside)	$46.40	$371.20	$69.25	$554.00	$51.04	$76.34
1 Skilled Worker	57.10	456.80	85.90	687.20		
1 Laborer	44.40	355.20	66.25	530.00		
.5 Equip. Oper. (crane)	61.45	245.80	91.55	366.20		
.5 S.P. Crane, 4x4, 5 Ton		190.97		210.07	6.82	7.50
28 L.H., Daily Totals		$1619.97		$2347.47	$57.86	$83.84

Crew B-21A	Hr.	Daily	Hr.	Daily	Bare Costs	Incl. O&P
1 Labor Foreman (outside)	$46.40	$371.20	$69.25	$554.00	$54.82	$81.78
1 Laborer	44.40	355.20	66.25	530.00		
1 Plumber	67.70	541.60	101.05	808.40		
1 Plumber Apprentice	54.15	433.20	80.80	646.40		
1 Equip. Oper. (crane)	61.45	491.60	91.55	732.40		
1 S.P. Crane, 4x4, 12 Ton		432.65		475.92	10.82	11.90
40 L.H., Daily Totals		$2625.45		$3747.11	$65.64	$93.68

Crew B-21B	Hr.	Daily	Hr.	Daily	Bare Costs	Incl. O&P
1 Labor Foreman (outside)	$46.40	$371.20	$69.25	$554.00	$48.21	$71.91
3 Laborers	44.40	1065.60	66.25	1590.00		
1 Equip. Oper. (crane)	61.45	491.60	91.55	732.40		
1 Hyd. Crane, 12 Ton		475.80		523.38	11.90	13.08
40 L.H., Daily Totals		$2404.20		$3399.78	$60.10	$84.99

Crew B-21C	Hr.	Daily	Hr.	Daily	Bare Costs	Incl. O&P
1 Labor Foreman (outside)	$46.40	$371.20	$69.25	$554.00	$48.28	$72.01
4 Laborers	44.40	1420.80	66.25	2120.00		
1 Equip. Oper. (crane)	61.45	491.60	91.55	732.40		
1 Equip. Oper. (oiler)	52.50	420.00	78.25	626.00		
2 Cutting Torches		25.90		28.49		
2 Sets of Gases		347.20		381.92		
1 Lattice Boom Crane, 90 Ton		1713.00		1884.30	37.25	40.98
56 L.H., Daily Totals		$4789.70		$6327.11	$85.53	$112.98

Crew B-22	Hr.	Daily	Hr.	Daily	Bare Costs	Incl. O&P
1 Labor Foreman (outside)	$46.40	$371.20	$69.25	$554.00	$51.73	$77.35
1 Skilled Worker	57.10	456.80	85.90	687.20		
1 Laborer	44.40	355.20	66.25	530.00		
.75 Equip. Oper. (crane)	61.45	368.70	91.55	549.30		
.75 S.P. Crane, 4x4, 5 Ton		286.46		315.11	9.55	10.50
30 L.H., Daily Totals		$1838.36		$2635.61	$61.28	$87.85

Crew B-22A	Hr.	Daily	Hr.	Daily	Bare Costs	Incl. O&P
1 Labor Foreman (outside)	$46.40	$371.20	$69.25	$554.00	$50.75	$75.84
1 Skilled Worker	57.10	456.80	85.90	687.20		
2 Laborers	44.40	710.40	66.25	1060.00		
1 Equipment Operator, Crane	61.45	491.60	91.55	732.40		
1 S.P. Crane, 4x4, 5 Ton		381.95		420.14		
1 Butt Fusion Mach., 4"-12" diam.		420.75		462.82	20.07	22.07
40 L.H., Daily Totals		$2832.70		$3916.57	$70.82	$97.91

For customer support on your Concrete & Masonry Costs with RSMeans Data, call 800.448.8182.

425

Crew No.	Bare Costs		Incl. Subs O&P		Cost Per Labor-Hour	
Crew B-22B	Hr.	Daily	Hr.	Daily	Bare Costs	Incl. O&P
1 Labor Foreman (outside)	$46.40	$371.20	$69.25	$554.00	$50.75	$75.84
1 Skilled Worker	57.10	456.80	85.90	687.20		
2 Laborers	44.40	710.40	66.25	1060.00		
1 Equip. Oper. (crane)	61.45	491.60	91.55	732.40		
1 S.P. Crane, 4x4, 5 Ton		381.95		420.14		
1 Butt Fusion Mach., 8"-24" diam.		1086.00		1194.60	36.70	40.37
40 L.H., Daily Totals		$3497.95		$4648.35	$87.45	$116.21
Crew B-22C	Hr.	Daily	Hr.	Daily	Bare Costs	Incl. O&P
1 Skilled Worker	$57.10	$456.80	$85.90	$687.20	$50.75	$76.08
1 Laborer	44.40	355.20	66.25	530.00		
1 Butt Fusion Mach., 2"-8" diam.		134.95		148.44	8.43	9.28
16 L.H., Daily Totals		$946.95		$1365.65	$59.18	$85.35
Crew B-23	Hr.	Daily	Hr.	Daily	Bare Costs	Incl. O&P
1 Labor Foreman (outside)	$46.40	$371.20	$69.25	$554.00	$44.80	$66.85
4 Laborers	44.40	1420.80	66.25	2120.00		
1 Drill Rig, Truck-Mounted		768.40		845.24		
1 Flatbed Truck, Gas, 3 Ton		850.05		935.05	40.46	44.51
40 L.H., Daily Totals		$3410.45		$4454.30	$85.26	$111.36
Crew B-23A	Hr.	Daily	Hr.	Daily	Bare Costs	Incl. O&P
1 Labor Foreman (outside)	$46.40	$371.20	$69.25	$554.00	$49.93	$74.47
1 Laborer	44.40	355.20	66.25	530.00		
1 Equip. Oper. (medium)	59.00	472.00	87.90	703.20		
1 Drill Rig, Truck-Mounted		768.40		845.24		
1 Pickup Truck, 3/4 Ton		112.20		123.42	36.69	40.36
24 L.H., Daily Totals		$2079.00		$2755.86	$86.63	$114.83
Crew B-23B	Hr.	Daily	Hr.	Daily	Bare Costs	Incl. O&P
1 Labor Foreman (outside)	$46.40	$371.20	$69.25	$554.00	$49.93	$74.47
1 Laborer	44.40	355.20	66.25	530.00		
1 Equip. Oper. (medium)	59.00	472.00	87.90	703.20		
1 Drill Rig, Truck-Mounted		768.40		845.24		
1 Pickup Truck, 3/4 Ton		112.20		123.42		
1 Centr. Water Pump, 6"		235.25		258.77	46.49	51.14
24 L.H., Daily Totals		$2314.25		$3014.64	$96.43	$125.61
Crew B-24	Hr.	Daily	Hr.	Daily	Bare Costs	Incl. O&P
1 Cement Finisher	$51.80	$414.40	$75.90	$607.20	$50.30	$74.60
1 Laborer	44.40	355.20	66.25	530.00		
1 Carpenter	54.70	437.60	81.65	653.20		
24 L.H., Daily Totals		$1207.20		$1790.40	$50.30	$74.60
Crew B-25	Hr.	Daily	Hr.	Daily	Bare Costs	Incl. O&P
1 Labor Foreman (outside)	$46.40	$371.20	$69.25	$554.00	$48.56	$72.43
7 Laborers	44.40	2486.40	66.25	3710.00		
3 Equip. Oper. (medium)	59.00	1416.00	87.90	2109.60		
1 Asphalt Paver, 130 H.P.		2143.00		2357.30		
1 Tandem Roller, 10 Ton		246.80		271.48		
1 Roller, Pneum. Whl., 12 Ton		349.90		384.89	31.13	34.25
88 L.H., Daily Totals		$7013.30		$9387.27	$79.70	$106.67
Crew B-25B	Hr.	Daily	Hr.	Daily	Bare Costs	Incl. O&P
1 Labor Foreman (outside)	$46.40	$371.20	$69.25	$554.00	$49.43	$73.72
7 Laborers	44.40	2486.40	66.25	3710.00		
4 Equip. Oper. (medium)	59.00	1888.00	87.90	2812.80		
1 Asphalt Paver, 130 H.P.		2143.00		2357.30		
2 Tandem Rollers, 10 Ton		493.60		542.96		
1 Roller, Pneum. Whl., 12 Ton		349.90		384.89	31.11	34.22
96 L.H., Daily Totals		$7732.10		$10361.95	$80.54	$107.94

Crew No.	Bare Costs		Incl. Subs O&P		Cost Per Labor-Hour	
Crew B-25C	Hr.	Daily	Hr.	Daily	Bare Costs	Incl. O&P
1 Labor Foreman (outside)	$46.40	$371.20	$69.25	$554.00	$49.60	$73.97
3 Laborers	44.40	1065.60	66.25	1590.00		
2 Equip. Oper. (medium)	59.00	944.00	87.90	1406.40		
1 Asphalt Paver, 130 H.P.		2143.00		2357.30		
1 Tandem Roller, 10 Ton		246.80		271.48	49.79	54.77
48 L.H., Daily Totals		$4770.60		$6179.18	$99.39	$128.73
Crew B-25D	Hr.	Daily	Hr.	Daily	Bare Costs	Incl. O&P
1 Labor Foreman (outside)	$46.40	$371.20	$69.25	$554.00	$49.82	$74.30
3 Laborers	44.40	1065.60	66.25	1590.00		
2.125 Equip. Oper. (medium)	59.00	1003.00	87.90	1494.30		
.125 Truck Driver (heavy)	51.30	51.30	76.70	76.70		
.125 Truck Tractor, 6x4, 380 H.P.		62.39		68.63		
.125 Dist. Tanker, 3000 Gallon		41.76		45.94		
1 Asphalt Paver, 130 H.P.		2143.00		2357.30		
1 Tandem Roller, 10 Ton		246.80		271.48	49.88	54.87
50 L.H., Daily Totals		$4985.06		$6458.35	$99.70	$129.17
Crew B-25E	Hr.	Daily	Hr.	Daily	Bare Costs	Incl. O&P
1 Labor Foreman (outside)	$46.40	$371.20	$69.25	$554.00	$50.03	$74.61
3 Laborers	44.40	1065.60	66.25	1590.00		
2.250 Equip. Oper. (medium)	59.00	1062.00	87.90	1582.20		
.25 Truck Driver (heavy)	51.30	102.60	76.70	153.40		
.25 Truck Tractor, 6x4, 380 H.P.		124.79		137.27		
.25 Dist. Tanker, 3000 Gallon		83.53		91.88		
1 Asphalt Paver, 130 H.P.		2143.00		2357.30		
1 Tandem Roller, 10 Ton		246.80		271.48	49.96	54.96
52 L.H., Daily Totals		$5199.51		$6737.52	$99.99	$129.57
Crew B-26	Hr.	Daily	Hr.	Daily	Bare Costs	Incl. O&P
1 Labor Foreman (outside)	$46.40	$371.20	$69.25	$554.00	$49.23	$73.32
6 Laborers	44.40	2131.20	66.25	3180.00		
2 Equip. Oper. (medium)	59.00	944.00	87.90	1406.40		
1 Rodman (reinf.)	58.90	471.20	88.05	704.40		
1 Cement Finisher	51.80	414.40	75.90	607.20		
1 Grader, 30,000 Lbs.		1073.00		1180.30		
1 Paving Mach. & Equip.		2503.00		2753.30	40.64	44.70
88 L.H., Daily Totals		$7908.00		$10385.60	$89.86	$118.02
Crew B-26A	Hr.	Daily	Hr.	Daily	Bare Costs	Incl. O&P
1 Labor Foreman (outside)	$46.40	$371.20	$69.25	$554.00	$49.23	$73.32
6 Laborers	44.40	2131.20	66.25	3180.00		
2 Equip. Oper. (medium)	59.00	944.00	87.90	1406.40		
1 Rodman (reinf.)	58.90	471.20	88.05	704.40		
1 Cement Finisher	51.80	414.40	75.90	607.20		
1 Grader, 30,000 Lbs.		1073.00		1180.30		
1 Paving Mach. & Equip.		2503.00		2753.30		
1 Concrete Saw		112.85		124.14	41.92	46.11
88 L.H., Daily Totals		$8020.85		$10509.74	$91.15	$119.43
Crew B-26B	Hr.	Daily	Hr.	Daily	Bare Costs	Incl. O&P
1 Labor Foreman (outside)	$46.40	$371.20	$69.25	$554.00	$50.04	$74.53
6 Laborers	44.40	2131.20	66.25	3180.00		
3 Equip. Oper. (medium)	59.00	1416.00	87.90	2109.60		
1 Rodman (reinf.)	58.90	471.20	88.05	704.40		
1 Cement Finisher	51.80	414.40	75.90	607.20		
1 Grader, 30,000 Lbs.		1073.00		1180.30		
1 Paving Mach. & Equip.		2503.00		2753.30		
1 Concrete Pump, 110' Boom		493.65		543.01	42.39	46.63
96 L.H., Daily Totals		$8873.65		$11631.82	$92.43	$121.16

Crew No.	Bare Costs		Incl. Subs O&P		Cost Per Labor-Hour	

Crew B-26C

	Hr.	Daily	Hr.	Daily	Bare Costs	Incl. O&P
1 Labor Foreman (outside)	$46.40	$371.20	$69.25	$554.00	$48.25	$71.86
6 Laborers	44.40	2131.20	66.25	3180.00		
1 Equip. Oper. (medium)	59.00	472.00	87.90	703.20		
1 Rodman (reinf.)	58.90	471.20	88.05	704.40		
1 Cement Finisher	51.80	414.40	75.90	607.20		
1 Paving Mach. & Equip.		2503.00		2753.30		
1 Concrete Saw		112.85		124.14	32.70	35.97
80 L.H., Daily Totals		$6475.85		$8626.24	$80.95	$107.83

Crew B-27

	Hr.	Daily	Hr.	Daily	Bare Costs	Incl. O&P
1 Labor Foreman (outside)	$46.40	$371.20	$69.25	$554.00	$44.90	$67.00
3 Laborers	44.40	1065.60	66.25	1590.00		
1 Berm Machine		253.55		278.90	7.92	8.72
32 L.H., Daily Totals		$1690.35		$2422.91	$52.82	$75.72

Crew B-28

	Hr.	Daily	Hr.	Daily	Bare Costs	Incl. O&P
2 Carpenters	$54.70	$875.20	$81.65	$1306.40	$51.27	$76.52
1 Laborer	44.40	355.20	66.25	530.00		
24 L.H., Daily Totals		$1230.40		$1836.40	$51.27	$76.52

Crew B-29

	Hr.	Daily	Hr.	Daily	Bare Costs	Incl. O&P
1 Labor Foreman (outside)	$46.40	$371.20	$69.25	$554.00	$48.28	$72.01
4 Laborers	44.40	1420.80	66.25	2120.00		
1 Equip. Oper. (crane)	61.45	491.60	91.55	732.40		
1 Equip. Oper. (oiler)	52.50	420.00	78.25	626.00		
1 Gradall, 5/8 C.Y.		850.65		935.72	15.19	16.71
56 L.H., Daily Totals		$3554.25		$4968.11	$63.47	$88.72

Crew B-30

	Hr.	Daily	Hr.	Daily	Bare Costs	Incl. O&P
1 Equip. Oper. (medium)	$59.00	$472.00	$87.90	$703.20	$53.87	$80.43
2 Truck Drivers (heavy)	51.30	820.80	76.70	1227.20		
1 Hyd. Excavator, 1.5 C.Y.		695.80		765.38		
2 Dump Trucks, 12 C.Y., 400 H.P.		1158.70		1274.57	77.27	85.00
24 L.H., Daily Totals		$3147.30		$3970.35	$131.14	$165.43

Crew B-31

	Hr.	Daily	Hr.	Daily	Bare Costs	Incl. O&P
1 Labor Foreman (outside)	$46.40	$371.20	$69.25	$554.00	$46.86	$69.93
3 Laborers	44.40	1065.60	66.25	1590.00		
1 Carpenter	54.70	437.60	81.65	653.20		
1 Air Compressor, 250 cfm		202.85		223.13		
1 Sheeting Driver		7.45		8.20		
2 -50' Air Hoses, 1.5"		45.60		50.16	6.40	7.04
40 L.H., Daily Totals		$2130.30		$3078.69	$53.26	$76.97

Crew B-32

	Hr.	Daily	Hr.	Daily	Bare Costs	Incl. O&P
1 Laborer	$44.40	$355.20	$66.25	$530.00	$55.35	$82.49
3 Equip. Oper. (medium)	59.00	1416.00	87.90	2109.60		
1 Grader, 30,000 Lbs.		1073.00		1180.30		
1 Tandem Roller, 10 Ton		246.80		271.48		
1 Dozer, 200 H.P.		1520.00		1672.00	88.74	97.62
32 L.H., Daily Totals		$4611.00		$5763.38	$144.09	$180.11

Crew B-32A

	Hr.	Daily	Hr.	Daily	Bare Costs	Incl. O&P
1 Laborer	$44.40	$355.20	$66.25	$530.00	$54.13	$80.68
2 Equip. Oper. (medium)	59.00	944.00	87.90	1406.40		
1 Grader, 30,000 Lbs.		1073.00		1180.30		
1 Roller, Vibratory, 25 Ton		672.35		739.59	72.72	80.00
24 L.H., Daily Totals		$3044.55		$3856.28	$126.86	$160.68

Crew B-32B

	Hr.	Daily	Hr.	Daily	Bare Costs	Incl. O&P
1 Laborer	$44.40	$355.20	$66.25	$530.00	$54.13	$80.68
2 Equip. Oper. (medium)	59.00	944.00	87.90	1406.40		
1 Dozer, 200 H.P.		1520.00		1672.00		
1 Roller, Vibratory, 25 Ton		672.35		739.59	91.35	100.48
24 L.H., Daily Totals		$3491.55		$4347.98	$145.48	$181.17

Crew B-32C

	Hr.	Daily	Hr.	Daily	Bare Costs	Incl. O&P
1 Labor Foreman (outside)	$46.40	$371.20	$69.25	$554.00	$52.03	$77.58
2 Laborers	44.40	710.40	66.25	1060.00		
3 Equip. Oper. (medium)	59.00	1416.00	87.90	2109.60		
1 Grader, 30,000 Lbs.		1073.00		1180.30		
1 Tandem Roller, 10 Ton		246.80		271.48		
1 Dozer, 200 H.P.		1520.00		1672.00	59.16	65.08
48 L.H., Daily Totals		$5337.40		$6847.38	$111.20	$142.65

Crew B-33A

	Hr.	Daily	Hr.	Daily	Bare Costs	Incl. O&P
1 Equip. Oper. (medium)	$59.00	$472.00	$87.90	$703.20	$54.83	$81.71
.5 Laborer	44.40	177.60	66.25	265.00		
.25 Equip. Oper. (medium)	59.00	118.00	87.90	175.80		
1 Scraper, Towed, 7 C.Y.		129.30		142.23		
1.25 Dozers, 300 H.P.		2231.25		2454.38	168.61	185.47
14 L.H., Daily Totals		$3128.15		$3740.61	$223.44	$267.19

Crew B-33B

	Hr.	Daily	Hr.	Daily	Bare Costs	Incl. O&P
1 Equip. Oper. (medium)	$59.00	$472.00	$87.90	$703.20	$54.83	$81.71
.5 Laborer	44.40	177.60	66.25	265.00		
.25 Equip. Oper. (medium)	59.00	118.00	87.90	175.80		
1 Scraper, Towed, 10 C.Y.		161.65		177.82		
1.25 Dozers, 300 H.P.		2231.25		2454.38	170.92	188.01
14 L.H., Daily Totals		$3160.50		$3776.19	$225.75	$269.73

Crew B-33C

	Hr.	Daily	Hr.	Daily	Bare Costs	Incl. O&P
1 Equip. Oper. (medium)	$59.00	$472.00	$87.90	$703.20	$54.83	$81.71
.5 Laborer	44.40	177.60	66.25	265.00		
.25 Equip. Oper. (medium)	59.00	118.00	87.90	175.80		
1 Scraper, Towed, 15 C.Y.		178.85		196.74		
1.25 Dozers, 300 H.P.		2231.25		2454.38	172.15	189.37
14 L.H., Daily Totals		$3177.70		$3795.11	$226.98	$271.08

Crew B-33D

	Hr.	Daily	Hr.	Daily	Bare Costs	Incl. O&P
1 Equip. Oper. (medium)	$59.00	$472.00	$87.90	$703.20	$54.83	$81.71
.5 Laborer	44.40	177.60	66.25	265.00		
.25 Equip. Oper. (medium)	59.00	118.00	87.90	175.80		
1 S.P. Scraper, 14 C.Y.		2424.00		2666.40		
.25 Dozer, 300 H.P.		446.25		490.88	205.02	225.52
14 L.H., Daily Totals		$3637.85		$4301.27	$259.85	$307.23

Crew B-33E

	Hr.	Daily	Hr.	Daily	Bare Costs	Incl. O&P
1 Equip. Oper. (medium)	$59.00	$472.00	$87.90	$703.20	$54.83	$81.71
.5 Laborer	44.40	177.60	66.25	265.00		
.25 Equip. Oper. (medium)	59.00	118.00	87.90	175.80		
1 S.P. Scraper, 21 C.Y.		2656.00		2921.60		
.25 Dozer, 300 H.P.		446.25		490.88	221.59	243.75
14 L.H., Daily Totals		$3869.85		$4556.48	$276.42	$325.46

Crew B-33F

	Hr.	Daily	Hr.	Daily	Bare Costs	Incl. O&P
1 Equip. Oper. (medium)	$59.00	$472.00	$87.90	$703.20	$54.83	$81.71
.5 Laborer	44.40	177.60	66.25	265.00		
.25 Equip. Oper. (medium)	59.00	118.00	87.90	175.80		
1 Elev. Scraper, 11 C.Y.		1059.00		1164.90		
.25 Dozer, 300 H.P.		446.25		490.88	107.52	118.27
14 L.H., Daily Totals		$2272.85		$2799.78	$162.35	$199.98

Crew B-33G

	Hr.	Daily	Hr.	Daily	Bare Costs	Incl. O&P
1 Equip. Oper. (medium)	$59.00	$472.00	$87.90	$703.20	$54.83	$81.71
.5 Laborer	44.40	177.60	66.25	265.00		
.25 Equip. Oper. (medium)	59.00	118.00	87.90	175.80		
1 Elev. Scraper, 22 C.Y.		1895.00		2084.50		
.25 Dozer, 300 H.P.		446.25		490.88	167.23	183.96
14 L.H., Daily Totals		$3108.85		$3719.38	$222.06	$265.67

Crew B-33H

	Hr.	Daily	Hr.	Daily	Bare Costs	Incl. O&P
.5 Laborer	$44.40	$177.60	$66.25	$265.00	$54.83	$81.71
1 Equipment Operator (med.)	59.00	472.00	87.90	703.20		
.25 Equipment Operator (med.)	59.00	118.00	87.90	175.80		
1 S.P. Scraper, 44 C.Y.		4695.00		5164.50		
.25 Dozer, 410 H.P.		701.75		771.92	385.48	424.03
14 L.H., Daily Totals		$6164.35		$7080.43	$440.31	$505.74

Crew B-33J

	Hr.	Daily	Hr.	Daily	Bare Costs	Incl. O&P
1 Equipment Operator (med.)	$59.00	$472.00	$87.90	$703.20	$59.00	$87.90
1 S.P. Scraper, 14 C.Y.		2424.00		2666.40	303.00	333.30
8 L.H., Daily Totals		$2896.00		$3369.60	$362.00	$421.20

Crew B-33K

	Hr.	Daily	Hr.	Daily	Bare Costs	Incl. O&P
1 Equipment Operator (med.)	$59.00	$472.00	$87.90	$703.20	$54.83	$81.71
.25 Equipment Operator (med.)	59.00	118.00	87.90	175.80		
.5 Laborer	44.40	177.60	66.25	265.00		
1 S.P. Scraper, 31 C.Y.		3707.00		4077.70		
.25 Dozer, 410 H.P.		701.75		771.92	314.91	346.40
14 L.H., Daily Totals		$5176.35		$5993.63	$369.74	$428.12

Crew B-34A

	Hr.	Daily	Hr.	Daily	Bare Costs	Incl. O&P
1 Truck Driver (heavy)	$51.30	$410.40	$76.70	$613.60	$51.30	$76.70
1 Dump Truck, 8 C.Y., 220 H.P.		407.60		448.36	50.95	56.05
8 L.H., Daily Totals		$818.00		$1061.96	$102.25	$132.75

Crew B-34B

	Hr.	Daily	Hr.	Daily	Bare Costs	Incl. O&P
1 Truck Driver (heavy)	$51.30	$410.40	$76.70	$613.60	$51.30	$76.70
1 Dump Truck, 12 C.Y., 400 H.P.		579.35		637.28	72.42	79.66
8 L.H., Daily Totals		$989.75		$1250.89	$123.72	$156.36

Crew B-34C

	Hr.	Daily	Hr.	Daily	Bare Costs	Incl. O&P
1 Truck Driver (heavy)	$51.30	$410.40	$76.70	$613.60	$51.30	$76.70
1 Truck Tractor, 6x4, 380 H.P.		499.15		549.07		
1 Dump Trailer, 16.5 C.Y.		138.35		152.19	79.69	87.66
8 L.H., Daily Totals		$1047.90		$1314.85	$130.99	$164.36

Crew B-34D

	Hr.	Daily	Hr.	Daily	Bare Costs	Incl. O&P
1 Truck Driver (heavy)	$51.30	$410.40	$76.70	$613.60	$51.30	$76.70
1 Truck Tractor, 6x4, 380 H.P.		499.15		549.07		
1 Dump Trailer, 20 C.Y.		153.55		168.91	81.59	89.75
8 L.H., Daily Totals		$1063.10		$1331.57	$132.89	$166.45

Crew B-34E

	Hr.	Daily	Hr.	Daily	Bare Costs	Incl. O&P
1 Truck Driver (heavy)	$51.30	$410.40	$76.70	$613.60	$51.30	$76.70
1 Dump Truck, Off Hwy., 25 Ton		1427.00		1569.70	178.38	196.21
8 L.H., Daily Totals		$1837.40		$2183.30	$229.68	$272.91

Crew B-34F

	Hr.	Daily	Hr.	Daily	Bare Costs	Incl. O&P
1 Truck Driver (heavy)	$51.30	$410.40	$76.70	$613.60	$51.30	$76.70
1 Dump Truck, Off Hwy., 35 Ton		945.65		1040.21	118.21	130.03
8 L.H., Daily Totals		$1356.05		$1653.82	$169.51	$206.73

Crew B-34G

	Hr.	Daily	Hr.	Daily	Bare Costs	Incl. O&P
1 Truck Driver (heavy)	$51.30	$410.40	$76.70	$613.60	$51.30	$76.70
1 Dump Truck, Off Hwy., 50 Ton		1979.00		2176.90	247.38	272.11
8 L.H., Daily Totals		$2389.40		$2790.50	$298.68	$348.81

Crew B-34H

	Hr.	Daily	Hr.	Daily	Bare Costs	Incl. O&P
1 Truck Driver (heavy)	$51.30	$410.40	$76.70	$613.60	$51.30	$76.70
1 Dump Truck, Off Hwy., 65 Ton		1938.00		2131.80	242.25	266.48
8 L.H., Daily Totals		$2348.40		$2745.40	$293.55	$343.18

Crew B-34I

	Hr.	Daily	Hr.	Daily	Bare Costs	Incl. O&P
1 Truck Driver (heavy)	$51.30	$410.40	$76.70	$613.60	$51.30	$76.70
1 Dump Truck, 18 C.Y., 450 H.P.		753.60		828.96	94.20	103.62
8 L.H., Daily Totals		$1164.00		$1442.56	$145.50	$180.32

Crew B-34J

	Hr.	Daily	Hr.	Daily	Bare Costs	Incl. O&P
1 Truck Driver (heavy)	$51.30	$410.40	$76.70	$613.60	$51.30	$76.70
1 Dump Truck, Off Hwy., 100 Ton		2769.00		3045.90	346.13	380.74
8 L.H., Daily Totals		$3179.40		$3659.50	$397.43	$457.44

Crew B-34K

	Hr.	Daily	Hr.	Daily	Bare Costs	Incl. O&P
1 Truck Driver (heavy)	$51.30	$410.40	$76.70	$613.60	$51.30	$76.70
1 Truck Tractor, 6x4, 450 H.P.		608.95		669.85		
1 Lowbed Trailer, 75 Ton		258.10		283.91	108.38	119.22
8 L.H., Daily Totals		$1277.45		$1567.36	$159.68	$195.92

Crew B-34L

	Hr.	Daily	Hr.	Daily	Bare Costs	Incl. O&P
1 Equip. Oper. (light)	$55.50	$444.00	$82.70	$661.60	$55.50	$82.70
1 Flatbed Truck, Gas, 1.5 Ton		198.50		218.35	24.81	27.29
8 L.H., Daily Totals		$642.50		$879.95	$80.31	$109.99

Crew B-34M

	Hr.	Daily	Hr.	Daily	Bare Costs	Incl. O&P
1 Equip. Oper. (light)	$55.50	$444.00	$82.70	$661.60	$55.50	$82.70
1 Flatbed Truck, Gas, 3 Ton		850.05		935.05	106.26	116.88
8 L.H., Daily Totals		$1294.05		$1596.66	$161.76	$199.58

Crew B-34N

	Hr.	Daily	Hr.	Daily	Bare Costs	Incl. O&P
1 Truck Driver (heavy)	$51.30	$410.40	$76.70	$613.60	$55.15	$82.30
1 Equip. Oper. (medium)	59.00	472.00	87.90	703.20		
1 Truck Tractor, 6x4, 380 H.P.		499.15		549.07		
1 Flatbed Trailer, 40 Ton		188.45		207.29	42.98	47.27
16 L.H., Daily Totals		$1570.00		$2073.16	$98.13	$129.57

Crew B-34P

	Hr.	Daily	Hr.	Daily	Bare Costs	Incl. O&P
1 Pipe Fitter	$68.35	$546.80	$102.00	$816.00	$58.72	$87.63
1 Truck Driver (light)	48.80	390.40	73.00	584.00		
1 Equip. Oper. (medium)	59.00	472.00	87.90	703.20		
1 Flatbed Truck, Gas, 3 Ton		850.05		935.05		
1 Backhoe Loader, 48 H.P.		216.20		237.82	44.43	48.87
24 L.H., Daily Totals		$2475.45		$3276.07	$103.14	$136.50

Crew B-34Q

Crew No.	Hr.	Daily	Hr.	Daily	Bare Costs	Incl. O&P
1 Pipe Fitter	$68.35	$546.80	$102.00	$816.00	$59.53	$88.85
1 Truck Driver (light)	48.80	390.40	73.00	584.00		
1 Equip. Oper. (crane)	61.45	491.60	91.55	732.40		
1 Flatbed Trailer, 25 Ton		137.20		150.92		
1 Dump Truck, 8 C.Y., 220 H.P.		407.60		448.36		
1 Hyd. Crane, 25 Ton		586.70		645.37	47.15	51.86
24 L.H., Daily Totals		$2560.30		$3377.05	$106.68	$140.71

Crew B-34R

Crew No.	Hr.	Daily	Hr.	Daily	Bare Costs	Incl. O&P
1 Pipe Fitter	$68.35	$546.80	$102.00	$816.00	$59.53	$88.85
1 Truck Driver (light)	48.80	390.40	73.00	584.00		
1 Equip. Oper. (crane)	61.45	491.60	91.55	732.40		
1 Flatbed Trailer, 25 Ton		137.20		150.92		
1 Dump Truck, 8 C.Y., 220 H.P.		407.60		448.36		
1 Hyd. Crane, 25 Ton		586.70		645.37		
1 Hyd. Excavator, 1 C.Y.		832.65		915.91	81.84	90.02
24 L.H., Daily Totals		$3392.95		$4292.97	$141.37	$178.87

Crew B-34S

Crew No.	Hr.	Daily	Hr.	Daily	Bare Costs	Incl. O&P
2 Pipe Fitters	$68.35	$1093.60	$102.00	$1632.00	$62.36	$93.06
1 Truck Driver (heavy)	51.30	410.40	76.70	613.60		
1 Equip. Oper. (crane)	61.45	491.60	91.55	732.40		
1 Flatbed Trailer, 40 Ton		188.45		207.29		
1 Truck Tractor, 6x4, 380 H.P.		499.15		549.07		
1 Hyd. Crane, 80 Ton		1458.00		1603.80		
1 Hyd. Excavator, 2 C.Y.		942.70		1036.97	96.51	106.16
32 L.H., Daily Totals		$5083.90		$6375.13	$158.87	$199.22

Crew B-34T

Crew No.	Hr.	Daily	Hr.	Daily	Bare Costs	Incl. O&P
2 Pipe Fitters	$68.35	$1093.60	$102.00	$1632.00	$62.36	$93.06
1 Truck Driver (heavy)	51.30	410.40	76.70	613.60		
1 Equip. Oper. (crane)	61.45	491.60	91.55	732.40		
1 Flatbed Trailer, 40 Ton		188.45		207.29		
1 Truck Tractor, 6x4, 380 H.P.		499.15		549.07		
1 Hyd. Crane, 80 Ton		1458.00		1603.80	67.05	73.75
32 L.H., Daily Totals		$4141.20		$5338.16	$129.41	$166.82

Crew B-34U

Crew No.	Hr.	Daily	Hr.	Daily	Bare Costs	Incl. O&P
1 Truck Driver (heavy)	$51.30	$410.40	$76.70	$613.60	$53.40	$79.70
1 Equip. Oper. (light)	55.50	444.00	82.70	661.60		
1 Truck Tractor, 220 H.P.		310.80		341.88		
1 Flatbed Trailer, 25 Ton		137.20		150.92	28.00	30.80
16 L.H., Daily Totals		$1302.40		$1768.00	$81.40	$110.50

Crew B-34V

Crew No.	Hr.	Daily	Hr.	Daily	Bare Costs	Incl. O&P
1 Truck Driver (heavy)	$51.30	$410.40	$76.70	$613.60	$56.08	$83.65
1 Equip. Oper. (crane)	61.45	491.60	91.55	732.40		
1 Equip. Oper. (light)	55.50	444.00	82.70	661.60		
1 Truck Tractor, 6x4, 450 H.P.		608.95		669.85		
1 Equipment Trailer, 50 Ton		207.25		227.97		
1 Pickup Truck, 4x4, 3/4 Ton		176.75		194.43	41.37	45.51
24 L.H., Daily Totals		$2338.95		$3099.84	$97.46	$129.16

Crew B-34W

Crew No.	Hr.	Daily	Hr.	Daily	Bare Costs	Incl. O&P
5 Truck Drivers (heavy)	$51.30	$2052.00	$76.70	$3068.00	$53.92	$80.50
2 Equip. Opers. (crane)	61.45	983.20	91.55	1464.80		
1 Equip. Oper. (mechanic)	61.50	492.00	91.65	733.20		
1 Laborer	44.40	355.20	66.25	530.00		
4 Truck Tractors, 6x4, 380 H.P.		1996.60		2196.26		
2 Equipment Trailers, 50 Ton		414.50		455.95		
2 Flatbed Trailers, 40 Ton		376.90		414.59		
1 Pickup Truck, 4x4, 3/4 Ton		176.75		194.43		
1 S.P. Crane, 4x4, 20 Ton		574.35		631.78	49.15	54.07
72 L.H., Daily Totals		$7421.50		$9689.01	$103.08	$134.57

Crew B-35

Crew No.	Hr.	Daily	Hr.	Daily	Bare Costs	Incl. O&P
1 Labor Foreman (outside)	$46.40	$371.20	$69.25	$554.00	$56.75	$84.76
1 Skilled Worker	57.10	456.80	85.90	687.20		
2 Welders	67.70	1083.20	101.05	1616.80		
1 Laborer	44.40	355.20	66.25	530.00		
1 Equip. Oper. (crane)	61.45	491.60	91.55	732.40		
1 Equip. Oper. (oiler)	52.50	420.00	78.25	626.00		
2 Welder, Electric, 300 amp		215.10		236.61		
1 Hyd. Excavator, .75 C.Y.		701.80		771.98	16.37	18.01
56 L.H., Daily Totals		$4094.90		$5754.99	$73.12	$102.77

Crew B-35A

Crew No.	Hr.	Daily	Hr.	Daily	Bare Costs	Incl. O&P
1 Labor Foreman (outside)	$46.40	$371.20	$69.25	$554.00	$53.42	$79.79
2 Laborers	44.40	710.40	66.25	1060.00		
1 Skilled Worker	57.10	456.80	85.90	687.20		
1 Welder (plumber)	67.70	541.60	101.05	808.40		
1 Equip. Oper. (crane)	61.45	491.60	91.55	732.40		
1 Equip. Oper. (oiler)	52.50	420.00	78.25	626.00		
1 Welder, Gas Engine, 300 amp		148.75		163.63		
1 Crawler Crane, 75 Ton		1967.00		2163.70	37.78	41.56
56 L.H., Daily Totals		$5107.35		$6795.32	$91.20	$121.35

Crew B-36

Crew No.	Hr.	Daily	Hr.	Daily	Bare Costs	Incl. O&P
1 Labor Foreman (outside)	$46.40	$371.20	$69.25	$554.00	$50.64	$75.51
2 Laborers	44.40	710.40	66.25	1060.00		
2 Equip. Oper. (medium)	59.00	944.00	87.90	1406.40		
1 Dozer, 200 H.P.		1520.00		1672.00		
1 Aggregate Spreader		59.95		65.94		
1 Tandem Roller, 10 Ton		246.80		271.48	45.67	50.24
40 L.H., Daily Totals		$3852.35		$5029.82	$96.31	$125.75

Crew B-36A

Crew No.	Hr.	Daily	Hr.	Daily	Bare Costs	Incl. O&P
1 Labor Foreman (outside)	$46.40	$371.20	$69.25	$554.00	$53.03	$79.05
2 Laborers	44.40	710.40	66.25	1060.00		
4 Equip. Oper. (medium)	59.00	1888.00	87.90	2812.80		
1 Dozer, 200 H.P.		1520.00		1672.00		
1 Aggregate Spreader		59.95		65.94		
1 Tandem Roller, 10 Ton		246.80		271.48		
1 Roller, Pneum. Whl., 12 Ton		349.90		384.89	38.87	42.76
56 L.H., Daily Totals		$5146.25		$6821.11	$91.90	$121.81

For customer support on your Concrete & Masonry Costs with RSMeans Data, call 800.448.8182.

429

Crew No.	Bare Costs		Incl. Subs O&P		Cost Per Labor-Hour	
Crew B-36B	Hr.	Daily	Hr.	Daily	Bare Costs	Incl. O&P
1 Labor Foreman (outside)	$46.40	$371.20	$69.25	$554.00	$52.81	$78.76
2 Laborers	44.40	710.40	66.25	1060.00		
4 Equip. Oper. (medium)	59.00	1888.00	87.90	2812.80		
1 Truck Driver (heavy)	51.30	410.40	76.70	613.60		
1 Grader, 30,000 Lbs.		1073.00		1180.30		
1 F.E. Loader, Crl, 1.5 C.Y.		668.35		735.18		
1 Dozer, 300 H.P.		1785.00		1963.50		
1 Roller, Vibratory, 25 Ton		672.35		739.59		
1 Truck Tractor, 6x4, 450 H.P.		608.95		669.85		
1 Water Tank Trailer, 5000 Gal.		154.10		169.51	77.53	85.28
64 L.H., Daily Totals		$8341.75		$10498.33	$130.34	$164.04
Crew B-36C	Hr.	Daily	Hr.	Daily	Bare Costs	Incl. O&P
1 Labor Foreman (outside)	$46.40	$371.20	$69.25	$554.00	$54.94	$81.93
3 Equip. Oper. (medium)	59.00	1416.00	87.90	2109.60		
1 Truck Driver (heavy)	51.30	410.40	76.70	613.60		
1 Grader, 30,000 Lbs.		1073.00		1180.30		
1 Dozer, 300 H.P.		1785.00		1963.50		
1 Roller, Vibratory, 25 Ton		672.35		739.59		
1 Truck Tractor, 6x4, 450 H.P.		608.95		669.85		
1 Water Tank Trailer, 5000 Gal.		154.10		169.51	107.33	118.07
40 L.H., Daily Totals		$6491.00		$7999.94	$162.28	$200.00
Crew B-36D	Hr.	Daily	Hr.	Daily	Bare Costs	Incl. O&P
1 Labor Foreman (outside)	$46.40	$371.20	$69.25	$554.00	$55.85	$83.24
3 Equip. Oper. (medium)	59.00	1416.00	87.90	2109.60		
1 Grader, 30,000 Lbs.		1073.00		1180.30		
1 Dozer, 300 H.P.		1785.00		1963.50		
1 Roller, Vibratory, 25 Ton		672.35		739.59	110.32	121.36
32 L.H., Daily Totals		$5317.55		$6546.98	$166.17	$204.59
Crew B-37	Hr.	Daily	Hr.	Daily	Bare Costs	Incl. O&P
1 Labor Foreman (outside)	$46.40	$371.20	$69.25	$554.00	$46.58	$69.49
4 Laborers	44.40	1420.80	66.25	2120.00		
1 Equip. Oper. (light)	55.50	444.00	82.70	661.60		
1 Tandem Roller, 5 Ton		258.75		284.63	5.39	5.93
48 L.H., Daily Totals		$2494.75		$3620.22	$51.97	$75.42
Crew B-37A	Hr.	Daily	Hr.	Daily	Bare Costs	Incl. O&P
2 Laborers	$44.40	$710.40	$66.25	$1060.00	$45.87	$68.50
1 Truck Driver (light)	48.80	390.40	73.00	584.00		
1 Flatbed Truck, Gas, 1.5 Ton		198.50		218.35		
1 Tar Kettle, T.M.		156.70		172.37	14.80	16.28
24 L.H., Daily Totals		$1456.00		$2034.72	$60.67	$84.78
Crew B-37B	Hr.	Daily	Hr.	Daily	Bare Costs	Incl. O&P
3 Laborers	$44.40	$1065.60	$66.25	$1590.00	$45.50	$67.94
1 Truck Driver (light)	48.80	390.40	73.00	584.00		
1 Flatbed Truck, Gas, 1.5 Ton		198.50		218.35		
1 Tar Kettle, T.M.		156.70		172.37	11.10	12.21
32 L.H., Daily Totals		$1811.20		$2564.72	$56.60	$80.15
Crew B-37C	Hr.	Daily	Hr.	Daily	Bare Costs	Incl. O&P
2 Laborers	$44.40	$710.40	$66.25	$1060.00	$46.60	$69.63
2 Truck Drivers (light)	48.80	780.80	73.00	1168.00		
2 Flatbed Trucks, Gas, 1.5 Ton		397.00		436.70		
1 Tar Kettle, T.M.		156.70		172.37	17.30	19.03
32 L.H., Daily Totals		$2044.90		$2837.07	$63.90	$88.66

Crew No.	Bare Costs		Incl. Subs O&P		Cost Per Labor-Hour	
Crew B-37D	Hr.	Daily	Hr.	Daily	Bare Costs	Incl. O&P
1 Laborer	$44.40	$355.20	$66.25	$530.00	$46.60	$69.63
1 Truck Driver (light)	48.80	390.40	73.00	584.00		
1 Pickup Truck, 3/4 Ton		112.20		123.42	7.01	7.71
16 L.H., Daily Totals		$857.80		$1237.42	$53.61	$77.34
Crew B-37E	Hr.	Daily	Hr.	Daily	Bare Costs	Incl. O&P
3 Laborers	$44.40	$1065.60	$66.25	$1590.00	$49.33	$73.62
1 Equip. Oper. (light)	55.50	444.00	82.70	661.60		
1 Equip. Oper. (medium)	59.00	472.00	87.90	703.20		
2 Truck Drivers (light)	48.80	780.80	73.00	1168.00		
4 Barrels w/ Flasher		16.60		18.26		
1 Concrete Saw		112.85		124.14		
1 Rotary Hammer Drill		52.25		57.48		
1 Hammer Drill Bit		25.25		27.77		
1 Loader, Skid Steer, 30 H.P.		179.50		197.45		
1 Conc. Hammer Attach.		118.50		130.35		
1 Vibrating Plate, Gas, 18"		31.90		35.09		
2 Flatbed Trucks, Gas, 1.5 Ton		397.00		436.70	16.68	18.34
56 L.H., Daily Totals		$3696.25		$5150.03	$66.00	$91.96
Crew B-37F	Hr.	Daily	Hr.	Daily	Bare Costs	Incl. O&P
3 Laborers	$44.40	$1065.60	$66.25	$1590.00	$45.50	$67.94
1 Truck Driver (light)	48.80	390.40	73.00	584.00		
4 Barrels w/ Flasher		16.60		18.26		
1 Concrete Mixer, 10 C.F.		147.15		161.87		
1 Air Compressor, 60 cfm		153.85		169.24		
1 -50' Air Hose, 3/4"		7.15		7.87		
1 Spade (Chipper)		8.55		9.40		
1 Flatbed Truck, Gas, 1.5 Ton		198.50		218.35	16.62	18.28
32 L.H., Daily Totals		$1987.80		$2758.98	$62.12	$86.22
Crew B-37G	Hr.	Daily	Hr.	Daily	Bare Costs	Incl. O&P
1 Labor Foreman (outside)	$46.40	$371.20	$69.25	$554.00	$46.58	$69.49
4 Laborers	44.40	1420.80	66.25	2120.00		
1 Equip. Oper. (light)	55.50	444.00	82.70	661.60		
1 Berm Machine		253.55		278.90		
1 Tandem Roller, 5 Ton		258.75		284.63	10.67	11.74
48 L.H., Daily Totals		$2748.30		$3899.13	$57.26	$81.23
Crew B-37H	Hr.	Daily	Hr.	Daily	Bare Costs	Incl. O&P
1 Labor Foreman (outside)	$46.40	$371.20	$69.25	$554.00	$46.58	$69.49
4 Laborers	44.40	1420.80	66.25	2120.00		
1 Equip. Oper. (light)	55.50	444.00	82.70	661.60		
1 Tandem Roller, 5 Ton		258.75		284.63		
1 Flatbed Truck, Gas, 1.5 Ton		198.50		218.35		
1 Tar Kettle, T.M.		156.70		172.37	12.79	14.07
48 L.H., Daily Totals		$2849.95		$4010.95	$59.37	$83.56

For customer support on your Concrete & Masonry Costs with RSMeans Data, call 800.448.8182.

Crew No.	Bare Costs		Incl. Subs O&P		Cost Per Labor-Hour	
Crew B-37I	Hr.	Daily	Hr.	Daily	Bare Costs	Incl. O&P
3 Laborers	$44.40	$1065.60	$66.25	$1590.00	$49.33	$73.62
1 Equip. Oper. (light)	55.50	444.00	82.70	661.60		
1 Equip. Oper. (medium)	59.00	472.00	87.90	703.20		
2 Truck Drivers (light)	48.80	780.80	73.00	1168.00		
4 Barrels w/ Flasher		16.60		18.26		
1 Concrete Saw		112.85		124.14		
1 Rotary Hammer Drill		52.25		57.48		
1 Hammer Drill Bit		25.25		27.77		
1 Air Compressor, 60 cfm		153.85		169.24		
1 -50' Air Hose, 3/4"		7.15		7.87		
1 Spade (Chipper)		8.55		9.40		
1 Loader, Skid Steer, 30 H.P.		179.50		197.45		
1 Conc. Hammer Attach.		118.50		130.35		
1 Concrete Mixer, 10 C.F.		147.15		161.87		
1 Vibrating Plate, Gas, 18"		31.90		35.09		
2 Flatbed Trucks, Gas, 1.5 Ton		397.00		436.70	22.33	24.56
56 L.H., Daily Totals		$4012.95		$5498.40	$71.66	$98.19
Crew B-37J	Hr.	Daily	Hr.	Daily	Bare Costs	Incl. O&P
1 Labor Foreman (outside)	$46.40	$371.20	$69.25	$554.00	$46.58	$69.49
4 Laborers	44.40	1420.80	66.25	2120.00		
1 Equip. Oper. (light)	55.50	444.00	82.70	661.60		
1 Air Compressor, 60 cfm		153.85		169.24		
1 -50' Air Hose, 3/4"		7.15		7.87		
2 Concrete Mixers, 10 C.F.		294.30		323.73		
2 Flatbed Trucks, Gas, 1.5 Ton		397.00		436.70		
1 Shot Blaster, 20"		208.70		229.57	22.10	24.31
48 L.H., Daily Totals		$3297.00		$4502.70	$68.69	$93.81
Crew B-37K	Hr.	Daily	Hr.	Daily	Bare Costs	Incl. O&P
1 Labor Foreman (outside)	$46.40	$371.20	$69.25	$554.00	$46.58	$69.49
4 Laborers	44.40	1420.80	66.25	2120.00		
1 Equip. Oper. (light)	55.50	444.00	82.70	661.60		
1 Air Compressor, 60 cfm		153.85		169.24		
1 -50' Air Hose, 3/4"		7.15		7.87		
2 Flatbed Trucks, Gas, 1.5 Ton		397.00		436.70		
1 Shot Blaster, 20"		208.70		229.57	15.97	17.57
48 L.H., Daily Totals		$3002.70		$4178.97	$62.56	$87.06
Crew B-38	Hr.	Daily	Hr.	Daily	Bare Costs	Incl. O&P
1 Labor Foreman (outside)	$46.40	$371.20	$69.25	$554.00	$49.94	$74.47
2 Laborers	44.40	710.40	66.25	1060.00		
1 Equip. Oper. (light)	55.50	444.00	82.70	661.60		
1 Equip. Oper. (medium)	59.00	472.00	87.90	703.20		
1 Backhoe Loader, 48 H.P.		216.20		237.82		
1 Hyd. Hammer (1200 lb.)		177.25		194.97		
1 F.E. Loader, W.M., 4 C.Y.		759.00		834.90		
1 Pvmt. Rem. Bucket		63.80		70.18	30.41	33.45
40 L.H., Daily Totals		$3213.85		$4316.68	$80.35	$107.92
Crew B-39	Hr.	Daily	Hr.	Daily	Bare Costs	Incl. O&P
1 Labor Foreman (outside)	$46.40	$371.20	$69.25	$554.00	$46.58	$69.49
4 Laborers	44.40	1420.80	66.25	2120.00		
1 Equip. Oper. (light)	55.50	444.00	82.70	661.60		
1 Air Compressor, 250 cfm		202.85		223.13		
2 Breakers, Pavement, 60 lb.		107.20		117.92		
2 -50' Air Hoses, 1.5"		45.60		50.16	7.41	8.15
48 L.H., Daily Totals		$2591.65		$3726.82	$53.99	$77.64

Crew No.	Bare Costs		Incl. Subs O&P		Cost Per Labor-Hour	
Crew B-40	Hr.	Daily	Hr.	Daily	Bare Costs	Incl. O&P
1 Pile Driver Foreman (outside)	$57.90	$463.20	$89.50	$716.00	$57.11	$87.06
4 Pile Drivers	55.90	1788.80	86.40	2764.80		
2 Equip. Oper. (crane)	61.45	983.20	91.55	1464.80		
1 Equip. Oper. (oiler)	52.50	420.00	78.25	626.00		
1 Crawler Crane, 40 Ton		1231.00		1354.10		
1 Vibratory Hammer & Gen.		2298.00		2527.80	55.14	60.65
64 L.H., Daily Totals		$7184.20		$9453.50	$112.25	$147.71
Crew B-40B	Hr.	Daily	Hr.	Daily	Bare Costs	Incl. O&P
1 Labor Foreman (outside)	$46.40	$371.20	$69.25	$554.00	$48.92	$72.97
3 Laborers	44.40	1065.60	66.25	1590.00		
1 Equip. Oper. (crane)	61.45	491.60	91.55	732.40		
1 Equip. Oper. (oiler)	52.50	420.00	78.25	626.00		
1 Lattice Boom Crane, 40 Ton		2127.00		2339.70	44.31	48.74
48 L.H., Daily Totals		$4475.40		$5842.10	$93.24	$121.71
Crew B-41	Hr.	Daily	Hr.	Daily	Bare Costs	Incl. O&P
1 Labor Foreman (outside)	$46.40	$371.20	$69.25	$554.00	$45.91	$68.49
4 Laborers	44.40	1420.80	66.25	2120.00		
.25 Equip. Oper. (crane)	61.45	122.90	91.55	183.10		
.25 Equip. Oper. (oiler)	52.50	105.00	78.25	156.50		
.25 Crawler Crane, 40 Ton		307.75		338.52	6.99	7.69
44 L.H., Daily Totals		$2327.65		$3352.13	$52.90	$76.18
Crew B-42	Hr.	Daily	Hr.	Daily	Bare Costs	Incl. O&P
1 Labor Foreman (outside)	$46.40	$371.20	$69.25	$554.00	$49.78	$74.67
4 Laborers	44.40	1420.80	66.25	2120.00		
1 Equip. Oper. (crane)	61.45	491.60	91.55	732.40		
1 Equip. Oper. (oiler)	52.50	420.00	78.25	626.00		
1 Welder	60.30	482.40	93.30	746.40		
1 Hyd. Crane, 25 Ton		586.70		645.37		
1 Welder, Gas Engine, 300 amp		148.75		163.63		
1 Horz. Boring Csg. Mch.		329.75		362.73	16.64	18.31
64 L.H., Daily Totals		$4251.20		$5950.52	$66.43	$92.98
Crew B-43	Hr.	Daily	Hr.	Daily	Bare Costs	Incl. O&P
1 Labor Foreman (outside)	$46.40	$371.20	$69.25	$554.00	$48.92	$72.97
3 Laborers	44.40	1065.60	66.25	1590.00		
1 Equip. Oper. (crane)	61.45	491.60	91.55	732.40		
1 Equip. Oper. (oiler)	52.50	420.00	78.25	626.00		
1 Drill Rig, Truck-Mounted		768.40		845.24	16.01	17.61
48 L.H., Daily Totals		$3116.80		$4347.64	$64.93	$90.58
Crew B-44	Hr.	Daily	Hr.	Daily	Bare Costs	Incl. O&P
1 Pile Driver Foreman (outside)	$57.90	$463.20	$89.50	$716.00	$56.10	$85.56
4 Pile Drivers	55.90	1788.80	86.40	2764.80		
2 Equip. Oper. (crane)	61.45	983.20	91.55	1464.80		
1 Laborer	44.40	355.20	66.25	530.00		
1 Crawler Crane, 40 Ton		1231.00		1354.10		
1 Lead, 60' High		211.80		232.98		
1 Hammer, Diesel, 15K ft.-lbs.		624.45		686.89	32.30	35.53
64 L.H., Daily Totals		$5657.65		$7749.57	$88.40	$121.09
Crew B-45	Hr.	Daily	Hr.	Daily	Bare Costs	Incl. O&P
1 Equip. Oper. (medium)	$59.00	$472.00	$87.90	$703.20	$55.15	$82.30
1 Truck Driver (heavy)	51.30	410.40	76.70	613.60		
1 Dist. Tanker, 3000 Gallon		334.10		367.51		
1 Truck Tractor, 6x4, 380 H.P.		499.15		549.07	52.08	57.29
16 L.H., Daily Totals		$1715.65		$2233.38	$107.23	$139.59

Crews - Standard

Crew No.	Bare Costs		Incl. Subs O&P		Cost Per Labor-Hour	

Crew B-46

	Hr.	Daily	Hr.	Daily	Bare Costs	Incl. O&P
1 Pile Driver Foreman (outside)	$57.90	$463.20	$89.50	$716.00	$50.48	$76.84
2 Pile Drivers	55.90	894.40	86.40	1382.40		
3 Laborers	44.40	1065.60	66.25	1590.00		
1 Chain Saw, Gas, 36" Long		41.65		45.81	.87	.95
48 L.H., Daily Totals		$2464.85		$3734.22	$51.35	$77.80

Crew B-47

	Hr.	Daily	Hr.	Daily	Bare Costs	Incl. O&P
1 Blast Foreman (outside)	$46.40	$371.20	$69.25	$554.00	$48.77	$72.73
1 Driller	44.40	355.20	66.25	530.00		
1 Equip. Oper. (light)	55.50	444.00	82.70	661.60		
1 Air Track Drill, 4"		1127.00		1239.70		
1 Air Compressor, 600 cfm		426.55		469.20		
2 -50' Air Hoses, 3"		76.70		84.37	67.93	74.72
24 L.H., Daily Totals		$2800.65		$3538.88	$116.69	$147.45

Crew B-47A

	Hr.	Daily	Hr.	Daily	Bare Costs	Incl. O&P
1 Drilling Foreman (outside)	$46.40	$371.20	$69.25	$554.00	$53.45	$79.68
1 Equip. Oper. (heavy)	61.45	491.60	91.55	732.40		
1 Equip. Oper. (oiler)	52.50	420.00	78.25	626.00		
1 Air Track Drill, 5"		1127.00		1239.70	46.96	51.65
24 L.H., Daily Totals		$2409.80		$3152.10	$100.41	$131.34

Crew B-47C

	Hr.	Daily	Hr.	Daily	Bare Costs	Incl. O&P
1 Laborer	$44.40	$355.20	$66.25	$530.00	$49.95	$74.47
1 Equip. Oper. (light)	55.50	444.00	82.70	661.60		
1 Air Compressor, 750 cfm		596.30		655.93		
2 -50' Air Hoses, 3"		76.70		84.37		
1 Air Track Drill, 4"		1127.00		1239.70	112.50	123.75
16 L.H., Daily Totals		$2599.20		$3171.60	$162.45	$198.22

Crew B-47E

	Hr.	Daily	Hr.	Daily	Bare Costs	Incl. O&P
1 Labor Foreman (outside)	$46.40	$371.20	$69.25	$554.00	$44.90	$67.00
3 Laborers	44.40	1065.60	66.25	1590.00		
1 Flatbed Truck, Gas, 3 Ton		850.05		935.05	26.56	29.22
32 L.H., Daily Totals		$2286.85		$3079.05	$71.46	$96.22

Crew B-47G

	Hr.	Daily	Hr.	Daily	Bare Costs	Incl. O&P
1 Labor Foreman (outside)	$46.40	$371.20	$69.25	$554.00	$47.67	$71.11
2 Laborers	44.40	710.40	66.25	1060.00		
1 Equip. Oper. (light)	55.50	444.00	82.70	661.60		
1 Air Track Drill, 4"		1127.00		1239.70		
1 Air Compressor, 600 cfm		426.55		469.20		
2 -50' Air Hoses, 3"		76.70		84.37		
1 Gunite Pump Rig		321.75		353.93	61.00	67.10
32 L.H., Daily Totals		$3477.60		$4422.80	$108.68	$138.21

Crew B-47H

	Hr.	Daily	Hr.	Daily	Bare Costs	Incl. O&P
1 Skilled Worker Foreman (out)	$59.10	$472.80	$88.90	$711.20	$57.60	$86.65
3 Skilled Workers	57.10	1370.40	85.90	2061.60		
1 Flatbed Truck, Gas, 3 Ton		850.05		935.05	26.56	29.22
32 L.H., Daily Totals		$2693.25		$3707.86	$84.16	$115.87

Crew B-48

	Hr.	Daily	Hr.	Daily	Bare Costs	Incl. O&P
1 Labor Foreman (outside)	$46.40	$371.20	$69.25	$554.00	$49.86	$74.36
3 Laborers	44.40	1065.60	66.25	1590.00		
1 Equip. Oper. (crane)	61.45	491.60	91.55	732.40		
1 Equip. Oper. (oiler)	52.50	420.00	78.25	626.00		
1 Equip. Oper. (light)	55.50	444.00	82.70	661.60		
1 Centr. Water Pump, 6"		235.25		258.77		
1 -20' Suction Hose, 6"		25.50		28.05		
1 -50' Discharge Hose, 6"		18.10		19.91		
1 Drill Rig, Truck-Mounted		768.40		845.24	18.70	20.57
56 L.H., Daily Totals		$3839.65		$5315.98	$68.57	$94.93

Crew B-49

	Hr.	Daily	Hr.	Daily	Bare Costs	Incl. O&P
1 Labor Foreman (outside)	$46.40	$371.20	$69.25	$554.00	$52.25	$78.46
3 Laborers	44.40	1065.60	66.25	1590.00		
2 Equip. Oper. (crane)	61.45	983.20	91.55	1464.80		
2 Equip. Oper. (oilers)	52.50	840.00	78.25	1252.00		
1 Equip. Oper. (light)	55.50	444.00	82.70	661.60		
2 Pile Drivers	55.90	894.40	86.40	1382.40		
1 Hyd. Crane, 25 Ton		586.70		645.37		
1 Centr. Water Pump, 6"		235.25		258.77		
1 -20' Suction Hose, 6"		25.50		28.05		
1 -50' Discharge Hose, 6"		18.10		19.91		
1 Drill Rig, Truck-Mounted		768.40		845.24	18.57	20.42
88 L.H., Daily Totals		$6232.35		$8702.15	$70.82	$98.89

Crew B-50

	Hr.	Daily	Hr.	Daily	Bare Costs	Incl. O&P
2 Pile Driver Foremen (outside)	$57.90	$926.40	$89.50	$1432.00	$54.27	$82.68
6 Pile Drivers	55.90	2683.20	86.40	4147.20		
2 Equip. Oper. (crane)	61.45	983.20	91.55	1464.80		
1 Equip. Oper. (oiler)	52.50	420.00	78.25	626.00		
3 Laborers	44.40	1065.60	66.25	1590.00		
1 Crawler Crane, 40 Ton		1231.00		1354.10		
1 Lead, 60' High		211.80		232.98		
1 Hammer, Diesel, 15K ft.-lbs.		624.45		686.89		
1 Air Compressor, 600 cfm		426.55		469.20		
2 -50' Air Hoses, 3"		76.70		84.37		
1 Chain Saw, Gas, 36" Long		41.65		45.81	23.32	25.66
112 L.H., Daily Totals		$8690.55		$12133.37	$77.59	$108.33

Crew B-51

	Hr.	Daily	Hr.	Daily	Bare Costs	Incl. O&P
1 Labor Foreman (outside)	$46.40	$371.20	$69.25	$554.00	$45.47	$67.88
4 Laborers	44.40	1420.80	66.25	2120.00		
1 Truck Driver (light)	48.80	390.40	73.00	584.00		
1 Flatbed Truck, Gas, 1.5 Ton		198.50		218.35	4.14	4.55
48 L.H., Daily Totals		$2380.90		$3476.35	$49.60	$72.42

Crew B-52

	Hr.	Daily	Hr.	Daily	Bare Costs	Incl. O&P
1 Carpenter Foreman (outside)	$56.70	$453.60	$84.60	$676.80	$50.76	$75.55
1 Carpenter	54.70	437.60	81.65	653.20		
3 Laborers	44.40	1065.60	66.25	1590.00		
1 Cement Finisher	51.80	414.40	75.90	607.20		
.5 Rodman (reinf.)	58.90	235.60	88.05	352.20		
.5 Equip. Oper. (medium)	59.00	236.00	87.90	351.60		
.5 Crawler Loader, 3 C.Y.		573.00		630.30	10.23	11.26
56 L.H., Daily Totals		$3415.80		$4861.30	$61.00	$86.81

Crew B-53

	Hr.	Daily	Hr.	Daily	Bare Costs	Incl. O&P
1 Equip. Oper. (light)	$55.50	$444.00	$82.70	$661.60	$55.50	$82.70
1 Trencher, Chain, 12 H.P.		158.60		174.46	19.82	21.81
8 L.H., Daily Totals		$602.60		$836.06	$75.33	$104.51

Crew B-54

Crew No.	Hr.	Daily	Hr.	Daily	Bare Costs	Incl. O&P
1 Equip. Oper. (light)	$55.50	$444.00	$82.70	$661.60	$55.50	$82.70
1 Trencher, Chain, 40 H.P.		450.70		495.77	56.34	61.97
8 L.H., Daily Totals		$894.70		$1157.37	$111.84	$144.67

Crew B-54A

Crew No.	Hr.	Daily	Hr.	Daily	Bare Costs	Incl. O&P
.17 Labor Foreman (outside)	$46.40	$63.10	$69.25	$94.18	$57.17	$85.19
1 Equipment Operator (med.)	59.00	472.00	87.90	703.20		
1 Wheel Trencher, 67 H.P.		1140.00		1254.00	121.79	133.97
9.36 L.H., Daily Totals		$1675.10		$2051.38	$178.96	$219.16

Crew B-54B

Crew No.	Hr.	Daily	Hr.	Daily	Bare Costs	Incl. O&P
.25 Labor Foreman (outside)	$46.40	$92.80	$69.25	$138.50	$56.48	$84.17
1 Equipment Operator (med.)	59.00	472.00	87.90	703.20		
1 Wheel Trencher, 150 H.P.		1238.00		1361.80	123.80	136.18
10 L.H., Daily Totals		$1802.80		$2203.50	$180.28	$220.35

Crew B-54C

Crew No.	Hr.	Daily	Hr.	Daily	Bare Costs	Incl. O&P
1 Laborer	$44.40	$355.20	$66.25	$530.00	$51.70	$77.08
1 Equipment Operator (med.)	59.00	472.00	87.90	703.20		
1 Wheel Trencher, 67 H.P.		1140.00		1254.00	71.25	78.38
16 L.H., Daily Totals		$1967.20		$2487.20	$122.95	$155.45

Crew B-54D

Crew No.	Hr.	Daily	Hr.	Daily	Bare Costs	Incl. O&P
1 Laborer	$44.40	$355.20	$66.25	$530.00	$51.70	$77.08
1 Equipment Operator (med.)	59.00	472.00	87.90	703.20		
1 Rock Trencher, 6" Width		434.20		477.62	27.14	29.85
16 L.H., Daily Totals		$1261.40		$1710.82	$78.84	$106.93

Crew B-54E

Crew No.	Hr.	Daily	Hr.	Daily	Bare Costs	Incl. O&P
1 Laborer	$44.40	$355.20	$66.25	$530.00	$51.70	$77.08
1 Equipment Operator (med.)	59.00	472.00	87.90	703.20		
1 Rock Trencher, 18" Width		1015.00		1116.50	63.44	69.78
16 L.H., Daily Totals		$1842.20		$2349.70	$115.14	$146.86

Crew B-55

Crew No.	Hr.	Daily	Hr.	Daily	Bare Costs	Incl. O&P
2 Laborers	$44.40	$710.40	$66.25	$1060.00	$45.87	$68.50
1 Truck Driver (light)	48.80	390.40	73.00	584.00		
1 Truck-Mounted Earth Auger		394.15		433.57		
1 Flatbed Truck, Gas, 3 Ton		850.05		935.05	51.84	57.03
24 L.H., Daily Totals		$2345.00		$3012.62	$97.71	$125.53

Crew B-56

Crew No.	Hr.	Daily	Hr.	Daily	Bare Costs	Incl. O&P
1 Laborer	$44.40	$355.20	$66.25	$530.00	$49.95	$74.47
1 Equip. Oper. (light)	55.50	444.00	82.70	661.60		
1 Air Track Drill, 4"		1127.00		1239.70		
1 Air Compressor, 600 cfm		426.55		469.20		
1 -50' Air Hose, 3"		38.35		42.19	99.49	109.44
16 L.H., Daily Totals		$2391.10		$2942.69	$149.44	$183.92

Crew B-57

Crew No.	Hr.	Daily	Hr.	Daily	Bare Costs	Incl. O&P
1 Labor Foreman (outside)	$46.40	$371.20	$69.25	$554.00	$50.77	$75.71
2 Laborers	44.40	710.40	66.25	1060.00		
1 Equip. Oper. (crane)	61.45	491.60	91.55	732.40		
1 Equip. Oper. (light)	55.50	444.00	82.70	661.60		
1 Equip. Oper. (oiler)	52.50	420.00	78.25	626.00		
1 Crawler Crane, 25 Ton		1152.00		1267.20		
1 Clamshell Bucket, 1 C.Y.		69.25		76.17		
1 Centr. Water Pump, 6"		235.25		258.77		
1 -20' Suction Hose, 6"		25.50		28.05		
20 -50' Discharge Hoses, 6"		362.00		398.20	38.42	42.26
48 L.H., Daily Totals		$4281.20		$5662.40	$89.19	$117.97

Crew B-58

Crew No.	Hr.	Daily	Hr.	Daily	Bare Costs	Incl. O&P
2 Laborers	$44.40	$710.40	$66.25	$1060.00	$48.10	$71.73
1 Equip. Oper. (light)	55.50	444.00	82.70	661.60		
1 Backhoe Loader, 48 H.P.		216.20		237.82		
1 Small Helicopter, w/ Pilot		2101.00		2311.10	96.55	106.21
24 L.H., Daily Totals		$3471.60		$4270.52	$144.65	$177.94

Crew B-59

Crew No.	Hr.	Daily	Hr.	Daily	Bare Costs	Incl. O&P
1 Truck Driver (heavy)	$51.30	$410.40	$76.70	$613.60	$51.30	$76.70
1 Truck Tractor, 220 H.P.		310.80		341.88		
1 Water Tank Trailer, 5000 Gal.		154.10		169.51	58.11	63.92
8 L.H., Daily Totals		$875.30		$1124.99	$109.41	$140.62

Crew B-59A

Crew No.	Hr.	Daily	Hr.	Daily	Bare Costs	Incl. O&P
2 Laborers	$44.40	$710.40	$66.25	$1060.00	$46.70	$69.73
1 Truck Driver (heavy)	51.30	410.40	76.70	613.60		
1 Water Tank Trailer, 5000 Gal.		154.10		169.51		
1 Truck Tractor, 220 H.P.		310.80		341.88	19.37	21.31
24 L.H., Daily Totals		$1585.70		$2184.99	$66.07	$91.04

Crew B-60

Crew No.	Hr.	Daily	Hr.	Daily	Bare Costs	Incl. O&P
1 Labor Foreman (outside)	$46.40	$371.20	$69.25	$554.00	$51.45	$76.71
2 Laborers	44.40	710.40	66.25	1060.00		
1 Equip. Oper. (crane)	61.45	491.60	91.55	732.40		
2 Equip. Oper. (light)	55.50	888.00	82.70	1323.20		
1 Equip. Oper. (oiler)	52.50	420.00	78.25	626.00		
1 Crawler Crane, 40 Ton		1231.00		1354.10		
1 Lead, 60' High		211.80		232.98		
1 Hammer, Diesel, 15K ft.-lbs.		624.45		686.89		
1 Backhoe Loader, 48 H.P.		216.20		237.82	40.78	44.85
56 L.H., Daily Totals		$5164.65		$6807.40	$92.23	$121.56

Crew B-61

Crew No.	Hr.	Daily	Hr.	Daily	Bare Costs	Incl. O&P
1 Labor Foreman (outside)	$46.40	$371.20	$69.25	$554.00	$47.02	$70.14
3 Laborers	44.40	1065.60	66.25	1590.00		
1 Equip. Oper. (light)	55.50	444.00	82.70	661.60		
1 Cement Mixer, 2 C.Y.		112.35		123.58		
1 Air Compressor, 160 cfm		212.30		233.53	8.12	8.93
40 L.H., Daily Totals		$2205.45		$3162.72	$55.14	$79.07

Crew B-62

Crew No.	Hr.	Daily	Hr.	Daily	Bare Costs	Incl. O&P
2 Laborers	$44.40	$710.40	$66.25	$1060.00	$48.10	$71.73
1 Equip. Oper. (light)	55.50	444.00	82.70	661.60		
1 Loader, Skid Steer, 30 H.P.		179.50		197.45	7.48	8.23
24 L.H., Daily Totals		$1333.90		$1919.05	$55.58	$79.96

For customer support on your Concrete & Masonry Costs with RSMeans Data, call 800.448.8182.

433

Crews - Standard

Crew B-62A

Crew No.	Bare Costs Hr.	Daily	Incl. Subs O&P Hr.	Daily	Cost Per Labor-Hour Bare Costs	Incl. O&P
2 Laborers	$44.40	$710.40	$66.25	$1060.00	$48.10	$71.73
1 Equip. Oper. (light)	55.50	444.00	82.70	661.60		
1 Loader, Skid Steer, 30 H.P.		179.50		197.45		
1 Trencher Attachment		66.25		72.88	10.24	11.26
24 L.H., Daily Totals		$1400.15		$1991.93	$58.34	$83.00

Crew B-63

Crew No.	Bare Costs Hr.	Daily	Incl. Subs O&P Hr.	Daily	Cost Per Labor-Hour Bare Costs	Incl. O&P
4 Laborers	$44.40	$1420.80	$66.25	$2120.00	$46.62	$69.54
1 Equip. Oper. (light)	55.50	444.00	82.70	661.60		
1 Loader, Skid Steer, 30 H.P.		179.50		197.45	4.49	4.94
40 L.H., Daily Totals		$2044.30		$2979.05	$51.11	$74.48

Crew B-63B

Crew No.	Bare Costs Hr.	Daily	Incl. Subs O&P Hr.	Daily	Cost Per Labor-Hour Bare Costs	Incl. O&P
1 Labor Foreman (inside)	$44.90	$359.20	$67.00	$536.00	$47.30	$70.55
2 Laborers	44.40	710.40	66.25	1060.00		
1 Equip. Oper. (light)	55.50	444.00	82.70	661.60		
1 Loader, Skid Steer, 78 H.P.		446.30		490.93	13.95	15.34
32 L.H., Daily Totals		$1959.90		$2748.53	$61.25	$85.89

Crew B-64

Crew No.	Bare Costs Hr.	Daily	Incl. Subs O&P Hr.	Daily	Cost Per Labor-Hour Bare Costs	Incl. O&P
1 Laborer	$44.40	$355.20	$66.25	$530.00	$46.60	$69.63
1 Truck Driver (light)	48.80	390.40	73.00	584.00		
1 Power Mulcher (small)		201.00		221.10		
1 Flatbed Truck, Gas, 1.5 Ton		198.50		218.35	24.97	27.47
16 L.H., Daily Totals		$1145.10		$1553.45	$71.57	$97.09

Crew B-65

Crew No.	Bare Costs Hr.	Daily	Incl. Subs O&P Hr.	Daily	Cost Per Labor-Hour Bare Costs	Incl. O&P
1 Laborer	$44.40	$355.20	$66.25	$530.00	$46.60	$69.63
1 Truck Driver (light)	48.80	390.40	73.00	584.00		
1 Power Mulcher (Large)		345.35		379.88		
1 Flatbed Truck, Gas, 1.5 Ton		198.50		218.35	33.99	37.39
16 L.H., Daily Totals		$1289.45		$1712.23	$80.59	$107.01

Crew B-66

Crew No.	Bare Costs Hr.	Daily	Incl. Subs O&P Hr.	Daily	Cost Per Labor-Hour Bare Costs	Incl. O&P
1 Equip. Oper. (light)	$55.50	$444.00	$82.70	$661.60	$55.50	$82.70
1 Loader-Backhoe, 40 H.P.		267.55		294.31	33.44	36.79
8 L.H., Daily Totals		$711.55		$955.90	$88.94	$119.49

Crew B-67

Crew No.	Bare Costs Hr.	Daily	Incl. Subs O&P Hr.	Daily	Cost Per Labor-Hour Bare Costs	Incl. O&P
1 Millwright	$58.75	$470.00	$84.90	$679.20	$57.13	$83.80
1 Equip. Oper. (light)	55.50	444.00	82.70	661.60		
1 R.T. Forklift, 5,000 Lb., diesel		272.85		300.13	17.05	18.76
16 L.H., Daily Totals		$1186.85		$1640.93	$74.18	$102.56

Crew B-67B

Crew No.	Bare Costs Hr.	Daily	Incl. Subs O&P Hr.	Daily	Cost Per Labor-Hour Bare Costs	Incl. O&P
1 Millwright Foreman (inside)	$59.25	$474.00	$85.60	$684.80	$59.00	$85.25
1 Millwright	58.75	470.00	84.90	679.20		
16 L.H., Daily Totals		$944.00		$1364.00	$59.00	$85.25

Crew B-68

Crew No.	Bare Costs Hr.	Daily	Incl. Subs O&P Hr.	Daily	Cost Per Labor-Hour Bare Costs	Incl. O&P
2 Millwrights	$58.75	$940.00	$84.90	$1358.40	$57.67	$84.17
1 Equip. Oper. (light)	55.50	444.00	82.70	661.60		
1 R.T. Forklift, 5,000 Lb., diesel		272.85		300.13	11.37	12.51
24 L.H., Daily Totals		$1656.85		$2320.14	$69.04	$96.67

Crew B-68A

Crew No.	Bare Costs Hr.	Daily	Incl. Subs O&P Hr.	Daily	Cost Per Labor-Hour Bare Costs	Incl. O&P
1 Millwright Foreman (inside)	$59.25	$474.00	$85.60	$684.80	$58.92	$85.13
2 Millwrights	58.75	940.00	84.90	1358.40		
1 Forklift, Smooth Floor, 8,000 Lb.		283.25		311.57	11.80	12.98
24 L.H., Daily Totals		$1697.25		$2354.78	$70.72	$98.12

Crew B-68B

Crew No.	Bare Costs Hr.	Daily	Incl. Subs O&P Hr.	Daily	Cost Per Labor-Hour Bare Costs	Incl. O&P
1 Millwright Foreman (inside)	$59.25	$474.00	$85.60	$684.80	$62.79	$92.40
2 Millwrights	58.75	940.00	84.90	1358.40		
2 Electricians	63.70	1019.20	94.65	1514.40		
2 Plumbers	67.70	1083.20	101.05	1616.80		
1 R.T. Forklift, 5,000 Lb., gas		283.30		311.63	5.06	5.56
56 L.H., Daily Totals		$3799.70		$5486.03	$67.85	$97.96

Crew B-68C

Crew No.	Bare Costs Hr.	Daily	Incl. Subs O&P Hr.	Daily	Cost Per Labor-Hour Bare Costs	Incl. O&P
1 Millwright Foreman (inside)	$59.25	$474.00	$85.60	$684.80	$62.35	$91.55
1 Millwright	58.75	470.00	84.90	679.20		
1 Electrician	63.70	509.60	94.65	757.20		
1 Plumber	67.70	541.60	101.05	808.40		
1 R.T. Forklift, 5,000 Lb., gas		283.30		311.63	8.85	9.74
32 L.H., Daily Totals		$2278.50		$3241.23	$71.20	$101.29

Crew B-68D

Crew No.	Bare Costs Hr.	Daily	Incl. Subs O&P Hr.	Daily	Cost Per Labor-Hour Bare Costs	Incl. O&P
1 Labor Foreman (inside)	$44.90	$359.20	$67.00	$536.00	$48.27	$71.98
1 Laborer	44.40	355.20	66.25	530.00		
1 Equip. Oper. (light)	55.50	444.00	82.70	661.60		
1 R.T. Forklift, 5,000 Lb., gas		283.30		311.63	11.80	12.98
24 L.H., Daily Totals		$1441.70		$2039.23	$60.07	$84.97

Crew B-68E

Crew No.	Bare Costs Hr.	Daily	Incl. Subs O&P Hr.	Daily	Cost Per Labor-Hour Bare Costs	Incl. O&P
1 Struc. Steel Foreman (inside)	$60.80	$486.40	$94.10	$752.80	$60.40	$93.46
3 Struc. Steel Workers	60.30	1447.20	93.30	2239.20		
1 Welder	60.30	482.40	93.30	746.40		
1 Forklift, Smooth Floor, 8,000 Lb.		283.25		311.57	7.08	7.79
40 L.H., Daily Totals		$2699.25		$4049.97	$67.48	$101.25

Crew B-68F

Crew No.	Bare Costs Hr.	Daily	Incl. Subs O&P Hr.	Daily	Cost Per Labor-Hour Bare Costs	Incl. O&P
1 Skilled Worker Foreman (out)	$59.10	$472.80	$88.90	$711.20	$57.77	$86.90
2 Skilled Workers	57.10	913.60	85.90	1374.40		
1 R.T. Forklift, 5,000 Lb., gas		283.30		311.63	11.80	12.98
24 L.H., Daily Totals		$1669.70		$2397.23	$69.57	$99.88

Crew B-68G

Crew No.	Bare Costs Hr.	Daily	Incl. Subs O&P Hr.	Daily	Cost Per Labor-Hour Bare Costs	Incl. O&P
2 Structural Steel Workers	$60.30	$964.80	$93.30	$1492.80	$60.30	$93.30
1 R.T. Forklift, 5,000 Lb., gas		283.30		311.63	17.71	19.48
16 L.H., Daily Totals		$1248.10		$1804.43	$78.01	$112.78

Crew B-69

Crew No.	Bare Costs Hr.	Daily	Incl. Subs O&P Hr.	Daily	Cost Per Labor-Hour Bare Costs	Incl. O&P
1 Labor Foreman (outside)	$46.40	$371.20	$69.25	$554.00	$48.92	$72.97
3 Laborers	44.40	1065.60	66.25	1590.00		
1 Equip. Oper. (crane)	61.45	491.60	91.55	732.40		
1 Equip. Oper. (oiler)	52.50	420.00	78.25	626.00		
1 Hyd. Crane, 80 Ton		1458.00		1603.80	30.38	33.41
48 L.H., Daily Totals		$3806.40		$5106.20	$79.30	$106.38

Crew B-69A

Crew No.	Bare Costs Hr.	Daily	Incl. Subs O&P Hr.	Daily	Cost Per Labor-Hour Bare Costs	Incl. O&P
1 Labor Foreman (outside)	$46.40	$371.20	$69.25	$554.00	$48.40	$71.97
3 Laborers	44.40	1065.60	66.25	1590.00		
1 Equip. Oper. (medium)	59.00	472.00	87.90	703.20		
1 Concrete Finisher	51.80	414.40	75.90	607.20		
1 Curb/Gutter Paver, 2-Track		1231.00		1354.10	25.65	28.21
48 L.H., Daily Totals		$3554.20		$4808.50	$74.05	$100.18

For customer support on your Concrete & Masonry Costs with RSMeans Data, call 800.448.8182.

Crew No.		Bare Costs		Incl. Subs O&P		Cost Per Labor-Hour	

Crew B-69B

	Hr.	Daily	Hr.	Daily	Bare Costs	Incl. O&P
1 Labor Foreman (outside)	$46.40	$371.20	$69.25	$554.00	$48.40	$71.97
3 Laborers	44.40	1065.60	66.25	1590.00		
1 Equip. Oper. (medium)	59.00	472.00	87.90	703.20		
1 Cement Finisher	51.80	414.40	75.90	607.20		
1 Curb/Gutter Paver, 4-Track		801.05		881.15	16.69	18.36
48 L.H., Daily Totals		$3124.25		$4335.56	$65.09	$90.32

Crew B-70

	Hr.	Daily	Hr.	Daily	Bare Costs	Incl. O&P
1 Labor Foreman (outside)	$46.40	$371.20	$69.25	$554.00	$50.94	$75.96
3 Laborers	44.40	1065.60	66.25	1590.00		
3 Equip. Oper. (medium)	59.00	1416.00	87.90	2109.60		
1 Grader, 30,000 Lbs.		1073.00		1180.30		
1 Ripper, Beam & 1 Shank		91.60		100.76		
1 Road Sweeper, S.P., 8' wide		723.65		796.01		
1 F.E. Loader, W.M., 1.5 C.Y.		441.40		485.54	41.60	45.76
56 L.H., Daily Totals		$5182.45		$6816.22	$92.54	$121.72

Crew B-70A

	Hr.	Daily	Hr.	Daily	Bare Costs	Incl. O&P
1 Laborer	$44.40	$355.20	$66.25	$530.00	$56.08	$83.57
4 Equip. Oper. (medium)	59.00	1888.00	87.90	2812.80		
1 Grader, 40,000 Lbs.		1219.00		1340.90		
1 F.E. Loader, W.M., 2.5 C.Y.		638.30		702.13		
1 Dozer, 80 H.P.		405.85		446.44		
1 Roller, Pneum. Whl., 12 Ton		349.90		384.89	65.33	71.86
40 L.H., Daily Totals		$4856.25		$6217.15	$121.41	$155.43

Crew B-71

	Hr.	Daily	Hr.	Daily	Bare Costs	Incl. O&P
1 Labor Foreman (outside)	$46.40	$371.20	$69.25	$554.00	$50.94	$75.96
3 Laborers	44.40	1065.60	66.25	1590.00		
3 Equip. Oper. (medium)	59.00	1416.00	87.90	2109.60		
1 Pvmt. Profiler, 750 H.P.		3483.00		3831.30		
1 Road Sweeper, S.P., 8' wide		723.65		796.01		
1 F.E. Loader, W.M., 1.5 C.Y.		441.40		485.54	83.00	91.30
56 L.H., Daily Totals		$7500.85		$9366.45	$133.94	$167.26

Crew B-72

	Hr.	Daily	Hr.	Daily	Bare Costs	Incl. O&P
1 Labor Foreman (outside)	$46.40	$371.20	$69.25	$554.00	$51.95	$77.45
3 Laborers	44.40	1065.60	66.25	1590.00		
4 Equip. Oper. (medium)	59.00	1888.00	87.90	2812.80		
1 Pvmt. Profiler, 750 H.P.		3483.00		3831.30		
1 Hammermill, 250 H.P.		857.40		943.14		
1 Windrow Loader		1461.00		1607.10		
1 Mix Paver, 165 H.P.		2172.00		2389.20		
1 Roller, Pneum. Whl., 12 Ton		349.90		384.89	130.05	143.06
64 L.H., Daily Totals		$11648.10		$14112.43	$182.00	$220.51

Crew B-73

	Hr.	Daily	Hr.	Daily	Bare Costs	Incl. O&P
1 Labor Foreman (outside)	$46.40	$371.20	$69.25	$554.00	$53.77	$80.16
2 Laborers	44.40	710.40	66.25	1060.00		
5 Equip. Oper. (medium)	59.00	2360.00	87.90	3516.00		
1 Road Mixer, 310 H.P.		1919.00		2110.90		
1 Tandem Roller, 10 Ton		246.80		271.48		
1 Hammermill, 250 H.P.		857.40		943.14		
1 Grader, 30,000 Lbs.		1073.00		1180.30		
.5 F.E. Loader, W.M., 1.5 C.Y.		220.70		242.77		
.5 Truck Tractor, 220 H.P.		155.40		170.94		
.5 Water Tank Trailer, 5000 Gal.		77.05		84.75	71.08	78.19
64 L.H., Daily Totals		$7990.95		$10134.29	$124.86	$158.35

Crew B-74

	Hr.	Daily	Hr.	Daily	Bare Costs	Incl. O&P
1 Labor Foreman (outside)	$46.40	$371.20	$69.25	$554.00	$53.67	$80.06
1 Laborer	44.40	355.20	66.25	530.00		
4 Equip. Oper. (medium)	59.00	1888.00	87.90	2812.80		
2 Truck Drivers (heavy)	51.30	820.80	76.70	1227.20		
1 Grader, 30,000 Lbs.		1073.00		1180.30		
1 Ripper, Beam & 1 Shank		91.60		100.76		
2 Stabilizers, 310 H.P.		2808.00		3088.80		
1 Flatbed Truck, Gas, 3 Ton		850.05		935.05		
1 Chem. Spreader, Towed		85.40		93.94		
1 Roller, Vibratory, 25 Ton		672.35		739.59		
1 Water Tank Trailer, 5000 Gal.		154.10		169.51		
1 Truck Tractor, 220 H.P.		310.80		341.88	94.46	103.90
64 L.H., Daily Totals		$9480.50		$11773.83	$148.13	$183.97

Crew B-75

	Hr.	Daily	Hr.	Daily	Bare Costs	Incl. O&P
1 Labor Foreman (outside)	$46.40	$371.20	$69.25	$554.00	$54.01	$80.54
1 Laborer	44.40	355.20	66.25	530.00		
4 Equip. Oper. (medium)	59.00	1888.00	87.90	2812.80		
1 Truck Driver (heavy)	51.30	410.40	76.70	613.60		
1 Grader, 30,000 Lbs.		1073.00		1180.30		
1 Ripper, Beam & 1 Shank		91.60		100.76		
2 Stabilizers, 310 H.P.		2808.00		3088.80		
1 Dist. Tanker, 3000 Gallon		334.10		367.51		
1 Truck Tractor, 6x4, 380 H.P.		499.15		549.07		
1 Roller, Vibratory, 25 Ton		672.35		739.59	97.83	107.61
56 L.H., Daily Totals		$8503.00		$10536.42	$151.84	$188.15

Crew B-76

	Hr.	Daily	Hr.	Daily	Bare Costs	Incl. O&P
1 Dock Builder Foreman (outside)	$57.90	$463.20	$89.50	$716.00	$56.98	$86.98
5 Dock Builders	55.90	2236.00	86.40	3456.00		
2 Equip. Oper. (crane)	61.45	983.20	91.55	1464.80		
1 Equip. Oper. (oiler)	52.50	420.00	78.25	626.00		
1 Crawler Crane, 50 Ton		1541.00		1695.10		
1 Barge, 400 Ton		869.15		956.07		
1 Hammer, Diesel, 15K ft.-lbs.		624.45		686.89		
1 Lead, 60' High		211.80		232.98		
1 Air Compressor, 600 cfm		426.55		469.20		
2 -50' Air Hoses, 3"		76.70		84.37	52.08	57.29
72 L.H., Daily Totals		$7852.05		$10387.42	$109.06	$144.27

Crew B-76A

	Hr.	Daily	Hr.	Daily	Bare Costs	Incl. O&P
1 Labor Foreman (outside)	$46.40	$371.20	$69.25	$554.00	$47.79	$71.29
5 Laborers	44.40	1776.00	66.25	2650.00		
1 Equip. Oper. (crane)	61.45	491.60	91.55	732.40		
1 Equip. Oper. (oiler)	52.50	420.00	78.25	626.00		
1 Crawler Crane, 50 Ton		1541.00		1695.10		
1 Barge, 400 Ton		869.15		956.07	37.66	41.42
64 L.H., Daily Totals		$5468.95		$7213.56	$85.45	$112.71

Crew B-77

	Hr.	Daily	Hr.	Daily	Bare Costs	Incl. O&P
1 Labor Foreman (outside)	$46.40	$371.20	$69.25	$554.00	$45.68	$68.20
3 Laborers	44.40	1065.60	66.25	1590.00		
1 Truck Driver (light)	48.80	390.40	73.00	584.00		
1 Crack Cleaner, 25 H.P.		53.00		58.30		
1 Crack Filler, Trailer Mtd.		170.95		188.04		
1 Flatbed Truck, Gas, 3 Ton		850.05		935.05	26.85	29.54
40 L.H., Daily Totals		$2901.20		$3909.40	$72.53	$97.73

For customer support on your Concrete & Masonry Costs with RSMeans Data, call 800.448.8182.

435

Crew No.	Bare Costs		Incl. Subs O&P		Cost Per Labor-Hour	
Crew B-78	Hr.	Daily	Hr.	Daily	Bare Costs	Incl. O&P
1 Labor Foreman (outside)	$46.40	$371.20	$69.25	$554.00	$45.47	$67.88
4 Laborers	44.40	1420.80	66.25	2120.00		
1 Truck Driver (light)	48.80	390.40	73.00	584.00		
1 Paint Striper, S.P., 40 Gallon		128.35		141.19		
1 Flatbed Truck, Gas, 3 Ton		850.05		935.05		
1 Pickup Truck, 3/4 Ton		112.20		123.42	22.72	24.99
48 L.H., Daily Totals		$3273.00		$4457.66	$68.19	$92.87
Crew B-78A	Hr.	Daily	Hr.	Daily	Bare Costs	Incl. O&P
1 Equip. Oper. (light)	$55.50	$444.00	$82.70	$661.60	$55.50	$82.70
1 Line Rem. (Metal Balls) 115 H.P.		996.25		1095.88	124.53	136.98
8 L.H., Daily Totals		$1440.25		$1757.47	$180.03	$219.68
Crew B-78B	Hr.	Daily	Hr.	Daily	Bare Costs	Incl. O&P
2 Laborers	$44.40	$710.40	$66.25	$1060.00	$45.63	$68.08
.25 Equip. Oper. (light)	55.50	111.00	82.70	165.40		
1 Pickup Truck, 3/4 Ton		112.20		123.42		
1 Line Rem.,11 H.P.,Walk Behind		114.75		126.22		
.25 Road Sweeper, S.P., 8' wide		180.91		199.00	22.66	24.92
18 L.H., Daily Totals		$1229.26		$1674.05	$68.29	$93.00
Crew B-78C	Hr.	Daily	Hr.	Daily	Bare Costs	Incl. O&P
1 Labor Foreman (outside)	$46.40	$371.20	$69.25	$554.00	$45.47	$67.88
4 Laborers	44.40	1420.80	66.25	2120.00		
1 Truck Driver (light)	48.80	390.40	73.00	584.00		
1 Paint Striper, T.M., 120 Gal.		603.10		663.41		
1 Flatbed Truck, Gas, 3 Ton		850.05		935.05		
1 Pickup Truck, 3/4 Ton		112.20		123.42	32.61	35.87
48 L.H., Daily Totals		$3747.75		$4979.89	$78.08	$103.75
Crew B-78D	Hr.	Daily	Hr.	Daily	Bare Costs	Incl. O&P
2 Labor Foremen (outside)	$46.40	$742.40	$69.25	$1108.00	$45.24	$67.53
7 Laborers	44.40	2486.40	66.25	3710.00		
1 Truck Driver (light)	48.80	390.40	73.00	584.00		
1 Paint Striper, T.M., 120 Gal.		603.10		663.41		
1 Flatbed Truck, Gas, 3 Ton		850.05		935.05		
3 Pickup Trucks, 3/4 Ton		336.60		370.26		
1 Air Compressor, 60 cfm		153.85		169.24		
1 -50' Air Hose, 3/4"		7.15		7.87		
1 Breaker, Pavement, 60 lb.		53.60		58.96	25.05	27.56
80 L.H., Daily Totals		$5623.55		$7606.78	$70.29	$95.08
Crew B-78E	Hr.	Daily	Hr.	Daily	Bare Costs	Incl. O&P
2 Labor Foremen (outside)	$46.40	$742.40	$69.25	$1108.00	$45.10	$67.31
9 Laborers	44.40	3196.80	66.25	4770.00		
1 Truck Driver (light)	48.80	390.40	73.00	584.00		
1 Paint Striper, T.M., 120 Gal.		603.10		663.41		
1 Flatbed Truck, Gas, 3 Ton		850.05		935.05		
4 Pickup Trucks, 3/4 Ton		448.80		493.68		
2 Air Compressors, 60 cfm		307.70		338.47		
2 -50' Air Hoses, 3/4"		14.30		15.73		
2 Breakers, Pavement, 60 lb.		107.20		117.92	24.28	26.71
96 L.H., Daily Totals		$6660.75		$9026.26	$69.38	$94.02

Crew No.	Bare Costs		Incl. Subs O&P		Cost Per Labor-Hour	
Crew B-78F	Hr.	Daily	Hr.	Daily	Bare Costs	Incl. O&P
2 Labor Foremen (outside)	$46.40	$742.40	$69.25	$1108.00	$45.00	$67.16
11 Laborers	44.40	3907.20	66.25	5830.00		
1 Truck Driver (light)	48.80	390.40	73.00	584.00		
1 Paint Striper, T.M., 120 Gal.		603.10		663.41		
1 Flatbed Truck, Gas, 3 Ton		850.05		935.05		
7 Pickup Trucks, 3/4 Ton		785.40		863.94		
3 Air Compressors, 60 cfm		461.55		507.70		
3 -50' Air Hoses, 3/4"		21.45		23.59		
3 Breakers, Pavement, 60 lb.		160.80		176.88	25.74	28.31
112 L.H., Daily Totals		$7922.35		$10692.58	$70.74	$95.47
Crew B-79	Hr.	Daily	Hr.	Daily	Bare Costs	Incl. O&P
1 Labor Foreman (outside)	$46.40	$371.20	$69.25	$554.00	$45.68	$68.20
3 Laborers	44.40	1065.60	66.25	1590.00		
1 Truck Driver (light)	48.80	390.40	73.00	584.00		
1 Paint Striper, T.M., 120 Gal.		603.10		663.41		
1 Heating Kettle, 115 Gallon		107.25		117.97		
1 Flatbed Truck, Gas, 3 Ton		850.05		935.05		
2 Pickup Trucks, 3/4 Ton		224.40		246.84	44.62	49.08
40 L.H., Daily Totals		$3612.00		$4691.28	$90.30	$117.28
Crew B-79A	Hr.	Daily	Hr.	Daily	Bare Costs	Incl. O&P
1.5 Equip. Oper. (light)	$55.50	$666.00	$82.70	$992.40	$55.50	$82.70
.5 Line Remov. (Grinder) 115 H.P.		515.50		567.05		
1 Line Rem. (Metal Balls) 115 H.P.		996.25		1095.88	125.98	138.58
12 L.H., Daily Totals		$2177.75		$2655.32	$181.48	$221.29
Crew B-79B	Hr.	Daily	Hr.	Daily	Bare Costs	Incl. O&P
1 Laborer	$44.40	$355.20	$66.25	$530.00	$44.40	$66.25
1 Set of Gases		173.60		190.96	21.70	23.87
8 L.H., Daily Totals		$528.80		$720.96	$66.10	$90.12
Crew B-79C	Hr.	Daily	Hr.	Daily	Bare Costs	Incl. O&P
1 Labor Foreman (outside)	$46.40	$371.20	$69.25	$554.00	$45.31	$67.64
5 Laborers	44.40	1776.00	66.25	2650.00		
1 Truck Driver (light)	48.80	390.40	73.00	584.00		
1 Paint Striper, T.M., 120 Gal.		603.10		663.41		
1 Heating Kettle, 115 Gallon		107.25		117.97		
1 Flatbed Truck, Gas, 3 Ton		850.05		935.05		
3 Pickup Trucks, 3/4 Ton		336.60		370.26		
1 Air Compressor, 60 cfm		153.85		169.24		
1 -50' Air Hose, 3/4"		7.15		7.87		
1 Breaker, Pavement, 60 lb.		53.60		58.96	37.71	41.48
56 L.H., Daily Totals		$4649.20		$6110.76	$83.02	$109.12
Crew B-79D	Hr.	Daily	Hr.	Daily	Bare Costs	Incl. O&P
2 Labor Foremen (outside)	$46.40	$742.40	$69.25	$1108.00	$45.45	$67.84
5 Laborers	44.40	1776.00	66.25	2650.00		
1 Truck Driver (light)	48.80	390.40	73.00	584.00		
1 Paint Striper, T.M., 120 Gal.		603.10		663.41		
1 Heating Kettle, 115 Gallon		107.25		117.97		
1 Flatbed Truck, Gas, 3 Ton		850.05		935.05		
4 Pickup Trucks, 3/4 Ton		448.80		493.68		
1 Air Compressor, 60 cfm		153.85		169.24		
1 -50' Air Hose, 3/4"		7.15		7.87		
1 Breaker, Pavement, 60 lb.		53.60		58.96	34.75	38.22
64 L.H., Daily Totals		$5132.60		$6788.18	$80.20	$106.07

Crew No.	Bare Costs		Incl. Subs O&P		Cost Per Labor-Hour	
Crew B-79E	Hr.	Daily	Hr.	Daily	Bare Costs	Incl. O&P
2 Labor Foremen (outside)	$46.40	$742.40	$69.25	$1108.00	$45.24	$67.53
7 Laborers	44.40	2486.40	66.25	3710.00		
1 Truck Driver (light)	48.80	390.40	73.00	584.00		
1 Paint Striper, T.M., 120 Gal.		603.10		663.41		
1 Heating Kettle, 115 Gallon		107.25		117.97		
1 Flatbed Truck, Gas, 3 Ton		850.05		935.05		
5 Pickup Trucks, 3/4 Ton		561.00		617.10		
2 Air Compressors, 60 cfm		307.70		338.47		
2 -50' Air Hoses, 3/4"		14.30		15.73		
2 Breakers, Pavement, 60 lb.		107.20		117.92	31.88	35.07
80 L.H., Daily Totals		$6169.80		$8207.66	$77.12	$102.60

Crew No.	Bare Costs		Incl. Subs O&P		Cost Per Labor-Hour	
Crew B-80	Hr.	Daily	Hr.	Daily	Bare Costs	Incl. O&P
1 Labor Foreman (outside)	$46.40	$371.20	$69.25	$554.00	$48.77	$72.80
1 Laborer	44.40	355.20	66.25	530.00		
1 Truck Driver (light)	48.80	390.40	73.00	584.00		
1 Equip. Oper. (light)	55.50	444.00	82.70	661.60		
1 Flatbed Truck, Gas, 3 Ton		850.05		935.05		
1 Earth Auger, Truck-Mtd.		202.55		222.81	32.89	36.18
32 L.H., Daily Totals		$2613.40		$3487.46	$81.67	$108.98

Crew No.	Bare Costs		Incl. Subs O&P		Cost Per Labor-Hour	
Crew B-80A	Hr.	Daily	Hr.	Daily	Bare Costs	Incl. O&P
3 Laborers	$44.40	$1065.60	$66.25	$1590.00	$44.40	$66.25
1 Flatbed Truck, Gas, 3 Ton		850.05		935.05	35.42	38.96
24 L.H., Daily Totals		$1915.65		$2525.05	$79.82	$105.21

Crew No.	Bare Costs		Incl. Subs O&P		Cost Per Labor-Hour	
Crew B-80B	Hr.	Daily	Hr.	Daily	Bare Costs	Incl. O&P
3 Laborers	$44.40	$1065.60	$66.25	$1590.00	$47.17	$70.36
1 Equip. Oper. (light)	55.50	444.00	82.70	661.60		
1 Crane, Flatbed Mounted, 3 Ton		238.75		262.63	7.46	8.21
32 L.H., Daily Totals		$1748.35		$2514.22	$54.64	$78.57

Crew No.	Bare Costs		Incl. Subs O&P		Cost Per Labor-Hour	
Crew B-80C	Hr.	Daily	Hr.	Daily	Bare Costs	Incl. O&P
2 Laborers	$44.40	$710.40	$66.25	$1060.00	$45.87	$68.50
1 Truck Driver (light)	48.80	390.40	73.00	584.00		
1 Flatbed Truck, Gas, 1.5 Ton		198.50		218.35		
1 Manual Fence Post Auger, Gas		54.40		59.84	10.54	11.59
24 L.H., Daily Totals		$1353.70		$1922.19	$56.40	$80.09

Crew No.	Bare Costs		Incl. Subs O&P		Cost Per Labor-Hour	
Crew B-81	Hr.	Daily	Hr.	Daily	Bare Costs	Incl. O&P
1 Laborer	$44.40	$355.20	$66.25	$530.00	$51.57	$76.95
1 Equip. Oper. (medium)	59.00	472.00	87.90	703.20		
1 Truck Driver (heavy)	51.30	410.40	76.70	613.60		
1 Hydromulcher, T.M., 3000 Gal.		256.85		282.54		
1 Truck Tractor, 220 H.P.		310.80		341.88	23.65	26.02
24 L.H., Daily Totals		$1805.25		$2471.22	$75.22	$102.97

Crew No.	Bare Costs		Incl. Subs O&P		Cost Per Labor-Hour	
Crew B-81A	Hr.	Daily	Hr.	Daily	Bare Costs	Incl. O&P
1 Laborer	$44.40	$355.20	$66.25	$530.00	$46.60	$69.63
1 Truck Driver (light)	48.80	390.40	73.00	584.00		
1 Hydromulcher, T.M., 600 Gal.		118.15		129.97		
1 Flatbed Truck, Gas, 3 Ton		850.05		935.05	60.51	66.56
16 L.H., Daily Totals		$1713.80		$2179.02	$107.11	$136.19

Crew No.	Bare Costs		Incl. Subs O&P		Cost Per Labor-Hour	
Crew B-82	Hr.	Daily	Hr.	Daily	Bare Costs	Incl. O&P
1 Laborer	$44.40	$355.20	$66.25	$530.00	$49.95	$74.47
1 Equip. Oper. (light)	55.50	444.00	82.70	661.60		
1 Horiz. Borer, 6 H.P.		184.15		202.57	11.51	12.66
16 L.H., Daily Totals		$983.35		$1394.17	$61.46	$87.14

Crew No.	Bare Costs		Incl. Subs O&P		Cost Per Labor-Hour	
Crew B-82A	Hr.	Daily	Hr.	Daily	Bare Costs	Incl. O&P
2 Laborers	$44.40	$710.40	$66.25	$1060.00	$49.95	$74.47
2 Equip. Opers. (light)	55.50	888.00	82.70	1323.20		
2 Dump Truck, 8 C.Y., 220 H.P.		815.20		896.72		
1 Flatbed Trailer, 25 Ton		137.20		150.92		
1 Horiz. Dir. Drill, 20k lb. Thrust		544.10		598.51		
1 Mud Trailer for HDD, 1500 Gal.		312.15		343.37		
1 Pickup Truck, 4x4, 3/4 Ton		176.75		194.43		
1 Flatbed Trailer, 3 Ton		71.15		78.27		
1 Loader, Skid Steer, 78 H.P.		446.30		490.93	78.21	86.04
32 L.H., Daily Totals		$4101.25		$5136.34	$128.16	$160.51

Crew No.	Bare Costs		Incl. Subs O&P		Cost Per Labor-Hour	
Crew B-82B	Hr.	Daily	Hr.	Daily	Bare Costs	Incl. O&P
2 Laborers	$44.40	$710.40	$66.25	$1060.00	$49.95	$74.47
2 Equip. Opers. (light)	55.50	888.00	82.70	1323.20		
2 Dump Truck, 8 C.Y., 220 H.P.		815.20		896.72		
1 Flatbed Trailer, 25 Ton		137.20		150.92		
1 Horiz. Dir. Drill, 30k lb. Thrust		647.65		712.41		
1 Mud Trailer for HDD, 1500 Gal.		312.15		343.37		
1 Pickup Truck, 4x4, 3/4 Ton		176.75		194.43		
1 Flatbed Trailer, 3 Ton		71.15		78.27		
1 Loader, Skid Steer, 78 H.P.		446.30		490.93	81.45	89.59
32 L.H., Daily Totals		$4204.80		$5250.24	$131.40	$164.07

Crew No.	Bare Costs		Incl. Subs O&P		Cost Per Labor-Hour	
Crew B-82C	Hr.	Daily	Hr.	Daily	Bare Costs	Incl. O&P
2 Laborers	$44.40	$710.40	$66.25	$1060.00	$49.95	$74.47
2 Equip. Opers. (light)	55.50	888.00	82.70	1323.20		
2 Dump Truck, 8 C.Y., 220 H.P.		815.20		896.72		
1 Flatbed Trailer, 25 Ton		137.20		150.92		
1 Horiz. Dir. Drill, 50k lb. Thrust		824.05		906.46		
1 Mud Trailer for HDD, 1500 Gal.		312.15		343.37		
1 Pickup Truck, 4x4, 3/4 Ton		176.75		194.43		
1 Flatbed Trailer, 3 Ton		71.15		78.27		
1 Loader, Skid Steer, 78 H.P.		446.30		490.93	86.96	95.66
32 L.H., Daily Totals		$4381.20		$5444.28	$136.91	$170.13

Crew No.	Bare Costs		Incl. Subs O&P		Cost Per Labor-Hour	
Crew B-82D	Hr.	Daily	Hr.	Daily	Bare Costs	Incl. O&P
1 Equip. Oper. (light)	$55.50	$444.00	$82.70	$661.60	$55.50	$82.70
1 Mud Trailer for HDD, 1500 Gal.		312.15		343.37	39.02	42.92
8 L.H., Daily Totals		$756.15		$1004.97	$94.52	$125.62

Crew No.	Bare Costs		Incl. Subs O&P		Cost Per Labor-Hour	
Crew B-83	Hr.	Daily	Hr.	Daily	Bare Costs	Incl. O&P
1 Tugboat Captain	$59.00	$472.00	$87.90	$703.20	$51.70	$77.08
1 Tugboat Hand	44.40	355.20	66.25	530.00		
1 Tugboat, 250 H.P.		726.10		798.71	45.38	49.92
16 L.H., Daily Totals		$1553.30		$2031.91	$97.08	$126.99

Crew No.	Bare Costs		Incl. Subs O&P		Cost Per Labor-Hour	
Crew B-84	Hr.	Daily	Hr.	Daily	Bare Costs	Incl. O&P
1 Equip. Oper. (medium)	$59.00	$472.00	$87.90	$703.20	$59.00	$87.90
1 Rotary Mower/Tractor		371.15		408.26	46.39	51.03
8 L.H., Daily Totals		$843.15		$1111.46	$105.39	$138.93

Crew No.	Bare Costs		Incl. Subs O&P		Cost Per Labor-Hour	
Crew B-85	Hr.	Daily	Hr.	Daily	Bare Costs	Incl. O&P
3 Laborers	$44.40	$1065.60	$66.25	$1590.00	$48.70	$72.67
1 Equip. Oper. (medium)	59.00	472.00	87.90	703.20		
1 Truck Driver (heavy)	51.30	410.40	76.70	613.60		
1 Telescoping Boom Lift, to 80'		387.75		426.52		
1 Brush Chipper, 12", 130 H.P.		366.05		402.65		
1 Pruning Saw, Rotary		26.40		29.04	19.50	21.46
40 L.H., Daily Totals		$2728.20		$3765.02	$68.20	$94.13

Crew B-86

Crew No.	Bare Costs Hr.	Bare Costs Daily	Incl. Subs O&P Hr.	Incl. Subs O&P Daily	Cost Per Labor-Hour Bare Costs	Cost Per Labor-Hour Incl. O&P
1 Equip. Oper. (medium)	$59.00	$472.00	$87.90	$703.20	$59.00	$87.90
1 Stump Chipper, S.P.		189.20		208.12	23.65	26.02
8 L.H., Daily Totals		$661.20		$911.32	$82.65	$113.92

Crew B-86A

Crew No.	Bare Costs Hr.	Bare Costs Daily	Incl. Subs O&P Hr.	Incl. Subs O&P Daily	Cost Per Labor-Hour Bare Costs	Cost Per Labor-Hour Incl. O&P
1 Equip. Oper. (medium)	$59.00	$472.00	$87.90	$703.20	$59.00	$87.90
1 Grader, 30,000 Lbs.		1073.00		1180.30	134.13	147.54
8 L.H., Daily Totals		$1545.00		$1883.50	$193.13	$235.44

Crew B-86B

Crew No.	Bare Costs Hr.	Bare Costs Daily	Incl. Subs O&P Hr.	Incl. Subs O&P Daily	Cost Per Labor-Hour Bare Costs	Cost Per Labor-Hour Incl. O&P
1 Equip. Oper. (medium)	$59.00	$472.00	$87.90	$703.20	$59.00	$87.90
1 Dozer, 200 H.P.		1520.00		1672.00	190.00	209.00
8 L.H., Daily Totals		$1992.00		$2375.20	$249.00	$296.90

Crew B-87

Crew No.	Bare Costs Hr.	Bare Costs Daily	Incl. Subs O&P Hr.	Incl. Subs O&P Daily	Cost Per Labor-Hour Bare Costs	Cost Per Labor-Hour Incl. O&P
1 Laborer	$44.40	$355.20	$66.25	$530.00	$56.08	$83.57
4 Equip. Oper. (medium)	59.00	1888.00	87.90	2812.80		
2 Feller Bunchers, 100 H.P.		1257.60		1383.36		
1 Log Chipper, 22" Tree		555.00		610.50		
1 Dozer, 105 H.P.		640.80		704.88		
1 Chain Saw, Gas, 36" Long		41.65		45.81	62.38	68.61
40 L.H., Daily Totals		$4738.25		$6087.35	$118.46	$152.18

Crew B-88

Crew No.	Bare Costs Hr.	Bare Costs Daily	Incl. Subs O&P Hr.	Incl. Subs O&P Daily	Cost Per Labor-Hour Bare Costs	Cost Per Labor-Hour Incl. O&P
1 Laborer	$44.40	$355.20	$66.25	$530.00	$56.91	$84.81
6 Equip. Oper. (medium)	59.00	2832.00	87.90	4219.20		
2 Feller Bunchers, 100 H.P.		1257.60		1383.36		
1 Log Chipper, 22" Tree		555.00		610.50		
2 Log Skidders, 50 H.P.		1826.70		2009.37		
1 Dozer, 105 H.P.		640.80		704.88		
1 Chain Saw, Gas, 36" Long		41.65		45.81	77.17	84.89
56 L.H., Daily Totals		$7508.95		$9503.13	$134.09	$169.70

Crew B-89

Crew No.	Bare Costs Hr.	Bare Costs Daily	Incl. Subs O&P Hr.	Incl. Subs O&P Daily	Cost Per Labor-Hour Bare Costs	Cost Per Labor-Hour Incl. O&P
1 Equip. Oper. (light)	$55.50	$444.00	$82.70	$661.60	$52.15	$77.85
1 Truck Driver (light)	48.80	390.40	73.00	584.00		
1 Flatbed Truck, Gas, 3 Ton		850.05		935.05		
1 Concrete Saw		112.85		124.14		
1 Water Tank, 65 Gal.		102.90		113.19	66.61	73.27
16 L.H., Daily Totals		$1900.20		$2417.98	$118.76	$151.12

Crew B-89A

Crew No.	Bare Costs Hr.	Bare Costs Daily	Incl. Subs O&P Hr.	Incl. Subs O&P Daily	Cost Per Labor-Hour Bare Costs	Cost Per Labor-Hour Incl. O&P
1 Skilled Worker	$57.10	$456.80	$85.90	$687.20	$50.75	$76.08
1 Laborer	44.40	355.20	66.25	530.00		
1 Core Drill (Large)		121.60		133.76	7.60	8.36
16 L.H., Daily Totals		$933.60		$1350.96	$58.35	$84.44

Crew B-89B

Crew No.	Bare Costs Hr.	Bare Costs Daily	Incl. Subs O&P Hr.	Incl. Subs O&P Daily	Cost Per Labor-Hour Bare Costs	Cost Per Labor-Hour Incl. O&P
1 Equip. Oper. (light)	$55.50	$444.00	$82.70	$661.60	$52.15	$77.85
1 Truck Driver (light)	48.80	390.40	73.00	584.00		
1 Wall Saw, Hydraulic, 10 H.P.		86.40		95.04		
1 Generator, Diesel, 100 kW		521.85		574.03		
1 Water Tank, 65 Gal.		102.90		113.19		
1 Flatbed Truck, Gas, 3 Ton		850.05		935.05	97.58	107.33
16 L.H., Daily Totals		$2395.60		$2962.92	$149.72	$185.18

Crew B-89C

Crew No.	Bare Costs Hr.	Bare Costs Daily	Incl. Subs O&P Hr.	Incl. Subs O&P Daily	Cost Per Labor-Hour Bare Costs	Cost Per Labor-Hour Incl. O&P
1 Cement Finisher	$51.80	$414.40	$75.90	$607.20	$51.80	$75.90
1 Masonry cut-off saw, gas		58.15		63.97	7.27	8.00
8 L.H., Daily Totals		$472.55		$671.16	$59.07	$83.90

Crew B-90

Crew No.	Bare Costs Hr.	Bare Costs Daily	Incl. Subs O&P Hr.	Incl. Subs O&P Daily	Cost Per Labor-Hour Bare Costs	Cost Per Labor-Hour Incl. O&P
1 Labor Foreman (outside)	$46.40	$371.20	$69.25	$554.00	$49.15	$73.35
3 Laborers	44.40	1065.60	66.25	1590.00		
2 Equip. Oper. (light)	55.50	888.00	82.70	1323.20		
2 Truck Drivers (heavy)	51.30	820.80	76.70	1227.20		
1 Road Mixer, 310 H.P.		1919.00		2110.90		
1 Dist. Truck, 2000 Gal.		303.25		333.57	34.72	38.19
64 L.H., Daily Totals		$5367.85		$7138.88	$83.87	$111.54

Crew B-90A

Crew No.	Bare Costs Hr.	Bare Costs Daily	Incl. Subs O&P Hr.	Incl. Subs O&P Daily	Cost Per Labor-Hour Bare Costs	Cost Per Labor-Hour Incl. O&P
1 Labor Foreman (outside)	$46.40	$371.20	$69.25	$554.00	$53.03	$79.05
2 Laborers	44.40	710.40	66.25	1060.00		
4 Equip. Oper. (medium)	59.00	1888.00	87.90	2812.80		
2 Graders, 30,000 Lbs.		2146.00		2360.60		
1 Tandem Roller, 10 Ton		246.80		271.48		
1 Roller, Pneum. Whl., 12 Ton		349.90		384.89	48.98	53.87
56 L.H., Daily Totals		$5712.30		$7443.77	$102.01	$132.92

Crew B-90B

Crew No.	Bare Costs Hr.	Bare Costs Daily	Incl. Subs O&P Hr.	Incl. Subs O&P Daily	Cost Per Labor-Hour Bare Costs	Cost Per Labor-Hour Incl. O&P
1 Labor Foreman (outside)	$46.40	$371.20	$69.25	$554.00	$52.03	$77.58
2 Laborers	44.40	710.40	66.25	1060.00		
3 Equip. Oper. (medium)	59.00	1416.00	87.90	2109.60		
1 Roller, Pneum. Whl., 12 Ton		349.90		384.89		
1 Road Mixer, 310 H.P.		1919.00		2110.90	47.27	52.00
48 L.H., Daily Totals		$4766.50		$6219.39	$99.30	$129.57

Crew B-90C

Crew No.	Bare Costs Hr.	Bare Costs Daily	Incl. Subs O&P Hr.	Incl. Subs O&P Daily	Cost Per Labor-Hour Bare Costs	Cost Per Labor-Hour Incl. O&P
1 Labor Foreman (outside)	$46.40	$371.20	$69.25	$554.00	$50.45	$75.28
4 Laborers	44.40	1420.80	66.25	2120.00		
3 Equip. Oper. (medium)	59.00	1416.00	87.90	2109.60		
3 Truck Drivers (heavy)	51.30	1231.20	76.70	1840.80		
3 Road Mixers, 310 H.P.		5757.00		6332.70	65.42	71.96
88 L.H., Daily Totals		$10196.20		$12957.10	$115.87	$147.24

Crew B-90D

Crew No.	Bare Costs Hr.	Bare Costs Daily	Incl. Subs O&P Hr.	Incl. Subs O&P Daily	Cost Per Labor-Hour Bare Costs	Cost Per Labor-Hour Incl. O&P
1 Labor Foreman (outside)	$46.40	$371.20	$69.25	$554.00	$49.52	$73.89
6 Laborers	44.40	2131.20	66.25	3180.00		
3 Equip. Oper. (medium)	59.00	1416.00	87.90	2109.60		
3 Truck Drivers (heavy)	51.30	1231.20	76.70	1840.80		
3 Road Mixers, 310 H.P.		5757.00		6332.70	55.36	60.89
104 L.H., Daily Totals		$10906.60		$14017.10	$104.87	$134.78

Crew B-90E

Crew No.	Bare Costs Hr.	Bare Costs Daily	Incl. Subs O&P Hr.	Incl. Subs O&P Daily	Cost Per Labor-Hour Bare Costs	Cost Per Labor-Hour Incl. O&P
1 Labor Foreman (outside)	$46.40	$371.20	$69.25	$554.00	$50.26	$74.96
4 Laborers	44.40	1420.80	66.25	2120.00		
3 Equip. Oper. (medium)	59.00	1416.00	87.90	2109.60		
1 Truck Driver (heavy)	51.30	410.40	76.70	613.60		
1 Road Mixer, 310 H.P.		1919.00		2110.90	26.65	29.32
72 L.H., Daily Totals		$5537.40		$7508.10	$76.91	$104.28

Crew B-91

Crew No.	Bare Costs Hr.	Bare Costs Daily	Incl. Subs O&P Hr.	Incl. Subs O&P Daily	Cost Per Labor-Hour Bare Costs	Cost Per Labor-Hour Incl. O&P
1 Labor Foreman (outside)	$46.40	$371.20	$69.25	$554.00	$52.81	$78.76
2 Laborers	44.40	710.40	66.25	1060.00		
4 Equip. Oper. (medium)	59.00	1888.00	87.90	2812.80		
1 Truck Driver (heavy)	51.30	410.40	76.70	613.60		
1 Dist. Tanker, 3000 Gallon		334.10		367.51		
1 Truck Tractor, 6x4, 380 H.P.		499.15		549.07		
1 Aggreg. Spreader, S.P.		859.10		945.01		
1 Roller, Pneum. Whl., 12 Ton		349.90		384.89		
1 Tandem Roller, 10 Ton		246.80		271.48	35.77	39.34
64 L.H., Daily Totals		$5669.05		$7558.35	$88.58	$118.10

Crews - Standard

Crew No.	Bare Costs		Incl. Subs O&P		Cost Per Labor-Hour	
Crew B-91B	Hr.	Daily	Hr.	Daily	Bare Costs	Incl. O&P
1 Laborer	$44.40	$355.20	$66.25	$530.00	$51.70	$77.08
1 Equipment Oper. (med.)	59.00	472.00	87.90	703.20		
1 Road Sweeper, Vac. Assist.		879.45		967.39	54.97	60.46
16 L.H., Daily Totals		$1706.65		$2200.59	$106.67	$137.54
Crew B-91C	Hr.	Daily	Hr.	Daily	Bare Costs	Incl. O&P
1 Laborer	$44.40	$355.20	$66.25	$530.00	$46.60	$69.63
1 Truck Driver (light)	48.80	390.40	73.00	584.00		
1 Catch Basin Cleaning Truck		542.60		596.86	33.91	37.30
16 L.H., Daily Totals		$1288.20		$1710.86	$80.51	$106.93
Crew B-91D	Hr.	Daily	Hr.	Daily	Bare Costs	Incl. O&P
1 Labor Foreman (outside)	$46.40	$371.20	$69.25	$554.00	$51.23	$76.42
5 Laborers	44.40	1776.00	66.25	2650.00		
5 Equip. Oper. (medium)	59.00	2360.00	87.90	3516.00		
2 Truck Drivers (heavy)	51.30	820.80	76.70	1227.20		
1 Aggreg. Spreader, S.P.		859.10		945.01		
2 Truck Tractors, 6x4, 380 H.P.		998.30		1098.13		
2 Dist. Tankers, 3000 Gallon		668.20		735.02		
2 Pavement Brushes, Towed		176.70		194.37		
2 Rollers Pneum. Whl., 12 Ton		699.80		769.78	32.71	35.98
104 L.H., Daily Totals		$8730.10		$11689.51	$83.94	$112.40
Crew B-92	Hr.	Daily	Hr.	Daily	Bare Costs	Incl. O&P
1 Labor Foreman (outside)	$46.40	$371.20	$69.25	$554.00	$44.90	$67.00
3 Laborers	44.40	1065.60	66.25	1590.00		
1 Crack Cleaner, 25 H.P.		53.00		58.30		
1 Air Compressor, 60 cfm		153.85		169.24		
1 Tar Kettle, T.M.		156.70		172.37		
1 Flatbed Truck, Gas, 3 Ton		850.05		935.05	37.92	41.72
32 L.H., Daily Totals		$2650.40		$3478.96	$82.83	$108.72
Crew B-93	Hr.	Daily	Hr.	Daily	Bare Costs	Incl. O&P
1 Equip. Oper. (medium)	$59.00	$472.00	$87.90	$703.20	$59.00	$87.90
1 Feller Buncher, 100 H.P.		628.80		691.68	78.60	86.46
8 L.H., Daily Totals		$1100.80		$1394.88	$137.60	$174.36
Crew B-94A	Hr.	Daily	Hr.	Daily	Bare Costs	Incl. O&P
1 Laborer	$44.40	$355.20	$66.25	$530.00	$44.40	$66.25
1 Diaphragm Water Pump, 2"		87.70		96.47		
1 -20' Suction Hose, 2"		3.55		3.90		
2 -50' Discharge Hoses, 2"		8.00		8.80	12.41	13.65
8 L.H., Daily Totals		$454.45		$639.17	$56.81	$79.90
Crew B-94B	Hr.	Daily	Hr.	Daily	Bare Costs	Incl. O&P
1 Laborer	$44.40	$355.20	$66.25	$530.00	$44.40	$66.25
1 Diaphragm Water Pump, 4"		106.35		116.99		
1 -20' Suction Hose, 4"		17.25		18.98		
2 -50' Discharge Hoses, 4"		25.60		28.16	18.65	20.52
8 L.H., Daily Totals		$504.40		$694.12	$63.05	$86.77
Crew B-94C	Hr.	Daily	Hr.	Daily	Bare Costs	Incl. O&P
1 Laborer	$44.40	$355.20	$66.25	$530.00	$44.40	$66.25
1 Centrifugal Water Pump, 3"		74.40		81.84		
1 -20' Suction Hose, 3"		8.75		9.63		
2 -50' Discharge Hoses, 3"		9.00		9.90	11.52	12.67
8 L.H., Daily Totals		$447.35		$631.37	$55.92	$78.92

Crew No.	Bare Costs		Incl. Subs O&P		Cost Per Labor-Hour	
Crew B-94D	Hr.	Daily	Hr.	Daily	Bare Costs	Incl. O&P
1 Laborer	$44.40	$355.20	$66.25	$530.00	$44.40	$66.25
1 Centr. Water Pump, 6"		235.25		258.77		
1 -20' Suction Hose, 6"		25.50		28.05		
2 -50' Discharge Hoses, 6"		36.20		39.82	37.12	40.83
8 L.H., Daily Totals		$652.15		$856.64	$81.52	$107.08
Crew C-1	Hr.	Daily	Hr.	Daily	Bare Costs	Incl. O&P
3 Carpenters	$54.70	$1312.80	$81.65	$1959.60	$52.13	$77.80
1 Laborer	44.40	355.20	66.25	530.00		
32 L.H., Daily Totals		$1668.00		$2489.60	$52.13	$77.80
Crew C-2	Hr.	Daily	Hr.	Daily	Bare Costs	Incl. O&P
1 Carpenter Foreman (outside)	$56.70	$453.60	$84.60	$676.80	$53.32	$79.58
4 Carpenters	54.70	1750.40	81.65	2612.80		
1 Laborer	44.40	355.20	66.25	530.00		
48 L.H., Daily Totals		$2559.20		$3819.60	$53.32	$79.58
Crew C-2A	Hr.	Daily	Hr.	Daily	Bare Costs	Incl. O&P
1 Carpenter Foreman (outside)	$56.70	$453.60	$84.60	$676.80	$52.83	$78.62
3 Carpenters	54.70	1312.80	81.65	1959.60		
1 Cement Finisher	51.80	414.40	75.90	607.20		
1 Laborer	44.40	355.20	66.25	530.00		
48 L.H., Daily Totals		$2536.00		$3773.60	$52.83	$78.62
Crew C-3	Hr.	Daily	Hr.	Daily	Bare Costs	Incl. O&P
1 Rodman Foreman (outside)	$60.90	$487.20	$91.05	$728.40	$55.10	$82.31
4 Rodmen (reinf.)	58.90	1884.80	88.05	2817.60		
1 Equip. Oper. (light)	55.50	444.00	82.70	661.60		
2 Laborers	44.40	710.40	66.25	1060.00		
3 Stressing Equipment		56.85		62.53		
.5 Grouting Equipment		123.33		135.66	2.82	3.10
64 L.H., Daily Totals		$3706.57		$5465.79	$57.92	$85.40
Crew C-4	Hr.	Daily	Hr.	Daily	Bare Costs	Incl. O&P
1 Rodman Foreman (outside)	$60.90	$487.20	$91.05	$728.40	$59.40	$88.80
3 Rodmen (reinf.)	58.90	1413.60	88.05	2113.20		
3 Stressing Equipment		56.85		62.53	1.78	1.95
32 L.H., Daily Totals		$1957.65		$2904.14	$61.18	$90.75
Crew C-4A	Hr.	Daily	Hr.	Daily	Bare Costs	Incl. O&P
2 Rodmen (reinf.)	$58.90	$942.40	$88.05	$1408.80	$58.90	$88.05
4 Stressing Equipment		75.80		83.38	4.74	5.21
16 L.H., Daily Totals		$1018.20		$1492.18	$63.64	$93.26
Crew C-5	Hr.	Daily	Hr.	Daily	Bare Costs	Incl. O&P
1 Rodman Foreman (outside)	$60.90	$487.20	$91.05	$728.40	$58.64	$87.58
4 Rodmen (reinf.)	58.90	1884.80	88.05	2817.60		
1 Equip. Oper. (crane)	61.45	491.60	91.55	732.40		
1 Equip. Oper. (oiler)	52.50	420.00	78.25	626.00		
1 Hyd. Crane, 25 Ton		586.70		645.37	10.48	11.52
56 L.H., Daily Totals		$3870.30		$5549.77	$69.11	$99.10
Crew C-6	Hr.	Daily	Hr.	Daily	Bare Costs	Incl. O&P
1 Labor Foreman (outside)	$46.40	$371.20	$69.25	$554.00	$45.97	$68.36
4 Laborers	44.40	1420.80	66.25	2120.00		
1 Cement Finisher	51.80	414.40	75.90	607.20		
2 Gas Engine Vibrators		54.30		59.73	1.13	1.24
48 L.H., Daily Totals		$2260.70		$3340.93	$47.10	$69.60

For customer support on your Concrete & Masonry Costs with RSMeans Data, call 800.448.8182.

439

Crew No.	Bare Costs		Incl. Subs O&P		Cost Per Labor-Hour	
Crew C-6A	**Hr.**	**Daily**	**Hr.**	**Daily**	**Bare Costs**	**Incl. O&P**
2 Cement Finishers	$51.80	$828.80	$75.90	$1214.40	$51.80	$75.90
1 Concrete Vibrator, Elec, 2 HP		45.80		50.38	2.86	3.15
16 L.H., Daily Totals		$874.60		$1264.78	$54.66	$79.05

Crew No.					Bare Costs	Incl. O&P
Crew C-7	**Hr.**	**Daily**	**Hr.**	**Daily**	**Bare Costs**	**Incl. O&P**
1 Labor Foreman (outside)	$46.40	$371.20	$69.25	$554.00	$47.97	$71.39
5 Laborers	44.40	1776.00	66.25	2650.00		
1 Cement Finisher	51.80	414.40	75.90	607.20		
1 Equip. Oper. (medium)	59.00	472.00	87.90	703.20		
1 Equip. Oper. (oiler)	52.50	420.00	78.25	626.00		
2 Gas Engine Vibrators		54.30		59.73		
1 Concrete Bucket, 1 C.Y.		45.90		50.49		
1 Hyd. Crane, 55 Ton		990.15		1089.17	15.14	16.66
72 L.H., Daily Totals		$4543.95		$6339.78	$63.11	$88.05

Crew No.					Bare Costs	Incl. O&P
Crew C-7A	**Hr.**	**Daily**	**Hr.**	**Daily**	**Bare Costs**	**Incl. O&P**
1 Labor Foreman (outside)	$46.40	$371.20	$69.25	$554.00	$46.38	$69.24
5 Laborers	44.40	1776.00	66.25	2650.00		
2 Truck Drivers (heavy)	51.30	820.80	76.70	1227.20		
2 Conc. Transit Mixers		1176.30		1293.93	18.38	20.22
64 L.H., Daily Totals		$4144.30		$5725.13	$64.75	$89.46

Crew No.					Bare Costs	Incl. O&P
Crew C-7B	**Hr.**	**Daily**	**Hr.**	**Daily**	**Bare Costs**	**Incl. O&P**
1 Labor Foreman (outside)	$46.40	$371.20	$69.25	$554.00	$47.79	$71.29
5 Laborers	44.40	1776.00	66.25	2650.00		
1 Equipment Operator, Crane	61.45	491.60	91.55	732.40		
1 Equipment Oiler	52.50	420.00	78.25	626.00		
1 Conc. Bucket, 2 C.Y.		55.65		61.22		
1 Lattice Boom Crane, 165 Ton		2403.00		2643.30	38.42	42.26
64 L.H., Daily Totals		$5517.45		$7266.92	$86.21	$113.55

Crew No.					Bare Costs	Incl. O&P
Crew C-7C	**Hr.**	**Daily**	**Hr.**	**Daily**	**Bare Costs**	**Incl. O&P**
1 Labor Foreman (outside)	$46.40	$371.20	$69.25	$554.00	$48.30	$72.04
5 Laborers	44.40	1776.00	66.25	2650.00		
2 Equipment Operators (med.)	59.00	944.00	87.90	1406.40		
2 F.E. Loaders, W.M., 4 C.Y.		1518.00		1669.80	23.72	26.09
64 L.H., Daily Totals		$4609.20		$6280.20	$72.02	$98.13

Crew No.					Bare Costs	Incl. O&P
Crew C-7D	**Hr.**	**Daily**	**Hr.**	**Daily**	**Bare Costs**	**Incl. O&P**
1 Labor Foreman (outside)	$46.40	$371.20	$69.25	$554.00	$46.77	$69.77
5 Laborers	44.40	1776.00	66.25	2650.00		
1 Equip. Oper. (medium)	59.00	472.00	87.90	703.20		
1 Concrete Conveyer		206.25		226.88	3.68	4.05
56 L.H., Daily Totals		$2825.45		$4134.07	$50.45	$73.82

Crew No.					Bare Costs	Incl. O&P
Crew C-8	**Hr.**	**Daily**	**Hr.**	**Daily**	**Bare Costs**	**Incl. O&P**
1 Labor Foreman (outside)	$46.40	$371.20	$69.25	$554.00	$48.89	$72.53
3 Laborers	44.40	1065.60	66.25	1590.00		
2 Cement Finishers	51.80	828.80	75.90	1214.40		
1 Equip. Oper. (medium)	59.00	472.00	87.90	703.20		
1 Concrete Pump (Small)		423.65		466.01	7.57	8.32
56 L.H., Daily Totals		$3161.25		$4527.61	$56.45	$80.85

Crew No.					Bare Costs	Incl. O&P
Crew C-8A	**Hr.**	**Daily**	**Hr.**	**Daily**	**Bare Costs**	**Incl. O&P**
1 Labor Foreman (outside)	$46.40	$371.20	$69.25	$554.00	$47.20	$69.97
3 Laborers	44.40	1065.60	66.25	1590.00		
2 Cement Finishers	51.80	828.80	75.90	1214.40		
48 L.H., Daily Totals		$2265.60		$3358.40	$47.20	$69.97

Crew No.	Bare Costs		Incl. Subs O&P		Cost Per Labor-Hour	
Crew C-8B	**Hr.**	**Daily**	**Hr.**	**Daily**	**Bare Costs**	**Incl. O&P**
1 Labor Foreman (outside)	$46.40	$371.20	$69.25	$554.00	$47.72	$71.18
3 Laborers	44.40	1065.60	66.25	1590.00		
1 Equip. Oper. (medium)	59.00	472.00	87.90	703.20		
1 Vibrating Power Screed		87.65		96.42		
1 Roller, Vibratory, 25 Ton		672.35		739.59		
1 Dozer, 200 H.P.		1520.00		1672.00	57.00	62.70
40 L.H., Daily Totals		$4188.80		$5355.20	$104.72	$133.88

Crew No.					Bare Costs	Incl. O&P
Crew C-8C	**Hr.**	**Daily**	**Hr.**	**Daily**	**Bare Costs**	**Incl. O&P**
1 Labor Foreman (outside)	$46.40	$371.20	$69.25	$554.00	$48.40	$71.97
3 Laborers	44.40	1065.60	66.25	1590.00		
1 Cement Finisher	51.80	414.40	75.90	607.20		
1 Equip. Oper. (medium)	59.00	472.00	87.90	703.20		
1 Shotcrete Rig, 12 C.Y./hr.		269.20		296.12		
1 Air Compressor, 160 cfm		212.30		233.53		
4 -50' Air Hoses, 1"		32.20		35.42		
4 -50' Air Hoses, 2"		115.80		127.38	13.11	14.43
48 L.H., Daily Totals		$2952.70		$4146.85	$61.51	$86.39

Crew No.					Bare Costs	Incl. O&P
Crew C-8D	**Hr.**	**Daily**	**Hr.**	**Daily**	**Bare Costs**	**Incl. O&P**
1 Labor Foreman (outside)	$46.40	$371.20	$69.25	$554.00	$49.52	$73.53
1 Laborer	44.40	355.20	66.25	530.00		
1 Cement Finisher	51.80	414.40	75.90	607.20		
1 Equipment Oper. (light)	55.50	444.00	82.70	661.60		
1 Air Compressor, 250 cfm		202.85		223.13		
2 -50' Air Hoses, 1"		16.10		17.71	6.84	7.53
32 L.H., Daily Totals		$1803.75		$2593.65	$56.37	$81.05

Crew No.					Bare Costs	Incl. O&P
Crew C-8E	**Hr.**	**Daily**	**Hr.**	**Daily**	**Bare Costs**	**Incl. O&P**
1 Labor Foreman (outside)	$46.40	$371.20	$69.25	$554.00	$47.82	$71.10
3 Laborers	44.40	1065.60	66.25	1590.00		
1 Cement Finisher	51.80	414.40	75.90	607.20		
1 Equipment Oper. (light)	55.50	444.00	82.70	661.60		
1 Shotcrete Rig, 35 C.Y./hr.		301.05		331.15		
1 Air Compressor, 250 cfm		202.85		223.13		
4 -50' Air Hoses, 1"		32.20		35.42		
4 -50' Air Hoses, 2"		115.80		127.38	13.58	14.94
48 L.H., Daily Totals		$2947.10		$4129.89	$61.40	$86.04

Crew No.					Bare Costs	Incl. O&P
Crew C-9	**Hr.**	**Daily**	**Hr.**	**Daily**	**Bare Costs**	**Incl. O&P**
1 Cement Finisher	$51.80	$414.40	$75.90	$607.20	$49.02	$72.78
2 Laborers	44.40	710.40	66.25	1060.00		
1 Equipment Oper. (light)	55.50	444.00	82.70	661.60		
1 Grout Pump, 50 C.F./hr.		190.35		209.38		
1 Air Compressor, 160 cfm		212.30		233.53		
2 -50' Air Hoses, 1"		16.10		17.71		
2 -50' Air Hoses, 2"		57.90		63.69	14.90	16.38
32 L.H., Daily Totals		$2045.45		$2853.11	$63.92	$89.16

Crew No.					Bare Costs	Incl. O&P
Crew C-10	**Hr.**	**Daily**	**Hr.**	**Daily**	**Bare Costs**	**Incl. O&P**
1 Laborer	$44.40	$355.20	$66.25	$530.00	$49.33	$72.68
2 Cement Finishers	51.80	828.80	75.90	1214.40		
24 L.H., Daily Totals		$1184.00		$1744.40	$49.33	$72.68

Crew No.					Bare Costs	Incl. O&P
Crew C-10B	**Hr.**	**Daily**	**Hr.**	**Daily**	**Bare Costs**	**Incl. O&P**
3 Laborers	$44.40	$1065.60	$66.25	$1590.00	$47.36	$70.11
2 Cement Finishers	51.80	828.80	75.90	1214.40		
1 Concrete Mixer, 10 C.F.		147.15		161.87		
2 Trowels, 48" Walk-Behind		188.60		207.46	8.39	9.23
40 L.H., Daily Totals		$2230.15		$3173.72	$55.75	$79.34

Crews - Standard

Crew No.	Hr.	Daily	Hr.	Daily	Bare Costs	Incl. O&P
Crew C-10C						
1 Laborer	$44.40	$355.20	$66.25	$530.00	$49.33	$72.68
2 Cement Finishers	51.80	828.80	75.90	1214.40		
1 Trowel, 48" Walk-Behind		94.30		103.73	3.93	4.32
24 L.H., Daily Totals		$1278.30		$1848.13	$53.26	$77.01
Crew C-10D	Hr.	Daily	Hr.	Daily	Bare Costs	Incl. O&P
1 Laborer	$44.40	$355.20	$66.25	$530.00	$49.33	$72.68
2 Cement Finishers	51.80	828.80	75.90	1214.40		
1 Vibrating Power Screed		87.65		96.42		
1 Trowel, 48" Walk-Behind		94.30		103.73	7.58	8.34
24 L.H., Daily Totals		$1365.95		$1944.55	$56.91	$81.02
Crew C-10E	Hr.	Daily	Hr.	Daily	Bare Costs	Incl. O&P
1 Laborer	$44.40	$355.20	$66.25	$530.00	$49.33	$72.68
2 Cement Finishers	51.80	828.80	75.90	1214.40		
1 Vibrating Power Screed		87.65		96.42		
1 Cement Trowel, 96" Ride-On		171.05		188.16	10.78	11.86
24 L.H., Daily Totals		$1442.70		$2028.97	$60.11	$84.54
Crew C-10F	Hr.	Daily	Hr.	Daily	Bare Costs	Incl. O&P
1 Laborer	$44.40	$355.20	$66.25	$530.00	$49.33	$72.68
2 Cement Finishers	51.80	828.80	75.90	1214.40		
1 Telescoping Boom Lift, to 60'		292.45		321.69	12.19	13.40
24 L.H., Daily Totals		$1476.45		$2066.09	$61.52	$86.09
Crew C-11	Hr.	Daily	Hr.	Daily	Bare Costs	Incl. O&P
1 Struc. Steel Foreman (outside)	$62.30	$498.40	$96.40	$771.20	$59.78	$91.78
6 Struc. Steel Workers	60.30	2894.40	93.30	4478.40		
1 Equip. Oper. (crane)	61.45	491.60	91.55	732.40		
1 Equip. Oper. (oiler)	52.50	420.00	78.25	626.00		
1 Lattice Boom Crane, 150 Ton		2324.00		2556.40	32.28	35.51
72 L.H., Daily Totals		$6628.40		$9164.40	$92.06	$127.28
Crew C-12	Hr.	Daily	Hr.	Daily	Bare Costs	Incl. O&P
1 Carpenter Foreman (outside)	$56.70	$453.60	$84.60	$676.80	$54.44	$81.22
3 Carpenters	54.70	1312.80	81.65	1959.60		
1 Laborer	44.40	355.20	66.25	530.00		
1 Equip. Oper. (crane)	61.45	491.60	91.55	732.40		
1 Hyd. Crane, 12 Ton		475.80		523.38	9.91	10.90
48 L.H., Daily Totals		$3089.00		$4422.18	$64.35	$92.13
Crew C-13	Hr.	Daily	Hr.	Daily	Bare Costs	Incl. O&P
1 Struc. Steel Worker	$60.30	$482.40	$93.30	$746.40	$58.43	$89.42
1 Welder	60.30	482.40	93.30	746.40		
1 Carpenter	54.70	437.60	81.65	653.20		
1 Welder, Gas Engine, 300 amp		148.75		163.63	6.20	6.82
24 L.H., Daily Totals		$1551.15		$2309.63	$64.63	$96.23
Crew C-14	Hr.	Daily	Hr.	Daily	Bare Costs	Incl. O&P
1 Carpenter Foreman (outside)	$56.70	$453.60	$84.60	$676.80	$53.39	$79.54
5 Carpenters	54.70	2188.00	81.65	3266.00		
4 Laborers	44.40	1420.80	66.25	2120.00		
4 Rodmen (reinf.)	58.90	1884.80	88.05	2817.60		
2 Cement Finishers	51.80	828.80	75.90	1214.40		
1 Equip. Oper. (crane)	61.45	491.60	91.55	732.40		
1 Equip. Oper. (oiler)	52.50	420.00	78.25	626.00		
1 Hyd. Crane, 80 Ton		1458.00		1603.80	10.13	11.14
144 L.H., Daily Totals		$9145.60		$13057.00	$63.51	$90.67

Crew No.	Hr.	Daily	Hr.	Daily	Bare Costs	Incl. O&P
Crew C-14A						
1 Carpenter Foreman (outside)	$56.70	$453.60	$84.60	$676.80	$54.68	$81.58
16 Carpenters	54.70	7001.60	81.65	10451.20		
4 Rodmen (reinf.)	58.90	1884.80	88.05	2817.60		
2 Laborers	44.40	710.40	66.25	1060.00		
1 Cement Finisher	51.80	414.40	75.90	607.20		
1 Equip. Oper. (medium)	59.00	472.00	87.90	703.20		
1 Gas Engine Vibrator		27.15		29.86		
1 Concrete Pump (Small)		423.65		466.01	2.25	2.48
200 L.H., Daily Totals		$11387.60		$16811.88	$56.94	$84.06
Crew C-14B	Hr.	Daily	Hr.	Daily	Bare Costs	Incl. O&P
1 Carpenter Foreman (outside)	$56.70	$453.60	$84.60	$676.80	$54.57	$81.36
16 Carpenters	54.70	7001.60	81.65	10451.20		
4 Rodmen (reinf.)	58.90	1884.80	88.05	2817.60		
2 Laborers	44.40	710.40	66.25	1060.00		
2 Cement Finishers	51.80	828.80	75.90	1214.40		
1 Equip. Oper. (medium)	59.00	472.00	87.90	703.20		
1 Gas Engine Vibrator		27.15		29.86		
1 Concrete Pump (Small)		423.65		466.01	2.17	2.38
208 L.H., Daily Totals		$11802.00		$17419.08	$56.74	$83.75
Crew C-14C	Hr.	Daily	Hr.	Daily	Bare Costs	Incl. O&P
1 Carpenter Foreman (outside)	$56.70	$453.60	$84.60	$676.80	$52.29	$77.96
6 Carpenters	54.70	2625.60	81.65	3919.20		
2 Rodmen (reinf.)	58.90	942.40	88.05	1408.80		
4 Laborers	44.40	1420.80	66.25	2120.00		
1 Cement Finisher	51.80	414.40	75.90	607.20		
1 Gas Engine Vibrator		27.15		29.86	.24	.27
112 L.H., Daily Totals		$5883.95		$8761.86	$52.54	$78.23
Crew C-14D	Hr.	Daily	Hr.	Daily	Bare Costs	Incl. O&P
1 Carpenter Foreman (outside)	$56.70	$453.60	$84.60	$676.80	$54.35	$81.07
18 Carpenters	54.70	7876.80	81.65	11757.60		
2 Rodmen (reinf.)	58.90	942.40	88.05	1408.80		
2 Laborers	44.40	710.40	66.25	1060.00		
1 Cement Finisher	51.80	414.40	75.90	607.20		
1 Equip. Oper. (medium)	59.00	472.00	87.90	703.20		
1 Gas Engine Vibrator		27.15		29.86		
1 Concrete Pump (Small)		423.65		466.01	2.25	2.48
200 L.H., Daily Totals		$11320.40		$16709.48	$56.60	$83.55
Crew C-14E	Hr.	Daily	Hr.	Daily	Bare Costs	Incl. O&P
1 Carpenter Foreman (outside)	$56.70	$453.60	$84.60	$676.80	$53.34	$79.52
2 Carpenters	54.70	875.20	81.65	1306.40		
4 Rodmen (reinf.)	58.90	1884.80	88.05	2817.60		
3 Laborers	44.40	1065.60	66.25	1590.00		
1 Cement Finisher	51.80	414.40	75.90	607.20		
1 Gas Engine Vibrator		27.15		29.86	.31	.34
88 L.H., Daily Totals		$4720.75		$7027.86	$53.64	$79.86
Crew C-14F	Hr.	Daily	Hr.	Daily	Bare Costs	Incl. O&P
1 Labor Foreman (outside)	$46.40	$371.20	$69.25	$554.00	$49.56	$73.02
2 Laborers	44.40	710.40	66.25	1060.00		
6 Cement Finishers	51.80	2486.40	75.90	3643.20		
1 Gas Engine Vibrator		27.15		29.86	.38	.41
72 L.H., Daily Totals		$3595.15		$5287.06	$49.93	$73.43

Crew C-14G

Crew No.	Hr.	Daily	Hr.	Daily	Bare Costs	Incl. O&P
1 Labor Foreman (outside)	$46.40	$371.20	$69.25	$554.00	$48.91	$72.19
2 Laborers	44.40	710.40	66.25	1060.00		
4 Cement Finishers	51.80	1657.60	75.90	2428.80		
1 Gas Engine Vibrator		27.15		29.86	.48	.53
56 L.H., Daily Totals		$2766.35		$4072.67	$49.40	$72.73

Crew C-14H

Crew No.	Hr.	Daily	Hr.	Daily	Bare Costs	Incl. O&P
1 Carpenter Foreman (outside)	$56.70	$453.60	$84.60	$676.80	$53.53	$79.68
2 Carpenters	54.70	875.20	81.65	1306.40		
1 Rodman (reinf.)	58.90	471.20	88.05	704.40		
1 Laborer	44.40	355.20	66.25	530.00		
1 Cement Finisher	51.80	414.40	75.90	607.20		
1 Gas Engine Vibrator		27.15		29.86	.57	.62
48 L.H., Daily Totals		$2596.75		$3854.67	$54.10	$80.31

Crew C-14L

Crew No.	Hr.	Daily	Hr.	Daily	Bare Costs	Incl. O&P
1 Carpenter Foreman (outside)	$56.70	$453.60	$84.60	$676.80	$51.19	$76.28
6 Carpenters	54.70	2625.60	81.65	3919.20		
4 Laborers	44.40	1420.80	66.25	2120.00		
1 Cement Finisher	51.80	414.40	75.90	607.20		
1 Gas Engine Vibrator		27.15		29.86	.28	.31
96 L.H., Daily Totals		$4941.55		$7353.06	$51.47	$76.59

Crew C-14M

Crew No.	Hr.	Daily	Hr.	Daily	Bare Costs	Incl. O&P
1 Carpenter Foreman (outside)	$56.70	$453.60	$84.60	$676.80	$53.08	$79.03
2 Carpenters	54.70	875.20	81.65	1306.40		
1 Rodman (reinf.)	58.90	471.20	88.05	704.40		
2 Laborers	44.40	710.40	66.25	1060.00		
1 Cement Finisher	51.80	414.40	75.90	607.20		
1 Equip. Oper. (medium)	59.00	472.00	87.90	703.20		
1 Gas Engine Vibrator		27.15		29.86		
1 Concrete Pump (Small)		423.65		466.01	7.04	7.75
64 L.H., Daily Totals		$3847.60		$5553.88	$60.12	$86.78

Crew C-15

Crew No.	Hr.	Daily	Hr.	Daily	Bare Costs	Incl. O&P
1 Carpenter Foreman (outside)	$56.70	$453.60	$84.60	$676.80	$51.31	$76.28
2 Carpenters	54.70	875.20	81.65	1306.40		
3 Laborers	44.40	1065.60	66.25	1590.00		
2 Cement Finishers	51.80	828.80	75.90	1214.40		
1 Rodman (reinf.)	58.90	471.20	88.05	704.40		
72 L.H., Daily Totals		$3694.40		$5492.00	$51.31	$76.28

Crew C-16

Crew No.	Hr.	Daily	Hr.	Daily	Bare Costs	Incl. O&P
1 Labor Foreman (outside)	$46.40	$371.20	$69.25	$554.00	$48.89	$72.53
3 Laborers	44.40	1065.60	66.25	1590.00		
2 Cement Finishers	51.80	828.80	75.90	1214.40		
1 Equip. Oper. (medium)	59.00	472.00	87.90	703.20		
1 Gunite Pump Rig		321.75		353.93		
2 -50' Air Hoses, 3/4"		14.30		15.73		
2 -50' Air Hoses, 2"		57.90		63.69	7.03	7.74
56 L.H., Daily Totals		$3131.55		$4494.94	$55.92	$80.27

Crew C-16A

Crew No.	Hr.	Daily	Hr.	Daily	Bare Costs	Incl. O&P
1 Laborer	$44.40	$355.20	$66.25	$530.00	$51.75	$76.49
2 Cement Finishers	51.80	828.80	75.90	1214.40		
1 Equip. Oper. (medium)	59.00	472.00	87.90	703.20		
1 Gunite Pump Rig		321.75		353.93		
2 -50' Air Hoses, 3/4"		14.30		15.73		
2 -50' Air Hoses, 2"		57.90		63.69		
1 Telescoping Boom Lift, to 60'		292.45		321.69	21.45	23.59
32 L.H., Daily Totals		$2342.40		$3202.64	$73.20	$100.08

Crew C-17

Crew No.	Hr.	Daily	Hr.	Daily	Bare Costs	Incl. O&P
2 Skilled Worker Foremen (out)	$59.10	$945.60	$88.90	$1422.40	$57.50	$86.50
8 Skilled Workers	57.10	3654.40	85.90	5497.60		
80 L.H., Daily Totals		$4600.00		$6920.00	$57.50	$86.50

Crew C-17A

Crew No.	Hr.	Daily	Hr.	Daily	Bare Costs	Incl. O&P
2 Skilled Worker Foremen (out)	$59.10	$945.60	$88.90	$1422.40	$57.55	$86.56
8 Skilled Workers	57.10	3654.40	85.90	5497.60		
.125 Equip. Oper. (crane)	61.45	61.45	91.55	91.55		
.125 Hyd. Crane, 80 Ton		182.25		200.47	2.25	2.48
81 L.H., Daily Totals		$4843.70		$7212.02	$59.80	$89.04

Crew C-17B

Crew No.	Hr.	Daily	Hr.	Daily	Bare Costs	Incl. O&P
2 Skilled Worker Foremen (out)	$59.10	$945.60	$88.90	$1422.40	$57.60	$86.62
8 Skilled Workers	57.10	3654.40	85.90	5497.60		
.25 Equip. Oper. (crane)	61.45	122.90	91.55	183.10		
.25 Hyd. Crane, 80 Ton		364.50		400.95		
.25 Trowel, 48" Walk-Behind		23.57		25.93	4.73	5.21
82 L.H., Daily Totals		$5110.98		$7529.98	$62.33	$91.83

Crew C-17C

Crew No.	Hr.	Daily	Hr.	Daily	Bare Costs	Incl. O&P
2 Skilled Worker Foremen (out)	$59.10	$945.60	$88.90	$1422.40	$57.64	$86.68
8 Skilled Workers	57.10	3654.40	85.90	5497.60		
.375 Equip. Oper. (crane)	61.45	184.35	91.55	274.65		
.375 Hyd. Crane, 80 Ton		546.75		601.42	6.59	7.25
83 L.H., Daily Totals		$5331.10		$7796.07	$64.23	$93.93

Crew C-17D

Crew No.	Hr.	Daily	Hr.	Daily	Bare Costs	Incl. O&P
2 Skilled Worker Foremen (out)	$59.10	$945.60	$88.90	$1422.40	$57.69	$86.74
8 Skilled Workers	57.10	3654.40	85.90	5497.60		
.5 Equip. Oper. (crane)	61.45	245.80	91.55	366.20		
.5 Hyd. Crane, 80 Ton		729.00		801.90	8.68	9.55
84 L.H., Daily Totals		$5574.80		$8088.10	$66.37	$96.29

Crew C-17E

Crew No.	Hr.	Daily	Hr.	Daily	Bare Costs	Incl. O&P
2 Skilled Worker Foremen (out)	$59.10	$945.60	$88.90	$1422.40	$57.50	$86.50
8 Skilled Workers	57.10	3654.40	85.90	5497.60		
1 Hyd. Jack with Rods		36.70		40.37	.46	.50
80 L.H., Daily Totals		$4636.70		$6960.37	$57.96	$87.00

Crew C-18

Crew No.	Hr.	Daily	Hr.	Daily	Bare Costs	Incl. O&P
.125 Labor Foreman (outside)	$46.40	$46.40	$69.25	$69.25	$44.62	$66.58
1 Laborer	44.40	355.20	66.25	530.00		
1 Concrete Cart, 10 C.F.		116.95		128.65	12.99	14.29
9 L.H., Daily Totals		$518.55		$727.89	$57.62	$80.88

Crew C-19

Crew No.	Hr.	Daily	Hr.	Daily	Bare Costs	Incl. O&P
.125 Labor Foreman (outside)	$46.40	$46.40	$69.25	$69.25	$44.62	$66.58
1 Laborer	44.40	355.20	66.25	530.00		
1 Concrete Cart, 18 C.F.		138.95		152.85	15.44	16.98
9 L.H., Daily Totals		$540.55		$752.10	$60.06	$83.57

Crew C-20

Crew No.	Hr.	Daily	Hr.	Daily	Bare Costs	Incl. O&P
1 Labor Foreman (outside)	$46.40	$371.20	$69.25	$554.00	$47.40	$70.54
5 Laborers	44.40	1776.00	66.25	2650.00		
1 Cement Finisher	51.80	414.40	75.90	607.20		
1 Equip. Oper. (medium)	59.00	472.00	87.90	703.20		
2 Gas Engine Vibrators		54.30		59.73		
1 Concrete Pump (Small)		423.65		466.01	7.47	8.21
64 L.H., Daily Totals		$3511.55		$5040.15	$54.87	$78.75

Crews - Standard

Crew C-21	Hr.	Daily	Hr.	Daily	Bare Costs	Incl. O&P
1 Labor Foreman (outside)	$46.40	$371.20	$69.25	$554.00	$47.40	$70.54
5 Laborers	44.40	1776.00	66.25	2650.00		
1 Cement Finisher	51.80	414.40	75.90	607.20		
1 Equip. Oper. (medium)	59.00	472.00	87.90	703.20		
2 Gas Engine Vibrators		54.30		59.73		
1 Concrete Conveyer		206.25		226.88	4.07	4.48
64 L.H., Daily Totals		$3294.15		$4801.01	$51.47	$75.02

Crew C-22	Hr.	Daily	Hr.	Daily	Bare Costs	Incl. O&P
1 Rodman Foreman (outside)	$60.90	$487.20	$91.05	$728.40	$59.19	$88.47
4 Rodmen (reinf.)	58.90	1884.80	88.05	2817.60		
.125 Equip. Oper. (crane)	61.45	61.45	91.55	91.55		
.125 Equip. Oper. (oiler)	52.50	52.50	78.25	78.25		
.125 Hyd. Crane, 25 Ton		73.34		80.67	1.75	1.92
42 L.H., Daily Totals		$2559.29		$3796.47	$60.94	$90.39

Crew C-23	Hr.	Daily	Hr.	Daily	Bare Costs	Incl. O&P
2 Skilled Worker Foremen (out)	$59.10	$945.60	$88.90	$1422.40	$57.48	$86.30
6 Skilled Workers	57.10	2740.80	85.90	4123.20		
1 Equip. Oper. (crane)	61.45	491.60	91.55	732.40		
1 Equip. Oper. (oiler)	52.50	420.00	78.25	626.00		
1 Lattice Boom Crane, 90 Ton		1713.00		1884.30	21.41	23.55
80 L.H., Daily Totals		$6311.00		$8788.30	$78.89	$109.85

Crew C-23A	Hr.	Daily	Hr.	Daily	Bare Costs	Incl. O&P
1 Labor Foreman (outside)	$46.40	$371.20	$69.25	$554.00	$49.83	$74.31
2 Laborers	44.40	710.40	66.25	1060.00		
1 Equip. Oper. (crane)	61.45	491.60	91.55	732.40		
1 Equip. Oper. (oiler)	52.50	420.00	78.25	626.00		
1 Crawler Crane, 100 Ton		2310.00		2541.00		
3 Conc. Buckets, 8 C.Y.		337.95		371.75	66.20	72.82
40 L.H., Daily Totals		$4641.15		$5885.15	$116.03	$147.13

Crew C-24	Hr.	Daily	Hr.	Daily	Bare Costs	Incl. O&P
2 Skilled Worker Foremen (out)	$59.10	$945.60	$88.90	$1422.40	$57.48	$86.30
6 Skilled Workers	57.10	2740.80	85.90	4123.20		
1 Equip. Oper. (crane)	61.45	491.60	91.55	732.40		
1 Equip. Oper. (oiler)	52.50	420.00	78.25	626.00		
1 Lattice Boom Crane, 150 Ton		2324.00		2556.40	29.05	31.95
80 L.H., Daily Totals		$6922.00		$9460.40	$86.53	$118.26

Crew C-25	Hr.	Daily	Hr.	Daily	Bare Costs	Incl. O&P
2 Rodmen (reinf.)	$58.90	$942.40	$88.05	$1408.80	$47.42	$73.15
2 Rodmen Helpers	35.95	575.20	58.25	932.00		
32 L.H., Daily Totals		$1517.60		$2340.80	$47.42	$73.15

Crew C-27	Hr.	Daily	Hr.	Daily	Bare Costs	Incl. O&P
2 Cement Finishers	$51.80	$828.80	$75.90	$1214.40	$51.80	$75.90
1 Concrete Saw		112.85		124.14	7.05	7.76
16 L.H., Daily Totals		$941.65		$1338.54	$58.85	$83.66

Crew C-28	Hr.	Daily	Hr.	Daily	Bare Costs	Incl. O&P
1 Cement Finisher	$51.80	$414.40	$75.90	$607.20	$51.80	$75.90
1 Portable Air Compressor, Gas		38.80		42.68	4.85	5.34
8 L.H., Daily Totals		$453.20		$649.88	$56.65	$81.23

Crew C-29	Hr.	Daily	Hr.	Daily	Bare Costs	Incl. O&P
1 Laborer	$44.40	$355.20	$66.25	$530.00	$44.40	$66.25
1 Pressure Washer		97.35		107.08	12.17	13.39
8 L.H., Daily Totals		$452.55		$637.09	$56.57	$79.64

Crew C-30	Hr.	Daily	Hr.	Daily	Bare Costs	Incl. O&P
1 Laborer	$44.40	$355.20	$66.25	$530.00	$44.40	$66.25
1 Concrete Mixer, 10 C.F.		147.15		161.87	18.39	20.23
8 L.H., Daily Totals		$502.35		$691.87	$62.79	$86.48

Crew C-31	Hr.	Daily	Hr.	Daily	Bare Costs	Incl. O&P
1 Cement Finisher	$51.80	$414.40	$75.90	$607.20	$51.80	$75.90
1 Grout Pump		321.75		353.93	40.22	44.24
8 L.H., Daily Totals		$736.15		$961.13	$92.02	$120.14

Crew C-32	Hr.	Daily	Hr.	Daily	Bare Costs	Incl. O&P
1 Cement Finisher	$51.80	$414.40	$75.90	$607.20	$48.10	$71.08
1 Laborer	44.40	355.20	66.25	530.00		
1 Crack Chaser Saw, Gas, 6 H.P.		73.25		80.58		
1 Vacuum Pick-Up System		74.95		82.44	9.26	10.19
16 L.H., Daily Totals		$917.80		$1300.22	$57.36	$81.26

Crew D-1	Hr.	Daily	Hr.	Daily	Bare Costs	Incl. O&P
1 Bricklayer	$53.70	$429.60	$80.90	$647.20	$48.70	$73.38
1 Bricklayer Helper	43.70	349.60	65.85	526.80		
16 L.H., Daily Totals		$779.20		$1174.00	$48.70	$73.38

Crew D-2	Hr.	Daily	Hr.	Daily	Bare Costs	Incl. O&P
3 Bricklayers	$53.70	$1288.80	$80.90	$1941.60	$50.15	$75.50
2 Bricklayer Helpers	43.70	699.20	65.85	1053.60		
.5 Carpenter	54.70	218.80	81.65	326.60		
44 L.H., Daily Totals		$2206.80		$3321.80	$50.15	$75.50

Crew D-3	Hr.	Daily	Hr.	Daily	Bare Costs	Incl. O&P
3 Bricklayers	$53.70	$1288.80	$80.90	$1941.60	$49.94	$75.20
2 Bricklayer Helpers	43.70	699.20	65.85	1053.60		
.25 Carpenter	54.70	109.40	81.65	163.30		
42 L.H., Daily Totals		$2097.40		$3158.50	$49.94	$75.20

Crew D-4	Hr.	Daily	Hr.	Daily	Bare Costs	Incl. O&P
1 Bricklayer	$53.70	$429.60	$80.90	$647.20	$49.15	$73.83
2 Bricklayer Helpers	43.70	699.20	65.85	1053.60		
1 Equip. Oper. (light)	55.50	444.00	82.70	661.60		
1 Grout Pump, 50 C.F./hr.		190.35		209.38	5.95	6.54
32 L.H., Daily Totals		$1763.15		$2571.78	$55.10	$80.37

Crew D-5	Hr.	Daily	Hr.	Daily	Bare Costs	Incl. O&P
1 Bricklayer	53.70	429.60	80.90	647.20	53.70	80.90
8 L.H., Daily Totals		$429.60		$647.20	$53.70	$80.90

Crew D-6	Hr.	Daily	Hr.	Daily	Bare Costs	Incl. O&P
3 Bricklayers	$53.70	$1288.80	$80.90	$1941.60	$48.94	$73.71
3 Bricklayer Helpers	43.70	1048.80	65.85	1580.40		
.25 Carpenter	54.70	109.40	81.65	163.30		
50 L.H., Daily Totals		$2447.00		$3685.30	$48.94	$73.71

Crew D-7	Hr.	Daily	Hr.	Daily	Bare Costs	Incl. O&P
1 Tile Layer	$51.70	$413.60	$75.55	$604.40	$46.65	$68.17
1 Tile Layer Helper	41.60	332.80	60.80	486.40		
16 L.H., Daily Totals		$746.40		$1090.80	$46.65	$68.17

Crew D-8	Hr.	Daily	Hr.	Daily	Bare Costs	Incl. O&P
3 Bricklayers	$53.70	$1288.80	$80.90	$1941.60	$49.70	$74.88
2 Bricklayer Helpers	43.70	699.20	65.85	1053.60		
40 L.H., Daily Totals		$1988.00		$2995.20	$49.70	$74.88

For customer support on your Concrete & Masonry Costs with RSMeans Data, call 800.448.8182.

443

Crew D-9

Crew No.	Bare Costs Hr.	Bare Costs Daily	Incl. Subs O&P Hr.	Incl. Subs O&P Daily	Cost Per Labor-Hour Bare Costs	Cost Per Labor-Hour Incl. O&P
3 Bricklayers	$53.70	$1288.80	$80.90	$1941.60	$48.70	$73.38
3 Bricklayer Helpers	43.70	1048.80	65.85	1580.40		
48 L.H., Daily Totals		$2337.60		$3522.00	$48.70	$73.38

Crew D-10

Crew No.	Bare Costs Hr.	Bare Costs Daily	Incl. Subs O&P Hr.	Incl. Subs O&P Daily	Cost Per Labor-Hour Bare Costs	Cost Per Labor-Hour Incl. O&P
1 Bricklayer Foreman (outside)	$55.70	$445.60	$83.90	$671.20	$53.64	$80.55
1 Bricklayer	53.70	429.60	80.90	647.20		
1 Bricklayer Helper	43.70	349.60	65.85	526.80		
1 Equip. Oper. (crane)	61.45	491.60	91.55	732.40		
1 S.P. Crane, 4x4, 12 Ton		432.65		475.92	13.52	14.87
32 L.H., Daily Totals		$2149.05		$3053.51	$67.16	$95.42

Crew D-11

Crew No.	Bare Costs Hr.	Bare Costs Daily	Incl. Subs O&P Hr.	Incl. Subs O&P Daily	Cost Per Labor-Hour Bare Costs	Cost Per Labor-Hour Incl. O&P
1 Bricklayer Foreman (outside)	$55.70	$445.60	$83.90	$671.20	$51.03	$76.88
1 Bricklayer	53.70	429.60	80.90	647.20		
1 Bricklayer Helper	43.70	349.60	65.85	526.80		
24 L.H., Daily Totals		$1224.80		$1845.20	$51.03	$76.88

Crew D-12

Crew No.	Bare Costs Hr.	Bare Costs Daily	Incl. Subs O&P Hr.	Incl. Subs O&P Daily	Cost Per Labor-Hour Bare Costs	Cost Per Labor-Hour Incl. O&P
1 Bricklayer Foreman (outside)	$55.70	$445.60	$83.90	$671.20	$49.20	$74.13
1 Bricklayer	53.70	429.60	80.90	647.20		
2 Bricklayer Helpers	43.70	699.20	65.85	1053.60		
32 L.H., Daily Totals		$1574.40		$2372.00	$49.20	$74.13

Crew D-13

Crew No.	Bare Costs Hr.	Bare Costs Daily	Incl. Subs O&P Hr.	Incl. Subs O&P Daily	Cost Per Labor-Hour Bare Costs	Cost Per Labor-Hour Incl. O&P
1 Bricklayer Foreman (outside)	$55.70	$445.60	$83.90	$671.20	$52.16	$78.28
1 Bricklayer	53.70	429.60	80.90	647.20		
2 Bricklayer Helpers	43.70	699.20	65.85	1053.60		
1 Carpenter	54.70	437.60	81.65	653.20		
1 Equip. Oper. (crane)	61.45	491.60	91.55	732.40		
1 S.P. Crane, 4x4, 12 Ton		432.65		475.92	9.01	9.91
48 L.H., Daily Totals		$2936.25		$4233.52	$61.17	$88.20

Crew D-14

Crew No.	Bare Costs Hr.	Bare Costs Daily	Incl. Subs O&P Hr.	Incl. Subs O&P Daily	Cost Per Labor-Hour Bare Costs	Cost Per Labor-Hour Incl. O&P
3 Bricklayers	$53.70	$1288.80	$80.90	$1941.60	$51.20	$77.14
1 Bricklayer Helper	43.70	349.60	65.85	526.80		
32 L.H., Daily Totals		$1638.40		$2468.40	$51.20	$77.14

Crew E-1

Crew No.	Bare Costs Hr.	Bare Costs Daily	Incl. Subs O&P Hr.	Incl. Subs O&P Daily	Cost Per Labor-Hour Bare Costs	Cost Per Labor-Hour Incl. O&P
1 Welder Foreman (outside)	$62.30	$498.40	$96.40	$771.20	$59.37	$90.80
1 Welder	60.30	482.40	93.30	746.40		
1 Equip. Oper. (light)	55.50	444.00	82.70	661.60		
1 Welder, Gas Engine, 300 amp		148.75		163.63	6.20	6.82
24 L.H., Daily Totals		$1573.55		$2342.82	$65.56	$97.62

Crew E-2

Crew No.	Bare Costs Hr.	Bare Costs Daily	Incl. Subs O&P Hr.	Incl. Subs O&P Daily	Cost Per Labor-Hour Bare Costs	Cost Per Labor-Hour Incl. O&P
1 Struc. Steel Foreman (outside)	$62.30	$498.40	$96.40	$771.20	$59.64	$91.34
4 Struc. Steel Workers	60.30	1929.60	93.30	2985.60		
1 Equip. Oper. (crane)	61.45	491.60	91.55	732.40		
1 Equip. Oper. (oiler)	52.50	420.00	78.25	626.00		
1 Lattice Boom Crane, 90 Ton		1713.00		1884.30	30.59	33.65
56 L.H., Daily Totals		$5052.60		$6999.50	$90.22	$124.99

Crew E-3

Crew No.	Bare Costs Hr.	Bare Costs Daily	Incl. Subs O&P Hr.	Incl. Subs O&P Daily	Cost Per Labor-Hour Bare Costs	Cost Per Labor-Hour Incl. O&P
1 Struc. Steel Foreman (outside)	$62.30	$498.40	$96.40	$771.20	$60.97	$94.33
1 Struc. Steel Worker	60.30	482.40	93.30	746.40		
1 Welder	60.30	482.40	93.30	746.40		
1 Welder, Gas Engine, 300 amp		148.75		163.63	6.20	6.82
24 L.H., Daily Totals		$1611.95		$2427.63	$67.16	$101.15

Crew E-3A

Crew No.	Bare Costs Hr.	Bare Costs Daily	Incl. Subs O&P Hr.	Incl. Subs O&P Daily	Cost Per Labor-Hour Bare Costs	Cost Per Labor-Hour Incl. O&P
1 Struc. Steel Foreman (outside)	$62.30	$498.40	$96.40	$771.20	$60.97	$94.33
1 Struc. Steel Worker	60.30	482.40	93.30	746.40		
1 Welder	60.30	482.40	93.30	746.40		
1 Welder, Gas Engine, 300 amp		148.75		163.63		
1 Telescoping Boom Lift, to 40'		281.90		310.09	17.94	19.74
24 L.H., Daily Totals		$1893.85		$2737.72	$78.91	$114.07

Crew E-4

Crew No.	Bare Costs Hr.	Bare Costs Daily	Incl. Subs O&P Hr.	Incl. Subs O&P Daily	Cost Per Labor-Hour Bare Costs	Cost Per Labor-Hour Incl. O&P
1 Struc. Steel Foreman (outside)	$62.30	$498.40	$96.40	$771.20	$60.80	$94.08
3 Struc. Steel Workers	60.30	1447.20	93.30	2239.20		
1 Welder, Gas Engine, 300 amp		148.75		163.63	4.65	5.11
32 L.H., Daily Totals		$2094.35		$3174.03	$65.45	$99.19

Crew E-5

Crew No.	Bare Costs Hr.	Bare Costs Daily	Incl. Subs O&P Hr.	Incl. Subs O&P Daily	Cost Per Labor-Hour Bare Costs	Cost Per Labor-Hour Incl. O&P
2 Struc. Steel Foremen (outside)	$62.30	$996.80	$96.40	$1542.40	$60.03	$92.24
5 Struc. Steel Workers	60.30	2412.00	93.30	3732.00		
1 Equip. Oper. (crane)	61.45	491.60	91.55	732.40		
1 Welder	60.30	482.40	93.30	746.40		
1 Equip. Oper. (oiler)	52.50	420.00	78.25	626.00		
1 Lattice Boom Crane, 90 Ton		1713.00		1884.30		
1 Welder, Gas Engine, 300 amp		148.75		163.63	23.27	25.60
80 L.H., Daily Totals		$6664.55		$9427.13	$83.31	$117.84

Crew E-6

Crew No.	Bare Costs Hr.	Bare Costs Daily	Incl. Subs O&P Hr.	Incl. Subs O&P Daily	Cost Per Labor-Hour Bare Costs	Cost Per Labor-Hour Incl. O&P
3 Struc. Steel Foremen (outside)	$62.30	$1495.20	$96.40	$2313.60	$59.96	$92.17
9 Struc. Steel Workers	60.30	4341.60	93.30	6717.60		
1 Equip. Oper. (crane)	61.45	491.60	91.55	732.40		
1 Welder	60.30	482.40	93.30	746.40		
1 Equip. Oper. (oiler)	52.50	420.00	78.25	626.00		
1 Equip. Oper. (light)	55.50	444.00	82.70	661.60		
1 Lattice Boom Crane, 90 Ton		1713.00		1884.30		
1 Welder, Gas Engine, 300 amp		148.75		163.63		
1 Air Compressor, 160 cfm		212.30		233.53		
2 Impact Wrenches		90.30		99.33	16.91	18.60
128 L.H., Daily Totals		$9839.15		$14178.39	$76.87	$110.77

Crew E-7

Crew No.	Bare Costs Hr.	Bare Costs Daily	Incl. Subs O&P Hr.	Incl. Subs O&P Daily	Cost Per Labor-Hour Bare Costs	Cost Per Labor-Hour Incl. O&P
1 Struc. Steel Foreman (outside)	$62.30	$498.40	$96.40	$771.20	$60.03	$92.24
4 Struc. Steel Workers	60.30	1929.60	93.30	2985.60		
1 Equip. Oper. (crane)	61.45	491.60	91.55	732.40		
1 Equip. Oper. (oiler)	52.50	420.00	78.25	626.00		
1 Welder Foreman (outside)	62.30	498.40	96.40	771.20		
2 Welders	60.30	964.80	93.30	1492.80		
1 Lattice Boom Crane, 90 Ton		1713.00		1884.30		
2 Welder, Gas Engine, 300 amp		297.50		327.25	25.13	27.64
80 L.H., Daily Totals		$6813.30		$9590.75	$85.17	$119.88

Crew E-8

Crew No.	Bare Costs Hr.	Bare Costs Daily	Incl. Subs O&P Hr.	Incl. Subs O&P Daily	Cost Per Labor-Hour Bare Costs	Cost Per Labor-Hour Incl. O&P
1 Struc. Steel Foreman (outside)	$62.30	$498.40	$96.40	$771.20	$59.73	$91.67
4 Struc. Steel Workers	60.30	1929.60	93.30	2985.60		
1 Welder Foreman (outside)	62.30	498.40	96.40	771.20		
4 Welders	60.30	1929.60	93.30	2985.60		
1 Equip. Oper. (crane)	61.45	491.60	91.55	732.40		
1 Equip. Oper. (oiler)	52.50	420.00	78.25	626.00		
1 Equip. Oper. (light)	55.50	444.00	82.70	661.60		
1 Lattice Boom Crane, 90 Ton		1713.00		1884.30		
4 Welder, Gas Engine, 300 amp		595.00		654.50	22.19	24.41
104 L.H., Daily Totals		$8519.60		$12072.40	$81.92	$116.08

For customer support on your Concrete & Masonry Costs with RSMeans Data, call 800.448.8182.

Crew No.	Bare Costs		Incl. Subs O&P		Cost Per Labor-Hour	
Crew E-9	Hr.	Daily	Hr.	Daily	Bare Costs	Incl. O&P
2 Struc. Steel Foremen (outside)	$62.30	$996.80	$96.40	$1542.40	$59.96	$92.17
5 Struc. Steel Workers	60.30	2412.00	93.30	3732.00		
1 Welder Foreman (outside)	62.30	498.40	96.40	771.20		
5 Welders	60.30	2412.00	93.30	3732.00		
1 Equip. Oper. (crane)	61.45	491.60	91.55	732.40		
1 Equip. Oper. (oiler)	52.50	420.00	78.25	626.00		
1 Equip. Oper. (light)	55.50	444.00	82.70	661.60		
1 Lattice Boom Crane, 90 Ton		1713.00		1884.30		
5 Welder, Gas Engine, 300 amp		743.75		818.13	19.19	21.11
128 L.H., Daily Totals		$10131.55		$14500.03	$79.15	$113.28
Crew E-10	Hr.	Daily	Hr.	Daily	Bare Costs	Incl. O&P
1 Welder Foreman (outside)	$62.30	$498.40	$96.40	$771.20	$61.30	$94.85
1 Welder	60.30	482.40	93.30	746.40		
1 Welder, Gas Engine, 300 amp		148.75		163.63		
1 Flatbed Truck, Gas, 3 Ton		850.05		935.05	62.42	68.67
16 L.H., Daily Totals		$1979.60		$2616.28	$123.72	$163.52
Crew E-11	Hr.	Daily	Hr.	Daily	Bare Costs	Incl. O&P
2 Painters, Struc. Steel	$47.20	$755.20	$75.80	$1212.80	$48.58	$75.14
1 Building Laborer	44.40	355.20	66.25	530.00		
1 Equip. Oper. (light)	55.50	444.00	82.70	661.60		
1 Air Compressor, 250 cfm		202.85		223.13		
1 Sandblaster, Portable, 3 C.F.		83.85		92.23		
1 Set Sand Blasting Accessories		15.55		17.11	9.45	10.39
32 L.H., Daily Totals		$1856.65		$2736.88	$58.02	$85.53
Crew E-11A	Hr.	Daily	Hr.	Daily	Bare Costs	Incl. O&P
2 Painters, Struc. Steel	$47.20	$755.20	$75.80	$1212.80	$48.58	$75.14
1 Building Laborer	44.40	355.20	66.25	530.00		
1 Equip. Oper. (light)	55.50	444.00	82.70	661.60		
1 Air Compressor, 250 cfm		202.85		223.13		
1 Sandblaster, Portable, 3 C.F.		83.85		92.23		
1 Set Sand Blasting Accessories		15.55		17.11		
1 Telescoping Boom Lift, to 60'		292.45		321.69	18.58	20.44
32 L.H., Daily Totals		$2149.10		$3058.57	$67.16	$95.58
Crew E-11B	Hr.	Daily	Hr.	Daily	Bare Costs	Incl. O&P
2 Painters, Struc. Steel	$47.20	$755.20	$75.80	$1212.80	$46.27	$72.62
1 Building Laborer	44.40	355.20	66.25	530.00		
2 Paint Sprayer, 8 C.F.M.		88.40		97.24		
1 Telescoping Boom Lift, to 60'		292.45		321.69	15.87	17.46
24 L.H., Daily Totals		$1491.25		$2161.74	$62.14	$90.07
Crew E-12	Hr.	Daily	Hr.	Daily	Bare Costs	Incl. O&P
1 Welder Foreman (outside)	$62.30	$498.40	$96.40	$771.20	$58.90	$89.55
1 Equip. Oper. (light)	55.50	444.00	82.70	661.60		
1 Welder, Gas Engine, 300 amp		148.75		163.63	9.30	10.23
16 L.H., Daily Totals		$1091.15		$1596.43	$68.20	$99.78
Crew E-13	Hr.	Daily	Hr.	Daily	Bare Costs	Incl. O&P
1 Welder Foreman (outside)	$62.30	$498.40	$96.40	$771.20	$60.03	$91.83
.5 Equip. Oper. (light)	55.50	222.00	82.70	330.80		
1 Welder, Gas Engine, 300 amp		148.75		163.63	12.40	13.64
12 L.H., Daily Totals		$869.15		$1265.63	$72.43	$105.47
Crew E-14	Hr.	Daily	Hr.	Daily	Bare Costs	Incl. O&P
1 Welder Foreman (outside)	$62.30	$498.40	$96.40	$771.20	$62.30	$96.40
1 Welder, Gas Engine, 300 amp		148.75		163.63	18.59	20.45
8 L.H., Daily Totals		$647.15		$934.83	$80.89	$116.85

Crew No.	Bare Costs		Incl. Subs O&P		Cost Per Labor-Hour	
Crew E-16	Hr.	Daily	Hr.	Daily	Bare Costs	Incl. O&P
1 Welder Foreman (outside)	$62.30	$498.40	$96.40	$771.20	$61.30	$94.85
1 Welder	60.30	482.40	93.30	746.40		
1 Welder, Gas Engine, 300 amp		148.75		163.63	9.30	10.23
16 L.H., Daily Totals		$1129.55		$1681.22	$70.60	$105.08
Crew E-17	Hr.	Daily	Hr.	Daily	Bare Costs	Incl. O&P
1 Struc. Steel Foreman (outside)	$62.30	$498.40	$96.40	$771.20	$61.30	$94.85
1 Structural Steel Worker	60.30	482.40	93.30	746.40		
16 L.H., Daily Totals		$980.80		$1517.60	$61.30	$94.85
Crew E-18	Hr.	Daily	Hr.	Daily	Bare Costs	Incl. O&P
1 Struc. Steel Foreman (outside)	$62.30	$498.40	$96.40	$771.20	$60.44	$92.84
3 Structural Steel Workers	60.30	1447.20	93.30	2239.20		
1 Equipment Operator (med.)	59.00	472.00	87.90	703.20		
1 Lattice Boom Crane, 20 Ton		1526.00		1678.60	38.15	41.97
40 L.H., Daily Totals		$3943.60		$5392.20	$98.59	$134.81
Crew E-19	Hr.	Daily	Hr.	Daily	Bare Costs	Incl. O&P
1 Struc. Steel Foreman (outside)	$62.30	$498.40	$96.40	$771.20	$59.37	$90.80
1 Structural Steel Worker	60.30	482.40	93.30	746.40		
1 Equip. Oper. (light)	55.50	444.00	82.70	661.60		
1 Lattice Boom Crane, 20 Ton		1526.00		1678.60	63.58	69.94
24 L.H., Daily Totals		$2950.80		$3857.80	$122.95	$160.74
Crew E-20	Hr.	Daily	Hr.	Daily	Bare Costs	Incl. O&P
1 Struc. Steel Foreman (outside)	$62.30	$498.40	$96.40	$771.20	$59.72	$91.59
5 Structural Steel Workers	60.30	2412.00	93.30	3732.00		
1 Equip. Oper. (crane)	61.45	491.60	91.55	732.40		
1 Equip. Oper. (oiler)	52.50	420.00	78.25	626.00		
1 Lattice Boom Crane, 40 Ton		2127.00		2339.70	33.23	36.56
64 L.H., Daily Totals		$5949.00		$8201.30	$92.95	$128.15
Crew E-22	Hr.	Daily	Hr.	Daily	Bare Costs	Incl. O&P
1 Skilled Worker Foreman (out)	$59.10	$472.80	$88.90	$711.20	$57.77	$86.90
2 Skilled Workers	57.10	913.60	85.90	1374.40		
24 L.H., Daily Totals		$1386.40		$2085.60	$57.77	$86.90
Crew E-24	Hr.	Daily	Hr.	Daily	Bare Costs	Incl. O&P
3 Structural Steel Workers	$60.30	$1447.20	$93.30	$2239.20	$59.98	$91.95
1 Equipment Operator (med.)	59.00	472.00	87.90	703.20		
1 Hyd. Crane, 25 Ton		586.70		645.37	18.33	20.17
32 L.H., Daily Totals		$2505.90		$3587.77	$78.31	$112.12
Crew E-25	Hr.	Daily	Hr.	Daily	Bare Costs	Incl. O&P
1 Welder Foreman (outside)	$62.30	$498.40	$96.40	$771.20	$62.30	$96.40
1 Cutting Torch		12.95		14.24	1.62	1.78
8 L.H., Daily Totals		$511.35		$785.45	$63.92	$98.18
Crew E-26	Hr.	Daily	Hr.	Daily	Bare Costs	Incl. O&P
1 Struc. Steel Foreman (outside)	$62.30	$498.40	$96.40	$771.20	$61.64	$94.84
1 Struc. Steel Worker	60.30	482.40	93.30	746.40		
1 Welder	60.30	482.40	93.30	746.40		
.25 Electrician	63.70	127.40	94.65	189.30		
.25 Plumber	67.70	135.40	101.05	202.10		
1 Welder, Gas Engine, 300 amp		148.75		163.63	5.31	5.84
28 L.H., Daily Totals		$1874.75		$2819.03	$66.96	$100.68

Crew No.	Bare Costs		Incl. Subs O&P		Cost Per Labor-Hour	

Crew E-27	Hr.	Daily	Hr.	Daily	Bare Costs	Incl. O&P
1 Struc. Steel Foreman (outside)	$62.30	$498.40	$96.40	$771.20	$59.72	$91.59
5 Struc. Steel Workers	60.30	2412.00	93.30	3732.00		
1 Equip. Oper. (crane)	61.45	491.60	91.55	732.40		
1 Equip. Oper. (oiler)	52.50	420.00	78.25	626.00		
1 Hyd. Crane, 12 Ton		475.80		523.38		
1 Hyd. Crane, 80 Ton		1458.00		1603.80	30.22	33.24
64 L.H., Daily Totals		$5755.80		$7988.78	$89.93	$124.82

Crew F-3	Hr.	Daily	Hr.	Daily	Bare Costs	Incl. O&P
4 Carpenters	$54.70	$1750.40	$81.65	$2612.80	$56.05	$83.63
1 Equip. Oper. (crane)	61.45	491.60	91.55	732.40		
1 Hyd. Crane, 12 Ton		475.80		523.38	11.90	13.08
40 L.H., Daily Totals		$2717.80		$3868.58	$67.94	$96.71

Crew F-4	Hr.	Daily	Hr.	Daily	Bare Costs	Incl. O&P
4 Carpenters	$54.70	$1750.40	$81.65	$2612.80	$55.46	$82.73
1 Equip. Oper. (crane)	61.45	491.60	91.55	732.40		
1 Equip. Oper. (oiler)	52.50	420.00	78.25	626.00		
1 Hyd. Crane, 55 Ton		990.15		1089.17	20.63	22.69
48 L.H., Daily Totals		$3652.15		$5060.36	$76.09	$105.42

Crew F-5	Hr.	Daily	Hr.	Daily	Bare Costs	Incl. O&P
1 Carpenter Foreman (outside)	$56.70	$453.60	$84.60	$676.80	$55.20	$82.39
3 Carpenters	54.70	1312.80	81.65	1959.60		
32 L.H., Daily Totals		$1766.40		$2636.40	$55.20	$82.39

Crew F-6	Hr.	Daily	Hr.	Daily	Bare Costs	Incl. O&P
2 Carpenters	$54.70	$875.20	$81.65	$1306.40	$51.93	$77.47
2 Building Laborers	44.40	710.40	66.25	1060.00		
1 Equip. Oper. (crane)	61.45	491.60	91.55	732.40		
1 Hyd. Crane, 12 Ton		475.80		523.38	11.90	13.08
40 L.H., Daily Totals		$2553.00		$3622.18	$63.83	$90.55

Crew F-7	Hr.	Daily	Hr.	Daily	Bare Costs	Incl. O&P
2 Carpenters	$54.70	$875.20	$81.65	$1306.40	$49.55	$73.95
2 Building Laborers	44.40	710.40	66.25	1060.00		
32 L.H., Daily Totals		$1585.60		$2366.40	$49.55	$73.95

Crew G-1	Hr.	Daily	Hr.	Daily	Bare Costs	Incl. O&P
1 Roofer Foreman (outside)	$50.20	$401.60	$81.35	$650.80	$44.99	$72.92
4 Roofers Composition	48.20	1542.40	78.15	2500.80		
2 Roofer Helpers	35.95	575.20	58.25	932.00		
1 Application Equipment		194.80		214.28		
1 Tar Kettle/Pot		209.95		230.94		
1 Crew Truck		168.15		184.97	10.23	11.25
56 L.H., Daily Totals		$3092.10		$4713.79	$55.22	$84.17

Crew G-2	Hr.	Daily	Hr.	Daily	Bare Costs	Incl. O&P
1 Plasterer	$49.85	$398.80	$74.25	$594.00	$46.27	$68.95
1 Plasterer Helper	44.55	356.40	66.35	530.80		
1 Building Laborer	44.40	355.20	66.25	530.00		
1 Grout Pump, 50 C.F./hr.		190.35		209.38	7.93	8.72
24 L.H., Daily Totals		$1300.75		$1864.18	$54.20	$77.67

Crew G-2A	Hr.	Daily	Hr.	Daily	Bare Costs	Incl. O&P
1 Roofer Composition	$48.20	$385.60	$78.15	$625.20	$42.85	$67.55
1 Roofer Helper	35.95	287.60	58.25	466.00		
1 Building Laborer	44.40	355.20	66.25	530.00		
1 Foam Spray Rig, Trailer-Mtd.		530.15		583.16		
1 Pickup Truck, 3/4 Ton		112.20		123.42	26.76	29.44
24 L.H., Daily Totals		$1670.75		$2327.78	$69.61	$96.99

Crew G-3	Hr.	Daily	Hr.	Daily	Bare Costs	Incl. O&P
2 Sheet Metal Workers	$65.45	$1047.20	$98.70	$1579.20	$54.92	$82.47
2 Building Laborers	44.40	710.40	66.25	1060.00		
32 L.H., Daily Totals		$1757.60		$2639.20	$54.92	$82.47

Crew G-4	Hr.	Daily	Hr.	Daily	Bare Costs	Incl. O&P
1 Labor Foreman (outside)	$46.40	$371.20	$69.25	$554.00	$45.07	$67.25
2 Building Laborers	44.40	710.40	66.25	1060.00		
1 Flatbed Truck, Gas, 1.5 Ton		198.50		218.35		
1 Air Compressor, 160 cfm		212.30		233.53	17.12	18.83
24 L.H., Daily Totals		$1492.40		$2065.88	$62.18	$86.08

Crew G-5	Hr.	Daily	Hr.	Daily	Bare Costs	Incl. O&P
1 Roofer Foreman (outside)	$50.20	$401.60	$81.35	$650.80	$43.70	$70.83
2 Roofers Composition	48.20	771.20	78.15	1250.40		
2 Roofer Helpers	35.95	575.20	58.25	932.00		
1 Application Equipment		194.80		214.28	4.87	5.36
40 L.H., Daily Totals		$1942.80		$3047.48	$48.57	$76.19

Crew G-6A	Hr.	Daily	Hr.	Daily	Bare Costs	Incl. O&P
2 Roofers Composition	$48.20	$771.20	$78.15	$1250.40	$48.20	$78.15
1 Small Compressor, Electric		39.30		43.23		
2 Pneumatic Nailers		55.40		60.94	5.92	6.51
16 L.H., Daily Totals		$865.90		$1354.57	$54.12	$84.66

Crew G-7	Hr.	Daily	Hr.	Daily	Bare Costs	Incl. O&P
1 Carpenter	$54.70	$437.60	$81.65	$653.20	$54.70	$81.65
1 Small Compressor, Electric		39.30		43.23		
1 Pneumatic Nailer		27.70		30.47	8.38	9.21
8 L.H., Daily Totals		$504.60		$726.90	$63.08	$90.86

Crew H-1	Hr.	Daily	Hr.	Daily	Bare Costs	Incl. O&P
2 Glaziers	$52.65	$842.40	$78.40	$1254.40	$56.48	$85.85
2 Struc. Steel Workers	60.30	964.80	93.30	1492.80		
32 L.H., Daily Totals		$1807.20		$2747.20	$56.48	$85.85

Crew H-2	Hr.	Daily	Hr.	Daily	Bare Costs	Incl. O&P
2 Glaziers	$52.65	$842.40	$78.40	$1254.40	$49.90	$74.35
1 Building Laborer	44.40	355.20	66.25	530.00		
24 L.H., Daily Totals		$1197.60		$1784.40	$49.90	$74.35

Crew H-3	Hr.	Daily	Hr.	Daily	Bare Costs	Incl. O&P
1 Glazier	$52.65	$421.20	$78.40	$627.20	$47.35	$71.03
1 Helper	42.05	336.40	63.65	509.20		
16 L.H., Daily Totals		$757.60		$1136.40	$47.35	$71.03

Crew H-4	Hr.	Daily	Hr.	Daily	Bare Costs	Incl. O&P
1 Carpenter	$54.70	$437.60	$81.65	$653.20	$51.44	$77.05
1 Carpenter Helper	42.05	336.40	63.65	509.20		
.5 Electrician	63.70	254.80	94.65	378.60		
20 L.H., Daily Totals		$1028.80		$1541.00	$51.44	$77.05

Crew J-1

Crew J-1	Hr.	Daily	Hr.	Daily	Bare Costs	Incl. O&P
3 Plasterers	$49.85	$1196.40	$74.25	$1782.00	$47.73	$71.09
2 Plasterer Helpers	44.55	712.80	66.35	1061.60		
1 Mixing Machine, 6 C.F.		113.35		124.69	2.83	3.12
40 L.H., Daily Totals		$2022.55		$2968.28	$50.56	$74.21

Crew J-2

Crew J-2	Hr.	Daily	Hr.	Daily	Bare Costs	Incl. O&P
3 Plasterers	$49.85	$1196.40	$74.25	$1782.00	$48.79	$72.41
2 Plasterer Helpers	44.55	712.80	66.35	1061.60		
1 Lather	54.10	432.80	79.00	632.00		
1 Mixing Machine, 6 C.F.		113.35		124.69	2.36	2.60
48 L.H., Daily Totals		$2455.35		$3600.28	$51.15	$75.01

Crew J-3

Crew J-3	Hr.	Daily	Hr.	Daily	Bare Costs	Incl. O&P
1 Terrazzo Worker	$51.75	$414.00	$75.60	$604.80	$47.65	$69.63
1 Terrazzo Helper	43.55	348.40	63.65	509.20		
1 Floor Grinder, 22" Path		96.05		105.66		
1 Terrazzo Mixer		162.90		179.19	16.18	17.80
16 L.H., Daily Totals		$1021.35		$1398.85	$63.83	$87.43

Crew J-4

Crew J-4	Hr.	Daily	Hr.	Daily	Bare Costs	Incl. O&P
2 Cement Finishers	$51.80	$828.80	$75.90	$1214.40	$49.33	$72.68
1 Laborer	44.40	355.20	66.25	530.00		
1 Floor Grinder, 22" Path		96.05		105.66		
1 Floor Edger, 7" Path		44.05		48.45		
1 Vacuum Pick-Up System		74.95		82.44	8.96	9.86
24 L.H., Daily Totals		$1399.05		$1980.95	$58.29	$82.54

Crew J-4A

Crew J-4A	Hr.	Daily	Hr.	Daily	Bare Costs	Incl. O&P
2 Cement Finishers	$51.80	$828.80	$75.90	$1214.40	$48.10	$71.08
2 Laborers	44.40	710.40	66.25	1060.00		
1 Floor Grinder, 22" Path		96.05		105.66		
1 Floor Edger, 7" Path		44.05		48.45		
1 Vacuum Pick-Up System		74.95		82.44		
1 Floor Auto Scrubber		179.55		197.51	12.33	13.56
32 L.H., Daily Totals		$1933.80		$2708.46	$60.43	$84.64

Crew J-4B

Crew J-4B	Hr.	Daily	Hr.	Daily	Bare Costs	Incl. O&P
1 Laborer	$44.40	$355.20	$66.25	$530.00	$44.40	$66.25
1 Floor Auto Scrubber		179.55		197.51	22.44	24.69
8 L.H., Daily Totals		$534.75		$727.51	$66.84	$90.94

Crew J-6

Crew J-6	Hr.	Daily	Hr.	Daily	Bare Costs	Incl. O&P
2 Painters	$46.45	$743.20	$68.90	$1102.40	$48.20	$71.69
1 Building Laborer	44.40	355.20	66.25	530.00		
1 Equip. Oper. (light)	55.50	444.00	82.70	661.60		
1 Air Compressor, 250 cfm		202.85		223.13		
1 Sandblaster, Portable, 3 C.F.		83.85		92.23		
1 Set Sand Blasting Accessories		15.55		17.11	9.45	10.39
32 L.H., Daily Totals		$1844.65		$2626.47	$57.65	$82.08

Crew J-7

Crew J-7	Hr.	Daily	Hr.	Daily	Bare Costs	Incl. O&P
2 Painters	$46.45	$743.20	$68.90	$1102.40	$46.45	$68.90
1 Floor Belt Sander		50.20		55.22		
1 Floor Sanding Edger		25.20		27.72	4.71	5.18
16 L.H., Daily Totals		$818.60		$1185.34	$51.16	$74.08

Crew K-1

Crew K-1	Hr.	Daily	Hr.	Daily	Bare Costs	Incl. O&P
1 Carpenter	$54.70	$437.60	$81.65	$653.20	$51.75	$77.33
1 Truck Driver (light)	48.80	390.40	73.00	584.00		
1 Flatbed Truck, Gas, 3 Ton		850.05		935.05	53.13	58.44
16 L.H., Daily Totals		$1678.05		$2172.26	$104.88	$135.77

Crew K-2

Crew K-2	Hr.	Daily	Hr.	Daily	Bare Costs	Incl. O&P
1 Struc. Steel Foreman (outside)	$62.30	$498.40	$96.40	$771.20	$57.13	$87.57
1 Struc. Steel Worker	60.30	482.40	93.30	746.40		
1 Truck Driver (light)	48.80	390.40	73.00	584.00		
1 Flatbed Truck, Gas, 3 Ton		850.05		935.05	35.42	38.96
24 L.H., Daily Totals		$2221.25		$3036.66	$92.55	$126.53

Crew L-1

Crew L-1	Hr.	Daily	Hr.	Daily	Bare Costs	Incl. O&P
1 Electrician	$63.70	$509.60	$94.65	$757.20	$65.70	$97.85
1 Plumber	67.70	541.60	101.05	808.40		
16 L.H., Daily Totals		$1051.20		$1565.60	$65.70	$97.85

Crew L-2

Crew L-2	Hr.	Daily	Hr.	Daily	Bare Costs	Incl. O&P
1 Carpenter	$54.70	$437.60	$81.65	$653.20	$48.38	$72.65
1 Carpenter Helper	42.05	336.40	63.65	509.20		
16 L.H., Daily Totals		$774.00		$1162.40	$48.38	$72.65

Crew L-3

Crew L-3	Hr.	Daily	Hr.	Daily	Bare Costs	Incl. O&P
1 Carpenter	$54.70	$437.60	$81.65	$653.20	$59.64	$89.16
.5 Electrician	63.70	254.80	94.65	378.60		
.5 Sheet Metal Worker	65.45	261.80	98.70	394.80		
16 L.H., Daily Totals		$954.20		$1426.60	$59.64	$89.16

Crew L-3A

Crew L-3A	Hr.	Daily	Hr.	Daily	Bare Costs	Incl. O&P
1 Carpenter Foreman (outside)	$56.70	$453.60	$84.60	$676.80	$59.62	$89.30
.5 Sheet Metal Worker	65.45	261.80	98.70	394.80		
12 L.H., Daily Totals		$715.40		$1071.60	$59.62	$89.30

Crew L-4

Crew L-4	Hr.	Daily	Hr.	Daily	Bare Costs	Incl. O&P
2 Skilled Workers	$57.10	$913.60	$85.90	$1374.40	$52.08	$78.48
1 Helper	42.05	336.40	63.65	509.20		
24 L.H., Daily Totals		$1250.00		$1883.60	$52.08	$78.48

Crew L-5

Crew L-5	Hr.	Daily	Hr.	Daily	Bare Costs	Incl. O&P
1 Struc. Steel Foreman (outside)	$62.30	$498.40	$96.40	$771.20	$60.75	$93.49
5 Struc. Steel Workers	60.30	2412.00	93.30	3732.00		
1 Equip. Oper. (crane)	61.45	491.60	91.55	732.40		
1 Hyd. Crane, 25 Ton		586.70		645.37	10.48	11.52
56 L.H., Daily Totals		$3988.70		$5880.97	$71.23	$105.02

Crew L-5A

Crew L-5A	Hr.	Daily	Hr.	Daily	Bare Costs	Incl. O&P
1 Struc. Steel Foreman (outside)	$62.30	$498.40	$96.40	$771.20	$61.09	$93.64
2 Structural Steel Workers	60.30	964.80	93.30	1492.80		
1 Equip. Oper. (crane)	61.45	491.60	91.55	732.40		
1 S.P. Crane, 4x4, 25 Ton		1155.00		1270.50	36.09	39.70
32 L.H., Daily Totals		$3109.80		$4266.90	$97.18	$133.34

For customer support on your Concrete & Masonry Costs with RSMeans Data, call 800.448.8182.

447

Crew L-5B	Hr.	Daily	Hr.	Daily	Bare Costs	Incl. O&P
1 Struc. Steel Foreman (outside)	$62.30	$498.40	$96.40	$771.20	$62.33	$94.01
2 Structural Steel Workers	60.30	964.80	93.30	1492.80		
2 Electricians	63.70	1019.20	94.65	1514.40		
2 Steamfitters/Pipefitters	68.35	1093.60	102.00	1632.00		
1 Equip. Oper. (crane)	61.45	491.60	91.55	732.40		
1 Equip. Oper. (oiler)	52.50	420.00	78.25	626.00		
1 Hyd. Crane, 80 Ton		1458.00		1603.80	20.25	22.27
72 L.H., Daily Totals		$5945.60		$8372.60	$82.58	$116.29

Crew L-6	Hr.	Daily	Hr.	Daily	Bare Costs	Incl. O&P
1 Plumber	$67.70	$541.60	$101.05	$808.40	$66.37	$98.92
.5 Electrician	63.70	254.80	94.65	378.60		
12 L.H., Daily Totals		$796.40		$1187.00	$66.37	$98.92

Crew L-7	Hr.	Daily	Hr.	Daily	Bare Costs	Incl. O&P
2 Carpenters	$54.70	$875.20	$81.65	$1306.40	$53.04	$79.11
1 Building Laborer	44.40	355.20	66.25	530.00		
.5 Electrician	63.70	254.80	94.65	378.60		
28 L.H., Daily Totals		$1485.20		$2215.00	$53.04	$79.11

Crew L-8	Hr.	Daily	Hr.	Daily	Bare Costs	Incl. O&P
2 Carpenters	$54.70	$875.20	$81.65	$1306.40	$57.30	$85.53
.5 Plumber	67.70	270.80	101.05	404.20		
20 L.H., Daily Totals		$1146.00		$1710.60	$57.30	$85.53

Crew L-9	Hr.	Daily	Hr.	Daily	Bare Costs	Incl. O&P
1 Labor Foreman (inside)	$44.90	$359.20	$67.00	$536.00	$50.19	$75.58
2 Building Laborers	44.40	710.40	66.25	1060.00		
1 Struc. Steel Worker	60.30	482.40	93.30	746.40		
.5 Electrician	63.70	254.80	94.65	378.60		
36 L.H., Daily Totals		$1806.80		$2721.00	$50.19	$75.58

Crew L-10	Hr.	Daily	Hr.	Daily	Bare Costs	Incl. O&P
1 Struc. Steel Foreman (outside)	$62.30	$498.40	$96.40	$771.20	$61.35	$93.75
1 Structural Steel Worker	60.30	482.40	93.30	746.40		
1 Equip. Oper. (crane)	61.45	491.60	91.55	732.40		
1 Hyd. Crane, 12 Ton		475.80		523.38	19.82	21.81
24 L.H., Daily Totals		$1948.20		$2773.38	$81.17	$115.56

Crew L-11	Hr.	Daily	Hr.	Daily	Bare Costs	Incl. O&P
2 Wreckers	$44.40	$710.40	$67.40	$1078.40	$51.44	$77.26
1 Equip. Oper. (crane)	61.45	491.60	91.55	732.40		
1 Equip. Oper. (light)	55.50	444.00	82.70	661.60		
1 Hyd. Excavator, 2.5 C.Y.		1567.00		1723.70		
1 Loader, Skid Steer, 78 H.P.		446.30		490.93	62.92	69.21
32 L.H., Daily Totals		$3659.30		$4687.03	$114.35	$146.47

Crew M-1	Hr.	Daily	Hr.	Daily	Bare Costs	Incl. O&P
3 Elevator Constructors	$90.30	$2167.20	$133.80	$3211.20	$85.79	$127.11
1 Elevator Apprentice	72.25	578.00	107.05	856.40		
5 Hand Tools		50.50		55.55	1.58	1.74
32 L.H., Daily Totals		$2795.70		$4123.15	$87.37	$128.85

Crew M-3	Hr.	Daily	Hr.	Daily	Bare Costs	Incl. O&P
1 Electrician Foreman (outside)	$65.70	$525.60	$97.65	$781.20	$67.62	$100.41
1 Common Laborer	44.40	355.20	66.25	530.00		
.25 Equipment Operator (med.)	59.00	118.00	87.90	175.80		
1 Elevator Constructor	90.30	722.40	133.80	1070.40		
1 Elevator Apprentice	72.25	578.00	107.05	856.40		
.25 S.P. Crane, 4x4, 20 Ton		143.59		157.95	4.22	4.65
34 L.H., Daily Totals		$2442.79		$3571.75	$71.85	$105.05

Crew M-4	Hr.	Daily	Hr.	Daily	Bare Costs	Incl. O&P
1 Electrician Foreman (outside)	$65.70	$525.60	$97.65	$781.20	$66.92	$99.38
1 Common Laborer	44.40	355.20	66.25	530.00		
.25 Equipment Operator, Crane	61.45	122.90	91.55	183.10		
.25 Equip. Oper. (oiler)	52.50	105.00	78.25	156.50		
1 Elevator Constructor	90.30	722.40	133.80	1070.40		
1 Elevator Apprentice	72.25	578.00	107.05	856.40		
.25 S.P. Crane, 4x4, 40 Ton		190.45		209.50	5.29	5.82
36 L.H., Daily Totals		$2599.55		$3787.09	$72.21	$105.20

Crew Q-1	Hr.	Daily	Hr.	Daily	Bare Costs	Incl. O&P
1 Plumber	$67.70	$541.60	$101.05	$808.40	$60.92	$90.92
1 Plumber Apprentice	54.15	433.20	80.80	646.40		
16 L.H., Daily Totals		$974.80		$1454.80	$60.92	$90.92

Crew Q-1A	Hr.	Daily	Hr.	Daily	Bare Costs	Incl. O&P
.25 Plumber Foreman (outside)	$69.70	$139.40	$104.00	$208.00	$68.10	$101.64
1 Plumber	67.70	541.60	101.05	808.40		
10 L.H., Daily Totals		$681.00		$1016.40	$68.10	$101.64

Crew Q-1C	Hr.	Daily	Hr.	Daily	Bare Costs	Incl. O&P
1 Plumber	$67.70	$541.60	$101.05	$808.40	$60.28	$89.92
1 Plumber Apprentice	54.15	433.20	80.80	646.40		
1 Equip. Oper. (medium)	59.00	472.00	87.90	703.20		
1 Trencher, Chain Type, 8' D		1894.00		2083.40	78.92	86.81
24 L.H., Daily Totals		$3340.80		$4241.40	$139.20	$176.72

Crew Q-2	Hr.	Daily	Hr.	Daily	Bare Costs	Incl. O&P
2 Plumbers	$67.70	$1083.20	$101.05	$1616.80	$63.18	$94.30
1 Plumber Apprentice	54.15	433.20	80.80	646.40		
24 L.H., Daily Totals		$1516.40		$2263.20	$63.18	$94.30

Crew Q-3	Hr.	Daily	Hr.	Daily	Bare Costs	Incl. O&P
1 Plumber Foreman (inside)	$68.20	$545.60	$101.80	$814.40	$64.44	$96.17
2 Plumbers	67.70	1083.20	101.05	1616.80		
1 Plumber Apprentice	54.15	433.20	80.80	646.40		
32 L.H., Daily Totals		$2062.00		$3077.60	$64.44	$96.17

Crew Q-4	Hr.	Daily	Hr.	Daily	Bare Costs	Incl. O&P
1 Plumber Foreman (inside)	$68.20	$545.60	$101.80	$814.40	$64.44	$96.17
1 Plumber	67.70	541.60	101.05	808.40		
1 Welder (plumber)	67.70	541.60	101.05	808.40		
1 Plumber Apprentice	54.15	433.20	80.80	646.40		
1 Welder, Electric, 300 amp		107.55		118.31	3.36	3.70
32 L.H., Daily Totals		$2169.55		$3195.91	$67.80	$99.87

Crew Q-5	Hr.	Daily	Hr.	Daily	Bare Costs	Incl. O&P
1 Steamfitter	$68.35	$546.80	$102.00	$816.00	$61.52	$91.83
1 Steamfitter Apprentice	54.70	437.60	81.65	653.20		
16 L.H., Daily Totals		$984.40		$1469.20	$61.52	$91.83

Crews - Standard

Crew Q-6

Crew No.	Bare Costs Hr.	Daily	Incl. Subs O&P Hr.	Daily	Cost Per Labor-Hour Bare Costs	Incl. O&P
2 Steamfitters	$68.35	$1093.60	$102.00	$1632.00	$63.80	$95.22
1 Steamfitter Apprentice	54.70	437.60	81.65	653.20		
24 L.H., Daily Totals		$1531.20		$2285.20	$63.80	$95.22

Crew Q-7

Crew No.	Bare Costs Hr.	Daily	Incl. Subs O&P Hr.	Daily	Cost Per Labor-Hour Bare Costs	Incl. O&P
1 Steamfitter Foreman (inside)	$68.85	$550.80	$102.75	$822.00	$65.06	$97.10
2 Steamfitters	68.35	1093.60	102.00	1632.00		
1 Steamfitter Apprentice	54.70	437.60	81.65	653.20		
32 L.H., Daily Totals		$2082.00		$3107.20	$65.06	$97.10

Crew Q-8

Crew No.	Bare Costs Hr.	Daily	Incl. Subs O&P Hr.	Daily	Cost Per Labor-Hour Bare Costs	Incl. O&P
1 Steamfitter Foreman (inside)	$68.85	$550.80	$102.75	$822.00	$65.06	$97.10
1 Steamfitter	68.35	546.80	102.00	816.00		
1 Welder (steamfitter)	68.35	546.80	102.00	816.00		
1 Steamfitter Apprentice	54.70	437.60	81.65	653.20		
1 Welder, Electric, 300 amp		107.55		118.31	3.36	3.70
32 L.H., Daily Totals		$2189.55		$3225.51	$68.42	$100.80

Crew Q-9

Crew No.	Bare Costs Hr.	Daily	Incl. Subs O&P Hr.	Daily	Cost Per Labor-Hour Bare Costs	Incl. O&P
1 Sheet Metal Worker	$65.45	$523.60	$98.70	$789.60	$58.90	$88.83
1 Sheet Metal Apprentice	52.35	418.80	78.95	631.60		
16 L.H., Daily Totals		$942.40		$1421.20	$58.90	$88.83

Crew Q-10

Crew No.	Bare Costs Hr.	Daily	Incl. Subs O&P Hr.	Daily	Cost Per Labor-Hour Bare Costs	Incl. O&P
2 Sheet Metal Workers	$65.45	$1047.20	$98.70	$1579.20	$61.08	$92.12
1 Sheet Metal Apprentice	52.35	418.80	78.95	631.60		
24 L.H., Daily Totals		$1466.00		$2210.80	$61.08	$92.12

Crew Q-11

Crew No.	Bare Costs Hr.	Daily	Incl. Subs O&P Hr.	Daily	Cost Per Labor-Hour Bare Costs	Incl. O&P
1 Sheet Metal Foreman (inside)	$65.95	$527.60	$99.50	$796.00	$62.30	$93.96
2 Sheet Metal Workers	65.45	1047.20	98.70	1579.20		
1 Sheet Metal Apprentice	52.35	418.80	78.95	631.60		
32 L.H., Daily Totals		$1993.60		$3006.80	$62.30	$93.96

Crew Q-12

Crew No.	Bare Costs Hr.	Daily	Incl. Subs O&P Hr.	Daily	Cost Per Labor-Hour Bare Costs	Incl. O&P
1 Sprinkler Installer	$66.50	$532.00	$99.35	$794.80	$59.85	$89.42
1 Sprinkler Apprentice	53.20	425.60	79.50	636.00		
16 L.H., Daily Totals		$957.60		$1430.80	$59.85	$89.42

Crew Q-13

Crew No.	Bare Costs Hr.	Daily	Incl. Subs O&P Hr.	Daily	Cost Per Labor-Hour Bare Costs	Incl. O&P
1 Sprinkler Foreman (inside)	$67.00	$536.00	$100.10	$800.80	$63.30	$94.58
2 Sprinkler Installers	66.50	1064.00	99.35	1589.60		
1 Sprinkler Apprentice	53.20	425.60	79.50	636.00		
32 L.H., Daily Totals		$2025.60		$3026.40	$63.30	$94.58

Crew Q-14

Crew No.	Bare Costs Hr.	Daily	Incl. Subs O&P Hr.	Daily	Cost Per Labor-Hour Bare Costs	Incl. O&P
1 Asbestos Worker	$60.95	$487.60	$93.25	$746.00	$54.85	$83.90
1 Asbestos Apprentice	48.75	390.00	74.55	596.40		
16 L.H., Daily Totals		$877.60		$1342.40	$54.85	$83.90

Crew Q-15

Crew No.	Bare Costs Hr.	Daily	Incl. Subs O&P Hr.	Daily	Cost Per Labor-Hour Bare Costs	Incl. O&P
1 Plumber	$67.70	$541.60	$101.05	$808.40	$60.92	$90.92
1 Plumber Apprentice	54.15	433.20	80.80	646.40		
1 Welder, Electric, 300 amp		107.55		118.31	6.72	7.39
16 L.H., Daily Totals		$1082.35		$1573.11	$67.65	$98.32

Crew Q-16

Crew No.	Bare Costs Hr.	Daily	Incl. Subs O&P Hr.	Daily	Cost Per Labor-Hour Bare Costs	Incl. O&P
2 Plumbers	$67.70	$1083.20	$101.05	$1616.80	$63.18	$94.30
1 Plumber Apprentice	54.15	433.20	80.80	646.40		
1 Welder, Electric, 300 amp		107.55		118.31	4.48	4.93
24 L.H., Daily Totals		$1623.95		$2381.51	$67.66	$99.23

Crew Q-17

Crew No.	Bare Costs Hr.	Daily	Incl. Subs O&P Hr.	Daily	Cost Per Labor-Hour Bare Costs	Incl. O&P
1 Steamfitter	$68.35	$546.80	$102.00	$816.00	$61.52	$91.83
1 Steamfitter Apprentice	54.70	437.60	81.65	653.20		
1 Welder, Electric, 300 amp		107.55		118.31	6.72	7.39
16 L.H., Daily Totals		$1091.95		$1587.51	$68.25	$99.22

Crew Q-17A

Crew No.	Bare Costs Hr.	Daily	Incl. Subs O&P Hr.	Daily	Cost Per Labor-Hour Bare Costs	Incl. O&P
1 Steamfitter	$68.35	$546.80	$102.00	$816.00	$61.50	$91.73
1 Steamfitter Apprentice	54.70	437.60	81.65	653.20		
1 Equip. Oper. (crane)	61.45	491.60	91.55	732.40		
1 Hyd. Crane, 12 Ton		475.80		523.38		
1 Welder, Electric, 300 amp		107.55		118.31	24.31	26.74
24 L.H., Daily Totals		$2059.35		$2843.28	$85.81	$118.47

Crew Q-18

Crew No.	Bare Costs Hr.	Daily	Incl. Subs O&P Hr.	Daily	Cost Per Labor-Hour Bare Costs	Incl. O&P
2 Steamfitters	$68.35	$1093.60	$102.00	$1632.00	$63.80	$95.22
1 Steamfitter Apprentice	54.70	437.60	81.65	653.20		
1 Welder, Electric, 300 amp		107.55		118.31	4.48	4.93
24 L.H., Daily Totals		$1638.75		$2403.51	$68.28	$100.15

Crew Q-19

Crew No.	Bare Costs Hr.	Daily	Incl. Subs O&P Hr.	Daily	Cost Per Labor-Hour Bare Costs	Incl. O&P
1 Steamfitter	$68.35	$546.80	$102.00	$816.00	$62.25	$92.77
1 Steamfitter Apprentice	54.70	437.60	81.65	653.20		
1 Electrician	63.70	509.60	94.65	757.20		
24 L.H., Daily Totals		$1494.00		$2226.40	$62.25	$92.77

Crew Q-20

Crew No.	Bare Costs Hr.	Daily	Incl. Subs O&P Hr.	Daily	Cost Per Labor-Hour Bare Costs	Incl. O&P
1 Sheet Metal Worker	$65.45	$523.60	$98.70	$789.60	$59.86	$89.99
1 Sheet Metal Apprentice	52.35	418.80	78.95	631.60		
.5 Electrician	63.70	254.80	94.65	378.60		
20 L.H., Daily Totals		$1197.20		$1799.80	$59.86	$89.99

Crew Q-21

Crew No.	Bare Costs Hr.	Daily	Incl. Subs O&P Hr.	Daily	Cost Per Labor-Hour Bare Costs	Incl. O&P
2 Steamfitters	$68.35	$1093.60	$102.00	$1632.00	$63.77	$95.08
1 Steamfitter Apprentice	54.70	437.60	81.65	653.20		
1 Electrician	63.70	509.60	94.65	757.20		
32 L.H., Daily Totals		$2040.80		$3042.40	$63.77	$95.08

Crew Q-22

Crew No.	Bare Costs Hr.	Daily	Incl. Subs O&P Hr.	Daily	Cost Per Labor-Hour Bare Costs	Incl. O&P
1 Plumber	$67.70	$541.60	$101.05	$808.40	$60.92	$90.92
1 Plumber Apprentice	54.15	433.20	80.80	646.40		
1 Hyd. Crane, 12 Ton		475.80		523.38	29.74	32.71
16 L.H., Daily Totals		$1450.60		$1978.18	$90.66	$123.64

Crew Q-22A

Crew No.	Bare Costs Hr.	Daily	Incl. Subs O&P Hr.	Daily	Cost Per Labor-Hour Bare Costs	Incl. O&P
1 Plumber	$67.70	$541.60	$101.05	$808.40	$56.92	$84.91
1 Plumber Apprentice	54.15	433.20	80.80	646.40		
1 Laborer	44.40	355.20	66.25	530.00		
1 Equip. Oper. (crane)	61.45	491.60	91.55	732.40		
1 Hyd. Crane, 12 Ton		475.80		523.38	14.87	16.36
32 L.H., Daily Totals		$2297.40		$3240.58	$71.79	$101.27

For customer support on your Concrete & Masonry Costs with RSMeans Data, call 800.448.8182.

449

Crew No.	Bare Costs		Incl. Subs O&P		Cost Per Labor-Hour	

Crew Q-23	Hr.	Daily	Hr.	Daily	Bare Costs	Incl. O&P
1 Plumber Foreman (outside)	$69.70	$557.60	$104.00	$832.00	$65.47	$97.65
1 Plumber	67.70	541.60	101.05	808.40		
1 Equip. Oper. (medium)	59.00	472.00	87.90	703.20		
1 Lattice Boom Crane, 20 Ton		1526.00		1678.60	63.58	69.94
24 L.H., Daily Totals		$3097.20		$4022.20	$129.05	$167.59

Crew R-1	Hr.	Daily	Hr.	Daily	Bare Costs	Incl. O&P
1 Electrician Foreman	$64.20	$513.60	$95.40	$763.20	$59.53	$88.46
3 Electricians	63.70	1528.80	94.65	2271.60		
2 Electrician Apprentices	50.95	815.20	75.70	1211.20		
48 L.H., Daily Totals		$2857.60		$4246.00	$59.53	$88.46

Crew R-1A	Hr.	Daily	Hr.	Daily	Bare Costs	Incl. O&P
1 Electrician	$63.70	$509.60	$94.65	$757.20	$57.33	$85.17
1 Electrician Apprentice	50.95	407.60	75.70	605.60		
16 L.H., Daily Totals		$917.20		$1362.80	$57.33	$85.17

Crew R-1B	Hr.	Daily	Hr.	Daily	Bare Costs	Incl. O&P
1 Electrician	$63.70	$509.60	$94.65	$757.20	$55.20	$82.02
2 Electrician Apprentices	50.95	815.20	75.70	1211.20		
24 L.H., Daily Totals		$1324.80		$1968.40	$55.20	$82.02

Crew R-1C	Hr.	Daily	Hr.	Daily	Bare Costs	Incl. O&P
2 Electricians	$63.70	$1019.20	$94.65	$1514.40	$57.33	$85.17
2 Electrician Apprentices	50.95	815.20	75.70	1211.20		
1 Portable cable puller, 8000 lb.		102.65		112.92	3.21	3.53
32 L.H., Daily Totals		$1937.05		$2838.51	$60.53	$88.70

Crew R-2	Hr.	Daily	Hr.	Daily	Bare Costs	Incl. O&P
1 Electrician Foreman	$64.20	$513.60	$95.40	$763.20	$59.81	$88.90
3 Electricians	63.70	1528.80	94.65	2271.60		
2 Electrician Apprentices	50.95	815.20	75.70	1211.20		
1 Equip. Oper. (crane)	61.45	491.60	91.55	732.40		
1 S.P. Crane, 4x4, 5 Ton		381.95		420.14	6.82	7.50
56 L.H., Daily Totals		$3731.15		$5398.55	$66.63	$96.40

Crew R-3	Hr.	Daily	Hr.	Daily	Bare Costs	Incl. O&P
1 Electrician Foreman	$64.20	$513.60	$95.40	$763.20	$63.45	$94.33
1 Electrician	63.70	509.60	94.65	757.20		
.5 Equip. Oper. (crane)	61.45	245.80	91.55	366.20		
.5 S.P. Crane, 4x4, 5 Ton		190.97		210.07	9.55	10.50
20 L.H., Daily Totals		$1459.97		$2096.67	$73.00	$104.83

Crew R-4	Hr.	Daily	Hr.	Daily	Bare Costs	Incl. O&P
1 Struc. Steel Foreman (outside)	$62.30	$498.40	$96.40	$771.20	$61.38	$94.19
3 Struc. Steel Workers	60.30	1447.20	93.30	2239.20		
1 Electrician	63.70	509.60	94.65	757.20		
1 Welder, Gas Engine, 300 amp		148.75		163.63	3.72	4.09
40 L.H., Daily Totals		$2603.95		$3931.22	$65.10	$98.28

Crew R-5	Hr.	Daily	Hr.	Daily	Bare Costs	Incl. O&P
1 Electrician Foreman	$64.20	$513.60	$95.40	$763.20	$55.87	$83.45
4 Electrician Linemen	63.70	2038.40	94.65	3028.80		
2 Electrician Operators	63.70	1019.20	94.65	1514.40		
4 Electrician Groundmen	42.05	1345.60	63.65	2036.80		
1 Crew Truck		168.15		184.97		
1 Flatbed Truck, 20,000 GVW		204.05		224.46		
1 Pickup Truck, 3/4 Ton		112.20		123.42		
.2 Hyd. Crane, 55 Ton		198.03		217.83		
.2 Hyd. Crane, 12 Ton		95.16		104.68		
.2 Earth Auger, Truck-Mtd.		40.51		44.56		
1 Tractor w/Winch		377.65		415.42	13.59	14.95
88 L.H., Daily Totals		$6112.55		$8658.52	$69.46	$98.39

Crew R-6	Hr.	Daily	Hr.	Daily	Bare Costs	Incl. O&P
1 Electrician Foreman	$64.20	$513.60	$95.40	$763.20	$55.87	$83.45
4 Electrician Linemen	63.70	2038.40	94.65	3028.80		
2 Electrician Operators	63.70	1019.20	94.65	1514.40		
4 Electrician Groundmen	42.05	1345.60	63.65	2036.80		
1 Crew Truck		168.15		184.97		
1 Flatbed Truck, 20,000 GVW		204.05		224.46		
1 Pickup Truck, 3/4 Ton		112.20		123.42		
.2 Hyd. Crane, 55 Ton		198.03		217.83		
.2 Hyd. Crane, 12 Ton		95.16		104.68		
.2 Earth Auger, Truck-Mtd.		40.51		44.56		
1 Tractor w/Winch		377.65		415.42		
3 Cable Trailers		194.25		213.68		
.5 Tensioning Rig		55.75		61.33		
.5 Cable Pulling Rig		306.65		337.32	19.91	21.91
88 L.H., Daily Totals		$6669.20		$9270.84	$75.79	$105.35

Crew R-7	Hr.	Daily	Hr.	Daily	Bare Costs	Incl. O&P
1 Electrician Foreman	$64.20	$513.60	$95.40	$763.20	$45.74	$68.94
5 Electrician Groundmen	42.05	1682.00	63.65	2546.00		
1 Crew Truck		168.15		184.97	3.50	3.85
48 L.H., Daily Totals		$2363.75		$3494.17	$49.24	$72.80

Crew R-8	Hr.	Daily	Hr.	Daily	Bare Costs	Incl. O&P
1 Electrician Foreman	$64.20	$513.60	$95.40	$763.20	$56.57	$84.44
3 Electrician Linemen	63.70	1528.80	94.65	2271.60		
2 Electrician Groundmen	42.05	672.80	63.65	1018.40		
1 Pickup Truck, 3/4 Ton		112.20		123.42		
1 Crew Truck		168.15		184.97	5.84	6.42
48 L.H., Daily Totals		$2995.55		$4361.59	$62.41	$90.87

Crew R-9	Hr.	Daily	Hr.	Daily	Bare Costs	Incl. O&P
1 Electrician Foreman	$64.20	$513.60	$95.40	$763.20	$52.94	$79.24
1 Electrician Lineman	63.70	509.60	94.65	757.20		
2 Electrician Operators	63.70	1019.20	94.65	1514.40		
4 Electrician Groundmen	42.05	1345.60	63.65	2036.80		
1 Pickup Truck, 3/4 Ton		112.20		123.42		
1 Crew Truck		168.15		184.97	4.38	4.82
64 L.H., Daily Totals		$3668.35		$5379.98	$57.32	$84.06

Crew R-10	Hr.	Daily	Hr.	Daily	Bare Costs	Incl. O&P
1 Electrician Foreman	$64.20	$513.60	$95.40	$763.20	$60.17	$89.61
4 Electrician Linemen	63.70	2038.40	94.65	3028.80		
1 Electrician Groundman	42.05	336.40	63.65	509.20		
1 Crew Truck		168.15		184.97		
3 Tram Cars		219.60		241.56	8.08	8.89
48 L.H., Daily Totals		$3276.15		$4727.73	$68.25	$98.49

For customer support on your Concrete & Masonry Costs with RSMeans Data, call 800.448.8182.

Crew No.	Bare Costs		Incl. Subs O&P		Cost Per Labor-Hour	

Crew R-11	Hr.	Daily	Hr.	Daily	Bare Costs	Incl. O&P
1 Electrician Foreman	$64.20	$513.60	$95.40	$763.20	$60.69	$90.26
4 Electricians	63.70	2038.40	94.65	3028.80		
1 Equip. Oper. (crane)	61.45	491.60	91.55	732.40		
1 Common Laborer	44.40	355.20	66.25	530.00		
1 Crew Truck		168.15		184.97		
1 Hyd. Crane, 12 Ton		475.80		523.38	11.50	12.65
56 L.H., Daily Totals		$4042.75		$5762.74	$72.19	$102.91

Crew R-12	Hr.	Daily	Hr.	Daily	Bare Costs	Incl. O&P
1 Carpenter Foreman (inside)	$55.20	$441.60	$82.40	$659.20	$51.90	$77.75
4 Carpenters	54.70	1750.40	81.65	2612.80		
4 Common Laborers	44.40	1420.80	66.25	2120.00		
1 Equip. Oper. (medium)	59.00	472.00	87.90	703.20		
1 Steel Worker	60.30	482.40	93.30	746.40		
1 Dozer, 200 H.P.		1520.00		1672.00		
1 Pickup Truck, 3/4 Ton		112.20		123.42	18.55	20.40
88 L.H., Daily Totals		$6199.40		$8637.02	$70.45	$98.15

Crew R-13	Hr.	Daily	Hr.	Daily	Bare Costs	Incl. O&P
1 Electrician Foreman	$64.20	$513.60	$95.40	$763.20	$61.55	$91.52
3 Electricians	63.70	1528.80	94.65	2271.60		
.25 Equip. Oper. (crane)	61.45	122.90	91.55	183.10		
1 Equipment Oiler	52.50	420.00	78.25	626.00		
.25 Hydraulic Crane, 33 Ton		245.79		270.37	5.85	6.44
42 L.H., Daily Totals		$2831.09		$4114.27	$67.41	$97.96

Crew R-15	Hr.	Daily	Hr.	Daily	Bare Costs	Incl. O&P
1 Electrician Foreman	$64.20	$513.60	$95.40	$763.20	$62.42	$92.78
4 Electricians	63.70	2038.40	94.65	3028.80		
1 Equipment Oper. (light)	55.50	444.00	82.70	661.60		
1 Telescoping Boom Lift, to 40'		281.90		310.09	5.87	6.46
48 L.H., Daily Totals		$3277.90		$4763.69	$68.29	$99.24

Crew R-15A	Hr.	Daily	Hr.	Daily	Bare Costs	Incl. O&P
1 Electrician Foreman	$64.20	$513.60	$95.40	$763.20	$55.98	$83.32
2 Electricians	63.70	1019.20	94.65	1514.40		
2 Common Laborers	44.40	710.40	66.25	1060.00		
1 Equip. Oper. (light)	55.50	444.00	82.70	661.60		
1 Telescoping Boom Lift, to 40'		281.90		310.09	5.87	6.46
48 L.H., Daily Totals		$2969.10		$4309.29	$61.86	$89.78

Crew R-18	Hr.	Daily	Hr.	Daily	Bare Costs	Incl. O&P
.25 Electrician Foreman	$64.20	$128.40	$95.40	$190.80	$55.89	$83.05
1 Electrician	63.70	509.60	94.65	757.20		
2 Electrician Apprentices	50.95	815.20	75.70	1211.20		
26 L.H., Daily Totals		$1453.20		$2159.20	$55.89	$83.05

Crew R-19	Hr.	Daily	Hr.	Daily	Bare Costs	Incl. O&P
.5 Electrician Foreman	$64.20	$256.80	$95.40	$381.60	$63.80	$94.80
2 Electricians	63.70	1019.20	94.65	1514.40		
20 L.H., Daily Totals		$1276.00		$1896.00	$63.80	$94.80

Crew R-21	Hr.	Daily	Hr.	Daily	Bare Costs	Incl. O&P
1 Electrician Foreman	$64.20	$513.60	$95.40	$763.20	$63.71	$94.67
3 Electricians	63.70	1528.80	94.65	2271.60		
.1 Equip. Oper. (medium)	59.00	47.20	87.90	70.32		
.1 S.P. Crane, 4x4, 25 Ton		115.50		127.05	3.52	3.87
32.8 L.H., Daily Totals		$2205.10		$3232.17	$67.23	$98.54

Crew R-22	Hr.	Daily	Hr.	Daily	Bare Costs	Incl. O&P
.66 Electrician Foreman	$64.20	$338.98	$95.40	$503.71	$58.30	$86.62
2 Electricians	63.70	1019.20	94.65	1514.40		
2 Electrician Apprentices	50.95	815.20	75.70	1211.20		
37.28 L.H., Daily Totals		$2173.38		$3229.31	$58.30	$86.62

Crew R-30	Hr.	Daily	Hr.	Daily	Bare Costs	Incl. O&P
.25 Electrician Foreman (outside)	$65.70	$131.40	$97.65	$195.30	$51.98	$77.40
1 Electrician	63.70	509.60	94.65	757.20		
2 Laborers (Semi-Skilled)	44.40	710.40	66.25	1060.00		
26 L.H., Daily Totals		$1351.40		$2012.50	$51.98	$77.40

Crew R-31	Hr.	Daily	Hr.	Daily	Bare Costs	Incl. O&P
1 Electrician	$63.70	$509.60	$94.65	$757.20	$63.70	$94.65
1 Core Drill, Electric, 2.5 H.P.		62.65		68.92	7.83	8.61
8 L.H., Daily Totals		$572.25		$826.12	$71.53	$103.26

Crew W-41E	Hr.	Daily	Hr.	Daily	Bare Costs	Incl. O&P
.5 Plumber Foreman (outside)	$69.70	$278.80	$104.00	$416.00	$58.78	$87.72
1 Plumber	67.70	541.60	101.05	808.40		
1 Laborer	44.40	355.20	66.25	530.00		
20 L.H., Daily Totals		$1175.60		$1754.40	$58.78	$87.72

The table below lists both the RSMeans® historical cost index based on Jan. 1, 1993 = 100 as well as the computed value of an index based on Jan. 1, 2021 costs. Since the Jan. 1, 2021 figure is estimated, space is left to write in the actual index figures as they become available through the quarterly *RSMeans Construction Cost Indexes.*

To compute the actual index based on Jan. 1, 2021 = 100, divide the historical cost index for a particular year by the actual Jan. 1, 2021 construction cost index. Space has been left to advance the index figures as the year progresses.

Year	Historical Cost Index Jan. 1, 1993 = 100		Current Index Based on Jan. 1, 2021 = 100		Year	Historical Cost Index Jan. 1, 1993 = 100	Current Index Based on Jan. 1, 2021 = 100		Year	Historical Cost Index Jan. 1, 1993 = 100	Current Index Based on Jan. 1, 2021 = 100	
	Est.	Actual	Est.	Actual		Actual	Est.	Actual		Actual	Est.	Actual
Oct 2021					July 2004	143.7	60.7		1984	82.0	34.6	
July 2021					2003	132.0	55.8		1983	80.2	33.9	
April 2021					2002	128.7	54.4		1982	76.1	32.2	
Jan 2021	236.7		100.0		2001	125.1	52.9		1981	70.0	29.6	
2020		234.6	99.1		2000	120.9	51.1		1980	62.9	26.6	
2019		232.2	98.1		1999	117.6	49.7		1979	57.8	24.4	
2018		222.9	94.2		1998	115.1	48.6		1978	53.5	22.6	
2017		213.6	90.2		1997	112.8	47.7		1977	49.5	20.9	
2016		207.3	87.6		1996	110.2	46.6		1976	46.9	19.8	
2015		206.2	87.1		1995	107.6	45.5		1975	44.8	18.9	
2014		204.9	86.6		1994	104.4	44.1		1974	41.4	17.5	
2013		201.2	85.0		1993	101.7	43.0		1973	37.7	15.9	
2012		194.6	82.2		1992	99.4	42.0		1972	34.8	14.7	
2011		191.2	80.8		1991	96.8	40.9		1971	32.1	13.6	
2010		183.5	77.5		1990	94.3	39.8		1970	28.7	12.1	
2009		180.1	76.1		1989	92.1	38.9		1969	26.9	11.4	
2008		180.4	76.2		1988	89.9	38.0					
2007		169.4	71.6		1987	87.7	37.0					
2006		162.0	68.4		1986	84.2	35.6					
2005		151.6	64.0		1985	82.6	34.9					

Adjustments to Costs

The "Historical Cost Index" can be used to convert national average building costs at a particular time to the approximate building costs for some other time.

Example:

Estimate and compare construction costs for different years in the same city.

To estimate the national average construction cost of a building in 1970, knowing that it cost $900,000 in 2021:

INDEX in 1970 = 28.7

INDEX in 2021 = 236.7

Note: The city cost indexes for Canada can be used to convert U.S. national averages to local costs in Canadian dollars.

Time Adjustment Using the Historical Cost Indexes:

$$\frac{\text{Index for Year A}}{\text{Index for Year B}} \times \text{Cost in Year B} = \text{Cost in Year A}$$

$$\frac{\text{INDEX 1970}}{\text{INDEX 2021}} \times \text{Cost 2021} = \text{Cost 1970}$$

$$\frac{28.7}{236.7} \times \$900,000 = .121 \times \$900,000 = \$108,900$$

The construction cost of the building in 1970 was $108,900.

Example:

To estimate and compare the cost of a building in Toronto, ON in 2021 with the known cost of $600,000 (US$) in New York, NY in 2021:

INDEX Toronto = 112.5

INDEX New York = 134.1

$$\frac{\text{INDEX Toronto}}{\text{INDEX New York}} \times \text{Cost New York} = \text{Cost Toronto}$$

$$\frac{112.5}{134.1} \times \$600,000 = .839 \times \$600,000 = \$503,400$$

The construction cost of the building in Toronto is $503,400 (CN$).

*Historical Cost Index updates and other resources are provided on the following website:
rsmeans.com/2021books

How to Use the City Cost Indexes

What you should know before you begin

RSMeans City Cost Indexes (CCI) are an extremely useful tool for when you want to compare costs from city to city and region to region.

This publication contains average construction cost indexes for 731 U.S. and Canadian cities covering over 930 three-digit zip code locations, as listed directly under each city.

Keep in mind that a City Cost Index number is a percentage ratio of a specific city's cost to the national average cost of the same item at a stated time period.

In other words, these index figures represent relative construction factors (or, if you prefer, multipliers) for material and installation costs, as well as the weighted average for Total In Place costs for each CSI MasterFormat division. Installation costs include both labor and equipment rental costs. When estimating equipment rental rates only for a specific location, use 01 54 33 EQUIPMENT RENTAL COSTS in the Reference Section.

The 30 City Average Index is the average of 30 major U.S. cities and serves as a national average. This national average represents the baseline from which all locations are calculated.

Index figures for both material and installation are based on the 30 major city average of 100 and represent the cost relationship as of July 1, 2020. The index for each division is computed from representative material and labor quantities for that division. The weighted average for each city is a weighted total of the components listed above it. It does not include relative productivity between trades or cities.

As changes occur in local material prices, labor rates, and equipment rental rates (including fuel costs), the impact of these changes should be accurately measured by the change in the City Cost Index for each particular city (as compared to the 30 city average).

Therefore, if you know (or have estimated) building costs in one city today, you can easily convert those costs to expected building costs in another city.

In addition, by using the Historical Cost Index, you can easily convert national average building costs at a particular time to the approximate building costs for some other time. The City Cost Indexes can then be applied to calculate the costs for a particular city.

Quick calculations

Location Adjustment Using the City Cost Indexes:

$$\frac{\text{Index for City A}}{\text{Index for City B}} \times \text{Cost in City B} = \text{Cost in City A}$$

Time Adjustment for the National Average Using the Historical Cost Index:

$$\frac{\text{Index for Year A}}{\text{Index for Year B}} \times \text{Cost in Year B} = \text{Cost in Year A}$$

Adjustment from the National Average:

$$\frac{\text{Index for City A}}{100} \times \text{National Average Cost} = \text{Cost in City A}$$

Since each of the other RSMeans data sets contains many different items, any *one* item multiplied by the particular city index may give incorrect results. However, the larger the number of items compiled, the closer the results should be to actual costs for that particular city.

The City Cost Indexes for Canadian cities are calculated using Canadian material and equipment prices and labor rates in Canadian dollars. When compared to the baseline mentioned previously, the resulting index enables the user to convert baseline costs to local costs. Therefore, indexes for Canadian cities can be used to convert U.S. national average prices to local costs in Canadian dollars.

How to use this section

1. Compare costs from city to city.

In using the RSMeans Indexes, remember that an index number is not a fixed number but a ratio: It's a percentage ratio of a building component's cost at any stated time to the national average cost of that same component at the same time period. Put in the form of an equation:

$$\frac{\text{Specific City Cost}}{\text{National Average Cost}} \times 100 = \text{City Index Number}$$

Therefore, when making cost comparisons between cities, do not subtract one city's index number from the index number of another city and read the result as a percentage difference. Instead, divide one city's index number by that of the other city. The resulting number may then be used as a multiplier to calculate cost differences from city to city.

The formula used to find cost differences between cities for the purpose of comparison is as follows:

$$\frac{\text{City A Index}}{\text{City B Index}} \times \text{City B Cost (Known)} = \text{City A Cost (Unknown)}$$

In addition, you can use RSMeans CCI to calculate and compare costs division by division between cities using the same basic formula. (Just be sure that you're comparing similar divisions.)

2. Compare a specific city's construction costs with the national average.

When you're studying construction location feasibility, it's advisable to compare a prospective project's cost index with an index of the national average cost.

For example, divide the weighted average index of construction costs of a specific city by that of the 30 City Average, which = 100.

$$\frac{\text{City Index}}{100} = \% \text{ of National Average}$$

As a result, you get a ratio that indicates the relative cost of construction in that city in comparison with the national average.

3. Convert U.S. national average to actual costs in Canadian City.

$$\frac{\text{Index for Canadian City}}{100} \times \text{National Average Cost} = \text{Cost in Canadian City in \$ CAN}$$

4. Adjust construction cost data based on a national average.

When you use a source of construction cost data which is based on a national average (such as RSMeans cost data), it is necessary to adjust those costs to a specific location.

$$\frac{\text{City Index}}{100} \times \frac{\text{Cost Based on}}{\text{National Average Costs}} = \frac{\text{City Cost}}{\text{(Unknown)}}$$

5. When applying the City Cost Indexes to demolition projects, use the appropriate division installation index. For example, for removal of existing doors and windows, use the Division 8 (Openings) index.

What you might like to know about how we developed the Indexes

The information presented in the CCI is organized according to the Construction Specifications Institute (CSI) MasterFormat 2018 classification system.

To create a reliable index, RSMeans researched the building type most often constructed in the United States and Canada. Because it was concluded that no one type of building completely represented the building construction industry, nine different types of buildings were combined to create a composite model.

The exact material, labor, and equipment quantities are based on detailed analyses of these nine building types, and then each quantity is weighted in proportion to expected usage. These various material items, labor hours, and equipment rental rates are thus combined to form a composite building representing as closely as possible the actual usage of materials, labor, and equipment in the North American building construction industry.

The following structures were chosen to make up that composite model:

1. Factory, 1 story
2. Office, 2–4 stories
3. Store, Retail
4. Town Hall, 2–3 stories
5. High School, 2–3 stories
6. Hospital, 4–8 stories
7. Garage, Parking
8. Apartment, 1–3 stories
9. Hotel/Motel, 2–3 stories

For the purposes of ensuring the timeliness of the data, the components of the index for the composite model have been streamlined. They currently consist of:

- specific quantities of 66 commonly used construction materials;
- specific labor-hours for 21 building construction trades; and
- specific days of equipment rental for 6 types of construction equipment (normally used to install the 66 material items by the 21 trades.) Fuel costs and routine maintenance costs are included in the equipment cost.

Material and equipment price quotations are gathered quarterly from cities in the United States and Canada. These prices and the latest negotiated labor wage rates for 21 different building trades are used to compile the quarterly update of the City Cost Index.

The 30 major U.S. cities used to calculate the national average are:

Atlanta, GA	Memphis, TN
Baltimore, MD	Milwaukee, WI
Boston, MA	Minneapolis, MN
Buffalo, NY	Nashville, TN
Chicago, IL	New Orleans, LA
Cincinnati, OH	New York, NY
Cleveland, OH	Philadelphia, PA
Columbus, OH	Phoenix, AZ
Dallas, TX	Pittsburgh, PA
Denver, CO	St. Louis, MO
Detroit, MI	San Antonio, TX
Houston, TX	San Diego, CA
Indianapolis, IN	San Francisco, CA
Kansas City, MO	Seattle, WA
Los Angeles, CA	Washington, DC

What the CCI does not indicate

The weighted average for each city is a total of the divisional components weighted to reflect typical usage. It does not include the productivity variations between trades or cities.

In addition, the CCI does not take into consideration factors such as the following:

- managerial efficiency
- competitive conditions
- automation
- restrictive union practices
- unique local requirements
- regional variations due to specific building codes

City Cost Indexes - V2

Table 1

DIVISION		UNITED STATES 30 CITY AVERAGE MAT.	INST.	TOTAL	ANNISTON 362 MAT.	INST.	TOTAL	BIRMINGHAM 350-352 MAT.	INST.	TOTAL	BUTLER 369 MAT.	INST.	TOTAL	DECATUR 356 MAT.	INST.	TOTAL	DOTHAN 363 MAT.	INST.	TOTAL
015433	CONTRACTOR EQUIPMENT		100.0	100.0		100.5	100.5		102.7	102.7		98.2	98.2		100.5	100.5		98.2	98.2
0241, 31 - 34	SITE & INFRASTRUCTURE, DEMOLITION	100.0	100.0	100.0	89.8	87.0	87.9	88.9	90.9	90.3	101.8	83.2	89.0	81.6	86.8	85.1	99.5	83.1	88.2
0310	Concrete Forming & Accessories	100.0	100.0	100.0	88.9	66.7	70.0	94.3	67.4	71.4	84.5	69.1	71.4	94.5	65.7	70.0	94.3	67.3	71.3
0320	Concrete Reinforcing	100.0	100.0	100.0	88.6	69.5	79.4	92.3	69.5	81.3	93.7	71.5	83.0	86.4	68.4	77.7	93.7	69.6	82.1
0330	Cast-in-Place Concrete	100.0	100.0	100.0	83.3	67.0	77.1	101.0	68.9	88.9	81.2	67.3	76.0	93.8	67.0	83.7	81.2	67.3	76.0
03	CONCRETE	100.0	100.0	100.0	91.4	69.0	81.4	97.0	69.8	84.8	92.5	70.5	82.6	90.0	68.3	80.3	91.8	69.4	81.7
04	MASONRY	100.0	100.0	100.0	96.7	62.5	75.8	89.8	62.6	73.2	101.5	62.5	77.7	87.1	62.6	72.1	102.9	62.5	78.3
05	METALS	100.0	100.0	100.0	101.4	94.5	99.3	98.7	92.8	96.8	100.3	95.9	99.0	100.9	93.8	98.7	100.4	95.1	98.8
06	WOOD, PLASTICS & COMPOSITES	100.0	100.0	100.0	85.3	67.5	76.0	93.4	67.7	80.0	79.6	70.1	74.6	97.4	65.6	80.7	92.5	67.5	79.4
07	THERMAL & MOISTURE PROTECTION	100.0	100.0	100.0	95.0	62.3	80.9	94.7	66.1	82.4	95.0	65.7	82.4	93.6	65.3	81.5	95.0	64.7	82.0
08	OPENINGS	100.0	100.0	100.0	94.4	68.2	88.0	102.3	68.7	94.1	94.5	70.6	88.6	106.7	67.3	97.1	94.5	68.6	88.2
0920	Plaster & Gypsum Board	100.0	100.0	100.0	86.0	67.2	73.7	97.5	67.2	77.6	84.0	69.9	74.8	96.9	65.2	76.1	93.4	67.2	76.2
0950, 0980	Ceilings & Acoustic Treatment	100.0	100.0	100.0	85.8	67.2	74.1	96.2	67.2	78.0	85.8	69.9	75.8	94.3	65.2	76.1	85.8	67.2	74.1
0960	Flooring	100.0	100.0	100.0	83.5	68.4	79.1	95.8	68.4	87.8	88.2	68.4	82.4	89.0	68.4	83.0	93.3	68.4	86.0
0970, 0990	Wall Finishes & Painting/Coating	100.0	100.0	100.0	87.4	53.7	67.2	91.6	53.7	68.9	87.4	45.5	62.3	81.3	60.5	68.8	87.4	78.1	81.8
09	FINISHES	100.0	100.0	100.0	84.4	65.7	74.3	94.3	65.9	78.9	87.4	66.3	74.9	89.3	65.3	76.3	89.9	68.4	78.3
COVERS	DIVS. 10 - 14, 25, 28, 41, 43, 44, 46	100.0	100.0	100.0	100.0	84.1	96.2	100.0	84.7	96.4	100.0	85.2	96.5	100.0	83.9	96.2	100.0	85.0	96.5
21, 22, 23	FIRE SUPPRESSION, PLUMBING & HVAC	100.0	100.0	100.0	102.3	51.9	82.0	100.8	65.0	86.4	98.5	64.4	84.7	100.8	64.9	86.4	98.5	63.8	84.5
26, 27, 3370	ELECTRICAL, COMMUNICATIONS & UTIL.	100.0	100.0	100.0	99.2	57.7	78.6	98.6	65.4	82.2	101.0	57.8	79.7	94.2	65.2	79.9	99.9	73.3	86.7
MF2018	WEIGHTED AVERAGE	100.0	100.0	100.0	97.1	66.6	83.9	98.2	70.9	86.4	97.1	69.7	85.2	96.9	70.2	85.4	97.2	71.6	86.1

Table 2 — ALABAMA

DIVISION		EVERGREEN 364 MAT.	INST.	TOTAL	GADSDEN 359 MAT.	INST.	TOTAL	HUNTSVILLE 357-358 MAT.	INST.	TOTAL	JASPER 355 MAT.	INST.	TOTAL	MOBILE 365-366 MAT.	INST.	TOTAL	MONTGOMERY 360-361 MAT.	INST.	TOTAL
015433	CONTRACTOR EQUIPMENT		98.2	98.2		100.5	100.5		100.5	100.5		100.5	100.5		98.2	98.2		103.5	103.5
0241, 31 - 34	SITE & INFRASTRUCTURE, DEMOLITION	102.3	83.2	89.1	87.5	87.0	87.2	81.3	86.7	85.0	87.4	87.0	87.1	95.1	83.2	86.9	93.5	92.2	92.6
0310	Concrete Forming & Accessories	80.9	67.0	69.0	86.0	67.2	70.0	94.5	65.3	69.7	91.7	67.1	70.8	93.6	67.0	71.0	97.8	67.4	71.9
0320	Concrete Reinforcing	93.8	71.5	83.0	91.6	69.5	80.9	86.4	72.8	79.8	86.4	69.5	78.2	91.3	71.5	81.8	99.5	69.6	85.1
0330	Cast-in-Place Concrete	81.2	67.2	75.9	93.8	67.2	83.8	91.3	66.7	82.1	104.0	67.2	90.1	85.2	67.2	78.4	85.1	68.9	79.0
03	CONCRETE	92.9	69.5	82.4	94.3	69.3	83.1	88.8	68.8	79.9	97.7	69.3	84.9	87.9	69.6	79.7	90.8	69.8	81.4
04	MASONRY	101.5	62.5	77.7	85.8	62.5	71.6	88.5	62.0	72.3	83.5	62.5	70.7	100.5	62.5	77.3	99.7	62.6	77.0
05	METALS	100.4	95.7	98.9	98.7	95.0	97.5	100.9	95.5	99.2	98.6	94.9	97.5	102.5	95.8	100.4	101.5	92.8	98.8
06	WOOD, PLASTICS & COMPOSITES	75.5	67.5	71.3	87.0	67.5	76.8	97.4	65.3	80.6	94.4	67.5	80.3	91.1	67.5	78.8	95.4	67.7	80.9
07	THERMAL & MOISTURE PROTECTION	95.0	64.7	82.0	93.9	65.4	81.6	93.6	65.0	81.3	93.9	65.2	81.6	94.6	64.7	81.7	93.3	66.5	81.8
08	OPENINGS	94.5	69.1	88.3	103.4	68.6	94.9	106.4	68.2	97.1	103.3	68.6	94.9	97.0	69.1	90.2	95.7	68.7	89.2
0920	Plaster & Gypsum Board	83.5	67.2	72.8	89.3	67.2	74.8	96.9	65.0	76.0	93.3	67.2	76.2	90.6	67.2	75.3	93.0	67.2	76.1
0950, 0980	Ceilings & Acoustic Treatment	85.8	67.2	74.1	87.9	67.2	74.9	95.7	65.0	76.4	87.9	67.2	74.9	92.9	67.2	76.8	94.8	67.2	77.5
0960	Flooring	86.1	68.4	80.9	85.3	68.4	80.4	89.0	68.4	83.0	87.4	68.4	81.9	92.8	68.4	85.7	92.5	68.4	85.5
0970, 0990	Wall Finishes & Painting/Coating	87.4	45.5	62.3	81.3	53.7	64.7	81.3	60.6	68.9	81.3	53.7	64.7	90.5	45.5	63.5	92.1	53.7	69.1
09	FINISHES	86.7	64.8	74.9	86.0	65.7	75.0	89.6	65.1	76.3	87.1	65.7	75.5	90.7	64.8	76.7	93.1	65.9	78.3
COVERS	DIVS. 10 - 14, 25, 28, 41, 43, 44, 46	100.0	84.9	96.4	100.0	84.1	96.3	100.0	83.7	96.2	100.0	84.1	96.2	100.0	84.9	96.5	100.0	84.7	96.4
21, 22, 23	FIRE SUPPRESSION, PLUMBING & HVAC	98.5	58.1	82.2	103.7	65.5	88.3	100.8	63.8	85.9	103.7	64.9	88.1	100.6	60.4	84.4	100.6	63.3	85.6
26, 27, 3370	ELECTRICAL, COMMUNICATIONS & UTIL.	98.6	57.8	78.4	94.2	65.4	80.0	95.0	63.4	79.4	93.9	57.7	76.0	101.6	60.3	81.2	101.5	73.2	87.5
MF2018	WEIGHTED AVERAGE	96.8	67.8	84.3	97.2	70.8	85.8	96.9	69.9	85.2	97.6	69.6	85.5	97.8	68.7	85.2	98.0	71.8	86.6

Table 3 — ALABAMA / ALASKA

DIVISION		PHENIX CITY 368 MAT.	INST.	TOTAL	SELMA 367 MAT.	INST.	TOTAL	TUSCALOOSA 354 MAT.	INST.	TOTAL	ANCHORAGE 995-996 MAT.	INST.	TOTAL	FAIRBANKS 997 MAT.	INST.	TOTAL	JUNEAU 998 MAT.	INST.	TOTAL
015433	CONTRACTOR EQUIPMENT		98.2	98.2		98.2	98.2		100.5	100.5		107.6	107.6		111.2	111.2		107.6	107.6
0241, 31 - 34	SITE & INFRASTRUCTURE, DEMOLITION	105.9	83.1	90.2	99.3	83.1	88.2	81.8	87.0	85.4	124.0	115.8	118.4	123.2	121.2	121.8	139.7	115.8	123.3
0310	Concrete Forming & Accessories	88.7	68.6	71.6	85.9	67.2	70.0	94.4	67.1	71.2	112.9	113.4	113.3	120.7	112.7	113.9	124.0	113.4	115.0
0320	Concrete Reinforcing	93.7	68.2	81.4	93.7	69.5	82.1	86.4	69.5	78.2	143.1	118.2	131.1	145.0	118.2	132.1	146.2	118.2	132.7
0330	Cast-in-Place Concrete	81.2	66.9	75.8	81.2	67.2	75.9	95.1	67.2	84.6	117.0	114.9	116.2	116.0	112.4	114.6	116.9	114.9	116.2
03	CONCRETE	95.8	69.6	84.1	91.3	69.3	81.4	90.6	69.3	81.0	112.6	114.1	113.3	105.1	113.1	108.7	119.7	114.1	117.2
04	MASONRY	101.5	61.9	77.3	105.9	62.5	79.4	87.4	62.5	72.2	185.2	116.7	143.4	187.5	115.0	143.3	171.5	116.7	138.0
05	METALS	100.3	94.6	98.5	100.3	95.0	98.7	100.1	95.0	98.5	126.3	103.8	119.3	122.7	105.5	117.3	121.4	103.8	115.9
06	WOOD, PLASTICS & COMPOSITES	84.9	69.6	76.9	81.5	67.5	74.2	97.4	67.5	81.7	102.3	111.1	106.9	117.3	111.0	114.0	117.0	111.1	113.9
07	THERMAL & MOISTURE PROTECTION	95.4	65.4	82.5	94.8	65.3	82.2	93.7	65.4	81.5	177.8	113.6	150.2	188.1	112.6	155.7	191.7	113.6	158.2
08	OPENINGS	94.4	69.4	88.3	94.4	68.6	88.1	106.4	68.6	97.2	131.0	113.6	126.8	133.2	112.9	128.2	131.7	113.6	127.3
0920	Plaster & Gypsum Board	87.4	69.4	75.6	85.4	67.2	73.5	96.9	67.2	77.4	139.6	111.3	121.0	165.5	111.3	129.9	141.3	111.3	121.6
0950, 0980	Ceilings & Acoustic Treatment	85.8	69.4	75.5	85.8	67.2	74.9	95.7	67.2	77.8	120.9	111.3	114.9	107.6	111.3	109.9	116.2	111.3	113.1
0960	Flooring	90.0	68.4	83.7	88.5	68.4	82.7	89.0	68.4	83.0	114.5	116.7	115.1	117.5	116.7	117.3	126.7	116.7	123.8
0970, 0990	Wall Finishes & Painting/Coating	87.4	79.8	82.8	87.4	53.7	67.2	81.3	53.7	64.7	108.6	117.6	114.0	109.0	113.4	111.6	108.9	117.6	114.1
09	FINISHES	88.8	69.6	78.4	87.4	65.7	75.6	89.6	65.7	76.7	123.7	114.5	118.7	125.5	113.5	119.0	125.5	114.5	119.5
COVERS	DIVS. 10 - 14, 25, 28, 41, 43, 44, 46	100.0	84.8	96.4	100.0	84.1	96.3	100.0	84.1	96.3	100.0	110.1	102.4	100.0	109.3	102.2	100.0	110.1	102.4
21, 22, 23	FIRE SUPPRESSION, PLUMBING & HVAC	98.5	63.0	84.2	98.5	63.8	84.5	100.8	64.3	86.1	101.2	104.1	102.4	101.2	106.5	103.3	101.8	104.1	102.7
26, 27, 3370	ELECTRICAL, COMMUNICATIONS & UTIL.	100.5	68.2	84.5	99.6	73.3	86.6	94.7	65.4	80.2	115.8	107.9	111.9	125.5	107.9	116.8	109.2	107.9	108.5
MF2018	WEIGHTED AVERAGE	97.7	70.8	86.1	96.9	71.2	85.8	96.9	70.5	85.5	119.7	110.4	115.7	120.2	111.0	116.2	119.7	110.4	115.7

For customer support on your Concrete & Masonry Costs with RSMeans Data, call 800.448.8182.

455

ALASKA / ARIZONA

DIVISION		KETCHIKAN 999 MAT.	INST.	TOTAL	CHAMBERS 865 MAT.	INST.	TOTAL	FLAGSTAFF 860 MAT.	INST.	TOTAL	GLOBE 855 MAT.	INST.	TOTAL	KINGMAN 864 MAT.	INST.	TOTAL	MESA/TEMPE 852 MAT.	INST.	TOTAL
015433	CONTRACTOR EQUIPMENT		111.2	111.2		85.9	85.9		85.9	85.9		87.5	87.5		85.9	85.9		87.5	87.5
0241, 31 - 34	SITE & INFRASTRUCTURE, DEMOLITION	176.0	121.3	138.4	69.8	87.5	82.0	89.3	87.6	88.2	103.1	88.7	93.2	69.8	87.6	82.0	93.2	88.9	90.2
0310	Concrete Forming & Accessories	112.4	113.3	113.2	96.6	68.8	73.0	102.5	68.9	74.0	86.2	69.0	71.6	94.7	64.9	69.3	89.0	70.1	72.9
0320	Concrete Reinforcing	108.2	118.2	113.0	99.5	73.6	87.0	99.3	73.6	86.9	104.3	73.6	89.5	99.6	73.6	87.0	105.0	73.6	89.8
0330	Cast-in-Place Concrete	234.6	113.8	189.0	88.4	67.6	80.6	88.5	67.8	80.7	79.6	67.4	75.0	88.1	67.8	80.4	80.3	67.6	75.5
03	CONCRETE	177.3	113.8	148.8	93.4	69.2	82.6	114.1	69.3	94.0	98.0	69.3	85.1	93.1	67.5	81.6	89.7	69.9	80.8
04	MASONRY	194.7	116.6	147.1	94.9	60.0	73.6	95.0	60.7	74.1	100.5	59.9	75.8	94.9	60.1	73.6	100.7	60.0	75.9
05	METALS	122.8	105.4	117.4	101.2	70.2	91.5	101.7	70.7	92.1	102.3	71.4	92.7	101.9	70.6	92.2	102.7	71.8	93.0
06	WOOD, PLASTICS & COMPOSITES	107.8	111.0	109.5	97.3	70.4	83.3	104.0	70.4	86.4	80.2	70.6	75.2	92.3	65.0	78.0	83.8	72.0	77.6
07	THERMAL & MOISTURE PROTECTION	194.3	112.5	159.2	99.3	70.0	86.7	101.2	70.4	88.0	99.0	68.7	86.0	99.2	69.6	86.5	98.7	69.1	86.0
08	OPENINGS	128.9	113.5	125.2	105.8	66.8	96.3	105.9	70.0	97.2	95.3	66.9	88.4	106.0	65.4	96.1	95.4	67.9	88.7
0920	Plaster & Gypsum Board	148.8	111.3	124.2	94.9	70.0	78.6	98.4	70.0	79.8	85.5	70.0	75.3	87.4	64.4	72.3	88.9	71.5	77.5
0950, 0980	Ceilings & Acoustic Treatment	101.1	111.3	107.5	116.5	70.0	87.4	117.1	70.0	87.6	103.7	70.0	82.6	117.1	64.4	84.1	103.7	71.5	83.5
0960	Flooring	117.2	116.7	117.1	89.1	61.5	81.1	91.6	62.7	83.2	100.0	61.5	88.7	87.7	61.5	80.1	101.7	62.7	90.3
0970, 0990	Wall Finishes & Painting/Coating	109.0	117.6	114.1	84.7	56.9	68.0	84.7	56.9	68.0	86.9	56.9	68.9	84.7	56.9	68.0	86.9	56.9	68.9
09	FINISHES	125.7	114.4	119.6	95.2	66.0	79.4	98.3	66.3	80.9	98.1	66.2	80.8	93.8	62.8	77.1	97.9	67.2	81.3
COVERS	DIVS. 10 - 14, 25, 28, 41, 43, 44, 46	100.0	109.8	102.3	100.0	84.0	96.2	100.0	84.0	96.2	100.0	84.4	96.3	100.0	83.4	96.1	100.0	84.5	96.4
21, 22, 23	FIRE SUPPRESSION, PLUMBING & HVAC	98.6	104.1	100.8	98.5	75.7	89.3	101.4	76.1	91.2	96.3	75.8	88.0	98.5	76.1	89.5	101.5	75.8	91.1
26, 27, 3370	ELECTRICAL, COMMUNICATIONS & UTIL.	125.5	107.9	116.8	101.5	64.1	83.0	100.6	60.8	80.9	97.3	64.6	81.1	101.5	64.6	83.2	95.4	64.6	80.1
MF2018	WEIGHTED AVERAGE	130.1	110.9	121.8	98.2	70.4	86.2	102.4	70.4	88.6	98.2	70.7	86.3	98.2	69.8	85.9	98.0	71.1	86.4

ARIZONA / ARKANSAS

DIVISION		PHOENIX 850,853 MAT.	INST.	TOTAL	PRESCOTT 863 MAT.	INST.	TOTAL	SHOW LOW 859 MAT.	INST.	TOTAL	TUCSON 856 - 857 MAT.	INST.	TOTAL	BATESVILLE 725 MAT.	INST.	TOTAL	CAMDEN 717 MAT.	INST.	TOTAL
015433	CONTRACTOR EQUIPMENT		94.1	94.1		85.9	85.9		87.5	87.5		87.5	87.5		86.8	86.8		86.8	86.8
0241, 31 - 34	SITE & INFRASTRUCTURE, DEMOLITION	92.8	96.1	95.0	77.0	87.7	84.3	105.4	88.7	94.0	88.6	88.9	88.8	71.6	82.8	79.3	77.4	82.8	81.1
0310	Concrete Forming & Accessories	92.8	70.9	74.1	98.4	71.0	75.1	92.7	69.0	72.6	89.4	77.4	79.2	85.5	59.9	63.7	77.9	60.4	63.0
0320	Concrete Reinforcing	99.7	73.6	87.1	99.3	73.6	86.9	105.0	73.6	89.8	87.1	73.7	80.6	83.0	67.4	75.5	91.3	68.5	80.3
0330	Cast-in-Place Concrete	84.5	69.9	79.0	88.4	67.9	80.7	79.7	67.4	75.0	82.6	67.6	77.0	68.7	73.2	70.4	74.9	73.3	74.3
03	CONCRETE	92.6	71.0	82.9	99.0	70.3	86.1	100.2	69.3	86.3	88.2	73.2	81.5	71.2	66.6	69.1	75.0	67.0	71.4
04	MASONRY	93.7	60.6	73.5	95.0	60.1	73.7	100.6	59.9	75.8	87.5	60.6	71.1	96.4	61.0	74.8	102.4	61.8	77.6
05	METALS	104.2	72.4	94.3	101.8	70.9	92.2	102.1	71.4	92.5	103.4	72.0	93.6	95.5	75.7	89.4	101.7	76.0	93.7
06	WOOD, PLASTICS & COMPOSITES	88.5	72.3	80.0	98.9	73.2	85.5	87.8	70.6	78.8	84.0	81.8	82.9	92.8	61.3	76.3	86.3	61.8	73.5
07	THERMAL & MOISTURE PROTECTION	99.2	69.9	86.6	99.8	70.5	87.3	99.3	68.1	86.1	99.6	70.3	87.0	104.4	61.1	85.8	98.1	61.8	82.5
08	OPENINGS	101.2	70.6	93.7	105.9	68.3	96.7	94.7	68.5	88.3	92.0	76.3	88.1	99.8	59.4	90.0	103.1	60.0	92.6
0920	Plaster & Gypsum Board	96.9	71.5	80.2	95.1	72.9	80.5	90.9	70.0	77.2	93.2	81.6	85.6	89.0	60.8	70.5	82.7	61.3	68.7
0950, 0980	Ceilings & Acoustic Treatment	116.2	71.5	88.2	115.2	72.9	88.7	103.7	70.0	82.6	105.0	81.6	90.3	92.4	60.8	72.6	90.9	61.3	72.4
0960	Flooring	102.6	62.7	90.9	90.1	64.0	82.5	103.6	61.5	91.3	93.3	65.3	85.1	84.9	70.4	80.7	85.7	71.8	81.7
0970, 0990	Wall Finishes & Painting/Coating	94.3	56.7	71.8	84.7	56.9	68.0	86.9	56.9	68.9	87.5	56.9	69.2	88.6	52.4	66.9	92.7	54.0	69.5
09	FINISHES	102.7	67.7	83.8	95.6	68.2	80.8	100.1	66.2	81.8	96.3	73.5	84.0	83.6	60.9	71.3	83.8	61.7	71.8
COVERS	DIVS. 10 - 14, 25, 28, 41, 43, 44, 46	100.0	85.6	96.6	100.0	84.3	96.3	100.0	84.4	96.3	100.0	85.6	96.6	100.0	78.8	95.0	100.0	78.8	95.0
21, 22, 23	FIRE SUPPRESSION, PLUMBING & HVAC	96.7	76.9	88.7	101.4	75.8	91.1	96.3	75.8	88.0	101.4	73.7	90.2	97.3	51.7	78.9	97.1	57.2	81.0
26, 27, 3370	ELECTRICAL, COMMUNICATIONS & UTIL.	97.6	62.7	80.3	100.4	64.6	82.7	95.1	64.6	80.0	97.2	59.7	78.6	96.2	58.2	77.4	94.1	56.0	75.2
MF2018	WEIGHTED AVERAGE	98.4	72.1	87.0	99.9	71.1	87.5	98.4	70.8	86.5	96.9	71.8	86.1	92.4	62.9	79.6	94.1	64.1	81.1

ARKANSAS

DIVISION		FAYETTEVILLE 727 MAT.	INST.	TOTAL	FORT SMITH 729 MAT.	INST.	TOTAL	HARRISON 726 MAT.	INST.	TOTAL	HOT SPRINGS 719 MAT.	INST.	TOTAL	JONESBORO 724 MAT.	INST.	TOTAL	LITTLE ROCK 720 - 722 MAT.	INST.	TOTAL
015433	CONTRACTOR EQUIPMENT		86.8	86.8		86.8	86.8		86.8	86.8		86.8	86.8		109.9	109.9		92.6	92.6
0241, 31 - 34	SITE & INFRASTRUCTURE, DEMOLITION	71.1	82.8	79.1	76.5	82.6	80.7	76.1	82.8	80.7	80.3	82.7	81.9	95.2	99.4	98.1	83.6	91.3	88.9
0310	Concrete Forming & Accessories	80.5	60.4	63.4	103.6	59.9	66.4	90.9	60.3	64.8	75.1	59.8	62.1	89.2	60.3	64.6	101.6	60.5	66.6
0320	Concrete Reinforcing	83.0	65.1	74.4	84.0	64.9	74.8	82.6	68.5	75.8	89.6	68.4	79.4	80.2	66.9	73.8	85.2	67.5	76.7
0330	Cast-in-Place Concrete	68.7	73.3	70.4	78.5	74.2	76.9	76.2	73.2	75.0	76.6	72.6	75.1	74.8	74.2	74.6	76.2	76.1	76.2
03	CONCRETE	70.9	66.4	68.9	78.2	66.5	72.9	77.5	66.9	72.7	78.0	66.5	72.8	75.2	68.0	71.9	79.7	67.7	74.3
04	MASONRY	87.2	61.8	71.7	94.6	60.6	73.9	96.5	61.8	75.3	76.7	58.3	65.5	88.6	60.6	71.5	91.6	61.5	73.3
05	METALS	95.5	74.8	89.1	97.9	74.2	90.5	96.7	75.8	90.2	101.7	75.5	93.5	92.4	90.0	91.7	98.0	74.2	90.6
06	WOOD, PLASTICS & COMPOSITES	88.2	61.8	74.4	114.0	61.8	86.7	99.8	61.8	79.9	83.2	61.8	72.0	97.2	62.1	78.8	107.3	62.0	83.6
07	THERMAL & MOISTURE PROTECTION	105.2	61.8	86.6	105.8	60.9	86.5	104.7	61.8	86.3	98.3	59.8	81.8	110.5	60.9	89.2	100.1	62.4	83.9
08	OPENINGS	99.8	59.1	89.9	101.8	58.6	91.3	100.6	59.9	90.7	103.1	59.5	92.4	105.4	59.8	94.3	95.5	59.1	86.6
0920	Plaster & Gypsum Board	88.4	61.3	70.7	94.6	61.3	72.8	93.7	61.3	72.5	81.0	61.3	68.1	102.6	61.3	75.5	109.0	61.3	77.7
0950, 0980	Ceilings & Acoustic Treatment	92.4	61.3	72.9	94.3	61.3	73.7	94.3	61.3	73.7	90.9	61.3	74.7	97.0	61.3	74.7	97.4	61.3	74.8
0960	Flooring	81.9	71.8	79.0	91.9	70.4	85.6	87.5	71.8	82.9	84.6	67.7	79.7	61.3	70.4	64.0	93.8	80.4	89.9
0970, 0990	Wall Finishes & Painting/Coating	88.6	54.0	67.9	88.6	51.6	66.5	88.6	54.0	67.9	92.7	51.9	68.3	78.4	53.0	63.2	91.9	53.0	68.6
09	FINISHES	82.8	61.7	71.3	87.1	61.0	73.0	85.8	61.7	72.7	83.6	60.3	71.0	82.4	61.4	71.0	95.4	63.4	78.1
COVERS	DIVS. 10 - 14, 25, 28, 41, 43, 44, 46	100.0	78.8	95.0	100.0	78.8	95.0	100.0	78.8	95.0	100.0	78.5	94.9	100.0	77.6	94.7	100.0	79.4	95.1
21, 22, 23	FIRE SUPPRESSION, PLUMBING & HVAC	97.4	61.6	83.0	101.2	48.3	79.9	97.3	49.5	78.0	97.1	49.0	77.7	101.8	50.8	81.2	100.8	48.4	79.7
26, 27, 3370	ELECTRICAL, COMMUNICATIONS & UTIL.	91.2	51.9	71.7	94.0	58.3	76.3	95.0	55.4	75.4	95.8	62.8	79.5	99.8	60.0	80.1	102.3	59.6	81.2
MF2018	WEIGHTED AVERAGE	91.3	64.2	79.6	95.1	61.9	80.8	93.7	62.3	80.1	93.4	62.5	80.1	94.8	65.7	82.2	96.0	63.5	81.9

For customer support on your Concrete & Masonry Costs with RSMeans Data, call 800.448.8182.

ARKANSAS / CALIFORNIA

DIVISION		PINE BLUFF 716			RUSSELLVILLE 728			TEXARKANA 718			WEST MEMPHIS 723			ALHAMBRA 917 - 918			ANAHEIM 928		
		MAT.	INST.	TOTAL	MAT.	INST.	TOTAL	MAT.	INST.	TOTAL	MAT.	INST.	TOTAL	MAT.	INST.	TOTAL	MAT.	INST.	TOTAL
015433	CONTRACTOR EQUIPMENT		86.8	86.8		86.8	86.8		88.0	88.0		109.9	109.9		92.5	92.5		97.5	97.5
0241, 31 - 34	SITE & INFRASTRUCTURE, DEMOLITION	82.8	82.6	82.7	72.8	82.6	79.5	93.4	84.6	87.4	101.1	99.2	99.8	99.1	101.8	101.0	100.7	102.6	102.1
0310	Concrete Forming & Accessories	74.8	60.1	62.3	86.6	59.6	63.6	81.6	60.2	63.4	95.6	60.1	65.4	115.5	135.6	132.6	102.0	139.4	133.3
0320	Concrete Reinforcing	91.2	67.4	79.7	83.5	68.3	76.2	90.8	68.5	80.0	80.2	60.4	70.6	98.9	133.2	115.4	93.2	133.1	112.5
0330	Cast-in-Place Concrete	76.6	73.0	75.2	71.9	72.5	72.2	83.5	73.0	79.6	78.4	73.7	76.6	82.3	126.4	99.0	88.3	129.7	104.0
03	CONCRETE	78.7	66.6	73.3	74.0	66.3	70.5	77.3	66.8	72.6	81.4	66.5	74.7	92.6	130.8	109.8	96.9	133.8	113.5
04	MASONRY	109.6	60.6	79.7	93.9	58.3	72.2	89.2	60.8	71.9	77.5	58.2	65.8	113.1	137.8	128.2	77.2	136.8	113.6
05	METALS	102.5	75.3	94.0	95.5	75.4	89.2	94.8	75.8	88.9	91.4	87.4	90.2	79.2	113.8	90.0	107.0	115.7	109.7
06	WOOD, PLASTICS & COMPOSITES	82.8	61.8	71.8	94.6	61.8	77.4	91.2	61.8	75.8	104.4	62.1	82.3	101.6	133.4	118.2	96.3	138.5	118.4
07	THERMAL & MOISTURE PROTECTION	98.4	61.2	82.4	105.5	59.8	85.9	99.1	61.0	82.8	111.1	59.8	89.1	103.4	128.8	114.3	109.4	132.9	119.5
08	OPENINGS	104.3	59.2	93.3	99.8	59.5	90.0	109.1	60.0	97.1	102.9	57.8	91.9	88.1	133.7	99.2	103.9	136.2	111.8
0920	Plaster & Gypsum Board	80.7	61.3	68.0	89.0	61.3	70.8	84.2	61.3	69.2	104.6	61.3	76.2	96.7	134.7	121.6	111.5	139.7	130.0
0950, 0980	Ceilings & Acoustic Treatment	90.9	61.3	72.4	92.4	61.3	72.9	96.6	61.3	74.5	95.6	61.3	74.1	115.1	134.7	127.4	106.8	139.7	127.4
0960	Flooring	84.4	71.8	80.7	84.4	67.7	79.5	86.5	70.8	81.9	63.8	67.7	65.0	104.7	119.9	109.1	98.9	119.9	105.0
0970, 0990	Wall Finishes & Painting/Coating	92.7	51.6	68.1	88.6	51.9	66.6	92.7	54.0	69.5	78.4	51.9	62.5	102.8	119.8	113.0	90.1	119.8	107.9
09	FINISHES	83.5	61.3	71.5	83.7	60.3	71.1	86.5	61.4	72.9	83.7	60.6	71.2	103.9	130.6	118.4	100.3	133.8	118.4
COVERS	DIVS. 10 - 14, 25, 28, 41, 43, 44, 46	100.0	78.7	95.0	100.0	78.5	94.9	100.0	78.7	95.0	100.0	79.3	95.1	100.0	116.6	103.9	100.0	117.7	104.2
21, 22, 23	FIRE SUPPRESSION, PLUMBING & HVAC	101.0	51.7	81.1	97.4	48.5	77.7	101.0	54.9	82.4	98.0	61.5	83.3	96.9	128.4	109.6	100.8	128.4	112.0
26, 27, 3370	ELECTRICAL, COMMUNICATIONS & UTIL.	94.2	57.6	76.1	94.0	55.4	74.9	95.7	57.6	76.9	101.1	62.8	82.1	121.7	131.9	126.7	91.7	114.7	103.1
MF2018	WEIGHTED AVERAGE	96.2	62.8	81.8	92.5	61.3	79.0	95.1	63.7	81.5	94.2	67.5	82.7	97.4	127.0	110.2	99.5	125.9	110.9

CALIFORNIA

DIVISION		BAKERSFIELD 932 - 933			BERKELEY 947			EUREKA 955			FRESNO 936 - 938			INGLEWOOD 903 - 905			LONG BEACH 906 - 908		
		MAT.	INST.	TOTAL	MAT.	INST.	TOTAL	MAT.	INST.	TOTAL	MAT.	INST.	TOTAL	MAT.	INST.	TOTAL	MAT.	INST.	TOTAL
015433	CONTRACTOR EQUIPMENT		100.3	100.3		98.8	98.8		95.3	95.3		95.5	95.5		95.6	95.6		95.6	95.6
0241, 31 - 34	SITE & INFRASTRUCTURE, DEMOLITION	100.0	108.7	106.0	115.4	103.4	107.2	112.0	100.4	104.1	102.2	100.9	101.3	87.5	99.7	95.9	94.5	99.7	98.1
0310	Concrete Forming & Accessories	100.8	138.4	132.8	115.8	173.3	164.7	111.2	160.0	152.7	100.3	159.0	150.3	105.7	136.0	131.5	99.8	136.0	130.6
0320	Concrete Reinforcing	95.8	133.1	113.8	87.7	134.6	110.4	101.3	134.7	117.4	80.1	133.7	106.0	99.7	133.2	115.9	98.9	133.2	115.5
0330	Cast-in-Place Concrete	84.7	130.1	101.9	110.8	136.5	120.5	95.9	132.9	109.9	91.7	132.5	107.1	81.8	129.1	99.6	93.1	129.1	106.7
03	CONCRETE	90.9	133.4	109.9	107.8	151.7	127.5	107.5	144.6	124.1	93.6	143.8	116.1	90.5	132.1	109.2	99.9	132.1	114.3
04	MASONRY	91.2	136.3	118.7	120.6	152.8	140.3	99.7	157.9	135.2	98.9	143.6	126.2	72.7	137.9	112.5	81.7	137.9	116.0
05	METALS	101.8	113.6	105.5	107.1	119.4	110.9	106.7	119.1	110.6	102.1	117.7	107.0	87.5	116.3	96.5	87.4	116.3	96.4
06	WOOD, PLASTICS & COMPOSITES	92.7	137.5	116.1	109.5	180.0	146.4	110.1	164.4	138.5	98.7	164.4	133.1	99.0	133.8	117.2	91.7	133.8	113.7
07	THERMAL & MOISTURE PROTECTION	108.0	123.5	114.6	114.9	155.3	132.2	112.9	153.3	130.3	98.9	133.7	113.8	105.2	129.9	115.8	105.5	129.9	116.0
08	OPENINGS	93.1	135.2	103.4	90.7	165.9	109.0	103.3	146.8	113.9	95.0	150.6	108.6	87.5	134.0	98.8	87.4	134.0	98.8
0920	Plaster & Gypsum Board	97.5	138.4	124.4	110.3	181.9	157.3	116.3	166.2	149.0	92.4	166.2	140.8	103.4	134.7	123.9	98.6	134.7	122.3
0950, 0980	Ceilings & Acoustic Treatment	109.8	138.4	127.7	105.7	181.9	153.4	111.3	166.2	145.7	102.6	166.2	142.4	102.5	134.7	122.7	102.5	134.7	122.7
0960	Flooring	99.9	119.9	105.7	121.5	149.0	129.5	102.5	149.0	116.1	99.3	122.1	106.0	113.7	119.9	115.5	110.4	119.9	113.2
0970, 0990	Wall Finishes & Painting/Coating	92.6	108.4	102.1	108.4	168.0	144.1	91.7	135.6	118.0	97.9	122.3	112.5	108.9	119.8	115.4	108.9	119.8	115.4
09	FINISHES	100.1	131.9	117.3	109.8	169.8	142.3	105.5	157.2	133.5	97.8	150.6	126.4	105.1	131.0	119.1	104.0	131.0	118.6
COVERS	DIVS. 10 - 14, 25, 28, 41, 43, 44, 46	100.0	127.0	106.4	100.0	132.5	107.7	100.0	129.6	107.0	100.0	129.5	107.0	100.0	117.6	104.1	100.0	117.6	104.1
21, 22, 23	FIRE SUPPRESSION, PLUMBING & HVAC	100.9	127.1	111.5	97.6	167.8	125.9	97.1	133.2	111.7	100.9	131.4	113.2	96.1	128.5	109.2	96.1	127.5	108.8
26, 27, 3370	ELECTRICAL, COMMUNICATIONS & UTIL.	105.5	109.7	107.6	100.1	155.8	127.7	98.1	130.8	114.3	96.2	108.6	102.3	99.5	131.9	115.5	99.3	131.9	115.4
MF2018	WEIGHTED AVERAGE	99.0	124.8	110.1	103.3	151.9	124.3	102.6	137.6	117.7	96.5	131.2	112.7	93.6	127.4	108.2	95.2	127.2	109.0

CALIFORNIA

DIVISION		LOS ANGELES 900 - 902			MARYSVILLE 959			MODESTO 953			MOJAVE 935			OAKLAND 946			OXNARD 930		
		MAT.	INST.	TOTAL	MAT.	INST.	TOTAL	MAT.	INST.	TOTAL	MAT.	INST.	TOTAL	MAT.	INST.	TOTAL	MAT.	INST.	TOTAL
015433	CONTRACTOR EQUIPMENT		104.7	104.7		95.3	95.3		95.3	95.3		95.5	95.5		98.8	98.8		94.5	94.5
0241, 31 - 34	SITE & INFRASTRUCTURE, DEMOLITION	94.0	110.4	105.3	108.4	100.0	102.6	103.7	100.4	101.4	94.8	100.6	98.8	121.8	103.4	109.2	102.4	98.8	99.9
0310	Concrete Forming & Accessories	103.2	140.0	134.5	100.4	159.0	150.2	96.1	159.3	149.9	109.7	138.1	133.8	103.2	173.4	162.9	101.9	139.4	133.8
0320	Concrete Reinforcing	98.1	131.4	114.2	101.3	133.5	116.9	105.0	133.8	118.9	96.3	133.0	114.0	89.7	137.2	112.6	94.5	133.1	113.1
0330	Cast-in-Place Concrete	84.6	131.3	102.1	107.1	132.4	116.7	95.9	132.6	109.7	79.3	128.8	98.0	105.1	136.5	117.0	92.4	129.1	106.2
03	CONCRETE	96.2	134.4	113.3	108.1	143.7	124.0	99.1	143.9	119.2	85.7	132.9	106.9	108.9	152.2	128.3	93.5	133.6	111.5
04	MASONRY	88.1	138.0	118.5	100.7	141.0	125.3	99.7	144.6	127.1	97.0	136.3	120.9	129.2	152.8	143.6	99.8	136.5	122.2
05	METALS	95.8	116.4	102.2	106.2	116.6	109.4	102.9	117.2	107.4	99.4	114.8	104.2	102.5	120.4	108.1	97.3	115.4	103.0
06	WOOD, PLASTICS & COMPOSITES	107.3	139.0	123.9	96.4	164.4	132.0	91.2	164.4	129.5	98.7	137.4	118.9	95.6	180.0	139.8	93.3	138.6	117.0
07	THERMAL & MOISTURE PROTECTION	101.7	132.5	114.9	112.3	137.8	123.3	111.9	139.8	123.9	104.8	121.8	112.1	113.2	155.3	131.3	108.1	131.7	118.2
08	OPENINGS	98.2	137.6	107.8	102.6	148.4	113.8	101.4	150.9	113.4	90.3	135.2	101.3	90.8	166.5	109.2	92.8	136.3	103.4
0920	Plaster & Gypsum Board	98.4	139.7	125.5	109.4	166.2	146.6	111.5	166.2	147.4	102.3	138.4	126.0	104.5	181.9	155.3	96.0	139.7	124.7
0950, 0980	Ceilings & Acoustic Treatment	109.6	139.7	128.5	110.6	166.2	145.4	106.8	166.2	144.0	104.9	138.4	125.9	107.7	181.9	154.2	104.7	139.7	126.6
0960	Flooring	110.9	119.9	113.5	98.1	114.0	102.7	98.4	133.0	108.5	100.7	115.5	105.0	114.4	149.0	124.5	93.1	119.9	100.9
0970, 0990	Wall Finishes & Painting/Coating	108.1	119.8	115.1	91.7	135.6	118.0	91.7	135.6	118.0	84.5	108.6	99.0	108.4	168.0	144.1	84.5	114.6	102.5
09	FINISHES	105.8	134.1	121.2	102.6	150.7	128.6	101.6	154.0	129.9	98.7	131.1	116.3	107.9	170.0	141.5	95.9	133.3	116.1
COVERS	DIVS. 10 - 14, 25, 28, 41, 43, 44, 46	100.0	118.9	104.5	100.0	129.6	107.0	100.0	129.6	107.0	100.0	113.0	103.1	100.0	132.5	107.7	100.0	117.9	104.2
21, 22, 23	FIRE SUPPRESSION, PLUMBING & HVAC	100.4	128.6	111.8	97.1	126.1	108.8	100.8	131.4	113.2	97.0	127.0	109.1	101.3	167.9	128.2	100.8	128.5	112.0
26, 27, 3370	ELECTRICAL, COMMUNICATIONS & UTIL.	98.1	131.9	114.8	95.3	116.3	105.6	97.4	110.2	103.7	94.8	109.7	102.2	99.4	163.0	130.9	100.6	117.4	108.9
MF2018	WEIGHTED AVERAGE	98.4	129.3	111.8	101.8	130.7	114.3	100.8	132.1	114.3	95.6	123.6	107.7	103.9	153.2	125.2	98.2	125.8	110.1

For customer support on your Concrete & Masonry Costs with RSMeans Data, call 800.448.8182.

457

CALIFORNIA

DIVISION		PALM SPRINGS 922			PALO ALTO 943			PASADENA 910 - 912			REDDING 960			RICHMOND 948			RIVERSIDE 925		
		MAT.	INST.	TOTAL	MAT.	INST.	TOTAL	MAT.	INST.	TOTAL	MAT.	INST.	TOTAL	MAT.	INST.	TOTAL	MAT.	INST.	TOTAL
015433	CONTRACTOR EQUIPMENT		96.3	96.3		98.8	98.8		92.5	92.5		95.3	95.3		98.8	98.8		96.3	96.3
0241, 31 - 34	SITE & INFRASTRUCTURE, DEMOLITION	91.8	100.8	98.0	111.6	103.4	106.0	95.9	101.8	100.0	124.5	100.2	107.8	120.5	103.4	108.8	99.2	100.8	100.3
0310	Concrete Forming & Accessories	98.4	132.1	127.0	101.1	163.9	154.6	102.8	135.6	130.7	102.3	155.1	147.3	119.0	173.0	165.0	102.4	139.4	133.9
0320	Concrete Reinforcing	106.5	132.9	119.3	87.7	136.9	111.4	99.8	133.2	115.9	134.5	133.8	134.2	87.7	136.9	111.4	103.5	133.1	117.8
0330	Cast-in-Place Concrete	84.2	129.6	101.3	93.7	136.4	109.8	78.1	126.4	96.3	117.0	132.6	122.9	107.8	136.4	118.6	91.6	129.7	106.0
03	CONCRETE	91.3	130.4	108.9	97.6	147.8	120.2	88.3	130.8	107.4	118.6	142.0	129.1	110.9	152.0	129.3	97.3	133.8	113.7
04	MASONRY	75.2	134.1	111.1	102.6	149.2	131.0	99.1	137.8	122.7	120.2	144.6	135.1	120.4	149.2	138.0	76.2	136.5	113.0
05	METALS	107.6	115.2	110.0	100.0	119.7	106.2	79.2	113.8	90.0	103.2	117.3	107.6	100.1	119.8	106.2	107.0	115.7	109.7
06	WOOD, PLASTICS & COMPOSITES	90.6	129.0	110.7	92.8	167.7	132.0	86.2	133.4	110.9	104.7	158.6	132.9	113.2	180.0	148.2	96.3	138.5	118.4
07	THERMAL & MOISTURE PROTECTION	108.7	129.3	117.5	112.4	149.9	128.5	102.9	128.3	113.8	126.7	139.9	132.4	113.1	152.8	130.1	109.6	132.8	119.6
08	OPENINGS	100.0	131.3	107.6	90.8	155.9	106.7	88.1	133.7	99.2	116.4	147.7	124.0	90.8	165.8	109.1	102.6	136.2	110.8
0920	Plaster & Gypsum Board	106.1	129.9	121.7	102.8	169.2	146.4	91.4	134.7	119.8	120.9	160.2	146.7	110.7	181.9	157.4	110.6	139.7	129.7
0950, 0980	Ceilings & Acoustic Treatment	103.6	129.9	120.1	106.3	169.2	145.7	115.1	134.7	127.4	138.0	160.2	151.9	106.3	181.9	153.7	111.5	139.7	129.1
0960	Flooring	101.0	115.5	105.2	112.9	149.0	123.5	98.2	119.9	104.6	88.8	126.7	99.9	124.1	149.0	131.3	102.4	119.9	107.5
0970, 0990	Wall Finishes & Painting/Coating	88.7	114.6	104.2	108.4	168.0	144.1	102.8	119.8	113.0	101.6	135.6	122.0	108.4	168.0	144.1	88.7	119.8	107.3
09	FINISHES	98.7	126.7	113.9	106.3	162.6	136.7	101.1	130.7	117.1	108.9	149.7	131.0	111.2	169.8	142.9	102.0	133.8	119.2
COVERS	DIVS. 10 - 14, 25, 28, 41, 43, 44, 46	100.0	116.7	103.9	100.0	131.2	107.4	100.0	116.6	103.9	100.0	129.0	106.8	100.0	132.5	107.6	100.0	120.4	104.8
21, 22, 23	FIRE SUPPRESSION, PLUMBING & HVAC	97.1	125.9	108.7	97.6	164.8	124.7	96.9	128.4	109.6	101.1	131.4	113.3	97.6	162.4	123.7	100.8	128.4	112.0
26, 27, 3370	ELECTRICAL, COMMUNICATIONS & UTIL.	94.4	106.7	100.5	99.3	170.0	134.3	118.7	131.9	125.2	99.5	121.7	110.5	99.8	140.3	119.8	91.5	116.0	103.6
MF2018	WEIGHTED AVERAGE	97.3	121.9	108.0	99.4	150.8	121.6	95.4	127.0	109.0	107.8	132.6	118.6	102.8	148.2	122.4	99.4	126.0	110.9

CALIFORNIA

DIVISION		SACRAMENTO 942,956 - 958			SALINAS 939			SAN BERNARDINO 923 - 924			SAN DIEGO 919 - 921			SAN FRANCISCO 940 - 941			SAN JOSE 951		
		MAT.	INST.	TOTAL	MAT.	INST.	TOTAL	MAT.	INST.	TOTAL	MAT.	INST.	TOTAL	MAT.	INST.	TOTAL	MAT.	INST.	TOTAL
015433	CONTRACTOR EQUIPMENT		97.6	97.6		95.5	95.5		96.3	96.3		103.0	103.0		111.2	111.2		97.0	97.0
0241, 31 - 34	SITE & INFRASTRUCTURE, DEMOLITION	99.6	108.5	105.7	115.6	100.9	105.5	78.2	100.8	93.7	104.5	109.5	107.9	121.3	114.9	116.9	135.4	95.7	108.2
0310	Concrete Forming & Accessories	101.9	157.3	149.1	105.9	157.9	150.1	106.5	139.4	134.5	104.0	129.4	125.6	107.4	168.5	159.4	102.5	173.6	163.0
0320	Concrete Reinforcing	83.3	133.8	107.7	95.0	134.3	113.9	103.5	133.1	117.8	96.0	131.1	112.9	108.0	135.5	117.5	92.7	137.4	114.3
0330	Cast-in-Place Concrete	88.9	133.2	105.6	91.2	132.9	107.0	63.3	129.7	88.4	89.5	125.9	103.3	117.4	133.9	123.6	111.4	135.9	120.6
03	CONCRETE	100.4	143.0	119.5	103.5	143.5	121.5	72.1	133.8	99.8	97.1	127.6	110.8	120.5	149.1	133.3	105.0	152.5	126.3
04	MASONRY	103.5	144.6	128.6	96.9	150.1	129.4	83.0	136.5	115.6	90.7	132.6	116.3	136.8	153.9	147.2	128.4	156.1	145.3
05	METALS	98.2	112.2	102.6	102.0	118.3	107.1	107.0	115.7	109.7	95.8	115.0	101.8	107.2	124.6	112.7	101.1	125.7	108.8
06	WOOD, PLASTICS & COMPOSITES	90.8	161.6	127.8	98.7	161.4	131.5	100.4	138.5	120.3	95.5	126.9	111.9	101.5	174.3	139.5	103.1	179.8	143.2
07	THERMAL & MOISTURE PROTECTION	122.2	141.7	130.6	105.5	145.8	122.8	107.8	132.8	118.6	108.1	120.8	113.5	119.0	156.1	134.9	108.5	160.6	130.8
08	OPENINGS	103.6	149.3	114.7	93.9	155.7	108.9	100.0	137.8	109.2	100.1	127.5	106.8	99.4	162.2	114.7	92.6	164.4	110.6
0920	Plaster & Gypsum Board	98.9	163.1	141.0	97.2	163.1	140.4	112.3	139.7	130.3	97.5	127.4	117.1	104.0	175.9	149.9	107.1	181.9	156.2
0950, 0980	Ceilings & Acoustic Treatment	106.3	163.1	141.9	104.9	163.1	141.4	106.8	139.7	127.4	126.2	127.4	127.0	119.8	175.9	154.9	107.6	181.9	154.2
0960	Flooring	112.7	126.7	116.8	95.0	142.9	109.0	104.6	119.9	109.0	108.0	119.9	111.5	111.2	142.9	120.5	92.1	149.0	108.7
0970, 0990	Wall Finishes & Painting/Coating	105.2	135.6	123.4	85.4	168.0	134.9	88.7	119.8	107.3	103.5	119.8	113.3	109.1	176.8	149.7	92.0	168.0	137.5
09	FINISHES	104.6	151.3	129.9	98.3	157.7	130.4	100.2	133.8	118.4	108.8	126.5	118.4	109.1	165.9	139.8	100.9	170.0	138.3
COVERS	DIVS. 10 - 14, 25, 28, 41, 43, 44, 46	100.0	129.8	107.0	100.0	129.3	106.9	100.0	117.7	104.2	100.0	116.1	103.8	100.0	128.9	106.8	100.0	132.1	107.6
21, 22, 23	FIRE SUPPRESSION, PLUMBING & HVAC	101.2	130.5	113.0	97.0	136.9	113.1	97.1	128.4	109.7	100.7	127.0	111.3	101.2	179.4	132.8	100.8	173.4	130.1
26, 27, 3370	ELECTRICAL, COMMUNICATIONS & UTIL.	95.3	121.6	108.3	95.7	132.6	114.0	94.4	111.9	103.0	105.3	103.3	104.3	100.0	186.3	142.7	100.5	188.8	144.2
MF2018	WEIGHTED AVERAGE	101.0	133.1	114.9	99.2	137.8	115.9	95.0	125.4	108.1	100.3	121.3	109.4	107.6	159.0	129.8	103.0	158.4	126.9

CALIFORNIA

DIVISION		SAN LUIS OBISPO 934			SAN MATEO 944			SAN RAFAEL 949			SANTA ANA 926 - 927			SANTA BARBARA 931			SANTA CRUZ 950		
		MAT.	INST.	TOTAL	MAT.	INST.	TOTAL	MAT.	INST.	TOTAL	MAT.	INST.	TOTAL	MAT.	INST.	TOTAL	MAT.	INST.	TOTAL
015433	CONTRACTOR EQUIPMENT		95.5	95.5		98.8	98.8		98.0	98.0		96.3	96.3		95.5	95.5		97.0	97.0
0241, 31 - 34	SITE & INFRASTRUCTURE, DEMOLITION	107.4	100.6	102.7	118.1	103.5	108.1	111.8	108.2	109.4	90.2	100.8	97.5	102.4	100.6	101.2	135.0	95.6	107.9
0310	Concrete Forming & Accessories	111.6	139.5	135.4	108.1	173.7	163.9	114.7	173.0	164.3	107.0	139.4	134.5	102.5	139.4	133.9	102.6	158.0	149.7
0320	Concrete Reinforcing	96.3	133.1	114.0	87.7	137.6	111.8	88.4	134.8	110.8	107.1	133.1	119.7	94.5	133.1	113.1	114.4	134.3	124.0
0330	Cast-in-Place Concrete	98.1	129.0	109.8	104.3	136.7	116.5	121.4	135.5	126.7	80.8	129.7	99.3	92.0	129.0	106.0	110.7	134.6	119.7
03	CONCRETE	101.1	133.6	115.7	107.3	152.5	127.6	131.4	151.0	140.2	88.9	133.8	109.1	93.4	133.6	111.4	107.6	144.4	124.1
04	MASONRY	98.5	135.2	120.9	120.1	159.0	143.8	97.5	155.8	133.0	72.5	136.8	111.8	97.2	135.2	120.4	132.2	150.2	143.2
05	METALS	100.1	115.4	104.9	99.9	121.2	106.5	101.2	115.6	105.7	107.0	115.7	109.7	97.9	115.3	103.3	108.6	122.6	112.9
06	WOOD, PLASTICS & COMPOSITES	100.9	138.6	120.6	101.7	180.0	142.7	101.5	179.8	142.5	102.4	138.5	121.3	93.3	138.6	117.0	103.1	161.5	133.6
07	THERMAL & MOISTURE PROTECTION	105.6	130.8	116.4	112.8	160.0	133.1	117.9	155.8	134.2	109.1	132.9	119.3	105.0	131.0	116.2	108.0	148.5	125.4
08	OPENINGS	92.1	135.8	102.8	90.8	166.5	109.2	101.3	165.1	116.8	99.3	136.2	108.3	93.6	137.8	104.4	93.9	155.7	108.9
0920	Plaster & Gypsum Board	102.8	139.7	127.0	108.7	181.9	156.7	111.1	181.9	157.5	113.4	139.7	130.7	96.0	139.7	124.7	114.8	163.1	146.5
0950, 0980	Ceilings & Acoustic Treatment	104.9	139.7	126.7	106.3	181.9	153.7	115.9	181.9	157.2	106.8	139.7	127.4	104.7	139.7	126.6	108.1	163.1	142.6
0960	Flooring	101.5	119.9	106.9	116.9	149.0	126.3	129.4	142.9	133.3	105.2	119.9	109.5	94.1	119.9	101.7	96.2	142.9	109.8
0970, 0990	Wall Finishes & Painting/Coating	84.5	114.6	102.5	108.4	168.0	144.1	104.6	168.0	142.6	88.7	119.8	107.3	84.5	114.6	102.5	92.0	168.0	137.5
09	FINISHES	100.1	133.3	118.0	108.7	170.2	142.0	113.0	168.6	143.1	101.7	133.8	119.0	96.3	133.3	116.3	103.3	157.8	132.8
COVERS	DIVS. 10 - 14, 25, 28, 41, 43, 44, 46	100.0	126.9	106.3	100.0	132.6	107.7	100.0	131.9	107.5	100.0	117.7	104.2	100.0	117.9	104.2	100.0	129.6	107.0
21, 22, 23	FIRE SUPPRESSION, PLUMBING & HVAC	97.0	128.5	109.7	97.6	169.3	126.5	97.6	179.6	130.6	97.1	128.4	109.7	100.8	128.5	112.0	100.8	137.0	115.4
26, 27, 3370	ELECTRICAL, COMMUNICATIONS & UTIL.	94.8	113.6	104.1	99.3	178.8	138.6	96.2	125.7	110.8	94.4	114.7	104.4	93.9	111.7	102.7	99.7	132.6	116.0
MF2018	WEIGHTED AVERAGE	98.4	125.5	110.1	101.9	156.6	125.5	105.0	150.3	124.6	97.1	125.7	109.5	97.4	125.0	109.4	104.9	138.1	119.2

CALIFORNIA / COLORADO

DIVISION		SANTA ROSA 954 MAT.	INST.	TOTAL	STOCKTON 952 MAT.	INST.	TOTAL	SUSANVILLE 961 MAT.	INST.	TOTAL	VALLEJO 945 MAT.	INST.	TOTAL	VAN NUYS 913-916 MAT.	INST.	TOTAL	ALAMOSA 811 MAT.	INST.	TOTAL
015433	CONTRACTOR EQUIPMENT		95.8	95.8		95.3	95.3		95.3	95.3		98.0	98.0		92.5	92.5		87.8	87.8
0241, 31 - 34	SITE & INFRASTRUCTURE, DEMOLITION	104.1	100.3	101.5	103.4	100.4	101.3	131.6	100.0	109.9	99.9	108.5	105.8	113.5	101.8	105.5	135.8	81.2	98.3
0310	Concrete Forming & Accessories	99.5	172.4	161.5	100.8	157.1	148.7	104.3	150.2	143.4	103.8	172.1	161.9	110.3	135.6	131.8	100.4	68.5	73.3
0320	Concrete Reinforcing	102.2	134.8	117.9	105.0	133.8	118.9	134.5	133.6	134.1	89.5	134.7	111.3	99.8	133.2	115.9	107.3	66.6	87.7
0330	Cast-in-Place Concrete	105.2	134.1	116.1	93.4	132.5	108.2	106.4	132.5	116.3	96.7	134.7	111.1	82.4	126.4	99.0	98.6	70.9	88.1
03	CONCRETE	108.0	150.6	127.1	98.2	142.9	118.3	120.9	139.8	129.4	105.0	150.3	125.3	102.2	130.8	115.0	112.5	69.6	93.2
04	MASONRY	98.8	154.8	132.9	99.7	144.6	127.1	118.6	141.0	132.3	75.8	154.8	124.0	113.1	137.8	128.2	128.3	58.5	85.7
05	METALS	107.4	120.3	111.4	103.2	117.2	107.5	102.2	116.7	106.7	101.1	115.1	105.5	78.4	113.8	89.4	103.2	77.4	95.2
06	WOOD, PLASTICS & COMPOSITES	92.3	179.6	138.0	97.4	161.4	130.9	106.6	152.4	130.6	89.3	179.8	136.6	95.6	133.4	115.4	93.9	71.0	81.9
07	THERMAL & MOISTURE PROTECTION	109.5	155.2	129.1	112.3	138.9	123.7	129.0	136.4	132.2	115.8	155.0	132.6	104.2	128.3	114.5	108.4	66.6	90.5
08	OPENINGS	100.9	165.7	116.7	101.4	149.2	113.0	117.4	141.8	123.3	102.9	164.4	117.9	88.0	133.7	99.1	96.4	69.9	89.9
0920	Plaster & Gypsum Board	108.3	181.9	156.6	111.5	163.1	145.4	121.4	153.8	142.7	105.4	181.9	155.6	94.9	134.7	121.0	83.8	70.3	75.0
0950, 0980	Ceilings & Acoustic Treatment	106.8	181.9	153.9	114.7	163.1	145.0	128.9	153.8	144.5	117.3	181.9	157.8	112.6	134.7	126.4	112.0	70.3	85.9
0960	Flooring	101.5	136.3	111.7	98.4	133.0	108.5	89.3	126.7	100.2	122.7	149.0	130.4	101.6	119.9	107.0	106.8	66.9	95.1
0970, 0990	Wall Finishes & Painting/Coating	88.7	168.0	136.2	91.7	132.0	115.8	101.6	135.6	122.0	105.6	168.0	143.0	102.8	119.8	113.0	97.5	76.1	84.7
09	FINISHES	100.9	167.2	136.8	103.4	151.9	129.6	108.0	145.9	128.5	109.3	169.6	141.9	103.3	130.7	118.1	102.6	69.0	84.4
COVERS	DIVS. 10 - 14, 25, 28, 41, 43, 44, 46	100.0	131.1	107.3	100.0	129.2	106.9	100.0	128.3	106.7	100.0	131.6	107.4	100.0	116.6	103.9	100.0	84.8	96.4
21, 22, 23	FIRE SUPPRESSION, PLUMBING & HVAC	97.1	179.8	130.4	100.8	131.4	113.2	97.3	131.4	111.0	101.3	147.6	120.0	96.9	128.4	109.6	97.1	69.8	86.1
26, 27, 3370	ELECTRICAL, COMMUNICATIONS & UTIL.	94.7	125.7	110.0	97.4	112.8	105.0	99.8	121.7	110.6	92.9	126.7	109.6	118.6	131.9	125.2	95.5	60.9	78.4
MF2018	WEIGHTED AVERAGE	101.2	149.8	122.2	101.0	131.9	114.3	107.3	131.0	117.5	100.5	143.4	119.0	98.5	127.0	110.8	103.3	69.2	88.6

COLORADO

DIVISION		BOULDER 803 MAT.	INST.	TOTAL	COLORADO SPRINGS 808-809 MAT.	INST.	TOTAL	DENVER 800-802 MAT.	INST.	TOTAL	DURANGO 813 MAT.	INST.	TOTAL	FORT COLLINS 805 MAT.	INST.	TOTAL	FORT MORGAN 807 MAT.	INST.	TOTAL
015433	CONTRACTOR EQUIPMENT		89.3	89.3		87.3	87.3		96.4	96.4		87.8	87.8		89.3	89.3		89.3	89.3
0241, 31 - 34	SITE & INFRASTRUCTURE, DEMOLITION	94.7	87.8	90.0	97.2	83.4	87.7	101.1	97.6	98.7	129.1	81.3	96.3	107.4	87.9	94.1	97.0	87.3	90.4
0310	Concrete Forming & Accessories	103.7	69.1	74.3	93.6	68.5	72.3	102.6	68.7	73.8	106.6	68.3	74.0	101.6	68.4	73.4	104.2	68.3	73.6
0320	Concrete Reinforcing	102.3	66.8	85.2	101.5	69.1	85.8	100.3	69.1	85.2	107.3	66.6	87.7	102.4	66.8	85.2	102.6	66.8	85.3
0330	Cast-in-Place Concrete	113.5	71.4	97.6	116.6	71.5	99.6	118.1	72.5	100.9	113.4	70.9	97.4	128.2	71.0	106.6	111.4	70.2	95.8
03	CONCRETE	107.8	70.0	90.8	111.4	70.2	92.9	115.2	70.6	95.1	114.2	69.5	94.1	119.2	69.6	96.9	116.2	69.2	93.2
04	MASONRY	104.1	63.9	79.6	107.5	62.1	79.8	113.5	62.7	82.5	116.0	59.5	81.6	120.9	61.9	84.9	119.6	63.7	85.5
05	METALS	93.9	76.5	88.5	97.1	77.6	91.0	99.6	77.3	92.6	103.2	77.4	95.2	95.2	76.5	89.4	93.6	76.6	88.3
06	WOOD, PLASTICS & COMPOSITES	102.9	70.5	86.0	91.8	70.7	80.7	107.2	70.7	88.1	103.4	71.0	86.4	100.2	70.5	84.7	102.9	70.5	86.0
07	THERMAL & MOISTURE PROTECTION	106.3	71.0	91.2	107.3	70.2	91.4	104.5	72.1	90.6	108.3	66.9	90.5	106.7	70.2	91.0	106.3	69.3	90.4
08	OPENINGS	95.0	69.7	88.8	99.1	70.3	92.1	102.4	70.3	94.6	102.9	69.9	94.9	95.0	69.7	88.8	95.0	69.7	88.8
0920	Plaster & Gypsum Board	123.3	70.3	88.5	108.5	70.3	83.5	119.1	70.3	87.1	97.1	70.3	79.5	116.5	70.3	86.2	123.3	70.3	88.5
0950, 0980	Ceilings & Acoustic Treatment	97.4	70.3	80.4	107.6	70.3	84.2	111.0	70.3	85.5	112.0	70.3	84.9	97.4	70.3	80.4	97.4	70.3	80.4
0960	Flooring	107.5	73.2	97.5	98.7	71.3	90.7	103.3	74.7	95.0	112.3	65.8	98.7	103.5	74.7	95.1	108.0	73.2	97.8
0970, 0990	Wall Finishes & Painting/Coating	96.2	76.1	84.2	95.9	80.7	86.8	105.9	74.7	87.2	97.5	76.1	84.7	96.2	76.1	84.2	96.2	76.1	84.2
09	FINISHES	102.3	70.4	85.1	100.2	70.1	83.9	104.6	70.2	86.0	105.2	68.9	85.6	100.7	70.2	84.2	102.4	69.9	84.8
COVERS	DIVS. 10 - 14, 25, 28, 41, 43, 44, 46	100.0	84.5	96.4	100.0	84.1	96.3	100.0	84.3	96.3	100.0	84.7	96.4	100.0	83.8	96.2	100.0	83.8	96.2
21, 22, 23	FIRE SUPPRESSION, PLUMBING & HVAC	97.2	71.9	87.0	101.1	71.5	89.2	100.9	74.7	90.4	97.1	56.8	80.9	101.0	71.5	89.1	97.2	70.9	86.6
26, 27, 3370	ELECTRICAL, COMMUNICATIONS & UTIL.	99.0	79.8	89.5	102.1	73.9	88.1	103.7	79.8	91.8	95.1	53.5	74.5	99.0	79.8	89.5	99.3	75.2	87.4
MF2018	WEIGHTED AVERAGE	99.2	73.7	88.2	101.8	72.3	89.1	104.0	75.1	91.5	103.6	65.5	87.1	102.8	73.2	90.0	99.9	72.5	88.0

COLORADO

DIVISION		GLENWOOD SPRINGS 816 MAT.	INST.	TOTAL	GOLDEN 804 MAT.	INST.	TOTAL	GRAND JUNCTION 815 MAT.	INST.	TOTAL	GREELEY 806 MAT.	INST.	TOTAL	MONTROSE 814 MAT.	INST.	TOTAL	PUEBLO 810 MAT.	INST.	TOTAL
015433	CONTRACTOR EQUIPMENT		90.9	90.9		89.3	89.3		90.9	90.9		89.3	89.3		89.4	89.4		87.8	87.8
0241, 31 - 34	SITE & INFRASTRUCTURE, DEMOLITION	145.1	88.7	106.3	107.6	87.6	93.9	129.0	89.1	101.6	94.0	87.8	89.8	138.5	85.4	102.0	120.8	81.5	93.8
0310	Concrete Forming & Accessories	97.0	68.1	72.4	96.3	68.7	72.8	105.4	69.3	74.7	99.2	68.4	73.0	96.6	68.3	72.5	102.6	68.8	73.9
0320	Concrete Reinforcing	106.1	66.8	87.2	102.6	66.8	85.3	106.5	66.6	87.3	102.3	66.8	85.2	106.0	66.8	87.1	102.6	69.0	86.4
0330	Cast-in-Place Concrete	98.6	69.9	87.7	111.5	71.4	96.4	109.2	72.2	95.2	107.1	71.0	93.5	98.6	70.4	88.0	97.9	71.9	88.1
03	CONCRETE	117.9	69.1	96.0	117.2	69.8	95.9	110.7	70.4	92.6	102.7	69.6	87.9	108.4	69.3	90.9	100.8	70.5	87.2
04	MASONRY	102.2	63.6	78.6	122.7	62.7	86.1	136.2	61.7	90.8	113.7	63.2	82.9	108.4	63.6	81.1	98.7	60.9	75.6
05	METALS	102.9	77.3	94.9	93.8	76.5	88.4	104.6	77.3	96.1	95.2	76.6	89.4	102.0	77.6	94.4	106.4	78.9	97.8
06	WOOD, PLASTICS & COMPOSITES	88.9	70.7	79.4	94.1	70.5	81.8	101.4	70.7	85.3	97.2	70.5	83.2	89.7	70.8	79.8	96.6	71.0	83.2
07	THERMAL & MOISTURE PROTECTION	108.2	68.0	90.9	107.5	70.6	91.6	107.2	69.3	90.9	105.9	70.5	90.7	108.4	68.0	91.0	106.8	68.6	90.4
08	OPENINGS	101.9	69.8	94.1	95.0	69.7	88.8	102.6	69.8	94.6	95.0	69.7	88.8	103.1	69.9	95.0	98.1	70.5	91.4
0920	Plaster & Gypsum Board	125.3	70.3	89.3	113.4	70.3	85.1	138.6	70.3	93.8	115.1	70.3	85.7	82.7	70.3	74.6	88.1	70.3	76.4
0950, 0980	Ceilings & Acoustic Treatment	111.4	70.3	85.7	97.4	70.3	80.4	111.4	70.3	85.7	97.4	70.3	80.4	112.0	70.3	85.9	119.7	70.3	88.8
0960	Flooring	105.8	73.2	96.3	101.0	73.2	92.9	111.6	69.6	99.3	102.3	72.6	93.7	109.5	69.8	97.9	108.1	72.6	97.7
0970, 0990	Wall Finishes & Painting/Coating	97.4	76.1	84.7	96.2	76.1	84.2	97.4	76.1	84.7	96.2	76.1	84.2	97.5	76.1	84.7	97.5	76.1	84.7
09	FINISHES	108.5	70.1	87.7	100.3	70.1	84.0	110.2	69.9	88.4	99.4	69.8	83.4	103.2	69.5	85.0	103.6	70.1	85.5
COVERS	DIVS. 10 - 14, 25, 28, 41, 43, 44, 46	100.0	84.1	96.3	100.0	84.1	96.3	100.0	84.9	96.5	100.0	83.8	96.2	100.0	84.4	96.3	100.0	84.8	96.4
21, 22, 23	FIRE SUPPRESSION, PLUMBING & HVAC	97.1	56.7	80.8	97.2	71.3	86.7	100.9	73.3	89.8	101.0	71.4	89.1	97.1	56.8	80.9	100.9	71.6	89.1
26, 27, 3370	ELECTRICAL, COMMUNICATIONS & UTIL.	93.0	53.5	73.4	99.3	77.6	88.5	94.8	52.5	73.9	99.0	79.8	89.5	94.8	53.5	74.3	95.5	63.3	79.6
MF2018	WEIGHTED AVERAGE	103.6	66.6	87.6	101.5	73.0	89.2	105.6	70.0	90.2	99.9	73.3	88.4	102.3	66.3	86.7	101.6	70.7	88.2

COLORADO / CONNECTICUT

DIVISION		SALIDA 812 MAT.	INST.	TOTAL	BRIDGEPORT 066 MAT.	INST.	TOTAL	BRISTOL 060 MAT.	INST.	TOTAL	HARTFORD 061 MAT.	INST.	TOTAL	MERIDEN 064 MAT.	INST.	TOTAL	NEW BRITAIN 060 MAT.	INST.	TOTAL
015433	CONTRACTOR EQUIPMENT		89.4	89.4		94.2	94.2		94.2	94.2		100.4	100.4		94.6	94.6		94.2	94.2
0241, 31 - 34	SITE & INFRASTRUCTURE, DEMOLITION	128.8	85.0	98.7	107.9	95.6	99.4	107.1	95.6	99.2	102.3	104.7	103.9	104.5	96.2	98.8	107.2	95.6	99.2
0310	Concrete Forming & Accessories	105.4	68.4	74.0	104.3	115.7	114.0	104.3	117.0	115.1	102.9	117.4	115.2	104.0	117.0	115.0	104.8	117.0	115.2
0320	Concrete Reinforcing	105.7	66.6	86.9	116.0	144.7	129.8	116.0	144.7	129.8	110.4	144.7	126.9	116.0	144.7	129.8	116.0	144.7	129.8
0330	Cast-in-Place Concrete	113.0	70.5	97.0	99.8	126.6	110.0	93.5	126.6	106.0	95.3	128.4	107.8	90.0	126.6	103.8	95.1	126.6	107.0
03	CONCRETE	109.0	69.4	91.3	101.9	123.9	111.8	99.0	124.5	110.5	100.2	125.2	111.4	97.3	124.5	109.5	99.7	124.5	110.9
04	MASONRY	136.5	59.5	89.5	111.8	130.5	123.2	103.7	130.5	120.1	107.3	130.6	121.5	103.4	130.5	119.9	105.4	130.5	120.8
05	METALS	101.7	77.5	94.1	103.5	118.7	108.2	103.5	118.7	108.2	108.2	117.6	111.1	100.3	118.6	106.0	99.4	118.7	105.4
06	WOOD, PLASTICS & COMPOSITES	97.9	70.8	83.8	106.2	112.7	109.6	106.2	114.5	110.6	97.6	114.8	106.6	106.2	114.5	110.6	106.2	114.5	110.6
07	THERMAL & MOISTURE PROTECTION	107.2	66.9	89.9	98.9	121.6	108.7	99.1	120.6	108.3	103.5	121.6	111.3	99.1	120.6	108.3	99.1	120.6	108.3
08	OPENINGS	96.4	69.9	90.0	96.6	119.9	102.3	96.6	121.6	102.7	98.3	121.7	104.0	98.8	120.9	104.2	96.6	121.6	102.7
0920	Plaster & Gypsum Board	82.9	70.3	74.7	122.9	112.9	116.3	122.9	114.8	117.6	110.1	114.8	113.2	125.4	114.8	118.4	122.9	114.8	117.6
0950, 0980	Ceilings & Acoustic Treatment	112.0	70.3	85.9	102.9	112.9	109.2	102.9	114.8	110.3	103.3	114.8	110.5	111.0	114.8	113.4	102.9	114.8	110.3
0960	Flooring	115.2	66.7	101.1	96.3	123.6	104.2	96.3	123.6	104.2	100.3	126.0	107.8	96.3	118.8	102.8	96.3	123.6	104.2
0970, 0990	Wall Finishes & Painting/Coating	97.5	76.1	84.7	90.7	126.8	112.3	90.7	130.8	114.7	98.9	130.8	118.0	90.7	126.8	112.3	90.7	130.8	114.7
09	FINISHES	103.9	68.8	84.9	100.3	117.7	109.8	100.4	119.2	110.6	101.8	119.9	111.6	102.7	118.0	111.0	100.4	119.2	110.6
COVERS	DIVS. 10 - 14, 25, 28, 41, 43, 44, 46	100.0	84.5	96.3	100.0	112.2	102.9	100.0	112.4	102.9	100.0	113.1	103.1	100.0	112.4	102.9	100.0	112.4	102.9
21, 22, 23	FIRE SUPPRESSION, PLUMBING & HVAC	97.1	70.9	86.6	101.0	119.0	108.3	101.0	119.0	108.3	100.9	119.0	108.2	97.2	119.0	106.0	101.0	119.0	108.3
26, 27, 3370	ELECTRICAL, COMMUNICATIONS & UTIL.	95.0	60.9	78.1	94.1	105.2	99.6	94.1	103.9	99.0	97.1	110.8	103.9	94.1	101.9	98.0	94.2	103.9	99.0
MF2018	WEIGHTED AVERAGE	102.9	69.8	88.6	100.9	116.7	107.7	100.1	116.9	107.4	101.7	118.7	109.0	98.8	116.4	106.4	99.7	116.9	107.1

CONNECTICUT

DIVISION		NEW HAVEN 065 MAT.	INST.	TOTAL	NEW LONDON 063 MAT.	INST.	TOTAL	NORWALK 068 MAT.	INST.	TOTAL	STAMFORD 069 MAT.	INST.	TOTAL	WATERBURY 067 MAT.	INST.	TOTAL	WILLIMANTIC 062 MAT.	INST.	TOTAL
015433	CONTRACTOR EQUIPMENT		94.6	94.6		94.6	94.6		94.2	94.2		94.2	94.2		94.2	94.2		94.2	94.2
0241, 31 - 34	SITE & INFRASTRUCTURE, DEMOLITION	107.2	96.2	99.7	98.6	96.2	97.0	107.7	95.6	99.4	108.4	95.6	99.6	107.6	95.6	99.3	107.8	95.4	99.3
0310	Concrete Forming & Accessories	104.0	116.9	115.0	104.0	116.9	115.0	104.3	115.7	114.0	104.3	116.0	114.3	104.3	117.0	115.1	104.3	116.9	115.0
0320	Concrete Reinforcing	116.0	144.7	129.8	90.9	144.7	116.6	116.0	144.6	129.8	116.0	144.7	129.8	116.0	144.7	129.8	116.0	144.7	129.8
0330	Cast-in-Place Concrete	96.7	124.7	107.3	82.4	124.7	98.3	98.2	125.8	108.7	99.8	126.0	109.7	99.8	126.6	110.0	93.2	124.7	105.1
03	CONCRETE	114.3	123.8	118.6	87.5	123.8	103.8	101.2	123.7	111.3	101.9	123.9	111.8	101.9	124.5	112.1	98.9	123.8	110.0
04	MASONRY	104.0	130.5	120.2	102.3	130.5	119.5	103.4	129.7	119.4	104.3	129.7	119.8	104.3	130.5	120.3	103.6	130.5	120.0
05	METALS	99.7	118.6	105.6	99.3	118.6	105.3	103.5	118.6	108.2	103.5	119.0	108.3	103.5	118.6	108.2	103.2	118.5	108.0
06	WOOD, PLASTICS & COMPOSITES	106.2	114.5	110.6	106.2	114.5	110.6	106.2	112.7	109.6	106.2	112.7	109.6	106.2	114.5	110.6	106.2	114.5	110.6
07	THERMAL & MOISTURE PROTECTION	99.2	119.4	107.9	99.0	120.3	108.1	99.2	121.3	108.7	99.1	121.3	108.6	99.1	119.6	107.9	99.3	120.0	108.2
08	OPENINGS	96.6	120.9	102.5	98.9	120.1	104.1	96.6	119.9	102.3	96.6	119.9	102.3	96.6	120.9	102.5	98.9	120.9	104.3
0920	Plaster & Gypsum Board	122.9	114.8	117.6	122.9	114.8	117.6	122.9	112.9	116.3	122.9	112.9	116.3	122.9	114.8	117.6	122.9	114.8	117.6
0950, 0980	Ceilings & Acoustic Treatment	102.9	114.8	110.3	101.4	114.8	109.8	102.9	112.9	109.2	102.9	112.9	109.2	102.9	114.8	110.3	101.4	114.8	109.8
0960	Flooring	96.3	123.6	104.2	96.3	121.2	103.5	96.3	123.6	104.2	96.3	123.6	104.2	96.3	123.6	104.2	96.3	115.5	101.9
0970, 0990	Wall Finishes & Painting/Coating	90.7	124.4	110.9	90.7	126.8	112.3	90.7	126.8	112.3	90.7	126.8	112.3	90.7	126.8	112.3	90.7	126.8	112.3
09	FINISHES	100.4	118.5	110.2	99.5	118.4	109.7	100.4	117.7	109.8	100.5	117.7	109.8	100.3	118.8	110.3	100.2	117.4	109.5
COVERS	DIVS. 10 - 14, 25, 28, 41, 43, 44, 46	100.0	112.4	102.9	100.0	112.4	102.9	100.0	112.2	102.9	100.0	112.4	102.9	100.0	112.4	102.9	100.0	112.4	102.9
21, 22, 23	FIRE SUPPRESSION, PLUMBING & HVAC	101.0	119.0	108.3	97.2	119.0	106.0	101.0	119.0	108.3	101.0	119.0	108.3	101.0	119.0	108.3	101.0	118.9	108.2
26, 27, 3370	ELECTRICAL, COMMUNICATIONS & UTIL.	94.1	106.2	100.1	91.9	103.9	97.8	94.1	107.1	100.6	94.1	156.4	125.0	93.8	107.3	100.4	94.1	107.1	100.5
MF2018	WEIGHTED AVERAGE	101.5	117.0	108.2	96.7	116.6	105.3	100.4	116.9	107.5	100.5	123.9	110.6	100.5	117.2	107.7	100.3	116.9	107.5

D.C. / DELAWARE / FLORIDA

DIVISION		WASHINGTON 200 - 205 MAT.	INST.	TOTAL	DOVER 199 MAT.	INST.	TOTAL	NEWARK 197 MAT.	INST.	TOTAL	WILMINGTON 198 MAT.	INST.	TOTAL	DAYTONA BEACH 321 MAT.	INST.	TOTAL	FORT LAUDERDALE 333 MAT.	INST.	TOTAL
015433	CONTRACTOR EQUIPMENT		104.0	104.0		115.6	115.6		117.0	117.0		115.8	115.8		98.2	98.2		91.4	91.4
0241, 31 - 34	SITE & INFRASTRUCTURE, DEMOLITION	105.0	95.4	98.4	102.7	103.6	103.3	102.0	106.0	104.7	98.9	104.5	102.8	117.0	82.6	93.4	94.9	72.4	79.4
0310	Concrete Forming & Accessories	100.6	73.0	77.1	100.0	100.2	100.2	96.2	99.9	99.4	98.0	100.4	100.0	98.6	61.6	67.1	93.8	67.7	71.6
0320	Concrete Reinforcing	115.4	90.3	103.3	100.7	116.3	108.2	98.8	116.3	107.2	96.7	116.4	106.2	90.5	61.5	76.5	88.7	57.6	73.7
0330	Cast-in-Place Concrete	98.3	79.7	91.3	106.2	108.3	107.0	88.6	106.3	95.3	103.4	108.3	105.2	85.5	65.0	77.8	88.0	72.3	82.1
03	CONCRETE	103.7	79.2	92.7	98.6	106.8	102.3	92.1	106.2	98.4	96.6	106.9	101.2	88.7	64.5	77.8	90.6	69.1	81.0
04	MASONRY	99.1	87.1	91.8	103.9	101.7	102.5	102.6	101.6	102.0	99.8	101.7	100.9	87.0	60.0	70.5	90.9	72.8	79.9
05	METALS	104.5	96.2	101.9	105.3	123.5	111.0	107.4	125.8	113.1	105.5	123.6	111.2	100.0	89.2	96.7	96.8	87.7	94.0
06	WOOD, PLASTICS & COMPOSITES	100.7	71.4	85.4	96.1	98.1	97.1	91.0	97.8	94.6	89.5	98.1	94.0	93.5	60.3	76.1	79.1	62.3	70.3
07	THERMAL & MOISTURE PROTECTION	100.8	86.0	94.5	105.4	112.4	108.4	109.9	111.2	110.5	104.9	112.4	108.1	100.8	64.5	85.2	105.5	68.1	89.5
08	OPENINGS	100.8	74.8	94.5	91.1	109.9	95.7	91.1	109.7	95.6	88.6	109.9	93.8	92.9	59.8	84.8	94.3	59.9	85.9
0920	Plaster & Gypsum Board	118.0	70.6	86.9	103.9	97.9	100.0	99.8	97.9	98.6	100.9	97.9	99.0	90.9	59.8	70.5	111.4	61.8	78.9
0950, 0980	Ceilings & Acoustic Treatment	108.3	70.6	84.7	105.5	97.9	100.7	100.8	97.9	99.0	101.9	97.9	99.4	86.1	59.8	69.7	94.4	61.8	74.0
0960	Flooring	94.7	76.7	89.5	91.5	106.8	95.9	91.9	106.8	96.2	99.5	106.8	101.6	104.4	61.0	91.8	102.1	61.0	90.1
0970, 0990	Wall Finishes & Painting/Coating	101.5	70.0	82.6	91.4	117.9	107.3	84.1	117.9	104.4	94.5	117.9	108.5	99.0	62.1	76.9	90.6	58.4	71.3
09	FINISHES	101.8	72.3	85.8	97.1	102.3	99.9	92.3	102.1	97.6	98.6	102.3	100.6	94.9	60.9	76.5	97.1	64.8	79.6
COVERS	DIVS. 10 - 14, 25, 28, 41, 43, 44, 46	100.0	95.5	99.0	100.0	100.0	100.0	100.0	99.3	99.8	100.0	106.3	101.5	100.0	82.0	95.8	100.0	88.0	97.3
21, 22, 23	FIRE SUPPRESSION, PLUMBING & HVAC	100.8	89.4	96.2	99.4	120.9	108.0	99.6	120.8	108.2	100.7	120.9	108.9	100.7	73.7	89.8	100.7	76.0	90.7
26, 27, 3370	ELECTRICAL, COMMUNICATIONS & UTIL.	96.4	103.9	100.1	98.2	109.9	104.0	98.0	109.9	103.9	97.4	109.9	103.6	98.0	61.4	79.9	95.1	67.5	81.4
MF2018	WEIGHTED AVERAGE	101.3	88.0	95.5	99.5	110.4	104.2	98.7	110.6	103.8	99.1	110.7	104.1	97.2	69.0	85.0	96.5	72.2	86.0

460

For customer support on your Concrete & Masonry Costs with RSMeans Data, call 800.448.8182.

City Cost Indexes - V2

FLORIDA

DIVISION		FORT MYERS 339,341			GAINESVILLE 326,344			JACKSONVILLE 320,322			LAKELAND 338			MELBOURNE 329			MIAMI 330 - 332,340		
		MAT.	INST.	TOTAL	MAT.	INST.	TOTAL	MAT.	INST.	TOTAL	MAT.	INST.	TOTAL	MAT.	INST.	TOTAL	MAT.	INST.	TOTAL
015433	CONTRACTOR EQUIPMENT		98.2	98.2		98.2	98.2		98.2	98.2		98.2	98.2		98.2	98.2		95.7	95.7
0241, 31 - 34	SITE & INFRASTRUCTURE, DEMOLITION	106.7	82.7	90.2	125.2	82.4	95.8	117.0	82.2	93.1	108.7	82.7	90.9	124.6	82.6	95.7	94.8	80.7	85.1
0310	Concrete Forming & Accessories	89.8	62.0	66.1	93.1	55.5	61.1	98.5	61.8	67.3	86.0	63.2	66.6	94.6	62.3	67.1	101.3	64.3	69.9
0320	Concrete Reinforcing	89.7	75.0	82.6	96.0	59.8	78.5	90.5	59.7	75.7	91.9	75.1	83.8	91.5	67.1	79.7	98.0	57.6	78.5
0330	Cast-in-Place Concrete	91.8	64.1	81.3	98.2	64.6	85.5	86.4	64.5	78.1	93.9	65.1	83.0	103.1	65.0	88.7	85.2	69.6	79.4
03	CONCRETE	91.0	66.7	80.1	99.5	61.3	82.3	89.1	64.0	77.9	92.6	67.6	81.4	99.7	65.8	84.5	89.6	66.5	79.2
04	MASONRY	85.7	57.3	68.4	99.7	59.2	75.0	86.8	57.7	69.1	101.1	60.0	76.0	84.3	60.0	69.4	94.7	62.6	75.1
05	METALS	98.8	93.7	97.2	98.8	87.5	95.3	98.6	87.0	95.0	98.7	94.2	97.3	108.4	91.4	103.1	97.0	85.5	93.4
06	WOOD, PLASTICS & COMPOSITES	75.9	61.0	68.1	86.6	52.7	68.9	93.5	61.3	76.6	71.2	62.7	66.8	88.6	61.0	74.2	94.1	62.5	77.6
07	THERMAL & MOISTURE PROTECTION	105.3	61.3	86.4	101.2	61.8	84.3	101.1	62.3	84.5	105.2	62.2	86.7	101.3	63.3	85.0	104.8	64.8	87.6
08	OPENINGS	95.7	63.3	87.8	92.6	55.2	83.5	92.9	59.9	84.9	95.6	64.2	88.0	92.2	61.5	84.7	96.7	60.1	87.8
0920	Plaster & Gypsum Board	107.3	60.6	76.6	87.9	52.0	64.4	90.9	60.8	71.1	105.0	62.3	77.0	87.9	60.6	70.0	98.6	61.8	74.5
0950, 0980	Ceilings & Acoustic Treatment	89.2	60.6	71.3	80.5	52.0	62.7	86.1	60.8	70.3	89.2	62.3	72.3	85.5	60.6	69.9	95.1	61.8	74.2
0960	Flooring	99.0	73.7	91.6	101.9	59.8	89.6	104.4	61.0	91.8	97.1	61.0	86.6	102.2	61.0	90.2	100.9	61.6	89.5
0970, 0990	Wall Finishes & Painting/Coating	95.1	62.5	75.5	99.0	62.5	77.1	99.0	62.5	77.1	95.1	62.5	75.5	99.0	79.5	87.3	96.8	58.4	73.8
09	FINISHES	96.6	63.9	78.9	93.4	56.2	73.3	95.0	61.5	76.8	95.8	62.3	77.7	94.4	63.2	77.5	96.9	62.6	78.3
COVERS	DIVS. 10 - 14, 25, 28, 41, 43, 44, 46	100.0	80.8	95.5	100.0	79.3	95.1	100.0	80.3	95.4	100.0	82.2	95.8	100.0	82.1	95.8	100.0	85.2	96.5
21, 22, 23	FIRE SUPPRESSION, PLUMBING & HVAC	97.6	57.0	81.2	97.8	60.6	82.8	100.7	64.0	85.9	97.6	58.4	81.8	100.7	74.1	90.0	96.3	67.9	84.8
26, 27, 3370	ELECTRICAL, COMMUNICATIONS & UTIL.	96.3	64.2	80.4	98.0	56.4	77.4	97.7	64.4	81.3	94.8	56.4	75.8	99.0	66.5	82.9	97.7	77.1	87.5
MF2018	WEIGHTED AVERAGE	96.4	66.7	83.6	98.3	63.8	83.4	97.0	66.9	84.0	97.1	66.3	83.8	99.9	70.6	87.2	96.2	70.4	85.1

FLORIDA

DIVISION		ORLANDO 327 - 328,347			PANAMA CITY 324			PENSACOLA 325			SARASOTA 342			ST. PETERSBURG 337			TALLAHASSEE 323		
		MAT.	INST.	TOTAL	MAT.	INST.	TOTAL	MAT.	INST.	TOTAL	MAT.	INST.	TOTAL	MAT.	INST.	TOTAL	MAT.	INST.	TOTAL
015433	CONTRACTOR EQUIPMENT		103.5	103.5		98.2	98.2		98.2	98.2		98.2	98.2		98.2	98.2		103.5	103.5
0241, 31 - 34	SITE & INFRASTRUCTURE, DEMOLITION	114.1	91.6	98.7	130.1	82.6	97.5	129.9	82.6	97.4	116.8	82.7	93.4	110.9	82.6	91.5	110.1	91.9	97.6
0310	Concrete Forming & Accessories	103.9	61.6	67.9	97.6	64.3	69.2	95.6	61.5	66.6	95.9	63.0	67.9	93.8	61.1	66.0	101.8	62.1	68.1
0320	Concrete Reinforcing	96.5	65.0	81.3	94.5	65.8	80.7	96.9	66.7	82.4	92.3	72.6	82.8	91.9	75.0	83.8	103.1	59.9	82.3
0330	Cast-in-Place Concrete	105.9	66.6	91.0	90.8	64.9	81.0	112.2	64.7	94.3	101.2	65.0	87.5	94.9	64.9	83.6	88.6	66.5	80.3
03	CONCRETE	99.2	65.5	84.0	98.1	66.4	83.9	107.2	65.3	88.4	94.9	67.0	82.4	94.3	66.6	81.8	93.2	64.8	80.4
04	MASONRY	93.2	60.0	73.0	91.2	59.4	71.8	109.8	58.9	78.7	91.4	60.0	72.3	139.1	57.9	89.6	91.5	59.5	72.0
05	METALS	95.2	88.0	93.0	99.7	90.5	96.8	100.7	90.7	97.6	102.2	92.5	99.1	99.6	93.8	97.8	99.9	86.1	95.6
06	WOOD, PLASTICS & COMPOSITES	94.0	60.6	76.5	92.2	64.3	77.6	90.7	61.3	75.3	93.5	62.7	77.4	80.7	60.0	69.9	98.6	61.5	79.2
07	THERMAL & MOISTURE PROTECTION	108.2	65.6	89.9	101.5	62.5	84.7	101.4	61.6	84.3	100.3	62.2	83.9	105.5	61.1	86.4	96.2	63.0	82.0
08	OPENINGS	97.7	60.7	88.7	91.1	63.0	84.2	91.1	61.5	83.9	97.7	63.2	89.3	94.5	62.7	86.7	97.8	60.1	88.6
0920	Plaster & Gypsum Board	92.7	59.8	71.1	90.1	63.9	72.9	98.9	60.8	73.9	98.3	62.3	74.6	109.3	59.6	76.6	101.5	60.8	74.8
0950, 0980	Ceilings & Acoustic Treatment	99.0	59.8	74.5	85.5	63.9	72.0	85.5	60.8	70.0	91.1	62.3	73.0	90.6	59.6	71.1	99.9	60.8	75.4
0960	Flooring	98.8	61.0	87.7	104.1	72.0	94.7	99.9	61.0	88.6	105.4	52.4	89.9	101.0	57.4	88.3	101.4	60.1	89.4
0970, 0990	Wall Finishes & Painting/Coating	94.5	62.1	75.1	99.0	62.5	77.1	99.0	62.5	77.1	95.1	62.5	75.6	95.1	62.5	75.5	99.7	62.5	77.4
09	FINISHES	98.0	61.1	78.0	96.0	65.1	79.3	95.7	61.2	77.0	98.5	60.8	78.1	98.0	60.0	77.4	100.0	61.4	79.1
COVERS	DIVS. 10 - 14, 25, 28, 41, 43, 44, 46	100.0	82.6	95.9	100.0	79.4	95.2	100.0	78.7	95.0	100.0	80.9	95.5	100.0	78.7	95.0	100.0	78.9	95.0
21, 22, 23	FIRE SUPPRESSION, PLUMBING & HVAC	96.4	56.0	80.1	100.7	64.2	86.0	100.7	62.0	85.1	100.3	56.2	82.5	100.7	58.4	83.6	96.5	64.2	83.5
26, 27, 3370	ELECTRICAL, COMMUNICATIONS & UTIL.	98.3	63.2	80.9	96.9	56.4	76.9	100.2	52.3	76.5	97.5	56.4	77.1	95.3	59.6	77.6	103.3	56.4	80.1
MF2018	WEIGHTED AVERAGE	97.9	66.3	84.3	98.7	67.2	85.1	101.2	65.3	85.7	99.1	65.3	84.5	100.4	65.8	85.4	98.1	66.7	84.6

FLORIDA / GEORGIA

DIVISION		TAMPA 335 - 336,346			WEST PALM BEACH 334,349			ALBANY 317,398			ATHENS 306			ATLANTA 300 - 303,399			AUGUSTA 308 - 309		
		MAT.	INST.	TOTAL	MAT.	INST.	TOTAL	MAT.	INST.	TOTAL	MAT.	INST.	TOTAL	MAT.	INST.	TOTAL	MAT.	INST.	TOTAL
015433	CONTRACTOR EQUIPMENT		98.2	98.2		91.4	91.4		92.3	92.3		91.0	91.0		97.5	97.5		91.0	91.0
0241, 31 - 34	SITE & INFRASTRUCTURE, DEMOLITION	111.3	84.6	93.0	91.1	72.4	78.2	104.9	74.5	84.0	101.9	88.8	92.9	99.5	97.2	97.9	95.4	89.4	91.3
0310	Concrete Forming & Accessories	97.0	63.4	68.4	97.7	67.4	71.9	90.2	67.7	71.0	91.9	42.0	49.5	97.2	75.8	78.9	93.1	73.6	76.5
0320	Concrete Reinforcing	88.7	75.2	82.2	91.2	54.8	73.7	91.8	71.8	82.1	92.3	60.2	76.8	94.5	71.8	83.6	92.6	70.8	82.1
0330	Cast-in-Place Concrete	92.8	65.2	82.4	83.7	72.2	79.4	86.0	69.7	79.8	103.9	69.5	90.9	107.3	72.4	94.1	98.1	70.8	87.8
03	CONCRETE	93.0	67.8	81.7	87.5	68.5	79.0	84.0	70.8	78.1	97.8	56.1	79.1	100.4	74.4	88.7	90.9	72.8	82.8
04	MASONRY	91.4	60.0	72.3	90.4	70.0	78.0	98.5	69.3	80.7	80.0	68.2	72.8	92.1	70.3	78.8	94.1	69.4	79.0
05	METALS	98.8	94.4	97.4	95.7	86.3	92.8	105.9	98.3	103.5	95.3	77.3	89.7	96.3	83.2	92.2	95.0	83.4	91.4
06	WOOD, PLASTICS & COMPOSITES	85.1	62.7	73.4	84.1	62.3	72.7	80.5	68.0	73.9	94.7	34.5	63.2	104.6	78.6	91.0	96.2	76.2	85.7
07	THERMAL & MOISTURE PROTECTION	105.8	62.2	87.1	105.2	68.0	89.2	99.5	68.8	86.3	94.3	65.5	82.0	95.9	73.9	86.5	94.0	71.0	84.1
08	OPENINGS	95.6	64.2	88.0	93.8	59.3	85.4	85.4	70.3	81.7	91.4	49.5	81.2	99.6	76.7	94.0	91.4	75.1	87.5
0920	Plaster & Gypsum Board	112.2	62.3	79.5	114.9	61.8	80.1	101.3	67.7	79.3	92.6	33.3	53.7	99.3	78.3	85.5	93.4	76.1	82.0
0950, 0980	Ceilings & Acoustic Treatment	94.4	62.3	74.3	89.2	61.8	72.0	83.6	67.7	73.7	102.6	33.3	59.2	93.9	78.3	84.1	103.3	76.1	86.2
0960	Flooring	102.1	61.0	90.1	104.0	52.4	88.9	103.5	71.1	94.1	93.9	71.1	87.3	95.6	72.5	88.9	94.1	71.1	87.4
0970, 0990	Wall Finishes & Painting/Coating	95.1	62.5	75.5	90.6	57.8	70.9	85.1	95.0	91.0	89.7	89.5	89.6	96.3	96.7	96.5	89.7	88.4	89.0
09	FINISHES	99.7	62.3	79.5	96.6	63.0	78.4	94.1	70.7	81.4	95.5	50.2	71.0	95.4	77.5	85.7	95.2	74.8	84.1
COVERS	DIVS. 10 - 14, 25, 28, 41, 43, 44, 46	100.0	81.0	95.5	100.0	87.9	97.2	100.0	84.3	96.3	100.0	80.5	95.4	100.0	86.3	96.8	100.0	85.4	96.6
21, 22, 23	FIRE SUPPRESSION, PLUMBING & HVAC	100.7	58.5	83.6	97.6	66.1	84.9	100.7	69.4	88.1	97.2	63.1	83.5	101.0	72.0	89.3	101.0	66.4	87.1
26, 27, 3370	ELECTRICAL, COMMUNICATIONS & UTIL.	95.0	63.2	79.3	95.6	64.9	80.4	95.2	62.0	78.8	99.2	61.5	80.6	98.9	72.7	85.9	99.7	68.5	84.3
MF2018	WEIGHTED AVERAGE	98.0	67.4	84.8	95.1	69.0	83.8	96.5	72.2	86.0	95.9	63.8	82.0	98.7	76.7	89.2	96.4	73.5	86.5

For customer support on your Concrete & Masonry Costs with RSMeans Data, call 800.448.8182. 461

GEORGIA

DIVISION		COLUMBUS 318-319			DALTON 307			GAINESVILLE 305			MACON 310-312			SAVANNAH 313-314			STATESBORO 304		
		MAT.	INST.	TOTAL	MAT.	INST.	TOTAL	MAT.	INST.	TOTAL	MAT.	INST.	TOTAL	MAT.	INST.	TOTAL	MAT.	INST.	TOTAL
015433	CONTRACTOR EQUIPMENT		92.3	92.3		106.1	106.1		91.0	91.0		101.8	101.8		97.7	97.7		94.0	94.0
0241, 31 - 34	SITE & INFRASTRUCTURE, DEMOLITION	104.8	74.8	84.2	101.6	93.8	96.2	101.6	88.7	92.7	106.3	88.8	94.3	106.2	83.8	90.8	102.8	75.0	83.7
0310	Concrete Forming & Accessories	90.1	70.3	73.3	83.5	62.2	65.4	96.0	39.9	48.3	89.8	69.2	72.3	96.2	71.6	75.3	78.1	50.2	54.3
0320	Concrete Reinforcing	91.7	71.8	82.1	91.8	57.7	75.3	92.1	60.2	76.7	92.9	71.8	82.7	97.8	70.9	84.8	91.4	67.5	79.8
0330	Cast-in-Place Concrete	85.7	70.0	79.8	100.9	68.0	88.5	109.2	69.0	94.0	84.5	70.9	79.4	86.8	71.3	81.0	103.7	68.6	90.5
03	CONCRETE	83.8	72.2	78.6	97.0	65.3	82.8	99.6	55.0	79.6	83.4	72.0	78.3	84.5	72.8	79.2	97.2	61.9	81.4
04	MASONRY	98.2	69.4	80.7	81.1	66.6	72.2	88.5	67.3	75.5	112.4	68.8	85.8	92.4	70.3	78.9	83.2	68.1	74.0
05	METALS	105.5	99.0	103.5	96.2	92.0	94.9	94.5	77.3	89.2	100.7	99.2	100.2	101.8	95.8	99.9	99.8	96.9	98.9
06	WOOD, PLASTICS & COMPOSITES	80.5	71.4	75.7	77.4	63.0	69.9	99.0	32.0	64.0	85.3	70.1	77.3	92.0	73.3	82.2	70.3	45.5	57.4
07	THERMAL & MOISTURE PROTECTION	99.4	70.1	86.8	96.1	65.7	83.0	94.3	63.4	81.0	98.0	71.2	86.5	97.4	70.6	85.9	94.6	62.8	81.0
08	OPENINGS	85.4	72.8	82.3	92.9	64.5	86.0	91.4	48.1	80.8	85.2	72.0	82.0	94.9	73.6	89.7	93.9	57.4	85.0
0920	Plaster & Gypsum Board	101.3	71.3	81.6	82.3	62.5	69.3	94.3	30.7	52.6	104.0	69.9	81.6	103.3	72.9	83.4	84.2	44.6	58.2
0950, 0980	Ceilings & Acoustic Treatment	83.6	71.3	75.9	118.4	62.5	83.4	102.6	30.7	57.6	79.2	69.9	73.4	98.2	72.9	82.4	113.4	44.6	70.3
0960	Flooring	103.5	71.1	94.1	94.8	71.1	87.9	95.5	71.1	88.4	81.4	71.1	78.4	97.9	72.5	90.5	111.4	71.1	99.6
0970, 0990	Wall Finishes & Painting/Coating	85.1	85.9	85.6	80.3	66.7	72.2	89.7	89.5	89.6	87.1	95.0	91.8	86.9	85.3	85.9	88.1	66.7	75.3
09	FINISHES	94.0	71.7	81.9	104.6	64.0	82.7	96.0	48.6	70.3	83.4	71.8	77.1	96.9	73.1	84.0	107.4	54.2	78.6
COVERS	DIVS. 10 - 14, 25, 28, 41, 43, 44, 46	100.0	84.7	96.4	100.0	81.0	95.5	100.0	80.0	95.3	100.0	84.4	96.3	100.0	85.4	96.6	100.0	81.6	95.7
21, 22, 23	FIRE SUPPRESSION, PLUMBING & HVAC	100.7	66.3	86.8	97.3	57.9	81.4	97.2	62.6	83.3	100.7	71.9	89.1	100.8	64.3	86.1	98.0	66.3	85.2
26, 27, 3370	ELECTRICAL, COMMUNICATIONS & UTIL.	95.3	70.8	83.2	108.0	59.9	84.2	99.2	70.7	85.1	93.9	62.8	78.5	99.2	63.8	81.7	99.7	63.4	81.7
MF2018	WEIGHTED AVERAGE	96.4	73.4	86.4	97.7	68.0	84.9	96.5	64.3	82.6	95.3	74.4	86.3	97.3	72.8	86.7	97.9	67.1	84.6

GEORGIA / HAWAII / IDAHO

DIVISION		GEORGIA VALDOSTA 316			WAYCROSS 315			HAWAII HILO 967			HONOLULU 968			STATES & POSS., GUAM 969			IDAHO BOISE 836 - 837		
		MAT.	INST.	TOTAL	MAT.	INST.	TOTAL	MAT.	INST.	TOTAL	MAT.	INST.	TOTAL	MAT.	INST.	TOTAL	MAT.	INST.	TOTAL
015433	CONTRACTOR EQUIPMENT		92.3	92.3		92.3	92.3		96.0	96.0		100.7	100.7		159.0	159.0		92.3	92.3
0241, 31 - 34	SITE & INFRASTRUCTURE, DEMOLITION	114.8	74.5	87.1	111.1	74.1	85.7	146.1	100.4	114.1	153.4	108.5	122.6	192.5	97.3	127.1	84.6	88.8	87.5
0310	Concrete Forming & Accessories	80.0	44.0	49.4	81.9	61.5	64.5	106.1	124.1	121.4	123.2	124.2	124.0	108.4	63.3	70.1	97.3	80.4	82.9
0320	Concrete Reinforcing	93.8	60.7	77.8	93.8	56.6	75.8	140.0	122.0	131.3	160.1	122.1	141.7	251.6	43.8	151.4	101.2	76.0	89.0
0330	Cast-in-Place Concrete	84.2	69.5	78.7	95.1	68.5	85.1	196.9	125.0	169.7	153.7	126.2	143.3	170.6	104.3	145.6	89.2	93.5	90.8
03	CONCRETE	88.2	58.1	74.7	91.1	64.8	79.3	153.2	123.2	139.7	146.4	123.5	136.1	156.1	74.9	119.7	97.6	84.3	91.7
04	MASONRY	104.4	69.3	83.0	105.1	68.1	82.5	140.4	124.4	130.7	132.4	124.5	127.6	200.5	48.8	108.0	125.0	84.2	100.1
05	METALS	105.0	93.4	101.4	104.0	86.8	98.7	110.0	107.8	109.4	124.9	106.4	119.1	141.9	81.4	123.0	107.7	80.4	99.2
06	WOOD, PLASTICS & COMPOSITES	67.9	36.3	51.3	69.6	61.2	65.2	108.3	124.7	116.9	135.2	124.8	129.8	119.7	63.6	90.4	89.5	79.4	84.2
07	THERMAL & MOISTURE PROTECTION	99.7	65.4	84.9	99.4	64.9	84.6	127.6	118.4	123.6	144.9	119.6	134.1	151.5	70.0	116.5	99.1	85.4	93.2
08	OPENINGS	82.2	50.5	74.5	82.5	60.7	77.2	114.7	122.3	116.6	128.7	122.4	127.2	118.9	52.8	102.8	97.1	74.0	91.4
0920	Plaster & Gypsum Board	94.2	35.1	55.4	94.2	60.8	72.3	122.9	125.3	124.5	169.2	125.3	140.4	249.0	53.0	120.4	94.3	79.1	84.3
0950, 0980	Ceilings & Acoustic Treatment	82.3	35.1	52.7	80.9	60.8	68.3	125.1	125.3	125.2	132.7	125.3	128.1	231.1	53.0	119.5	112.5	79.1	91.6
0960	Flooring	96.9	71.1	89.4	98.3	71.1	90.4	105.9	140.4	116.0	122.8	140.4	127.9	123.3	47.3	101.1	91.8	86.5	90.3
0970, 0990	Wall Finishes & Painting/Coating	85.1	93.1	89.9	85.1	66.7	74.1	96.1	137.0	120.6	109.0	137.0	125.8	101.6	39.6	64.5	89.3	45.7	63.2
09	FINISHES	91.8	51.9	70.2	91.4	63.4	76.3	111.9	129.0	121.1	126.8	129.0	128.0	186.6	58.2	117.1	96.1	77.7	86.1
COVERS	DIVS. 10 - 14, 25, 28, 41, 43, 44, 46	100.0	80.9	95.5	100.0	81.2	95.6	100.0	112.1	102.9	100.0	112.3	102.9	100.0	92.9	98.3	100.0	86.9	96.9
21, 22, 23	FIRE SUPPRESSION, PLUMBING & HVAC	100.7	69.1	88.0	98.4	60.5	83.1	101.1	108.9	104.2	101.1	108.9	104.3	104.1	52.5	83.2	101.2	71.5	89.2
26, 27, 3370	ELECTRICAL, COMMUNICATIONS & UTIL.	93.8	56.7	75.5	97.1	63.4	80.4	105.8	122.0	113.8	106.7	122.0	114.3	150.4	49.2	100.3	97.2	69.3	83.4
MF2018	WEIGHTED AVERAGE	96.7	65.4	83.2	96.6	66.9	83.8	115.6	117.2	116.3	120.3	117.8	119.2	138.0	63.4	105.7	100.9	78.2	91.1

IDAHO / ILLINOIS

DIVISION		COEUR D'ALENE 838			IDAHO FALLS 834			LEWISTON 835			POCATELLO 832			TWIN FALLS 833			ILLINOIS BLOOMINGTON 617		
		MAT.	INST.	TOTAL	MAT.	INST.	TOTAL	MAT.	INST.	TOTAL	MAT.	INST.	TOTAL	MAT.	INST.	TOTAL	MAT.	INST.	TOTAL
015433	CONTRACTOR EQUIPMENT		87.3	87.3		92.3	92.3		87.3	87.3		92.3	92.3		92.3	92.3		98.8	98.8
0241, 31 - 34	SITE & INFRASTRUCTURE, DEMOLITION	84.4	83.1	83.5	83.3	88.6	86.9	91.3	84.4	86.5	86.1	88.8	87.9	93.0	88.6	90.0	94.4	92.2	92.9
0310	Concrete Forming & Accessories	103.6	78.3	82.1	90.7	75.5	77.8	109.1	81.3	85.5	97.4	79.9	82.5	98.5	79.8	82.6	80.3	111.8	107.1
0320	Concrete Reinforcing	109.3	99.8	104.7	103.2	74.2	89.2	109.3	100.0	104.8	101.6	76.1	89.3	103.6	77.0	90.7	95.4	97.5	96.4
0330	Cast-in-Place Concrete	96.5	81.5	90.8	85.0	80.1	83.1	100.3	81.9	93.4	91.7	93.3	92.3	94.2	81.6	89.4	94.4	109.3	100.0
03	CONCRETE	103.8	83.3	94.6	89.5	77.2	84.0	107.5	84.8	97.1	96.7	84.0	91.0	104.6	80.1	93.6	91.5	109.3	99.5
04	MASONRY	126.7	83.6	100.4	120.2	84.2	98.2	127.1	85.3	101.6	122.5	87.2	101.0	125.4	81.8	98.8	115.8	116.7	116.4
05	METALS	101.4	88.5	97.4	115.9	79.1	104.4	100.8	89.8	97.4	116.0	80.1	104.8	116.0	80.0	104.8	93.0	121.1	101.7
06	WOOD, PLASTICS & COMPOSITES	90.7	77.1	83.6	82.7	73.6	78.0	96.9	80.1	88.1	89.5	79.4	84.2	90.7	79.4	84.7	76.5	109.2	93.6
07	THERMAL & MOISTURE PROTECTION	152.8	80.4	121.7	97.7	72.2	86.8	153.1	82.5	122.8	98.3	74.7	88.2	99.2	80.7	91.2	95.5	107.9	100.8
08	OPENINGS	114.1	75.0	104.6	99.8	68.6	92.2	107.0	81.1	100.7	97.7	68.2	90.5	100.5	64.3	91.7	91.4	114.3	96.9
0920	Plaster & Gypsum Board	170.7	76.8	109.1	83.2	73.2	76.6	172.4	79.9	111.7	85.0	79.1	81.1	87.2	79.1	81.9	85.4	109.8	101.4
0950, 0980	Ceilings & Acoustic Treatment	146.0	76.8	102.6	113.9	73.2	88.4	146.0	79.9	104.6	119.7	79.1	94.2	116.5	79.1	93.0	84.1	109.8	100.2
0960	Flooring	129.9	73.5	113.5	91.5	73.5	86.2	133.5	78.0	117.3	95.2	78.0	90.2	96.5	82.3	92.3	85.5	115.3	94.2
0970, 0990	Wall Finishes & Painting/Coating	106.4	70.3	84.7	89.3	37.6	58.3	106.4	72.1	85.8	89.2	40.7	60.2	89.3	39.9	59.7	85.6	128.3	111.2
09	FINISHES	159.1	76.3	114.3	94.3	71.2	81.8	160.4	79.2	116.5	97.5	75.8	85.7	98.1	76.6	86.4	86.1	114.2	101.3
COVERS	DIVS. 10 - 14, 25, 28, 41, 43, 44, 46	100.0	88.2	97.2	100.0	86.3	96.8	100.0	97.4	99.4	100.0	86.9	96.9	100.0	86.9	96.9	100.0	105.5	101.3
21, 22, 23	FIRE SUPPRESSION, PLUMBING & HVAC	100.6	81.5	92.9	102.3	78.2	92.6	102.4	85.6	95.6	100.9	73.0	89.7	100.9	71.5	89.0	96.8	100.4	98.2
26, 27, 3370	ELECTRICAL, COMMUNICATIONS & UTIL.	88.4	78.5	83.5	88.0	68.2	78.2	86.8	84.7	85.8	94.3	67.5	81.1	89.1	70.7	80.0	94.7	87.0	90.8
MF2018	WEIGHTED AVERAGE	108.1	81.5	96.6	100.2	76.9	90.1	108.4	84.9	98.2	101.8	77.8	91.4	102.9	76.9	91.7	94.8	105.4	99.4

For customer support on your Concrete & Masonry Costs with RSMeans Data, call 800.448.8182.

City Cost Indexes - V2

ILLINOIS

DIVISION		CARBONDALE 629 MAT.	INST.	TOTAL	CENTRALIA 628 MAT.	INST.	TOTAL	CHAMPAIGN 618-619 MAT.	INST.	TOTAL	CHICAGO 606-608 MAT.	INST.	TOTAL	DECATUR 625 MAT.	INST.	TOTAL	EAST ST. LOUIS 620-622 MAT.	INST.	TOTAL
015433	CONTRACTOR EQUIPMENT		107.2	107.2		107.2	107.2		99.7	99.7		97.0	97.0		99.7	99.7		107.2	107.2
0241, 31 - 34	SITE & INFRASTRUCTURE, DEMOLITION	103.0	92.8	96.0	103.4	94.7	97.4	103.2	93.2	96.3	105.6	100.0	101.7	96.7	93.4	94.4	106.0	94.3	97.9
0310	Concrete Forming & Accessories	88.6	104.8	102.4	90.5	109.9	107.0	86.9	112.4	108.6	100.5	159.0	150.3	91.5	116.6	112.9	86.0	112.8	108.8
0320	Concrete Reinforcing	90.9	98.5	94.6	90.9	99.1	94.8	95.4	96.8	96.1	104.8	151.0	127.1	89.1	96.9	92.9	90.8	105.1	97.7
0330	Cast-in-Place Concrete	92.1	98.4	94.5	92.5	114.7	100.9	109.5	106.3	108.3	120.6	154.2	133.3	100.5	113.9	105.6	94.1	117.7	103.0
03	CONCRETE	82.4	103.1	91.7	82.9	111.3	95.6	103.5	108.3	105.7	112.4	155.4	131.7	92.9	112.9	101.9	83.9	114.7	97.7
04	MASONRY	81.6	106.8	97.0	81.6	117.8	103.7	141.6	120.2	128.5	102.7	164.4	140.4	77.1	120.0	103.3	81.9	121.5	106.1
05	METALS	98.5	127.7	107.6	98.6	130.8	108.7	93.0	117.8	100.7	95.7	147.2	111.8	102.5	119.3	107.8	99.7	134.5	110.6
06	WOOD, PLASTICS & COMPOSITES	85.6	102.2	94.3	88.4	107.4	98.3	83.4	110.1	97.4	108.0	158.3	134.3	87.7	116.1	102.6	82.6	109.5	96.6
07	THERMAL & MOISTURE PROTECTION	90.6	96.7	93.2	90.7	106.5	97.5	96.3	110.4	102.3	94.7	149.2	118.1	95.9	111.7	102.7	90.7	109.1	98.6
08	OPENINGS	87.1	112.0	93.2	87.1	114.8	93.9	91.9	113.4	97.1	101.2	168.5	117.6	97.9	116.8	102.5	87.2	115.3	94.0
0920	Plaster & Gypsum Board	86.8	102.6	97.2	87.9	107.9	101.0	87.1	110.7	102.6	96.0	159.9	138.0	89.8	116.9	107.6	85.4	110.1	101.6
0950, 0980	Ceilings & Acoustic Treatment	79.6	102.6	94.0	79.6	107.9	97.4	84.1	110.7	100.8	89.6	159.9	133.7	86.6	116.9	105.6	79.6	110.1	98.7
0960	Flooring	116.3	111.3	114.8	117.3	111.3	115.5	88.9	111.3	95.4	95.4	161.3	114.6	104.6	114.0	107.3	115.2	113.5	114.7
0970, 0990	Wall Finishes & Painting/Coating	99.9	99.9	99.9	99.9	106.3	103.7	85.6	109.4	99.9	93.9	162.4	135.0	92.8	112.2	104.4	99.9	110.9	106.5
09	FINISHES	91.4	106.2	99.4	91.9	109.8	101.6	88.0	112.2	101.1	94.8	160.6	130.4	92.0	116.2	105.1	91.0	112.6	102.7
COVERS	DIVS. 10 - 14, 25, 28, 41, 43, 44, 46	100.0	101.3	100.3	100.0	100.2	100.1	100.0	106.1	101.4	100.0	127.2	106.4	100.0	106.2	101.5	100.0	107.4	101.8
21, 22, 23	FIRE SUPPRESSION, PLUMBING & HVAC	96.7	101.7	98.7	96.7	93.2	95.3	96.8	101.9	98.9	100.4	134.9	114.3	100.5	95.9	98.7	100.5	96.7	99.0
26, 27, 3370	ELECTRICAL, COMMUNICATIONS & UTIL.	91.9	103.8	97.8	92.9	102.3	97.6	97.4	91.9	94.6	96.1	136.4	116.0	94.9	99.9	97.4	92.6	105.6	99.1
MF2018	WEIGHTED AVERAGE	92.7	105.2	98.1	92.9	106.7	98.9	98.4	106.3	101.8	100.5	144.4	119.5	96.7	107.6	101.4	94.1	109.8	100.9

ILLINOIS

DIVISION		EFFINGHAM 624 MAT.	INST.	TOTAL	GALESBURG 614 MAT.	INST.	TOTAL	JOLIET 604 MAT.	INST.	TOTAL	KANKAKEE 609 MAT.	INST.	TOTAL	LA SALLE 613 MAT.	INST.	TOTAL	NORTH SUBURBAN 600-603 MAT.	INST.	TOTAL
015433	CONTRACTOR EQUIPMENT		99.7	99.7		98.8	98.8		90.4	90.4		90.4	90.4		98.8	98.8		90.4	90.4
0241, 31 - 34	SITE & INFRASTRUCTURE, DEMOLITION	101.0	92.0	94.8	96.7	92.1	93.5	101.7	94.8	97.0	94.9	91.3	93.7	96.1	92.9	93.9	101.0	92.9	96.1
0310	Concrete Forming & Accessories	95.9	110.3	108.2	86.8	112.8	108.9	100.1	157.8	149.2	92.5	139.9	132.8	101.1	120.7	117.7	99.2	152.6	144.6
0320	Concrete Reinforcing	92.1	89.3	90.7	94.9	100.6	97.6	111.9	138.0	124.5	112.6	128.7	120.4	95.1	126.5	110.3	111.9	142.0	126.4
0330	Cast-in-Place Concrete	100.2	105.4	102.1	97.2	104.9	100.1	109.6	146.6	123.5	102.1	130.3	112.8	97.1	116.9	104.6	109.6	146.8	123.7
03	CONCRETE	93.6	105.6	99.0	94.4	108.8	100.9	104.1	149.8	124.6	97.9	134.3	114.2	95.3	121.3	106.9	104.1	148.2	123.9
04	MASONRY	85.7	110.5	100.8	116.0	117.4	116.8	101.8	159.4	136.9	98.3	140.9	124.3	116.0	124.3	121.1	98.8	152.2	131.4
05	METALS	99.5	112.4	103.6	93.0	123.5	102.5	93.8	138.8	107.8	93.8	130.6	105.3	93.0	139.3	107.5	94.8	138.9	108.6
06	WOOD, PLASTICS & COMPOSITES	90.3	110.1	100.6	83.3	111.4	98.0	101.1	158.5	131.1	93.0	139.8	117.5	99.4	119.0	109.6	100.0	152.9	127.7
07	THERMAL & MOISTURE PROTECTION	95.4	104.2	99.2	95.7	105.1	99.7	99.3	145.0	118.9	98.4	133.2	113.3	95.9	117.0	105.0	99.7	141.0	117.4
08	OPENINGS	93.1	110.3	97.3	91.4	112.9	96.6	99.2	163.6	114.9	92.5	150.5	106.6	91.4	128.4	100.4	99.2	161.8	114.5
0920	Plaster & Gypsum Board	89.3	110.7	103.3	87.1	112.1	103.5	90.0	160.4	136.2	89.4	141.2	123.4	94.5	119.9	111.1	94.3	154.7	133.9
0950, 0980	Ceilings & Acoustic Treatment	79.6	110.7	99.1	84.1	112.1	101.6	89.3	160.4	133.9	89.3	141.2	121.9	84.1	119.9	106.5	89.3	154.7	130.1
0960	Flooring	105.9	111.3	107.4	88.6	115.3	96.4	94.2	154.7	111.9	91.1	143.8	106.5	95.5	119.1	102.4	94.6	152.2	111.4
0970, 0990	Wall Finishes & Painting/Coating	92.8	104.1	99.6	85.6	93.8	90.5	83.9	169.6	135.2	83.9	129.2	111.0	85.6	126.7	110.2	85.4	156.1	127.8
09	FINISHES	90.6	110.5	101.4	87.3	111.7	100.5	91.4	159.6	128.3	90.1	140.1	117.1	90.3	119.9	106.3	92.1	153.8	125.5
COVERS	DIVS. 10 - 14, 25, 28, 41, 43, 44, 46	100.0	104.7	101.1	100.0	105.7	101.3	100.0	123.1	105.4	100.0	111.6	102.7	100.0	106.7	101.6	100.0	117.7	104.2
21, 22, 23	FIRE SUPPRESSION, PLUMBING & HVAC	96.7	99.5	97.9	96.8	100.7	98.4	100.4	131.4	112.9	96.6	123.4	107.4	96.8	122.3	107.1	100.3	129.7	112.2
26, 27, 3370	ELECTRICAL, COMMUNICATIONS & UTIL.	93.3	103.7	98.5	95.4	83.2	89.4	95.2	133.5	114.2	91.1	131.9	111.3	93.0	131.9	112.3	95.1	130.5	112.6
MF2018	WEIGHTED AVERAGE	95.3	104.9	99.4	95.5	104.8	99.5	98.5	140.3	116.5	95.1	129.5	110.0	95.8	122.2	107.2	98.5	137.3	115.3

ILLINOIS

DIVISION		PEORIA 615-616 MAT.	INST.	TOTAL	QUINCY 623 MAT.	INST.	TOTAL	ROCK ISLAND 612 MAT.	INST.	TOTAL	ROCKFORD 610-611 MAT.	INST.	TOTAL	SOUTH SUBURBAN 605 MAT.	INST.	TOTAL	SPRINGFIELD 626-627 MAT.	INST.	TOTAL
015433	CONTRACTOR EQUIPMENT		98.8	98.8		99.7	99.7		98.8	98.8		98.8	98.8		90.4	90.4		105.4	105.4
0241, 31 - 34	SITE & INFRASTRUCTURE, DEMOLITION	97.7	93.0	94.5	99.7	92.4	94.7	94.9	90.9	92.1	97.1	94.1	95.0	101.0	93.8	96.0	102.5	102.4	102.4
0310	Concrete Forming & Accessories	90.6	117.9	113.8	93.5	113.8	110.8	88.4	95.3	94.3	94.3	130.0	124.6	99.2	152.6	144.6	90.9	119.9	115.6
0320	Concrete Reinforcing	92.5	103.8	98.0	91.7	82.1	87.0	94.9	94.2	94.6	87.6	132.9	109.4	111.9	142.0	126.4	88.3	101.9	94.9
0330	Cast-in-Place Concrete	94.3	114.9	102.1	100.4	102.1	101.0	95.1	94.2	94.8	96.5	129.2	108.8	109.6	146.7	123.6	95.6	113.1	102.2
03	CONCRETE	91.7	115.2	102.3	93.2	104.9	98.5	92.3	96.0	94.0	92.3	130.9	109.6	104.1	148.1	123.8	89.3	114.9	100.8
04	MASONRY	114.6	122.5	119.4	109.1	108.5	108.7	115.8	96.1	103.8	89.4	142.0	121.5	98.8	152.1	131.3	89.0	124.3	110.5
05	METALS	95.7	125.1	104.8	99.6	110.1	102.9	93.0	115.9	100.2	95.7	144.2	110.8	94.8	138.7	108.5	100.1	120.3	106.4
06	WOOD, PLASTICS & COMPOSITES	91.8	115.9	104.4	87.5	116.1	102.5	85.0	94.2	89.8	91.7	126.6	110.0	100.0	152.9	127.7	87.1	119.0	103.8
07	THERMAL & MOISTURE PROTECTION	96.4	112.7	103.4	95.4	102.6	98.5	95.7	93.9	94.9	99.1	129.3	112.1	99.7	141.0	117.4	98.1	116.3	105.9
08	OPENINGS	96.6	120.9	102.5	93.8	111.3	98.1	91.4	99.5	93.3	96.6	138.1	106.7	99.2	161.8	114.5	98.8	119.9	103.9
0920	Plaster & Gypsum Board	91.9	116.6	108.1	87.9	116.9	106.9	87.1	94.4	91.9	91.9	127.6	115.3	94.3	154.7	133.9	90.4	119.5	109.5
0950, 0980	Ceilings & Acoustic Treatment	89.2	116.6	106.4	79.6	116.9	103.0	84.1	94.4	90.5	89.2	127.6	113.3	89.3	154.7	130.3	88.5	119.5	108.0
0960	Flooring	92.3	121.0	100.7	104.6	108.4	105.7	89.8	90.3	89.9	92.3	124.0	101.5	94.6	152.2	111.4	105.3	128.7	112.1
0970, 0990	Wall Finishes & Painting/Coating	85.6	140.2	118.3	92.8	106.2	100.8	85.6	91.0	88.8	85.6	144.2	120.8	85.4	156.1	127.8	98.4	114.0	107.7
09	FINISHES	90.3	121.0	106.9	90.0	113.2	102.5	87.6	93.7	90.9	90.3	130.0	111.8	92.1	153.8	125.5	95.5	121.7	109.7
COVERS	DIVS. 10 - 14, 25, 28, 41, 43, 44, 46	100.0	106.8	101.6	100.0	104.5	101.1	100.0	92.3	98.2	100.0	113.3	103.1	100.0	117.7	104.2	100.0	107.9	101.9
21, 22, 23	FIRE SUPPRESSION, PLUMBING & HVAC	100.6	102.2	101.2	96.7	98.0	97.2	96.8	92.9	95.2	100.7	115.5	106.7	100.3	129.7	112.2	100.5	102.5	101.3
26, 27, 3370	ELECTRICAL, COMMUNICATIONS & UTIL.	96.3	93.5	94.9	91.5	79.2	85.4	88.9	89.7	89.3	96.5	128.3	112.3	95.1	130.5	112.6	100.9	91.4	96.2
MF2018	WEIGHTED AVERAGE	97.4	109.9	102.8	96.1	101.1	98.3	94.5	95.5	94.9	96.3	126.3	109.3	98.5	137.2	115.3	97.8	110.4	103.2

For customer support on your Concrete & Masonry Costs with RSMeans Data, call 800.448.8182.

463

DIVISION		INDIANA																
		ANDERSON 460			BLOOMINGTON 474			COLUMBUS 472			EVANSVILLE 476-477			FORT WAYNE 467-468			GARY 463-464		
		MAT.	INST.	TOTAL	MAT.	INST.	TOTAL	MAT.	INST.	TOTAL	MAT.	INST.	TOTAL	MAT.	INST.	TOTAL	MAT.	INST.	TOTAL
015433	CONTRACTOR EQUIPMENT		92.7	92.7		79.6	79.6		79.6	79.6		108.7	108.7		92.7	92.7		92.7	92.7
0241, 31 - 34	SITE & INFRASTRUCTURE, DEMOLITION	99.7	86.5	90.6	87.3	85.1	85.8	83.6	85.1	84.6	92.8	112.6	106.4	100.5	86.4	90.9	100.3	90.2	93.4
0310	Concrete Forming & Accessories	94.5	78.9	81.3	103.2	80.5	83.9	96.8	79.5	82.1	95.6	80.1	82.4	92.9	73.8	76.7	94.6	109.1	107.0
0320	Concrete Reinforcing	102.1	83.9	93.3	89.6	84.9	87.3	90.0	84.9	87.6	98.1	80.4	89.6	102.1	77.2	90.1	102.1	112.5	107.1
0330	Cast-in-Place Concrete	102.2	75.4	92.1	99.8	75.1	90.5	99.4	73.3	89.5	95.4	83.2	90.8	108.6	74.3	95.7	106.8	109.8	107.9
03	CONCRETE	94.9	79.2	87.8	99.2	79.1	90.2	98.4	78.0	89.2	99.3	81.5	91.3	97.8	75.3	87.7	97.0	109.9	102.8
04	MASONRY	83.7	75.3	78.6	90.4	73.1	79.8	90.2	73.1	79.7	85.8	78.7	81.4	87.6	71.2	77.6	85.1	107.9	99.0
05	METALS	102.9	89.4	98.7	100.8	75.6	92.9	100.8	75.1	92.8	93.4	84.6	90.7	102.9	86.8	97.8	102.9	106.8	104.1
06	WOOD, PLASTICS & COMPOSITES	92.8	79.4	85.8	110.3	81.3	95.1	104.0	80.1	91.5	91.9	79.4	85.4	92.7	74.1	83.0	90.6	107.5	99.4
07	THERMAL & MOISTURE PROTECTION	107.9	75.3	93.9	95.3	77.3	87.6	94.7	76.6	86.9	99.8	81.7	92.0	107.6	76.0	94.0	106.5	103.5	105.2
08	OPENINGS	92.9	77.4	89.1	97.3	78.9	92.8	93.8	78.2	90.0	91.7	76.9	88.1	92.9	71.3	87.6	92.9	112.6	97.7
0920	Plaster & Gypsum Board	100.4	79.3	86.5	95.4	81.7	86.4	92.3	80.4	84.5	90.9	78.7	82.9	99.8	73.8	82.8	94.2	108.2	103.3
0950, 0980	Ceilings & Acoustic Treatment	96.0	79.3	85.5	78.5	81.7	80.5	78.5	80.4	79.7	82.1	78.7	80.0	96.0	73.8	82.1	96.0	108.2	103.6
0960	Flooring	94.1	76.0	88.8	100.5	81.8	95.1	95.2	81.8	91.3	95.4	72.3	88.6	94.1	70.6	87.2	94.1	107.6	98.0
0970, 0990	Wall Finishes & Painting/Coating	92.0	67.9	77.5	84.7	79.7	81.7	84.7	79.7	81.7	90.9	83.6	86.5	92.0	71.0	79.4	92.0	117.3	107.1
09	FINISHES	92.8	77.2	84.3	91.3	80.8	85.6	89.3	80.1	84.3	90.2	78.8	84.1	92.6	72.9	81.9	91.8	109.5	101.4
COVERS	DIVS. 10 - 14, 25, 28, 41, 43, 44, 46	100.0	89.9	97.6	100.0	89.3	97.5	100.0	89.2	97.5	100.0	91.6	98.0	100.0	90.2	97.7	100.0	102.9	100.7
21, 22, 23	FIRE SUPPRESSION, PLUMBING & HVAC	100.6	76.7	90.9	100.2	77.6	91.1	96.4	76.9	88.6	100.5	78.2	91.5	100.6	71.6	88.9	100.6	103.1	101.6
26, 27, 3370	ELECTRICAL, COMMUNICATIONS & UTIL.	89.4	84.9	87.2	100.2	85.0	92.7	99.6	87.1	93.4	96.2	83.2	89.7	90.1	75.0	82.6	100.6	108.1	104.3
MF2018	WEIGHTED AVERAGE	96.8	80.4	89.7	98.2	79.7	90.2	96.5	79.4	89.1	96.0	83.2	90.5	97.5	75.9	88.2	98.3	105.8	101.5

DIVISION		INDIANA																
		INDIANAPOLIS 461 - 462			KOKOMO 469			LAFAYETTE 479			LAWRENCEBURG 470			MUNCIE 473			NEW ALBANY 471		
		MAT.	INST.	TOTAL	MAT.	INST.	TOTAL	MAT.	INST.	TOTAL	MAT.	INST.	TOTAL	MAT.	INST.	TOTAL	MAT.	INST.	TOTAL
015433	CONTRACTOR EQUIPMENT		87.9	87.9		92.7	92.7		79.6	79.6		98.9	98.9		91.0	91.0		88.9	88.9
0241, 31 - 34	SITE & INFRASTRUCTURE, DEMOLITION	99.5	93.7	95.5	95.4	86.5	89.3	84.3	85.1	84.8	81.6	99.5	93.8	87.1	85.5	86.0	79.5	87.3	84.9
0310	Concrete Forming & Accessories	97.4	85.0	86.8	97.6	77.3	80.3	93.8	81.9	83.7	92.7	77.6	79.8	93.1	78.5	80.7	92.1	76.2	78.6
0320	Concrete Reinforcing	101.0	85.3	93.4	92.3	85.2	88.9	89.6	85.2	87.4	88.8	78.0	83.6	99.1	83.8	91.7	90.2	80.4	85.5
0330	Cast-in-Place Concrete	99.3	85.1	94.0	101.2	81.0	93.6	99.9	78.4	91.8	93.5	72.7	85.6	104.9	74.6	93.5	96.5	72.6	87.5
03	CONCRETE	97.7	84.5	91.8	91.7	80.6	86.7	98.6	80.9	90.6	91.8	76.6	85.0	97.5	78.7	89.0	97.2	76.1	87.7
04	MASONRY	94.5	79.0	85.0	83.4	76.0	78.8	95.3	76.2	83.6	75.0	73.0	73.8	92.1	75.4	81.9	81.4	70.3	74.7
05	METALS	99.7	75.4	92.1	99.2	89.9	96.3	99.1	75.7	91.8	95.5	86.1	92.6	102.5	89.3	98.4	97.6	82.0	92.8
06	WOOD, PLASTICS & COMPOSITES	96.5	85.6	90.8	96.0	76.7	85.9	100.7	83.1	91.5	89.9	77.9	83.6	101.3	79.0	89.6	92.5	77.1	84.4
07	THERMAL & MOISTURE PROTECTION	97.6	81.3	90.6	106.9	76.4	93.8	94.7	78.9	87.9	99.7	75.9	89.5	97.7	76.7	88.7	88.3	73.3	81.9
08	OPENINGS	103.5	81.2	98.1	88.2	76.3	85.3	92.3	79.9	89.3	93.5	74.1	88.8	90.8	77.2	87.4	91.3	73.2	86.9
0920	Plaster & Gypsum Board	87.9	85.3	86.2	105.1	76.5	86.3	89.9	83.6	85.7	70.8	78.0	75.5	90.9	79.3	83.3	88.3	77.1	81.0
0950, 0980	Ceilings & Acoustic Treatment	97.6	85.3	89.9	96.6	76.5	84.0	74.7	83.6	80.2	86.5	78.0	81.2	78.5	79.3	79.0	82.1	77.1	79.0
0960	Flooring	95.9	81.8	91.8	98.2	87.9	95.2	94.0	81.8	90.4	69.7	81.8	73.2	94.8	76.0	89.3	92.6	51.5	80.6
0970, 0990	Wall Finishes & Painting/Coating	99.6	79.7	87.7	92.0	67.2	77.2	84.7	82.3	83.2	85.5	71.8	77.3	84.7	67.9	74.6	90.9	64.4	75.0
09	FINISHES	94.1	84.1	88.7	94.6	78.1	85.6	87.7	82.1	84.7	80.7	77.9	79.2	88.7	76.8	82.3	89.3	70.1	78.9
COVERS	DIVS. 10 - 14, 25, 28, 41, 43, 44, 46	100.0	92.3	98.2	100.0	89.7	97.6	100.0	89.4	97.5	100.0	88.8	97.4	100.0	88.9	97.4	100.0	88.6	97.3
21, 22, 23	FIRE SUPPRESSION, PLUMBING & HVAC	100.5	78.5	91.6	96.8	77.1	88.9	96.4	77.9	88.9	97.4	73.8	87.9	100.2	76.5	90.7	96.7	76.7	88.7
26, 27, 3370	ELECTRICAL, COMMUNICATIONS & UTIL.	102.5	87.1	94.9	93.3	77.5	85.5	99.1	80.3	89.8	94.2	73.9	84.2	92.2	75.7	84.0	94.9	74.8	84.9
MF2018	WEIGHTED AVERAGE	99.5	82.9	92.3	94.9	79.8	88.4	96.2	79.9	89.1	93.0	78.3	86.6	96.6	78.9	89.0	94.3	76.3	86.5

DIVISION		INDIANA									IOWA								
		SOUTH BEND 465 - 466			TERRE HAUTE 478			WASHINGTON 475			BURLINGTON 526			CARROLL 514			CEDAR RAPIDS 522 - 524		
		MAT.	INST.	TOTAL	MAT.	INST.	TOTAL	MAT.	INST.	TOTAL	MAT.	INST.	TOTAL	MAT.	INST.	TOTAL	MAT.	INST.	TOTAL
015433	CONTRACTOR EQUIPMENT		108.4	108.4		108.7	108.7		108.7	108.7		97.0	97.0		97.0	97.0		93.8	93.8
0241, 31 - 34	SITE & INFRASTRUCTURE, DEMOLITION	100.0	95.3	96.8	94.3	112.8	107.0	93.8	113.1	107.1	98.1	88.8	91.7	86.9	89.6	88.7	100.4	89.1	92.6
0310	Concrete Forming & Accessories	96.2	77.8	80.5	96.8	78.6	81.3	97.7	81.6	84.0	98.0	90.7	91.8	83.6	80.8	81.2	104.8	85.9	88.7
0320	Concrete Reinforcing	100.3	82.9	91.9	98.1	84.0	91.3	90.8	85.0	88.0	94.3	93.2	93.8	95.0	81.8	88.6	95.0	99.3	97.1
0330	Cast-in-Place Concrete	102.3	80.1	93.9	92.4	78.9	87.3	100.5	85.8	95.0	105.0	52.5	85.2	105.0	78.9	95.1	105.3	84.4	97.4
03	CONCRETE	101.5	80.9	92.2	102.3	80.0	92.3	108.0	83.9	97.2	96.3	78.4	88.2	94.9	81.1	88.7	96.4	88.4	92.8
04	MASONRY	81.6	75.4	77.8	93.6	75.4	82.5	85.9	78.3	81.2	101.6	68.9	81.7	103.1	69.4	82.6	107.5	82.7	92.4
05	METALS	106.8	102.5	105.5	94.1	86.1	91.6	88.6	87.1	88.1	86.0	97.3	89.5	86.1	93.7	88.4	88.5	102.5	92.8
06	WOOD, PLASTICS & COMPOSITES	97.3	77.1	86.7	94.5	78.4	86.1	94.8	80.9	87.5	93.7	94.8	94.3	77.6	85.4	81.7	102.0	85.6	93.4
07	THERMAL & MOISTURE PROTECTION	102.2	79.6	92.5	99.9	79.9	91.3	99.8	82.7	92.5	104.1	73.6	91.0	104.3	75.3	91.9	105.2	81.4	95.0
08	OPENINGS	92.9	76.5	88.9	92.2	76.6	88.4	89.1	80.3	87.0	94.3	92.3	93.8	98.5	81.1	94.3	99.0	87.8	96.3
0920	Plaster & Gypsum Board	88.4	76.6	80.7	90.9	77.7	82.2	90.9	80.2	83.9	99.9	95.0	96.7	95.9	85.4	89.0	105.4	85.7	92.5
0950, 0980	Ceilings & Acoustic Treatment	96.2	76.6	83.9	82.1	77.7	79.3	77.6	80.2	79.2	103.1	95.0	98.0	103.1	85.4	92.0	105.6	85.7	93.1
0960	Flooring	92.3	87.3	90.8	95.4	76.6	89.9	96.4	81.8	92.1	95.3	65.9	86.7	89.3	75.9	85.4	109.7	87.9	103.3
0970, 0990	Wall Finishes & Painting/Coating	95.9	83.4	88.4	90.9	79.1	83.8	90.9	86.5	88.3	88.7	80.6	83.8	88.7	80.6	83.8	90.6	72.3	79.6
09	FINISHES	93.1	79.8	85.9	90.2	78.0	83.6	89.7	81.9	85.5	95.9	86.2	90.7	92.1	80.3	85.7	101.4	84.7	92.4
COVERS	DIVS. 10 - 14, 25, 28, 41, 43, 44, 46	100.0	92.0	98.1	100.0	91.1	97.9	100.0	91.8	98.1	100.0	93.8	98.5	100.0	89.7	97.6	100.0	93.1	98.4
21, 22, 23	FIRE SUPPRESSION, PLUMBING & HVAC	100.5	75.9	90.6	100.5	78.1	91.5	96.7	80.0	90.0	96.9	79.8	90.0	96.9	74.1	87.7	100.7	84.4	94.1
26, 27, 3370	ELECTRICAL, COMMUNICATIONS & UTIL.	103.5	88.8	96.2	94.7	85.3	90.0	95.1	83.8	89.5	99.7	69.6	84.8	100.3	74.1	87.3	97.7	80.2	89.1
MF2018	WEIGHTED AVERAGE	99.6	83.4	92.6	96.9	82.9	90.8	95.1	84.8	90.7	95.8	81.1	89.5	95.5	79.3	88.5	98.3	86.6	93.2

464

For customer support on your Concrete & Masonry Costs with RSMeans Data, call 800.448.8182.

		IOWA																
	DIVISION	COUNCIL BLUFFS			CRESTON			DAVENPORT			DECORAH			DES MOINES			DUBUQUE		
		515			508			527 - 528			521			500 - 503,509			520		
		MAT.	INST.	TOTAL	MAT.	INST.	TOTAL	MAT.	INST.	TOTAL	MAT.	INST.	TOTAL	MAT.	INST.	TOTAL	MAT.	INST.	TOTAL
015433	CONTRACTOR EQUIPMENT		93.2	93.2		97.0	97.0		97.0	97.0		97.0	97.0		103.5	103.5		92.7	92.7
0241, 31 - 34	SITE & INFRASTRUCTURE, DEMOLITION	105.5	86.2	92.2	92.0	90.6	91.0	98.8	92.7	94.6	96.7	88.6	91.1	98.1	101.0	100.1	98.2	86.5	90.2
0310	Concrete Forming & Accessories	82.7	76.4	77.3	77.2	83.5	82.6	104.3	100.5	101.1	95.1	69.1	73.0	96.1	87.1	88.4	84.5	83.2	83.4
0320	Concrete Reinforcing	96.9	84.5	91.0	94.0	81.8	88.1	95.0	99.9	97.4	94.3	81.2	88.0	103.3	99.6	101.5	93.7	80.8	87.5
0330	Cast-in-Place Concrete	109.4	88.2	101.4	107.8	82.2	98.1	101.4	96.5	99.6	102.2	72.9	91.1	91.9	92.7	92.2	103.1	83.8	95.8
03	CONCRETE	98.3	82.9	91.4	95.5	83.5	90.1	94.6	99.6	96.8	94.2	73.7	85.0	95.4	91.5	93.6	93.0	83.8	88.9
04	MASONRY	107.3	78.0	89.4	104.4	76.5	87.4	103.1	95.1	98.3	123.7	66.3	88.7	89.6	87.2	88.1	108.6	70.0	85.1
05	METALS	93.1	96.0	94.0	86.4	94.3	88.9	88.5	108.4	94.7	86.1	92.5	88.1	92.0	96.4	93.4	87.1	93.8	89.2
06	WOOD, PLASTICS & COMPOSITES	76.1	75.1	75.6	69.3	85.4	77.7	102.0	100.3	101.1	90.1	68.5	78.8	90.4	85.6	87.9	78.2	84.2	81.3
07	THERMAL & MOISTURE PROTECTION	104.4	75.6	92.1	106.0	78.5	94.2	104.6	93.8	100.0	104.3	66.8	88.2	108.6	85.5	92.9	104.7	76.9	92.8
08	OPENINGS	98.1	80.0	93.7	107.9	82.8	101.8	99.0	101.9	99.7	97.2	75.0	91.8	100.6	87.8	97.5	98.1	84.2	94.7
0920	Plaster & Gypsum Board	95.9	75.0	82.2	91.6	85.4	87.5	105.4	100.7	102.3	98.8	67.9	78.5	84.6	85.4	85.1	95.9	84.3	88.3
0950, 0980	Ceilings & Acoustic Treatment	103.1	75.0	85.5	88.4	85.4	86.5	105.6	100.7	102.5	103.1	67.9	81.0	91.0	85.4	87.5	103.1	84.3	91.3
0960	Flooring	88.3	85.3	87.4	79.3	65.9	75.4	97.8	90.6	95.7	94.8	65.9	86.4	92.2	92.0	92.2	100.5	71.3	92.0
0970, 0990	Wall Finishes & Painting/Coating	85.6	57.0	68.4	82.2	80.6	81.2	88.7	93.0	91.3	88.7	80.6	83.8	93.1	82.9	87.0	89.8	79.3	83.5
09	FINISHES	93.0	75.6	83.6	83.5	80.1	81.7	98.0	97.9	97.9	95.6	69.5	81.5	90.3	87.4	88.7	96.7	80.6	88.0
COVERS	DIVS. 10 - 14, 25, 28, 41, 43, 44, 46	100.0	90.5	97.8	100.0	92.1	98.1	100.0	97.3	99.4	100.0	89.2	97.5	100.0	94.7	98.8	100.0	91.9	98.1
21, 22, 23	FIRE SUPPRESSION, PLUMBING & HVAC	100.7	79.6	92.2	96.8	76.0	88.4	100.7	95.6	98.6	96.9	70.9	86.4	100.5	85.0	94.3	100.7	76.3	90.9
26, 27, 3370	ELECTRICAL, COMMUNICATIONS & UTIL.	102.1	82.0	92.2	93.3	74.1	83.8	95.9	88.1	92.0	97.7	45.5	71.9	103.8	91.4	97.6	100.9	77.0	89.1
MF2018	WEIGHTED AVERAGE	98.8	81.9	91.5	95.2	81.1	89.1	97.3	96.6	97.0	96.7	71.0	85.6	97.3	90.0	94.2	97.3	80.6	90.1

		IOWA																
	DIVISION	FORT DODGE			MASON CITY			OTTUMWA			SHENANDOAH			SIBLEY			SIOUX CITY		
		505			504			525			516			512			510 - 511		
		MAT.	INST.	TOTAL	MAT.	INST.	TOTAL	MAT.	INST.	TOTAL	MAT.	INST.	TOTAL	MAT.	INST.	TOTAL	MAT.	INST.	TOTAL
015433	CONTRACTOR EQUIPMENT		97.0	97.0		97.0	97.0		92.7	92.7		93.2	93.2		97.0	97.0		97.0	97.0
0241, 31 - 34	SITE & INFRASTRUCTURE, DEMOLITION	99.5	87.5	91.2	99.5	88.5	92.0	98.1	83.7	88.2	103.2	85.6	91.1	108.1	88.5	94.7	110.2	90.7	96.8
0310	Concrete Forming & Accessories	77.8	75.1	75.5	82.2	68.7	70.7	92.7	83.9	85.2	84.7	74.3	75.8	85.4	36.4	43.7	104.8	75.2	79.6
0320	Concrete Reinforcing	94.0	81.1	87.8	93.9	81.2	87.8	94.3	93.3	93.8	96.9	81.7	89.6	96.9	81.1	89.3	95.0	97.9	96.4
0330	Cast-in-Place Concrete	101.2	39.9	78.0	101.2	68.3	88.8	105.7	61.2	88.9	105.8	79.3	95.8	103.7	53.1	84.6	104.4	87.3	97.9
03	CONCRETE	91.1	64.9	79.4	91.4	71.9	82.6	95.7	78.4	88.0	95.9	78.3	88.0	94.9	51.9	75.6	95.8	84.3	90.7
04	MASONRY	103.4	49.8	70.7	117.4	65.1	85.5	104.3	52.9	73.0	106.8	71.4	85.2	126.4	50.0	79.8	101.1	68.4	81.2
05	METALS	86.5	92.2	88.3	86.5	92.5	88.4	86.0	98.0	89.7	92.1	93.8	92.6	86.2	91.6	87.9	88.5	101.2	92.4
06	WOOD, PLASTICS & COMPOSITES	69.8	85.4	78.0	74.4	68.5	71.3	86.8	94.6	90.9	78.2	75.8	77.0	79.2	32.9	55.0	102.0	74.6	87.7
07	THERMAL & MOISTURE PROTECTION	105.4	62.2	86.8	104.9	67.3	88.8	104.9	67.1	88.7	103.7	70.0	89.3	104.0	52.7	82.0	104.6	71.6	90.4
08	OPENINGS	102.1	73.1	95.1	94.6	75.0	89.9	98.5	89.4	96.3	90.2	73.6	86.2	95.4	44.3	82.9	99.0	80.9	94.6
0920	Plaster & Gypsum Board	91.6	85.4	87.5	91.6	67.9	76.1	96.2	95.0	95.4	95.9	75.7	82.6	95.9	31.4	53.6	105.4	74.2	84.9
0950, 0980	Ceilings & Acoustic Treatment	88.4	85.4	86.5	88.4	67.9	75.6	103.1	95.0	98.0	103.1	75.7	85.9	103.1	31.4	58.2	105.6	74.2	86.0
0960	Flooring	80.5	65.9	76.3	82.6	65.9	77.7	103.7	65.9	92.7	89.1	70.5	83.7	90.2	65.9	83.2	97.9	71.3	90.1
0970, 0990	Wall Finishes & Painting/Coating	82.2	77.4	79.3	82.2	80.6	81.2	89.8	80.6	84.3	85.6	80.6	82.6	88.7	79.6	83.2	88.7	67.8	76.1
09	FINISHES	85.3	74.9	79.7	85.8	69.2	76.8	97.7	81.5	88.9	93.1	74.0	82.8	95.4	44.5	67.8	99.5	73.1	85.2
COVERS	DIVS. 10 - 14, 25, 28, 41, 43, 44, 46	100.0	85.6	96.6	100.0	88.9	97.4	100.0	87.3	97.0	100.0	83.7	96.2	100.0	80.0	95.3	100.0	90.5	97.8
21, 22, 23	FIRE SUPPRESSION, PLUMBING & HVAC	96.8	68.0	85.2	96.8	74.0	87.6	96.9	69.6	85.9	96.9	79.9	90.0	96.9	68.1	85.3	100.7	78.8	91.9
26, 27, 3370	ELECTRICAL, COMMUNICATIONS & UTIL.	98.7	66.8	82.9	97.9	45.5	72.0	99.6	68.3	84.1	97.7	75.5	86.7	97.7	45.5	71.9	97.7	74.1	86.0
MF2018	WEIGHTED AVERAGE	95.0	71.0	84.6	95.0	71.2	84.7	96.4	75.7	87.5	96.1	78.7	88.6	96.9	60.1	81.0	98.0	80.3	90.3

		IOWA						KANSAS										
	DIVISION	SPENCER			WATERLOO			BELLEVILLE			COLBY			DODGE CITY			EMPORIA		
		513			506 - 507			669			677			678			668		
		MAT.	INST.	TOTAL	MAT.	INST.	TOTAL	MAT.	INST.	TOTAL	MAT.	INST.	TOTAL	MAT.	INST.	TOTAL	MAT.	INST.	TOTAL
015433	CONTRACTOR EQUIPMENT		97.0	97.0		97.0	97.0		100.6	100.6		100.6	100.6		100.6	100.6		98.8	98.8
0241, 31 - 34	SITE & INFRASTRUCTURE, DEMOLITION	108.1	87.3	93.9	105.1	90.4	95.0	110.4	88.3	95.2	103.8	88.8	93.5	106.8	89.1	94.6	102.5	85.9	91.1
0310	Concrete Forming & Accessories	92.5	36.1	44.5	94.4	70.1	73.7	90.9	53.1	58.7	93.9	59.1	64.3	87.1	60.8	64.7	82.1	67.5	69.7
0320	Concrete Reinforcing	96.9	81.0	89.3	94.6	81.1	88.1	96.1	99.7	97.8	96.5	99.7	98.0	94.0	99.5	96.7	94.8	99.8	97.2
0330	Cast-in-Place Concrete	103.7	62.8	88.3	108.4	84.8	99.5	113.4	81.1	101.2	108.8	83.5	99.2	110.7	87.9	102.1	109.6	83.6	99.8
03	CONCRETE	95.4	55.1	77.3	97.0	78.2	88.6	107.6	72.4	91.8	104.3	76.0	91.6	105.5	78.2	93.3	100.3	79.8	91.1
04	MASONRY	126.4	50.0	79.8	104.1	76.4	87.2	91.2	58.0	70.9	97.2	61.9	75.6	111.1	59.8	79.8	96.9	63.3	76.4
05	METALS	86.2	91.3	87.8	88.8	94.5	90.6	93.4	93.6	94.7	90.6	98.3	93.1	92.2	98.0	94.0	93.1	98.6	94.8
06	WOOD, PLASTICS & COMPOSITES	86.7	32.9	58.6	89.3	65.8	77.0	85.8	49.3	66.7	92.8	55.3	73.2	84.5	56.2	69.7	76.8	66.4	71.4
07	THERMAL & MOISTURE PROTECTION	104.9	53.2	82.7	105.2	77.7	93.4	91.3	60.3	78.0	97.1	62.5	82.2	97.0	66.5	83.9	89.4	71.9	81.9
08	OPENINGS	106.1	44.3	91.0	95.1	74.5	90.0	93.7	60.6	85.6	97.5	63.9	89.3	97.5	64.3	89.4	91.8	72.0	87.0
0920	Plaster & Gypsum Board	96.2	31.4	53.7	99.5	65.2	77.0	85.3	48.3	61.0	97.1	54.4	69.1	90.6	55.3	67.4	82.2	65.9	71.5
0950, 0980	Ceilings & Acoustic Treatment	103.1	31.4	58.2	89.7	65.2	74.3	82.2	48.3	60.9	79.7	54.4	63.9	79.7	55.3	64.4	82.2	65.9	71.9
0960	Flooring	93.1	65.9	85.2	87.9	79.9	85.5	89.6	65.2	82.5	87.7	65.2	81.1	83.8	66.5	78.7	85.0	65.2	79.2
0970, 0990	Wall Finishes & Painting/Coating	88.7	54.5	68.2	82.2	84.6	83.6	89.0	98.1	94.4	94.9	98.1	96.8	94.9	98.1	96.8	89.0	98.1	94.4
09	FINISHES	96.2	40.6	66.1	89.0	72.4	80.0	86.5	58.6	71.4	86.6	63.1	73.9	84.6	64.3	73.7	83.7	69.6	76.1
COVERS	DIVS. 10 - 14, 25, 28, 41, 43, 44, 46	100.0	86.3	97.8	100.0	90.5	97.8	100.0	79.5	95.2	100.0	81.5	95.6	100.0	85.5	96.6	100.0	81.3	95.6
21, 22, 23	FIRE SUPPRESSION, PLUMBING & HVAC	96.9	68.1	85.3	100.6	81.0	92.7	96.7	68.0	85.1	96.7	67.7	85.0	100.5	76.3	90.7	96.7	70.0	85.9
26, 27, 3370	ELECTRICAL, COMMUNICATIONS & UTIL.	99.2	45.5	72.6	94.9	64.2	79.7	99.8	60.4	80.3	96.8	65.4	81.2	94.6	71.2	83.0	97.4	64.4	81.1
MF2018	WEIGHTED AVERAGE	98.3	59.9	81.7	96.5	78.4	88.7	96.9	69.1	84.9	96.5	71.7	85.7	98.0	74.8	88.0	95.2	74.0	86.0

For customer support on your Concrete & Masonry Costs with RSMeans Data, call 800.448.8182.

465

KANSAS

DIVISION		FORT SCOTT 667			HAYS 676			HUTCHINSON 675			INDEPENDENCE 673			KANSAS CITY 660 - 662			LIBERAL 679		
		MAT.	INST.	TOTAL	MAT.	INST.	TOTAL	MAT.	INST.	TOTAL	MAT.	INST.	TOTAL	MAT.	INST.	TOTAL	MAT.	INST.	TOTAL
015433	CONTRACTOR EQUIPMENT		99.7	99.7		100.6	100.6		100.6	100.6		100.6	100.6		97.3	97.3		100.6	100.6
0241, 31 - 34	SITE & INFRASTRUCTURE, DEMOLITION	99.2	86.0	90.1	108.8	88.6	94.9	88.5	88.9	88.8	107.2	89.0	94.7	94.0	86.8	89.1	108.5	88.6	94.8
0310	Concrete Forming & Accessories	99.4	79.8	82.7	91.4	57.6	62.7	81.6	54.7	58.7	102.5	65.4	70.9	95.9	99.3	98.8	87.5	56.8	61.4
0320	Concrete Reinforcing	94.1	96.0	95.0	94.0	99.7	96.8	94.0	99.7	96.7	93.5	96.0	94.7	91.3	104.3	97.6	95.4	99.4	97.3
0330	Cast-in-Place Concrete	101.6	78.9	93.1	85.8	81.5	84.2	79.5	83.4	81.0	111.1	83.5	100.7	87.1	96.7	90.7	85.8	80.1	83.7
03	CONCRETE	96.0	83.1	90.2	96.7	74.6	86.8	80.2	73.9	77.4	106.7	78.2	93.9	88.6	99.7	93.6	98.5	73.7	87.4
04	MASONRY	99.4	53.8	71.6	108.5	58.1	77.8	97.2	61.9	75.6	94.6	61.9	74.6	98.5	98.1	98.3	109.0	51.4	73.8
05	METALS	93.1	96.9	94.3	90.2	98.3	92.8	90.0	97.8	92.5	89.9	97.0	92.1	101.0	105.9	102.5	90.5	97.5	92.7
06	WOOD, PLASTICS & COMPOSITES	96.1	87.4	91.5	89.8	55.3	71.8	79.2	49.5	63.7	103.9	63.5	82.8	92.1	99.8	96.1	85.1	55.3	69.5
07	THERMAL & MOISTURE PROTECTION	90.5	69.5	81.5	97.4	61.0	81.8	95.8	61.6	81.1	97.2	71.5	86.1	89.9	97.8	93.3	97.6	58.2	80.7
08	OPENINGS	91.8	82.7	89.6	97.4	63.9	89.3	97.4	60.7	88.5	95.6	67.5	88.7	93.0	98.5	94.3	97.5	63.9	89.3
0920	Plaster & Gypsum Board	87.9	87.4	87.5	94.9	54.4	68.3	89.2	48.5	62.5	106.2	62.9	77.8	81.9	100.1	93.9	91.5	54.4	67.2
0950, 0980	Ceilings & Acoustic Treatment	82.2	87.4	85.4	79.7	54.4	63.9	79.7	48.5	60.1	79.7	62.9	69.2	82.2	100.1	93.4	79.7	54.4	63.9
0960	Flooring	99.0	63.8	88.7	86.4	65.2	80.2	80.7	65.2	76.2	92.1	63.8	83.9	79.7	96.0	84.4	84.0	65.2	78.5
0970, 0990	Wall Finishes & Painting/Coating	90.5	73.8	80.5	94.9	98.1	96.8	94.9	98.1	96.8	94.9	98.1	96.8	96.5	104.4	101.2	94.9	98.1	96.8
09	FINISHES	88.7	77.4	82.6	86.4	62.1	73.3	82.1	59.7	70.0	89.3	68.1	77.8	84.2	99.3	92.3	85.5	61.6	72.6
COVERS	DIVS. 10 - 14, 25, 28, 41, 43, 44, 46	100.0	82.2	95.8	100.0	80.2	95.3	100.0	80.8	95.5	100.0	82.4	95.9	100.0	95.6	99.0	100.0	79.6	95.2
21, 22, 23	FIRE SUPPRESSION, PLUMBING & HVAC	96.7	64.2	83.6	96.7	65.3	84.0	96.7	67.7	85.0	96.7	67.1	84.8	100.4	103.0	101.4	96.7	66.0	84.3
26, 27, 3370	ELECTRICAL, COMMUNICATIONS & UTIL.	96.7	65.4	81.2	96.0	65.4	80.9	92.5	58.7	75.7	94.0	70.1	82.2	101.6	97.1	99.4	94.6	65.4	80.1
MF2018	WEIGHTED AVERAGE	95.2	73.9	86.0	96.0	70.4	84.9	91.9	69.7	82.3	96.4	73.6	86.6	96.4	99.2	97.6	96.0	69.5	84.6

DIVISION		KANSAS SALINA 674			TOPEKA 664 - 666			WICHITA 670 - 672			KENTUCKY ASHLAND 411 - 412			BOWLING GREEN 421 - 422			CAMPTON 413 - 414		
		MAT.	INST.	TOTAL	MAT.	INST.	TOTAL	MAT.	INST.	TOTAL	MAT.	INST.	TOTAL	MAT.	INST.	TOTAL	MAT.	INST.	TOTAL
015433	CONTRACTOR EQUIPMENT		100.6	100.6		104.7	104.7		106.1	106.1		95.7	95.7		88.9	88.9		95.0	95.0
0241, 31 - 34	SITE & INFRASTRUCTURE, DEMOLITION	97.5	88.7	91.5	98.6	96.9	97.4	95.0	98.7	97.6	112.9	76.0	87.6	80.0	87.5	85.1	87.9	87.8	87.8
0310	Concrete Forming & Accessories	83.6	61.6	64.9	91.8	68.0	71.6	91.0	57.3	62.3	87.2	86.3	86.5	87.8	81.4	82.3	89.9	78.4	80.1
0320	Concrete Reinforcing	93.5	99.6	96.4	92.7	101.1	96.8	87.4	101.4	94.1	91.8	91.3	91.5	89.0	79.2	84.3	89.8	88.7	89.2
0330	Cast-in-Place Concrete	96.1	84.2	91.6	96.5	88.8	93.6	89.9	78.8	85.7	87.8	90.9	89.0	87.1	68.3	80.0	97.2	65.4	85.2
03	CONCRETE	93.3	77.4	86.1	91.4	82.0	87.2	87.6	73.6	81.3	92.9	90.0	91.6	91.5	76.9	84.9	94.9	76.1	86.5
04	MASONRY	126.2	55.1	82.8	94.2	65.1	76.4	95.3	53.2	69.6	92.5	88.2	89.9	94.7	69.6	79.4	91.3	52.4	67.6
05	METALS	92.0	98.6	94.1	97.3	98.6	97.7	94.1	96.6	94.9	96.7	106.5	99.8	98.4	85.1	94.2	97.6	88.6	94.8
06	WOOD, PLASTICS & COMPOSITES	80.8	60.7	70.3	90.6	66.7	78.1	96.3	55.1	74.8	74.3	84.7	79.7	86.3	83.8	85.0	84.5	85.3	84.9
07	THERMAL & MOISTURE PROTECTION	96.5	61.8	81.6	94.1	84.1	89.8	95.3	60.9	80.5	90.5	85.1	88.2	88.3	77.9	83.9	99.9	64.8	84.8
08	OPENINGS	97.4	67.3	90.1	103.2	74.8	96.3	100.9	64.6	92.1	90.5	86.1	89.4	91.3	79.5	88.4	92.5	84.1	90.4
0920	Plaster & Gypsum Board	89.2	59.9	70.0	94.3	65.8	75.6	93.9	53.9	67.7	56.9	84.7	75.2	84.1	84.0	84.0	84.1	84.7	84.5
0950, 0980	Ceilings & Acoustic Treatment	79.7	59.9	67.3	95.5	65.8	76.9	91.9	53.9	68.1	77.8	84.7	82.1	82.1	84.0	83.3	82.1	84.7	83.7
0960	Flooring	82.1	66.5	77.6	92.3	66.5	84.7	93.0	66.9	85.4	74.8	77.5	75.6	90.4	60.3	81.6	92.8	61.4	83.6
0970, 0990	Wall Finishes & Painting/Coating	94.9	98.1	96.8	95.3	98.1	97.0	96.4	57.3	73.0	91.6	86.1	88.3	90.9	66.9	76.5	90.9	51.3	67.2
09	FINISHES	83.3	65.3	73.6	93.6	69.5	80.6	92.7	57.6	73.7	76.9	84.6	81.1	88.0	76.0	81.5	88.8	72.9	80.2
COVERS	DIVS. 10 - 14, 25, 28, 41, 43, 44, 46	100.0	85.5	96.6	100.0	85.9	96.7	100.0	85.4	96.6	100.0	83.6	96.1	100.0	89.3	97.5	100.0	87.6	97.1
21, 22, 23	FIRE SUPPRESSION, PLUMBING & HVAC	100.5	69.4	87.9	100.6	74.5	90.1	100.3	70.6	88.3	96.4	81.0	90.2	100.5	77.9	91.4	96.7	72.8	87.1
26, 27, 3370	ELECTRICAL, COMMUNICATIONS & UTIL.	94.4	73.4	84.0	102.0	72.0	87.2	99.5	73.4	86.6	92.7	85.0	88.9	95.2	76.5	85.9	92.9	84.9	89.0
MF2018	WEIGHTED AVERAGE	96.8	73.3	86.6	98.1	78.0	89.4	96.5	72.2	86.0	93.6	86.3	90.5	95.2	78.4	87.9	94.9	76.3	86.9

KENTUCKY

DIVISION		CORBIN 407 - 409			COVINGTON 410			ELIZABETHTOWN 427			FRANKFORT 406			HAZARD 417 - 418			HENDERSON 424		
		MAT.	INST.	TOTAL	MAT.	INST.	TOTAL	MAT.	INST.	TOTAL	MAT.	INST.	TOTAL	MAT.	INST.	TOTAL	MAT.	INST.	TOTAL
015433	CONTRACTOR EQUIPMENT		95.0	95.0		98.9	98.9		88.9	88.9		99.9	99.9		95.0	95.0		108.7	108.7
0241, 31 - 34	SITE & INFRASTRUCTURE, DEMOLITION	92.1	88.2	89.4	83.1	98.8	93.9	73.8	86.5	82.5	90.3	97.6	95.3	85.7	88.8	87.8	82.2	111.3	102.2
0310	Concrete Forming & Accessories	86.2	73.1	75.1	85.1	68.2	70.7	81.8	69.7	71.5	101.4	75.9	79.7	85.9	79.3	80.3	93.3	74.2	77.0
0320	Concrete Reinforcing	87.1	87.9	87.5	88.5	74.9	81.9	89.4	79.8	84.7	103.5	80.1	92.2	90.2	88.2	89.2	89.1	80.3	84.9
0330	Cast-in-Place Concrete	90.3	69.6	82.5	93.0	78.2	87.4	78.7	64.9	73.5	90.5	74.1	84.3	93.4	66.9	83.4	77.0	81.1	78.5
03	CONCRETE	84.3	75.0	80.2	93.2	73.8	84.5	83.5	70.4	77.6	88.8	76.4	83.2	91.7	76.9	85.1	88.6	78.1	83.9
04	MASONRY	90.6	57.5	70.4	105.7	69.0	83.3	78.7	59.2	66.8	88.7	71.2	78.0	90.2	54.4	68.4	98.1	75.5	84.3
05	METALS	92.5	87.7	91.0	95.5	86.5	92.7	97.5	84.0	93.3	95.2	84.2	91.7	97.6	88.6	94.8	88.3	85.3	87.4
06	WOOD, PLASTICS & COMPOSITES	74.1	74.9	74.5	81.5	65.9	73.3	80.3	71.4	75.7	100.2	75.0	87.0	80.7	85.3	83.1	89.1	73.2	80.8
07	THERMAL & MOISTURE PROTECTION	104.5	66.4	88.1	99.9	69.5	86.9	87.7	65.4	78.2	102.1	72.9	89.5	99.8	66.4	85.4	99.1	79.0	90.5
08	OPENINGS	87.1	67.9	82.4	94.3	69.5	88.2	91.3	70.3	86.2	98.7	75.7	93.1	92.8	84.0	90.6	89.5	75.6	86.1
0920	Plaster & Gypsum Board	93.4	74.1	80.7	67.4	65.8	66.3	83.0	71.2	75.3	99.3	74.1	82.7	83.0	84.7	84.1	87.2	72.3	77.4
0950, 0980	Ceilings & Acoustic Treatment	79.5	74.1	76.1	86.5	65.8	73.5	82.1	71.2	75.3	96.5	74.1	82.4	82.1	84.7	83.7	77.6	72.3	74.3
0960	Flooring	88.3	61.4	80.4	67.0	79.6	70.7	87.5	70.4	82.5	95.8	75.0	89.7	90.9	61.4	82.3	94.4	72.6	88.0
0970, 0990	Wall Finishes & Painting/Coating	85.4	59.0	69.6	85.5	67.7	74.8	90.9	64.9	75.3	96.1	87.4	90.9	90.9	51.3	67.2	90.9	82.6	85.9
09	FINISHES	85.6	69.6	76.9	79.6	69.8	74.3	86.6	68.7	76.9	95.7	77.0	85.6	88.0	73.4	80.1	87.9	74.8	80.8
COVERS	DIVS. 10 - 14, 25, 28, 41, 43, 44, 46	100.0	89.0	97.4	100.0	87.8	97.1	100.0	86.1	96.7	100.0	91.1	97.9	100.0	88.3	97.2	100.0	52.8	88.9
21, 22, 23	FIRE SUPPRESSION, PLUMBING & HVAC	97.2	70.6	86.5	97.5	72.4	87.3	96.9	74.3	87.8	100.7	77.2	91.2	96.8	73.9	87.5	96.9	75.6	88.3
26, 27, 3370	ELECTRICAL, COMMUNICATIONS & UTIL.	89.3	84.9	87.1	96.0	69.8	83.0	92.8	75.0	84.0	100.1	75.0	87.7	92.9	84.9	89.0	94.7	75.0	84.9
MF2018	WEIGHTED AVERAGE	91.8	75.0	84.6	94.9	75.0	86.3	91.8	73.4	83.8	96.8	78.6	88.9	94.3	77.1	86.9	92.7	78.7	86.7

KENTUCKY

DIVISION		LEXINGTON 403 - 405			LOUISVILLE 400 - 402			OWENSBORO 423			PADUCAH 420			PIKEVILLE 415 - 416			SOMERSET 425 - 426		
		MAT.	INST.	TOTAL	MAT.	INST.	TOTAL	MAT.	INST.	TOTAL	MAT.	INST.	TOTAL	MAT.	INST.	TOTAL	MAT.	INST.	TOTAL
015433	CONTRACTOR EQUIPMENT		95.0	95.0		95.3	95.3		108.7	108.7		108.7	108.7		95.7	95.7		95.0	95.0
0241, 31 - 34	SITE & INFRASTRUCTURE, DEMOLITION	94.6	90.7	91.9	88.4	97.1	94.4	92.7	112.7	106.5	84.9	111.5	103.2	124.2	75.3	90.6	78.2	88.3	85.1
0310	Concrete Forming & Accessories	100.4	75.0	78.8	101.1	81.5	84.4	91.5	78.4	80.4	89.1	78.2	79.8	97.5	82.4	84.7	87.1	73.8	75.8
0320	Concrete Reinforcing	95.5	86.8	91.3	87.9	86.7	87.3	89.1	79.8	84.6	89.6	78.5	84.3	92.3	91.2	91.8	89.4	88.1	88.8
0330	Cast-in-Place Concrete	92.4	83.9	89.2	87.0	69.7	80.5	89.8	83.9	87.6	82.0	77.8	80.4	96.5	85.8	92.5	77.0	84.4	79.8
03	CONCRETE	87.4	80.6	84.3	85.1	78.5	82.1	100.4	81.0	91.7	93.0	78.5	86.5	106.4	86.5	97.5	78.7	80.5	79.5
04	MASONRY	89.2	71.8	78.6	87.3	70.6	77.1	90.5	80.0	84.1	93.3	75.9	82.7	89.7	79.8	83.6	84.9	60.6	70.1
05	METALS	94.9	88.7	93.0	97.1	86.1	93.6	89.9	87.3	89.1	86.8	85.8	86.5	96.6	106.3	99.6	97.5	88.3	94.7
06	WOOD, PLASTICS & COMPOSITES	92.1	73.2	82.2	97.8	84.1	90.6	86.7	77.5	81.9	84.1	78.7	81.3	85.8	84.7	85.2	81.4	74.9	78.0
07	THERMAL & MOISTURE PROTECTION	104.9	78.0	93.4	101.8	75.1	90.3	99.8	82.0	92.2	99.2	78.9	90.5	91.4	76.9	85.2	99.1	68.7	86.0
08	OPENINGS	87.4	76.6	84.8	88.7	79.1	86.3	89.5	78.4	86.8	88.8	78.1	86.2	91.0	82.9	89.1	91.9	74.2	87.6
0920	Plaster & Gypsum Board	102.8	72.3	82.8	94.0	84.0	87.4	85.8	76.7	79.8	85.2	77.9	80.4	60.6	84.7	76.4	83.0	74.1	77.1
0950, 0980	Ceilings & Acoustic Treatment	82.0	72.3	75.9	99.0	84.0	89.6	77.6	76.7	77.1	77.6	77.9	77.8	77.8	84.7	82.1	82.1	74.1	77.1
0960	Flooring	93.4	63.8	84.8	96.2	60.1	85.6	93.8	60.3	84.0	92.7	72.6	86.8	79.2	61.4	74.0	91.2	61.4	82.5
0970, 0990	Wall Finishes & Painting/Coating	85.4	80.5	82.5	93.7	66.8	77.6	90.9	86.0	87.9	90.9	69.6	78.2	91.6	87.4	89.0	90.9	64.9	75.3
09	FINISHES	89.1	72.8	80.3	95.2	76.1	84.9	88.1	75.3	81.1	87.3	76.2	81.3	79.5	79.3	79.4	87.3	70.3	78.1
COVERS	DIVS. 10 - 14, 25, 28, 41, 43, 44, 46	100.0	90.7	97.8	100.0	89.8	97.8	100.0	93.4	98.4	100.0	86.4	96.8	100.0	45.6	87.2	100.0	89.2	97.5
21, 22, 23	FIRE SUPPRESSION, PLUMBING & HVAC	101.0	76.9	91.2	100.7	78.4	91.7	100.5	77.9	91.4	96.9	74.5	87.9	96.4	78.4	89.2	96.9	70.5	86.2
26, 27, 3370	ELECTRICAL, COMMUNICATIONS & UTIL.	91.5	76.3	84.0	95.5	76.5	86.1	94.7	73.8	84.4	96.7	74.4	85.7	95.3	85.0	90.2	93.3	84.9	89.2
MF2018	WEIGHTED AVERAGE	94.2	78.8	87.5	94.9	79.6	88.3	95.3	81.8	89.4	93.0	80.0	87.4	96.1	82.1	90.1	92.1	76.6	85.4

LOUISIANA

DIVISION		ALEXANDRIA 713 - 714			BATON ROUGE 707 - 708			HAMMOND 704			LAFAYETTE 705			LAKE CHARLES 706			MONROE 712		
		MAT.	INST.	TOTAL	MAT.	INST.	TOTAL	MAT.	INST.	TOTAL	MAT.	INST.	TOTAL	MAT.	INST.	TOTAL	MAT.	INST.	TOTAL
015433	CONTRACTOR EQUIPMENT		88.0	88.0		92.5	92.5		86.6	86.6		86.6	86.6		86.1	86.1		88.0	88.0
0241, 31 - 34	SITE & INFRASTRUCTURE, DEMOLITION	99.2	84.7	89.2	101.4	92.8	95.5	99.5	82.6	87.9	101.1	84.5	89.7	101.8	84.2	89.7	99.2	84.6	89.2
0310	Concrete Forming & Accessories	77.1	61.1	63.5	94.2	71.5	74.9	74.4	54.3	57.3	91.4	65.5	69.4	91.9	69.1	72.5	76.7	60.6	63.0
0320	Concrete Reinforcing	92.6	53.8	73.9	88.3	53.8	71.7	86.3	51.8	69.6	87.5	51.8	70.3	87.5	53.8	71.3	91.6	53.8	73.3
0330	Cast-in-Place Concrete	87.2	67.1	79.6	92.2	69.5	83.6	87.0	61.5	78.9	86.6	67.5	79.3	91.1	68.4	82.6	87.2	66.3	79.3
03	CONCRETE	82.0	62.6	73.3	89.2	68.3	79.8	84.3	58.6	72.8	85.3	64.3	75.9	87.5	66.6	78.1	81.8	62.2	73.0
04	MASONRY	106.0	64.5	80.7	86.4	62.3	71.7	90.4	62.2	73.2	90.4	65.3	75.1	89.9	66.8	75.8	100.9	63.2	77.9
05	METALS	93.5	70.3	86.2	93.7	73.6	87.4	85.6	68.0	80.1	84.9	68.3	79.7	84.9	69.2	80.0	93.4	70.3	86.2
06	WOOD, PLASTICS & COMPOSITES	86.2	60.1	72.6	96.1	75.0	85.0	75.4	52.3	63.3	95.7	65.7	80.0	94.0	69.5	81.2	85.6	60.1	72.2
07	THERMAL & MOISTURE PROTECTION	99.6	66.2	85.2	98.0	68.1	85.2	97.3	63.3	82.7	97.9	67.1	84.7	97.6	68.2	85.0	99.5	65.6	85.0
08	OPENINGS	110.5	59.5	98.1	97.9	71.9	91.5	96.2	52.8	85.7	99.6	59.8	89.9	99.6	63.2	90.8	110.5	58.0	97.7
0920	Plaster & Gypsum Board	81.3	59.6	67.1	98.4	74.6	82.8	100.4	51.6	68.4	109.1	65.3	80.4	109.1	69.3	83.0	81.0	59.6	67.0
0950, 0980	Ceilings & Acoustic Treatment	92.2	59.6	71.8	100.3	74.6	84.2	103.0	51.6	70.8	99.8	65.3	78.2	100.5	69.3	81.0	92.2	59.6	71.8
0960	Flooring	84.6	67.6	79.6	93.6	67.6	86.0	92.7	67.6	85.3	101.3	67.6	91.5	101.3	67.6	91.5	84.2	67.6	79.4
0970, 0990	Wall Finishes & Painting/Coating	92.7	60.4	73.3	95.8	60.4	74.6	94.9	57.0	72.2	94.9	56.9	72.1	94.9	60.4	74.2	92.7	62.2	74.4
09	FINISHES	84.9	61.9	72.5	96.6	70.1	82.3	95.8	56.4	74.4	98.7	65.0	80.5	98.9	68.1	82.2	84.8	61.8	72.3
COVERS	DIVS. 10 - 14, 25, 28, 41, 43, 44, 46	100.0	82.7	95.9	100.0	85.3	96.5	100.0	82.3	95.8	100.0	84.8	96.4	100.0	85.7	96.6	100.0	82.3	95.8
21, 22, 23	FIRE SUPPRESSION, PLUMBING & HVAC	101.0	63.0	85.6	100.9	61.5	85.0	97.3	59.8	82.2	101.1	61.2	85.0	101.1	64.2	86.2	101.0	61.9	85.2
26, 27, 3370	ELECTRICAL, COMMUNICATIONS & UTIL.	92.7	54.9	74.0	103.0	59.3	81.4	95.9	67.0	81.6	96.8	61.4	79.3	96.5	66.1	81.4	94.2	58.5	76.5
MF2018	WEIGHTED AVERAGE	96.1	64.7	82.5	97.0	68.3	84.6	93.2	63.4	80.3	95.1	65.9	82.5	95.3	68.4	83.7	96.0	64.6	82.4

LOUISIANA / MAINE

DIVISION		NEW ORLEANS 700 - 701			SHREVEPORT 710 - 711			THIBODAUX 703			AUGUSTA 043			BANGOR 044			BATH 045		
		MAT.	INST.	TOTAL	MAT.	INST.	TOTAL	MAT.	INST.	TOTAL	MAT.	INST.	TOTAL	MAT.	INST.	TOTAL	MAT.	INST.	TOTAL
015433	CONTRACTOR EQUIPMENT		88.6	88.6		94.3	94.3		86.6	86.6		99.3	99.3		94.2	94.2		94.2	94.2
0241, 31 - 34	SITE & INFRASTRUCTURE, DEMOLITION	102.0	93.6	96.2	101.8	94.0	96.5	101.9	84.3	89.8	90.4	101.7	98.2	92.8	93.0	92.9	90.1	91.5	91.1
0310	Concrete Forming & Accessories	94.6	69.8	73.5	93.5	62.4	67.1	85.2	62.3	65.7	101.2	78.6	82.0	95.5	78.4	81.0	90.7	78.1	80.0
0320	Concrete Reinforcing	87.7	53.2	71.0	93.0	52.8	73.6	86.3	51.8	69.7	97.7	80.9	89.6	88.8	80.9	85.0	87.9	80.6	84.4
0330	Cast-in-Place Concrete	83.7	71.5	79.1	90.3	68.3	82.0	93.5	65.4	82.9	90.3	111.5	98.3	71.4	110.3	86.1	71.4	110.4	86.1
03	CONCRETE	88.9	67.4	79.3	88.9	63.4	77.4	89.0	62.2	77.0	96.1	90.9	93.8	87.5	90.5	88.9	87.5	90.4	88.8
04	MASONRY	92.9	63.2	74.8	88.9	63.4	73.5	112.5	61.7	81.5	93.0	93.1	93.1	108.3	92.7	98.7	114.5	90.6	99.9
05	METALS	95.9	61.8	85.2	95.5	68.7	87.1	85.6	68.2	80.2	115.0	92.3	107.9	98.7	93.8	97.2	97.0	93.2	95.8
06	WOOD, PLASTICS & COMPOSITES	94.8	72.6	83.2	97.7	62.3	79.2	82.7	63.3	72.5	98.5	76.0	86.7	94.6	75.9	84.8	87.7	75.9	81.5
07	THERMAL & MOISTURE PROTECTION	96.1	68.6	84.3	98.0	67.1	84.8	97.5	64.3	83.3	110.2	100.7	106.1	107.7	100.0	104.4	107.6	99.0	103.9
08	OPENINGS	98.7	65.0	90.5	104.5	57.6	93.1	100.5	54.8	89.4	102.9	79.1	97.1	96.6	82.0	93.1	96.6	78.1	92.1
0920	Plaster & Gypsum Board	101.0	72.1	82.0	97.8	61.6	74.0	102.1	62.9	76.3	107.5	75.0	86.2	110.3	75.0	87.2	105.2	75.0	85.4
0950, 0980	Ceilings & Acoustic Treatment	95.8	72.1	80.9	100.9	61.6	76.3	103.0	62.9	77.9	97.1	75.0	83.3	80.2	75.0	77.0	79.5	75.0	76.7
0960	Flooring	107.9	68.9	96.6	92.9	67.6	85.5	98.4	67.0	89.3	91.0	109.8	96.5	84.2	109.8	91.7	82.6	105.5	89.3
0970, 0990	Wall Finishes & Painting/Coating	99.9	59.9	75.9	94.6	62.2	75.2	95.3	57.0	72.4	95.3	89.8	92.0	89.1	98.4	94.7	89.1	84.7	86.5
09	FINISHES	100.1	69.0	83.3	95.3	63.2	77.9	97.8	62.6	78.8	96.6	84.6	90.1	88.9	85.4	87.0	87.5	83.1	85.1
COVERS	DIVS. 10 - 14, 25, 28, 41, 43, 44, 46	100.0	85.4	96.6	100.0	83.2	96.0	100.0	83.3	96.1	100.0	100.7	100.2	100.0	100.3	100.1	100.0	96.2	99.1
21, 22, 23	FIRE SUPPRESSION, PLUMBING & HVAC	101.0	61.4	85.0	100.8	62.5	85.3	97.3	59.6	82.1	101.2	75.0	90.6	101.3	74.7	90.6	97.5	74.9	88.4
26, 27, 3370	ELECTRICAL, COMMUNICATIONS & UTIL.	100.3	69.4	85.0	101.4	65.6	83.7	94.8	67.0	81.0	99.2	78.4	88.9	98.4	70.9	84.8	96.8	78.4	87.7
MF2018	WEIGHTED AVERAGE	97.7	68.3	85.0	97.7	66.8	84.4	95.5	65.0	82.3	101.6	86.0	94.9	97.5	84.5	91.9	96.2	84.6	91.2

For customer support on your Concrete & Masonry Costs with RSMeans Data, call 800.448.8182.

467

MAINE

DIVISION		HOULTON 047 MAT.	INST.	TOTAL	KITTERY 039 MAT.	INST.	TOTAL	LEWISTON 042 MAT.	INST.	TOTAL	MACHIAS 046 MAT.	INST.	TOTAL	PORTLAND 040-041 MAT.	INST.	TOTAL	ROCKLAND 048 MAT.	INST.	TOTAL
015433	CONTRACTOR EQUIPMENT		94.2	94.2		94.2	94.2		94.2	94.2		94.2	94.2		99.4	99.4		94.2	94.2
0241, 31 - 34	SITE & INFRASTRUCTURE, DEMOLITION	91.9	91.5	91.7	81.0	91.5	88.2	90.1	93.0	92.1	91.3	91.5	91.5	90.6	102.0	98.4	87.7	91.5	90.3
0310	Concrete Forming & Accessories	100.0	78.1	81.3	89.1	78.3	79.9	101.8	78.4	81.9	96.4	78.1	80.8	102.8	78.5	82.2	97.8	78.1	81.1
0320	Concrete Reinforcing	88.8	80.6	84.9	87.5	80.7	84.2	109.1	80.9	95.5	88.8	80.6	84.9	108.3	80.9	95.1	88.8	80.6	84.9
0330	Cast-in-Place Concrete	71.4	109.3	85.7	71.9	110.4	86.4	72.9	110.3	87.0	71.4	110.3	86.1	86.7	111.4	96.0	72.9	110.4	87.0
03	CONCRETE	88.6	90.0	89.2	81.9	90.5	85.8	87.5	90.5	88.9	88.0	90.4	89.1	95.9	90.8	93.6	85.3	90.4	87.6
04	MASONRY	91.5	90.6	91.0	111.6	90.6	98.8	92.2	92.7	92.5	91.5	90.6	91.0	98.3	92.7	94.9	86.2	90.6	88.9
05	METALS	97.3	93.1	96.0	91.6	93.4	92.1	102.4	93.9	99.8	97.3	93.2	96.0	108.6	92.4	103.5	97.1	93.2	95.9
06	WOOD, PLASTICS & COMPOSITES	99.2	75.9	87.0	89.7	75.9	82.4	101.5	75.9	88.1	95.6	75.9	85.3	101.7	76.0	88.2	96.7	75.9	85.8
07	THERMAL & MOISTURE PROTECTION	107.8	99.0	104.0	108.1	99.0	104.2	107.5	100.0	104.3	107.7	99.0	104.0	110.7	101.0	106.6	107.4	99.0	103.8
08	OPENINGS	96.7	78.1	92.2	96.6	81.5	92.9	99.6	82.0	95.3	96.7	78.1	92.2	97.0	82.0	93.4	96.6	78.1	92.1
0920	Plaster & Gypsum Board	112.0	75.0	87.8	100.3	75.0	83.7	115.4	75.0	88.9	110.9	75.0	87.4	107.5	75.0	86.2	110.9	75.0	87.4
0950, 0980	Ceilings & Acoustic Treatment	79.5	75.0	76.7	90.6	75.0	80.9	89.7	75.0	80.5	79.5	75.0	76.7	96.1	75.0	82.9	79.5	75.0	76.7
0960	Flooring	85.3	105.5	91.2	89.0	105.5	93.8	86.8	109.8	93.5	84.7	105.5	90.8	90.7	109.8	96.3	85.0	105.5	91.0
0970, 0990	Wall Finishes & Painting/Coating	89.1	84.7	86.5	79.1	97.3	90.0	89.1	98.4	94.7	89.1	84.7	86.5	95.4	98.4	97.2	89.1	84.7	86.5
09	FINISHES	89.3	83.1	86.0	91.9	84.5	87.9	92.3	85.4	88.5	88.9	83.1	85.8	95.2	85.5	89.9	88.7	83.1	85.7
COVERS	DIVS. 10 - 14, 25, 28, 41, 43, 44, 46	100.0	93.1	98.4	100.0	96.2	99.1	100.0	100.3	100.1	100.0	93.1	98.4	100.0	100.6	100.1	100.0	96.2	99.1
21, 22, 23	FIRE SUPPRESSION, PLUMBING & HVAC	97.5	74.9	88.4	97.6	74.9	88.4	101.3	74.8	90.6	97.5	74.9	88.4	100.7	74.8	90.3	97.5	74.9	88.4
26, 27, 3370	ELECTRICAL, COMMUNICATIONS & UTIL.	100.1	78.4	89.4	93.6	78.4	86.1	100.1	73.9	87.1	100.1	78.4	89.4	103.2	73.8	88.7	100.0	78.4	89.3
MF2018	WEIGHTED AVERAGE	95.9	84.4	91.0	94.4	84.9	90.3	98.0	84.9	92.4	95.8	84.5	90.9	100.6	85.6	94.1	95.0	84.6	90.5

MAINE / MARYLAND

DIVISION		WATERVILLE 049 MAT.	INST.	TOTAL	ANNAPOLIS 214 MAT.	INST.	TOTAL	BALTIMORE 210-212 MAT.	INST.	TOTAL	COLLEGE PARK 207-208 MAT.	INST.	TOTAL	CUMBERLAND 215 MAT.	INST.	TOTAL	EASTON 216 MAT.	INST.	TOTAL
015433	CONTRACTOR EQUIPMENT		94.2	94.2		105.0	105.0		102.6	102.6		106.0	106.0		100.4	100.4		100.4	100.4
0241, 31 - 34	SITE & INFRASTRUCTURE, DEMOLITION	91.8	91.5	91.6	98.7	94.3	95.7	99.8	95.2	96.7	101.7	90.8	94.2	89.9	87.3	88.1	96.5	85.6	89.0
0310	Concrete Forming & Accessories	90.1	78.1	79.9	102.6	75.4	79.5	103.2	75.3	79.5	84.9	73.6	75.3	94.6	80.5	82.6	92.3	71.4	74.5
0320	Concrete Reinforcing	88.8	80.6	84.9	104.8	90.4	97.9	113.9	90.5	102.6	110.6	97.8	104.4	95.2	87.4	91.4	94.3	83.1	88.9
0330	Cast-in-Place Concrete	71.4	110.3	86.1	114.5	77.1	100.3	123.0	78.2	106.1	103.7	73.5	92.3	95.0	85.2	91.3	105.5	63.2	89.5
03	CONCRETE	89.1	90.4	89.6	103.1	79.9	92.7	113.1	79.9	98.2	103.1	79.4	92.5	90.3	84.6	87.8	98.3	72.1	86.5
04	MASONRY	101.8	90.6	95.0	103.3	73.4	85.1	106.0	75.2	87.2	110.9	71.6	86.9	103.3	88.9	94.5	118.6	56.5	80.7
05	METALS	97.2	93.2	96.0	104.8	102.3	104.0	103.2	97.6	101.4	90.7	110.5	96.9	100.2	103.7	101.3	100.4	100.9	100.6
06	WOOD, PLASTICS & COMPOSITES	87.0	75.9	81.2	99.2	74.7	86.4	105.6	74.5	89.4	76.9	73.1	74.9	89.3	78.2	83.5	86.6	78.1	82.1
07	THERMAL & MOISTURE PROTECTION	107.7	99.0	104.0	99.9	80.7	91.7	99.6	81.4	91.7	101.9	79.2	92.2	97.6	81.2	90.6	97.8	71.7	86.6
08	OPENINGS	96.7	78.1	92.2	102.1	81.0	97.0	100.8	81.0	96.0	92.7	82.1	90.1	97.7	82.5	94.0	96.1	81.0	92.4
0920	Plaster & Gypsum Board	105.2	75.0	85.4	102.9	74.2	84.1	104.9	73.9	84.6	109.2	72.6	85.2	108.0	78.1	88.4	108.0	77.9	88.3
0950, 0980	Ceilings & Acoustic Treatment	79.5	75.0	76.7	94.3	74.2	81.7	106.9	73.9	86.2	113.8	72.6	88.0	107.3	78.1	89.0	107.3	77.9	88.9
0960	Flooring	82.3	105.5	89.1	94.7	75.9	89.2	96.5	75.9	90.5	88.1	74.8	84.2	89.4	94.3	90.8	88.5	72.2	83.7
0970, 0990	Wall Finishes & Painting/Coating	89.1	84.7	86.5	96.0	69.4	80.1	97.7	69.4	80.7	100.3	69.4	81.8	91.8	80.6	85.1	91.8	69.4	78.4
09	FINISHES	87.6	83.1	85.2	95.2	74.3	83.9	99.8	74.2	85.9	99.0	72.6	84.7	98.5	83.0	90.1	98.7	71.5	83.9
COVERS	DIVS. 10 - 14, 25, 28, 41, 43, 44, 46	100.0	93.0	98.3	100.0	89.5	97.5	100.0	89.7	97.6	100.0	82.3	95.8	100.0	91.4	98.0	100.0	83.8	96.2
21, 22, 23	FIRE SUPPRESSION, PLUMBING & HVAC	97.5	74.9	88.4	100.8	82.9	93.6	100.8	81.6	93.0	97.1	83.4	91.6	96.8	70.7	86.3	96.8	69.4	85.8
26, 27, 3370	ELECTRICAL, COMMUNICATIONS & UTIL.	100.1	78.4	89.4	101.1	85.6	93.4	96.7	86.9	91.8	96.4	103.9	100.1	97.2	79.9	88.6	96.7	60.0	78.6
MF2018	WEIGHTED AVERAGE	96.2	84.5	91.2	101.4	83.4	93.6	102.3	83.1	94.0	97.5	85.8	92.5	97.1	83.1	91.0	98.9	72.6	87.5

MARYLAND / MASSACHUSETTS

DIVISION		ELKTON 219 MAT.	INST.	TOTAL	HAGERSTOWN 217 MAT.	INST.	TOTAL	SALISBURY 218 MAT.	INST.	TOTAL	SILVER SPRING 209 MAT.	INST.	TOTAL	WALDORF 206 MAT.	INST.	TOTAL	BOSTON 020-022, 024 MAT.	INST.	TOTAL
015433	CONTRACTOR EQUIPMENT		100.4	100.4		100.4	100.4		100.4	100.4		97.9	97.9		97.9	97.9		105.8	105.8
0241, 31 - 34	SITE & INFRASTRUCTURE, DEMOLITION	84.4	86.1	85.6	88.9	87.5	87.9	96.4	85.3	88.8	88.8	82.9	84.8	95.0	82.8	86.6	92.0	105.4	101.2
0310	Concrete Forming & Accessories	99.5	89.0	90.6	93.6	78.0	80.3	109.4	48.7	57.8	93.3	72.9	75.9	101.4	72.8	77.1	104.7	136.7	131.9
0320	Concrete Reinforcing	94.3	111.8	102.8	95.2	87.4	91.4	94.3	62.1	78.8	109.4	97.7	103.7	110.1	97.7	104.1	117.1	149.5	132.8
0330	Cast-in-Place Concrete	85.4	69.5	79.4	90.5	85.3	88.5	105.5	61.1	88.7	106.2	73.9	94.0	118.9	73.7	101.8	98.2	142.4	114.9
03	CONCRETE	83.2	87.3	85.0	86.8	83.5	85.3	99.4	57.5	80.6	100.9	79.0	91.1	111.1	78.9	96.6	103.4	140.4	120.0
04	MASONRY	103.2	64.3	79.5	109.7	88.9	97.0	117.7	53.1	78.3	110.2	71.9	86.9	94.8	71.9	80.8	107.9	144.2	130.1
05	METALS	100.5	112.7	104.3	100.3	103.9	101.4	100.5	91.6	97.7	95.2	106.5	98.7	95.2	106.0	98.6	102.4	134.6	112.4
06	WOOD, PLASTICS & COMPOSITES	95.0	97.9	96.5	88.2	74.4	81.0	107.9	49.6	77.4	84.3	72.4	78.1	92.4	72.4	81.9	104.9	135.7	121.0
07	THERMAL & MOISTURE PROTECTION	97.3	77.5	88.8	97.8	84.5	92.1	98.2	67.0	84.8	104.3	83.9	95.5	104.9	83.9	95.9	108.4	138.0	121.1
08	OPENINGS	96.1	99.5	96.9	96.1	79.8	92.1	96.3	59.6	87.4	84.8	81.7	84.1	85.4	81.7	84.5	100.3	143.3	110.8
0920	Plaster & Gypsum Board	110.0	98.3	102.3	108.0	74.2	85.8	118.0	48.7	72.5	116.0	72.6	87.5	118.5	72.6	88.4	105.5	136.3	125.7
0950, 0980	Ceilings & Acoustic Treatment	107.3	98.3	101.6	110.1	74.2	87.6	107.3	48.7	70.6	121.4	72.6	90.9	121.4	72.6	90.9	94.0	136.3	120.5
0960	Flooring	90.9	72.2	86.4	88.9	94.3	90.5	94.5	72.2	88.0	94.0	74.8	88.4	97.6	74.8	91.0	96.8	162.4	115.9
0970, 0990	Wall Finishes & Painting/Coating	91.8	69.4	78.4	91.8	69.4	78.4	91.8	69.4	78.4	108.0	69.4	84.9	108.0	69.4	84.9	101.6	154.9	133.5
09	FINISHES	98.8	85.0	91.4	98.8	79.5	88.4	101.8	53.9	75.9	99.6	72.0	84.7	101.3	72.1	85.5	98.5	143.8	123.0
COVERS	DIVS. 10 - 14, 25, 28, 41, 43, 44, 46	100.0	53.7	89.1	100.0	91.0	97.9	100.0	79.6	95.2	100.0	80.8	95.5	100.0	78.8	95.0	100.0	117.4	104.1
21, 22, 23	FIRE SUPPRESSION, PLUMBING & HVAC	96.8	75.1	88.0	100.6	88.3	95.6	96.8	67.5	85.0	97.1	83.8	91.7	97.1	83.8	91.7	96.7	127.1	109.0
26, 27, 3370	ELECTRICAL, COMMUNICATIONS & UTIL.	98.2	83.7	91.0	97.0	79.9	88.5	95.7	57.9	77.0	93.8	103.9	98.8	91.8	103.9	97.8	98.3	129.3	113.7
MF2018	WEIGHTED AVERAGE	96.1	83.0	90.5	97.7	86.2	92.8	99.3	65.0	84.5	96.7	84.9	91.6	97.4	84.8	92.0	100.2	132.9	114.3

		MASSACHUSETTS																
	DIVISION	BROCKTON			BUZZARDS BAY			FALL RIVER			FITCHBURG			FRAMINGHAM			GREENFIELD		
		023			025			027			014			017			013		
		MAT.	INST.	TOTAL	MAT.	INST.	TOTAL	MAT.	INST.	TOTAL	MAT.	INST.	TOTAL	MAT.	INST.	TOTAL	MAT.	INST.	TOTAL
015433	CONTRACTOR EQUIPMENT		97.1	97.1		97.1	97.1		98.0	98.0		94.2	94.2		95.9	95.9		94.2	94.2
0241, 31 - 34	SITE & INFRASTRUCTURE, DEMOLITION	91.0	96.4	94.7	80.6	96.0	91.2	90.2	96.5	94.5	82.8	95.9	91.8	79.6	95.4	90.4	86.5	94.5	92.0
0310	Concrete Forming & Accessories	101.9	122.8	119.7	99.4	122.0	118.7	101.9	122.5	119.4	94.5	115.2	112.1	102.5	122.7	119.7	92.6	119.0	115.1
0320	Concrete Reinforcing	105.1	144.2	124.0	84.3	121.3	102.1	105.1	121.3	112.9	85.2	135.8	109.6	85.2	144.0	113.5	88.8	120.8	104.3
0330	Cast-in-Place Concrete	84.5	133.0	102.8	70.2	132.7	93.8	81.7	133.3	101.2	77.1	132.7	98.1	77.1	132.9	98.2	79.2	117.8	93.8
03	CONCRETE	91.6	129.8	108.7	77.7	125.4	99.1	89.0	125.8	105.5	78.6	124.6	99.2	81.3	129.7	103.0	82.0	118.5	98.4
04	MASONRY	104.3	136.3	123.8	96.8	136.3	120.9	104.9	136.2	124.0	98.7	128.1	116.7	105.3	132.1	121.7	103.3	118.6	112.6
05	METALS	100.9	129.6	109.9	95.5	119.2	102.9	100.9	119.8	106.8	100.3	121.8	107.0	100.3	128.7	109.2	102.9	113.1	106.1
06	WOOD, PLASTICS & COMPOSITES	101.3	122.2	112.2	97.4	122.2	110.4	101.3	122.4	112.4	95.2	112.6	104.3	102.9	122.0	112.9	92.8	122.6	108.4
07	THERMAL & MOISTURE PROTECTION	105.5	127.4	114.9	104.4	124.1	112.8	105.4	124.0	113.3	102.4	120.0	109.9	102.6	125.1	112.2	102.5	109.9	105.7
08	OPENINGS	97.8	130.5	105.8	93.9	119.4	100.1	97.8	120.5	103.4	99.6	123.0	105.3	90.2	130.3	99.9	99.8	121.0	105.0
0920	Plaster & Gypsum Board	92.9	122.6	112.4	87.2	122.6	110.5	92.9	122.6	112.4	108.7	112.8	111.4	112.4	122.6	119.1	109.7	123.0	118.4
0950, 0980	Ceilings & Acoustic Treatment	98.6	122.6	113.7	80.4	122.6	106.9	98.6	122.6	113.7	86.4	112.8	103.0	86.4	122.6	109.1	94.7	123.0	112.5
0960	Flooring	89.9	158.8	110.0	87.4	158.8	108.2	88.7	158.8	109.1	89.1	158.8	109.4	90.7	158.8	110.6	88.3	136.7	102.5
0970, 0990	Wall Finishes & Painting/Coating	87.0	138.9	118.1	87.0	138.9	118.1	87.0	138.9	118.1	84.9	138.9	117.2	85.8	138.9	117.6	84.9	114.9	102.9
09	FINISHES	92.8	131.5	113.7	86.3	131.5	110.7	92.5	131.6	113.7	90.1	125.8	109.5	90.9	131.3	112.8	92.4	123.0	108.9
COVERS	DIVS. 10 - 14, 25, 28, 41, 43, 44, 46	100.0	112.8	103.0	100.0	112.8	103.0	100.0	113.3	103.1	100.0	106.3	101.5	100.0	112.4	102.9	100.0	105.4	101.3
21, 22, 23	FIRE SUPPRESSION, PLUMBING & HVAC	101.7	104.3	102.8	97.2	99.5	98.1	101.7	104.0	102.6	97.6	100.5	98.8	97.6	116.3	105.2	97.6	98.2	97.8
26, 27, 3370	ELECTRICAL, COMMUNICATIONS & UTIL.	100.8	99.8	100.3	98.0	96.4	97.2	100.6	101.2	100.9	99.8	100.2	100.0	97.0	119.8	108.3	99.8	101.5	100.7
MF2018	WEIGHTED AVERAGE	98.9	117.8	107.1	93.3	114.1	102.3	98.5	116.0	106.1	95.4	113.2	103.1	94.9	122.4	106.8	96.7	109.5	102.3

		MASSACHUSETTS																
	DIVISION	HYANNIS			LAWRENCE			LOWELL			NEW BEDFORD			PITTSFIELD			SPRINGFIELD		
		026			019			018			027			012			010 - 011		
		MAT.	INST.	TOTAL	MAT.	INST.	TOTAL	MAT.	INST.	TOTAL	MAT.	INST.	TOTAL	MAT.	INST.	TOTAL	MAT.	INST.	TOTAL
015433	CONTRACTOR EQUIPMENT		97.1	97.1		96.6	96.6		94.2	94.2		98.0	98.0		94.2	94.2		94.2	94.2
0241, 31 - 34	SITE & INFRASTRUCTURE, DEMOLITION	87.4	96.2	93.4	92.0	96.1	94.8	91.1	96.3	94.7	88.6	96.5	94.0	92.1	94.6	93.8	91.5	95.0	93.9
0310	Concrete Forming & Accessories	92.7	122.5	118.0	103.8	123.2	120.3	100.4	125.2	121.5	101.9	122.6	119.5	100.3	106.6	105.6	100.6	120.3	117.4
0320	Concrete Reinforcing	84.3	121.3	102.2	105.5	141.8	123.0	106.3	141.6	123.3	105.1	121.3	112.9	88.2	117.5	102.3	106.3	120.9	113.3
0330	Cast-in-Place Concrete	77.1	132.9	98.1	89.2	133.1	105.8	81.1	136.1	101.8	72.0	133.3	95.1	88.5	116.0	98.9	84.4	119.7	97.7
03	CONCRETE	83.3	125.6	102.3	94.9	129.6	110.5	87.0	131.3	106.9	85.8	125.8	103.8	88.0	111.6	98.6	88.6	119.7	102.6
04	MASONRY	103.4	136.3	123.4	110.9	138.8	127.9	97.8	139.9	123.5	103.3	134.8	122.5	98.4	113.3	107.5	98.1	121.9	112.6
05	METALS	97.1	119.7	104.1	103.3	128.9	111.3	103.3	125.6	110.2	100.9	119.8	106.8	103.0	111.1	105.6	106.3	113.2	108.4
06	WOOD, PLASTICS & COMPOSITES	89.7	122.2	106.7	103.7	122.2	113.4	102.6	122.2	112.8	101.3	122.4	112.4	102.6	107.1	104.9	102.6	122.6	113.0
07	THERMAL & MOISTURE PROTECTION	104.9	126.5	114.2	103.2	128.1	113.9	102.9	129.3	114.2	105.3	123.6	113.1	103.0	106.3	104.4	102.9	111.4	106.6
08	OPENINGS	94.4	120.6	100.8	93.7	129.7	102.5	100.9	129.7	107.9	97.8	124.9	104.4	100.9	111.6	103.5	100.9	121.0	105.8
0920	Plaster & Gypsum Board	83.2	122.6	109.1	115.1	122.6	120.0	115.1	122.6	120.0	92.9	122.6	112.4	115.1	107.1	109.8	115.1	123.0	120.3
0950, 0980	Ceilings & Acoustic Treatment	89.7	122.6	110.3	96.2	122.6	112.8	96.2	122.6	112.8	98.6	122.6	113.7	96.2	107.1	103.0	96.2	123.0	113.0
0960	Flooring	84.8	158.8	106.4	91.3	158.8	111.0	91.3	158.8	111.0	88.7	158.8	109.1	91.6	131.5	103.3	90.7	136.7	104.1
0970, 0990	Wall Finishes & Painting/Coating	87.0	138.9	118.1	85.0	138.9	117.3	84.9	138.9	117.2	87.0	138.9	118.1	84.9	114.9	102.9	85.7	114.9	103.2
09	FINISHES	87.6	131.5	111.3	94.5	131.4	114.5	94.5	132.8	115.2	92.4	131.6	113.6	94.6	112.2	104.1	94.4	123.8	110.3
COVERS	DIVS. 10 - 14, 25, 28, 41, 43, 44, 46	100.0	112.8	103.0	100.0	112.9	103.0	100.0	114.6	103.4	100.0	113.3	103.1	100.0	102.9	100.7	100.0	106.5	101.5
21, 22, 23	FIRE SUPPRESSION, PLUMBING & HVAC	101.7	108.6	104.5	101.0	120.9	108.8	101.0	126.6	111.3	101.7	104.0	102.6	101.0	95.2	98.7	101.0	99.9	100.6
26, 27, 3370	ELECTRICAL, COMMUNICATIONS & UTIL.	98.8	101.2	100.0	99.0	125.6	112.2	99.4	124.4	111.8	101.6	103.2	102.4	99.4	101.5	100.5	99.5	102.9	101.2
MF2018	WEIGHTED AVERAGE	96.0	116.9	105.1	99.4	124.9	110.4	98.4	126.4	110.5	98.1	116.5	106.0	98.6	105.1	101.4	99.1	110.8	104.2

		MASSACHUSETTS			MICHIGAN													
	DIVISION	WORCESTER			ANN ARBOR			BATTLE CREEK			BAY CITY			DEARBORN			DETROIT		
		015 - 016			481			490			487			481			482		
		MAT.	INST.	TOTAL	MAT.	INST.	TOTAL	MAT.	INST.	TOTAL	MAT.	INST.	TOTAL	MAT.	INST.	TOTAL	MAT.	INST.	TOTAL
015433	CONTRACTOR EQUIPMENT		94.2	94.2		106.7	106.7		94.4	94.4		106.7	106.7		106.7	106.7		97.5	97.5
0241, 31 - 34	SITE & INFRASTRUCTURE, DEMOLITION	91.4	95.9	94.5	82.0	90.0	87.5	93.9	79.0	83.7	72.7	88.3	83.4	81.7	90.2	87.5	100.3	102.4	101.8
0310	Concrete Forming & Accessories	101.0	124.0	120.5	98.0	105.4	104.3	100.4	79.4	82.6	98.1	79.2	82.0	97.9	106.1	104.9	98.6	104.1	103.3
0320	Concrete Reinforcing	106.3	148.7	126.8	99.0	104.8	101.8	90.4	81.5	86.1	99.0	103.9	101.3	99.0	106.1	102.4	96.9	102.4	99.6
0330	Cast-in-Place Concrete	83.9	132.8	102.4	85.2	99.1	90.5	83.2	93.3	87.0	81.6	85.2	83.0	83.4	100.0	89.7	101.5	99.3	100.7
03	CONCRETE	88.4	130.9	107.4	88.0	104.0	95.2	84.4	84.4	84.4	86.3	87.1	86.7	87.1	104.9	95.1	99.6	101.5	100.5
04	MASONRY	97.7	134.7	120.3	96.0	100.1	98.5	93.9	77.4	83.8	95.6	79.8	85.9	95.9	123.5	112.7	101.0	98.3	99.4
05	METALS	106.3	127.9	113.1	99.3	118.8	105.4	105.5	83.2	98.5	99.9	114.3	104.4	99.4	120.7	106.1	100.8	93.0	98.4
06	WOOD, PLASTICS & COMPOSITES	103.1	124.1	114.1	90.9	107.1	99.3	93.0	78.6	85.5	90.9	79.1	84.7	90.9	107.1	99.3	93.7	106.0	100.1
07	THERMAL & MOISTURE PROTECTION	102.9	123.0	111.6	104.6	100.1	102.6	96.3	78.8	88.8	102.1	82.1	93.5	103.1	107.8	105.1	102.2	103.7	102.9
08	OPENINGS	100.9	132.7	108.6	94.1	102.1	96.1	86.7	75.5	84.0	94.1	83.6	91.6	94.1	102.1	96.1	97.0	100.9	97.9
0920	Plaster & Gypsum Board	115.1	124.6	121.3	93.9	107.1	102.6	87.6	75.0	79.3	93.9	78.4	83.7	93.9	107.1	102.6	91.5	106.1	101.1
0950, 0980	Ceilings & Acoustic Treatment	96.2	124.6	114.0	80.2	107.1	97.1	87.9	75.0	79.8	80.9	78.4	79.3	80.2	107.1	97.1	84.2	106.1	97.9
0960	Flooring	91.3	160.0	111.4	92.8	110.8	98.0	92.1	67.2	84.8	92.8	85.5	90.7	92.0	104.9	95.8	94.9	104.8	97.8
0970, 0990	Wall Finishes & Painting/Coating	84.9	138.9	117.2	83.4	100.4	93.6	86.5	72.6	78.2	83.4	82.4	82.8	83.4	98.6	92.5	84.8	97.3	92.3
09	FINISHES	94.5	132.8	115.2	90.5	106.0	98.9	88.7	76.0	81.8	90.2	80.0	84.7	90.3	105.2	98.3	94.0	103.9	99.4
COVERS	DIVS. 10 - 14, 25, 28, 41, 43, 44, 46	100.0	107.6	101.8	100.0	101.8	100.4	100.0	90.2	97.7	100.0	95.1	98.8	100.0	102.2	100.5	100.0	101.0	100.2
21, 22, 23	FIRE SUPPRESSION, PLUMBING & HVAC	101.0	105.1	102.7	100.7	94.1	98.0	100.6	79.9	92.3	100.7	80.2	92.4	100.7	103.1	101.6	100.5	101.5	100.9
26, 27, 3370	ELECTRICAL, COMMUNICATIONS & UTIL.	99.5	106.4	102.9	97.2	105.3	101.2	95.6	74.8	85.3	96.3	79.4	88.0	97.2	100.1	98.6	99.8	101.6	100.7
MF2018	WEIGHTED AVERAGE	99.1	118.6	107.5	96.3	102.0	98.7	95.7	79.4	88.6	95.7	85.3	91.2	96.1	106.0	100.4	99.5	100.9	100.1

MICHIGAN

DIVISION		FLINT 484 - 485			GAYLORD 497			GRAND RAPIDS 493,495			IRON MOUNTAIN 498 - 499			JACKSON 492			KALAMAZOO 491		
		MAT.	INST.	TOTAL	MAT.	INST.	TOTAL	MAT.	INST.	TOTAL	MAT.	INST.	TOTAL	MAT.	INST.	TOTAL	MAT.	INST.	TOTAL
015433	CONTRACTOR EQUIPMENT		106.7	106.7		101.9	101.9		98.3	98.3		89.3	89.3		101.9	101.9		94.4	94.4
0241, 31 - 34	SITE & INFRASTRUCTURE, DEMOLITION	70.3	88.9	83.1	87.6	76.8	80.2	93.3	87.7	89.5	96.4	85.6	89.0	111.0	78.4	88.6	94.2	79.0	83.8
0310	Concrete Forming & Accessories	101.5	84.9	87.4	98.6	71.4	75.4	97.0	78.1	80.9	92.0	75.8	78.3	95.2	79.4	81.8	100.4	79.8	82.8
0320	Concrete Reinforcing	99.0	104.2	101.5	84.3	88.6	86.4	93.6	81.5	87.8	84.3	80.9	82.7	81.9	103.9	92.5	90.4	80.0	85.4
0330	Cast-in-Place Concrete	85.8	88.7	86.9	83.0	75.3	80.1	87.5	91.2	88.9	98.3	64.9	85.7	82.9	90.0	85.6	84.9	93.4	88.1
03	CONCRETE	88.5	90.9	89.6	81.2	77.5	79.5	93.3	83.1	88.7	89.5	73.7	82.4	76.6	88.7	82.0	87.4	84.4	86.1
04	MASONRY	96.1	89.0	91.7	104.3	69.7	83.2	92.1	76.8	82.8	90.8	76.4	82.0	84.9	83.3	83.9	92.6	81.2	85.6
05	METALS	99.4	114.9	104.2	107.2	109.2	107.9	102.5	82.5	96.2	106.6	89.5	101.3	107.5	112.5	109.0	105.5	83.1	98.5
06	WOOD, PLASTICS & COMPOSITES	94.9	84.1	89.2	86.6	70.9	78.4	93.2	76.7	84.6	84.7	76.2	80.3	85.3	77.1	81.0	93.0	78.6	85.5
07	THERMAL & MOISTURE PROTECTION	102.4	86.2	95.4	95.0	71.4	84.9	98.0	73.4	87.4	98.4	71.4	86.8	94.3	84.6	90.1	96.3	80.1	89.4
08	OPENINGS	94.1	85.5	92.0	86.0	75.8	83.5	101.8	76.2	95.6	92.5	67.3	86.3	85.3	83.7	84.9	86.7	76.7	84.3
0920	Plaster & Gypsum Board	95.6	83.5	87.7	87.1	69.4	75.5	95.3	73.4	80.9	52.9	76.3	68.3	85.7	75.7	79.1	87.6	75.0	79.3
0950, 0980	Ceilings & Acoustic Treatment	80.2	83.5	82.3	86.0	69.4	75.6	98.2	73.4	82.7	85.8	76.3	79.9	86.0	75.7	79.5	87.9	75.0	79.8
0960	Flooring	92.8	88.8	91.6	85.5	81.7	84.4	94.8	78.6	90.0	102.3	86.8	97.8	84.0	75.1	81.4	92.1	73.5	86.6
0970, 0990	Wall Finishes & Painting/Coating	83.4	80.7	81.8	82.6	77.7	79.7	97.1	77.7	85.5	99.1	66.3	79.4	82.6	89.3	86.6	86.5	78.7	81.8
09	FINISHES	89.8	84.6	87.0	87.8	73.2	79.9	95.4	77.6	85.8	88.8	76.8	82.3	88.9	78.6	83.4	88.7	78.0	82.9
COVERS	DIVS. 10 - 14, 25, 28, 41, 43, 44, 46	100.0	97.2	99.3	100.0	94.4	98.7	100.0	100.9	100.2	100.0	93.0	98.4	100.0	97.7	99.5	100.0	101.8	100.4
21, 22, 23	FIRE SUPPRESSION, PLUMBING & HVAC	100.7	85.9	94.7	97.0	75.3	88.3	100.6	81.8	93.0	96.9	80.0	90.1	97.0	81.7	90.9	100.6	80.5	92.5
26, 27, 3370	ELECTRICAL, COMMUNICATIONS & UTIL.	97.2	92.8	95.0	93.8	73.5	83.7	100.7	86.4	93.6	99.4	77.7	88.7	97.2	99.2	98.2	95.4	77.2	86.4
MF2018	WEIGHTED AVERAGE	95.9	90.9	93.8	94.6	78.1	87.5	98.9	82.2	91.6	96.5	78.9	88.9	94.1	88.0	91.4	96.0	80.9	89.5

MICHIGAN / MINNESOTA

DIVISION		LANSING 488 - 489			MUSKEGON 494			ROYAL OAK 480,483			SAGINAW 486			TRAVERSE CITY 496			BEMIDJI 566		
		MAT.	INST.	TOTAL	MAT.	INST.	TOTAL	MAT.	INST.	TOTAL	MAT.	INST.	TOTAL	MAT.	INST.	TOTAL	MAT.	INST.	TOTAL
015433	CONTRACTOR EQUIPMENT		110.1	110.1		94.4	94.4		87.2	87.2		106.7	106.7		89.3	89.3		95.8	95.8
0241, 31 - 34	SITE & INFRASTRUCTURE, DEMOLITION	92.8	97.2	95.8	92.0	78.9	83.0	86.2	88.5	87.8	73.7	88.3	83.7	81.9	84.6	83.8	95.2	92.3	93.2
0310	Concrete Forming & Accessories	95.5	82.2	84.2	101.0	78.5	81.9	93.7	104.1	102.6	98.0	82.3	84.6	92.0	69.8	73.1	87.3	84.8	85.1
0320	Concrete Reinforcing	100.8	103.9	102.3	91.1	81.6	86.5	89.7	99.8	94.5	99.0	103.9	101.3	85.6	76.5	81.2	97.1	100.7	98.8
0330	Cast-in-Place Concrete	94.0	88.4	91.9	82.9	91.6	86.2	74.6	93.5	81.7	84.2	85.2	84.6	76.7	77.0	76.8	99.7	95.0	97.9
03	CONCRETE	94.2	89.4	92.0	82.9	83.5	83.2	75.7	99.0	86.1	87.5	88.5	88.0	73.8	74.4	74.0	91.3	92.5	91.9
04	MASONRY	89.3	85.2	86.8	91.4	78.0	83.2	89.8	95.4	93.2	97.6	79.8	86.7	89.1	71.7	78.5	100.5	103.0	102.0
05	METALS	98.3	111.4	102.4	103.1	83.5	97.0	102.8	90.0	98.8	99.4	114.1	104.0	106.5	87.3	100.5	89.2	118.8	98.4
06	WOOD, PLASTICS & COMPOSITES	89.7	80.4	84.8	90.9	77.4	83.8	86.4	107.1	97.2	88.0	83.5	85.6	84.7	69.5	76.7	70.2	80.1	75.4
07	THERMAL & MOISTURE PROTECTION	101.0	85.2	94.2	95.3	72.9	85.7	100.6	96.8	99.0	103.2	82.6	94.3	97.4	71.1	86.1	106.3	90.0	99.3
08	OPENINGS	102.5	83.4	97.9	86.0	76.7	83.8	93.9	99.8	95.4	92.3	86.0	90.8	92.5	63.0	85.3	99.4	103.7	100.4
0920	Plaster & Gypsum Board	86.8	79.8	82.2	69.5	73.8	72.3	91.0	107.1	101.6	93.9	82.9	86.7	52.9	69.4	63.7	98.3	80.1	86.3
0950, 0980	Ceilings & Acoustic Treatment	84.2	79.8	81.5	87.9	73.8	79.0	79.6	107.1	96.9	80.2	82.9	81.9	85.8	69.4	75.5	131.6	80.1	99.3
0960	Flooring	97.2	93.6	96.2	90.7	84.1	88.8	89.8	101.8	93.3	92.8	85.5	90.7	102.3	81.7	96.3	92.2	104.1	95.7
0970, 0990	Wall Finishes & Painting/Coating	96.8	77.7	85.4	84.7	79.0	81.3	84.9	89.3	87.6	83.4	82.4	82.8	99.1	37.6	62.2	85.6	96.5	92.1
09	FINISHES	91.9	83.3	87.2	85.2	79.2	82.0	89.4	102.5	96.5	90.2	82.6	86.1	87.7	68.2	77.1	101.1	89.0	94.5
COVERS	DIVS. 10 - 14, 25, 28, 41, 43, 44, 46	100.0	97.0	99.3	100.0	101.5	100.3	100.0	95.3	98.9	100.0	95.6	99.0	100.0	91.6	98.0	100.0	97.5	99.4
21, 22, 23	FIRE SUPPRESSION, PLUMBING & HVAC	100.5	85.9	94.6	100.4	81.5	92.8	97.1	96.6	96.9	100.7	79.7	92.2	96.9	75.2	88.1	97.1	83.6	91.7
26, 27, 3370	ELECTRICAL, COMMUNICATIONS & UTIL.	101.3	94.5	97.9	95.8	70.2	83.2	99.0	99.3	99.2	95.6	85.2	90.5	95.3	73.4	84.5	103.3	96.9	100.2
MF2018	WEIGHTED AVERAGE	98.1	90.5	94.8	94.5	79.7	88.1	94.1	96.9	95.3	95.6	86.7	91.8	93.5	75.2	85.6	96.8	94.6	95.8

MINNESOTA

DIVISION		BRAINERD 564			DETROIT LAKES 565			DULUTH 556 - 558			MANKATO 560			MINNEAPOLIS 553 - 555			ROCHESTER 559		
		MAT.	INST.	TOTAL	MAT.	INST.	TOTAL	MAT.	INST.	TOTAL	MAT.	INST.	TOTAL	MAT.	INST.	TOTAL	MAT.	INST.	TOTAL
015433	CONTRACTOR EQUIPMENT		98.4	98.4		95.8	95.8		104.2	104.2		98.4	98.4		108.3	108.3		99.5	99.5
0241, 31 - 34	SITE & INFRASTRUCTURE, DEMOLITION	95.6	97.0	96.6	93.4	92.5	92.8	100.9	103.6	102.8	92.2	97.9	96.1	96.0	107.6	104.0	98.0	96.1	96.7
0310	Concrete Forming & Accessories	88.3	85.4	85.8	83.8	80.4	81.0	100.3	100.0	100.0	97.8	101.7	101.1	102.4	116.9	114.7	104.9	105.9	105.7
0320	Concrete Reinforcing	95.9	100.8	98.3	97.1	100.6	98.8	108.1	101.2	104.7	95.8	110.2	102.7	94.0	108.1	100.8	100.3	110.5	105.2
0330	Cast-in-Place Concrete	108.4	99.7	105.1	96.8	98.1	97.3	102.2	103.0	102.5	99.8	110.4	103.8	90.7	118.0	101.0	97.5	102.5	99.4
03	CONCRETE	95.1	94.4	94.8	88.9	91.6	90.1	102.0	102.3	102.1	90.9	107.2	98.2	95.9	116.4	105.1	93.3	106.5	99.2
04	MASONRY	127.3	112.3	118.2	126.0	108.9	115.6	100.4	112.3	107.7	114.9	125.9	121.6	109.4	121.3	116.7	101.3	114.3	109.2
05	METALS	90.2	118.7	99.1	89.2	118.0	98.2	97.5	120.1	104.6	90.1	123.8	100.6	96.1	125.4	105.2	96.6	126.9	106.1
06	WOOD, PLASTICS & COMPOSITES	84.4	77.1	80.6	66.9	71.9	69.5	97.6	97.7	97.6	95.5	92.2	93.7	103.6	114.6	109.3	104.7	104.6	104.6
07	THERMAL & MOISTURE PROTECTION	103.8	98.2	101.4	106.1	94.0	100.9	98.0	106.4	101.6	104.3	100.4	102.6	100.1	118.1	107.8	104.4	98.6	101.9
08	OPENINGS	86.3	102.0	90.1	99.4	99.2	99.3	103.1	108.5	104.4	90.7	111.8	95.8	98.9	123.7	104.9	98.2	120.4	103.6
0920	Plaster & Gypsum Board	85.8	77.2	80.2	98.0	71.7	80.7	93.3	98.0	96.4	89.8	92.7	91.7	96.1	115.0	108.5	101.9	105.3	104.1
0950, 0980	Ceilings & Acoustic Treatment	60.2	77.2	70.8	131.6	71.7	94.0	92.0	98.0	95.8	60.2	92.7	80.5	101.2	115.0	109.8	94.5	105.3	101.3
0960	Flooring	90.9	104.1	94.8	90.9	81.6	88.2	94.8	103.5	97.4	93.0	88.8	91.8	101.9	121.5	107.6	94.9	88.8	93.1
0970, 0990	Wall Finishes & Painting/Coating	80.1	96.5	90.0	85.6	79.4	81.9	91.4	108.9	101.9	90.0	98.8	95.6	99.5	127.3	116.2	84.4	102.5	95.2
09	FINISHES	81.5	89.0	85.6	100.6	79.8	89.3	93.3	101.5	97.8	83.2	98.8	91.6	98.7	118.9	109.6	92.8	102.8	98.2
COVERS	DIVS. 10 - 14, 25, 28, 41, 43, 44, 46	100.0	99.3	99.8	100.0	98.5	99.6	100.0	101.2	100.3	100.0	105.4	101.3	100.0	106.7	101.6	100.0	102.0	100.5
21, 22, 23	FIRE SUPPRESSION, PLUMBING & HVAC	96.2	87.4	92.7	97.1	86.4	92.8	100.5	96.4	98.8	96.2	95.6	96.0	100.6	113.3	105.7	100.6	101.7	101.1
26, 27, 3370	ELECTRICAL, COMMUNICATIONS & UTIL.	101.4	101.9	101.7	103.1	68.4	86.0	101.4	101.9	101.7	106.9	99.5	103.2	104.2	115.4	109.8	101.4	99.5	100.4
MF2018	WEIGHTED AVERAGE	95.5	97.9	96.5	97.6	90.3	94.4	99.9	103.9	101.6	95.5	105.0	99.6	99.7	116.6	107.0	98.4	106.1	101.7

MINNESOTA / MISSISSIPPI

DIVISION		SAINT PAUL 550 - 551			ST. CLOUD 563			THIEF RIVER FALLS 567			WILLMAR 562			WINDOM 561			BILOXI 395		
		MAT.	INST.	TOTAL	MAT.	INST.	TOTAL	MAT.	INST.	TOTAL	MAT.	INST.	TOTAL	MAT.	INST.	TOTAL	MAT.	INST.	TOTAL
015433	CONTRACTOR EQUIPMENT		104.2	104.2		98.4	98.4		95.8	95.8		98.4	98.4		98.4	98.4		98.7	98.7
0241, 31 - 34	SITE & INFRASTRUCTURE, DEMOLITION	97.6	104.3	102.2	91.4	98.3	96.2	94.1	92.1	92.8	90.3	97.0	94.9	84.1	96.2	92.4	101.8	83.7	89.4
0310	Concrete Forming & Accessories	102.2	120.4	117.7	85.2	119.7	114.5	88.3	84.1	84.7	84.9	85.2	85.2	89.9	81.2	82.5	86.3	68.0	70.7
0320	Concrete Reinforcing	105.3	110.8	108.0	95.9	110.5	102.9	97.4	100.6	98.9	95.5	110.1	102.6	95.5	109.2	102.1	89.6	64.5	77.5
0330	Cast-in-Place Concrete	102.5	119.7	109.0	95.6	117.5	103.9	98.8	92.0	96.2	97.1	78.5	90.1	84.0	82.8	83.5	121.8	66.9	101.0
03	CONCRETE	101.6	119.0	109.4	86.8	117.9	100.7	90.1	91.1	90.5	86.7	88.7	87.6	78.0	88.2	82.6	98.8	68.6	85.3
04	MASONRY	98.7	123.0	113.5	110.2	119.5	115.8	100.5	103.0	102.0	113.7	112.3	112.9	125.9	89.1	103.5	90.1	63.9	74.1
05	METALS	97.4	126.6	106.5	90.9	125.4	101.7	89.3	117.8	98.2	90.0	123.4	100.4	89.9	121.5	99.8	90.1	90.6	90.3
06	WOOD, PLASTICS & COMPOSITES	101.3	118.2	110.1	81.1	117.8	100.3	71.6	80.1	76.1	80.7	77.2	78.9	86.1	77.2	81.4	83.0	68.6	75.5
07	THERMAL & MOISTURE PROTECTION	101.5	119.2	109.1	104.0	112.7	107.7	107.2	87.9	98.9	103.7	97.9	101.2	103.7	81.5	94.2	101.4	63.5	85.1
08	OPENINGS	97.9	127.9	105.2	91.1	127.7	100.0	99.4	103.7	100.4	88.2	101.1	91.3	91.6	101.1	93.9	96.1	63.1	88.1
0920	Plaster & Gypsum Board	92.5	119.0	109.9	85.8	119.0	107.6	98.3	80.1	86.3	85.8	77.3	80.2	85.8	77.3	80.2	112.4	68.3	83.5
0950, 0980	Ceilings & Acoustic Treatment	93.3	119.0	109.4	60.2	119.0	97.1	131.6	80.1	99.3	60.2	77.3	70.9	60.2	77.3	70.9	90.4	68.3	76.6
0960	Flooring	93.5	121.5	101.7	87.7	119.6	97.0	91.9	81.6	88.9	89.2	81.6	87.0	91.8	81.6	88.8	93.8	64.3	85.2
0970, 0990	Wall Finishes & Painting/Coating	90.6	131.3	115.0	90.8	127.3	112.7	85.6	79.4	81.9	85.6	79.4	81.9	85.6	98.8	93.5	80.7	48.3	61.3
09	FINISHES	93.1	121.8	108.6	80.9	120.7	102.5	101.0	83.2	91.3	81.0	83.3	82.2	81.2	82.6	82.0	92.5	65.4	77.8
COVERS	DIVS. 10 - 14, 25, 28, 41, 43, 44, 46	100.0	107.0	101.6	100.0	106.1	101.4	100.0	97.4	99.4	100.0	99.2	99.8	100.0	96.2	99.1	100.0	82.9	96.0
21, 22, 23	FIRE SUPPRESSION, PLUMBING & HVAC	100.5	117.9	107.5	100.0	109.3	103.8	97.1	83.3	91.5	96.2	100.6	98.0	96.2	80.7	90.0	101.1	56.6	83.1
26, 27, 3370	ELECTRICAL, COMMUNICATIONS & UTIL.	102.1	119.9	110.9	101.4	119.7	110.6	100.9	68.4	84.8	101.4	80.6	91.1	106.9	99.4	103.2	100.2	56.1	78.4
MF2018	WEIGHTED AVERAGE	99.3	119.2	107.9	94.9	115.9	104.0	96.4	89.4	93.4	93.7	96.6	94.9	94.0	91.7	93.0	97.1	66.6	83.9

MISSISSIPPI

DIVISION		CLARKSDALE 386			COLUMBUS 397			GREENVILLE 387			GREENWOOD 389			JACKSON 390 - 392			LAUREL 394		
		MAT.	INST.	TOTAL	MAT.	INST.	TOTAL	MAT.	INST.	TOTAL	MAT.	INST.	TOTAL	MAT.	INST.	TOTAL	MAT.	INST.	TOTAL
015433	CONTRACTOR EQUIPMENT		98.7	98.7		98.7	98.7		98.7	98.7		98.7	98.7		103.9	103.9		98.7	98.7
0241, 31 - 34	SITE & INFRASTRUCTURE, DEMOLITION	99.6	82.0	87.5	100.1	82.8	88.2	105.8	83.7	90.6	102.6	81.8	88.3	97.2	92.6	94.0	105.3	82.1	89.4
0310	Concrete Forming & Accessories	83.7	42.6	48.7	75.8	44.7	49.4	80.2	62.0	64.7	94.0	42.7	50.4	87.3	67.4	70.3	75.9	57.6	60.4
0320	Concrete Reinforcing	99.6	63.4	82.1	96.0	63.5	80.3	100.1	63.7	82.5	99.6	63.5	82.1	110.9	64.6	83.4	116.5	31.2	65.1
0330	Cast-in-Place Concrete	100.3	57.4	84.1	124.2	59.6	99.8	103.3	66.1	89.2	107.8	57.2	88.7	105.0	67.6	90.9	121.5	58.8	97.8
03	CONCRETE	93.3	53.5	75.4	100.3	55.3	80.1	98.6	65.5	83.8	99.6	53.5	78.9	96.4	68.4	83.8	102.3	55.3	81.2
04	MASONRY	91.9	50.0	66.4	114.1	52.7	76.6	137.0	63.5	92.2	92.5	49.9	66.5	97.1	64.3	77.1	110.5	52.6	75.2
05	METALS	92.3	86.7	90.6	87.2	89.2	87.8	93.4	90.3	92.4	92.3	86.5	90.5	96.6	88.6	94.1	87.3	75.3	83.6
06	WOOD, PLASTICS & COMPOSITES	78.7	42.9	59.9	69.9	43.7	56.2	74.8	60.7	67.4	92.4	42.9	66.5	88.3	67.8	77.5	70.9	62.3	66.4
07	THERMAL & MOISTURE PROTECTION	97.7	50.1	77.3	101.3	53.5	80.8	98.1	62.7	83.0	98.2	52.3	78.4	99.8	64.7	84.7	101.5	55.6	81.8
08	OPENINGS	94.9	46.4	83.1	95.8	46.8	83.8	94.6	56.3	85.3	94.9	46.4	83.1	99.2	60.4	89.8	93.1	49.9	82.6
0920	Plaster & Gypsum Board	91.5	41.9	59.0	101.8	42.7	63.0	91.0	60.2	70.8	101.7	41.9	62.5	95.9	67.2	77.1	101.8	61.9	75.6
0950, 0980	Ceilings & Acoustic Treatment	89.2	41.9	59.5	84.0	42.7	58.1	91.3	60.2	71.8	89.2	41.9	59.5	93.6	67.2	77.1	84.0	61.9	70.1
0960	Flooring	99.6	64.3	89.3	87.5	64.3	80.7	98.0	64.3	88.2	105.4	64.3	93.4	91.5	65.6	84.0	86.2	64.3	79.9
0970, 0990	Wall Finishes & Painting/Coating	90.9	44.7	63.2	80.7	44.7	59.1	90.9	56.2	70.1	90.9	44.7	63.2	87.0	56.2	68.5	80.7	44.7	59.1
09	FINISHES	92.8	46.5	67.7	87.7	47.6	66.0	93.3	61.6	76.1	96.2	46.5	69.3	92.5	66.0	78.2	87.9	58.6	72.0
COVERS	DIVS. 10 - 14, 25, 28, 41, 43, 44, 46	100.0	45.4	87.2	100.0	46.4	87.4	100.0	81.9	95.7	100.0	45.4	87.1	100.0	83.2	96.0	100.0	32.4	84.1
21, 22, 23	FIRE SUPPRESSION, PLUMBING & HVAC	99.1	48.9	78.8	99.0	50.8	79.5	100.9	56.5	83.0	99.1	49.2	79.0	101.2	60.0	84.6	99.0	45.4	77.4
26, 27, 3370	ELECTRICAL, COMMUNICATIONS & UTIL.	97.4	40.1	69.0	98.0	53.0	75.7	97.4	56.1	77.0	97.4	37.5	67.8	102.6	56.1	79.6	99.3	55.8	77.8
MF2018	WEIGHTED AVERAGE	95.7	53.8	77.6	96.7	57.1	79.6	99.4	65.2	84.6	97.0	53.5	78.2	98.6	67.9	85.3	96.8	56.4	79.4

MISSISSIPPI / MISSOURI

DIVISION		MCCOMB 396			MERIDIAN 393			TUPELO 388			BOWLING GREEN 633			CAPE GIRARDEAU 637			CHILLICOTHE 646		
		MAT.	INST.	TOTAL	MAT.	INST.	TOTAL	MAT.	INST.	TOTAL	MAT.	INST.	TOTAL	MAT.	INST.	TOTAL	MAT.	INST.	TOTAL
015433	CONTRACTOR EQUIPMENT		98.7	98.7		98.7	98.7		98.7	98.7		104.0	104.0		104.0	104.0		98.6	98.6
0241, 31 - 34	SITE & INFRASTRUCTURE, DEMOLITION	93.0	81.9	85.4	97.4	83.8	88.0	97.2	81.9	86.7	88.2	86.1	86.8	90.2	86.3	87.5	101.7	84.8	90.1
0310	Concrete Forming & Accessories	75.0	43.9	48.6	73.3	68.1	68.9	80.7	44.1	49.6	91.6	90.7	90.8	84.2	81.4	81.8	77.7	92.3	90.1
0320	Concrete Reinforcing	97.2	32.7	66.1	96.0	64.6	80.8	97.4	63.3	81.0	100.6	92.0	96.4	101.9	84.9	93.7	100.7	97.8	99.3
0330	Cast-in-Place Concrete	107.8	57.0	88.6	115.3	66.9	97.1	100.3	58.3	84.4	82.7	91.5	86.0	81.8	84.7	82.9	97.5	82.5	91.8
03	CONCRETE	89.9	48.6	71.3	94.4	68.7	82.8	92.8	54.5	75.6	87.9	92.7	90.1	87.1	85.0	86.1	96.7	90.7	94.0
04	MASONRY	115.5	49.5	75.3	89.7	63.9	74.0	126.1	51.0	80.3	111.9	93.8	100.9	109.4	80.1	91.5	101.3	88.4	93.4
05	METALS	87.4	74.3	83.3	88.4	90.8	89.1	92.2	86.2	90.4	88.9	115.3	97.1	90.0	111.5	96.7	82.3	108.6	90.5
06	WOOD, PLASTICS & COMPOSITES	69.9	45.2	57.0	67.6	68.6	68.1	75.4	43.7	58.8	90.7	91.2	90.9	83.0	79.6	81.2	80.0	93.7	87.2
07	THERMAL & MOISTURE PROTECTION	100.8	52.3	80.0	101.0	63.9	85.1	97.7	50.5	77.4	96.6	94.7	95.8	96.1	83.0	90.5	94.1	88.5	91.7
08	OPENINGS	95.8	39.7	82.1	95.5	64.2	87.9	94.9	49.5	83.8	98.2	94.6	97.3	98.2	78.0	93.3	86.4	92.6	87.9
0920	Plaster & Gypsum Board	101.8	44.3	64.1	101.8	68.3	79.8	91.0	42.7	59.3	95.9	91.5	93.0	95.4	79.6	85.0	103.7	93.7	97.1
0950, 0980	Ceilings & Acoustic Treatment	84.0	44.3	59.1	84.7	68.3	74.4	89.2	42.7	60.1	89.6	91.5	90.8	89.6	79.6	83.3	88.5	93.7	91.7
0960	Flooring	87.5	64.3	80.7	86.1	64.3	79.8	98.2	64.3	88.3	95.6	92.4	94.7	92.1	84.6	89.9	92.4	95.1	93.2
0970, 0990	Wall Finishes & Painting/Coating	80.7	44.7	59.1	80.7	56.2	66.0	90.9	43.1	62.3	95.1	101.2	98.7	95.1	76.7	84.1	90.4	98.3	95.1
09	FINISHES	87.1	47.7	65.8	87.2	66.3	75.9	92.3	47.4	68.0	96.5	91.9	94.0	95.4	80.8	87.5	95.9	93.7	94.7
COVERS	DIVS. 10 - 14, 25, 28, 41, 43, 44, 46	100.0	48.2	87.8	100.0	82.9	96.0	100.0	46.4	87.4	100.0	96.8	99.3	100.0	95.6	99.0	100.0	96.6	99.2
21, 22, 23	FIRE SUPPRESSION, PLUMBING & HVAC	99.0	48.3	78.6	101.1	60.1	84.5	99.2	50.2	79.4	96.9	94.8	96.1	100.7	98.2	99.7	97.3	95.2	96.4
26, 27, 3370	ELECTRICAL, COMMUNICATIONS & UTIL.	96.8	53.9	75.6	99.3	58.1	78.9	97.2	53.0	75.3	99.4	74.0	86.8	99.4	94.8	97.1	94.2	73.0	83.7
MF2018	WEIGHTED AVERAGE	95.1	53.8	77.3	95.4	67.8	83.5	97.2	56.4	79.6	95.6	92.3	94.2	96.4	90.6	93.9	93.7	90.7	92.4

MISSOURI

DIVISION		COLUMBIA 652			FLAT RIVER 636			HANNIBAL 634			HARRISONVILLE 647			JEFFERSON CITY 650 - 651			JOPLIN 648		
		MAT.	INST.	TOTAL	MAT.	INST.	TOTAL	MAT.	INST.	TOTAL	MAT.	INST.	TOTAL	MAT.	INST.	TOTAL	MAT.	INST.	TOTAL
015433	CONTRACTOR EQUIPMENT		107.2	107.2		104.0	104.0		104.0	104.0		98.6	98.6		112.3	112.3		102.0	102.0
0241, 31 - 34	SITE & INFRASTRUCTURE, DEMOLITION	93.0	91.4	91.9	90.7	85.8	87.3	86.0	85.9	85.9	93.6	85.9	88.3	93.0	99.4	97.4	103.0	89.7	93.9
0310	Concrete Forming & Accessories	77.7	89.1	87.4	98.4	85.0	87.0	89.9	79.6	81.1	75.1	95.5	92.5	90.8	78.5	80.4	88.0	75.8	77.6
0320	Concrete Reinforcing	87.1	100.0	93.3	101.9	98.0	100.0	100.1	91.9	96.1	100.3	105.7	102.9	90.9	89.1	90.0	104.0	102.3	103.2
0330	Cast-in-Place Concrete	83.7	87.0	84.9	85.3	86.7	85.8	78.3	89.5	82.5	99.9	95.7	98.3	88.8	85.6	87.6	105.6	74.7	93.9
03	CONCRETE	79.4	91.9	85.0	90.8	89.5	90.2	84.5	87.0	85.6	93.2	98.2	95.4	85.6	84.6	85.2	96.3	81.2	89.5
04	MASONRY	146.8	91.4	113.0	109.3	73.2	87.3	102.9	90.6	95.4	95.6	95.3	95.4	99.2	91.5	94.5	94.5	80.6	86.0
05	METALS	94.1	119.5	102.0	88.8	116.6	97.5	88.9	114.8	97.0	82.7	113.1	92.2	93.6	112.1	99.4	85.2	106.6	91.8
06	WOOD, PLASTICS & COMPOSITES	71.4	86.7	79.4	99.3	86.5	92.6	88.8	78.3	83.3	77.0	95.4	86.6	89.1	73.1	80.7	90.2	75.2	82.3
07	THERMAL & MOISTURE PROTECTION	92.1	89.7	91.0	96.8	86.8	92.5	96.4	89.2	93.3	93.3	96.2	94.5	99.2	88.9	94.8	93.7	80.7	88.2
08	OPENINGS	94.3	88.7	93.0	98.2	93.8	97.1	98.2	80.6	93.9	86.1	99.4	89.4	94.5	78.4	90.6	87.4	78.5	85.2
0920	Plaster & Gypsum Board	72.2	86.7	81.7	101.9	86.7	91.9	95.7	78.3	84.3	100.0	95.3	96.9	80.3	72.4	75.1	110.4	74.6	86.9
0950, 0980	Ceilings & Acoustic Treatment	79.8	86.7	84.1	89.6	86.7	87.8	89.6	78.3	82.5	88.5	95.3	92.8	84.2	72.4	76.8	89.1	74.6	80.0
0960	Flooring	87.3	95.9	89.8	98.9	80.7	93.6	94.8	92.4	94.1	87.9	96.1	90.3	93.5	68.2	86.1	116.9	71.5	103.6
0970, 0990	Wall Finishes & Painting/Coating	94.1	82.0	86.8	95.1	68.3	79.1	95.1	89.6	91.8	94.6	102.7	99.4	95.1	82.0	87.2	90.0	72.2	79.3
09	FINISHES	80.9	89.1	85.4	98.4	81.7	89.4	96.1	81.9	88.4	93.7	96.2	95.1	88.4	76.1	81.7	102.5	74.7	87.4
COVERS	DIVS. 10 - 14, 25, 28, 41, 43, 44, 46	100.0	96.9	99.3	100.0	94.7	98.8	100.0	94.4	98.7	100.0	98.0	99.5	100.0	97.6	99.4	100.0	93.4	98.4
21, 22, 23	FIRE SUPPRESSION, PLUMBING & HVAC	100.9	101.2	101.0	96.9	92.2	95.0	96.9	93.1	95.4	97.2	96.1	96.7	101.0	95.6	98.8	101.1	69.9	88.5
26, 27, 3370	ELECTRICAL, COMMUNICATIONS & UTIL.	94.2	82.5	88.3	103.4	94.8	99.1	98.3	74.0	86.3	99.9	97.5	98.7	100.9	82.5	91.8	92.5	74.7	83.7
MF2018	WEIGHTED AVERAGE	95.7	94.5	95.2	96.6	90.6	94.0	94.5	88.5	91.9	93.2	97.4	95.0	95.7	90.0	93.2	95.2	80.1	88.7

MISSOURI

DIVISION		KANSAS CITY 640 - 641			KIRKSVILLE 635			POPLAR BLUFF 639			ROLLA 654 - 655			SEDALIA 653			SIKESTON 638		
		MAT.	INST.	TOTAL	MAT.	INST.	TOTAL	MAT.	INST.	TOTAL	MAT.	INST.	TOTAL	MAT.	INST.	TOTAL	MAT.	INST.	TOTAL
015433	CONTRACTOR EQUIPMENT		105.1	105.1		94.1	94.1		96.4	96.4		107.2	107.2		97.3	97.3		96.4	96.4
0241, 31 - 34	SITE & INFRASTRUCTURE, DEMOLITION	95.5	100.3	98.8	89.1	81.1	83.6	76.6	84.6	82.1	91.4	90.2	90.6	89.1	85.5	86.6	79.9	85.2	83.6
0310	Concrete Forming & Accessories	87.2	100.5	98.5	82.2	75.6	76.6	82.5	76.3	77.2	84.9	90.5	89.6	82.9	75.4	76.5	83.2	76.4	77.4
0320	Concrete Reinforcing	92.2	107.0	99.4	100.8	79.4	90.5	103.9	72.4	88.7	87.6	88.9	88.2	86.4	105.1	95.4	103.2	72.4	88.4
0330	Cast-in-Place Concrete	103.6	100.5	102.4	85.3	78.7	82.8	65.6	79.8	71.0	85.6	90.6	87.5	89.5	77.9	85.1	70.1	79.8	73.8
03	CONCRETE	94.3	102.1	97.8	102.5	78.6	91.8	78.9	78.1	78.6	81.1	91.8	85.9	94.1	82.7	89.0	82.3	78.2	80.4
04	MASONRY	101.2	100.7	100.9	116.0	81.3	94.8	107.7	71.0	85.3	117.2	81.8	95.6	124.2	78.2	96.2	107.1	71.0	85.1
05	METALS	91.9	110.2	97.6	88.6	98.7	91.7	89.1	95.3	91.0	93.4	113.4	99.7	92.4	110.5	98.1	89.5	95.4	91.3
06	WOOD, PLASTICS & COMPOSITES	85.1	100.4	93.1	76.7	73.9	75.2	75.7	77.2	76.5	78.4	92.3	85.7	73.0	74.1	73.5	77.4	77.2	77.3
07	THERMAL & MOISTURE PROTECTION	92.2	103.0	96.9	101.9	86.7	95.4	100.5	79.2	91.4	92.3	88.7	90.8	96.9	83.8	91.3	100.7	78.4	91.1
08	OPENINGS	96.9	100.2	97.7	103.7	75.9	96.9	104.6	72.4	96.8	94.3	87.7	92.7	99.4	83.8	95.6	104.6	72.4	96.8
0920	Plaster & Gypsum Board	110.5	100.5	103.9	91.1	73.7	79.7	91.4	77.2	82.1	73.6	92.4	86.0	68.5	73.7	71.9	93.4	77.2	82.7
0950, 0980	Ceilings & Acoustic Treatment	90.7	100.5	96.9	88.4	73.7	79.2	89.6	77.2	81.8	79.8	92.4	87.7	79.8	73.7	76.0	89.6	77.2	81.8
0960	Flooring	93.1	99.9	95.1	73.8	92.0	79.1	87.1	80.7	85.2	91.0	92.0	91.3	71.2	69.6	70.7	87.7	80.7	85.6
0970, 0990	Wall Finishes & Painting/Coating	96.2	111.5	105.4	90.9	74.6	81.1	90.2	63.9	74.4	94.1	85.9	89.2	94.1	98.3	96.6	90.2	63.9	74.4
09	FINISHES	96.8	101.5	99.4	95.1	77.8	85.7	94.9	75.2	84.2	82.3	90.3	86.6	80.2	75.9	77.9	95.6	75.5	84.7
COVERS	DIVS. 10 - 14, 25, 28, 41, 43, 44, 46	100.0	99.4	99.9	100.0	93.9	98.6	100.0	93.0	98.4	100.0	96.8	99.2	100.0	92.0	98.1	100.0	93.0	98.4
21, 22, 23	FIRE SUPPRESSION, PLUMBING & HVAC	100.8	101.4	101.0	97.0	92.5	95.2	97.0	90.4	94.3	97.1	93.9	95.8	97.0	90.4	94.4	97.0	90.4	94.3
26, 27, 3370	ELECTRICAL, COMMUNICATIONS & UTIL.	101.7	99.7	100.7	98.3	73.9	86.2	98.6	94.8	96.7	93.0	76.4	84.8	93.8	97.4	95.6	97.8	94.8	96.3
MF2018	WEIGHTED AVERAGE	97.5	101.8	99.3	98.0	83.6	91.8	94.4	84.3	90.0	93.4	90.6	92.2	95.6	88.1	92.4	94.9	84.4	90.4

MISSOURI / MONTANA

DIVISION		SPRINGFIELD 656 - 658			ST. JOSEPH 644 - 645			ST. LOUIS 630 - 631			BILLINGS 590 - 591			BUTTE 597			GREAT FALLS 594		
		MAT.	INST.	TOTAL	MAT.	INST.	TOTAL	MAT.	INST.	TOTAL	MAT.	INST.	TOTAL	MAT.	INST.	TOTAL	MAT.	INST.	TOTAL
015433	CONTRACTOR EQUIPMENT		99.7	99.7		98.6	98.6		107.9	107.9		96.1	96.1		95.8	95.8		95.8	95.8
0241, 31 - 34	SITE & INFRASTRUCTURE, DEMOLITION	91.6	89.3	90.0	97.5	84.2	88.3	96.9	98.8	98.2	91.9	90.6	91.0	98.2	90.3	92.8	101.9	90.6	94.1
0310	Concrete Forming & Accessories	90.7	78.3	80.2	86.8	93.4	92.4	94.6	102.5	101.3	99.1	69.4	73.8	85.7	69.1	71.6	99.4	69.2	73.7
0320	Concrete Reinforcing	84.0	99.6	91.5	97.5	108.6	102.9	91.4	112.4	101.5	90.1	81.0	85.7	97.7	80.6	89.4	90.1	81.0	85.7
0330	Cast-in-Place Concrete	90.9	77.0	85.7	98.1	96.5	97.5	92.4	104.8	97.1	112.2	70.3	96.4	123.5	70.1	103.3	130.5	71.1	108.1
03	CONCRETE	90.5	82.7	87.0	92.3	98.0	94.8	92.0	106.0	98.3	95.7	72.6	85.4	99.0	72.3	87.1	104.1	72.8	90.0
04	MASONRY	92.8	84.5	87.7	96.4	93.2	94.5	87.6	107.7	99.8	131.1	83.1	101.8	126.6	76.1	95.8	131.5	80.3	100.3
05	METALS	98.5	104.9	100.5	88.5	113.9	96.4	94.3	121.5	102.8	104.6	89.4	99.9	98.9	88.5	95.7	102.1	89.4	98.1
06	WOOD, PLASTICS & COMPOSITES	80.0	76.3	78.1	90.2	93.5	91.9	92.1	100.1	96.3	89.4	66.5	77.5	75.0	66.5	70.6	90.9	66.5	78.1
07	THERMAL & MOISTURE PROTECTION	96.1	78.5	88.6	93.6	92.7	93.2	95.3	105.6	99.7	108.7	72.2	93.0	108.3	69.8	91.7	108.9	72.1	93.1
08	OPENINGS	101.7	88.2	98.4	90.3	99.9	92.7	100.0	106.8	101.7	98.4	66.4	90.6	96.8	66.4	89.4	99.6	71.0	92.6
0920	Plaster & Gypsum Board	74.8	76.0	75.6	111.9	93.4	99.8	98.1	100.5	99.7	117.1	66.1	83.7	116.3	66.1	83.4	126.4	66.1	86.9
0950, 0980	Ceilings & Acoustic Treatment	79.8	76.0	77.4	94.9	93.4	94.0	86.3	100.5	95.2	110.0	66.1	82.5	117.4	66.1	85.3	119.3	66.1	86.0
0960	Flooring	90.3	71.5	84.8	97.5	99.9	98.2	97.5	97.1	97.4	90.6	77.1	86.7	87.9	79.4	85.4	94.6	77.1	89.5
0970, 0990	Wall Finishes & Painting/Coating	88.5	104.5	98.1	90.4	103.0	97.9	99.3	106.8	103.8	87.6	88.4	88.1	86.1	64.5	73.2	86.1	91.1	89.1
09	FINISHES	84.4	80.0	82.0	99.4	95.7	97.4	97.2	101.5	99.5	97.8	72.2	84.0	98.3	70.1	83.0	102.4	72.3	86.1
COVERS	DIVS. 10 - 14, 25, 28, 41, 43, 44, 46	100.0	94.3	98.7	100.0	97.3	99.4	100.0	101.8	100.4	100.0	92.3	98.2	100.0	92.3	98.2	100.0	92.0	98.1
21, 22, 23	FIRE SUPPRESSION, PLUMBING & HVAC	100.9	74.0	90.1	101.1	90.6	96.8	100.7	104.5	102.3	100.8	74.4	90.2	100.8	68.3	87.7	100.8	74.0	90.0
26, 27, 3370	ELECTRICAL, COMMUNICATIONS & UTIL.	97.4	67.1	82.4	99.9	80.0	90.1	103.2	96.7	100.0	99.2	72.4	85.9	105.1	70.5	88.0	98.7	73.4	86.1
MF2018	WEIGHTED AVERAGE	96.5	81.4	90.0	96.0	93.3	94.9	97.5	104.6	100.6	101.4	77.2	90.9	101.3	74.4	89.7	102.8	77.1	91.7

For customer support on your Concrete & Masonry Costs with RSMeans Data, call 800.448.8182.

MONTANA

DIVISION		HAVRE 595 MAT.	INST.	TOTAL	HELENA 596 MAT.	INST.	TOTAL	KALISPELL 599 MAT.	INST.	TOTAL	MILES CITY 593 MAT.	INST.	TOTAL	MISSOULA 598 MAT.	INST.	TOTAL	WOLF POINT 592 MAT.	INST.	TOTAL
015433	CONTRACTOR EQUIPMENT		95.8	95.8		100.9	100.9		95.8	95.8		95.8	95.8		95.8	95.8		95.8	95.8
0241, 31 - 34	SITE & INFRASTRUCTURE, DEMOLITION	105.1	90.1	94.8	92.1	98.1	96.2	87.8	90.2	89.5	94.1	90.1	91.4	81.0	90.3	87.4	111.6	90.1	96.8
0310	Concrete Forming & Accessories	78.0	67.7	69.2	102.5	68.9	74.0	89.3	68.8	71.9	97.3	67.9	72.3	89.3	69.3	72.2	89.4	67.8	71.1
0320	Concrete Reinforcing	98.5	75.1	87.2	101.6	80.6	91.5	100.3	84.7	92.8	98.1	75.2	87.0	99.4	84.5	92.2	99.5	74.5	87.5
0330	Cast-in-Place Concrete	133.0	68.3	108.6	95.2	71.2	86.1	107.2	69.5	93.0	117.4	68.3	98.9	91.0	70.1	83.1	131.5	67.3	107.3
03	CONCRETE	106.6	70.2	90.3	96.9	72.5	86.0	89.2	72.7	81.8	96.1	70.3	84.5	78.4	73.1	76.0	110.1	69.8	92.0
04	MASONRY	127.6	77.2	96.9	115.5	75.1	90.9	125.5	79.2	97.3	133.6	77.2	99.2	153.1	75.9	106.0	134.8	77.2	99.7
05	METALS	95.1	86.4	92.4	100.9	86.7	96.5	95.0	90.2	93.5	94.3	86.6	91.9	95.5	89.9	93.8	94.4	86.3	91.9
06	WOOD, PLASTICS & COMPOSITES	65.7	66.5	66.2	94.6	66.8	80.1	78.8	66.5	72.4	87.5	66.5	76.5	78.8	66.5	72.4	77.9	66.5	72.0
07	THERMAL & MOISTURE PROTECTION	108.6	63.7	89.3	104.8	70.4	90.0	107.8	70.7	91.9	108.2	65.0	89.7	107.4	71.9	92.2	109.2	64.9	90.2
08	OPENINGS	97.3	65.1	89.5	96.1	66.6	88.9	97.3	67.3	90.0	96.8	65.1	89.1	96.8	67.3	89.6	96.8	65.0	89.1
0920	Plaster & Gypsum Board	111.7	66.1	81.8	114.1	66.1	82.6	116.3	66.1	83.4	124.7	66.1	86.3	116.3	66.1	83.4	118.8	66.1	84.2
0950, 0980	Ceilings & Acoustic Treatment	117.4	66.1	85.3	122.3	66.1	87.1	117.4	66.1	85.3	112.9	66.1	83.6	117.4	66.1	85.3	112.9	66.1	83.6
0960	Flooring	85.5	86.8	85.9	97.2	79.4	92.0	89.6	86.8	88.8	94.5	86.8	92.2	89.6	79.4	86.7	91.0	86.8	89.8
0970, 0990	Wall Finishes & Painting/Coating	86.1	64.5	73.2	93.9	57.2	71.9	86.1	86.7	86.4	86.1	64.5	73.2	86.1	86.7	86.4	86.1	64.5	73.2
09	FINISHES	97.7	70.9	83.2	105.0	69.3	85.6	98.2	73.8	85.0	100.2	70.9	84.3	97.7	72.5	84.1	99.7	70.9	84.1
COVERS	DIVS. 10 - 14, 25, 28, 41, 43, 44, 46	100.0	84.7	96.4	100.0	92.6	98.2	100.0	85.4	96.6	100.0	84.7	96.4	100.0	92.2	98.2	100.0	84.7	96.4
21, 22, 23	FIRE SUPPRESSION, PLUMBING & HVAC	97.0	65.6	84.3	100.9	67.9	87.6	97.0	66.1	84.5	97.0	71.7	86.8	100.8	68.4	87.7	97.0	71.6	86.7
26, 27, 3370	ELECTRICAL, COMMUNICATIONS & UTIL.	98.7	66.7	82.8	105.3	70.9	88.3	102.3	65.6	84.1	98.7	71.7	85.3	103.2	69.1	86.3	98.7	71.7	85.3
MF2018	WEIGHTED AVERAGE	100.2	72.5	88.3	101.2	74.7	89.7	98.0	74.1	87.7	99.1	74.6	88.5	98.8	74.8	88.5	101.3	74.5	89.7

NEBRASKA

DIVISION		ALLIANCE 693 MAT.	INST.	TOTAL	COLUMBUS 686 MAT.	INST.	TOTAL	GRAND ISLAND 688 MAT.	INST.	TOTAL	HASTINGS 689 MAT.	INST.	TOTAL	LINCOLN 683 - 685 MAT.	INST.	TOTAL	MCCOOK 690 MAT.	INST.	TOTAL
015433	CONTRACTOR EQUIPMENT		92.2	92.2		98.8	98.8		98.8	98.8		98.8	98.8		104.9	104.9		98.8	98.8
0241, 31 - 34	SITE & INFRASTRUCTURE, DEMOLITION	97.6	92.3	94.0	100.4	86.8	91.1	105.6	87.3	93.0	103.8	86.8	92.1	95.3	96.6	96.2	100.5	86.8	91.1
0310	Concrete Forming & Accessories	83.9	53.2	57.8	95.7	72.2	75.7	95.2	67.4	71.5	98.9	69.9	74.2	96.8	73.9	77.3	89.2	53.8	59.0
0320	Concrete Reinforcing	110.1	83.2	97.1	100.7	81.9	91.6	100.1	80.4	90.6	100.1	72.5	86.8	95.3	80.7	88.3	102.7	72.7	88.2
0330	Cast-in-Place Concrete	104.2	78.3	94.4	104.6	78.3	94.7	110.7	75.6	97.5	110.7	70.8	95.6	87.5	81.9	85.4	112.6	70.7	96.8
03	CONCRETE	113.9	68.0	93.3	100.2	77.2	89.9	104.8	73.8	90.9	105.1	71.9	90.2	93.8	78.9	87.1	103.2	64.6	85.9
04	MASONRY	109.4	74.2	87.9	115.8	74.2	90.4	109.1	73.0	87.1	119.3	70.9	89.8	98.4	77.2	85.5	103.6	74.2	85.7
05	METALS	96.7	83.0	92.4	89.2	95.8	91.3	91.0	94.8	92.2	91.7	91.5	91.6	98.6	93.4	97.0	91.7	91.5	91.6
06	WOOD, PLASTICS & COMPOSITES	78.2	47.0	61.9	92.3	72.3	81.8	91.4	64.0	77.0	95.5	69.6	81.9	100.2	72.6	85.7	85.8	47.9	66.0
07	THERMAL & MOISTURE PROTECTION	101.1	72.3	88.7	101.4	76.7	90.8	101.5	77.2	91.1	101.6	74.1	89.8	100.3	79.0	91.1	96.1	72.6	86.0
08	OPENINGS	90.9	55.9	82.4	91.7	71.0	86.7	91.7	66.3	85.5	91.7	67.2	85.8	105.8	68.3	96.7	91.3	53.7	82.1
0920	Plaster & Gypsum Board	75.5	45.9	56.1	84.4	71.8	76.2	83.5	63.3	70.3	85.2	69.1	74.6	90.3	71.8	78.2	87.1	46.8	60.7
0950, 0980	Ceilings & Acoustic Treatment	88.5	45.9	61.8	84.8	71.8	76.7	84.8	63.3	71.3	84.8	69.1	74.9	106.7	71.8	84.8	84.1	46.8	60.7
0960	Flooring	90.9	83.3	88.7	83.7	83.3	83.6	83.4	75.3	81.1	84.8	77.5	82.7	94.9	82.6	91.3	89.6	83.3	87.8
0970, 0990	Wall Finishes & Painting/Coating	156.3	49.2	92.1	74.2	57.2	64.0	74.2	59.5	65.4	74.2	57.2	64.0	94.7	75.6	83.2	85.6	42.9	60.0
09	FINISHES	91.0	56.8	72.5	85.1	72.5	78.3	85.1	67.1	75.4	85.8	69.7	77.1	96.5	75.4	85.1	88.2	56.6	71.1
COVERS	DIVS. 10 - 14, 25, 28, 41, 43, 44, 46	100.0	85.1	96.5	100.0	88.0	97.2	100.0	88.1	97.2	100.0	87.7	97.1	100.0	89.7	97.6	100.0	85.3	96.5
21, 22, 23	FIRE SUPPRESSION, PLUMBING & HVAC	97.1	71.4	86.7	97.0	71.7	86.8	100.8	80.2	92.5	97.0	71.0	86.5	100.6	80.3	92.4	96.8	71.7	86.7
26, 27, 3370	ELECTRICAL, COMMUNICATIONS & UTIL.	92.7	63.2	78.1	96.1	78.4	87.3	94.9	66.0	80.6	94.3	76.5	85.5	105.4	66.0	85.9	94.4	63.3	79.0
MF2018	WEIGHTED AVERAGE	98.4	70.4	86.3	96.0	77.7	88.1	97.4	76.2	88.2	97.1	75.2	87.7	99.8	79.3	90.9	95.9	70.2	84.8

NEBRASKA / NEVADA

DIVISION		NORFOLK 687 MAT.	INST.	TOTAL	NORTH PLATTE 691 MAT.	INST.	TOTAL	OMAHA 680 - 681 MAT.	INST.	TOTAL	VALENTINE 692 MAT.	INST.	TOTAL	CARSON CITY 897 MAT.	INST.	TOTAL	ELKO 898 MAT.	INST.	TOTAL
015433	CONTRACTOR EQUIPMENT		88.5	88.5		98.8	98.8		94.2	94.2		91.9	91.9		95.7	95.7		92.3	92.3
0241, 31 - 34	SITE & INFRASTRUCTURE, DEMOLITION	82.6	85.9	84.9	102.3	86.6	91.5	89.9	95.2	93.6	85.3	91.3	89.4	80.5	93.7	89.6	65.3	88.0	80.9
0310	Concrete Forming & Accessories	82.3	71.4	73.0	91.8	72.8	75.6	92.3	78.3	80.4	80.6	51.4	55.7	100.8	84.7	87.1	105.3	92.7	94.6
0320	Concrete Reinforcing	100.8	63.3	82.7	102.2	77.7	90.4	103.7	81.0	92.7	102.7	63.2	83.6	116.4	117.8	117.1	117.6	109.1	113.5
0330	Cast-in-Place Concrete	105.3	68.1	91.2	112.6	58.0	92.0	86.7	79.6	84.0	99.4	51.6	81.4	97.1	84.3	92.3	90.9	69.3	82.8
03	CONCRETE	98.8	69.4	85.6	103.3	69.7	88.2	94.2	79.5	87.6	100.8	54.7	80.1	106.5	90.2	99.2	97.7	87.2	92.7
04	MASONRY	123.9	74.1	93.5	93.1	73.7	81.2	96.8	81.6	87.5	104.8	73.6	85.8	116.8	67.2	86.5	123.4	64.9	87.7
05	METALS	92.5	78.1	88.0	91.0	93.4	91.8	98.6	84.8	94.3	102.5	77.5	94.7	109.5	95.4	105.1	113.1	91.6	106.4
06	WOOD, PLASTICS & COMPOSITES	76.8	71.8	74.2	87.9	74.2	80.8	92.6	78.0	85.0	73.6	45.7	59.0	87.1	85.6	86.3	96.2	98.8	97.6
07	THERMAL & MOISTURE PROTECTION	100.3	74.7	89.3	96.0	74.8	86.9	95.9	82.2	90.0	95.8	70.6	85.0	114.3	80.1	99.6	110.4	70.2	93.2
08	OPENINGS	92.9	66.1	86.4	90.7	70.6	85.8	99.7	79.0	94.7	92.7	51.4	82.7	101.1	83.8	96.9	102.4	89.2	99.2
0920	Plaster & Gypsum Board	84.4	71.8	76.1	87.1	73.9	78.4	92.2	77.9	82.9	88.7	45.0	60.0	98.5	85.2	89.7	103.2	99.0	100.5
0950, 0980	Ceilings & Acoustic Treatment	99.4	71.8	82.1	84.1	73.9	77.7	93.6	77.9	83.8	98.4	45.0	64.9	107.1	85.2	93.3	105.3	99.0	101.4
0960	Flooring	102.7	83.3	97.0	90.8	75.3	86.3	95.0	90.4	93.7	115.7	83.3	106.3	99.0	64.1	88.8	99.1	64.1	88.9
0970, 0990	Wall Finishes & Painting/Coating	133.5	57.2	87.8	85.6	56.1	67.9	103.2	58.1	76.1	154.7	58.1	96.8	91.6	85.1	87.7	87.4	75.6	80.3
09	FINISHES	101.7	72.2	85.7	88.6	71.7	79.5	96.7	78.2	86.7	108.1	56.8	80.4	98.3	80.4	88.7	95.4	86.8	90.8
COVERS	DIVS. 10 - 14, 25, 28, 41, 43, 44, 46	100.0	86.9	96.9	100.0	87.9	97.2	100.0	89.3	97.5	100.0	83.9	96.2	100.0	106.3	101.5	100.0	86.2	96.8
21, 22, 23	FIRE SUPPRESSION, PLUMBING & HVAC	96.6	70.8	86.2	100.6	73.4	89.6	100.6	83.9	93.9	96.3	70.1	85.7	100.9	81.4	93.0	99.0	73.3	88.6
26, 27, 3370	ELECTRICAL, COMMUNICATIONS & UTIL.	95.2	78.4	86.8	92.9	66.0	79.6	102.9	83.1	93.1	90.6	63.2	77.1	100.9	92.0	96.5	97.5	85.0	91.3
MF2018	WEIGHTED AVERAGE	97.4	74.5	87.5	96.1	74.9	86.9	98.6	83.0	91.9	98.0	67.4	84.8	103.2	85.6	95.6	101.5	81.8	93.0

For customer support on your Concrete & Masonry Costs with RSMeans Data, call 800.448.8182.

473

DIVISION		NEVADA									NEW HAMPSHIRE								
		ELY 893			LAS VEGAS 889-891			RENO 894-895			CHARLESTON 036			CLAREMONT 037			CONCORD 032-033		
		MAT.	INST.	TOTAL	MAT.	INST.	TOTAL	MAT.	INST.	TOTAL	MAT.	INST.	TOTAL	MAT.	INST.	TOTAL	MAT.	INST.	TOTAL
015433	CONTRACTOR EQUIPMENT		92.3	92.3		92.3	92.3		92.3	92.3		94.2	94.2		94.2	94.2		98.6	98.6
0241, 31 - 34	SITE & INFRASTRUCTURE, DEMOLITION	70.7	89.1	83.3	73.8	92.7	86.8	70.9	89.8	83.9	81.6	93.4	89.7	75.5	93.4	87.8	91.0	102.3	98.7
0310	Concrete Forming & Accessories	97.9	97.8	97.8	99.1	112.9	110.9	94.5	84.5	85.9	86.6	82.9	83.5	93.3	83.0	84.5	99.3	94.4	95.1
0320	Concrete Reinforcing	116.2	109.3	112.9	107.1	125.5	116.0	109.9	123.8	116.6	87.5	84.4	86.0	87.5	84.4	86.0	93.9	84.5	89.4
0330	Cast-in-Place Concrete	97.6	91.3	95.2	94.5	109.4	100.1	103.2	82.2	95.3	86.0	111.8	95.7	79.0	111.8	91.4	104.7	113.8	108.2
03	CONCRETE	106.0	97.3	102.1	101.0	113.3	106.5	105.1	90.5	98.6	90.3	93.5	91.7	83.0	93.5	87.7	102.2	99.3	100.9
04	MASONRY	129.0	71.7	94.1	115.5	105.1	109.2	122.4	67.1	88.7	92.3	95.4	94.2	93.1	95.4	94.5	105.2	98.9	101.4
05	METALS	113.0	94.4	107.2	122.1	105.3	116.8	114.8	98.7	109.8	99.2	90.8	96.6	99.2	90.8	96.6	106.7	89.8	101.4
06	WOOD, PLASTICS & COMPOSITES	86.5	100.9	94.0	85.0	112.0	99.1	81.0	85.3	83.2	87.6	80.6	83.9	94.7	80.6	87.3	98.2	94.4	96.2
07	THERMAL & MOISTURE PROTECTION	110.9	88.0	101.1	125.1	103.2	115.7	110.4	78.5	96.7	107.6	102.2	105.3	107.4	102.2	105.2	111.3	107.1	109.5
08	OPENINGS	102.3	90.3	99.4	101.3	115.0	104.7	100.2	84.9	96.5	96.6	79.8	92.5	97.6	79.8	93.3	97.4	87.4	95.0
0920	Plaster & Gypsum Board	98.1	101.1	100.1	92.4	112.5	105.6	87.8	85.2	86.1	100.3	79.9	86.9	100.8	79.9	87.1	100.2	94.1	96.2
0950, 0980	Ceilings & Acoustic Treatment	105.3	101.1	102.7	111.7	112.5	112.2	108.5	85.2	93.9	90.6	79.9	83.9	90.6	79.9	83.9	92.6	94.1	93.5
0960	Flooring	96.6	64.1	87.1	88.5	102.0	92.4	93.3	64.1	84.8	87.5	107.9	93.4	90.0	107.9	95.2	96.0	107.9	99.5
0970, 0990	Wall Finishes & Painting/Coating	87.4	106.7	99.0	89.9	118.5	107.1	87.4	85.1	86.1	79.1	85.7	83.1	79.1	85.7	83.1	95.0	85.7	89.4
09	FINISHES	94.4	93.1	93.7	93.0	112.2	103.4	92.6	80.2	85.9	90.6	87.2	88.8	90.9	87.2	88.9	95.0	95.9	95.5
COVERS	DIVS. 10 - 14, 25, 28, 41, 43, 44, 46	100.0	67.0	92.2	100.0	106.4	101.5	100.0	105.6	101.3	100.0	84.4	96.3	100.0	84.4	96.3	100.0	106.8	101.6
21, 22, 23	FIRE SUPPRESSION, PLUMBING & HVAC	99.0	91.6	96.0	101.1	103.6	102.1	100.9	81.4	93.0	97.6	78.3	89.8	97.6	78.3	89.8	101.2	85.4	94.8
26, 27, 3370	ELECTRICAL, COMMUNICATIONS & UTIL.	97.8	89.5	93.7	102.1	109.9	105.9	98.1	92.0	95.1	94.6	72.4	83.7	94.6	72.4	83.7	95.9	72.4	84.3
MF2018	WEIGHTED AVERAGE	102.9	89.5	97.1	104.2	107.0	105.4	102.8	85.6	95.4	95.6	85.6	91.3	94.8	85.7	90.8	100.8	91.3	96.7

DIVISION		NEW HAMPSHIRE															NEW JERSEY		
		KEENE 034			LITTLETON 035			MANCHESTER 031			NASHUA 030			PORTSMOUTH 038			ATLANTIC CITY 082,084		
		MAT.	INST.	TOTAL	MAT.	INST.	TOTAL	MAT.	INST.	TOTAL	MAT.	INST.	TOTAL	MAT.	INST.	TOTAL	MAT.	INST.	TOTAL
015433	CONTRACTOR EQUIPMENT		94.2	94.2		94.2	94.2		99.1	99.1		94.2	94.2		94.2	94.2		91.8	91.8
0241, 31 - 34	SITE & INFRASTRUCTURE, DEMOLITION	89.5	93.4	92.2	75.6	92.4	87.1	87.4	102.4	97.7	91.2	93.6	92.8	85.0	94.3	91.4	92.0	97.2	95.6
0310	Concrete Forming & Accessories	91.8	83.2	84.5	103.8	77.3	81.3	100.2	94.8	95.6	100.8	94.7	95.7	88.2	94.3	93.4	112.0	145.5	140.5
0320	Concrete Reinforcing	87.5	84.4	86.0	88.2	84.4	86.4	110.5	84.6	98.0	109.3	84.6	97.4	87.5	84.6	86.1	82.0	137.7	108.9
0330	Cast-in-Place Concrete	86.4	111.9	96.0	77.5	103.4	87.3	98.7	114.3	104.6	81.8	113.4	93.7	77.5	113.6	91.2	79.8	135.5	100.8
03	CONCRETE	90.0	93.7	91.6	82.4	88.0	85.0	101.6	99.7	100.7	90.3	99.4	94.4	82.5	99.4	90.1	87.8	139.2	110.9
04	MASONRY	94.9	95.4	95.2	105.1	80.6	90.1	100.7	98.9	99.6	97.7	98.8	98.4	93.2	98.3	96.3	106.3	140.0	126.9
05	METALS	99.9	91.2	97.2	99.9	90.8	97.1	107.4	90.3	102.0	106.0	91.8	101.5	101.5	93.2	98.9	104.0	116.0	107.7
06	WOOD, PLASTICS & COMPOSITES	93.0	80.6	86.5	105.7	80.6	92.6	100.4	94.5	97.3	104.3	94.4	99.1	89.1	94.4	91.9	117.5	147.1	133.0
07	THERMAL & MOISTURE PROTECTION	108.2	102.2	105.6	107.6	95.5	102.4	112.5	107.1	110.2	108.7	106.1	107.6	108.1	105.7	107.1	100.7	136.4	116.0
08	OPENINGS	95.3	83.1	92.3	98.5	79.8	94.0	96.3	92.5	95.4	99.5	90.1	97.2	100.0	81.1	95.4	96.7	142.0	107.7
0920	Plaster & Gypsum Board	100.5	79.9	87.0	115.8	79.9	92.2	99.2	94.1	95.8	110.7	94.1	99.8	100.3	94.1	96.2	109.6	148.2	134.9
0950, 0980	Ceilings & Acoustic Treatment	90.6	79.9	83.9	90.6	79.9	83.9	89.3	94.1	92.3	100.9	94.1	96.6	91.4	94.1	93.1	89.1	148.2	126.1
0960	Flooring	89.6	107.9	95.0	100.4	107.9	102.6	95.0	110.1	99.4	93.3	110.1	98.2	87.7	110.1	94.2	102.7	158.5	119.0
0970, 0990	Wall Finishes & Painting/Coating	79.1	98.4	90.7	79.1	85.7	83.1	96.3	118.0	109.3	79.1	118.0	102.4	79.1	97.4	90.1	77.1	147.2	119.1
09	FINISHES	92.5	88.6	90.4	96.0	83.5	89.2	94.3	99.9	97.3	97.4	99.9	98.7	91.5	97.4	94.7	94.4	149.6	124.3
COVERS	DIVS. 10 - 14, 25, 28, 41, 43, 44, 46	100.0	90.3	97.7	100.0	91.1	97.9	100.0	107.0	101.7	100.0	106.9	101.6	100.0	106.6	101.5	100.0	116.5	103.9
21, 22, 23	FIRE SUPPRESSION, PLUMBING & HVAC	97.6	78.4	89.8	97.6	70.6	86.7	101.2	85.5	94.9	101.4	85.5	95.0	101.4	85.0	94.8	100.4	130.0	112.4
26, 27, 3370	ELECTRICAL, COMMUNICATIONS & UTIL.	94.6	72.4	83.7	95.4	47.3	71.6	97.4	78.5	88.0	96.3	78.5	87.5	95.0	77.6	86.4	92.2	138.0	114.9
MF2018	WEIGHTED AVERAGE	96.1	86.2	91.8	96.1	77.7	88.1	100.5	93.0	97.3	99.3	92.3	96.3	96.5	91.5	94.3	97.8	132.6	112.8

DIVISION		NEW JERSEY																	
		CAMDEN 081			DOVER 078			ELIZABETH 072			HACKENSACK 076			JERSEY CITY 073			LONG BRANCH 077		
		MAT.	INST.	TOTAL	MAT.	INST.	TOTAL	MAT.	INST.	TOTAL	MAT.	INST.	TOTAL	MAT.	INST.	TOTAL	MAT.	INST.	TOTAL
015433	CONTRACTOR EQUIPMENT		91.8	91.8		94.2	94.2		94.2	94.2		94.2	94.2		91.8	91.8		91.4	91.4
0241, 31 - 34	SITE & INFRASTRUCTURE, DEMOLITION	92.8	97.5	96.0	107.6	98.9	101.6	113.1	98.9	103.3	108.7	98.9	101.9	98.3	98.8	98.6	103.5	97.6	99.5
0310	Concrete Forming & Accessories	101.8	146.6	139.9	96.0	148.3	140.5	109.0	148.4	142.5	96.0	148.1	140.3	100.0	148.3	141.1	100.6	134.2	129.2
0320	Concrete Reinforcing	107.7	137.9	122.2	72.3	152.4	110.9	72.3	152.4	110.9	72.3	152.4	110.9	93.8	152.4	122.1	72.3	151.9	110.7
0330	Cast-in-Place Concrete	77.4	135.7	99.4	82.7	135.9	102.8	71.1	137.8	96.2	80.8	137.7	102.3	64.5	135.9	91.5	71.7	128.9	93.3
03	CONCRETE	88.5	139.8	111.5	85.9	143.3	111.7	83.0	140.0	110.4	84.3	143.8	111.0	80.6	143.2	108.7	84.6	134.2	106.9
04	MASONRY	95.5	138.5	121.8	98.0	141.7	124.6	115.6	141.7	131.5	102.2	141.7	126.3	92.5	141.7	122.5	107.4	129.8	121.1
05	METALS	110.3	116.4	112.2	101.5	125.2	108.9	103.1	125.3	110.0	101.6	125.0	108.9	107.9	122.3	112.4	101.6	120.6	107.6
06	WOOD, PLASTICS & COMPOSITES	104.7	149.8	128.3	93.1	149.8	122.8	109.1	149.8	130.4	93.1	149.8	122.8	94.2	149.8	123.3	95.4	134.7	115.9
07	THERMAL & MOISTURE PROTECTION	100.5	135.9	115.7	102.2	140.6	118.6	102.5	140.8	118.9	102.0	133.1	115.3	101.7	140.6	118.4	101.8	126.2	112.3
08	OPENINGS	98.8	143.4	109.7	102.7	144.8	113.0	101.2	144.8	111.8	100.6	144.8	111.4	99.2	144.8	110.3	95.3	134.1	104.7
0920	Plaster & Gypsum Board	105.8	151.0	135.4	118.5	151.0	139.8	125.5	151.0	142.2	118.5	151.0	139.8	121.8	151.0	140.9	120.4	135.5	130.3
0950, 0980	Ceilings & Acoustic Treatment	98.1	151.0	131.2	83.8	151.0	125.9	85.2	151.0	126.4	83.8	151.0	125.9	93.5	151.0	129.5	83.8	135.5	116.2
0960	Flooring	98.9	158.5	116.3	82.4	183.6	112.0	87.4	183.6	115.5	82.4	183.6	112.0	83.3	183.6	112.6	83.6	168.8	108.5
0970, 0990	Wall Finishes & Painting/Coating	77.1	147.2	119.1	82.6	142.9	118.7	82.6	142.9	118.7	82.6	142.9	118.7	82.7	142.9	118.8	82.7	147.2	121.4
09	FINISHES	94.8	150.4	124.9	90.5	154.7	125.2	93.8	155.2	127.1	90.4	154.7	125.2	93.3	155.2	126.8	91.5	142.7	119.2
COVERS	DIVS. 10 - 14, 25, 28, 41, 43, 44, 46	100.0	115.8	103.7	100.0	125.9	106.1	100.0	125.9	106.1	100.0	125.9	106.1	100.0	125.9	106.1	100.0	113.2	103.1
21, 22, 23	FIRE SUPPRESSION, PLUMBING & HVAC	100.7	132.5	113.6	100.4	135.3	114.5	100.7	135.2	114.6	100.4	135.3	114.5	100.7	135.2	114.7	100.4	125.0	110.3
26, 27, 3370	ELECTRICAL, COMMUNICATIONS & UTIL.	95.8	141.5	118.4	94.7	144.4	119.3	95.2	144.4	119.5	94.7	143.1	118.7	98.7	143.1	120.6	94.5	125.1	109.6
MF2018	WEIGHTED AVERAGE	98.9	133.7	114.0	97.6	137.5	114.8	98.9	137.6	115.6	97.4	137.1	114.6	97.7	137.1	114.7	97.1	126.6	109.9

For customer support on your Concrete & Masonry Costs with RSMeans Data, call 800.448.8182.

NEW JERSEY

DIVISION		NEW BRUNSWICK 088 - 089 MAT.	INST.	TOTAL	NEWARK 070 - 071 MAT.	INST.	TOTAL	PATERSON 074 - 075 MAT.	INST.	TOTAL	POINT PLEASANT 087 MAT.	INST.	TOTAL	SUMMIT 079 MAT.	INST.	TOTAL	TRENTON 085 - 086 MAT.	INST.	TOTAL
015433	CONTRACTOR EQUIPMENT		91.4	91.4		99.5	99.5		94.2	94.2		91.4	91.4		94.2	94.2		95.5	95.5
0241, 31 - 34	SITE & INFRASTRUCTURE, DEMOLITION	105.8	98.0	100.4	114.9	107.9	110.1	110.5	98.8	102.5	107.2	97.6	100.6	110.3	98.9	102.4	90.8	103.4	99.5
0310	Concrete Forming & Accessories	105.4	147.6	141.3	99.8	148.4	141.2	98.2	148.0	140.6	99.0	134.2	129.0	98.9	148.2	140.9	98.9	133.8	128.6
0320	Concrete Reinforcing	83.0	151.9	116.2	97.7	152.4	124.1	93.8	152.4	122.1	83.0	151.9	116.2	72.3	152.4	110.9	107.7	110.7	109.2
0330	Cast-in-Place Concrete	98.6	132.0	111.3	93.5	138.9	110.6	82.2	137.6	103.1	98.6	130.8	110.8	68.7	137.7	94.8	94.5	129.4	107.7
03	CONCRETE	104.2	141.4	120.9	93.7	144.3	116.4	88.7	143.8	113.4	103.8	134.9	117.7	80.2	143.9	108.8	95.4	127.2	109.7
04	MASONRY	104.4	135.2	123.2	102.5	141.7	126.4	98.5	141.7	124.8	92.6	129.8	115.3	101.0	141.7	125.8	102.5	129.9	119.2
05	METALS	104.1	121.0	109.3	109.6	123.9	114.0	102.9	125.0	109.8	104.1	120.7	109.2	101.5	125.2	108.9	109.6	106.4	108.6
06	WOOD, PLASTICS & COMPOSITES	110.5	149.7	131.0	96.4	149.9	124.4	95.9	149.8	124.1	101.9	134.7	119.1	97.2	149.8	124.7	96.7	134.7	116.6
07	THERMAL & MOISTURE PROTECTION	101.0	136.2	116.1	104.1	141.9	120.3	102.3	133.1	115.5	101.0	128.4	112.8	102.5	140.8	119.0	100.9	132.6	114.5
08	OPENINGS	92.1	144.8	104.9	101.2	144.9	111.8	105.7	144.8	115.2	93.8	138.5	104.7	106.6	144.8	115.9	97.2	125.2	104.1
0920	Plaster & Gypsum Board	108.2	151.0	136.3	112.0	151.0	137.6	121.8	151.0	140.9	102.8	135.5	124.3	120.4	151.0	140.5	97.7	135.5	122.5
0950, 0980	Ceilings & Acoustic Treatment	89.1	151.0	127.9	92.7	151.0	129.2	93.5	151.0	129.5	89.1	135.5	118.2	83.8	151.0	125.9	92.7	135.5	119.5
0960	Flooring	100.3	183.6	124.6	97.6	183.6	122.7	83.3	183.6	112.6	97.6	158.5	115.4	83.6	183.6	112.8	104.3	168.8	123.2
0970, 0990	Wall Finishes & Painting/Coating	77.1	142.9	116.5	96.1	142.9	124.2	82.6	142.9	118.7	77.1	147.2	119.1	82.6	142.9	118.7	92.4	147.2	125.3
09	FINISHES	94.8	154.7	127.2	98.0	155.3	129.0	93.4	154.7	126.6	93.3	141.0	119.1	91.5	154.7	125.7	97.0	142.8	121.8
COVERS	DIVS. 10 - 14, 25, 28, 41, 43, 44, 46	100.0	125.7	106.1	100.0	126.2	106.2	100.0	125.9	106.1	100.0	105.9	101.4	100.0	125.9	106.1	100.0	113.3	103.1
21, 22, 23	FIRE SUPPRESSION, PLUMBING & HVAC	100.4	127.8	111.5	100.8	134.2	114.3	100.7	135.1	114.6	100.4	125.0	110.3	100.4	132.3	113.3	100.8	124.6	110.4
26, 27, 3370	ELECTRICAL, COMMUNICATIONS & UTIL.	92.7	128.7	110.5	103.3	143.1	123.0	98.7	144.4	121.3	92.2	125.1	108.5	95.2	144.4	119.5	100.5	123.9	112.1
MF2018	WEIGHTED AVERAGE	99.7	132.2	113.8	101.8	137.9	117.4	99.3	137.3	115.7	99.1	126.5	110.9	97.7	136.9	114.6	100.5	124.4	110.8

NEW JERSEY / NEW MEXICO

DIVISION		VINELAND 080,083 MAT.	INST.	TOTAL	ALBUQUERQUE 870 - 872 MAT.	INST.	TOTAL	CARRIZOZO 883 MAT.	INST.	TOTAL	CLOVIS 881 MAT.	INST.	TOTAL	FARMINGTON 874 MAT.	INST.	TOTAL	GALLUP 873 MAT.	INST.	TOTAL
015433	CONTRACTOR EQUIPMENT		91.8	91.8		105.2	105.2		105.2	105.2		105.2	105.2		105.2	105.2		105.2	105.2
0241, 31 - 34	SITE & INFRASTRUCTURE, DEMOLITION	96.7	97.3	97.1	88.4	94.4	92.5	106.6	94.4	98.2	93.9	94.4	94.2	94.6	94.4	94.5	103.2	94.4	97.1
0310	Concrete Forming & Accessories	95.7	133.1	127.5	103.1	65.0	70.7	95.9	65.0	69.6	95.9	64.9	69.5	103.2	65.0	70.7	103.2	65.0	70.7
0320	Concrete Reinforcing	82.0	125.1	102.8	96.3	69.5	83.4	109.2	69.5	90.0	110.4	69.5	90.7	105.3	69.5	85.1	100.8	69.5	85.7
0330	Cast-in-Place Concrete	86.0	129.1	102.3	90.4	68.9	82.2	93.4	68.9	84.1	93.3	68.8	84.1	91.2	68.9	82.8	85.8	68.9	79.4
03	CONCRETE	92.2	129.2	108.8	94.1	68.5	82.6	116.6	68.5	95.0	104.3	68.4	88.2	97.6	68.5	84.5	104.4	68.5	88.3
04	MASONRY	94.6	129.9	116.1	105.7	62.4	79.3	107.4	62.4	80.0	107.4	62.4	80.0	112.7	62.4	82.0	101.8	62.4	77.8
05	METALS	103.9	111.5	106.3	106.2	89.8	101.1	104.7	89.8	100.0	104.3	89.6	99.7	103.9	89.8	99.5	103.0	89.8	98.9
06	WOOD, PLASTICS & COMPOSITES	98.3	133.3	116.6	104.5	64.9	83.8	89.5	64.9	76.6	89.5	64.9	76.6	104.6	64.9	83.8	104.6	64.9	83.8
07	THERMAL & MOISTURE PROTECTION	100.4	131.1	113.6	98.9	72.0	87.4	103.9	72.0	90.2	102.5	72.0	89.4	99.1	72.0	87.4	100.3	72.0	88.1
08	OPENINGS	93.4	131.8	102.8	98.9	65.1	90.7	96.5	65.1	88.9	96.7	65.1	89.0	101.1	65.1	92.3	101.1	65.1	92.3
0920	Plaster & Gypsum Board	101.1	134.0	122.7	126.4	63.9	85.4	82.9	63.9	70.5	82.9	63.9	70.5	111.9	63.9	80.4	111.9	63.9	80.4
0950, 0980	Ceilings & Acoustic Treatment	89.1	134.0	117.3	111.4	63.9	81.6	112.0	63.9	81.9	112.0	63.9	81.9	108.5	63.9	80.6	108.5	63.9	80.6
0960	Flooring	96.9	158.5	114.9	88.4	65.0	81.6	96.5	65.0	87.3	96.5	65.0	87.3	89.8	65.0	82.6	89.8	65.0	82.6
0970, 0990	Wall Finishes & Painting/Coating	77.1	147.2	119.1	92.0	51.1	67.5	89.3	51.1	66.4	89.3	51.1	66.4	86.4	51.1	65.3	86.4	51.1	65.3
09	FINISHES	92.0	140.1	118.0	99.0	63.3	79.6	97.5	63.3	78.9	96.2	63.3	78.4	96.6	63.3	78.5	97.8	63.3	79.1
COVERS	DIVS. 10 - 14, 25, 28, 41, 43, 44, 46	100.0	113.2	103.1	100.0	86.4	96.8	100.0	86.4	96.8	100.0	86.4	96.8	100.0	86.4	96.8	100.0	86.4	96.8
21, 22, 23	FIRE SUPPRESSION, PLUMBING & HVAC	100.4	124.7	110.2	101.4	68.1	87.9	98.7	68.1	86.3	98.7	67.7	86.2	101.2	68.1	87.9	98.9	68.1	86.5
26, 27, 3370	ELECTRICAL, COMMUNICATIONS & UTIL.	92.2	138.0	114.9	86.6	69.5	78.1	89.7	69.5	79.7	87.8	69.5	78.7	84.9	69.5	77.3	84.3	69.5	77.0
MF2018	WEIGHTED AVERAGE	97.2	126.6	109.9	98.9	71.6	87.1	101.4	71.6	88.5	99.1	71.5	87.2	99.3	71.6	87.3	99.2	71.6	87.3

NEW MEXICO

DIVISION		LAS CRUCES 880 MAT.	INST.	TOTAL	LAS VEGAS 877 MAT.	INST.	TOTAL	ROSWELL 882 MAT.	INST.	TOTAL	SANTA FE 875 MAT.	INST.	TOTAL	SOCORRO 878 MAT.	INST.	TOTAL	TRUTH/CONSEQUENCES 879 MAT.	INST.	TOTAL
015433	CONTRACTOR EQUIPMENT		81.1	81.1		105.2	105.2		105.2	105.2		109.7	109.7		105.2	105.2		81.1	81.1
0241, 31 - 34	SITE & INFRASTRUCTURE, DEMOLITION	95.6	74.7	81.2	93.4	94.4	94.1	96.5	94.4	95.0	95.6	102.8	100.5	90.0	94.4	93.0	108.8	74.7	85.4
0310	Concrete Forming & Accessories	92.5	64.0	68.3	103.2	65.0	70.7	95.9	65.0	69.6	101.9	65.1	70.6	103.2	65.0	70.7	100.9	64.0	69.5
0320	Concrete Reinforcing	106.0	69.3	88.3	102.6	69.5	86.6	110.4	69.5	90.7	102.1	69.5	86.4	104.5	69.5	87.6	98.0	69.4	84.2
0330	Cast-in-Place Concrete	88.2	62.1	78.3	88.7	68.9	81.2	93.3	68.9	84.1	94.2	70.0	85.1	86.9	68.9	80.1	95.3	62.1	82.7
03	CONCRETE	82.0	65.2	74.5	95.2	68.5	83.2	105.2	68.5	88.7	99.1	68.8	85.5	94.1	68.5	82.6	87.3	65.2	77.4
04	MASONRY	103.0	60.7	77.2	102.1	62.4	77.9	118.1	62.4	84.2	98.8	62.5	76.7	102.0	62.4	77.9	98.6	62.1	76.4
05	METALS	103.1	82.1	96.5	102.7	89.8	98.7	105.6	89.8	100.7	101.0	87.7	96.8	103.0	89.8	98.9	102.5	82.1	96.2
06	WOOD, PLASTICS & COMPOSITES	79.9	63.9	71.5	104.6	64.9	83.8	89.5	64.9	76.6	101.3	64.9	82.3	104.6	64.9	83.8	96.9	63.9	79.6
07	THERMAL & MOISTURE PROTECTION	91.8	67.2	81.2	98.6	72.0	87.2	102.7	72.0	89.5	100.5	73.4	88.8	98.6	72.0	87.2	89.2	67.6	79.9
08	OPENINGS	92.1	64.6	85.4	97.7	65.1	89.8	96.5	65.1	88.8	99.1	65.2	90.9	97.6	65.1	89.7	91.1	64.6	84.6
0920	Plaster & Gypsum Board	82.0	63.9	70.1	111.9	63.9	80.4	82.9	63.9	70.5	125.2	63.9	85.0	111.9	63.9	80.4	113.4	63.9	80.9
0950, 0980	Ceilings & Acoustic Treatment	96.8	63.9	76.2	108.5	63.9	80.6	112.0	63.9	81.9	109.4	63.9	80.9	108.5	63.9	80.6	98.5	63.9	76.9
0960	Flooring	127.1	65.0	109.0	89.8	65.0	82.6	96.5	65.0	87.3	98.9	65.0	89.0	89.8	65.0	82.6	118.3	65.0	102.8
0970, 0990	Wall Finishes & Painting/Coating	78.5	51.1	62.1	86.4	51.1	65.3	89.3	51.1	66.4	96.7	51.1	69.4	86.4	51.1	65.3	79.1	51.1	62.3
09	FINISHES	104.6	62.5	81.8	96.4	63.3	78.5	96.3	63.3	78.4	103.8	63.3	81.9	96.3	63.3	78.4	107.6	62.5	83.2
COVERS	DIVS. 10 - 14, 25, 28, 41, 43, 44, 46	100.0	84.1	96.3	100.0	86.4	96.8	100.0	86.4	96.8	100.0	86.5	96.8	100.0	86.4	96.8	100.0	84.1	96.2
21, 22, 23	FIRE SUPPRESSION, PLUMBING & HVAC	101.7	67.8	88.0	98.9	68.1	86.5	100.9	68.1	87.7	101.2	68.1	87.8	98.9	68.1	86.5	98.8	67.8	86.3
26, 27, 3370	ELECTRICAL, COMMUNICATIONS & UTIL.	90.0	80.7	85.4	86.2	69.5	77.9	88.9	69.5	79.3	100.7	69.4	85.2	84.7	69.5	77.2	88.1	69.4	78.9
MF2018	WEIGHTED AVERAGE	96.8	70.0	85.2	97.4	71.6	86.3	100.7	71.6	88.1	100.5	72.2	88.3	97.1	71.6	86.1	96.8	68.6	84.6

City Cost Indexes - V2

NEW MEXICO / NEW YORK

DIVISION		TUCUMCARI 884 MAT.	INST.	TOTAL	ALBANY 120-122 MAT.	INST.	TOTAL	BINGHAMTON 137-139 MAT.	INST.	TOTAL	BRONX 104 MAT.	INST.	TOTAL	BROOKLYN 112 MAT.	INST.	TOTAL	BUFFALO 140-142 MAT.	INST.	TOTAL
015433	CONTRACTOR EQUIPMENT		105.2	105.2		117.0	117.0		115.8	115.8		102.9	102.9		107.5	107.5		101.3	101.3
0241, 31-34	SITE & INFRASTRUCTURE, DEMOLITION	93.5	94.4	94.1	78.5	107.4	98.4	94.1	87.0	89.2	98.8	107.1	104.5	117.8	117.5	117.6	99.2	103.8	102.4
0310	Concrete Forming & Accessories	95.9	64.9	69.5	99.3	104.4	103.7	100.5	92.2	93.4	94.9	183.3	170.1	106.9	183.3	171.9	103.4	115.3	113.6
0320	Concrete Reinforcing	108.2	69.5	89.5	99.2	112.1	105.4	96.5	106.7	101.4	95.6	173.5	133.2	97.9	231.2	162.2	95.5	113.4	104.2
0330	Cast-in-Place Concrete	93.3	68.8	84.1	76.1	115.5	91.0	105.2	103.6	104.6	82.3	167.1	114.3	107.2	165.5	129.2	117.3	122.0	119.1
03	CONCRETE	103.6	68.4	87.8	89.6	110.6	99.0	94.1	101.0	97.2	86.3	175.5	126.3	105.6	183.9	140.7	107.1	116.9	111.5
04	MASONRY	119.1	62.4	84.5	97.2	116.3	108.9	109.7	103.6	106.0	88.5	184.6	147.1	119.1	184.6	159.1	110.8	123.5	118.5
05	METALS	104.3	89.6	99.7	103.1	123.4	109.4	96.0	135.0	108.2	92.9	173.2	117.9	103.9	173.4	125.6	93.9	107.2	98.0
06	WOOD, PLASTICS & COMPOSITES	89.5	64.9	76.6	97.4	100.6	99.1	102.1	89.0	95.2	93.5	182.4	140.0	105.3	182.1	145.5	100.9	114.0	107.8
07	THERMAL & MOISTURE PROTECTION	102.5	72.0	89.4	107.7	109.7	108.6	110.1	93.1	102.8	102.8	163.8	129.0	110.3	163.2	133.0	104.0	111.2	107.1
08	OPENINGS	96.4	65.1	88.8	96.4	101.9	97.7	90.5	94.1	91.4	92.5	191.0	116.5	87.9	190.9	113.0	100.0	109.5	102.3
0920	Plaster & Gypsum Board	82.9	63.9	70.5	105.1	100.4	102.0	113.6	88.6	97.2	92.2	184.6	152.8	106.0	184.6	157.6	119.9	114.3	116.2
0950, 0980	Ceilings & Acoustic Treatment	112.0	63.9	81.9	97.3	100.4	99.3	102.4	88.6	93.8	80.3	184.6	145.7	96.1	184.6	151.6	102.1	114.3	109.7
0960	Flooring	96.5	65.0	87.3	92.6	107.6	97.0	103.0	97.8	101.5	99.4	182.7	123.7	110.0	182.7	131.2	94.1	119.1	101.4
0970, 0990	Wall Finishes & Painting/Coating	89.3	51.1	66.4	93.9	99.4	97.2	84.5	101.3	94.5	104.7	164.2	140.4	109.6	164.2	142.3	93.8	114.4	106.1
09	FINISHES	96.1	63.3	78.3	94.8	103.9	99.7	96.7	93.4	94.9	93.8	181.8	141.4	107.8	181.6	147.7	101.6	116.4	109.6
COVERS	DIVS. 10 - 14, 25, 28, 41, 43, 44, 46	100.0	86.4	96.8	100.0	100.9	100.2	100.0	98.0	99.5	100.0	137.9	108.9	100.0	137.2	108.8	100.0	106.3	101.5
21, 22, 23	FIRE SUPPRESSION, PLUMBING & HVAC	98.7	67.7	86.2	101.2	108.0	103.9	101.6	98.6	100.4	100.4	171.5	129.1	100.5	171.5	129.1	100.9	102.0	101.3
26, 27, 3370	ELECTRICAL, COMMUNICATIONS & UTIL.	89.7	69.5	79.7	101.8	107.2	104.5	100.2	97.3	98.7	93.6	183.2	137.9	100.2	183.2	141.3	99.9	104.3	102.1
MF2018	WEIGHTED AVERAGE	99.7	71.5	87.6	98.3	109.4	103.1	98.5	100.5	99.3	94.8	171.2	127.8	102.7	173.1	133.1	100.9	109.8	104.8

NEW YORK

DIVISION		ELMIRA 148-149 MAT.	INST.	TOTAL	FAR ROCKAWAY 116 MAT.	INST.	TOTAL	FLUSHING 113 MAT.	INST.	TOTAL	GLENS FALLS 128 MAT.	INST.	TOTAL	HICKSVILLE 115,117,118 MAT.	INST.	TOTAL	JAMAICA 114 MAT.	INST.	TOTAL
015433	CONTRACTOR EQUIPMENT		118.7	118.7		107.5	107.5		107.5	107.5		110.8	110.8		107.5	107.5		107.5	107.5
0241, 31-34	SITE & INFRASTRUCTURE, DEMOLITION	97.9	87.0	90.4	120.6	117.5	118.5	120.6	117.5	118.5	70.0	96.4	88.1	111.0	116.0	114.4	114.9	117.5	116.7
0310	Concrete Forming & Accessories	83.6	94.4	92.8	92.0	183.3	169.7	96.2	183.3	170.3	82.5	96.2	94.2	88.1	153.9	144.1	96.2	183.3	170.3
0320	Concrete Reinforcing	97.6	105.5	101.4	97.9	231.2	162.2	99.6	231.2	163.1	95.2	107.4	101.1	97.9	169.7	132.5	97.9	231.2	162.2
0330	Cast-in-Place Concrete	103.2	102.9	103.1	116.0	165.5	134.7	116.0	165.5	134.7	73.3	107.2	86.1	98.4	155.6	120.0	107.2	165.5	129.2
03	CONCRETE	92.8	101.7	96.8	111.7	183.9	144.1	112.2	183.9	144.4	78.7	103.3	89.7	97.4	156.5	123.9	104.9	183.9	140.4
04	MASONRY	106.2	103.1	104.3	124.2	184.6	161.0	117.4	184.6	158.4	102.3	107.9	105.7	114.2	168.6	147.4	122.3	184.6	160.3
05	METALS	92.4	136.3	106.1	104.0	173.4	125.6	104.0	173.4	125.6	96.4	122.5	104.6	105.5	147.9	118.7	104.0	173.4	125.6
06	WOOD, PLASTICS & COMPOSITES	81.9	92.7	87.6	86.5	182.1	136.5	91.8	182.1	139.0	83.3	92.9	88.3	82.6	151.9	118.9	91.8	182.1	139.0
07	THERMAL & MOISTURE PROTECTION	110.8	92.9	103.1	110.2	163.4	133.0	110.2	163.4	133.1	101.0	102.3	101.5	109.8	155.1	129.2	110.0	163.4	132.9
08	OPENINGS	96.8	95.8	96.6	86.7	190.9	112.1	86.7	190.9	112.1	90.2	95.4	91.5	87.1	159.2	104.7	86.7	190.9	112.1
0920	Plaster & Gypsum Board	110.2	92.6	98.6	92.7	184.6	153.0	95.6	184.6	154.0	90.9	92.8	92.1	92.5	153.6	132.6	95.6	184.6	154.0
0950, 0980	Ceilings & Acoustic Treatment	109.1	92.6	98.7	83.4	184.6	146.8	83.4	184.6	146.8	88.5	92.8	91.2	82.6	153.6	127.1	83.4	184.6	146.8
0960	Flooring	86.8	97.8	90.0	104.5	182.7	127.3	106.1	182.7	128.4	83.8	102.8	89.4	103.4	180.3	125.9	106.1	182.7	128.4
0970, 0990	Wall Finishes & Painting/Coating	90.7	88.9	89.6	109.6	164.2	142.3	109.6	164.2	142.3	85.3	99.4	93.8	109.6	164.2	142.3	109.6	164.2	142.3
09	FINISHES	97.2	94.1	95.5	101.7	181.6	144.9	102.5	181.6	145.3	86.6	97.1	92.3	100.3	159.5	132.3	102.1	181.6	145.1
COVERS	DIVS. 10 - 14, 25, 28, 41, 43, 44, 46	100.0	98.2	99.6	100.0	137.2	108.8	100.0	137.2	108.8	100.0	91.9	98.1	100.0	131.3	107.4	100.0	137.2	108.8
21, 22, 23	FIRE SUPPRESSION, PLUMBING & HVAC	97.3	94.4	96.1	96.6	171.5	126.8	96.6	171.5	126.8	97.5	104.1	100.1	100.5	155.3	122.6	96.6	171.5	126.8
26, 27, 3370	ELECTRICAL, COMMUNICATIONS & UTIL.	97.6	100.3	98.9	106.4	183.2	144.4	106.4	183.2	144.4	95.8	101.1	98.4	99.7	145.8	122.5	98.9	183.2	140.6
MF2018	WEIGHTED AVERAGE	96.9	100.4	98.4	102.7	173.1	133.1	102.5	173.1	133.0	92.8	103.2	97.3	100.5	151.7	122.6	100.9	173.1	132.1

NEW YORK

DIVISION		JAMESTOWN 147 MAT.	INST.	TOTAL	KINGSTON 124 MAT.	INST.	TOTAL	LONG ISLAND CITY 111 MAT.	INST.	TOTAL	MONTICELLO 127 MAT.	INST.	TOTAL	MOUNT VERNON 105 MAT.	INST.	TOTAL	NEW ROCHELLE 108 MAT.	INST.	TOTAL
015433	CONTRACTOR EQUIPMENT		89.2	89.2		107.5	107.5		107.5	107.5		107.5	107.5		102.9	102.9		102.9	102.9
0241, 31-34	SITE & INFRASTRUCTURE, DEMOLITION	99.4	87.2	91.0	137.1	112.4	120.1	119.0	117.5	118.0	132.6	112.4	118.7	104.6	102.9	103.5	103.8	102.9	103.2
0310	Concrete Forming & Accessories	83.7	86.7	86.2	83.9	126.0	119.7	101.0	183.3	171.0	92.0	126.0	120.9	84.9	134.4	127.0	99.9	131.0	126.4
0320	Concrete Reinforcing	97.8	105.9	101.7	95.6	152.8	123.2	97.9	231.2	162.2	94.8	152.8	122.8	94.5	172.2	132.0	94.6	172.1	132.0
0330	Cast-in-Place Concrete	107.1	101.1	104.8	99.0	140.7	114.7	110.6	165.5	131.3	92.6	140.8	110.8	91.9	145.2	112.1	91.9	144.9	111.9
03	CONCRETE	95.9	95.3	95.6	98.8	135.6	115.3	108.1	183.9	142.1	94.1	135.6	112.7	95.0	145.5	117.7	94.5	143.8	116.6
04	MASONRY	116.0	100.3	106.4	118.4	148.9	137.0	115.6	184.6	157.6	110.9	148.9	134.1	94.5	151.4	129.2	94.5	151.4	129.2
05	METALS	90.0	100.3	93.2	104.8	133.9	113.9	103.9	173.4	125.6	104.8	134.0	113.9	92.7	168.0	116.1	92.9	167.2	116.1
06	WOOD, PLASTICS & COMPOSITES	80.7	83.3	82.1	84.7	119.9	103.1	98.6	182.1	142.3	92.9	119.8	107.0	83.3	128.0	106.7	100.9	124.4	113.2
07	THERMAL & MOISTURE PROTECTION	110.3	91.1	102.0	123.0	139.6	130.2	110.2	163.4	133.0	122.7	139.6	130.0	103.6	141.4	119.8	103.8	139.0	118.9
08	OPENINGS	96.7	89.6	95.0	92.4	137.8	103.5	86.7	190.9	112.1	88.5	137.7	100.5	92.5	162.7	109.6	92.5	160.7	109.2
0920	Plaster & Gypsum Board	98.3	82.9	88.2	91.3	120.7	110.6	100.9	184.6	155.8	91.9	120.7	110.8	88.3	128.7	114.8	100.4	125.0	116.5
0950, 0980	Ceilings & Acoustic Treatment	104.6	82.9	91.0	75.7	120.7	103.9	83.4	184.6	146.8	75.7	120.7	103.9	78.4	128.7	109.9	78.4	125.0	107.6
0960	Flooring	89.2	97.8	91.7	100.9	153.2	116.2	107.9	182.7	129.7	103.4	153.2	117.9	90.8	171.4	114.3	98.9	156.5	115.7
0970, 0990	Wall Finishes & Painting/Coating	92.2	95.8	94.4	111.2	124.7	119.3	109.6	164.2	142.3	111.2	117.5	115.0	103.2	164.2	139.8	103.2	164.2	139.8
09	FINISHES	95.3	88.9	91.8	98.5	130.1	115.6	103.6	181.6	145.8	98.9	129.2	115.3	90.9	143.1	119.1	94.8	137.9	118.1
COVERS	DIVS. 10 - 14, 25, 28, 41, 43, 44, 46	100.0	91.9	98.1	100.0	115.1	103.6	100.0	137.2	108.8	100.0	115.1	103.6	100.0	123.2	105.5	100.0	107.8	101.8
21, 22, 23	FIRE SUPPRESSION, PLUMBING & HVAC	97.1	89.2	93.9	97.4	128.3	109.8	100.5	171.5	129.1	97.4	132.3	111.5	96.8	146.3	116.7	96.8	146.0	116.6
26, 27, 3370	ELECTRICAL, COMMUNICATIONS & UTIL.	96.6	91.4	94.0	97.2	112.0	104.5	99.3	183.2	140.8	97.2	112.0	104.5	91.7	164.2	127.6	91.7	136.4	113.9
MF2018	WEIGHTED AVERAGE	97.1	92.3	95.0	101.2	128.8	113.1	102.2	173.1	132.8	99.8	129.5	112.7	94.9	147.0	117.4	95.3	141.4	115.2

For customer support on your Concrete & Masonry Costs with RSMeans Data, call 800.448.8182.

NEW YORK

DIVISION		NEW YORK 100 - 102			NIAGARA FALLS 143			PLATTSBURGH 129			POUGHKEEPSIE 125 - 126			QUEENS 110			RIVERHEAD 119		
		MAT.	INST.	TOTAL	MAT.	INST.	TOTAL	MAT.	INST.	TOTAL	MAT.	INST.	TOTAL	MAT.	INST.	TOTAL	MAT.	INST.	TOTAL
015433	CONTRACTOR EQUIPMENT		105.6	105.6		89.2	89.2		92.5	92.5		107.5	107.5		107.5	107.5		107.5	107.5
0241, 31 - 34	SITE & INFRASTRUCTURE, DEMOLITION	106.6	112.7	110.8	101.5	88.1	92.3	107.7	93.5	98.0	133.6	111.4	118.3	114.2	117.5	116.5	112.0	115.7	114.6
0310	Concrete Forming & Accessories	103.6	186.4	174.0	83.6	108.9	105.1	88.0	88.0	88.0	83.9	161.5	149.9	88.3	183.3	169.1	93.3	154.5	145.4
0320	Concrete Reinforcing	107.3	175.3	140.1	96.5	106.7	101.4	99.5	106.7	103.0	95.6	152.4	123.0	99.6	231.2	163.1	99.7	241.9	168.3
0330	Cast-in-Place Concrete	95.3	168.6	123.0	110.8	119.4	114.0	89.6	96.6	92.2	95.8	130.2	108.8	101.8	165.5	125.9	100.1	155.7	121.1
03	CONCRETE	99.3	177.7	134.5	98.2	111.8	104.3	91.1	94.2	92.5	96.2	147.9	119.4	100.6	183.9	138.0	98.0	169.2	130.0
04	MASONRY	99.0	186.2	152.2	123.1	119.9	121.1	96.8	93.3	94.7	110.0	133.7	124.5	109.4	184.6	155.3	119.6	168.6	149.5
05	METALS	105.2	174.6	126.8	92.5	100.8	95.0	100.7	97.6	99.7	104.9	133.6	113.8	103.9	173.4	125.6	106.0	174.2	127.3
06	WOOD, PLASTICS & COMPOSITES	99.3	186.1	144.7	80.6	104.6	93.2	90.5	85.2	87.7	84.7	173.3	131.1	82.8	182.1	134.7	88.5	151.9	121.7
07	THERMAL & MOISTURE PROTECTION	105.0	168.7	132.3	110.4	105.1	108.1	116.3	92.8	106.2	123.0	138.8	129.8	109.7	163.4	132.7	111.3	154.4	129.8
08	OPENINGS	96.2	197.3	120.8	96.7	100.8	97.7	97.6	91.2	96.0	92.5	164.3	110.0	86.7	190.9	112.1	87.1	177.0	109.0
0920	Plaster & Gypsum Board	99.4	188.3	157.7	98.3	104.8	102.6	111.6	84.5	93.8	91.3	175.6	146.6	92.5	184.6	152.9	93.8	153.6	133.1
0950, 0980	Ceilings & Acoustic Treatment	98.1	188.3	154.6	104.6	104.8	104.7	104.9	84.5	92.1	75.7	175.6	138.1	83.4	184.6	146.8	83.3	153.6	127.4
0960	Flooring	99.8	184.4	124.5	89.2	106.9	94.4	106.3	100.2	104.5	100.9	151.4	115.6	103.4	182.7	126.6	104.5	180.3	126.6
0970, 0990	Wall Finishes & Painting/Coating	104.0	169.0	143.0	92.2	106.1	100.6	106.7	90.6	97.1	111.2	117.5	115.0	109.6	164.2	142.3	109.6	164.2	142.3
09	FINISHES	99.6	185.1	145.9	95.4	108.5	102.5	98.2	89.8	93.7	98.3	157.8	130.5	100.6	181.6	144.4	100.9	159.5	132.6
COVERS	DIVS. 10 - 14, 25, 28, 41, 43, 44, 46	100.0	141.5	109.8	100.0	97.5	99.4	100.0	87.9	97.2	100.0	110.5	102.5	100.0	137.2	108.8	100.0	128.5	106.7
21, 22, 23	FIRE SUPPRESSION, PLUMBING & HVAC	99.4	174.6	129.7	97.1	100.2	98.4	97.4	95.6	96.6	97.4	114.6	104.3	100.5	171.5	129.1	100.8	157.4	123.6
26, 27, 3370	ELECTRICAL, COMMUNICATIONS & UTIL.	99.3	185.3	141.9	95.4	95.2	95.3	93.4	88.3	90.9	97.2	116.7	106.9	99.7	183.2	141.0	101.1	146.0	123.3
MF2018	WEIGHTED AVERAGE	100.3	174.2	132.2	98.0	103.4	100.3	97.7	92.8	95.6	100.3	131.7	113.9	100.5	173.1	131.8	101.3	156.9	125.3

NEW YORK

DIVISION		ROCHESTER 144 - 146			SCHENECTADY 123			STATEN ISLAND 103			SUFFERN 109			SYRACUSE 130 - 132			UTICA 133 - 135		
		MAT.	INST.	TOTAL	MAT.	INST.	TOTAL	MAT.	INST.	TOTAL	MAT.	INST.	TOTAL	MAT.	INST.	TOTAL	MAT.	INST.	TOTAL
015433	CONTRACTOR EQUIPMENT		117.0	117.0		110.8	110.8		102.9	102.9		102.9	102.9		110.8	110.8		110.8	110.8
0241, 31 - 34	SITE & INFRASTRUCTURE, DEMOLITION	89.1	104.7	99.8	79.9	97.2	91.8	109.3	107.1	107.8	100.9	100.7	100.8	92.9	96.5	95.4	71.8	96.0	88.4
0310	Concrete Forming & Accessories	103.4	98.9	99.6	100.9	104.1	103.6	84.3	183.5	168.7	93.5	135.6	129.4	99.6	93.3	94.2	100.5	92.8	93.9
0320	Concrete Reinforcing	96.8	105.7	101.1	94.2	112.0	102.8	95.6	205.9	148.8	94.6	144.8	118.8	97.5	106.8	102.0	97.5	105.7	101.5
0330	Cast-in-Place Concrete	99.2	107.0	102.1	88.5	113.5	97.9	91.9	167.1	120.3	89.1	133.7	106.0	97.7	107.2	101.3	89.3	106.6	95.9
03	CONCRETE	98.0	104.1	100.0	91.9	109.9	100.0	96.7	180.6	134.4	91.9	136.1	111.7	97.0	101.9	99.2	95.1	101.2	97.8
04	MASONRY	104.3	106.9	105.9	99.6	116.2	109.7	100.6	184.6	151.8	94.3	136.2	119.9	100.9	108.0	105.2	92.5	107.5	101.6
05	METALS	101.2	121.3	107.4	100.7	125.7	108.5	91.0	173.5	116.7	91.0	130.1	103.2	99.6	121.8	106.5	97.6	121.3	105.0
06	WOOD, PLASTICS & COMPOSITES	104.7	97.0	100.6	104.9	100.3	102.5	82.0	182.4	134.5	93.0	137.7	116.4	98.9	89.3	93.9	98.9	88.4	93.4
07	THERMAL & MOISTURE PROTECTION	117.5	102.1	110.9	102.6	108.5	105.2	103.1	163.8	129.2	103.6	135.8	117.4	105.3	99.4	102.8	94.9	99.1	96.7
08	OPENINGS	100.6	98.4	100.1	95.3	101.7	96.9	92.5	191.0	116.5	92.5	145.7	105.5	92.5	92.4	92.5	95.1	91.7	94.2
0920	Plaster & Gypsum Board	106.4	96.8	100.1	102.1	100.4	101.0	88.5	184.6	151.5	91.4	138.7	122.4	102.4	89.1	93.7	102.4	88.1	93.0
0950, 0980	Ceilings & Acoustic Treatment	104.8	96.8	99.8	99.2	100.4	100.0	80.3	184.6	145.7	78.4	138.7	116.2	102.4	89.1	94.0	102.4	88.1	93.0
0960	Flooring	92.2	107.5	96.7	91.4	107.6	96.2	94.5	182.7	120.2	94.6	163.7	114.8	91.9	96.7	93.3	89.6	98.1	92.1
0970, 0990	Wall Finishes & Painting/Coating	93.7	99.7	97.3	85.3	99.4	93.8	104.7	164.2	140.4	103.2	120.9	113.8	87.1	102.6	96.4	81.0	102.6	93.9
09	FINISHES	98.8	100.4	99.7	93.6	103.7	99.0	92.8	181.8	141.0	92.1	140.1	118.1	95.3	94.1	94.6	93.6	93.9	93.8
COVERS	DIVS. 10 - 14, 25, 28, 41, 43, 44, 46	100.0	99.7	99.9	100.0	100.1	100.0	100.0	137.9	108.9	100.0	108.9	102.1	100.0	98.5	99.7	100.0	98.0	99.5
21, 22, 23	FIRE SUPPRESSION, PLUMBING & HVAC	100.9	91.1	96.9	101.2	107.9	103.9	100.4	171.5	129.1	96.8	117.8	105.2	101.2	95.8	99.0	101.2	101.7	101.4
26, 27, 3370	ELECTRICAL, COMMUNICATIONS & UTIL.	102.6	96.5	99.6	99.8	107.2	103.5	93.6	183.2	137.9	97.3	110.4	103.8	100.1	106.8	103.4	98.0	106.8	102.4
MF2018	WEIGHTED AVERAGE	100.8	101.2	101.0	97.9	108.6	102.5	96.5	171.9	129.1	94.9	125.4	108.1	98.7	101.6	99.9	96.8	102.5	99.3

DIVISION		NEW YORK WATERTOWN 136			WHITE PLAINS 106			YONKERS 107			NORTH CAROLINA ASHEVILLE 287 - 288			CHARLOTTE 281 - 282			DURHAM 277		
		MAT.	INST.	TOTAL	MAT.	INST.	TOTAL	MAT.	INST.	TOTAL	MAT.	INST.	TOTAL	MAT.	INST.	TOTAL	MAT.	INST.	TOTAL
015433	CONTRACTOR EQUIPMENT		110.8	110.8		102.9	102.9		102.9	102.9		97.0	97.0		100.2	100.2		102.2	102.2
0241, 31 - 34	SITE & INFRASTRUCTURE, DEMOLITION	79.3	96.3	90.9	99.0	103.0	101.7	106.6	103.0	104.1	96.5	76.0	82.4	99.5	83.5	88.5	101.8	84.5	89.9
0310	Concrete Forming & Accessories	83.9	96.0	94.2	98.8	145.4	138.4	99.0	145.5	138.6	93.8	59.6	64.7	98.3	64.7	69.8	101.5	59.6	65.9
0320	Concrete Reinforcing	98.2	106.8	102.3	94.6	172.3	132.1	98.4	172.3	134.0	93.6	62.9	78.8	96.6	68.3	83.0	95.4	63.6	80.1
0330	Cast-in-Place Concrete	104.1	109.4	106.1	81.6	145.4	105.7	91.3	145.4	111.8	102.3	70.1	90.1	104.4	72.6	92.4	115.8	70.1	98.5
03	CONCRETE	107.5	103.8	105.8	86.0	150.5	114.9	94.6	150.6	119.7	89.6	65.6	78.8	92.6	69.5	82.3	99.0	65.7	84.1
04	MASONRY	93.7	111.5	104.6	93.9	151.4	129.0	96.6	151.4	130.0	85.3	62.6	71.4	87.3	65.2	73.8	86.9	62.6	72.0
05	METALS	97.7	121.5	105.1	92.5	168.3	116.2	101.3	168.4	122.3	101.0	89.1	97.3	101.9	89.1	97.9	117.4	89.3	108.6
06	WOOD, PLASTICS & COMPOSITES	78.9	91.9	85.7	98.9	142.5	121.7	99.8	142.5	121.7	91.4	57.0	73.4	92.6	63.2	77.3	96.9	57.0	76.1
07	THERMAL & MOISTURE PROTECTION	95.2	101.2	97.8	103.4	144.0	120.8	103.8	144.6	121.3	98.7	61.5	82.7	91.9	66.2	80.9	101.7	61.4	84.4
08	OPENINGS	95.1	95.4	95.1	92.5	170.7	111.6	96.2	170.7	114.3	89.3	57.6	81.6	99.2	61.5	90.1	95.9	57.8	86.6
0920	Plaster & Gypsum Board	93.1	91.7	92.2	94.8	143.6	126.8	99.7	143.6	128.5	108.8	55.9	74.1	100.2	62.3	75.3	95.8	55.9	69.6
0950, 0980	Ceilings & Acoustic Treatment	102.4	91.7	95.7	78.4	143.6	119.3	98.1	143.6	126.6	89.6	55.9	68.5	92.1	62.3	73.4	98.1	55.9	71.6
0960	Flooring	82.6	98.1	87.1	97.1	180.3	121.4	96.5	182.7	121.7	98.2	63.7	88.1	97.2	68.4	88.8	104.5	63.7	92.6
0970, 0990	Wall Finishes & Painting/Coating	81.0	101.0	93.0	103.2	164.2	139.8	103.2	164.2	139.8	98.6	54.4	72.1	95.0	66.8	78.1	100.1	54.4	72.7
09	FINISHES	90.9	96.4	93.9	92.9	153.5	125.7	98.7	153.9	128.6	96.0	59.3	76.2	95.1	65.3	79.0	96.7	59.3	76.5
COVERS	DIVS. 10 - 14, 25, 28, 41, 43, 44, 46	100.0	99.6	99.9	100.0	127.5	106.5	100.0	134.8	108.2	100.0	84.2	96.3	100.0	85.0	96.5	100.0	84.2	96.3
21, 22, 23	FIRE SUPPRESSION, PLUMBING & HVAC	101.2	89.9	96.6	101.0	146.4	119.3	101.0	146.4	119.3	101.8	58.8	84.5	101.0	63.4	85.8	101.8	58.8	84.5
26, 27, 3370	ELECTRICAL, COMMUNICATIONS & UTIL.	100.0	88.7	94.4	91.7	164.2	127.6	97.4	164.2	130.5	99.9	55.7	78.0	101.6	84.2	93.0	95.4	55.2	75.5
MF2018	WEIGHTED AVERAGE	98.4	98.9	98.6	94.9	149.8	118.6	99.1	150.1	121.1	97.0	64.6	83.0	98.4	72.1	87.0	101.2	65.2	85.6

NORTH CAROLINA

DIVISION		ELIZABETH CITY 279 MAT.	INST.	TOTAL	FAYETTEVILLE 283 MAT.	INST.	TOTAL	GASTONIA 280 MAT.	INST.	TOTAL	GREENSBORO 270,272-274 MAT.	INST.	TOTAL	HICKORY 286 MAT.	INST.	TOTAL	KINSTON 285 MAT.	INST.	TOTAL
015433	CONTRACTOR EQUIPMENT		106.2	106.2		102.2	102.2		97.0	97.0		102.2	102.2		102.2	102.2		102.2	102.2
0241, 31 - 34	SITE & INFRASTRUCTURE, DEMOLITION	105.9	85.8	92.1	95.7	84.2	87.8	96.2	76.0	82.3	101.6	84.5	89.8	94.9	84.2	87.6	93.7	84.2	87.2
0310	Concrete Forming & Accessories	84.6	61.5	64.9	93.0	58.1	63.3	101.1	59.8	66.0	101.3	59.6	65.8	89.0	59.7	64.1	84.9	58.1	62.1
0320	Concrete Reinforcing	93.4	66.5	80.4	97.2	63.6	81.0	93.9	63.7	79.3	94.2	63.6	79.5	93.6	63.6	79.1	93.1	63.6	78.9
0330	Cast-in-Place Concrete	116.0	70.1	98.7	107.3	68.4	92.6	100.0	70.8	89.0	114.9	70.7	98.2	102.3	69.9	90.1	98.8	67.8	87.0
03	CONCRETE	98.8	67.0	84.5	91.6	64.4	79.4	88.5	66.0	78.4	98.4	65.9	83.8	89.3	65.6	78.7	86.5	64.2	76.5
04	MASONRY	99.1	58.5	74.4	88.2	58.5	70.1	89.4	62.6	73.0	83.7	62.6	70.8	75.0	62.5	67.4	81.5	58.5	67.5
05	METALS	103.3	91.4	99.6	121.6	89.4	111.5	101.7	89.5	97.9	109.7	89.4	103.4	101.1	88.7	97.2	99.9	89.3	96.6
06	WOOD, PLASTICS & COMPOSITES	77.7	61.2	69.1	89.8	57.0	72.7	100.8	57.0	77.9	96.6	57.0	75.9	84.9	57.0	70.3	81.0	57.0	68.4
07	THERMAL & MOISTURE PROTECTION	100.7	59.8	83.2	99.6	61.4	81.5	99.0	61.4	82.9	101.3	61.4	84.2	99.1	61.4	82.9	98.9	59.6	82.0
08	OPENINGS	93.3	60.7	85.4	89.4	57.8	81.7	92.5	57.8	84.1	95.9	57.8	86.6	89.4	57.8	81.7	89.4	57.8	81.7
0920	Plaster & Gypsum Board	89.7	59.6	70.0	112.1	55.9	75.2	114.7	55.9	76.1	96.8	55.9	70.0	108.8	55.9	74.1	108.7	55.9	74.1
0950, 0980	Ceilings & Acoustic Treatment	98.1	59.6	74.0	92.8	55.9	69.7	94.7	55.9	70.4	98.1	55.9	71.6	89.6	55.9	68.5	94.7	55.9	70.4
0960	Flooring	96.3	63.7	86.8	98.3	63.7	88.2	101.6	63.7	90.5	104.5	63.7	92.6	98.1	63.7	88.1	95.2	63.7	86.0
0970, 0990	Wall Finishes & Painting/Coating	100.1	54.4	72.7	98.6	54.4	72.1	98.6	54.4	72.1	100.1	54.4	72.7	98.6	54.4	72.1	98.6	54.4	72.1
09	FINISHES	93.8	60.9	76.0	97.2	58.3	76.2	98.9	59.3	77.5	96.8	59.3	76.6	96.2	59.3	76.2	96.3	58.3	75.8
COVERS	DIVS. 10 - 14, 25, 28, 41, 43, 44, 46	100.0	85.3	96.5	100.0	82.9	96.0	100.0	84.3	96.3	100.0	84.2	96.3	100.0	84.3	96.3	100.0	82.8	96.0
21, 22, 23	FIRE SUPPRESSION, PLUMBING & HVAC	97.9	55.8	80.9	101.6	56.7	83.5	101.8	57.8	84.1	101.7	58.8	84.4	98.0	57.8	81.8	98.0	55.6	80.9
26, 27, 3370	ELECTRICAL, COMMUNICATIONS & UTIL.	95.1	66.6	81.0	99.9	55.2	77.8	99.4	84.2	91.9	94.7	55.7	75.4	97.9	84.2	91.1	97.7	53.7	76.0
MF2018	WEIGHTED AVERAGE	98.1	66.6	84.5	100.5	64.0	84.7	97.7	68.5	85.1	99.7	65.3	84.8	95.2	69.0	83.9	94.9	63.5	81.3

NORTH CAROLINA / NORTH DAKOTA

DIVISION		MURPHY 289 MAT.	INST.	TOTAL	RALEIGH 275 - 276 MAT.	INST.	TOTAL	ROCKY MOUNT 278 MAT.	INST.	TOTAL	WILMINGTON 284 MAT.	INST.	TOTAL	WINSTON-SALEM 271 MAT.	INST.	TOTAL	BISMARCK 585 MAT.	INST.	TOTAL
015433	CONTRACTOR EQUIPMENT		97.0	97.0		106.9	106.9		102.2	102.2		97.0	97.0		102.2	102.2		100.9	100.9
0241, 31 - 34	SITE & INFRASTRUCTURE, DEMOLITION	97.1	75.7	82.4	101.1	93.2	95.7	103.8	84.1	90.3	97.5	75.7	82.5	102.0	84.5	90.0	101.0	99.3	99.8
0310	Concrete Forming & Accessories	101.9	57.9	64.5	99.4	59.3	65.3	92.2	59.0	64.0	95.3	58.1	63.6	103.1	59.6	66.1	107.3	78.9	83.1
0320	Concrete Reinforcing	93.1	61.3	77.8	95.7	63.6	80.2	93.4	63.6	79.0	94.3	63.6	79.5	94.2	63.6	79.5	94.4	100.2	97.2
0330	Cast-in-Place Concrete	105.9	67.7	91.5	119.2	70.7	100.0	113.5	69.1	96.7	101.9	67.8	89.0	117.8	70.1	99.8	99.4	89.1	95.5
03	CONCRETE	92.5	63.7	79.6	99.4	65.6	84.2	99.3	65.0	83.9	89.6	64.2	78.2	99.9	65.7	84.5	96.5	86.6	92.1
04	MASONRY	78.1	58.5	66.2	85.1	61.4	70.6	76.9	61.2	67.3	75.8	58.5	65.2	84.0	62.6	70.9	102.0	84.1	91.1
05	METALS	98.8	88.5	95.6	102.0	87.0	97.3	102.5	88.5	98.1	100.6	89.3	97.1	106.8	89.3	101.4	93.6	94.5	93.8
06	WOOD, PLASTICS & COMPOSITES	101.6	56.9	78.2	93.7	57.2	74.6	86.5	57.0	71.1	93.2	57.0	74.3	96.6	57.0	75.9	103.2	75.4	88.6
07	THERMAL & MOISTURE PROTECTION	99.0	59.6	82.1	94.6	62.0	80.6	101.3	60.9	83.9	98.7	59.6	81.9	101.3	61.4	84.2	107.9	86.4	98.6
08	OPENINGS	89.3	57.2	81.5	98.0	57.9	88.2	92.6	57.8	84.1	89.5	57.8	81.7	95.9	57.8	86.6	103.4	84.6	98.8
0920	Plaster & Gypsum Board	113.3	55.7	75.5	91.5	55.9	68.1	90.9	55.9	67.9	110.2	55.9	74.6	96.8	55.9	70.0	91.1	74.9	80.5
0950, 0980	Ceilings & Acoustic Treatment	89.6	55.7	68.4	97.4	55.9	71.4	94.9	55.9	70.5	92.8	55.9	69.4	98.1	55.9	71.6	113.7	74.9	89.4
0960	Flooring	101.9	63.7	90.8	96.2	63.7	86.7	99.9	63.7	89.4	98.9	63.7	88.6	104.5	63.7	92.6	88.9	54.9	79.0
0970, 0990	Wall Finishes & Painting/Coating	98.6	54.4	72.1	96.7	54.4	71.4	100.1	54.4	72.7	98.6	54.4	72.1	100.1	54.4	72.7	89.1	61.0	72.3
09	FINISHES	97.9	58.2	76.4	95.1	59.1	75.6	94.4	59.0	75.3	97.1	58.3	76.1	96.8	59.3	76.6	97.5	73.2	84.3
COVERS	DIVS. 10 - 14, 25, 28, 41, 43, 44, 46	100.0	82.8	96.0	100.0	84.1	96.3	100.0	83.8	96.2	100.0	82.8	96.0	100.0	84.2	96.3	100.0	94.3	98.6
21, 22, 23	FIRE SUPPRESSION, PLUMBING & HVAC	98.0	55.6	80.9	100.9	58.2	83.7	97.9	57.0	81.4	101.8	56.7	83.6	101.7	58.8	84.4	100.5	78.1	91.5
26, 27, 3370	ELECTRICAL, COMMUNICATIONS & UTIL.	100.6	53.7	77.4	100.4	54.2	77.5	96.7	55.2	76.2	100.2	53.7	77.2	94.7	55.7	75.4	98.8	76.2	87.6
MF2018	WEIGHTED AVERAGE	96.1	62.6	81.6	99.0	65.3	84.4	97.1	64.5	83.0	96.6	63.1	82.1	99.5	65.3	84.7	99.1	83.1	92.2

NORTH DAKOTA

DIVISION		DEVILS LAKE 583 MAT.	INST.	TOTAL	DICKINSON 586 MAT.	INST.	TOTAL	FARGO 580 - 581 MAT.	INST.	TOTAL	GRAND FORKS 582 MAT.	INST.	TOTAL	JAMESTOWN 584 MAT.	INST.	TOTAL	MINOT 587 MAT.	INST.	TOTAL
015433	CONTRACTOR EQUIPMENT		95.8	95.8		95.8	95.8		100.9	100.9		95.8	95.8		95.8	95.8		95.8	95.8
0241, 31 - 34	SITE & INFRASTRUCTURE, DEMOLITION	104.7	90.7	95.1	113.1	90.6	97.7	101.2	99.0	99.7	109.3	91.4	97.0	103.8	90.7	94.8	106.7	91.2	96.0
0310	Concrete Forming & Accessories	106.2	69.3	74.8	93.9	69.1	72.8	96.6	69.8	73.8	98.7	76.2	79.6	95.8	69.2	73.2	93.5	69.4	73.0
0320	Concrete Reinforcing	97.8	100.2	99.0	98.7	100.1	99.4	97.4	100.8	99.1	96.3	100.3	98.2	98.4	100.7	99.5	99.7	100.1	99.9
0330	Cast-in-Place Concrete	116.6	78.4	102.2	105.5	78.3	95.3	96.6	84.9	92.2	105.5	84.9	97.7	115.2	78.4	101.3	105.5	78.6	95.4
03	CONCRETE	101.9	78.6	91.5	101.2	78.5	91.0	98.7	81.1	90.8	98.7	84.1	92.1	100.4	78.7	90.7	97.2	78.8	88.9
04	MASONRY	115.6	82.4	95.4	116.1	79.8	93.9	98.2	91.2	94.0	109.1	85.6	94.7	128.5	91.1	105.7	109.0	79.1	90.8
05	METALS	92.6	95.3	93.5	92.5	94.7	93.2	98.0	95.1	97.1	92.6	96.1	93.7	92.9	95.9	93.7	92.9	95.0	93.7
06	WOOD, PLASTICS & COMPOSITES	98.8	65.2	81.2	84.4	65.2	74.4	91.8	65.5	78.0	89.6	72.6	80.7	86.6	65.2	75.4	84.0	65.2	74.2
07	THERMAL & MOISTURE PROTECTION	106.3	81.0	95.4	106.9	81.4	95.9	104.3	85.9	96.4	106.5	84.2	96.9	106.1	83.8	96.5	106.2	81.9	95.8
08	OPENINGS	100.6	79.0	95.3	100.6	79.0	95.3	100.8	79.1	95.5	99.2	83.0	95.3	100.6	79.0	95.3	99.4	79.0	94.4
0920	Plaster & Gypsum Board	102.2	64.8	77.6	93.1	64.8	74.5	87.1	64.8	72.4	94.3	72.3	79.9	94.0	64.8	74.8	93.1	64.8	74.5
0950, 0980	Ceilings & Acoustic Treatment	109.4	64.8	81.4	109.4	64.8	81.4	97.9	64.8	77.1	109.4	72.3	86.2	109.4	64.8	81.4	109.4	64.8	81.4
0960	Flooring	96.3	54.9	84.2	89.3	54.9	79.2	100.3	54.9	87.1	91.2	54.9	80.6	90.0	54.9	79.8	89.0	54.9	79.0
0970, 0990	Wall Finishes & Painting/Coating	83.5	53.7	65.6	83.5	53.7	65.6	94.7	64.1	76.4	83.5	62.9	71.2	83.5	53.7	65.6	83.5	54.7	66.3
09	FINISHES	97.1	65.2	79.8	94.7	65.2	78.7	97.0	66.6	80.5	94.9	71.3	82.1	94.1	65.2	78.5	93.8	65.3	78.4
COVERS	DIVS. 10 - 14, 25, 28, 41, 43, 44, 46	100.0	82.8	95.9	100.0	82.7	95.9	100.0	91.8	98.1	100.0	90.8	97.8	100.0	82.7	95.9	100.0	91.2	97.9
21, 22, 23	FIRE SUPPRESSION, PLUMBING & HVAC	96.9	74.8	88.0	96.9	69.9	86.0	100.6	70.7	88.5	100.7	78.0	91.5	96.9	68.9	85.6	100.7	68.8	87.8
26, 27, 3370	ELECTRICAL, COMMUNICATIONS & UTIL.	94.2	70.0	82.2	100.7	65.3	83.2	100.6	66.8	83.9	96.8	70.1	83.6	94.2	65.7	80.1	99.2	69.9	84.7
MF2018	WEIGHTED AVERAGE	98.6	77.8	89.6	99.2	75.7	89.0	99.6	78.9	90.6	98.8	81.0	91.1	98.7	76.9	89.3	98.7	76.5	89.1

For customer support on your Concrete & Masonry Costs with RSMeans Data, call 800.448.8182.

		NORTH DAKOTA			OHIO														
		WILLISTON			AKRON			ATHENS			CANTON			CHILLICOTHE			CINCINNATI		
	DIVISION	588			442 - 443			457			446 - 447			456			451 - 452		
		MAT.	INST.	TOTAL	MAT.	INST.	TOTAL	MAT.	INST.	TOTAL	MAT.	INST.	TOTAL	MAT.	INST.	TOTAL	MAT.	INST.	TOTAL
015433	CONTRACTOR EQUIPMENT		95.8	95.8		87.0	87.0		83.5	83.5		87.0	87.0		94.0	94.0		96.6	96.6
0241, 31 - 34	SITE & INFRASTRUCTURE, DEMOLITION	106.6	91.5	96.2	96.4	91.2	92.8	108.7	82.7	90.8	96.5	91.1	92.8	95.0	92.1	93.0	93.3	98.5	96.9
0310	Concrete Forming & Accessories	100.7	76.2	79.8	102.6	79.9	83.3	94.3	76.1	78.8	102.6	72.0	76.6	97.7	78.6	81.4	100.4	74.9	78.7
0320	Concrete Reinforcing	100.6	100.1	100.4	92.2	87.2	89.8	93.1	85.0	89.2	92.2	71.5	82.2	90.0	84.6	87.4	95.5	83.7	89.8
0330	Cast-in-Place Concrete	105.5	84.9	97.7	107.2	85.5	99.0	111.7	92.1	104.3	108.2	84.5	99.2	101.4	88.8	96.6	96.8	76.5	89.1
03	CONCRETE	98.6	84.0	92.1	101.3	82.8	93.0	103.4	83.1	94.3	101.8	76.2	90.3	97.9	83.7	91.5	96.0	77.2	87.6
04	MASONRY	103.2	82.2	90.4	99.8	85.7	91.2	79.7	93.2	87.9	100.5	78.5	87.1	86.3	86.7	86.6	91.7	79.2	84.1
05	METALS	92.7	95.8	93.7	94.3	79.0	89.5	102.3	79.1	95.1	94.3	72.4	87.5	94.1	88.4	92.3	96.3	81.9	91.8
06	WOOD, PLASTICS & COMPOSITES	91.5	72.6	81.6	104.1	78.3	90.6	87.1	71.5	78.9	104.5	69.3	86.1	99.3	75.3	86.8	101.3	73.6	86.8
07	THERMAL & MOISTURE PROTECTION	106.5	83.3	96.5	104.1	88.1	97.3	102.9	87.4	96.2	105.0	84.9	96.4	104.3	84.4	95.8	102.1	80.9	93.0
08	OPENINGS	100.7	83.0	96.4	107.3	80.0	100.6	97.4	73.0	91.5	101.1	67.2	92.9	88.4	75.2	85.2	98.6	73.2	92.4
0920	Plaster & Gypsum Board	94.3	72.3	79.9	98.1	77.9	84.8	94.2	70.8	78.8	99.2	68.6	79.1	97.5	75.2	82.9	96.5	73.2	81.2
0950, 0980	Ceilings & Acoustic Treatment	109.4	72.3	86.2	93.2	77.9	83.6	102.8	70.8	82.7	93.2	68.6	77.8	95.0	75.2	82.6	88.5	73.2	78.9
0960	Flooring	92.3	54.9	81.4	91.6	79.5	88.1	124.9	71.8	109.4	91.8	70.7	85.6	101.5	71.8	92.8	101.5	79.3	95.0
0970, 0990	Wall Finishes & Painting/Coating	83.5	60.4	69.7	97.9	86.5	91.1	104.9	84.9	92.9	97.9	77.3	85.5	102.2	84.9	91.8	101.5	73.0	84.4
09	FINISHES	95.1	71.2	82.1	95.5	80.3	87.3	102.9	75.6	88.1	95.7	71.5	82.6	99.4	77.2	87.4	97.2	75.1	85.3
COVERS	DIVS. 10 - 14, 25, 28, 41, 43, 44, 46	100.0	93.0	98.4	100.0	91.6	98.0	100.0	90.3	97.7	100.0	90.2	97.7	100.0	89.4	97.5	100.0	88.1	97.2
21, 22, 23	FIRE SUPPRESSION, PLUMBING & HVAC	96.9	74.7	87.9	100.5	83.7	93.7	96.7	77.3	88.9	100.5	75.5	90.4	97.2	88.0	93.5	100.4	78.1	91.4
26, 27, 3370	ELECTRICAL, COMMUNICATIONS & UTIL.	97.1	75.1	86.2	99.2	78.8	89.1	100.7	86.4	93.6	98.5	81.9	90.3	99.8	86.4	93.2	99.1	74.0	86.7
MF2018	WEIGHTED AVERAGE	97.8	80.7	90.4	99.7	83.0	92.5	99.2	81.8	91.7	99.2	77.5	89.8	96.3	85.3	91.5	98.0	79.2	89.9

		OHIO																	
		CLEVELAND			COLUMBUS			DAYTON			HAMILTON			LIMA			LORAIN		
	DIVISION	441			430 - 432			453 - 454			450			458			440		
		MAT.	INST.	TOTAL	MAT.	INST.	TOTAL	MAT.	INST.	TOTAL	MAT.	INST.	TOTAL	MAT.	INST.	TOTAL	MAT.	INST.	TOTAL
015433	CONTRACTOR EQUIPMENT		91.8	91.8		93.7	93.7		87.4	87.4		94.0	94.0		86.8	86.8		87.0	87.0
0241, 31 - 34	SITE & INFRASTRUCTURE, DEMOLITION	95.5	95.2	95.3	102.9	94.1	96.8	91.4	91.5	91.5	91.4	91.7	91.6	102.5	82.4	88.7	95.7	91.4	92.7
0310	Concrete Forming & Accessories	101.3	89.9	91.6	100.6	76.3	80.0	99.8	74.8	78.5	99.9	74.6	78.4	94.3	73.7	76.7	102.6	72.0	76.6
0320	Concrete Reinforcing	92.8	90.7	91.8	104.6	78.8	92.2	95.5	75.4	85.8	95.5	73.6	84.9	93.1	75.6	84.6	92.2	87.5	89.9
0330	Cast-in-Place Concrete	106.9	95.9	102.7	102.9	86.0	96.5	86.8	77.9	83.5	93.1	79.5	87.9	102.6	86.1	96.4	102.0	88.3	96.8
03	CONCRETE	101.7	91.7	97.2	100.1	80.2	91.2	89.3	76.0	83.4	92.2	76.8	85.3	96.3	78.5	88.3	98.9	80.3	90.5
04	MASONRY	104.6	97.3	100.1	87.9	83.9	85.5	85.4	74.1	78.5	85.7	77.8	80.9	108.9	74.9	88.2	96.4	91.0	93.1
05	METALS	95.9	83.8	92.1	103.2	79.6	95.9	95.6	75.5	89.4	95.7	84.3	92.1	102.3	79.0	95.1	94.9	79.8	90.2
06	WOOD, PLASTICS & COMPOSITES	98.4	87.9	92.9	100.3	75.1	87.1	103.2	74.5	88.2	102.2	74.5	87.7	87.0	72.6	79.4	104.1	67.7	85.0
07	THERMAL & MOISTURE PROTECTION	100.9	96.7	99.1	92.7	83.8	88.9	108.1	76.3	94.4	104.4	76.8	92.5	102.5	79.9	92.8	104.9	87.6	97.5
08	OPENINGS	100.2	86.0	96.7	97.8	73.6	91.9	95.1	72.2	89.5	93.0	72.3	87.9	97.4	71.0	91.0	101.1	74.2	94.6
0920	Plaster & Gypsum Board	98.1	87.5	91.1	92.7	74.5	80.7	99.0	74.3	82.8	99.0	74.3	82.8	94.2	71.9	79.5	98.1	66.9	77.6
0950, 0980	Ceilings & Acoustic Treatment	91.2	87.5	88.8	88.8	74.5	79.8	95.7	74.3	82.3	95.0	74.3	82.1	102.8	71.9	83.4	93.2	66.9	76.7
0960	Flooring	92.5	87.3	91.0	96.0	75.3	89.9	105.2	69.0	94.7	102.5	74.9	94.5	124.2	73.6	109.4	91.8	87.3	90.5
0970, 0990	Wall Finishes & Painting/Coating	101.9	89.3	94.3	99.5	77.7	86.4	102.2	67.4	81.3	102.2	67.9	81.6	104.9	72.8	85.7	97.9	86.5	91.1
09	FINISHES	95.4	89.2	92.0	93.2	75.7	83.8	100.6	72.6	85.5	99.6	73.9	85.7	102.3	73.0	86.4	95.6	75.4	84.7
COVERS	DIVS. 10 - 14, 25, 28, 41, 43, 44, 46	100.0	97.4	99.4	100.0	88.7	97.3	100.0	87.2	97.0	100.0	87.1	97.0	100.0	91.6	98.0	100.0	86.5	96.8
21, 22, 23	FIRE SUPPRESSION, PLUMBING & HVAC	100.5	91.3	96.8	100.4	89.8	96.1	101.2	77.8	91.7	101.0	71.4	89.1	96.7	87.4	93.0	100.5	84.6	94.1
26, 27, 3370	ELECTRICAL, COMMUNICATIONS & UTIL.	99.1	92.0	95.6	101.2	81.0	91.2	97.6	72.9	85.4	97.8	72.7	85.4	101.0	72.9	87.1	98.6	75.3	87.1
MF2018	WEIGHTED AVERAGE	99.4	91.5	96.0	99.2	83.2	92.3	96.9	76.6	88.1	96.9	76.7	88.2	99.6	79.0	90.7	98.7	81.8	91.4

		OHIO																	
		MANSFIELD			MARION			SPRINGFIELD			STEUBENVILLE			TOLEDO			YOUNGSTOWN		
	DIVISION	448 - 449			433			455			439			434 - 436			444 - 445		
		MAT.	INST.	TOTAL	MAT.	INST.	TOTAL	MAT.	INST.	TOTAL	MAT.	INST.	TOTAL	MAT.	INST.	TOTAL	MAT.	INST.	TOTAL
015433	CONTRACTOR EQUIPMENT		87.0	87.0		87.4	87.4		87.4	87.4		91.1	91.1		91.1	91.1		87.0	87.0
0241, 31 - 34	SITE & INFRASTRUCTURE, DEMOLITION	91.5	91.1	91.2	94.9	88.0	90.2	91.7	91.5	91.6	139.6	95.6	109.4	99.5	88.3	91.8	96.2	91.0	92.6
0310	Concrete Forming & Accessories	90.5	70.9	73.8	98.8	76.5	79.8	99.8	74.8	78.5	98.9	76.3	79.7	103.1	81.7	84.9	102.6	74.0	78.3
0320	Concrete Reinforcing	84.0	75.3	79.8	98.8	75.3	87.4	95.5	75.4	85.8	96.1	95.6	95.9	107.2	78.9	93.5	92.2	84.2	88.3
0330	Cast-in-Place Concrete	99.2	84.3	93.6	89.0	84.7	87.4	89.2	77.9	84.9	96.7	86.0	92.7	97.5	86.8	93.4	106.1	83.9	97.7
03	CONCRETE	93.3	76.3	85.7	88.1	79.1	84.1	90.4	76.0	84.0	91.9	82.8	87.8	96.1	83.2	90.3	100.8	79.1	91.0
04	MASONRY	98.8	85.2	90.5	90.5	86.1	87.8	85.4	74.1	78.5	78.7	87.3	83.9	96.6	88.0	91.3	100.2	83.0	89.7
05	METALS	95.2	74.6	88.7	102.2	77.5	94.5	95.6	75.5	89.3	98.4	82.6	93.5	103.0	84.2	97.1	94.4	77.6	89.2
06	WOOD, PLASTICS & COMPOSITES	89.7	67.7	78.2	95.0	74.9	84.5	104.4	74.5	88.8	89.8	73.8	81.5	99.8	81.1	90.0	104.1	72.1	87.3
07	THERMAL & MOISTURE PROTECTION	103.3	85.4	95.7	90.0	84.9	87.8	108.0	76.3	94.4	101.6	84.2	94.2	90.8	87.1	89.3	105.2	84.8	96.4
08	OPENINGS	102.3	68.4	94.1	90.9	72.4	86.4	93.4	71.9	88.2	91.1	76.7	87.6	93.4	78.3	89.7	101.1	77.7	95.4
0920	Plaster & Gypsum Board	91.2	66.9	75.2	91.7	74.5	80.4	99.0	74.3	82.8	89.5	72.8	78.6	93.7	80.8	85.2	98.1	71.4	80.6
0950, 0980	Ceilings & Acoustic Treatment	93.9	66.9	77.0	95.0	74.5	82.1	95.7	74.3	82.3	92.0	72.8	80.0	95.0	80.8	86.1	93.2	71.4	79.6
0960	Flooring	86.4	88.4	87.0	96.4	88.4	94.1	105.2	69.0	94.7	125.5	90.8	115.4	96.9	90.1	94.9	91.8	85.9	90.1
0970, 0990	Wall Finishes & Painting/Coating	97.9	74.1	83.6	101.1	74.1	84.9	102.2	67.4	81.3	113.5	82.9	95.2	101.1	82.9	90.2	97.9	76.2	84.9
09	FINISHES	92.9	73.6	82.4	95.1	78.0	85.9	100.6	72.6	85.5	111.6	78.8	93.9	95.9	83.1	89.0	95.6	75.7	84.8
COVERS	DIVS. 10 - 14, 25, 28, 41, 43, 44, 46	100.0	89.7	97.6	100.0	88.4	97.3	100.0	87.1	97.0	100.0	91.8	98.1	100.0	86.7	96.9	100.0	90.0	97.6
21, 22, 23	FIRE SUPPRESSION, PLUMBING & HVAC	96.7	83.3	91.3	96.6	88.7	93.4	101.2	77.8	91.7	97.1	87.1	93.1	100.4	89.2	95.9	100.5	81.7	92.9
26, 27, 3370	ELECTRICAL, COMMUNICATIONS & UTIL.	96.6	87.4	92.0	97.1	87.4	92.3	97.6	77.2	87.5	91.5	102.5	96.9	102.3	99.6	101.0	98.6	71.6	85.2
MF2018	WEIGHTED AVERAGE	96.6	81.1	89.9	95.5	83.6	90.3	96.9	77.2	88.4	97.1	87.4	92.9	98.9	87.7	94.1	99.1	79.7	90.7

For customer support on your Concrete & Masonry Costs with RSMeans Data, call 800.448.8182.

479

OHIO / OKLAHOMA

DIVISION		ZANESVILLE 437-438 MAT.	INST.	TOTAL	ARDMORE 734 MAT.	INST.	TOTAL	CLINTON 736 MAT.	INST.	TOTAL	DURANT 747 MAT.	INST.	TOTAL	ENID 737 MAT.	INST.	TOTAL	GUYMON 739 MAT.	INST.	TOTAL
015433	CONTRACTOR EQUIPMENT		87.4	87.4		77.7	77.7		76.9	76.9		76.9	76.9		76.9	76.9		76.9	76.9
0241, 31 - 34	SITE & INFRASTRUCTURE, DEMOLITION	98.0	87.9	91.0	94.4	89.1	90.8	95.7	87.7	90.2	92.9	85.0	87.5	97.9	87.7	90.9	99.9	86.8	90.9
0310	Concrete Forming & Accessories	95.2	75.5	78.5	90.9	53.4	59.0	89.5	53.4	58.8	83.1	52.9	57.4	93.5	53.6	59.6	97.0	53.1	59.7
0320	Concrete Reinforcing	98.2	88.2	93.4	88.2	66.3	77.6	88.7	66.3	77.9	92.1	60.6	76.9	88.1	66.3	77.6	88.7	60.3	75.0
0330	Cast-in-Place Concrete	93.8	83.1	89.8	93.3	67.5	83.5	90.1	67.5	81.6	90.3	67.3	81.6	90.1	67.6	81.6	90.2	66.9	81.4
03	CONCRETE	91.7	80.5	86.7	86.4	60.5	74.8	85.9	60.5	74.5	85.3	59.2	73.6	86.4	60.7	74.8	89.0	59.1	75.6
04	MASONRY	88.9	83.0	85.3	96.7	56.7	72.3	123.2	56.7	82.7	88.5	59.6	70.8	104.6	56.7	75.4	98.4	53.1	70.8
05	METALS	103.7	84.7	97.8	98.0	61.0	86.5	98.1	61.0	86.6	93.2	58.6	82.4	99.6	61.2	87.6	98.6	57.3	85.8
06	WOOD, PLASTICS & COMPOSITES	89.7	74.9	82.0	99.0	52.5	74.7	98.0	52.5	74.2	86.4	52.5	68.7	101.9	52.5	76.1	106.1	52.5	78.1
07	THERMAL & MOISTURE PROTECTION	90.1	81.2	86.3	101.2	62.7	84.7	101.3	62.7	84.7	97.7	62.0	82.4	101.5	62.7	84.8	101.8	59.4	83.6
08	OPENINGS	90.9	77.3	87.6	103.4	53.3	91.2	103.4	53.3	91.2	95.9	51.9	85.2	104.6	54.5	92.4	103.6	51.9	91.0
0920	Plaster & Gypsum Board	87.5	74.5	78.9	89.9	51.9	65.0	89.6	51.9	64.9	79.2	51.9	61.3	90.5	51.9	65.2	90.5	51.9	65.2
0950, 0980	Ceilings & Acoustic Treatment	95.0	74.5	82.1	93.5	51.9	67.5	93.5	51.9	67.5	87.1	51.9	65.1	93.5	51.9	67.5	93.5	51.9	67.5
0960	Flooring	94.5	71.8	87.9	86.9	54.7	77.5	85.8	54.7	76.7	92.3	49.3	79.8	87.5	71.6	82.9	89.1	54.7	79.0
0970, 0990	Wall Finishes & Painting/Coating	101.1	84.9	91.4	84.3	43.5	59.9	84.3	43.5	59.9	91.5	43.5	62.8	84.3	43.5	59.9	84.3	40.2	57.9
09	FINISHES	94.2	75.1	83.9	88.1	51.4	68.2	88.0	51.4	68.2	87.3	50.3	67.3	88.6	54.9	70.4	89.5	52.3	69.3
COVERS	DIVS. 10 - 14, 25, 28, 41, 43, 44, 46	100.0	87.5	97.1	100.0	77.4	94.7	100.0	77.4	94.7	100.0	77.4	94.7	100.0	77.4	94.7	100.0	77.4	94.7
21, 22, 23	FIRE SUPPRESSION, PLUMBING & HVAC	96.6	85.6	92.2	97.3	65.5	84.4	97.3	65.5	84.4	97.4	65.4	84.5	101.1	65.5	86.7	97.3	65.2	84.3
26, 27, 3370	ELECTRICAL, COMMUNICATIONS & UTIL.	97.3	86.4	91.9	95.2	70.0	82.7	96.1	70.0	83.2	97.6	66.6	82.3	96.1	70.0	83.2	97.5	63.2	80.5
MF2018	WEIGHTED AVERAGE	96.1	83.0	90.4	95.9	63.8	82.0	97.3	63.7	82.8	93.8	62.7	80.4	97.8	64.2	83.3	97.0	61.7	81.7

OKLAHOMA

DIVISION		LAWTON 735 MAT.	INST.	TOTAL	MCALESTER 745 MAT.	INST.	TOTAL	MIAMI 743 MAT.	INST.	TOTAL	MUSKOGEE 744 MAT.	INST.	TOTAL	OKLAHOMA CITY 730-731 MAT.	INST.	TOTAL	PONCA CITY 746 MAT.	INST.	TOTAL
015433	CONTRACTOR EQUIPMENT		77.7	77.7		76.9	76.9		88.0	88.0		88.0	88.0		87.3	87.3		76.9	76.9
0241, 31 - 34	SITE & INFRASTRUCTURE, DEMOLITION	94.0	89.1	90.7	86.3	86.9	86.7	87.6	84.5	85.5	88.7	84.5	85.8	94.3	99.0	97.5	93.5	87.3	89.2
0310	Concrete Forming & Accessories	96.7	53.6	60.0	81.0	39.8	46.0	95.5	53.1	59.5	100.4	56.6	63.1	99.6	62.7	68.2	90.4	53.3	58.8
0320	Concrete Reinforcing	88.3	66.3	77.7	91.7	60.4	76.6	90.3	66.3	78.7	91.2	65.7	78.9	98.4	66.4	83.0	91.2	66.3	79.2
0330	Cast-in-Place Concrete	87.2	67.5	79.7	79.2	66.8	74.5	83.0	68.3	77.4	84.0	68.9	78.3	91.6	72.8	84.5	92.8	67.4	83.2
03	CONCRETE	83.0	60.6	73.0	76.5	53.1	66.0	81.0	61.7	72.3	82.6	63.3	74.0	90.4	66.7	79.8	87.4	60.5	75.3
04	MASONRY	98.4	56.7	73.0	106.3	56.6	76.0	91.3	56.8	70.2	108.2	48.3	71.6	98.8	57.5	73.6	83.8	56.7	67.3
05	METALS	103.4	61.2	90.3	93.2	57.4	82.0	93.1	75.7	87.7	94.7	74.9	88.5	96.6	64.0	86.4	93.1	60.9	83.1
06	WOOD, PLASTICS & COMPOSITES	105.0	52.5	77.6	83.7	34.9	58.2	100.4	52.6	75.4	105.5	56.8	80.0	100.9	64.2	81.7	95.3	52.5	72.9
07	THERMAL & MOISTURE PROTECTION	101.2	62.7	84.7	97.3	59.3	81.0	97.8	62.0	82.5	98.1	59.9	81.7	93.9	65.8	81.8	97.9	62.9	82.9
08	OPENINGS	106.4	54.5	93.8	95.9	42.2	82.8	95.9	53.4	85.5	97.1	55.7	87.0	99.8	60.9	90.3	95.9	53.3	85.5
0920	Plaster & Gypsum Board	92.8	51.9	66.0	78.7	33.8	49.2	84.9	51.9	63.3	87.8	56.2	67.1	93.9	63.7	74.1	83.8	51.9	62.9
0950, 0980	Ceilings & Acoustic Treatment	102.4	51.9	70.8	87.1	33.8	53.7	87.1	51.9	65.1	97.9	56.2	71.8	90.7	63.7	73.8	87.1	51.9	65.1
0960	Flooring	89.3	71.6	84.2	91.2	54.7	80.5	98.6	49.3	84.2	101.2	31.7	80.9	89.6	71.6	84.3	95.6	54.7	83.7
0970, 0990	Wall Finishes & Painting/Coating	84.3	43.5	59.9	91.5	40.2	60.8	91.5	40.2	60.8	91.5	40.2	60.8	90.1	43.5	62.2	91.5	43.5	62.8
09	FINISHES	91.1	54.9	71.5	86.4	40.7	61.6	89.3	50.0	68.0	93.0	49.5	69.5	91.1	61.9	75.3	89.0	52.3	69.1
COVERS	DIVS. 10 - 14, 25, 28, 41, 43, 44, 46	100.0	77.4	94.7	100.0	75.5	94.2	100.0	77.8	94.8	100.0	78.7	95.0	100.0	79.5	95.2	100.0	77.4	94.7
21, 22, 23	FIRE SUPPRESSION, PLUMBING & HVAC	101.1	65.5	86.7	97.4	60.5	82.5	97.4	60.6	82.5	101.2	62.1	85.4	101.0	65.9	86.8	97.4	60.6	82.5
26, 27, 3370	ELECTRICAL, COMMUNICATIONS & UTIL.	97.5	66.6	82.2	96.2	64.5	80.5	97.4	64.5	81.2	95.9	86.7	91.4	102.4	70.0	86.3	95.9	63.2	79.7
MF2018	WEIGHTED AVERAGE	98.1	63.9	83.3	93.2	58.3	78.1	93.5	63.0	80.3	96.1	65.7	83.0	97.6	67.8	84.8	93.9	61.8	80.0

OKLAHOMA / OREGON

DIVISION		POTEAU 749 MAT.	INST.	TOTAL	SHAWNEE 748 MAT.	INST.	TOTAL	TULSA 740-741 MAT.	INST.	TOTAL	WOODWARD 738 MAT.	INST.	TOTAL	BEND 977 MAT.	INST.	TOTAL	EUGENE 974 MAT.	INST.	TOTAL
015433	CONTRACTOR EQUIPMENT		86.8	86.8		76.9	76.9		88.0	88.0		76.9	76.9		95.5	95.5		95.5	95.5
0241, 31 - 34	SITE & INFRASTRUCTURE, DEMOLITION	74.4	80.3	78.4	96.4	87.3	90.1	94.7	84.2	87.5	96.1	87.7	90.3	112.5	95.7	100.9	102.9	95.7	98.0
0310	Concrete Forming & Accessories	88.1	52.9	58.1	83.0	53.3	57.7	100.2	57.0	63.4	89.6	42.1	49.2	103.1	101.8	102.0	99.5	102.1	101.7
0320	Concrete Reinforcing	92.2	66.2	79.7	91.2	62.4	77.3	91.4	66.3	79.3	88.1	66.3	77.6	90.5	114.3	102.0	94.4	114.4	104.0
0330	Cast-in-Place Concrete	83.0	68.2	77.4	95.7	67.4	85.0	91.5	70.8	83.7	90.1	67.4	81.6	118.3	101.0	111.8	114.6	101.1	109.5
03	CONCRETE	83.0	61.5	73.3	88.9	59.8	75.9	87.6	64.3	77.1	86.1	55.4	72.3	110.9	103.2	107.5	102.0	103.4	102.6
04	MASONRY	91.5	56.8	70.3	107.7	56.7	76.6	92.2	57.8	71.2	92.3	56.7	70.6	101.7	102.7	102.3	98.6	102.7	101.1
05	METALS	93.2	75.4	87.6	93.1	59.5	82.6	97.9	76.1	91.1	98.2	60.9	86.6	107.3	98.1	104.4	108.0	98.5	105.0
06	WOOD, PLASTICS & COMPOSITES	91.5	52.6	71.2	86.2	52.5	68.6	104.6	56.8	79.6	98.1	37.3	66.3	95.9	102.1	99.1	91.5	102.1	97.0
07	THERMAL & MOISTURE PROTECTION	97.9	62.0	82.5	97.9	61.6	82.3	98.1	64.9	83.9	101.4	61.2	84.2	118.6	102.9	111.8	117.7	105.5	112.5
08	OPENINGS	95.9	53.4	85.5	95.9	52.4	85.3	98.8	56.9	88.6	103.4	44.9	89.2	97.0	105.0	99.0	97.3	105.0	99.2
0920	Plaster & Gypsum Board	81.8	51.9	62.2	79.2	51.9	61.3	87.8	56.2	67.0	89.6	36.3	54.6	116.1	102.2	107.0	113.9	102.2	106.2
0950, 0980	Ceilings & Acoustic Treatment	87.1	51.9	65.1	87.1	51.9	65.1	97.9	56.2	71.8	93.5	36.3	57.7	90.0	102.2	97.6	90.7	102.2	97.9
0960	Flooring	94.9	49.3	81.6	92.3	54.7	81.3	100.0	62.4	89.0	85.8	52.1	76.0	101.3	106.4	102.8	99.7	106.4	101.7
0970, 0990	Wall Finishes & Painting/Coating	91.5	43.5	62.8	91.5	40.2	60.8	91.5	54.9	69.6	84.3	43.5	59.9	93.9	76.5	83.5	93.9	66.6	77.6
09	FINISHES	87.0	50.4	67.2	87.5	51.0	67.8	92.9	57.0	73.5	88.0	42.0	63.1	100.7	99.9	100.3	99.0	98.8	98.9
COVERS	DIVS. 10 - 14, 25, 28, 41, 43, 44, 46	100.0	77.7	94.7	100.0	77.4	94.7	100.0	78.6	95.0	100.0	75.8	94.3	100.0	102.5	100.6	100.0	102.5	100.6
21, 22, 23	FIRE SUPPRESSION, PLUMBING & HVAC	97.4	60.6	82.5	97.4	65.4	84.5	101.2	61.3	85.1	97.3	65.5	84.4	97.2	102.3	99.3	101.2	117.3	107.7
26, 27, 3370	ELECTRICAL, COMMUNICATIONS & UTIL.	96.0	64.5	80.4	97.6	70.0	83.9	97.6	69.9	83.3	97.4	70.0	83.8	98.3	96.1	97.2	97.6	96.1	96.8
MF2018	WEIGHTED AVERAGE	93.0	62.6	79.9	95.3	63.3	81.5	96.9	65.5	83.3	95.9	61.2	80.9	102.3	100.5	101.5	101.6	103.7	102.5

OREGON

DIVISION		KLAMATH FALLS 976			MEDFORD 975			PENDLETON 978			PORTLAND 970 - 972			SALEM 973			VALE 979		
		MAT.	INST.	TOTAL	MAT.	INST.	TOTAL	MAT.	INST.	TOTAL	MAT.	INST.	TOTAL	MAT.	INST.	TOTAL	MAT.	INST.	TOTAL
015433	CONTRACTOR EQUIPMENT		95.5	95.5		95.5	95.5		93.1	93.1		95.5	95.5		101.4	101.4		93.1	93.1
0241, 31 - 34	SITE & INFRASTRUCTURE, DEMOLITION	116.6	95.7	102.2	110.8	95.7	100.4	110.6	89.2	95.9	105.6	95.7	98.8	98.5	104.0	102.3	97.5	89.2	91.8
0310	Concrete Forming & Accessories	96.1	101.4	100.6	94.9	101.6	100.6	97.2	101.7	101.1	101.0	102.0	101.9	99.2	102.2	101.7	103.8	100.4	100.9
0320	Concrete Reinforcing	90.5	114.2	101.9	92.0	114.3	102.7	89.7	114.4	101.6	95.1	114.5	104.4	102.2	114.5	108.1	87.6	114.2	100.4
0330	Cast-in-Place Concrete	118.4	95.7	109.8	118.3	100.9	111.7	119.2	96.2	110.5	117.8	101.1	111.5	108.3	103.0	106.3	93.7	96.4	94.7
03	CONCRETE	113.8	101.2	108.1	108.3	103.1	106.0	94.7	101.5	97.8	103.7	103.4	103.6	99.7	104.0	101.8	78.8	100.9	88.7
04	MASONRY	114.7	102.7	107.4	95.2	102.7	99.8	105.6	104.9	105.2	100.5	104.8	103.1	101.3	104.9	103.5	103.5	104.9	104.3
05	METALS	107.3	97.8	104.3	107.6	97.9	104.5	115.5	98.7	110.2	109.4	98.6	106.0	116.4	97.2	110.4	115.3	97.2	109.7
06	WOOD, PLASTICS & COMPOSITES	86.2	102.1	94.5	85.1	102.1	94.0	88.7	102.2	95.8	92.7	102.1	97.6	85.5	102.4	94.3	97.7	102.2	100.1
07	THERMAL & MOISTURE PROTECTION	118.8	98.4	110.0	118.4	98.0	109.7	110.5	95.7	104.1	117.7	103.4	111.6	114.0	104.8	110.0	109.9	91.8	102.1
08	OPENINGS	97.0	105.0	99.0	99.6	105.0	100.9	93.1	105.0	96.0	95.3	105.0	97.7	102.7	105.1	103.3	93.1	92.3	92.9
0920	Plaster & Gypsum Board	110.2	102.2	104.9	109.6	102.2	104.7	96.3	102.2	100.2	113.5	102.2	106.1	109.9	102.2	104.8	103.4	102.2	102.6
0950, 0980	Ceilings & Acoustic Treatment	97.6	102.2	100.5	102.6	102.2	102.3	65.8	102.2	88.6	92.8	102.2	98.7	105.1	102.2	103.3	65.8	102.2	88.6
0960	Flooring	98.3	106.4	100.7	97.8	106.4	100.3	67.3	106.4	78.7	97.6	106.4	100.1	104.8	106.4	101.9	69.4	106.4	80.2
0970, 0990	Wall Finishes & Painting/Coating	93.9	76.8	83.7	93.9	76.8	83.7	85.0	76.8	80.1	93.8	76.8	83.6	95.5	74.5	82.9	85.0	76.5	79.9
09	FINISHES	101.3	99.9	100.6	101.4	99.9	100.6	71.3	100.0	86.8	98.9	99.9	99.5	100.6	99.9	100.2	71.9	100.0	87.1
COVERS	DIVS. 10 - 14, 25, 28, 41, 43, 44, 46	100.0	102.4	100.6	100.0	102.4	100.6	100.0	105.2	101.2	100.0	102.6	100.6	100.0	103.1	100.7	100.0	102.6	100.6
21, 22, 23	FIRE SUPPRESSION, PLUMBING & HVAC	97.2	102.3	99.2	101.2	108.1	104.0	100.0	108.8	103.5	101.2	111.1	105.2	100.2	108.2	103.4	100.0	69.8	87.8
26, 27, 3370	ELECTRICAL, COMMUNICATIONS & UTIL.	97.2	79.6	88.5	100.6	79.6	90.2	90.3	93.3	91.8	97.9	110.5	104.1	101.5	96.1	98.8	90.3	65.5	78.0
MF2018	WEIGHTED AVERAGE	103.3	97.8	100.9	103.1	99.3	101.4	98.3	101.0	99.5	102.0	104.7	103.2	103.3	102.8	103.0	96.0	87.8	92.4

PENNSYLVANIA

DIVISION		ALLENTOWN 181			ALTOONA 166			BEDFORD 155			BRADFORD 167			BUTLER 160			CHAMBERSBURG 172		
		MAT.	INST.	TOTAL	MAT.	INST.	TOTAL	MAT.	INST.	TOTAL	MAT.	INST.	TOTAL	MAT.	INST.	TOTAL	MAT.	INST.	TOTAL
015433	CONTRACTOR EQUIPMENT		110.8	110.8		110.8	110.8		108.0	108.0		110.8	110.8		110.8	110.8		110.0	110.0
0241, 31 - 34	SITE & INFRASTRUCTURE, DEMOLITION	91.6	95.0	93.9	94.9	95.1	95.0	102.8	91.1	94.8	90.1	92.9	92.0	85.8	94.8	92.0	87.6	92.2	90.7
0310	Concrete Forming & Accessories	98.9	109.3	107.7	82.5	87.6	86.9	81.1	78.4	78.8	85.0	93.1	91.9	83.9	92.0	90.8	87.4	74.0	76.0
0320	Concrete Reinforcing	97.5	112.8	104.9	94.5	109.6	101.8	93.7	100.7	97.1	96.5	100.8	98.6	95.1	121.2	107.7	98.9	106.9	102.7
0330	Cast-in-Place Concrete	88.5	103.6	94.2	98.6	90.6	95.6	107.2	82.3	97.8	94.2	87.1	91.5	87.1	97.0	90.8	93.8	90.0	92.4
03	CONCRETE	91.8	108.8	99.4	87.2	93.8	90.2	100.9	85.1	93.8	93.5	93.4	93.5	79.4	100.0	88.6	97.4	87.0	92.7
04	MASONRY	96.6	98.8	97.9	100.3	89.8	93.9	119.2	76.9	93.4	97.9	78.6	86.1	102.6	91.8	96.0	98.3	75.8	84.6
05	METALS	99.8	122.3	106.8	93.8	118.2	101.4	98.8	111.2	102.7	97.5	111.5	101.9	93.5	123.0	102.7	99.2	116.8	104.7
06	WOOD, PLASTICS & COMPOSITES	98.2	110.8	104.8	75.5	86.2	81.1	78.1	78.9	78.5	81.3	97.8	90.0	77.1	90.9	84.3	83.6	73.0	78.0
07	THERMAL & MOISTURE PROTECTION	105.3	112.6	108.4	103.3	95.0	99.7	99.6	85.1	93.4	105.1	85.4	96.6	102.7	94.0	99.0	96.8	79.3	89.3
08	OPENINGS	92.5	108.5	96.4	86.3	89.6	87.1	93.6	83.6	91.1	92.4	93.0	92.5	86.3	99.3	89.4	89.0	78.4	86.4
0920	Plaster & Gypsum Board	99.7	111.2	107.3	91.2	85.9	87.7	98.2	78.5	85.3	91.6	97.9	95.7	91.2	90.7	90.9	115.9	72.4	87.3
0950, 0980	Ceilings & Acoustic Treatment	92.2	111.2	104.1	95.4	85.9	89.4	113.3	78.5	91.5	95.5	97.9	97.0	96.1	90.7	92.7	96.9	72.4	81.5
0960	Flooring	91.9	97.7	93.6	85.4	103.7	90.7	92.2	98.9	94.1	86.1	98.9	89.8	86.4	100.5	90.5	91.4	74.4	86.4
0970, 0990	Wall Finishes & Painting/Coating	87.1	108.2	99.7	82.8	112.0	100.3	89.7	100.6	96.2	87.1	100.6	95.2	82.8	101.1	93.8	87.1	95.9	92.4
09	FINISHES	92.4	107.2	100.4	90.4	92.5	91.5	101.6	83.9	92.0	90.4	95.4	93.1	90.3	94.0	92.3	93.8	75.7	84.0
COVERS	DIVS. 10 - 14, 25, 28, 41, 43, 44, 46	100.0	103.5	100.8	100.0	98.6	99.7	100.0	94.9	98.8	100.0	97.7	99.5	100.0	99.7	99.9	100.0	92.6	98.3
21, 22, 23	FIRE SUPPRESSION, PLUMBING & HVAC	101.2	116.9	107.5	100.5	90.6	96.5	93.1	80.5	88.0	97.4	86.3	92.9	96.7	92.5	95.0	95.1	85.5	91.2
26, 27, 3370	ELECTRICAL, COMMUNICATIONS & UTIL.	99.4	99.5	99.5	90.7	113.3	101.9	94.7	113.3	103.9	94.1	113.3	103.6	91.2	103.7	97.4	93.6	82.7	88.2
MF2018	WEIGHTED AVERAGE	97.5	108.1	102.1	94.2	97.5	95.6	98.0	90.1	94.6	95.6	95.0	95.3	92.1	98.7	95.0	95.4	86.0	91.3

PENNSYLVANIA

DIVISION		DOYLESTOWN 189			DUBOIS 158			ERIE 164 - 165			GREENSBURG 156			HARRISBURG 170 - 171			HAZLETON 182		
		MAT.	INST.	TOTAL	MAT.	INST.	TOTAL	MAT.	INST.	TOTAL	MAT.	INST.	TOTAL	MAT.	INST.	TOTAL	MAT.	INST.	TOTAL
015433	CONTRACTOR EQUIPMENT		89.5	89.5		108.0	108.0		110.8	110.8		108.0	108.0		112.2	112.2		110.8	110.8
0241, 31 - 34	SITE & INFRASTRUCTURE, DEMOLITION	104.5	82.9	89.6	107.6	91.5	96.6	91.8	94.9	93.9	98.7	93.2	94.9	88.7	98.1	95.1	84.6	93.9	91.0
0310	Concrete Forming & Accessories	81.3	122.1	116.1	80.5	80.8	80.7	98.1	87.4	89.0	87.9	86.7	86.8	100.5	89.6	91.2	79.0	85.6	84.6
0320	Concrete Reinforcing	94.3	143.7	118.1	93.0	113.7	103.0	96.5	109.3	102.7	93.0	120.0	106.5	103.9	112.4	108.0	94.7	107.7	101.0
0330	Cast-in-Place Concrete	83.6	124.7	99.1	103.3	93.2	99.5	96.9	91.0	94.7	99.4	96.4	98.3	92.5	104.9	96.2	83.6	91.8	86.7
03	CONCRETE	87.3	126.3	104.8	102.9	92.2	98.1	86.4	93.8	89.7	96.1	97.3	96.6	92.6	99.2	95.6	84.4	93.0	88.3
04	MASONRY	100.2	129.3	117.9	120.1	89.5	101.4	88.2	90.7	89.7	130.3	88.0	104.5	93.7	92.9	93.2	110.1	85.8	95.3
05	METALS	97.3	121.9	105.0	98.8	116.7	104.4	94.0	117.3	101.3	98.8	120.7	105.6	106.0	118.6	109.9	99.5	117.8	105.2
06	WOOD, PLASTICS & COMPOSITES	76.8	121.6	100.2	76.9	78.9	78.0	94.9	85.7	90.1	85.0	84.1	84.5	100.3	86.4	93.0	75.1	84.1	79.8
07	THERMAL & MOISTURE PROTECTION	101.9	125.8	112.2	99.9	91.2	96.2	103.6	90.7	98.1	99.9	91.8	96.2	98.9	108.2	102.9	104.5	96.2	100.9
08	OPENINGS	94.9	128.3	103.0	93.6	86.6	91.9	86.4	90.7	87.5	93.5	95.6	94.0	100.6	87.8	97.4	93.0	89.9	92.2
0920	Plaster & Gypsum Board	90.1	122.2	111.2	96.8	78.5	84.8	99.7	85.4	90.3	98.6	83.9	89.0	122.6	85.9	98.5	90.4	83.7	86.0
0950, 0980	Ceilings & Acoustic Treatment	91.6	122.2	110.8	113.3	78.5	91.5	92.2	85.4	87.9	112.7	83.9	94.6	102.3	85.9	92.0	92.9	83.7	87.2
0960	Flooring	75.9	132.9	92.6	91.9	98.9	93.9	92.5	103.7	95.8	95.9	74.4	89.6	95.2	94.1	94.9	83.1	85.2	83.7
0970, 0990	Wall Finishes & Painting/Coating	86.7	136.8	116.7	89.7	101.1	96.6	92.9	93.5	93.3	89.7	101.1	96.6	93.7	89.5	91.2	87.1	99.1	94.3
09	FINISHES	84.1	125.4	106.4	101.8	85.3	92.9	93.5	90.5	91.9	102.3	85.2	93.0	99.2	89.6	94.0	88.3	86.6	87.4
COVERS	DIVS. 10 - 14, 25, 28, 41, 43, 44, 46	100.0	114.9	103.5	100.0	96.7	99.2	100.0	98.9	99.7	100.0	98.8	99.7	100.0	98.6	99.7	100.0	98.5	99.6
21, 22, 23	FIRE SUPPRESSION, PLUMBING & HVAC	96.7	126.3	108.7	93.1	83.4	89.2	100.5	93.8	97.8	93.1	84.7	89.7	96.4	98.5	97.2	97.4	91.0	94.8
26, 27, 3370	ELECTRICAL, COMMUNICATIONS & UTIL.	93.5	127.5	110.3	95.2	113.3	104.1	92.2	90.5	91.4	95.2	113.3	104.2	100.6	89.0	94.9	94.8	86.7	90.8
MF2018	WEIGHTED AVERAGE	94.7	122.6	106.7	98.5	94.0	96.6	94.0	94.6	94.3	98.0	95.8	97.1	98.5	97.0	97.9	95.1	92.5	94.0

For customer support on your Concrete & Masonry Costs with RSMeans Data, call 800.448.8182.

481

PENNSYLVANIA

DIVISION		INDIANA 157			JOHNSTOWN 159			KITTANNING 162			LANCASTER 175 - 176			LEHIGH VALLEY 180			MONTROSE 188		
		MAT.	INST.	TOTAL	MAT.	INST.	TOTAL	MAT.	INST.	TOTAL	MAT.	INST.	TOTAL	MAT.	INST.	TOTAL	MAT.	INST.	TOTAL
015433	CONTRACTOR EQUIPMENT		108.0	108.0		108.0	108.0		110.8	110.8		110.0	110.0		110.8	110.8		110.8	110.8
0241, 31 - 34	SITE & INFRASTRUCTURE, DEMOLITION	96.9	92.1	93.6	103.3	93.4	96.5	88.3	94.7	92.7	79.7	94.3	89.7	88.4	94.3	92.4	87.2	94.0	91.8
0310	Concrete Forming & Accessories	81.7	89.8	88.6	80.5	87.3	86.3	83.9	86.8	86.3	89.5	87.8	88.0	91.8	106.3	104.1	80.0	86.1	85.2
0320	Concrete Reinforcing	92.4	121.2	106.3	93.7	121.1	106.9	95.1	121.0	107.6	98.5	112.5	105.2	94.7	102.9	98.6	99.2	110.7	104.7
0330	Cast-in-Place Concrete	97.5	93.6	96.0	108.1	89.8	101.2	90.5	94.0	91.8	79.8	96.7	86.2	90.5	95.3	92.3	88.7	89.3	88.9
03	CONCRETE	93.7	97.7	95.5	101.9	95.3	98.9	81.9	96.6	88.5	85.7	96.6	90.6	90.8	102.8	96.2	89.4	92.9	91.0
04	MASONRY	115.5	89.6	99.7	116.4	89.2	99.8	104.9	85.1	92.8	104.0	89.5	95.1	96.6	92.5	94.1	96.5	87.7	91.1
05	METALS	98.9	120.5	105.6	98.9	120.6	105.6	93.6	121.8	102.4	99.2	121.3	106.1	99.4	117.1	104.9	97.6	118.5	104.1
06	WOOD, PLASTICS & COMPOSITES	78.9	90.8	85.1	76.9	86.1	81.7	77.1	86.3	81.9	86.8	86.2	86.5	88.8	108.8	99.2	76.0	84.1	80.2
07	THERMAL & MOISTURE PROTECTION	99.3	92.5	96.4	99.6	91.9	96.3	102.9	90.9	97.7	96.1	105.1	100.0	105.0	106.9	105.8	104.6	86.9	97.0
08	OPENINGS	93.6	94.9	93.9	93.5	92.3	93.2	86.3	96.8	88.8	89.0	87.8	88.7	92.9	102.7	95.3	89.8	88.5	89.5
0920	Plaster & Gypsum Board	98.5	90.7	93.4	96.7	85.9	89.6	91.2	86.0	87.8	118.7	85.9	97.2	92.4	109.1	103.3	91.1	83.7	86.3
0950, 0980	Ceilings & Acoustic Treatment	113.3	90.7	99.1	112.7	85.9	95.9	96.1	86.0	89.8	96.9	85.9	90.0	92.9	109.1	103.1	95.5	83.7	88.1
0960	Flooring	92.7	98.9	94.5	91.9	103.7	95.3	86.4	98.9	90.0	92.5	94.4	93.0	88.8	89.8	89.1	83.8	88.9	88.2
0970, 0990	Wall Finishes & Painting/Coating	89.7	101.1	96.6	89.7	112.0	103.1	82.8	101.1	93.8	87.1	89.5	88.6	87.1	96.2	92.5	87.1	99.1	94.3
09	FINISHES	101.4	92.2	96.4	101.2	92.2	96.4	89.9	89.6	90.0	93.9	88.3	90.9	90.5	102.3	96.9	89.5	89.0	89.2
COVERS	DIVS. 10 - 14, 25, 28, 41, 43, 44, 46	100.0	98.0	99.5	100.0	98.2	99.6	100.0	97.8	99.5	100.0	96.5	99.2	100.0	102.7	100.6	100.0	99.1	99.8
21, 22, 23	FIRE SUPPRESSION, PLUMBING & HVAC	93.1	81.5	88.4	93.1	88.3	91.1	96.7	86.6	92.6	95.1	96.0	95.5	97.4	112.0	103.3	97.4	90.3	94.5
26, 27, 3370	ELECTRICAL, COMMUNICATIONS & UTIL.	95.2	113.3	104.2	95.2	113.3	104.2	90.7	113.3	101.9	94.8	96.4	95.6	94.8	124.6	109.5	94.1	91.7	92.9
MF2018	WEIGHTED AVERAGE	96.8	96.2	96.5	98.0	97.2	97.7	92.6	96.6	94.3	94.2	96.5	95.2	95.6	107.5	100.8	94.6	93.3	94.0

PENNSYLVANIA

DIVISION		NEW CASTLE 161			NORRISTOWN 194			OIL CITY 163			PHILADELPHIA 190 - 191			PITTSBURGH 150 - 152			POTTSVILLE 179		
		MAT.	INST.	TOTAL	MAT.	INST.	TOTAL	MAT.	INST.	TOTAL	MAT.	INST.	TOTAL	MAT.	INST.	TOTAL	MAT.	INST.	TOTAL
015433	CONTRACTOR EQUIPMENT		110.8	110.8		95.3	95.3		110.8	110.8		100.4	100.4		96.9	96.9		110.0	110.0
0241, 31 - 34	SITE & INFRASTRUCTURE, DEMOLITION	86.2	95.3	92.5	94.7	91.9	92.8	84.8	93.1	90.5	97.9	102.1	100.8	102.7	93.3	96.3	82.6	92.7	89.5
0310	Concrete Forming & Accessories	83.9	95.9	94.1	81.6	120.7	114.9	83.9	89.4	88.6	95.6	140.6	133.9	98.9	99.1	99.1	80.3	84.4	83.8
0320	Concrete Reinforcing	94.1	99.4	96.6	97.4	143.7	119.7	95.1	90.4	92.8	107.8	151.5	128.9	92.6	125.8	108.6	97.7	108.7	103.0
0330	Cast-in-Place Concrete	87.9	96.9	91.3	83.6	122.4	98.3	85.4	90.6	87.4	93.9	133.6	108.9	106.7	100.9	104.5	85.0	92.5	87.8
03	CONCRETE	79.7	98.0	87.9	86.6	124.8	103.8	78.2	91.5	84.1	100.3	138.9	117.6	101.5	104.3	102.8	89.1	93.0	90.8
04	MASONRY	102.7	94.8	97.9	111.2	125.7	120.0	101.0	85.0	91.3	99.9	135.2	121.4	105.5	100.7	102.6	97.5	83.9	89.2
05	METALS	93.6	115.1	100.3	102.9	122.0	108.8	93.6	112.7	99.5	106.5	124.1	112.0	100.3	108.9	103.0	99.4	118.5	105.4
06	WOOD, PLASTICS & COMPOSITES	77.1	96.4	87.2	74.4	121.5	99.1	77.1	90.9	84.3	93.8	142.0	119.0	100.3	99.2	99.7	75.6	82.3	79.1
07	THERMAL & MOISTURE PROTECTION	102.8	94.3	98.8	108.8	124.3	115.4	102.6	87.6	96.2	103.7	137.2	118.1	99.9	99.6	99.8	96.3	95.2	95.8
08	OPENINGS	86.3	97.0	88.9	86.6	128.3	96.7	86.3	92.3	87.7	97.7	143.0	108.8	97.0	104.7	98.9	89.0	89.6	89.2
0920	Plaster & Gypsum Board	91.2	96.4	94.6	87.2	122.2	110.2	91.2	90.7	90.9	103.0	143.0	129.3	100.4	99.1	99.5	113.3	81.9	92.7
0950, 0980	Ceilings & Acoustic Treatment	96.1	96.4	96.3	95.1	122.2	112.1	96.1	90.7	92.7	99.7	143.0	126.9	106.5	99.1	101.9	96.9	81.9	87.5
0960	Flooring	86.4	103.4	91.3	87.1	132.9	100.5	86.4	98.9	90.0	98.4	149.0	113.2	99.8	106.7	101.8	88.3	98.9	91.4
0970, 0990	Wall Finishes & Painting/Coating	82.8	109.2	98.6	85.2	136.8	116.2	82.8	101.1	93.8	95.6	168.5	139.3	95.3	110.6	104.5	87.1	99.1	94.3
09	FINISHES	90.4	98.7	94.9	87.4	124.4	107.4	90.2	92.3	91.4	98.7	145.6	124.1	102.5	101.3	101.9	92.1	87.7	89.7
COVERS	DIVS. 10 - 14, 25, 28, 41, 43, 44, 46	100.0	100.3	100.1	100.0	113.6	103.2	100.0	98.2	99.6	100.0	119.6	104.6	100.0	101.4	100.3	100.0	98.3	99.6
21, 22, 23	FIRE SUPPRESSION, PLUMBING & HVAC	96.7	96.0	96.4	95.7	124.4	107.3	96.7	89.8	93.9	96.4	136.9	112.7	96.5	98.6	97.3	95.1	91.2	93.5
26, 27, 3370	ELECTRICAL, COMMUNICATIONS & UTIL.	91.2	100.8	95.9	93.4	136.7	114.9	92.7	103.7	98.2	99.2	157.6	128.1	96.8	111.5	104.1	93.2	87.9	90.6
MF2018	WEIGHTED AVERAGE	92.2	98.9	95.1	95.1	123.4	107.3	92.1	94.4	93.1	99.7	136.9	115.8	99.3	102.6	100.7	94.0	92.5	93.3

PENNSYLVANIA

DIVISION		READING 195 - 196			SCRANTON 184 - 185			STATE COLLEGE 168			STROUDSBURG 183			SUNBURY 178			UNIONTOWN 154		
		MAT.	INST.	TOTAL	MAT.	INST.	TOTAL	MAT.	INST.	TOTAL	MAT.	INST.	TOTAL	MAT.	INST.	TOTAL	MAT.	INST.	TOTAL
015433	CONTRACTOR EQUIPMENT		117.0	117.0		110.8	110.8		110.0	110.0		110.8	110.8		110.8	110.8		108.0	108.0
0241, 31 - 34	SITE & INFRASTRUCTURE, DEMOLITION	99.9	105.0	103.4	92.1	94.6	93.8	82.5	93.8	90.3	86.2	94.2	91.7	94.0	93.5	93.7	97.4	93.0	94.4
0310	Concrete Forming & Accessories	98.4	89.2	90.6	99.0	89.9	91.3	84.0	88.0	87.4	85.8	87.2	87.0	92.3	81.7	83.2	74.5	91.8	89.2
0320	Concrete Reinforcing	98.8	147.8	122.4	97.5	112.5	104.8	95.8	109.9	102.6	97.8	110.9	104.1	100.5	106.9	103.6	93.0	121.3	106.7
0330	Cast-in-Place Concrete	74.9	97.3	83.4	92.4	93.5	92.8	89.1	90.8	89.7	87.1	90.8	88.5	92.9	91.1	92.2	97.5	96.3	97.0
03	CONCRETE	85.9	103.6	93.8	93.6	96.4	94.9	93.7	94.1	93.9	88.2	94.0	90.8	92.7	90.9	91.9	93.3	99.6	96.1
04	MASONRY	99.8	92.0	95.0	97.0	94.1	95.2	102.4	92.9	96.6	94.4	92.3	93.1	98.1	77.6	85.6	133.8	94.1	109.6
05	METALS	103.2	135.3	113.2	101.9	120.9	107.8	97.4	118.8	104.1	99.5	119.0	105.6	99.2	117.4	104.9	98.6	121.2	105.6
06	WOOD, PLASTICS & COMPOSITES	94.0	86.1	89.9	98.2	88.5	93.1	83.4	86.2	84.8	82.3	84.1	83.2	84.9	82.3	83.5	70.4	90.8	81.1
07	THERMAL & MOISTURE PROTECTION	109.9	106.2	108.3	105.1	92.7	99.8	104.3	105.9	105.0	104.7	87.2	97.2	97.3	88.5	93.6	99.2	94.6	97.2
08	OPENINGS	90.8	103.1	93.8	92.5	92.7	92.5	89.7	89.6	89.6	93.0	91.1	92.5	89.1	86.6	88.5	93.5	97.4	94.4
0920	Plaster & Gypsum Board	96.3	85.9	89.5	102.4	88.2	93.1	92.2	85.9	88.1	90.9	83.7	86.2	112.5	81.9	92.4	94.7	90.7	92.1
0950, 0980	Ceilings & Acoustic Treatment	86.7	85.9	86.2	102.4	88.2	93.5	92.3	85.9	88.3	91.6	83.7	86.7	93.7	81.9	86.3	112.7	90.7	98.9
0960	Flooring	91.9	94.4	92.6	91.9	96.9	93.4	89.3	96.4	91.4	86.6	89.8	87.5	89.2	98.9	92.1	88.7	105.0	93.5
0970, 0990	Wall Finishes & Painting/Coating	84.1	108.2	98.5	87.1	108.2	99.7	87.1	112.0	102.0	87.1	99.1	94.3	87.1	99.1	94.3	89.7	109.2	101.4
09	FINISHES	88.4	90.7	89.7	95.2	92.6	93.8	89.7	91.3	90.6	89.2	88.2	88.7	92.9	86.1	89.2	99.6	95.3	97.3
COVERS	DIVS. 10 - 14, 25, 28, 41, 43, 44, 46	100.0	98.9	99.8	100.0	99.6	99.9	100.0	96.9	99.3	100.0	100.1	100.0	100.0	94.2	98.6	100.0	99.5	99.9
21, 22, 23	FIRE SUPPRESSION, PLUMBING & HVAC	101.1	114.5	106.5	101.2	96.0	99.1	97.4	96.3	97.0	97.4	93.3	95.7	95.1	83.2	90.3	93.1	88.9	91.4
26, 27, 3370	ELECTRICAL, COMMUNICATIONS & UTIL.	99.5	96.5	98.0	99.5	92.0	95.8	93.3	113.3	103.2	94.8	136.5	115.4	93.6	88.5	91.1	92.7	113.3	102.9
MF2018	WEIGHTED AVERAGE	97.3	104.7	100.5	98.3	96.8	97.7	95.2	99.1	96.9	95.0	100.9	97.6	95.0	89.3	92.5	97.1	99.2	98.0

For customer support on your Concrete & Masonry Costs with RSMeans Data, call 800.448.8182.

City Cost Indexes - V2

PENNSYLVANIA

DIVISION		WASHINGTON 153			WELLSBORO 169			WESTCHESTER 193			WILKES-BARRE 186 - 187			WILLIAMSPORT 177			YORK 173 - 174		
		MAT.	INST.	TOTAL	MAT.	INST.	TOTAL	MAT.	INST.	TOTAL	MAT.	INST.	TOTAL	MAT.	INST.	TOTAL	MAT.	INST.	TOTAL
015433	CONTRACTOR EQUIPMENT		108.0	108.0		110.8	110.8		95.3	95.3		110.8	110.8		110.8	110.8		110.0	110.0
0241, 31 - 34	SITE & INFRASTRUCTURE, DEMOLITION	97.5	93.9	95.0	93.5	93.4	93.5	100.7	92.8	95.3	84.2	94.6	91.4	85.2	94.4	91.5	84.7	94.3	91.3
0310	Concrete Forming & Accessories	81.9	98.1	95.7	84.2	81.0	81.5	88.6	121.9	117.0	88.8	89.9	89.7	88.6	86.9	87.2	84.0	87.3	86.8
0320	Concrete Reinforcing	93.0	121.5	106.8	95.8	110.6	102.9	96.5	143.7	119.3	96.5	112.5	104.2	99.7	112.5	105.9	100.5	112.5	106.3
0330	Cast-in-Place Concrete	97.5	96.9	97.3	93.4	84.6	90.1	92.7	124.5	104.7	83.6	93.4	87.3	78.6	91.6	83.5	85.4	96.6	89.6
03	CONCRETE	93.8	102.8	97.8	95.9	89.0	92.8	94.2	126.1	108.5	85.3	96.4	90.3	80.9	94.5	87.0	90.3	96.4	93.0
04	MASONRY	114.3	96.7	103.5	102.6	78.0	87.6	105.6	129.3	120.1	110.4	94.1	100.5	89.7	91.9	91.1	99.5	89.5	93.4
05	METALS	98.6	122.4	106.0	97.5	117.7	103.8	102.9	122.0	108.8	97.6	120.9	104.9	99.2	120.9	106.0	100.9	121.2	107.2
06	WOOD, PLASTICS & COMPOSITES	79.0	98.8	89.3	80.7	81.5	81.1	81.9	121.5	102.6	85.1	88.5	86.8	81.0	86.2	83.7	79.4	85.7	82.7
07	THERMAL & MOISTURE PROTECTION	99.3	96.9	98.3	105.4	82.1	95.4	109.3	121.4	114.5	104.5	92.7	99.4	96.6	91.3	94.3	96.4	105.0	100.1
08	OPENINGS	93.5	103.6	95.9	92.3	86.7	91.0	86.6	128.3	96.7	89.8	92.7	90.5	89.1	91.4	89.7	89.0	87.4	88.6
0920	Plaster & Gypsum Board	98.4	98.9	98.7	90.8	81.1	84.4	87.8	122.2	110.4	91.9	88.2	89.5	113.3	85.9	95.3	113.3	85.4	95.0
0950, 0980	Ceilings & Acoustic Treatment	112.7	98.9	104.1	92.3	81.1	85.3	95.1	122.2	112.1	95.5	88.2	90.9	96.9	85.9	90.0	96.2	85.4	89.4
0960	Flooring	92.8	105.0	96.4	85.6	98.9	89.5	90.0	132.9	102.5	87.5	96.9	90.3	88.1	94.4	89.9	89.8	94.4	91.1
0970, 0990	Wall Finishes & Painting/Coating	89.7	109.2	101.4	87.1	99.1	94.3	85.2	136.8	116.2	87.1	108.2	99.7	87.1	108.2	99.7	87.1	89.5	88.6
09	FINISHES	101.3	100.3	100.7	89.8	86.1	87.8	88.8	125.3	108.6	90.4	92.6	91.6	92.7	90.0	91.2	94.7	88.0	90.0
COVERS	DIVS. 10 - 14, 25, 28, 41, 43, 44, 46	100.0	100.4	100.1	100.0	94.1	98.6	100.0	114.7	103.5	100.0	99.6	99.9	100.0	96.3	99.1	100.0	96.4	99.2
21, 22, 23	FIRE SUPPRESSION, PLUMBING & HVAC	93.1	92.5	92.8	97.4	83.2	91.7	95.7	126.3	108.0	97.4	96.0	96.8	95.1	93.4	94.4	100.9	96.0	98.9
26, 27, 3370	ELECTRICAL, COMMUNICATIONS & UTIL.	94.7	113.3	103.9	94.1	91.7	92.9	93.3	127.5	110.2	94.8	92.0	93.4	94.1	85.1	89.6	95.6	83.9	89.8
MF2018	WEIGHTED AVERAGE	96.6	101.9	98.9	96.2	89.3	93.2	96.1	123.2	107.8	94.9	96.8	95.7	92.8	94.2	93.4	96.2	94.6	95.5

DIVISION		PUERTO RICO SAN JUAN 009			RHODE ISLAND NEWPORT 028			PROVIDENCE 029			SOUTH CAROLINA AIKEN 298			BEAUFORT 299			CHARLESTON 294		
		MAT.	INST.	TOTAL	MAT.	INST.	TOTAL	MAT.	INST.	TOTAL	MAT.	INST.	TOTAL	MAT.	INST.	TOTAL	MAT.	INST.	TOTAL
015433	CONTRACTOR EQUIPMENT		83.7	83.7		96.9	96.9		102.4	102.4		101.8	101.8		101.8	101.8		101.8	101.8
0241, 31 - 34	SITE & INFRASTRUCTURE, DEMOLITION	129.7	83.1	97.7	86.0	95.7	92.7	89.3	104.5	99.8	133.3	83.3	99.0	128.2	83.3	97.4	112.3	83.5	92.5
0310	Concrete Forming & Accessories	86.6	21.0	30.8	101.8	120.8	118.0	101.6	121.0	118.1	97.7	62.5	67.8	96.2	38.1	46.8	95.2	62.9	67.7
0320	Concrete Reinforcing	179.7	18.0	101.7	105.1	121.3	112.9	99.2	121.3	109.9	94.4	57.8	76.8	93.5	63.6	79.1	93.4	64.2	79.3
0330	Cast-in-Place Concrete	97.1	31.1	72.2	68.7	116.8	86.9	94.2	118.3	103.3	89.6	65.3	80.4	89.5	65.3	80.4	105.1	65.6	90.2
03	CONCRETE	97.5	25.0	65.0	84.3	119.2	99.9	98.1	119.7	107.8	99.0	64.3	83.4	96.5	54.3	77.5	93.4	65.7	81.0
04	MASONRY	87.9	23.3	48.5	97.5	121.7	112.3	106.6	121.8	115.9	85.7	62.1	71.3	102.1	62.1	77.7	103.5	65.6	80.4
05	METALS	127.6	38.7	99.9	100.9	115.9	105.6	105.3	114.5	108.2	102.4	87.6	97.8	102.4	89.7	98.4	104.4	90.8	100.2
06	WOOD, PLASTICS & COMPOSITES	84.8	20.0	50.9	101.2	119.8	110.9	103.2	120.1	112.0	94.6	64.5	78.9	92.7	31.3	60.6	91.2	64.5	77.3
07	THERMAL & MOISTURE PROTECTION	136.9	26.4	89.5	105.0	117.2	110.2	109.2	118.1	113.1	96.7	61.5	81.6	96.3	58.5	80.1	95.3	63.5	81.7
08	OPENINGS	155.1	18.6	121.8	97.8	120.6	103.4	99.4	120.7	104.6	94.1	60.2	85.8	94.1	43.2	81.7	97.6	62.7	89.1
0920	Plaster & Gypsum Board	157.4	18.0	66.0	91.6	120.2	110.4	105.2	120.2	115.0	91.2	63.6	73.1	95.8	29.5	52.3	96.5	63.6	74.9
0950, 0980	Ceilings & Acoustic Treatment	226.3	18.0	95.8	84.2	120.2	106.8	96.9	120.2	111.5	80.1	63.6	69.8	85.8	29.5	50.5	85.8	63.6	71.9
0960	Flooring	212.2	28.4	158.5	88.7	125.9	99.5	89.1	125.9	99.8	101.5	85.7	96.9	103.0	72.3	94.1	102.7	80.0	96.1
0970, 0990	Wall Finishes & Painting/Coating	183.5	23.4	87.6	87.0	116.8	104.9	93.8	116.8	107.6	90.9	62.9	74.1	90.9	54.2	68.9	90.9	68.0	77.1
09	FINISHES	201.9	22.4	104.8	88.9	121.8	106.7	95.1	122.0	109.6	93.4	66.9	79.0	95.4	44.1	67.6	93.4	66.7	78.9
COVERS	DIVS. 10 - 14, 25, 28, 41, 43, 44, 46	99.3	25.1	81.8	100.0	107.8	101.8	100.0	108.3	101.9	100.0	83.3	96.1	100.0	76.1	94.4	100.0	83.3	96.1
21, 22, 23	FIRE SUPPRESSION, PLUMBING & HVAC	117.7	18.0	77.5	100.9	111.4	105.1	101.2	111.4	105.3	97.8	52.6	79.6	97.8	58.2	81.8	101.6	58.3	84.1
26, 27, 3370	ELECTRICAL, COMMUNICATIONS & UTIL.	117.5	24.1	71.3	101.3	97.6	99.4	102.1	97.6	99.8	95.2	59.6	77.6	98.3	63.9	81.3	96.9	56.8	77.1
MF2018	WEIGHTED AVERAGE	124.9	28.3	83.1	97.0	112.6	103.8	101.0	112.6	106.3	98.1	65.2	83.9	98.9	61.5	82.7	99.4	67.0	85.4

DIVISION		SOUTH CAROLINA COLUMBIA 290 - 292			FLORENCE 295			GREENVILLE 296			ROCK HILL 297			SPARTANBURG 293			SOUTH DAKOTA ABERDEEN 574		
		MAT.	INST.	TOTAL	MAT.	INST.	TOTAL	MAT.	INST.	TOTAL	MAT.	INST.	TOTAL	MAT.	INST.	TOTAL	MAT.	INST.	TOTAL
015433	CONTRACTOR EQUIPMENT		105.4	105.4		101.8	101.8		101.8	101.8		101.8	101.8		101.8	101.8		95.8	95.8
0241, 31 - 34	SITE & INFRASTRUCTURE, DEMOLITION	112.3	92.3	98.6	122.4	83.2	95.5	117.2	83.5	94.0	114.1	83.2	92.9	117.0	83.5	94.0	100.1	90.9	93.8
0310	Concrete Forming & Accessories	98.1	62.9	68.1	81.5	62.8	65.6	95.0	63.0	67.7	92.9	62.2	66.8	98.3	63.0	68.2	100.0	66.5	71.5
0320	Concrete Reinforcing	95.8	64.2	80.6	93.0	64.2	79.1	92.9	61.5	77.7	93.7	62.7	78.8	92.9	64.2	79.1	92.9	70.1	81.9
0330	Cast-in-Place Concrete	106.4	66.0	91.2	89.5	65.3	80.4	89.5	65.6	80.5	89.5	64.8	80.2	89.5	65.6	80.5	111.2	75.5	97.7
03	CONCRETE	93.4	65.7	81.0	91.4	65.5	79.8	90.4	65.3	79.2	88.6	64.9	78.0	90.6	65.8	79.5	100.0	71.2	87.1
04	MASONRY	95.8	65.6	77.4	85.8	65.5	73.4	83.5	65.6	72.5	109.3	61.3	80.0	85.8	65.6	73.5	118.1	69.8	88.6
05	METALS	101.5	88.3	97.4	103.2	90.0	99.1	103.2	90.5	99.3	102.4	89.3	98.3	103.2	90.9	99.4	91.7	81.5	88.6
06	WOOD, PLASTICS & COMPOSITES	95.4	64.5	79.3	75.9	64.5	69.9	91.1	64.5	77.2	89.2	64.5	76.3	95.5	64.5	79.3	98.9	62.7	79.9
07	THERMAL & MOISTURE PROTECTION	91.1	64.3	79.6	95.6	63.5	81.9	95.6	63.5	81.8	95.5	56.7	78.8	95.7	63.5	81.9	102.9	73.5	90.3
08	OPENINGS	99.8	62.7	90.7	94.2	62.7	86.5	94.1	62.4	86.4	94.1	61.3	86.1	94.1	62.7	86.5	96.9	55.4	86.8
0920	Plaster & Gypsum Board	94.4	63.6	74.2	85.3	63.6	71.1	89.8	63.6	72.6	89.2	63.6	72.4	92.0	63.6	73.4	99.6	62.2	75.1
0950, 0980	Ceilings & Acoustic Treatment	88.5	63.6	72.9	81.4	63.6	70.2	80.1	63.6	69.8	80.1	63.6	69.8	80.1	63.6	69.8	103.6	62.2	77.7
0960	Flooring	92.8	80.0	89.1	93.7	80.0	89.7	100.3	80.0	94.4	99.2	72.3	91.3	101.7	80.0	95.4	93.9	41.8	78.7
0970, 0990	Wall Finishes & Painting/Coating	92.4	68.0	77.8	90.9	68.0	77.1	90.9	68.0	77.1	90.9	62.9	74.1	90.9	68.0	77.1	83.5	31.8	52.5
09	FINISHES	92.2	66.7	78.4	89.4	66.7	77.1	91.2	66.7	77.9	90.5	64.4	76.4	91.9	66.7	78.3	94.1	59.6	75.4
COVERS	DIVS. 10 - 14, 25, 28, 41, 43, 44, 46	100.0	83.3	96.1	100.0	83.3	96.1	100.0	83.4	96.1	100.0	83.1	96.0	100.0	83.4	96.1	100.0	86.0	96.7
21, 22, 23	FIRE SUPPRESSION, PLUMBING & HVAC	101.0	56.5	83.0	101.6	56.5	83.4	101.6	55.5	83.0	97.8	52.3	79.4	101.6	55.5	83.0	100.6	51.4	80.8
26, 27, 3370	ELECTRICAL, COMMUNICATIONS & UTIL.	100.3	60.4	80.5	95.2	60.4	78.0	97.0	83.5	90.3	97.0	83.5	90.3	97.0	83.5	90.3	99.8	64.8	82.4
MF2018	WEIGHTED AVERAGE	98.8	67.6	85.3	97.4	67.0	84.2	97.5	70.0	85.6	97.3	68.2	84.7	97.7	70.2	85.8	99.1	66.7	85.1

SOUTH DAKOTA

DIVISION		MITCHELL 573			MOBRIDGE 576			PIERRE 575			RAPID CITY 577			SIOUX FALLS 570 - 571			WATERTOWN 572		
		MAT.	INST.	TOTAL	MAT.	INST.	TOTAL	MAT.	INST.	TOTAL	MAT.	INST.	TOTAL	MAT.	INST.	TOTAL	MAT.	INST.	TOTAL
015433	CONTRACTOR EQUIPMENT		95.8	95.8		95.8	95.8		100.9	100.9		95.8	95.8		102.0	102.0		95.8	95.8
0241, 31 - 34	SITE & INFRASTRUCTURE, DEMOLITION	96.3	90.2	92.1	96.3	90.2	92.1	99.5	98.4	98.7	98.4	90.5	92.9	93.7	101.1	98.8	96.2	90.2	92.1
0310	Concrete Forming & Accessories	98.9	42.8	51.2	87.8	43.2	49.9	99.0	46.2	54.1	108.3	55.8	63.7	101.8	84.8	87.4	83.7	72.1	73.8
0320	Concrete Reinforcing	92.4	67.3	80.3	94.8	67.2	81.5	94.8	97.5	96.1	86.9	97.6	92.0	97.3	99.8	98.5	89.8	67.2	78.9
0330	Cast-in-Place Concrete	108.0	50.1	86.2	108.0	74.2	95.3	112.8	77.8	99.6	107.2	75.4	95.2	88.9	82.6	86.5	108.0	74.1	95.2
03	CONCRETE	97.7	51.2	76.8	97.3	59.7	80.4	102.3	67.3	86.6	97.2	70.9	85.4	94.3	86.9	91.0	96.4	72.7	85.8
04	MASONRY	104.5	74.2	86.1	112.4	69.8	86.4	105.3	74.4	86.5	111.7	72.7	87.9	106.1	79.2	89.7	139.5	70.6	97.5
05	METALS	90.7	80.4	87.5	90.8	80.7	87.6	94.2	88.6	92.4	93.4	90.7	92.6	94.8	93.0	94.3	90.7	80.6	87.6
06	WOOD, PLASTICS & COMPOSITES	97.7	32.5	63.6	84.4	32.2	57.1	103.8	33.0	66.8	104.4	46.3	74.0	98.4	84.0	90.9	79.7	71.7	75.5
07	THERMAL & MOISTURE PROTECTION	102.7	70.1	88.7	102.6	71.7	89.4	103.0	72.7	90.0	103.3	77.2	92.1	103.3	86.1	95.9	102.4	73.2	89.8
08	OPENINGS	96.1	38.2	82.0	98.6	37.5	83.7	99.5	62.4	90.5	100.6	69.8	93.1	103.8	90.9	100.6	96.1	59.2	87.1
0920	Plaster & Gypsum Board	98.5	31.1	54.3	92.8	30.9	52.2	95.0	31.4	53.3	99.4	45.3	63.9	91.5	83.8	86.5	90.8	71.4	78.1
0950, 0980	Ceilings & Acoustic Treatment	100.4	31.1	57.0	103.6	30.9	58.0	109.6	31.4	60.6	104.9	45.3	67.6	105.6	83.8	91.9	100.4	71.4	82.3
0960	Flooring	93.5	74.8	88.1	89.0	44.3	76.0	99.9	29.8	79.4	93.2	69.2	86.2	95.8	77.2	90.3	87.6	41.8	74.2
0970, 0990	Wall Finishes & Painting/Coating	83.5	34.8	54.3	83.5	35.7	54.8	93.6	118.9	108.8	83.5	118.9	104.7	96.5	118.9	109.90	83.5	31.8	52.5
09	FINISHES	92.9	46.2	67.6	91.6	42.0	64.8	99.8	50.0	72.8	94.1	64.2	77.9	97.4	87.0	91.8	90.1	64.3	76.1
COVERS	DIVS. 10 - 14, 25, 28, 41, 43, 44, 46	100.0	77.8	94.8	100.0	77.7	94.7	100.0	85.6	96.7	100.0	86.1	96.7	100.0	91.4	98.0	100.0	81.4	95.6
21, 22, 23	FIRE SUPPRESSION, PLUMBING & HVAC	96.8	46.7	76.6	96.8	65.8	84.3	100.6	75.1	90.3	100.6	74.2	90.0	100.5	72.2	89.1	96.8	49.6	77.8
26, 27, 3370	ELECTRICAL, COMMUNICATIONS & UTIL.	98.3	61.0	79.9	99.8	38.3	69.3	102.9	45.9	74.7	96.7	45.9	71.6	102.6	66.7	84.8	97.6	61.0	79.5
MF2018	WEIGHTED AVERAGE	96.6	59.6	80.6	97.1	60.7	81.4	100.2	68.9	86.7	98.7	71.0	86.7	99.3	82.1	91.8	97.7	66.6	84.3

TENNESSEE

DIVISION		CHATTANOOGA 373 - 374			COLUMBIA 384			COOKEVILLE 385			JACKSON 383			JOHNSON CITY 376			KNOXVILLE 377 - 379		
		MAT.	INST.	TOTAL	MAT.	INST.	TOTAL	MAT.	INST.	TOTAL	MAT.	INST.	TOTAL	MAT.	INST.	TOTAL	MAT.	INST.	TOTAL
015433	CONTRACTOR EQUIPMENT		103.4	103.4		98.2	98.2		98.2	98.2		104.2	104.2		97.4	97.4		97.4	97.4
0241, 31 - 34	SITE & INFRASTRUCTURE, DEMOLITION	106.1	91.8	96.3	89.8	82.3	84.6	95.5	79.0	84.2	99.4	91.6	94.0	112.8	81.3	91.2	91.9	82.1	85.2
0310	Concrete Forming & Accessories	101.1	62.0	67.9	80.5	60.3	63.3	80.7	31.2	38.6	87.8	39.7	46.9	85.3	59.8	63.6	99.9	63.1	68.6
0320	Concrete Reinforcing	96.3	68.3	82.8	87.4	61.9	75.1	87.4	61.6	75.0	87.4	69.2	78.6	96.9	64.3	81.2	96.3	67.8	82.5
0330	Cast-in-Place Concrete	98.2	63.8	85.2	89.4	63.2	79.5	101.3	56.2	84.3	98.9	57.7	83.3	79.0	58.0	71.1	92.3	65.1	82.0
03	CONCRETE	95.3	65.4	81.9	91.9	63.3	79.1	101.8	47.6	77.5	92.6	53.3	75.0	104.8	61.7	85.5	92.8	66.2	80.9
04	MASONRY	100.5	56.9	73.9	115.5	53.2	77.5	110.6	37.8	66.2	115.8	43.8	71.9	115.3	43.9	71.7	78.5	56.3	65.0
05	METALS	94.4	90.6	93.2	93.2	87.6	91.5	93.3	86.6	91.2	95.8	90.3	94.1	91.7	88.7	90.8	94.8	90.0	93.3
06	WOOD, PLASTICS & COMPOSITES	108.9	62.7	84.7	67.8	61.2	64.4	68.0	28.4	47.3	82.4	38.3	59.3	80.3	65.3	72.4	97.4	63.3	79.5
07	THERMAL & MOISTURE PROTECTION	100.7	62.2	84.2	95.5	60.3	80.4	96.0	50.5	76.5	97.7	52.4	78.2	95.9	57.1	79.3	93.8	62.6	80.4
08	OPENINGS	101.7	61.3	91.8	91.0	51.3	81.3	91.0	34.2	77.1	97.4	43.8	84.3	98.0	62.1	89.3	95.5	56.6	86.0
0920	Plaster & Gypsum Board	81.2	62.3	68.8	88.2	60.8	70.2	88.2	27.1	48.1	90.2	37.2	55.4	101.5	65.0	77.5	108.9	62.9	78.7
0950, 0980	Ceilings & Acoustic Treatment	99.0	62.3	76.0	80.9	60.8	68.3	80.9	27.1	47.1	88.7	37.2	56.4	94.8	65.0	76.1	96.1	62.9	75.3
0960	Flooring	98.9	61.4	87.9	83.3	51.5	74.0	83.4	47.7	73.0	82.7	56.1	75.0	92.8	56.1	82.1	98.3	56.1	86.0
0970, 0990	Wall Finishes & Painting/Coating	92.3	56.7	71.0	81.9	53.8	65.0	81.9	53.8	65.0	84.0	60.6	70.0	89.1	56.7	69.7	89.1	56.7	69.7
09	FINISHES	94.8	61.0	76.5	87.8	57.7	71.5	88.3	34.7	59.3	88.1	43.5	64.0	98.2	58.9	76.9	92.3	60.9	75.3
COVERS	DIVS. 10 - 14, 25, 28, 41, 43, 44, 46	100.0	81.3	95.6	100.0	81.0	95.5	100.0	73.5	93.8	100.0	75.0	94.1	100.0	78.1	94.9	100.0	82.1	95.8
21, 22, 23	FIRE SUPPRESSION, PLUMBING & HVAC	101.1	56.7	83.2	98.3	70.2	87.0	98.3	63.9	84.5	101.0	58.6	83.9	100.8	53.0	81.5	100.8	62.9	85.5
26, 27, 3370	ELECTRICAL, COMMUNICATIONS & UTIL.	101.8	79.8	90.9	94.9	48.7	72.0	96.3	59.0	77.8	101.1	63.3	82.4	92.9	50.5	72.0	97.9	57.7	78.0
MF2018	WEIGHTED AVERAGE	99.0	68.7	85.9	95.2	64.6	82.0	96.6	56.1	79.1	98.1	60.0	81.6	99.2	60.5	82.5	95.7	66.0	82.9

TENNESSEE / TEXAS

DIVISION		MCKENZIE 382			MEMPHIS 375,380 - 381			NASHVILLE 370 - 372			ABILENE 795 - 796			AMARILLO 790 - 791			AUSTIN 786 - 787		
		MAT.	INST.	TOTAL	MAT.	INST.	TOTAL	MAT.	INST.	TOTAL	MAT.	INST.	TOTAL	MAT.	INST.	TOTAL	MAT.	INST.	TOTAL
015433	CONTRACTOR EQUIPMENT		98.2	98.2		99.2	99.2		106.9	106.9		88.0	88.0		94.3	94.3		92.5	92.5
0241, 31 - 34	SITE & INFRASTRUCTURE, DEMOLITION	95.1	79.2	84.2	90.0	91.3	90.9	103.9	99.0	100.5	93.8	84.9	87.7	93.9	94.4	94.2	97.0	91.0	92.9
0310	Concrete Forming & Accessories	89.4	33.4	41.8	98.0	72.1	76.0	102.1	68.6	73.6	99.0	60.4	66.2	99.3	54.0	60.8	98.7	54.8	61.3
0320	Concrete Reinforcing	87.6	62.5	75.5	104.0	71.5	88.3	102.4	69.0	86.3	97.1	50.5	74.6	97.8	50.4	74.9	92.2	45.2	69.5
0330	Cast-in-Place Concrete	99.1	57.3	83.3	93.9	69.8	84.8	91.2	70.5	83.4	86.0	66.6	78.7	83.9	67.6	77.8	93.6	67.1	83.6
03	CONCRETE	100.6	49.2	77.5	96.7	72.0	85.6	94.5	70.4	83.7	86.5	64.1	75.3	93.3	59.0	77.9	89.2	58.2	75.3
04	MASONRY	114.4	43.2	71.0	93.6	60.0	73.1	92.1	63.9	74.9	101.3	63.3	78.1	98.4	61.7	76.0	98.8	63.3	77.1
05	METALS	93.3	87.2	91.4	84.5	83.1	84.1	107.0	85.7	100.4	108.3	70.2	96.4	103.5	68.6	92.6	102.5	65.4	91.0
06	WOOD, PLASTICS & COMPOSITES	77.8	30.6	53.1	100.2	73.0	86.0	105.3	67.9	85.7	103.7	62.1	81.9	107.1	53.2	78.9	96.9	54.8	74.9
07	THERMAL & MOISTURE PROTECTION	96.1	51.4	76.9	93.2	67.3	82.1	99.2	68.9	86.2	100.4	64.7	85.1	102.3	63.8	85.8	96.3	65.1	82.9
08	OPENINGS	91.0	35.8	77.5	99.2	66.9	91.4	100.8	66.5	92.5	101.7	57.9	91.0	103.1	53.0	90.9	101.3	49.9	88.8
0920	Plaster & Gypsum Board	91.0	29.3	50.5	95.7	72.4	80.4	90.3	66.9	75.0	101.7	61.6	75.4	113.8	52.3	73.4	88.4	53.9	65.8
0950, 0980	Ceilings & Acoustic Treatment	80.9	29.3	48.6	101.5	72.4	83.2	89.3	66.9	75.3	97.3	61.6	74.9	95.4	52.3	68.4	99.4	53.9	70.9
0960	Flooring	86.2	51.5	76.1	98.1	56.1	85.8	96.8	64.1	87.2	96.5	71.2	89.1	96.3	67.8	88.0	94.5	60.9	84.7
0970, 0990	Wall Finishes & Painting/Coating	81.9	42.0	58.0	90.6	74.8	81.1	97.7	67.9	79.9	91.5	57.0	70.8	94.5	54.3	70.4	94.3	42.1	63.0
09	FINISHES	89.4	36.0	60.5	97.5	69.4	82.3	95.3	67.3	80.1	92.2	62.1	75.9	97.7	56.1	75.2	94.8	54.1	72.8
COVERS	DIVS. 10 - 14, 25, 28, 41, 43, 44, 46	100.0	74.3	94.0	100.0	85.3	96.5	100.0	85.1	96.5	100.0	79.6	95.2	100.0	79.4	95.1	100.0	79.1	95.1
21, 22, 23	FIRE SUPPRESSION, PLUMBING & HVAC	98.3	55.8	81.2	100.7	74.2	90.0	101.1	82.1	93.4	101.2	50.4	80.7	100.9	52.4	81.4	100.8	59.6	84.2
26, 27, 3370	ELECTRICAL, COMMUNICATIONS & UTIL.	96.1	53.3	74.9	101.6	64.8	83.4	99.6	62.0	81.0	97.6	55.2	76.6	100.3	60.6	80.6	99.4	59.4	79.6
MF2018	WEIGHTED AVERAGE	96.7	54.7	78.6	96.5	72.5	86.1	100.0	74.5	89.0	99.0	61.6	82.9	99.9	61.8	83.5	98.6	62.3	82.9

TEXAS

DIVISION		BEAUMONT 776 - 777			BROWNWOOD 768			BRYAN 778			CHILDRESS 792			CORPUS CHRISTI 783 - 784			DALLAS 752 - 753		
		MAT.	INST.	TOTAL	MAT.	INST.	TOTAL	MAT.	INST.	TOTAL	MAT.	INST.	TOTAL	MAT.	INST.	TOTAL	MAT.	INST.	TOTAL
015433	CONTRACTOR EQUIPMENT		92.1	92.1		88.0	88.0		92.1	92.1		88.0	88.0		97.9	97.9		107.6	107.6
0241, 31 - 34	SITE & INFRASTRUCTURE, DEMOLITION	90.8	88.8	89.4	102.1	84.8	90.3	81.3	88.6	86.3	104.0	84.9	90.8	141.2	79.9	99.1	108.1	99.1	101.9
0310	Concrete Forming & Accessories	102.4	57.2	63.9	97.6	58.5	64.4	80.7	56.1	59.8	97.5	58.7	64.5	106.9	54.1	62.0	96.4	62.9	67.9
0320	Concrete Reinforcing	93.9	59.3	77.2	84.6	48.7	67.3	96.3	48.7	73.3	97.3	48.7	73.8	82.4	50.1	66.8	103.3	51.8	78.5
0330	Cast-in-Place Concrete	98.6	63.2	85.2	107.8	56.3	88.3	79.0	56.8	70.6	88.2	56.4	76.2	114.2	67.3	96.5	100.1	69.8	88.7
03	CONCRETE	95.5	60.6	79.8	95.6	56.9	78.3	79.7	56.1	69.1	95.0	57.0	77.9	96.2	59.9	79.9	97.2	64.7	82.6
04	MASONRY	99.2	63.9	77.7	126.7	58.2	84.9	139.3	58.1	89.8	104.9	57.8	76.1	81.5	61.7	69.4	102.7	60.0	76.6
05	METALS	101.1	74.6	92.8	103.1	68.7	92.3	101.0	69.9	91.3	105.5	68.8	94.0	98.8	83.9	94.1	103.4	80.7	96.3
06	WOOD, PLASTICS & COMPOSITES	111.4	57.3	83.1	102.8	60.9	80.9	76.5	57.2	66.4	103.0	60.9	81.0	121.9	53.6	86.1	97.9	64.6	80.5
07	THERMAL & MOISTURE PROTECTION	92.8	64.7	80.7	92.9	61.1	79.2	86.2	62.1	75.8	101.0	61.0	83.8	99.3	63.1	83.8	90.4	67.8	80.7
08	OPENINGS	94.6	55.8	85.1	100.6	55.7	89.6	95.4	53.9	85.3	96.1	55.7	86.3	101.5	51.5	89.3	96.6	60.0	87.7
0920	Plaster & Gypsum Board	106.6	56.8	73.9	96.9	60.5	73.0	91.6	56.7	68.7	100.7	60.5	74.3	90.6	52.8	65.8	93.3	63.6	73.8
0950, 0980	Ceilings & Acoustic Treatment	116.3	56.8	79.0	92.5	60.5	72.4	105.7	56.7	75.0	93.5	60.5	72.8	103.2	52.8	71.6	109.6	63.6	80.9
0960	Flooring	120.5	77.1	107.8	76.0	52.1	69.0	89.7	63.9	82.1	94.7	52.1	82.3	111.3	73.7	100.3	97.7	69.8	89.5
0970, 0990	Wall Finishes & Painting/Coating	95.9	51.3	69.2	89.5	49.2	65.3	93.2	54.7	70.1	91.5	49.2	66.1	106.8	45.6	70.1	102.3	54.3	73.5
09	FINISHES	104.1	60.0	80.3	84.0	56.4	69.1	90.6	56.9	72.4	91.9	56.4	72.7	105.7	56.4	79.0	101.5	63.2	80.8
COVERS	DIVS. 10 - 14, 25, 28, 41, 43, 44, 46	100.0	80.3	95.4	100.0	79.1	95.1	100.0	79.8	95.2	100.0	79.1	95.1	100.0	78.4	94.9	100.0	81.3	95.6
21, 22, 23	FIRE SUPPRESSION, PLUMBING & HVAC	101.0	60.8	84.8	97.0	47.3	77.0	97.2	59.5	82.0	97.4	51.6	78.9	101.0	60.8	84.8	101.0	62.1	85.3
26, 27, 3370	ELECTRICAL, COMMUNICATIONS & UTIL.	99.6	61.0	80.5	99.4	44.4	72.2	98.0	61.0	79.7	97.6	53.7	75.9	96.7	66.1	81.6	98.2	60.5	79.5
MF2018	WEIGHTED AVERAGE	99.2	64.9	84.4	99.1	57.2	81.0	96.3	62.4	81.7	98.7	59.4	81.7	100.2	64.6	84.8	100.1	67.4	86.0

TEXAS

DIVISION		DEL RIO 788			DENTON 762			EASTLAND 764			EL PASO 798 - 799,885			FORT WORTH 760 - 761			GALVESTON 775		
		MAT.	INST.	TOTAL	MAT.	INST.	TOTAL	MAT.	INST.	TOTAL	MAT.	INST.	TOTAL	MAT.	INST.	TOTAL	MAT.	INST.	TOTAL
015433	CONTRACTOR EQUIPMENT		87.2	87.2		97.4	97.4		88.0	88.0		94.3	94.3		94.3	94.3		104.9	104.9
0241, 31 - 34	SITE & INFRASTRUCTURE, DEMOLITION	120.3	82.8	94.6	101.8	79.3	86.3	104.9	84.8	91.1	89.7	94.0	92.7	97.7	94.4	95.4	108.5	87.3	93.9
0310	Concrete Forming & Accessories	102.5	49.3	57.2	106.3	58.9	65.9	98.7	58.6	64.6	100.5	57.8	64.2	97.1	60.7	66.1	89.8	53.8	59.2
0320	Concrete Reinforcing	83.0	45.3	64.8	86.1	48.9	68.1	84.9	48.7	67.4	99.2	50.5	75.7	93.6	51.4	73.2	95.7	56.9	77.0
0330	Cast-in-Place Concrete	122.9	55.8	97.6	83.3	57.5	73.5	113.9	56.4	92.2	76.1	66.4	72.4	96.9	67.4	85.7	104.9	57.8	87.1
03	CONCRETE	115.8	51.8	87.1	75.3	58.5	67.8	100.3	57.0	80.9	90.0	60.2	76.6	89.5	62.1	77.2	96.9	57.8	79.3
04	MASONRY	95.8	58.0	72.8	135.3	60.1	89.4	95.1	58.2	72.6	96.8	63.3	76.3	93.8	61.2	73.9	96.0	58.2	72.9
05	METALS	98.6	65.4	88.2	102.6	84.1	96.9	102.9	68.7	92.2	103.7	67.8	92.5	105.4	69.1	94.1	102.6	89.1	98.4
06	WOOD, PLASTICS & COMPOSITES	103.1	48.6	74.6	115.7	61.0	87.1	109.5	60.9	84.1	94.1	58.4	75.4	96.5	62.3	78.6	92.5	53.9	72.3
07	THERMAL & MOISTURE PROTECTION	96.4	60.4	81.0	91.4	62.4	79.0	93.2	61.1	79.4	94.6	61.5	80.4	90.9	65.1	79.8	85.6	62.4	75.7
08	OPENINGS	96.7	46.0	84.4	118.8	55.4	103.4	70.6	55.7	67.0	96.3	53.7	85.9	101.0	58.2	90.6	99.4	53.9	88.3
0920	Plaster & Gypsum Board	86.9	47.8	61.3	101.8	60.5	74.7	96.9	60.5	73.0	112.0	57.6	76.3	106.5	61.6	77.1	99.1	53.2	69.0
0950, 0980	Ceilings & Acoustic Treatment	98.6	47.8	66.8	96.3	60.5	73.8	92.5	60.5	72.4	92.4	57.6	70.6	100.9	61.6	76.3	110.8	53.2	74.7
0960	Flooring	95.7	52.1	83.0	72.0	52.1	66.2	95.3	52.1	82.7	97.6	70.6	89.7	95.3	61.5	85.5	106.0	63.9	93.7
0970, 0990	Wall Finishes & Painting/Coating	95.1	40.5	62.4	100.1	46.5	68.0	90.8	49.2	65.9	98.2	49.2	68.8	100.8	54.4	73.0	106.0	51.1	73.1
09	FINISHES	98.2	48.2	71.2	82.8	56.2	68.4	89.9	56.4	71.8	96.8	59.1	76.4	97.6	60.0	77.3	101.0	54.6	75.9
COVERS	DIVS. 10 - 14, 25, 28, 41, 43, 44, 46	100.0	77.2	94.6	100.0	79.4	95.1	100.0	79.1	95.1	100.0	78.9	95.0	100.0	80.2	95.3	100.0	79.7	95.2
21, 22, 23	FIRE SUPPRESSION, PLUMBING & HVAC	97.1	56.6	80.8	97.0	53.3	79.4	97.0	47.3	77.0	100.9	63.7	85.9	100.8	56.8	83.1	97.1	59.6	82.0
26, 27, 3370	ELECTRICAL, COMMUNICATIONS & UTIL.	98.6	57.6	78.3	101.7	56.5	79.3	99.3	56.4	78.1	100.4	49.6	75.3	100.8	59.8	80.5	99.5	61.1	80.5
MF2018	WEIGHTED AVERAGE	100.7	58.2	82.3	98.9	61.6	82.8	95.7	58.9	79.8	98.3	63.3	83.2	99.0	63.9	83.9	98.8	64.0	83.8

TEXAS

DIVISION		GIDDINGS 789			GREENVILLE 754			HOUSTON 770 - 772			HUNTSVILLE 773			LAREDO 780			LONGVIEW 756		
		MAT.	INST.	TOTAL	MAT.	INST.	TOTAL	MAT.	INST.	TOTAL	MAT.	INST.	TOTAL	MAT.	INST.	TOTAL	MAT.	INST.	TOTAL
015433	CONTRACTOR EQUIPMENT		87.2	87.2		97.5	97.5		102.2	102.2		92.1	92.1		87.2	87.2		88.8	88.8
0241, 31 - 34	SITE & INFRASTRUCTURE, DEMOLITION	105.5	83.1	90.1	97.4	81.8	86.7	107.7	94.9	98.9	97.2	88.6	91.3	99.5	83.3	88.3	95.1	87.7	90.0
0310	Concrete Forming & Accessories	100.2	49.4	57.0	87.0	55.5	60.2	94.8	59.5	64.8	88.7	52.2	57.7	102.5	53.7	61.0	82.4	58.7	62.2
0320	Concrete Reinforcing	83.7	46.2	65.6	103.8	48.9	77.3	95.5	50.6	73.8	96.5	48.7	73.5	83.0	50.1	67.1	102.7	48.4	76.5
0330	Cast-in-Place Concrete	104.1	56.1	86.0	102.5	57.4	85.5	97.5	68.6	86.6	108.7	56.7	89.1	87.8	62.8	78.4	109.5	56.8	95.5
03	CONCRETE	93.4	52.2	74.9	93.2	56.8	76.9	94.9	62.2	80.2	103.4	54.3	81.3	78.5	57.2	74.1	109.5	56.8	85.9
04	MASONRY	103.2	58.1	75.7	163.6	60.1	100.5	94.2	63.7	75.6	136.9	58.1	88.8	89.3	62.4	72.9	159.1	56.8	96.7
05	METALS	98.0	66.4	88.2	100.8	82.5	95.1	105.6	75.0	96.1	100.9	69.8	91.2	101.2	68.5	91.0	93.6	67.4	85.4
06	WOOD, PLASTICS & COMPOSITES	102.4	48.6	74.3	84.2	56.4	69.7	98.0	59.8	78.0	86.5	52.1	68.5	103.1	53.5	77.1	77.8	61.0	69.0
07	THERMAL & MOISTURE PROTECTION	96.9	61.1	81.5	89.4	61.3	77.3	87.8	68.4	79.5	87.3	61.5	76.2	95.7	63.2	81.7	90.6	60.2	77.6
08	OPENINGS	95.9	46.8	83.9	92.3	53.2	82.8	106.3	56.9	94.3	95.4	50.5	84.4	96.9	50.4	85.5	83.5	55.3	76.6
0920	Plaster & Gypsum Board	86.1	47.8	61.0	79.3	55.6	63.7	100.1	58.7	72.9	95.5	51.5	66.6	88.1	52.8	64.9	77.5	60.5	66.3
0950, 0980	Ceilings & Acoustic Treatment	98.6	47.8	66.8	103.3	55.6	73.4	103.7	58.7	75.5	105.7	51.5	71.7	103.0	52.8	71.6	96.3	60.5	73.8
0960	Flooring	95.4	52.1	82.8	92.3	52.1	80.5	105.4	75.2	96.6	94.3	52.1	82.0	95.6	60.9	85.5	98.2	52.1	84.7
0970, 0990	Wall Finishes & Painting/Coating	95.1	42.1	63.3	93.4	49.2	66.9	106.5	58.0	77.4	93.2	54.7	70.1	95.1	42.1	63.3	84.8	46.5	61.9
09	FINISHES	96.9	48.4	70.6	95.3	53.8	72.9	105.6	62.0	82.0	93.3	51.9	70.9	97.7	53.3	73.7	99.7	56.2	76.1
COVERS	DIVS. 10 - 14, 25, 28, 41, 43, 44, 46	100.0	77.5	94.7	100.0	79.0	95.1	100.0	82.1	95.8	100.0	79.2	95.1	100.0	78.0	94.8	100.0	75.1	94.1
21, 22, 23	FIRE SUPPRESSION, PLUMBING & HVAC	97.2	58.5	81.6	97.3	55.4	80.4	101.0	63.8	85.9	97.2	59.5	82.0	100.9	56.8	83.1	97.2	55.0	80.2
26, 27, 3370	ELECTRICAL, COMMUNICATIONS & UTIL.	95.9	53.5	74.9	95.4	56.5	76.1	102.0	65.1	83.7	98.0	56.9	77.7	98.7	59.1	79.1	95.5	49.0	72.5
MF2018	WEIGHTED AVERAGE	97.3	58.3	80.5	99.6	61.3	83.1	101.3	67.3	86.6	99.9	60.7	83.0	97.6	60.9	81.8	99.8	59.2	82.3

For customer support on your Concrete & Masonry Costs with RSMeans Data, call 800.448.8182.

485

TEXAS

DIVISION		LUBBOCK 793 - 794			LUFKIN 759			MCALLEN 785			MCKINNEY 750			MIDLAND 797			ODESSA 797		
		MAT.	INST.	TOTAL	MAT.	INST.	TOTAL	MAT.	INST.	TOTAL	MAT.	INST.	TOTAL	MAT.	INST.	TOTAL	MAT.	INST.	TOTAL
015433	CONTRACTOR EQUIPMENT		100.1	100.1		88.8	88.8		98.0	98.0		97.5	97.5		100.1	100.1		88.0	88.0
0241, 31 - 34	SITE & INFRASTRUCTURE, DEMOLITION	118.4	83.8	94.7	90.0	87.1	88.0	145.4	80.0	100.5	93.8	81.8	85.5	120.9	83.8	95.4	94.0	85.0	87.8
0310	Concrete Forming & Accessories	99.3	53.1	60.0	85.7	52.4	57.4	106.9	49.9	58.4	86.0	55.5	60.1	103.8	60.4	66.9	99.0	60.4	66.2
0320	Concrete Reinforcing	98.3	50.4	75.2	104.6	63.6	84.8	82.6	50.0	66.9	103.8	48.9	77.3	99.4	50.2	75.7	97.1	50.3	74.5
0330	Cast-in-Place Concrete	86.2	66.7	78.8	106.7	55.7	87.4	124.0	57.8	99.0	96.1	57.4	81.5	91.6	64.3	81.3	86.0	65.7	78.4
03	CONCRETE	85.3	59.3	73.6	101.4	56.3	81.2	103.7	54.6	81.7	88.3	56.8	74.2	89.4	61.7	77.0	86.5	61.3	75.2
04	MASONRY	100.6	61.6	76.9	120.5	58.0	82.3	96.1	61.7	75.1	176.4	60.1	105.4	117.8	59.2	82.1	101.3	61.6	77.1
05	METALS	112.0	84.9	103.5	100.9	71.4	91.7	98.5	83.2	93.7	100.7	82.5	95.0	110.1	84.7	102.1	107.5	69.9	95.8
06	WOOD, PLASTICS & COMPOSITES	104.8	52.1	77.2	84.8	52.2	67.8	120.0	48.7	82.8	83.0	56.4	69.1	110.2	62.2	85.1	103.7	62.1	81.9
07	THERMAL & MOISTURE PROTECTION	90.3	63.4	78.7	90.4	59.6	77.2	99.2	62.4	83.4	89.2	61.3	77.2	90.5	63.4	78.9	100.4	63.6	84.6
08	OPENINGS	108.7	52.0	94.9	63.4	54.4	61.2	100.6	47.1	87.6	92.3	53.2	82.8	110.2	57.8	97.4	101.7	57.7	91.0
0920	Plaster & Gypsum Board	101.9	51.3	68.7	76.1	51.5	60.0	91.5	47.8	62.9	79.3	55.6	63.7	103.0	61.6	75.9	101.7	61.6	75.4
0950, 0980	Ceilings & Acoustic Treatment	97.9	51.3	68.7	91.2	51.5	66.3	103.0	47.8	68.4	103.3	55.6	73.4	94.7	61.6	74.0	97.3	61.6	74.9
0960	Flooring	90.1	67.7	83.6	131.5	52.1	108.3	110.8	73.7	100.0	91.7	52.1	80.2	91.3	60.9	82.4	96.5	66.5	87.7
0970, 0990	Wall Finishes & Painting/Coating	102.4	52.2	72.3	84.8	49.2	63.5	106.8	40.5	67.1	93.4	49.2	66.9	102.4	49.2	70.5	91.5	52.2	67.9
09	FINISHES	94.7	55.2	73.3	107.5	51.7	77.3	106.1	53.0	77.4	95.0	53.8	72.7	94.8	59.3	75.6	92.2	60.7	75.2
COVERS	DIVS. 10 - 14, 25, 28, 41, 43, 44, 46	100.0	78.9	95.0	100.0	79.5	95.2	100.0	77.7	94.7	100.0	79.0	95.1	100.0	80.0	95.3	100.0	79.7	95.2
21, 22, 23	FIRE SUPPRESSION, PLUMBING & HVAC	100.5	50.6	80.4	97.2	55.5	80.4	97.2	47.1	77.0	97.3	57.5	81.2	96.7	50.6	78.1	101.2	50.6	80.8
26, 27, 3370	ELECTRICAL, COMMUNICATIONS & UTIL.	96.4	52.2	74.5	96.7	59.1	78.1	96.5	30.4	63.8	95.5	56.5	76.2	96.4	52.6	74.7	97.7	52.5	75.3
MF2018	WEIGHTED AVERAGE	100.4	60.7	83.3	96.6	60.5	81.0	100.9	55.2	81.2	99.5	61.8	83.2	100.9	61.7	84.0	98.9	60.8	82.5

TEXAS

DIVISION		PALESTINE 758			SAN ANGELO 769			SAN ANTONIO 781 - 782			TEMPLE 765			TEXARKANA 755			TYLER 757		
		MAT.	INST.	TOTAL	MAT.	INST.	TOTAL	MAT.	INST.	TOTAL	MAT.	INST.	TOTAL	MAT.	INST.	TOTAL	MAT.	INST.	TOTAL
015433	CONTRACTOR EQUIPMENT		88.8	88.8		88.0	88.0		95.0	95.0		88.0	88.0		88.8	88.8		88.8	88.8
0241, 31 - 34	SITE & INFRASTRUCTURE, DEMOLITION	95.3	87.7	90.1	98.5	84.9	89.1	97.5	95.9	96.4	87.3	84.4	85.3	84.7	87.8	86.8	94.3	87.8	89.8
0310	Concrete Forming & Accessories	76.2	58.7	61.3	97.9	49.7	56.9	101.3	55.7	62.5	101.9	49.3	57.2	93.9	59.0	64.2	87.6	58.8	63.1
0320	Concrete Reinforcing	102.0	48.7	76.3	84.5	50.2	67.9	89.1	47.3	68.9	84.7	48.4	67.2	101.8	51.2	77.4	102.7	51.2	77.9
0330	Cast-in-Place Concrete	97.5	56.3	81.9	101.7	62.4	86.8	83.8	69.4	78.3	83.3	55.9	72.9	98.2	64.8	85.6	117.2	56.3	94.2
03	CONCRETE	102.4	56.9	82.0	91.2	55.3	75.1	86.3	59.6	74.3	77.8	52.5	66.5	94.1	60.4	79.0	109.0	57.4	85.5
04	MASONRY	115.0	56.8	79.5	123.2	58.2	83.6	89.0	63.4	73.4	134.3	58.1	87.8	181.0	56.8	105.3	169.7	56.8	100.9
05	METALS	100.7	67.5	90.3	103.3	69.4	92.7	103.4	64.2	91.2	102.9	67.2	91.8	93.5	68.8	85.8	100.5	68.5	90.5
06	WOOD, PLASTICS & COMPOSITES	75.0	61.0	67.7	103.1	48.6	74.6	103.8	54.9	78.2	112.5	48.6	79.1	91.3	61.0	75.5	86.7	61.0	73.3
07	THERMAL & MOISTURE PROTECTION	90.9	60.4	77.8	92.7	60.9	79.1	88.7	67.6	79.6	92.3	60.1	78.5	90.3	61.9	78.1	90.8	60.4	77.7
08	OPENINGS	63.3	55.7	61.5	100.6	49.3	88.1	102.3	51.7	90.0	67.3	48.9	62.8	83.5	56.3	76.8	63.3	56.3	61.6
0920	Plaster & Gypsum Board	75.6	60.5	65.7	96.9	47.8	64.7	92.1	53.9	67.0	96.9	47.8	64.7	82.6	60.5	68.1	76.1	60.5	65.8
0950, 0980	Ceilings & Acoustic Treatment	91.2	60.5	71.9	92.5	47.8	64.5	93.2	53.9	68.5	92.5	47.8	64.5	96.3	60.5	73.8	91.2	60.5	71.9
0960	Flooring	121.6	52.1	101.3	76.1	52.1	69.1	105.7	73.9	96.4	96.9	52.1	83.8	107.4	60.9	93.8	134.1	52.1	110.2
0970, 0990	Wall Finishes & Painting/Coating	84.8	49.2	63.5	89.5	49.2	65.3	101.1	45.6	67.8	90.8	42.1	61.6	84.8	49.2	63.5	84.8	49.2	63.5
09	FINISHES	105.2	56.5	78.8	83.8	49.2	65.1	106.7	57.7	80.2	89.1	48.4	67.1	102.5	58.2	78.5	108.6	56.5	80.4
COVERS	DIVS. 10 - 14, 25, 28, 41, 43, 44, 46	100.0	79.2	95.1	100.0	77.5	94.7	100.0	79.7	95.2	100.0	77.8	94.8	100.0	79.2	95.1	100.0	79.2	95.1
21, 22, 23	FIRE SUPPRESSION, PLUMBING & HVAC	97.2	54.1	79.9	97.0	49.8	78.0	100.9	60.0	84.4	97.0	50.6	78.3	97.2	53.0	79.4	97.2	55.6	80.4
26, 27, 3370	ELECTRICAL, COMMUNICATIONS & UTIL.	93.3	46.6	70.2	103.2	52.6	78.1	100.2	61.5	81.0	100.4	50.1	75.5	96.6	52.5	74.8	95.5	51.7	73.9
MF2018	WEIGHTED AVERAGE	95.9	58.9	80.0	98.7	57.3	80.8	98.9	63.8	83.7	94.1	56.4	77.8	99.1	60.4	82.4	100.1	60.2	82.8

DIVISION		TEXAS VICTORIA 779			WACO 766 - 767			WAXAHACHIE 751			WHARTON 774			WICHITA FALLS 763			UTAH LOGAN 843		
		MAT.	INST.	TOTAL	MAT.	INST.	TOTAL	MAT.	INST.	TOTAL	MAT.	INST.	TOTAL	MAT.	INST.	TOTAL	MAT.	INST.	TOTAL
015433	CONTRACTOR EQUIPMENT		103.5	103.5		88.0	88.0		97.5	97.5		104.9	104.9		88.0	88.0		91.5	91.5
0241, 31 - 34	SITE & INFRASTRUCTURE, DEMOLITION	113.6	85.0	93.9	96.0	84.6	88.2	95.6	81.7	86.0	119.1	87.0	97.1	96.8	85.0	88.7	97.2	86.8	90.0
0310	Concrete Forming & Accessories	89.8	49.6	55.6	100.2	60.4	66.3	86.0	58.9	63.0	84.0	51.3	56.1	100.2	60.5	66.4	100.2	70.4	74.9
0320	Concrete Reinforcing	91.9	48.5	71.0	84.3	47.7	66.7	103.8	48.9	77.3	84.3	48.5	72.9	84.3	50.4	68.0	101.6	87.8	94.9
0330	Cast-in-Place Concrete	117.8	57.7	95.1	90.1	65.7	80.9	101.5	57.2	84.8	121.1	56.8	96.8	96.3	71.1	86.7	85.1	75.0	81.3
03	CONCRETE	104.9	54.4	82.2	84.2	60.8	73.7	92.2	58.3	77.0	108.9	54.8	84.6	87.1	63.2	76.4	105.3	75.5	91.9
04	MASONRY	113.9	58.2	79.9	93.2	62.4	74.4	164.3	60.0	100.7	97.3	58.1	73.4	93.7	62.0	74.4	108.5	63.1	80.8
05	METALS	101.0	85.7	96.2	105.4	68.1	93.7	100.8	82.2	95.0	102.6	85.2	97.2	105.3	70.2	94.4	107.5	84.3	100.3
06	WOOD, PLASTICS & COMPOSITES	95.1	48.7	70.9	110.5	62.1	85.2	83.0	61.1	71.6	85.0	50.8	67.1	110.5	62.1	85.2	83.0	70.7	76.5
07	THERMAL & MOISTURE PROTECTION	88.6	59.5	76.1	93.1	64.9	81.0	89.3	61.8	77.5	85.9	62.0	75.7	93.1	63.8	80.5	100.9	68.7	87.0
08	OPENINGS	99.2	49.1	87.0	78.2	57.5	73.2	92.3	55.8	83.4	99.4	50.3	87.4	78.2	57.9	73.2	92.4	73.1	87.7
0920	Plaster & Gypsum Board	94.7	47.8	63.9	97.0	61.6	73.8	79.8	60.5	67.1	94.3	50.0	65.2	97.0	61.6	73.8	83.4	70.2	74.7
0950, 0980	Ceilings & Acoustic Treatment	112.1	47.8	71.8	93.1	61.6	73.4	105.2	60.5	73.4	110.8	50.0	72.7	93.1	61.6	73.4	113.9	70.2	86.5
0960	Flooring	104.3	52.1	89.0	96.3	69.9	88.6	91.7	52.1	80.2	103.0	63.5	91.5	96.9	74.3	90.3	96.5	62.9	86.7
0970, 0990	Wall Finishes & Painting/Coating	106.2	54.7	75.3	90.8	57.0	70.6	93.4	49.2	66.9	106.0	53.1	74.3	92.8	54.3	69.7	89.3	60.8	72.3
09	FINISHES	98.3	49.9	72.1	89.5	61.8	74.5	95.7	56.6	74.5	100.1	52.9	74.6	89.9	62.4	75.0	97.3	67.7	81.3
COVERS	DIVS. 10 - 14, 25, 28, 41, 43, 44, 46	100.0	79.1	95.1	100.0	79.6	95.2	100.0	79.5	95.2	100.0	79.4	95.1	100.0	79.6	95.2	100.0	85.6	96.6
21, 22, 23	FIRE SUPPRESSION, PLUMBING & HVAC	97.2	59.4	81.9	100.8	59.7	84.2	97.3	55.4	80.4	97.1	59.0	81.7	100.8	54.3	82.1	100.9	74.6	90.3
26, 27, 3370	ELECTRICAL, COMMUNICATIONS & UTIL.	104.2	52.0	78.3	103.3	55.4	79.6	95.4	56.5	76.2	103.2	57.0	80.3	105.3	54.7	80.3	94.6	70.9	82.9
MF2018	WEIGHTED AVERAGE	101.0	60.8	83.6	95.8	63.2	81.7	99.5	62.0	83.3	101.0	62.1	84.2	96.5	62.5	81.8	100.7	74.0	89.2

For customer support on your Concrete & Masonry Costs with RSMeans Data, call 800.448.8182.

		UTAH												VERMONT					
		OGDEN			PRICE			PROVO			SALT LAKE CITY			BELLOWS FALLS			BENNINGTON		
DIVISION		842,844			845			846 - 847			840 - 841			051			052		
		MAT.	INST.	TOTAL	MAT.	INST.	TOTAL	MAT.	INST.	TOTAL	MAT.	INST.	TOTAL	MAT.	INST.	TOTAL	MAT.	INST.	TOTAL
015433	CONTRACTOR EQUIPMENT		91.5	91.5		90.6	90.6		90.6	90.6		91.5	91.5		94.2	94.2		94.2	94.2
0241, 31 - 34	SITE & INFRASTRUCTURE, DEMOLITION	85.1	86.8	86.3	94.4	84.4	87.5	93.5	85.3	87.9	84.9	86.7	86.1	86.8	93.3	91.3	86.2	93.3	91.1
0310	Concrete Forming & Accessories	100.2	70.4	74.9	102.9	63.3	69.2	102.1	70.4	75.2	102.6	70.4	75.2	99.6	98.4	98.6	96.7	98.2	98.0
0320	Concrete Reinforcing	101.3	87.8	94.7	108.9	80.0	95.0	109.9	87.8	99.2	103.4	87.8	95.9	82.2	82.8	82.5	82.2	82.8	82.5
0330	Cast-in-Place Concrete	86.4	75.0	82.1	85.2	70.8	79.7	85.2	75.0	81.3	94.4	75.0	87.1	86.5	111.7	96.0	86.5	111.6	96.0
03	CONCRETE	94.3	75.5	85.9	106.5	69.4	89.9	104.9	75.5	91.7	114.1	75.5	96.8	86.9	100.2	92.9	86.7	100.1	92.7
04	MASONRY	102.8	63.1	78.6	114.0	66.3	84.9	114.2	63.1	83.0	116.8	63.1	84.0	98.8	91.9	94.6	108.9	91.9	98.5
05	METALS	108.1	84.3	100.7	104.6	80.6	97.1	105.6	84.3	98.9	111.8	84.3	103.2	100.2	90.8	97.2	100.1	90.6	97.2
06	WOOD, PLASTICS & COMPOSITES	83.0	70.7	76.5	86.4	61.8	73.6	84.7	70.7	77.4	85.0	70.7	77.5	106.0	103.2	104.5	102.9	103.2	103.1
07	THERMAL & MOISTURE PROTECTION	99.7	68.7	86.4	102.7	67.0	87.4	102.8	68.7	88.1	107.6	68.7	90.9	101.4	87.7	95.5	101.3	87.7	95.5
08	OPENINGS	92.4	73.1	87.7	96.1	73.4	90.5	96.1	73.1	90.5	94.2	73.1	89.0	100.6	94.0	99.0	100.6	94.0	99.0
0920	Plaster & Gypsum Board	83.4	70.2	74.7	86.6	61.0	69.8	84.3	70.2	75.0	92.8	70.2	77.9	113.9	103.1	106.8	113.0	103.1	106.5
0950, 0980	Ceilings & Acoustic Treatment	113.9	70.2	86.5	113.9	61.0	80.8	113.9	70.2	86.5	106.8	70.2	83.8	81.4	103.1	95.0	81.4	103.1	95.0
0960	Flooring	94.6	62.9	85.3	97.7	53.5	84.8	97.5	62.9	87.4	98.6	62.9	88.1	90.3	110.7	96.3	89.4	110.7	95.6
0970, 0990	Wall Finishes & Painting/Coating	89.3	60.8	72.3	89.3	57.8	70.4	89.3	60.8	72.3	92.4	62.1	74.2	82.8	96.7	91.1	82.8	96.7	91.1
09	FINISHES	95.5	67.7	80.4	98.5	60.2	77.8	98.0	67.7	81.6	97.0	67.8	81.2	88.5	101.2	95.4	88.1	101.2	95.2
COVERS	DIVS. 10 - 14, 25, 28, 41, 43, 44, 46	100.0	85.6	96.6	100.0	84.6	96.4	100.0	85.6	96.6	100.0	85.6	96.6	100.0	100.5	100.1	100.0	100.4	100.1
21, 22, 23	FIRE SUPPRESSION, PLUMBING & HVAC	100.9	74.6	90.3	98.9	62.2	84.1	100.9	74.6	90.3	101.2	74.6	90.5	97.6	83.5	91.9	97.6	83.5	91.9
26, 27, 3370	ELECTRICAL, COMMUNICATIONS & UTIL.	95.0	70.9	83.1	98.9	66.2	82.7	95.1	71.6	83.4	97.5	68.6	83.2	104.3	76.6	90.6	104.3	52.4	78.6
MF2018	WEIGHTED AVERAGE	98.6	74.0	88.0	101.1	68.6	87.1	101.1	74.0	89.4	103.3	73.7	90.5	97.1	90.6	94.3	97.5	87.2	93.0

		VERMONT																	
		BRATTLEBORO			BURLINGTON			GUILDHALL			MONTPELIER			RUTLAND			ST. JOHNSBURY		
DIVISION		053			054			059			056			057			058		
		MAT.	INST.	TOTAL	MAT.	INST.	TOTAL	MAT.	INST.	TOTAL	MAT.	INST.	TOTAL	MAT.	INST.	TOTAL	MAT.	INST.	TOTAL
015433	CONTRACTOR EQUIPMENT		94.2	94.2		99.4	99.4		94.2	94.2		99.4	99.4		94.2	94.2		94.2	94.2
0241, 31 - 34	SITE & INFRASTRUCTURE, DEMOLITION	87.7	93.3	91.6	91.6	103.4	99.7	85.8	93.0	90.7	90.0	102.5	98.6	90.6	94.3	93.2	85.8	92.4	90.3
0310	Concrete Forming & Accessories	99.9	98.4	98.6	98.8	81.9	84.5	96.0	92.7	93.2	101.0	99.2	99.5	100.2	81.8	84.6	93.2	92.7	92.8
0320	Concrete Reinforcing	81.3	82.8	82.0	105.8	82.8	94.7	82.9	82.8	82.8	96.1	82.8	89.7	102.8	82.8	93.1	81.3	82.7	82.0
0330	Cast-in-Place Concrete	89.3	111.7	97.7	107.3	113.9	109.8	83.9	103.5	91.3	99.6	113.9	105.0	84.8	112.9	95.4	83.9	103.5	91.3
03	CONCRETE	88.9	100.2	93.9	104.9	93.4	99.8	84.4	94.8	89.1	100.2	101.2	100.6	89.7	93.1	91.2	84.0	94.8	88.8
04	MASONRY	108.6	91.9	98.4	109.6	94.2	100.2	108.9	77.8	90.0	108.7	94.2	99.9	89.7	94.1	92.4	136.9	77.8	100.9
05	METALS	100.1	90.8	97.2	108.3	89.2	102.4	100.2	90.5	97.1	106.7	89.2	101.3	106.5	90.6	101.6	100.2	90.4	97.1
06	WOOD, PLASTICS & COMPOSITES	106.4	103.2	104.7	99.7	79.8	89.3	101.1	103.2	102.2	101.3	103.3	102.3	106.6	79.7	92.5	95.1	103.2	99.4
07	THERMAL & MOISTURE PROTECTION	101.5	87.7	95.6	107.3	96.8	102.8	101.2	81.3	92.6	110.1	99.3	105.5	101.6	95.8	99.1	101.0	81.3	92.6
08	OPENINGS	100.6	94.2	99.1	103.1	81.5	97.9	100.6	94.4	99.1	103.6	94.5	101.4	103.7	81.5	98.3	100.6	94.4	99.1
0920	Plaster & Gypsum Board	113.9	103.1	106.8	115.4	79.0	91.5	120.4	103.1	109.0	117.5	103.1	108.0	114.4	79.0	91.1	122.4	103.1	109.7
0950, 0980	Ceilings & Acoustic Treatment	81.4	103.1	95.0	93.3	79.0	84.3	81.4	103.1	95.0	83.9	103.1	95.9	87.6	79.0	82.2	81.4	103.1	95.0
0960	Flooring	90.6	110.7	96.4	96.9	110.7	100.9	93.4	110.7	98.4	97.4	110.7	101.3	90.3	110.7	96.3	96.7	110.7	100.8
0970, 0990	Wall Finishes & Painting/Coating	82.8	96.7	91.1	95.3	104.6	100.9	82.8	104.6	95.9	98.3	104.6	102.1	82.8	104.6	95.9	82.8	104.6	95.9
09	FINISHES	88.7	101.2	95.4	97.0	88.9	92.6	90.2	98.5	94.7	95.4	102.7	99.4	90.2	88.8	89.4	91.5	98.5	95.3
COVERS	DIVS. 10 - 14, 25, 28, 41, 43, 44, 46	100.0	100.5	100.1	100.0	98.9	99.7	100.0	95.7	99.0	100.0	101.4	100.3	100.0	98.6	99.7	100.0	95.7	99.0
21, 22, 23	FIRE SUPPRESSION, PLUMBING & HVAC	97.6	83.5	91.9	101.2	68.1	87.8	97.6	59.5	82.3	97.4	68.1	85.6	101.4	68.0	87.9	97.6	59.5	82.3
26, 27, 3370	ELECTRICAL, COMMUNICATIONS & UTIL.	104.3	76.6	90.6	104.0	56.4	80.5	104.3	52.4	78.6	102.8	56.4	79.9	103.4	56.4	80.6	104.3	52.4	78.6
MF2018	WEIGHTED AVERAGE	97.8	90.7	94.7	103.1	82.3	94.1	97.3	79.2	89.5	101.2	86.0	94.6	99.4	81.6	91.7	98.7	79.2	90.3

		VERMONT			VIRGINIA														
		WHITE RIVER JCT.			ALEXANDRIA			ARLINGTON			BRISTOL			CHARLOTTESVILLE			CULPEPER		
DIVISION		050			223			222			242			229			227		
		MAT.	INST.	TOTAL	MAT.	INST.	TOTAL	MAT.	INST.	TOTAL	MAT.	INST.	TOTAL	MAT.	INST.	TOTAL	MAT.	INST.	TOTAL
015433	CONTRACTOR EQUIPMENT		94.2	94.2		103.4	103.4		102.2	102.2		102.2	102.2		106.2	106.2		102.2	102.2
0241, 31 - 34	SITE & INFRASTRUCTURE, DEMOLITION	90.2	92.5	91.8	114.5	86.0	94.9	125.0	83.8	96.7	108.9	84.0	91.8	113.5	85.7	94.4	112.0	83.5	92.4
0310	Concrete Forming & Accessories	93.7	93.2	93.2	94.9	72.8	76.1	93.1	61.8	66.5	87.7	59.2	63.4	85.9	60.9	64.6	82.9	68.7	70.8
0320	Concrete Reinforcing	82.2	82.7	82.5	86.9	88.5	87.7	98.0	86.2	92.3	98.0	70.4	84.7	97.4	71.9	85.1	98.0	88.5	93.4
0330	Cast-in-Place Concrete	89.3	104.3	94.9	107.2	76.7	95.7	104.3	75.2	93.3	103.9	49.2	83.3	108.1	73.3	94.9	106.8	74.5	94.6
03	CONCRETE	90.5	95.2	92.6	99.4	78.3	89.9	103.8	72.4	89.7	100.2	59.5	81.9	100.7	68.9	86.5	97.9	75.6	87.9
04	MASONRY	122.1	79.1	95.9	91.9	72.8	80.3	105.7	71.4	84.8	95.2	55.0	70.7	120.7	55.3	80.8	108.0	70.4	85.0
05	METALS	100.2	90.4	97.1	105.9	102.2	104.8	104.4	101.1	103.4	103.2	93.4	100.2	103.5	96.4	101.3	103.6	101.1	102.8
06	WOOD, PLASTICS & COMPOSITES	99.2	103.2	101.3	94.6	71.7	82.6	90.7	57.7	73.4	81.8	57.7	69.2	79.9	58.3	68.6	78.3	67.7	72.8
07	THERMAL & MOISTURE PROTECTION	101.5	81.9	93.1	100.6	79.9	91.7	102.5	77.4	91.8	101.8	60.7	84.1	101.4	68.1	87.1	101.6	77.6	91.3
08	OPENINGS	100.6	94.4	99.1	95.5	76.0	90.8	93.8	67.8	87.4	96.4	64.7	88.7	94.8	65.3	87.6	95.1	73.8	89.9
0920	Plaster & Gypsum Board	111.6	103.1	106.0	101.6	71.0	81.5	98.1	56.6	70.9	94.2	56.6	69.5	94.2	56.6	69.5	94.4	66.9	76.3
0950, 0980	Ceilings & Acoustic Treatment	81.4	103.1	95.0	99.0	71.0	81.5	96.5	56.6	71.5	95.8	56.6	71.2	95.8	56.6	71.2	96.5	66.9	77.9
0960	Flooring	88.4	110.7	94.9	99.0	75.3	92.1	97.5	74.8	90.8	94.1	55.0	82.7	92.5	55.0	81.6	92.5	71.0	86.2
0970, 0990	Wall Finishes & Painting/Coating	82.8	104.6	95.9	111.0	70.1	86.5	111.0	70.1	86.5	97.7	75.0	84.1	97.7	59.1	74.6	111.0	69.8	86.4
09	FINISHES	88.0	98.8	93.9	98.3	72.2	84.2	98.0	63.5	79.3	94.7	59.8	75.8	94.3	58.2	74.8	94.9	68.4	80.6
COVERS	DIVS. 10 - 14, 25, 28, 41, 43, 44, 46	100.0	96.1	99.1	100.0	88.8	97.4	100.0	79.8	95.2	100.0	82.2	95.8	100.0	83.8	96.2	100.0	87.6	97.1
21, 22, 23	FIRE SUPPRESSION, PLUMBING & HVAC	97.6	60.2	82.5	101.5	86.1	95.3	101.5	85.6	95.1	97.7	49.7	78.3	97.7	86.0	93.0	97.7	83.2	91.8
26, 27, 3370	ELECTRICAL, COMMUNICATIONS & UTIL.	104.3	52.3	78.6	95.4	100.8	98.0	93.4	102.2	97.7	95.0	35.0	65.3	95.0	72.0	83.6	97.0	97.3	97.1
MF2018	WEIGHTED AVERAGE	98.7	79.6	90.4	100.0	84.8	93.4	101.0	81.6	92.6	98.5	59.4	81.6	99.8	74.2	88.7	99.1	82.0	91.7

VIRGINIA

DIVISION		FAIRFAX 220-221			FARMVILLE 239			FREDERICKSBURG 224-225			GRUNDY 246			HARRISONBURG 228			LYNCHBURG 245		
		MAT.	INST.	TOTAL	MAT.	INST.	TOTAL	MAT.	INST.	TOTAL	MAT.	INST.	TOTAL	MAT.	INST.	TOTAL	MAT.	INST.	TOTAL
015433	CONTRACTOR EQUIPMENT		102.2	102.2		106.2	106.2		102.2	102.2		102.2	102.2		102.2	102.2		102.2	102.2
0241, 31 - 34	SITE & INFRASTRUCTURE, DEMOLITION	123.2	83.8	96.1	109.8	84.6	92.5	111.5	83.4	92.2	106.7	82.5	90.1	120.0	83.8	95.2	107.5	84.0	91.4
0310	Concrete Forming & Accessories	86.3	61.6	65.3	102.5	60.2	66.5	86.3	63.3	66.7	91.5	58.3	63.2	81.7	61.2	64.3	87.7	60.6	64.6
0320	Concrete Reinforcing	98.0	88.5	93.4	95.4	64.9	80.7	98.7	77.4	88.5	96.7	42.2	70.4	98.0	80.5	89.6	97.4	70.7	84.5
0330	Cast-in-Place Concrete	104.3	75.3	93.4	107.7	83.8	98.6	105.9	72.5	93.3	103.9	49.3	83.3	104.3	75.5	93.4	103.9	74.1	92.7
03	CONCRETE	103.4	72.7	89.6	102.5	70.9	88.3	97.6	70.6	85.5	98.8	53.9	78.7	101.2	71.3	87.8	98.7	68.9	85.3
04	MASONRY	105.6	72.0	85.1	105.6	56.1	75.4	106.9	81.6	91.5	97.1	56.0	72.0	104.5	71.4	84.3	112.5	55.3	77.6
05	METALS	103.7	101.8	103.1	101.3	89.6	97.7	103.6	97.9	101.8	103.2	74.2	94.2	103.5	99.0	102.1	103.4	95.4	100.9
06	WOOD, PLASTICS & COMPOSITES	81.8	57.1	68.9	98.5	58.3	77.4	81.8	62.4	71.7	85.4	57.7	70.9	77.0	57.9	67.0	81.8	57.9	69.3
07	THERMAL & MOISTURE PROTECTION	102.2	77.2	91.5	103.8	69.7	89.1	101.6	79.5	92.1	101.8	59.1	83.4	102.0	76.8	91.2	101.6	68.2	87.3
08	OPENINGS	93.8	68.0	87.5	94.3	53.2	84.3	94.8	67.5	88.2	96.4	44.4	83.7	95.1	67.0	88.2	95.1	64.8	87.7
0920	Plaster & Gypsum Board	94.4	56.0	69.2	106.4	56.6	73.7	94.4	61.5	72.8	94.2	56.6	69.5	94.2	56.8	69.6	94.2	56.8	69.6
0950, 0980	Ceilings & Acoustic Treatment	96.5	56.0	71.1	90.1	56.6	69.1	96.5	61.5	74.5	95.8	56.6	71.2	95.8	56.8	71.4	95.8	56.8	71.4
0960	Flooring	94.3	76.4	89.1	95.3	50.0	83.5	94.3	69.1	86.9	95.3	55.0	83.5	92.3	76.9	87.8	94.1	55.0	82.7
0970, 0990	Wall Finishes & Painting/Coating	111.0	70.1	86.5	93.3	54.5	70.0	111.0	54.3	77.0	97.7	29.7	57.0	111.0	58.9	79.8	97.7	75.0	84.1
09	FINISHES	96.5	63.6	78.7	94.4	58.0	74.7	95.4	62.4	77.5	94.9	54.9	73.2	95.4	62.6	77.7	94.5	59.6	75.6
COVERS	DIVS. 10 - 14, 25, 28, 41, 43, 44, 46	100.0	87.0	96.9	100.0	83.7	96.2	100.0	74.3	93.9	100.0	82.4	95.8	100.0	82.4	95.9	100.0	82.0	95.8
21, 22, 23	FIRE SUPPRESSION, PLUMBING & HVAC	97.7	84.0	92.2	98.0	52.9	79.8	97.7	76.1	89.0	97.7	63.1	83.7	97.7	75.0	88.5	97.7	86.0	93.0
26, 27, 3370	ELECTRICAL, COMMUNICATIONS & UTIL.	95.9	102.2	99.0	89.8	69.9	79.9	93.6	102.1	97.8	95.0	69.8	82.6	95.2	78.3	86.8	95.9	65.3	80.7
MF2018	WEIGHTED AVERAGE	99.9	81.7	92.0	98.5	66.0	84.5	98.6	79.9	90.5	98.4	63.2	83.2	99.4	75.6	89.1	99.1	73.1	87.9

VIRGINIA

DIVISION		NEWPORT NEWS 236			NORFOLK 233-235			PETERSBURG 238			PORTSMOUTH 237			PULASKI 243			RICHMOND 230-232		
		MAT.	INST.	TOTAL	MAT.	INST.	TOTAL	MAT.	INST.	TOTAL	MAT.	INST.	TOTAL	MAT.	INST.	TOTAL	MAT.	INST.	TOTAL
015433	CONTRACTOR EQUIPMENT		106.3	106.3		107.9	107.9		106.2	106.2		106.2	106.2		102.2	102.2		108.0	108.0
0241, 31 - 34	SITE & INFRASTRUCTURE, DEMOLITION	109.0	85.8	93.0	109.6	92.1	97.6	112.4	85.7	94.1	107.7	85.4	92.4	106.1	82.8	90.1	101.8	92.3	95.3
0310	Concrete Forming & Accessories	101.5	60.5	66.6	103.2	60.2	66.6	93.8	61.0	65.9	89.0	60.2	64.5	91.5	59.0	63.8	96.3	60.9	66.2
0320	Concrete Reinforcing	95.3	66.8	81.5	105.0	66.8	86.5	94.9	71.9	83.8	94.9	66.8	81.3	96.7	80.0	88.6	104.0	71.9	88.5
0330	Cast-in-Place Concrete	104.6	76.1	93.8	116.2	77.0	101.4	111.1	76.3	98.0	103.6	64.7	88.9	103.9	80.3	95.0	88.4	77.3	84.2
03	CONCRETE	99.6	68.8	85.8	104.1	68.8	88.2	105.0	70.0	89.3	98.2	64.7	83.2	98.8	71.7	86.7	90.5	70.1	81.4
04	MASONRY	100.4	55.3	72.9	104.4	55.3	74.5	114.9	55.3	78.5	106.8	55.2	75.3	91.0	53.5	68.1	103.3	55.3	74.1
05	METALS	103.8	94.0	100.7	105.2	91.0	100.8	101.4	96.4	99.9	102.7	93.6	99.9	103.3	94.8	100.6	105.9	93.6	102.0
06	WOOD, PLASTICS & COMPOSITES	96.9	54.5	76.8	99.2	58.3	77.8	86.7	58.3	71.8	82.2	58.3	69.7	85.4	57.7	70.9	96.5	58.3	76.5
07	THERMAL & MOISTURE PROTECTION	103.7	66.3	87.6	101.2	68.8	87.3	103.7	68.5	88.6	103.7	66.6	87.8	101.8	68.6	87.5	98.8	69.7	86.3
08	OPENINGS	94.6	63.8	87.1	95.3	64.2	87.7	94.0	65.3	87.0	94.7	63.7	87.2	96.4	56.0	86.6	101.6	65.3	92.8
0920	Plaster & Gypsum Board	107.9	56.8	74.3	99.3	56.8	71.4	100.4	56.6	71.6	101.4	56.6	72.0	94.2	56.6	69.5	98.9	56.8	71.3
0950, 0980	Ceilings & Acoustic Treatment	95.8	56.8	71.4	90.7	56.8	69.4	92.0	56.6	69.8	95.8	56.6	71.2	95.8	56.6	71.2	93.6	56.8	70.5
0960	Flooring	95.3	55.0	83.5	95.9	55.0	83.9	91.3	55.0	80.7	88.3	55.0	78.6	95.3	55.0	83.5	91.1	55.0	80.6
0970, 0990	Wall Finishes & Painting/Coating	93.3	58.0	72.1	97.7	58.9	74.4	93.3	59.1	72.8	93.3	58.9	72.7	97.7	46.0	66.7	92.5	59.1	72.5
09	FINISHES	95.8	58.2	75.4	94.5	58.2	74.9	93.0	58.2	74.2	92.9	58.2	74.1	94.9	56.0	73.8	93.7	58.2	74.5
COVERS	DIVS. 10 - 14, 25, 28, 41, 43, 44, 46	100.0	83.8	96.2	100.0	83.4	96.1	100.0	83.8	96.2	100.0	81.5	95.6	100.0	81.5	95.7	100.0	83.5	96.1
21, 22, 23	FIRE SUPPRESSION, PLUMBING & HVAC	101.8	65.5	87.1	101.1	65.0	86.5	98.0	86.0	93.1	101.8	65.5	87.1	97.7	63.9	84.1	101.1	86.0	95.0
26, 27, 3370	ELECTRICAL, COMMUNICATIONS & UTIL.	91.9	72.1	82.1	95.9	62.3	79.3	92.0	72.0	82.1	90.6	62.3	76.6	95.0	80.3	87.7	98.6	72.0	85.5
MF2018	WEIGHTED AVERAGE	99.5	69.5	86.5	100.7	68.3	86.7	99.4	74.4	88.6	99.0	67.4	85.3	98.1	69.7	85.8	99.6	74.6	88.8

VIRGINIA / WASHINGTON

DIVISION		ROANOKE 240-241			STAUNTON 244			WINCHESTER 226			CLARKSTON 994			EVERETT 982			OLYMPIA 985		
		MAT.	INST.	TOTAL	MAT.	INST.	TOTAL	MAT.	INST.	TOTAL	MAT.	INST.	TOTAL	MAT.	INST.	TOTAL	MAT.	INST.	TOTAL
015433	CONTRACTOR EQUIPMENT		102.2	102.2		106.2	106.2		102.2	102.2		88.4	88.4		97.7	97.7		100.2	100.2
0241, 31 - 34	SITE & INFRASTRUCTURE, DEMOLITION	106.9	84.1	91.2	110.1	84.4	92.5	118.7	83.5	94.5	102.2	83.1	89.1	93.2	103.3	100.1	94.8	105.9	102.4
0310	Concrete Forming & Accessories	99.0	60.8	66.5	91.1	60.1	64.8	84.5	63.9	67.0	102.5	62.0	68.0	114.5	105.5	106.8	98.0	106.2	105.0
0320	Concrete Reinforcing	97.7	70.7	84.7	97.4	69.3	83.8	97.4	76.4	87.3	106.6	99.4	103.1	107.5	113.7	110.5	120.0	113.5	116.9
0330	Cast-in-Place Concrete	118.0	83.8	105.1	108.1	83.6	98.8	104.3	61.2	88.0	87.0	79.6	84.2	100.9	111.7	105.0	88.8	114.3	98.5
03	CONCRETE	103.3	72.4	89.4	100.2	71.5	87.3	100.7	66.8	85.5	90.0	75.1	83.3	94.5	108.6	100.8	93.0	109.7	100.5
04	MASONRY	98.4	57.1	73.2	107.8	56.0	76.2	101.8	66.8	80.5	100.6	87.8	92.8	109.1	104.3	106.2	106.7	102.2	104.0
05	METALS	105.7	95.3	102.4	103.5	90.8	99.5	103.6	96.1	101.3	92.9	88.6	91.6	116.2	99.5	111.0	117.2	98.0	111.2
06	WOOD, PLASTICS & COMPOSITES	95.7	57.9	75.9	85.4	58.3	71.2	79.9	62.6	70.9	95.5	56.5	75.1	112.1	105.1	108.4	92.4	105.4	99.2
07	THERMAL & MOISTURE PROTECTION	101.6	70.4	88.2	101.4	69.6	87.7	102.1	74.2	90.1	157.0	79.1	123.6	113.5	107.0	110.7	109.4	107.5	108.6
08	OPENINGS	95.5	64.8	88.0	95.1	54.1	85.1	96.5	65.8	89.0	116.6	64.4	103.9	105.6	107.8	106.2	109.3	108.0	109.0
0920	Plaster & Gypsum Board	101.6	56.8	72.2	94.2	56.6	69.5	94.4	61.6	72.9	143.7	55.3	85.7	109.5	105.5	106.9	105.5	105.5	105.5
0950, 0980	Ceilings & Acoustic Treatment	99.0	56.8	72.5	95.8	56.6	71.2	96.5	61.6	74.6	91.6	55.3	68.8	104.4	105.5	105.1	114.7	105.5	108.9
0960	Flooring	99.0	56.7	86.7	94.9	55.0	83.2	93.7	71.0	87.1	85.8	73.9	82.3	109.8	101.4	107.1	99.9	91.6	97.5
0970, 0990	Wall Finishes & Painting/Coating	97.7	75.0	84.1	97.7	29.2	56.7	111.0	73.9	88.8	83.5	68.7	74.6	95.4	92.6	93.7	92.6	95.3	94.3
09	FINISHES	97.3	60.3	77.2	94.7	55.3	73.4	95.9	65.6	79.5	105.6	63.1	82.6	106.2	103.0	104.5	102.6	101.5	102.0
COVERS	DIVS. 10 - 14, 25, 28, 41, 43, 44, 46	100.0	82.3	95.8	100.0	83.8	96.2	100.0	78.4	94.9	100.0	93.5	98.5	100.0	102.5	100.6	100.0	103.6	100.9
21, 22, 23	FIRE SUPPRESSION, PLUMBING & HVAC	101.5	67.2	87.6	97.7	56.8	81.2	97.7	77.5	89.6	97.5	76.8	89.2	101.2	102.5	101.7	101.1	108.1	103.9
26, 27, 3370	ELECTRICAL, COMMUNICATIONS & UTIL.	95.0	59.7	77.5	94.1	77.8	86.0	93.9	88.4	91.2	91.6	92.1	91.8	105.7	105.9	105.8	103.5	105.6	104.6
MF2018	WEIGHTED AVERAGE	100.5	69.1	87.0	99.0	67.8	85.5	99.2	76.4	89.3	99.9	79.5	91.1	104.5	104.2	104.4	103.9	105.2	104.5

DIVISION		WASHINGTON																	
		RICHLAND 993			SEATTLE 980 - 981,987			SPOKANE 990 - 992			TACOMA 983 - 984			VANCOUVER 986			WENATCHEE 988		
		MAT.	INST.	TOTAL	MAT.	INST.	TOTAL	MAT.	INST.	TOTAL	MAT.	INST.	TOTAL	MAT.	INST.	TOTAL	MAT.	INST.	TOTAL
015433	CONTRACTOR EQUIPMENT		88.4	88.4		100.2	100.2		88.4	88.4		97.7	97.7		94.1	94.1		97.7	97.7
0241, 31 - 34	SITE & INFRASTRUCTURE, DEMOLITION	104.7	83.9	90.5	97.8	105.0	102.8	104.2	83.9	90.3	96.3	103.3	101.1	106.3	90.8	95.6	104.9	99.9	101.4
0310	Concrete Forming & Accessories	102.7	82.2	85.2	110.0	105.8	106.4	108.0	81.7	85.7	104.5	106.2	105.9	104.9	97.1	98.3	105.9	73.7	78.5
0320	Concrete Reinforcing	102.2	99.3	100.8	103.9	111.8	107.7	102.8	100.1	101.5	106.2	113.7	109.8	107.1	113.7	110.3	107.1	91.4	99.5
0330	Cast-in-Place Concrete	87.2	85.3	86.5	106.5	110.7	108.1	90.6	85.1	88.5	103.8	112.4	107.0	115.9	100.3	110.0	105.9	87.4	98.9
03	CONCRETE	89.5	86.2	88.0	101.4	108.1	104.4	91.7	86.1	89.2	96.3	109.1	102.0	106.2	100.0	103.8	104.1	81.8	94.1
04	MASONRY	101.7	85.8	92.0	114.3	102.8	107.3	102.3	100.4	101.2	108.1	105.1	106.5	108.7	104.8	106.3	111.4	90.2	98.5
05	METALS	93.3	89.6	92.2	115.8	100.2	110.9	95.4	90.3	93.8	118.1	99.5	112.3	115.6	100.0	110.7	115.4	86.1	106.2
06	WOOD, PLASTICS & COMPOSITES	95.7	80.5	87.8	108.8	105.2	106.9	104.2	80.5	91.8	100.4	105.1	102.9	93.9	96.2	95.1	102.3	70.8	85.8
07	THERMAL & MOISTURE PROTECTION	158.7	87.1	128.0	112.4	106.5	109.9	155.4	88.0	126.5	113.3	108.5	111.2	113.7	100.9	108.2	112.9	84.3	100.6
08	OPENINGS	114.7	76.6	105.4	105.7	106.7	106.0	115.3	77.4	106.0	106.4	107.8	106.7	102.5	101.2	102.2	105.9	70.3	97.2
0920	Plaster & Gypsum Board	143.7	80.0	101.9	104.6	105.5	105.2	132.3	80.0	98.0	106.7	105.5	105.9	104.8	96.6	99.4	110.3	70.3	84.1
0950, 0980	Ceilings & Acoustic Treatment	98.0	80.0	86.7	107.2	105.5	106.1	94.4	80.0	85.4	107.6	105.5	106.3	106.2	96.6	100.2	100.4	70.3	81.5
0960	Flooring	86.1	81.8	84.9	106.6	98.7	104.3	85.3	101.4	90.0	102.0	101.4	101.9	107.8	102.2	106.2	105.2	73.9	96.1
0970, 0990	Wall Finishes & Painting/Coating	83.5	75.8	78.9	108.1	91.2	98.0	83.7	79.1	80.9	95.4	92.6	93.7	97.6	79.1	86.5	95.4	67.5	78.7
09	FINISHES	107.4	80.5	92.8	106.5	102.3	104.2	104.8	84.8	94.0	104.5	103.3	103.8	103.0	95.7	99.1	105.2	71.9	87.2
COVERS	DIVS. 10 - 14, 25, 28, 41, 43, 44, 46	100.0	97.3	99.4	100.0	102.7	100.6	100.0	97.2	99.3	100.0	102.9	100.7	100.0	98.1	99.5	100.0	95.3	98.9
21, 22, 23	FIRE SUPPRESSION, PLUMBING & HVAC	101.4	111.4	105.4	101.2	117.2	107.6	101.3	84.2	94.4	101.3	108.1	104.0	101.3	107.5	103.8	97.5	87.6	93.5
26, 27, 3370	ELECTRICAL, COMMUNICATIONS & UTIL.	89.1	101.3	95.2	103.1	114.5	108.7	88.0	83.5	85.8	105.5	105.7	105.6	111.7	109.1	110.4	106.3	91.0	98.7
MF2018	WEIGHTED AVERAGE	100.7	93.0	97.4	105.4	108.4	106.7	101.0	86.8	94.9	104.9	105.7	105.2	106.2	102.2	104.4	105.0	85.6	96.6

DIVISION		WASHINGTON			WEST VIRGINIA														
		YAKIMA 989			BECKLEY 258 - 259			BLUEFIELD 247 - 248			BUCKHANNON 262			CHARLESTON 250 - 253			CLARKSBURG 263 - 264		
		MAT.	INST.	TOTAL	MAT.	INST.	TOTAL	MAT.	INST.	TOTAL	MAT.	INST.	TOTAL	MAT.	INST.	TOTAL	MAT.	INST.	TOTAL
015433	CONTRACTOR EQUIPMENT		97.7	97.7		102.2	102.2		102.2	102.2		102.2	102.2		107.1	107.1		102.2	102.2
0241, 31 - 34	SITE & INFRASTRUCTURE, DEMOLITION	98.7	101.7	100.8	100.1	84.5	89.4	100.6	84.5	89.5	106.8	84.7	91.6	99.1	94.7	96.1	105.7	84.7	91.8
0310	Concrete Forming & Accessories	104.8	102.1	102.5	85.3	84.3	84.5	88.1	84.1	84.7	87.4	87.3	87.3	97.9	88.8	90.2	84.5	87.4	87.0
0320	Concrete Reinforcing	106.7	100.2	103.6	95.7	85.1	90.6	96.3	78.6	87.8	96.9	86.3	91.8	101.7	85.3	93.8	96.9	96.2	96.6
0330	Cast-in-Place Concrete	110.8	87.3	101.9	100.0	89.1	95.9	101.6	89.0	96.8	101.3	90.8	97.3	96.8	90.5	94.4	111.0	86.6	101.8
03	CONCRETE	100.9	96.1	98.7	91.9	87.3	89.8	94.8	86.0	90.8	97.6	89.4	94.0	91.2	89.7	90.5	101.6	89.7	96.3
04	MASONRY	101.5	102.1	101.9	91.7	85.4	87.8	92.6	85.4	88.2	104.0	88.3	94.5	87.2	90.5	89.2	108.0	88.3	96.0
05	METALS	116.1	91.1	108.3	98.0	102.6	99.5	103.5	100.1	102.4	103.7	103.3	103.6	96.8	100.9	98.1	103.7	106.8	104.7
06	WOOD, PLASTICS & COMPOSITES	100.8	105.1	103.0	80.4	84.4	82.5	84.0	84.4	84.2	83.1	86.8	85.0	92.2	88.6	90.3	79.1	86.8	83.1
07	THERMAL & MOISTURE PROTECTION	113.5	95.8	105.9	102.6	83.1	94.2	101.5	83.1	93.6	101.8	86.7	95.3	97.9	87.6	93.5	101.7	86.4	95.2
08	OPENINGS	105.9	100.8	104.6	94.4	84.1	91.7	96.9	81.6	93.2	96.9	84.7	94.0	95.0	85.4	92.6	96.9	86.9	94.5
0920	Plaster & Gypsum Board	106.3	105.5	105.8	86.6	84.1	85.0	93.3	84.1	87.2	93.9	86.5	89.1	92.6	88.1	89.7	91.7	86.5	88.3
0950, 0980	Ceilings & Acoustic Treatment	101.9	105.5	104.1	81.0	84.1	82.9	93.3	84.1	87.5	95.8	86.5	90.0	90.6	88.1	89.1	95.8	86.5	90.0
0960	Flooring	103.1	81.8	96.9	89.1	98.4	91.8	91.7	98.4	93.7	91.4	94.7	92.4	93.4	98.4	94.9	90.3	94.7	91.6
0970, 0990	Wall Finishes & Painting/Coating	95.4	79.1	85.6	90.2	87.7	88.7	97.7	85.5	90.4	97.7	89.0	92.5	92.3	89.2	90.4	97.7	89.0	92.5
09	FINISHES	103.6	95.9	99.4	86.6	87.4	87.0	92.7	87.2	89.7	93.8	88.9	91.2	92.6	90.7	91.6	93.1	88.9	90.8
COVERS	DIVS. 10 - 14, 25, 28, 41, 43, 44, 46	100.0	100.4	100.1	100.0	91.6	98.0	100.0	91.6	98.0	100.0	92.9	98.3	100.0	93.4	98.5	100.0	92.9	98.3
21, 22, 23	FIRE SUPPRESSION, PLUMBING & HVAC	101.3	113.5	106.2	98.0	89.4	94.6	97.7	80.1	90.6	97.7	91.1	95.0	101.1	91.3	97.1	97.7	91.3	95.1
26, 27, 3370	ELECTRICAL, COMMUNICATIONS & UTIL.	108.2	101.3	104.8	91.9	81.0	86.5	94.2	81.0	87.6	95.3	90.7	93.0	98.9	86.4	92.7	95.3	90.7	93.0
MF2018	WEIGHTED AVERAGE	105.0	101.6	103.5	95.2	87.7	91.9	97.3	85.2	92.1	96.5	88.8	92.9	96.6	91.1	94.2	99.3	91.0	95.7

DIVISION		WEST VIRGINIA																	
		GASSAWAY 266			HUNTINGTON 255 - 257			LEWISBURG 249			MARTINSBURG 254			MORGANTOWN 265			PARKERSBURG 261		
		MAT.	INST.	TOTAL	MAT.	INST.	TOTAL	MAT.	INST.	TOTAL	MAT.	INST.	TOTAL	MAT.	INST.	TOTAL	MAT.	INST.	TOTAL
015433	CONTRACTOR EQUIPMENT		102.2	102.2		102.2	102.2		102.2	102.2		102.2	102.2		102.2	102.2		102.2	102.2
0241, 31 - 34	SITE & INFRASTRUCTURE, DEMOLITION	104.2	84.7	90.8	105.0	85.9	91.9	116.4	84.5	94.5	103.8	84.9	90.9	101.5	85.8	90.7	110.4	85.8	93.5
0310	Concrete Forming & Accessories	86.7	84.5	84.8	98.2	88.2	89.7	84.8	84.1	84.2	85.3	76.5	77.8	84.8	87.7	87.3	89.4	89.3	89.3
0320	Concrete Reinforcing	96.9	87.3	92.3	97.1	90.3	93.8	96.9	78.6	88.1	95.7	90.6	93.2	96.9	96.3	96.6	96.3	86.2	91.4
0330	Cast-in-Place Concrete	106.0	88.9	99.6	109.0	94.2	103.4	101.7	89.0	96.9	104.7	84.8	97.2	101.3	92.9	98.1	103.5	92.8	99.5
03	CONCRETE	98.1	87.6	93.4	97.1	91.7	94.7	104.2	86.0	96.0	95.3	83.1	89.8	94.5	92.0	93.4	99.7	91.0	95.8
04	MASONRY	108.8	88.2	96.2	89.7	92.1	91.2	96.4	85.4	89.7	92.9	80.5	85.3	126.8	88.3	103.4	82.5	85.5	84.3
05	METALS	103.6	103.3	103.5	100.5	105.2	102.0	103.6	100.2	102.5	98.4	102.5	99.7	103.7	107.1	104.8	104.4	103.2	104.0
06	WOOD, PLASTICS & COMPOSITES	81.9	83.2	82.6	92.8	87.3	89.9	79.5	84.4	82.1	80.4	75.9	78.1	79.5	86.8	83.3	83.5	89.2	86.5
07	THERMAL & MOISTURE PROTECTION	101.5	86.3	95.0	102.8	86.9	96.0	102.5	83.1	94.2	102.8	77.1	91.8	101.5	87.3	95.4	101.8	86.8	95.4
08	OPENINGS	95.3	82.9	92.3	93.7	85.8	91.8	96.9	81.6	93.2	96.1	71.1	90.0	98.0	86.9	95.3	95.9	83.4	92.8
0920	Plaster & Gypsum Board	92.8	82.8	86.3	92.0	87.0	88.7	91.7	84.1	86.7	86.6	75.3	79.2	91.7	86.5	88.3	94.2	89.0	90.8
0950, 0980	Ceilings & Acoustic Treatment	95.8	82.8	87.7	81.0	87.0	84.8	95.8	84.1	88.5	81.0	75.3	77.4	95.8	86.5	90.0	95.8	89.0	91.5
0960	Flooring	91.3	98.4	93.3	96.7	100.4	97.8	90.5	98.4	92.8	89.1	94.4	90.6	90.5	94.7	91.7	94.4	98.4	95.6
0970, 0990	Wall Finishes & Painting/Coating	97.7	89.2	92.6	90.2	89.2	89.6	97.7	61.4	75.9	90.2	79.7	83.9	97.7	89.0	92.5	97.7	89.2	92.6
09	FINISHES	93.3	87.6	90.2	89.5	90.5	90.1	94.2	84.5	89.0	86.8	79.7	83.0	92.7	88.9	90.7	94.8	91.1	92.8
COVERS	DIVS. 10 - 14, 25, 28, 41, 43, 44, 46	100.0	92.5	98.2	100.0	93.1	98.4	100.0	91.6	98.0	100.0	77.1	94.6	100.0	92.9	98.3	100.0	93.1	98.4
21, 22, 23	FIRE SUPPRESSION, PLUMBING & HVAC	97.7	81.5	91.2	101.8	91.6	97.7	97.7	89.4	94.3	98.0	77.1	89.6	97.7	91.3	95.1	101.4	91.2	97.3
26, 27, 3370	ELECTRICAL, COMMUNICATIONS & UTIL.	95.3	86.4	90.9	95.0	89.8	92.4	92.1	81.0	86.6	96.8	74.0	85.5	95.5	90.7	93.1	95.4	85.0	90.3
MF2018	WEIGHTED AVERAGE	98.7	87.3	93.7	97.8	91.7	95.1	99.0	86.8	93.8	96.5	80.8	89.7	99.3	91.4	95.9	98.9	90.0	95.1

For customer support on your Concrete & Masonry Costs with RSMeans Data, call 800.448.8182.

489

WEST VIRGINIA / WISCONSIN

DIVISION		PETERSBURG 268 MAT.	INST.	TOTAL	ROMNEY 267 MAT.	INST.	TOTAL	WHEELING 260 MAT.	INST.	TOTAL	BELOIT 535 MAT.	INST.	TOTAL	EAU CLAIRE 547 MAT.	INST.	TOTAL	GREEN BAY 541-543 MAT.	INST.	TOTAL
015433	CONTRACTOR EQUIPMENT		102.2	102.2		102.2	102.2		102.2	102.2		96.6	96.6		98.1	98.1		95.8	95.8
0241, 31 - 34	SITE & INFRASTRUCTURE, DEMOLITION	100.8	85.6	90.4	103.7	85.6	91.3	111.0	85.7	93.7	96.1	98.3	97.6	96.4	97.7	97.3	99.9	93.7	95.7
0310	Concrete Forming & Accessories	88.6	85.3	85.8	84.0	85.2	85.0	91.5	87.1	87.7	99.0	93.1	94.0	100.2	98.1	98.4	112.4	102.8	104.3
0320	Concrete Reinforcing	96.3	87.3	91.9	96.9	90.9	94.0	95.7	96.3	96.0	102.7	136.5	119.0	93.7	113.9	103.5	91.9	113.9	102.5
0330	Cast-in-Place Concrete	101.3	90.7	97.3	106.0	82.4	97.1	103.5	93.0	99.6	101.9	96.5	99.8	97.8	103.1	99.8	101.1	103.0	101.8
03	CONCRETE	94.6	88.6	91.9	97.9	86.3	92.7	99.7	91.8	96.2	98.8	102.0	100.2	93.6	102.8	97.7	96.9	105.0	100.5
04	MASONRY	99.1	88.3	92.5	96.8	88.2	91.5	107.0	86.8	94.7	95.1	109.8	104.0	92.1	111.9	104.1	122.5	111.0	115.5
05	METALS	103.8	103.3	103.6	103.8	104.4	104.0	104.5	107.1	105.3	96.7	113.6	102.0	91.7	108.5	96.9	94.3	108.9	98.8
06	WOOD, PLASTICS & COMPOSITES	84.3	84.4	84.4	78.5	84.4	81.6	85.4	86.8	86.1	95.7	90.9	93.2	101.8	96.5	99.0	112.8	102.9	107.6
07	THERMAL & MOISTURE PROTECTION	101.6	81.9	93.2	101.7	81.8	93.1	102.0	86.0	95.1	105.0	93.5	100.1	103.7	105.4	104.5	106.2	105.9	106.0
08	OPENINGS	98.0	80.5	93.8	98.0	81.4	93.9	96.6	86.9	94.2	97.9	105.5	99.7	103.0	96.1	101.3	98.8	105.8	100.5
0920	Plaster & Gypsum Board	93.9	84.1	87.5	91.4	84.1	86.6	94.2	86.5	89.2	90.9	91.3	91.2	99.7	96.9	97.9	98.3	103.5	101.7
0950, 0980	Ceilings & Acoustic Treatment	95.8	84.1	88.5	95.8	84.1	88.5	95.8	86.5	90.0	87.4	91.3	89.9	94.1	96.9	95.9	86.3	103.5	97.1
0960	Flooring	92.3	94.7	93.0	90.2	94.4	91.5	95.3	94.7	95.1	95.4	113.0	100.6	82.3	117.9	92.7	97.7	117.9	103.6
0970, 0990	Wall Finishes & Painting/Coating	97.7	89.0	92.5	97.7	89.0	92.5	97.7	89.0	92.5	88.9	98.7	94.7	80.1	83.6	82.2	87.9	106.2	98.9
09	FINISHES	93.5	87.5	90.3	92.8	87.5	89.9	90.5	88.5	89.5	92.5	96.7	94.8	89.0	99.9	94.9	92.4	106.0	99.7
COVERS	DIVS. 10 - 14, 25, 28, 41, 43, 44, 46	100.0	92.5	98.2	100.0	92.5	98.2	100.0	91.6	98.0	100.0	98.2	99.6	100.0	96.7	99.2	100.0	98.1	99.5
21, 22, 23	FIRE SUPPRESSION, PLUMBING & HVAC	97.7	81.9	91.3	97.7	81.6	91.2	101.5	90.5	97.1	98.4	91.2	95.5	100.6	90.3	96.5	100.9	97.4	99.5
26, 27, 3370	ELECTRICAL, COMMUNICATIONS & UTIL.	98.0	74.0	86.1	97.4	74.0	85.8	93.2	90.7	91.9	98.8	81.1	90.1	102.4	102.9	102.6	97.9	102.9	100.4
MF2018	WEIGHTED AVERAGE	98.3	85.6	92.8	98.5	85.3	92.8	100.1	90.9	96.1	97.8	97.3	97.6	97.4	100.4	98.7	99.4	103.1	101.0

WISCONSIN

DIVISION		KENOSHA 531 MAT.	INST.	TOTAL	LA CROSSE 546 MAT.	INST.	TOTAL	LANCASTER 538 MAT.	INST.	TOTAL	MADISON 537 MAT.	INST.	TOTAL	MILWAUKEE 530,532 MAT.	INST.	TOTAL	NEW RICHMOND 540 MAT.	INST.	TOTAL
015433	CONTRACTOR EQUIPMENT		94.6	94.6		98.1	98.1		96.6	96.6		101.6	101.6		90.8	90.8		98.4	98.4
0241, 31 - 34	SITE & INFRASTRUCTURE, DEMOLITION	103.0	96.2	98.3	90.1	97.7	95.3	94.8	97.7	96.8	94.9	107.7	103.7	91.0	97.2	95.3	93.0	96.7	95.5
0310	Concrete Forming & Accessories	110.2	113.8	113.3	84.6	98.0	96.0	98.2	92.0	92.9	103.6	106.2	105.8	102.6	116.9	114.8	93.8	89.2	89.9
0320	Concrete Reinforcing	102.5	113.9	108.0	93.5	113.9	103.3	103.9	102.1	103.0	103.4	114.0	108.5	102.0	116.7	109.1	91.0	109.7	100.0
0330	Cast-in-Place Concrete	110.9	105.5	108.8	87.9	103.0	93.6	101.3	96.0	99.3	93.7	108.9	99.4	94.5	118.1	103.4	101.8	99.0	100.7
03	CONCRETE	103.7	110.6	106.8	85.0	102.8	92.9	98.5	95.4	97.1	99.9	108.2	103.6	100.0	116.2	107.3	91.1	96.6	93.6
04	MASONRY	93.0	115.8	106.9	91.3	111.9	103.8	95.1	109.8	104.1	94.5	119.5	109.7	97.4	121.2	111.9	117.0	109.9	112.7
05	METALS	97.8	106.1	100.4	91.6	108.4	96.8	94.0	100.2	95.9	100.8	104.6	102.0	95.7	98.9	96.7	92.0	105.5	96.2
06	WOOD, PLASTICS & COMPOSITES	105.7	115.2	110.7	84.0	96.5	90.5	94.8	90.9	92.8	98.9	103.0	101.0	102.3	115.6	109.3	89.7	87.1	88.4
07	THERMAL & MOISTURE PROTECTION	105.7	109.9	107.5	103.1	104.8	103.8	104.8	93.2	99.8	106.0	111.2	108.2	104.4	115.3	109.1	104.0	100.1	102.3
08	OPENINGS	92.2	112.7	97.2	102.9	96.2	101.3	93.8	88.5	92.6	101.5	105.7	102.5	102.5	113.5	105.2	88.5	87.0	88.1
0920	Plaster & Gypsum Board	78.8	116.3	103.4	95.1	96.9	96.3	89.5	91.3	90.7	100.5	103.5	102.5	92.9	116.0	108.0	85.8	87.5	86.9
0950, 0980	Ceilings & Acoustic Treatment	87.4	116.3	105.5	92.9	96.9	95.4	82.3	91.3	88.0	95.7	103.5	100.6	88.8	116.0	105.8	58.9	87.5	76.8
0960	Flooring	113.2	117.9	114.6	76.3	117.9	88.4	95.1	105.9	98.2	95.0	117.9	101.7	101.5	118.1	106.4	91.2	109.3	96.5
0970, 0990	Wall Finishes & Painting/Coating	99.7	122.4	113.3	80.1	81.1	80.7	88.9	93.0	91.4	90.9	104.7	99.2	100.8	122.2	113.6	90.8	77.8	83.0
09	FINISHES	96.7	116.2	107.2	85.9	99.7	93.4	90.9	94.5	92.9	95.8	108.0	102.4	96.3	118.1	108.1	81.9	91.4	87.0
COVERS	DIVS. 10 - 14, 25, 28, 41, 43, 44, 46	100.0	99.8	100.0	100.0	96.7	99.2	100.0	98.2	99.6	100.0	103.4	100.8	100.0	104.2	101.0	100.0	90.3	97.7
21, 22, 23	FIRE SUPPRESSION, PLUMBING & HVAC	100.8	97.3	99.4	100.6	90.3	96.5	94.7	85.0	90.8	96.1	100.6	97.9	96.1	117.1	104.6	96.2	84.8	91.6
26, 27, 3370	ELECTRICAL, COMMUNICATIONS & UTIL.	100.0	102.9	101.4	102.7	102.9	102.8	98.5	81.7	90.2	98.9	102.8	100.9	97.8	102.8	100.3	100.8	81.7	91.4
MF2018	WEIGHTED AVERAGE	99.2	106.2	102.2	95.7	100.3	97.7	95.8	92.9	94.6	98.6	106.4	101.9	97.9	111.7	103.9	94.9	92.9	94.0

WISCONSIN

DIVISION		OSHKOSH 549 MAT.	INST.	TOTAL	PORTAGE 539 MAT.	INST.	TOTAL	RACINE 534 MAT.	INST.	TOTAL	RHINELANDER 545 MAT.	INST.	TOTAL	SUPERIOR 548 MAT.	INST.	TOTAL	WAUSAU 544 MAT.	INST.	TOTAL
015433	CONTRACTOR EQUIPMENT		95.8	95.8		96.6	96.6		96.6	96.6		95.8	95.8		98.4	98.4		95.8	95.8
0241, 31 - 34	SITE & INFRASTRUCTURE, DEMOLITION	91.1	92.9	92.3	85.7	98.6	94.5	96.7	99.8	98.8	103.5	92.4	95.9	89.7	96.4	94.3	86.9	93.5	91.5
0310	Concrete Forming & Accessories	92.5	92.4	92.4	89.7	92.7	92.2	99.4	110.8	109.1	89.7	91.5	91.2	91.6	85.6	86.5	91.9	94.3	94.0
0320	Concrete Reinforcing	92.0	105.5	98.6	104.0	102.2	103.1	102.7	113.9	108.1	92.2	102.3	97.1	91.0	109.5	99.9	92.2	113.7	102.6
0330	Cast-in-Place Concrete	93.7	95.7	94.5	86.9	94.7	89.9	99.9	105.3	101.9	106.3	94.4	101.8	95.9	94.7	95.5	87.4	89.2	88.1
03	CONCRETE	86.7	96.2	91.0	86.2	95.5	90.4	97.9	109.2	103.0	98.4	94.8	96.8	81.9	93.5	89.3	81.9	96.3	88.4
04	MASONRY	105.8	111.3	109.1	94.0	109.8	103.6	95.3	115.8	107.8	125.9	109.1	115.7	116.3	107.6	111.0	105.3	109.1	107.6
05	METALS	92.2	104.0	95.9	94.7	104.3	97.7	98.3	106.1	100.8	92.0	102.8	95.4	93.0	105.8	97.0	91.9	107.6	96.8
06	WOOD, PLASTICS & COMPOSITES	89.0	91.0	90.1	83.8	90.9	87.5	96.1	111.2	104.0	85.9	91.0	88.6	87.8	82.9	85.2	88.4	93.9	91.3
07	THERMAL & MOISTURE PROTECTION	105.1	85.3	96.6	104.2	99.4	102.1	105.1	109.1	106.8	106.0	83.6	96.4	103.7	94.7	99.8	104.9	95.2	100.7
08	OPENINGS	95.4	92.8	94.8	94.0	96.0	94.5	97.9	110.4	101.0	95.4	88.5	93.8	87.9	91.0	88.6	95.6	96.3	95.8
0920	Plaster & Gypsum Board	86.1	91.3	89.5	83.2	91.3	88.5	90.9	112.1	104.8	86.1	91.3	89.5	86.1	83.1	84.1	86.1	94.3	91.5
0950, 0980	Ceilings & Acoustic Treatment	86.3	91.3	89.4	86.1	91.3	89.4	87.4	112.1	102.9	86.3	91.3	89.4	60.2	83.1	74.5	86.3	94.3	91.3
0960	Flooring	89.2	109.5	95.1	91.1	109.3	96.4	95.4	117.9	102.0	88.5	109.3	94.6	92.3	117.4	99.7	89.1	109.3	95.0
0970, 0990	Wall Finishes & Painting/Coating	85.6	106.2	98.0	88.9	93.0	91.4	88.8	121.1	108.2	85.6	76.2	79.9	80.1	100.7	92.5	85.6	79.3	81.8
09	FINISHES	87.5	97.3	92.8	89.2	95.4	92.6	92.5	113.6	103.9	88.3	93.4	91.1	81.5	92.5	87.5	87.2	95.4	91.7
COVERS	DIVS. 10 - 14, 25, 28, 41, 43, 44, 46	100.0	81.9	95.7	100.0	98.3	99.6	100.0	99.4	99.9	100.0	96.0	99.1	100.0	89.1	97.4	100.0	96.4	99.1
21, 22, 23	FIRE SUPPRESSION, PLUMBING & HVAC	97.1	84.4	92.0	94.7	91.3	93.4	100.6	99.2	100.0	97.1	84.4	92.0	96.2	86.8	92.5	97.1	85.0	92.2
26, 27, 3370	ELECTRICAL, COMMUNICATIONS & UTIL.	101.3	81.2	91.3	101.8	90.5	96.2	99.6	102.9	101.2	100.8	75.2	88.1	105.0	94.9	100.0	102.3	75.2	88.9
MF2018	WEIGHTED AVERAGE	95.2	92.7	94.1	94.2	96.5	95.2	98.5	106.2	101.8	98.0	91.1	95.0	94.6	94.6	94.6	94.5	92.9	93.8

WYOMING

DIVISION		CASPER 826 MAT.	INST.	TOTAL	CHEYENNE 820 MAT.	INST.	TOTAL	NEWCASTLE 827 MAT.	INST.	TOTAL	RAWLINS 823 MAT.	INST.	TOTAL	RIVERTON 825 MAT.	INST.	TOTAL	ROCK SPRINGS 829-831 MAT.	INST.	TOTAL
015433	CONTRACTOR EQUIPMENT		98.5	98.5		92.3	92.3		92.3	92.3		92.3	92.3		92.3	92.3		92.3	92.3
0241, 31 - 34	SITE & INFRASTRUCTURE, DEMOLITION	98.8	96.8	97.4	91.5	87.2	88.5	83.3	87.2	86.0	97.1	87.2	90.3	90.7	87.2	88.3	87.0	87.2	87.1
0310	Concrete Forming & Accessories	99.6	64.4	69.7	103.4	63.8	69.7	92.8	64.2	68.5	97.4	64.2	69.1	91.6	64.1	68.2	99.2	64.0	69.2
0320	Concrete Reinforcing	105.9	81.4	94.1	97.3	81.5	89.7	104.5	81.6	93.5	104.2	81.6	93.3	105.2	81.6	93.8	105.2	80.9	93.5
0330	Cast-in-Place Concrete	104.7	79.4	95.1	98.7	78.0	90.9	99.7	78.0	91.5	99.8	78.0	91.5	99.7	77.9	91.5	99.7	77.9	91.4
03	CONCRETE	109.2	73.0	93.0	101.7	72.3	88.6	101.8	72.5	88.7	116.6	72.5	96.9	110.7	72.4	93.5	102.3	72.3	88.8
04	MASONRY	96.8	65.0	77.4	98.5	66.5	79.0	95.5	68.0	78.7	95.5	68.0	78.7	95.5	68.0	78.7	152.5	61.1	96.8
05	METALS	101.5	79.0	94.5	103.8	80.4	96.5	100.0	80.5	93.9	100.0	80.5	93.9	100.1	80.3	93.9	100.9	79.7	94.2
06	WOOD, PLASTICS & COMPOSITES	94.7	62.2	77.7	94.4	61.4	77.2	83.1	61.9	72.0	87.7	61.9	74.2	81.8	61.9	71.4	92.4	61.9	76.4
07	THERMAL & MOISTURE PROTECTION	109.2	67.6	91.3	105.5	67.5	89.2	106.5	67.2	89.6	108.0	67.2	90.5	107.4	70.4	91.5	106.6	68.3	90.2
08	OPENINGS	109.2	67.1	99.0	107.0	66.7	97.2	111.0	65.5	99.9	110.6	65.5	99.6	110.8	65.5	99.8	111.4	66.3	100.4
0920	Plaster & Gypsum Board	96.9	61.1	73.4	85.8	60.6	69.3	82.8	61.1	68.6	83.1	61.1	68.7	82.8	61.1	68.6	94.7	61.1	72.7
0950, 0980	Ceilings & Acoustic Treatment	119.8	61.1	83.0	107.8	60.6	78.2	110.7	61.1	79.6	110.7	61.1	79.6	110.7	61.1	79.6	110.7	61.1	79.6
0960	Flooring	103.9	72.8	94.8	102.9	72.8	94.1	96.6	67.6	88.1	99.7	67.6	90.4	95.9	67.6	87.7	102.4	58.3	89.5
0970, 0990	Wall Finishes & Painting/Coating	98.3	58.1	74.2	97.7	58.1	74.0	94.3	74.6	82.5	94.3	74.6	82.5	94.3	57.6	72.3	94.3	74.6	82.5
09	FINISHES	103.7	64.6	82.6	99.4	64.1	80.3	95.0	65.2	78.9	97.2	65.2	79.9	95.6	63.3	78.1	98.3	63.3	79.4
COVERS	DIVS. 10 - 14, 25, 28, 41, 43, 44, 46	100.0	88.6	97.3	100.0	87.9	97.2	100.0	98.4	99.6	100.0	98.4	99.6	100.0	87.2	97.0	100.0	85.0	96.5
21, 22, 23	FIRE SUPPRESSION, PLUMBING & HVAC	100.9	74.6	90.3	101.2	74.6	90.5	99.1	71.8	88.1	99.1	71.8	88.1	99.1	71.8	88.1	101.1	71.8	89.3
26, 27, 3370	ELECTRICAL, COMMUNICATIONS & UTIL.	97.0	62.0	79.7	95.2	67.7	81.6	94.0	60.0	77.2	94.0	60.0	77.2	94.0	64.2	79.3	92.7	64.8	78.9
MF2018	WEIGHTED AVERAGE	102.5	72.3	89.4	101.0	72.4	88.7	99.4	71.3	87.3	101.9	71.3	88.7	100.8	71.4	88.1	103.3	70.6	89.1

WYOMING / CANADA

DIVISION		SHERIDAN 828 MAT.	INST.	TOTAL	WHEATLAND 822 MAT.	INST.	TOTAL	WORLAND 824 MAT.	INST.	TOTAL	YELLOWSTONE NAT'L PA 821 MAT.	INST.	TOTAL	BARRIE, ONTARIO MAT.	INST.	TOTAL	BATHURST, NEW BRUNSWICK MAT.	INST.	TOTAL
015433	CONTRACTOR EQUIPMENT		92.3	92.3		92.3	92.3		92.3	92.3		92.3	92.3		97.9	97.9		97.4	97.4
0241, 31 - 34	SITE & INFRASTRUCTURE, DEMOLITION	91.1	87.2	88.4	87.6	87.2	87.3	84.9	87.2	86.5	85.0	87.2	86.5	116.9	91.6	99.5	104.6	87.4	92.8
0310	Concrete Forming & Accessories	101.8	64.0	69.7	94.8	63.9	68.5	94.9	64.2	68.8	95.0	64.1	68.7	125.7	80.9	87.6	104.9	56.8	63.9
0320	Concrete Reinforcing	105.2	81.6	93.8	104.5	80.8	93.1	105.2	81.6	93.8	107.0	79.0	93.5	171.9	84.6	129.8	137.9	56.0	98.4
0330	Cast-in-Place Concrete	103.1	77.9	93.6	104.1	77.9	94.2	99.7	78.0	91.5	99.7	77.9	91.5	155.6	80.1	127.1	112.6	54.5	90.6
03	CONCRETE	110.8	72.4	93.6	107.0	72.2	91.4	102.0	72.6	88.8	102.2	72.0	88.7	141.5	84.1	114.7	116.6	56.9	89.8
04	MASONRY	95.8	68.0	78.9	95.8	56.3	71.7	95.5	68.0	78.7	95.5	68.0	78.8	169.0	87.4	119.2	164.9	56.2	98.6
05	METALS	103.6	80.3	96.4	99.9	79.3	93.5	100.1	80.5	94.0	100.7	79.0	94.0	111.5	90.6	105.0	113.4	72.2	100.6
06	WOOD, PLASTICS & COMPOSITES	96.1	61.9	78.2	85.2	61.9	73.0	85.3	61.9	73.1	85.3	61.9	73.1	116.6	79.7	97.3	95.0	57.0	75.2
07	THERMAL & MOISTURE PROTECTION	107.9	68.0	90.8	106.7	63.9	88.3	106.5	67.2	89.6	106.0	67.2	89.3	114.2	83.8	101.1	110.4	56.8	87.4
08	OPENINGS	111.6	65.5	100.4	109.6	65.5	98.8	111.2	65.5	100.1	104.3	64.6	94.6	90.4	79.6	87.8	84.2	50.3	75.9
0920	Plaster & Gypsum Board	104.5	61.1	76.0	82.8	61.1	68.6	82.8	61.1	68.6	83.0	61.1	68.6	151.7	79.4	104.2	119.4	56.0	77.8
0950, 0980	Ceilings & Acoustic Treatment	114.9	61.1	81.2	110.7	61.1	79.6	110.7	61.1	79.6	111.3	61.1	79.8	95.4	79.4	85.3	115.9	56.0	78.4
0960	Flooring	101.0	57.5	88.3	98.2	67.6	89.3	98.2	67.6	89.3	98.2	67.6	89.3	121.5	85.1	110.8	101.2	40.5	83.4
0970, 0990	Wall Finishes & Painting/Coating	96.6	58.1	73.5	94.3	55.6	71.1	94.3	74.6	82.5	94.3	57.6	72.3	102.9	81.6	90.1	108.6	46.6	71.5
09	FINISHES	103.0	61.8	80.7	95.7	63.1	78.1	95.4	65.2	79.1	95.6	63.3	78.1	112.9	81.9	96.1	109.1	52.7	78.6
COVERS	DIVS. 10 - 14, 25, 28, 41, 43, 44, 46	100.0	87.9	97.2	100.0	87.9	97.2	100.0	98.4	99.6	100.0	98.4	99.6	139.3	62.8	121.3	131.2	55.8	113.4
21, 22, 23	FIRE SUPPRESSION, PLUMBING & HVAC	99.1	71.8	88.1	99.1	71.8	88.1	99.1	71.8	88.1	99.1	71.8	88.1	106.2	91.1	100.1	106.2	63.0	88.8
26, 27, 3370	ELECTRICAL, COMMUNICATIONS & UTIL.	95.9	64.2	80.2	94.0	64.2	79.3	94.0	64.2	79.3	93.2	64.2	78.9	117.2	81.6	99.6	117.0	55.0	86.3
MF2018	WEIGHTED AVERAGE	102.4	71.1	88.9	100.1	69.9	87.0	99.5	71.9	87.6	98.9	71.4	87.0	118.0	85.2	103.8	112.7	60.8	90.3

CANADA

DIVISION		BRANDON, MANITOBA MAT.	INST.	TOTAL	BRANTFORD, ONTARIO MAT.	INST.	TOTAL	BRIDGEWATER, NOVA SCOTIA MAT.	INST.	TOTAL	CALGARY, ALBERTA MAT.	INST.	TOTAL	CAP-DE-LA-MADELEINE, QUEBEC MAT.	INST.	TOTAL	CHARLESBOURG, QUEBEC MAT.	INST.	TOTAL
015433	CONTRACTOR EQUIPMENT		99.6	99.6		97.9	97.9		97.3	97.3		122.7	122.7		98.1	98.1		98.1	98.1
0241, 31 - 34	SITE & INFRASTRUCTURE, DEMOLITION	124.4	89.9	100.7	116.1	91.9	99.4	100.2	89.3	92.7	126.4	112.5	116.9	96.2	90.6	92.4	96.2	90.6	92.4
0310	Concrete Forming & Accessories	149.5	64.9	77.5	129.0	87.6	93.8	97.4	66.7	71.3	125.3	92.0	96.9	136.3	77.8	86.5	136.3	77.8	86.5
0320	Concrete Reinforcing	167.0	53.1	112.0	160.7	83.3	123.3	136.9	46.4	93.2	133.9	78.7	107.3	136.9	70.5	104.8	136.9	70.5	104.8
0330	Cast-in-Place Concrete	108.0	68.4	93.0	129.0	99.6	117.9	132.5	65.1	107.0	134.5	101.3	122.0	104.1	85.4	97.0	104.1	85.4	97.0
03	CONCRETE	125.5	65.1	98.4	128.4	91.4	111.8	126.4	63.6	98.3	136.8	93.7	117.5	114.1	79.8	98.7	114.1	79.8	98.7
04	MASONRY	221.4	59.4	122.6	173.0	91.5	123.3	167.9	64.2	104.6	231.4	86.3	142.9	169.1	74.8	111.5	169.1	74.8	111.5
05	METALS	126.0	77.2	110.8	111.4	91.2	105.1	110.6	74.9	99.5	128.4	98.4	119.1	109.7	84.0	101.7	109.7	84.0	101.7
06	WOOD, PLASTICS & COMPOSITES	150.7	65.9	106.3	120.4	86.6	102.7	86.3	66.5	75.9	99.8	91.9	95.7	132.8	78.0	104.1	132.8	78.0	104.1
07	THERMAL & MOISTURE PROTECTION	126.8	66.8	101.0	120.1	89.0	106.8	114.8	65.8	93.8	136.1	94.8	118.4	113.9	81.9	100.2	113.9	81.9	100.2
08	OPENINGS	100.2	58.9	90.1	87.7	85.3	87.1	82.7	60.3	77.3	82.4	81.2	82.1	89.2	71.1	84.8	89.2	71.1	84.8
0920	Plaster & Gypsum Board	114.9	64.8	82.0	114.7	86.4	96.1	120.4	65.7	84.5	132.6	90.9	105.2	143.9	77.4	100.3	143.9	77.4	100.3
0950, 0980	Ceilings & Acoustic Treatment	122.1	64.8	86.2	102.7	86.4	92.5	102.7	65.7	79.5	146.1	90.9	111.5	102.7	77.4	86.9	102.7	77.4	86.9
0960	Flooring	133.4	60.4	112.1	115.3	85.1	106.5	96.7	57.4	85.2	114.1	81.3	111.6	115.3	83.8	106.1	115.3	83.8	106.1
0970, 0990	Wall Finishes & Painting/Coating	114.4	52.6	77.4	107.2	89.5	96.6	107.2	57.7	77.6	107.8	101.3	103.9	107.2	81.4	91.7	107.2	81.4	91.7
09	FINISHES	124.1	63.1	91.1	109.7	87.6	97.8	104.5	64.3	82.8	127.8	91.5	108.2	113.1	79.6	95.0	113.1	79.6	95.0
COVERS	DIVS. 10 - 14, 25, 28, 41, 43, 44, 46	131.2	58.0	114.0	131.2	64.7	115.5	131.2	58.7	114.1	131.2	88.4	121.1	131.2	73.1	117.5	131.2	73.1	117.5
21, 22, 23	FIRE SUPPRESSION, PLUMBING & HVAC	106.5	76.8	94.5	106.2	93.8	101.2	106.2	77.4	94.6	106.6	86.0	98.3	106.6	82.0	96.7	106.6	82.0	96.7
26, 27, 3370	ELECTRICAL, COMMUNICATIONS & UTIL.	116.4	62.1	89.5	114.5	81.1	98.0	118.8	57.7	88.6	114.0	90.4	102.3	113.2	64.2	89.0	113.2	64.2	89.0
MF2018	WEIGHTED AVERAGE	122.8	68.9	99.5	115.2	88.8	103.8	113.3	68.7	94.0	123.3	91.7	109.7	112.8	78.2	97.8	112.8	78.2	97.8

For customer support on your Concrete & Masonry Costs with RSMeans Data, call 800.448.8182.

491

CANADA

DIVISION		CHARLOTTETOWN, PRINCE EDWARD ISLAND			CHICOUTIMI, QUEBEC			CORNER BROOK, NEWFOUNDLAND			CORNWALL, ONTARIO			DALHOUSIE, NEW BRUNSWICK			DARTMOUTH, NOVA SCOTIA		
		MAT.	INST.	TOTAL	MAT.	INST.	TOTAL	MAT.	INST.	TOTAL	MAT.	INST.	TOTAL	MAT.	INST.	TOTAL	MAT.	INST.	TOTAL
015433	CONTRACTOR EQUIPMENT		115.1	115.1		97.9	97.9		97.7	97.7		97.9	97.9		98.0	98.0		96.8	96.8
0241, 31 - 34	SITE & INFRASTRUCTURE, DEMOLITION	132.1	101.4	111.0	101.7	90.5	94.0	127.6	87.3	99.9	114.3	91.4	98.6	99.9	87.8	91.6	116.0	88.9	97.4
0310	Concrete Forming & Accessories	125.5	52.1	63.0	138.3	89.8	97.0	122.9	74.1	81.4	126.3	81.0	87.8	104.3	57.0	64.0	111.1	66.7	73.3
0320	Concrete Reinforcing	153.8	45.3	101.4	104.0	93.7	99.0	153.1	47.5	102.1	160.7	83.0	123.2	140.0	56.0	99.5	160.1	46.4	105.2
0330	Cast-in-Place Concrete	150.0	56.3	114.6	106.0	93.8	101.4	125.7	61.8	101.5	116.0	90.7	106.4	109.0	54.7	88.5	121.3	65.0	100.0
03	CONCRETE	147.7	54.0	105.6	107.7	92.0	100.6	160.3	65.9	118.0	122.2	85.2	105.6	120.0	57.1	91.7	140.5	63.6	106.0
04	MASONRY	189.1	53.8	106.6	167.5	89.8	120.1	216.3	72.6	128.7	172.0	83.8	118.2	170.1	56.2	100.6	231.9	64.2	129.6
05	METALS	133.6	75.8	115.6	113.2	91.5	106.4	126.4	73.9	110.0	111.2	89.9	104.6	104.6	72.8	94.6	126.8	74.6	110.5
06	WOOD, PLASTICS & COMPOSITES	102.2	51.7	75.8	133.0	90.7	110.9	125.7	80.1	101.8	118.2	80.5	98.5	93.2	57.1	74.3	112.1	66.4	88.2
07	THERMAL & MOISTURE PROTECTION	135.4	56.3	101.5	111.0	95.6	104.4	130.8	64.1	102.2	119.9	83.9	104.5	119.6	56.8	92.6	128.1	65.8	101.3
08	OPENINGS	85.8	45.0	75.9	87.9	76.3	85.1	106.0	65.3	96.1	89.2	79.3	86.8	85.7	50.3	77.1	91.6	60.3	83.9
0920	Plaster & Gypsum Board	126.1	49.9	76.1	138.9	90.5	107.2	148.0	79.7	103.1	172.0	80.2	111.7	127.7	56.0	80.6	144.2	65.7	92.7
0950, 0980	Ceilings & Acoustic Treatment	131.4	49.9	80.3	114.6	90.5	99.5	122.7	79.7	95.7	106.6	80.2	90.0	105.0	56.0	74.3	131.0	65.7	90.1
0960	Flooring	119.9	53.9	100.6	117.8	83.8	107.9	114.3	48.3	95.0	115.3	83.9	106.1	103.2	61.9	91.1	108.9	57.4	93.9
0970, 0990	Wall Finishes & Painting/Coating	108.7	38.8	66.8	108.6	103.3	105.4	114.3	55.0	78.7	107.2	83.4	93.0	110.3	46.6	72.2	114.3	57.7	80.4
09	FINISHES	123.0	51.3	84.2	116.0	91.2	102.6	123.8	68.7	94.0	118.7	82.0	98.8	109.3	57.1	81.1	122.4	64.2	90.9
COVERS	DIVS. 10 - 14, 25, 28, 41, 43, 44, 46	131.2	56.4	113.6	131.2	78.2	118.7	131.2	58.4	114.1	131.2	62.4	115.0	131.2	55.9	113.4	131.2	58.7	114.1
21, 22, 23	FIRE SUPPRESSION, PLUMBING & HVAC	106.5	57.4	86.7	106.2	80.3	95.7	106.5	64.1	89.4	106.6	91.8	100.7	106.3	63.1	88.8	106.5	77.4	94.7
26, 27, 3370	ELECTRICAL, COMMUNICATIONS & UTIL.	119.8	46.7	83.6	113.5	84.3	99.0	113.5	50.7	82.4	114.6	82.1	98.5	117.6	52.0	85.1	117.6	57.7	88.0
MF2018	WEIGHTED AVERAGE	124.1	58.7	95.8	112.5	86.9	101.4	127.3	66.7	101.1	115.3	85.5	102.4	112.4	61.0	90.2	124.0	68.6	100.1

CANADA

DIVISION		EDMONTON, ALBERTA			FORT MCMURRAY, ALBERTA			FREDERICTON, NEW BRUNSWICK			GATINEAU, QUEBEC			GRANBY, QUEBEC			HALIFAX, NOVA SCOTIA		
		MAT.	INST.	TOTAL	MAT.	INST.	TOTAL	MAT.	INST.	TOTAL	MAT.	INST.	TOTAL	MAT.	INST.	TOTAL	MAT.	INST.	TOTAL
015433	CONTRACTOR EQUIPMENT		123.6	123.6		100.2	100.2		115.0	115.0		98.1	98.1		98.1	98.1		112.9	112.9
0241, 31 - 34	SITE & INFRASTRUCTURE, DEMOLITION	126.6	114.2	118.1	120.8	92.5	101.4	116.1	102.5	106.8	96.0	90.6	92.3	96.4	90.6	92.4	101.8	102.3	102.1
0310	Concrete Forming & Accessories	128.7	92.0	97.4	126.2	85.9	91.9	128.1	57.7	68.2	136.3	77.7	86.4	136.3	77.6	86.4	125.4	85.1	91.1
0320	Concrete Reinforcing	133.5	78.7	107.1	148.7	78.6	114.9	138.3	56.2	98.7	144.7	70.5	108.9	144.7	70.5	108.9	153.3	75.3	115.7
0330	Cast-in-Place Concrete	143.8	101.3	127.8	171.2	95.1	142.5	111.1	56.8	90.6	102.7	85.4	96.1	106.1	85.3	98.3	97.9	82.3	92.0
03	CONCRETE	141.3	93.7	119.9	146.3	88.3	120.3	127.7	58.6	96.7	114.5	79.7	98.9	116.1	79.7	99.7	122.5	83.4	105.0
04	MASONRY	230.8	86.3	142.7	214.4	83.6	134.6	187.9	57.6	108.4	168.9	74.8	111.5	169.2	74.8	111.6	184.4	83.7	122.9
05	METALS	131.9	98.3	121.4	138.5	89.6	123.2	133.8	81.1	117.4	109.7	83.9	101.6	109.9	83.8	101.8	133.5	95.0	121.5
06	WOOD, PLASTICS & COMPOSITES	99.1	91.9	95.3	114.1	85.6	99.1	104.4	57.5	79.8	132.8	78.0	104.1	132.8	78.0	104.1	101.6	85.6	93.2
07	THERMAL & MOISTURE PROTECTION	141.3	94.8	121.3	128.8	89.4	111.9	135.5	58.3	102.4	113.9	81.9	100.2	113.9	80.5	99.5	137.8	85.9	115.5
08	OPENINGS	81.1	81.2	81.1	89.2	77.7	86.4	88.1	49.5	78.7	89.2	66.8	83.8	89.2	66.8	83.8	89.1	77.0	86.2
0920	Plaster & Gypsum Board	131.9	90.9	105.0	115.2	85.0	95.4	128.8	56.0	81.0	114.1	77.4	90.0	114.1	77.4	90.0	122.7	85.0	97.9
0950, 0980	Ceilings & Acoustic Treatment	141.7	90.9	109.9	111.7	85.0	95.0	132.7	56.0	84.6	102.7	77.4	86.9	102.7	77.4	86.9	129.1	85.0	101.4
0960	Flooring	126.3	81.3	113.1	115.3	81.3	105.4	120.7	64.6	104.3	115.3	83.8	106.1	115.3	83.8	106.1	113.9	78.1	103.4
0970, 0990	Wall Finishes & Painting/Coating	110.4	101.3	105.0	107.3	85.9	94.5	109.9	59.4	79.7	107.2	81.4	91.7	107.2	81.4	91.7	109.5	87.5	96.3
09	FINISHES	126.8	91.5	107.7	113.1	85.4	98.1	126.3	59.3	90.1	108.9	79.6	93.0	108.9	79.6	93.0	121.1	84.7	101.4
COVERS	DIVS. 10 - 14, 25, 28, 41, 43, 44, 46	131.2	89.6	121.4	131.2	86.5	120.7	131.2	56.8	113.7	131.2	73.1	117.5	131.2	73.1	117.5	131.2	65.6	115.7
21, 22, 23	FIRE SUPPRESSION, PLUMBING & HVAC	106.4	86.0	98.2	106.7	90.7	100.2	106.7	71.8	92.6	106.6	82.0	96.7	106.2	82.0	96.4	106.5	78.9	95.4
26, 27, 3370	ELECTRICAL, COMMUNICATIONS & UTIL.	123.4	90.4	107.1	108.5	76.3	92.6	122.1	68.5	95.6	113.2	64.2	89.0	113.9	64.2	89.3	123.7	87.2	105.6
MF2018	WEIGHTED AVERAGE	125.3	91.9	110.9	123.9	86.2	107.6	121.9	67.8	98.5	112.5	78.0	97.6	112.7	77.9	97.7	120.5	84.9	105.1

CANADA

DIVISION		HAMILTON, ONTARIO			HULL, QUEBEC			JOLIETTE, QUEBEC			KAMLOOPS, BRITISH COLUMBIA			KINGSTON, ONTARIO			KITCHENER, ONTARIO		
		MAT.	INST.	TOTAL	MAT.	INST.	TOTAL	MAT.	INST.	TOTAL	MAT.	INST.	TOTAL	MAT.	INST.	TOTAL	MAT.	INST.	TOTAL
015433	CONTRACTOR EQUIPMENT		110.1	110.1		98.1	98.1		98.1	98.1		101.3	101.3		100.1	100.1		99.4	99.4
0241, 31 - 34	SITE & INFRASTRUCTURE, DEMOLITION	105.3	100.8	102.2	96.0	90.6	92.3	96.5	90.6	92.5	119.7	94.1	102.1	114.3	95.0	101.0	93.6	96.1	95.3
0310	Concrete Forming & Accessories	132.8	91.0	97.2	136.3	77.7	86.4	136.3	77.8	86.5	123.0	80.8	87.1	126.5	81.1	87.8	118.4	84.7	89.7
0320	Concrete Reinforcing	134.4	98.4	117.0	144.7	70.5	108.9	136.9	70.5	104.8	107.3	73.8	91.1	94.7	98.2	96.4	94.7	98.2	96.4
0330	Cast-in-Place Concrete	106.1	97.9	103.0	102.7	85.4	96.1	106.9	85.4	98.8	92.6	89.6	91.5	116.0	90.7	106.4	110.3	93.3	103.9
03	CONCRETE	120.1	95.2	108.9	114.5	79.7	98.9	115.4	79.8	99.4	126.4	83.2	107.0	124.3	85.3	106.8	102.2	90.4	96.9
04	MASONRY	189.2	96.3	132.5	168.9	74.8	111.5	169.3	74.8	111.6	177.4	82.2	119.3	179.6	83.9	121.2	150.0	94.5	116.1
05	METALS	132.0	102.4	122.7	109.9	83.9	101.8	109.9	84.0	101.8	111.9	86.2	103.9	112.6	89.9	105.6	122.8	96.1	114.5
06	WOOD, PLASTICS & COMPOSITES	110.7	90.7	100.2	132.8	78.0	104.1	132.8	78.0	104.1	100.5	79.6	89.5	118.2	80.6	98.6	109.9	83.0	95.8
07	THERMAL & MOISTURE PROTECTION	132.8	97.6	117.7	113.9	81.9	100.1	113.9	81.9	100.1	129.8	79.5	108.2	119.9	85.0	104.9	115.3	94.3	106.3
08	OPENINGS	85.5	89.8	86.5	89.2	66.8	83.8	89.2	71.1	84.8	86.3	77.2	84.1	89.2	79.0	86.7	80.3	83.9	81.2
0920	Plaster & Gypsum Board	124.4	90.0	101.8	114.1	77.4	90.0	143.9	77.4	100.3	100.1	78.8	86.2	175.4	80.3	113.0	104.8	82.7	90.3
0950, 0980	Ceilings & Acoustic Treatment	137.4	90.0	107.7	102.7	77.4	86.9	102.7	77.4	86.9	102.7	78.8	94.8	119.3	80.3	94.8	106.7	82.7	91.7
0960	Flooring	121.1	91.0	112.3	115.3	83.8	106.1	115.3	83.8	106.1	114.8	48.0	95.3	115.3	83.9	106.1	102.9	91.1	99.5
0970, 0990	Wall Finishes & Painting/Coating	107.2	99.2	102.4	107.2	81.4	91.7	107.2	81.4	91.7	107.2	74.4	87.5	107.2	77.1	89.2	101.7	89.8	94.6
09	FINISHES	123.1	91.8	106.1	108.9	79.6	93.0	113.1	79.6	95.0	109.3	74.3	90.4	122.2	81.3	100.1	103.2	85.9	93.9
COVERS	DIVS. 10 - 14, 25, 28, 41, 43, 44, 46	131.2	87.6	120.9	131.2	73.1	117.5	131.2	73.1	117.5	131.2	81.6	119.5	131.2	62.4	115.0	131.2	85.2	120.3
21, 22, 23	FIRE SUPPRESSION, PLUMBING & HVAC	106.8	87.7	99.1	106.2	82.0	96.4	106.2	82.0	96.4	106.2	85.2	97.7	106.6	91.9	100.7	105.5	87.7	98.3
26, 27, 3370	ELECTRICAL, COMMUNICATIONS & UTIL.	115.5	99.8	107.7	113.5	64.2	90.0	113.9	64.2	89.3	117.3	73.0	95.4	114.6	80.8	97.9	111.3	97.5	104.5
MF2018	WEIGHTED AVERAGE	119.2	94.5	108.6	112.6	78.0	97.7	113.0	78.2	97.9	115.6	81.6	100.9	116.5	85.5	103.1	109.9	91.2	101.8

City Cost Indexes - V2

CANADA

DIVISION		LAVAL, QUEBEC MAT.	INST.	TOTAL	LETHBRIDGE, ALBERTA MAT.	INST.	TOTAL	LLOYDMINSTER, ALBERTA MAT.	INST.	TOTAL	LONDON, ONTARIO MAT.	INST.	TOTAL	MEDICINE HAT, ALBERTA MAT.	INST.	TOTAL	MONCTON, NEW BRUNSWICK MAT.	INST.	TOTAL
015433	CONTRACTOR EQUIPMENT		98.1	98.1		100.2	100.2		100.2	100.2		113.0	113.0		100.2	100.2		97.4	97.4
0241, 31 - 34	SITE & INFRASTRUCTURE, DEMOLITION	96.4	90.7	92.5	113.8	93.1	99.6	113.8	92.5	99.2	104.4	101.5	102.4	112.6	92.6	98.8	104.1	89.2	93.8
0310	Concrete Forming & Accessories	136.5	78.4	87.0	128.1	86.0	92.3	125.8	76.9	84.2	131.9	85.8	92.7	128.0	76.8	84.4	137.9	66.7	103.6
0320	Concrete Reinforcing	144.7	71.2	109.2	148.7	78.6	114.9	148.7	78.5	114.9	124.4	97.2	111.3	148.7	78.5	114.9	137.9	66.7	103.6
0330	Cast-in-Place Concrete	106.1	86.0	98.5	128.4	95.1	115.9	119.2	91.7	108.8	118.5	97.3	110.5	119.2	91.6	108.8	108.4	65.3	92.1
03	CONCRETE	116.1	80.3	100.0	126.5	88.3	109.4	122.0	83.0	104.5	124.5	92.6	110.1	122.2	82.9	104.6	114.6	66.0	92.8
04	MASONRY	169.2	75.4	112.0	188.1	83.6	124.4	168.3	77.6	113.0	200.2	95.4	136.3	168.3	77.6	113.0	164.6	71.7	107.9
05	METALS	109.8	84.2	101.8	132.7	89.6	119.3	111.9	89.4	104.9	132.0	104.1	123.3	112.1	89.4	105.0	113.4	85.0	104.5
06	WOOD, PLASTICS & COMPOSITES	132.9	78.6	104.5	117.9	85.6	101.0	114.1	76.4	94.4	113.7	83.9	98.1	117.9	76.4	96.2	95.0	63.0	78.3
07	THERMAL & MOISTURE PROTECTION	114.7	82.4	100.8	126.1	89.4	110.4	122.4	85.2	106.4	128.3	95.3	114.1	128.5	85.2	109.9	114.6	69.3	95.2
08	OPENINGS	89.2	67.5	83.9	89.2	77.7	86.4	89.2	72.7	85.2	81.8	85.0	82.6	89.2	72.7	85.2	84.2	59.8	78.3
0920	Plaster & Gypsum Board	114.4	78.1	90.6	106.2	85.0	92.3	101.6	75.6	84.6	128.9	82.9	98.7	103.9	75.6	85.4	119.4	62.2	81.9
0950, 0980	Ceilings & Acoustic Treatment	102.7	78.1	87.3	111.7	85.0	95.0	102.7	75.6	85.8	135.0	82.9	102.4	102.7	75.6	85.8	115.9	62.2	82.2
0960	Flooring	115.3	84.7	106.4	115.3	81.3	105.4	115.3	81.3	105.4	115.7	91.1	108.5	115.3	81.3	105.4	101.2	63.3	90.1
0970, 0990	Wall Finishes & Painting/Coating	107.2	82.1	92.2	107.2	93.8	99.2	107.3	73.1	86.8	109.6	96.1	101.5	107.2	73.1	86.8	108.6	79.9	91.4
09	FINISHES	108.9	80.2	93.4	111.0	86.3	97.6	108.3	77.1	91.4	123.5	87.5	104.0	108.5	77.1	91.5	109.1	65.1	85.3
COVERS	DIVS. 10 - 14, 25, 28, 41, 43, 44, 46	131.2	73.7	117.6	131.2	85.3	120.4	131.2	83.5	119.9	131.2	87.1	120.8	131.2	82.3	119.7	131.2	58.2	114.0
21, 22, 23	FIRE SUPPRESSION, PLUMBING & HVAC	105.8	82.7	96.5	106.6	87.6	98.9	106.6	87.6	98.9	106.6	86.3	98.4	106.2	84.5	97.4	106.2	70.5	91.8
26, 27, 3370	ELECTRICAL, COMMUNICATIONS & UTIL.	114.3	64.8	89.8	110.1	76.3	93.4	107.7	76.3	92.2	110.1	98.2	104.2	107.7	76.3	92.2	120.9	85.5	103.4
MF2018	WEIGHTED AVERAGE	112.6	78.6	97.9	118.9	85.7	104.6	113.6	82.6	100.2	119.2	92.9	107.8	113.7	81.9	100.0	113.0	73.2	95.8

CANADA

DIVISION		MONTREAL, QUEBEC MAT.	INST.	TOTAL	MOOSE JAW, SASKATCHEWAN MAT.	INST.	TOTAL	NEW GLASGOW, NOVA SCOTIA MAT.	INST.	TOTAL	NEWCASTLE, NEW BRUNSWICK MAT.	INST.	TOTAL	NORTH BAY, ONTARIO MAT.	INST.	TOTAL	OSHAWA, ONTARIO MAT.	INST.	TOTAL
015433	CONTRACTOR EQUIPMENT		116.2	116.2		96.5	96.5		96.8	96.8		97.4	97.4		97.4	97.4		99.4	99.4
0241, 31 - 34	SITE & INFRASTRUCTURE, DEMOLITION	114.3	102.2	106.0	113.6	87.1	99.6	109.9	88.9	95.5	104.6	87.4	92.8	125.0	90.6	101.4	103.8	96.3	98.6
0310	Concrete Forming & Accessories	132.9	90.6	96.9	108.5	53.6	61.8	111.0	66.7	73.3	104.9	57.0	64.1	150.5	78.6	89.4	124.9	88.2	93.6
0320	Concrete Reinforcing	124.6	93.8	109.8	105.0	59.8	83.1	153.1	46.4	101.6	137.9	56.0	98.4	181.1	82.6	133.5	150.0	99.2	125.5
0330	Cast-in-Place Concrete	122.7	97.5	113.2	115.6	62.0	95.4	121.3	65.0	100.0	112.6	54.6	90.7	116.8	78.2	102.3	127.5	102.1	118.0
03	CONCRETE	126.5	94.2	112.0	106.5	58.7	85.0	139.6	63.6	105.5	116.6	57.0	89.8	141.6	79.8	113.8	124.7	95.2	111.5
04	MASONRY	185.6	89.9	127.2	166.9	55.2	98.8	215.9	64.2	123.4	164.9	56.2	98.6	222.8	80.2	135.8	152.8	97.7	119.2
05	METALS	142.7	101.5	129.8	108.6	74.0	97.8	124.4	74.6	108.8	113.4	72.5	100.6	125.2	89.4	114.0	113.1	96.8	108.0
06	WOOD, PLASTICS & COMPOSITES	109.7	91.4	100.1	97.1	52.4	73.7	112.1	66.4	88.2	95.0	57.0	75.2	154.0	79.2	114.8	118.1	86.7	101.7
07	THERMAL & MOISTURE PROTECTION	126.1	97.1	113.7	113.1	58.4	89.6	128.1	65.8	101.3	114.6	56.8	89.8	134.2	80.5	111.1	116.2	99.6	109.1
08	OPENINGS	87.6	78.6	85.4	85.5	50.5	77.0	91.6	60.3	83.9	84.2	50.3	75.9	98.4	77.0	93.2	85.1	87.2	85.6
0920	Plaster & Gypsum Board	126.6	90.5	102.9	97.5	51.3	67.2	141.8	65.7	91.9	119.4	56.0	77.8	138.3	78.8	99.3	108.5	86.6	94.1
0950, 0980	Ceilings & Acoustic Treatment	139.7	90.5	108.9	102.7	51.3	70.5	122.1	65.7	86.8	115.9	56.0	78.4	122.1	78.8	95.0	102.2	86.6	92.4
0960	Flooring	118.9	86.8	109.5	105.0	52.3	89.6	108.9	57.4	93.9	101.2	61.9	89.7	133.4	83.9	119.0	105.9	93.2	102.2
0970, 0990	Wall Finishes & Painting/Coating	115.1	103.3	108.0	107.2	59.4	78.6	114.3	57.7	80.4	108.6	46.6	71.5	114.3	82.8	95.4	101.7	103.0	102.5
09	FINISHES	124.7	92.2	107.1	104.6	53.7	77.0	119.9	64.2	89.8	109.1	57.1	80.9	127.2	80.4	101.8	104.1	90.2	96.6
COVERS	DIVS. 10 - 14, 25, 28, 41, 43, 44, 46	131.2	79.8	119.1	131.2	55.5	113.4	131.2	58.7	114.1	131.2	55.8	113.4	131.2	61.2	114.7	131.2	85.9	120.5
21, 22, 23	FIRE SUPPRESSION, PLUMBING & HVAC	106.8	80.5	96.2	106.6	68.4	91.2	106.5	77.4	94.7	106.2	63.0	88.8	106.5	90.0	99.8	105.5	87.8	98.3
26, 27, 3370	ELECTRICAL, COMMUNICATIONS & UTIL.	118.0	84.3	101.4	116.3	55.4	86.2	113.9	57.7	86.1	116.4	55.0	86.0	115.2	82.0	98.7	112.2	100.3	106.3
MF2018	WEIGHTED AVERAGE	122.1	89.4	108.0	111.0	62.4	90.0	121.9	68.6	98.9	112.8	61.4	90.5	125.0	83.4	107.0	112.4	93.6	104.3

CANADA

DIVISION		OTTAWA, ONTARIO MAT.	INST.	TOTAL	OWEN SOUND, ONTARIO MAT.	INST.	TOTAL	PETERBOROUGH, ONTARIO MAT.	INST.	TOTAL	PORTAGE LA PRAIRIE, MANITOBA MAT.	INST.	TOTAL	PRINCE ALBERT, SASKATCHEWAN MAT.	INST.	TOTAL	PRINCE GEORGE, BRITISH COLUMBIA MAT.	INST.	TOTAL
015433	CONTRACTOR EQUIPMENT		112.7	112.7		97.9	97.9		97.9	97.9		100.1	100.1		96.5	96.5		101.3	101.3
0241, 31 - 34	SITE & INFRASTRUCTURE, DEMOLITION	108.0	101.0	103.2	116.9	91.5	99.4	116.1	91.3	99.1	114.6	90.2	97.8	109.6	87.2	94.2	122.9	94.1	103.1
0310	Concrete Forming & Accessories	126.4	85.0	91.2	125.7	77.3	84.5	129.0	79.6	87.0	128.2	64.5	74.0	108.5	53.4	61.6	114.6	76.1	81.9
0320	Concrete Reinforcing	134.5	97.2	116.5	171.9	84.6	129.7	160.7	83.1	123.2	148.7	53.1	102.6	109.6	59.7	85.5	107.3	73.8	91.1
0330	Cast-in-Place Concrete	120.6	99.1	112.5	155.6	74.7	125.1	129.0	79.8	110.4	119.2	67.9	99.8	104.8	61.9	88.6	116.0	89.7	106.1
03	CONCRETE	126.4	92.9	111.4	141.5	78.3	113.1	128.4	80.9	107.1	115.1	64.8	92.5	102.1	58.5	82.5	136.7	81.1	111.8
04	MASONRY	189.1	95.7	132.1	169.0	85.3	117.9	173.0	86.2	120.1	171.7	58.5	102.7	166.2	55.2	98.5	179.3	82.2	120.1
05	METALS	132.7	105.0	124.0	111.5	90.4	104.9	111.4	90.1	104.7	112.1	77.4	101.3	108.7	73.8	97.8	111.9	86.3	103.9
06	WOOD, PLASTICS & COMPOSITES	106.3	82.7	94.0	116.6	76.2	95.5	120.4	78.0	98.2	117.9	65.9	90.7	97.1	52.4	73.7	100.5	73.1	86.1
07	THERMAL & MOISTURE PROTECTION	141.4	95.9	121.9	114.2	81.3	100.0	120.1	85.8	105.4	113.6	66.4	93.4	113.0	57.4	89.1	123.8	78.8	104.5
08	OPENINGS	90.5	84.3	89.0	90.4	76.4	87.0	87.7	78.6	85.5	89.2	58.9	81.8	84.5	50.5	76.2	86.3	73.7	83.2
0920	Plaster & Gypsum Board	124.0	81.8	96.3	151.7	75.8	101.9	114.7	77.6	90.4	103.6	64.8	78.2	97.5	51.3	67.2	100.1	72.2	81.8
0950, 0980	Ceilings & Acoustic Treatment	145.7	81.8	105.6	95.4	75.8	83.1	102.7	77.6	87.0	102.7	64.8	79.0	102.7	51.3	70.5	102.7	72.2	83.6
0960	Flooring	114.4	86.8	106.4	121.5	85.1	110.8	115.3	83.9	106.1	115.3	60.4	99.3	105.0	52.3	89.6	110.5	65.7	97.4
0970, 0990	Wall Finishes & Painting/Coating	109.6	91.3	98.6	102.9	81.6	90.1	107.2	84.8	93.8	107.3	52.6	74.6	107.2	50.6	73.3	107.2	74.4	87.5
09	FINISHES	126.0	85.5	104.1	112.9	79.3	94.7	109.7	81.0	94.2	108.4	62.8	83.8	104.6	52.7	76.5	108.1	73.6	89.4
COVERS	DIVS. 10 - 14, 25, 28, 41, 43, 44, 46	131.2	85.0	120.3	139.3	61.7	121.0	131.2	62.6	115.0	131.2	57.7	113.9	131.2	55.5	113.4	131.2	80.9	119.3
21, 22, 23	FIRE SUPPRESSION, PLUMBING & HVAC	106.7	87.9	99.1	106.2	89.9	99.7	106.2	93.2	101.0	106.2	76.3	94.1	106.6	61.7	88.5	106.2	85.2	97.7
26, 27, 3370	ELECTRICAL, COMMUNICATIONS & UTIL.	111.3	98.1	104.8	118.5	80.8	99.9	114.5	81.6	98.2	116.0	53.6	85.1	116.3	55.4	86.2	114.5	73.0	94.0
MF2018	WEIGHTED AVERAGE	120.6	93.0	108.7	118.1	83.5	103.2	115.2	85.2	102.3	113.6	67.5	93.6	110.2	60.7	88.8	116.5	81.0	101.2

For customer support on your Concrete & Masonry Costs with RSMeans Data, call 800.448.8182.

493

CANADA

DIVISION		QUEBEC CITY, QUEBEC MAT.	INST.	TOTAL	RED DEER, ALBERTA MAT.	INST.	TOTAL	REGINA, SASKATCHEWAN MAT.	INST.	TOTAL	RIMOUSKI, QUEBEC MAT.	INST.	TOTAL	ROUYN-NORANDA, QUEBEC MAT.	INST.	TOTAL	SAINT HYACINTHE, QUEBEC MAT.	INST.	TOTAL
015433	CONTRACTOR EQUIPMENT		117.5	117.5		100.2	100.2		124.0	124.0		98.1	98.1		98.1	98.1		98.1	98.1
0241, 31 - 34	SITE & INFRASTRUCTURE, DEMOLITION	113.4	102.2	105.7	112.6	92.6	98.8	126.0	113.0	117.1	96.3	90.6	92.4	96.0	90.6	92.3	96.4	90.6	92.4
0310	Concrete Forming & Accessories	129.6	90.7	96.5	140.1	76.8	86.3	131.7	88.2	94.7	136.3	89.8	96.8	136.3	77.7	86.4	136.3	77.7	86.4
0320	Concrete Reinforcing	121.9	93.9	108.4	148.7	78.5	114.9	121.5	84.6	103.7	103.3	93.7	98.7	144.7	70.5	108.9	144.7	70.5	108.9
0330	Cast-in-Place Concrete	129.8	97.6	117.6	119.2	91.6	108.8	151.6	94.6	130.1	108.0	93.8	102.7	102.7	85.4	96.1	106.1	85.4	98.3
03	CONCRETE	129.2	94.4	113.6	122.9	82.9	105.0	141.3	90.6	118.5	111.4	92.0	102.7	114.5	79.7	98.9	116.1	79.7	99.8
04	MASONRY	189.0	89.9	128.5	168.3	77.6	113.0	221.2	83.1	137.0	168.7	89.8	120.6	168.9	74.8	111.5	169.2	74.8	111.6
05	METALS	136.5	103.3	126.2	112.1	89.4	105.0	137.4	97.7	125.0	109.4	91.6	103.9	109.9	83.9	101.8	109.9	83.9	101.8
06	WOOD, PLASTICS & COMPOSITES	112.5	91.4	101.4	117.9	76.4	96.2	105.7	89.2	97.1	132.8	90.7	110.8	132.8	78.0	104.1	132.8	78.0	104.1
07	THERMAL & MOISTURE PROTECTION	123.0	97.1	111.9	138.2	85.2	115.5	148.5	83.4	120.5	113.9	95.6	106.0	113.9	81.9	100.1	114.4	81.9	100.4
08	OPENINGS	88.5	85.7	87.9	89.2	72.7	85.2	88.2	77.0	85.4	88.8	76.3	85.8	89.2	66.8	83.8	89.2	66.8	83.8
0920	Plaster & Gypsum Board	123.0	90.5	101.7	103.9	75.6	85.4	142.0	88.1	106.6	143.7	90.5	108.8	113.9	77.4	90.0	113.9	77.4	90.0
0950, 0980	Ceilings & Acoustic Treatment	134.2	90.5	106.8	102.7	75.6	85.8	156.4	88.1	113.6	102.1	90.5	94.8	102.1	77.4	86.6	102.1	77.4	86.6
0960	Flooring	115.6	86.8	107.2	118.3	81.3	107.5	125.4	91.1	115.4	116.4	83.8	106.9	115.3	83.8	106.1	115.3	83.8	106.1
0970, 0990	Wall Finishes & Painting/Coating	115.9	103.3	108.3	107.2	73.1	86.8	109.6	85.1	95.0	110.0	103.3	105.9	107.2	81.4	91.7	107.2	81.4	91.7
09	FINISHES	123.5	92.2	106.6	109.4	77.1	91.9	134.2	89.4	109.9	113.5	91.2	101.4	108.7	79.6	93.0	108.7	79.6	93.0
COVERS	DIVS. 10 - 14, 25, 28, 41, 43, 44, 46	131.2	79.8	119.1	131.2	82.3	119.7	131.2	67.6	116.2	131.2	78.3	118.7	131.2	73.1	117.5	131.2	73.1	117.5
21, 22, 23	FIRE SUPPRESSION, PLUMBING & HVAC	106.5	80.5	96.0	106.2	84.5	97.4	106.4	82.8	96.9	106.2	80.3	95.8	106.2	82.0	96.4	102.4	82.0	94.1
26, 27, 3370	ELECTRICAL, COMMUNICATIONS & UTIL.	124.2	84.3	104.5	107.7	76.3	92.2	121.0	88.1	104.7	113.9	84.3	99.2	113.9	64.2	89.3	114.5	64.2	89.6
MF2018	WEIGHTED AVERAGE	122.2	89.9	108.2	114.2	81.9	100.2	127.0	88.5	110.3	112.3	86.9	101.4	112.4	78.0	97.6	111.8	78.0	97.2

CANADA

DIVISION		SAINT JOHN, NEW BRUNSWICK MAT.	INST.	TOTAL	SARNIA, ONTARIO MAT.	INST.	TOTAL	SASKATOON, SASKATCHEWAN MAT.	INST.	TOTAL	SAULT STE MARIE, ONTARIO MAT.	INST.	TOTAL	SHERBROOKE, QUEBEC MAT.	INST.	TOTAL	SOREL, QUEBEC MAT.	INST.	TOTAL
015433	CONTRACTOR EQUIPMENT		97.4	97.4		97.9	97.9		97.0	97.0		97.9	97.9		98.1	98.1		98.1	98.1
0241, 31 - 34	SITE & INFRASTRUCTURE, DEMOLITION	104.8	89.2	94.1	114.7	91.4	98.7	111.8	90.0	96.8	105.1	91.0	95.4	96.4	90.6	92.4	96.5	90.6	92.5
0310	Concrete Forming & Accessories	128.1	61.3	71.2	127.5	86.1	92.3	108.3	87.3	90.4	115.0	83.4	88.1	136.3	77.7	86.4	136.3	77.8	86.5
0320	Concrete Reinforcing	137.9	66.7	103.6	114.6	84.4	100.1	112.2	84.4	98.8	103.1	83.2	93.5	144.7	70.5	108.9	136.9	70.5	104.8
0330	Cast-in-Place Concrete	110.9	65.3	93.7	119.0	92.1	108.8	112.0	89.5	103.5	106.9	78.5	96.2	106.1	85.4	98.3	106.9	85.4	98.8
03	CONCRETE	117.5	64.8	93.8	117.5	88.3	104.4	106.6	87.8	98.1	102.6	82.3	93.5	116.1	79.7	99.8	115.4	79.8	99.4
04	MASONRY	188.3	71.5	117.1	185.9	88.7	126.6	179.7	83.0	120.7	169.9	89.2	120.7	169.2	74.8	111.6	169.3	74.8	111.6
05	METALS	113.3	84.9	104.5	111.3	90.5	104.8	105.4	87.1	99.7	110.5	93.9	105.3	109.7	83.9	101.6	109.9	84.0	101.8
06	WOOD, PLASTICS & COMPOSITES	120.8	59.5	88.7	119.3	85.6	101.7	93.8	88.3	90.9	105.1	84.9	94.5	132.8	78.0	104.1	132.8	78.0	104.1
07	THERMAL & MOISTURE PROTECTION	115.0	69.2	95.3	120.2	89.1	106.8	116.7	81.0	101.4	118.9	85.1	104.4	113.9	81.9	100.2	113.9	81.9	100.1
08	OPENINGS	84.1	56.9	77.5	90.7	82.2	88.6	85.5	76.5	83.3	82.5	83.3	82.7	89.2	66.8	83.8	89.2	71.1	84.8
0920	Plaster & Gypsum Board	133.4	58.5	84.3	139.9	85.4	104.1	116.5	88.1	97.9	106.4	84.7	92.1	113.9	77.4	90.0	143.7	77.4	100.2
0950, 0980	Ceilings & Acoustic Treatment	120.4	58.5	81.6	107.8	85.4	93.8	124.2	88.1	101.6	102.7	84.7	91.4	102.1	77.4	86.6	102.1	77.4	86.6
0960	Flooring	113.3	63.3	98.7	115.3	91.8	108.4	108.0	91.1	103.1	107.9	88.7	102.3	115.3	83.8	106.1	115.3	83.8	106.1
0970, 0990	Wall Finishes & Painting/Coating	108.6	79.9	91.4	107.2	95.9	100.4	110.3	85.1	95.2	107.2	88.9	96.2	107.2	81.4	91.7	107.2	81.4	91.7
09	FINISHES	115.6	63.1	87.2	114.5	88.5	100.5	114.4	88.7	100.5	105.7	85.3	94.6	108.7	79.6	93.0	112.9	79.6	94.9
COVERS	DIVS. 10 - 14, 25, 28, 41, 43, 44, 46	131.2	58.1	114.0	131.2	63.9	115.3	131.2	65.9	115.8	131.2	83.8	120.0	131.2	73.1	117.5	131.2	73.1	117.5
21, 22, 23	FIRE SUPPRESSION, PLUMBING & HVAC	106.2	72.0	92.4	106.2	98.9	103.2	106.0	82.6	96.5	106.2	87.8	98.8	106.6	82.0	96.7	106.2	82.0	96.4
26, 27, 3370	ELECTRICAL, COMMUNICATIONS & UTIL.	123.5	85.5	104.7	117.6	83.6	100.8	118.6	88.0	103.5	116.2	82.0	99.3	113.9	64.2	89.3	113.9	64.2	89.3
MF2018	WEIGHTED AVERAGE	115.6	72.9	97.1	115.5	89.4	104.2	112.0	85.1	100.4	110.6	86.4	100.1	112.8	78.0	97.7	112.9	78.2	97.9

CANADA

DIVISION		ST. CATHARINES, ONTARIO MAT.	INST.	TOTAL	ST JEROME, QUEBEC MAT.	INST.	TOTAL	ST. JOHN'S, NEWFOUNDLAND MAT.	INST.	TOTAL	SUDBURY, ONTARIO MAT.	INST.	TOTAL	SUMMERSIDE, PRINCE EDWARD ISLAND MAT.	INST.	TOTAL	SYDNEY, NOVA SCOTIA MAT.	INST.	TOTAL
015433	CONTRACTOR EQUIPMENT		97.3	97.3		98.1	98.1		120.5	120.5		97.3	97.3		96.8	96.8		96.8	96.8
0241, 31 - 34	SITE & INFRASTRUCTURE, DEMOLITION	93.8	92.8	93.1	96.0	90.6	92.3	113.1	111.0	111.6	93.9	92.4	92.8	119.4	86.4	96.7	106.1	88.9	94.3
0310	Concrete Forming & Accessories	116.1	90.6	94.4	136.3	77.7	86.4	127.4	83.3	89.8	111.3	85.6	89.4	111.3	51.5	60.4	111.0	66.7	73.3
0320	Concrete Reinforcing	95.5	98.3	96.8	144.7	70.5	108.9	168.7	80.9	126.4	96.2	96.8	96.5	151.1	45.2	100.0	153.1	46.4	101.6
0330	Cast-in-Place Concrete	105.3	95.0	101.4	102.7	85.4	96.1	135.2	96.3	120.5	106.2	91.5	100.7	115.2	53.1	91.7	93.6	65.0	82.8
03	CONCRETE	99.9	93.7	97.1	114.5	79.7	98.9	143.9	88.2	118.9	100.1	89.9	95.5	149.5	52.1	105.8	126.7	63.6	98.4
04	MASONRY	149.6	97.0	117.5	168.9	74.8	111.5	212.1	88.8	136.9	149.7	93.2	115.2	215.2	53.7	116.7	213.6	64.2	122.5
05	METALS	113.0	96.0	107.7	109.9	83.9	101.8	139.5	95.2	125.7	112.5	95.4	107.1	124.4	68.0	106.8	124.4	74.6	108.8
06	WOOD, PLASTICS & COMPOSITES	107.2	90.4	98.4	132.8	78.0	104.1	105.1	81.6	92.8	102.4	84.9	93.2	112.8	51.1	80.5	112.1	66.4	88.2
07	THERMAL & MOISTURE PROTECTION	115.3	97.6	107.7	113.9	81.9	100.1	137.8	92.7	118.5	114.5	92.6	105.1	127.6	55.6	96.7	128.1	65.8	101.3
08	OPENINGS	79.8	88.3	81.9	89.2	66.8	83.8	85.6	72.7	82.5	80.4	83.7	81.2	102.3	44.7	88.3	91.6	60.3	83.9
0920	Plaster & Gypsum Board	98.8	90.3	93.2	113.9	77.4	90.0	146.3	80.6	103.2	97.6	84.7	89.1	143.0	49.9	81.9	141.8	65.7	91.9
0950, 0980	Ceilings & Acoustic Treatment	102.2	90.3	94.8	102.1	77.4	86.6	144.0	80.6	104.3	97.1	84.7	89.3	122.1	49.9	76.9	122.1	65.7	86.8
0960	Flooring	101.7	88.1	97.8	115.3	83.8	106.1	118.7	50.0	98.7	99.9	88.7	96.6	109.0	53.9	92.9	108.9	57.4	93.9
0970, 0990	Wall Finishes & Painting/Coating	101.7	99.6	100.4	107.2	81.4	91.7	111.8	96.7	102.8	101.7	90.7	95.2	114.3	38.8	69.0	114.3	57.7	80.4
09	FINISHES	101.0	91.2	95.7	108.7	79.6	93.0	131.0	78.4	102.5	99.0	86.5	92.2	121.1	50.8	83.1	119.9	64.2	89.8
COVERS	DIVS. 10 - 14, 25, 28, 41, 43, 44, 46	131.2	66.2	115.9	131.2	73.1	117.5	131.2	65.4	115.7	131.2	85.1	120.3	131.2	55.0	113.2	131.2	58.7	114.1
21, 22, 23	FIRE SUPPRESSION, PLUMBING & HVAC	105.5	87.1	98.1	106.2	82.0	96.4	106.3	81.4	96.3	105.9	85.4	97.7	106.5	57.3	86.6	106.5	77.4	94.7
26, 27, 3370	ELECTRICAL, COMMUNICATIONS & UTIL.	112.8	98.5	105.7	114.6	64.2	89.6	122.9	78.3	100.8	110.7	98.3	104.6	112.8	46.7	80.1	113.9	57.7	86.1
MF2018	WEIGHTED AVERAGE	108.0	92.1	101.1	112.5	78.0	97.6	126.2	85.2	108.5	107.7	90.3	100.2	124.4	56.4	95.0	120.1	68.6	97.9

For customer support on your Concrete & Masonry Costs with RSMeans Data, call 800.448.8182.

City Cost Indexes - V2

		CANADA																
	DIVISION	THUNDER BAY, ONTARIO			TIMMINS, ONTARIO			TORONTO, ONTARIO			TROIS RIVIERES, QUEBEC			TRURO, NOVA SCOTIA			VANCOUVER, BRITISH COLUMBIA		
		MAT.	INST.	TOTAL	MAT.	INST.	TOTAL	MAT.	INST.	TOTAL	MAT.	INST.	TOTAL	MAT.	INST.	TOTAL	MAT.	INST.	TOTAL
015433	CONTRACTOR EQUIPMENT		97.3	97.3		97.9	97.9		112.2	112.2		97.6	97.6		97.3	97.3		130.1	130.1
0241, 31 - 34	SITE & INFRASTRUCTURE, DEMOLITION	98.3	92.7	94.4	116.1	91.0	98.8	107.8	101.6	103.6	106.3	90.3	95.3	100.4	89.3	92.7	115.2	116.1	115.8
0310	Concrete Forming & Accessories	124.9	89.4	94.7	129.0	78.7	86.2	131.3	98.5	103.4	160.0	77.8	90.1	97.4	66.7	71.3	132.4	89.3	95.7
0320	Concrete Reinforcing	85.3	98.1	91.5	160.7	82.6	123.0	134.4	100.9	118.2	153.1	70.5	113.3	136.9	46.4	93.2	130.1	91.4	111.4
0330	Cast-in-Place Concrete	116.0	94.3	107.8	129.0	78.3	109.8	106.9	109.2	107.8	96.9	85.3	92.5	133.9	65.1	107.9	121.8	98.7	113.0
03	CONCRETE	107.6	92.8	101.0	128.4	79.8	106.6	120.3	103.1	112.6	129.2	79.7	107.0	127.1	63.6	98.6	130.1	94.0	113.9
04	MASONRY	150.2	96.7	117.6	173.0	80.2	116.4	189.2	104.2	137.4	218.0	74.8	130.6	168.0	64.2	104.7	180.3	93.1	127.1
05	METALS	113.0	95.2	107.5	111.4	89.7	104.6	132.5	106.3	124.3	123.6	83.7	111.2	110.6	74.9	99.5	132.2	106.9	124.3
06	WOOD, PLASTICS & COMPOSITES	118.1	88.6	102.7	120.4	79.2	98.9	113.5	97.0	104.9	170.1	77.9	121.9	86.3	66.5	75.9	108.4	88.8	98.1
07	THERMAL & MOISTURE PROTECTION	115.6	95.2	106.9	120.1	80.5	103.1	139.9	105.1	125.0	127.1	81.9	107.7	114.8	65.8	93.8	145.8	90.2	121.9
08	OPENINGS	79.1	86.9	81.0	87.7	77.0	85.1	84.1	95.7	87.0	100.2	71.1	93.1	82.7	60.3	77.3	84.7	86.0	85.0
0920	Plaster & Gypsum Board	123.9	88.5	100.7	114.7	78.8	91.2	123.9	96.6	106.0	171.2	77.4	109.7	120.4	65.7	84.5	135.1	87.5	103.9
0950, 0980	Ceilings & Acoustic Treatment	97.1	88.5	91.7	102.7	78.8	87.8	143.2	96.6	114.0	120.8	77.4	93.6	102.7	65.7	79.5	147.4	87.5	109.9
0960	Flooring	105.9	94.3	102.5	115.3	83.9	106.1	115.9	96.6	110.3	133.4	83.8	118.9	96.7	57.4	85.2	130.0	84.8	116.8
0970, 0990	Wall Finishes & Painting/Coating	101.7	92.0	95.9	107.2	82.8	92.6	106.0	103.0	104.2	114.3	81.4	94.5	107.2	57.7	77.6	105.8	95.5	99.6
09	FINISHES	104.8	90.5	97.1	109.7	80.4	93.8	123.6	98.7	110.1	130.7	79.6	103.0	104.5	64.3	82.8	131.7	89.0	108.6
COVERS	DIVS. 10 - 14, 25, 28, 41, 43, 44, 46	131.2	66.3	115.9	131.2	61.3	114.7	131.2	90.4	121.6	131.2	73.0	117.5	131.2	58.7	114.1	131.2	86.6	120.7
21, 22, 23	FIRE SUPPRESSION, PLUMBING & HVAC	105.5	87.2	98.1	106.2	90.0	99.7	106.8	95.0	102.0	106.5	82.0	96.6	106.2	77.4	94.6	106.8	86.4	98.6
26, 27, 3370	ELECTRICAL, COMMUNICATIONS & UTIL.	111.3	97.8	104.6	116.2	82.0	99.3	113.7	100.0	106.9	114.2	64.2	89.5	113.6	57.7	85.9	118.8	79.5	99.4
MF2018	WEIGHTED AVERAGE	109.3	91.6	101.6	115.4	83.4	101.6	119.3	99.9	111.0	122.8	78.1	103.5	112.8	68.7	93.7	121.7	91.7	108.7

		CANADA																	
	DIVISION	VICTORIA, BRITISH COLUMBIA			WHITEHORSE, YUKON			WINDSOR, ONTARIO			WINNIPEG, MANITOBA			YARMOUTH, NOVA SCOTIA			YELLOWKNIFE, NWT		
		MAT.	INST.	TOTAL	MAT.	INST.	TOTAL	MAT.	INST.	TOTAL	MAT.	INST.	TOTAL	MAT.	INST.	TOTAL	MAT.	INST.	TOTAL
015433	CONTRACTOR EQUIPMENT		104.8	104.8		139.6	139.6		97.3	97.3		122.1	122.1		96.8	96.8		126.5	126.5
0241, 31 - 34	SITE & INFRASTRUCTURE, DEMOLITION	123.6	99.0	106.7	134.2	120.5	124.8	90.4	92.7	92.0	112.7	109.6	110.6	109.7	88.9	95.4	143.2	116.8	125.0
0310	Concrete Forming & Accessories	114.0	86.6	90.7	136.3	56.3	68.2	124.9	87.2	92.8	131.3	63.3	73.4	153.1	46.4	101.6	136.0	64.1	101.3
0320	Concrete Reinforcing	109.7	91.2	100.8	164.6	61.7	114.9	93.6	97.1	95.3	125.0	58.4	92.9	120.0	65.0	99.2	172.6	86.5	140.1
0330	Cast-in-Place Concrete	116.0	94.5	107.9	150.5	72.3	121.0	107.9	95.7	103.3	141.6	70.1	114.6	120.0	65.0	99.2	172.6	86.5	140.1
03	CONCRETE	138.8	90.4	117.1	156.1	64.8	115.2	101.4	92.2	97.2	137.1	66.2	105.3	139.0	63.6	105.1	162.8	78.1	124.8
04	MASONRY	182.8	93.0	128.0	256.6	56.9	134.8	149.8	96.0	116.9	211.7	62.8	120.9	215.8	64.2	123.3	244.6	67.6	136.6
05	METALS	106.7	91.9	102.1	143.5	93.1	127.8	113.0	95.5	107.6	139.8	84.8	122.7	124.4	74.6	108.8	138.9	90.0	123.7
06	WOOD, PLASTICS & COMPOSITES	98.2	86.1	91.9	120.5	55.1	86.3	118.1	86.0	101.3	105.3	64.1	83.8	112.1	66.4	88.2	129.9	76.2	101.8
07	THERMAL & MOISTURE PROTECTION	126.3	87.3	109.6	143.9	62.8	109.1	115.4	94.7	106.5	140.8	67.9	109.5	128.1	65.8	101.3	142.4	77.5	114.5
08	OPENINGS	87.0	80.6	85.4	99.6	52.4	88.1	79.0	85.8	80.6	87.8	57.8	80.5	91.6	60.3	83.9	94.1	64.6	86.9
0920	Plaster & Gypsum Board	109.6	85.5	93.8	180.6	52.4	96.5	109.0	85.8	93.8	138.3	62.3	88.4	141.8	65.7	91.9	189.8	74.7	114.2
0950, 0980	Ceilings & Acoustic Treatment	104.4	85.5	92.6	170.0	52.4	96.3	97.1	85.8	90.0	148.7	62.3	94.6	122.1	65.7	86.8	168.4	74.7	109.6
0960	Flooring	113.6	65.7	99.7	124.5	53.8	103.8	105.9	91.8	101.8	121.0	65.6	104.8	108.9	57.4	93.9	124.3	80.9	111.6
0970, 0990	Wall Finishes & Painting/Coating	110.3	95.5	101.5	117.1	51.7	77.9	101.7	92.5	96.2	106.0	51.1	73.1	114.3	57.7	80.4	120.3	73.9	92.5
09	FINISHES	111.8	84.1	96.8	148.1	55.1	97.7	102.4	88.5	94.9	131.5	62.8	94.3	119.9	64.2	89.8	150.9	75.3	110.0
COVERS	DIVS. 10 - 14, 25, 28, 41, 43, 44, 46	131.2	63.2	115.2	131.2	59.8	114.4	131.2	65.7	115.8	131.2	61.0	114.6	131.2	58.7	114.1	131.2	62.0	114.9
21, 22, 23	FIRE SUPPRESSION, PLUMBING & HVAC	106.3	84.7	97.6	106.9	69.9	92.0	105.5	87.6	98.3	106.8	61.8	88.7	106.5	77.4	94.7	107.6	86.2	99.0
26, 27, 3370	ELECTRICAL, COMMUNICATIONS & UTIL.	116.5	79.0	98.0	138.0	56.2	97.5	116.0	98.1	107.1	123.2	61.6	92.7	113.9	57.7	86.1	132.2	76.3	104.5
MF2018	WEIGHTED AVERAGE	116.8	86.4	103.7	136.0	68.6	106.9	108.6	91.2	101.1	125.8	68.4	101.0	121.9	68.6	98.8	135.0	81.1	111.7

Costs shown in RSMeans cost data publications are based on national averages for materials and installation. To adjust these costs to a specific location, simply multiply the base cost by the factor and divide by 100 for that city. The data is arranged alphabetically by state and postal zip code numbers. For a city not listed, use the factor for a nearby city with similar economic characteristics.

STATE/ZIP	CITY	MAT.	INST.	TOTAL
ALABAMA				
350-352	Birmingham	98.2	70.9	86.4
354	Tuscaloosa	96.9	70.5	85.5
355	Jasper	97.6	69.6	85.5
356	Decatur	96.9	70.2	85.4
357-358	Huntsville	96.9	69.9	85.2
359	Gadsden	97.2	70.8	85.8
360-361	Montgomery	98.0	71.8	86.6
362	Anniston	97.1	66.6	83.9
363	Dothan	97.2	71.6	86.1
364	Evergreen	96.8	67.8	84.3
365-366	Mobile	97.8	68.7	85.2
367	Selma	96.9	71.2	85.8
368	Phenix City	97.7	70.8	86.1
369	Butler	97.1	69.7	85.2
ALASKA				
995-996	Anchorage	119.7	110.4	115.7
997	Fairbanks	120.2	111.0	116.2
998	Juneau	119.7	110.4	115.7
999	Ketchikan	130.1	110.9	121.8
ARIZONA				
850,853	Phoenix	98.4	72.1	87.0
851,852	Mesa/Tempe	98.0	71.1	86.4
855	Globe	98.2	70.7	86.3
856-857	Tucson	96.9	71.8	86.1
859	Show Low	98.4	70.8	86.5
860	Flagstaff	102.4	70.4	88.6
863	Prescott	99.9	71.1	87.5
864	Kingman	98.2	69.8	85.9
865	Chambers	98.2	70.4	86.2
ARKANSAS				
716	Pine Bluff	96.2	62.8	81.8
717	Camden	94.1	64.1	81.1
718	Texarkana	95.1	63.7	81.5
719	Hot Springs	93.4	62.5	80.1
720-722	Little Rock	96.0	63.5	81.9
723	West Memphis	94.2	67.5	82.7
724	Jonesboro	94.8	65.7	82.2
725	Batesville	92.4	62.9	79.6
726	Harrison	93.7	62.3	80.1
727	Fayetteville	91.3	64.2	79.6
728	Russellville	92.5	61.3	79.0
729	Fort Smith	95.1	61.9	80.8
CALIFORNIA				
900-902	Los Angeles	98.4	129.3	111.8
903-905	Inglewood	93.6	127.4	108.2
906-908	Long Beach	95.2	127.2	109.0
910-912	Pasadena	95.4	127.0	109.0
913-916	Van Nuys	98.5	127.0	110.8
917-918	Alhambra	97.4	127.0	110.2
919-921	San Diego	100.3	121.3	109.4
922	Palm Springs	97.3	121.9	108.0
923-924	San Bernardino	95.0	125.4	108.1
925	Riverside	99.4	126.0	110.9
926-927	Santa Ana	97.1	125.7	109.5
928	Anaheim	99.5	125.9	110.9
930	Oxnard	98.2	125.8	110.1
931	Santa Barbara	97.4	125.0	109.4
932-933	Bakersfield	99.0	124.8	110.1
934	San Luis Obispo	98.4	125.5	110.1
935	Mojave	95.6	123.6	107.7
936-938	Fresno	98.6	131.2	112.7
939	Salinas	99.2	137.8	115.9
940-941	San Francisco	107.6	159.0	129.8
942,956-958	Sacramento	101.0	133.1	114.9
943	Palo Alto	99.4	150.8	121.6
944	San Mateo	101.9	156.6	125.5
945	Vallejo	100.5	143.4	119.0
946	Oakland	103.9	153.2	125.2
947	Berkeley	103.3	151.9	124.3
948	Richmond	102.8	148.2	122.4
949	San Rafael	105.0	150.3	124.6
950	Santa Cruz	104.9	138.1	119.2

STATE/ZIP	CITY	MAT.	INST.	TOTAL
CALIFORNIA (CONT'D)				
951	San Jose	103.0	158.4	126.9
952	Stockton	101.0	131.9	114.3
953	Modesto	100.8	132.1	114.3
954	Santa Rosa	101.2	149.8	122.2
955	Eureka	102.6	137.6	117.7
959	Marysville	101.8	130.7	114.3
960	Redding	107.8	132.6	118.6
961	Susanville	107.3	131.0	117.5
COLORADO				
800-802	Denver	104.0	75.1	91.5
803	Boulder	99.2	73.7	88.2
804	Golden	101.5	73.0	89.2
805	Fort Collins	102.8	73.2	90.0
806	Greeley	99.9	73.3	88.4
807	Fort Morgan	99.9	72.5	88.0
808-809	Colorado Springs	101.8	72.3	89.1
810	Pueblo	101.6	70.7	88.2
811	Alamosa	103.3	69.2	88.6
812	Salida	102.9	69.8	88.6
813	Durango	103.6	65.5	87.1
814	Montrose	102.3	66.3	86.7
815	Grand Junction	105.6	70.0	90.2
816	Glenwood Springs	103.6	66.6	87.6
CONNECTICUT				
060	New Britain	99.7	116.9	107.1
061	Hartford	101.7	118.7	109.0
062	Willimantic	100.3	116.9	107.5
063	New London	96.7	116.6	105.3
064	Meriden	98.8	116.4	106.4
065	New Haven	101.5	117.0	108.2
066	Bridgeport	100.9	116.7	107.7
067	Waterbury	100.5	117.2	107.7
068	Norwalk	100.4	116.9	107.5
069	Stamford	100.5	123.9	110.6
D.C.				
200-205	Washington	101.3	88.0	95.5
DELAWARE				
197	Newark	98.7	110.6	103.8
198	Wilmington	99.1	110.7	104.1
199	Dover	99.5	110.4	104.2
FLORIDA				
320,322	Jacksonville	97.0	66.9	84.0
321	Daytona Beach	97.2	69.0	85.0
323	Tallahassee	98.1	66.7	84.6
324	Panama City	98.7	67.2	85.1
325	Pensacola	101.2	65.3	85.7
326,344	Gainesville	98.3	63.8	83.4
327-328,347	Orlando	97.9	66.3	84.3
329	Melbourne	99.9	70.6	87.2
330-332,340	Miami	96.2	70.4	85.1
333	Fort Lauderdale	96.5	72.2	86.0
334,349	West Palm Beach	95.1	69.0	83.8
335-336,346	Tampa	98.0	67.4	84.8
337	St. Petersburg	100.4	65.8	85.4
338	Lakeland	97.1	66.3	83.8
339,341	Fort Myers	96.4	66.7	83.6
342	Sarasota	99.1	65.3	84.5
GEORGIA				
300-303,399	Atlanta	98.7	76.7	89.2
304	Statesboro	97.9	67.1	84.6
305	Gainesville	96.5	64.3	82.6
306	Athens	95.9	63.8	82.0
307	Dalton	97.7	68.0	84.9
308-309	Augusta	96.4	73.5	86.5
310-312	Macon	95.3	74.4	86.3
313-314	Savannah	97.3	72.8	86.7
315	Waycross	96.6	66.9	83.8
316	Valdosta	96.7	65.4	83.2
317,398	Albany	96.5	72.2	86.0
318-319	Columbus	96.4	73.4	86.4

Location Factors - Commercial - V2

STATE/ZIP	CITY	MAT.	INST.	TOTAL
HAWAII				
967	Hilo	115.6	117.2	116.3
968	Honolulu	120.3	117.8	119.2
STATES & POSS.				
969	Guam	138.0	63.4	105.7
IDAHO				
832	Pocatello	101.8	77.8	91.4
833	Twin Falls	102.9	76.9	91.7
834	Idaho Falls	100.2	76.9	90.1
835	Lewiston	108.4	84.9	98.2
836-837	Boise	100.9	78.2	91.1
838	Coeur d'Alene	108.1	81.5	96.6
ILLINOIS				
600-603	North Suburban	98.5	137.3	115.3
604	Joliet	98.5	140.3	116.5
605	South Suburban	98.5	137.2	115.3
606-608	Chicago	100.5	144.4	119.5
609	Kankakee	95.1	129.5	110.0
610-611	Rockford	96.3	126.3	109.3
612	Rock Island	94.5	95.5	94.9
613	La Salle	95.8	122.2	107.2
614	Galesburg	95.5	104.8	99.5
615-616	Peoria	97.4	109.9	102.8
617	Bloomington	94.8	105.4	99.4
618-619	Champaign	98.4	106.3	101.8
620-622	East St. Louis	94.1	109.8	100.9
623	Quincy	96.1	101.1	98.3
624	Effingham	95.3	104.9	99.4
625	Decatur	96.7	107.6	101.4
626-627	Springfield	97.8	110.4	103.2
628	Centralia	92.9	106.7	98.9
629	Carbondale	92.7	105.2	98.1
INDIANA				
460	Anderson	96.8	80.4	89.7
461-462	Indianapolis	99.5	82.9	92.3
463-464	Gary	98.3	105.8	101.5
465-466	South Bend	99.6	83.4	92.6
467-468	Fort Wayne	97.5	75.9	88.2
469	Kokomo	94.9	79.8	88.4
470	Lawrenceburg	93.0	78.3	86.6
471	New Albany	94.3	76.3	86.5
472	Columbus	96.5	79.4	89.1
473	Muncie	96.6	78.9	89.0
474	Bloomington	98.2	79.7	90.2
475	Washington	95.1	84.8	90.7
476-477	Evansville	96.0	83.2	90.5
478	Terre Haute	96.9	82.9	90.8
479	Lafayette	96.2	79.9	89.1
IOWA				
500-503,509	Des Moines	97.3	90.0	94.2
504	Mason City	95.0	71.2	84.7
505	Fort Dodge	95.0	71.0	84.6
506-507	Waterloo	96.5	78.4	88.7
508	Creston	95.2	81.1	89.1
510-511	Sioux City	98.0	80.3	90.3
512	Sibley	96.9	60.1	81.0
513	Spencer	98.3	59.9	81.7
514	Carroll	95.5	79.3	88.5
515	Council Bluffs	98.8	81.9	91.5
516	Shenandoah	96.1	78.7	88.6
520	Dubuque	97.3	80.6	90.1
521	Decorah	96.7	71.0	85.6
522-524	Cedar Rapids	98.3	86.6	93.2
525	Ottumwa	96.4	75.7	87.5
526	Burlington	95.8	81.1	89.5
527-528	Davenport	97.3	96.6	97.0
KANSAS				
660-662	Kansas City	96.4	99.2	97.6
664-666	Topeka	98.1	78.0	89.4
667	Fort Scott	95.2	73.9	86.0
668	Emporia	95.2	74.0	86.0
669	Belleville	96.9	69.1	84.9
670-672	Wichita	96.5	72.2	86.0
673	Independence	96.4	73.6	86.6
674	Salina	96.8	73.3	86.6
675	Hutchinson	91.9	69.7	82.3
676	Hays	96.0	70.4	84.9
677	Colby	96.5	71.7	85.7

STATE/ZIP	CITY	MAT.	INST.	TOTAL
KANSAS (CONT'D)				
678	Dodge City	98.0	74.8	88.0
679	Liberal	96.0	69.5	84.6
KENTUCKY				
400-402	Louisville	94.9	79.6	88.3
403-405	Lexington	94.2	78.8	87.5
406	Frankfort	96.8	78.6	88.9
407-409	Corbin	91.8	75.0	84.6
410	Covington	94.9	75.0	86.3
411-412	Ashland	93.6	86.3	90.5
413-414	Campton	94.9	76.3	86.9
415-416	Pikeville	96.1	82.1	90.1
417-418	Hazard	94.3	77.1	86.9
420	Paducah	93.0	80.0	87.4
421-422	Bowling Green	95.2	78.4	87.9
423	Owensboro	95.3	81.8	89.4
424	Henderson	92.7	78.7	86.7
425-426	Somerset	92.1	76.6	85.4
427	Elizabethtown	91.8	73.4	83.8
LOUISIANA				
700-701	New Orleans	97.7	68.3	85.0
703	Thibodaux	95.5	65.0	82.3
704	Hammond	93.2	63.4	80.3
705	Lafayette	95.1	65.9	82.5
706	Lake Charles	95.3	68.4	83.7
707-708	Baton Rouge	97.0	68.3	84.6
710-711	Shreveport	97.7	66.8	84.4
712	Monroe	96.0	64.6	82.4
713-714	Alexandria	96.1	64.7	82.5
MAINE				
039	Kittery	94.4	84.9	90.3
040-041	Portland	100.6	85.6	94.1
042	Lewiston	98.0	84.9	92.4
043	Augusta	101.6	86.0	94.9
044	Bangor	97.5	84.5	91.9
045	Bath	96.2	84.6	91.2
046	Machias	95.8	84.5	90.9
047	Houlton	95.9	84.4	91.0
048	Rockland	95.0	84.6	90.5
049	Waterville	96.2	84.5	91.2
MARYLAND				
206	Waldorf	97.4	84.8	92.0
207-208	College Park	97.5	85.8	92.5
209	Silver Spring	96.7	84.9	91.6
210-212	Baltimore	102.3	83.1	94.0
214	Annapolis	101.4	83.4	93.6
215	Cumberland	97.1	83.1	91.0
216	Easton	98.9	72.6	87.5
217	Hagerstown	97.7	86.2	92.8
218	Salisbury	99.3	65.0	84.5
219	Elkton	96.1	83.0	90.5
MASSACHUSETTS				
010-011	Springfield	99.1	110.8	104.2
012	Pittsfield	98.6	105.1	101.4
013	Greenfield	96.7	109.5	102.3
014	Fitchburg	95.4	113.2	103.1
015-016	Worcester	99.1	118.6	107.5
017	Framingham	94.9	122.4	106.8
018	Lowell	98.4	126.4	110.5
019	Lawrence	99.4	124.9	110.4
020-022, 024	Boston	100.2	132.9	114.3
023	Brockton	98.9	117.8	107.1
025	Buzzards Bay	93.3	114.1	102.3
026	Hyannis	96.0	116.9	105.1
027	New Bedford	98.1	116.3	106.0
MICHIGAN				
480,483	Royal Oak	94.1	96.9	95.3
481	Ann Arbor	96.3	102.0	98.7
482	Detroit	99.5	100.9	100.1
484-485	Flint	95.9	90.9	93.8
486	Saginaw	95.6	86.7	91.8
487	Bay City	95.7	85.3	91.2
488-489	Lansing	98.1	90.5	94.8
490	Battle Creek	95.7	79.4	88.6
491	Kalamazoo	96.0	80.9	89.5
492	Jackson	94.1	88.0	91.4
493,495	Grand Rapids	98.9	82.2	91.6
494	Muskegon	94.5	79.7	88.1

STATE/ZIP	CITY	MAT.	INST.	TOTAL	STATE/ZIP	CITY	MAT.	INST.	TOTAL
MICHIGAN (CONT'D)					**NEW HAMPSHIRE (CONT'D)**				
496	Traverse City	93.5	75.2	85.6	032-033	Concord	100.8	91.3	96.7
497	Gaylord	94.6	78.1	87.5	034	Keene	96.1	86.2	91.8
498-499	Iron Mountain	96.5	78.9	88.9	035	Littleton	96.1	77.7	88.1
					036	Charleston	95.6	85.6	91.3
MINNESOTA					037	Claremont	94.8	85.7	90.8
550-551	Saint Paul	99.3	119.2	107.9	038	Portsmouth	96.5	91.5	94.3
553-555	Minneapolis	99.7	116.6	107.0					
556-558	Duluth	99.9	103.9	101.6	**NEW JERSEY**				
559	Rochester	98.4	106.1	101.7	070-071	Newark	101.8	137.9	117.4
560	Mankato	95.5	105.0	99.6	072	Elizabeth	98.9	137.6	115.6
561	Windom	94.0	91.7	93.0	073	Jersey City	97.7	137.1	114.7
562	Willmar	93.7	96.6	94.9	074-075	Paterson	99.3	137.3	115.7
563	St. Cloud	94.9	115.9	104.0	076	Hackensack	97.4	137.1	114.6
564	Brainerd	95.5	97.9	96.5	077	Long Branch	97.1	126.6	109.9
565	Detroit Lakes	97.6	90.3	94.4	078	Dover	97.6	137.5	114.8
566	Bemidji	96.8	94.6	95.8	079	Summit	97.7	136.9	114.6
567	Thief River Falls	96.4	89.4	93.4	080,083	Vineland	97.2	126.6	109.9
					081	Camden	98.9	133.7	114.0
MISSISSIPPI					082,084	Atlantic City	97.8	132.6	112.8
386	Clarksdale	95.7	53.8	77.6	085-086	Trenton	100.5	124.4	110.8
387	Greenville	99.4	65.2	84.6	087	Point Pleasant	99.1	126.5	110.9
388	Tupelo	97.2	56.4	79.6	088-089	New Brunswick	99.7	132.2	113.8
389	Greenwood	97.0	53.5	78.2					
390-392	Jackson	98.6	67.9	85.3	**NEW MEXICO**				
393	Meridian	95.4	67.8	83.5	870-872	Albuquerque	98.9	71.6	87.1
394	Laurel	96.8	56.4	79.4	873	Gallup	99.2	71.6	87.3
395	Biloxi	97.1	66.6	83.9	874	Farmington	99.3	71.6	87.3
396	McComb	95.1	53.8	77.3	875	Santa Fe	100.5	72.2	88.3
397	Columbus	96.7	57.1	79.6	877	Las Vegas	97.4	71.6	86.3
					878	Socorro	97.1	71.6	86.1
MISSOURI					879	Truth/Consequences	96.8	68.6	84.6
630-631	St. Louis	97.5	104.6	100.6	880	Las Cruces	96.8	70.0	85.2
633	Bowling Green	95.6	92.3	94.2	881	Clovis	99.1	71.5	87.2
634	Hannibal	94.5	88.5	91.9	882	Roswell	100.7	71.6	88.1
635	Kirksville	98.0	83.6	91.8	883	Carrizozo	101.4	71.6	88.5
636	Flat River	96.6	90.6	94.0	884	Tucumcari	99.7	71.5	87.6
637	Cape Girardeau	96.4	90.6	93.9					
638	Sikeston	94.9	84.4	90.4	**NEW YORK**				
639	Poplar Bluff	94.4	84.3	90.0	100-102	New York	100.3	174.2	132.2
640-641	Kansas City	97.5	101.8	99.3	103	Staten Island	96.5	171.9	129.1
644-645	St. Joseph	96.0	93.3	94.9	104	Bronx	94.8	171.2	127.8
646	Chillicothe	93.7	90.7	92.4	105	Mount Vernon	94.9	147.0	117.4
647	Harrisonville	93.2	97.4	95.0	106	White Plains	94.9	149.8	118.6
648	Joplin	95.2	80.1	88.7	107	Yonkers	99.1	150.1	121.1
650-651	Jefferson City	95.7	90.0	93.2	108	New Rochelle	95.3	141.4	115.2
652	Columbia	95.7	94.5	95.2	109	Suffern	94.9	125.4	108.1
653	Sedalia	95.6	88.1	92.4	110	Queens	100.5	173.1	131.8
654-655	Rolla	93.4	90.6	92.2	111	Long Island City	102.2	173.1	132.8
656-658	Springfield	96.5	81.4	90.0	112	Brooklyn	102.7	173.1	133.1
					113	Flushing	102.5	173.1	133.0
MONTANA					114	Jamaica	100.9	173.1	132.1
590-591	Billings	101.4	77.2	90.9	115,117,118	Hicksville	100.5	151.7	122.6
592	Wolf Point	101.3	74.5	89.7	116	Far Rockaway	102.7	173.1	133.1
593	Miles City	99.1	74.6	88.5	119	Riverhead	101.3	156.9	125.3
594	Great Falls	102.8	77.1	91.7	120-122	Albany	98.3	109.4	103.1
595	Havre	100.2	72.5	88.3	123	Schenectady	97.9	108.6	102.5
596	Helena	101.2	74.7	89.7	124	Kingston	101.2	128.8	113.1
597	Butte	101.3	74.4	89.7	125-126	Poughkeepsie	100.3	131.7	113.9
598	Missoula	98.8	74.8	88.5	127	Monticello	99.8	129.5	112.7
599	Kalispell	98.0	74.1	87.7	128	Glens Falls	92.8	103.2	97.3
					129	Plattsburgh	97.7	92.8	95.6
NEBRASKA					130-132	Syracuse	98.7	101.6	99.9
680-681	Omaha	98.6	83.0	91.9	133-135	Utica	96.8	102.5	99.3
683-685	Lincoln	99.8	79.3	90.9	136	Watertown	98.4	98.9	98.6
686	Columbus	96.0	77.7	88.1	137-139	Binghamton	98.5	100.5	99.3
687	Norfolk	97.4	74.5	87.5	140-142	Buffalo	100.9	108.8	104.8
688	Grand Island	97.4	76.2	88.2	143	Niagara Falls	98.0	103.4	100.3
689	Hastings	97.1	75.2	87.7	144-146	Rochester	100.8	101.2	101.0
690	McCook	95.9	70.2	84.8	147	Jamestown	97.1	92.3	95.0
691	North Platte	96.1	74.9	86.9	148-149	Elmira	96.9	100.4	98.4
692	Valentine	98.0	67.4	84.8					
693	Alliance	98.4	70.4	86.3	**NORTH CAROLINA**				
					270,272-274	Greensboro	99.7	65.3	84.8
NEVADA					271	Winston-Salem	99.5	65.3	84.7
889-891	Las Vegas	104.2	107.0	105.4	275-276	Raleigh	99.0	65.3	84.4
893	Ely	102.9	89.5	97.1	277	Durham	101.2	65.2	85.6
894-895	Reno	102.8	85.6	95.4	278	Rocky Mount	97.1	64.5	83.0
897	Carson City	103.2	85.6	95.6	279	Elizabeth City	98.1	66.6	84.5
898	Elko	101.5	81.8	93.0	280	Gastonia	97.7	68.5	85.1
					281-282	Charlotte	98.4	72.1	87.0
NEW HAMPSHIRE					283	Fayetteville	100.5	64.0	84.7
030	Nashua	99.3	92.3	96.3	284	Wilmington	96.6	63.1	82.1
031	Manchester	100.5	93.0	97.3	285	Kinston	94.9	63.5	81.3

STATE/ZIP	CITY	MAT.	INST.	TOTAL
NORTH CAROLINA (CONT'D)				
286	Hickory	95.2	69.0	83.9
287-288	Asheville	97.0	64.6	83.0
289	Murphy	96.1	62.6	81.6
NORTH DAKOTA				
580-581	Fargo	99.6	78.9	90.6
582	Grand Forks	98.8	81.0	91.1
583	Devils Lake	98.6	77.8	89.6
584	Jamestown	98.7	76.9	89.3
585	Bismarck	99.1	83.1	92.2
586	Dickinson	99.2	75.7	89.0
587	Minot	98.7	76.5	89.1
588	Williston	97.8	80.7	90.4
OHIO				
430-432	Columbus	99.2	83.2	92.3
433	Marion	95.5	83.6	90.3
434-436	Toledo	98.9	87.7	94.1
437-438	Zanesville	96.1	83.0	90.4
439	Steubenville	97.1	87.4	92.9
440	Lorain	98.7	81.8	91.4
441	Cleveland	99.4	91.5	96.0
442-443	Akron	99.7	83.0	92.5
444-445	Youngstown	99.1	79.7	90.7
446-447	Canton	99.2	77.5	89.8
448-449	Mansfield	96.6	81.1	89.9
450	Hamilton	96.9	76.7	88.2
451-452	Cincinnati	98.0	79.2	88.9
453-454	Dayton	96.9	76.6	88.1
455	Springfield	96.9	77.2	88.4
456	Chillicothe	96.3	85.3	91.5
457	Athens	99.2	81.8	91.7
458	Lima	99.6	79.0	90.7
OKLAHOMA				
730-731	Oklahoma City	97.6	67.8	84.8
734	Ardmore	95.9	63.8	82.0
735	Lawton	98.1	63.9	83.3
736	Clinton	97.3	63.7	82.8
737	Enid	97.8	64.2	83.3
738	Woodward	95.9	61.2	80.9
739	Guymon	97.0	61.7	81.7
740-741	Tulsa	96.9	65.5	83.3
743	Miami	93.5	63.0	80.3
744	Muskogee	96.1	65.7	83.0
745	McAlester	93.2	58.3	78.1
746	Ponca City	93.9	61.8	80.0
747	Durant	93.8	62.7	80.4
748	Shawnee	95.3	63.3	81.5
749	Poteau	93.0	62.6	79.9
OREGON				
970-972	Portland	102.0	104.7	103.2
973	Salem	103.3	102.8	103.0
974	Eugene	101.6	103.7	102.5
975	Medford	103.1	99.3	101.4
976	Klamath Falls	103.3	97.8	100.9
977	Bend	102.3	100.5	101.5
978	Pendleton	98.3	101.0	99.5
979	Vale	96.0	87.8	92.4
PENNSYLVANIA				
150-152	Pittsburgh	99.3	102.6	100.7
153	Washington	96.6	101.9	98.9
154	Uniontown	97.1	99.2	98.0
155	Bedford	98.0	90.1	94.6
156	Greensburg	98.0	95.8	97.1
157	Indiana	96.8	96.2	96.5
158	Dubois	98.5	94.0	96.6
159	Johnstown	98.0	97.2	97.7
160	Butler	92.1	98.7	95.0
161	New Castle	92.2	98.9	95.1
162	Kittanning	92.6	96.6	94.3
163	Oil City	92.1	94.4	93.1
164-165	Erie	94.0	94.6	94.3
166	Altoona	94.2	97.5	95.6
167	Bradford	95.6	95.0	95.3
168	State College	95.2	99.1	96.9
169	Wellsboro	96.2	89.3	93.2
170-171	Harrisburg	98.5	97.0	97.9
172	Chambersburg	95.4	86.0	91.3
173-174	York	96.2	94.6	95.5
175-176	Lancaster	94.2	96.5	95.2

STATE/ZIP	CITY	MAT.	INST.	TOTAL
PENNSYLVANIA (CONT'D)				
177	Williamsport	92.8	94.2	93.4
178	Sunbury	95.0	89.3	92.5
179	Pottsville	94.0	92.5	93.3
180	Lehigh Valley	95.6	107.5	100.8
181	Allentown	97.5	108.1	102.1
182	Hazleton	95.1	92.5	94.0
183	Stroudsburg	95.0	100.9	97.6
184-185	Scranton	98.3	96.8	97.7
186-187	Wilkes-Barre	94.9	96.8	95.7
188	Montrose	94.6	93.3	94.0
189	Doylestown	94.7	122.6	106.7
190-191	Philadelphia	99.7	136.9	115.8
193	Westchester	96.1	123.2	107.8
194	Norristown	95.1	123.4	107.3
195-196	Reading	97.3	104.7	100.5
PUERTO RICO				
009	San Juan	124.9	28.3	83.1
RHODE ISLAND				
028	Newport	97.0	112.6	103.8
029	Providence	101.0	113.3	106.3
SOUTH CAROLINA				
290-292	Columbia	98.8	67.6	85.3
293	Spartanburg	97.7	70.2	85.8
294	Charleston	99.4	67.0	85.4
295	Florence	97.4	67.0	84.2
296	Greenville	97.5	70.0	85.6
297	Rock Hill	97.3	68.2	84.7
298	Aiken	98.1	65.2	83.9
299	Beaufort	98.9	61.5	82.7
SOUTH DAKOTA				
570-571	Sioux Falls	99.3	82.1	91.8
572	Watertown	97.7	66.6	84.3
573	Mitchell	96.6	59.6	80.6
574	Aberdeen	99.1	66.7	85.1
575	Pierre	100.2	68.9	86.7
576	Mobridge	97.1	60.7	81.4
577	Rapid City	98.7	71.0	86.7
TENNESSEE				
370-372	Nashville	100.0	74.5	89.0
373-374	Chattanooga	99.0	68.7	85.9
375,380-381	Memphis	96.5	72.5	86.1
376	Johnson City	99.2	60.5	82.5
377-379	Knoxville	95.7	66.0	82.9
382	McKenzie	96.7	54.7	78.6
383	Jackson	98.1	60.0	81.6
384	Columbia	95.2	64.6	82.0
385	Cookeville	96.6	56.1	79.1
TEXAS				
750	McKinney	99.5	61.8	83.2
751	Waxahachie	99.5	62.0	83.3
752-753	Dallas	100.1	67.4	86.0
754	Greenville	99.6	61.3	83.1
755	Texarkana	99.1	60.4	82.4
756	Longview	99.8	59.2	82.3
757	Tyler	100.1	60.2	82.8
758	Palestine	95.9	58.9	80.0
759	Lufkin	96.6	60.5	81.0
760-761	Fort Worth	99.0	63.9	83.9
762	Denton	98.9	61.6	82.8
763	Wichita Falls	96.5	62.5	81.8
764	Eastland	95.7	58.9	79.8
765	Temple	94.1	56.4	77.8
766-767	Waco	95.8	63.2	81.7
768	Brownwood	99.1	57.2	81.0
769	San Angelo	98.7	57.3	80.8
770-772	Houston	101.3	67.3	86.6
773	Huntsville	99.9	60.7	83.0
774	Wharton	101.0	62.1	84.2
775	Galveston	98.8	64.0	83.8
776-777	Beaumont	99.2	64.9	84.4
778	Bryan	96.3	62.4	81.7
779	Victoria	101.0	60.8	83.6
780	Laredo	97.6	60.9	81.8
781-782	San Antonio	98.9	63.8	83.7
783-784	Corpus Christi	100.2	64.6	84.8
785	McAllen	100.9	55.2	81.2
786-787	Austin	98.6	62.3	82.9

For customer support on your Concrete & Masonry Costs with RSMeans Data, call 800.448.8182.

499

STATE/ZIP	CITY	MAT.	INST.	TOTAL	STATE/ZIP	CITY	MAT.	INST.	TOTAL
TEXAS (CONT'D)					**WISCONSIN (CONT'D)**				
788	Del Rio	100.7	58.2	82.3	538	Lancaster	95.8	92.9	94.6
789	Giddings	97.3	58.3	80.5	539	Portage	94.2	96.5	95.2
790-791	Amarillo	99.9	61.8	83.5	540	New Richmond	94.9	92.9	94.0
792	Childress	98.7	59.4	81.7	541-543	Green Bay	99.4	103.1	101.0
793-794	Lubbock	100.4	60.7	83.3	544	Wausau	94.5	92.9	93.8
795-796	Abilene	99.0	61.6	82.9	545	Rhinelander	98.0	91.1	95.0
797	Midland	100.9	61.7	84.0	546	La Crosse	95.7	100.3	97.7
798-799,885	El Paso	98.3	63.3	83.2	547	Eau Claire	97.4	100.4	98.7
					548	Superior	94.6	94.6	94.6
UTAH					549	Oshkosh	95.2	92.7	94.1
840-841	Salt Lake City	103.3	73.7	90.5					
842,844	Ogden	98.6	74.0	88.0	**WYOMING**				
843	Logan	100.7	74.0	89.2	820	Cheyenne	101.0	72.4	88.7
845	Price	101.1	68.6	87.1	821	Yellowstone Nat'l Park	98.9	71.4	87.0
846-847	Provo	101.1	74.0	89.4	822	Wheatland	100.1	69.9	87.0
					823	Rawlins	101.9	71.3	88.7
VERMONT					824	Worland	99.5	71.9	87.6
050	White River Jct.	98.7	79.6	90.4	825	Riverton	100.8	71.4	88.1
051	Bellows Falls	97.1	90.6	94.3	826	Casper	102.5	72.3	89.4
052	Bennington	97.5	87.2	93.0	827	Newcastle	99.4	71.3	87.3
053	Brattleboro	97.8	90.7	94.7	828	Sheridan	102.4	71.1	88.9
054	Burlington	103.1	82.3	94.1	829-831	Rock Springs	103.3	70.6	89.1
056	Montpelier	101.2	86.0	94.6					
057	Rutland	99.4	81.6	91.7	**CANADIAN FACTORS (reflect Canadian currency)**				
058	St. Johnsbury	98.7	79.2	90.3					
059	Guildhall	97.3	79.2	89.5	**ALBERTA**				
						Calgary	123.3	91.7	109.7
VIRGINIA						Edmonton	125.3	91.9	110.9
220-221	Fairfax	99.9	81.7	92.0		Fort McMurray	123.9	86.2	107.6
222	Arlington	101.0	81.6	92.6		Lethbridge	118.9	85.7	104.6
223	Alexandria	100.0	84.8	93.4		Lloydminster	113.6	82.6	100.2
224-225	Fredericksburg	98.6	79.9	90.5		Medicine Hat	113.7	81.9	100.0
226	Winchester	99.2	76.4	89.3		Red Deer	114.2	81.9	100.2
227	Culpeper	99.1	82.0	91.7					
228	Harrisonburg	99.4	75.6	89.1	**BRITISH COLUMBIA**				
229	Charlottesville	99.8	74.2	88.7		Kamloops	115.6	81.6	100.9
230-232	Richmond	99.6	74.6	88.8		Prince George	116.5	81.0	101.2
233-235	Norfolk	100.7	68.3	86.7		Vancouver	121.7	91.7	108.7
236	Newport News	99.5	69.5	86.5		Victoria	116.8	86.4	103.7
237	Portsmouth	99.0	67.4	85.3					
238	Petersburg	99.4	74.4	88.6	**MANITOBA**				
239	Farmville	98.5	66.0	84.5		Brandon	122.8	68.9	99.5
240-241	Roanoke	100.5	69.1	87.0		Portage la Prairie	113.6	67.5	93.6
242	Bristol	98.5	59.4	81.6		Winnipeg	125.8	68.4	101.0
243	Pulaski	98.1	69.7	85.8					
244	Staunton	99.0	67.8	85.5	**NEW BRUNSWICK**				
245	Lynchburg	99.1	73.1	87.9		Bathurst	112.7	60.8	90.3
246	Grundy	98.4	63.2	83.2		Dalhousie	112.4	61.0	90.2
						Fredericton	121.9	67.8	98.5
WASHINGTON						Moncton	113.0	73.2	95.8
980-981,987	Seattle	105.4	108.4	106.7		Newcastle	112.8	61.4	90.5
982	Everett	104.5	104.2	104.4		St. John	115.6	72.9	97.1
983-984	Tacoma	104.9	105.7	105.2					
985	Olympia	103.9	105.2	104.5	**NEWFOUNDLAND**				
986	Vancouver	106.2	102.2	104.4		Corner Brook	127.3	66.7	101.1
988	Wenatchee	105.0	85.6	96.6		St. Johns	126.2	85.2	108.5
989	Yakima	105.0	101.6	103.5					
990-992	Spokane	101.0	86.8	94.9	**NORTHWEST TERRITORIES**				
993	Richland	100.7	93.0	97.4		Yellowknife	135.0	81.1	111.7
994	Clarkston	99.9	79.5	91.1					
					NOVA SCOTIA				
WEST VIRGINIA						Bridgewater	113.3	68.7	94.0
247-248	Bluefield	97.3	85.2	92.1		Dartmouth	124.0	68.6	100.1
249	Lewisburg	99.0	86.8	93.8		Halifax	120.5	84.9	105.1
250-253	Charleston	96.6	91.1	94.2		New Glasgow	121.9	68.6	98.9
254	Martinsburg	96.5	80.8	89.7		Sydney	120.1	68.6	97.9
255-257	Huntington	97.8	91.7	95.1		Truro	112.8	68.7	93.7
258-259	Beckley	95.2	87.7	91.9		Yarmouth	121.9	68.6	98.8
260	Wheeling	100.1	90.9	96.1					
261	Parkersburg	98.9	90.0	95.1	**ONTARIO**				
262	Buckhannon	98.7	90.5	95.1		Barrie	118.0	85.2	103.8
263-264	Clarksburg	99.3	91.0	95.7		Brantford	115.2	88.8	103.8
265	Morgantown	99.3	91.4	95.9		Cornwall	115.3	85.5	102.4
266	Gassaway	98.7	87.3	93.7		Hamilton	119.2	94.5	108.6
267	Romney	98.5	85.3	92.8		Kingston	116.5	85.5	103.1
268	Petersburg	98.3	85.6	92.8		Kitchener	109.9	91.2	101.8
						London	119.2	92.9	107.8
WISCONSIN						North Bay	125.0	83.4	107.0
530,532	Milwaukee	97.9	111.7	103.9		Oshawa	112.4	93.6	104.3
531	Kenosha	99.2	106.2	102.2		Ottawa	120.6	93.0	108.7
534	Racine	98.5	106.2	101.8		Owen Sound	118.1	83.5	103.2
535	Beloit	97.8	97.3	97.6		Peterborough	115.2	85.2	102.3
537	Madison	98.6	106.4	101.9		Sarnia	115.5	89.4	104.2

STATE/ZIP	CITY	MAT.	INST.	TOTAL
ONTARIO (CONT'D)				
	Sault Ste. Marie	110.6	86.4	100.1
	St. Catharines	108.0	92.1	101.1
	Sudbury	107.7	90.3	100.2
	Thunder Bay	109.3	91.6	101.6
	Timmins	115.4	83.4	101.6
	Toronto	119.3	99.9	111.0
	Windsor	108.6	91.2	101.1
PRINCE EDWARD ISLAND				
	Charlottetown	124.1	58.7	95.8
	Summerside	124.4	56.4	95.0
QUEBEC				
	Cap-de-la-Madeleine	112.8	78.2	97.8
	Charlesbourg	112.8	78.2	97.8
	Chicoutimi	112.5	86.9	101.4
	Gatineau	112.5	78.0	97.6
	Granby	112.7	77.9	97.7
	Hull	112.6	78.0	97.7
	Joliette	113.0	78.2	97.9
	Laval	112.6	78.6	97.9
	Montreal	122.1	89.4	108.0
	Quebec City	122.2	89.9	108.2
	Rimouski	112.3	86.9	101.4
	Rouyn-Noranda	112.4	78.0	97.6
	Saint-Hyacinthe	111.8	78.0	97.2
	Sherbrooke	112.8	78.0	97.7
	Sorel	112.9	78.2	97.9
	Saint-Jerome	112.5	78.0	97.6
	Trois-Rivieres	122.8	78.1	103.5
SASKATCHEWAN				
	Moose Jaw	111.0	62.4	90.0
	Prince Albert	110.2	60.7	88.8
	Regina	127.0	88.5	110.3
	Saskatoon	112.0	85.1	100.4
YUKON				
	Whitehorse	136.0	68.6	106.9

R011105-05   Tips for Accurate Estimating

1. Use pre-printed or columnar forms for orderly sequence of dimensions and locations and for recording telephone quotations.

2. Use only the front side of each paper or form except for certain pre-printed summary forms.

3. Be consistent in listing dimensions: For example, length x width x height. This helps in rechecking to ensure that, the total length of partitions is appropriate for the building area.

4. Use printed (rather than measured) dimensions where given.

5. Add up multiple printed dimensions for a single entry where possible.

6. Measure all other dimensions carefully.

7. Use each set of dimensions to calculate multiple related quantities.

8. Convert foot and inch measurements to decimal feet when listing. Memorize decimal equivalents to .01 parts of a foot (1/8″ equals approximately .01′).

9. Do not "round off" quantities until the final summary.

10. Mark drawings with different colors as items are taken off.

11. Keep similar items together, different items separate.

12. Identify location and drawing numbers to aid in future checking for completeness.

13. Measure or list everything on the drawings or mentioned in the specifications.

14. It may be necessary to list items not called for to make the job complete.

15. Be alert for: Notes on plans such as N.T.S. (not to scale); changes in scale throughout the drawings; reduced size drawings; discrepancies between the specifications and the drawings.

16. Develop a consistent pattern of performing an estimate. For example:
    a. Start the quantity takeoff at the lower floor and move to the next higher floor.
    b. Proceed from the main section of the building to the wings.
    c. Proceed from south to north or vice versa, clockwise or counterclockwise.
    d. Take off floor plan quantities first, elevations next, then detail drawings.

17. List all gross dimensions that can be either used again for different quantities, or used as a rough check of other quantities for verification (exterior perimeter, gross floor area, individual floor areas, etc.).

18. Utilize design symmetry or repetition (repetitive floors, repetitive wings, symmetrical design around a center line, similar room layouts, etc.). Note: Extreme caution is needed here so as not to omit or duplicate an area.

19. Do not convert units until the final total is obtained. For instance, when estimating concrete work, keep all units to the nearest cubic foot, then summarize and convert to cubic yards.

20. When figuring alternatives, it is best to total all items involved in the basic system, then total all items involved in the alternates. Therefore you work with positive numbers in all cases. When adds and deducts are used, it is often confusing whether to add or subtract a portion of an item; especially on a complicated or involved alternate.

R011105-50   Metric Conversion Factors

Description: This table is primarily for converting customary U.S. units in the left hand column to SI metric units in the right hand column. In addition, conversion factors for some commonly encountered Canadian and non-SI metric units are included.

	If You Know		Multiply By		To Find
Length	Inches	x	25.4[a]	=	Millimeters
	Feet	x	0.3048[a]	=	Meters
	Yards	x	0.9144[a]	=	Meters
	Miles (statute)	x	1.609	=	Kilometers
Area	Square inches	x	645.2	=	Square millimeters
	Square feet	x	0.0929	=	Square meters
	Square yards	x	0.8361	=	Square meters
Volume	Cubic inches	x	16,387	=	Cubic millimeters
(Capacity)	Cubic feet	x	0.02832	=	Cubic meters
	Cubic yards	x	0.7646	=	Cubic meters
	Gallons (U.S. liquids)[b]	x	0.003785	=	Cubic meters[c]
	Gallons (Canadian liquid)[b]	x	0.004546	=	Cubic meters[c]
	Ounces (U.S. liquid)[b]	x	29.57	=	Milliliters[c, d]
	Quarts (U.S. liquid)[b]	x	0.9464	=	Liters[c, d]
	Gallons (U.S. liquid)[b]	x	3.785	=	Liters[c, d]
Force	Kilograms force[d]	x	9.807	=	Newtons
	Pounds force	x	4.448	=	Newtons
	Pounds force	x	0.4536	=	Kilograms force[d]
	Kips	x	4448	=	Newtons
	Kips	x	453.6	=	Kilograms force[d]
Pressure,	Kilograms force per square centimeter[d]	x	0.09807	=	Megapascals
Stress,	Pounds force per square inch (psi)	x	0.006895	=	Megapascals
Strength	Kips per square inch	x	6.895	=	Megapascals
(Force per unit area)	Pounds force per square inch (psi)	x	0.07031	=	Kilograms force per square centimeter[d]
	Pounds force per square foot	x	47.88	=	Pascals
	Pounds force per square foot	x	4.882	=	Kilograms force per square meter[d]
Flow	Cubic feet per minute	x	0.4719	=	Liters per second
	Gallons per minute	x	0.0631	=	Liters per second
	Gallons per hour	x	1.05	=	Milliliters per second
Bending	Inch-pounds force	x	0.01152	=	Meter-kilograms force[d]
Moment	Inch-pounds force	x	0.1130	=	Newton-meters
Or Torque	Foot-pounds force	x	0.1383	=	Meter-kilograms force[d]
	Foot-pounds force	x	1.356	=	Newton-meters
	Meter-kilograms force[d]	x	9.807	=	Newton-meters
Mass	Ounces (avoirdupois)	x	28.35	=	Grams
	Pounds (avoirdupois)	x	0.4536	=	Kilograms
	Tons (metric)	x	1000	=	Kilograms
	Tons, short (2000 pounds)	x	907.2	=	Kilograms
	Tons, short (2000 pounds)	x	0.9072	=	Megagrams[e]
Mass per	Pounds mass per cubic foot	x	16.02	=	Kilograms per cubic meter
Unit	Pounds mass per cubic yard	x	0.5933	=	Kilograms per cubic meter
Volume	Pounds mass per gallon (U.S. liquid)[b]	x	119.8	=	Kilograms per cubic meter
	Pounds mass per gallon (Canadian liquid)[b]	x	99.78	=	Kilograms per cubic meter
Temperature	Degrees Fahrenheit	(F-32)/1.8		=	Degrees Celsius
	Degrees Fahrenheit	(F+459.67)/1.8		=	Degrees Kelvin
	Degrees Celsius	C+273.15		=	Degrees Kelvin

[a]The factor given is exact
[b]One U.S. gallon = 0.8327 Canadian gallon
[c]1 liter = 1000 milliliters = 1000 cubic centimeters
 1 cubic decimeter = 0.001 cubic meter

[d]Metric but not SI unit
[e]Called "tonne" in England and
 "metric ton" in other metric countries

R011105-60 Weights and Measures

Measures of Length
1 Mile = 1760 Yards = 5280 Feet
1 Yard = 3 Feet = 36 inches
1 Foot = 12 Inches
1 Mil = 0.001 Inch
1 Fathom = 2 Yards = 6 Feet
1 Rod = 5.5 Yards = 16.5 Feet
1 Hand = 4 Inches
1 Span = 9 Inches
1 Micro-inch = One Millionth Inch or 0.000001 Inch
1 Micron = One Millionth Meter + 0.00003937 Inch

Surveyor's Measure
1 Mile = 8 Furlongs = 80 Chains
1 Furlong = 10 Chains = 220 Yards
1 Chain = 4 Rods = 22 Yards = 66 Feet = 100 Links
1 Link = 7.92 Inches

Square Measure
1 Square Mile = 640 Acres = 6400 Square Chains
1 Acre = 10 Square Chains = 4840 Square Yards =
   43,560 Sq. Ft.
1 Square Chain = 16 Square Rods = 484 Square Yards =
   4356 Sq. Ft.
1 Square Rod = 30.25 Square Yards = 272.25 Square Feet = 625 Square
   Lines
1 Square Yard = 9 Square Feet
1 Square Foot = 144 Square Inches
An Acre equals a Square 208.7 Feet per Side

Cubic Measure
1 Cubic Yard = 27 Cubic Feet
1 Cubic Foot = 1728 Cubic Inches
1 Cord of Wood = 4 x 4 x 8 Feet = 128 Cubic Feet
1 Perch of Masonry = 16½ x 1½ x 1 Foot = 24.75 Cubic Feet

Avoirdupois or Commercial Weight
1 Gross or Long Ton = 2240 Pounds
1 Net or Short Ton = 2000 Pounds
1 Pound = 16 Ounces = 7000 Grains
1 Ounce = 16 Drachms = 437.5 Grains
1 Stone = 14 Pounds

Power
1 British Thermal Unit per Hour = 0.2931 Watts
1 Ton (Refrigeration) = 3.517 Kilowatts
1 Horsepower (Boiler) = 9.81 Kilowatts
1 Horsepower (550 ft-lb/s) = 0.746 Kilowatts

Shipping Measure
For Measuring Internal Capacity of a Vessel:
   1 Register Ton = 100 Cubic Feet

For Measurement of Cargo:
   Approximately 40 Cubic Feet of Merchandise is considered a Shipping
   Ton, unless that bulk would weigh more than 2000 Pounds, in which case
   Freight Charge may be based upon weight.

40 Cubic Feet = 32.143 U.S. Bushels = 31.16 Imp. Bushels

Liquid Measure
1 Imperial Gallon = 1.2009 U.S. Gallon = 277.42 Cu. In.
1 Cubic Foot = 7.48 U.S. Gallons

R011110-10 Architectural Fees

Tabulated below are typical percentage fees by project size, for good professional architectural service. Fees may vary from those listed depending upon degree of design difficulty and economic conditions in any particular area.

Rates can be interpolated horizontally and vertically. Various portions of the same project requiring different rates should be adjusted proportionately. For alterations, add 50% to the fee for the first $500,000 of project cost and add 25% to the fee for project cost over $500,000.

Architectural fees tabulated below include Structural, Mechanical and Electrical Engineering Fees. They do not include the fees for special consultants such as kitchen planning, security, acoustical, interior design, etc.

Civil Engineering fees are included in the Architectural fee for project sites requiring minimal design such as city sites. However, separate Civil Engineering fees must be added when utility connections require design, drainage calculations are needed, stepped foundations are required, or provisions are required to protect adjacent wetlands.

Building Types	Total Project Size in Thousands of Dollars						
	100	250	500	1,000	5,000	10,000	50,000
Factories, garages, warehouses, repetitive housing	9.0%	8.0%	7.0%	6.2%	5.3%	4.9%	4.5%
Apartments, banks, schools, libraries, offices, municipal buildings	12.2	12.3	9.2	8.0	7.0	6.6	6.2
Churches, hospitals, homes, laboratories, museums, research	15.0	13.6	12.7	11.9	9.5	8.8	8.0
Memorials, monumental work, decorative furnishings	—	16.0	14.5	13.1	10.0	9.0	8.3

R011110-30 Engineering Fees

Typical **Structural Engineering Fees** based on type of construction and total project size. These fees are included in Architectural Fees.

Type of Construction	Total Project Size (in thousands of dollars)			
	$500	$500-$1,000	$1,000-$5,000	Over $5000
Industrial buildings, factories & warehouses	Technical payroll times 2.0 to 2.5	1.60%	1.25%	1.00%
Hotels, apartments, offices, dormitories, hospitals, public buildings, food stores		2.00%	1.70%	1.20%
Museums, banks, churches and cathedrals		2.00%	1.75%	1.25%
Thin shells, prestressed concrete, earthquake resistive		2.00%	1.75%	1.50%
Parking ramps, auditoriums, stadiums, convention halls, hangars & boiler houses		2.50%	2.00%	1.75%
Special buildings, major alterations, underpinning & future expansion		Add to above 0.5%	Add to above 0.5%	Add to above 0.5%

For complex reinforced concrete or unusually complicated structures, add 20% to 50%.

Typical **Mechanical and Electrical Engineering Fees** are based on the size of the subcontract. The fee structure for both is shown below. These fees are included in Architectural Fees.

Type of Construction	Subcontract Size							
	$25,000	$50,000	$100,000	$225,000	$350,000	$500,000	$750,000	$1,000,000
Simple structures	6.4%	5.7%	4.8%	4.5%	4.4%	4.3%	4.2%	4.1%
Intermediate structures	8.0	7.3	6.5	5.6	5.1	5.0	4.9	4.8
Complex structures	10.1	9.0	9.0	8.0	7.5	7.5	7.0	7.0

For renovations, add 15% to 25% to applicable fee.

R012153-60 Security Factors

Contractors entering, working in, and exiting secure facilities often lose productive time during a normal workday. The recommended allowances in this section are intended to provide for the loss of productivity by increasing labor costs. Note that different costs are associated with searches upon entry only and searches upon entry and exit. Time spent in a queue is unpredictable and not part of these allowances. Contractors should plan ahead for this situation.

Security checkpoints are designed to reflect the level of security required to gain access or egress. An extreme example is when contractors, along with any materials, tools, equipment, and vehicles, must be physically searched and have all materials, tools, equipment, and vehicles inventoried and documented prior to both entry and exit.

Physical searches without going through the documentation process represent the next level and take up less time.

Electronic searches—passing through a detector or x-ray machine with no documentation of materials, tools, equipment, and vehicles—take less time than physical searches.

Visual searches of materials, tools, equipment, and vehicles represent the next level of security.

Finally, access by means of an ID card or displayed sticker takes the least amount of time.

Another consideration is if the searches described above are performed each and every day, or if they are performed only on the first day with access granted by ID card or displayed sticker for the remainder of the project. The figures for this situation have been calculated to represent the initial check-in as described and subsequent entry by ID card or displayed sticker for up to 20 days on site. For the situation described above, where the time period is beyond 20 days, the impact on labor cost is negligible.

There are situations where tradespeople must be accompanied by an escort and observed during the work day. The loss of freedom of movement will slow down productivity for the tradesperson. Costs for the observer have not been included. Those costs are normally born by the owner.

For customer support on your Concrete & Masonry Costs with RSMeans Data, call 800.448.8182.

505

R012909-80   Sales Tax by State

State sales tax on materials is tabulated below (5 states have no sales tax). Many states allow local jurisdictions, such as a county or city, to levy additional sales tax.

Some projects may be sales tax exempt, particularly those constructed with public funds.

State	Tax (%)	State	Tax (%)	State	Tax (%)	State	Tax (%)
Alabama	4	Illinois	6.25	Montana	0	Rhode Island	7
Alaska	0	Indiana	7	Nebraska	5.5	South Carolina	6
Arizona	5.6	Iowa	6	Nevada	6.85	South Dakota	4.5
Arkansas	6.5	Kansas	6.5	New Hampshire	0	Tennessee	7
California	7.25	Kentucky	6	New Jersey	6.625	Texas	6.25
Colorado	2.9	Louisiana	4.45	New Mexico	5.125	Utah	6.10
Connecticut	6.35	Maine	5.5	New York	4	Vermont	6
Delaware	0	Maryland	6	North Carolina	4.75	Virginia	5.3
District of Columbia	6	Massachusetts	6.25	North Dakota	5	Washington	6.5
Florida	6	Michigan	6	Ohio	5.75	West Virginia	6
Georgia	4	Minnesota	6.875	Oklahoma	4.5	Wisconsin	5
Hawaii	4	Mississippi	7	Oregon	0	Wyoming	4
Idaho	6	Missouri	4.225	Pennsylvania	6	Average	5.11 %

Sales Tax by Province (Canada)

GST - a value-added tax, which the government imposes on most goods and services provided in or imported into Canada. PST - a retail sales tax, which five of the provinces impose on the prices of most goods and some

services. QST - a value-added tax, similar to the federal GST, which Quebec imposes. HST - Three provinces have combined their retail sales taxes with the federal GST into one harmonized tax.

Province	PST (%)	QST(%)	GST(%)	HST(%)
Alberta	7	0	5	0
British Columbia	7	0	5	0
Manitoba	7	0	5	0
New Brunswick	0	0	0	15
Newfoundland	0	0	0	15
Northwest Territories	0	0	5	0
Nova Scotia	0	0	0	15
Ontario	0	0	0	13
Prince Edward Island	0	0	0	15
Quebec	9.975	0	5	0
Saskatchewan	6	0	5	0
Yukon	0	0	5	0

R012909-85   Unemployment Taxes and Social Security Taxes

State unemployment tax rates vary not only from state to state, but also with the experience rating of the contractor. The federal unemployment tax rate is 6.2% of the first $7,000 of wages. This is reduced by a credit of up to 5.4% for timely payment to the state. The minimum federal unemployment tax is 0.6% after all credits.

Social security (FICA) for 2021 is estimated at time of publication to be 7.65% of wages up to $137,700.

R012909-86   Unemployment Tax by State

Information is from the U.S. Department of Labor, state unemployment tax rates.

State	Tax (%)	State	Tax (%)	State	Tax (%)	State	Tax (%)
Alabama	6.80	Illinois	6.93	Montana	6.30	Rhode Island	9.49
Alaska	5.4	Indiana	7.4	Nebraska	5.4	South Carolina	5.46
Arizona	12.76	Iowa	7.5	Nevada	5.4	South Dakota	9.35
Arkansas	14.3	Kansas	7.6	New Hampshire	7.5	Tennessee	10.0
California	6.2	Kentucky	9.3	New Jersey	5.8	Texas	6.5
Colorado	8.15	Louisiana	6.2	New Mexico	5.4	Utah	7.1
Connecticut	6.8	Maine	5.46	New York	9.1	Vermont	7.7
Delaware	8.20	Maryland	7.50	North Carolina	5.76	Virginia	6.21
District of Columbia	7	Massachusetts	12.65	North Dakota	10.74	Washington	7.73
Florida	5.4	Michigan	10.3	Ohio	9	West Virginia	8.5
Georgia	5.4	Minnesota	9.1	Oklahoma	5.5	Wisconsin	12.0
Hawaii	5.6	Mississippi	5.6	Oregon	5.4	Wyoming	8.8
Idaho	5.4	Missouri	8.37	Pennsylvania	11.03	Median	7.40%

R012909-90   Overtime

One way to improve the completion date of a project or eliminate negative float from a schedule is to compress activity duration times. This can be achieved by increasing the crew size or working overtime with the proposed crew.

To determine the costs of working overtime to compress activity duration times, consider the following examples. Below is an overtime efficiency and cost chart based on a five, six, or seven day week with an eight through twelve hour day. Payroll percentage increases for time and one half and double times are shown for the various working days.

Days per Week	Hours per Day	Production Efficiency					Payroll Cost Factors	
		1st Week	2nd Week	3rd Week	4th Week	Average 4 Weeks	@ 1-1/2 Times	@ 2 Times
	8	100%	100%	100%	100%	100%	1.000	1.000
	9	100	100	95	90	96	1.056	1.111
5	10	100	95	90	85	93	1.100	1.200
	11	95	90	75	65	81	1.136	1.273
	12	90	85	70	60	76	1.167	1.333
	8	100	100	95	90	96	1.083	1.167
	9	100	95	90	85	93	1.130	1.259
6	10	95	90	85	80	88	1.167	1.333
	11	95	85	70	65	79	1.197	1.394
	12	90	80	65	60	74	1.222	1.444
	8	100	95	85	75	89	1.143	1.286
	9	95	90	80	70	84	1.183	1.365
7	10	90	85	75	65	79	1.214	1.429
	11	85	80	65	60	73	1.240	1.481
	12	85	75	60	55	69	1.262	1.524

For customer support on your Concrete & Masonry Costs with RSMeans Data, call 800.448.8182.

507

R013113-40   Builder's Risk Insurance

Builder's risk insurance is insurance on a building during construction. Premiums are paid by the owner or the contractor. Blasting, collapse and underground insurance would raise total insurance costs.

R013113-50   General Contractor's Overhead

There are two distinct types of overhead on a construction project: Project overhead and main office overhead. Project overhead includes those costs at a construction site not directly associated with the installation of construction materials. Examples of project overhead costs include the following:

1. Superintendent
2. Construction office and storage trailers
3. Temporary sanitary facilities
4. Temporary utilities
5. Security fencing
6. Photographs
7. Cleanup
8. Performance and payment bonds

The above project overhead items are also referred to as general requirements and therefore are estimated in Division 1. Division 1 is the first division listed in the CSI MasterFormat but it is usually the last division estimated. The sum of the costs in Divisions 1 through 49 is referred to as the sum of the direct costs.

All construction projects also include indirect costs. The primary components of indirect costs are the contractor's main office overhead and profit. The amount of the main office overhead expense varies depending on the following:

1. Owner's compensation
2. Project managers' and estimators' wages
3. Clerical support wages
4. Office rent and utilities
5. Corporate legal and accounting costs
6. Advertising
7. Automobile expenses
8. Association dues
9. Travel and entertainment expenses

These costs are usually calculated as a percentage of annual sales volume. This percentage can range from 35% for a small contractor doing less than $500,000 to 5% for a large contractor with sales in excess of $100 million.

R013113-55   Installing Contractor's Overhead

Installing contractors (subcontractors) also incur costs for general requirements and main office overhead.

Included within the total incl. overhead and profit costs is a percent mark-up for overhead that includes:

1. Compensation and benefits for office staff and project managers
2. Office rent, utilities, business equipment, and maintenance
3. Corporate legal and accounting costs
4. Advertising
5. Vehicle expenses (for office staff and project managers)
6. Association dues
7. Travel, entertainment
8. Insurance
9. Small tools and equipment

R013113-60   Workers' Compensation Insurance Rates by Trade

The table below tabulates the national averages for workers' compensation insurance rates by trade and type of building. The average "Insurance Rate" is multiplied by the "% of Building Cost" for each trade. This produces the "Workers' Compensation" cost by % of total labor cost, to be added for each trade by building type to determine the weighted average workers' compensation rate for the building types analyzed.

Trade	Insurance Rate (% Labor Cost)			% of Building Cost			Workers' Compensation		
	Range		Average	Office Bldgs.	Schools & Apts.	Mfg.	Office Bldgs.	Schools & Apts.	Mfg.
Excavation, Grading, etc.	2.1 % to	16.5%	9.3%	4.8%	4.9%	4.5%	0.45%	0.46%	0.42%
Piles & Foundations	3.1 to	28.0	15.6	7.1	5.2	8.7	1.11	0.81	1.36
Concrete	2.5 to	24.9	13.7	5.0	14.8	3.7	0.69	2.03	0.51
Masonry	3.2 to	53.5	28.4	6.9	7.5	1.9	1.96	2.13	0.54
Structural Steel	3.3 to	30.8	17.1	10.7	3.9	17.6	1.83	0.67	3.01
Miscellaneous & Ornamental Metals	2.3 to	22.8	12.5	2.8	4.0	3.6	0.35	0.50	0.45
Carpentry & Millwork	2.6 to	29.1	15.8	3.7	4.0	0.5	0.58	0.63	0.08
Metal or Composition Siding	4.2 to	135.3	69.7	2.3	0.3	4.3	1.60	0.21	3.00
Roofing	4.2 to	105.1	54.7	2.3	2.6	3.1	1.26	1.42	1.70
Doors & Hardware	3.1 to	29.1	16.1	0.9	1.4	0.4	0.14	0.23	0.06
Sash & Glazing	2.6 to	20.5	11.5	3.5	4.0	1.0	0.40	0.46	0.12
Lath & Plaster	2.1 to	33.7	17.9	3.3	6.9	0.8	0.59	1.24	0.14
Tile, Marble & Floors	1.9 to	17.4	9.7	2.6	3.0	0.5	0.25	0.29	0.05
Acoustical Ceilings	1.7 to	21.5	11.6	2.4	0.2	0.3	0.28	0.02	0.03
Painting	2.8 to	37.7	20.3	1.5	1.6	1.6	0.30	0.32	0.32
Interior Partitions	2.6 to	29.1	15.8	3.9	4.3	4.4	0.62	0.68	0.70
Miscellaneous Items	1.6 to	97.7	49.7	5.2	3.7	9.7	2.58	1.84	4.82
Elevators	1.0 to	8.9	5.0	2.1	1.1	2.2	0.11	0.06	0.11
Sprinklers	1.4 to	14.2	7.8	0.5	—	2.0	0.04	—	0.16
Plumbing	1.4 to	13.2	7.3	4.9	7.2	5.2	0.36	0.53	0.38
Heat., Vent., Air Conditioning	2.8 to	15.8	9.3	13.5	11.0	12.9	1.26	1.02	1.20
Electrical	1.2 to	10.6	5.9	10.1	8.4	11.1	0.60	0.50	0.65
Total	1.0 % to	135.3%	68.1	100.0%	100.0%	100.0%	17.36%	16.05%	19.81%
			Overall Weighted Average	17.74%					

Workers' Compensation Insurance Rates by States

The table below lists the weighted average Workers' Compensation base rate for each state with a factor comparing this with the national average of 15.1%.

State	Weighted Average	Factor	State	Weighted Average	Factor	State	Weighted Average	Factor
Alabama	17.2%	159	Kentucky	17.5%	162	North Dakota	8.3%	77
Alaska	11.9	110	Louisiana	25.5	236	Ohio	6.4	59
Arizona	11.1	103	Maine	13.5	125	Oklahoma	13.3	123
Arkansas	7.3	68	Maryland	15.2	141	Oregon	10.8	100
California	32.6	302	Massachusetts	12.0	111	Pennsylvania	26.1	242
Colorado	8.1	75	Michigan	5.8	54	Rhode Island	11.0	102
Connecticut	19.8	183	Minnesota	20.7	192	South Carolina	24.6	228
Delaware	18.2	169	Mississippi	13.7	127	South Dakota	11.3	105
District of Columbia	11.3	105	Missouri	21.4	198	Tennessee	8.8	81
Florida	12.5	116	Montana	12.2	113	Texas	8.0	74
Georgia	50.0	463	Nebraska	16.6	154	Utah	9.2	85
Hawaii	14.0	130	Nevada	12.0	111	Vermont	13.6	126
Idaho	12.9	119	New Hampshire	12.7	118	Virginia	9.9	92
Illinois	27.9	258	New Jersey	21.5	199	Washington	10.4	96
Indiana	4.9	45	New Mexico	20.7	192	West Virginia	6.7	62
Iowa	15.1	140	New York	25.9	240	Wisconsin	16.0	148
Kansas	8.8	81	North Carolina	17.4	161	Wyoming	7.0	65
			Weighted Average for U.S. is	15.1% of payroll = 100%				

The weighted average skilled worker rate for 35 trades is 17.74%. For bidding purposes, apply the full value of Workers' Compensation directly to total labor costs, or if labor is 38%, materials 42% and overhead and profit 20% of total cost, carry 38/80 x 17.74% = 8.43% of cost (before overhead and profit) into overhead. Rates vary not only from state to state but also with the experience rating of the contractor.

Rates are the most current available at the time of publication.

R013113-80 Performance Bond

This table shows the cost of a Performance Bond for a construction job scheduled to be completed in 12 months. Add 1% of the premium cost per month for jobs requiring more than 12 months to complete. The rates are "standard" rates offered to contractors that the bonding company considers financially sound and capable of doing the work. Preferred rates are offered by some bonding companies based upon financial strength of the contractor. Actual rates vary from contractor to contractor and from bonding company to bonding company. Contractors should prequalify through a bonding agency before submitting a bid on a contract that requires a bond.

Contract Amount		Building Construction Class B Projects			Highways & Bridges					
					Class A New Construction			Class A-1 Highway Resurfacing		
First $ 100,000 bid		$25.00 per M			$15.00 per M			$9.40 per M		
Next 400,000 bid		$ 2,500	plus $15.00	per M	$ 1,500	plus $10.00	per M	$ 940	plus $7.20	per M
Next 2,000,000 bid		8,500	plus 10.00	per M	5,500	plus 7.00	per M	3,820	plus 5.00	per M
Next 2,500,000 bid		28,500	plus 7.50	per M	19,500	plus 5.50	per M	15,820	plus 4.50	per M
Next 2,500,000 bid		47,250	plus 7.00	per M	33,250	plus 5.00	per M	28,320	plus 4.50	per M
Over 7,500,000 bid		64,750	plus 6.00	per M	45,750	plus 4.50	per M	39,570	plus 4.00	per M

General Requirements R0154 Construction Aids

R015423-10 Steel Tubular Scaffolding

On new construction, tubular scaffolding is efficient up to 60' high or five stories. Above this it is usually better to use a hung scaffolding if construction permits. Swing scaffolding operations may interfere with tenants. In this case, the tubular is more practical at all heights.

In repairing or cleaning the front of an existing building the cost of tubular scaffolding per S.F. of building front increases as the height increases above the first tier. The first tier cost is relatively high due to leveling and alignment.

The minimum efficient crew for erecting and dismantling is three workers. They can set up and remove 18 frame sections per day up to 5 stories high. For 6 to 12 stories high, a crew of four is most efficient. Use two or more on top and two on the bottom for handing up or hoisting. They can also set up and remove 18 frame sections per day. At 7' horizontal spacing, this will run about 800 S.F. per day of erecting and dismantling. Time for placing and removing planks must be added to the above. A crew of three can place and remove 72 planks per day up to 5 stories. For over 5 stories, a crew of four can place and remove 80 planks per day.

The table below shows the number of pieces required to erect tubular steel scaffolding for 1000 S.F. of building frontage. This area is made up of a scaffolding system that is 12 frames (11 bays) long by 2 frames high.

For jobs under twenty-five frames, add 50% to rental cost. Rental rates will be lower for jobs over three months duration. Large quantities for long periods can reduce rental rates by 20%.

Description of Component	Number of Pieces for 1000 S.F. of Building Front	Unit
5' Wide Standard Frame, 6'-4" High	24	Ea.
Leveling Jack & Plate	24	
Cross Brace	44	
Side Arm Bracket, 21"	12	
Guardrail Post	12	
Guardrail, 7' section	22	
Stairway Section	2	
Stairway Starter Bar	1	
Stairway Inside Handrail	2	
Stairway Outside Handrail	2	
Walk-Thru Frame Guardrail	2	

Scaffolding is often used as falsework over 15' high during construction of cast-in-place concrete beams and slabs. Two foot wide scaffolding is generally used for heavy beam construction. The span between frames depends upon the load to be carried with a maximum span of 5'.

Heavy duty shoring frames with a capacity of 10,000#/leg can be spaced up to 10' O.C. depending upon form support design and loading.

Scaffolding used as horizontal shoring requires less than half the material required with conventional shoring.

On new construction, erection is done by carpenters.

Rolling towers supporting horizontal shores can reduce labor and speed the job. For maintenance work, catwalks with spans up to 70' can be supported by the rolling towers.

R015433-10 Contractor Equipment

Rental Rates shown elsewhere in the book pertain to late model high quality machines in excellent working condition, rented from equipment dealers. Rental rates from contractors may be substantially lower than the rental rates from equipment dealers depending upon economic conditions; for older, less productive machines, reduce rates by a maximum of 15%. Any overtime must be added to the base rates. For shift work, rates are lower. Usual rule of thumb is 150% of one shift rate for two shifts; 200% for three shifts.

For periods of less than one week, operated equipment is usually more economical to rent than renting bare equipment and hiring an operator.

Costs to move equipment to a job site (mobilization) or from a job site (demobilization) are not included in rental rates, nor in any Equipment costs on any Unit Price line items or crew listings. These costs can be found elsewhere. If a piece of equipment is already at a job site, it is not appropriate to utilize mob/demob costs in an estimate again.

Rental rates vary throughout the country with larger cities generally having lower rates. Lease plans for new equipment are available for periods in excess of six months with a percentage of payments applying toward purchase.

Rental rates can also be treated as reimbursement costs for contractor-owned equipment. Owned equipment costs include depreciation, loan payments, interest, taxes, insurance, storage, and major repairs.

Monthly rental rates vary from 2% to 5% of the cost of the equipment depending on the anticipated life of the equipment and its wearing parts. Weekly rates are about 1/3 the monthly rates and daily rental rates are about 1/3 the weekly rates.

The hourly operating costs for each piece of equipment include costs to the user such as fuel, oil, lubrication, normal expendables for the equipment, and a percentage of the mechanic's wages chargeable to maintenance. The hourly operating costs listed do not include the operator's wages.

The daily cost for equipment used in the standard crews is figured by dividing the weekly rate by five, then adding eight times the hourly operating cost to give the total daily equipment cost, not including the operator. This figure is in the right hand column of the Equipment listings under Equipment Cost/Day.

Pile Driving rates shown for the pile hammer and extractor do not include leads, cranes, boilers or compressors. Vibratory pile driving requires an added field specialist during set-up and pile driving operation for the electric model. The hydraulic model requires a field specialist for set-up only. Up to 125 reuses of sheet piling are possible using vibratory drivers. For normal conditions, crane capacity for hammer type and size is as follows.

Crane Capacity	Hammer Type and Size		
	Air or Steam	Diesel	Vibratory
25 ton	to 8,750 ft.-lb.		70 H.P.
40 ton	15,000 ft.-lb.	to 32,000 ft.-lb.	170 H.P.
60 ton	25,000 ft.-lb.		300 H.P.
100 ton		112,000 ft.-lb.	

Cranes should be specified for the job by size, building and site characteristics, availability, performance characteristics, and duration of time required.

Backhoes & Shovels rent for about the same as equivalent size cranes but maintenance and operating expenses are higher. The crane operator's rate must be adjusted for high boom heights. Average adjustments: for 150' boom add 2% per hour; over 185', add 4% per hour; over 210', add 6% per hour; over 250', add 8% per hour and over 295', add 12% per hour.

Tower Cranes of the climbing or static type have jibs from 50' to 200' and capacities at maximum reach range from 4,000 to 14,000 pounds. Lifting capacities increase up to maximum load as the hook radius decreases.

Typical rental rates, based on purchase price, are about 2% to 3% per month.

Erection and dismantling run between 500 and 2000 labor hours. Climbing operation takes 10 labor hours per 20' climb. Crane dead time is about 5 hours per 40' climb. If crane is bolted to side of the building add cost of ties and extra mast sections. Climbing cranes have from 80' to 180' of mast while static cranes have 80' to 800' of mast.

Truck Cranes can be converted to tower cranes by using tower attachments. Mast heights over 400' have been used.

A single 100' high material **Hoist and Tower** can be erected and dismantled in about 400 labor hours; a double 100' high hoist and tower in about 600 labor hours. Erection times for additional heights are 3 and 4 labor hours per vertical foot respectively up to 150', and 4 to 5 labor hours per vertical foot over 150' high. A 40' high portable Buck hoist takes about 160 labor hours to erect and dismantle. Additional heights take 2 labor hours per vertical foot to 80' and 3 labor hours per vertical foot for the next 100'. Most material hoists do not meet local code requirements for carrying personnel.

A 150' high **Personnel Hoist** requires about 500 to 800 labor hours to erect and dismantle. Budget erection time at 5 labor hours per vertical foot for all trades. Local code requirements or labor scarcity requiring overtime can add up to 50% to any of the above erection costs.

Earthmoving Equipment: The selection of earthmoving equipment depends upon the type and quantity of material, moisture content, haul distance, haul road, time available, and equipment available. Short haul cut and fill operations may require dozers only, while another operation may require excavators, a fleet of trucks, and spreading and compaction equipment. Stockpiled material and granular material are easily excavated with front end loaders. Scrapers are most economically used with hauls between 300' and 1-1/2 miles if adequate haul roads can be maintained. Shovels are often used for blasted rock and any material where a vertical face of 8' or more can be excavated. Special conditions may dictate the use of draglines, clamshells, or backhoes. Spreading and compaction equipment must be matched to the soil characteristics, the compaction required and the rate the fill is being supplied.

R015433-15 Heavy Lifting

Hydraulic Climbing Jacks

The use of hydraulic heavy lift systems is an alternative to conventional type crane equipment. The lifting, lowering, pushing, or pulling mechanism is a hydraulic climbing jack moving on a square steel jackrod from 1-5/8" to 4" square, or a steel cable. The jackrod or cable can be vertical or horizontal, stationary or movable, depending on the individual application. When the jackrod is stationary, the climbing jack will climb the rod and push or pull the load along with itself. When the climbing jack is stationary, the jackrod is movable with the load attached to the end and the climbing jack will lift or lower the jackrod with the attached load. The heavy lift system is normally operated by a single control lever located at the hydraulic pump.

The system is flexible in that one or more climbing jacks can be applied wherever a load support point is required, and the rate of lift synchronized.

Economic benefits have been demonstrated on projects such as: erection of ground assembled roofs and floors, complete bridge spans, girders and trusses, towers, chimney liners and steel vessels, storage tanks, and heavy machinery. Other uses are raising and lowering offshore work platforms, caissons, tunnel sections and pipelines.

For customer support on your Concrete & Masonry Costs with RSMeans Data, call 800.448.8182.

511

R015436-50 Mobilization

Costs to move rented construction equipment to a job site from an equipment dealer's or contractor's yard (mobilization) or off the job site (demobilization) are not included in the rental or operating rates, nor in the equipment cost on a unit price line or in a crew listing. These costs can be found consolidated in the Mobilization section of the data and elsewhere in particular site work sections. If a piece of equipment is already on the job site, it is not appropriate to include mob/demob costs in a new estimate that requires use of that equipment. The following table identifies approximate sizes of rented construction equipment that would be hauled on a towed trailer. Because this listing is not all-encompassing, the user can infer as to what size trailer might be required for a piece of equipment not listed.

3-ton Trailer	20-ton Trailer	40-ton Trailer	50-ton Trailer
20 H.P. Excavator	110 H.P. Excavator	200 H.P. Excavator	270 H.P. Excavator
50 H.P. Skid Steer	165 H.P. Dozer	300 H.P. Dozer	Small Crawler Crane
35 H.P. Roller	150 H.P. Roller	400 H.P. Scraper	500 H.P. Scraper
40 H.P. Trencher	Backhoe	450 H.P. Art. Dump Truck	500 H.P. Art. Dump Truck

R024119-10   Demolition Defined

Whole Building Demolition - Demolition of the whole building with no concern for any particular building element, component, or material type being demolished. This type of demolition is accomplished with large pieces of construction equipment that break up the structure, load it into trucks and haul it to a disposal site, but disposal or dump fees are not included. Demolition of below-grade foundation elements, such as footings, foundation walls, grade beams, slabs on grade, etc., is not included. Certain mechanical equipment containing flammable liquids or ozone-depleting refrigerants, electric lighting elements, communication equipment components, and other building elements may contain hazardous waste, and must be removed, either selectively or carefully, as hazardous waste before the building can be demolished.

Foundation Demolition - Demolition of below-grade foundation footings, foundation walls, grade beams, and slabs on grade. This type of demolition is accomplished by hand or pneumatic hand tools, and does not include saw cutting, or handling, loading, hauling, or disposal of the debris.

Gutting - Removal of building interior finishes and electrical/mechanical systems down to the load-bearing and sub-floor elements of the rough building frame, with no concern for any particular building element, component, or material type being demolished. This type of demolition is accomplished by hand or pneumatic hand tools, and includes loading into trucks, but not hauling, disposal or dump fees, scaffolding, or shoring. Certain mechanical equipment containing flammable liquids or ozone-depleting refrigerants, electric lighting elements, communication equipment components, and other building elements may contain hazardous waste, and must be removed, either selectively or carefully, as hazardous waste, before the building is gutted.

Selective Demolition - Demolition of a selected building element, component, or finish, with some concern for surrounding or adjacent elements, components, or finishes (see the first Subdivision (s) at the beginning of appropriate Divisions). This type of demolition is accomplished by hand or pneumatic hand tools, and does not include handling, loading,

storing, hauling, or disposal of the debris, scaffolding, or shoring. "Gutting" methods may be used in order to save time, but damage that is caused to surrounding or adjacent elements, components, or finishes may have to be repaired at a later time.

Careful Removal - Removal of a piece of service equipment, building element or component, or material type, with great concern for both the removed item and surrounding or adjacent elements, components or finishes. The purpose of careful removal may be to protect the removed item for later re-use, preserve a higher salvage value of the removed item, or replace an item while taking care to protect surrounding or adjacent elements, components, connections, or finishes from cosmetic and/or structural damage. An approximation of the time required to perform this type of removal is 1/3 to 1/2 the time it would take to install a new item of like kind. This type of removal is accomplished by hand or pneumatic hand tools, and does not include loading, hauling, or storing the removed item, scaffolding, shoring, or lifting equipment.

Cutout Demolition - Demolition of a small quantity of floor, wall, roof, or other assembly, with concern for the appearance and structural integrity of the surrounding materials. This type of demolition is accomplished by hand or pneumatic hand tools, and does not include saw cutting, handling, loading, hauling, or disposal of debris, scaffolding, or shoring.

Rubbish Handling - Work activities that involve handling, loading or hauling of debris. Generally, the cost of rubbish handling must be added to the cost of all types of demolition, with the exception of whole building demolition.

Minor Site Demolition - Demolition of site elements outside the footprint of a building. This type of demolition is accomplished by hand or pneumatic hand tools, or with larger pieces of construction equipment, and may include loading a removed item onto a truck (check the Crew for equipment used). It does not include saw cutting, hauling, or disposal of debris, and, sometimes, handling or loading.

R024119-20   Dumpsters

Dumpster rental costs on construction sites are presented in two ways.

The cost per week rental includes the delivery of the dumpster; its pulling or emptying once per week, and its final removal. The assumption is made that the dumpster contractor could choose to empty a dumpster by simply bringing in an empty unit and removing the full one. These costs also include the disposal of the materials in the dumpster.

The Alternate Pricing can be used when actual planned conditions are not approximated by the weekly numbers. For example, these lines can be used when a dumpster is needed for 4 weeks and will need to be emptied 2 or 3 times per week. Conversely the Alternate Pricing lines can be used when a dumpster will be rented for several weeks or months but needs to be emptied only a few times over this period.

R031113-10 Wall Form Materials

Aluminum Forms

Approximate weight is 3 lbs. per S.F.C.A. Standard widths are available from 4″ to 36″ with 36″ most common. Standard lengths of 2′, 4′, 6′ to 8′ are available. Forms are lightweight and fewer ties are needed with the wider widths. The form face is either smooth or textured.

Metal Framed Plywood Forms

Manufacturers claim over 75 reuses of plywood and over 300 reuses of steel frames. Many specials such as corners, fillers, pilasters, etc. are available. Monthly rental is generally about 15% of purchase price for first month and 9% per month thereafter with 90% of rental applied to purchase for the first month and decreasing percentages thereafter. Aluminum framed forms cost 25% to 30% more than steel framed.

After the first month, extra days may be prorated from the monthly charge. Rental rates do not include ties, accessories, cleaning, loss of hardware or freight in and out. Approximate weight is 5 lbs. per S.F. for steel; 3 lbs. per S.F. for aluminum.

Forms can be rented with option to buy.

Plywood Forms, Job Fabricated

There are two types of plywood used for concrete forms.

1. Exterior plyform is completely waterproof. This is face oiled to facilitate stripping. Ten reuses can be expected with this type with 25 reuses possible.
2. An overlaid type consists of a resin fiber fused to exterior plyform. No oiling is required except to facilitate cleaning. This is available in both high density (HDO) and medium density overlaid (MDO). Using HDO, 50 reuses can be expected with 200 possible.

Plyform is available in 5/8″ and 3/4″ thickness. High density overlaid is available in 3/8″, 1/2″, 5/8″ and 3/4″ thickness.

5/8″ thick is sufficient for most building forms, while 3/4″ is best on heavy construction.

Plywood Forms, Modular, Prefabricated

There are many plywood forming systems without frames. Most of these are manufactured from 1-1/8″ (HDO) plywood and have some hardware attached. These are used principally for foundation walls 8′ or less high. With care and maintenance, 100 reuses can be attained with decreasing quality of surface finish.

Steel Forms

Approximate weight is 6-1/2 lbs. per S.F.C.A. including accessories. Standard widths are available from 2″ to 24″, with 24″ the most common. Standard lengths are from 2′ to 8′, with 4′ the most common. Forms are easily ganged into modular units.

Forms are usually leased for 15% of the purchase price per month prorated daily over 30 days.

Rental may be applied to sale price, and usually rental forms are bought. With careful handling and cleaning 200 to 400 reuses are possible.

Straight wall gang forms up to 12′ x 20′ or 8′ x 30′ can be fabricated. These crane handled forms usually lease for approx. 9% per month.

Individual job analysis is available from the manufacturer at no charge.

R031113-40 Forms for Reinforced Concrete

Design Economy

Avoid many sizes in proportioning beams and columns.

From story to story avoid changing column dimensions. Gain strength by adding steel or using a richer mix. If a change in size of column is necessary, vary one dimension only to minimize form alterations. Keep beams and columns the same width.

From floor to floor in a multi-story building vary beam depth, not width, as that will leave the slab panel form unchanged. It is cheaper to vary the strength of a beam from floor to floor by means of a steel area than by 2″ changes in either width or depth.

Cost Factors

Material includes the cost of lumber, cost of rent for metal pans or forms if used, nails, form ties, form oil, bolts and accessories.

Labor includes the cost of carpenters to make up, erect, remove and repair, plus common labor to clean and move. Having carpenters remove forms minimizes repairs.

Improper alignment and condition of forms will increase finishing cost. When forms are heavily oiled, concrete surfaces must be neutralized before finishing. Special curing compounds will cause spillages to spall off in first frost. Gang forming methods will reduce costs on large projects.

Materials Used

Boards are seldom used unless their architectural finish is required. Generally, steel, fiberglass and plywood are used for contact surfaces. Labor on plywood is 10% less than with boards. The plywood is backed up with

2 x 4′s at 12″ to 32″ O.C. Walers are generally 2 - 2 x 4′s. Column forms are held together with steel yokes or bands. Shoring is with adjustable shoring or scaffolding for high ceilings.

Reuse

Floor and column forms can be reused four or possibly five times without excessive repair. Remember to allow for 10% waste on each reuse.

When modular sized wall forms are made, up to twenty uses can be expected with exterior plyform.

When forms are reused, the cost to erect, strip, clean and move will not be affected. 10% replacement of lumber should be included and about one hour of carpenter time for repairs on each reuse per 100 S.F.

The reuse cost for certain accessory items normally rented on a monthly basis will be lower than the cost for the first use.

After the fifth use, new material required plus time needed for repair prevent the form cost from dropping further; it may go up. Much depends on care in stripping, the number of special bays, changes in beam or column sizes and other factors.

Costs for multiple use of formwork may be developed as follows:

2 Uses	3 Uses	4 Uses
$\dfrac{(\text{1st Use + Reuse})}{2} = \text{avg. cost/2 uses}$	$\dfrac{(\text{1st Use + 2 Reuses})}{3} = \text{avg. cost/3 uses}$	$\dfrac{(\text{1st use + 3 Reuses})}{4} = \text{avg. cost/4 uses}$

R031113-60 Formwork Labor-Hours

Item	Unit	Hours Required			Total Hours	Multiple Use		
		Fabricate	Erect & Strip	Clean & Move	1 Use	2 Use	3 Use	4 Use
Beam and Girder, interior beams, 12" wide	100 S.F.	6.4	8.3	1.3	16.0	13.3	12.4	12.0
Hung from steel beams		5.8	7.7	1.3	14.8	12.4	11.6	11.2
Beam sides only, 36" high		5.8	7.2	1.3	14.3	11.9	11.1	10.7
Beam bottoms only, 24" wide		6.6	13.0	1.3	20.9	18.1	17.2	16.7
Box out for openings		9.9	10.0	1.1	21.0	16.6	15.1	14.3
Buttress forms, to 8' high		6.0	6.5	1.2	13.7	11.2	10.4	10.0
Centering, steel, 3/4" rib lath			1.0		1.0			
3/8" rib lath or slab form	▼		0.9		0.9			
Chamfer strip or keyway	100 L.F.		1.5		1.5	1.5	1.5	1.5
Columns, fiber tube 8" diameter			20.6		20.6			
12"			21.3		21.3			
16"			22.9		22.9			
20"			23.7		23.7			
24"			24.6		24.6			
30"	▼		25.6		25.6			
Columns, round steel, 12" diameter			22.0		22.0	22.0	22.0	22.0
16"			25.6		25.6	25.6	25.6	25.6
20"			30.5		30.5	30.5	30.5	30.5
24"	▼		37.7		37.7	37.7	37.7	37.7
Columns, plywood 8" x 8"	100 S.F.	7.0	11.0	1.2	19.2	16.2	15.2	14.7
12" x 12"		6.0	10.5	1.2	17.7	15.2	14.4	14.0
16" x 16"		5.9	10.0	1.2	17.1	14.7	13.8	13.4
24" x 24"		5.8	9.8	1.2	16.8	14.4	13.6	13.2
Columns, steel framed plywood 8" x 8"			10.0	1.0	11.0	11.0	11.0	11.0
12" x 12"			9.3	1.0	10.3	10.3	10.3	10.3
16" x 16"			8.5	1.0	9.5	9.5	9.5	9.5
24" x 24"			7.8	1.0	8.8	8.8	8.8	8.8
Drop head forms, plywood		9.0	12.5	1.5	23.0	19.0	17.7	17.0
Coping forms		8.5	15.0	1.5	25.0	21.3	20.0	19.4
Culvert, box			14.5	4.3	18.8	18.8	18.8	18.8
Curb forms, 6" to 12" high, on grade		5.0	8.5	1.2	14.7	12.7	12.1	11.7
On elevated slabs		6.0	10.8	1.2	18.0	15.5	14.7	14.3
Edge forms to 6" high, on grade	100 L.F.	2.0	3.5	0.6	6.1	5.6	5.4	5.3
7" to 12" high	100 S.F.	2.5	5.0	1.0	8.5	7.8	7.5	7.4
Equipment foundations		10.0	18.0	2.0	30.0	25.5	24.0	23.3
Flat slabs, including drops		3.5	6.0	1.2	10.7	9.5	9.0	8.8
Hung from steel		3.0	5.5	1.2	9.7	8.7	8.4	8.2
Closed deck for domes		3.0	5.8	1.2	10.0	9.0	8.7	8.5
Open deck for pans		2.2	5.3	1.0	8.5	7.9	7.7	7.6
Footings, continuous, 12" high		3.5	3.5	1.5	8.5	7.3	6.8	6.6
Spread, 12" high		4.7	4.2	1.6	10.5	8.7	8.0	7.7
Pile caps, square or rectangular		4.5	5.0	1.5	11.0	9.3	8.7	8.4
Grade beams, 24" deep		2.5	5.3	1.2	9.0	8.3	8.0	7.9
Lintel or Sill forms		8.0	17.0	2.0	27.0	23.5	22.3	21.8
Spandrel beams, 12" wide		9.0	11.2	1.3	21.5	17.5	16.2	15.5
Stairs			25.0	4.0	29.0	29.0	29.0	29.0
Trench forms in floor		4.5	14.0	1.5	20.0	18.3	17.7	17.4
Walls, Plywood, at grade, to 8' high		5.0	6.5	1.5	13.0	11.0	9.7	9.5
8' to 16'		7.5	8.0	1.5	17.0	13.8	12.7	12.1
16' to 20'		9.0	10.0	1.5	20.5	16.5	15.2	14.5
Foundation walls, to 8' high		4.5	6.5	1.0	12.0	10.3	9.7	9.4
8' to 16' high		5.5	7.5	1.0	14.0	11.8	11.0	10.6
Retaining wall to 12' high, battered		6.0	8.5	1.5	16.0	13.5	12.7	12.3
Radial walls to 12' high, smooth		8.0	9.5	2.0	19.5	16.0	14.8	14.3
2' chords		7.0	8.0	1.5	16.5	13.5	12.5	12.0
Prefabricated modular, to 8' high		—	4.3	1.0	5.3	5.3	5.3	5.3
Steel, to 8' high		—	6.8	1.2	8.0	8.0	8.0	8.0
8' to 16' high		—	9.1	1.5	10.6	10.3	10.2	10.2
Steel framed plywood to 8' high		—	6.8	1.2	8.0	7.5	7.3	7.2
8' to 16' high	▼	—	9.3	1.2	10.5	9.5	9.2	9.0

For customer support on your Concrete & Masonry Costs with RSMeans Data, call 800.448.8182.

515

R032110-10   Reinforcing Steel Weights and Measures

Bar Designation No.**	Nominal Weight Lb./Ft.	U.S. Customary Units Nominal Dimensions*			SI Units Nominal Dimensions*			
		Diameter in.	Cross Sectional Area, in.²	Perimeter in.	Nominal Weight kg/m	Diameter mm	Cross Sectional Area, cm²	Perimeter mm
3	.376	.375	.11	1.178	.560	9.52	.71	29.9
4	.668	.500	.20	1.571	.994	12.70	1.29	39.9
5	1.043	.625	.31	1.963	1.552	15.88	2.00	49.9
6	1.502	.750	.44	2.356	2.235	19.05	2.84	59.8
7	2.044	.875	.60	2.749	3.042	22.22	3.87	69.8
8	2.670	1.000	.79	3.142	3.973	25.40	5.10	79.8
9	3.400	1.128	1.00	3.544	5.059	28.65	6.45	90.0
10	4.303	1.270	1.27	3.990	6.403	32.26	8.19	101.4
11	5.313	1.410	1.56	4.430	7.906	35.81	10.06	112.5
14	7.650	1.693	2.25	5.320	11.384	43.00	14.52	135.1
18	13.600	2.257	4.00	7.090	20.238	57.33	25.81	180.1

* The nominal dimensions of a deformed bar are equivalent to those of a plain round bar having the same weight per foot as the deformed bar.
** Bar numbers are based on the number of eighths of an inch included in the nominal diameter of the bars.

R032110-20   Metric Rebar Specification - ASTM A615-81

Grade 300 (300 MPa* = 43,560 psi; +8.7% vs. Grade 40)				
Grade 400 (400 MPa* = 58,000 psi; −3.4% vs. Grade 60)				
Bar No.	Diameter mm	Area mm²	Equivalent in.²	Comparison with U.S. Customary Bars
10M	11.3	100	.16	Between #3 & #4
15M	16.0	200	.31	#5 (.31 in.²)
20M	19.5	300	.47	#6 (.44 in.²)
25M	25.2	500	.78	#8 (.79 in.²)
30M	29.9	700	1.09	#9 (1.00 in.²)
35M	35.7	1000	1.55	#11 (1.56 in.²)
45M	43.7	1500	2.33	#14 (2.25 in.²)
55M	56.4	2500	3.88	#18 (4.00 in.²)

* MPa = megapascals

R032110-40   Weight of Steel Reinforcing Per Square Foot of Wall (PSF)

Reinforced Weights: The table below suggests the weights per square foot for reinforcing steel in walls. Weights are approximate and will be the same for all grades of steel bars. For bars in two directions, add weights for each size and spacing.

C/C Spacing in Inches	#3 Wt. (PSF)	#4 Wt. (PSF)	#5 Wt. (PSF)	#6 Wt. (PSF)	#7 Wt. (PSF)	#8 Wt. (PSF)	#9 Wt. (PSF)	#10 Wt. (PSF)	#11 Wt. (PSF)
2"	2.26	4.01	6.26	9.01	12.27				
3"	1.50	2.67	4.17	6.01	8.18	10.68	13.60	17.21	21.25
4"	1.13	2.01	3.13	4.51	6.13	8.10	10.20	12.91	15.94
5"	.90	1.60	2.50	3.60	4.91	6.41	8.16	10.33	12.75
6"	.752	1.34	2.09	3.00	4.09	5.34	6.80	8.61	10.63
8"	.564	1.00	1.57	2.25	3.07	4.01	5.10	6.46	7.97
10"	.451	.802	1.25	1.80	2.45	3.20	4.08	5.16	6.38
12"	.376	.668	1.04	1.50	2.04	2.67	3.40	4.30	5.31
18"	.251	.445	.695	1.00	1.32	1.78	2.27	2.86	3.54
24"	.188	.334	.522	.751	1.02	1.34	1.70	2.15	2.66
30"	.150	.267	.417	.600	.817	1.07	1.36	1.72	2.13
36"	.125	.223	.348	.501	.681	.890	1.13	1.43	1.77
42"	.107	.191	.298	.429	.584	.753	.97	1.17	1.52
48"	.094	.167	.261	.376	.511	.668	.85	1.08	1.33

R032110-50   Minimum Wall Reinforcement Weight (PSF)

This table lists the approximate minimum wall reinforcement weights per S.F. according to the specification of .12% of gross area for vertical bars and .20% of gross area for horizontal bars.

Location	Wall Thickness	Bar Size	Horizontal Steel Spacing C/C	Sq. In. Req'd per S.F.	Total Wt. per S.F.	Bar Size	Vertical Steel Spacing C/C	Sq. In. Req'd per S.F.	Total Wt. per S.F.	Horizontal & Vertical Steel Total Weight per S.F.
Both Faces	10"	#4	18"	.24	.89#	#3	18"	.14	.50#	1.39#
	12"	#4	16"	.29	1.00	#3	16"	.17	.60	1.60
	14"	#4	14"	.34	1.14	#3	13"	.20	.69	1.84
	16"	#4	12"	.38	1.34	#3	11"	.23	.82	2.16
	18"	#5	17"	.43	1.47	#4	18"	.26	.89	2.36
One Face	6"	#3	9"	.15	.50	#3	18"	.09	.25	.75
	8"	#4	12"	.19	.67	#3	11"	.12	.41	1.08
	10"	#5	15"	.24	.83	#4	16"	.14	.50	1.34

R032110-70   Bend, Place and Tie Reinforcing

Placing and tying by rodmen for footings and slabs run from nine hrs. per ton for heavy bars to fifteen hrs. per ton for light bars. For beams, columns, and walls, production runs from eight hrs. per ton for heavy bars to twenty hrs. per ton for light bars. The overall average for typical reinforced concrete buildings is about fourteen hrs. per ton. These production figures include the time for placing accessories and usual inserts, but not their material cost (allow 15% of the cost of delivered bent rods). Equipment handling is necessary for the larger-sized bars so that installation costs for the very heavy bars will not decrease proportionately.

Installation costs for splicing reinforcing bars include allowance for equipment to hold the bars in place while splicing as well as necessary scaffolding for iron workers.

R032110-80   Shop-Fabricated Reinforcing Steel

The material prices for reinforcing, shown in the unit cost sections of the data set, are for 50 tons or more of shop-fabricated reinforcing steel and include:

1. Mill base price of reinforcing steel
2. Mill grade/size/length extras
3. Mill delivery to the fabrication shop
4. Shop storage and handling
5. Shop drafting/detailing
6. Shop shearing and bending
7. Shop listing
8. Shop delivery to the job site

Both material and installation costs can be considerably higher for small jobs consisting primarily of smaller bars, while material costs may be slightly lower for larger jobs.

For customer support on your Concrete & Masonry Costs with RSMeans Data, call 800.448.8182.

517

R032205-30 Common Stock Styles of Welded Wire Fabric

This table provides some of the basic specifications, sizes, and weights of welded wire fabric used for reinforcing concrete.

New Designation		Old Designation		Steel Area per Foot				Approximate Weight per 100 S.F.	
Spacing — Cross Sectional Area (in.) — (Sq. in. 100)		Spacing — Wire Gauge (in.) — (AS & W)		Longitudinal		Transverse			
				in.	cm	in.	cm	lbs	kg
Rolls	6 x 6 — W1.4 x W1.4	6 x 6 — 10 x 10		.028	.071	.028	.071	21	9.53
	6 x 6 — W2.0 x W2.0	6 x 6 — 8 x 8	1	.040	.102	.040	.102	29	13.15
	6 x 6 — W2.9 x W2.9	6 x 6 — 6 x 6		.058	.147	.058	.147	42	19.05
	6 x 6 — W4.0 x W4.0	6 x 6 — 4 x 4		.080	.203	.080	.203	58	26.91
	4 x 4 — W1.4 x W1.4	4 x 4 — 10 x 10		.042	.107	.042	.107	31	14.06
	4 x 4 — W2.0 x W2.0	4 x 4 — 8 x 8	1	.060	.152	.060	.152	43	19.50
	4 x 4 — W2.9 x W2.9	4 x 4 — 6 x 6		.087	.227	.087	.227	62	28.12
	4 x 4 — W4.0 x W4.0	4 x 4 — 4 x 4		.120	.305	.120	.305	85	38.56
Sheets	6 x 6 — W2.9 x W2.9	6 x 6 — 6 x 6		.058	.147	.058	.147	42	19.05
	6 x 6 — W4.0 x W4.0	6 x 6 — 4 x 4		.080	.203	.080	.203	58	26.31
	6 x 6 — W5.5 x W5.5	6 x 6 — 2 x 2	2	.110	.279	.110	.279	80	36.29
	4 x 4 — W1.4 x W1.4	4 x 4 — 4 x 4		.120	.305	.120	.305	85	38.56

NOTES: 1. Exact W—number size for 8 gauge is W2.1
      2. Exact W—number size for 2 gauge is W5.4

The above table was compiled with the following excerpts from the WRI Manual of Standard Practices, 7th Edition, Copyright 2006. Reproduced with permission of the Wire Reinforcement Institute, Inc.:

1. Chapter 3, page 7, Table 1 Common Styles of Metric Wire Reinforcement (WWR) With Equivalent US Customary Units
2. Chapter 6, page 19, Table 5 Customary Units
3. Chapter 6, Page 23, Table 7 Customary Units (in.) Welded Plain Wire Reinforcement
4. Chapter 6, Page 25 Table 8 Wire Size Comparison
5. Chapter 9, Page 30, Table 9 Weight of Longitudinal Wires Weight (Mass) Estimating Tables
6. Chapter 9, Page 31, Table 9M Weight of Longitudinal Wires Weight (Mass) Estimating Tables
7. Chapter 9, Page 32, Table 10 Weight of Transverse Wires Based on 62″ lengths of transverse wire (60″ width plus 1″ overhand each side)
8. Chapter 9, Page 33, Table 10M Weight of Transverse Wires

R033053-10 Spread Footings

General: A spread footing is used to convert a concentrated load (from one superstructure column, or substructure grade beams) into an allowable area load on supporting soil.

Because of punching action from the column load, a spread footing is usually thicker than strip footings which support wall loads. One or two story commercial or residential buildings should have no less than 1' thick spread footings. Heavier loads require no less than 2' thick. Spread footings may be square, rectangular or octagonal in plan.

Spread footings tend to minimize excavation and foundation materials, as well as labor and equipment. Another advantage is that footings and soil conditions can be readily examined. They are the most widely used type of footing, especially in mild climates and for buildings of four stories or under. This is because they are usually more economical than other types, if suitable soil and site conditions exist.

They are used when suitable supporting soil is located within several feet of the surface or line of subsurface excavation. Suitable soil types include sands and gravels, gravels with a small amount of clay or silt, hardpan, chalk, and rock. Pedestals may be used to bring the column base load down to the top of the footing. Alternately, undesirable soil between the underside of the footing and the top of the bearing level can be removed and replaced with lean concrete mix or compacted granular material.

Depth of footing should be below topsoil, uncompacted fill, muck, etc. It must be lower than frost penetration but should be above the water table. It must not be at the ground surface because of potential surface erosion. If the ground slopes, approximately three horizontal feet of edge protection must remain. Differential footing elevations may overlap soil stresses or cause excavation problems if clear spacing between footings is less than the difference in depth.

Other footing types are usually used for the following reasons:

A. Bearing capacity of soil is low.

B. Very large footings are required, at a cost disadvantage.

C. Soil under footing (shallow or deep) is very compressible, with probability of causing excessive or differential settlement.

D. Good bearing soil is deep.

E. Potential for scour action exists.

F. Varying subsoil conditions within building perimeter.

Cost of spread footings for a building is determined by:

1. The soil bearing capacity.
2. Typical bay size.
3. Total load (live plus dead) per S.F. for roof and elevated floor levels.
4. The size and shape of the building.
5. Footing configuration. Does the building utilize outer spread footings or are there continuous perimeter footings only or a combination of spread footings plus continuous footings?

Soil Bearing Capacity in Kips per S.F.

Bearing Material	Typical Allowable Bearing Capacity
Hard sound rock	120 KSF
Medium hard rock	80
Hardpan overlaying rock	24
Compact gravel and boulder-gravel; very compact sandy gravel	20
Soft rock	16
Loose gravel; sandy gravel; compact sand; very compact sand-inorganic silt	12
Hard dry consolidated clay	10
Loose coarse to medium sand; medium compact fine sand	8
Compact sand-clay	6
Loose fine sand; medium compact sand-inorganic silts	4
Firm or stiff clay	3
Loose saturated sand-clay; medium soft clay	2

R033053-50 Industrial Chimneys

Foundation requirements in C.Y. of concrete for various sized chimneys.

Size Chimney	2 Ton Soil	3 Ton Soil	Size Chimney	2 Ton Soil	3 Ton Soil	Size Chimney	2 Ton Soil	3 Ton Soil
75' x 3'-0"	13 C.Y.	11 C.Y.	160' x 6'-6"	86 C.Y.	76 C.Y.	300' x 10'-0"	325 C.Y.	245 C.Y.
85' x 5'-6"	19	16	175' x 7'-0"	108	95	350' x 12'-0"	422	320
100' x 5'-0"	24	20	200' x 6'-0"	125	105	400' x 14'-0"	520	400
125' x 5'-6"	43	36	250' x 8'-0"	230	175	500' x 18'-0"	725	575

R033053-60 Maximum Depth of Frost Penetration in Inches

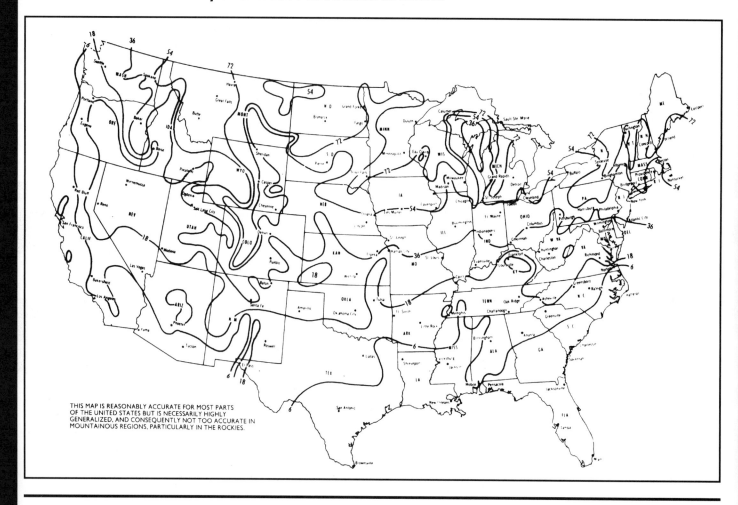

THIS MAP IS REASONABLY ACCURATE FOR MOST PARTS OF THE UNITED STATES BUT IS NECESSARILY HIGHLY GENERALIZED, AND CONSEQUENTLY NOT TOO ACCURATE IN MOUNTAINOUS REGIONS, PARTICULARLY IN THE ROCKIES.

R033105-10 Proportionate Quantities

The tables below show both quantities per S.F. of floor areas as well as form and reinforcing quantities per C.Y. Unusual structural requirements would increase the ratios below. High strength reinforcing would reduce the steel weights. Figures are for 3000 psi concrete and 60,000 psi reinforcing unless specified otherwise.

Type of Construction	Live Load	Span	Per S.F. of Floor Area				Per C.Y. of Concrete		
			Concrete	Forms	Reinf.	Pans	Forms	Reinf.	Pans
Flat Plate	50 psf	15 Ft.	.46 C.F.	1.06 S.F.	1.71 lb.		62 S.F.	101 lb.	
		20	.63	1.02	2.40		44	104	
		25	.79	1.02	3.03		35	104	
	100	15	.46	1.04	2.14		61	126	
		20	.71	1.02	2.72		39	104	
		25	.83	1.01	3.47		33	113	
Flat Plate (waffle construction) 20" domes	50	20	.43	1.00	2.10	.84 S.F.	63	135	53 S.F.
		25	.52	1.00	2.90	.89	52	150	46
		30	.64	1.00	3.70	.87	42	155	37
	100	20	.51	1.00	2.30	.84	53	125	45
		25	.64	1.00	3.20	.83	42	135	35
		30	.76	1.00	4.40	.81	36	160	29
Waffle Construction 30" domes	50	25	.69	1.06	1.83	.68	42	72	40
		30	.74	1.06	2.39	.69	39	87	39
		35	.86	1.05	2.71	.69	33	85	39
		40	.78	1.00	4.80	.68	35	165	40
Flat Slab (two way with drop panels)	50	20	.62	1.03	2.34		45	102	
		25	.77	1.03	2.99		36	105	
		30	.95	1.03	4.09		29	116	
	100	20	.64	1.03	2.83		43	119	
		25	.79	1.03	3.88		35	133	
		30	.96	1.03	4.66		29	131	
	200	20	.73	1.03	3.03		38	112	
		25	.86	1.03	4.23		32	133	
		30	1.06	1.03	5.30		26	135	
One Way Joists 20" Pans	50	15	.36	1.04	1.40	.93	78	105	70
		20	.42	1.05	1.80	.94	67	120	60
		25	.47	1.05	2.60	.94	60	150	54
	100	15	.38	1.07	1.90	.93	77	140	66
		20	.44	1.08	2.40	.94	67	150	58
		25	.52	1.07	3.50	.94	55	185	49
One Way Joists 8" x 16" filler blocks	50	15	.34	1.06	1.80	.81 Ea.	84	145	64 Ea.
		20	.40	1.08	2.20	.82	73	145	55
		25	.46	1.07	3.20	.83	63	190	49
	100	15	.39	1.07	1.90	.81	74	130	56
		20	.46	1.09	2.80	.82	64	160	48
		25	.53	1.10	3.60	.83	56	190	42
One Way Beam & Slab	50	15	.42	1.30	1.73		84	111	
		20	.51	1.28	2.61		68	138	
		25	.64	1.25	2.78		53	117	
	100	15	.42	1.30	1.90		84	122	
		20	.54	1.35	2.69		68	154	
		25	.69	1.37	3.93		54	145	
	200	15	.44	1.31	2.24		80	137	
		20	.58	1.40	3.30		65	163	
		25	.69	1.42	4.89		53	183	
Two Way Beam & Slab	100	15	.47	1.20	2.26		69	130	
		20	.63	1.29	3.06		55	131	
		25	.83	1.33	3.79		43	123	
	200	15	.49	1.25	2.70		41	149	
		20	.66	1.32	4.04		54	165	
		25	.88	1.32	6.08		41	187	

For customer support on your Concrete & Masonry Costs with RSMeans Data, call 800.448.8182.

521

R033105-10   Proportionate Quantities (cont.)

Item	Size	Forms	Reinforcing	Minimum	Maximum
			4000 psi Concrete and 60,000 psi Reinforcing—Form and Reinforcing Quantities per C.Y.		
	10″ x 10″	130 S.F.C.A.	#5 to #11	220 lbs.	875 lbs.
	12″ x 12″	108	#6 to #14	200	955
	14″ x 14″	92	#7 to #14	190	900
	16″ x 16″	81	#6 to #14	187	1082
	18″ x 18″	72	#6 to #14	170	906
	20″ x 20″	65	#7 to #18	150	1080
Columns	22″ x 22″	59	#8 to #18	153	902
(square tied)	24″ x 24″	54	#8 to #18	164	884
	26″ x 26″	50	#9 to #18	169	994
	28″ x 28″	46	#9 to #18	147	864
	30″ x 30″	43	#10 to #18	146	983
	32″ x 32″	40	#10 to #18	175	866
	34″ x 34″	38	#10 to #18	157	772
	36″ x 36″	36	#10 to #18	175	852
	38″ x 38″	34	#10 to #18	158	765
	40″ x 40″	32	#10 to #18	143	692

Item	Size	Form	Spiral	Reinforcing	Minimum	Maximum
	12″ diameter	34.5 L.F.	190 lbs.	#4 to #11	165 lbs.	1505 lb.
		34.5	190	#14 & #18	—	1100
	14″	25	170	#4 to #11	150	970
		25	170	#14 & #18	800	1000
	16″	19	160	#4 to #11	160	950
		19	160	#14 & #18	605	1080
	18″	15	150	#4 to #11	160	915
		15	150	#14 & #18	480	1075
	20″	12	130	#4 to #11	155	865
		12	130	#14 & #18	385	1020
	22″	10	125	#4 to #11	165	775
		10	125	#14 & #18	320	995
	24″	9	120	#4 to #11	195	800
		9	120	#14 & #18	290	1150
Columns	26″	7.3	100	#4 to #11	200	729
(spirally reinforced)		7.3	100	#14 & #18	235	1035
	28″	6.3	95	#4 to #11	175	700
		6.3	95	#14 & #18	200	1075
	30″	5.5	90	#4 to #11	180	670
		5.5	90	#14 & #18	175	1015
	32″	4.8	85	#4 to #11	185	615
		4.8	85	#14 & #18	155	955
	34″	4.3	80	#4 to #11	180	600
		4.3	80	#14 & #18	170	855
	36″	3.8	75	#4 to #11	165	570
		3.8	75	#14 & #18	155	865
	40″	3.0	70	#4 to #11	165	500
		3.0	70	#14 & #18	145	765

R033105-10 Proportionate Quantities (cont.)

3000 psi Concrete and 60,000 psi Reinforcing—Form and Reinforcing Quantities per C.Y.

Item	Type	Loading	Height	C.Y./L.F.	Forms/C.Y.	Reinf./C.Y.
Retaining Walls	Cantilever	Level Backfill	4 Ft.	0.2 C.Y.	49 S.F.	35 lbs.
			8	0.5	42	45
			12	0.8	35	70
			16	1.1	32	85
			20	1.6	28	105
		Highway Surcharge	4	0.3	41	35
			8	0.5	36	55
			12	0.8	33	90
			16	1.2	30	120
			20	1.7	27	155
		Railroad Surcharge	4	0.4	28	45
			8	0.8	25	65
			12	1.3	22	90
			16	1.9	20	100
			20	2.6	18	120
	Gravity, with Vertical Face	Level Backfill	4	0.4	37	None
			7	0.6	27	
			10	1.2	20	
		Sloping Backfill	4	0.3	31	
			7	0.8	21	
			10	1.6	15	↓

		Live Load in Kips per Linear Foot							
	Span	Under 1 Kip		2 to 3 Kips		4 to 5 Kips		6 to 7 Kips	
		Forms	Reinf.	Forms	Reinf.	Forms	Reinf.	Forms	Reinf.
Beams	10 Ft.	—	—	90 S.F.	170 #	85 S.F.	175 #	75 S.F.	185 #
	16	130 S.F.	165 #	85	180	75	180	65	225
	20	110	170	75	185	62	200	51	200
	26	90	170	65	215	62	215	—	—
	30	85	175	60	200	—	—	—	—

Item	Size	Type	Forms per C.Y.	Reinforcing per C.Y.
Spread Footings	Under 1 C.Y.	1,000 psf soil	24 S.F.	44 lbs.
		5,000	24	42
		10,000	24	52
	1 C.Y. to 5 C.Y.	1,000	14	49
		5,000	14	50
		10,000	14	50
	Over 5 C.Y.	1,000	9	54
		5,000	9	52
		10,000	9	56
Pile Caps (30 Ton Concrete Piles)	Under 5 C.Y.	shallow caps	20	65
		medium	20	50
		deep	20	40
	5 C.Y. to 10 C.Y.	shallow	14	55
		medium	15	45
		deep	15	40
	10 C.Y. to 20 C.Y.	shallow	11	60
		medium	11	45
		deep	12	35
	Over 20 C.Y.	shallow	9	60
		medium	9	45
		deep	10	40

R033105-10   Proportionate Quantities (cont.)

3000 psi Concrete and 60,000 psi Reinforcing — Form and Reinforcing Quantities per C.Y.						
Item	Size	Pile Spacing	50 T Pile	100 T Pile	50 T Pile	100 T Pile
Pile Caps (Steel H Piles)	Under 5 C.Y.	24" O.C.	24 S.F.	24 S.F.	75 lbs.	90 lbs.
		30"	25	25	80	100
		36"	24	24	80	110
	5 C.Y. to 10 C.Y.	24"	15	15	80	110
		30"	15	15	85	110
		36"	15	15	75	90
	Over 10 C.Y.	24"	13	13	85	90
		30"	11	11	85	95
		36"	10	10	85	90

		8" Thick		10" Thick		12" Thick		15" Thick	
	Height	Forms	Reinf.	Forms	Reinf.	Forms	Reinf.	Forms	Reinf.
Basement Walls	7 Ft.	81 S.F.	44 lbs.	65 S.F.	45 lbs.	54 S.F.	44 lbs.	41 S.F.	43 lbs.
	8		44		45		44		43
	9		46		45		44		43
	10		57		45		44		43
	12		83		50		52		43
	14		116		65		64		51
	16				86		90		65
	18	↓	↓	↓		↓	106	↓	70

R033105-20   Materials for One C.Y. of Concrete

This is an approximate method of figuring quantities of cement, sand and coarse aggregate for a field mix with waste allowance included.

With crushed gravel as coarse aggregate, to determine barrels of cement required, divide 10 by total mix; that is, for 1:2:4 mix, 10 divided by 7 = 1-3/7 barrels.

If the coarse aggregate is crushed stone, use 10-1/2 instead of 10 as given for gravel.

To determine tons of sand required, multiply barrels of cement by parts of sand and then by 0.2; that is, for the 1:2:4 mix, as above, 1-3/7 x 2 x .2 = .57 tons.

Tons of crushed gravel are in the same ratio to tons of sand as parts in the mix, or 4/2 x .57 = 1.14 tons.

1 bag cement = 94#	1 C.Y. sand or crushed gravel = 2700#	1 C.Y. crushed stone = 2575#
4 bags = 1 barrel	1 ton sand or crushed gravel = 20 C.F.	1 ton crushed stone = 21 C.F.

Average carload of cement is 692 bags; of sand or gravel is 56 tons.

Do not stack stored cement over 10 bags high.

R033105-30   Metric Equivalents of Cement Content for Concrete Mixes

94 Pound Bags per Cubic Yard	Kilograms per Cubic Meter	94 Pound Bags per Cubic Yard	Kilograms per Cubic Meter
1.0	55.77	7.0	390.4
1.5	83.65	7.5	418.3
2.0	111.5	8.0	446.2
2.5	139.4	8.5	474.0
3.0	167.3	9.0	501.9
3.5	195.2	9.5	529.8
4.0	223.1	10.0	557.7
4.5	251.0	10.5	585.6
5.0	278.8	11.0	613.5
5.5	306.7	11.5	641.3
6.0	334.6	12.0	669.2
6.5	362.5	12.5	697.1

a. If you know the cement content in pounds per cubic yard, multiply by .5933 to obtain kilograms per cubic meter.

b. If you know the cement content in 94 pound bags per cubic yard, multiply by 55.77 to obtain kilograms per cubic meter.

R033105-40 Metric Equivalents of Common Concrete Strengths (to convert other psi values to megapascals, multiply by 0.006895)

U.S. Values psi	SI Values Megapascals	Non-SI Metric Values kgf/cm²*
2000	14	140
2500	17	175
3000	21	210
3500	24	245
4000	28	280
4500	31	315
5000	34	350
6000	41	420
7000	48	490
8000	55	560
9000	62	630
10,000	69	705

* kilograms force per square centimeter

R033105-50 Quantities of Cement, Sand and Stone for One C.Y. of Concrete per Various Mixes

This table can be used to determine the quantities of the ingredients for smaller quantities of site mixed concrete.

Concrete (C.Y.)	Mix = 1:1:1-3/4 Cement (sacks)	Sand (C.Y.)	Stone (C.Y.)	Mix = 1:2:2.25 Cement (sacks)	Sand (C.Y.)	Stone (C.Y.)	Mix = 1:2.25:3 Cement (sacks)	Sand (C.Y.)	Stone (C.Y.)	Mix = 1:3:4 Cement (sacks)	Sand (C.Y.)	Stone (C.Y.)
1	10	.37	.63	7.75	.56	.65	6.25	.52	.70	5	.56	.74
2	20	.74	1.26	15.50	1.12	1.30	12.50	1.04	1.40	10	1.12	1.48
3	30	1.11	1.89	23.25	1.68	1.95	18.75	1.56	2.10	15	1.68	2.22
4	40	1.48	2.52	31.00	2.24	2.60	25.00	2.08	2.80	20	2.24	2.96
5	50	1.85	3.15	38.75	2.80	3.25	31.25	2.60	3.50	25	2.80	3.70
6	60	2.22	3.78	46.50	3.36	3.90	37.50	3.12	4.20	30	3.36	4.44
7	70	2.59	4.41	54.25	3.92	4.55	43.75	3.64	4.90	35	3.92	5.18
8	80	2.96	5.04	62.00	4.48	5.20	50.00	4.16	5.60	40	4.48	5.92
9	90	3.33	5.67	69.75	5.04	5.85	56.25	4.68	6.30	45	5.04	6.66
10	100	3.70	6.30	77.50	5.60	6.50	62.50	5.20	7.00	50	5.60	7.40
11	110	4.07	6.93	85.25	6.16	7.15	68.75	5.72	7.70	55	6.16	8.14
12	120	4.44	7.56	93.00	6.72	7.80	75.00	6.24	8.40	60	6.72	8.88
13	130	4.82	8.20	100.76	7.28	8.46	81.26	6.76	9.10	65	7.28	9.62
14	140	5.18	8.82	108.50	7.84	9.10	87.50	7.28	9.80	70	7.84	10.36
15	150	5.56	9.46	116.26	8.40	9.76	93.76	7.80	10.50	75	8.40	11.10
16	160	5.92	10.08	124.00	8.96	10.40	100.00	8.32	11.20	80	8.96	11.84
17	170	6.30	10.72	131.76	9.52	11.06	106.26	8.84	11.90	85	9.52	12.58
18	180	6.66	11.34	139.50	10.08	11.70	112.50	9.36	12.60	90	10.08	13.32
19	190	7.04	11.98	147.26	10.64	12.36	118.76	9.84	13.30	95	10.64	14.06
20	200	7.40	12.60	155.00	11.20	13.00	125.00	10.40	14.00	100	11.20	14.80
21	210	7.77	13.23	162.75	11.76	13.65	131.25	10.92	14.70	105	11.76	15.54
22	220	8.14	13.86	170.05	12.32	14.30	137.50	11.44	15.40	110	12.32	16.28
23	230	8.51	14.49	178.25	12.88	14.95	143.75	11.96	16.10	115	12.88	17.02
24	240	8.88	15.12	186.00	13.44	15.60	150.00	12.48	16.80	120	13.44	17.76
25	250	9.25	15.75	193.75	14.00	16.25	156.25	13.00	17.50	125	14.00	18.50
26	260	9.64	16.40	201.52	14.56	16.92	162.52	13.52	18.20	130	14.56	19.24
27	270	10.00	17.00	209.26	15.12	17.56	168.76	14.04	18.90	135	15.02	20.00
28	280	10.36	17.64	217.00	15.68	18.20	175.00	14.56	19.60	140	15.68	20.72
29	290	10.74	18.28	224.76	16.24	18.86	181.26	15.08	20.30	145	16.24	21.46

For customer support on your Concrete & Masonry Costs with RSMeans Data, call 800.448.8182.

525

R033105-65   Field-Mix Concrete

Presently most building jobs are built with ready-mixed concrete except at isolated locations and some larger jobs requiring over 10,000 C.Y. where land is readily available for setting up a temporary batch plant.

The most economical mix is a controlled mix using local aggregate proportioned by trial to give the required strength with the least cost of material.

R033105-70   Placing Ready-Mixed Concrete

For ground pours allow for 5% waste when figuring quantities.

Prices in the front of the data set assume normal deliveries. If deliveries are made before 8 A.M. or after 5 P.M. or on Saturday afternoons add 30%. Negotiated discounts for large volumes are not included in prices in front of data set.

For the lower floors without truck access, concrete may be wheeled in rubber-tired buggies, conveyer handled, crane handled or pumped. Pumping is economical if there is top steel. Conveyers are more efficient for thick slabs.

At higher floors the rubber-tired buggies may be hoisted by a hoisting tower and wheeled to the location. Placement by a conveyer is limited to three floors and is best for high-volume pours. Pumped concrete is best when the building has no crane access. Concrete may be pumped directly as high as thirty-six stories using special pumping techniques. Normal maximum height is about fifteen stories.

The best pumping aggregate is screened and graded bank gravel rather than crushed stone.

Pumping downward is more difficult than pumping upward. The horizontal distance from pump to pour may increase preparation time prior to pour. Placing by cranes, either mobile, climbing or tower types, continues as the most efficient method for high-rise concrete buildings.

Reference Tables

R033105-80 Slab on Grade

General: Ground slabs are classified on the basis of use. Thickness is generally controlled by the heaviest concentrated load supported. If load area is greater than 80 sq. in., soil bearing may be important. The base granular fill must be a uniformly compacted material of limited capillarity, such as gravel or crushed rock. Concrete is placed on this surface of the vapor barrier on top of the base.

Ground slabs are either single or two course floors. Single course floors are widely used. Two course floors have a subsequent wear resistant topping.

Reinforcement is provided to maintain tightly closed cracks.

Control joints limit crack locations and provide for differential horizontal movement only. Isolation joints allow both horizontal and vertical differential movement.

Use of Table: Determine the appropriate type of slab (A, B, C, or D) by considering the type of use or amount of abrasive wear of traffic type.

Determine thickness by maximum allowable wheel load or uniform load, opposite 1st column, thickness. Increase the controlling thickness if details require, and select either plain or reinforced slab thickness and type.

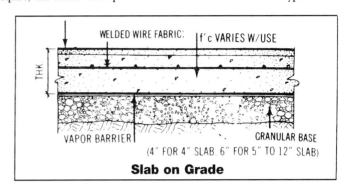

Slab on Grade

Thickness and Loading Assumptions by Type of Use

SLAB THICKNESS (IN.)	TYPE	A — Non — Little — Foot Only — Load* (K)	B — Light — Light — Pneumatic Wheels — Load* (K)	C — Normal — Moderate — Solid Rubber Wheels — Load* (K)	D — Heavy — Severe — Steel Tires — Load* (K)	◄ Slab I.D. / ◄ Industrial / ◄ Abrasion / ◄ Type of Traffic — Max. Uniform Load to Slab ▼ (PSF)
4"	Reinf. / Plain	4K				100
5"	Reinf. / Plain	6K	4K			200
6"	Reinf. / Plain		8K	6K	6K	500 to 800
7"	Reinf. / Plain			9K	8K	1,500
8"	Reinf. / Plain				11K	
10"	Reinf. / Plain				14K	* Max. Wheel Load in Kips (incl. impact)
12"	Reinf. / Plain					
DESIGN ASSUMPTIONS	Concrete, Chuted	f'c = 3.5 KSI	4 KSI	4.5 KSI	Slab @ 3.5 KSI	ASSUMPTIONS BY SLAB TYPE
	Toppings			1" Integral	1" Bonded	
	Finish	Steel Trowel	Steel Trowel	Steel Trowel	Screed & Steel Trowel	
	Compacted Granular Base	4" deep for 4" slab thickness / 6" deep for 5" slab thickness & greater				
	Vapor Barrier	6 mil polyethylene				ASSUMPTIONS FOR ALL SLAB TYPES
	Forms & Joints	Allowances included				
	Rein-forcement	WWF as required ≥ 60,000 psi				

For customer support on your Concrete & Masonry Costs with RSMeans Data, call 800.448.8182.

527

R033105-85   Lift Slabs

The cost advantage of the lift slab method is due to placing all concrete, reinforcing steel, inserts and electrical conduit at ground level and in reduction of formwork. Minimum economical project size is about 30,000 S.F. Slabs may be tilted for parking garage ramps.

It is now used in all types of buildings and has gone up to 22 stories high in apartment buildings. The current trend is to use post-tensioned flat plate slabs with spans from 22' to 35'. Cylindrical void forms are used when deep slabs are required. One pound of prestressing steel is about equal to seven pounds of conventional reinforcing.

To be considered cured for stressing and lifting, a slab must have attained 75% of design strength. Seven days are usually sufficient with four to five days possible if high early strength cement is used. Slabs can be stacked using two coats of a non-bonding agent to insure that slabs do not stick to each other. Lifting is done by companies specializing in this work. Lift rate is 5' to 15' per hour with an average of 10' per hour. Total areas up to 33,000 S.F. have been lifted at one time. 24 to 36 jacking columns are common. Most economical bay sizes are 24' to 28' with four to fourteen stories most efficient. Continuous design reduces reinforcing steel cost. Use of post-tensioned slabs allows larger bay sizes.

R033543-10   Polished Concrete Floors

A polished concrete floor has a glossy mirror-like appearance and is created by grinding the concrete floor with finer and finer diamond grits, similar to sanding wood, until the desired level of reflective clarity and sheen are achieved. The technical term for this type of polished concrete is bonded abrasive polished concrete. The basic piece of equipment used in the polishing process is a walk-behind planetary grinder for working large floor areas. This grinder drives diamond-impregnated abrasive discs, which progress from coarse- to fine-grit discs.

The process begins with the use of very coarse diamond segments or discs bonded in a metallic matrix. These segments are coarse enough to allow the removal of pits, blemishes, stains, and light coatings from the floor surface in preparation for final smoothing. The condition of the original concrete surface will dictate the grit coarseness of the initial grinding step which will generally end up being a three- to four-step process using ever finer grits. The purpose of this initial grinding step is to remove surface coatings and blemishes and to cut down into the cream for very fine aggregate exposure, or deeper into the fine aggregate layer just below the cream layer, or even deeper into the coarse aggregate layer. These initial grinding steps will progress up to the 100/120 grit. If wet grinding is done, a waste slurry is produced that must be removed between grit changes and disposed of properly. If dry grinding is done, a high performance vacuum will pick up the dust during grinding and collect it in bags which must be disposed of properly.

The process continues with honing the floor in a series of steps that progresses from 100-grit to 400-grit diamond abrasive discs embedded in a plastic or resin matrix. At some point during, or just prior to, the honing step, one or two coats of stain or dye can be sprayed onto the surface to give color to the concrete, and two coats of densifier/hardener must be applied to the floor surface and allowed to dry. This sprayed-on densifier/hardener will penetrate about 1/8" into the concrete to make the surface harder, denser and more abrasion-resistant.

The process ends with polishing the floor surface in a series of steps that progresses from resin-impregnated 800-grit (medium polish) to 1500-grit (high polish) to 3000-grit (very high polish), depending on the desired level of reflective clarity and sheen.

The Concrete Polishing Association of America (CPAA) has defined the flooring options available when processing concrete to a desired finish. The first category is aggregate exposure, the grinding of a concrete surface with bonded abrasives, in as many abrasive grits necessary, to achieve one of the following classes:

©Concrete Polishing Association of America "Glossary."

A. Cream – very little surface cut depth; little aggregate exposure

B. Fine aggregate (salt and pepper) – surface cut depth of 1/16"; fine aggregate exposure with little or no medium aggregate exposure at random locations

C. Medium aggregate – surface cut depth of 1/8"; medium aggregate exposure with little or no large aggregate exposure at random locations

D. Large aggregate – surface cut depth of 1/4"; large aggregate exposure with little or no fine aggregate exposure at random locations

The second CPAA defined category is reflective clarity and sheen, the polishing of a concrete surface with the minimum number of bonded abrasives as indicated to achieve one of the following levels:

1. Ground – flat appearance with none to very slight diffused reflection; none to very low reflective sheen; using a minimum total of 4 grit levels up to 100-grit

2. Honed – matte appearance with or without slight diffused reflection; low to medium reflective sheen; using a minimum total of 5 grit levels up to 400-grit

3. Semi-polished – objects being reflected are not quite sharp and crisp but can be easily identified; medium to high reflective sheen; using a minimum total of 6 grit levels up to 800-grit

4. Highly-polished – objects being reflected are sharp and crisp as would be seen in a mirror-like reflection; high to highest reflective sheen; using a minimum total of up to 8 grit levels up to 1500-grit or 3000-grit

The CPAA defines reflective clarity as the degree of sharpness and crispness of the reflection of overhead objects when viewed 5' above and perpendicular to the floor surface. Reflective sheen is the degree of gloss reflected from a surface when viewed at least 20' from and at an angle to the floor surface. These terms are relatively subjective. The final outcome depends on the internal makeup and surface condition of the original concrete floor, the experience of the floor polishing crew, and the expectations of the owner. Before the grinding, honing, and polishing work commences on the main floor area, it might be beneficial to do a mock-up panel in the same floor but in an out of the way place to demonstrate the sequence of steps with increasingly fine abrasive grits and to demonstrate the final reflective clarity and reflective sheen. This mock-up panel will be within the area of, and part of, the final work.

R034105-30 Prestressed Precast Concrete Structural Units

Type	Location	Depth	Span in Ft.		Live Load Lb. per S.F.
Double Tee 8' to 10'	Floor	28" to 34"	60 to 80		50 to 80
	Roof	12" to 24"	30 to 50		40
	Wall	Width 8'	Up to 55' high		Wind
Multiple Tee 8'	Roof	8" to 12"	15 to 40		40
	Floor	8" to 12"	15 to 30		100
Plank or	Roof or Floor		Roof	Floor	40 for Roof
		4"	13	12	
		6"	22	18	
		8"	26	25	
		10"	33	29	100 for Floor
		12"	42	32	
Single Tee 8' to 10'	Roof	28"	40		
		32"	80		
		36"	100		40
		48"	120		
AASHO Girder	Bridges	Type 4	100		
		5	110		Highway
		6	125		
Box Beam 4'	Bridges	15" 27" 33"	40 to 100		Highway

The majority of precast projects today utilize double tees rather than single tees because of speed and ease of installation. As a result casting beds at manufacturing plants are normally formed for double tees. Single tee projects will therefore require an initial set up charge to be spread over the individual single tee costs.

For floors, a 2" to 3" topping is field cast over the shapes. For roofs, insulating concrete or rigid insulation is placed over the shapes.

Member lengths up to 40' are standard haul, 40' to 60' require special permits and lengths over 60' must be escorted. Excessive width and/or length can add up to 100% on hauling costs.

Large heavy members may require two cranes for lifting which would increase erection costs by about 45%. An eight man crew can install 12 to 20 double tees, or 45 to 70 quad tees or planks per day.

Grouting of connections must also be included.

Several system buildings utilizing precast members are available. Heights can go up to 22 stories for apartment buildings. The optimum design ratio is 3 S.F. of surface to 1 S.F. of floor area.

For customer support on your Concrete & Masonry Costs with RSMeans Data, call 800.448.8182.

529

R034136-90   Prestressed Concrete, Post-Tensioned

In post-tensioned concrete the steel tendons are tensioned after the concrete has reached about 3/4 of its ultimate strength. The cableways are grouted after tensioning to provide bond between the steel and concrete. If bond is to be prevented, the tendons are coated with a corrosion-preventative grease and wrapped with waterproofed paper or plastic. Bonded tendons are usually used when ultimate strength (beams & girders) are controlling factors.

High strength concrete is used to fully utilize the steel, thereby reducing the size and weight of the member. A plasticizing agent may be added to reduce water content. Maximum size aggregate ranges from 1/2″ to 1-1/2″ depending on the spacing of the tendons.

The types of steel commonly used are bars and strands. Job conditions determine which is best suited. Bars are best for vertical prestresses since they are easy to support. The trend is for steel manufacturers to supply a finished package, cut to length, which reduces field preparation to a minimum.

Bars vary from 3/4″ to 1-3/8″ diameter. The table below gives time in labor-hours per tendon for placing, tensioning and grouting (if required) a 75′ beam. Tendons used in buildings are not usually grouted; tendons for bridges usually are grouted. For strands the table indicates the labor-hours per pound for typical prestressed units 100′ long. Simple span beams usually require one-end stressing regardless of lengths. Continuous beams are usually stressed from two ends. Long slabs are poured from the center outward and stressed in 75′ increments after the initial 150′ center pour.

Labor Hours per Tendon and per Pound of Prestressed Steel						
Length	100′ Beam		75′ Beam		100′ Slab	
Type Steel	Strand		Bars		Strand	
Diameter	0.5″		3/4″	1-3/8″	0.5″	0.6″
Number	4	12	1	1	1	1
Force in Kips	100	300	42	143	25	35
Preparation & Placing Cables	3.6	7.4	0.9	2.9	0.9	1.1
Stressing Cables	2.0	2.4	0.8	1.6	0.5	0.5
Grouting, if required	2.5	3.0	0.6	1.3		
Total Labor Hours	8.1	12.8	2.3	5.8	1.4	1.6
Prestressing Steel Weights (Lbs.)	215	640	115	380	53	74
Labor-hours per Lb. Bonded	0.038	0.020	0.020	0.015		
Non-bonded					0.026	0.022

Flat Slab construction — 4000 psi concrete with span-to-depth ratio between 36 and 44. Two way post-tensioned steel averages 1.0 lb. per S.F. for 24′ to 28′ bays (usually strand) and additional reinforcing steel averages .5 lb. per S.F.

Pan and Joist construction — 4000 psi concrete with span-to-depth ratio between 28 to 30. Post-tensioned steel averages .8 lb. per S.F. and reinforcing steel about 1.0 lb. per S.F. Placing and stressing average 40 hours per ton of total material.

Beam construction — 4000 to 5000 psi concrete. Steel weights vary greatly.

Labor cost per pound goes down as the size and length of the tendon increase. The primary economic consideration is the cost per kip for the member.

Post-tensioning becomes feasible for beams and girders over 30′ long; for continuous two-way slabs over 20′ clear; and for transferring upper building loads over longer spans at lower levels. Post-tension suppliers will provide engineering services at no cost to the user. Substantial economies are possible by using post-tensioned Lift Slabs.

Concrete   R0345 Precast Architectural Concrete

R034513-10   Precast Concrete Wall Panels

Panels are either solid or insulated with plain, colored or textured finishes. Transportation is an important cost factor. Prices shown in the unit cost section of the data set are based on delivery within 50 miles of a plant including fabricators' overhead and profit. Engineering data is available from fabricators to assist with construction details. Usual minimum job size for economical use of panels is about 5000 S.F. Small jobs can double the prices shown. For large, highly repetitive jobs, deduct up to 15% from the prices shown.

2″ thick panels cost about the same as 3″ thick panels, and maximum panel size is less. For building panels faced with granite, marble or stone, add the material prices from those unit cost sections to the plain panel price shown. There is a growing trend toward aggregate facings and broken rib finishes rather than plain gray concrete panels.

No allowance has been made in the unit cost section for supporting steel framework. On one story buildings, panels may rest on grade beams and require only wind bracing and fasteners. On multi-story buildings panels can span from column to column and floor to floor. Plastic-designed steel-framed structures may have large deflections which slow down erection and raise costs.

Large panels are more economical than small panels on a S.F. basis. When figuring areas include all protrusions, returns, etc. Overhangs can triple erection costs. Panels over 45′ have been produced. Larger flat units should be prestressed. Vacuum lifting of smooth finish panels eliminates inserts and can speed erection.

Concrete   R0347 Site-Cast Concrete

R034713-20   Tilt Up Concrete Panels

The advantage of tilt up construction is in the low cost of forms and the placing of concrete and reinforcing. Panels up to 75′ high and 5-1/2″ thick have been tilted using strongbacks. Tilt up has been used for one to five story buildings and is well-suited for warehouses, stores, offices, schools and residences.

The panels are cast in forms on the floor slab. Most jobs use 5-1/2″ thick solid reinforced concrete panels. Sandwich panels with a layer of insulating materials are also used. Where dampness is a factor, lightweight aggregate is used. Optimum panel size is 300 to 500 S.F.

Slabs are usually poured with 3000 psi concrete which permits tilting seven days after pouring. Slabs may be stacked on top of each other and are separated from each other by either two coats of bond breaker or a film of polyethylene. Use of high early-strength cement allows tilting two days after a pour. Tilting up is done with a roller outrigger crane with a capacity of at least 1-1/2 times the weight of the panel at the required reach. Exterior precast columns can be set at the same time as the panels; interior precast columns can be set first and the panels clipped directly to them. The use of cast-in-place concrete columns is diminishing due to shrinkage problems. Structural steel columns are sometimes used if crane rails are planned. Panels can be clipped to the columns or lowered between the flanges. Steel channels with anchors may be used as edge forms for the slab. When the panels are lifted the channels form an integral steel column to take structural loads. Roof loads can be carried directly by the panels for wall heights to 14′.

Requirements of local building codes may be a limiting factor and should be checked. Building floor slabs should be poured first and should be a minimum of 5″ thick with 100% compaction of soil or 6″ thick with less than 100% compaction.

Setting times as fast as nine minutes per panel have been observed, but a safer expectation would be four panels per hour with a crane and a four-man setting crew. If a crane erects from inside a building, some provision must be made to get the crane out after walls are erected. Good yarding procedure is important to minimize delays. Equalizing three-point lifting beams and self-releasing pick-up hooks speed erection. If panels must be carried to their final location, setting time per panel will be increased and erection costs may approach the erection cost range of architectural precast wall panels. Placing panels into slots formed in continuous footers will speed erection.

Reinforcing should be with #5 bars with vertical bars on the bottom. If surface is to be sandblasted, stainless steel chairs should be used to prevent rust staining.

Use of a broom finish is popular since the unavoidable surface blemishes are concealed.

Precast columns run from three to five times the C.Y. price of the panels only.

Concrete   R0352 Lightweight Concrete Roof Insulation

R035216-10   Lightweight Concrete

Lightweight aggregate concrete is usually purchased ready mixed, but it can also be field mixed.

Vermiculite or Perlite comes in bags of 4 C.F. under various trade names. The weight is about 8 lbs. per C.F. For insulating roof fill use 1:6 mix. For a structural deck use 1:4 mix over gypsum boards, steeltex, steel centering, etc., supported by closely spaced joists or bulb trees. For structural slabs use 1:3:2 vermiculite sand concrete over steeltex, metal lath, steel centering, etc., on joists spaced 2′-0″ O.C. for maximum L.L. of 80 P.S.F. Use same mix

for slab base fill over steel flooring or regular reinforced concrete slab when tile, terrazzo or other finish is to be laid over.

For slabs on grade use 1:3:2 mix when tile, etc., finish is to be laid over. If radiant heating units are installed use a 1:6 mix for a base. After coils are in place, cover with a regular granolithic finish (mix 1:3:2) to a minimum depth of 1-1/2″ over top of units.

Reinforce all slabs with 6 x 6 or 10 x 10 welded wire mesh.

R040130-10   Cleaning Face Brick

On smooth brick a person can clean 70 S.F. an hour; on rough brick 50 S.F. per hour. Use one gallon muriatic acid to 20 gallons of water for 1000

S.F. Do not use acid solution until wall is at least seven days old, but a mild soap solution may be used after two days.

Time has been allowed for clean-up in brick prices.

R040513-10   Cement Mortar (material only)

Type N - 1:1:6 mix by volume. Use everywhere above grade except as noted below. - 1:3 mix using conventional masonry cement which saves handling two separate bagged materials.

Type M - 1:1/4:3 mix by volume, or 1 part cement, 1/4 (10% by wt.) lime, 3 parts sand. Use for heavy loads and where earthquakes or hurricanes may occur. Also for reinforced brick, sewers, manholes and everywhere below grade.

Mix Proportions by Volume and Compressive Strength of Mortar

| Where Used | Allowable Proportions by Volume | | | | | Compressive Strength @ 28 days |
	Mortar Type	Portland Cement	Masonry Cement	Hydrated Lime	Masonry Sand	
Plain Masonry	M	1	1	—	6	
		1	—	1/4	3	2500 psi
	S	1/2	1	—	4	
		1	—	1/4 to 1/2	4	1800 psi
	N	—	1	—	3	
		1	—	1/2 to 1-1/4	6	750 psi
	O	—	1	—	3	
		1	—	1-1/4 to 2-1/2	9	350 psi
	K	1	—	2-1/2 to 4	12	75 psi
Reinforced Masonry	PM	1	1	—	6	2500 psi
	PL	1	—	1/4 to 1/2	4	2500 psi

Note: The total aggregate should be between 2.25 to 3 times the sum of the cement and lime used.

The labor cost to mix the mortar is included in the productivity and labor cost of unit price lines in unit cost sections for brickwork, blockwork and stonework.

The material cost of mixed mortar is included in the material cost of those same unit price lines and includes the cost of renting and operating a 10 C.F. mixer at the rate of 200 C.F. per day.

There are two types of mortar color used. One type is the inert additive type with about 100 lbs. per M brick as the typical quantity required. These colors are also available in smaller-batch-sized bags (1 lb. to 15 lb.) which can be placed directly into the mixer without measuring. The other type is premixed and replaces the masonry cement. Dark green color has the highest cost.

R040519-50   Masonry Reinforcing

Horizontal joint reinforcing helps prevent wall cracks where wall movement may occur and in many locations is required by code. Horizontal joint reinforcing is generally not considered to be structural reinforcing and an unreinforced wall may still contain joint reinforcing.

Reinforcing strips come in 10' and 12' lengths and in truss and ladder shapes, with and without drips. Field labor runs between 2.7 to 5.3 hours per 1000 L.F. for wall thicknesses up to 12".

The wire meets ASTM A82 for cold drawn steel wire and the typical size is 9 ga. sides and ties with 3/16" diameter also available. Typical finish is mill galvanized with zinc coating at .10 oz. per S.F. Class I (.40 oz. per S.F.) and Class III (.80 oz. per S.F.) are also available, as is hot dipped galvanizing at 1.50 oz. per S.F.

R042110-10   Economy in Bricklaying

Have adequate supervision. Be sure bricklayers are always supplied with materials so there is no waiting. Place experienced bricklayers at corners and openings.

Use only screened sand for mortar. Otherwise, labor time will be wasted picking out pebbles. Use seamless metal tubs for mortar as they do not leak or catch the trowel. Locate stack and mortar for easy wheeling.

Have brick delivered for stacking. This makes for faster handling, reduces chipping and breakage, and requires less storage space. Many dealers will deliver select common in 2′ x 3′ x 4′ pallets or face brick packaged. This affords quick handling with a crane or forklift and easy tonging in units of ten, which reduces waste.

Use wider bricks for one wythe wall construction. Keep scaffolding away from the wall to allow mortar to fall clear and not stain the wall.

On large jobs develop specialized crews for each type of masonry unit.

Consider designing for prefabricated panel construction on high rise projects.

Avoid excessive corners or openings. Each opening adds about 50% to the labor cost for area of opening.

Bolting stone panels and using window frames as stops reduce labor costs and speed up erection.

R042110-20   Common and Face Brick

Common building brick manufactured according to ASTM C62 and facing brick manufactured according to ASTM C216 are the two standard bricks available for general building use.

Building brick is made in three grades: SW, where high resistance to damage caused by cyclic freezing is required; MW, where moderate resistance to cyclic freezing is needed; and NW, where little resistance to cyclic freezing is needed. Facing brick is made in only the two grades SW and MW. Additionally, facing brick is available in three types: FBS, for general use; FBX, for general use where a higher degree of precision and lower permissible variation in size than FBS are needed; and FBA, for general use to produce characteristic architectural effects resulting from non-uniformity in size and texture of the units.

In figuring the material cost of brickwork, an allowance of 25% mortar waste and 3% brick breakage was included. If bricks are delivered palletized

with 280 to 300 per pallet, or packaged, allow only 1-1/2% for breakage. Packaged or palletized delivery is practical when a job is big enough to have a crane or other equipment available to handle a package of brick. This is so on all industrial work but not always true on small commercial buildings.

The use of buff and gray face is increasing, and there is a continuing trend to the Norman, Roman, Jumbo and SCR brick.

Common red clay brick for backup is not used that often. Concrete block is the most usual backup material with occasional use of sand lime or cement brick. Building brick is commonly used in solid walls for strength and as a fire stop.

Brick panels built on the ground and then crane erected to the upper floors have proven to be economical. This allows the work to be done under cover and without scaffolding.

R042110-50   Brick, Block & Mortar Quantities

Running Bond					C.F. of Mortar per M Bricks, Waste Included		For Other Bonds Standard Size Add to S.F. Quantities in Table to Left		
Number of Brick per S.F. of Wall - Single Wythe with 3/8″ Joints									
Type Brick	Nominal Size (incl. mortar) L H W			Modular Coursing	Number of Brick per S.F.	3/8″ Joint	1/2″ Joint	Bond Type	Description	Factor
Standard	8 x 2-2/3 x 4			3C=8″	6.75	8.1	10.3	Common	full header every fifth course	+20%
Economy	8 x 4 x 4			1C=4″	4.50	9.1	11.6		full header every sixth course	+16.7%
Engineer	8 x 3-1/5 x 4			5C=16″	5.63	8.5	10.8	English	full header every second course	+50%
Fire	9 x 2-1/2 x 4-1/2			2C=5″	6.40	550 # Fireclay	—	Flemish	alternate headers every course	+33.3%
Jumbo	12 x 4 x 6 or 8			1C=4″	3.00	22.5	29.2		every sixth course	+5.6%
Norman	12 x 2-2/3 x 4			3C=8″	4.50	11.2	14.3	Header = W x H exposed		+100%
Norwegian	12 x 3-1/5 x 4			5C=16″	3.75	11.7	14.9	Rowlock = H x W exposed		+100%
Roman	12 x 2 x 4			2C=4″	6.00	10.7	13.7	Rowlock stretcher = L x W exposed		+33.3%
SCR	12 x 2-2/3 x 6			3C=8″	4.50	21.8	28.0	Soldier = H x L exposed		—
Utility	12 x 4 x 4			1C=4″	3.00	12.3	15.7	Sailor = W x L exposed		-33.3%

Concrete Blocks Nominal Size		Approximate Weight per S.F.		Blocks per 100 S.F.	Mortar per M block, waste included	
		Standard	Lightweight		Partitions	Back up
2″	x 8″ x 16″	20 PSF	15 PSF	113	27 C.F.	36 C.F.
4″		30	20		41	51
6″		42	30		56	66
8″		55	38		72	82
10″		70	47		87	97
12″		85	55		102	112

Brick & Mortar Quantities
©Brick Industry Association. 2009 Feb. Technical Notes on
Brick Construction 10:
    Dimensioning and Estimating Brick Masonry. Reston (VA): BIA. Table 1
    Modular Brick Sizes and Table 4 Quantity Estimates for Brick Masonry.

For customer support on your Concrete & Masonry Costs with RSMeans Data, call 800.448.8182.

533

R042210-20 Concrete Block

The material cost of special block such as corner, jamb and head block can be figured at the same price as ordinary block of equal size. Labor on specials is about the same as equal-sized regular block.

Bond beams and 16″ high lintel blocks are more expensive than regular units of equal size. Lintel blocks are 8″ long and either 8″ or 16″ high.

Use of a motorized mortar spreader box will speed construction of continuous walls.

Hollow non-load-bearing units are made according to ASTM C129 and hollow load-bearing units according to ASTM C90.

R050521-20 Welded Structural Steel

Usual weight reductions with welded design run 10% to 20% compared with bolted or riveted connections. This amounts to about the same total cost compared with bolted structures since field welding is more expensive than bolts. For normal spans of 18′ to 24′ figure 6 to 7 connections per ton.

Trusses — For welded trusses add 4% to weight of main members for connections. Up to 15% less steel can be expected in a welded truss compared to one that is shop bolted. Cost of erection is the same whether shop bolted or welded.

General — Typical electrodes for structural steel welding are E6010, E6011, E60T and E70T. Typical buildings vary between 2# to 8# of weld rod per

ton of steel. Buildings utilizing continuous design require about three times as much welding as conventional welded structures. In estimating field erection by welding, it is best to use the average linear feet of weld per ton to arrive at the welding cost per ton. The type, size and position of the weld will have a direct bearing on the cost per linear foot. A typical field welder will deposit 1.8# to 2# of weld rod per hour manually. Using semiautomatic methods can increase production by as much as 50% to 75%.

R051223-10 Structural Steel

The bare material prices for structural steel, shown in the unit cost sections of the data set, are for 100 tons of shop-fabricated structural steel and include:
1. Mill base price of structural steel
2. Mill scrap/grade/size/length extras
3. Mill delivery to a metals service center (warehouse)
4. Service center storage and handling
5. Service center delivery to a fabrication shop
6. Shop storage and handling
7. Shop drafting/detailing
8. Shop fabrication
9. Shop coat of primer paint
10. Shop listing
11. Shop delivery to the job site

In unit cost sections of the data set that contain items for field fabrication of steel components, the bare material cost of steel includes:
1. Mill base price of structural steel
2. Mill scrap/grade/size/length extras
3. Mill delivery to a metals service center (warehouse)
4. Service center storage and handling
5. Service center delivery to the job site

R051223-35 Common Steel Sections

The upper portion of this table shows the name, shape, common designation and basic characteristics of commonly used steel sections. The lower portion explains how to read the designations used for the above illustrated common sections.

Shape & Designation	Name & Characteristics	Shape & Designation	Name & Characteristics
W	W Shape Parallel flange surfaces	MC	Miscellaneous Channel Infrequently rolled by some producers
S	American Standard Beam (I Beam) Sloped inner flange	L	Angle Equal or unequal legs, constant thickness
M	Miscellaneous Beams Cannot be classified as W, HP or S; infrequently rolled by some producers	T	Structural Tee Cut from W, M or S on center of web
C	American Standard Channel Sloped inner flange	HP	Bearing Pile Parallel flanges and equal flange and web thickness

Common drawing designations follow:

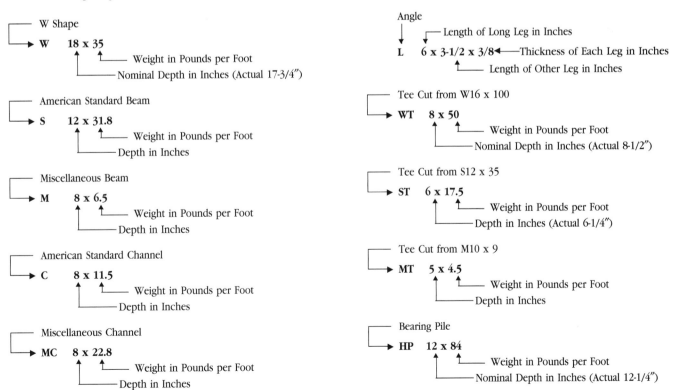

W Shape
W   18 x 35
— Weight in Pounds per Foot
— Nominal Depth in Inches (Actual 17-3/4")

American Standard Beam
S   12 x 31.8
— Weight in Pounds per Foot
— Depth in Inches

Miscellaneous Beam
M   8 x 6.5
— Weight in Pounds per Foot
— Depth in Inches

American Standard Channel
C   8 x 11.5
— Weight in Pounds per Foot
— Depth in Inches

Miscellaneous Channel
MC   8 x 22.8
— Weight in Pounds per Foot
— Depth in Inches

Angle
L   6 x 3-1/2 x 3/8 ◄—Thickness of Each Leg in Inches
— Length of Long Leg in Inches
— Length of Other Leg in Inches

Tee Cut from W16 x 100
WT   8 x 50
— Weight in Pounds per Foot
— Nominal Depth in Inches (Actual 8-1/2")

Tee Cut from S12 x 35
ST   6 x 17.5
— Weight in Pounds per Foot
— Depth in Inches (Actual 6-1/4")

Tee Cut from M10 x 9
MT   5 x 4.5
— Weight in Pounds per Foot
— Depth in Inches

Bearing Pile
HP   12 x 84
— Weight in Pounds per Foot
— Nominal Depth in Inches (Actual 12-1/4")

For customer support on your Concrete & Masonry Costs with RSMeans Data, call 800.448.8182.

535

R051223-45   Installation Time for Structural Steel Building Components

The following tables show the expected average installation times for various structural steel shapes. Table A presents installation times for columns, Table B for beams, Table C for light framing and bolts, and Table D for structural steel for various project types.

Table A		
Description	**Labor-Hours**	**Unit**
Columns		
Steel, Concrete Filled		
3-1/2" Diameter	.933	Ea.
6-5/8" Diameter	1.120	Ea.
Steel Pipe		
3" Diameter	.933	Ea.
8" Diameter	1.120	Ea.
12" Diameter	1.244	Ea.
Structural Tubing		
4" x 4"	.966	Ea.
8" x 8"	1.120	Ea.
12" x 8"	1.167	Ea.
W Shape 2 Tier		
W8 x 31	.052	L.F.
W8 x 67	.057	L.F.
W10 x 45	.054	L.F.
W10 x 112	.058	L.F.
W12 x 50	.054	L.F.
W12 x 190	.061	L.F.
W14 x 74	.057	L.F.
W14 x 176	.061	L.F.

Table B				
Description	**Labor-Hours**	**Unit**	**Labor-Hours**	**Unit**
Beams, W Shape				
W6 x 9	.949	Ea.	.093	L.F.
W10 x 22	1.037	Ea.	.085	L.F.
W12 x 26	1.037	Ea.	.064	L.F.
W14 x 34	1.333	Ea.	.069	L.F.
W16 x 31	1.333	Ea.	.062	L.F.
W18 x 50	2.162	Ea.	.088	L.F.
W21 x 62	2.222	Ea.	.077	L.F.
W24 x 76	2.353	Ea.	.072	L.F.
W27 x 94	2.581	Ea.	.067	L.F.
W30 x 108	2.857	Ea.	.067	L.F.
W33 x 130	3.200	Ea.	.071	L.F.
W36 x 300	3.810	Ea.	.077	L.F.

Table C		
Description	**Labor-Hours**	**Unit**
Light Framing		
Angles 4" and Larger	.055	lbs.
Less than 4"	.091	lbs.
Channels 8" and Larger	.048	lbs.
Less than 8"	.072	lbs.
Cross Bracing Angles	.055	lbs.
Rods	.034	lbs.
Hanging Lintels	.069	lbs.
High Strength Bolts in Place		
3/4" Bolts	.070	Ea.
7/8" Bolts	.076	Ea.

Table D				
Description	**Labor-Hours**	**Unit**	**Labor-Hours**	**Unit**
Apartments, Nursing Homes, etc.				
1-2 Stories	4.211	Piece	7.767	Ton
3-6 Stories	4.444	Piece	7.921	Ton
7-15 Stories	4.923	Piece	9.014	Ton
Over 15 Stories	5.333	Piece	9.209	Ton
Offices, Hospitals, etc.				
1-2 Stories	4.211	Piece	7.767	Ton
3-6 Stories	4.741	Piece	8.889	Ton
7-15 Stories	4.923	Piece	9.014	Ton
Over 15 Stories	5.120	Piece	9.209	Ton
Industrial Buildings				
1 Story	3.478	Piece	6.202	Ton

R051223-50   Subpurlins

Bulb tee subpurlins are structural members designed to support and reinforce a variety of roof deck systems such as precast cement fiber roof deck tiles, monolithic roof deck systems, and gypsum or lightweight concrete over formboard. Other uses include interstitial service ceiling systems, wall panel systems, and joist anchoring in bond beams. See the Unit Price section for pricing on a square foot basis at 32-5/8" O.C. Maximum span is based on a 3-span condition with a total allowable vertical load of 40 psf.

R051223-80 Dimensions and Weights of Sheet Steel

Gauge No.	Approximate Thickness					Weight		
	Inches (in fractions)	Inches (in decimal parts)		Millimeters				per Square Meter in Kg.
	Wrought Iron	Wrought Iron	Steel	Steel		per S.F. in Ounces	per S.F. in Lbs.	
0000000	1/2"	.5	.4782	12.146		320	20.000	97.650
000000	15/32"	.46875	.4484	11.389		300	18.750	91.550
00000	7/16"	.4375	.4185	10.630		280	17.500	85.440
0000	13/32"	.40625	.3886	9.870		260	16.250	79.330
000	3/8"	.375	.3587	9.111		240	15.000	73.240
00	11/32"	.34375	.3288	8.352		220	13.750	67.130
0	5/16"	.3125	.2989	7.592		200	12.500	61.030
1	9/32"	.28125	.2690	6.833		180	11.250	54.930
2	17/64"	.265625	.2541	6.454		170	10.625	51.880
3	1/4"	.25	.2391	6.073		160	10.000	48.820
4	15/64"	.234375	.2242	5.695		150	9.375	45.770
5	7/32"	.21875	.2092	5.314		140	8.750	42.720
6	13/64"	.203125	.1943	4.935		130	8.125	39.670
7	3/16"	.1875	.1793	4.554		120	7.500	36.320
8	11/64"	.171875	.1644	4.176		110	6.875	33.570
9	5/32"	.15625	.1495	3.797		100	6.250	30.520
10	9/64"	.140625	.1345	3.416		90	5.625	27.460
11	1/8"	.125	.1196	3.038		80	5.000	24.410
12	7/64"	.109375	.1046	2.657		70	4.375	21.360
13	3/32"	.09375	.0897	2.278		60	3.750	18.310
14	5/64"	.078125	.0747	1.897		50	3.125	15.260
15	9/128"	.0713125	.0673	1.709		45	2.813	13.730
16	1/16"	.0625	.0598	1.519		40	2.500	12.210
17	9/160"	.05625	.0538	1.367		36	2.250	10.990
18	1/20"	.05	.0478	1.214		32	2.000	9.765
19	7/160"	.04375	.0418	1.062		28	1.750	8.544
20	3/80"	.0375	.0359	.912		24	1.500	7.324
21	11/320"	.034375	.0329	.836		22	1.375	6.713
22	1/32"	.03125	.0299	.759		20	1.250	6.103
23	9/320"	.028125	.0269	.683		18	1.125	5.490
24	1/40"	.025	.0239	.607		16	1.000	4.882
25	7/320"	.021875	.0209	.531		14	.875	4.272
26	3/160"	.01875	.0179	.455		12	.750	3.662
27	11/640"	.0171875	.0164	.417		11	.688	3.357
28	1/64"	.015625	.0149	.378		10	.625	3.052

For customer support on your Concrete & Masonry Costs with RSMeans Data, call 800.448.8182.

537

R053100-10 Decking Descriptions

General - All Deck Products

A steel deck is made by cold forming structural grade sheet steel into a repeating pattern of parallel ribs. The strength and stiffness of the panels are the result of the ribs and the material properties of the steel. The deck lengths can be varied to suit job conditions, but because of shipping considerations, are usually less than 40 feet. Standard deck width varies with the product used but full sheets are usually 12″, 18″, 24″, 30″, or 36″. Deck is typically furnished in a standard width with the ends cut square. Any cutting for width, such as at openings or for angular fit, is done at the job site.

The deck is typically attached to the building frame with arc puddle welds, self-drilling screws, or powder or pneumatically driven pins. Sheet to sheet fastening is done with screws, button punching (crimping), or welds.

Composite Floor Deck

After installation and adequate fastening, a floor deck serves several purposes. It (a) acts as a working platform, (b) stabilizes the frame, (c) serves as a concrete form for the slab, and (d) reinforces the slab to carry the design loads applied during the life of the building. Composite decks are distinguished by the presence of shear connector devices as part of the deck. These devices are designed to mechanically lock the concrete and deck together so that the concrete and the deck work together to carry subsequent floor loads. These shear connector devices can be rolled-in embossments, lugs, holes, or wires welded to the panels. The deck profile can also be used to interlock concrete and steel.

Composite deck finishes are either galvanized (zinc coated) or phosphatized/painted. Galvanized deck has a zinc coating on both the top and bottom surfaces. The phosphatized/painted deck has a bare (phosphatized) top surface that will come into contact with the concrete. This bare top surface can be expected to develop rust before the concrete is placed. The bottom side of the deck has a primer coat of paint.

A composite floor deck is normally installed so the panel ends do not overlap on the supporting beams. Shear lugs or panel profile shapes often prevent a tight metal to metal fit if the panel ends overlap; the air gap caused by overlapping will prevent proper fusion with the structural steel supports when the panel end laps are shear stud welded.

Adequate end bearing of the deck must be obtained as shown on the drawings. If bearing is actually less in the field than shown on the drawings, further investigation is required.

Roof Deck

A roof deck is not designed to act compositely with other materials. A roof deck acts alone in transferring horizontal and vertical loads into the building frame. Roof deck rib openings are usually narrower than floor deck rib openings. This provides adequate support of the rigid thermal insulation board.

A roof deck is typically installed to endlap approximately 2″ over supports. However, it can be butted (or lapped more than 2″) to solve field fit problems. Since designers frequently use the installed deck system as part of the horizontal bracing system (the deck as a diaphragm), any fastening substitution or change should be approved by the designer. Continuous perimeter support of the deck is necessary to limit edge deflection in the finished roof and may be required for diaphragm shear transfer.

Standard roof deck finishes are galvanized or primer painted. The standard factory applied paint for roof decks is a primer paint and is not intended to weather for extended periods of time. Field painting or touching up of abrasions and deterioration of the primer coat or other protective finishes is the responsibility of the contractor.

Cellular Deck

A cellular deck is made by attaching a bottom steel sheet to a roof deck or composite floor deck panel. A cellular deck can be used in the same manner as a floor deck. Electrical, telephone, and data wires are easily run through the chase created between the deck panel and the bottom sheet.

When used as part of the electrical distribution system, the cellular deck must be installed so that the ribs line up and create a smooth cell transition at abutting ends. The joint that occurs at butting cell ends must be taped or otherwise sealed to prevent wet concrete from seeping into the cell. Cell interiors must be free of welding burrs, or other sharp intrusions, to prevent damage to wires.

When used as a roof deck, the bottom flat plate is usually left exposed to view. Care must be maintained during erection to keep good alignment and prevent damage.

A cellular deck is sometimes used with the flat plate on the top side to provide a flat working surface. Installation of the deck for this purpose requires special methods for attachment to the frame because the flat plate, now on the top, can prevent direct access to the deck material that is bearing on the structural steel. It may be advisable to treat the flat top surface to prevent slipping.

A cellular deck is always furnished galvanized or painted over galvanized.

Form Deck

A form deck can be any floor or roof deck product used as a concrete form. Connections to the frame are by the same methods used to anchor floor and roof decks. Welding washers are recommended when welding a deck that is less than 20 gauge thickness.

A form deck is furnished galvanized, prime painted, or uncoated. A galvanized deck must be used for those roof deck systems where a form deck is used to carry a lightweight insulating concrete fill.

R061110-30   Lumber Product Material Prices

The price of forest products fluctuates widely from location to location and from season to season depending upon economic conditions. The bare material prices in the unit cost sections of the data set show the National Average material prices in effect Jan. 1 of this data year. It must be noted that lumber prices in general may change significantly during the year.

Availability of certain items depends upon geographic location and must be checked prior to firm-price bidding.

R061636-20   Plywood

There are two types of plywood used in construction: interior, which is moisture-resistant but not waterproofed, and exterior, which is waterproofed.

The grade of the exterior surface of the plywood sheets is designated by the first letter: A, for smooth surface with patches allowed; B, for solid surface with patches and plugs allowed; C, which may be surface plugged or may have knot holes up to 1″ wide; and D, which is used only for interior type plywood and may have knot holes up to 2-1/2″ wide. "Structural Grade" is specifically designed for engineered applications such as box beams. All CC & DD grades have roof and floor spans marked on them.

Underlayment-grade plywood runs from 1/4″ to 1-1/4″ thick. Thicknesses 5/8″ and over have optional tongue and groove joints which eliminate the need for blocking the edges. Underlayment 19/32″ and over may be referred to as Sturd-i-Floor.

The price of plywood can fluctuate widely due to geographic and economic conditions.

Typical uses for various plywood grades are as follows:

AA-AD Interior — cupboards, shelving, paneling, furniture

BB Plyform — concrete form plywood

CDX — wall and roof sheathing

Structural — box beams, girders, stressed skin panels

AA-AC Exterior — fences, signs, siding, soffits, etc.

Underlayment — base for resilient floor coverings

Overlaid HDO — high density for concrete forms & highway signs

Overlaid MDO — medium density for painting, siding, soffits & signs

303 Siding — exterior siding, textured, striated, embossed, etc.

R092000-50 Lath, Plaster and Gypsum Board

Gypsum board lath is available in 3/8″ thick x 16″ wide x 4′ long sheets as a base material for multi-layer plaster applications. It is also available as a base for either multi-layer or veneer plaster applications in 1/2″ and 5/8″ thick–4′ wide x 8′, 10′ or 12′ long sheets. Fasteners are screws or blued ring shank nails for wood framing and screws for metal framing.

Metal lath is available in diamond mesh patterns with flat or self-furring profiles. Paper backing is available for applications where excessive plaster waste needs to be avoided. A slotted mesh ribbed lath should be used in areas where the span between structural supports is greater than normal. Most metal lath comes in 27″ x 96″ sheets. Diamond mesh weighs 1.75, 2.5 or 3.4 pounds per square yard, slotted mesh lath weighs 2.75 or 3.4 pounds per square yard. Metal lath can be nailed, screwed or tied in place.

Many **accessories** are available. Corner beads, flat reinforcing strips, casing beads, control and expansion joints, furring brackets and channels are some examples. Note that accessories are not included in plaster or stucco line items.

Plaster is defined as a material or combination of materials that when mixed with a suitable amount of water, forms a plastic mass or paste. When applied to a surface, the paste adheres to it and subsequently hardens, preserving in a rigid state the form or texture imposed during the period of elasticity.

Gypsum plaster is made from ground calcined gypsum. It is mixed with aggregates and water for use as a base coat plaster.

Vermiculite plaster is a fire-retardant plaster covering used on steel beams, concrete slabs and other heavy construction materials. Vermiculite is a group name for certain clay minerals, hydrous silicates or aluminum, magnesium and iron that have been expanded by heat.

Perlite plaster is a plaster using perlite as an aggregate instead of sand. Perlite is a volcanic glass that has been expanded by heat.

Gauging plaster is a mix of gypsum plaster and lime putty that when applied produces a quick drying finish coat.

Veneer plaster is a one or two component gypsum plaster used as a thin finish coat over special gypsum board.

Keenes cement is a white cementitious material manufactured from gypsum that has been burned at a high temperature and ground to a fine powder. Alum is added to accelerate the set. The resulting plaster is hard and strong and accepts and maintains a high polish, hence it is used as a finishing plaster.

Stucco is a Portland cement based plaster used primarily as an exterior finish.

Plaster is used on both interior and exterior surfaces. Generally it is applied in multiple-coat systems. A three-coat system uses the terms scratch, brown and finish to identify each coat. A two-coat system uses base and finish to describe each coat. Each type of plaster and application system has attributes that are chosen by the designer to best fit the intended use.

Gypsum Plaster Quantities for 100 S.Y.	2 Coat, 5/8″ Thick		3 Coat, 3/4″ Thick		
	Base	Finish	Scratch	Brown	Finish
	1:3 Mix	2:1 Mix	1:2 Mix	1:3 Mix	2:1 Mix
Gypsum plaster	1,300 lb.		1,350 lb.	650 lb.	
Sand	1.75 C.Y.		1.85 C.Y.	1.35 C.Y.	
Finish hydrated lime		340 lb.			340 lb.
Gauging plaster		170 lb.			170 lb.

Vermiculite or Perlite Plaster Quantities for 100 S.Y.	2 Coat, 5/8″ Thick		3 Coat, 3/4″ Thick		
	Base	Finish	Scratch	Brown	Finish
Gypsum plaster	1,250 lb.		1,450 lb.	800 lb.	
Vermiculite or perlite	7.8 bags		8.0 bags	3.3 bags	
Finish hydrated lime		340 lb.			340 lb.
Gauging plaster		170 lb.			170 lb.

Stucco–Three-Coat System Quantities for 100 S.Y.	On Wood Frame	On Masonry
Portland cement	29 bags	21 bags
Sand	2.6 C.Y.	2.0 C.Y.
Hydrated lime	180 lb.	120 lb.

Finishes | R0966 Terrazzo Flooring

R096613-10   Terrazzo Floor

The table below lists quantities required for 100 S.F. of 5/8″ terrazzo topping, either bonded or not bonded.

Description	Bonded to Concrete 1-1/8″ Bed, 1:4 Mix	Not Bonded 2-1/8″ Bed and 1/4″ Sand
Portland cement, 94 lb. Bag	6 bags	8 bags
Sand	10 C.F.	20 C.F.
Divider strips, 4′ squares	50 L.F.	50 L.F.
Terrazzo fill, 50 lb. Bag	12 bags	12 bags
15 Lb. tarred felt		1 C.S.F.
Mesh 2 x 2 #14 galvanized		1 C.S.F.
Crew J-3	0.77 days	0.87 days

2′ x 2′ panels require 1.00 L.F. divider strip per S.F.

3′ x 3′ panels require 0.67 L.F. divider strip per S.F.

4′ x 4′ panels require 0.50 L.F. divider strip per S.F.

5′ x 5′ panels require 0.40 L.F. divider strip per S.F.

6′ x 6′ panels require 0.33 L.F. divider strip per S.F.

Special Construction | R1311 Swimming Pools

R131113-20   Swimming Pools

Pool prices given per square foot of surface area include pool structure, filter and chlorination equipment, pumps, related piping, ladders/steps, maintenance kit, skimmer and vacuum system. Decks and electrical service to equipment are not included.

Residential in-ground pool construction can be divided into two categories: vinyl lined and gunite. Vinyl lined pool walls are constructed of different materials including wood, concrete, plastic or metal. The bottom is often graded with sand over which the vinyl liner is installed. Vermiculite or soil cement bottoms may be substituted for an added cost.

Gunite pool construction is used both in residential and municipal installations. These structures are steel reinforced for strength and finished with a white cement limestone plaster.

Municipal pools will have a higher cost because plumbing codes require more expensive materials, chlorination equipment and higher filtration rates.

Municipal pools greater than 1,800 S.F. require gutter systems to control waves. This gutter may be formed into the concrete wall. Often a vinyl/stainless steel gutter or gutter/wall system is specified, which will raise the pool cost.

Competition pools usually require tile bottoms and sides with contrasting lane striping, which will also raise the pool cost.

For customer support on your Concrete & Masonry Costs with RSMeans Data, call 800.448.8182.

541

R312316-40 Excavating

The selection of equipment used for structural excavation and bulk excavation or for grading is determined by the following factors.

1. Quantity of material
2. Type of material
3. Depth or height of cut
4. Length of haul
5. Condition of haul road
6. Accessibility of site
7. Moisture content and dewatering requirements
8. Availability of excavating and hauling equipment

Some additional costs must be allowed for hand trimming the sides and bottom of concrete pours and other excavation below the general excavation.

Number of B.C.Y. per truck = 1.5 C.Y. bucket x 8 passes = 12 loose C.Y.

$$= 12 \text{ x } \frac{100}{118} = 10.2 \text{ B.C.Y. per truck}$$

Truck Haul Cycle:

Load truck, 8 passes	=	4 minutes
Haul distance, 1 mile	=	9 minutes
Dump time	=	2 minutes
Return, 1 mile	=	7 minutes
Spot under machine	=	1 minute
		23 minute cycle

Add the mobilization and demobilization costs to the total excavation costs. When equipment is rented for more than three days, there is often no mobilization charge by the equipment dealer. On larger jobs outside of urban areas, scrapers can move earth economically provided a dump site or fill area and adequate haul roads are available. Excavation within sheeting bracing or cofferdam bracing is usually done with a clamshell and production

When planning excavation and fill, the following should also be considered.

1. Swell factor
2. Compaction factor
3. Moisture content
4. Density requirements

A typical example for scheduling and estimating the cost of excavation of a 15′ deep basement on a dry site when the material must be hauled off the site is outlined below.

Assumptions:

1. Swell factor, 18%
2. No mobilization or demobilization
3. Allowance included for idle time and moving on job
4. No dewatering, sheeting, or bracing
5. No truck spotter or hand trimming

Fleet Haul Production per day in B.C.Y.

$$4 \text{ trucks x } \frac{50 \text{ min. hour}}{23 \text{ min. haul cycle}} \text{ x 8 hrs. x 10.2 B.C.Y.}$$

$$= 4 \text{ x 2.2 x 8 x 10.2} = 718 \text{ B.C.Y./day}$$

is low, since the clamshell may have to be guided by hand between the bracing. When excavating or filling an area enclosed with a wellpoint system, add 10% to 15% to the cost to allow for restricted access. When estimating earth excavation quantities for structures, allow work space outside the building footprint for construction of the foundation and a slope of 1:1 unless sheeting is used.

R312316-45 Excavating Equipment

The table below lists theoretical hourly production in C.Y./hr. bank measure for some typical excavation equipment. Figures assume 50 minute hours, 83% job efficiency, 100% operator efficiency, 90° swing and properly sized hauling units, which must be modified for adverse digging and loading conditions. Actual production costs in the front of the data set average about 50% of the theoretical values listed here.

Equipment	Soil Type	B.C.Y. Weight	% Swell	1 C.Y.	1-1/2 C.Y.	2 C.Y.	2-1/2 C.Y.	3 C.Y.	3-1/2 C.Y.	4 C.Y.
Hydraulic Excavator	Moist loam, sandy clay	3400 lb.	40%	85	125	175	220	275	330	380
"Backhoe"	Sand and gravel	3100	18	80	120	160	205	260	310	365
15' Deep Cut	Common earth	2800	30	70	105	150	190	240	280	330
	Clay, hard, dense	3000	33	65	100	130	170	210	255	300
	Moist loam, sandy clay	3400	40	170 (6.0)	245 (7.0)	295 (7.8)	335 (8.4)	385 (8.8)	435 (9.1)	475 (9.4)
Power Shovel	Sand and gravel	3100	18	165 (6.0)	225 (7.0)	275 (7.8)	325 (8.4)	375 (8.8)	420 (9.1)	460 (9.4)
Optimum Cut (Ft.)	Common earth	2800	30	145 (7.8)	200 (9.2)	250 (10.2)	295 (11.2)	335 (12.1)	375 (13.0)	425 (13.8)
	Clay, hard, dense	3000	33	120 (9.0)	175 (10.7)	220 (12.2)	255 (13.3)	300 (14.2)	335 (15.1)	375 (16.0)
	Moist loam, sandy clay	3400	40	130 (6.6)	180 (7.4)	220 (8.0)	250 (8.5)	290 (9.0)	325 (9.5)	385 (10.0)
Drag Line	Sand and gravel	3100	18	130 (6.6)	175 (7.4)	210 (8.0)	245 (8.5)	280 (9.0)	315 (9.5)	375 (10.0)
Optimum Cut (Ft.)	Common earth	2800	30	110 (8.0)	160 (9.0)	190 (9.9)	220 (10.5)	250 (11.0)	280 (11.5)	310 (12.0)
	Clay, hard, dense	3000	33	90 (9.3)	130 (10.7)	160 (11.8)	190 (12.3)	225 (12.8)	250 (13.3)	280 (12.0)

Equipment	Soil Type	B.C.Y. Weight	% Swell	Wheel Loaders				Track Loaders		
				3 C.Y.	4 C.Y.	6 C.Y.	8 C.Y.	2-1/4 C.Y.	3 C.Y.	4 C.Y.
	Moist loam, sandy clay	3400	40	260	340	510	690	135	180	250
Loading Tractors	Sand and gravel	3100	18	245	320	480	650	130	170	235
	Common earth	2800	30	230	300	460	620	120	155	220
	Clay, hard, dense	3000	33	200	270	415	560	110	145	200
	Rock, well-blasted	4000	50	180	245	380	520	100	130	180

For customer support on your Concrete & Masonry Costs with RSMeans Data, call 800.448.8182.

543

R312319-90 Wellpoints

A single stage wellpoint system is usually limited to dewatering an average 15' depth below normal ground water level. Multi-stage systems are employed for greater depth with the pumping equipment installed only at the lowest header level. Ejectors with unlimited lift capacity can be economical when two or more stages of wellpoints can be replaced or when horizontal clearance is restricted, such as in deep trenches or tunneling projects, and where low water flows are expected. Wellpoints are usually spaced on 2-1/2' to 10' centers along a header pipe. Wellpoint spacing, header size, and pump size are all determined by the expected flow as dictated by soil conditions.

In almost all soils encountered in wellpoint dewatering, the wellpoints may be jetted into place. Cemented soils and stiff clays may require sand wicks about 12" in diameter around each wellpoint to increase efficiency and eliminate weeping into the excavation. These sand wicks require 1/2 to 3 C.Y. of washed filter sand and are installed by using a 12" diameter steel casing and hole puncher jetted into the ground 2' deeper than the wellpoint. Rock may require predrilled holes.

Labor required for the complete installation and removal of a single stage wellpoint system is in the range of 3/4 to 2 labor-hours per linear foot of header, depending upon jetting conditions, wellpoint spacing, etc.

Continuous pumping is necessary except in some free draining soil where temporary flooding is permissible (as in trenches which are backfilled after each day's work). Good practice requires provision of a stand-by pump during the continuous pumping operation.

Systems for continuous trenching below the water table should be installed three to four times the length of expected daily progress to ensure uninterrupted digging, and header pipe size should not be changed during the job.

For pervious free draining soils, deep wells in place of wellpoints may be economical because of lower installation and maintenance costs. Daily production ranges between two to three wells per day, for 25' to 40' depths, to one well per day for depths over 50'.

Detailed analysis and estimating for any dewatering problem is available at no cost from wellpoint manufacturers. Major firms will quote "sufficient equipment" quotes or their affiliates will offer lump sum proposals to cover complete dewatering responsibility.

Description for 200' System with 8" Header		Quantities
Equipment & Material	Wellpoints 25' long, 2" diameter @ 5' O.C.	40 Each
	Header pipe, 8" diameter	200 L.F.
	Discharge pipe, 8" diameter	100 L.F.
	8" valves	3 Each
	Combination jetting & wellpoint pump (standby)	1 Each
	Wellpoint pump, 8" diameter	1 Each
	Transportation to and from site	1 Day
	Fuel for 30 days x 60 gal./day	1800 Gallons
	Lubricants for 30 days x 16 lbs./day	480 Lbs.
	Sand for points	40 C.Y.
Labor	Technician to supervise installation	1 Week
	Labor for installation and removal of system	300 Labor-hours
	4 Operators straight time 40 hrs./wk. for 4.33 wks.	693 Hrs.
	4 Operators overtime 2 hrs./wk. for 4.33 wks.	35 Hrs.

R312323-30 Compacting Backfill

Compaction of fill in embankments, around structures, in trenches, and under slabs is important to control settlement. Factors affecting compaction are:

1. Soil gradation
2. Moisture content
3. Equipment used
4. Depth of fill per lift
5. Density required

Production Rate:

$$\frac{1.75' \text{ plate width x 50 F.P.M. x 50 min./hr. x .67' lift}}{27 \text{ C.F. per C.Y.}} = 108.5 \text{ C.Y./hr.}$$

Production Rate for 4 Passes:

$$\frac{108.5 \text{ C.Y.}}{4 \text{ passes}} = 27.125 \text{ C.Y./hr. x 8 hrs.} = 217 \text{ C.Y./day}$$

Example:

Compact granular fill around a building foundation using a 21" wide x 24" vibratory plate in 8" lifts. Operator moves at 50 F.P.M. working a 50 minute hour to develop 95% Modified Proctor Density with 4 passes.

Earthwork   R3141 Shoring

R314116-40   Wood Sheet Piling

Wood sheet piling may be used for depths to 20' where there is no ground water. If moderate ground water is encountered Tongue & Groove

sheeting will help to keep it out. When considerable ground water is present, steel sheeting must be used.

For estimating purposes on trench excavation, sizes are as follows:

Depth	Sheeting	Wales	Braces	B.F. per S.F.
To 8'	3 x 12's	6 x 8's, 2 line	6 x 8's, @ 10'	4.0 @ 8'
8' x 12'	3 x 12's	10 x 10's, 2 line	10 x 10's, @ 9'	5.0 average
12' to 20'	3 x 12's	12 x 12's, 3 line	12 x 12's, @ 8'	7.0 average

Sheeting to be toed in at least 2' depending upon soil conditions. A five person crew with an air compressor and sheeting driver can drive and brace 440 SF/day at 8' deep, 360 SF/day at 12' deep, and 320 SF/day at 16' deep.

For normal soils, piling can be pulled in 1/3 the time to install. Pulling difficulty increases with the time in the ground. Production can be increased by high pressure jetting.

R314116-45   Steel Sheet Piling

Limiting weights are 22 to 38#/S.F. of wall surface with 27#/S.F. average for usual types and sizes. (Weights of piles themselves are from 30.7#/L.F. to 57#/L.F. but they are 15" to 21" wide.) Lightweight sections 12" to 28" wide from 3 ga. to 12 ga. thick are also available for shallow excavations. Piles may be driven two at a time with an impact or vibratory hammer (use vibratory to pull) hung from a crane without leads. A reasonable estimate of the life of steel sheet piling is 10 uses with up to 125 uses possible if a vibratory hammer is used. Used piling costs from 50% to 80% of new piling depending on location and market conditions. Sheet piling and H piles

can be rented for about 30% of the delivered mill price for the first month and 5% per month thereafter. Allow 1 labor-hour per pile for cleaning and trimming after driving. These costs increase with depth and hydrostatic head. Vibratory drivers are faster in wet granular soils and are excellent for pile extraction. Pulling difficulty increases with the time in the ground and may cost more than driving. It is often economical to abandon the sheet piling, especially if it can be used as the outer wall form. Allow about 1/3 additional length or more for toeing into ground. Add bracing, waler and strut costs. Waler costs can equal the cost per ton of sheeting.

Earthwork   R3145 Vibroflotation & Densification

R314513-90   Vibroflotation and Vibro Replacement Soil Compaction

Vibroflotation is a proprietary system of compacting sandy soils in place to increase relative density to about 70%. Typical bearing capacities attained will be 6000 psf for saturated sand and 12,000 psf for dry sand. Usual range is 4000 to 8000 psf capacity. Costs in the front of the data set are for a vertical foot of compacted cylinder 6' to 10' in diameter.

Vibro replacement is a proprietary system of improving cohesive soils in place to increase bearing capacity. Most silts and clays above or below the water table can be strengthened by installation of stone columns.

The process consists of radial displacement of the soil by vibration. The created hole is then backfilled in stages with coarse granular fill which is thoroughly compacted and displaced into the surrounding soil in the form of a column.

The total project cost would depend on the number and depth of the compacted cylinders. The installing company guarantees relative soil density of the sand cylinders after compaction and the bearing capacity of the soil after the replacement process. Detailed estimating information is available from the installer at no cost.

R316326-60   Caissons

The three principal types of caissons are:

(1) Belled Caissons, which except for shallow depths and poor soil conditions, are generally recommended. They provide more bearing than shaft area. Because of its conical shape, no horizontal reinforcement of the bell is required.

(2) Straight Shaft Caissons are used where relatively light loads are to be supported by caissons that rest on high value bearing strata. While the shaft is larger in diameter than for belled types this is more than offset by the saving in time and labor.

(3) Keyed Caissons are used when extremely heavy loads are to be carried. A keyed or socketed caisson transfers its load into rock by a combination of end-bearing and shear reinforcing of the shaft. The most economical shaft often consists of a steel casing, a steel wide flange core and concrete. Allowable compressive stresses of .225 f'c for concrete, 16,000 psi for the wide flange core, and 9,000 psi for the steel casing are commonly used. The usual range of shaft diameter is 18″ to 84″. The number of sizes specified for any one project should be limited due to the problems of casing and auger storage. When hand work is to be performed, shaft diameters should not be less than 32″. When inspection of borings is required a minimum shaft diameter of 30″ is recommended. Concrete caissons are intended to be poured against earth excavation so permanent forms, which add to cost, should not be used if the excavation is clean and the earth is sufficiently impervious to prevent excessive loss of concrete.

Soil Conditions for Belling		
Good	**Requires Handwork**	**Not Recommended**
Clay	Hard Shale	Silt
Sandy Clay	Limestone	Sand
Silty Clay	Sandstone	Gravel
Clayey Silt	Weathered Mica	Igneous Rock
Hard-pan		
Soft Shale		
Decomposed Rock		

R347216-10   Single Track R.R. Siding

The costs for a single track RR siding in the Unit Price section include the components shown in the table below.

Description of Component	Qty. per L.F. of Track	Unit
Ballast, 1-1/2″ crushed stone	.667	C.Y.
6″ x 8″ x 8′-6″ Treated timber ties, 22″ O.C.	.545	Ea.
Tie plates, 2 per tie	1.091	Ea.
Track rail	2.000	L.F.
Spikes, 6″, 4 per tie	2.182	Ea.
Splice bars w/ bolts, lock washers & nuts, @ 33′ O.C.	.061	Pair
Crew B-14 @ 57 L.F./Day	.018	Day

R347216-20   Single Track, Steel Ties, Concrete Bed

The costs for a R.R. siding with steel ties and a concrete bed in the Unit Price section include the components shown in the table below.

Description of Component	Qty. per L.F. of Track	Unit
Concrete bed, 9′ wide, 10″ thick	.278	C.Y.
Ties, W6x16 x 6′-6″ long, @ 30″ O.C.	.400	Ea.
Tie plates, 4 per tie	1.600	Ea.
Track rail	2.000	L.F.
Tie plate bolts, 1″, 8 per tie	3.200	Ea.
Splice bars w/bolts, lock washers & nuts, @ 33′ O.C.	.061	Pair
Crew B-14 @ 22 L.F./Day	.045	Day

Change Orders

Change Order Considerations

A change order is a written document usually prepared by the design professional and signed by the owner, the architect/engineer, and the contractor. A change order states the agreement of the parties to: an addition, deletion, or revision in the work; an adjustment in the contract sum, if any; or an adjustment in the contract time, if any. Change orders, or "extras", in the construction process occur after execution of the construction contract and impact architects/engineers, contractors, and owners.

Change orders that are properly recognized and managed can ensure orderly, professional, and profitable progress for everyone involved in the project. There are many causes for change orders and change order requests. In all cases, change orders or change order requests should be addressed promptly and in a precise and prescribed manner. The following paragraphs include information regarding change order pricing and procedures.

The Causes of Change Orders

Reasons for issuing change orders include:

- Unforeseen field conditions that require a change in the work
- Correction of design discrepancies, errors, or omissions in the contract documents
- Owner-requested changes, either by design criteria, scope of work, or project objectives
- Completion date changes for reasons unrelated to the construction process
- Changes in building code interpretations, or other public authority requirements that require a change in the work
- Changes in availability of existing or new materials and products

Procedures

Properly written contract documents must include the correct change order procedures for all parties—owners, design professionals, and contractors—to follow in order to avoid costly delays and litigation.

Being "in the right" is not always a sufficient or acceptable defense. The contract provisions requiring notification and documentation must be adhered to within a defined or reasonable time frame.

The appropriate method of handling change orders is by a written proposal and acceptance by all parties involved. Prior to starting work on a project, all parties should identify their authorized agents who may sign and accept change orders, as well as any limits placed on their authority.

Time may be a critical factor when the need for a change arises. For such cases, the contractor might be directed to proceed on a "time and materials" basis, rather than wait for all paperwork to be processed—a delay that could impede progress. In this situation, the contractor must still follow the prescribed change order procedures including, but not limited to, notification and documentation.

Lack of documentation can be very costly, especially if legal judgments are to be made, and if certain field personnel are no longer available. For time and material change orders, the contractor should keep accurate daily records of all labor and material allocated to the change.

Owners or awarding authorities who do considerable and continual building construction (such as the federal government) realize the inevitability of change orders for numerous reasons, both predictable and unpredictable. As a result, the federal government, the American Institute of Architects (AIA), the Engineers Joint Contract Documents Committee (EJCDC), and other contractor, legal, and technical organizations have developed standards and procedures to be followed by all parties to achieve contract continuance and timely completion, while being financially fair to all concerned.

Pricing Change Orders

When pricing change orders, regardless of their cause, the most significant factor is when the change occurs. The need for a change may be perceived in the field or requested by the architect/engineer *before* any of the actual installation has begun, or may evolve or appear *during* construction when the item of work in question is partially installed. In the latter cases, the original sequence of construction is disrupted, along with all contiguous and supporting systems. Change orders cause the greatest impact when they occur *after* the installation has been completed and must be uncovered, or even replaced. Post-completion changes may be caused by necessary design changes, product failure, or changes in the owner's requirements that are not discovered until the building or the systems begin to function.

Specified procedures of notification and record keeping must be adhered to and enforced regardless of the stage of construction: *before, during,* or *after* installation. Some bidding documents anticipate change orders by requiring that unit prices including overhead and profit percentages—for additional as well as deductible changes—be listed. Generally these unit prices do not fully take into account the ripple effect, or impact on other trades, and should be used for general guidance only.

When pricing change orders, it is important to classify the time frame in which the change occurs. There are two basic time frames for change orders: *pre-installation change orders,* which occur before the start of construction, and *post-installation change orders,* which involve reworking after the original installation. Change orders that occur between these stages may be priced according to the extent of work completed using a combination of techniques developed for pricing *pre-* and *post-installation* changes.

Factors To Consider When Pricing Change Orders

As an estimator begins to prepare a change order, the following questions should be reviewed to determine their impact on the final price.

General

- *Is the change order work* pre-installation *or* post-installation?

  Change order work costs vary according to how much of the installation has been completed. Once workers have the project scoped in their minds, even though they have not started, it can be difficult to refocus. Consequently they may spend more than the normal amount of time understanding the change. Also, modifications to work in place, such as trimming or refitting, usually take more time than was initially estimated. The greater the amount of work in place, the more reluctant workers are to change it. Psychologically they may resent the change and as a result the rework takes longer than normal. Post-installation change order estimates must include demolition of existing work as required to accomplish the change. If the work is performed at a later time, additional obstacles, such as building finishes, may be present which must be protected. Regardless of whether the change occurs

pre-installation or post-installation, attempt to isolate the identifiable factors and price them separately. For example, add shipping costs that may be required pre-installation or any demolition required post-installation. Then analyze the potential impact on productivity of psychological and/or learning curve factors and adjust the output rates accordingly. One approach is to break down the typical workday into segments and quantify the impact on each segment.

Change Order Installation Efficiency

The labor-hours expressed (for new construction) are based on average installation time, using an efficiency level. For change order situations, adjustments to this efficiency level should reflect the daily labor-hour allocation for that particular occurrence.

- *Will the change substantially delay the original completion date?*

A significant change in the project may cause the original completion date to be extended. The extended schedule may subject the contractor to new wage rates dictated by relevant labor contracts. Project supervision and other project overhead must also be extended beyond the original completion date. The schedule extension may also put installation into a new weather season. For example, underground piping scheduled for October installation was delayed until January. As a result, frost penetrated the trench area, thereby changing the degree of difficulty of the task. Changes and delays may have a ripple effect throughout the project. This effect must be analyzed and negotiated with the owner.

- *What is the net effect of a deduct change order?*

In most cases, change orders resulting in a deduction or credit reflect only bare costs. The contractor may retain the overhead and profit based on the original bid.

Materials

- *Will you have to pay more or less for the new material, required by the change order, than you paid for the original purchase?*

The same material prices or discounts will usually apply to materials purchased for change orders as new construction. In some

instances, however, the contractor may forfeit the advantages of competitive pricing for change orders. Consider the following example:

A contractor purchased over $20,000 worth of fan coil units for an installation and obtained the maximum discount. Some time later it was determined the project required an additional matching unit. The contractor has to purchase this unit from the original supplier to ensure a match. The supplier at this time may not discount the unit because of the small quantity, and he is no longer in a competitive situation. The impact of quantity on purchase can add between 0% and 25% to material prices and/or subcontractor quotes.

- *If materials have been ordered or delivered to the job site, will they be subject to a cancellation charge or restocking fee?*

Check with the supplier to determine if ordered materials are subject to a cancellation charge. Delivered materials not used as a result of a change order may be subject to a restocking fee if returned to the supplier. Common restocking charges run between 20% and 40%. Also, delivery charges to return the goods to the supplier must be added.

Labor

- *How efficient is the existing crew at the actual installation?*

Is the same crew that performed the initial work going to do the change order? Possibly the change consists of the installation of a unit identical to one already installed; therefore, the change should take less time. Be sure to consider this potential productivity increase and modify the productivity rates accordingly.

- *If the crew size is increased, what impact will that have on supervision requirements?*

Under most bargaining agreements or management practices, there is a point at which a working foreman is replaced by a nonworking foreman. This replacement increases project overhead by adding a nonproductive worker. If additional workers are added to accelerate the project or to perform changes while maintaining the schedule, be sure to add additional supervision time if warranted. Calculate the

hours involved and the additional cost directly if possible.

- *What are the other impacts of increased crew size?*

The larger the crew, the greater the potential for productivity to decrease. Some of the factors that cause this productivity loss are: overcrowding (producing restrictive conditions in the working space) and possibly a shortage of any special tools and equipment required. Such factors affect not only the crew working on the elements directly involved in the change order, but other crews whose movements may also be hampered. As the crew increases, check its basic composition for changes by the addition or deletion of apprentices or nonworking foreman, and quantify the potential effects of equipment shortages or other logistical factors.

- *As new crews, unfamiliar with the project, are brought onto the site, how long will it take them to become oriented to the project requirements?*

The orientation time for a new crew to become 100% effective varies with the site and type of project. Orientation is easiest at a new construction site and most difficult at existing, very restrictive renovation sites. The type of work also affects orientation time. When all elements of the work are exposed, such as concrete or masonry work, orientation is decreased. When the work is concealed or less visible, such as existing electrical systems, orientation takes longer. Usually orientation can be accomplished in one day or less. Costs for added orientation should be itemized and added to the total estimated cost.

- *How much actual production can be gained by working overtime?*

Short term overtime can be used effectively to accomplish more work in a day. However, as overtime is scheduled to run beyond several weeks, studies have shown marked decreases in output. The following chart shows the effect of long term overtime on worker efficiency. If the anticipated change requires extended overtime to keep the job on schedule, these factors can be used as a guide to predict the impact on time and cost. Add project overhead, particularly supervision, that may also be incurred.

548

Days per Week	Hours per Day	Production Efficiency					Payroll Cost Factors	
		1st Week	2nd Week	3rd Week	4th Week	Average 4 Weeks	@ 1-1/2 Times	@ 2 Times
5	8	100%	100%	100%	100%	100%	100%	100%
	9	100	100	95	90	96	1.056	1.111
	10	100	95	90	85	93	1.100	1.200
	11	95	90	75	65	81	1.136	1.273
	12	90	85	70	60	76	1.167	1.333
6	8	100	100	95	90	96	1.083	1.167
	9	100	95	90	85	93	1.130	1.259
	10	95	90	85	80	88	1.167	1.333
	11	95	85	70	65	79	1.197	1.394
	12	90	80	65	60	74	1.222	1.444
7	8	100	95	85	75	89	1.143	1.286
	9	95	90	80	70	84	1.183	1.365
	10	90	85	75	65	79	1.214	1.429
	11	85	80	65	60	73	1.240	1.481
	12	85	75	60	55	69	1.262	1.524

Effects of Overtime

Caution: Under many labor agreements, Sundays and holidays are paid at a higher premium than the normal overtime rate.

The use of long-term overtime is counterproductive on almost any construction job; that is, the longer the period of overtime, the lower the actual production rate. Numerous studies have been conducted, and while they have resulted in slightly different numbers, all reach the same conclusion. The figure above tabulates the effects of overtime work on efficiency.

As illustrated, there can be a difference between the *actual* payroll cost per hour and the *effective* cost per hour for overtime work. This is due to the reduced production efficiency with the increase in weekly hours beyond 40. This difference between actual and effective cost results from overtime work over a prolonged period. Short-term overtime work does not result in as great a reduction in efficiency and, in such cases, effective cost may not vary significantly from the actual payroll cost. As the total hours per week are increased on a regular basis, more time is lost due to fatigue, lowered morale, and an increased accident rate.

As an example, assume a project where workers are working 6 days a week, 10 hours per day. From the figure above (based on productivity studies), the average effective productive hours over a 4-week period are:

$$0.875 \times 60 = 52.5$$

Depending upon the locale and day of week, overtime hours may be paid at time and a half or double time. For time and a half, the overall (average) *actual* payroll cost (including regular and overtime hours) is determined as follows:

$$\frac{40 \text{ reg. hrs.} + (20 \text{ overtime hrs.} \times 1.5)}{60 \text{ hrs.}} = 1.167$$

Based on 60 hours, the payroll cost per hour will be 116.7% of the normal rate at 40 hours per week. However, because the effective production (efficiency) for 60 hours is reduced to the equivalent of 52.5 hours, the effective cost of overtime is calculated as follows:

For time and a half:

$$\frac{40 \text{ reg. hrs.} + (20 \text{ overtime hrs.} \times 1.5)}{52.5 \text{ hrs.}} = 1.33$$

The installed cost will be 133% of the normal rate (for labor).

Thus, when figuring overtime, the actual cost per unit of work will be higher than the apparent overtime payroll dollar increase, due to the reduced productivity of the longer work week. These efficiency calculations are true only for those cost factors determined by hours worked. Costs that are applied weekly or monthly, such as equipment rentals, will not be similarly affected.

Equipment

- *What equipment is required to complete the change order?*

Change orders may require extending the rental period of equipment already on the job site, or the addition of special equipment brought in to accomplish the change work. In either case, the additional rental charges and operator labor charges must be added.

Summary

The preceding considerations and others you deem appropriate should be analyzed and applied to a change order estimate. The impact of each should be quantified and listed on the estimate to form an audit trail.

Change orders that are properly identified, documented, and managed help to ensure the orderly, professional, and profitable progress of the work. They also minimize potential claims or disputes at the end of the project.

Back by customer demand!

You asked and we listened. For customer convenience and estimating ease, we have made the 2021 Project Costs available for download at **RSMeans.com/2021books**. You will also find sample estimates, an RSMeans data overview video, and a book registration form to receive quarterly data updates throughout 2021.

Estimating Tips

- The cost figures available in the download were derived from hundreds of projects contained in the RSMeans database of completed construction projects. They include the contractor's overhead and profit. The figures have been adjusted to January of the current year.

- These projects were located throughout the U.S. and reflect a tremendous variation in square foot (S.F.) costs. This is due to differences, not only in labor and material costs, but also in individual owners' requirements. For instance, a bank in a large city would have different features than one in a rural area. This is true of all the different types of buildings analyzed. Therefore, caution should be exercised when using these Project Costs. For example, for courthouses, costs in the database are local courthouse costs and will not apply to the larger, more elaborate federal courthouses.

- None of the figures "go with" any others. All individual cost items were computed and tabulated separately. Thus, the sum of the median figures for plumbing, HVAC, and electrical will not normally total up to the total mechanical and electrical costs arrived at by separate analysis and tabulation of the projects.

- Each building was analyzed as to total and component costs and percentages. The figures were arranged in ascending order with the results tabulated as shown. The 1/4 column shows that 25% of the projects had lower costs and 75% had higher. The 3/4 column shows that 75% of the projects had lower costs and 25% had higher. The median column shows that 50% of the projects had lower costs and 50% had higher.

- Project Costs are useful in the conceptual stage when no details are available. As soon as details become available in the project design, the square foot approach should be discontinued and the project should be priced as to its particular components. When more precision is required, or for estimating the replacement cost of specific buildings, the current edition of *Square Foot Costs with RSMeans data* should be used.

- In using the figures in this section, it is recommended that the median column be used for preliminary figures if no

additional information is available. The median figures, when multiplied by the total city construction cost index figures (see City Cost Indexes) and then multiplied by the project size modifier at the end of this section, should present a fairly accurate base figure, which would then have to be adjusted in view of the estimator's experience, local economic conditions, code requirements, and the owner's particular requirements. There is no need to factor in the percentage figures, as these should remain constant from city to city.

- The editors of this data would greatly appreciate receiving cost figures on one or more of your recent projects, which would then be included in the averages for next year. All cost figures received will be kept confidential, except that they will be averaged with other similar projects to arrive at square foot cost figures for next year.

See the website above for details and the discount available for submitting one or more of your projects.

Same Data. Simplified.

Enjoy the convenience and efficiency of accessing your costs anywhere:

- **Skip the multiplier** by setting your location
- **Quickly search,** edit, favorite and share costs
- **Stay on top of price changes** with automatic updates

Discover more at rsmeans.com/online

50 17 00 | Project Costs

		UNIT	UNIT COSTS			% OF TOTAL		
			1/4	MEDIAN	3/4	1/4	MEDIAN	3/4
01	0000	**Auto Sales with Repair**	S.F.							01
	0100	Architectural		107	120	130	58%	64%	67%
	0200	Plumbing		8.95	9.40	12.55	4.84%	5.20%	6.80%
	0300	Mechanical		12	16.10	17.75	6.40%	8.70%	10.15%
	0400	Electrical		18.45	23	28.50	9.05%	11.70%	15.90%
	0500	Total Project Costs		180	187	193			
02	0000	**Banking Institutions**	S.F.							02
	0100	Architectural		161	198	241	59%	65%	69%
	0200	Plumbing		6.50	9.10	12.60	2.12%	3.39%	4.19%
	0300	Mechanical		12.95	17.90	21	4.41%	5.10%	10.75%
	0400	Electrical		31.50	38	58.50	10.45%	13.05%	15.90%
	0500	Total Project Costs		268	300	370			
03	0000	**Court House**	S.F.							03
	0100	Architectural		85	167	167	54.50%	58.50%	58.50%
	0200	Plumbing		3.22	3.22	3.22	2.07%	2.07%	2.07%
	0300	Mechanical		20	20	20	12.95%	12.95%	12.95%
	0400	Electrical		26	26	26	16.60%	16.60%	16.60%
	0500	Total Project Costs		155	286	286			
04	0000	**Data Centers**	S.F.							04
	0100	Architectural		192	192	192	68%	68%	68%
	0200	Plumbing		10.55	10.55	10.55	3.71%	3.71%	3.71%
	0300	Mechanical		27	27	27	9.45%	9.45%	9.45%
	0400	Electrical		25.50	25.50	25.50	9%	9%	9%
	0500	Total Project Costs		284	284	284			
05	0000	**Detention Centers**	S.F.							05
	0100	Architectural		178	188	200	52%	53%	60.50%
	0200	Plumbing		18.80	23	27.50	5.15%	7.10%	7.25%
	0300	Mechanical		24	34	41	7.55%	9.50%	13.80%
	0400	Electrical		39	46.50	60.50	10.90%	14.85%	17.95%
	0500	Total Project Costs		300	320	375			
06	0000	**Fire Stations**	S.F.							06
	0100	Architectural		101	130	190	46%	54.50%	61.50%
	0200	Plumbing		10.25	14.05	18.35	4.59%	5.60%	6.30%
	0300	Mechanical		15.10	22	29.50	6.05%	8.25%	10.20%
	0400	Electrical		23.50	30	40	10.75%	12.55%	14.95%
	0500	Total Project Costs		211	241	335			
07	0000	**Gymnasium**	S.F.							07
	0100	Architectural		89	118	118	57%	64.50%	64.50%
	0200	Plumbing		2.19	7.15	7.15	1.58%	3.48%	3.48%
	0300	Mechanical		3.35	30	30	2.42%	14.65%	14.65%
	0400	Electrical		11	21.50	21.50	7.95%	10.35%	10.35%
	0500	Total Project Costs		139	206	206			
08	0000	**Hospitals**	S.F.							08
	0100	Architectural		108	178	193	43%	47.50%	48%
	0200	Plumbing		7.95	15.15	33	6%	7.45%	7.65%
	0300	Mechanical		52.50	59	77	14.20%	17.95%	23.50%
	0400	Electrical		24	48	62	10.95%	13.75%	16.85%
	0500	Total Project Costs		253	375	405			
09	0000	**Industrial Buildings**	S.F.							09
	0100	Architectural		59	116	235	52%	56%	82%
	0200	Plumbing		1.70	4.62	13.35	1.57%	2.11%	6.30%
	0300	Mechanical		4.86	9.25	44	4.77%	5.55%	14.80%
	0400	Electrical		7.40	8.45	70.50	7.85%	13.55%	16.20%
	0500	Total Project Costs		88	133	435			
10	0000	**Medical Clinics & Offices**	S.F.							10
	0100	Architectural		90.50	122	163	48.50%	55.50%	62.50%
	0200	Plumbing		9	13.50	21.50	4.47%	6.65%	8.75%
	0300	Mechanical		14.60	22.50	41.50	7.80%	10.90%	16.05%
	0400	Electrical		19.90	24.50	38	8.90%	11.40%	14.05%
	0500	Total Project Costs		168	224	295			

50 17 00 | Project Costs

			UNIT	UNIT COSTS			% OF TOTAL			
				1/4	MEDIAN	3/4	1/4	MEDIAN	3/4	
11	0000	**Mixed Use**	S.F.							11
	0100	Architectural		92	130	198	45.50%	52.50%	61.50%	
	0200	Plumbing		6.25	11.45	12.15	3.31%	3.47%	4.18%	
	0300	Mechanical		15.25	25	46	4.68%	13.60%	17.05%	
	0400	Electrical		16.15	36	53.50	8.30%	11.40%	15.65%	
	0500	Total Project Costs		190	335	340				
12	0000	**Multi-Family Housing**	S.F.							12
	0100	Architectural		77.50	105	155	54.50%	61.50%	66.50%	
	0200	Plumbing		6.90	11.10	15.10	5.30%	6.85%	8%	
	0300	Mechanical		7.15	9.55	27.50	4.92%	6.90%	10.40%	
	0400	Electrical		10	15.70	22.50	6.20%	8%	10.25%	
	0500	Total Project Costs		128	210	253				
13	0000	**Nursing Home & Assisted Living**	S.F.							13
	0100	Architectural		72.50	94.50	119	51.50%	55.50%	63.50%	
	0200	Plumbing		7.80	11.75	12.90	6.25%	7.40%	8.80%	
	0300	Mechanical		6.40	9.45	18.50	4.04%	6.70%	9.55%	
	0400	Electrical		10.60	16.70	23.50	7%	10.75%	13.10%	
	0500	Total Project Costs		123	161	194				
14	0000	**Office Buildings**	S.F.							14
	0100	Architectural		93	130	179	54.50%	61%	69%	
	0200	Plumbing		5.15	8.10	15.15	2.70%	3.78%	5.85%	
	0300	Mechanical		10.10	17.15	26.50	5.60%	8.20%	11.10%	
	0400	Electrical		12.90	22	34	7.75%	10%	12.70%	
	0500	Total Project Costs		159	202	285				
15	0000	**Parking Garage**	S.F.							15
	0100	Architectural		32	39	40.50	70%	79%	88%	
	0200	Plumbing		1.05	1.10	2.06	2.05%	2.70%	2.83%	
	0300	Mechanical		.82	1.25	4.76	2.11%	3.62%	3.81%	
	0400	Electrical		2.80	3.08	6.45	5.30%	6.35%	7.95%	
	0500	Total Project Costs		39	47.50	51.50				
16	0000	**Parking Garage/Mixed Use**	S.F.							16
	0100	Architectural		103	113	115	61%	62%	65.50%	
	0200	Plumbing		3.33	4.36	6.65	2.47%	2.72%	3.66%	
	0300	Mechanical		14.20	16	23	7.80%	13.10%	13.60%	
	0400	Electrical		14.90	21.50	22	8.20%	12.65%	18.15%	
	0500	Total Project Costs		169	176	182				
17	0000	**Police Stations**	S.F.							17
	0100	Architectural		117	131	165	49%	56.50%	61%	
	0200	Plumbing		15.45	18.55	18.65	5.05%	5.55%	9.05%	
	0300	Mechanical		35	48.50	50.50	13%	14.55%	16.55%	
	0400	Electrical		26.50	29	30.50	9.15%	12.10%	14%	
	0500	Total Project Costs		219	270	305				
18	0000	**Police/Fire**	S.F.							18
	0100	Architectural		114	114	350	55.50%	66%	68%	
	0200	Plumbing		9.15	9.45	35	5.45%	5.50%	5.55%	
	0300	Mechanical		13.95	22	80	8.35%	12.70%	12.80%	
	0400	Electrical		15.90	20.50	91.50	9.50%	11.75%	14.55%	
	0500	Total Project Costs		167	173	630				
19	0000	**Public Assembly Buildings**	S.F.							19
	0100	Architectural		116	160	240	57.50%	61.50%	66%	
	0200	Plumbing		6.15	9	13.50	2.60%	3.36%	4.79%	
	0300	Mechanical		12.95	23	35.50	6.55%	8.95%	12.45%	
	0400	Electrical		19.15	26	41.50	8.60%	10.75%	13%	
	0500	Total Project Costs		186	255	375				
20	0000	**Recreational**	S.F.							20
	0100	Architectural		111	176	238	49.50%	59%	65.50%	
	0200	Plumbing		8.60	14.85	25.50	3.12%	4.76%	7.40%	
	0300	Mechanical		13.25	20	32	4.98%	6.60%	11.55%	
	0400	Electrical		19.15	27.50	40	7.15%	8.95%	10.75%	
	0500	Total Project Costs		198	296	445				

For customer support on your Concrete & Masonry Costs with RSMeans Data, call 800.448.8182.

553

50 17 00 | Project Costs

		UNIT	UNIT COSTS			% OF TOTAL		
			1/4	MEDIAN	3/4	1/4	MEDIAN	3/4
21	0000 **Restaurants**	S.F.						
	0100 Architectural		127	195	250	59%	60%	63.50%
	0200 Plumbing		13.90	32	40.50	7.35%	7.75%	8.95%
	0300 Mechanical		15	17.70	37.50	6.50%	8.15%	11.15%
	0400 Electrical		14.90	24.50	48.50	7.10%	10.30%	11.60%
	0500 Total Project Costs		210	305	420			
22	0000 **Retail**	S.F.						
	0100 Architectural		56.50	88.50	182	54.50%	60%	64.50%
	0200 Plumbing		7.05	9.85	12.25	4.18%	5.45%	8.45%
	0300 Mechanical		6.60	9.35	17.20	4.98%	6.15%	7.05%
	0400 Electrical		10.50	21.50	31.50	7.90%	11.25%	12.45%
	0500 Total Project Costs		86	153	292			
23	0000 **Schools**	S.F.						
	0100 Architectural		98	127	173	51%	56%	60.50%
	0200 Plumbing		7.75	10.70	15.65	3.66%	4.59%	7%
	0300 Mechanical		18.70	26.50	38.50	8.95%	12%	14.55%
	0400 Electrical		18.45	25.50	33	9.45%	11.25%	13.30%
	0500 Total Project Costs		178	227	310			
24	0000 **University, College & Private School Classroom & Admin Buildings**	S.F.						
	0100 Architectural		126	160	194	50.50%	55%	59.50%
	0200 Plumbing		8.50	11.50	19.90	3.02%	4.46%	7.35%
	0300 Mechanical		25	37.50	46.50	9.95%	11.95%	14.70%
	0400 Electrical		20	28.50	35.50	7.70%	9.90%	11.55%
	0500 Total Project Costs		219	293	380			
25	0000 **University, College & Private School Dormitories**	S.F.						
	0100 Architectural		81.50	143	152	53%	61.50%	68%
	0200 Plumbing		10.80	15.25	17.50	6.35%	6.65%	8.95%
	0300 Mechanical		4.84	20.50	32.50	4.13%	9%	11.80%
	0400 Electrical		5.75	19.95	30.50	4.75%	7.35%	10.60%
	0500 Total Project Costs		121	229	271			
26	0000 **University, College & Private School Science, Eng. & Lab Buildings**	S.F.						
	0100 Architectural		140	166	195	48.50%	56.50%	58%
	0200 Plumbing		9.65	14.65	24.50	3.29%	3.77%	5%
	0300 Mechanical		37	69	71	11.70%	19.40%	23.50%
	0400 Electrical		29	38	39	9%	12.05%	13.15%
	0500 Total Project Costs		293	315	370			
27	0000 **University, College & Private School Student Union Buildings**	S.F.						
	0100 Architectural		111	292	292	54.50%	54.50%	59.50%
	0200 Plumbing		16.80	16.80	25	3.13%	4.27%	11.45%
	0300 Mechanical		32	51.50	51.50	9.60%	9.60%	14.55%
	0400 Electrical		28	48.50	48.50	9.05%	12.80%	13.15%
	0500 Total Project Costs		219	535	535			
28	0000 **Warehouses**	S.F.						
	0100 Architectural		47.50	72.50	132	60.50%	67%	71.50%
	0200 Plumbing		2.48	5.30	10.20	2.82%	3.72%	5%
	0300 Mechanical		2.93	16.70	26	4.56%	8.15%	10.70%
	0400 Electrical		6.15	20	33.50	7.75%	10.10%	18.30%
	0500 Total Project Costs		71	113	228			

Square Foot Project Size Modifier

One factor that affects the S.F. cost of a particular building is the size. In general, for buildings built to the same specifications in the same locality, the larger building will have the lower S.F. cost. This is due mainly to the decreasing contribution of the exterior walls plus the economy of scale usually achievable in larger buildings. The Area Conversion Scale shown below will give a factor to convert costs for the typical size building to an adjusted cost for the particular project.

The Square Foot Base Size lists the median costs, most typical project size in our accumulated data, and the range in size of the projects.

The Size Factor for your project is determined by dividing your project area in S.F. by the typical project size for the particular Building Type. With this factor, enter the Area Conversion Scale at the appropriate Size Factor and determine the appropriate Cost Multiplier for your building size.

Example: Determine the cost per S.F. for a 107,200 S.F. Multi-family housing.

$$\frac{\text{Proposed building area} = 107,200 \text{ S.F.}}{\text{Typical size from below} = 53,600 \text{ S.F.}} = 2.00$$

Enter Area Conversion Scale at 2.0, intersect curve, read horizontally the appropriate cost multiplier of .94. Size adjusted cost becomes .94 x $210.00 = $197.40 based on national average costs.

Note: For Size Factors less than .50, the Cost Multiplier is 1.1
For Size Factors greater than 3.5, the Cost Multiplier is .90

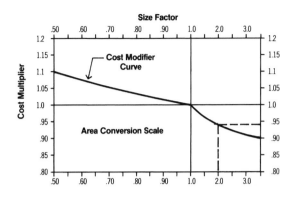

System	Median Cost (Total Project Costs)	Typical Size Gross S.F. (Median of Projects)	Typical Range (Low – High) (Projects)
Auto Sales with Repair	$187.00	24,900	4,680 – 29,253
Banking Institutions	300.00	9,300	3,267 – 38,148
Court House	286.00	47,600	24,680 – 70,500
Data Centers	284.00	14,400	14,369 – 14,369
Detention Centers	320.00	37,800	12,257 – 183,339
Fire Stations	241.00	12,500	6,300 – 49,577
Gymnasium	206.00	52,400	22,844 – 81,992
Hospitals	375.00	87,100	22,428 – 410,273
Industrial Buildings	133.00	21,100	5,146 – 200,625
Medical Clinics & Offices	224.00	22,500	2,273 – 327,000
Mixed Use	335.00	28,500	7,200 – 188,944
Multi-Family Housing	210.00	53,600	2,472 – 1,161,450
Nursing Home & Assisted Living	161.00	38,200	1,515 – 242,555
Office Buildings	202.00	20,600	1,115 – 930,000
Parking Garage	47.5	151,800	99,884 – 287,040
Parking Garage/Mixed Use	176.00	254,200	5,302 – 318,033
Police Stations	270.00	28,500	15,377 – 88,625
Police/Fire	173.00	44,300	8,600 – 50,279
Public Assembly Buildings	255.00	21,000	2,232 – 235,301
Recreational	296.00	27,700	1,000 – 223,787
Restaurants	305.00	6,000	3,975 – 42,000
Retail	153.00	26,000	4,035 – 84,270
Schools	227.00	69,900	1,344 – 410,848
University, College & Private School Classroom & Admin Buildings	293.00	48,200	8,806 – 196,187
University, College & Private School Dormitories	229.00	39,600	1,500 – 126,889
University, College & Private School Science, Eng. & Lab Buildings	315.00	60,000	5,311 – 117,643
University, College & Private School Student Union Buildings	535.00	47,300	42,075 – 50,000
Warehouses	113.00	11,100	640 – 303,750

For customer support on your Concrete & Masonry Costs with RSMeans Data, call 800.448.8182.

555

A	Area Square Feet; Ampere	Brk., brk	Brick	Csc	Cosecant
AAFES	Army and Air Force Exchange Service	brkt	Bracket	C.S.F.	Hundred Square Feet
		Brs.	Brass	CSI	Construction Specifications Institute
ABS	Acrylonitrile Butadiene Stryrene; Asbestos Bonded Steel	Brz.	Bronze		
		Bsn.	Basin	CT	Current Transformer
A.C., AC	Alternating Current; Air-Conditioning; Asbestos Cement; Plywood Grade A & C	Btr.	Better	CTS	Copper Tube Size
		BTU	British Thermal Unit	Cu	Copper, Cubic
		BTUH	BTU per Hour	Cu. Ft.	Cubic Foot
		Bu.	Bushels	cw	Continuous Wave
ACI	American Concrete Institute	BUR	Built-up Roofing	C.W.	Cool White; Cold Water
ACR	Air Conditioning Refrigeration	BX	Interlocked Armored Cable	Cwt.	100 Pounds
ADA	Americans with Disabilities Act	°C	Degree Centigrade	C.W.X.	Cool White Deluxe
AD	Plywood, Grade A & D	c	Conductivity, Copper Sweat	C.Y.	Cubic Yard (27 cubic feet)
Addit.	Additional	C	Hundred; Centigrade	C.Y./Hr.	Cubic Yard per Hour
Adh.	Adhesive	C/C	Center to Center, Cedar on Cedar	Cyl.	Cylinder
Adj.	Adjustable	C-C	Center to Center	d	Penny (nail size)
af	Audio-frequency	Cab	Cabinet	D	Deep; Depth; Discharge
AFFF	Aqueous Film Forming Foam	Cair.	Air Tool Laborer	Dis., Disch.	Discharge
AFUE	Annual Fuel Utilization Efficiency	Cal.	Caliper	Db	Decibel
AGA	American Gas Association	Calc	Calculated	Dbl.	Double
Agg.	Aggregate	Cap.	Capacity	DC	Direct Current
A.H., Ah	Ampere Hours	Carp.	Carpenter	DDC	Direct Digital Control
A hr.	Ampere-hour	C.B.	Circuit Breaker	Demob.	Demobilization
A.H.U., AHU	Air Handling Unit	C.C.A.	Chromate Copper Arsenate	d.f.t.	Dry Film Thickness
A.I.A.	American Institute of Architects	C.C.F.	Hundred Cubic Feet	d.f.u.	Drainage Fixture Units
AIC	Ampere Interrupting Capacity	cd	Candela	D.H.	Double Hung
Allow.	Allowance	cd/sf	Candela per Square Foot	DHW	Domestic Hot Water
alt., alt	Alternate	CD	Grade of Plywood Face & Back	DI	Ductile Iron
Alum.	Aluminum	CDX	Plywood, Grade C & D, exterior glue	Diag.	Diagonal
a.m.	Ante Meridiem			Diam., Dia	Diameter
Amp.	Ampere	Cefi.	Cement Finisher	Distrib.	Distribution
Anod.	Anodized	Cem.	Cement	Div.	Division
ANSI	American National Standards Institute	CF	Hundred Feet	Dk.	Deck
		C.F.	Cubic Feet	D.L.	Dead Load; Diesel
APA	American Plywood Association	CFM	Cubic Feet per Minute	DLH	Deep Long Span Bar Joist
Approx.	Approximate	CFRP	Carbon Fiber Reinforced Plastic	dlx	Deluxe
Apt.	Apartment	c.g.	Center of Gravity	Do.	Ditto
Asb.	Asbestos	CHW	Chilled Water; Commercial Hot Water	DOP	Dioctyl Phthalate Penetration Test (Air Filters)
A.S.B.C.	American Standard Building Code				
Asbe.	Asbestos Worker	C.I., CI	Cast Iron	Dp., dp	Depth
ASCE	American Society of Civil Engineers	C.I.P., CIP	Cast in Place	D.P.S.T.	Double Pole, Single Throw
A.S.H.R.A.E.	American Society of Heating, Refrig. & AC Engineers	Circ.	Circuit	Dr.	Drive
		C.L.	Carload Lot	DR	Dimension Ratio
ASME	American Society of Mechanical Engineers	CL	Chain Link	Drink.	Drinking
		Clab.	Common Laborer	D.S.	Double Strength
ASTM	American Society for Testing and Materials	Clam	Common Maintenance Laborer	D.S.A.	Double Strength A Grade
		C.L.F.	Hundred Linear Feet	D.S.B.	Double Strength B Grade
Attchmt.	Attachment	CLF	Current Limiting Fuse	Dty.	Duty
Avg., Ave.	Average	CLP	Cross Linked Polyethylene	DWV	Drain Waste Vent
AWG	American Wire Gauge	cm	Centimeter	DX	Deluxe White, Direct Expansion
AWWA	American Water Works Assoc.	CMP	Corr. Metal Pipe	dyn	Dyne
Bbl.	Barrel	CMU	Concrete Masonry Unit	e	Eccentricity
B&B, BB	Grade B and Better; Balled & Burlapped	CN	Change Notice	E	Equipment Only; East; Emissivity
		Col.	Column	Ea.	Each
B&S	Bell and Spigot	CO₂	Carbon Dioxide	EB	Encased Burial
B.&W.	Black and White	Comb.	Combination	Econ.	Economy
b.c.c.	Body-centered Cubic	comm.	Commercial, Communication	E.C.Y	Embankment Cubic Yards
B.C.Y.	Bank Cubic Yards	Compr.	Compressor	EDP	Electronic Data Processing
BE	Bevel End	Conc.	Concrete	EIFS	Exterior Insulation Finish System
B.F.	Board Feet	Cont., cont	Continuous; Continued, Container	E.D.R.	Equiv. Direct Radiation
Bg. cem.	Bag of Cement	Corkbd.	Cork Board	Eq.	Equation
BHP	Boiler Horsepower; Brake Horsepower	Corr.	Corrugated	EL	Elevation
		Cos	Cosine	Elec.	Electrician; Electrical
B.I.	Black Iron	Cot	Cotangent	Elev.	Elevator; Elevating
bidir.	bidirectional	Cov.	Cover	EMT	Electrical Metallic Conduit; Thin Wall Conduit
Bit., Bitum.	Bituminous	C/P	Cedar on Paneling		
Bit., Conc.	Bituminous Concrete	CPA	Control Point Adjustment	Eng.	Engine, Engineered
Bk.	Backed	Cplg.	Coupling	EPDM	Ethylene Propylene Diene Monomer
Bkrs.	Breakers	CPM	Critical Path Method		
Bldg., bldg	Building	CPVC	Chlorinated Polyvinyl Chloride	EPS	Expanded Polystyrene
Blk.	Block	C.Pr.	Hundred Pair	Eqhv.	Equip. Oper., Heavy
Bm.	Beam	CRC	Cold Rolled Channel	Eqlt.	Equip. Oper., Light
Boil.	Boilermaker	Creos.	Creosote	Eqmd.	Equip. Oper., Medium
bpm	Blows per Minute	Crpt.	Carpet & Linoleum Layer	Eqmm.	Equip. Oper., Master Mechanic
BR	Bedroom	CRT	Cathode-ray Tube	Eqol.	Equip. Oper., Oilers
Brg., brng.	Bearing	CS	Carbon Steel, Constant Shear Bar Joist	Equip.	Equipment
Brhe.	Bricklayer Helper			ERW	Electric Resistance Welded
Bric.	Bricklayer				

E.S.	Energy Saver	H	High Henry	Lath.	Lather
Est.	Estimated	HC	High Capacity	Lav.	Lavatory
esu	Electrostatic Units	H.D., HD	Heavy Duty; High Density	lb.; #	Pound
E.W.	Each Way	H.D.O.	High Density Overlaid	L.B., LB	Load Bearing; L Conduit Body
EWT	Entering Water Temperature	HDPE	High Density Polyethylene Plastic	L. & E.	Labor & Equipment
Excav.	Excavation	Hdr.	Header	lb./hr.	Pounds per Hour
excl	Excluding	Hdwe.	Hardware	lb./L.F.	Pounds per Linear Foot
Exp., exp	Expansion, Exposure	H.I.D., HID	High Intensity Discharge	lbf/sq.in.	Pound-force per Square Inch
Ext., ext	Exterior; Extension	Help.	Helper Average	L.C.L.	Less than Carload Lot
Extru.	Extrusion	HEPA	High Efficiency Particulate Air	L.C.Y.	Loose Cubic Yard
f.	Fiber Stress		Filter	Ld.	Load
F	Fahrenheit; Female; Fill	Hg	Mercury	LE	Lead Equivalent
Fab., fab	Fabricated; Fabric	HIC	High Interrupting Capacity	LED	Light Emitting Diode
FBGS	Fiberglass	HM	Hollow Metal	L.F.	Linear Foot
F.C.	Footcandles	HMWPE	High Molecular Weight	L.F. Hdr	Linear Feet of Header
f.c.c.	Face-centered Cubic		Polyethylene	L.F. Nose	Linear Foot of Stair Nosing
f'c.	Compressive Stress in Concrete;	HO	High Output	L.F. Rsr	Linear Foot of Stair Riser
	Extreme Compressive Stress	Horiz.	Horizontal	Lg.	Long; Length; Large
F.E.	Front End	H.P., HP	Horsepower; High Pressure	L & H	Light and Heat
FEP	Fluorinated Ethylene Propylene	H.P.F.	High Power Factor	LH	Long Span Bar Joist
	(Teflon)	Hr.	Hour	L.H.	Labor Hours
F.G.	Flat Grain	Hrs./Day	Hours per Day	L.L., LL	Live Load
F.H.A.	Federal Housing Administration	HSC	High Short Circuit	L.L.D.	Lamp Lumen Depreciation
Fig.	Figure	Ht.	Height	lm	Lumen
Fin.	Finished	Htg.	Heating	lm/sf	Lumen per Square Foot
FIPS	Female Iron Pipe Size	Htrs.	Heaters	lm/W	Lumen per Watt
Fixt.	Fixture	HVAC	Heating, Ventilation & Air-	LOA	Length Over All
FJP	Finger jointed and primed		Conditioning	log	Logarithm
Fl. Oz.	Fluid Ounces	Hvy.	Heavy	L-O-L	Lateralolet
Flr.	Floor	HW	Hot Water	long.	Longitude
Flrs.	Floors	Hyd.; Hydr.	Hydraulic	L.P., LP	Liquefied Petroleum; Low Pressure
FM	Frequency Modulation;	Hz	Hertz (cycles)	L.P.F.	Low Power Factor
	Factory Mutual	I.	Moment of Inertia	LR	Long Radius
Fmg.	Framing	IBC	International Building Code	L.S.	Lump Sum
FM/UL	Factory Mutual/Underwriters Labs	I.C.	Interrupting Capacity	Lt.	Light
Fdn.	Foundation	ID	Inside Diameter	Lt. Ga.	Light Gauge
FNPT	Female National Pipe Thread	I.D.	Inside Dimension; Identification	L.T.L.	Less than Truckload Lot
Fori.	Foreman, Inside	I.F.	Inside Frosted	Lt. Wt.	Lightweight
Foro.	Foreman, Outside	I.M.C.	Intermediate Metal Conduit	L.V.	Low Voltage
Fount.	Fountain	In.	Inch	M	Thousand; Material; Male;
fpm	Feet per Minute	Incan.	Incandescent		Light Wall Copper Tubing
FPT	Female Pipe Thread	Incl.	Included; Including	M²CA	Meters Squared Contact Area
Fr	Frame	Int.	Interior	m/hr.; M.H.	Man-hour
F.R.	Fire Rating	Inst.	Installation	mA	Milliampere
FRK	Foil Reinforced Kraft	Insul., insul	Insulation/Insulated	Mach.	Machine
FSK	Foil/Scrim/Kraft	I.P.	Iron Pipe	Mag. Str.	Magnetic Starter
FRP	Fiberglass Reinforced Plastic	I.P.S., IPS	Iron Pipe Size	Maint.	Maintenance
FS	Forged Steel	IPT	Iron Pipe Threaded	Marb.	Marble Setter
FSC	Cast Body; Cast Switch Box	I.W.	Indirect Waste	Mat; Mat'l.	Material
Ft., ft	Foot; Feet	J	Joule	Max.	Maximum
Ftng.	Fitting	J.I.C.	Joint Industrial Council	MBF	Thousand Board Feet
Ftg.	Footing	K	Thousand; Thousand Pounds;	MBH	Thousand BTU's per hr.
Ft lb.	Foot Pound		Heavy Wall Copper Tubing, Kelvin	MC	Metal Clad Cable
Furn.	Furniture	K.A.H.	Thousand Amp. Hours	MCC	Motor Control Center
FVNR	Full Voltage Non-Reversing	kcmil	Thousand Circular Mils	M.C.F.	Thousand Cubic Feet
FVR	Full Voltage Reversing	KD	Knock Down	MCFM	Thousand Cubic Feet per Minute
FXM	Female by Male	K.D.A.T.	Kiln Dried After Treatment	M.C.M.	Thousand Circular Mils
Fy.	Minimum Yield Stress of Steel	kg	Kilogram	MCP	Motor Circuit Protector
g	Gram	kG	Kilogauss	MD	Medium Duty
G	Gauss	kgf	Kilogram Force	MDF	Medium-density fibreboard
Ga.	Gauge	kHz	Kilohertz	M.D.O.	Medium Density Overlaid
Gal., gal.	Gallon	Kip	1000 Pounds	Med.	Medium
Galv., galv	Galvanized	KJ	Kilojoule	MF	Thousand Feet
GC/MS	Gas Chromatograph/Mass	K.L.	Effective Length Factor	M.F.B.M.	Thousand Feet Board Measure
	Spectrometer	K.L.F.	Kips per Linear Foot	Mfg.	Manufacturing
Gen.	General	Km	Kilometer	Mfrs.	Manufacturers
GFI	Ground Fault Interrupter	KO	Knock Out	mg	Milligram
GFRC	Glass Fiber Reinforced Concrete	K.S.F.	Kips per Square Foot	MGD	Million Gallons per Day
Glaz.	Glazier	K.S.I.	Kips per Square Inch	MGPH	Million Gallons per Hour
GPD	Gallons per Day	kV	Kilovolt	MH, M.H.	Manhole; Metal Halide; Man-Hour
gpf	Gallon per Flush	kVA	Kilovolt Ampere	MHz	Megahertz
GPH	Gallons per Hour	kVAR	Kilovar (Reactance)	Mi.	Mile
gpm, GPM	Gallons per Minute	KW	Kilowatt	MI	Malleable Iron; Mineral Insulated
GR	Grade	KWh	Kilowatt-hour	MIPS	Male Iron Pipe Size
Gran.	Granular	L	Labor Only; Length; Long;	mj	Mechanical Joint
Grnd.	Ground		Medium Wall Copper Tubing	m	Meter
GVW	Gross Vehicle Weight	Lab.	Labor	mm	Millimeter
GWB	Gypsum Wall Board	lat	Latitude	Mill.	Millwright
				Min., min.	Minimum, Minute

For customer support on your Concrete & Masonry Costs with RSMeans Data, call 800.448.8182.

557

Misc.	Miscellaneous	PCM	Phase Contrast Microscopy	SBS	Styrene Butadiere Styrene
ml	Milliliter, Mainline	PDCA	Painting and Decorating	SC	Screw Cover
M.L.F.	Thousand Linear Feet		Contractors of America	SCFM	Standard Cubic Feet per Minute
Mo.	Month	P.E., PE	Professional Engineer;	Scaf.	Scaffold
Mobil.	Mobilization		Porcelain Enamel;	Sch., Sched.	Schedule
Mog.	Mogul Base		Polyethylene; Plain End	S.C.R.	Modular Brick
MPH	Miles per Hour	P.E.C.I.	Porcelain Enamel on Cast Iron	S.D.	Sound Deadening
MPT	Male Pipe Thread	Perf.	Perforated	SDR	Standard Dimension Ratio
MRGWB	Moisture Resistant Gypsum	PEX	Cross Linked Polyethylene	S.E.	Surfaced Edge
	Wallboard	Ph.	Phase	Sel.	Select
MRT	Mile Round Trip	P.I.	Pressure Injected	SER, SEU	Service Entrance Cable
ms	Millisecond	Pile.	Pile Driver	S.F.	Square Foot
M.S.F.	Thousand Square Feet	Pkg.	Package	S.F.C.A.	Square Foot Contact Area
Mstz.	Mosaic & Terrazzo Worker	Pl.	Plate	S.F. Flr.	Square Foot of Floor
M.S.Y.	Thousand Square Yards	Plah.	Plasterer Helper	S.F.G.	Square Foot of Ground
Mtd., mtd., mtd	Mounted	Plas.	Plasterer	S.F. Hor.	Square Foot Horizontal
Mthe.	Mosaic & Terrazzo Helper	plf	Pounds Per Linear Foot	SFR	Square Feet of Radiation
Mtng.	Mounting	Pluh.	Plumber Helper	S.F. Shlf.	Square Foot of Shelf
Mult.	Multi; Multiply	Plum.	Plumber	S4S	Surface 4 Sides
MUTCD	Manual on Uniform Traffic Control	Ply.	Plywood	Shee.	Sheet Metal Worker
	Devices	p.m.	Post Meridiem	Sin.	Sine
M.V.A.	Million Volt Amperes	Pntd.	Painted	Skwk.	Skilled Worker
M.V.A.R.	Million Volt Amperes Reactance	Pord.	Painter, Ordinary	SL	Saran Lined
MV	Megavolt	pp	Pages	S.L.	Slimline
MW	Megawatt	PP, PPL	Polypropylene	Sldr.	Solder
MXM	Male by Male	P.P.M.	Parts per Million	SLH	Super Long Span Bar Joist
MYD	Thousand Yards	Pr.	Pair	S.N.	Solid Neutral
N	Natural; North	P.E.S.B.	Pre-engineered Steel Building	SO	Stranded with oil resistant inside
nA	Nanoampere	Prefab.	Prefabricated		insulation
NA	Not Available; Not Applicable	Prefin.	Prefinished	S-O-L	Socketolet
N.B.C.	National Building Code	Prop.	Propelled	sp	Standpipe
NC	Normally Closed	PSF, psf	Pounds per Square Foot	S.P.	Static Pressure; Single Pole; Self-
NEMA	National Electrical Manufacturers	PSI, psi	Pounds per Square Inch		Propelled
	Assoc.	PSIG	Pounds per Square Inch Gauge	Spri.	Sprinkler Installer
NEHB	Bolted Circuit Breaker to 600V.	PSP	Plastic Sewer Pipe	spwg	Static Pressure Water Gauge
NFPA	National Fire Protection Association	Pspr.	Painter, Spray	S.P.D.T.	Single Pole, Double Throw
NLB	Non-Load-Bearing	Psst.	Painter, Structural Steel	SPF	Spruce Pine Fir; Sprayed
NM	Non-Metallic Cable	P.T.	Potential Transformer		Polyurethane Foam
nm	Nanometer	P. & T.	Pressure & Temperature	S.P.S.T.	Single Pole, Single Throw
No.	Number	Ptd.	Painted	SPT	Standard Pipe Thread
NO	Normally Open	Ptns.	Partitions	Sq.	Square; 100 Square Feet
N.O.C.	Not Otherwise Classified	Pu	Ultimate Load	Sq. Hd.	Square Head
Nose.	Nosing	PVC	Polyvinyl Chloride	Sq. In.	Square Inch
NPT	National Pipe Thread	Pvmt.	Pavement	S.S.	Single Strength; Stainless Steel
NQOD	Combination Plug-on/Bolt on	PRV	Pressure Relief Valve	S.S.B.	Single Strength B Grade
	Circuit Breaker to 240V.	Pwr.	Power	sst, ss	Stainless Steel
N.R.C., NRC	Noise Reduction Coefficient/	Q	Quantity Heat Flow	Sswk.	Structural Steel Worker
	Nuclear Regulator Commission	Qt.	Quart	Sswl.	Structural Steel Welder
N.R.S.	Non Rising Stem	Quan., Qty.	Quantity	St.; Stl.	Steel
ns	Nanosecond	Q.C.	Quick Coupling	STC	Sound Transmission Coefficient
NTP	Notice to Proceed	r	Radius of Gyration	Std.	Standard
nW	Nanowatt	R	Resistance	Stg.	Staging
OB	Opposing Blade	R.C.P.	Reinforced Concrete Pipe	STK	Select Tight Knot
OC	On Center	Rect.	Rectangle	STP	Standard Temperature & Pressure
OD	Outside Diameter	recpt.	Receptacle	Stpi.	Steamfitter, Pipefitter
O.D.	Outside Dimension	Reg.	Regular	Str.	Strength; Starter; Straight
ODS	Overhead Distribution System	Reinf.	Reinforced	Strd.	Stranded
O.G.	Ogee	Req'd.	Required	Struct.	Structural
O.H.	Overhead	Res.	Resistant	Sty.	Story
O&P	Overhead and Profit	Resi.	Residential	Subj.	Subject
Oper.	Operator	RF	Radio Frequency	Subs.	Subcontractors
Opng.	Opening	RFID	Radio-frequency Identification	Surf.	Surface
Orna.	Ornamental	Rgh.	Rough	Sw.	Switch
OSB	Oriented Strand Board	RGS	Rigid Galvanized Steel	Swbd.	Switchboard
OS&Y	Outside Screw and Yoke	RHW	Rubber, Heat & Water Resistant;	S.Y.	Square Yard
OSHA	Occupational Safety and Health		Residential Hot Water	Syn.	Synthetic
	Act	rms	Root Mean Square	S.Y.P.	Southern Yellow Pine
Ovhd.	Overhead	Rnd.	Round	Sys.	System
OWG	Oil, Water or Gas	Rodm.	Rodman	t.	Thickness
Oz.	Ounce	Rofc.	Roofer, Composition	T	Temperature; Ton
P.	Pole; Applied Load; Projection	Rofp.	Roofer, Precast	Tan	Tangent
p.	Page	Rohe.	Roofer Helpers (Composition)	T.C.	Terra Cotta
Pape.	Paperhanger	Rots.	Roofer, Tile & Slate	T & C	Threaded and Coupled
P.A.P.R.	Powered Air Purifying Respirator	R.O.W.	Right of Way	T.D.	Temperature Difference
PAR	Parabolic Reflector	RPM	Revolutions per Minute	TDD	Telecommunications Device for
P.B., PB	Push Button	R.S.	Rapid Start		the Deaf
Pc., Pcs.	Piece, Pieces	Rsr	Riser	T.E.M.	Transmission Electron Microscopy
P.C.	Portland Cement; Power Connector	RT	Round Trip	temp	Temperature, Tempered, Temporary
P.C.F.	Pounds per Cubic Foot	S.	Suction; Single Entrance; South	TFFN	Nylon Jacketed Wire

Abbreviations

TFE	Tetrafluoroethylene (Teflon)	U.L., UL	Underwriters Laboratory	w/	With
T. & G.	Tongue & Groove;	Uld.	Unloading	W.C., WC	Water Column; Water Closet
	Tar & Gravel	Unfin.	Unfinished	W.F.	Wide Flange
Th., Thk.	Thick	UPS	Uninterruptible Power Supply	W.G.	Water Gauge
Thn.	Thin	URD	Underground Residential	Wldg.	Welding
Thrded	Threaded		Distribution	W. Mile	Wire Mile
Tilf.	Tile Layer, Floor	US	United States	W-O-L	Weldolet
Tilh.	Tile Layer, Helper	USGBC	U.S. Green Building Council	W.R.	Water Resistant
THHN	Nylon Jacketed Wire	USP	United States Primed	Wrck.	Wrecker
THW.	Insulated Strand Wire	UTMCD	Uniform Traffic Manual For Control	WSFU	Water Supply Fixture Unit
THWN	Nylon Jacketed Wire		Devices	W.S.P.	Water, Steam, Petroleum
T.L., TL	Truckload	UTP	Unshielded Twisted Pair	WT., Wt.	Weight
T.M.	Track Mounted	V	Volt	WWF	Welded Wire Fabric
Tot.	Total	VA	Volt Amperes	XFER	Transfer
T-O-L	Threadolet	VAT	Vinyl Asbestos Tile	XFMR	Transformer
tmpd	Tempered	V.C.T.	Vinyl Composition Tile	XHD	Extra Heavy Duty
TPO	Thermoplastic Polyolefin	VAV	Variable Air Volume	XHHW	Cross-Linked Polyethylene Wire
T.S.	Trigger Start	VC	Veneer Core	XLPE	Insulation
Tr.	Trade	VDC	Volts Direct Current	XLP	Cross-linked Polyethylene
Transf.	Transformer	Vent.	Ventilation	Xport	Transport
Trhv.	Truck Driver, Heavy	Vert.	Vertical	Y	Wye
Trlr	Trailer	V.F.	Vinyl Faced	yd	Yard
Trlt.	Truck Driver, Light	V.G.	Vertical Grain	yr	Year
TTY	Teletypewriter	VHF	Very High Frequency	Δ	Delta
TV	Television	VHO	Very High Output	%	Percent
T.W.	Thermoplastic Water Resistant	Vib.	Vibrating	~	Approximately
	Wire	VLF	Vertical Linear Foot	Ø	Phase; diameter
UCI	Uniform Construction Index	VOC	Volatile Organic Compound	@	At
UF	Underground Feeder	Vol.	Volume	#	Pound; Number
UGND	Underground Feeder	VRP	Vinyl Reinforced Polyester	<	Less Than
UHF	Ultra High Frequency	W	Wire; Watt; Wide; West	>	Greater Than
U.I.	United Inch			Z	Zone

Index

For customer support on your Concrete & Masonry Costs with RSMeans Data, call 800.448.8182.

For customer support on your Concrete & Masonry Costs with RSMeans Data, call 800.448.8182.

Index

For customer support on your Concrete & Masonry Costs with RSMeans Data, call 800.448.8182.

565

For customer support on your Concrete & Masonry Costs with RSMeans Data, call 800.448.8182.

For customer support on your Concrete & Masonry Costs with RSMeans Data, call 800.448.8182.

For customer support on your Concrete & Masonry Costs with RSMeans Data, call 800.448.8182.

Index

Division Notes

	CREW	DAILY OUTPUT	LABOR-HOURS	UNIT	BARE COSTS				TOTAL INCL O&P
					MAT.	LABOR	EQUIP.	TOTAL	

Division Notes

	CREW	DAILY OUTPUT	LABOR-HOURS	UNIT	BARE COSTS				TOTAL INCL O&P
					MAT.	LABOR	EQUIP.	TOTAL	

Division Notes

	CREW	DAILY OUTPUT	LABOR-HOURS	UNIT	BARE COSTS				TOTAL INCL O&P
					MAT.	LABOR	EQUIP.	TOTAL	
Division Notes									

Division Notes

		CREW	DAILY OUTPUT	LABOR-HOURS	UNIT	BARE COSTS				TOTAL INCL O&P
						MAT.	LABOR	EQUIP.	TOTAL	
					CREW	OUTPUT	HOURS	UNIT	MAT.	LABOR	EQUIP.	TOTAL	TOTAL INCL O&P

A tradition of excellence in construction cost information
and services since 1942

For more information visit our website at RSMeans.com

Unit prices according to the latest MasterFormat®

Cost Data Selection Guide

The following table provides definitive information on the content of each cost data publication. The number of lines of data provided in each unit price or assemblies division, as well as the number of crews, is listed for each data set. The presence of other elements such as reference tables, square foot models, equipment rental costs, historical cost indexes, and city cost indexes, is also indicated. You can use the table to help select the RSMeans data set that has the quantity and type of information you most need in your work.

Unit Cost Divisions	Building Construction	Mechanical	Electrical	Commercial Renovation	Square Foot	Site Work Landsc.	Green Building	Interior	Concrete Masonry	Open Shop	Heavy Construction	Light Commercial	Facilities Construction	Plumbing	Residential
1	621	454	475	576	0	543	200	377	505	620	560	322	1092	462	219
2	825	347	157	781	0	972	181	466	221	824	739	550	1268	355	340
3	1745	341	232	1265	0	1537	1043	355	2274	1745	1930	538	2028	317	445
4	960	22	0	920	0	724	180	613	1158	928	614	532	1175	0	446
5	1895	158	155	1099	0	858	1793	1106	735	1895	1031	985	1912	204	752
6	2468	18	18	2127	0	110	589	1544	281	2464	123	2157	2141	22	2677
7	1593	215	128	1633	0	580	761	532	523	1590	26	1326	1693	227	1046
8	2140	80	3	2733	0	255	1138	1813	105	2142	0	2328	2966	0	1552
9	2125	86	45	1943	0	313	464	2216	424	2062	15	1779	2379	54	1544
10	1088	17	10	684	0	232	32	898	136	1088	34	588	1179	237	224
11	1095	199	166	539	0	135	56	923	29	1062	0	229	1115	162	108
12	539	0	2	297	0	219	147	1552	14	506	0	272	1571	23	216
13	745	149	157	252	0	370	124	250	77	721	271	109	761	115	103
14	273	36	0	223	0	0	0	257	0	273	0	12	293	16	6
21	127	0	41	37	0	0	0	293	0	127	0	121	665	685	259
22	1165	7543	160	1226	0	2010	1061	849	20	1154	2109	875	7505	9400	719
23	1263	7010	639	1033	0	250	864	775	38	1246	191	980	5244	2011	579
25	0	0	14	14	0	0	0	0	0	0	0	0	0	0	0
26	1513	491	10473	1293	0	859	645	1160	55	1439	648	1360	10254	399	636
27	95	0	467	105	0	0	0	71	0	95	45	67	389	0	56
28	143	79	223	124	0	0	28	97	0	127	0	70	209	57	41
31	1511	733	610	807	0	3261	284	7	1216	1456	3280	607	1568	660	616
32	906	49	8	944	0	4523	419	418	367	877	1942	496	1803	140	544
33	1319	1100	563	310	0	3142	33	0	253	596	3277	135	1790	2165	161
34	112	0	47	9	0	195	0	0	36	67	226	0	141	0	0
35	18	0	0	0	0	327	0	0	0	18	442	0	84	0	0
41	63	0	0	34	0	8	0	22	0	62	31	0	69	14	0
44	75	79	0	0	0	0	0	0	0	0	0	0	75	75	0
46	23	16	0	0	0	274	261	0	0	23	264	0	33	33	0
48	8	0	36	2	0	0	21	0	0	8	15	8	21	0	8
Totals	26453	19222	14829	21010	0	21697	10324	16594	8467	25215	17813	16446	51423	17833	13297

Assem Div	Building Construction	Mechanical	Electrical	Commercial Renovation	Square Foot	Site Work Landscape	Assemblies	Green Building	Interior	Concrete Masonry	Heavy Construction	Light Commercial	Facilities Construction	Plumbing	Asm Div	Residential
A		15	0	190	165	579	599	0	0	537	573	155	24	0	1	378
B		0	0	848	2567	0	5673	56	329	1979	368	2108	174	0	2	211
C		0	0	647	956	0	1338	0	1645	146	0	846	251	0	3	591
D		1060	945	712	1866	72	2544	330	827	0	0	1353	1108	1090	4	851
E		0	0	85	262	0	302	0	5	0	0	259	5	0	5	391
F		0	0	0	114	0	143	0	0	0	0	114	0	0	6	357
G		527	447	318	312	3378	792	0	0	535	1349	205	293	677	7	307
															8	760
															9	80
															10	0
															11	0
															12	0
Totals		1602	1392	2800	6242	4029	11391	386	2806	3197	2290	5040	1855	1767		3926

Reference Section	Building Construction Costs	Mechanical	Electrical	Commercial Renovation	Square Foot	Site Work Landscape	Assem.	Green Building	Interior	Concrete Masonry	Open Shop	Heavy Construction	Light Commercial	Facilities Construction	Plumbing	Resi.
Reference Tables	yes	yes	yes	yes	no	yes	yes	yes	yes	yes	yes	yes	yes	yes	yes	yes
Models					111			25					50			28
Crews	584	584	584	564		584		584	584	584	562	584	562	564	584	560
Equipment Rental Costs	yes	yes	yes	yes		yes		yes	yes	yes	yes	yes	yes	yes	yes	yes
Historical Cost Indexes	yes	yes	yes	yes	yes	yes	yes	yes	yes	yes	yes	yes	yes	yes	yes	no
City Cost Indexes	yes	yes	yes	yes	yes	yes	yes	yes	yes	yes	yes	yes	yes	yes	yes	yes

2021 Training Class Options   📞 877.620.6245

RSMeans offers training classes in a variety of formats—eLearning training modules, instructor-led classes, on-site training, and virtual training classes. Due to our current Covid-19 situation, RSMeans has adjusted our current training offerings and instructor-led classes for 2021 have not been determined. Please visit our website at https://www.rsmeans.com/products/training/seminars for our current schedule of class offerings.

Training classes that are offered in one or more of our formats are: Building Systems and the Construction Process; Construction Cost Estimating: Concepts & Practice; Introduction to Estimating; Scope of Work for Facilities Estimating; Facilities Construction Estimating; Maintenance & Repair Estimating for Facilities; Mechanical & Electrical Estimating; RSMeans CostWorks CD; and RSMeans Online Training.

If you have any questions or you would like to register for any class or purchase a training module, call us at 877-620.6245.

Facilities Construction Estimating

In this two-day course, professionals working in facilities management can get help with their daily challenges to establish budgets for all phases of a project.

Some of what you'll learn:
- Determining the full scope of a project
- Understanding of RSMeans data and what is included in prices
- Identifying appropriate factors to be included in your estimate
- Creative solutions to estimating issues
- Organizing estimates for presentation and discussion
- Special estimating techniques for repair/remodel and maintenance projects
- Appropriate use of contingency, city cost indexes, and reference notes
- Techniques to get to the correct estimate quickly

Who should attend: facility managers, engineers, contractors, facility tradespeople, planners, and project managers.

Construction Cost Estimating: Concepts and Practice

This one- or two-day introductory course to improve estimating skills and effectiveness starts with the details of interpreting bid documents and ends with the summary of the estimate and bid submission.

Some of what you'll learn:
- Using the plans and specifications to create estimates
- The takeoff process—deriving all tasks with correct quantities
- Developing pricing using various sources; how subcontractor pricing fits in
- Summarizing the estimate to arrive at the final number
- Formulas for area and cubic measure, adding waste and adjusting productivity to specific projects
- Evaluating subcontractors' proposals and prices
- Adding insurance and bonds
- Understanding how labor costs are calculated
- Submitting bids and proposals

Who should attend: project managers, architects, engineers, owners' representatives, contractors, and anyone who's responsible for budgeting or estimating construction projects.

Assessing Scope of Work for Facilities Construction Estimating

This two-day practical training program addresses the vital importance of understanding the scope of projects in order to produce accurate cost estimates for facilities repair and remodeling.

Some of what you'll learn:
- Discussions of site visits, plans/specs, record drawings of facilities, and site-specific lists
- Review of CSI divisions, including means, methods, materials, and the challenges of scoping each topic
- Exercises in scope identification and scope writing for accurate estimating of projects
- Hands-on exercises that require scope, takeoff, and pricing

Who should attend: corporate and government estimators, planners, facility managers, and others who need to produce accurate project estimates.

Maintenance & Repair Estimating for Facilities

This two-day course teaches attendees how to plan, budget, and estimate the cost of ongoing and preventive maintenance and repair for existing buildings and grounds.

Some of what you'll learn:
- The most financially favorable maintenance, repair, and replacement scheduling and estimating
- Auditing and value engineering facilities
- Preventive planning and facilities upgrading
- Determining both in-house and contract-out service costs
- Annual, asset-protecting M&R plan

Who should attend: facility managers, maintenance supervisors, buildings and grounds superintendents, plant managers, planners, estimators, and others involved in facilities planning and budgeting.

Mechanical & Electrical Estimating

This two-day course teaches attendees how to prepare more accurate and complete mechanical/electrical estimates, avoid the pitfalls of omission and double-counting, and understand the composition and rationale within the RSMeans mechanical/electrical database.

Some of what you'll learn:
- The unique way mechanical and electrical systems are interrelated
- M&E estimates—conceptual, planning, budgeting, and bidding stages
- Order of magnitude, square foot, assemblies, and unit price estimating
- Comparative cost analysis of equipment and design alternatives

Who should attend: architects, engineers, facilities managers, mechanical and electrical contractors, and others who need a highly reliable method for developing, understanding, and evaluating mechanical and electrical contracts.

Building Systems and the Construction Process

This one-day course was written to assist novices and those outside the industry in obtaining a solid understanding of the construction process - from both a building systems and construction administration approach.

Some of what you'll learn:
- Various systems used and how components come together to create a building
- Start with foundation and end with the physical systems of the structure such as HVAC and Electrical
- Focus on the process from start of design through project closeout

This training session requires you to bring a laptop computer to class.

Who should attend: building professionals or novices to help make the crossover to the construction industry; suited for anyone responsible for providing high level oversight on construction projects.

Practical Project Management for Construction Professionals

In this two-day course you will acquire the essential knowledge and develop the skills to effectively and efficiently execute the day-to-day responsibilities of the construction project manager.

Some of what you'll learn:
- General conditions of the construction contract
- Contract modifications: change orders and construction change directives
- Negotiations with subcontractors and vendors
- Effective writing: notification and communications
- Dispute resolution: claims and liens

Who should attend: architects, engineers, owners' representatives, and project managers.

RSMeans data Training

Training for our Online Estimating Solution

Construction estimating is vital to the decision-making process at each state of every project. Our online solution works the way you do. It's systematic, flexible, and intuitive. In this one-day class you will see how you can estimate any phase of any project faster and better.

Some of what you'll learn:
- Customizing our online estimating solution
- Making the most of RSMeans "Circle Reference" numbers
- How to integrate your cost data
- Generating reports, exporting estimates to MS Excel, sharing, collaborating, and more

Also offered as a self-paced or on-site training program!

Training for our CD Estimating Solution

This one-day course helps users become more familiar with the functionality of the CD. Each menu, icon, screen, and function found in the program is explaine in depth. Time is devoted to hands-on estimating exercises.

Some of what you'll learn:
- Searching the database using all navigation methods
- Exporting RSMeans data to your preferred spreadsheet format
- Viewing crews, assembly components, and much more
- Automatically regionalizing the database

This training session requires you to bring a laptop computer to class.

When you register for this course you will receive an outline for your laptop requirements.

Also offered as a self-paced or on-site training program!

Site Work Estimating with RSMeans data

This one-day program focuses directly on site work costs. Accurately scoping, quantifying, and pricing site preparation, underground utility work, and improvements to exterior site elements are often the most difficult estimating tasks on any project. Some of what you'll learn:
- Evaluation of site work and understanding site scope including: site clearing, grading, excavation, disposal and trucking of materials, backfill and compaction, underground utilities, paving, sidewalks, and seeding & planting.
- Unit price site work estimates—Correct use of RSMeans site work cost data to develop a cost estimate.
- Using and modifying assemblies—Save valuable time when estimating site work activities using custom assemblies.

Who should attend: engineers, contractors, estimators, project managers, owners' representatives, and others who are concerned with the proper preparation and/or evaluation of site work estimates.

Please bring a laptop with capability to access the internet.

Facilities Estimating Using the CD

This two-day class combines hands-on skill-building with best estimating practices and real-life problems. You will learn key concepts, tips, pointers, and guidelines to save time and avoid cost oversights and errors.

Some of what you'll learn:
- Estimating process concepts
- Customizing and adapting RSMeans cost data
- Establishing scope of work to account for all known variables
- Budget estimating: when, why, and how
- Site visits: what to look for and what you can't afford to overlook
- How to estimate repair and remodeling variables

This training session requires you to bring a laptop computer to class.

Who should attend: facility managers, architects, engineers, contractors, facility tradespeople, planners, project managers, and anyone involved with JOC, SABRE, or IDIQ.

Registration Information

How to register

By Phone
Register by phone at 877.620.6245

Online
Register online at
RSMeans.com/products/services/training

Note: Purchase Orders or Credits Cards are required to register.

Two-day seminar registration fee - $1,300*.

One-Day Construction Cost Estimating or Building Systems and the Construction Process - $825*.

Two-day virtual training classes - $825*.

Three-day virtual training classes - $995*.

Instructor-led Government pricing

All federal government employees save off the regular seminar price. Other promotional discounts cannot be combined with the government discount. Call 781.422.5115 for government pricing.

CANCELLATION POLICY FOR INSTRUCTOR-LED CLASSES:

If you are unable to attend a seminar, substitutions may be made at any time before the session starts by notifying the seminar registrar at 781.422.5115 or your sales representative.
If you cancel twenty-one (21) days or more prior to the seminar, there will be no penalty and your registration fees will be refunded. These cancellations must be received by the seminar registrar or your sales representative and will be confirmed to be eligible for cancellation.
If you cancel fewer than twenty-one (21) days prior to the seminar, you will forfeit the registration fee.
In the unfortunate event of an RSMeans cancellation, RSMeans will work with you to reschedule your attendance in the same seminar at a later date or will

fully refund your registration fee. RSMeans cannot be responsible for any non-refundable travel expenses incurred by you or another as a result of your registration, attendance at, or cancellation of an RSMeans seminar.
Any on-demand training modules are not eligible for cancellation, substitution, transfer, return, or refund.

AACE approved courses

Many seminars described and offered here have been approved for 14 hours (1.4 recertification credits) of credit by the AACE International Certification Board toward meeting the continuing education requirements for recertification as a Certified Cost Engineer/Certified Cost Consultant.

AIA Continuing Education

We are registered with the AIA Continuing Education System (AIA/CES) and are committed to developing quality learning activities in accordance with the CES criteria. Many seminars meet the AIA/CES criteria for Quality Level 2. AIA members may receive 14 learning units (LUs) for each two-day RSMeans course.

Daily course schedule

The first day of each seminar session begins at 8:30 a.m. and ends at 4:30 p.m. The second day begins at 8:00 a.m. and ends at 4:00 p.m. Participants are urged to bring a hand-held calculator since many actual problems will be worke out in each session.

Continental breakfast

Your registration includes the cost of a continental breakfast and a morning and afternoon refreshment break. These informal segments allow you to discuss topics of mutual interest with other seminar attendees. (You are free to make your own lunch and dinner arrangements.)

Hotel/transportation arrangements

We arrange to hold a block of rooms at most host hotels. To take advantage of special group rates when making your reservation, be sure to mention that you are attending the RSMeans Institute data seminar. You are, of course, free to stay at the lodging place of your choice. (Hotel reservations and transportation arrangements should be made directly by seminar attendees.)

Important

Class sizes are limited, so please register as soon as possible.

*Note: Pricing subject to change.**

Orthodontic Treatment of Impacted Teeth

To my wife Sheila, to our children and grandchildren,
and to the memories of our parents and my sister.

Orthodontic Treatment of Impacted Teeth

Third Edition

Adrian Becker

BDS, LDS RCS, DDO RCPS

Clinical Associate Professor Emeritus
Department of Orthodontics
Hebrew University–Hadassah School of Dental Medicine
Jerusalem
Israel

A John Wiley & Sons, Ltd., Publication

This edition first published 2012
© 2012 by Adrian Becker

First published in 1998 by Martin Dunitz and in 2007 by Informa UK Ltd

Wiley-Blackwell is an imprint of John Wiley & Sons, formed by the merger of Wiley's global Scientific, Technical and Medical business with Blackwell Publishing.

Registered office: John Wiley & Sons, Ltd, The Atrium, Southern Gate, Chichester, West Sussex, PO19 8SQ, UK

Editorial offices: 9600 Garsington Road, Oxford, OX4 2DQ, UK
          The Atrium, Southern Gate, Chichester, West Sussex, PO19 8SQ, UK
          2121 State Avenue, Ames, Iowa 50014-8300, USA

For details of our global editorial offices, for customer services and for information about how to apply for permission to reuse the copyright material in this book please see our website at www.wiley.com/wiley-blackwell.

The right of the author to be identified as the author of this work has been asserted in accordance with the UK Copyright, Designs and Patents Act 1988.

Library of Congress Cataloging-in-Publication Data
Becker, Adrian.
  Orthodontic treatment of impacted teeth / Adrian Becker. – 3rd ed.
        p. ; cm.
  Includes bibliographical references and index.
  ISBN-13: 978-1-4443-3675-7 (hard cover : alk. paper)
  ISBN-10: 1-4443-3675-4 (hard cover : alk. paper)
  I. Title.
  [DNLM: 1. Tooth, Impacted–surgery.   2. Orthodontics, Corrective–methods. WU 101.5]
  LC classification not assigned
  617.6'43–dc23
                        2011034239

A catalogue record for this book is available from the British Library.

Wiley also publishes its books in a variety of electronic formats. Some content that appears in print may not be available in electronic books.

Set in 10/12 pt Minion by Toppan Best-set Premedia Limited, Hong Kong
Printed and bound in Singapore by Markono Print Media Pte Ltd

1   2012

Contents

Preface to the First Edition

There can be little question that the treatment of impacted teeth has caught the imagination of many in the dental profession. The challenge has, over the years, been taken up by the general practitioner and by a number of dental specialists, including the paedodontist, the periodontist, the orthodontist and, most of all, the oral and maxillofacial surgeon. Each of these professionals has much 'input' to offer in the resolution of the immediate problem and each is able to show some fine results. However, no single individual on this specialist list can completely and successfully treat more than a few of these cases without the assistance of one or more of others of his/her colleagues on that list. Thus, the type of treatment prescribed may depend upon which of these dental specialists sees the patient first and the level of his/her experience with the problem in his/her field. Such treatment may involve surgical exposure and packing, orthodontic space opening, perhaps autotransplantation, or a surgical dento-alveolar set-down procedure, or even just an abnormally angulated prosthetic crown reconstruction.

Experience has come to show that the orthodontic/surgical modality has the potential to achieve the most satisfactory results in the long term. Despite this, many orthodontists have ignored or abrogated their responsibility towards the subject of impacted teeth to others, accounting for the popularity of other modalities of treatment. The subject has become something of a Cinderella of dentistry.

Within the orthodontic/surgical modality, much room exists for debate as to what should be done first and to what lengths each of the two specialties represented should go in the zealous pursuit of its allotted portion of the procedure. The literature offers scant information and guidance to resolve these issues, leaving the practitioner to fend for himself/herself, with a problem that has ramifications in several different specialist realms.

This book discusses the many aspects of impacted teeth, including their prevalence, aetiology, diagnosis, treatment timing, treatment and prognosis. Since these aspects differ between incisors and canines, and between these and the other teeth, a separate chapter is devoted to each. The material presented is based on the findings of clinical research that has been carried out in Jerusalem by a small group of clinicians over the past 15 years or so, at the Hebrew University – Hadassah School of Dental Medicine, founded by the Alpha Omega Fraternity and from the gleanings of clinical experience in the treatment of many hundreds of my patients, young and old.

An overall and recommended approach to the treatment of impacted teeth is presented and emphasis is placed on the periodontal prognosis of the results. Among the many other aspects of this book, the intention has been to propose ideas and principles that may be used to resolve even the most difficult impaction, employing orthodontic auxiliaries of many different types and designs. None of these is specific to any particular orthodontic appliance system or treatment 'philosophy', notwithstanding the author's own personal preferences, which will become obvious from many of the illustrations. These auxiliaries may be used with equal facility in virtually any appliance system with which the reader may be fluent. The only limitation in the use of these ideas and principles are those imposed on the reader by his/her own imagination and willingness to adapt.

The orthodontic manufacturers' catalogues are replete with the more commonly and routinely used attachments, archwires and auxiliaries, which are offered to the profession with the aim of streamlining the busy practice. These items have not been tailored to the demands of the clinical issues that are raised in this book. These issues, by their very nature, are exceptional, problematic and often unique, while occurring alongside and in addition to the routine. Among the more common limitations self-imposed by many orthodontists has been the disturbing trend to rely so completely upon the use of preformed and pre-welded attachments that they have forgotten the arts of welding and soldering and no longer carry the necessary modest equipment. This then restricts one's practice to using only what is available and sufficiently commonly used to make it commercially worthwhile for the manufacturer to produce. By consenting to this unhealthy situation, the orthodontist is agreeing to work with 'one hand tied behind his/her back' and treatment results with inevitably suffer.

I acknowledge and am grateful for the help given me by several colleagues in the preparation of this book. An excellent professional relationship has been established, and has withstood the test of time, with two senior members of the Department of Oral and Maxillofacial Surgery at Hadassah, with whom a *modus operandi* has been developed, in the treatment of our patients. Professor Arye Shteyer, Head of the Department, and, subsequently, Professor Joshua Lustmann have educated me in the finer points of surgical procedure and care while, at the same time, have demonstrated a respect and understanding of the needs of the orthodontist at the time of surgery. I am grateful to them for their collaboration in the writing of Chapter 3.

Dr Ilana Brin read the original manuscript and made some useful suggestions, which have been included in the text. I am grateful to Dr Alexander Vardimon for his comments regarding the use of magnets and to Dr Tom

Weinberger for the discussions that we have had regarding several issues realised in the book. My wife, Sheila, read the earlier drafts and made many important recommendations and corrections. More than anyone else, she encouraged me to keep writing during the many months when other and more pressing responsibilities could have been used as justifiable excuses for putting the project aside.

My colleagues Dr Monica Barzel, Dr Yocheved ben Bassat, Dr Gabi Engel, Dr Doron Harary, Dr Tom Weinberger and Professor Yerucham Zilberman, and my former graduate students Dr Yossi Abed, Dr Dror Eisenbud, Dr Sylvia Geron, Dr Immanuel Gillis, Dr Raffi Romano and Dr Nir Shpack, have provided me with several of the illustrations included here and I am indebted to them.

I am grateful, too, to Ms Alison Campbell, Commissioning Editor at Martin Dunitz Publishers and to Dr Joanna Battagel, Technical Editor, for their constructive and professional critique of the manuscript, which contributed so much to its ultimate format. I also thank Naomi and Dudley Rogg, of the British Hernia Centre, for the computer and office facilities that they placed at my disposal during my short sabbatical in London in the latter stages of the preparation of the work for publication.

Permission to use illustrations from my own articles that were published in various learned journals was granted by the publishers of those journals or by the owners of the copyright, as follows:

Figure 5.13 was reprinted from Pertz B, Becker A, Chosak A, The repositioning of a traumatically-intruded mature rooted permanent incisor with a removable appliance. *J. Pedodont* 1982; 6: 343–354, with kind permission of the Journal of Pedodontics Inc.

Figure 5.4 and 5.12 were reprinted from Becker A, Stern N, Zelcer Z, Utilization of a dilacerated incisor tooth as its own space maintainer. *J Dent* 1976; 4: 263–264, with kind permission from Elsevier Science Ltd, The Boulevard, Langford Lane, Kidlington OX5 1GB, UK.

Figures 9.8–9.14 were reprinted from Becker A, Shteyer, A, Bimstein, E, Lustmann, J, Cleidocranial dysplasia: part 2 – a Treatment Protocol for the Orthodontic and Surgical Modality, *Am J Orthod Dentofac* Orthop 1997; 111: 173–183, with kind permission of Mosby-Year Book Inc., St Louis, MO, USA.

Figure 6.35 was reprinted from Kornhauser, S, Abed, Y, Harary, D, Becker, A, The resolution of palatally-impacted canines using palatalocclusal force from a buccal auxiliary, *Am J Orthod Dentofac* Orthop 1996; 110: 528–534, with kind permission of Mosby-Year Book Inc., St Louis, MO, USA.

I am very thankful for their cooperation and for their agreement.

Adrian Becker
Jerusalem

Preface to the Second Edition

In the nine years that have elapsed since the publication of the first edition of this book, much has changed in the field of orthodontics in general and, perhaps even more so, as it relates to the treatment of impacted teeth. The advances in imaging, particularly cone beam computerized tomography, have made accurate positional diagnosis of an impacted tooth virtually foolproof, enabling the application of appropriately directed traction to resolve even the most difficult cases. Temporary orthodontic implants have provided the opportunity to resolve the impaction, in many cases without the need for an orthodontic appliance and before orthopaedic treatment *per se* is begun. They have opened up a whole new area to exploit for mechanotherapeutic solutions to many of the problems we face.

The first edition was based on the findings of clinical research that was carried out over a long period of time in Jerusalem during the 1980s and 1990s. In much the same way, this second edition documents the findings of ongoing and evidence-based studies carried out by largely the same small group of clinical investigators, since then. Most of these published articles were the product of an excellent working collaboration with Dr Stella Chaushu, a former student of mine and now Senior Lecturer in the Department of Orthodontics. Her industrious and intellectual qualities have contributed to the output of a large number of valuable published studies in just a few short years.

Under the leadership of Professor Refael Zeltser, chairperson of the Department of Oral and Maxillo-facial Surgery at the Hebrew University – Hadassah School of Dental Medicine in Jerusalem, a whole generation of young surgeons has grown up who exhibit the ability to appreciate and value the finer points of cooperation with the orthodontist. Dr Eran Regev and Dr Nardi Casap in Jerusalem, Dr Gavriel Chaushu, the chairperson at the parallel department of the Sourasky Hospital in Tel Aviv, and Dr Harvey Samen in private practice, have worked closely with me in the treatment of our patients. Many of these cases are illustrated in the pages of this book. I derive considerable satisfaction from seeing the surgical expertise learned from and handed down by Professors Arye Shteyer and Joshua Lustmann being practised by these highly professional colleagues, on a day-by-day basis. Their awareness and perception of the significance of their work in determining the long-term outcome have helped me to aim for the highest quality results and the well-being of the patient. They deserve my gratitude.

In the preparation of this book, I have called upon and am grateful for the expertise of a small number of people, who have provided me with authoritative and essential information that has permitted me to make the text more comprehensive and more complete. In particular, I mention Dr James Mah and Dr David Hatcher in California, with regard to cone beam CT imaging and Dr Joe Noar in London, with regard to the use of magnets.

I have given and continue to give courses and lectures on the subject of impacted teeth in many places all over the world which, in the past few years, have been presented in collaboration with Dr Stella Chaushu. It is at these meetings that I come across some of the most interesting and rare material. I am indebted to several individual members of these audiences who frequently approach us during a coffee break, radiograph in hand, with some truly remarkable conditions, several of which have been included in this book, together with appropriate recognition.

My colleagues in the Orthodontic Department in Jerusalem have often become the sounding board for many of the ideas that are presented herein and I am thankful to them for the discussions that we have had. I appreciate their taking the stand of devil's advocate in these situations, forcing me to justify or to discard. Nevertheless, none of this would ever have been published had I not spent so many years teaching the students on our postgraduate orthodontics specialty course. These future orthodontic standard bearers are privileged to learn from the various individual teaching preferences of mentors who rely on years of experience in practice, particularly when it comes to this bracket or that, this treatment philosophy or that and this orthodontic guru or that. Additionally, they have learned to look for and even demand clinical ideas and treatment policies that have a proven evidence-based, track record to commend them and to justify their use. I know of no other postgraduate orthodontic course, worldwide, in which the subject of impacted teeth is explicitly taught in a comprehensive and integrative manner, including a designated weekly clinical session. It was this more than any other factor which encouraged me to embark on this mammoth task.

The future of our profession and the long-term superior care of the even younger generation of our patients is in the hands of these aspiring orthodontists. I am grateful to them for having, perhaps unwittingly, cajoled me into writing this text. I hope that it will be a source of information for them as they undertake the challenge of some of the more difficult, unconventional and unusual cases that they will inevitably come across in practice and for which they will be expected to find appropriate therapeutic answers.

I wish to thank the following publishers of two articles, as follows:

Several of the illustrations comprising Figure 7.8 were reprinted from the *World Journal of Orthodontics. Vol. 5. The Role of Digital Volume Tomography in the Imaging of Impacted Teeth,* by Adrian Becker and Stella Chaushu. 2004. with permission from Quintessence Publishing Co, Inc.

Several of the illustrations comprising Figure 11.9 were reprinted from *Healthy periodontium with bone and soft tissue regeneration following orthodontic-surgical retrieval of teeth impacted within cysts,* by Adrian Becker & Stella Chaushu, in *Biological Mechanisms of Tooth Movement and Craniofacial Adaptation.* Proceedings of the Fourth International Conference, 2004, pp. 155–162. Z Davidovitch and J Mah, editors. Sponsored by the Harvard Society for the Advancement of Orthodontics. Reproduced with permission.

Adrian Becker
Jerusalem, Israel

Preface to the Third Edition

Only 14 years have passed since the publication of the first edition of this book and much has changed in orthodontics, in general and in the context of the treatment of impacted teeth, in particular. The subject material that appeared in that small monograph has developed several fold, in the light of research and the advent of new technology. These two factors have encouraged the orthodontic specialist to be more discerning in the diagnosis of pathology and more innovative and resourceful in the application of directional traction. Mistaken positional diagnosis and surgical blunders have become less common and consequent failure to resolve the impaction less frequent. At the same time, they have permitted the orthodontist to become more adventurous and to successfully apply his/her knowledge and experience to the treatment of cases where previously the tooth would have been scheduled for extraction. If this third edition may yet contribute to the furtherance of this favorable trend in any way, I will consider that my mission will have been accomplished.

It was the aim in each of the earlier editions of this book to present reasoned principles of treatment for tooth impaction, illustrated by examples from real life. Following these principles to their logical conclusion, Chapter 15 has been added in the present edition to illustrate how some extreme examples or cases with concurrent complicating factors may be resolved, several of which involve the expertise of colleagues in our sister specialties. Oddities, such as the "banana" third molar, with its impacting influence on its immediate neighbor, are also new to this edition.

Failure has intrigued me for a long time and, while Chapter 12 was new to the second edition, it has been enlarged now in the third. The recognition and importance of invasive cervical root resorption (ICRR) as a cause of failure to resolve an affected impacted tooth seems to be hardly known within the profession. There is a section added herein which discusses the etiology of this pathological entity, its disease process, its potency as a factor for failure and speculates on accepted standard procedures that may predispose to its occurrence.

To write a textbook or to update an edition may take several years. Once it is finished, it has to go through the many months of the publishing process, with questions and corrections, proofreading and amendments. In the meantime, what was written becomes progressively obsolete – new ideas are put forward in the journals, some are disciplined studies and others just innovative clinical methods learned in the very singular one-on-one situation in the orthodontic operatory between orthodontist and patient. In order to provide at least a partial answer to this, I have set up an internet website at www.dr-adrianbecker.com, in which regular updates on clinical research and technique, vignettes describing individual conditions or just a customized approach to the treatment of a specific case, are published with the aim of complementing the book. The site also features a "troubleshooting impacted teeth" page for individual clinical consultations – open to anyone, whether orthodontist, patient or concerned parent. Details of the patient and his/her condition will need to be filled in and existing radiographs, CBCT and other relevant information uploaded. A report is returned to the sender within a few days with suggestions and recommendations for treatment.

The clinical research on which this text is largely based has been the product of long-term cooperation with Professor Stella Chaushu, PhD, DMD, MSc, Chairperson of the Orthodontic Department in Jerusalem, to whom are due my special thanks. I am grateful to my co-authors who have advised me in my writing of several of the chapters herein and to a number of my colleagues who have sent me illustrative material which I have included, with their permission. I would also like to recognize Mr. Israel Vider, director of the Dent-Or Imaging Center in Jerusalem, for his CT imaging expertise, his assistance in granting me access to his technical laboratory and for his work on several of the illustrations that are published in this edition.

Adrian Becker
Jerusalem, October 2011

1

General Principles Related to the Diagnosis and Treatment of Impacted Teeth

In order for us to understand what an impacted tooth is and whether and when it should be treated, it is necessary first to define our perception of normal development of the dentition as a whole and the time-frame within which it operates.

Dental age

A patient's growth and development may be faster or slower than average and we may assess his age in line with this development [1]. Thus, a child may be tall for his age, so that his morphological age may be considered to be advanced. By studying radiographs of the progress of ossification of the epiphyseal cartilages of the bones in the hands of a young patient (the carpal index) and comparing this with average data values for children of his age, we are in a position to assess the child's skeletal age. Similarly, there is a sexual age assessment related to the appearance of primary and secondary sexual features, a mental age assessment (intelligence quotient, or IQ, tests), an assessment for behaviour and another to measure the child's self-concept. These indices are used to complement the chronological age, which is calculated directly from the child's birth date, to give further information regarding a particular child's growth and development.

Dental age is another of these parameters, and it is a particularly relevant and important assessment, used in advising proper orthodontic treatment timing. Schour and Massler [2], Moorrees et al. [3, 4], Nolla [5], Demerjian et al. [6] and Koyoumdjisky-Kaye et al. [7] have drawn up tables and diagrammatic charts of stages of development of the teeth, from initiation of the calcification process through to the completion of the root apex of each of the teeth, together with the average chronological ages at which each stage occurs.

Eruption of each of the various groups of teeth is expected at a particular time, but this may be influenced by local factors, which may cause premature or delayed eruption with a wide time-span discrepancy. For this reason, eruption time is an unreliable method of assessing dental age.

With few exceptions, mainly related to frank pathology, root development proceeds in a fairly constant manner usually regardless of tooth eruption or the fate of the deciduous predecessor. It therefore follows that the use of tooth development as the basis for dental age assessment, as determined by an examination of periapical or panoramic X-rays, is a far more accurate tool.

Thus, we may find that a child of 11–12 years of age has four erupted first permanent molars and all the permanent incisors only, with deciduous canines and molars completing the erupted dentition. If the practitioner were merely to run to the eruption chart, he would note that at this age all the permanent canines and premolars should have erupted, and he would conclude that the 12 deciduous teeth have been retained beyond their due time. The treatment

that would then appear to be the logical follow-up of this observation requires the extraction of these 12 deciduous teeth! However, there are two possibilities in this situation and, in order to prevent unnecessary harm being inflicted on the child and his parents, the radiographs must be carefully studied to distinguish one context from the other.

In the event that the radiographs show the unerupted permanent canines and premolars having completed most of their expected root length, then the child's dental age corresponds with his chronological age (Figure 1.1). The deciduous teeth have not shed naturally, due to insufficient resorption of their roots. As such, we have to presume that they are the impediment to the normal eruption of the permanent teeth. Their permanent successors may then strictly be defined as having delayed eruption. Under these circumstances it would be logical to extract the deciduous teeth on the grounds that their continued presence defines them as over-retained.

The second possibility is that the radiographs reveal relatively little root development, more closely corresponding, perhaps, to the picture of the 9-year-old child on the tooth development chart (Figure 1.2). The child's birth certificate

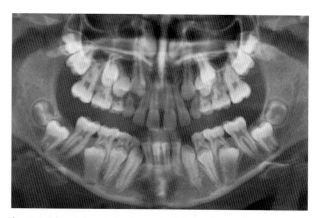

Fig. 1.1 Advanced root development of the canines and premolars in a 10-year-old child defines these teeth as exhibiting delayed eruption. The overall dental age is 12–13 years, with very late developing second permanent molars, particularly on the mandibular right side.

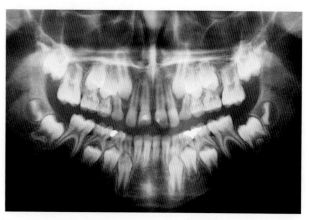

Fig. 1.2 A 12-year-old patient with root development defining dental age as 9 years. Extraction of deciduous teeth is contraindicated.

may indicate that he is 12 years of age, and this may well be supported by his body size and development, and by his intelligence. Nevertheless, his dentition is that of a child three years younger, determining his dental age as 9 years and diagnosing a late-developing dentition. Extraction in these circumstances would be the wrong line of treatment, since it is to be expected that these teeth will shed normally at the appropriate dental age and early extraction may lead to the undesired sequelae characteristic of early extraction, performed for any other reason.

From this discussion, we are now in a position to define the terms that we shall use throughout this text. The first refers to a retained deciduous tooth, which has a positive connotation and which may be defined as a tooth that remains in place beyond its normal, chronological shedding time due to the absence or retarded development of the permanent successor. By contrast and with a negative connotation, an over-retained deciduous tooth is one whose unerupted permanent successor exhibits a root development in excess of three-quarters of its expected final length (Figure 1.3). Thus, a radiograph of the permanent successor is needed to determine the status of the deciduous tooth and, by implication, its treatment.

A permanent tooth with delayed eruption is an unerupted tooth whose root is developed in excess of this length and whose spontaneous eruption may be expected in time. A tooth which is not expected to erupt in a reasonable time in these circumstances is termed an impacted tooth.

Dental age is not assessed with reference to a single tooth only, since some variation is found within the different groups of teeth. An all-round assessment must be made and, only then, can a definitive determination be offered. However, in doing this one should be wary of including the maxillary lateral incisors, the mandibular second premolars and the third molars, the timing of whose development is not always in line with that of the remaining teeth [8, 9]. These are the same teeth that are most frequently congenitally missing in cases of partial anodontia (oligodontia). Indeed, reduced size, poorly contoured crown form and late development of these teeth are all considered microforms of missing teeth [9–11]. It is important to note that this variation in the timing of their development is only ever expressed in lateness, and they are not to be found in a more advanced state of development than the other teeth. If the individual dental ages of any of these variable groups of teeth is advanced, then so too is the entire dentition in which they occur.

Assessing dental age

When studying full-mouth periapical radiographs or a panoramic film several criteria can be used in the estimation of tooth development. The first radiographic signs of the presence of a tooth are seen shortly after the initiation of calcification of the cusp tips. Thereafter, one may attempt to delineate the completed crown formation and various degrees of root formation (usually expressed in fractions), through to the fully closed root apex. By and large, orthodontic treatment is performed on a relatively older section of the child population and, as such, the stages of root formation are usually the only factors which remain relevant.

The stage of tooth development that is easiest to define is that relating to the closure of the root apex. For as long as the dental papilla is discernible at the root end, the apex is open and still developing. Once fully closed, the papilla disappears and a continuous lamina dura is seen to intimately follow the root outline. The accuracy with which one may assess fractions of an unmeasurable and merely 'expected' final root length is far less reliable and much more subject to individual observer variation.

Root development of the permanent teeth is completed approximately 2.5–3 years after normal eruption [5]. This allows us to conclude that, at the age of 9 years, the mandibular incisors (which erupt at age 6) will be the first teeth to exhibit closed apices and that these will usually be closely followed by the four first permanent molars. At 9.5 years, the mandibular lateral incisors will complete, while at 10 years and 11 years, respectively, the roots of the maxillary central and normally developing lateral incisors will be fully formed. This being so, when presented with a set of radiographs, we may proceed to assess dental age by following a simple line of investigation, which uses the dental age of 9 years as its starting point and then progresses forward or traces its steps back, depending on its findings.

If the mandibular central incisor roots are complete, we may presume the patient is at least 9 years old (dental age) and we may then advance, checking for closed apices of first molars (9–9.5 years), mandibular lateral incisors (9.5 years), maxillary central incisors (10 years), normally developing

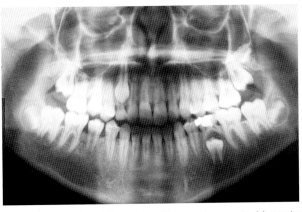

Fig. 1.3 The mandibular left second deciduous molar is retained (extraction contraindicated), since the root development of its successor is inadequate for normal eruption. The right maxillary deciduous canine, in contrast, is over-retained (extraction advised), since the long root of its successor illustrates delayed eruption.

maxillary lateral incisors (11 years), mandibular canines and first premolars (12–13 years), maxillary first premolars (13–14 years), normally developing second premolars and maxillary canines (14–15 years) and second molars (15 years).

By this method, we may arrive at a tentative determination for dental age on the basis of the last tooth in this sequence which has a closed apex (Figure 1.4). It is now important to relate the actual development of the remaining teeth in the sequence to their expected development that may be derived from the wall chart or from tables that have been presented in the literature. This may then provide corroborative evidence in support of an overall and definitive dental age determination.

When the dental age is less than 9 years, none of the permanent teeth will have completed their root development and the clinician will have no choice but to rely on an estimation of the degree of root development, degree of crown completion and, in the very young, initiation of crown calcification (Figure 1.5). This is most conveniently done by working backwards from the expected development at age 9 years and comparing the dental development

status of the patient to this, beginning with the mandibular central incisors and the first permanent molars. Thus, at dental age 6 years, one would find one-half to two-thirds root length of these teeth and this could be confirmed by studying the development of the other teeth. At the same time, one should expect unerupted maxillary central incisors with half root length, mandibular canines with one-third root length, first premolars with one-quarter root length, and so on.

As pointed out earlier, variation occurs, and this may lead to certain apparent contradictions. In such cases, excluding the affected maxillary lateral incisors, mandibular second premolars or third molars from the calculation will usually simplify the procedure and contribute to its accuracy. As we have noted, early development of these teeth in relation to the development of the remainder of the dentition does not appear to occur. Individual variability is expressed only in terms of degrees of lateness. This means that the developmental status of these teeth may be used as corroborative evidence for the determination of dental age, provided that their own timing is first confirmed as being in line with the remainder of the dentition.

Unusually small teeth, coniform premolars and mandibular incisors, and peg-shaped lateral incisors are most often seen developing very much later than normally shaped and sized teeth of the same series, sometimes as much as three or four years later, and should not be included in the overall estimation. Thus, in diagnosing dental age for a patient with an abnormality of this nature, one may present a determination for the dentition as a whole, with the added notation that this individual tooth may have a much lower dental age. Typically, we may occasionally examine a 14-year-old patient who has a complete permanent dentition, including the second molars, with the exception that a mandibular second deciduous molar is present. The radiographs (Figure 1.6) show the apices of

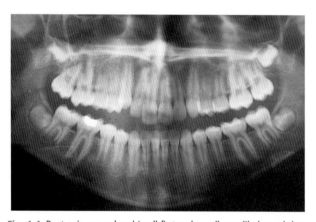

Fig. 1.4 Root apices are closed in all first molars, all mandibular and three of the maxillary incisors, excluding the left lateral incisor.

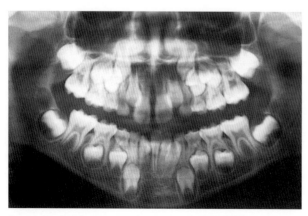

Fig. 1.5 No closed apices. Dental age assessment 7.5 years.

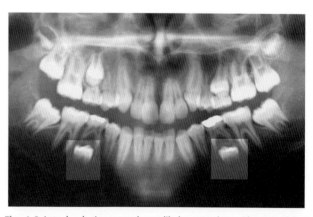

Fig. 1.6 Late-developing second mandibular premolars with retained (not over-retained) deciduous second molars in a child with dental age 11–12 years. The contrast and brightness of the picture have been adjusted in the relevant areas to clearly show the stage of development of these tooth buds.

the first molars, central and lateral incisors, mandibular canines and premolars to be closed, while the maxillary canines and the second molars are almost closed. However, the unerupted mandibular second premolar has an open root apex and development equivalent to about a quarter of its expected eventual length, or less. On the basis of this information, we may assess the dental age of the dentition as a whole to be 14 years. At the same time, we would have to note that the dental age of the unerupted second premolar is approximately 7 years. Having made this determination, we may now confidently say that the second premolar, individually, does not exhibit delayed eruption and the deciduous second molar is not over-retained in the context of the terminology used here. Accordingly, it would not be appropriate to extract the deciduous tooth at this time, but to wait at least a further few years, at which time the tooth may be expected to shed normally. To summarize this discussion, it is essential to differentiate between four different conditions that may exist when we encounter a dentition which includes certain deciduous teeth inconsistent with the patient's chronological age. Because the ensuing classification of these conditions is treatment oriented, the labelling of a patient within one of these groupings indicates the treatment that is required:

1 *A late-developing dentition.* The dental age of the patient lags behind the chronological age, as witnessed radiographically by less root formation than is to be expected at a given age, in the entire dentition. Typically, this will be evident clinically by the continued and symmetrical presence of all the deciduous molars and canines on each side of each jaw. The extraction of deciduous teeth is contraindicated, since the teeth are expected to exfoliate normally when the appropriate dental age is reached.
2 *Over-retained deciduous teeth.* The dental age of the patient may be positively correlated with the chronological age, but the radiograph shows an individual permanent tooth or teeth with well-developed roots, which remain unerupted. This tends to be localized in a single area and may be due to an ectopic siting of the permanent tooth bud, which has stimulated the resorption of only a portion of the root of its deciduous predecessor, but shedding has not occurred due to the persistence of the remaining part of the root or of a second and unresorbed root. Nevertheless, the condition may occasionally be found symmetrically in a single dental arch or in both arches. Extraction of the over-retained tooth or teeth is indicated.
3 *A normal dental age, with single or multiple late-developing permanent teeth.* This condition is commonly found in relation to the maxillary lateral incisor and the mandibular second premolar teeth, and extraction of the deciduous predecessor is to be avoided. Normal shedding of the tooth is to be expected when the root of the

permanent tooth reaches two-thirds to three-quarters of its expected length.
4 *A combination of the above.* Sometimes one may see features of each of the above three alternatives in a single dentition.

The importance of interpreting the differential diagnosis for a given patient cannot be over-emphasized, since it has far-reaching effects on all the aspects of diagnosis, treatment planning and treatment timing for cases with impacted teeth.

When is a tooth considered to be impacted?

From the work of Grøn [12] we learn that, under normal circumstances, a tooth erupts with a developing root and with approximately three-quarters of its final root length. The mandibular central incisors and first molars have marginally less root development and the mandibular canines and second molars marginally more when they erupt. We may therefore take this as a diagnostic baseline from which to assess the eruption of teeth in general. Thus, should an erupted tooth have less root development (Figure 1.7), it would be appropriate to label this tooth as prematurely erupted. This will usually be the consequence of the early loss of a deciduous tooth, particularly one whose extraction was dictated by deep caries with resultant periapical pathology.

At the opposite end of the scale, we find the unerupted tooth which exhibits a more completely developed root. The normal eruption process of this tooth must be presumed to have been impeded by one of several aetiological possibilities. These include such factors as a failure of resorption of the roots of a deciduous tooth, an abnormal eruptive path, a supernumerary tooth, dental crowding, a much enlarged dental follicle/dentigerous cyst, other forms of soft tissue pathology or a disturbance in the eruption mechanism of the tooth. However, a thickened

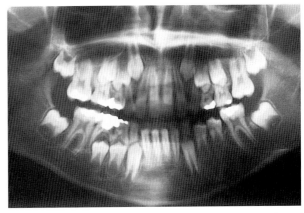

Fig. 1.7 The left mandibular premolars are prematurely erupted, with insufficient root development.

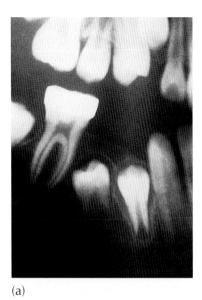

(a)

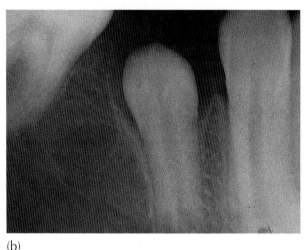

(b)

Fig. 1.8 (a) The right mandibular second premolar was extracted at age 8.5 years. (b) Seen at age 11, the root of the unerupted first premolar is almost completed.

post-extraction or post-trauma repair of the mucosa (Figure 1.8) should not be overlooked as a potent cause of non-eruption.

Not infrequently, and particularly in the mandibular premolar region, there may be a history of very early extraction of one or both deciduous molars. Delayed or non-eruption of the premolars will occur due to a thickened mucosa overlying the teeth. It is usually possible to palpate these teeth, their distinct outline clearly seen bulging the gum for a period of a year or more, although eruption may not occur.

Impacted teeth and local space loss

A time lapse exists between the performance of a surgical procedure to remove the cause of an impaction and the full eruption of the impacted tooth into its place in the dental arch. The extent of this timespan is dependent on several factors, such as the initial distance between the tooth and the occlusal plane, the stage of development of the particular tooth, the age of the patient and the manner in which hard and soft tissue may be laid down in the healing wound. During this period, therefore, local changes in the erupted dentition may occur as the result of the break in integrity of the dental arch caused by the surgical procedure, such as space loss and tipping of the adjacent erupted teeth. This intervention is no less susceptible to the drifting of neighbouring teeth than is any other factor that may produce interproximal loss of dental tissue.

With an odontome or supernumerary tooth in the path of an unerupted permanent tooth, vertical (and sometimes mesial, distal, buccal or lingual) displacement of the permanent tooth is likely to be considerable. It would be convenient if removal of the space-occupying body could be performed leaving the deciduous teeth intact, since the deciduous tooth would maintain arch integrity during the extended period needed for the permanent tooth to erupt normally. Unfortunately, in order to gain access to perform the desired surgery, one or more deciduous teeth often need to be extracted. This being so, and having regard for the long distance that a displaced permanent tooth may have to travel before it erupts into the mouth, space maintenance should be regarded as essential in most cases, particularly in the posterior area. It should be the first orthodontic procedure to be considered, preferably in advance of the surgical procedure, and it should be retained until full eruption of the permanent tooth has occurred.

Impacted teeth are often associated with a lack of space in the immediate area. This is frequently due to the drifting of adjacent teeth, although crowding of the dentition in general may be the prime cause. In such cases, the spontaneous eruption of an impacted tooth is unlikely to occur unless adequate or, preferably, excessive space is provided. It would be convenient if excision of the associated pathological entity could be comfortably delayed until this time to bring about the desired eruption and permit this corrective treatment to be attempted when the root development of the unerupted tooth is adequate. However, the surgeon will insist on removing most forms of pathology as soon as a tentative diagnosis is reached, in order to obtain examinable biopsy material for the establishment of a definitive diagnosis. Odontomes and supernumerary teeth are generally considered to be exceptions to this rule and the timing of their removal may be considered more leisurely.

Whose problem?

Patients do not go to their dentist complaining of an impacted tooth. They are frequently unaware that this abnormality exists, since there is no pain, discomfort or swelling. Nor is it obvious to the layman that there is a

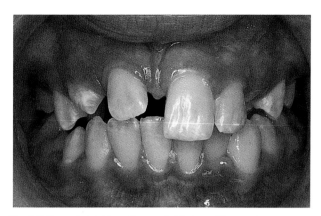
Fig. 1.9 Unerupted right maxillary central incisor with space loss.

missing tooth, since the deciduous predecessor may not shed naturally in these circumstances. The vast majority of impacted teeth come to light by chance, in routine dental examination, and are not the result of a patient's direct complaint. As a general rule, it is the paedodontist or general dental practitioner who, during a routine dental examination, discovers and records the existence of an over-retained deciduous tooth. A periapical radiograph will then confirm the diagnosis.

There are two principal exceptions whereby an abnormal appearance may motivate the patient to seek professional advice. The first of these usually brings the patient to the office at the age of 8–10 years, when a single maxillary central incisor will have erupted a year or so earlier and the parent points out that the erupting lateral incisor of the opposite side has not left enough space for the expected eruption of the second central incisor (Figure 1.9). Often, the deciduous central incisor is over-retained. In this situation, the parent has recognized the abnormality, but will not generally have the technical understanding to suggest the possibility of impaction of the unerupted central incisor.

The second exception occurs with a 14–15-year-old patient who requests the restoration of an unsightly carious lesion on an over-retained maxillary deciduous canine. Generally speaking, the patient will be unaware that this is not a permanent tooth and it will require suitable professional advice to point out that restoration is probably not the appropriate line of treatment, but rather extraction and resolution of the impaction of the permanent canine.

A very small percentage of cases may initially be seen by their general dental practitioner because of symptoms related to relatively rare complications of impacted teeth. Among these symptoms are mobility or migration of adjacent teeth (due to extensive root resorption), painless bony expansion (dentigerous or radicular cyst) or perhaps pain and/or discharge (non-vital over-retained deciduous tooth or infected cyst, with communication to the oral cavity) [13].

Initially, the practitioner should ascertain whether there is a good chance that resolution will be spontaneous once the aetiological factor has been removed or whether active appliance therapy will be needed. To be able to do this, the exact position, long-axis angulation and rotational status of the tooth have to be accurately visualized and an assessment of space in the arch needs to be made. Following this initial assessment, the paedodontist or general dental practitioner now has to decide who should treat the problem.

Many dentists will prefer not to accept responsibility for the case and will refer the patient to an oral and maxillofacial surgeon on the premise that surgery will be needed. Many surgeons will agree that the problem is essentially surgical in nature and will proceed to remove over-retained deciduous teeth, clear away other possible aetiological factors, such as supernumerary teeth, odontomes, cysts and tumours, and will also expose the impacted permanent tooth. If the impacted tooth is buccally located, the surgical flap may be apically repositioned to prevent primary closure and to maintain subsequent visual contact with the impacted tooth after healing has occurred. This will have the effect of encouraging eruption in many cases. Until healing (by 'secondary intention') has occurred, the wound will usually be packed with a proprietary zinc oxide/eugenol-based periodontal pack (e.g. CoePack®) or a gauze strip impregnated with Whitehead's varnish, over a period of a few weeks. Careful placement and wedging of the pack between an impacted tooth and its neighbour is used by surgeons to help free the tooth to erupt naturally when the pack is later removed. Often, in the more difficult impactions, wider surgical exposure is undertaken, which includes fairly radical bone resection, both around the crown and down to the cemento-enamel junction, with complete removal of the dental follicle. The principal aims of this procedure are to clear away all possible impediments to eruption and to ensure that subsequent healing of the soft tissues does not cover the tooth again.

Following a period of many months and (for some of the more awkwardly positioned teeth) sometimes extending into years, the surgeon, family dentist or paedodontist will usually then follow up the spontaneous eruption of the impacted tooth until it reaches the occlusal level. If, at that time, alignment is poor or the tooth still has not erupted, the patient will be referred to the orthodontist.

They may alternatively and preferably refer the patient directly to an orthodontist in the first place. Certainly, the orthodontist cannot directly influence the position of the impacted tooth until appropriate access has been provided surgically and an attachment has been placed on the tooth. Nevertheless, with proper planning and management, including referral for surgical exposure at the appropriate stage in the treatment, a much higher level of quality care may be provided and in a very much shorter timeframe. This will be discussed in the ensuing chapters of this book.

The timing of the surgical intervention

From the above discussion, we see that the timing and nature of the surgical procedure are determined by the degree of development of the teeth concerned, at the time of the initial diagnosis. At an early stage, a radiographic survey of a very young child may reveal pathology, such as a supernumerary tooth, an odontome, a cyst or benign tumour, which appears likely to prevent the normal and spontaneous eruption of a neighbouring tooth.

At this stage, it would be inappropriate to expose the crown of an immature tooth from every point of view. In the first place, one would not want to encourage the tooth to erupt before an adequate (half to two-thirds) root length has been produced. Second, at that early stage of its development, the tooth cannot be considered as impacted and, given time and freedom to manoeuvre, will probably erupt by itself. Early exposure risks the possibility of damage to the crown and to the subsequent root development of the tooth.

Nevertheless, with the discovery of the pathological condition (Figure 1.10), the potential for impaction exists and leaving the condition untreated will worsen the prognosis. Accordingly, removal of the pathological entity, without disturbing the adjacent permanent teeth or their follicular crypts, should be the aim of any treatment at that time. It may then reasonably be expected that normal development and eruption will eventually occur. Whilst this is an obviously desirable course of action, access to the targeted area may be thwarted by the presence and closeness of adjacent developing structures and delay may still be advised.

The second scenario occurs when the condition is only discovered much later. In this case (Figure 1.11), it may be seen that the superiorly displaced central incisors have fully developed, if angulated, roots and the adjacent lateral incisors have erupted with almost the full length of their roots completed. The central incisors may justifiably be defined as impacted, and the aims of surgical treatment become two-fold: first, to eliminate the pathology, and then to create optimal conditions for the eruption of the permanent tooth, which is already late. This will usually involve exposure of the crown of the tooth. For many teeth, given adequate space in the dental arch and little or no displacement of the impacted tooth, spontaneous eruption may be expected [14, 15]. As we shall see in subsequent chapters, there are several situations and tooth types where this may not occur, or may not occur in a reasonable time-frame, often due to severe displacement of the affected tooth. For these cases, the natural eruptive potential of the tooth is supplemented and, if necessary, diverted mechanically, with the use of an orthodontic appliance.

Patient motivation and the orthodontic option

Angle's class 2 malocclusion is present in between one-fifth and one-quarter of the child population in most countries of the western world [16, 17]. However, even a cursory analysis of the patient load of any given orthodontic practice will reveal that around three-quarters of the patients are being treated for this malocclusion. The reason for this has to do with the fact that a patient's appearance is adversely affected to a greater extent by this condition than by most others. In other words, appearance plays an extremely large part in the initiative and motivation on the part of the parent of this young patient to seek treatment.

A significant section of the remaining quarter of the patients in this hypothetical orthodontic practice are being treated for various less unsightly conditions (crowding, single ectopic teeth, open bites or class 3 relationships). This leaves only a few patients in this practice sample who have been referred for strictly health reasons, which may not be obvious to the patient.

Appearance is not a problem for this small group of patients, who will have agreed to orthodontic treatment only after they have been motivated by the careful and

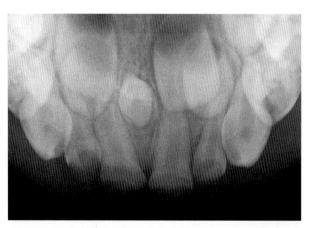

Fig. 1.10 A midline supernumerary tooth (mesiodens) discovered in routine periapical radiographic view of the maxillary incisor area in a 4-year-old child.

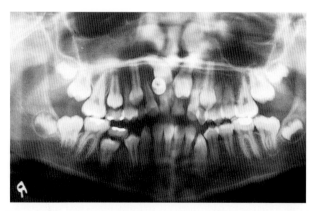

Fig. 1.11 The panoramic view shows erupted maxillary permanent lateral incisors and over-retained deciduous central incisors. The unerupted permanent central incisors can be seen superior to the two unerupted supernumerary teeth. (Courtesy of Dr I. Gillis.)

persuasive explanations of a dentist, orthodontist, periodontist, prosthodontist or oral surgeon, regarding the ills that are otherwise likely to befall them and their dentition.

Most impactions are symptomless and, aside from maxillary central incisors, do not usually present an obviously abnormal appearance. Accordingly, motivation for treatment in these cases is minimal, and much time has to be spent with the patient before he/she agrees to treatment. The story does not end there, since these patients may often require periodic 'pep talks' to maintain their cooperation and the resolve to complete the treatment. Many of them will not maintain the required standard of oral hygiene, and, while it is difficult to justify continuing treatment in these circumstances, it is just as difficult to remove appliances from a patient in the middle of treatment, when impacted teeth have partially erupted and large spaces are present in the dental arch. For these reasons, while ambitious and innovative treatment plans may be suggested, it is essential to take motivation into account before advising lengthy and complicated treatment, since the risk of non-completion may be high.

References

1. Krogman WM. Biological timing and the dentofacial complex. J Dent Child 1968; 35: 175–185.
2. Schour I, Massler M. The development of the human dentition. J Am Dent Assoc 1941; 28: 1153–1160.
3. Moorrees CFA, Fanning EA, Grøn A-M, Lebret L. The timing of orthodontic treatment in relation to tooth formation. Trans Eur Orthod Soc 1962; 38: 1–14.
4. Moorrees CFA, Fanning EA, Hunt EE Jr. Age variation of formation stages for ten permanent teeth. J Dent Res 1963; 42: 1490–1502.
5. Nolla CM. The development of permanent teeth. J Dent Child 1960; 27: 254–266.
6. Demerjian A, Goldstein H, Tanner JM. A new system of dental age assessment. Hum Biol 1973; 45: 211–227.
7. Koyoumdjisky-Kaye E, Baras M, Grover NB. Stages in the emergence of the dentition: an improved classification and its application to Israeli children. Growth 1977; 41: 285–296.
8. Garn SM, Lewis AB, Vicinus JH. Third molar polymorphism and its significance to dental genetics. J Dent Res 1963; 42: 1344–1363.
9. Sofaer JA. Dental morphologic variation and the Hardy-Weinberg law. J Dent Res 1970; 49 (Suppl): 1505.
10. Gràhnen H. Hypodontia in the permanent dentition. A clinical and genetic investigation. Odontol Revy 1956; 79 (Suppl 3): 1–100.
11. Alvesalo L, Portin P. The inheritance pattern of missing, peg-shaped and strongly mesio-distally reduced upper lateral incisors. Acta Odontol Scand 1969; 27: 563–575.
12. Grøn A-M. Prediction of tooth emergence. J Dent Res 1962; 41: 573–585.
13. Shafer WG, Hine MK, Levy BM. A Textbook of Oral Pathology, 4th edn. Philadelphia: WB Saunders, 1983.
14. DiBiase DD. The effects of variations in tooth morphology and position on eruption. Dent Pract Dent Rec 1971; 22: 95–108.
15. Mitchell L, Bennett TG. Supernumerary teeth causing delayed eruption – a retrospective study. Br J Orthod 1992; 19: 41–46.
16. Brin I, Becker A, Shalhav M. Position of the maxillary permanent canine in relation to anomalous or missing lateral incisors: a population study. Eur J Orthod 1986; 8: 12–16.
17. Massler M, Frankel JM. Prevalence of malocclusion in children aged 14–18 yrs. Am J Orthod 1951; 37: 751–760.

2

Radiographic Methods Related to the Diagnosis of Impacted Teeth

(In Collaboration with Stella Chaushu)

It is not the purpose of this chapter to present a complete manual on dental radiography, but rather to highlight concisely those techniques and methods that are useful in the clinical setting, as it pertains to impacted teeth. The methods offered have two main aims [1, 2]. The first relates to the furnishing of qualitative information regarding normal and abnormal conditions that may be associated with unerupted teeth. Thus, the different ways of radiologically displaying and recognizing pathological entities, such as supernumerary teeth, enlarged eruption follicles, odontomes, root resorption and other pathological entities, are discussed and compared. The second aim is to describe the various radiological techniques that the clinician may find helpful in accurately pinpointing the position of a clinically invisible, unerupted tooth in the three planes of space. The relative merits of these techniques are discussed and indications for their use are suggested in relation to the different groups of teeth concerned.

Qualitative radiography

Periapical radiographs

The first, simplest and most informative X-ray film is the periapical view. This view is oriented to pass through the minimum of surrounding tissue, in order to give accuracy and quality of resolution. It is generally aimed to be perpendicular to an imaginary plane which bisects the angle between the long axis of an erupted tooth and the plane of the film, to produce the minimum of distortion. The periapical film is designed to view the tooth itself from the angle of best advantage, unrelated to its position in space.

From this view, it will be immediately obvious if there is an impacted tooth and if its stage of development is similar to that of its erupted antimere, with at least two-thirds of its root length. The presence and size of a follicle will be obvious, and crown or root resorption, root pattern and integrity will be possible to ascertain. The presence and description of hard tissue obstruction will be evident, allowing the observer to distinguish connate, incisiform and barrel-shaped supernumeraries, and odontomes of the complex or compound composite type. Similarly, it will show soft tissue lesions, such as cysts. The great clarity that the view offers is superior to other views and should always be used as the initial film of a suspected impacted tooth in a radiographic examination. As with any radiographic film, however, the periapical view is two-dimensional, and thus can give no information in the bucco-lingual plane. Overlapping structures cannot be differentiated on a single film as to which is lingual and which buccal.

For this film to give the most advantageous view of the teeth in the maxillary arch and in the mandibular anterior segment, the central ray of the periapical view is oblique, and will vary between 20° and 55° to the occlusal plane [3] depending on the region to be X-rayed. Given this oblique direction, any attempt to estimate the height of an impacted

tooth or its bucco-lingual location, without additional information, must fail.

When performing periapical radiography on the posterior teeth in the mandibular arch, however, the most advantageous direction has the central ray very close to the horizontal and as such also offers a true lateral view of these teeth. Thus, not only will the observer see the most precise detail of the tooth and its surrounding tissues, it will also be possible to accurately assess its height in the jaw.

Occlusal radiographs

Mandibular arch

In the mandibular arch, this view is properly executed by tipping the patient's head backwards and pointing the X-ray tube at right-angles to a film, held between the teeth, in the occlusal plane (Figure 2.1). The head will need to be tipped back to permit the positioning of the X-ray tube under the chin. In the lower canine/premolar region, the occlusal view is a 'true' occlusal view and should depict all the posterior standing teeth in cross-section, and as such should also provide bucco-lingual positional information on the tooth and any associated structures in a plane at right-angles to that seen on the periapical film. Due to the thickness of bone traversed, detail is much poorer, unless there is expansion owing to a large cyst or a bucco-lingually displaced tooth.

In order to produce a true occlusal view in the anterior region of the mandibular arch (Figure 2.1), the head will need to be tipped back further and the tube pointed at the symphysis menti, at an angle of 110° to the horizontal, in line with the long axes of the incisor teeth. To achieve the same for the molar teeth, the 90° angle to the horizontal will need to be augmented by a 15° medial tilt of the tube, to compensate for the characteristic slight lingual tipping of these teeth [3]. This means that, ideally, the film should

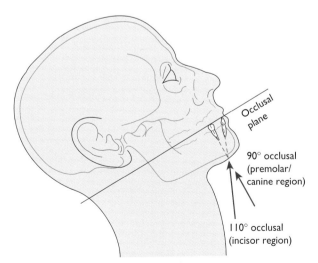

Fig. 2.1 The angle of the central ray in a true occlusal view of the lower jaw depends on the area of interest. Reproduced from previous edition with the kind permission of Informa Healthcare – Books.

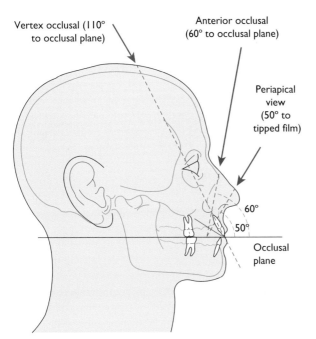

Vertex occlusal (110° to occlusal plane)

Anterior occlusal (60° to occlusal plane)

Periapical view (50° to tipped film)

60°

50°

Occlusal plane

Fig. 2.2 A diagram showing incisor inclination, film position and central X-ray beam, differentiating the periapical view, the anterior (oblique) occlusal view and the true vertex occlusal view. Reproduced from previous edition with the kind permission of Informa Healthcare – Books.

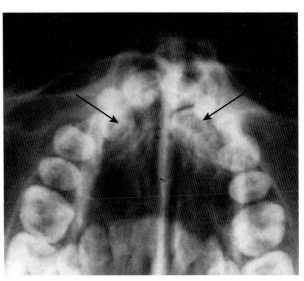

Fig. 2.3 A true vertex occlusal film using Ong's projection, showing two palatal canines. The right canine is close to the arch and almost vertical. The crown of the left canine reaches the midline suture, while its root apex is close to the line of the arch.

be performed individually for each side, in order to capture each molar in its long axis and its true occlusal view.

Maxillary arch

Maxillary anterior occlusal

In the maxillary arch, the nose and forehead interfere with the positioning of the X-ray tube, close to the area to be viewed. The best that can be achieved by positioning the tube close to the face is an oblique, anterior, maxillary occlusal view of the teeth, which is perhaps better described as a high or steeply angled periapical view (Figure 2.2). The view will 'shorten' the apparent length of the roots, but it will be a far cry from the cross-sectional view that is so easy to achieve in the mandibular arch. Since the central ray passes through cancellous rather than the compact bone that is found in the mandible, detail is usually good, although not as clear as with the periapical view.

True (vertex) occlusal

A true occlusal view of the anterior maxilla is a view in which the central ray of the X-ray beam runs parallel to the long axis of the central incisors (Figure 2.3). This is only possible when the cone is placed over the vertex of the skull, to produce the vertex occlusal film. Since the beam has to travel a great distance through the cranium and its contents, the base of the skull and the maxilla, there is a considerable loss in clarity. An excellent alternative method of producing this view, with the film positioned extra-orally, has been described [4]. Notwithstanding, a very long exposure is required, and a fast film should be used in a cassette with

intensifying screens. For all these reasons, the method has never been popular. It is, therefore, almost with a collective sigh of relief among professionals that the method has been totally superseded by the introduction of volumetric cone beam computerized tomography (CBCT) scanning. This imaging modality, which can give the same and much more information with little or no increase in radiation dosage, has developed considerable sophistication within a very short space of time and is discussed at the end of this chapter.

Nevertheless, in this view (Figure 2.3), all the anterior teeth will be seen in their cross-sectional aspect as small circles with a tiny concentric circle in the centre, denoting the pulp chamber and root canal. No information is available regarding the relative height of the object in the alveolus and it certainly cannot be used for fine detail. A single tooth which is palatal to the line of the arch will appear within this arc of small circles. If the tooth is at an angle, not parallel with its neighbours, it will show up in its elliptical, oblique cross-section, representing a tilted long axis. In the event that the tooth is horizontal across the palate, its full length will be obvious on this view, together with the exact mesio-distal and bucco-lingual orientation of both the root and the crown, in the horizontal plane.

The difference may not seem to be very great between the two types of occlusal film, but it should be appreciated that from the vantage point of an anterior occlusal film, the anterior teeth will be foreshortened, but they will still have appreciable length. In this situation, a high and mesially placed labial canine could give virtually the same picture as

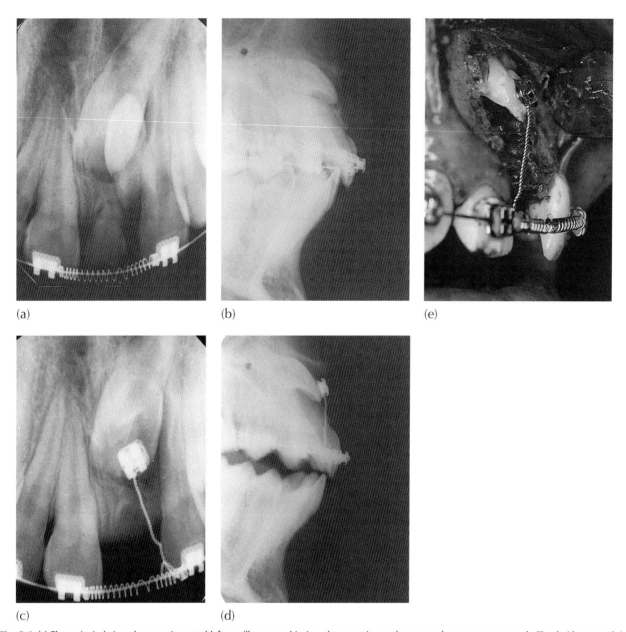

(a) (b) (e)

(c) (d)

Fig. 2.4 (a) The periapical view shows an impacted left maxillary central incisor, due to an inverted, unerupted, supernumerary tooth. The deciduous tooth is over-retained. Accurate diagnosis of the height of the impacted tooth in the alveolus is not possible from this view. (b) The anterior maxilla seen on a lateral cephalometric radiograph shows the high impacted central incisor, facing the labial sulcus. (c) and (d) The same views as (a) and (b) after removal of the supernumerary tooth and bracket bonding on the exposed incisor. (e) A parallel intra-oral photographic view. The film has been laterally inverted to simplify comparison (Courtesy of Dr D. Harary.)

a low and mesially placed palatal canine. This could not happen in a vertex occlusal projection.

Extra-oral radiographs

The panoramic view, while not showing detail to the same degree as a periapical film, has the advantage of simply and quickly offering a good scan of teeth and jaws, from the temporo-mandibular (TM) joint on one side to the TM joint on the other. It is probably true to say that, today, orthodontists are in general agreement that this film gives the most qualitative information to act as a starting point from which to proceed to other forms of radiography, in line with the demands of the particular situation in any given case.

True and oblique extra-oral views (Figure 2.4a–e) and the variously angulated oblique occlusal films all provide information that may be used to complement the periapical film, particularly when tooth displacement is severe. However, the use of any oblique film for the accurate localization of a buried tooth may frequently be misleading, be

it a single periapical, an occlusal or a lateral jaw film. This being so, two incipient dangers exist. First, a surgical procedure may be misdirected and a flap opened on the wrong side of the alveolar process. Second, misinterpretation of the tooth's position may lead the operator to consider there to be a very favourable prognosis for biomechanical resolution when, in fact, the tooth may be in a completely intractable position. Thus, the choice of treatment will be inappropriate.

Three-dimensional diagnosis of tooth position

As dentists, we are used to seeing periapical films of individual teeth and, provided that the teeth concerned are erupted and in the line of the arch, these films have many advantages. However, in this view, the X-ray tube is not directed in either the true horizontal, true vertical or true lateral plane. Aside from radiography of the mandibular posterior teeth, the tube is always tipped at an angle to one or more of these planes. For an erupted tooth, this is unimportant, since the third dimension is supplied by direct vision within the mouth. Thus, while it gives a good two-dimensional representation of the tooth, this view has limited value when visualization of an unerupted tooth is required in the three planes of space.

Parallax method [5]

By following the principles involved in binocular vision, two periapical views of the same object and taken from slightly different angles can provide depth to the flat, two-dimensional picture depicted by each of the films individually (Figure 2.5). This is of considerable help with distinguishing the buccal or lingual displacement of the canine which is low down and fairly close to the line of the arch, and is performed in the following manner (Figure 2.6):

1. A periapical-sized film is placed in the mouth, with the patient's finger holding it against the palatal aspect of the area where the tooth would normally be situated. The X-ray tube is directed at right angles (ortho-radial)

to a tangent to the line of the arch at this point, as for any periapical view, and at the appropriate angle to the horizontal plane for the tooth in question (50° for the central incisor, etc.)

2. A second film is placed in the mouth in the identical position but, on this occasion, the X-ray tube is shifted (rotated) mesially or distally round the arch, but held at the same angle to the horizontal plane and directed at the mesially or distally adjacent tooth. To achieve this, the tube should describe an arc of between 30° and 45° of a circle whose centre is somewhere in the middle of the palate.

There should be no problem identifying which of the two films is the ortho-radial view and which was taken from the distally deviated aspect by studying the relative distortion of the erupted teeth on the two films. However, by radiographically 'labelling' the deviated film with the placement of a paper clip in one corner or by using a different film size for the deviated view, such as an occlusal-sized film, this distinction will be simplified.

Let us assume that a right unerupted canine is palatally placed (Figure 2.6), and then this tooth will be close to the

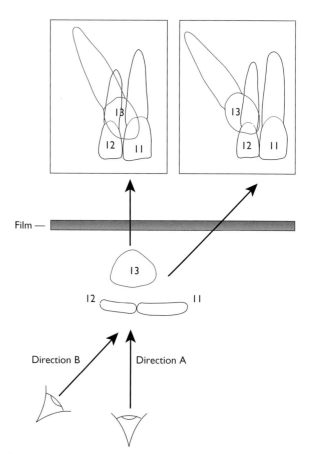

Fig. 2.6 A diagrammatic representation of the parallax method. If the observer's eye peers along the axis of the X-ray beam in each case, the image on the film will be easy to reconstruct. Reproduced from previous edition with the kind permission of Informa Healthcare – Books.

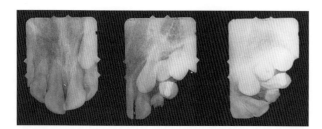

Fig. 2.5 The left periapical view, oriented for the central incisors, shows the crown of the canine superimposed on the distal half of the central incisor root. The middle film, rotated 30° to the left, shows the canine overlapping only the lateral incisor root. By rotating the central beam a further 30°, superimposition of the canine over the lateral incisor root has been eliminated. The canine is palatally displaced.

middle of the picture obtained in both films. However, in the first picture (direction B), where the tube is directed over the designated canine area of the ridge, the lateral incisor root will be on the right of the picture. If the canine is also mesially displaced, there will be some overlap of the canine crown and the lateral incisor root. On the second picture, taken from the front (direction A), the right lateral incisor root and the crown of the palatal canine will be in the middle of the picture, superimposed on one another, to a much greater degree.

Jacobs [7, 8] enjoins the observer to use the right eye in place of the X-ray tube and suggests the useful exercise of holding up two fingers vertically, at eye level, with one finger obscuring the other. If the observer now closes this eye and opens the other, his/her new vantage point for inspection will have resulted in a visual separation of the two fingers. Through the left eye, the obscured finger will have 'moved' to the left of the forward finger, to become partially visible. Transferring this to the radiographic context, in the second picture, the tooth furthest from the tube (i.e. the palatal tooth) will 'move' in the same direction that the X-ray tube has travelled from the first exposure.

This method has its limitation, although it is very useful in cases where there is a minimal height discrepancy between the erupted and unerupted adjacent teeth. However, when the canine is high and the periapical view shows no superimposition of the canine with the roots of the erupted teeth, or where the superimposition is only in the apical area, then the overall picture may be very misleading and a different method of localization should be used. The periapical view is directed from above the occlusal plane and in an oblique downward and medial direction, which distances the palatal canine from the roots of the other teeth and makes it appear higher than the anatomy of the maxilla would allow. While it may prove useful in locating the position of the crown of the impacted tooth, it is not adequate to the task of accurately placing the root apex and, thereby, defining the orientation of its long axis. These are important parameters when assessing treatment difficulty and prognosis during the treatment planning stage and critical for the successful resolution of an impacted tooth, as we shall see in the following chapters.

Vertical parallax may sometimes be a useful variant of the same technique, in which two films are taken of the area, with the central ray of one periapical film being more steeply angled in the vertical plane than the other. In this manner, the separation of the images in the more steeply angled (above the occlusal plane) film will result in a palatal tooth being more superiorly related *vis-à-vis* the target tooth than in the regular film.

Unfortunately, the parallax method in general offers a relatively low degree of reliability. In a study to evaluate the usefulness of its two variants [6], six experienced orthodontists were given the case records of 39 patients with ectopic canines. The cases were evaluated twice, once using films

that showed vertical parallax and once with films that featured horizontal parallax, although the parallax pairs were not revealed to the examiners as being of the same individual. In 83% of cases the correct positional diagnosis was made with the horizontal method, while only 68% of cases were correctly diagnosed with the vertical method. These results expose the method as being too crude, or the experts insufficiently discerning, for it to be relied on with any degree of confidence. Thus, while often useful to obtain an initial overall impression, the method should certainly be backed up by more reliable diagnostic radiographic methods before a final treatment plan is presented to the patient.

In the incisor region, an unerupted permanent incisor may be associated with one or two supernumerary teeth (mesiodentes). The parallax method is insufficiently clear in these cases, due to the presence of two or three hard tissue entities in the bone, superimposed on the outline of the roots of the deciduous teeth and at varying heights in the alveolus.

The question arises whether the parallax principles may be applied to other types of film combinations, possibly with a greater degree of reliability. A vertical imaging discrepancy between teeth in the line of the arch and those that are buccally or palatally displaced can be created between the panoramic view and the periapical/anterior occlusal views (Figure 2.7). The panoramic view is a rotational tomograph, with the cone of the machine pointing upwards with a very small 7° tilt from below the occlusal plane, as it circles around the head of the patient. Because this view is recorded when the film is on the buccal side of the teeth and the cone directed from the palatal aspect, this is equivalent to a 7° tilt of the X-ray cone, when translated into a buccal-to-palatal direction.

By contrast, the direction of the central ray in an anterior occlusal (6°–65°) or periapical view (5°–55°) is angulated much more steeply to the film. These will both show superimposition of an ectopic tooth over the tooth in the line of the arch, but a degree of vertical discrepancy between these films and the panoramic view will reveal the position of the displaced tooth.

The same panoramic view will project the anterior midline area in its postero-anterior aspect, with the X-ray beam hitting this area when the cone is at the back of the patient's head. The canine and premolar regions will be projected from an increasingly angulated viewpoint, as the X-ray cone moves from the back to the side of the head. The molar and retromolar areas will be projected from the side on the same revolving film as the consequence of the further rotation of the X-ray beam.

If all the teeth are in the same approximate semicircular line of the arch, then their mesio-distal relationships will be fairly accurately represented on the film. However, a palatally displaced canine or premolar tooth is imaged when the X-ray cone is at a point in its arc of circle just

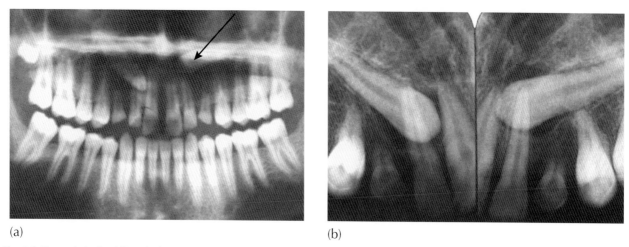

(a)   (b)

Fig. 2.7 The vertical tube shift method using a panoramic film and periapical views. (a) The panoramic film shows the left canine very high and above the root apices of the incisors (arrow). The right canine superimposes on the apical third of the adjacent incisor. (b) The periapical views show the left and right canines overlapping one-third and two-thirds of the incisor roots, respectively. Both canines are labial.

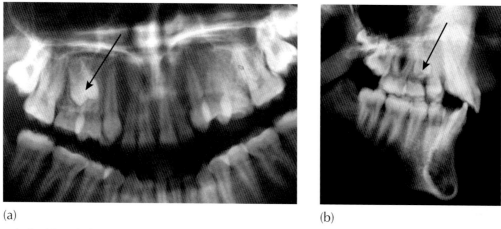

(a)   (b)

Fig. 2.8 The lateral tube shift method using a panoramic film and a lateral cephalogram. (a) In the panoramic view, the X-ray cone projects this image in the premolar area when it is behind the ear of the opposite side and therefore provides an oblique lateral view. This gives the misleading impression that the unerupted right second premolar is rotated. (b) The lateral cephalometric view of the same patient shows only a very mild mesial displacement of the second premolar, with a minimal rotation of its palatal cusp in a mesial direction. Since this film is a true lateral view, this is the true mesio-distal position of the tooth.

behind the ear on the opposite side. Viewed from this position, the palatally placed tooth will be 'thrown' mesial to its true mesio-distal position and will be shown superimposed more mesially on other structures than would be evident from its appearance on a lateral cephalogram [9]. Accordingly, a panoramic film (an oblique lateral view) and a lateral cephalogram (a true lateral view) may be used together to determine the bucco-lingual location in the canine or premolar regions, in a similar manner to the use of two periapical views in Clark's parallax method [5]. Obviously, this is dependent on the individual teeth being clearly discernible on the cephalogram, in which unavoidable superimposition in the anterior region may sometimes invalidate the method (Figure 2.8).

One further clue to the position of an ectopic canine is governed by the physical principle that objects projected optically on a screen become markedly larger as the distance between the object and the screen increases, or the distance between the object and the source decreases, with the degree of enlargement being directly proportional to the square of the object–screen distance.

The panoramic X-ray machine is normally adjusted so that its circling of the jaws maintains a fixed distance of the cone to the dental arch, whose perimeter falls within the focal trough of the machine. Teeth which are palatal to the line of the dental arch become enlarged, because they are further from the film and closer to the cone.

The mesio-distal width of a maxillary permanent canine is approximately 90% of the width of the maxillary central incisors. With a normally located canine, the distance between it and the film may be slightly larger than that of the central incisor, due to the form of the arch in that area.

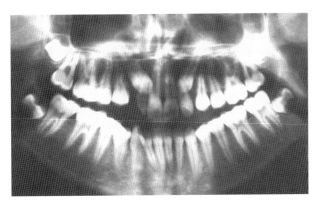

Fig. 2.9 A panoramic film showing two unerupted maxillary canines. Note the contrast in image size of the two canines. By direct measurement of the crown of the right canine, the mesio-distal width of its crown is considerably more than 15% larger than that of the right central incisor and the left is approximately the same width as the left central incisor. Since each shows superimposition on the middle portion of the root of its immediate neighbour, the right canine is palatal and the left buccal.

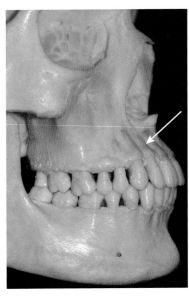

Fig. 2.10 On the dry skull, the roots of the maxillary incisor teeth can be seen to tip palatally at a significant angle to the vertical, creating a depression in the bone at the level of their apices. A canine impacted labially in this depression will be more distant from the panoramic film than the incisor crowns and will therefore cast a much enlarged image on the film. The use of the panoramic view for positional diagnosis at this relative height would therefore be incorrect.

Thus, in these cases, it is common to see similar mesio-distal widths of these two teeth on the panoramic film. A buccally displaced canine, on the other hand, will generally reflect the true width difference between the two teeth, because its distance from the film is similar to that of the central incisor (Figure 2.9).

This principle was used in an investigation of this phenomenon, which revealed that when the mesio-distal width of the crown of an unerupted canine (as it appears and is measured directly on the panoramic film) was 1.15 times larger (i.e. 15% greater) than that of the adjacent central incisor (the canine-to-incisor index), then the canine was palatally displaced [10, 11]. This was found to be reliable in 100% of cases in which the canine was seen on the film to be superimposed on the coronal or middle portions of the root of the adjacent incisor.

Earlier studies that attempted to diagnose canine position on the panoramic film using the principle of differential enlargement revealed only an 80–89% degree of reliability of diagnosis [12, 13]. This was due to the inclusion of cases where the image of the canine was superimposed on the apical portion of the root of the incisor. The anatomy of the anterior portion of the maxilla is responsible for this aberration. Erupted permanent incisor teeth do not stand vertically upright, but their roots tip palatally at a significant angle to the vertical (Figure 2.10). This means that the root apices are considerably more distant from the film drum of the panoramic machine than are the crowns. If a canine is located high up on the labial side of the root apices, in the labial alveolar depression in the incisor region inferior to the nose, the tooth may still be considerably more distant from the film than the crowns of the incisors. Thus, the image of the canine crown will be enlarged to a greater extent than those of the incisor crowns and will appear disproportionately large on the film.

Accordingly, the 1.15 canine-to-incisor index formula excludes all canines whose superimposition on the incisor root is high in the apical area. If the method is restricted to those cases in which the canine traverses the root of the incisor inferior to its apical third, then its use in determining the bucco-lingual positioning of the crown of an impacted tooth is valid, without the need to resort to other views.

Radiographic views at right angles

Radiographic views may be taken at right-angles to one another in various ways but, for the method to be of value, it must be possible to determine the exact orientation in space of both the film and the central ray [1, 2]. The observer must be in a position to deduce these from observation of other structures on the film, whose locations are known. Thus, if one begins with a periapical view, it becomes necessary to provide another view which is at 90° to it, in order to satisfy the minimum geometric conditions. However, having done this, it must be possible to reconstruct mentally the exact orientation of this second view at a later date, by looking at the film alone and without necessarily having prior knowledge of exactly how the tube and film were placed. This is obviously very confusing and completely impractical.

Standardization

Standardization of views within the confines of a strict adherence to the planes of space is required. Performed in

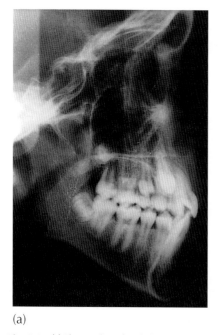
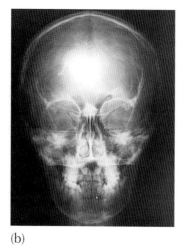
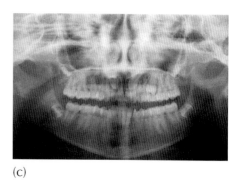

(a)   (b)   (c)

Fig. 2.11 (a) The true lateral cephalometric radiograph shows both canines superimposed at a higher level than the other teeth. Their axial inclination on the antero-posterior plane is favourable, with the crowns and apices apparently normally located. (b) The postero-anterior cephalometric radiograph shows the two canines similarly angulated, with their apices in the line of the arch and their crowns close to the midline. From these two films, we may conclude that the apices are ideally placed and that the long axes of the teeth have a downward, mesial and palatal inclination. (c) The panoramic view of the same patient. The appearance of canines close to the midline is very similar to that seen on the posterior-anterior cephalometric radiograph.

this manner, each two-dimensional view obtained becomes simple for the observer to appreciate and, when information from the other views at right-angles is merged with it, the composite three-dimensional picture is easy to mentally reconstruct. A true lateral view (Figure 2.11a) will give exact information regarding both the antero-posterior and vertical location of an object, relative to other structures that may be seen both on that radiograph and clinically. It will not give any clue to the bucco-lingual (transverse plane) picture. A true occlusal view will provide positional information in both the antero-posterior and the transverse planes, but not in the vertical plane. The third possibility is the true antero-posterior view (Figure 2.11b, c), which defines the height (vertical plane) and the bucco-lingual relationship only. By combining the information provided by any two of these three films, three-dimensional localization may be accurately determined.

Translating these principles into radiographic practice presents some difficulties. However, these are not insurmountable and, insofar as they provide the clinician with accurate positional visualization of the unerupted tooth, they may be entirely worthwhile.

For most orthodontic cases, a lateral cephalometric radiograph (a cephalogram) is a prerequisite whose primary purpose is the routine measurement of angles and planes. However, this film potentially contains much useful positional information regarding the location and angulation of unerupted teeth. The film represents a true lateral view

of the skull and, for present purposes, of the jaws and the anterior maxilla in particular (Figure 2.11a). Although there are many superimposed structures in this area, the outline of a canine may be clearly seen. The direction of the long axis of the tooth in the anterior–posterior and vertical planes may be defined, together with the mesio-distal position of both crown and apex.

In the mandibular posterior area, we have pointed out that the routine periapical radiograph is also a true lateral view, with the X-ray tube pointing at right-angles across the body of the mandible and in the horizontal plane. The height and mesio-distal position of a buried tooth may then be accurately defined. The occlusal radiograph of this area is directed perpendicular to the occlusal plane and adds the bucco-lingual dimension to complete the three-dimensional picture. Accordingly, these two views will provide accurate localization of the position of unerupted teeth in this area (Figure 2.12).

If a cephalometric radiograph is not available, the same view of the anterior maxilla may be obtained on a small, occlusal-sized film. This film is held vertically against the cheek and parallel to the sagittal plane of the skull. The X-ray tube is directed horizontally above and parallel to the occlusal plane from the opposite side of the face and at right-angles to the film. The result is called the tangential view and has the advantage of simplicity. This view is particularly useful in monitoring progress in the resolution of impacted incisors, during active treatment.

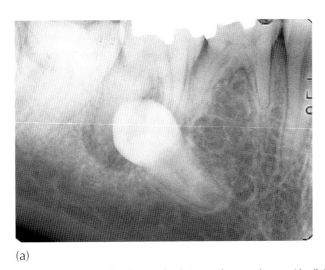

(a)

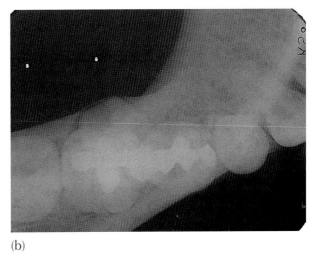

(b)

Fig. 2.12 The true lateral and true occlusal views, taken together, provide all the information needed for an accurate positional assessment of crown and root in the three planes of space. (a) The periapical view (a true lateral in this case) of an impacted mandibular right second premolar shows the tooth to be tipped 60° distally from the vertical, with its incomplete apex at the correct height and mesio-distal location. (b) The true occlusal view shows the crown of the tooth to be lingual to the molar and the apex to be in the bucco-lingual line of the arch. The long axis of the tooth, proceeding from its ideally sited apex, can be described as rising at a 30° angle in a distal and lingual direction, to overlap the molar roots on the lingual side.

At the age that most patients first present with an impacted central incisor, around 8–10 years, the permanent canine teeth are unerupted and are located both well forward and high in the anterior maxilla. Thus, on the lateral cephalometric or tangential view, right and left canines will be impossible to distinguish from one another. The roots of the incisors, at the same height as the canines, as well as the superimposed images of the more inferiorly placed crowns of the erupted incisors and deciduous canines, will all be impossible to differentiate from one another and from any supernumerary teeth that may also be present. For this reason, the lateral view may be of limited value in cases where there is obstructive impaction, with minimal displacement. When gross displacement is present, however, the outline of the altered axial inclination and height of the tooth are usually possible to delineate, despite the considerable superimposition of other teeth.

Nowhere is this view a greater asset than when a dilacerate central incisor is present, since it separates out this malformed tooth superiorly from the root apices of the other teeth and from the permanent canines, because of its relative height (Figure 2.13). Furthermore, its morphology may be seen to best advantage from this aspect, which allows definitive and accurate diagnosis of the condition to be made, together with its precise relations *vis-à-vis* surrounding structures. The lateral cephalogram/tangential view should be considered an essential requirement in radiographically recording the dilacerated central incisor.

For maxillary canines, the lateral view is extremely useful. It should be remembered that most impacted maxillary canines are diagnosed in the full permanent dentition when all the other teeth have erupted. This permits clear radio-

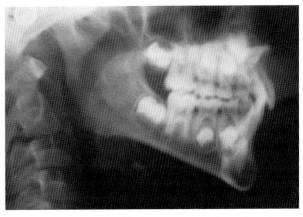

Fig. 2.13 A dilacerated central incisor seen in a lateral cephalometric film.

graphic imaging of the canine, when it is sited at a higher level than the other teeth.

A postero-anterior cephalometric film is used less routinely in orthodontics, but it offers the clinician the opportunity to view the maxilla in a different plane, the true postero-anterior view (Figure 2.11b), which is at right-angles to the lateral cephalogram. The overlap of structures of the base of the skull and the maxilla renders detail of individual teeth less clear, but a good postero-anterior radiograph will show the height of both the crown and the root of a markedly displaced tooth, as with the lateral film. This view also shows whether the root apex of an ectopic posterior tooth is in the line of the arch and how far the crown is deflected in the palatal direction. The bucco-lingual tilt of the long axis of the tooth will be plainly visible (Figure 2.11b). However, the view is less practical in the mandible,

where the left and right sides of its V-shaped body converge as they proceed forward towards the anterior midline and are thus oblique to the central ray. There is usually excessive overlap, more radio-opaque bone and difficulty in discerning even markedly bucco-lingually displaced teeth. For structures close to the midline, the panoramic view offers a very similar representation (Figure 2.11b, c) and a much clearer picture. Since this view is a rotational tomograph, it eliminates all structures that are either lingually or buccally outside the narrow focal trough at which it is aimed.

An occlusal projection of the anterior maxilla (Figure 2.3) offers the possibility of viewing in the third plane of space, at right-angles to each of the two earlier radiographs, and recording the position of the displaced incisor or canine without overlap. However, for it to be of greatest value, it is important to project the X-ray beam through the long axis of the maxillary teeth, as we have just described.

Any two of these three views (the lateral cephalogram or tangential view, the postero-anterior cephalogram and the true occlusal) will provide complete information regarding every aspect of the height, bucco-lingual and mesio-distal location of the crown and the root, and the degree of tilt of the long axis of the impacted tooth and its relation with neighbouring teeth. The postero-anterior cephalogram and the occlusal views, however, are not always as clear as is desirable, and they may need to be repeated or discarded. The lateral cephalometric or the tangential view in a case of bilateral canine impaction may create confusion, since one canine will be superimposed on the other and distinguishing them may be a problem, although other views will usually facilitate differentiation. Two identically oriented and superimposed canines (Figure 2.11) will obviously not need to be differentiated. It is emphasized that many of the radiographs that we take for other reasons should always be scrutinized for useful information regarding positions of unerupted teeth but, particularly regarding cephalograms, whose principal aim is to measure and compare angles and distances on tracings of specific landmarks on the film; often these potentially valuable views are filed away without sufficient thought.

From these projections, it is very easy to build up a three-dimensional picture of the exact position and angulation of the impacted tooth and to define the type of movement that will be necessary to bring the tooth into alignment. When a composite mental reconstruction of the position of the unerupted tooth in space is built, the design of the appliance needed to resolve the impaction is simplified and fewer surprises are likely to be encountered. It is, however, a *sine qua non* in all these cases to examine a periapical view of the tooth to eliminate the possibility of local pathology, which could be missed on the extra-oral views.

The foregoing description has covered the various methods available for envisioning the anatomical form and the three-dimensional location of unerupted teeth using plain film radiography. We have seen that, in order to achieve an adequate picture of the relationships between the crowns of these teeth and the surrounding anatomic structures, including adjacent teeth, information gleaned from several different types of view needs to be put together to make up the complex picture. Nevertheless, even when enough information is available, mistakes and misdiagnoses are sometimes made by experienced orthodontists [6], occasionally with serious repercussions for the patient.

Is the same degree of precision in root apex location and three-dimensional root orientation essential for both the orthodontist and the surgeon? From the point of view of the oral surgeon, diagnosis of the position of the crown, buccal or palatal to the line of the arch, is generally all that is needed, regardless of whether the tooth is to be exposed for orthodontic alignment or extracted. The position of the root apex and the orientation of the long axis of the tooth are irrelevant for surgical exposure. If the tooth is to be extracted in one piece, careful dissection of the tissue surrounding the crown and dislodging it with an elevator or extraction forceps will deliver the tooth, together with its root. Even if the root portion is sectioned and scheduled for extraction after removal of the crown portion, its general orientation and apex position can be determined by the anatomy and general orientation of the crown. For these reasons, many surgeons will rely solely on the tube shift parallax method of positional diagnosis for exposure or extraction of impacted teeth, and will do so with considerable confidence.

From the point of view of the orthodontist, however, while the position of the crown is important, the position of the apex and the orientation of the long axis of the tooth are crucial. When the root apex of the impacted tooth is displaced, re-siting it in its correct position is fraught with difficulty technically and can only be done once the crown has been brought into its place in the arch and ligated into the main archwire. As we shall discuss in Chapter 6, root movement is the most damaging movement to the supporting tissues and has been shown to be one of the principal factors that most undermine the periodontal prognosis of the treated result [14–16]. An orthodontist cannot be expected to be confident in his/her ability to bring a tooth into full alignment if an accurate three-dimensional diagnosis of its position is not available. In the event, a surgical exposure may be attempted from the wrong side of the alveolus and unnecessary damage will ensue. Alternatively, the impacted tooth may be drawn in the wrong direction and will be brought into contact with the root of an adjacent tooth, which may lead to resorption of that root or to the blocking of further progress. The reader is referred to Chapter 12 for an illustrated description of failures and how most of these were due to positional misdiagnosis.

Unfortunately, the above-mentioned plane film methods exhibit many shortcomings. This is particularly true in relation to the bucco-lingual plane. Undoubtedly, the most difficult aspect to define is the relative proximity of the

impacted tooth to the root of an adjacent tooth on which its image is superimposed in the bucco-lingual plane, as seen on a periapical, anterior occlusal, panoramic or cephalometric film. Whether there is a small distance between them or whether the crown of the impacted tooth lies in a resorbed crater on the palatal or labial aspect of the root of an incisor may be impossible to determine using plane film radiography. As a result, an undiagnosed and severely resorbed tooth, with a poor long-term prognosis, may be mistakenly included as an integral but 'weak link' in the final scheme of the dentition in a projected treatment plan [17].

The relative accuracy of positional diagnosis using plain film radiography is, therefore, inadequate in many instances. While this is so, there can be no question that a good number of cases continue to be successfully treated, despite a lack of adequate imaging documentation that would be needed to make even an approximate positional diagnosis. In some cases, no serious attempt at definitive diagnosis of the position of the impacted tooth is made until the unsuspecting and potentially unfortunate patient is on the operating table.

Computerized tomography

The use of CT scanning was first proposed in the present context in the late 1980s [18, 19] to identify the exact position of the palatally impacted canine, particularly when root resorption of the lateral incisor is suspected [20]. At that time, while its excellent potential for diagnosis of the position of impacted and supernumerary teeth was recognized, the large dosage of radiation that routine CT imaging required was difficult to justify for all but the most complex and exceptional cases. Nevertheless, because of the relative inadequacy of plain film radiography, many of these cases escape the discovery that they are difficult and exceptional, a revelation that can only be ascertained when a CT scan has been performed.

It has become quite clear, since writing the first edition of this text, that CT has found and established an important place in the planning of treatment of impacted teeth. Accurate three-dimensional (3-D) localization of the impacted tooth is immediately available from the 3-D views or by following the outlines of the individual teeth in successive 'cuts' on the films. In this way, the exact relationships between the impacted teeth and their adjacent teeth can be seen along the entire lengths of the crowns and roots of each.

Using this modality, it has become possible to improve the overall assessment of cases in which the impaction may best be resolved with orthodontic treatment and to sufficiently separate them from those where the tooth is in an intractable position. Trial-and-error is slowly becoming a practice of the past [21, 22] since it is now possible to present a 3-D radiographic image of what the surgical field will look like when an impacted tooth is uncovered by the oral surgeon. This helps to eliminate positional misdiagnosis and the undertaking of treatment for those relatively few cases in which the position and proximity of other teeth make it impossible to arrive at a successful conclusion to the treatment.

Similarly, the axial (horizontal) and transaxial (vertical) 'slices' made at predetermined intervals provide information in the bucco-lingual plane, which is largely impossible to discern with routine plane radiography. These views contribute materially to the evaluation of the prognosis of the intended treatment outcome. Thus, the bucco-lingual proximity of teeth and the existence and extent of oblique root resorption all become assessable, and these are important factors in deciding whether to undertake treatment and in determining choice of teeth for extraction.

The prevalence of resorption of the roots of incisor teeth in association with an impacted canine was investigated by plain film radiography in a study performed in 1988 [19] and found to affect 12% of the individuals in the sample. When the same investigators repeated their study 12 years later, using spiral CT scanning [23], the number of affected individuals increased to 48%! There can be little doubt that this was due to this vastly improved diagnostic tool and to the fact that resorption of the buccal or palatal aspects of the roots of the incisor teeth cannot be seen on regular film. It is only when the buccal or palatal resorption has become sufficiently extensive to cause a change in the shape of the mesio-distal profile of the root that it may be suspected. Until then, this type of resorption may go undiagnosed on plain radiography films.

CT offers advantages in assessing the proximity of the impacted tooth to an adjacent pathological entity and in evaluating aberration in the shape and appearance of the crowns and roots of teeth that are suspected of having become damaged or have suffered from abnormal development due to past trauma [24].

Conventional spiral CT machines, as used in routine hospital practice for imaging various parts of the body, expose the body to an X-ray beam in the form of a progressive spiral, encircling the body over a specific, defined area. This submits the patient to a high dose of ionizing radiation and is a subject for concern when considering its use in the present context. Just how large this dose is was evaluated by Dula et al. [25, 26] using what is referred to as a hypothetical mortality risk. In this assessment, the mortality risk associated with routine dental radiographs ranged between 0.05 and 0.3×10^{-6} units, depending on the type and number of films performed, while a CT scan of the dental area alone was assessed at 28.2×10^{-6} for the maxilla and 18.2×10^{-6} for the mandible.

Cone beam computerized tomography

More recently, digital volume tomographic (DVT) machines (Table 2.1), which use the cone beam principle (NewTom 9000TM and NewTom 3GTM, Italy; i-CATTM, Imaging

Table 2.1 A comparison matrix of the important features of different cone beam computerized tomography machines. (Courtesy of Dr D Hatcher.)

	NewTom 9000®	New Tom 3G®	CB MercuRay®	Iluma DentalCAT™	i-CAT™
Sensor type	Image intensifier	Image intensifier	Image intensifier	Amorphous silicone flat panel	Amorphous silicone flat panel
Gray levels (bits)	8	12	12	14	14
Voxel size (mm³)	0.29	0.2–0.4	0.2–0.4	0.1–0.4	0.1–0.4
Collimation	Limited	Limited	Limited	No	Yes
Scan time(s)	70	40	10	20 or 40	10, 20, or 40
Dose type	Pulsed	Pulsed	Continuous	Continuous	Pulsed
Frames/revolution	360	360	300	?	150, 300, or 600
Effective dose (μSv)	50	44.7	487 or 869	?	68.7
Field of view	6 in diameter	4, 6, or 9 in diameter	4, 6, or 9 in diameter	17 cm	6–22 cm
Reconstruction shape	Sphere	Sphere	Sphere	Cylinder	Cylinder

Sensor type. Image intensifiers are bulky, require a large space and structurally robust support. Image intensifiers have a lower signal/noise ratio than flat panel sensors. Signal is good and noise is bad.

Gray levels. The number of gray levels is directly proportional to the image quality. 8 bits = 2^8 = 256 shades of gray, 12 bits = 2^{12} = 4096 shades of gray and 14 bits = 2^{14} = 16384 shades of gray. The increase in the number of gray levels will give greater access to soft tissue detail.

Voxel size. The reconstructed data are presented as voxel data. A voxel is the smallest element of a three-dimensional image. Each voxel possesses the following attributes: size, location, and gray level. Cone beam CT voxels are nearly isotopic, i.e. they have equal x, y, and z dimensions.

Collimation. Collimation allows the operator to reduce the field of view (FOV) to the area of interest. A small FOV has the advantages of reducing the effective dose and reducing scatter (noise).

Scan time. The main advantage for a short scan time is the potential for the reduction of motion artifact. Motion is a significant contributor to noise.

Dose: pulsed or continuous. A pulsed dose results in a lower effective dose. Effective dose is a method for assessing radiation burden to the patient and allows a direct comparison between different radiation sources. It considers the absorbed dose in key tissue in and around the field of view. The absorbed dose is weighted by percent of body being imaged and by the sensitivity of the tissues being measured. In the case of the CB MercuRay, two different mA settings (10 and 15 mA) were used in the studies. For comparison, in maxillo-mandibular study by fan beam CT, the dose was reported to be 2100 μSv, and in a maxillary study was reported in the 1400 μSv.

Frames/revolution. The frames are the number of acquired images, known as the raw data. The frame number is directly proportional to the signal generation.

Field of view (FOV). FOV refers to the dimensions of the anatomic region being imaged. Using the image intensifier the FOV is linked to the voxel size. A larger FOV requires a larger voxel and therefore a lower resolution. The selected voxel size is independent of the FOV when using a flat panel sensor; therefore, a large FOV with a small voxel size can be used.

Reconstruction shape. A cylindrical reconstruction fits the human head better than a spherical shape.

Image quality. Image quality is related to several factors including: signal, noise, voxel size, and gray levels. The best-quality image has high signal, low noise, small voxel size, and a large number of gray levels.

Sciences International, USA; CB MercuRayTM, Hitachi Medical Corporation, Japan; 3D AccuitomoTM, J. Morita Mfg Corporation, Japan; Iluma DentalCATTM, IMTEC Imaging, USA; Promax 3D, Planmeca, Finland), have become available, in which an X-ray source and an image intensifier perform a single 360° rotation around the patient's head during which raw data are acquired and, on every 10 steps of the rotation, a different projection of the skull is taken, starting from the anterior midline position. The projection is captured by an image intensifier and stored on a hard drive unit, and the raw data are transferred into axial (horizontal or parallel to the occlusal plane) views, which form their primary reconstruction. Subsequently, these are secondarily reconstructed into views in other directions, which include panoramic views at various depths and diverse (coronal, radial, sagittal) transaxial views. Furthermore, because of a 1:1 size relationship of the images to reality, valid measurement may be made directly on the films themselves [27]. As with the conventional spiral CT machines, 3-D views from any aspect may also be reconstructed, although the resolution and definition of soft tissues are not quite as clear.

With the more recent advances in the development of these machines, companies now offer the possibility of serially moving through the consecutive horizontal or vertical 'slices', on a personal computer screen and under the control of the mouse. In this way the intimate details of the crown and root relationships of the impacted tooth and adjacent structures may be followed and resorbed areas clearly recognized. Similarly, the 3-D views of the impacted tooth within the erupted dental arch may be shown as a continuous animated movie, showing the teeth (stripped of all soft tissue and bone) from all angles. As a result of these sophisticated features and improvements with the cone beam method, a renewed investigation of cases of maxillary canine impaction in relation to detectable root resorption recorded 66.7% resorption in lateral incisors and 11.1% in central incisors [27] (Figures 2.14 and 2.15).

One of the highlights of this machine is that the operator may use the program to trace the course of the inferior

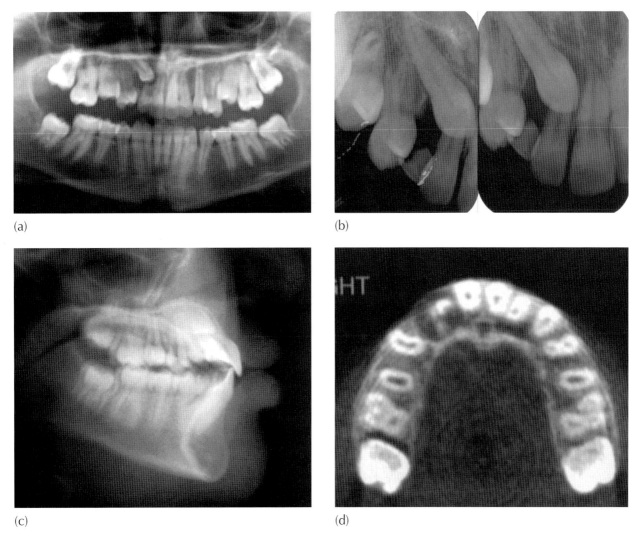

Fig. 2.14 A comparison of information available from plain film radiographs with that from cone beam computerized tomography. (a–c) The panoramic, periapical and lateral cephalometric radiographs show in impacted maxillary canine, in close relation to the roots of the lateral incisor and the first premolar. There is evidence of root resorption of the lateral incisor. The canine has been diagnosed as buccal, but its relation with the premolar is undiagnosed. Its apex appears superior to that of the premolar. (d) A horizontal (axial) cut shows the tip of the crown of the canine at the level of the middle of the incisor roots and in the line of the arch. (e) A higher parallel axial cut shows the horizontal cut obliquely imaging the canine, indicating its crown to be buccal and its root oriented across the ridge to the palatal of the premolar roots and at the same level. (f, g) Vertical (transaxial) views show the labio-lingual and height relation between the canine and it immediate neighbours and the site adds severity of the resorption of the incisor root. (h, i) The three-dimensional view of the area shows an accurate visualization of what the surgical field will look like, when the exposure is undertaken.

alveolar canal on the imaged panoramic view with a red marker and then generate vertical or horizontal cuts which will depict the inferior dental canal as a red dot or line. This is normally used in planning the placement of dental implants. It may also be used in the present context to define the relationship between the inferior dental nerve and a severely and deeply infra-occluded/impacted molar or premolar, whose roots are in close proximity to it. In addition, accurate 1:1 measurement may be made on the films to assist in overall orientation [28].

The Israeli group published the first clinical article that describes CBCT and its use in the accurate positional diag-

nosis of impacted teeth [29]. In that article, the enormous advantages to be gained with this method are highlighted, together with examples whereby seemingly impossible positions of teeth may be clearly identified and appropriate biomechanical solutions devised to successfully overcome them. The information obtainable with plane film radiography is very poor by comparison and is the reason for failure in many of the more unusual cases [30–32].

The greatest advantage that the cone beam volumetric machine has over conventional CT machines is that its radiation dosage is only a fraction of that emitted by the conventional machine. According to one published source,

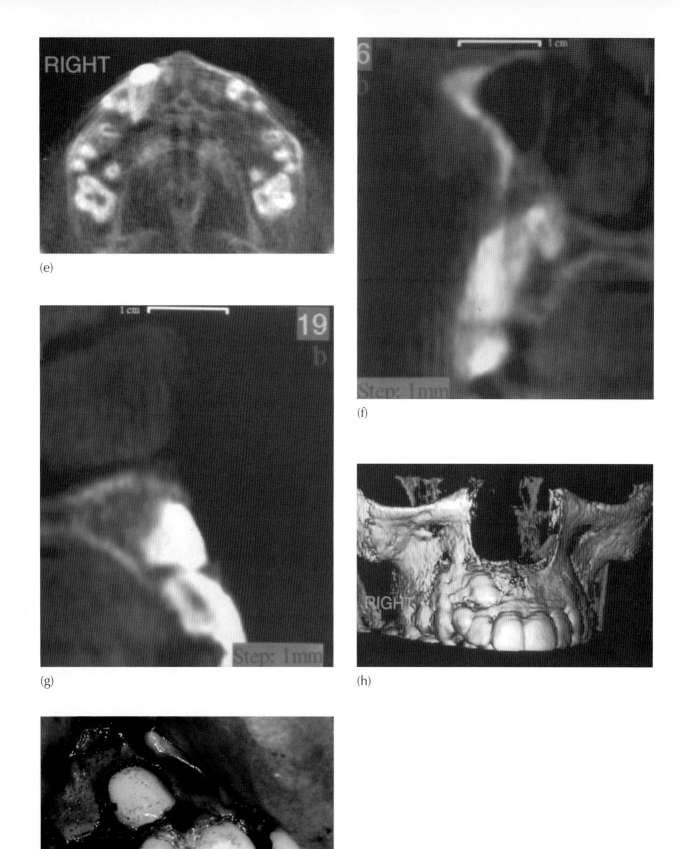

(e)

(f)

(g)

(h)

(i)

Fig. 2.14 (*Continued*)

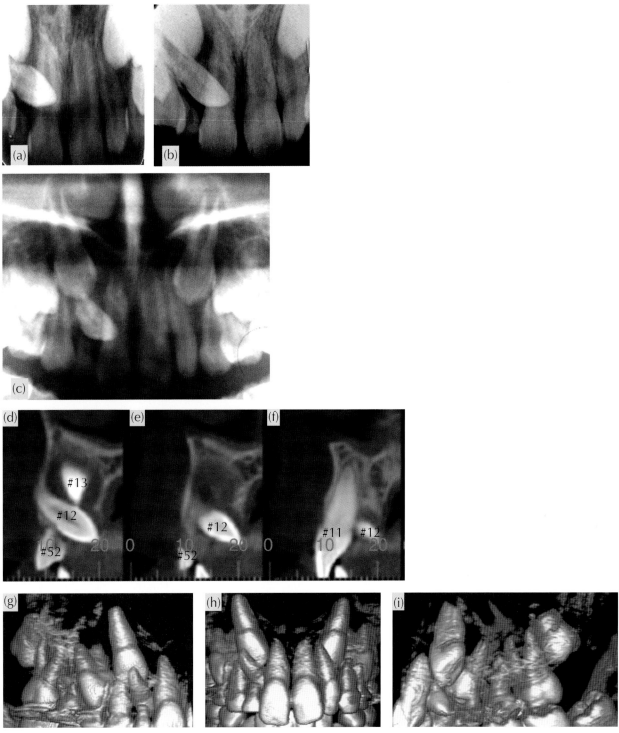

Fig. 2.15 A case in which maxillary canines and the right lateral incisor were impacted, but the bucco-lingual relations and axial inclinations between these teeth need to be evaluated. (Courtesy of Dr M. Yitschaky.) (a) A periapical view of the maxillary midline area showing the impacted lateral incisor. (b) A more steeply angled periapical view confirms the right lateral incisor to be palatal to the root of the central incisor, using the principles of vertical parallax. (c) Using Chaushu's method for this anterior portion of a panoramic film of the same patient, the canines are also palatally displaced. However, these views are inadequate to determine the complicated bucco-lingual relations and axial inclinations of the six anterior teeth. (d–f) Consecutive transaxial cuts show the vertical and bucco-lingual relationship between the right permanent canine (#13) and the adjacent lateral incisor (#12), deciduous canine (#53) and central incisor (#11). (g–i) show 3-D views from the right side, the front and the palatal aspect, to determine the root of the lateral incisor to be protruding labially and its crown palatal to the central incisor. The canine is palatal to the root but labial to the crown of the lateral incisor. The left permanent canine is also palatally displaced.

the DVT machine irradiates the patient at approximately 20% of the regular CT machine [33], while another source [34] has quoted the radiation dose for the mandible as 8580 mR for the conventional machine, as against 140 mR for the DVT machine.

An issue of the *Journal of the Californian Dental Association* reported on the 2002 Inaugural Conference of the (California) Coast Conference on Orthodontic Advances in Science and Technology, and the articles in that volume were devoted to the various aspects of cone beam volume tomography. In these reports, the level of the patient's radiation dose is variously described as 'similar to standard dental radiology' [35], 'an absorbed dose that is comparable to other dental surveys' [36] and 'is 50.3 μSv (NewTom 9000), while that of a dental panoramic film ranges from 2.9–9.6 μSv and a complete mouth series ranges from approximately 33–84 μSv and 14–100 μSv' [37].

What do these figures mean to the lay public? With our responsibility as dentists to convey information in a manner understandable to those requesting our treatment and in order to obtain informed consent, it is imperative to present the issue in its context, without blinding the patient with scientific data. Thus, it may be more pertinent to use the comparison that (a) the average man-in-the-street receives a daily dose of about 8 μSv per day or 2700 μSv per year from the environment [38] and (b) flying from New York to Tokyo by the transpolar route exposes the passenger to ionizing (cosmic) X-rays of approximately 150 μSv and from New York to Seattle of approximately 60 μSv [39]. This may be put more explicitly for the benefit of the clinician in practice in Table 2.2, which presents helpful comparisons of dosages for commonly used dental and medical radiographs with CBCT and spiral CT, as compared to the natural background radiation to which we are passively subjected in our daily lives and, at the same time, couched in terms readily comprehensible to the lay public.

In the past and with the availability of only the very high radiation dosage, traditional spiral CT machines, there was

little justification for the routine use of this method in the diagnosis of impacted teeth in any but the clinically most difficult cases. Nevertheless, given the amount of information that may be provided by CT methodology and the reduction in radiation that is now possible with these newer machines, the significant reduction in the risk/benefit ratio has brought the method to within much more acceptable limits. For plain film radiography to provide a comparable level of positional information, a number of different views would need to be taken and, together, the accumulated level of radiation that these would generate is of the same order as that emitted by the new CT machines. Panoramic radiographs are already produced by the CBCT machines, among the full range of views available. The CT is generally focused on a rotating narrow band of tissue, the focal trough, at a fixed distance from the beam source, at the same time eliminating all structures that are in front of or behind this trough. Thus, while a traditional panoramic film, which is generally focused on a broader band of tissue, will reveal a labially or palatally placed ectopic supernumerary tooth, it will be understood that if this body were to lie outside this trough, either in a CBCT or in a panoramic view, it will not appear on the image.

The more recent versions of CBCT machines have the capacity to produce standardized cephalometric views generated from the same scan that will have produced the panoramic view and the 3-D views. This has led to a broadening of the services and to a streamlining of the radiographic records available to the orthodontist, which has in turn led serendipitously to the discovery of incidental findings that had not shown in the patient's plane film records in up to 30% of the orthodontic cases seen in one study [41]. More pertinently, in another study of patients with impacted maxillary canines, 43.7% of the initially proposed treatment plans, based initially on plain film radiographs, were amended in the light of subsequent findings seen on the CBCT images [42]. The main reason for the revised treatment plans was linked to the demonstrable superiority of CBCT over plane film radiography in relation to the bucco-lingual dimension in general and to the many cases where existing incisor root resorption was diagnosed on the CBCT but missed on the plane films.

CBCT represents state-of-the-art technology with direct relevance to the determination of macroscopic anatomy and accurate positional diagnosis of impacted teeth. The machinery is not beyond the financial means of most radiology institutes, small diagnostic centres and dental school radiology departments, its usefulness to the orthodontist and surgeon is manifest, its level of emitted ionizing radiation is low and the cost to the patient affordable. Accordingly, it must now be considered a recommended procedure for many of the cases that are discussed within the context of this book.

But there is also an inherent danger with this type of comprehensive imaging. The means of presentation of the

Table 2.2 Effective radiation doses (after ref. [40], reproduced with permission)

Examination	Exposure (μSv)	Equivalent natural background radiation
Dental		
Panoramic	3–11	1/2–1 day
Cephalogram	5–7	1/2–1 day
Occlusal film	5	1/2 day
Bitewing	1–4	1/2 day
Full-mouth series	30–170	4–21 days
TMJ series	20–30	3–4 days
CBCT exam	40–135	4–17 days
Medical		
Chest X-ray	100	10–12 days
Mammogram	700	88 days
Medical CT	8,000	1,000 days (3 years)

results of the CBCT scan are very attractive to the layman and several of the films may be undertaken to impress the orthodontic patient, who may request a copy of the before-and-after portfolio as a souvenir of his/her orthodontic treatment and outcome. In today's world, this can easily become part of the 'hard sell' and a means of attracting new patients. So the danger is that the stage may be set for the production of a whole gamut of films for the sake of 'completeness', many of which may be superfluous to the clinical needs of the patient, and to achieve this the patient will be subjected to a large overdose of ionizing radiation.

What is ALARA?

The risks attached to any form of radiation are only seen many years later in its stochastic effects, which include a higher susceptibility of the individual to various forms of cancer. It is known that these effects are amplified with increased exposure and that children are more susceptible than adults. Yet it is the children and young adults who are the main targets of the population for the provision of orthodontic treatment. It is therefore incumbent on the practitioner to reduce this exposure to the minimum, while deriving a maximum of information, adequate to the problem in hand. This is ALARA – as low as reasonably achievable.

As we have already noted, there is ample documentation that cone beam computerized tomography will irradiate the patient less – possibly far less – than a conventional CT machine and that the many different CBCT machines available on the market today have widely differing outputs of radiation [43–45] for similar results. It stands to reason, therefore, that when a CT scan is justified, it is because plane film radiography cannot maximize the information needed for that patient. It should also be performed using the CBCT machine with the lowest radiation specifications, and the irradiated area should be limited as much as possible.

References

1. Seward GR. Radiology in general practice IX. Unerupted maxillary canines, central incisors and supernumeraries. Br Dent J 1968; 115: 85–91.
2. Hunter SB. The radiographic assessment of the unerupted maxillary canine. Br Dent J 1981; 150: 151–155.
3. Mason RA. A Guide to Dental Radiography, 2nd ed. Bristol: Wright PSG, 1982.
4. Ong A. An alternative technique to the vertex/true occlusal view. Am J Orthod Dentofac Orthop 1994; 106: 621–626.
5. Clark CA. A method of ascertaining the relative position of unerupted teeth by means of film radiographs. Proc Roy Soc Med (Sec. Odont) 1910; 3: 87–90.
6. Armstrong C, Johnston C, Burden D, Stevenson M. Localizing ectopic maxillary canines – horizontal or vertical parallax? Eur J Orthod 2003; 25: 585–589.
7. Jacobs SG. Localisation of the unerupted maxillary canine. Aust Orthod J 1986; 9: 313–316.
8. Jacobs SG. Exercises in the localisation of unerupted teeth. Aust Orthod J 1987; 10: 33–35, 58–60.
9. Nohadani N, Pohl Y, Ruf S. Displaced premolars in panoramic radiography—fact or fallacy? Angle Orthod 2008, 78: 309–316.
10. Chaushu S, Chaushu G, Becker A. The use of panoramic radiographs to localize maxillary palatal canines. Oral Surg Oral Med Oral Pathol Oral Radiol Endod 1999; 88: 511–516.
11. Chaushu S, Chaushu G, Becker A. Reliability of a method for the localization of displaced maxillary canines using a single panoramic radiograph. Clin Orthod Res 1999; 2: 194–199.
12. Wolf JE, Mattila K. Localization of impacted maxillary canines by panoramic tomography. Dentomaxillofac Radiol 1979; 8: 85–91.
13. Fox NA, Fletcher GA, Horner K. Localising maxillary canines using dental panoramic tomography. Br Dent J 1995; 179: 416–420.
14. Kohavi D, Zilberman Y, Becker A. Periodontal status following the alignment of buccally ectopic maxillary canine teeth. Am J Orthod 1984; 85: 78–82.
15. Becker A, Kohavi D, Zilberman Y. Periodontal status following the alignment of palatally impacted canine teeth. Am J Orthod 1983; 84: 332–336.
16. Kohavi D, Becker A, Zilberman Y. Surgical exposure, orthodontic movement and final tooth position as factors in periodontal breakdown of treated palatally impacted canines. Am J Orthod 1984; 85: 72–77.
17. Becker A, Chaushu S. Long-term follow-up of severely resorbed maxillary incisors following resolution of etiologically-associated canine impaction. Am J Orthod Dentofacial Orthop 2005; 127: 650–654.
18. Ericson S, Kurol J. CT diagnosis of ectopically erupting maxillary canines – a case report. Eur J Orthod 1988; 10: 115–120.
19. Ericson S, Kurol J. Resorption of maxillary lateral incisors caused by ectopic eruption of canines. Am J Orthod Dentofacial Orthop 1988; 94: 503–513.
20. Ericson S, Kurol J. Radiographic examination of ectopically erupting maxillary canines. Am J Orthod Dentofacial Orthop 1987; 91: 483–492.
21. Odegaard J. The treatment of a Class I malocclusion with two horizontally impacted maxillary canines. Am J Orthod Dentofacial Orthop 1997; 111: 357–365.
22. Becker A. Comment about making outcome of treatment more predictable. Am J Orthod Dentofacial Orthop 1997; 112: 17A–19A.
23. Ericson S, Kurol PJ. Resorption of incisors after ectopic eruption of maxillary canines: a CT study. Angle Orthod 2000; 70: 415–423.
24. Bodner L, Bar Ziv J, Becker A. Image accuracy of plain film radiography and computerized tomography in assessing morphological abnormality of impacted teeth. Am J Orthod Dentofacial Orthop 2001; 120: 623–628.
25. Dula K, Mini R, van der Stelt PF et al. Hypothetical mortality risk associated with spiral computed tomography of the maxilla and mandible. Eur J Oral Sci 1996; 104: 503–510.
26. Dula K, Mini R, van der Stelt PF, Buser D. The radiographic assessment of implant patients: decision-making criteria. Int J Oral Maxillofac Implants 2001; 16: 80–89.
27. Walker L, Enciso R, Hatcher DC, Mah J. Three-dimensional analysis of impacted canines using volumetric imaging. Am J Orthod Dentofacial Orthop 2005; 125: 418–423.
28. Mah J, Hatcher D. Three-dimensional craniofacial imaging. Am J Orthod Dentofacial Orthop 2004; 126: 308–309.
29. Chaushu S, Chaushu G, Becker A. The role of digital volume tomography in the imaging of impacted teeth. World J Orthod 2004; 5: 120–132.
30. Mah JK, Alexandroni S. Cone-beam computed tomography in the management of impacted canines. Semin Orthod 2010; 16: 199–204.
31. Becker A, Chaushu S, Casap-Caspi N. CBCT and the orthosurgical management of impacted teeth. J Am Dent Assoc 2010;141(10 suppl):14S-18S.
32. Becker A, Chaushu G, Chaushu A. An analysis of failure in the treatment of impacted maxillary canines. American Journal of Orthodontics & Dentofacial Orthopedics 2010; 137: 743–754.
33. Mozzo P, Procacci C, Tacconi A, Martini PT, Andreis IA. A new volumetric CT machine for dental imaging based on the cone-beam technique: preliminary results. Eur J Radiol 1998; 8: 1558–1564.
34. Mah JK, Danforth RA, Bumann A, Hatcher D. Radiation absorbed in maxillofacial imaging with a new dental computed tomography device. Oral Surg Oral Med Oral Pathol Oral Radiol Endod 2003; 96: 508–513.

35. Danforth RA, Dus I, Mah J. 3-D volume imaging for dentistry: a new dimension. J Calif Dent Assoc 2003; 31: 817–823.
36. Hatcher DC, Dial C, Mayorga C. Cone beam CT for pre-surgical assessment of implant sites. J Calif Dent Assoc 2003; 31: 825–833.
37. Mah J, Enciso R, Jorgensen M. Management of impacted cuspids using 3-D volumetric imaging. J Calif Dent Assoc 2003; 31: 835–841.
38. Health Protection Agency. Ionising Radiation Exposure of the UK Population: Review, Ref: HPA-RPD-001. Radiation Protection Division, Chilton, Oxon, UK: 2005.
39. Barish RJ. In-flight radiation exposure during pregnancy. Obstet Gynecol 2004; 103: 1326–1330.
40. Mah J. Cone Beam CT Radiography: A Certification Program for Dentists. Appendix I as part of a CE Course Manual entitled " Copyright 2010. eDental Academy, Las Vegas, NV.
41. Cha J-Y, Mah J, Sinclair P. Incidental findings in the maxillofacial area with 3-dimensional cone-beam imaging. Am. J Orthod Dentofac Orthop, 2007; 132: 7–14.
42. Bjerklin K, Ericson S. How a computerized tomography examination changed the treatment plans of 80 children with retained and ectopically positioned maxillary canines. Angle Orthod, 2006; 76: 43–51.
43. Tsiklakis K, Dontaa C, Gavala S, Karayianni K, Kamenopoulou V, Hourdakis CJ. Dose reduction in maxillofacial imaging using low dose Cone Beam CT. Eur J Radiol 2005, 56: 413–417.
44. Brooks SL. CBCT Dosimetry: orthodontic considerations. Semin Orthod 2009, 15: 14–18.
45. Silva MAG, Wolf U, Heinicke F, Bumann A, Visser H, Hirsch E. Cone-beam computed tomography for routine orthodontic treatment planning: a radiation dose evaluation. Am J Orthod Dentofac Orthop 2008; 133: 640.

3

Surgical Exposure of Impacted Teeth

(In Collaboration with Arye Shteyer and Joshua Lustmann)

Aims of surgery for impacted teeth

In the past, the decision on how a particular impacted tooth should be treated was usually made by the oral surgeon and, by and large, the alternatives were decided and stage-managed by him. This situation has changed in recent years.

Prior to the 1950s, few orthodontists were prepared to adapt their skills and their ingenuity to the task of resolving the impaction of maxillary canines and incisors. Accordingly, the orthodontists themselves referred patients to the oral surgeon, who would decide if the impacted tooth could be brought into the dental arch. Where the circumstances were potentially favourable, the tooth would be surgically exposed and, when the surgical field was displayed fully, the surgeon would make his assessment of the prognosis of the case, and decide and act solely in accordance with his own judgement. In this way, many potentially retrievable impacted teeth were scheduled for extraction.

There are no surgical methods, other than transplantation, by which positive and active alignment of an impacted tooth may be carried out. The best a surgeon can do is to provide the optimal environment for normal and unhindered eruption and then hope that the tooth will oblige. In the past and with this in mind, therefore, those teeth that were considered worth trying to recover were widely exposed and packed with some form of surgical or periodontal pack to protect the wound during the healing phase and to prevent re-healing of the tissues over the tooth. For a variety of reasons, several other steps were taken, depending on the preferences and beliefs of the operator, with the aim of providing that 'extra something' that would improve the chances of spontaneous eruption still further. These measures were often very empirical in nature and included one or more of the following:

- clearing the follicular sac completely, down to the cemento-enamel junction (CEJ) area;
- clearing the bone around the tooth, down to the CEJ area, to dissect out and free the entire crown and the coronal portion of the root of the impacted tooth;
- 'loosening' the tooth by subluxating it with an elevator;
- bone channelling in the desired direction of movement of the tooth;
- packing gauze or hot gutta percha into the area of the CEJ, under pressure, in order to apply force to deflect the eruption path of the tooth in a particular direction.

In those years, few patients were referred to the orthodontist until full eruption had been achieved, and the tooth then needed only to be moved horizontally into line with its neighbours. Up to that point, the problem was considered to be within the realm of the oral surgeon. In many cases, 'success' in achieving the eruption of the tooth was pyrrhic and often subordinated to failure of a different kind, namely the periodontal condition of the newly erupted tooth and its survival potential – its prognosis. This was the inevitable result of the aggressive and overenthusiastic surgical techniques that had been used, specifically those listed above, which typically left the tooth with an elongated clinical crown, a lack of attached gingiva and a reduced alveolar crest height [1–6].

Surgical intervention without orthodontic treatment

We come across cases in which the only clinical problem relates to the impacted tooth, the occlusion and alignment being otherwise acceptable. For these patients, the following question needs to be addressed: what surgical methods are available that may be expected to provide a more or less complete solution without orthodontic assistance? To be in a position to answer this question, it is necessary to provide a description of the position of the teeth that will respond to this kind of treatment.

Exposure only

A superficially placed tooth, palpable beneath the bulging gum, is an obvious candidate. This type of tooth may be seen in the maxillary canine area (Figure 3.1a), but also in the mandibular premolar area (see Figure 1.8) and the maxillary central incisor area, usually where very early extraction of the deciduous predecessor was performed while the immature permanent tooth bud was still deep in the bone and unready for eruption. Healing occurred, the gum closed over and the permanent teeth were unable to penetrate the thickened mucosa [7, 8]. Removing the fibrous mucosal covering or incising and resuturing it to leave the incisal edges exposed (Figure 3.2) will generally lead to a fairly rapid eruption of the soft tissue impacted tooth, particularly in the maxillary incisor area. The more the tooth bulges the soft tissue, the less likely is a reburial of the tooth in healing soft tissue and the faster is the eruption.

Exposure with pack

Taking this one step further, we will understand that a less superficial tooth requires a more radical exposure procedure and may need a pack to prevent the tissues from re-healing over the tooth. While the surgeon may be rewarded with spontaneous eruption, this will take longer and a compromised periodontal result should be expected (Figure 3.3).

We have defined over-retained deciduous teeth as teeth still present in the mouth when their permanent successors have reached a stage of root development that is compatible with their full eruption. These deciduous teeth may then be considered as obstructing the normal development which would be expected to proceed in their absence. The deciduous teeth should be extracted, but provision should be made to encourage the permanent teeth to erupt quickly.

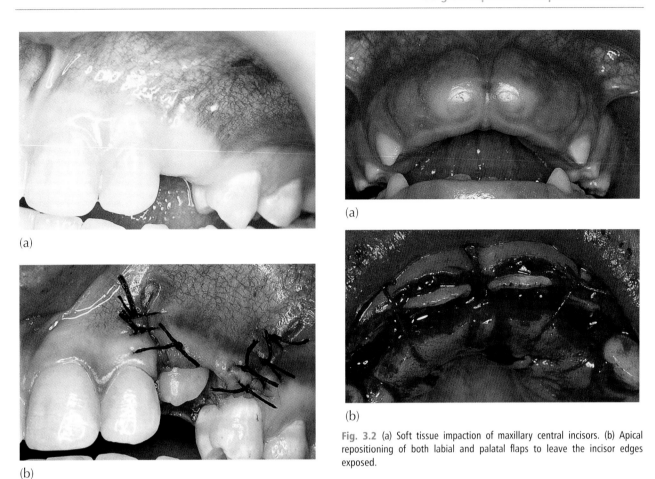

(a)

(b)

Fig. 3.2 (a) Soft tissue impaction of maxillary central incisors. (b) Apical repositioning of both labial and palatal flaps to leave the incisor edges exposed.

(c)

Fig. 3.1 (a) A 16-year-old female exhibits an unerupted maxillary left canine, which has been present in this position for two years and has not progressed. (b) The tooth was exposed and the flap, which consisted of attached gingiva, was apically repositioned. (c) At nine months post-surgery, the tooth has erupted normally. (Courtesy of Professor L. Shapira.)

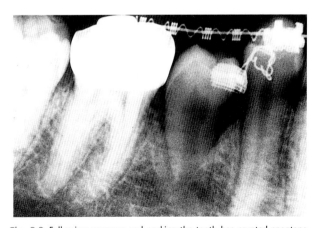

Fig. 3.3 Following exposure and packing the tooth has erupted spontaneously, but the bone level is compromised.

Many of these permanent teeth with delayed eruption are abnormally low in the alveolus and are in danger of becoming reburied by the healing tissue of the evacuated socket of the deciduous tooth. Accordingly, the crowns of the teeth should be exposed to their widest diameter and a surgical or periodontal pack placed over them and sutured in place for 2–3 weeks. This will encourage epithelialization down the sides of the socket and, generally, prevent the reformation of bone over the unerupted tooth.

Most surgeons and periodontists today will use a proprietary pack, such as CoePak™, to maintain the opening which, at the same time, acts as a dressing for the wound. A careful assessment of the space requirement should be made in these cases and consideration given to the need for space maintenance. It should be remembered that space loss in the mixed dentition may often be very rapid, and the erupting tooth may be arrested in its progress by its proximal contact with the adjacent teeth. Alternatively, particularly with regard to maxillary canines, placement of a removable acrylic plate, which is prepared before surgery, can be used to hold in a small pack over the exposed tooth [9].

Exposing and packing has recently been reintroduced and recommended for the treatment of severely palatally displaced maxillary canines [10]. When this is done, spontaneous resolution of even quite severe displacements has been claimed to occur in most cases, in the months that follow. This takes the form of at least partial eruption through the surgically created and pack-maintained opening and permits relatively easy access for attachment bonding and subsequent alignment of the tooth when appliance therapy is later initiated.

Exposure with pressure pack

Mesial impaction of a mandibular second permanent molar beneath the distal bulbosity of the first permanent molar is analogous to the more common mesial impaction of a third molar beneath the distal of the second. In either case, and in its mildest form, it is a condition that may sometimes respond to surgical intervention and packing only. This involves exposure of the occlusal surface of the tooth and the deliberate wedging of the pack in the area between the two teeth and leaving it there for 2–3 weeks. During this time, the pressure will often succeed in eliciting a distal movement of the impacted molar, which may then erupt more freely when the pack is removed. The degree of control available to the operator in judging the amount of pressure applied and the extent to which the pack interferes periodontally is minimal, although damage to the periodontium of the two adjacent teeth is possible. Success in bringing about an improved position of the tooth may thus not be matched by the health of its supporting structures. Others have used brass wire [11] or elastic separators to apply a similar disimpacting force.

The surgical elimination of pathology

Soft tissue lesions

In Chapters 8 and 11 we shall refer more specifically to benign tumours and cysts. Surgery is the only treatment that is indicated for these conditions, in the first instance. This should be performed without delay, if only for reasons of obtaining biopsy material to confirm the innocence of a tentative diagnosis. Orthodontic treatment should be suggested then, but begun only after a filling-in of bone has occurred. At that point, there will be an improvement of the positions of the grossly displaced teeth, together with an improvement of the bony defect that will be evident in the anatomy of the alveolar bone in the area. However, this may take many months to occur. In the interim, the preparation of the patient for the proposed orthodontic treatment may be undertaken, which must begin with seeing positive results from a preventive dental health programme aimed at eliminating marginal gingival inflammation and reducing the caries incidence for that patient.

Hard tissue obstruction

Obstructive impaction invites the logical step of removing the offending body causing the non-eruption. This is performed by the surgeon and, on many occasions, it is without recourse to orthodontic assistance and enjoys varying degrees of success. In Chapter 5 we shall refer to the reliability of spontaneous eruption, following the various surgical procedures involved in the treatment of impacted incisors. For the present discussion, we must recognize that there is a significant number of cases in which eruption does not occur in a reasonable time-frame.

Following the removal of the obstruction, be it a supernumerary tooth, an odontome, residual deciduous roots or an infra-occluded primary tooth, the position of most unerupted teeth improves with time. However, many of these teeth do not erupt without assistance due to their severe displacement, which is a result of the existence of the erstwhile obstruction and the healing tissues.

A hard tissue body is generally made up of the dental tissues and, with its accompanying dental follicle, occupies much space. This causes a gross displacement of the developing tooth bud of the normal tooth, which is true in terms of both overall distance from its normal location and the usually marked deflection of the orientation of its long axis. Thus, the root or the crown of the tooth may be deflected mesially, distally, lingually or buccally, or displaced superiorly (in the upper jaw) or inferiorly (in the lower), compromising its chances for spontaneous eruption. Abnormally shaped roots may develop in the cramped circumstances in which they find themselves between the displacing influence of the pathological entity and the adjacent teeth, on the one hand, and the floor of the nose or lower border of the mandible [12], on the other. Teeth with abnormally

shaped roots may have deviated eruption paths and do not always erupt spontaneously, although they may be successfully erupted with orthodontic appliances, provided their periodontal ligament (PDL) is normal.

Non-eruption of an impacted tooth disturbs the eruption pattern of the adjacent teeth, which then assume abnormal relationships to one another, usually characterized by space reduction and tipping. This then provides a secondary physical impediment to the eruption of the impacted tooth.

Infra-occlusion

As we shall discuss in Chapter 8, infra-occluded permanent teeth are usually ankylosed to the surrounding bone and as such cannot respond to orthodontic traction. In many cases, the ankylosed area of root is minute and may be easily broken by a deliberate, but gentle luxation of the tooth. This is usually performed with an elevator or extraction forceps and is done in such a way as to loosen the rigid connection of the bony union, which is unbending. The tooth is not removed from its socket, nor is the principal aim even to tear the periodontal fibres. The purpose is to bring the tooth to a higher degree of mobility, beyond that characteristic of a normal tooth.

Unfortunately, the fate of the tooth that has undergone this procedure is usually a rehealing and reattachment of the ankylotic connection, leading to a return to the original situation. Accordingly, this approach can only be successful if a continuously active traction force is applied to the tooth from the time of its luxation. This force may then act to modify the rehealing of bone due to a localized microcosm of distraction osteogenesis [13, 14] that it causes. If the range of force is small and loses its potency between visits for adjustment, re-ankylosis will result and the tooth will not move. Thus, to be effective, it must be of sufficient magnitude to cause distraction and of sufficient range to remain active between one visit for adjustment and the

next. The risk is that a poor biomechanical auxiliary, insufficient force levels or missed appointments may cause the exercise to founder, due to the re-establishment of the ankylosis bridge.

The principles of the surgical exposure of impacted teeth

In general there are two basic approaches to surgically exposing impacted teeth, described below.

The open eruption technique

Historically, the first method used to uncover impacted teeth left the tooth exposed to the oral environment, while surrounded by freshly trimmed soft tissue of the palate or labial oral mucosa, following the removal of the mucosa and bone actually covering the tooth. This is known as the open eruption technique and it may be performed in two ways.

The window technique involves the surgical removal of a circular section of the overlying mucosa and the thin bony covering. For most labially displaced teeth, due to their height, this entire surgical procedure would most likely only be possible above the level of attached gingiva (Figure 3.4), in the mobile area of the oral mucosa. Notwithstanding, it is clear that this is the simplest, most conservative and most direct manner to expose a tooth which is palpable immediately under the oral mucosa and it may often be accomplished with surface anaesthetic spray only. An attachment may then be bonded to the tooth and orthodontically encouraged eruption may proceed without delay, to complete its alignment within a very short time. While this obviously represents a significant advantage in the treatment of a young patient, the long-term outcome of the procedure will be characterized by a muco-gingival attachment on the labial side of the tooth which is not of attached gustatory epithelium, but

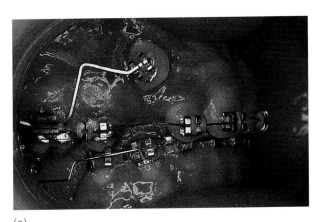
(a)

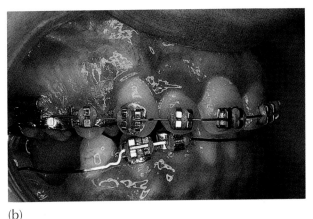
(b)

Fig. 3.4 (a) A high buccal canine exposed by circular incision of the sulcus mucosa. (b) Following alignment, the oral mucosa is attached directly to the gingiva. (Courtesy of Dr G. Engel.)

rather a mobile, thin, oral mucosa that does not function well as a marginal tissue, as has been widely documented in the periodontal literature. The only situation in which this exposure procedure is clinically advantageous is when there is a very wide band of attached gingiva and where a labially impacted tooth is situated well down in this band, such that a simple removal of the tissue overlying the crown will still leave 1–2 mm of bound epithelial attachment inferior to the free, movable, oral mucosal lining of the sulcus.

By contrast, the palatal mucosa is very thick and tightly bound down to the underlying bone. Thus, no parallel precautions need to be made to ensure a good attachment for the final periodontal status of a palatally impacted tooth, following its eruption into the palate. When the window technique is used on the palatal side, the cut edges of the wound need to be substantially trimmed back and the dental follicle removed to prevent re-closure of the very considerable width of palatal soft tissue over the exposed tooth (see Figure 6.29). For a deeply buried palatal canine, the exposure will additionally need to be maintained using a surgical pack.

The apically repositioned flap is an alternative way of performing an open exposure technique on the buccal side. It is aimed at improving the periodontal outcome by ensuring that attached gingiva covers the labial aspect of the erupted tooth in the final instance. This is done by raising a labial flap, taken from the crest of the ridge, and relocating it higher up on the crown of the newly exposed tooth. This method, a recognized and accepted procedure in periodontics, was first described in the context of surgical and orthodontic treatment of unerupted labially displaced teeth by Vanarsdall and Corn [15]. In their method, and in the absence of the deciduous canine, a muco-gingival flap, which incorporates attached gingiva, is raised from the crest of the ridge (Figure 3.1). If a deciduous canine is present, the flap is designed to include the entire area of buccal gingiva that invests it, and the deciduous tooth itself is extracted. In either case, the flap is detached from the underlying hard tissue some way up into the sulcus, to expose the canine. The flap is then sutured to the labial side of the exposed crown of the permanent canine, to cover the denuded periosteum and overlie its cervical area, while the remainder of the crown remains exposed. Subsequent eruption of the tooth is accompanied by the healing gingival tissue and, when the tooth takes up its final position in the arch, it will be found to be invested with a good width of attached gingiva.

This particular method of exposure is best suited to buccally/labially impacted teeth which are situated above the band of attached gingiva, but which are not displaced mesially or distally from their place in the dental arch. If the case presents with more than a minor degree of horizontal displacement in the sagittal plane, a raised and full thickness soft tissue flap will denude the alveolar bone cov-

ering the adjacent tooth to an unacceptable degree, contraindicating the use of this surgical modality. In order to overcome this, a partial thickness surgical flap may be raised, which leaves the donor area invested with a connective tissue cover [16], which will heal over by epithelial proliferation.

With any form of open exposure surgery, the tooth acquires a new gingival margin which comprises the cut edge of gingival tissue, which will heal in this position and will move with the tooth as it is drawn down into its place in the arch. While the periodontal parameters may be very satisfactory, the appearance of the tissues surrounding the aligned tooth at the end of treatment by this method lacks a completely natural look and it is usually possible to distinguish the previously affected tooth with ease, even several years later (Figure 3.8).

The closed eruption technique

The alternative approach to surgical exposure, the closed eruption technique, has an attachment placed at the time of the exposure and the tissues fully replaced and sutured to their former place, to re-cover the impacted tooth. This was described by Hunt [17] and McBride [18], although it seems to have been in use earlier [19, 20], and it is a procedure that may be used regardless of the height or mesiodistal displacement of the tooth. For a buccally impacted tooth, a surgical flap is raised from the attached gingiva at the crest of the ridge, with suitable vertical releasing cuts, and elevated as high as is necessary to expose the unerupted tooth. An attachment is then bonded and the flap fully sutured back to its former place. The twisted stainless steel ligature wire or gold chain, which is preferred by some clinicians, which has been tied or linked to the attachment, is then drawn inferiorly and through the sutured edges of the fully replaced flap. The surgical wound is, therefore, completely closed and the exposed tooth and its new attachment are sealed off from the oral environment. Spontaneous eruption is less likely to occur than when the tooth remains exposed following apical repositioning, and active orthodontic force will probably need to be applied to the tooth to bring about its eruption. Traction is then applied to the twisted stainless steel ligature or gold chain to bring about the full eruption of the tooth [18, 21, 22]. In this method, the tooth progresses towards and through the area of the attached gingiva several weeks or months after complete healing of the repositioned surgical flap has occurred and it creates its own portal through which it exits the tissues and erupts into the mouth. As such, it very closely simulates normal eruption and the clinical outcome will usually be difficult to distinguish from any normally and spontaneously erupting tooth, in terms of its clinical appearance and objective periodontal parameters.

A modification of the closed eruption technique has been described by Crescini et al. specifically in respect of maxillary permanent canines [23]. In this procedure, a full

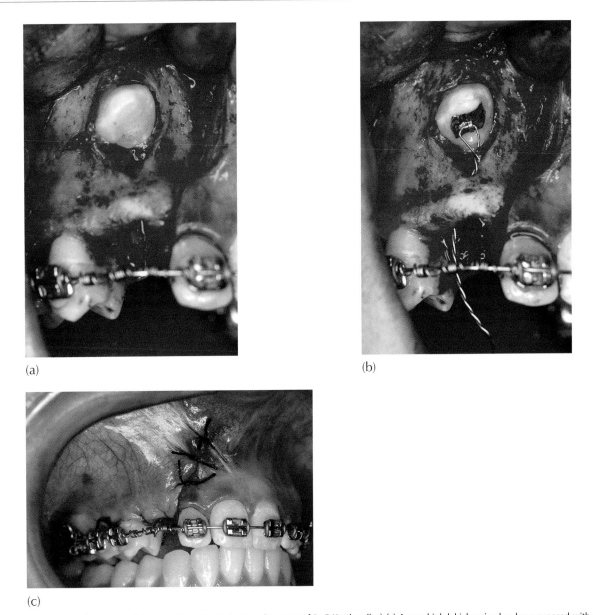

(a)

(b)

(c)

Fig. 3.5 Crescini's tunnel variation of the closed eruption technique. (Courtesy of Dr E Ketzhandler.) (a) A very high labial canine has been exposed with a full flap exposure, which included the gingival margin of the extracted deciduous canine. The bridge of buccal bone is left intact. (b) An attachment is bonded to the palatal aspect of the permanent canine and its pigtail ligature is directed through the socket vacated by the extracted deciduous tooth. (c) The flap is sutured to its former place and vertical traction will draw the tooth down, maintaining alveolar bone on its labial side.

buccal flap is raised from the attached gingival at the neck of the deciduous canine and adjacent teeth, to expose the surface of alveolar bone up to and including that covering the labially impacted buccal canine. The buccal crown surface of the canine is exposed and the deciduous canine extracted. The twisted steel ligature or gold chain linked to the eyelet, which is now bonded to the tooth, is threaded into the apical area of the recently vacated socket of the deciduous canine and drawn downwards to exit through its coronal end. No buccal bone is removed beyond that immediately overlying the crown of the exposed canine.

The flap is now sutured back to its former position, leaving only the end of the ligature/gold chain visible through the socket of the deciduous canine. The aim of this aptly named 'tunnel' technique is to mimic even further the natural eruption process by applying extrusive force to move the impacted canine directly through the socket of the recently extracted deciduous canine (Figure 3.5). Furthermore, by retaining the buccal bridge of bone during surgery, the final outcome will show the aligned tooth to have an excellent bony support, in terms of both its width and level.

Each method has its advantages and its drawbacks from the points of view of efficacy of treatment, postsurgical recover and the overall treatment outcome in relation to aesthetics, periodontal prognosis and stability of the final result.

Efficacy of treatment

Should the orthodontist be present at the time of surgery?

The greatest inconvenience of the closed eruption technique is that it is preferable that the orthodontist be present in the operating room for bonding of the attachment before the flap is sutured back to its former place. It is true that many oral surgeons bond the attachments themselves. However, since the surgeon bonds orthodontic attachments far less frequently than does the orthodontist, the chances of bond failure are relatively increased – the more so since the surgeon, when working with only chair-side nursing assistance, will need to undertake both the task of maintaining a dry and uncontaminated tooth surface in a very haemorrhagic field and, at the same time, that of performing the bonding procedure. It should be emphasized that in the case of the closed eruption technique, bond failure will dictate the need for a second surgical intervention. If the attachment is to be bonded at a later visit, the orthodontist does not need to be present at the surgeon's side for an open exposure case. However, this means that the surgeon must expose the tooth much more widely, place surgical packs and aim for healing by 'secondary intention' only, with attendant negative periodontal implications.

Of far greater importance and directly related to ensuring successful resolution of the impaction, the orthodontist is able to see the exact position of the crown, the direction of the long axis and the deduced location of the root apex, by being present. The height of the tooth and its relation to adjacent roots may all be noted, and the orthodontist may confirm or change the original plan for the strategy of its resolution, by direct vision, in the light of what he/she now sees 'in the flesh'. The orthodontist will be in a position to decide exactly where the attachment should be placed from a mechano-therapeutic point of view and will bond it there. The orthodontist is also the best person to fabricate and place a suitable and efficient auxiliary to apply a directional force of optimal magnitude and a wide range of movement, and to do so at the time of or, preferably, immediately prior to surgery. It is not fair to expect the oral surgeon to be aware of how different attachment positions may affect the orthodontic or periodontic prognosis; nor should it be expected that he/she will be sufficiently experienced with the bonding technique to do this. Indeed, without the presence of the orthodontist, the surgeon may carry out the exposure of the tooth and place a bracket in the most convenient location that may seem to him/her to be entirely appropriate. At a subsequent visit, the application of traction by the orthodontist may need to be made in a particular direction which, because of an incorrectly placed attachment, is impossible to attain, and the tooth may cause damage to adjacent structures by being drawn in an unfavourable direction.

It becomes evident that the inconvenience caused to the orthodontist by his/her being present at the exposure is handsomely rewarded in the long run by far greater control of the destiny of the impacted tooth, including efficacy and predictability of treatment and the quality of outcome.

The reliability of bonding

For the patient who has had an open exposure procedure, the reliability of bonding at a subsequent visit is, paradoxically, much poorer than when the attachment bonding is performed at the time of surgery [24] for the following reasons.

During a closed surgery procedure, a wide tissue flap is usually raised, which provides good visibility and access, especially to a deeply buried tooth (Figure 3.6). The margins of the wide flap are distant from the tooth, enabling better control of moisture and bleeding in the immediate area. The orthodontist can bond the attachment efficiently, while the surgeon and the nurse maintain haemostasis and the necessary dry field. In contrast, an open exposure involves raising a small flap in the area immediately surrounding the tooth and maintaining the patency of the opening (usually with the help of a periodontal pack) until an attachment is bonded at a later date. At that visit, secondary healing will have occurred, and the newly epithelialized cut surface will be very sensitive to any form of manipulation. Accordingly, the patient will have avoided brushing the area and a degree of inflammation will be present, due to the accumulation of plaque and the freshness of the healing wound. Prophying the tooth under these conditions of restricted access and fragile haemorrhagic tissue is not conducive to successful bonding. In addition, the presence of eugenol from the periodontal pack may inhibit composite polymerization and thereby weaken the bond strength.

When orthodontic brackets are bonded in day-to-day practice, the teeth are first cleaned, using a rubber cup and pumice. The aim of this procedure is to remove extraneous materials, which include soft plaque, dried saliva, organic and chemical staining, and deposits which adhere or adsorb to the enamel prisms and which may prevent penetration of the etchant. Once these are removed, the enamel surface becomes vulnerable to the orthophosphoric acid liquid or gel, which is the key to a successful adhesion of the attachment. In contrast, newly exposed impacted teeth are completely free of these extraneous materials. Their only covering is Nasmyth's membrane, which is made up of the enamel cuticle and the reduced enamel epithelium and is about 1 micron thick. This appears to present no barrier whatsoever to the etching effect that is achieved by the application of orthophosphoric acid [24]. Accordingly, there is no advantage to be gained by pumicing these teeth

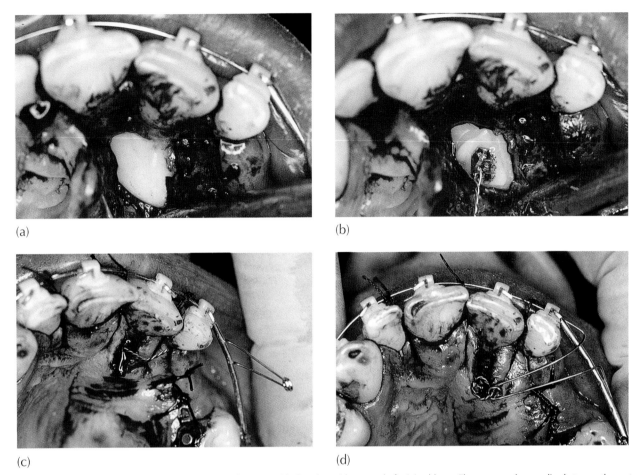

(a)   (b)

(c)   (d)

Fig. 3.6 (a) The crown of an impacted canine is exposed using a wide flap, but with removal of minimal bone. The unexposed crown lies between the roots of the central incisors, having traversed the midline suture. (b) An attachment is bonded, while haemostasis is maintained by the surgeon. (c) The flap has been divided to accommodate the ligature pigtail in its desired position, before being fully replaced and sutured. (d) The labial spring auxiliary loop, seen in its passive position in (c), has been turned inwards towards the palate and secured to the stainless steel ligature pigtail.

as part of the bonding procedure. Rather, the reverse is the case. To permit the introduction of a hand piece and rubber cup or a small electric toothbrush or hand brush, exposure has to be considerably broader for prophylaxis to be effective. It is difficult to control these implements during the brushing exercise and, as a direct consequence, the brush or cup traumatizes the exposed bone and soft tissues. This generates renewed bleeding, while giving rise to a dispersal of the pumice over the immediate surgical field. Prophylaxis is therefore completely superfluous.

A significant problem with the closed eruption technique is sometimes caused by a poor choice of orthodontic attachment. Since the mid-buccal position of this tooth is easy to expose and to bond to, the orthodontist may be tempted to use a conventional orthodontic bracket in this instance [25] and there is no contradiction in doing this with an open eruption technique.

However, due to the buccal prominence of the tooth, the lack of buccal bone, and the relative tightness of the replaced flap, damage may be caused to this muco-gingival tissue by

the bulk of wide and high profile conventional brackets (see Figure 4.3), with the closed eruption technique. This may lead to a breakdown of the overlying tissue, to cause a dehiscence or even 'buttonholing' (Figure 3.7) of the mucosa. In such circumstances, any attachment placed on the tooth should be as small as is practical and with a minimum height profile, in order that it will cause as little adverse effect on the gingival tissues as possible, on its way through. This will be dealt with more fully in the next chapter.

Duration of the surgical procedure

When the surgical field is opened, there is considerable bleeding at the cut edges of the surgical flap and of the whole area that has been exposed by its reflection. The exposed area needs to be wide enough to afford the surgeon good visibility to perform the needed episode accurately and efficiently and, in order to fulfil any task, whether this be to remove a supernumerary tooth, dissect away bone, free an impacted tooth from its surrounding tissues or

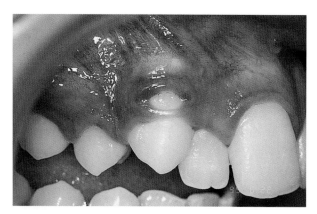

Fig. 3.7 Buttonholing.

bond an attachment, good haemostasis must be achieved. This is usually done using high-power suction, packs of various sorts and the application of pressure. If the duration of the surgically inflicted wound is excessive, the surface of exposed bone will become desiccated, leading to cell death and it may take several weeks or months of healing before the necrotic section of bone is resorbed and replaced with healthy new bone. More importantly, desiccation of exposed roots of teeth, periodontal ligament and cementum may occur, which will damage these tissues – a factor that may sometimes be compounded by the over-generous use of acid etchant. Adverse changes may then occur in these tissues which could result in impairment of the eruption mechanism of the tooth. This may be irreversible and result in failure to elicit eruption of the impacted tooth, even when traction is applied. The reader is referred to the sections on ankylosis and invasive cervical root resorption in Chapter 7 for a description of these phenomena. It is therefore important to select a surgical procedure or technique that may be completed in as short a time as possible.

On the face of it, it would seem that an open exposure should take less time than a closed procedure. However, the results of a study performed by our group have indicated quite the opposite [26, 27]. It seems that the wide tissue flap raised in the closed procedure improves visibility, permitting easier and quicker exposure of the impacted tooth, thus shortening its overall duration. Suturing the full flap back to its former place is also considerably quicker and neater than suturing an apically repositioned flap into its new position with accompanying pack placement.

Initiation of traction

During a closed surgical technique procedure, the orthodontist may or may not be present but it is imperative that an attachment be bonded at the time. Absence places the onus on the surgeon to do this. It is obviously propitious to apply the eruptive force to the impacted tooth immediately, taking full advantage of the prevailing anaesthesia.

In contrast, when open surgery is performed, the presence of the orthodontist is unnecessary, the intention being to prepare the stage for the placement of an attachment at a future date by the orthodontist in his/her office. This means that the surgeon must complete the exposure in such a manner as to be sure that the tissues will not heal over and make the tooth inaccessible in the few post-surgical weeks until an attachment is bonded in the orthodontist's office and traction may begin. Since orthodontic procedures in general do not require local anaesthesia, the orthodontist is unlikely to offer this to the patient at this orthodontic appointment, and additional delay in the application of force is inevitable due to the general sensitivity of the area to even gentle manipulation. It is therefore beneficial that an attachment be placed at surgery to maintain the treatment momentum.

Speed of eruption

When traction is applied to a palatally impacted canine in the closed eruption technique, the tooth may move rapidly and create a very palpable bulge beneath the thick mucosa of the palate, but will often experience difficulty in erupting through it. In these circumstances, it is recommended that a small circular incision be made around the crown tip of the impacted tooth and the tissue removed, to re-expose the tooth to a level not exceeding the greatest circumference of its crown. Further traction will then erupt the tooth very rapidly. Delay in performing this simple procedure will simply cause the anchor teeth to intrude and the overall arch form to become disrupted.

The final treatment outcome

Over the years, several groups of researchers in a number of countries have studied the post-treatment pulpal and periodontal status following the orthodontic resolution of impacted teeth, particularly in relation to maxillary canines and the open exposure technique. A Norwegian group [28] found an increased depth of periodontal pockets on the distal of the impacted teeth and bone loss on the mesial. In a group of patients with impacted canines, treated by undefined 'conservative' surgical procedures, a Seattle group [29] found attachment loss, reduced alveolar bone height and frequent instances of pulp obliteration, discoloration and misalignment, with the previously impacted canine being identifiable in 75% of the treated cases, presumably because of marred appearance. In relation to the outcome of the treatment of palatal canines with the closed eruption technique, a Jerusalem group of researchers found an excellent appearance, with slightly deeper pockets and a 4% loss of alveolar bone support [1]. In relation to buccally ectopic maxillary canines. They also found a minor reduction in the width of the attached gingiva, but otherwise a good periodontal outcome [3]. Further corroboration of the good clinical periodontal results to be seen when the closed eruption technique is used, in both buccal and

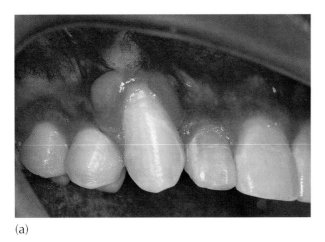
(a)

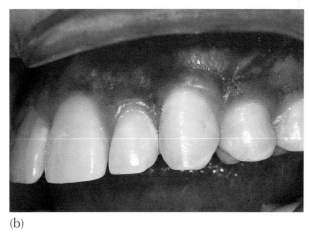
(b)

Fig. 3.8 Treatment for the right buccally impacted maxillary canine was performed using an open exposure technique and an apically repositioned flap. (a) The post-treatment outcome shows a thick band of attached gingiva, but a long clinical crown with an unaesthetic lumpy appearance of the gingival margin. (b) The normally erupted canine of the left side is shown for comparison.

palatal canine cases, comes from a study carried out in the UK [21].

Comparison of open (specifically the apically repositioned flap exposure) and closed procedures was undertaken by the Seattle group, who used a mixed sample of incisors and canines in their study [30]. With regard to those treated with the open surgical approach, they found poorer results in both periodontic measurement and aesthetic assessment. For buccal canines, clinical crown length was increased and the gingival margins were often uneven and of poor appearance (Figure 3.8). Loss of attachment and alveolar bone were noted, together with frequent vertical relapse of the erupted tooth, after the completion of treatment. They reasoned that during the tissue healing that occurs after the surgical repositioning, horizontal mucosal lines are formed which undergo stretching and distortion during the incisal movement of the tooth. Once orthodontic control is released, vertical relapse occurs due to a contraction of these extended mucosal lines. By contrast, they found that clinical crown length and gingival appearance in the closed eruption group were similar to those of the control side in the closed eruption cases, with a completely normal periodontal attachment. No post-treatment relapse in incisor position was seen.

Impacted incisors are seen far less frequently in the orthodontic office than are canines, accounting for a relative absence of reports regarding their treatment outcomes. However, given differences in their varied aetiologies, there was reason to believe that there could be differences in the end result of their treatment. The first clear-cut, 'maxillary incisors only' investigations were conducted by the Jerusalem group [26, 27], who found poor periodontal and aesthetic results in the open eruption group, with increased pocket depth and a 10% loss of alveolar bone height. The clinical crowns were elongated and the band of attached

gingiva reduced. Those treated by the closed surgical procedure showed only minimal changes, with greater bone support, a lesser increase in clinical crown length and better appearance than in the open surgery group. Crown length and attached gingiva were closely similar to those of the unaffected side, while the bone support level was reduced by between 5% and 6%.

In a subsequent open vs. closed exposure study, using similar impacted incisor cases matched for age, aetiology and location of tooth, the results strongly favoured the closed surgical exposure method, with the open exposure cases showing statistically poorer outcomes, characterized by increased crown length and more significant loss of bone support [30].

The evidence presented here indicates that a closed surgical exposure approach produces good, predictable, long-term periodontal and aesthetic results, and has advantages over the apically repositioned flap method [2, 32]. These have been attributed to the close similarity of the conditions brought about by full flap closure to those associated with normal tooth eruption [10, 23].

The Cochrane Data Base Collaboration offers systematic reviews of primary research in human health care and health policy and the clinical researchers investigate the effects of interventions for prevention, treatment and rehabilitation in various fields. All the existing primary research that meets certain criteria on a topic is searched for and collated, and then assessed using stringent guidelines, to establish whether or not there is conclusive evidence about a specific treatment. The reviews are updated regularly, ensuring that treatment decisions can be based on the most current and reliable evidence. This is obviously the ideal framework through which to objectively adjudicate the rival claims of the open exposure advocates and those favouring closed exposure, which may be tainted with

personal subjective preference or bias. A Cochrane Collaboration Systematic Review was recently undertaken in the UK [33] in which only those factors directly related to the surgical aspect were examined and no regard was paid to the age of the patient, the presenting malocclusion or the type of active orthodontic treatment undertaken.

It is clear from what has been described above and what will be discussed in later chapters that the quality of the outcome is only partly due to the surgical procedure and that these other factors are intimately involved in the outcome. Indeed, some may be just as potent in dictating the result. Without doubt, a number of these factors may be neutralized in a very large sample because randomization may have distributed them equally between the two groups. Nevertheless, there are several which cannot be discounted [1–3]. Among these one must look at three distinct sources, namely factors related to the patient, those related to the oral surgeon and those related to the orthodontist.

Patients present with a wide variation in positional diagnosis of the tooth and its proximity to adjacent teeth, some of which are not amenable to open surgery procedures. There is great variation in the level of oral hygiene among our patients and poor oral hygiene has a direct and negative relationship on the quality of outcome, even in the best treatment plan.

No two oral surgeons work in the same manner, and often exhibit differing preferences in relation to flap design, amount of bone and soft tissue removal, whether or not to remove the follicle in its entirety, pack size and the amount of pressure applied by the pack. Some surgeons are known to 'assist' in the eruption by radical exposure and by pushing an elevator down the PDL of the tooth, to be sure that it is mobilized, particularly when the orthodontist is not watching! When the surgeon is called on to place the attachment, his/her bonding skill and knowledge of the most advantageous bonding location cannot match those of the orthodontist, raising the spectre of detachment, repeat surgery and inappropriate traction direction. Any one of these factors may alter periodontal health and gingival architecture.

Orthodontists also work in different ways, with widely differing appliance methods and customized traction device designs. There will be a periodontal price to pay for inappropriate directional traction, excessive traction forces and treatment inefficiency, not to mention varying quality standards of case completion.

In short, the number of factors involved is so great and so diverse that it is not possible to arrive at a 'squeaky clean' sample that will permit the comparisons necessary to achieve a neat and authoritative answer to the question of surgical exposure preference – Cochrane Collaboration notwithstanding. This leaves the reader to rely on what is available, which is a large body of published studies based on retrospective samples, from which to draw some appropriate, if empirical, conclusions.

Partial and full flap closure on the palatal side

Often, impacted teeth that are located on the palatal side are palpable immediately beneath the palatal mucosa, which is firmly bound down to the underlying bone. The surgical removal of a circular section of the overlying mucosa (see Figure 6.29) and the thin bony covering, and leaving the tooth exposed, is tempting and has obvious advantages. In particular, the newly exposed tooth will be favourably invested with attached gingiva when it finally erupts. However, the palatal mucosal covering is very thick and will leave a broad cut surface, which will tend to close over unless its edges are substantially trimmed back and the dental follicle removed. Thus, for a deeply placed tooth, the exposure will additionally need to be maintained using a surgical pack. This type of surgical approach will, therefore, leave the palatal side of the tooth with a soft tissue deficiency and a long clinical crown, at the completion of the orthodontic alignment and in the long term, even though the character of its surrounding tissue will be desirable attached gingiva (Figure 3.9). This method has been favoured and promoted in relation to palatally impacted canines, since it has been suggested that the canines will improve their position in many instances and erupt autonomously in the palate in time [34]. This study retrospectively investigated the post-treatment periodontal status of a group of patients who had been successfully treated by this method. However, there is no published controlled study that looks into the reliability and predictability of this treatment option protocol.

As we have described for the buccal side, full flap closure on the palatal side allows the tooth to be exposed with the minimum of tissue removal and consequent surgical trauma and it also requires the bonding of an attachment on the exposed tooth prior to suturing. When this is done and given appropriate orthodontic mechanics, the final result will show that the bone support for the tooth, as well as the health and appearance of the muco-gingival tissues, is at its best, as will be demonstrated in the following chapters. The accumulated evidence that has been presented in the various clinical comparisons of surgical methods of exposure [1, 2, 4–6, 23, 26–28, 30, 31, 35] provides the evidence base to recommend the full flap closure approach over any other, in most situations, in relation to a palatally impacted tooth. This refers to both qualitative clinical appearance assessment of the crown length and gingival architecture and quantitatively to objective parameters evaluated in a periodontal examination.

Buccally accessible impacted teeth

Indications for apically repositioned flap exposure

On the basis of their results for exposure on the buccal side of the ridge, the Seattle group questioned the justification

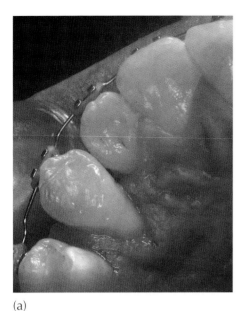

(a)

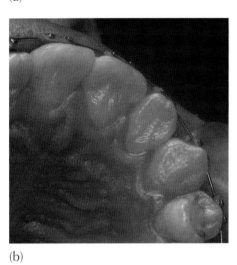

(b)

Fig. 3.9 Treatment for the right palatally impacted canine was performed with an open-exposure technique. (a) The post-treatment result shows attached gingiva of the palatal tissues covering most of the root, although the clinical crown length extends well down on the palatal side of the tooth, leaving several millimetres of root exposed on that side. The bone level is expected to be 8–10% defective compared to the untreated side. (b) The normally erupted left canine is shown for comparison.

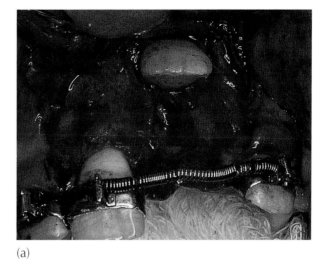

(a)

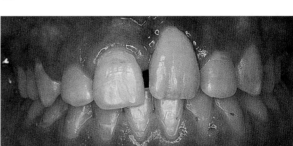

(b)

Fig. 3.10 A case treated by the author in the mid-1970s, before the era of the acid-etch technique. (a) A left impacted maxillary central incisor has been exposed and the entire follicular sac removed, prior to cementing a band. (b) At two years post-treatment, the gingival contour is poor, the clinical crown is long and there has been positional deterioration of the previously aligned tooth.

for the continued use of the apically repositioned flap method [30]. From the above discussion, it is clear that there are grounds for their scepticism in relation to teeth that are more severely displaced. Nevertheless, it should be realized that for the more trivial cases, where the tooth is mesio-distally fairly close to its final position and bulging the oral mucosa at its junction with the attached gingiva, the application of the apically repositioned flap method may eliminate the need for subsequent orthodontic treatment, as we have noted earlier in this chapter and in Figure 3.1, while producing a good periodontal result. Experience

shows that many of these teeth never come down fully to the occlusal level, and those that do erupt well may take many months, sometimes stretching to a year or more, to do so. This appears to be due to the relapse tendency engendered by the distorted mucosal lines in the muco-gingival area produced as a result of the surgery [10]. In the final outcome of this form of surgical exposure, an unaesthetic gingival contour may sometimes result (Figure 3.10). Grafting may be needed to improve this [15–16, 36–38].

When left untreated, buccally palpable unerupted teeth may take many months to break through the mucosa and reach their final positions. After an apically repositioned flap is performed, eruption is speeded up. Additionally, with the sutured soft tissue applying some pressure on the buccal side of the tooth, and assuming there to be space in the immediate vicinity, buccal displacement may be spontaneously reduced.

If the unerupted tooth is very high, the surgical flap, which must include attached gingiva on the crest of the ridge or at the free gingiva of the deciduous tooth up to the

depth of the sulcus, will be excessively large. Under these circumstances, the procedure is not recommended, since the apically repositioned flap would leave a wide area of periosteum of the labial bony plate unnecessarily exposed to the oral environment. In order to cover this area, grafts would need to be taken from elsewhere in the oral cavity. A useful alternative for these very high teeth is to use a combination of two distinct techniques [39], particularly for conditions like the labial dilacerate central incisor. Initially, a closed eruption procedure would be used and the tooth brought down until it bulges the labial mucosa, well above the attached gingiva and is in danger of erupting through this loose and thin oral epithelium. At this point, a second procedure is undertaken in which an apically repositioned flap, taken from the crest of the ridge and incorporating attached gingiva, is raised and re-exposes the impacted tooth from below. The flap is then taken over the incisal edge/occlusal tip of the tooth and sutured on the labial side. The tooth may now continue to be drawn occlusally, completely encompassed by firm gingival tissue. At the same time, it should be recognized that the well-sutured flap will additionally apply pressure on the labial side of the buccally/labially displaced tooth that it is partially covering and will be a positive influence in moving it lingually towards the general line of the dental arch (see Chapter 15).

While the closed eruption exposure method is to be preferred in most situations, an important advantage of the apically repositioned flap method is that the buccally impacted canine is exposed to the oral environment and remains accessible for attachment bonding. In some cases, where orthodontic treatment is not needed for other problems, the progress of the tooth may be followed for many months, until full eruption has occurred without the use of appliances (Figure 3.1). In others, an attachment may be bonded by the orthodontist at any appropriate later date and active extrusion subsequently undertaken.

Indications for circular incision (window) exposure

Perhaps the only situation in which the circular incision is appropriate for a buccally sited tooth is when the tooth is correctly mesio-distally located, the band of attached gingiva is very broad and the tooth can be palpated occlusal to the muco-gingival junction. In this case, and provided that at least 2 mm of attached gingiva remain intact apical to the crown after removal of the cut tissue over the tooth, this procedure provides the simplest appropriate approach [10].

The relief of crowding to reduce canine displacement

If the displacement of the canine has been due to crowding, then it follows that spontaneous improvement of the position of the canine may well occur if the crowding is eliminated.

Time may not be on the side of the clinician opting for this approach, since the tooth may erupt through the oral mucosa if delay is incurred. Nevertheless, for the case in which this approach is to be used, a full case analysis is required, leading to a diagnosis and treatment plan for the overall malocclusion. If the crowding is to be dispersed by distal movement of the molars, it will take longer before space is available in the canine region. Considerable delay must be expected while the treatment is proceeding before spontaneous improvement of the canine position may be seen. On the other hand, a premolar extraction will provide immediate relief of the crowding and an excellent opportunity for a self-correction of the buccal displacement and, with it, the disappearance of the potential periodontal hazard.

A buccal approach may sometimes be preferred in the treatment of a palatally impacted maxillary canine, provided that its palatal displacement from the line of the arch is fairly minimal. By using the buccal approach with an impacted tooth that is vertically fairly close to the CEJs of the adjacent teeth, there is a danger that interproximal bone will be destroyed in the process. The greater the palatal displacement of the tooth, the greater will be the bony defect. Nevertheless, the approach may be advised for those teeth that are marginally palatally displaced and situated higher in the maxilla, since it offers the prospect of a more direct route for traction of the tooth to the labial archwire. Impaction of the maxillary canine which is close to the line of the dental arch may occur when the canine has a mesial tip and when its mesial crown incline comes into direct contact with the distal side of the root of the lateral incisor. This is a fairly easy impaction to treat, and may be self-correcting, when the crown of the lateral incisor is tipped mesially to close anterior spacing and to provide room for the canine in the arch. By doing this, the root apex of the incisor moves distally and favourably alters the orientation of its long axis with that of the canine. If the tooth still shows no sign of erupting, a labial surgical approach to the very mildly palatally displaced canine will often be the most suitable and, in terms of the traction, the most direct.

A conservative attitude to the dental follicle

The dental sac or follicle develops from a mesodermal condensation of cells on the outer surface of the external enamel epithelium of the enamel organ of a forming tooth, into a fibrovascular capsule. The follicle has an inner vascular plexus through which the enamel organ is supplied with nutrients during growth, and an outer vascular plexus which is concerned with enlarging the bony crypt in which the tooth germ lies. It achieves this by its inherent capability to resorb the alveolar bone. The follicle encompasses the entire tooth germ and, as Hertwig's root sheath proliferates as an apically directed extension of the follicle, the root

portion of the tooth is formed. Later, the outer surface of this sheath eventually becomes the periodontal membrane, connecting the cementum covering of the developing root to the developing alveolar bone.

The enamel cuticle covering the crown is made up of a keratinous deposit from the ameloblasts and the reduced enamel epithelium, and is continuous with Hertwig's epithelial root sheath. This separates the crown of the tooth from the follicle, from which the root develops and cementum forms. It is because there is this separation between follicle and enamel that cementum is not formed on the crown of the tooth.

Eruptive movements that accompany the resorption of the bony crypt bring the tooth follicle into close proximity to the oral mucosa. Their epithelia fuse and the tooth thus breaks through an epithelium-lined opening. As eruption proceeds, the remainder of the follicle everts and comes to be turned 'inside out', with the reduced enamel epithelium becoming the gingival cuff and most superficial point of attachment. When the tooth erupts in this normal manner, it is accompanied by a relatively generous band of gingival tissue, which is rounded and appears a little swollen initially. The most coronal portion of this gingival tissue is the free gingiva, and, at its base, the attachment is direct to the cervical area of the enamel of the crown of the tooth, several millimetres coronal to the CEJ. This represents a very special type of attachment, in that the gingival band immediately adjacent to the crown adheres directly to the enamel through the agency of hemidesmosomes (cells that originate in the reduced enamel epithelium). Over the subsequent period of 3–4 years, this transforms into the junctional epithelium which constitutes the initial form of attachment of the gingiva to the tooth, on the cervical enamel area [40].

It is in the patient's interest for us, when treating an impacted tooth, to try to create conditions that will replicate this natural eruption process as closely as possible, and this obliges us to treat the follicle conservatively. If the follicular tissue is totally removed down to the CEJ, then we can no longer expect there to be a gingival attachment direct to the enamel of the crown of the tooth. The attachment that will result will be on cementum, at or somewhere down the root surface, beyond the CEJ. This will result in an elongated clinical crown and a compromised periodontal result (Figure 3.10).

Point to remember

When an impacted wisdom tooth is removed surgically, the dental follicle is always carefully dissected out to prevent the possible later occurrence of cysts that may arise from residual follicle epithelium. Without the tooth, therefore, the residual follicle has no function, merely a potential nuisance value. However, when a tooth is exposed only, its surrounding follicle has an important function to fulfil – a function which is identical with that of a tooth which

erupts normally and which is integral to the establishment of a normal biological support system. This difference in attitude to the follicle of a tooth committed for extraction compared with that of a tooth that is to be exposed and subsequently to be erupted into the mouth is basic and important to understand.

From studies of the causation of external crown resorption of long-standing buried teeth, it has been found that pathological changes occur in the follicle surrounding its crown (see Figure 6.12) over a period of many years, changes which have brought the enamel surface into direct contact with the surrounding tissues [41]. It is easy to draw a parallel between this condition and the artificially produced environment of an impacted tooth that has been surgically exposed and, without extrusive force being applied, has subsequently become reburied in the tissues. If, for whatever reason, the tooth does not erupt spontaneously, there will be long-term direct contact between the tissues and the enamel of the tooth.

In clearing the tissues surrounding an impacted tooth for the purpose of creating an opening adequate in size which will not eventually close over, the surgeon will, generally, deliberately and completely remove the follicle surrounding the tooth (Figure 3.10). The oral epithelium grows down the sides of the opening, into the area that has been surgically cleared of follicular tissue. It will grow down as far as the deepest point where instrumentation has occurred which, in this case, will be at least to the CEJ or possibly a little way down the root surface, depending on how carefully the surgery was performed. This is considerably further apically than one would expect to find in a tooth whose recent eruption has occurred spontaneously, and a compromised gingival attachment will result. The subsequently erupted tooth will have a longer clinical crown and reduced alveolar crest height, in sharp contrast to a tooth that has erupted recently and autonomously.

The application of orthodontic traction only requires an opening in the follicle which is large enough for the attachment to be placed, while the rest of the follicle may be left intact. The surgical flap may then be fully sutured back and the wound completely closed. Orthodontic traction brings the tooth towards the oral cavity and the intact remains of the follicle fuse with the oral mucosa, to mimic normal eruption. This leads to the establishment of a normal gingival attachment [23].

It is clear, then, that a new look must be taken at the surgical plan for the exposure of unerupted teeth. If bonding will not take place at surgery, then a wider exposure must be performed and a surgical pack may need to be placed. The principal reason for this is to prevent the reclosing of the wound. Thus, even though one must stress the importance of avoiding overzealous surgical removal of the follicle and of damaging the CEJ area, by the forceful placement of the pack, a poorer periodontal result should be expected. Attachment bonding must subsequently be performed at a

convenient time after pack removal. However, at this time, the healing and swollen gingival tissue surrounding the exposed tooth will be tender, will be covered with plaque and will bleed with minimum provocation, since effective toothbrushing in this delicate and sensitive area is unlikely to have been possible. These are not conditions that are conducive to reliable attachment bonding, despite the ease of access.

A wide flap design has the advantage of displaying the form of the area of bone covering the tooth, which is helpful in identification of the exact site of the tooth. A canine tooth buried in a bony crypt in the palate will alter the shape of the palate inferiorly by creating a distinct bulge of thinned bone, and this will be all the more obvious if much of the surrounding bone is also visible. This is just as true in the labial plate of the maxilla and in the buccal or lingual plate of the mandible, where the tooth in question may be a maxillary central incisor, a second premolar or any other tooth. The distancing of the edges and underside of the flap from the field of operation is important, if contamination with blood is to be avoided during bonding. This is most easily performed when the flap design is generous.

Once the bony surface has been exposed and the location of the buried tooth identified, the thin overlying bone may be lifted off very easily. The surgeon will generally use a sharp chisel with light hand pressure to cut open the bony crypt and remove the superficial part of its wall. The bone is often paper-thin and can be cut with a sharp scalpel. Immediately beneath the bone, the dental follicle will be seen to glisten in the beam of the operating lamp. A window should be cut in the follicle to fully match the extent of the very minimal bony opening that has already been achieved, in order to see the orientation of the tooth, as it lies in its crypt.

As we shall describe in later chapters, it is important to place the orthodontic attachment as close as possible to the mid-buccal position of the crown of the tooth, so that traction will tend to reduce any existing rotation, thereby reducing the amount of mechano-therapy to which the tooth will need to be subjected. For this reason, where a rotated tooth is exposed, the bony opening should be extended around the crown of the tooth, towards the mid-buccal area of the crown, provided that this can be done with ease and with the infliction of little or no further surgical damage. In this instance, flap replacement may be completed and the pigtail ligature, tied into the newly placed attachment, drawn in the direction of the proposed target site in the dental arch.

During exposure of the crown of a tooth, instrumentation of the enamel surface is not detrimental to the eruption process or to the quality of the treated result. However, exposure and instrumentation of the root surface is potentially damaging. Exposing root surface presupposes that the natural attachment of the tooth at the CEJ will have been ruptured, and renewed attachment will probably only be established more apically. Additionally, periodontal fibres are severed and cementum exposed, subjected to drying (suction and air syringe) and contact with foreign substances (etchant, bonding materials). This can lead to the later initiation of a resorption process on the root surface, to ankylosis and to failure of eruption in extreme instances. More common sequelae include seriously reduced bone support, long clinical crowns and poor gingival attachment and quality.

The *raison d'être* of the closed eruption technique has been called into question by Kokich, who prefers an open surgical approach for the exposure of canines that are deeply impacted in the palate. The method he describes demands the excision of enough bone until the cavity created is larger than the crown of the tooth, at its greatest mesio-distal diameter, and extending from crown tip to CEJ and, at the same time, the follicle is removed in its entirety. The rationale for this procedure is that while contact of the follicle of an advancing unerupted tooth with alveolar bone causes resorption of the bone, as seen in the normal, unaided eruption of teeth, the proximity of bare enamel to alveolar bone does not physiologically initiate resorption 'since there are no cells in the enamel to resorb the bone'. The contention is that 'resorption will eventually occur through pressure necrosis, but it will occur slowly'. Hence, when an impacted tooth is located in mid-palate, the advice has been to perform an open exposure and maintain its patency pending natural, spontaneous eruption, which may or may not occur. The confident claim is that 'these palatally displaced canines will erupt on their own . . . in about 6 to 8 months' [10]. This hypothesis has not been tested on a random sample of impacted canine cases and, more importantly, neither has the periodontal outcome of the orthodontically successful outcome of such a sample been evaluated.

Reparatory bone deposition begins in the organizing blood clot within a few weeks following surgery. This being so, it follows that, unless the widest part of the crown of the impacted tooth has been drawn fully outside the bony plate in this very short time period, bone must be expected to re-form over parts of the crown of the tooth and that, according to the hypothesis, this will cause the tooth to stop moving. This would obviously not augur well for patients with teeth which are deeply impacted in bone, or others for whom several weeks or months may have elapsed after the surgical exposure and before orthodontic traction is applied. Chapter 12 describes many of the more common reasons why failure to resolve the impactions may occur and how these may be avoided. Cases illustrated there show resolution of the impaction being successful as much as a year or more after the surgical exposure had been performed and, sometimes, after an initially failed treatment by another practitioner. Before treatment was started in these failed transfer cases, much mature bone would have been laid down as a potential stumbling block in the path

of the impacted tooth, yet the second attempt at treatment was successful, rapid and did not need re-exposure of the tooth.

Three decades ago, it was shown that the presence of an intact follicle is a prerequisite for the process of normal and spontaneous eruption [42]. Nevertheless, experience has taught us that light orthodontic traction is capable of encouraging the resorption of bone needed for the eruption of an exposed tooth, even in the absence of a follicle. Anecdotal clinical evidence contradicting Kokich's view is provided in the section dealing with impacted maxillary canines that are associated with root resorption of their immediate neighbours (Chapter 7). In the more extreme examples of this anomaly, the canine crown and the resorbed incisor root are intimately related and situated in the middle of the ridge, surrounded by bone on all sides [39]. The exposure has to be carefully planned to avoid surgical trauma to the incisor root area. Wide clearance of bone and dental follicle to the full width and length of the crown of the canine, down to the CEJ, is obviously out of the question. Nevertheless, these teeth can be routinely drawn through the surrounding bone and the impaction resolved with light forces which are suitably directed, as with any other impacted tooth, and with considerable speed in most cases. Similarly, in Chapter 14, which describes the treatment of patients with cleidocranial dysplasia (CCD), surgical exposure has to be performed on multiple impacted teeth deeply displaced low down in the basal bone. For reasons outlined in that chapter, exposure of the canine and premolar teeth is typically aimed at the buccal aspect of the teeth, without the deliberate wide removal of bone surrounding the remainder of the crown of the tooth and without exposing the occlusal surface of the crown of the tooth superiorly. This seems to have no noticeable retarding effect on the eruptive response of the tooth to occlusally directed light forces, which is rapid and generates eruption within a short time. This is despite the fact that it may often have to resorb a thick layer of bone on its way, and despite the fact that alveolar bone in CCD patients is considered to be particularly dense and the largely acellular cementum on the roots of their teeth is associated with slower resorption [43].

Individual clinical case reports illustrating the eruption of teeth through bone grafts appear in the literature from time to time. In one of these, a palatally impacted maxillary canine was diagnosed in a unilateral cleft palate patient, located superiorly to the reparative autogenous alveolar bone graft that had been placed some years earlier. The patient underwent a closed eruption procedure and the tooth was erupted through the bone graft [44]. In another case report [45], an impacted canine was exposed and then successfully drawn through a synthetic bone substitute, which had earlier been grafted to fill an area where a significant volume of buccal cortical bone had been lost, due to the surgical removal of multiple odontomes.

In each of these difficult and extreme scenarios, a successful outcome of the treatment will almost always show good clinical and radiographic features, notwithstanding the need to erupt the teeth through surrounding alveolar bone. In terms of clinical appearance, the periodontal condition and an absence of any signs of pathology, it is also difficult to distinguish the treated teeth from other unaffected teeth in the mouth.

The tunnel method [23] mentioned above deliberately aims to bring a large canine down through the much narrower socket recently vacated by the extraction of its deciduous predecessor. This cannot be done without the resorption of bone lining the socket. Furthermore, given the time involved in bringing a severely displaced canine into its place in the arch and however rapidly this may be achieved, the lower part of the socket will surely have undergone healing. Therefore, the eruptive progress of this tooth cannot go far along this vacated socket before healing of the extraction will have brought about the deposition of new bone directly in its path.

In the closed eruption approach described in the present chapter, we have recommended reflecting a wide soft tissue flap, but only opening the dental follicle itself to a very minimal degree – enough only to permit the maintenance of haemostasis while the bonding of a small attachment is performed. The remainder of the follicle survives intact, which means that all other parts of the crown of the tooth are invested in follicular tissue, which will initiate bone resorption when traction is applied. The break in the integrity of the follicle exists solely at the site of the attachment, where only a minimum of bone has been removed. It is open to speculation whether repair of the follicular tissue may occur over a low-profile attachment and its integrity fully restored around the steel ligature to which traction is applied. This ideal scenario would appear to be most unlikely, although if it were indeed true, then the ideal eruption environment will have been re-established and light directional traction should bring about successful eruption of the tooth, with the reward of an excellent gingival appearance and periodontal health. Regardless of the reasons why a conservative attitude to the dental follicle works so well, the fact remains that post-orthodontic corrective periodontal flap and graft procedures are largely superfluous in these cases. For the most part, the previously impacted teeth are impossible to distinguish from their unaffected neighbours and antimeres, clinically, radiographically and aesthetically (Figures 3.11 and 3.12).

Quality-of-life issues following surgical exposure

When a patient undergoes traditional surgery for an inguinal hernia, he must be informed that, following his stay in hospital, his previously active lifestyle will need to be curtailed. He will suffer pain and swelling at the

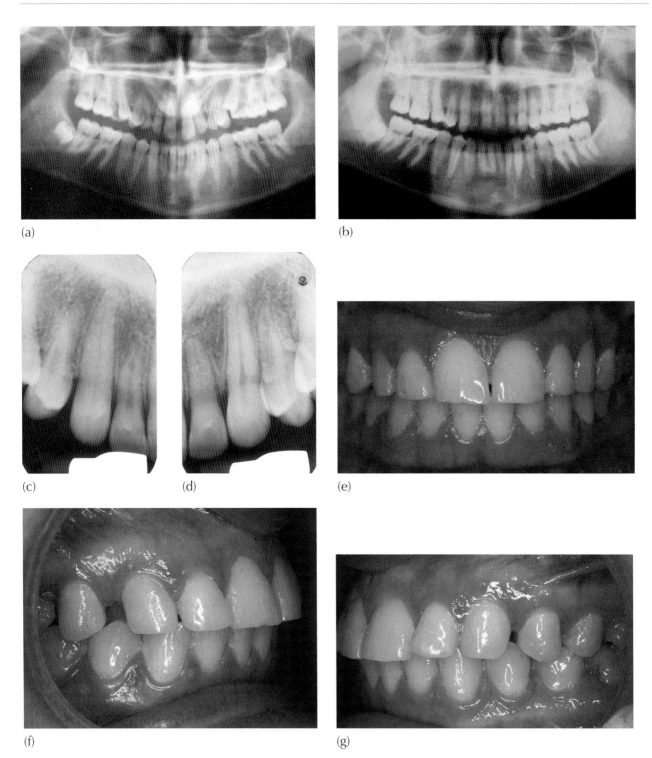

Fig. 3.11 A case of bilateral palatal impaction of maxillary canine treatment with the closed eruption surgical technique. (a) Panoramic view to show canines almost contacting in the midline. (b–d) Panoramic and periapical views of the post-treatment result. (e–g) Centre, right and left clinical views of the treated result, 14 years post-treatment, show the treated teeth to be indistinguishable from their neighbouring teeth in terms of their crown length, gingival contour and excellent appearance. (h–k) Clinical views of the buccal and palatal sides of the two canines, 14 years post-treatment, to show normal gingival contour, normal crown length and no recession. No reparative periodontal procedures were performed after the original closed surgical exposure.

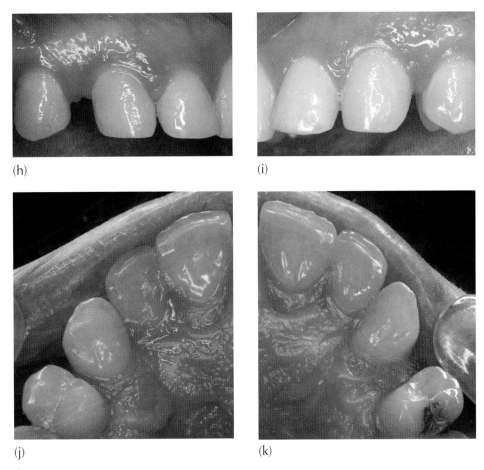

(h)

(i)

(j)

(k)

Fig. 3.11 (*Continued*)

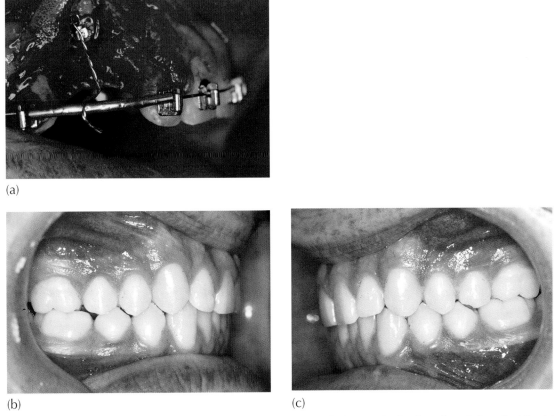

(a)

(b)

(c)

Fig. 3.12 (a) an impacted right maxillary canine treated with full flap closure on the buccal side, using the closed eruption technique. (b, c) The right and left sides are indistinguishable at the completion of treatment.

operation site. He will be unable to undertake heavy work, his posture may be initially affected, driving a car will be difficult and he will have pain when passing urine in the early postoperative period. Under such circumstances, the character of his entire lifestyle will need to be much subdued and his quality of life will suffer.

In the present context, the (usually young) patient who is about to undergo surgical exposure of an impacted tooth needs to be informed how the procedure will affect his daily life, in terms of pain, function, speech and the several other aspects that surround the oral cavity, including the risks and benefits of the intended treatment. The patient is often apprehensive at the thought of surgery, particularly if he is a young and healthy individual with little or no previous experience of surgical procedures. The incidence and magnitude of these possible sequelae are all part of the post-surgical follow-up, which the patient and his parent must be aware of, and constitute information that they must legally have in order for them to sign a statement of informed consent. While this is true of all types of orthodontic treatment, it is particularly so when surgery is involved.

Oddly, while a number of articles have recently appeared in oral surgery journals regarding these parameters in relation to the extraction of third molars, there is a significant paucity of published work that relates to quality-of-life issues related to the surgical exposure of impacted teeth. The result has been that the information available to both clinicians and patients often comes from one-time anecdotal episodes and hearsay from individuals who have themselves experienced some form of oral surgery. These reports are notoriously unreliable and rarely offer appropriate comparisons with the surgical exposure at hand.

It was for this reason that prospective clinical studies were undertaken in Israel to quantitatively assess the various aspects of quality of life following the performance of both open and closed surgery [46–48].

Two groups of patients were assembled for study. One group included young patients who were scheduled for open surgery and the second group for closed surgery of an impacted tooth. Each patient was presented with seven identical questionnaires on the day the exposure was performed and was instructed to complete one of the questionnaires on each post-treatment day, for seven days. From the answers, information was collected regarding pain, oral function, general disability, limitation in eating, absence from school and related parameters. The results for the group of patients who had had open exposure were then analysed and compared with those for the patients who had undergone a closed exposure technique.

In general, it was found that recovery from an open eruption exposure required five days vs. three days for a closed procedure. The prolonged recovery was particularly observed in relation to a higher level of pain, difficulty in eating and swallowing, and to the use of analgesics. There

was much greater discomfort with the open exposure in association with a palatally impacted canine, especially if bone removal had been performed. However, it is noteworthy that exposure of impacted teeth with a buccal approach resulted in a high level of discomfort, regardless of the surgical method employed. It may be speculated that this is due to the fact that paranasal and oral musculature is severed during buccal procedures, and the healing surgical flap is sited in highly mobile oral mucosa.

We may summarize the advantages and disadvantages of complete flap closure (healing by primary intention) compared with the alternative open exposure techniques, in which the opening in the tissue over the impacted tooth is maintained by reducing the size of the flap and packing the wound or by repositioning the flap more apically (healing by secondary intention), in the following manner.

Primary full flap closure
Advantages

- rapid healing
- less discomfort
- good postoperative haemostasis
- less impediment to function
- conservative bone removal
- immediate traction possible
- high degree of reliability of bonding
- possible in close proximity to resorbing root area

Disadvantages

- presence of orthodontist advised
- bond failure dictates re-exposure

Open exposure
Advantages

- orthodontist's presence unnecessary
- bond failure – needs no surgery

Disadvantages

- greater risk of infection
- greater discomfort
- interference with function
- wider bone exposure
- bad taste and smell in mouth
- possibility of reclosure of exposure
- bonding reliability poorer
- delayed initiation of traction
- poorer periodontal condition
- extra visits to change packs

Cooperation between surgeon and orthodontist

From the discussion in this chapter, it is hoped that the reader will have come to realize the narrow limitations of

the surgeon's ability to materially assist these cases and to appreciate that the inclusion of orthodontic procedures offers most cases a better chance of success. Today, orthodontists have come to play a more important role in the initial stages of the treatment of impacted teeth, by providing the traction that is necessary to encourage this eruption and, in many cases, to do so successfully in teeth that were previously felt to have a poor prognosis for eruption.

The status of an impacted tooth today is largely dependent on the ability and the ingenuity of the orthodontist to apply light traction in an appropriate direction and with efficient means, once the tooth has been made accessible by the oral surgeon. If orthodontic treatment is available to the patient, there is little merit in the surgeon offering any of the other procedures listed at the beginning of this chapter, since there is no evidence that suggests that these procedures may improve the opportunity for orthodontic resolution, without causing concurrent harm.

We may therefore conclude that, with respect to the treatment of impacted teeth, the aims of the oral surgeon should be limited to:

- the provision of access to the buried tooth;
- the clearing of any obstruction in the tooth's eruptive path, such as supernumerary teeth, odontomes or thickened overlying mucosa;
- taking an active part with the orthodontist in bonding an attachment to the exposed teeth at surgery by maintaining haemostasis, which is so critical in ensuring success.

The single most important aim of the surgical episode is to provide access to a tooth which is otherwise buried. This enables the orthodontist to provide the means by which force may be applied to the tooth in question, through several subsequent visits over a longish time span, and in as simple a manner as possible. For this to happen, an attachment has to be securely bonded and a firm ligature or other form of intermediary drawn to the exterior, to which steel wires, super-elastic nickel–titanium wires, elastic ligatures or an auxiliary spring may be tied. A sharing of the responsibility for the successful execution of this procedure should be undertaken by the oral surgeon and the orthodontist, with one specialist complementing the other in applying their very special skills to the resolution of the immediate task. Together, they possess all the tools that are needed to complete the job for which, separately, each is perhaps ill-equipped. Their combined efforts should be geared to achieving this.

Bonding an attachment to the tooth a few weeks after the surgery has been performed has the advantage of not requiring the presence of the orthodontist at the surgeon's side. By doing this, the surgeon must expose the tooth much more widely, place surgical packs and aim for healing 'by secondary intention', as has been pointed out above and as will be explained in greater detail in later chapters.

Additionally, the reliability of the bonding at this later date is much poorer than when performed at the time of surgery [24].

The team approach to attachment bonding

This episode primarily represents an adjunctive surgical procedure, whose aim is to provide a site for the application of an orthodontic force-delivery system. As such, it should be carried out on the surgeon's territory and not in the orthodontic clinic. The orthodontic treatment will have been initiated and orthodontic appliances will, most often, be in place before the surgical exposure is attempted. Orthodontic procedures that need to be carried out during the surgical episode are few and relatively simple and can all be performed in the oral surgeon's operatory. However, the orthodontist should prepare a small tray of instruments and materials which are not normally available in the operating room, together with a customized auxiliary spring, which may have been fabricated at a previous visit for the purpose of applying a directional force to the impacted tooth, such as a ballista [49], a flexible palatal arch [50–51] or a labial auxiliary arch [52] (see Chapter 6). The instrument tray should contain the following items:

Instruments

- a fine wire bending plier (e.g. Begg plier)
- a fine wire-cutter
- reverse-action bracket-holding tweezers which are closed when not held and release when the handles are lightly squeezed
- a ligature director
- a mosquito or Matthieu forceps
- a fine scaler

Materials

- etching gel
- composite bonding material, preferably a light-curing material, which is probably the most manageable type of bonding method that can be used in these special circumstances, although a no-mix paste and catalyst may serve very well
- applicators (wood sticks, fine brushes, etc.)

Attachments

- eyelets welded to thin band material, backed with stainless steel mesh; these should be cut and trimmed into patches of various sizes, but not larger than the base of a small bracket
- cut lengths of dead soft stainless steel ligature wire of gauge 0.011 in or 0.012 in
- elastic thread and elastic chain

It has been the author's practice to tie the labial auxiliary arch into the orthodontic brackets immediately prior to the surgical exposure since, in its relaxed position, the active

loop stands well away from the immediate surgical field and does not interfere with the work of the surgeon. Other customized auxiliaries may be similarly tied in or placed on the instrument tray, in readiness for placement at the end of the procedure.

The surgeon reflects a muco-periosteal flap over the impacted tooth and removes the intervening bone, which is usually very thin and easy to peel with a scalpel blade. If a supernumerary tooth or odontome is present, this is removed first. The dental follicle is removed from the target area immediately overlying the crown and the resultant exposure is not widened more than is necessary to satisfy two basic requirements (Figure 3.6a). These are: (a) to provide enough enamel surface to accept a small attachment, and (b) to do so in an area wide enough for adequate haemostasis to allow the bonding procedure to take place without fear of contamination.

The surgeon then moves to the other side of the operating couch in order to take over the entire responsibility for maintaining the enamel surface free of blood and saliva throughout the critical bonding phase – which is, after all, the point of the entire exercise!

Under the conditions of exposed and oozing soft tissue and bone surfaces, the surgeon will generally need to use a regular suction tip and a second and very fine tip in the form of a cannula no. 14 or 16, in order to maintain a bloodless field of operation for the bonding procedure. Occasionally, a persistent bleeding point from the bone surface may require the surgeon to use either pressure from a blunt instrument or the application of bone wax to occlude the tiny vessel. Soft tissue bleeding may be controlled with electro-cautery, a hot burnisher or, occasionally, ligation of the vessel. Bleeding does not occur in the follicular space, but seepage from adjacent areas may be present and is best arrested with the use of light pressure from a strip of gauze, which may be left in place until suturing is ready to begin – but it must not be forgotten! Then, armed with a retractor in one hand and alternating the suction tips as necessary with the other, the surgeon will maintain the access and haemostasis to the immediate area of the newly exposed and impacted tooth.

The orthodontist proceeds directly to rinsing the tooth surface with atomized water spray from a standard triple syringe or, if preferred, with sterile saline from a large syringe, through a wide bore needle, to disperse any blood from the tooth surface. The saline is evacuated through the broad suction tip, operated by the surgeon. The fine suction tip is then substituted and is made to hover over the entire exposed crown, close to the tooth surface, with the aim of drawing air over the clean enamel. This achieves effective drying, and the use of sterile saline as a rinsing agent does not appear to undermine the reliability of the bonded union.

Liquid etchants should not be used in the exposed surgical field [22] since it is difficult to limit their spread to the exposed soft tissues and bone surfaces and, perhaps more important, to the area of the CEJ, the PDL and cementum. It should be applied by the orthodontist as a gel on the end of a fine wooden toothpick applicator (non-medicated), left in place for 30 seconds and then drawn off by the surgeon through the fine suction tip, before the surface is rinsed again with saline to remove the last traces of acid.

Continuation of use of the fine tip will draw air over the surface of the crown of the tooth until drying is achieved, when the typical white matte appearance of the etched surface will quickly appear. The surface is now ready for bonding.

Many practitioners may feel concern about the adequacy of the desiccation and may also prefer to be sure that no salt crystals remain from the dried saline. Experience shows the concern to be groundless. Nevertheless, to allay these fears, a final rinse with atomized water from the triple syringe may be followed by a fine compressed airstream to achieve the appropriate degree of dryness of the enamel surface. For this to be successful, the compressed airstream must be very gentle to avoid splashing up blood from the surgical area, which will contaminate the enamel and cause bond failure. The use of a hair dryer has the advantage of providing a gentle stream of warm air, which may be more effective in drying the etched enamel surface and is a method favoured by some clinicians.

The prepared eyelet attachment has a pliable base. An attachment of appropriate size should be selected and adapted between the plier and the gloved hand of the orthodontist, to conform to the target bonding site. A cut length of 0.011 in (0.275 mm) or 0.012 in (0.3 mm) dead soft stainless steel ligature wire is threaded through the eyelet and, with the use of the mosquito or Matthieu forceps, is twisted into a medium-tight and firm pigtail, which should swing freely in the eyelet.

Strictly, any type of bonding agent may be used, including light-activated and chemically-activated systems. Attachments will be discussed in more detail in the next chapter. Nevertheless, one or two points regarding bonding under conditions of surgical exposure are pertinent in the present context.

In the first edition of this book, we expressed a preference for a chemically activated 'no-mix' system. In this system, the attachment is seized in the reverse-action bonding tweezers and its mesh base painted with the catalyst. A small quantity of bonding paste is placed on the attachment, which is delivered to the operating area. The etched enamel is checked for dryness once again and then painted with the liquid catalyst. The attachment is pressed firmly into place on the crown of the tooth until paste oozes from underneath the mesh base. Even without waiting for initial setting, the viscosity of the paste will hold the attachment in place without the need for any support, and careful fine-tip suction in the surrounding area is probably all that is needed to guarantee bonding success.

The mid-buccal aspect of the tooth possesses a smooth curved surface, to which the base of a standard bracket, button or eyelet may be easily and reliably bonded. The contour of the base closely conforms to the tooth surface at this site, which offers virtually ideal conditions, provided that good haemostasis is obtained. However, following placement of the attachment, several minutes of continued isolation of the area are needed to permit uncontaminated setting, which often requires the deft use of retractors and fine suction to prevent incipient haemorrhage.

When the available bonding site is elsewhere on the tooth, the base of the attachment is unsuited and the undulating palatal surface of a maxillary canine, in particular, makes adequate adaptation quite impossible. The attachment base then has a two- or three-point contact with the tooth only. Under these circumstances, chemical bonding has strong disadvantages, since the thickness of the 'no-mix' bonding paste will be excessive in large areas and the liquid catalyst will not penetrate sufficiently to harden all the composite material. The bond will be undermined and failure is possible [24]. Using a light-cured adhesive, the full thickness of the material will be hardened more satisfactorily and the bond will be more reliable. Given the high-intensity LED curing lights now available, full hardening of the adhesive takes place within seconds, which is a boon in this delicate procedure. Furthermore, the units are wire-less and do not entail dragging a cable from an electric socket across the sterile operating field, which makes light curing the method of choice for bonding attachments to impacted teeth, at the time of surgery.

Many operators prefer to use mosquito or Matthieu forceps to carry the attachment to its place and to hold it there until setting has occurred. Unfortunately, the freeing of the attachment from the forceps is achieved by changing the hand grip and unlocking the ratchet that holds the handles closed. These manoeuvres produce considerable jolting and jarring of the attachment, which may easily lead to loss of control in its accurate placement. Accordingly, it is better to use the reverse-action bonding tweezers, which may be much more gently disengaged once the attachment has been placed and left unsupported during the curing process. The viscosity of the bonding paste should be adequate to prevent any movement. If continuous pressure is desired during the setting period, a ligature director may be substituted for the tweezers and placed with its notch engaged astride the eyelet loop and pressing against it, while the light is used. To free the ligature director once setting is complete, it is merely withdrawn in the direction of its long axis, without generating any undue lateral jarring. The newly bonded attachment should always be tested for strength by firmly pulling at the pigtail ligature before the flap is sutured.

As part of the original orthodontic treatment plan, an accurate radiographic assessment of the position of the impacted tooth will have been made and an approach to its orthodontic resolution formulated. With the impacted tooth now in full view, the orthodontist must re-evaluate the earlier decision and confirm or revise the traction direction accordingly. If this is to be performed in a direct line to the prepared place in the dental arch, then the pigtail ligature will be swivelled on the eyelet until it points in that direction. The surgeon will then suture the flap back over the wire, leaving its end freely protruding through the cut and sutured edges.

As we shall discuss with regard to a palatally impacted maxillary canine (Chapter 6), sometimes the direction of the traction cannot be pointed straight to the labial archwire because of the proximity of the roots of adjacent teeth. In this case, the wire may initially need to be drawn vertically downwards, towards the tongue or in a posterior direction. To achieve this, the pigtail cannot be drawn through the sutured edges of the flap, but rather taken through the middle of the palatal area. This means the reflected flap must be divided into two, one on either side of the pigtail, to accommodate this (Figure 3.6c). A better alternative is to pass the pigtail through a small pinhole in the palatal mucosa, prior to the replacement and suturing of the flap. When suturing is completed and the palatal area completely closed off, the orthodontist should shorten the pigtail and turn it up into a hook or circle, to be attached to an active palatal arch, ballista or auxiliary archwire.

The replacement of the flap will hide the impacted tooth from sight once again and it will become evident in later stages that it is prudent to photograph the tooth and its attachment before closure is performed (Figure 3.6b). By recording its position in this way, subsequent decisions related to the direction of orthodontic traction may be made more reliably when the patient returns for routine orthodontic adjustment and further activation of the traction mechanism at future orthodontic visits.

The application of traction should be immediate, regardless of which method is used. It will be appreciated that later manipulation of the ligature pigtail, as it passes through the soft tissues, is very unpleasant and even painful for the patient. While subsequent manipulation may only be necessary for two or three additional adjustment visits before the tooth is erupted and the pigtail free of the soft tissue, there is much to be said for the first of these being fully exploited with the application of appropriate traction while a local anaesthetic is operational, i.e. at the time of surgical exposure. If an auxiliary labial archwire or a 'ballista' spring has been tied into the arch in their passive mode prior to the surgery, as recommended above, then lightly pushing the loop from its vertical, inactive position towards the mid-palate and turning the pigtail ligature around it will provide appropriate extrusive force which will be active over a wide range of movement, and will remain active for several weeks. Similarly, an auxiliary palatal arch may be slotted into the palatal horizontal molar tubes, and then

raised into the pigtail ligature and held there by similar means (see Figure 6.34). Whichever of these devices is used, this orthodontic manoeuvre should take no more than a minute or two and can be done while the surgical instruments are being cleared away.

The reliability of the bonding procedure under the circumstances described here has been shown to be extremely high [24]. Nevertheless, in the past, bonding in the presence of an open and bleeding wound, involving both soft and hard tissues, has been strongly contraindicated, since it was thought to be inconsistent with the attainment of a dry and uncontaminated field. This attitude, on the part of the orthodontist, was probably nurtured more out of a reticence to be present at the surgical episode than out of any experience of a high incidence of failure in attachment bonding at that time.

It is important to emphasize the need to properly adapt the base of the attachment to the shape of the recipient surface of the crown of the tooth. Thus, the use of standard brackets with 'anatomic' bases, as supplied by the manufacturer, was shown in that study [24] to fare considerably better in the mid-buccal position of the impacted tooth (80.6%) than on any other surface, particularly the palatal surface, where the chances of its survival were 58.3% – a failure rate of almost 1 in 2! Better still, and at a 96.7% level of reliability, a small attachment (such as an eyelet) on a pliable base, properly and individually adapted to the form of the recipient site, will allow the orthodontist to work with the greatest degree of confidence.

In Chapter 12, we shall refer to a wide range of reasons why failure to resolve an impaction occurs and among these, ankylosis is arguably the most commonly claimed by the often embarrassed orthodontist [53]. Why ankylosis should occur is largely unknown, but there may be one or two well-intentioned approaches that the surgeon may use with the best of intentions and to facilitate different aspects of the exposure procedure, but which could contribute to this unfortunate pathological sequela.

Thus, overzealous and wide exposure of the tooth, down to the CEJ and beyond, will cause tearing of the fibres of the periodontal ligament and exposure of the surface of the cementum. It most often heals without further complications. Notwithstanding, an aggressive resorption process may be initiated in the exposed cervical area of the root due to chronic inflammation in the granulation tissue which is in contact with the dentine, which may prevent the marginal gingival epithelium from forming a protective cervical cell layer in an angular defect.

Pushing elevators down the side of the tooth aimed at ensuring a good degree of mobility causes damage to the cementum which may then heal with an ankylotic union.

During the bonding process, the necessary drying of the enamel surface will also cause desiccation of these sensitive tissues. Should liquid etchant reach the area, then chemical damage will also occur, which will cause cell death. Many dental practitioners will use liquid etchant to control bleeding from the gingiva into a prepared carious cavity to facilitate placement of a dental restoration. If this method is used for haemostasis in the present context, the damage that occurs will not be revealed for some time. However, there is reason to believe that this may be a potent cause of ankylosis or of aggressive cervical root resorption, both of which will effectively prevent further movement of the affected tooth (see Chapter 7).

From the above account, it will be appreciated that the presence of the orthodontist at the surgical intervention has much to commend it. In the first place, the orthodontist is able to see the exact position of the crown, the direction of the long axis and the deduced location of the root apex. The height of the tooth and its relation to adjacent roots may all be noted and the orthodontist may plan the strategy of its resolution by direct visualization. The orthodontist will be in a position to decide exactly where he/she would like to see the attachment placed from the mechano-therapeutic point of view and will bond it there. The orthodontist is also the best person to fabricate and place a suitable and efficient auxiliary to apply a directional force of optimal magnitude and a wide range of movement and to place it at the time of surgery.

It is not fair to expect the oral surgeon to be aware of how different attachment positions may affect the orthodontic or periodontic prognosis; nor should it be expected of him/her to be sufficiently experienced with the bonding technique to do this. For most oral surgeons, bonding is not a procedure that they routinely carry out. The presence of the orthodontist allows for bonding to be performed efficiently, with the surgeon and the nurse maintaining haemostasis and the necessary dry field. So, if a surgeon were to take exception to the present recommendation that the orthodontist be present at the exposure, with the words 'even the lowliest oral surgeon can place a bracket' or that it is 'a waste of time' [54], that oral surgeon would be sorely missing the point and the wider context of ensuring quality care and overall treatment success.

The ultimate responsibility for the success of the case, from the initiation of orthodontic treatment up to the point where the impacted tooth is brought into full alignment, rests firmly on the shoulders of the orthodontist. It would seem irresponsible to abrogate this crucial stage of the treatment, where a force is to be applied to the newly exposed impacted tooth and where so much is at stake that affects the future of the case, to another party. Absenting himself from the procedure, as has been advocated by many orthodontists and surgeons alike, leaves the surgeon to make orthodontic decisions for which he is not equipped, thereby endangering the outcome and inviting legal proceedings, from which the orthodontist will not be immune [55].

References

1. Becker A, Kohavi D, Zilberman Y. Periodontal status following the alignment of palatally impacted canine teeth. Am J Orthod 1983; 84: 332–336.

2. Kohavi D, Becker A, Zilberman Y. Surgical exposure, orthodontic movement and final tooth position as factors in periodontal breakdown of treated palatally impacted canines. Am J Orthod 1984; 85: 72–77.

3. Kohavi D, Zilberman Y, Becker A. Periodontal status following the alignment of buccally ectopic maxillary canine teeth. Am J Orthod 1984; 85: 78–82.

4. Boyd R. Clinical assessment of injuries in orthodontic movement of impacted teeth. I. Methods of attachment. Am J Orthod 1982; 82: 478–486.

5. Boyd R. Clinical assessment of injuries in orthodontic movement of impacted teeth. II. Surgical recommendations. Am J Orthod 1984; 86: 407–418.

6. Odenrick L, Modeer T. Periodontal status following surgical-orthodontic alignment of impacted teeth. Acta Odontol Scand 1978; 36: 233–236.

7. Di Biase DD. The effects of variations in tooth morphology and position on eruption. Dent Pract Dent Rec 1971; 22: 95–108.

8. Andreasen JO, Andreasen FM. Textbook and Color Atlas of Traumatic Injuries to the Teeth. Copenhagen: Munksgaard, 1994.

9. Korbendau J-M, Guyomard F. Chirurgie parodontale orthodontique. Velizy-Villacoublay Cedex, France: Editions CdP, 1998.

10. Kokich VG. Surgical and orthodontic management of impacted maxillary canines. Am J Orthod Dentofacial Orthop 2004; 126: 278–283.

11. Moro N, Murakami T, Tanaka T, Ohto C. Uprighting of impacted lower third molars using brass ligature wire. Aust Orthod J 2002; 18: 35–38.

12. Becker A, Shochat S. Submergence of a deciduous tooth, its ramifications on the dentition and treatment of the resulting malocclusion. Am J Orthod 1982; 81: 240–244.

13. Ilizarov G, Devyatov A, Kamerin V. Plastic reconstruction of longitudinal bone defects by means of compression and subsequent distraction. Acta Chir Plast 1980; 22: 32–46.

14. Altuna G, Walker DA, Freeman E. Rapid orthopedic lengthening of the mandible in primates by sagittal split osteotomy and distraction osteogenesis: a pilot study. Int J Adult Orthod Orthognath Surg 1995; 10: 59–64.

15. Vanarsdall RL, Corn H. Soft-tissue management of labially positioned unerupted teeth. Am J Orthod 1977; 72: 53–64.

16. Vanarsdall RL. Efficient management of unerupted teeth: a time-tested treatment modality. Semin Orthod. 2010, 16: 212–221.

17. Hunt NP. Direct traction applied to unerupted teeth using the acid-etch technique. Br J Orthod 1977; 4: 211–212.

18. McBride LJ. Traction – a surgical/orthodontic procedure. Am J Orthod 1979; 76: 287–299.

19. Gensior AM, Strauss RE. The direct bonding technique applied to the management of the maxillary impacted canine. J Am Dent Assoc 1974; 89: 1332–1337.

20. Nielsen LI, Prydso U, Winkler T. Direct bonding on impacted teeth. Am J Orthod 1975; 68: 666–670.

21. McDonald F, Yap WL. The surgical exposure and application of direct traction of unerupted teeth. Am J Orthod 1986; 89: 331–340.

22. Kokich VG, Mathews DP. Surgical and orthodontic management of impacted teeth. Dent Clin North Am 1993; 37: 181–204.

23. Crescini A, Clauser C, Giorgetti R, Cortellini P, Pini Prato GP. Tunnel traction of intraosseous impacted maxillary canines: a three-year periodontal follow-up. Am J Orthod Dentofac Orthop 1994; 105: 61–72.

24. Becker A, Shpack N, Shteyer A. Attachment bonding to impacted teeth at the time of surgical exposure. Eur J Orthod 1996; 18: 457–463.

25. Wong-Lee TK, Wong FCK. Maintaining an ideal tooth–gingiva relationship when exposing and aligning an impacted tooth. Br J Orthod 1985; 12: 189–192.

26. Becker A, Brin I, Ben-Bassat Y, Zilberman Y, Chaushu S. Closed-eruption surgical technique for impacted maxillary incisors: a post-orthodontic periodontal evaluation. Am J Orthod Dentofac Orthop, 2002; 122: 9–14.

27. Chaushu S, Brin I, Ben-Bassat Y, Zilberman Y, Becker A. Periodontal status following surgical-orthodontic alignment of impacted central incisors by an open-eruption technique. Eur J Orthod 2003; 25: 579–584.

28. Wisth PJ, Nordervall K, Boe OE. Periodontal status of orthodontically treated impacted maxillary canines. Angle Orthod 1976; 46: 69–76.

29. Woloshyn H, Artun J, Kennedy DB, Joondeph DR. Pulpal and periodontal reactions to orthodontic alignment of palatally impacted canines. Angle Orthod 1994; 64: 257–264.

30. Vermette ME, Kokich VG, Kennedy DB. Uncovering labially impacted teeth: apically positioned flap and closed-eruption technique. Angle Orthod 1995; 65: 23–32.

31. Chaushu S, Dykstein N, Ben-Bassat Y, Becker A. Periodontal status of impacted maxillary incisors uncovered by two different surgical techniques. J Oral Maxillofac Surg, 2009; 67: 120–124.

32. Becker A. An interview with Adrian Becker. World J Orthod 2004; 5: 277–282.

33. Parkin N, Benson PE, Thind B, Shah A. Open versus closed surgical exposure of canine teeth that are displaced in the roof of the mouth. Cochrane Database of Systematic Reviews 2008, 4. CD006966. DOI: 10.1002/14651858.CD006966.pub2.

34. Schmidt AD, Kokich VG. Periodontal response to early uncovering, autonomous eruption, and orthodontic alignment of palatally impacted maxillary canines. Am J Orthod Dentofac Orthop 2007, 131: 449–455.

35. Heaney TG, Atherton JD. Periodontal problems associated with the surgical exposure of unerupted teeth. Br J Orthod 1976; 3: 79–85.

36. Vanarsdall RL, Corn H. Soft tissue management of labially positioned unerupted teeth. Am J Orthod Dentofac Orthop 2004; 125: 284–293.

37. Vanarsdall RL Jr. An interview with Robert L. Vanarsdall, Jr. World J Orthod 2004; 5: 74–76.

38. Vanarsdall RJ. Periodontal/orthodontic interrelationships. In Graber TM, Vanarsdall RL, eds. Orthodontics. Current Principles and Techniques, 4th edn. St Louis, MO: Mosby, 2000: 822–836.

39. Becker A. Extreme tooth impaction and its resolution. Semin. Orthod, 2010, 16: 222–233.

40. Ten Cate AR. Oral Histology: Development, Structure and Function, 4th edn. St Louis, MO: Mosby, 1994: 270.

41. Blackwood HJJ. Resorption of enamel and dentine in the unerupted tooth. Oral Surg Oral Med Oral Pathol 1958; 11: 79–85.

42. Cahill DR, Marks SC Jr. Tooth eruption: evidence for the central role of the dental follicle. J Oral Pathol. 1980; 9: 189–200.

43. Hu JC-C, Nurko C, Sun X et al. Characteristics of cementum in cleidocranial dysplasia. J Hard Tissue Biol 2002; 11: 9–15.

44. Becker A, Caspi N, Chaushu S. Conventional wisdom and the surgical exposure of impacted teeth. Orthod Craniofac Res, 2009; 12: 82–93.

45. Danan M, Zenou A, Bouaziz-Attal A-S, Dridi S-M. Orthodontic traction of an impacted canine through a synthetic bone substitute. J Clin Orthod 2004; 38: 39–44.

46. Chaushu G, Becker A, Zeltser R, Branski S, Chaushu S. Patients' perceptions of recovery after exposure of impacted teeth with a closed-eruption technique. Am J Orthod Dentofac Orthop 2004; 125: 690–696.

47. Chaushu S, Becker A, Zeltser R, Vasker N, Chaushu G. Patients' perception of recovery after surgical exposure of impacted maxillary teeth treated with an open-eruption surgical-orthodontic technique. Eur J Orthod 2004; 26: 591–596.

48. Chaushu G, Becker A, Chaushu S. Patients' perception of recovery after exposure of impacted teeth: a comparison of closed versus open-eruption techniques. J Oral Maxillofac Surg 2005; 63: 323–329.

49. Jacoby H. The ballista spring system for impacted teeth. Am J Orthod 1979; 75: 143–151.

50. Becker A, Zilberman Y. A combined fixed-removable approach to the treatment of impacted maxillary canines. J Clin Orthod 1975; 9: 162–169.

51. Becker A, Zilberman Y. The palatally impacted canine: a new approach to its treatment. Am J Orthod 1978; 74: 422–429.

52. Kornhauser S, Abed Y, Harari D, Becker A. The resolution of palatally-impacted canines using palatal-occlusal force from a buccal auxilliary. Am J Orthod Dentofacial Orthop 1996; 110: 528–534.

53. Becker A, Chaushu G, Chaushu S. An analysis of failure in the treatment of impacted maxillary canines. Am J Orthod Dentofac Orthop 2010; 137: 743–754.

54. Haskell R. Book review. Br J Oral Maxillofac Surg 1999; 37: 157–158.

55. Haney E, Gansky SA, Lee JS, Johnson E, Maki K, Miller AJ, Huang JC. Comparative analysis of traditional radiographs and cone-beam computed tomography volumetric images in the diagnosis and treatment planning of maxillary impacted canines. Am J Orthod Dentofac Orthop 2010: 137(5): 590–597.

4

Treatment Components and Strategy

Orthodontic Treatment of Impacted Teeth, Third Edition. Adrian Becker.
© 2012 Adrian Becker. Published 2012 by Blackwell Publishing Ltd.

Orthodontic treatment duration

The opening of adequate space in the arch may initiate movement in an unimpeded impacted tooth, which may start moving in the appropriate direction, sometimes quite quickly [1]. By the time the space is of a suitable size and arrangements for the surgery have been made, a new periapical radiograph may show much positive change in its position and lead the clinician to believe that spontaneous eruption will render the surgery unnecessary. If the eruption is likely to occur imminently, or at least within a reasonable period, there is merit in waiting for this to occur.

If, on the other hand, eruption will take many months, then the orthodontist must weigh the benefits of avoiding surgery against the drawbacks involved in leaving orthodontic appliances in place for all this time or removing them before all the teeth are in their appropriate places. Orthodontic appliances raise the level of vulnerability of the teeth to caries and of the periodontium to inflammation – the longer they are in place, the greater is the risk. Removing the appliances before time runs the different risk of having to replace them later to correct a malposition of the newly erupted and erstwhile impacted tooth or to accept a compromised, inadequate, outcome. In order to solve this dilemma, the clinician may elect to advise surgical exposure and orthodontic traction to expedite the eruption of the tooth and complete the treatment in a very much shorter time.

When orthodontic treatment has provided space and surgery is undertaken to remove a physical obstacle, a similar dilemma may arise. In the absence of the obstacle, the impaction is potentially resolvable, without further treatment. However, the surgical intervention involved in removing the obstacle offers the opportunity of access to the unerupted tooth. Subsequent healing of the wound will deny that access and, in the event that eruption does not take place, a second surgical intervention in the same area will be necessary and much time will have been wasted confirming that spontaneous eruption will not occur. Clearly, then, the time factor must not be ignored. Orthodontic appliances are in place and, perhaps, the space in the arch is unsightly. Without question, orthodontically aided eruption will speed up the resolution enormously and, this being so, the patient's best interests are to be served by including this option among the factors to be considered at the planning stage.

When the existence of an impaction is only a small part of an overall complex malocclusion, the time factor becomes more critical. It is quite conceivable and reasonable to estimate that a given overall problem alone may require two years of treatment. The complete alignment of an awkwardly placed impacted tooth may add a further year or more to this [2–4]. To permit the luxury of a wait-and-see period is to add this to an already extended three-year period, during which appliances are worn. While the ortho-dontist may well be rewarded by a much improved position of the impacted tooth, a deteriorating state of oral health, due to poor oral hygiene, may deprive the achievement of all meaningful content.

It becomes relevant to refer to our original definitions in Chapter 1, in which we noted that 'A permanent tooth with delayed eruption is an unerupted tooth whose root is developed in excess of this length and whose spontaneous eruption may be expected, in time. A tooth which is not expected to erupt in a reasonable time, in these circumstances, is termed an impacted tooth.' Thus, in the present context and while the tooth may be expected to erupt spontaneously 'in time', this period may not be considered 'reasonable' in relation to the prospect of detrimental iatrogenic effects on the remainder of the dentition engendered by this extra and often considerable waiting period. This then redefines the tooth as impacted, and a proactive surgical exposure should be considered preferable.

The anchor unit

At this juncture, it is not the intention to go into the details of appliance therapy. This will come later, as the different groups of impacted teeth that are seen in practice are dealt with. However, some general principles are in order.

For most malocclusions, quality treatment is best provided by the use of one or other of the recognized fixed appliance treatment techniques. If the dental arches are correctly related and adequate space is present, then the teeth are initially 'levelled' to a labial archwire of standardized archform and a given coefficient of elasticity. Later, heavier round or rectangular archwires are substituted to perform root movements which will pave the way to achieving an optimal result. Incorrectly related dental arches will benefit from the use of other appliances, such as headgears, functional appliances or intermaxillary elastics, prior to or in addition to the fixed appliances, while space may be provided by the extraction of teeth or by lengthening the arches mesio-distally or expanding them laterally.

When dealing with a malocclusion that incorporates an impacted tooth, modifications must be made in this procedure. Unlike other teeth in the mouth, the impacted tooth may be severely displaced from its normal position in all three planes, and much anchorage will be expended in bringing it into alignment. Accordingly, a rigid anchor base must be developed against which to pit the forces required to reduce the impaction.

At the age at which an impacted maxillary canine is treated, the full permanent dentition (with the exception of third molars) is usually present. Accordingly, a fully multi-bracketed appliance should normally be placed and the entire dentition treated with the use of light archwires through the stages of levelling and the opening of adequate space in the arch for the impacted tooth. A heavy and more

rigid archwire is then placed into the brackets on all the teeth of the fully aligned and complete dental arch. The aim of this is to provide a solid anchorage base [5] which will not allow distortion of the archwire to occur as the result of the forces that will eventually be applied to the impacted tooth after its exposure. The effect on the anchor unit of forces that are designed to resolve a grossly displaced canine should not be underestimated, particularly if they are applied for an extended duration.

By contrast, at the time an impacted upper central incisor needs to be treated, only first permanent molars and three permanent incisor teeth are present in the maxillary arch. Accordingly, alternative means of making the appliance system rigid must be employed before light forces may be applied to the impacted tooth, in order not to compromise the remainder of the dentition.

Attachments

To be in the position of being able to influence the future development of an impacted tooth, it is necessary to place some form of attachment on the tooth. These attachments have changed over the years, reflecting the advances made in the field of dental materials.

Lasso wires

In the years prior to the mid-1960s, a lasso wire (Figure 4.1) twisted tightly around the neck of the canine was employed widely and was used in our earlier cases, in the initial stages. It will be readily appreciated that the shape of the crown of a tooth is such that its narrowest diameter is at the cemento-enamel junction (CEJ), which is where the lasso wire will inevitably settle. This will result in irritation of the gingiva and prevent reattachment of the healing tissues in this vital

area. It has also been reported that external resorption and ankylosis have been produced in the area of the CEJ following employment of this method [6]. Given the excellent alternatives that are available today, the lasso wire is obsolete.

Threaded pins

Several systems of threaded pins (Figure 4.2) have been available for many years. Their specific purpose is to provide retention for an amalgam or composite core, to allow the provision of a cast crown in a severely broken down tooth. These threaded pins may also be used to provide the attachment for an impacted tooth. This is a method which was used in the past [7, 8] and has been totally superseded. Its disadvantages include the fact that it is dentally invasive, necessitating a subsequent restoration. Given the difficulties of access to many impacted teeth and the desirability of limiting surgical exposure as much as possible, the orientation of the long axis of the tooth may be difficult to determine and the drilled hole may inadvertently enter the pulp – unerupted teeth often have large pulp chambers! Even in the most favourable of circumstances, it seems that this unnecessarily aggressive method produces avoidable damage to a virgin tooth, when there are eminently suitable, non-invasive methods and efficacious alternatives, as we shall discuss below. Nevertheless, the method was still in use and apparently recommended until quite recently [5, 9].

Orthodontic bands

Preformed orthodontic bands largely replaced the lasso wire, and clinical experience with them showed them to be considerably more compatible with ensuring the health of the periodontal tissues. As with the lasso wire, however, the

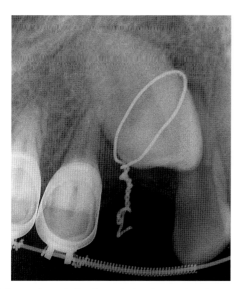

Fig. 4.1 Lasso wire encircling the neck of an impacted canine.

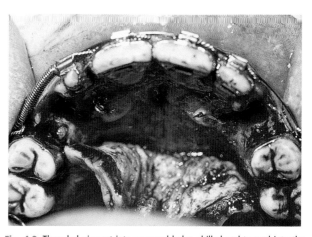

Fig. 4.2 Threaded pins set into prepared holes, drilled and tapped into the enamel and dentine of the surgically exposed canines.

use of a band dictated the very wide surgical clearance of tissue on all sides of the tooth (see Figure 6.35c), in order to permit the introduction of the band, to adequately control haemorrhage around the crown and to avoid contamination from oozing blood inside the cement-filled band, at the time of placement.

Bonded attachments

Since the introduction of enamel bonding, all the above-mentioned methods have become obsolete. The employment of the acid-etch composite bonding technique to the crown of a tooth has much merit [10–12], notably in terms of simplicity and reliability of the bond. Its greatest advantage is that it requires relatively little exposed surface of enamel to be successful, a fact which may contribute much to the subsequent periodontal health of the treated result. It is, presently, without doubt the method of choice from almost every point of view, and should replace other methods in virtually all circumstances.

Standard orthodontic brackets

As far as the actual choice of type of attachment to be placed on impacted teeth is concerned, there are several salient points to consider regarding the impacted tooth when compared with an erupted tooth that needs to be brought into its position in the dental arch. The wide array of orthodontic brackets, advertised in the catalogues of the various orthodontic manufacturing companies, represent sophisticated designs of attachment that enable the orthodontist to perform any type of movement on a tooth in all three planes. Since many or perhaps most impacted teeth are in need of a wide variety of movements, it may seem logical to place a sophisticated orthodontic bracket on the affected tooth, from the outset.

That part of the movement of the impacted tooth, from its initial ectopic position until it reaches the main archwire, represents resolution of the impaction. During this entire period, which is the most difficult part of the treatment of this displaced tooth, it is not possible to achieve more than tipping, extrusion and some rotation until the bracket arrives at and fully engages the main archwire. In other words, the value of the bracket up to that point is no greater than that of a simple eyelet [13]. Indeed, on several counts, the potential of the eyelet outweighs that of the conventional bracket during the resolution stage.

The base of a conventional bracket is wide, rigid and difficult to adapt to the shape of another part of the tooth's surface other than the mid-buccal, for which it has been designed. Thus, composite bonding elsewhere on a tooth is very likely to lead to failure [13]. Orthodontic brackets are highly specialized, each having a slot milled to a very precise blueprint, specific to the particular tooth for which it is intended. The mesio-distal angulation differs between one tooth and another, the 'in–out' bucco-lingual depth of the slot will vary, the torque angulation will not be the same

for the individual teeth, and the height at which the bracket should be placed on an incisor will not be the same as that on the canine. These are the principles on which the so-called 'straight wire' appliances are built. However, it should be immediately obvious that all this highly sophisticated programmed engineering is only meaningful if the bracket is bonded in its appropriate, predetermined position on the crown of the tooth. We shall see in later chapters that it is very frequently impossible or unwise, at the time of the surgical exposure of an impacted tooth, to bond an attachment in this position on the crown. This site on the crown of the tooth may not be accessible due to its relationship to the root of an adjacent tooth, or an excessive amount of soft and hard tissue might need to be surgically removed, thereby producing unnecessary damage in order to provide this access.

The standard orthodontic bracket in any technique is relatively large, possesses a wide, high and sharp profile, and, even when placed in alternative positions on the tooth, by force of circumstance at the time of surgery, it may find itself deeply sited in the surgical wound. The bracket's shear bulk creates irritation as the tooth is later drawn through the soft tissues, particularly the mucosa (Figure 4.3). A ligature wire or elastic thread tied to it must also originate deep in the wound and will be stretched across the replaced flap tissue towards the labial archwire. This increases the possibility of impingement with the investing tissues and leads to inflammation and to probable permanent periodontal damage.

As the displaced tooth moves towards its place in the arch, exuberant gingival tissue bunches up in front of it, which will also lead to impingement by a conventional orthodontic bracket. The existence of the exuberant gingival tissue in advance of the tooth can often cause 'pinching' between it and the teeth in the arch immediately adjacent

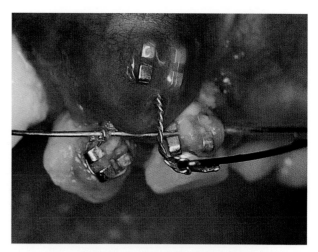

Fig. 4.3 As the impacted tooth is about to erupt, the high-profile Siamese edgewise bracket has fenestrated the swollen gingival tissue.

Fig. 4.4 Eyelets welded to a pliable band material base, backed by steel mesh.

to it. This is less likely to occur if a deliberately generous space is provided in the arch for the tooth. This precaution may avoid unnecessary periodontal damage.

A simple eyelet

An eyelet, welded to orthodontic band material with a mesh backing (Figure 4.4), is soft and easy to contour, permitting its adaptation to the bonding surface to be more intimate, which makes for superior retentive properties. Its relatively small size and low profile makes the mid-buccal position of several of the more awkwardly placed teeth considerably more accessible when compared to the placing of a conventional bracket. Its modest, low-profile dimensions are also less irritant to the surrounding tissues, particularly during the critical phase as it breaks through epithelial tissues in the final stages of its eruption into the oral cavity.

For these reasons, a small eyelet is recommended as the initial attachment, placed at the time of surgery and removed only when the tooth has progressed to the point where it is in close proximity to the archwire. At that time, it should be substituted for the same type of sophisticated bracket which is being used on the other teeth, thereby heralding the initiation of the more intricate root manipulations of the tooth (rotating, uprighting and torqueing).

Intermediaries

We have seen in Chapter 3 that there are important periodontal advantages to be gained by full closure of the surgical flap at the end of the visit during which the surgical exposure is performed. Thus, the impacted tooth is recovered by the surgical flap and is lost from sight, unless the impacted tooth is fairly superficially placed. The only manner in which contact may be maintained with it is through the some form of ligature wire, gold chain or elastic thread, which is attached to the eyelet, before it is bonded to the tooth. This may be termed an intermediary or connector.

Since elastic thread can only be tied once, it is not recommended for an attachment that is not clearly visible and accessible in the mouth. Gold chain has found a surprising degree of acclaim and acceptance worldwide in view of the fact that it appears to be unnecessarily sophisticated, expensive and not widely available, although it is undoubtedly suitable and sufficiently strong for the purpose. There is perhaps only one practical drawback to its use, which relates to its physical properties. If a closed surgical approach is used, following bonding of its attachment base to a tooth, the end of the chain needs to be held in a locking tweezers or artery forceps until it is ligated to its active traction element, be it a spring or elastic thread. If this is not done, then the fine-linked chain may collapse down and slip between the recently sutured edges of the flaps and be lost from sight. This may also happen when an open surgical approach is performed, with the collapsed chain falling between the wound edges and into the cervical area of the newly exposed tooth. It may also occur during later visits for re-ligation of the still only partially erupted tooth. Subsequent search for the lost chain is very uncomfortable for the patient and may even require a reopening of the healing soft tissue cover.

The use of a stainless steel ligature is far easier from every point of view, and is readily at hand in both the orthodontic and the surgical operatory. Such a ligature is passed through the eyelet and twisted into a long braid with an artery forceps before bonding is undertaken. The braided wire or pigtail hangs loosely in the eyelet until bonding and suturing have been completed, and it should be of sufficient substance for it to be rolled up into a loop, which will not easily be unravelled by application of the extrusive force. On the other hand, it must not be so thick that the effort needed to form the loop will seriously test the bond strength of the newly placed attachment. In practice, the use of a dead soft stainless steel ligature wire of 0.011 in or 0.012 in gauge is generally the most suitable.

It has been recommended that the pigtail be braided in such a way that each two or three turns of the braid are followed by a small loop, then two or three more turns, another loop, and so on. In this way, the braid comprises a convenient chain of loops, which may be shortened as necessary by cutting off the excess, while exploiting the loop closest to the gingival tissue [14]. However, 'rolling up' the terminal loop of a simply twisted stainless steel ligature (Figure 4.5) as the tooth progresses is simpler and 'user-friendly'.

Elastic ties and modules versus auxiliary springs

At first glance, elastic ties of one sort or another present the orthodontist with the most convenient means of applying light forces to a tooth, with a good range of action. However,

their use is more disappointing than one may initially realize.

The manufacturer's spool of elastic thread usually comes in the form of fine hollow tubing, which is easier to tie than a solid thread. Most orthodontists tie the thread with a simple knot which, when tying string, will not unravel. This is done very empirically and there can be no control on the amount of force applied. Furthermore, when tying elastomeric thread, the knot tends to loosen and much of the original force of the tie will then be lost. When under tension, all materials used to make this elastic thread suffer a high degree of force decay, which is very rapid and very significant. The force levels of chains of various lengths are known to decay to below that required for tooth movement, in a period of between one and three weeks, depending upon the amount of tension initially applied [15, 16].

A shorter piece of stretched elastic (Figure 4.5) will have a very short range, and runs the risk of applying an initial

excessive amount of pressure if the tie is good – or no effective pressure if the tie loosens. The immature periodontal membrane of the recently exposed tooth and the strength of its bonded attachment could be severely tested. In the case of an unerupted tooth close to the line of the arch, traction applied directly from its attachment to the archwire will generally be very inefficient, requiring frequent changes and producing only a very slow response. It is impossible to measure or control such a force, for all practical purposes.

It is prudent to use more distant sites from which to apply traction to the unerupted tooth in order to include a greater length of stretched elastic thread to increase the range of the traction force and, thereby, its effectiveness in moving the tooth over a longer period of time. To do this, the elastic thread needs to be drawn to the target area on the archwire, through the agency of a loop bent into the archwire at that point. The thread may then be tied back to the hook on the molar tube of the same side, with care being taken to insert a stop in the archwire, mesial to the tube, in order to prevent mesial movement of the molar.

The 'slingshot' elastic is an excellent alternative for the situation in which space for an unerupted or partially erupted and buccally, palatally or superiorly displaced tooth has been opened. A cut piece of steel tubing may be placed on the archwire between the brackets in order to maintain the space, and an elastic chain or E-link is stretched across the gap between the brackets on the two adjacent teeth. The middle of the chain is then seized with an artery forceps or Howe plier and stretched over the attachment on the ectopic tooth or on the twisted steel ligature coming through the sutured edge of the replaced surgical flap (Figure 4.6a, b).

As a general rule, elastic thread should only be used as a go-between, connecting the non-elastic steel pigtail to a

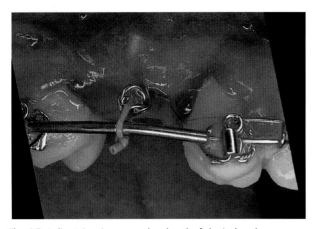

Fig. 4.5 A direct tie using a very short length of elastic thread.

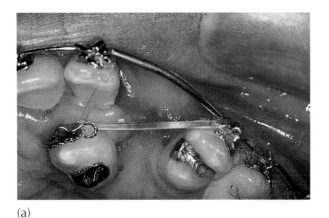

(a)

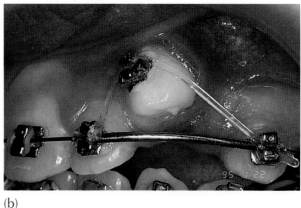

(b)

Fig. 4.6 (a) The slingshot elastic. A palatally impacted canine has erupted into the palate (see Chapter 6). The elastic module, placed between the bracket of lateral incisor and first premolar, is stretched towards the canine and tied into the buccal eyelet. The steel tube on the archwire maintains the space. (b) The slingshot used on a buccal canine.

similarly non-elastic and heavy archwire. If a lighter arch-wire is used, then the tie should be made with a steel liga-ture – the archwire providing the elastic displacement. Nickel–titanium alloy wires may be used with great effect in this context, but the distortion of the archwire will bring about alteration in both the horizontal and vertical planes, to produce unwanted change in the form of the dental arch and an uneven occlusal plane. So, if a single super-elastic archwire is tied into the brackets on each of the teeth in the levelled and aligned arch and then into an attachment bonded to a severely displaced impacted tooth, control of overall archform will be lost. This will be evident in the three planes of space, causing tipping movements of indi-vidual teeth, alterations in the occlusal plane, asymmetric skewing of the shape of the arch and loss of occlusal con-tacts. The adjacent teeth will be relatively intruded and will be displaced buccally or palatally and tipped towards the space reserved for the impacted tooth. Super-elastic wires should not be used in circumstances of severe displacement without a heavy base arch in place to resist these unwanted movements of the anchor teeth. However, it should be clearly understood that for a nickel–titanium archwire to develop adequate vertically directed eruptive force, it must be free to slide in the bracket slots of the other teeth to which it is ligatured (Figure 4.7a, b). The presence of a heavy base arch tied in with elastomeric ligatures will con-siderably increase the friction and binding of the super-elastic archwire in the brackets. This may not be evident when the last elastomeric is placed and it is difficult to check. Accordingly, the pressure from the deflection that was applied to fully engage the super-elastic wire in the slots may be nullified by the inability of the wire to slide freely through the brackets.

The combined use of both a flexible archwire and an elastic thread tie [6] is counterproductive, since the elastic-ity of the one that exerts the stronger force will be effectively neutralized and offer no physical advantage over a steel ligature, while the displacement of the weaker element will be the only factor which is active in moving the teeth.

In general, orthodontists use elastic ligatures and chains to move teeth by first elongating the material and drawing the dental elements towards one another. The range of elas-ticity in this direction is limited and, as pointed out above, decays rapidly. However, the lateral displacement of an elongated elastic thread produces a potentially greater range of movement, within suitable orthodontic force levels, than does a longitudinal displacement. This principle may be applied to moving teeth which lie at a distance from the main arch more efficiently and with controlled and measurable forces (Figure 4.7a, b).

Given a little thought in the planning of their use, elastic ties, nickel–titanium auxiliary archwires, chains and modules are extremely helpful in many situations incurred by the presence of impacted teeth. However, properly designed springs, always auxiliary to a heavy base arch, are usually more efficient, since their ability to deliver a meas-ured and controlled force is good, the force decay is lower, the variety of metallic alloys that is available for spring fabrication is broad, their range of action may be very wide and their direction is accurate. These will be illustrated in the discussion of cases as they pertain to the individual groups of teeth, in the succeeding chapters.

Thus far, the discussion has centred on maintenance of a steady force through as wide a range as possible, but what force values are appropriate for application to an impacted tooth? If we are to apply traction to a tooth through its long axis, pure extrusion is produced and it implies that there is no resistance from the bone of the coronally divergent socket. The force is applied to the tooth, transferred directly to the supporting fibres of the PDL and as such it requires to be minimal – of the order of 10–15 g – because resistance is small. If greater force is applied, the tooth will become excessively loose and the extrusion achieved will bring with it relatively little supporting alveolar bone.

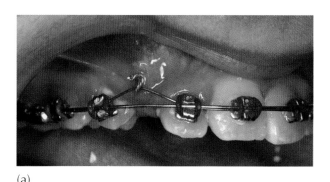

(a)

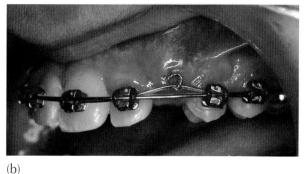

(b)

Fig. 4.7 (a, b) The use of nickel–titanium auxiliary wire as the active element in applying eruptive force to the unerupted canines, by being thread through the 'rolled-up' stainless steel pigtails which are ligated to the eyelets. There is a heavy 0.020 in gauge base arch. The low-profile eyelets, which were bonded at exposure in the closed surgery procedure, can be seen through the translucency of the healthy and uninflamed gingiva.

If there is a modicum of tip introduced into this movement, then the tooth will be brought into close proximity with the bony socket walls, interjecting resistance. Compression of the fibres on the pressure side and stretching on the tension side will generate hyalinization and undermining resorption of the alveolar bone. The force required to elicit eruption will be in the range of 20–40 g/cm^2 of root surface.

Soft tissue resistance also has to be accounted for. With a simple window technique exposure, the crown of the tooth is freed to erupt directly to its place, with little or no interference from the soft tissues, which means that the full traction force is transmitted to the PDL at the cementum/alveolar bone interface. By contrast, a closed eruption technique will leave the tooth covered by a soft tissue flap, which will have been firmly sutured into its former place. Some of the applied traction will be dissipated in overcoming the tension of this soft tissue flap and thus must be increased to reach the threshold needed for tooth movement.

When traction is applied to a tooth following an apically repositioned flap procedure, tension is created in the tissues by their being sutured superiorly to the labial side of the tooth. This tension is eruptively directed and may then magnify the applied force, with the effect being cumulative. For this reason, it is sometimes advisable to rely only on pressure from the sutured flap in the first instance, leaving the application of biomechanical traction to a subsequent visit. This is particularly relevant in cases where the tooth is high and labially or buccally displaced.

Temporary anchorage devices (TADs)

As a means of obtaining skeletal anchorage, a simple titanium screw temporary implant is often used in routine orthodontics to act as a 'stake in the ground', against which to tie elastic modules and chains for the horizontal movement of teeth in various directions [17]. Treating impacted teeth has largely to do with the facility to develop vertical eruptive forces and to bring them to bear on 'vertically challenged' teeth. At best, there is very little bone in the same jaw coronal to an impacted tooth at the outset, and certainly less or none after the tooth has been surgically exposed. This leaves precious little opportunity to place a screw device as the base from which traction may be applied to it and only the smallest distance from which to seek to obtain any appreciable range of action. In this limited sense, therefore, using a TAD in the same jaw as the impacted tooth is largely inappropriate. Some of these screws are designed with two slots at right-angles to one another, which serve as the means for driving the screw into the bone but which may also be used as an orthodontic bracket slot, into which a rectangular archwire may be ligated, in the usual manner. Other designs include a slot for an ortho-

dontic archwire in the neck of the screw, beneath the screw head. A short length of rectangular 0.019 × 0.025 in wire may be fabricated into a custom-made, self-supported spring and rigidly tied into the slot on this screw which may be placed at some distance and in a more convenient mesial, distal or apical location position in relation to the impacted tooth. It may then be used to apply traction to the tooth and to erupt it vertically to a considerable degree (see Chapter 10).

Using a TAD in the opposite jaw as a direct anchor, in the absence of an orthodontic appliance, has the obvious advantage that there can be no adverse, particularly intrusive, movement of teeth adjacent to the impacted tooth, since these are not included as anchor units. Nevertheless, this demands a patient compliant in the placement of intermaxillary elastics from the device to the pigtail ligature hook which extends from the bonded orthodontic bracket/button/eyelet on the impacted tooth. These two points of elastic ensnarement are not always easily accessible for the patient or for a dedicated parent and may prove to be impractical.

To circumvent this problem, it becomes necessary to place a full orthodontic appliance in the opposing dental arch, which is secured by ligation to a TAD on the affected side. This provides an implant-supported anchor arch configuration. Intermaxillary traction may then be applied from any conveniently located hook or bracket on that appliance directly to the attachment on the impacted tooth, without fear that the teeth in the anchor arch will over-erupt. This assumes that the attachment hook on the impacted tooth is accessible for the patient and not painful to manipulate.

Impacted teeth such as second molars, which are largely inaccessible for the patient, require a more circuitous approach. This dictates setting up the same implant-supported configuration in the opposing anchor arch but, in this case, a full appliance also needs to be placed in the affected arch. An accessory archwire, custom-designed eruption spring or elastic chain is actively ligated by the orthodontist, between a convenient location on the main arch to the attachment on the impacted tooth, to erupt the tooth. To overcome the tendency of the reactive force to intrude the teeth on that side of the dental arch, vertical elastics are prescribed to be placed by the patient between convenient and easily accessible hooks or buttons between upper and lower appliances. An indirect anchorage system is thus created in which active extrusive forces are applied and controlled by the orthodontist. Loss of anchorage in the same jaw (intrusion of the teeth) is combatted by the patient placing intermaxillary vertical elastics and loss of anchorage in the opposite jaw (extrusion of the teeth) is obviated by ligation to a TAD in that jaw (Figure 4.8a–c).

Titanium screws have been reported as having a success rate in excess of 80% [18], although this author is unable

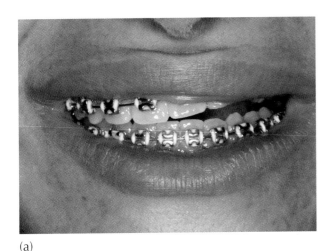

(a)

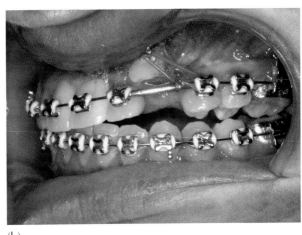

(b)

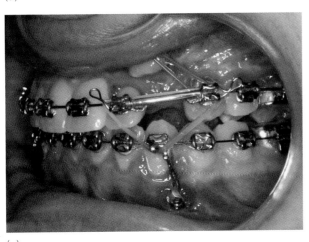

(c)

Fig. 4.8 An indirect anchorage system. (a) Extra-oral view to show tipped occlusal plane due to anchorage loss during the attempted active eruption of the left maxillary canine. (b) Intra-oral view of the same case shows the exposed canine ligated with elastic ligature to the first premolar. The space is held open by a steel tube tied between the incisor and premolar brackets. The extrusive force has resulted in the lateral open bite and cant in the occlusal plane. (c) Vertical inter-maxillary bite-closing elastics (blue) are used to support the anchorage of the maxillary arch. A titanium screw TAD is tied to the mandibular canine bracket with an elastic chain to prevent unwanted reactive eruption of the lower teeth.

level of success. Thus, while they are very easily and rapidly replaced in the event of failure, such failure creates a nuisance in the smooth running of the treatment. In cases where considerable movement is needed on a fairly long-term basis, a good alternative is the use of a malleable titanium plate onlay which may be adapted to the shape of an area of bone surface, such as in the palate or the inferior surface of the zygomatic process of the maxilla [19]. Titanium screws are then used to secure the plate to the strategically selected area, and the flap sutured to leave only the extremity of the plate exposed at one end, for use as an elastic attachment device (Figure 4.9a–d). The zygomatic plate TAD appears to be much more successful in terms of its fairly long-term usefulness and experience shows that is has a much lower failure rate, although scientific data are at present unavailable.

The phenomenon of non-eruption of a tooth in one jaw is often accompanied by over-eruption of its antagonist in the opposing jaw, particularly in the molar region. The ostensibly successful resolution of an impacted mandibular second molar may thus be prejudiced by the tooth being prevented from reaching the occlusal plane, because of its elongated opposite number. To treat the over-erupted tooth, a simple titanium screw implant may be inserted on the palatal side of the alveolus adjacent to the second molar and an elastic chain stretched from this TAD to the zygomatic plate, across the occlusal surface of the tooth. In this manner, intrusive force is applied by the chain and a rapid reduction in the height of the tooth may be achieved. This then permits the vertical elastic from the plate to the mandibular second molar to erupt the tooth to its ideal height in relation to the occlusal plane

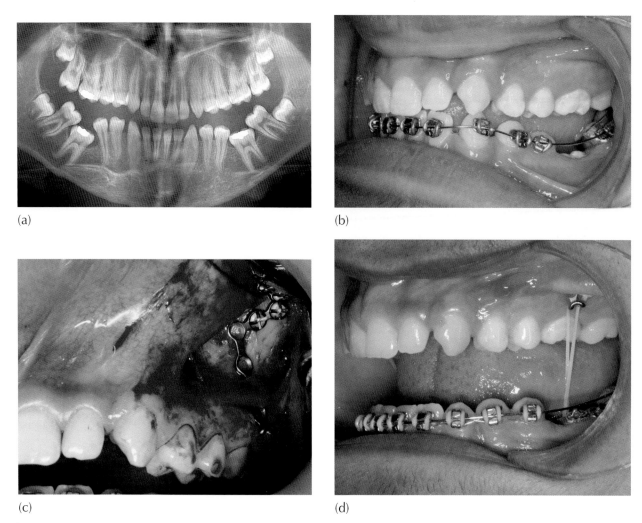

(a)

(b)

(c)

(d)

Fig. 4.9 The zygomatic plate. (a) Panoramic view of bilateral infra-occluded first permanent mandibular molars. These teeth are almost certainly ankylosed and require to be subluxated if they may be made to erupt. (b) Initial mandibular aligning appliance. (c) The zygomatic plate affixed firmly with titanium screws to the inferior surface of the zygomatic process of the maxilla. The free end of the plate is drawn through a separate cut in the attached gingiva, while the larger flap will be fully replaced and sutured. (d) The mandibular dentition has been realigned, anterior spaces closed and space created in the molar region. The patient places the latex elastic between an attachment on the molar band and the hooked end of the zygomatic plate.

(Figure 4.10a–k), without the need to involve conventional multi-bracketed orthodontics in the eruption process.

Infra-occluded deciduous teeth

Infra-occluded deciduous teeth (Figure 4.11) may be exploited in a variety of situations and may be used alone or incorporated into an appliance system, to enhance anchorage. These teeth cannot be moved orthodontically due to an ankylotic union with the alveolar bone, thus providing the opportunity for skeletal anchorage for the movement of other teeth. In much the same way as with temporary anchorage devices, a rectangular cross-section tube or bracket bonded to the buccal surface may serve as the origin for a self-supported rectangular wire spring to

provide the necessary light extrusive force for an impacted tooth.

Magnets

Rare earth magnets were developed more than 40 years ago, but only recently has this reached the point where their reduction in size, with the introduction of the lanthanide alloys, has provided the possibility of applying suitable forces that may be exploited in the present context. Various authors have presented successful clinical results of the treatment of impacted teeth in humans [20–23] using magnetic forces. Using rare earth magnets, the forces are generated along the line of the magnetic plane and, therefore, it is possible to prescribe tooth movement in all three planes

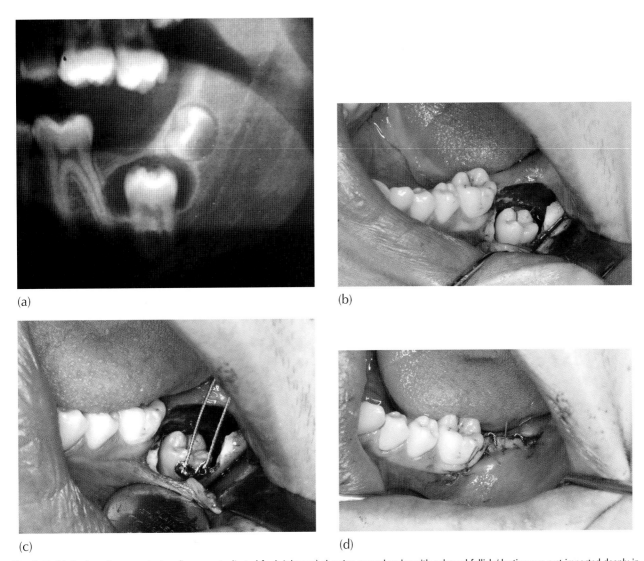

(a)

(b)

(c)

(d)

Fig. 4.10 (a) Section of panoramic view (lower part adjusted for brightness) showing second molar with enlarged follicle/dentigerous cyst impacted deeply in the mandible, despite almost complete root closure and in close relation to inferior alveolar canal.

(b) Surgical exposure of crown of the second molar, following enucleation of the third molar.

(c) Two bonded eyelet attachments with steel ligature pigtails.

(d) Full flap closure, leaving only the twisted steel pigtails turned into hooks and easily accessible to the patient.

(e) Post-surgical periapical radiograph.

(f) The maxillary second molar is over-erupted (note the prominent mesial marginal ridge). An elastic chain is drawn across its occlusal surface from a palatal screw TAD to the intra-oral extremity of the zygomatic plate implant. The naturally bonded button is to prevent slippage of the elastic chain off the occlusal surface.

(g) The vertical intermaxillary elastic is placed by the patient from zygomatic plate to one of the two available steel hooks emanating through the sutured exposure site in the mandible.

(h) The lower second molar has erupted and the upper has become markedly intruded (reverse mesial marginal ridge).

(i) At completion, the maxillary second molar has been markedly intruded and the mandibular second molar erupted to the height of its mesial neighbour. Orthodontic treatment *per se* was not performed and no other orthodontic appliances were used.

(j) The periapical view of the day treatment was completed, showing normal bony picture.

(k) The panoramic view on the same day.

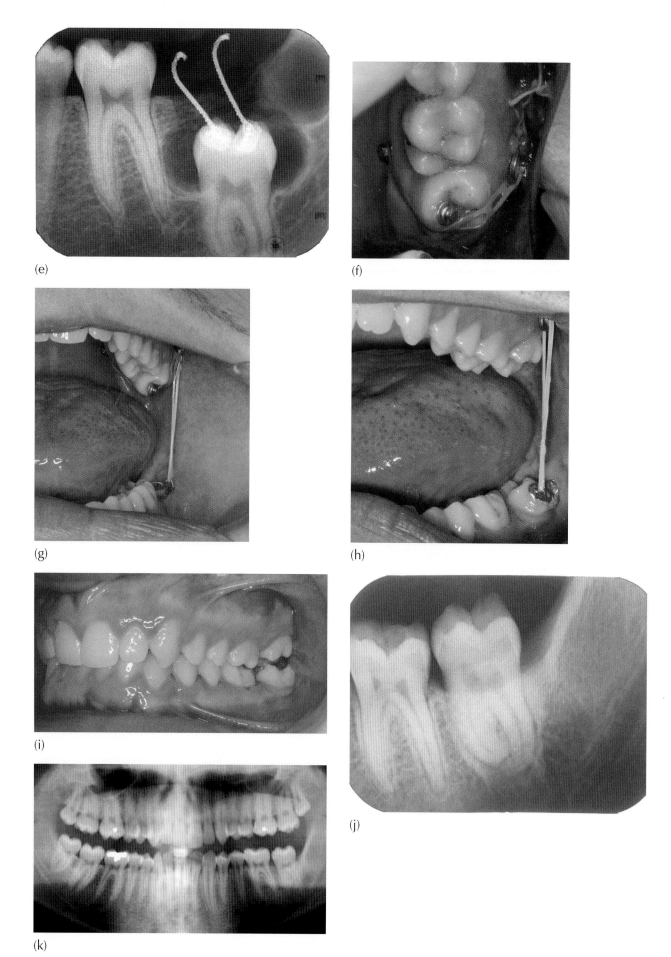

(e)

(f)

(g)

(h)

(i)

(j)

(k)

Fig. 4.10 (*Continued*)

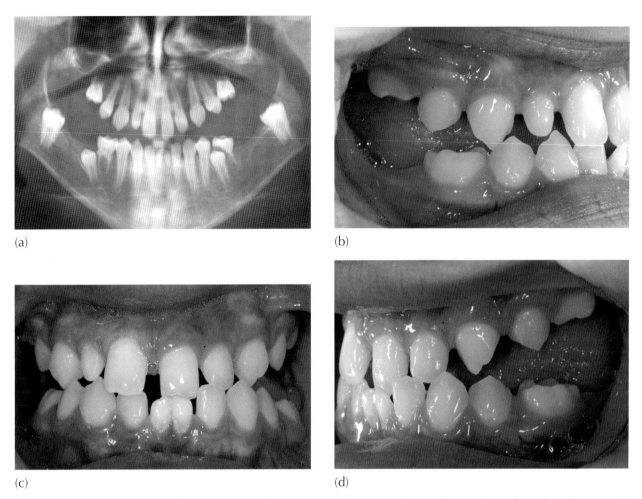

(a)

(b)

(c)

(d)

Fig. 4.11 (a) The patient is a 12-year-old male, congenitally missing 10 of his 12 permanent molars, maxillary second premolars and mandibular central incisors. The maxillary second deciduous molars are severely infra-occluded and the right mandibular deciduous second molar is mildly infra-occluded, with clear signs of ankylosis. The bilaterally solitary permanent molars were unerupted and buried in the anterior ramus at the height of the occlusal plane.

(b–d) Intra-oral views show a distinct class 3 relation and a large posterior open bite on both sides.

(e) Occlusal view to show artificial acrylic teeth on a lingual arch soldered to first premolar bands, replacing the extracted deciduous lower incisors. The anchorage for the resolution of the right impacted molar was based on the infra-occluded deciduous second molar and for the left impacted molar on the temporary anchorage screw device. In the upper jaw treatment, a protraction face mask was used with elastic traction direct to attachments placed on the severely infra-occluded deciduous molars, followed by space closing mechanics using an upper multi-bracketed appliance.

(f) A short piece of wire has been bonded into place between the temporary anchorage screw device and the second premolar tooth to provide anchorage for the mesial bodily traction of the unerupted molar.

(g) The most recent panoramic view shows the previously unerupted molars have been drawn bodily about 0mm on each side in a mesial direction along the occlusal plane, with an accompanying and considerable regeneration of alveolar bone. There has been marked vertical growth of the jaws during the treatment period, as witnessed by the further infra-occlusion of the deciduous second molars.

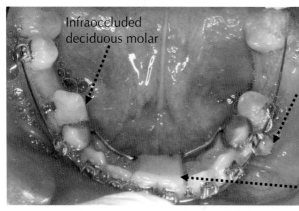

Infraoccluded deciduous molar

Temporary anchorage screw device

Artificial teeth on soldered lingual arch

(e)

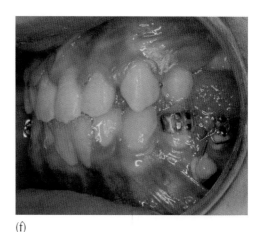

(f)

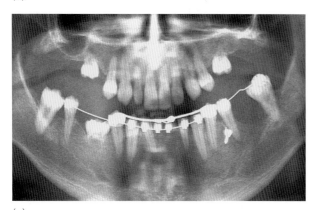

(g)

Fig. 4.11 (*Continued*)

[24–25]. In addition, these magnets corrode significantly in the intra-oral environment, and have to be carefully coated in order to render them safe. A parylene coating has been shown to seal them successfully and, when embedded in acrylic appliances, these magnets can be isolated from the intra-oral environment and protected from heavy forces [25].

There are, however, a number of significant problems with their use. Attracting forces that exist between the two magnets are in inverse proportion to the square of the distance between them. This means that when used to move displaced or ectopically positioned teeth, the magnet that is sited on the appliance must be placed close to the magnet which has been bonded to the displaced tooth, otherwise the force between them will be too low. In addition, if the magnets are not sited ideally on top of each other, there is a dramatic drop in force level [26].

The notion that traction may be applied without the need to trail a wire through the soft tissues of the palate appeals to authors. They speculate that this may improve the final periodontal condition of the teeth, since 'eruption simulates a normal eruption process'. However, it should be remembered that:

- the tooth must nevertheless initially be exposed surgically;

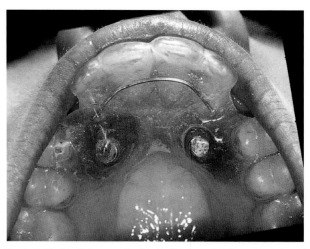

Fig. 4.12 The bonded magnet 'backpack'. (Courtesy of Professor A. Vardimon.)

- the magnet must be bonded to it;
- the flap must be partially or fully replaced and healing must occur;
- the tooth must then pass through the tissues with this relatively large magnetic 'backpack' (Figure 4.12);

each of which presents obstacles that signify a departure from any similarity with normal eruption. Even if the idea is 'attractive' [27], the use of magnets for impacted teeth is still in its early developmental stages and the methods that have been described show a number of technological disadvantages. The size of the magnets and the inverse square rule of their force of attraction are the most pertinent. Presently, it remains a method that cannot unequivocally replace the more traditional and conventional methods described above [28–31].

References

1. Olive RJ. Factors influencing the non-surgical eruption of palatally impacted canines. Aust Orthod J 2005; 21: 95–101.
2. Iramaneerat S, Cunningham SJ, Horrocks EN. The effect of two alternative methods of canine exposure upon subsequent duration of orthodontic treatment. Int J Paediatr Dent 1998; 8: 123–129.
3. Becker A. Alternative methods of canine exposure and subsequent duration of treatment. Int J Paediatr Dent 1998; 8: 298–299 [letter to the editor].
4. Becker A, Chaushu S. Success rate and duration of orthodontic treatment for adult patients with palatally impacted maxillary canines. Am J Orthod Dentofacial Orthop 2003; 124: 509–514.
5. Kokich VG, Mathews DP. Surgical and orthodontic management of impacted teeth. Dent Clin North Am 1993; 37: 181–214.
6. Shapira Y, Kuftinec MM. Treatment of impacted cuspids: the hazard lasso. Angle Orthod 1981; 51: 203–207.
7. Kettle MA. Treatment of the unerupted maxillary canine. Dent Pract Dent Rec 1958; 8: 245–255.
8. Becker A, Zilberman Y. The palatally impacted canine: a new approach to its treatment. Am J Orthod 1978; 74: 422–429.
9. Kokich VG. Surgical and orthodontic management of impacted maxillary canines. Am J Orthod Dentofacial Orthop 2004; 126: 278–283.
10. Gensior AM, Strauss RE. The direct bonding technique applied to the management of the maxillary impacted canine. J Am Dent Assoc 1974; 89: 1332–1337.
11. Nielsen LI, Prydso U, Winkler T. Direct bonding on impacted teeth. Am J Orthod 1975; 68: 666–670.
12. Hunt NP. Direct traction applied to unerupted teeth using the acid-etch technique. Br J Orthod 1977; 4: 211–212.
13. Becker A, Shpack N, Shteyer A. Attachment bonding to impacted teeth at the time of surgical exposure. Eur J Orthod 1996; 18: 457–463.
14. Ziegler TF. A modified technique for ligating impacted canines. Am J Orthod Dentofacial Orthop 1977; 72: 665–670.
15. Lu TC, Wang WN, Tarng TH, Chen JW. Force decay of elastomeric chains – a serial study. Part 2. Am J Orthod Dentofacial Orthop 1993; 104: 373–377.
16. Storie DJ, Regennitter F, von Fraunhofer JA. Characteristics of a fluoride-releasing elastomeric chain. Angle Orthod 1994; 64: 199–210.
17. Vachiramon A, Urata M, Kyung HM, Yamashita DD, Yen SL. Clinical applications of orthodontic microimplant anchorage in craniofacial patients. Cleft Palate Craniofac J. 2009; 46: 136–146.
18. Kuroda S, Sugawara Y, Kuroda S, Sugawara Y, Kyung H-M, Takano-Yamamoto T. Clinical use of miniscrew implants as orthodontic anchorage: success rates and postoperative discomfort. American Journal of Orthodontics and Dentofacial Orthopaedics 2007; 131: 9–15.
19. Erverdi N, Usumez S, Solak A. New generation open-bite treatment with zygomatic anchorage. Angle Orthod. 2006 May; 76(3): 519–526.
20. Vardimon AD, Graber TM, Voss LR. Hygienic magnetic technique to align impacted teeth. Presented at the 87th annual session of the American Association of Orthodontists, Montreal, Canada, 1987.
21. Sandler PJ, Meghji S, Murray AM et al. Magnets and orthodontics. Br J Orthod 1989; 16: 243–249.
22. Darendeliler MA, Friedli JM. Treatment of an impacted canine with magnets. J Clin Orthod 1994; 28: 639–643.
23. Vardimon AD, Graber TM, Drescher D, Bourauel C. Rare earth magnets and impaction. Am J Orthod Dentofacial Orthop 1991; 100: 494–512.
24. Mancini GP, Noar JH, Evans RD. Neodymium iron boron magnets for tooth extrusion. Eur J Orthod 1999; 21: 541–550.
25. Noar JH, Wahab A, Evans RD, Wojcik AG. The durability of parylene coatings on neodymium iron boron magnets. Eur J Orthod 1999; 21: 685–693.
26. Noar JH, Shell N, Hunt NP. The performance of bonded magnets used in the treatment of anterior open bite. Am J Orthod Dentofacial Orthop 1996; 109: 549–557.
27. Sandler JP. An attractive solution to unerupted teeth. Am J Orthod Dentofacial Orthop 1991; 100: 489–493.
28. Ingervall B. The use of magnets in orthodontic therapy: panel discussion. Eur J Orthod 1993; 15: 421–424.
29. Gianelly A. The use of magnets in orthodontic therapy: panel discussion. Eur J Orthod 1993; 15: 421–424.
30. Rygh P. The use of magnets in orthodontic therapy: panel discussion. Eur J Orthod 1993; 15: 421–4.
31. Vardimon AD. The use of magnets in orthodontic therapy: panel discussion. Eur J Orthod 1993; 15: 421–424.

5

Maxillary Central Incisors

Orthodontic Treatment of Impacted Teeth, Third Edition. Adrian Becker.
© 2012 Adrian Becker. Published 2012 by Blackwell Publishing Ltd.

Aetiology

At the age of about 6 years, in most children, a sudden and dramatic change occurs in the anterior part of the dentition, with the shedding of the deciduous incisor teeth and the appearance of the permanent incisor teeth. As has been described in Chapter 1, the first to erupt in the young child are usually the mandibular central incisors, although the first permanent molars may sometimes precede them. The mandibular lateral and the maxillary central incisors then erupt shortly after, at 6.5–7 years. Under normal circumstances, the maxillary lateral incisors are the last of the incisors to erupt, completing the anterior dentition with their appearance about a year after the eruption of the adjacent central incisors. Further changes in the dentition, in terms of shedding of deciduous teeth and eruption of their permanent successors, do not occur until approximately the age of 9.5–10 years. This means that there is a two-year period of relative stability, which is known as the mixed dentition stage.

The spectacle of erupted lateral incisors associated with the non-appearance of one or both of the central incisors should always be deemed abnormal, whether or not a deciduous central incisor is still present, and further investigation should be undertaken to ascertain the reason for the aberration.

Congenital absence of a maxillary permanent central incisor, given the presence of permanent lateral incisors, is exceptionally rare, although it has been reported. In this situation, the shape of the single central incisor makes it impossible to distinguish whether it belongs to the right or left side. The patient's appearance is abnormal (Figure 5.1a–c) and rather reminiscent of a 'dental cyclops'! However, the abnormality in the appearance is also due to other clinical features and malformation of other elements of the craniofacial complex. The patients have other midline anomalies, including an indistinct philtrum of the upper lip, with an absence of the typical Cupid's bow and no midline frenulum, a mid-palatal ridge, and nasal obstruction or septum deviation. They exhibit a short anterior

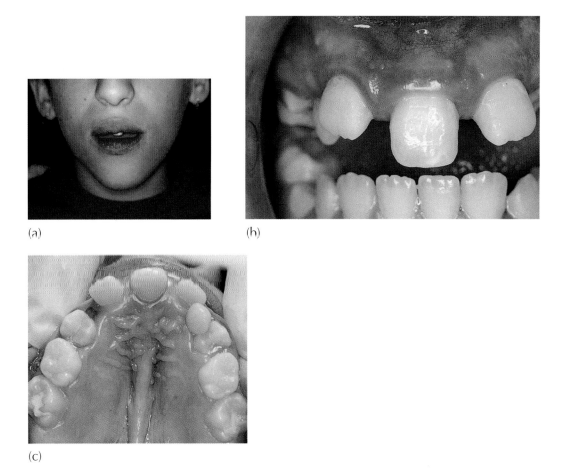

(a)   (b)

(c)

Fig. 5.1 (a) Abnormal lip morphology, absence of philtrum and midline position of single central incisor, in a case of holoprosencephaly. (b) Intra-oral view of same patient to show 'square' anatomy of incisor and indeterminate right/left designation. The lateral incisors are laterally flared as part of an otherwise normal 'ugly duckling' stage of development. (c) A view of the palate to show bilateral submucous clefting. (Courtesy of Dr S. Geron.)

cranial base and maxilla, together with retrognathic and posteriorly inclined maxilla and mandible. The sella turcica also has an abnormal shape. Accordingly, the condition should not be treated as a simple, local and isolated congenital absence, as would be the case in relation to a missing maxillary lateral incisor or mandibular second premolar. This patient exhibits a mild form of holoprosencephaly (Figure 5.1).

At 5–6 weeks of human embryonic development, the forebrain is formed and the face begins to take shape. In holoprosencephaly, the forebrain fails to develop into two cerebral hemispheres and the structures that are normally paired into identical right and left units are allowed to merge. This causes defects in the development of the face and in brain structure and function which, in less severe cases, may result in normal or near-normal brain development and facial deformities that may affect the eyes, nose and upper lip. However, in most cases of holoprosencephaly, the malformations are so severe that the foetus dies *in utero* [1].

Trauma to the anterior maxilla in a young patient is probably an almost daily occurrence in many schools, playgrounds and sports activities. Trauma at home should not be forgotten and may occasionally be the first clue of child abuse. By and large, the incidents are trivial and characterized by much bleeding from lacerations of the lip and gingivae which may occasionally divert attention from underlying damage to the incisor teeth. These may have been traumatically intruded and lost from sight or even completely avulsed.

A careful history should be elicited to eliminate the remote possibility of avulsion of the permanent central incisor which may have been completely overlooked. This may happen if the tooth had been in the very earliest stages of its eruption and, therefore, its absence went unnoticed at the time. A newly erupting central incisor is invested with a rudimentary, underdeveloped periodontal ligament (PDL) such that a relatively light and unfortunately well-directed blow may easily bring about its loss.

Obstructive causes

Supernumerary teeth

When an existing permanent central incisor does not erupt and the diagnosis of impaction is made, the most common aetiological factor is the presence of one or more midline supernumerary teeth [2–4] (see Figures 1.10 and 1.11). In a study of a sample of schoolchildren [5] the prevalence of supernumerary teeth was found to be between 1.5% and 3.5%. It has also been shown in a different sample group [6] that between 28% and 60% of cases with supernumerary teeth will have resultant eruption disturbances of the adjacent teeth. By arbitrarily integrating these two studies and assuming that they represented random samples of patients attending a children's hospital dental department, it would seem that between 0.42% and 2.1% of child dental

patients suffer from impacted central incisor teeth from this cause. For the most part, supernumerary teeth develop on the palatal side of the permanent incisors and as such occupy space within the narrow alveolar ridge. This results in a labial and superior displacement of the permanent incisors, which includes their roots.

Odontome

Among the other and rarer causes of obstruction, which may equally prevent the eruption of a normal central incisor, are odontomes. These are very variable in size and type, but, whether they are of the complex or the composite type, they usually have a broader and wider cross-section and their presence will be more likely to impede the eruption of an incisor than a supernumerary tooth.

Ectopic position of the tooth bud

The development of a tooth bud in an abnormal position or in an abnormal angulation may have no apparent cause and may thus be attributed to traumatic or genetic factors (Figure 5.2). As a result of the displacement, normally placed adjacent teeth may provide the physical obstacle to the normal eruption of such a tooth. Alternatively, other physical obstacles, such as the above-mentioned supernumerary teeth or odontomes, may be the more likely reason for a secondary displacement of this tooth. While the early removal of this obvious aetiological factor may be strongly indicated, this will not necessarily affect the position of the tooth bud, which will probably continue to develop in its existing location.

Variation in the position of a developing tooth will produce a concomitant variation in its eruption path. When the eruption path is very slightly deflected, the tooth will usually erupt, but with an abnormal angulation of its long axis, indicating the path along which it will have travelled. As its vertical development proceeds, its relationship to the deciduous predecessor is more lateral, medial, palatal or

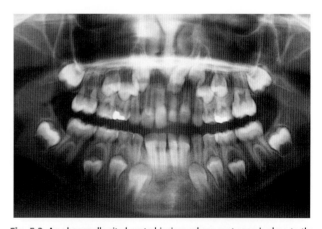

Fig. 5.2 An abnormally sited central incisor, whose root apex is close to the canine area. The shape of the root also appears to be abnormal.

labial to it. This will bring about a partial and oblique resorption along one side of the root of the deciduous incisor. In due time, the further progress of the permanent incisor and the oblique resorption pattern bring it into contact with the crown of the deciduous tooth on that side.

The permanent tooth may remain impacted if insufficient space exists. Alternatively, the tooth may finally erupt adjacent to the over-retained deciduous tooth and many months later than expected, possibly into a cross-bite relationship, or it may be proclined labially, or a diastema may be produced in the midline. This situation may not be self-correcting after the belated shedding or extraction of the stubborn deciduous tooth.

Should the position of the developing tooth be more markedly displaced, which is unusual, its potential eruption path will be in a more obtuse direction and little or no resorption of the deciduous tooth will occur. In these circumstances, eruptive movements are minimal and the permanent tooth remains in a more or less unchanged position over a long period. Finally, when positive intervention is undertaken, removal of the cause will need to be supplemented with active appliance therapy for its resolution.

Traumatic causes

Obstruction due to soft tissue repair

The traumatic, sudden and very early loss of the deciduous incisor teeth is most often the result of an accident that delivered a blow to dislodge the tooth, usually while the child was at play. The possibility also exists for this to have been caused by a dental extraction, prompted by the presence of deep caries or following the sequelae of an earlier trauma. This typically happens at the age of 3 or 4 years, at a time when the permanent incisor is not ready to erupt and a healing-over of the macerated gingival tissue occurs without the early eruption of the tooth. In time, changes take place in the connective tissue overlying the developing permanent incisor, which prevent the tooth from penetrating the mucosa [7, 8]. By the age of 7 or 8 years, one may see and be able to palpate the bulging profile of the central incisor (see Figure 3.2a).

Dilaceration

In the early stages of their development, high in the anterior part of the maxilla, the permanent central incisors are situated lingual and superior to the apices of the deciduous incisors. Their positions change as the result of their labially and inferiorly directed migration, which is part of their normal further development and, as they do so, an oblique resorption of the roots of their deciduous predecessors is initiated.

During this critical period, children are frequently involved in traumatic episodes of various degrees of intensity and a blow may be inflicted on the deciduous maxillary incisors from the front. A blow of sufficient magnitude will produce a lingual avulsion of the crowns of the incisors and

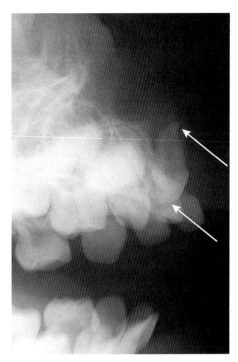

Fig. 5.3 The tangential view shows severe labial displacement of the root of the deciduous incisor. (Courtesy of Professor J. Lustmann.)

a concomitant labial displacement of the roots, which will sometimes cause a fracture of the labial plate (Figure 5.3). In this situation, the roots of the teeth and the sharp edge of bone will be palpable in the labial sulcus, although the overlying mucosa will usually remain intact. Since the unerupted permanent incisors are superiorly and lingually placed in relation to the resorbing root ends of the deciduous incisors, the successional teeth will usually not be displaced and will often continue unhindered in their normal developmental path.

The roots of incisor teeth are tapered apically, which means that a blow directed along their long axis from below will drive the tooth upwards into a similarly tapered socket. It follows that the alveolar process fracture that may occur will cause a part of the labial plate of bone to become displaced labially and the teeth to be displaced superiorly (intruded). Under these circumstances, trauma to the anterior part of the deciduous dentition may cause damage to the growing tooth buds of the permanent dentition, and this may alter the further development of the permanent teeth in many ways. The anomalies that may occur vary from a discoloured hypoplastic mark on the enamel, through anatomic malformations of the crown or root, to non-proliferation of the root, to pulp death and sequestration of the tooth bud [8]. The alterations in the form of the crown and the root of these teeth are collectively referred to as dilacerations. These are the result of recovery from the blow that will have disturbed the natural development, resulting in a variable combination of a hypocalcified,

Fig. 5.4 An extracted dilacerated incisor. Reprinted from ref. [37], with permission.

altered, attenuated and directionally reoriented form of the crown and/or root. The final shape and appearance of the injured tooth will be dictated by the direction and force of the blow and by the particular stage of development of the unerupted permanent tooth. Each these anomalies will be unique in its form and no two trauma incidents will produce the same result.

In direct contrast, there is one type of dilaceration, which we shall refer to here as the 'classic' dilaceration, which consistently reproduces an almost identically shaped tooth (Figure 5.4), whose unerupted location is always the same. Its causation has, over the past six or more decades, been the subject of much discussion among paediatric dentists and orthodontists.

When the trauma is directed in the long axis of the deciduous tooth, it may be transmitted superiorly to the developing permanent tooth. According to this hypothesis, the force delivered by the violent blow to the deciduous incisor is transmitted to the resorbing root apex, which momentarily establishes a point of impact with the palatal side of the crown of the permanent incisor. This has been believed to cause the partially developed and unerupted permanent incisor tooth an immediate upward and labial rotation, in its crypt. It is then assumed that any further root development that occurs in the post-trauma period will continue in the same direction as before, producing a bizarre angle between the pre- and post-trauma portions of the tooth – the 'classic' dilacerated central incisor – with labial displacement of the crown portion. The site of this junction will depend on the stage of development of the root at the time of trauma, and with it the prognosis of any proposed orthodontic treatment for the tooth. This reconstructed scenario for the causation of this classic form of dilaceration is extremely well known and probably represents a majority opinion, if not consensus, within the profession.

Palatal, rather than labial, displacement of the crown *vis-à-vis* the post-traumatic root portion may occur in rare instances as a variation of the above theme and is due to a more palatal position of the tip of the developing permanent incisor and the apex of the deciduous incisor root, at

impact. This is a more likely variation in the very young, during the early calcification of the tooth crown when its palatal location is normal (Figure 5.5).

In an attempt to determine whether there was an evidence base for accepting the above hypothesis, a protocol was instituted some years ago in the emergency clinic of the Department of Paediatric Dentistry at the Hebrew University–Hadassah School of Dental Medicine, Jerusalem. Children who attended the clinic having suffered trauma to their anterior deciduous teeth were subjected to routine clinical and radiographic examination, vitality testing and emergency pulpal and restorative treatments, as indicated. Until the new protocol was established, the radiographic examination had generally been limited to a periapical film of the area. It is obvious that this radiographic view is not suitable for showing any labial and superior displacement of the crown of the unerupted permanent central incisor within the crypt. Accordingly, the new protocol included a tangential radiographic view of the anterior maxilla, which has been described in Chapter 2 as being a view taken from the side, across the occlusal plane, and identical to a restricted area of a lateral cephalometric radiograph. It is obtained on a small, occlusal-sized film by holding the film vertically against the cheek and parallel to the mid-sagittal plane of the skull. Any alteration in the labio-lingual and vertical orientation of the crown of an incisor will be immediately discernible on this film.

The protocol was run for several years and did not discover a single case where the crown of the tooth had been displaced in the immediate post-trauma period in this way. However, in the succeeding years among the small but significant number of patients who turned up in the Department of Orthodontics with a fully developed, classic, dilacerated central incisor, several had been seen in the emergency clinic of the Department of Paediatric Dentistry following trauma in their earlier years. Since records existed from that time, the radiographs that had been taken when the protocol was in force were re-examined with specific regard to these patients. No displacement of the crowns of the unerupted permanent central incisors could be discerned on the immediate post-trauma films, effectively discrediting immediate tooth germ displacement as a cogent explanation for the aetiology of the phenomenon.

A developmental origin has been suggested as an alternative cause for the classic dilacerations [9], with the contention that the active process of the development of cysts, odontomes or supernumerary teeth may produce this phenomenon by displacement of the crown of the tooth or by interference and redirection of its root. In a retrospective study of the phenomenon [10] no history of trauma could be elicited in 70% of the patients in the sample, nor was there macroscopic or microsopic evidence of trauma, or the existence of a cyst, odontome or extra tooth in these patients. No case was found with both central incisors affected and there had been no damage to neighbouring

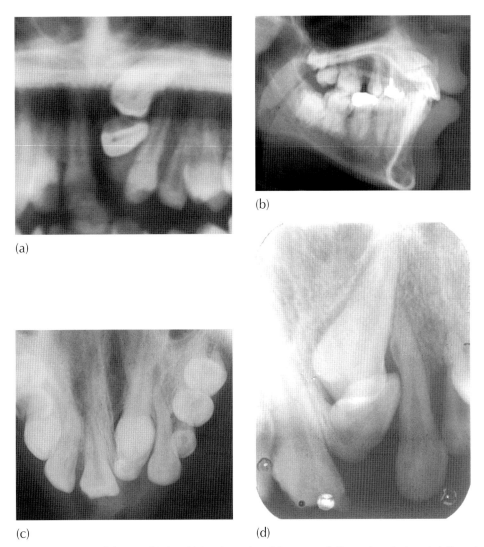

(a)

(b)

(c)   (d)

Fig. 5.5 A palatally-oriented dilaceration of the central incisor. (a) Anterior section of the panoramic film to show the impacted dilacerate left central incisor with its labial surface facing inferiorly. The unerupted permanent left canine, which is transposed with the mesially-tipped lateral incisor, is seen to be directly above it. The deciduous canine with an intact root is seen distal to the lateral incisor. From this view, it is impossible to tell whether the crown of the central incisor points labially or palatally. (b) The lateral cephalometric view shows the outline of a screw post in the restored right central incisor. Above it, the profile of the palatally directed crown of the central incisor can be clearly seen, with its root portion curving labially and superiorly in a tight quarter-circle, encompassing the more superiorly placed and labially directed permanent canine. (c) The anterior occlusal view shows the typical form of the canine superimposed and largely masking the crown of the incisor. The apical portion of the incisor root is viewed in its cross-section anterior to the tip of the canine crown. (d) The periapical view provides a separation of the two teeth although, taken in isolation, it may give the inexperienced diagnostician the misleading impression that the crown of the incisor is labial.

teeth, which could be expected to occur in at least a few instances had trauma been the cause.

These cases also failed to show two distinct and angulated portions to the root, but rather a continuous and tight curve (Figure 5.4), quite different from those in which trauma, as an aetiological factor, had resulted in any of the other forms of dilaceration. The conclusion of the study was that a fairly high proportion of dilacerations occur due to an ectopic siting of the tooth germ, whose root development is deformed by its proximity to and the anatomy of the palatal vault in the immediate vicinity.

These explanations are unsatisfactory on several counts. Different cases show an almost identical and very typical anatomy of the tooth, which affects maxillary central incisors exclusively – no case of bilateral occurrence or parallel incidence in lateral incisors has been reported in the literature. As if to illustrate the exception that proves the rule, the records of a single case of bilateral occurrence were sent to the author (Figure 5.6) a few years ago and it remains the sole case that the author has seen. The morphology of the crown of the affected tooth is normal and is an exact mirror-image of its erupted antimere and the coronal

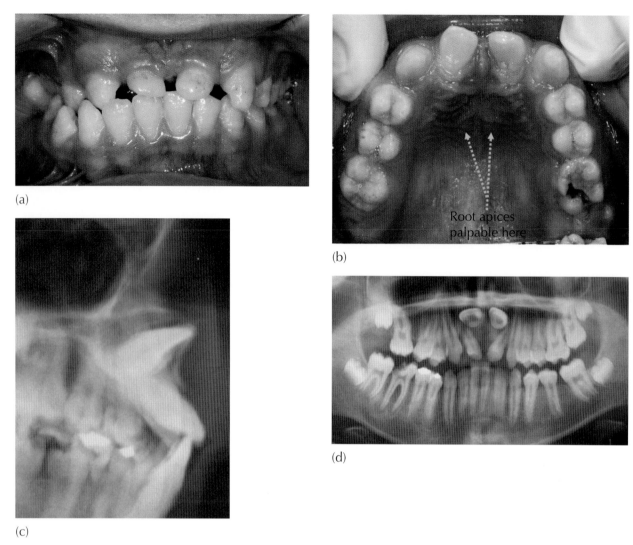

Fig. 5.6 An extreme rarity: bilateral classic dilacerations of the central incisors. (Courtesy of Dr. Srdan Marelic.)
(a) Mesial drifting of the erupted lateral incisors has largely closed off the space for the unerupted central incisors. The intra-oral anterior view of the dentition.
(b) The intra-oral occlusal view.
(c) The cephalogram.
(d) Panoramic view.

portion of the root shows initial normal development. The remainder of the root develops strictly in the labio-lingual plane along a tight and continuous circular path, rather than two straight portions of root at an angle to one another. The crown is grossly displaced in the bucco-lingual and vertical planes, while exhibiting minimal or no mesio-distal rotation.

Not only is the anatomy typical, but the position and orientation of the tooth are also unique. The crown of the tooth is upturned and displaced high in the sulcus on the labial side of the alveolus, with its palatal aspect palpable on the labial side of the sulcus close to the root of the nose. Often, the root apex of the tooth is palpable on the palatal side of the alveolus (Figure 5.5) and may be the sign that

influences an unwary surgeon to mistakenly approach exposure of the tooth on the wrong side (see Figure 12.12).

It has been suggested that the loss of a deciduous incisor may lead to scarring along the eruption path of the permanent incisor, which deflects the developing tooth labially [8]. This runs counter to Stewart's observation that no history of early traumatic loss of the deciduous tooth had occurred in 70% of the cases in his series.

The Jerusalem hypothesis

It is possible to read a completely different aetiological interpretation into these very constantly occurring features, and it is pertinent to begin by questioning the reliability of a child's or parent's memory regarding traumatic injury of

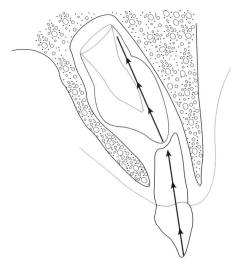

Fig. 5.7 A diagram to show how a vertically directed force through the deciduous incisor is transmitted to the labial aspect of the mineralizing root of the unerupted permanent incisor. Reproduced from previous edition with the kind permission of Informa Healthcare – Books.

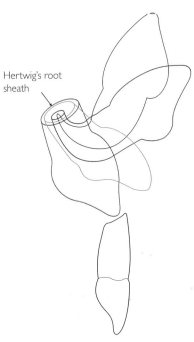

Hertwig's root sheath

Fig. 5.8 A diagrammatic illustration of the progressive alteration in the orientation of a dilacerated incisor, during the disturbed root formation. Note that the position of Hertwig's proliferating root sheath remains unaltered. Reproduced from previous edition with the kind permission of Informa Healthcare – Books.

the front teeth. Severe trauma is rare and always remembered, but non-disfiguring trauma (i.e. trauma that causes no loss, fracture or displacement of the deciduous anterior teeth) occurs quite frequently in young children, is rarely noted and almost never remembered in later years. Indeed, following a relatively minor blow, the child may have come home in tears, but the incident will often have been forgotten the next day, at most with some minor, residual, temporary soreness or tenderness.

Abrupt and vertically directed force through the long axis of the deciduous tooth will bring about the transference of the impact to the intimately related, unerupted, permanent central incisor. Because the long axis of the permanent incisor has a more labially tipped orientation, the force will be transmitted in an oblique line which runs through the incisal edge and the most superior point on the labial side of the newly forming root, close to or at the root mineralization interface (Figure 5.7). This most recently formed extremity of the partially developed root has a circular knife edge of calcified dentine. Thus, the intrusive blow will be delivered directly to the sensitive cells of Hertwig's root sheath at this narrow rim. The intrusive force is thus concentrated along the knife edge root extremity and will inflict considerable damage on cells of the formative root sheath, with relatively low force values. It will be appreciated that precision in direction may be more critical than force magnitude.

It is entirely possible that the root sheath may only partly recover from the blow, which may result in an attenuated rate of production of dentine on the labial side of the tooth. With the remainder of the root-forming system continuing to produce dentine unscathed, undeterred and unabated, it follows that the final shape of the root of this tooth will conform to a continuous labially directed curve (Figure

5.8), until apexification is achieved. Furthermore, since the dental papilla base of Hertwig's root sheath maintains its position within the alveolar process fairly constantly – against the eruptive force of the developing tooth – and provides the platform from which the downward development of the root is normally directed, the crown of the incisor moves labially and superiorly for as long as this bucco-lingually unequal mineralization gradient of the developing root continues. In other words, dilaceration of this classic type is an anomaly which is traumatic in origin but developmental in its expression.

This hypothesis provides an explanation for the typical appearance of the dilacerated tooth as well as its final unerupted position under the nose. Furthermore, since it may occur with a relatively minor trauma, this could account for the high proportion of cases with no apparent history of trauma experience and no premature loss of or damage to the adjacent teeth. Its relative rarity may be due to the positional relationship between the deciduous and permanent incisor at the time of injury and a very specific directional relationship with the traumatic force vector. This may account for an absence of bilaterally affected cases, for its non-occurrence among lateral incisors and for an absence of any association with supernumerary teeth, cysts and odontomes. It also provides an explanation as to why there is no obvious displacement visible on the tangential radiographs that were taken immediately following the traumatic episode (Figure 5.9) and why the tooth is never rotated in the horizontal plane.

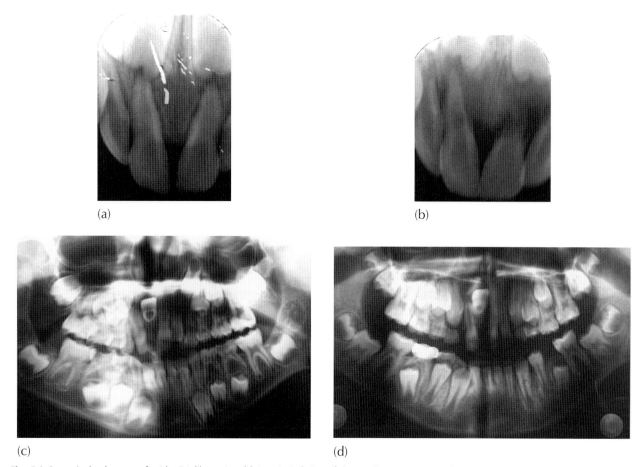

(a)

(b)

(c)

(d)

Fig. 5.9 Dynamic development of a 'classic' dilaceration. (a) A periapical view of the maxillary incisor area of a 3.11-year-old child taken in the emergency room on the day that trauma had occurred. Note there is no apparent displacement of either central incisor. The left side appears to be the worst affected by the trauma. (b) A six-month follow-up film shows considerable inflammatory resorption of the root of the left deciduous central incisor and less on the right side, but there is a concomitant subtle change in the relative positions of the unerupted permanent incisors. On the right side, the incisal edge is more superiorly located in relation to the left incisor and to the root of its deciduous predecessor. (c) The initial diagnosis of dilaceration of the right central incisor was made on the basis of this panoramic view, taken at age 7.6 years (i.e. 3.7 years post-trauma). Note the anatomical labial surface of the crown faces superiorly, with the orientation of the crown portion being slightly above the horizontal. The pulp chamber is very large, indicating that root development is continuing. (d) A new panoramic view taken 2.4 years later shows the tooth to be higher still, to one side of the anteriorly nasal spine and tipped further posteriorly. Its palatal surface is now facing anteriorly and its pulp chamber is much smaller.

Arrested root development

When a pre-school child suffers a very severe blow to the maxillary anterior dentition, it is likely to result in the loss of the deciduous incisors, but it may also cause fracture of one or both jaws. The root-producing ring of cells that comprises Hertwig's sheath may have been so seriously damaged as to have effectively caused serious disruption or actual cessation of any further root development. These teeth may lose their eruption potential, while the adjacent teeth will continue to erupt, bringing with them vertical proliferation of alveolar bone. Only at a much later stage will this phenomenon be discovered when the affected tooth or teeth do not erupt and an area of vertically deficient and bucco-palatally narrow, edentulous alveolar ridge becomes evident.

Radiographs will usually reveal a normal axial inclination and an undisturbed sagittal and lateral location of the crown of the tooth. In the vertical plane, it will be situated very high up in the premaxilla, with minimal or no root formation, depending on how much root had already developed at the time of the accident (Figure 5.10).

Acute traumatic intrusion (intrusive luxation)

Traumatic injury occurs in young children as the result of play-related activities in school or at home, in accidents involving a fall or, occasionally, as the result of deliberate physical violence [8]. The effects on the teeth range from a transitory pulp inflammation, through the various types of fracture of the crown of the tooth or of its root, to, in the severest cases, avulsion of the entire tooth. Intrusion of one

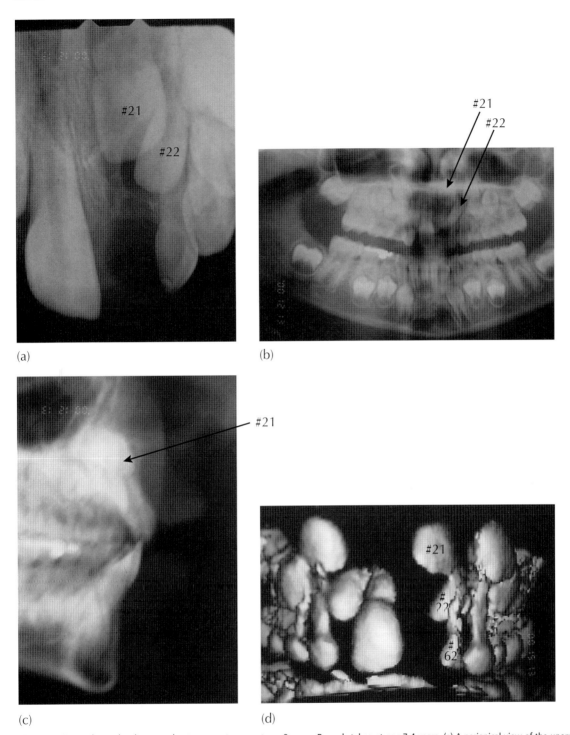

(a)

(b)

(c)

(d)

Fig. 5.10 Arrested root development due to severe trauma at age 3 years. Records taken at age 7.4 years. (a) A periapical view of the unerupted, but normally oriented, left central incisor (#21) shows a marked height discrepancy compared to the erupting right incisor. Note the late developing lateral incisor (#22) is inferiorly placed, tipping mesially and more advanced in terms of its eruption status. The affected incisor shows little or no root development. (b, c) The panoramic and cephalometric views show the incisor to be close to the floor of the nose and a normal crown orientation. (d) The anterior view seen in this three-dimensional image from the cone beam computerized tomography (CT) scan. (e, f) Two paraxial (vertical) cuts from the CT scan illustrate the vertical relationship between the unerupted central and lateral incisors and the deciduous lateral incisor (#62). They also show the labial cervical area of the crown coming to an abrupt step where cessation of crown formation occurred and where the minimal root development is almost at a right-angle to the general orientation of the crown and in close relation with the floor of the nose.

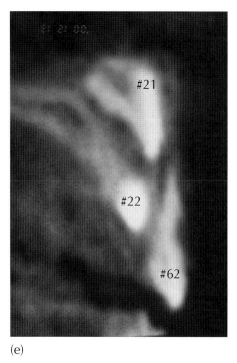

(e)

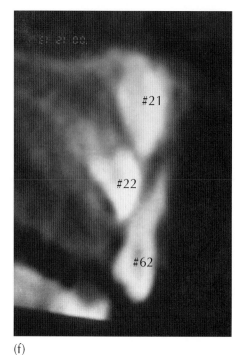

(f)

Fig. 5.10 (*Continued*)

or more of the incisor teeth is often associated with a fracture or comminution of the labial plate of bone and a tearing of the periodontal fibres.

A child may present at the emergency clinic with what gives the appearance of a total avulsion, since the tooth is not visible in its former place, the gingivae are lacerated and there is a considerable amount of blood clot. A periapical radiograph of the area will reveal a superior displacement of the tooth into the alveolar bone, without necessarily producing a fracture of either the crown or the root. The labial plate is displaced labially, although it is most often held closed by the injured but uninterrupted band of labial gingiva and oral mucosa. The integrity of the blood supply to the soft tissues is usually intact, and this allows good and rapid soft tissue healing.

In a sense this tooth has been totally avulsed, but in an atypical manner. It is completely displaced from its socket, with total severance of its attachment apparatus and disruption of its vital supply lines. When compared with the typical avulsed case, however, it has one enormous advantage, and that is that the tooth has not been allowed to dry; it has not usually been in contact with any form of contaminated material and it is not necessary to store it in saliva or milk or other recommended isotonic medium, before restoring it to its rightful place. It is situated in an area initially surrounded by a coagulating haematoma and later by organizing blood clot, and it must be assumed that the damaged periodontal fibres in this situation fare considerably better than do those of the replanted tooth, which

has spent some time out of the mouth. This is dealt with in detail in Chapter 13.

Diagnosis

History

A missing permanent central incisor is likely to be the presenting symptom of a child when the contralateral central incisor has been present in the mouth for several months. Urgency on the part of the parent may be sharpened when the lateral incisor of the same side erupts and clearly reduces the size of the place for the absent tooth. Ideally, many children will have been seen by the paediatric dentist for routine dental checks in the past and will have been referred by the dentist for the orthodontic opinion, but many others may be self-referred directly to the orthodontist. With the notable exception of the acute traumatic impaction cases referred to above, urgency does not exist in the present context in contrast to most other forms of presenting symptom, such as pain and swelling.

When seeing a patient who attends for the first time exhibiting absence of an erupted permanent central incisor, the general medical history should be recorded carefully. It must be borne in mind that surgical intervention is very likely to be needed as an essential part of the treatment that is to be provided. Accordingly, such aspects as previous illnesses, particularly rheumatic heart disease, drugs being taken and bleeding tendencies, together with any other

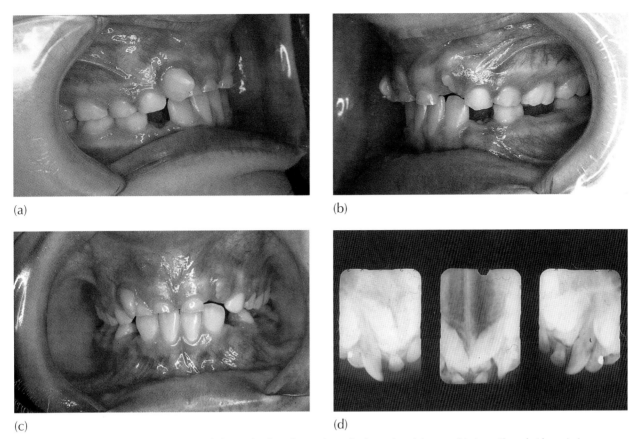

(a)

(b)

(c)

(d)

Fig. 5.11 (a–c) Clinical views of a patient with a bulging ridge form due to obstructive impaction of the central incisors. Three deciduous incisors are over-retained and the permanent lateral incisors are erupting rotated and with a mesio-labial angulation. The ridge bulges due to the presence of dental obstruction. (d) Periapical views of the anterior region show displaced long axis orientation of the central incisors and the existence of several supernumerary teeth.

important and relevant items of information, require to be elicited at the outset.

Questions should be asked, with particular emphasis placed on the possibility of an episode of past trauma. The parent should be questioned carefully to discover whether the child is generally accident-prone. Direct mention of bicycle accidents, falling from a chair, ladder or tree, or being hit in the mouth during play should be made and all relevant answers should be carefully recorded, together with the approximate dates of their occurrence. The possibility of well-hidden child abuse should be considered and, when there is suspicion of this, a report should be made to the local police.

Clinical examination

Much of the patient's dental experience and history is possible to ascertain from a glance at the dentition itself. The existence of sealants and restorations, the absence of teeth, gingival inflammation and the level of oral hygiene will often tell a great deal about past attitudes of both the patient and the attending dentist to prevention and to restorative procedures. At the time the patient attends for treatment, the presence or absence of the deciduous incisor is generally irrelevant. The central incisor of the opposite side and the lateral incisor of the same side will usually be seen to be tipped towards one another, and there will usually be insufficient space at the occlusal level for the placement of the unerupted tooth. Widely apically divergent long axes of the two adjacent teeth will suggest the presence of an unseen and undiagnosed space-occupying physical obstruction.

Palpation

In the obstructed cases, the unerupted tooth itself is often high on the labial side of the alveolar ridge and there may be additional and smaller irregularities bulging the alveolus more inferiorly. These are best identified by palpation. There will almost always be a labio-lingual widening of the ridge (Figure 5.11). If the ridge area is relatively thin inferiorly, it will indicate that teeth are not present at this level (Figure 5.12).

The importance of palpation of the area is not to be underestimated since, if it is not performed sufficiently thoroughly, an important diagnosis may be missed. The presence of a dilacerated central incisor will only be revealed by clinical examination if palpation is made very high in the labial sulcus. Normally, the superior midline area is

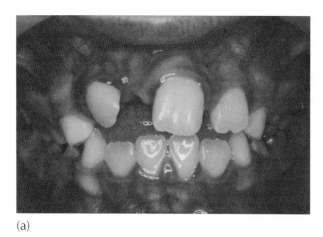

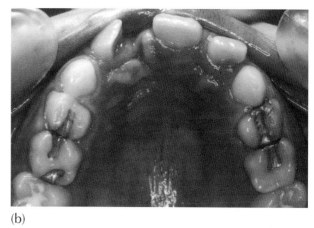

(a)   (b)

Fig. 5.12 (a, b) Frontal and occlusal clinical views of a patient with a dilacerate central incisor, illustrating a bucco-lingually and vertically deficient ridge, due to the absence of teeth in the immediate area – the 'hourglass' ridge.

delineated by the prominence of the anterior nasal spine, on each side of which a shallow depression will be felt. The palatal surface of a dilacerated incisor crown faces forwards and produces a hard swelling in the place where the shallow depression is expected. By superiorly retracting the upper lip, the oral mucosa may be seen to move freely over the stretched area, which will emphasize the outline of the cingulum of the tooth.

Palpation of a dilacerated central incisor may often be made in two places. With the abnormal position of the coronal portion of the tooth, the further development of the root portion may be along an axis which is tipped more lingually and, in the later stages of root closure, the apex may become palpable as a small, hard lump in the palate. This is a feature which few clinicians seem to look for and is a more consistent finding than may be realized [11]. Mistaken diagnosis of crown location is an unusual event since both the orthodontist and the oral surgeon need to undertake a clinical examination and review the radiographs before treatment and it is difficult to imagine that the same mistake could be made twice. Notwithstanding, as we shall see in Chapter 12, such occurrences are seen from time to time (Figure 12.12).

Radiographic examination

In Chapter 2, the different methods of radiographically viewing unerupted teeth were discussed in general and it was pointed out that a periapical view provides essential qualitative information and as such, should be the first step in this part of the examination.

In the case of an unerupted central incisor, this view will generally show associated pathology with great clarity, including hard tissue obstructions (supernumerary teeth of the carious types, odontomes), soft tissue lesions (cysts, tumours), and abnormal root and crown morphology of the unerupted tooth. From this alone, it will usually be possible to establish the reason for the failure of the tooth to erupt.

If supernumerary teeth or odontomes are seen on the film, the information that will then be required relates to their size, number and mesio-distal relationship to the midline and the incisor teeth, all of which will be obvious from this view. However, their labio-lingual orientation in relation to the adjacent erupted teeth will not be obvious from this one film. Since the periapical view is obliquely angled to the horizontal plane, a labial supernumerary tooth will appear lower in the vertical plane than a palatal supernumerary which is situated at the same height. Accordingly, the assessment of height is directly related to the labio-lingual position of the tooth, and this provides the basis for the vertical tube shift method of locating the bucco-lingual relationship of an impacted tooth to an erupted tooth, described in Chapter 2.

For patients in whom the aetiology is the presence of unerupted supernumerary teeth and at the age at which most patients will attend for treatment, a true lateral, tangential view is not helpful, due to the superimposition of central and lateral incisors, deciduous and unerupted permanent canines, and the supernumerary teeth. A second periapical view, directed from a more distal vantage point, will usually help to localize the relative position of the unerupted teeth, using the principles of parallax. Similarly, a routine anterior (oblique) occlusal film will help to separate out the images of the unerupted teeth, using the same parallax principles, this time in the vertical plane. This will provide the information needed to compute the relative heights and labio-lingual relations of the individual structures.

A good vertex occlusal film, directed through the long axes of the anterior teeth, will provide unequivocal evidence of labio-lingual tooth position, particularly if there is marked displacement. However, the relative lack of contrast in this view and the long exposure needed weigh heavily against its risk/benefit efficacy and have contributed to obsolescence of the method and to its demise.

Dilacerated central incisor teeth with labial displacement have a very special and characteristic appearance on a periapical radiograph. We have already described how the crown and the developed part of the root become rotated labially and superiorly following the trauma to the deciduous incisor. The long axis of the coronal part lies in the direct line of the X-ray beam, which is pointed at a periapical film of the area and, accordingly, will show up as a cross-sectional view of the crown, superimposed on and concentric with a cross-sectional view of the widest part of the root. It will be readily understood, therefore, that the labial surface will be seen to face superiorly and the cingulum will be clearly outlined inferiorly (Figure 5.9c, d). The apical (post-trauma) portion of the root, on the other hand, progressively turns in a tight labial and vertical arc as development proceeds, and ends up with a 90° or more angulation to the coronal portion. The root apex will be seen as a very short 'tail' extending superiorly above the image of the tooth's labial surface. The picture is reminiscent of a scorpion viewed from the front (Figure 5.13a).

Although this is clearly recognizable, the periapical film gives only an indication of its height in the alveolus, while the detail of its curved axis and its general apico-incisal orientation cannot be defined. Confirmation of the diagnosis and the degree of its severity may then be positively completed using a tangential or lateral skull radiograph (Figure 5.13b). This will give information that will help to build up a more comprehensive picture of the tooth, particularly regarding details of its morphology, height and the overall orientation of its long axes.

Computerized tomography in these impacted incisor cases in general is helpful for locating the exact three-dimensional positions of these teeth, while eliminating the superimpositions inevitably seen in plain film radiography [12]. It will also facilitate pinpointing the location of the initial curvature of the root of a dilacerate tooth, so that the orthodontist may decide whether an apicoectomy will later be necessary as root torque is being performed. Similarly, the relative position of the impacted tooth *vis-à-vis* supernumerary teeth will help the surgeon to decide where to open the initial flap and to more easily identify one from another and from the permanent incisor. It will also assist the orthodontist in assessing the relative difficulty of the case in attempting to resolve the impaction and to enable the design of appliance auxiliaries to suit the task.

Treatment timing

In Chapter 1, we discussed the occurrence of a chance pathological finding during routine X-ray examination. Obstructions and any other form of pathology should be eliminated wherever possible before they have the chance

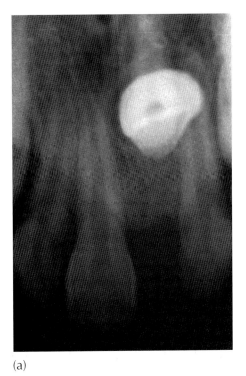

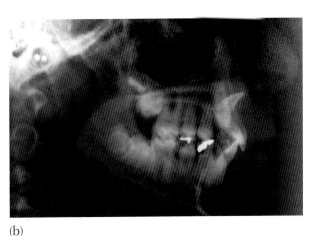

(a)   (b)

Fig. 5.13 (a) The periapical view of a dilacerate central incisor typically shows the coronal portion to be viewed through its long axis, with the labial surface facing superiorly and cingulum area clearly depicted inferiorly. The pulp chamber is depicted in cross-section as a circle in the middle part of the root, while the apical portion points vertically upwards and is visible superior to the labial surface of the crown (the 'scorpion' appearance). (b) The lateral cephalogram shows the lateral profile of the typical 'classic' dilacerate tooth, with its incisal edge at the level of the anterior nasal spine. The palatal aspect of the tooth faces labially and the labial aspect superiorly and posteriorly. The exact configuration of the obviously shortened root is obscured by the superimposition of adjacent teeth.

to create delayed eruption, in order to obviate the need for orthodontic treatment. Very young children are encouraged to first visit the paediatric dentist as early as 3 years of age. The practitioner may perform a minimal radiological examination which includes a pair of bite wings to identify interproximal caries and, often, a maxillary anterior periapical view as part of the initial induction protocol. This film will show the developing central incisor teeth separated by a small midline space, with the midline suture running between them and the lateral incisors overlapping the distal third of the central incisor crowns, on either side. Occasionally, this view will also reveal the existence of a midline supernumerary tooth (mesiodens). As we pointed out at the beginning of this chapter, the presence of the extra tooth in this position will not always cause an eruption disturbance, although the risk is high.

However, it will undoubtedly trigger a round of consultations between the parents and paediatric dentist on the one hand, and an orthodontist and an oral surgeon on the other, in search of an appropriate preventive treatment regimen to avoid later problems. The orthodontist will express the preference for removal of the mesiodens in order to permit the adjacent and unerupted permanent incisor teeth to develop normally, to erupt in their due time without assistance and into their ideal location. The oral surgeon will caution against early surgery because of the possibility of collateral damage being unintentionally inflicted on the dental follicles of the immature permanent incisors, given such a constricted and dentally crowded area from which to pluck the unwanted extra tooth. In addition, this exercise will not be possible to achieve without the simultaneous extraction of the deciduous incisors in order to gain access to the area. If we are to assume that the time is not ripe for the permanent incisors to erupt, then the patient will be anteriorly edentulous for an extended period and healing of the tissues following these deciduous extractions will probably result in a thickened scar tissue which will later obstruct the normal eruption of the permanent incisor teeth, when their time is due.

By and large, the oral surgeon's opinion should take precedence in this scenario, although there are times and situations when the removal should be undertaken, particularly as the child reaches the age when the teeth would normally erupt, as indicated by their root development.

There is, however, no preventive treatment to be recommended once trauma has generated a dilaceration and, regardless of the steady progression of severity of the crown displacement, the patient will need to wait until corrective treatment is appropriate.

When a patient presents with a single central incisor and both lateral incisors erupted, the normal eruption date of the second central incisor will have passed. The impacted tooth will be seen on the periapical radiograph to have at least two-thirds of its eventual root, which is the developmental landmark that determines that a tooth should be erupted. Orthodontic and surgical treatment is, therefore, indicated at that time, both for obstructive impactions and for the dilacerated tooth.

Often, at this early stage one may be able to foresee an obvious need for orthodontic treatment for other reasons, treatment that may not normally be advised until three or four years later. It is not reasonable to delay the resolution of an impacted central incisor for this period of time merely in the interests of trying to achieve a single-phase orthodontic treatment plan in the full permanent dentition. The alignment of the impacted tooth should be undertaken and executed efficiently, avoiding unnecessary attention to other details. Root uprighting and torqueing should be performed as indicated, but an ideal, final and artistic alignment is not the aim at this stage, and extending this first phase of treatment in order to achieve this is contraindicated.

Once appliances are removed following the restoration of normality, there will be natural spontaneous changes in the alignment of these and adjacent teeth during the many months that follow and before the permanent canines come into their place. It must be pointed out to the parent that these changes are expected, normal and not a matter for concern. The parent and patient should be advised that further treatment will most probably be necessary at a later date to treat the rest of the overall malocclusion, and that retention of the aligned incisors against these physiological movements may often not be advised.

In this regard, a further factor has come to light following a clinical study by the Jerusalem group [13]. It will be realized that the advice and treatment proffered in this book are based on many years' clinical experience, with many hundreds of treated impaction cases. Patterns sometimes emerge that would never be noticed under the more usual and more random conditions of orthodontic practice. This is particularly so in relation to impacted central incisors, which are too often treated at a young age by the paediatric dentist, general practitioner or oral surgeon, who rarely follow up their cases to the same degree as do orthodontists.

Impacted central incisors are usually first diagnosed at 7–8 years of age and they are best treated at that time, which is approximately four years before the maxillary canines are due to erupt. It had been noticed that there is frequent disturbance of eruption of the canine on the side of the formerly impacted central incisor, and a large study of unilaterally affected impacted incisor cases was set up to investigate this. Its experimental group involved patients who had been treated for an impacted central incisor and had been under follow-up care until the eruption time of the canines.

On the side where the maxillary incisor had erupted normally, 4.7% of the neighbouring canines showed eruption disturbances or displacement. On the side previously affected by impaction of the central incisor, 41.3% of the

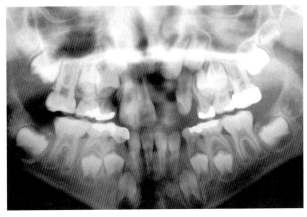

(a)

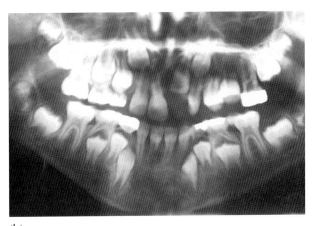

(b)

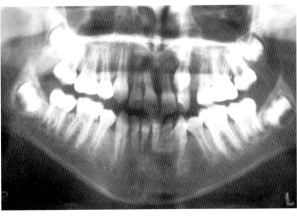

(c)

Fig. 5.14 The development of maxillary canine ectopia adjacent to an impacted central incisor. (a) An 8-year-old boy has an impacted maxillary left central incisor. The contralateral central incisor and both lateral incisors are erupted. The unerupted canines are similarly normally located and the deciduous canines had been extracted earlier. (b) At 9.7 years and without treatment, the left central incisor has improved its position and the right canine has progressed to a limited degree both vertically and slightly mesially. The left canine has also progressed vertically, but it shows considerable mesial migration, superimposing the lateral incisor root. In view of the degree of crowding evident in this case, the three deciduous first molars and all four first premolars were extracted. A short first phase treatment was instituted to expose and erupt the incisor only. (c) At 12.3 years and in common with successful serial extraction procedures, the crowding in three quadrants of the mouth has spontaneously resolved, with the normal eruption of the canines and second premolars. On the side of the resolved incisor impaction, the canine and lateral incisor are in an incomplete transposition relationship, with a strongly buccal and mesial displacement of the canine.

canines showed abnormality, with 30.2% buccally displaced, 9.5% palatally displaced and 1.6% with complete canine–lateral incisor transposition. Half the buccally displaced canines were pseudo-transposed with the adjacent lateral incisor [13] (Figure 5.14).

This study emphasizes the need to thoroughly examine the positions of adjacent teeth, particularly canines, whenever an impacted incisor is present, and to warn parents of the distinct possibility of later canine eruption disturbance. Just as important, it should serve to counsel the orthodontist to make the first phase of treatment as short as possible, without unnecessary attention being paid to meticulous alignment detail at that stage.

Attitudes to treatment

There has been a more or less standard protocol of treatment available for several decades in the orthodontic profession in Europe regarding central incisors that are anatomically normal in their development, but impacted. The recommendations advice is: (a) that adequate space be prepared for the tooth in the arch; and (b) that the cause of the non-eruption (usually a supernumerary tooth) be eliminated. The impacted central incisor teeth may then be expected to erupt spontaneously [14–17].

Studies that have been made of patients who have undergone this type of treatment have shown disappointing results with regard to three important parameters:

1. Non-eruption. Spontaneous eruption has been variously reported as occurring in 54–78% of cases [7, 17, 18, 19], which represents a low degree of reliability.
2. Delay in eruption. Even when eruption occurs, the average time for the affected tooth to make its appearance in the mouth is between 16 and 20 months [7, 17, 19]. This is an unacceptably long period of time, when one considers that the patient will be without a front tooth or teeth for so many months. Additionally, 25% of the patients required two surgical episodes, followed by a waiting period of 2.5–3 years before the tooth erupted! [17]. In a retrospective study of cases in which (a) space had been provided, (b) the supernumerary tooth had been removed, and (c) a stainless steel crown had been cemented to the impacted tooth, eruption was found to have occurred in 96% of the patients. However, this took on average three years! [20].
3. Alignment. The third parameter that showed disappointing results relates to the adequacy of spontaneous alignment. Mitchell and Bennett [17] found that 36% of the teeth in their sample failed to erupt and 41% of the remainder required orthodontic assistance to rectify incomplete results. This means that 62% of the whole sample of cases needed mechano-therapy at that stage. Gardiner [21] also found that spontaneous alignment occurred in only a minority of patients. Obviously, the criteria for deciding what constitutes an acceptable 'alignment' varies from one clinician to another and, one may be permitted to speculate, depends on whether the treatment is being carried out in a community health (managed care) clinic, a hospital orthodontic department (from which the material for these studies was collected) or a private orthodontic practice. Ashkenazi et al. [22], in their more recent study, have found that in 64% of the cases where supernumerary teeth had been removed without other forms of treatment, the impacted incisors failed to erupt and in 9% there was only partial eruption. In 17% of their cases, eruption was successful but into an ectopic location, aggregating to 90% of the cases in need of orthodontic treatment to bring these teeth into acceptable locations.

Some workers [23–25] recognized the need for affirmative action to control (or actively encourage) incisor eruption and they devised methods to perform this using wire loops and pinning, and even advocated the passing of a wire through a drilled hole in the incisal edge, in those early days.

Mills warned against exposure of the crown of the permanent tooth during the procedure to remove the supernumerary tooth, pointing out that periodontal prognosis of the final result would be compromised [16]. Beyond the use of a simple removable appliance to make space in the arch for the unerupted incisor, he displayed a reticence to use mechano-therapy and seemed to have influenced opinion in Britain, where there appears to be a wide consensus that the use of appliances in bringing down impacted central incisors is to be avoided. His reasons for this are as follows:

- these teeth often erupt spontaneously, without help
- loss of labial bony plate
- poor gingival margin, with less attached gingiva
- gingival level discrepancy

Little objective and evidence-based research was offered in support of these contentions, and it is equally open to speculation that most or all of these factors could be the result of overenthusiastic or otherwise poor surgical technique [26, 27]. This is discussed fully in Chapters 3 and 6.

Few children are brought to the orthodontist before 10 or 11 years of age, yet the marred appearance of the child with a single erupted central incisor generally encourages the parent to seek treatment much earlier. An orthodontist's lack of concern for a rapid solution in response to the parent's disquiet for the child's compromised appearance is insensitive, if not callous. This is particularly disturbing in the present context, since simple and effective means of achieving this are freely available.

Treatment of impacted central incisors

Following determination of the overall orthodontic diagnosis, a problem list should be drawn up. Occasionally, there may be just one item on the list, namely the impacted tooth. More often, the presence of mandibular incisor crowding, posterior cross-bite or a class 2 relationship may also be noted, and the clinician must then decide which of these should be treated in this early treatment phase and which left until later. As a general rule, treatment priority should be given to the unerupted incisor, and all other orthodontic procedures will be delayed until the incisor has been brought into alignment. However, an anterior or posterior cross-bite and malalignment of the erupted adjacent incisors will usually be treated at the same time, either because it is simple, of short duration, convenient and advantageous to do so or because their treatment may provide the needed space for the impacted incisor tooth. By

and large, however, most other elements of an existing overall malocclusion will be left until the eruption of the full permanent dentition.

An orthodontic appliance for use in the early mixed dentition

As has been pointed out, opening the space for the unerupted tooth prior to removal of the supernumerary requires some form of orthodontic appliance. The simple removable plate is manifestly unsuitable, since the only positive influence it is capable of producing is a distal tipping of the adjacent teeth. While this creates space in the arch, it does so only at the coronal level. The effect is quite the opposite more apically, where the roots of the tipped teeth will be moved towards one another and directly into the path of the (hopefully) erupting impacted tooth. Vertical force control on the impacted tooth is difficult to achieve with a removable appliance. Corrective rotatory movement of the finally erupted tooth is rarely adequate, while uprighting and torqueing root movements are quite impossible. A practitioner using this method would essentially be placed in the position of an observer, being able to exercise virtually no control over eruption, which is the most important aim of the exercise.

It is true that in the event that the tooth erupts (a) spontaneously, (b) into good alignment and (c) within a reasonably short period of time, there will have been much to gain by this relative inactivity. However, an impacted tooth initially shares the limited labio-lingual width of the alveolar ridge with a supernumerary tooth, which is likely to have prevented its eruption. Sharing this space usually causes the root of the central incisor to become displaced in either or both the labio-lingual and mesio-distal planes of space. It is therefore likely that, in the final analysis, the erupting tooth will require root movement in a labio-lingual (torqueing) and/or a mesio-distal (uprighting) direction. This may be accompanied by the need for significant rotational movement.

It becomes evident that a specific, purpose-designed type of appliance is indicated, one that has the potential to deal efficiently with all these eventualities. Given the significantly wide spectrum of cases where spontaneous eruption cannot be expected to result in a satisfactory alignment or where eruption has failed, it is essential to seek an alternative method of mechano-therapy which provides simple and rapid solutions to all the movements required.

A technique must be employed that provides satisfactory answers to the following four aspects:

1. The appliance should have the capability to level and rotate the erupted incisor teeth rapidly and, with controlled crown and root movements, open adequate space to accommodate the impacted tooth. This space is demanded at both the occlusal level and for the entire length of the roots of the adjacent teeth.

2. The surgical exposure of the crown of the impacted tooth, together with the bonding of an attachment, must be performed in a manner that will achieve a good periodontal prognosis, as described in Chapter 3. Therefore, the appliance must hold the space during and after the surgical procedure, while not hindering the surgeon.

3. Light and controlled extrusive forces must be generated from the appliance to be effective over a long range of movement to bring the tooth down to the occlusal level, and attention must be paid to providing anchorage which is adequate for the purpose.

4. Final detailing of the position of the impacted tooth and its erupted neighbours must be completed without changing to another appliance, including movement of the crowns and roots of each of these teeth in all three planes.

5. While it is important that the appliance has the potential to achieve a good finish, we have argued against aiming for the meticulous placement of teeth in the young child – the pursuit of perfection and a 'final' alignment of the anterior dentition are contraindicated at such an early stage of dental development. This is particularly so when the permanent canines are very high and in close relation with the apical third of the lateral incisor roots.

Typically, our young patient has two molars and three (or fewer) incisor teeth of the erupted maxillary permanent dentition. Strictly and in order to eliminate the long span between the permanent molars and lateral incisors, it is possible to include the deciduous canines and molars into a fully bonded appliance scheme with the use of regular brackets. Because of the anatomy of their buccal surfaces, this presents several difficulties. In the first place, no company presently manufactures bracket bases suitable for deciduous teeth, and bonding brackets designed for permanent premolars and canines will leave large voids and ledges that will have to be filled with composite bonding material. In order to improve the fit of the bases, the buccal surfaces would have to be fashioned to a more accommodating form which, given the short crown height characteristic of these deciduous teeth, would be challenging to achieve satisfactorily. The diminutive crown height of the deciduous teeth restricts the options for determining bracket height. This has repercussions on the subsequent fabrication of archwires, with bends and offsets included in each of the series of archwires to ensure passivity, because there is rarely any need or intention to move them. Accordingly, most practitioners side-step them and band only the molars, with bonded brackets on the incisors, leaving a long span of unsupported wire in the buccal region – the 2 × 4 appliance.

With short inter-bracket distances in the anterior region, a light flexible wire will be needed to perform the initial levelling, aligning and space adjustment. But this wire is completely unsuitable in an unsupported state for the long

span between lateral incisors and the first molars. It is impossible to avoid distortion and there can be no vertically extrusive force directed from the molar tubes, even when the archwire gauge is substantially increased. Anchorage for the extrusive movement of the impacted tooth is therefore poorly exploited and is limited to the adjacent incisors, which progressively intrude as the treatment proceeds. Conversely, using a much heavier archwire to achieve an acceptable degree of support in the posterior region and an extrusive component anteriorly is not compatible with the very specific finer movements needed for the levelling, alignment, rotating, uprighting and torqueing that may be necessary for the initially malaligned incisor teeth.

Johnson's (modified) twin-wire arch

An updated version of the obsolete and largely forgotten Johnson twin-wire appliance [28, 29] presents certain unique features which make it especially suited and efficient in the many treatment aspects of this specific problem in the mixed dentition stage, and particularly regarding anchorage in the vertical plane.

The appliance is based on fixed molar bands, interconnected by a soldered palatal arch (Figure 5.15). Long, narrow gauge (0.020 in internal diameter) tubes slide freely, but accurately and without lateral 'play' in the round molar buccal tubes (0.036 in), and are made to extend anteriorly to the deciduous canine area. An initial anterior, multi-strand, sectional 0.036 in round molar tubes is used. The anterior sectional archwire in this case is a single 0.016 in wire, since only one erupted permanent incisor is present and the usual multi-strand wire is not needed. The alignment of the buccal tubes shows a downward tip as they proceed mesially, to encourage open bite closure and to aid the mechanically assisted eruption of the impacted tooth.

Wire (0.0175 in) or nickel–titanium wire (0.016 in) is held in the long narrow tubes by a friction fit, created by placing three or four bends in the multi-strand wire (more difficult to do effectively with the nickel–titanium wire) and then drawing it through the tube between two pliers, one of which should be grooved. When the appliance was in general use several decades ago, as its name suggests, two fine 0.010 in stainless steel wires were used in the anterior portion to provide improved flexibility and springiness, which today's nickel–titanium and multi-strand stainless steel wires achieve much better. The first step in the construction of the appliance requires that an impression of the dental arch of the patient is taken, with properly adapted, plain molar bands in place. The bands are removed carefully from the teeth and re-set into the impression before pouring.

On the model, a palatal arch is fabricated and soldered to the lingual side of the molar bands to provide resistance to the extrusive forces that will later be needed. In a case requiring more than minimal anchorage, it is advisable to

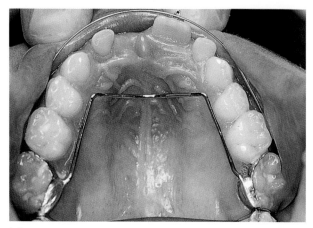

(a)

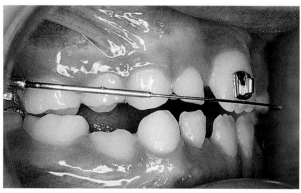

(b)

Fig. 5.15 (a) An occlusal view of Johnson's (modified) twin wire arch, to show the soldered palatal arch. (b) 0.020 in round tube sections are slotted into the 0.036 in round molar tubes. The anterior sectional archwire in this case is a single 0.016 in wire, since only one erupted permanent incisor is present and the usual multi-stranded wire is not needed. The alignment of the buccal tubes shows a downward tip as they proceed mesially, to encourage open bite closure and to aid the mechanically assisted eruption of the impacted tooth. Reprinted from ref. [30], with permission.

add an acrylic (Nance) button to the palatal arch. Buccal tubes are best soldered to ensure precise alignment with the anterior bracket position. Slight upward or downward tipping of the molar tubes to suit the needs of an individual case (Figure 5.13b) will create a significant intrusive or extrusive vertical force component on the anterior teeth. The labial archwire is constructed on the model, initially using the multi-strand or nickel–titanium wire in the buccal tubes.

The appliance is transferred to the mouth and cemented in place, brackets are bonded to the anterior teeth and the prepared initial archwire is placed. The brackets may be of virtually any type, although there are several advantages to Begg brackets in this situation as their vertical slot makes them particularly suited to the light vertical traction that may need to be applied to encourage eruption of the impacted tooth (Figure 5.16).

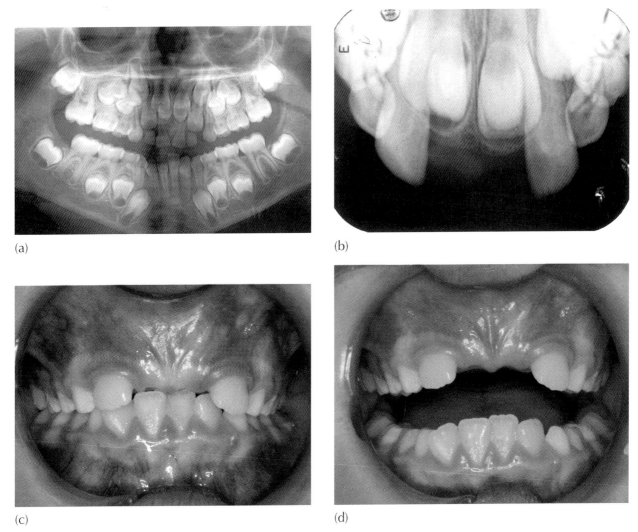

Fig. 5.16 Impacted central incisors due to unerupted supernumerary teeth.

(a) Panoramic view one year pre-treatment shows the permanent lateral incisor well advanced in their eruptive progress. The two supernumerary teeth are superimposed on the unerupted central incisors.

(b) Periapical radiograph taken on the same date.

(c) Intra-oral anterior view shows unerupted lateral incisors and over-erupted mandibular incisors in contact with upper gingiva.

(d) With the teeth apart, the degree of mandibular incisor over-eruption is evident.

(e) Three-dimensional view seen on CBCT image shows the central incisor to be very high, with supernumerary laterally and palatally situated.

(f) A transaxial (vertical) CBCT slice cuts through the central incisor, supernumerary, the lateral incisor and the mandibular lateral incisor in the superior-inferior direction.

(g) The modified Johnson twin-arch 2 × 4 (2 × 2) appliance in place, using Begg brackets on the lateral incisors.

(h) At surgery, the supernumerary teeth are identified on the palatal side and extracted to leave empty sockets.

(i) A minimal area of the labial surface of the incisors is exposed, without removing surrounding bone. Removal of bone inferior to the incisors would have left a deep defect, by including the sockets of the extracted supernumerary teeth.

(j) Attachment bonding.

(k) The lock pins in the Begg brackets are released to permit raising the archwire and ensnarement by the twisted ligatures.

(l) The look pins are re-engaged to apply vertical traction to the incisor teeth.

(m) The full flap is sutured back to its former place.

(n) Close to the time the teeth are erupted, a similar Johnson 2 × 4 appliance is placed in the mandibular dentition to align and intrude the lower incisors.

(o) Anterior intra-oral view, at the end of phase 1.

(p) Note the root of the left lateral incisor is displaced palatally by the labially palpable and unerupted canine. This will be addressed in phase 2 of treatment.

(q) Panoramic view at completion of phase 1.

(r) Periapical view of the central incisors at the end of treatment.

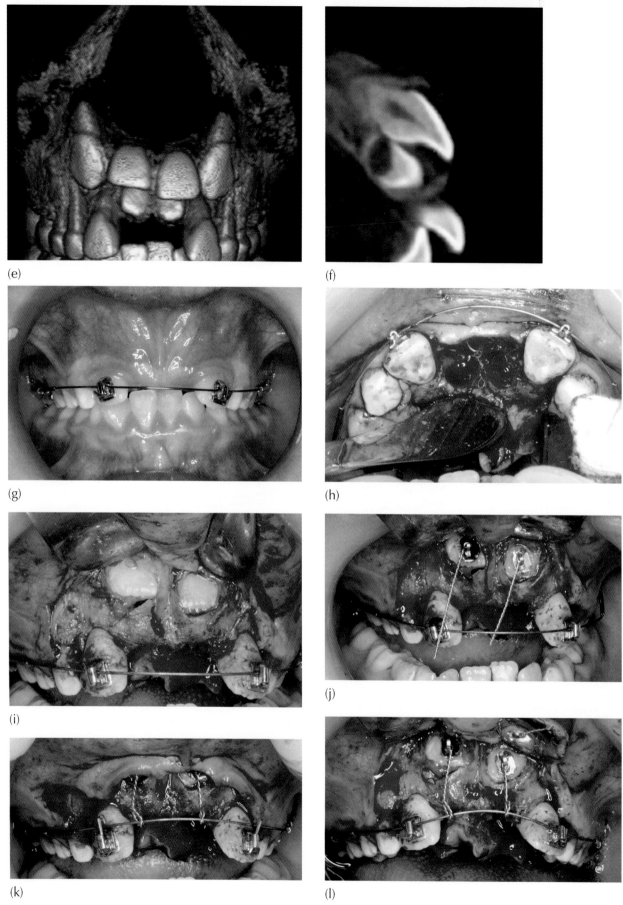

(e)

(f)

(g)

(h)

(i)

(j)

(k)

(l)

Fig. 5.16 (*Continued*)

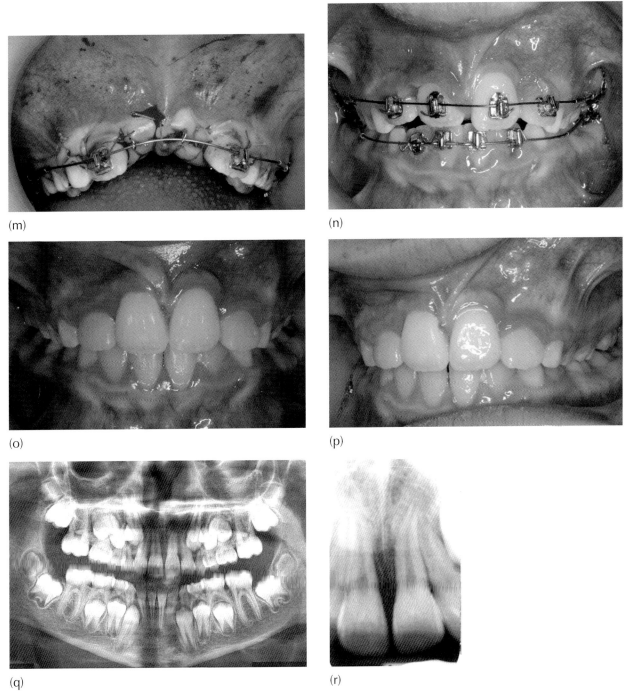

Fig. 5.16 (*Continued*)

By the second or third visit, some weeks later, alignment of the three erupted incisor teeth will usually have been achieved and the multi-strand/nickel–titanium wire is discarded and replaced by a plain round stainless steel 0.018 in or 0.020 in wire, which is similarly inserted in the long narrow gauge buccal tubes. An expanded coil spring is threaded onto it and tied into the anterior portion, compressed between the teeth on either side of the impacted tooth (Figure 5.17).

Space is gained very rapidly, with the movement being at least partly achieved by tipping, depending on which brackets are used. Subsequent uprighting will need to be performed with vertically offset and compensated bends or using auxiliary springs. Once there is adequate room for the unerupted tooth and uprighting has been achieved, a carefully measured piece of stainless steel tubing should be threaded onto the archwire, the length of the tubing being cut so that it fits exactly between the brackets of the teeth adjacent to the impacted tooth. The presence of the tubing between the adjacent teeth ensures maintenance of the required space and contributes to the rigidity of the archwire. This in turn provides a firm platform from which light force may be applied to the unerupted tooth. Together with the rigid buccal tube arms and the soldered palatal arch, this constitutes a very significant anchorage unit in this biomechanical system.

At this point the patient is ready for the surgical episode, at which an over-retained deciduous incisor and any buried supernumerary teeth are removed.

The permanent incisor is exposed and an attachment placed, without removing any part of the orthodontic appliance.

Erupting the impacted tooth ideally requires relatively little force, but a mechanism with a long range of action. As we have discussed in Chapter 4, directly ligating with elastic thread between the twisted pigtail ligature that was tied into the eyelet at the time of surgery and the main archwire is a very poor method, since force is high and uncontrolled and its range minimal. A more appropriate system is to stretch an elastic chain horizontally between the brackets on the teeth on either side of the affected tooth and to raise the middle of the chain and ensnare it over the rolled-up pigtail ligature. This produces a much more controllable force with a good range of movement.

If the patient is seen every 3 to 4 weeks for adjustment, a previously obstructed tooth will usually erupt very rapidly and within a few months will be visible, providing the young patient with a more acceptable appearance which will enhance his self-confidence and self-image. The favourable time factor achieved by this approach is an advantage that should not be underestimated and one that needs to be taken into account, even when treating the younger patient. An obstructed incisor will take less time

to resolve than a dilacerate tooth, simply because it has a shorter distance to travel and considerably less root movement to experience, but the treatment principles are the same.

At the point when the tooth reaches the occlusal level, a reassessment is made as to whether uprighting, torqueing or rotation of the tooth is needed, and whether it should be done at this stage or in the second phase of treatment, assuming that this will be necessary. If so, the eyelet is removed and a bracket similar to that on the other teeth is placed in its ideal position. Finishing is then achieved in the appropriate manner. In the dilacerate incisor case illustrated in Figure 5.17, the final photographs were taken at the end of phase 2 of treatment, at age 13 years and after considerable labial root torque had been achieved, both before and following the apicoectomy and root canal treatment.

The construction, placement and activation procedure of the appliance is very simple and does not require a high level of expertise. The laboratory stage of appliance construction requires accurate soldering of the palatal arch and careful alignment of the buccal tubes, which a good orthodontic technician should master very quickly, although the orthodontist may prefer to do this himself. In the mouth, cementation of the bands and bonding of the brackets is routine, and the application of the prepared archwire presents a neat and robust appliance in the long span between the molar tubes and the incisor brackets, while anteriorly providing light and gentle vertically directed forces of good range, to give rapid results. The presence of the palatal arch will ensure that undesired movement of the adjacent incisors and the distant anchor molars cannot occur.

In common with other patients with unerupted incisors, the patient with arrested root development will usually seek advice and treatment before the age of 9 years. By contrast in these cases, however, more than one and occasionally all the maxillary incisors may have been affected, since the condition is usually the result of a severe blow to the area at a very young age (Figure 5.18).

While the principles of treatment are the same as for the patients we have already discussed, it is clear that in the most severe of cases, when all four incisors are affected, a different labial arch mechanism is needed. The same basic appliance, with palatal arch soldered to molar bands, is cemented into place before the surgical exposure is performed. This type of case is much more demanding on molar anchorage, and an acrylic (Nance) button on the palatal arch should be considered mandatory.

All four impacted incisors are exposed and eyelets bonded to them, with the pigtail ligatures trailing through the sutured edges of the flap. In this case, a heavy buccal archwire of 0.036 in gauge is slotted into the round tubes on the molar bands, and is held at a minimal distance labial

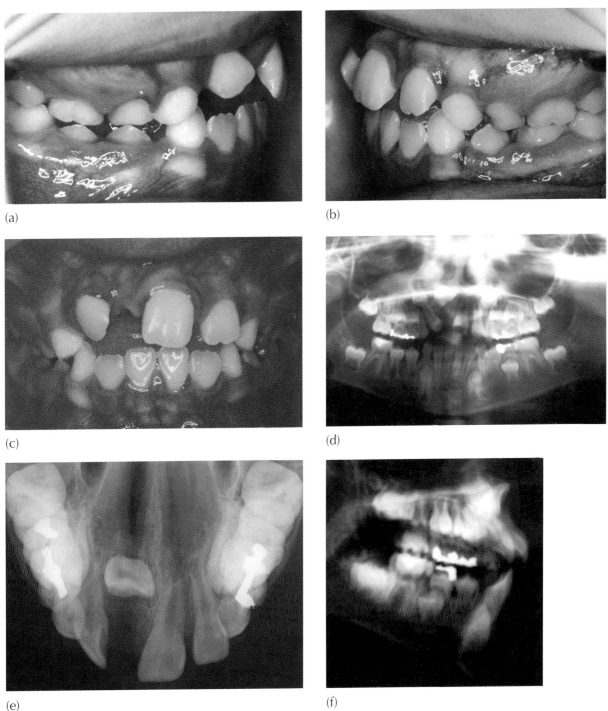

(a)

(b)

(c)

(d)

(e)

(f)

Fig. 5.17 An 8-year-old female with a dilacerate maxillary incisor and considerable caries experience treated entirely by postgraduate students. (a–c) Clinical intra-oral views of the dentition. Space has been lost due to tipping of the adjacent incisors. (d–f) Panoramic, occlusal and lateral cephalometric radiographs show the typical configuration and location of the affected tooth, adjacent to the anterior nasal spine. (g–i) A modified Johnson's twin-wire arch has achieved a reopening of the space with an open coil spring. The space opening has been deliberately exaggerated and the spring is subsequently replaced by a measured and contoured steel tube to act as a space maintainer. This tube and the buccal tube arms, together with the rigid soldered palatal arch, constitute the very significant anchorage unit in this biomechanical system. (j–n) In preparation for force application, a T-pin is inserted from the occlusal into the vertical slot of the bracket of the two adjacent incisors. The surgical flap is taken from the crest of the ridge in order for it to include attached gingiva and is reflected high into the sulcus to expose only the palatal aspect of the incisor, which faces forwards. An eyelet attachment is bonded and the flap fully sutured to its former place, with the pigtail ligature exiting through its sutured edge and turned upwards into the form of a hook. The mid-point of a clear (and barely visible) elastomeric module, stretched between the two T-pins, is gently raised and ensnared in the hook, to provide an easily measurable, light traction force to the tooth. (o, p) Occasionally, as was seen in this case, providing the tooth with adequate torque will bring its forward-facing root apex through the labial plate of bone, to become excessively prominent in the sulcus. Radiographs of the completed treatment show the affected tooth after elective root treatment and apicoectomy. (q–u) The final treatment outcome of the overall malocclusion. The clinical crown is slightly longer than that of its antimere and is the only clue to identifying the affected tooth. There were no differences between the two teeth regarding any other periodontal parameters. A removable retainer is worn at night.

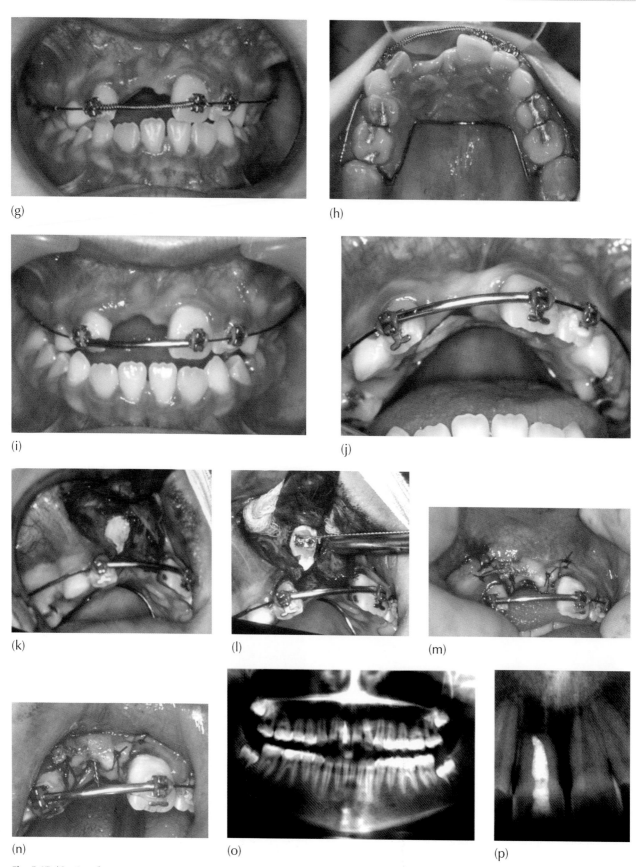

(g)

(h)

(i)

(j)

(k)

(l)

(m)

(n)

(o)

(p)

Fig. 5.17 *(Continued)*

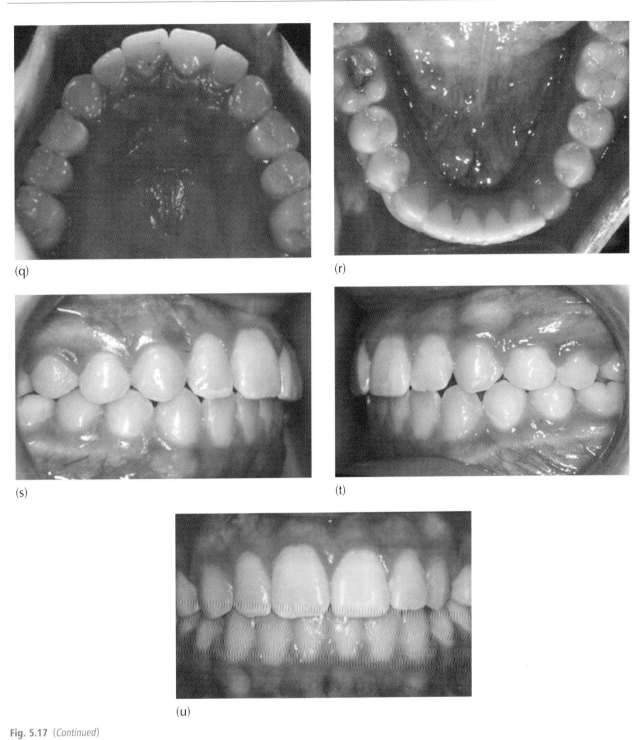

(q)

(r)

(s)

(t)

(u)

Fig. 5.17 (*Continued*)

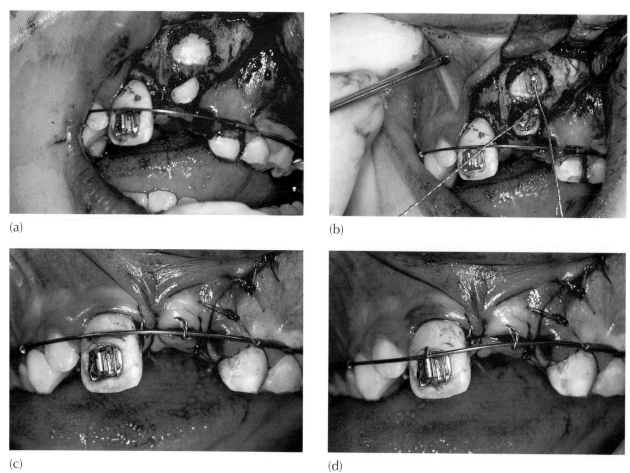

(a)

(b)

(c)

(d)

Fig. 5.18 The initial diagnostic radiographs and computerized tomographic imaging of this patient, at age 7.4 years, were presented in Figure 5.10. This figure displays the surgical exposure of the affected central incisor showing its extreme height, with the lateral incisor inferior to it and directly in its potential eruption path. (a–d) Surgical exposure displays the affected central incisor to be very high and the lateral incisor inferior to it and directly in its potential eruption path. Eyelets are bonded and the pigtails separated and labelled for subsequent identification. The surgical flap is fully replaced and only the pigtail of the lateral incisor attachment ensnared into the archwire of the modified Johnson's twin arch, which is initially free of its engagement in the sole existing bracket, on the only fully erupted incisor. By pinning the archwire back into the vertical slot of the Begg bracket, extrusive force is applied to the lateral incisor. The pigtail from the central incisor lies freely unattached. (e, f) Periapical and tangential views show the enormous height discrepancy at the time of exposure. (g–j) Clinical views showing the erupted lateral incisor and the use of a coil spring to move it distally, reopening space for the central incisor, to which traction is then applied. (k) Once erupted, the central incisor was torqued and uprighted, using auxiliary springs with the very lightest of forces, suitable to the severely reduced root length. In contrast to using 'straight wire' mechanics, with tip and torque built into the bracket angulation, the force exerted by these auxiliaries (when placed in Begg or Tip-Edge® brackets) can be easily measured, controlled and tailored to the needs of the moment. (l–n) Completion of the first phase has provided the child with a front tooth 14 months later, at age 8.6 years. Note the periodontal condition on the labial side of this tooth, which is due to the coronally sited junction of the incompletely formed crown and its right-angle dilaceration with the very short root. No retainers were used subsequently. (o–q) At age 11.1, the full permanent dentition has established and there is slight positional deterioration in the alignment of the incisor while its periodontal condition remains unchanged. The present aim is to maintain the tooth until it can be replaced by an implant, and to delay second phase orthodontic treatment until closer to that time. (r–u) the patient seen at 18 years in these intra-oral frontal and occlusal views, with new periapical and panoramic views of the dentition still showing the trauma-driven developmental cervical defect. Alveolar bone height is good and she has now been referred for an elective replacement of the left central incisor with an implant borne restoration.

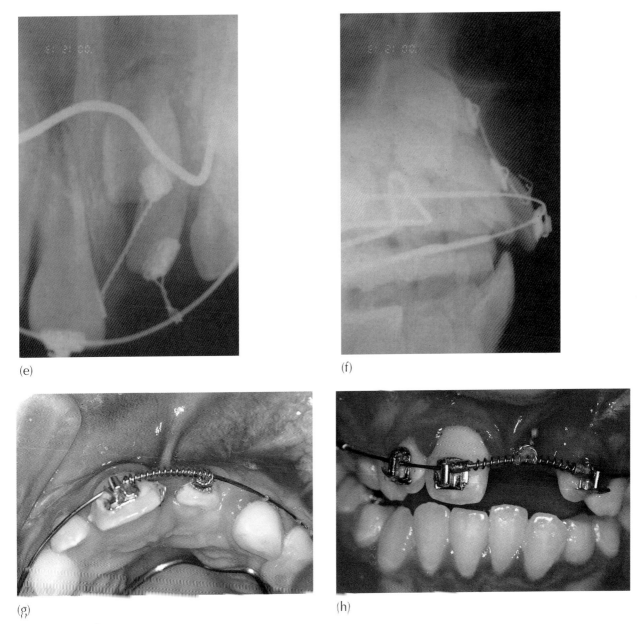

(e)

(f)

(g)

(h)

Fig. 5.18 (*Continued*)

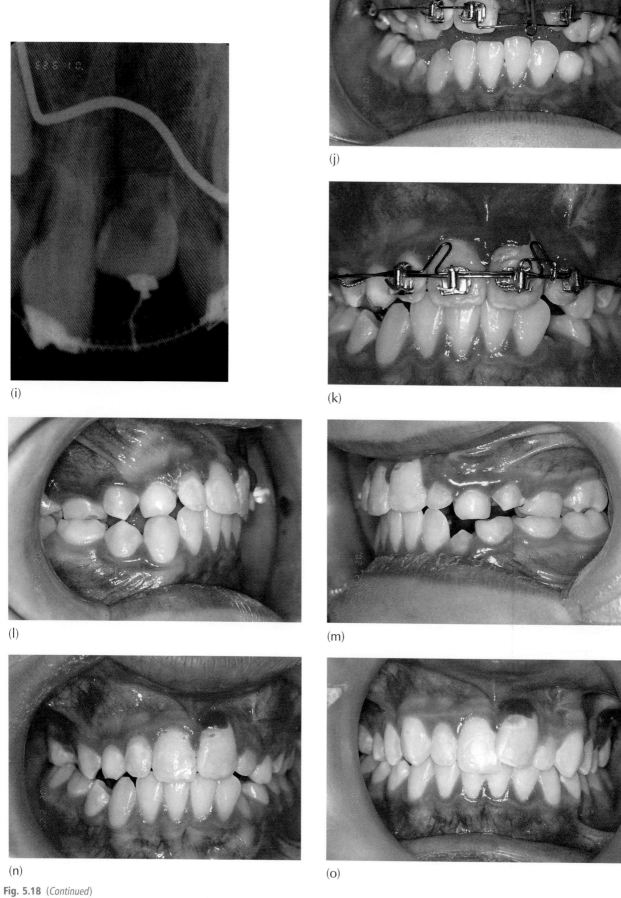

(i)

(j)

(k)

(l)

(m)

(n)

(o)

Fig. 5.18 (*Continued*)

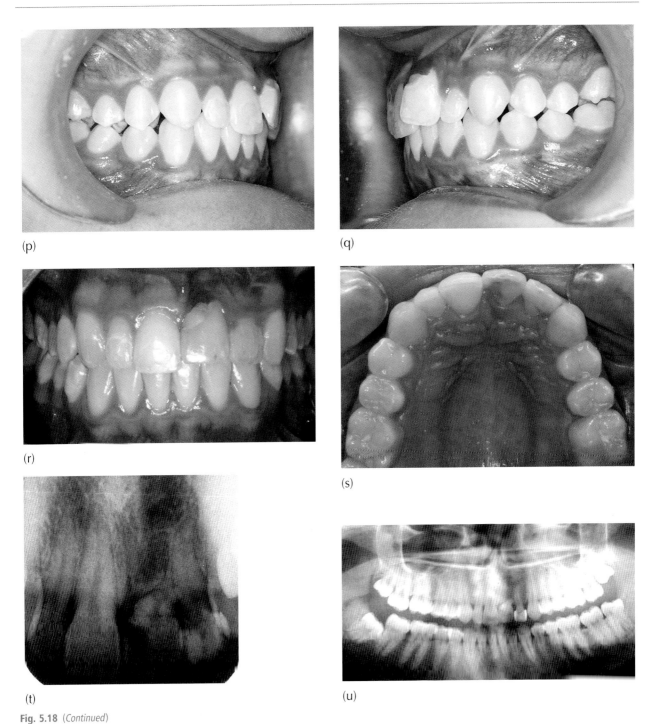

(p)

(q)

(r)

(s)

(t)

(u)

Fig. 5.18 *(Continued)*

to the line of the arch with bayonet bend stops at the molar tubes (Figure 5.19). The anterior portion of the heavy labial arch is raised with gentle finger pressure and ensnared by the four pigtail ligatures, thereby applying a gentle extrusive force to the four unerupted teeth.

At the completion of this phase of treatment, and for all of the types of problem described here, it is good practice to reassess the overall orthodontic condition. Further treatment may be advised at this stage, as mentioned earlier. More frequently, however, the appliance is removed and the

patient is placed on recall over a period of several years, until the eruption of the full permanent dentition, when a new clinical assessment is made and an overall treatment plan developed for the entire dentition. In the meantime, the night-time wear of a simple removable retainer is usually advised in order to hold the achieved alignment. Delaying further treatment affords the orthodontist the opportunity to monitor the survival and progress of the tooth/teeth and the ability to predict its/their long-term prognosis with greater reliability. This is a wise precaution

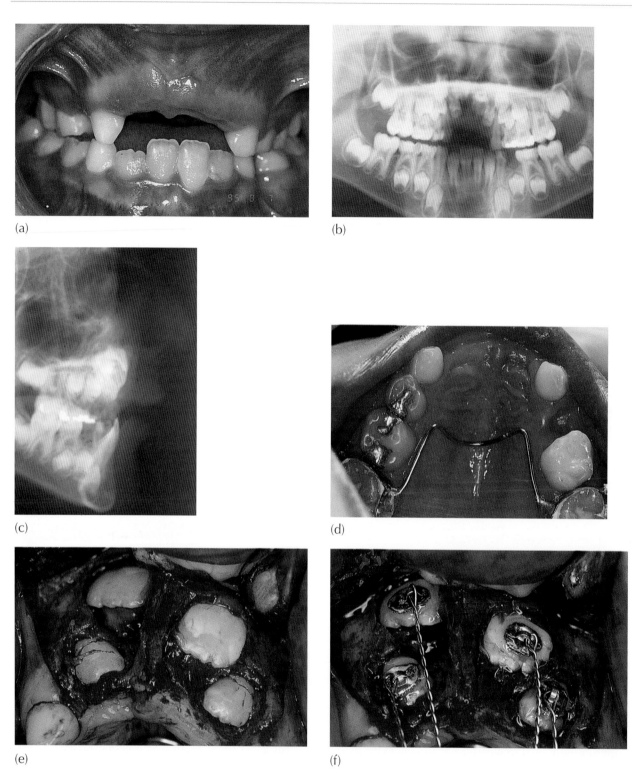

(a)

(b)

(c)

(d)

(e)

(f)

Fig. 5.19 (a) This 9-year-old child has lost alveolar bone height following traumatic avulsion of his anterior deciduous teeth at age 2 years in an accident. (b, c) The panoramic and lateral cephalometric films show very little root development of all maxillary incisors. The central incisors are at the level of the anterior nasal spine. (d) View of soldered palatal arch. (e) The incisors are surgically exposed (note the large areas of hypoplastic enamel). (f) Eyelet attachments are bonded. (g) After suturing, a self-supporting 'stopped' labial archwire is placed into the molar tubes. Displaced superiorly by gentle finger pressure, the anterior part of the archwire is ensnared by all four steel pigtails, to deliver extrusive force. (h, i) Tangential and anterior occlusal radiographic views immediately post-surgery. (j) A clinical view at completion of the first phase of treatment. (k, l) Periapical and tangential radiographs at completion of treatment. Note poor development, anatomic form and life expectancy of the teeth at the end of treatment. Uprighting of the roots was considered inappropriate.

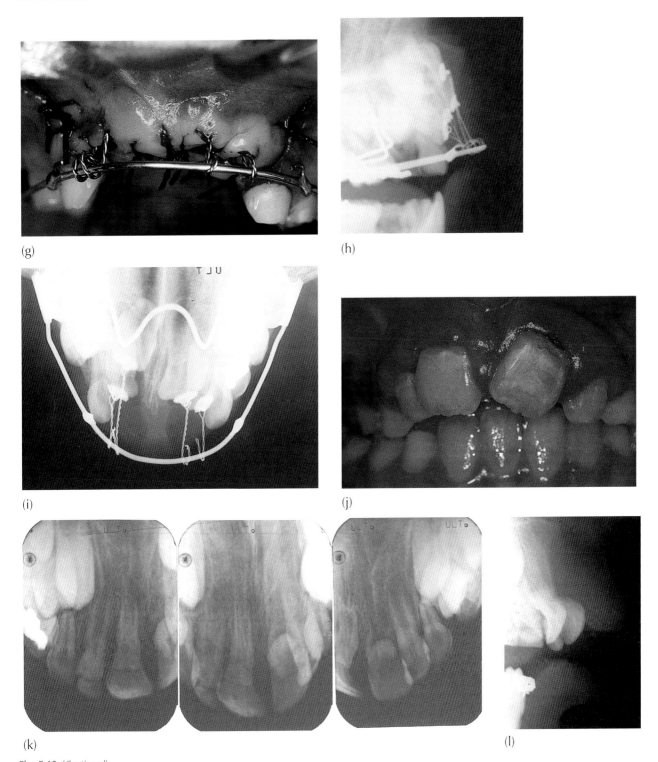

(g)

(h)

(i)

(j)

(k)

(l)

Fig. 5.19 (Continued)

before committing to overall orthodontic treatment that may often require irreversible steps to be taken, such as extractions.

Prognosis

The obstructed impaction

The prognosis of the result depends on several factors.

Root length

For teeth whose impaction is due to obstruction, the root length is usually normal, although in some cases it may be somewhat shorter because of the cramped circumstances in which the root developed prior to the treatment. Nevertheless, mild shortening will not normally adversely affect prognosis.

Surgical exposure

Overly radical exposure, taken down to and sometimes beyond the CEJ, with complete removal of the follicle is an unnecessary and harmful step which will inevitably have a long-term debilitating effect on the periodontium. Even though the eruption of the impacted tooth will bring with it new alveolar bone, this will not be sufficient to establish a bone level similar to that of the adjacent teeth and there will be an increased crown length relative to its unaffected and normally erupted antimere.

Type and height of periodontal attachment

If the surgeon opened a window in the oral mucosa directly over the impacted tooth, above the attached gingiva, then the prognosis of the result will be relatively poorer. Steps should be taken to properly manage the muco-gingival soft tissue by raising a full flap from the crest of the ridge at the time of the surgical exposure and by fully suturing the wound, at the end, wherever appropriate. If this is done, the tooth will be erupted through the attached gingiva and the periodontal result will be good.

It has already been pointed out that in cases of obstructive incisor impaction the long axis of the unerupted tooth is abnormally oriented due to the presence of the extraneous displacing hard tissue. The mesiodens or odontome are usually palatally placed and the root of the incisor consequently displaced labially. Thus, when the tooth is drawn vertically downward following a closed exposure procedure, it will tend to break through the oral mucosa above the attached gingiva and the newly erupted incisor will exhibit a long clinical crown because of the relative labially prominent cervical region and a thin and delicate mucosal attachment.

In order to avoid this unfortunate sequela, traction may be applied in a downward and palatal direction in an attempt to draw it though attached gingiva. However, this will concurrently tip its root labially, thereby increasing the amount of subsequent root torque that will be necessary. A better approach is to draw the tooth down until it bulges the labial oral mucosa and then to pre-empt its eruption by performing an apically repositioned flap at this late stage. The flap is raised from the crest of the ridge immediately inferior to the unerupted tooth and this formerly attached gingival tissue is sutured over the labial (superior) aspect of the tooth. Further traction will bring the tooth down and into alignment with the repositioned tissue. Performing this procedure as the initial exposure may be preferred if the vertical displacement of the tooth is relatively mild but, for a higher displacement, the two-stage procedure described here will produce a much better outcome.

A dilacerate incisor will require a similar surgical initiative. A closed exposure procedure is probably the only reasonable surgical approach to take in these cases in the first instance, because of its extreme ectopic location in the root of the nose. As the tooth is drawn vertically downwards, its incisal edge bulges the oral mucosa covering the labial side of the alveolar ridge and the new horizontal orientation of its crown may be clearly outlined through the almost transparent mucosa. Here, too, one should wait until the tooth is well down and close to eruption before performing a similar apically repositioned flap. In this case in particular, the tightness of the suturing of the tissue above the crown of the tooth will exert considerable pressure to assist in the further traction of the tooth into its place. When the crown has been brought down to occlusal level, the extreme degree of labial root torque that is needed to properly and artistically reposition the crown of the tooth will become apparent.

The question then arises as to how much of this root torque should be performed in the very young patient and how much should be left until the phase 2 of the treatment in the permanent dentition. It is quite clear that any appreciable torque will tip the curved root end forward causing it to protrude labially and to alter the contour of the mucosa covering the alveolar process. Yet for most cases this will still not be enough to adequately reorient the crown. Nevertheless, in the early mixed dentition this should be accepted as representing the completion of phase 1. The appliances should be removed and retainers placed. Over the next year or so, there may be some relapse of the achieved root torque. This is due to the final and natural apexification of the root end being resisted by the labial periosteum, which causes a reversal movement affecting the tooth axis.

It becomes apparent therefore that, from the orthodontic viewpoint, the problem is considerably more challenging.

In the young patient, a careful study of the tangential radiograph may reveal the shape and orientation of the root, although this will be much more accurately assessed on a computerized tomography (CT) series. The more apical the dilaceration *vis-à-vis* the coronal third of the root, the better the prognosis. Similarly, if the dilaceration is in the crown of the tooth, the prognosis improves the closer it is to the incisal edge.

A dilaceration situated apical to the coronal third of the root provides no serious impediment to orthodontic alignment and it should have an excellent prognosis. The crown of the tooth should be surgically exposed and an eyelet bonded to it. In most cases, the only surface of the crown that is available for the eyelet is the palatal surface. This surface faces the operator when the tooth is first exposed, while the anatomically labial surface of the tooth is inaccessible, facing superiorly and posteriorly and buried in the hard tissue adjacent to the anterior nasal spine. A stainless steel ligature is threaded through the eyelet, twisted into a pigtail and drawn downwards, to be ligated with an upward stretched horizontal elastomeric module, as described above and in Chapter 4. Care should be taken not to apply much pressure initially, since the tension introduced by suturing the surgical flap may itself apply a considerable downward force to the tooth in the first few weeks.

As the crown of the tooth responds to the force it rotates downwards, causing the incisal edge of the tooth to become labially prominent, outlined beneath the oral mucosa on the labial side of the alveolar ridge. At the same time, the root apex rotates forward towards the labial plate of bone. For the most part, these cases may be completed without the root apex ever protruding excessively in the labial sulcus.

If the apex becomes prominent and further labial root movement is still desirable, this will necessitate a surgical intervention, performed with the aim of amputating the root apex of this tooth. The labially directed portion of the root, corresponding to the post-traumatic developmental portion, is sectioned in a line that is continuous with the labial side of the main coronal portion of the root of the tooth. The pulp is extirpated and the root canal is obliterated using a combined conventional (coronal) and a surgical retrograde endodontic approach, wherever possible (Figure 5.17). Further extrusion and appropriate root torqueing of the incisor may necessitate a second root-shortening procedure, but, if the first procedure was delayed as late as possible and the root apex amputation was properly designed, this may usually be avoided. This is most advantageously done in the final stage of treatment in the full permanent dentition, at the dental age of 12–13 years. Further root torque will need to be initiated until the crown orientation is ideal, at which point endodontic treatment and apicoectomy of the prominent root will

need to be performed. In these cases, the outcome will display a shortened root, but the normal crown and soft tissue appearance should be excellent and largely indistinguishable from the adjacent incisor. The degree by which the final prognosis of the short-rooted central incisor will be compromised depends largely on how much root remains after the amputation has been performed. The desired site of the amputation is entirely dependent on the location of the dilaceration, and it eliminates a majority of that portion of the root apical to it, which had developed after the traumatic episode. Thus, the closer the dilaceration is to the coronal portion of the root, the shorter will be the final root length at the completion of treatment and the poorer its prognosis.

Treatment duration

As we have seen, treatment of impacted central incisor teeth is generally undertaken in the early to middle period of the mixed dentition, which means that it is classed as a phase 1 treatment, with its aims usually limited to resolution of the incisor impacted. Phase 2 becomes appropriate only in the fully erupted permanent dentition, at around the age of 12–13 years. Naturally, the later this treatment is embarked upon, the greater is the likelihood that the two phases will be incorporated into one long period of comprehensive biomechanical therapy.

In an unpublished study [31], the duration and the rate of success of this phase 1 treatment was evaluated in relation to both obstructed and classic dilacerate impacted incisors. The study sample included 59 patients, 31 associated with supernumerary or odontome obstruction and 28 with dilaceration. Three stages of treatment were defined:

T1 – the pre-surgical period between the application of fixed orthodontic appliances and the referral for surgery

T2 – the period between the surgical exposure and the engagement of the orthodontic bracket of the impacted tooth in the labial archwire

T3 – the period from the engagement in the archwire to completion of phase 1.

For the group as a whole, T1 was found to be 5.2 months, with a range of ±4 months, the impaction resolution period T2 took 8 months (±5 months) and the T3 to completion was 6.3 months (±4.5 months). While each stage may individually not appear very long, in total they amounted to 19.5 months (±9 months), which is a considerable period. Taken individually, the duration of treatment in the obstructed cases was 17 months (±7 months) and that of the dilacerate cases 22 months (±17 months), largely due to the greater complexity of treatment of the dilacerate

incisors, particularly in the T3 period, where much labial root torque needs to be achieved. This is also reflected in the success rate of the two constituent diagnoses, with one failure (97%) in the obstructed group and five (82%) in the dilacerate group.

Relative bone height of the crestal alveolus

In Chapter 6, we will refer to the fact that when teeth are supra-erupted, their vertical movement is accompanied by a vertical increase in their supporting alveolar bone. Thus, when the impaction of a tooth is resolved by augmenting the natural eruptive force following the removal of the causative agent, it will be seen that the bone support of that tooth will be greater than that around normally erupting adjacent teeth [27, 32–35]. However, this positive response on the part of the alveolar bone to the extrusive forces is dependent on the amount of pressure applied being within relatively narrow limits. A periapical radiograph taken at that time will show radiolucent areas where new bone is being laid down – new bone does not show up on X-ray. A similar view taken 4–6 months after cessation of this movement, when the bone will have matured and calcified, will show the excellence of its regeneration

In the event that excessive extrusive force is brought to bear on these teeth, eruption will also occur rapidly, but without a regeneration of alveolar bone. The result will be characterized by the tooth having a long clinical crown and considerable mobility. The periapical radiograph performed even six months later in this case will show a much reduced bone level around the newly and apparently successfully resolved impaction. The prognosis of such a tooth will be impaired.

Preservation of vitality

During the surgical procedure, removal of awkwardly placed supernumerary teeth may lead to an unavoidable devitalization of the impacted tooth. In the hands of a competent surgeon this is quite rare and it is more likely that excessive extrusive force, which compromises its bony support, may be a factor in bringing about the demise of its pulpal tissue.

Oral hygiene

During the initial phase of eruption of an impacted tooth, the surrounding gingiva is sensitive, tender and bleeds very easily. This will usually make the younger patient very apprehensive of brushing the area regularly and to an adequate standard of cleanliness. Secondary inflammation of the gingiva and a concurrent adverse effect on the regeneration of bone will be the inevitable result [36].

For all these reasons, the surgeon should be meticulous in the proper planning and execution of the surgical technique and the orthodontist in the application of extrusive forces whose magnitude is difficult to control. Elastic ligation thread is used widely for applying traction to impacted teeth, by tying it directly and tightly between the attachment and a relatively rigid archwire. It is exceptionally difficult to accurately judge the amount of force being applied by this method and so, when it is the only practical one available, great care should be taken not to tie too tightly. Wherever possible, alternative methods should be used, as described in Chapter 4.

As the impacted teeth respond to the traction force, the orthodontist should be perceptive, aware and responsive to their changing relationship to the soft tissues, altering force directions as necessary and, if needed, requesting mucogingival surgery to cause the tooth to exit the tissues in the best possible location from both the periodontal and aesthetic points of view.

Vertical 'box' elastics are often used in routine cases in orthodontics to enhance intercuspation at the close of treatment. These elastics are small and can produce forces which are very much in excess of that desirable for a single impacted tooth. This will be increased further during mouth opening. It is difficult to measure or control the forces applied in this way. Nevertheless, this is a valuable tool and should be used with only the very lightest and largest of elastics. It should also be remembered that, as orthodontists, we are apt to apply more than adequate extrusive forces by the downward deflection of an archwire and we then place a 'box' elastic, in addition, just to be sure! The aggregated force may thus become very much in excess of the physiological limit. The instructions to the patient who is prescribed this form of vertical intermaxillary anchorage reinforcement should insist on full time wear, including at meal times and removed only for toothbrushing, in order to maintain a light continuous force.

Impaction due to past trauma

Soft tissue obstruction

When deciduous incisor teeth have been lost early and well ahead of the time that the incisors are ready to erupt, the mucosa will generally heal over with a thickened attached gingival tissue, which is often fairly fibrous. This tissue may then act as a barrier, preventing the normal and timely eruption of the permanent incisors, although they are at the appropriate developmental stage for eruption. The only treatment needed to overcome the resistance of the thickened mucosa to eruption is surgical, involving an incision in the thickened mucosa, enough to expose the incisal edges of the already bulging crowns of the teeth. This is described in Chapter 3 and the prognosis is usually excellent.

The benefits of salvaging severely compromised impacted teeth

It is clear that the long-term prognosis of many dilacerated teeth and teeth with arrested root development is poor, and their extraction and replacement may be a part of the long-term treatment strategy, as noted in the treatments described in the case reports above (Figures 5.17–5.19). This being so, the most pertinent question that needs to be asked regarding these teeth is whether there is any justification to expend energy on treating them at all. The answer is not necessarily the same for every case, and each must be considered on its own merits. No decision should be made until the practitioner first considers the following points in relation to the particular patient concerned:

1. A permanent artificial solution cannot be considered much before early adulthood, whether by conventional prosthodontic treatment or implant-borne restoration.
2. Any tissue-borne form of temporary replacement (partial 'flipper' denture) will be far less satisfactory from every point of view. Even when this solution is acceptable to the patient, long-term wear will lead to deterioration in the health of the palatal mucosa and of the gingiva adjacent to any teeth which are in contact with the removable denture – not to mention the possibility of enamel caries of the teeth concerned.
3. Any tooth-borne, resin-bonded (Maryland) bridge replacement may require some preparation of the adjacent teeth and may be unreliable in the long term [37]. It is unlikely that this restoration will remain for more than a few years, and an implant-borne permanent substitute will be preferred in early adulthood, which means that the earlier invasive tooth preparation will need to be reversed by restorative treatment.
4. Should it be decided to extract the dilacerate tooth, the already deficient alveolar ridge area will become even more deficient both vertically and in its labio lingual width, making the case unsuitable for an implant and unaesthetic for a conventional bridge.
5. Orthodontic alignment of the dilacerate tooth will bring with it much alveolar bone to enhance both ridge width and vertical height to normal dimensions.
6. The retention of even a very short-rooted tooth will preserve the normal shape and architecture of the alveolar ridge.
7. The original trauma in these cases may have caused the teeth to develop close to the nasal floor and this may have been the cause of stunting of the root growth and loss of their eruptive potential. When these teeth are drawn down towards the line of the arch and assuming they still have open apices, they will often rapidly develop long spindly roots, which may improve their prognosis (Figure 5.19). With suitable restorative enhancement of their crowns, these rescued teeth may then usually be maintained well into the second and sometimes third decade of life.

Under these conditions, orthodontic alignment of the impacted and dilacerated tooth will usually be preferable. For the most part, however, this procedure must be viewed as providing only a temporary solution and, at a later time, when growth has ceased and conditions are more favourable, some form of permanent restoration will need to be considered. In the meantime, however, this decision will have been made much easier and will include a wider choice of prosthodontic modality options, and the result will be much more satisfactory in the long term because of the enhancement of the bony ridge that will have accompanied the eruption and retention of the damaged tooth.

Apical root dilacerations

In the very young patient, treatment of the dilacerated incisor follows much the same lines as described for the obstructed incisor. Before this is done, however, a careful study of the tangential radiograph may reveal the shape and orientation of the root, although this will be much more accurately assessed on a CT series – the more apical the dilaceration vis-à-vis the coronal third of the root, the better the prognosis. Similarly, if the dilaceration is in the crown of the tooth, the prognosis improves the closer it is to the incisal edge.

A dilaceration situated apical to the coronal third of the root provides no serious impediment to orthodontic alignment and it should have an excellent prognosis. The crown of the tooth should be surgically exposed and an eyelet bonded to it. In most cases, the only surface of the crown available for the eyelet is the palatal surface. This surface faces the operator when the tooth is first exposed, while the anatomically labial surface of the tooth is inaccessible, facing superiorly and posteriorly and buried in the hard tissue adjacent to the anterior nasal spine.

A stainless steel ligature is threaded through the eyelet, twisted into a pigtail and is drawn downwards, to be ligated with an upward stretched horizontal elastomeric module, as described above and in Chapter 4. Care should be taken not to apply much pressure initially, since the tension introduced by suturing the surgical flap may itself apply a considerable downward force to the tooth, in the first few weeks.

As the crown of the tooth responds to the force it rotates downwards, causing the incisal edge of the tooth to become labially prominent, outlined beneath the oral mucosa on the labial side of the alveolar ridge. At the same time, the root apex rotates forward towards the labial plate of bone. For the most part, these cases may be completed without the root apex ever protruding excessively in the labial sulcus.

If the apex becomes prominent and further labial root movement is still desirable, this will necessitate a surgical

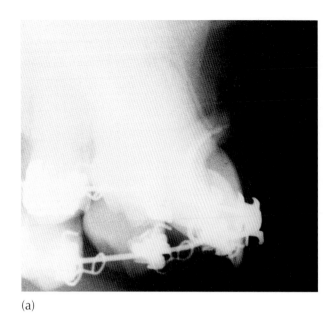

(a)

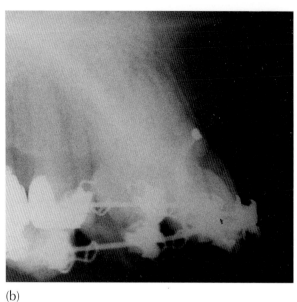

(b)

Fig. 5.20 Tangential views of the completed alignment of a dilacerate tooth with root canal filling (a) prior to and (b) after apicoectomy and retrograde amalgam filling. (Courtesy of Professor I. Heling and Dr M. Morag.) This is the same case as in Figure 5.17.

intervention, performed with the aim of amputating the root apex of this tooth. The labially-directed portion of the root, corresponding to the post-traumatic developmental portion, is sectioned in a line which is continuous with the labial side of the main coronal portion of the root of the tooth. The pulp is extirpated and the root canal is obliterated using a combined conventional (coronal) and a surgical retrograde endodontic approach, wherever possible (Figure 5.20). Further extrusion and appropriate root torqueing of the incisor may necessitate a second root-shortening procedure, but, if the first procedure was delayed as late as possible and the root apex amputation was properly designed, this may usually be avoided.

The degree by which the final prognosis of the short-rooted central incisor will be compromised depends largely on how much root remains after the amputation has been performed. The site of the amputation is entirely dependent on the location of the dilaceration and eliminates a majority of that portion of the root apical to it, which had developed after the traumatic episode. Thus, the closer the dilaceration is to the coronal portion of the root, the shorter will be the final root length at the completion of treatment and the poorer its prognosis.

Crown dilaceration

Dilaceration may also occur in the crown portion of the tooth, although this is an extremely unusual sequel to trauma. In this eventuality, the tooth itself may be mildly labially displaced. When the crown is surgically exposed, an attachment is best placed on the labial surface, which is

almost certainly accessible. In this way, the continued downward-directed orthodontic traction will bring the root portion of the tooth from its more palatally displaced location close to its normal position and its long axis within normal limits. This will be due to the lingual tipping effect of the attachment on the labial surface. The tooth will erupt with the more incisal section of its crown labially tipped and the post-traumatic section in an acceptable angulation.

Restorative treatment is indicated after grinding off that portion of the crown that represents the pre-trauma dental developmental portion. The future of the tooth will then be founded on the majority portion of the tooth, which developed subsequent to the traumatic episode. Root canal treatment may be required, and the ideal contour of the tooth restored using a crown and core restoration. If the dilaceration site is found to be very incisally placed, then a more modest composite material restoration may be used, with vitality maintained. In either case, the initial crown reconstruction that is placed should be made in a line continuous with the long axis of the root portion and an orthodontic bracket bonded to its appropriate mid-labial position, ready for finishing. Final root alignment, in both the labio-lingual and mesio-distal planes, is then achieved in the usual manner.

Dilaceration of the coronal third of the root

The critical portion of the tooth for the dilaceration to occur is the coronal part of the root, close to the CEJ. In this situation, the prognosis of the aligned tooth is extremely

poor, since the majority of its root – that relating to the post-trauma period of development – will need to be amputated during the procedure, leaving the tooth with a non-viable coronal remnant of the root. Such a tooth will need to be extracted. Nevertheless, it will be necessary to reopen the space in the dental arch for some form of artificial replacement. Accordingly, and regardless of the prognosis, the same preparatory orthodontic procedure is still advisable.

A modified Johnson's twin-wire arch may be placed and space opened up as accurately as possible, to provide exactly the right amount of space, by comparing it with its erupted antimere. At this point, the dilacerate tooth should be exposed and, if its condition is confirmed as hopeless, it may be extracted – but not discarded! Instead, its contorted root is carefully severed from its perfectly shaped crown with a high-speed diamond burr, and the pulp chamber cleaned and filled with a composite filling material. The newly prepared natural crown may then be bonded to the two adjacent teeth, to act as its own space maintainer [38] until a more satisfactory permanent replacement may be made which, in view of the patient's age, may not be for several years (Figure 5.21).

Alternatively, and provided the apical portion of the root is substantial, it may be advantageous to treat it in the manner of a tooth whose root has been fractured below the level of the crest of the interproximal alveolar bone. To achieve this, it will be necessary to remove the crown of the tooth at the time of surgical exposure and perform an immediate root canal filling. A fixed, threaded post is prepared and a small hole is bored through its coronal end. The post is then firmly cemented. The cut root face and much of the coronal part of the post are covered with a composite filling material, leaving the tip of the post exposed. A stainless steel ligature wire is passed through the prepared hole and lightly twisted into a pigtail, with the help of artery forceps. In the absence of the acutely angled

crown portion, the remainder of the tooth presents a less complicated impaction, whose orthodontic resolution (so-called 'forced eruption' in perio-prosthesis parlance) is straightforward.

The prepared tooth is erupted into the mouth until the post and the restoration covering the root surface become apparent at the gingival level. The orientation of the root of the tooth is then reassessed by palpation and by taking new radiographs – a periapical for the mesio-distal inclination and a tangential for the bucco-lingual relationship. The latter view will be considerably easier to discern than before, since the root canal filling will act as an excellent radio-opaque marker for the root orientation despite the superimposition of other teeth.

The patient may then be referred for any necessary and appropriate muco-gingival surgery by a competent periodontist, followed by the construction of a good quality temporary acrylic crown, which is placed over the existing post. A general dentist or a specialist prosthodontist is familiar with the need for 'correcting' an abnormal root orientation by constructing the artificial crown in line with the crowns of the adjacent teeth, regardless of the root axis. This may be a sensible compromise in the more minor non-impacted displacement cases, since orthodontic root movement may then be avoided. However, in dilaceration cases, considerable root movement is needed and this is most suitably performed with the existing orthodontic appliance. For this to be made possible, the temporary crown must be placed in a proper axial relationship to the recently confirmed orientation of the root. The desired orientation of this intended reconstruction of the crown of the tooth will be at odds with the alignment of its neighbours, and this is not always an easy message to convey to the prosthodontist! Once the temporary artificial restoration is cemented securely in place, a bracket is placed in the usual manner. Crown alignment, root uprighting and root torqueing are then undertaken.

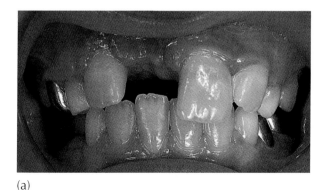
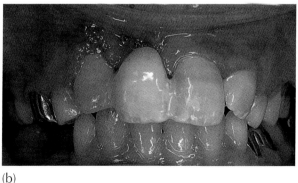

(a)   (b)

Fig. 5.21 (a) Precisely measured reopening of the space has been performed, using a removable appliance. (b) The prepared crown of the dilacerate central incisor, shown in Figure 5.4, is bonded between the etched interproximal surfaces of the adjacent teeth.

It may be seen that the point beyond which a root amputation should be avoided is when less than one-third of the root will remain after treatment. Crown amputation may be used as a viable alternative up to that point, but it must be remembered that the remaining root portion, which may be as little as one-half to two-thirds its original length, will be narrower in both mesio-distal and bucco-lingual dimensions. Additionally, this root stump will need to be erupted to the point where its amputated surface is just subgingival, because this will become the margin or shoulder of the future post and crown restoration. The diameter of this stump will be considerably smaller than is normal for a central incisor, which means that the emergence profile of the restored tooth will markedly compromise the appearance. These considerations illustrate the difficulties that exist in the placement of a satisfactory crown.

It is emphasized that, once the space in the arch has been gained and the time has come for surgical exposure, accurate diagnosis of the exact location of the dilaceration is critical if a valid decision is to be made. As pointed out earlier, this is not always possible from the radiographs, due to the superimposition of other unerupted teeth and neighbouring roots on the tangential view; the periapical view can contribute nothing in this respect. However, the use of cone beam volumetric CT can easily supply the needed answer [12].

If CT is unavailable and it is still impossible to locate the dilaceration, then an attachment should be placed on the lingual side of the tooth as before and the initial traction applied to bring the crown of the tooth occlusally. A decision to amputate the root portion of the tooth must in any case be delayed until the root is palpably bulging into the labial sulcus. At each subsequent visit, the sulcus should be carefully checked and any palpable change in the position of the root apex should be compared with the downward progress and eruption status of the crown and its angulation. Progress radiographs should be taken at appropriate stages, until the exact location of the dilaceration can be pinpointed or otherwise clinically diagnosed, particularly in relation to the long axes of the two parts of the tooth. Once the accurate diagnosis has been established, a reassessment of the treatment approach should be made to decide whether the line of treatment is indeed appropriate or whether the crown portion should be amputated and the direction of traction altered accordingly.

References

1. Kjaer I, Becktor KB, Lisson J, Gormsen C, Russell BG. Face, palate, and craniofacial morphology in patients with a solitary median maxillary central incisor. Eur J Orthod 2001; 23: 63–73.
2. Howard RD. The unerupted incisor. Dent Pract Dent Rec 1967; 17: 332–342.
3. Brin I, Zilberman Y, Azaz B. The unerupted maxillary central incisor: review of its etiology and treatment. J Dent Child 1982; 43: 352–356.
4. Zilberman Y, Malron M, Shteyer A. Assessment of 100 children in Jerusalem with supernumerary teeth in the premaxillary region. J Dent Child 1992; 59: 44–47.
5. Brook AH. Dental anomalies of number, form and size: their prevalence in British schoolchildren. J Int Assoc Dent Child 1974; 5: 37–53.
6. Tay F, Pang A, Yuen S. Unerupted maxillary anterior supernumerary teeth: report of 204 cases. J Dent Child 1984; 51: 289–294.
7. Di Biase DD. The effects of variations in tooth morphology and position on eruption. Dent Pract Dent Rec 1971; 22: 95–108.
8. Andreasen JO, Andreasen FM. Textbook and Color Atlas of Traumatic Injuries to the Teeth. Copenhagen: Munksgaard, 1994.
9. Howe GL. Minor Oral Surgery, 2nd edn. Bristol: Wright, 1971: 135–137.
10. Stewart DJ. Dilacerate unerupted maxillary central incisors. Br Dent J 1978; 145: 229–233.
11. Seward GR. Radiology in general dental practice. IX-unerupted maxillary canines, central incisors and supernumeraries. Br Dent J 1968; 115: 85–91.
12. Chaushu S, Chaushu G, Becker A. The role of digital volume tomography in the imaging of impacted teeth. World J Orthod 2004; 5: 120–132.
13. Chaushu S, Zilberman Y, Becker A. Maxillary incisor impaction and its relationship to canine displacement. Am J Orthod Dentofacial Orthop 2003; 124: 144–150.
14. Battagel J. The case for early assessment: 2: treatment with specialist support. Dent Update 1985; 12: 293–298.
15. Houston WJB, Tulley WJ. A Textbook of Orthodontics. Bristol: Wright, 1986: 126–131.
16. Mills JRE. Principles and Practice of Orthodontics, 2nd edn. Edinburgh: Churchill Livingstone, 1987.
17. Mitchell L, Bennett TG. Supernumerary teeth causing delayed eruption – a retrospective study. Br J Orthod 1992; 19: 41–46.
18. Witsenberg B, Boering G. Eruption of impacted permanent upper incisor teeth after removal of supernumerary teeth. J Oral Surg 1981; 10: 423–431.
19. Ashkenazi M, Greenberg BP, Chodik G, Rakocz M. Postoperative prognosis of unerupted teeth after removal of supernumerary teeth or odontomas Am J Orthod Dentofac Orthop 2007; 131: 614–619.
20. Bodenham RS. The treatment and prognosis of unerupted maxillary incisors, associated with the presence of supernumerary teeth. Br Dent J 1967; 123: 173–177.
21. Munns D. Unerupted incisors. Br J Orthod 1981; 8: 39–42.
22. Gardiner JH. Supernumerary teeth. Dent Pract Dent Rec 1961; 12: 63–73.
23. Day RCB. Supernumerary teeth in the premaxillary region. Br Dent J 1964; 116: 304–308.
24. Kettle MA. Unerupted upper incisors. Trans Eur Orthod Soc 1958; 34: 388–395.
25. Hotz R. Orthodontia in Everyday Practice. Berne: Huber, 1961.
26. Becker A, Kohavi D, Zilberman Y. Periodontal status following the alignment of palatally impacted canine teeth. Am J Orthod 1983; 84: 332–336.
27. Kohavi D, Becker A, Zilberman Y. Surgical exposure, orthodontic movement and final tooth position as factors in periodontal breakdown of treated palatally impacted canines. Am J Orthod 1984; v85: 72–77.
28. Johnson JE. A new orthodontic mechanism: the twin wire alignment appliance. Int J Orthod 1934; 20: 946–963.
29. Shepard ES. Technique and Treatment with the Twin-wire Appliance. St Louis, MO: CV Mosby, 1961.
30. Peretz B, Becker A, Chosak A. The repositioning of a traumatically-intruded mature root permanent incisor with a removable appliance. Journal of Pedodontics, 1982.
31. Chaushu S, Becker T, Becker A. Duration and Success Rate of Orthodontic Treatment for Impacted Maxillary Central Incisors 2010, unpublished.
32. Ingber SJ. Forced eruption. Part I. A method of treating isolated one and two wall infrabony osseous defects – rationale and case report. J Periodontol 1974; 45: 199–206.
33. Ingber SJ. Forced eruption. Part II. A method of treating non-restorable teeth – periodontal and restorative considerations. J Periodontol 1976; 47: 203–216.

34. Stern N, Becker A. Forced eruption: biological and clinical considerations. J Oral Rehabil 1980; 7: 395–402.

35. Melsen B. Tissue reaction following application of extrusive and intrusive forces to teeth in adult monkeys. Am J Orthod 1986; 89: 469–475.

36. Bimstein E, Becker A. Malocclusion, periodontal health and orthodontic intervention. In Bimstein E, Needleman HL, Karimbux N, van Dyke TE, eds. Periodontal Health and Diseases in Children Adolescents and Young Adults. London: Martin Dunitz, 2001: 251–274.

37. Boyer DB, Williams VD, Thayer KE, Denehey GE, Diaz-Arnold AM. Analysis of debond rates of resin-bonded prostheses. J Dent Res 1993; 72: 1244–1248.

38. Becker A, Stern N, Zelcer Z. Utilization of a dilacerated incisor tooth as its own space maintainer. J Dent 1976; 4: 263–264.

6

Palatally Impacted Canines

Orthodontic Treatment of Impacted Teeth, Third Edition. Adrian Becker.
© 2012 Adrian Becker. Published 2012 by Blackwell Publishing Ltd.

Prevalence

In any population, the prevalence of palatally impacted maxillary canines is low, but it seems to have a variable distribution with regard to ethnic origin. The lowest frequency reported in the literature relates to Japan [1], where the anomaly occurred in only 0.27% of the sample population. Some very early studies by Cramer [2] among white Americans and Mead [3] in an undefined sample found 1.4% and 1.57%, respectively. A study of a large series of full mouth dental radiographs among patients in the USA revealed a figure of 0.92% [4], while Brin et al. [5], in a study of an Israeli population, found a 1.5% prevalence. Higher figures for the anomaly have been found in more recent surveys: 1.8% was reported in the study by Thilander and Jacobson [6] of an Icelandic population, and 2.4% in an Italian sample [7].

Montelius [8] was the first to indicate a difference between Caucasian and Oriental populations, finding a frequency of 1.7% for Chinese and 5.9% for Caucasians. However, since he did not distinguish between buccal and palatal impaction in his study, little useful information may be gleaned from these figures in the immediate context. More recently, the work of Oliver et al. [9] indirectly indicated that Asians may suffer from buccally impacted canines more frequently than from palatal canines. While this appears to be supported by various case reports that have appeared in the literature from the Far East, no definitive study has been undertaken to investigate this possibility.

A strong prevalence of impacted canines is found among females, with a ratio of 2.3:1 in the above-mentioned group of American patients [4]. 2.5:1 in an Israeli orthodontic group [10] and 3:1 in each of a Welsh orthodontic group [9], a US orthodontic sample [11] and an Italian sample [7].

However, some confusion exists with regard to these figures, since a subsequent random Israeli population study [5] showed an approximately equal male–female occurrence of the anomaly. Furthermore, Oliver et al [9] indicated that, although a higher female incidence was present in their study of Welsh patients, this reflected the trend for more females to seek orthodontic treatment in the UK.

Appearance is rarely marred by the presence of an over-retained deciduous canine, since there is a complete and uninterrupted display of teeth, and any abnormalities are usually not disfiguring. That being the case, and if we are to accept an 'appearance/aesthetics' motivation for girls seeking treatment more frequently than boys, then the diagnosis of an impacted maxillary canine alone does not seem to be an adequate reason for the preponderance of females seeking treatment. Motivation for treatment may therefore rather depend on the ability and persuasiveness of a particular practitioner in pointing out the potential hazards of non-treatment. There may be no basis to expect that this would convince more female patients than males to accept treatment.

Aetiology

There is no single cause for the palatal displacement of the maxillary canine tooth. Space-occupying, extraneous entities of dental origin (Figure 6.1) will undoubtedly produce abnormal positioning of an unerupted permanent maxillary canine, but they are comparatively rare in the canine area. The fact that the majority of impacted canines occur in their absence compels us to look elsewhere for the main causes of impaction. Nevertheless, when a supernumerary tooth (Figure 6.1a) or an odontome (Figure 6.1b) appears in the area, they disturb the position and orientation of a developing canine tooth and deflect the eruptive force to express itself in a futile direction. Their timely removal may sometimes bring about spontaneous re-orientation and resolution, although this is by no means certain and active orthodontic intervention is usually necessary in the final instance.

One other factor may be found in the context of premolar root morphology. An example of this is seen in Figure 6.1c, d, in which the buccal and/or palatal roots or a fused single root develops directly in the path of the unerupted canine. The canine may then become impacted, deflected or halted, but the physical presence of the canine may act as an impediment in the way of the developing root end, which may also serve to cause a mesial curving of the interfering root.

In order to resolve this impaction, the premolar root may be uprighted distally and rotated mesially, with a modicum of palatal root torque, thereby moving its root apex distally and palatally to distance it from the canine. The canine may then be exposed and mechanically erupted in the usual manner and into its ideal position. Finally, the premolar should be returned to its upright position and re-rotated as much as possible, using light force, and the effort should be discontinued immediately it meets with resistance that will be caused by contact between the mesially curved premolar root end and the canine root.

Other treatment options include:

1. Root treating and apicoectomizing the premolar and extracting the deciduous canine, which should all be done at the same time as exposure of the canine.
2. Extracting the canine and leaving the short rooted deciduous canine in place.
3. Extracting the premolar and the deciduous canine, which is an approach to be considered if the overall malocclusion may be treatment planned as a 2- or 4-unit extraction case.

To return to the more usual forms and to explain the mechanism of palatal displacement of the maxillary canine, some of the hypotheses that have been put forward have been intimately involved with aberration in the normal process by which the maxillary anterior teeth erupt. For this reason, an understanding of normal development in this area is important.

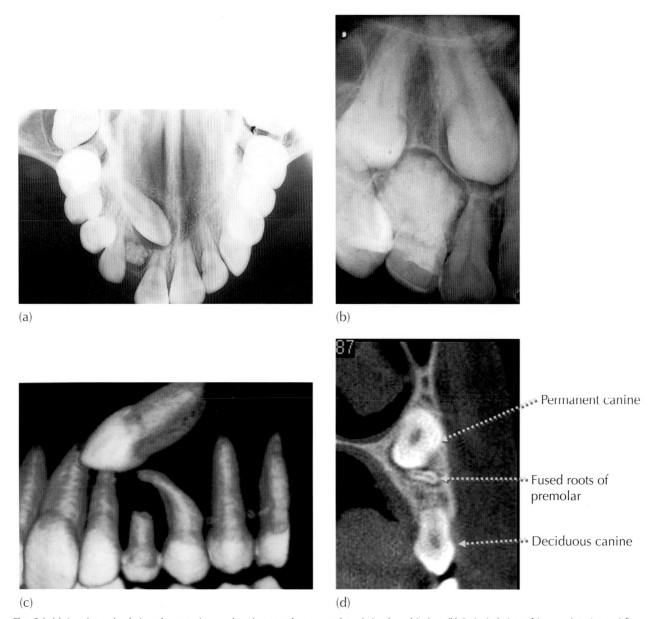

(a)

(b)

(c)

(d)

Fig. 6.1 (a) Anterior occlusal view shows an impacted canine, an odontome and a missing lateral incisor. (b) Periapical view of impacted canine and first premolar associated with an odontome and over-retained deciduous first molar. (c, d) A three-dimensional CBCT view showing the apical third of the root of a normally sited first premolar to be turned 90° to the mesial, directly in the path of the unerupted canine, which is being deflected further mesially and prevented from erupting. The transaxial (vertical) CBCT slice across the canine in the area of the CEJ and longitudinally through the deciduous canine. It also cuts across the mesio-distally oriented horizontal portion of the root close to the apex of the premolar. (Courtesy of Dr N. Dykstein.)

Normal development

In the middle period of the deciduous dentition, a periapical radiograph of the premaxillary region will show the fully completed deciduous incisor roots. It will show the overlapping shadows of the permanent central and lateral incisors more or less in the same vertical plane (Figure 6.2a) and at the level of the apical area of the roots of the deciduous incisors, with the canines being sited higher up. The overlap of the permanent teeth crowns is due to the fact that these relatively wide permanent teeth are all contained in a narrow area and, at this time, are initially located palatally in the alveolus. The developmental position of the lateral incisors is palatal in relation to both the central incisors and the permanent canines. For these reasons, the periapical view described above gives the appearance of severe crowding.

During the early eruptive movements of the central incisors, a progressive resorption of the roots of the deciduous incisors occurs. The permanent incisors migrate slowly across from the palatal side of the arch to the labial as they proceed in their downward path, until the teeth erupt into

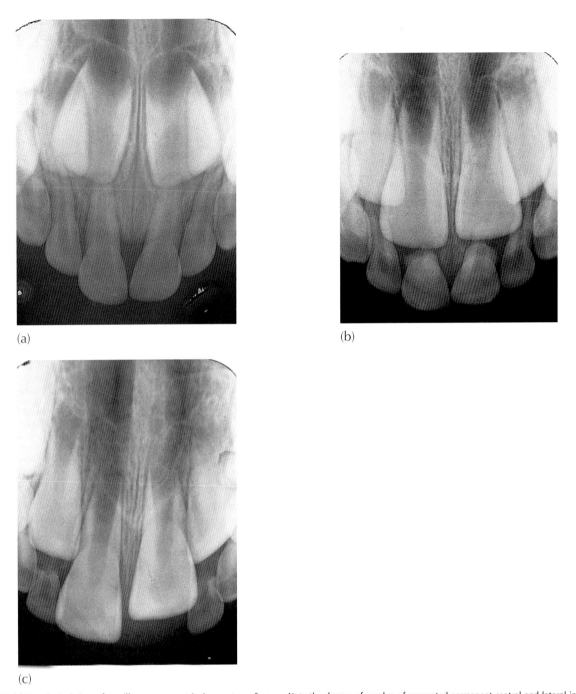

(a)

(b)

(c)

Fig. 6.2 (a) A periapical view of maxillary permanent incisors, at age 3 years. Note the degree of overlap of unerupted permanent central and lateral incisors. (b) The same patient at 5 years. The permanent central incisors have migrated inferiorly and labially relative to the lateral incisors. Note the reduced degree of incisor overlap. (c) The central incisors are erupting at age 6.5 years. Note how the lateral incisors have migrated labially into the arch to eliminate the overlap completely. (Courtesy of Professor B Peretz.)

a more labial perimeter than was defined by the deciduous incisor teeth before their shedding. During this process, the wide crown portion of the central incisors will have moved downwards and labially, ahead of the lateral incisors (Figure 6.2b). As this occurs, the progressively narrower cemento-

enamel junction (CEJ) area and the root portion of the central incisor come to lie mesial to the unerupted lateral incisor crowns. This leads to the fairly rapid provision of space at this level in the alveolus [12]. The lateral incisor migrates labially into this area as it begins its downward

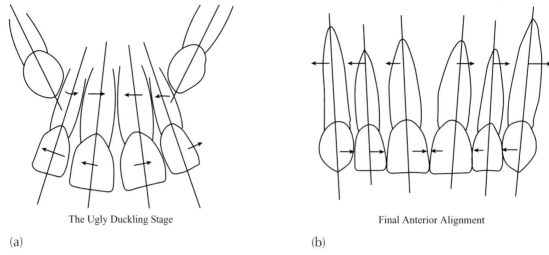

The Ugly Duckling Stage

Final Anterior Alignment

(a)

(b)

Fig. 6.3 (a) In the early stages the unerupted canines are mesially directed, restricting the incisor roots to the area of the midline, thereby flaring their crowns distally, to create spacing. (b) As the canines descend towards the crown, down the distal aspects of the lateral incisor roots, their influence is reversed. They now apply a mesially directed force on the distal surfaces of the lateral incisor crowns, which closes off the anterior spaces and encourages a distal root flaring.

eruption path. Additionally, the downward eruption movement distances it from the permanent canine crown, providing more space for it to move labially, following closely after the central incisor.

With the eruption of the central incisors, the lateral incisor crowns move from a lingual relationship into a direct distal relationship with the central incisor roots, initially at a higher level (Figure 6.2c). As this occurs, the presence of the lateral incisor crowns displaces the developing roots of the central incisors towards one another, since these are at the same level within the alveolar bone. With the central incisor apices held together in this way, the crowns of these teeth are flared distally. A developmentally normal median diastema is thus produced, which has been termed the 'ugly duckling' [13].

A year or so later, the lateral incisors will have descended along the distal side of the central incisor roots, to release their 'hold' on the narrowed inter-radicular width of the central incisor roots, allowing the roots to drift apart. The lateral incisors continue to move inferiorly along their eruptive path, progressively reducing their constricting influence on the central incisor roots until they reach the distal side of the necks of the central incisor crowns. At this point, their presence and continued downward migration serve to provide a mesially directed force to the crowns of these teeth, moving them towards one another and partially closing off the median diastema. The long axes of the central incisor teeth will also have changed, with the roots becoming more parallel. The lateral incisor long axes, however, are relatively flared in the coronal direction, with their root apices close to those of the central incisors.

A periapical view of the area at this time will show the unerupted permanent canine crowns of each side pointing mesially towards the lateral incisor apical area. They appear to be the containing influence that causes the apical convergence of the incisor roots and the reason that the median diastema has not completely closed. Subsequent follow-up radiographs of the area will show the permanent canines altering their relationship as they move downwards along the distal side of the root of the lateral incisors. Having initially constricted the roots of the lateral incisors mesially, the downward descent of the canines alters their influence on the lateral incisor crowns, to tip them mesially. This brings about an uprighting of the long axes of the incisors on each side and a closure of the anterior spacing (Figure 6.3). The canines' own long axes become more vertical as they progress and as the roots of the deciduous canines become resorbed. With the shedding of the deciduous canines, they finally erupt with a slight mesial inclination, taking up their place in the arch by moving the crowns of the incisors towards the midline, to close off the diastema completely [14]. As all this occurs, the long axes of the incisor teeth change from being apically convergent to become more parallel and even slightly divergent.

Throughout the period of their downward progress, the permanent canines are conspicuously palpable on the buccal side of the alveolar ridge, from as early as 2 or 3 years prior to their normal eruption, which normally occurs at the age of 11–13 years.

Theories regarding the causes of palatal displacement

Long path of eruption

From the days of Broadbent, in the 1940s, the most common reason given for palatal displacement of the permanent maxillary canine was the fact that it had a long and tortuous eruption path, beginning close to the floor of the orbit. It was considered that, compared with other permanent teeth,

this tooth had much further to travel before it erupted into the mouth and that it therefore had a greater chance of 'losing its way'. This has been standard teaching for many decades.

Crowding

Hitchin [15] considered that crowding of the dentition was the reason for this condition, although he offered no evidence to support his contention.

In general, crowding of the dentition results in the exaggerated displacement of a tooth from its developmental position in the arch. The developmental position of the maxillary lateral incisor is lingual to the line of the arch, as we have described above. Thus, when crowding affects the early mixed dentition, there will be insufficient space for the lateral incisor to migrate labially between the root of the erupting or newly erupted central incisor and the deciduous canine teeth, which is the manner in which it normally comes into the dental arch. For the most part, therefore, it continues to develop downwards, but in a lingual position, and erupts lingual to the adjacent teeth.

A parallel environment is created when a second deciduous maxillary molar is extracted before its due time and the first permanent molar drifts mesially into the available space. Similarly to the lateral incisor, the developing second premolar develops palatally to the line of the arch, and its continued development and eruptive path will be in an exaggerated palatal direction in much the same way.

We have pointed out above that the normal eruption path of the permanent canine is buccal to the line of the arch, and we also know that the lateral incisor and first premolar, the teeth immediately adjacent to the canine, erupt before the canine. Thus, in the presence of crowding, there will be reduced space in the arch in the canine area and the close proximity of these adjacent teeth will prevent the canine from moving into the arch. The vertical development of the maxillary permanent canine will therefore be accompanied by its buccal displacement, to give the typical picture seen in the class 1 crowded case (Figure 6.4). Whether the tooth eventually erupts or remains impacted is irrelevant, although buccal impaction is uncommon in Caucasian population groups. It is therefore quite clear that

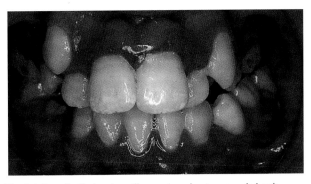

Fig. 6.4 Buccally displaced maxillary canines due to a crowded arch.

the cause of this type of displacement of the canine is completely different from that involved with palatal displacement. The two conditions are different entities. They should never be confused, nor should they be lumped together to form an experimental group for clinical research, as if to offer a homogeneous sample of impacted teeth. For the purposes of study, it is far more logical to combine all palatally displaced canines, whether they are unerupted or erupted, since they share a common aetiology, although their clinical presentation may be different.

In a series of clinical research studies, Jacoby [16], Becker [17] and Brin et al. [5] pointed out that the likelihood of palatal displacement is much reduced where crowding is present. They have shown it to be a far more prevalent occurrence when there is excessive space in the dental arch.

Non-resorption of the root of the deciduous canine

Lappin [18] considered that it is the failure of the root of the deciduous canine to resorb that causes a palatal deflection of the eruption path of the permanent canine, leading to its impaction. Here, too, one may draw a parallel with other teeth. In cases where a second deciduous molar is over-retained, owing to the presence of a malposed premolar tooth germ, one may often see on the periapical or panoramic radiograph that one of the roots has totally resorbed, while the second root is only partially so. The long spicule of unresorbed root that may be present retains the tooth against natural shedding, while the fully developed and unerupted second premolar is situated immediately beneath the crown of the deciduous tooth in the area previously occupied by the resorbed portion of the roots.

From this type of clinical evidence, which is seen so widely and frequently in practice, it is generally considered that the presence and advancing eruption of the permanent tooth provides the stimulus for the resorption, and a portion of root distant from the unerupted permanent tooth may be unaffected by this process. On the basis of this, Lappin's view would appear to be 'putting the cart before the horse'. Nevertheless, and in support of his argument, subsequent studies [19–22] have shown the spontaneous eruption of previously impacted canines in many cases, following the extraction of deciduous canines. This will be discussed at length later in this chapter, under the heading 'Preventive treatment and its timing'.

Trauma

In a clinical report, Brin et al. [23] have illustrated how trauma, which leads to a cessation in the development of a lateral incisor root, may be associated with palatal canine impaction. They explain this by assuming:

- that the traumatic episode may have caused movement of the lateral incisor, or
- by conduction, movement of the unerupted canine itself, or

- in terms of the guidance theory, that this could be explained as being due to the shortness of the lateral incisor root, whose development ceased as a result of the trauma.

Soft tissue pathology

A further alternative could place the blame on the possible presence of chronic irritation, residual infection or granuloma around the apex of a non-vital deciduous canine tooth (Figure 6.5). Deciduous maxillary canines may often be affected by interproximal, usually distal, caries. The lesions are frequently left untreated in the belief that the teeth are about to be shed. The tooth loses its vitality when caries reaches the pulp and a chronic periapical area develops. This soft tissue lesion by itself is a potent cause for deflection of the path of a developing unerupted maxillary canine (Figure 6.6). In rare instances, it may develop into a radicular cyst or it may initiate cystic change in the follicle

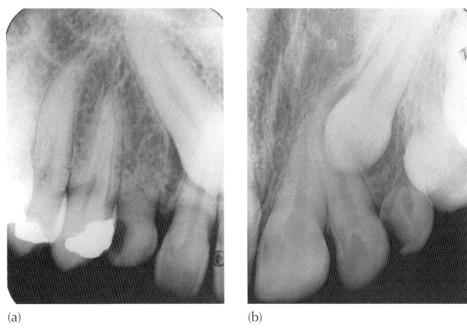

(a)   (b)

Fig. 6.5 (a, b) Periapical views of bilaterally impacted canines, each associated with a non-vital deciduous canine.

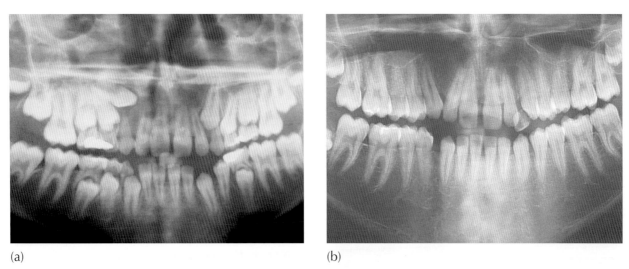

(a)   (b)

Fig. 6.6 (a) Panoramic view of a patient in the mixed dentition stage with a markedly displaced and unerupted right maxillary canine. The immediate area shows a large area of bone loss involving the canine and first premolar, associated with the non-vital deciduous first molar. The chronic periapical abscess represents a soft tissue obstruction that has deflected the eruption path of the permanent canine. The deciduous canine has an obliterated pulp. (b) Following extraction of the deciduous canine and deciduous first molar, there has been spontaneous resolution, with eruption of the teeth. For no apparent reason, the deciduous canine of the opposite side was overlooked! (Courtesy of Dr A. Renert.)

of the canine, either of which will alter the path of eruption of the canine, or prevent its further eruptive progress. This will be discussed in Chapter 11.

The guidance theory

Miller [24] and Bass [25] reported that there appeared to be an unusually high prevalence of congenitally missing lateral incisors associated with palatally impacted canine teeth. They theorized that, under these circumstances, the permanent canine lacks the guidance normally afforded by the distal aspect of the lateral incisor root. As pointed out earlier, in relation to normal development, the canine initially has a strong mesial developmental path, which alters early on, with the canine being guided downwards, apparently along the distal aspect of the lateral incisor root. They concluded that, in the absence of this guiding influence, the canine continues in its initial mesial and palatal path. The tooth then becomes impacted in the palatal area, posterior to the central incisors, and fails to erupt in its due time, if at all.

Miller's concept was founded on information gleaned from the study of six such cases. He assumed that since a peg-shaped or otherwise abnormally small lateral incisor develops a root of more or less normal length, such a tooth would provide the required guidance for the normal eruption of its adjacent canine. He therefore rationalized that these anomalous teeth could not be an aetiological factor in canine impaction.

Following the treatment of several hundred cases of this type by the present author, a different pattern of association seemed apparent. Palatal impaction of the maxillary canine appeared to be intimately bound up with the occurrence of anomalous lateral incisors and less with the congenitally missing teeth. Furthermore, a stereotype of the maxillary impacted canine patient could be offered (Figure 6.7), in which the patient is frequently a 15-year-old female, with well-aligned and normally related dental arches, slight spacing and no real malocclusion. Characteristically, the teeth are small, the lateral incisors particularly so, the incisors lack their normal rounded contour, there may be missing teeth, dental development is late and the patient's motivation for treatment is low.

A series of clinical research studies followed, in which a sample of patients who were successfully treated for a palatally displaced canine was investigated. In the first study [10] a wide and highly significant discrepancy in the numbers of normal, small and peg-shaped lateral incisors adjacent to an affected canine was found, compared with the published data for normal populations (Figure 6.8). In the interests of accuracy, a random study was later performed by the same research group [5] to quantify the various types of lateral incisors found within the general population of the same geographic area, while using the same definitions of anomaly. In the general population, 93% of all lateral incisor teeth were of normal shape and size, compared with only

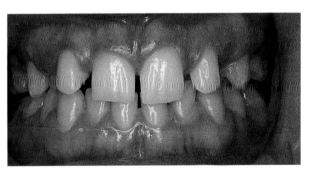

Fig. 6.7 Late-developing dentition showing spacing, small peg-shaped lateral incisors, teeth of poor anatomic contour and minor class 1 malocclusion.

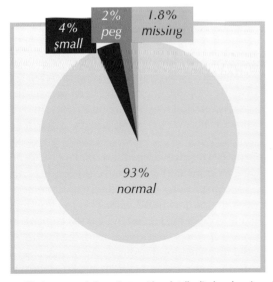
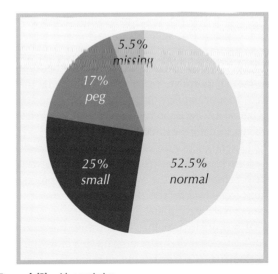

Fig. 6.8 Lateral incisor anomaly in patients with palatally displaced canines. From ref. [5], with permission.

52% in the palatal canine sample. In the random population sample, missing lateral incisors were found in approximately 1.8% of the cases [26], which contrasted markedly with the 5.5% of this anomaly among the impacted canine cases [10] or three times as frequent (Figure 6.8). These figures are valid for the Israeli population sample studied. However, congenital absence of maxillary lateral incisors in a meta-analysis of the collected data from a large number of different population studies was found to be lower, at 1.55% for males and 1.78% for females [27].

These results clearly support Miller and Bass regarding the part played by the lateral incisor as a guide in the normal eruption of the permanent canine. Without this guidance, normal eruption is compromised fivefold. However, Miller's rationalization regarding the positive role of anomalous lateral incisors appears very much misplaced. Small lateral incisors were seen in only 4% of the random sample, while the palatal canine cases showed this anomaly to be six times as frequent (25%). Furthermore, only 2% of the general population had peg-shaped incisors, while 17% (nine times the frequency) were seen among the palatal canine cases. Similar results have since been shown in confirmatory studies that have examined Welsh [9] and west of Scotland [28] samples. If small or peg-shaped lateral incisors provide similar conditions to normally shaped lateral incisors in their role in the normal eruption of the canines, then one would expect figures for the association of these anomalous teeth with palatal canines to be much lower than the 5.5% figure for missing lateral incisors. Their presence is shown here to offer more than merely a loss of guidance to the developing permanent canine. The fact that they are associated with a greater prevalence of impaction compared to that of the missing lateral suggests that an additional, obstructive role is played by these teeth.

In the first study [10] a hypothesis was presented based on the fact that the anomalous small and peg-shaped lateral incisors develop very much later than normal lateral incisors. While no figures are available for the extent of this delay, it seems clear from clinical observation that it may be as much as three years – and this for teeth whose calcification normally begins at age 10–12 months!

If we are now to relate this to the 'guidance theory of impaction', we may postulate that at the critical time that the permanent canine requires the guidance, the root of the anomalous lateral incisor is too rudimentarily developed to provide it. Thus, initially the situation is parallel to that seen in congenital lateral incisor absence. The result is that the canine develops mesially and palatally and usually in a downwards direction, into the vertical alveolar process, where it proceeds towards the palatal periosteum. This describes the first stage of palatal displacement.

The palatal periosteum may then halt further progress of the tooth, or it may alter the eruption path to a more horizontal direction, across the palate. In either instance, this may then be defined as a first-stage palatal impaction.

In particular circumstances, it seems that the palatal periosteum may deflect the developing canine from its first-stage displacement, in a downward direction. The alveolar process in the canine region is V-shaped in cross-section, such that, with continued vertical movement, the progressively narrowing alveolus will tend to guide the aberrant canine in a buccal/labial direction. These corrective movements of the palatally displaced canine are the characteristic feature of what may be termed the first stage of palatal displacement with secondary correction (Figure 6.9).

In cases of congenital absence of the lateral incisor, a canine that was not palpable buccally at any point in its earlier development may often be seen to finally erupt more mesially than normal and in the line of the arch. In the presence of an over-retained deciduous lateral incisor or canine only, the corrective movements of the canine lead to the initiation of root resorption of these deciduous teeth. Following their shedding, the permanent canine may then erupt into the line of the arch and, often, in a more mesial location in the place of the absent lateral incisor. If a late-developing lateral incisor is present, it will now lie directly in the path of the displaced canine. The physical presence of the lateral incisor will bring an abrupt stop to these corrective movements, and any further vertical development of the canine may only then be on the palatal side of the dental arch, completing the second stage of palatal displacement. In summary, therefore, the 'guidance theory' comprises five elements:

1. Normal eruption. It adopts Broadbent's original view that, given the timely and normal development of a lateral incisor, guidance for the canine is provided by the presence of a normally developing lateral incisor and a buccal path of eruption is to be expected, with the tooth palpable in the buccal sulcus early on.
2. First-stage impaction. It offers an explanation for the absence of guidance at a critical time in the normal development of the permanent canine, which leads to a deflection of the developmental path of the tooth, causing it to move palatally. This etiologic factor may be created by a congenitally missing lateral incisor or by a late-developing, anomalous lateral incisor. In the event that no vertical movement of the canine into the alveolar process occurs, the result may be a horizontal palatal impaction.
3. First-stage impaction with secondary correction. It goes on to explain the corrective influence of the vertical alveolar process, which redirects the canine on a more favourable downward path. This scenario may be difficult to diagnose accurately, and the clinician must draw his or her own conclusions from the further progress of the impacted tooth, which may be palpable, low down on the palatal side, before it finally erupts close to the line of the arch. The tooth may then spontaneously

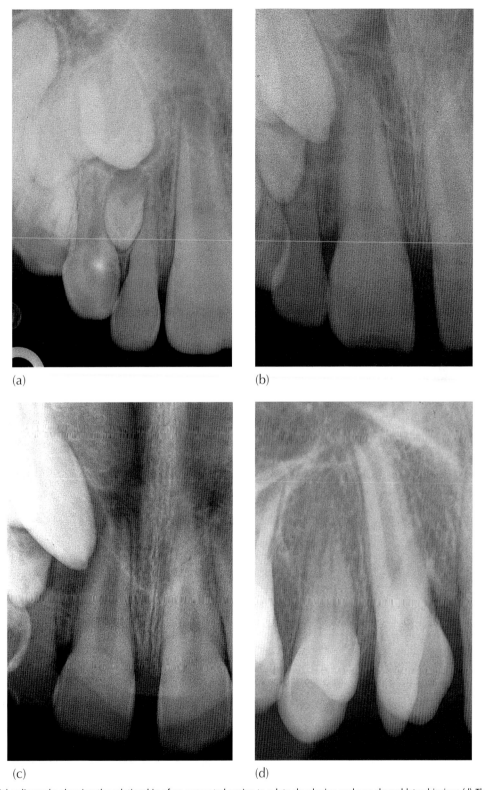

(a)

(b)

(c)

(d)

Fig. 6.9 (a–c) Serial radiographs showing the relationship of an unerupted canine to a late-developing and peg-shaped lateral incisor. (d) The two teeth have erupted and are superimposed on one another. At clinical examination the erupted canine was found to be on the palatal side of the lateral incisor.

move more buccally and mesially, in the absence of a lateral incisor, to reach the occlusal plane in a close-to-normal bucco-lingual position.

4. Second-stage impaction. Self-correction is prevented by the presence of an anomalous and late-developing lateral incisor, redeflecting the tooth further palatally. This may be termed second-stage displacement, and is an aetiological factor that is not seen when the lateral incisor is absent.

5. Second-stage impaction with secondary correction. As we shall see later, in the discussion of treatment timing, extraction of an over-retained deciduous canine, or even the anomalous lateral incisor itself, may often lead to spontaneous eruption of the impacted tooth.

There are several elements in this process which clearly indicate that the erupting canine is strongly influenced in the progress of its eruptive movements by environmental conditions, which determine the degree of success of its final eruption status, including its final erupted location, or its impaction. Small, peg-shaped, and missing teeth are more frequent findings among females than among males, in the approximate ratio of 2–3:1, as discussed earlier in this chapter. Furthermore, the maxillary permanent canine erupts earlier in females, which could mean that earlier lateral incisor guidance will be necessary for its normal eruption. These facts provide the hypothesis with some support in explaining why palatal canines are more frequent in females and why anomalous lateral incisors are a more powerful causal agent than congenitally absent lateral incisors.

It is quite clear that heredity plays an important role in this hypothesis. The assumption is that the genetically determined factors (small, peg-shaped, missing lateral incisors, spaced dentitions, etc.) provide the environment that leads to a loss of guidance of the canine, its abnormal palatal path, and impaction.

Heredity

Given the strong hereditary influence in palatal canine displacement, there are those who believe that heredity is the direct cause and dismiss other relationships as secondary or as similarly linked hereditary factors. In other words, the palatal canine is another link in the chain of genetically linked phenomena.

In a study of the families of children affected by palatally displaced maxillary canines [29] a search was made among the parents and the siblings for the related anomalies to which we have referred above. The prevalence of small, peg-shaped and missing lateral incisors, late-developing dentitions and other missing teeth among these close relatives was very high, in addition to palatally impacted canines. This evidence seems to favour heredity as the causal agent for these associated phenomena. The guidance theory contends that their presence creates an environment

favourable to the development of palatally displaced canines and, as was to be expected, the lateral incisor phenomena were found to occur in an unusually high proportion of these cases. The view that these phenomena are each genetically determined and frequently occur together, including the canine displacement [30–32], is equally tenable, but would appear to be an oversimplification. The fact that the extraction of adjacent deciduous canines and anomalous lateral incisors and orthodontic space opening greatly improve the chances of canine eruption would lead us to believe that local factors cannot be ignored as exerting a powerful influence on the aetiology of canine impaction.

Other forms of maxillary canine positional anomaly do occur from time to time, and are difficult to equate with the more usual pattern of palatal or buccal displacement. Genetic factors seem likely to be the governing factors, in which the entire tooth is located in an abnormal position. In these cases, it seems that the original site or orientation of the anlage from which the tooth developed was abnormal. Thus, we occasionally come across patients in whom there is adequate space in the arch for the ideal eruption and alignment of the maxillary permanent canines, yet these teeth erupt buccally ectopic (Figures 6.10 and 6.11). Additionally, the eruption occurs relatively high in the alveolus and the tooth has no mesio-distal contact with its immediate neighbours and, therefore, cannot be considered to have been guided into this position.

On the other hand, this may be looked at and perhaps explained in the parlance and context of the guidance theory. As we have illustrated above, buccally displaced canines (BDC) are usually found in crowded dentitions. Nevertheless, there is a small but significant percentage of BDC cases where there is no crowding to account for the buccal displacement. A study was undertaken by the Jerusalem group to investigate the features of dentitions of BDC cases where no crowding was present and compared them with BDC cases with crowding and cases in which the canines had erupted normally. The results revealed reduced dimensions of the maxillary incisors in BDC in dentitions

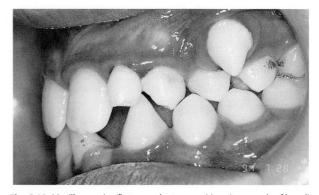

Fig. 6.10 Maxillary canine/first premolar transposition. An example of hereditary primary tooth germ displacement.

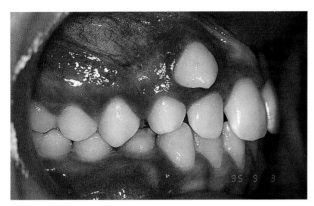

Fig. 6.11 Despite the absence of crowding, the canine has erupted in an abnormal location. Is this evidence of a lack of guidance on the part of the adjacent peg-shaped lateral incisor or hereditary primary tooth displacement?

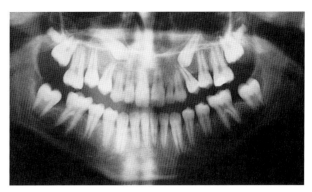

Fig. 6.12 A case of bilateral hereditary primary tooth germ displacement.

in which no crowding existed in comparison to the other two groups. More specifically, the lateral incisor was the only tooth which was consistently smaller when compared to both crowded BDC and normally erupted canine cases [33]. As pointed out earlier in this chapter, small lateral incisors develop very late, growing adequate root length at a time which is too late for it to influence the developing canine. Thus, it may be reasonably postulated that the lack of guidance from the anomalous adjacent lateral incisor provides a cogent alternative explanation for the buccal displacement of the canines in non-crowded dentitions, in a similar way to what occurs in the palatally displaced canine cases. The canine simply takes a buccal path instead.

It becomes clear from a careful reading of this section that there are both undeniable hereditary and environmental/developmental factors at work in the aetiology of maxillary canine impaction, neither of which may be discounted and both influencing the expression of the aberrant eruption. For a balanced, updated, evaluative and critical discourse on the place of both these elements as causative agents, the reader is referred to a recently published review (34).

Peck et al. [35] have studied maxillary permanent canine/first premolar transposition (Figure 6.11), and have found a strong hereditary influence in its aetiology. They point out that this very specific type of canine anomaly cannot be construed as deriving from guidance from the lateral incisor, and cannot be influenced in any way by the size, form or timing of the development of that tooth. With somewhat questionable justification, they then extend their reasoning to cover all maxillary canine displacements. They go on to claim that their findings, specifically *vis-à-vis* transposed canines, represent conclusive evidence that the aetiology of palatal impaction is also under total genetic control (Figure 6.12).

Canine/first premolar transposition and palatal displacement of the canine are both aberrations of tooth position, but there is no reason for assuming that the aetiology of the one is the same as for the other, and there can be no basis for comparing transposition with palatal displacement. This does not, therefore, present a valid argument with which to refute the 'guidance theory' [36].

We may prefer to define canine/first premolar transposition as primary tooth germ displacement [37]. In other words, their site of development is not in its expected location in the jaw and in relation to the other teeth. In the developed dentition, this is reflected in an abnormal position of the root apex, which is usually assessed clinically by the mental exercise of extending the orientation of the long axis of the tooth crown. This helps us to distinguish them from other and more common forms of displacement, which have an environmentally influenced aetiology.

As a result of crowding, the position of the more frequent buccally placed canine is dictated by the amount of space available for it in the arch and by direct interproximal contacts with the adjacent teeth. The vast majority of buccally and palatally displaced canines show the root apex to be ideally placed, in the line of the arch [36]. Experience shows that extrusion and tipping of the crown into its place in the arch is usually sufficient to resolve the malalignment, with little or no root torque being required in most cases.

Other associated clinical features

Repeated studies have found that palatal canine patients have dentitions characterized by their small teeth [5, 10, 16, 35, 36]. In the general population, the individual teeth of males are larger than those of females, but a study of patients with palatal canines discovered that the individual teeth in the affected males were found to be significantly smaller than in unaffected males [38]. However, there was no difference in the size of teeth between affected and unaffected females. Oddly, the teeth of affected males were similarly sized to those of affected and unaffected (control) groups of females.

Both male and female patients with palatal canines often feature missing teeth, such as third molars, maxillary lateral

incisors, mandibular and maxillary second premolars, and mandibular central incisors [5, 7, 24, 25, 30, 33, 34].

From the literature, we learn that small and missing teeth in a dentition have been shown to be associated with late development [39–41], a fact that has largely escaped attention. Scant notice was taken within the profession of the shrewd observation made by Newcomb half a century ago [41] that 'with few exceptions . . . potential impaction of permanent teeth is seen in patients exhibiting moderate to severe retardation of dental maturation . . . [and] . . . a slow rate of permanent teeth formation'. He considered 'it would be useful . . . to correlate dental and bone ages' among these patients. Newcomb based his conclusions on dental age as determined by the eruption status of the dentition. However, no specific study was ever undertaken to investigate this connection until recently.

In a more recent study that attempted to investigate this factor, the 55 cases with palatal canines which constituted the experimental group were assessed for dental age on the basis of the root development of the dentition, as seen on the radiographs and using the principles outlined in Chapter 1. This is a more accurate method of age assessment than eruption status, which may be influenced by local factors. In approximately half of the cases, development was seen to be in line with the norms for their ages, while in the remainder, significant developmental delay was seen. None of the affected cases showed advanced dental development, underlining the absence of a normal distribution for dental development and a strong tendency for lateness [42].

As seen in the discussion in relation to tooth size, a sexual dimorphism in the pattern of delayed dental development was found. A dental age significantly younger than the chronological age was noted among affected males more than twice as frequently than among affected females. In the males, late dental development was observed in half the individuals with palatal canines, which was accompanied by the presence of smaller than average teeth and a high frequency of lateral incisor anomaly. The other half of the males showed a timely developed dentition, a statistically non-significant increase in the incidence of lateral incisor anomaly, and mesio-distal width reduction only in the maxillary central incisors and first molars. This latter male sub-group therefore resembled the unaffected cases that made up the control group in this study. Among the females, late dental age was accompanied by a slight increase in lateral incisor anomaly, although overall tooth size was not affected. With such contrasting and partially conflicting findings regarding tooth size and retarded dental development seen in males and females who exhibit palatally displaced maxillary canines, investigations of tooth size, congenital absence and dental age which involve a combined male–female group of patients, will produce confusing results. This combination may obscure important differences that exist between the sexes.

Infra-occluded deciduous mandibular molars have also been found in larger numbers in cases with palatal canines [7, 42]. The explanation may be viewed from the same two distinct standpoints. The first assumes that the canine aberration is totally hereditary and is linked with the associated hereditary factors of lateral incisor anomaly, late dental development, small and missing teeth – to which it is now proposed to add infra-occluded deciduous molars as an additional hereditary factor. The alternative standpoint notes that over-retained and infra-occluded deciduous molars are often found in situations where their permanent successors are small or congenitally absent. In both these situations, resorption of the roots of the deciduous molars may be partial, sometimes involving only one of the widely divergent roots. This results in over-retention of the deciduous molar and, under these circumstances, it is more likely to become infra-occluded than one that sheds normally and in its due time. Dentitions with small or missing premolars are often associated with lateral incisor anomalies, which have been shown to be allied with palatal canines, as we have discussed earlier in this chapter. Thus, the association between infra-occluded mandibular molars and palatal canines is an indirect one, with linked hereditary factors bringing about changes elsewhere, which generate the guidance factor that causes the canine aberration.

There are more compelling arguments that seem to favour the guidance theory. For around half a century, it has been accepted that small and peg-shaped incisors represent a weak or partial expression (incomplete penetrance or microform) of congenital absence [43–48]. It follows, therefore, that if palatal displacement of a maxillary canine is under hereditary control, then one would expect to see this form of ectopia more frequently associated with congenital absence of the adjacent lateral incisor than with a lateral incisor of reduced size. Not only does this not occur, but the reverse appears to be true, with a significantly higher proportion of palatal canines involved with lateral incisors of reduced size [9, 10, 28, 36, 42].

In order to confirm or negate this reported tendency and to provide a firm scientific base for this apparent paradox, a study was designed in which the sample consisted of patients who were taken serially from the files of the Orthodontic Departments in the Universities of Jerusalem and Tel Aviv and in orthodontic private practice in Israel [49]. From a patient base of approximately 12 000 consecutively treated patients, only those exhibiting an anterior maxilla with the following pre-treatment conditions:

- a missing lateral incisor on one side
- an anomalous (i.e. peg-shaped or reduced) lateral incisor on the other, and
- a unilateral palatally displaced canine

were included in the study. Given such rigorous inclusion criteria, 19 patients remained to form the experimental group. The null hypothesis of the investigation was that if:

(a) reduced or peg-shaped lateral incisors are hereditary and represent a microform/incomplete penetrance/partial expression of congenital absence and (b) palatal displacement of maxillary canines is a hereditary condition and associated with the lateral incisor anomaly, then it is logical to assume that the palatally displaced canine will occur far more frequently in association with congenital absence of the lateral incisor than with the dimensionally diminutive tooth. Not only was this not so, but the results of the study showed that in an overwhelming majority of the cases (84%), the palatal canine was found on the side of the anomalous lateral incisor, with only three cases (16%) found on the side of the missing tooth. From this, it was concluded that environmental factors are strongly bound up with the causation of palatal displacement of the maxillary canine and explained in terms of a second stage impaction, as described above.

In the previous chapter, relating to impacted maxillary incisors, we pointed out that an early first phase of treatment is indicated in order to resolve the impaction of the affected tooth. In the long-term follow-up of many of these maxillary incisor impaction cases, it was observed that there seemed to be a substantial number of patients in whom there was a serious disturbance in the eruption of the canine of the same side. Accordingly, a study was undertaken to monitor the further development of children who had been treated for maxillary incisor impaction in the months and years following the resolution of the impaction in their first phase of treatment. Abnormality of position and disturbance of eruption were seen in the canine on the same side as the previously impacted central incisor in 43% of the cases. This was in contrast to the contralateral canine, where the rate of anomaly was only 4.7%. The abnormality in the affected side canine was expressed in several different ways, namely palatal impaction, buccal ectopia and pseudo-transposition with the lateral incisor [50]. The frequent occurrence of canine aberration only on the side where the central incisor had been impacted indicates clear environmental influence (see Figure 6.20). In a radiographic study [51] of a series of 122 Israeli patients with multiple congenitally missing teeth, 20.4% of the maxillary canines were congenitally absent and 42.4% were mesially displaced, of which 5.7% became impacted and mostly adjacent to a missing lateral incisor. Of the remainder, 5.6% were distally displaced and only 26.4% were positioned in their correct locations, mainly adjacent to a lateral incisor. It was concluded that displaced and impacted maxillary canines were very frequently found in this highly special group of patients, although inadequate sample size prevented the drawing of meaningful conclusions.

In summary, it may be learned from all this that the causation of palatal displacement or impaction of the maxillary canine is not due to any single factor. The simultaneously occurring factors related to or causing the canine impaction may be hereditary in nature, such as anomalous or missing lateral incisors, late dentitions and infra-occluded deciduous molars, which themselves are under total genetic control. Equally, however, the path of eruption of the canine may be influenced both favourably and unfavourably by conditions and events that are environmental in nature and include local therapeutic countermeasures, such as orthodontic space opening, prophylactic extraction of the deciduous canine or of a minuscule lateral incisor, or the existence and/or treatment of an adjacent impacted central incisor, or due to a soft tissue lesion or hard tissue body. So, while the aetiological stage may be set by either genetic or environmental factors, there is strong evidence that casts doubt on the simplistic and dogmatic view that the palatal canine itself is solely under genetic control. Unquestionably, maxillary canine eruption is influenced by and responds to altered conditions within its immediate environment. At the time of writing, therefore, 'there is currently too little robust statistical or genetic evidence to definitively ascribe malposition of the permanent canine as an isolated disorder of either genetics or environment' [49, 50].

Complications of the untreated impacted canine

Morbidity of the deciduous canine

Early morbidity of the deciduous canine is common for two reasons. First, its root may become markedly resorbed, even when its unerupted successor is quite distant from it, causing considerable mobility and eventual shedding without the possibility of replacement by the permanent tooth. This creates a problem in terms of restoration, since the space is usually too small for a satisfactory replacement either by the misplaced permanent canine or by some form of artificial fixed bridge pontic or implant.

The second reason that such a tooth may not survive relates to its relatively high susceptibility to interproximal (particularly distal) caries. As we have pointed out, it is still common to see a fairly extensive distal cavity in this tooth at around the age of 11 or 12 onwards, which may have been deliberately left untreated by a general practitioner who was unaware of the likelihood or existence of impaction of its permanent successor.

Cystic change

Loss of vitality may occur very early on in the carious process in the deciduous canine teeth, owing to the narrow width of the hard structures of these teeth and the relatively large pulp. Necrosis of the pulp and periapical pathology may then be asymptomatic. Under these circumstances, there may be a direct interconnection between the apical pathology and the follicular sac surrounding the impacted canine. This may stimulate an enlargement of the follicular sac, which is clearly seen on a periapical radiograph. It may also undergo cystic change, to produce a dentigerous cyst

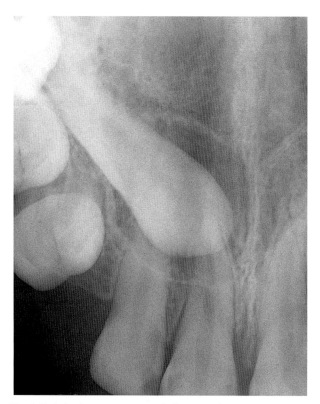

Fig. 6.13 A dentigerous cyst surrounds the crown of an impacted canine.

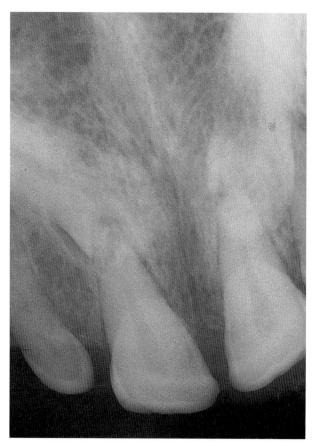

Fig. 6.14 Periapical view of maxillary incisor area in a 63-year-old female, showing advanced replacement resorption of the crowns of two impacted canines. The follicles of both teeth are almost completely absent and the teeth are very radiolucent, with poor definition.

(Figure 6.13). This may also occur without any relation to pathosis of the deciduous canine. In strictly radiological terms, an enlargement of the follicular sac to beyond 2 or 3 mm is generally considered to represent cystic change. In rare cases, these cysts may expand at the expense of surrounding maxillary bone and displace the canine higher and higher in the maxilla. Alternatively, the chronic periapical lesion on the deciduous canine may itself become cystic – a radicular cyst – and its subsequent enlargement may displace the adjacent teeth, including the palatal canine. The subject of cysts in relation to impacted teeth will be described separately in Chapter 11.

Crown resorption

The reduced enamel epithelium surrounding the completed crown of a tooth separates the crown of the tooth from the surrounding tissues. This intact epithelial covering may degenerate with age if the tooth remains unerupted, and its integrity may be lost. This allows bone and connective tissue to come into direct contact with the crown of the tooth. In time, osteoclastic activity will lead to resorption of the enamel and its replacement by bone – a process known as replacement resorption. Over a long period, repeated radiographs of the tooth will show a poorer definition of the profile of the crown, with the enamel becoming less and less contrasting in its opacity, highlighting this bone-for-enamel substitution (Figure 6.14). Subsequent surgical exposure of the crown of this tooth will show a pitted surface, which is difficult to separate from the surrounding bone, and sparse soft tissues.

This condition seems more likely to occur in adult patients in whom the impaction has been left untreated over two or three decades [52] and is almost certainly the reason why, when attempting treatment of impacted teeth in adults in the fourth or fifth decade of life, the chances that the tooth will not respond to orthodontic force may be high [53].

Resorption of the roots of the incisors

The proximity of the follicular sac of an unerupted permanent tooth to the roots of its deciduous predecessor appears to be the trigger that initiates the process of root resorption, probably as the result of pressure. The progress of this resorption process is then maintained by the further advance of the eruption of the permanent tooth, which moves into new areas vacated by the resorbing root. This is part of the normal process of transition from the deciduous to the permanent dentition.

Little is known about the reasons for the resorption of the roots of deciduous teeth that leads to their eventual

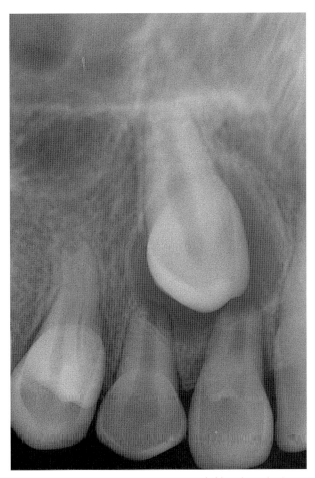

Fig. 6.15 The impacted canine crown is surrounded by a large dentigerous cyst, with associated root resorption of both the deciduous canine (to be expected) and the permanent lateral incisor (pathological).

shedding and why this does not normally occur with the roots of permanent teeth. Histologically, there is no way to tell the difference between the root tissue of a deciduous and that of a permanent tooth. Under certain conditions, however, the presence of an uncrupted permanent canine tooth may be associated with the resorption of the root of the adjacent lateral (Figure 6.15) or central incisor [53–59]. Furthermore, and in a manner similar to that seen with deciduous teeth, the progress of this undesirable phenomenon depends on further eruptive movements on the part of the impacted tooth.

In this context, it is perhaps pertinent to comment that the maxillary canine and, occasionally, the third molars are almost the only permanent teeth whose eruptive movements, both successful and unsuccessful, may cause resorption of the roots of neighbouring teeth to any significant degree. They are also the only permanent teeth that normally develop in close relationship with the developing apical areas of the roots of other permanent teeth, while the premolar teeth develop in a restricted area, encompassed by the roots of the deciduous molars and at a distance from

other permanent teeth. However, in rare instances an aberrant and deeply located second premolar may cause the resorption of the mesial root of the first molar (see Figure 7.2). The central incisors and first molars erupt before their adjacent neighbours, and the lateral incisor is related to the neck area of the crown of the central incisor. The canine, however, is closely related to the roots of the adjacent lateral incisor and first premolar, while it is still fairly high in the maxilla during most of its eruption period. Similarly, unerupted third molars in crowded positions in the ramus or tuberosity areas may come into close relation with the root portion of the second molars, where similar damage may occur.

Diagnosis

Unerupted permanent maxillary canines cause the patient relatively few problems, unlike impacted mandibular third molars. A retained deciduous canine may have a relatively poor appearance compared with a properly aligned permanent canine, but most patients are frequently unaware of the presence of and do not seek treatment for the over-retained deciduous canine. The discovery of palatal impaction is therefore usually made by the general dentist during routine dental examination.

Inspection

The maxillary permanent canine normally erupts at a dental age of about 11 years. Its non-appearance at this age should invite clinical inspection and radiographic investigation, especially if its antimere is present.

The maxillary incisor teeth are normally spaced and flared laterally until the age of 10 years, as described earlier in this chapter. Should this situation still be true by 11 or 12 years, the clinician should be suspicious, since this means that there is a detail missing from the mechanism that smoothly transfers Broadbent's ugly duckling stage into the final adult alignment, with interproximal incisor contacts. Indeed, a resultant persistent median diastema may be the factor that brings the patient to the office to request treatment, unaware of the impacted canine.

It is unlikely that a missing lateral incisor or a frankly peg-shaped incisor will be overlooked. Nevertheless, care should be taken to examine the size and shape of existing lateral incisors. Central and lateral incisors whose crowns have mesio-distal straight or slightly tapering sides and lack the classical proximal contour are usually small teeth and often develop late. Some of these are frankly peg-shaped, a condition defined by their widest mesio-distal dimension being at the CEJ.

Furthermore, the discovery of a late-developing dentition and a dentition in which there are missing teeth, other than the lateral incisors, should also be treated with a degree of caution. We have reported above that all these factors have been linked with palatally displaced canines, and this

possibility should be thoroughly investigated, both at the time when the phenomena are first noticed and in subsequent follow-up examinations that have been scheduled to oversee the smooth changeover from the mixed to the permanent dentition.

Abnormally positioned unerupted canines frequently affect the positions of neighbouring teeth, particularly lateral incisors. We have already pointed out that the root apex of a palatally impacted canine is usually in the line of the arch, with the crown mesially displaced, in addition to its palatal tilt. This brings it into close relation with the palatal side of the lateral incisor, often displacing its root labially. Clinically, this will be identified by the root being very prominent on the labial side of the ridge or by a lingual tilt of the crown of the tooth, sometimes into a cross-bite relationship. In contrast, a lateral incisor whose root orientation indicates a strong palatally directed apical displacement suggests that the unerupted canine is labially placed and, at least to a degree, also somewhat mesially directed (Figure 6.16).

Palpation

We have pointed out above that, under conditions of normal development, the permanent canine is palpable buccally above the deciduous canine for two or three years prior to its eruption. The buccal aspect of the alveolus should be palpated above the attached gingiva and up to the reflection of the oral mucosa. A wide convex contour of the bone is indicative of the canine, immediately beneath. Care should be taken not to confuse this with the narrower profile of the root of the deciduous canine. In the event that this contour is concave, the palatal side of the alveolar process should be palpated to see if there is a clue to its location there. The deciduous canine should always be tested for mobility. If this test is even mildly positive, it will suggest that the permanent canine is fairly close to the desired eruption path and that severe displacement is unlikely. In this situation, the unerupted canine may not be palpable on either side of the alveolar ridge.

When there is a strong palatally directed apical displacement of the root of the lateral incisor, we have just pointed out that it is likely that the canine is also mesially directed. In this case, the unerupted cuspid will usually be palpable high in the labial sulcus and much closer to the central incisor root than to its normal position in the arch, and the contour of the alveolus, superior to the deciduous canine, may be concave. Given the unusual distancing of the permanent canine from its normal position, its deciduous predecessor will have most of its root intact and will consequently exhibit no pre-shedding mobility.

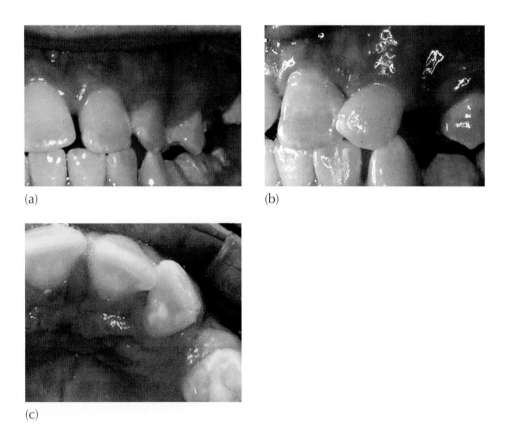

(a)

(b)

(c)

Fig. 6.16 Clues to the position of the unerupted canine. (a, b) The unerupted left canine presents a palpable bulge overlying the root of the lateral incisors, seen from the front and the lateral aspects. At the same time, the lateral incisor root appears to be displaced distally and palatally. (c) From the occlusal view, compared to the orientation of the central incisor, the lateral incisor root has a distinctly palatal axial inclination, which is less obvious in the other two views.

Radiography

As we shall see later in this chapter, to plan the strategy of mechano-therapy properly for a particular case and to obtain a pre-treatment assessment of the periodontal prognosis of the treated result, it is essential to know the exact positions of both the crown and the root apex of the unerupted tooth. A single periapical radiograph is essential to identify pathology, such as root resorption, obstruction and cystic change, but it should be supplemented by other films that will help to precisely locate the unseen tooth.

The use of a second periapical radiograph in the parallax method has the advantage of simplicity of technique and provides both the orthodontist and the surgeon with important information regarding positioning, although the exact locations of crown and apex are difficult to compute from these pictures. A true lateral view (as seen on the lateral cephalogram or on a tangential film) paired with a vertex occlusal or postero-anterior cephalometric view is technically more difficult to obtain, but will provide accurate three-dimensional positional information of the unerupted tooth in its most comprehensible form. However, these films involve a high radiation dosage for a relatively poor return on the amount of information provided.

The central portion of a panoramic radiograph shows the incisor region in the postero-anterior view and will indicate a palatal displacement as an overlap of the impacted canine with the roots of the incisors. This is by far the most popular method used today. The reader is referred to Chapter 2 for a description of how this film may be used by itself, or in combination with a periapical film, with an occlusal film or with a lateral cephalogram, to extract the maximum information regarding the position of an impacted canine.

We have already pointed out that plain film radiography cannot provide reliable information in the bucco-lingual plane and, therefore, incisor root resorption may occur and remain undiagnosed until well advanced. Additionally, the bucco-lingual distance that exists between the impacted tooth and its neighbour is very difficult to assess from these films and this may be an important factor in planning the strategy for the biomechanical resolution of the impaction.

For both these reasons and certainly in relation to all of the more involved cases, it is recommended that the immediate area be subjected to a computerized tomographic examination, using a cone beam volumetric scanning machine, to provide the maximum positional information while, at the same time, reducing the dosage of ionizing radiation to the absolute minimum.

Treatment timing

In normal circumstances, by the age of 9–10, it is usually possible to palpate a normally developing maxillary permanent canine tooth on the buccal side of the alveolus, high above its deciduous predecessor. In the presence of crowding, and particularly after the eruption of the first premolar, the bulging of the unerupted canine is emphasized. The greater the degree of crowding the greater will be this displacement and the more palpable will the canine become, as its eruptive process brings it further and further down on the facial side of the dental arch. It follows, too, that the greater the buccal displacement the greater the risk that it will erupt through oral mucosa, higher up the alveolar process, rather than through attached gingiva.

In the event that the tooth is not palpable at this age, radiographs should be taken to assist in locating the tooth accurately and to secure other information regarding the presence, size, shape, position and state of development of individual unerupted teeth and any pathology. In a patient younger than 9 years, the radiographs will not usually show abnormality in the position of the unerupted canine teeth, even if the canines are not palpable and even if they are destined subsequently to become palatally displaced. Many of these non-palpable canines will finally erupt into good positions in the dental arch in their due time, provided that there is little or no mesial and palatal displacement of the crown of the unerupted tooth. It may be argued that even canines with an initial mild palatal displacement will achieve spontaneous eruption and alignment despite a first-stage displacement, if they undergo secondary correction (see 'The guidance theory' in the section on Aetiology above). Other canines, however, will not erupt, and their positions may worsen in time, as may be seen in follow up radiographs. If it were possible to distinguish between the two early enough, a line of preventive treatment might be advised.

Preventive treatment and its timing

Using panoramic radiographs of young patients in the mixed dentition, Lindauer et al. [21] were able, to a relatively low degree of reliability (78%) only, to predict palatal impaction on the basis of canine overlap of the root of the lateral incisor.

Extraction as a means of prevention: deciduous canines

As we indicated earlier in this chapter, several authorities [20, 22] prescribe the extraction of the deciduous canine teeth in an attempt to encourage the permanent canines to erupt. They have recommended seeing the patient and diagnosing the palatal positioning before the age of 11 years, and have shown that extraction performed at this time offers a good prognosis for the natural eruption of the canine, with 78% of the canines in their sample erupting into a clinically correct position. Caution must be advised in interpreting these results, however, since the authors did not study an untreated control group and thus are not in a position to determine just how many of these teeth would have erupted without this preventive treatment [21] (Figures 6.17 and 6.18).

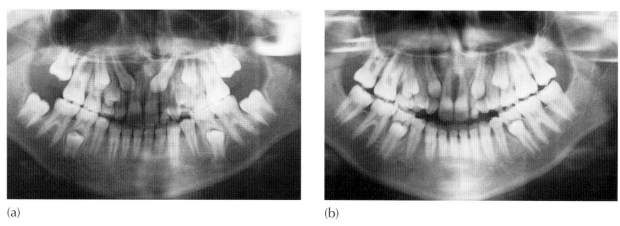

(a)   (b)

Fig. 6.17 (a) A case diagnosed from this panoramic view as having bilateral palatal canine displacement and referred for extraction of the deciduous canines. (b) A year later, a repeat film shows great improvement in the position of both canines and normal eruption of the canines is imminent, despite the fact that the deciduous canines have not been extracted.

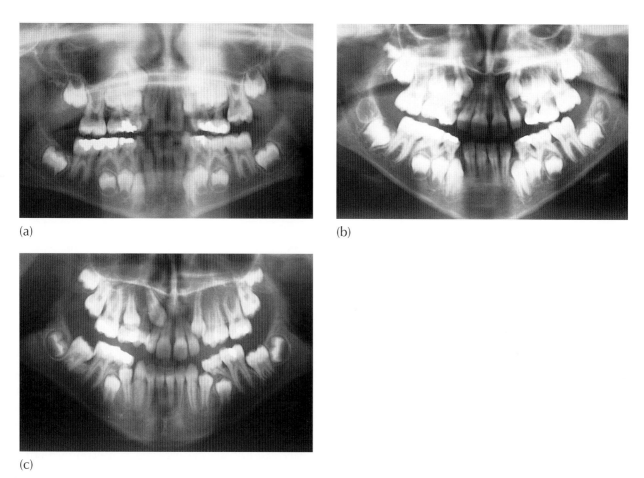

(a)   (b)

(c)

Fig. 6.18 (a) A case of early crowding treated by extraction of four deciduous canines, to relieve crowding at age 8 years. No hint of impending palatal displacement of the canine is discernible on this panoramic film. (b) One year later, the incisors are aligned and spaced. Extraction of the four deciduous first molars (the second stage of serial extraction) was advised. (c) The panoramic view taken a year later reveals the maxillary right canine in a palatally displaced location, despite early extraction of the deciduous canine. Treatment of this case may be seen in Figure 6.41.

From their study, Ericson and Kurol concluded that the prognosis becomes less favourable as the palatally displaced canine's medial overlap of the lateral incisor root increases and as the angle between the long axis of the canine and the mid-sagittal plane widens. They also noted that, if positional improvement of the canine was not evident within 12 months of the prophylactic extraction, it was unlikely that improvement would occur.

From this discussion and from some considerable anecdotal clinical experience, we may assume that, under certain circumstances, the extraction of a maxillary deciduous canine may be a useful measure in the prevention of incipient canine impaction. To achieve maximum reliability, the following conditions should be met before extraction is advised:

1. The diagnosis of palatal displacement must be made as early as possible.
2. The patient must be in the 10–13-year age range, preferably with a delayed dental age.
3. Accurate identification of the position of the apex should be made and confirmed to be in the line of the arch.
4. Medial overlap of the unerupted canine cusp tip should be less than half-way across the root of the lateral incisor, on the panoramic view.
5. The angulation of the long axis should be less than 55° to the mid-sagittal plane.

4 and 5 on this list represent conditions that, if not fulfilled, may still lead to spontaneous eruption and alignment, so that while the chances are reduced, extraction may still be worth considering.

Notwithstanding, in two studies carried out in Italy by the same researchers, their earlier paper [60] found no statistically significant differences between those cases in which the deciduous canines had been extracted and those in which no extractions were performed. Contradictory figures were presented in their second paper [61] in which deciduous canine extraction produced 65% success against 36% for non-extraction cases.

One further factor that is frequently missed when studying the radiograph and, therefore, not normally taken into account as a causative agent is the occurrence of non-symptomatic soft tissue pathology, specifically chronic periapical abscesses associated with non-vital deciduous teeth and enlarged dental follicles or early dentigerous cysts. These may lead to quite severe displacement and impaction of the unerupted permanent canine. Their elimination may often have a favourable outcome even when displacement is quite extreme (Figure 6.6). The reader is referred to Chapter 11 for a description of the treatment of teeth impacted in positions of extreme displacement by dentigerous cysts.

Given that there is no truly reliable method of early detection of a potential palatal displacement [21], a claim that it was the pre-emptive extraction of the deciduous canine that had elicited the normal eruption of a permanent canine must be viewed with some reservation on the basis of the present state of our knowledge. Clinical experience would lead us to be encouraged by the procedure in many cases, but an accurate assessment of its efficacy has still to be determined.

In the mixed dentition period, the unerupted maxillary canine is often held too far mesially by the mesio-distally wide crown of the unerupted first premolar immediately distal to it. If clinical experience is to be countenanced in these days of evidence-based treatment decisions, the author has found that, together with the extraction of the deciduous canine, there is merit in the simultaneous extraction of the adjacent first deciduous molar. The rationale for this 'pearl' is that loss of the deciduous molar encourages a very rapid eruption of the first premolar. With its eruption, the large crown of the tooth erupts and a much narrower cervical root is substituted at the level of and distal to the unerupted canine. This creates a potential void distal to the canine, which appears to encourage the latter to drop back distally into the space that has relatively suddenly become available. There is reason to believe that this may redirect a potentially wayward canine and encourage its more normal eruption.

Extraction as a means of prevention: first premolars
Within the minority group of patients with impacted canines who are considered to be extraction cases, usually because of incisor crowding, a class 2 relation or bimaxillary protrusion, the choice of teeth for extraction usually devolves on the first or second premolar teeth. The reasons for this choice are bound up with the history of orthodontics itself. This offers much potential benefit to the displaced canine, since the proximity of these teeth to the canine facilitates the immediate provision of space close by. It also affords considerable opportunity for a spontaneous improvement in the canine position (Figure 6.19) during the early levelling and aligning phase of the mechano-therapy.

Extraction as a means of prevention: lateral incisors
We have noted above that many of the impacted cases that we see are associated with anomalous lateral incisors. At the end of the treatment it is often necessary to alter the shape of these teeth by prosthetic crowning, laminates or composite build-ups to make them aesthetically acceptable, particularly those that are peg-shaped. We have pointed out earlier that palatal canine cases generally have spaced dentitions, comprising small teeth, such that crowding and the need for extractions in the overall treatment are unusual. Nevertheless, if extraction has to be made to treat the overall malocclusion, consideration should be given to the extraction of these malformed lateral incisors as an alternative to the conventional but healthy and anatomically perfect first premolars.

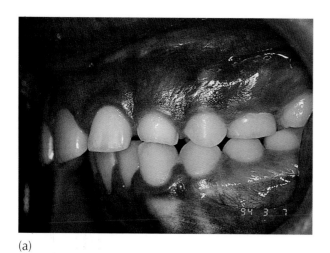
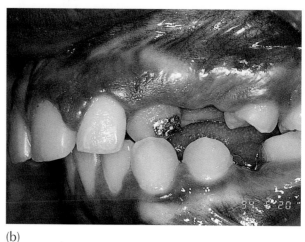

(a) (b)

Fig. 6.19 (a) The left side of a case with bilateral maxillary palatal canine impaction. The maxillary deciduous canine, deciduous second molars and first premolars were extracted and an attachment placed on each of the exposed canines. No active orthodontic treatment was commenced. (b) The same patient seen 14 weeks later. Both canines and second premolars have erupted spontaneously and to a similar degree.

We have described how the guidance theory of eruption of the canine offers a cogent argument of how palatal displacement of the canine may occur. In those cases described as a first-stage displacement, it was pointed out how the vertical wall of the alveolar process on the palatal side steers the vertically and palatally directed eruption on a more buccal course, to produce a secondary correction. This continues until the developing canine comes up against the root of a late-developing lateral incisor, which forms a barrier to its further progress. A second-stage impaction is thus created.

Logically, the removal of this barrier should lead to a natural improvement in the position and eruption status of the impacted canine – the second-stage impaction with secondary correction. In practice, clinical experience shows this to be largely true (Figure 6.20).

Extraction of the lateral incisor is not a suitable procedure in most cases, but in those patients where it is indicated, treatment time may often be very brief. However, a normally sized canine adjacent to a central incisor may create a marred and unsatisfactory appearance, particularly if the central incisor has a poor profile (frequently seen in these cases). Furthermore, by lining up the canine and first premolar in place of the lateral incisor and canine, respectively, a discrepancy between upper and lower tooth sizes may compromise the occlusion.

Orthodontic space opening

The only preventive measures that we have described up to this point have involved the extraction of teeth adjacent to the impacted, namely the deciduous canine, the lateral incisor or the first premolar, in the hope that the impaction will resolve spontaneously. An alternative and sometimes supplementary line of preventive treatment involves the generous opening of space for the teeth, using orthodontic appliances. One of the primary functions of orthodontic treatment preparatory to the treatment of impacted teeth is the creation of space in the dental arches for the impacted teeth. When this is done, unerupted teeth may often begin to improve their positions, as will be seen on repeat radiographs, and may often erupt without surgical intervention (Figure 6.21). This is clearly due to an alteration in the relation of the canine crown to the roots of the incisors and a concomitant alteration in the guidance influence of these teeth. It provides further evidence to support the guidance theory of impaction [60–62].

Rapid maxillary expansion

Over the past few years, there has been speculation regarding the efficacy of a rapid maxillary expander as a means of prevention. There would not seem to be any logical reason to suppose that skeletal mid-palatal suture-splitting expansion should provide the impetus for the spontaneous correction of an incipient canine impaction – a latero-lateral response to a sagittal problem. Nevertheless, a study by the same Italian researchers has shown that the method used on 7.6–9.6 year olds will increase the chances for eruption from 13.6% for an untreated control group to 65.7% for the group treated with rapid maxillary expansion [63, 64]. The diagnostic parameter used by the authors for confirming impending impaction was a reduction in the distance between the unerupted canine and the midline on a postero-anterior cephalogram in these very young patients!

We may summarize that, on the basis of the evidence from the many studies presented here, there are several steps that may be taken in order to reduce the chances of palatal impaction of maxillary permanent canines, provided that the patient is seen early enough. These include the extraction of deciduous canines, rapid maxillary expansion and distal movement of posterior teeth to create exces-

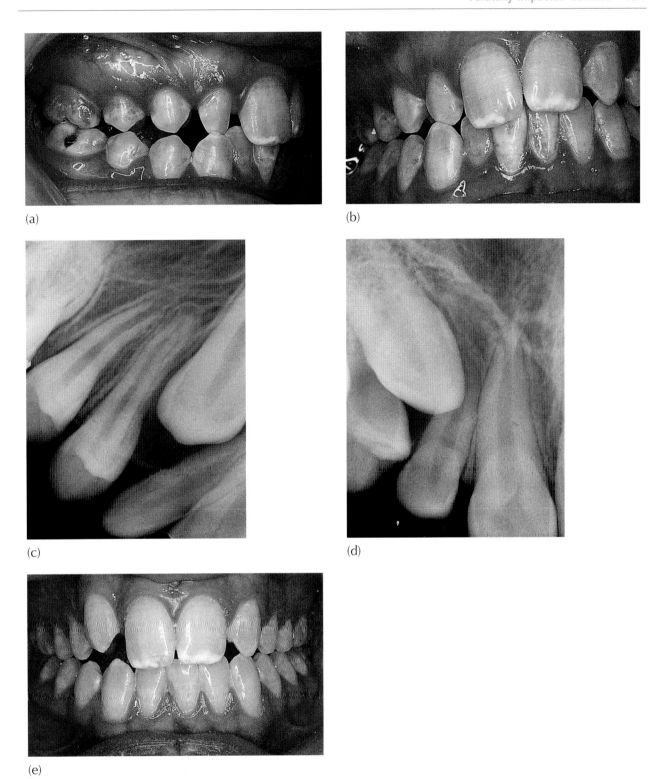

(a)

(b)

(c)

(d)

(e)

Fig. 6.20 (a, b) A palatally impacted right canine is adjacent to the peg-shaped right lateral incisor, while the opposite canine has erupted in place of the congenitally absent lateral incisor. (c, d) The periapical films used to diagnose the palatal position of the canine by parallax. At the time of extraction, the palatal position of the canine was confirmed clinically. (e) The canine has erupted on the buccal side.

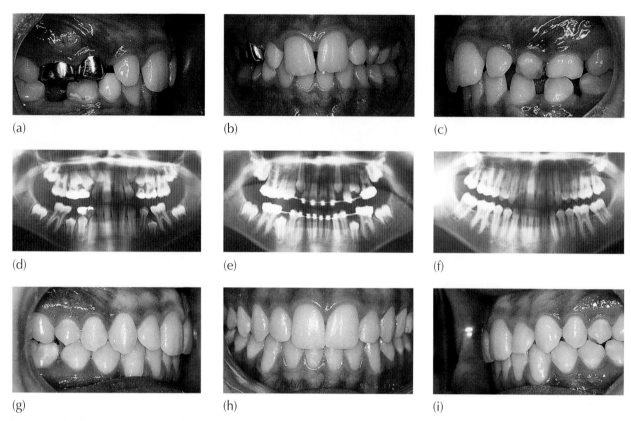

(a) (b) (c)

(d) (e) (f)

(g) (h) (i)

Fig. 6.21 (a–c) A class 2, division 2 case with crowding in the maxillary arch and severe space loss due to early extraction in the mandibular arch. (d) A panoramic view shows a palatally displaced right maxillary canine. (e) A similar radiograph taken following distal movement of all four molars and space reopening. Note improved positions and prospects of all the unerupted teeth, particularly the canine. (f–i) The final dental alignment and occlusion. The right maxillary canine erupted unaided.

sive space for the canine. The exploitation of two or more of these methods in a specific case will, in all probability, be synergistic.

Timing of mechano-therapy

Most cases are not identified early enough to take advantage of these preventive steps, and will usually be seen for the first time by the orthodontist only after the initial and, by then, more obvious diagnosis has been made by the generalist or paedodontist. A coexisting malocclusion will have often been the reason for the patient requesting treatment, and the impacted tooth may have been discovered only as the result of the orthodontist's routine clinical and radiographic examination.

The patient is generally in the full permanent dentition stage, with the exception of the deciduous canine of the affected side. Sometimes, the remainder of the dentition is in a close-to-ideal alignment and inter-arch relation, as has been pointed out earlier, although a minor degree of local tooth malalignment may often be seen. This generally includes a laterally flattened or collapsed archform (Figure 6.22) and space loss in the immediate area, with space opening more mesially [9]. In only about 15% of cases [16, 17] is actual crowding present.

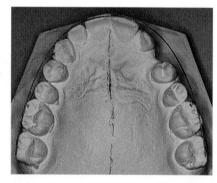

Fig. 6.22 A preformed archwire blank laid over the occlusal surface of the teeth closely conforms to this dental arch, except in the area of the impacted canine, which shows lateral flattening.

The periapical radiographs should be carefully scrutinized to discover any evidence of resorption of the lateral incisor roots. Should this be seen, orthodontic treatment, designed to rapidly deflect the developing canine away from the incisor, should be undertaken as soon as possible. If the resorption is advanced, and in the relatively unlikely event that this is an extraction case, consideration should be given to extraction of this affected lateral incisor. A reasoned

approach to the problem of incisor root resorption seen in association with an impacted maxillary canine will be discussed at length in Chapter 7.

For the most part, however, there is no reason to hurry into treatment simply because palatal displacement has been diagnosed. The patient must first be prepared for the treatment to be undertaken, initially by explaining the nature and ramifications of the problem using the radiographs and plaster models as visual aids. The principal function of this exercise is to overcome complacency that most of these patients have, since their problem is one in which the facial appearance and oral function are rarely compromised. The aim is to inspire motivation and the necessary future cooperation.

The dentition must then be protected against the incipient dangers that the placement of appliances is likely to generate. The necessary measures include a high level of oral hygiene and the use of appropriate fluoridating procedures, at home and in the dental office, as well as the treatment of any carious or periodontal lesions.

After a period of a few weeks, during which the patient will, it is to be hoped, have undertaken these oral hygiene responsibilities on a regular basis, an oral examination should show pink, firm and stippled gums and an absence of plaque on the teeth. This being so, the time will be ripe to begin orthodontic treatment. In the non-cooperative patient, treatment should be delayed until the above conditions are fulfilled. Should the dental awareness of the patient be too low for this ever to occur, alternative treatment modalities should be considered, particularly prosthetic replacement, although for these to be successful in the long term they may be just as reliant on oral hygiene as is orthodontic treatment. Nevertheless, an operative decision may be delayed for quite a long time, so long as there are no signs of morbidity, particularly root resorption. This means that periodic radiographic monitoring will need to be carried out on an infrequent but regular basis. Postponement for a few months or even a year is rarely a problem in straightforward orthodontic terms, and if it serves to bring the patient round to the ways of proper home care, then the time spent will have been worthwhile.

Unfortunately, as already mentioned, the dental development is often delayed, which is why these patients reach the permanent dentition stage with the canine impaction diagnosis made only at the age of 14 or 15 years. Thus, from a social point of view, the patient may be less inclined to wear appliances if further postponement is entailed.

From the strictly developmental point of view, the best time for therapeutic intervention is when the root of any affected tooth is of a length that is seen at the time of normal eruption. For the canine, this is a little in excess of three-quarters of the potential root length, which is virtually always present by the time the diagnosis of palatal displacement may be determined.

General principles of mechano-therapy

When a patient arrives at the orthodontist's office and a palatally displaced and unerupted canine is diagnosed, treatment must be planned in a disciplined manner. We have seen in Chapter 5, in regard to the impacted maxillary central incisor, that orthodontic preparation of the case is required and that surgical intervention is not to be undertaken in any haphazard or unplanned manner. By and large, appliances that are to be used to disimpact, erupt and align these teeth are, with very little modification, the same appliances that are used to align the other teeth. For impacted incisor cases, the patient is in the early mixed dentition stage and the time is not ripe for the treatment of the overall orthodontic problem. Thus, a first phase of treatment is planned to deal with the incisor anomaly only, leaving the remainder of the malocclusion to be treated in a second-phase intervention much later. This is not so in the present context where, with the exception of the impacted canine, the full permanent dentition is usually erupted. For this reason, the local anomaly and the overall malocclusion are usually dealt with together in one full and comprehensive orthodontic treatment plan.

A diagnosis of the overall malocclusion needs to be made and a problem list set out, which includes the palatal canine. The problem list is then sorted into a treatment priority list, in which alignment of the impacted canine should precede many of the other items, but only after space has been made for it in the dental arch. The same principles that were used in the planning of treatment in Chapter 5 are applicable in the present context, although we shall now be dealing with the treatment of the entire dentition and not merely the area immediately adjacent to the impacted tooth.

These principles need to be adapted to the new circumstances, and may therefore be presented as follows:

1. The appliance should have the capability to level and align all the erupted teeth in the same jaw rapidly and, with controlled crown and root movements, to open adequate space to accommodate the impacted tooth. As we saw when dealing with impacted incisors, this space is required both at the occlusal level and between the roots of the adjacent teeth for their entire length. This stage requires the use of fine levelling and aligning archwires initially, and space subsequently gained with the use of a more substantial base arch and sliding mechanics.

2. With the initial alignment achieved and no further movement of individual erupted teeth needed, these teeth are transformed into a composite and rigid anchorage unit, in which each tooth plays an integral part. This is done with a heavier wire, whose gauge is as large as the bracket will take, in order to allow as little 'play' of wire within the bracket as possible, thereby maximizing the anchorage value of each erupted tooth.

3. The surgical exposure of the crown of the impacted tooth should be performed in a manner that will achieve a good periodontal prognosis of the treated result. For a closed procedure, an attachment is bonded to it and the flap fully closed, with only a fine ligature wire leading through the gingival tissue to the recovered tooth. Alternatively, an apically repositioned flap procedure or window technique may be indicated and a dressing placed as required, while the placement of an attachment may be performed then or at a later date.

4. Using an auxiliary means of traction from the now rigid orthodontic appliance, a gentle and continuous light force, with a wide range of activity, is applied to the tooth and is aimed at erupting the impacted tooth along a path that is free of obstruction from neighbouring teeth [65–68]. When the impacted canine is located mesial to the lateral incisor, there is no direct path from the canine to the space created for it in the arch, since the root of the lateral incisor stands directly between the two. This means that the tooth will need to be diverted in a different direction first, to circumvent the obstruction, and, only then, drawn along a new and unimpeded path to its place. This will be discussed later on in this chapter.

5. There should be final detailing of the position of the formerly impacted tooth, together with that of all the other teeth in both jaws. A class 2 or class 3 dental relation will usually be reduced at this point.

These principles are by no means immutable and the orthodontist should always be prepared to re-evaluate and adapt them in the light of other findings in a particular situation and to suit a particular case. There is one specific palatal canine scenario in which a race against time exists in relation to the timing of treatment for class 2 and class 3 cases which have a skeletal component. Many of these cases require to be treated with the use of orthopaedic/functional appliances whose aim is to realize as much of the growth potential of the deficient jaw as is possible. In these cases, there are several potentially conflicting factors which need to be carefully managed and prioritized in relation to treatment sequence. These are as follows:

1. The results in orthopaedic/functional treatment are best realized during the growth period.
2. Maxillary canine impaction is much more frequent in females who complete their growth much earlier than males.
3. Patients with palatally impacted canines frequently exhibit an overall late dental development, which could mean that menarche and the cessation of growth could occur before the eruption of the full permanent dentition (i.e. a mature 15–16-year-old girl with a dental development more akin to age 12).
4. The resolution of the impacted canine may take considerable time and is not dependent on whether further growth may be expected.

Thus orthopaedic/functional treatment will be advised first, aimed at reducing the skeletal discrepancy while there is the promise of growth. The canine impaction would then be treated only when the dental arches had been brought to a class 1 relation and in the late mixed/early permanent dentition period – but in a 16+-year-old female.

At this juncture, it is important to recognize two very important exceptions to the order of things that we have outlined above. The priority rating for an impacted canine which is causing resorption of the roots of adjacent teeth will overrule space considerations and will make treatment of the canine a matter for immediate attention. In Chapter 7, the discussion will concentrate entirely on these cases in which the progress of the resorption often proceeds at an alarming rate and in which time spent preparing space can lead to loss of the incisor tooth. Once incisor root resorption has been positively identified, therefore, the full focus of treatment should be directed at distancing the impacted tooth from the adjacent root as quickly as possible. In the absence of space in the arch, as will be seen in the examples illustrated in that chapter, the tooth should be exposed as early as possible and traction applied to erupt the tooth either into the palate or into the labial sulcus. In these locations, the tooth is out of harm's way and further resorption of the incisor root will be drastically reduced or entirely eliminated [69] even when subsequent orthodontic movement to modify the position and angulation of the damaged incisor will be performed with forces directed from the appliance. Space may then be provided in the normal way and the canine realigned in due course.

In order to provide space for an impacted tooth, there is a sequence of tasks that orthodontists normally go through with fixed orthodontic appliances that constitutes a largely immutable routine. The sequence first concerns levelling and alignment, which are achieved with the use of light, springy archwires. Only when a heavier and more rigid base arch can be inserted do we begin to create space, using compressed open coil springs and sliding mechanics. If all of this is performed on round cross-section archwires, the adjacent teeth are tipped mesially and/or distally, labially and/or lingually, intruded and/or extruded and rotated. With few exceptions, for every orthodontic bracket in use today, the horizontal slot generates mesio-distal uprighting movements which, in some cases, may move the root apex several millimetres mesially and/or distally. If rectangular archwires are used at any stage in this initial sequence, then torqueing movements of the root apex in a lingual and/or labial direction will also be introduced.

Under the heading of 'Diagnosis' earlier in this chapter, there is a paragraph describing clinical clues that may be present to indicate the position of the impacted tooth, by virtue of the displacing effect that the tooth has on the positions of the neighbouring teeth, particularly lateral incisors. On the radiographs of the area, one may usually discern a cause-and-effect relation between the angle of the

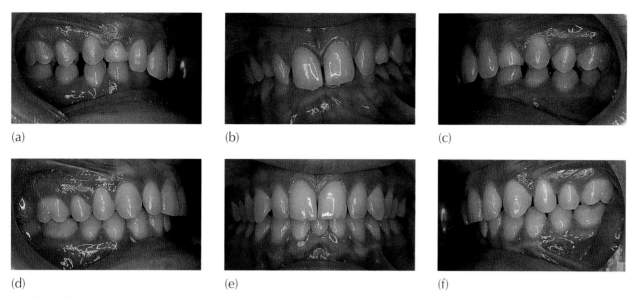

Fig. 6.23 (a–c) Inadequate space for unerupted permanent canines with inter-incisal spacing. (d–f) The permanent canines in place at the completion of treatment.

unerupted canine and that of the lateral incisor. It follows, therefore, that virtually every 'corrective' root movement performed during the levelling and alignment stage will have the effect of forcing the root of the incisor against the crown of the impacted canine. In some cases, this may encourage a realignment of the orientation of the long axis of the canine, which may then rekindle its eruptive potential and spontaneous resolution of the impaction may occur, obviating the need for exposure surgery. Equally, it may further displace the canine, which may not be a serious consequence, since exposure surgery had been the planned line of treatment anyway.

The third and most dangerous possibility is that the clash between the two teeth may result in resorption of the root of the incisor. This distressing outcome will paradoxically be seen alongside the 'success' of the incisor uprighting effort. Radiographic monitoring of the initial few months of treatment, using simple periapical films, should certainly be considered, but unexpected difficulty and duration of what is normally a simple and rapid movement should alert the practitioner to this possible eventuality.

There are essentially four ways in which space may be provided for the palatally displaced canine:

1. Existing incisor spacing is due to failure in the progress from the ugly duckling stage of development to completion of the natural alignment [13, 14] and it may be closed off by moving the lateral incisor mesially to increase the space in the canine area. It was pointed out at the beginning of this chapter that the final stage in anterior space closure occurs when the canine erupts and influences the lateral incisor to move mesially. It was also pointed out that impacted maxillary canine teeth

are often intimately linked with small and peg-shaped lateral incisors and with small teeth in general. These are the reasons that it is common to find anterior spacing in these patients (Figure 6.23).

2. Improving archform. When the maxillary permanent canine erupts normally, it does so along a more buccal path than the deciduous canine and slightly buccal to the lateral incisor and first premolar, earning the title 'cornerstone of the arch'. Comparing the two sides of the maxillary arch in a unilaterally affected patient, we have already pointed out that in the canine areas there is a much narrower maxillary width on the side of the deciduous canine than on the normal side. Exploiting improvement in the form of the arch in this region, which is almost automatically achieved with any fixed orthodontic appliance and preformed arches, will add two or three millimetres of space for the displaced tooth (Figure 6.24).

3. Increasing arch length. Studying the plaster casts from the occlusal aspect, it may often be noted that there is mesio-palatal rotation of the first molars. Additionally, mesio-buccal rotation of the first premolar is a frequent occurrence in these cases. Correcting these rotations can provide the much-needed millimetre or more of space on each side. If crowding is mild, the use of a headgear or class 2 traction against open coil springs is recommended in order to move the maxillary molars distally. This will provide the extra space which may then be concentrated in the canine area, using a multi-bracketed appliance system (Figure 6.25).

4. Extraction of teeth. When crowding is more severe, particularly where there is also a class 2 dental relation that is to be treated with the use of intermaxillary elastics, the

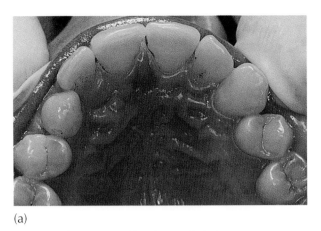

(a)

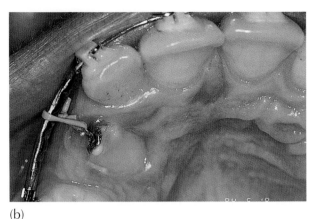

(b)

Fig. 6.24 (a, b) Improving archform has provided adequate space.

extraction of a premolar tooth on each side of the dental arch in the maxilla and usually in the mandible will be required. With the extraction of the premolar, space for the impacted canine is immediately and very locally available, and so appliance therapy is not needed to provide the space. Second, following the loss of the first premolars, alignment, levelling and rotation of the remaining teeth are very much simplified. Third, with a local anaesthetic already covering the area and a surgical wound inevitable, it is logical to extract the deciduous canine and expose the impacted canine at the same time to reduce the number of surgical interventions and post-surgical discomfort to a minimum. Thus, in extraction cases, it may sometimes be recommended that the surgical exposure be undertaken prior to the placement of an appliance (Figure 6.19).

When orthodontic appliances are placed at the beginning of treatment, the achievement of a good archform is an important first step in the maxillary arch. After the initial levelling wires, a coil spring is placed on a more substantial archwire of idealized form to increase the canine space by moving the lateral incisor mesially and the first premolar distally, until interproximal contacts are established elsewhere in the maxillary arch. This will usually provide more than enough space for the unerupted canine.

A heavier-gauge archwire is then firmly ligated into the maxillary appliance and the space for the canine must be retained. In the various edgewise and prescription pre-torqued bracket techniques, as heavy a rectangular base arch as possible should be used. In the latest Tip-Edge® Plus™ technique, a round 0.020 in or 0.022 in wire or, preferably, a rectangular 0.0215 × 0.028 in wire should be used as the base arch, with the addition of an auxiliary nickel–titanium archwire threaded in the horizontal channel behind the main bracket slot, or uprighting springs and torqueing auxiliaries to act as 'brakes' if necessary.

The space that has been reopened for the canine may be maintained using the same coil spring, which will need to be deactivated. However, it is difficult to adjust the spring to maintain the space accurately, and one will usually find that the space will increase or decrease slightly over the succeeding months. Furthermore, a coil spring quickly fills with food particles and is impossible to clean effectively. A much better alternative involves using a measured and slightly curved length of stainless steel tubing, which is threaded on the archwire and tied between the brackets of the premolar and lateral incisor, in place of the coil spring. This adds a great deal of rigidity to the archwire in the area of greatest importance and helps in resisting distortion, thereby providing an excellent and firm base from which to apply force to the impacted canine. Many of these canines have to be moved over a long distance to bring them into the arch, and several will require root movement of the different types before they may be properly brought into position and the case completed. This inevitably expends anchorage. The measures and precautions that we have described will contribute much to the preservation of anchorage.

The need for classification of the palatal canine

During the orthodontic treatment of a patient, as with any other prescribed form of medical or dental treatment, attention is paid to achieving the maximum benefit that the approach has to offer, while sustaining the minimum possible adverse collateral changes to the health of the dentition and its supporting tissues that may be caused by the treatment. To this end, the orthodontist must ensure an adequate level of oral hygiene before and during the period when the procedures are performed and the forces generated by the appliances must be within certain limits, compatible with physiological tooth movement, so that permanent and irreversible damage is not inflicted on the dentition.

In an extraction case, the decision regarding which teeth to extract is usually made on strictly strategic criteria, insofar as certain teeth require to be brought to particular

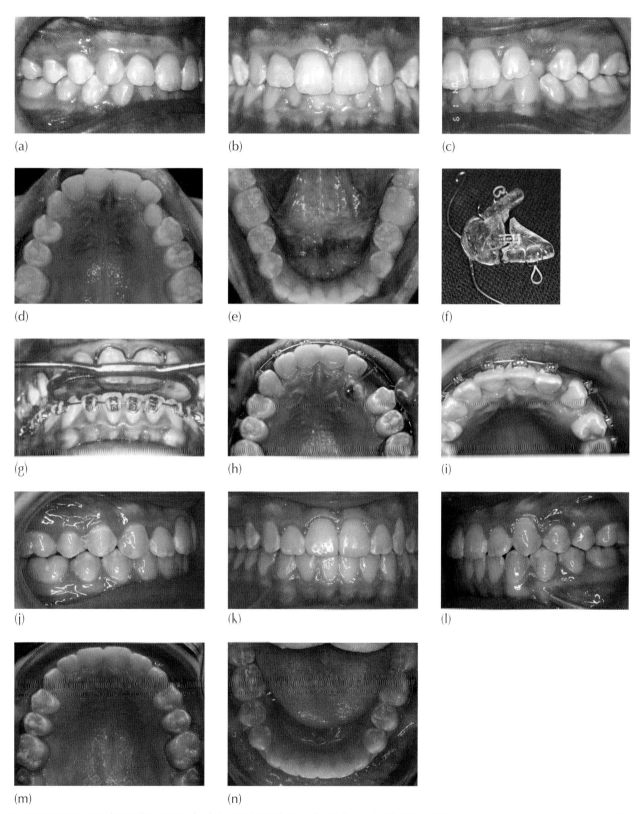

Fig. 6.25 Creating space by distal movement. (a–e) The initial clinical views of a mildly crowded dentition with incisor retroclination and a deep incisor overbite. The left deciduous canine is in cross-bite. (f, g) A safety, removable, acrylic high-pull headgear appliance with the facility to move posterior teeth selectively distally (with two jackscrews), to torque the incisors (Bass torqueing spring), in addition to the overall distalizing effect. Note the facility to use inter-maxillary elastic from the Adams clasps on the molars, to aid in the distal movement of the mandibular dentition. (h, i) More than adequate space has been created for the canine to be aligned. Note how ill-advised labial traction from a palatal eyelet has introduced a degree of undesired rotation into the canine as it achieves its place in the arch. This needs to be redressed in the final stage. (j–n) The treated result seen five months post-treatment.

places and appropriate anchor teeth need to be chosen in order to achieve this. Given a good prognosis of each of the teeth in the mouth, with no severely carious teeth and excellent periodontal health, the criteria of orthodontic treatment strategy are the guide to this extraction decision. However, when a tooth or teeth are present whose long-term prognosis is in doubt, such as a molar tooth that is in need of root canal treatment and a post-and-crown restoration, this becomes an additional factor that must influence the choice of tooth for extraction. It would not make long-term sense if the immediate and beautiful orthodontic results were to be predicated on teeth that would not be present in the mouth a few years later, while healthy teeth of excellent prognosis had been sacrificed in the name of strategic convenience.

Maxillary canine impaction in a case where teeth require to be extracted as part of the overall orthodontic treatment presents a similar dilemma. The canine is a tooth that has an important role in the establishment of a good functional occlusion. It has a long root and contributes much to the patient's appearance, particularly the smile. These are factors which makes it a very valuable tooth and one worth the expenditure of considerable effort to bring it into position. Its substitution by a first premolar is not usually desirable. It is inappropriate to automatically and blindly extract an impacted canine with a good prognosis in preference to an erupted first premolar, just as it is inappropriate to extract a first molar rather than a first premolar simply because the molar has a small occlusal filling. But what if the long-term prognosis of the canine is poor because of its initial intractable position or as a consequence of its having been through the processes of surgical exposure and orthodontic alignment over the period of many months or years that were spent in its meticulous alignment? In such a case, perhaps it would have been better to remove that particular canine at the outset and to have brought the first premolar to its place [70].

Since extraction cases are very much in the minority among patients with palatally impacted canines [16, 17] every effort must be made to bring the canine into the arc, and to do this in a manner that will provide it with its best possible periodontal prognosis. The clinician must carefully assess the several factors that affect the prognosis of the results of treatment in each individual case prior to the beginning of treatment, in order to be in a position to make the optimal decision regarding the choice of tooth for extraction.

At the time that the patient's records are being studied in order to formulate a suitable treatment plan, it would be helpful if there were a way in which it is possible to assess the long-term prognosis of an impacted canine before treatment is started. It is therefore crucial to seek a key that may be available to help us decide which canines will be adversely affected in periodontal terms and which will take an inordinately long time to resolve their impaction:

- as a result of the creation of surgical access to them
- by the relative difficulty in orthodontically moving them into alignment.

From the surgical aspect, which we have discussed in Chapter 3, minimal exposure and full flap closure (with attendant attachment bonding) is the preferred line of treatment, aimed at primary healing, for the majority of cases.

Does a tooth that requires a whole range of different types of orthodontic movement pay a periodontic penalty, in the final analysis, in comparison with one that is more simply aligned?

Teeth that are mechanically erupted bring with them a generous amount of alveolar bone. It has been shown in studies in Israel [71–73] that the assisted eruption of buried teeth with the use of orthodontic appliances produces a collar of alveolar bone around the erupted tooth that is greater than that seen on normally erupted adjacent teeth. These studies have shown this to be true only where surgical exposure was conservative and did not involve removal of the entire follicular sac. Radical surgery leads to less bone support than is present in a normally erupted tooth and considerably less than the minimally exposed impacted tooth.

The most likely explanation for this is to be found in the procedure that prosthodontists call 'forced eruption' [74, 75]. When one side of a tooth is fractured or destroyed by caries to below the height of the crestal bone, an infra-bony pocket is produced. The treatment that is prescribed to eliminate this is to mechanically erupt the tooth away from the bone margin and to thus orthodontically reverse the relationship between the prepared crown shoulder or cavity margin and the interproximal alveolar bone. At the same time, the other sides of the same tooth, whose relationship with the bone was normal to begin with, are extruded together with their adjacent alveolar bone. This generates excessive bone in the latter areas, extending more coronally than is normally seen, which sometimes needs to be reduced by periodontal/osseous surgery.

In relation to the resolution of impaction, it has been shown [71–73] that, in contrast with extrusive movement, teeth that undergo root uprighting and torqueing movements end up with a significantly lower crestal bone level than untreated controls. The histogram in Figure 6.26 shows the influence of the various combinations of conservative vs. radical surgical exposure and extrusion/tipping vs. root movement orthodontic forces on the relative bone support of impacted canine teeth. These results are an indication of the periodontal prognosis of the teeth concerned.

One last factor, which is often ignored or simply overlooked, relates to root resorption that may occur in the impacted tooth during the extended period that may be involved in its alignment. In orthodontic treatment generally resorption of the root apices of teeth may sometimes

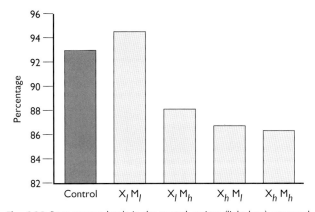

Fig. 6.26 Bone support levels in the treated canines (light bars) compared with the normally erupted opposite canines (dark bar); X_l, minimal surgery and primary closure; X_h, surgery in which the follicular sac was removed down to the CEJ; M_l, orthodontic tipping, extrusion and rotation movements only; M_h, root uprighting and torqueing movements. (Adapted from ref. 72.)

be seen. The reasons for this are not clear and although some clues are available, there is no known key that may be used to predict with confidence those patients in whom resorption will occur. What is found, however, is that the resorption process almost invariably stops when orthodontic tooth movement is completed.

In most patients for whom major orthodontic movements have been carried out, resorption is exceptionally small and of no clinical significance. Nevertheless, there are occasional cases where these movements may account for the loss of 3 or 4 mm of the original root length. Although there are dissenting reports [76] regarding resorption and treatment duration, the overriding opinion is that there is a linear relation between resorption and treatment duration [77–80]. In the absence of definitive information regarding the mechanisms involved, therefore, it would seem wise that, for those patients who are prone to root resorption, orthodontic treatment should be kept to a minimum in terms of duration and complexity. The optimal result for those individuals may not even come close to the ideal.

For the resorption-susceptible case, it could conceivably happen that all the precautions that we have described may be followed scrupulously and a good periodontal result may be obtained. However, unusually severe resorption may account for a final root length of, say, 12 mm. A 2 mm difference in height between crestal bone and the CEJ will give a relative bone support in this patient of only 83%, and the long-term prognosis must be judged accordingly. Thus, for cases in which there is a more compromised bone loss in the cervical area, with a greater CEJ-to-crestal bone height difference, the relative importance of root resorption as a factor in long-term tooth survival increases.

Monitoring for early indications of root resorption may be performed during the progress of the mechano-therapy of a given impacted canine using periapical radiographs. However, since the position and, more specifically, the ori-

entation of the tooth is changing during the orthodontic treatment, comparison with earlier films may be difficult. It should also be remembered that even when marked resorption is noticed, it is unlikely that the orthodontic treatment will be stopped much before full eruption of the impacted tooth has been accomplished. Less would render the canine valueless for all practical purposes and under any circumstances.

The efficacy of radiographic monitoring is only relevant at the point when the crown of the tooth has been brought into its place in the arch, and a decision has to be made whether root uprighting and torqueing movements are merely desirable in the interests of a meticulous alignment or essential, the absence of which would be to the long-term detriment of the treatment result. If a periapical radiograph taken at this juncture shows that significant resorption is evident, its severity must be offset against the relative importance of producing these root movements.

A classification of palatally impacted canines

All forms of surgery inflict trauma which, in the present context, may have lasting effects on the success of the outcome in terms of its periodontal prognosis. In order to simplify the discussion and to exclude complicating factors, we shall assume that the most appropriate surgical technique has been chosen for exposure of a given impacted maxillary canine and that this procedure is performed with a high level of expertise. We have concluded that: trauma generated by surgery is greater when access is difficult; and orthodontic alignment is more complex when the impacted tooth has a greater displacement, particularly if the root apex is not in the line of the arch. Yet it is these two factors, both of which relate to the position of the tooth, that will later dictate the quality of the supporting structures of the treated result. Accordingly, it becomes clear that the patient's best interests are served if an accurate visualization of the canine location of the impacted tooth is made at the time of treatment planning. It follows that if palatal canines are classified in relation to their position in the maxilla, they will essentially be grouped in accordance with the prognosis of their therapeutic outcome.

The classification that is offered here is based on two variables:

- the transverse relationship of the crown of the tooth to the line of the dental arch, which may be close or distant (nearer the midline);
- the height of the crown of the tooth in relation to the occlusal plane, which may be defined as high or low.

Determination of the location of the crown of the impacted tooth is achieved by employing the radiographic methods that we have outlined earlier. More than anywhere else, the use of computerized tomography (CT) scanning has made this exercise much simpler and more relevant. It provides

the clinician with a reconstruction of the three-dimensional image of the impacted tooth and its environs, with no effort and no possibility of error. This positional determination may be subsequently confirmed by direct vision at the time of surgical exposure.

In this section, several cases will be presented to illustrate the salient clinical features of the impacted canines in each of the classification groups. An approach to treatment will be discussed within each group and how this needs to be adapted to suit the conditions seen in each. Where relevant, cases that were treated inappropriately will also be presented, with the aim of revealing how the shortcomings of the results occurred and to discuss what alternatives could have been employed to prevent the undesired sequelae.

Group 1

- proximity to line of arch: close
- position in maxilla: low.

Typically, palatal canines that are close to the line of the arch and low in the maxilla suggest a good prognosis, insofar as the tooth is usually palpable in the palate and readily accessible to surgery (Figure 6.27). These canines represent by far the most common form of palatal impaction. In its simplest form, the canine is opposite the space and is not rotated. The root apex is usually in its correct location and root movements are rarely necessary.

Surgery

If the canine is only mildly displaced to the palatal, it may be approached from the occluso-buccal (Figure 6.27d), in which case little bone removal is needed to reach it. A labial flap is raised from the crest of the ridge or from the gingival margin of the deciduous canine, which would be extracted. The edge of this flap would then comprise thick attached gingival tissue. For a more palatally located canine and following the reflection of a palatal flap (i.e. a flap of thick firmly bound mucosa), the canine is immediately obvious under its bulging but thin covering of bone on the inner surface of the alveolar ridge. Minimal removal of eggshell-thin bone is needed to reach the follicular sac, and access for bonding an attachment to the tooth is good. After suturing of the full flap, the pigtail ligature is drawn through the sutured edge in the direction of the main archwire.

Planning the orthodontic strategy

With the tooth immediately opposite its place in the arch, orthodontic alignment requires some extrusion, but principally a buccal tipping movement. Thus, direct force application between pigtail and archwire is the most appropriate (Figure 6.27e). Properly planned surgery will result in the intermediary, be it twisted steel ligature or gold chain, emanating from the sutured edge of the replaced flap. Traction applied to this intermediary will encourage the impacted tooth to erupt through attached gingiva, closely resembling a naturally erupting tooth, to provide it with a normal periodontal environment.

Problems that may be encountered

In the simplest group 1 case, the eyelet may be sufficient to complete all the movements required. Alternatively, and if surgery has exposed a sufficiently long clinical crown, a conventional bracket may be placed immediately. However, it should be remembered that, as the tooth moves buccally, it gathers gingival tissue ahead of it and, if oral hygiene is not excellent, the exuberant soft tissue will become inflamed and may impinge on the bracket. Undoubtedly, wider exposure of the crown will eliminate this, but will compromise the periodontal tissues in the final analysis. Thus, it is wiser to use an eyelet initially and replace it as necessary when the tooth reaches the main archwire.

Complications

Group 1 canines in their initial positions may be complicated by rotation, mesial crown displacement or palatal root displacement.

Rotation

The type of rotation that the canine generally presents is a mesio-palatal, with the buccal surface of the tooth facing mesially towards the root of the lateral incisor. This means that during treatment the appliance must incorporate a rotational mechanism to bring the tooth into alignment. The simplest manner in which to do this is to initially place the eyelet on the anatomic labial surface of the crown of the canine, which faces anteriorly, towards the lateral incisor and orienting it vertically in the long axis of the tooth (Figure 6.28).

A full-arch nickel–titanium auxiliary wire, together with heavy main arch, is the method of choice [81]. This fine and highly elastic wire is threaded directly through the vertically oriented eyelet, which introduces a strong deflection into this auxiliary archwire. This will then exert a force couple that will both bring the tooth to its place in the arch and, at the same time, bring about a very efficient correctional rotation (Figure 6.28). It is, however, essential to check that the ligation of the auxiliary into the other brackets of the appliance is not too tight so as to limit its movement, since the efficiency of the appliance is dependent on this free-sliding attribute.

A good alternative involves the use of a 'slingshot' elastic (see Figure 4.7) or elastic thread tied between the eyelet and the cut length of stainless steel tube that has been threaded on to the main archwire, for use as the canine space maintainer and to add rigidity to the base arch. While the canine is being moved towards the line of the arch, it is also being rotated about its long axis in a corrective mesio-buccal rotatory movement. Since the stainless steel tube space maintainer will not allow individual movement of the adjacent teeth, the direction of rotation may be changed or

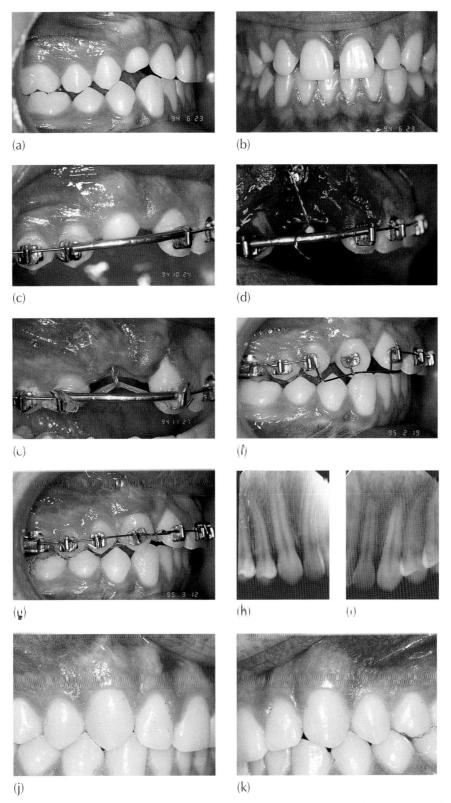

Fig. 6.27 (a, b) Intra-oral views of the initial condition. (c) Space was opened using a coil spring and sliding mechanics. An over-sized stainless steel tube was cut to measure, curved and placed in the archwire to maintain space and to increase base arch rigidity. (d) Exposure and attachment bonding – a closed eruption procedure. The pigtail ligature was drawn downward and shaped over the archwire to allow for replacement of the flap without impingement. Traction was applied immediately. (e) Two weeks post-surgery, a new 'slingshot' elastic module is stretched between the brackets of the lateral incisor and first premolar. Its middle portion is raised to engage the pigtail hook with a controlled and measurable light extrusive force. (f, g) Three months post-surgery. An inferiorly and laterally offset light wire arch is substituted and the tooth ligated with steel ligature wire to achieve full eruption. An orthodontic bracket now replaces the eyelet. (h, i) Periapical view, showing comparable supporting bone levels in the treated and untreated canines. (j, k) The gingival appearance shows comparable gingival levels on the treated versus the untreated side. It is unlikely that an open exposure procedure could produce a result of this calibre.

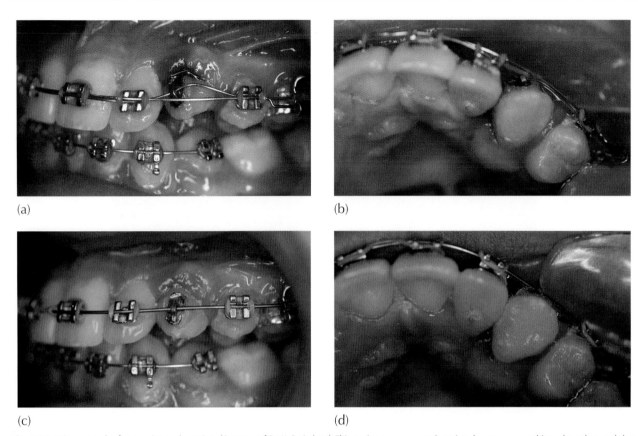

(a)

(b)

(c)

(d)

Fig. 6.28 Using an eyelet for eruption and rotation. (Courtesy of Dr H Corimlow.) This canine was a group 2 canine, but was erupted into the palate and then drawn towards the labial archwire. (a, b) With the canine only partially erupted but markedly rotated, the original vertically oriented eyelet (bonded at the time of exposure) is exploited to rotate the tooth, using a fine (0.014 in) nickel–titanium auxiliary archwire, under the main 0.020 in stainless steel base arch. (c, d) Four weeks later, the auxiliary wire is removed and the base arch threaded directly through the eyelet. At the next visit a bracket will be substituted to effect appropriate finishing.

increased to fit other types of rotated palatal canines, by tying the elastic thread from the eyelet directly to the premolar or lateral incisor teeth.

Mesial crown displacement

This is very commonly seen in conjunction with the mesio-lingual rotation that we have just described. Whether or not the rotation is present, the proximity of the anatomic labial surface of the canine to the lateral incisor creates constraints on the placing of a bracket at the mid-buccal position of the canine crown. As a general rule, this space is too small for the placement of any of the conventional brackets, which are much bulkier. To overcome this, many practitioners bond the conventional bracket, with its rigid and contoured base, on the irregular palatal surface of the tooth, to which it is totally unsuited and to which the reliability of the bond is much lower, increasing the risk of detachment to almost one case in two [82].

A further drawback to using a regular bracket on the palatal aspect of the tooth is that traction applied directly between it and the archwire will substantially increase the existing and adverse rotation of the tooth. This will be very

difficult to correct later and will significantly add to the amount of mechano-therapeutic manipulation that the tooth must undergo. As a result, the periodontal prognosis of the tooth will be compromised unnecessarily.

Traction from an eyelet placed in the ideal mid-buccal position on the tooth, even if it is more incisally located because of the physical limitations imposed by the proximity of the lateral incisor, will bring about a corrective rotational movement as the tooth is drawn towards the target area. As with a bracket, palatal bracket-siting of an eyelet risks a complication (detachment) and, while it solves one problem (the impaction), it creates another (increased rotation), which may be just as formidable.

It is relatively easy to bond an eyelet close to the ideal mid-buccal position of the exposed tooth and to draw elastic thread from it to the rigid tubing that has been placed on the archwire to maintain the canine space in the arch. It may be advantageous to tie the elastic thread to the bracket of the first premolar to increase the mesio-buccal rotatory component of the traction.

The premolar will not close down the canine space because of the presence of the stainless steel tubing. When

the period of traction becomes extended, it may slowly bring about adverse changes in the dental midline by tipping of the incisors in that direction. In this case, additional precautions will need to be taken to protect the anchorage. Super-elastic auxiliary archwires threaded through a vertically oriented eyelet are a good alternative and will achieve a similar result. They will usually work with greater efficiency, provided the distortion of the wire is not so great as to cause binding in the brackets of the adjacent teeth.

Palatal root displacement

If the root apex of the canine is palatally displaced, in addition to the palatal crown displacement, the crown will first need to be aligned in the manner that we have just described. This includes the correction of any possible rotation and mesial crown displacement. With the canine crown in place and the main archwire firmly ligated into the newly substituted, conventional bracket, the palatal inclination of the crown-to-root orientation of the tooth will dictate that its palatal surface bulges inferiorly, while the buccal surface tips labially and superiorly.

The heavy archwire is now needed to serve as the base arch to a labial root-torqueing auxiliary. The heavy base arch provides the fulcrum about which the auxiliary will buccally rotate the root apex. Employing a full rectangular arch, which is torqued in stages on the canine only, the long-axis inclination will be seen to improve. However, the equal and opposite reactive force is on the roots of the adjacent teeth, which provide the anchorage for this difficult movement. These adjacent teeth are being torqued lingually at each torque-adjusting stage and then buccally as the torqueing force is expended – a classic example of 'round-tripping' which, when considerable torque needs to be applied to the canine, transfers an equal and opposite torqueing force on the adjacent teeth. This, in the case of a small lateral incisor or one with a partially resorbed root that is associated with the impaction, may undermine the long-term prognosis of the incisor.

Regardless of the type of orthodontic brackets employed, there are advantages to using a torqueing auxiliary that derives its anchorage from the narrowed archform of the heavy main archwire. This will avoid distortion of the dental arch, and will not create unwanted side-effects on the adjacent teeth.

Group 2

- proximity to line of arch: close
- position in maxilla: forward, low, and mesial to lateral incisor root.

The root apex of the canine in this group is usually to be found in its correct mesio-distal location in the line of the arch, and at more or less the correct height. The crown of the tooth, however, is tilted mesially (forward) and in close

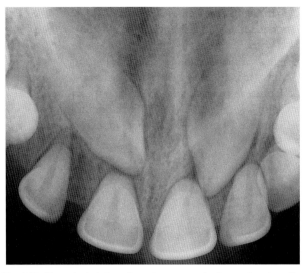

Fig. 6.29 The periapical view of an extreme example of group 2 canines. The left canine is located between lateral and central incisor roots and the right canine is mesial to the central incisor root. Reproduced with permission from Gill and Naini, forthcoming, with permission.

association with the palatal aspect of the root of the lateral incisor (Figure 6.29) and often sited between the roots of lateral and central incisors [66]. The tooth is not always palpable on the palatal side.

Surgery

Surgical exposure in this group is complicated by the often unavoidable simultaneous exposure of the roots of those adjacent teeth. Aggressive surgical techniques may occasionally open the way for the orthodontist to bond on the labial surface, but not without considerable damage to the adjacent exposed incisor roots by the radical removal of bone.

With the position of the crown of the impacted canine situated mesially to the root of the lateral incisor, several operative problems present themselves. In the first place, surgical exposure has to be carefully undertaken so as not to damage the roots of the incisors. The temptation to expose too widely should be resisted and only enough of the most conveniently accessible surface of the tooth should be uncovered to permit bonding. The palatal surgical flap should be replaced in its entirety in order to provide maximum protection and tissue reattachment for the exposed incisor roots and the area of exposed bone, and to re-establish a normal periodontium. However, to provide an exit for the twisted ligature pigtail that is tied to the attachment, a small piercing should be made in the flap immediately opposite the attachment on the tooth, using a wide-bore needle. Alternatively, the flap may be divided into two halves by a slit that is made with electro-cautery rather than a scalpel in order to prevent bleeding at the cut edge. The pigtail is passed through the pierced hole or the slit, and the flap is sutured back into place (Figure 6.30).

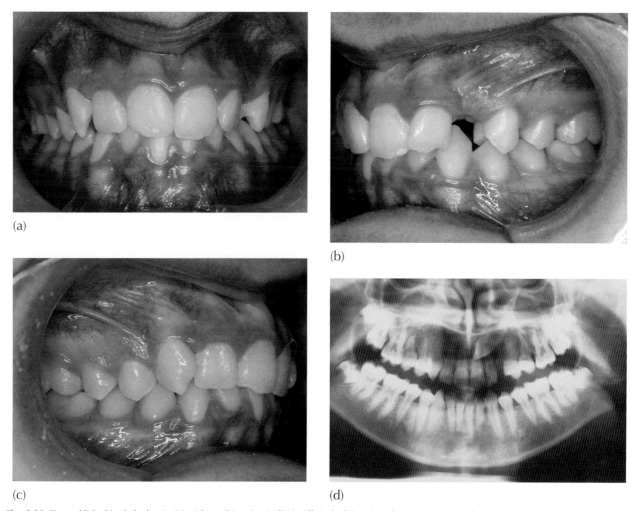

(a)

(b)

(c)   (d)

Fig. 6.30 First published in *Orthodontics Principles and Practice, Daljit S. Gill, Farhad B. Naini editors, Impacted Teeth and their Orthodontic Management by Adrian Becker & Stella Chaushu, 2011, Wiley-Blackwell, Oxford, UK*, with permission of the publisher.

(a–c) Intra-oral views of a class 1 malocclusion with retroclined and crowded incisors and an impacted left maxillary canine.

(d) The initial panoramic view showing the canine to be displaced almost to the midline of the palate.

(e) The cephalogram confirms palatal displacement.

(f) The auxiliary stainless steel archwire intended to disimpact the canine.

(g) Following alignment and space opening, the auxiliary archwire is ligated in, over the space maintaining steel tube place between lateral incisor and first premolar. The auxiliary is in its vertical passive mode.

(h) The occlusal view immediately pre-surgery, showing the canine bulge in the palate.

(i) A palatal flap is raised from the cervical margins of the teeth, up to the midline, leaving the incisive canal bundle intact. The canine is minimally exposed, revealing its proximity to the midline. (Surgery by Dr Harvey Samen.)

(j) An eyelet attachment is bonded by the orthodontist on the anatomically labial aspect of the crown of the canine.

(k) The twisted steel ligature pigtail is pushed through the palatal flap at a point immediately opposite the eyelet.

(l) The flap is fully replaced and sutured, to leave no unprotected wound. The pigtail ligature can be seen to exit the palatal tissue immediately opposite the eyelet.

(m) The pigtail is shortened and bent to form a firm hook around the vertical loop of the auxiliary archwire, which has been turned inwards and raised flush with the palatal tissue. This applies measurable extrusive force to the impacted tooth.

(n) Several weeks later, the fully healed palatal tissue shows a very much increased and palpable bulge outlining the tooth beneath.

(o) A small incision in the thick mucosa around the tip of the crown of the tooth has permitted the tooth to erupt. Labial traction is now applied using elastic thread to draw the tooth to the main archwire.

(p) The occlusal view shows that much labial root torque is needed.

(q) An accessory torqueing arch has been placed to move the root apex buccally.

(r–u) The finished case showing a shorter clinical crown than on the untreated side.

(v–x) Panoramic, lateral skull and periapical films of the completed case.

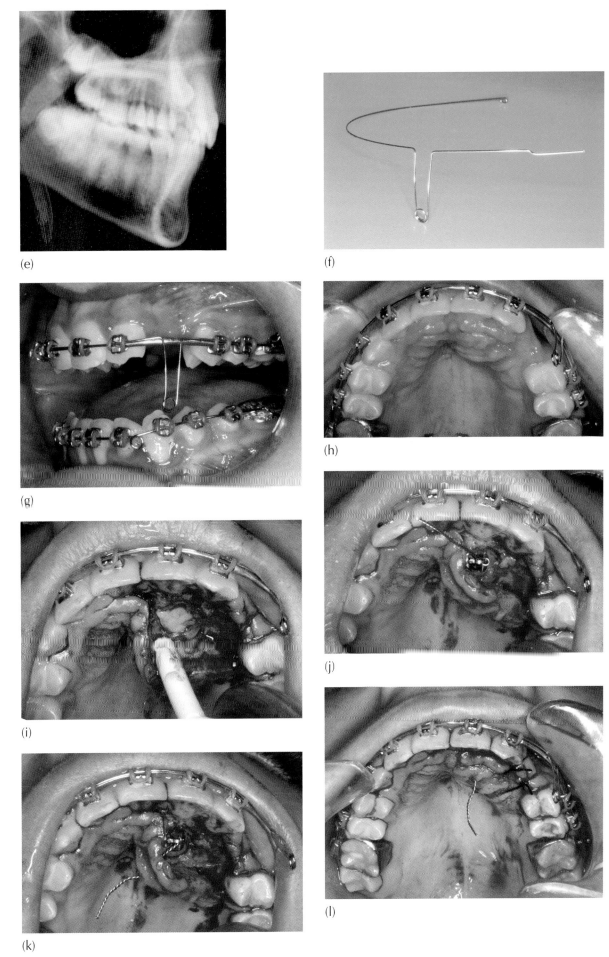

(e)

(f)

(g)

(h)

(i)

(j)

(k)

(l)

Fig. 6.30 (*Continued*)

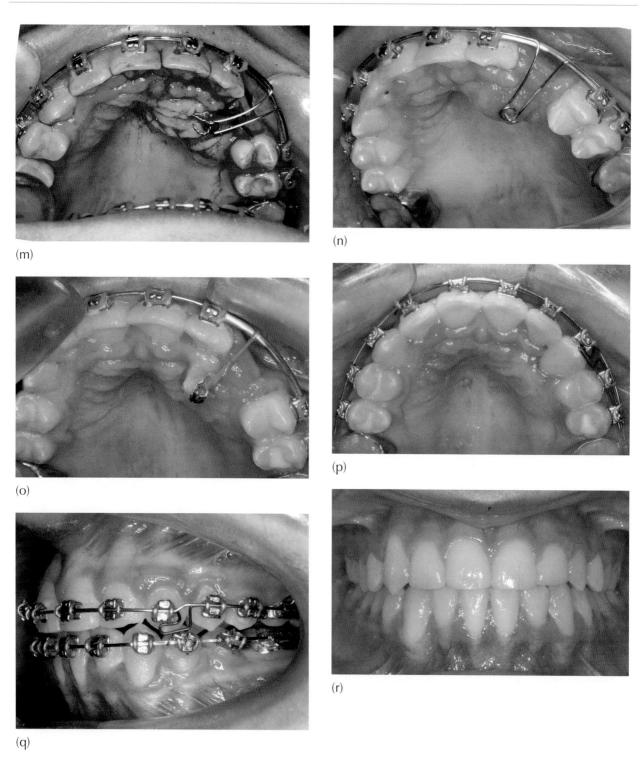

(m)

(n)

(o)

(p)

(q)

(r)

Fig. 6.30 (*Continued*)

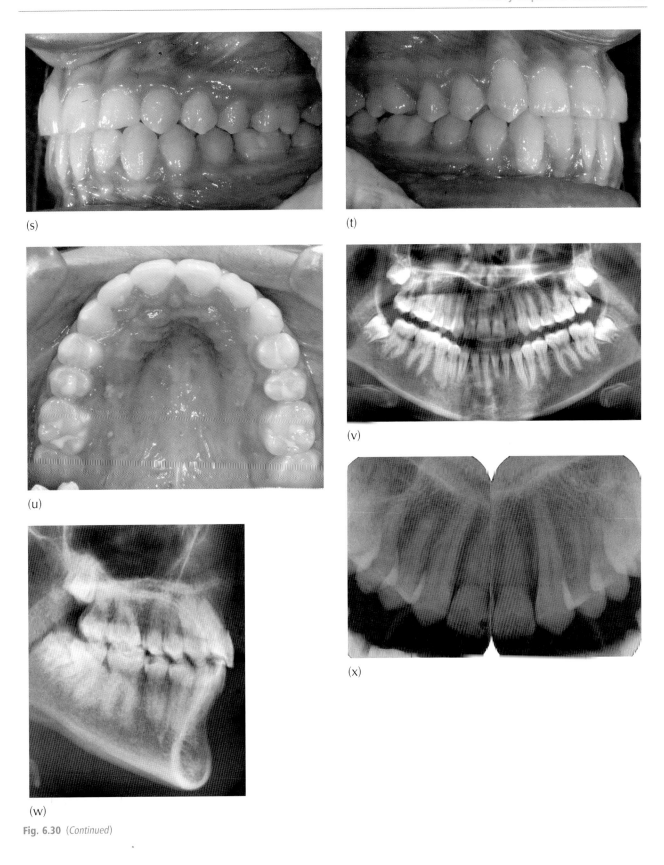

(s)

(t)

(u)

(v)

(w)

(x)

Fig. 6.30 *(Continued)*

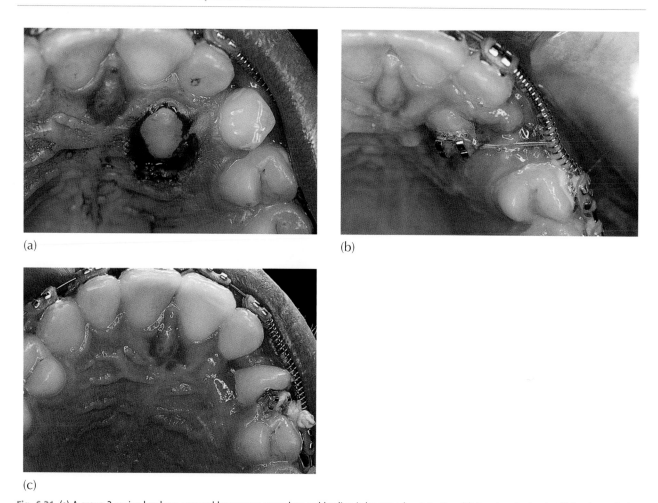

(a)

(b)

(c)

Fig. 6.31 (a) A group 3 canine has been exposed by an open procedure and healing is by secondary intention. (b) An edgewise bracket (!) has been sited on the palatal aspect of the canine. The tooth is being drawn from this palatal attachment directly to a flexible labial archwire (!). (c) The tooth has reached the archwire, and is now rotated a further 30–40°. Note the swollen appearance of the gingiva and its poor contour.

The mesio-palatal rotation of the canine that is usually present in these cases places the labial surface of the canine in a completely inaccessible position. This means that only the palatal or distal aspect of the canine is available for siting the attachment.

Planning the orthodontic strategy

From an eyelet placed on the palatal side of the tooth, direct traction to the labial archwire is sometimes possible. However, if traction is performed in this manner, the attached surface leads the way, and it will inevitably cause the canine to 'roll' over the root of the lateral incisor, to increase the existing rotation (Figure 6.31) and to risk damage to the incisor. Furthermore, the operator could well be faced with a 180° rotation to perform once the tooth reaches the archwire! Few will dispute that this task is formidable, but before the adventurous clinician even begins to argue that it is not insurmountable, the following three questions should be considered.

1. How long will the derotation prolong the appliance therapy?
2. Will the rotational relapse factor be possible to overcome?

3. What will be the prediction for the health of the periodontium at the completion of correction of what will have been an iatrogenic rotation?

For many of the impacted teeth in group 2, the intimate relation between canine crown and lateral incisor root will block movement of the canine when traction direct to the archwire is applied. The inexperienced operator may then respond by increasing the pressure applied to the tooth and, within a fairly short period of time, there will be signs of loss of anchorage on the other teeth, characterized by the production of a cross-bite tendency, a midline shift from the affected side and tooth mobility, not to mention the possibility of damage to the lateral incisor root and failure to bring the canine down.

A completely different approach is needed, in which the tooth must first be moved in a different direction to free it from potential entanglement with the incisor roots. The most practical manner of doing this is to draw it vertically downwards (towards the tongue), erupting it into the palate and thereby bringing it to face its target location. In this way, the tooth will have a clear path to its place in the arch without any interposing adjacent roots and the group 2

canine will have been converted to a group 1 canine. It may then be moved directly across the line of the arch, towards the labial archwire, in a second movement.

It is for this reason that the pigtail ligature wire must be drawn through the fully replaced surgical flap and not through its sutured edge. The site of passage of the ligature wire through the flap should be immediately opposite the buried tooth, in order that vertical traction may be applied with ease and relative comfort for the patient. Traction in this direction from a palatally sited eyelet will not cause a rotation of the impacted tooth as it progresses.

Three types of maxillary spring auxiliary may be used to bring about this desired movement that is needed initially, in this first stage of the resolution. These are the ballista [67], the active palatal arch [65, 66] and the light auxiliary labial arch [68]. With each of these methods it is essential that a heavy base arch be tied into the brackets of all the teeth on the labial side and to hold the opened space for the canine in the arch, to resist secondary distortion of the occlusal plane and archform and to provide a base from which to apply the force to the tooth.

The ballista [67] (Figure 6.32) is a unilateral spring of rectangular wire, which is tied into one of the rectangular molar tubes. It proceeds forwards until it is opposite the canine space. At this point, it is bent vertically downwards towards the lower jaw and terminates in a small loop. With light finger pressure, the vertical portion is turned upwards and inwards, across the canine space, and ensnared in this active mode by turning over the pigtail ligature around its terminal loop to hold it close to the palatal mucosa. In this way, torque is introduced into the horizontal part of the rectangular wire, which is the source of the vertical traction. Its equal and opposite reactive force is thus transferred to the anchor molar.

The elasticity of the ballista spring exerts pressure for it to return to its original vertical position, which in turn applies extrusive force to the unerupted tooth. If the impacted tooth is fairly resistant to movement or if the distance that the tooth needs to be moved is great, lingual molar root torque may occur, representing a loss of anchorage. To overcome this, a rectangular main arch or a soldered palatal arch may be used.

The active palatal arch (Figure 6.33) consists of a fine 0.024 in (0.6 mm) palatal archwire carrying an omega loop on each side. The wire is slotted into a soldered horizontal 0.024 in (0.6 mm) tube on the palatal side of the maxillary

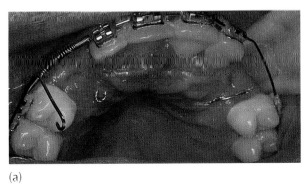
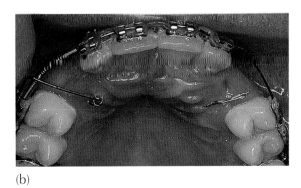

(a)   (b)

Fig. 6.32 (a) A ballista in its passive mode, pointing downwards. (b) Using light finger pressure, the looped end of the spring is turned inwards and upwards towards the palate, where it is ensnared by the stainless steel pigtail from the unerupted canine.

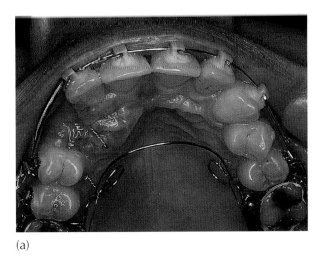
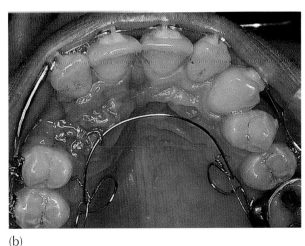

(a)   (b)

Fig. 6.33 (a) The active palatal arch in its passive mode, lying several millimetres away from the palate as seen from the occlusal. (b) The same view after the active palatal arch has been gently raised towards the palate and ensnared by the pigtail hooks, thereby applying vertically extrusive traction to the unseen canines.

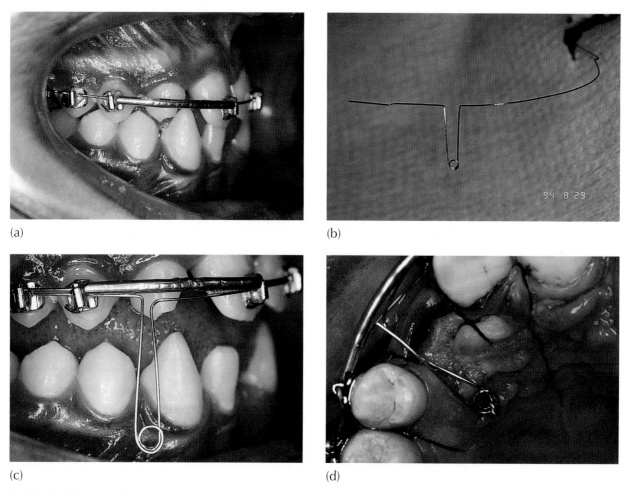

(a)

(b)

(c)

(d)

Fig. 6.34 (a) Initial treatment has created space and a heavy base arch, carrying a stainless steel tube space maintainer, in place. (b) A typical auxiliary labial archwire of 0.016 in gauge with its vertical loop and terminal helix. (c) The auxiliary labial archwire was tied into the brackets piggyback style over the heavy base arch, immediately prior to surgical exposure. (d) Following full flap closure, the vertical loop was gently raised and turned inwards, with its helix secured into the terminal hook of the pigtail.

molars. It may optionally be further secured by a steel liga-ture tie and distanced from the palate by finger pressure. By elevating the downward-activated palatal archwire and hooking the pigtail ligature around it, the unerupted tooth comes to be erupted through the palatal tissue, in a direc-tion slightly away from the teeth.

The auxiliary labial wire [68] is a third possibility (Figure 6.34); it requires no advance preparation of any sort, such as the soldering of lingual tubes. It is most conveniently fashioned from an archform blank of 0.014 in or 0.016 in diameter round wire by forming a vertical loop in the area of the impacted canine. This loop has a small terminal helix. The auxiliary is tied into all the brackets of the arch, in piggyback style over a heavy main arch, with the extremi-ties slotted into a spare tube on the molars or left free distal to the second premolar brackets. In a similar manner to the ballista, the vertical loop is activated by raising it palatally across the canine space, and ensnaring it in the pigtail liga-ture in the palate. The auxiliary labial wire draws its activa-

tion from its curved archform, which does not therefore transfer torque to the molar.

This is a particularly useful method for use with a bilat-eral impaction, when two different loops will need to be inserted into the archform. Used without a base arch, as has been recommended elsewhere [83, 84], it will extrude the adjacent teeth and thereby alter the occlusal plane. It will move the molars buccally and additionally will alter the horizontal arch form in the incisor area. A base arch is therefore mandatory.

In the construction of the ballista and auxiliary labial wire, it is important to calculate the length of the active arm in advance. This may be done by pinpointing the intended location of the eyelet on the tooth and its projected exit from the palatal flap, either on the plaster cast or directly in the mouth, by palpation of the bulge in the palate. Alternatively, the location may be estimated from the radio-graphs and, particularly, from the CBCT which will provide greater accuracy. The distance between this point and the

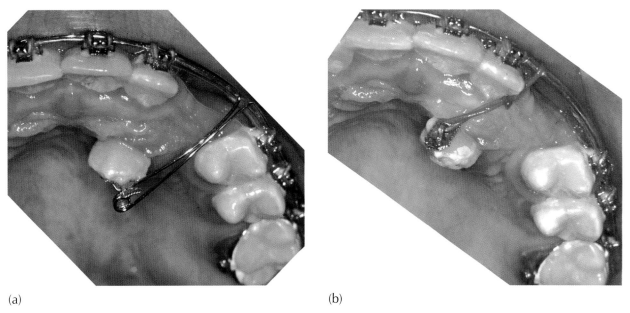

(a)   (b)

Fig. 6.35 (a, b) With the eruption of the canine into the mid-palate, the eyelet position has to be changed to enable labial traction that encourages simultaneous corrective rotation.

labial archwire represents the optimum length of the active arm. If the arm is made shorter or longer, it will be difficult to approximate it to the palatal tissue and the arm will stand uncomfortably away from the palatal mucosa, to interfere with tongue activity. If it is made too long, it will draw the tooth away towards the **midline** as it vertically erupts it and, with it, the apex of the canine will be similarly drawn down and palatally, adversely altering the tooth's centre of resistance away from the line of the arch. This will mean that its final alignment will require more labial root torque than should otherwise have been necessary. It is therefore preferable to err on the side of a shorter active arm.

A mandibular removable appliance represents a method that requires no maxillary appliance at all in the first instance [85]. Vertical extrusive force is derived from a removable appliance in the opposing jaw through the agency of a simple rubber band. This elastic is applied by the patient between the clasps of the mandibular appliance and the pigtail ligature in the palate. In order for this to be efficient, several clasps used for retention of the plate on the teeth need to be included in the design and the form of the teeth must lend itself to good retention, otherwise the plate will be easily dislodged by the vertical intermaxillary elastic force. A maxillary orthodontic appliance is later needed to move the tooth into the line of the arch, achieve levelling and aligning of all the teeth and any needed finishing, but only when the tooth has erupted through the palatal mucosa.

A much better and more reliable method of applying intermaxillary force rests with the use of a simple temporary anchorage device. A screw implant may be placed in the mandibular jaw in the inter-radicular bony area between the canine and first premolar. Traction in the form of

simple latex elastic rings may be applied from the implant device directly to the pigtail ligature by the patient. However, if a maxillary appliance is present from the outset, it is more effective if the elastic is engaged in a hook attachment on the appliance, while the orthodontist controls the traction applied to the impacted canine by ligating it to the archwire. This represents indirect traction from the implant to reinforce the anchorage of the whole appliance on that side of the maxilla to permit the safe and controlled application of an extrusive force to the canine.

Regardless of which method is employed, the successful end-result will find the newly erupted tooth, surrounded by a wide rim of palatal mucosa and bone, in the middle of the palate (Figure 6.35). The more the tooth is erupted, the easier it will be to place an attachment on its buccal surface to enable the tooth to be moved buccally, without the bracket impinging on the gingiva, yet encouraging a simultaneous corrective rotation. However, an excessively erupted tooth will lead to occlusal trauma as it moves across the line of mandibular teeth.

For the first stage of the two-stage manoeuvre, the position of the eyelet is not critical. Therefore, bonding is performed to the most convenient surface available, since no adverse rotation of the tooth will occur while it is being moved vertically downwards. The tooth is cleared of the lateral incisor root and moved until it has an unobstructed path to the labial archwire. A second eyelet is bonded to the tooth, this time on the mid-buccal aspect, which will have become accessible as the result of the initial orthodontic movement. The second stage of traction may then commence, with the application of force from the second eyelet directly to the labial archwire.

The point should be made that in this case, and in any other group or situation, direct traction to the archwire should only be performed from an attachment sited in the mid-buccal position of the tooth. Traction applied at any other site will engender an unwanted rotation as the tooth progresses towards its place in the arch – a rotation that will require correction in an extra and superfluous phase of orthodontics.

Problems that may be encountered

As the tooth initially progresses in the eruption path that has been planned for it, the thick and resistant palatal tissue bulges more and more, but may not allow the tooth to erupt through it (Figure 6.36). Delay or increasing trac-

tion forces will not improve the chances of progress of the impacted canine, but the tissue resistance will encourage the forces of the auxiliary to be expressed against the anchor unit, causing the adjacent teeth to intrude, to generate an open bite and a serious disruption in archform. This obstacle will necessitate a very limited and superficial surgical removal of the thick mucosa immediately over the crown of the tooth. The anterior palatine artery is located in this immediate area and care should be taken not to sever it, in what is otherwise a very simple procedure. This may be done without releasing the traction mechanism and then left another two or three weeks, during which time the tooth will be seen to advance at considerable speed.

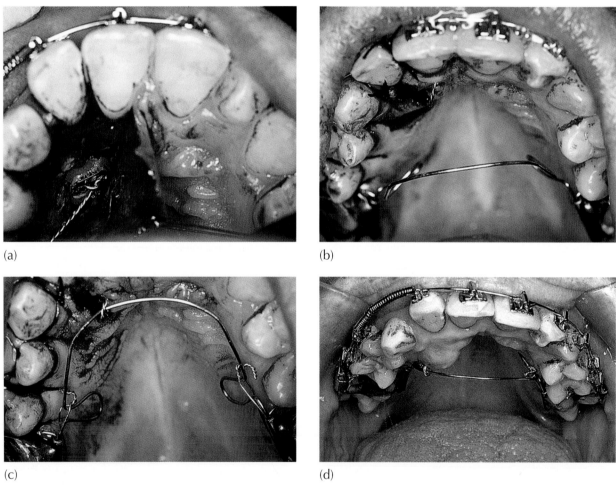

(a) (b)

(c) (d)

Fig. 6.36 (a) Minimal exposure and eyelet attachment bonding of the palatal aspect of a group 2 canine in a 17-year-old female. The tip of the cusp has not been exposed. (b) Suturing was completed with the pigtail emerging through the divided flap. The palatal arch is in its passive state. (c) The palatal arch fully tied in and active. (d) Seen three months later, after two intervening visits for adjustment, the archwire has become distanced from the palate and the canine can be seen to bulge the contour of the palate almost to the occlusal level. (e) The deciduous canine was extracted at the same time as the minimal re-exposure was performed, preparatory to buccal movement. A second eyelet is bonded slightly mesial to the mid-buccal position and elastic ligation is drawing the tooth directly to the archwire, with a favourable rotation vector from the second eyelet. (f) At 13 months post-surgery the canine is in the arch and a bracket is substituted for the eyelet. (g, h) The gingival health of the treated canine is good, but its clinical crown is longer than the untreated left canine. (i) The periapical view. Note the resorbed root apex of the right lateral incisor. (j, k) Post-treatment periapical views to show comparable bone support of treated and untreated canines.

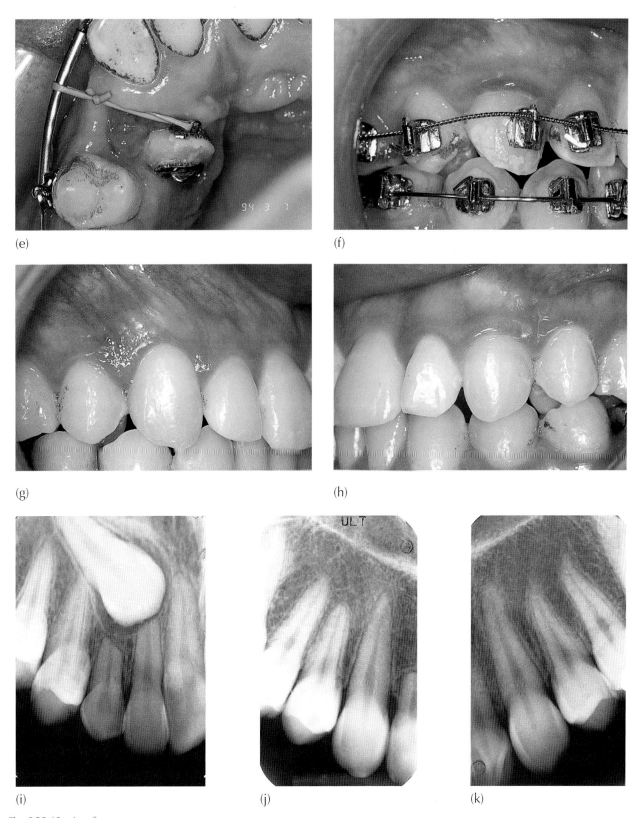

(e)

(f)

(g)

(h)

(i)

(j)

(k)

Fig. 6.36 (*Continued*)

Once the tooth is well erupted at the level of the occlusal plane, the new eyelet attachment is bonded to its anatomically buccal surface (Figure 6.36e). It is then drawn in a direct line to the labial archwire and to the place where space will have been provided for it in the initial orthodontic phase of the treatment. Initial rotation of the canine will still be present when this second phase of the traction begins, but it will correct steadily as the traction proceeds, with the attachment and buccal surface leading the way.

Even with good oral hygiene during the traction period, much exuberant gingival tissue precedes the canine during its migration towards the line of the arch. The premature bonding of a conventional bracket may result in tissue impingement, particularly as the tooth comes into close relation with the adjacent teeth. The orthodontist should not relinquish the use of the eyelet until no further buccal movement is needed.

Complications

The initial location of the tooth may be complicated by one or both of the following conditions:

1. Rotation. As with the group 1 canine, a mesio-lingual rotation is common and is largely corrected during the second stage of resolution, when traction is made from the second eyelet on the buccal surface of the tooth.
2. Palatally displaced root. Occasionally a tooth in this group may present with the root palatally displaced, in addition to the crown (i.e. a palatal translation of the entire tooth). Its occurrence will complicate the appliance work by requiring the introduction of buccal root torque and often some mesio-distal root uprighting. These forces may only be applied once the crown has been engaged by the main labial archwire.

It will be appreciated that a group 2 canine has to be approached with a good measure of preparedness. From the surgical point of view, access is not difficult, but exposure must be performed carefully to avoid exposing and damaging the roots of the adjacent teeth. The orthodontic appliance may need to execute as many as five different types of movement, involving both crown tipping and root movement (i.e. vertical extrusion to move the tooth away from the palate), buccal tipping to the line of the arch, rotation, mesio-distal root uprighting and buccal root torque. From the periodontal point of view, the prognosis of the result is dependent on the smooth execution of each of the surgical and orthodontic operative procedures. Mismanagement of any part of the orthodontic and the surgical stages may not be germane to the question of whether the final alignment of the tooth will be technically successful, but it will be critical in determining the final bone level (Figure 6.36j, k), periodontal condition, clinical crown length, gingival architecture and natural appearance that may be achieved (Figure 6.36g, h).

Group 3

- proximity to line of arch: close
- position in maxilla: high.

The root apex of the canine in this group is situated very high in the maxilla, although most often in the general bucco-lingual line of the arch and in its correct location in the antero-posterior plane. The crown is high and only relatively mildly displaced palatally; it is not usually palpable.

Surgical and orthodontic strategy

Access to this tooth may be from either the buccal or the palatal side, with advantages and disadvantages to both approaches, since there is a significant thickness of bone both on the buccal side and the palatal side of the tooth. Considerable bone removal is needed to reach it from either side, with similar difficulty in the application of an attachment. The orthodontic treatment entailed in aligning such a tooth involves principally extrusion, together with a buccal tipping movement.

The buccal approach

From the buccal side, the tooth is approached as for a buccal impaction, described in Chapter 3. The surgeon may have greater difficulty in locating the impacted tooth and will be more dependent on the radiographs.

The apically repositioned flap [86, 87] would, at first sight, appear to be a good approach, since it offers suitable access for the application of orthodontic force direct to the archwire. It also ensures that an adequate band of attached gingiva be raised as a part of a split-thickness flap above the level of the impacted tooth and then accompany that tooth in its subsequent downward path.

From the surgical point of view, however, this approach takes no account of the three-dimensional location of the canine. The method may be very suitable for a buccal canine whose vertical displacement is relatively minor, but in the case of a more superiorly displaced palatal canine, this type of exposure would leave a considerable expanse of periosteum and alveolar bone open to the oral environment. The flap would need to be sutured several millimetres lateral to the crown of the tooth unless considerably more bone were to be removed on the buccal side of the crown of the tooth to enable the flap to bridge the large gap between the labial mucosa and the palatal tooth. Furthermore, the maintenance of the exposure of a canine crown palatal to the line of the arch would secondarily cause a denuding of the interproximal areas of the roots of the adjacent lateral incisor and first premolar teeth. Postsurgical discomfort with this method is more severe and more prolonged than with a closed exposure procedure [88–90]. This method is only suitable for cases of very minor palatal displacement.

Full labial flap reflection after reducing the size of the flap and its subsequent partial replacement over the exposed

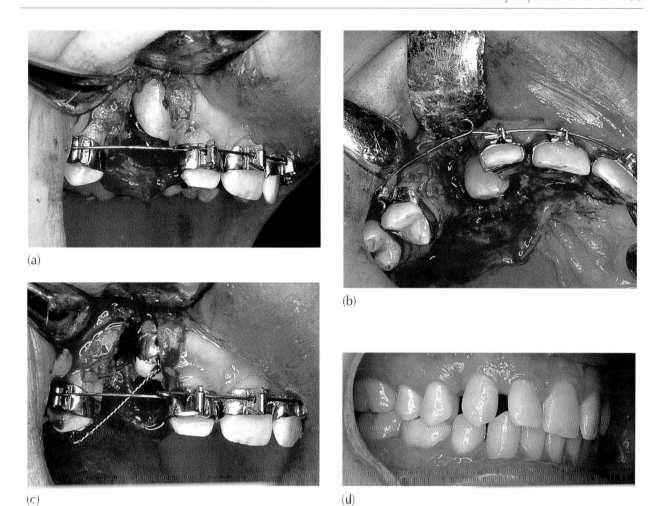

(a)

(b)

(c)

(d)

Fig. 6.37 A case treated by the author c. 1972, using the approach recommended by Johnston (ref. 11), Lewis (ref. 91) and von der Heydt (ref. 92). (a, b) The group 3 canine has been exposed from the buccal side and bone has been channelled to clear a direct path to the archwire. (c) With the band cemented to the canine the radical nature of the surgery may be seen. (d) The aligned canine shows a healthy gingival appearance, but a long clinical crown and a lack of normal bony contour, with deficient interdental papillae. (Reprinted from Kornhauser, S, Abed, Y, Harary D and Becker, A (1996), The Resolution of Palatally-impacted canines using palatal-occlusal force from a buccal auxiliary, Am. J. Orthod. Dentofac. Orthop. 110: 528–534, with permission of Elsevier.)

tooth, together with the placement of a surgical pack to cover the exposed tooth and surrounding tissues during healing, has been used for many years and was widely advocated [11, 18, 91, 92] for mildly palatally impacted canines. It shares features with the apically repositioned flap procedure described above. In the case of a group 3 canine, however, it must be remembered that a considerable thickness of alveolar bone is present both inferiorly and inferobuccally to the canine, which must be traversed by the tooth. In order to overcome this physical impediment, the above authors have recommended the surgical channelling of bone to free a path in the direction of the dental arch (Figure 6.37). Experience with this procedure shows that while it lives up to its expectations regarding the provision of access and the enablement of direct traction, it does so only by the planned sacrifice of much of the bone of the immediate area of the alveolar process! The treated result then leaves an aligned canine with an unacceptably reduced

bone support, poor gingival contour and a poorer periodontal prognosis than could be achieved by other means.

The tunnel approach

An excellent modification of this method has been described [93] in which the buccal plate of bone is preserved, while the impacted tooth is drawn through a tunnel in the bone provided by the vacated socket of the simultaneously extracted deciduous canine. This method is particularly suited to the group 3 canine. A full labial flap is reflected to include attached gingiva from the crest of the ridge and the impacted tooth exposed from its buccal aspect, leaving the buccal plate inferior to it intact. The deciduous canine is extracted and its socket extended and widened sufficiently to allow the passage of a fine wire through it as far as the impacted tooth (Figure 6.38). It is not necessary, however, to widen it to the full diameter of the crown of the canine, which must pass along this path. Much time will elapse

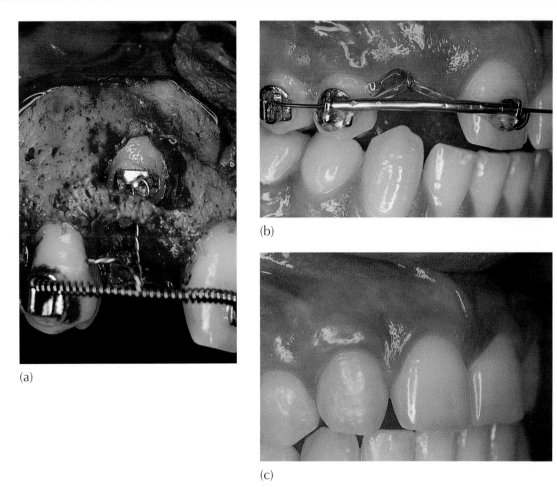

(a)

(b)

(c)

Fig. 6.38 (a) The 'tunnel' approach of Crescini. Note the preservation of the buccal plate inferior to the tooth. The stainless steel pigtail is drawn inferiorly through the vacated socket of the deciduous canine. (b) At 2.5 months post-surgery, the 'slingshot' elastic module has brought the canine into a buccally palpable position. (c) One year after completion of treatment. Note the gingival height, wide attached gingival band and good bony contour. The lateral incisor is congenitally missing.

before a markedly displaced canine reaches the coronal end of this 'eruption tunnel' and much reparatory bone will be laid down in its path, well ahead of its final and unimpeded eruption. Nevertheless, the progress of the biomechanically encouraged eruption may be achieved with no more difficulty than that of impacted teeth in more favourable locations and with unimpeded paths of eruption.

Given the narrow eruption path planned for this tooth in the present context, it is obvious that no attachment other than a small eyelet is suitable for bonding at the time of exposure. The eyelet is threaded with 0.011 in or 0.012 in soft ligature wire, ready for bonding to the newly exposed impacted canine. Following appropriate acid etching of the enamel surface, the attachment is loaded with the bonding agent and its steel ligature pigtail is lightly curved and threaded into the immediate area of the exposed tooth and on into the prepared tunnel until it emerges from the occlusal end of the deciduous canine socket. At this point, the eyelet attachment is pressed firmly into place on the impacted tooth and cured. Feeding a gold chain down

the tunnel is impossible, since the individual links adhere to the bloody walls of the prepared deciduous tooth socket, unless the chain is led down using a wire ligature threaded through the terminal link as a pathfinder.

The surgical flap is fully sutured to its former place and visual contact with the impacted tooth is lost. Control of future movement of the canine is exercised through the application of force to the steel pigtail ligature or gold chain, whose extremity may be seen to extend through the sutured edges of the flap within the deciduous canine socket at the crest of the ridge. Any excess in length of the ligature is cut and it is fashioned into a small hook, to which elastic traction may be applied (Figure 6.38b). The surgical method affords good access to the canine and a minor degree of difficulty in threading the ligature through the vacated deciduous canine socket. Orthodontic traction may be efficiently applied and the treatment result shows a good bony profile and an uncompromised periodontal result, similar to that seen on normally erupted teeth [93] (Figure 6.38c).

The palatal approach

If the canine is more palatally displaced, surgery approached from the buccal side will be more radical, involving the removal of comparatively large quantities of labial bone and, in these circumstances, a palatal approach is to be preferred. Following the raising of a palatal flap, the canine will be revealed high up, palatal to the roots of the adjacent teeth, which occasionally may themselves become denuded in the exposure process. Inferior and lateral to the canine is the vertical wall of the alveolar process. Bonding of the eyelet attachment is performed in the usual manner to the most conveniently accessible site, which is the palatal side of the tooth, although the buccal surface, close to the tip, is occasionally sufficiently approachable (Figure 6.39).

In this situation, many surgeons will remove a part of the flap in order to leave the impacted tooth in visual contact with the exterior and will place a pack to cover the open area. If the stainless steel pigtail ligature is drawn towards the line of the arch and sutured into place so that its extremity comes through the deficient part of the flap, the application of orthodontic forces will give rise to significant danger of irritation and infection of the area. This is caused by the tooth being drawn buccally and vertically downwards by its ligation to the labial archwire. The resultant direction of this force will cause the impacted canine to be drawn laterally against the alveolus and its healing granulation tissue. The exposed tooth will become reburied in these tissues (Figure 6.40) as it proceeds downwards and buccally. This leads to inflammation, false pocketing, acute pain and the likely occurrence of an acute lateral periodontal abscess.

It is therefore advised that orthodontic strategy for group 3 canines be approached in the same manner as with group 2 canines, by dividing their resolution into two distinct stages (Figure 6.41). Because of the considerable risk of exposing the root surfaces of the adjacent teeth, a closed eruption surgical procedure is to be preferred, since it requires a much more limited entry into the follicle and exposure of enamel surface – enough only to provide a bonding site for a small attachment under conditions that permit adequate haemostasis. More radical removal of

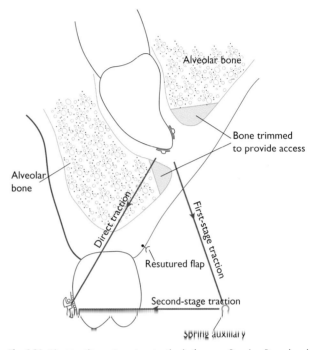

Fig. 6.39 Direct traction vs. two-stage traction in the group 3 canine. Reproduced from previous edition with the kind permission of Informa Healthcare – Books.

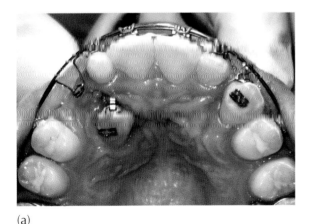

(a)

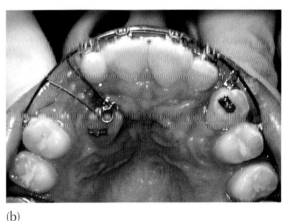

(b)

Fig. 6.40 Acute periodontal pain from prematurely attempted buccal movement in a group 3 canine. (a) The patient had bilaterally impacted palatal canines and these were drawn vertically downwards using an active palatal arch (see Figure 6.32). The left side canine erupted ahead of the right and the active palatal arch was discarded. New eyelets were bonded and traction was applied in a buccal direction on each. The left canine moved rapidly into place, but the insufficiently erupted right canine became partially buried in the vertical alveolar process and caused acute inflammation and pain. The swelling and redness can be clearly seen. The remedy is to irrigate the area, prescribe mouthwashes and return to the vertical eruption stage. An auxiliary labial archwire was placed, seen here in its passive mode, unattached to the eyelets. (b) The auxiliary labial archwire in its active mode to elicit further vertical eruption prior to the second attempt at buccal movement.

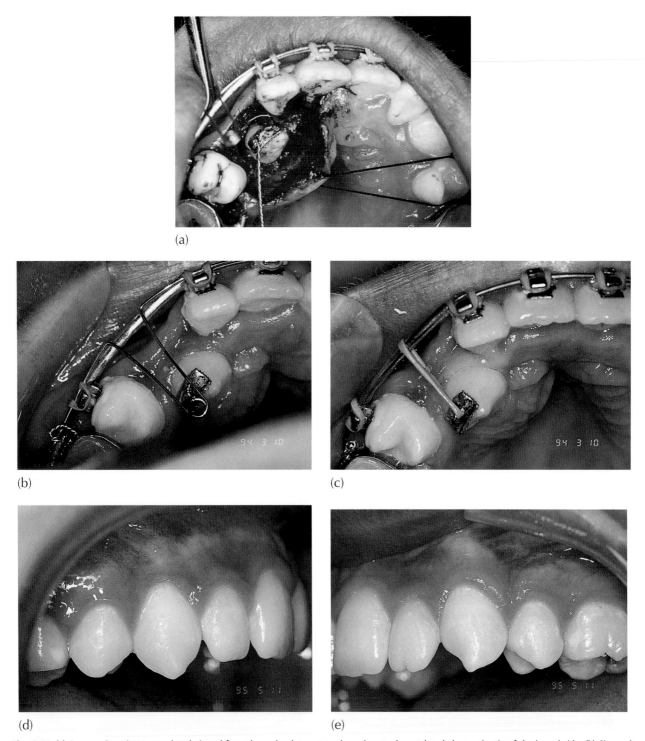

Fig. 6.41 (a) A group 3 canine exposed and viewed from the occlusal aspect to show the attachment bonded near the tip of the buccal side. (b) Six weeks later and without further adjustment, the canine has erupted through the closed flap. (c) Buccal traction to the buccally placed eyelet. (d, e) The buccal and palatal views of gingival tissues on the treated side, 16 months post-surgery. (f, g) The same views of the normal side. (h, i) Periapical views of the treated (right) and untreated sides, showing comparable bone support levels.

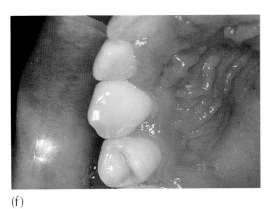
(f)

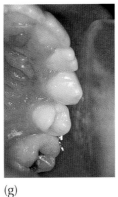
(g)

(h)

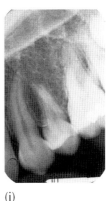
(i)

Fig. 6.41 *(Continued)*

bone around the mesial and distal curvatures of the crown and enucleation of the follicle are both unnecessary and harmful to the final periodontal outcome and prognosis of the aligned tooth [94].

Orthodontic traction is first applied in the lingual and vertically downward direction, to erupt the tooth into the palate, palatal to the line of the arch, and brought down to the occlusal level. As it comes down it is accompanied by a wide collar of newly formed alveolar bone. As was noted in relation to the group 2 canines, in many cases the palatal tissue is very resistant and bulges more and more as the tooth progresses, not allowing the tooth to erupt – a situation that demands the very limited and superficial surgical removal of the thick mucosa immediately over the crown of the tooth.

At this point, an additional eyelet should be placed on the buccal aspect, as for a group 2 case, and the direction of traction altered to a pure buccal tipping movement to bring it into the arch. It is often possible to bond a conventional bracket to complete this second stage if there is enough gingival clearance on the buccal surface of the tooth, although this may not be necessary. Since neither rotation nor ectopic root apex position is common in these cases, the second stage tipping movement generally brings the canine into its desired position and inclination. The canine that is located in the position that we have described here presents different problems from the group 2 cases. Direct traction is, under these circumstances, technically possible, but periodontally hazardous. The most direct surgical remedy (from the buccal side) may be too radical and leaves the tooth relatively unsupported by bone, in the final analysis, unless the 'tunnel' approach [93] is used. However, for the palatal approach, careful two-stage orthodontic movement will require minimal surgery and avoid undesirable periodontal sequelae. As pointed out earlier in regard to the use of an auxiliary labial wire or ballista, it is important not to make the active arm too long and to thereby draw the root apex of the impacted tooth palatally. In such a case, a tooth which might otherwise have had a normally located root apex will now iatrogenically require labial root torque in its final alignment.

The principal feature that distinguishes the group 3 case from the simple and straightforward group 1 case is its relative height in the alveolus. The root apex is usually in the line of the arch and the crown is only relatively minimally displaced palatally, which often means that the tooth cannot be palpated. The group 1 canine requires a minimal degree of vertical development in the mechano-therapy and mainly a buccal tipping movement from its more severe palatal position. The group 3 canine, on the other hand, has primarily to be extruded vertically. If only a very minor buccal component is needed, then a buccal approach to surgery, using the 'tunnel' method, may be the best way to go with the promise of a superior periodontal outcome.

Group 4

- proximity to line of arch: distant
- position in maxilla: high.

When the crown of the palatally displaced canine is not directly related to the roots of the incisors, it generally points medially and approaches or even crosses the mid palatal suture (Figure 6.42). It is not always palpable in the palate.

Surgery

These teeth are generally at some distance from the adjacent teeth and little bone removal is needed to expose them, with scant danger of exposing the roots of other teeth. There is usually reasonably good access for the placement of a bonded attachment, although the immediate exposed surface is unlikely to be the buccal aspect of the tooth.

Planning the orthodontic strategy

Since there is normal positioning of the root apex in most of these cases, all that would appear necessary is to draw the tooth directly to the labial archwire. If the long axis of the tooth is close to the horizontal plane, it would be inappropriate to do this since the direction of this force would

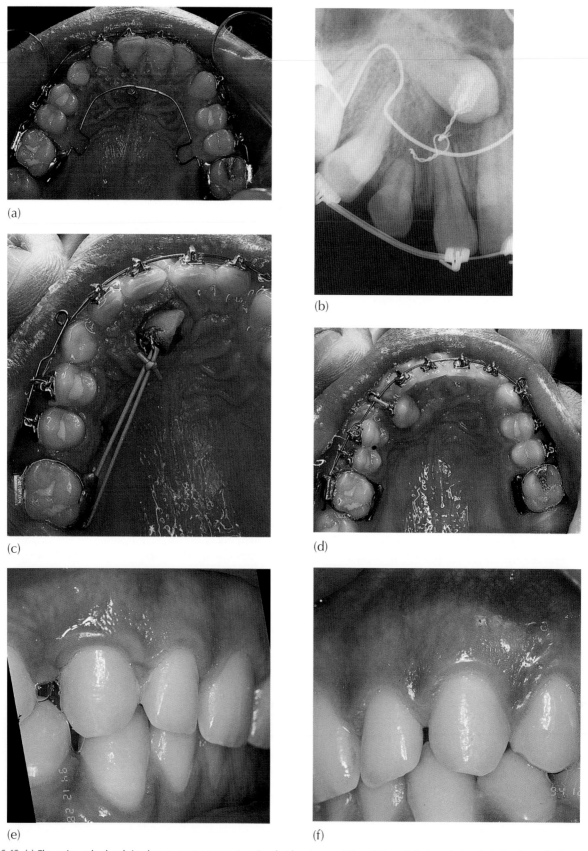

Fig. 6.42 (a) The active palatal arch in place to erupt a group 4 canine that has traversed the midline. (b) Post-surgical periapical radiograph shows space opening and an active palatal arch ligated to the bonded attachment. (c) After three months of traction, minimal re-exposure of the now very superficial and palpable canine was performed. A posterior component was achieved using elastic thread to the lingual tube of the molar of that side. (d) Disto-buccal followed by purely buccal traction was also performed. Distal uprighting and buccal root torque were later needed. (e, f) Intra-oral views to compare the buccal gingival health and clinical crown length of the treated (right) and untreated canines. (g, h) The same on the palatal side.

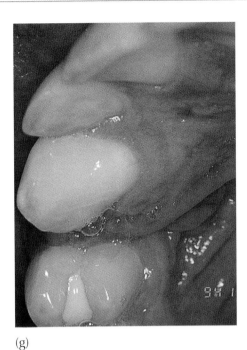

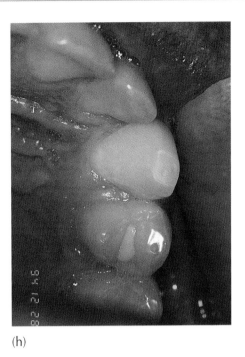

(g)

(h)

Fig. 6.42 (Continued)

be virtually coincident with its long axis. This makes the mechanics highly inefficient and little progress in resolving the impaction will be seen. The procedure will become excessively taxing on anchorage and will give rise to a reactive movement of the entire maxillary dentition to the opposite side.

With the tooth close to the horizontal, a more cautious approach should be embarked upon. A wide downward tipping movement will be achieved more efficiently with the same vertical directional approach that has been described for use in group 2 and 3 cases with the use of spring auxiliaries. In their tied-in and activated positions, the ballista and the light auxiliary labial arch mechanisms lie across the palate, parallel to and closely mimicking the orientation of the long axis of the unerupted and horizontal canine. As these mechanisms work themselves out they move in the same downward and buccal arc of circle that is needed for the resolution of the canine impaction.

Were the fulcrum for this wide tipping movement of the canine to be at the root end, it would follow that the root apex position would be unchanged when the crown finally reached its destination, vertically oriented and in the line of the arch. Unfortunately, however, the fulcrum is often some short way along the apical portion of the root; therefore, during the alignment of the tooth there will be a concomitant, but relatively minor, palatal displacement of the root apex of the canine. Thus, some buccal root torque of the group 4 canine will later be necessary. It also follows that in the unusual situation where there is a palatal displacement of the root apex at the outset much torque will be required, and this presents a major clinical problem.

Problems

Torque cannot be usefully applied until the crown of the tooth has reached its place in the arch and its newly bonded conventional orthodontic bracket firmly engaged by the labial archwire. At that juncture, the crown of the canine is at the occlusal level and has a strong buccal tilt. This places the palatally displaced root low in the palate, with its profile clearly outlined under the muco-periosteum. That critical portion of the lingual side of the root closest to the crown of the canine has a markedly convex shape, which dictates a distinct bulging of the mucosa covering it. In the clinical context, some dehiscence of this lingual area of the CEJ is usually present. Additionally, the cingulum area of the palatal aspect of the crown is very prominent and is likely to interfere with the occlusion.

It is important to recognize that in group 4 and many group 2 cases, considerable lateral movement of the canine is required, most of which involves tipping. Quite frequently, buccal root torque is also needed, and this may be quite substantial, particularly in the present group. Accordingly, there is an equal and opposite reactive force acting over a long period on the entire anchor unit, which, if properly planned and prepared, will respond only minimally. This will be expressed as a movement of the dental midline to the opposite side and a cross-bite tendency on the same side. In order to minimize this, a heavy base arch should be used and its form altered to compensate for the expected movements. In the most extreme circumstances, the use of intermaxillary S-elastics will be a necessary adjunct, i.e. from the buccal of the lower molar to the lingual of the upper molar on the side where a cross-bite

tendency has occurred, and from the lingual of the lower to the buccal of the upper on the other side. Thought should be given to the inclusion of temporary anchorage devices in the general scheme of anchorage preparation. A titanium screw may be placed in the same dental arch for direct traction application or in the opposing arch for use with intermaxillary elastics. A bilaterally affected case provides the opportunity for nullifying loss of anchorage by pitting one side against the other, such that midline and archform alterations need not occur.

To summarize the group 4 cases, the clinician must be alert to difficulties in the mechanics that may be inherent due to the initial location of the canine. Care must be taken to preserve orthodontic anchorage by properly planning the mechano-therapeutic strategy of reducing the canine displacement. The practical limitations imposed may lead to adverse effects on the periodontal status of the lingual aspect of the tooth, where occlusal interference may be present in the interim until the root position is corrected. Finally, following long, meticulous and successful treatment of the group 4 canines, there appears to be a strong relapse tendency that will result in the canine crown dropping back a millimetre or two into an edge-to-edge relationship with the opposing teeth and sometimes into a renewed cross-bite relationship. This seems to occur even after a fairly lengthy period of retention and the tooth may require permanent splinting. It is also related to the tendency to underestimate the amount of labial root torque that is need in these cases and an under-torqued canine will be very prone to positional relapse into an edge-to-edge relation with the lower canine.

The Seattle group [95, 96] has recommended open exposure and packing of canines before any orthodontic treatment is initiated. They do not give guidelines as to which cases are most suitable, nor have they investigated the success rate of this procedure, although it seems likely to be most appropriate for those canines in groups 2 and 4. In their view, the canines will erupt autonomously into the palate in the succeeding several months, thereby saving valuable time in appliance therapy when this is initiated at a later date. In order for this to be successful, it is incumbent on the surgeon to ensure that the exposure wound will not later heal over and this will make it mandatory for the exposure to be wide and fully down to the CEJ, with removal of the follicle in its entirety. Moreover, it has been shown that canines may take a very long time to erupt adequately into the mouth – of the order of 20 months or so [60] – offsetting any advantages that a shorter period of time with orthodontic appliances might bring. From the periodontal investigation of these cases the researchers found loss of bone support on the affected side compared with the controls and 79% of these teeth could be identified as having been previously impacted. It remains to be seen whether the same level of outcome can be achieved using this method as with the closed exposure approach.

Group 5

- canine root apex mesial to that of lateral incisor or distal to that of first premolar.

This tooth should be considered as transposed. To be completely consistent with the definition of transposition, the canine apex should be in the line of the arch in the place of the root apex of the adjacent tooth, but it is independent of mesio-distal or bucco-lingual crown location. However, partial transpositions or pseudo-transpositions are more common, in which the apices are displaced to a more limited extent and the order of the crowns of the teeth has reversed.

The canine–first premolar (CPm1) transposition must be considered as a three-dimensional phenomenon. The premolar may often be found in its ideal erupted location in the arch adjacent to the second premolar, but with a strong mesial displacement of its root, and also with a palatal displacement of its long axis when viewed from the occlusal aspect. On the buccal aspect of the sulcus above the premolars, there is usually a bulge which clearly identifies the position of the canine. In many cases, this tooth will erupt spontaneously, although it will emerge high in the oral mucosa (Figure 6.43).

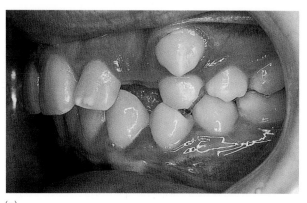
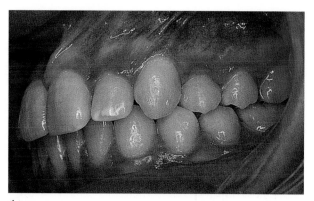

(a) (b)

Fig. 6.43 (a, b) A maxillary canine/first premolar transposition, treated to reverse the transposition.

Among the canine–lateral incisor (I2C) transpositions, the phenomenon is rarely limited to a two-dimensional model since this would mean that, while the positions of the crowns and/or roots of the two teeth involved are reversed, there is no bucco-lingual displacement from the line of the arch of either tooth.

The more typical and frequent presentation of this transposition involves the interrelationship between the canine and both incisors mesial to it, and it affects their alignment in all three dimensions (Figure 6.44). There are discrepancies in the mesio-distal, bucco-lingual and height locations of the crowns and roots of the teeth concerned. The root of the lateral incisor is distally and palatally displaced under the influence of a more mesially sited and mesially tipped canine, which is located superiorly and labially to it. This strictly defines the canine as a labial rather than palatal impaction in relation to the lateral incisor. However, the further eruptive movement of the tooth may occasionally bring the crown tip further mesially and on the palatal side of the root of the central incisor. Thus, the canine traverses the alveolar ride obliquely with its crown on the palatal side of the central incisor, while its root is labial to that of the lateral incisor. This is not an uncommon configuration and it is difficult to diagnose using plane film radiography only. Without the benefit of CBCT imaging, this may nevertheless be achieved in the following manner:

1. Viewed clinically from the front of the patient, the orientation of the long axis of the lateral incisor will usually exhibit a distal displacement of the root in the apical (crown-to-root) direction. The canine is not palpable on the labial or palatal sides of the ridge.
2. Viewed from the occlusal aspect, with the patient's head tipped back, the long axis of the lateral incisor also has a posterior displacement in the apical direction and the outline of the root may often be seen to bulge immediately beneath the palatal tissue. Its orientation is widely divergent from that of the long axis of the adjacent more upright central incisor.
3. A pair of periapical views employing Clark's tube-shift method will place the canine crown palatal to the root of the central incisor on which it is superimposed.
4. A panoramic view will show the canine 'riding high' over the lateral incisor and, usually, partially superimposed on it. The root apices of the canine will be above and in line with or possibly distal to that of the incisor, but more superiorly located.

The classic picture of the arrangement of the maxillary incisors in a class 2 division 2 malocclusion is depicted by central incisors tipped lingually and lateral incisors tipped labially. This is an incisor arrangement that seems to be associated with this particular scenario within the group 5 canine, presumably because of the wide divergence of the central and lateral incisor roots. In view of the very different eruption times of the incisors and canines, it would seem

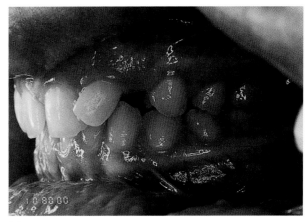

(a)

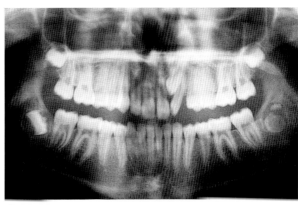

(b)

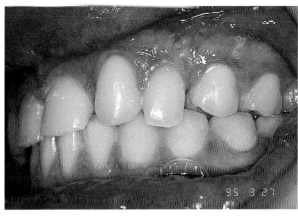

(c)

Fig. 6.44 (a, b) Canine/lateral incisor transposition seen intra-orally and on panoramic radiograph. (c) The completed alignment of the teeth in the transposed order. Grinding of the incisor edge of the lateral incisor needs to be performed to avoid occlusal interference and to improve appearance.

that the complexity of this impaction is the result of the incisor presentation and not its cause.

On the CBCT, the various views clearly show the interrelations between the teeth with great clarity and the orthodontist should be in a position to devise a method to resolve this very awkward entanglement.

Surgery

Surgical exposure of the CPm1 impacted canine is much simpler because of its favourable buccal position above the premolar teeth. Since the tooth is high in the oral mucosa, an apically repositioned flap is the most appropriate means by which the tooth will be rendered accessible and a periodontist will probably be the most suitable surgeon to perform this.

It would appear that there are two possible approaches to provide surgical access to the I2C transposed canine: buccally or palatally. We have seen that for a palatally impacted canine the exposure must be performed on the palatal side of the ridge and the tooth drawn palatally away from the incisor roots and towards its place in the arch. The crown of the canine for the present canine–lateral incisor (I2C) situation is palatal to the crown of the central incisor, but it would be a serious mistake to use a palatal surgical approach. There can be no question that a labial approach to the canine is the simpler way. Palatal access is especially difficult because the tooth is very high up on the palatal side of the root of the incisor, much bone will need to be removed to reach it and attachment bonding will be fraught with difficulty. However, a palatal approach would be fatal to the outcome of the case because traction applied to the canine from a palatal auxiliary will draw the root of the canine against the labial side of the lateral incisor root and displace its apex palatally while, at the same time, obstructing the progress of the canine. The most likely outcome of such a manoeuvre would be the loss of both these entangled teeth (Figure 12.16).

The only way to achieve success in these very special cases is to expose the tooth on the labial side, using a wide flap raised from the area of attached gingiva. Since the canine is high and close to the root apex of the central incisor, only a small portion of the more distally presenting surface of the canine should be exposed. This surface is readily accessible with little or no bone removal. The follicle is conservatively opened, wide enough for the bonding of a small attachment at that site. The tooth cannot be drawn directly down to the archwire, but must be directed disto-labially for it to have a clear path of exit from the palatal to the labial side of the ridge between the central and lateral incisors, respectively. To achieve this, the stainless steel pigtail or gold chain will need to be drawn horizontally and distally through a small puncture in the oral mucosa portion of the fully replaced surgical flap. It cannot, therefore, be drawn through attached gingiva in the first instance.

In the subsequent weeks of successful activation, the tooth comes to bulge the oral mucosa as it moves labially and distally around the lateral incisor, towards its place in the arch. If care is taken not to apply too effective a labial traction component, the tooth may remain beneath the mucosa for much of its early movement. When appropriate, surgical apical repositioning of a flap taken from the attached mucosa over the alveolar ridge may be used to re-expose the tooth and continue its traction to its final destination, encompassed by firm gustatory tissue. However, the response of these teeth to an efficient spring auxiliary may lead to early eruption of the tooth in the labial sulcus, often in a matter of weeks.

Planning the orthodontic strategy

The canines in group 5 may be offered four possible lines of treatment that are appropriate procedures for their resolution:

- to resolve the transposition to the ideal relationship (Figures 6.43 and 6.46);
- to move the premolar mesially (or incisor distally) into the canine location and align the canine between the two premolars (or between central and lateral incisors) (Figure 6.44);
- to use the canine for auto-transplantation into a prepared socket in its ideal site;
- to extract the severely displaced canine, incisor or premolar, depending on which has the least chance to be aligned with a good prognosis and leaving the deciduous canine in place.

In the scenario described above for the I2C transposed lateral incisor–canine, with the canine labial to the lateral incisor and palatal to the central incisor, many practitioners facing this scenario will elect to accept the transposition of the canine and lateral incisor and prefer to align these teeth in their transposed order (Figure 6.44). On the other hand, a three-dimensional approach to its resolution is necessary if the orthodontist is to aim for complete correction of the impaction and the transposition (Figures 6.45 and 6.46). Prior to the surgical exposure, an auxiliary labial arch similar to the one illustrated in Figures 6.30 and 6.34 is prepared for use on the labial side of the arch. Its loop is formed in the horizontal plane and tied piggyback into the brackets over the heavy main arch. At the conclusion of the surgical episode, the horizontal loop is pushed superiorly and mesially and ensnared in the pigtail ligature, as close to the mucosal flap as possible, to apply a traction force. This force draws the tooth back along its original path between the two incisors, tipping it labially and distally as it goes. It is important to recognize that ligating the lateral incisor bracket into the archwire will tend to upright this tooth and, if the archwire is rectangular in cross-section, to apply labial root torque to it. This will immediately close the window of opportunity for the canine, by reducing the mesio-distal and labio-lingual space between the roots of the two adjacent incisors. It is far better to leave the lateral incisor without a bracket at this stage. Once the canine has moved labially and is palpable on the labial side of the ridge, the bracket is placed on the lateral incisor and this tooth may then be uprighted and labially root-torqued to clear the way for the canine to be drawn distally and into its place.

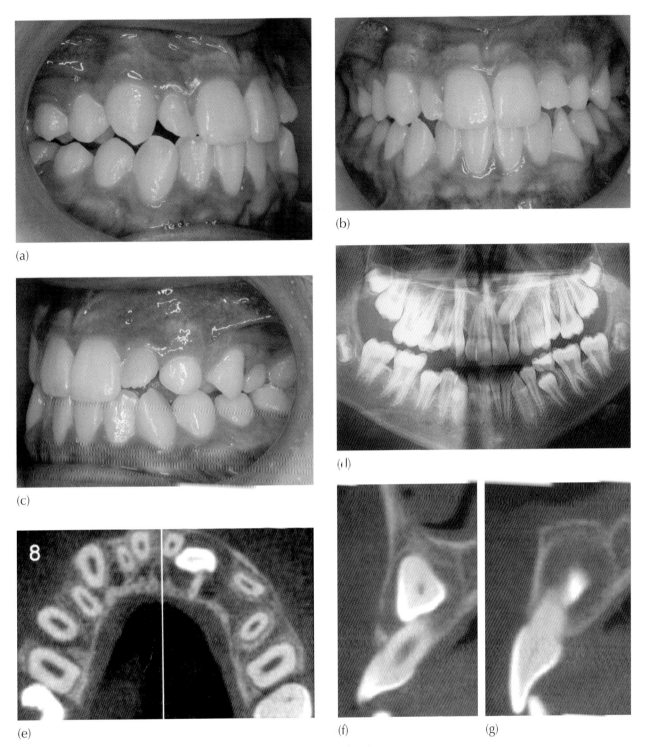

Fig. 6.45 (a–c) Intra-oral views of a 12-year-old male with left maxillary impacted canine.

(d) Panoramic view of a 12-year-old male with left maxillary impacted canine and two over-retained second deciduous molars.

(e) An axial cut from the CBCT at a level high on the roots of the anterior teeth shows the lateral incisor root displaced palatally, with the crown of the canine labial to it but palatal to the root of the central incisor.

(f) This transaxial cut shows the canine to be located labial to the root of the lateral incisor.

(g) The transaxial cut along the central incisor shows the canine crown on its palatal side.

(h) A three-dimensional CBCT view from the palatal side shows the canine to be labial to the lateral incisor and encroaching on the palatal side of the central incisor.

(i) Surgery must be approached from the labial side, care being taken not to proceed too far towards the crown tip of the canine in order not to damage the central incisor root.

(j) Attachment bonding.

(k) The flap is fully closed and sutured.

(l) Traction is applied in a labial direction by engaging the auxiliary looped archwire.

(m) Due to missed appointments, the patient was seen only six weeks later. The tooth has emerged through the oral mucosa!

(n) The left side in the closing stages of treatment.

(o–q) The completed case. The left canine has a slightly longer clinical crown than the unaffected right canine. This is more common for labial impactions and when the tooth is drawn through oral mucosa, as occurred in this case.

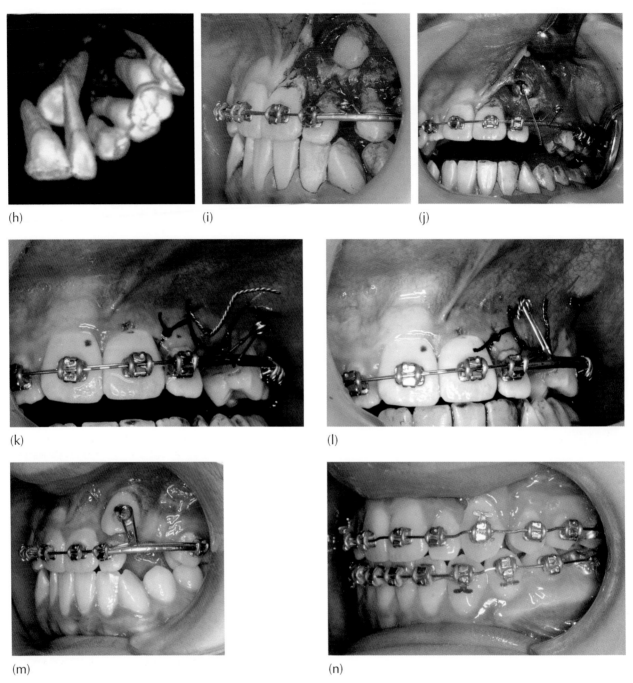

(h) (i) (j)

(k) (l)

(m) (n)

Fig. 6.45 *(Continued)*

After a relatively short period of time, often within a month or so, the canine bulges the labial tissue as the tooth is drawn labially, indicating that it has moved clear of any contact with the root of the lateral incisor and care should be exercised not to draw the tooth too far labially, since this will reduce the width of its labial bone cover and risk a long clinical crown. Therefore, at this point mesial uprighting of the lateral incisor should be undertaken to permit distal and palatal movement of the canine into its designated location. The final alignment manoeuvres needed include uprighting of the tooth, with labial root-torque of the lateral incisor and palatal root-torque to the canine, which can usually be performed with considerable efficacy using a simple reciprocal torqueing auxiliary.

Problems that may be encountered

Care should be taken in the construction of the auxiliary looped arches in these cases, so that when the loop is

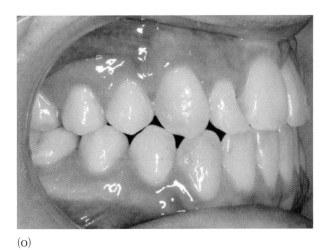

(o)

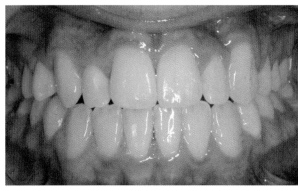

(p)

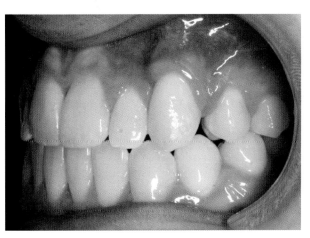

(q)

Fig. 6.45 *(Continued)*

directed superiorly (following flap replacement and suturing) to ensnare the twisted steel ligature and apply the required traction, it is not excessively long. Ideally, its length should be a couple of millimetres short of the height of the sulcus at that site, otherwise the patient will suffer unnecessary after-pain and swelling due to ulceration of this very mobile and delicate oral mucosal tissue at the corner of the mouth.

If the CPm1 transposed canine tooth is palatal to the line of the arch, the secondary effect of root contact with the premolar will rotate the canine's root apex both mesially and palatally across the palate, in a wide-sweeping movement. The tooth will become 'laid out' immediately beneath the periosteum and the long profile of its root will be palpable under the palatal mucosa. Dehiscence of the cervical area of the palatal surface of the root will occur. The amount of unavoidable labial root-torque that will then be required will be extreme and beyond therapeutic reason. If the canine position is buccal to the root of the adjacent tooth and the tooth is brought buccally to the first premolar or the lateral incisor, further buccal displacement

of its root will occur, with gross dehiscence of the buccal periodontium.

Plane film radiography in these cases can easily distinguish which of the teeth is buccal and which lingual in relation to their crowns. But there is great difficulty in determining the interrelation between the teeth for the full length of their roots and in measuring the exact proximity of the root of one to the root of the other. Without this information, the choice of which treatment approach to follow cannot be made with any degree of certainty regarding its outcome. The information needed is now easy to obtain using cone beam volumetric CT scanning techniques.

While each of the above four treatment possibilities will be recommended in specific instances, it becomes clear that the most likely and practical course, in many cases, will be to recognize and accept the transposition. The canine should be brought into the dental arch with the first and second premolars (or between central and lateral incisors) as its mesial and distal neighbours, respectively (Figure 6.44).

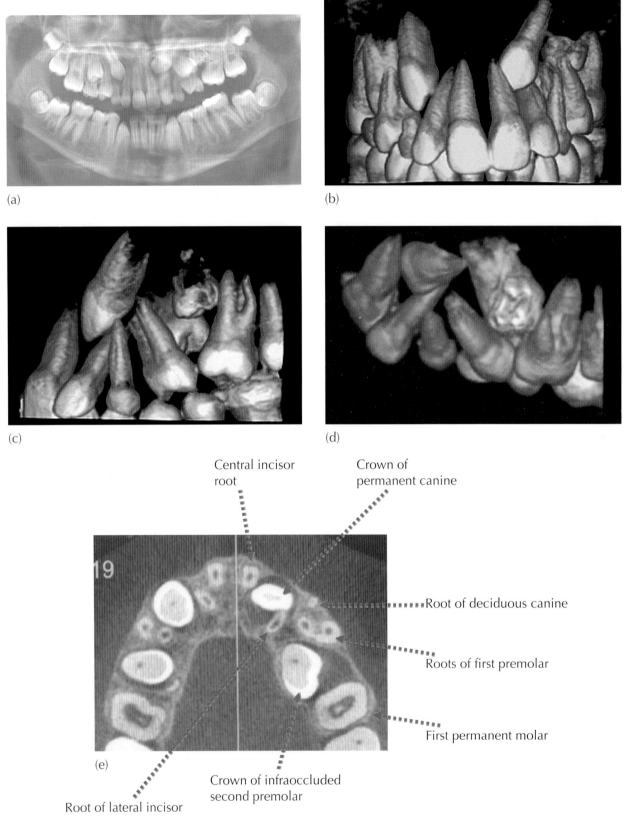

Central incisor
root

Crown of
permanent canine

19

Root of deciduous canine

Roots of first premolar

First permanent molar

Root of lateral incisor

Crown of infraoccluded
second premolar

(e)

Fig. 6.46 A group 5 canine with obvious transposition between the left lateral incisor and the adjacent canine.

(a) The panoramic view taken from the CBCT. The canine is high and although its apex is slightly distal to that of the lateral incisor, its crown is clearly mesial. The lateral incisor crown is close to its normal place, but its root orientation shows a marked distal displacement.

(b) The three-dimensional view from the front shows the mesial edge of the canine crown palatal to the central incisor and the distal edge labial to the lateral incisor. It will be clear that surgical exposure from the palatal side followed by traction to the palatal will torque the root of the lateral incisor even more palatally.

(c) Seen from the left side, these interrelations are very clear and the deciduous canine is unresorbed. Incidentally, there is a distally tipped first premolar and mesially tipped first molar which are due to the severely infra-occluded and almost totally resorbed deciduous second molar. The impacted second premolar can be seen on the palatal side of these teeth.

(d) Seen from above, the strongly divergent long axes of the three anterior teeth are evident. The canine is firmly wedged between the central incisor and lateral incisor roots.

(e) The axial cut across the maxilla at this level shows the relationship between the crown of the canine and the roots of the central and lateral incisors, close to their apices.

'Machismo' may be the driving force behind the decision to place the teeth in their correct order since, after all, we are orthodontists and this is the sort of challenge for which we have been trained. It is difficult to pass over an opportunity to display our initiative, dexterity and clinical excellence. Sometimes this is justified – but not often.

It should always be remembered that the reversal of a transposition of necessity dictates that the two teeth have to pass by one another in an alveolar process whose bucco-lingual width is suitable for just one of them. True, when teeth are moved buccally or lingually on the ridge, there is a concomitant bucco-lingual expansion of the alveolar bone. Clinically, however, there is loss in bone height and a dehiscence may occur, particularly if oral hygiene is inadequate or, paradoxically, when toothbrushing is compulsively aggressive, as seen in some individuals. Furthermore, the biomechanics are difficult to perform with adequate root control, and root proximity may occur during the exercise. This could seriously compromise bone support on that surface of the two roots and some loss of attachment or root resorption may occur.

Group 6

- erupting in the line of the arch, in place of and resorbing the roots of the incisors.

The teeth that fall into this last category of impacted canines are only marginally displaced buccally or palatally, since most of them do not merely generate resorption, but actually move into the newly resorbed area where resorption of the incisor root has occurred. They may frequently, therefore, be difficult to palpate in the clinical examination. Similarly, the plane film radiographic methods used in determination of the exact three-dimensional location of the tooth will be very difficult to interpret because of the minimal bucco-palatal discrepancy of the two structures. The difficulty in positional diagnosis is markedly increased if there is little or no radiographic superimposition of the image of the crown of the impacted tooth on the remainder of the resorbed incisor root, which occurs when the resorption begins at the root end due to a superiorly placed and axially oriented canine.

Why unerupted permanent teeth, in particular the maxillary canine, which are in close proximity to the roots of their neighbours cause resorption of these roots is unclear. Nevertheless, a cause-and-effect relationship is present and it has been conclusively shown that the resorption largely ceases when the impacted tooth is removed from the area of the affected roots [69].

The type of root resorption that is seen frequently in the routine orthodontic treatment of patients without impacted teeth, although worrying, is fairly insidious in nature. While it is a cause for wariness and caution in advising treatment, and for careful monitoring during treatment, it is not usually more than a marginal phenomenon, which stops when active tooth movement stops. The finding before or during treatment may make the case for shortening the duration of appliance therapy and limiting its goals to the essentials only. However, when the cause of the resorption is very specifically related to the proximity of an impacted tooth, the character or type of resorption appears to be different, insofar as its conduct is more aggressive [69, 97]. It follows that, with the very survival of one or more teeth at stake, early diagnosis is crucial in order to initiate appropriate treatment. The condition constitutes one of the very few situations in which orthodontic treatment may be considered a quasi-emergency.

Given the very special nature of this type of canine impaction, the difficulty in gaining an accurate picture of its extent, its rate of progress and the implications for its treatment, the next chapter will be given over to a comprehensive discussion of this important topic, which haunts even the most proficient orthodontic practitioner.

Treatment duration

It is probably stating the obvious that orthodontic treatment of a malocclusion with an impacted canine will take longer than the same malocclusion in a fully erupted dentition. Nevertheless, the presence of an impacted canine will clearly involve an extra dimension in the treatment which, by and large, cannot usually be addressed at the same time as other required orthodontic movements.

As we have seen here, the initial part of treatment of these cases is the same as with any other case, in that alignment and levelling need to be achieved and this requires the use of one of today's sophisticated fixed orthodontic appliance systems. At this point, passive and heavy archwires are inserted to fill the bracket slots in order to convert the whole appliance into a multiple anchor unit, incorporating all the teeth in one or both jaws. Further progress in the treatment of the overall malocclusion is placed on hold, while purpose-designed auxiliaries are introduced into the system aimed solely at reducing the impaction of the ectopic tooth/teeth. Thus, therapeutic attention is diverted away from the overall malocclusion and becomes focused solely on the impacted tooth and its resolution, until it is brought into alignment with the other teeth. Once the canine becomes an integral part of the alignment scheme, renewed attention is directed to bringing the overall malocclusion to its successful conclusion.

There are then both subjective factors that influence treatment duration (i.e. factors inherent in the treatment provider) and objective factors (i.e. factors related to the three-dimensional geometry of the canine ectopy).

Subjectively, a more inventive orthodontist with a flair for original design based on sound mechanical principles will usually be able to fabricate a device that is particularly suited to a given situation, while another may rely on standard "cookbook" solutions that require more frequent and more numerous adjustments. The more efficient the auxiliary, in terms of its traction force level and range of action,

the fewer the visits for adjustment will be and the shorter the treatment duration. Objectively, there are factors that make for extended treatment times. These include the height of the impaction [98–100], the horizontal position of the crown of the canine to the adjacent teeth and the maxillary dental midline [99, 101], the sector of overlap of the incisors, as seen on a panoramic radiograph [62, 99] and the angulation of the tooth [98, 99].

Studies described in the literature have shown that orthodontic cases that exhibit impacted canines have been variously reported with treatment times of 26.3 months [101], 19.7 months [102], 22 months [103] and 25.8 months for unilateral cases and 32.4 months for bilateral cases [98]. While these treatment periods do not seem unduly long compared with many routine malocclusion cases, it should be remembered that many impacted canine cases have largely normal alignment and jaw relations, with adequate space within the dental arches and little in the way of malocclusion, apart from the aberrant canine. This being so, most of the treatment period will be concentrated on canine resolution, while the biomechanics needed for the remainder of the treatment will be minimal, straightforward and of short duration.

References

1. Takahama Y, Aiyama Y. Maxillary canine impaction as a possible microform of cleft lip and palate. Eur J Orthod 1982; 4: 275–277.
2. Cramer HC. Dental survey of one thousand adult males: a statistical study correlated with physical and laboratory findings. J Am Dent Assoc 1929; 16: 122.
3. Mead SV. Incidence of impacted teeth. Int J Orthod 1930; 16: 885–890.
4. Dachi SF, Howell FV. A survey of 3874 routine full-mouth radiographs. II. A study of impacted teeth. Oral Surg Oral Med Oral Pathol 1961; 14: 1165–1169.
5. Brin I, Becker A, Shalhav M. Position of the maxillary permanent canine in relation to anomalous or missing lateral incisors: a population study. Eur J Orthod 1986; 8: 12–16.
6. Thilander B, Jacobson SO. Local factors in impaction of maxillary canines. Acta Odont Scand 1968; 26: 145–168.
7. Sacerdoti R, Baccetti T. Dentoskeletal features associated with unilateral or bilateral palatal displacement of maxillary canines. Angle Orthod 2004; 74: 725–732.
8. Montelius GA. Impacted teeth. A comparative study of Chinese and Caucasian dentitions. J Dent Res 1932; 12: 931–938.
9. Oliver RG, Mannion JE, Robinson JM. Morphology of the maxillary lateral incisor in cases of unilateral impaction of the maxillary canine. Br J Orthod 1989; 16: 9–16.
10. Becker A, Smith P, Behar R. The incidence of anomalous lateral incisors in relation to palatally-displaced cuspids. Angle Orthod 1981; 51: 24–29.
11. Johnston WD. Treatment of palatally impacted canine teeth. Am J Orthod 1969; 56: 589–596.
12. Friel ES. Migration of teeth. Dent Rec 1949; 69: 74.
13. Broadbent BH. Ontogenic development of occlusion. Angle Orthod 1941; 1: 45.
14. Becker A. The median diastema. Dent Clin North Am 1978; 22: 685 710.
15. Hitchin AD. The impacted maxillary canine. Br Dent J 1956; 100: 1–12.
16. Jacoby H. The aetiology of maxillary canine impactions. Am J Orthod 1983; 84: 125–132.
17. Becker A. Aetiology of maxillary canine impactions. Am J Orthod 1984; 86: 437–438.
18. Lappin MM. Practical management of the impacted maxillary canine. Am J Orthod 1951; 37: 769–778.
19. Howard RD. The unerupted incisor. A study of the postoperative eruptive history of incisors delayed in their eruption by supernumerary teeth. Dent Pract Dent Rec 1967; 17: 332–341.
20. Ericson S, Kurol J. Early treatment of palatally erupting maxillary canines by extraction of the primary canines. Eur J Orthod 1988; 10: 283–295.
21. Lindauer SJ, Rubinstein LK, Hang WM et al. Canine impaction identified early with panoramic radiographs. J Am Dent Assoc 1992; 123: 91–97.
22. Power SM, Short MBE. An investigation into the response of palatally displaced canines to the removal of deciduous canines and an assessment of factors contributing to favourable eruption. Br J Orthod 1993; 20: 215–223.
23. Brin I, Solomon Y, Zilberman Y. Trauma as a possible etiologic factor in maxillary canine impaction. Am J Orthod Dentofacial Orthop 1993; 104: 132–137.
24. Miller BH. The influence of congenitally missing teeth on the eruption of the upper canine. Dent Pract Dent Rec 1963; 13: 497–504.
25. Bass TB. Observations on the misplaced upper canine tooth. Dent Pract Dent Rec 1967; 18: 25–33.
26. Chosack A, Eidelman E, Cohen T. Hypodontia: a polygenic trait – a family study among Israeli Jews. J Dent Res 1975; 54: 16–19.
27. Polder BJ, Van't Hof MA, Van der Linden FP, Kuijpers-Jagtman AM. A meta-analysis of the prevalence of dental agenesis of permanent teeth. Community Dent Oral Epidemiol 2004; 32: 217–226.
28. Mossey PA, Campbell HM, Luffingham JK. The palatal canine and the adjacent lateral incisor: a study of a West of Scotland population. Br J Orthod 1994; 21: 169–174.
29. Zilberman Y, Cohen B, Becker A. Familial trends in palatal canines, anomalous lateral incisors and related phenomena. Eur J Orthod 1990; 12: 135–139.
30. Bjerklin K, Kurol J, Valentin J. Ectopic eruption of maxillary first permanent molars and association with other tooth and developmental disturbances. Eur J Orthod 1992; 14: 369–375.
31. Peck S, Peck L, Kataja M. The palatally displaced canine as a dental anomaly of genetic origin. Angle Orthod 1994; 64: 249–256.
32. Peck S, Peck L, Kataja M. Palatal canine displacement: guidance theory or an anomaly of genetic origin? Sense and nonsense regarding palatal canines. Angle Orthod 1995; 65: 13–17.
33. Chaushu S, Bongart M, Aksoy A, Ben Bassat Y. Becker Buccal Ectopia of Maxillary Canines in the Absence of Crowding. Am J Orthod Dentofacial Orthop 2009; v136: v218–223.
34. Rutledge MS, Hartsfield, JK Jr., Genetic Factors in the aetiology of Palatally Displaced Canines, Semin Orthod 2010; 16: 165–171.
35. Peck L, Peck S, Attia Y. Maxillary canine-first premolar transposition, associated dental anomalies and genetic basis. Angle Orthod 1993; 63: 99–109.
36. Becker A. Palatal canine displacement: guidance theory or an anomaly of genetic origin? Angle Orthod 1995; 65: 95–98.
37. Ely NJ, Sherriff M, Cobourne MT. Dental transposition as a disorder of genetic origin. Eur J Orthod 2006; 28: 145–151.
38. Becker A, Sharabi S, Chaushu S. Maxillary tooth size variation in dentitions with palatal canine displacement. Eur J Orthod 2002; 24: 313–318.
39. Garn SM, Lewis AB, Vicinus JH. Third molar polymorphism and its significance to dental genetics. J Dent Res 1963; 42: 1344–1363.
40. Sofaer JA. Dental morphologic variation and the Hardy–Weinberg law. J Dent Res 1970; 49 (Suppl): 1505–1508.
41. Newcomb MR. Recognition and interception of aberrant canine eruption. Angle Orthod 1959; 29: 161–168.
42. Becker A, Chaushu S. Dental age in maxillary canine ectopia. Am J Orthod Dentofacial Orthop 2000; 17: 657–662.
43. Baccetti T. A controlled study of associated dental anomalies. Angle Orthod 1998; 68: 471–474.
44. Grahnen H. Hypodontia in the permanent dentition. A clinical and genetic investigation. Odontol Revy 1956; 79 (Suppl 3): 1–100.
45. Alvesalo L, Portin P. The inheritance pattern of missing, peg-shaped and strongly mesio-distally reduced upper lateral incisors. Acta Odontol Scand 1969; 27: 563–575.
46. Garn SM, Lewis AB. The gradient and the pattern of crown-size reduction in simple hypodontia. Angle Orthod 1970; 40: 51–58.

47. Brook AH. A unifying aetiological explanation for anomalies of human tooth number and size. Arch Oral Biol 1984; 29: 373–378.

48. Pinho T, Tavares P, Maciel P, Pollmann C. Developmental absence of maxillary lateral incisors in the Portuguese population. Eur J Orthod 2005; 27: 443–449.

49. Becker A, Gillis I, Shpack N. The aetiology of palatal displacement of maxillary canines. Clin Orthod Res 1999; 2: 62–66.

50. Chaushu S, Zilberman Y, Becker A. Maxillary incisor impaction and its relation to canine displacement. Am J Orthod Dentofacial Orthop 2003; 124: 144–150.

51. Ben Bassat Y, Brin I. Maxillary canines in patients with multiple congenitally missing teeth: a roentgenographic study. Semin Orthod 2010; 16: 193–198.

52. Azaz B, Shteyer A. Resorption of the crown in impacted maxillary canine. A clinical, radiographic and histologic study. Int J Oral Surg 1978; 7: 167–171.

53. Becker A, Chaushu S. Success rate and duration of orthodontic treatment for adult patient with palatally impacted maxillary canines. Am J Orthod Dentofacial Orthop 2003; 124: 509–514.

54. Ericson S, Kurol J. Incisor resorption caused by maxillary cuspids. A radiographic study. Angle Orthod 1987; 57: 332–345.

55. Ericson S, Kurol J. Radiographic examination of ectopically erupting maxillary canines. Am J Orthod Dentofacial Orthop 1987; 91: 483–492.

56. Ericson S, Kurol J. Resorption of maxillary lateral incisors caused by ectopic eruption of the canines. A clinical and radiographic analysis of predisposing factors. Am J Orthod Dentofacial Orthop 1988; 94: 503–513.

57. Ericson S, Kurol J. CT diagnosis of ectopically erupting maxillary canines – a case report. Eur J Orthod 1988; 10: 115–120.

58. Ericson S, Kurol PJ. Resorption of incisors after ectopic eruption of maxillary canines: a CT study. Angle Orthod 2000; 70: 415–423.

59. Ericson S, Kurol J. Incisor root resorptions due to ectopic maxillary canines imaged by computerized tomography: a comparative study in extracted teeth. Angle Orthod 2000; 70: 276–283.

60. Leonardi M, Armi P, Franchi L, Baccetti T. Two interceptive approaches to palatally displaced canines: a prospective longitudinal study. AO 2004; 74: 581–586.

61. Baccetti T, Leonardi M, Armi P. A randomized clinical study of two interceptive approaches to palatally displaced canines. Eur J Orthod 2008; 30: 381–385.

62. Olive RJ. Orthodontic treatment of palatally impacted maxillary canines. Aust Orthod J 2002; 18: 64–70.

63. Baccetti T, Mucedero M, Leonardi M, Cozza P. Interceptive treatment of palatal impaction of maxillary canines with rapid maxillary expansion: a randomized clinical trial. Am J Orthod Dentofac Orthop 2009; 136: 657–665.

64. Baccetti T. Risk indicators and interceptive treatment alternatives for palatally displaced canines. Semin Orthod 2010; 16: 182–192.

65. Becker A, Zilberman Y. A combined fixed–removable approach to the treatment of impacted maxillary canines J Clin Orthod 1975; 9: 162–169.

66. Becker A, Zilberman Y. The palatally impacted canine: a new approach to its treatment. Am J Orthod 1978; 74: 422–429.

67. Jacoby H. The ballista spring system for impacted teeth. Am J Orthod 1979; 75: 143–151.

68. Kornhauser S, Abed Y, Harari D, Becker A. The resolution of palatally-impacted canines using palatal-occlusal force from a buccal auxiliary. Am J Orthod Dentofacial Orthop 1996; 110: 528–534.

69. Becker A, Chaushu S. Long-term follow-up of severely resorbed maxillary incisors following resolution of etiologically-associated canine impaction. Am J Orthod Dentofacial Orthop 2005; 127: 650–654.

70. Freeman RS. Adult treatment with removal of all four permanent canines. Am J Orthod Dentofacial Orthop 1994; 106: 549–554.

71. Becker A, Kohavi D, Zilberman Y. Periodontal status following the alignment of palatally impacted canine teeth. Am J Orthod 1983; 84: 332–336.

72. Kohavi D, Becker A, Zilberman Y. Surgical exposure, orthodontic movement and final tooth position as factors in periodontal breakdown of treated palatally impacted canines. Am J Orthod 1984; 185: 72–77.

73. Kohavi D, Zilberman Y, Becker A. Periodontal status following the alignment of buccally ectopic maxillary canine teeth. Am J Orthod 1984; 85: 78–82.

74. Ingber SJ. Forced eruption. Part II. A method of treating non-restorable teeth – periodontal and restorative considerations. J Periodont 1974; 45: 199–206.

75. Stern N, Becker A. Forced eruption: biological and clinical considerations. J Oral Rehab 1980; 7: 395–402.

76. Beck BW, Harris EF. Apical root resorption in orthodontically treated subjects: analysis of edgewise and light wire mechanics. Am J Orthod Dentofacial Orthop 1994; 106: 350–361.

77. Hendrix I, Carels C, Kuijpers-Jagtman AM, Van'T Hof M. A radiographic study of posterior apical root resorption in orthodontic patients. Am J Orthod Dentofacial Orthop 1994; 105: 345–349.

78. Sameshima GT, Sinclair PM. Predicting and preventing root resorption: Part II. Treatment factors. Am J Orthod Dentofacial Orthop 2001; 119: 511–515.

79. Sameshima GT, Sinclair PM. Characteristics of patients with severe root resorption. Orthod Craniofac Res 2004; 7: 108–114.

80. Segal GR, Schiffman PH, Tuncay OC. Meta-analysis of the treatment-related factors of external apical root resorption. Orthod Craniofac Res 2004; 7: 71–78.

81. Sandler PJ, Murray AM, Di Biase D. Piggyback archwires. Clin Orthod Res 1999; 2: 99–104.

82. Becker A, Shpack N, Shteyer A. Attachment bonding to impacted teeth at the time of surgical exposure. Eur J Orthod 1996; 18: 457–463.

83. Proffit WR. Contemporary Orthodontics. St Louis, MO: Mosby Year Book, 1992.

84. Kokich VG, Mathews DP. Surgical and orthodontic management of impacted teeth. Dent Clin North Am 1993; 37: 181–204.

85. Orton HS, Garvey MT, Pearson MH. Extrusion of the ectopic maxillary canine using a lower removable appliance. Am J Orthod 1995; 107: 349–359.

86. Vanarsdall RL, Corn H. Soft-tissue management of labially positioned unerupted teeth. Am J Orthod 1977; 72: 53–64.

87. Vanarsdall RL.Jr. Efficient management of unerupted teeth: a time-tested treatment modality. Semin Orthod 2010; 16: 212–221.

88. Chaushu G, Becker A, Zeltser R, Branski S, Chaushu S. Patients' perceptions of recovery after exposure of impacted teeth with a closed-eruption technique. Am J Orthod Dentofacial Orthop 2004; 125: 690–696.

89. Chaushu S, Becker A, Zeltser R, Vasker N, Chaushu G. Patients' perceptions of recovery after surgical exposure of impacted maxillary teeth treated with an open-eruption surgical-orthodontic technique. Eur J Orthod 2004; 26: 591–596.

90. Chaushu S, Becker A, Zeltser R et al. Patients' perception of recovery after exposure of impacted teeth: a comparison of closed versus open-eruption techniques. J Oral Maxillofac Surg 2005; 63: 323–329.

91. Lewis PD. Pre-orthodontic surgery in the treatment of impacted canines. Am J Orthod 1971; 60: 382–397.

92. von der Heydt K. The surgical uncovering and orthodontic positioning of unerupted maxillary canines. Am J Orthod 1975; 68: 256–276.

93. Crescini A, Clauser C, Giorgetti R et al. Tunnel traction of intraosseous impacted maxillary canines: a three-year periodontal follow-up. Am J Orthod Dentofacial Orthop 1994; 105: 61–72.

94. Becker A. Extreme tooth impaction and its resolution. Semin Orthod 2010; 16: 222–233.

95. Schmidt A, Kokich V. Periodontal response to early uncovering, autonomous eruption and orthodontic alignment of palatally impacted maxillary canine. Am J. Orthod. Dentofacial Orthop 2007; 131: 449–455.

96. Kokich VG. Preorthodontic uncovering and autonomous eruption of palatally impacted maxillary canines. Semin Orthod 2010; 16: 205–211.

97. Brin I, Becker A, Zilberman Y. Resorbed lateral incisors adjacent to impacted canines have normal crown size. Am J Orthod 1993; 104: 60–66.

98. Stewart JA, Heo G, Glover KE, Williamson PC, Lam EW, Major PW. Factors that relate to treatment duration for patients with palatally

impacted maxillary canines. Am J Orthod Dentofacial Orthop 2001; 119: 216–225.

99. Zuccati G, Ghobadlu J, Nieri M, Clauser C. Factors associated with the duration of forced eruption of impacted maxillary canines: a retrospective study. Am J Orthod Dentofacial Orthop 2006; 130: 349–356.

100. Crescini A, Nieri M, Buti J, Baccetti T, Pini Prato GP. Orthodontic and periodontal outcomes of treated impacted maxillary canines. Angle Orthod 2007; 77: 571–577.

101. Fleming PS, Scott P, Heidari N, di Biase AT. Influence of radiographic position of ectopic canines on the duration of orthodontic treatment. Angle Orthod 2009; 79: 442–446.

102. Becker A, Chaushu S. Success rate and duration of orthodontic treatment for adult patients with palatally impacted maxillary canines. Am J Orthod Dentofacial Orthop 2003; 124: 509–514.

103. Baccetti T, Crescini A, Nieri M, Rotundo R, Pini Prato GP. Orthodontic treatment of impacted maxillary canines: an appraisal of prognostic factors. Progress in Orthodontics 2007; 8: 6–15.

7

Impacted Teeth and Resorption of the Roots of Adjacent Teeth

Orthodontic Treatment of Impacted Teeth, Third Edition. Adrian Becker.
© 2012 Adrian Becker. Published 2012 by Blackwell Publishing Ltd.

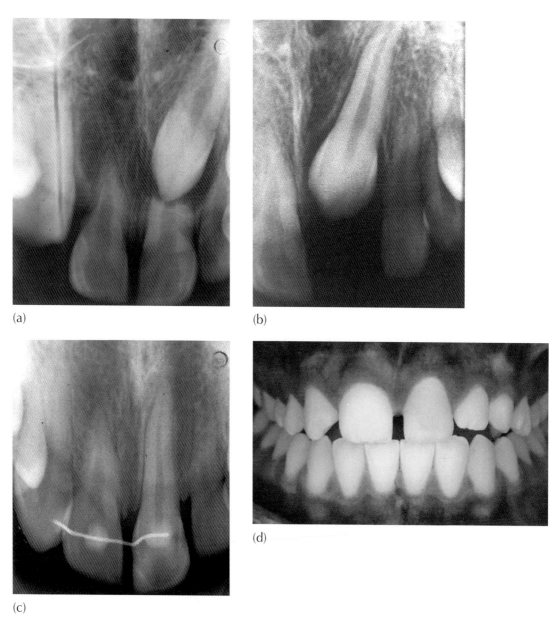

(a)

(b)

(c)

(d)

Fig. 7.1 An extreme case of resorption. (Courtesy of Dr P. Svanholt.) (a) Most of the root of the left central incisor has resorbed in response to the advancing unerupted canine in an 8.5-year-old female. On the right side, the central incisor shows considerable root resorption, which has arrested with the eruption of the right canine. The right lateral incisor is congenitally absent. (b) 16 months later the incisor is shed, following continued resorption. (c) The canine has erupted spontaneously in the position of the lost incisor. (d) Prosthodontic reshaping and restoration has made the canine more acceptable in its new role as an incisor.

One of the essential characteristics of the deciduous dentition is that the roots of the teeth physiologically resorb with the development and progress of eruption of their permanent successors. This leads to a smooth transfer through the mixed dentition and on to the permanent dentition in the adolescent. In contrast, resorption of the roots of permanent teeth does not normally occur even when there are unerupted or crowded teeth in the close vicinity of these roots. Nevertheless, in relatively exceptional circumstances, pathological resorption of the roots of certain teeth can and does occur (Figure 7.1). The 'aggressor' teeth concerned are principally the maxillary canines and, to a much lesser extent, the mandibular third molars, which cause resorp-

tion of the roots of the maxillary lateral/central incisors and the mandibular second molars, respectively (Figure 7.2). Occasionally, too, one finds an aberrant mandibular second premolar developing low in the alveolus and with a strong distal tilt, with signs of resorption of the mesial root of the first molar (Figure 7.3). In each of these circumstances the crown of the impacted tooth is in close relation with the root of its neighbour – a situation not normally seen in other areas, where unerupted teeth are to be found at the level of the cervical area of the crowns of the adjacent teeth. Nevertheless, why resorption of the roots occurs routinely and physiologically in deciduous teeth and rarely and pathologically in these exceptional permanent teeth is not

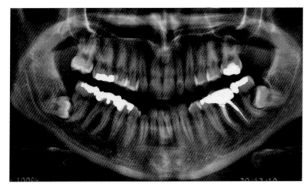

Fig. 7.2 Bilaterally impacted mandibular third molars associated with resorption of the roots of the adjacent second molars in an adult.

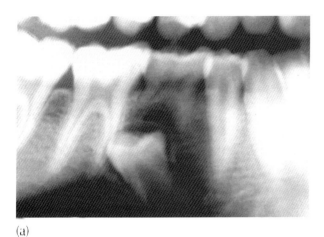

(a)

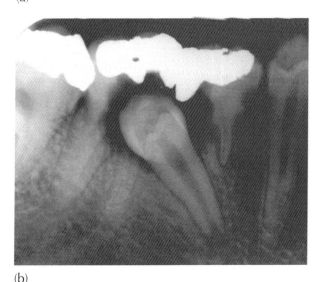

(b)

Fig. 7.3 (a) A section of the panoramic view of a female patient aged 12 years, showing a late-developing second premolar with a distal inclination. (b) A periapical view taken 2.5 years later. The second premolar has caused large-scale destruction of the first permanent molar.

understood. Since resorption of the roots of permanent teeth has wide-ranging clinical implications, a discussion of the various aspects of this phenomenon is clearly warranted. The present chapter will deal mainly with the resorption of the roots of erupted permanent teeth by unerupted adjacent teeth in its most common location and where its clinical implications are also the most significant, namely the permanent canine/incisor area of the maxilla.

Prevalence

Based on the use of plane film radiography, which was the standard of care in the late 1980s, early studies of the incidence of discernible root resorption of lateral and central incisors adjacent to an impacted maxillary canine revealed that 12% of the cases in the sample studied were affected [1–4]. While this finding in itself was alarming, the authors of the papers and others were at pains to point out that because of the limitations of plane film radiography, this figure was probably understating the extent of the problem [5–6]. It will be readily understood that an unerupted palatal canine is to be found closer to the palatal surface of the incisor roots than to any other surface of the root and, should there be root resorption of the incisor, it is logical to assume that it will be palatal surfaces that are most likely to be affected. By the same token, a buccal canine may cause similar damage to the labial surface of the root of the adjacent incisor. Furthermore, because the canine generally approaches the root of the incisor from above and along a path which is at an angle to the long axis of the resorbing incisor, the resorption process usually affects the incisor roots obliquely. However, the labial and palatal surfaces of the roots of the teeth are largely impossible to image using traditional plane film radiography. It is these palatal and labial root surfaces which, if affected by resorption, are least likely to be recognized in plane film radiography until the extent of the lesion is sufficiently well advanced for it to alter the mesio-distal profile of the root or the relative radiolucency of its projected shadow.

In line with technical progress in the development of diagnostic radiology, computerized tomography (CT) presented the window of opportunity that could overcome this bucco-lingual imaging 'blind spot', as discussed in Chapter 2. Accordingly, when a similar study of root resorption was undertaken by the same researchers [7], this time using spiral CT scanning in place of the earlier plane film radiography, it revealed signs of root resorption in almost half the cases affected by an aberrant/ectopic and unerupted maxillary canine. The actual figures in this study showed 38% of lateral incisors and 9% of central incisors affected to a greater or lesser extent, or 47% of affected individuals, in total. In a more recent study, in which the imaging modality used was cone beam three-dimensional volumetric CT, the figures were higher still, with 66.7% of adjacent lateral incisors and 11.1% of adjacent central incisors affected (Figure 7.4). All those cases

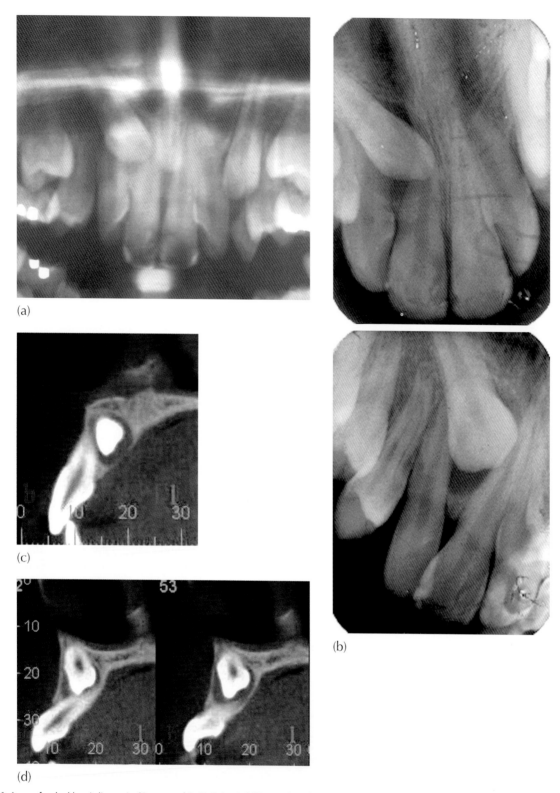

Fig. 7.4 A case for double misdiagnosis. (Courtesy of Dr N. Dykstein.) (a) A section of a panoramic view showing the right impacted canine superimposed on the lateral and central incisors. There is no apparent resorption of the incisor roots and the canine is palatal according to Chaushu's method (see Chapter 2). (b) Two periapical views, with lateral tube shift intended for positional diagnosis, show the crown of the canine to be palatal to the central incisor root and no apparent root resorption. The canine is clearly palatally impacted. (c) From the CBCT images, a single paraxial cut of the canine and central incisor relation shows marked oblique and palatal resorption of the incisor root, and the canine can be confirmed to be on the palatal side of the central incisor. (d) From the CBCT images, two parallel paraxial (vertical antero-posterior) cuts, at an interval of 1.5 mm, of the canine and lateral incisor relation show marked oblique and labial resorption of the incisor root. The canine is on the labial side of the lateral incisor.

which exhibited central incisor resorption also showed resorption of the lateral incisors and in all the cases in the study close proximity between the impacted tooth and the incisor root was found [8].

Of course, in a significant proportion of the impacted canine patients in whom root resorption has occurred, these canines will eventually erupt and the roots of the incisors may suffer little or no further resorption in the long term (see Figure 6.34i, j). The eruption may be generated by a spontaneous change in the orientation of the canine in relation to the affected incisor root, as has been shown in anecdotal case reports [9] or during orthodontic treatment to create space for the canine by an appliance-generated movement of adjacent teeth [10, 11] as described in Chapter 6 (see Figure 6.19) or by the prophylactic extraction of deciduous [5, 12, 13] or permanent teeth (see Figures 6.19, 6.20).

Aetiology, diagnosis and prevention

Over the years, clinicians have searched for clues, specific tell-tale features, which may indicate a high risk for incisor root resorption associated with an impacted canine. Obviously, the mobility of an incisor, in the absence of any signs of premature contact or symptoms of pain, would offer such an indication *after* the fact. However, the quest was for an associated anomaly or phenomenon in the dentition that could predict the occurrence of incisor root resorption *before* it occurred.

Two studies that have investigated sexual dimorphism as a factor in the prevalence of this phenomenon have found that it occurs 3–4 times more frequently in females than in males [3, 8]. In relation to those cases in which the resorption has been extensive, accounting for more than a third of the incisor roots, both published individual and multiple case reports in the orthodontic literature and the author's clinical experience indicate that, in its severest form, the phenomenon may be almost completely limited to females [14]. However, caution should be exercised in this regard since no formal study has been made of these more advanced resorption cases in contrast to those that may be minimally affected or where the extent of the damage is clinically insignificant.

Suspicion regarding the presence of an enlarged dental follicle surrounding the impacted canine is another factor that has been investigated. Two studies have been performed by the same Swedish group and they found there to be no apparent cause-and-effect relation between an enlarged follicle and root resorption [3, 15]. Notwithstanding, other clinicians have expressed concern regarding a possible cause-and-effect relation between an enlarged follicle and root resorption [16, 17].

In light of the potential seriousness of the condition in the more extreme expression of its occurrence and its very much higher prevalence among females, this concern should be heeded, care taken and caution advised in the

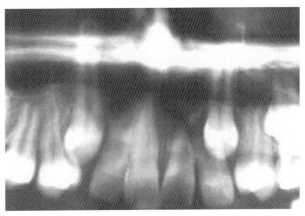

Fig. 7.5 An enlarged follicle associated with severe resorption of the left lateral and central incisor teeth, in a 10-year-old female patient.

clinical context, despite the absence of confirmatory investigative studies. Good anecdotal examples of this may be seen in Figures 7.5, 6.15 and 6.36 and 7.10b, where unambiguous evidence of advanced root resorption of the incisor is seen adjacent to an impacted canine which is encompassed by an enlarged dental follicle. By and large, the overwhelming majority of cases where resorption has been confirmed show only a very mild and clinically insignificant degree of root loss, which at least partly explains why the condition requires the use of CT for the diagnosis to be determined definitively. However, it should also be remembered that all severely affected cases began the pathological process to this advanced state with very small initial resorption lesions which, had they been diagnosed earlier, may have markedly improved the prognosis.

As we have pointed out in Chapter 6, a number of studies relating to the aetiology of palatal canine impaction have been published over the past 20 years, and these have established a firm link between palatally displaced maxillary canines and anomalous lateral incisors. However, when the prevalence of resorption of an incisor adjacent to a palatal canine was investigated, anomalous incisors appeared to be relatively immune to resorption. The phenomenon was found to be much more frequently associated with a normal-sized adjacent lateral incisor [6]. The conclusion drawn from this study was that the practitioner would be well advised to investigate the possible existence of resorption in any impacted canine patient who shows no lateral incisor anomaly. This is, therefore, a factor to be taken into consideration in deciding the strategy of the orthodontic treatment and the timing and sequencing of the individual tasks within the overall treatment plan.

Treatment

When the diagnosis of impacted canine-related resorption of the incisor root is made, a lack of positive action on the part of the orthodontist or surgeon carries with it the

danger of further resorption of the root of the incisor, for which these two practitioners may justifiably be held responsible. Furthermore, for as long as the impacted tooth remains in close proximity to the resorbing root, its continued destructive nature remains undeterred by any initial orthodontic alignment and space-gaining procedures that may be under way. The resorption process seen in many of these cases is rapid and it has been shown conclusively that it will advance significantly during the preparatory orthodontic phase [14]. On the other hand, and as has been pointed out earlier, the creation of space in the arch by moving the adjacent teeth, one or more of which are themselves being actively resorbed, might be a factor that may change the course of the canine, causing it to move away from the area and spontaneously erupt (Figure 7.6).

Treatment options

Three possible lines of treatment are available, and each has its advantages, disadvantages, unanswered questions and manner in which it influences the outcome.

Extract the impacted tooth

First principles in medicine dictate that, in order to successfully treat a disease, it is first necessary to remove the cause. In the present context, the cause of the resorption is the unerupted tooth, and its extraction carries with it the bonus of eliminating what may be a problematic impaction from the point of view of the provision of surgical access and the mechano-therapeutic difficulty. If this apparently logical line of treatment is followed, then a healthy canine will be extracted and the incisor with the resorbed root will be left in place. There is uncertainty as to whether the resorption process will continue and, in this compromised state, it may be reasonable to assume that the tooth has a reduced long-term prognosis. Given that it is unlikely that the incisor is in its ideal position and will thus need to be orthodontically moved into alignment, we need to know if the orthodontic treatment prescribed to achieve this will generate further resorption, potentially adversely affecting its prognosis still further. If an implant may be planned as the ultimate substitute for the extracted canine, considerable root movement of the resorbed lateral incisor will probably be mandatory in order to provide adequate space between the roots of the adjacent teeth. Alternatively, it may be wiser to attempt space closure. This would increase the length of time that orthodontic forces will need to be applied to this endangered tooth in its capacity as an anchor unit, and it may finally need to be splinted. Although it could be argued that perhaps a fixed cast bridge might reduce the amount of orthodontic movement to which the tooth will need to be subjected, it should not be forgotten that a resorbed lateral incisor will not make a good abutment. With a premolar in the canine position, the appearance may be compromised, and canine protection of the occlusion would not be present.

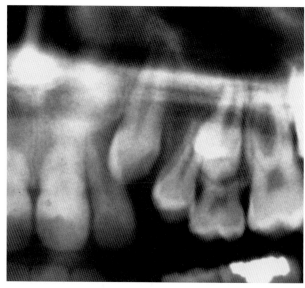

(a)

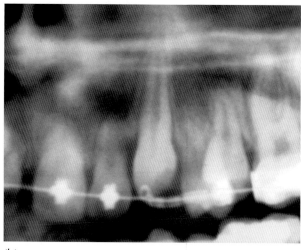

(b)

Fig. 7.6 Root resorption, space opening and spontaneous eruption. (a) The left impacted canine has caused a marked oblique resorption, which has shortened the root in general and its distal side in particular. The central incisors have short rounded roots. (b) Following the orthodontic creation of space, the canine is seen to be erupting spontaneously.

Leaving the deciduous canine in its place, in the hope of its enjoying long-term survival in the present situation, is a real possibility. Given the degree of displacement of the aberrant canine, it is quite likely that the root of the deciduous canine has remained untouched by the normal resorption process affecting deciduous teeth, and, in these cases, the tooth may last for many years before implant and/or prosthetic replacement is needed. However, if its position needs to be altered, to enable the placement of an appearance improving restoration, then perhaps the required orthodontic movement may trigger the physiological resorption that has not occurred.

Extract the resorbed tooth

Extracting the 'victim' tooth may appear unfair, but must be considered the pragmatic choice if it is to be assumed that it has a poor prognosis. By its extraction, the treatment of the impaction is much simplified. Whether the deciduous canine will also be lost in the overall long-term planning strategy or not, the permanent canine would be aligned in the lateral incisor site, with the accompanying poor appearance and lack of a canine-protected occlusion [18]. To address the appearance, some orthodontists, prosthodontists and 'aesthetic' dentists might prefer aligning the impacted canine in its normal place in the arch, performing a one-tooth implant to replace the lateral incisor and to achieve a more normal, uncompromised appearance [19, 20].

Non-extraction

This is perhaps the most difficult and most heroic line of treatment, the more so since the literature offers little evidence for us to believe that the outcome can provide hope for a stable future. In the first place, without the expedient of extraction, the canine impaction is much more difficult to resolve since the tooth will need to be surgically exposed, attachment bonded and drawn away from the resorbing root in a separate orthodontic manoeuvre and before the more usual directional forces may be applied to bring it to its place in the arch. Second, the lateral incisor has a shortened root and will almost certainly require to be moved orthodontically. It may need to be splinted for the duration of its presumably reduced lifespan. Third, this incisor may become non-vital as a complication of the surgical exposure and may subsequently require root canal treatment. These factors appear likely to reduce its long-term prognosis still further. On the other side of the scale, it has been noted in the previous chapter that most palatally impacted canines and a minority section of the buccally ectopic canines are to be found in conditions of spacing and the mildest malocclusion, which may generally be considered to be non-extraction cases. Therefore, this approach potentially offers the prospect of the greatest gain for the patient and, if carefully managed, perhaps provides a way in which our worst fears will not be realized.

It has been proposed that there is a fourth alternative in this situation – a 'masterly inactivity' or 'wait-and-see' option. An anecdotal series of three cases has been reported in which severe resorption of incisor roots had been noted adjacent to impacted maxillary canines, in each case [9]. The patients were followed up clinically and radiographically for a long period on a half-yearly recall basis, without treatment of any sort being provided. The canines eventually erupted, although the progress of the spontaneous resolution of the impaction generated further root resorption. It is a *sine qua non* that additional damage to the roots of the adjacent incisors was fully anticipated at the outset in these three patients, although the full extent of its expression could not be predicted. It is therefore reasonable to assume that the significant risk that these teeth would be lost during this observation period was accepted as a conscious decision. The dilemma that faced the clinicians was whether more damage would have been inflicted iatrogenically had the impacted teeth been treated by the orthodontic/surgical modality.

It is germane to attempt to answer this question in relation to group 6 impacted canines in general. To do so, it becomes necessary to examine the anatomical, surgical and orthodontic contexts of the dilemma and discuss the several factors that need to be carefully evaluated.

The anatomical context

The first factor addresses the anatomical relationship between the impacted tooth, the resorbing tooth and their location in the alveolar process. In general, and having been the first to erupt some 3–4 years earlier, the adjacent incisor tooth is sited in mid-alveolus in the bucco-lingual plane. The impacted canine generally follows a high and mesially angulated path towards the root of the incisor. The approach of a palatal canine is on the distal or disto-palatal aspect of the root of the incisor, some way down the root from the apex (Figure 7.6a), while a buccal canine will be on the disto-labial side. Resorption occurs in the immediate area of the proximity of the two, and will only be discernible radiographically if it also affects the distal profile of the root of the incisor and is not hidden by the superimposition of the two teeth.

In the most extreme cases, which are the context of the present discussion, the canine does not 'side-swipe' the root of the incisor, but rather comes down from above its apex, obliquely resorbing most of the full width of the root as it progresses inferiorly into the area vacated by the resorbing root (Figure 7.1) The canine is also thus situated in mid-alveolus. In one scenario, further progress finds the incisor root being resorbed obliquely with the destruction proceeding coronally at an alarming rate, until the crown of the still unerupted canine reaches one side of the cemento-enamel junction (CEJ) of the incisor. At this point, no further resorption can take place, since the enamel of the tooth crown will not resorb but will act to deflect the progress of the canine to one side, where it may then erupt. The incisor does not then shed naturally, since the oblique resorptive process will have left a longish spicule of root which will hold the tooth in the alveolus, although the tooth may (but not necessarily) become a little mobile. The incisor crown is also likely to be displaced in the opposite direction to the canine's new eruptive path due to the eruptive force of the latter acting obliquely on one side of the incisor's CEJ.

The surgical context

If we are to try to change the course of what appears to be inevitable, it is logical to attempt to distance the canine

from the resorption area as early as possible, which is immediately after the diagnosis has been established. In the orthodontic/surgical modality of treatment, this means that surgical access must be provided and an attachment bonded in order that orthodontic forces may be applied direct to the tooth to effectively draw it away from the immediate area by diverting it from its present path.

Because of the intimate relationship between the crown of the impacted tooth and the resorbing root apex of the incisor, accuracy of positional diagnosis is vital, if simultaneous collateral surgical damage to the root area of the incisor tooth is to be avoided. A wide flap is reflected, the crown of the impacted tooth is identified immediately beneath its thin bony covering and only the smallest opening is made in the bony crypt to expose that area of the crown of the impacted tooth most superficially accessible and most distant from the resorption site. No attempt should be made to seek out the root apex of the incisor, or even to gently probe the area (Figure 7.7).

It will be appreciated that a good radiographic diagnostic technique, specifically CT views employing a cone beam volumetric machine, should be considered an essential diagnostic aid that will contribute enormously to determining the preferred exposure site as accurately as possible. From these, the direction in which the tooth must be moved away from the sensitive resorption area will be decided. This may be in a labial direction or in a palatal direction, depending on the labio-lingual location of the crown tip vis-à-vis the residual root end.

Exposure of the crown, clearance of the follicle and bone removal to the maximum width of the crown and down to the CEJ in this situation, as recommended by Kokich for the more accessible palatal canines [19], is strongly contraindicated in these cases, since it will surely inflict unnecessary trauma to the resorption area and lead to devitalization of the incisor.

In these highly sensitive situations it is recommended that the area of crown to be exposed should be on an aspect of the crown as far as possible from the site of resorption. It should be exposed minimally, with an opening only large enough for placement of the small eyelet, while permitting adequate haemostasis during the bonding procedure (Figure 7.7j, k).

An eyelet attachment is then bonded to the tooth in the most accessible site available on the crown surface, and the twisted steel ligature is made to point in the direction that traction will be applied. The full flap is then sutured back to its former place, leaving only the ligature wire peeking through it, either at the sutured edge or, preferably, to piercing through the flap in the direction that traction has been determined. Primary closure is essential in these cases to protect the wound from damage or infection of the vital tissue at the root end of the lateral incisor. If, following the earlier resorption, the canine crown tip has crossed over to the labial side of the incisor roots, a labial resolution will be required (Figure 7.8). In this case, labial traction will be needed, and this will mean that a labial surgical approach will be preferred, with the attachment bonded in a convenient place on the crown and its ligature protruding through the flap, high in the labial sulcus. If the tooth is palatal, a palatal approach will be dictated, with the palatal aspect of the tooth being exposed and an attachment bonded there. Traction will need to be directed in the horizontal plane in a posterior horizontal direction.

The orthodontic context

As we have discussed in the previous chapter, directional traction must be designed to meet the demands of each individual case. Nowhere is this more crucial than for the cases that we are discussing here.

Ballista springs [21] and light auxiliary arches [22] are particularly suited to the provision of force in a palatal direction. They are specifically useful when designed to produce forces in a labial or vertically downward direction, but may be directionally modified to include apical or mesio-distal force components which are sometimes helpful in elevating a canine over the root apex of an incisor or to draw it in a wide labial sweeping movement around the root of the adjacent tooth.

When a canine needs to be moved posteriorly in the horizontal plane, a trans-palatal arch carrying three or four soldered loops may be used. Elastic thread may then be tied between the steel pigtail wire, which is ligated to the eyelet attachment on the canine, and that soldered loop which is most suitably placed to provide the optimum directional pull. Exploiting the lingual cleat of a molar band may sometimes be adequate in many of these cases. Using these simple and mainly custom-designed accessories, rapid movement of the crown of the impacted canine may be effected to widely separate it from the resorption area.

It is quite clear that the approach of Kokich [19] recommended for the open surgical exposure of an impacted canine uncomplicated by resorption of the root of the incisor cannot be used when the canine crown is in close association with a resorbing incisor root. In that approach, bone and follicular tissue are completely eliminated around the full width of the crown of the buried tooth, down to the CEJ, and a large circular opening is made in the flap, with placement of a periodontal dressing to maintain patency. To attempt this in the present context would seriously endanger the vitality and survival of the resorbing incisor.

Acceptance of the rationale for Kokich's method (see description in Chapter 3) precludes any remedy for the case where canine-related incisor root resorption is present, and one can only counsel 'not to interfere with Mother Nature' in its hoped-for trade-off between the possibility of spontaneous eruption and the near certainty of additional collateral resorption [19] as illustrated in a series of three cases [9].

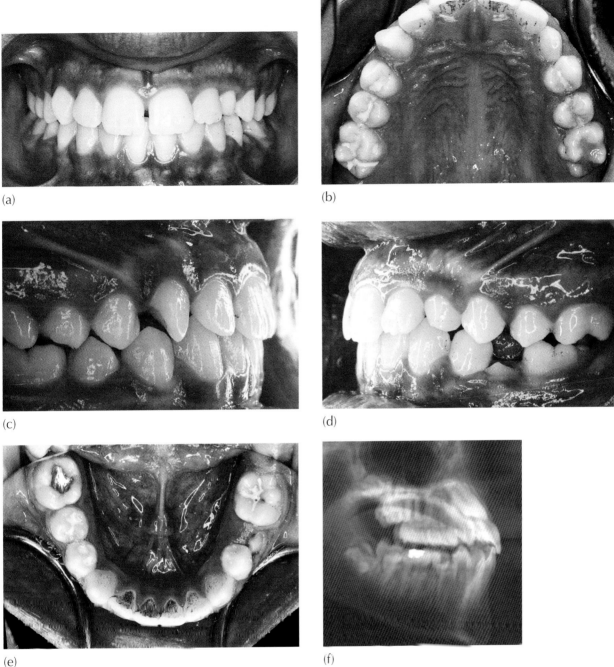

(a)

(b)

(c)

(d)

(e)

(f)

Fig. 7.7 This case was treated before the introduction of cone beam computerized tomography. (a–e) Initial clinical intra-oral views of the dentition. Note deciduous left maxillary canine in place and space loss in the left mandibular quadrant following early extraction of the deciduous second molar. (f) Section of the lateral cephalogram to show a single labially ectopic maxillary canine. (g) A section of the panoramic view shows the canine to superimpose on the lateral incisor root. It appears that most of the root of the lateral incisor has been resorbed. (h, i) The periapical views confirm labial impaction. (j) Surgical exposure is minimal to reveal only the disto-labial aspect of the tooth. The camera angle shows the canine to be in the same long axis of the incisor, obviously moving into the freshly resorbing area as it develops. (k) An eyelet is bonded to the disto-labial corner of the crown of the tooth. Its pigtail ligature is seen to hang loosely. Traction cannot be made direct to the archwire, for fear of resorbing more of the incisor root. (l, m) An open surgical exposure is strongly contraindicated here, since it would leave the sensitive apical area of the incisor exposed. Full flap closure is made to cover the entire exposed area, with the pigtail ligature taken through the oral mucosa, high up in the sulcus. In this situation, it cannot be taken through attached gingiva. The long loop of the light auxiliary labial archwire lies horizontally, in its passive state, ready for activation. (n, o) The loop is flexed upwards and engaged by the pigtail ligature, which ensnares it. This produces a light and highly controllable labial force to move the unseen canine. (p) At one month post-surgery, the tooth has exited the oral mucosa. (q) At two months post-surgery, the tooth has cleared the root of the lateral incisor and a distal direction of traction is applied, with a measured piece of cut tubing holding the space between incisor and premolar. The disto-labial position of the eyelet permits this movement without incurring unwanted rotation. (r, s) At three months post-surgery, the tooth has moved opposite its place in the arch and a regular bracket is substituted, in preparation for the finishing phase of treatment. (t) At five months post-surgery, finishing procedures are in progress. (u) Periapical radiograph checking for the status of root resorption indicates considerably more root than realized at the outset. The appearance is typical of an oblique resorption, in which the labial side of the tooth has been 'shaved' off, leaving a long spicule of palatal root. (v–z) The final clinical result. In view of the severity of the resorption process, meticulous attention to root uprighting and torqueing was considered inappropriate at the time. (a'–f') Periapical and panoramic views of the immediate area, at the end of treatment. (g'–i') Patient seen 7.6 years post-treatment. (j', k'). Periapical and panoramic views of the dentition 7.6 years post-treatment, showing no change in the condition of the root of the resorbed lateral incisor. There is no hypermobility of the tooth and the removable retainer was discarded three years ago.

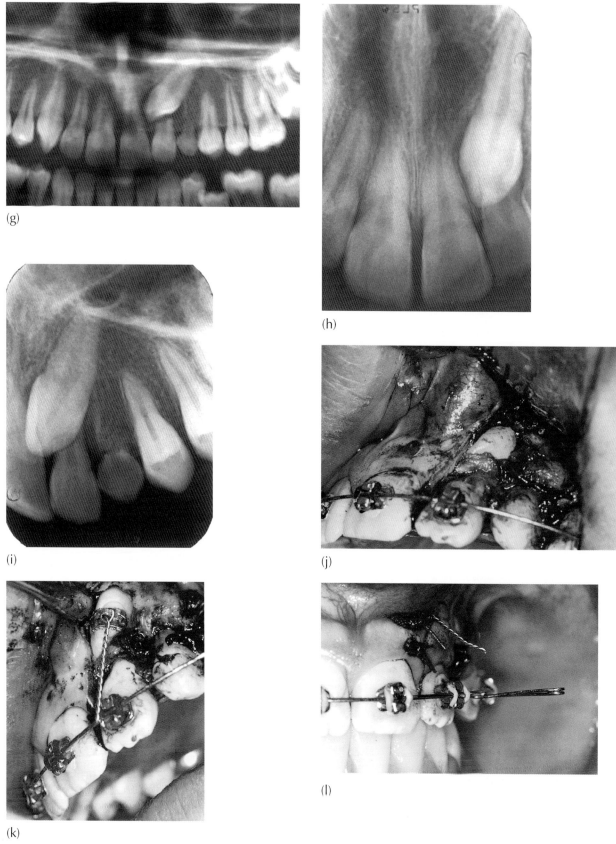

(g)

(h)

(i)

(j)

(k)

(l)

Fig. 7.7 (*Continued*)

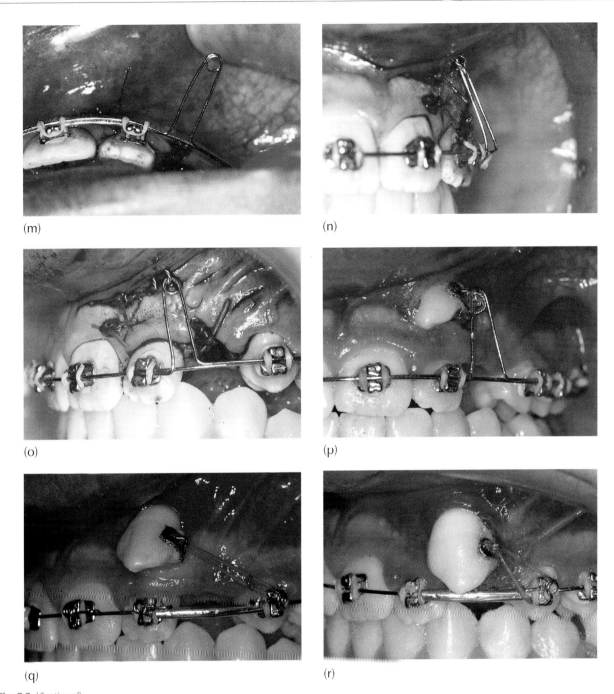

(m)

(n)

(o)

(p)

(q)

(r)

Fig. 7.7 (Continued)

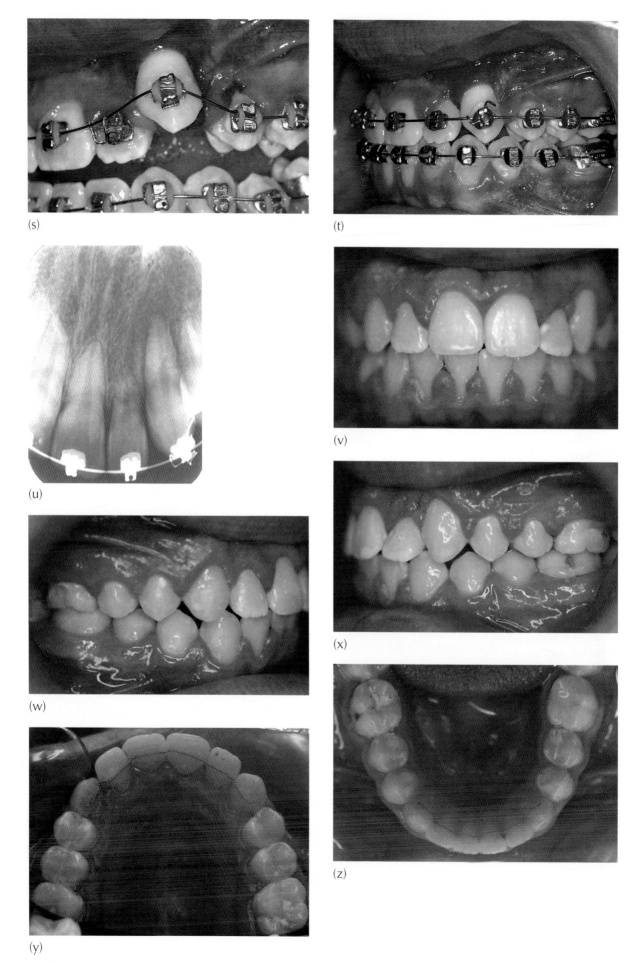

(s)

(t)

(u)

(v)

(w)

(x)

(y)

(z)

Fig. 7.7 (*Continued*)

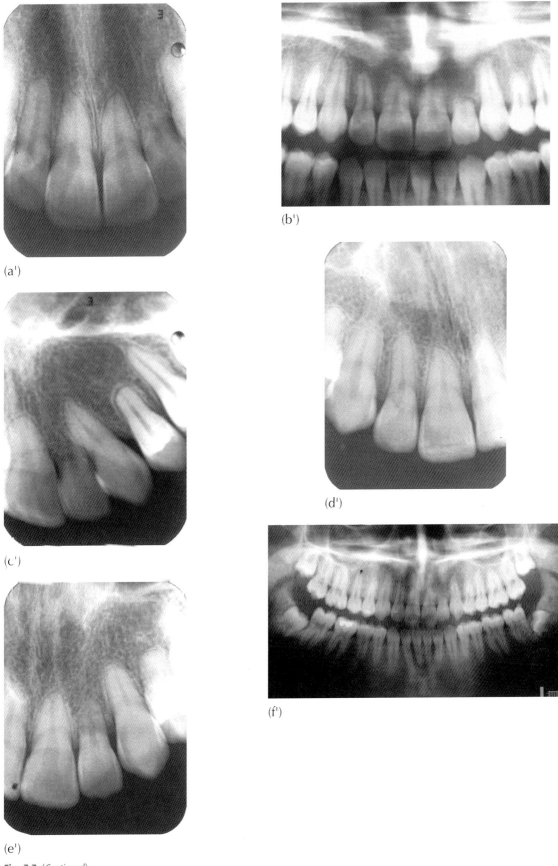

(a')

(b')

(c')

(d')

(e')

(f')

Fig. 7.7 (Continued)

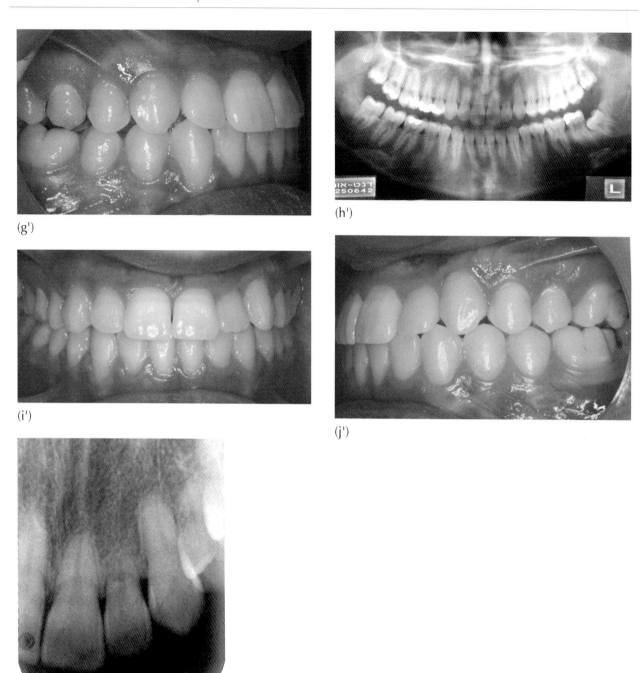

(g')

(h')

(i')

(j')

(k')

Fig. 7.7 (Continued)

Clinical experience with a disproportionately large number of these cases over the years, employing the closed exposure approach to surgery, with the application of directional force that has been described in this text and supported by the findings of an investigative study [14], has permitted different conclusions to be drawn. During the past 35 years or so, a conscious and standardized effort has been made during the surgical episode to expose only the smallest area of available enamel surface for these teeth and then to apply immediate and efficient light directional orthodontic traction. The patients have been rewarded with an eruption process that has occurred with considerable speed, and with periodontal and aesthetic outcomes that largely make the previously impacted teeth indistinguishable from their normally erupting antimeres and from other untreated teeth (Figures 7.7 and 7.8).

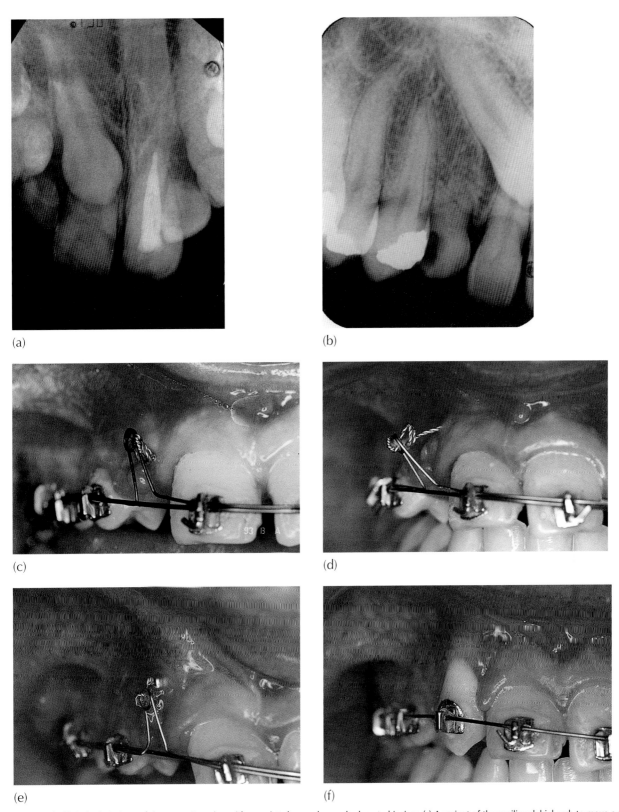

(a)

(b)

(c)

(d)

(e)

(f)

Fig. 7.8 (a, b) Periapical views of the group 6 canine with associated severely resorbed central incisor. (c) A variant of the auxiliary labial arch to move a tooth buccally. The loop is drawn upwards and into the sulcus to ensnare the pigtail. (d) Four weeks later, the pigtail has elongated, indicating progress of the canine. (e) After several adjustments over a three-month period, the eyelet attachment of the canine becomes visible. (f, g) Clinical and periapical views of the final stage of treatment. Note cessation of root resorption of the incisor and the gingival condition of the canine. (h, i) Lateral and anterior views one year after completion of treatment.

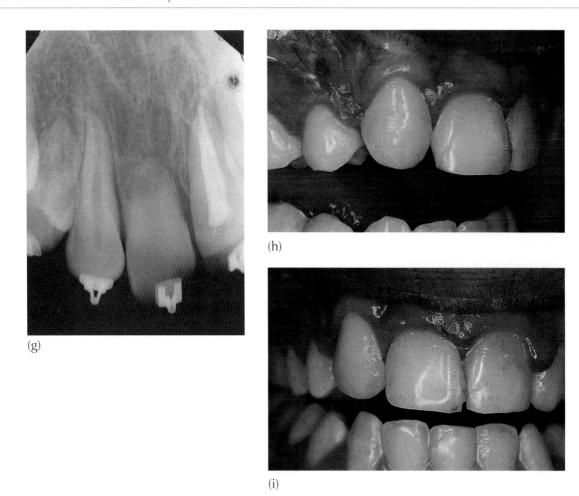

(g)

(h)

(i)

Fig. 7.8 (*Continued*)

Evidence-based answers to questions regarding canine-related incisor root resorption

When faced with the challenge of the patient whose impacted maxillary canines are associated with extensive resorption of the root of the adjacent incisor, several questions arise. These may be enumerated as follows:

1. Will the resorption process continue after the canine impaction is resolved?
2. After resolution of the impaction, will the resorbed tooth suffer further resorption if it is subsequently moved orthodontically into its ideal position, including the likely need for root movement?
3. Will the introduction of calcium hydroxide into the root canal of the resorbed tooth [23] be a necessary preventive remedy to help to eliminate further resorption?
4. Will rigid splinting of the severely resorbed incisor be an essential element of a successful outcome?
5. Can the patient reasonably expect a long-term prognosis for the resorbed incisor?

With no guidance in the literature until recently, other than anecdotal reports, the practitioner needs to make decisions that are based on clinical experience with lesser problems or what may appear to be common sense. In the more severe resorption cases, it may be that there is little to lose, since non-intervention may well result in exfoliation of the resorbed tooth and eruption of the impacted tooth into its place (Figure 7.1). As succinctly stated in correspondence with the author: 'it is from the treatment or non-treatment of these extreme and apparently hopeless situations that we may sometimes find answers' (Kokich, personal communication). If enough of these cases may be found to establish a database, then evidence-based and reliable conclusions may be reached that may set the foundations for valid treatment protocols to guide us when facing similar challenges. With this in mind, and to fully address the above five questions, a study was undertaken of severe resorption of the roots of 21 incisor teeth, each due to an associated impacted maxillary canine, in a group of 12 patients [14]. Radiographs of each of the cases were analysed and compared at four distinct stages, namely, T1 prior to the commencement of orthodontic treatment, T2 at the time the canine was distanced from the resorption site, T3 at completion of the overall orthodontic treatment and T4 at a follow-up

visit, which was at least one year later and up to 26 years post-treatment.

The conclusions from this study may be listed as follows:

- Resorption of the roots stops when the canine impaction has been resolved.
- Subsequent orthodontic movement of the resorption-affected teeth does not generate further resorption
- The incisor teeth with severely resorbed roots have a high survival rate.
- Prophylactic root canal therapy is contraindicated [23].
- The teeth remain vital.
- The teeth retain their colour and appearance in the long term.
- The teeth show a very low degree of mobility and an improvement in bone support, following post-treatment retention.
- Splinting of the resorbed teeth is not usually necessary, although it is often recommended as the means to retain the alignment of the teeth at the completion of the orthodontic treatment.

These results may therefore lead us to conclude that a tooth that has suffered advanced root resorption *from this cause* can usually be treated conservatively with a fair prognosis (Figure 7.10). It is emphasized that these conclusions do not apply in cases where there has been root resorption from other causes. Reducing the time taken in the first stages of treatment and avoiding the customary pre-surgical preparation of space in the arch for the impacted tooth are considered by the authors to be crucial. They strongly advise that distancing the canine from the resorption area should take precedence over all other orthodontic movements, even before space opening and even if this means leaving the newly disimpacted tooth in limbo for several months while space for it is subsequently being prepared in the dental arch (Figures 7.8, 7.9).

Finally, they issue a fairly optimistic assessment of the longevity of the resorbed teeth and encourage the orthodontist not to underestimate the prognosis of even those with very short roots. At the end of treatment, when the appliances are removed, the teeth may exhibit a fairly high degree of mobility, and this will undoubtedly be so for the more seriously affected incisors. However, after a few months of retention the teeth become very firm. It should be remembered that the resorption process is usually oblique, with a strong vertical component, which means that a long spicule of root may often remain firmly ensconced in the alveolar bone.

Resorption in relation to vitality of the dental pulp

It is pertinent at this point to discuss the clinical histopathological implications of root resorption that has involved the pulp chamber. When an impacted canine is situated in close relation to the apex of the incisor root, resorption begins at the apex and the pulpal tissues are immediately involved, but the vitality of the tooth is unaffected. The advance of the impacted canine goes hand-in-hand with the resorption process of the incisor, which may continue until much of the root is destroyed, but the pulp remains vital. Similarly, when the canine is associated with a severe resorption on one side of the root of the incisor, be it mesial or distal, palatal or buccal, the pulpal tissues will inevitably become involved with the resorption process at some point.

Does this mean that the tooth will lose its vitality? Is it incumbent on the practitioner to perform root canal treatment? If this were so, then it would be necessary to restore the resorbed area of root with some form of filling material in addition to the normal procedures involved in the endodontic treatment.

The situation may be likened to that of a shedding deciduous tooth. For both the deciduous tooth and the resorbing permanent incisor, the resorptive cellular layer or resorptive front fenestrates through to the inner pulpal surface of the root and meets the odontoblasts that line that surface. The periodontal and pulpal tissues merge and metaplastic changes occur in the pulpal tissue. Its vitality remains unchallenged, with neither inflammation nor pain.

This is one reason why, when resorption is so severe as to penetrate the pulp in the apical or lateral areas, the process is asymptomatic and root canal therapy is not indicated. Once the impacted tooth associated with the resorption process is moved away, the resorption will generally cease and the affected tooth can usually be moved orthodontically without any significant danger of renewed resorption. Dental traumatologists recommend elective root canal treatment as a means of limiting or preventing the resorption that occurs frequently following trauma to incisor teeth [23], but there is no evidence that the same treatment will minimize or eliminate the resorption caused by the proximity of an uncrupted canine.

When the canine has caused root resorption of the lateral incisor close to the CEJ and to the gingival crevice, there is the danger of perforation, recession and secondary caries in the resorbed area. It would seem to be desirable to orthodontically extrude the tooth to expose the resorbed area and then to restore it, with or without endodontic treatment, as indicated by the depth of the lesion. However, this type of resorption is almost always oblique and extends a long way up the root, as dictated by the eruption path of the canine. Thus, if the resorbed root area is large, then the tooth will most likely need to be extracted. If it is small, then a flap operation will be needed to make good the defect, using a glass ionomer restoration. Since reattachment to the restoration cannot occur, the incisor will need to be actively erupted and its crown shortened, until the most apical extremity of the restoration becomes supra-crestal and a gingival crevice of normal depth is attained.

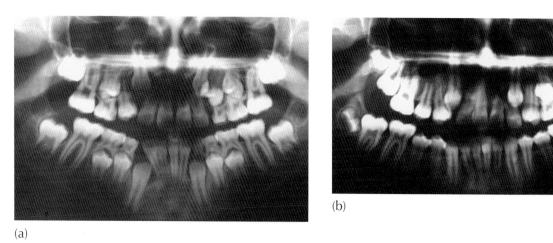

(a)

(b)

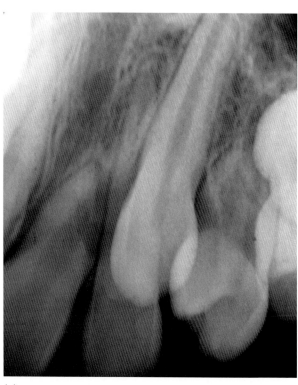

(c)

Fig. 7.9 Management of the rapid progress of resorption. (a) At age 8.5 years, the panoramic view shows a normally developing dentition, with enlarged follicles around the crowns of the maxillary canines, which are otherwise in a normal relationship with the lateral incisors at this age. No resorption is visible. Since treatment was not appropriate at this time, the patient was advised to return for follow-up a year later. (b, c) The patient returned two years later and new panoramic and periapical views show marked resorption of the central incisor and signs of an oblique resorption of the lateral incisors. (d, e) The patient was sent for CBCT imaging, which demonstrated the loss of root contour (resorption) in the horizontal and vertical planes. (f) Following diagnosis and treatment planning, a soldered trans-palatal arch with a Nance acrylic button was placed for the purposes of anchorage. (g–j) Two maxillary premolars and the second deciduous molar were extracted at the time of surgical exposure of both maxillary canines. The right canine was an uncomplicated impaction. The left canine was exposed minimally on its buccal aspect and an eyelet with a pigtail steel ligature was placed. The full surgical flap was sutured back and immediate traction made to the hook on the molar band. (k) At one month post-surgery, distal movement of the tooth has been rapid and the eyelet can be seen through the translucent gingiva. (l) At four months post-surgery, the remainder of the teeth have been bonded and the canine has reached its place in the arch. (m) A radiograph taken at this stage to monitor further resorption shows no further deterioration. (n–p) The alignment and occlusion on the day the fixed appliances were removed. (q) At one year post-treatment, the periapical radiograph shows complete elimination of the resorptive process, with a return of the lamina dura and good bony fill-in and texture. (r, s) At one year post-treatment, the clinical view comparing the canine on the side involved with the resorption and that of the opposite side, where a simple routine impaction was treated, shows them to be indistinguishable from one other and also from any normally erupting canine. The gingival contour is excellent, the clinical crown is of normal length, there is a good width of attached gingiva on each and the appearance is close to ideal – a tribute to accurate positional diagnosis (imaging), skilled execution of the closed eruption technique (surgery), with carefully planned directional force application (orthodontics).

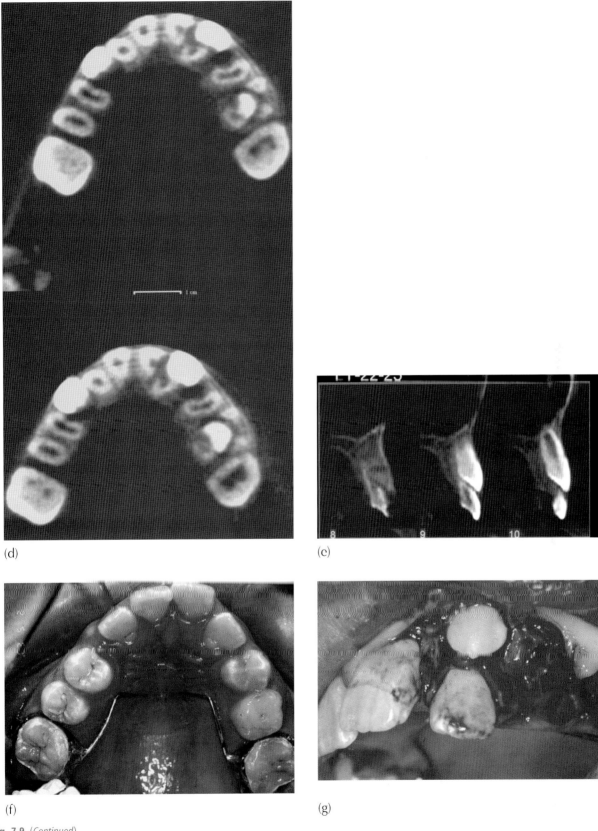

(d)

(e)

(f)

(g)

Fig. 7.9 (*Continued*)

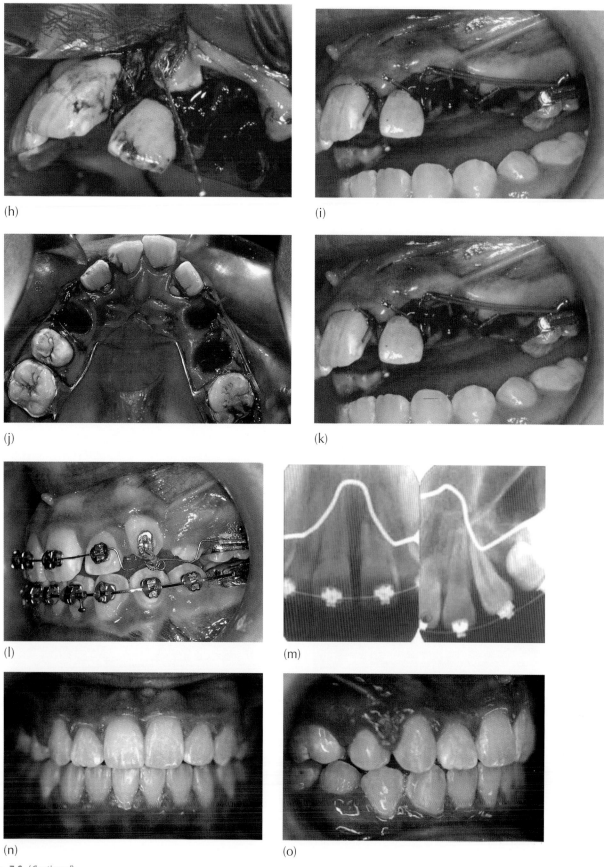

(h)

(i)

(j)

(k)

(l)

(m)

(n)

(o)

Fig. 7.9 (Continued)

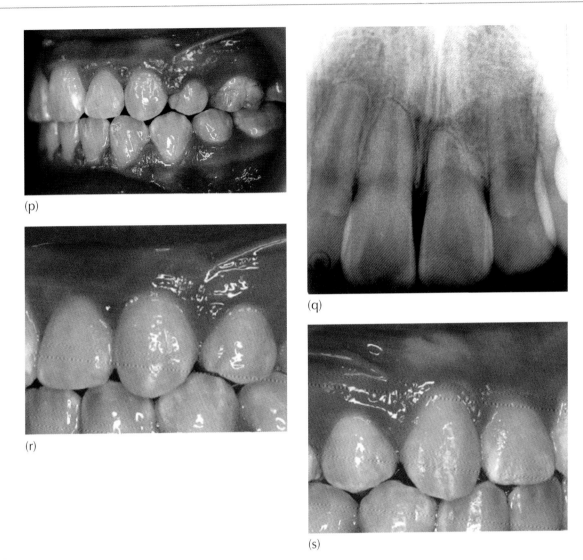

(p)

(q)

(r)

(s)

Fig. 7.9 (Continued)

In summary, therefore, it is important to realize that even severely resorbed teeth need not be extracted and may well be treated conservatively to successfully occupy an important place in the scheme of the dentition for a long time, often for years to come. For this to happen, accurate 3-D localization of the tooth and its surrounding structures is mandatory, in order to guide the surgeon in identifying the exact position of the impacted tooth *vis-à-vis* the resorbing root face. Accurate positioning and designing of the soft tissue flap, minimal precision bone removal and attachment bonding then follow and a closed exposure procedure is absolutely essential if the vitality of the resorbed tooth is to be retained, if the canine can be diverted from its progressively worsening threat to its immediate neighbours and if both teeth are to be profitably salvaged from this debacle. Emphasis is placed on the surgery being undertaken very early in the treatment in order to permit immediate orthodontic traction to distance the tooth from the resorbing incisor root, in a planned direction. This will effectively and almost completely eliminate the resorptive

process. The orthodontist's attention should then turn to treating the existing overall malocclusion and creating space for the canine in the dental arch, which may take several months. During this time, the canine remains in a 'neutral' location on the buccal or palatal side, depending on the escape route that will have been taken in its movement out of harm's way. Following full correction of the canine ectopy and 'fine-tuning' of the alignment of all the teeth, reparative periodontal procedures are generally superfluous and the long-term prognosis of the canine will be excellent. In many cases, even those with a much reduced root length, the adjacent incisor usually enjoys a fair to good prognosis and, for the most part, the tooth will last the patient well into adulthood. The gingival architecture, crown lengths and dental alignment should be such that even a dental colleague will be able to recognize neither the previously impacted canine nor the incisor that was affected by root resorption. Appearance of the anterior dentition will be unsurpassed in comparison to any other form of treatment (Figures 7.9 and 7.10).

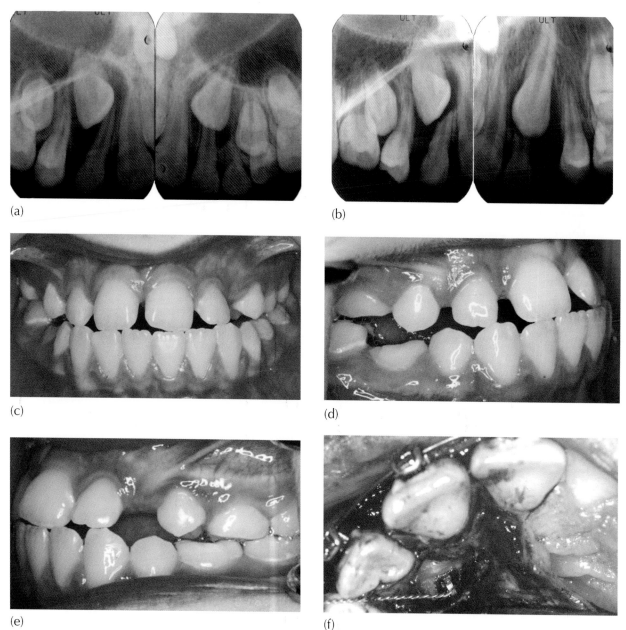

Fig. 7.10 Enlarged dental follicle and root resorption. (a) Periapical radiographs of the canine areas, taken at the initial consultation of a 10-year-old female. The films show enlarged follicles around the unerupted canines. Aside from this and the obvious crowding, it was considered too early for orthodontic treatment *per se*, and the patient was placed on one year recall and sent for extraction of the left deciduous canine and first molar teeth in the hope that this might encourage the canine to improve its position. (b) Thirteen months later, the new periapical views show extremely aggressive root resorption of both lateral incisors. Orthodontic treatment was advised urgently. (c–e) The initial clinical photographs show a class 3 relation of the teeth on a mild skeletal class 3 base. Crowding was present in the maxilla but not in the mandible, and it was considered a non-extraction case with the possible option of dental compensation or later surgical advancement of the maxilla. (f) The right canine was palatally impacted, and it was exposed and attachment bonded immediately after placement of the fixed appliance. Note that this was done before space was made available in the arch in order to separate the canine from the incisor and, hopefully, thereby to prevent further resorption. (g) An active palatal arch was placed immediately after suturing of the closed exposure full flap and is seen in its passive state at a distance from the palatal mucosa. The pigtail ligature protrudes through the middle of the sutured flap, opposite the buried tooth. (h) The active palatal arch has been raised up close to the palatal mucosa arch and held in place by ensnaring it in the pigtail ligature. This produces a vertical force on the unerupted canine, away from the roots of the incisor. (i) The left canine was buccally impacted and treated with a closed surgical procedure. This picture shows the latter stages of its buccal and distal resolution, using a long rectangular lever arm. In the immediate post-surgical stage, a labial auxiliary arch was used, as in (j–l). The completed case shows the dental compensation for the skeletal class 3 relationship and all teeth in place. Overall gingival condition and contour are good, but the left maxillary canine shows an increased crown length. (m–o) At 3.5 years post-treatment, the gingival condition has remained largely unchanged. A bonded multi-strand wire splint functions as an orthodontic retainer and splint. (p) The monitoring periapical views taken following resolution of the impactions show severe resorption of both lateral incisors. There is no apparent bone support for the right and only a millimetre or so for the left lateral incisor. Should these teeth be removed? (q) Similar films at the time that orthodontic movement of the lateral incisors had been completed show minor improvement in the bone 'support' and obliteration of the pulp on the right side. (r) At debonding of the appliances and placement of the bonded multi-strand wire splint, corresponding to nine months after cessation of incisor movement, a new lamina dura and bone condensation have appeared around the root ends of the affected teeth. (s) At 3.5 years post-treatment, further bone support has been generated.

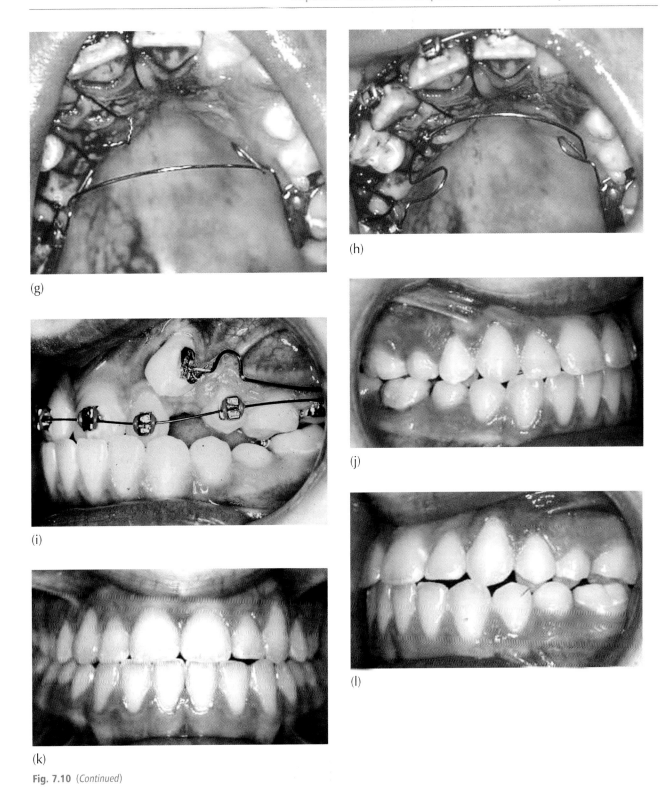

(g)

(h)

(i)

(j)

(k)

(l)

Fig. 7.10 (*Continued*)

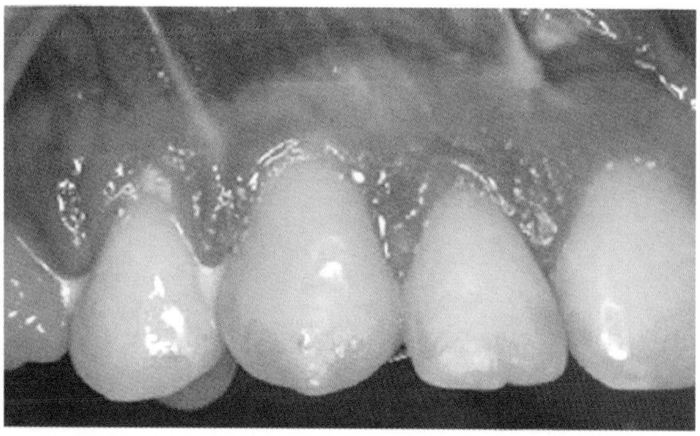

(m)

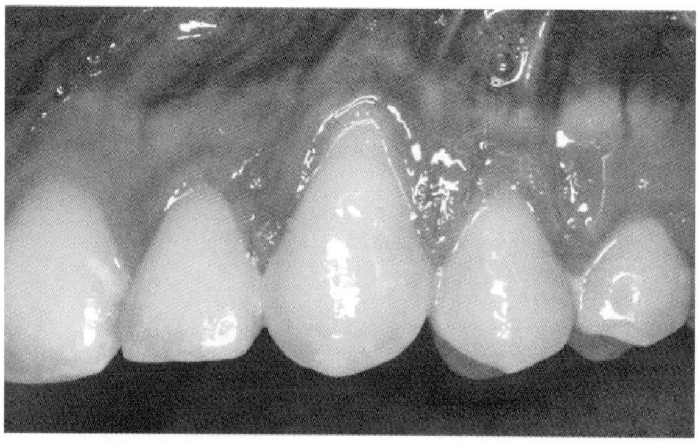

(n)

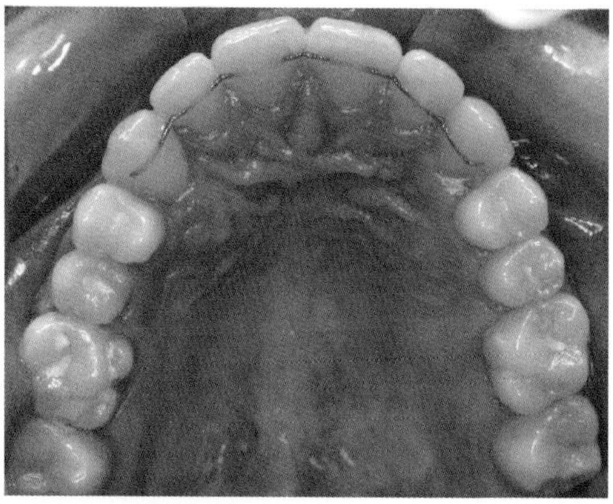

(o)

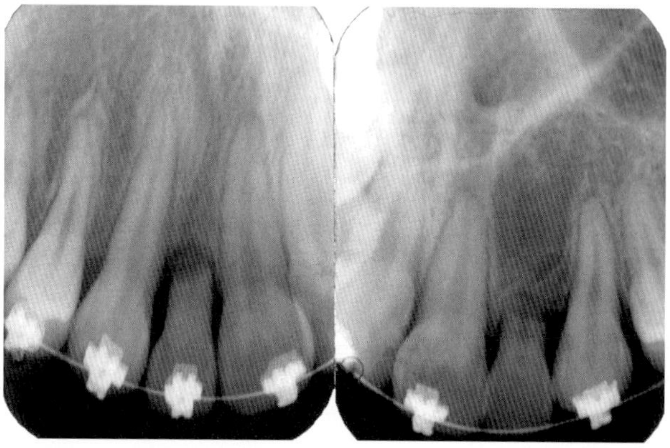

(p)

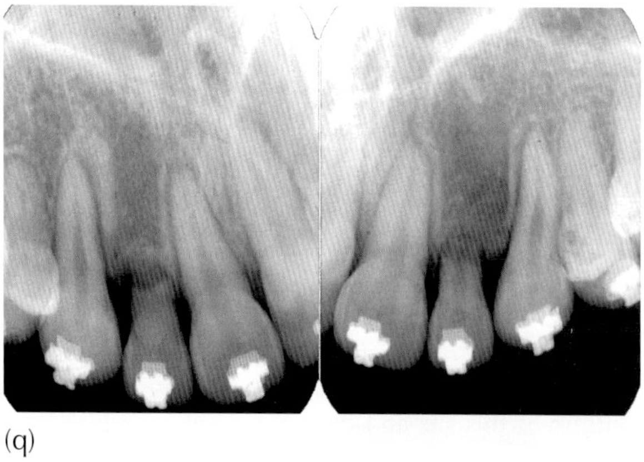

(q)

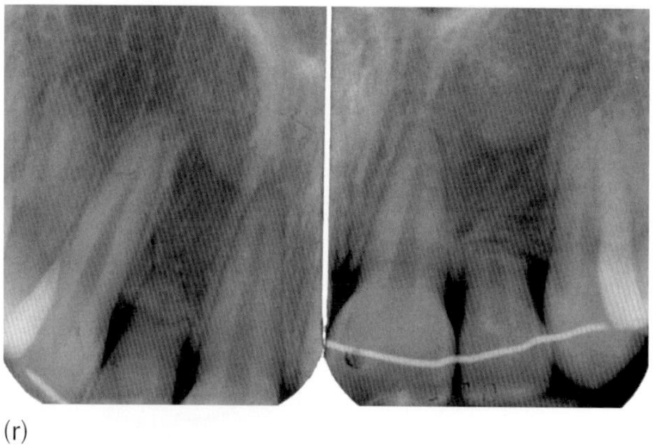

(r)

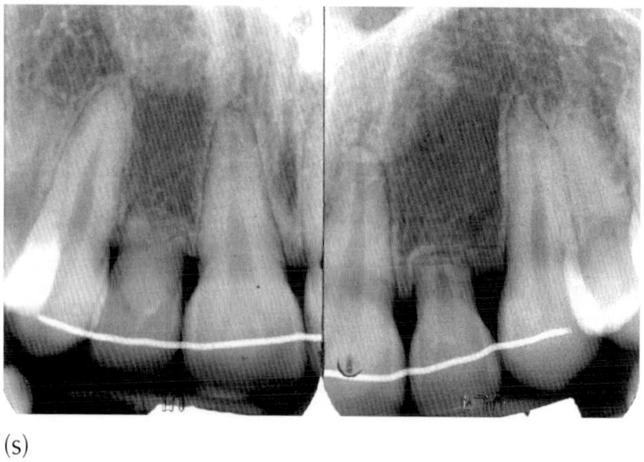

(s)

Fig. 7.10 (*Continued*)

Invasive cervical root resorption

Quite unrelated to the presence of an adjacent impacted tooth, resorption of the cervical area on one side of the root of a tooth has been referred to in the literature and is described as a specific, recognizable and recurring entity in its own right, apparently unrelated to the factors that cause the more usually diagnosed resorption. Potential predisposing factors have been identified as trauma (15.1%), intracoronal restoration (14.4%), surgery (5.4%) and intracoronal bleaching (3.9%), while a significant number (16.4%) of the affected teeth showed no identifiable cause, but the most common link of this unusual phenomenon was found to be a history of having had past orthodontic treatment. This treatment had been provided for patients who comprised 24.1% of the affected teeth in the study sample [24–27].

By themselves, none of these factors or the condition itself would be relevant or appear to merit discussion in a book on impacted teeth. However, among the orthodontically treated cases in that study, arrested eruption was seen in five of the teeth concerned. Indeed, a number of instances of resistant impacted teeth have been seen by the present author in association with invasive cervical resorption. The resorption referred to here specifically affects the impacted tooth itself and not the adjacent incisor.

The diagnosis of invasive cervical root resorption can easily be missed or mistaken for the 'interproximal cervical burnout' seen on periapical and bitewing radiographs (Figure 7.11). A typical narrative then follows. Presuming a healthy periodontium and root surface integrity, the tooth

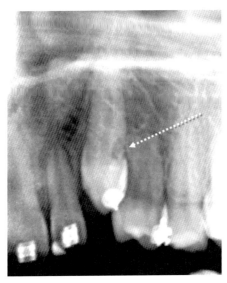

Fig. 7.11 An impacted canine has resisted attempts to mechanically erupt it. The periapical film shows a distal resorptive lesion burrowing into the root of the tooth in its cervical region. The point of entry and the loss of integrity of the lamina dura are clearly seen.

is exposed, an attachment bonded and traction applied in the usual manner. The tooth does not respond and the orthodontist applies greater traction force. This is repeated for several more visits and increased traction force. The tooth still does not respond, but the neighbouring teeth show signs of anchorage loss, with the occurrence of intrusion and the creation of an open bite (Figure 7.12). The orthodontist then assumes the tooth to be ankylosed and refers the patient back to the surgeon. The surgeon's intention is to re-expose the tooth and free the presumed ankylotic connection by seizing the tooth with a forceps and applying a force sufficient to break this direct connection to the bone – a luxation. However, with the tooth re-exposed and in full view, it is found to be mobile – sometimes even excessively mobile, if the orthodontist had indeed increased the traction force, in the usually many months of unproductive traction.

If we can assume that the direction of applied orthodontic force had been appropriate to the resolution of the impaction and that ankylosis is clearly not a factor, then there must be a flaw in the periodontium which does not permit the tooth to respond, since a healthy, complete periodontal ligament is a precondition for tooth eruption. It is in these circumstances that a tentative diagnosis of invasive cervical resorption may be determined by default. Identification is not easy because it entails subgingival probing and periapical radiography at different angles. Many of these lesions occur in buccal or lingual areas of the root surface which do not show up in plane film radiography. However, a careful examination of the relative radiolucency of the crown of the tooth may provide the clue to the presence of the condition (Figure 7.13). It may be seen as an atypical caries-like shadow, ballooning into the dentine, undermining the enamel of the crown of the tooth and, because the tooth is still unerupted, it is easily passed over or presumed to be an artefact. Unless the tooth is extracted, only a periodontal surgical flap will reveal the lesion. It seems that the invasive nature of this condition destroys the normal structure and integrity of the periodontium at this site and does not permit the histological changes normally associated with the application of orthodontic force to occur and to generate the desired tooth movement.

Cervical root resorption is initiated by a local inflammatory process, indicating that infection, or perhaps trauma, is a necessary aetiological factor. A defective junctional epithelium has also been blamed, although assumptions are largely speculative [28]. An experimental model for cervical root resorption was developed to study the significance of the junctional epithelium in the prevention of cervical resorption [29]. The researchers reflected a muco-periosteal flap on the mesial and distal sides of the canines in the upper and lower jaws, respectively. The crestal alveolar bone with adhering periodontal membrane and superficial cementum and dentin was removed with a round burr, to

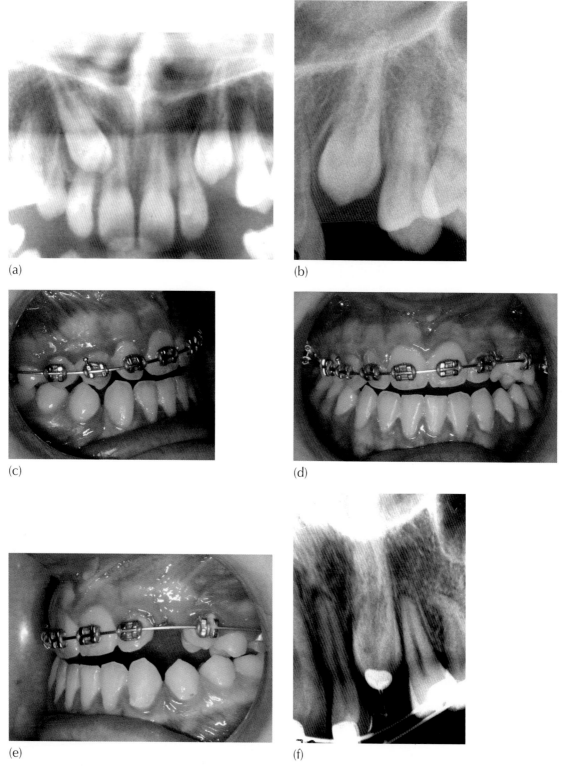

Fig. 7.12 (a) Anterior section of the pre-treatment panoramic view shows the right canine to be palatally impacted and superimposed on the anomalous lateral incisor. The left canine has a complete root, there is no space problem to account for its non-eruption and it is in a direct line to its appointed place in the arch. The cervical resorption on the mesial aspect has affected approximately half the root length, although unnoticed at the time of commencement of treatment.
(b) Periapical view of the left canine clearly shows the lesion and the disappearance of the lamina dura
(c–e) Intra-oral views of the dentition after 28 months of treatment. An orthodontic appliance is present in the maxilla, with molar bands and brackets on the first premolars, incisors and right canine, whose palatal impaction has been resolved. The left canine has not responded to treatment. The application of traction to the left canine has caused intrusion of the adjacent teeth and an open bite which is larger on the left side, indicating loss of anchorage. The second premolars are unconnected to the orthodontic appliance and have remained in occlusion.
(f) Periapical view taken 28 months into treatment shows the progress of the resorptive process into the crown of the canine and extending widely to involve most of the root.
(g–i) Successive axial (horizontal) cuts of the CBCT scan, taken 28 months into treatment, show the lingual and mesial radiolucency characteristic of resorption process. The slice represented in slice (g) is completely encircled in crown enamel, but slices (h) and (i) show the external portal of entry of the lesion on the palatal side.
(j) The three-dimensional view from the buccal aspect shows the canine to have a complete and unblemished root surface.
(k) From the palatal side, the three-dimensional view clearly shows the resorption process to have affected the entire palatal aspect of the tooth as far as the apex.

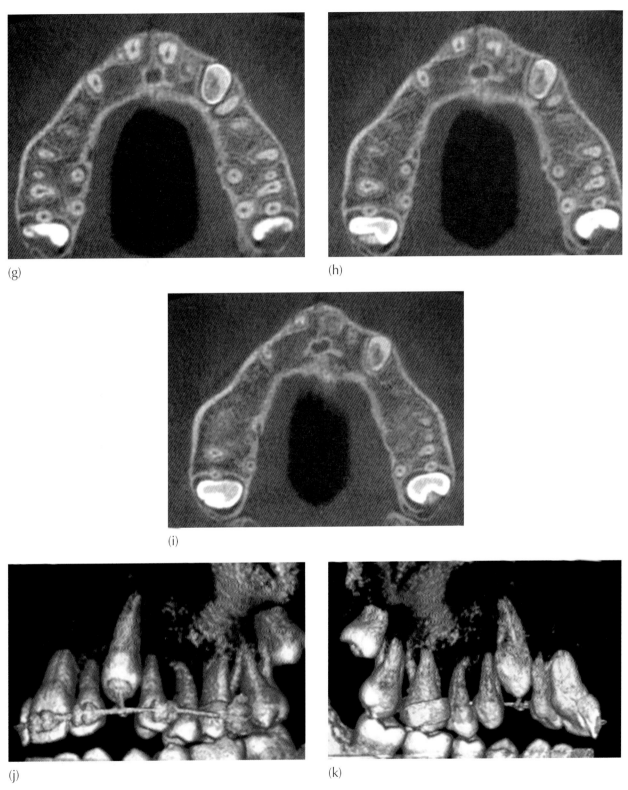

(g)

(h)

(i)

(j)

(k)

Fig. 7.12 *(Continued)*

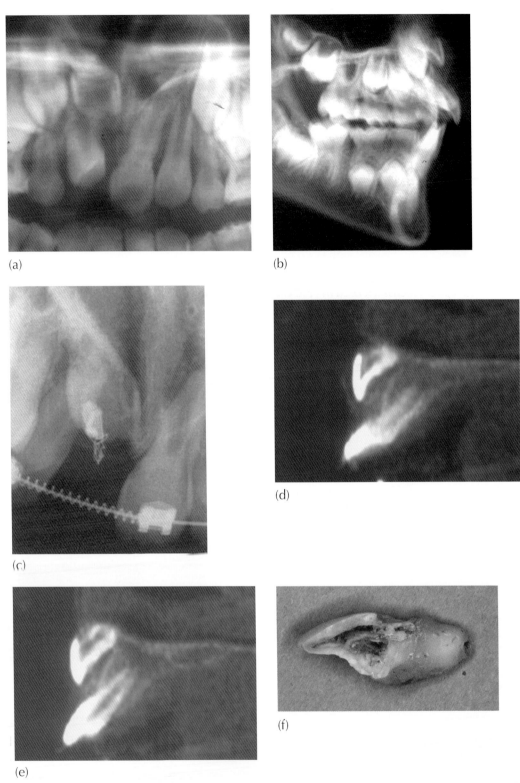

Fig. 7.13 Invasive cervical resorption with a history of early trauma. (Courtesy of Dr Joel Becker.) (a, b) The pre-treatment panoramic and cephalometric views (anterior portions presented only) show an impacted central incisor with a short root, very high in the anterior maxilla, at the level of the anterior nasal spine and floor of the nose. The crown appears 'empty' and there is a break in its labial cervical contour. (c) After space had been made and an attachment placed at surgery, traction was applied to no avail. (d, e) Frames from the paraxial views of the CBCT show the break in the enamel layer of the gingival third of the crown and the low density of the dentine area, compared with the similar view of the erupted incisor. The root of the tooth is not visible because of the angulation of the tooth *vis-à-vis* the vertical cut. (f) The sectioned tooth clearly demonstrates the labial cervical crown/root defect. The dentine is completely resorbed and the enamel remains as an unsupported hollow shell. There is a thin and continuous partition of secondary dentine separating the resorbed area from the pulpal tissue, which was vital.

a depth of 2 mm in an apical direction. In half the teeth in the sample, the exposed dentine was covered with thin polycarbonate foil. Eight weeks later, the dentine surface had not developed an epithelial cover but exhibited numerous resorption cavities associated with moderately chronically inflamed and cell-rich granulation tissue. Actively resorbing odontoclasts and osteoclasts were seen along the dentine and crestal bone surfaces, respectively. During the same period, the dentine surfaces of the non-foiled teeth were all covered by a dense squamous epithelium. The originally denuded dentine surfaces that did not come in contact with mucosal or periodontal connective tissue showed cementum repair in the more apical areas of the root. The reparative connective tissue that appeared under the epithelial coverage was cell-rich and presented a mild chronic inflammation.

In the cervical area, where the damage was adjacent to the oral environment, repair was characterized by epithelial proliferation down the side of the root to epithelialize the exposed dentine and thus to produce a long junctional epithelium. However, before the proliferation was fully accomplished to completely cover all the dentine surface, there were sites of initiation of root resorption which were later inhibited once epithelium covered them. The authors of the study concluded that chronic inflammation in the granulation tissue, when in contact with the dentine, prevents the marginal gingival epithelium from forming a protective cervical cell layer in an angular defect and thus generates the resorption process.

While this study helps us to understand why the lesion is essentially cervical, it should be remembered that, once initiated, the unchecked progress of the resorption process will see it extend into both the root and the crown. In the crown it may progress without symptoms of pain to destroy virtually the entire dentine content, with the exception of the layer of pre-dentine immediately covering the pulp tissue (Figure 7.13f). The pulp remains protected by the pre-dentine until late in the process because pre-dentine is thought to possess an enzyme inhibitor (anti-invasion factor) present during the organic phase of pre-dentine formation. The resorption starts at what becomes a portal of entry on the root surface and continues to mushroom into the dentin (Figure 7.11). When the pre-dentine is reached, the resorption proceeds laterally and in an apical and coronal direction, progressively enveloping the root canal [30, 31]. The anti-resorptive effect of the pre-dentine and of the outer surface of the enamel causes the resorption to stop, leaving a narrow layer of dentine and pre-dentine around the pulp and giving it an irregular radiographic appearance. The enamel is thinned down, leaving only an outer, rodless, translucent layer and producing a characteristic 'pink tooth' due to the vascular pulpal and resorptive tissue [32].

Unless there is access to the oral environment from a periodontal defect or generalized periodontal disease, the portal of entry is not to be found in a more apically location along the root, since an area of incomplete epithelial coverage that is isolated from direct access to the mouth will typically repair with a cementum deposit on the root surface. However, a resorptive lesion with a cervical portal of entry, will spread progressively coronally and apically to eliminate much of the root substance (Figures 7.12 and 7.14).

A majority of the existing literature on the invasive cervical root resorption refers to its occurrence in relation to non-vital teeth which have undergone intracoronal bleaching. It seems that there is a high prevalence of invasive cervical resorption in previously bleached teeth in the years that follow, although the cause of the problem is from within. Non-vital bleaching is performed by sealing a bleaching agent into the prepared pulp chamber and the most coronal portion of the root canal (the remainder of the root canal will have already been sealed off by the root canal filling). The bleaching agent may leak through the more coronal accessory root canals of the tooth and its effect on the immediately adjacent cervical area is to generate a resorptive process, which is typically annular and uniform around the neck of the tooth [33]. Invasive cervical root resorption from this cause, therefore, is indeed beyond the context of this book. However, a successfully treated impacted tooth in the years that follow may sometimes become relatively infra-occluded and it is important to make the differential diagnosis of its cause. One of the possibilities is post-bleaching root resorption of the tooth, if it had been root-treated as described above. Another is vertical relapse that occurs when apically repositioned flaps are used to expose an impacted tooth [34]. This is due to a realignment of the mucosal lines between the flap and the adjacent mucosa, which was discussed in Chapter 3. Strictly, invasive cervical resorption of a vital tooth must be considered to be a third possibility, although this could have been the result of the treatment, just as it could have been the cause of the impaction.

Clinical examination

Invasive cervical root resorption in an unerupted permanent tooth is characterized by the absence of any response on the part of the tooth to orthodontic traction force and by the tell-tale radiolucent area in the cervical area of the tooth, as may often be seen on a periapical or panoramic radiograph. As we have noted, imaging and definitive diagnosis may be difficult to establish, particularly in an early lesion, and a cone beam CT will assist materially in recognizing the defect and mapping its extent. Because definitive diagnosis is difficult to establish, a thorough clinical examination assumes greater importance. Indirect clinical recognition of the phenomenon may often be derived from the clinical examination and is a very important first step in these cases. It may contribute much to the practitioner's initial diagnosis of the case and to informing the patient at an early stage that there may be a questionable prognosis for the tooth.

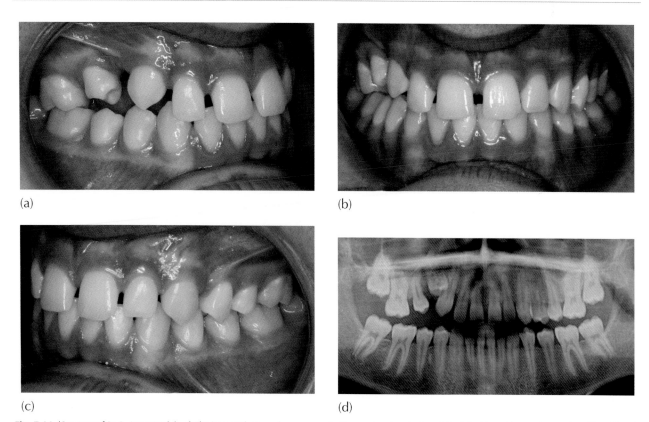

(a)

(b)

(c)

(d)

Fig. 7.14 (Courtesy of Dr B. Jesperson) (a–c) The intra-oral views show a normal alignment and occlusion of the left side, with anterior spacing. The right side shows a normal molar relation, but strongly mesial tipping of the second premolar and distal tipping of the canine, with under-eruption of both teeth. There are large spaces on either side of these teeth and a localized lateral open bite. (d) The impacted first premolar is seen with a very 'woolly' appearance of its crown and root, indicating very advanced involvement of the resorption process, undermining the enamel of the crown and much of the root. There is a loss of integrity of the lamina dura on the entire mesial side. Although apparently unassociated, the central incisors have very short roots in comparison with the lateral incisors.

It is common knowledge that when a deciduous molar tooth is extracted and space is lost in the arch, this will have occurred by the adjacent teeth mildly tipping into the area, from both the distal and the mesial sides. The successional premolar will either become impacted between the two adjacent tipped teeth or will find a pathway to erupt buccally or lingually displaced from the line of the arch. In contrast, the teeth adjacent to a markedly infra-occluded deciduous second molar will characteristically show much more pronounced tipping and their occlusal level has been shown to also be infra-occluded relative to the other teeth in the same jaw, but less so than the affected deciduous molar [35–37]. There is also usually a deviation of the dental midline to that side, even in a spaced dentition. In similar manner, the teeth adjacent to a site where an impacted tooth has been prevented from erupting due to invasive cervical resorption will show the same pronounced degree of tipping and the same picture of relative infra-occlusion, as with those adjacent to an infra-occluded deciduous molar. So, from a clinical point of view, the orthodontist should look for the following signs:

- exaggerated tipping of the teeth adjacent to the impacted tooth (Figure 7.14);
- relative height deficiency of the adjacent teeth, often with a localized lateral open bite;
- deviation of the midline to that side.

If the affected tooth had erupted prior to the onset of the cervical root resorption, the tooth will often be seen, in a growing child, to infra-occlude progressively further and further in relation to the occlusal level, to create an open bite (Figure 7.15).

On the radiograph, the signs usually seen are:

- a radiolucency in the cervical area of the tooth which may range from a small area that can be confused with typical interproximal radiographic 'burn-out', to a 'woolly' area affecting one side of the tooth and stretching into the crown and longitudinally further down the root;
- loss of the lamina dura in the immediate area;
- the root apices of the teeth on either side of the affected tooth are excessively distanced from one another and

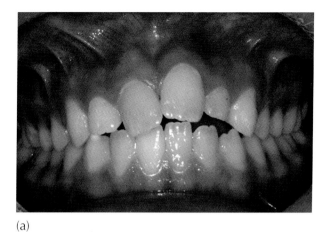

(a)

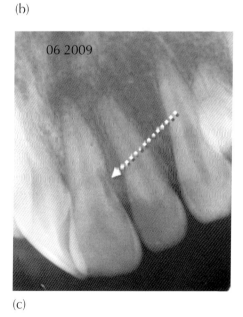

(b)

(c)

Fig. 7.15 (Courtesy of Dr M. Friedman.)

(a) The left central incisor is relatively under-erupted in comparison with the right central incisor and an open bite is present on that side.

(b) This picture was taken 14 months later and shows a very much increased infra-occlusion of the left central incisor and a much larger open bite, with mesial tipping of the right central incisor across the midline.

(c) A periapical view of the left central incisor shows a small resorption lesion, with a very small portal of entry at the cervical level (see arrow). The lesion may be seen to have progressed coronally and apically parallel with the pulp chamber, with a narrow wall separating the two.

imagining their corrective uprighting will create space for almost two teeth.

Given the histopathological background to the possible causation of this very aggressive condition, it is time to look again at some of the time-honoured customs that are often routinely practised during the most crucial event in the treatment of the impacted tooth, namely the surgical exposure. It is conceivable that some of these customs might be precipitating factors that initiate the lesion. Thus, it is unlikely that a surgeon will deliberately remove bony tissue to such an extent that there is exposure of the cervical area of the root surface. Nevertheless and in the interests of making sure that the tissues do not subsequently heal over, this may be the unintended consequence. The risk may be increased if a surgical pack is pushed forcefully down the periodontal space. Old habits and idiosyncrasies in the exposure protocol die hard and the surgeon must be made aware of this possible source of danger.

When the surgical field has been laid open the surgeon will usually check the tooth for mobility. The sole purpose of this test is to see if the tooth is ankylosed or not. However, many surgeons will actually aim to mechanically increase this mobility by passing an elevator beyond the CEJ and along the root surface to act as a lever. This exercise cannot be condemned strongly enough, because it damages the periodontium and cementum and may cause an ankylosis. This methodology has been used to deliberately ankylose a deciduous maxillary canine, to prepare it to be used as an absolute anchor unit against which forces may be applied to move other teeth [38]. In the light of the studies referred to above [24–27], there may be some basis to further speculate that this site-specific, acute trauma to the periodontium and cementum may alternatively initiate the process of invasive and equally site-specific cervical resorption by denuding the dentine surface and preventing the cementum or the oral epithelium from forming a protective cervical cell layer in an angular defect. During many years of orthodontic practice, the author has found that the temptation among many of the excellent surgeons with whom he has worked to 'loosen' the impacted tooth has been seen to be rife and it is only his restraining presence that has often prevented this from happening.

Similarly, with the use of orthophosphoric acid etchant in the form of a liquid rather than a gel, it is difficult to limit the spread of the caustic material into the surrounding vital tissues that will have been exposed at surgery. Should the etchant come into contact with the denuded tooth beyond the CEJ, it is likely that inflamed granulation tissue, rather than cementum or oral epithelium, will encompass the tooth surface during the healing period and the stage will be set for the initiation of an invasive cervical resorption lesion.

Performed in expert hands, surgery of this kind need not take long to complete. However, time is often wasted with the surgical field wide open while suitable implements or bonding agents are searched for and, unless continuously irrigated with saline, the exposed tissues will become subject to desiccation. This becomes more acute if efficient suction is still being used by the attending staff to stave off the bleeding. Consequent cell death in these sensitive areas could create an environment conducive to the initiation of an ICRR process.

It is known that acid etchant is a good haemostatic agent and has been recommended for use by paediatric dentists to control gingival haemorrhage to permit an uncontaminated restoration to be placed. Anecdotally, the author has been in irregular correspondence with one particular orthodontist regarding an abnormally disappointing record of failure in treatment of a series of impacted teeth, in which the most frequently diagnosed cause was invasive cervical root resorption. In a quasi-detective investigation of this strange phenomenon, it was found that the surgeon had been routinely flooding the exposed area with orthophosphoric acid to provide both a bloodless field and, at the same time, a suitably enamel-etched bonding surface for placing an attachment on the impacted tooth!

By not being present at the surgical episode, the orthodontist loses control of the dynamics of the treatment, as pointed out in Chapter 3. Nevertheless and despite this apparently elementary safety precaution, the overriding majority of orthodontists still refer their patients, unaccompanied, to the oral surgeon or periodontist for the exposure. It is worth remembering that it is not the surgeon but the orthodontist who will be responsible for achieving a successful outcome.

The notion is offered that the presence of the orthodontist during this critical and highly sensitive surgical procedure in the patient's treatment will contribute much to the success of the treatments provided. It should be remembered that the role of the surgeon is to provide therapeutic access to a tooth whose future would otherwise be highly questionable. With the orthodontist on hand to answer questions regarding type of surgical exposure needed, the aspect of the tooth to be exposed, the degree of exposure required, as well as to take an active part in the bonding of the attachment and application of traction on the spot, control of the treatment is retained in the hands of the person ultimately responsible for its outcome.

Treatment of this atypical form of resorption is possible, depending on its location and extent. It usually consists of gaining surgical access to the lesion, removing the fibrovascular tissue and repairing the defect with restorative material. It has also been recommended [27] to apply a topical 90% aqueous solution of trichloracetic acid, curettage, endodontic therapy where necessary and restoration with a glass ionomer cement. With the histopathological character of the lesion eliminated, the tooth may then be expected to respond to traction (Figure 7.16). An infrabony periodontal pocket will remain and will require

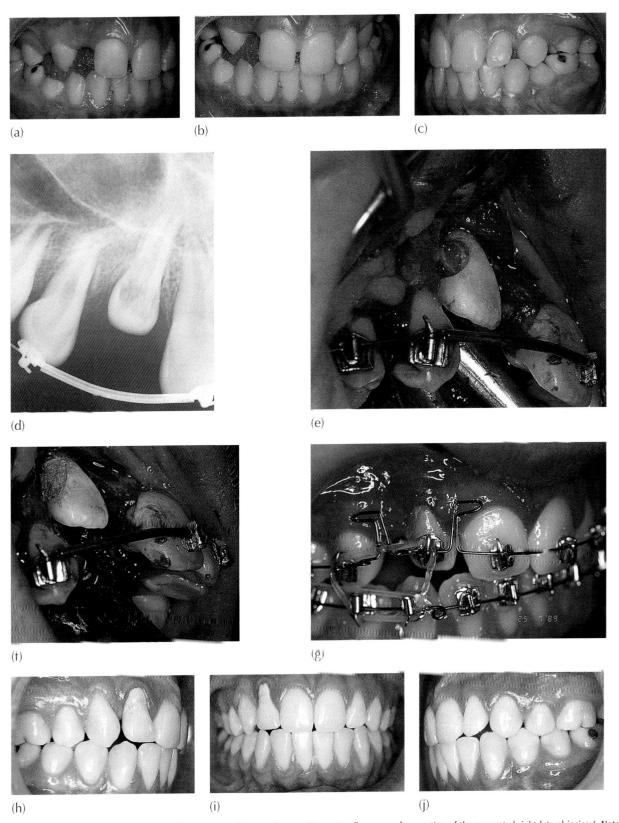

Fig. 7.16 (a–c) Maxillary first premolars had been extracted by another practitioner 'to allow space for eruption of the unerupted right lateral incisor'. Note the relative lack of vertical development and tipping of adjacent teeth, reminiscent of an association with infra-occlusion. (Compare with Figure 7.14.)

(d) The pre-surgical periapical radiographic view shows a dark shadow in the distal half of the crown of the incisor and this extends to the cervical margin on the distal side of the tooth.

(e) Following orthodontic space regaining, surgical exposure reveals the cervical resorption defect. Soft resorption-replacement tissue had prevented the eruption.

(f) The resorption area was debrided and the exposed pulp extirpated and root-filled temporarily with calcium hydroxide. After amalgam restoration placement, an eyelet was bonded and immediate traction applied. (Surgery by Prof. A. Shteyer, endodontics by Prof. I. Heling.)

(g) The final stages of eruption mechanics.

(h–j) The completed orthodontic result, after definitive root canal therapy was completed and a porcelain crown placed.

(k, l) The clinical and periapical radiographic views of the tooth.

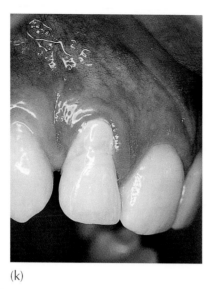

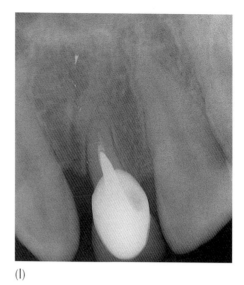

(k)   (l)

Fig. 7.16 (*Continued*)

periodontal treatment and a considerable degree of supra-eruption (forced eruption) of the tooth to bring the affected root area to the surface. At best, then, this will leave the tooth with a short root and a long clinical crown. When an invasive cervical resorption lesion is extensive or occurs in a severely displaced impacted tooth, access for the necessary treatment may be inadequate and extraction is likely to be the only reasonable alternative.

Pre-eruptive invasive coronal resorption

An invasive resorptive process can only attack the crown of an unerupted tooth as its primary source through an enamel defect. This type of resorption is also very aggressive and may consume all the dentine within the crown, as it painlessly mushrooms beneath the enamel outer layer [39]. It may also resorb away some or most of the enamel shell itself (Figure 7.17). In these circumstances, it is easy to understand the concern and even panic of the orthodontist or the surgeon, when faced with this chance finding on the radiographic film. On presumptions of serious doubt for the chances of long-term survival of this tooth and the likelihood of its early demise, such teeth are often extracted or an ill-advised immediate and risky restorative procedure undertaken under the existing, far from ideal, conditions of partial eruption or impaction. Because the affected tooth is unerupted, its root development is incomplete, with wide open apical areas and broad, apically- divergent root canals. Nevertheless, the pulp tissue is entirely normal, with no inflammation and no stimulation for the formation of reparative secondary dentine.

In these circumstances, it is easy for the orthodontist or the surgeon, when faced with this chance finding on the radiographic film, to come to hasty and mistaken conclusions. On presumptions of serious doubt for the chances of long-term survival of this tooth and the likelihood of its early demise, such teeth are often extracted or an ill-advised

immediate and highly risky endodontic and restorative procedure undertaken under the existing, far from ideal, location of under-eruption or impaction. Even when extensive destruction of the intracoronal dentine has occurred, pulpal involvement is unlikely and one may see the continuation of normal apexification in a diagnosed, coronally resorbed, unerupted tooth. As we have pointed out in relation to invasive cervical root resorption, the resorptive process stops short at the predentine, presumably for the same reason. It is usually possible to see the predentine bridge on a radiograph (Figure 7.18).

It should be understood that these teeth are not likely to be impacted and they will usually erupt normally, regardless of the extent of the resorption and provided their dental follicle has remained intact. Nevertheless, treatment should be aimed at exposing and erupting the tooth as soon as possible after diagnosis in order to eliminate further damage. It is therefore logical to eliminate the urgency of the situation by simply exposing the crown of the tooth with an open surgical procedure, aimed at cutting off the blood supply to the resorptive tissue within the crown. This will deprive the tissue of its lifeline, causing it to become necrotic. Without question, an under-erupted tooth in that condition, and with the passage of time, will most likely and secondarily develop caries within the large dentine defect, making the need for restorative treatment still essential. A good temporary restoration may therefore be placed, without excavating the deeper part of the cavity, to seal off the lesion from the opportunistic initiation of caries. This effectively defuses and freezes the issue for the immediate to middle future, thereby making restorative treatment far less urgent.

Permanent restorative procedures may be delayed until the tooth has fully erupted and when conditions for this will be optimal. Root canal therapy is not normally required if all the diseased material is carefully eliminated, but this may leave insufficient dental hard tissue to be retentive of

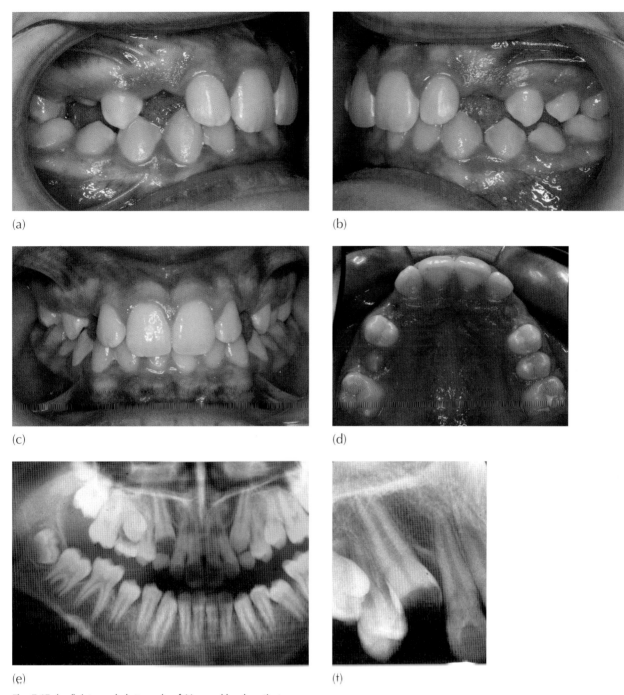

(a)

(b)

(c)

(d)

(e)

(f)

Fig. 7.17 (a–d) intra-oral photographs of 11 year old male patient.
(e, f) Panoramic and periapical views to show crown resorption of maxillary right permanent canine.
(g) The exposed crown showing extreme degree of resorption.
(h) An eyelet attachment in place, with twisted pigtail ligature.
(i) The extrusion auxiliary arch in its passive mode.
(j) The extrusion auxiliary arch engaged in the pigtail ligature, exerting extrusive force.
(k) Seen from the occlusal aspect.
(l) The tooth has erupted rapidly, seen from the occlusal.
(m) The buccal view.
(n–q) The competed case after root canal treatment and rehabilitation.
(r) Periapical radiograph of completed case.

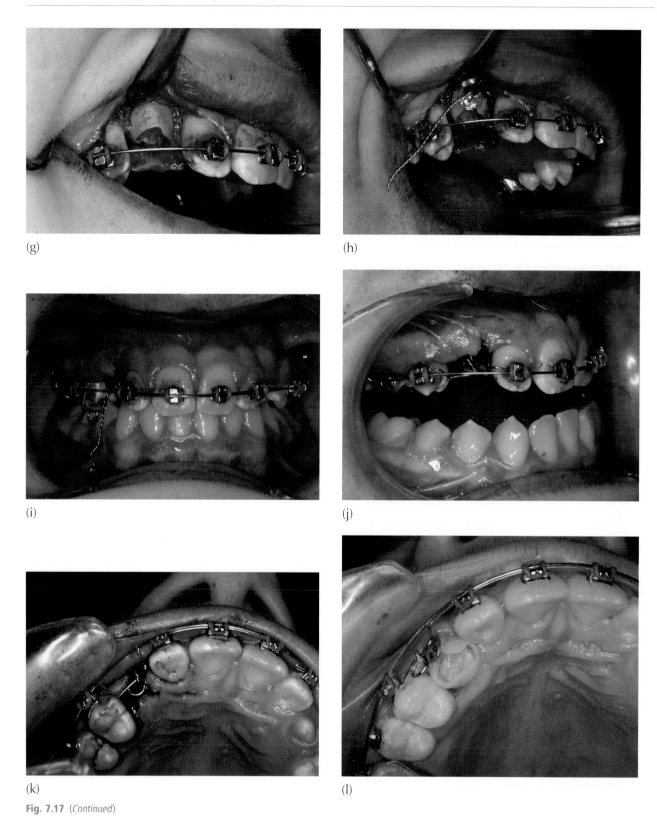

(g)

(h)

(i)

(j)

(k)

(l)

Fig. 7.17 (*Continued*)

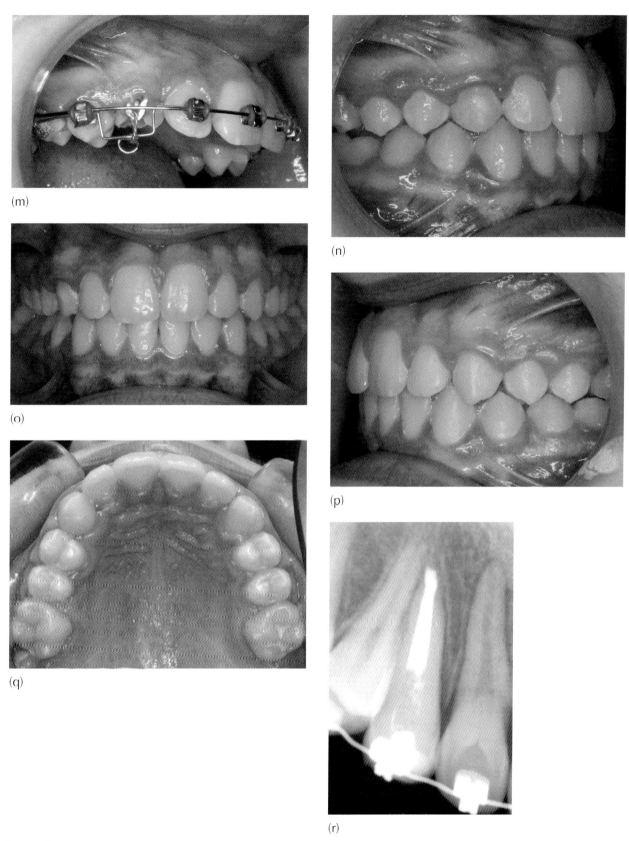

(m)

(n)

(o)

(p)

(q)

(r)

Fig. 7.17 *(Continued)*

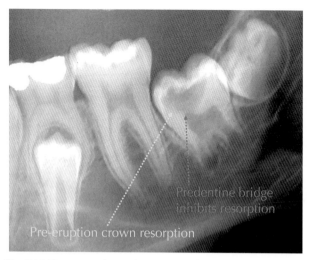

Fig. 7.18 The unerupted second permanent molar has an extensive area of coronal resorption, which is separated from the wide pulp chamber by a narrow dentine bridge. Access to this lesion is extremely limited, the pulp chamber is wide and the roots are only partly developed.

the restoration. In this circumstance, an elective root canal procedure will need to be undertaken, to facilitate placement of a post and core reconstruction, in order to provide the foundation for good crown rehabilitation.

References

1. Ericson S, Kurol J. Incisor resorption caused by maxillary cuspids. A radiographic study. Angle Orthod 1987; 57: 332–345.
2. Ericson S, Kurol J. Radiographic examination of ectopically erupting maxillary canines. Am J Orthod Dentofacial Orthop 1987; 91: 483–492.
3. Ericson S, Kurol J. Resorption of maxillary lateral incisors caused by ectopic eruption of the canines. A clinical and radiographic analysis of predisposing factors. Am J Orthod Dentofacial Orthop 1988; 94: 503–513.
4. Ericson S, Kurol J. CT diagnosis of ectopically erupting maxillary canines – a case report. Eur J Orthod 1988; 10: 115–120.
5. Ericson S, Kurol J. Early treatment of palatally erupting maxillary canines by extraction of the primary canines. Eur J Orthod 1988; 10: 283–295.
6. Brin I, Becker A, Zilberman Y. Resorbed lateral incisors adjacent to impacted canines have normal crown size. Am J Orthod 1993; 104: 60–66.
7. Ericson S, Kurol J. Resorption of incisors after ectopic eruption of maxillary canines: a CT study. Angle Orthod 2000; 70: 415–423.
8. Walker L, Enciso R, Mah J. Three-dimensional localization of maxillary canines with cone-beam computed tomography. Am J Orthod Dentofacial Orthop 2005; 128: 418–423.
9. Savage RR, Kokich VG Sr. Restoration and retention of maxillary anteriors with severe root resorption. J Am Dent Assoc 2002; 133: 67–71.
10. Olive R. Factors influencing the non-surgical eruption of palatally impacted canines. Aust Orthod J 2005; 21: 95–101.
11. Leonardi M, Armi P, Franchi L, Baccetti T. Two interceptive approaches to palatally displaced canines: A prospective longitudinal study. Angle Orthod. 2004; 75: 581–586.
12. Lindauer SJ, Rubenstein LK, Hang WM, Andersen WC, Isaacson RJ. Canine impaction identified early with panoramic radiographs. J Am Dent Assoc 1992; 123: 91–92, 95–97.
13. Power SM, Short MBE. An investigation into the response of palatally displaced canines to the removal of deciduous canines and an assessment of factors contributing to favourable eruption. Br J Orthod 1993; 20: 215–223.
14. Becker A, Chaushu S. Long-term follow-up of severely resorbed maxillary incisors following resolution of etiologically-associated canine impaction. American Journal of Orthodontics and Dentofacial Orthopedics 2005, 127: 650–654.
15. Ericson S, Kurol J, Falahat B. Does the canine dental follicle cause resorption of permanent tooth roots? A computed tomographic study of erupting maxillary canines. Angle Orthod 2002; 72: 95–104.
16. Vanarsdall RL. What every orthodontist should know about impacted teeth. Presented at the 106th American Association of Orthodontists' Annual Session, Las Vegas, May 2006. Audio 28:186B.
17. Becker A, Chaushu S. Impacted canines and associated incisor root resorption: can it be a win–win situation? Presented at the 106th American Association of Orthodontists' Annual Session, Las Vegas, May 2006. Audio recording 28:186A.
18. 16. Becker A, Chaushu S. Long-term follow-up of severely resorbed maxillary incisors following resolution of etiologically-associated canine impaction. Am J Orthod Dentofacial Orthop 2005; 127: 650–654.
19. Kokich VG Jr, Kinzer GA. Managing congenitally missing lateral incisors. Part I: canine substitution. J Esthet Restor Dent 2005; 17: 5–10.
20. Kokich VG. Surgical and orthodontic management of impacted maxillary canines. Am J Orthod Dentofacial Orthop 2004; 126: 278–283. Kinzer GA, Kokich VO Jr. Managing congenitally missing lateral incisors. Part III: single-tooth implants. J Esthet Restor Dent 2005; 17: 202–210.Kokich VG. Maxillary lateral incisor implants: planning with the aid of orthodontics. J Oral Maxillofac Surg 2004; 62: 48–56.
21. Jacoby H. The ballista spring system for impacted teeth. Am J Orthod 1979; 75: 143–151.
22. Kornhauser S, Abed Y, Harari D, Becker A. The resolution of palatally-impacted canines using palatal-occlusal force from a buccal auxiliary. Am J Orthod Dentofacial Orthop 1996; 110: 528–534.
23. Farhad A, Mohammadi Z. Calcium hydroxide: a review. Int Dent J 2005; 55: 293–301.
24. Heithersay GS. Clinical, radiologic, and histopathologic features of invasive cervical resorption. Quintessence Int 1999; 30: 27–37.
25. Heithersay GS. Invasive cervical resorption: an analysis of potential predisposing factors. Quintessence Int 1999; 30: 83–95.
26. Heithersay GS. Treatment of invasive cervical resorption: an analysis of results using topical application of trichloracetic acid, curettage, and restoration. Quintessence Int 1999; 30: 96–110.
27. Heithersay GS. Invasive cervical resorption following trauma. Aust Endod J 1999; 25: 79–85.
28. Patel S, Kanagasingam S, Pitt Ford T. External cervical resorption: a review. Journal of Endodontics, 2009; 35: 616–625,
29. Brosjö M, Andersson K, Dug J-O, Lindskog S. An experimental model for cervical resorption in monkeys. Dental Traumatology, 1990; 6: 118–120.
30. Hiremath H, Yakub S, Metgud S, Bhagwat S, Kulkarni S. Invasive cervical resorption: a case report. Journal of Endodontics, 2007; 33: 999–1003.
31. Frank AL, Torabinejad M. Diagnosis and treatment of extracanal invasive resorption. Journal of Endodontics 1998; 24: 500–504.
32. Iqbal MK. Clinical and scanning electron microscopic features of invasive cervical resorption in a maxillary molar. Oral Surg Oral Med Oral Pathol Oral Radiol Endod 2007; 103: e49–e54.
33. Friedman S, Rotstein I, Libfeld H, Stabholz A, Heling I. Incidence of external root resorption and esthetic results in 58 bleached pulpless teeth. Dental Traumatology 1988; 4: 23–26.
34. Vermette ME, Kokich VG, Kennedy DB. Uncovering labially impacted teeth: apically positioned flap and closed-eruption techniques. Angle Orthod. 1995; 65: 23–32.
35. Becker A, Karnei-R'em RM. The effects of infraocclusion: part 1 – tilting of the adjacent teeth and space loss. American Journal of Orthodontics 1992; 102: 257–264.
36. Becker A, Karnei-R'em RM. The effects of infraocclusion: part 2 – the type of movement of the adjacent teeth and their vertical development. American Journal of Orthodontics 1992; 102: 302–309.
37. Becker A, Karnei-R'em RM, Steigman S. The effects of infraocclusion: part 3 – dental arch length and the midline. American Journal of Orthodontics 1992; 201: 427–433.
38. Kokich VG, Shapiro PA, Oswald R, Koskinen-Moffett L, Clarren SK. Ankylosed teeth as abutments for maxillary protraction: a case report. Am J Orthod 1985; 88: 303–307.
39. Holan G, Eidelman E, Mass E. Pre-eruptive coronal resorption of permanent teeth: report of three cases and their treatments. Pediatric Dentistry, 1994; 16: 373–377.

8

Other Single Teeth

Orthodontic Treatment of Impacted Teeth, Third Edition. Adrian Becker.
© 2012 Adrian Becker. Published 2012 by Blackwell Publishing Ltd.

Aside from third molars, the maxillary canines and central incisors are the principal teeth that may become impacted but, from time to time, other teeth may also be affected. For some of these teeth, familiar patterns emerge, typically affecting the same tooth and with the same aetiology in many of the cases. In others, unusual pathology is involved, which may affect any tooth or group of teeth and is, therefore, quite non-specific. Nevertheless, even with a widely heterogeneous group, trends may be recognized and treatment protocols may be suggested to cover a good proportion of them.

Before moving on to buccally displaced canines, there is a small group of maxillary canines which are neither buccally nor palatally displaced, but are in the line of the arch. According to the panoramic or periapical view, the orientation of their long axes is mildly mesially inclined. By virtue of this combination of location and angulation, the crown comes to be jammed at an angle against the distal aspect of the root of the lateral incisor, whose long axis may be oriented distally. From there, the canine appears to be unable to free itself to erupt, unless the incisor itself is tipped mesially during the initial space-opening procedure.

This standard preparatory orthodontic movement reduces the angulation between the long axes of canine and incisor to a marked degree, but it also moves the apex of the lateral incisor distally, providing the impetus to secondarily tip the crown of the canine distally and inferiorly and reduce the angulation still further. This will often be sufficient to elicit resolution of the impaction and spontaneous eruption of the tooth into its normal place in the arch.

In the event that progress is not achieved within a few months, simple surgery, attachment bonding and traction from an auxiliary nickel–titanium wire or elastic ligature will be sufficient to resolve the problem. The type of surgery advised will depend on the height of the impacted tooth *vis-à-vis* the gingival tissue. If there is a thick band of attached gingiva above the height of the unerupted tooth, then a simple window exposure will be ideal, since the tooth will be erupted through this band and a width of attached gingiva will remain on the labial side of the tooth in the long term. Should this not be the case, then an apically repositioned flap will ensure that the tooth will have a normal gingival attachment [1].

Equally, a full-flap closed procedure will achieve largely the same result as the apically repositioned flap procedure and, for these relatively minor impactions, there would appear to be very little to choose between the two procedures in terms of the periodontal outcome.

Buccally displaced maxillary canines (BDC)

In Chapter 6, it was recorded that palatal impaction of maxillary canines occurs in 1–2% of most Caucasian populations studied. They exceed the prevalence of buccally impacted maxillary canines by a ratio of 2 or 3:1 [2, 3] while, paradoxically, buccally ectopic maxillary canines which have erupted into the mouth represent one of the most frequently encountered conditions in orthodontic practice. In contrast to Caucasian populations, one study [4] found that Orientals suffer more buccal impactions, while the frequency of palatal impaction in that ethnic group is very low.

Tooth size and arch length play important roles in the difference between palatal and buccal impaction, insofar as the dentition with palatal impaction is characterized by an excess of space in the dental arches for the most part [5–7], while the buccal impaction cases show marked crowding [5, 8]. In males this seems to be more due to a deficiency in length of the dental arch, while in females it was found to be more related to larger than average teeth [9]. We have pointed out that the developmental location of the unerupted maxillary canine is slightly buccal to the general line of the dental arch and that the two adjacent teeth erupt ahead of the canine. This being so, and in the presence of any crowding, the space for the canine between these two teeth will be reduced, and this will cause the canine to exaggerate the buccal tendency of its eruptive path (see Chapter 6).

Dental age also appears to be associated with these two very different phenomena. While patients with palatal canines showed a strong tendency for delayed dental development, a similar study of a large sample of patients with buccally ectopic canines showed a normal dental age distribution [10]. There was only a small variation on either side, when dental age and chronological age were compared. The dental age of patients with palatally impacted canines showed two distinct and separate trends. In half the cases, the dental age was normal, and in the other half, the dental development was very late, by as much as two years. Advanced dental development was not seen in any of the cases within the palatal canine group [11].

Ectopic canines in the absence of crowding

Among the individuals that we see with buccal displacement of the canines, there is a small but significant number in whom there is no crowding to account for this phenomenon. In order to shed more light on this, the clinical features of the dentitions of a large sample of such cases were compared to those of a similar sample of BDC cases with crowding and another sample of cases in whom the canine had erupted into its place [12].

Dental age vs. chronological age and mesio-distal tooth dimensions showed no difference between the patients in the three groups. There was a single significant finding in this study, which related to the adjacent lateral incisor tooth. This tooth exhibited a much increased prevalence of anomaly and delayed development, in exactly the same way as has been recorded in relation to palatal canine displacement. Both the Guidance Theory and hereditary primary

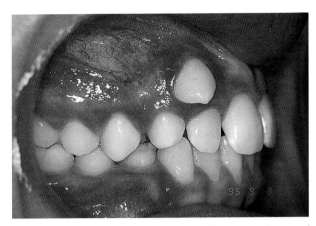

Fig. 8.1 The canine has developed in an abnormal location – primary tooth germ displacement or lack of guidance from the adjacent peg-shaped lateral incisor?

displacement of the tooth germ offer cogent aetiological explanations for the occurrence of this phenomenon, as pointed out in Chapter 6 (Figure 8.1).

In the absence of crowding, the canine may erupt higher up in the area of the sulcus oral mucosa, which creates a poor gingival attachment. From the periodontal point of view, having only thin oral epithelium covering the root leaves the patient with a delicate and easily traumatized attachment apparatus. Different surgical approaches have been described in Chapter 3 to resolve the problem.

Buccally impacted canines with mesial displacement

Most buccally displaced canines have a slight mesial displacement, overlapping the distal side of the lateral incisor. These are so common in orthodontics that we may consider them routine, whether they are erupted or unerupted. However, on occasion one may find that the degree of displacement of the unerupted canine crown is quite extreme, overlapping very high up on the mesial side of the root of the lateral incisor. These canines will progress more mesially and inferiorly in the early teen years and may be found at the level of the apical third of the root of the incisor in the bony depression that is located in the height of the sulcus. Although these are often palpable, the unusual height and mesial displacement of the tooth in the sulcus may lead to an erroneous initial clinical diagnosis.

In the clinical intra-oral examination, clues to their position may be found by studying the orientation of the adjacent incisors (Figure 8.2). Because the canines occupy bucco-lingual space in the alveolus, which is very narrow superiorly, their presence causes a palatal displacement of the root apex of the lateral and, occasionally, the central incisors. Tipping the patient's head backwards and studying the anterior part of the maxilla from the occlusal aspect, the astute clinician will note that the lateral incisor has a strong palatal inclination, occasionally with its root promi-

nently outlined beneath the palatal mucosa. The tooth clearly needs considerable labial root torque.

Diagnosis of labial location of the canine may not be easy to confirm radiographically. At this relative height, the panoramic view shows it to traverse the apical areas of the lateral and, sometimes, central incisors. Nevertheless, confirmation of labial location may be deduced given the presence of the tooth as it is imaged superimposed on the root of this lateral incisor and, from the clinical examination, the fact of palatally prominent root. Although the impacted canine is labial to the incisor roots, it is more distant from the film than are the incisor crowns, given its height and situation in the labial depression of the anterior maxilla and hence its radiographic image on the panoramic view will be enlarged, relative to those of the incisors. Canines in this location are precisely the teeth for which the method described for labio-lingual determination of canine position using a single panoramic view [13, 14] does not apply. Periapical or occlusal radiography of these teeth will superimpose them on the incisor roots at a lower level due to the acute angulation of the X-ray cone. Thus, the combination of a periapical or occlusal view with the panoramic view will facilitate diagnosis using vertical tube shift information (Figure 8.2c, d). Antero-posterior information will be seen on the lateral cephalogram, which will show the location of the crowns of these teeth relative to the incisor roots and to the anterior nasal spine.

When a mesially directed and labially impacted canine is present in a patient whose maxillary incisor arrangement accords with the archetypal class 2 division malocclusion, the situation may become a very complex canine/lateral incisor partial transposition. In this scenario, the unerupted canine lies high on the labial side of the root of the lateral incisor and, as pointed out above, the root of this lateral incisor is tipped palatally and its crown proclined labially. The long axes of the central incisors, on the other hand, show lingually tipped crowns and a labial orientation of their roots, which are often palpably outlined in the labial sulcus.

With the lateral incisor roots palpable on the palatal side and the central incisor roots on the labial, there is much space in the bucco-lingual plane between the root apices of these teeth. It will be appreciated that in these circumstances, the impacted canine may progress further mesially and migrate to the palatal side of the central incisor roots (Figure 8.3a–d). With the canine root labial to the root of the lateral incisor and its crown palatal to the central incisor, orthodontic resolution and alignment may be impossible.

If positional diagnosis is based solely on the buccal object rule using two periapical films at different angles, the palatal location of the canine crowns to the central incisors will be confirmed, but without clinical observation or more sophisticated imaging the relation between canine and lateral incisor will be missed. The unsuspecting orthodontist, who routinely relies solely on Clark's tube-shift method

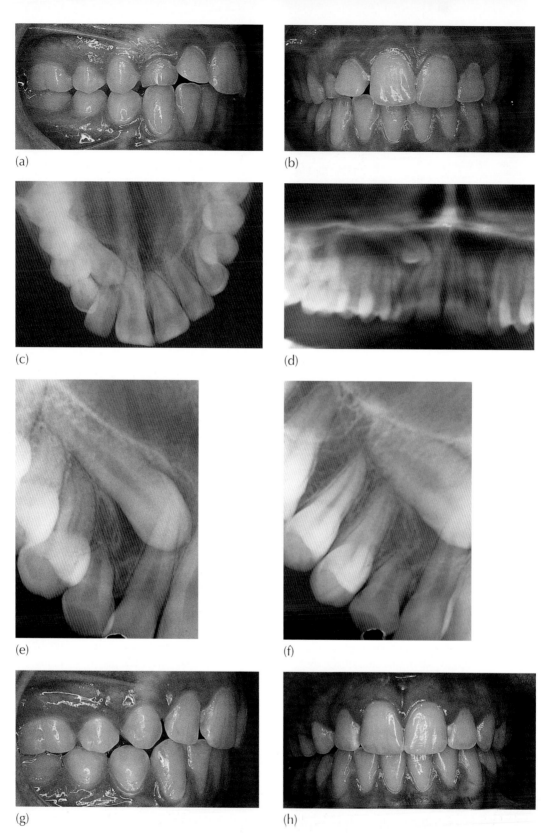

Fig. 0.2 (a, b) Clinical views showing an over-retained deciduous right maxillary canine. Note the labial and distal tipping of the right lateral incisor crown and the palatal root position.

(c) Anterior occlusal view showing superimposition of canine crown and lateral incisor root on right side. From this view, the canine could be inferior to the lateral's root on the palatal side or superior to the root on the labial.

(d) A section of the panoramic view to show the severe mesial displacement and unusual height of the canine. Taken together with the anterior occlusal view, the vertical tube shift created shows the canine to be buccal.

(e, f) Mesio-distal tube shift periapical views confirm the buccal diagnosis.

(g, h) Extraction of the right deciduous and permanent canines only, together with maxillary arch mechano-therapy, has achieved space closure and good intercuspation in class 1 on the left and class 2 on the right, with midline correction.

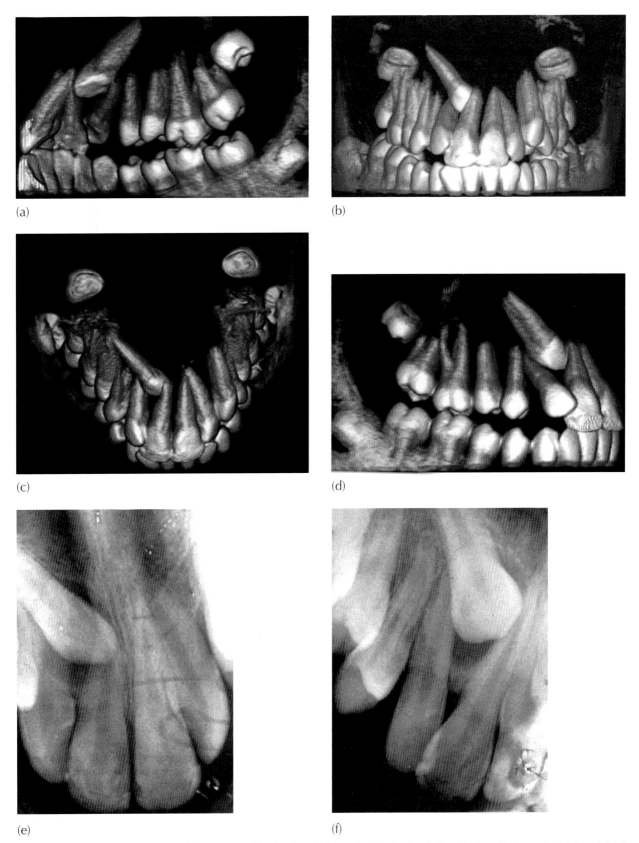

(a)

(b)

(c)

(d)

(e)

(f)

Fig. 8.3 (a–d) Three-dimensional CT views of the anterior maxilla, showing a high buccal canine that has displaced the lateral incisor root palatally and distally and which has progressed mesially to traverse the arch on the palatal side of the central incisor. (e, f) The two tube shift periapical views confirm the palatal location of the canine crown *vis-à-vis* the central incisor, but give no hint as to the complexity of the case. A palatal approach to surgical exposure would be disastrous (see Figures 12.16, 12.17 and 12.18). (Courtesy of Dr N. Dykstein.)

(Figure 8.3e, f), will surely offer inappropriate treatment in this case because it will lead him/her to think of a palatal surgical and orthodontic approach to the crown of the canine. This would be totally misleading since the subsequent orthodontic manipulation of the canine in a palatal direction will hinge the canine root around the root of the lateral incisor, inflicting damage to that tooth. At the same time, the apex of the canine will be swung anteriorly to clash against the labial plate, from where there would be no reasonable possibility of success. The only available approach would be to expose the canine crown from the labial side and apply traction in a labial and distal direction to bring the tooth clear of the central incisor. It would then need to be drawn labially and distally around the root of the lateral incisor and into its place in the arch.

Nevertheless, a treatment decision will still be necessary since the canine will not remain static, but will continue on its ectopic path. The treatment options are:

1. To leave the canine in place – with suitable long-term radiographic follow-up planned; this policy precludes any possibility of orthodontic correction of the incisor relationship and it always carries with it the risk that neglect on the part of the patient will lead to further complications in later life.
2. To extract the impacted tooth and align the lateral incisor.
3. To extract the lateral incisor, align the canine into its normal place and prepare for an implant to replace the missing incisor.
4. To extract the lateral incisor, align the canine into the place of the missing incisor and draw the posterior teeth mesially to eliminate the spaces.
5. To align the teeth in their transposed order.
6. To expose the tooth and bring it into designated place in the arch, while realigning the lateral incisor in its place, with carefully planned directional orthodontic treatment.

Regardless of its relation to the central incisor, a labial surgical and orthodontic approach is essential because of the relationship of the canine to the lateral incisor. Surgical access and orthodontic traction need to be negotiated very delicately between the roots of the central and lateral incisors and, above all, no attempt at root uprighting or torque of the incisors should be undertaken until the canine is well clear of both (Figure 8.3). This, therefore, rules out the early use of rectangular archwires and it is often wiser not to ligate the lateral incisor into the arch until the later stages of treatment. In these cases, the canine has to be drawn through between the incisor roots at a level high in the sulcus, above to the labial archwire, which makes it difficult to erupt it through attached gingiva (Figure 6.45).

The three-dimensional relationship of the teeth to one another and the possible existence of incisor root resorption are very difficult diagnoses to make, particularly in the bucco-lingual plane, as pointed out in Chapter 2, and cone beam computerized tomography should be performed to provide the relevant and much needed accurate information.

For most other buccally impacted maxillary canines surgical access is good, but the ability to provide a satisfactory orthodontic strategy to reduce the impaction and still provide for a good periodontal prognosis may be poor. This is because the high buccal canine tooth must be brought buccally, inferiorly and distally in a manner that circumvents the root of the adjacent incisor. This involves its being drawn in a semicircular flanking movement around the lateral incisor root, in an area where the alveolar bone is too narrow to allow one root to pass by another. As the canine is moved labially, the bony alveolar plate responds and becomes more prominent as it remodels labially.

This bony remodelling does not add width to the same degree as the dental movement, so that the root of the tooth loses some of its labial bone and soft tissue support, and a long clinical crown often results. The prospects for mucogingival surgery, performed at the time of exposure while possible [15, 16] are very limited for the high buccal canine and a long clinical crown with a relatively poor periodontal outcome are to be expected [17].

Given these drawbacks, the more severely displaced buccal canines of this type may occasionally need to be extracted, and, as far as possible, the deciduous canine left in place. In this instance, it is recommended to provide additional space mesially and distally, to allow for its crown to be prosthetically enlarged in anticipation of a later implant restoration when the deciduous tooth is finally lost. If the deciduous canine has a poor prognosis, an early decision regarding space closure or space opening should be made. Where appropriate, controlled orthodontic space closure may then be carried out, with or without a compensating extraction on the opposite, unaffected side. Alternatively, orthodontic preparation of the case for an implant-borne replacement crown will need to be undertaken as part of the overall orthodontic treatment.

Buccally impacted canines with distal displacement

It is unusual to come across a buccally located canine which is also displaced distally. These are almost invariably transposed with the first premolar tooth, to a greater or lesser extent (Figure 8.4). Many of these transposed canines erupt high in the sulcus where they are invested with oral epithelium rather than with attached gingiva. Others do not erupt and may need to be surgically exposed, depending on the orthodontic treatment plan. However, in these cases the orthodontist will frequently decide to align the teeth in their transposed order. By and large, these transposition patients present with the first premolar crown in its normal place, but its roots are displaced mesially, giving the tooth a distally inclined long axis. The tooth may be quickly

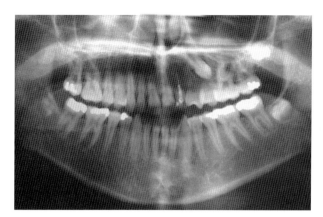

Fig. 8.4 A distally displaced maxillary canine, transposed with the first premolar.

tipped mesially, uprighting it into the site vacated by extraction of the deciduous canine. Under these circumstances, tipping the first premolar into the canine location reciprocally creates space for the canine in the first premolar place and the canine will usually respond with autonomous eruptive movement.

Many of these teeth may then come down without the application of orthodontic forces, although full eruption may take many months, sometimes extending into a year or more. Nevertheless, and even though there may be no apparent reason for this natural process to be retarded, a good number remain unerupted or in their partially erupted state for a very long time and hardly progress. Therefore, following the mesial movement of the first premolar in the earlier part of the treatment, these teeth should be exposed and actively drawn to the labial archwire, using light elastic traction or an auxiliary nickel–titanium wire placed piggyback style over the existing heavier base arch. Leaving an orthodontic appliance in place for all this time without exploiting it to erupt and align the canine is counterproductive, insofar as it prolongs and runs the risk of producing the deleterious side-effects of caries and periodontal inflammation and is to be discouraged. Thus, when the impacted tooth shows no signs of erupting or does not fully erupt within a reasonable period of time after space has been prepared for it, orthodontic traction and alignment should be initiated.

If the tooth remains unerupted and situated above the level of the line of the attached gingiva, the window technique exposure will leave it with an undesirable labial attachment consisting of oral mucosa. Consequently, an apically repositioned flap or a fully closed flap exposure will be more appropriate, and the tooth brought down by direct traction to the labial archwire of the appliance. Many clinicians tend to judge severity of transposition by the relative positions of the crowns of the two affected teeth, when the only valid determinant of severity is in the relative positions of their root apices. As the result, the inexperienced practitioner may be unwisely drawn into attempting to bring the

teeth to their ideally corrected order – a task that requires considerable clinical skill and an extended treatment time.

If the displaced canine can be moved mesially while it is still relatively high, then the danger of root proximity of the affected teeth, as they are drawn past each other across the narrow alveolar width, is reduced. However, while a closed surgical exposure may be preferred with a high canine, this can only be conveniently used when the direction of traction is vertically downwards. It is not easy to adapt the method when the direction of traction has more than a minor degree of mesial or distal component, because the elastic thread or closed coil spring that may be used will cause soft tissue impingement.

Nevertheless, if full correction of the transposition is to be undertaken, it is wise initially to place an eyelet attachment on the mesial aspect of the exposed canine, so that mesial traction will not generate a mesio-lingual rotational component on the tooth as it moves forward. Once the crown of the tooth has been moved into its correct location, the eyelet is removed and a bracket of the type used on the other teeth is placed in its ideal position on the tooth to complete the remaining movements needed, usually mesial uprighting, palatal root torqueing and a degree of derotation.

Mandibular canines

Mandibular impacted canines are rarely seen and, as the result, the more bizarre cases are published as single case reports in the literature [18, 19]. The very few numerically significant case series that have been published have gathered the cases from many centres and individual practitioners, on an international multicentre basis [20].

Impacted mandibular canines are usually chance findings and are not discovered because the condition is almost always symptomless. The over-retained deciduous canine may not raise the suspicions of a dental practitioner until well into the second decade of life and often when later. It seems possible that females are more frequently affected, although the evidence for this is tenuous, given the lack of large sample studies. The frequency with which this phenomenon occurs has been quoted as being 20 times less frequent than the parallel condition in the maxillary arch. While maxillary impacted canines crossing the mid-palatal suture has not been reported, mandibular canines do cross the symphysis (Figure 8.5) and have been reported to have reached as far as the permanent molar on the opposite side [21, 22].

They may sometimes be located on the lingual side of the alveolar process, when they will appear as a palpable hard swelling under the lingual mucosa. They are more frequently to be found buccally ectopic or in the general line of the arch. They travel relatively large intra-osseous distances and may become embedded in the chin prominence.

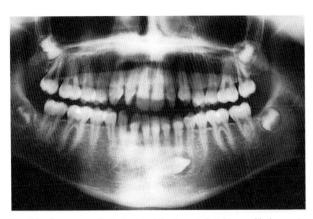

Fig. 8.5 The crown of the horizontally impacted right mandibular canine overlies the root of the erupted mandibular canine of the opposite side. (Courtesy of Dr T. Weinberger.)

The first clue to the existence of an impacted canine is the over-retention and lack of mobility of its deciduous predecessor and this should stimulate the clinician to perform a radiographic examination.

Regarding aetiology, there may be obvious local factors to which the condition may be attributed, namely supernumerary teeth, odontomes (Figures 8.6 and 8.7) and an enlarged dental follicle. However, it is important to note that soft tissue lesions, such as expanding radicular cysts that may have developed from non-vital deciduous first molars or canines, are potent displacing agents for developing adjacent teeth (see Figure 11.7). Nevertheless, for most of these cases, including the extreme examples, there is no apparent local cause and it seems likely that a hereditary primary tooth germ displacement may account for the abnormal angulation of the long axis of the tooth [23]. If this angulation is between 30° and 50°, there is a good chance that the tooth will migrate across the midline in a relatively short period of time, while an angle in excess of 50° will make this eventuality virtually certain.

Although there is some anecdotal evidence alleging a remarkably high speed of migration of these teeth, this is difficult to assess since these cases are usually seen *post-factum* and, even when seen before much of the movement has occurred, most patients will be advised to have interceptive or corrective treatment. Few will be merely kept under observation (i.e., supervised neglect!), given that a worsening of the situation seems inevitable.

Interceptive treatment should generally be instituted as soon as the anomaly is discovered. The extraction of the deciduous canine may exert a positive influence to alter the orientation of the aberrant permanent canine. By also extracting the first deciduous molar, the first premolar may often be influenced to erupt early, assuming it has about half of its expected root development. This will have the effect of providing more space in the alveolar bone distal to the canine because, with the eruption of the tooth, the broader diameter of the crown will have given way to the

narrower root diameter adjacent to the canine. Removing a non-vital deciduous canine or first molar, particularly if these had been associated with an unresolved granuloma or with cystic change, may encourage a dramatic improvement (see Figure 11.7).

When a canine has progressed beyond the mesial side of the lateral incisor, the majority of authorities have advised extraction of the canine itself, leaving the deciduous canine in place. Alternatively, if the case is considered to be an orthodontic extraction case, this tooth will be tagged for extraction rather than the more usual choice of a premolar.

Corrective orthodontic treatment may seem to be the best choice, but this is likely to be disappointing, particularly if the tooth is on the labial side and has migrated mesially, which essentially means that it has developed into a transposition. Good 3-D imaging of the area will be necessary in order to decide if orthodontic treatment can provide a good answer. The periapical radiograph will most often provide adequate qualitative information regarding the mandibular canine, unless it is very deeply displaced. This is because it may not be possible to insert the film sufficiently deeply into the lingual sulcus. In these cases, a panoramic view may provide a better view of the tooth and its mesio-distal orientation. A true occlusal view will be required to provide the third dimension needed to accurately localize the tooth and it is important to remember that the central ray of the X-ray machine should pass along the long axes of the mandibular incisor teeth for this view to be of value. In the midline region, the missing aspect will be depicted on the lateral cephalogram that will have been taken to assist in the diagnosis and treatment planning of what initially appeared to be routine orthodontic treatment (Figure 8.7). In this type of case, extraction may sometimes be the only practical line of treatment. Given the presence of the roots of adjacent teeth immediately superior to it and the narrow dimensions of the mandibular body in this area, there may be inadequate room for successful orthodontic manoeuvre, particularly when a partial or complete transposition of teeth is to be corrected. A parallel anomaly occurring in the maxillary arch will usually be much more amenable to treatment, because the impacted tooth may be temporarily moved from the narrow alveolar ridge inwards into palatal bone to permit the movement of an adjacent tooth.

For most of the impacted mandibular canines, however, the radiographic evaluation will indicate a reasonable prospect for orthodontic alignment. In line with the general principles set out in Chapter 4, an orthodontic appliance is placed and space is prepared in the arch to accommodate the tooth before its exposure is undertaken.

Lingually impacted mandibular canines will almost invariably be found with their root apex in their normal location and the lingual displacement of the crowns of the teeth is due to an abnormal orientation of the long axis – a

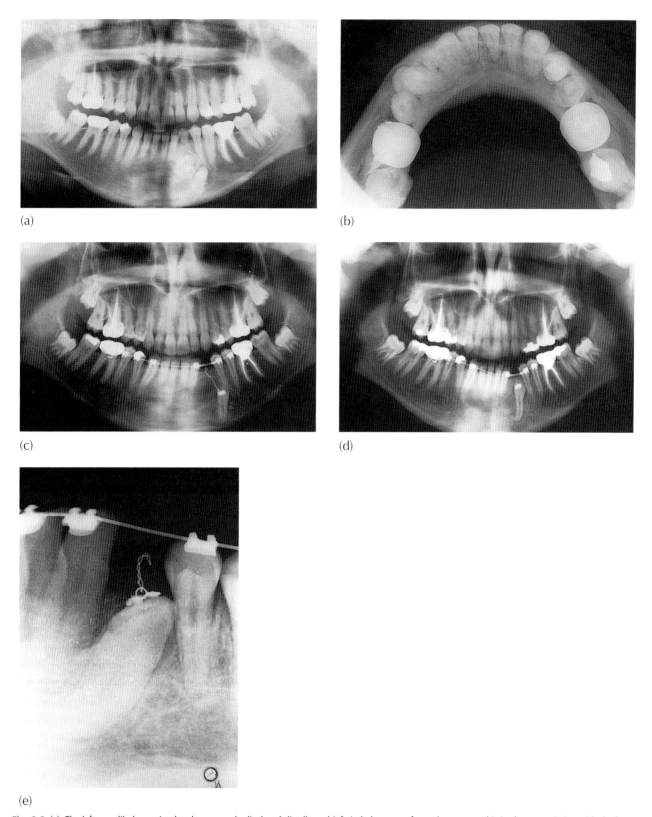

(a)

(b)

(c)

(d)

(e)

Fig. 8.6 (a) The left mandibular canine has been grossly displaced distally and inferiorly because of an odontome, and is in close association with the lower border of the mandible.

(b) A true occlusal view of the canine/premolar area.

(c) After alignment and space-opening, surgical removal of the over-retained deciduous tooth and odontome has permitted attachment placement.

(d) Rapid improvement in position has occurred. Note the deleterious effects on archform and midline due to the use of a base arch of inadequate gauge.

(e) A periapical view in the latter stages of resolution. Note that the tooth has responded well, despite having been covered with repaired and calcified bone after attachment bonding at the time of the closed exposure.

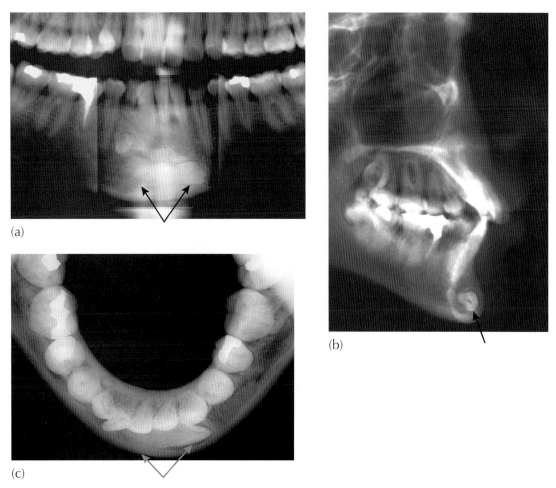

(a)

(b)

(c)

Fig. 8.7 (a) Panoramic view of the mandible with mental region – the contrast of the mid-section over the chin has been altered to show the horizontally impacted canine, with compound odontome. (b) Cephalometric radiograph shows the canine to be completely horizontal and at 90° to the mid-sagittal plane and the film. (c) Occlusal view of the anterior mandible.

tipping displacement. Therefore, orthodontic realignment involves merely a corrective tipping movement of the crown in a buccal, extrusive and, possibly, distal direction. The tooth should be exposed, an attachment bonded to the buccal aspect and the wound fully closed with the sutured flap, unless the tooth is very superficial. In this way, traction from the attachment direct to the labial archwire is all that is needed to bring it to its place. The wire ligature pigtail, tied to the bonded attachment at the time of surgery, is rolled downwards to form a loop, close to the sutured gingival tissues. An elastic chain is placed across the span between first premolar and lateral incisor, and its middle portion is stretched downwards with a haemostat or ligature director and ensnared in the rolled down pigtail. This provides a light, easily measurable and vertically directed force on the impacted tooth, with a wide range of action. Alternatively, an auxiliary nickel–titanium wire may be passed through this rolled-down pigtail and through the brackets of a number of teeth on each side, under the main arch, to achieve the same effect.

Migration, transmigration and transposition

Mupparapu [21] has attempted to classify buccal canines – teeth that are remarkably prone to the most bizarre eruptive movements – into five types, as follows:

1. Mesio-angular, lying inferiorly to the front teeth and, with its crown crossing the midline and termed a transmigrated tooth.
2. Impacted horizontally below the apices of the incisors.
3. Erupted mesially or distally to the contralateral canine.
4. Impacted horizontally below the apices of the contralateral canine and premolars.
5. Vertical, coinciding with the midline. It is the only type which may be declared a true transposition *ab initio*, rather than as the result of migration.

A special case, therefore, must be made for the buccally displaced mandibular canine, migrated, transmigrated or transposed mesial to the lateral incisor. Although very uncommon, it is the most frequent form of transposition

in the mandible, aside from third molars. The crown of the canine will need to be moved buccally in order to sidestep the lateral incisor root, before being moved towards the archwire. As with the parallel situation in relation to the buccally displaced maxillary canine, the orthodontic and periodontic prognoses of treatment for these teeth deteriorate in inverse relation to the amount and type of mechanotherapy used. Nevertheless, in a minority of these patients, full resolution of the transposition may be successfully achieved provided the cases are carefully selected, taking into consideration the periodontal prognosis, in addition to the biomechanics.

In theory, it is possible to apply appropriate tipping and bodily movements to move the tooth back from whence it has clearly come. However, the more horizontal the tooth, the greater will be the need for a large component of force being applied through its long axis – a horizontal 'intru-

sion', which is clearly futile. The tooth will also need to be drawn below the incisor apices with its crown exposed in highly mobile mucosa and in the deepest part of the labial sulcus. Even the more amenable mesially migrated canines will need to be drawn distally and occlusally, but also laterally in order to skirt the roots of one or more incisors. Once the crown of the tooth has been brought to its place in the arch, the canine root must be distally uprighted and finally lingually torqued to an appreciable degree. Given in the best circumstances, it will be appreciated that it is unlikely there will be much bone covering the root on the labial side of the treated tooth, its clinical crown will be long and the marginal soft tissue on the labial side will be largely devoid of attached gingiva. Treatment will have been inordinately long to achieve an acceptable orthodontic result, but it will be accompanied by a poor periodontal outcome (Figure 8.8).

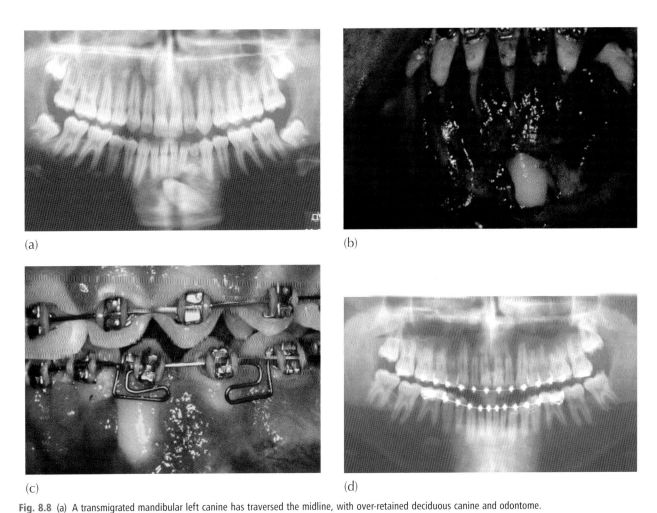

(a)   (b)

(c)   (d)

Fig. 8.8 (a) A transmigrated mandibular left canine has traversed the midline, with over-retained deciduous canine and odontome.
(b) At surgical exposure, the canine is located inferior to the right central incisor and is situated below the depth of the sulcus.
(c) In the final stages of orthodontic treatment aimed at completing the uprighting of its root and lingual root torque, the clinical crown is very long due to gingival recession and there is deep periodontal pocketing on the mesial side of the canine.
(d) Root paralleling of the teeth is good, but apical root resorption is present on all teeth carrying orthodontic attachments. There is crestal bone loss in the immediate area of the affected canine. The periodontal prognosis of this tooth contrasts sharply with the success of the orthodontic treatment.

There remain three additional and alternative lines of treatment for the non-crowded case. The clinician may:

1. extract the canine, leaving the deciduous canine in its place, provided its root is of reasonable length and prognosis;
2. extract an incisor and align the canine in its place, leaving the deciduous canine in place;
3. align the two teeth in the transposed relationship which, in the mandibular arch, may offer the optimal solution [24–27].

In the presence of crowding, extraction of the deciduous canine and the adjacent permanent incisor, or of the deciduous canine and the permanent canine, should be considered in addition (or in preference) to a more conventional choice. The space provided may then be used for the relief of crowding as an integral part of a comprehensive orthodontic treatment programme that includes other aspects of the malocclusion.

One final thought in regard to transmigrated mandibular canine relates to its innervation. It should be remembered that, regardless of the distance it travels, the tooth takes with it the blood vessels and nerve supply with which it started out life. This needs to be taken into account when considering the surgical exposure or removal of the tooth under local anaesthetic cover.

Mandibular second premolars

Crowding and space loss

Perhaps the most common cause of impaction of the second mandibular premolar is the early extraction of its deciduous predecessor, although this has become less common with the decline of caries in the Western world. With the loss of the second deciduous molar, the adjacent permanent molar will usually tip mesially and 'roll' lingually. Additionally, there will be a degree of distal drifting of the first deciduous molar of the same side, but total elimination of the space for the second premolar does not often occur. The result will be that this successional tooth will be blocked from erupting into the line of the dental arch. Given that its early developmental position is slightly lingual to the line of the arch, and that it is prevented from migrating upwards in the normal manner, it either will move more lingually and erupt on the lingual side, or it may remain impacted and beneath the 'pitched roof' formed by the two adjacent erupted and tilted teeth.

The radiographic method for these cases is very similar to that described for mandibular canine teeth. The periapical film is used to provide detail but, in the mandibular premolar area, it also provides a lateral horizontal view in this area. In theory, therefore, it may be supplemented by an occlusal view to provide the third dimension and enable accurate localization. Unfortunately, the occlusal view has the X-ray beam passing through the full thickness of the body of the mandible and, unless the tooth is markedly displaced to the lingual or buccal side it will not be possible to differentiate it from the mass of bone. If its presence can be confirmed in the periapical view and there is no clear view of the outline of the tooth on this film, it will be safe to assume the tooth to be close to the line of the arch and undeviated buccally or lingually.

Alignment requires space, and this may be achieved by re-siting the drifted teeth back in their former or improved positions using a fixed orthodontic appliance with a coil spring. This may often require intermaxillary (class 3) traction to reinforce the anchorage of the lower jaw and to prevent undesirable incisor proclination.

Alternatively, extraction may sometimes be necessary, in which case the impacted tooth or its immediate premolar neighbour may be the tooth that will be sacrificed along with a matching tooth in the other three quadrants of the mouth, in order to treat the overall malocclusion. Given space by distal movement of the molar and/or by mesial movement of a distally tipped first premolar or by extraction of the adjacent premolar, an impacted premolar tooth will normally erupt with considerable speed without further assistance.

From the periodontal point of view, surgical removal of unerupted mandibular second premolars, which may be needed in an extraction case, may leave a marked bony defect in the area, even after the excess space has been closed and adjacent teeth have been fully uprighted. This may be accompanied by a deep mucosal fold or cleft in the interproximal area in the site where the extraction had been made. This may disappear once space closure has been completed, although it may persist and thus prevent the regeneration of bone in the interproximal area, to cause a periodontal defect.

Abnormal premolar orientation

The second deciduous molar of the lower jaw has much to answer for in relation to the non-eruption of its permanent successor, not merely when it is prematurely lost due to the ravages of caries, but also when its presence is abnormally prolonged. The second premolar tooth germ is not always in its ideal developmental position, directly between the mesial and distal roots of the deciduous molar. Indeed, an abnormal angulation or location seems to be a frequent finding.

The premolar may often be tipped more distally, initiating resorption of only the distal root, leaving the mesial root of the deciduous molar largely intact. This will lead to over-retention of the deciduous tooth, often despite the complete disappearance of the distal root and much of the coronal dentine. A periapical radiograph will show the long-rooted premolar very superiorly positioned, almost

inside the distal part of the crown of the deciduous tooth, whose long and thin spicule of the mesial root remains, grimly resisting exfoliation. A parallel scenario may occur with resorption of the mesial root due to mesial tilt of the second premolar from early on in its development, although it seems to enjoy a lesser frequency. In either of these cases, as long as the degree of tilting is relatively slight and the tooth is relatively high up in the alveolus, the extraction of the deciduous tooth will usually suffice to achieve the rapid and trouble-free eruption of the premolar tooth. Space is never a problem in these cases, since the second premolar has a smaller mesio-distal crown width than its healthy predecessor.

A premolar tooth which has a stronger distal tilt is usually situated more apically, and the distal-occlusal aspect of its crown is in close relation with the mesial root of the first permanent molar. The second deciduous molar is usually over-retained at the time of detection and more than adequately holding the space in the arch (Figures 8.9 and 8.10).

In terms of aetiology, it has been found that an exaggerated disto-angular malposition of the unerupted mandibular second premolar is associated with agenesis of its antimere [28] and with retarded dental development. In Chapter 6, we have alluded to the existence of a connection between second premolar anomaly and palatally displaced canines, which are similarly affected by late development. It seems that individuals with both maxillary canine and mandibular second premolar anomalies suffer greater delay in dental development [29].

In these cases, a space-holding device should be used when the deciduous molar is removed to prevent tipping of the permanent molar, and an attempt may also need to be made to upright the premolar. An appropriate space-maintaining device may be designed in many ways, but, classically, a buccal and lingual bar may be soldered to two bands to form a simplified fixed bridge, which is then cemented to these teeth. Alternatively a single rigid bar, with terminal loops or a mesh pad at each extremity, may be bonded to the buccal surface of the first molar and first premolar. This is a fairly good alternative provided it is well clear of the occlusion, although it may still become debonded by occlusal forces transferred through bulky and hard foods. Because of its small size, the debonded bar with terminal loops on a mesh pad presents a potential hazard, since it may be ingested or, worse, inhaled by the patient.

At surgery, only the mesial and occlusal aspects of the impacted and distally tipped premolar tooth are exposed and, where possible, an eyelet should be bonded to this area of the crown of the tooth, carrying a twisted steel pigtail ligature. The tooth is fairly deep down and an open exposure is likely to leave the mesial root of the first molar exposed and devoid of attachment. For this reason, the flap should be completely sutured back into its place and the

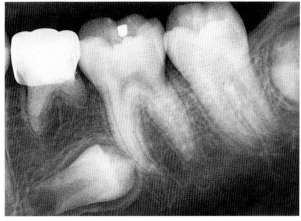

(a)

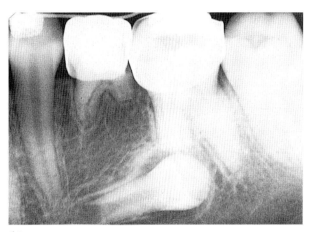

(b)

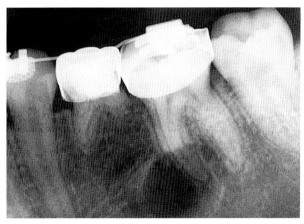

(c)

Fig. 8.9 (a) A late-developing left second premolar, horizontally oriented. (b) A year later, the tooth has moved distally to overlap the mesial root of the first permanent molar. (c) Extraction reveals some resorption of the mesial root of the molar. (Courtesy of Professor Y. Zilberman.)

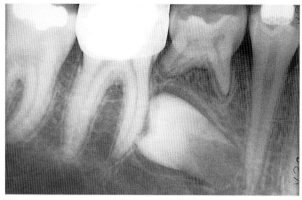

(a)

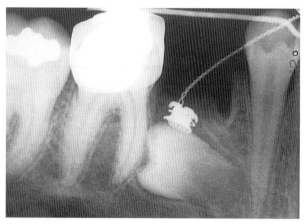

(b)

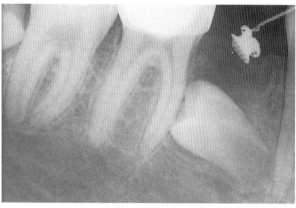

(c)

Fig. 8.10 (a–c) Serial periapical views of a failed attempt to bond an edgewise bracket to an inadequately developed second mandibular premolar. (Courtesy of Dr D. Harary.)

stainless steel ligature wire pigtail, ligated through the bonded eyelet, becomes the means of applying force to the unerupted tooth.

An elastic chain may be stretched between a hook on the fixed band of erupted first premolar and first molar, parallel to and overlying the rigid bar. Once in place, the middle of the elastic chain is drawn downwards with artery forceps and ensnared in the pigtail ligature to apply a vertical erupting force to the impacted tooth. The greater the degree of movement required, the more substantial must be the anchor base, and, where indicated, a fixed lingual arch to the opposite molar may be advisable.

This region of the mouth does not provide easy access to permit acid etch bonding and, particularly when the orthodontist is not present to do this part of the surgical procedure, eyelet attachment may not be possible (Figure 8.10). When this is the case, more radical surgical compromises may have to be made in order to salvage the tooth. Wider exposure is indicated and a buccal or lingual extension to the exposure may be needed, depending on the orientation of the tooth (Figure 8.11). This will then hopefully provide the needed access for bonding the attachment, with a satisfactory degree of reliability.

Because of the difficulty in bonding, it is common practice to use an open exposure procedure, leaving the tooth exposed but covered with a surgical pack, particularly distally where the pack is wedged between the premolar crown and molar root. The pack is designed to remain in place for two or three weeks to prevent healing over of the gingival tissues. The deliberate wedging of gauze into the distal area helps to divert the eruption path of the premolar in a more mesial direction; subsequent eruption may then occur spontaneously. As was pointed out in Chapter 3, however, these procedures will make the establishment of an ideal periodontium unlikely for both the molar and the premolar because the open surgery will have left the mesial side of the mesial root of the molar exposed to the detriment of its final periodontal status. It is to be expected that the clinical crown of the premolar will, in the final erupted position, be longer, the gingival attachment and architecture compromised, the bone support reduced and the prognosis poorer than normal.

While there are many impacted second premolars that may be treated in this way, success cannot be expected in some of the more extreme cases and the tooth has to be extracted.

On occasion, extraction of the tooth is advised for other reasons, both objective and subjective. Care should be taken to follow up these cases once diagnosis has been established, since there may be a danger of resorption of the mesial root of the first molar, if the tooth is left untreated without suitable and long-term radiographic supervision (Figure 7.3).

A lingually ectopic second premolar that is low down in the floor of the mouth and has migrated mesially in relation

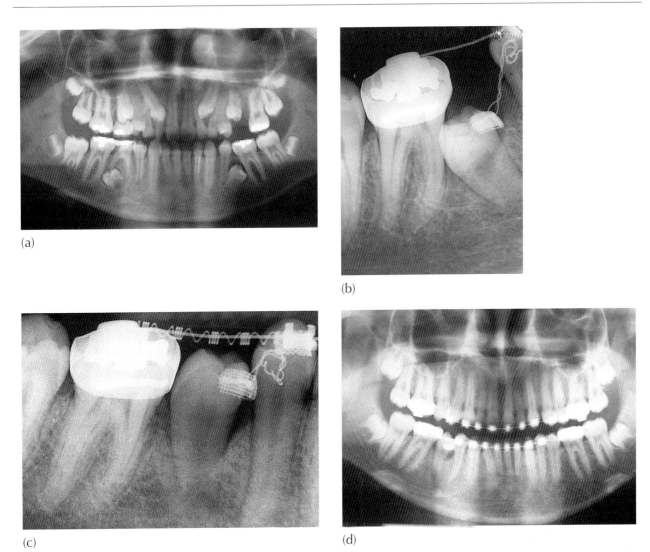

Fig. 8.11 (a) The left second premolar is impacted due to loss of space in the dental arch due to extraction of the second deciduous molar three years earlier. Both mandibular second premolars are very late-developing, relative to the other permanent teeth and both are strongly distally tipped. Note the associated bilateral palatally impacted maxillary canines.
(b) A periapical view shows bonding of an attachment to the right second premolar was achieved only after wide exposure.
(c) A similar follow-up radiograph shows improvement in the position of the tooth, but crestal bone height is deficient.
(d) A panoramic view of the final stages of treatment clearly shows the reduced periodontal prognosis of the right mandibular second premolar. (Courtesy of Dr L. Rothstein.)

to the adjacent teeth is less commonly seen than the distally drifted teeth. In either case, direct traction to the labial archwire will not be practical. In these situations, the use of the same auxiliary labial arch – always as an addition to the main base arch, as noted above (Figure 8.12) and as recommended for many palatal canines – has excellent application and can be used with great efficacy to provide the vertical and distal eruptive components (Figure 8.13).

Maxillary second premolars

The most common reason that these teeth become impacted, as with the mandibular second premolar, is related to space loss in the dental arch following the early extraction of the deciduous second molar and the drifting of the two adjacent teeth, particularly the first permanent molar. It is also true that mesial movement of the first permanent molar in the maxilla is more rapid than occurs in the mandible, it more completely closes off the space and it does so by mesial tipping and a mesio-lingual 'rolling', even in the late mixed dentition stage. When, therefore, space has been lost, the maxillary second premolar is most often to be seen developing with its root apex in the line of the arch and its crown deflected palatally, palpable on the palatal side of the alveolar process. As with the mandibular second premolar and the lateral incisors of

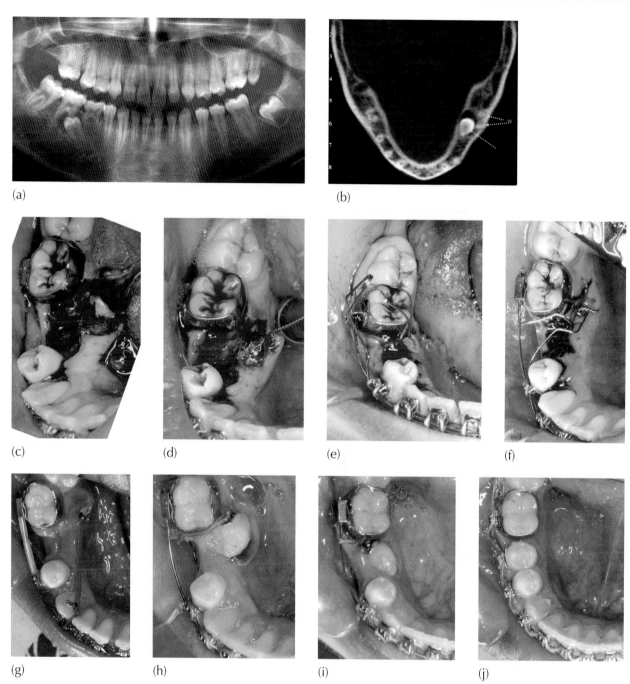

Fig. 8.12 (a) Panoramic view showing impacted right second mandibular premolar tipped 60° distally, overlapping the mesial root of the first molar. Note the missing maxillary right second and all four third molars, peg-shaped left second maxillary molar.

(b) Axial slice from the CBCT shows the second premolar situated lingual to the roots of the first molar (arrowed).

(c) At surgery, the deciduous molar is removed and the premolar exposed deep in the lingual sulcus in the floor of the mouth. (Surgery by Prof. R. Seltzer.)

(d) An attachment is bonded.

(e) The auxiliary labial arch is shown in its vertically passive mode and the pigtail ligature may be seen on the lingual side of the alveolus.

(f) The auxiliary labial arch is engaged in the pigtail ligature after the full flap is replaced and sutured, exerting a vertical and slightly lingually directed force, to draw the tooth away from the molar tooth.

(g) The tooth bulges the lingual mucosa more superiorly re-exposing the eyelet. Mesial traction is applied to the labial archwire at a point between canine and lateral incisor, to circumvent the mesial root of the molar tooth.

(h) Elastic traction is applied direct to the archwire to de-rotate the tooth as it moves buccally.

(i) The premolar moves into the arch, with much correctional rotation.

(j) Completion of the alignment has been achieved and an orthodontic bracket has been substituted for the initial eyelet, to permit control of root movement.

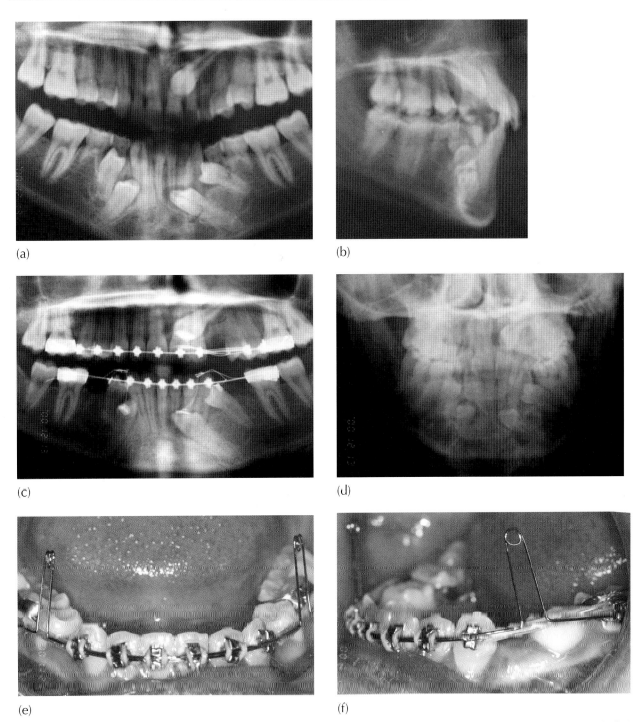

(a)

(b)

(c)

(d)

(e)

(f)

Fig. 8.13 (a–d) The initial radiographs including panoramic, lower occlusal views and partial views of lateral and postero-anterior cephalograms. (e, f) The auxiliary arch is placed in passive mode prior to surgery. (g, h) The mandibular premolars are exposed and bonded with eyelet attachments. Plastic tubing threaded onto the base arch holds the regained canine space. Because of their superficial positions, the second premolars were left partially exposed, while the first premolars were fully recovered with the surgical flap. (Surgery by Prof. J. Lustmann.) (i) Occlusal view of the lower jaw one week after the exposure of the teeth. The loops on the auxiliary arch are seen in their active mode, hooked round the pigtail ligatures of the first premolars. The more superficial second premolars are ligated with elastic ligature direct to the buccal hooks on the molar bands. Plastic tubing threaded onto the base arch holds the premolar space bilaterally. (j, k) Occlusal and panoramic radiographic views of the postoperative condition. (l–n) The finished case 25 months after completion of treatment. (o) Panoramic view taken 25 months post-treatment. Note the abnormal crown–root angulations and root dilacerations of the treated mandibular premolars. (p) Partial view of post-treatment lateral cephalogram.

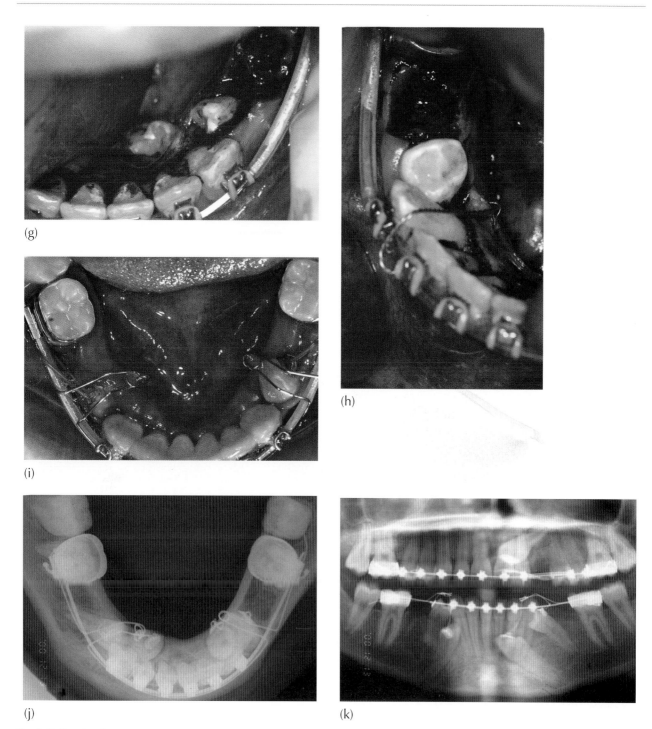

(g)

(h)

(i)

(j)

(k)

Fig. 8.13 (*Continued*)

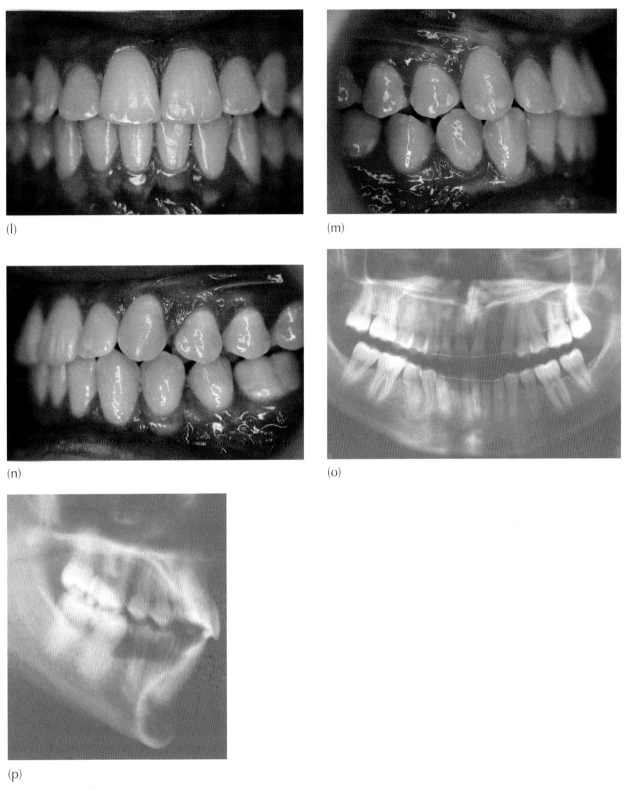

(l)

(m)

(n)

(o)

(p)

Fig. 8.13 (*Continued*)

both jaws, the developmental position of the maxillary second premolar is lingual to the line of the arch and any physical limitation in the mesio-distal width of its normal eruption path will tend to deflect the tooth in a more palatal direction.

Palatal displacement can occur with over-retention of the deciduous second molar; it can also occur when there has been a more palatal orientation of the tooth bud of the unerupted premolar in its early development. The outcome may be its eruption or partial eruption at the palatal aspect of the cervical margin of the over-retained deciduous tooth. Alternatively, it may become more horizontally oriented, to remain unerupted and, occasionally, very palpable close to the mid-palatal suture.

Because of the oblique angle of the X-ray tube in the vertical plane in periapical radiography of this area, the palatally tipped and unerupted second premolar will be viewed almost through the long axis, which may give the appearance of a short and underdeveloped root and, with a more exaggerated palatal tip, will be depicted on the film as an ellipse. This being the case, it is probable that further X-ray films are not necessary, particularly if the tooth is palpable in the palatal area. The conventional (oblique) occlusal film or a second, laterally shifted periapical film will offer the opportunity to add three-dimensional information. Both a postero-anterior cephalogram and the true vertex view would be more decisive, but the relatively high dose of radiation needed for these views should be considered excessive in relation to the information that they may provide in this situation. Nevertheless, if these views are available, they must be studied for any relevant information.

In Chapter 2, we pointed out that a panoramic view of the maxillary second premolar area is taken when the X-ray beam is directed from behind the patient's ear of the opposite side, which means that a palatally displaced second premolar will be imaged further mesially in relation to its erupted neighbours. If, at the same time, the lateral cephalometric view shows the tooth to be in its normal position, then the diagnosis of palatal displacement will have been confirmed in what is essentially another form of the tube shift parallax method (buccal object rule).

Infra-occlusion of deciduous teeth and its influence on premolar successors

Infra-occlusion of deciduous molars in either jaw is a relatively common occurrence and the affected teeth are known to shed quite normally in most cases, with only a relatively minor delay in their exfoliation time [30, 31]. The cause of the phenomenon is a local ankylosis at some point or area of the root of the tooth. The tooth will not then alter its position or develop vertically along with the adjacent teeth, which continue their passive eruption as vertical growth of the alveolus progresses. The result is that the affected tooth becomes relatively 'submerged' or infra-occluded.

Attempts to move these teeth and to overcome the infra-occlusion by orthodontic means will fail and it is rarely possible to identify the site or extent of the ankylotic connection. Indeed, it is rarely a task that the orthodontist will need to or be called on to perform for as long as an unerupted permanent premolar is present. On the other hand, such a tooth may be used as an anchorage for the movement of other teeth without fear of unwanted movement, since it behaves very much like a temporary anchorage device.

When the infra-occlusion is very marked, an extreme vertical displacement of the apically placed successor will also occur, as has been discussed earlier in this chapter. Indeed, in the mandible the apex of the root of the developing premolar may even cause a palpable prominence in the otherwise smooth profile of its lower border [32]. In these cases, extraction of the infra-occluded tooth should be made in line with the appropriate development of the root of the permanent successor, or earlier if the location of the premolar is abnormal, since the deciduous tooth may have been the cause of the ectopic location.

It is often quite impractical to attempt to bond an attachment to many of these grossly displaced premolar teeth. However, their position generally improves quite dramatically once the deciduous tooth has been removed, usually in a reasonably short time. No studies or even case series have appeared of these extreme cases in the literature and opinions are strongly influenced by single published case reports or individual practitioner clinical experience. However, there would appear to be merit to a wait-and-see policy and some cause for optimism in these very special circumstances (Figure 8.14) following space opening and extraction of the infra-occluded deciduous tooth.

Fortunately with infra-occluded teeth, actual space loss within the mandibular dental arch is very minimal, despite the obvious fact that the teeth adjacent to an infra-occluded tooth are often severely tipped towards it. The reason for this is that there is a displacement of the roots of the teeth immediately adjacent to the infra-occluded tooth, away from the affected tooth, with little or no approximation of their crowns [33–35]. Once the infra-occluded tooth is removed, space loss may be quite rapid and a space maintainer should be placed. No attempt should be made in the first instance to uncover the very deeply placed premolar but, following the removal of the deciduous tooth, the patient should be followed over a long period of time, with an occasional periapical radiograph taken to check for eruption progress.

Infra-occluded deciduous teeth are associated with a lack of alveolar bone height in the immediate area. The height from the inferior border of the mandible to the occlusal table is significantly reduced, when compared with the normal, unaffected, opposite side. The height of the

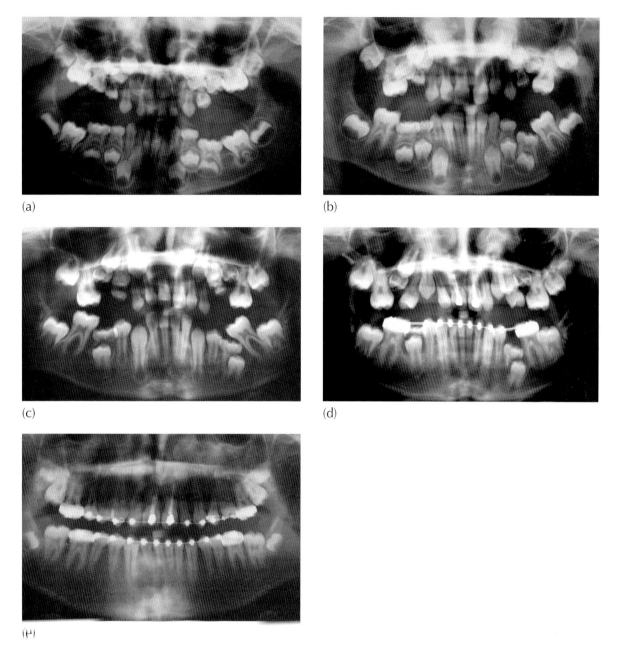

Fig. 8.14 (a) At age 6 years, all the deciduous molars appear infra-occluded to a greater or lesser degree, as can be seen in relation to the erupted deciduous canines. (b) At age 8 years, the first molars have erupted with mesial inclinations and the deciduous teeth have further infra-occluded. (c) At age 10 years, the infra-occlusion has increased and the permanent molars have developed a strong mesial inclination. Note that the maxillary second premolar teeth both are mesially displaced and have caused the direct external resorption of the maxillary second deciduous molars. (d) Nine months later, from treatment being performed on the upper right deciduous molar, orthodontic treatment was begun to upright the molars with simple distal tipping, using a fixed lower and removable upper appliance, supported by an integral headgear. The right second deciduous molars have further resorbed and their degree of infra-occlusion has lessened, presumably due to the resorption of the causative ankylotic site. The four deciduous molars were finally extracted a few months later, when adequate space had been made. All the premolar teeth erupted spontaneously over a two-year period, while space was maintained and without the need to expose them. (e) At age 13 years, all teeth have erupted and orthodontic treatment has recommenced to bring about full alignment, particularly the de-rotation of the maxillary left second premolar. Full bony regeneration and alveolar height have been achieved. (f–j) Clinical views of the condition immediately prior to the commencement of orthodontic space opening procedures at age 10.6 years. Note severe bony defects in premolar areas, constituting a marked lateral open bite. (k–o) Clinical views of the condition at age 13 years and after eruption of the premolars. The maxillary right second premolar has a 110° rotation which is being treated in a final finishing stage of orthodontic treatment. Note full regeneration of alveolar bone height, establishment of occlusion and normal appearance of the teeth and gingivae. (p–r) Intra-oral photographs at age 16 years, two years post-treatment. (s) Panoramic view two years post-treatment.

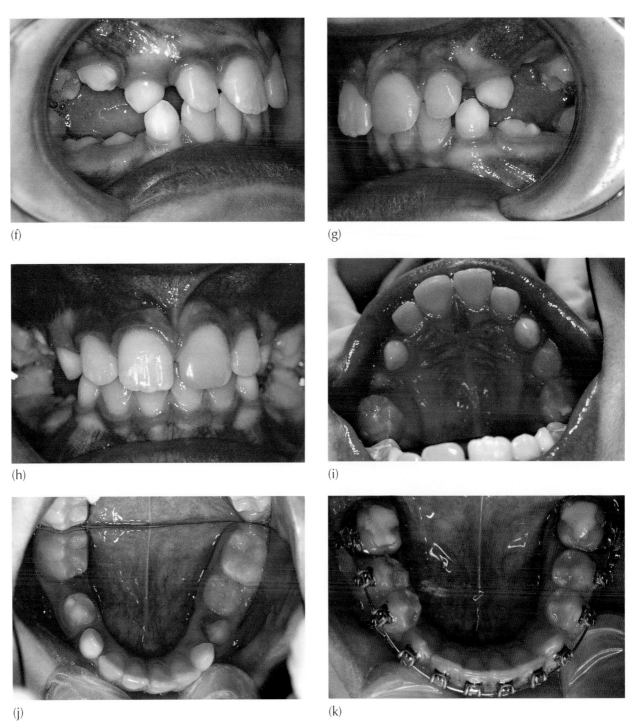

(f)

(g)

(h)

(i)

(j)

(k)

Fig. 8.14 (*Continued*)

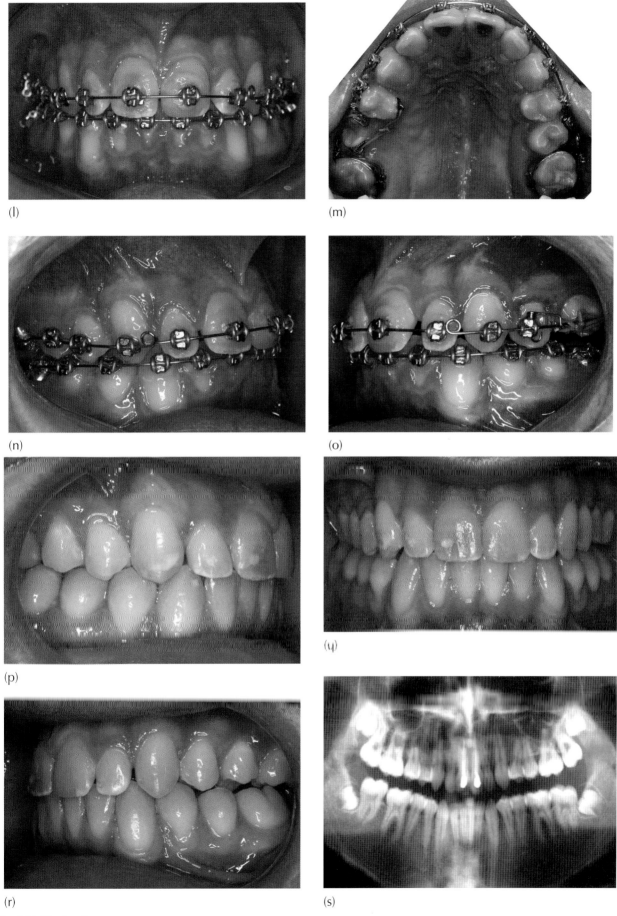

(l)

(m)

(n)

(o)

(p)

(q)

(r)

(s)

Fig. 8.14 (*Continued*)

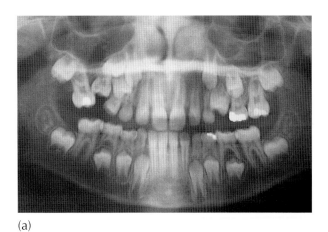
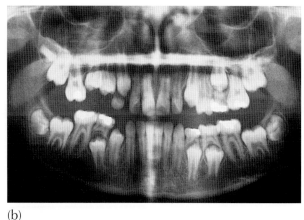

(a)   (b)

Fig. 8.15 (a) Characteristic extreme tipping of the teeth adjacent to an infra-occluded maxillary deciduous second molar. The second premolar is displaced superiorly and mesially, exhibiting a marked distal orientation. (b) Space regaining and maintenance, with extraction of the infra-occluded deciduous molar, has resulted in spontaneous correction of tooth position. Eruption is imminent.

teeth immediately adjacent to a severely infra-occluded deciduous molar is also reduced when compared with the height of unaffected contralateral teeth. This phenomenon has been blamed on the inclusion, by the infra-occluded tooth, of a vertical component in the normally horizontal trans-septal fibres, which inhibits their vertical development [34].

Extraction of the infra-occluded tooth initially leads to healing and bone reorganization which, in the absence of permanent teeth, will not increase the vertical height of the body of the mandible in that area. If a permanent tooth is present and begins to erupt, alveolar bone development accompanies the eruption and the vertical bony deficiency is eventually made good. However, this may be partial only and a lateral open bite may sometimes persist.

Second premolars in either jaw occasionally become impacted in a situation complicated by the severe infra-occlusion of the second primary molar (Figure 8.15). When infra-occlusion occurs in the very young child, the relative submergence of the tooth will occur rapidly in line with the vertical growth of the alveolar ridges, until the deciduous tooth is lost from sight beneath the gingiva. The erupting first permanent molar then migrates mesially to an excessive degree and tips in an exaggerated manner, greater than would normally be seen as the result of the early extraction of a deciduous tooth. This is seemingly due to the influence of the infra-occluded tooth in the vertical plane [33–35]. Similarly, the deciduous or permanent tooth immediately mesial to the affected tooth tips strongly distally, such that the long axes of the two adjacent teeth converge coronally at an angle of almost 90°, instead of being parallel. When the practitioner comes across this type of unusually severe convergence, an infra-occluded deciduous tooth should always be suspected.

The infra-occluded tooth is now firmly locked in by the reduced space in the arch and remains there during the entire period of the resorption of its root, which normally would precede its shedding. In time, the entire dentinal contents of the crown become resorbed and the empty enamel shell remains sequestrated *in situ*. In the maxilla, the unerupted second premolar develops in these cramped circumstances and, with its further root growth, becomes displaced usually mesio-palatally, with its root oriented mesially. It may eventually side-step the empty enamel crown and erupt into the palate, but usually it will remain high up in the maxilla and close to the floor of the maxillary sinus. In a panoramic radiograph, the tooth may frequently be seen to superimpose on the unerupted maxillary canine of that side.

Treatment of the problem is surprisingly simple (Figure 8.16). The molar must be moved distally to its ideal location to reopen the space in the arch. Since the tooth has a strong mesial tip, a removable appliance carrying a finger spring to distalize the molar is probably the most efficient appliance available for this task, particularly in the mixed dentition stage, and will usually take no more than three or four months to achieve its goal.

The patient is referred to the surgeon for removal of the resorbed remains of the deciduous second molar. For this, the removable appliance may be taken out of the mouth, its acrylic base trimmed to be clear of the surgical field and, at the conclusion of the surgical procedure, replaced to retain the molar position. In the longer term, it is preferable to place a soldered lingual arch in the maxillary arch, based on two molar bands. The removable appliance itself is perfectly adequate as a space maintainer, but compliance may be a problem, particularly in the immediate postsurgical period.

With the older patient, in the permanent dentition stage, a fully-bracketed fixed appliance should be used in order

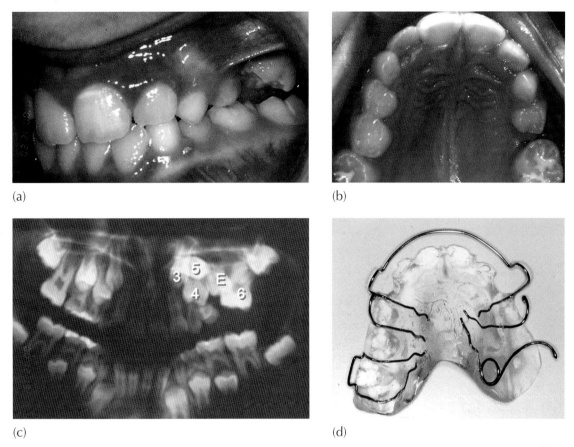

Fig. 8.16 (a) The left first molar is strongly tipped mesially into the place of the second deciduous molar and out of occlusion. The first deciduous molar is distally tipped.
(b) The occlusal view shows some distal drift of the deciduous first molar, canine and permanent lateral incisor.
(c) The complex interrelations between the first permanent molar (6), the intra-occluded second deciduous molar (E) and the unerupted second premolar (5), first premolar (4) and permanent canine (3). Note the almost 90° angle described by the long axes of the first permanent and deciduous molars.
(d) A simple removable appliance which is retained by clasps on the right permanent molar, both deciduous first molars and a labial arch. The spring is designed to move the molar distally.
(e) The imprint of the spring in the palatal soft tissue can be seen, following removal of the first and second deciduous molars.
(f) A soldered lingual arch holds the molar in its place in this occlusal view, taken 18 months later. The second premolar is now erupting into its place. No other orthodontic treatment was performed in this case up to this point.
(g) The view from the left side on the same day.

to combine this local problem with the treatment of the overall malocclusion (Figure 8.17).

Following the removal of the impediment and free of the physical constraints, the second premolar will rapidly respond spontaneously and, within a period of between several months to a year or so, it is highly predictable that it will erupt unaided into the location prepared for it in the arch. For this reason, the second premolar does not need to have an attachment placed on it, nor is it usually necessary to even expose it. In contrast to other impacted teeth, there appears to be little benefit in going through the elaborate orthodontic and surgical preparations that we have described in relation to incisor and canine teeth. Much surgical damage will be inflicted by the effort and difficulty involved in exposing and bonding an attachment to a tooth

in this location and the exercise would most often be superfluous.

Maxillary first molars

In the early mixed dentition, one may occasionally see the erupting maxillary molar caught by the distal bulbosity of the adjacent deciduous second molar (Figure 8.18). This is usually an early sign of crowding of the dentition, although it may simply be due to an abnormal mesial tilt of the first molar. Clinically, the essential diagnostic criterion is that the marginal ridges of the two adjacent teeth are at different levels, with that of the deciduous tooth being more occlusally placed. In the severer cases, the mesial marginal ridge of the permanent molar is unseen beneath the area of the

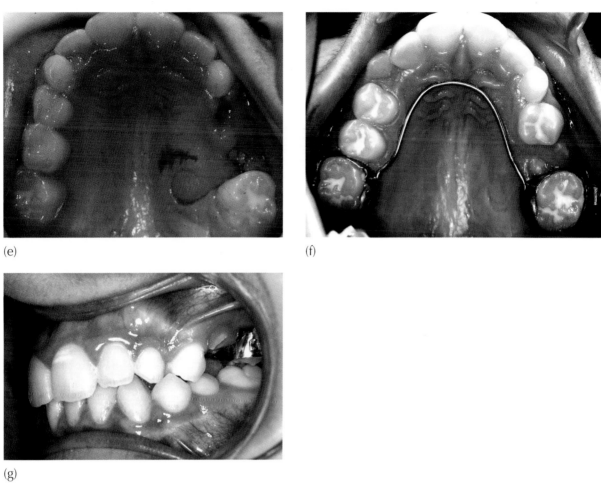

(e)

(f)

(g)

Fig. 8.16 (*Continued*)

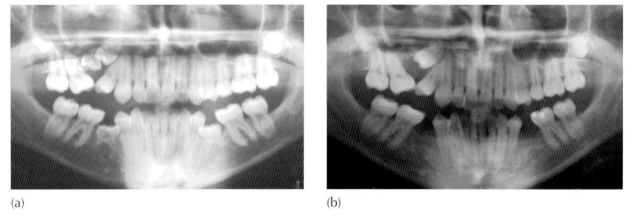

(a)

(b)

Fig. 8.17 (a) The panoramic view of a 17-year-old female with missing mandibular second premolars and third molars and with infra-occlusion of three premolars. In the upper right side, the vertical and mesial displacement of the second premolar is evident.

(b) The patient underwent the surgical removal of the deciduous second molars before being referred to the author for treatment.

(c–e) The initial intra-oral views.

(f) The bracket placement jigs demonstrate the 90° angulation between the long axes of first premolar and molar.

(g) Five months later, levelling, alignment and space-opening has been prepared in the maxilla and the second premolar may already be palpated as a bulge in the buccal area above the archwire.

(h) Seven further months later and without surgical exposure, the second premolar has erupted and is in its place in the arch.

(i–m) Intra-oral views of the completed orthodontic treatment on the day appliances were removed.

(n) Panoramic radiograph of the treated result on the same day.

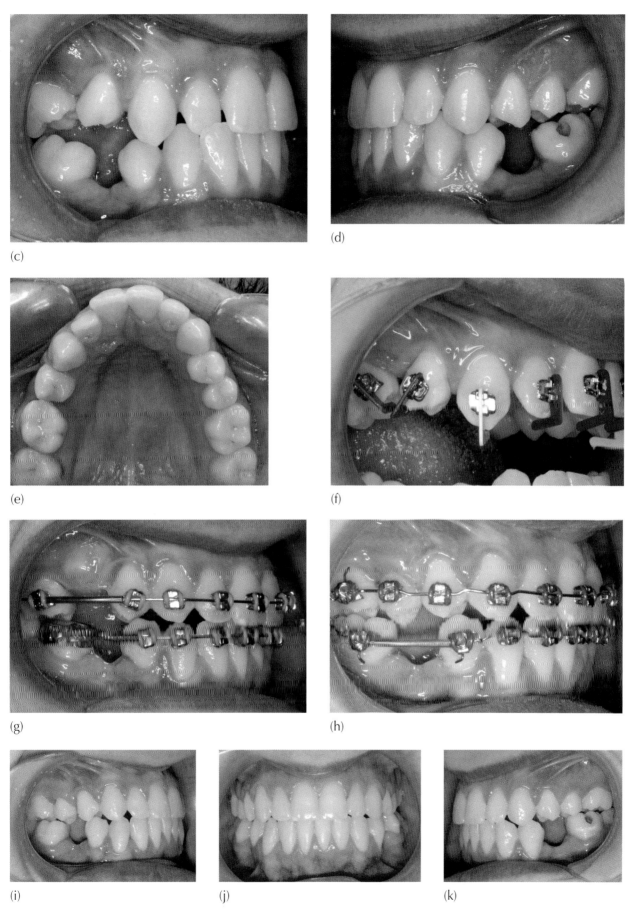

(c)

(d)

(e)

(f)

(g)

(h)

(i)

(j)

(k)

Fig. 8.17 (*Continued*)

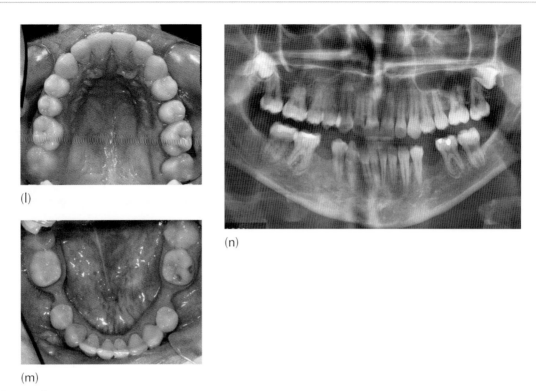

(l)

(n)

(m)

Fig. 8.17 (Continued)

distal CEJ of the deciduous tooth. At best, the distal cusps may be in occlusion with the mandibular molars, but usually the impaction prevents adequate eruption for this to occur. In rare instances, the permanent molar is completely prevented from erupting and does not break through the mucosa.

A periapical or panoramic radiograph of the area will usually show that most or all of the distal root of the deciduous second molar has been resorbed, and the general long axis and path of eruption of the permanent tooth to be tipped too far mesially. To leave a partially erupted molar tooth without treatment is to invite caries in the large stagnation area that has been created between it and the deciduous molar, which serves to compound the existing potential gingival/pulpal problem in the area of contact. In general, the mesial and palatal roots of the deciduous second molar remain intact, which generally suffices to retain this tooth in its place.

If the deciduous second molar is shed prematurely or is extracted, the permanent molar will move rapidly forward, significantly closing off the space vacated by the lost tooth within a few short months. The tooth also erupts occlusally at the same time, and its movement has a significant forward translation component, in addition to its mesial tip. With its centre of rotation more mesially placed on the ridge, orthodontic treatment to push the tooth back to its ideal position after full eruption will generate an excessive distal tip to the molar, which will leave the roots too far mesially. Second, the achieved space will need to be maintained for around five years, until all the successional teeth have erupted. Regardless of how well designed and constructed the proposed space-maintaining device may be, its placement in a 7- or 8-year-old patient must provide some concern *vis-à-vis* its deleterious effect on the long-term prognosis of the first molar.

Many original ideas and gadgets have been suggested to disimpact the ectopic first molar from beneath the distal bulbosity of the deciduous second molar, and the simplest of these has been to use an elastic separating ring or other form of orthodontic separator, which is normally used to create spaces to facilitate banding. This may be successful in some of the very mild cases, although what frequently happens is that a space between the teeth is produced within a week or so after placement of the ring. The elastic ring is then removed to allow the permanent tooth to erupt occlusally. In many cases, the molar tips right back into its previously impacted position and nothing will have been achieved.

This type of relapse may occur with any method that is only concerned with tipping the tooth distally, without providing for a retention period. The retention period is essential in order to permit additional eruption of the affected molar, while preventing re-impaction. A fixed device has several apparent inherent advantages, but it must be remembered that the patient is a young child, who may be relatively unwilling to cooperate in its construction and placement. Furthermore, most orthodontists do not have a selection of preformed deciduous second molar bands, and

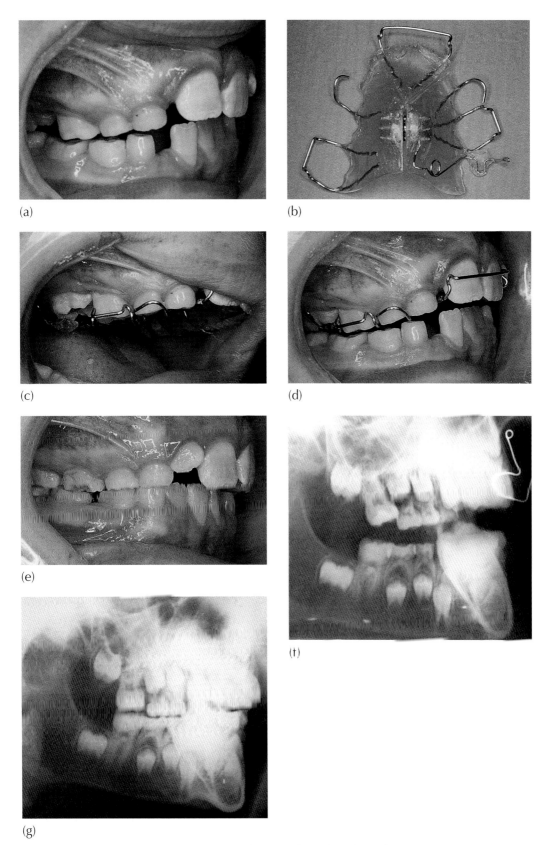

Fig. 8.18 (a) Incomplete eruption of the maxillary first permanent molar, due to abnormal angulation of its long axis. It has become impacted beneath the distal bulbosity of the deciduous second molar. (b) The removable appliance carries five retention clasps and a palatal 'finger' spring that traverses the occlusal surface of the partially erupted first molar. An acrylic button is placed on that portion of the spring that comes into contact with the occlusal surface of the first molar. (c) The incompletely seated appliance shows the spring in its passive mode, distal to the molar. (d) The patient has brought the spring mesially, prior to fully seating the appliance. (e) The molar is disimpacted and in full class 2 occlusion. (f, g) Lateral oblique extra-oral films before and after treatment show the advanced resorption of the second deciduous molar.

the placement of a band on the permanent molar is impossible in all but the simplest of cases. Suitable conditions for composite bonding are also exceptionally difficult to attain in the molar area of such a young child and results are unreliable. Nevertheless, appliances have been designed which are based on a fixed band on the deciduous second molar only, or on an additional fixed band on the second deciduous molar of the opposite side, through the agency of a soldered palatal arch. A soldered spring is formed on a model, which fits into the most convenient and deepest occlusal pit of the ectopic molar and activated prior to cementation of the appliance [36]. It is quite difficult to readjust this spring in the mouth, which may make this a one-time activation appliance, unless the whole cementation process is repeated.

At first glance, a removable appliance seems inappropriate, since there appears to be no way of passing a cantilever spring across the interproximal area, mesial to the tooth, to apply the needed distal force. However, the removable appliance offers excellent ways of overcoming the drawbacks described above (Figure 8.18) [37]. Good retention is necessary, and this is best supplied by placing an Adams clasp on the second deciduous molar of that side, on the erupted first permanent molar on the opposite side and on the two central incisors. Additionally, a three-quarter circumferential clasp placed on the first deciduous molar usually offers excellent retentive support. On the plaster model, a simple cantilever spring is drawn across the erupted portion of the occlusal surface of the impacted molar, mesial to the palatal and between the two buccal cusps. In the area of the mesial occlusal pit, a dab of acrylic is cured onto the spring to adapt it to that surface of the tooth. Activation of the spring is made after the patient has experienced the passive appliance for a couple of weeks or so and is taught how to place the appliance with the spring in its appropriate place. The tooth will generally move distally and disimpact from the deciduous molar within a few weeks. At that time the spring is altered in shape to enter the newly accessible interproximal area between the two teeth and, once there, to maintain the molar position while allowing its eruption. When the permanent molar has reached occlusal level, the appliance may be discarded since, at this point, the presence of the second deciduous molars will prevent relapse.

In the more severe cases of this type, the resorption process will have completely eliminated the distal root of the deciduous molar and, if careless probing is performed with a sharp explorer, perforation at the distal aspect of the deciduous molar is very likely to occur. Clinically, the sharp enamel edge at the cervical margin of the crown and the abrupt absence of continuity may be quite obvious in the same way as is seen with any deciduous tooth immediately prior to its normal shedding. In this area, former pulpal tissue will have merged with the surrounding gingival tissue. Inflammation of this tissue at this juncture will, therefore, no longer give rise to a pulpitis and it will repair symptom-free, as with any other injured gingival tissue.

In the cases of this type with which we have had experience there have been no adverse clinical symptoms related to the deciduous second molar tooth; nor has periapical radiolucency been evident on X-ray to suggest pulp death. Extraction has not been required and the tooth has remained to act as a natural space maintainer until shedding has occurred at a much later stage, closer to the normal shedding time and late enough to find the second premolar erupting rapidly thereafter. This is preferable from every point of view to extraction and replacement by an artificial space maintainer.

To find the parallel situation in the first permanent molar of the mandibular arch is very rare. However, management would be very similar to what has been described for the maxillary first molar.

Mandibular second molars

Impaction of the mandibular second molar is uncommon, but when it occurs it is very similar in its appearance to that so frequently associated with third molars. Unlike the maxillary first molar, the mesially impacted mandibular second molar is frequently unerupted and it usually comes to light in a routine dental examination, when it shows up on the bite-wing radiographs. A periapical film or a panoramic scan will show detail of the tooth from crown to apex and its relationship with the unerupted third molar. The tooth may also have a buccal or lingual tilt, which will generally be revealed by palpation, although an occlusal view will serve to confirm. Clues to a lingual inclination may also be seen on the panoramic view, since it is easy to distinguish between the form of the lingual cusps and that of the buccal cusps. It is also unusual to see a vertical discrepancy of any magnitude in the superimposition of the buccal cusps and the lingual cusps of each molar tooth. Thus, when the lingual cusps are seen to be much lower than the superimposed buccal cusps, one may conclude that the tooth has a strong lingual inclination.

There are several aetiological possibilities and these may be listed as follows:

1. Wide molar crown contour: The teeth of some patients may sometimes show unusually wide crowns sited on narrow roots, which results in a deep concavity on both mesial and distal aspects of the teeth concerned. In the molar region, this presents the unerupted distally-adjacent molar with a potential cul-de-sac beneath the distal bulbosity, into which it may migrate in its mesial and vertical eruption path (Figure 8.19a).
2. Abnormal mesial angulation of the second molar tooth germ: The crypt of the developing second molar may sometimes be seen, *ab initio*, to have a mesial orienta-

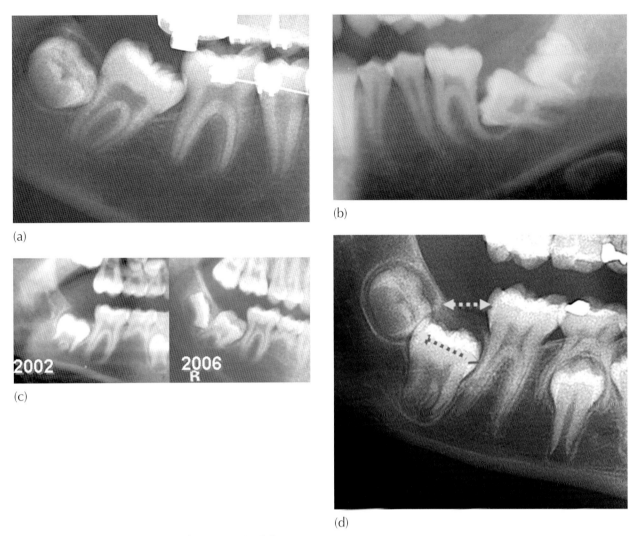

(a)

(b)

(c)   2002   2006
R

(d)

Fig. 8.19 (a) Bulbous crown of second molar entrapped by similar contour of first molar.
(b) Abnormal mesial angulation probably the result of abnormal crypt orientation.
(c) a developing second molar mesio-angular impaction radiographed in 2002 and then in 2006 to show influence of third molar developing with a mesial angulation in the vertical ramus and its subsequent mesial migration.
(d) Ramal development of the third molar has reduced the width of the eruption path of the second molar, to cause its vertical impaction.
(e) A dentigerous cyst encompasses the crown of the second molar.
(f) Two panoramic views of the same patient taken a year apart. The right mandibular second molar is relatively more inferiorly placed and a gap has developed between it and the first molar. There are clear signs of ankylosis.
(g) An Odontome has displaced the molar to the lower border of the mandible, from where it is palpable externally.
(h) CBCT images show the close proximity and encirclement of the inferior alveolar nerve bundle by the roots of the second molar.

tion. From this compromised position, the tooth will generally continue to progress along an eruption path that has been thus determined (Figure 8.19b).

3. Posterior crowding, with the third molar developing in the vertical ramus: Teeth normally develop in the alveolar bone of the horizontal body of the mandible. When crowding exists, it may be present in the incisor region, with the teeth developing, erupting and competing for space on the anterior rim of the two dental arches between the deciduous canines of each side. In the retromolar area of the mandible, it will be recognized by the third molar being located in the vertical ramus at a higher level than would be expected and with a marked mesial rotation of the tooth bud. Any expression of eruptive potential will be directed mesially, towards the second molar, effectively pushing this tooth to tip mesially and become lodged below the distal bulbosity of the first molar (Figure 8.19c, d).

4. Mesial/distal root length differential: It has been noted that mesial angulation of what appears to have been a normally erupting second molar is seen frequently in these cases and it seems to occur quite late in its eventually unsuccessful eruptive progress. As this edition goes to press, an association has been recognized between the

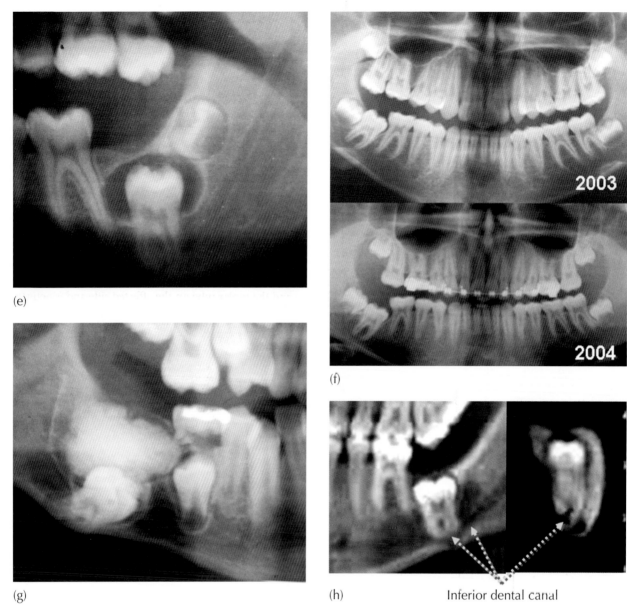

(e)

(f)

(g)

(h)   Inferior dental canal

Fig. 8.19 *(Continued)*

development of a mesial root which is shorter than the distal root. Similarly, distally angulated second molars are associated with a shorter distal root. Assuming that the two roots develop at the same time and at the same rate, it is reasonable to speculate that a longer mesial or distal root will continue to grow after the other has apexified, thereby contributing a factor for change in the orientation of the tooth in the opposite direction [38].

5. Heredity: This same study found an autosomal dominant hereditary factor, which was noted in comparing two different population samples [38].

There is almost certainly a considerable overlap in the five aetiological factors of this list so far and it may be that a given case could be associated with two or more of these

potential causes combined. This may make difficult the intellectual exercise of deciding which was the precipitating factor.

6. Enlarged follicle/dentigerous cyst: As we have noted in regard to maxillary canines (see Chapter 6), an enlarged follicle or dentigerous cyst will often be associated with non-eruption. This is because the intra-cyst pressure overcomes the natural eruption force of the tooth (Figure 8.19e).

7. Ankylosis: As with any other impacted tooth, this diagnosis is a rare and unlikely aetiological factor. Its presence will mean that orthodontic force to erupt the tooth will be unproductive and this may serve to confirm a tentative diagnosis (Figure 8.19f).

8. Other local pathology (Figure 8.19g).
9. Root entanglement with the inferior alveolar nerve (figure 8.19h).

Local treatment

If the third molar is implicated in the impaction, or if a mesial and superior positioning of the third molar secondarily prevents the eruption of the second molar, then extraction of the third molar is obligatory. Following this step and in the absence of further treatment, the second molar will frequently be seen to improve its position and its spontaneous eruption is certainly possible. On the other hand, this would pass up the opportunity to place an attachment on the second molar at the time of the third molar enucleation and mandate a second exposure if the tooth did not erupt.

Corrective orthodontic treatment is nevertheless often needed and, if it is to be performed on a local basis only in the absence of any other form of treatment for a concurrent malocclusion, then the first part of the treatment involves the surgical removal of the overlying mucosa, to expose the occlusal surface. It would be advantageous if the buccal aspect were also to be exposed, but this is not always possible due to the oblique line on the outer surface of the mandible that runs backward and upward from each mental tubercle. This oblique ridge reduces depth of the buccal sulcus, making it shallower as it proceeds past the molar region to become continuous with the anterior border of the ramus. Exposure of the buccal surface of an unerupted second molar therefore is difficult at best and frequently impossible. As the result, the only accessible surface may be the occlusal, for which a simple approach was described above in relation to the maxillary first molar, where a removable appliance was described and which may be employed here, with the same design and the same mechanical principles [37].

Alternative fixed appliance methods abound, but their design is too frequently restricted to just two or three teeth, which may seriously undermine the anchorage. Uprighting a second molar is notorious for causing the anchor teeth to move mesially or for the adjacent molar and premolars to be intruded and tipped buccally into a cross bite relationship with the uppers. Every effort should be made to counter these tendencies.

Ideally, teeth on both sides of the mandibular dental arch should be included in the anchorage unit [38] in order to minimize unwanted movement of other teeth in the arch, particularly to eliminate the possible occurrence of lower incisor crowding. A minimum suggested anchorage unit should include fixed bands on both first molars, joined by a soldered lingual arch, and orthodontic brackets on the premolars of the affected side. A useful alternative to the soldered lingual arch, particularly when there is bilateral impaction of the second molars, is to place a bonded multistrand wire retainer on the lingual side of the incisors and canines and extend it to the mesial occlusal pit of the first premolar. In this way, the anchor unit comprises a sectional arch on each side engaging first molars and both premolars, together with the bonded retainer joining first premolar to first premolar. This brings the 12 teeth anterior to the second molars into the anchor unit without the need for anterior brackets.

When the second molar is only mildly impacted and partially erupted beneath the distal bulbosity of the first molar, it is frequently possible to place a bonded tube or other attachment on its buccal surface and simply include it in the initial levelling arch that is used in the early stages of treatment. For the more severe cases, however, creative and problem-specific mechanisms will be needed to resolve the problem.

The active element may be constructed in one of many ways:

1. A free-sliding sectional wire is slotted into the brackets and the molar tube on the affected side and is activated by an elastic module, with the distal end of the sectional wire fashioned into a small hook, which latches onto an eyelet attachment on the buccal or occlusal surface of the impacted tooth. This will tip the molar distally until the buccal surface rises sufficiently for a buccal tube to be placed and the remainder of the treatment completed with a buccal aligning archwire.
2. A similar method can be employed using a compressed coil spring (Figure 8.20).
3. A rectangular section arch can be used, containing an expanded open loop which is tied into the brackets after being compressed against the attachment on the molar.
4. A large and stiff open loop of rectangular wire is placed in the distal end of a buccal tube or wide Siamese bracket on the molar and is designed to widely encircle the impacted tooth, with a small helix at its extremity. By tying a stainless steel ligature between a bonded eyelet or button on the occlusal of this tooth and the small terminal loop, distal pressure is brought to bear on the impacted molar [39].
5. A complete round wire loop with a distal helix may be slotted into buccal and lingual horizontal tubes on the molar band. Activation is made as for 4, by tying a steel ligature between the helix and an attachment on the tooth (Figure 8.21).

With a more deeply sited tooth, one which may be seen on a radiograph to be well down in the junction between body and ramus, two essential problems exist. First, access in this area is difficult to obtain in order that bonding an attachment may be undertaken, but, with good teamwork on the part of the surgeon and the orthodontist, it can usually be done. The more difficult part of the problem involves the ability to conjure up a biomechanical method for applying appropriately directed traction, together with a solid anchor base that will resist unwanted movement.

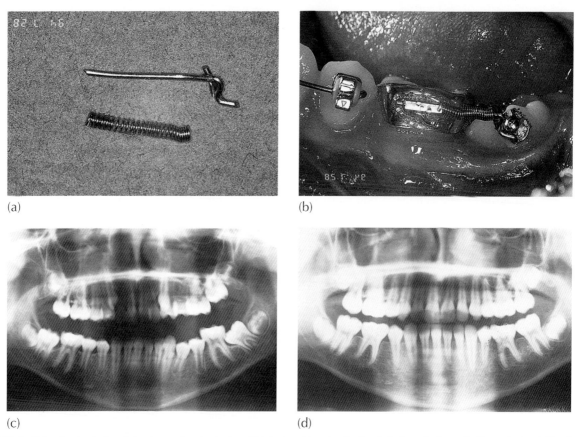

Fig. 8.20 (a, b) A coil spring is threaded onto a sectional archwire, which is slotted into the distal end of a buccal tube on the first molar band. The distal end of the wire carries a welded stop or cross-piece, which permits compression of the coil spring against a button or eyelet attachment on the second molar. A lingual arch and fully bracketed appliance is present for additional anchorage. (c, d) Panoramic views before and after treatment.

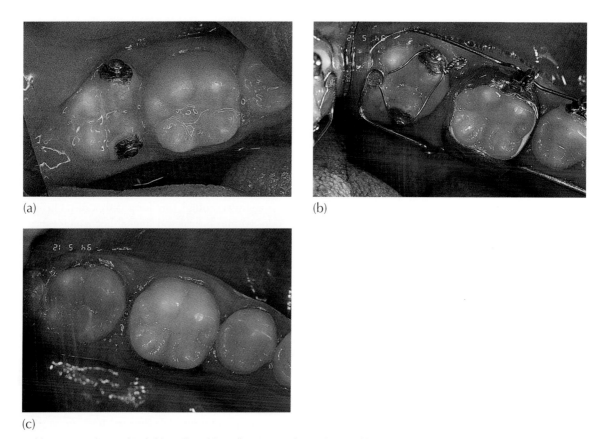

Fig. 8.21 (a) Button attachments bonded buccally and lingually to impacted second molar. (b) A wire loop carrying a distal helix is slotted into buccal and lingual tubes on the molar band. The wire loop is compressed by tying steel ligatures between the buttons and distal helix. (c) The final result.

Applying forces from adjacent teeth only will intrude these teeth and generate a cant in the occlusal plane. Furthermore, the use of loop mechanics built into the archwire, as the means of applying uprighting, levelling or aligning forces, is extremely limited because of the shallowness of the buccal sulcus in the second molar area. However, the lingual sulcus is very much deeper and, with a little ingenuity, it may be exploited to accommodate a suitably designed uprighting, levelling and/or aligning spring.

Applying forces from teeth in the maxillary arch will increase what is probably already a considerable over-eruption of the opposing molar, caused by the long-time absence of its antagonist. In the past, the only method was to use a high-pull headgear to maintain the vertical position of the molar while applying intermaxillary vertical traction to the impacted tooth (Figure 8.22).

Today, this would be approached using a temporary micro-screw in the maxilla or a zygomatic plate. The

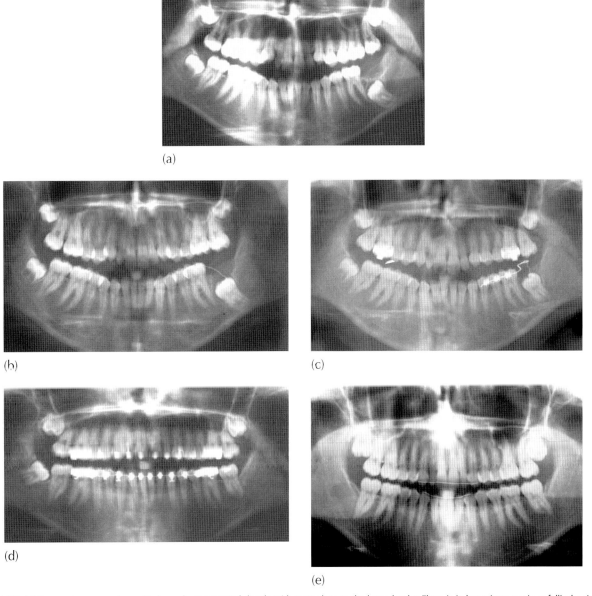

(a)

(b)

(c)

(d)

(e)

Fig. 8.22 (a) A second permanent mandibular molar is impacted deeply with roots close to the lower border. There is indeterminate cystic or follicular tissue superiorly placed to it, which appears to have been the causative factor. (b) Following removal of the abnormal tissue, an attachment has been bonded to the tooth, with a minimal closed surgical procedure and without the removal of overlying bone or follicular tissue. The bony crypt is largely intact and the pigtail ligature is seen to be temporarily secured to the first molar. (c) An *en bloc*, high-pull, integral and removable headgear plate (not seen in the film) is the source of anchorage for the vertical elastics that are joined between the upper molar band and the sectional traction mechanism on the lower left side. Bone can be seen to have regenerated superior to the crown of the impacted tooth, similar to that of a normally erupting tooth. (d) The tooth has erupted and full multi-bracketed appliances are in place to treat the remaining malocclusion. (e) The case five years after completion of treatment, with bonded maxillary and mandibular 3-3 retainers still in place.

advantages of a temporary anchorage device (TAD) include the fact that an orthodontic appliance in the affected jaw is, at least theoretically, unnecessary. However, this presupposes that the application of the vertical force will raise the tooth and align it in its ideal position (Figure 8.23). In practice, simple extrusion is rarely adequate because the tooth usually needs other force components to upright it. This means that an orthodontic appliance will be needed to complement the TAD in order to produce these aligning movements once the tooth is erupted and accessible for the placement of a suitable attachment. A second problem involves the patient's ability to attach an intermaxillary elastic ring between the bone anchor and the hooked ligature wire that has been tied into the eyelet on the impacted second molar.

The twisted ligature hook emanating from the impacted molar is not easily accessible for the patient to perform direct traction. It should be appreciated and accepted that not every patient is sufficiently dextrous or skilful at placing the elastic ring so far at the back of the mouth. It is usually considerably easier to teach the patient to apply vertical force from a convenient hook or bracket on one of the more mesial mandibular teeth to the TAD. This supports the anchorage, while active extrusion may be applied by the orthodontist direct from a suitably configured short rectangular sectional archwire to the pigtail hooks attached to the impacted molar.

Treatment as part of a comprehensive orthodontic treatment plan

For most cases, some form of overall malocclusion is present, for which treatment needs to be prescribed. The resolution of the impacted tooth should be integrated into the general treatment plan. If resolution of the overall malocclusion demands a reduction in the number of teeth, consideration should be given to the extraction of this impacted tooth, together with appropriate balancing and compensating extractions in the other quadrants of the mouth (Figure 8.24) [40–45]. Extraction of mandibular second molars is not a frequent extraction of choice, but under the circumstances of a very difficult impaction should be considered. This enables the dispersal of mild crowding to be effected with great facility in a distal direction, without the need for the extensive root uprighting movements that are seen with premolar extractions. However, the orthodontist will be counting on the favourable eruption of the third molars (which will be significantly earlier in these cases) and their spontaneous alignment. Should this not occur, a further period of treatment will need to be initiated, three or four years later in the young adult stage, aimed at uprighting the mesially tipped third molars from a position that may be reminiscent of the initial position of the extracted second molar! In this situation, the above procedures may be applied to

bring about third molar uprighting, in exactly the same way as with second molars.

Perhaps the best way to visualize the potential influence of the extraction of different groups of teeth is on the panoramic radiograph. With a mesially tipped second molar in an extraction case, the loss of a first or second premolar and the mesial movement of the first molar may often free the tooth to erupt, but the tipping will remain and, in a few cases, may even become worse, with the tooth appearing to 'fall flat on its face'. Uprighting will then need to be undertaken, using the existing multi-bracketed appliances.

Where the mandibular arch, mesial to the first molars, is well aligned or slightly crowded, but there is a mild class 2 relationship or crowding of the maxillary dentition, all the upper and the posterior lower teeth will probably need distal movement, rather than extractions made. An extraoral headgear will be needed to take the entire maxillary dentition posteriorly and to provide the space for the alignment of its individual teeth. Class 3 intermaxillary elastics, supported by the headgear, may then be used against a full lower fixed appliance, whose purpose is to align the teeth and move the mandibular first molars distally. The force will be transferred through interproximal contacts to the second molar and, if this is tipped less than approximately 20° and has a relatively high contact with the distal of the first molar crown, above the CEJ area, the tooth will ultimately tip distally and upright, while the first molar will itself be tipped distally. Once the second molars have erupted and the distalizing force is discontinued, the first molar will spontaneously upright.

Second molars that are tipped more than 20° and/or are located at or below the CEJ and the distal bulbosity of the first molar will not spontaneously upright in this way and will need to be treated more aggressively. This will necessitate surgical exposure and attachment bonding to the impacted tooth and the inclusion of this unit into a full mandibular fixed appliance. Initial levelling and aligning archwires extended to this tooth will usually suffice to bring it into alignment. In more extreme cases, the employment of a temporary implant should be considered, as detailed above.

Maxillary second molars

In the posterior region of the maxilla crowding will be seen in the tuberosity area. Unerupted second and third molars will be seen to be 'stacked' almost vertically above the erupted first molar and above one another, rather than in a more horizontal, distally directed arrangement. Their root apices are mesial and their crown point distal, usually with a buccal component – clearly due to a basal shortness of the maxilla, with disproportionately large teeth.

In this situation, the second molars will sometimes fail to erupt, which may delay the completion of orthodontic treatment. Provided there is no other obstacle most of these teeth will spontaneously erupt after the covering and

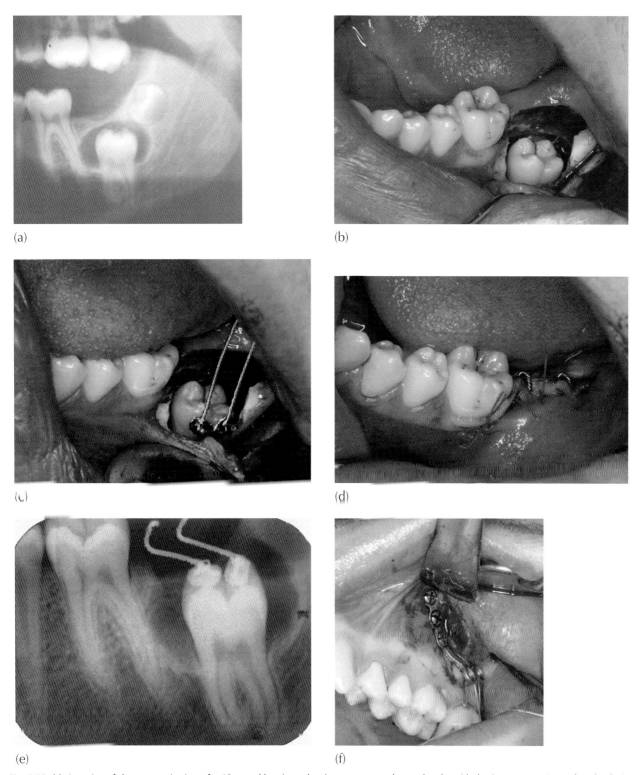

(a)　(b)

(c)　(d)

(e)　(f)

Fig. 8.23 (a) A section of the panoramic view of a 19-year-old patient, showing an unerupted second molar with dentigerous cyst. A very late-developing third molar is present and the maxillary second molar is over-erupted.

(b) The tooth is exposed to show the size of the cyst. The third molar was removed at the same visit.

(c) Two attachments bonded since access is good and a large surface is available.

(d) The exposure site is fully sutured closed with the two pigtail ligatures fashioned into hooks for elastic traction.

(e) Post-surgical periapical radiograph.

(f) The zygomatic plate is placed under local anaesthetic cover.

(g) A palatal screw temporary anchorage device is placed.

(h) An elastic chain is drawn between the palatal TAD to the zygomatic plate, across the occlusal surface of the over-erupted tooth and secured by a bonded button on the occlusal surface. An inter-maxillary elastic is placed by the patient between the zygomatic plate and the molar pigtail ligature hooks.

(i) At the end of treatment, lasting eight months, the maxillary second molar is well intruded and the mandibular second molar has been erupted to the occlusal plane.

(j) Periapical view of the mandibular second molar post-treatment.

(k) Section of the post-treatment panoramic view to show maxillary and mandibular molars of affected side.

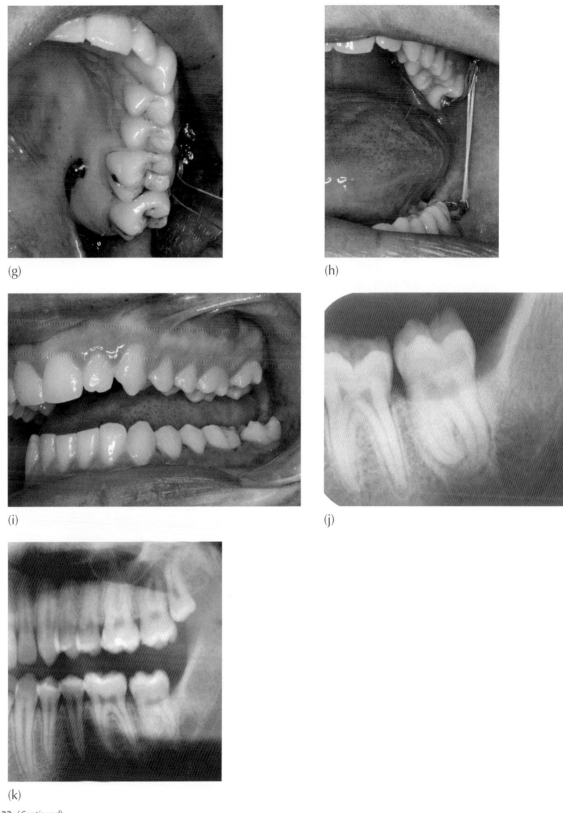

(g)

(h)

(i)

(j)

(k)

Fig. 8.23 (*Continued*)

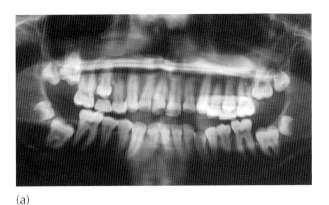

(a)

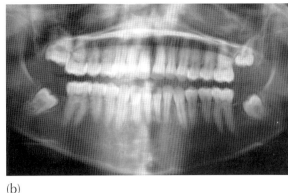

(b)

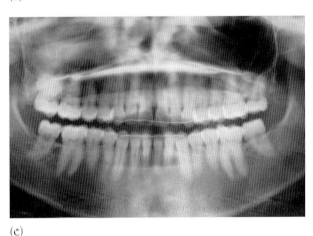

(c)

Fig. 8.24 (a) Age 13.2 years all four second molars are in abnormal positions in a class 2, division 1 extraction case. Extraction of three second molars and right maxillary third molar performed. (b) Age 16.4 years follow-up radiograph at the end of orthodontic treatment, to assess the third molars (c) Age 19.3 years the three-year post-treatment follow-up panoramic view shows the third molars in ideal alignment and fully juxtaposed with the first molars. Unfortunately, there are few reliable keys or guidelines by which this excellent result could have been predicted at the outset.

usually thickened mucosa is cut away. They will, however, often erupt buccal to the line of the arch into a buccal cross-bite relationship with the lower second molar and will thus need to be aligned as a last stage in the overall orthodontic treatment.

The occasional second molar may have a mesial tip with its mesial curvature locked in the distal concavity of the first molar, just beneath the surface. Following exposure, this may sometimes need the placement of a simple elastic separator ring for a few days to help, but these teeth usually come in well.

There is one final condition that is seen on the panoramic radiograph that causes the maxillary second molar to become impacted. This is related to an anomaly in the development of the adjacent third molar. This tooth is characterized by having a banana-shaped crown, little or no root development and it is located inferiorly to the crown of the second molar, thereby acting as an obstruction to the latter's eruption (Figure 8.25). In fact, the tooth is probably both inferior and palatal to the second molar, but nevertheless obstructing its path of eruption. Identifying this tooth at the time of exposure and removing it is disconcerting for the surgeon,

because of its unexpected position occlusal and lingual to the second molar. Nevertheless, following its extraction, the second molar will erupt very rapidly (Figure 8.23).

Mandibular third molars

In the past, the 'prophylactic' removal of third molars was carried out almost blindly and, in many cases, in the belief that this would prevent secondary incisor crowding in the post-retention period [46–48]. Today, there are positive indications that an effort should be made to bring these teeth into the dental arch and make them useful functional dental units. This is particularly pertinent in the latter stages of the treatment of a premolar extraction case when excess space is closed by drawing the first and second molars mesially. This potentially opens up space at the distal end of the arch, which may sometimes be adequate to accommodate an erstwhile impacted third molar. In these instances, successful conservative treatment of the impaction, by orthodontic alignment, provides the maxillary third molar with an antagonist. This in turn allows both teeth to assume the same functional rating as any other tooth in the dentition.

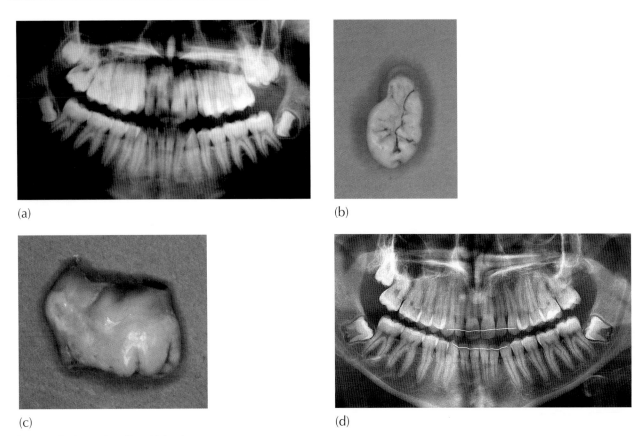

(a)

(b)

(c)

(d)

Fig. 8.25 The 'banana' maxillary third molar.
(a) In the panoramic view of a 13-year-old female, the left maxillary third molar is superimposed on the impacted second molar and is more occlusally located. The other second molars have erupted.
(b, c) The extracted 'banana' molar from the occlusal and buccal aspects.
(d) The second molar erupted autonomously four months after the extraction. This panoramic view was taken many months later, at the completion of the orthodontic treatment.

The decision to upright a mesially inclined and impacted third molar must be made taking into consideration its prospective final position in relation to the ascending ramus of the mandible and its thick soft tissue investment. Should the final position of the tooth be at the expense of a localized resorption of the anterior border of the ramus or in the exuberant and inflamed soft tissue covering it, then the effort will have been wasted. Given adequate space at the end of the arch, however, there are often good indications for this treatment.

Treatment is very similar to that described for mandibular second molars above. Distally inclined and impacted third molars are much more of a problem, since control of root movement is poor at the end of the arch and extraction is often to be advised.

Impaction and crown resorption

In teeth that have remained unerupted for many years, the outer enamel epithelium of the dental follicle surrounding the crown of the tooth may occasionally break down and direct contact between bone and enamel occurs. In time,

this may lead to resorption of the enamel, often with the laying down of bone in the resorption lacunae. Any tooth may be affected, although it tends to occur in much older individuals whose dentition is established, and the unerupted tooth may have meanwhile been replaced prosthetically. Rarely, it may occur in a young person, for whom there is a relatively greater importance to bringing the tooth into the dental arch. The first clue of the existence of this pathological process is seen radiographically, with difficulty in distinguishing the outline of the dental follicle. This usually signifies the presence of a replacement resorption process of the enamel, with bone substituting for the resorbed enamel. In its more advanced stages, the radiographic appearance of the tooth shows a loss of the sharp outline of the crown and, later still, a reduction in the radiopacity of the crown, associated with a steady decalcification of the enamel [49, 50] and its replacement with trabecular bone (Figure 8.26).

In order to move this tooth orthodontically, the entire crown area must be dissected free of the bone and a pack inserted to prevent the healing bone from again coming into contact with the enamel surface. A preformed crown

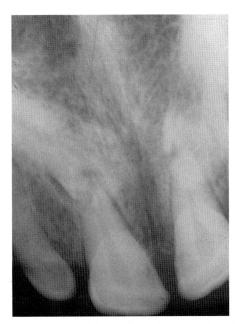

Fig. 8.26 A periapical view of the maxillary incisor region of an elderly lady, to show two impacted canines in an advanced stage of replacement resorption. The follicle is still visible in a localized area of the distal aspect of the right canine while elsewhere it is absent. The general vague outline of the two teeth may be seen, but the texture of each has given way to bony trabeculation.

would be a better alternative, but access to the tooth is too compromised to allow its proper adaptation and cementation. At the same time, orthodontic force should be applied to the tooth and its activity maintained by frequent re-ligation. Success is far from guaranteed, because it may be difficult to fully separate the crown from the tissues and there may be secondary deterioration of tissues surrounding the root, as the tooth will also have been impacted for so many years.

This is a very different entity from invasive coronal resorption, which was described at the end of Chapter 7. The tooth that has undergone invasive coronal resorption appears to have a normal dental follicle, a normal periodontal ligament and, while the initiation of the resorption is through a fault in the enamel itself, the resorption mushrooms out in the dentine and only affects the enamel last. There is no substitution of the resorbed tooth substance with bone, but rather with soft and vascular resorptive tissue. Consequently, there is no problem with orthodontic alignment provided there is an adequate area of residual enamel or new restorative material on which to bond an attachment.

Infra-occlusion of permanent teeth

There is considerable confusion in terminology in relation to teeth that do not erupt to reach the occlusal plane and different designations are often applied to the same condition. Some authorities use a specific term to describe a very specific condition, while others will use the same term indicating a much more general connotation. There are also language differences which, when translated literally, alter their meaning and an author's geographical location may determine the particular word used in a journal report.

Thus, clinicians and researchers from the Netherlands and Denmark refer to the condition in which there is arrested eruption occurring before the tooth penetrates the oral mucosa, as 'primary retention' [51, 52]. Elsewhere, the same condition will be classified under 'impacted' or, specifically in the French- and Italian-speaking world, as 'included'. 'Secondary retention' in these same countries will mean cessation of eruption of a tooth after emergence without a physical barrier or ectopic position of the tooth, due to ankylosis [53, 54] which has been termed submergence or infra-occlusion elsewhere. The only recognizable similarity between the two forms of retention is that they both largely affect molars. These researchers restrict their definition of 'impaction' to mean a cessation of the eruption of a tooth caused by a clinically or radiographically detectable physiological barrier in the eruption path, or due to an abnormal position of the unerupted tooth.

Primary retention is more frequently seen than secondary retention and there is an equal frequency of occurrence between the sexes. Primary retention occurs more often in the maxilla than in the mandible, while the opposite is seen in secondary retention. Primary retention may be seen in dentitions that have normal development of the second molar, as well as in dentitions with abnormal eruption patterns in the second molar region (late development and second molar agenesis), with or without generalized malocclusion. Secondary retention can occur in dentitions with normal development of the second molar independent of the presence or absence of generalized malocclusion. Furthermore, primary retention is an acquired condition, judging by phenotypic differences between the primary and secondary retention, in that variations of normal eruption of the second molar are seen in primary retention [52].

Over the past two decades or so, the Danish research group has come to implicate a defective nerve supply of the tooth as possibly instrumental in causing primary and second retention. Innervation of the dental follicle appears to be an important factor for the continued development of the teeth [55] and different groups of teeth have different sources of innervation, with one source covering a specific field. It is pointed out that neural pathways can spread viral or bacterial infections, particularly Herpes zoster and the mumps virus, but also in other conditions. Children suffer many undiagnosed acute infections in their younger years and neural spread would seem to be a possible cause of a compromised nerve supply, with consequent effects on the teeth [56].

Unilateral occurrence of the primary or secondary retention phenomena suggests infection as the cause, while bilateral occurrence suggests a genetic aetiology, either directly

on the eruption mechanism or indirectly by affecting the nerve supply.

The term 'primary failure of eruption' (PFE) was coined in 1981 by Proffit and Vig [57] and suggests a disturbance of the eruption mechanism that caused the complete or partial failure of a non-ankylosed tooth to erupt, although the molecular basis for this failure was unknown (Figure 8.27). This phenomenon is characterized by the following:

1. The posterior teeth are more frequently affected, that is, the first and second molars are more frequently affected than premolars and canines.
2. If a tooth in a further anterior position presents an eruption disturbance, the posterior teeth are usually affected as well.
3. The affected teeth resorb the alveolar bone above the crown, but erupt only partially or fail to erupt.
4. Both deciduous and permanent teeth can be affected.
5. The condition is usually asymmetrical.
6. Primarily non-ankylosed teeth tend to become ankylosed as soon as orthodontic forces are applied.

Figure 8.28 shows panoramic radiographs of three children in one family, together with those of the father and the paternal grandmother, to show the differing familial expression of the phenomenon for each of these affected persons.

Research on the molecular basis for PFE has recently placed the aetiological blame for PFE on mutations in the parathyroid hormone receptor 1(PTH1R) gene, which will provide the key to distinguish PFE from isolated ankylosis and secondary retention [58–62].

Differentiating the aetiology is of importance because it gives the clinician the ability to choose appropriate treatment. Teeth suffering from secondary retention, ankylosis and PFE will not be amenable to orthodontic treatment, but in PFE only the adjacent under-erupted teeth will also not respond to orthodontic traction, whereas the teeth adjacent to an isolated under-erupted tooth will. Nevertheless, in the clinical context at the time of writing, the time has yet to come when the diagnosis will be routinely determined by an appropriate genetic test.

With continued vertical growth of the adjacent teeth and alveolar bone, ankylosed teeth become relatively lower and lower in the alveolus, until the gingival tissue grows over the tooth and they may even sometimes become covered by bone. Given that these teeth cannot be orthodontically moved, their 'absolute' anchorage potential may be exploited to orthodontically alter the position of neighbouring teeth in much the same way as any other ankylosed tooth, or an osseo-integrated implant may be used. Having no effect on the infra-occluded teeth, orthodontic extrusive forces will tend to intrude the adjacent teeth towards the level of the infra-occluded teeth.

Nevertheless, it is not easy to give up on a tooth, particularly a single first permanent molar, in otherwise ideal circumstances [63]. In these situations and under local anaesthetic, an attempt may be made to forcibly erupt an infra-occluded tooth following surgical luxation. Adequate space must be made for it and the tooth is held in extraction forceps and a rocking force applied aimed at fracturing the ankylotic junction and achieving a considerable degree of mobility of the tooth (Figure 8.29). This mobility will be achieved easily only when the area of ankylosis is small. In the presence of a wider area of ankylosis, the application of force is much more likely to cause root fracture. Accordingly, a good quality periapical radiograph should be carefully studied for signs and extent of any abnormality, ahead of time, since in the vast majority of cases of ankylosis the lesion and its location are radiographically undetectable.

Healing is rapid and generally quite painless, but the ankylosis will soon become re-established if delay is incurred before traction is applied. Therefore, a force level must be maintained, not permitted to lessen and renewed on a regular (ideally weekly) basis. This means that the timing for this treatment must not span a period when the patient or the practitioner will be on vacation for any length of time. Should it happen that the patient is not seen for a few weeks and the force value drops, then re-ankylosis will occur and the tooth will resist further movement. In this event, a second surgical subluxation will be required to remobilize the tooth. Attempts at orthodontic correction are, therefore, frequently disappointing.

Nevertheless, and despite the advice of Proffit and Vig, the parent and the orthodontist may not be prepared to accept the 'no treatment' verdict for a single affected tooth in a young patient. If the parents understand the poor prognosis, treatment may be attempted employing the partial surgical luxation technique referred to above, provided that attendance for frequent activation is assured. Beyond the treatment of single accessible teeth, attempts to resolve the impaction of multiple teeth whose roots have undergone ankylosis will generally fail (Figure 8.30).

If strictly dental anchorage is used and the anchor teeth are in the same jaw as the infra-occluded tooth, the reactive force will likely lead to their intrusion. If intermaxillary forces are used, then the anchor teeth in the opposing jaw may become over-erupted and a canting of the occlusal plane may well occur. Skeletal anchorage, using a temporary screw implant or bone plate, should always be considered in these cases in order to avoid deleterious effects on the remainder of the dentition. This type of movement is more a microcosm of distraction osteogenesis than orthodontic movement *per se* and as such requires heavier and continuous forces to be applied.

In many of these cases, the use of an orthodontic implant as the platform from which forces are applied, particularly a mini-plate secured with screws to the zygomatic process of the maxilla, or titanium mini-screws, may be very advantageous, because this source of anchorage will have no ill-effects on the dentition as a whole. Accordingly, there is a

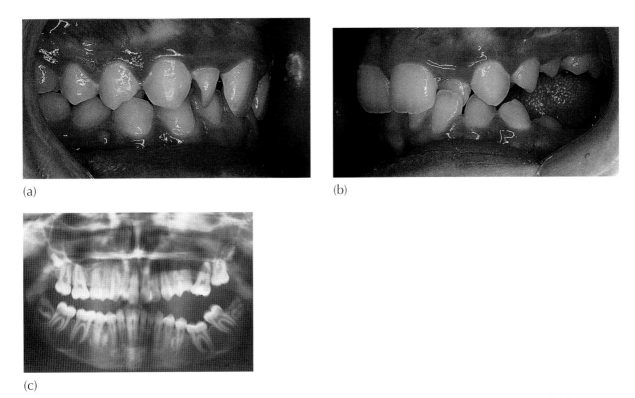

(a)

(b)

(c)

Fig. 8.27 (a) A normal occlusion of the posterior teeth is present on the right side. (b) A severe lateral open bite has developed on the left side due to primary failure of eruption. (c) The panoramic radiograph shows very marked infra occlusion of the left mandibular first molar and, to a lesser extent, also of the left maxillary second premolar and first molar.

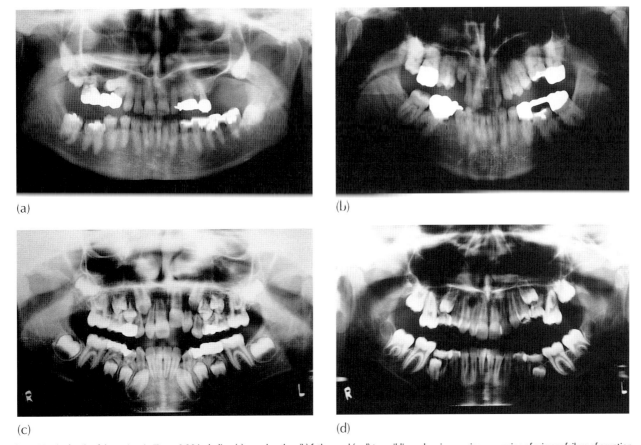

(a)

(b)

(c)

(d)

Fig. 8.28 The family of the patient in Figure 8.28 including (a) grandmother, (b) father, and (c, d) two siblings showing varying expression of primary failure of eruption.

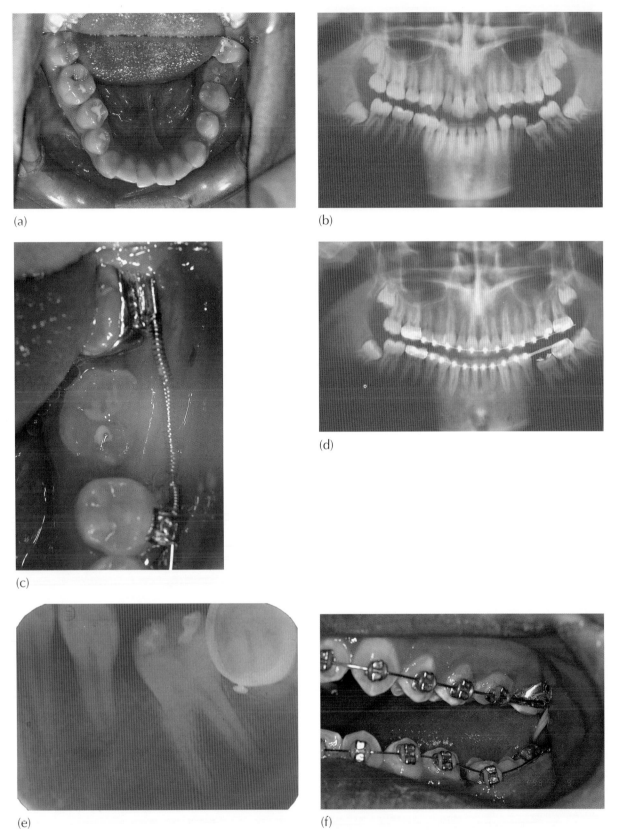

Fig. 8.29 (a) An occlusal view of the mandibular dentition to show the severely infra-occluded left first molar, with tipping of the adjacent teeth. (b) Panoramic view of the initially diagnosed condition. (c) A clinical occlusal view of the achieved space opening. (d) Panoramic view of the achieved space opening. (e) Periapical view of the infra-occluded molar after bonding of two eyelets to the available occlusal surface. (f) Eight weeks after partial surgical luxation the tooth has erupted sufficiently and a regular bracket is bonded to the now available buccal surface. (g, h) Post-treatment panoramic and periapical views after extrusion, alignment and mesial root uprighting. (i, j) Clinical views of the treated tooth at two months post-treatment. (k–m) The occlusion at two months post-treatment.

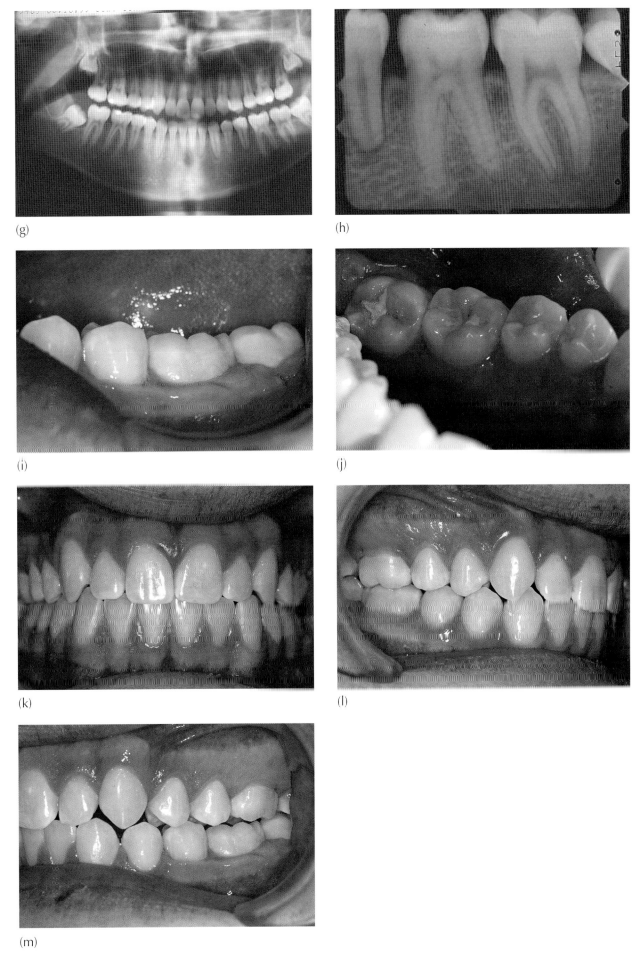

(g)

(h)

(i)

(j)

(k)

(l)

(m)

Fig. 8.29 *(Continued)*

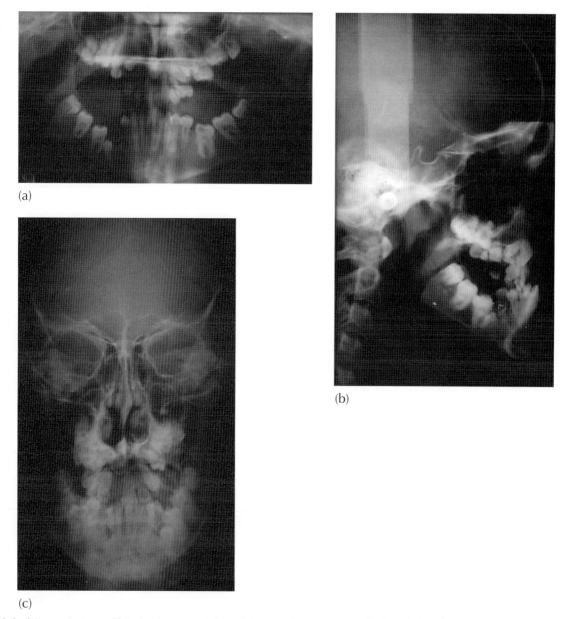

(a)

(b)

(c)

Fig.8.30 (a–c) Panoramic view, and lateral and postero-anterior cephalograms of an extreme case of primary failure of eruption.

case to be made for the use of implant anchorage in most cases where ankylotic teeth are treated in this way.

On the basis of the above argument, it is logical to expect that, once successful treatment has been completed in the younger patient and the infra-occluded tooth is brought to the occlusal level, a renewed ankylosis will occur and the treated tooth will steadily and progressively infra-occlude again. Anecdotally, in several of the cases treated by the author, this has not happened, encouraging the belief that perhaps re-ankylosis does not always result and the tooth maintains its occlusal height over the long term. Given this favourable outcome, it should become possible to subsequently include orthodontic movement of this tooth as part of an overall orthodontic treatment plan for a given patient.

The trend in orthodontics today is to attempt to incorporate all malrelations of the jaws, occlusal interferences, individual tooth malalignments and space problems in one all-encompassing episode of treatment. By doing so, treatment is streamlined in terms of appliance efficiency, quality of results and overall treatment duration. Controlled studies, from the Universities of South Carolina [64], Washington [65], Michigan [66], Florida [67] and Oxford [68], have indicated that there is little long-term advantage to be gained from treating a class 2 case in a two-phase treatment.

Provided the optimum timing for the treatment of the impacted tooth coincides with the establishment of the full permanent dentition, therefore, it makes sense to integrate

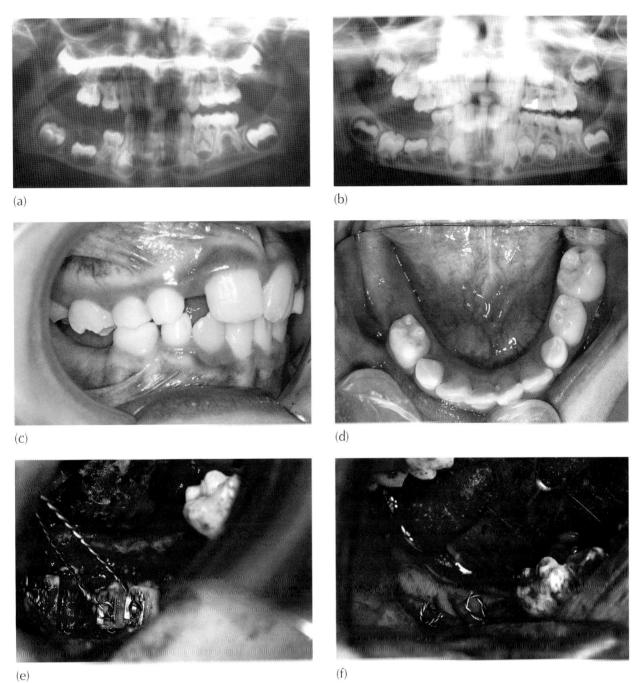

Fig. 8.31 (a) This 4-year-old child was seen in the Department of Oral Surgery with a cystic lesion around the right deciduous second molar, whose development appears to have become arrested at an early stage. The adjacent deciduous first molar is infra-occluded. There appears to be a similar disturbance of the follicle of the first permanent molar. Surgical removal of the cystic lesion was undertaken and the pathological report yielded a diagnosis of ameloblastic fibro-odontoma. No teeth were extracted at that time. (b) Two years later healing had occurred and there were no signs of recurrence. The first deciduous molar had erupted normally and both the second deciduous molar and first permanent molar showed the beginnings of root development. However, they showed no signs of eruptive movement, in contrast to the other molars. The second premolar had failed to develop. (c, d) Right lateral and occlusal clinical views of the edentulous ridge immediately before surgery at 8 years of age. (e) Two eyelets are bonded to the buccal surface of the molar. (f) A closed eruption procedure leaves only the two pigtail ligatures exposed. (g) The zygomatic implant is screwed into place. A pigtail ligature will be placed here too. (h) Vertically extrusive latex elastics are placed by the parent and subsequently the child between the hooked pigtail ligatures. (i) The panoramic view at this stage. Note how the developing root and the lack of eruption have caused an alteration in the shape of the lower border of the mandible. (j, k) The tooth responded slowly to the force, showing early signs of erupting into the mouth within about half a year. (l, m) Full eruption was achieved in a further nine months. Treatment was stopped at this point. The eyelets were detached and, under local anaesthetic, the zygomatic implant was removed. (n) Initiation of orthodontic treatment with fixed appliances for the first time. (o-q) The final result of the orthodontic treatment. The first mandibular right premolar is located in the place of the second premolar. The patient will be retained until an implant may be placed. (r) Post-treatment panoramic radiograph showing normal root development of the previously affected teeth and adequate space preparation for a future implant to replace the missing premolar.

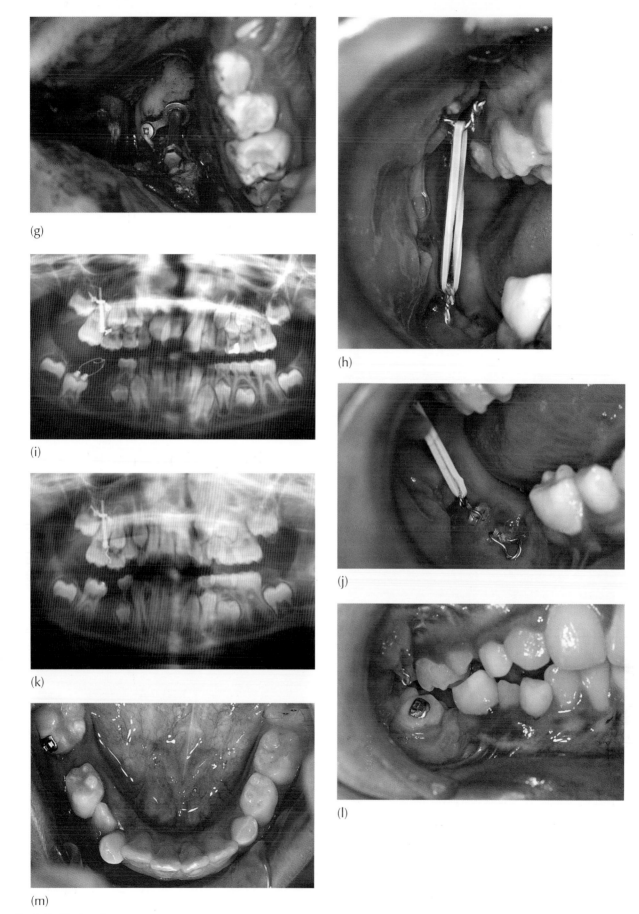

(g)

(h)

(i)

(j)

(k)

(l)

(m)

Fig. 8.31 (*Continued*)

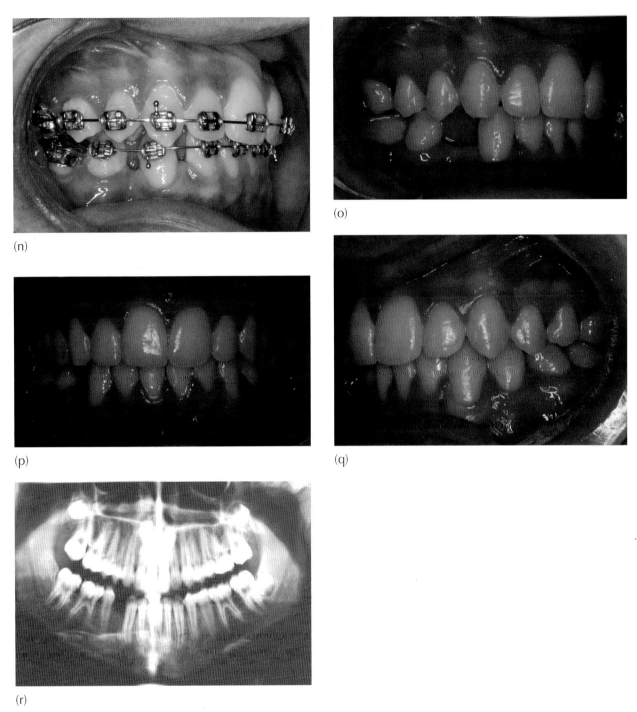

(n)

(o)

(p)

(q)

(r)

Fig. 8.31 (*Continued*)

it into the overall treatment plan. While this is fine for the maxillary canine, which is often among the last permanent teeth to make up the full complement of erupted permanent teeth (apart from the third molars), this would be totally inappropriate for a maxillary central incisor or for the permanent molars, which make up a small but significant number of affected cases.

Thus, due to developments in the area of implantology, one of the more valuable approaches that now presents itself is the ability to divorce the treatment of an impacted tooth from the treatment of a malocclusion. Using only an implant in the opposite jaw and an elastic band, it is now possible to resolve an impaction and bring about full eruption of a tooth without the use of an orthodontic appliance and without causing ill-effects either to the adjacent teeth or to the teeth in the opposing jaw. Being no longer dependent on the presence of other teeth to influence the decision, the practitioner may choose the optimum timing for treatment of the impaction, based solely and objectively on considerations related to the impacted tooth itself, without reference to the other teeth, which may then be treated at a different time (Figure 8.31).

Iatrogenics

Aside from the naturally occurring ankylosis, there may be reason to suspect the possibility of an iatrogenic variety. It was pointed out in Chapter 3 that it is not uncommon to find an oral surgeon widely exposing the crown of the impacted tooth down to the CEJ and beyond, with consequent instrumentation of the cementum of the root surface of the tooth and a drying out of the PDL. We have also discussed the accidental spillage of liquid etchant over the root surface during the bonding of an attachment. Surgical trauma, desiccation and chemical burns are potent causes for both ankylosis and invasive cervical root resorption and this would never be discovered until many months later, when attempts at moving the tooth have failed.

Aggressive surgical practices or careless bonding procedures have no place in the treatment of impacted teeth, since these will undermine the highly sensitive eruptive mechanism of the tooth from which it may not recover. The best way to reduce the chances of morbidity is for the orthodontist to be present and an active participant in the critical surgical episode.

References

1. Kokich VG. Surgical and orthodontic management of impacted maxillary canines. Am J Orthod Dentofacial Orthop 2004; 126: 278–283.
2. Nordenram A, Stromberg C. Positional variations of the impacted upper canine: a clinical and radiologic study. Oral Surg Oral Med Oral Pathol 1966; 22: 711–714.
3. Fournier A, Turcotte JY, Bernard C. Orthodontic considerations in the treatment of maxillary impacted canines. Am J Orthod 1982; 81: 236–239.
4. Oliver RG, Mannion JE, Robinson JM. Morphology of the maxillary lateral incisor in cases of unilateral impaction of the maxillary canine. Br J Orthod 1989; 16: 9–16.
5. Jacoby H. The etiology of maxillary canine impactions. Am J Orthod 1983; 84: 125–132.
6. Zilberman Y, Cohen B, Becker A. Familial trends in palatal canines, anomalous lateral incisors, and related phenomena. Eur J Orthod 1990; 12: 135–139.
7. Peck S, Peck L, Kataja M. Prevalence of tooth agenesis and peg-shaped maxillary lateral incisor associated with palatally displaced canine (PDC) anomaly. Am J Orthod Dentofacial Orthop 1996; 110: 441–443.
8. Becker A. Etiology of maxillary canine impactions. Am J Orthod 1984; 86: 437–438.
9. Chaushu S, Sharabi S, Becker A. Tooth size in dentitions with buccal canine ectopia. Eur J Orthod 2003; 25: 485–491.
10. Becker A, Chaushu S. Dental age in maxillary canine ectopia. Am J Orthod Dentofacial Orthop 2000; 117: 657–662.
11. Chaushu S, Sharabi S, Becker A. Dental morphologic characteristics of normal versus delayed developing dentitions with palatally displaced canines. Am J Orthod Dentofacial Orthop 2002; 121: 339–346.
12. Chaushu S, Bongart M, Aksoy A, Ben Bassat Y, Becker A buccal ectopia of maxillary canines in the absence of crowding. Am J Orthod Dentofacial Orthop, 2009; 136: 218–223.
13. Chaushu S, Chaushu G, Becker A. The use of panoramic radiographs to localize maxillary palatal canines. Oral Surg Oral Med Oral Pathol Oral Radiol Endod 1999; 88: 511–516.
14. Chaushu S, Chaushu G, Becker A. Reliability of a method for the localization of displaced maxillary canines using a single panoramic radiograph. Clin Orthod Res 1999; 2: 194–199.
15. Vanarsdall RL, Corn H. Soft tissue management of labially positioned unerupted teeth. Am J Orthod 1977; 72: 53–61.
16. Vanarsdall RJ. Periodontal/orthodontic interrelationships. In Graber TM, Vanarsdall RL, eds. Orthodontics. Current Principles and Techniques, 4th edn. St Louis, MO: Mosby, 2000: 801–838.
17. Kohavi D, Zilberman Y, Becker A. Periodontal status following the alignment of buccally ectopic maxillary canine teeth. Am J Orthod 1984; 85: 78–82.
18. Kuftinec MM, Shapira Y, Nahlieli O. A case report. Bilateral transmigration of impacted mandibular canines. J Am Dent Assoc 1995; 126: 1022–1024.
19. Shapira Y. Bilateral transposition of mandibular canines and lateral incisors: orthodontic management of a case. Br J Orthod 1978; 5: 207–209.
20. Joshi MR. Transmigrant mandibular canines: a record of 28 cases and a retrospective review of the literature. Angle Orthod 2001; 71: 12–22.
21. Mupparapu M. Patterns of intra-osseous transmigration and ectopic eruption of mandibular canines: review of literature and report of nine additional cases. Dentomaxillofac Radiol 2002; 31: 355–360.
22. Peck S. On the phenomenon of intraosseous migration of nonerupting teeth. Am J Orthod Dentofacial Orthop 1998; 113: 515–517.
23. Nodine, A.M. Aberrant teeth, their history, causes and treatment. Dent Items of Interest 1943. 65: 440–451.
24. Shapira Y. Transposition of canines. J Am Dent Assoc.1980; 100: 710–712.
25. Shapira Y, Kuftinec MM. Orthodontic management of mandibular canine–incisor transposition. Am J Orthod 1983; 83: 271–276.
26. Brezniak N, ben-Yehuda A, Shapira Y. Unusual mandibular canine transposition: a case report. Am J Orthod Dentofacial Orthop 1993; 104: 91–94.
27. Shapira Y, Kuftinec MM. Intraosseous transmigration of mandibular canines – review of the literature and treatment options. Compend Contin Educ Dent 1995; 16: 1014, 1018–1020, 1022–1024.
28. Shalish M, Peck S, Wasserstein A, Peck L. Malposition of unerupted mandibular second premolar associated with agenesis of its antimere. Am J Orthod Dentofacial Orthop. 2001; 121: 53–55.
29. Shalish M, Chaushu S, Wasserstein A. Malposition of unerupted mandibular second premolar in children with palatally displaced canines. Angle Orthod. 2009; 79: 796–799.
30. Kurol J. Infra-occlusion of primary molars. An epidemiological, familial, longitudinal, clinical and histological study. Swed Dent J Suppl 1984; 21: 1–67.
31. Kurol J, Thilander B. Infra-occlusion of primary molars and the effect on occlusal development, a longitudinal study. Eur J Orthod 1984; 6: 277–293.

32. Becker A, Shochat S. Submergence of a deciduous tooth, its ramifications on the dentition and treatment of the resulting malocclusion. Am J Orthod 1982; 81: 240–244.

33. Becker A, Karnei-R'em RM. The effects of infra-occlusion: part 1 – tilting of the adjacent teeth and space loss. Am J Orthod 1992; 102: 257–264.

34. Becker A, Karnei-R'em RM. The effects of infra-occlusion: part 2 – the type of movement of the adjacent teeth and their vertical development. Am J Orthod 1992; 102: 302–309.

35. Becker A, Karnei-R'em RM, Steigman S. The effects of infra-occlusion: part 3 – dental arch length and the midline. Am J Orthod 1992; 201: 427–433.

36. Proffit WR. Contemporary Orthodontics. St Louis, MO: Mosby Year Book, 1992.

37. Becker A. The correction of mesially angulated semi-impacted molar teeth by simple orthodontic means. Isr J Dent Med 1977; 2: 17–22.

38. Shapira Y, Finkelstein T, Shpack N, Lai YH, Kuftinec MM, Vardimon A. Mandibular second molar impaction – Part 1: Genetic traits and characteristics. Am J Orthod Dentofacial Orthop 2011; 140: 32–37.

39. Majourau A, Norton LA. Uprighting impacted second molars with segmented springs. Am J Orthod Dentofacial Orthop 1995; 107: 235–238.

40. Wilson HE. The extraction of second permanent molars as a therapeutic measure. Trans Eur Orthod Soc 1966; 42: 141–145.

41. Richardson ME, Richardson A. Lower third molar development subsequent to second molar extraction. Am J Orthod Dentofacial Orthop 1993; 104: 566–574.

42. Cavanaugh JJ. Third molar changes following second molar extraction. Angle Orthod 1985; 55: 70–76.

43. Gaumond G. Second molar germectomy and third molar eruption. Angle Orthod 1985; 55: 788.

44. Gooris CGM, Artun J, Joondeph DR. Eruption of mandibular third molars after second molar extractions: a radiographic study. Am J Orthod Dentofacial Orthop 1990; 98: 161–167.

45. Staggers JA. A comparison of results of second molar and first premolar extraction treatment. Am J Orthod Dentofacial Orthop 1990; 98: 430–436.

46. Richardson ME. The role of the third molar as the cause of late lower arch crowding: a review. Am J Orthod Dentofacial Orthop 1989; 95: 79–83.

47. Southard TE. Third molars and incisor crowding: when removal is unwarranted. J Am Dent Assoc 1992; 123: 75–79.

48. Zachrisson BU. Mandibular third molars and late lower arch crowding – the evidence base. World J Orthod 2005; 6: 180–186.

49. Blackwood HJJ. Resorption of enamel and dentine in the unerupted tooth. Oral Surg Oral Med Oral Pathol 1958; 11: 79–85.

50. Azaz B, Shteyer A. Resorption of the crown in impacted maxillary canine. A clinical, radiographic and histologic study. Int J Oral Surg 1978; 7: 167–171.

51. Raghoebar GM, Boering G, Vissink A, et al. Eruption disturbances of permanent molars: a review. J Oral Pathol Med 1991, 20: 159–166.

52. Kjaer I. Phenotypic classification of 90 dentitions with arrested eruption of first permanent mandibular or maxillary molars. Seminars in Orthodontics, 201; 16: 172–179.

53. Raghoebar GM, Boering G, Jansen HWB, et al. Secondary retention of permanent molars: a histologic study. J Oral Pathol Med 1989; 18: 427–431.

54. Raghoebar GM, Boering G Vissink A, Stegenga B. Eruption disturbances of permanent molars: a review. Journal of Oral Pathology and Medicine, 1991; 20: 159–166.

55. Christensen LR, Janas MS, Mollgaard K, Kjaer I. An immunocytochemical study of the innervation of developing human fetal teeth using protein gene product 9.5 (PGP9.5). Archives of Oral Biology 1993; 38: 1113–1120.

56. Becktor KB, Bangstrup MI, Rolling S, Kjaer I. Unilateral primary or secondary retention of permanent teeth and dental malformations. Eur. J Orthod 2002; 24: 205–214.

57. Proffit WR, Vig KWL. Primary failure of eruption: a possible cause of posterior open bite. Am J Orthod 1981; 80: 173–190.

58. Frazier-Bowers SA, Koehler KE, Ackerman JL, et al. Primary failure of eruption: further characterization of a rare eruption disorder. Am J Orthod Dentofacial Orthop 2007; 131: 578.e1–11.

59. Decker E, Stellzig-Eisenhauer A, Fiebig BS, et al. *PTHR1* loss-of-function mutations in familial, nonsyndromic primary failure of tooth eruption. Am J Hum Genet 2008; 83: 781–786.

60. Frazier-Bowers SA, Simmons D, Wright JT, Proffit WR, Ackerman JL. Primary failure of eruption and PTH1R: The importance of a genetic diagnosis for orthodontic treatment planning. Am J Orthod Dentofacial Orthop 2010; 127: 160.e1–7.

61. Stellzig-Eisenhauer A, Decker E, Meyer-Marcotty P et al. Primary failure of eruption (PFE) – clinical and molecular genetics analysis. J Orofac Orthop 2010; 71. 6–16.

62. Frazier-Bowers SA, Chaitanya PP, Mahaney MC. The aetiology of eruption disorders – further evidence of a 'genetic paradigm'. Seminars in Orthodontics 2010; 16: 180–185.

63. Chaushu S, Becker A, Chaushu G. Orthosurgical treatment with lingual orthodontics of an infra-occluded maxillary first molar in an adult. Am J Orthod Dentofacial Orthop 2004; 125: 379–387.

64. Tulloch JFS, Proffit WR, Phillips C. Outcomes in a 2-phase randomized clinical trial of early class II treatment. Am J Orthod Dentofacial Orthop 2004; 125: 657–667.

65. King GJ, McGorray SP, Wheeler TT, Dolce C, Taylor M. Comparison of peer assessment ratings (PAR) from 1-phase and 2-phase treatment protocols for class II malocclusions. Am J Orthod Dentofacial Orthop 2003; 123: 489–496.

66. Livieratos FA, Johnston LE Jr. A comparison of one-stage and two-stage nonextraction alternatives in matched class II samples. Am J Orthod Dentofacial Orthop 1995; 108: 118–131.

67. Dolce C, Schader RE, McGorray SP, Wheeler TT. Centrographic analysis of 1-phase versus 2-phase treatment for class II malocclusion. Am J Orthod Dentofacial Orthop 2005; 128: 195–200.

68. McKnight MM, Daniels CP, Johnston LE Jr. A retrospective study of two-stage treatment outcomes assessed with two modified PAR indices. Angle Orthod 1998; 68: 521–524.

9
Impacted Teeth in the Adult Patient

(In Collaboration with Stella Chaushu)

Neglect and disguise

A small but significant number of untreated impacted teeth will eventually find some way of erupting into the mouth without treatment, although this may be many years after their normal eruption time and then often into an ectopic eruption site. This is particularly true of maxillary canines [1] and, at least to a degree, in contradiction to the popular view that eruption potential is lost when the root apex closes [2]. Nevertheless, a good proportion will remain unerupted and asymptomatic for many years. Prosthodontists are aware of the occasional patient complaining of the eruption of a tooth under a denture often many years after the patient had become otherwise edentulous.

For the most part, during the childhood of the particular adult patient concerned, advice was probably sought and rejected, with the reasons for this being very varied. The patient may have been an orthodontically unmanageable child at the appropriate age; perhaps the dentist or orthodontist was insufficiently convincing in the task of informing the parent of the consequences of non-treatment; or the parents' level of dental awareness was inadequate, the idea that surgery would be needed was possibly abhorrent to the parents or simply the cost and duration of the proposed treatment were unacceptable.

Just occasionally, a surgical exposure procedure would have been carried out at the appropriate time, but failed to elicit eruption and was not then followed up. Some impacted teeth, particularly maxillary canines, may simply have never been diagnosed. One further possibility that is not unfamiliar is that a dentist succumbed to the pleadings of the parent to 'do something temporary to make it look good', until they would be ready for the definitive treatment – a time that never arrived!

The impacted maxillary central incisor

It may be difficult for the orthodontist to imagine the situation where a patient has reached adulthood still with an impacted central incisor. This will have been obvious from around the age of 7 years, but the patient only sought treatment in his/her twenties or even later. This type of neglect is indeed unusual and its prevalence seems likely to vary from country to country, in inverse proportion to the level of dental awareness in the population. A country that offers its citizens some form of national dental insurance may be expected to have a lower prevalence among its adults, since one would hope that treatment would have been carried out at the appropriate time, given the relative freedom from financial constraints in a welfare state. Cost, however, is not the only factor, probably not even the dominant one.

Whatever the reason, the adult patient will usually present with the incisor anomaly unsuccessfully disguised in one of three ways:

1. A retained deciduous tooth may have been enlarged with the addition of composite material, although this will probably only have improved its length. Any increase in its width will be limited by the reduced mesio-distal space available to the tooth, the result of the marked mesial tipping of the adjacent lateral incisor and the central incisor of the opposite side.
2. This reduced space may have been maintained with a 'flipper' (spoon) partial denture, carrying a single and poorly matched small tooth (Figure 9.1a, b).
3. The lateral incisor may have been enlarged, with the use of composite material, in an attempt to simulate the shape of the impacted central incisor (Figure 9.2a).

There are serious drawbacks with each of these treatment alternatives, which focus principally on the very poor appearance of the results. The absence or reduction in size of a central incisor is always obvious, as is any significant shift in a maxillary dental midline, even to the casual observer. The tipping of the two teeth adjacent to the impacted incisor is too severe for this to escape notice and the angle of the lateral incisor is too acute for its long axis

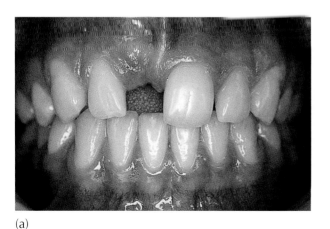
(a)

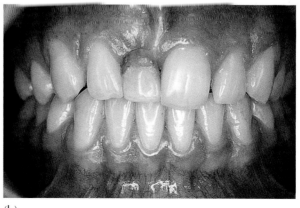
(b)

Fig. 9.1 Impacted right maxillary central incisor (a) replaced by poorly matched artificial tooth on 'flipper' (spoon) denture (b).

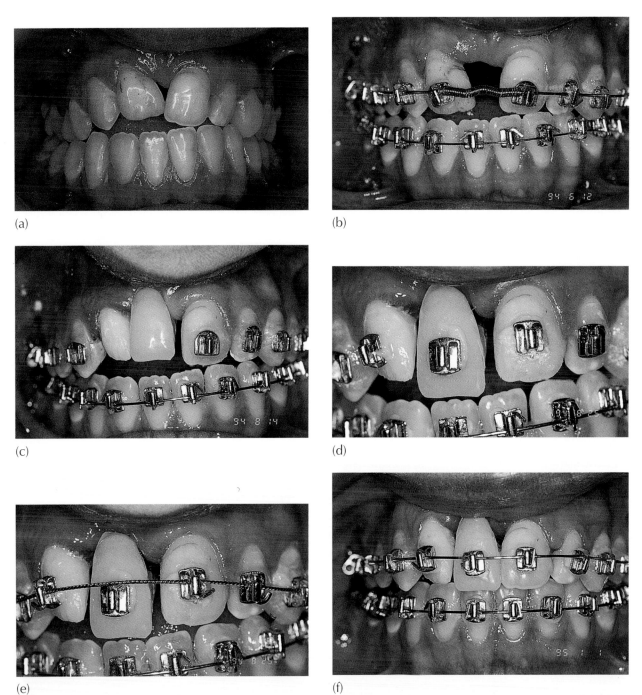

(a)   (b)   (c)   (d)   (e)   (f)

Fig. 9.2 (a) Impacted right maxillary central incisor. The right lateral incisor crown has been enlarged by a composite build-up. (b) Initial stages of alignment and space-opening. The composite build-up of the maxillary right lateral incisor is still in place. (c) Space has been reopened, the composite build-up of the lateral incisor has been removed and the artificial central incisor is in place, attached to the soldered palatal arch closing the anterior open bite. (d) The soldered arch has been deflected slightly vertically downwards to elongate the artificial tooth. (e) Placement of the archwire shows the degree of vertical displacement. The artificial tooth is now fully engaged in the archwire. (f) The extrusive force of the palatal archwire closes off the open bite. (g) The periapical view shows the impacted tooth to be dilacerated. The palatal arch is clearly seen, with the forward-pointing loop used to carry the radiolucent artificial tooth. An orthodontic bracket is also attached to this radiologically invisible tooth. (h) Pre-treatment tangential radiograph. (i) Bonding of an eyelet to the anatomically palatal aspect of the incisor crown. (j) The elastic chain is gently raised and ensnared in the pigtail to provide immediate and controlled vertical traction. (k) Post-surgical periapical view to show the bonded eyelet and pigtail ligature. (l) The tangential view post-surgery showing the length of the unseen part of the ligature and the relative heights of the tooth, the ligature extremity, and the occlusal plane. (m) The impacted tooth has erupted five months later: note the reduction of the cervical portion of the artificial tooth to allow for further progress. Traction was made to a newly placed labial attachment at this juncture. (n) Periapical view of the dilacerate incisor at the completion of treatment. (o) The orthodontic result: note the gingival appearance of the treated and untreated maxillary central incisors.

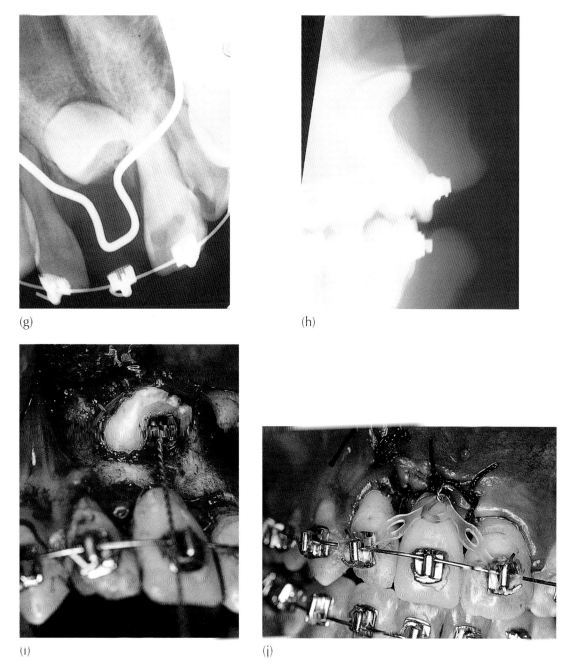

(g) (h)

(i) (j)

Fig. 9.2 (*Continued*)

to be visually 'realigned' by composite additions or by reshaping. The narrowness of the neck of the tooth makes an aesthetically convincing reconstruction, as a central incisor, difficult to achieve.

Management

It is quite clear that in these circumstances the first and most important prerequisite to any form of treatment for the missing tooth is to provide the maxillary dental arch

with an ideal shape in each of the three planes of space. In practical terms, this means:

- Levelling and aligning the entire dental arch. All ectopically placed teeth will need to be brought into an ideal archform, teeth will need to be aligned in a single, uniform occlusal plane and all rotations dealt with.
- Reopening a space of suitable mesio-distal width in order to accommodate the impacted tooth in the arch. Correcting the palatal inclination of the canine and tipping the lateral incisor of the same side and the central

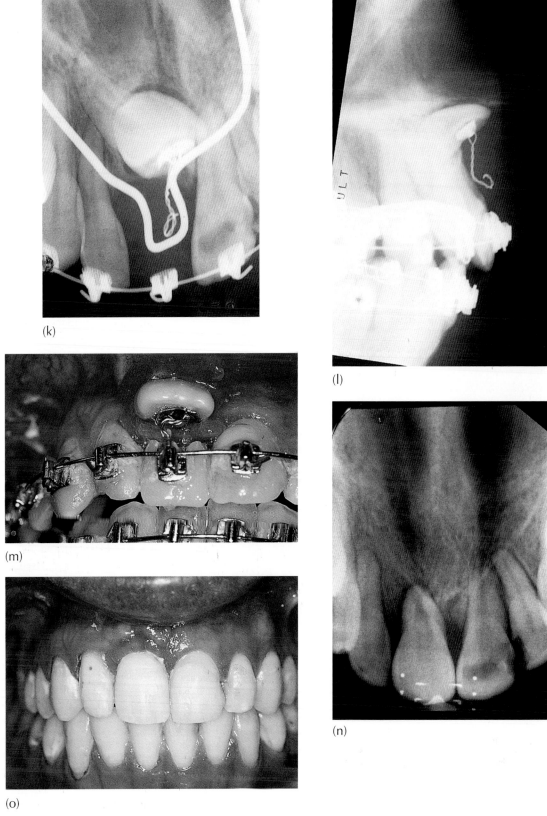

(k)

(l)

(m)

(n)

(o)

Fig. 9.2 (Continued)

and lateral incisor of the opposite side will usually provide adequate space, although distal movement, extraction or interproximal enamel stripping may need to be considered.

- Correcting the dental midline, to be continuous with the lower and with the midline of the face. This is normally achieved as a result of the reopening of space, but it may require the use of coil springs or anterior, oblique, intermaxillary elastics as part of a more comprehensive orthodontic appliance programme. The use of anterior intermaxillary elastics in the adult is often a difficult request to put before the patient for obvious social reasons. The judicious placement of temporary anchorage devices to permit forces to be applied unilaterally within the same arch is clearly to be preferred.
- Closing down an anterior open bite and bringing the teeth into occlusion. This may sometimes be achieved by properly aligning the molar tubes and by altering bracket height on the anterior teeth, thereby bringing about the desired extrusion of the teeth – and unpopular anterior vertical elastics are of material help in this situation. Often, however, an increased height of the lower third of the face and excessive vertical exposure of the maxillary anterior teeth and gingivae (gummy smile) may dictate the intrusion of the posterior teeth rather than extruding the anteriors. This may be achieved using zygomatic plates screwed into the inferior aspect of the zygomatic process of the maxilla, opposite the molar teeth. In this scenario, it should be understood that a force applied from a zygomatic plate, high in the sulcus, to the buccal tube on the molar will produce a rolling-out of the molar. It is therefore mandatory to use a transpalatal bar, preferably soldered from molar to molar. This bar should be prepared in such a manner as to be positioned a couple of millimetres away from the palatal mucosa to permit molar intrusion without palatal impingement.

Once these aims have been achieved, the patient is ready for that stage in treatment when all resources will need to be concentrated on the impacted tooth. The entire dental arch must be consolidated into a compound and united anchorage unit, to which the unerupted tooth will be drawn. In Chapters 6 and 7, it was noted that teeth that have been impacted for many years sometimes undergo pathological change which prevents their eruption [3], even when all other factors are favourable. It is by no means always possible to diagnose pathological change from a radiograph, unless there is a loss of the follicular sac and actual enamel resorption has become evident over wide areas of the surface of the unerupted tooth. It is therefore true to say that, whenever an adult patient presents for the treatment of an impacted tooth, a calculated risk is taken in offering this kind of treatment to resolve the impaction.

In the most adverse of circumstances, the central incisor tooth will have to be extracted and, perhaps, replaced with an implant-borne restoration. In that event, the preparation of the dental arch described above will have provided optimal clinical conditions to accept the implant or other form of artificial restoration of the space. An ideal pontic width is present, the roots of the adjacent teeth have been uprighted to make sufficient inter-radicular space for implant placement, all other teeth are aligned and the occlusion is good. However, the surgical removal of a grossly displaced impacted tooth, high above its normal position, will leave a considerable and unsightly bony defect. This will be difficult to conceal in the gingival area around a fixed prosthesis and will not lend itself to the placing of an implant without suitable and prior osseous ridge reconstruction [4–7].

The patient must be brought into the decision-making process from the outset and should be informed of the potential advantages of each of the stages of treatment. This is best done using a set of plaster models of the patient's teeth, together with a wax duplicate set-up, to show a scheme of the proposed treatment result. Prognosis for the success of the pre-surgical stages of the treatment is excellent, but for the alignment of the impacted tooth it is not so certain. Offering the treatment plan to the patient is, therefore, probably best accomplished if it is based on explaining the benefits of the limited objectives, i.e. aligning the teeth for the purposes of achieving improved conditions for the construction of a conventional prosthodontic or implant-borne replacement. The added bonus, which will be derived from success in the resolution of the impaction, may then be properly brought into perspective to provide the desirable added incentive. By whatever means the value of the treatment is explained, care should be taken to fully inform the patient that the possibility of failure to bring the impacted tooth into the arch is real, but that contingency plans are available in this disappointing eventuality.

The need for temporary prostheses during the treatment

The impacted central incisor

For the adult patient, planned orthodontic space opening for an unerupted anterior tooth is a daunting prospect. All central and lateral incisors and maxillary canines require some form of immediate temporary prosthetic replacement until such time as the permanent tooth comes into its place. With some patients, particularly those more concerned with their appearance or those who have a broad smile or a wide dental display which is evident in facial expression and social intercourse, there may be a need to artificially replace even premolar teeth.

Perhaps the most popular solution for an impacted incisor is to trim an artificial acrylic tooth to a suitable shape and size, bond a bracket to it and ligate it into the archwire of the appliance. In the early stages, the archwire

is probably nickel–titanium and of round cross-section, which means that the new artificial pontic is very unstable and will rotate around or slide along the archwire. Placing a V-bend in the wire or welding a small piece of wire in the vertical position will prevent the rotation but complicate use of the archwire. It will also need to be copied into each of the subsequent archwires until a rectangular cross-section wire is used.

An artificial acrylic tooth may be bonded to the mesial surface of the contralateral central or ipsilateral lateral incisor, although this assumes that the archwire is absolutely passive when ligated into the bracket on these teeth. A minor rotatory movement will cause a major lateral or medial swing of the pontic, while an uprighting component will markedly intrude or extrude it at its opposite end.

Alternatively, the artificial restoration may take the form of a removable plate carrying a single tooth – a 'flipper' (spoon) denture. The average adult patient may have considerable difficulty becoming accustomed to it, even if its retention is adequate initially. Of greater concern, however, is the fact that the adjacent teeth and many other teeth need to be moved during the orthodontic treatment, which will rapidly make this artificial denture ill-fitting. Furthermore, the close adaptation of the acrylic base to the contour of the other teeth in the jaw may actually interfere with the planned orthodontic movement. For this type of artificial replacement to be successful, Adams clasps may have to be used on the second molars; alternatively, a modified circumferential clasp may hook over the buccal tubes of the first molars. These teeth are often excluded from the planned dental movements and may sometimes be helpful in retaining such a plate, although the distance between the clasps and an incisor pontic may be the cause of an unacceptable degree of instability.

Clearly, more satisfactory alternative methods of artificial replacement are essential to the successful pursuit of treatment for the adult patient, and these must provide an answer to the several shortcomings of the 'flipper' denture. Indeed, given a little thought in their design, and rather than their playing the role of the villain of the piece, assistance in the application of force to the impacted tooth may be derived from the method of artificial replacement, which may actually contribute to the smooth running of the active orthodontic appliance.

The active removable plate

If a removable plate is to be worn to hold the artificial tooth in place, it makes good sense to augment that plate with active elements that will also produce tooth movement. Looked at in a different way, this means designing a simple removable orthodontic appliance carrying springs of one sort or another, which are aimed at realigning the teeth to reopen the anterior space and, at the same time, to fill that space with an artificial tooth.

This method has some important advantages. The active plate is straightforward and easy to use, requiring very limited expertise in adjustment of the forces applied and in their direction. Because the plate is removable and generally adjusted at the chair side out of the mouth, the size and shape of the artificial tooth may be easily altered in size or speedily replaced as space opening occurs, to maintain appearance.

However, the active removable appliance is unable to produce more than tipping movements of the adjacent teeth. Methods have been described where removable appliances have been used to produce the extrusive movements needed to resolve the impaction of teeth [8, 9]. Their effectiveness in this capacity is limited and they cannot be expected to perform the labio-lingual or mesio-distal root movements or rotatory movements that are often needed to fully align these aberrant teeth. In cases where removable appliances are used to resolve the initial impaction and to erupt the tooth, therefore, they must be followed by a fixed appliance to bring about the successful completion of a second phase of treatment, which is aimed at the fine functional and aesthetic positioning of the teeth in all three planes of space.

The soldered palatal arch

In the adult patient, the scope of orthodontic correction that is planned tends to be more localized and less comprehensive, particularly when a single and grossly displaced tooth is present. The first maxillary molar teeth are most commonly used as anchor teeth for the fixed appliance, and their orthodontic movement is not usually required. This being so, the buccal aspects of these teeth and buccal/labial aspects of the teeth more anteriorly placed will be used to carry the orthodontic attachments, archwires and auxiliaries. This leaves the palatal side of the teeth and the palate area free, and available to serve the interests of the patient's appearance.

A soldered palatal arch, based on the molar bands, can provide the orthodontic appliance with an excellent anchorage base, at the same time as acting as the vehicle for a satisfactory prosthetic replacement. Several approaches are available. They depend on the adaptation of well-fitting preformed orthodontic bands to the molar teeth and their accurate transference to a plaster working model of the jaw.

On the working model, a palatal arch is fabricated and soldered on the palatal side of the molar bands. A small wire extension may then be soldered or bent into the anterior portion of the palatal arch, extending towards the space in the arch and terminating immediately palatal to the position of the missing tooth, with a configuration that will mechanically retain an artificial acrylic tooth. The exact location of the artificial tooth should be decided in accord-

ance with the projected treatment goals of the case and not necessarily in line with the adjacent natural teeth. Thus, if an overjet is to be closed or a cross-bite treated, the siting of the artificial tooth should be made according to the intended final, post-treatment position of the adjacent teeth. An occluded plaster cast of the opposite jaw is therefore necessary to assist in its accurate placement.

This is the simplest approach of this type and offers the patient a good artificial replacement, which is well tolerated (Figures 9.2a, b). It also allows the adjacent teeth to be aligned without hindrance, while actually enhancing the anchorage value of the molars during retraction of a procumbent labial segment. The anchor molars cannot be rotated or tipped easily when using horizontal, intramaxillary elastics, due to the stabilizing effect of the rigid soldered palatal arch.

A significant and valuable refinement of this approach involves bonding a conventional bracket to the artificial tooth, as with the other teeth. This makes the artificial tooth aesthetically compromised to a similar degree as the other teeth and, therefore, less recognizable as other than a part of the natural dentition. Since this tooth is rigidly attached to the molar teeth and at a fixed distance from them, this method has much more to offer. Its integration into the appliance system makes alignment and levelling more accurate and more rapid. Additionally, the need for elastic traction to reduce the overjet will be eliminated, since the use of the initial fine gauge levelling and aligning wires in the early weeks of treatment will perform this without any further modification. The distance and relationship between molars and the artificial incisor is fixed to the ideal length and position by the palatal arch. Thus, a progression of ideal wire archforms will align all other teeth within that arch. Essentially, by linking the archwire to the fixed pontic in its normal overjet and overbite location, the first stage of mechano therapy, which generally deals with initial levelling and alignment only, now comes to include automatic overjet and overbite reduction. The overall length of the heavy palatal arch provides it with a degree of elasticity despite its heavy gauge. Thus, while carrying a temporary prosthetic replacement, it may be used to widen or constrict the dental arch. In the present context, however, it has one other possible function which is less obvious, but most helpful. The palatal arch has the potential to provide the vertical component of force that is needed to close an anterior open bite and, subsequently, the vertical traction needed to resolve the incisor impaction (Figures 9.2 d–g).

The impacted maxillary canine

Aside from third molars, and in common with the younger patient, the tooth most frequently found to be impacted in the adult is the maxillary canine. The principles of diagnosis, treatment planning and appliance therapy in the adult are no different from those in the child, although certain demands are made by the adult patient, which may make treatment methods less routine and more individualized.

Palatally impacted canines are frequently associated with only minor malocclusions and, as we have already pointed out in Chapter 6, in dentitions in which the dental age is often very late. Therefore, we may occasionally see a case of impaction which has eluded diagnosis until a much later age, and the circumstance that led to the discovery was the exfoliation of the deciduous canine or, sometimes, a routine examination by the general dentist, which revealed the buried tooth. Additionally, the increase in demand for orthodontic treatment among adults in recent years may change the attitude of some patients who had strongly opposed orthodontic treatment in adolescence and who may be more inclined to reconsider it later.

The question therefore arises as to whether we can expect that treating impacted teeth in adults will produce the same results, in the same period of time and with a similar degree of confidence in achieving a successful outcome. Little has been published in the orthodontic and surgical journals to enlighten us regarding these extremely important points, since the number of cases is very small and most orthodontists have limited experience with them. Nevertheless, two articles have appeared which have reported conflicting findings.

According to one study, the duration of such treatment in the over-25 years is likely to be significantly longer than for the child patient [10]. A more recent investigation found the opposite, namely that treatment was longer in younger patients with impacted canines than in an older group [11]. However, the patients who took part in the latter study were all under the age of 20 years, and among these, the younger individuals in the group had, on average, more serious impactions than did the older ones. Thus, given the fact that the 'adult' patients were little more than adolescents and barely definable as adults, both chronologically and dentally, as well as the small difference in age and a difference in the objective complexity of the treatment of the younger vs. the older patients, it seems that little may be concluded from this study regarding the effect of age. Aside from these two papers and until quite recently, the available articles on adult patients have been anecdotal case presentations, which can give no indication regarding these basic questions.

Accordingly, a study was initiated [12] in which 19 adults, whose ages ranged between 20 and 47 years (mean age 28.8 years) and who had been treated for 23 impacted maxillary canines, took part. They were compared with a control group of younger patients, ranging from 12 to 16 years of age (mean age 13.7 years). The control group had been objectively selected on a matched case-by-case basis to

ensure a similar degree of complexity of the impaction and with similar positions of the canine location in three planes of space, using the classification described in Chapter 6 above. Successful treatment was possible in approximately 70% of the adult canines and in 100% of the younger controls. While the duration of the overall treatment of the general dental malocclusion did not differ significantly between the two groups, the part of the treatment that was strictly concerned with the impacted canines took considerably longer in the adult group, both for the simpler cases and for the more complicated cases, when compared with their matched younger counterparts. All the failed cases occurred in the over-30-year-old patients. The conclusions that were drawn were that the prognosis for successful reduction and alignment of impacted canines among adults was lower in general and that it worsens with increasing age. Additionally, it was noted that a successful outcome of the part of the treatment concerned with the resolution of the canine impaction and its alignment in the adult should be expected to take considerably longer – in contrast to the remainder of the orthodontic treatment for the overall malocclusion.

To camouflage the absence in the arch of an impacted canine, an artificial acrylic tooth may be bonded to the mesial surface of the first premolar (see Figure 10.11), although this assumes that the archwire is passively ligated into the premolar bracket. A minor rotatory movement will cause a major lateral or medial swing of the pontic, while an uprighting component will markedly intrude it into the gingival tissues or extrude it into premature occlusion at its opposite end.

In a case with impacted incisor teeth, the presence of a palatal arch does not encroach on the area where surgical exposure will be performed and where postsurgical swelling is likely to occur, provided the anterior portion of the palatal arch is not brought too far forward. A cut-back design is usually most appropriate. For the surgical episode involved in the exposure of a palatally displaced maxillary canine, a wide area of palatal mucosa may need to be reflected back, and this, together with the possible sequel of even a minimal degree of postsurgical oedema, effectively disqualifies the use of a rigid palatal arc. A trans-palatal bar, such as a Goshgarian or a simple 'across-the-palate' soldered arch, is usually sufficiently posteriorly located from the surgical site to be used in these circumstances. It cannot be used for prosthetic replacement and its only functions are to enhance the anchorage and to maintain arch width. However, in combination with a buccal arm, it may be very useful and have definite indications.

In the illustrated case (Figure 9.3), the adapted molar bands were transferred to a plaster working model and a trans-palatal arch was soldered to the palatal side of the bands. To the buccal side of the bands and gingival to the buccal tubes a heavy buccal arm was soldered, which

extended vertically upwards into the sulcus. The arm was then fabricated to follow the depth of the sulcus anteriorly until it reached the canine area, where it again dipped inferiorly to terminate in a loop in the canine site. An artificial acrylic tooth was cured into this retention loop, in the place of the missing permanent canine.

Immediately after extraction of the natural deciduous tooth, the unit, which comprised two molar bands, a palatal arch and a buccal extension arm, was cemented into place, followed by appropriate orthodontic attachments on the first premolar teeth. After placement of a sectional arch on each side, the impacted canine was exposed surgically in the normal way. Traction was then applied behind the façade of the buccally retained artificial tooth.

As with the placement of a bracket on the artificial central incisor, described above, it is advantageous to design the integration of these prosthetic expedients into the orthodontic appliance system in such a way that they may materially contribute to the efficiency of the appliance. Thus, after cementation of the bands carrying these additions, the buccal arm may be displaced further buccally, so that its passive position stands a few millimetres buccal to its original location. If it is then tied directly to the impacted canine, using a steel ligature, which will draw the displaced buccal arm and artificial tooth back into alignment with the adjacent teeth. The energy stored by this long, elastic and now deflected buccal arm will thus provide the traction needed to draw the impacted tooth towards its place in the arch.

Supplementary clinical concerns

The basic premise for the use of these palatal and buccal arches for both impacted incisors and canines has been that the first molar tooth does not require to be orthodontically moved. Needless to say, there are cases in which movement of the first molars is an essential part of the orthodontic strategy of the treatment of a particular adult. These may include cases where there is a pronounced rotation or a palatal or buccal displacement of this tooth, but they may also include the premolar extraction cases where closure of excess space from the distal will be needed.

In these cases, several options are still available to allow the smooth pursuit of orthodontic treatment under aesthetically acceptable conditions. In the first place, a single buccally displaced or palatally displaced molar tooth, which is planned to be used as an anchor unit, may be tipped into its place using a removable appliance. This appliance will need to carry some form of buccal or palatal spring, which will be used to move the tooth in the appropriate direction, suitable clasps to retain the appliance firmly in position and, possibly, carry an artificial tooth to replace the impacted incisor tooth in the interim. Alternatively, an existing

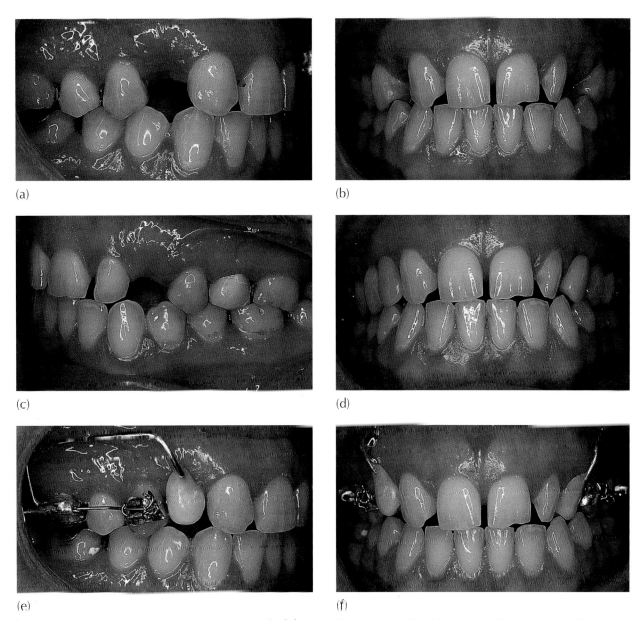

(a)

(b)

(c)

(d)

(e)

(f)

Fig. 9.2 (a–c) A 47 year-old female has a maxillary left canine that is impacted adjacent to a peg-shaped lateral incisor. This case was treated by the author in the early 1970s. The erupted right canine is in the place of the congenitally absent lateral incisor. (d) The patient wears a removable partial plate to fill the canine sites on each side. (e–g) The molar bands are interconnected by a soldered 'across the palatal' heavy arch (not seen). A high buccal arm is soldered on each molar band, carrying an artificial canine tooth, to replace the discarded partial plate. (h) A small hook is cured into the left artificial canine, and the buccal arm is deflected buccally and inferiorly at the time of surgical exposure of the impacted canine. (i) Ligating the impacted tooth to the artificial tooth applies extrusive and buccal traction. (j–l) The left canine has been brought into its place and a fixed partial prosthesis fills the gap on the right side.

'flipper' denture may be augmented to incorporate the same clasp and spring elements.

When more extensive movement of the anchor teeth is required, this is usually enacted in a preparatory orthodontic treatment phase, which is aimed at producing good alignment by uprighting, rotating and torqueing the teeth, while limiting the appliance work and movement of the anterior teeth to levelling and aligning. During this procedure, it is important to enable the patient who has been wearing a removable artificial prosthesis to continue to do so until such time as the alignment stage is complete. Similarly, an over-retained deciduous incisor tooth should be allowed to remain until the planned temporary prosthetic rehabilitation becomes practical. With the satisfactory completion of the preparatory orthodontic treatment phase and the teeth having been brought into good alignment, the precise reopening of the space for the missing tooth is undertaken.

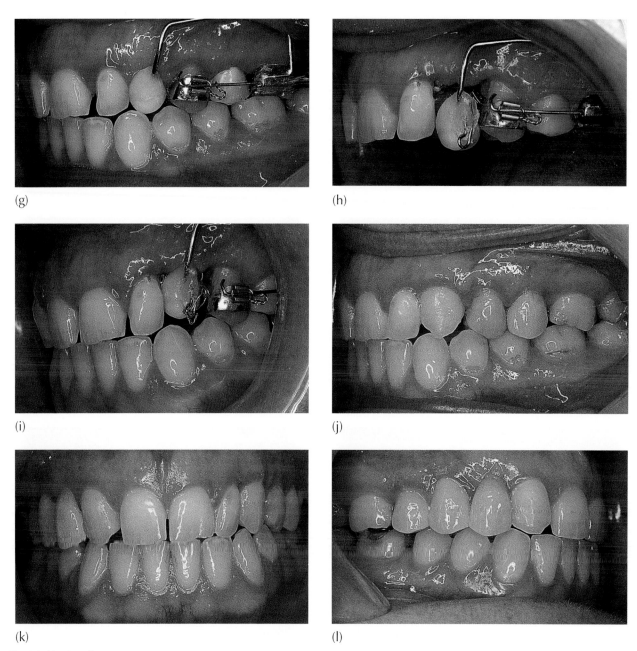

(g)

(h)

(i)

(j)

(k)

(l)

Fig. 9.3 (*Continued*)

At this point, the palatal arch is constructed on a plaster model, into which the molar bands have been accurately seated and the artificial tooth is sited, as described above. The patient's 'flipper' denture is now discarded, or the over-retained deciduous tooth is extracted, and the palatal arch carrying the artificial incisor tooth is inserted by re-cementing the molar bands to their former place.

When moving teeth mesio-distally along an archwire with a multi-bracketed fixed appliance, the establishment of interproximal contacts between the teeth enables a high degree of control of individual tooth position. Once this is

achieved, uprighting and torqueing movements may be carried out, with care being taken to see that the spaces do not reopen. The desired treatment result includes closed contacts.

In the situation where a tooth is unerupted and space needs to be made for it, the orthodontist should err on the side of reopening excess space and then maintaining it until the tooth reaches its place. However, to do so, the space needs to be maintained during the many further months of treatment, when the adjacent teeth will be altering their relationships to one another. While attention is diverted to

the details of treatment in these other areas, changes may inadvertently alter the space. The original coil spring which may have opened the space requires to be deactivated, since it will otherwise continue to increase the dimensions of the space. However, the placement of active root-uprighting springs will tend to close the space by compressing the spring. The size of the artificial incisor tooth and its mesio-distal siting may be very useful in holding the achieved space and in ideally placing the adjacent teeth, regardless of their locations and without the need for the coil spring or the ligation of groups of teeth. The natural teeth are swiftly brought into interproximal contact with the artificial tooth and with each other, and this status is then simple to maintain.

Tooth transposition and temporary prosthetic replacement

Rarely, one or more of the impacted teeth are also transposed. For the most part, it is preferable to align the teeth in their transposed positions rather than try to re-transpose them to their ideal positions, for reasons discussed in Chapter 6. Nevertheless, there are situations in which this may be the preferred line of treatment.

Given the high frequency of missing lateral incisors in cases of transposition of maxillary canine/first premolar transposition, treatment of the transposition may have to consider the artificial replacement of the missing teeth during the appliance therapy. The use of a palatal arch to augment the anchorage value of the molar teeth and to provide the vehicle for the artificial replacement teeth is an effective way to deal with the problem (Figure 9.4).

By tying in a labial archwire from molar tube to molar tube, passing through a bracket on each of the artificial lateral incisors, these four strategic attachments on the perimeter of the arch are defined in relation to one another. Since the incisor brackets are also rigidly fixed to the molar bands on the lingual side. The long span of labial archwire that intervenes between molar tube and incisor bracket is therefore well supported in terms of anchorage and may be used to slide the more buccal of the transposed teeth (usually the canine) in the mesio-distal plane. At the same time, the more lingual of the transposed teeth must be moved further lingually to allow its neighbour to pass by. Finally, it must be moved in the opposite mesio-distal direction and back in the line of the arch. To achieve this, the more lingual tooth may be ligated to several different and strategically planned loops and cross-pieces that will have been prepared on the palatal arch, ahead of time, using elastic thread. Once again, positive use is made of the palatal arch as an integral part of the orthodontic appliance system, together with its function as a buttress for anchorage. It does not serve merely as a means of supporting an artificial tooth.

The unerupted third molar as a potential bridge abutment or antagonist for an unopposed tooth

In the previous chapter, we discussed the disimpaction of molar teeth, which were prevented from erupting partially or fully by their relationship with an immediate mesial neighbour. It is pertinent in the context of the present chapter to discuss a related scenario which commonly presents.

Following the extraction of posterior teeth in the adult patient, the establishment of a 'free-end edentulous saddle' makes oral rehabilitation problematic. This may sometimes find a potentially convenient solution in the discovery of an unerupted third molar. However, the absence of standing posterior teeth creates mechano-therapeutic difficulties in providing the vertical traction which is aimed at enhancing the eruptive potential of an unerupted third molar. Elastic traction of the tooth to the opposing jaw is a useful method, and has been referred to above [9], but the use of removable appliances in the adult patient is unreliable. Adults have far greater difficulty becoming accustomed to the bulk of the removable plate and its interference with masticatory and articular function. The use of a fixed mandibular appliance offers a much more satisfactory and dimensionally modest alternative, which interferes neither with eating nor with speech. Furthermore, if the mandibular appliance is to be used only as a source of anchorage, brackets and archwires may be dispensed with, making it very inconspicuous indeed.

The second molar in the opposite arch is unopposed. It is the tooth which faces the potential eruption site and the proposed final position of the unerupted tooth. Using this tooth as the sole source of elastic traction will cause its highly undesirable over-eruption [13]. The premolars in the same quadrant will may be in occlusion further forward and these teeth may be included in the anchor unit if they are rigidly linked together, to prevent or limit their reactive intrusion.

Using a plaster model of the patient's opposing jaw, a length of 0.024 in (0.6 mm) stainless steel wire is adapted to the general form of the buccal surfaces of the teeth, from second molar to the second or, preferably, first premolar (Figure 9.5). The wire extends very slightly mesial to the most anterior tooth and a few millimetres distal to the second molar, where it is bent into the form of a hook. Those parts of the wire immediately overlying the tooth surfaces should also include small retention loops and/or welded mesh pads.

Returning to the patient, the buccal surfaces of the teeth are etched and the wire bonded to them using a composite material. Transbond™ or equivalent lingual retainer bonding agent is probably the easiest to use, is adequate for the task and is easy to remove at the conclusion of the treatment.

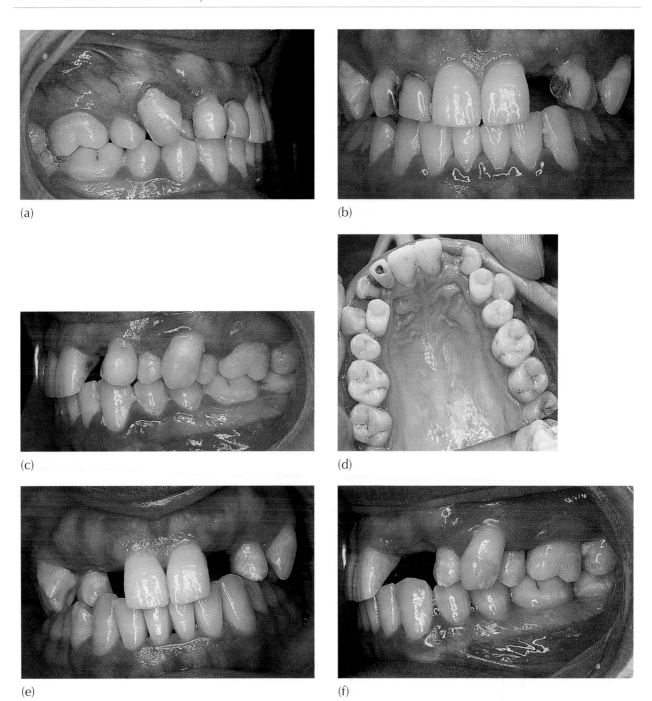

(a)

(b)

(c)

(d)

(e)

(f)

Fig. 9.4 (a–c) A 27-year-old female with congenitally absent maxillary lateral incisors and maxillary canine/first premolar bilateral transposition treated by the author in the 1970s, before the advent of bonded brackets. The deciduous canines and right lateral incisor are still present. (d) Occlusal view of maxillary arch. (e–g) Intra oral views after extraction of deciduous teeth. (h–j) Treatment progress seen from the right side. (k, l) Use of a palatal arch as support for lateral incisor pontics and also to move premolars through varying use of elastic thread. (m) Occlusal view of complete maxillary dentition, with wire splint from first premolar, including lateral incisor pontics. (n–p) Intra-oral views of the completed orthodontic result. The patient was referred for permanent prosthodontic treatment of the lateral incisor problem.

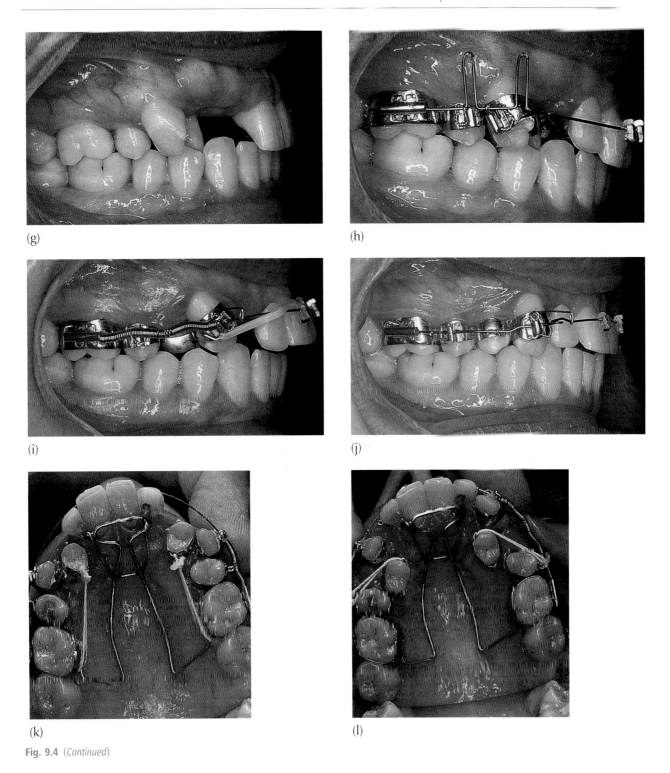

(g)

(h)

(i)

(j)

(k)

(l)

Fig. 9.4 (*Continued*)

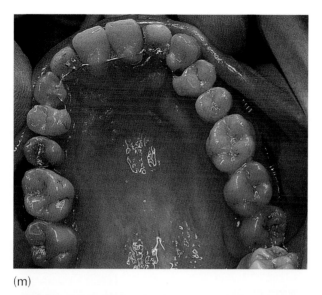

(m)

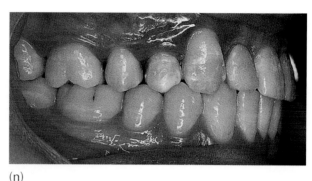

(n)

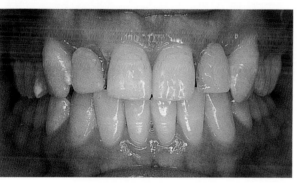

(o)

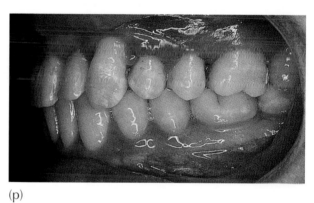

(p)

Fig. 9.4 *(Continued)*

A small, custom-made hook or button is prepared and bonded to the unerupted tooth following its exposure. At the same visit as these procedures are performed, the patient is taught to place a small latex elastic (3/8 in medium or 5/16 in light gauge) on the hook and to draw it round the wire extension distal to the opposing second molar, and then forward to engage the small protrusion of the wire mesial to the first premolar. This forms an L-shaped configuration to the elastic, which has a dual purpose: the overall length provides a light force of excellent range while, at the same time, making the manipulation of the elastic very easy for the patient. The vertical traction that is applied to the tooth may be altered to include a horizontal component by altering the position of the distal end of the bonded wire, thereby also directing the tooth mesially, distally, lingually or buccally. An easier alternative in this case would be to place a mini-screw in bone on the buccal side of the alveolus just distal to the roots of the second molar tooth and to use that as the source of anchorage for the vertical

elastic. However, while this particular treatment was carried out several years before temporary implant devices were in routine use, both approaches to treatment are entirely valid and today provide both operator and patient with a choice.

Implant anchorage

When an adult patient requires oral rehabilitation following the loss of one or more teeth, the prosthetist/prosthodontist will often favour artificial replacement with an osseo-integrated implant. If that same patient also has an impacted tooth, then the implant may be exploited to assist in its resolution. A successful implant can be used to provide 'absolute' anchorage, since it forms an osseo-integrated union with the bone and, like an ankylosed tooth, will not respond to orthodontic forces [14] (Figure 9.6).

Alternatively, a temporary orthodontic implant, in the form of a mini-screw which can be used in a variety of places in either jaw (see Figure 10.16), or a zygomatic plate screwed into the inferior surface of the zygomatic arch of

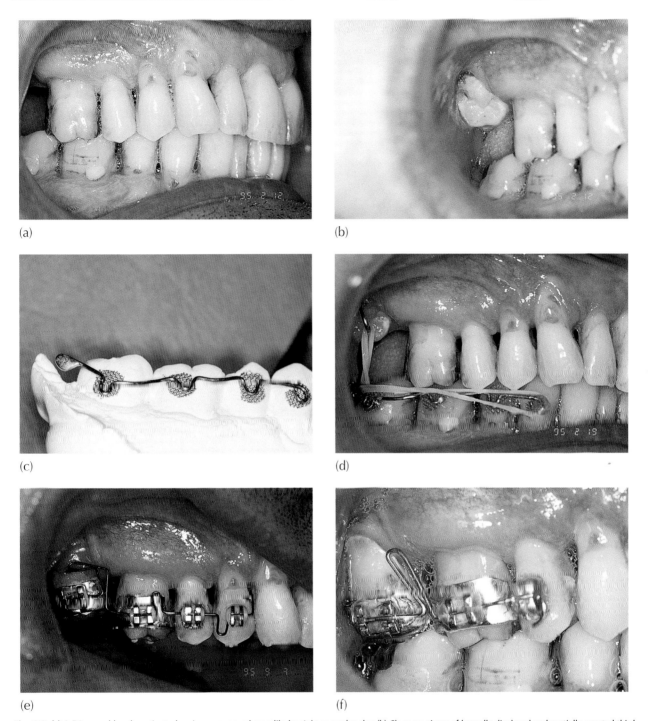

Fig. 9.5 (a) A 54-year-old male patient, showing unopposed mandibular right second molar. (b) Close-up views of buccally displaced and partially erupted third molar. (c) A 0.024 in round wire has been adapted to buccal surfaces of the premolar and molar teeth on the mandibular model. Note the retention loops and welded mesh pads. The distal extremity is in the shape of a hook, which has been covered with solder for patient comfort. (d) The L-shaped elastic configuration for ease of placement and wide range of action. (e) Following eruption, a partially bonded appliance is used to upright the third molar. Additional anchorage is derived from a soldered palatal arch from first molar to first molar. (f) The final stage of treatment.

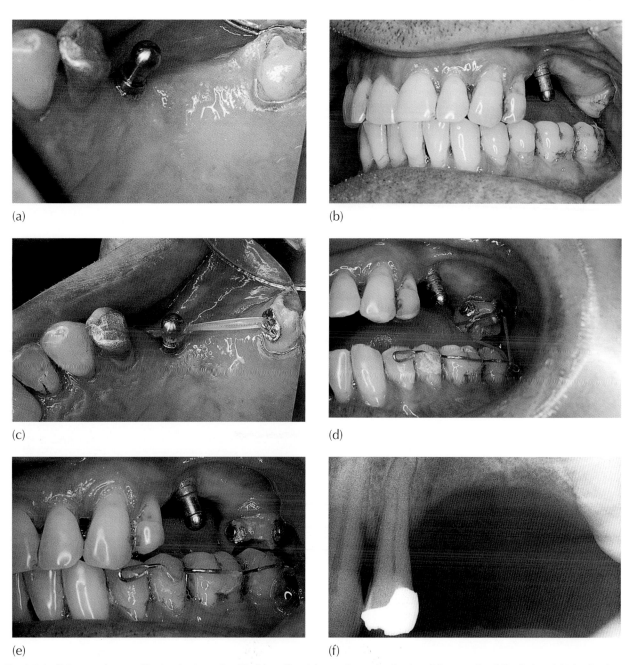

Fig. 9.6 (a, b) An osseo-integrated implant has been placed in the maxillary left second premolar site. A partially erupted and distally tipped third molar shows only its mesial surface and mesial occlusal cusps. (c) An elastomeric chain module applies force to the implant post via the bonded eyelet on the molar. (d) Vertical force is applied to a mandibular bonded buccal bar to achieve occlusal contact. (e) Fully occluding third molar. Note the use of three button attachments to provide vertical force with buccal or lingual vectors, as needed. The mesial button is to prevent the elastic from impinging on the gingiva. (f, g) Pre- and post-treatment radiographs. (h) Lateral view of the prosthodontic reconstruction.

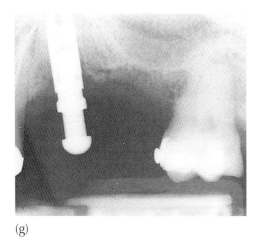

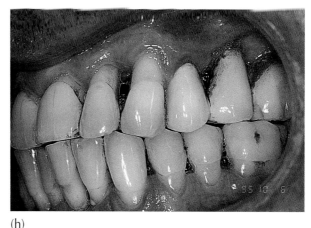

(g)   (h)

Fig. 9.6 (*Continued*)

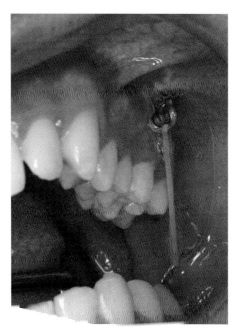

Fig 9.7 A plate orthodontic implant has been screwed onto the surface of the inferior aspect of the zygomatic arch, with its looped extension passing through the oral mucosa and lying adjacent to the first or second molar. The patient places vertical elastics from this plate to the hooked end of the pigtail ligature which is tied to an attachment bonded to an impacted third molar (not seen). No other orthodontic appliance is presently being used to elevate this tooth.

the maxilla (Figures 9.7 and 9.9), may be used as a platform from which elastic traction is applied to individual teeth, with or without the need for the use of complex orthodontic appliances. An implant may also be placed in the midline of the palatal vault – a midplant.

However, unlike the above-mentioned devices, this is unsuitable for direct traction to an impacted tooth, because of its non-strategic position *vis-à-vis* the tooth. It may nevertheless be used to reinforce the anchorage of an orthodontic appliance that has been customized to open space within the dental arch and to apply the needed directional extrusive force to the impacted tooth (Figure 9.8).

At the completion of their specific tasks, these non-integrated devices are removed very simply, provided they have not been in place for more than a half year or so. Those that are present for much longer may become progressively osseo-integrated in many instances and will be more difficult to remove. The use of both temporary and osseo-integrated implants in this manner offers considerable opportunity for the development of novel ways of applying forces in general, particularly for partially dentate adults, for replacing extra-oral anchorage in non-growing patients [15] (Figure 9.9) and in lingual orthodontics, as will be seen in the next chapter.

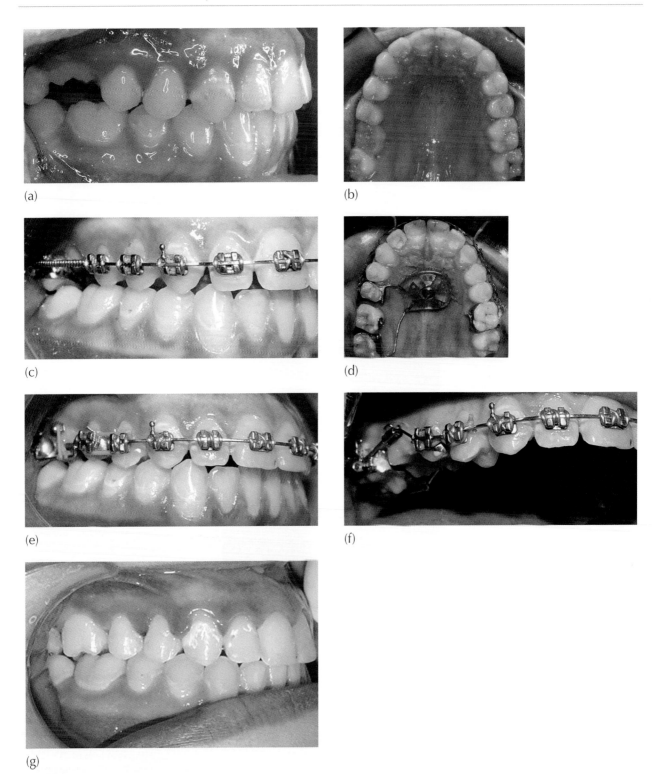

Fig. 9.8 Anchorage derived from a mid-plant. (a, b) Buccal and occlusal intra-oral views of an infra-occluded maxillary first molar. (c) A regular fixed appliance is used with a coil spring to open the space between second premolar and second molar. (d, e) The space is opened and the mid-plant is placed, with soldered arms connecting it to the second premolar and second molar. Buccal and lingual attachments are bonded to the infra-occluded molar prior to the surgical procedure to loosen the tooth from its ankylotic attachment. Elastic ties apply extrusive force from the buccal eyelet to the vertically offset archwire. (f) The buccal surface has become more exposed as the tooth erupts, and a new horizontal tube is substituted for the eyelet. (g) The molar tooth has reached the occlusal level and intercuspates well with the opposing teeth.

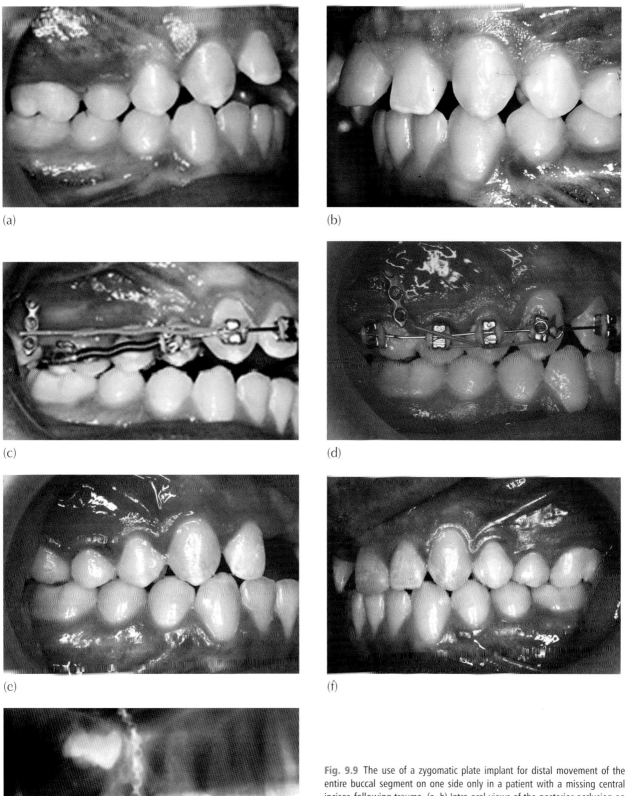

(a)

(b)

(c)

(d)

(e)

(f)

(g)

Fig. 9.9 The use of a zygomatic plate implant for distal movement of the entire buccal segment on one side only in a patient with a missing central incisor, following trauma. (a, b) Intra-oral views of the posterior occlusion on each side. (c) A compressed open coil spring is placed between first premolar and first molar, with the intention of moving first and second molars distally. The canine is tied back with an elastic power chain to the inferior extremity of the plate implant, to provide anchorage against the reactive force of the coil spring. (d) With the molars in class 1 intercuspation, the implant is used to tie elastic elements to retract the premolars and canine and to support the anchorage for the uprighting of the canine. (e, f) The completed result. (g) A section of the panoramic radiograph shows the plate implant.

References

1. Thilander B, Jacobson SO. Local factors in impaction of maxillary canines. Acta Odont Scand 1968; 26: 145–168.
2. Kokich VG, Mathews DP. Surgical and orthodontic management of impacted teeth. Dent Clin North Am 1993; 37: 181–204.
3. Azaz B, Shteyer A. Resorption of the crown in impacted maxillary canine. A clinical, radiographic and histologic study. Int J Oral Surg 1978; 7: 167–171.
4. Sailer HF. Two new methods combining osteotomies and endosseous titanium screw implants for the narrow maxillary ridge and the atrophic lateral mandible. Abstract presented at the Third International Congress on Preprosthetic Surgery, 'The edentulous jaw'. Arnhem, the Netherlands, 1989: 62–63.
5. Richardson D, Cawood JI. Anterior maxillary osteoplasty to broaden the narrow maxillary ridge. Int J OralMaxillofac Surg 1991; 20: 342–348.
6. Simion M, Baldoni M, Zaffe D. Jawbone enlargement using immediate implant placement associated with a split-crest technique and guided tissue regeneration. Int J Periodont Restor Dent 1992; 12: 463–473.
7. Lustmann J, Lewinstein I. Interpositional bone grafting technique to widen narrow maxillary ridge. Int J Oral Maxillofac Implants 1995; 10: 568–577.
8. Fournier A, Turcotte J, Bernard C. Orthodontic considerations in the treatment of maxillary impacted canines. Am J Orthod 1982; 81: 236–239.
9. Orton HS, Garvey MT, Pearson MH. Extrusion of the ectopic maxillary canine using a lower removable appliance. Am J Orthod Dentofacial Orthop 1995; 107: 349–359.
10. Harzer W, Seifert D, Mahdi Y. The orthodontic classification of impacted canines with special reference to the age at treatment, the angulation and dynamic occlusion. Fortsch Kieferorthop 1994; 55: 47–53.
11. Stewart JA, Heo G, Glover KE et al. Factors that relate to treatment duration for patients with palatally impacted maxillary canines. Am J Orthod Dentofacial Orthop 2001; 119: 216–225.
12. Becker A, Chaushu S. Success rate and duration of orthodontic treatment for adult patients with palatally impacted maxillary canines. Am J Orthod Dentofacial Orthop 2003; 124: 509–514.
13. Peck L, Peck S, Attia Y. Maxillary canine-first premolar transposition, associated dental anomalies and genetic basis. Angle Orthod 1993; 63: 99–109.
14. Stern N, Becker A. Forced eruption: biological and clinical considerations. J Oral Rehabil 1980; 7: 395–402.
15. Roberts WE, Smith RK, Zilberman Y, Mozsary PG, Smith RS. Osseous adaptation to continuous loading of rigid endosseous implants. Am J Orthod 1984; 86: 95–111.

10

Lingual Appliances, Implants and Impacted Teeth

(Stella Chaushu and Gabriel Chaushu)

Orthodontic Treatment of Impacted Teeth, Third Edition. Adrian Becker.
© 2012 Adrian Becker. Published 2012 by Blackwell Publishing Ltd.

The context of impacted canines *vis-à-vis* the lingual appliance

There are four major areas of concern in the ortho-surgical approach for the resolution of impacted teeth in adults, namely, the need to wear orthodontic appliances, prognosis, duration of treatment and anchorage. While the objective treatment difficulties are considerable, the adult patient may reject the whole plan of treatment because of the need to wear unaesthetic fixed orthodontic appliances for long periods. Among the 'invisible' appliances, the lingual orthodontic appliance is, at present, the only viable alternative to the traditional labial appliance which may be efficiently used to treat such complex conditions in adults.

In a Medline search of the English-language orthodontic literature, only one article was found that describes the use of lingual appliances in the treatment of impacted teeth in adults [1]. This is rather surprising in the light of the growing demand for facial and dental aesthetics by adult patients, and in view of the fact that lingual orthodontics has become established as a well-recognized and widely accepted discipline. It is nevertheless understandable because, in the treatment of the cases under discussion, an orthodontic appliance may need to execute as many as five different movements on the impacted tooth, involving vertical extrusion, tipping to the line of the arch, rotation, mesio-distal root uprighting and buccal root-torque (see Chapter 6). Achieving these with a lingual appliance is still considered to be more difficult by most clinicians, who would undoubtedly prefer to treat cases requiring these complex manipulations with the more familiar labial appliances.

In Chapter 9 it was pointed out that the prognosis for the success of the orthodontic resolution of the impacted canine in an adult is lower than in the young patient and that it worsens with advancing age. Furthermore, when such treatment is undertaken, its successful completion should be expected to take considerably longer than in younger patients [2]. For this reason, it is important to find creative ways to shorten the whole treatment, especially that part of treatment related to the canine impaction.

Differences in treatment approach engendered by the use of lingual appliances

Changes and adaptations need to be made to the protocols for the treatment of impacted teeth, which have been suggested in earlier chapters of this book in line with the demands of each of the various stages of treatment. Several problematic areas arise, and it is necessary to show how these may be overcome using the lingual appliance.

Following accurate positional diagnosis and a carefully planned strategy for erupting the impacted tooth, the lingual appliance will be bonded into place. Its aims will initially be directed at levelling, alignment and space-

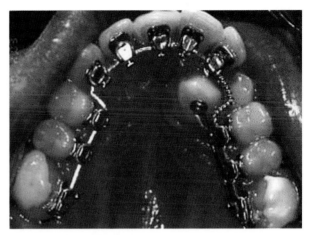

Fig. 10.1 Space opening with an open coil spring.

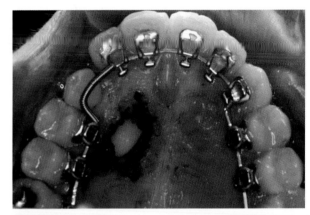

Fig. 10.2 Space maintenance with a closed coil spring. Note open surgical exposure of the canine.

opening for the impacted tooth. These goals may be realized in the present context with the help of copper–nickel–titanium (CuNiTi) archwires initially, followed by a heavier steel archwire, an open coil spring and sliding mechanics (Figure 10.1).

The space achieved must then be maintained until the impacted tooth has been initially aligned in the arch. With labial appliances, the use of a gently curved stainless steel tube threaded on the wire has been recommended to act as the space maintainer (see Chapter 6). However, this is not feasible in the lingual appliances, because of the need for a premolar offset. Therefore, in lingual treatment, space is usually maintained with a closed coiled spring (Figure 10.2) or with a pair of offsets distal to the lateral incisor and mesial to the first premolar, or by 'figure-of-eight' ligation of the teeth on either side of the space. At this stage, the patient is referred for surgical exposure of the impacted tooth.

In Chapter 3, the advantages of a closed eruption technique have been discussed and it was shown that it provides

a better periodontal and aesthetic result when compared to the open eruption technique [3, 4]

The closed eruption technique also results in less postoperative discomfort for exposures in the palate [5]. However, a closed eruption procedure for a palatal canine generally requires an intra-sulcular incision along the cervical margins on the palatal side of all the teeth, with the flap raised from first molar forward. In the presence of a lingual appliance, this procedure is clearly difficult to perform, since the brackets and their hook attachments are adjacent to and extend deeper than the cervical margins, obstructing access. Removing the archwire is mandatory prior to surgery, which complicates its postsurgical replacement and the application of traction.

The alternative approaches are:

- to perform closed surgery before appliance placement, leaving the ligature unattached and free in the palate until traction can be initiated, several weeks or months later;
- to perform an open surgical exposure, accepting the disadvantages of its postsurgical discomfort and post-treatment outcome.

Canine traction, eruption and alignment

In cases in which traditional labial orthodontic appliances are employed, direct traction to the archwire is often the most efficient line of treatment and is best achieved using elastic ties from the impacted tooth across the line of the arch to the labial side. However, with lingual appliances the distance to the lingual archwire is very short and direct traction is rarely appropriate, except in the early eruption phase of the traction. Following eruption, the lingual wire becomes an obstacle in the way of further progress of the canine. A buccal offset with a helix may be incorporated in the canine area of the lingual archwire to increase this distance and the range of the elastic (Figure 10.3). A palatal offset may also be used, in which the archwire is designed

to circumnavigate the canine on its medial side and close to the palatal midline raphe, while labial/buccal traction will need to be made to a small buccal attachment on a posterior tooth (Figure 10.4).

In group 2 canine cases, the intimate relation between the canine crown and lateral incisor root will block the canine movement if direct traction is applied, while in group 3 canines the height of the tooth may contraindicate the use of direct buccal traction. Therefore, the canine must be erupted first in a vertically downward and somewhat palatal movement in order to free it from its entanglement with the incisor roots, as has been amply demonstrated in earlier chapters. An appropriate canine auxiliary should be prepared to be placed at the time of surgery as in a labial approach, such as a suitably modified full auxiliary arch [6]. Again, due to the considerably shorter distance between the impacted tooth and the lingual archwire, the range of action will be significantly decreased and there is the risk of inadvertently applying excessive extrusive forces (Figure 10.5).

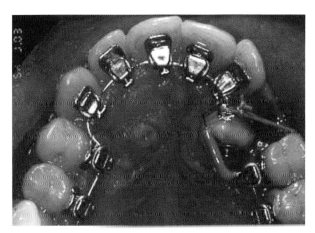

Fig. 10.4 Elastic thread tied between the buccal eyelet of the impacted canine and an attachment bonded on the buccal aspect of the first molar (unseen). A palatal offset has been inserted in the lingual archwire

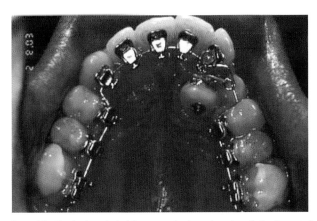

Fig. 10.3 Elastic thread tied between the buccal eyelet of the impacted canine and a loop in the buccal offset of the lingual archwire.

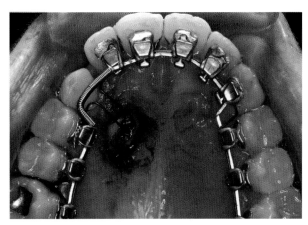

Fig. 10.5 Canine auxiliary ligated to canine eyelet under main lingual arch.

Therefore, the auxiliary should be made from lighter wires or its activation range should be decreased. This spring is inserted piggyback under a lingual heavy rectangular base arch – the latter having been placed to consolidate the anchor unit, while the spring provides a light extrusive force to the canine. In this way, unwanted movement on the adjacent teeth will be avoided. Alternatively, a light active palatal arch (see Chapter 6) may be used in combination with lingual appliances, although it requires double molar tubes. Preferably, these should be welded to preformed orthodontic bands, although the bands themselves may not be acceptable by the patient for aesthetic reasons, despite the fact that they are so distally sited in the mouth.

Once the canine has erupted in the palate, it must be moved buccally towards its place in the arch and the same means may be used as described above. Vertical offsets, designed to erupt the tooth, are limited by the likelihood of impingement by the occlusion of the lower teeth. However, during movement towards the buccal, occlusal interferences with the opposing teeth, which are sometimes encountered when a labial appliance is used, are obviated by the bite-opening effect of the lingual appliances, which eliminates these interferences and which facilitates canine migration at this stage.

The short inter-bracket span characteristic of lingual appliances is another difference which has clinical implications and will demand the much wider use of super-elastic archwires. With such short inter-bracket spans, even these wires are sometimes too stiff to be fully engaged in the bracket slots, and may often be tied tightly into the initial canine attachment only when this tooth has reached a position which is relatively close to its place in the arch. This is done to avoid the application of excessive force on the resolving impacted tooth and also to minimize the reactive forces on the adjacent teeth (Figure 10.6).

When the canine reaches its place in the arch a bracket has to be bonded. In traditional labial orthodontics,

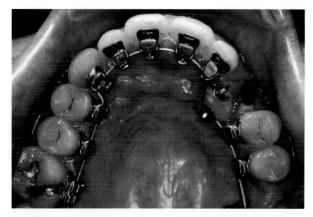

Fig. 10.6 Nickel–titanium archwire inserted through the palatal eyelet.

bonding of a bracket at its ideal height on the buccal aspect of the canine is usually impeded at this stage by exuberant gingival tissue, which accumulates on the buccal aspect during the canine's migration towards the buccal, even with good oral hygiene. In contrast, bonding of a lingual bracket is much easier, since the clinical crown on the palatal side is usually fairly long, particularly if an open exposure was performed for this tooth.

Finishing procedures

Finishing procedures with lingual appliances are similar to those with labial appliances. However, since torqueing auxiliaries are largely unsuitable and certainly difficult to design and insert, torque has to be introduced in the rectangular archwire or in the bracket base. As with the labial appliance, torqueing a canine with a rectangular archwire might take a long time because of the small range of deflection possible with the wire and the need for a number of progressive activations. It should be clearly understood that these torqueing deflections will generate small but undesirable reciprocal torqueing movements of the adjacent anchor teeth at each activation, which correct themselves as each activation works itself out. This is generally referred to as 'round-tripping' the anchor teeth, a phenomenon which has been blamed as a cause of root resorption generally. So, while torqueing auxiliaries are highly recommended in labial appliances to avoid this potentially harmful side-effect, since these do not derive their anchorage from a reverse torque of the adjacent teeth their use with lingual appliances is presently limited.

An additional aspect of lingual treatment, already referred to, is the fact that many of the patients concerned have very high aesthetic demands and will not tolerate the extraction space of the deciduous canine without some form of temporary pontic placement. For this reason, the extraction of the deciduous canine has to be delayed until a very late stage of the treatment. If it is necessary to remove the deciduous tooth at an earlier stage, an artificial tooth must be bonded to the adjacent teeth during the intervening period (Figure 10.7). This may also be a problem with an adult undergoing any form of orthodontic treatment, although the likelihood is much less when the patient has agreed to wear labial appliances, particularly if these are all metal.

Anchorage considerations

One of the most important principles of mechano-therapy in the treatment of cases with impacted canines is to establish a firm anchorage unit. This is obviously true for both labial and lingual techniques. Because impacted teeth are

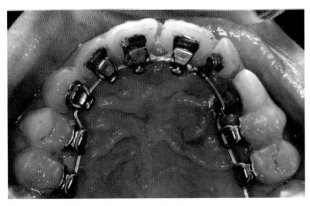

Fig. 10.7 Aesthetic pontic bonded to the first premolar. Note the impacted canine has been brought to the line of the arch, beneath the pontic.

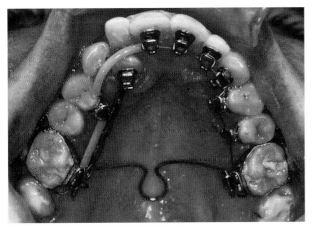

Fig. 10.8 Trans-palatal arch integrated into lingual appliance design.

much more resistant to movement in adults than in children, the effort to bring them into alignment will be reflected in a greater loss of anchorage. This may be further undermined and the problem compounded if, in addition, the anchor teeth are periodontally involved, with loss of bone support.

The use of most traditional means of enhancing the anchorage in children, such as palatal arches (Figure 10.8), consolidation of anchor units with heavy rectangular arches and intermaxillary traction, are appropriate for adults, while others, such as extra-oral appliances and lip bumpers, are taboo for adults in general, and particularly for the patient who wants the appliance to be completely invisible and who is averse to anyone knowing that the treatment is being performed at all. New opportunities have been opened for adults with the introduction of temporary and osseo-integrated anchorage devices, which have been developed and used in cases in which a large amount of tooth

movement is required in the absence of adequate alternative anchorage.

Mini-/micro-screws have many advantages. They are inexpensive, small, simple to place, immediately loadable and well tolerated by patients [7–11]. Their main disadvantage is their proximity to the roots, which may be damaged during placement of the screws or when the adjacent teeth are displaced [12].

Integrating implants with lingual appliances

A well thought-out strategy for the placement of implants offers the possibility of facilitating and shortening treatment while, at the same time, decreasing the deleterious effects of anchor loss. This may be achieved in the following ways:

1. A titanium mini-screw may be inserted on one side of the posterior area of the palate, to provide anchorage for erupting a group 2 or group 3 canine (see Chapter 6) into the palate (Figure 10.9a). To generate the extrusive forces, a 0.016 in stainless steel ballista spring with a loop at its end may be inserted through the internal slot of the mini-screw (Figure 10.9b). The spring is activated by making a permanent bend in it, down towards the tongue, and then tying it to the pigtail ligature wire of the impacted tooth (Figure 10.9c). The elasticity of the spring exerts pressure for it to return to its more vertical resting shape and position, thereby applying extrusive force to the uncrupted tooth. The ballista spring should be ligated into place at the end of the surgical procedure, while the local anaesthetic is still effective.

2. A micro-screw may be inserted into the bone on the labial side of the alveolar ridge [1] to provide the necessary anchorage for moving the tooth buccally towards its place in the arch, once it has been erupted into the palate. Elastic thread, tied from an attachment bonded on the buccal aspect of the canine to the micro-screw, will create a buccally directed force and a moment for correcting its rotation. The archwire should be offset in the canine area to permit the canine movement and a helix added in the offset will direct the buccal migration more mesially or more distally, as needed (Figure 10.10a). If the canine is located in a direct line open to the implant, then a useful alternative would involve cutting out the short span of archwire between the distal of the lateral incisor and the mesial of the first premolar to leave a larger and a smaller sectional archwire in place and the canine area unencumbered (Figure 10.10b). A continuous archwire may be reinserted once the canine is close to its place, through the palatal eyelet or a newly bonded lingual bracket.

3. In cases where direct traction of the impacted tooth to the archwire is possible from the beginning of treatment,

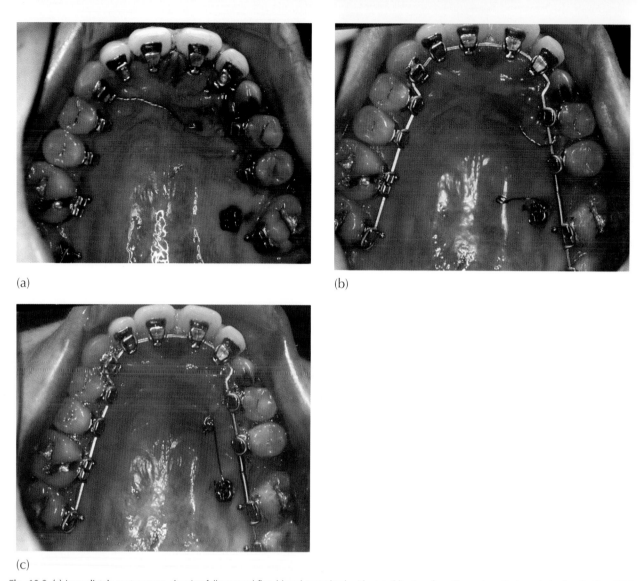

(a)

(b)

(c)

Fig. 10.9 (a) Immediately post-surgery, showing fully sutured flap (closed procedure) with pigtail ligature from the canine attachment piercing the flap. Mini-screw inserted in the palatal mucosa near the space of the missing first molar. Ballista spring tied into mini-screw. (b) Passive state. (c) Activated by ligation to pigtail ligature, applying traction to the unerupted canine.

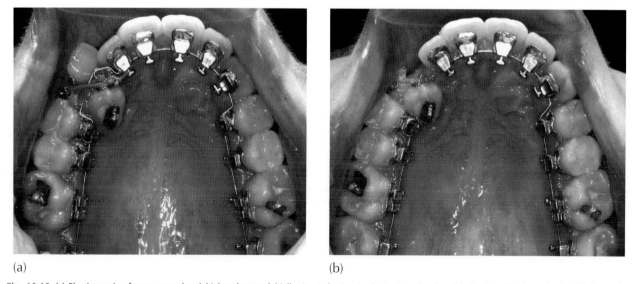

(a)

(b)

Fig. 10.10 (a) Elastic traction from a secondary labial eyelet to a labially situated micro-implant, via a directional helix placed in the archwire. (b) The canine is in a direct line to its place. The lingual archwire has been divided into two, leaving the canine area unencumbered.

a micro-implant may be introduced in the buccal bone at the same appointment during which the surgical exposure is performed (Figure 10.11a). An elastic force from the impacted canine to the micro-screw may then be applied immediately, while the area is still anesthetized (Figure 10.11b). If the deciduous canine has to be extracted, an aesthetic pontic needs to be fabricated and the elastic passed under the pontic. In these cases, it is recommended to delay the insertion of the implant until a period of healing of the extraction space has occurred, since the chances of failure are greater when an implant is inserted into a fresh extraction site. Thereafter, the canine may be moved under the palatal mucosa towards its place in the arch. In order to move it further buccally and also to extrude it, a titanium–molybdenum alloy

(TMA™) spring with an artificial tooth may be connected to the implant and activated buccally and vertically, as necessary (Figure 10.12).

It will be seen that a group 2 canine requires two implants, one in the palate to erupt and move the canine away from the lateral incisor root and the other in the buccal alveolar plate for moving the tooth buccally into the dental arch. Group 1 cases, in which direct traction is possible, require only one implant placed in the buccal plate. The most important advantage of using orthodontic implants is that treatment of the impaction may be initiated before levelling and alignment and opening of adequate space in the arch and continued in an entirely independent manner. Hence, the clinician has two possible options:

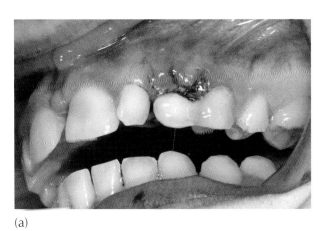
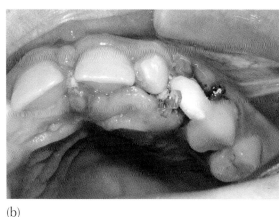

(a)

(b)

Fig. 10.11 (a) Micro-implant inserted during the appointment for canine exposure. (b) A clear chain will be drawn beneath the pontic to the micro-implant, to apply buccal traction to the impacted canine.

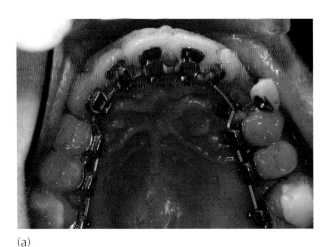
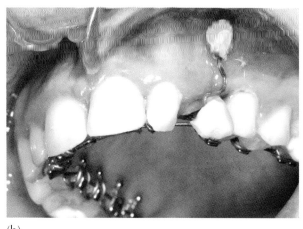

(a)

(b)

Fig. 10.12 (a, b) The titanium–molybdenum alloy (TMA™) spring in its passive mode, bonded to the implant and carrying an artificial tooth. Ligation to the canine will draw the tooth vertically and buccally.

1. To erupt the canine concomitantly with and as an integral part of the orthodontic treatment of the other teeth. This is relevant when space opening in the dental arch is needed first. In this case, it would be wise to erupt the canine only partially (in contrast to what is recommended for labial treatment – see Chapter 6), and then to continue its buccal movement underneath the palatal mucosa until it passes the lingual archwire to the buccal side. This will avoid archwire interferences and the need to fabricate offsets.

2. To perform the entire stage of treatment required for treating the impaction before placing any orthodontic appliance. This approach is suitable in cases in which there is adequate space for the canine in the arch, but also in many other cases, in the interests of reducing the time that orthodontic appliances need to be in the mouth.

The use of implants in this way is valid for both labial and lingual treatments and may significantly shorten the period the patient has to wear either unaesthetic labial or uncomfortable lingual orthodontic appliances. The idea of completing the resolution of the impaction before placing appliances is particularly advantageous in lingual treatment, since the problem of interference of canine movement by the archwire is circumvented before appliances are placed. Orthodontic treatment in general is a discipline in which a force system is set up to move certain teeth in predetermined directions and in a particular way, while at the same time preventing the reactive forces that are transferred to the anchor teeth from causing unwanted deterioration in their positions. Thus, implant anchorage, which is absolute and not relative anchorage, greatly simplifies the approach, because forces may be applied solely to the impacted tooth, thereby avoiding ill-effects on the other dentition and the need for a cumbersome orthodontic system. Implants may be placed in a wide variety of positions, both within and well outside the dental arch, and as such may be carefully planned and strategically chosen for optimal direction and range of force application. The combination of lingual appliances and implant anchorage is useful in the ortho-surgical treatment of impacted teeth in the adult patients who would otherwise reject treatment altogether because of the need to wear an unaesthetic labial appliance over a long period of time. This combination has the ability to provide answers to three major drawbacks of the ortho-surgical treatment of impacted teeth in the adult patient, namely, unimpaired orofacial appearance, length of time that fixed appliances need to be worn and anchorage.

Case report

A 24-year-old woman presented with a palatally impacted left maxillary canine, an adjacent over-retained deciduous canine and a missing maxillary left first molar (Figure 10.13a, b). Lingual brackets (Ormco 0.018 in slot) were bonded on all the teeth in the upper arch, with the aim of improving the anterior alignment, correcting the midline deviation and opening adequate space for the impacted canine and the missing first molar. Since the impacted tooth was classified as a group 2 canine, it needed to be moved in two stages: first, to erupt it into the palate to the occlusal level, in order to distance it from the roots of the lateral incisor, and second, to move it buccally into its place in the arch.

To provide the extrusive forces, a 0.016 in stainless steel ballista spring with a loop at its end was inserted through the internal slot of a Spider Screw® orthodontic implant (11 mm length and 2.3 mm diameter; HDC Company, Sarcedo, Italy) [13] as described above. In this case, the mini-screw was inserted through the palatal mucosa near the space of the missing first molar, into an area of bone that would later become the molar implant site and where there was no possibility of damaging adjacent teeth (Figure 10.13c, d).

In the succeeding months, the canine was erupted using a force applied from the implant, through the agency of the ballista spring. At the same time, routine orthodontic treatment was under way on the other teeth using forces provided from a conventional lingual orthodontic appliance (Ormco 0.018 in slot), based on all the other teeth. The two systems operated independently in the early stages and there was no initial interaction between them.

When enough space had been created in the arch, an additional eyelet was placed on the buccal aspect of the canine and the direction of traction altered to a pure buccal tipping movement to bring the tooth into the arch. A new, round, lingual archwire with a buccally directed offset and a loop was fabricated. Ligation with elastic thread tied between the buccal eyelet of the impacted tooth and the loop produced the buccally directed force at this stage (Figure 10.13e).

When the canine approached its place in the arch, the deciduous canine was extracted and the rectangular wire changed to a 0.014 in nickel–titanium wire, inserted through the palatal eyelet of the impacted tooth. In addition, a clear elastic thread was inserted through the buccal eyelet and ligated to an attachment bonded on the buccal side of the upper molar (Figure 10.13f). Subsequently, the lingual eyelet was changed to a lingual bracket and a 0.016 × 0.022 in CuNiTi archwire was reinserted. Active treatment was completed in 19 months (Figures 10.13g–i). The spider screw was subsequently removed without complications. No treatment was performed in the lower arch.

At the end of the orthodontic phase, the patient was referred for placement of a single tooth implant in the reopened extraction site of the maxillary first molar.

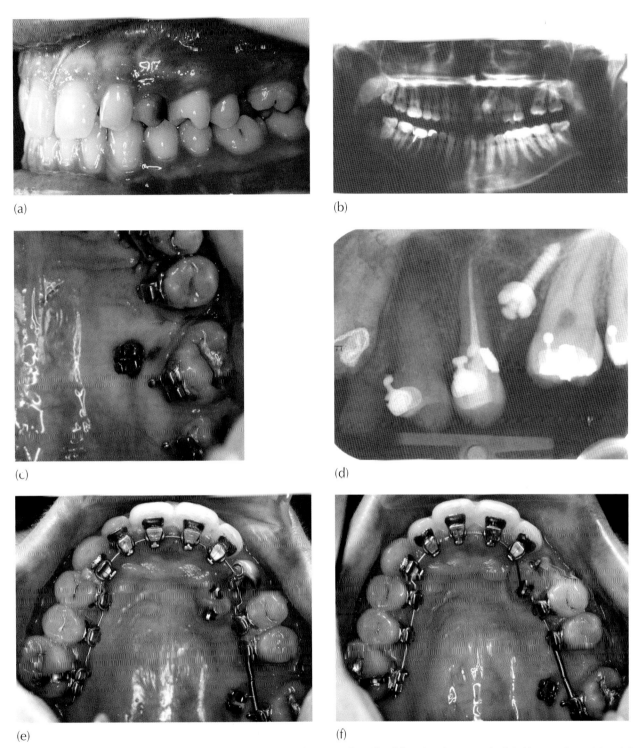

(a)

(b)

(c)

(d)

(e)

(f)

Fig. 10.13 Pre-treatment clinical (a) and panoramic (b) views showing the impacted maxillary left canine, the over-retained deciduous canine and the missing maxillary first molar. Mini-screw inserted through the palatal mucosa and into the bone opposite the space of the missing first molar: (c) clinical view, (d) radiographic view. Canine traction: (e) elastic traction to buccal offset in archwire, (f) engagement of nickel–titanium wire in palatal eyelet and elastic chain in buccal eyelet, drawn to buccal attachment on second molar. Post-treatment intra-oral views: (g) frontal, (h) left side, (i) post-treatment panoramic view showing the impacted canine aligned in the arch.

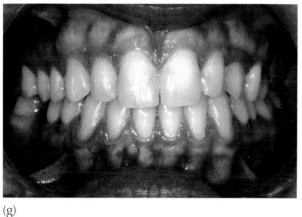

(g)

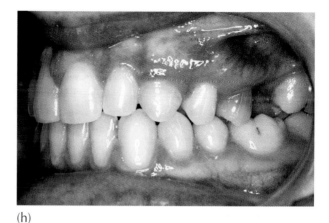

(h)

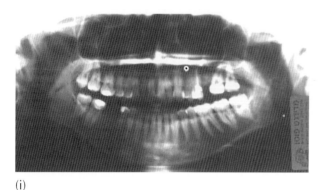

(i)

Fig. 10.13 (*Continued*)

References

1. Park HS, Kwon OW, Sung JH. Micro-implant anchorage for forced eruption of impacted canines. J Clin Orthod 2004; 38: 297–302.
2. Becker A, Chaushu S. Success rate and duration of orthodontic treatment for adult patients with palatally impacted maxillary canines. Am J Orthod Dentofacial Orthop 2003; 124: 509–514.
3. Becker A, Brin I, Ben-Bassat Y, Zilberman Y, Chaushu S. Closed-eruption surgical technique for impacted maxillary incisors: a postorthodontic periodontal evaluation. Am J Orthod Dentofacial Orthop 2002; 122: 9–14.
4. Chaushu S, Brin I, Ben-Bassat Y, Zilberman Y, Becker A. Periodontal status following surgical-orthodontic alignment of impacted central incisors with an open-eruption technique. Eur J Orthod 2003; 25: 579–584.
5. Chaushu S, Becker A, Zeltser R, Branski S, Chaushu G. Patients' perception of recovery after exposure of impacted teeth: a comparison of closed- versus open-eruption techniques. J Oral Maxillofac Surg 2005; 63: 323–329.
6. Kornhauser S, Abed Y, Harari D, Becker A. The resolution of palatally impacted canines using palatal-occlusal force from a buccal auxiliary. Am J Orthod Dentofacial Orthop 1996; 110: 528–534.
7. Roberts WE, Marshall KJ, Mozsary PG. Rigid endosseous implant utilized as anchorage to protract molars and close an atrophic extraction site. Angle Orthod 1990; 60: 135–152.
8. Wehrbein H, Merz BR, Diedrich P, Glatzmaier J. The use of palatal implants for orthodontic anchorage. Design and clinical application of the orthosystem. Clin Oral Implants Res 1996; 7: 410–416.
9. Kanomi R. Mini-implant for orthodontic anchorage. J Clin Orthod 1997; 31: 763–767.
10. Melsen B, Costa A. Immediate loading of implants used for orthodontic anchorage. Clin Orthod Res 2000; 3: 23–28.
11. Lee JS, Park HS. Micro-implant anchorage for lingual treatment of a skeletal Class II malocclusion. J Clin Orthod 2001; 35: 643–647.
12. De Clerck H, Geerinckx V, Siciliano S. The zygoma anchorage system. J Clin Orthod 2002; 36: 455–459.
13. Maino BG, Bednar J, Pagin P, Mura P. The spider screw for skeletal anchorage. J Clin Orthod 2003; 37: 90–97.

11

Rescuing Teeth Impacted in Dentigerous Cysts

Orthodontic Treatment of Impacted Teeth, Third Edition. Adrian Becker.
© 2012 Adrian Becker. Published 2012 by Blackwell Publishing Ltd.

Cysts are found in a variety of tissues and in many sites in the human body. They are fluid-filled, epithelium-lined, balloon-like lesions, which generally enlarge progressively and painlessly due to hydrostatic pressure from within. If they are developing in a homogeneous medium, then the laws of physics determine that they will be spherical in shape. If the medium is bone, then this will resorb in response to pressure from within, thereby progressively permitting the enlargement. Simultaneously, there is a reactive bone apposition process on the outer side, which causes the bone to become expanded. This apposition is slower than the resorption from within, and so the bony expansion is accompanied by a thinning of the bony walls of the cyst, which may eventually become paper-thin. In the final instance, the cyst will resorb the last remnant of the hard tissue to become fluctuant beneath the skin or, in the context of this book, the oral mucosa.

Dentigerous cysts

A dentigerous cyst is a specialized type of cyst insofar as it is located around the crown of an unerupted tooth and arises as an expansion of what is normally the very narrow space between the inner and outer enamel epithelia of the dental follicle, in which the completed crown of the tooth has developed. These two layers meet at the cemento-enamel junction (CEJ) at the neck of the developing tooth [1]. The cystic area enlarges due to the production of fluid by the epithelium, forcing the layers apart into its typical spherical shape, except where the crown of the tooth protrudes into the cyst, with its long axis traversing the centre of the lesion. In its early stages, therefore, it is difficult to distinguish a dentigerous cyst from a benign enlargement of the dental follicle, and between that and the normal follicle of the tooth. Furthermore, when eruptive movements of the tooth bring it through the bone and in close proximity with the oral mucosa the early dentigerous cyst may be fluctuant to palpation, is termed an eruption cyst and usually resolves spontaneously by rupture into the oral cavity. The tooth then erupts normally, its gingival attachment becomes normal and no subsequent signs of the earlier pathology remain.

However, when the cyst becomes larger, the pressure from within overcomes the tooth's inherent eruptive force potential to stop its normal eruption, and it may even cause the tooth to 'back up' along its former eruption path, displacing it apically, while still remaining in the middle of the cyst (Figure 11.1). It

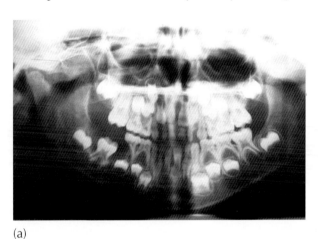

(a)

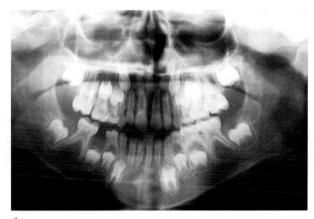

(b)

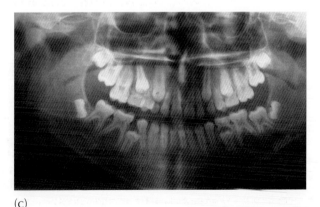

(c)

Fig. 11.1 A dentigerous cyst surrounds the crown of the mandibular right first molar has arrested eruption and has apparently slowed down root development. (b, c) The follow-up radiographs, taken two and five years later, show space closure having occurred from both the distal and the mesial, impacting the second premolar. The bizarre root morphology of the molar is the same on the affected and unaffected sides. Nevertheless, for what seemed a short root of questionable prospects for further growth in its initial position, growth has rebounded very impressively.

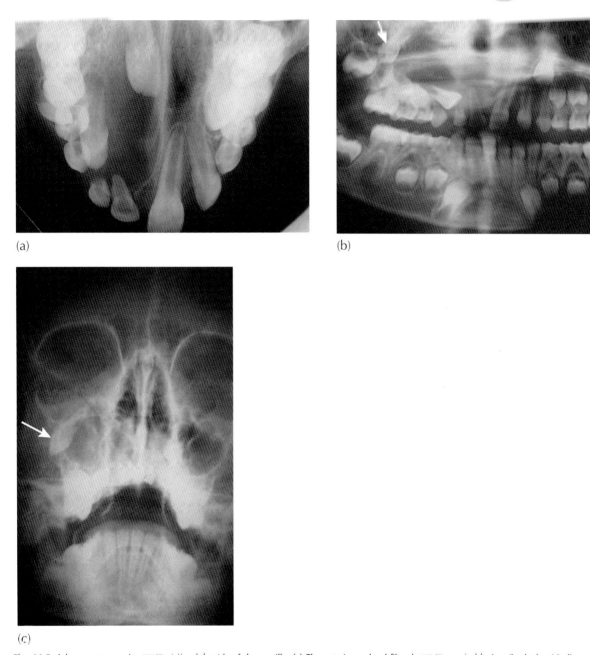

(a)

(b)

(c)

Fig 11.2 A large cyst occupies much of the right side of the maxilla. (a) The anterior occlusal film shows the central incisor 'backed up' in line with its long axis. (b) The initial panoramic film shows the lateral incisor and both premolars laid out horizontally in the floor of the cyst. The canine has become displaced posteriorly into the molar region (arrow). The location and orientation of these teeth defines the perimeter of the cyst. Because the central incisor has been displaced superiorly and posteriorly, and is distant from the film, its image is beyond the focal trough of the film and hence has largely disappeared. (c) A Waters projection radiograph of the skull shows the degree of buccal displacement of the canine (arrow), indicating the lateral boundary of the cyst.

seems, therefore, that the term 'eruption cyst' may represent a state of pressure equilibrium between the forces of eruption and the opposing force of intra-cystic pressure. With a smaller eruption cyst, the forces of eruption might prevail, as we have pointed out above. However, a larger eruption cyst may require surgical incision to bring about decompression, which will then permit the tooth to erupt spontaneously and quite quickly in the younger patient.

Regarding the affected tooth itself, its long axis is generally perpendicular to the epithelial outer wall of the cyst at its CEJ (Figure 11.2a). A good portion of the apical section of its root remains firmly in bone, but the pressure-generated enlargement of the cyst will cause its lining to circumferentially and progressively outline more and more of the otherwise bared coronal portion of the root, by a resorptive squeezing out of the crestal area of bone.

With further expansion, the lateral aspects of the cyst come up against adjacent unerupted teeth, which become pushed aside as their supporting bone becomes thinner and as the cyst lining comes into direct contact with their

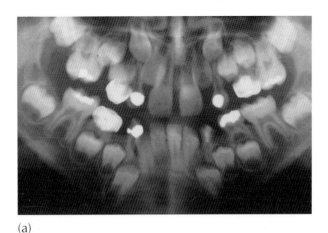
(a)

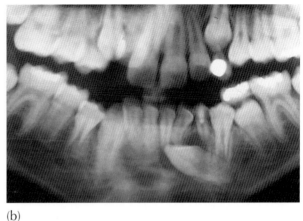
(b)

Fig. 11.3 (a) An incomplete root canal treatment has been performed in the mandibular left deciduous canine, as seen on this panoramic view. (b) A routine follow-up panoramic radiograph, taken prior to impending orthodontic treatment, reveals the development of a dentigerous cyst around, and displacement of, the permanent canine. (Courtesy of Dr M. Barzel.)

periodontal ligament (PDL). Each of these teeth then becomes profiled in the walls of the cyst, with only a thin epithelial lining to cover one side of the bared tooth along its full length (Figure 11.2b, c). The other side of the displaced adjacent tooth is invested by bone, with PDL intervening over the root area and its own dental follicle intervening in the crown area. The degree of displacement of these teeth will depend on the size of the cyst, but they lie in and are oriented parallel to the walls of the cyst, in contrast to the perpendicular orientation of the causal tooth.

Histologically, the cyst lining is squamous or stratified squamous epithelium and as such is similar to the follicle from which it originated. Thus, the only criterion for distinguishing between an enlarged follicular sac and a dentigerous cyst is size, as seen on a periapical radiograph. Accordingly, a workable definition for a dentigerous cyst is when the distance between the crown of the tooth and the dental sac is larger than 2.5–3 mm on the film. This is the definition most frequently used in radiology and surgery. However, we may like to refine this definition with more discerning biological criteria, which take into account the implications of the increased hydrostatic pressure within the lesion. Thus, should the eruptive progress of the affected tooth be halted or reversed, or should the radiolucent outline of the lesion appear to connect with the tooth some way down the root, apical to the CEJ, these might be considered more valid diagnostic signs.

The aetiology of a dentigerous cyst is strictly unknown, although it is associated with chronic local inflammation, as may be seen when the deciduous predecessor has become non-vital and an apical granuloma has initiated cystic change in the follicular sac of the unerupted permanent successor. An incomplete root canal treatment may leave the deciduous tooth symptomless, but the chronic periapical lesion may remain unresolved and this may act

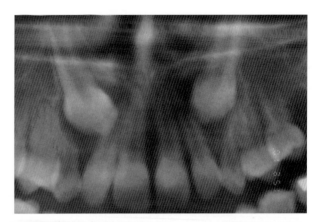

Fig. 11.4 The anterior portion of a panoramic view showing cystic enlargement of the follicles of the two maxillary permanent canines.

as the irritant that induces a cyst to develop from the follicle of the permanent successor (Figure 11.3). Unerupted permanent teeth, particularly maxillary canines, often exhibit enlarged follicles, and these may sometimes become further enlarged (Figure 11.4). This seems to occur frequently when there is a close association with the roots of the lateral incisors, although no specific explanation is forthcoming.

Radicular cysts

After root formation of a tooth is completed, Hertwig's root sheath disappears, leaving behind only the epithelial rests of Malassez in the immediate area. These remnants may sometimes be stimulated by the presence of a periapical granuloma to become cystic [2]. The radicular cyst is therefore to be found at the apex of a non-vital tooth, and it too may grow at the expense of the surrounding bone, to displace and partly envelop unerupted adjacent teeth, with the crowns and roots covered by the cyst epithelium. In this

case, all the displaced teeth will be found in the walls of the cyst and oriented at a tangent to the cyst lining that covers them. No tooth will have its entire crown enveloped within the cyst, and the cyst lining will not be attached to the cervical area of any one of the teeth involved.

Histologically and clinically, the radicular cyst is similar to the dentigerous cyst and the clues to its identity are largely clinical (Figure 11.5). Accordingly, the following discussion of treatment and outcome will not differentiate between the two in any substantial manner.

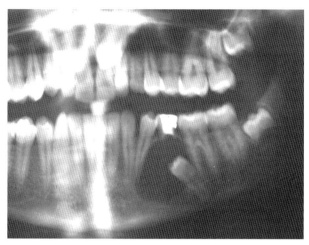

(a)

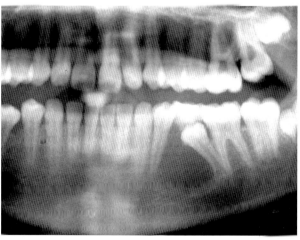

(b)

Fig. 11.5 This figure shows two similar situations arising from different causes. (a) The crown and coronal part of the root of the second premolar are encompassed within the cyst and the orientation of the tooth is towards the middle of the cyst. The asymmetry may be due to the relative ease with which the first premolar root has been moved away from the lesion, while the more robust, two-rooted molar has withstood the cyst pressure. This is a dentigerous cyst. (b) The root of the erupted first premolar has been displaced mesially as with the above case, but the second premolar has been pushed distally against the molar and there has been distal root movement away from the cyst. There has been total stripping of bone support of the entire mesial side, while a thin wedge of bone can be seen between the molar and the affected premolar. This is a radicular cyst, presumably due to an incomplete root treatment on the second deciduous molar.

Treatment principles

Surgery

Professional intervention may influence the future of the displaced tooth, depending on the type of treatment employed to resolve the cyst. The cyst may be opened and its lining completely shelled out in a procedure called enucleation [3]. The excised tissue is then sent for histological examination. Following the procedure, the crown of the tooth is left fully exposed within the cystic cavity and the tooth itself exhibits a high degree of mobility, due to its very rudimentary and reduced supporting periodontal attachment. A complete and hermetic closure of the surgical flap is usually attempted, sealing the former cyst cavity from the exterior to protect the exposed bone and soft tissue from infection. The aim is that this will fill with a blood clot and healing is by primary intention.

If the cyst is very large, then the chances of infection and breakdown of the clot become significant and, even in successful cases, the associated unerupted tooth becomes deeply buried in the newly forming bone, and it will need to erupt a considerable distance through this repairing bony tissue. In such circumstances, spontaneous eruptive movement of the tooth may be slight and a tooth grossly displaced by a large cyst may remain in an inaccessible position, following a filling-in of the surrounding tissues. Its extraction may then be unavoidable [4]. The prognosis for its eruption would appear to be in inverse proportion to its distance from the oral cavity. In the event that the blood clot breaks down due to infection and the area is left to heal by secondary intention, the entire former cystic area remains exposed to the exterior and will require some form of protective dressing or pack. In this situation, most of the teeth involved will need to be extracted, leaving a large defect of the basal bone in the final analysis, which may alter the shape and contour of the patient's face. This will have marked functional, cosmetic and psychological consequences in the long term [4], with the need for artificial replacement of the missing teeth and, possibly, even a maxillofacial prosthesis to overcome the basal defect. Accordingly, enucleation is generally recommended for relatively small cysts.

For larger cysts, a two-stage procedure is often advised, in which the first stage is aimed at decompression and drainage of the cyst without removing its epithelial lining. Several months later, when the cyst has reduced in size, the lesion will be reopened and the remaining epithelium shelled out, in what has been termed the 'curative enucleation of the cyst lining' [4]. There can be little doubt that total removal is not always easy to achieve, since the exercise may threaten the unerupted teeth that are lodged in the walls of the cyst (Figure 11.6). These teeth have a very limited, rudimentary and tenuous periodontal attachment to the adjacent bone. Thus, the teeth themselves may be inadvertently plucked from their places and, conversely, some detached areas of epithelium may remain within the healing tissues, which may later regenerate.

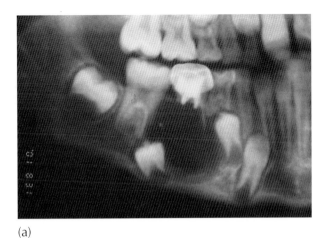

(a)

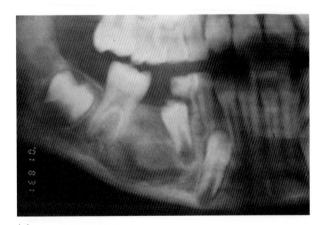

(b)

(c)

Fig. 11.6 (a) A radicular cyst, due to unresolved periapical pathology after root canal treatment of a second deciduous molar. The very rudimentarily developed second premolar has become severely displaced inferiorly in the lower border of the mandible. (b) The same view immediately after enucleation of the entire cyst lining. (c) The same view six months later, showing bony fill-in.

Figure 11.6a shows a portion of a panoramic view of the right side of the lower jaw of a 7-year-old patient, in which a large cyst is present. It seems likely that this is a radicular cyst related to the non-vital second deciduous molar, which has pushed aside the developing and largely rootless un-erupted second premolar, whose long axis is parallel to the wall of the cyst. This patient was treated by extraction of the deciduous tooth, together with excision and complete enucleation of the cyst. In the accompanying transfer letter, the surgeon noted that he had found it necessary to remove the second premolar, and the second film (Figure 11.6b), taken postoperatively, shows the absence of this tooth.

From the surgeon's perspective, three points are noteworthy in the management of this patient:

1. The very underdeveloped second premolar had a minimal attachment to its surroundings, as seen on the original film.
2. The cyst lining itself probably constituted a significant part of the attachment of this tooth.
3. During a surgical procedure of this type, visibility inside a cyst cavity is poor.

The surgeon may have judged that a tooth in this position, with this orientation, with an obvious absence of bony support and in the absence of a developed root had a poor prognosis and therefore removed it. Indeed, had the tooth remained in place, there is room to question whether it would have continued to develop a normal root and if it could ever have been expected to erupt into the mouth, either spontaneously or with orthodontic help. However, its root shows the very earliest stages of its development, is wide open and is vital, indicating potential for the genesis of a reasonable root, which will generally occur in time, as illustrated above (Figure 11.1).

One remaining possibility is that under these exceptional circumstances it is conceivable that this very early developing tooth, with its very tenuous attachment to its surroundings, had been dislodged inadvertently during the removal of the cyst lining!

It should be remembered that the lining of a dentigerous cyst arises from the inner and outer enamel epithelia of the tooth germ. In the long term and under normal circumstances, this epithelial tissue of the follicle eventually fuses with the epithelium of the oral mucosa, opens and everts to form the normal gingival attachment of the healthy tooth (see Chapter 6). During cyst formation the lining usually remains unaltered, and opening it to the oral environment, with a marsupialization procedure, will allow it to become contiguous with the oral mucosa. In time, a metaplasia of its epithelium will occur, as is the case with every normally erupting tooth, following which the former cyst lining will become histologically indistinguishable from the remainder of the oral mucosa.

Marsupialization is the surgical treatment option that is to be preferred for the larger lesions of this type [5–8]. This

Involves opening the cyst into the oral cavity at its most superficial point and maintaining the patency of this orifice over a long period of time.

The cut linings of the cyst and the oral mucosa fuse to become continuous with one another. The teeth involved directly or indirectly with the lesion are deliberately left undisturbed and remain covered by the thin epithelial continuum of the cyst lining.

In time, the lined cyst cavity becomes smaller and smaller as bony regeneration occurs behind the epithelium, to fill the defect in from the bottom up. As it does so, the causal tooth is generally carried bodily forward, as if on the crest of a wave, in the vanguard of the naturally occurring regeneration.

Understanding this natural process was, no doubt, the stimulus that caused periodontists to invent guided tissue regeneration in which bone will regenerate under an artificial membrane, by deliberately preventing the encroachment of epithelium.

Of much greater significance, the areas of root surface that had earlier become denuded by the enlarging cyst, with only the thinnest epithelial covering of cyst lining, now become infused with new bony support. Bone levels return to the more normal values seen in other unerupted teeth due to the osteogenesis that lifts the epithelial lining away from the root surface. In this way, teeth that previously appeared to be in hopeless situations may now take on a new lease of life, provided that orthodontic treatment may later be instituted to complete any final eruptive assistance they may need and then to align them. However, any attempt to apply orthodontic traction to teeth before complete resolution of the cyst and before the maximum amount of spontaneous eruption has occurred will extrude them ahead of the advancing bone, thereby weakening their bony and periodontal support and prejudicing their longevity.

Spontaneous resolution of the impaction may be expected to occur to a significant degree when the cyst is eliminated in this way, and several truly remarkable cases have been reported in the literature [5, 6]. From our experience in Jerusalem, these case reports are by no means exceptional and, with rational and careful management, have been repeated with a high degree of reliability in the most extreme cases that have come under our care [8].

It is therefore to be concluded that the cyst must be treated first and that monitoring of the healing process should be instituted until the bone has completed the reparative fill-in of the bony defect. Only at that point, which will be many months later, should an assessment be made of how much improvement has occurred naturally, how much more may be expected and how much orthodontic treatment is needed to improve the positions of the teeth. Not only, therefore, is there no value in bonding an attachment to the tooth at the time that the marsupialization is undertaken initially, but the procedure is likely to be quite harmful to the natural recuperation and positional

improvement of the target teeth and for the health and integrity of the surrounding tissues – not least the cyst lining.

From the point of view of the oral and maxillofacial surgeon, treatment of this cyst is a priority in order to confirm the relatively innocent diagnosis. It must be remembered that, until a biopsy and pathological investigation are performed, the diagnosis is only tentative. The more sinister alternative diagnoses are fortunately extremely rare; nevertheless, the surgeon cannot take the chance involved in delaying the performance of the necessary diagnostic procedures until the patient is ready to accept orthodontic treatment. Therefore, the first stage of surgical treatment, aimed at eliminating the pathological entity, must be regarded as obligatory, and must include histopathological examination of biopsy material and careful follow-up.

Orthodontics

From the orthodontic point of view, the patient should be prepared for treatment, with an understanding of the demands of oral hygiene and the need to wear appliances and to expose the impacted teeth. These requirements may often be fulfilled quite quickly, but it is unfair to coerce young patients into a hurried decision before they are ready and is usually counterproductive. It would be logical to infer from this discussion that, although the orthodontist has much to contribute to the outcome of the treatment, this contribution relates to the later stages and not to the immediate prospects. It would therefore be wise to include the orthodontist in the decision-making team, but inappropriate for him/her to become actively involved in the early stages of treatment, until much of the bony filling-in has occurred and the tooth has migrated down ahead of this.

The prognosis of teeth which have been severely displaced by cysts

Given the presence of defined pathology and an extreme displacement of the affected tooth or teeth in relation to dentigerous cysts in general, the orthodontist and the oral surgeon need answers to the following five questions, before being equipped to formulate a treatment protocol:

1. Following treatment to eliminate the cyst, how much improvement in the realignment of the severely displaced teeth can be expected to occur spontaneously?
2. Can these teeth be subsequently brought into the arch?
3. Might there be disturbances or anomalies in the calcification or form of these teeth that may occur as the result of the circumstances and the site of their development, which may limit the quality of the treatment outcome?
4. Will they be accompanied by a good periodontium and bone support?

5. Can this be done in such a way as to make them indistinguishable from any other teeth, at the end of treatment?

The short answer to four of these questions is in the emphatic affirmative, as the following discussion will attempt to illustrate. Perhaps the most dramatic part of this is the incredible potential for spontaneous resolution, with an amazing response of the body to regenerate lost bone and to restore normal anatomy (Figure 11.7).

Figure 11.8 shows clinical and radiographic views of an 8-year-old child with a cyst of dental origin in the left man-

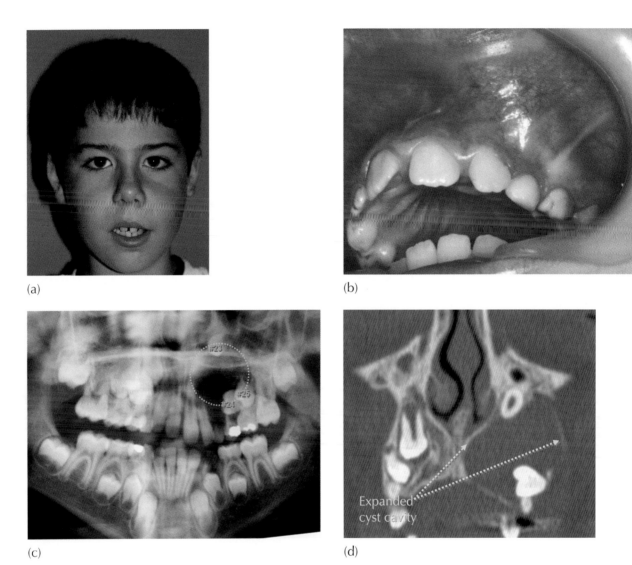

(a)

(b)

(c)

(d)

Fig. 11.7 Extreme displacement of teeth and its spontaneous resolution.

(a) The left side of the patient's face shows considerable swelling when first seen.

(b) The left side of the maxilla in the canine premolar area shows bone expansion and a blue discoloration of the overlying mucosa, with fluctuant mucosal swelling.

(c) In the initial panoramic view, a large radiolucent cystic area may be clearly discerned and demarcated approximately with the orange broken line circle, which has pushed aside the lateral incisor and the two premolars (#24, #25) and has displaced the maxillary canine (#23) very high and close to the inferior border of the orbit. The origin of the cyst is likely to have been caused by the presence of the heavily filled and poorly restored maxillary first deciduous molar.

(d) Gross maxillary expansion and the degree to which displacement of the canine and premolar has occurred, may be easily seen in this coronal 'slice' from the CBCT.

(e) The canine can be seen to protrude anteriorly from the base of the nose.

(f, g) More than 3 ml of fluid is drawn from the cyst.

(h) A large opening for continued drainage is made, through which the unerupted first premolar can be discerned.

(i) The stomium is loosely packed with an antiseptic-impregnated ribbon pack.

(j) The panoramic view taken 26 month later shows a total reversal of the condition, with the maxillary left canine and premolars erupting into ideal locations and a regeneration of the lost bone on that side of the maxilla. The only orthodontic treatment provided in this case was the placement of the soldered palatal arch space maintainer. An eruptive cyst is now seen surrounding the crown of the unerupted right canine.

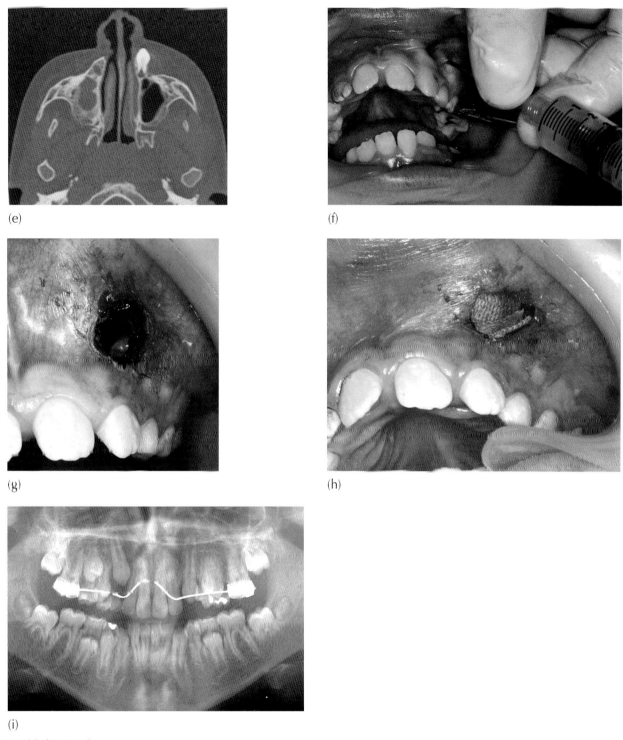

(e)

(f)

(g)

(h)

(i)

Fig. 11.7 (*Continued*)

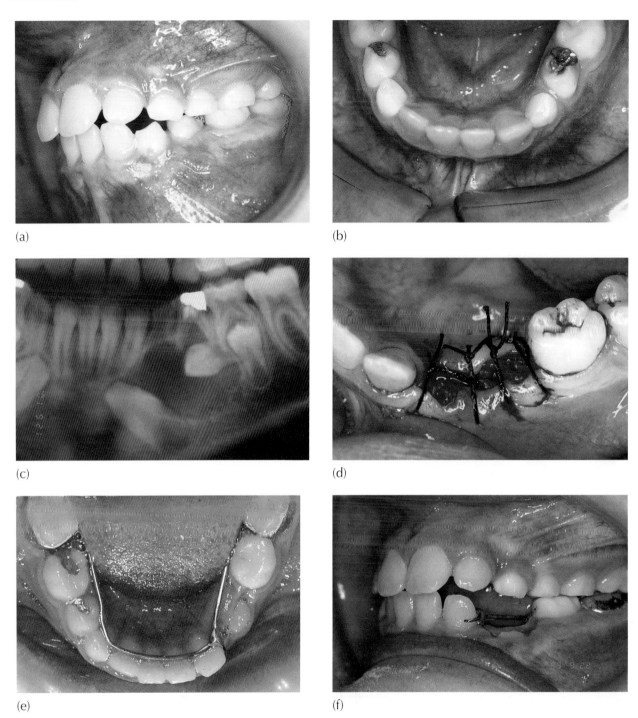

(a)

(b)

(c)

(d)

(e)

(f)

Fig. 11.8 (a, b) Clinical lateral and occlusal views of the left side to show expansion of the alveolus adjacent to the deciduous canine and first molar. The deciduous molar has a large restoration. (c) The panoramic view shows a large cystic lesion apparently centred on the horizontally displaced, unerupted premolar, 'pushing' the unerupted canine mesially. The premolar has a 110° tip, and the canine 30°, from their normal angulation. (d) The deciduous canine and first molar are extracted and the wound packed and sutured. (e, f) A fixed lingual arch space maintainer is placed. (g, h) Thirteen months later, the permanent canine and first premolar have erupted into their normal locations. (i) A panoramic view taken at the same time shows the canine to have self-corrected by 30°, while the premolar has uprighted spontaneously by 115° and now sports a 5° distal tip.

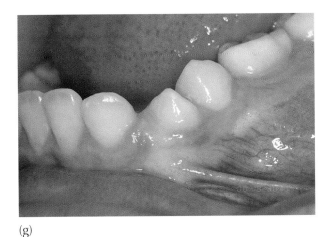

(g)

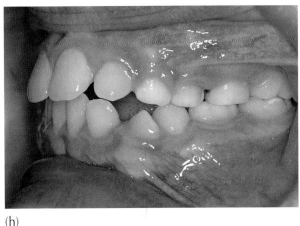

(h)

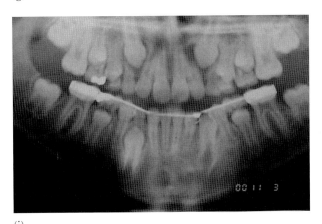

(i)

Fig. 11.8 (*Continued*)

dibular canine/premolar area. In terms of aetiology, it may be reasoned that this is a radicular cyst originating from the non-vital deciduous first molar, which has displaced the first premolar and canine downwards and mesially to a very marked degree. Equally, it may be argued that the chronic periapical granuloma of the deciduous tooth has stimulated the follicular sac of the first premolar to enlarge into a dentigerous cyst and, because of the resistance of the erupted deciduous tooth, the premolar and canine have rotated downwards and mesially. The long axis of the premolar appears to traverse the middle of the cyst. As already noted, this argument is academic, since the approach to treatment will be the same in either case. Treatment was provided in the form of extraction of the deciduous canine and deciduous first molar. This effectively resulted in the surgical marsupialization of the cyst by virtue of the fact that, as expected, the cyst lining was ruptured during the extraction. The remainder of the cyst lining was left undisturbed. A pack was placed in the wound and the socket margins partially approximated with sutures. The sutures and pack were removed a week or so later and, at the same time, a lingual arch space maintainer was placed.

In follow-up visits over the succeeding few months, the two displaced teeth improved their positions and erupted into the mouth very rapidly and without any further assistance. A comparison of the panoramic radiographs taken at the time of surgery and at the 13-month post-surgery follow-up shows the amount of spontaneous correction that occurred, with the regeneration of alveolar bone into the void previously occupied by the cyst. In the pre-surgical film, the canine shows a mesial tipping displacement of about 30°, and the first premolar about 110° from their normal axial inclinations. Ten months later finds the long axis of the first premolar fully corrected and that of the canine to exhibit a 5° over-correction and a distal tilt! All this had occurred spontaneously, strictly due to a reversal of the progress of the expanding pathological entity that had caused the initial displacement and without the need for any orthodontic correction [7].

The larger the cyst, the greater is the displacement and, with the defusing of the lesion, there is a great deal of migration of the affected teeth in a very short time and it is directly related to the contraction of the previously cystic area. There can be no question that corrective treatment of

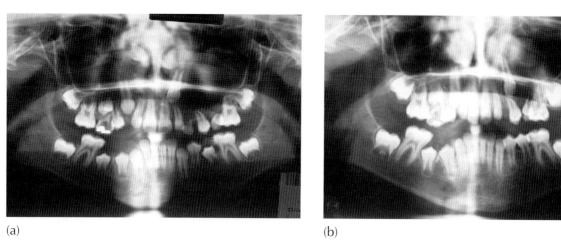

(a)

(b)

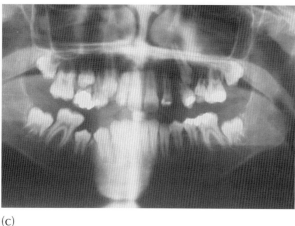

(c)

Fig. 11.9 (a) The initial film taken in January 2007 is of an 8-year-old patient with a history of extensive dental caries and early extraction of many deciduous teeth. The left side of the maxilla exhibits a large cyst extending from the lateral incisor to the first molar, which has apparently displaced the canine against the mesial extremity wall of the lesion. It seems likely that this developed from apical pathology associated with one of the extracted deciduous molars. Surgical treatment of the cyst was performed a month or two later.
(b) The film taken 7–8 months later shows considerable reduction in the size of the lesion and a distinct migration of the canine distally.
(c) At approximately one year post-surgery, the canine is seen to have moved much further distally and occupying a location between the two premolars. Note that the development of the roots of the canine and premolars has continued uninterrupted.

these cysts of dental origin depends largely on surgical intervention and that the displacement of teeth trapped within or adjacent to them may sometimes be completely reversed. Notwithstanding, total spontaneous resolution is not always the outcome and it does not always bring the teeth back to their ideal place in the dental arch (Figure 11.9).

Integrating spontaneous resolution into a combined treatment regimen

Trauma to the anterior teeth is common in the very young and may result in the loss of vitality of a deciduous incisor, as indicated by a deleterious colour change of the crown of the tooth. Should this tooth remain untreated in the long term, the chronic granuloma present may itself produce a radicular cyst, or it may stimulate the follicle of the devel-

oping central incisor to become cystic, as we have described above. Clinically, an expansion of the alveolus is seen both in the labial sulcus and on one side of the anterior portion of the palatal vault, where its normally concave anatomic form has become convex (Figure 11.10a, b). In the panoramic radiograph (Figure 11.10c), the unerupted central incisor shows a marked height discrepancy, with its long axis remaining in the vertical plane. This tooth is bare of bone on its distal side and is in contact with continuous bone from crown to developing root end on its mesial side. This is due to hydrostatic pressure from within the cyst, which has pressed the tooth against the bony wall of its perimeter, suggesting the diagnosis of radicular rather than dentigerous cyst.

The dilaceration of the root of the central incisor is also indicative of its lying in the wall of the cyst. The canine is seen on the panoramic view to be markedly displaced

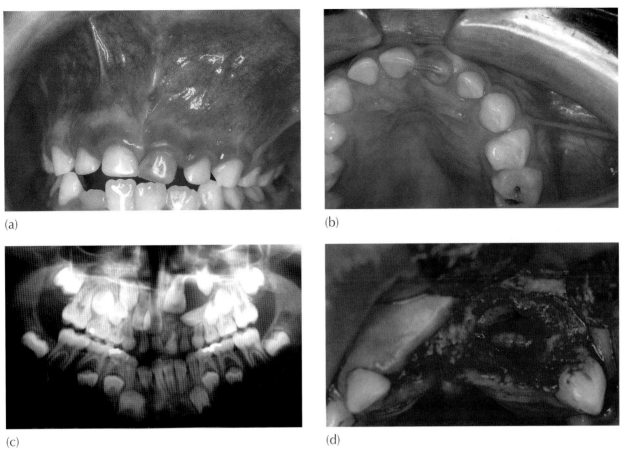

(a)

(b)

(c)

(d)

Fig. 11.10 (a, b) Clinical anterior and occlusal views of a 6-year-old child, showing a discoloured deciduous central incisor and considerable alveolar expansion in the labial sulcus and palatal area on either side of the teeth. (c) The panoramic view at age 6 years shows the left incisors and canine arranged in the form of a circle, apparently 'chasing each other's tails'. The first premolar has also been affected by the presence of the cyst whose perimeter may be fairly accurately assessed, using these four teeth as landmark signs. (d) All four deciduous incisors were removed and the cyst cavity opened to the exterior. (e, f) These intra cystic photographs were taken with a camera-mounted optic probe, which was passed into the cyst cavity to show the glistening, epithelium-covered incisors in the cyst wall. (g) The edges of the oral mucosa flaps, which are also the deciduous tooth sockets, were approximated after including a small pack and drain to permit further drainage and healing of the oral mucosa to the cyst lining. (h) The follow-up radiograph taken 26 months later shows that the incisors and canine have moved forwards and downwards and have come together very close to their normal locations. This, together with the bony trabeculation seen on the film, indicates the elimination of the cystic cavity. Note that the central and lateral incisors have dilacerated roots in their apical areas. (i, j) There has been spontaneous eruption of the affected incisors, a little later than normal, but the orientation of their long axes means that considerable root torque will be necessary. (k, l) At the age of 9 years, a 2 × 4 appliance (modified Johnson, twin arch) was placed to align these teeth and to improve the parallelism of the roots. (m) At the end of the short first phase of treatment, it was noted that further labial root torque of the lateral incisor was not possible due to the apical section of the dilacerate root of the central incisor and also by the mesio-labial location of the erupting canine. (n–r) The final photographic records of the treated case, at age 12 years, show an excellent position, crown height, gingival level and appearance of the two incisors. The quality of the canine gingival appearance, gingival level and crown height are poorer than for the canine of the unaffected side. However, the lateral incisor needs further labial root torque and the canine needs palatal root torque, which could not be completed adequately because of the root dilacerations. (s, t) The occlusal views show the bonded multi-strand wire splints. (u–w) The intra-oral clinical views seen two years post-treatment. (x–z) Close-up views of the anterior region to show the gingival condition and appearance at two years post-treatment. (a') Two-year follow-up periapical radiographs to show the excellent bone resolution and the root entanglement that limits the root movement possible.

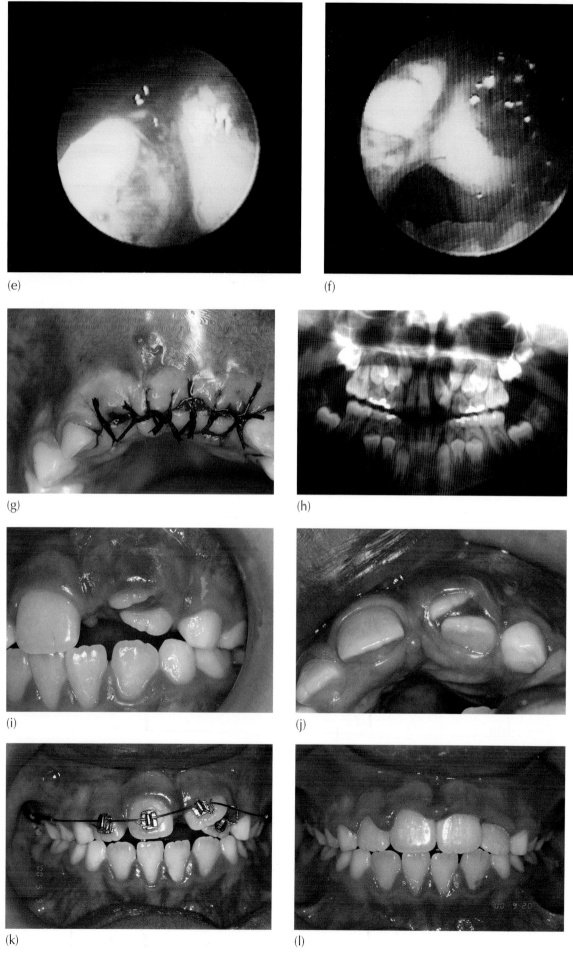

Fig. 11.10 (*Continued*)

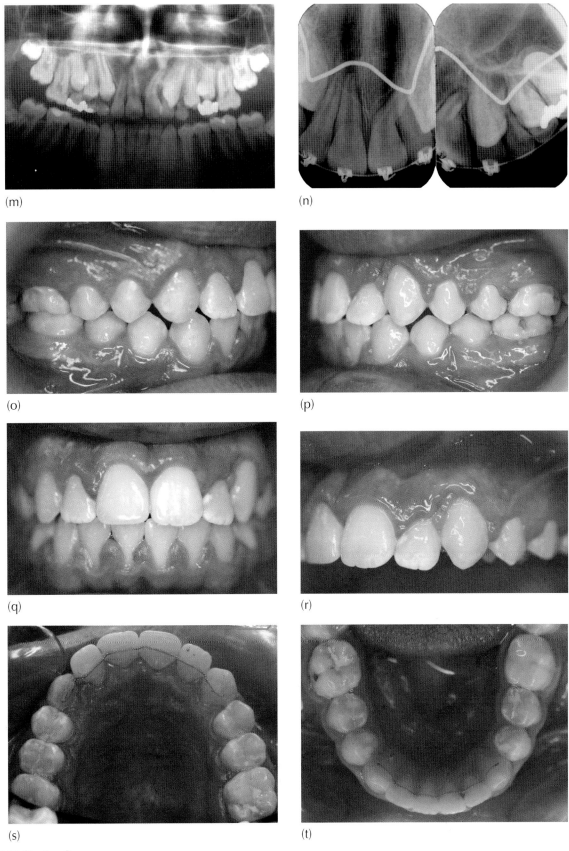

Fig. 11.10 (*Continued*)

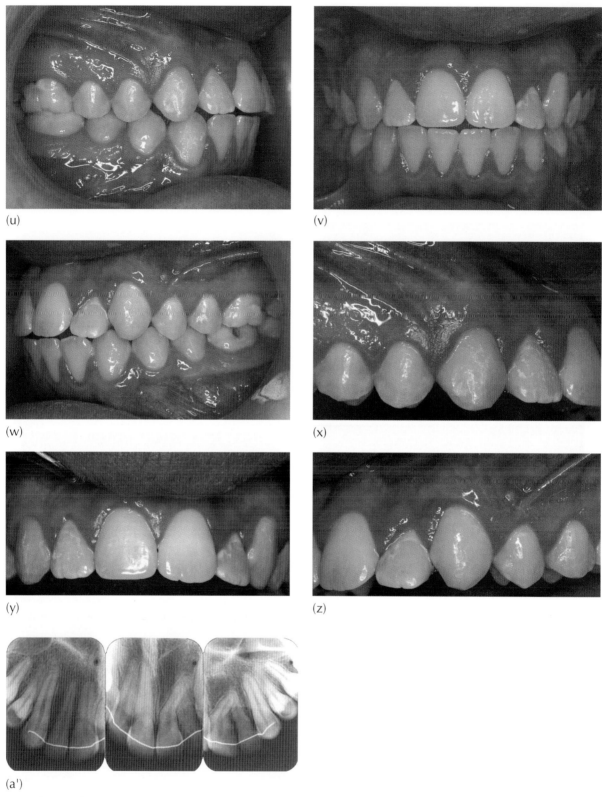

(u)

(v)

(w)

(x)

(y)

(z)

(a')

Fig. 11.10 (*Continued*)

upwards and backwards with a distal tip, and the lateral incisor has adopted a severe horizontal posture inferiorly. The unerupted developing premolars are also mildly affected. The marked variation in the long axes of the three anterior teeth, as though they are lying in a circle, is indicative of the fact that they are to be found in the wall of the cyst and, as such, they outline its extent. Using these indicators, it is now possible to estimate the shape and size of the cyst cavity and to define its borders.

In this case, the non-vital deciduous central and vital deciduous lateral incisors were extracted, which was all that was needed to produce a marsupialization of the cyst. The extractions immediately brought about spontaneous drainage of the cyst, and the edges of the torn lining of the cyst healed and became contiguous with the oral epithelium. The insertion of a very small camera, mounted on a long optic probe, into the cyst cavity allowed us to see and record the appearance of the teeth, as they lay glistening under the epithelium, in the walls of the cyst (Figure 11.10e, f).

No further treatment was provided at that time and the patient was followed up periodically. The two permanent incisors of that side began to erupt, after 26 months, into abnormal positions. A simple partially bracketed appliance (a modified Johnson's twin-arch appliance – see Chapter 5) was placed to draw the teeth to their normal positions.

In Figure 11.10l it can be seen that the central incisor is in an excellent position, with a short clinical crown and even an excess of gingival tissue. While the lateral incisor crown has been brought into alignment, the orientation of its long axis indicates a strong palatal displacement requiring considerable root torque, which may usually be corrected at this stage or in the full permanent dentition, when a second phase treatment will be initiated. However, the radiographs reveal a marked dilaceration of the root of the central incisor. If we study the initial panoramic view of the case, it will be seen that the tooth had been displaced very high in the alveolus and its root was developing in close relation to the floor of the nasal cavity. It is reasonable to assume that its development in this site and subsequent deliverance from this extreme displacement in the wall of the cyst was the cause of the dilaceration. This does not account for the lesser dilaceration of the lateral incisor.

The orientation of the apical portion of the root is distal and labial to the root of the lateral incisor, as can be determined from the two periapical films, employing the tube shift method (Figure 11.10n). This implies that labial root torque of the lateral incisor must not be attempted before this anomaly is remedied or circumvented, if this is possible and judged justifiable. The following options are available:

1. Leaving the situation without further treatment and the lateral incisor in this position.
2. Rotating the central incisor disto-labially, which effectively rotates the abnormally shaped apical portion labially and mesially. However, this will then involve crowning the tooth to achieve a good appearance.
3. Performing root canal treatment and an apicoectomy, thereby eliminating the root apex, but compromising the health and appearance of the tooth.
4. Over-uprighting the central incisor root mesially, thereby moving the apical portion mesially and superiorly. This may then be reversed, when the lateral incisor has been adequately (or possibly excessively) torqued, to permit the root apex of the central incisor to be re-sited palatal to the lateral incisor root. This was the treatment actually attempted, in the full permanent dentition, as a second-phase procedure. Unfortunately, the degree of the dilacerations of both the incisors did not permit full torque of the root of the three teeth involved (Figure 11.10o–s).

Because dentigerous and radicular cysts are not normally associated with pain, they may become very large before the patient seeks treatment, and the presenting symptom may be facial swelling, as seen in Figure 11.11a–g. In this patient, there was marked swelling on both sides of the alveolar processes, over-retention of the deciduous teeth in the area and non-eruption of the permanent teeth that would normally be erupting at this time. The maxillary right deciduous central incisor was discoloured and non-vital. The radiographs of the area (Figure 11.11h–l) indicate severe displacement of all the teeth on that side of the maxilla, from the anterior midline to the first permanent molar, as well as an upward extension of the cyst, raising the floor of the maxillary antrum. The pattern of the displacement clearly indicates the extent of the cyst. The maxillary central incisor of that side has been pushed superiorly and mesially across the midline. The root apices of the lateral incisor and, to a lesser extent, the two premolars have been pushed inferiorly and distally to cause these teeth to lie almost horizontally above their deciduous predecessors in the floor of the cyst, while the canine has been displaced superiorly and laterally in close relation to the floor of the orbit. Given the orientation of these teeth, together with the presence of a non-vital deciduous incisor, it seems likely that the cyst was radicular, but this is by no means certain. With much of the right side of the patient's maxilla virtually devoid of bone, the likelihood of a minor blow to the face causing a pathological fracture and becoming the presenting symptom was quite high.

Under the influence of simple infiltration local anaesthesia, the fluid was first drained from the cyst using a syringe and wide bore cannula needle, easily passed directly through the buccal sulcular area of the bony alveolar ridge, which had become expanded and extremely thin due to the size of the cyst. A relatively large quantity of brownish liquid was withdrawn from the cyst into the syringe. A circular incision was made in the immediate area over the cyst, to produce a hole of about 5 mm in diameter, communicating

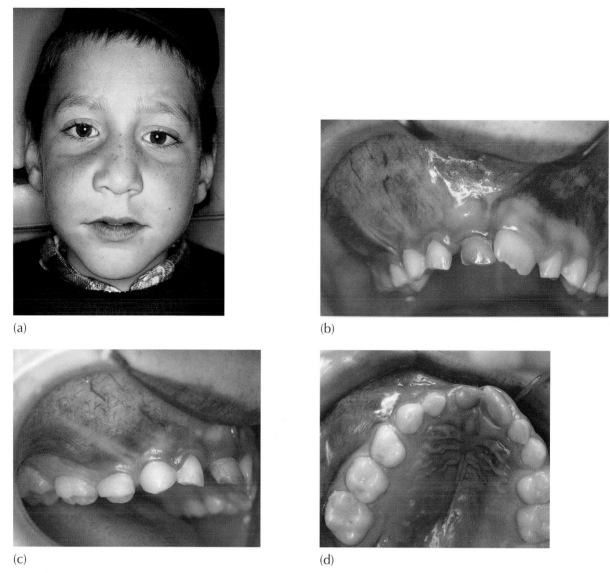

(a)

(b)

(c)

(d)

Fig. 11.11 (a) A 7.6-year-old boy presented because of swelling of the right side of the face. (b–d) Clinical views of the right side of the maxilla show extensive hard swelling of the buccal alveolus and palatal areas, extending from the first molar to the anterior midline. (e–g) Clinical views of the dentition showing normal buccal and antero-posterior relations, with a discoloured maxillary right deciduous central incisor. (h–k) Oblique occlusal, panoramic, Waters' view, and lateral and postero-anterior cephalograms show the extent of the cyst and the extreme degree to which the involved teeth have been dispersed in its walls. (l) The barrel of the syringe contains a brownish liquid, drawn off from the cyst through the cannula. (m, n) A 10 mm diameter stomium is established. (o) At the 39 months follow-up visit, at age 11 years, the hard swelling in the face has disappeared. (p, q) Panoramic and oblique occlusal films show that the affected teeth have come together, indicating elimination of the cystic area and its replacement by bone. The bony trabeculation can be seen around the teeth on the occlusal view. (s, t) Clinical views of the affected area from the right and the front. The two premolars are erupting and the stomium made into the former cystic area has remained patent. (u–w) Treatment initiated at age 11 years, aimed at levelling, alignment and the creation of space for the three unerupted teeth, including midline correction. (x) Eight months later, a pre-surgical panoramic view shows the widely divergent long axes and locations of the affected teeth. (y) The lateral incisor is seen 'peeking' through the stomium. (z, a') Exposure and eyelet bonding, after full flap reflection. (b') The flap is fully sutured back, with the pigtail ligatures temporarily held in place by turning them over the steel tube that has been placed on the archwire as a space maintainer. (c') The pigtail ligatures are shortened and formed into hooks, as close as possible to the sutured flap, and elastic ties apply traction to the archwire. Note that the lateral incisor is being drawn downwards through attached gingiva, rather than through the oral mucosa of the stomium, which was later sutured closed. (c') A year later, the teeth have been actively erupted, the central incisor has been brought to abut the midline and the lateral incisor lies horizontally, with its apex distal to that of the canine, whose root lies very buccally prominent. (d'–f') These clinical photographs show the extreme degree of root torque that is needed to align the lateral incisor, canine and first premolar, and the auxiliaries that were used to achieve this. There is a combined orientation discrepancy between the long axes of the lateral incisor and the canine, requiring approximately 90° of aggregated reciprocal torque! (h'–j') The final result shows excellent alignment, axial orientations and occlusion, with only a mild degree of lateral incisor and canine clinical crown elongation. The periodontal condition is good, and an experienced practitioner would be hard-pressed to guess that these teeth had even been impacted, let alone affected to the degree that the initial films show. (k') At two years post-treatment, periapical radiographs of the anterior teeth show a lack of bony support around the two previously affected incisors, a dilacerate central incisor root and obliteration of the pulp of the canine. All other periodontal parameters were normal.

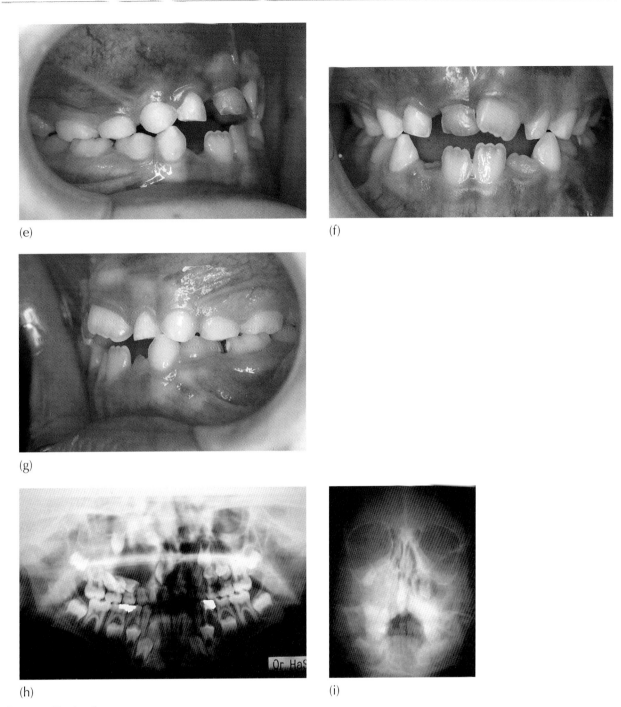

(e)

(f)

(g)

(h)

(i)

Fig. 11.11 (*Continued*)

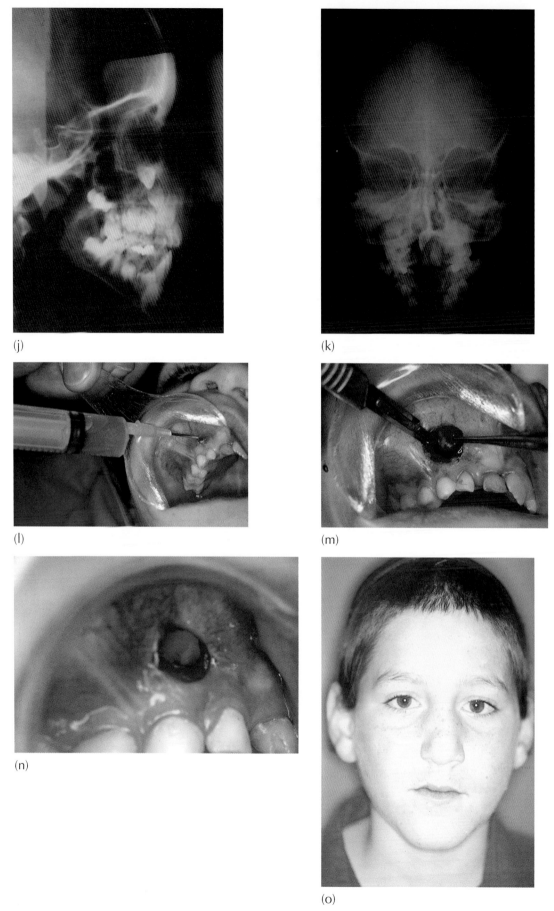

(j)

(k)

(l)

(m)

(n)

(o)

Fig. 11.11 (*Continued*)

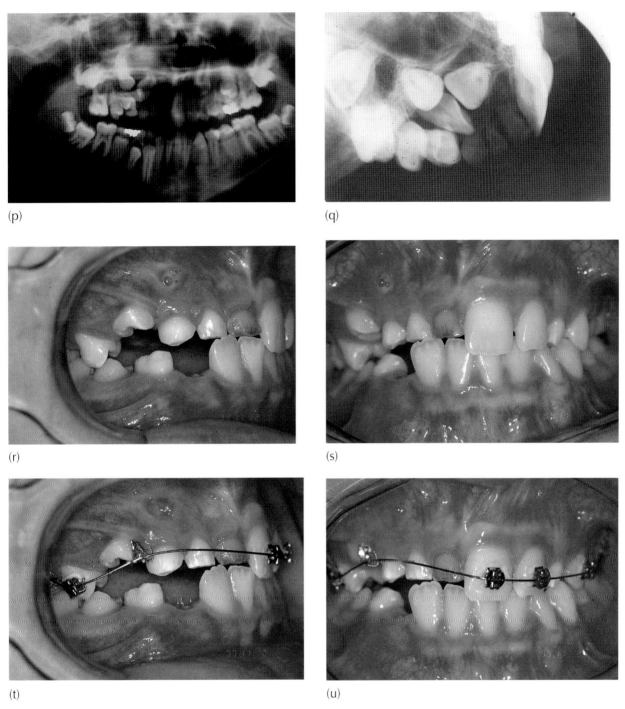

(p)

(q)

(r)

(s)

(t)

(u)

Fig. 11.11 *(Continued)*

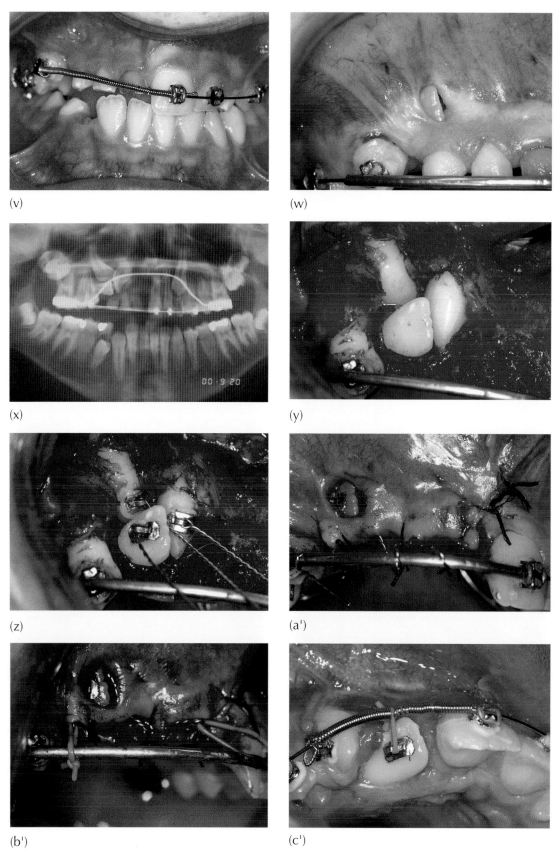

(v)

(w)

(x)

(y)

(z)

(a')

(b')

(c')

Fig. 11.11 (Continued)

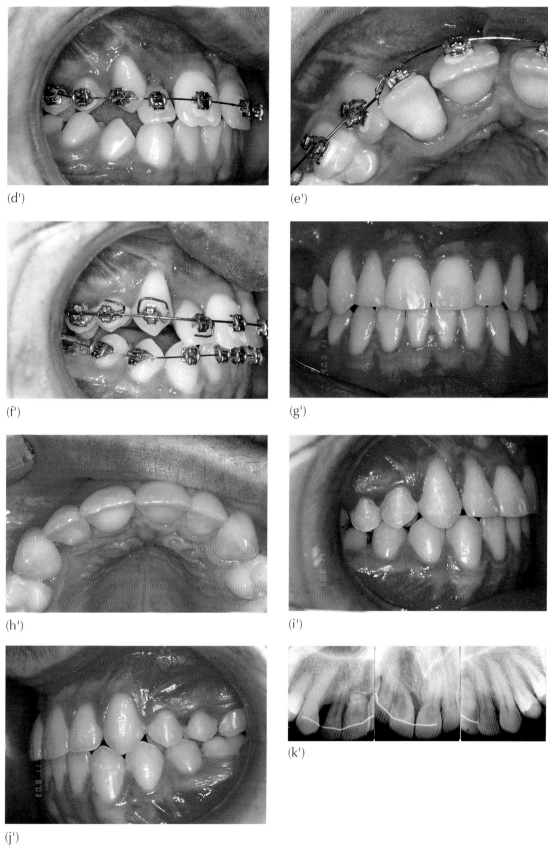

(d')

(e')

(f')

(g')

(h')

(i')

(j')

(k')

Fig. 11.11 (*Continued*)

directly with the cyst (Figure 11.11m–o). The oral mucosa and cyst lining were left to heal and to become contiguous, while the hole itself remained patent. The excised tissue was sent for pathological examination and found to be innocent and consistent with benign cystic epithelium.

In the 39 months that constituted the follow-up period, the widely dispersed teeth in the walls of the cyst were seen to improve their positions and to come together, indicating the disappearance of the cyst and a considerable filling-in of the bony defect (Figure 11.11p–t).

At this point, a fixed orthodontic bonded appliance was placed on the erupted teeth, to act as a composite anchorage unit (Figure 11.11u, v) for the extrusion and alignment of the teeth within the former cyst cavity. The central incisor of the healthy left side of the maxilla had traversed the midline and the first premolar had partially erupted, mesio-lingually rotated and mesially displaced. A simple coil spring was used after initial levelling to move the premolar distally and correct its rotation, while reciprocally moving the left central incisor back to the midline (Figure 11.11w, x). This effectively opened more than adequate space to accommodate the three impacted teeth. While this preliminary orthodontic space regaining and alignment was taking place, the lateral incisor in the floor of the former cyst area had suddenly become visible through the still patent stomium that had earlier been surgically opened to drain the cyst (Figure 11.11y). The relatively simple and minor surgical exposure of the crowns of the three teeth was then undertaken on an ambulatory basis under local anaesthetic in the Department of Oral and Maxillofacial Surgery, and small eyelets bonded in strategic positions, with twisted steel pigtail ligatures threaded into the eyelets (Figure 11.11z, a'). The surgical area was then fully closed and the pigtail ligatures were taken through the sutured edges of the flap (Figure 11.11b'). Directional elastic traction was applied between these ligatures and the main archwire (Figure 11.11c') before the patient was released from the operatory. The elastic traction was subsequently renewed periodically until the teeth had fully erupted and been brought into approximate alignment. At that stage, sophisticated brackets of the same type used on the other teeth were substituted for the eyelets and the treatment was continued until full alignment was achieved and, it should particularly be noted, until the extreme amount of root uprighting and root torque were completed (Figure 11.11d'–k').

Orthodontic brackets have been developed over the years with the incorporation of various highly sophisticated slot angle prescriptions and ligature tie alternatives, to make them as versatile as possible in aligning, rotating, tipping, uprighting and torqueing the teeth. This has allowed the orthodontist to cater to well over 90% of cases without having to change attachments and to achieve excellent alignment, offset, tip, torque and inter-arch relationships. The prerequisites for these brackets to achieve their goal are that they be placed in the mid-buccal position of the teeth and, at the end of the treatment, that a full-sized archwire be fully tied into them for them to realize their maximum programmed potential.

When the ideal bonding site of a tooth is not accessible, the route taken by most orthodontic practitioners is to place their regular 'straight wire' bracket at a different site on the tooth, where it can be used to achieve some alignment and levelling. In this position, however, it cannot be used to realize the full potential of the prescription slot – a prescription that is able to perform any other type of movement with considerable accuracy and efficiency – and nor would one want it to, unless a series of complicated bends are introduced into the archwire by way of compensation. Beyond a minor degree of rotation, height discrepancy and ectopia, it is difficult to ligate the archwire into the bracket, and this stage of the treatment becomes very clumsy and inefficient, slowing down the momentum of the desired correction.

It is obvious that brackets cannot be placed in their ideal positions on teeth which are severely rotated or partially erupted; nor can an archwire be easily and usefully tied into the bracket of a tooth which is grossly tipped or ectopically placed. In these cases, an initial temporary attachment should be preferred, or the archwire modified with a series of offsets and loops.

Eyelets and brackets

In Chapter 4, we discussed the use of eyelets in relation to impacted and markedly displaced teeth. Using eyelets as initial attachments on the impacted teeth provides for superior management of the gross movements of the teeth, movements that are needed in the early stages of their resolution. It simplifies ligation of teeth in the most extreme positions of displacement in these situations and makes directional traction much more efficient. Threading a super-elastic auxiliary archwire through one and sometimes a pair of eyelets makes large-scale extrusion, tipping, rotation and uprighting movements extremely easy over a broad amplitude of positional dislocation. Within a relatively few short appointments, the formerly impacted teeth may be brought into general alignment over large distances in this context, to a point prior to the detailing and finishing procedures that will be necessary to achieve excellence.

At this stage, in the patient who is being treated for the resolution of a dentigerous cyst, new radiographs of the involved teeth should be evaluated to check the form of their roots. We have noted above that teeth which have developed their roots while being displaced by the progressive enlargement of a dentigerous cyst are likely to have abnormal root form. The discovery of this phenomenon in a particular tooth or teeth will obviously modify the amount and direction of root movement that is indicated for the patient. Once this has been determined, the eyelets should

be removed and regular orthodontic brackets, of the type used on the other teeth, should be substituted in preparation for the final stage of treatment aimed at the fine-tuning of the alignment and occlusion of the dentition.

In essence, the eyelets are responsible for the resolution of the impaction and the gross movement of bringing the severely displaced teeth into the orthodontic 'ball park'. When all that remains to be performed are the final millimetre or two of levelling and crown alignment, the extensive root uprighting and torque, and the additional few degrees of tooth rotation, this can only be achieved with a sophisticated bracket system, which should then be applied.

Alternative surgical approaches are presently in use in many centres, particularly where the orthodontist is not involved in the decision-making process. Perhaps the most common approach has been described by two oral surgeons in a study of 40 dentigerous cyst cases [4]. Apart from one instance, all their cases were treated by enucleation. Their treatment sequence was aspiration, followed by an incisional biopsy for pathological examination. The cysts were subsequently enucleated, and extraction of the affected impacted tooth was considered to have been indicated in 34 of their 40 cases. In one isolated case, decompression was performed first and then excision of the cyst done several months later. The histopathological reports indicated only innocent diagnoses, and postsurgical bone regeneration occurred in all cases within 6–12 months. Grafts were not needed in any of the cases.

In the view of these authors, 'cyst enucleation . . . in extensive cysts will lead to loss of several teeth' and 'when the teeth involved with the cyst are extracted (especially in children) function, cosmetic and psychological consequences may follow'. While they mention that marsupialization is one of the treatment options, it was not the modality chosen in any of the individuals in this series of cases. Their clear preference is for 'curative' enucleation of the lining.

As a general rule, surgeons will prefer not to run the risk of introducing infection, which could track along a stainless steel ligature wire that is ligated to an attachment on the impacted tooth, when the wound is a relatively large one. Many are uncomfortable with the idea of leaving the cyst lining in place and prefer to shell it out completely on the premise that it is pathological tissue.

These two surgical concerns are addressed in the approach to treatment described in this chapter. In the first place, by virtue of the marsupialization procedure that has been performed many months earlier, the large surface area of exposed bone that would be engendered by a 'curative enucleation of the lining' is avoided. Instead, an epithelial-lined cavity of ever-reducing size is present, with no exposure of the underlying tissues to the exterior. At the later stage, when the teeth themselves are finally exposed and attachment bonded, the exposure is extremely minimal, less than in a routine impacted tooth exposure. There are no large wounds whatsoever in this approach and the risk of infection is insignificant.

The literature reports only the rarest incidences of neoplastic change in the lining of a dentigerous cyst, and these have been cases in which the histopathological examination has been performed on excisional biopsy material taken at the time of the initial treatment of the cyst. In the present approach, a small portion of the lining is sent for microscopic examination and the patient is followed up clinically over the succeeding months until full resolution of the cyst cavity has occurred. It would appear that the chances of undiagnosed neoplastic alteration occurring in the former cystic epithelium are extremely small, probably no greater than finding it in another unrelated area of the oral mucosa.

The rate with which neoplastic change might occur in a dentigerous cyst has never been investigated. An early histological study of a sample of 52 patients with widened follicles, as distinct from dentigerous cysts, around their impacted maxillary canines showed a zero incidence of neoplastic change [9]. However, there are no studies investigating neoplastic change in large samples of patients with clearly defined dentigerous cysts and as such there can be no foundation for always assuming the worst. The fact of the acceptance for publication in the dental literature of only occasional, single individual, case reports of neoplasia in these cases speaks more of typical editorial eye-catching sensationalism and, therefore, of extreme rarity. With no other information available, these articles arouse readership interest based on emotional appeal and exert an exaggerated influence on surgical decision-making. There is a great deal of highly emotive, alarmist and even incendiary input influencing the decision of which surgical route to take on the part of both the patient and the surgeon.

Hence the surgeon will sleep better if he/she performs a 'curative enucleation' of the lining of the cyst, along with removing the tooth or teeth involved in the lesion. Legally, this degree of caution cannot be easily faulted. But clinically in some of these cases, this relatively radical surgical procedure will leave a large bony defect, which may also adversely affect the patient's facial appearance. Bony reconstruction and implant-borne restorations will be necessary in the long term.

But the clinical experience of experienced surgeons tells us that neoplastic alteration of the epithelial cyst lining is extremely rare and, although it is essential to biopsy that part of the cyst epithelium removed in establishing adequate drainage, experience in the overriding majority of these cases must justify the use of the more conservative marsupialization option whenever possible. This naturally presupposes careful radiography of the healing process and clinical follow-up of the appearance of the epithelial tissue in the ensuing months during which the lesion is resolving.

By the nature of their work, surgeons do not have the same follow-up as do orthodontists. The orthodontist, on

the other hand, maintains considerable control over frequent periodic visits. He/she will place a space maintainer, with the patient on close recall for clinical and regular radiographic examination and later for orthodontic appliance adjustment. A follow-up periapical film taken a month later should reveal rapid resolution and healing and a dramatic improvement of the position and orientation of the affected tooth. In the extremely unlikely event that the lesion becomes more sinister, the change will be seen on this film and another taken three months later. For these reasons, the parents must be told of the importance of follow-up with the argument that marsupialization surgery is simple and minimal, it will most likely save teeth, it will generate a normal ridge bony contour and cause no facial disfigurement. This is in direct contrast to enucleation surgery which is more radical, will likely necessitate the loss of tooth or teeth, create a large bony defect with possible consequent facial change and require long term space maintenance, ridge enhancement bony procedures, implants and crowns.

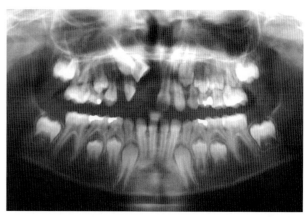

Fig. 11.12 The same patient illustrated in Figure 11.2, seen 20 months after the initial decompression surgery. This panoramic radiographic view shows the coming together of the three teeth that were grossly displaced by the cyst and some incidental improvement in the positions of the premolars on the same side.

Conclusions

In regard to the marsupialization of dentigerous and radicular cysts, the anecdotal evidence that has been presented above and from other cases treated by the author points to several important, if tentative, conclusions that may be made at this stage. These are as follows:

1. The cystic area shrinks rapidly as bone fills in behind the cyst lining, to repair the initial bony defect.
2. The teeth that have been displaced by the lesion come together within the repairing defect, their orientation improves and there are positive movements in the direction of eruption, and actual eruption may occur in some instances over time.
3. Alveolar bone regenerates around the roots of the teeth.
4. The root form of teeth impacted within cysts may be dilacerate.
5. Treated in the following protocol, there is rarely any need to extract the involved teeth, and their eventual prognosis is excellent.

The following is the suggested protocol for a conservative approach to the resolution of teeth impacted within dentigerous cysts:

- Simple decompression surgery – marsupialization and biopsy of the removed epithelial tissue.
- Awaiting bony fill-in, which will be seen by periodic radiographic monitoring, indicated by the coming together of the teeth within the former cyst location – a year or more after the decompression, depending on the size of the lesion (Figure 11.12).

- Orthodontic aligning of the erupted teeth, levelling and preparation of the anchor unit.
- Surgery to expose impacted teeth and bond eyelet attachments.
- Traction to main arch with extrusion, tipping, rotation and some root uprighting movements.
- Check root morphology of involved teeth.
- Substituting standard brackets for controlled root uprighting and torqueing and finishing procedures.

References

1. Shafer WG, Hine MK, Levy BM. Oral Pathology. Philadelphia: Saunders, 1983.
2. Shear M. Cysts of the Oral Regions, 2nd edn. Bristol: Wright, 1983: 114–141.
3. Archer WA. Oral Surgery, 4th edn. Philadelphia: Saunders, 1966.
4. Motamedi MHK, Talesh KT. Management of extensive dentigerous cysts. Br Dent J 2005; 198: 203–206.
5. Fearne J, Lee RT. Favourable spontaneous eruption of severely displaced maxillary canines with associated follicular disturbance. Br J Orthod 1988; 15: 93–98.
6. Sain DR, Hollis WA, Togrye AR. Correction of a superiorly displaced impacted canine due to a large dentigerous cyst. Am J Orthod Dentofacial Orthop 1992; 102: 270–276.
7. Miyawaki S, Hyomoto M, Tsubouchi J, Kirita T, Sugimura M. Eruption speed and rate of angulation change of a cyst-associated mandibular second premolar after marsupialization of a dentigerous cyst. Am J Orthod Dentofacial Orthop 1999; 116: 578–584.
8. Becker A, Chaushu A. Healthy periodontium with bone and soft tissue regeneration following the orthodonticsurgical retrieval of teeth impacted within cysts. In Davidovitch Z, Mah J, eds. Biological Mechanisms and Craniofacial Adaptation. Boston, MA: Harvard Society for the Advancement of Orthodontics, 2004: 155–162.
9. Olow-Nordenram M, Anneroth G. Eruption of maxillary canines. Scand J Dent Res. 1982; 90: 1–8.

12

The Anatomy of Failure

Orthodontic Treatment of Impacted Teeth, Third Edition. Adrian Becker.
© 2012 Adrian Becker. Published 2012 by Blackwell Publishing Ltd.

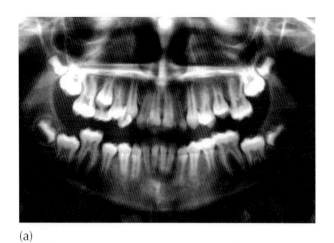

(a)

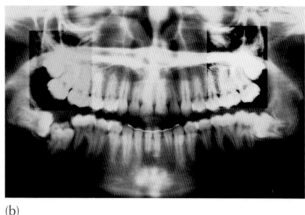

(b)

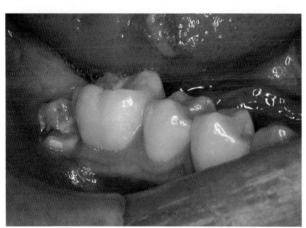

(c)

Fig. 12.1 (a) A pre-treatment panoramic view of a patient whose mandibular right second molar had not erupted at the completion of treatment. (b) A panoramic view of the same patient seven years later and five years after the completion of the treatment shows the mandibular right second molar impacted between the two adjacent molars. (c) A month after the surgical removal of the unerupted third molar and exposure of the impacted second molar, the second molar is erupting spontaneously. The impaction is due to the obstruction created by the distal anatomy of the first molar and the large and horizontal third molar.

There are many factors complicating the treatment of impacted teeth that are not present in routine general orthodontics. In the first place, the affected tooth is not visible, and is only imaged using clinical and radiographic aids. It thus cannot be examined for abnormality in the same manner or with the same degree of thoroughness as a normally erupted tooth. The precise, three-dimensional location of a normally erupted tooth is obvious from a clinical examination. So too is its degree of rotation, the orientation of its long axis and its relation to the erupted adjacent teeth, as well as minor flaws in the morphological features of its crown, surface imperfections in the smooth outline or colour of the enamel. Exactly what types of corrective movement its orthodontic treatment will demand can be seen by direct vision, and biomechanical planning is straightforward. These luxuries are not available when dealing with an impacted tooth.

One of the most constant features of the normal growing child is the natural and spontaneous eruption of teeth. In

the deciduous dentition, this occurs between the ages of 6 months to 2.5 years and these teeth shed normally between 6 and 10 years later, to be replaced within just a few months and in rapid succession by the permanent teeth. This innate attribute is so universal that a single tooth failing to erupt, when all other factors are apparently favourable, should raise the suspicions of the discerning orthodontist. The very fact of its non-eruption raises questions as to why this should be, and the answers range from the common and the obvious to the unusual and least expected.

Determining aetiology must not be looked on as a mere theoretical exercise, but rather an essential prerequisite that provides the basis for the treatment plan for each individual case. Perhaps the tooth is quite normal, but there is some local impediment blocking its path (Figures 12.1 and 12.2), or perhaps the location of the tooth germ is ectopic. Alternatively, the cause may lie in a local abnormality of the tooth follicle itself, or perhaps the patient suffers from a general pathological condition that, among other features,

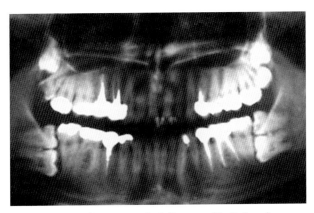

Fig. 12.2 A cause of incisor crowding? (Courtesy of Dr N. Casap.)

adversely affects tooth eruption. Accurate positional diagnosis is often fraught with difficulty and mistakes may be made in locating the tooth – even by experts [1, 2]. As the result, a tooth in an intractable position may be thought to have a good treatment prognosis, and an inappropriate, ill-advised and ill-fated course of treatment will be prescribed.

In many cases all that may be required is for space to be made and this will sometimes reawaken dormant eruptive movements in the tooth, which may then respond by improving its position and even open the window of opportunity to its spontaneous eruption, with the passage of time [3] (see Figure 6.21). At the other end of the scale, a complicated directional traction strategy may be needed to bring the tooth into its place in the arch, while avoiding the roots of adjacent teeth. This we have seen in Chapters 6 and 7 in relation to maxillary canines which are located mesial to the root of adjacent incisors or where the tooth is associated with the resorption of those roots.

An open surgical exposure of the impacted tooth may close over in the succeeding days and weeks and make later attachment bonding unreliable or impossible to achieve. When bonding is performed by the surgeon as an integral task during an open or closed exposure, an attachment may be placed in an inappropriate position on the tooth surface or the pigtail ligature wire or gold chain may have been drawn through the tissues in the wrong direction for traction to resolve the impaction. Alternatively, the bond may fail and, without further surgery, suitable conditions for rebonding may be limited or unattainable (see Figure 8.10).

We have seen in the preceding chapters that the successful outcome of treatment of a difficult impaction, from the orthodontic and surgical points of view, may founder in the long term because of a lack of periodontal support of the treated tooth – a situation that could have turned out very differently had certain precautions been taken in diagnosis, treatment planning, surgical exposure or biomechanics. From the foregoing, we come to appreciate that there are many different areas where opinions regarding attitudes

to treatment and where levels of expertise may vary among the three principal specialists involved with the treatment: orthodontist, oral radiologist, oral and maxillofacial surgeon. Taken from the standpoint of how to serve the best interests of our patients, it helps us to understand that communication and rapport between these specialists is of paramount importance. Given the broad spectrum of concerns that are so intimately bound up with the whole approach to this relatively limited phenomenon, it should not be surprising that the chances of error are legion, and with them the distinct likelihood of embarrassing failure. The purpose of the present chapter, therefore, is to analyse the possible causes of failure that occur from time to time in clinical practice and to group them in relation to the various aspects involved in each scenario.

In a recently published study [4], the Jerusalem group collected a sample of 28 young patients for whom treatment for the resolution of their impacted canines had failed. These patients had been referred to the authors by a large number of different orthodontists individually and over a period of several years, with the intention and in the hope that the cases could be salvaged. These 28 patients had 37 failed canines in total and had been in treatment for an average of 26 months for the sample, with a range of 7–72 months, before being referred and, for 10 of these canines, surgical exposure had been performed three times! The referring orthodontists were questioned as to why they thought the case had failed and a large majority had assumed it to be due to ankylosis, while most of the others had missed diagnosing root resorption of the adjacent teeth. A few had blamed bond failure, an intractable location of the tooth and failure to adequately surgically expose the impacted tooth.

These cases were then re-evaluated, many with the help of a CBCT examination and re-treated in accordance with the new findings. The outcome saw 28 of the original 37 failed teeth successfully aligned in the dental arch. Of the remaining cases, two individuals had refused the revised treatment. Of the seven canines with confirmed ankylosis, three were successfully realigned following surgical luxation.

In their conclusions, the authors of this study noted that there are many aspects and minutiae in the treatment of impacted maxillary canine that may cause the treatment to founder. They specifically had the following comments to make:

1. Diagnosis of the location of the tooth and its immediate relationship with the roots of the adjacent teeth is generally treated with cavalier and often negligent simplicity, even though modern technology has provided the tools to achieve this with great accuracy in all three dimensions.
2. With inappropriate positional diagnosis, it follows that traction will be applied in the wrong direction.

3. A lack of appreciation of the considerable anchorage requirements of the case and the need to exploit all available means of enhancing them will inevitably lead to inefficient mechano-therapy and unnecessarily longer treatment.
4. Ankylosis might have afflicted the impacted tooth either *a priori* or as the result of the earlier surgical or orthodontic manoeuvres.

Age

In Chapter 6, we referred to the fact that the chances for orthodontic traction having a positive effect on an impacted tooth are extremely high in the young patient, but that with advancing age, notably in the over-thirties, the risk of non-movement becomes quite significant [5]. By contrast, the compliance factor among adults undergoing orthodontic treatment towards the relatively undemanding tasks which are in the patient's realm of responsibility within the treatment protocol (i.e. maintaining a high level of oral hygiene, attending pre-arranged treatment visits, taking proper care of the appliances, placing elastics), is usually considerably better. Thus, while tissue response to orthodontic forces in the child is very positive, the endeavour may be more likely to founder due to a lack of compliance.

In the adult, the ravages of time will often be evidenced by a loss of attachment and bone support, which have been due to large measure by chronic inflammation. But once these factors have been brought under control by appropriate periodontal treatment, routine orthodontic treatment may be recommended with considerable confidence and with a high degree of predictability. But this is true only in regard to the root of a tooth. For the unerupted crown of the tooth this situation may be quite different, since the dental follicle undergoes deterioration in time and direct contact between tooth enamel and the encroaching bony tissues may occur, which will effectively eliminate the orthodontic therapeutic option. Since this phenomenon is age-related, it will rarely be encountered in the young patient, but must be a factor to be considered in planning the treatment of an adult, particularly in the over-thirties age group.

Abnormal morphology of the impacted and adjacent teeth

Unusually large or small teeth and abnormalities of crown form or root configuration, regardless of their cause, are findings which may affect the decision whether or not to attempt to bring the tooth into the arch. Uncomplicated abnormal morphology presents no impediment to orthodontic movement, since the roots of the teeth are generally invested with normal periodontal tissues and their crowns protected within a normal follicle in which they developed, however abnormally. The unusual form of the teeth may affect the position of the centre of rotation and the centre of resistance, which in turn may affect the direction that

orthodontic traction will need to be applied. However, the teeth will respond. In many of these cases, as discussed in Chapter 5 with reference to dilacerate incisors, the orthodontic/surgical modality of treatment may provide an optimal result in aligning these abnormal teeth and in carrying the young patient through the years of childhood and adolescent growth, during which time prosthetic/implant substitution is generally contraindicated. Minor artificial modifications in crown or root form may be necessary during this time but, with this treatment modality, the child is free of iatrogenic, periodontal disease-producing or caries-generating prosthetic devices, when the dentition is at its most vulnerable. At least as important, alveolar bone in the area proliferates quite normally with the resolution of the impaction and maintains its height throughout growth, in parallel with the other teeth. It would be a serious mistake to indiscriminately extract these teeth simply because of a diagnosis of morphological abnormality, and the practitioner is strongly advised to consider all the alternatives before recommending extraction.

Fixed bridges and implants have a limited life expectancy, and failure may occur much earlier in some instances. This, together with the atrophy of alveolar bone in the immediate area, which inevitably occurs following an extraction, must be considered to be a failure to utilize naturally occurring and available raw material (i.e. the imperfectly formed tooth). Nevertheless, there are cases in which the location of the dilacerations and the degree of the tooth's distortion leave little option but extraction (Figure 12.3).

Rather than the canine itself being the cause of the problem, it may occasionally occur that the root morphology of the first premolar is the hidden impediment. The adjacent first premolar normally erupts before the canine and has a buccal and a palatal root, which may sometimes be fused. These roots, particularly the palatal one, may develop in the direct eruption path of the canine which may happen when the premolar erupts with a distal crown tip or when the tooth is rotated mesio-buccally. The root itself may be deformed and exhibit a mesially directed dilaceration of its root apex, as has been described in Chapter 6 (Fig. 6.1c, d). In plane film radiography, the individual roots of the premolar are sometimes not possible to distinguish and superimposition on the canine leaves the bucco-lingual interrelation of the two teeth difficult to ascertain. CBCT imaging of these cases will often clarify the problem and permit the determination of a suitable direction for the application of traction to resolve it. In the case illustrated in Figure 6.1, the root orientation of the premolar must be over-uprighted in a distal direction and then the tooth should be rotated mesio-palatally to free the canine from interference. Once the canine has been brought into alignment, the first premolar may then be re-uprighted and re-rotated into a more optimal position, having due care to avoid a clash of roots in the final analysis.

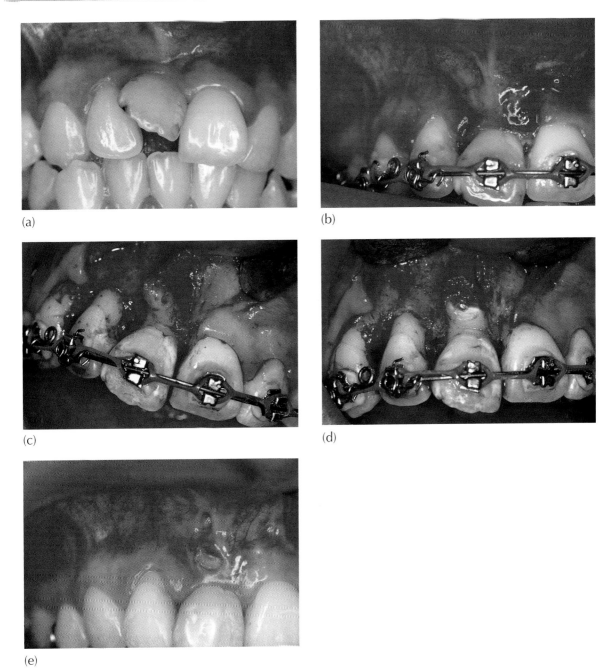

(a)

(b)

(c)

(d)

(e)

Fig. 12.3 A 15-year-old female presented with a dilacerate but erupted right maxillary central incisor. (a) The initial malalignment. (b) Following alignment and levelling, the root apex can be seen to be pointing labially under a thin cover of overlying oral mucosa. (c) When the flap was raised over the area, with the view to perform an apicoectomy, it was seen that there was no labial bone covering the short and malformed root. (d) The root apex was trimmed back and a retrograde filling placed. (e) At the two-month follow-up visit, the root apex had fenestrated the oral mucosa and the tooth was extracted.

Ankylosis and invasive cervical resorption

These two phenomena are difficult to diagnose in their early stages, even with good radiographic technique. The presence of either will result in failure of the affected tooth to respond to the applied orthodontic force because of a loss of integrity of the normal periodontal tissue in one or more, often very small, locations on the root surface

(Figures 12.4–12.7). This has been discussed fully in Chapter 7 (Figures 7.11–7.13).

Wildly ectopic teeth

Although wildly ectopic teeth are seen extremely rarely, there would appear to be few limitations on where teeth may be found in the tooth-bearing areas and beyond, in

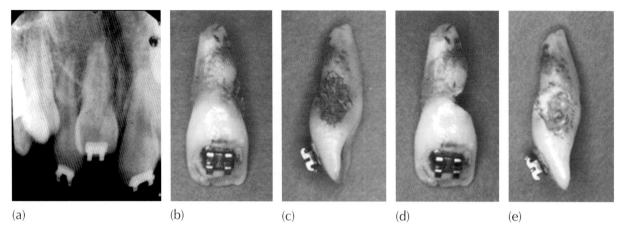

(a)   (b)   (c)   (d)   (e)

Fig. 12.4 Invasive cervical resorption. (a) The orthodontic attempt to erupt the left central incisor had failed and the tooth was extracted. A close look at the distal side of the cervical area of the tooth shows a defective outline. (b, c) A large area of inflammatory soft tissue in the cervical area of the extracted tooth on the distal side. (d, e) With the soft tissue removed with a scaler, the full extent of the erosive lesion can be seen.

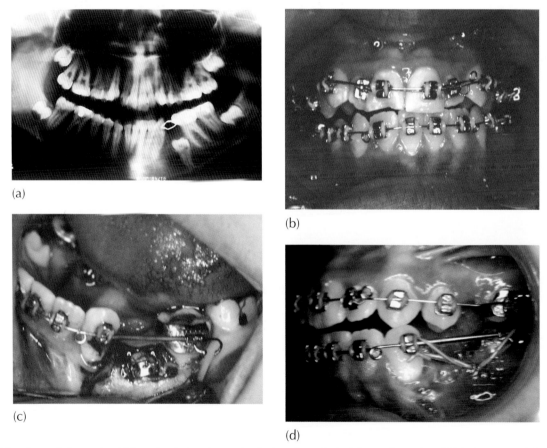

(a)

(b)

(c)

(d)

Fig. 12.5 Invasive cervical resorption. (a) A panoramic view shows a deeply situated mandibular left second premolar, which had not responded to extraction of the adjacent first premolar. (b) The treatment plan involved the extraction of a premolar in each of the other quadrants and placement of a fixed appliance. (c) The premolar was exposed and an attachment bonded. Traction was applied immediately. (d) The left side, seen one month post-surgery. (e) Despite the application of light forces, the impacted tooth did not respond and the adjacent teeth had intruded. (f, g) These light forces, over a long period, had generated a left-side open bite and a strong cant of the occlusal plane. (h) A new periapical view at this stage showed a radiolucent area within the crown of the tooth, which turned out to be advanced invasive cervical resorption on the lingual side, which was not seen at the time of surgery, although it was obvious on the original panoramic view and had been overlooked! The original location of the premolar, with its developing apex on the lower border of the mandible and with no obstruction superiorly, had raised the suspicion of the author at the outset.

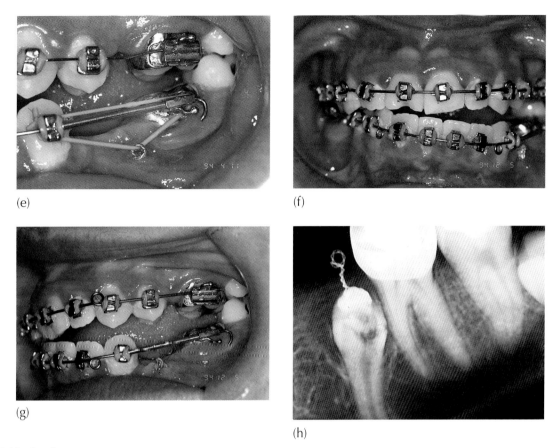

(e)

(f)

(g)

(h)

Fig. 12.5 (*Continued*)

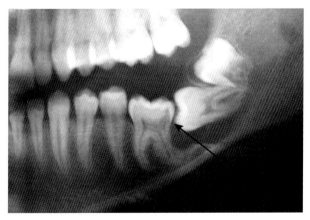

Fig. 12.6 This patient was seen by the oral surgeon and referred for an orthodontic opinion as to which tooth should be extracted. The surgeon suggested the following options: (a) to extract the third molar, expose the first and second molars, drain the cyst, and leave the remainder to the orthodontist; (b) to extract the second molar, together with cyst enucleation, and to free the first molar to erupt; or (c) to extract the first molar, curette the cyst lining, and then upright the second and third molars. Are all these equal possibilities? The missing element is the diagnosis. The first molar was infra-occluded due to invasive cervical resorption (arrow) and must therefore be the tooth to be sacrificed, together with defusing of the cyst. The possibility of pathologic fracture of the mandible was known and accepted, and the molar therefore removed piecemeal. Enucleation of the cyst was impossible, due to poor access and the danger of damaging the second molar – it was marsupialized. An attachment was placed on the second molar and occlusally directed traction was applied from a zygomatic arch implant.

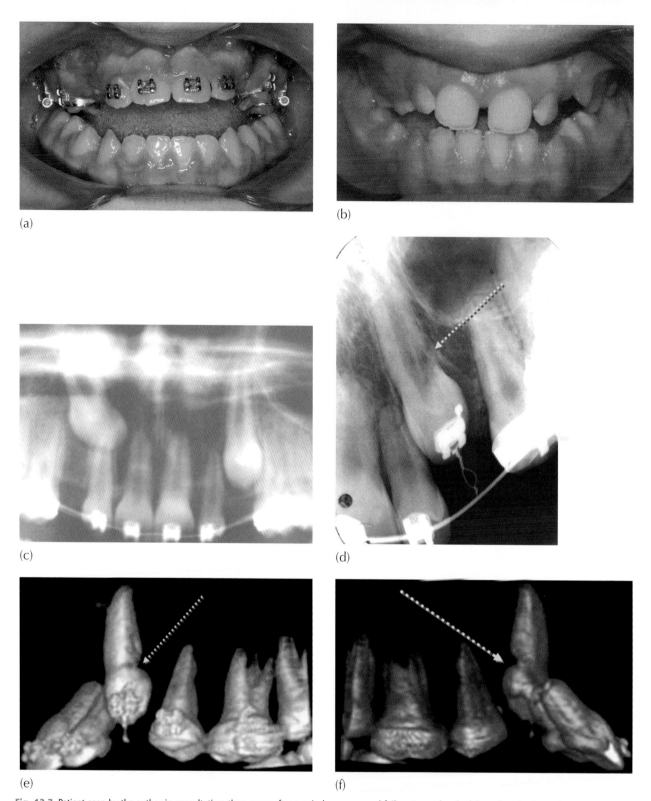

Fig. 12.7 Patient seen by the author in consultation, three years after surgical exposure and failure to resolve the bilateral canine impaction.

(a) Intra-oral view at consultation: patient complaining of an open bite.

(b) Intra-oral view from the initial treatment records, five years earlier.

(c) Anterior section of panoramic view taken prior to surgical exposure showing canines with no apparent pathological involvement.

(d) Periapical view taken at time of consultation visit shows aggressive cervical root resorption on the distal aspect of the root of the left canine.

(e) A buccal three-dimensional view from the CBCT series shows the lesion extending from the distal to lingual sides of the root.

(f) A lingual three-dimensional view from the CBCT series shows the lesion extending from the distal to lingual sides of the root.

(g) A single axial (horizontal) slice from the CBCT to show the right canine lying horizontally above the roots of the incisors. The cross-section of the left canine shows a distinct break in the continuity of the root on the distal side. It is possible to discern the narrow pulp chamber outlined by more radiolucent predentine and the resorbed area of root around it.

(h) Two transaxial (vertical) slices from the CBCT show the palatal point of entry of the resorption process and its mushrooming extension inwards.

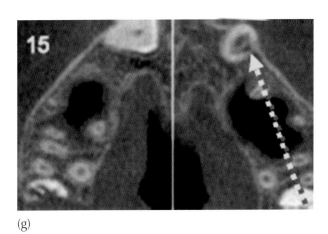

(g)

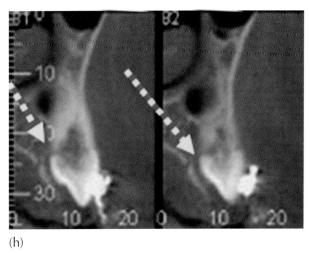

(h)

Fig. 12.7 *(Continued)*

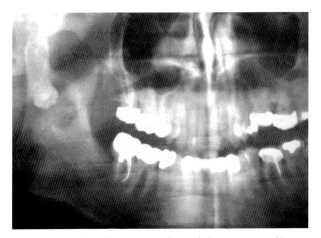

Fig. 12.8 A 'condylar' third molar complete with dentigerous cyst. (Courtesy of Dr A. Renert.)

remote areas of the jaws generally. Several of these are worth exhibiting here, for curiosity value only (Figures 12.8–12.12). Satisfactory imaging of many of these teeth may be achieved with the use of straightforward plane film radiography because of the wide separation of these teeth from the remainder of the dentition. There may be little or no added value in the use of CBCT in many of these cases. As with teeth of abnormal form, their surrounding tissues are generally quite normal and the potential that they may have for responding to the application of orthodontic forces may be excellent. Nevertheless, the question that needs to be answered is whether the estimated length of time involved in their treatment and the periodontal prognosis of the outcome are sufficiently favourable to make the orthodontic/surgical modality superior to other therapeutic options. The answer will usually lean towards extraction or, occasionally, towards leaving the tooth untreated because of difficulty and possible complications in extraction.

Incorrect positional diagnosis

Inadequate or inappropriate use of imaging techniques may sometimes give the operator a false impression of the position of an unerupted tooth. Often, films exist, but are simply not considered as contributory because positional diagnosis was not the original intention for obtaining the particular view concerned. As a routine, most orthodontists insist that the initial records of all their new patients include a panoramic, lateral cephalogram and sometimes a postero-anterior cephalogram. The panoramic films are used to ensure the presence of all the teeth and to register their state of development, and the cephalograms are studied and evaluated by the measurement of angles and distances between anatomic and dental structures. An undetermined but significant proportion of our orthodontic and surgical colleagues fail to exploit all the information that is contained in these films, particularly in regard to three-dimensional diagnosis of impacted tooth position (Figure 12.13). The patient is then further irradiated in the search of information already available.

We have discussed wildly ectopic teeth as one of the factors that a patient may present to us for which the orthodontic/surgical modality may not provide the best treatment solution. However, to arrive at this decision a careful analysis of the orientation of the tooth in its ectopic location needs to be made, and this must be done by the orthodontist.

One of the most important guiding principles relates to the accurate diagnosis of the location of the root apex of the aberrant tooth. If the apex is in a fairly normal position, the prognosis will generally be fairly good, because the main part of the treatment for that tooth will involve a tipping movement, which is easy to apply and rapid in its response. However, whenever the root apex is also displaced, particularly when it is located on the same side of the arch as the displaced crown, very considerable

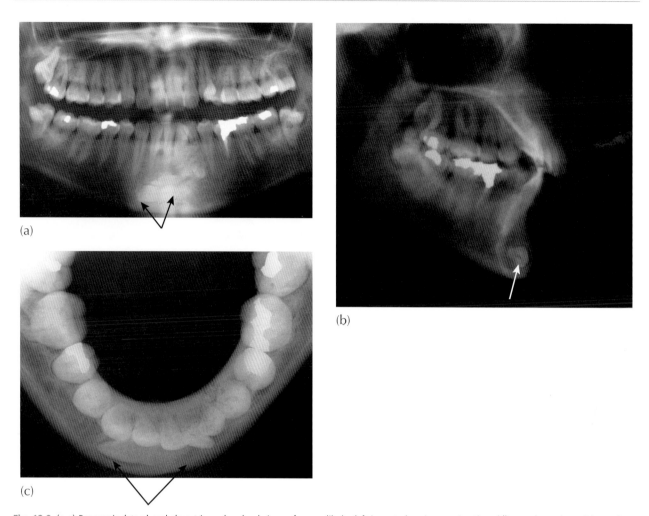

Fig. 12.9 (a–c) Panoramic, lateral cephalometric, and occlusal views of a mandibular left impacted canine crossing the midline and exactly at right-angles to the mid-palatal and antero-posterior planes (arrows). The direction of the X-ray for the cephalogram is in the long axis of the tooth and depicts it in cross-section.

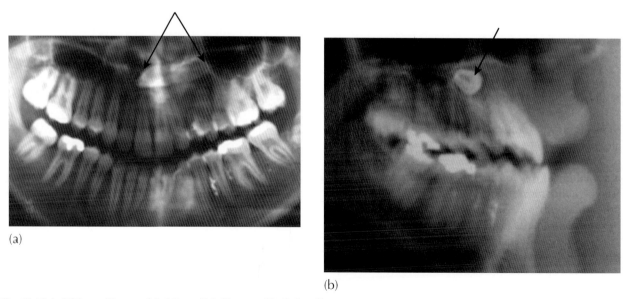

Fig. 12.10 (a, b) The maxillary parallel of Figure 12.9. (Courtesy of Dr M. Barzel.)

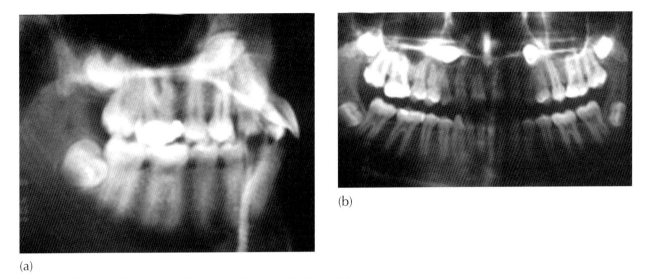

Fig. 12.11 (a, b) Bilateral 'high-flying' maxillary canines. (Courtesy of Dr P. Teper-Adler.)

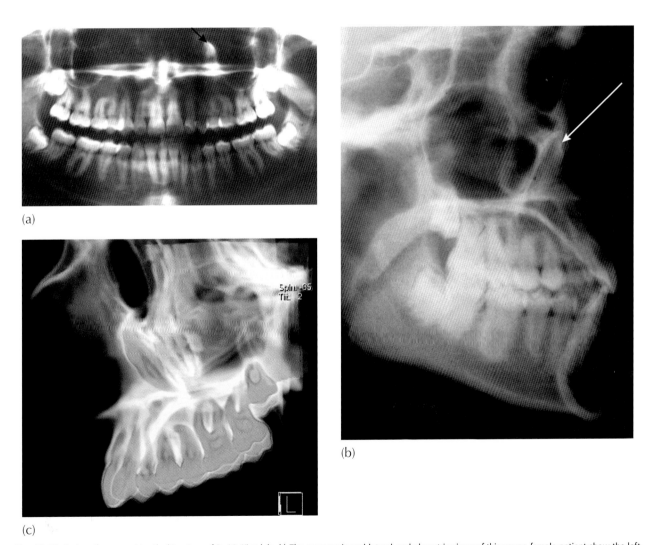

Fig. 12.12 An 'eye-for-an-eye' tooth. (Courtesy of Dr M. King.) (a, b) The panoramic and lateral cephalometric views of this young female patient show the left maxillary canine (arrows) to be in close relation with the floor of the orbit of that side. It was important to establish its exact location in the three planes of space and whether further 'eruptive' movements in the present direction would threaten the eye. (c–e) Three-dimensional computerized tomographic views of the tooth provide accurate positional information with which to consult the ophthalmologist.

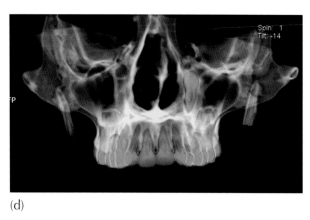

(d)

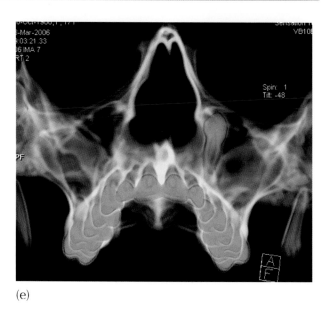

(e)

Fig. 12.12 (*Continued*)

torqueing and/or uprighting movement will be required. Root movements are largely impossible to accomplish before the bonded attachment on the crown of the tooth has reached and been ligated into the main archwire. Therefore, during the mechano-therapy employed to bring the tooth to the line of the arch, the orientation of the impacted tooth will generally change unfavourably to cause the root to bulge the palatal or buccal mucosa to an exaggerated degree and lead to a dehiscence of its cervical area. The severity of this bulging and, by extension, its dehiscence will depend on the distance that separates the root apex from the general line of the arch – its normal place. In the more extreme cases, this will require a disagreeably wide angle of torque correction (Figure 12.14), which will translate into many months or years of treatment, and the periodontal implications of the treated result will often leave much to be desired.

In Chapter 8, the buccally impacted maxillary canine with mesial displacement was highlighted, and it was explained why its positional diagnosis was particularly prone to error when using plane film radiography. When these teeth are more severely mesially displaced, a periapical or anterior occlusal radiographic view will depict them superimposed on the mesial aspect of the root of the lateral incisor and on the distal aspect of the root of the central incisor. Being sited in the depression between the two roots, it may not produce an adequately convincing differential overlap of the incisors that is needed for the lateral tube shift method in determining its labio-lingual positional diagnosis. Because of the angle of the central ray from the periapical or anterior occlusal vantage points, this labially displaced tooth will appear to be more occlusal than its true position, superimposing on the middle portion of the roots of the adjacent teeth. However, on a panoramic film, the relative height of the tooth will show it to be more apically superimposed on the roots of the incisors (for a description

of vertical tube shift, see Chapter 2). Given the frequency with which misdiagnosis appears to be made in these cases, the attempt to surgically expose a labially impacted canine may be mistakenly made from the palatal side! The surgeon may then report the case as having been a very difficult one or that the tooth was impossible to find (Figure 12.15) – this would be the good news! A more persistent surgeon might succeed in exposing it, but fail to bond an attachment to it, or perhaps the lateral incisor may be subject to a totally superfluous and iatrogenic vital apicoectomy on the way (Figure 12.16)! Compounding the problem, the orthodontist may follow the surgeon's lead by attempting to draw the tooth from the palatal side, which may then result in one of three possible ill-fated outcomes: (a) the tooth will not move at all, (b) it may cause resorption of the labial and interproximal aspects of the incisor roots, against which it is being drawn, or (c) it may erupt through the palatal mucosa, with the orientation of its strongly tipped long axis clearly indicating that the root apex is on the labial side (Figure 12.17)!

Surgical exposure without prior orthodontic planning

For the surgical approach to most impacted canines, the essential design and extent of a palatal surgical flap in a closed eruption case differ very little, regardless of the position of the canine. The same is true of those canines to be exposed in an open 'window' technique. For this reason, and once the surgeon knows that the tooth is on the palatal side, he/she may consider further radiographic refinement of the positional diagnosis to be unnecessary and will simply rely on the two, very conveniently acquired, periapical views used in Clark's tube shift method. After all, and in contrast to the needs of the orthodontist, knowing the position of the root apex and orientation of the root portion

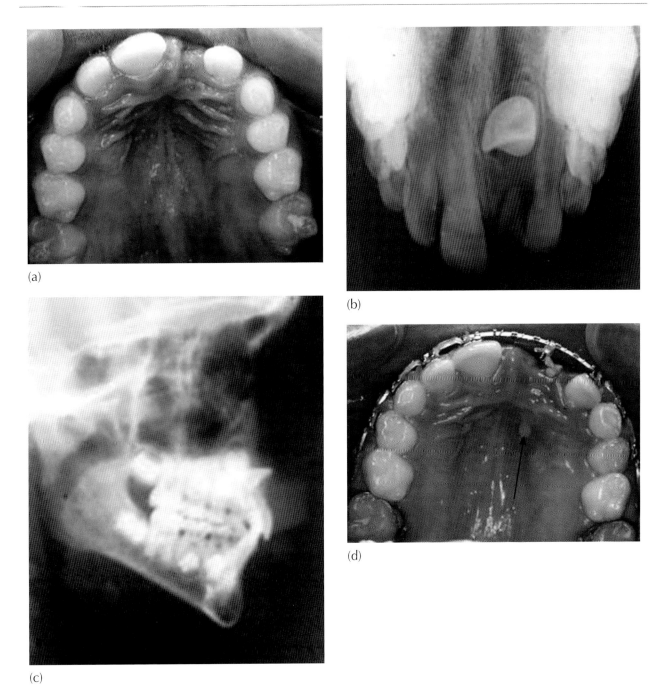

Fig. 12.13 A case of mistaken identity. (Courtesy of Dr N. Shpack.) (a) An occlusal view of a patient with an unerupted maxillary left central incisor. (b) The anterior occlusal view of the maxilla shows the typical appearance of a classical dilacerate central incisor, with the crown viewed in its long axis and the apical portion of the root pointing superiorly. (c) The cephalogram confirms the position of the crown high up and labial, adjacent to the root of the nose. (d) Excessive space had been deliberately made and a rectangular archwire placed, ready to be used as the base from which to apply traction to the impacted tooth. After the second surgical exposure an elastic ligature was tied to the labial archwire. However, the site of the first surgical exposure was mistaken and the decision was made through inadequate attention being given to information that was readily available from the radiographs. The exposed root apex of the tooth (arrow) can be clearly seen in the mid-palate. This tooth was subsequently extracted.

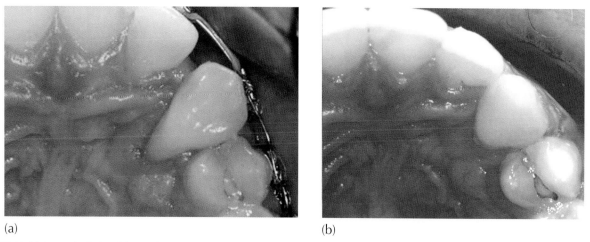

(a)

(b)

Fig. 12.14 (a) A severely displaced (group 4) canine crown has been brought to the line of the arch and needs much labial root torque. Considering the length of the root of a canine, its apex must be moved buccally through approximately 10 mm of alveolar bone! (b) Root torque was achieved with a torqueing auxiliary and took 17 months to complete. (Courtesy of Professor S. Chaushu.)

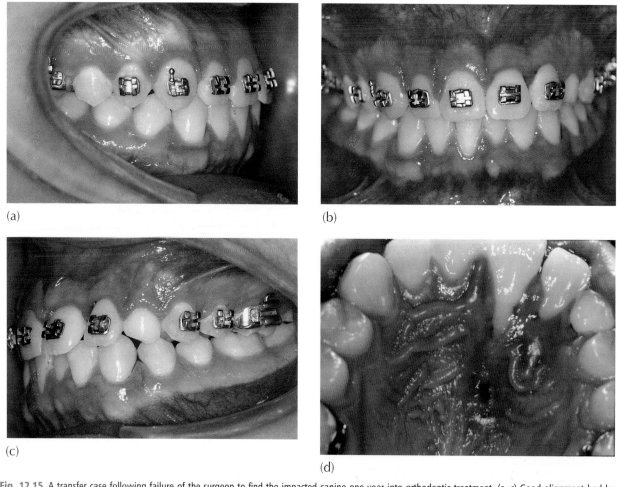

(a)

(b)

(c)

(d)

Fig. 12.15 A transfer case following failure of the surgeon to find the impacted canine one year into orthodontic treatment. (a–c) Good alignment had been achieved by the previous practitioner, but all potential clinical signs of canine location were eradicated. (d) A palatal approach had been chosen for the surgical exposure. The canine was not found. This is the residual soft tissue and bony defects can be clearly seen. (e, f) Intra-oral views taken from the original pre-treatment records show labial tipping of the crown of the left maxillary central incisor and palatal tipping of its root – an indication that the canine is on the labial side of the incisor root. (g, h) The anterior section of the original panoramic view and an anterior occlusal view show differing levels of superimposition of the canine crown over the incisor root. The vertical tube shift method confirms the diagnosis of a labially displaced canine. (i, j) A vertical (paraxial) cut and a three-dimensional view from the newly ordered CBCT to show the labial position of the canine. (k) Surgical access to the canine was simple and shows the canine crown to be sited between the roots of the two central incisors. (l) The full flap has been sutured back in place. The pigtail ligature exits through the flap opposite the tooth. Note the horizontal passive position of the auxiliary arch. (m) The horizontal loop of the auxiliary arch is flexed vertically upwards and a little mesially to be engaged by the pigtail ligature and to impart a labially direct force with a distal component in order to circumvent the central incisor root. (n–r) Intra-oral views of the case 2.3 years post-treatment. The clinical crown is slightly elongated and the palatal view shows the central incisor defect. There is no difference between the appearance and clinical condition of right and left canine. (s, t) Anterior portion of panoramic film and a periapical view, 2.3 years post-treatment.

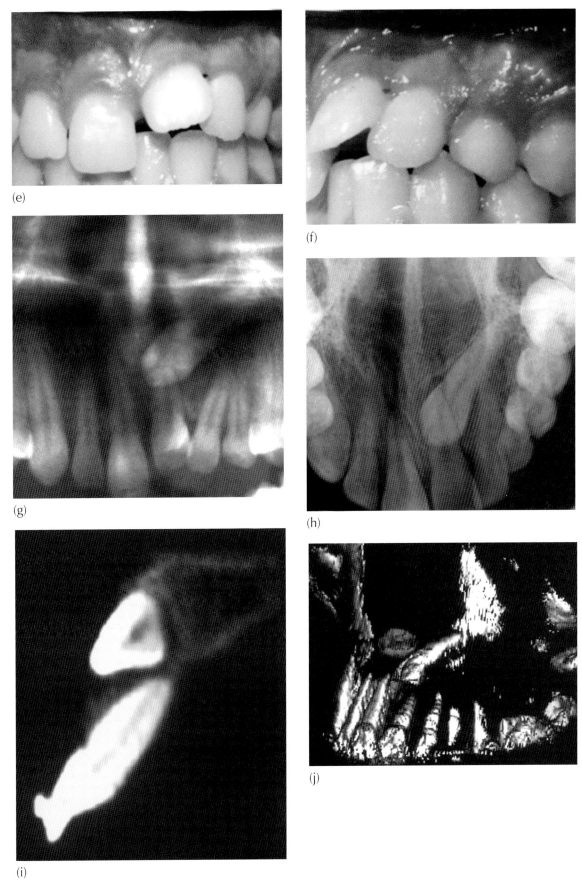

(e)

(f)

(g)

(h)

(i)

(j)

Fig. 12.15 *(Continued)*

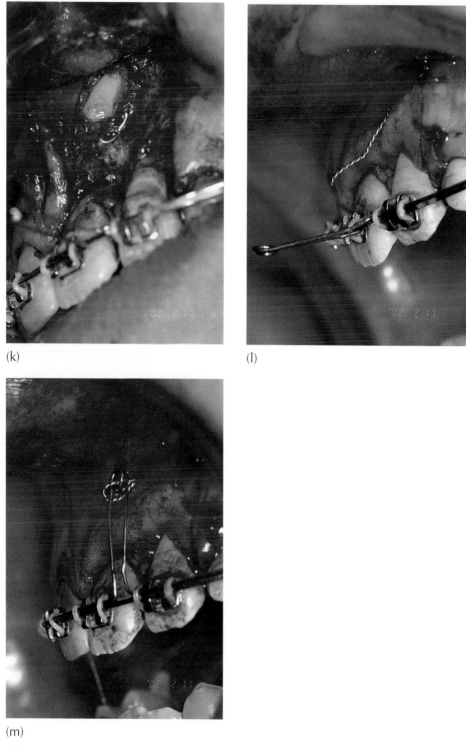

(k)

(l)

(m)

Fig. 12.15 (*Continued*)

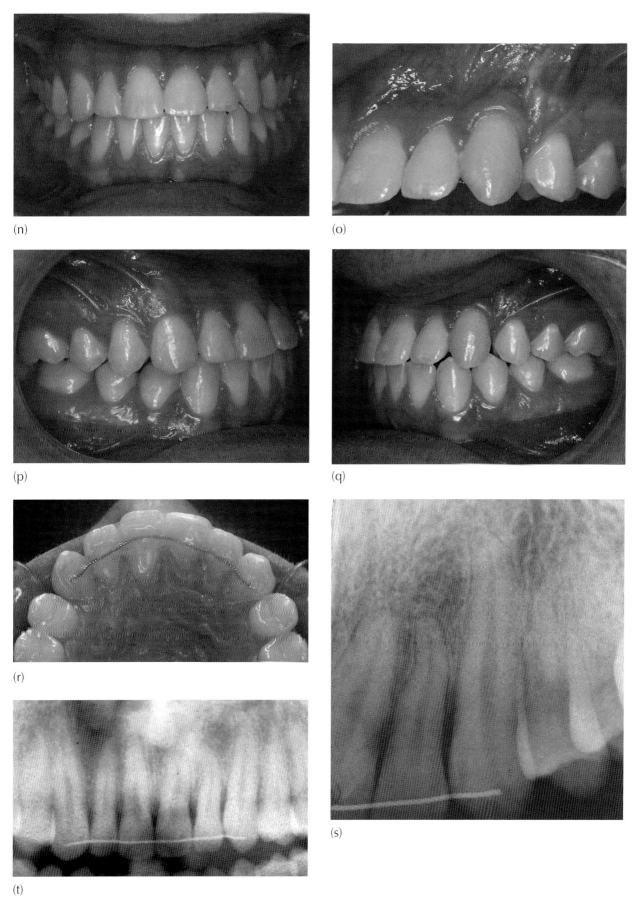

(n)

(o)

(p)

(q)

(r)

(s)

(t)

Fig. 12.15 (*Continued*)

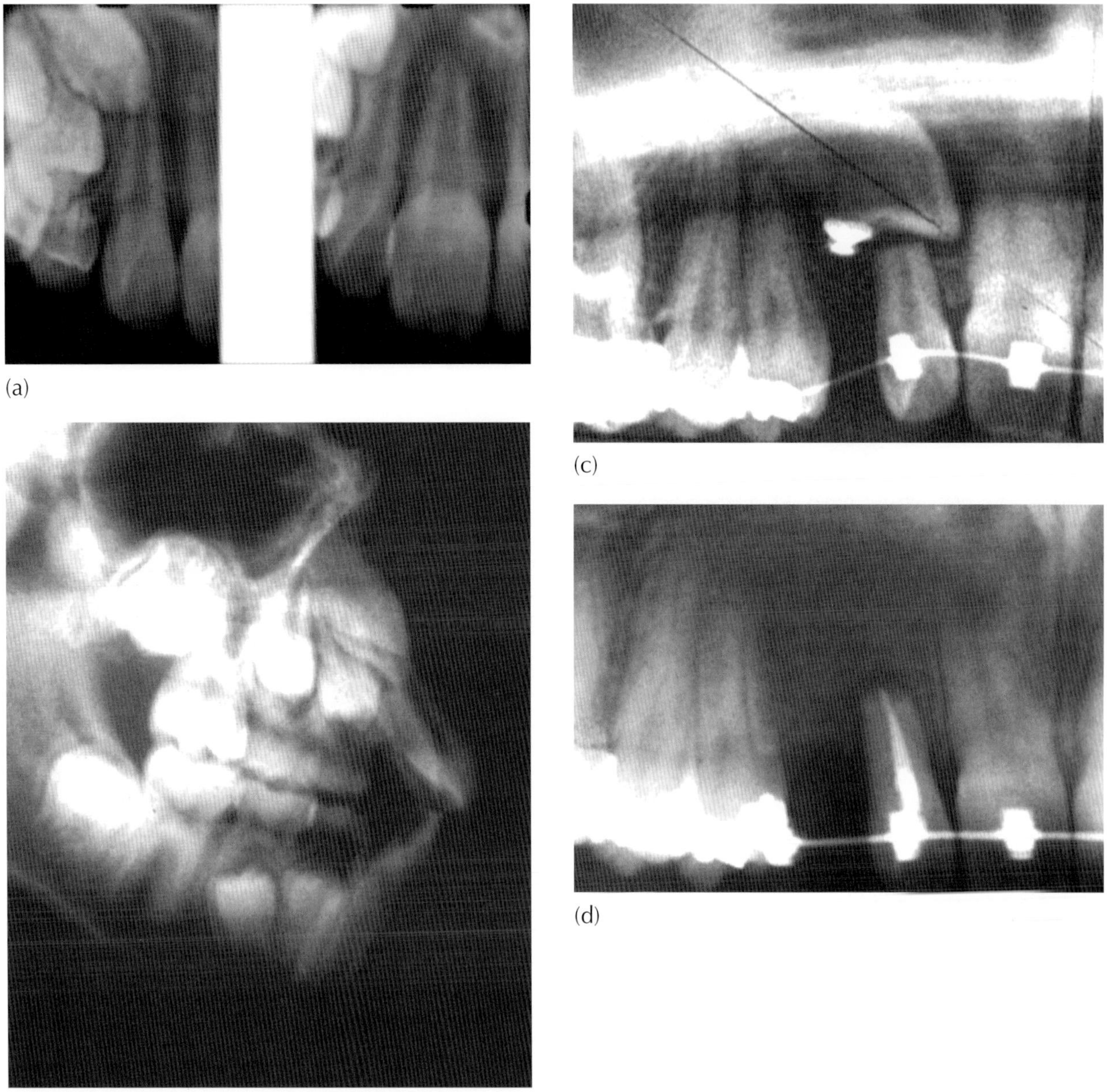

(a)

(b)

(c)

(d)

Fig. 12.16 A miscarriage of justice! (a) Two adjacent views from the initial periapical survey radiographs. Using the buccal object rule, the right maxillary canine is labial to the lateral incisor root. (b) The initial lateral cephalogram shows both canines to be labial to the incisor roots. (c) A panoramic film taken seven months post-surgery. Exposure was mistakenly performed from the palatal side (!) and a button bonded to the inferior (anatomically palatal) aspect of the canine. Surgical access was 'very difficult'. The lateral incisor has a shortened root – the immediately pre- and post-surgical periapical radiographs were 'lost'. (d) Months later, the surgeon extracted the canine (!), and endodontic treatment of the incisor was prescribed in order 'to stop further root resorption'. Subsequently the lateral incisor was extracted because it was considered to have a poor prognosis and implants placed to replace the missing teeth. The orthodontist was successfully sued by the patient for negligence 'for having caused the entire problem'.

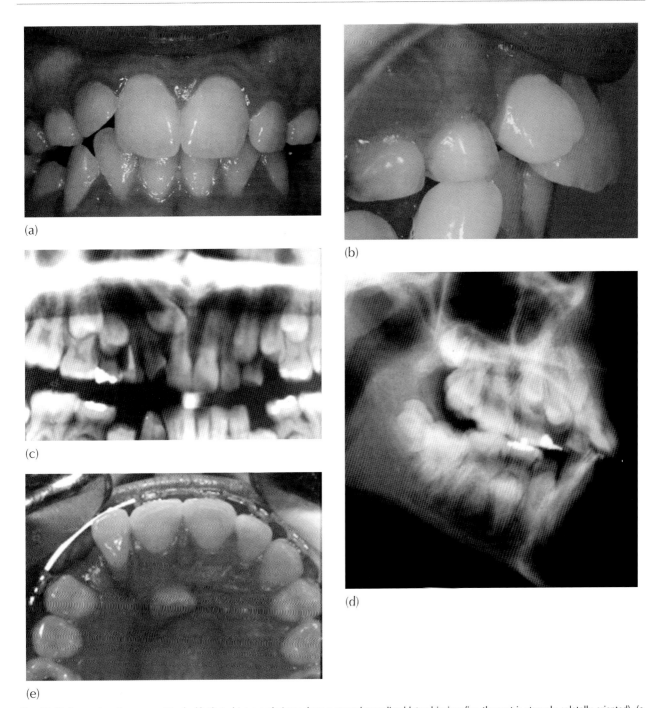

(a)

(b)

(c)

(d)

(e)

Fig. 12.17 Exposed on the wrong side. (a, b) Clinical intra-oral views show a strongly proclined lateral incisor (i.e. the root is strongly palatally oriented). (c, d) The panoramic and lateral cephalometric views show the canine to be sited between the incisor roots. Combining the clinical image of lateral incisor orientation with that seen on the cephalogram places the canine labial to the lateral incisor root. (e) Surgical access was attempted from the palatal side, despite the fact that much tissue was removed, the impacted canine was very difficult to find (!). An attachment was placed by the surgeon and the orthodontist applied traction from the palatal side. The canine is seen here having been drawn from the labial side, between roots of the central and lateral incisors, and erupted into the palate! There is a severe periodontal defect on the lateral incisor. The lateral incisor and canine were subsequently extracted and the case completed with implants and crowns.

is less crucial for the surgeon, who has only to locate the crown of the tooth in order to approach it with surgical precision.

The palatal canine is often palpable in the palate and, when the flap is reflected, the position of the canine is usually self-evident by the manner in which its crypt bulges the bone overlying it. While this *modus operandi* is shared by a large number of surgeons, it does not fully take into account those canines in group 3, which are to be found higher up and whose position may not be obvious after initial flap reflection. Thus, when surgical exposure is attempted with merely a pair of periapical views to hand, much surgical rummaging in the palatal bone may be necessary before the exact location of the canine crown is found.

The intention may be to perform an open exposure and to maintain the opening with the placement of a pack in the hope that the impacted tooth will oblige and erupt spontaneously. In the more favourable cases this will take many months, but there is a distinct possibility that the tooth will not erupt, and access to it may have become severely limited by palatal soft tissue overgrowth in the interim. In due time, a clinical and radiographic reassessment may show little or no progress, and the conclusion reached, in retrospect, might well be that perhaps the earlier procedure had been too conservative. A second attempt at surgical exposure may be advised, this time with a more radical bone-clearing procedure, a wider opening and a larger pack. This may or may not finally elicit the eruption of the impacted canine. However, under the most favourable circumstances, and even taking into consideration the alveolar bone that will undoubtedly be regenerated along with the eruption process, the finally erupted and aligned tooth will suffer from a compromised bone support and a very long clinical crown, with a good section of the exposed cervical third of its root. This will be accompanied by a similarly compromised condition of the palatal surface of the root of the neighbouring incisor (Figure 12.18).

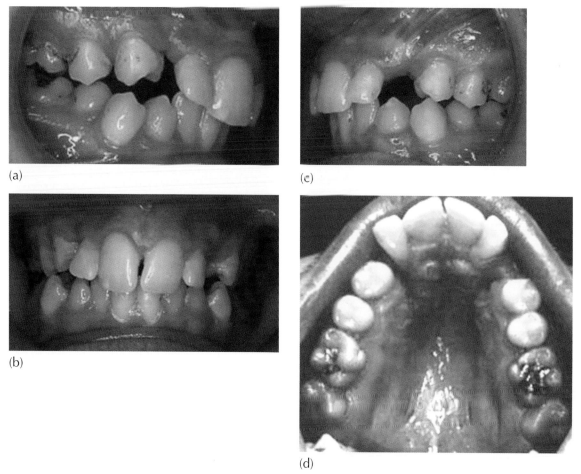

(a)

(c)

(b)

(d)

Fig. 12.18 Three failed surgical interventions. (a–d) The intra-oral photographic records show the missing deciduous canines (extracted during the first procedure) and severe soft tissue and bony defects on the palatal sides of the lateral incisors, following three 'open-and-pack' interventions, over a three-year period. Neither canine had shown any signs of erupting and the periodontal prognosis of the lateral incisors was considered to be poor. (e) A periapical view shows the canines to be associated with resorption of the incisor roots. (f–i) The case was treated as an extraction case, with removal of the periodontally involved lateral incisors, and compensated by first premolars in the lower jaw.

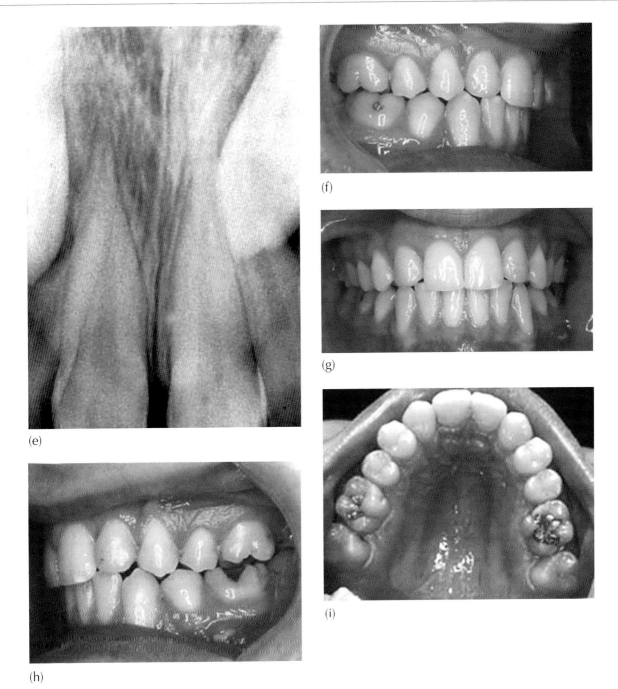

(e)

(f)

(g)

(h)

(i)

Fig. 12.18 *(Continued)*

It seems clear that the choice in these situations is difficult. On the one hand, we seek to make a net gain to avoid early active orthodontic intervention by substituting it with an extended surgical alternative. But in so doing, we undertake to bring about a spontaneous eruption of the impacted canine, using a surgical procedure that will compromise the final outcome of the overall treatment of the tooth, in terms of its final periodontal condition. At the same time, we must not forget that an integral part of this deal is that we run the risk that the tooth will still not respond. There are no published figures or available evidence to indicate what degree of reliability may be attributed to this method that may justify its routine use in generating spontaneous eruption, thus to overcome our natural and healthy reserve regarding its efficacy in a sufficient proportion of cases. By contrast, the use of an orthodontic force applied from a suitable auxiliary, with an appropriate direction, range of activity and energy level, must accord the treatment a much greater level of reliability and the possibility of limiting the surgical exposure to the least damaging of the surrounding structures, both directly and indirectly. By taking this proactive stance and with proper management, therefore, the chances of failure must be very considerably reduced.

Resorption of the root of an adjacent tooth

Chapter 7 is devoted in its entirety to this subject. It is emphasized that the occurrence of this phenomenon in the incisor area is much more common (66.7% of adjacent lateral incisors and 11.1% of adjacent central incisors) than once thought [6] and, in its severest form, it develops rapidly and may irretrievably destroy much of the root of an adjacent central or lateral incisor in a relatively short time. The young patient with a mixed dentition may be seen on a regular basis, watching the development of the early malocclusion in anticipation of starting the orthodontic treatment in the full permanent dentition. Even with yearly radiographic monitoring of the immediate area, resorption may cause much root shortening in the intervening months before it comes to the attention of the orthodontist (Figure 12.19).

In these cases, therefore, suspicious positioning of an unerupted canine vis-à-vis the lateral incisor root should encourage the orthodontist to recommend a closer follow-up procedure. A single periapical view (not a panoramic view) taken 4–6 months after an initial base-line film should show the presence or absence of downward eruption progress of the canine, and comparison should be made with its antimere. A lack of progress, strong overlap of the incisor roots or alteration in root form or length should be treated with suspicion.

When the condition is mild, traction on the impacted tooth will distance it from the resorbing area and the resorption will effectively stop [7]. However, when the resorption has become extensive, accurate relative posi-

tional diagnosis of the impacted canine and the affected root end must be established for two reasons: the exact direction that traction must take needs to be determined; and on the basis of this, the site of the surgical exposure will be chosen such that the attachment is placed on a surface that faces the direction of future traction. Exposure should be as conservative as possible, sufficient only to accept the full base of a small attachment, while haemostasis is maintained. Meticulous closure of the surgical flap is then performed.

Failure to observe these precautions will lead to failure in one or more aspects of the treatment. The application of orthodontic force to the impacted canine in an ill-determined direction vis-à-vis the resorbing root end may bring the two in closer contact and thereby encourage further resorption (Figures 12.20 and 12.21). An open surgical exposure in this situation and the clearance of tissue around the crown of the tooth, aimed at freeing the crown from all potential contact with bone, are to be avoided. Either of these two procedures will risk exposing the resorption front at the root face to the oral environment, which will lead to the devitalization of the incisor and further endanger its viability, whether or not a surgical pack has been placed.

In the case illustrated in Figure 12.22, the influence of the unerupted canines in producing root resorption of the lateral incisor root was not apparent in the initial panoramic view taken at 11 years of age. A repeat film taken two years later, immediately prior to the commencement of treatment for the patients class 3 malocclusion, showed oblique distal resorption of the roots of both lateral incisors, clearly in association with the proximity of the unerupted canines. On the other hand, the left central incisor had lost almost half its root length due to resorption from no obvious cause, since the unerupted canine was quite distant from it.

Orthodontic treatment was begun and, before adequate space was provided in the arch, the left canine was exposed with traction being applied directly to the labial archwire. From that height, this caused the canine to move down and directly towards the root of the lateral incisor. The follow-up radiograph showed the resultant severe resorption of the lateral incisor root. At this point, the patient was referred to the author for continuation of treatment and, on the first visit, an auxiliary labial arch was placed to change the direction of the traction to the labial. The rest of the treatment was completed with the now buccally displaced canine remaining 'in limbo' until space became available for its alignment. Root uprighting and torqueing procedures were avoided and the treatment was completed with simple levelling and aligning movement only and in a very short time period.

At 17 years of age, the follow-up radiograph, taken one year after bonded splints had been placed at the end of treatment, revealed severe resorption of the roots of the

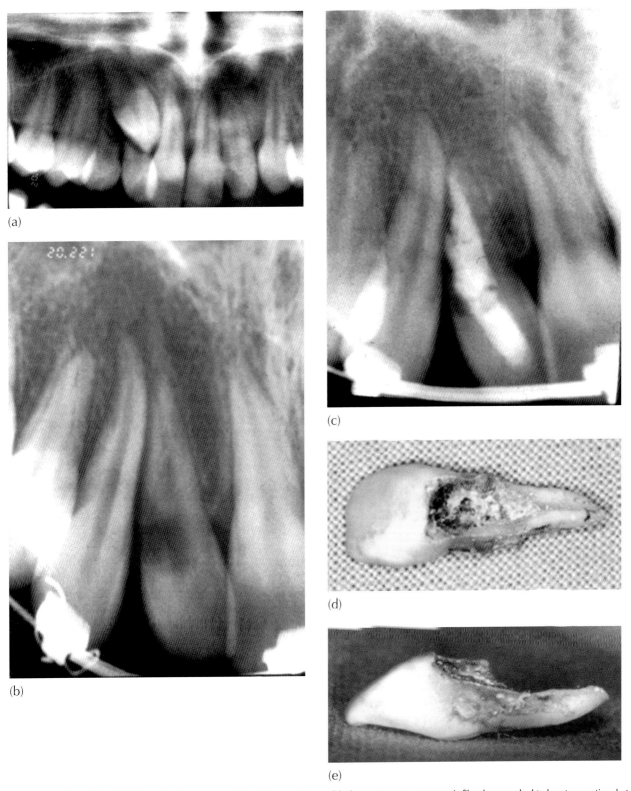

Fig. 12.19 The bucco-lingual blind spot in the pre-computerized tomography era. (a) The pre-treatment panoramic film shows undoubted root resorption, but a continuation of the root under the superimposition of the canine. (b) Following careful surgical exposure and orthodontic treatment, the canine is distanced from its intimate relationship with the root of the incisor, to reveal an area of root resorption at and just apical to the CEJ. Because of the danger of exposure of this area to the oral environment, the patient was referred for endodontic treatment of this tooth. (c) Obturating the canal with root canal filling material proved impossible due to obvious lack of a wall to the pulp chamber. The tooth was extracted. (d, e) Views of the extracted tooth from the labial and mesial sides show a severe longitudinal resorption of the entire labial side of the root of the tooth, involving the pulp. This was not possible to diagnose sufficiently well from the radiographs in hand, and ill-advised treatment was undertaken which, in today's world of CBCT, would have been avoided.

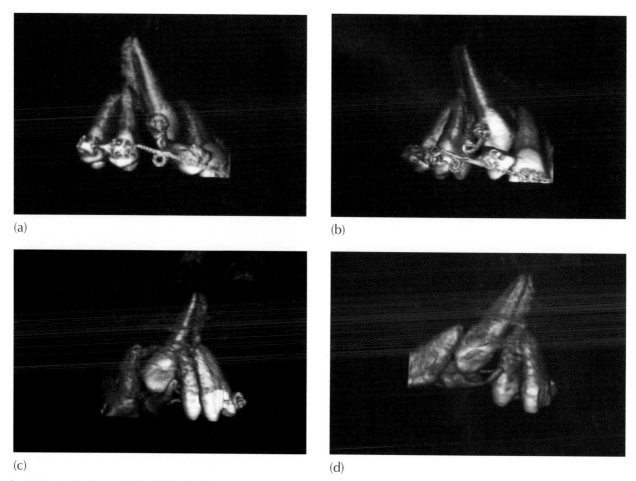

(a)

(b)

(c)

(d)

Fig. 12.20 A canine that resisted applied traction due to inappropriate directional traction. Treatment was begun before accurate positional diagnosis had been made. (a, b) Cone beam tomography presents two aspects from the labial side in a three-dimensional reconstruction. The canine is mesial and labial to the lateral incisor root, which is almost horizontally displaced. The crown tip of the canine appears to be in a resorption trough in the lateral incisor root. It is being inappropriately pulled labially and inferiorly, against the cervical part of the root of the incisor. (c, d) The two views from the palatal side show that most of the incisor root has disappeared and that only a thin sliver of its palatal side remains. Further traction to the labial archwire will almost certainly result in total root resorption. The patient would have been better served by the canine being drawn labially and superiorly, using an auxiliary labial archwire, as seen in Figure 12.16m.

other three maxillary incisors, having lost between one half and two-thirds of their initial root length. The left lateral incisor had lost almost its entire root but, with the splint in place, there was no obvious mobility. With the cessation of orthodontic movement, it is estimated that further resorption will not occur and it is hoped the patient may look forward to several years of stability before artificial replacements become necessary [8]. In stark contrast, none of the other teeth in the mouth were affected by root resorption!

This raises an interesting point. On the one hand, resorption of the root of the tooth immediately adjacent to an impacted canine is well known and the effect is clearly a local one. On the other hand, the fact that there are cases in which there may be generalized resorption of the roots of all the teeth either with or without orthodontic treatment has also been extensively documented and its aetiology has been variously ascribed to genetic, humoral, hormonal or idiopathic factors. It is therefore an enigma,

as seen in this case, that three incisor teeth with no obvious proximity to the impacted canine are resorbed to more or less the same degree as the immediately adjacent lateral incisor, while the remainder of the dentition is totally unaffected.

Iatrogenic damage may also be inflicted because of inadequate attention to detail and specifically to an incomplete examination of the diagnostic radiographic material. It was pointed out in Chapter 6 that a displaced lateral incisor is usually due to the proximity and angulation of an adjacent impacted canine. Reopening space with simple mechanics brings the root of the incisor into direct contact with the unerupted crown of the canine. When the angle between the long axes of the two teeth is small, this will often bring about improvement in the angulation of the impacted tooth and subsequent spontaneous eruption, as has been pointed out by other workers [9–11]. However, when the angle of orientation of the teeth approaches 90°, then

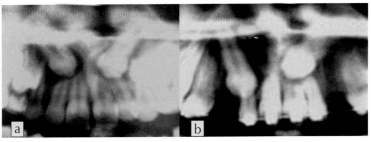

Jan 1997   Sept 1998

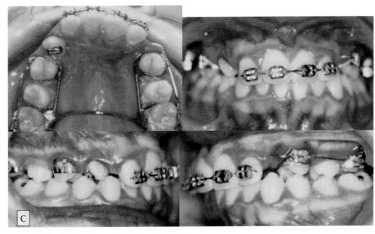

Clinical views in Oct 1998

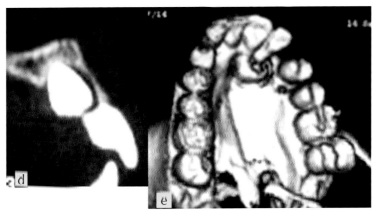

Fig. 12.21 A transfer case first seen by the author in October 1999. The patient spent eight months in Jerusalem during her father's sabbatical leave. Accompanying radiographic records were received as follows. (a) Anterior portion of initial panoramic view (January 1997) at commencement of treatment, shows bilateral maxillary canine impaction and severe resorption of all four incisor roots. (b) Follow-up view (September 1998) taken after alignment, levelling and space opening, shows improvement in position of right canine and worsening position of left canine, with considerable additional resorption of the roots. Surgical exposure and attachment bonding was performed at this time. (c) The clinical condition at the patient's first visit in October 1999 shows a palatal arch soldered to molar and premolar bands and a sectional arch linking the incisors only. The right canine had erupted with its palatal eyelet attachment in place. The ligature wire attached to the Edgewise bracket on the unerupted left canine is visible exiting the palatal mucosa and a Ballista spring is present, tied into molar and premolar brackets. The following are signs of failing anchorage: the burrowing of the ballista into the palatal mucosa, intrusion of the left premolar, extrusion of the incisors. (d) From the newly commissioned CBCT, a paraxial cut shows the severity of the resorption of the root of the central incisor and the location of the canine tip immediately superior to the resorbed root end. (e) An oblique occlusal three-dimensional view of the palatal side of the left incisor area shows how the canine crown is immediately above the resorbed root end. The vertically directed traction force has clearly extruded the incisors (all four are linked by the sectional archwire), it has largely resisted the force of the ballista, causing it to burrow into the mucosa and brought about the infra-occlusion of the premolar. Additionally, it has aggravated the resorption of the incisor roots. (f) Traction force was directed posteriorly and laterally to a transpalatal arch with soldered loops in November 1999. (g) The canine was drawn away from the incisor root and erupted in the mid-palate in March 1999. (h) The bonded bracket on the palatal side of the canine was substituted with an eyelet on the labial side in May 1999 and drawn towards its place in the arch, while a coil spring was used to reopen space by moving the four incisors to the right. (i) A view from the left side at that stage. (j) Photographs of the case 6.7 years post-treatment. (k) Periapical radiographs taken one year post-treatment in June 2001. (l) Periapical radiographs taken of the case 10 years post-treatment in September 2009 show no pathology, arrest of the resorption process, excellent regeneration of bone, intact lamina dura and pulp obliteration of and caused by a vital left lateral incisor.

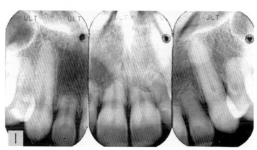

Fig. 12.21 (*Continued*)

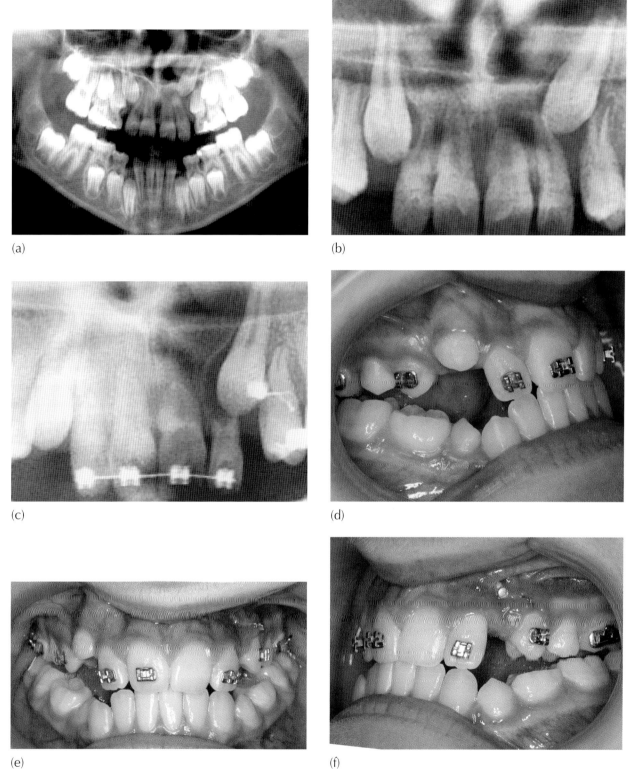

Fig. 12.22 (a) Panoramic view at age 11 years showing no abnormality.

(b) Panoramic view at age 13, with early oblique root resorption on the distal side of the root of the lateral incisors. The left central incisor has lost almost half its root to resorption.

(c) Following exposure and traction of the canine, there is resorption loss of almost the entire lateral incisor root.

(d–f) The situation at the first transfer visit. The open surgery exposure had almost completely closed over, with the head of the bonded button attachment still visible and part of the steel ligature wire re-buried in oral mucosa.

(g) At the same visit the appliance was refurbished and an auxiliary labial arch placed to draw the canine labially away from the incisor.

(h) Full appliances placed and space opening begins.

(i–m) The completed treatment with lingual twistflex wire bonded retainers in place. Note that labial root torque of the resorbed lateral incisor was considered inappropriate.

(n) The one-year follow-up panoramic view shows the severe degree of resorption of the maxillary incisor teeth and the normal roots of the other teeth.

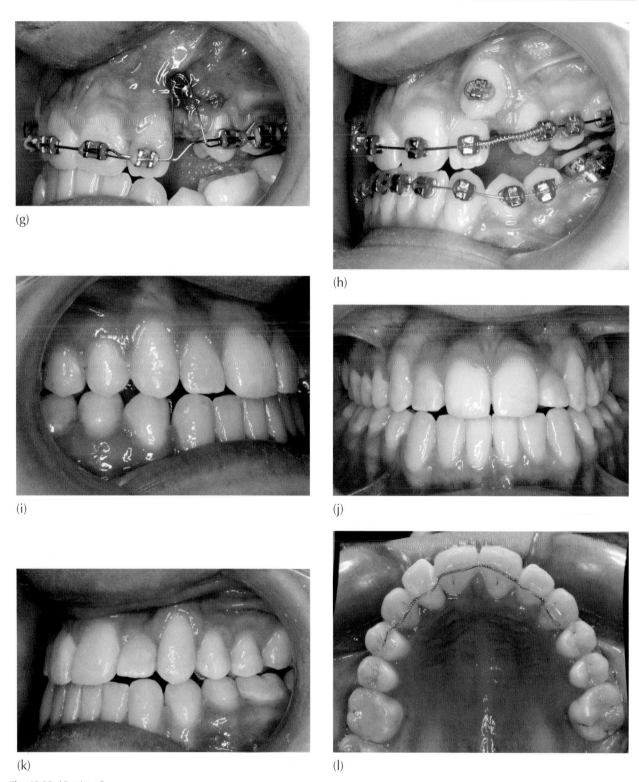

(g)

(h)

(i)

(j)

(k)

(l)

Fig. 12.22 (*Continued*)

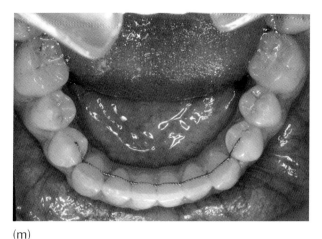

(m)

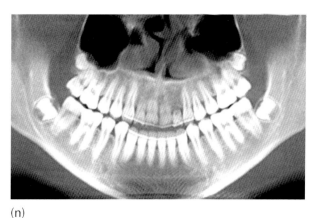

(n)

Fig. 12.22 *(Continued)*

uprighting the incisor will bring its root apex into contact with the crown tip of the canine and develop a force that attempts to force the canine to 'back up' along its long axis. This will be a very potent cause of resorption of the incisor root apex. In the case illustrated (Figure 12.23), it will also be seen that negligent positioning of the bracket on the first premolar has brought the root of that tooth mesially and into contact with the inferior aspect of the impacted canine. This further limits the canine's room for manoeuvre and effectively blocks the desired pathway for its eventual resolution.

In this case, a re-evaluation of the biomechanics was made and the bracket of the lateral incisor was immediately removed. The bracket on the first premolar was debonded and replaced by another whose function was to upright the roots distally and, at the same time, to rotate the tooth in a mesio-lingual direction to thereby distance the palatal root from contact with the canine. Once these movements were completed, the canine was exposed on the labial side as high as possible above the resorbing root area of the lateral incisor. Traction was made using an auxiliary labial arch to bring the canine labially high in the sulcus and above the lateral incisor root. Once clear of this entanglement, the labial auxiliary arch was discarded and the canine drawn occlusally towards the archwire. Only at this point does it become appropriate to rebond a bracket on the lateral incisor. In the final stages of the treatment, as the canine comes down to the alveolar junction of the oral mucosa with the attached gingiva, a partial thickness flap will be raised from the attached gingiva on the crest of the ridge, to be apically repositioned over the labial side of the tooth.

Poor anchorage

Anchorage is an important consideration in all forms of orthodontic treatment where forces are applied to teeth. The reactive force is distributed to all the other teeth that share in supporting the fixed or removable appliance. When the unit to be moved comprises a single tooth that may require levelling and tipping only, there will be no perceivable counter-movements on the anchor teeth. However, when an impacted tooth needs to be drawn over a long distance from its initial location to its place in the arch, and then requires to be subjected to significant root uprighting and torqueing movements, loss of anchorage may often be noticeable, specifically altering intermaxillary relations and creating premature contacts and functional mandibular shifts. Since extrusion is one of the main force vectors needed in the treatment of impacted teeth, open bites will often result due to the reactive intrusive forces brought to bear over longish periods on the other teeth in the same jaw. Many practitioners will employ 'up-and-down' inter-maxillary elastics from the outset as a means of reinforcing the anchorage of the affected arch. However, a mandibular dentition is not immune to adverse movement from consistently applied intermaxillary forces. The long-term use of these vertical elastics will cause an over-eruption of the teeth in the opposing arch and a consequent cant in the occlusal plane, with similar occlusal disturbances. The way to avoid these unfortunate sequelae is to use anchorage elements that are not dependent on other teeth, such as extra-oral headgears and temporary or osseo-integrated implants, as described in earlier chapters.

When the side-effect that has occurred is a vertical loss of anchorage, as expressed in the establishment of an open bite, an 'up-and-down' elastic placed on a mini-implant in the opposing jaw will be most useful in controlling this unintended consequence. Alternatively, the problem may be conveniently ignored in some cases while vertical traction is being applied to the impacted tooth. The degree of intrusion that may occur is not without limit, and, once the impacted tooth has been aligned in its place in the dental arch, the situation is easily reversible. In this way, and strictly in the final stages of treatment only, the use of

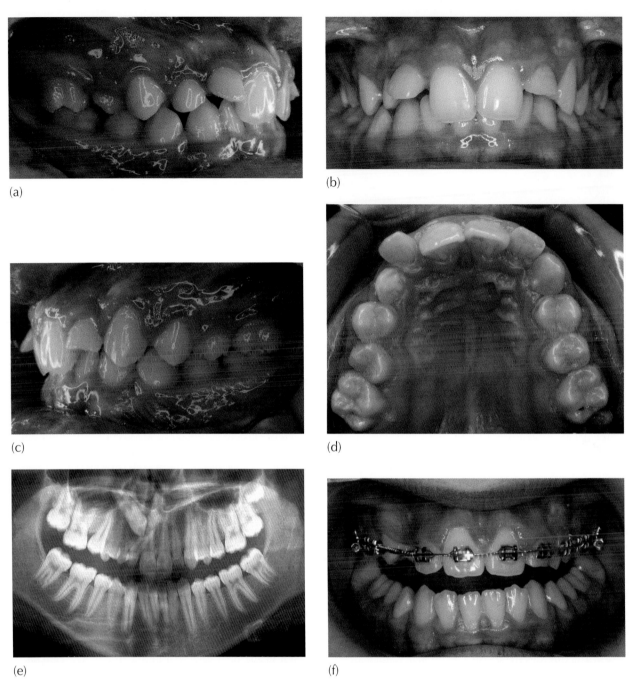

(a)

(b)

(c)

(d)

(e)

(f)

Fig. 12.23 (a–d) Intra-oral photographs to show the initial alignment of the anterior teeth. The disto-labially flared right lateral incisor indicates the expected orientation of the impacted canine. Note the normal axial angulation of the right maxillary first premolar.

(e) Initial panoramic view confirming the relationship of impacted canine to incisor root.

(f, g) Space is made using routine biomechanics, including levelling and alignment. Poor bracket position has displaced the first premolar root in an excessively mesial direction.

(h) Periapical radiograph reveals canine-related, extensive resorption of the root of the lateral incisor.

(i) A transaxial slice from the CBCT to show the bucco-lingual and vertical resorption damage and the presence of an enlarged follicle around the unerupted canine. A palatal surgical and orthodontic approach would be technically very difficult and be more damaging to the incisor.

(j) A three-dimensional view from the CBCT also indicates the mesial movement of the roots of the first premolar due to the improperly placed Edgewise bracket (compare a and g, above). The palatal root of the premolar introduces a new factor in the impaction of the canine.

(k) A transaxial slice from the CBCT to show the premolar palatal root actually contacting the root of the canine and negating a labial approach to the canine impaction. In order to resolve this impaction, the premolar root must be over-uprighted distally and rotated mesio-palatally. This will permit a labial approach both for surgical exposure and orthodontic resolution of the canine.

(l) Re-bracketing the case. The red bracket placement jigs illustrate the angulation built into the orientation of the slot.

(m) The initial 0.016 in nickel–titanium archwire is tied in.

(n) The occlusal view shows the rotation component achieved by correct bracket placement.

(o) Surgical exposure and eyelet attachment bonding are performed with the prior placement of the auxiliary archwire ligated in its passive mode.

(p) Activation of the horizontal loop of the auxiliary arch is ensured by its vertical ensnarement in the twisted wire ligature emanating through the sutured flap, immediately opposite the re-covered canine.

(q) Ten weeks later, the canine breaks through the oral mucosa, high in the sulcus.

(r) The auxiliary arch is discarded in favour of vertical traction with elastic thread. An orthodontic bracket may now be re-bonded to the lateral incisor.

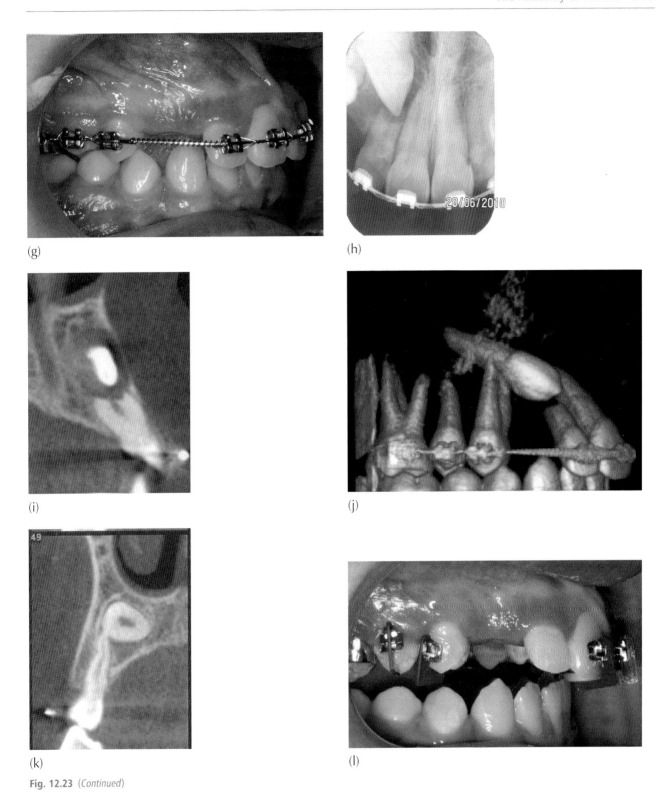

(g)

(h)

(i)

(j)

(k)

(l)

Fig. 12.23 (*Continued*)

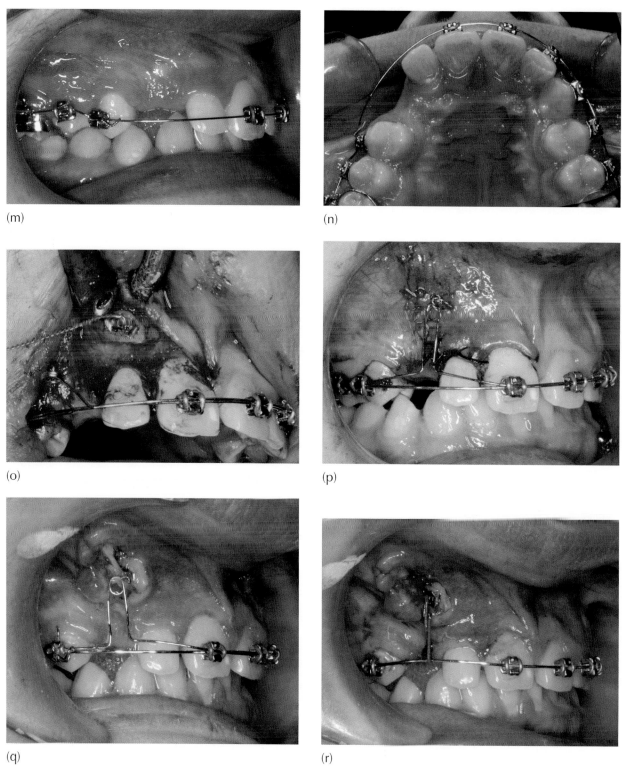

(m)

(n)

(o)

(p)

(q)

(r)

Fig. 12.23 (*Continued*)

vertical elastics against two heavy base arches of slightly exaggerated, corrective archform will rapidly achieve this goal. Notwithstanding, this is a form of 'round-tripping' that many will frown on.

It was pointed out in Chapter 6 that, following a closed eruption exposure in the palate, vertical traction of the tooth will cause the thick palatal mucosa to bulge more and more occlusally, with the shape of the tooth clearly outlined beneath. Yet the tissue is sometimes too resistant to permit eruption. In this situation, a small window of palatal mucosa needs to be opened and the tooth will then respond very quickly. Failure to do this will give rise to unwanted movement of the anchor unit (loss of anchorage) and palatal arches may become buried in the palatal tissue (Figure 12.24).

A corollary to the use of the labial auxiliary arch in the treatment of a palatal canine is the lack of an appreciation of the three-dimensional picture. A group 2 or a group 3 canine usually needs to be brought down on the palatal side of the arch and erupted through the palatal mucosa, as we have seen above and in several instances illustrated in Chapter 6. This movement is essential in order to take the canine on a vertically circuitous route around the roots of the neighbouring incisors and to provide an obstacle-free path to its place in the arch. However, there is also a height factor, since the tooth may often have been erupted on the palatal side of the line of the dental arch in a high-vaulted palate.

Drawing this tooth labially with traction applied from the main archwire, before it has been adequately erupted vertically, may now cause the tooth to become re-buried in the palatal mucosa of the medial wall of the alveolar ridge. This will cause an acute inflammation of these tissues, including swelling and much pain. The mucosal tissue will frequently grow to obscure the tooth, the bonded attachment and the elastic thread (Figure 12.25). The immediate treatment required is to eliminate the lateral traction element, which usually means cutting the elastic tie to the labial archwire, together with irrigation of the area with an atomized water spray and prescribing antiseptic mouthwashes. The symptoms will disappear and the inflammation will subside within a few days.

In the longer term, traction should be renewed within a week or so of the acute episode by reinstating the auxiliary labial arch to reapply vertically extrusive traction for a further period until the canine is at the level of the occlusal plane. At that point, traction towards the labial archwire will usually not then produce further problems.

A second corollary can produce the opposite effect, due to a lack of appreciation of the efficacy and range of a well-activated auxiliary labial arch. It was pointed out above that when the buried canine bulges the palatal tissue and does not succeed to break through this thick tissue, it is essential to cut a small window in the mucosa to free the tooth to erupt under the influence of this spring device. Much of

this impressive bulge comprises palatal mucosa and the tooth itself still needs much extrusion. The auxiliary arch is therefore left in place for a further period of activity to overcome this vertical discrepancy. In some cases, however, the speed and extent of the extrusion may be so exaggerated as to markedly over-erupt the aberrant tooth until it actually interferes with the patient's occlusion (Figure 12.26).

Inefficient appliances

The various orthodontic techniques in general use today are extremely efficient in moving erupted teeth around the dental arches, to correct abnormal intermaxillary dental relations and to align crowded teeth to a high standard of finish. The individual brackets have been specifically engineered to achieve this and, given proper professional attention and good patient compliance, most treatment plans can be easily completed within a two-year period. These same fixed appliances are, however, not so efficient when it comes to focusing all one's energies into the singular effort to resolve the impaction of a severely displaced tooth, and it may take much time to bring the tooth from its ectopic location to an advantageous position close to the main archwire. It is only at that point that a tooth may be reckoned to be in the orthodontic 'ball-park'.

This is the arbitrary point at which further treatment of the erstwhile severely ectopic tooth may be combined with that of all the other teeth, and where the practitioner may be permitted the luxury of relating to the patient as 'any other routine case'. For many of us, elastic thread tied to the most gingival link of a gold chain and stretched to a rigid archwire is the manner in which we apply traction to the impacted tooth. As we have explained in Chapter 4, this is a very poor way to achieve good and efficient results, because of the small distances across which the elastic thread is drawn. Much clinic time is wasted in the short time-lapses that are necessary between appointments to change the elastic thread and to maintain the extrusive momentum. The practitioner's ability to control the force delivered and to differentiate between a knotted elastic tie which produces a high force from one that generates no force at all is very limited. The patient may quickly become disconcerted by largely superfluous attrition of his time, which may exhaust his cooperation, courting the possibility of failure by default. Using a spring with a wide range of activity and a measurable force level produces much more rapid results, with excellent control of the force levels being applied and the freedom to allow several weeks between visits, without loss of activity in the spring.

To summarize the causes of failure in the resolution of impacted teeth, the orthodontist should consider the following points:

1. Patient-dependent factors
   (a) abnormal morphology of impacted tooth
   (b) age

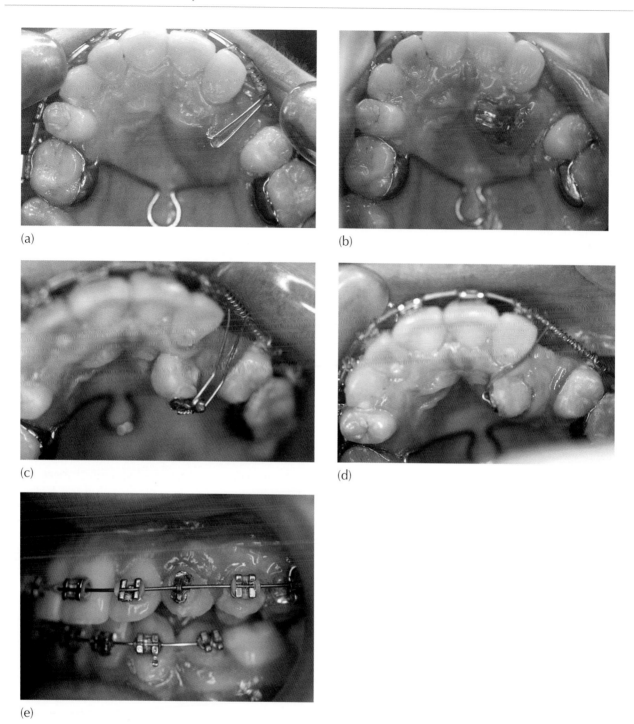

(a)

(b)

(c)

(d)

(e)

Fig. 12.24 Lost anchorage. (a) Following a closed surgical exposure of this palatal (group 2) canine, traction was applied and the tooth quickly became palpably evident beneath the palatal mucosa. Despite many months of further traction, the tooth did not break through the mucosa but, instead, the soldered palatal arch became buried in the palate, indicating loss of anchorage. (b) A simple incision and removal of palatal tissue over the crown tip reveals the original bonded attachment. (c) A month after the re-application of light occlusal force, the tooth has erupted well. (d) A new eyelet is placed in a more advantageous position and elastic traction direct to the archwire incorporates a rotatory component. (e) Six months later, the archwire is engaged through the eyelet, as the tooth completes its rotation and is brought to its place in the arch.

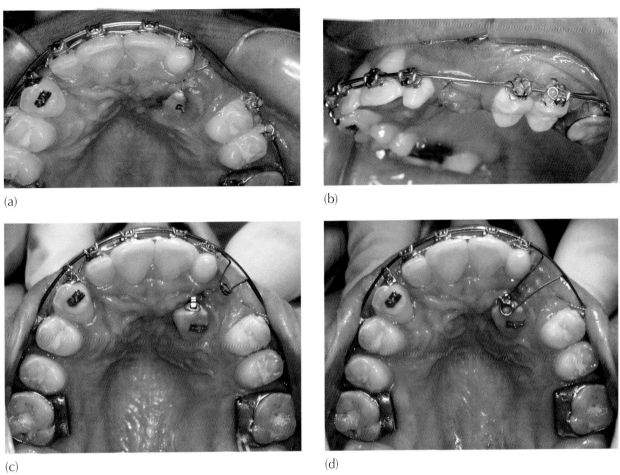

(a)

(b)

(c)

(d)

Fig. 12.25 Inadequate vertical extrusion and iatrogenic damage. (a, b) Viewed from the occlusal aspect and from the left side, the left maxillary canine was mechanically erupted into the palate. It was prematurely drawn labially and became buried in the medial side of the alveolar ridges to produce acute pain, inflammation and swelling. The tooth, the attachment and the traction elastic were buried in the inflammatory tissue. (c) Ten days later, the inflammation and swelling have receded to reveal the formerly impacted canine. An auxiliary labial arch is tied into the brackets piggyback fashion and the position of its active loop section may be seen in its unligated passive mode. (d) In its active extrusive mode, the ligated active loop is now drawing the canine vertically downwards towards the occlusal plane.

(c) pathology of the impacted tooth

(d) grossly ectopic teeth

(e) resorption of the root of an adjacent tooth

(f) lack of compliance (missed appointments, inadequate oral hygiene, etc.)

2. Orthodontist-dependent factors

    (a) mistaken positional diagnosis and inappropriate directional force

    (b) missed diagnosis of resorption of the root of an adjacent tooth

    (c) poor anchorage

    (d) inefficient appliance

    (e) inadequate torque

3. Surgeon-dependent factors

    (a) mistaken positional diagnosis – exposure on wrong side – rummaging exposure

    (b) injury to impacted tooth

(c) injury to adjacent tooth

(d) soft tissue damage

(e) surgery without orthodontic planning

It is paradoxical that, on the one hand, orthodontics has reached the pinnacle of excellence in the treatment of malocclusion with the highest level of predictability and confidence, which is the envy of virtually every other specialty within dentistry or medicine. The wide spectrum of malocclusions that can be treated well and within a fairly short space of time, with the continuing evolution of more efficient appliances, is truly impressive. On the other hand, when an impacted tooth is one of the elements of this malocclusion, the confidence of the orthodontist is often shaken to the core and there is a degree of apprehension and uncertainty as to whether a successful result may be achieved. The clinician may decide not to accept the patient

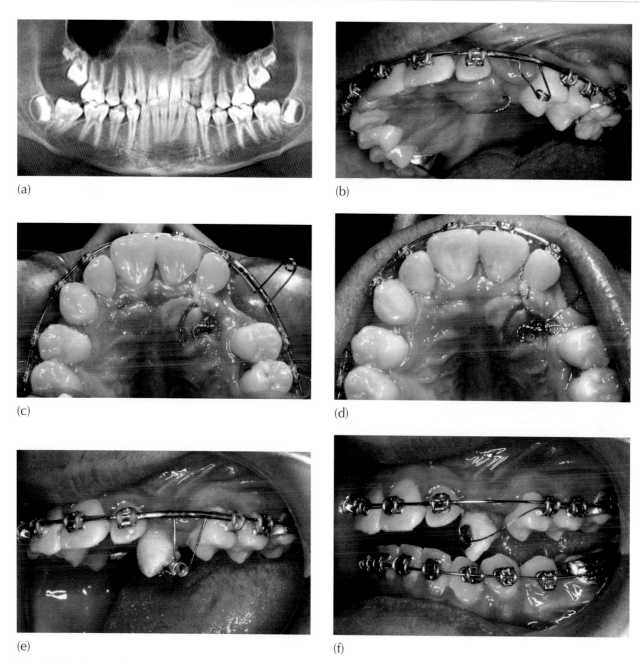

(a)

(b)

(c)

(d)

(e)

(f)

Fig. 12.26 Excessive vertical extrusion and iatrogenic damage.

(a) The initial panoramic view showing the maxillary left canine impacted high in the palate due to an adjacent supernumerary tooth. The deciduous canine has a full-length root.

(b) Following a closed surgical exposure, eyelet bonding and immediate traction with a labial auxiliary arch, the tooth has come down and now bulges the thick palatal mucosa, but is unable to penetrate it. The auxiliary arch is seen with its loop in the passive mode (i.e. lying vertically), freed from the twisted steel ligature.

(c) A simple minimal re-exposure of the cusp tip is performed, with the loop of the auxiliary arch in its passive mode.

(d) The loop of the auxiliary arch is re-ligated in its active mode (i.e. horizontal) to erupt the crown of the tooth.

(e) Three weeks later, an unexpected, highly unusual and rapid over-eruption has occurred, bringing the tooth down below the occlusal level.

(f) The view from the left side to show severe occlusal interference. An auxiliary nickel–titanium wire is placed in the brackets under the main arch, to apply a corrective rotational, labial and intrusive force to the tooth.

(g) The tooth has a long clinical crown, with root exposure on the palatal side, and now requires considerable labial root torque. Both of these handicaps are due to the excessive extrusion.

(h) The final stage of treatment to effect the labial root torque is best achieved using a Begg-type torqueing auxiliary.

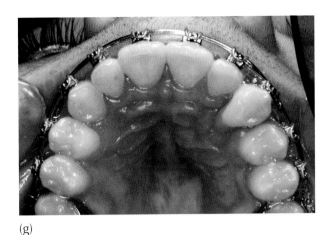

(g)

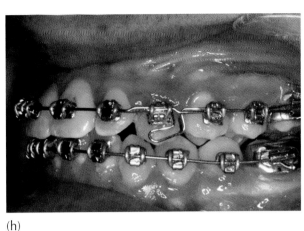

(h)

Fig. 12.26 (*Continued*)

for treatment and suggest that the parent seek treatment with another practitioner, who may be more experienced and have greater expertise in this specific area of orthodontics. In the final instance, treatment is most often undertaken accompanied by some form of let-out clause or warning that this particular element of the treatment could fail.

Expert opinion and second opinions during treatment

When treatment of an impacted tooth has failed, is in the process of failing or has not shown the expected resolution after a long period, the patient or the orthodontist treating the case may want to consult another orthodontist. Alternatively, the request may come from a lawyer representing a patient who feels aggrieved at what he perceives as failure and the possibility of negligence that he assumes must have caused it (Figure 12.27). Regardless of the source of the referral, the principal aim must be the well-being of the patient and a satisfactory resolution of the problem.

In general, the patient will arrive for the consultation appointment, together with some radiographs of the initial condition, which may or may not be adequate for accurate positional diagnosis. The orthodontist should perform a careful examination and take note of every detail that may appear relevant. The clinical examination will usually reveal fixed appliances in place on the teeth, which will probably have achieved levelling and alignment of the teeth. Some form of traction applied to a gold chain or wire ligature will be present.

If the exposure had been with the closed approach, then careful palpation of the expected location of the impacted tooth should be made, to try to assess its position, and it is especially relevant to attempt to define the direction that force has been applied to the buried tooth, by observing the

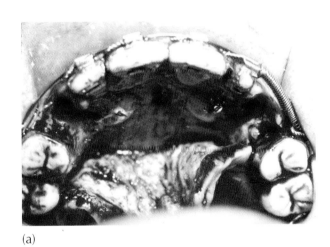

(a)

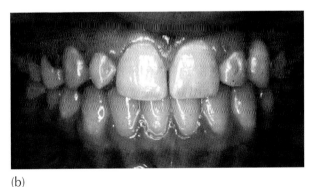

(b)

Fig. 12.27 A patient treated by the author c. 1971, using threaded pin attachments for maxillary bilaterally impacted canines. (a) At surgery, the threaded pins were screwed into prepared drill holes in the canine crowns. There was severe, unexplained post-surgical pain, which lasted several days. (b) The orthodontic treatment was successfully completed, but the right canine was markedly discoloured and found to be non-vital. It transpired that the threaded pin had inadvertently entered the pulp chamber during surgery! It should be remembered that the pulps of unerupted teeth in young patients are large, that under the conditions pertaining during a surgical procedure, a clear view of the crown anatomy is not easy to obtain, and directional control for accurate drilling is poor.

traction mechanism. The amount of space prepared in the dental arch for this tooth should then be measured, to see if this is adequate to accommodate the impacted tooth, although this will only be a cause of non eruption if the tooth is close to the line of the arch. The existing films should be reviewed to see if they are adequate for the task of accurately locating the tooth in three planes of space, and a note made of the dates on which they were taken. If these films are older than a year or so, progress films should be made, and any supplementary views required for positive positional diagnosis ordered.

If the diagnosis is not completely clear, or the tooth is in a position and orientation in which three-dimensional relations are difficult to mentally reconstruct, then a CBCT examination should be requested. At the same time, clinical photographs should be taken of the case, without disturbing the appliances. If the patient has not brought the original radiographs, clinical photographs, and plaster casts to the first appointment, every effort should be made to obtain them from the treating orthodontist, even if they are old records. It should be remembered that clinical signs seen (or missed!) in the original malocclusion models or intraoral photographs may show displacement of the crown or root of an adjacent tooth, which may help in diagnosing the initial position of the tooth concerned. These signs will have been largely eradicated by the initial levelling and aligning procedures that will have been completed by the time the second orthodontist sees the patient (Figure 12.16). Second, the active movement of teeth adjacent to an impacted tooth, whether to bring about alignment and levelling or to reopen space, is known to be often associated with renewed eruption activity of the impacted tooth. This may finally be successful in generating eruption or, conversely, in directing the tooth further along the wrong path.

So, comparing the old films and the old photographs with new films and a clinical examination may usually lead to a better understanding of the dynamics surrounding the impaction and to the chance of a more successful resolution of the problem.

Before a new treatment plan may be formulated, new casts of the teeth will be needed, and this should be done when all the ordered radiographs have been assembled and after the archwires have been removed. The archwires should then be replaced without renewed activation until a definite line of treatment is devised and until it is decided whether the original practitioner will continue with the case or, for whatever reason, the case will transfer to the new orthodontist or elsewhere, for this part of the treatment.

References

1. Jacobs SG. Localisation of the unerupted maxillary canine. Aust Orthod J 1986; 9: 313–216.
2. Armstrong C, Johnston C, Burden D, Stevenson M. Localizing ectopic maxillary canines – horizontal or vertical parallax? Eur J Orthod 2003; 25: 585–589.
3. Olive RJ. Orthodontic treatment of palatally impacted maxillary canines. Aust Orthod J 2002; 18: 64–70.
4. Becker A, Chaushu G, Chaushu A. An analysis of failure in the treatment of impacted maxillary canines. American Journal of Orthodontics & Dentofacial Orthopedics, 2010; 137: 743–754.
5. Becker A, Chaushu S. Success rate and duration of orthodontic treatment for adult patients with palatally impacted maxillary canines. Am J Orthod Dentofacial Orthop 2003; 124: 509–514.
6. Walker L, Enciso R, Mah J. Three-dimensional localization of maxillary canines with cone-beam computed tomography. Am J Orthod Dentofacial Orthop 2005; 128: 418–423.
7. Becker A, Chaushu S. Long-term follow-up of severely resorbed maxillary incisors following resolution of etiologically-associated canine impaction. Am J Orthod Dentofacial Orthop 2005; 127: 650–654.
8. Becker A, Chaushu S. Long-term follow-up of severely resorbed maxillary incisors following resolution of etiologically associated canine impaction. American Journal of Orthodontics and Dentofacial Orthopedics 2005, 127: 650–654,
9. Ling KK, Ho CT, Kravchuk O, Olive RJ. Comparison of surgical and non-surgical methods of treating palatally impacted canines I Periodontal and pulpal outcomes. Aust Orthod J. 2007; 23: 1–7.
10. Ling KK, Ho CT, Kravchuk O, Olive RJ. Comparison of surgical and non-surgical methods of treating palatally impacted canines. II. Aesthetic outcomes. Aust Orthod J 2007; 23: 8–15.
11. Olive RJ. Factors influencing the non-surgical eruption of palatally impacted canines. Aust Orthod J. 2005; 2195–2201.

13

Traumatic Impaction

(In Collaboration with Stella Chaushu)

Orthodontic Treatment of Impacted Teeth, Third Edition. Adrian Becker.
© 2012 Adrian Becker. Published 2012 by Blackwell Publishing Ltd.

Acute traumatic intrusion

The waiting areas attached to the offices of local dentists, to emergency rooms of general hospitals and to departments of paediatric dentistry of dental schools are places where one will often see young children who have suffered trauma to the lower part of the face and jaws, which may have resulted in fracture or avulsion of the front teeth. By and large, the traumatic incident will have occurred an hour or two earlier as the result of an accidental blow to the mouth during innocent play, a fall or sporting activity, although the attending practitioner should be watchful and alert to the possibility of child abuse. The treatment for the dental condition and any soft tissue lacerations involves the considerable skills of the paediatric dentist, the endodontist and the oral surgeon, and is not within the scope and context of this book. Nevertheless, there is one small corner of the field of traumatic injury in which emergency adjunctive orthodontic treatment may sometimes be of value, and this relates to those injuries in which there has been displacement of the damaged teeth, particularly intrusive luxation. This may be alternatively referred to as the acute, traumatically induced impaction of a previously erupted tooth, and it requires special attention. It will be readily understood that, however successful endodontic and restorative treatments may be in providing for a renewed healthy retention of the tooth in its surroundings, it is with regard to our ability to realign it to its former place in the dental arch that a favourable outcome must be judged and success measured.

This type of injury occurs as the result of a severe blow in the general orientation of the long axis of the tooth, which will drive the tooth upwards into the alveolar process. This results in injury to the periodontal ligament, involving a severance of the gingival and periodontal fibres, which will usually be accompanied by varying degrees of crown fracture and fracture and comminution of the bone lining the socket. In the longer term it will be a major factor in the occurrence of root resorption [1]. The insult to the pulp will cause pulp necrosis in virtually every case where the apex is closed and in about half the cases when the root apex is still open [2]. In the latter, there will often be arrest in further root development [1].

Intrusive luxation may find its resolution in one of three ways:

- spontaneous re-eruption
- manipulative/surgical repositioning and splinting
- orthodontic reduction

Spontaneous re-eruption

Following the vertical displacement, an affected tooth may re-erupt and eventually return to its original position [3], particularly if its root apex is still open [1] (Figure 13.1), although some will re-erupt even after root closure. Should the teeth remain intruded after several weeks of follow-up, corrective relocation will be needed. For these cases, some emergency treatment and initial restorative procedures will already have been carried out by the paediatric dentist, endodontist or oral surgeon, and the patient will not be in pain at the time that this assistance is required. However, whether the relocation of the intruded tooth is achieved by orthodontic movement or surgical manipulation, additional insult is borne by the tooth, which would be avoided if the potential for spontaneous re-eruption could be realized [1].

Manipulative/surgical repositioning and splinting

For immediate repositioning of the tooth to be successful, the tooth needs to be gently manoeuvred into its former place and splinted there for several weeks, in the hope that the severed gingival and periodontal fibres will heal to maintain the tooth in its place for the long term. Concern for the future of the tooth is expressed on three levels, namely:

- restoration of the periodontal attachment, assuring re-establishment of the integrity between the tooth and its supporting tissues;
- healing of the pulp tissues or of the periapical environment after pulp extirpation and root canal therapy (the reader is referred to appropriate texts in paediatric dentistry, endodontics and oral surgery for details of these procedures);
- the degree of root resorption that is virtually an inevitable sequela

Orthodontic reduction

If extrusive forces are applied to an intruded tooth which has few intact periodontal fibres immediately following the traumatic episode, it is highly likely that the tooth will be exfoliated within a very short time. Thus, before initiating traction, it is essential to wait for at least a couple of weeks to allow for healing and the re-establishment of some measure of periodontal support that will hopefully occur with the organization of the blood clot and reattachment of the periodontal fibres.

For this tooth to 'take' and be successful, the desired union of tooth to the surrounding bone is by periodontal fibre healing alone or by healing with surface resorption. According to Andreasen and Andreasen [4], healing without surface resorption is probably not a possibility in the clinical situation since it needs to be completed totally without injury to the innermost layer of the periodontal ligament. However, healing with surface resorption will leave the luxated tooth attached to the socket with a normal periodontal ligament and new cementum. Such a tooth will respond to orthodontic forces.

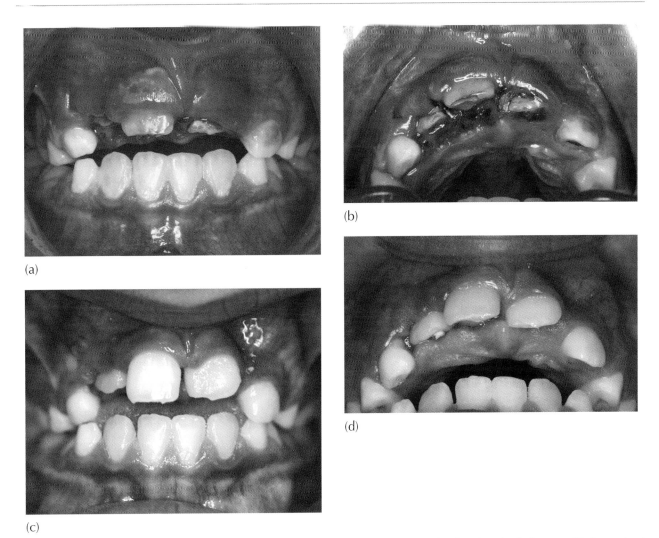

(a)

(b)

(c)

(d)

Fig. 13.1 An 8-year-old female following trauma to the lower face. (a, b) Seen four days post-trauma, there is an intrusive displacement of both central and right lateral incisors. (c, d) At seven weeks post-trauma, the teeth have partially re-erupted without orthodontic treatment.

If, on the other hand, healing is by replacement resorption, there is a direct union between the root and the surrounding bone. Repair will be counted as successful, but the tooth will then never be amenable to orthodontic forces. Sometimes a transient replacement resorption will occur and the tooth may then regain a normal attachment. This is by no means certain, and it is probably more likely that areas of ankylosis will occur over the root surface and the tooth will then remain intruded permanently. In this condition, the tooth is a liability to the dentition; nor is it useful as a foundation for a lasting prosthodontic restoration. An intruded and ankylotic tooth in a growing child will become more and more infra-occluded in relation to the adjacent teeth, and its accompanying bone will also be lacking when compared to the normal vertical development of the alveolar bone surrounding the unaffected teeth. For this reason, there is only limited value in maintaining the tooth as a

means of preserving alveolar bone and under these conditions its extraction may often be preferred. Orthodontic intervention at the appropriate time may offer the only viable treatment option, which, together with certain relatively minor restorative procedures, may produce a good result, with a fair prognosis. If the tooth is still completely sub-gingival, then the labial gingival soft tissue will need to be pared back or apically repositioned until 2 mm of the incisal edge of the tooth is revealed. Light extrusive force must be applied after the time that the periodontal fibres have begun to reunite and in the earlier stages of organization of the blood clot, but before the deposition of bone (i.e. 10–28 days post-trauma).

In a meta-analysis designed to evaluate the orthodontic modality of treatment of these cases [2] and which included material taken from other published studies of individual cases and case series reports, there was a 90% rate of

successful repositioning of the intruded teeth. As pointed out above, all those teeth with closed apices lost their vitality, together with approximately half those with an open apex. External root resorption was diagnosed in 54.8% of the involved teeth. Marginal bone loss was rarely seen in the patients, which is in sharp contrast to the findings of other studies [1]. Inflammatory root resorption occurred as a late complication in teeth with closed apices, and pulp obliteration was seen in those teeth that remained vital. The 'take-home message' from that study was that orthodontic reduction was the most reliable of the three alternatives and was found to be kinder to both hard and soft tissues than the surgical option. It provided the opportunity for a superior outcome in terms of fewer teeth lost and potential complications.

This view is not universally held, since other studies have found little difference in the long-term results between the orthodontic and surgical options, and it has been argued that the extra clinic time and attention needed for orthodontic reduction of the traumatic intrusion does not justify any potential minor difference in the long-term results [1].

Orthodontic treatment considerations

Special care must be exercised when using fixed appliances in this situation [4]. First, for any extrusive movement of a single tooth or group of teeth, some kind of resistant framework needs to be bonded to the adjacent teeth, which acts as a multiple anchor unit from which force would be applied to the intruded tooth or teeth. This may take the form of a few brackets and an archwire, although the composite bonding of a customized rigid wire directly to the labial enamel of these teeth (illustrated below) would be more satisfactory from many points of view. Unfortunately, these adjacent teeth will themselves have almost certainly been traumatized at the time of the accident, and using them in this manner may lead to further damage even at the light force levels involved. Second, if the intended extrusion is initiated later than two months post-trauma, ankylosis may have affected the intruded tooth [4]. Active extrusive forces generated by the appliance will then be to no avail. Instead, the reactive forces will be absorbed by the anchor teeth and these will become intruded.

Many children attending for routine comprehensive orthodontic treatment, particularly in the presence of an enlarged overjet, have a history of non-displacement trauma to the maxillary incisor teeth, which have then undergone root canal therapy. The advice usually given is that orthodontic treatment should be delayed for several months, until there is some radiographic evidence of repair. However, following intrusive trauma, the possibility of the occurrence of ankylosis (replacement resorption) is significant and will be evident within two months. Teeth that had completed root development at the time of the accident are

generally scheduled for root treatment in the first week after the traumatic incident. Thus, these exceptional circumstances dictate that the orthodontic extrusion of these teeth must begin at the latest six weeks or so after the traumatic episode, although the 10–28-day time-frame is to be preferred [2]. The risk of failure of extrusion due to replacement resorption is high and absolute, which is why treatment should begin within this time. The risk of an orthodontically induced complication of the root treatment is much lower and of less therapeutic significance.

Once the tooth is brought into alignment it may be retained and splinted to its immediate neighbours, using a short length of multi-strand wire, which is bonded to the labial surface of the three teeth for a few weeks only. It has been considered important not to cover the wire completely with composite material, but to place a small quantity of composite material across the wire over each tooth and to leave broad interproximal areas of exposed and flexible wire. Rigid bonding for long periods is contraindicated, since it seems to lead to a greater incidence of pulp necrosis and pulp obliteration [5, 6]. The use of a multi-strand wire splint allows a degree of movement, which is similar in extent to that seen in physiological mobility [7–10] and thus it may be argued that it can be safely used for considerably longer if desired. However, newer evidence appears to find no difference in outcome between rigid and flexible splinting, nor does it find that the length of time that stabilization is present alters the prognosis [1].

The indications for the different types of orthodontic appliance

The removable appliance

Simple removable appliances (Figure 13.2) are most suitable since they need apply no force to the adjacent teeth, the anchorage being derived from their broad contact with the palatal mucosa, via the acrylic base [11, 12]. A small button attachment is bonded on the labial side of the tooth and the labial bow of the removable appliance is divided at the midline. One arm of this divided labial bow may then be activated vertically downward against the button while the other half of the labial bow is removed.

Treatment generally proceeds rapidly, with the tooth appearing in the mouth and at the level of its neighbours within a few weeks, depending on the amount of extrusion required. This method has the advantage of simplicity and, since it is independent of support from other anterior teeth which may have been damaged in the trauma incident, its use will incur fewer dangers than other methods. However, since the activation depends on the patient placing the activated labial bow over the labial button attachment, clumsiness or lack of care may result in the tooth being displaced labially or palatally or the activation being nullified.

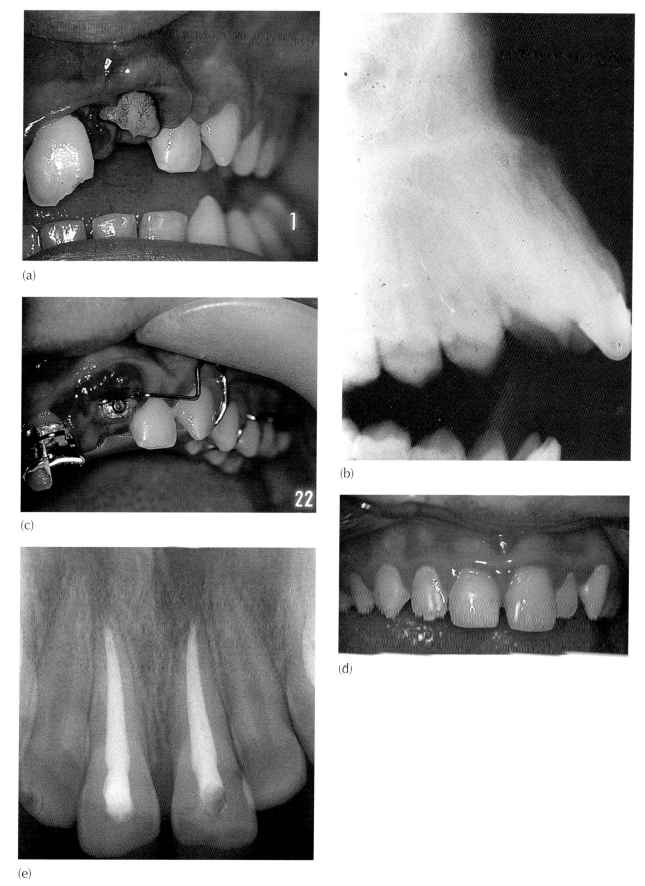

Fig. 13.2 The removable appliance. (a) Acute intrusion of a maxillary left and crown fracture of the maxillary right central incisors. (b) The tangential radiograph shows intrusive and buccal displacement. (c) At 14 days post-trauma, a button attachment is bonded to the intruded incisor. The labial arch is activated to extrude the tooth and the patient is instructed in accurate placement and care. (d, e) The tooth has re-erupted. Both central incisors have had root canal treatment and crown restoration. (Courtesy of Professor B. Peretz.)

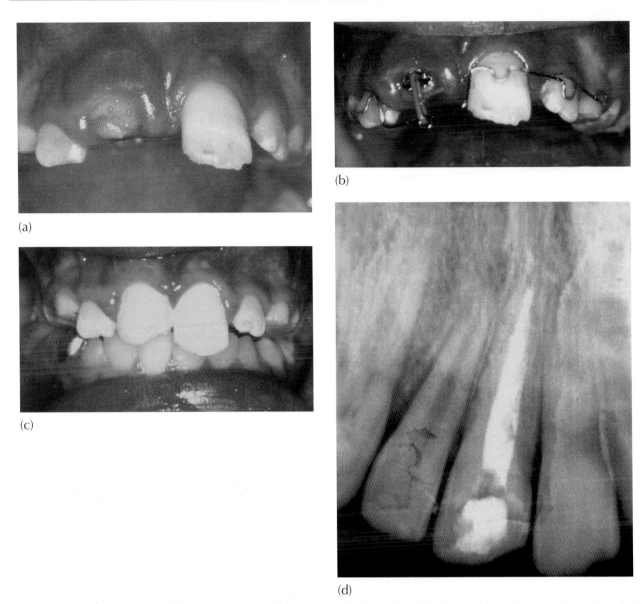

Fig. 13.3 The wire frame. (a) As the result of a fall in an 8-year-old female patient disabled with cerebral palsy, the right maxillary central incisor has suffered intrusive avulsion and its incisal edge is barely visible at the gingival margin. (b) During the intravenous (propofol) sedation session, endodontic treatment was undertaken, while the wire frame was fashioned chair side and bonded to the adjacent teeth. After debridement, an eyelet was bonded to the visible incisal edge and light elastic traction was applied. (c) A month later, the tooth was fully erupted and the frame was removed. No further retention was used. (d) The periapical view of the root-filled tooth shows evidence of resorption of the root.

The bonded wire frame

Fractured incisors are very common in special needs children [13–18], particularly those afflicted by cerebral palsy. Depending on the area of the brain affected, the child may be unable to walk or may do so only with great difficulty and will consequently be more prone to accidental falling. With similarly affected upper limbs and slower reactions, he may be unable to protect himself with his hands and the result will often be that the nose, lips and teeth – particularly protruding teeth – will absorb the brunt of the trauma as he/she hits the ground. Most young children, healthy or disabled, will need sedation or general anaesthesia to enable the successful performance of the variety of different treatments that are likely to be required in the emergency session that may follow the traumatic incident. While undergoing gingival debridement, suturing of a lacerated tongue or lip tissues or an initial pulp treatment the bonded wire frame (Figure 13.3) may be fabricated at the chair side, even if these are being carried out under a general anaesthetic. It may also be made on a plaster model

cast from a snap impression taken at the start of this multidisciplinary session and while the other procedures are being performed. A length of 0.020 in steel wire may be adapted and bonded to the labial surfaces of two or three teeth either side of the intruded tooth. The middle portion of this wire is formed into a vertically offset section with a loop opposite the intruded tooth and 3 or 4 mm vertically below it. Tying elastic thread lightly between an eyelet bonded to the tooth and this loop will apply gentle extrusive pressure to the tooth (Figure 13.3). Directly bonding the wire to the adjacent teeth transfers the reactive force to these teeth with relatively little chance of overtaxing them, since they are united into a rigid anchor unit and the force is divided more or less equally between them. This method has the advantage that it may be constructed and placed as part of the single session with simplicity and requires neither sophisticated orthodontic brackets nor any great skill.

A self-supported labial arch on fixed molar bands

A third possible appliance design involves the placement of molar bands with a soldered palatal arch, preferably (although not compulsorily) including an acrylic Nance button in the palate. Round 0.036 in (0.9 mm) tubes are soldered to the buccal side of the bands and a heavy self-supporting 0.036 in labial archwire is fashioned to include adjustment loops immediately mesial to the tubes to act as stops. These loop stops hold the archwire a millimetre or two labial to the anterior region and, in its passive position, is at or slightly below the occlusal level. With light finger pressure the anterior portion may be raised and tied with steel ligature wires to the button or eyelet attachments to the labial side of one or more intruded teeth to generate extrusive traction, whose force is measurable and controllable and its range adjustable (Figure 13.4)

This method is to be preferred over the use of a regular fixed orthodontic appliance, with brackets on each of the anterior teeth and the use of progressively heavier wires as the teeth come into alignment, since the need for initial levelling will take unaffordable time and it is unnecessarily sophisticated in these emergency conditions. The self-supported arch is also a method that is particularly suited to the resolution of multiple intrusion cases. Once the tooth erupts, root canal therapy is usually needed [2] and a permanent restoration may be placed, followed by a short period of retention.

Parents of very young children who have recently experienced this type of acute trauma are often confused by the need for the multidisciplinary treatments and they may have approached their insurance company, or the insurance company representing the school where the accident occurred, seeking financial recompense. Accordingly, much

time may be lost in the interim, during which the various aspects of the treatment slowly become understood and accepted and informed consent granted. Frequently, therefore, orthodontics is delayed and ankylosis may have begun. The recommended approach in these cases is to place appliances as planned and to spend a few weeks during which light force is applied to the tooth to see if it will respond. If there are several teeth involved, and if the active extrusive mechanism is a shared one, it should be remembered that it requires just one ankylosed tooth to prevent the others from re-erupting. Freeing one or another of a group of teeth or applying individual extrusive force delivery elements, therefore, may permit one of the teeth to respond. If there should be no change in the position of one or more of the teeth, then the very rudimentary and poorly developed ankylotic connection should be gently broken, using an extraction forceps under local anaesthetic cover, and the extrusive force immediately reapplied.

The palato-labial partial avulsion

A child may sustain a severe blow to the premaxillary area so as to displace a maxillary central incisor in such a way that the crown is tipped inwards and the root, still covered by the labial oral mucosa, protrudes through the labial plate of alveolar bone – the so-called lateral avulsion. The patient is unable to close the teeth together due to prematurity of the displaced tooth. In the heat of the moment and lacking suitable direction, the parents of a child do not always attend the appropriate clinic or the most knowledgeable dental practitioner. Emergency treatment indicated for this case is to manipulate the tooth to its original site under local anaesthetic, and to splint it in place.

The patient seen in Figure 13.5 was treated by grinding the incisor to reduce the occlusal interference and given a bite plate to disarticulate the teeth! By the time the parent was finally referred to a trained paediatric dentist, several days had elapsed. The paediatric dentist referred the patient on to an oral surgeon, with the request that manipulative relocation and splinting be undertaken. The oral surgeon considered that reduction without evacuation of the several days-old blood clot was no longer appropriate, and evacuation of the blood clot would have dictated an open flap procedure which, in turn, would have relegated the prognosis of the incisor to an unacceptably low level.

Orthodontic treatment was prescribed to re-site the tooth by applying labial tipping and then palatal root torque. Since this involved its being moved through freshly organizing blood clot, the treatment proceeded with great speed. The tooth maintained its vitality, as evidenced by positive pulp testing and subsequent pulp obliteration. Finally, the ground-down crown was restored with composite material, and follow-up periapical radiography has

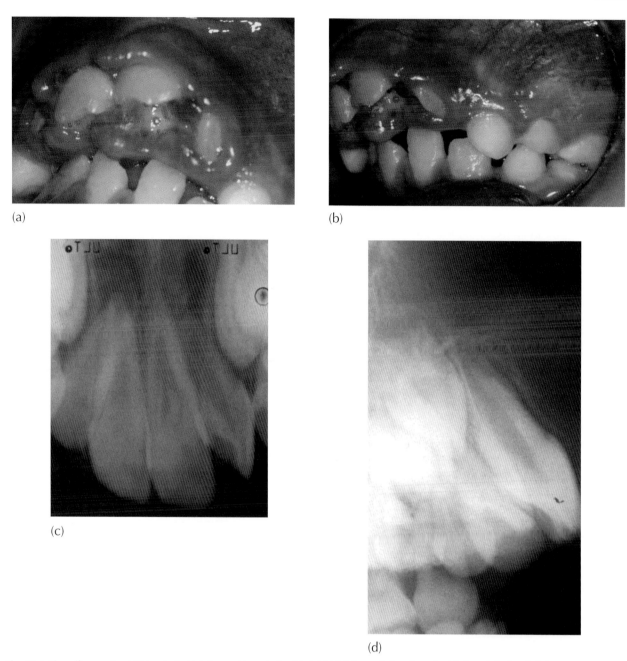

(a)

(b)

(c)

(d)

Fig. 13.4 The self-supporting labial arch. (a, b) A 9-year-old cerebral palsied child fell forwards from her wheelchair and displaced all four maxillary incisors superiorly upwards into their sockets. (c) A periapical view of the anterior teeth shows the degree of the displacement and the absence of root fracture. (d) A tangential view of the anterior maxilla shows the incisor teeth to have been displaced upwards and labially, with the labial plate of bone visibly displaced. (e, f) Molar bands were adapted and an impression taken from which this simple appliance was constructed, involving a soldered palatal arch and a labial archwire of 0.036 in gauge, which slots into soldered round molar tubes of the same diameter. This preparation was done during the same intravenous (propofol) sedation session as was exploited for surgical debridement of the gingival tissue and root canal therapy. (g, h) Eyelet attachments are composite bonded to the labial surfaces of the intruded incisor teeth, after some gingival reduction, and the removable heavy archwire is inserted into the molar tubes. The front of the labial arch is gently raised and tied with steel ligatures up to the eyelet attachments. (i) A month later, the teeth have all re-erupted. A periapical view of the left lateral incisor shows inflammatory root resorption. (j) At six months post-treatment, the lateral incisor has been removed and there is early cervical root resorption of the right lateral incisor. (k) The intra-oral view at six months shows the central incisors in good positions. The cervical root resorption lesion of the right lateral incisor is clearly seen.

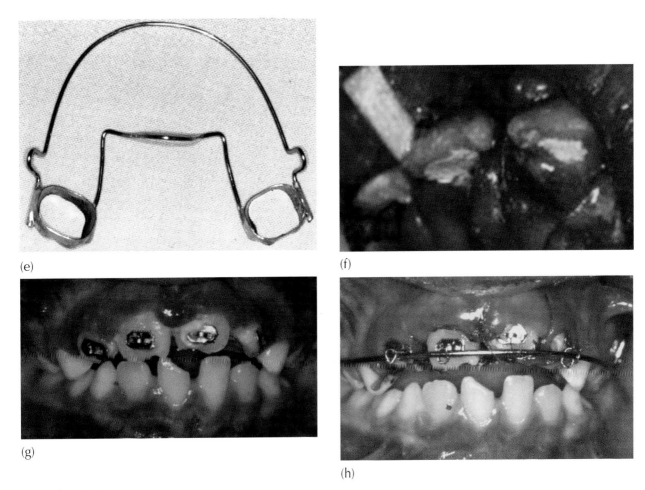

(e)

(f)

(g)

(h)

Fig. 13.4 (Continued)

not shown signs of root resorption, while the pulp chamber has become totally obliterated.

It is clear that the paediatric dentist, rather than the orthodontist, is the person who has to face the challenges that the patient who has suffered intrusive luxation or gross displacement of a maxillary incisor tooth presents and to accept responsibility for the conduct of the treatment. It is the paediatric dentist who must surely be the main contractor in the care of the injured child, bringing in a surgical, endodontic or orthodontic 'subcontractor' to perform specific parts of the multidisciplinary treatment, as and when necessary. In spite of this, there appears to be an unexplained over-enthusiasm among paediatric dentists to undertake the limited orthodontic procedures described here, which may sometimes be quite complicated. At the same time, there is a reticence on the part of orthodontists to accept these patients for treatment because of the very real risk that the injured tooth may become resorbed, ankylosed or lost during the treatment. Such sequelae are rarely encountered in routine orthodontic treatment and, when they are diagnosed, they come as an unwelcome surprise to the orthodontist. Orthodontists see these as signs of treatment failure, and many will thus prefer to shy away from undertaking orthodontic treatment on traumatized teeth when the chances of these phenomena occurring may be high. This represents an unfortunate and misplaced evasion of responsibility on the part of the orthodontist towards the patient. The experienced orthodontist is the most skilled professional trained and able to apply appropriate directional forces to resolve intruded and other displaced teeth and to do so with speed, with suitable force levels and with least discomfort to the patient. In this respect, he/she has an important role to play as an occasional but integral member of the multidisciplinary dental trauma team in the early stages of the emergency treatment in cases of traumatic impaction.

One further point leads on from this specific issue. Judging by what we see in our orthodontic practices, there appears to be a very significant minority of children who show signs of past trauma to their anterior teeth to a greater or lesser extent, which has obviously occurred over the few years between their eruption and the age at which they are ready for orthodontic treatment. But of the remainder, the apparently silent majority who show no

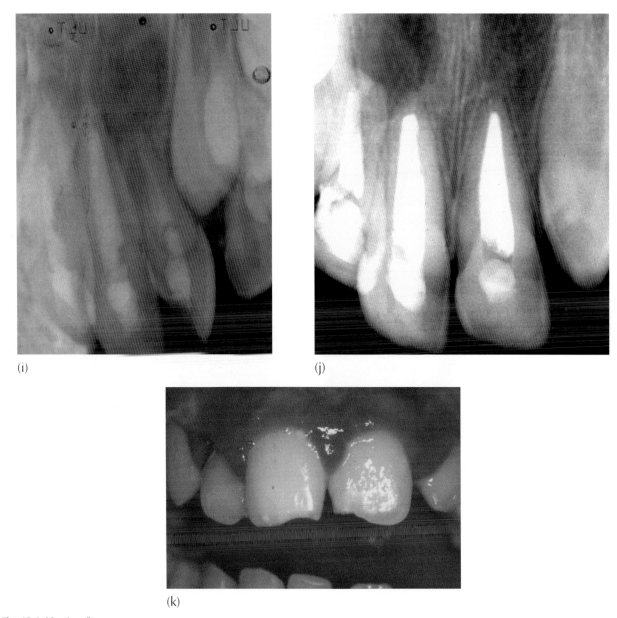

(i)

(j)

(k)

Fig. 13.4 (*Continued*)

outward signs of trauma, there is an unknown number who have suffered a relatively minor blow at one time or another and who woke up the following morning with no pain or discomfort.

Most of these children will soon have no memory of the traumatic incident, yet this may be the cause of the silent, degenerating health and possibly a slow and asymptomatic death of the pulp. There is reason to believe that this may be a more common sequence of events than is commonly thought. Suffice it to say that when orthodontic treatment is initiated and even minimal force is applied to such a

tooth, the hyperemia that normally affects the pulp in the first few days merely serves to reawaken the vital remnants of degenerated pulp tissue, and a cycle of events occurs which may then lead to discoloration of the crown and, from there, to endodontic therapy [19, 20]. The common accusation of the general dentist or the endodontist may then be that the orthodontist has applied excessive force with the appliance and that this was the cause of the pulp pathology. From these ill-considered and irresponsible comments to the law courts may sometimes be a very short path!

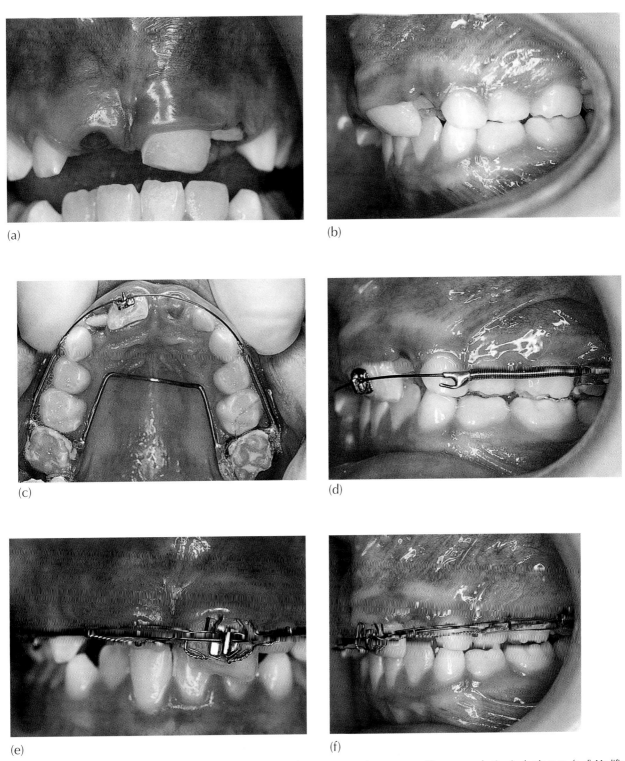

(a)

(b)

(c)

(d)

(e)

(f)

Fig. 13.5 (a, b) Front and left views, showing palatally displaced central incisor, one week post-trauma. The crown reduction is clearly seen. (c, d) Modified Johnson's twin-arch appliance in place. Buccal coil springs apply labially directed force on the single 0.018 in archwire, which engages the bracket of the displaced tooth. (e, f) A torqueing auxiliary is laced down to the main archwire and tied back to the molar tubes. (g–i) Twelve months after completion of treatment: front and left views of the occlusion and a close-up view of the maxillary central incisor teeth. (j, k) Periapical and tangential pre-treatment views. (l, m) Tangential and periapical views four weeks later. (n) Periapical view at two years post-treatment showing obliterated pulp – evidence of maintained pulp vitality.

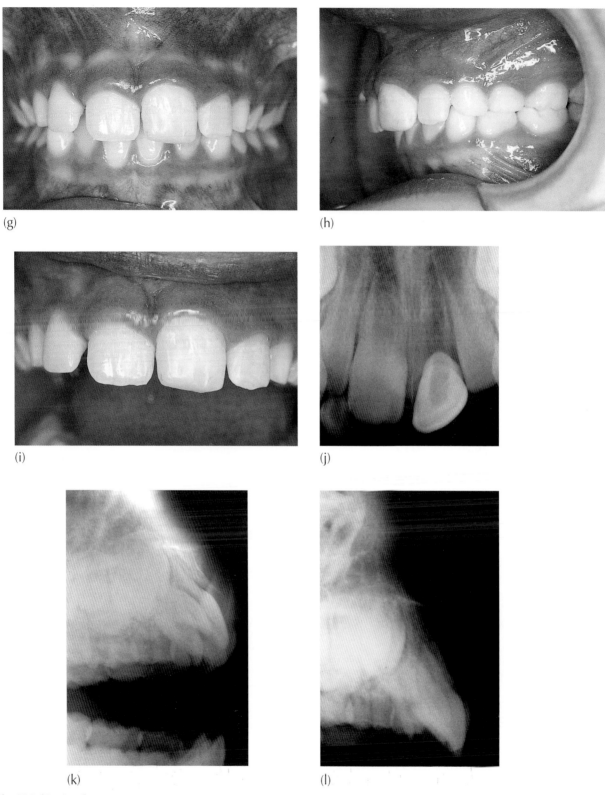

Fig. 13.5 (*Continued*)

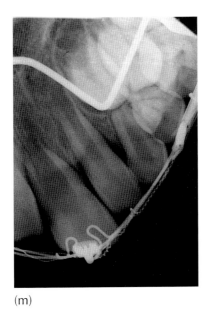

(m)

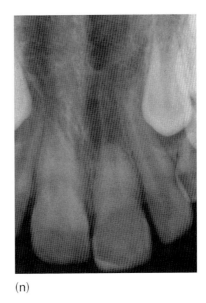

(n)

Fig. 13.5 (*Continued*)

References

1. Andreasen JO, Bakland LK, Andreasen FM. Traumatic intrusion of permanent teeth. Part 3. A clinical study of the effect of treatment variables such as treatment delay, method of repositioning, type of splint, length of splinting and antibiotics on 140 teeth. Dent Traumatol 2006; 22: 99–111.

2. Chaushu S, Shapira J, Heling I, Becker A. Emergency orthodontic treatment following the traumatic intrusive luxation of maxillary incisor teeth. Am J Orthod Dentofacial Orthop 2004; 126: 162–172.

3. Shapira J, Regev L, Liebfeld H. Re-eruption of completely intruded immature permanent incisors. Endod Dent Traumatol 1986; 2: 113–116.

4. Andreasen JO, Andreasen FM. Textbook and Color Atlas of Traumatic Injuries to the Teeth. Copenhagen: Munksgaard, 1994.

5. Andreasen FM, Vestergaard Pedersen B. Prognosis of luxated permanent teeth – the development of pulp necrosis. Endod Dent Traumatol 1985; 1: 207–220.

6. Rock WP, Grundy MC. The effect of luxation and subluxation upon the prognosis of traumatized incisor teeth. J Dent 1981; 9: 224–230.

7. Zachrisson BU. Clinical experience with direct-bonded orthodontic retainers. Am J Orthod 1977; 71: 440–448.

8. Becker A. Periodontal splinting with multistrand wire following orthodontic realignment of migrated teeth: report of 38 cases. Int J Adult Orthod Orthogn Surg 1987; 2: 99–109.

9. Becker A, Goultschin J. The multistrand retainer and splint. Am J Orthod 1984; 85: 470–474.

10. Dahl EH, Zachrisson B. Long-term experience with direct-bonded lingual retainers. J Clin Orthod 1991; 25: 619–630.

11. Peretz B, Becker A, Chosak A. The repositioning of a traumatically-intruded mature rooted permanent incisor with a removable appliance. J Pedodont 1982; 6: 343–354.

12. Mamber EK. Treatment of intruded permanent incisors: a multidisciplinary approach. Endod Dent Traumatol 1994; 10: 98–104.

13. Becker A, Shapira J. Orthodontics for the handicapped child. Eur. J. Orthod 1996; 18: 55–67.

14. Becker A, Shapira J, Chaushu S. Orthodontic treatment for disabled children: motivation, expectation and satisfaction. European Journal of Orthodontics. 2000; 22: 151–158.

15. Chaushu S, Gozal D, Becker A. Intravenous sedation: an adjunct to enable orthodontic treatment for children with disabilities. European Journal of Orthodontics 2002; 24: 81–89.

16. Chaushu S, Shapira J, Heling I, Becker A. Emergency orthodontic treatment following the traumatic intrusive luxation of maxillary incisor teeth. Am J Orthod Dentofacial Orthop 2004; 126: 162–172.

17. Becker A, Chaushu S, Shapira J. Orthodontic treatment for the special needs child. Seminars in Orthodontics 2004; 10: 281–292.

18. Becker A, Shapira J, Chaushu S. Orthodontic treatment for the special needs child. Progress in Orthodontics, 2009, 10: 34–47.

19. Brin I, Ben Bassat Y, Heling I, Engelberg A. The influence of orthodontic treatment on previously traumatized permanent incisors. Eur J Orthod 1991; 13: 372–377.

20. Brin I, Ben-Bassat Y, Heling I, Brezniak N. Profile of an orthodontic patient at risk of dental trauma. Endod Dent Traumatol 2000; 16: 111–115.

14

Cleidocranial Dysplasia

Clinical features and dental characteristics

The cleidocranial dysplasia patient is typically of short stature, with a brachycephalic skull and bossing of the parietal and frontal bones. There is hypoplasia of the mid-face, giving the misleading appearance of mandibular prognathism. The skull sutures and fontanelles exhibit delayed closure and secondary centres of ossification occur in these areas, with the formation of wormian bones. The development of the clavicles is defective and ranges from a small medial gap to total absence in severe cases [1–7]. The patient usually has a narrow chest and sloping shoulders.

According to Stewart and Prescott [8] more than 100 other anomalies have been associated with these major clinical features of the condition. Cleidocranial dysplasia does not affect the sufferer mentally or intellectually and, from this aspect, he/she is completely normal. The palate is narrow and occasionally high, and there is a marked degree of lateness in the development of the deciduous dentition, while eruption is generally normal. There is rarely any alteration in the number of the deciduous teeth, although on occasion, erupted supernumerary/supplemental deciduous teeth may be seen in the incisor area. The permanent molars usually erupt late, but spontaneously, while the remainder of the permanent dentition (i.e. the successional teeth – incisor, canines and premolars) exhibit very delayed or non-eruption. Additionally, supernumerary teeth typically develop in the successional teeth areas and much less frequently in the molar areas in numbers that may vary from none to around 12 in general. While greater numbers are uncommon in these cases, the highest recorded number reported in the literature was 63 [9]. Apart from barrel-shaped teeth and the rare occurrence of peg-shaped teeth in the maxillary incisor area only, the supernumerary teeth take the form of premolars in the premolar area, canines in the canine area and incisors in the incisor area. They may therefore be more appropriately referred to as supplemental teeth.

At the other end of the scale, the author's experience with a large number of these cases has revealed two patients with the correct number of teeth, another with congenitally missing maxillary lateral incisors and yet another with a missing third molar.

Equally affecting males and females, cleidocranial dysplasia is an autosomal dominant inherited disease [4] and there is a high incidence of new mutations at around 20–40% of all cases [5]. Part of the RUNX family of transcription factors, RUNX2 (Runt-related transcription factor 2) is known to be the gene directly involved. It encodes a nuclear protein, 521 amino acids, (56648 Da) with a Runt DNA-binding domain. RUNX2 is essential for osteoblastic differentiation and skeletal morphogenesis, acting as a scaffold for nucleic acids and regulatory factors involved in skeletal gene expression.

RUNX2 is the only gene specific to the aetiology of CCD, with sequence analysis showing 60–70% of CCD patients present with a missense or nonsense mutations, small insertions or deletions and exon skipping. Among the remaining 30–40% of patients with mutations, 13% may be found by other means of testing (karyotype for visible deletions, insertions and rearrangements involving the RUNX2 locus in chromosome 6p21, qPCR, Real-time PCR, FISH, etc.) to have large deletions of the gene and of the genes following. Mutations in the RUNX2 gene span the whole gene and are highly penetrant.

There is no clear genotype/phenotype correlation in CCD. Although, when mutations of RUNX2 are found, they are pathognomonic of the disease and phenotypes of other diseases are not associated with them [10].

Approximately one-third of the cases we have seen have been sporadic occurrences, with no family history of the condition. Most of the cases that have been under our care come from families where a parent was affected and thus diagnosis was usually (but not always) made at birth. For most of the other cases, a tentative or initial diagnosis was suspected only several years later by the child's paediatrician or orthopaedist, although the discovery may sometimes be made at a routine paediatric dental examination. Corroborative evidence from a clinical examination and a wider radiological examination was then obtained to establish the definitive diagnosis. During the physical examination, the clinician should attempt to confirm as many of the features described above as possible. In particular, the patient should be asked to approximate the shoulders to confirm the clavicle anomaly (Figure 14.1). Palpation should also be made of the areas between the parietal bones on the crown of the skull and between the frontal bones at the upper forehead/hairline region. In both of these midline areas, a smooth and wide hollow, concavity or furrow (Figure 14.2) may be clearly felt, in contrast to the convex contour of the skull of a normal child. Radiological examination should include views of the clavicles (Figure 14.3), the fontanelles, which may be seen on lateral and postero anterior cephalometric films (Figures 14.4 a, b) and an initial panoramic film of the jaws (Figure 14.6).

Aside from the principal bony defects of the cranium and clavicles, a good proportion of CCD individuals suffer other skeletal and orthopaedic anomalies. According to Cooper et al. in 2001, 57% suffer from flat feet, 28% from knock-knees and 18% from scoliosis, while joint dislocation of the shoulder and elbow may also occur [11]. Additionally, infections of the upper respiratory tract are common, specifically the sinuses and the ears, where conductive hearing loss occurs in 39% of the cases. Treatment for these associated conditions is indicated and should be performed by the paediatric specialist concerned. For the most part, the signs and symptoms of the condition are very distinct and, these predispositions aside, entirely benign. They are in no way progressive, the patient is not physically or mentally disabled and, in general, other body systems are not adversely affected.

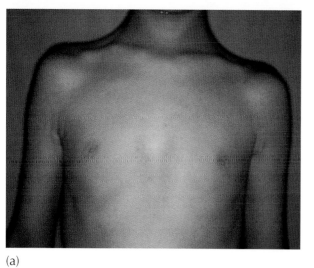

(a)

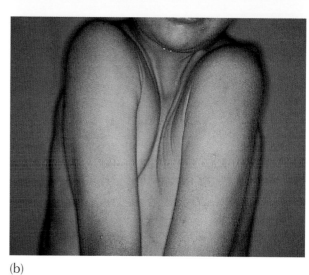

(b)

Fig. 14.1 (a, b) The approximated shoulders of a cleidocranial dysplasia patient.

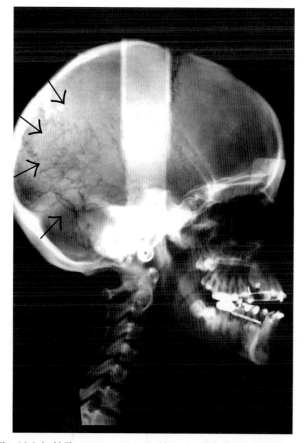

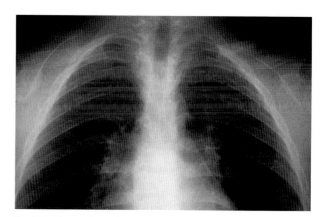

Fig. 14.2 Frontal midline furrow passing through the hairline.

Fig. 14.3 Chest radiograph to show incomplete clavicles.

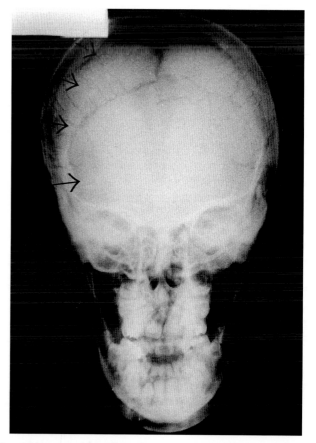

Fig. 14.4 (a, b) The postero-anterior and lateral cephalograms show abnormal cranial form, open fontanelles and numerous wormian bones (arrowed).

At present, there is no way to change the underlying inherited condition, treatment cannot therefore be advised for the primary condition [5, 6] and its diagnosis does little more than label the child as an oddity. The only situation where treatment is needed is when one or more of the associated consequences become burdensome. By and large, it is the oral and dental disability of the condition that undoubtedly presents its most serious ramifications, since this affects the vertical and horizontal growth of the face and oral structures, including alveolar bone and teeth. From student to experienced practitioner in the dental profession, and out of all proportion to the rarity of the condition, its clinical features are surprisingly well known, again reflecting its curiosity value rather than any ability on the part of the profession to have been able to promote change and correction in the past. For many decades, the profession has stood in awe of the overwhelming number of the dental problems that these cases present, unable to offer satisfactory answers.

The dental characteristics typically include over-retention of the deciduous dentition, non-eruption of the permanent dentition and the presence of many supernumerary teeth (Figure 14.5). Nevertheless, there is no dental discomfort or disturbance, unless the deciduous teeth become decayed. These teeth are small relative to the growing face and are not visible below the upper lip, particularly when an often seen anterior open bite is present. In many cases, in the rest position it is the tongue that is visible between the lips (Figure 14.6). The horizontal growth pattern generally produces a mandible of normal length and an underdeveloped maxilla which therefore produces a skeletal and dental class 3 relationship. However, this relationship is not always present initially in the younger patient, but may gather momentum during the adolescent growth spurt to a greater or lesser degree [12] (Figure 14.7). This trend is by no means certain, and there are cases of good class 1 relationships and even the occasional class 2 case that may be diagnosed, as they complete the development of their facial pattern, at the end of the adolescent growth period. The vertical growth of the alveolar processes is generally deficient, which leaves the patient with very shallow labial and lingual sulci in both jaws. Taken together, these features give the patient an edentulous appearance, which may often be the presenting symptom.

Some light has been shed on a possible reason for the non-resorption and over-retention of the deciduous teeth and the non-eruption of the permanent teeth in these cases [13]. It has been proposed that decreased root surface resorption occurs in the deciduous teeth due to a thin uniform layer of resorption-resistant acellular cementum covering virtually the entire root. In the permanent teeth, very little cellular cementum is seen on the roots of the teeth. However, secondary deposition of reparative cementum is found in areas of focal resorptive defects, and this

may be blamed for the non-eruption of the permanent teeth.

Given that:

1. the child with cleidocranial dysplasia has little by way of a tangible complaint involving pathology (no pain, no swelling, no difficulty in functioning), yet
2. the dentist has diagnosed a benign condition of extraordinary therapeutic magnitude,
3. the dentist has no available guidelines on how to even begin to approach the resolution of the problem,
4. a practitioner seeing a case for the first time is unable to predict treatment results, and
5. the degree of facial deformity is usually of insufficient consequence to demand surgical modification, at least in the younger patient,

it is entirely understandable that a responsible clinician will hesitate before undertaking treatment.

The options are:

- not to offer treatment at all;
- to suggest the more radical approach of extraction of many teeth, followed by prosthetic replacement; or
- to advise an orthodontic–surgical treatment procedure, with an unknown level of confidence in the ability to achieve the desired outcome.

However, non-treatment becomes less of an option as the patient grows older. Because of considerable occlusal attrition and caries there is a progressive morbidity of the deciduous dentition, which starts in the early teens and gathers pace over just a few years. Root canal treatment is often needed and restoration becomes difficult. The patient's appearance suffers further, with a reduced lower face height [14], impaired masticatory function and continuing facial growth contributing to the increasing over-closed appearance. Treatment is needed to provide an efficient masticatory apparatus and improvements in the dental appearance and the facial proportions.

Treatment modalities

These goals may be realized in several ways. The methods that have been proposed over a period of many years have most often reflected the particular area of dentistry in which the treating dentist has specialized. Therefore, to a degree, the mode of treatment may depend on whose door the patient first knocks!

Prosthodontics

The most popular approach has been to provide the patient with removable partial or full prostheses, which fulfil all the immediate needs of the patient. This approach has been suggested by many only after the removal of all the deciduous teeth and the unerupted supernumerary and permanent teeth [9, 15, 16]. Given that the alveolar bone height

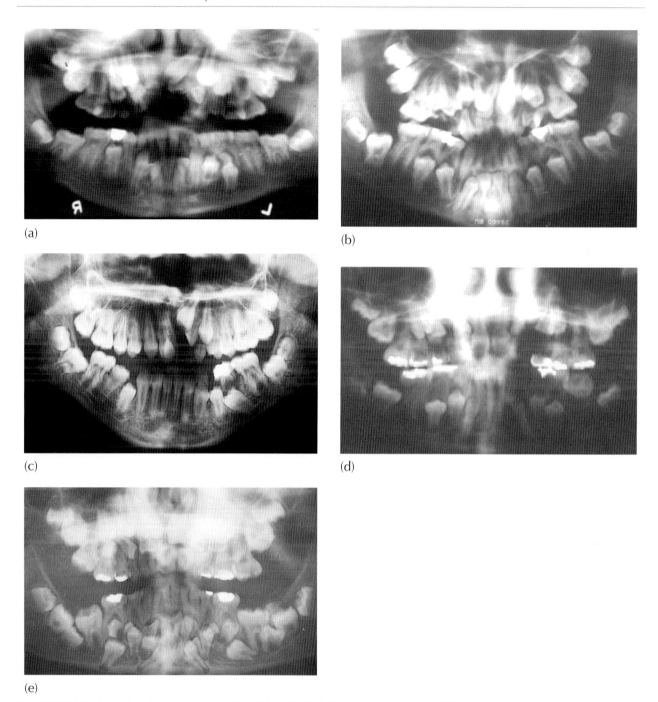

(a)

(b)

(c)

(d)

(e)

Fig. 14.5 Variation in number of supernumerary teeth in cleidocranial dysplasia patients. (a) A 14-year-old female with eight extra teeth. (b) A 14-year-old female with five extra teeth. (c) A 13-year-old male with one mesiodens. (d) A 12-year-old male with a missing tooth and no supernumeraries. (e) A 14-year-old female with a full deciduous dentition and two erupted supernumerary deciduous maxillary lateral incisors. Only one erupted (deeply carious) permanent molar is present. There are 27 supernumerary teeth, in addition to the 31 unerupted permanent teeth – a total of 81 teeth! Note the presence of supplemental molar teeth.

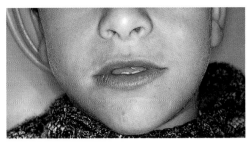

Fig. 14.6 The tongue is visible at rest, protruding between the lips.

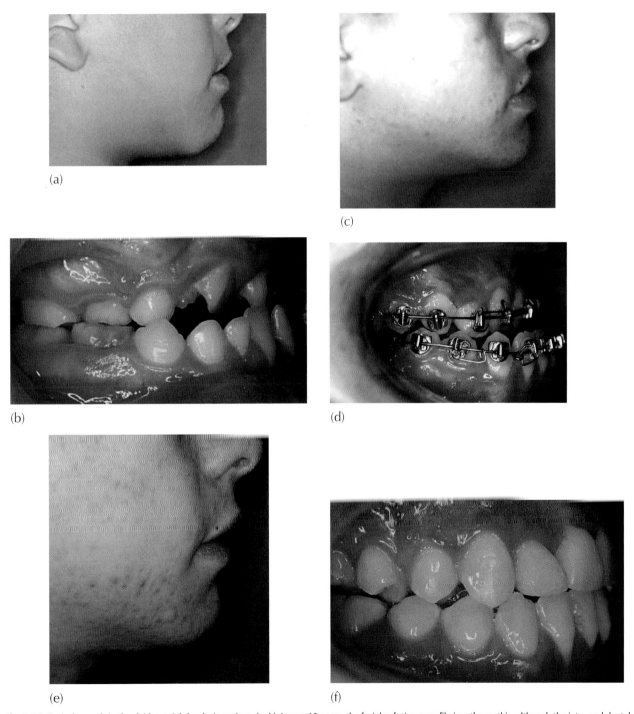

(a)

(c)

(b)

(d)

(e)

(f)

Fig. 14.7 Typical growth in the cleidocranial dysplasia patient. (a, b) At age 10 years, the facial soft tissue profile is orthognathic, although the intra-oral dental and dental base relations are class 3. (c, d) At age 17 years, there is a clear underdevelopment of the mid-face, with an accompanying mandibular prognathism and a marked worsening of the class 3 dental and skeletal relations. The patient is being prepared for orthognathic surgery. (Courtesy of Dr D. Harary.) (e, f) At age 17.5 years and following surgical correction.

in these cases is very limited, with shallow sulci, the dentist should ponder the implications of the construction of replacement tissue-borne prostheses 10 years later, after additional and iatrogenic ridge resorption has occurred.

Others have advised retaining the standing teeth and the construction of prostheses around them [8, 16–18]. A further refinement recommends the exploitation of any standing teeth, together with the surgical exposure of the more superficial unerupted teeth, to serve as supports for an over-denture [18–20]. By making it tooth-borne, the denture is less likely to cause further ridge resorption, but the supporting teeth will deteriorate quickly under these circumstances, both from caries of their crowns and their root surfaces and from loss of their periodontal attachment.

Implant-based prosthodontics is largely ruled out because implants cannot be placed in a jaw that is full of unerupted teeth, without the removal of a very large number of these teeth. Extensive extraction will take its toll with the postsurgical resorption of additional alveolar bone from the already severely reduced bony ridge. For the most part, therefore, the relative absence of suitable implant sites and the thinness of the mandible itself will rule out the use of implants.

Surgical relocation

In the search for a non-prosthetic method that utilizes the existing teeth, surgical removal of the supernumerary teeth, followed by careful dissection of the unerupted teeth and their repositioning or transplantation into artificially prepared sockets, has been proposed [5, 21, 22]. However, the long-term results of this method, in the context of its use in cleidocranial dysplasia, and with the presence of multiple impactions and deficient alveolar bone width, have not been investigated. This is an important area for study, since it would give the profession some inkling as to the fate of these teeth in later years. Accordingly, the practitioner seeking appropriate answers is left to ponder whether the transplanted teeth will erupt autonomously and behave like normal teeth or perhaps undergo ankylosis and root resorption in common with other transplanted teeth.

These methods all suffer one serious drawback, namely that the results thus achieved deteriorate fairly rapidly and their prognoses are relatively poor. When one considers that treatment for the condition needs to be carried out in the patient's second decade of life, these modalities must be considered to be of limited value and essentially inadequate to the aim of lasting through to old age.

Orthodontics and surgery

Before the mid-1960s, while some limited positive results were obtained with orthodontics, the idea was considered to be fanciful and impractical, and was widely derided. Nevertheless, it caught the imagination of a small number of clinicians, and, particularly in the late 1970s and early 1980s, several publications appeared advocating a surgical and orthodontic method. Methods were designed to bring about the eruption of the teeth by extracting the deciduous teeth, surgically removing the unerupted supernumerary teeth and exposing the buried permanent teeth, with or without the use of a surgical pack, depending on the depth of the individual teeth within the tissues [14, 23–25].

The patient was then seen in routine follow-up visits, until the teeth erupted or had reached a sufficiently accessible position, occlusal to the healed gingival tissues, for the application of orthodontic bands or bonded attachments. In other words, assisted eruption was provided only for those teeth that had already partially erupted. For CCD, characterized as it is by a lessened power of eruption, many months will pass before teeth appear and some of the more deeply sited teeth will never erupt. Additional surgical exposure is needed for some of these, but still with no guarantee of success.

Immediate bonding and ligation at the time of surgery for these cases was introduced in the literature in the 1980s, when Trimble et al. [26] and Davies et al. [27] each showed a single case in which this was done. The advantage of being able to apply forces to the most intractably impacted teeth is well illustrated in these two cases.

The results and prognosis that may be achieved by a method involving surgical removal of the unwanted deciduous and supernumerary teeth, followed by the orthodontically assisted eruption and alignment of the natural permanent teeth, must be viewed as warranting exploitation. What, after all, could be better than to restore the dentition with the patient's own teeth and with normal alveolar bone support, through the medium of a healthy periodontal ligament?

Since the orthodontic literature records few attempts to standardize orthodontic treatment strategy beyond the above-mentioned single case reports, it must be concluded that the orthodontic option had not been exercised for many cases and that there seem to be few centres anywhere in the world where a significant number of patients had been treated. For these reasons, the present state of opinion regarding recommended or appropriate procedure is difficult to assess accurately.

Nevertheless, within this modality, three courses of action have been suggested over the years, each based on the experience of the treatment of several cases and each with its own relative merits. These will be referred to as:

- the Toronto–Melbourne approach
- the Belfast–Hamburg approach
- the Jerusalem approach.

The Toronto–Melbourne approach
This method was originated by a team from Toronto [14] and was later further developed in Melbourne [24]. Surgical procedures are performed in a stage-by-stage series under

endotracheal general anaesthesia, with the degree of root development of the permanent teeth dictating the timing of each stage. Thus, initially, the deciduous incisor teeth are extracted at 6 years of age, followed by the deciduous canines and molars at 9–10 years. Supernumerary teeth overlying the crypts of the unerupted permanent teeth are removed, together with substantial amounts of bone to uncover the crowns of the permanent teeth to their maximum diameter. The teeth are left widely exposed. The Melbourne team prefers to expose the incisors at a separate and additional surgical episode, and this is done after the first molar bands are placed following the late eruption of these teeth, which may reach full expression only at about the age of 10–11 years. Surgical packs are used to maintain the patency of the surgical exposure and to safeguard access for eventual bonding of the teeth.

The expectation is that, following the removal of the obstructive elements (i.e. the deciduous and supernumerary teeth, together with a liberal amount of bone and soft tissue), the teeth will then erupt of their own accord to a varying degree and over an extended time-frame. When convenient, orthodontic brackets are bonded to individual teeth and these are drawn to a light archwire, which spans the unsupported premolar/canine areas, from the banded molars to one or more anteriorly erupted incisors. Teeth are then drawn to the archwire, depending on their becoming accessible to bracket bonding.

Smylski et al. [14] and Hall and Hyland [24] do not propose any special or purpose-designed appliances to deal with the vertical traction that is needed in every area of the mouth, but appear to rely on conventional methods used in routine orthodontic treatment.

Limitations

In this method, the patient is under treatment for many years, starting from a very early age and requiring several recommended and fairly extensive surgical interventions, followed by several smaller ones for individual teeth. The age of the patient in the early stages and the scope of the surgery are the major determinants as to whether all these interventions should be carried out under general anaesthesia.

The deciduous anterior teeth are removed at an early stage in order to encourage the eruption of permanent incisors. Nevertheless, in their subsequent recommendation to fully expose the permanent incisor teeth in a distinct and separate surgical stage, Smylski et al. [14] and Hall and Hyland [24] recognize that spontaneous eruption does not always occur. This means that the patient is anteriorly edentulous for some considerable time. This would seem a high price to pay for what may be undue optimism regarding the potential in cleidocranial dysplasia for normal eruption. In two of the three cases described by Smylski et al. [14] unerupted supernumerary teeth were not present in the anterior segments and the permanent incisor teeth res-

ponded to simple exposure and packing. However, there are many cases where spontaneous eruption does not occur, this being one of the diagnostic criteria of the condition, which may be associated specifically with the frequent presence of supernumerary teeth in this region.

The placement of attachments to the deeply sited permanent teeth is not performed at the time of surgery, but some time later, after full healing (by secondary intention) has occurred and the surgical packs have been removed. Thus, at each surgical stage, valuable time is lost between the exposure and the force application needed to encourage the eruption of the teeth.

The Belfast–Hamburg approach

Simultaneously, but quite independently, Richardson and Swinson [28] of Belfast and Behlfelt [29] of Hamburg proposed a diametrically opposite method of treatment of the teeth in cleidocranial dysplasia. They recognized that, while there is the need for extensive surgery in these cases, this could be completed at one time, including the extraction of all deciduous and supernumerary teeth and the exposure of all unerupted permanent teeth. This is carried out under general anaesthesia, under operating theatre conditions, and with surgical packs placed over the remaining teeth to encourage epithelialization of the exposed tissue, which is the essence of healing by secondary intention.

During the succeeding weeks, these surgical packs remain in place and are perhaps changed over a further quite short period, until brackets may be conveniently bonded to the exposed teeth. This can then be done under what the proponents consider to be more reliable conditions for bonding than those pertaining during the surgical procedure.

Whether or not eruption of these teeth occurs without assistance is the subject of some debate, with one source insisting that, while there is apparent improvement, this is due to the radical loss of surrounding soft and hard tissue during the surgical procedure rather than actual vertical dental change [30].

Nevertheless, even with the most favourable and optimistic assessment, there can be no doubt that the eruption will be neither sufficient nor reliable enough to eliminate the need for extrusive mechanics. As with the Toronto–Melbourne approach, appliances consist of molar bands and bonded brackets, with long spans of unsupported and relatively fine archwire used to vertically develop the partially erupted teeth.

Limitations

By recommending all extractions and exposures at one time, the Belfast–Hamburg surgical policy has clear advantages from the patient's point of view, although a balance has to be struck in terms of timing. The earlier-developing permanent teeth, particularly the incisors, should not be exposed too late in their development to lose any eruptive potential that they may have, while the later-developing

teeth should not be exposed too early when their roots are insufficiently developed. Accordingly, the Belfast team [28] recommends that the one-time, comprehensive, surgical intervention be performed at age 12–14 years.

The immediate advantage of this policy is very clear and encouraging, although its drawbacks are of considerable consequence and not so obvious. By delaying treatment until this late age, the teeth of the normal series will have been held deep down in basal bone by the supernumerary teeth, particularly in the lateral incisor/canine/premolar area, for an extended period of time. Their roots will have reached an advanced stage of development in these cramped circumstances, which is likely to exaggerate the tendency for a stunted, tortuous and distorted root morphology [31]. Removal of the unwanted extra teeth at this late stage will relieve the impaction of the permanent teeth of the normal series, but it will do so at a time when they exhibit even less potential for spontaneous eruption, particularly in the incisor region, since the root apices will already have been completed.

During growth in a normal child and with the eruption of permanent teeth, the vertical development of the alveolar processes that occurs makes a significant contribution to the height of the lower face. It also leads to the establishment of deep vestibular and lingual sulci, with a clear differentiation of wide zones of oral mucosa and attached gingiva. In the untreated cleidocranial dysplasia patient, vertical growth of the alveolar bone appears to be markedly diminished. This brings about the typically reduced height of the lower third of the face that is so frequently a feature of the condition.

Thus, with the late removal of the unwanted deciduous and supernumerary teeth at a time when most of the patient's growth has already occurred, the ultimate vertical alveolar growth that accompanies the erupting permanent teeth will be correspondingly less, leaving a shallower sulcus, an absence or reduced width of attached gingiva and an incompletely vertically developed lower third of the face.

Furthermore, and in addition to removing the unwanted supernumerary teeth, it is necessary to gain access to the canine and premolar teeth of the normal series and to expose them widely. When the procedure is performed at this late stage, these target teeth are very deeply situated, often with their developing root apices close to the lower border of the mandible or the floor of the nose and maxillary sinus. This necessitates the removal of considerable quantities of bone [14] and, as recommended by several authors, the placement of a surgical pack over and around the crowns and necks of the teeth to prevent bony healing-over and to encourage spontaneous eruption. This packing procedure will markedly delay healing and is designed to prevent the reparative filling-in of bone. It is difficult under these circumstances to avoid pushing the pack into the area of the cemento-enamel junction (CEJ), which will inevitably lead to a poor periodontal prognosis for the finally erupted tooth, with an exposed CEJ and lessened bone support [32].

The frequent need to change packs over a long period incurs pain, discomfort and nuisance, difficulty in maintaining oral hygiene and a limitation of normal function, with a prolonged bad taste and odour in the mouth due to the unhygienic circumstances. From the surgeon's point of view, this entails seeing the patient for many time-consuming appointments. There is no active encouragement of eruption until brackets may be successfully bonded and traction applied, in a case already afflicted by slow or non-eruption as a characteristic of the disease. Thus, at an age when facial appearance is very important, the patient will spend an unacceptably long time without teeth. Furthermore, bone regeneration will have been retarded by the use of a method involving healing by secondary intention [33]. Eruption is thus delayed, and a growing over of the soft tissues, to re-cover the deeper and newly exposed teeth, may still occur.

From the discussion of these two approaches, it becomes clear that treatment could be vastly improved if the placement of attachments and the initial application of extrusive traction could be included as two of the functions to be addressed during the surgical procedure. This creates several formidable obstacles, the largest of which is the ability of the surgeon to create a series of microenvironments whereby a small area of the crown of each of the teeth is exposed and conditions of haemostasis and isolation prepared, to permit the delicate attachment bonding procedure.

There can be no question that the skills needed to achieve this type of complicated treatment protocol require more than one pair of hands and more than one qualified operator in the operating theatre. The ideal situation is the interdisciplinary cooperation of the surgeon with the orthodontist at this critical time. As pointed out in Chapter 3 and in several subsequent chapters, there are enormous advantages to be gained by the presence of the orthodontist at this crucial stage, which has a long-term bearing on the efficacy of the later orthodontic resolution of impacted teeth in general. If this is true of the treatment of a single impacted tooth, then its benefit is exponentially more valuable in relation to multiple impactions, many of which are often located deep in basal bone.

Without the placement of attachments at the time of surgery, access to the unerupted teeth must be guaranteed by the surgeon performing wide opening and radical bone resection, with the placement of surgical packs. With attachment placement, a conservative surgical policy is possible – only enough bone is removed to allow access for the placement of a small eyelet attachment on the minimally exposed tooth surface. The surgery may then be aimed at preserving rather than removing bone, since the presence of bone in the eruption path does not hinder mechanically encouraged eruption of the teeth, in these cases or in

general. Its lack would be a greater drawback in terms of the eventual degree of bone support and thus of the peri odontal prognosis of the erupted teeth [32]

It has been reported [14] that denser alveolar bone is present in cleidocranial dysplasia. We have not found abnormal bone in any of the cases in our care, although the observer could understandably be misled by the fact that cortical bone is found, to the relative exclusion of spongiosum. It should be remembered that the impaction of many teeth within the jaws takes up much of the volume within the body of the mandible where spongiosum would normally be present. Thus, while cortical bone encompasses all these teeth and is present in normal amounts, spongiosum is sparse.

The Jerusalem approach

This method [34, 35] was presented for the first time at the same forum as the Belfast–Hamburg approach described above [36]. Its *modus operandi* is quite different from either of the two other approaches. The Jerusalem approach is based on a rationale that is related to the abnormal dental development of the patient and on the factors that produce it.

This comprehensive approach to treatment addresses the following points:

- recognition of the clinical features of the facial, oral and dento-alveolar structures in the disease;
- the surgical measures that are required to provide access to the areas concerned;
- the need for an orthodontic strategy to enable the application of extrusive mechanics to the buried teeth in an efficient and reliable manner;
- attending to the patient's psychological well-being by focusing the earliest stages of treatment on the resolution of the incisor impactions.

Recognition of the clinical features

Cleidocranial dysplasia patients exhibit each of the following features to a variable degree:

1. non-resorption of deciduous teeth roots;
2. the presence of supernumerary teeth, markedly displacing the developing permanent teeth and providing a physical barrier to their eruption;
3. lessened eruptive force, although eruptive movements are evident;
4. poor vertical development of alveolar bone, as witnessed by a shallow sulcus, a reduced height of the lower face and a class 3 skeletal tendency, due to an underdeveloped maxilla and to a counter-clockwise mandibular rotation;
5. late but normal and unhindered eruption of first and, much later, second permanent molars in both arches;
6. late dental development, as judged by the root development of the permanent teeth, whether erupted or unerupted – a 12-year-old patient will typically show a dental age more appropriate to that of a 9-year-old [24, 35, 37].

Surgical therapeutic measures

The timing regarding the actual exposure of the permanent teeth is critical, and only two interventions are planned at distinct points in time, depending on the extent of root development, as follows.

Intervention 1

At the dental age of 7–8 years, the anterior deciduous teeth, together with all the supernumerary teeth, in the anterior and, as far as reasonably possible, posterior areas are extracted. However, only the crowns of the anterior permanent teeth whose roots are sufficiently developed (two-thirds of their expected length) are surgically exposed. Attachments are placed immediately and surgical flaps fully closed. Given the usual lateness in development of the dentition in these cases, the chronological age of the patient at this stage is usually around 10–12 years. At this time, the canine and premolar teeth are at an early stage of development, with their roots less than half their expected final length. The surgical intervention in the posterior region is therefore limited to removal of supernumerary teeth. Actual exposure of the developmentally immature posterior teeth of the permanent series is not undertaken and, most importantly, their dental follicles are left intact until later. If there are no supernumerary teeth in the posterior areas, the deciduous teeth are left in place.

Intervention 2

The dental age of 10–11 years (chronological age 13–15 years) is the most appropriate time for the second intervention, because the root development of the posterior successional teeth will be sufficiently well advanced and eruption and alignment of the incisor teeth will have been achieved. This intervention involves the exposure of the crowns of the canines and premolars in both dental arches and the immediate placement of orthodontic attachments. The special requirements of the surgical procedure relate to the conservation of bone, in general and of the cortical part of the bone in particular. Removal of the unerupted supernumerary teeth with a minimum of buccal plate of bone creates enough space around the crowns of the impacted permanent teeth of the normal series to allow the immediate bonding of small eyelet attachments. The lingual plate is left intact and at its original height. Maxillary second premolars may require a palatal approach, in which case the buccal plate should be left intact. Bone that covers the occlusal surface of the crowns of deeply impacted teeth is not removed, and the dental follicle is untouched except for the small window opening in the immediate area of attachment bonding.

A wide soft tissue flap exposing the surgical field is advised to enable good vision and access and to help in

maintaining the conservative attitude to the removal of bone. The partial-thickness muco-gingival flaps are finally replaced intact and sutured back, without the use of packs, in the manner of primary soft tissue closure [32, 36–40].

Immediately following the first intervention, it becomes necessary to supplement the eruptive force of the incisors. In this way, the vertical migration of the teeth that rapidly occurs brings with it a pronounced vertical development of the alveolar bone [14]. This will have been planned for the stage of dental development when root length is between half and three-quarters of its final expected form, which corresponds to the stage of development at which teeth normally erupt [41]. Similarly, occlusally directed forces are applied to the posterior teeth immediately following the second intervention.

Orthodontic strategy

In the broad overview, the provision of space within the arch is made by appliance-generated antero-posterior expansion of posterior-versus-anterior erupted teeth [21], while the removal of deciduous and supernumerary teeth provides space in the vertical plane. In this way, and while space is being provided, self-realization of any eruptive potential that the permanent teeth may possess is permitted, to present the opportunity for them to migrate towards the occlusal plane and to take up a more normal developmental position within the alveolus. This seems to occur to a varying degree and has the advantage of allowing the roots to develop in uncramped circumstances, thereby leading to the acquisition of more normal root morphology [33].

However, it is important to emphasize that no reliance is placed on spontaneous eruption of these teeth [14], although should this occur, it is only to be welcomed and will simplify the treatment plan. However, the present approach has been formulated to combat the worst eventuality: non-eruption. From the point of view of the orthodontic mechano-therapy, achieving efficient force application in an appropriate direction for each tooth, requires examination of the following points.

- There must be a sufficient number of erupted anchor teeth in the mouth to act as a base from which forces may be generated. As we have already pointed out, the permanent molars usually erupt without help, and one or two incisors may also be visible.
- One has to design a rigid appliance frame that will endure chewing and other functional and para-functional movements that may be expected to occur during daily oral function, considering the long spans of free, unattached and unprotected archwire mesial to the two erupted anchor molar teeth in each jaw.
- Individual and groups of unerupted teeth must be subjected to light continuous extrusive forces.
- Appliance design has to feature sufficient versatility to enable it to:

- apply vertical extrusive forces to erupt the impacted teeth rapidly;
- open spaces between recently erupted teeth, to provide room for other unerupted teeth and to establish interproximal contacts and archform;
- bring these teeth into occlusion and to upright their roots.

And all this with only minor alterations!

The patient's psychological well-being

In Chapter 5, we pointed out that it is unacceptable to leave even the youngest patient without front teeth for an extended period of time, and that it is important to make the child aware that efforts are being made to rectify such a situation speedily. The physical obstacles to eruption (i.e. the deciduous and supernumerary teeth) must be removed in order to facilitate the eruption of the anterior teeth. Proper timing is critical. This should only be done at the age when the permanent incisor teeth indicate adequate root development for eruption and only when an appliance is in place to actively supplement their limited eruption potential.

Dental crowding

When studying the radiographs of an untreated cleidocranial dysplasia patient, one is immediately struck by the intra-bony crowding provided by the large number of unerupted permanent teeth (those of the normal series and the supernumeraries). During surgery, and after all of the superfluous deciduous and supernumerary teeth have been removed, the surgeon and the orthodontist will view the open surgical field and, given the relatively underdeveloped alveolar processes, find it difficult to see how it is possible to fit all the remaining permanent teeth into the dental arch and in full alignment.

On the basis of this 'spot' diagnosis, the orthodontist will be tempted to advise the oral surgeon to take advantage of the prevailing general anaesthetic to remove a premolar tooth in each quadrant of the mouth, in what would appear to be a logical step necessary to reduce the apparent crowding. However, for most cases, this step would be regretted later when the subsequent size and form of the alveolar processes, which may be developed as a by-product of the mechanical eruption of the teeth, become evident. Initially, the appliance-generated eruption of the anterior teeth brings the teeth into the mouth with a pronounced lingual tipping of their long axes. This is due to the influence of purely vertical forces that will have been brought to bear on the permanent incisors, whose developmental position is very much lingually placed and apical to the recently extracted deciduous incisors.

For this reason, the permanent incisors must be tipped labially to create a normal archform and to provide a more

procumbent support for the lips. This will contribute much additional space in the dental arches for the premolar and canine teeth, and will be instrumental in significantly eliminating the dental crowding and the patient's edentulous appearance. It is only after complete or almost complete eruption has occurred that a decision should be made as to whether extractions are needed. Our limited experience has shown that, in these younger patients, extractions usually are not needed, and that adequate space for alignment of all the teeth anterior to the first molars may easily be provided.

Retention of the treated result

Once the permanent teeth have all reached their final positions in the arch, the removal of the fixed appliances will not usually be accompanied by a loss of vertical height of these teeth, despite the fact that their vertical positions will have changed so dramatically over the treatment period. The lateral width of the two arches will have been set initially by the first permanent molars and the over-retained deciduous teeth. The use of fixed lingual arches during treatment will have allowed good control against any change in this dimension. Therefore, the orthodontist may confidently expect no natural post-treatment alteration of the arch width. However, in the event of maxillary narrowness, rapid suture-splitting expansion may be performed early on in the treatment and its results maintained throughout the treatment, initially by a transpalatal soldered bar and, thereafter, by the archform of the successive labial archwires. Subsequent post-treatment loss of width is generally marginal.

The only dilemmas of any consequence in the context of retention relate to the labio-lingual post-treatment position of the incisors of both jaws and to those teeth that have undergone rotational orthodontic movement during treatment.

It is axiomatic to say that proper labio-lingual positioning of the anterior teeth in any patient is dependent on the muscular balance between the lips and the tongue. Teeth placed too far labially or lingually will inevitably be pushed by the lips or tongue in the opposite direction when all retaining devices are removed. During the orthodontic treatment of the normal patient and in the interests of stability, the positions of adjacent teeth are often used as a measure against which the displaced teeth should be moved to achieve the desired alignment. Alternatively, cephalometric standards may be preferred against which to compare the dentition, such as the lower incisal edges vis-à-vis the A–Po line [42].

With the cleidocranial dysplasia patient, there is no scientific way to judge the 'biologically correct' and therefore stable position of the incisors. Nor are there any published cephalometric data on a large group of treated and post-retention cleidocranial dysplasia patients to help establish such norms. By no means may the cephalometric values of these patients be compared with the average values found in the various growth studies that have been carried out with samples of normal patients.

Consequently, the use of Holdaway or Ricketts analyses and a growth prediction analysis, as proposed elsewhere [24], is invalid and highly misleading. The orthodontist can never be sure of the stability of the final result in this aspect of the treatment, and some form of long-term retention will usually be advisable. This being so, we have adopted the view that the incisors should be brought well forward, extruded below the upper lip, so that 2 or 3 mm of their incisal edges are clearly visible at rest, and the recommendations of Zachrisson regarding the so-called 'smile line' be followed [43, 44] to slightly over-compensate to some degree for the years that the patient has lived with very short and largely unseen teeth. When only deciduous teeth were present, or in the initial stages of treatment, the patient's social interaction with others would have been dictated by a desire to mask the missing anterior teeth, and he/she may have adopted unnaturally unwelcoming and unsmiling facial expressions and a reserved attitude. Once dental alignment is complete and appliances removed, a positive and dramatic psychological change in the patient's attitude to life seems to occur and, from then on, many treated cleidocranial dysplasia patients seem to have a permanent smile on their faces, consciously and deliberately displaying their newfound teeth! They adopt an optimistic outlook on life and become much more open in their day-to-day contact with others.

After a short period of time with conventional removable retainers, our practice has been to prepare and apply fixed multi-stranded bonded retainers to the maxillary and mandibular six or eight anterior teeth [45–48]. These will then hold the labio-lingual positions of all the anterior teeth, as well as preventing rotational relapse. The conventional removable retainers may then be discarded.

The Jerusalem approach in clinical practice

Patients who are suspected to be suffering from cleidocranial dysplasia are referred to us through different agencies, including the various medical specialties and general dental practitioners or dental specialists. A small proportion also arrive on their own initiative, requesting advice and help in the search for a solution to the presence of 'very small teeth' or to their 'toothless' appearance. CCD individuals vary widely in the expression of the phenotypal features, with some exhibiting a broad spectrum of the associated characteristics and others with relatively few. Some may even escape detection and diagnosis until early adolescence.

In order to confirm that the patient does indeed suffer from cleidocranial dysplasia, our diagnostic routine has come to include:

- a family history to check for parents or other relatives who may have been diagnosed with cleidocranial dysplasia;
- a clinical examination in search of the general characteristics of the condition, which takes in the form of the cranium, face and clavicles, including the mobility of the shoulders;
- an intra-oral examination to relate the eruption status of the dentition vis-à-vis the patient's chronological age;
- a radiographic evaluation, which plays a critical role in the confirmation of the clinical diagnosis, includes a chest X-ray and antero-posterior and lateral skull radiographs, which are performed in a cephalostat. At the same time, a panoramic radiograph is studied and supplemented with periapical and occlusal views, as required.

Following the establishment of the tentative diagnosis, a genetic analysis should be undertaken. It is recommended that a small blood sample be taken and forwarded to a laboratory specializing in dealing with the gene specific to CCD, namely RUNX2. The laboratory will extract the DNA from the blood sample and then perform the sequencing. There is a 60–70% likelihood of finding the gene although, as has been pointed out at the beginning of this chapter, its absence does not mean that the disease is not present.

Once the diagnosis has been confirmed, genetic counselling is offered to the parents regarding their own future offspring, but more particularly regarding future offspring of the affected child. An important part of the geneticist's examination will include gathering information about relatives and the possibility that other, more distant family members may be similarly affected.

Treatment of the oro-facial characteristics of the condition will involve the talents of a team of three dental specialists: the paedodontist, the orthodontist and the oral and maxillofacial surgeon. They will need to work in close collaboration and the first stage may begin immediately.

Stage 1: Assuring the health of the dentition

Treatment of the cleidocranial dysplasia patient will necessitate the wearing of orthodontic appliances for several years. Therefore, an essential requirement in all cases is that the health of the dentition be guaranteed by proper oral hygiene instruction, with follow-up to check that an adequate level of compliance is attained. Appropriate use of fissure sealants and fluoride applications is recommended. Carious teeth will need to be treated, but, in order for the paedodontist to be in a position to decide on the type of restoration indicated, the timing of the extraction of the remaining deciduous teeth will need to be determined at the outset.

Stage 2: Vertical correction in the incisor region

It has already been emphasized that there is a lag in the dental development of about three years in relation to the child's chronology. Thus, for most cleidocranial dysplasia cases, it will not be until about 10 years of age that the spontaneous and unimpaired eruption of all first permanent molars will herald the dental age of 7–8 years. Sometimes, one or more of the permanent incisors will also have erupted, but the following description of the technique will assume the least favourable initial scenario and we shall assume that no permanent anterior teeth are present.

Orthodontics

Plain orthodontic bands are fitted on the erupted first molars (Figure 14.8) and an impression is taken of each dental arch. The bands are then carefully removed from the teeth, re-seated in the impressions and a model is cast. Heavy soldered palatal and lingual arches are prepared on the two models, and single soldered buccal tubes (0.036 in round) are oriented such that they run mesially, close to the buccal side of the deciduous teeth, parallel to the occlusal plane. A lingually displaced first permanent molar may need to be aligned buccally with a removable appliance first in order to be able to align the buccal tubes accurately as described, since the efficient working of the appliance depends on this.

A heavy 'incisor-erupting' archwire is prepared for each arch in advance (Figure 14.8f), and its function is to achieve a correction in the vertical plane. This archwire is made of 0.036 in round wire, which slots into the buccal molar tubes up to a predetermined bayonet bend on each side. This holds the wire 2–3 mm labial to the anterior teeth and 3–4 mm gingival to the occlusal plane. In the canine area, an S-shaped hook is soldered, with its mesially pointing extremity on the occlusal side and the distally pointing extremity gingival. In the midline area, a small fine-wire frame is also soldered and points towards the sulcus depth.

An 'incisor-aligning' archwire is also prepared in advance, although it will not be put to use until all the permanent incisors have been fully erupted. This consists of 0.020 in internal-diameter tube (0.036 in external diameter) side-pieces, which freely slide accurately into the 0.036 in buccal tubes, without allowing lateral or vertical play. The tubes are cut so that their mesial extremity is in line with the distal of the deciduous canine. A length of 0.0155 in or 0.0175 in multi-stranded wire or of 0.016 in or 0.018 in nickel–titanium wire is then drawn into the buccal side-piece tubes, and is made to 'friction-fit' by incorporating three or four sharp bends in that part of the wire. This updated version of Johnson's twin-wire arch [49, 50] is placed to one side until it is needed at that later stage, when it will be used to achieve correction of the alignment of the teeth in the horizontal plane. The reader is referred to Chapter 5 for a full description of the Johnson twin-wire appliance.

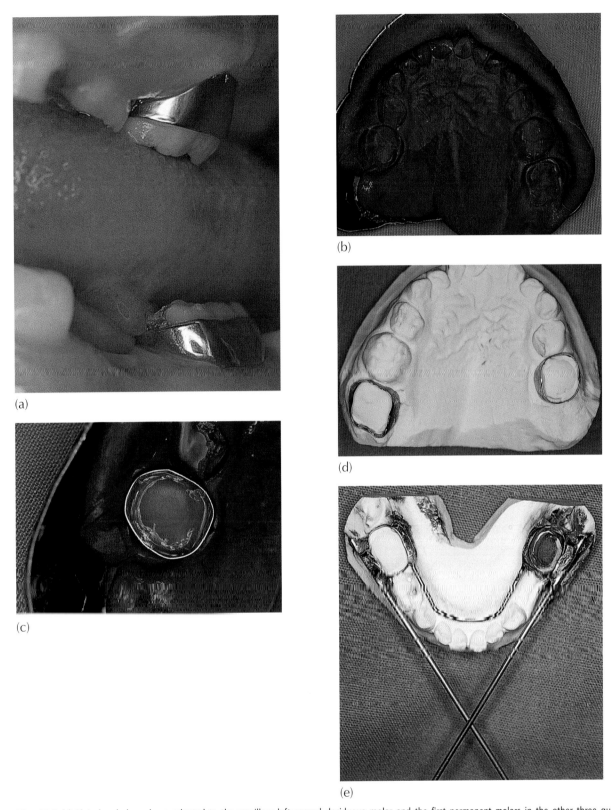

Fig. 14.8 (a) Plain bands have been adapted to the maxillary left second deciduous molar and the first permanent molars in the other three quadrants. (b) Upper (shown here) and lower compound impressions are made. (c) The bands are removed from the mouth and carefully replaced in the impression. The bands are then partially filled with wax, before a model is poured. (d) The cast model with accurately located and stabilized molar bands. (e) The occlusal view of the mandibular model shows right and left tubes converging in the midline and the lingual arch in place. (f) The heavy 'incisor-erupting' archwire is slotted into the molar tubes. Note the S-shaped hook soldered in the canine area and the anteriorly soldered fine-wire frame. (g) Disassembled mandibular appliance ready for intra-oral placement. (h, i) The appliances cemented in the mouth.

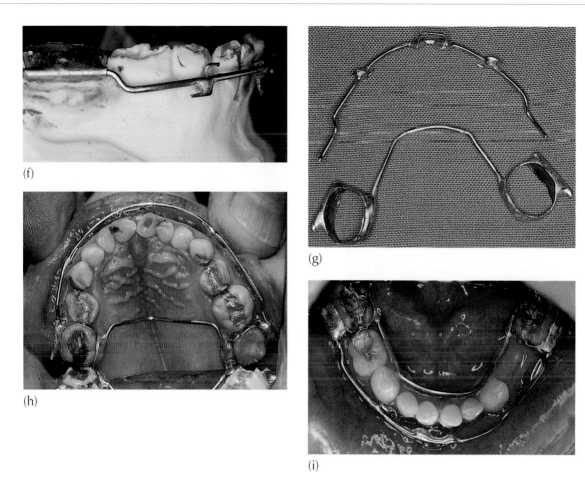

Fig. 14.8 (Continued)

Surgery

The patient is now ready for the first surgical intervention (Figures 14.8 and 14.9), which must be performed under endotracheal general anaesthesia. At the completion of this episode the patient will have lost all of the anterior deciduous teeth and will also have had the unerupted supernumerary teeth removed in both arches. Fine pigtail ligatures will have been placed in the eyelet attachments bonded to the incisor teeth (ideally by the orthodontist) and, following suturing of the flaps back to their original place, these will be the only link between the invisible unerupted permanent teeth and the exterior (Figure 14.10). Deciduous teeth not associated with supernumerary teeth are generally left until the next surgical stage. Under these circumstances, this surgical procedure will generally take between 3 and 5 hours, depending on the number of supernumerary teeth removed and on the number of unerupted permanent teeth that need to be exposed, with the placement of attachments.

Orthodontics

Still under the endotracheal anaesthesia in the operating theatre, the orthodontist replaces the prepared 'incisor-erupting' archwire in the buccal tubes. The archwire may be secured by drawing an elastic chain between the distal

of the buccal tubes to the mesially pointing extremity of the S-hook, although this is rarely necessary. The anterior portion of the archwire is raised with light finger pressure and engaged by looping the pigtail ligatures around it. Since these pigtails are tied directly to the buried incisor teeth, this displacement of the archwire elicits a vertical extruding force which is shared by the unerupted teeth (Figure 14.12c). The extrusive force and its range are very easily measured and, therefore, controllable to within appropriate levels.

This force generates a rapid response of the teeth, as witnessed by elongation of the pigtail ligatures over a period of a few weeks. By displacing the archwire apically and then rolling up the pigtail around it, extrusive pressure may be simply and efficiently reapplied over several visits, until eruption occurs.

This procedure is performed simultaneously in both arches in order that the two may then subsequently be treated more efficiently. To support the orthodontic anchorage, a single, large and fine-gauge elastic should be placed in the form of a rectangle, engaging the distally-pointing, soldered hooks in the canine area on each side of the archwires in both jaws. When the elastic is placed, its midline portion is laid over the soldered vertical frame of the archwire in order that tissue impingement may be avoided

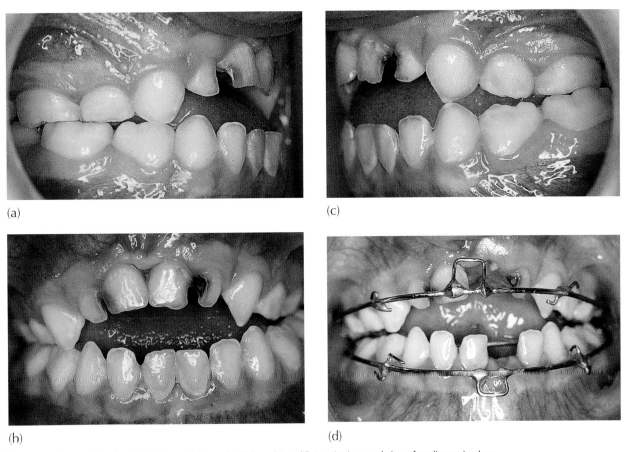

(a)

(c)

(b)

(d)

Fig. 14.9 A 14-year-old patient. (a–c) Intra-oral views of initial condition. (d) Anterior intra-oral view of appliances in place.

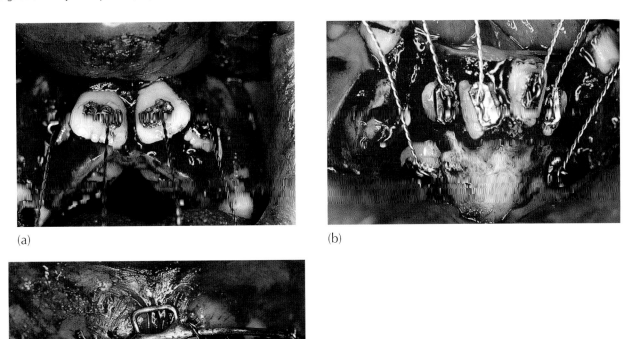

(a)

(b)

(c)

Fig. 14.10 The same patient as in Figure 14.9. (a) Four maxillary incisors have been exposed and attachments bonded. (b) Six mandibular anterior teeth have been exposed and attachments bonded. The exposed chin button and the lingual arch indicate the depth of these teeth. (c) The 'incisor-erupting' archwire is replaced when full-flap suturing is completed, and the steel ligature pigtails are made to ensnare the archwire, which has been displaced superiorly by light finger pressure.

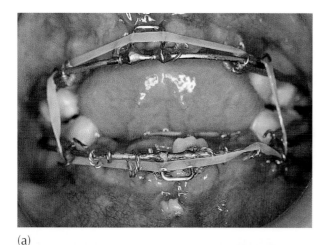

(a)

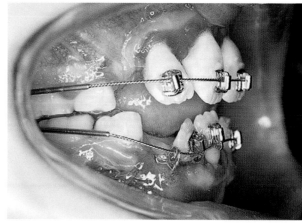

(a)

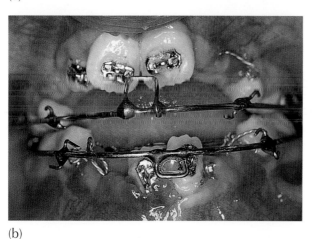

(b)

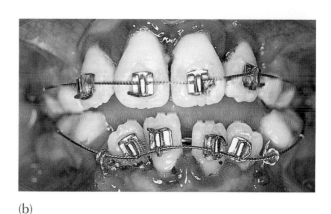

(b)

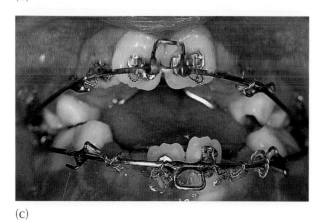

(c)

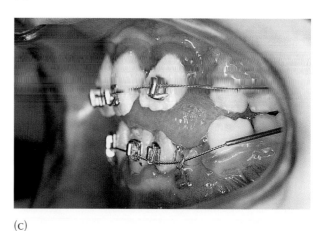

(c)

Fig. 14.11 The same patient as in Figures 14.9 and 14.10. (a) At five weeks post-surgery, a very light anterior 'box elastic' is placed on the distally pointing element of the S-hooks in the canine areas to enhance the anchorage. Note the use of the midline frames to prevent tissue impingement. (b) At nine weeks post surgery, five incisors have erupted and the archwires have been disengaged to increase their deflection. (c) Re-engaging the archwires illustrates their range of effectiveness.

Fig. 14.12 The same patient as in Figures 14.9–14.11. (a–c) At 6.5 months post-surgery, all incisors have erupted. Conventional brackets have been substituted for the eyelet attachments, and the 'incisor-aligning' archwires are in place. Note the extrusive component generated by these archwires.

Stage 3: horizontal correction in the incisor region

Orthodontics

The incisors erupt relatively quickly and with a strong lingual inclination, in general. At that point, their eyelets should be replaced by the orthodontic bracket of the orthodontist's choice, which should be sited in the routine manner (Figure 14.12). The prepared 'incisor-

(Figure 14.11). This anterior vertical elastic provides an intermaxillary vertical force to each archwire. This means that the forces used to erupt teeth in one jaw are anchored by the reactive force that is producing the eruption of the teeth in the opposite jaw.

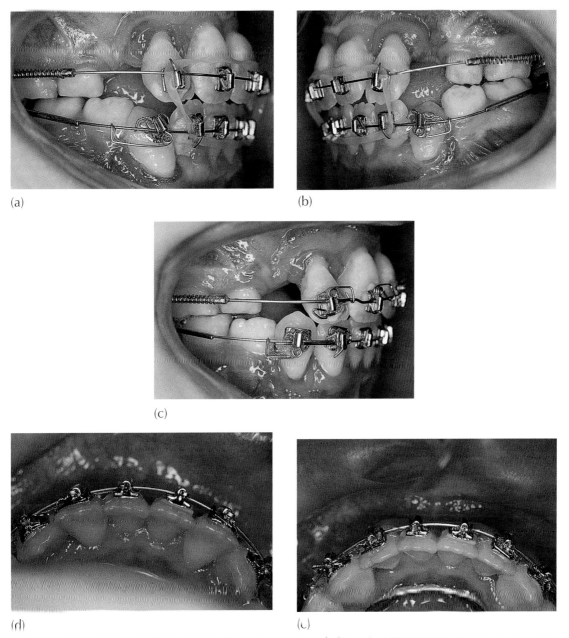

(a)　　　　　　(b)

(c)

(d)　　　　　　(c)

Fig. 14.13　The same patient as in Figures 14.9–14.11. (a, b) At 12 months post surgery, the incisors have been moved labially, aligned and a positive overbite overjet relationship has been established. In this case, the mandibular canines were included in this stage of treatment. Note the severe mesial root displacement of these teeth. (c) The occlusal view of the mandibular anterior teeth indicates gross labio lingual displacement of the roots of the lateral incisors and canines. (d, c) Following four months of further treatment, the root displacements have been corrected. This was done during a waiting period for adequate premolar development, to allow the initiation of the second surgical intervention.

aligning' archwire is then ligated into place, where its first task will be to undertake the levelling and alignment phase of treatment (i.e. incisor height, rotation and uprighting).

With proper buccal tube orientation at the outset, the long buccal tube side-pieces maintain the achieved vertical extrusion, while providing stability and resistance to distortion. The light anterior vertical elastics may be attached to the lateral incisor brackets to continue the vertical extrusive force, if and when necessary.

As levelling proceeds, the light wire middle portion of the 'incisor-aligning' archwire may be substituted for heavier and thicker stainless steel wires (Figure 14.13) until an 0.018 in or 0.020 in wire is in place, using the same side-pieces. At this point, an expanded coil spring is placed on the side-pieces, which now have a 'stop' soldered or welded to their mesial end. When the side-pieces are replaced in the buccal tubes, the coil spring is compressed between the buccal tube and the mesial stop, which displaces the archwire forwards. The archwire is ligated into

the anterior brackets under pressure. The compressed coil spring is now producing an antero-posterior expansion force, acting between the first molars and the incisors. This rapidly tips the incisors to a more normal labial inclination and, at the same time lengthens the arch in the buccal region to subsequently accept the unerupted canine and premolar teeth. Archform will be greatly enhanced and the patient's appearance will, for the first time, begin to improve markedly.

An estimate of treatment time up to this point (stages 2 and 3) would be between 9 and 18 months. New radiographs should now be taken of the unerupted canines and premolars (Figure 14.14) to assess the degree of their root development and to reassess their vertical location within the alveolus. Absence or earlier removal of supernumerary teeth in this area, together with an increase of space in the arch and the passage of time, may have led to an improvement in their position, which should be recorded.

Stage 4: vertical correction in the posterior region

Surgery

The second surgical intervention (Figure 14.15a–d) is performed at dental age 10–11 years and will leave the patient devoid of any remaining deciduous teeth that had been left undisturbed in the first intervention. The surgical flaps will have been replaced to completely cover the eyelet attachments bonded to the buccal side of the unerupted permanent teeth, although it is unlikely that full closure of the sockets of the deciduous teeth will be possible. Stainless steel pigtail ligatures will be visible emanating superiorly through these sockets.

Strictly speaking, the scope of this surgical procedure, potentially involving 12 teeth (eight premolars and four canines in the four quadrants) is large enough to warrant a repetition of endotracheal anaesthesia and operating theatre conditions. However, in the more favourable cases, spontaneous eruptive movement of the unerupted teeth will have occurred since treatment was initiated and several teeth may have partially erupted. This may encourage the oral surgeon to prefer to perform the remaining exposures under local anaesthetic one or two quadrants at a time and with an open exposure technique.

Orthodontics

The buccal side-pieces of the modified Johnson twin wire arch, with its more rigid 0.020 in middle section, are now used as a rigid beam, from which elastic thread may be tied, under pressure, to the rolled-up pigtail ligatures of the unerupted premolar and canine teeth. Re-ligation will be needed at frequent intervals because of the relatively poor range of action of the elastic thread. Alternatively, this composite archwire may be discarded in favour of a plain 0.018

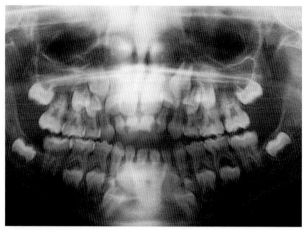

(a)

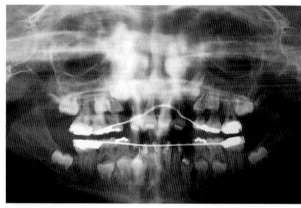

(b)

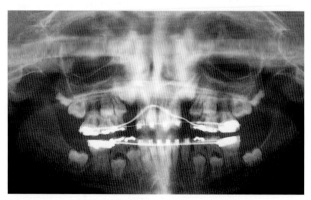

(c)

Fig. 14.14 The same patient as in Figures 14.9–14.13. (a) A panoramic view taken 10 months pre-surgery. (b) A similar view six months post-surgery. (c) A similar view 21 months post-surgery.

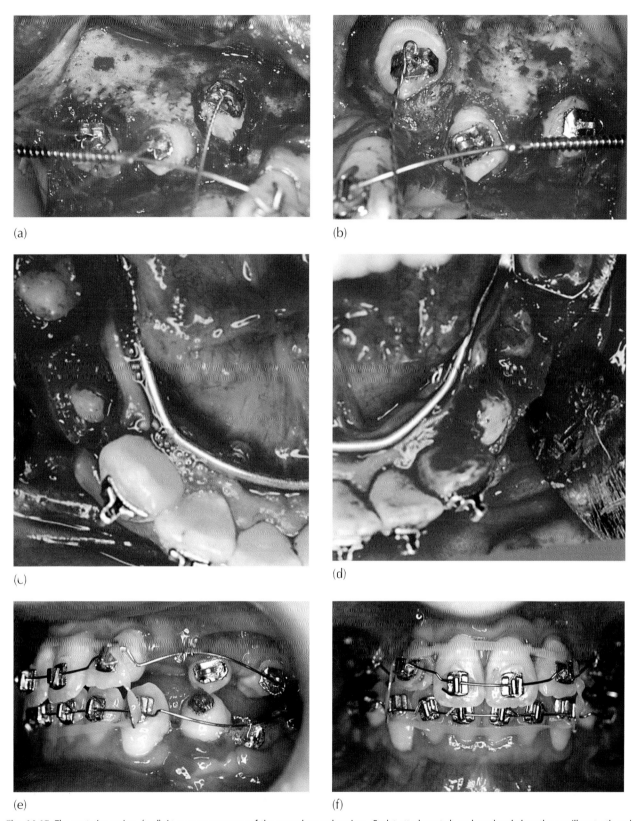

(a)

(b)

(c)

(d)

(e)

(f)

Fig. 14.15 The posterior region. (a–d) At surgery, exposure of the premolars and canines. Eyelet attachments have been bonded on the maxillary teeth and full closure was performed. (e, f) Four months post-surgery, vertically offset 0.018 in round archwires are used to erupt the teeth, with anterior vertical inter-maxillary elastics maintaining the vertical dimension and reinforcing the anchorage in both arches. (g–i) Immediately prior to appliance removal. (j–p) Immediately after appliance removal, with 3–3 twistflex splints placed the same day. Note the pitting of the enamel of the incisors and several other teeth. (q–u) At 18 months follow-up. (v, w) Panoramic views at appliance removal and at 18 months after completion of treatment, respectively. Note the steady progress in the spontaneous second molar eruption and development of third molars. (x, y) Profile views of the face before and after treatment (5.11 years), showing an atypical exaggerated growth of the mid-face.

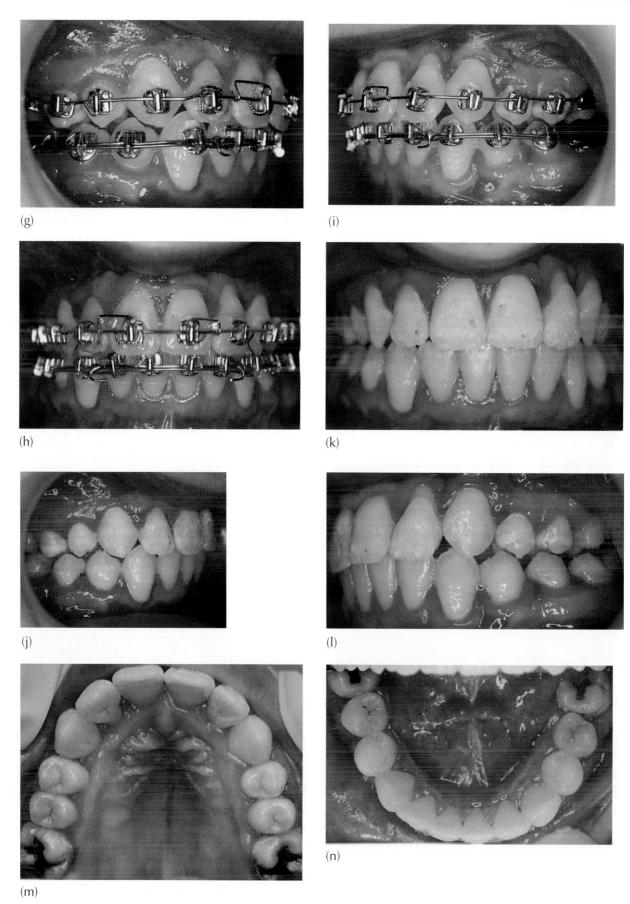

(g)

(i)

(h)

(k)

(j)

(l)

(m)

(n)

Fig. 14.15 (Continued)

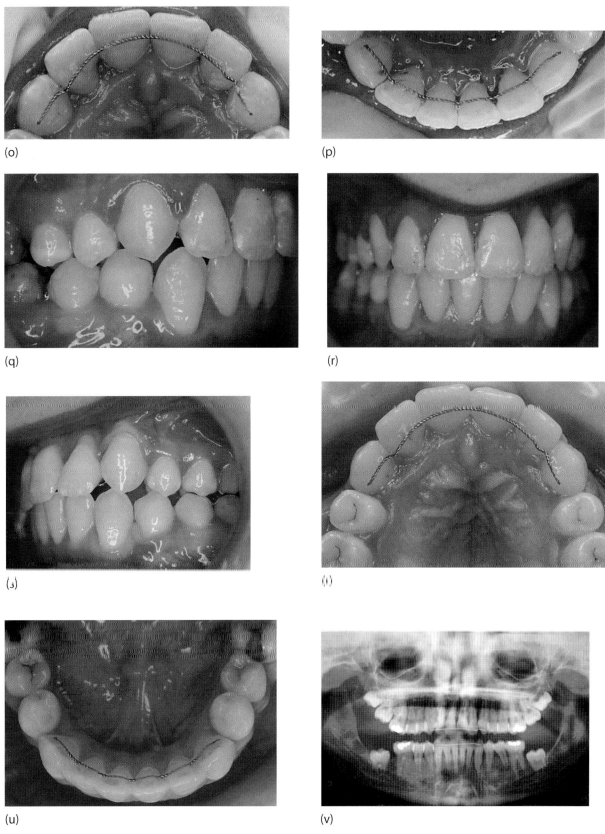

(o)

(p)

(q)

(r)

(s)

(t)

(u)

(v)

Fig. 14.15 (*Continued*)

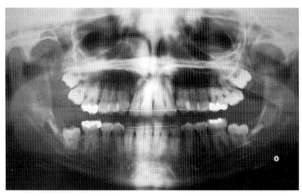

(w)

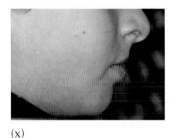

(x)

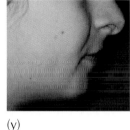

(y)

Fig. 14.15 (*Continued*)

in or 0.020 in round archwire, stretching from molar to molar (Figure 14.15e). Its long span, between lateral incisor and first molar, is flexible in the vertical plane, and the pigtail ligatures may be turned over to engage the wire under light extrusive tension (Figure 14.15e). Stage 4 is expected to be completed within 5–12 months.

Stage 5: correction of the axial orientation of the roots of the teeth

Orthodontics

Much mesio-distal uprighting of the anterior teeth may have already occurred in the earlier stages of treatment, although root-torqueing movements will usually need to await the use of a full rectangular arch. However, given the amount of root displacement seen in these cases, it is most likely that stage 5 will see these movements still being applied to the anterior teeth, while uprighting and rotating movements will need to be initiated in the premolar/canine regions, following the substitution of the eyelets with standard brackets (Figure 14.15g–i).

At this stage, the palatal and lingual heavy arches are no longer needed, while rectangular tubes on the molars now become an essential aid to the remaining finishing procedures. The molar bands are removed and new bands adapted, with rectangular tubes. Molar expansion or rota-

tion may now be performed, if desired, together with the uprighting and torqueing movement needed to produce optimal alignment. Standard finishing procedures that are *de rigueur* in the protocol in any routine orthodontic case now become possible and are carried out in the usual way.

It is important to note that in many cases the teeth had earlier been markedly displaced by the physical presence of supernumerary teeth and, as a result, may need degrees of uprighting and torqueing movements almost never seen in unaffected individuals. These movements take much time even when an efficient root-moving mechanism is used because of the distances that the root apices have to travel through alveolar bone in order for the tooth to become optimally sited. Under these circumstances, the orthodontist is encouraged to use auxiliary springs on rigid base arches, since they possess a greater range of action with more controllable forces, while the base arch holds the overall archform.

Treatment follow-up and the eruption of second molars

For most patients, the stages of whose treatment have been undertaken at the recommended ages and developmental milestones, the protocol described above will be completed before the eruption of the second molars. It was pointed out at the beginning of this chapter that, in cleidocranial dysplasia, permanent molars generally erupt under their own steam and in accordance with the patient's dental age. This is largely true of both the first and second molars. This being so, we would not expect to see eruption of the second molars before the age of 15–16 years, with a normal range of +2 years – hopefully after completion of the overall treatment of the remainder of the teeth. A new panoramic radiographic view at this time (Figure 14.15v, w) will usually show the eruption of the second molars to have progressed markedly, with the teeth heading towards their place in the arch. The film should also be scanned for signs of the possible early development of new supernumerary teeth. If all appears well, orthodontic treatment should not be prolonged until the second molars erupt and the appliances should be removed.

The second molars may still take many months, stretching into a year or two, before intra-oral signs of eruption are noted. For as long as unerupted supernumerary teeth do not interfere with the eruption of teeth of the normal series and do not threaten to disturb an existing dental alignment and occlusion, the decision whether or not to extract them should be made by a competent oral surgeon and not by the orthodontist. Therefore, if newly developing supernumerary teeth are identified on this new film, a surgical opinion should be sought regarding the need and the timing for their removal.

Supplemental molars

From time to time, supplemental molars are seen in a small number of CCD cases and these may occur in what appears to be a simple distal extension of the dental lamina, to produce a fourth molar which is located distal to the unerupted third molar, in either jaw. Their presence will not adversely affect the autonomous eruption of the second molars. However, supernumerary molars may occasionally be seen lateral, medial or superior to the unerupted second molar, when it is almost certain to impede eruption. The difference between the supernumerary tooth and the normal second molar will usually be decided by the root development of the teeth. Supernumerary teeth usually develop much later than teeth of the normal series and, as such, will have rudimentary root development which lags far behind the dental age of the rest of the dentition. Furthermore, when fully developed, the supernumerary teeth will often exhibit considerable root stunting.

When eruption of the second molars is obstructed by these unerupted teeth, surgical removal is justified, following which some improvement in the eruption status of the second molars is very likely. It is therefore wise to review the case on a half-yearly recall basis, for at least a year or more, before advising further treatment to augment the limited eruptive potential of the teeth.

Retention

Retention procedures are similar to those used in any orthodontic case. It should be remembered that beautifully aligned teeth need to be retained long term. All teeth move during the life of normal individuals and of CCD patients, whether or not orthodontic treatment has been carried out. The degree of immediate post-treatment movement is always considerably greater than after some months in retention and it is also greater for teeth that have been moved over large distances or had been initially rotated. For this and other reasons, the long term use of anterior lingual bonded retainers is advised [46–48].

Treatment experience

To date, over 40 cleidocranial dysplasia patients have been treated or are still in the various stages of comprehensive orthodontic/surgical treatment at our centre (Figures 14.16, 14.17). It is from their treatment that the Jerusalem approach has been formulated and refined over the years.

In the published reports that described the treatment of the first 16 patients in this series [34, 35] three had no supernumerary teeth and one had a congenitally missing premolar. Seven patients had four supernumerary teeth or fewer and four others had six, eight, 10 and 27,

respectively. Treatment of the remaining case was initiated elsewhere, and information was not available for that patient.

Gathering information from all the cases that we have seen over the years, most supernumerary teeth were found to be recognizable supplemental teeth, similar to those adjacent to them (i.e. with the characteristic morphology of incisors, canines, premolars or molars). In the anterior region of the maxilla, occasional barrel-shaped supernumerary teeth are relatively frequent. The presence of supplemental teeth has been helpful in permitting a choice of teeth for extraction, since a displaced tooth may be removed and the better placed adjacent tooth aligned in its place without regard to distinguishing which is the abnormal tooth, provided the root development is potentially adequate.

Without exception, the dental age range of our cases lagged between 2.5 and 4 years behind their chronological ages. Spontaneously erupted first permanent molars have been seen in all but two cases, which were associated with supernumerary molar teeth. Unlike other teeth, the permanent molars appear to erupt spontaneously in cleidocranial dysplasia, and it seems reasonable to assume that, in the absence of these obstructions, the molars of these two patients would also have erupted normally. Following the removal of the supernumerary teeth, in both of these cases the unerupted molars were exposed and a surgical pack placed to encourage healing by secondary intention and to maintain the patency of the exposure. The teeth erupted spontaneously after this preparatory surgical procedure and both patients were then treated with the above protocol. A single erupted maxillary first premolar was present in three patients whose deciduous predecessor had had an apical abscess.

The shape of the well-developed permanent teeth, while clearly recognizable and classifiable into their different tooth types, showed labial concavities in the incisors, broad mesial and distal ridges on the labial and lingual aspects of both canines and incisors, and mesio-distally wide and bucco lingually narrow widths of the lower second premolars. The roots of the supernumerary teeth were relatively short and the orientation of their long axes was often significantly divergent from the overall orientation of the crown in both the mesio-distal and bucco-lingual planes. This created the need for a more controlled periapical X-ray monitoring of the uprighting and torqueing movements of these teeth than may be usual.

In the above treatment protocol, the ability to use inter-maxillary vertical elastics on the previously impacted teeth was emphasized. Their value is seen in the augmentation of the eruptive forces, to improve anchorage and to encourage vertical alveolar growth that normally accompanies erupting teeth. Therefore, efficiency is not served by treating the mandibular anterior teeth before the maxillary anterior

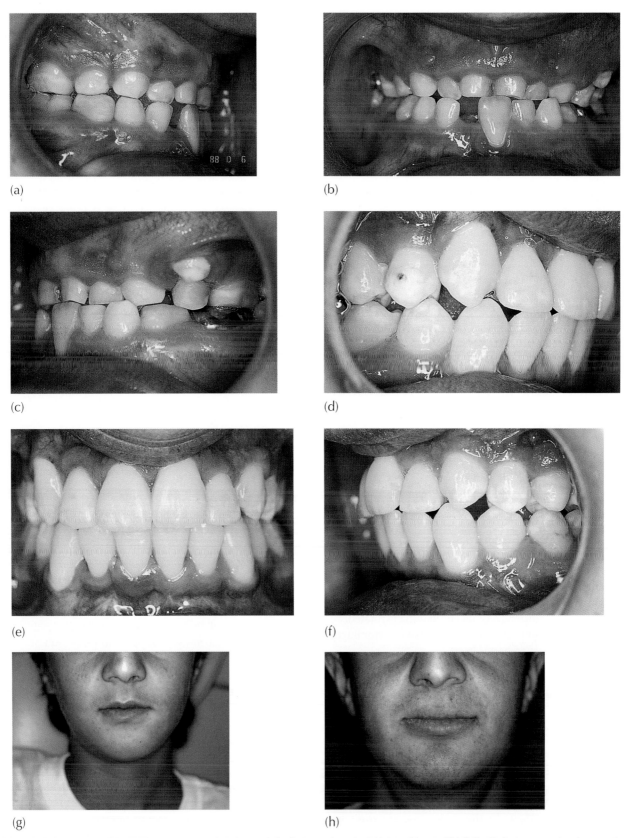

Fig. 14.16 (a–c) Intra oral views of a class 1 case of cleidocranial dysplasia, to show the initial condition and (d–f) the final outcome, using the Jerusalem approach to treatment, as described here. A class 3 relationship did not develop. (g–j) The *en face* and profile comparisons of the facial appearance before and after treatment.

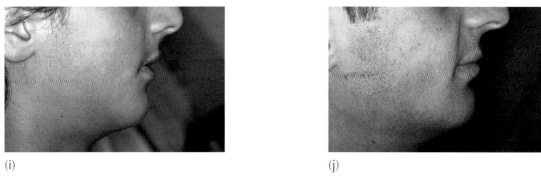

(i)

(j)

Fig. 14.16 (*Continued*)

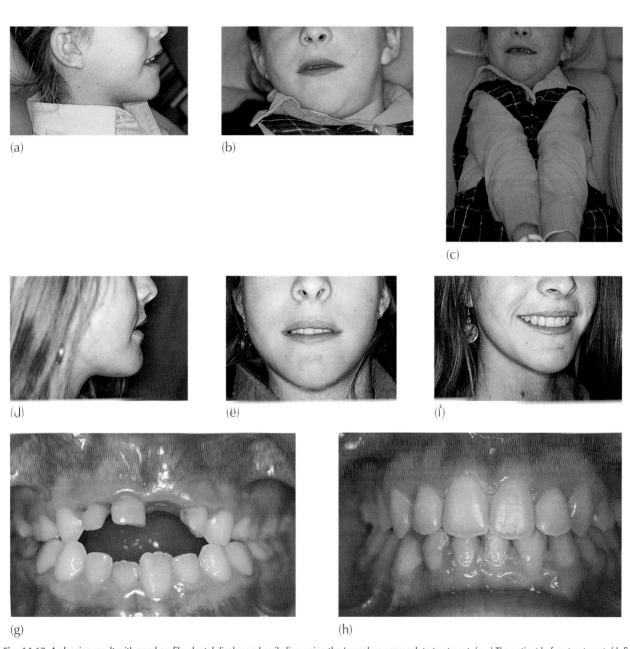

(a)

(b)

(c)

(d)

(e)

(f)

(g)

(h)

Fig. 14.17 A pleasing result with good profile, dental display and smile line, using the Jerusalem approach to treatment. (a–c) The patient before treatment. (d–f) At one year post-retention. (g) The dentition prior to treatment, at age 10. (h) The dentition a year after the completion of treatment at age 15 years. During the treatment the patient was involved in a car accident and fractured her mandible, with the fracture line passing through the unerupted right canine socket, causing ankylosis. This tooth was extracted and the remaining teeth aligned across the midline. All the mandibular incisors had been fractured and restored.

teeth, merely because there is a time lapse of a year or so between the normal eruption times of these teeth. The treatment of the anterior areas of both jaws should begin at the same time. Nevertheless, it is essential that the forces produced by these vertical 'up-and-down' elastics be kept within very minimal values, avoiding the temptation to use elastics which are too heavy.

Excessive extrusive forces will bring about rapid eruption, but the clinical crown length of the erupting teeth will be long, owing to a relatively lesser generation of alveolar bone. The teeth will be very mobile, the cervical root area will be exposed and sensitive and they may lose their vitality.

In the first surgical intervention, the aim is to remove all the supernumerary teeth, since these seem to be the principal factor in the displacement of the adjacent permanent teeth more deeply into basal bone. The continued presence of these extra teeth will prevent the teeth of the normal series from expressing what little eruptive potential they may have and benefit immensely is attached to their early removal. Furthermore, late-developing supernumerary teeth increase in size as they complete their crowns and their roots grow, thereby occupying more space in the alveolus. This results in a progressive displacement of apically placed permanent teeth, pushing them deeper into basal bone and away from the oral cavity.

Our experience has been that, when supernumerary teeth are not present in certain areas, the normal permanent teeth in those areas are not severely displaced and, while they may still not erupt spontaneously, their development is fairly normal. Occasionally, their presence may stimulate the shedding of the deciduous predecessor and they may, in time, erupt. For this reason, in those areas where supernumerary teeth are not found, deciduous teeth are not removed in the first surgical intervention.

Normal development and maximum eruption potential are probably best realized when the integrity of the dental follicles is maintained until two-thirds to three-quarters of the root length has developed. For this reason, the canine and premolar teeth are not exposed when the supernumerary teeth are removed in the first surgical phase of the treatment. Trauma to developing teeth has been shown to cause damage to both developing roots and the enamel of their completed crowns [51–53] and surgical trauma is no exception.

Both in this text and elsewhere [32, 39, 54–57] we have been at pains to point out our opposition to the accepted and established practice of wide surgical exposure of unerupted teeth in normal cases with isolated impacted teeth [58–61]. Typically, in cleidocranial dysplasia there is underdevelopment of the maxilla in the antero-posterior plane and of both jaws in the vertical plane [1, 4, 5]. A very large number of unerupted teeth are present within the bone, largely eliminating the presence of spongiosum. This being so, the wide removal of the cortical plate [14, 24, 28, 29] would appear to be wasteful and compromising.

In the Jerusalem approach, access to the teeth is gained by a minimal opening in the cortical plate immediately overlying the teeth, together with a small window in the dental follicle. The size of the opening is determined by two factors:

- exposing a surface large enough to accommodate a small eyelet attachment;
- enlarging this to the minimum size that will allow the surgeon to achieve haemostasis for long enough to allow the bonding procedure to take place in a contamination-free micro-environment.

There is no reason to remove further bone and certainly not to reduce the vertical height of the adjacent cortical plate. The complete surgical flaps are sutured back to close off the surgical field fully, and healing is by primary intention, which should offer a healthier and more rapid period of healing and promote a speedier and more generous response on the part of the alveolar bone [62].

Deeply displaced teeth may thus be drawn occlusally through the overlying bone, which offers no real resistance to the mechanically assisted eruptive force. New alveolar bone accompanies the erupting teeth as they progress towards the occlusal plane in a manner similar to that occurring with normal, unassisted eruption in the normal patient [32, 39]. This will enhance the vertical height of the alveolar processes of the two jaws, incidentally deepening the labial and lingual sulci and improving the overall facial proportions. The final periodontal status of the teeth will also be more normal [32, 37, 55, 56] and the appearance will make it difficult for the trained eye to distinguish between a previously impacted tooth and a normally erupted adjacent tooth. Even in the cleidocranial dysplasia case exhibiting impaction of the patient's entire complement of permanent teeth, it is now possible to make the teeth, their supporting tissues and the sulcus depth appear completely normal, as in any routine orthodontic case.

In both the Toronto–Melbourne and the Belfast–Hamburg approaches, the need to maintain patency of the exposure and visual contact with the unerupted teeth dictates that the bone level must be pared down to that of the deepest tooth. This clearly includes the reduction of both the lingual and buccal cortical plates to this level – and this in a patient whose alveolar processes are already of reduced height because of the syndrome, and whose teeth are deeply sited in basal bone due to displacement by supernumerary teeth.

Patient variation

Similarities are seen among all of the patients suffering from cleidocranial dysplasia, with a varying degree of

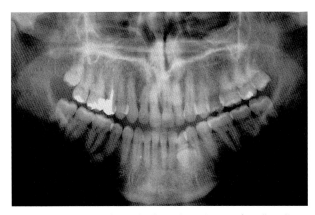

Fig. 14.18 Panoramic radiograph of a patient 12 years after all appliances were removed. A newly developing supernumerary premolar tooth is noted.

expression of the various characteristic features of the disease, together with many other sporadic phenomena that have been reported to occur with the condition. Erupted permanent incisors are sometimes seen and these may eliminate the need for the 'incisor-erupting' initial heavy archwire in one or both arches. Instead, the treatment may begin with the levelling phase using the modified Johnson twin-wire arch.

At the completion of all the treatment and the placement of retainers, new radiographs should be taken to check for the development of recurrent supernumerary teeth, which sometimes occurs in the 14–16-year-old patient (Figure 14.18). At this point, any decision regarding the extraction of such teeth is strictly a surgical one [36, 63]. When newly developing teeth are identified at this late stage, their extraction is no longer an orthodontic decision, since full alignment of the teeth has been achieved and suitable retention will prevent any adverse effects. Such factors as location and accessibility, incipient resorption of neighbouring erupted teeth and cysts will all influence the surgeon regarding the necessity and timing of their extraction.

The very young and the adult cases

While the parents of most cleidocranial dysplasia patients will seek advice and treatment from about 9 10 years of age, at a time when the initiation of the first phase of treatment is opportune, some children may present earlier. If the diagnosis has been made at or shortly after birth, the parents may wish to consult regarding prevention or simply to know what the future holds for their young child. With the accessibility of the internet, its ease of use and the body of information it can make available to the general public, they will often attend the first consultation visit already well informed of the implications of the condition and, as with

many of the rarer conditions and syndromes, may sometimes possess more knowledge of the condition than the first practitioner they consult.

Among the initial complaints about which the parents are concerned is the steadily progressive problem of appearance and its psychological effect on this child, whose intelligence is normal. The child is generally shorter, he may have a class 3 skeletal appearance and there is the continued presence of deciduous teeth, when the other children in the class at school are already erupting permanent teeth. However, a panoramic radiograph is likely to show that the 7- or 8-year-old child has a dental age which is three years younger and, as such, is too young for the first phase of treatment to begin.

For this patient, a pre-first phase treatment of an existing class 3 skeletal relation may be usefully initiated in the form of maxillary protraction, with or without preparatory rapid maxillary expansion (Figure 14.19). Since no permanent teeth are present at this dental age, it is probably most practical to use an acrylic cap splint cemented with glass ionomer cement or acid-etch and composite bonding agent.

A standard or, preferably, custom-made face mask is then provided, with elastic force to a hook cured into the buccal aspects on each side of the acrylic splint. Compliance is always a problem with these appliances, but it will usually be at its best in patients of this age and younger, and whose parents are understanding, encouraging and proactive. At the completion of this stage, the patient is reassessed regarding the timing for the first phase.

At the other end of the scale (Figure 14.20), we may see an untreated patient in his mid- to late teens, with a virtually complete deciduous dentition, few erupted permanent teeth and a strong class 3 skeletal relationship that clearly demands surgical reduction. Despite the obvious facial disfigurement, the presenting symptom may often be pain or swelling from the deciduous teeth that will likely have suffered a considerable degree of morbidity. This may have triggered the patient's initial request for emergency treatment for their immediate distress. Even following only a cursory clinical and radiological examination, the concerned dentist will then (hopefully) refer the patient on for orthodontic and surgical evaluation and treatment.

As a general rule, bringing the teeth into the dental arches and aligning them will precede the skeletal correction. In the child patient, the Jerusalem approach describes the division into two separate phases of treatment because of developmental determinants of the dentition. With the young adult patient this is irrelevant, since all the teeth will have passed the stage of development that is appropriate for eruption in the unaffected individual.

Nevertheless, erupting all the teeth in both arches in a single biomechanical treatment phase has its limitations.

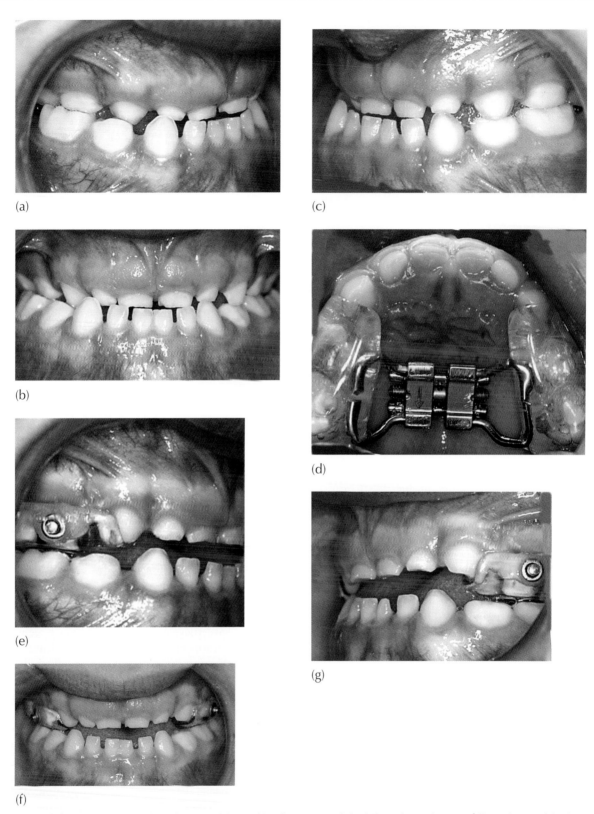

Fig. 14.19 Pre-first phase protraction orthopaedics at age 8.5 years (dental age 5.6 years). (a–c) The occlusion showing a full complement of deciduous teeth in a strong class 3 relationship. (d–g) An acrylic cap splint carrying a Hyrax® screw is bonded to the maxillary posterior teeth. This carries a small button cured into the acrylic, one each side for elastic attachment to the face mask (not shown). (h–k) The occlusal relationships have been slightly over-corrected. (l, m) The profile before and after protraction.

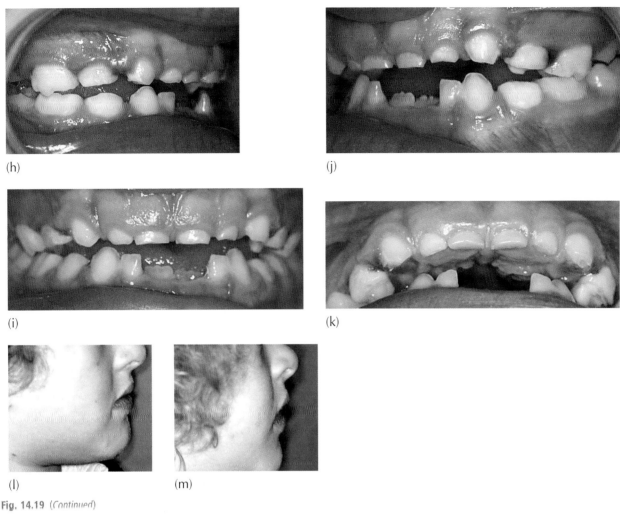

(h)

(j)

(i)

(k)

(l)

(m)

Fig. 14.19 *(Continued)*

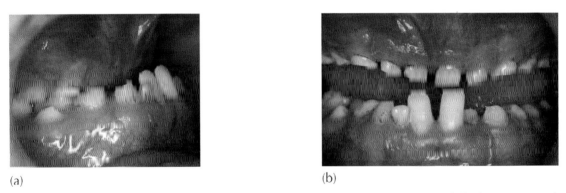

(a)

(b)

Fig. 14.20 Late treatment in a 15.7-year-old male. (a–e) Initial clinical views. The patient never brought his teeth into occlusion due to severe overclosure – he chewed on his tongue. In the lateral views in forced occlusion, the anterior maxilla is totally enveloped by the anterior portion of the mandible. Apart from the permanent molars, only two lower incisors of the permanent dentition have erupted. (f, g) Cephalogram and panoramic views show the extreme skeletal class 3 relation and the fully developed, but unerupted permanent dentition. Note supernumerary teeth in the maxillary incisor and canine regions, but congenitally absent mandibular third molars. (h) Intra-oral occlusal views of the appliances immediately prior to exposure surgery. (i, j) All the unerupted teeth are exposed only to the most minimal degree to permit bonding of the small eyelet attachments (not shown). (k) The post-surgical panoramic radiograph after extraction of 18 deciduous teeth, five supernumerary teeth, and the exposure of 16 unerupted permanent teeth with attachment bonding to 15 of them. The right mandibular second premolar was left uncovered and unattached, while the remainder of the teeth were covered fully by the surgical flaps in a closed exposure technique. Note that bone may be seen to surround and cover most of the unerupted teeth, except those that are more superficially placed or had been adjacent to the areas where the supernumerary teeth were removed. Ligation to the vertically activated labial arches was effected immediately. (l–n) Two years later, all teeth had been erupted into the class 3 relation and treatment was discontinued. (o–q) The panoramic, profile and cephalometric views prior to appliance removal. (r–u) Two years later, maxillary (Le Fort 1) advancement surgery was performed. (Courtesy of Dr E. Regev and Dr D. Harary.) The outcome may be seen in these intra-oral clinical and panoramic radiographic films, profile photograph and cephalogram. Note the excellent appearance of the teeth and supporting tissues – hardly what may be expected in a case with previously multiple impacted teeth!

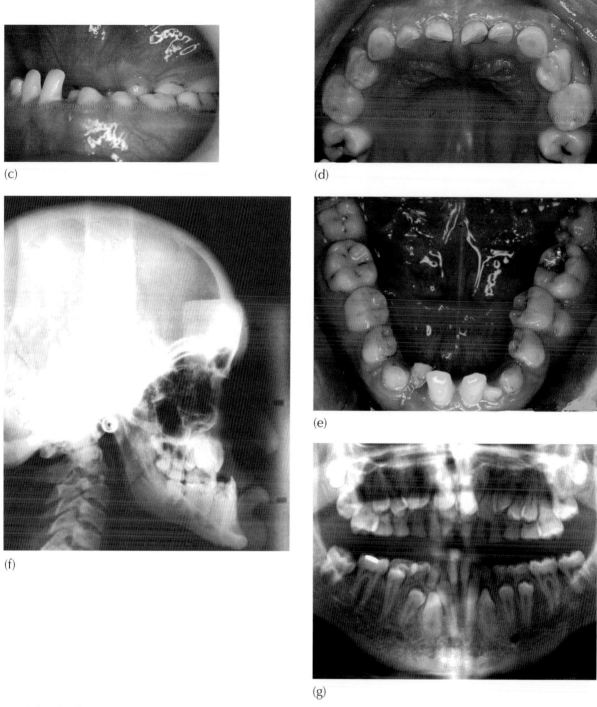

(c)

(d)

(e)

(f)

(g)

Fig. 14.20 (*Continued*)

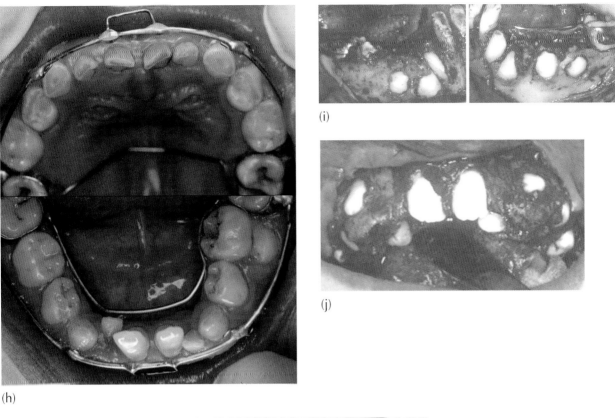

(h)

(i)

(j)

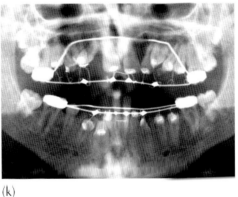

(k)

Fig. 14.20 (continued)

Accordingly, there is merit in following the same overall two-stage procedure as recommended for the younger person, with the exception that the phases will follow on, one after the other. Thus, for practical/operative rather than biological/developmental reasons, it remains convenient and makes clinical sense to divide the objectives of treatment of the multiple impactions into anterior and posterior areas of activity.

The fact of closure of the root apices of these teeth does not appear to impede their eruption once their own innate, minimally existing eruption potential is augmented with extraneous forces provided by the orthodontic appliances.

Appliance-driven eruption of the teeth during the active phase of their root development is generally accompanied by an excellent regenerative alveolar bone reaction. This contributes much to an increased height of the alveolar ridges, to a deepening of the labial and lingual sulci and to an enhancement of the reduced height of the lower third of the face, as has already been pointed out. When the patient reports for treatment only after his growth is completed, and while resolution of the impactions may generally be completely successful, there will be a much less favourable contribution of the alveolar bone to the overall result.

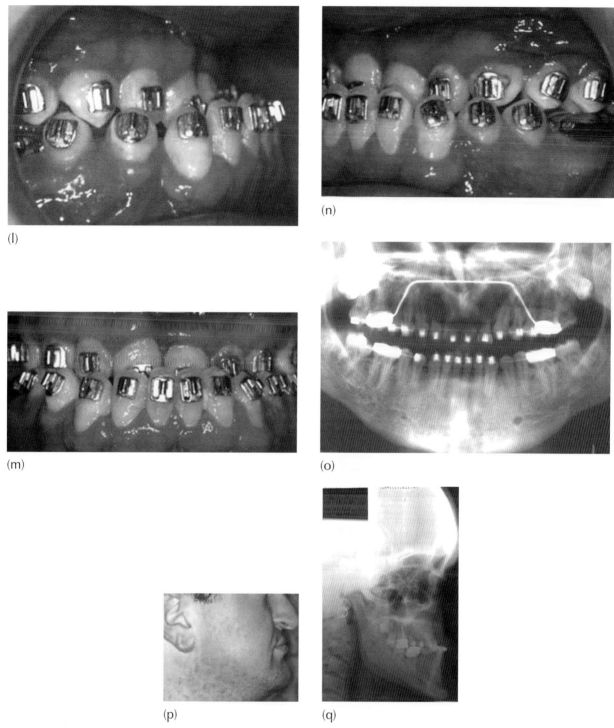

(l)

(n)

(m)

(o)

(p)

(q)

Fig. 14.20 (*Continued*)

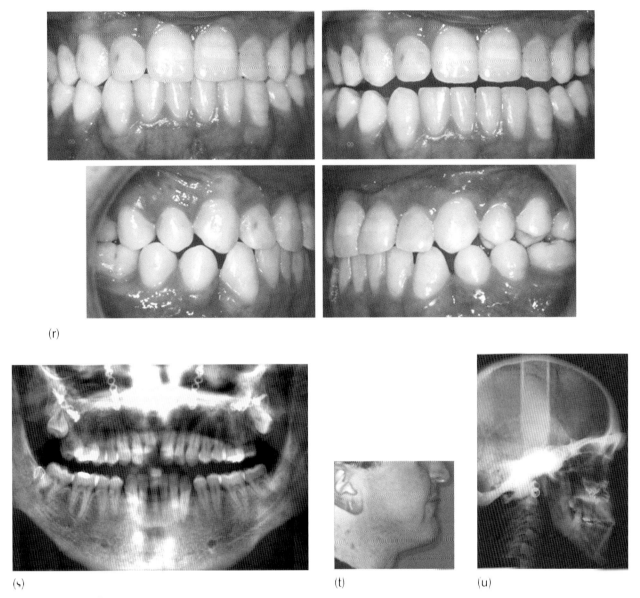

(r)

(s)   (t)   (u)

Fig. 14.20 (Continued)

For the most part, the overall active treatment to resolve the multiple tooth impaction and achieve good alignment should take between three and five years, depending on its complexity and, generally, excluding the interlude that may occur between first and second phases.

At the conclusion of the orthodontic treatment, the teeth will be fully erupted and aligned, with tip, rotation, uprighting and torque all corrected. At this stage, a renewed look must be taken at the jaw relations. New records need to be assessed and the orthosurgical aspects of the condition evaluated, as with any other non-cleidocranial dysplasia orthosurgical case. By and large, it will be the maxilla that needs to be brought forward, but mandibular surgery is sometimes needed either because of occasional true man-dibular prognathism or in order to create a harmonious relationship of the jaws, in the presence of other abnormal skeletal parameters.

Considerable jaw growth occurs during and around the adolescent growth spurt, which sometimes alters the jaw relation into a more class 3 relation, and this generally occurs during the mid-treatment second phase period. In this situation, the orthodontist should continue on the steady path of eruption resolution until all the teeth are in the arch, with the exception of the third molars. Only at this stage should the case be re-evaluated for orthosurgery. In those cases where the growth has brought about a marked skeletal class 3 relation, the changes that occur in the jaw relation are usually far too major for

a protraction headgear or a chin cap to ameliorate this developing skeletal pattern, even with excellent compliance. For these patients, surgery will be the only satisfactory answer.

References

1. Kulliala E, Taskinen PJ. Cleidocranial dystosis: report of 6 typical cases and 1 atypical case. J Oral Surg 1962; 15: 808–822.
2. Bixler D. Hereditable disorders affecting cementum and the periodontal structures. In Stewart RE, Prescott GH, eds. Oral Facial Genetics. St Louis, MO: Mosby, 1976: 282–284.
3. Cohen MM Jr. Dysmorphic syndromes with craniofacial manifestations. In Stewart RE, Prescott GH, eds. Oral Facial Genetics. St Louis, MO: Mosby, 1976: 566–567.
4. Zegarelli EV, Kutscher AH, Hyman GA. Diagnosis of Diseases of the Mouth and Jaws. Philadelphia: Lea & Febiger, 1978: 137.
5. Shafer WG, Hine MK, Levy BM. A Textbook of Oral Pathology, 4th edn. Philadelphia: Saunders, 1983: 678–680.
6. Tachdjian MO. Pediatric Orthopedics, 2nd edn. Philadelphia: Saunders, 1990: 840–844.
7. Gorlin RJ, Cohen MM Jr, Levin LS. Syndromes of the Head and Neck, 3rd edn. New York: Oxford University Press, 1990: 249–253.
8. Stewart RE, Prescott GH, eds. Oral Facial Genetics. St Louis, MO: Mosby, 1976.
9. Yamamoto H, Sakae T, Davies JE. Cleidocranial dysplasia: a light microscope, electron microscope and crystallographic study. Oral Surg Oral Med Oral Pathol 1989; 68: 195–200.
10. Mendoza-Londono R, Lee B, Cleidocranial Dysostosis. NCBI Resources, Gene Reviews, Pagon RA, Bird TC, Dolan CR, et al., editors, 1993.
11. Cooper SC, Flaitz CM, Johnston DA, et al. A natural history of cleidocranial dysplasia. Am J Med Genet. 2001; 104: 1–6.
12. Ishii K, Nielsen IL, Vargervik K. Characteristics of jaw growth in cleidocranial dysplasia. Cleft Palate Craniofac J 1998; 35: 161–166.
13. Hu JCC, Nurko C, Sun X et al. Characteristics of cementum in cleidocranial dysplasia. J Hard Tissue Biol 2002; 11: 9–15.
14. Smylski PT, Woodside DG, Harnett BE. Surgical and orthodontic treatment of cleidocranial dysostosis. Int J Oral Surg 1974; 3: 380–385.
15. Winther JE, Khan MW. Cleidocranial dysostosis: report of 4 cases Dent Pract 1972; 22: 215–219.
16. Kelly L, Nakamoto RY. Cleidocranial dysostosis – a prosthodontic problem. J Pros Dent 1974; 31: 518–526.
17. Frommer HH, Lapeyrolerie FM. Two case reports of cleidocranial dysostosis. New York J Dent 1964; 34: 103–107.
18. Hitchin AD, Fairley JM. Dental management in cleidocranial dysostosis. Br J Oral Surg 1974; 12: 46–55.
19. Weintraub GS, Yasilove IL. Prosthodontic therapy for cleidocranial dysostosis. Report of a case. J Am Dent Assoc 1978; 96: 301–305.
20. Probster L, Bachmann R, Weber H. Custom-made resin-bonded attachments supporting a removable partial denture using the spark erosion technique: a case report. Quintessence Int 1991; 22: 349–354.
21. Muller EE. Transplantation of teeth in cleidocranial dysostosis. In Husted E, Hjorting-Hansen E, eds. Oral Surgery: Transactions of the 2nd Congress of the International Association of Oral Surgeons. Copenhagen: Munksgaard, 1967: 375–379.
22. Oksala E, Fagerstrom G. A two-stage sutotransplantation of 14 teeth in a patient with cleidocranial dysostosis. Suom Hammaslaak Toim 1971; 67: 333–338.
23. Elomaa E, Elomaa M. Orthodontic treatment of a case of cleidocranial dysostosis. Suom Hammaslaak Toim 1967; 67: 139–151.
24. Hall RK, Hyland AL. Combined surgical and orthodontic management of the oral abnormalities in children with cleidocranial dysplasia. Int J Oral Surg 1978; 7: 267–273.
25. Frame K, Evans RIW. Progressive development of supernumerary teeth in cleidocranial dysplasia. Br J Orthod 1989; 16: 103–106.
26. Trimble LD, West RA, McNeill RW. Cleidocranial dysplasia: comprehensive treatment of the dentofacial abnormalities. J Am Dent Assoc 1982; 105: 661–666.
27. Davies TM, Lewis DH, Gillbe GV. The surgical and orthodontic management of unerupted teeth in cleidocranial dysostosis. Br J Orthod 1987; 14: 43–47.
28. Richardson A, Swinson T. Combined orthodontic and surgical approach to cleido-cranial dysostosis. Trans Eur Orthod Soc 1987; 63: 23 [abstract].
29. Behlfelt K. Cleido-cranial dysplasia: diagnosis and treatment concept. Trans Eur Orthod Soc 1987; 63: 25 [abstract].
30. Miller R, Sakamoto E, Zell A et al. Cleidocranial dysostosis. A multidisciplinary approach to treatment. J Am Dent Assoc 1978; 96: 296–300.
31. Becker A, Shochat S. Submergence of a deciduous tooth, its ramifications on the dentition and treatment of the resulting malocclusion. Am J Orthod 1982; 81: 240–244.
32. Kohavi D, Becker A, Zilberman Y. Surgical exposure, orthodontic movement and final tooth position as factors in periodontal breakdown of treated palatally impacted canines. Am J Orthod 1984; 85: 72–77.
33. Howe GL. Minor Oral Surgery, 2nd edn. Bristol: Wright, 1971: 135–137.
34. Becker A, Lustmann J, Shteyer A. Cleidocranial dysplasia: part 1 – general principles of the orthodontic and surgical treatment modality. Am J Orthod Dentofacial Orthop 1997; 111: 28–33.
35. Becker A, Shteyer A, Bimstein E, Lustmann J. Cleidocranial dysplasia: part 2 – a treatment protocol for the orthodontic and surgical modality. Am J Orthod Dentofacial Orthop 1997; 111: 173–183.
36. Becker A, Shteyer A. A surgical and orthodontic approach to the dentition in cleidocranial dysostosis. Trans Eur Orthod Soc 1987; 63: 121 [abstract].
37. Seow WK, Hertzberg J. Dental development and molar root length in children with cleidocranial dysplasia. Pediatr Dent 1995; 17: 101–105.
38. Becker A, Zilberman Y. The palatally impacted canine: a new approach to its treatment. Am J Orthod 1978; 74: 422–429.
39. Becker A, Kohavi D, Zilberman Y. Periodontal status following the alignment of palatally impacted canine teeth. Am J Orthod 1983; 84: 332–336.
40. Vermette ME, Kokich VG, Kennedy DB. Uncovering labially impacted teeth: apically repositioned flap and closed-eruption techniques. Angle Orthod 1995; 65: 23–32.
41. Gron A. Prediction of tooth emergence. J Dent Res 1962; 41: 573–585.
42. Ricketts RM. Perspectives in the clinical application of cephalometrics. The first fifty years. Angle Orthod 1981; 51; 115–150.
43. Zachrisson BU. Premolar extraction and smile esthetics. Am J Orthod Dentofacial Orthop 2003; 124: 11A–12A.
44. Zachrisson BU. Incisal edge recontouring in orthodontic finishing. World J Orthod 2005; 6: 398–405.
45. Zachrisson BU. Clinical experience with direct-bonded orthodontic retainers. Am J Orthod 1977; 71: 440–448.
46. Becker A, Goultschin J. The multistrand retainer and splint. Am J Orthod 1984; 81: 470–474.
47. Becker A. Periodontal splinting with multistrand wire following orthodontic realignment of migrated teeth: report of 38 cases. Int J Adult Orthod Orthogn Surg 1987; 2: 99–109.
48. Becker A, Chaushu S. Non-invasive periodontal splinting with multistrand wire following the orthodontic realignment of periodontally migrated teeth. Orthodontics, 2004; 1: 159–167.
49. Johnson JE. A new orthodontic mechanism: the twin wire alignment appliance. Int J Orthod 1934; 20: 946–963.
50. Shepard ES. Technique and Treatment with the Twin-Wire Appliance. St Louis, MO: Mosby, 1961.
51. Brin I, Ben Bassat Y, Fuks A, Zilberman Y. Trauma to the primary incisors and its effect on the permanent incisors. Pediatr Dent 1984; 6: 78–82.
52. Ben Bassat Y, Brin I, Fuks A, Zilberman Y. Effect of trauma to the primary incisors on permanent successors in different developmental stages. Pediatr Dent 1985; 7: 37–40.
53. Zilberman Y, Fuks A, Ben Bassat Y et al. Effect of trauma to primary incisors on root development of their permanent successors. Pediatr Dent 1986; 8: 289–293.
54. Becker A. Early treatment for impacted maxillary incisors. Am J Orthod Dentofacial Orthop 2002; 121: 586–587.

55. Becker A, Brin I, Ben-Bassat Y, Zilberman Y, Chaushu S. Periodontal status following surgical-orthodontic alignment of impacted maxillary incisors by a closed eruption technique. Am J Orthod Dentofacial Orthop 2002; 122: 9–14.

56. Chaushu S, Brin I, Ben-Bassat Y, Zilberman Y, Becker A. Periodontal status following surgical-orthodontic alignment of impacted central incisors by an open-eruption technique. Eur J Orthod 2003; 25: 579–584.

57. Becker A. An interview with Adrian Becker. World J Orthod 2004; 5: 277–282.

58. Lappin MM. Practical management of the impacted maxillary canine. Am J Orthod 1951; 37: 769–778.

59. Lewis PD. Preorthodontic surgery in the treatment of impacted canines. Am J Orthod 1971; 60: 382–397.

60. von der Heydt K. The surgical uncovering and orthodontic positioning of unerupted maxillary canines. Am J Orthod 1975; 68: 256–276.

61. Kokich VG. Surgical and orthodontic management of impacted maxillary canines. Am J Orthod Dentofacial Orthop 2004; 126: 278–283.

62. Laskin D. Oral and Maxillofacial Surgery, Vol 2. St Louis, MO: Mosby, 1985: 44–47.

63. Becker A, Bimstein E, Shteyer A. Interdisciplinary treatment of multiple unerupted supernumerary teeth. Am J Orthod 1982; 81: 417–422.

15

Extreme Impactions, Unusual Phenomena and Difficult Decisions

Orthodontic Treatment of Impacted Teeth, Third Edition. Adrian Becker.
© 2012 Adrian Becker. Published 2012 by Blackwell Publishing Ltd.

In this chapter, the intention is to present a few cases featuring impacted teeth in a variety of difficult, interdisciplinary or extreme situations. For these cases, there are no hard and-fast rules, the literature offers the practitioner little assistance and, despite all the advice that may be sought and received, the orthodontist largely remains alone in a diagnostic or treatment planning wilderness without a compass. Options may be few, but they also may be so many that each potential scenario must be acted out in the fertile imagination of the orthodontist to find the direction that will offer relative success, cause the least collateral damage and carry with it the least risk of failure. Some may require the active participation of specialists in fields other than orthodontics and oral surgery, particularly the paediatric dentist, the endodontist and the periodontist. By their very nature, these cases are frequently one of a kind and thus determining the diagnosis, the treatment options, chosen treatment approach and prognosis may never be evidence-based, but largely dependent on the logic acquired from the collected clinical experience of the operators.

This is both the strength and the weakness of published single case reports. There is often much to learn from them individually but, whatever that may be, it cannot be used to draw conclusions in relation to other apparently or arguably similar cases in the future. Without doubt legitimate criticism may be levelled by any dentist regarding the treatment decisions that were made in the following individual presentations. Moreover, an alternative approach to the same case might be preferred, a choice that will be influenced or biased by that orthodontist's positive or negative experience with the same or contrary modalities of treatment.

The clinical cases shown here are mostly finished cases with long-term post-treatment follow-up and they are offered here specifically as they relate to impacted teeth. Others are in the final stages of their overall active treatment as this book goes to print but, in all cases, the principal issue under discussion (i.e. resolution of the impaction) will have been fully and successfully addressed.

Case 1: Monster tooth, supernumerary tooth, impacted central incisor and the maxillary midline

The existence of an unusually large maxillary central incisor with talon cusp or cusps, otherwise known as dens evaginatus, is rare and usually published in the literature as a single case report [1–3]. Since it takes up more than the space of a normal central incisor and its location is at the front of the mouth, the dens evaginatus is unsightly and disfiguring. To reduce it in size with the view to reshaping it to more normal proportions is difficult or impossible, because of a large pulp chamber and/or a very broad cross-section in the cervical region. In such circumstances, it is often extracted and replaced prosthetically, or it may be enlarged and reshaped to make it resemble two teeth, a central and lateral incisor, while sacrificing the adjacent normal and healthy lateral incisor.

In the present case of a 7-year-old healthy girl (Figure 15.1a-e), the right central incisor was much enlarged with a Y-shaped crown cross-section, due to the talon cusp and a much enlarged pulp chamber and root cross-section at the CEJ. The dentition comprised the four erupted first molars, mandibular central incisors and rotated erupting mandibular lateral incisors, associated with mild dental crowding. On the maxillary left side, there was an atypical incisiform 'central incisor', which was assumed to be a supernumerary tooth. The dens evaginatus occupied most of the space for both central and lateral incisor of that side. The remainder of the dentition comprised healthy and restored deciduous teeth in normal intermaxillary occlusal relations.

Periapical and panoramic radiographs (Figure 15.1f, g) revealed the unerupted teeth superimposed on one another with inadequate differentiation for appropriate diagnosis. Accordingly a CBCT was performed (Figure 15.1h, i) from which it was concluded that the unerupted teeth in the maxillary incisor region included two normal lateral incisors and an unerupted, normally shaped, left maxillary central incisor. This lent credence to the assertion that the erupted 'central incisor' was a supernumerary tooth and that the dens evaginatus represented a fusion between a right central incisor and an additional supernumerary tooth.

The shape and size of the dens evaginatus, together with its pulp chamber dimension and cross-sectional breadth at the CEJ, determined that the limited alteration of crown shape possible would not provide a satisfactory answer to the overall space problem, nor would it contribute to improving the patient's appearance. Thus, its extraction was unavoidable. This created the unbalanced situation in which an incisor would be missing on the right side, while an extra incisor was present on the other.

The treatment proposed, therefore, was to move the erupted and presumed supernumerary tooth across from the left to fill the place of the extracted dens evaginatus on the right side and thus to permit the eruption of the impacted normal left central incisor tooth into its designated place. Once this was completed, the deciduous canines in both jaws would be extracted to provide space temporarily to permit the alignment of the four incisors in both jaws.

A modified form of the 2 × 4 appliance (in fact, a 2 × 1 appliance) was placed, with a palatal arch soldered to two molar bands and the typical modified Johnson twin arch set-up described in Chapter 5, with a Tip-Edge® bracket placed on the single erupted permanent supernumerary incisor and an archwire of 0.016 in round steel wire inserted into the buccal arms of the composite archwire which, in turn, would be inserted into the molar tubes (Figure 15.1j–m).

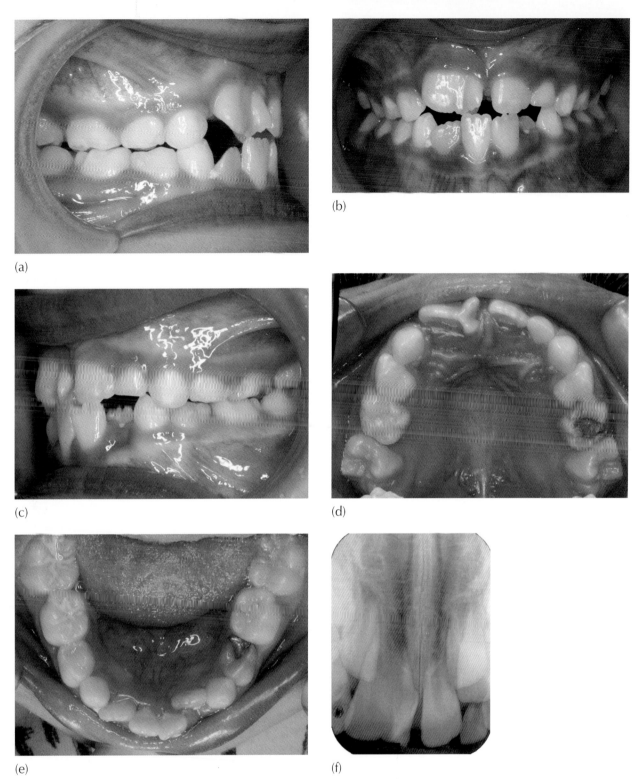

Fig. 15.1 (a–e) The clinical intra-oral views. (f, g) Periapical and panoramic radiographic views. (h) CBCT axial cut through the roots of the erupted teeth. #D.E is the dens evaginatus and #S the erupted supernumerary tooth. (i) Right and left CBCT transaxial views. (j–m) The modified Johnson 2 × 4 (2 × 1) appliance in place. (n) Following the extraction of the dens evaginatus, a coil spring is compressed between the soldered stop on the mesial end of the left buccal tube arm and the bracket on the supernumerary tooth. (o) With the TipEdge bracket, tipping of the left supernumerary across the facial midline takes only six weeks. (p) Four months later, the left central incisor has erupted and is incorporated in the appliance, with a coil spring moving it to the midline, while the supernumerary tooth is being uprighted in the right central incisor position. (q) The condition prior to lateral incisor eruption. (r) The lateral incisors erupt spontaneously and are incorporated in the appliance. (s–u) The completion of phase 1 treatment. (v, w) Panoramic and periapical views clearly show the midline suture on the right side of the supernumerary tooth, together with an enlarged right canine dental follicle/early dentigerous cyst. (x) The midline raphe is seen to deviate sharply to the right in this pre-phase 2 occlusal view of the palate, as indicated by the arrows. The incisive papilla has also been displaced. The dental follicle of the unerupted right canine has enlarged into a small dentigerous cyst and can be seen to have expanded the alveolar ridge. (y) The pre-phase 2 view of the patient from the front shows a pleasing appearance, a normal dental smile line and an apparently natural dental midline.

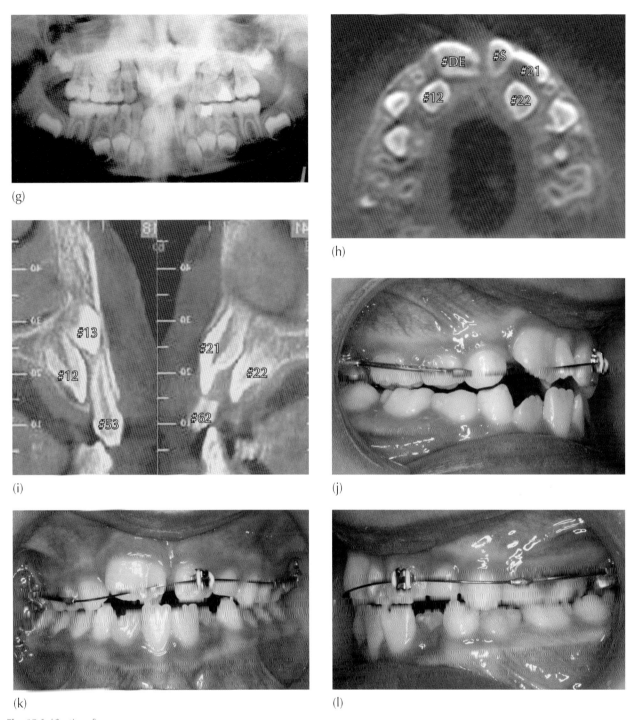

Fig. 15.1 *(Continued)*

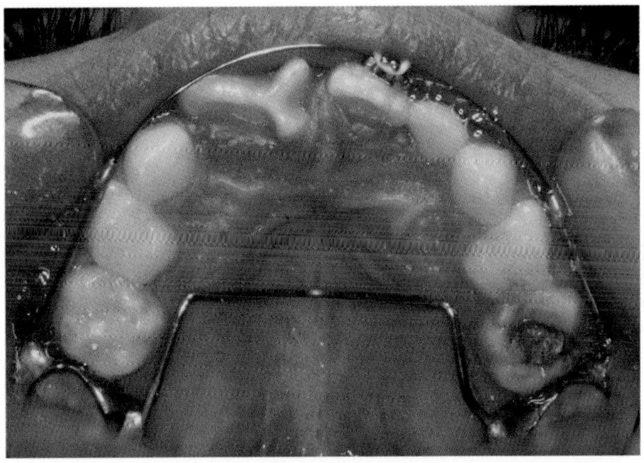

(m)

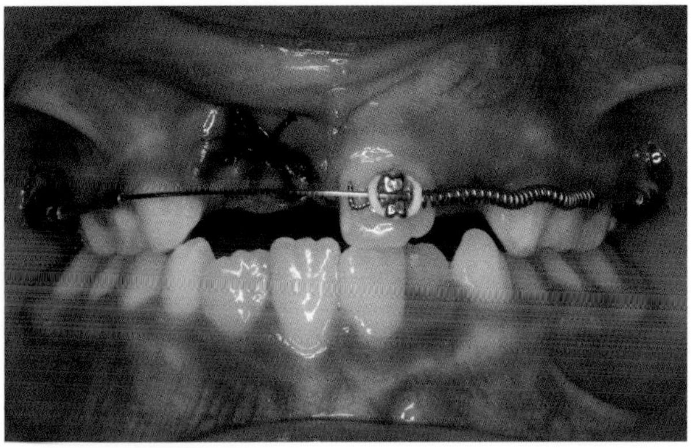

(n)

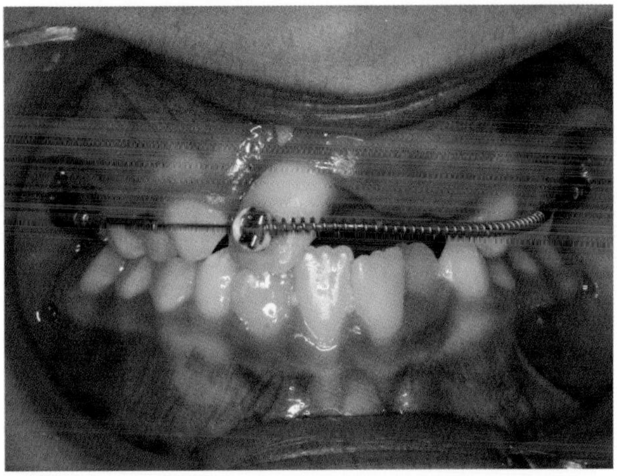

(o)

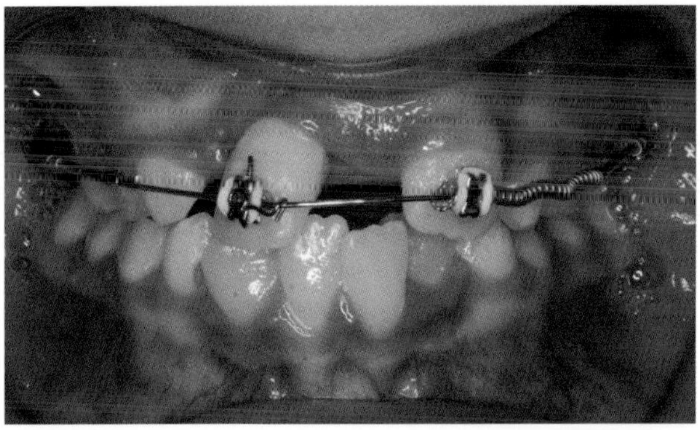

(p)

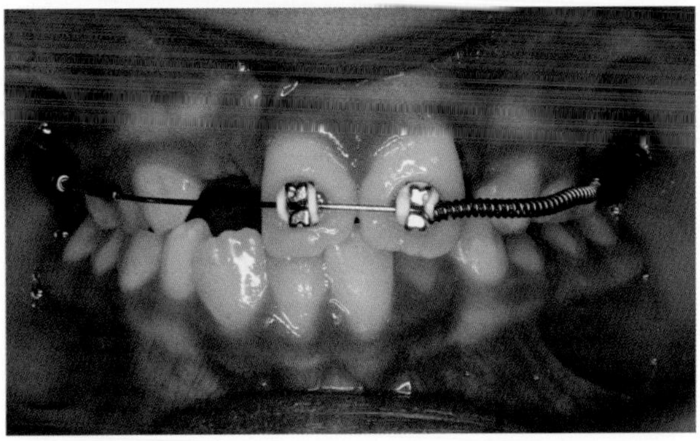

(q)

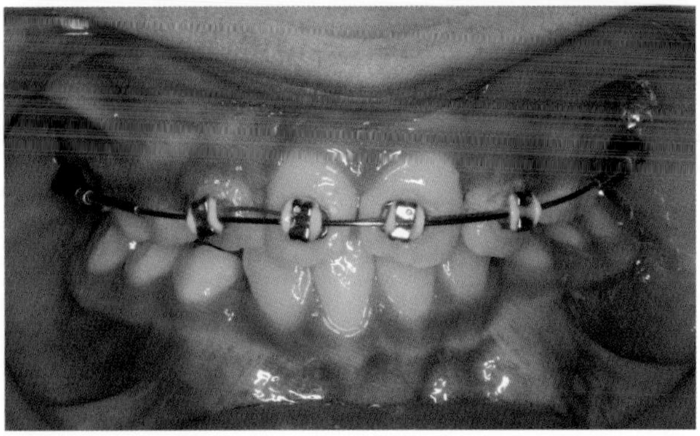

(r)

Fig. 15.1 (Continued)

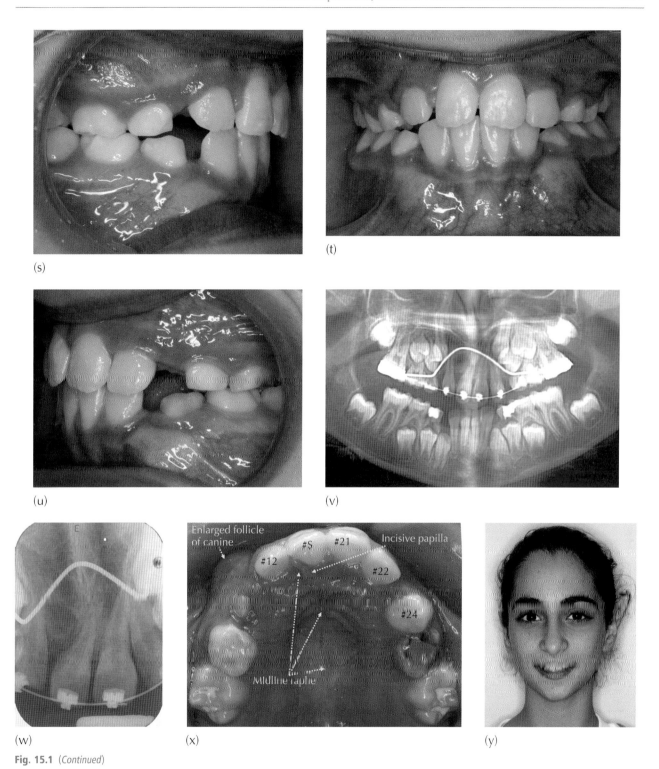

Fig. 15.1 (*Continued*)

The orientation of the soldered molar tubes and these buccal arms was tipped slightly downwards on each side to encourage incisor eruption and bite closure and a welded stop was placed on the buccal arms of each side mesial to the molar tubes to maintain the initial arch length and to prevent unwanted sliding of the buccal arms through the tubes.

The abnormal right dens evaginus central incisor was extracted at the next visit and, on the same day, a coil spring was compressed between the soldered stop at the mesial end of the left buccal arm and the bracket on the left erupted incisiform supernumerary tooth (Figure 15.1n). Six weeks later, the tooth was seen to have tipped across the midline and into proximal contact with the deciduous canine of the opposite side (Figure 15.1o). The impacted left central incisor erupted rapidly and autonomously five months after placement of the coil spring, driven by the sudden provision of space in the arch. This tooth was bracketed and a coil spring again placed on the arch to move it to the midline. An auxiliary spring was placed in the vertical slot of the Tip-Edge® bracket of the supernumerary tooth to upright its root across the facial midline (Figure 15.1p, q).

During the subsequent 12 months of treatment, the four deciduous canines were extracted and the lateral incisors erupted without further assistance. Brackets were placed on them and the teeth were aligned and moved towards and slightly across the midline. Further root uprighting was then initiated (Figure 15.1r). Phase 1 treatment was completed seven months later, with satisfactory alignment of the teeth (Figure 15.1s–u). The root configuration of the substitute 'right central incisor', formerly the supernumerary tooth (Figure 15.1v, w), strengthens the assumption that it was indeed a supernumerary tooth.

It should be clearly understood that moving a tooth across the left side of the maxilla to the right, across the midline, does not infer that the tooth actually traverses the midline palatal suture. What happens is that the bone on each side of the suture is remodelled and moves together with the tooth, so that the suture remains on the mesial side of the left incisor, as can be seen on the radiographs and the radiograph (Figure 15.1v–x).

The total treatment time was 26 months, a removable maxillary Hawley retainer is being worn nightly and phase 2 treatment will be considered in the full permanent dentition.

Case 2: Bilaterally impacted maxillary canines in a patient suffering with aggressive juvenile periodontitis

The patient was a 15-year-old girl, who had been referred to the author for the treatment of her bilaterally impacted maxillary canine teeth. No mention was made in the referral letter to the effect that the patient suffered from perio-

dontal disease and there were no symptoms that might have indicated this.

An examination of the initial panoramic view (Figure 15.2a) indicated a routine palatal impaction of both maxillary canines, both classified as group 1 type (see Chapter 6). Clinically, there was a close to normal occlusion with excellent alignment in both jaws, good intercuspation and a normal overbite and overjet.

The complicating factor here was that the patient suffered from aggressive juvenile periodontitis, with the loss of much alveolar bone in the molar and incisor regions of both jaws. The patient returned 18 months later having completed her periodontal treatment, which included the grafting and integration of bovine bone in the more severely affected locations mesial to the first molars (periodontic treatment by Professor Ayala Stabholz). These may be clearly seen as small radio-opacities on the new film after a long period of post-treatment follow-up (Figure 15.2b). Note also the spontaneous closure of the spaces between the second premolars and the first molars which had been present before treatment began.

Pre-surgical orthodontic preparation for this almost normal occlusion (Figure 15.2c–e) involved levelling, aligning and space-opening and lasted just four months, before closed exposures were performed on both canines (Figure 15.2f–h). A measured length of steel tube was threaded over the base arch, with the purpose of holding the distance between the premolar and lateral incisor brackets. The right side is shown here, to illustrate the use of an elastic e-link stretched between the occlusally inserted power pins into the vertical slots of the Tip-Edge® brackets of the lateral incisor and first premolar, which was then raised and engaged in the pigtail ligature close to the sutured flap. In this way, renewable vertical traction was applied to the impacted teeth to bring about their eruption (Figure 15.2i–k). The final result, shown here eight years after completion of the orthodontic treatment (Figure 15.2l–n), shows excellent alignment and inter-arch relations. The clinical appearance of the teeth and the new periapical radiographs offered no signs or clues that would indicate that the canines had previously been palatally impacted (Figure 15.2o–t). While still discernible in the radiographs, the bone grafts have become progressively more integrated into the trabecular picture. The orthodontic treatment time for this case was 18 months and the patient is followed up by the periodontist on a regular basis.

Case 3: Labially impacted maxillary canine at the level of the nasal floor

The patient, a girl aged 11 years in the late mixed dentition stage, was referred to the author with accompanying radiographs from which bilateral labially impacted upper canines had been diagnosed, one of which was extremely high in the maxilla.

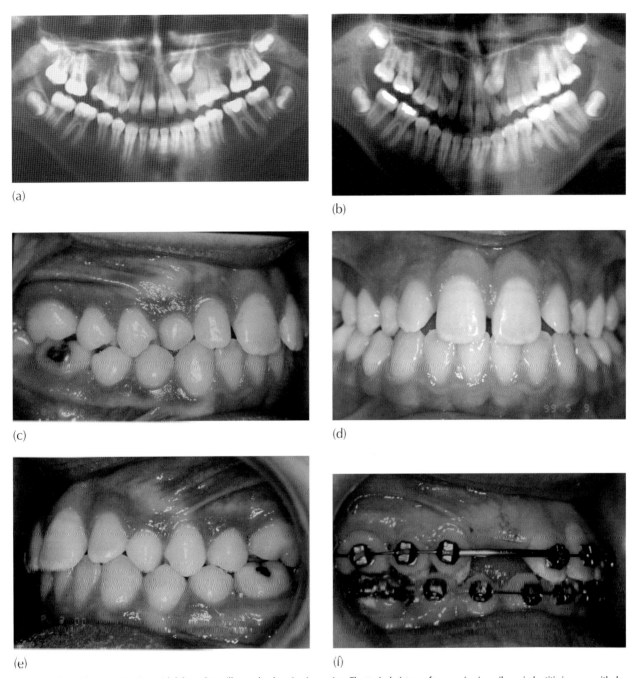

Fig. 15.2 (a) Initial panoramic view with bilateral maxillary palatal canine impaction. The typical picture of aggressive juvenile periodontitis is seen, with deep vertical periodontal defects and severe bone loss in the four molar and maxillary incisor regions.

(b) A new panoramic film taken after the successful completion of periodontal treatment shows the radio-opaque areas where bovine bone was used to regenerate bone in the defective areas. (Periodontics by Professor Ayala Stabholz.)

(c–e) The post-periodontic intra-oral appearance of the malocclusion.

(f–h) After extraction of the maxillary deciduous canine, orthodontic alignment, levelling and space opening, the canine spaces are held open with the use of cut lengths of stainless steel tube of broad gauge (0.036 in internal diameter) threaded onto the main arch.

(i–k) At surgery, the orthodontist has bonded an eyelet attachment to the exposed canine and traction is applied using an elastic chain stretched between the inverted power pins on the adjacent teeth. This is raised in its middle portion to engage the twisted pigtail ligature that has been turned into a small hook close to the sutured tissues. (Surgery by Dr H.P. Samen.)

(l–n) At 8 years following the completion of orthodontic treatment, there is an excellent alignment and appearance of the gingival tissue around the previously impacted canines. There are no signs that these had been impacted.

(o–t) Periapical views of the areas that were most seriously affected by the condition, seen eight years post-treatment.

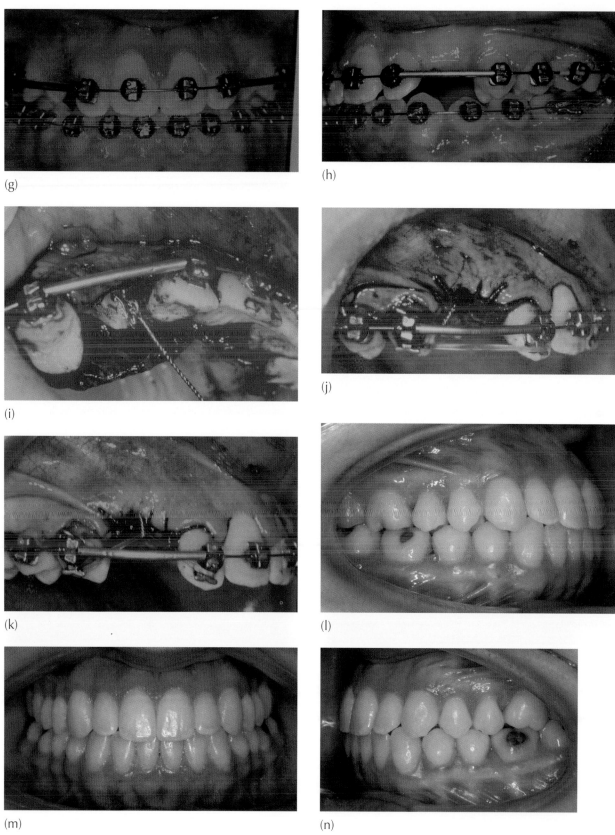

(g)

(h)

(i)

(j)

(k)

(l)

(m)

(n)

Fig. 15.2 (Continued)

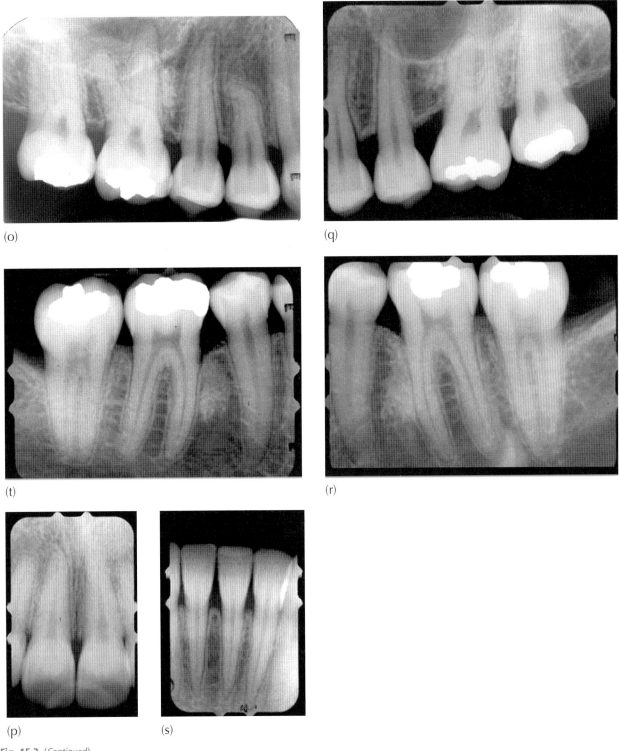

(o)

(q)

(t)

(r)

(p)

(s)

Fig. 15.2 (*Continued*)

The existing occlusion was almost ideal, with good general dental alignment, a class 1 occlusal relationship of the molars and normal incisor overbite and overjet (Figure 15.3a-c). The panoramic radiograph (Figure 15.3d) showed the presence of a lingual arch space maintainer, which had been placed following the extraction of deciduous teeth. It showed the presence of all the permanent teeth, together with the about-to-be-shed maxillary left first and second deciduous molars and the mandibular left second deciduous molar. The two maxillary deciduous canines were also present with virtually complete and unresorbed roots. The maxillary left first premolar root was mesially displaced and in partial transposition with the left canine, which was extremely high and lying horizontally in the palatal plane on a line shared by the floor of the nose and the floor of the maxillary sinus. The right canine was also very high, in

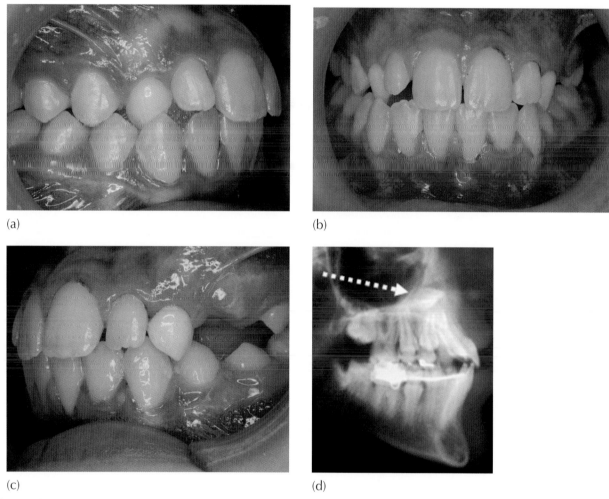

(a)

(b)

(c)

(d)

Fig. 15.3 (First published in *Seminars in Orthodontics*, reprinted with permission.)
(a–c) The intra-oral clinical views of the patient when first seen by the author.
(d, e) The initial panoramic view and cephalogram to show the position and orientation of the left canine (arrow). The right canine is also impacted on the labial side of the arch.
(f–h) CBCT three-dimensional views and a transaxial view to show the extreme height of the canine, adjacent to the nasal cavity and maxillary sinus, high above the sulcus height.
(i–k) Closed exposure technique used to expose the canines on both sides, with eyelet bonding. At the completion of the surgical procedure, the pigtail ligature was shortened and formed into a small hook and an elastic ligature was used initially to apply immediate traction. (Surgery by Dr E. Regev.)
(l–n) On the right side an auxiliary NiTi wire has been used to draw down the canine, which is about to erupt through attached gingiva. On the left side, an offset 0.016 in round wire with steel ligature has been used progressively, until the tooth is about to erupt through the oral mucosa. At this point an apically reposition flap has been used to place attached gingiva on the labial aspect of the canine before continuing the traction process.
(o–q) The final result shows some labial recession on all the teeth due to obsessive toothbrushing, although the canines are worst affected, which is typical of labial canines.
(r, s) Post-treatment periapical and panoramic views of the teeth show normal bone pattern, but marked apical root resorption of several teeth. This is not expected to alter in the future, following the cessation of orthodontic movement.

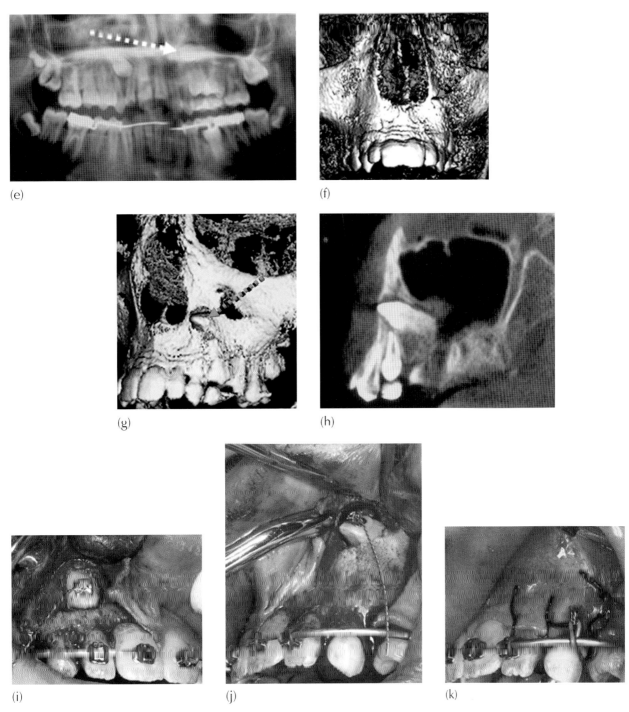

(e)

(f)

(g)

(h)

(i)

(j)

(k)

Fig. 15.3 (*Continued*)

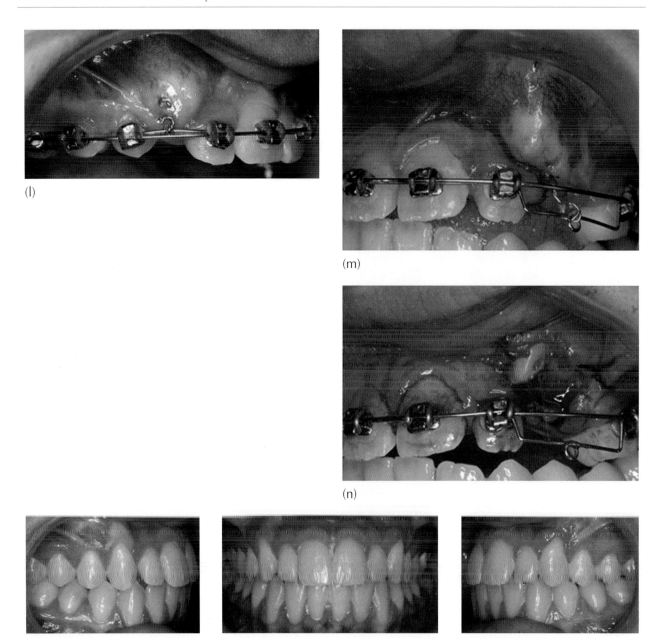

(l)

(m)

(n)

(o) (p) (q)

Fig. 15.3 (*Continued*)

comparison with any 'regular' impacted canine, although not reaching the severity of the left canine. The a–p and vertical orientation of the two canines was clearly depicted in the lateral cephalogram, as well as the mesial displacement of the root of the first premolar (Figure 15.3e). In the CBCT 3-D views and the transaxial slice shown here, the relative difficulty of the impactions will be clearly appreciated (Figure 15.3f–h).

Orthodontic alignment, levelling and space-opening were completed very quickly and measured stainless steel tube lengths were threaded on to the heavy 0.020 in main arch to maintain canine spaces and add rigidity to the anchor unit.

Labial surgical flaps were raised from the attached gingiva around the deciduous canines on each side of the maxilla under local anaesthetic cover (Figure 15.3i, j). These were reflected high into the depth of the sulcus area on both sides but, on the left side, exposure of the canine was only achieved about 10 mm above the extremity of the sulcus. Small eyelet attachments were bonded, with the twisted pigtail ligature drawn vertically downward, closely adapted to and lying over the exposed alveolar bone. The surgical flaps were then fully replaced and sutured, with the terminal hook of the pigtail ligatures emanating from the sutured edges on each side. The deciduous canines were not extracted at this stage.

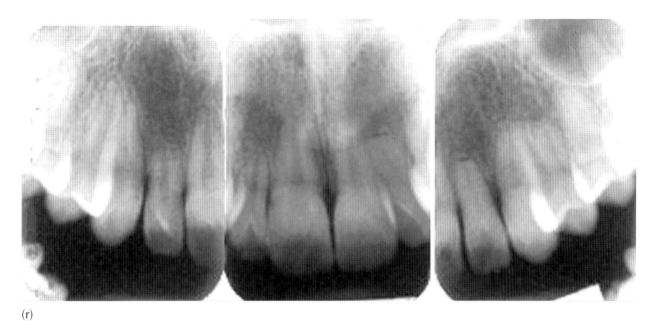

(r)

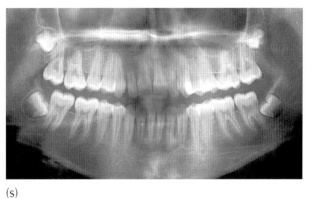

(s)

Fig. 15.3 (Continued)

Orthodontic traction was provided immediately following the suturing of the flaps with elastic thread ties (Figure 15.3k). Subsequently, traction was variously applied with NiTi auxiliary wires and with offset light wire auxiliary arches. On the right side, it was found possible to erupt the canine through the attached gingiva and directly into its designated location (Figure 15.3l). On the left side, the canine bulged the oral mucosa above the attached gingiva (Figure 15.3m) and a secondary surgical procedure was undertaken to apically reposition the attached gingiva over the tooth (Figure 15.3n).

At the end of treatment, the alignment of the teeth was excellent and, due to the compulsive aggressive toothbrushing, there was a general displacement of the gingival tissues in an apical direction. This was more noticeable over the two canines, which is largely to be expected with labially displaced canines (Figure 15.3o–q). Post-treatment radiographs (Figure 15.3r, s) show significant root resorption of the lateral incisors on both sides and of the central incisors and the left canine to a lesser degree.

Case 4: Impacted mandibular molars and premolars with over-eruption of the opposing teeth

This patient was seen first by his dentist at the age of 8.6 years, when it was noted that all the teeth distal to the deciduous canine in the lower right side of the jaw were unerupted. The deciduous molars on that side had been extracted some time earlier, together with those on the maxillary opposite side. The other three first permanent molars were erupted and the deciduous canines were still present. As the result of the extractions, the left maxillary molar had drifted mesially, reducing the extraction space and preventing the second premolar from erupting fully between it and the erupted first premolar. The maxillary posterior teeth, together with the alveolar ridge of the right side, had markedly over-erupted due to the absence of antagonists and the permanent molar was in direct occlusal contact with the mucosa covering its lower counterpart.

Radiographic records from that time consisted only of a panoramic film (Figure 15.4a), which showed that all the teeth were present, were situated low down in a narrowed body of the mandible and fully enclosed within alveolar bone, with no apparent follicular abnormality. The roots of the first molar were short, their apices were distally hooked, with almost completed apexification, and their apices were abnormally close to the lower border of the mandible. These observations label the first molar as impacted due to an unknown causal agent which had clearly overcome the natural eruption potential of the tooth. The radiographic appearance of the adjacent premolars and second molar was normal for the age of the patient, with early root development only and, as such, would not have been expected to erupt for some time. No treatment was performed at that time.

The child was referred to the author three years later, at the age of 11.6 years, when the situation of the first mandibular permanent molar had not changed and both premolars and second molar also remained unerupted, although their root development was now beyond that expected of normal unerupted teeth. However, the crowns of the first premolar and second molar could be palpated immediately beneath the overlying mucosa and, before treatment was commenced, the second molar had just broken through the tissues. Elsewhere, the premolars and permanent second molars had all erupted or were palpably close to eruption (Figure 15.4b–g).

It was obvious from the outset that any attempt to raise these teeth using forces derived from intra- or inter-arch dental anchorage would seriously worsen the occlusal plane in both jaws, particularly since the maxillary right side was already severely vertically compromised and was itself in need of intrusive mechanics to bring about some semblance of normality. Furthermore, without knowledge of the cause of the non-eruption of the molar, it was by no means certain that it would respond to extrusive mechanics. The differential diagnosis was principally between primary failure of eruption, ankylosis and invasive cervical root resorption, none of which could be confirmed. Accordingly, therapeutic diagnosis was the only remaining option.

The non-eruption of the premolars was considered to be due in part, at least, to the very early extraction and thickened reparative mucosal covering, having regard to their being prominently outlined in the gingiva. Nevertheless, the second premolar was also deeply impacted and this may have been secondary to the more severe impaction of its distal neighbour [4]. Accordingly, it was felt that the only serious problem related to the first molar and, to a lesser extent, the second premolar, while the first premolar and second molar would be expected to erupt following simple elimination of the mucosa covering the occlusal surfaces of these teeth.

Treatment was planned with the simultaneous use of two separate and unconnected systems. Skeletal anchorage from a zygomatic plate was to be employed to apply vertical traction to the impacted first molar and second premolar teeth in the attempt to overcome the non-eruption. At the same time, the plate would also be used as the anchor base from which to apply intrusive forces to the maxillary teeth of the right side. The second independent system was the placement of traditional orthodontic appliances for the purpose of levelling, alignment and treating the inter-maxillary relationship of the erupted teeth and, after their hopefully successful eruption, to the impacted teeth.

The surgical procedure was performed under intravenous sedation [5], during which all the unerupted teeth on the right side of the mandible were exposed. The superficially located second molar and both premolars were treated by open exposure without attachment bonding. The first molar and second premolar were much more deeply placed and small attachments were bonded to their buccal surfaces, before the teeth were re-covered by full replacement of the surgical flap, to leave only the two twisted steel ligatures visible above the sutured mucosa (Figure 15.4h, i). These were fashioned into two small hooks close to their exit from the tissues.

During the same surgical episode, the zygomatic plate was placed via an incision at the height of the sulcus, into the exposed inferior surface of the zygomatic arch (Figure 15.4j), with its free end drawn through attached gingiva close to the buccal side of the first permanent molar.

At the next visit, sutures were removed and the patient was instructed in the placement of latex elastics between the hooked end of the zygomatic plate and the pigtail hooks in the lower jaw. Over the period of several months, traction was applied continuously and the first molar and second premolar teeth slowly responded. In the meantime, the first premolar and second molar teeth erupted normally without assistance. A fixed multi-bracketed appliance was subsequently placed in the maxillary arch to align the teeth (Figure 15.4k). A mandibular fixed appliance was placed only when the affected molar had fully erupted (Figure 15.4l–n). The treatment was brought to its successful conclusion at 24 months post-surgery, having achieved levelled right and left sides of the occlusal plane in both arches (Figure 15.4o–t).

Case 5: Severe trauma in infancy causing damage to anterior tooth buds

The patient was a 10-year-old female who attended with the complaint that she had missing maxillary anterior teeth. She had suffered the ravages of caries in the deciduous dentition, the restorative and preventive treatment for which had been grossly neglected. From the patient's history it was learned that the child had suffered severe trauma from a fall at the age of 4 years, when the four deciduous maxillary incisors had been lost. Since that time, she had been without her front teeth.

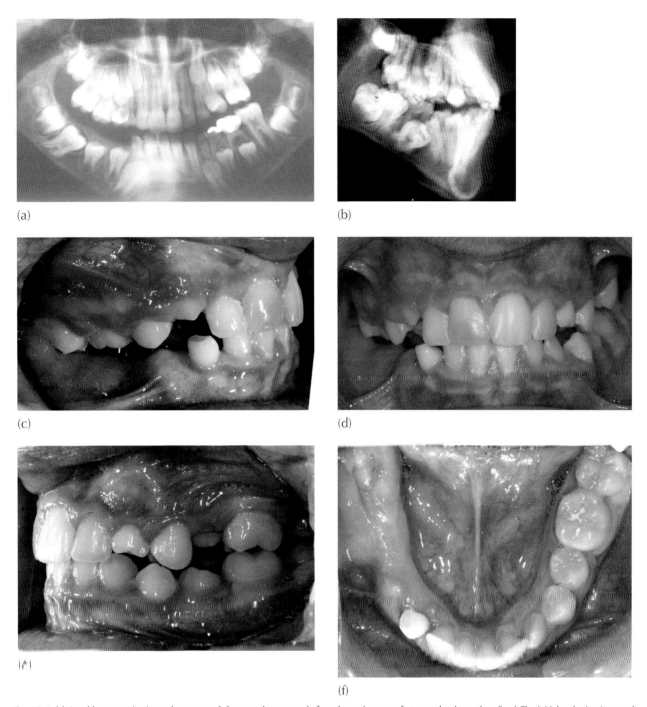

(a)

(b)

(c)

(d)

|᷿|

(f)

Fig. 15.4 (a) An old panoramic view taken at age 8.6 years, three years before the patient was first seen by the author. (b–g) The initial orthodontic records taken at age 11.6 years. (h, i) At surgery, the mandibular canine, premolars and molars were exposed. Attachments were placed on second premolar and first molar, which were subsequently recovered with the soft tissue flap, to leave only their pigtail ligatures visible intra-orally. The other three teeth remained exposed in an open procedure. (j) The zygomatic plate was placed at the same visit. (k) Vertical elastic traction was applied by the patient from the pigtail ligature hooks to the TAD, simultaneously with the placement of an orthodontic appliance in the maxilla. (l–n) The closing stages of appliance therapy. (o–q) The immediate post-treatment outcome. Note the leveled occlusal plane achieved by the intrusion of the teeth in the right side of the maxilla. (r–t) lateral cephalogram and panoramic view, with a periapical film of the immediate area, taken 24 months after initiation of treatment.

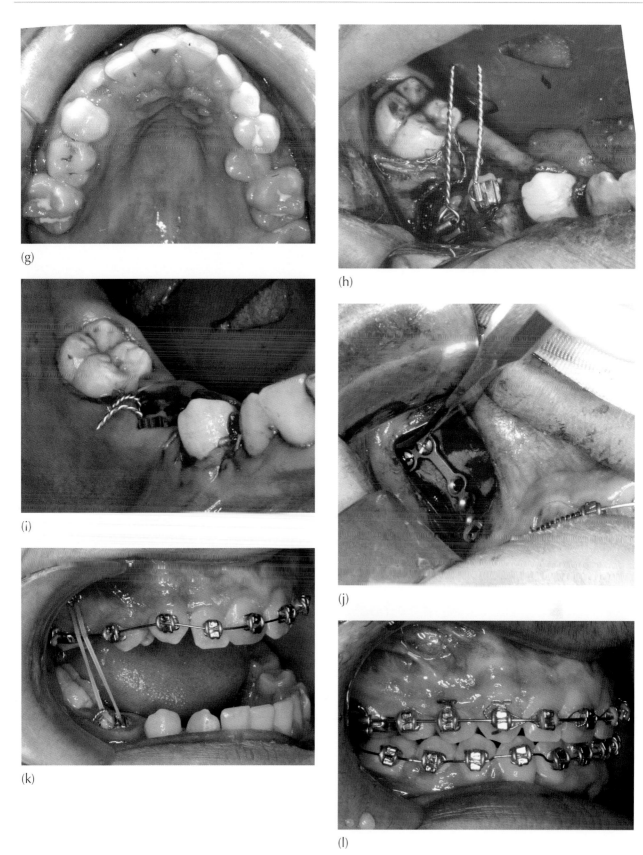

(g)

(h)

(i)

(j)

(k)

(l)

Fig. 15.4 (*Continued*)

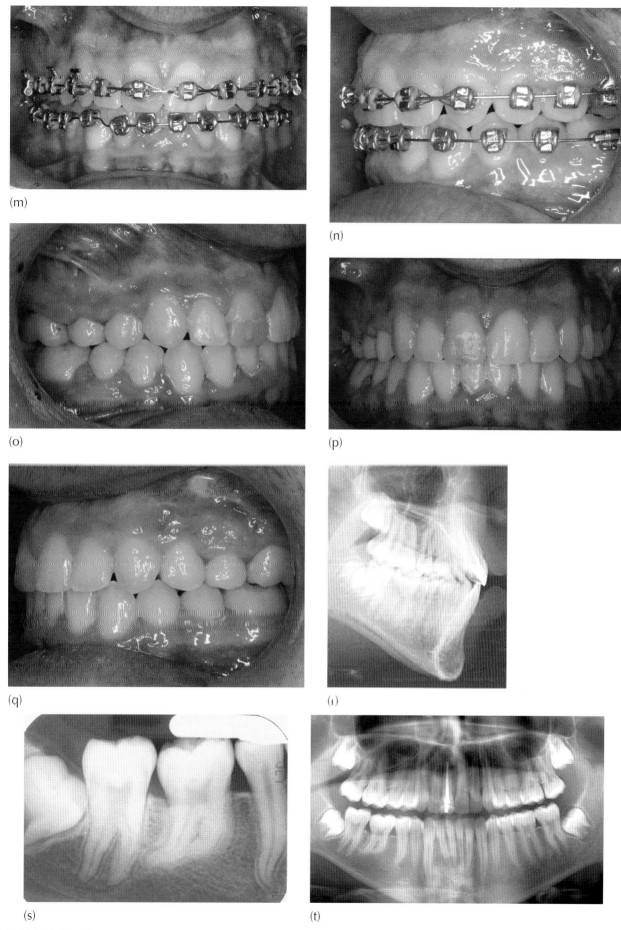

(m)

(n)

(o)

(p)

(q)

(r)

(s)

(t)

Fig. 15.4 (*Continued*)

At examination, the mandibular counterparts of the missing incisors were over-erupted and retroclined, the maxillary arch was narrowed and a left side unilateral crossbite was present, with accompanying functional shift into full closure. The dental arches were restricted, with obvious potential crowding. The anterior maxillary bony ridge was very thin and underdeveloped, as is to be expected in the absence of teeth (Figure 15.5a-e).

The radiographic records that accompanied the child included periapical, panoramic, cephalometric films and a CBCT scan (Figure 15.5f–j). These showed the presence of all permanent teeth, with anomalous development of the maxillary incisors and canines. The roots of the incisor teeth were very rudimentary, of a length to be expected at 3–4 years and with wide, open apices. These teeth were of abnormal form and they were located very high in the anterior maxilla close to the nasal floor. From the point of view of the child's age, the eruption time of these teeth was long overdue, but from the length of their roots and their height in the alveolus, normal eruption could not be expected for 4–5 years, although the likelihood of this occurring at all was in considerable doubt. By contrast, the developmental status of her unaffected teeth, erupted and unerupted, was closely identified with the child's chronological age.

She had been taken to several orthodontists for treatment of the unerupted incisors, each of whom had been at a loss to prescribe appropriate treatment.

When finally seen by the author, the conclusions drawn in relation to these teeth were that their development had been seriously hindered by the trauma that had occurred many years earlier, root development seemed to be adversely influenced by the proximity to the nasal floor and it could not be assumed that further root length development would occur, particularly given the apparent lack of eruptive potential of these teeth. To extract them, with the view to temporary prosthodontic replacement and, subsequently, with implant-supported crowns, would have caused very much more alveolar bone resorption and a large anterior bony defect. In light of the age of the child and her dental history, long-term wear of artificial prostheses would likely seriously reduce the prognosis for the survival of the other teeth. The prostheses would need to be present for 10 or more years, for the majority of which she would only be seen periodically by her dentist and therefore at considerable risk to develop further caries and gingival inflammation.

On the other hand, a successful plan that attempted to enhance the attenuated innate eruptive force these teeth would bring with it the benefits of providing the child with her own natural and adequate, if temporary, anterior dental rehabilitation. Furthermore, this would be accompanied by excellent natural regeneration, contributing materially to the replacement and reconstitution of the defective alveolar bone height. True, the prognosis of these teeth was unknown

and the presence of orthodontic appliances in the mouth was also a risk factor for caries and gingival inflammation. Nevertheless, the expected treatment duration would be relatively short and the child would be under frequent and routine professional supervision, thereby reducing potential collateral damage to a minimum.

Molar bands with a soldered palatal arch were cemented and a removable self-supporting labial arch slotted into the round buccal tubes on the bands, as with the first phase of CCD treatment (see Chapter 14). At surgery, the four maxillary incisors were exposed, bonded with small eyelet attachments and the full flap sutured back to its former place, with the twisted steel pigtail ligatures emanating from the sutured edge (Figure 15.5k–m). In subsequent visits, the pigtail ligatures were rolled up higher as the teeth responded and a mandibular Johnson modified appliance was placed with the intention of intruding and proclining the lower incisors (Figure 15.5n–p).

Once the incisors had erupted, the eyelet attachments were substituted by orthodontic brackets on the incisors, which were then aligned and proclined labially (Figure 15.5q, r). At this point, sufficient space had been provided for the maxillary canine teeth which were impacted in the line of the arch, but which had also suffered anomalous development, as the result of the trauma, with an abnormal bucco-lingual crown angle and form. A second surgical procedure was performed to expose and bond attachments to both maxillary canines (Figure 15.5s, t) and this was followed by full alignment of the remaining teeth (Figure 15.5u). In this final stage, mandibular second premolars were extracted to provide space to eliminate the lower crowding and to permit the necessary anchorage for class 2 traction to control the overjet that was produced. Thus, in the final occlusion, the molars were brought to a full unit class 3 intercuspation, while the canines were normally related. In this manner, the orthodontic treatment was completed, with a good smile line relationship between the lower lip and the incisal edges (Figure 15.5v–y).

Treatment duration was 26 months, with bonded 3-3 twistflex retainers in place in both arches. The post-treatment radiographs (Figure 15.5z–za) show considerable root growth which, it was believed, would have been a most unlikely outcome in the pre-treatment location of these severely damaged teeth. It is reasonable to speculate that this length of root may have been the reward generated by the extrusion of the teeth away from their anatomically limiting nasal floor.

The follow-up records taken at age 17 years (i.e. four years post-treatment; Figure 15.5zb, zc) show further improvement in the root length, with the exception of the left central incisor, which required root canal treatment about nine months after the completion of the orthodontic treatment. The health of the teeth and surrounding tissues was good and the prognosis of the teeth appeared to be fairly good in the medium term. Nevertheless, there was a

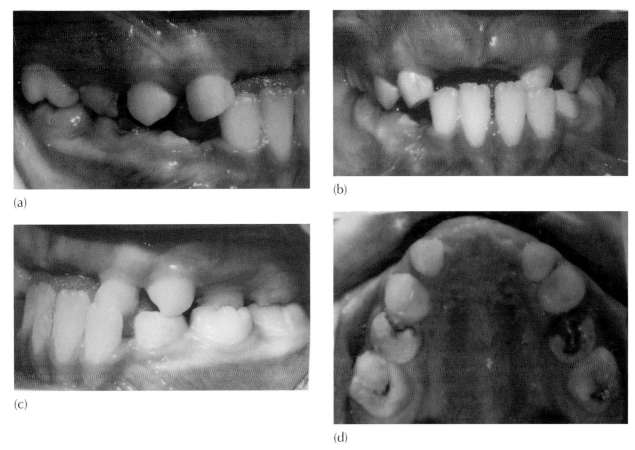

(a)

(b)

(c)

(d)

Fig. 15.5 (a–e) Intra-oral photographs transferred with the patient to the author, showing the initial condition. Note missing incisors and defective alveolar ridge, marked potential crowding, left side cross-bite and poor overall dental status.

(f–h) Initial cephalometric, panoramic and maxillary anterior occlusal radiographs.

(i, j) Transaxial and anterior coronal 'slices' from the CBCT to show incisor crown anomaly, very early root development and location of the teeth in relation to the nasal floor.

(k, l) Bonding of eyelets to the exposed incisor crowns.

(m) The surgical flap is fully sutured with only the twisted steel pigtail ligatures engaging the raised self supporting labial arch.

(n) An occlusal view of the active extrusion appliance one month post-surgery.

(o) The anterior occlusal radiographic view shows the soldered palatal arch and the bonded eyelet attachments with their pigtail ligatures attached to the removable, self-supported, labial arch.

(p) A month later, the mandibular Johnson modified appliance is placed. The anterior portion of the composite archwire may be seen to be exerting an intrusive influence on the anterior teeth.

(q, r) Regular orthodontic (TipEdge plus) brackets are placed on the newly erupted teeth, the teeth are aligned and brought forward with buccal coil springs, to provide space for the canines.

(s, t) The canines are exposed and bonded.

(u) The final alignment showing lingual torqueing auxiliaries on the two canines.

(v–x) The final alignment and occlusion. Note the anomalous long clinical crowns of the incisors and the crown/root angle of the canines and the mottled hypoplastic enamel.

(y) The normal appearance of the patient's smile and the excellent smile line.

(z, za) Panoramic and periapical radiographs show the remarkable growth of the incisor roots.

(zb, zc) Panoramic and periapical views taken four years post-treatment show further incisor root growth and the root canal treatment in the left central incisor.

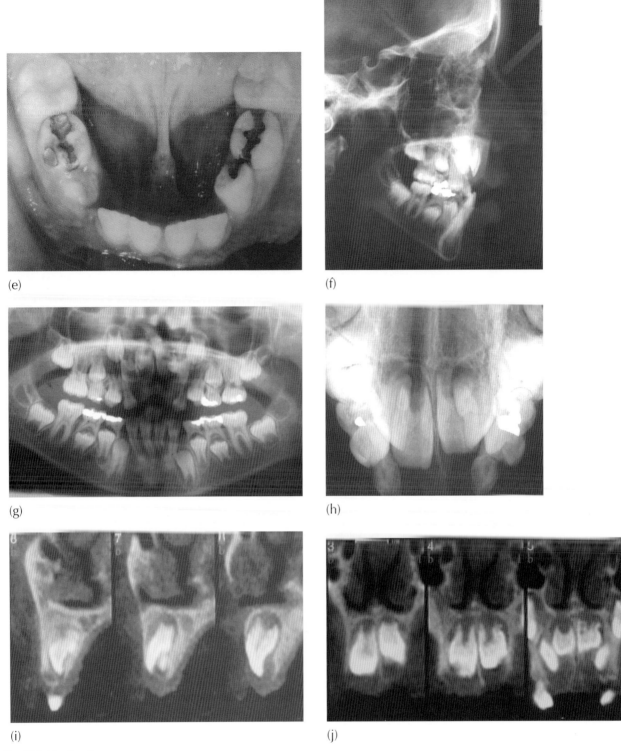

(e)

(f)

(g)

(h)

(i)

(j)

Fig. 15.5 (Continued)

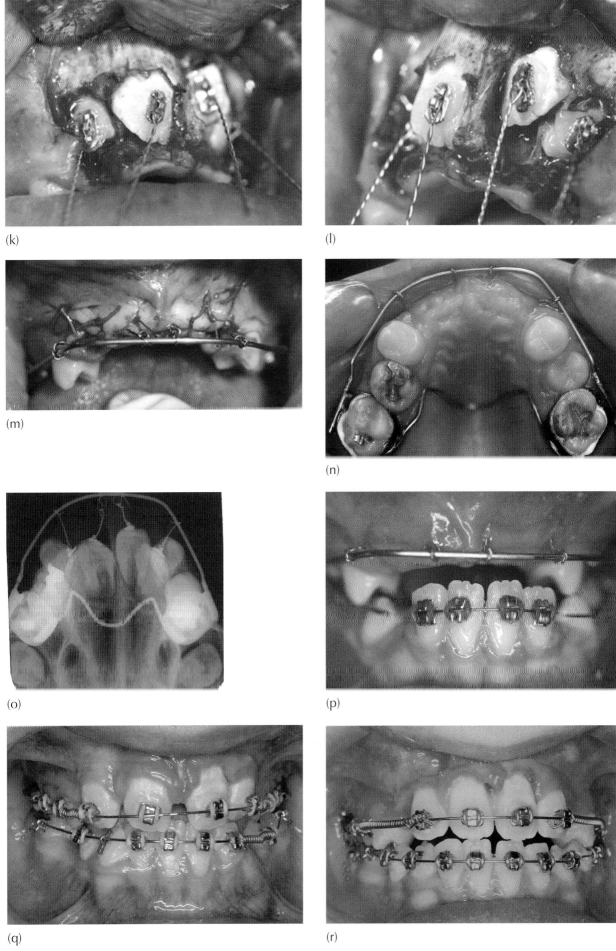

(k)

(l)

(m)

(n)

(o)

(p)

(q)

(r)

Fig. 15.5 *(Continued)*

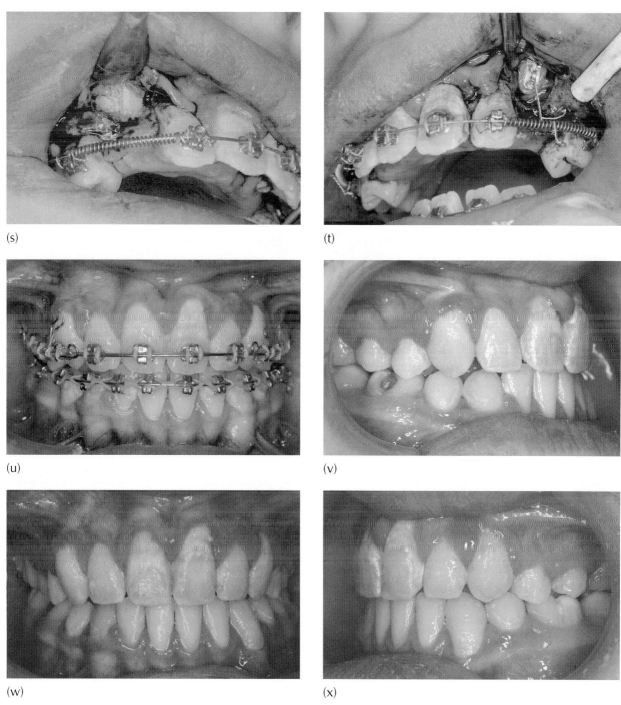

(s)

(t)

(u)

(v)

(w)

(x)

Fig. 15.5 *(Continued)*

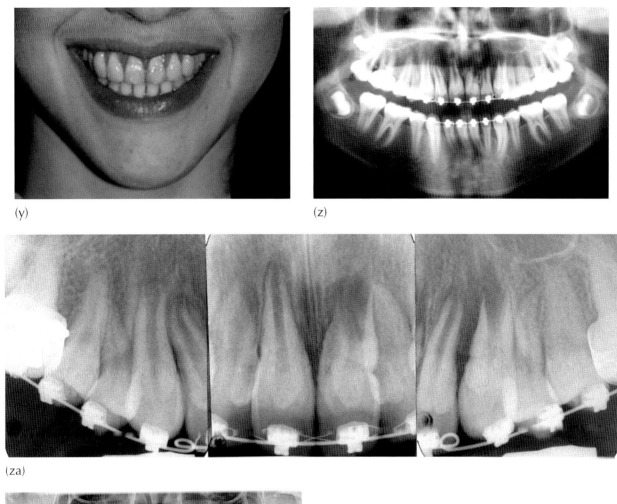

(y)

(z)

(za)

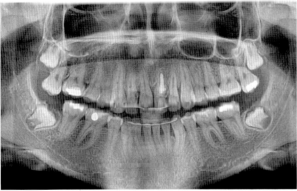

(zb)

Fig. 15.5 *(Continued)*

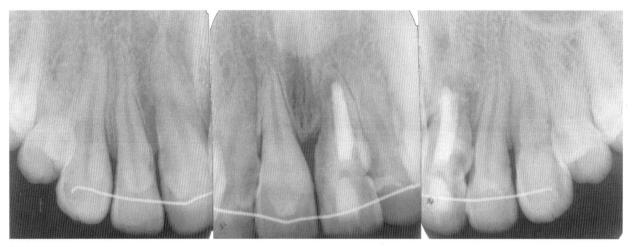

(zc)

Fig. 15.5 *(Continued)*

mildly marred appearance due to enamel hypoplasia and long clinical crowns of the affected teeth, despite the reshaping that was performed at that time.

Case 6: Buccal to the lateral incisor and palatal to the central incisor

The 14-year-old male patient attended with what seemed a very minor problem. The dental arches in both jaws were well aligned and the occlusion textbook perfect. The posterior teeth met in a fully interdigitated class 1 occlusion, the incisor relations were ideal and there was a modicum of spacing generally in the dentition. The right maxillary canine had clearly recently shown its first signs of eruption into its place in the dental arch. The only flaw was to be found in the left anterior maxilla, where the canine had not erupted and the long axis of the lateral incisor appeared to be proceeding apically in a slowly rising distal and palatal direction, such that the root could be envisaged low down on the palatal surface, lingual to the line of the arch in the canine location (Figure 15.6a–e). Additionally, the tooth was rotated 30° mesio-labially, overlapping the adjacent central incisor.

In a situation like this, the clinician should always suspect a labial canine, with the strong possibility of a canine-lateral incisor transposition. The scenario indicates a labially placed space-occupying body within the alveolus, the presence of which displaces the lateral incisor root palatally and distally. The displacing factor is usually an unerupted labial canine, which may be palpated on the labial side of the ridge, above the lateral incisor. However, on occasion, futile and vertically downward eruptive movement of a relatively high labial canine brings it more mesially and on the lingual side of a vertical root of the central incisor. Thus, the canine may end up straddling the ridge labial to the lateral incisor and lingual to the central incisor, in partial or complete

transposition with the lateral incisor (see Figures 7.3, 8.3, 12.16 and 12.19). Under these circumstances, the canine will not be palpable labially or palatally.

The radiographs of the patient under discussion in the present context showed a full complement of teeth, including the unerupted third molars (Figure 15.6f, g). The right maxillary canine was unerupted, but could be seen to have an unimpeded line of approach to its place in the arch. The left maxillary canine was noted on the panoramic view to have migrated mesially and its crown tip was superimposed on the middle root area of the central incisor. On this film, the lateral incisor was tipped in the mesio-distal plane at an angle of 45° with its partially resorbed apex displaced into the area of the alveolar ridge reserved for the canine. From the CBCT, it was confirmed that the body of the canine was labial to the lateral incisor and its crown tip was palatal to the root of the central incisor (Figure 15.6h, i). Furthermore, root resorption of the lateral incisor root was identified as an oblique defect in the integrity of the normal root outline.

A maxillary orthodontic appliance was placed on all the teeth from the first molars forward on each side, with the exception of the two canines and the ectopically oriented left lateral incisor. Immediately prior to the surgical exposure, a prepared auxiliary arch of 0.016 in gauge was tied into the brackets with elastic modules, in piggy-back fashion over the 0.020 in steel base arch, with its active loop lying horizontally, in its passive mode (Figure 15.6j). Surgical exposure was undertaken from the labial side of the ridge under local anaesthetic and an eyelet attachment was bonded by the author at the time (Figure 15.6k, l). Access to the canine was difficult, since coincidental exposure of the apically and widely diverging roots of the two incisors needed to be explicitly avoided. The pigtail ligature from the bonded attachment was directed labially and slightly distally and pushed through the oral mucosa area of the surgical flap at the height of the canine. Once the

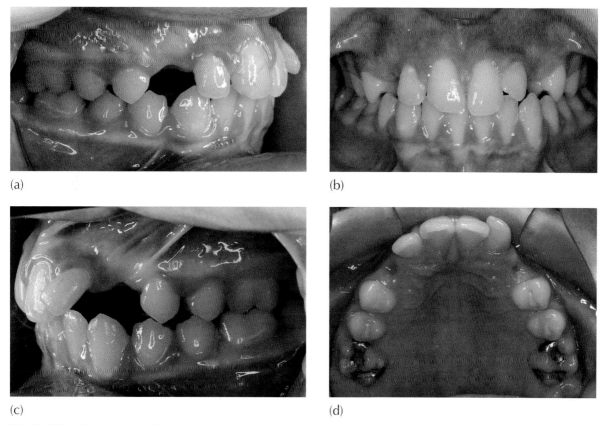

(a)　　　　　　　　　　　　　　(b)

(c)　　　　　　　　　　　　　　(d)

Fig. 15.6 (a–e) The initial intra-oral condition.

(f) The relevant section of the cephalogram shows no A-P or height difference between the two superimposed maxillary canines.

(g) The panoramic view confirms a full complement of teeth, with the maxillary right third molar 'banana'-shaped (see Chapter 8). The left maxillary canine has tipped mesially and its tip superimposes on the distal aspect of the central incisor roots. The left lateral incisor root apex is tipped distally into the canine location, in partial transposition with the canine. This view gives no bucco-lingual information.

(h) The anterior three-dimensional view taken from the CBCT reconstructions provides the essential missing bucco-lingual information needed to plan the treatment. The canine is labial to the root of the lateral incisor and palatal to the root of the central incisor.

(i) The angle that exists between the adjacent surfaces, as indicated by the green dotted line drawn on the three-dimensional view from the left side, provides the 'window of opportunity' through which the canine may be accurately drawn high in a labial direction to resolve the impaction.

(j) The auxiliary labial arch in its passive mode is ligated piggyback over the main arch, immediately before the surgical exposure begins. Note the absence of a bonded bracket on the left lateral incisor.

(k) With the full flap exposure, the crown of the canine is exposed at a distance from its tip, which is buried between the roots of the incisors. The surgeon refrains from wider exposure and bone removal to protect the vitality of the teeth and to avoid exposing the root surface. (Surgery by Dr H. Samen.)

(l) A small eyelet is bonded. The loop and terminal helix of the auxiliary archwire is seen in its passive horizontal mode.

(m) Full replacement and suturing of the flap is performed, with the pigtail ligature pushed through the flap, high in the sulcus. The loop of the auxiliary archwire is turned upwards and engaged tightly in the shortened end of the pigtail ligature, thereby transferring labial traction force on the hidden canine. Note that the auxiliary loop is too high in the sulcus and caused subsequent pain and swelling of the lip tissue, necessitating removal of the pigtail ligature after three months, by which time the canine was palpable on the labial side.

(n) The canine was now well clear of its earlier relationship with the palatally displaced root of the lateral incisor and a mesial uprighting spring was placed in the lateral incisor bracket.

(o) With the lateral incisor fully uprighted, the canine was re-exposed and a new ligature applied.

(p) The flap was again fully sutured back to its former place, with elastic thread drawing the pigtail distally towards the canine location. There are auxiliaries mesially uprighting the incisor and distally uprighting the first premolar, to clear a path to accommodate the distally and vertically moving canine.

(q) With the canine close to its intended location, a bracket is substituted for the eyelet in the closing stages of the case.

(r–t) The alignment of the teeth on the day of debanding and prior to the placement of removable finishing and retaining appliances.

(u) Periapical views of the anterior maxilla.

(v) Panoramic view shows good parallelism of the teeth concerned. Note the 'banana'-shaped right maxillary third molar.

(w) The post-treatment cephalogram.

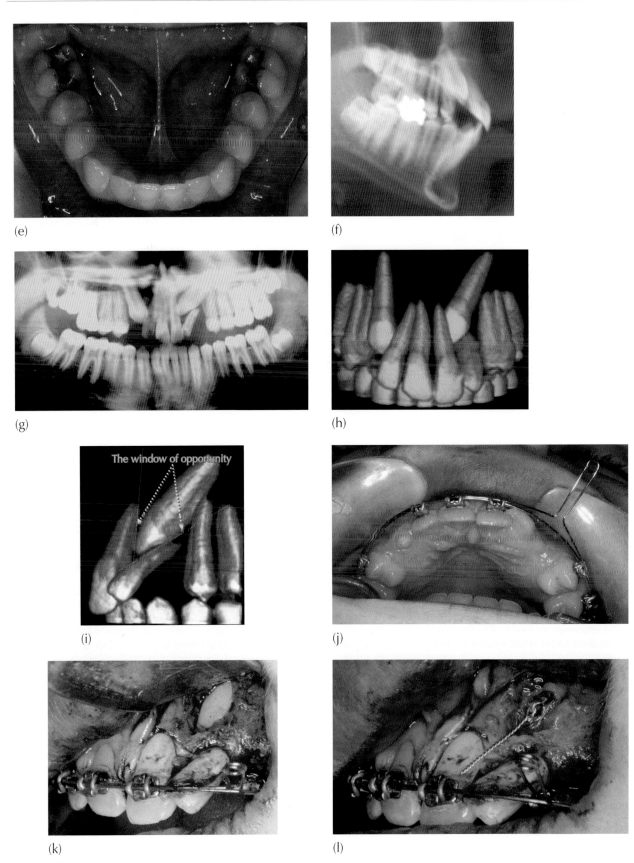

(e)

(f)

(g)

(h)

(i)

(j)

(k)

(l)

Fig. 15.6 (Continued)

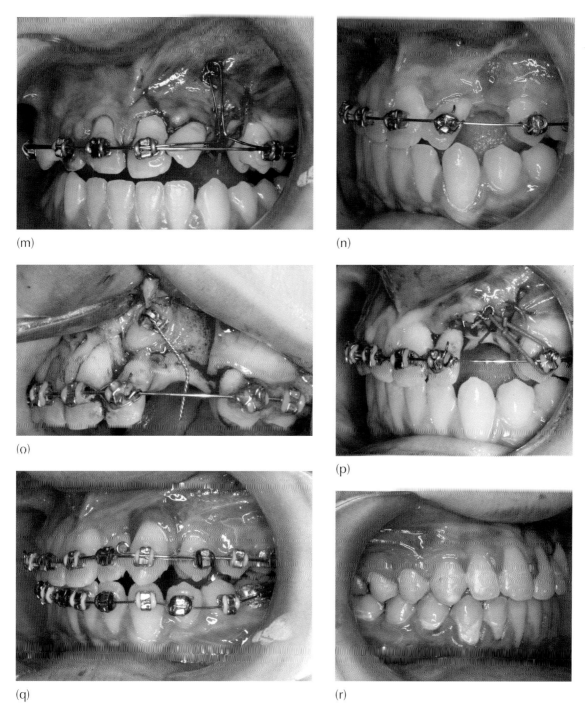

(m)

(n)

(o)

(p)

(q)

(r)

Fig. 15.6 *(Continued)*

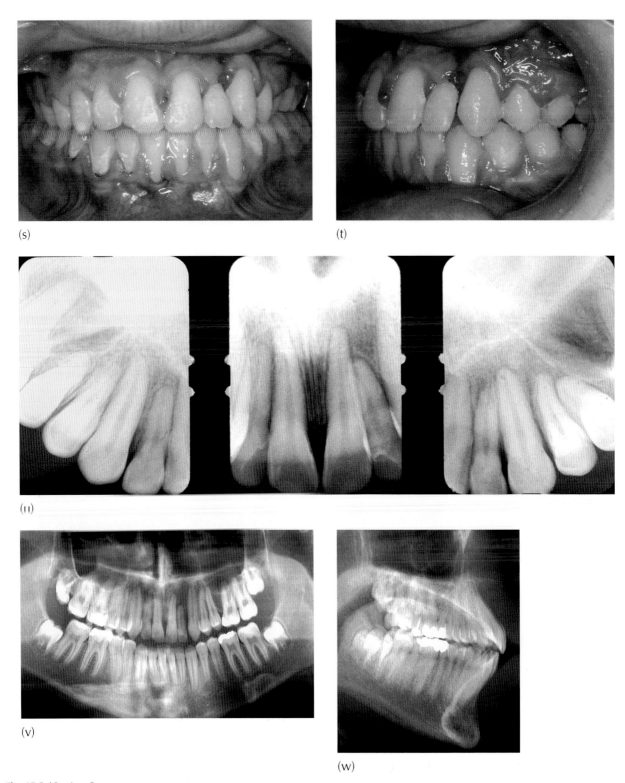

(s)

(t)

(u)

(v)

(w)

Fig. 15.6 (*Continued*)

attached gingival flap was fully sutured to its former place, the pigtail ligature could be seen to pierce the oral mucosa, high up in the sulcus and was the only means of communication with the impacted tooth. The loop of the auxiliary archwire was then raised, turned towards the pigtail ligature and ensnared by turning the pigtail around its terminal loop, as close as possible to the oral mucosa (Figure 15.6m). This had the effect of applying traction to the canine, high up and in a labial direction, through the 'window of opportunity' provided by the widely divergent roots of the two incisors in the mesio-distal and bucco-lingual planes (Figure 15.6i).

In the subsequent 2–3 months the patient had considerable discomfort due to ulceration of the highly mobile oral mucosa from what transpired to be excessive length of the loop of the auxiliary archwire. The inflamed tissue grew over the loop and the auxiliary archwire had to be removed. However, at this point the canine was palpable on the buccal side of the ridge, which meant that it had moved from the palatal side of the central incisor and was also now clear of the root of the lateral incisor. Accordingly, a bracket was placed on the lateral incisor. It was aligned and its root mesially uprighted with relative ease, using an auxiliary uprighting spring (Figure 15.6n). The canine was then re-exposed with an apically repositioned flap and its ligation renewed to permit distally and vertically downward traction towards its place in the arch (Figure 15.6o, p). With the eruption of the canine into its appropriate location, a

bracket was substituted for the eyelet attachment (Figure 15.6q) and, as expected in the finishing stage, much labial root torque of the lateral incisor and lingual root torque of the canine were then applied.

After 25 months of treatment, the fixed appliances were removed (Figure 15.6r–t) and removable finishing and retaining appliances were placed. The radiographs show good root parallelism and good intra- and inter-maxillary relationships of the teeth in the immediate area of concern (Figure 15.6u–w). The lateral incisor, which had suffered pre-treatment root resorption, remained with the same degree of root shortening, while some minimal and clinically insignificant resorptive blunting was noted of the apices of the maxillary central incisors.

References

1. Dankner E, Harari D, Rotstein I, Dens evaginatus of anterior teeth. Literature review and radiographic survey of 15,000 teeth. Oral Surg Oral Med Oral Pathol Oral Radiol Endod, 1996; 81; 472–476.
2. Danesh G, Schrijnemakers T, Lippod C, Shafer E. A fused maxillary central incisor with dens evaginatus as a talon cusp. Angle Orthod 2007, 77: 176–180.
3. Abbott PV, Labial and palatal 'talon cusps' on the same tooth: a case report. Oral Surg Oral Med Oral Pathol Oral Radiol Endod 1998; 85: 726–730.
4. Becker A, Karnei-R'em RM: The effects of infraocclusion: part 2 – the type of movement of the adjacent teeth and their vertical development. American Journal of Orthodontics 1992; 102: 302–309.
5. Chaushu S, Gozal D, Becker A. Intravenous sedation: an adjunct to enable orthodontic treatment for children with disabilities. European Journal of Orthodontics 2002; 24: 81–89.

Index

Note: Page numbers in *italic* refer to tables and the legends of illustrations.

Orthodontic Treatment of Impacted Teeth, Third Edition. Adrian Becker.
© 2012 Adrian Becker. Published 2012 by Blackwell Publishing Ltd.

premolars
  first 322
  impaction of maxillary canines 111
  resorption 45, 173–210, 340–7, 348–50
    anatomy 179
    incisors 340
      anomalies and 177
      computerized tomography 21, 22
      maxillary canine impaction 124–5, 132–3, 134, 135, 169–70
      invasive cervical 197–206, 324–5, 326
      mandibular second premolar impaction 223
      maxillary canine impaction 138–9, 169–70
      prevalence 175–7
      trauma 358, 360
      treatment 177–87
    surface damage 44
    torque, see torque, on roots
    see also palatal root displacement
rotations
  correction 135
  maxillary canines, palatal impaction
    Group 1 140–2
    Group 2 148, 154
'round-tripping'
  canines 143
  lingual appliances and 286
rubber bands, to mandibular removable appliance 151
RUNX2 gene 371, 382

sacs, see dental follicles
saline, rinsing 30
'scorpion' appearance, dilaceration of maxillary central incisors 83
screws
  mini-screws, lingual appliances with 287
  temporary anchorage devices 62–3
second intention, healing 36
second molars
  cleidocranial dysplasia 392
  deciduous 234, 235–8
  mandibular 240–9
    failed treatment 320
    pressure packs 32
  maxillary 249
second opinions 355–6
secondary retention 251
self-supported arch, for traumatic intrusion 363, 364–6
separators, maxillary first molar impaction 238
sexual dimorphism, see gender differences
shoulders, cleidocranial dysplasia 371, 372
size, see small teeth; tooth size
skeletal anchorage, traction for infra-occlusion 260

skull, cleidocranial dysplasia 371, 372
'slingshot' elastic ties 60
small teeth 322
  cleidocranial dysplasia 373
  maxillary canine impaction and 121, 122–3
smile line, treated cleidocranial dysplasia 381
soft tissues
  cleidocranial dysplasia 378
  contact with enamel cuticle 43
  devitalized maxillary canine on 117
  effect of brackets 58
  resistance 62, 351, 354
    management 30, 31
    maxillary canine impaction treatment 152, 159
    trauma 73, 104
soldered palatal arch
  Johnson's twin-wire arch 88, 92
  for maxillary central incisor impaction (adults) 268–9
soldered springs, maxillary first molar impaction 238
space closure, root resorption and 178
space-holding devices, for mandibular second premolar impaction 223
space loss
  local 6, 7, 7, 32
  second premolar impaction
    mandibular 222
    maxillary 225–30
  see also crowding
space maintenance
  adults 272–3
  maxillary central incisors 107
space opening 321
  cleidocranial dysplasia 380
  infra-occlusion 254
  maxillary canine impaction
    prevention 130
    treatment 133, 134–6, 138
  maxillary central incisor impaction (adults) 264, 265–7
  for root resorption 178
Spider Screw*.110 implant 290
splinting
  cleidocranial dysplasia 397, 398
  trauma 358, 360
spontaneous eruption 56
  in cleidocranial dysplasia 376
spontaneous resolution
  impaction from cysts 299, 300–16
  traumatic intrusion 358
spoon denture ('flipper' partial denture) 263, 268
springs 61, 351
  cleidocranial dysplasia 387–8, 392
  for Johnson's twin-wire arch 92, 93
  lingual appliances and 285–6, 289
  mandibular second molar impaction 243, 244

maxillary canine impaction
  palatal, Group 2 149
  treatment 136
maxillary first molar impaction 238
space maintenance, adults 273
for temporary anchorage devices 62
see also ballista springs
stainless steel wires
  Johnson's twin-wire arch 88, 92
  see also ligature wires
standardization, radiographs at right angles 17–18
steel
  tubing, with elastic ties 60
  see also ligature wires; stainless steel wires
'straight-wire' brackets 58
suction 50
super-elastic wires 61
  lingual appliances and 286
  maxillary canines, palatal impaction, Group 1 143
supernumerary teeth 8, 72
  central 84
  cleidocranial dysplasia 371, 374, 392, 393, 396
    recurrent 397
  dilaceration of maxillary central incisors 74–5
  impaction of maxillary canines 111
  maxillary central incisors 407–12
  periapical radiographs 11, 13
  radiography 82
  results of removal 86
  on tooth eruption 6, 8, 89
supplemental molars, cleidocranial dysplasia 393
surgeons 30, 49
  attachment bonding by 46, 49
  cooperation with orthodontists 48–52
  three-dimensional diagnosis and 20
  views on mesiodens 84
surgical exposure 29–54, 50
  alone 30
  cleidocranial dysplasia 45, 376–9, 384, 388, 396
  cysts 297–9
  duration of procedures 57–8
  effect on outcomes 43–4
  invasive cervical root resorption and 204
  for mandibular second premolar impaction 224
  maxillary canines 134
    alone 30, 31
    buccal approach 42
      palatal impaction (Group 3) 154–5
    buccally-displaced 216, 217
      misdiagnosis 330, 331, 332–5, 336, 337
    discomfort after 48
    Kokich technique 44
    and packing 32